HANDBOOK OF

Aqueous Solubility Data

HANDBOOK OF

Aqueous Solubility Data

Samuel H. Yalkowsky
Yan He

CRC PRESS

Boca Raton   London   New York   Washington, D.C.

Library of Congress Cataloging-in-Publication Data

Handbook of aqueous solubility data : an extensive compilation of aqueous solubility
data for organic compounds extracted from the AQUASOL dATAbASE / Samuel H.
Yalkowsky and Yan He, editors.
     p. cm.
     Includes bibliographical references and index.
     ISBN 0-8493-1532-8 (alk. paper)
     1. Organic compounds—Solubility—Handbooks, manuals, etc. I. Yalkowsky, Samuel H.
(Samuel Hyman), 1942- II. He, Yan, 1970-

QD257.7 .H33 2003
547′.13422—dc21
                                          2002191165

Visit the CRC Press Web site at www.crcpress.com

Authors

Samuel Yalkowsky earned his B.S. in pharmacy from Columbia University in 1965 and his Ph.D. in pharmaceutical chemistry from the University of Michigan in 1969. He is a professor of pharmaceutical sciences at the University of Arizona. His work has led to over two hundred scientific publications, four patents, and four books. His most recent book *Solubility and Solubilization in Aqueous Media*, ACS-Oxford, was published in 1999. Dr. Yalkowsky co-authored two papers that won the Ebert Prize for the best scientific paper in the *Journal of Pharmaceutical Sciences*. He received the University of Arizona Graduate Teaching and Mentoring Award in 2001.

Yan He earned her M.S. degree in pharmaceutical sciences from the University of Arizona in 1999. She is a research specialist in the pharmaceutical sciences at the University of Arizona and is currently pursuing her Ph.D. degree. Her research interests include the development of formulations for low water solubility anticancer agents. She also conducts basic research on the relationships between chemical structure and physical properties of organic compounds.

Acknowledgment

We would like to thank all the people who helped us in completing this book. Our deepest appreciation goes to Jingsong Zhang, who helped us manage the data and generate the final output. We thank Dr. Julianne M. Braun for her assistance on the alphabetization of chemical names. We would also like to express our gratitude to Dr. Wei-Youh Kuu and Dr. Rose-Marie Dannenfelser for their contributions to this work.

Introduction

The *Handbook of Aqueous Solubility Data* is an extensive compilation of published data for the solubility of a wide variety of organic nonelectrolytes and unionized weak electrolytes in water. It includes data for pharmaceuticals, pollutants, nutrients, herbicides, pesticides, and agricultural, industrial, and energy-related compounds. This handbook contains over 16,000 solubility records for more than 4000 compounds. These data were extracted from about 1800 scientific references contained in the AQUASOL dATAbASE.

Each compound is identified by a sequential number along with molecular formula, compound name, synonyms, molecular weight, Chemical Abstracts Service Registry Number, melting point, and boiling point if available. For user convenience, all solubility data are converted to moles per liter and grams per liter. Also, reported numerical temperature values are converted to Celsius. The following symbols are included in the temperature field when nonnumerical temperature descriptors are reported:

amb	ambient temperature
c	cold water
h	hot water
rt	room temperature
ns	temperature not stated

Each record has a 5-point evaluation for reporting of the data and a reference code for the citation. Comments are included when necessary. The following alternatives are used in the comments field:

EFG	estimated from graph
LCST	lower critical solution temperature
UCST	upper critical solution temperature

Solubility Data

The compounds are sorted by their molecular formula using the Hill system (number of carbons, number of hydrogens, and then alphabetically by element). Each compound can contain up to five synonyms. This is followed by the Chemical Abstracts Service Registry Number (RN), melting point (MP) in Celsius, molecular weight (MW), and boiling point (BP) in Celsius. Multiple values are presented whenever available. These are sorted by temperature and then by reference source.

Citations

The reference citation is given as a four-character code in which the first character is alphabetic, referring to the first author's last name, and the next three are numeric. The complete reference citation is provided in the Reference section.

Evaluation

As listed in the Table of the Explanation of Evaluation Scores, a five-point evaluation is provided for the quality of the reporting of temperature (T), purity of solute (P), equilibration time/agitation (E), analysis (A), and accuracy and/or precision (A).

Explanation of Evaluation Scores

Parameter		Score		
		0	1	2
T	Temperature	Not given, ambient, or room temp	Given with no range	Given with range
P	Purity of solute	Not stated or as received	Stated with no range or as received	Stated with range or altered with range or calculated
E	Equilibration time/agitation	Not stated	Stated briefly	Described in detail
A	Analysis	Not stated	Stated briefly or stated in other paper	Described in detail
A	Accuracy and/or precision	1 significant figure or range > 20%	2 significant figures or range 5–20%	3 significant figures or range 0–5%

Indices

Entries in the indices are referenced to the compound sequential numbers, not to page numbers. The alphabetization of Index 3 was performed by an Excel macro, which can be found at http://academic.athens.tec.ga.us/jbraun/excel.

Separate indices referring to the compound sequential numbers are provided for:

Index 1: Molecular Formula
Index 2: Chemical Abstracts Service Registry Number (RN)
Index 3: Names and Synonyms

Table of Contents

Solubility Data 1

References 1275

Index 1: Molecular Formula 1347

Index 2: Chemical Abstracts Service Registry Number (RN)...................................1369

Index 3: Names and Synonyms 1401

Solubility Data

1. CHBrCl$_2$
Bromodichloromethane
Dichlorobromomethane
BDCM

RN: 75-27-4   **MP** (°C): -55
MW: 163.83   **BP** (°C): 87

Solubility (Moles/L)	Solubility (Grams/L)	Temp (°C)	Ref (#)	Evaluation (T P E A A)	Comments
1.851E-02	3.032E+00	30	M300	1 1 2 2 2	
1.812E-02	2.968E+00	30	M311	1 1 2 2 2	

2. CHBr$_2$Cl
Chlorodibromomethane
Dibromochloromethane
CDBM

RN: 124-48-1   **MP** (°C): -22
MW: 208.29   **BP** (°C): 119.5

Solubility (Moles/L)	Solubility (Grams/L)	Temp (°C)	Ref (#)	Evaluation (T P E A A)	Comments
5.041E-03	1.050E+00	30	M300	1 1 2 2 2	
1.205E-02	2.509E+00	30	M311	1 1 2 2 2	

3. CHBr$_3$
Bromoform
Tribromomethane
Methyl Tribromide

RN: 75-25-2   **MP** (°C): 7.5
MW: 252.75   **BP** (°C): 149

Solubility (Moles/L)	Solubility (Grams/L)	Temp (°C)	Ref (#)	Evaluation (T P E A A)	Comments
1.187E-02	3.001E+00	15	G029	1 0 2 2 2	
3.957E-03	1.000E+00	20	F300	1 0 0 0 0	
<7.91E-04	<2.00E-01	25	B019	1 0 1 2 0	*sic*
1.262E-02	3.190E+00	30	F300	1 0 0 0 2	
1.258E-02	3.180E+00	30	G029	1 0 2 2 2	
1.555E-02	3.931E+00	30	M311	1 1 2 2 2	
1.256E-02	3.174E+00	30	V009	1 0 0 0 2	
1.227E-02	3.100E+00	ns	O006	0 0 0 0 2	

4.   CHClF$_2$
Chlorodifluoromethane
Freon 22
Halocarbon 22
RN: 75-45-6   **MP** (°C):   -146
MW: 86.47   **BP** (°C):   -40.8

Solubility (Moles/L)	Solubility (Grams/L)	Temp (°C)	Ref (#)	Evaluation (T P E A A)	Comments
3.018E-01	2.610E+01	21	M065	1 0 2 1 2	

5.   CHCl$_3$
Chloroform
Trichloromethane
Methyl Trichloride
Formyl Trichloride
RN: 67-66-3   **MP** (°C):   -63
MW: 119.38   **BP** (°C):   61

Solubility (Moles/L)	Solubility (Grams/L)	Temp (°C)	Ref (#)	Evaluation (T P E A A)	Comments
8.896E-02	1.062E+01	0	H101	2 0 0 0 2	
7.077E-02	8.448E+00	15	G029	1 0 2 2 2	
7.134E-02	8.517E+00	15	J036	1 2 0 0 2	
6.648E-02	7.937E+00	20	E019	1 0 1 1 0	
6.785E-02	8.100E+00	20	F300	1 0 0 0 1	
6.886E-02	8.220E+00	20	H101	2 0 0 0 2	
6.869E-02	8.200E+00	20	M133	1 0 0 0 2	
6.827E-02	8.150E+00	20	M368	1 0 0 0 1	
6.648E-02	7.937E+00	20	N034	1 0 0 0 0	
6.869E-02	8.200E+00	20	P046	1 0 0 0 0	
6.750E-02	8.058E+00	20	P073	1 0 0 1 2	
3.504E-02	4.182E+00	22	H072	1 0 1 1 2	
7.472E-02	8.920E+00	25	B019	1 0 1 2 0	
6.050E-02	7.222E+00	25	B173	2 0 2 2 2	
6.660E-02	7.950E+00	25	F071	1 1 2 1 2	
6.648E-02	7.937E+00	25	G056	1 0 0 0 2	
6.813E-02	8.133E+00	25	L319	1 0 2 1 2	
6.618E-02	7.900E+00	25	M037	1 1 0 0 1	
6.648E-02	7.937E+00	25	O026	1 2 0 1 0	
7.472E-02	8.920E+00	25	R321	1 2 1 1 1	
6.236E-02	7.444E+00	25.0	C055	1 2 1 0 1	
6.409E-02	7.651E+00	30	G029	1 0 2 2 2	
6.500E-02	7.760E+00	30	H101	2 0 0 0 2	
2.114E-02	2.524E+00	30	M311	1 1 2 2 2	
6.411E-02	7.653E+00	30	V009	1 0 0 0 2	
6.648E-02	7.937E+00	56.1	C055	2 2 1 0 0	
6.236E-02	7.444E+00	60	R321	1 2 1 1 1	

6.660E-02	7.950E+00	ns	H123	0 0 0 0 2	
4.168E-02	4.975E+00	ns	I306	0 0 0 0 0	
6.660E-02	7.950E+00	ns	M344	0 0 0 0 2	
6.830E-02	8.153E+00	ns	R028	0 0 0 0 2	

6.   CHI$_3$
Iodoform
Triiodomethane

RN: 75-47-8   **MP** (°C):   121.5
MW: 393.73   **BP** (°C):   218

Solubility (Moles/L)	Solubility (Grams/L)	Temp (°C)	Ref (#)	Evaluation (T P E A A)	Comments
3.000E-04	1.181E-01	25	V009	1 0 0 0 0	
2.540E-04	9.999E-02	rt	D021	0 0 1 1 0	

7.   CH$_2$BrCl
Bromochloromethane
Bromo-chloro-methane
Chlorobromomethane
CBM

RN: 74-97-5   **MP** (°C):   -86.5
MW: 129.39   **BP** (°C):   68.1

Solubility (Moles/L)	Solubility (Grams/L)	Temp (°C)	Ref (#)	Evaluation (T P E A A)	Comments
1.290E-01	1.669E+01	25	M342	1 0 1 1 2	
1.142E-01	1.478E+01	ns	O006	0 0 0 0 1	

8.   CH$_2$Br$_2$
Methylene Bromide
Dibrom-methan

RN: 74-95-3   **MP** (°C):   -52.7
MW: 173.85   **BP** (°C):   97

Solubility (Moles/L)	Solubility (Grams/L)	Temp (°C)	Ref (#)	Evaluation (T P E A A)	Comments
6.747E-02	1.173E+01	0	H101	2 0 0 0 2	
6.652E-02	1.156E+01	15	G029	1 0 2 2 2	
6.604E-02	1.148E+01	20	H101	2 0 0 0 2	
6.259E-02	1.088E+01	25	O006	1 0 0 0 1	
6.782E-02	1.179E+01	30	G029	1 0 2 2 2	
6.765E-02	1.176E+01	30	H101	2 0 0 0 2	
6.779E-02	1.179E+01	30	V009	1 0 0 0 2	
6.558E-02	1.140E+01	ns	F300	0 0 0 0 2	

9.   CH₂Cl₂

Methylene Chloride
Dichlor-methan
Dichloromethane
Methylene Dichloride
Methane Dichloride

RN: 75-09-2   **MP** (°C):   -95.1
MW: 84.93   **BP** (°C):   39.8

Solubility (Moles/L)	Solubility (Grams/L)	Temp (°C)	Ref (#)	Evaluation (T P E A A)	Comments
2.782E-01	2.363E+01	0	H101	2 0 0 0 2	
2.309E-01	1.961E+01	20	C057	1 0 0 0 2	
2.355E-01	2.000E+01	20	F300	1 0 0 0 0	
2.355E-01	2.000E+01	20	H101	2 0 0 0 2	
2.263E-01	1.922E+01	20	N034	1 0 0 0 2	
1.887E-01	1.603E+01	20	N038	1 0 0 1 2	
2.309E-01	1.961E+01	25	A094	1 0 0 0 1	
1.534E-01	1.303E+01	25	G056	1 0 0 0 2	
1.554E-01	1.320E+01	25	M037	1 1 0 0 2	
1.554E-01	1.320E+01	25	M133	1 0 0 0 2	
1.554E-01	1.320E+01	25	P046	1 0 0 0 0	
2.275E-01	1.932E+01	30	V009	1 0 0 0 2	
2.284E-01	1.940E+01	ns	H123	0 0 0 0 2	

10.   CH₂I₂

Methylene Iodide
Diiod-methan

RN: 75-11-6   **MP** (°C):   6.0
MW: 267.84   **BP** (°C):   181

Solubility (Moles/L)	Solubility (Grams/L)	Temp (°C)	Ref (#)	Evaluation (T P E A A)	Comments
3.110E-03	8.330E-01	25	A032	1 2 1 1 2	
4.624E-03	1.238E+00	30	G029	1 0 2 2 2	
4.594E-03	1.231E+00	30	V009	1 0 0 0 1	

11.   CH₂N₂

Cyanamide
Cyanamid

RN: 420-04-2   **MP** (°C):
MW: 42.04   **BP** (°C):

Solubility (Moles/L)	Solubility (Grams/L)	Temp (°C)	Ref (#)	Evaluation (T P E A A)	Comments
1.057E+01	4.444E+02	ns	N013	0 0 0 0 1	

12.   CH₃Br

Methyl Bromide
Bromomethane
Celfume

RN:	74-83-9	**MP** (°C):	-94
MW:	94.94	**BP** (°C):	3.56

Solubility (Moles/L)	Solubility (Grams/L)	Temp (°C)	Ref (#)	Evaluation (T P E A A)	Comments
2.748E-01	2.609E+01	10	H081	1 0 2 0 2	
1.893E-01	1.797E+01	17	H081	1 0 2 0 2	
1.893E-01	1.797E+01	17	M061	1 0 0 0 2	
1.933E-01	1.835E+01	19.9	G061	1 2 1 1 2	774.3mm Hg @ 25 °C
1.685E-01	1.600E+01	20	G080	1 0 0 0 1	
1.659E-01	1.575E+01	20	P081	1 0 0 0 1	
1.394E-01	1.323E+01	25	H081	1 0 2 0 2	
1.411E-01	1.340E+01	25	M161	1 0 0 0 2	
1.196E-01	1.136E+01	32	H081	1 0 2 0 2	
9.479E-03	9.000E-01	ns	N013	0 0 0 0 1	

13.   CH₃BrO₆S₂

Bromomethionic Acid
Methanedisulfonic Acid, Bromo-

RN:	187610-86-2	**MP** (°C):	
MW:	255.07	**BP** (°C):	

Solubility (Moles/L)	Solubility (Grams/L)	Temp (°C)	Ref (#)	Evaluation (T P E A A)	Comments
3.039E+00	7.752E+02	25	B077	1 2 0 0 2	

14.   CH₃Cl

Methyl Chloride
Chloromethane

RN:	74-87-3	**MP** (°C):	-97.0
MW:	50.49	**BP** (°C):	-23.7

Solubility (Moles/L)	Solubility (Grams/L)	Temp (°C)	Ref (#)	Evaluation (T P E A A)	Comments
1.531E+01	7.727E+02	0	M061	1 0 0 0 1	*sic*
1.436E-01	7.250E+00	20	M133	1 0 0 0 2	
9.069E-02	4.579E+00	20	N034	1 0 0 0 1	
1.436E-01	7.250E+00	20	P046	1 0 0 0 0	
1.059E-01	5.347E+00	24.9	G061	1 2 1 1 2	756.1mm Hg @ 25 °C
1.455E-01	7.346E+00	30	G056	1 0 0 0 2	
1.466E-01	7.400E+00	30	M037	1 1 0 0 1	

15.   CH$_3$ClO$_6$S$_2$
Chloromethionic Acid
Acide Chloromethionique
RN: 74692-14-1   **MP** (°C):
MW: 210.61   **BP** (°C):

Solubility (Moles/L)	Solubility (Grams/L)	Temp (°C)	Ref (#)	Evaluation (T P E A A)	Comments
1.540E+01	3.243E+03	25	B075	1 2 0 0 2	

16.   CH$_3$F
Fluoromethane
Methylfluoride
RN: 593-53-3   **MP** (°C):   -141.8
MW: 34.03   **BP** (°C):   -78.2

Solubility (Moles/L)	Solubility (Grams/L)	Temp (°C)	Ref (#)	Evaluation (T P E A A)	Comments
~7.05E-02	~2.40E+00	15	F300	1 0 0 0 0	
5.250E-02	1.787E+00	29.9	G061	1 2 1 1 2	766.8mm Hg @25 °C

17.   CH$_3$I
Iodomethane
Methyl-iodide
Halon 10001
Methyl Iodine
Methyliodide
RN: 74-88-4   **MP** (°C):   -64
MW: 141.94   **BP** (°C):   42

Solubility (Moles/L)	Solubility (Grams/L)	Temp (°C)	Ref (#)	Evaluation (T P E A A)	Comments
1.103E-01	1.565E+01	0	H101	2 0 0 0 2	
9.997E-02	1.419E+01	20	H101	2 0 0 0 2	
9.727E-02	1.381E+01	20	H127	1 0 0 0 1	
9.727E-02	1.381E+01	20	I316	0 0 0 0 1	
9.600E-02	1.363E+01	20	M171	1 0 0 0 2	
9.590E-02	1.361E+01	22	F001	1 0 1 2 2	
9.511E-02	1.350E+01	22	F300	1 0 0 0 2	
9.590E-02	1.361E+01	22	S006	1 0 0 0 2	
1.007E-01	1.429E+01	30	H101	2 0 0 0 2	
9.957E-02	1.413E+01	30	V009	1 0 0 0 2	
8.725E-03	1.238E+00	ns	O006	0 0 0 0 2	

18.   CH₃NO

Formaldehyde Oxime

Formaldehyd-oxim

RN: 75-17-2   **MP** (°C):

MW: 45.04   **BP** (°C):

Solubility (Moles/L)	Solubility (Grams/L)	Temp (°C)	Ref (#)	Evaluation (T P E A A)	Comments
3.774E+00	1.700E+02	20	F300	1 0 0 0 1	

19.   CH₃NO₂

Nitromethane

Nitrocarbol

NM

RN: 75-52-5   **MP** (°C):   -29

MW: 61.04   **BP** (°C):   101

Solubility (Moles/L)	Solubility (Grams/L)	Temp (°C)	Ref (#)	Evaluation (T P E A A)	Comments
1.421E+00	8.676E+01	20	C121	0 0 0 0 1	unit assumed, *sic*
1.627E+00	9.934E+01	25	F049	2 0 2 0 0	
1.802E+00	1.100E+02	25	M136	2 0 0 0 2	
1.802E+00	1.100E+02	25	M139	2 0 0 0 2	
3.039E-01	1.855E+01	ns	D348	0 0 2 2 2	

20.   CH₃N₅

5-Aminotetrazole

5-Amino-tetrazol

RN: 4418-61-5   **MP** (°C):   204

MW: 85.07   **BP** (°C):

Solubility (Moles/L)	Solubility (Grams/L)	Temp (°C)	Ref (#)	Evaluation (T P E A A)	Comments
1.411E-01	1.200E+01	18	F300	1 0 0 0 1	

21.   CH₄

Methane

Methan

RN: 74-82-8   **MP** (°C):   -183

MW: 16.04   **BP** (°C):   -161

Solubility (Moles/L)	Solubility (Grams/L)	Temp (°C)	Ref (#)	Evaluation (T P E A A)	Comments
2.468E-03	3.960E-02	0	F300	1 0 0 0 2	
2.210E-03	3.545E-02	4.99	C115	2 0 2 2 2	
1.926E-03	3.090E-02	9.99	C115	2 0 2 2 2	
1.633E-03	2.620E-02	14.99	C115	2 0 2 2 2	
1.567E-03	2.513E-02	19.8	G058	1 0 0 0 2	
1.511E-03	2.424E-02	19.99	C115	2 0 2 2 2	
1.446E-03	2.320E-02	20	F300	1 0 0 0 2	

1.381E-03	2.215E-02	24.99	C115	2 0 2 2 2
1.521E-03	2.440E-02	25	M001	2 1 2 2 2
1.521E-03	2.440E-02	25	M002	2 2 1 2 2
1.502E-03	2.410E-02	25	M040	1 0 0 1 2
1.550E-03	2.487E-02	25	M102	1 2 2 1 2
1.266E-03	2.030E-02	29.99	C115	2 0 2 2 2
1.189E-03	1.907E-02	34.99	C115	2 0 2 2 2
1.079E-03	1.732E-02	39.99	C115	2 0 2 2 2
1.056E-03	1.693E-02	40	S212	2 1 2 2 2
1.055E-03	1.693E-02	44.99	C115	2 0 2 2 2
8.477E-04	1.360E-02	50	F300	1 0 0 0 2
9.000E-04	1.444E-02	60	S212	2 1 2 2 2
8.000E-04	1.283E-02	80	S212	2 1 2 2 2
1.434E-03	2.300E-02	ns	M091	0 1 0 0 2
1.378E-03	2.210E-02	ns	S212	2 1 2 2 2

22. CH_4N_2O

Urea
Harnstoff
Uree

RN: 57-13-6   **MP** (°C):   132.7
MW: 60.06   **BP** (°C):

Solubility (Moles/L)	Solubility (Grams/L)	Temp (°C)	Ref (#)	Evaluation (T P E A A)	Comments
6.680E+00	4.012E+02	0	F300	1 0 0 0 2	
4.757E+00	2.857E+02	0	J021	1 0 0 0 2	
6.680E+00	4.012E+02	0	M043	1 0 0 0 1	
6.680E+00	4.012E+02	0	P023	1 2 1 1 2	
7.297E+00	4.382E+02	5	D041	1 0 0 0 1	
5.088E+00	3.056E+02	7	J021	1 0 0 0 2	
5.246E+00	3.151E+02	10	D020	1 2 1 1 2	
5.246E+00	3.151E+02	10	D060	2 2 1 1 2	
7.651E+00	4.595E+02	10	M043	1 0 0 0 1	
7.602E+00	4.565E+02	10	P023	1 2 1 1 2	
5.550E+00	3.333E+02	17	J021	1 0 0 0 2	
7.382E+00	4.433E+02	18.72	S131	2 2 1 1 2	recrystallized
5.536E+00	3.324E+02	20	C052	1 2 1 1 2	
5.617E+00	3.373E+02	20	J021	1 0 0 0 2	
8.529E+00	5.122E+02	20	M043	1 0 0 0 2	
8.517E+00	5.115E+02	20	P023	1 2 1 1 2	
7.594E+00	4.561E+02	21.59	S131	2 2 1 1 2	recrystallized
7.738E+00	4.647E+02	23.85	S131	2 2 1 1 2	recrystallized
5.874E+00	3.528E+02	25	D020	1 2 1 1 2	
9.058E+00	5.440E+02	25	D041	1 0 0 0 2	
5.874E+00	3.528E+02	25	D060	2 2 1 1 2	
8.326E+00	5.000E+02	25	M136	2 0 0 0 2	
7.910E+00	4.750E+02	26.83	S131	2 2 1 1 2	recrystallized

7.966E+00	4.784E+02	27.31	S131	2 2 1 1 2	recrystallized
9.566E+00	5.745E+02	30	M043	1 0 0 0 2	
9.596E+00	5.763E+02	30	P023	1 2 1 1 2	
8.171E+00	4.907E+02	30.38	S131	2 2 1 1 2	recrystallized
6.244E+00	3.750E+02	35	J021	1 0 0 0 2	
1.712E+01	1.028E+03	35	S200	1 0 0 0 2	loc. cit.
8.469E+00	5.086E+02	35.15	S131	2 2 1 1 2	recrystallized
8.465E+00	5.083E+02	35.42	S131	2 2 1 1 2	recrystallized
8.575E+00	5.150E+02	37.36	S131	2 2 1 1 2	recrystallized
1.038E+01	6.232E+02	39.7	P023	1 2 1 1 2	
6.392E+00	3.839E+02	40	D020	1 2 1 1 2	
6.392E+00	3.839E+02	40	D060	2 2 1 1 2	
1.037E+01	6.226E+02	40	M043	1 0 0 0 2	
1.837E+01	1.103E+03	40	S200	1 0 0 0 2	loc. cit.
8.822E+00	5.298E+02	41.11	S131	2 2 1 1 2	recrystallized
8.982E+00	5.394E+02	43.85	S131	2 2 1 1 2	recrystallized
8.967E+00	5.386E+02	43.94	S131	2 2 1 1 2	recrystallized
1.961E+01	1.178E+03	45	S200	1 0 0 0 2	loc. cit.
9.107E+00	5.469E+02	46.56	S131	2 2 1 1 2	recrystallized
1.119E+01	6.721E+02	50	P023	1 2 1 1 2	
2.109E+01	1.267E+03	50	S200	1 0 0 0 2	loc. cit.
1.122E+01	6.736E+02	50.6	P023	1 2 1 1 2	
9.560E+00	5.742E+02	54.77	S131	2 2 1 1 2	recrystallized
9.584E+00	5.756E+02	54.97	S131	2 2 1 1 2	recrystallized
2.283E+01	1.371E+03	55	S200	1 0 0 0 2	loc. cit.
9.649E+00	5.795E+02	55.88	S131	2 2 1 1 2	recrystallized
9.681E+00	5.814E+02	57.02	S131	2 2 1 1 2	recrystallized
9.806E+00	5.889E+02	59.13	S131	2 2 1 1 2	recrystallized
6.936E+00	4.166E+02	60	J021	1 0 0 0 2	
9.847E+00	5.914E+02	60	K013	1 0 1 1 2	
1.189E+01	7.143E+02	60	M043	1 0 0 0 2	
2.422E+01	1.455E+03	60	S200	1 0 0 0 2	loc. cit.
1.184E+01	7.110E+02	60.0	P023	1 2 1 1 2	
9.930E+00	5.963E+02	61.76	S131	2 2 1 1 2	recrystallized
1.005E+01	6.034E+02	63.79	S131	2 2 1 1 2	recrystallized
1.009E+01	6.060E+02	65	K013	1 0 1 1 2	
2.570E+01	1.543E+03	65	S200	1 0 0 0 2	loc. cit.
1.244E+01	7.468E+02	68.5	P023	1 2 1 1 2	
1.020E+01	6.127E+02	68.50	M059	1 1 2 1 2	
1.270E+01	7.629E+02	70	F300	1 0 0 0 2	
7.206E+00	4.328E+02	70	J021	1 0 0 0 2	
1.033E+01	6.206E+02	70	K013	1 0 1 1 2	
1.263E+01	7.588E+02	70	P023	1 2 1 1 2	
2.730E+01	1.640E+03	70	S200	1 0 0 0 2	loc. cit.
1.038E+01	6.231E+02	70.49	S131	2 2 1 1 2	recrystallized
1.048E+01	6.295E+02	73.11	S131	2 2 1 1 2	recrystallized
1.057E+01	6.345E+02	75	K013	1 0 1 1 2	
1.048E+01	6.296E+02	75.30	M059	1 1 2 1 2	
1.079E+01	6.480E+02	80	K013	1 0 1 1 2	
1.332E+01	8.000E+02	80	M043	1 0 0 0 2	

1.090E+01	6.546E+02	84.40	M059	1 1 2 1 2	
1.101E+01	6.610E+02	85	K013	1 0 1 1 2	
3.229E+01	1.939E+03	85	S200	1 0 0 0 2	loc. cit.
1.122E+01	6.738E+02	90	K013	1 0 1 1 2	
3.426E+01	2.058E+03	90	S200	1 0 0 0 2	loc. cit.
1.131E+01	6.791E+02	93.80	M059	1 1 2 1 2	
1.142E+01	6.858E+02	95	K013	1 0 1 1 2	
3.611E+01	2.169E+03	95	S200	1 0 0 0 2	loc. cit.
1.161E+01	6.975E+02	100	K013	1 0 1 1 2	
1.465E+01	8.795E+02	100	M043	1 0 0 0 2	
3.778E+01	2.269E+03	100	S200	1 0 0 0 2	loc. cit.
1.177E+01	7.066E+02	104.40	M059	1 1 2 1 2	
1.199E+01	7.199E+02	109.90	M059	1 1 2 1 2	
1.219E+01	7.321E+02	115.30	M059	1 1 2 1 2	
1.229E+01	7.383E+02	118.30	M059	1 1 2 1 2	
1.234E+01	7.411E+02	118.70	M059	1 1 2 1 2	
1.245E+01	7.479E+02	121.90	M059	1 1 2 1 2	
1.249E+01	7.503E+02	123.20	M059	1 1 2 1 2	
1.264E+01	7.592E+02	127.50	M059	1 1 2 1 2	
1.269E+01	7.619E+02	128.80	M059	1 1 2 1 2	
1.281E+01	7.694E+02	132.60	M059	1 1 2 1 2	
1.665E+01	1.000E+03	ns	B338	0 0 0 0 1	
1.332E+01	8.000E+02	ns	D072	0 0 0 0 0	

23. CH_4N_2S

Thiourea
Thiouree

RN: 62-56-6   **MP** (°C):   176
MW: 76.12   **BP** (°C):

Solubility (Moles/L)	Solubility (Grams/L)	Temp (°C)	Ref (#)	Evaluation (T P E A A)	Comments
6.136E-01	4.671E+01	0	M043	1 0 0 0 1	
9.731E-01	7.407E+01	10	M043	1 0 0 0 1	
1.118E+00	8.507E+01	10	O017	1 0 1 1 2	
1.206E+00	9.180E+01	13	F300	1 0 0 0 2	
1.206E+00	9.179E+01	13	O019	1 0 0 1 2	
1.383E+00	1.053E+02	15	O017	1 0 1 1 2	
1.573E+00	1.197E+02	20	M043	1 0 0 0 2	
1.544E+00	1.175E+02	20	O017	1 0 1 1 2	
1.085E+00	8.257E+01	25	I310	0 0 0 0 0	
1.759E+00	1.339E+02	25	O017	1 0 1 1 2	
2.199E+00	1.674E+02	30	M043	1 0 0 0 2	
3.093E+00	2.355E+02	40	M043	1 0 0 0 2	
5.455E+00	4.152E+02	60	M043	1 0 0 0 1	
7.617E+00	5.798E+02	80	M043	1 0 0 0 2	
9.250E+00	7.041E+02	100	M043	1 0 0 0 2	
7.882E-01	6.000E+01	ns	D072	0 0 0 0 0	

24.   CH₄N₄O₂

α-Nitroguanidine
Nitroguanidine
Nitroguanidin

RN: 556-88-7   **MP** (°C):   235
MW: 104.07   **BP** (°C):

Solubility (Moles/L)	Solubility (Grams/L)	Temp (°C)	Ref (#)	Evaluation (T P E A A)	Comments
2.597E-02	2.703E+00	19.5	D027	1 2 0 0 2	
1.173E-01	1.221E+01	25	D022	1 1 2 2 2	
4.228E-02	4.400E+00	25	F300	1 0 0 0 1	
4.305E-02	4.480E+00	29.87	M028	1 2 2 1 0	EFG
1.122E-01	1.167E+01	50	D027	1 2 0 0 2	
3.070E-01	3.195E+01	71.67	M028	1 2 2 1 0	EFG
5.695E-01	5.927E+01	83.98	M028	1 2 2 1 0	EFG
9.025E-01	9.392E+01	100	D027	1 2 0 0 2	
7.620E-01	7.930E+01	100	F300	1 0 0 0 2	

25.   CH₄O

Methanol
Methyl Alcohol

RN: 67-56-1   **MP** (°C):   -97.8
MW: 32.04   **BP** (°C):   64.7

Solubility (Moles/L)	Solubility (Grams/L)	Temp (°C)	Ref (#)	Evaluation (T P E A A)	Comments
1.689E+01	5.411E+02	ns	L003	0 0 2 1 2	

26.   CH₄O₆S₂

Methionic Acid
Acide Methionique
Methanedisulfonic Acid

RN: 503-40-2   **MP** (°C):   98.0
MW: 176.17   **BP** (°C):

Solubility (Moles/L)	Solubility (Grams/L)	Temp (°C)	Ref (#)	Evaluation (T P E A A)	Comments
1.395E+01	2.458E+03	25	B075	1 2 0 0 2	
4.035E+00	7.108E+02	25	B076	1 2 0 0 2	
4.862E+00	8.566E+02	25	F300	1 0 0 0 2	

27.   CH$_4$O$_6$S$_2$.H$_2$O
Methionic Acid (Monohydrate)
RN: 503-40-2   MP (°C):
MW: 194.18   BP (°C):

Solubility (Moles/L)	Solubility (Grams/L)	Temp (°C)	Ref (#)	Evaluation (T P E A A)	Comments
4.409E+00	8.562E+02	25	B076	1 2 0 0 2	

28.   CH$_5$N
Methylamine
Aminomethane
Carbinamine
Mercurialin
RN: 74-89-5   MP (°C):   -93.5
MW: 31.06   BP (°C):   -6.3

Solubility (Moles/L)	Solubility (Grams/L)	Temp (°C)	Ref (#)	Evaluation (T P E A A)	Comments
1.906E+01	5.920E+02	4.50	F300	1 0 0 0 2	
2.963E+01	9.202E+02	12.5	D041	1 0 0 0 2	
2.147E+01	6.667E+02	12.50	M081	1 0 0 0 2	
1.916E+01	5.951E+02	20	M081	1 0 0 0 2	
1.789E+01	5.556E+02	25	M081	1 0 0 0 2	
1.664E+01	5.169E+02	30	M081	1 0 0 0 2	
1.380E+01	4.286E+02	40	M081	1 0 0 0 1	
1.143E+01	3.548E+02	50	M081	1 0 0 0 1	
9.034E+00	2.806E+02	60	M081	1 0 0 0 1	

29.   CH$_5$N$_5$O$_2$
Nitroaminoguanidine
Hydrazinecarboximidamide, N-Nitro-
1-Amino-3-nitroguanidine
3-Amino-1-nitroguanidine
1-Amino-2-nitroguanidine
1-Nitro-3-aminoguanidine
RN: 18264-75-0   MP (°C):   185
MW: 119.08   BP (°C):

Solubility (Moles/L)	Solubility (Grams/L)	Temp (°C)	Ref (#)	Evaluation (T P E A A)	Comments
1.360E-02	1.619E+00	9.33	M047	2 2 1 1 0	EFG
2.254E-02	2.684E+00	20.96	M047	2 2 1 1 0	EFG
3.567E-02	4.248E+00	29.87	M047	2 2 1 1 0	EFG
4.384E-02	5.221E+00	34.53	M047	2 2 1 1 0	EFG
7.087E-02	8.440E+00	44.30	M047	2 2 1 1 0	EFG
9.318E-02	1.110E+01	49.42	M047	2 2 1 1 0	EFG

30.   CH$_5$O$_3$As
Methanearsonic Acid
MAA
Methylarsonsaeure

RN: 124-58-3   **MP** (°C):   132
MW: 139.97   **BP** (°C):

Solubility (Moles/L)	Solubility (Grams/L)	Temp (°C)	Ref (#)	Evaluation (T P E A A)	Comments
1.456E+00	2.038E+02	20	B200	1 0 0 0 2	
1.563E+00	2.188E+02	25	D305	1 0 0 0 1	

31.   CH$_5$As
Methylarsine
Methylarsin

RN: 593-52-2   **MP** (°C):   -143
MW: 91.97   **BP** (°C):   2

Solubility (Moles/L)	Solubility (Grams/L)	Temp (°C)	Ref (#)	Evaluation (T P E A A)	Comments
9.242E-04	8.500E-02	20	F300	1 0 0 0 1	

32.   CBrClF$_2$
Bromochlorodifluoromethane
Halon 1211
Chlorodifluorobromomethane
Bromochlorodifluoromethine

RN: 353-59-3   **MP** (°C):
MW: 165.37   **BP** (°C):

Solubility (Moles/L)	Solubility (Grams/L)	Temp (°C)	Ref (#)	Evaluation (T P E A A)	Comments
9.555E-05	1.580E-02	0	G055	1 2 2 2 1	

33.   CBr$_3$F
Tribromo-fluoro-methane
Methane, Tribromofluoro-
Fluorotribromomethane

RN: 353-54-8   **MP** (°C):
MW: 270.74   **BP** (°C):

Solubility (Moles/L)	Solubility (Grams/L)	Temp (°C)	Ref (#)	Evaluation (T P E A A)	Comments
1.477E-03	3.998E-01	25	O006	1 0 0 0 1	

34.   CBr$_4$
Carbon Tetrabromide
Tetrabromomethane
RN: 558-13-4   **MP** (°C):   89
MW: 331.65   **BP** (°C):

Solubility (Moles/L)	Solubility (Grams/L)	Temp (°C)	Ref (#)	Evaluation (T P E A A)	Comments
7.235E-04	2.399E-01	30	G029	1 0 2 2 1	
6.998E-04	2.321E-01	30	V009	1 0 0 0 0	

35.   CClN
Cyanogen Chloride
Chlorcyan
RN: 506-77-4   **MP** (°C):   -6
MW: 61.47   **BP** (°C):   13.8

Solubility (Moles/L)	Solubility (Grams/L)	Temp (°C)	Ref (#)	Evaluation (T P E A A)	Comments
9.761E-01	6.000E+01	0	F300	1 0 0 0 0	

36.   CClN$_3$O$_6$
Chlorotrinitromethane
Chlor-trinitro-methan
RN: 1943-16-4   **MP** (°C):
MW: 185.48   **BP** (°C):

Solubility (Moles/L)	Solubility (Grams/L)	Temp (°C)	Ref (#)	Evaluation (T P E A A)	Comments
1.186E-02	2.200E+00	20	F300	1 0 0 0 1	

37.   CCl$_2$F$_2$
Dichlorodifluoromethane
Difluorodichloromethane
Freon 12
RN: 75-71-8   **MP** (°C):   -158
MW: 120.91   **BP** (°C):   -29.8

Solubility (Moles/L)	Solubility (Grams/L)	Temp (°C)	Ref (#)	Evaluation (T P E A A)	Comments
1.544E-02	1.867E+00	21	M065	1 0 2 1 2	
2.316E-03	2.800E-01	25	M133	1 0 0 0 2	
2.316E-03	2.800E-01	25	P046	1 0 0 0 0	
2.315E-03	2.799E-01	25	R048	1 0 0 0 1	

38.   CCl₃F
Trichlorofluoromethane
Fluorotrichloromethane
Freon 11

RN: 75-69-4   **MP** (°C):   -111
MW: 137.37   **BP** (°C):   23.7

Solubility (Moles/L)	Solubility (Grams/L)	Temp (°C)	Ref (#)	Evaluation (T P E A A)	Comments
1.020E-02	1.401E+00	20	H041	1 0 0 0 2	
8.008E-03	1.100E+00	20	M133	1 0 0 0 2	
8.008E-03	1.100E+00	20	P046	1 0 0 0 0	
1.020E-02	1.401E+00	21	H041	1 0 0 0 2	
8.013E-03	1.101E+00	25	H041	1 0 0 0 2	
7.999E-03	1.099E+00	25	R048	1 0 0 0 1	
7.997E-03	1.099E+00	27	H041	1 0 0 0 2	
7.853E-03	1.079E+00	30	H041	1 0 0 0 2	
9.892E-03	1.359E+00	31	H041	1 0 0 0 2	
4.152E-03	5.703E-01	50	H041	1 0 0 0 2	
2.258E-03	3.102E-01	75	H041	1 0 0 0 2	

39.   CCl₃NO₂
Chloropicrin
Chlorpikrin

RN: 76-06-2   **MP** (°C):   -64
MW: 164.38   **BP** (°C):   112

Solubility (Moles/L)	Solubility (Grams/L)	Temp (°C)	Ref (#)	Evaluation (T P E A A)	Comments
1.381E-02	2.270E+00	0	M161	1 0 0 0 2	
1.396E-02	2.295E+00	20	C121	1 0 0 0 1	unit assumed, *sic*
1.186E-02	1.950E+00	20	G080	1 0 0 0 1	
9.718E-03	1.597E+00	20	M061	1 0 0 0 1	
1.214E-02	1.996E+00	20	P081	1 0 0 0 0	
9.874E-03	1.623E+00	25	F300	1 0 0 0 2	
1.217E-02	2.000E+00	ns	N013	0 0 0 0 2	

40.   CCl₄
Carbon Tetrachloride
Tetrachloromethane
Methane Tetrachloride

RN: 56-23-5   **MP** (°C):   -23
MW: 153.82   **BP** (°C):   76.7

Solubility (Moles/L)	Solubility (Grams/L)	Temp (°C)	Ref (#)	Evaluation (T P E A A)	Comments
6.306E-03	9.700E-01	0	H101	2 0 0 0 1	
5.002E-03	7.694E-01	15	G029	1 0 2 2 1	
5.002E-03	7.694E-01	15	J036	1 2 0 0 2	
5.197E-03	7.994E-01	20	C121	1 0 0 0 0	unit assumed, *sic*

5.201E-03	8.000E-01	20	H101	2 0 0 0 1	
5.201E-03	8.000E-01	20	M040	1 0 0 1 2	
5.103E-03	7.850E-01	20	M133	1 0 0 0 2	
5.200E-03	7.999E-01	20	M312	1 0 0 0 2	
4.612E-03	7.095E-01	20	N038	1 0 0 1 2	
5.103E-03	7.850E-01	20	P046	1 0 0 0 0	
6.494E-03	9.990E-01	25	B019	1 0 1 2 0	
4.920E-03	7.568E-01	25	B173	2 0 2 2 2	
5.000E-03	7.691E-01	25	G038	1 2 2 2 1	
5.000E-03	7.691E-01	25	G053	2 1 2 1 1	
5.197E-03	7.994E-01	25	G056	1 0 0 0 2	
5.197E-03	7.994E-01	25	L319	1 0 2 1 1	
5.201E-03	8.000E-01	25	M037	1 1 0 0 0	
5.197E-03	7.994E-01	25	M061	1 0 0 0 0	
1.820E-03	2.800E-01	25	M161	1 0 0 0 1	
5.006E-03	7.700E-01	25	M368	1 0 0 0 1	
1.038E-02	1.597E+00	25	N034	1 0 0 0 1	*sic*
5.556E-03	8.546E-01	25	S133	1 1 1 1 1	
5.262E-03	8.093E-01	30	G029	1 0 2 2 1	
5.526E-03	8.500E-01	30	H101	2 0 0 0 1	
5.296E-03	8.146E-01	30	V009	1 0 0 0 1	
5.201E-03	8.000E-01	ns	F071	0 1 2 1 2	
5.201E-03	8.000E-01	ns	H080	0 0 0 0 2	
3.249E-03	4.998E-01	ns	I306	0 0 0 0 0	
5.201E-03	8.000E-01	ns	M344	0 0 0 0 2	

41. CF_4
Carbon Tetrafluoride
Tetrafluoromethane

RN: 75-73-0   **MP** (°C):   -184
MW: 88.00   **BP** (°C):   -128

Solubility (Moles/L)	Solubility (Grams/L)	Temp (°C)	Ref (#)	Evaluation (T P E A A)	Comments
2.319E-04	2.041E-02	19.99	C115	2 0 2 2 2	
2.083E-04	1.833E-02	24.99	C115	2 0 2 2 2	
2.111E-04	1.858E-02	25	D055	1 0 0 0 1	
1.940E-04	1.707E-02	29.99	C115	2 0 2 2 2	

42.   COS
Carbonyl Sulfide
Kohlenoxidsulfid
RN: 463-58-1   **MP** (°C):   -138
MW: 60.07   **BP** (°C):   -50

Solubility (Moles/L)	Solubility (Grams/L)	Temp (°C)	Ref (#)	Evaluation (T P E A A)	Comments
6.259E-02	3.760E+00	0	F300	1 0 0 0 2	
2.081E-02	1.250E+00	25	F300	1 0 0 0 2	

43.   CO₂
Carbon Dioxide
Carbonic Acid Gas
Carbonic Anhydride
RN: 124-38-9   **MP** (°C):   -57
MW: 44.01   **BP** (°C):

Solubility (Moles/L)	Solubility (Grams/L)	Temp (°C)	Ref (#)	Evaluation (T P E A A)	Comments
8.641E-02	3.803E+00	16	B109	1 0 0 0 2	unit assumed, *sic*
8.377E-02	3.687E+00	17	B109	1 0 0 0 2	unit assumed, *sic*
8.641E-02	3.803E+00	18	B109	1 0 0 0 2	unit assumed, *sic*
8.123E-02	3.575E+00	18	B109	1 0 0 0 2	unit assumed, *sic*
7.886E-02	3.471E+00	19	B109	1 0 0 0 2	unit assumed, *sic*
7.654E-02	3.369E+00	20	B109	1 0 0 0 2	unit assumed, *sic*
7.432E-02	3.271E+00	21	B109	1 0 0 0 2	unit assumed, *sic*
7.427E-02	3.269E+00	21	B109	1 0 0 0 2	unit assumed, *sic*
7.213E-02	3.174E+00	22	B109	1 0 0 0 2	unit assumed, *sic*
6.582E-02	2.897E+00	25	B109	1 0 0 0 2	unit assumed, *sic*
3.360E-02	1.479E+00	25	H124	1 0 0 1 2	
6.204E-02	2.730E+00	27	B109	1 0 0 0 2	unit assumed, *sic*
6.127E-02	2.696E+00	28	B109	1 0 0 0 2	unit assumed, *sic*
5.714E-02	2.515E+00	30	B109	1 0 0 0 2	unit assumed, *sic*

44.   CS₂
Carbon Disulfide
Carbon Disulphide
Schwefelkohlenstoff
RN: 75-15-0   **MP** (°C):   -112
MW: 76.14   **BP** (°C):   46.5

Solubility (Moles/L)	Solubility (Grams/L)	Temp (°C)	Ref (#)	Evaluation (T P E A A)	Comments
2.679E-02	2.040E+00	0	F300	1 0 0 0 2	
3.257E-02	2.480E+00	0	H101	2 0 0 0 2	
2.883E-02	2.195E+00	20	C121	0 0 0 0 1	unit assumed, *sic*
2.351E-02	1.790E+00	20	F300	1 0 0 0 2	
2.850E-02	2.170E+00	20	G080	1 0 0 0 1	
2.844E-02	2.165E+00	20	M061	1 0 0 0 2	

3.850E-02	2.931E+00	20	N038	1 0 0 1 2	
2.889E-02	2.200E+00	22	P076	1 2 1 1 1	
3.746E-02	2.852E+00	25	L319	1 0 2 1 1	
2.036E-02	1.550E+00	30	F300	1 0 0 0 2	
2.889E-02	2.200E+00	32	M161	1 0 0 0 1	
2.627E-02	2.000E+00	ns	N013	0 0 0 0 2	

45.   C$_2$HBrClF$_3$
Halothane
2-Bromo-2-chloro-1,1,1-trifluoroethane
Fluothane
RN: 151-67-7   **MP** (°C):   <25
MW: 197.39   **BP** (°C):   50.2

Solubility (Moles/L)	Solubility (Grams/L)	Temp (°C)	Ref (#)	Evaluation (T P E A A)	Comments
1.742E-02	3.438E+00	ns	R028	0 0 0 0 2	

46.   C$_2$HCl$_3$
Trichloroethylene
Trichloroethene
Trichloro-ethylene
Ethinyl Trichloride
Acetylene Trichloride
1,1,2-Trichloroethylene
RN: 79-01-6   **MP** (°C):   -87
MW: 131.39   **BP** (°C):   86.7

Solubility (Moles/L)	Solubility (Grams/L)	Temp (°C)	Ref (#)	Evaluation (T P E A A)	Comments
8.372E-03	1.100E+00	20	M133	1 0 0 0 2	
9.654E-03	1.268E+00	20	P041	1 0 0 0 1	
8.372E-03	1.100E+00	20	P046	1 0 0 0 0	
7.603E-03	9.990E-01	25	A094	1 0 0 0 1	
1.120E-02	1.472E+00	25	B173	2 0 2 2 2	
8.363E-03	1.099E+00	25	G056	1 0 0 0 2	
8.372E-03	1.100E+00	25	M037	1 1 0 0 1	
1.040E-02	1.366E+00	25	M342	1 0 1 1 2	
8.372E-03	1.100E+00	25	M368	1 0 0 0 1	
8.363E-03	1.099E+00	25	N034	1 0 0 0 1	
3.032E-02	3.984E+00	25	N309	1 0 0 0 1	*sic*
5.656E-03	7.431E-01	30	M311	1 1 2 2 2	
9.274E-03	1.219E+00	37	P041	1 0 0 0 1	
8.363E-03	1.099E+00	ns	O006	0 0 0 0 1	

47.   C$_2$HCl$_3$O.H$_2$O
Chloral (Monhydrate)
Chloral-hydrat
RN: 302-17-0   **MP** (°C):   57.0
MW: 165.40   **BP** (°C):

Solubility (Moles/L)	Solubility (Grams/L)	Temp (°C)	Ref (#)	Evaluation (T P E A A)	Comments
2.056E+00	3.400E+02	0	F300	1 0 0 0 2	
4.837E+00	8.000E+02	11.30	F300	1 0 0 0 2	
5.629E+00	9.310E+02	38.10	F300	1 0 0 0 2	
4.794E+00	7.930E+02	rt	D021	0 0 1 1 2	

48.   C$_2$HCl$_3$O$_2$
Trichloroacetic Acid
TCA
RN: 76-03-9   **MP** (°C):   57.5
MW: 163.39   **BP** (°C):   196.5

Solubility (Moles/L)	Solubility (Grams/L)	Temp (°C)	Ref (#)	Evaluation (T P E A A)	Comments
3.338E+00	5.455E+02	25	B185	1 0 0 0 2	
5.685E+00	9.289E+02	25	B200	1 0 0 0 2	
2.146E+00	3.506E+02	25	F018	1 0 0 0 1	
4.024E+00	6.575E+02	25	K040	1 2 1 2 2	
1.000E+01	1.634E+03	ns	M163	0 0 0 0 0	EFG
2.146E+00	3.506E+02	ns	N013	0 0 0 0 1	

49.   C$_2$HCl$_5$
Pentachloroethane
Pentachloro-ethane
Pentalin
Pentachlorethane
Ethane Pentachloride
RN: 76-01-7   **MP** (°C):   -29
MW: 202.30   **BP** (°C):   161

Solubility (Moles/L)	Solubility (Grams/L)	Temp (°C)	Ref (#)	Evaluation (T P E A A)	Comments
2.322E-03	4.698E-01	20	V009	1 0 0 0 1	
2.470E-03	4.998E-01	25	G056	1 0 0 0 2	
2.472E-03	5.000E-01	25	M037	1 1 0 0 1	
2.373E-03	4.800E-01	ns	H123	0 0 0 0 2	
2.322E-03	4.698E-01	ns	O006	0 0 0 0 1	

50. C_2H_2
Acetylene
Acetylen

RN: 74-86-2   **MP** (°C):   -81
MW: 26.04   **BP** (°C):

Solubility (Moles/L)	Solubility (Grams/L)	Temp (°C)	Ref (#)	Evaluation (T P E A A)	Comments
7.796E-02	2.030E+00	0	F300	1 0 0 0 2	*sic*
4.609E-02	1.200E+00	20	F300	1 0 0 0 2	*sic*
1.862E+01	4.848E+02	25	M101	1 0 0 0 2	
1.959E-02	5.100E-01	60	F300	1 0 0 0 1	*sic*

51. $C_2H_2Br_4$
sym-Tetrabromoethane
1,1,2,2-Tetrabrom-aethan
Acetylene Tetrabromide
1,1,2,2-Tetrabromoethane
Tetrabromoacetylene

RN: 79-27-6   **MP** (°C):   0
MW: 345.67   **BP** (°C):   151

Solubility (Moles/L)	Solubility (Grams/L)	Temp (°C)	Ref (#)	Evaluation (T P E A A)	Comments
1.880E-03	6.500E-01	30	F300	1 0 0 0 1	
1.879E-03	6.496E-01	30	O006	1 0 0 0 1	

52. $C_2H_2Cl_2$
cis-Acetylene Dichloride
cis-1,2-Dichloroethylene
cis-Dichlorethylene

RN: 156-59-2   **MP** (°C):   -80
MW: 96.94   **BP** (°C):   60

Solubility (Moles/L)	Solubility (Grams/L)	Temp (°C)	Ref (#)	Evaluation (T P E A A)	Comments
3.610E-02	3.500E+00	25	M037	1 1 0 0 1	

53. $C_2H_2Cl_2$
Vinylidene Chloride
1,1-Dichloroethylene
RN: 75-35-4   **MP** (°C): -122.0
MW: 96.94   **BP** (°C): 31.7

Solubility (Moles/L)	Solubility (Grams/L)	Temp (°C)	Ref (#)	Evaluation (T P E A A)	Comments
2.470E-02	2.394E+00	15	D086	1 0 2 2 1	
2.624E-02	2.544E+00	17	D086	1 0 2 2 2	
4.126E-03	4.000E-01	20	M133	1 0 0 0 2	
4.126E-03	4.000E-01	20	P046	1 0 0 0 0	
2.572E-02	2.494E+00	20.5	D086	1 0 2 2 1	
2.316E-02	2.245E+00	25	D086	1 0 2 2 2	
2.470E-02	2.394E+00	28.5	D086	1 0 2 2 1	
2.624E-02	2.544E+00	29.5	D086	1 0 2 2 2	
2.302E-02	2.232E+00	30	M311	1 1 2 2 2	
2.264E-02	2.195E+00	38.5	D086	1 0 2 2 1	
2.162E-02	2.096E+00	45	D086	1 0 2 2 1	
2.367E-02	2.295E+00	51	D086	1 0 2 2 1	
2.162E-02	2.096E+00	55	D086	1 0 2 2 1	
2.470E-02	2.394E+00	60	D086	1 0 2 2 1	
2.316E-02	2.245E+00	65	D086	1 0 2 2 2	
3.034E-02	2.941E+00	71	D086	1 0 2 2 2	
2.572E-02	2.494E+00	74.5	D086	1 0 2 2 1	
3.034E-02	2.941E+00	81	D086	1 0 2 2 2	
3.803E-02	3.686E+00	85.5	D086	1 0 2 2 1	
3.598E-02	3.488E+00	90.5	D086	1 0 2 2 1	

54. $C_2H_2Cl_2$
trans-Acetylene Dichloride
trans-1,2-Dichloroethylene
trans-Dichlorethylene
RN: 156-60-5   **MP** (°C): -50
MW: 96.94   **BP** (°C): 48

Solubility (Moles/L)	Solubility (Grams/L)	Temp (°C)	Ref (#)	Evaluation (T P E A A)	Comments
6.499E-02	6.300E+00	25	M037	1 1 0 0 1	

55. $C_2H_2Cl_3As$
Chlorovinyldichloroarsine
Chlorvinylarsin-dichlorid
RN: 541-25-3   **MP** (°C):
MW: 207.32   **BP** (°C):

Solubility (Moles/L)	Solubility (Grams/L)	Temp (°C)	Ref (#)	Evaluation (T P E A A)	Comments
2.412E-03	5.000E-01	20	F300	1 0 0 0 0	

56.   C₂H₂Cl₄
1,1,2,2-Tetrachloroethane
sym-Tetrachloroethane
RN: 79-34-5   **MP** (°C):   -36
MW: 167.85   **BP** (°C):   146.5

Solubility (Moles/L)	Solubility (Grams/L)	Temp (°C)	Ref (#)	Evaluation (T P E A A)	Comments
1.924E-02	3.230E+00	20	C094	1 0 0 0 2	
1.758E-02	2.951E+00	23.5	S171	2 1 2 2 2	
1.770E-02	2.971E+00	25	B173	2 0 2 2 2	
1.782E-02	2.991E+00	25	F050	1 0 0 0 0	
1.728E-02	2.900E+00	25	M037	1 1 0 0 1	
1.737E-02	2.915E+00	30	M311	1 1 2 2 2	

57.   C₂H₂Cl₄
1,1,1,2-Tetrachloroethane
Ethane, 1,1,1,2-Tetrachloro-
F 130α
TCA
HCC 130α
RN: 630-20-6   **MP** (°C):   -44
MW: 167.85   **BP** (°C):

Solubility (Moles/L)	Solubility (Grams/L)	Temp (°C)	Ref (#)	Evaluation (T P E A A)	Comments
7.141E-03	1.199E+00	0	V009	1 0 0 0 2	
6.487E-03	1.089E+00	20	V009	1 0 0 0 2	
1.723E-02	2.892E+00	25	G056	1 0 0 0 2	
<1.66E-02	<2.79E+00	25.50	O005	2 0 2 2 1	
6.843E-03	1.149E+00	35	V009	1 0 0 0 2	
7.438E-03	1.248E+00	50	V009	1 0 0 0 2	

58.   C₂H₂O₄
Oxalic Acid
Oxalsaeure
RN: 144-62-7   **MP** (°C):   189
MW: 90.04   **BP** (°C):

Solubility (Moles/L)	Solubility (Grams/L)	Temp (°C)	Ref (#)	Evaluation (T P E A A)	Comments
3.683E-01	3.316E+01	0	C066	1 0 1 1 2	
3.665E-01	3.300E+01	0	L041	1 0 0 1 1	
3.756E-01	3.382E+01	0	M043	1 0 0 0 1	
4.907E-01	4.418E+01	4.99	A339	2 0 2 2 2	
5.912E-01	5.323E+01	9.99	A339	2 0 2 2 2	
6.287E-01	5.660E+01	10	M043	1 0 0 0 1	
7.752E-01	6.979E+01	14.99	A339	2 0 2 2 2	

7.441E-01	6.700E+01	15	F066	2 2 2 2 1
7.464E-01	6.720E+01	15	F300	1 0 0 0 2
7.775E-01	7.000E+01	15	L041	1 0 0 1 1
9.468E-01	8.524E+01	19.99	A339	2 0 2 2 2
9.219E-01	8.300E+01	20	F066	2 2 2 2 1
9.219E-01	8.300E+01	20	F300	1 0 0 0 1
9.552E-01	8.600E+01	20	L041	1 0 0 1 1
9.636E-01	8.676E+01	20	M043	1 0 0 0 1
8.836E-01	7.956E+01	20	M171	1 0 0 0 1
1.146E+00	1.032E+02	24.99	A339	2 0 2 2 2
1.088E+00	9.800E+01	25	F066	2 2 2 2 1
1.378E+00	1.240E+02	25	F317	2 1 1 1 2
2.480E+00	2.233E+02	25	H084	1 0 0 0 2
2.409E+00	2.169E+02	25	K040	1 0 2 1 2
1.317E+00	1.186E+02	29.99	A339	2 0 2 2 2
1.407E+00	1.266E+02	30	M043	1 0 0 0 2
1.623E+00	1.461E+02	34.99	A339	2 0 2 2 2
1.710E+00	1.540E+02	35	L041	1 0 0 1 2
1.903E+00	1.713E+02	39.99	A339	2 0 2 2 2
1.973E+00	1.776E+02	40	M043	1 0 0 0 2
2.199E+00	1.979E+02	44.99	A339	2 0 2 2 2
2.527E+00	2.275E+02	49.99	A339	2 0 2 2 2
2.150E+00	1.935E+02	50	C066	1 0 1 1 2
2.821E+00	2.540E+02	50	L041	1 0 0 1 2
2.867E+00	2.581E+02	54.99	A339	2 0 2 2 2
3.121E+00	2.810E+02	59.99	A339	2 0 2 2 2
3.410E+00	3.070E+02	60	M043	1 0 0 0 2
3.661E+00	3.296E+02	64.99	A339	2 0 2 2 2
4.121E+00	3.710E+02	65	L041	1 0 0 1 2
3.583E+00	3.226E+02	80	C066	1 0 1 1 2
5.084E+00	4.577E+02	80	M043	1 0 0 0 2
6.059E+00	5.455E+02	90	F300	1 0 0 0 2

59. $C_2H_2O_4.2H_2O$
Oxalic Acid Dihydrate
Ethanedioic Acid, Dihydrate
RN: 6153-56-6   **MP** (°C):   101
MW: 126.07   **BP** (°C):

Solubility (Moles/L)	Solubility (Grams/L)	Temp (°C)	Ref (#)	Evaluation (T P E A A)	Comments
1.443E-02	1.820E+00	23	C038	2 2 2 2 0	EFG, 0.1N HCl
1.070E-02	1.349E+00	30	C038	2 2 2 2 0	EFG, 0.1N HCl
7.234E-03	9.120E-01	35	C038	2 2 2 2 0	EFG, 0.1N HCl

60.   C₂H₃Br₃O

2,2,2-Tribromoethanol
2,2,2-Tribrom-aethanol

RN: 75-80-9   **MP** (°C):   80
MW: 282.77   **BP** (°C):   92

Solubility (Moles/L)	Solubility (Grams/L)	Temp (°C)	Ref (#)	Evaluation (T P E A A)	Comments
1.206E-01	3.410E+01	40	F300	1 0 0 0 2	

61.   C₂H₃Cl

Vinyl Chloride
Chloroethylene

RN: 75-01-4   **MP** (°C):   -153.0
MW: 62.50   **BP** (°C):   -13.3

Solubility (Moles/L)	Solubility (Grams/L)	Temp (°C)	Ref (#)	Evaluation (T P E A A)	Comments
9.600E-04	6.000E-02	10	M133	1 0 0 0 1	*sic*
9.600E-04	6.000E-02	10	P046	1 0 0 0 0	*sic*
1.506E-01	9.411E+00	15	D086	1 0 2 2 1	
1.576E-01	9.852E+00	16	D086	1 0 2 2 2	
1.081E-01	6.754E+00	20	N034	1 0 0 0 1	
1.451E-01	9.067E+00	20.5	D086	1 0 2 2 2	
<1.76E-02	<1.10E+00	25	I310	0 0 0 0 0	
1.396E-01	8.723E+00	26	D086	1 0 2 2 1	
1.411E-01	8.821E+00	29.5	D086	1 0 2 2 1	
1.490E-01	9.312E+00	35	D086	1 0 2 2 1	
1.411E-01	8.821E+00	41	D086	1 0 2 2 1	
1.396E-01	8.723E+00	46.5	D086	1 0 2 2 1	
6.717E-03	4.198E-01	50	M065	0 0 2 1 1	
1.506E-01	9.411E+00	55	D086	1 0 2 2 1	
1.459E-01	9.116E+00	65	D086	1 0 2 2 1	
1.553E-01	9.705E+00	72.5	D086	1 0 2 2 1	
1.584E-01	9.901E+00	80	D086	1 0 2 2 2	
1.772E-01	1.108E+01	85	D086	1 0 2 2 2	

62.   C₂H₃Cl₂NO₂

1,1-Dichloro-1-nitroethane
Dichloronitroethane
Ethide

RN: 594-72-9   **MP** (°C):
MW: 143.96   **BP** (°C):

Solubility (Moles/L)	Solubility (Grams/L)	Temp (°C)	Ref (#)	Evaluation (T P E A A)	Comments
3.456E-02	4.975E+00	20	C121	1 0 0 0 0	unit assumed, *sic*
1.732E-02	2.494E+00	20	M061	1 0 0 0 1	

63. $C_2H_3Cl_3$
1,1,1-Trichloroethane
1,1,1- Trichloroethane
Trichloroethane
1,1,1-Trichloethane

RN: 71-55-6   **MP** (°C):   -35
MW: 133.41   **BP** (°C):   74.1

Solubility (Moles/L)	Solubility (Grams/L)	Temp (°C)	Ref (#)	Evaluation (T P E A A)	Comments
1.190E-02	1.587E+00	0	V009	1 0 0 0 2	
1.342E-02	1.790E+00	3.5	C094	1 0 0 0 2	
1.019E-02	1.360E+00	20	C094	1 0 1 0 2	
3.358E-02	4.480E+00	20	G056	1 0 0 0 2	
3.598E-03	4.800E-01	20	M133	1 0 0 0 2	
9.895E-03	1.320E+00	20	M368	1 0 0 0 1	
3.598E-03	4.800E-01	20	P046	1 0 0 0 0	
9.882E-03	1.318E+00	20	V009	1 0 0 0 2	
8.797E-03	1.174E+00	23.5	S171	2 1 2 2 2	
5.244E-03	6.995E-01	25	A094	1 0 0 0 0	
1.000E-02	1.334E+00	25	B173	2 0 2 2 2	
3.284E-02	4.381E+00	25	N309	1 0 0 0 1	*sic*
9.732E-03	1.298E+00	25	O006	1 0 0 0 1	
3.597E-03	4.798E-01	30	M311	1 1 2 2 2	
9.433E-03	1.258E+00	35	V009	1 0 0 0 2	
9.583E-03	1.278E+00	50	V009	1 0 0 0 2	
5.397E-03	7.200E-01	ns	H123	0 0 0 0 2	

64. $C_2H_3Cl_3$
1,1,2-Trichloroethane
1,1,2-β-Trichloroethane

RN: 79-00-5   **MP** (°C):   -37
MW: 133.41   **BP** (°C):   113

Solubility (Moles/L)	Solubility (Grams/L)	Temp (°C)	Ref (#)	Evaluation (T P E A A)	Comments
3.477E-02	4.638E+00	0	V009	1 0 0 0 2	
3.254E-02	4.341E+00	20	V009	1 0 0 0 2	
3.804E-02	5.074E+00	25	C119	2 2 2 2 2	
3.298E-02	4.400E+00	25	M037	1 1 0 0 1	
3.272E-02	4.365E+00	30	M311	1 1 2 2 2	
3.417E-02	4.559E+00	35	V009	1 0 0 0 2	
3.967E-02	5.292E+00	55	V009	1 0 0 0 2	

65.   C₂H₃FO₂

Fluoroacetic Acid

Essigsaeurefluorid

RN: 144-49-0   **MP** (°C):

MW: 78.04   **BP** (°C):

Solubility (Moles/L)	Solubility (Grams/L)	Temp (°C)	Ref (#)	Evaluation (T P E A A)	Comments
6.407E-04	5.000E-02	20	F300	1 0 0 0 0	

66.   C₂H₃N

Acetonitrile

Acetonitril

RN: 75-05-8   **MP** (°C):   -45

MW: 41.05   **BP** (°C):   81.6

Solubility (Moles/L)	Solubility (Grams/L)	Temp (°C)	Ref (#)	Evaluation (T P E A A)	Comments
>1.95E+01	>8.00E+02	25	B019	1 0 1 2 0	

67.   C₂H₃N

Methylisocyanide

Methyl-isocyanid

RN: 593-75-9   **MP** (°C):

MW: 41.05   **BP** (°C):

Solubility (Moles/L)	Solubility (Grams/L)	Temp (°C)	Ref (#)	Evaluation (T P E A A)	Comments
2.217E+00	9.100E+01	15	F300	1 0 0 0 1	

68.   C₂H₃NS

Methyl Isothiocyanate

Isothiocyanatomethane

RN: 556-61-6   **MP** (°C):   35

MW: 73.12   **BP** (°C):   119

Solubility (Moles/L)	Solubility (Grams/L)	Temp (°C)	Ref (#)	Evaluation (T P E A A)	Comments
1.039E-01	7.600E+00	20	M161	1 0 0 0 1	
1.032E-01	7.543E+00	20	O300	1 0 0 0 1	
1.085E-01	7.937E+00	20	P081	1 0 0 0 0	

69. C_2H_4
Ethylene
Ethene
RN: 74-85-1   **MP** (°C):   -169
MW: 28.05   **BP** (°C):   102

Solubility (Moles/L)	Solubility (Grams/L)	Temp (°C)	Ref (#)	Evaluation (T P E A A)	Comments
7.129E+00	2.000E+02	0	R028	0 0 0 0 1	
3.240E+00	9.091E+01	25	R028	0 0 0 0 1	
3.187E+00	8.942E+01	30	C116	1 0 0 0 2	

70. C_2H_4BrCl
Ethylene Chlorobromide
1-Bromo-2-chloroethane
RN: 107-04-0   **MP** (°C):   -17
MW: 143.42   **BP** (°C):   106

Solubility (Moles/L)	Solubility (Grams/L)	Temp (°C)	Ref (#)	Evaluation (T P E A A)	Comments
4.778E-02	6.853E+00	20	C121	1 0 0 0 1	unit assumed, *sic*

71. $C_2H_4Br_2$
1,2-Dibromoethane
Ethylene Dibromide
Curafume
Haltox
1,2-Dibromaethan
RN: 106-93-4   **MP** (°C):   9.97
MW: 187.87   **BP** (°C):   131.7

Solubility (Moles/L)	Solubility (Grams/L)	Temp (°C)	Ref (#)	Evaluation (T P E A A)	Comments
1.777E-02	3.339E+00	0	V009	1 0 0 0 2	
2.078E-02	3.905E+00	15	G029	1 0 2 2 2	
1.874E-02	3.520E+00	20	C094	1 0 1 0 2	
2.279E-02	4.282E+00	20	C121	1 0 0 0 1	unit assumed, *sic*
1.794E-02	3.370E+00	20	G080	1 0 0 0 1	
2.300E-02	4.321E+00	20	M312	1 0 0 0 1	
1.592E-02	2.991E+00	20	P081	1 0 0 0 0	
2.142E-02	4.024E+00	20	V009	1 0 0 0 2	
2.210E-02	4.153E+00	25	O006	1 0 0 0 2	
2.294E-02	4.310E+00	30	F300	1 0 0 0 2	
2.284E-02	4.292E+00	30	G029	1 0 2 2 2	
2.279E-02	4.282E+00	30	M061	1 0 0 0 1	
2.289E-02	4.300E+00	30	M161	1 0 0 0 1	
2.390E-02	4.490E+00	35	V009	1 0 0 0 2	
2.817E-02	5.292E+00	50	V009	1 0 0 0 2	

72.   C$_2$H$_4$ClNO

Acetohydroxamic Acid Chloride
Acethydroximsaeure-chlorid
2-Chloroacetamide
Chloroacetamide
Chloressigsaeureamid
Essigsaeure-N-chloramid

RN: 79-07-2   **MP** (°C):   119.5
MW: 93.51   **BP** (°C):   225

Solubility (Moles/L)	Solubility (Grams/L)	Temp (°C)	Ref (#)	Evaluation (T P E A A)	Comments
9.624E-01	9.000E+01	24	F300	1 0 0 0 0	

73.   C$_2$H$_4$ClNO$_2$

1-Chloro-1-nitroethane
1-Chloronitroethane

RN: 598-92-5   **MP** (°C):
MW: 109.51   **BP** (°C):   124

Solubility (Moles/L)	Solubility (Grams/L)	Temp (°C)	Ref (#)	Evaluation (T P E A A)	Comments
3.638E-02	3.984E+00	20	C121	1 0 0 0 0	unit assumed, *sic*
3.638E-02	3.984E+00	20	M061	1 0 0 0 0	

74.   C$_2$H$_4$Cl$_2$

Ethylene Dichloride
1,2-Dichloraethan

RN: 107-06-2   **MP** (°C):   -35
MW: 98.96   **BP** (°C):

Solubility (Moles/L)	Solubility (Grams/L)	Temp (°C)	Ref (#)	Evaluation (T P E A A)	Comments
9.095E-02	9.000E+00	0	F300	1 0 0 0 0	
9.317E-02	9.220E+00	0	H101	2 0 0 0 2	
9.232E-02	9.136E+00	0	L103	1 0 0 0 2	unit assumed
8.745E-02	8.654E+00	0	V009	1 0 0 0 2	
8.735E-02	8.645E+00	15	G029	1 0 2 2 2	
8.539E-02	8.450E+00	20	C094	1 0 1 0 2	
8.716E-02	8.625E+00	20	C121	1 0 0 0 1	unit assumed, *sic*
8.716E-02	8.625E+00	20	D052	1 1 0 0 1	
8.716E-02	8.625E+00	20	G056	1 0 0 0 2	
8.781E-02	8.690E+00	20	H101	2 0 0 0 2	
8.706E-02	8.615E+00	20	L103	1 0 0 0 2	unit assumed
8.706E-02	8.615E+00	20	M061	1 0 0 0 2	
8.616E-02	8.527E+00	20	M062	1 0 0 0 1	
8.892E-02	8.800E+00	20	M133	1 0 0 0 2	
8.716E-02	8.625E+00	20	O006	1 0 0 0 1	

Solubility (Moles/L)	Solubility (Grams/L)	Temp (°C)	Ref (#)	Evaluation (T P E A A)	Comments
8.892E-02	8.800E+00	20	P046	1 0 0 0 0	
8.507E-02	8.419E+00	20	V009	1 0 0 0 2	
8.070E-02	7.986E+00	25	B173	2 0 2 2 2	
1.060E-01	1.049E+01	25	C119	2 2 2 2 2	
8.690E-02	8.600E+00	25	F300	1 0 0 0 2	
8.740E-02	8.649E+00	25	G038	1 2 2 2 2	
8.740E-02	8.649E+00	25	G053	2 1 2 1 2	
8.488E-02	8.400E+00	25	M037	1 1 0 0 1	
9.013E-02	8.920E+00	30	G029	1 0 2 2 1	
8.954E-02	8.861E+00	30	L103	1 0 0 0 2	unit assumed
3.543E-02	3.506E+00	30	M311	1 1 2 2 2	
8.964E-02	8.871E+00	35	V009	1 0 0 0 2	
1.030E-01	1.019E+01	56	V009	1 0 0 0 2	
8.716E-02	8.625E+00	72	B197	0 0 0 0 1	at bp of 72 °C
5.927E-02	5.865E+00	89.3	B197	0 0 0 0 1	at bp of 89.3 °C
4.327E-02	4.282E+00	92.3	B197	0 0 0 0 1	at bp of 92.3 °C
3.324E-02	3.289E+00	94	B197	0 0 0 0 1	at bp of 94 °C
1.312E-02	1.298E+00	98	B197	0 0 0 0 1	at bp of 98 °C
4.345E-02	4.300E+00	rt	M161	0 0 0 0 1	

75. $C_2H_4Cl_2$

Ethylidene Chloride
1,1-Dichloraethan
1,1-Dichloroethane

RN: 75-34-3   **MP** (°C):   -97
MW: 98.96   **BP** (°C):

Solubility (Moles/L)	Solubility (Grams/L)	Temp (°C)	Ref (#)	Evaluation (T P E A A)	Comments
6.669E-02	6.600E+00	0	F300	1 0 0 0 1	
6.629E-02	6.560E+00	0	H101	2 0 0 0 2	
5.967E-02	5.905E+00	0	V009	1 0 0 0 2	
5.558E-02	5.500E+00	20	F300	1 0 0 0 1	
5.558E-02	5.500E+00	20	H101	2 0 0 0 2	
5.087E-02	5.035E+00	20	V009	1 0 0 0 2	
5.110E-02	5.057E+00	25	G038	1 2 2 2 2	
5.110E-02	5.057E+00	25	G053	2 2 2 1 2	
5.457E-02	5.400E+00	30	F300	1 0 0 0 1	
4.885E-02	4.834E+00	30	M300	1 1 2 2 2	
4.637E-02	4.589E+00	30	M311	1 1 2 2 2	
5.397E-02	5.341E+00	30	N034	1 0 0 0 2	
4.847E-02	4.797E+00	35	V009	1 0 0 0 2	
5.217E-02	5.163E+00	50	V009	1 0 0 0 2	

76.   C$_2$H$_4$F$_2$
1,1-Difluoroethane
Ethylidene Fluoride
RN: 75-37-6   **MP** (°C):   -117
MW: 66.05   **BP** (°C):   -24.7

Solubility (Moles/L)	Solubility (Grams/L)	Temp (°C)	Ref (#)	Evaluation (T P E A A)	Comments
8.132E-02	5.371E+00	0	M065	0 0 2 1 2	

77.   C$_2$H$_4$N$_2$O$_2$
Oxamide
Oxalsaeure-diamid
RN: 471-46-5   **MP** (°C):
MW: 88.07   **BP** (°C):

Solubility (Moles/L)	Solubility (Grams/L)	Temp (°C)	Ref (#)	Evaluation (T P E A A)	Comments
4.201E-03	3.700E-01	7.30	F300	1 0 0 0 1	
7.040E-02	6.200E+00	100	F300	1 0 0 0 1	

78.   C$_2$H$_4$N$_4$
Dicyanodiamide
Dicyandiamid
Dicyandiamide
RN: 461-58-5   **MP** (°C):   210
MW: 84.08   **BP** (°C):

Solubility (Moles/L)	Solubility (Grams/L)	Temp (°C)	Ref (#)	Evaluation (T P E A A)	Comments
1.526E-01	1.283E+01	0	M043	1 0 0 0 1	
1.492E-01	1.254E+01	0.0	H037	1 2 2 1 2	
2.218E-01	1.865E+01	10	M043	1 0 0 0 1	
2.617E-01	2.200E+01	13	F300	1 0 0 0 1	
3.688E-01	3.101E+01	20	M043	1 0 0 0 1	
4.876E-01	4.100E+01	25	F300	1 0 0 0 1	
4.717E-01	3.966E+01	25.0	H037	1 2 2 1 2	
5.663E-01	4.762E+01	30	M043	1 0 0 0 1	
8.565E-01	7.201E+01	39.9	H037	1 2 2 1 2	
8.606E-01	7.236E+01	40	M043	1 0 0 0 1	
1.255E+00	1.055E+02	49.8	H037	1 2 2 1 2	
1.899E+00	1.597E+02	60	M043	1 0 0 0 1	
1.878E+00	1.579E+02	60.1	H037	1 2 2 1 2	
2.236E+00	1.880E+02	60.10	F300	1 0 0 0 2	
2.978E+00	2.504E+02	74.5	H037	1 2 2 1 2	
3.275E+00	2.754E+02	80	M043	1 0 0 0 1	

79.   C$_2$H$_4$N$_4$
Amitrole
3-Amino-1,2,4-triazole
3-Amino-s-triazole
ATA
Aminotriazole

RN: 61-82-5   **MP** (°C):   159.0
MW: 84.08   **BP** (°C):

Solubility (Moles/L)	Solubility (Grams/L)	Temp (°C)	Ref (#)	Evaluation (T P E A A)	Comments
2.602E+00	2.188E+02	23	M061	1 0 0 0 1	
2.602E+00	2.188E+02	25	B185	1 0 0 0 1	
2.602E+00	2.188E+02	25	B200	1 0 0 0 1	
2.602E+00	2.188E+02	25	I310	0 0 0 0 1	
3.330E+00	2.800E+02	25	M161	1 0 0 0 2	
2.602E+00	2.188E+02	ns	B100	0 0 0 0 1	
3.162E+00	2.659E+02	ns	M163	0 0 0 0 0	EFG

80.   C$_2$H$_4$N$_4$O$_2$S$_2$
2-Amino-1,3,4-thiadiazole-5-sulfonamide
5-Amino-1,3,4-thiadiazol-2-sulfonamide
5-Amino-1,3,4-thiadiazole-2-sulfonamide
CL 5343
Tio-urasin

RN: 14949-00-9   **MP** (°C):
MW: 180.21   **BP** (°C):

Solubility (Moles/L)	Solubility (Grams/L)	Temp (°C)	Ref (#)	Evaluation (T P E A A)	Comments
2.630E-02	4.739E+00	15	K024	1 2 1 1 2	

81.   C$_2$H$_4$O$_2$
Acetic Acid Glacial
Acetic Acid
Essigsaeure

RN: 64-19-7   **MP** (°C):   16.7
MW: 60.05   **BP** (°C):   118

Solubility (Moles/L)	Solubility (Grams/L)	Temp (°C)	Ref (#)	Evaluation (T P E A A)	Comments
1.004E+01	6.029E+02	25	H084	1 0 0 0 2	

82.   C$_2$H$_4$O$_3$
Glycolic Acid
Glykolsaeure
RN: 79-14-1   **MP** (°C):   80
MW: 76.05   **BP** (°C):

Solubility (Moles/L)	Solubility (Grams/L)	Temp (°C)	Ref (#)	Evaluation (T P E A A)	Comments
6.084E+00	4.627E+02	6.99	A340	2 0 2 2 2	
6.913E+00	5.258E+02	10.89	A340	2 0 2 2 2	
7.894E+00	6.004E+02	20.69	A340	2 0 2 2 2	
8.015E+00	6.096E+02	24.99	A340	2 0 2 2 2	
8.168E+00	6.212E+02	30.09	A340	2 0 2 2 2	
8.296E+00	6.309E+02	35.99	A340	2 0 2 2 2	
8.400E+00	6.388E+02	39.99	A340	2 0 2 2 2	
8.533E+00	6.489E+02	47.99	A340	2 0 2 2 2	
8.536E+00	6.492E+02	48.99	A340	2 0 2 2 2	
8.654E+00	6.582E+02	54.99	A340	2 0 2 2 2	
8.721E+00	6.632E+02	59.49	A340	2 0 2 2 2	
8.808E+00	6.698E+02	64.49	A340	2 0 2 2 2	
8.866E+00	6.743E+02	69.99	A340	2 0 2 2 2	
8.932E+00	6.793E+02	74.99	A340	2 0 2 2 2	
8.968E+00	6.820E+02	79.89	A340	2 0 2 2 2	
9.016E+00	6.857E+02	84.49	A340	2 0 2 2 2	
9.043E+00	6.877E+02	88.09	A340	2 0 2 2 2	

83.   C$_2$H$_5$Br
Bromoethane
Ethyl Bromide
Aethylbromid
RN: 74-96-4   **MP** (°C):   -119
MW: 108.97   **BP** (°C):   38.5

Solubility (Moles/L)	Solubility (Grams/L)	Temp (°C)	Ref (#)	Evaluation (T P E A A)	Comments
9.792E-02	1.067E+01	0	H101	2 0 0 0 2	
8.810E-02	9.600E+00	17.5	F001	1 0 1 2 2	
8.810E-02	9.600E+00	17.5	S006	1 0 0 0 2	
8.259E-02	9.000E+00	20	F300	1 0 0 0 0	
8.388E-02	9.140E+00	20	H101	2 0 0 0 2	
8.185E-02	8.920E+00	20	H127	1 0 0 0 0	
8.127E-02	8.856E+00	30	V009	1 0 0 0 1	

84. C_2H_5Cl
Ethyl Chloride
Aethylchlorid
Chloroethane
Monochloroethane
RN: 75-00-3   **MP** (°C):   -139.0
MW: 64.52   **BP** (°C):   12.3

Solubility (Moles/L)	Solubility (Grams/L)	Temp (°C)	Ref (#)	Evaluation (T P E A A)	Comments
6.975E-02	4.500E+00	0	M037	1 1 0 0 1	
6.898E-02	4.450E+00	0	V009	1 0 0 0 2	
7.865E-02	5.074E+00	20	G056	1 0 0 0 2	
8.846E-02	5.707E+00	20	N034	1 0 0 0 2	
8.900E-02	5.742E+00	ns	F001	0 0 1 2 2	
8.433E-02	5.440E+00	ns	R028	0 0 0 0 2	

85. C_2H_5I
Iodoethane
Ethyl Iodide
Aethyliodid
Iodaethan
RN: 75-03-6   **MP** (°C):   -108
MW: 155.97   **BP** (°C):   71

Solubility (Moles/L)	Solubility (Grams/L)	Temp (°C)	Ref (#)	Evaluation (T P E A A)	Comments
2.828E-02	4.410E+00	0	H101	2 0 0 0 2	
2.571E-02	4.010E+00	20	F300	1 0 0 0 2	
2.584E-02	4.030E+00	20	H101	2 0 0 0 2	
2.510E-02	3.915E+00	20	M171	1 0 0 0 2	
2.510E-02	3.915E+00	22.5	F001	1 0 1 2 2	
2.510E-02	3.915E+00	22.5	S006	1 0 0 0 2	
2.580E-02	4.024E+00	30	G029	1 0 2 2 2	
2.661E-02	4.150E+00	30	H101	2 0 0 0 2	
2.580E-02	4.023E+00	30	V009	1 0 0 0 2	

86. C_2H_5N
Ethylenimine
Aethylenimin
Aziridine
Ethyleneimine
Dimethyleneimine
RN: 151-56-4   **MP** (°C):   -78
MW: 43.07   **BP** (°C):   56

Solubility (Moles/L)	Solubility (Grams/L)	Temp (°C)	Ref (#)	Evaluation (T P E A A)	Comments
2.117E-01	9.116E+00	20	P315	1 2 1 2 2	

87.   C₂H₅NO

Acetamide
Acetamid
RN: 60-35-5   **MP** (°C):   81.0
MW: 59.07   **BP** (°C):   222.0

Solubility (Moles/L)	Solubility (Grams/L)	Temp (°C)	Ref (#)	Evaluation (T P E A A)	Comments
8.342E+00	4.927E+02	0	M022	1 0 0 0 2	
9.816E+00	5.798E+02	0	M043	1 0 0 0 2	
1.021E+01	6.030E+02	0.3	F300	1 0 0 0 2	
1.077E+01	6.364E+02	10	M043	1 0 0 0 2	
1.165E+01	6.880E+02	20	F300	1 0 0 0 2	
9.691E+00	5.724E+02	20	M022	1 0 0 0 2	
1.180E+01	6.970E+02	20	M043	1 0 0 0 2	
1.194E+01	7.050E+02	24.50	F300	1 0 0 0 2	
3.386E+01	2.000E+03	25	I310	0 0 0 0 0	
1.280E+01	7.561E+02	30	M043	1 0 0 0 2	
1.093E+01	6.455E+02	40	M022	1 0 0 0 2	
1.379E+01	8.148E+02	40	M043	1 0 0 0 2	
1.208E+01	7.138E+02	60	M022	1 0 0 0 2	
1.515E+01	8.947E+02	60	M043	1 0 0 0 2	
8.358E+00	4.937E+02	rt	D021	0 0 1 1 2	

88.   C₂H₅NO₂

Glycine
Glycin
Glycocoll
RN: 56-40-6   **MP** (°C):   245
MW: 75.07   **BP** (°C):

Solubility (Moles/L)	Solubility (Grams/L)	Temp (°C)	Ref (#)	Evaluation (T P E A A)	Comments
1.612E+00	1.210E+02	0	C347	2 0 2 2 0	EFG
1.668E+00	1.252E+02	0	D018	2 2 2 1 2	
1.656E+00	1.243E+02	0	M043	1 0 0 0 2	
1.905E+00	1.430E+02	10	C347	2 0 2 2 0	EFG
2.032E+00	1.525E+02	10	M043	1 0 0 0 2	
3.025E+00	2.271E+02	15	D349	2 1 1 2 2	
1.710E+00	1.284E+02	15	G081	1 0 1 1 2	
3.009E+00	2.259E+02	20	B032	1 2 2 2 2	
2.336E+00	1.754E+02	20	C347	2 0 2 2 0	EFG
3.180E+00	2.387E+02	20	D349	2 1 1 2 2	
2.447E+00	1.837E+02	20	M043	1 0 0 0 2	
2.616E+00	1.964E+02	21	P045	1 0 2 1 2	
3.316E+00	2.489E+02	25	B032	1 2 2 2 2	
2.885E+00	2.166E+02	25	C018	1 0 2 2 2	
3.329E+00	2.499E+02	25	D016	1 0 0 0 2	

2.691E+00	2.020E+02	25	D018	2 2 2 1 2	
2.663E+00	1.999E+02	25	D041	1 0 0 0 2	
3.325E+00	2.496E+02	25	D349	2 1 1 2 2	
2.886E+00	2.166E+02	25	E015	1 2 1 1 2	
2.660E+00	1.997E+02	25	F300	1 0 0 0 2	
2.664E+00	2.000E+02	25	G092	2 1 1 1 1	
2.664E+00	2.000E+02	25	G315	1 0 2 2 2	
2.526E+00	1.897E+02	25	K031	2 1 2 1 2	
2.886E+00	2.166E+02	25	M024	1 2 0 1 2	
3.334E+00	2.503E+02	25	M029	2 2 2 2 2	
2.760E+00	2.072E+02	25	N001	2 0 2 1 0	EFG
2.900E+00	2.177E+02	25	N012	2 0 2 1 2	
2.664E+00	2.000E+02	25	O316	1 0 1 2 2	
2.544E+00	1.910E+02	25	O316	1 0 1 2 2	
2.715E+00	2.038E+02	25	O317	1 0 1 2 2	
3.330E+00	2.500E+02	25.1	N024	2 0 2 2 2	
3.352E+00	2.516E+02	25.1	N025	2 0 2 2 2	
3.342E+00	2.509E+02	25.1	N026	2 0 2 2 2	
2.673E+00	2.006E+02	25.1	N027	1 1 2 2 2	
3.144E+00	2.360E+02	27	D036	2 1 2 2 2	
3.074E+00	2.308E+02	27	D036	2 1 2 2 2	
3.630E+00	2.725E+02	29.80	B032	1 2 2 1 2	
2.737E+00	2.054E+02	30	C347	2 0 2 2 0	EFG
2.832E+00	2.126E+02	30	M043	1 0 0 0 1	
3.109E+00	2.334E+02	40	C347	2 0 2 2 0	EFG
3.305E+00	2.481E+02	40	M043	1 0 0 0 1	
3.547E+00	2.662E+02	50	C347	2 0 2 2 0	EFG
3.816E+00	2.865E+02	50	D018	2 2 2 1 2	
3.745E+00	2.811E+02	50	F300	1 0 0 0 2	
3.921E+00	2.943E+02	60	C347	2 0 2 2 0	EFG
4.134E+00	3.103E+02	60	M043	1 0 0 0 1	
4.215E+00	3.164E+02	70	C347	2 0 2 2 0	EFG
4.863E+00	3.650E+02	75	D018	2 2 2 1 2	
4.693E+00	3.523E+02	75	D041	1 0 0 0 2	
4.693E+00	3.523E+02	75	F300	1 0 0 0 2	
4.517E+00	3.390E+02	80	C347	2 0 2 2 0	EFG
4.836E+00	3.631E+02	80	M043	1 0 0 0 1	
4.753E+00	3.568E+02	90	C347	2 0 2 2 0	EFG
5.353E+00	4.018E+02	99.99	P349	1 0 0 2 2	
4.911E+00	3.686E+02	100	C347	2 0 2 2 0	EFG
5.353E+00	4.018E+02	100	F300	1 0 0 0 2	
5.485E+00	4.118E+02	100	M043	1 0 0 0 1	
6.661E+00	5.000E+02	ns	D072	0 0 0 0 0	
4.499E+00	3.377E+02	rt	D021	0 0 1 1 2	

89. $C_2H_5NO_2$

Glycolamide
2-Hydroxyacetamide
2-Hydroxyacetimidic Acid
Glycolic Amide
Glycolic Acid Amide

RN: 598-42-5   **MP** (°C):
MW: 75.07   **BP** (°C):

Solubility (Moles/L)	Solubility (Grams/L)	Temp (°C)	Ref (#)	Evaluation (T P E A A)	Comments
5.509E+00	4.135E+02	25	M008	1 0 0 0 2	

90. $C_2H_5NO_2$

Methyl Carbamate
Carbamidsaeure-methyl Ester
Methyl Urethane

RN: 598-55-0   **MP** (°C):   52
MW: 75.07   **BP** (°C):   177

Solubility (Moles/L)	Solubility (Grams/L)	Temp (°C)	Ref (#)	Evaluation (T P E A A)	Comments
9.125E+00	6.850E+02	11	F300	1 0 0 0 2	
9.119E+00	6.845E+02	11	I314	0 0 0 0 2	
9.200E+00	6.906E+02	15.50	F001	1 0 1 0 2	
5.462E+00	4.100E+02	15.50	F300	1 0 0 0 1	

91. $C_2H_5NO_2$

Nitroethane
Nitroetan

RN: 79-24-3   **MP** (°C):   -50
MW: 75.07   **BP** (°C):   114

Solubility (Moles/L)	Solubility (Grams/L)	Temp (°C)	Ref (#)	Evaluation (T P E A A)	Comments
5.736E-01	4.306E+01	20	C121	1 0 0 0 1	unit assumed, *sic*
6.404E-01	4.807E+01	25	M346	2 1 1 1 2	

92. C$_2$H$_5$NS
Thiacetamide
Thioessigsaeureamid
Thioacetamide
Acetothioamide
Ethanethioamide

RN: 62-55-5   **MP** (°C):   113
MW: 75.13   **BP** (°C):

Solubility (Moles/L)	Solubility (Grams/L)	Temp (°C)	Ref (#)	Evaluation (T P E A A)	Comments
1.865E+00	1.402E+02	25	I310	0 0 0 0 1	

93. C$_2$H$_5$N.2H$_2$O
Ethyleneimine (Dihydrate)
Aziridine (Dihydrate)

RN: 151-56-4   **MP** (°C):
MW: 79.10   **BP** (°C):

Solubility (Moles/L)	Solubility (Grams/L)	Temp (°C)	Ref (#)	Evaluation (T P E A A)	Comments
6.840E-02	5.411E+00	20	P315	1 2 1 2 2	

94. C$_2$H$_5$N$_3$O$_2$
Biuret
Carbamylurea

RN: 108-19-0   **MP** (°C):
MW: 103.08   **BP** (°C):

Solubility (Moles/L)	Solubility (Grams/L)	Temp (°C)	Ref (#)	Evaluation (T P E A A)	Comments
1.164E-01	1.200E+01	0	F300	1 0 0 0 2	
1.475E-01	1.520E+01	15	F300	1 0 0 0 2	
3.104E+00	3.200E+02	106	F300	1 0 0 0 1	

95. C$_2$H$_5$N$_3$O$_2$
Methylnitrosourea
MNU
Nitrosomethylurea

RN: 684-93-5   **MP** (°C):   123
MW: 103.08   **BP** (°C):

Solubility (Moles/L)	Solubility (Grams/L)	Temp (°C)	Ref (#)	Evaluation (T P E A A)	Comments
1.400E-01	1.443E+01	24	M031	1 1 1 1 1	

96. $C_2H_5N_5O_3$
N-Methyl-N'-Nitro-N-Nitrosoguanidine
MNNG
1-Methyl-3-nitro-1-nitrosoguanidine
RN: 70-25-7 **MP** (°C): 118
MW: 147.09 **BP** (°C):

Solubility (Moles/L)	Solubility (Grams/L)	Temp (°C)	Ref (#)	Evaluation (T P E A A)	Comments
<3.38E-02	<4.98E+00	ns	I307	0 0 0 0 0	

97. $C_2H_5O_5P$
Phosphoacetic Acid
Phosphor Carboxymethyl-phosphonsaeure
Phosphonoacetic Acid
RN: 4408-78-0 **MP** (°C): 144.5
MW: 140.03 **BP** (°C):

Solubility (Moles/L)	Solubility (Grams/L)	Temp (°C)	Ref (#)	Evaluation (T P E A A)	Comments
2.799E+00	3.920E+02	0	F300	1 0 0 0 2	
2.800E+00	3.921E+02	0	N028	1 0 0 0 2	

98. $C_2H_5O_5As$
Arsonoacetic Acid
Arsono-essigsaeure
RN: 107-38-0 **MP** (°C): 152
MW: 183.98 **BP** (°C):

Solubility (Moles/L)	Solubility (Grams/L)	Temp (°C)	Ref (#)	Evaluation (T P E A A)	Comments
2.174E+00	4.000E+02	18	F300	1 0 0 0 1	

99. C_2H_6
Ethane
Aethan
RN: 74-84-0 **MP** (°C): -172
MW: 30.07 **BP** (°C): -88

Solubility (Moles/L)	Solubility (Grams/L)	Temp (°C)	Ref (#)	Evaluation (T P E A A)	Comments
2.587E-01	7.779E+00	0	C075	1 0 1 0 1	
4.157E-03	1.250E-01	0	F300	1 0 0 0 2	
3.601E-03	1.083E-01	4.99	C115	2 0 2 2 2	
2.903E-03	8.730E-02	9.99	C115	2 0 2 2 2	
2.465E-03	7.413E-02	14.99	C115	2 0 2 2 2	
2.222E-03	6.682E-02	19.8	G058	1 0 0 0 2	
2.129E-03	6.401E-02	19.99	C115	2 0 2 2 2	

1.929E-03	5.800E-02	20	F300	1 0 0 0 1
1.850E-03	5.563E-02	24.99	C115	2 0 2 2 2
2.009E-03	6.040E-02	25	M001	2 1 2 2 2
2.009E-03	6.040E-02	25	M002	2 2 1 2 2
1.760E-03	5.292E-02	25	M102	1 2 2 1 2
1.620E-03	4.871E-02	29.99	C115	2 0 2 2 2
7.981E-04	2.400E-02	60	F300	1 0 0 0 1

100. C_2H_6O
Methyl Ether
Dimethyl Ether
Dimethylaether
RN: 115-10-6   **MP** (°C):   -138
MW: 46.07   **BP** (°C):   -23.6

Solubility (Moles/L)	Solubility (Grams/L)	Temp (°C)	Ref (#)	Evaluation (T P E A A)	Comments
1.476E+00	6.800E+01	18	F300	1 0 0 0 1	
5.669E+00	2.612E+02	24	M065	1 0 2 1 2	

101. $C_2H_6O_2$
Ethylene Glycol
Glycol
1,2-Ethandiol
RN: 107-21-1   **MP** (°C):   -13
MW: 62.07   **BP** (°C):   197.6

Solubility (Moles/L)	Solubility (Grams/L)	Temp (°C)	Ref (#)	Evaluation (T P E A A)	Comments
6.710E+00	4.165E+02	4.50	C022	1 2 0 0 2	
5.562E-01	3.452E+01	25	B004	1 0 0 0 2	

102. $C_2H_6O_3S$
Methyl Methanesulphonate
Methyl Mesylate
Methanesulfonic Acid Methyl Ester
RN: 66-27-3   **MP** (°C):   20
MW: 110.13   **BP** (°C):   203

Solubility (Moles/L)	Solubility (Grams/L)	Temp (°C)	Ref (#)	Evaluation (T P E A A)	Comments
1.513E+00	1.667E+02	25	I310	0 0 0 0 1	

103.   C$_2$H$_6$O$_4$S
Dimethyl Sulfate
Sulfuric Acid Dimethyl Ester
RN: 77-78-1   **MP** (°C):   -27
MW: 126.13   **BP** (°C):   188

Solubility (Moles/L)	Solubility (Grams/L)	Temp (°C)	Ref (#)	Evaluation (T P E A A)	Comments
2.220E-01	2.800E+01	18	B078	1 0 0 0 1	
2.159E-01	2.724E+01	18	D049	1 2 0 0 1	

104.   C$_2$H$_7$N
Ethylamine
Aethylamin
RN: 75-04-7   **MP** (°C):   -81
MW: 45.08   **BP** (°C):   16.6

Solubility (Moles/L)	Solubility (Grams/L)	Temp (°C)	Ref (#)	Evaluation (T P E A A)	Comments
2.686E-02	1.211E+00	25	B004	1 0 0 0 2	

105.   C$_2$H$_7$NO$_3$S
Taurine
Taurin
RN: 107-35-7   **MP** (°C):   328
MW: 125.15   **BP** (°C):

Solubility (Moles/L)	Solubility (Grams/L)	Temp (°C)	Ref (#)	Evaluation (T P E A A)	Comments
2.999E-01	3.754E+01	0	M043	1 0 0 0 1	
4.523E-01	5.660E+01	10	M043	1 0 0 0 1	
4.842E-01	6.060E+01	12	F300	1 0 0 0 2	
3.919E-01	4.905E+01	15	G081	1 0 1 1 2	
6.448E-01	8.070E+01	20	F300	1 0 0 0 2	
6.463E-01	8.088E+01	20	M043	1 0 0 0 1	
4.700E-01	5.882E+01	24	D031	1 0 0 0 2	
7.580E-01	9.486E+01	25	D041	1 0 0 0 2	
8.815E-01	1.103E+02	30	M043	1 0 0 0 2	
1.149E+00	1.438E+02	40	M043	1 0 0 0 2	
1.719E+00	2.151E+02	60	M043	1 0 0 0 2	
1.985E+00	2.484E+02	70	F300	1 0 0 0 2	
2.105E+00	2.634E+02	75	D041	1 0 0 0 2	
2.217E+00	2.775E+02	80	M043	1 0 0 0 2	
2.506E+00	3.137E+02	100	M043	1 0 0 0 2	

106. $C_2H_7O_2As$
Cacodylic Acid
Dimethylarsinsaeure
Kakodylsaeure
Arsine Oxide, Hydroxydimethyl-
Cacodylic Acid

RN: 75-60-5   **MP** (°C):   195
MW: 138.00   **BP** (°C):

Solubility (Moles/L)	Solubility (Grams/L)	Temp (°C)	Ref (#)	Evaluation (T P E A A)	Comments
2.899E+00	4.001E+02	20	B200	1 0 0 0 2	
3.287E+00	4.536E+02	22	B185	1 0 0 0 1	
3.290E+00	4.540E+02	22	F300	1 0 0 0 2	
4.961E+00	6.845E+02	25	D305	1 0 0 0 2	
1.449E+01	2.000E+03	25	M161	1 0 0 0 0	

107. C_2H_7As
Ethylarsine
Aethylarsin
Arsen

RN: 593-59-9   **MP** (°C):
MW: 106.00   **BP** (°C):   36

Solubility (Moles/L)	Solubility (Grams/L)	Temp (°C)	Ref (#)	Evaluation (T P E A A)	Comments
1.226E-03	1.300E-01	19	F300	1 0 0 0 1	

108. $C_2Cl_2F_4$
1,2-Dichlorotetrafluoroethane
CFC-114
sym-Dichlorotetrafluoroethane
Halon 242

RN: 76-14-2   **MP** (°C):   -94
MW: 170.92   **BP** (°C):   3.8

Solubility (Moles/L)	Solubility (Grams/L)	Temp (°C)	Ref (#)	Evaluation (T P E A A)	Comments
7.605E-04	1.300E-01	25	R048	1 0 0 0 1	

109. $C_2Cl_3F_3$

1,1,2-Trichloro-1,2,2-trifluoroethane
Freon 113
Fluorocarbon 113
Halocarbon 113

RN: 76-13-1   **MP** (°C):   -36.4
MW: 187.38   **BP** (°C):   47.6

Solubility (Moles/L)	Solubility (Grams/L)	Temp (°C)	Ref (#)	Evaluation (T P E A A)	Comments
9.071E-04	1.700E-01	25	R048	1 0 0 0 1	

110. C_2Cl_4

Tetrachloroethylene
Ethylene Tetrachloride
Perchloroethylene
Tetrachloroethene
Tetrachloro-ethylene
PERC

RN: 127-18-4   **MP** (°C):   -22
MW: 165.83   **BP** (°C):   121

Solubility (Moles/L)	Solubility (Grams/L)	Temp (°C)	Ref (#)	Evaluation (T P E A A)	Comments
1.206E-03	2.000E-01	20	C094	1 0 1 0 2	
1.206E-03	2.000E-01	20	C121	0 0 0 0 0	unit assumed, *sic*
9.045E-04	1.500E-01	20	M133	1 0 0 0 2	
9.045E-04	1.500E-01	20	P046	1 0 0 0 0	
9.044E-04	1.500E-01	25	A094	1 0 0 0 1	
2.920E-03	4.842E-01	25	B173	2 0 2 2 2	
1.206E-03	2.000E-01	25	C119	2 2 2 2 2	
2.412E-03	4.000E-01	25	F071	1 1 2 1 2	
9.044E-04	1.500E-01	25	G056	1 0 0 0 2	
9.045E-04	1.500E-01	25	M037	1 1 0 0 1	
9.045E-04	1.500E-01	25	M368	1 0 0 0 1	
9.044E-04	1.500E-01	25	N034	1 0 0 0 1	
2.412E-03	4.000E-01	ns	M344	0 0 0 0 2	
9.044E-04	1.500E-01	ns	O006	0 0 0 0 1	

111. C_2Cl_6
Hexachloroethane
1,1,1,2,2,2-Hexachloroethane
Avlothane
Distopin
Distopan
Distokal

RN: 67-72-1   **MP** (°C):   187
MW: 236.74   **BP** (°C):   186.8

Solubility (Moles/L)	Solubility (Grams/L)	Temp (°C)	Ref (#)	Evaluation (T P E A A)	Comments
3.253E-05	7.700E-03	20	M339	2 2 2 2 1	
2.112E-04	5.000E-02	22.3	M037	1 1 0 0 0	

112. C_2N_2
Cyanogen
Dicyan

RN: 460-19-5   **MP** (°C):
MW: 52.04   **BP** (°C):

Solubility (Moles/L)	Solubility (Grams/L)	Temp (°C)	Ref (#)	Evaluation (T P E A A)	Comments
1.572E+01	8.182E+02	20	F300	1 0 0 0 1	

113. $C_2N_4S_2$
Cyanogen Azidodithiocarbonate

RN: **MP** (°C):
MW: 144.18   **BP** (°C):

Solubility (Moles/L)	Solubility (Grams/L)	Temp (°C)	Ref (#)	Evaluation (T P E A A)	Comments
1.040E-02	1.500E+00	0	A055	0 0 0 0 2	

114. $C_2N_6S_4$
Thioperoxydicarbonic Diazide
Azidoschwefel-kohlenstoff
Azidocarbonicdisulfide

RN: 148832-09-1   **MP** (°C):
MW: 236.32   **BP** (°C):

Solubility (Moles/L)	Solubility (Grams/L)	Temp (°C)	Ref (#)	Evaluation (T P E A A)	Comments
1.269E-03	3.000E-01	25	F300	1 0 0 0 0	

115. $C_3H_2Cl_2N_2O_2$
1,3-Dichlorohydantoin
2,4-Imidazolidinedione, 1,3-Dichloro-
RN: 2958-99-8   **MP** (°C):
MW: 168.97   **BP** (°C):

Solubility (Moles/L)	Solubility (Grams/L)	Temp (°C)	Ref (#)	Evaluation (T P E A A)	Comments
4.114E-02	6.951E+00	20	B080	1 0 1 1 0	
8.171E-02	1.381E+01	40	B080	1 0 1 1 1	

116. $C_3H_2N_2$
Malononitrile
Malonsaeure-dinitril
RN: 109-77-3   **MP** (°C):   32
MW: 66.06   **BP** (°C):   218.5

Solubility (Moles/L)	Solubility (Grams/L)	Temp (°C)	Ref (#)	Evaluation (T P E A A)	Comments
1.780E+00	1.176E+02	20	F300	1 0 0 0 2	

117. $C_3H_2N_2O_3$
Parabanic Acid
Parabansaeure
RN: 120-89-8   **MP** (°C):
MW: 114.06   **BP** (°C):

Solubility (Moles/L)	Solubility (Grams/L)	Temp (°C)	Ref (#)	Evaluation (T P E A A)	Comments
3.945E-01	4.500E+01	8	F300	1 0 0 0 1	

118. $C_3H_3Cl_3O_3$
β,β,β-Trichlorolactic Acid
β,β,β-Trichlor-milchsaeure
RN: 599-01-9   **MP** (°C):
MW: 193.41   **BP** (°C):

Solubility (Moles/L)	Solubility (Grams/L)	Temp (°C)	Ref (#)	Evaluation (T P E A A)	Comments
2.265E+00	4.380E+02	25	F300	1 0 0 0 2	

119. C_3H_3N
Acrylonitrile
Propenitrile
RN: 107-13-1   **MP** (°C):   -83.5
MW: 53.06   **BP** (°C):   77.3

Solubility (Moles/L)	Solubility (Grams/L)	Temp (°C)	Ref (#)	Evaluation (T P E A A)	Comments
1.266E+00	6.716E+01	0	D046	1 2 1 1 1	
1.266E+00	6.716E+01	0	D046	2 2 0 0 1	EFG
1.282E+00	6.803E+01	20	D046	1 2 1 1 1	
1.282E+00	6.803E+01	20	D046	2 2 0 0 1	EFG
1.298E+00	6.890E+01	25	D046	2 2 0 0 1	EFG
1.298E+00	6.890E+01	25	D046	1 2 1 1 1	
1.298E+00	6.890E+01	25	L096	1 2 0 2 1	
1.413E+00	7.500E+01	25	M161	1 0 0 0 1	
1.315E+00	6.977E+01	28	D046	2 2 0 0 1	EFG
1.347E+00	7.149E+01	36	D046	2 2 0 0 1	EFG
1.364E+00	7.236E+01	39	D046	2 2 0 0 1	EFG
1.388E+00	7.365E+01	41	D046	2 2 0 0 2	EFG
1.508E+00	8.004E+01	49	D046	2 2 0 0 1	EFG
1.508E+00	8.004E+01	53	D046	2 2 0 0 1	EFG
1.540E+00	8.173E+01	59	D046	2 2 0 0 1	EFG
1.603E+00	8.509E+01	63	D046	2 2 0 0 1	EFG
1.760E+00	9.338E+01	65	A324	2 2 2 1 2	
1.651E+00	8.759E+01	68	D046	2 2 0 0 0	EFG
1.721E+00	9.132E+01	72	D046	2 2 0 0 0	EFG
1.869E+00	9.918E+01	80	D046	2 2 0 0 0	EFG
1.974E+00	1.047E+02	85	D046	2 2 1 1 0	EFG
2.124E+00	1.127E+02	90	D046	2 2 1 1 0	EFG

120. $C_3H_3NOS_2$
Rhodanine
Rhodanin
RN: 141-84-4   **MP** (°C):   170
MW: 133.19   **BP** (°C):

Solubility (Moles/L)	Solubility (Grams/L)	Temp (°C)	Ref (#)	Evaluation (T P E A A)	Comments
1.689E-02	2.250E+00	25	F300	1 0 0 0 2	

121. $C_3H_3N_3O_3$
Cyamelide
Cyamelid
RN: 462-02-2   **MP** (°C):
MW: 129.08   **BP** (°C):

Solubility (Moles/L)	Solubility (Grams/L)	Temp (°C)	Ref (#)	Evaluation (T P E A A)	Comments
7.747E-04	1.000E-01	15	F300	1 0 0 0 0	

122. $C_3H_3N_3O_3$
Cyanuric Acid
Cyanursaeure
Isocyanuric Acid
Isocyanursaeure
RN: 108-80-5   **MP** (°C):
MW: 129.08   **BP** (°C):

Solubility (Moles/L)	Solubility (Grams/L)	Temp (°C)	Ref (#)	Evaluation (T P E A A)	Comments
2.300E-02	2.969E+00	2	B193	1 2 0 0 1	
3.874E-02	5.000E+00	20	F300	1 0 0 0 0	
3.874E-02	5.000E+00	20	F300	1 0 0 0 0	
2.009E-02	2.593E+00	25	B384	1 0 2 2 2	

123. $C_3H_3N_3S_3$
Trithiocyanuric Acid
s-Triazine-2,4,6-trithiol
Trimercapto-s-triazine
RN: 638-16-4   **MP** (°C):
MW: 177.27   **BP** (°C):

Solubility (Moles/L)	Solubility (Grams/L)	Temp (°C)	Ref (#)	Evaluation (T P E A A)	Comments
1.354E-03	2.399E-01	25	B384	1 0 2 2 2	

124. C_3H_4
Propyne
Methyl Acetylene
Methylacetylene
RN: 74-99-7   **MP** (°C):   -101
MW: 40.07   **BP** (°C):   -23.2

Solubility (Moles/L)	Solubility (Grams/L)	Temp (°C)	Ref (#)	Evaluation (T P E A A)	Comments
8.085E-02	3.239E+00	21	I011	1 2 2 1 2	
9.085E-02	3.640E+00	25	M001	2 1 2 2 2	
5.488E-02	2.199E+00	38	I011	1 2 2 1 1	
3.606E-02	1.445E+00	54	I011	1 2 2 1 1	
2.220E-02	8.895E-01	71	I011	1 2 2 1 1	
8.886E-03	3.560E-01	88	I011	1 2 2 1 1	

125.   C$_3$H$_4$ClN$_5$
Desethyl Simazine
Amino-2-chloro-6-ethylamino-s-triazine
6-Chloro-N-ethyl-1,3,5-triazine-2,4-diamine
RN: 1007-28-9   **MP** (°C):
MW: 145.55   **BP** (°C):

Solubility (Moles/L)	Solubility (Grams/L)	Temp (°C)	Ref (#)	Evaluation (T P E A A)	Comments
1.200E-03	1.747E-01	2	B193	1 1 0 0 0	

126.   C$_3$H$_4$Cl$_2$
1,2-Dichloropropene
Dichloropropylene
RN: 26952-23-8   **MP** (°C):
MW: 110.97   **BP** (°C):   92

Solubility (Moles/L)	Solubility (Grams/L)	Temp (°C)	Ref (#)	Evaluation (T P E A A)	Comments
2.427E-02	2.693E+00	20	C121	1 0 0 0 1	unit assumed, *sic*

127.   C$_3$H$_4$Cl$_2$
cis-1,3-Dichloropropene
1,3-Dichloropropropylene (cis)
cis-1,3-Dichloropropylene
cis 1,3-Dichloro-propene
cis-1,3-Dichloro-1-propene
(Z)-1,3-Dichloropropene
RN: 10061-01-5   **MP** (°C):
MW: 110.97   **BP** (°C):   108

Solubility (Moles/L)	Solubility (Grams/L)	Temp (°C)	Ref (#)	Evaluation (T P E A A)	Comments
2.433E-02	2.700E+00	20	G080	1 0 0 0 1	
9.651E-03	1.071E+00	30	M300	1 1 2 2 2	
8.211E-03	9.112E-01	30	M311	1 1 2 2 2	

128.   C$_3$H$_4$Cl$_2$
trans 1,3-Dichloro-propene
trans-1,3-Dichloro-1-propene
(E)-1,3-Dichloro-1-Propene
E-1,3-Dichloropropene
RN: 10061-02-6   **MP** (°C):
MW: 110.97   **BP** (°C):   111

Solubility (Moles/L)	Solubility (Grams/L)	Temp (°C)	Ref (#)	Evaluation (T P E A A)	Comments
2.523E-02	2.800E+00	20	G080	1 0 0 0 1	

129. $C_3H_4Cl_2$
trans-1,3-Dichloropropene
1,3-Dichloropropylene (trans)
trans-1,3-Dichloropropylene
1,3-Dichloropropene
RN: 542-75-6   **MP** (°C):
MW: 110.97   **BP** (°C):   112

Solubility (Moles/L)	Solubility (Grams/L)	Temp (°C)	Ref (#)	Evaluation (T P E A A)	Comments
2.703E-03	2.999E-01	20	C121	1 0 0 0 0	unit assumed, *sic*
9.011E-03	1.000E+00	20	M161	1 0 0 0 0	
1.071E-02	1.188E+00	30	M300	1 1 2 2 2	

130. $C_3H_4Cl_2O_2$
Dalapon
α,α-Dichlor-propionsaeure
RN: 75-99-0   **MP** (°C):
MW: 142.97   **BP** (°C):   187.5

Solubility (Moles/L)	Solubility (Grams/L)	Temp (°C)	Ref (#)	Evaluation (T P E A A)	Comments
3.511E+00	5.020E+02	25	M161	1 0 0 0 2	
3.511E+00	5.020E+02	ns	K138	0 0 0 0 1	

131. $C_3H_4N_2O$
Cyanoacetamide
Cyanessigsaeure-amid
RN: 107-91-5   **MP** (°C):
MW: 84.08   **BP** (°C):

Solubility (Moles/L)	Solubility (Grams/L)	Temp (°C)	Ref (#)	Evaluation (T P E A A)	Comments
1.546E+00	1.300E+02	20	F300	1 0 0 0 1	

132. $C_3H_4N_2O_2$
Hydantoin
2,4-Imidazolidinedione
RN: 461-72-3   **MP** (°C):   220
MW: 100.08   **BP** (°C):

Solubility (Moles/L)	Solubility (Grams/L)	Temp (°C)	Ref (#)	Evaluation (T P E A A)	Comments
2.944E+00	2.946E+02	100	F300	1 0 0 0 2	
3.970E-01	3.973E+01	ns	M025	0 2 0 1 2	

133. C₃H₄N₂O₃S

2-Imidazole Sulfonic Acid
Imidazol-sulfosaeure-(2)
RN: 53744-47-1 **MP** (°C):
MW: 148.14 **BP** (°C):

Solubility (Moles/L)	Solubility (Grams/L)	Temp (°C)	Ref (#)	Evaluation (T P E A A)	Comments
5.009E-01	7.420E+01	20	F300	1 0 0 0 2	

134. C₃H₄N₄O₂

Ammelide
2,4-Dihydroxy-6-amino-1,3,5-triazine
RN: 645-93-2 **MP** (°C):
MW: 128.09 **BP** (°C):

Solubility (Moles/L)	Solubility (Grams/L)	Temp (°C)	Ref (#)	Evaluation (T P E A A)	Comments
6.000E-04	7.685E-02	2	B193	1 2 0 0 0	

135. C₃H₄O

Acrolein
2-Propenal
Acrylaldehyde
RN: 107-02-8 **MP** (°C): -88.0
MW: 56.06 **BP** (°C): 52.5

Solubility (Moles/L)	Solubility (Grams/L)	Temp (°C)	Ref (#)	Evaluation (T P E A A)	Comments
8.690E+00	4.872E+02	0	B111	1 0 0 1 1	Quinol as a stabilizer
3.764E+00	2.110E+02	20	F300	1 0 0 0 2	
3.071E+00	1.722E+02	20	M161	1 0 0 0 1	
8.522E+00	4.778E+02	32.50	B111	1 0 0 1 2	Quinol as a stabilizer
8.429E+00	4.726E+02	44.40	B111	1 0 0 1 2	Quinol as a stabilizer
8.339E+00	4.675E+02	50	B111	1 0 0 1 2	Quinol as a stabilizer
8.288E+00	4.647E+02	53	B111	1 0 0 1 2	Quinol as a stabilizer
7.889E+00	4.423E+02	74.50	B111	1 0 0 1 2	Quinol as a stabilizer
7.338E+00	4.114E+02	82	B111	1 0 0 1 2	Quinol as a stabilizer
7.013E+00	3.932E+02	84	B111	1 0 0 1 2	Quinol as a stabilizer
6.597E+00	3.699E+02	87.80	B111	1 0 0 1 2	Quinol as a stabilizer
6.417E+00	3.598E+02	88	B111	1 0 0 1 2	Quinol as a stabilizer
5.096E+00	2.857E+02	ns	B185	0 0 0 0 1	
3.567E+00	2.000E+02	ns	B200	0 0 0 0 0	

136.   C₃H₄O₄

Malonic Acid
Acide Malonique
Malonsaeure

RN: 141-82-2   **MP** (°C):   135
MW: 104.06   **BP** (°C):

Solubility (Moles/L)	Solubility (Grams/L)	Temp (°C)	Ref (#)	Evaluation (T P E A A)	Comments
3.645E+00	3.793E+02	0	F300	1 0 0 0 2	
5.871E+00	6.110E+02	0	L041	1 0 0 1 2	
4.990E+00	5.192E+02	0	M043	1 0 0 0 2	
5.871E+00	6.110E+02	0	M051	1 0 0 0 2	
4.743E+00	4.936E+02	4.99	A339	2 0 2 2 2	
4.888E+00	5.087E+02	9.99	A339	2 0 2 2 2	
5.427E+00	5.648E+02	10	K077	1 2 2 2 2	average of 3
5.395E+00	5.614E+02	10	M043	1 0 0 0 2	
5.034E+00	5.238E+02	14.99	A339	2 0 2 2 2	
5.608E+00	5.836E+02	15	K077	1 2 2 2 2	
6.746E+00	7.020E+02	15	L041	1 0 0 1 2	
6.746E+00	7.020E+02	15	M051	1 0 0 0 2	
5.728E+00	5.961E+02	18	K077	1 2 2 2 2	
5.198E+00	5.409E+02	19.99	A339	2 0 2 2 2	
7.063E+00	7.350E+02	20	L041	1 0 0 1 2	
5.811E+00	6.047E+02	20	M043	1 0 0 0 2	
4.067E+00	4.232E+02	20	M171	1 0 0 0 2	
2.670E+00	2.778E+02	20	S006	1 0 0 0 2	
5.928E+00	6.169E+02	24	K077	1 2 2 2 2	
5.354E+00	5.571E+02	24.99	A339	2 0 2 2 2	
4.221E+00	4.393E+02	25	F300	1 0 0 0 2	
5.990E+00	6.233E+02	25	K077	1 2 2 2 2	
7.332E+00	7.630E+02	25	M051	1 0 0 0 2	
5.494E+00	5.717E+02	29.99	A339	2 0 2 2 2	
6.178E+00	6.429E+02	30	M043	1 0 0 0 2	
5.638E+00	5.867E+02	34.99	A339	2 0 2 2 2	
7.938E+00	8.260E+02	35	L041	1 0 0 1 2	
5.800E+00	6.035E+02	39.99	A339	2 0 2 2 2	
6.530E+00	6.795E+02	40	M043	1 0 0 0 2	
5.913E+00	6.153E+02	44.99	A339	2 0 2 2 2	
6.028E+00	6.273E+02	49.99	A339	2 0 2 2 2	
8.898E+00	9.260E+02	50	L041	1 0 0 1 2	
8.898E+00	9.260E+02	50	M051	1 0 0 0 2	
6.895E+00	7.175E+02	53	K077	1 2 2 2 2	
6.182E+00	6.433E+02	54.99	A339	2 0 2 2 2	
6.328E+00	6.585E+02	59.99	A339	2 0 2 2 2	
7.158E+00	7.449E+02	60	M043	1 0 0 0 2	
6.451E+00	6.713E+02	64.99	A339	2 0 2 2 2	
9.831E+00	1.023E+03	65	L041	1 0 0 1 2	

7.878E+00	8.198E+02	80	M043	1 0 0 0 2	
8.267E+00	8.603E+02	93	K077	1 2 2 2 2	
8.554E+00	8.901E+02	100	M043	1 0 0 0 2	
9.610E+00	1.000E+03	132	K077	1 2 2 2 2	
1.441E+01	1.500E+03	ns	D072	0 0 0 0 1	

137.   C₃H₅Br
Allyl Bromide
3-Bromopropene

RN: 106-95-6   **MP** (°C):   -119
MW: 120.98   **BP** (°C):   71.3

Solubility (Moles/L)	Solubility (Grams/L)	Temp (°C)	Ref (#)	Evaluation (T P E A A)	Comments
3.170E-02	3.835E+00	25	M342	1 0 1 1 2	

138.   C₃H₅Br₂Cl
1,2-Dibromo-3-chloropropane
1-Chloro-2,3-dibromopropane
Nemagon

RN: 96-12-8   **MP** (°C):
MW: 236.34   **BP** (°C):   196

Solubility (Moles/L)	Solubility (Grams/L)	Temp (°C)	Ref (#)	Evaluation (T P E A A)	Comments
5.204E-03	1.230E+00	20	G080	1 0 0 0 1	
4.227E-03	9.990E-01	20	P081	1 0 0 0 0	
4.227E-03	9.990E-01	ns	I316	0 0 0 0 0	
4.227E-03	9.990E-01	ns	M061	0 0 0 0 0	
4.231E-03	1.000E+00	rt	M161	0 0 0 0 0	

139.   C₃H₅Cl
Allyl Chloride
3-Chloro-1-propene

RN: 107-05-1   **MP** (°C):   -134
MW: 76.53   **BP** (°C):   44

Solubility (Moles/L)	Solubility (Grams/L)	Temp (°C)	Ref (#)	Evaluation (T P E A A)	Comments
4.687E-02	3.587E+00	20	G056	1 0 0 0 2	
1.305E-02	9.990E-01	ns	N034	0 0 0 0 0	

140.   C₃H₅ClO

Chloroacetone
1-Chloro-2-propanone
Chloraceton

RN: 78-95-5   **MP** (°C):   -44.5
MW: 92.53   **BP** (°C):   119.7

Solubility (Moles/L)	Solubility (Grams/L)	Temp (°C)	Ref (#)	Evaluation (T P E A A)	Comments
8.924E-01	8.257E+01	ns	N034	0 0 0 0 0	

141.   C₃H₅ClO

Epichlorohydrin
Epichloridrina

RN: 106-89-8   **MP** (°C):   -25.6
MW: 92.53   **BP** (°C):   117.9

Solubility (Moles/L)	Solubility (Grams/L)	Temp (°C)	Ref (#)	Evaluation (T P E A A)	Comments
6.577E-01	6.086E+01	0	L061	1 2 2 1 2	
6.615E-01	6.121E+01	10	L061	1 2 2 1 2	
6.501E-01	6.015E+01	20	I313	0 0 0 0 1	
6.692E-01	6.191E+01	30.20	L061	1 2 2 1 2	
7.568E-01	7.003E+01	52	L061	1 2 2 1 2	
8.421E-01	7.792E+01	65	L061	1 2 2 1 2	
9.232E-01	8.542E+01	72	L061	1 2 2 1 2	
1.024E+00	9.478E+01	80.20	L061	1 2 2 1 2	

142.   C₃H₅Cl₂NO₂

1,1-Dichloro-1-nitropropane
Propane, 1,1-Dichloro-1-nitro-

RN: 595-44-8   **MP** (°C):
MW: 157.98   **BP** (°C):   141

Solubility (Moles/L)	Solubility (Grams/L)	Temp (°C)	Ref (#)	Evaluation (T P E A A)	Comments
3.149E-02	4.975E+00	20	C121	1 0 0 0 0	unit assumed, *sic*

143.   C₃H₅Cl₃

1,2,3-Trichloropropane
Allyl Trichloride
Trichlorohydrin
Glycerol Trichlorohydrin

RN: 96-18-4   **MP** (°C):   -14
MW: 147.43   **BP** (°C):   156

Solubility (Moles/L)	Solubility (Grams/L)	Temp (°C)	Ref (#)	Evaluation (T P E A A)	Comments
1.289E-02	1.900E+00	ns	H123	0 0 0 0 2	

144.   C₃H₅IO₂

β-Iodopropionic Acid
β-Iod-propionsaeure

RN: 141-76-4   **MP** (°C):   81.5
MW: 199.98   **BP** (°C):

Solubility (Moles/L)	Solubility (Grams/L)	Temp (°C)	Ref (#)	Evaluation (T P E A A)	Comments
3.715E-01	7.430E+01	25	F300	1 0 0 0 2	

145.   C₃H₅N

Ethyl Isocyanide
Ethanc, Isocyano-

RN: 624-79-3   **MP** (°C):
MW: 55.08   **BP** (°C):

Solubility (Moles/L)	Solubility (Grams/L)	Temp (°C)	Ref (#)	Evaluation (T P E A A)	Comments
1.814E-02	9.990E-01	ns	L055	0 0 0 0 1	

146.   C₃H₅N

Propionitrile
Propionsaeure-nitril
n-Propioniitrile

RN: 107-12-0   **MP** (°C):   -93
MW: 55.08   **BP** (°C):   97

Solubility (Moles/L)	Solubility (Grams/L)	Temp (°C)	Ref (#)	Evaluation (T P E A A)	Comments
6.151E-02	3.388E+00	25	B004	1 0 0 0 2	

147.   C₃H₅NO

Acrylamide

2-Propenamide

RN: 79-06-1   **MP** (°C):   84

MW: 71.08   **BP** (°C):

Solubility (Moles/L)	Solubility (Grams/L)	Temp (°C)	Ref (#)	Evaluation (T P E A A)	Comments
4.299E+00	3.056E+02	0	M147	0 2 1 1 0	EFG
4.690E+00	3.333E+02	10	M147	0 2 1 1 0	EFG
5.220E+00	3.711E+02	20	M147	0 2 1 1 0	EFG
5.695E+00	4.048E+02	30	M147	0 2 1 1 0	EFG
6.075E+00	4.318E+02	40	M147	0 2 1 1 0	EFG
6.253E+00	4.444E+02	50	M147	0 2 1 1 0	EFG
6.625E+00	4.709E+02	60	M147	0 2 1 1 0	EFG
7.034E+00	5.000E+02	80	M147	0 2 1 1 0	EFG

148.   C₃H₅NO₃

Formylglycine

N-Formyl Glycine

RN: 2491-15-8   **MP** (°C):

MW: 103.08   **BP** (°C):

Solubility (Moles/L)	Solubility (Grams/L)	Temp (°C)	Ref (#)	Evaluation (T P E A A)	Comments
1.849E+00	1.906E+02	25	M024	1 2 0 1 2	
1.849E+00	1.906E+02	ns	M025	0 2 0 1 2	

149.   C₃H₅N₃O

Ethylnitrosocyanamide

ENC

RN: 38434-77-4   **MP** (°C):

MW: 99.09   **BP** (°C):

Solubility (Moles/L)	Solubility (Grams/L)	Temp (°C)	Ref (#)	Evaluation (T P E A A)	Comments
1.400E-01	1.387E+01	24	M031	1 1 1 1 1	

150. $C_3H_5N_3O_9$
Nitroglycerin
Nitroglycerol
RN: 55-63-0 **MP** (°C): 13.5
MW: 227.09 **BP** (°C): 256

Solubility (Moles/L)	Solubility (Grams/L)	Temp (°C)	Ref (#)	Evaluation (T P E A A)	Comments
5.629E-03	1.278E+00	15	L063	2 0 1 1 2	
7.926E-03	1.800E+00	20	F300	1 0 0 0 1	
6.069E-03	1.378E+00	20	L063	2 0 1 1 2	
5.504E-03	1.250E+00	25	P312	1 2 2 2 2	
6.595E-03	1.498E+00	30	L063	2 0 1 1 2	
7.342E-03	1.667E+00	40	L063	2 0 1 1 2	
8.570E-03	1.946E+00	50	L063	2 0 1 1 2	
1.041E-02	2.364E+00	60	L063	2 0 1 1 2	
1.265E-02	2.872E+00	70	L063	2 0 1 1 2	
1.518E-02	3.448E+00	80	L063	2 0 1 1 2	

151. $C_3H_5N_5O$
Ammeline
Ammelin
RN: 645-92-1 **MP** (°C):
MW: 127.11 **BP** (°C):

Solubility (Moles/L)	Solubility (Grams/L)	Temp (°C)	Ref (#)	Evaluation (T P E A A)	Comments
6.000E-04	7.626E-02	2	B193	1 1 0 0 0	
5.901E-04	7.500E-02	23	F300	1 0 0 0 1	
2.486E-03	3.160E-01	100	F300	1 0 0 0 2	

152. C_3H_6
Propylene
Methyl Ethylene
Propene
RN: 115-07-1 **MP** (°C): -185
MW: 42.08 **BP** (°C): -48

Solubility (Moles/L)	Solubility (Grams/L)	Temp (°C)	Ref (#)	Evaluation (T P E A A)	Comments
2.139E-02	9.000E-01	0	F300	1 0 0 0 1	
7.553E-03	3.178E-01	21	A052	1 1 1 2 2	smoothed
7.842E-03	3.300E-01	25	F300	1 0 0 0 1	
4.753E-03	2.000E-01	25	M001	2 1 2 2 2	
4.221E-03	1.776E-01	38	A052	1 1 1 2 1	smoothed
2.333E-03	9.818E-02	54	A052	1 1 1 2 1	smoothed
1.500E-03	6.312E-02	71	A052	1 1 1 2 1	smoothed
7.222E-04	3.039E-02	88	A052	1 1 1 2 1	smoothed

153.   C$_3$H$_6$
Cyclopropane
Trimethylene
RN: 75-19-4   **MP** (°C):   -127
MW: 42.08   **BP** (°C):   -33

Solubility (Moles/L)	Solubility (Grams/L)	Temp (°C)	Ref (#)	Evaluation (T P E A A)	Comments
2.461E-02	1.036E+00	5.05	Z008	2 1 2 2 2	at 97.26 kPa
1.281E-02	5.390E-01	20	R060	1 0 2 2 2	
1.754E-02	7.382E-01	21	I017	1 2 2 1 2	at 16.9 psia
1.103E-02	4.640E-01	25	R060	1 0 2 2 2	
9.315E-03	3.920E-01	30	R060	1 0 2 2 2	
8.983E-03	3.780E-01	31	R060	1 0 2 2 2	
7.723E-03	3.250E-01	35	R060	1 0 2 2 2	
1.083E-02	4.557E-01	38	I017	1 2 2 1 2	at 17.0 psia
6.844E-03	2.880E-01	39	R060	1 0 2 2 2	
5.917E-03	2.490E-01	45	R060	1 0 2 2 2	
8.386E-03	3.529E-01	71	I017	1 2 2 1 2	at 19.9 psia
3.999E-03	1.683E-01	104	I017	1 2 2 1 2	at 24.9 psia
5.896E+00	2.481E+02	ns	R028	0 0 0 0 1	

154.   C$_3$H$_6$BrCl
1-Bromo-3-chloropropane
ω-Chlorobromopropane
3-Bromopropyl chloride
3-Chloro-1-bromopropane
RN: 109-70-6   **MP** (°C):   -58.9
MW: 157.44   **BP** (°C):   143.3

Solubility (Moles/L)	Solubility (Grams/L)	Temp (°C)	Ref (#)	Evaluation (T P E A A)	Comments
1.420E-02	2.236E+00	25	M342	1 0 1 1 2	

155.   C$_3$H$_6$BrNO$_4$
Bronopol
2-Bromo-2-nitropropane-1,3-diol
RN: 52-51-7   **MP** (°C):   130
MW: 199.99   **BP** (°C):

Solubility (Moles/L)	Solubility (Grams/L)	Temp (°C)	Ref (#)	Evaluation (T P E A A)	Comments
1.000E+00	2.000E+02	22	M161	1 0 0 0 1	

156. $C_3H_6Br_2$
Trimethylene Bromide
1,3-Dibromopropane
RN: 109-64-8 **MP** (°C): -36
MW: 201.90 **BP** (°C): 167

Solubility (Moles/L)	Solubility (Grams/L)	Temp (°C)	Ref (#)	Evaluation (T P E A A)	Comments
8.406E-03	1.697E+00	20	C121	1 0 0 0 1	unit assumed, *sic*

157. $C_3H_6ClNO_2$
1-Chloro-1-nitropropane
Propane, 1-Chloro-1-nitro-
RN: 600-25-9 **MP** (°C):
MW: 123.54 **BP** (°C): 141

Solubility (Moles/L)	Solubility (Grams/L)	Temp (°C)	Ref (#)	Evaluation (T P E A A)	Comments
6.424E-02	7.937E+00	20	C121	1 0 0 0 0	unit assumed, *sic*
4.027E-02	4.975E+00	20	M061	1 0 0 0 0	

158. $C_3H_6ClNO_2$
1-Chloro-2-nitropropane
Propane, 1-Chloro-2-nitro-
RN: 37809-02-2 **MP** (°C):
MW: 123.54 **BP** (°C): 174

Solubility (Moles/L)	Solubility (Grams/L)	Temp (°C)	Ref (#)	Evaluation (T P E A A)	Comments
6.424E-02	7.937E+00	20	M061	1 0 0 0 0	

159. $C_3H_6Cl_2$
1,3-Dichloropropane
1,3-Dichlor-propan
RN: 142-28-9 **MP** (°C): -99
MW: 112.99 **BP** (°C): 120

Solubility (Moles/L)	Solubility (Grams/L)	Temp (°C)	Ref (#)	Evaluation (T P E A A)	Comments
2.559E-02	2.892E+00	20	C121	1 0 0 0 1	unit assumed, *sic*
2.416E-02	2.730E+00	25	F300	1 0 0 0 2	
2.430E-02	2.746E+00	25	G038	1 2 2 2 2	
2.430E-02	2.746E+00	25	G053	2 1 2 1 2	
9.027E-03	1.020E+00	30	M311	1 1 2 2 2	

160.   C$_3$H$_6$Cl$_2$
Propylene Dichloride
1,2-Dichlor-propan
1,2-Dichloropropane
Propylene Chloride
Dichloropropane
RN: 78-87-5   **MP** (°C):   -100.3
MW: 112.99   **BP** (°C):   96.8

Solubility (Moles/L)	Solubility (Grams/L)	Temp (°C)	Ref (#)	Evaluation (T P E A A)	Comments
3.160E-02	3.570E+00	20	C094	1 0 1 0 2	
2.383E-02	2.693E+00	20	C121	1 0 0 0 1	unit assumed, *sic*
2.390E-02	2.700E+00	20	F300	1 0 0 0 1	
2.390E-02	2.700E+00	20	M037	1 1 0 0 1	
2.383E-02	2.693E+00	20	M061	1 0 0 0 1	
2.295E-02	2.593E+00	20	M062	1 0 0 0 1	
2.390E-02	2.700E+00	20	M161	1 0 0 0 1	
2.500E-02	2.825E+00	20	M312	1 0 0 0 1	
2.383E-02	2.693E+00	20	N034	1 0 0 0 1	
2.478E-02	2.800E+00	25	F300	1 0 0 0 1	
2.480E-02	2.802E+00	25	G038	1 2 2 2 2	
2.480E-02	2.802E+00	25	G053	2 1 2 1 2	
2.295E-02	2.593E+00	25	G056	1 0 0 0 2	
2.142E-02	2.420E+00	30	M300	1 1 2 2 2	
1.831E-02	2.069E+00	30	M311	1 1 2 2 2	

161.   C$_3$H$_6$Cl$_2$O
1,3-Dichloro-2-propanol
1,3-Dichlor-propanol-(2)
RN: 96-23-1   **MP** (°C):   -4
MW: 128.99   **BP** (°C):   174.3

Solubility (Moles/L)	Solubility (Grams/L)	Temp (°C)	Ref (#)	Evaluation (T P E A A)	Comments
7.675E-01	9.900E+01	19	F300	1 0 0 0 1	
6.984E-01	9.008E+01	19	N034	1 0 0 0 1	
1.124E+00	1.450E+02	72	F300	1 0 0 0 2	

162.   C$_3$H$_6$N$_2$O$_2$
1-Acetylurea
Acetylharnstoff
RN: 591-07-1   **MP** (°C):   218
MW: 102.09   **BP** (°C):   185

Solubility (Moles/L)	Solubility (Grams/L)	Temp (°C)	Ref (#)	Evaluation (T P E A A)	Comments
1.273E-01	1.300E+01	15	F300	1 0 0 0 1	

163. $C_3H_6N_2O_2$
Malonic Acid Diamide
Malonsaeure-diamid
Malonamide
Malonodiamide
Propanediamide
RN: 108-13-4   **MP** (°C):   170
MW: 102.09   **BP** (°C):

Solubility (Moles/L)	Solubility (Grams/L)	Temp (°C)	Ref (#)	Evaluation (T P E A A)	Comments
7.513E-01	7.670E+01	8	F300	1 0 0 0 2	
7.830E-03	7.994E-01	ns	L055	0 0 0 0 1	

164. $C_3H_6N_2O_2$
Methylglyoxime
Methylglyoxim
RN: 1804-15-5   **MP** (°C):
MW: 102.09   **BP** (°C):

Solubility (Moles/L)	Solubility (Grams/L)	Temp (°C)	Ref (#)	Evaluation (T P E A A)	Comments
4.506E-01	4.600E+01	26	F300	1 0 0 0 1	
7.444E-01	7.600E+01	40	F300	1 0 0 0 1	

165. $C_3H_6N_2O_2$
Methylnitrosoacetamide
MNA
RN: 7417-67-6   **MP** (°C):
MW: 102.09   **BP** (°C):

Solubility (Moles/L)	Solubility (Grams/L)	Temp (°C)	Ref (#)	Evaluation (T P E A A)	Comments
1.700E-01	1.736E+01	24	M031	1 1 1 1 1	

166. $C_3H_6N_2O_3$
Hydantoic Acid
N-(Carboxymethyl)urea
N-Carbamoylglycine
Carbamoylglycine
Glycoluric Acid
RN: 462-60-2   **MP** (°C):
MW: 118.09   **BP** (°C):

Solubility (Moles/L)	Solubility (Grams/L)	Temp (°C)	Ref (#)	Evaluation (T P E A A)	Comments
2.549E-01	3.010E+01	20	F300	1 0 0 0 2	
3.290E-01	3.885E+01	25	M024	1 2 0 1 2	
3.290E-01	3.885E+01	ns	M025	0 2 0 1 2	

167.   C₃H₆N₂O₇

Glycerol 1,3-Dinitrate

Glycerol-α,α'-dinitrate

Glycerin-α,α'-dinitrat

RN:	623-87-0	**MP** (°C):	26	
MW:	182.09	**BP** (°C):	116	

Solubility (Moles/L)	Solubility (Grams/L)	Temp (°C)	Ref (#)	Evaluation (T P E A A)	Comments
3.993E-01	7.270E+01	20	D013	1 0 1 1 2	

168.   C₃H₆N₂O₇

Glycerol 1,2-Dinitrate

1,2,3-Propanetriol 1,2-dinitrate

1,2-Dinitroglycerol

RN:	131287-51-9	**MP** (°C):		
MW:	182.09	**BP** (°C):	106	

Solubility (Moles/L)	Solubility (Grams/L)	Temp (°C)	Ref (#)	Evaluation (T P E A A)	Comments
3.386E-01	6.165E+01	20	D013	1 0 1 1 2	

169.   C₃H₆N₂S

Ethylenethiourea

Mercaptoimidazoline

Mercozen

RN:	96-45-7	**MP** (°C):	203	
MW:	102.16	**BP** (°C):		

Solubility (Moles/L)	Solubility (Grams/L)	Temp (°C)	Ref (#)	Evaluation (T P E A A)	Comments
1.919E-01	1.961E+01	30	I310	0 0 0 0 0	
8.082E-01	8.257E+01	60	I310	0 0 0 0 0	
2.991E+00	3.056E+02	90	I310	0 0 0 0 1	

170.   C₃H₆N₄Hg

Methylmercuridicyanodiamide

Panogen

RN:	502-39-6	**MP** (°C):	156	
MW:	298.70	**BP** (°C):		

Solubility (Moles/L)	Solubility (Grams/L)	Temp (°C)	Ref (#)	Evaluation (T P E A A)	Comments
7.265E-02	2.170E+01	20	M061	1 0 0 0 2	
7.265E-02	2.170E+01	rt	M161	0 0 0 0 2	

171. $C_3H_6N_6$
Melamine
1,3,5-Triazine-2,4,6-triamine
Cymel
RN: 108-78-1   **MP** (°C):
MW: 126.12   **BP** (°C):

Solubility (Moles/L)	Solubility (Grams/L)	Temp (°C)	Ref (#)	Evaluation (T P E A A)	Comments
9.503E-03	1.199E+00	0	M043	1 0 0 0 1	
1.000E-02	1.261E+00	2	B193	1 1 0 0 1	
1.425E-02	1.797E+00	10	M043	1 0 0 0 1	
2.561E-02	3.230E+00	19.90	C023	2 2 0 1 2	
2.135E-02	2.693E+00	20	M043	1 0 0 0 1	
3.316E-02	4.182E+00	30	M043	1 0 0 0 1	
4.651E-02	5.865E+00	34.90	C023	2 2 0 1 2	
5.590E-02	7.050E+00	40	M043	1 0 0 0 1	
8.200E-02	1.034E+01	49.80	C023	2 2 0 1 2	
1.172E-01	1.478E+01	60	M043	1 0 0 0 1	
1.325E-01	1.672E+01	64.10	C023	2 2 0 1 2	
1.836E-01	2.315E+01	74.50	C023	2 2 0 1 2	
2.160E-01	2.724E+01	80	M043	1 0 0 0 1	
2.421E-01	3.054E+01	83.50	C023	2 2 0 1 2	
3.480E-01	4.389E+01	94.80	C023	2 2 0 1 2	
3.812E-01	4.807E+01	99	C023	2 2 0 1 2	
3.776E-01	4.762E+01	100	M043	1 0 0 0 1	

172. $C_3H_6N_6O_6$
Cyclonite
RDX
RN: 121-82-4   **MP** (°C):   205
MW: 222.12   **BP** (°C):

Solubility (Moles/L)	Solubility (Grams/L)	Temp (°C)	Ref (#)	Evaluation (T P E A A)	Comments
2.690E-04	5.975E-02	25	B173	2 0 2 2 2	

173. C_3H_6O
Propylene Oxide
Methyl Ethylene Oxide
RN: 75-56-9   **MP** (°C):   -112
MW: 58.08   **BP** (°C):   34.23

Solubility (Moles/L)	Solubility (Grams/L)	Temp (°C)	Ref (#)	Evaluation (T P E A A)	Comments
4.963E+00	2.883E+02	20	I313	0 0 0 0 2	
2.544E-01	1.478E+01	20	M065	1 0 2 1 1	*sic*
6.389E+00	3.711E+02	25	I313	0 0 0 0 1	

174.   C₃H₆O

Propaldehyde
Propyl Aldehyde
Propanal

RN: 123-38-6   **MP** (°C):   -81
MW: 58.08   **BP** (°C):

Solubility (Moles/L)	Solubility (Grams/L)	Temp (°C)	Ref (#)	Evaluation (T P E A A)	Comments
2.870E+00	1.667E+02	20	D041	1 0 0 0 0	
2.927E+00	1.700E+02	20	F300	1 0 0 0 1	
5.269E+00	3.060E+02	25	A049	1 0 0 0 2	
3.105E+00	1.803E+02	25	B060	2 0 1 1 1	
2.880E+00	1.673E+02	25	F044	1 0 0 0 2	

175.   C₃H₆O₂

Ethyl Formate
Ameisensaeure-aethyl Ester
Formic Acid Ethyl Ester

RN: 109-94-4   **MP** (°C):   -80
MW: 74.08   **BP** (°C):   53

Solubility (Moles/L)	Solubility (Grams/L)	Temp (°C)	Ref (#)	Evaluation (T P E A A)	Comments
1.094E+00	8.108E+01	5.0	K079	1 0 0 0 2	
1.139E+00	8.437E+01	15.9	K079	1 0 0 0 2	
1.350E+00	1.000E+02	18	F300	1 0 0 0 1	
1.350E+00	1.000E+02	22	S006	1 0 0 0 2	
1.194E+00	8.848E+01	30.2	K079	1 0 0 0 2	
1.239E+00	9.178E+01	38.0	K079	1 0 0 0 2	
1.283E+00	9.507E+01	45.1	K079	1 0 0 0 2	
1.339E+00	9.918E+01	50.0	K079	1 0 0 0 2	
1.383E+00	1.025E+02	55.5	K079	1 0 0 0 2	
1.517E+00	1.124E+02	63.9	K079	1 0 0 0 2	
1.639E+00	1.214E+02	70.0	K079	1 0 0 0 2	
1.778E+00	1.317E+02	75.5	K079	1 0 0 0 2	

176. C₃H₆O₂

176. $C_3H_6O_2$
Methyl Acetate
Essigsaeures Methyl
Methylacetat

RN:	79-20-9	**MP** (°C):	-98.0	
MW:	74.08	**BP** (°C):	56.9	

Solubility (Moles/L)	Solubility (Grams/L)	Temp (°C)	Ref (#)	Evaluation (T P E A A)	Comments
3.678E+00	2.725E+02	5.0	K079	1 0 0 0 2	
4.017E+00	2.976E+02	20	E002	1 0 0 0 2	
3.290E+00	2.437E+02	20	F001	1 0 1 2 2	
2.647E+00	1.961E+02	20	F300	1 0 0 0 2	
3.290E+00	2.437E+02	20	M171	1 0 0 0 2	
4.617E+00	3.420E+02	20	P040	1 0 0 0 2	
4.300E+00	3.185E+02	20	S006	1 0 0 0 1	
3.722E+00	2.757E+02	21.0	K079	1 0 0 0 2	
2.772E-02	2.054E+00	25	B004	1 0 0 0 2	*sic*
3.772E+00	2.794E+02	35.0	K079	1 0 0 0 2	
3.889E+00	2.881E+02	58.0	K079	1 0 0 0 2	
3.906E+00	2.893E+02	58.9	K079	1 0 0 0 2	
3.922E+00	2.906E+02	60.1	K079	1 0 0 0 2	
3.950E+00	2.926E+02	61.7	K079	1 0 0 0 2	
4.172E+00	3.091E+02	69.1	K079	1 0 0 0 2	
4.256E+00	3.153E+02	70.5	K079	1 0 0 0 2	
4.294E+00	3.181E+02	71.9	K079	1 0 0 0 2	
4.906E+00	3.634E+02	83.5	K079	1 0 0 0 2	
4.252E-02	3.150E+00	c	L055	0 0 0 0 2	

177. C₃H₆O₂

177. $C_3H_6O_2$
Propionic Acid
n-Propionic Acid

RN:	79-09-4	**MP** (°C):	-22	
MW:	74.08	**BP** (°C):	141	

Solubility (Moles/L)	Solubility (Grams/L)	Temp (°C)	Ref (#)	Evaluation (T P E A A)	Comments
2.733E-01	2.025E+01	25	B004	1 0 0 0 2	

178. C₃H₆O₂S₃

178. $C_3H_6O_2S_3$
α-Trimethylene Trisulphide Dioxide
1,3,5-Trithiane, 1,3-Dioxide, trans-

RN:	60077-04-5	**MP** (°C):		
MW:	170.27	**BP** (°C):		

Solubility (Moles/L)	Solubility (Grams/L)	Temp (°C)	Ref (#)	Evaluation (T P E A A)	Comments
9.817E-02	1.672E+01	25	B112	1 2 1 1 2	

179. $C_3H_6O_2S_3$
β-Trimethylene Trisulphide Dioxide
1,3,5-Trithiane, 1,3-Dioxide, cis-
RN: 60041-48-7   **MP** (°C):
MW: 170.27   **BP** (°C):

Solubility (Moles/L)	Solubility (Grams/L)	Temp (°C)	Ref (#)	Evaluation (T P E A A)	Comments
2.545E-01	4.334E+01	25	B112	1 2 1 1 2	

180. $C_3H_6O_3$
Hydracrylic Acid
Hydracrylsaeure
RN: 503-66-2   **MP** (°C):
MW: 90.08   **BP** (°C):

Solubility (Moles/L)	Solubility (Grams/L)	Temp (°C)	Ref (#)	Evaluation (T P E A A)	Comments
2.998E+00	2.701E+02	25	I307	0 0 0 0 1	

181. $C_3H_6O_3$
s-Trioxane
1,3,5-Trioxan
RN: 110-88-3   **MP** (°C):   64
MW: 90.08   **BP** (°C):   114.5

Solubility (Moles/L)	Solubility (Grams/L)	Temp (°C)	Ref (#)	Evaluation (T P E A A)	Comments
1.715E+00	1.544E+02	20.00	B394	1 1 2 2 2	
1.943E+00	1.750E+02	25	F300	1 0 0 0 2	
2.033E+00	1.831E+02	25.00	B394	1 1 2 2 2	
2.403E+00	2.165E+02	30.10	B394	1 1 2 2 2	
2.741E+00	2.469E+02	34.45	B394	1 1 2 2 2	
4.187E+00	3.772E+02	43.00	B394	1 1 2 2 2	
4.462E+00	4.019E+02	44.00	B394	1 1 2 2 2	
4.606E+00	4.149E+02	44.40	B394	1 1 2 2 2	
4.826E+00	4.348E+02	45.00	B394	1 1 2 2 2	
4.816E+00	4.338E+02	45.10	B394	1 1 2 2 2	
5.355E+00	4.824E+02	46.00	B394	1 1 2 2 2	
5.311E+00	4.784E+02	46.10	B394	1 1 2 2 2	
6.401E+00	5.766E+02	47.10	B394	1 1 2 2 2	
8.161E+00	7.351E+02	47.80	B394	1 1 2 2 2	
8.534E+00	7.687E+02	48.95	B394	1 1 2 2 2	
8.741E+00	7.874E+02	50.20	B394	1 1 2 2 2	
9.095E+00	8.192E+02	55.30	B394	1 1 2 2 2	

182.   C$_3$H$_6$O$_3$
DL-Glyceraldehyde
DL-Glycerin-aldehyd

RN: 56-82-6   **MP** (°C):   145
MW: 90.08   **BP** (°C):   150

Solubility (Moles/L)	Solubility (Grams/L)	Temp (°C)	Ref (#)	Evaluation (T P E A A)	Comments
3.233E-01	2.913E+01	18	D041	1 0 0 0 0	
3.242E-01	2.920E+01	18	F300	1 0 0 0 2	

183.   C$_3$H$_6$O$_3$S$_3$
β-Trimethylene Trisulphoxide
1,3,5-Trithiane, 1,3,5-Trioxide, (1α,3α,5β)-

RN: 60102-88-7   **MP** (°C):
MW: 186.27   **BP** (°C):

Solubility (Moles/L)	Solubility (Grams/L)	Temp (°C)	Ref (#)	Evaluation (T P E A A)	Comments
7.605E-02	1.417E+01	25	B112	1 2 1 1 2	

184.   C$_3$H$_6$O$_3$S$_3$
α-Trimethylene Trisulphoxide
1,3,5-Trithiane, 1,3,5-Trioxide, (1α,3α,5α)-

RN: 60102-87-6   **MP** (°C):
MW: 186.27   **BP** (°C):

Solubility (Moles/L)	Solubility (Grams/L)	Temp (°C)	Ref (#)	Evaluation (T P E A A)	Comments
7.184E-03	1.338E+00	25	B112	1 2 1 1 2	

185.   C$_3$H$_6$O$_3$S
1,3-Propane Sultone
1,2-Oxathiolane 2,2-dioxide
3-Hydroxy-1-propanesulfonic Acid γ-sultone

RN: 1120-71-4   **MP** (°C):   31
MW: 122.14   **BP** (°C):   112

Solubility (Moles/L)	Solubility (Grams/L)	Temp (°C)	Ref (#)	Evaluation (T P E A A)	Comments
8.187E-01	1.000E+02	ns	I307	0 0 0 0 2	

186.   C₃H₇Br

Isopropyl Bromide
Isopropylbromid
RN: 75-26-3   **MP** (°C):   -89
MW: 123.00   **BP** (°C):   59

Solubility (Moles/L)	Solubility (Grams/L)	Temp (°C)	Ref (#)	Evaluation (T P E A A)	Comments
3.398E-02	4.180E+00	0	H101	2 0 0 0 2	
2.340E-02	2.878E+00	18	F001	1 0 1 2 2	
2.602E-02	3.200E+00	20	F300	1 0 0 0 1	
2.585E-02	3.180E+00	20	H101	2 0 0 0 2	
2.592E-02	3.188E+00	30	V009	1 0 0 0 1	

187.   C₃H₇Br

Propyl Bromide
1-Bromopropane
Propylbromid
Bromopropane
RN: 106-94-5   **MP** (°C):   -110
MW: 123.00   **BP** (°C):   71

Solubility (Moles/L)	Solubility (Grams/L)	Temp (°C)	Ref (#)	Evaluation (T P E A A)	Comments
2.415E-02	2.970E+00	0	F300	1 0 0 0 2	
2.423E-02	2.980E+00	0	H101	2 0 0 0 2	
1.850E-02	2.275E+00	19.5	S006	1 0 0 0 2	
1.850E-02	2.275E+00	19.50	F001	1 0 1 0 2	
1.992E-02	2.450E+00	20	H101	2 0 0 0 2	
1.947E-02	2.394E+00	20	H127	1 0 0 0 1	
1.874E-02	2.305E+00	30	G029	1 0 2 2 2	
1.876E-02	2.307E+00	30	V009	1 0 0 0 2	
1.140E-01	1.402E+01	ns	H307	1 0 1 1 2	

188.   C₃H₇BrO

3-Bromo-1-propanol
3-Brom-propanol-(1)
RN: 627-18-9   **MP** (°C):
MW: 139.00   **BP** (°C):

Solubility (Moles/L)	Solubility (Grams/L)	Temp (°C)	Ref (#)	Evaluation (T P E A A)	Comments
1.022E+00	1.420E+02	20	F300	1 0 0 0 2	

189.   C₃H₇Cl
Chloropropane
Propyl Chloride
1-Chloropropane

RN: 540-54-5   **MP** (°C):   -123
MW: 78.54   **BP** (°C):   43.47

Solubility (Moles/L)	Solubility (Grams/L)	Temp (°C)	Ref (#)	Evaluation (T P E A A)	Comments
4.787E-02	3.760E+00	0	H101	2 0 0 0 2	
2.970E-02	2.333E+00	12.50	F001	1 0 1 0 2	
3.438E-02	2.700E+00	20	F300	1 0 0 0 1	
3.463E-02	2.720E+00	20	H101	2 0 0 0 2	
3.428E-02	2.693E+00	20	N034	1 0 0 0 1	
2.970E-02	2.333E+00	20	S006	1 0 0 0 2	
3.520E-02	2.765E+00	30	V009	1 0 0 0 2	

190.   C₃H₇Cl
Isopropyl Chloride
2-Chloropropane

RN: 75-29-6   **MP** (°C):   -117
MW: 78.54   **BP** (°C):   35

Solubility (Moles/L)	Solubility (Grams/L)	Temp (°C)	Ref (#)	Evaluation (T P E A A)	Comments
5.602E-02	4.400E+00	0	H101	2 0 0 0 2	
4.380E-02	3.440E+00	12.50	F001	1 0 1 0 2	
3.947E-02	3.100E+00	20	F300	1 0 0 0 1	
3.883E-02	3.050E+00	20	H101	2 0 0 0 2	
3.935E-02	3.090E+00	20	N034	1 0 0 0 1	
3.888E-02	3.054E+00	30	V009	1 0 0 0 1	

191.   C₃H₇ClO
3-Chloro-1-propanol
3-Chlor-propanol-(1)

RN: 627-30-5   **MP** (°C):
MW: 94.54   **BP** (°C):

Solubility (Moles/L)	Solubility (Grams/L)	Temp (°C)	Ref (#)	Evaluation (T P E A A)	Comments
2.644E+00	2.500E+02	20	F300	1 0 0 0 1	

192.   C$_3$H$_7$I
Isopropyl Iodide
2-Iodopropane
RN: 75-30-9   **MP** (°C):   -90
MW: 169.99   **BP** (°C):   89

Solubility (Moles/L)	Solubility (Grams/L)	Temp (°C)	Ref (#)	Evaluation (T P E A A)	Comments
9.824E-03	1.670E+00	0	H101	2 0 0 0 2	
8.236E-03	1.400E+00	20	F300	1 0 0 0 1	
8.236E-03	1.400E+00	20	H101	2 0 0 0 2	
7.889E-03	1.341E+00	30	V009	1 0 0 0 1	

193.   C$_3$H$_7$I
Iodopropane
n-Propyl Iodide
RN: 107-08-4   **MP** (°C):   -101
MW: 169.99   **BP** (°C):   101.5

Solubility (Moles/L)	Solubility (Grams/L)	Temp (°C)	Ref (#)	Evaluation (T P E A A)	Comments
6.706E-03	1.140E+00	0	H101	2 0 0 0 2	
5.100E-03	8.670E-01	20	F001	1 0 1 0 2	
5.118E-03	8.700E-01	20	F300	1 0 0 0 1	
6.294E-03	1.070E+00	20	H101	2 0 0 0 2	
5.100E-03	8.670E-01	20	M171	1 0 0 0 1	
5.100E-03	8.670E-01	20	S006	1 0 0 0 1	
6.258E-03	1.064E+00	23.5	S171	2 1 2 2 2	
6.112E-03	1.039E+00	30	G029	1 0 2 2 2	
6.094E-03	1.036E+00	30	V009	1 0 0 0 1	

194.   C$_3$H$_7$NO$_2$
DL-Alanine
DL-α-Alanine
DL-2-Aminopropionic Acid
RN: 302-72-7   **MP** (°C):   289
MW: 89.09   **BP** (°C):

Solubility (Moles/L)	Solubility (Grams/L)	Temp (°C)	Ref (#)	Evaluation (T P E A A)	Comments
1.212E+00	1.080E+02	0	D018	2 2 2 1 2	
1.212E+00	1.080E+02	0	F300	1 0 0 0 2	
1.212E+00	1.079E+02	0	M043	1 0 0 0 2	
1.361E+00	1.213E+02	10	M043	1 0 0 0 0	
1.523E+00	1.357E+02	20	M043	1 0 0 0 0	
1.557E+00	1.387E+02	21	P045	1 0 2 1 2	
1.659E+00	1.478E+02	25	C018	1 0 2 2 2	
1.596E+00	1.422E+02	25	D018	2 2 2 1 2	

1.598E+00	1.424E+02	25	D041	1 0 0 0 2
1.607E+00	1.432E+02	25	F300	1 0 0 0 2
1.900E+00	1.693E+02	25	J303	2 0 2 2 2
1.530E+00	1.363E+02	25	K031	2 1 2 1 2
2.024E+00	1.803E+02	30	J303	2 0 2 2 2
1.704E+00	1.518E+02	30	M043	1 0 0 0 0
2.307E+00	2.055E+02	40	J303	2 0 2 2 2
1.894E+00	1.687E+02	40	M043	1 0 0 0 0
2.134E+00	1.902E+02	50	D018	2 2 2 1 2
2.106E+00	1.876E+02	50	F300	1 0 0 0 2
2.591E+00	2.308E+02	50	J303	2 0 2 2 2
2.954E+00	2.632E+02	60	J303	2 0 2 2 2
2.337E+00	2.082E+02	60	M043	1 0 0 0 0
2.733E+00	2.435E+02	75	D018	2 2 2 1 2
2.734E+00	2.436E+02	75	D041	1 0 0 0 2
2.714E+00	2.418E+02	75	F300	1 0 0 0 2
2.842E+00	2.532E+02	80	M043	1 0 0 0 0
3.432E+00	3.057E+02	99.99	P349	1 0 0 2 2
3.431E+00	3.057E+02	100	F300	1 0 0 0 2
3.430E+00	3.056E+02	100	M043	1 0 0 0 2

195. $C_3H_7NO_2$
Lactamide
2-Hydroxypropionamide
RN: 2043-43-8   **MP** (°C):
MW: 89.09   **BP** (°C):

Solubility (Moles/L)	Solubility (Grams/L)	Temp (°C)	Ref (#)	Evaluation (T P E A A)	Comments
8.779E+00	7.822E+02	25	M008	1 0 0 0 2	

196. $C_3H_7NO_2$
Sarcosine
Sarkosin
RN: 107-97-1   **MP** (°C):   208
MW: 89.09   **BP** (°C):

Solubility (Moles/L)	Solubility (Grams/L)	Temp (°C)	Ref (#)	Evaluation (T P E A A)	Comments
5.151E-01	4.589E+01	20	D041	1 0 0 0 2	
3.367E+00	3.000E+02	20	F300	1 0 0 0 2	
4.807E+00	4.282E+02	20	P045	1 0 2 1 2	

197. $C_3H_7NO_2$
β-Alanine
β-Alanin
RN: 107-95-9   **MP** (°C):
MW: 89.09   **BP** (°C):

Solubility (Moles/L)	Solubility (Grams/L)	Temp (°C)	Ref (#)	Evaluation (T P E A A)	Comments
3.959E+00	3.528E+02	25	D041	1 0 0 0 2	
6.123E+00	5.455E+02	25	M024	1 2 0 1 2	

198. $C_3H_7NO_2$
α-Alanine
Alanine
2-Aminopropanoic Acid
2-Ammoniopropanoate
L-2-Aminopropionic Acid
RN: 56-41-7   **MP** (°C):   314.5-316.5
MW: 89.09   **BP** (°C):

Solubility (Moles/L)	Solubility (Grams/L)	Temp (°C)	Ref (#)	Evaluation (T P E A A)	Comments
1.192E+00	1.062E+02	0	C347	2 0 2 2 0	EFG
1.366E+00	1.217E+02	10	C347	2 0 2 2 0	EFG
1.640E+00	1.461E+02	15	D349	2 1 1 2 2	
1.744E+00	1.554E+02	20	B032	1 2 2 1 2	
1.535E+00	1.367E+02	20	C347	2 0 2 2 0	EFG
1.780E+00	1.586E+02	20	D349	2 1 1 2 2	
1.838E+00	1.638E+02	25	B032	1 2 2 1 2	
1.590E+00	1.417E+02	25	D005	2 2 1 1 2	
1.602E+00	1.427E+02	25	D041	1 0 0 0 2	
1.870E+00	1.666E+02	25	D349	2 1 1 2 2	
1.660E+00	1.479E+02	25	E015	1 2 1 1 1	
1.595E+00	1.421E+02	25	G092	2 1 1 1 1	
1.595E+00	1.421E+02	25	G315	1 0 2 2 2	
1.852E+00	1.650E+02	25	J303	2 0 2 2 2	
1.600E+00	1.426E+02	25	N001	2 0 2 1 0	EFG
1.630E+00	1.452E+02	25	N012	2 0 2 1 2	
1.555E+00	1.386E+02	25	O316	1 0 1 2 2	
1.598E+00	1.424E+02	25	O316	1 0 1 2 2	
1.623E+00	1.446E+02	25	O317	1 0 1 2 2	
1.871E+00	1.667E+02	25.1	N024	2 0 2 2 2	
1.871E+00	1.667E+02	25.1	N026	2 0 2 2 2	
1.606E+00	1.431E+02	25.1	N027	1 1 2 2 2	
1.695E+00	1.510E+02	27	D036	2 1 2 2 2	
1.704E+00	1.518E+02	27	D036	2 1 2 2 2	
1.940E+00	1.728E+02	29.80	B032	1 2 2 1 2	
1.657E+00	1.477E+02	30	C347	2 0 2 2 0	EFG

Solubility (Moles/L)	Solubility (Grams/L)	Temp (°C)	Ref (#)	Evaluation (T P E A A)	Comments
1.956E+00	1.743E+02	30	J303	2 0 2 2 2	
1.816E+00	1.618E+02	40	C347	2 0 2 2 0	EFG
2.192E+00	1.953E+02	40	J303	2 0 2 2 2	
1.931E+00	1.720E+02	45	F300	1 0 0 0 2	
1.932E+00	1.721E+02	50	C347	2 0 2 2 0	EFG
2.430E+00	2.165E+02	50	J303	2 0 2 2 2	
2.118E+00	1.887E+02	60	C347	2 0 2 2 0	EFG
2.706E+00	2.411E+02	60	J303	2 0 2 2 2	
2.333E+00	2.078E+02	70	C347	2 0 2 2 0	EFG
2.489E+00	2.218E+02	75	D041	1 0 0 0 2	
2.504E+00	2.230E+02	80	C347	2 0 2 2 0	EFG
2.668E+00	2.377E+02	90	C347	2 0 2 2 0	EFG
2.888E+00	2.573E+02	100	C347	2 0 2 2 0	EFG
1.587E+00	1.414E+02	rt	D021	0 0 1 1 2	

199. $C_3H_7NO_2$
2-Nitropropane
Nitroisopropane
Dimethylnitromethane

RN: 79-46-9   **MP** (°C):   -93
MW: 89.09   **BP** (°C):   120.3

Solubility (Moles/L)	Solubility (Grams/L)	Temp (°C)	Ref (#)	Evaluation (T P E A A)	Comments
1.876E-01	1.672E+01	20	C121	0 0 0 0 1	unit assumed, *sic*
1.874E-01	1.670E+01	20	F300	1 0 0 0 2	
2.376E-01	2.117E+01	20	H118	1 1 1 1 2	

200. $C_3H_7NO_2$
1-Nitropropane
n-Nitropropane

RN: 108-03-2   **MP** (°C):   -108
MW: 89.09   **BP** (°C):   131.6

Solubility (Moles/L)	Solubility (Grams/L)	Temp (°C)	Ref (#)	Evaluation (T P E A A)	Comments
1.550E-01	1.381E+01	20	C121	1 0 0 0 1	unit assumed, *sic*

201. $C_3H_7NO_2$
D-Alanine
D(-)-Alanine

RN: 338-69-2   **MP** (°C):   292
MW: 89.09   **BP** (°C):

Solubility (Moles/L)	Solubility (Grams/L)	Temp (°C)	Ref (#)	Evaluation (T P E A A)	Comments
1.265E+00	1.127E+02	0	M043	1 0 0 0 2	
1.396E+00	1.243E+02	10	M043	1 0 0 0 2	
1.530E+00	1.363E+02	20	D041	1 0 0 0 2	
1.531E+00	1.364E+02	20	M043	1 0 0 0 2	

1.589E+00	1.416E+02	25	D005	2 2 1 1 2
1.680E+00	1.497E+02	30	M043	1 0 0 0 2
1.839E+00	1.639E+02	40	M043	1 0 0 0 2
2.194E+00	1.955E+02	60	M043	1 0 0 0 2
2.590E+00	2.308E+02	80	M043	1 0 0 0 2
3.049E+00	2.717E+02	99.99	P349	1 0 0 2 2
3.049E+00	2.717E+02	100	M043	1 0 0 0 2

202. $C_3H_7NO_2$
Urethan
Carbamidsaeure-aethyl Ester
Eythyl Urethan
Urethane
Ethyl Carbamate
Carbamic Acid Ethyl Ester
RN: 51-79-6   **MP** (°C):   49
MW: 89.09   **BP** (°C):   183

Solubility (Moles/L)	Solubility (Grams/L)	Temp (°C)	Ref (#)	Evaluation (T P E A A)	Comments
2.918E+00	2.600E+02	11	F300	1 0 0 0 1	
5.393E+00	4.805E+02	15.5	F001	1 0 1 2 2	
2.245E+01	2.000E+03	25	I310	0 0 0 0 0	
5.074E+00	4.521E+02	25	P065	2 0 1 1 2	
1.800E+01	1.604E+03	37	H006	1 2 2 1 1	
8.901E+00	7.930E+02	40	F300	1 0 0 0 2	

203. $C_3H_7NO_2S$
Cysteine
2-Amino-3-mercaptopropanoic Acid
RN: 3374-22-9   **MP** (°C):   225
MW: 121.16   **BP** (°C):

Solubility (Moles/L)	Solubility (Grams/L)	Temp (°C)	Ref (#)	Evaluation (T P E A A)	Comments
2.773E-02	3.360E+00	20	P045	1 0 2 1 2	

204. $C_3H_7NO_3$
DL-Isoserine
DL-Isoserin
RN: 632-12-2   **MP** (°C):   235
MW: 105.09   **BP** (°C):

Solubility (Moles/L)	Solubility (Grams/L)	Temp (°C)	Ref (#)	Evaluation (T P E A A)	Comments
1.456E-01	1.530E+01	20	F300	1 0 0 0 2	

205.   C$_3$H$_7$NO$_3$
DL-Serine
DL-2-Amino-3-hydroxypropanoic Acid
RN: 302-84-1   **MP** (°C):   240
MW: 105.09   **BP** (°C):

Solubility (Moles/L)	Solubility (Grams/L)	Temp (°C)	Ref (#)	Evaluation (T P E A A)	Comments
2.778E-01	2.920E+01	10	F300	1 0 0 0 2	
3.787E-01	3.980E+01	20	F300	1 0 0 0 2	
4.548E-01	4.780E+01	25	D041	1 0 0 0 2	
4.805E-01	5.050E+01	25	J303	2 0 2 2 2	
7.403E-01	7.780E+01	40	J303	2 0 2 2 2	
8.916E-01	9.370E+01	50	F300	1 0 0 0 2	
1.261E+00	1.325E+02	60	J303	2 0 2 2 2	
1.533E+00	1.611E+02	75	D041	1 0 0 0 2	
1.532E+00	1.610E+02	75	F300	1 0 0 0 2	
2.320E+00	2.438E+02	99.99	P349	1 0 0 2 2	
2.320E+00	2.438E+02	100	F300	1 0 0 0 2	

206.   C$_3$H$_7$NO$_3$
Serine
2-Amino-3-hydroxypropanoic Acid
L(-)-Serin
RN: 56-45-1   **MP** (°C):   220
MW: 105.09   **BP** (°C):

Solubility (Moles/L)	Solubility (Grams/L)	Temp (°C)	Ref (#)	Evaluation (T P E A A)	Comments
4.530E-01	4.761E+01	15	D349	2 1 1 2 2	
1.903E+00	2.000E+02	20	D041	1 0 0 0 1	
4.610E-01	4.845E+01	20	D349	2 1 1 2 2	
9.512E-01	9.997E+01	20	F300	1 0 0 0 2	
3.405E+00	3.578E+02	20.00	B032	1 2 2 1 2	*sic*
4.700E-01	4.939E+01	25	D349	2 1 1 2 2	
2.807E+00	2.950E+02	25	G315	1 0 2 2 2	*sic*
4.013E+00	4.217E+02	25	J303	2 0 2 2 2	
4.043E+00	4.249E+02	25.00	B032	1 2 2 0 2	*sic*
3.578E+00	3.760E+02	27	D036	2 1 2 2 2	
4.690E+00	4.929E+02	29.80	B032	1 2 2 1 2	*sic*
5.633E+00	5.920E+02	40	J303	2 0 2 2 2	
7.574E+00	7.960E+02	60	J303	2 0 2 2 2	

207.   C₃H₇NO₃
D-Serine
D-2-Amino-3-hydroxypropanoic Acid
RN: 312-84-5   **MP** (°C):   220
MW: 105.09   **BP** (°C):

Solubility (Moles/L)	Solubility (Grams/L)	Temp (°C)	Ref (#)	Evaluation (T P E A A)	Comments
1.903E+00	2.000E+02	20	D041	1 0 0 0 0	
4.010E+00	4.214E+02	25	J303	2 0 2 2 2	
5.709E+00	6.000E+02	40	J303	2 0 2 2 2	
7.631E+00	8.020E+02	60	J303	2 0 2 2 2	

208.   C₃H₇NO₅
Glycerol-α-nitrate
Glycerin-α-nitrat
RN: 27321-61-5   **MP** (°C):
MW: 137.09   **BP** (°C):

Solubility (Moles/L)	Solubility (Grams/L)	Temp (°C)	Ref (#)	Evaluation (T P E A A)	Comments
3.004E+00	4.118E+02	15	F300	1 0 0 0 2	

209.   C₃H₇N₃O₂
Nitrosoethylurea
N-Nitroso-N-Ethylurea
RN: 759-73-9   **MP** (°C):   103
MW: 117.11   **BP** (°C):

Solubility (Moles/L)	Solubility (Grams/L)	Temp (°C)	Ref (#)	Evaluation (T P E A A)	Comments
1.096E-01	1.283E+01	rt	I306	0 0 0 0 1	

210.   C₃H₇N₃O₂
Glycocyamine
Guanidin-essigsaeure
Guanidineacetic Acid
RN: 352-97-6   **MP** (°C):   280
MW: 117.11   **BP** (°C):

Solubility (Moles/L)	Solubility (Grams/L)	Temp (°C)	Ref (#)	Evaluation (T P E A A)	Comments
3.825E-02	4.480E+00	15	D041	1 0 0 0 1	
3.074E-02	3.600E+00	15	F300	1 0 0 0 1	

211. $C_3H_7O_5P$
2-Carboxyethylphosphonic Acid
3-Phosphonopropionic Acid
RN: 5962-42-5   **MP** (°C):
MW: 154.06   **BP** (°C):

Solubility (Moles/L)	Solubility (Grams/L)	Temp (°C)	Ref (#)	Evaluation (T P E A A)	Comments
1.845E+00	2.842E+02	0	N028	1 0 0 0 2	
2.129E+00	3.280E+02	20	N028	1 0 0 0 2	

212. C_3H_8
Propane
Propan
RN: 74-98-6   **MP** (°C):   -187
MW: 44.10   **BP** (°C):   -42

Solubility (Moles/L)	Solubility (Grams/L)	Temp (°C)	Ref (#)	Evaluation (T P E A A)	Comments
3.460E-03	1.526E-01	4	K031	2 1 2 1 2	
2.472E-03	1.090E-01	10	F300	1 0 0 0 2	
2.721E-03	1.200E-01	18	M065	0 0 2 1 1	1 atm, *sic*
1.761E-03	7.765E-02	19.8	G058	1 0 0 0 2	
1.746E-03	7.700E-02	20	F300	1 0 0 0 1	
1.420E-03	6.261E-02	25	B342	1 1 2 1 1	
1.530E-03	6.747E-02	25	K031	2 1 2 1 2	
1.415E-03	6.240E-02	25	M001	2 1 2 2 2	
1.415E-03	6.240E-02	25	M002	2 1 2 2 2	
8.400E-04	3.704E-02	50	K031	2 1 2 1 2	
6.123E-04	2.700E-02	60	F300	1 0 0 0 1	

213. $C_3H_8NO_5P$
Glyphosate
N-(Phosphonomethyl)glycine
Bronco
RN: 1071-83-6   **MP** (°C):   230.0
MW: 169.07   **BP** (°C):

Solubility (Moles/L)	Solubility (Grams/L)	Temp (°C)	Ref (#)	Evaluation (T P E A A)	Comments
7.097E-02	1.200E+01	25	M161	1 0 0 0 1	
5.856E-02	9.901E+00	ns	B100	0 0 0 0 0	

214. C_3H_8O
n-Propyl Alcohol
Propanol
RN: 71-23-8   **MP** (°C):   -127.0
MW: 60.10   **BP** (°C):   97.2

Solubility (Moles/L)	Solubility (Grams/L)	Temp (°C)	Ref (#)	Evaluation (T P E A A)	Comments
3.132E+00	1.882E+02	ns	L003	0 0 2 1 2	

215. C_3H_8O
Isopropyl Alcohol
2-Propanol
RN: 67-63-0   **MP** (°C):   -88
MW: 60.10   **BP** (°C):   82.5

Solubility (Moles/L)	Solubility (Grams/L)	Temp (°C)	Ref (#)	Evaluation (T P E A A)	Comments
5.033E+00	3.025E+02	ns	L003	0 0 2 1 1	

216. $C_3H_8OS_2$
2,3-Dimercapto-1-propanol
Dimercaprol
RN: 59-52-9   **MP** (°C):
MW: 124.22   **BP** (°C):

Solubility (Moles/L)	Solubility (Grams/L)	Temp (°C)	Ref (#)	Evaluation (T P E A A)	Comments
5.963E-01	7.407E+01	20	D041	1 0 0 0 0	

217. $C_3H_8O_2$
Methylal
Formaldehyd-dimethyl-acetal
RN: 109-87-5   **MP** (°C):   -105
MW: 76.10   **BP** (°C):   41.5

Solubility (Moles/L)	Solubility (Grams/L)	Temp (°C)	Ref (#)	Evaluation (T P E A A)	Comments
3.208E+00	2.441E+02	16	B117	1 0 0 1 2	
3.022E+00	2.300E+02	20	F300	1 0 0 0 1	
3.022E+00	2.300E+02	20	F300	1 0 0 0 1	

218.   C₃H₈O₃

$C_3H_8O_3$

Glycerol
Glycerin

RN: 56-81-5   **MP** (°C):   20
MW: 92.10   **BP** (°C):

Solubility (Moles/L)	Solubility (Grams/L)	Temp (°C)	Ref (#)	Evaluation (T P E A A)	Comments
5.973E+00	5.501E+02	4.50	C022	1 2 0 0 2	
5.751E-01	5.296E+01	25	B004	1 0 0 0 2	

219.   C₃H₉N

C_3H_9N

Propylamine
Propylamin
n-Propylamine

RN: 107-10-8   **MP** (°C):   -83
MW: 59.11   **BP** (°C):   48

Solubility (Moles/L)	Solubility (Grams/L)	Temp (°C)	Ref (#)	Evaluation (T P E A A)	Comments
2.469E-02	1.459E+00	25	B004	1 0 0 0 2	

220.   C₃H₉N

C_3H_9N

Trimethylamine
N,N-Dimethylmethanamine

RN: 75-50-3   **MP** (°C):   -124.0
MW: 59.11   **BP** (°C):   3.2

Solubility (Moles/L)	Solubility (Grams/L)	Temp (°C)	Ref (#)	Evaluation (T P E A A)	Comments
>6.77E+00	>4.00E+02	20	F300	1 0 0 0 0	
6.936E+00	4.100E+02	25	A049	1 0 0 0 2	

221.   C₃H₉O₄P

$C_3H_9O_4P$

Trimethyl Phosphate
Phosphorsaeure-trimethyl Ester

RN: 512-56-1   **MP** (°C):
MW: 140.08   **BP** (°C):   197

Solubility (Moles/L)	Solubility (Grams/L)	Temp (°C)	Ref (#)	Evaluation (T P E A A)	Comments
3.569E+00	5.000E+02	25	F300	1 0 0 0 1	

222. $C_3H_{12}N_6O_3$
Guanidine Carbonate
Guanidin-carbonat
RN: 3425-08-9   **MP** (°C):   198
MW: 180.17   **BP** (°C):

Solubility (Moles/L)	Solubility (Grams/L)	Temp (°C)	Ref (#)	Evaluation (T P E A A)	Comments
1.850E+00	3.333E+02	24	F300	1 0 0 0 2	

223. $C_3Cl_3N_3O_3$
Trichloroisocyanuric Acid
Symclosene
RN: 87-90-1   **MP** (°C):   246.5
MW: 232.41   **BP** (°C):

Solubility (Moles/L)	Solubility (Grams/L)	Temp (°C)	Ref (#)	Evaluation (T P E A A)	Comments
3.439E-03	7.994E-01	20	B080	1 0 1 1 0	
2.311E-02	5.371E+00	40	B080	1 0 1 1 1	

224. C_4HI_4N
Iodol
2,3,4,5-Tetraiodpyrrol
RN: 87-58-1   **MP** (°C):
MW: 570.68   **BP** (°C):

Solubility (Moles/L)	Solubility (Grams/L)	Temp (°C)	Ref (#)	Evaluation (T P E A A)	Comments
3.505E-04	2.000E-01	15	F300	1 0 0 0 2	

225. C_4H_2
Butadiyne
Diacetylen
RN: 460-12-8   **MP** (°C):   -36.4
MW: 50.06   **BP** (°C):   10.3

Solubility (Moles/L)	Solubility (Grams/L)	Temp (°C)	Ref (#)	Evaluation (T P E A A)	Comments
1.998E-03	1.000E-01	25	F300	1 0 0 0 0	

226. $C_4H_2N_2O_4$
Alloxan
Alloxane
RN: 50-71-5   **MP** (°C):   256dec
MW: 142.07   **BP** (°C):

Solubility (Moles/L)	Solubility (Grams/L)	Temp (°C)	Ref (#)	Evaluation (T P E A A)	Comments
5.631E-02	8.000E+00	ns	D072	0 0 0 0 0	

227. $C_4H_3FN_2O_2$
5 Fluorouracil
5-Fluorouracil
Fluorouracil
5-Fluoro-2,4(1H,3H)-Pyrimidinedione
Fluroblastin
Fluororuracil
RN: 51-21-8   **MP** (°C):   281
MW: 130.08   **BP** (°C):

Solubility (Moles/L)	Solubility (Grams/L)	Temp (°C)	Ref (#)	Evaluation (T P E A A)	Comments
8.533E-02	1.110E+01	22	B321	1 0 2 2 2	pH 4.0
8.533E-02	1.110E+01	22	B332	1 1 0 0 1	pH 4.0
8.533E-02	1.110E+01	22	B388	1 0 2 2 2	
9.379E-02	1.220E+01	22	M317	1 1 1 1 1	
9.379E-02	1.220E+01	25	R023	1 0 0 0 2	
8.533E-02	1.110E+01	37	B332	1 1 0 0 1	pH 4.0

228. $C_4H_3N_2S$
2-Methyl-1,3,4-thiadiazole
Thiodiazolique Methyle
RN: 26584-42-9   **MP** (°C):
MW: 111.15   **BP** (°C):

Solubility (Moles/L)	Solubility (Grams/L)	Temp (°C)	Ref (#)	Evaluation (T P E A A)	Comments
7.918E-03	8.800E-01	37	D084	1 0 1 0 1	

229. $C_4H_3N_3O_5$
5-Nitrobarbituric Acid
Dilitursaeure
RN: 28176-10-5   **MP** (°C):   176
MW: 173.09   **BP** (°C):

Solubility (Moles/L)	Solubility (Grams/L)	Temp (°C)	Ref (#)	Evaluation (T P E A A)	Comments
5.200E-03	9.000E-01	25.60	F300	1 0 0 0 0	

230. $C_4H_4Br_2O_4$
meso-2,3-Dibromosuccinic Acid
meso-Dibrom-bernsteinsaeure
DL-2,3-Dibromosuccinic Acid
DL-Dibrom-bernsteinsaeure
RN: 526-78-3   **MP** (°C):   171
MW: 275.89   **BP** (°C):

Solubility (Moles/L)	Solubility (Grams/L)	Temp (°C)	Ref (#)	Evaluation (T P E A A)	Comments
7.249E-02	2.000E+01	17	F300	1 0 0 0 2	

231. $C_4H_4Cl_2N_2O_2$
1,3-Dichloro-5-methylhydantoin
2,4-Imidazolidinedione, 1,3-Dichloro-5-methyl-
Hydantoin, 1,3-Dichloro-5-methyl-
RN: 15216-12-3   **MP** (°C):
MW: 182.99   **BP** (°C):

Solubility (Moles/L)	Solubility (Grams/L)	Temp (°C)	Ref (#)	Evaluation (T P E A A)	Comments
1.634E-02	2.991E+00	20	B080	1 0 1 1 0	
4.498E-02	8.232E+00	40	B080	1 0 1 1 1	

232. $C_4H_4Cl_2O_4$
L-2,3-Dichlorosuccinic Acid
L(-)-Dichlor-bernsteinsaeure
D-2,3-Dichlorosuccinic Acid
D(+)-Dichlor-bernsteinsaeure
2,3-Dichlorosuccinic Acid
meso-2,3-Dichlorosuccinic Acid
RN: 19922-87-3   **MP** (°C):   168
MW: 186.98   **BP** (°C):

Solubility (Moles/L)	Solubility (Grams/L)	Temp (°C)	Ref (#)	Evaluation (T P E A A)	Comments
2.674E+00	5.000E+02	25	H090	0 1 1 1 1	
1.701E-02	3.180E+00	ns	H090	0 2 2 1 2	

233. $C_4H_4N_2$
Succinonitrile
Bersteinsaeure-dinitril
RN: 110-61-2   **MP** (°C):   57
MW: 80.09   **BP** (°C):   265

Solubility (Moles/L)	Solubility (Grams/L)	Temp (°C)	Ref (#)	Evaluation (T P E A A)	Comments
1.584E+00	1.269E+02	20	F300	1 0 0 0 2	

234.   C$_4$H$_4$N$_2$O
4(3H)-Pyrimidone
4-Hydroxypyrimidine
RN: 51953-17-4   **MP** (°C):   164
MW: 96.09   **BP** (°C):

Solubility (Moles/L)	Solubility (Grams/L)	Temp (°C)	Ref (#)	Evaluation (T P E A A)	Comments
2.813E+00	2.703E+02	20	B050	1 0 0 0 0	

235.   C$_4$H$_4$N$_2$O
2-Hydroxypyrimidine
2-Pyrimidinol
RN: 51953-13-0   **MP** (°C):
MW: 96.09   **BP** (°C):

Solubility (Moles/L)	Solubility (Grams/L)	Temp (°C)	Ref (#)	Evaluation (T P E A A)	Comments
3.252E+00	3.125E+02	20	B050	1 0 0 0 0	

236.   C$_4$H$_4$N$_2$OS
2-Thiouracil
Thiouracil
4(1H)-Pyrimidinone
RN: 141-90-2   **MP** (°C):   340
MW: 128.15   **BP** (°C):

Solubility (Moles/L)	Solubility (Grams/L)	Temp (°C)	Ref (#)	Evaluation (T P E A A)	Comments
4.679E-03	5.996E-01	20	D041	1 0 0 0 0	
5.530E-03	7.087E-01	25	G016	1 2 1 2 2	intrinsic
3.900E-03	4.998E-01	ns	I310	0 0 0 0 0	

237.   C$_4$H$_4$N$_2$O$_2$
4,6-Dihydroxypyrimidine
4,6-Pyrimidinediol
RN: 1193-24-4   **MP** (°C):   >300
MW: 112.09   **BP** (°C):

Solubility (Moles/L)	Solubility (Grams/L)	Temp (°C)	Ref (#)	Evaluation (T P E A A)	Comments
2.225E-02	2.494E+00	20	B050	1 0 0 0 0	

238. $C_4H_4N_2O_2$
Maleic Hydrazide
Dihydropyridazine-3,6-dione
RN: 123-33-1   **MP** (°C):
MW: 112.09   **BP** (°C):

Solubility (Moles/L)	Solubility (Grams/L)	Temp (°C)	Ref (#)	Evaluation (T P E A A)	Comments
3.554E-02	3.984E+00	20	B185	1 0 0 0 1	
5.321E-02	5.964E+00	25	B185	1 0 0 0 1	
5.353E-02	6.000E+00	25	B200	1 0 0 0 2	
5.321E-02	5.964E+00	25	M061	1 0 0 0 0	
5.353E-02	6.000E+00	25	M161	1 0 0 0 0	
5.321E-02	5.964E+00	ns	B100	0 0 0 0 0	
6.310E-03	7.072E-01	ns	M163	0 0 0 0 0	EFG
3.554E-02	3.984E+00	ns	N013	0 0 0 0 0	

239. $C_4H_4N_2O_2$
Uracil
2,4-Dihydroxypyrimidine
RN: 66-22-8   **MP** (°C):   335
MW: 112.09   **BP** (°C):

Solubility (Moles/L)	Solubility (Grams/L)	Temp (°C)	Ref (#)	Evaluation (T P E A A)	Comments
2.964E-02	3.322E+00	20	B050	1 0 0 0 0	
2.500E-02	2.802E+00	20	N019	2 2 1 2 2	
3.200E-02	3.587E+00	25	D041	1 0 0 0 1	
3.212E-02	3.600E+00	25	F300	1 0 0 0 1	
2.380E-02	2.668E+00	25	H061	1 2 2 0 2	
4.015E-02	4.500E+00	37	B390	1 0 2 2 1	
2.676E-02	3.000E+00	ns	B177	0 0 0 0 0	

240. $C_4H_4N_2O_3$
2,4,6-Trihydroxypyrimidine
2,4,6-Pyrimidinetriol
RN: 223674-01-9   **MP** (°C):
MW: 128.09   **BP** (°C):

Solubility (Moles/L)	Solubility (Grams/L)	Temp (°C)	Ref (#)	Evaluation (T P E A A)	Comments
5.170E-02	6.623E+00	20	B050	1 0 0 0 0	

241.   C₄H₄N₂O₃

Barbituric Acid
Barbitursaeure
RN: 67-52-7   **MP** (°C):   248
MW: 128.09   **BP** (°C):

Solubility (Moles/L)	Solubility (Grams/L)	Temp (°C)	Ref (#)	Evaluation (T P E A A)	Comments
1.483E-07	1.900E-05	37	B166	1 0 1 1 1	

242.   C₄H₄O₄

Maleic Acid
Maleinsaeure
RN: 110-16-7   **MP** (°C):   138
MW: 116.07   **BP** (°C):

Solubility (Moles/L)	Solubility (Grams/L)	Temp (°C)	Ref (#)	Evaluation (T P E A A)	Comments
2.431E+00	2.821E+02	0	M043	1 0 0 0 2	
2.607E+00	3.026E+02	4.99	A339	2 0 2 2 2	
2.880E+00	3.343E+02	9.99	A339	2 0 2 2 2	
2.872E+00	3.334E+02	10	F300	1 0 0 0 2	
2.872E+00	3.333E+02	10	M043	1 0 0 0 1	
3.094E+00	3.591E+02	14.99	A339	2 0 2 2 2	
3.312E+00	3.845E+02	19.99	A339	2 0 2 2 2	
3.547E+00	4.118E+02	20	M043	1 0 0 0 1	
6.789E+00	7.880E+02	22.5	G301	2 1 0 1 2	
3.592E+00	4.170E+02	24.99	A339	2 0 2 2 2	
3.797E+00	4.407E+02	25	D041	1 0 0 0 2	
3.797E+00	4.407E+02	25	F300	1 0 0 0 2	
3.797E+00	4.407E+02	25	W011	1 2 2 1 2	
3.823E+00	4.437E+02	29.99	A339	2 0 2 2 2	
4.081E+00	4.737E+02	30	M043	1 0 0 0 1	
4.117E+00	4.778E+02	34.99	A339	2 0 2 2 2	
4.300E+00	4.991E+02	39.99	A339	2 0 2 2 2	
4.608E+00	5.349E+02	40	M043	1 0 0 0 2	
4.561E+00	5.294E+02	40	W011	1 2 2 1 2	
4.562E+00	5.295E+02	44.99	A339	2 0 2 2 2	
4.677E+00	5.429E+02	49.99	A339	2 0 2 2 2	
4.842E+00	5.620E+02	54.99	A339	2 0 2 2 2	
5.031E+00	5.840E+02	59.99	A339	2 0 2 2 2	
5.516E+00	6.403E+02	60	M043	1 0 0 0 2	
5.151E+00	5.979E+02	60	W011	1 2 2 1 2	
5.166E+00	5.997E+02	64.99	A339	2 0 2 2 2	
6.366E+00	7.389E+02	80	M043	1 0 0 0 2	
6.864E+00	7.967E+02	97.5	D041	1 0 0 0 2	
6.866E+00	7.970E+02	97.5	F300	1 0 0 0 2	
6.866E+00	7.970E+02	97.5	W011	1 2 2 1 2	

243.   C$_4$H$_4$O$_4$
trans-Fumaric Acid
Fumaric Acid
Fumarsaeure
RN: 110-17-8   **MP** (°C):   287
MW: 116.07   **BP** (°C):

Solubility (Moles/L)	Solubility (Grams/L)	Temp (°C)	Ref (#)	Evaluation (T P E A A)	Comments
1.977E-02	2.295E+00	0	M043	1 0 0 0 1	
3.005E-02	3.488E+00	10	M043	1 0 0 0 1	
4.286E-02	4.975E+00	20	M043	1 0 0 0 1	
5.989E-02	6.951E+00	25	D041	1 0 0 0 1	
6.031E-02	7.000E+00	25	F300	1 0 0 0 0	
5.989E-02	6.951E+00	25	W011	1 2 2 1 1	
6.159E-02	7.149E+00	30	M043	1 0 0 0 1	
9.218E-02	1.070E+01	40	F300	1 0 0 0 2	
9.374E-02	1.088E+01	40	M043	1 0 0 0 1	
9.121E-02	1.059E+01	40	W011	1 2 2 1 2	
1.937E-01	2.248E+01	60	M043	1 0 0 0 1	
2.019E-01	2.344E+01	60	W011	1 2 2 1 1	
4.258E-01	4.943E+01	80	M043	1 0 0 0 1	
7.689E-01	8.925E+01	100	D041	1 0 0 0 1	
8.012E-01	9.300E+01	100	F300	1 0 0 0 1	
7.689E-01	8.925E+01	100	M043	1 0 0 0 1	
7.689E-01	8.925E+01	100	W011	1 2 2 1 1	

244.   C$_4$H$_4$S
Thiophene
Thiofuran
Thiacyclopentadiene
RN: 110-02-1   **MP** (°C):   -38.3
MW: 84.14   **BP** (°C):   84.4

Solubility (Moles/L)	Solubility (Grams/L)	Temp (°C)	Ref (#)	Evaluation (T P E A A)	Comments
3.583E-02	3.015E+00	25	K119	1 0 0 0 2	
3.583E-02	3.015E+00	25	P051	2 1 1 2 2	
3.583E-02	3.015E+00	25.00	P007	2 1 2 2 2	

245.   C$_4$H$_5$BrO$_4$

Bromosuccinic Acid

DL-Brombernsteinsaeure

RN: 923-06-8   **MP** (°C):

MW: 196.99   **BP** (°C):   161

Solubility (Moles/L)	Solubility (Grams/L)	Temp (°C)	Ref (#)	Evaluation (T P E A A)	Comments
6.092E-01	1.200E+02	15.5	F300	1 0 0 0 1	

246.   C$_4$H$_5$ClO$_2$

3-Chloroisocrotonic Acid

β-Chlor-isocrotonsaeure

RN: 6625-00-9   **MP** (°C):

MW: 120.54   **BP** (°C):

Solubility (Moles/L)	Solubility (Grams/L)	Temp (°C)	Ref (#)	Evaluation (T P E A A)	Comments
1.037E-01	1.250E+01	7	F300	1 0 0 0 2	
1.560E-01	1.880E+01	19	F300	1 0 0 0 2	

247.   C$_4$H$_5$ClO$_2$

2-Chlorocrotonic Acid

α-Chlor-crotonsaeure

RN: 600-13-5   **MP** (°C):

MW: 120.54   **BP** (°C):

Solubility (Moles/L)	Solubility (Grams/L)	Temp (°C)	Ref (#)	Evaluation (T P E A A)	Comments
1.742E-01	2.100E+01	19	F300	1 0 0 0 1	

248.   C$_4$H$_5$ClO$_2$

2-Chloroisocrotonic Acid

α-Chlor-isocrotonsaeure

RN: 24253-33-6   **MP** (°C):

MW: 120.54   **BP** (°C):

Solubility (Moles/L)	Solubility (Grams/L)	Temp (°C)	Ref (#)	Evaluation (T P E A A)	Comments
5.102E-01	6.150E+01	19	F300	1 0 0 0 2	

249.   C$_4$H$_5$ClO$_2$
3-Chlorocrotonic Acid
β-Chlor-crotonsaeure
RN: 6214-28-4   **MP** (°C):   94
MW: 120.54   **BP** (°C):   206

Solubility (Moles/L)	Solubility (Grams/L)	Temp (°C)	Ref (#)	Evaluation (T P E A A)	Comments
1.842E-01	2.220E+01	12.5	F300	1 0 0 0 2	
2.481E-01	2.990E+01	19	F300	1 0 0 0 2	

250.   C$_4$H$_5$ClO$_4$
L-Chlorosuccinic Acid
L(-)-Chlor-bernsteinsaeure
D-Chlorosuccinic Acid
D(+)-Chlor-bernsteinsaeure
RN: 16045-92-4   **MP** (°C):
MW: 152.54   **BP** (°C):

Solubility (Moles/L)	Solubility (Grams/L)	Temp (°C)	Ref (#)	Evaluation (T P E A A)	Comments
1.180E+00	1.800E+02	20	F300	1 0 0 0 1	
1.193E+00	1.820E+02	20	F300	1 0 0 0 2	

251.   C$_4$H$_5$F$_3$O
Fluroxene
2,2,2-(Trifluoroethoxy)ethene
Redeptin
Fluoromar
RN: 406-90-6   **MP** (°C):
MW: 126.08   **BP** (°C):   42.7

Solubility (Moles/L)	Solubility (Grams/L)	Temp (°C)	Ref (#)	Evaluation (T P E A A)	Comments
3.173E-05	4.000E-03	ns	R028	0 0 0 0 0	

252.   C$_4$H$_5$N
Pyrrole
Azole
Imidole
RN: 109-97-7   **MP** (°C):   -23
MW: 67.09   **BP** (°C):

Solubility (Moles/L)	Solubility (Grams/L)	Temp (°C)	Ref (#)	Evaluation (T P E A A)	Comments
7.098E-01	4.762E+01	rt	B099	0 2 0 0 0	

253.   C₄H₅N

Methacrylonitrile

2-Methyl-2-Propenenitrile

RN: 126-98-7   **MP** (°C):   -35.8

MW: 67.09   **BP** (°C):   90.3

Solubility (Moles/L)	Solubility (Grams/L)	Temp (°C)	Ref (#)	Evaluation (T P E A A)	Comments
3.692E-01	2.477E+01	25	L096	1 2 0 2 2	

254.   C₄H₅NO₂

Succinimide

2,5-Pyrrolidinedione

Butanimide

RN: 123-56-8   **MP** (°C):   126

MW: 99.09   **BP** (°C):   288

Solubility (Moles/L)	Solubility (Grams/L)	Temp (°C)	Ref (#)	Evaluation (T P E A A)	Comments
9.174E-01	9.091E+01	0	M043	1 0 0 0 1	
1.392E+00	1.379E+02	10	M043	1 0 0 0 1	
2.082E+00	2.063E+02	20	M043	1 0 0 0 1	
1.978E+00	1.960E+02	21	F300	1 0 0 0 2	
3.273E+00	3.243E+02	30	M043	1 0 0 0 1	
4.577E+00	4.536E+02	40	M043	1 0 0 0 1	
5.887E+00	5.833E+02	60	M043	1 0 0 0 2	
6.868E+00	6.805E+02	80	M043	1 0 0 0 2	
1.413E+00	1.400E+02	ns	D072	0 0 0 0 1	

255.   C₄H₅NO₂

Hymexazol

3-Hydroxy-5-methyl Isoxazole

5-Methyl-3(2H)-isoxazolone

Tachigaren

Isoxazolol, 5-Methyl-

RN: 10004-44-1   **MP** (°C):   86

MW: 99.09   **BP** (°C):

Solubility (Moles/L)	Solubility (Grams/L)	Temp (°C)	Ref (#)	Evaluation (T P E A A)	Comments
8.578E-01	8.500E+01	25	M161	1 0 0 0 2	
8.578E-01	8.500E+01	25	N306	1 0 0 0 1	

256.   C₄H₅NS

Allyl Isothiocyanate
Allyl Mustardiol
Allylsenfoel
RN: 57-06-7   **MP** (°C):   -8
MW: 99.16   **BP** (°C):   152

Solubility (Moles/L)	Solubility (Grams/L)	Temp (°C)	Ref (#)	Evaluation (T P E A A)	Comments
2.017E-02	2.000E+00	20	F300	1 0 0 0 0	

257.   C₄H₅N₃O

Cytosine
2-Oxy-4-amino Pyrimidine
2(1H)-Pyrimidinone, 4-Amino-
RN: 71-30-7   **MP** (°C):   320
MW: 111.10   **BP** (°C):

Solubility (Moles/L)	Solubility (Grams/L)	Temp (°C)	Ref (#)	Evaluation (T P E A A)	Comments
5.000E-02	5.555E+00	20	C017	2 0 0 1 0	EFG
6.877E-02	7.641E+00	25	D041	1 0 0 0 1	
7.200E-02	8.000E+00	25	F300	1 0 0 0 0	
6.580E-02	7.311E+00	25	H061	1 2 2 0 2	
6.500E-02	7.222E+00	25	R030	1 0 0 0 1	

258.   C₄H₅N₃OS

6-Amino-2-thiouracil
2-Mercapto-4-amino-6-hydroxypyrimidine
2-Thio-4-amino-6-hydroxypyrimidine
2-Mercapto-6-aminouracil
RN: 1004-40-6   **MP** (°C):
MW: 143.17   **BP** (°C):

Solubility (Moles/L)	Solubility (Grams/L)	Temp (°C)	Ref (#)	Evaluation (T P E A A)	Comments
1.790E-03	2.563E-01	25	G016	1 2 1 2 2	intrinsic

259.   C₄H₅N₃O₂

2-Methyl-4(5)-nitroimidazole
2-Methyl-5-nitroimidazole
Menidazole
RP 8532
L 581490

RN: 696-23-1   **MP** (°C):   257-258
MW: 127.10   **BP** (°C):

Solubility (Moles/L)	Solubility (Grams/L)	Temp (°C)	Ref (#)	Evaluation (T P E A A)	Comments
2.368E-02	3.010E+00	20	D344	1 1 2 2 2	
2.367E-02	3.009E+00	20	D344	1 1 2 2 2	
2.353E-02	2.991E+00	20	D344	1 1 2 2 2	
2.370E-02	3.012E+00	20	D344	1 1 2 2 2	

260.   C₄H₅N₃O₂

5-Aminouracil
5-Amino-uracil

RN: 932-52-5   **MP** (°C):   >300
MW: 127.10   **BP** (°C):

Solubility (Moles/L)	Solubility (Grams/L)	Temp (°C)	Ref (#)	Evaluation (T P E A A)	Comments
3.934E-03	5.000E-01	20	F300	1 0 0 0 0	
1.259E-01	1.600E+01	100	F300	1 0 0 0 1	

261.   C₄H₆

1-Butyne
Ethylacetylene
Ethylethyne

RN: 107-00-6   **MP** (°C):   -125.7
MW: 54.09   **BP** (°C):   8.1

Solubility (Moles/L)	Solubility (Grams/L)	Temp (°C)	Ref (#)	Evaluation (T P E A A)	Comments
5.306E-02	2.870E+00	25	M001	2 1 2 2 2	

262.   C₄H₆

1,3-Butadiene
Pyrrolylene

RN: 106-99-0   **MP** (°C):   -108.9
MW: 54.09   **BP** (°C):   -4.5

Solubility (Moles/L)	Solubility (Grams/L)	Temp (°C)	Ref (#)	Evaluation (T P E A A)	Comments
1.359E-02	7.350E-01	25	M001	2 1 2 2 2	

263.   C$_4$H$_6$BrNO$_4$
5-Bromo-5-nitro-1,3-dioxane
Bronidox
Microcide I
Bronidox L
1,3-Dioxane, 5-Bromo-5-nitro-
RN: 30007-47-7   **MP** (°C):   49-50
MW: 212.01   **BP** (°C):

Solubility (Moles/L)	Solubility (Grams/L)	Temp (°C)	Ref (#)	Evaluation (T P E A A)	Comments
2.706E-02	5.737E+00	25	L013	1 0 2 1 2	

264.   C$_4$H$_6$Cl$_2$O$_2$S
3,4-Dichlorotetrahydrothiophene Dioxide
3,4-Dichlorotetrahydrothiophene 1,1-dioxide
3,4-Dichlorosulfolane
DAC PRD
3,4-Dichlorothiolane 1,1-dioxide
RN: 3001-57-8   **MP** (°C):   130
MW: 189.06   **BP** (°C):

Solubility (Moles/L)	Solubility (Grams/L)	Temp (°C)	Ref (#)	Evaluation (T P E A A)	Comments
1.161E-02	2.195E+00	20	M061	1 0 0 0 1	

265.   C$_4$H$_6$N$_2$O$_2$
2,5-Piperazinedione
Diketopiperazine
RN: 106-57-0   **MP** (°C):
MW: 114.10   **BP** (°C):

Solubility (Moles/L)	Solubility (Grams/L)	Temp (°C)	Ref (#)	Evaluation (T P E A A)	Comments
1.232E-01	1.406E+01	20	B032	1 2 2 1 2	
1.253E-01	1.430E+01	20	M075	2 0 1 1 2	
1.475E-01	1.683E+01	25	B032	1 2 2 1 2	
1.754E-01	2.001E+01	29.80	B032	1 2 2 1 2	

266. $C_4H_6N_2S_4Zn$
Zineb
Zinc Ethylenebis(dithiocarbamate)
RN: 12122-67-7   **MP** (°C):
MW: 275.74   **BP** (°C):

Solubility (Moles/L)	Solubility (Grams/L)	Temp (°C)	Ref (#)	Evaluation (T P E A A)	Comments
3.627E-06	1.000E-03	20	M061	1 0 0 0 0	
3.627E-05	1.000E-02	rt	M161	0 0 0 0 1	

267. $C_4H_6N_4O_3$
Allantoin
Allantoine
RN: 97-59-6   **MP** (°C):   238
MW: 158.12   **BP** (°C):

Solubility (Moles/L)	Solubility (Grams/L)	Temp (°C)	Ref (#)	Evaluation (T P E A A)	Comments
3.303E-02	5.223E+00	20	D041	1 0 0 0 2	
4.755E-02	7.519E+00	c	D004	1 0 0 0 0	
2.040E-01	3.226E+01	h	D004	1 0 0 0 0	
2.530E-02	4.000E+00	ns	D072	0 0 0 0 1	

268. $C_4H_6N_4O_3S_2$
Acetazolamide
5-Acetamido-1,3,4-thiadiazole-2-sulfonamide
RN: 59-66-5   **MP** (°C):   258
MW: 222.25   **BP** (°C):

Solubility (Moles/L)	Solubility (Grams/L)	Temp (°C)	Ref (#)	Evaluation (T P E A A)	Comments
2.700E-03	6.001E-01	15	K024	1 2 1 1 2	
2.249E-03	4.998E-01	20	D041	1 0 0 0 0	
4.409E-03	9.799E-01	30	E049	2 0 2 2 2	
5.174E-03	1.150E+00	37	C054	2 0 2 1 2	
4.144E-03	9.210E-01	ns	I304	0 0 2 2 2	

269. C_4H_6O
α-Methylacrolein
α-Methyl-acrolein
RN: 78-85-3   **MP** (°C):
MW: 70.09   **BP** (°C):

Solubility (Moles/L)	Solubility (Grams/L)	Temp (°C)	Ref (#)	Evaluation (T P E A A)	Comments
8.089E-01	5.670E+01	20	F300	1 0 0 0 2	

270. C_4H_6O
Vinyl Ether
1,1'-Oxybisethene
Divinyl Ether
RN: 109-93-3   **MP** (°C):
MW: 70.09   **BP** (°C):   28.4

Solubility (Moles/L)	Solubility (Grams/L)	Temp (°C)	Ref (#)	Evaluation (T P E A A)	Comments
7.490E-02	5.250E+00	37	R047	1 0 0 0 2	
5.487E-01	3.846E+01	ns	R028	0 0 0 0 1	

271. C_4H_6O
trans-Crotonaldehyde
Crotonaldehyd
RN: 123-73-9   **MP** (°C):   -77
MW: 70.09   **BP** (°C):

Solubility (Moles/L)	Solubility (Grams/L)	Temp (°C)	Ref (#)	Evaluation (T P E A A)	Comments
2.140E+00	1.500E+02	20	F300	1 0 0 0 1	

272. $C_4H_6O_2$
Methyl Acrylate
Acrylic Acid Methyl Ester
2-Propenoic Acid Methyl Ester
RN: 96-33-3   **MP** (°C):   -76.5
MW: 86.09   **BP** (°C):   70

Solubility (Moles/L)	Solubility (Grams/L)	Temp (°C)	Ref (#)	Evaluation (T P E A A)	Comments
5.742E-01	4.943E+01	30	L096	1 2 0 2 1	

273. $C_4H_6O_2$
trans-Crotonic Acid
trans-Crotonsaeure
RN: 3724-65-0   **MP** (°C):
MW: 86.09   **BP** (°C):

Solubility (Moles/L)	Solubility (Grams/L)	Temp (°C)	Ref (#)	Evaluation (T P E A A)	Comments
9.989E-01	8.600E+01	25	F300	1 0 0 0 1	
4.600E+00	3.960E+02	40	F300	1 0 0 0 2	

274. $C_4H_6O_2$
Vinyl Acetate
Vinylacetate
RN: 108-05-4   **MP** (°C):   -100
MW: 86.09   **BP** (°C):   72

Solubility (Moles/L)	Solubility (Grams/L)	Temp (°C)	Ref (#)	Evaluation (T P E A A)	Comments
3.136E-01	2.700E+01	50	L097	1 1 1 1 1	

275. $C_4H_6O_2$
β-Butyrolacetone
3-Hydroxybutanoic Acid β-Lactone
RN: 3068-88-0   **MP** (°C):
MW: 86.09   **BP** (°C):

Solubility (Moles/L)	Solubility (Grams/L)	Temp (°C)	Ref (#)	Evaluation (T P E A A)	Comments
1.541E+00	1.327E+02	18	I313	0 0 0 0 2	

276. $C_4H_6O_2$
Diacetyl
2,3-Butanedione
RN: 431-03-8   **MP** (°C):
MW: 86.09   **BP** (°C):

Solubility (Moles/L)	Solubility (Grams/L)	Temp (°C)	Ref (#)	Evaluation (T P E A A)	Comments
2.323E+00	2.000E+02	15	F300	1 0 0 0 1	
2.323E+00	2.000E+02	20	D041	1 0 0 0 1	

277. $C_4H_6O_2$
Crotonic Acid
2-Butenoic Acid
3-Methylacrylic Acid
RN: 107-93-7   **MP** (°C):   73
MW: 86.09   **BP** (°C):

Solubility (Moles/L)	Solubility (Grams/L)	Temp (°C)	Ref (#)	Evaluation (T P E A A)	Comments
8.882E-01	7.647E+01	20	D041	1 0 0 0 2	

278.   C$_4$H$_6$O$_2$S$_4$
bis(Methylxanthogen) Disulfide
Dimethylxanthogen Disulfide
Methyl Dixanthogen
RN: 1468-37-7   **MP** (°C):   22.75
MW: 214.35   **BP** (°C):

Solubility (Moles/L)	Solubility (Grams/L)	Temp (°C)	Ref (#)	Evaluation (T P E A A)	Comments
1.150E-04	2.465E-02	25	H102	1 2 1 2 2	

279.   C$_4$H$_6$O$_3$
Acetic Anhydride
Essigsaeure-anhydrid
RN: 108-24-7   **MP** (°C):   -73
MW: 102.09   **BP** (°C):   139

Solubility (Moles/L)	Solubility (Grams/L)	Temp (°C)	Ref (#)	Evaluation (T P E A A)	Comments
1.175E+00	1.200E+02	20	F300	1 0 0 0 2	

280.   C$_4$H$_6$O$_4$
Methylmalonic Acid
Acide Methylmalonique
Methyl-malonsaeure
RN: 516-05-2   **MP** (°C):   129.5
MW: 118.09   **BP** (°C):

Solubility (Moles/L)	Solubility (Grams/L)	Temp (°C)	Ref (#)	Evaluation (T P E A A)	Comments
2.600E+00	3.070E+02	0	F300	1 0 0 0 2	
3.743E+00	4.420E+02	0	M051	1 0 0 0 2	
4.954E+00	5.850E+02	15	M051	1 0 0 0 2	
5.750E+00	6.790E+02	25	M051	1 0 0 0 2	
4.071E+00	4.808E+02	50	F300	1 0 0 0 2	
7.748E+00	9.150E+02	50	M051	1 0 0 0 2	

281. $C_4H_6O_4$
Succinic Acid
Bernsteinsaeure
RN: 110-15-6   **MP** (°C):   185
MW: 118.09   **BP** (°C):   235

Solubility (Moles/L)	Solubility (Grams/L)	Temp (°C)	Ref (#)	Evaluation (T P E A A)	Comments
2.363E-01	2.790E+01	0	L041	1 0 0 1 2	
2.273E-01	2.684E+01	0	M020	1 0 0 1 1	
2.306E-01	2.724E+01	0	M043	1 0 0 0 1	
2.892E-01	3.415E+01	4.99	A339	2 0 2 2 2	
3.616E-01	4.271E+01	9.99	A339	2 0 2 2 2	
3.569E-01	4.215E+01	10	M043	1 0 0 0 1	
3.854E-01	4.551E+01	11.85	L064	2 2 2 1 2	
4.518E-01	5.335E+01	14.99	A339	2 0 2 2 2	
4.102E-01	4.843E+01	15	F055	1 2 2 2 2	
4.149E-01	4.900E+01	15	L041	1 0 0 1 1	
4.149E-01	4.900E+01	15	M051	1 0 0 0 1	
4.912E-01	5.800E+01	17.50	F300	1 0 0 0 1	
4.974E-01	5.874E+01	18	L064	2 2 2 1 2	
5.661E-01	6.685E+01	19.99	A339	2 0 2 2 2	
5.392E-01	6.367E+01	20	D041	1 0 0 0 1	
5.019E-01	5.927E+01	20	F055	1 2 2 2 2	
5.420E-01	6.400E+01	20	F300	1 0 0 0 2	
4.912E-01	5.800E+01	20	L041	1 0 0 1 1	
5.466E-01	6.455E+01	20	M043	1 0 0 0 1	
5.510E-01	6.507E+01	20	M153	1 0 0 0 0	cal. from fitted equation
4.632E-01	5.470E+01	20	M171	1 0 0 0 1	
5.716E-01	6.750E+01	20	W026	1 0 1 1 1	average of 2
6.344E-01	7.492E+01	23.75	L064	2 2 2 1 2	
6.829E-01	8.064E+01	24.99	A339	2 0 2 2 2	
5.930E-01	7.003E+01	25	D061	1 0 0 0 2	
6.032E-01	7.124E+01	25	F055	1 2 2 2 2	
6.518E-01	7.697E+01	25	M020	1 0 0 1 2	
6.634E-01	7.834E+01	25	M153	1 0 0 0 0	cal. from fitted equation
7.402E-01	8.741E+01	28	D050	1 2 1 2 2	
8.003E-01	9.451E+01	29.99	A339	2 0 2 2 2	
8.047E-01	9.502E+01	30	M043	1 0 0 0 2	
8.047E-01	9.502E+01	30	M153	1 0 0 0 0	cal. from fitted equation
8.849E-01	1.045E+02	30	W026	1 0 1 1 2	average of 2
9.508E-01	1.123E+02	34.99	A339	2 0 2 2 2	
8.976E-01	1.060E+02	35	L041	1 0 0 1 2	
9.742E-01	1.150E+02	35	M153	1 0 0 0 0	cal. from fitted equation
1.145E+00	1.353E+02	39.99	A339	2 0 2 2 2	
1.149E+00	1.357E+02	40	B088	1 0 0 0 2	
1.181E+00	1.394E+02	40	M043	1 0 0 0 2	
1.168E+00	1.379E+02	40	M153	1 0 0 0 0	cal. from fitted equation
1.377E+00	1.627E+02	44.99	A339	2 0 2 2 2	
1.600E+00	1.889E+02	49.99	A339	2 0 2 2 2	

1.524E+00	1.800E+02	50	L041	1 0 0 1 2	
1.633E+00	1.929E+02	50	M020	1 0 0 1 2	
1.842E+00	2.175E+02	54.99	A339	2 0 2 2 2	
2.048E+00	2.418E+02	59.99	A339	2 0 2 2 2	
2.232E+00	2.636E+02	60	M043	1 0 0 0 2	
2.398E+00	2.832E+02	64.99	A339	2 0 2 2 2	
2.380E+00	2.810E+02	65	L041	1 0 0 1 2	
3.238E+00	3.824E+02	75	F300	1 0 0 0 2	
3.191E+00	3.768E+02	75	M020	1 0 0 1 2	
3.510E+00	4.145E+02	80	M043	1 0 0 0 2	
8.515E-01	1.006E+02	84.30	B118	1 0 0 0 2	unit assumed
4.636E+00	5.475E+02	100	D041	1 0 0 0 2	
4.738E+00	5.595E+02	100	M043	1 0 0 0 2	

282. $C_4H_6O_4$
Methyl Oxalate
Oxalic Acid Ethyl Ester
Oxalsaeure-monoaethyl Ester

RN: 553-90-2   **MP** (°C):   54.0
MW: 118.09   **BP** (°C):   163.5

Solubility (Moles/L)	Solubility (Grams/L)	Temp (°C)	Ref (#)	Evaluation (T P E A A)	Comments
3.006E-01	3.549E+01	0.1	K079	1 0 0 0 2	
6.900E-01	8.148E+01	11.1	K079	1 0 0 0 2	
1.029E+00	1.216E+02	19.5	K079	1 0 0 0 2	
5.106E-01	6.030E+01	25	F300	1 0 0 0 2	
1.489E+00	1.758E+02	27.1	K079	1 0 0 0 2	
1.867E+00	2.204E+02	31.9	K079	1 0 0 0 2	
2.978E+00	3.516E+02	44.4	K079	1 0 0 0 2	
3.372E+00	3.982E+02	49.2	K079	1 0 0 0 2	
3.589E+00	4.238E+02	51.0	K079	1 0 0 0 2	
3.839E+00	4.533E+02	53.0	K079	1 0 0 0 2	
4.783E+00	5.649E+02	75.0	K079	1 0 0 0 2	
4.939E+00	5.832E+02	79.3	K079	1 0 0 0 2	
5.678E+00	6.705E+02	96.1	K079	1 0 0 0 2	
4.929E-01	5.820E+01	rt	D021	0 0 1 1 2	

283.   C$_4$H$_6$O$_5$
Diglycolic Acid
Di-glykolsaeure
RN: 110-99-6   **MP** (°C):   148
MW: 134.09   **BP** (°C):

Solubility (Moles/L)	Solubility (Grams/L)	Temp (°C)	Ref (#)	Evaluation (T P E A A)	Comments
1.596E+00	2.140E+02	5.09	A340	2 0 2 2 2	
1.932E+00	2.590E+02	10.99	A340	2 0 2 2 2	
2.522E+00	3.382E+02	15.59	A340	2 0 2 2 2	
2.668E+00	3.577E+02	20.59	A340	2 0 2 2 2	
2.834E+00	3.801E+02	23.49	A340	2 0 2 2 2	
3.252E+00	4.361E+02	28.09	A340	2 0 2 2 2	
3.645E+00	4.887E+02	37.49	A340	2 0 2 2 2	
3.794E+00	5.087E+02	39.99	A340	2 0 2 2 2	
4.061E+00	5.445E+02	47.99	A340	2 0 2 2 2	
4.135E+00	5.545E+02	49.99	A340	2 0 2 2 2	
4.353E+00	5.837E+02	54.49	A340	2 0 2 2 2	
4.508E+00	6.044E+02	59.49	A340	2 0 2 2 2	
4.631E+00	6.209E+02	64.99	A340	2 0 2 2 2	
4.776E+00	6.404E+02	69.99	A340	2 0 2 2 2	
4.877E+00	6.540E+02	74.99	A340	2 0 2 2 2	
4.969E+00	6.663E+02	79.89	A340	2 0 2 2 2	
5.067E+00	6.794E+02	83.99	A340	2 0 2 2 2	
5.125E+00	6.872E+02	88.19	A340	2 0 2 2 2	

284.   C$_4$H$_6$O$_5$
DL-Malic Acid
Malic Acid
RN: 6915-15-7   **MP** (°C):   131.5
MW: 134.09   **BP** (°C):

Solubility (Moles/L)	Solubility (Grams/L)	Temp (°C)	Ref (#)	Evaluation (T P E A A)	Comments
3.512E+00	4.709E+02	0	M043	1 0 0 0 1	
3.820E+00	5.122E+02	10	M043	1 0 0 0 2	
4.158E+00	5.575E+02	20	M043	1 0 0 0 2	
4.401E+00	5.902E+02	26	D041	1 0 0 0 2	
4.415E+00	5.920E+02	26	F300	1 0 0 0 2	
5.605E+00	7.516E+02	30	D062	1 0 1 1 0	data given in normality
4.475E+00	6.000E+02	30	M043	1 0 0 0 2	
4.794E+00	6.429E+02	40	M043	1 0 0 0 2	
5.442E+00	7.297E+02	60	M043	1 0 0 0 2	
5.998E+00	8.043E+02	79	D041	1 0 0 0 2	
6.033E+00	8.089E+02	79	F300	1 0 0 0 2	
6.126E+00	8.214E+02	80	M043	1 0 0 0 2	

285.   C$_4$H$_6$O$_5$
D-Malic Acid
D(-)-Aepfelsaeure
RN: 636-61-3   **MP** (°C):   100
MW: 134.09   **BP** (°C):

Solubility (Moles/L)	Solubility (Grams/L)	Temp (°C)	Ref (#)	Evaluation (T P E A A)	Comments
3.397E+00	4.555E+02	4.99	A339	2 0 2 2 2	
3.542E+00	4.749E+02	9.99	A339	2 0 2 2 2	
3.695E+00	4.954E+02	14.99	A339	2 0 2 2 2	
3.878E+00	5.200E+02	19.99	A339	2 0 2 2 2	
4.030E+00	5.403E+02	24.99	A339	2 0 2 2 2	
4.146E+00	5.560E+02	29.99	A339	2 0 2 2 2	
4.282E+00	5.742E+02	34.99	A339	2 0 2 2 2	
4.441E+00	5.955E+02	39.99	A339	2 0 2 2 2	
4.544E+00	6.094E+02	44.99	A339	2 0 2 2 2	
4.719E+00	6.328E+02	49.99	A339	2 0 2 2 2	
4.840E+00	6.490E+02	54.99	A339	2 0 2 2 2	
4.976E+00	6.672E+02	59.99	A339	2 0 2 2 2	
5.119E+00	6.865E+02	64.99	A339	2 0 2 2 2	

286.   C$_4$H$_6$O$_6$
L-Tartaric Acid
L(+)-Weinsaeure
L(+)-Tartaric Acid
RN: 87-69-4   **MP** (°C):   169
MW: 150.09   **BP** (°C):

Solubility (Moles/L)	Solubility (Grams/L)	Temp (°C)	Ref (#)	Evaluation (T P E A A)	Comments
3.565E+00	5.350E+02	0	F300	1 0 0 0 2	
3.564E+00	5.349E+02	0	F302	1 0 0 0 2	
3.634E+00	5.455E+02	5	F302	1 0 0 0 2	
3.702E+00	5.556E+02	10	F302	1 0 0 0 2	
3.791E+00	5.690E+02	15	F302	1 0 0 0 2	
3.878E+00	5.820E+02	20	F300	1 0 0 0 2	
3.875E+00	5.816E+02	20	F302	1 0 0 0 2	
3.965E+00	5.951E+02	25	F302	1 0 0 0 2	
4.060E+00	6.094E+02	30	F302	1 0 0 0 2	
4.158E+00	6.241E+02	35	F302	1 0 0 0 2	
4.249E+00	6.377E+02	40	F302	1 0 0 0 2	
4.325E+00	6.491E+02	45	F302	1 0 0 0 2	
4.397E+00	6.600E+02	50	F300	1 0 0 0 1	
4.404E+00	6.610E+02	50	F302	1 0 0 0 2	
4.485E+00	6.732E+02	55	F302	1 0 0 0 2	
4.568E+00	6.855E+02	60	F302	1 0 0 0 2	
4.644E+00	6.970E+02	65	F302	1 0 0 0 2	

4.726E+00	7.093E+02	70	F302	1 0 0 0 2	
4.802E+00	7.207E+02	75	F302	1 0 0 0 2	
4.876E+00	7.319E+02	80	F302	1 0 0 0 2	
4.954E+00	7.436E+02	85	F302	1 0 0 0 2	
5.026E+00	7.543E+02	90	F302	1 0 0 0 2	
5.095E+00	7.647E+02	95	F302	1 0 0 0 2	
5.157E+00	7.740E+02	100	F300	1 0 0 0 2	
5.159E+00	7.743E+02	100	F302	1 0 0 0 2	

287. $C_4H_6O_6$
DL-Tartaric Acid
DL-Weinsaeure
Tartaric Acid (Racemic)

RN: 133-37-9   **MP** (°C):   206
MW: 150.09   **BP** (°C):

Solubility (Moles/L)	Solubility (Grams/L)	Temp (°C)	Ref (#)	Evaluation (T P E A A)	Comments
2.279E+00	3.421E+02	0	D039	2 2 1 2 0	EFG
5.630E-01	8.450E+01	0	D041	1 0 0 0 2	
5.084E-01	7.630E+01	0	F300	1 0 0 0 2	
5.049E-01	7.579E+01	0	M043	1 0 0 0 1	
2.333E+00	3.502E+02	10	D039	2 2 1 2 0	EFG
7.298E-01	1.095E+02	10	M043	1 0 0 0 2	
2.350E+00	3.528E+02	20	D039	2 2 1 2 0	EFG
1.138E+00	1.708E+02	20	D041	1 0 0 0 2	
1.139E+00	1.710E+02	20	F300	1 0 0 0 2	
1.016E+00	1.525E+02	20	M043	1 0 0 0 2	
2.459E+00	3.690E+02	25	D039	2 2 1 2 2	EFG
1.179E+00	1.770E+02	25	F017	1 0 0 0 2	
1.026E+01	1.540E+03	25	K040	1 0 2 1 2	
2.483E+00	3.726E+02	30	D039	2 2 1 2 0	EFG
1.341E+00	2.013E+02	30	M043	1 0 0 0 2	
2.563E+00	3.846E+02	40	D039	2 2 1 2 0	EFG
1.799E+00	2.701E+02	40	M043	1 0 0 0 2	
2.612E+00	3.921E+02	50	D039	2 2 1 2 0	EFG
2.687E+00	4.033E+02	60	D039	2 2 1 2 0	EFG
2.612E+00	3.921E+02	60	M043	1 0 0 0 2	
2.750E+00	4.128E+02	70	D039	2 2 1 2 0	EFG
2.811E+00	4.220E+02	80	D039	2 2 1 2 0	EFG
3.299E+00	4.952E+02	80	M043	1 0 0 0 2	
2.860E+00	4.292E+02	90	D039	2 2 1 2 0	EFG
2.920E+00	4.382E+02	100	D039	2 2 1 2 0	EFG
4.324E+00	6.490E+02	100	D041	1 0 0 0 2	
4.331E+00	6.500E+02	100	F300	1 0 0 0 1	
3.863E+00	5.798E+02	100	M043	1 0 0 0 2	

288. $C_4H_6O_6$
D-(-)-Tartaric Acid
D-(-)-Dihydroxysuccinic Acid
RN: 147-71-7   **MP** (°C):   173
MW: 150.09   **BP** (°C):   171

Solubility (Moles/L)	Solubility (Grams/L)	Temp (°C)	Ref (#)	Evaluation (T P E A A)	Comments
3.564E+00	5.349E+02	0	M043	1 0 0 0 2	
3.348E+00	5.024E+02	4.99	A339	2 0 2 2 2	
3.431E+00	5.149E+02	9.99	A339	2 0 2 2 2	
2.350E+00	3.528E+02	10	D020	1 2 1 1 2	
3.715E+00	5.575E+02	10	M043	1 0 0 0 2	
3.499E+00	5.251E+02	14.99	A339	2 0 2 2 2	
3.553E+00	5.332E+02	19.99	A339	2 0 2 2 2	
3.875E+00	5.816E+02	20	M043	1 0 0 0 2	
3.629E+00	5.447E+02	24.99	A339	2 0 2 2 2	
2.459E+00	3.691E+02	25	D020	1 2 1 1 2	
3.973E+00	5.963E+02	25	F017	1 0 0 0 2	
3.706E+00	5.562E+02	29.99	A339	2 0 2 2 2	
4.060E+00	6.094E+02	30	M043	1 0 0 0 2	
3.791E+00	5.690E+02	34.99	A339	2 0 2 2 2	
3.846E+00	5.773E+02	39.99	A339	2 0 2 2 2	
4.249E+00	6.377E+02	40	M043	1 0 0 0 2	
3.926E+00	5.892E+02	44.99	A339	2 0 2 2 2	
4.021E+00	6.036E+02	49.99	A339	2 0 2 2 2	
4.104E+00	6.160E+02	54.99	A339	2 0 2 2 2	
4.157E+00	6.238E+02	59.99	A339	2 0 2 2 2	
4.581E+00	6.875E+02	60	M043	1 0 0 0 2	
4.232E+00	6.352E+02	64.99	A339	2 0 2 2 2	
4.876E+00	7.319E+02	80	M043	1 0 0 0 2	
5.159E+00	7.743E+02	100	M043	1 0 0 0 2	

289. $C_4H_6O_6$
meso-Tartaric Acid
meso-Weinsaeure
RN: 147-73-9   **MP** (°C):   147
MW: 150.09   **BP** (°C):

Solubility (Moles/L)	Solubility (Grams/L)	Temp (°C)	Ref (#)	Evaluation (T P E A A)	Comments
2.239E+00	3.360E+02	0	F300	1 0 0 0 2	
3.702E+00	5.556E+02	15	D041	1 0 0 0 2	
3.731E+00	5.600E+02	15	F300	1 0 0 0 1	
3.731E+00	5.600E+02	20	F300	1 0 0 0 1	

290. C_4H_7Br
4-Bromo-1-butene
1-Bromo-3-butene
Homoallyl Bromide
4-Bromobutene-1
3-Butenyl Bromide
RN: 5162-44-7   **MP** (°C):
MW: 135.01   **BP** (°C):   98.5

Solubility (Moles/L)	Solubility (Grams/L)	Temp (°C)	Ref (#)	Evaluation (T P E A A)	Comments
5.660E-03	7.642E-01	25	M342	1 0 1 1 2	

291. $C_4H_7BrN_2O_2$
α-Bromo-methyl-acetic Ureide
Propanamide, N-(Aminocarbonyl)-2-bromo-
(2-Bromopropionyl)urea
α-Bromopropionylurea
RN: 14299-55-9   **MP** (°C):
MW: 195.02   **BP** (°C):

Solubility (Moles/L)	Solubility (Grams/L)	Temp (°C)	Ref (#)	Evaluation (T P E A A)	Comments
2.581E-01	5.033E+01	ns	F056	0 2 2 2 1	

292. $C_4H_7BrO_2$
α-Bromobutyric Acid
DL-2-Bromobutyric Acid
DL-Brombuttersaeure
RN: 80-58-0   **MP** (°C):   -4
MW: 167.01   **BP** (°C):   181

Solubility (Moles/L)	Solubility (Grams/L)	Temp (°C)	Ref (#)	Evaluation (T P E A A)	Comments
4.191E-01	7.000E+01	ns	F300	1 0 0 0 0	

293. C_4H_7Cl
1-Chloro-2-butene
1-Chloro-2-methylpropene-2
α-Methylallyl Chloride
RN: 591-97-9   **MP** (°C):
MW: 90.55   **BP** (°C):   84

Solubility (Moles/L)	Solubility (Grams/L)	Temp (°C)	Ref (#)	Evaluation (T P E A A)	Comments
1.103E-02	9.990E-01	ns	M061	0 0 0 0 0	

294.   C$_4$H$_7$Cl$_2$O$_4$P
Dichlorvos
O,O-Dimethyl O-2-Dichlorovinyl Phosphate
RN: 62-73-7   **MP** (°C):
MW: 220.98   **BP** (°C):   84

Solubility (Moles/L)	Solubility (Grams/L)	Temp (°C)	Ref (#)	Evaluation (T P E A A)	Comments
4.481E-02	9.901E+00	ns	M061	0 0 0 0 0	
4.525E-02	1.000E+01	rt	M161	0 0 0 0 1	

295.   C$_4$H$_7$Cl$_3$O
1,1,1-Trichloro-tert-butanol
Acetonchloroform
Chloreton
RN: 57-15-8   **MP** (°C):   98
MW: 177.46   **BP** (°C):   167

Solubility (Moles/L)	Solubility (Grams/L)	Temp (°C)	Ref (#)	Evaluation (T P E A A)	Comments
4.508E-02	8.000E+00	20	F300	1 0 0 0 0	

296.   C$_4$H$_7$N
n-Butyroniitrile
γ-Butyronitrile
Propyl cyanide
1-Cyanopropane
n-Butyronitrile
RN: 109-74-0   **MP** (°C):   -112
MW: 69.11   **BP** (°C):   115-117

Solubility (Moles/L)	Solubility (Grams/L)	Temp (°C)	Ref (#)	Evaluation (T P E A A)	Comments
5.446E-02	3.764E+00	25	B004	1 0 0 0 2	

297.   C$_4$H$_7$NO$_3$
N-Acetyl Glycine
Aceturic Acid
Glycin-N-acetat
Glycine-N-acetate
RN: 543-24-8   **MP** (°C):   206
MW: 117.11   **BP** (°C):

Solubility (Moles/L)	Solubility (Grams/L)	Temp (°C)	Ref (#)	Evaluation (T P E A A)	Comments
2.246E-01	2.630E+01	15	F300	1 0 0 0 2	

298. $C_4H_7NO_4$
Butanoic Acid, 4-Amino-2-hydroxy-4-oxo-
D-β-Malaminsaeure
R-β-Malaminsaeure
RN: 82310-91-6 **MP** (°C): 149
MW: 133.10 **BP** (°C):

Solubility (Moles/L)	Solubility (Grams/L)	Temp (°C)	Ref (#)	Evaluation (T P E A A)	Comments
2.903E-01	3.865E+01	18	L039	1 0 0 0 2	
5.255E-01	6.994E+01	18	L039	1 0 0 0 2	

299. $C_4H_7NO_4$
Iminodiacetic Acid
Imino-diessigsaeure
RN: 142-73-4 **MP** (°C): 247.5
MW: 133.10 **BP** (°C):

Solubility (Moles/L)	Solubility (Grams/L)	Temp (°C)	Ref (#)	Evaluation (T P E A A)	Comments
1.781E-01	2.370E+01	5	F300	1 0 0 0 2	

300. $C_4H_7NO_4$
L-Aspartic Acid
Aspartic Acid
L(+)-Asparaginsaeure
L-(+)-Asparaginic Acid
L-(+)-Aspartic Acid
RN: 56-84-8 **MP** (°C): 270.5
MW: 133.10 **BP** (°C):

Solubility (Moles/L)	Solubility (Grams/L)	Temp (°C)	Ref (#)	Evaluation (T P E A A)	Comments
1.675E-02	2.230E+00	0	D018	2 2 2 1 2	
3.170E-02	4.220E+00	20	B032	1 2 2 1 2	
3.770E-02	5.018E+00	25	B032	1 2 2 1 2	
4.030E-02	5.364E+00	25	D018	2 2 2 1 2	
3.738E-02	4.975E+00	25	D041	1 0 0 0 0	
3.805E-02	5.064E+00	25	G315	1 0 2 2 2	
3.719E-02	4.950E+00	25	J303	2 0 2 2 2	
3.644E-02	4.850E+00	27	D036	2 1 2 2 2	
4.469E-02	5.948E+00	29.80	B032	1 2 2 1 2	
6.348E-02	8.450E+00	40	J303	2 0 2 2 2	
9.304E-02	1.238E+01	50	D018	2 2 2 1 2	
1.232E-01	1.640E+01	60	J303	2 0 2 2 2	
1.985E-01	2.642E+01	75	D018	2 2 2 1 2	
2.100E-01	2.795E+01	75	D041	1 0 0 0 2	
2.885E-01	3.840E+01	99	M160	2 1 1 1 0	
3.750E-02	4.991E+00	ns	M025	0 2 0 1 2	

301.   C$_4$H$_7$NO$_4$
L-β-Malamidic Acid
L-β-Malaminsaeure
RN: 57229-74-0   **MP** (°C):   149
MW: 133.10   **BP** (°C):

Solubility (Moles/L)	Solubility (Grams/L)	Temp (°C)	Ref (#)	Evaluation (T P E A A)	Comments
5.242E-01	6.977E+01	18	L039	1 0 0 0 2	

302.   C$_4$H$_7$NO$_4$
DL-Aspartic Acid
DL-2-Aminobutanedioic Acid
RN: 617-45-8   **MP** (°C):
MW: 133.10   **BP** (°C):

Solubility (Moles/L)	Solubility (Grams/L)	Temp (°C)	Ref (#)	Evaluation (T P E A A)	Comments
2.367E-02	3.151E+00	0	D018	2 2 2 1 2	
6.081E-02	8.094E+00	25	D018	2 2 2 1 2	
6.110E-02	8.133E+00	25	D041	1 0 0 0 1	
1.544E-01	2.055E+01	50	D018	2 2 2 1 2	
3.437E-01	4.575E+01	75	D018	2 2 2 1 2	
3.434E-01	4.571E+01	75	D041	1 0 0 0 2	

303.   C$_4$H$_7$N$_2$O$_4$
Glycine Dipeptide
RN: **MP** (°C):
MW: 147.11   **BP** (°C):

Solubility (Moles/L)	Solubility (Grams/L)	Temp (°C)	Ref (#)	Evaluation (T P E A A)	Comments
1.418E+00	2.086E+02	20	B032	1 2 2 1 2	
1.534E+00	2.257E+02	25	B032	1 2 2 1 2	
1.540E+00	2.266E+02	25.1	N024	2 0 2 2 2	
1.546E+00	2.275E+02	25.1	N026	2 0 2 2 2	
1.647E+00	2.423E+02	29.80	B032	1 2 2 1 2	

304.   C$_4$H$_7$N$_3$O
Creatinine
Kreatinin
RN: 60-27-5   **MP** (°C):   220.5
MW: 113.12   **BP** (°C):

Solubility (Moles/L)	Solubility (Grams/L)	Temp (°C)	Ref (#)	Evaluation (T P E A A)	Comments
7.075E-01	8.004E+01	16	D041	1 0 0 0 1	
7.081E-01	8.010E+01	16	F300	1 0 0 0 2	

305.   C₄H₈

Isobutylene
2-Methylpropene

RN: 115-11-7   **MP** (°C):   -140.3
MW: 56.11   **BP** (°C):   -6.90

Solubility (Moles/L)	Solubility (Grams/L)	Temp (°C)	Ref (#)	Evaluation (T P E A A)	Comments
4.687E-03	2.630E-01	25	M001	2 1 2 2 2	

306.   C₄H₈

1-Butene
α-Butene
Ethylethylene
α-Butylene
1-Butylene
Butene-1

RN: 106-98-9   **MP** (°C):   -185
MW: 56.11   **BP** (°C):   -6.47

Solubility (Moles/L)	Solubility (Grams/L)	Temp (°C)	Ref (#)	Evaluation (T P E A A)	Comments
3.957E-03	2.220E-01	25	M001	2 1 2 2 2	
1.210E-02	6.791E-01	38	B123	1 2 1 1 2	
1.582E-02	8.876E-01	71	B123	1 2 1 1 2	
2.746E-02	1.541E+00	104	B123	1 2 1 1 2	
3.526E-02	1.979E+00	138	B123	1 2 1 1 2	
3.858E-02	2.165E+00	144.00	B123	1 2 1 1 2	

307.   C₄H₈Cl₂

2,3-Dichlorobutane
Butane, 2,3-Dichloro-

RN: 7581-97-7   **MP** (°C):   -80
MW: 127.01   **BP** (°C):   117

Solubility (Moles/L)	Solubility (Grams/L)	Temp (°C)	Ref (#)	Evaluation (T P E A A)	Comments
1.430E-02	1.817E+00	0	L103	1 0 0 0 2	unit assumed
4.422E-03	5.617E-01	20	L103	1 0 0 0 2	unit assumed
1.464E-03	1.860E-01	30	L103	1 0 0 0 2	unit assumed
1.755E-03	2.230E-01	40	L103	1 0 0 0 2	unit assumed

308.   C$_4$H$_8$Cl$_2$O
sym-Dichloroethyl Ether
2,2'-Dichlorodiethylether
RN: 111-44-4   **MP** (°C):   -50
MW: 143.01   **BP** (°C):   66

Solubility (Moles/L)	Solubility (Grams/L)	Temp (°C)	Ref (#)	Evaluation (T P E A A)	Comments
7.060E-02	1.010E+01	20	D052	1 1 0 0 0	
7.403E-02	1.059E+01	20	M062	1 0 0 0 2	

309.   C$_4$H$_8$Cl$_2$OS
β,β'-Dichlorodiethylsulfoxide
β,β'-Dichlor-diaethylsulfoxid
RN: 5819-08-9   **MP** (°C):
MW: 175.08   **BP** (°C):

Solubility (Moles/L)	Solubility (Grams/L)	Temp (°C)	Ref (#)	Evaluation (T P E A A)	Comments
6.854E-02	1.200E+01	20	F300	1 0 0 0 1	

310.   C$_4$H$_8$Cl$_2$O$_2$S
β,β'-Dichlorodiethylsulfone
β,β'-Dichlor-diaethylsulfon
RN: 471-03-4   **MP** (°C):
MW: 191.08   **BP** (°C):

Solubility (Moles/L)	Solubility (Grams/L)	Temp (°C)	Ref (#)	Evaluation (T P E A A)	Comments
3.140E-02	6.000E+00	20	F300	1 0 0 0 0	
1.256E-01	2.400E+01	100	F300	1 0 0 0 1	

311.   C$_4$H$_8$Cl$_2$S
Mustard Gas
Sulfure β'-Ethyl Dichlore
β,β'-Dichlor-diaethylsulfid
RN: 505-60-2   **MP** (°C):
MW: 159.08   **BP** (°C):

Solubility (Moles/L)	Solubility (Grams/L)	Temp (°C)	Ref (#)	Evaluation (T P E A A)	Comments
4.337E-03	6.900E-01	25	F300	1 0 0 0 1	
3.017E-03	4.800E-01	c	B079	0 0 1 1 1	

312.   C$_4$H$_8$Cl$_3$O$_4$P
Trichlorfon
O,O-Dimethyl (1-Hydroxy-2,2,2-trichloroethyl)phosphonate
RN: 52-68-6   **MP** (°C):   83.5
MW: 257.44   **BP** (°C):

Solubility (Moles/L)	Solubility (Grams/L)	Temp (°C)	Ref (#)	Evaluation (T P E A A)	Comments
5.982E-01	1.540E+02	25	M161	1 0 0 0 2	
4.255E-01	1.095E+02	ns	M061	0 0 0 0 2	

313.   C$_4$H$_8$N$_2$O$_2$
Dimethylglyoxime
Dimethylglyoxim
RN: 95-45-4   **MP** (°C):   240.5
MW: 116.12   **BP** (°C):

Solubility (Moles/L)	Solubility (Grams/L)	Temp (°C)	Ref (#)	Evaluation (T P E A A)	Comments
5.167E-03	6.000E-01	20	F300	1 0 0 0 0	
3.100E-02	3.600E+00	80	F300	1 0 0 0 1	
5.081E-02	5.900E+00	100	F300	1 0 0 0 1	

314.   C$_4$H$_8$N$_2$O$_2$
Succinamide
Bersteinsaeure-diamid
RN: 110-14-5   **MP** (°C):   260
MW: 116.12   **BP** (°C):

Solubility (Moles/L)	Solubility (Grams/L)	Temp (°C)	Ref (#)	Evaluation (T P E A A)	Comments
3.858E-02	4.480E+00	15	D041	1 0 0 0 1	
2.842E-02	3.300E+00	15	F300	1 0 0 0 1	
8.534E-01	9.910E+01	100	D041	1 0 0 0 2	
3.445E-04	4.000E-02	c	L055	0 0 0 0 2	
9.463E-03	1.099E+00	h	L055	0 0 0 0 1	

315.   C$_4$H$_8$N$_2$O$_3$
N-Nitroso-N-methylurethane
N-Nitroso-N-methyl-urethan
RN: 615-53-2   **MP** (°C):
MW: 132.12   **BP** (°C):

Solubility (Moles/L)	Solubility (Grams/L)	Temp (°C)	Ref (#)	Evaluation (T P E A A)	Comments
2.800E-01	3.699E+01	24	M031	1 1 1 1 1	

316. $C_4H_8N_2O_3$

Asparagine
L-Asparagine
L-Asparagin

RN: 70-47-3   **MP** (°C):   235
MW: 132.12   **BP** (°C):

Solubility (Moles/L)	Solubility (Grams/L)	Temp (°C)	Ref (#)	Evaluation (T P E A A)	Comments
6.509E-02	8.600E+00	0	F300	1 0 0 0 1	
2.180E-01	2.880E+01	15	D349	2 1 1 2 2	
1.759E-01	2.324E+01	20	B032	1 2 2 1 2	
2.210E-01	2.920E+01	20	D349	2 1 1 2 2	
1.589E-01	2.100E+01	20	F300	1 0 0 0 2	
8.477E-02	1.120E+01	21.5	P045	0 0 2 1 2	
2.226E-01	2.941E+01	25	B032	1 2 2 1 2	
2.260E-01	2.986E+01	25	D349	2 1 1 2 2	
1.709E-01	2.258E+01	25	G315	1 0 2 2 2	
1.900E-01	2.510E+01	25.1	N024	2 0 2 2 2	
1.900E-01	2.510E+01	25.1	N025	2 0 2 2 2	
1.900E-01	2.510E+01	25.1	N026	2 0 2 2 2	
1.853E-01	2.449E+01	25.1	N027	1 1 2 2 2	
1.918E-01	2.534E+01	27	D036	2 1 2 2 2	
2.233E-01	2.950E+01	27	D036	2 1 2 2 2	
2.777E-01	3.669E+01	29.80	B032	1 2 2 1 2	
2.604E+00	3.440E+02	98	F300	1 0 0 0 2	
1.817E-01	2.400E+01	ns	D072	0 0 0 0 1	
1.860E-01	2.457E+01	ns	M025	0 2 0 1 2	
1.774E-01	2.344E+01	rt	D021	0 0 1 1 2	

317. $C_4H_8N_2O_3$

α-Alanine Hydantoic Acid
Methylhydantoic Acid

RN: 77340-50-2   **MP** (°C):
MW: 132.12   **BP** (°C):

Solubility (Moles/L)	Solubility (Grams/L)	Temp (°C)	Ref (#)	Evaluation (T P E A A)	Comments
1.930E-01	2.550E+01	25	M024	1 2 0 1 2	
1.930E-01	2.550E+01	ns	M025	0 2 0 1 2	

318. $C_4H_8N_2O_3$
N-Glycylglycine
Diglycine
RN: 556-50-3   **MP** (°C):   215
MW: 132.12   **BP** (°C):

Solubility (Moles/L)	Solubility (Grams/L)	Temp (°C)	Ref (#)	Evaluation (T P E A A)	Comments
1.253E+00	1.656E+02	21	F300	1 0 0 0 2	
1.399E+00	1.848E+02	25	G092	2 1 1 1 1	
1.399E+00	1.848E+02	25	G315	1 0 2 2 2	
1.430E+00	1.890E+02	25.1	N027	1 2 2 2 2	
1.512E+00	1.998E+02	ns	M025	0 2 0 1 2	

319. $C_4H_8N_2O_3$
β-Alanine Hydantoic Acid
β-Uramidopropionic Acid
Glycine, N-(Aminocarbonyl)-N-methyl-
RN: 30565-25-4   **MP** (°C):
MW: 132.12   **BP** (°C):

Solubility (Moles/L)	Solubility (Grams/L)	Temp (°C)	Ref (#)	Evaluation (T P E A A)	Comments
1.580E-01	2.087E+01	25	M024	1 2 0 1 2	

320. $C_4H_8N_2O_3.H_2O$
L-Asparagine Monohydrate
Asparagine, Monohydrate, L-
RN: 5794-13-8   **MP** (°C):   234
MW: 150.14   **BP** (°C):

Solubility (Moles/L)	Solubility (Grams/L)	Temp (°C)	Ref (#)	Evaluation (T P E A A)	Comments
1.933E-01	2.902E+01	25	D041	1 0 0 0 2	
1.933E-01	2.902E+01	25	D041	1 0 0 0 2	
1.858E-01	2.790E+01	25	O316	1 0 1 2 2	
1.853E-01	2.781E+01	25	O316	1 0 1 2 2	
1.293E+00	1.941E+02	75	D041	1 0 0 0 2	

321. $C_4H_8N_4O_2$
N,N'-Dinitrosopiperazine
Dinitrosopiperazine
RN: 140-79-4   **MP** (°C):
MW: 144.13   **BP** (°C):

Solubility (Moles/L)	Solubility (Grams/L)	Temp (°C)	Ref (#)	Evaluation (T P E A A)	Comments
4.000E-02	5.765E+00	24	D083	2 0 0 0 1	

322.   C₄H₈O
Ethyl Vinyl Ether
Aethyl-vinyl-aether
RN: 109-92-2   **MP** (°C):   -115.0
MW: 72.11   **BP** (°C):   35

Solubility (Moles/L)	Solubility (Grams/L)	Temp (°C)	Ref (#)	Evaluation (T P E A A)	Comments
1.390E-01	1.002E+01	37	E028	1 1 2 2 2	

323.   C₄H₈O
Isobutyraldehyde
2-Methyl Propanal
RN: 78-84-2   **MP** (°C):   -66
MW: 72.11   **BP** (°C):   64

Solubility (Moles/L)	Solubility (Grams/L)	Temp (°C)	Ref (#)	Evaluation (T P E A A)	Comments
1.167E+00	8.413E+01	20	M146	1 2 2 2 2	
1.234E+00	8.900E+01	25	A049	1 0 0 0 0	

324.   C₄H₈O
2-Butyraldehyde
Butyraldehyde
Butyraldehyd
n-Butanal
RN: 123-72-8   **MP** (°C):   -96
MW: 72.11   **BP** (°C):   75

Solubility (Moles/L)	Solubility (Grams/L)	Temp (°C)	Ref (#)	Evaluation (T P E A A)	Comments
4.948E-01	3.568E+01	20	D041	1 0 0 0 1	
4.993E-01	3.600E+01	20	F300	1 0 0 0 1	
9.694E-01	6.990E+01	25	A049	1 0 0 0 2	
9.194E-01	6.629E+01	25	B060	2 0 1 1 1	
5.077E-01	3.661E+01	38	J020	2 2 2 1 1	

325. C$_4$H$_8$O
Methyl Ethyl Ketone
Butanon-(2)
RN: 78-93-3   **MP** (°C): -87
MW: 72.11   **BP** (°C): 80

Solubility (Moles/L)	Solubility (Grams/L)	Temp (°C)	Ref (#)	Evaluation (T P E A A)	Comments
1.015E+00	7.322E+01	20	A075	1 0 0 0 1	
2.827E+00	2.038E+02	20	D052	1 1 0 0 2	
2.922E+00	2.107E+02	20	E019	1 0 1 1 2	
2.399E+00	1.730E+02	20	F300	1 0 0 0 2	
2.977E+00	2.146E+02	20	G030	1 2 0 0 2	
5.020E+00	3.620E+02	20	P040	1 0 0 0 2	
2.931E+00	2.114E+02	25	A094	1 0 0 0 2	
3.302E+00	2.381E+02	25	A356	2 1 2 2 1	
2.931E+00	2.114E+02	25	B060	2 0 1 1 1	
3.130E+00	2.257E+02	25	F044	1 0 0 0 2	
2.824E+00	2.036E+02	25	G030	1 2 0 0 2	
2.657E+00	1.916E+02	25	J005	1 0 2 1 2	
6.112E+00	4.407E+02	25	K105	2 0 0 0 2	
2.912E+00	2.100E+02	25	M136	2 0 0 0 2	
2.912E+00	2.100E+02	25	M139	2 0 0 0 2	
2.720E+00	1.961E+02	25	N309	1 0 0 0 2	
2.756E+00	1.987E+02	25	O028	2 2 2 2 2	
2.556E+00	1.843E+02	25	P055	1 0 0 0 1	
2.774E+00	2.000E+02	25	R320	1 0 1 1 2	
2.690E+00	1.940E+02	30	G030	1 2 0 0 2	
1.703E+00	1.228E+02	30	R319	2 2 2 1 2	
2.900E+00	2.091E+02	35	A356	2 1 2 2 1	
2.969E+00	2.141E+02	35	C309	2 2 2 2 1	
2.538E+00	1.830E+02	38	J020	2 0 2 1 2	
7.726E-01	5.571E+01	40	A075	1 0 0 0 1	
2.723E+00	1.964E+02	45	A356	2 1 2 2 1	
2.723E+00	1.964E+02	45	A356	2 1 2 2 1	
2.615E+00	1.885E+02	45	C309	2 2 2 2 1	
6.257E+00	4.512E+02	45	K105	2 0 0 0 2	
6.855E-01	4.943E+01	60	A075	1 0 0 0 1	
6.319E+00	4.556E+02	60	K105	2 0 0 0 2	
6.352E-01	4.580E+01	70	A075	1 0 0 0 1	
3.453E+00	2.490E+02	70	P040	1 0 0 0 2	
2.219E+00	1.600E+02	90	F300	1 0 0 0 1	
3.627E+00	2.615E+02	100	P040	1 0 0 0 2	
6.844E+00	4.935E+02	140	P040	1 0 0 0 2	
3.334E+00	2.404E+02	ns	C309	2 2 2 2 1	

326.   C$_4$H$_8$O
Tetrahydrofuran
1,4-Epoxybutane
Butylene Oxide
RN: 109-99-9   **MP** (°C):   -108.0
MW: 72.11   **BP** (°C):   66.0

Solubility (Moles/L)	Solubility (Grams/L)	Temp (°C)	Ref (#)	Evaluation (T P E A A)	Comments
4.498E+00	3.243E+02	72.2	M347	2 2 2 1 2	
4.504E+00	3.248E+02	72.25	M347	2 2 2 1 2	
4.536E+00	3.271E+02	72.3	M347	2 2 2 1 2	
4.251E+00	3.065E+02	73.4	M347	2 2 2 1 2	
4.019E+00	2.898E+02	75.4	M347	2 2 2 1 2	
3.678E+00	2.652E+02	78.6	M347	2 2 2 1 2	
3.595E+00	2.593E+02	78.9	M347	2 2 2 1 2	
3.378E+00	2.436E+02	83.3	M347	2 2 2 1 2	
3.257E+00	2.349E+02	87.9	M347	2 2 2 1 2	
3.217E+00	2.320E+02	89.5	M347	2 2 2 1 2	
3.118E+00	2.248E+02	92.9	M347	2 2 2 1 2	
3.042E+00	2.194E+02	102.5	M347	2 2 2 1 2	
3.042E+00	2.194E+02	110.5	M347	2 2 2 1 2	
3.118E+00	2.248E+02	119.3	M347	2 2 2 1 2	
3.257E+00	2.349E+02	127.8	M347	2 2 2 1 2	
3.595E+00	2.593E+02	132.9	M347	2 2 2 1 2	
3.998E+00	2.883E+02	136.1	M347	2 2 2 1 2	
4.067E+00	2.933E+02	136.5	M347	2 2 2 1 2	
4.617E+00	3.329E+02	137.1	M347	2 2 2 1 2	
6.934E+00	5.000E+02	rt	B066	0 2 0 0 2	

327.   C$_4$H$_8$O$_2$
3-Hydroxytetrahydrofuran
(RS)-3-Hydroxytetrahydrofuran
Tetrahydro-3-furanol
(±)-3-Hydroxytetrahydrofuran
3-Hydroxyoxolane
RN: 453-20-3   **MP** (°C):   <25
MW: 88.11   **BP** (°C):

Solubility (Moles/L)	Solubility (Grams/L)	Temp (°C)	Ref (#)	Evaluation (T P E A A)	Comments
5.675E+00	5.000E+02	rt	B066	0 2 0 0 2	

328. $C_4H_8O_2$
Propyl Formate
Ameisensaeure-propylester
Propyl Methanoate
n-Propyl Formate
Propyl Formate
RN: 110-74-7   **MP** (°C):   -93
MW: 88.11   **BP** (°C):   81

Solubility (Moles/L)	Solubility (Grams/L)	Temp (°C)	Ref (#)	Evaluation (T P E A A)	Comments
4.222E-01	3.720E+01	-1.0	K079	1 0 0 0 2	
3.861E-01	3.402E+01	4.0	K079	1 0 0 0 2	
3.722E-01	3.280E+01	6.0	K079	1 0 0 0 2	
3.444E-01	3.035E+01	12.5	K079	1 0 0 0 2	
3.220E-01	2.837E+01	20	S006	1 0 0 0 2	
3.272E-01	2.883E+01	20.0	K079	1 0 0 0 2	
2.497E-01	2.200E+01	22	F300	1 0 0 0 1	
3.161E-01	2.785E+01	30.0	K079	1 0 0 0 2	
2.880E-01	2.537E+01	32.5	N014	1 1 1 0 2	
3.083E-01	2.717E+01	34.0	K079	1 0 0 0 2	
2.972E-01	2.619E+01	45.0	K079	1 0 0 0 2	

329. $C_4H_8O_2$
Methyl Propionate
Methylester Propanolc Acid
RN: 554-12-1   **MP** (°C):   -87.0
MW: 88.11   **BP** (°C):   79.7

Solubility (Moles/L)	Solubility (Grams/L)	Temp (°C)	Ref (#)	Evaluation (T P E A A)	Comments
1.083E+00	9.545E+01	-2.1	K079	1 0 0 0 2	
1.000E+00	8.811E+01	1.0	K079	1 0 0 0 2	
8.778E-01	7.734E+01	11.5	K079	1 0 0 0 2	
8.500E-01	7.489E+01	14.9	K079	1 0 0 0 2	
8.150E-01	7.181E+01	20	S006	1 0 0 0 2	
8.167E-01	7.195E+01	20.0	K079	1 0 0 0 2	
7.778E-01	6.853E+01	27.1	K079	1 0 0 0 2	
7.667E-01	6.755E+01	32.5	K079	1 0 0 0 2	
7.389E-01	6.510E+01	42.7	K079	1 0 0 0 2	

330.   C$_4$H$_8$O$_2$
Isobutyric Acid
Isobuttersaeure
RN: 79-31-2   **MP** (°C):   -47
MW: 88.11   **BP** (°C):   153.5

Solubility (Moles/L)	Solubility (Grams/L)	Temp (°C)	Ref (#)	Evaluation (T P E A A)	Comments
1.931E+00	1.701E+02	15.2	P060	1 0 0 0 2	
1.931E+00	1.701E+02	15.2	P060	1 2 0 0 2	
4.171E+00	3.675E+02	17	P060	1 0 0 0 2	
4.171E+00	3.675E+02	17	P060	1 2 0 0 2	
2.619E+00	2.308E+02	17.7	H068	2 0 0 0 1	
1.892E+00	1.667E+02	20	D041	1 0 0 0 0	
1.894E+00	1.669E+02	20	F300	1 0 0 0 2	
3.768E+00	3.320E+02	20.0	P060	1 0 0 0 2	
3.768E+00	3.320E+02	20.0	P060	1 2 0 0 2	
3.732E+00	3.289E+02	20.1	P060	1 0 0 0 2	
3.732E+00	3.289E+02	20.1	P060	1 2 0 0 2	
2.255E+00	1.987E+02	20.2	P060	1 2 0 0 2	
2.255E+00	1.987E+02	20.25	P060	1 0 0 0 2	
2.367E+00	2.085E+02	20.9	P060	1 0 0 0 2	
2.363E+00	2.082E+02	20.9	P060	1 2 0 0 2	
3.363E+00	2.963E+02	21.2	P060	1 2 0 0 2	
3.363E+00	2.963E+02	21.2	P060	1 0 0 0 2	
3.161E+00	2.785E+02	21.5	P060	1 2 0 0 2	
3.161E+00	2.785E+02	21.5	P060	1 0 0 0 2	
2.500E+00	2.203E+02	21.5	P060	1 2 0 0 2	
2.500E+00	2.203E+02	21.5	P060	1 0 0 0 2	
3.240E+00	2.855E+02	21.7	P060	1 2 0 0 2	
3.001E+00	2.644E+02	21.76	P060	1 0 0 0 2	
3.003E+00	2.645E+02	21.79	P060	1 0 0 0 2	
2.831E+00	2.495E+02	21.8	P060	1 2 0 0 2	
2.831E+00	2.495E+02	21.89	P060	1 0 0 0 2	
2.709E+00	2.387E+02	21.9	P060	1 0 0 0 2	
2.709E+00	2.387E+02	21.9	P060	1 2 0 0 2	

331.   C$_4$H$_8$O$_2$
Butyric Acid
Buttersaeure
n-Butyric Acid
RN: 107-92-6   **MP** (°C):   -7.9
MW: 88.11   **BP** (°C):   163.5

Solubility (Moles/L)	Solubility (Grams/L)	Temp (°C)	Ref (#)	Evaluation (T P E A A)	Comments
2.943E-02	2.593E+00	1.13	H068	2 0 0 0 1	
1.149E-01	1.012E+01	25	B004	1 0 0 0 2	

332. $C_4H_8O_2$
1,4-Dioxane
1,4-Dioxan
Dioxane
RN: 123-91-1   **MP** (°C):   11.8
MW: 88.11   **BP** (°C):   101

Solubility (Moles/L)	Solubility (Grams/L)	Temp (°C)	Ref (#)	Evaluation (T P E A A)	Comments
>9.08E+00	>8.00E+02	25	B019	1 0 1 2 0	

333. $C_4H_8O_2$
Ethyl Acetate
Athylacetat
Essigsaeureaethyl Ester
RN: 141-78-6   **MP** (°C):   -83
MW: 88.11   **BP** (°C):   77

Solubility (Moles/L)	Solubility (Grams/L)	Temp (°C)	Ref (#)	Evaluation (T P E A A)	Comments
1.097E+00	9.666E+01	0	B108	1 2 0 1 2	
9.941E-01	8.759E+01	0	B108	1 2 0 1 1	
1.069E+00	9.420E+01	0	G062	1 2 2 2 2	
1.032E+00	9.091E+01	0	M088	2 0 0 0 1	
1.144E+00	1.008E+02	0	M111	1 0 1 1 2	
1.156E+00	1.018E+02	0.0	K079	1 0 0 0 2	
9.333E-01	8.223E+01	10	G062	1 2 2 2 2	
1.001E+00	8.817E+01	10	M111	1 0 1 1 2	
9.944E-01	8.762E+01	10.0	K079	1 0 0 0 2	
8.698E-01	7.664E+01	15	M088	2 0 0 0 1	
9.419E-01	8.299E+01	15	M111	1 0 1 1 2	
8.329E-01	7.339E+01	17.0	G101	1 2 1 1 2	
8.718E-01	7.681E+01	20	A016	1 2 1 1 2	
8.795E-01	7.749E+01	20	B108	1 2 0 1 2	
8.212E-01	7.236E+01	20	B108	1 2 0 1 1	
7.346E-01	6.472E+01	20	D052	1 1 0 0 2	
9.556E-01	8.419E+01	20	E002	1 0 0 0 2	
7.310E-01	6.441E+01	20	F001	1 0 1 2 2	
8.932E-01	7.870E+01	20	F300	1 0 0 0 2	
8.920E-01	7.860E+01	20	M111	1 0 1 1 2	
7.300E-01	6.432E+01	20	M171	1 0 0 0 1	
7.732E-01	6.812E+01	20	M348	2 2 1 1 2	
9.200E-01	8.106E+01	20	S006	1 0 0 0 1	
8.778E-01	7.734E+01	20.0	K079	1 0 0 0 2	
8.708E-01	7.672E+01	20.40	A016	1 2 1 1 2	
8.417E-01	7.416E+01	25	A016	1 2 1 1 2	
9.084E-01	8.004E+01	25	A094	1 0 0 0 1	
8.243E-01	7.263E+01	25	A326	1 2 0 1 1	
8.243E-01	7.263E+01	25	A326	1 2 0 1 1	
5.396E-02	4.755E+00	25	B004	1 0 0 0 2	*sic*

9.084E-01	8.004E+01	25	B060	2 0 1 1 1
9.180E-01	8.088E+01	25	B092	2 1 1 1 2
9.080E-01	8.000E+01	25	B304	2 0 2 2 0
7.810E-01	6.881E+01	25	G062	1 2 2 2 2
7.977E-01	7.029E+01	25	L062	2 2 0 1 2
9.847E-01	8.676E+01	25	L319	1 0 2 1 2
8.485E-01	7.476E+01	25	M111	1 0 1 1 2
8.310E-01	7.322E+01	25	P055	1 0 0 0 1
8.222E-01	7.244E+01	25.0	K079	1 0 0 0 2
8.436E-01	7.433E+01	25.10	A016	1 2 1 1 2
7.653E-01	6.743E+01	27.0	G101	1 2 1 1 2
7.603E-01	6.699E+01	27.5	G101	1 2 1 1 2
8.115E-01	7.149E+01	30	A016	1 2 1 1 2
8.124E-01	7.158E+01	30	A016	1 2 1 1 2
7.524E-01	6.629E+01	30	M088	2 0 0 0 1
8.124E-01	7.158E+01	30	M111	1 0 1 1 2
7.524E-01	6.629E+01	30	S357	1 2 1 0 2
7.889E-01	6.951E+01	30.0	K079	1 0 0 0 2
7.800E-01	6.873E+01	34	A016	1 2 1 1 2
7.810E-01	6.881E+01	35	A016	1 2 1 1 2
7.791E-01	6.864E+01	35	M111	1 0 1 1 2
8.170E-01	7.198E+01	37	E028	1 0 1 1 2
7.077E-01	6.235E+01	37	G062	1 2 2 2 2
7.444E-01	6.559E+01	37.0	K079	1 0 0 0 2
7.425E-01	6.542E+01	38	J020	2 1 2 1 1
7.574E-01	6.673E+01	39.90	A016	1 2 1 1 2
7.504E-01	6.612E+01	40	A016	1 2 1 1 2
7.395E-01	6.516E+01	40	B108	1 2 0 1 2
7.524E-01	6.629E+01	40	M111	1 0 1 1 2
6.696E-01	5.900E+01	40	M348	2 2 1 1 2
7.278E-01	6.412E+01	40.0	K079	1 0 0 0 2
6.465E-01	5.696E+01	50	G062	1 2 2 2 2
6.722E-01	5.923E+01	50.0	K079	1 0 0 0 2
5.907E-01	5.204E+01	55	M348	2 2 1 1 2
7.820E-01	6.890E+01	60	B092	2 1 1 1 2
6.790E-01	5.983E+01	70	A326	1 2 0 1 1
5.549E-01	4.889E+01	70	M348	2 2 1 1 2
6.727E-01	5.927E+01	70.4	G101	1 2 1 1 1
1.600E-01	1.410E+01	ns	D348	0 0 2 2 2

334. C_4H_9Br
Isobutyl Bromide
1-Bromo-2-methylpropane
RN: 78-77-3   **MP** (°C):   -119
MW: 137.03   **BP** (°C):   91.5

Solubility (Moles/L)	Solubility (Grams/L)	Temp (°C)	Ref (#)	Evaluation (T P E A A)	Comments
3.700E-03	5.070E-01	18	F001	1 0 1 0 2	
3.722E-03	5.100E-01	18	F300	1 0 0 0 1	

335. C_4H_9Br
n-Butyl Bromide
Bromobutane
RN: 109-65-9   **MP** (°C):   -112
MW: 137.03   **BP** (°C):   101.3

Solubility (Moles/L)	Solubility (Grams/L)	Temp (°C)	Ref (#)	Evaluation (T P E A A)	Comments
4.300E-03	5.892E-01	16	F001	1 0 1 0 2	
4.300E-03	5.892E-01	17	S006	1 0 0 0 1	
<1.46E-03	<2.00E-01	25	B019	1 0 1 2 0	
4.500E-03	6.166E-01	25	K012	1 0 0 0 1	
6.340E-03	8.687E-01	25	M342	1 0 1 1 2	
4.434E-03	6.076E-01	30	G029	1 0 2 2 2	
4.500E-02	6.166E+00	ns	H307	1 0 1 1 2	

336. C_4H_9Cl
sec-Butyl Chloride
2-Chlorobutane
RN: 78-86-4   **MP** (°C):   -140
MW: 92.57   **BP** (°C):   68

Solubility (Moles/L)	Solubility (Grams/L)	Temp (°C)	Ref (#)	Evaluation (T P E A A)	Comments
1.079E-02	9.990E-01	25	N034	1 0 0 0 0	

337. C_4H_9Cl
Isobutyl Chloride
Isobutylchlorid
RN: 513-36-0   **MP** (°C):   -131
MW: 92.57   **BP** (°C):   68

Solubility (Moles/L)	Solubility (Grams/L)	Temp (°C)	Ref (#)	Evaluation (T P E A A)	Comments
1.000E-02	9.257E-01	12.5	F001	1 0 1 2 2	
9.722E-03	9.000E-01	12.50	F300	2 0 0 0 1	

338.   C₄H₉Cl
tert-Butyl Chloride
2-Chloro-2-methylpropane
RN: 507-20-0   **MP** (°C):   -26.5
MW: 92.57   **BP** (°C):   51.0

Solubility (Moles/L)	Solubility (Grams/L)	Temp (°C)	Ref (#)	Evaluation (T P E A A)	Comments
8.180E-02	7.572E+00	0.9	C064	2 2 1 1 2	
6.620E-02	6.128E+00	5.00	C064	2 2 1 1 2	
3.110E-02	2.879E+00	14.90	C064	2 2 1 1 2	

339.   C₄H₉Cl
n-Butyl Chloride
1-Chlorobutane
RN: 109-69-3   **MP** (°C):   -123.0
MW: 92.57   **BP** (°C):   78.5

Solubility (Moles/L)	Solubility (Grams/L)	Temp (°C)	Ref (#)	Evaluation (T P E A A)	Comments
7.200E-03	6.665E-01	12.5	F001	1 0 1 0 2	
7.130E-02	6.600E+00	12.50	F300	1 0 0 0 1	
8.000E-03	7.406E-01	25	K012	1 0 0 0 0	
9.430E-03	8.729E-01	25	M342	1 0 1 1 2	
7.557E-03	6.995E-01	ns	N034	0 0 0 0 1	

340.   C₄H₉I
Iodobutane
n-Butyl Iodide
RN: 542-69-8   **MP** (°C):   -103
MW: 184.02   **BP** (°C):   130.5

Solubility (Moles/L)	Solubility (Grams/L)	Temp (°C)	Ref (#)	Evaluation (T P E A A)	Comments
1.100E-03	2.024E-01	17.5	F001	1 0 1 0 2	
1.100E-03	2.024E-01	17.5	S006	1 0 0 0 1	
1.100E-03	2.024E-01	20	M171	1 0 0 0 1	
1.700E-03	3.128E-01	25	K012	1 0 0 0 1	

341. C$_4$H$_9$NO
Butyramide
n-Butyramide
RN: 541-35-5   **MP** (°C): 116
MW: 87.12   **BP** (°C): 216

Solubility (Moles/L)	Solubility (Grams/L)	Temp (°C)	Ref (#)	Evaluation (T P E A A)	Comments
1.960E+00	1.708E+02	6	H059	1 2 2 0 2	
2.190E+00	1.908E+02	16	H059	1 2 2 0 2	
2.640E+00	2.300E+02	25	H059	1 2 2 0 2	

342. C$_4$H$_9$NO
N,N-Dimethylacetamide
Acetdimethylamide
U-5954
RN: 127-19-5   **MP** (°C): -20
MW: 87.12   **BP** (°C): 163

Solubility (Moles/L)	Solubility (Grams/L)	Temp (°C)	Ref (#)	Evaluation (T P E A A)	Comments
6.071E+00	5.289E+02	4.50	C022	1 2 0 0 2	

343. C$_4$H$_9$NO$_2$
1-Nitrobutane
Butane, 1-Nitro-
RN: 627-05-4   **MP** (°C): -81
MW: 103.12   **BP** (°C): 152.5

Solubility (Moles/L)	Solubility (Grams/L)	Temp (°C)	Ref (#)	Evaluation (T P E A A)	Comments
3.500E-02	3.609E+00	25	K012	1 0 0 0 1	

344. C$_4$H$_9$NO$_2$
Propyl Carbamate
n-Propyl Carbamate
RN: 627-12-3   **MP** (°C): 60
MW: 103.12   **BP** (°C): 196

Solubility (Moles/L)	Solubility (Grams/L)	Temp (°C)	Ref (#)	Evaluation (T P E A A)	Comments
1.940E+00	2.001E+02	37	H006	1 2 2 1 2	

345.   C$_4$H$_9$NO$_2$
DL-α-Aminobutyric Acid
DL-2-Aminobutyric Acid
RN: 2835-81-6   **MP** (°C):   304
MW: 103.12   **BP** (°C):

Solubility (Moles/L)	Solubility (Grams/L)	Temp (°C)	Ref (#)	Evaluation (T P E A A)	Comments
2.121E+00	2.188E+02	20	D041	1 0 0 0 1	
1.615E+00	1.665E+02	25	K031	2 1 2 1 2	

346.   C$_4$H$_9$NO$_2$
α-Aminobutyric Acid
2-Aminobutanoic Acid
α-Amino-n-butyric Acid
Butanoic Acid
RN: 80-60-4   **MP** (°C):   304
MW: 103.12   **BP** (°C):

Solubility (Moles/L)	Solubility (Grams/L)	Temp (°C)	Ref (#)	Evaluation (T P E A A)	Comments
1.845E+00	1.902E+02	25	A048	1 1 1 1 2	form A
1.624E+00	1.674E+02	25	A048	1 1 1 1 2	form B
1.800E+00	1.856E+02	25	C018	1 0 2 2 2	
1.800E+00	1.856E+02	25	E015	1 2 1 1 2	
2.041E+00	2.105E+02	25	M029	2 2 2 2 2	
1.852E+00	1.910E+02	35	A048	1 1 1 1 2	form A
1.771E+00	1.826E+02	35	A048	1 1 1 1 2	form B
1.931E+00	1.991E+02	45	A048	1 1 1 1 2	form A
1.917E+00	1.977E+02	45	A048	1 1 1 1 2	form B

347.   C$_4$H$_9$NO$_2$
α-Aminoisobutyric Acid
α-Amino-isobuttersaeure
α- Aminoisobutyric Acid
2-Methylalanine
RN: 62-57-7   **MP** (°C):
MW: 103.12   **BP** (°C):

Solubility (Moles/L)	Solubility (Grams/L)	Temp (°C)	Ref (#)	Evaluation (T P E A A)	Comments
1.330E+00	1.371E+02	25	C018	1 0 2 2 2	
1.170E+00	1.206E+02	25	D041	1 0 0 0 2	
1.482E+00	1.528E+02	25	M029	2 2 2 2 2	
1.759E+00	1.814E+02	25	M097	2 2 2 2 2	

348. $C_4H_9NO_2$

β-Aminobutyric Acid

β-Amino-n-butyric Acid

RN: 2835-82-7   **MP** (°C):   193

MW: 103.12   **BP** (°C):

Solubility (Moles/L)	Solubility (Grams/L)	Temp (°C)	Ref (#)	Evaluation (T P E A A)	Comments
1.212E+01	1.250E+03	25	M029	2 2 2 2 2	

349. $C_4H_9NO_2$

γ-Aminobutyric Acid

γ-Amino-buttersaeure

γ-Amino-n-butyric Acid

RN: 56-12-2   **MP** (°C):

MW: 103.12   **BP** (°C):

Solubility (Moles/L)	Solubility (Grams/L)	Temp (°C)	Ref (#)	Evaluation (T P E A A)	Comments
1.261E+01	1.300E+03	25	M029	2 2 2 2 2	

350. $C_4H_9NO_3$

DL-Threonine

(±)-Threonine

RN: 80-68-2   **MP** (°C):   244

MW: 119.12   **BP** (°C):

Solubility (Moles/L)	Solubility (Grams/L)	Temp (°C)	Ref (#)	Evaluation (T P E A A)	Comments
1.405E+00	1.674E+02	25	D041	1 0 0 0 2	
2.979E+00	3.548E+02	80	D041	1 0 0 0 1	

351. $C_4H_9NO_3$

DL-allo-Threonine

DL-Allothreonine

RN: 144-98-9   **MP** (°C):

MW: 119.12   **BP** (°C):

Solubility (Moles/L)	Solubility (Grams/L)	Temp (°C)	Ref (#)	Evaluation (T P E A A)	Comments
1.024E+00	1.220E+02	25	D041	1 0 0 0 2	
1.987E+00	2.366E+02	80	D041	1 0 0 0 2	

352.   C₄H₉NO₃
Butyl Nitrate
N-Butyl Nitrate
RN: 928-45-0   **MP** (°C):
MW: 119.12   **BP** (°C):

Solubility (Moles/L)	Solubility (Grams/L)	Temp (°C)	Ref (#)	Evaluation (T P E A A)	Comments
6.500E-03	7.743E-01	25	K012	1 0 0 0 1	

353.   C₄H₉NO₃
L-Threonine
Threonine
RN: 72-19-5   **MP** (°C):   270
MW: 119.12   **BP** (°C):

Solubility (Moles/L)	Solubility (Grams/L)	Temp (°C)	Ref (#)	Evaluation (T P E A A)	Comments
7.606E-01	9.060E+01	20	B032	1 2 2 1 2	
8.139E-01	9.695E+01	25	B032	1 2 2 1 2	
7.346E-01	8.751E+01	25	G315	1 0 2 2 2	
8.202E-01	9.770E+01	25.1	N024	2 0 2 2 2	
8.227E-01	9.800E+01	25.1	N026	2 0 2 2 2	
7.493E-01	8.925E+01	25.1	N027	1 1 2 2 2	
8.168E-01	9.730E+01	27	D036	2 1 2 2 2	
8.695E-01	1.036E+02	29.80	B032	1 2 2 1 2	

354.   C₄H₉N₃O₂
Creatine
Kreatin
RN: 57-00-1   **MP** (°C):   219
MW: 131.14   **BP** (°C):

Solubility (Moles/L)	Solubility (Grams/L)	Temp (°C)	Ref (#)	Evaluation (T P E A A)	Comments
8.222E-02	1.078E+01	10	D041	1 0 0 0 2	
1.016E-01	1.332E+01	18	D041	1 0 0 0 2	
1.014E-01	1.330E+01	18	F300	1 0 0 0 2	

355. $C_4H_9O_5P$

γ-Phosphono-n-butyric Acid
4-Phosphonobutyric Acid
Phosphonic Acid, (3-Carboxypropyl)-
Butyric Acid, 4-Phosphono-
RN: 4378-43-2   **MP** (°C):
MW: 168.09   **BP** (°C):

Solubility (Moles/L)	Solubility (Grams/L)	Temp (°C)	Ref (#)	Evaluation (T P E A A)	Comments
1.739E+00	2.923E+02	0	N028	1 0 0 0 2	
2.068E+00	3.477E+02	20	N028	1 0 0 0 2	

356. C_4H_{10}

Butane
n-Butane
Diethyl
HC 600
Liquefied Petroleum Gas
R 600 (alkane)
RN: 106-97-8   **MP** (°C):   -138
MW: 58.12   **BP** (°C):   -0.5

Solubility (Moles/L)	Solubility (Grams/L)	Temp (°C)	Ref (#)	Evaluation (T P E A A)	Comments
3.138E-03	1.824E-01	3	R063	1 0 2 2 2	
3.210E-03	1.866E-01	4	K031	2 1 2 1 2	
2.622E-03	1.524E-01	6	R063	1 0 2 2 2	
2.314E-03	1.345E-01	9	R063	1 0 2 2 2	
1.886E-03	1.096E-01	14	R063	1 0 2 2 2	
1.461E-03	8.492E-02	19.8	G058	1 0 0 0 2	
1.260E-03	7.324E-02	25	K031	2 1 2 1 2	
1.056E-03	6.140E-02	25	M001	2 1 2 2 2	
1.056E-03	6.140E-02	25	M002	2 1 2 2 2	
1.056E-03	6.140F-02	25	M040	1 0 0 1 2	
2.773E-02	1.612E+00	38	R078	1 0 2 2 1	
6.600E-04	3.836E-02	50	K031	2 1 2 1 2	
1.159E-01	6.735E+00	71	R078	1 0 2 2 1	
4.596E-01	2.671E+01	104	R078	1 0 2 2 1	
1.370E+00	7.965E+01	138	R078	1 0 2 2 2	

357.   C₄H₁₀
Isobutane
1,1-Dimethylethane
2-Methylpropane
Trimethylmethane
Purifrigor Iso 3.5
R 600α
RN: 75-28-5   **MP** (°C):   -159
MW: 58.12   **BP** (°C):

Solubility (Moles/L)	Solubility (Grams/L)	Temp (°C)	Ref (#)	Evaluation (T P E A A)	Comments
~5.68E-03	~3.30E-01	17	F300	1 0 0 0 0	
8.413E-04	4.890E-02	25	M001	2 1 2 2 2	
8.413E-04	4.890E-02	25	M002	2 1 2 2 2	

358.   C₄H₁₀NO₃PS
Acephate
Orthene
Acetylphosphoramidothioic Acid O,S-Dimethyl Ester
RN: 30560-19-1   **MP** (°C):   85.5
MW: 183.17   **BP** (°C):

Solubility (Moles/L)	Solubility (Grams/L)	Temp (°C)	Ref (#)	Evaluation (T P E A A)	Comments
2.151E+00	3.939E+02	rt	M161	0 0 0 0 1	

359.   C₄H₁₀N₂O
N-Nitrosodiethylamine
Diethyl Nitrosamine
RN: 55-18-5   **MP** (°C):
MW: 102.14   **BP** (°C):

Solubility (Moles/L)	Solubility (Grams/L)	Temp (°C)	Ref (#)	Evaluation (T P E A A)	Comments
1.040E+00	1.062E+02	24	D083	2 0 0 0 2	

360.   C₄H₁₀O

$C_4H_{10}O$

n-Butyl Alcohol
Butanol-(1)
n-Butanol
1-Butanol
Butyl Alcohol
n-Butyl Acohol

RN: 71-36-3   **MP** (°C):   -90
MW: 74.12   **BP** (°C):   117

Solubility (Moles/L)	Solubility (Grams/L)	Temp (°C)	Ref (#)	Evaluation (T P E A A)	Comments
1.262E+00	9.355E+01	0	E029	1 2 0 1 2	
1.176E+00	8.717E+01	0	M095	2 2 1 2 2	
1.176E+00	8.717E+01	5	H003	1 2 1 1 2	
1.077E+00	7.987E+01	10	E029	1 2 0 1 2	
1.104E+00	8.181E+01	10	H003	1 2 1 1 2	
6.015E+00	4.459E+02	13.0	J012	1 2 0 1 2	
1.024E+00	7.587E+01	15	H003	1 2 1 1 2	
1.034E+00	7.664E+01	15	M095	2 2 1 2 2	
9.190E-01	6.812E+01	18	F001	1 0 1 0 2	
8.634E-01	6.400E+01	18	F300	1 0 0 0 1	
7.396E-01	5.482E+01	20	A075	1 0 0 0 1	
9.762E-01	7.236E+01	20	D040	2 2 1 1 2	
9.993E-01	7.407E+01	20	D052	1 1 0 0 0	
9.482E-01	7.029E+01	20	E029	1 2 0 1 2	
9.773E-01	7.244E+01	20	H003	1 2 1 1 2	
6.302E-01	4.671E+01	20	L084	1 1 1 1 1	
1.040E+00	7.709E+01	20	M312	1 0 0 0 1	
8.270E-01	6.130E+01	23	D063	1 0 1 2 2	
1.021E+00	7.567E+01	23.5	D063	1 0 0 2 2	
9.983E-01	7.400E+01	25	A049	1 0 1 0 0	
1.125E+00	8.341E+01	25	B019	1 0 1 2 0	
9.645E-01	7.149E+01	25	B060	2 0 1 1 1	
1.000E+00	7.412E+01	25	F044	1 0 0 0 0	EFG
8.708E-01	6.455E+01	25	F325	1 2 0 1 1	
9.200E-01	6.819E+01	25	G075	1 0 1 0 1	
9.237E-01	6.847E+01	25	H003	1 2 1 1 2	
9.307E-01	6.899E+01	25	H028	2 0 2 0 2	
1.070E+00	7.931E+01	25	K012	1 0 0 0 2	
9.700E-01	7.190E+01	25	K025	2 2 1 1 1	
8.867E-01	6.572E+01	25	L322	1 1 2 2 1	
8.904E-01	6.600E+01	25	M136	2 0 0 0 1	
8.904E-01	6.600E+01	25	M139	2 0 0 0 1	
8.826E-01	6.542E+01	25.0	P077	1 1 1 1 1	
8.234E-01	6.103E+01	26	O012	1 2 1 1 2	
8.826E-01	6.542E+01	27	R319	2 2 2 1 1	
5.976E+00	4.429E+02	29.82	J012	1 2 0 1 2	
8.944E-01	6.629E+01	30	D040	2 2 1 1 2	
8.897E-01	6.594E+01	30	E029	1 2 0 1 2	

8.920E-01	6.612E+01	30	F053	1 0 2 0 2
8.920E-01	6.612E+01	30	H003	1 2 1 1 2
8.838E-01	6.551E+01	30.0	H043	2 2 1 1 2
8.625E-01	6.393E+01	35	H003	1 2 1 1 2
9.061E-01	6.716E+01	38	J020	2 0 2 1 1
8.471E-01	6.279E+01	38	M125	1 1 1 1 1
5.933E-01	4.398E+01	40	A075	1 0 0 0 1
8.353E-01	6.191E+01	40	D040	2 2 1 1 2
8.495E-01	6.297E+01	40	E029	1 2 0 1 2
8.353E-01	6.191E+01	40	H003	1 2 1 1 2
8.234E-01	6.103E+01	45	M095	2 2 1 2 2
8.293E-01	6.147E+01	50	E029	1 2 0 1 2
8.186E-01	6.068E+01	50	H003	1 2 1 1 2
7.756E-01	5.749E+01	50	O012	1 2 1 1 2
5.837E+00	4.327E+02	58.50	J012	1 2 0 1 2
5.064E-01	3.754E+01	60	A075	1 0 0 0 1
8.258E-01	6.121E+01	60	E029	1 2 0 1 2
8.258E-01	6.121E+01	60	H003	1 2 1 1 2
5.064E-01	3.754E+01	70	A075	1 0 0 0 1
8.436E-01	6.253E+01	70	E029	1 2 0 1 2
8.850E-01	6.560E+01	70	F001	1 0 1 0 2
8.507E-01	6.306E+01	70	H003	1 2 1 1 2
6.669E-01	4.943E+01	75	L084	1 1 1 1 1
8.590E-01	6.367E+01	75	M095	2 2 1 2 1
8.708E-01	6.455E+01	80	E029	1 2 0 1 2
9.460E-01	7.012E+01	80	F001	1 0 1 0 2
8.696E-01	6.446E+01	80	H003	1 2 1 1 2
9.412E-01	6.977E+01	90	E029	1 2 0 1 2
1.054E+00	7.813E+01	90	F001	1 0 1 0 2
9.762E-01	7.236E+01	90	M095	2 2 1 2 1
1.084E+00	8.038E+01	97.90	H003	1 2 1 1 2
1.101E+00	8.164E+01	98.3	R072	2 2 2 1 2
4.900E+00	3.632E+02	100	E029	1 2 0 1 2
1.228E+00	9.102E+01	100	F001	1 0 1 0 2
1.204E+00	8.925E+01	105	M095	2 2 1 2 1
1.342E+00	9.950E+01	110	E029	1 2 0 1 2
1.473E+00	1.092E+02	110	F001	1 0 1 0 2
1.523E+00	1.129E+02	114.50	H003	1 2 1 1 2
1.600E+00	1.186E+02	116.90	H003	1 2 1 1 2
1.805E+00	1.338E+02	120	E029	1 2 0 1 2
2.223E+00	1.648E+02	123.30	H003	1 2 1 1 2
2.890E+00	2.142E+02	124.80	H003	1 2 1 1 2
2.567E+00	1.903E+02	125	E029	1 2 0 1 2
3.334E+00	2.471E+02	125.10	H003	1 2 1 1 2
3.148E+00	2.334E+02	125.20	H003	1 2 1 1 2
7.920E-01	5.871E+01	ns	D348	0 0 2 2 2
9.744E-01	7.222E+01	ns	L003	0 0 2 1 2
9.033E+00	6.695E+02	ns	M314	2 1 2 1 2

361.   C$_4$H$_{10}$O
sec-Butyl Alcohol
DL-sec-Butyl Alcohol
DL-Butanol-(2)
sec-DL-Butyl Alcohol
RN: 78-92-2   **MP** (°C):   -114
MW: 74.12   **BP** (°C):   99.5

Solubility (Moles/L)	Solubility (Grams/L)	Temp (°C)	Ref (#)	Evaluation (T P E A A)	Comments
2.602E+00	1.929E+02	10.04	M119	2 2 2 2 2	
3.222E+00	2.388E+02	20	A070	1 2 1 0 2	
1.499E+00	1.111E+02	20	D052	1 1 0 0 0	
2.106E+00	1.561E+02	20	E019	1 0 1 1 2	
1.497E+00	1.110E+02	20	F300	1 0 0 0 2	
2.230E+00	1.653E+02	20	M112	2 2 1 1 2	
2.267E+00	1.681E+02	20.04	M119	2 2 2 2 2	
1.348E+00	9.991E+01	25	B019	1 0 1 2 0	
1.057E+00	7.834E+01	25	B060	2 0 1 1 1	
1.699E+00	1.260E+02	25	B165	1 0 1 1 1	
2.048E+00	1.518E+02	27.04	M119	2 2 2 2 2	
2.556E+00	1.894E+02	40	A070	1 2 1 0 2	
1.821E+00	1.349E+02	40	M112	2 0 1 1 2	
1.749E+00	1.297E+02	40.04	M119	2 2 2 2 2	
1.573E+00	1.166E+02	50.04	M119	2 2 2 2 2	
2.167E+00	1.606E+02	60	A070	1 2 1 0 2	
1.657E+00	1.228E+02	60	M112	2 0 1 1 2	
1.531E+00	1.135E+02	60.04	M119	2 2 2 2 2	
1.541E+00	1.143E+02	70.04	M119	2 2 2 2 2	
2.167E+00	1.606E+02	80	A070	1 2 1 0 2	
1.657E+00	1.228E+02	80	M112	2 0 1 1 2	
1.636E+00	1.213E+02	80.04	M119	2 2 2 2 2	
1.760E+00	1.304E+02	85	M112	2 0 1 1 2	
5.107E-02	3.786E+00	87.30	B165	1 0 1 1 2	
1.810E+00	1.342E+02	90.04	M119	2 2 2 2 2	
2.087E+00	1.547E+02	100.04	M119	2 2 2 2 2	
2.602E+00	1.929E+02	110.04	M119	2 2 2 2 2	
1.901E+00	1.409E+02	ns	L003	0 0 2 1 2	

362.   C$_4$H$_{10}$O
Ethyl Ether
Diaethylaether
Diethyl Ether
RN: 60-29-7   **MP** (°C):   -116
MW: 74.12   **BP** (°C):   34.6

Solubility (Moles/L)	Solubility (Grams/L)	Temp (°C)	Ref (#)	Evaluation (T P E A A)	Comments
1.526E+00	1.131E+02	-3.8	H002	2 0 0 1 2	
1.410E+00	1.045E+02	0	H002	1 0 0 1 2	
1.662E+00	1.232E+02	0	K077	1 2 2 2 2	average of 3
1.338E+00	9.920E+01	7.5	K077	1 2 2 2 2	
1.263E+00	9.360E+01	8.5	K077	1 2 2 2 2	
1.118E+00	8.291E+01	10	H002	1 0 0 1 2	
1.115E+00	8.265E+01	10	K002	1 2 1 1 2	
1.105E+00	8.190E+01	12	K077	1 2 2 2 2	
9.796E-01	7.261E+01	15	F055	1 2 2 2 2	
1.133E+00	8.400E+01	15	F300	1 0 0 0 1	
9.893E-01	7.333E+01	15	H002	1 0 0 1 2	
9.843E-01	7.296E+01	15	K002	1 2 1 1 2	
8.430E+00	6.249E+02	15	M069	1 0 0 0 2	
1.137E+00	8.430E+01	15	T033	1 2 1 1 2	
1.029E+00	7.630E+01	16	K077	1 2 2 2 2	
8.837E-01	6.550E+01	19	K077	1 2 2 2 2	average
8.696E-01	6.446E+01	20	F055	1 2 2 2 2	
8.703E-01	6.451E+01	20	H002	1 0 0 1 2	
8.684E-01	6.437E+01	20	K002	1 2 1 1 2	
8.353E-01	6.191E+01	20	M345	2 1 1 1 1	
8.341E-01	6.183E+01	20	N038	1 0 0 1 2	
8.769E-03	6.500E-01	21	H337	1 0 1 0 2	*sic*
1.012E+00	7.502E+01	22	H072	1 0 1 1 2	
9.993E-01	7.407E+01	25	B019	1 0 1 2 0	
7.636E-01	5.660E+01	25	F055	1 2 2 2 2	
8.095E-01	6.000E+01	25	F300	1 0 0 0 1	
7.669E-01	5.684E+01	25	H002	1 0 0 1 2	
7.684E-01	5.696E+01	25	K002	1 2 1 1 2	
8.800E-01	6.523E+01	25	K012	1 0 0 0 1	
6.050E+00	4.484E+02	25	M069	1 0 0 0 2	
8.471E-01	6.279E+01	25	M345	2 1 1 1 1	
8.162E-01	6.050E+01	25	T033	1 2 1 1 2	
1.048E-02	7.770E-01	26	H337	1 0 1 0 2	*sic*
6.839E-01	5.069E+01	30	H002	1 0 0 1 2	
6.839E-01	5.069E+01	30	K002	1 2 1 1 2	
6.799E-01	5.040E+01	30	K077	1 2 2 2 2	
1.073E-02	7.950E-01	32	H337	1 0 1 0 2	*sic*
5.950E-01	4.410E+01	37	E022	1 0 1 1 0	
7.120E-01	5.278E+01	37	E028	1 0 1 1 2	

9.484E-03	7.030E-01	37	H337	1 0 1 0 2	*sic*
6.314E-01	4.680E+01	38	K077	1 2 2 2 2	
9.417E-03	6.980E-01	38.5	H337	1 0 1 0 2	*sic*
9.808E-03	7.270E-01	40	H337	1 0 1 0 2	*sic*
5.545E-01	4.110E+01	49	K077	1 2 2 2 2	
5.491E-01	4.070E+01	51.5	K077	1 2 2 2 2	
4.857E-01	3.600E+01	62.5	K077	1 2 2 2 2	
4.600E-01	3.410E+01	65	K077	1 2 2 2 2	
4.209E-01	3.120E+01	66.5	K077	1 2 2 2 2	
4.020E-01	2.980E+01	71	K077	1 2 2 2 2	
3.912E-01	2.900E+01	72	K077	1 2 2 2 2	
3.643E-01	2.700E+01	82	K077	1 2 2 2 2	
1.770E-01	1.312E+01	ns	D348	0 0 2 2 2	
9.412E-01	6.977E+01	ns	R028	0 0 0 0 1	
8.826E-01	6.542E+01	rt	B066	0 2 0 0 0	

363. $C_4H_{10}O$
Methyl Propyl Ether
1-Methoxypropane
RN: 557-17-5   **MP** (°C):   <25
MW: 74.12   **BP** (°C):   38.8

Solubility (Moles/L)	Solubility (Grams/L)	Temp (°C)	Ref (#)	Evaluation (T P E A A)	Comments
7.154E-01	5.303E+01	0	B002	2 1 1 2 2	
4.939E-01	3.661E+01	10	B002	2 1 1 2 2	
4.436E-01	3.288E+01	15	B002	2 1 1 2 2	
4.183E-01	3.101E+01	20	B002	2 1 1 2 2	
3.993E-01	2.960E+01	25	B002	2 1 1 2 2	

364. $C_4H_{10}O$
Isobutyl Alcohol
2-Methyl-1-propanol
RN: 78-83-1   **MP** (°C):   -108
MW: 74.12   **BP** (°C):   108

Solubility (Moles/L)	Solubility (Grams/L)	Temp (°C)	Ref (#)	Evaluation (T P E A A)	Comments
1.351E+00	1.001E+02	18	F001	1 0 1 2 2	
1.228E+00	9.100E+01	18	F300	1 0 0 0 0	
1.278E+00	9.471E+01	20	M146	1 2 2 2 2	
1.280E+00	9.488E+01	20	M312	1 0 0 0 1	
1.000E+00	7.416E+01	25	A037	2 2 2 2 2	
1.226E+00	9.091E+01	25	D052	1 1 0 0 2	
9.529E-01	7.063E+01	25	F050	1 0 0 0 1	
8.967E-01	6.647E+01	25	F317	2 1 1 1 2	
1.045E+00	7.749E+01	29.84	M114	2 2 1 1 1	
9.529E-01	7.063E+01	39.74	M114	2 2 1 1 1	
8.234E-01	6.103E+01	49.64	M114	2 2 1 1 1	
8.590E-01	6.367E+01	59.54	M114	2 2 1 1 1	

9.295E-01	6.890E+01	79.24	M114	2 2 1 1 1
9.645E-01	7.149E+01	89.14	M114	2 2 1 1 1
5.168E+00	3.831E+02	90.5	J017	1 0 1 2 2
5.033E+00	3.730E+02	91.0	J017	1 0 1 2 2
4.887E+00	3.622E+02	92.0	J017	1 0 1 2 2
4.871E+00	3.610E+02	92.1	J017	1 0 1 2 2
4.615E+00	3.421E+02	93.0	J017	1 0 1 2 2
4.135E+00	3.065E+02	94.3	J017	1 0 1 2 2
3.820E+00	2.832E+02	95.3	J017	1 0 1 2 2
1.215E+00	9.008E+01	99.04	M114	2 2 1 1 1
1.348E+00	9.991E+01	108.94	M114	2 2 1 1 2
1.708E+00	1.266E+02	118.74	M114	2 2 1 1 2
2.009E+00	1.489E+02	123.74	M114	2 2 1 1 2
2.239E+00	1.660E+02	125.64	M114	2 2 1 1 2
2.415E+00	1.790E+02	128.64	M114	2 2 1 1 2
2.637E+00	1.955E+02	130.64	M114	2 2 1 1 2
3.000E+00	2.224E+02	132.64	M114	2 2 1 1 2
3.527E+00	2.614E+02	134.14	M114	2 2 1 1 2
1.179E+00	8.740E+01	ns	L003	0 0 2 1 1

365. $C_4H_{10}O$
Methyl Isopropyl Ether
2-Methoxypropane
RN: 598-53-8   **MP** (°C): <25
MW: 74.12   **BP** (°C): 32

Solubility (Moles/L)	Solubility (Grams/L)	Temp (°C)	Ref (#)	Evaluation (T P E A A)	Comments
1.193E+00	8.842E+01	10	B002	2 1 1 2 2	
1.068E+00	7.919E+01	15	B002	2 1 1 2 2	
9.295E-01	6.890E+01	20	B002	2 1 1 2 2	
8.234E-01	6.103E+01	25	B002	2 1 1 2 2	
8.437E-01	6.254E+01	ns	J300	0 0 0 0 0	

366. $C_4H_{10}O$
tert-Butyl Alcohol
2-Methyl-2-propanol
tert-Butanol
RN: 75-65-0   **MP** (°C): 25.6
MW: 74.12   **BP** (°C): 82.41

Solubility (Moles/L)	Solubility (Grams/L)	Temp (°C)	Ref (#)	Evaluation (T P E A A)	Comments
8.712E-02	6.458E+00	79.40	B165	1 0 1 1 2	

367. $C_4H_{10}O_2S$

Diethyl Sulfone
Diaethylsulfon

RN: 597-35-3   **MP** (°C):   73
MW: 122.19   **BP** (°C):   248

Solubility (Moles/L)	Solubility (Grams/L)	Temp (°C)	Ref (#)	Evaluation (T P E A A)	Comments
1.105E+00	1.350E+02	16	F300	1 0 0 0 2	

368. $C_4H_{10}O_4$

DL-Threitol
DL-1,2,3,4-Butanetetrol

RN: 6968-16-7   **MP** (°C):   90
MW: 122.12   **BP** (°C):

Solubility (Moles/L)	Solubility (Grams/L)	Temp (°C)	Ref (#)	Evaluation (T P E A A)	Comments
7.353E+00	8.980E+02	25	C346	2 0 2 1 2	

369. $C_4H_{10}O_4$

Erythritol
Erythrit

RN: 149-32-6   **MP** (°C):   121.5
MW: 122.12   **BP** (°C):   330

Solubility (Moles/L)	Solubility (Grams/L)	Temp (°C)	Ref (#)	Evaluation (T P E A A)	Comments
3.118E+00	3.808E+02	rt	D021	0 0 1 1 2	
4.995E+00	6.100E+02	rt	F300	0 0 0 0 2	

370. $C_4H_{10}S$

Ethyl Sulfide
1,1'-Thiobisethane
Diethyl Thioether

RN: 352-93-2   **MP** (°C):   -100
MW: 90.19   **BP** (°C):   91

Solubility (Moles/L)	Solubility (Grams/L)	Temp (°C)	Ref (#)	Evaluation (T P E A A)	Comments
3.400E-02	3.066E+00	25	K012	1 0 0 0 1	

371. $C_4H_{11}N$
sec-Butylamine
DL-sec-Butylamine
DL-sec-Butylamin

RN: 13952-84-6   **MP** (°C):
MW: 73.14   **BP** (°C):   63

Solubility (Moles/L)	Solubility (Grams/L)	Temp (°C)	Ref (#)	Evaluation (T P E A A)	Comments
1.531E+00	1.120E+02	20	F300	1 0 0 0 2	

372. $C_4H_{11}N$
n-Butylamine
n-Butylamin
1-Aminobutane

RN: 109-73-9   **MP** (°C):   -50
MW: 73.14   **BP** (°C):   78

Solubility (Moles/L)	Solubility (Grams/L)	Temp (°C)	Ref (#)	Evaluation (T P E A A)	Comments
3.259E-02	2.384E+00	25	B004	1 0 0 0 2	

373. $C_4H_{11}NO_3$
Tromethamine
Tris-(hydroxymethyl)-amino-methan
Tris-(hydroxymethyl)-Aminomethane
2-Amino-2-(hydroxymethyl)-1,3-propanediol
Tris(hydroxymethyl)methylamine

RN: 77-86-1   **MP** (°C):   171.5
MW: 121.14   **BP** (°C):   219.5

Solubility (Moles/L)	Solubility (Grams/L)	Temp (°C)	Ref (#)	Evaluation (T P E A A)	Comments
4.564E+00	5.529E+02	15	E305	1 2 2 2 2	
5.766E+00	6.985E+02	25	E305	1 2 2 2 2	
7.160E+00	8.673E+02	35	E305	1 2 2 2 2	

374. $C_4H_{11}NO_8P_2$
Glyphosine
Polaris
N,N-bis(Phosphonomethyl)glycine

RN: 2439-99-8   **MP** (°C):
MW: 263.08   **BP** (°C):

Solubility (Moles/L)	Solubility (Grams/L)	Temp (°C)	Ref (#)	Evaluation (T P E A A)	Comments
9.427E-01	2.480E+02	20	M161	1 0 0 0 2	

375.   C₄Cl₆

Hexachloro-1,3-butadiene
Hexachlorobutadiene
RN: 87-68-3   **MP** (°C):   -19
MW: 260.76   **BP** (°C):   210

Solubility (Moles/L)	Solubility (Grams/L)	Temp (°C)	Ref (#)	Evaluation (T P E A A)	Comments
9.772E-06	2.548E-03	20	C113	1 0 2 1 2	
1.917E-05	5.000E-03	20	M068	1 0 0 0 0	
~7.67E-06	~2.00E-03	20	M133	1 0 0 0 0	
1.240E-05	3.233E-03	25	B173	2 0 2 2 2	
7.668E-04	2.000E-01	ns	M061	0 0 0 0 1	

376.   C₅H₂Cl₃NO

3,5,6-Trichloro-2-pyridinol
3,5,6-Trichloropyridinol
Hydroxy-3,5,6-trichloropyridine
Pyridinone, 3,5,6-trichloro-
RN: 6515-38-4   **MP** (°C):
MW: 198.44   **BP** (°C):

Solubility (Moles/L)	Solubility (Grams/L)	Temp (°C)	Ref (#)	Evaluation (T P E A A)	Comments
1.109E-03	2.200E-01	26.70	L095	2 2 1 1 2	
1.109E-03	2.200E-01	ns	K138	0 0 0 0 1	

377.   C₅H₂Cl₃NO

2,3,5-Trichloro-4-hydroxypyridine
Daxtrom
RN: 1970-40-7   **MP** (°C):   216
MW: 198.44   **BP** (°C):

Solubility (Moles/L)	Solubility (Grams/L)	Temp (°C)	Ref (#)	Evaluation (T P E A A)	Comments
2.871E-03	5.697E-01	25	M061	1 0 0 0 1	

378.   C₅H₄ClN₅

2-Chloroadenine
1H-Purin-6-amine, 2-Chloro-
6-Amino-2-chloropurine
2-Chloro-6-aminopurine
SQ 22982
RN: 1839-18-5   **MP** (°C):
MW: 169.57   **BP** (°C):

Solubility (Moles/L)	Solubility (Grams/L)	Temp (°C)	Ref (#)	Evaluation (T P E A A)	Comments
4.895E-05	8.300E-03	25	A336	2 2 2 2 2	

379.   C$_5$H$_4$N$_2$O$_4$
α,β-Imidazoledicarboxylic Acid
4,5-Imidazoledicarboxylic Acid
Imidazol-di-carbonsaeure-(4,5)

RN: 570-22-9   **MP** (°C):   288
MW: 156.10   **BP** (°C):

Solubility (Moles/L)	Solubility (Grams/L)	Temp (°C)	Ref (#)	Evaluation (T P E A A)	Comments
3.203E-03	5.000E-01	20	F300	1 0 0 0 1	
8.328E-03	1.300E+00	100	F300	1 0 0 0 1	

380.   C$_5$H$_4$N$_2$O$_4$
Orotic Acid
Vitamin B13
1,2,3,6-Tetrahydro-2,6-dioxo-4-pyrimidinecarboxylic Acid

RN: 65-86-1   **MP** (°C):   345.5
MW: 156.10   **BP** (°C):

Solubility (Moles/L)	Solubility (Grams/L)	Temp (°C)	Ref (#)	Evaluation (T P E A A)	Comments
1.163E-02	1.815E+00	18	B135	1 0 0 0 0	

381.   C$_5$H$_4$N$_2$O$_4$
5-Carboxyuracil
5-Uracilcarboxylic Acid
2,4-Dihydroxypyrimidine-5-carboxylic Acid
Uracil-carbonsaeure-(4)

RN: 23945-44-0   **MP** (°C):   283
MW: 156.10   **BP** (°C):

Solubility (Moles/L)	Solubility (Grams/L)	Temp (°C)	Ref (#)	Evaluation (T P E A A)	Comments
1.153E-02	1.800E+00	20	F300	1 0 0 0 1	
7.000E-03	1.093E+00	20	N019	2 2 1 2 2	

382.   C$_5$H$_4$N$_4$
Purine
7-Imidazo(4,5-d)pyrimidine

RN: 120-73-0   **MP** (°C):   216
MW: 120.11   **BP** (°C):

Solubility (Moles/L)	Solubility (Grams/L)	Temp (°C)	Ref (#)	Evaluation (T P E A A)	Comments
2.775E+00	3.333E+02	20	A018	1 0 1 1 0	

383.   C$_5$H$_4$N$_4$O
8-Hydroxypurine
9H-Purin-8-ol
RN: 51953-05-0   **MP** (°C):
MW: 136.11   **BP** (°C):

Solubility (Moles/L)	Solubility (Grams/L)	Temp (°C)	Ref (#)	Evaluation (T P E A A)	Comments
3.048E-02	4.149E+00	20	A022	1 0 0 0 0	

384.   C$_5$H$_4$N$_4$O
Hypoxanthine
Hypoxanthin
RN: 68-94-0   **MP** (°C):   150dec
MW: 136.11   **BP** (°C):

Solubility (Moles/L)	Solubility (Grams/L)	Temp (°C)	Ref (#)	Evaluation (T P E A A)	Comments
5.139E-03	6.995E-01	19	D041	1 0 0 0 0	
5.143E-03	7.000E-01	23	F300	1 0 0 0 1	
5.290E-03	7.200E-01	25	A337	1 0 2 2 2	
1.042E-01	1.418E+01	100	D004	1 0 0 0 0	
1.080E-01	1.470E+01	100	F300	1 0 0 0 2	
5.359E-03	7.294E-01	c	D004	1 0 0 0 0	

385.   C$_5$H$_4$N$_4$O
Allopurinol
1H-Pyrazolo(3,4-d)pyrimidin-4-ol
Lopurin
RN: 315-30-0   **MP** (°C):   >350
MW: 136.11   **BP** (°C):

Solubility (Moles/L)	Solubility (Grams/L)	Temp (°C)	Ref (#)	Evaluation (T P E A A)	Comments
2.535E-03	3.450E-01	15	C095	1 0 0 1 2	
3.673E-03	5.000E-01	22	B322	1 0 2 2 2	
3.526E-03	4.800E-01	25	B189	1 0 0 0 1	
4.180E-03	5.690E-01	25	C095	1 0 0 1 2	
6.502E-03	8.850E-01	35	C095	1 0 0 1 2	
7.964E-03	1.084E+00	40	C095	1 0 0 1 2	
3.526E-03	4.800E-01	ns	A351	0 0 1 1 1	
5.730E-03	7.800E-01	ns	H067	0 2 0 0 2	

386.   C$_5$H$_4$N$_4$O$_2$
Xanthine
2,6-Dioxopurine
1H-Purine-2,6-dione, 3,7-Dihydro-
RN: 69-89-6   **MP** (°C):   >300
MW: 152.11   **BP** (°C):

Solubility (Moles/L)	Solubility (Grams/L)	Temp (°C)	Ref (#)	Evaluation (T P E A A)	Comments
3.285E-03	4.998E-01	20	D041	1 0 0 0 0	
2.458E-04	3.739E-02	21	L015	1 0 1 1 2	
5.246E-04	7.980E-02	37	L015	1 0 1 1 2	
1.312E-02	1.996E+00	100	D041	1 0 0 0 0	

387.   C$_5$H$_4$N$_4$O$_2$.H$_2$O
Xanthine (Monohydrate)
RN: 69-89-6   **MP** (°C):   >150dec
MW: 170.13   **BP** (°C):

Solubility (Moles/L)	Solubility (Grams/L)	Temp (°C)	Ref (#)	Evaluation (T P E A A)	Comments
4.082E-04	6.944E-02	c	D004	1 0 0 0 0	
3.916E-03	6.662E-01	h	D004	1 0 0 0 0	

388.   C$_5$H$_4$N$_4$O$_3$
Uric Acid
Harnsaeure
RN: 69-93-2   **MP** (°C):
MW: 168.11   **BP** (°C):

Solubility (Moles/L)	Solubility (Grams/L)	Temp (°C)	Ref (#)	Evaluation (T P E A A)	Comments
1.190E-04	2.000E-02	0	M043	1 0 0 0 0	
7.110E-05	1.195E-02	2.6	M315	1 0 1 1 2	
1.029E-04	1.730E-02	5	R042	1 2 2 1 2	
1.050E-04	1.765E-02	9.3	M315	1 0 1 1 2	
2.379E-04	4.000E-02	10	M043	1 0 0 0 0	
1.326E-04	2.230E-02	14	B116	2 0 1 1 2	
1.190E-04	2.000E-02	20	D041	1 0 0 0 0	
3.569E-04	6.000E-02	20	M043	1 0 0 0 0	
6.610E-04	1.111E-01	22	M145	1 0 1 2 2	intrinsic
1.862E-04	3.130E-02	25	R042	1 2 2 1 2	
2.070E-04	3.480E-02	25.0	M315	1 0 1 1 2	
5.354E-04	9.000E-02	30	F300	1 0 0 0 2	
5.353E-04	8.999E-02	30	M043	1 0 0 0 0	
3.660E-04	6.153E-02	37.0	M315	1 0 1 1 2	
7.137E-04	1.200E-01	40	M043	1 0 0 0 1	
3.753E-04	6.310E-02	40	R042	1 2 2 1 2	

Solubility (Moles/L)	Solubility (Grams/L)	Temp (°C)	Ref (#)	Evaluation (T P E A A)	Comments
6.280E-04	1.056E-01	50.0	M315	1 0 1 1 2	
6.960E-04	1.170E-01	54	R042	1 2 2 1 2	
1.368E-03	2.299E-01	60	M043	1 0 0 0 1	
1.457E-03	2.450E-01	70	F300	1 0 0 0 2	
2.319E-03	3.898E-01	80	M043	1 0 0 0 1	
2.974E-04	5.000E-02	100	D041	1 0 0 0 0	
4.961E-03	8.340E-01	100	F300	1 0 0 0 0	
3.686E-03	6.196E-01	100	M043	1 0 0 0 1	

389. $C_5H_4N_4O_3 \cdot 2H_2O$
Uric Acid (Dihydrate)
RN: 69-93-2   **MP** (°C):
MW: 204.14   **BP** (°C):

Solubility (Moles/L)	Solubility (Grams/L)	Temp (°C)	Ref (#)	Evaluation (T P E A A)	Comments
9.620E-05	1.964E-02	2.6	M315	1 0 1 1 2	
1.420E-04	2.899E-02	9.3	M315	1 0 1 1 2	
3.390E-04	6.920E-02	25.0	M315	1 0 1 1 2	
6.560E-04	1.339E-01	37.0	M315	1 0 1 1 2	
1.440E-03	2.940E-01	50.0	M315	1 0 1 1 2	

390. $C_5H_4N_4S$
6-Mercaptopurine
6-Purinethiol
Mercaptopurlne
Purine-6-thiol
Leukeran
RN: 50-44-2   **MP** (°C):
MW: 152.18   **BP** (°C):

Solubility (Moles/L)	Solubility (Grams/L)	Temp (°C)	Ref (#)	Evaluation (T P E A A)	Comments
3.000E-04	4.565E-02	4.62	A034	1 1 2 2 0	EFG
8.148E-04	1.240E-01	25	N063	1 1 1 1 2	
4.500E-02	6.848E+00	29.87	A034	1 1 2 2 1	EFG
1.703E-03	2.591E-01	37	H046	1 1 1 1 2	
2.658E-03	4.045E-01	ns	N050	0 1 1 0 0	

391.   C$_5$H$_4$O$_2$
Furfural
2-Furaldehyde
Furfurol
RN: 98-01-1   **MP** (°C):   -36
MW: 96.09   **BP** (°C):   162

Solubility (Moles/L)	Solubility (Grams/L)	Temp (°C)	Ref (#)	Evaluation (T P E A A)	Comments
7.620E-01	7.322E+01	10	M099	1 2 0 1 1	
7.816E-01	7.510E+01	16	M099	1 2 0 1 2	
7.869E-01	7.561E+01	17	M099	1 2 0 1 2	
7.976E-01	7.664E+01	20	D052	1 1 0 0 0	
7.972E-01	7.660E+01	20	F300	1 0 0 0 2	
7.976E-01	7.664E+01	20	M099	1 2 0 1 1	
7.620E-01	7.322E+01	25	C056	1 2 1 1 1	
8.197E-01	7.877E+01	25	C329	1 2 1 1 1	average
7.709E-01	7.407E+01	25	H338	2 2 1 2 2	
7.976E-01	7.664E+01	25	H340	1 0 2 2 1	
7.441E-01	7.149E+01	25	L062	2 2 1 2 1	
7.709E-01	7.407E+01	25	L320	2 2 1 2 1	
8.242E-01	7.919E+01	25	M099	1 2 0 1 1	
8.347E-01	8.021E+01	27	M099	1 2 0 1 2	
8.347E-01	8.021E+01	27.20	M099	1 2 0 1 2	
8.312E-01	7.987E+01	27.50	M099	1 2 0 1 2	
8.418E-01	8.088E+01	30	M099	1 2 0 1 1	
8.488E-01	8.156E+01	35	H338	2 2 1 2 2	
8.506E-01	8.173E+01	35	L320	2 2 1 2 1	
9.029E-01	8.676E+01	38	G050	1 0 2 1 1	
8.619E-01	8.282E+01	39.50	E037	1 2 2 2 2	
9.029E-01	8.676E+01	40	M099	1 2 0 1 1	
9.289E-01	8.925E+01	44	M099	1 2 0 1 2	
9.804E-01	9.420E+01	50	M099	1 2 0 1 2	
1.023E+00	9.829E+01	52	G050	1 0 2 1 2	
9.306E-01	8.942E+01	53.10	E037	1 2 2 2 2	
4.982E+00	4.787E+02	53.30	E037	1 2 2 2 2	
1.090E+00	1.047E+02	60	M099	1 2 0 1 2	
1.107E+00	1.063E+02	61	M099	1 2 0 1 2	
1.156E+00	1.111E+02	66	G050	1 0 2 1 2	
1.156E+00	1.111E+02	66	M099	1 2 0 1 2	
1.214E+00	1.166E+02	70	M099	1 2 0 1 2	
4.895E+00	4.703E+02	73.60	E037	1 2 2 2 2	
1.318E+00	1.266E+02	79	G050	1 0 2 1 2	
1.342E+00	1.289E+02	80	M099	1 2 0 1 2	
1.361E+00	1.307E+02	85.80	E037	1 2 2 2 2	
1.482E+00	1.424E+02	90	M099	1 2 0 1 2	

1.512E+00	1.453E+02	92	M099	1 2 0 1 2
1.684E+00	1.618E+02	93	G050	1 0 2 1 2
4.721E+00	4.536E+02	95.90	E037	1 2 2 2 2
1.617E+00	1.554E+02	97.90	M099	1 2 0 1 2

392.   C$_5$H$_4$O$_2$S
3-Thenoic Acid
Thiophen-carbonsaeure-(3)
RN: 88-13-1   **MP** (°C):   137
MW: 128.15   **BP** (°C):

Solubility (Moles/L)	Solubility (Grams/L)	Temp (°C)	Ref (#)	Evaluation (T P E A A)	Comments
3.355E-02	4.300E+00	25	F300	1 0 0 0 1	

393.   C$_5$H$_4$O$_3$
Isopyromucic Acid
Isobrenzschleimsaeure
RN: 496-64-0   **MP** (°C):
MW: 112.09   **BP** (°C):

Solubility (Moles/L)	Solubility (Grams/L)	Temp (°C)	Ref (#)	Evaluation (T P E A A)	Comments
3.845E-01	4.310E+01	0	F300	1 0 0 0 2	

394.   C$_5$H$_4$O$_3$
2-Furoic Acid
Furan-carbon-saeure-(2)
RN: 88-14-2   **MP** (°C):   129.5
MW: 112.09   **BP** (°C):   231

Solubility (Moles/L)	Solubility (Grams/L)	Temp (°C)	Ref (#)	Evaluation (T P E A A)	Comments
2.227E-01	2.496E+01	5.99	A341	2 0 2 2 2	
2.243E-01	2.514E+01	6.99	A341	2 0 2 2 2	
2.332E-01	2.614E+01	10.49	A341	2 0 2 2 2	
2.498E-01	2.799E+01	10.99	A341	2 0 2 2 2	
2.543E-01	2.851E+01	11.99	A341	2 0 2 2 2	
3.310E-01	3.710E+01	15	F300	1 0 0 0 2	
2.606E-01	2.921E+01	15.99	A341	2 0 2 2 2	
3.385E-01	3.794E+01	20.99	A341	2 0 2 2 2	
4.216E-01	4.725E+01	24.99	A341	2 0 2 2 2	
4.665E-01	5.229E+01	27.99	A341	2 0 2 2 2	
5.182E-01	5.808E+01	28.99	A341	2 0 2 2 2	
6.448E-01	7.227E+01	33.99	A341	2 0 2 2 2	
6.677E-01	7.484E+01	35.99	A341	2 0 2 2 2	
7.816E-01	8.761E+01	37.99	A341	2 0 2 2 2	
1.120E+00	1.256E+02	41.99	A341	2 0 2 2 2	
1.229E+00	1.378E+02	43.99	A341	2 0 2 2 2	
1.444E+00	1.618E+02	46.64	A341	2 0 2 2 2	

2.159E+00	2.420E+02	49.99	A341	2 0 2 2 2
2.610E+00	2.926E+02	51.99	A341	2 0 2 2 2
2.768E+00	3.103E+02	53.99	A341	2 0 2 2 2
2.815E+00	3.155E+02	54.49	A341	2 0 2 2 2
3.221E+00	3.610E+02	54.99	A341	2 0 2 2 2
3.964E+00	4.443E+02	57.49	A341	2 0 2 2 2
4.219E+00	4.729E+02	60.04	A341	2 0 2 2 2
4.224E+00	4.735E+02	61.39	A341	2 0 2 2 2
4.940E+00	5.537E+02	62.99	A341	2 0 2 2 2
5.529E+00	6.197E+02	67.99	A341	2 0 2 2 2
1.838E+00	2.060E+02	100	F300	1 0 0 0 2

395. C_5H_5NO
2-Hydroxypyridine
2-Pyridinol

RN: 72762-00-6   **MP** (°C):   106
MW: 95.10   **BP** (°C):   280.5

Solubility (Moles/L)	Solubility (Grams/L)	Temp (°C)	Ref (#)	Evaluation (T P E A A)	Comments
5.258E+00	5.000E+02	20	B050	1 0 0 0 0	

396. C_5H_5NO
4-Hydroxypyridine
4-Pyridinol

RN: 626-64-2   **MP** (°C):   148
MW: 95.10   **BP** (°C):   232.5

Solubility (Moles/L)	Solubility (Grams/L)	Temp (°C)	Ref (#)	Evaluation (T P E A A)	Comments
5.258E+00	5.000E+02	20	B050	1 0 0 0 0	

397. C_5H_5NO
3-Hydroxypyridine
3-Pyridinol

RN: 109-00-2   **MP** (°C):   127.5
MW: 95.10   **BP** (°C):   152

Solubility (Moles/L)	Solubility (Grams/L)	Temp (°C)	Ref (#)	Evaluation (T P E A A)	Comments
3.392E-01	3.226E+01	20	B050	1 0 0 0 0	

398.   C₅H₅NO₂

398. $C_5H_5NO_2$

2,4-Dihydroxypyridine
3-Deazauracil
2,4-Pyridinediol

RN: 626-03-9   **MP** (°C):   278
MW: 111.10   **BP** (°C):

Solubility (Moles/L)	Solubility (Grams/L)	Temp (°C)	Ref (#)	Evaluation (T P E A A)	Comments
5.591E-02	6.211E+00	20	B050	1 0 0 0 0	

399. $C_5H_5N_3O$

Pyrazinamide
Pyrazine-2-carboxamide
Prazina

RN: 98-96-4   **MP** (°C):   190
MW: 123.12   **BP** (°C):

Solubility (Moles/L)	Solubility (Grams/L)	Temp (°C)	Ref (#)	Evaluation (T P E A A)	Comments
1.413E-01	1.740E+01	25	N041	2 0 1 1 0	EFG

400. $C_5H_5N_5$

Adenine
Adenin

RN: 73-24-5   **MP** (°C):   363
MW: 135.13   **BP** (°C):

Solubility (Moles/L)	Solubility (Grams/L)	Temp (°C)	Ref (#)	Evaluation (T P E A A)	Comments
4.719E-03	6.377E-01	17.5	S306	1 0 1 2 2	
6.328E-03	8.551E-01	18.8	S306	1 0 1 2 2	
6.494E-03	8.776E-01	19.2	S306	1 0 1 2 2	
7.382E-03	9.975E-01	19.7	S306	1 0 1 2 2	
7.000E-03	9.459E-01	20	C017	2 0 0 1 0	EFG
6.907E-03	9.333E-01	20.08	D307	2 1 2 2 2	
7.680E-03	1.038E+00	22.36	D307	2 1 2 2 2	
6.586E-03	8.900E-01	25	A337	1 0 2 2 2	
6.654E-03	8.992E-01	25	D041	1 0 0 0 0	
7.040E-03	9.513E-01	25	H061	1 2 2 0 2	
7.600E-03	1.027E+00	25	L080	2 1 2 1 2	
8.000E-03	1.081E+00	25	R039	2 2 2 2 1	
8.610E-03	1.163E+00	25.01	D307	2 1 2 2 2	
8.690E-03	1.174E+00	25.03	D307	2 1 2 2 2	
8.250E-03	1.115E+00	25.5	T008	1 1 2 2 2	
7.936E-03	1.072E+00	26.6	S306	1 0 1 2 2	
9.740E-03	1.316E+00	27.47	D307	2 1 2 2 2	
1.087E-02	1.469E+00	29.97	D307	2 1 2 2 2	
9.377E-03	1.267E+00	31.1	S306	1 0 1 2 2	
1.540E-02	2.081E+00	37	L042	2 0 2 2 2	pH 6.47

1.390E-02	1.878E+00	38	T008	1 1 2 2 2
1.514E-02	2.045E+00	44.0	S306	1 0 1 2 2
1.707E-02	2.307E+00	45.1	S306	1 0 1 2 2
1.862E-02	2.516E+00	45.5	S306	1 0 1 2 2
1.805E-01	2.439E+01	100	D041	1 0 0 0 0
6.808E-03	9.200E-01	c	D004	1 0 0 0 0
1.805E-01	2.439E+01	h	D004	1 0 0 0 0

401. $C_5H_5N_5O$
Guanine
2-Aminohypoxanthine
2-Amino-6-hydroxypurine
RN: 73-40-5   **MP** (°C):   >300
MW: 151.13   **BP** (°C):

Solubility (Moles/L)	Solubility (Grams/L)	Temp (°C)	Ref (#)	Evaluation (T P E A A)	Comments
1.920E-05	2.902E-03	15.02	D307	2 1 2 2 2	
6.000E-05	9.068E-03	20	C017	2 0 0 1 1	EFG
2.740E-05	4.141E-03	20.05	D307	2 1 2 2 2	
3.290E-05	4.972E-03	22.50	D307	2 1 2 2 2	
3.870E-05	5.849E-03	25.02	D307	2 1 2 2 2	
4.520E-05	6.831E-03	27.54	D307	2 1 2 2 2	
5.350E-05	8.085E-03	30.01	D307	2 1 2 2 2	
7.230E-05	1.093E-02	35.05	D307	2 1 2 2 2	
2.647E-04	4.000E-02	40	D041	1 0 0 0 0	
9.880E-05	1.493E-02	40.22	D307	2 1 2 2 2	

402. $C_5H_5N_5O$
Isoguanine
2-Hydroxy-6-aminopurine
RN: 3373-53-3   **MP** (°C):
MW: 151.13   **BP** (°C):

Solubility (Moles/L)	Solubility (Grams/L)	Temp (°C)	Ref (#)	Evaluation (T P E A A)	Comments
3.970E-04	6.000E-02	25	D041	1 0 0 0 0	
1.654E-03	2.499E-01	100	D041	1 0 0 0 1	

403. C_5H_5N_5O_2

$C_5H_5N_5O_2$

2,8-Dioxyadenine
2,8-Dihydroxyadenine
RN: 30377-37-8 **MP** (°C):
MW: 167.13 **BP** (°C):

Solubility (Moles/L)	Solubility (Grams/L)	Temp (°C)	Ref (#)	Evaluation (T P E A A)	Comments
1.316E-05	2.200E-03	25	B049	1 0 1 1 1	
8.556E-06	1.430E-03	37	P068	1 0 1 1 2	

404. C_5H_6Cl_2N_2

$C_5H_6Cl_2N_2$

3-Methyluracil
2,4(1H,3H)-Pyrimidinedione, 3-Methyl-
Uracil, 3-Methyl-
RN: 608-34-4 **MP** (°C):
MW: 165.02 **BP** (°C):

Solubility (Moles/L)	Solubility (Grams/L)	Temp (°C)	Ref (#)	Evaluation (T P E A A)	Comments
1.212E+00	2.000E+02	ns	B177	0 0 0 0 2	

405. C_5H_6Cl_2N_2O_2

$C_5H_6Cl_2N_2O_2$

Dantoin
1,3-Dichloro-5,5-Dimethyl-2,4-Imidazolidinedione
1,3-Dichloro-5,5-dimethylhydantoin
RN: 118-52-5 **MP** (°C): 132
MW: 197.02 **BP** (°C):

Solubility (Moles/L)	Solubility (Grams/L)	Temp (°C)	Ref (#)	Evaluation (T P E A A)	Comments
2.537E-03	4.998E-01	20	B080	1 0 1 1 0	
6.590E-03	1.298E+00	40	B080	1 0 1 1 1	

406. C_5H_6N_2OS

$C_5H_6N_2OS$

5-Methyl-2-thiouracil
4(1H)-Pyrimidinone, 2,3-Dihydro-5-methyl-2-thioxo-
2-Thiothymine
RN: 636-26-0 **MP** (°C): 284
MW: 142.18 **BP** (°C):

Solubility (Moles/L)	Solubility (Grams/L)	Temp (°C)	Ref (#)	Evaluation (T P E A A)	Comments
3.580E-03	5.090E-01	25	G016	1 2 1 2 2	intrinsic

407.   C₅H₆N₂OS

Methylthiouracil
6-Methyl-2-thiouracil
RN: 56-04-2   **MP** (°C):   330
MW: 142.18   **BP** (°C):

Solubility (Moles/L)	Solubility (Grams/L)	Temp (°C)	Ref (#)	Evaluation (T P E A A)	Comments
3.750E-03	5.332E-01	25	G016	1 2 1 2 2	intrinsic
7.026E-03	9.990E-01	c	I310	0 0 0 0 0	

408.   C₅H₆N₂O₂

1-Methyluracil
2,4(1H,3H)-Pyrimidinedione, 1-Methyl-
N1-Methyluracil
RN: 615-77-0   **MP** (°C):   179
MW: 126.12   **BP** (°C):

Solubility (Moles/L)	Solubility (Grams/L)	Temp (°C)	Ref (#)	Evaluation (T P E A A)	Comments
1.586E-01	2.000E+01	ns	B177	0 0 0 0 1	

409.   C₅H₆N₂O₂

Thymine
2,4-Dihydroxy-5-methylpyrimidine
5-Methyluracil
RN: 65-71-4   **MP** (°C):   316
MW: 126.12   **BP** (°C):

Solubility (Moles/L)	Solubility (Grams/L)	Temp (°C)	Ref (#)	Evaluation (T P E A A)	Comments
2.200E-02	2.775E+00	20	C017	2 0 0 1 1	EFG
2.379E-02	3.000E+00	23	F300	1 0 0 0 0	
3.552E-02	4.480E+00	25	D041	1 0 0 0 1	
2.780E-02	3.506E+00	25	H061	1 2 2 0 2	
3.030E-02	3.821E+00	25	L080	2 1 2 1 2	
2.860E-02	3.607E+00	25	R039	2 2 2 2 2	
2.740E-02	3.456E+00	25.5	T008	1 1 2 2 2	
3.500E-02	4.414E+00	30	L080	2 1 2 1 2	

410.   C$_5$H$_6$N$_2$O$_4$
5-Carboxymethylhydantoin
Hydantoin of Aspartic Acid
RN: 5427-26-9   **MP** (°C):   216
MW: 158.11   **BP** (°C):

Solubility (Moles/L)	Solubility (Grams/L)	Temp (°C)	Ref (#)	Evaluation (T P E A A)	Comments
7.050E-02	1.115E+01	ns	M025	0 2 0 1 2	

411.   C$_5$H$_6$O$_2$
α-Angelica Lactone
α-Angelica-lacton
RN: 591-12-8   **MP** (°C):   18
MW: 98.10   **BP** (°C):   56

Solubility (Moles/L)	Solubility (Grams/L)	Temp (°C)	Ref (#)	Evaluation (T P E A A)	Comments
4.689E-01	4.600E+01	15	F300	1 0 0 0 1	

412.   C$_5$H$_6$O$_4$
Citraconic Acid
Citraconsaeure
RN: 498-23-7   **MP** (°C):
MW: 130.10   **BP** (°C):

Solubility (Moles/L)	Solubility (Grams/L)	Temp (°C)	Ref (#)	Evaluation (T P E A A)	Comments
6.018E+00	7.830E+02	25	F300	1 0 0 0 2	

413.   C$_5$H$_6$O$_4$
Itaconic Acid
Itaconsaeure
RN: 97-65-4   **MP** (°C):   163
MW: 130.10   **BP** (°C):

Solubility (Moles/L)	Solubility (Grams/L)	Temp (°C)	Ref (#)	Evaluation (T P E A A)	Comments
4.281E-01	5.570E+01	10	F300	1 0 0 0 2	
5.891E-01	7.664E+01	20	D041	1 0 0 0 1	
5.903E-01	7.680E+01	20	F300	1 0 0 0 2	

414.   C$_5$H$_6$O$_4$
Mesaconic Acid
Mesaconsaeure
RN: 498-24-8   **MP** (°C):   204.5
MW: 130.10   **BP** (°C):

Solubility (Moles/L)	Solubility (Grams/L)	Temp (°C)	Ref (#)	Evaluation (T P E A A)	Comments
2.022E-01	2.630E+01	18	F300	1 0 0 0 2	
4.241E+00	5.518E+02	100	F300	1 0 0 0 2	

415.   C$_5$H$_7$NO$_2$
Ethyl Cyanoacetate
Cyanessigsaeure-aethyl Ester
RN: 105-56-6   **MP** (°C):
MW: 113.12   **BP** (°C):

Solubility (Moles/L)	Solubility (Grams/L)	Temp (°C)	Ref (#)	Evaluation (T P E A A)	Comments
1.768E-01	2.000E+01	25	F300	1 0 0 0 0	
7.072E-01	8.000E+01	80	F300	1 0 0 0 0	

416.   C$_5$H$_7$N$_2$O$_2$
6-Methyluracil
4-Methyl-uracil
RN: 626-48-2   **MP** (°C):   318dec
MW: 127.12   **BP** (°C):

Solubility (Moles/L)	Solubility (Grams/L)	Temp (°C)	Ref (#)	Evaluation (T P E A A)	Comments
5.506E-02	7.000E+00	22	F300	1 0 0 0 0	

417.   C$_5$H$_7$N$_3$O
5-Methylcytosine
Mec
RN: 554-01-8   **MP** (°C):   270
MW: 125.13   **BP** (°C):

Solubility (Moles/L)	Solubility (Grams/L)	Temp (°C)	Ref (#)	Evaluation (T P E A A)	Comments
3.441E-01	4.306E+01	25	D041	1 0 0 0 1	

418.   C₅H₇N₃O₂

Dimetridazole

1,2-Dimethyl-5-nitroimidazole

RN: 551-92-8   **MP** (°C):   137-139

MW: 141.13   **BP** (°C):

Solubility (Moles/L)	Solubility (Grams/L)	Temp (°C)	Ref (#)	Evaluation (T P E A A)	Comments
6.866E-02	9.690E+00	20	D344	1 1 2 2 2	
6.866E-02	9.690E+00	20	D344	1 1 2 2 2	
6.738E-02	9.509E+00	20	D344	1 1 2 2 2	
6.870E-02	9.696E+00	20	D344	1 1 2 2 2	

419.   C₅H₈

Cyclopentene

RN: 142-29-0   **MP** (°C):   -135

MW: 68.12   **BP** (°C):   44

Solubility (Moles/L)	Solubility (Grams/L)	Temp (°C)	Ref (#)	Evaluation (T P E A A)	Comments
2.411E-02	1.642E+00	24.8	L007	2 1 1 2 2	
7.854E-03	5.350E-01	25	M001	2 1 2 2 2	
2.411E-02	1.642E+00	25.1	L007	2 2 1 1 2	
2.562E-02	1.745E+00	34.8	L007	2 1 1 2 2	

420.   C₅H₈

Isoprene

2-Methyl-1,3-butadiene

RN: 78-79-5   **MP** (°C):   -120

MW: 68.12   **BP** (°C):   34.07

Solubility (Moles/L)	Solubility (Grams/L)	Temp (°C)	Ref (#)	Evaluation (T P E A A)	Comments
9.425E-03	6.420E-01	25	M001	2 1 2 2 2	

421.   C₅H₈

1-Pentyne

Pent-1-yne

RN: 627-19-0   **MP** (°C):   -106

MW: 68.12   **BP** (°C):   40

Solubility (Moles/L)	Solubility (Grams/L)	Temp (°C)	Ref (#)	Evaluation (T P E A A)	Comments
2.305E-02	1.570E+00	25	M001	2 1 2 2 2	
1.154E-02	7.861E-01	25	M342	1 0 1 1 2	

422.   C$_5$H$_8$
1,4-Pentadiene
Penta-1,4-diene
RN: 591-93-5   **MP** (°C):   -148
MW: 68.12   **BP** (°C):   26

Solubility (Moles/L)	Solubility (Grams/L)	Temp (°C)	Ref (#)	Evaluation (T P E A A)	Comments
8.191E-03	5.580E-01	25	M001	2 1 2 2 2	

423.   C$_5$H$_8$BrNO$_4$
5-Bromo-2-methyl-5-nitro-1,3-dioxane
Dioxane, 5-Bromo-2-methyl-5-nitro-
Nibroxane
RN: 53983-00-9   **MP** (°C):   72
MW: 226.03   **BP** (°C):

Solubility (Moles/L)	Solubility (Grams/L)	Temp (°C)	Ref (#)	Evaluation (T P E A A)	Comments
2.695E-02	6.093E+00	25	L013	1 0 2 1 2	

424.   C$_5$H$_8$N$_2$O$_2$
5,5'-Dimethylhydantoin
5,5-Dimethylhydantoin
5,5-Dimethyl-2,4-imidazolidinedione
5,5-Dimethylimidazolidine-2,4-dione
RN: 77-71-4   **MP** (°C):   177
MW: 128.13   **BP** (°C):

Solubility (Moles/L)	Solubility (Grams/L)	Temp (°C)	Ref (#)	Evaluation (T P E A A)	Comments
1.018E+00	1.304E+02	37	F183	1 0 1 1 1	intrinsic

425.   C$_5$H$_8$N$_2$O$_2$
5-Ethylhydantoin
Hydantoin of α-Aminobutyric Acid
RN: 15414-82-1   **MP** (°C):   119
MW: 128.13   **BP** (°C):

Solubility (Moles/L)	Solubility (Grams/L)	Temp (°C)	Ref (#)	Evaluation (T P E A A)	Comments
8.630E-01	1.106E+02	ns	M025	0 2 0 1 2	

426. $C_5H_8N_4O_3S_2$

Methazolamide
Acetamide, N-[5-(Aminosulfonyl)-3-methyl-1,3,4-thiadiazol-2(3H)-ylidene]-
N-(4-Methyl-2-sulfamoyl-D2-1,3,4-thiadiazolin-5-ylidene)acetamide
Neptazaneat
Metazolamide
Methenamide

RN: 554-57-4   **MP** (°C):   213
MW: 236.27   **BP** (°C):

Solubility (Moles/L)	Solubility (Grams/L)	Temp (°C)	Ref (#)	Evaluation (T P E A A)	Comments
2.000E-03	4.725E-01	15	K024	1 2 1 1 1	
1.481E-02	3.500E+00	ns	M032	0 0 0 0 2	

427. $C_5H_8N_4O_{12}$

Pentaerythritol Tetranitrate
Nitropentaerythritol
1,3-Propanediol, 2,2-bis[(nitrooxy)methyl]-, Dinitrate (Ester)

RN: 78-11-5   **MP** (°C):   140
MW: 316.14   **BP** (°C):

Solubility (Moles/L)	Solubility (Grams/L)	Temp (°C)	Ref (#)	Evaluation (T P E A A)	Comments
6.326E-06	2.000E-03	ns	M013	0 2 0 1 1	

428. C_5H_8O

Cyprethylene Ether

RN: **MP** (°C):
MW: 84.12   **BP** (°C):

Solubility (Moles/L)	Solubility (Grams/L)	Temp (°C)	Ref (#)	Evaluation (T P E A A)	Comments
9.435E-02	7.937E+00	27	K058	1 0 1 1 0	

429. C_5H_8O

α-Methylcrotonaldehyde
α-Methyl-crotonaldehyd

RN: 623-36-9   **MP** (°C):
MW: 84.12   **BP** (°C):

Solubility (Moles/L)	Solubility (Grams/L)	Temp (°C)	Ref (#)	Evaluation (T P E A A)	Comments
2.378E-01	2.000E+01	20	F300	1 0 0 0 1	

430.   C₅H₈O₂
Acetylacetone
2,4-Pentanedione
Acetylaceton
RN: 123-54-6   **MP** (°C):   -23
MW: 100.12   **BP** (°C):   140.5

Solubility (Moles/L)	Solubility (Grams/L)	Temp (°C)	Ref (#)	Evaluation (T P E A A)	Comments
1.678E+00	1.680E+02	19.0	N051	1 2 1 1 2	
1.703E+00	1.705E+02	19.5	N051	1 2 1 1 2	
1.089E+00	1.090E+02	20	F300	1 0 0 0 2	
1.706E+00	1.708E+02	25	B019	1 0 1 2 0	

431.   C₅H₈O₂
Methyl Methacrylate
Methacrylic Acid Methyl Ester
Methyl 2-Methyl-2-Propenoate
RN: 80-62-6   **MP** (°C):   -48
MW: 100.12   **BP** (°C):   100

Solubility (Moles/L)	Solubility (Grams/L)	Temp (°C)	Ref (#)	Evaluation (T P E A A)	Comments
1.563E-01	1.565E+01	20	L096	1 2 0 2 2	

432.   C₅H₈O₂
Ethyl Acrylate
Ethyl Propenoate
2-Propenoic Acid Ethyl Ester
RN: 140-88-5   **MP** (°C):   -71
MW: 100.12   **BP** (°C):   99.4

Solubility (Moles/L)	Solubility (Grams/L)	Temp (°C)	Ref (#)	Evaluation (T P E A A)	Comments
1.785E-01	1.787E+01	30	L096	1 2 0 2 2	

433. $C_5H_8O_3$
Levulinic Acid
Laevulinsaeure
4-Oxopentanoic Acid
3-Acetyl Propionic Acid
RN: 123-76-2   **MP** (°C):   37.2
MW: 116.12   **BP** (°C):   245

Solubility (Moles/L)	Solubility (Grams/L)	Temp (°C)	Ref (#)	Evaluation (T P E A A)	Comments
4.632E+00	5.378E+02	6.99	A340	2 0 2 2 2	
4.990E+00	5.795E+02	9.99	A340	2 0 2 2 2	
5.530E+00	6.422E+02	14.49	A340	2 0 2 2 2	
6.087E+00	7.068E+02	20.79	A340	2 0 2 2 2	
6.400E+00	7.431E+02	24.99	A340	2 0 2 2 2	
6.631E+00	7.700E+02	30.09	A340	2 0 2 2 2	

434. $C_5H_8O_3$
Dimethylpyruvic Acid
DL-Methyl-bernsteinsaeure
α-Ketoisovaleric Acid
RN: 759-05-7   **MP** (°C):
MW: 116.12   **BP** (°C):

Solubility (Moles/L)	Solubility (Grams/L)	Temp (°C)	Ref (#)	Evaluation (T P E A A)	Comments
3.450E+00	4.006E+02	20	F300	1 0 0 0 2	

435. $C_5H_8O_4$
Methylsuccinic Acid
Acide Methylsuccinique
1,2-Propanedicarboxylic Acid
RN: 498-21-5   **MP** (°C):   117.5
MW: 132.12   **BP** (°C):

Solubility (Moles/L)	Solubility (Grams/L)	Temp (°C)	Ref (#)	Evaluation (T P E A A)	Comments
5.041E+00	6.660E+02	15	M051	1 0 0 0 2	

436.   C₅H₈O₄

Ethylmalonic Acid
1,1-Propanedicarboxylic Acid
Aethylmalonsaeure
Mono-Ethyl Malonate
Malonic Acid Monoethyl Ester
Malonsaeure-monoaethyl Ester
RN: 601-75-2   **MP** (°C):   114
MW: 132.12   **BP** (°C):   160

Solubility (Moles/L)	Solubility (Grams/L)	Temp (°C)	Ref (#)	Evaluation (T P E A A)	Comments
2.619E+00	3.460E+02	0	F300	1 0 0 0 2	
3.996E+00	5.280E+02	0	M051	1 0 0 0 2	
4.814E+00	6.360E+02	15	M051	1 0 0 0 2	
5.389E+00	7.120E+02	25	M051	1 0 0 0 2	
3.626E+00	4.790E+02	50	F300	1 0 0 0 2	
6.873E+00	9.080E+02	50	M051	1 0 0 0 2	

437.   C₅H₈O₄

Glutaric Acid
Glutarsaeure
1,3-Propanedicarboxylic Acid
RN: 110-94-1   **MP** (°C):   96.5
MW: 132.12   **BP** (°C):

Solubility (Moles/L)	Solubility (Grams/L)	Temp (°C)	Ref (#)	Evaluation (T P E A A)	Comments
2.272E+00	3.002E+02	0	F300	1 0 0 0 2	
3.247E+00	4.290E+02	0	L041	1 0 0 1 2	
2.410E+00	3.183E+02	3.40	A031	1 2 2 2 2	
2.650E+00	3.501E+02	5.99	A341	2 0 2 2 2	
2.764E+00	3.651E+02	7.99	A341	2 0 2 2 2	
3.127E+00	4.131E+02	10.40	A031	1 2 2 2 2	
2.909E+00	3.843E+02	10.99	A341	2 0 2 2 2	
3.213E+00	4.245E+02	12.99	A341	2 0 2 2 2	
3.433E+00	4.536E+02	14	A031	1 2 2 2 0	
4.443E+00	5.870E+02	15	L041	1 0 0 1 2	
4.443E+00	5.870E+02	15	M051	1 0 0 0 2	
3.521E+00	4.652E+02	15.99	A341	2 0 2 2 2	
3.674E+00	4.854E+02	17.99	A341	2 0 2 2 2	
3.861E+00	5.100E+02	18	A031	1 2 2 2 2	
3.816E+00	5.041E+02	19.99	A341	2 0 2 2 2	
2.954E+00	3.902E+02	20	D041	1 0 0 0 1	
4.837E+00	6.390E+02	20	L041	1 0 0 1 2	
2.952E+00	3.900E+02	20	M171	1 0 0 0 2	
1.340E+00	1.770E+02	20	S006	1 0 0 0 2	
4.278E+00	5.652E+02	23.90	A031	1 2 2 2 2	

4.088E+00	5.401E+02	24.99	A341	2 0 2 2 2
4.653E+00	6.148E+02	28.30	A031	1 2 2 2 2
4.394E+00	5.805E+02	28.99	A341	2 0 2 2 2
4.503E+00	5.949E+02	30.99	A341	2 0 2 2 2
4.642E+00	6.133E+02	33.99	A341	2 0 2 2 2
6.033E+00	7.970E+02	35	L041	1 0 0 1 2
4.796E+00	6.336E+02	36.99	A341	2 0 2 2 2
4.894E+00	6.466E+02	38.99	A341	2 0 2 2 2
5.096E+00	6.732E+02	42.99	A341	2 0 2 2 2
5.131E+00	6.779E+02	43.99	A341	2 0 2 2 2
5.143E+00	6.795E+02	44.99	A341	2 0 2 2 2
5.246E+00	6.930E+02	46.99	A341	2 0 2 2 2
5.341E+00	7.057E+02	49.99	A341	2 0 2 2 2
7.244E+00	9.570E+02	50	L041	1 0 0 1 2
5.470E+00	7.227E+02	54.49	A341	2 0 2 2 2
5.640E+00	7.451E+02	55.99	A341	2 0 2 2 2
5.713E+00	7.548E+02	58.99	A341	2 0 2 2 2
5.729E+00	7.569E+02	61.09	A341	2 0 2 2 2
5.890E+00	7.782E+02	62.99	A341	2 0 2 2 2
4.032E+00	5.327E+02	65	F300	1 0 0 0 2
8.462E+00	1.118E+03	65	L041	1 0 0 1 2
6.038E+00	7.977E+02	68.99	A341	2 0 2 2 2

438. $C_5H_8O_4$
Dimethylmalonic Acid
Dimethyl-malonsaeure
Dimethyl-propanedioic Acid
RN: 595-46-0   **MP** (°C): 192
MW: 132.12   **BP** (°C):

Solubility (Moles/L)	Solubility (Grams/L)	Temp (°C)	Ref (#)	Evaluation (T P E A A)	Comments
6.812E-01	9.000E+01	13	F300	1 0 0 0 0	
1.968E+00	2.600E+02	100	F300	1 0 0 0 1	

439. $C_5H_9BrO_2$
α-Ethyl-β-bromo-propionic Ureide
RN: **MP** (°C):
MW: 181.04   **BP** (°C):

Solubility (Moles/L)	Solubility (Grams/L)	Temp (°C)	Ref (#)	Evaluation (T P E A A)	Comments
2.130E-01	3.855E+01	ns	F056	0 2 2 2 1	

440.   C$_5$H$_9$BrO$_2$
α-Bromo-methyl-ethyl-acetate
Ethyl DL-α-Bromopropionate
Propanoic Acid, 2-Bromo-, Ethyl Ester
Ethyl DL-2-Bromopropionate
RN: 535-11-5   **MP** (°C):
MW: 181.04   **BP** (°C):

Solubility (Moles/L)	Solubility (Grams/L)	Temp (°C)	Ref (#)	Evaluation (T P E A A)	Comments
2.780E-01	5.033E+01	ns	F057	0 2 2 2 1	

441.   C$_5$H$_9$NO$_2$
DL-Proline
Pyrrolidine-2-carboxylic Acid
RN: 609-36-9   **MP** (°C):   208
MW: 115.13   **BP** (°C):

Solubility (Moles/L)	Solubility (Grams/L)	Temp (°C)	Ref (#)	Evaluation (T P E A A)	Comments
1.217E+01	1.401E+03	20	J303	2 0 2 2 2	
1.146E+01	1.319E+03	25	J303	2 0 2 2 2	
1.425E+01	1.641E+03	40	J303	2 0 2 2 2	
1.708E+01	1.967E+03	50	J303	2 0 2 2 2	
2.082E+01	2.397E+03	60	J303	2 0 2 2 2	

442.   C$_5$H$_9$NO$_2$
L-Proline
2-Pyrrolidinecarboxylic Acid
RN: 147-85-3   **MP** (°C):
MW: 115.13   **BP** (°C):

Solubility (Moles/L)	Solubility (Grams/L)	Temp (°C)	Ref (#)	Evaluation (T P E A A)	Comments
5.374E+00	6.188E+02	25	D041	1 0 0 0 2	
6.653E+00	7.660E+02	27	D036	2 1 2 2 2	
6.123E+00	7.050E+02	65	D041	1 0 0 0 2	
6.691E+00	7.704E+02	99.99	P349	1 0 0 2 2	

443.   C$_5$H$_9$NO$_3$
Formyl-α-aminobutyric Acid
Butanoic Acid, 2-(Formylamino)-
RN: 106873-99-8   **MP** (°C):
MW: 131.13   **BP** (°C):

Solubility (Moles/L)	Solubility (Grams/L)	Temp (°C)	Ref (#)	Evaluation (T P E A A)	Comments
2.560E-01	3.357E+01	25	M024	1 2 0 1 2	
2.560E-01	3.357E+01	ns	M025	0 2 0 1 2	

444.   C$_5$H$_9$NO$_3$
L-Hydroxyproline
trans-4-Hydroxy-L-Proline
L-4-Hydroxyproline
(4S)-4-Hydroxy-L-proline
RN: 51-35-4   **MP** (°C):
MW: 131.13   **BP** (°C):

Solubility (Moles/L)	Solubility (Grams/L)	Temp (°C)	Ref (#)	Evaluation (T P E A A)	Comments
3.158E+00	4.141E+02	99.99	P349	1 0 0 2 2	

445.   C$_5$H$_9$NO$_4$
L-Glutamic Acid
L-2-Aminoglutaric Acid
L(+)-Glutaminsaeure
Glutamic Acid
L(+) Glutaminic Acid
RN: 56-86-0   **MP** (°C):   250
MW: 147.13   **BP** (°C):

Solubility (Moles/L)	Solubility (Grams/L)	Temp (°C)	Ref (#)	Evaluation (T P E A A)	Comments
4.866E-02	7.160E+00	20	B032	1 2 2 1 2	
4.486E-02	6.600E+00	21	F302	1 0 0 0 1	
5.825E-02	8.570E+00	25	B032	1 2 2 1 2	
5.822E-02	8.566E+00	25	D041	1 0 0 0 2	
5.845E-02	8.600E+00	25	F300	1 0 0 0 1	
7.262E-02	1.068E+01	25	G315	1 0 2 2 2	
5.614E-02	8.260E+00	27	D036	2 1 2 2 2	
6.980E-02	1.027E+01	29.80	B032	1 2 2 1 2	
1.454E-01	2.140E+01	50	F300	1 0 0 0 2	
3.562E-01	5.240E+01	75	D041	1 0 0 0 2	
3.561E-01	5.240E+01	75	F300	1 0 0 0 2	
8.346E-01	1.228E+02	100	F300	1 0 0 0 2	
4.078E-02	6.000E+00	ns	D072	0 0 0 0 0	

446.   C₅H₉NO₄

DL-Glutamic Acid
DL-2-Aminoglutaric Acid
RN: 617-65-2   **MP** (°C):   194
MW: 147.13   **BP** (°C):

Solubility (Moles/L)	Solubility (Grams/L)	Temp (°C)	Ref (#)	Evaluation (T P E A A)	Comments
5.601E-02	8.241E+00	0	D018	2 2 2 1 2	
1.750E-01	2.575E+01	25	D018	2 2 2 1 2	
1.368E-01	2.013E+01	25	D041	1 0 0 0 2	
5.131E-01	7.549E+01	50	D018	2 2 2 1 2	
7.206E-01	1.060E+02	75	D041	1 0 0 0 2	

447.   C₅H₉NO₄

D-Glutamic Acid
D-2-Aminoglutaric Acid
RN: 6893-26-1   **MP** (°C):   201
MW: 147.13   **BP** (°C):

Solubility (Moles/L)	Solubility (Grams/L)	Temp (°C)	Ref (#)	Evaluation (T P E A A)	Comments
2.337E-02	3.439E+00	0	D018	2 2 2 1 2	
2.303E-02	3.388E+00	0	M043	1 0 0 0 1	
3.381E-02	4.975E+00	10	M043	1 0 0 0 1	
1.004E-01	1.478E+01	20	D041	1 0 0 0 1	
4.859E-02	7.149E+00	20	M043	1 0 0 0 1	
4.472E-02	6.580E+00	21	P045	1 0 2 1 2	
5.981E-02	8.800E+00	25	D018	2 2 2 1 2	
6.729E-02	9.901E+00	30	M043	1 0 0 0 1	
1.004E-01	1.478E+01	40	M043	1 0 0 0 1	
1.481E-01	2.179E+01	50	D018	2 2 2 1 2	
2.107E-01	3.101E+01	60	M043	1 0 0 0 1	
4.148E-01	6.103E+01	80	M043	1 0 0 0 1	
8.347E-01	1.228E+02	100	M043	1 0 0 0 2	
5.850E-02	8.607E+00	ns	M025	0 2 0 1 2	

448.   C$_5$H$_{10}$
Cyclopentane
Pentamethylene
Exxsol Cyclopentane S
Zeonsolv HP
RN: 287-92-3   **MP** (°C):   -94.4
MW: 70.14   **BP** (°C):   49.3

Solubility (Moles/L)	Solubility (Grams/L)	Temp (°C)	Ref (#)	Evaluation (T P E A A)	Comments
4.826E-03	3.385E-01	4.8	L007	2 2 1 2 2	
4.826E-03	3.385E-01	5.1	L007	2 1 1 1 2	
4.870E-03	3.416E-01	14.8	L007	2 2 1 2 2	
4.870E-03	3.416E-01	15.2	L007	2 1 1 1 2	
4.873E-03	3.418E-01	24.8	L007	2 2 1 2 2	
2.338E-03	1.640E-01	25	G313	2 1 1 2 2	
2.281E-03	1.600E-01	25	K119	1 0 0 0 2	
2.224E-03	1.560E-01	25	M001	2 1 2 2 2	
2.224E-03	1.560E-01	25	M002	2 1 2 2 2	
2.281E-03	1.600E-01	25.0	P051	2 1 1 2 2	
2.281E-03	1.600E-01	25.00	P007	2 1 2 2 2	
4.873E-03	3.418E-01	25.1	L007	2 1 1 1 2	
5.252E-03	3.684E-01	34.8	L007	2 2 1 2 2	
5.252E-03	3.684E-01	35.2	L007	2 1 1 1 2	
2.324E-03	1.630E-01	40.1	P051	2 1 1 2 2	
2.324E-03	1.630E-01	40.10	P007	2 1 2 2 2	
4.867E-03	3.414E-01	44.8	L007	2 2 1 2 2	
2.566E-03	1.800E-01	55.7	P051	2 1 1 2 2	
2.566E-03	1.800E-01	55.70	P007	2 1 2 2 2	
4.220E-03	2.960E-01	99.1	P051	2 1 1 2 2	
4.220E-03	2.960E-01	99.10	P007	2 1 2 2 2	
5.304E-03	3.720E-01	118.0	P051	2 1 1 2 2	
5.304E-03	3.720E-01	118.00	P007	2 1 2 2 2	
8.712E-03	6.110E-01	137.3	P051	2 1 1 2 2	
8.712E-03	6.110E-01	137.30	P007	2 1 2 2 2	
1.129E-02	7.920E-01	153.1	P051	2 1 1 2 2	
1.129E-02	7.920E-01	153.10	P007	2 1 2 2 2	
2.224E-03	1.560E-01	ns	H123	0 0 0 0 2	

449.   C$_5$H$_{10}$
2-Pentene
1-Methyl-2-ethylethylene
sym-Methylethylethylene
β-Amylene
β-n-Amylene
3-Pentene
RN: 109-68-2   **MP** (°C):   -136
MW: 70.14   **BP** (°C):   36

Solubility (Moles/L)	Solubility (Grams/L)	Temp (°C)	Ref (#)	Evaluation (T P E A A)	Comments
2.894E-03	2.030E-01	25	M001	2 1 2 2 2	

450.   C$_5$H$_{10}$
1-Pentene
Propylethylene
α-n-Amylene
1-Methyl-3-butene
RN: 109-67-1   **MP** (°C):   -165
MW: 70.14   **BP** (°C):   30.1

Solubility (Moles/L)	Solubility (Grams/L)	Temp (°C)	Ref (#)	Evaluation (T P E A A)	Comments
2.609E-03	1.830E-01	23	C332	2 0 2 2 1	
2.110E-03	1.480E-01	25	M001	2 1 2 2 2	

451.   C$_5$H$_{10}$
3-Methyl-1-butene
2-Methyl-3-butene
3,3-Dimethylpropene
Isopropylethylene
RN: 563-45-1   **MP** (°C):   -168
MW: 70.14   **BP** (°C):   20

Solubility (Moles/L)	Solubility (Grams/L)	Temp (°C)	Ref (#)	Evaluation (T P E A A)	Comments
1.854E-03	1.300E-01	25	M001	2 1 2 2 2	

452. $C_5H_{10}Cl_3O_3P$
Diethyl Trichloromethyl Phosphonate
Phosphonic Acid, (Trichloromethyl)-, Diethyl Ester
Ro 3-0658
RN: 866-23-9   **MP** (°C):
MW: 255.47   **BP** (°C):

Solubility (Moles/L)	Solubility (Grams/L)	Temp (°C)	Ref (#)	Evaluation (T P E A A)	Comments
1.761E-02	4.500E+00	25	B070	1 2 0 1 1	

453. $C_5H_{10}N_2O$
N-Nitrosopiperidine
Pyridine, Hexahydro-N-nitroso
NPIP
RN: 100-75-4   **MP** (°C):   <25
MW: 114.15   **BP** (°C):

Solubility (Moles/L)	Solubility (Grams/L)	Temp (°C)	Ref (#)	Evaluation (T P E A A)	Comments
6.700E-01	7.648E+01	24	D083	2 0 0 0 1	

454. $C_5H_{10}N_2O_2S$
Methomyl
Nudrin
Lannate
RN: 16752-77-5   **MP** (°C):   78.5
MW: 162.21   **BP** (°C):

Solubility (Moles/L)	Solubility (Grams/L)	Temp (°C)	Ref (#)	Evaluation (T P E A A)	Comments
3.576E-01	5.800E+01	25	M161	1 0 0 0 1	

455. $C_5H_{10}N_2O_3$
Glycolylglycineamide
RN: **MP** (°C):
MW: 146.15   **BP** (°C):

Solubility (Moles/L)	Solubility (Grams/L)	Temp (°C)	Ref (#)	Evaluation (T P E A A)	Comments
5.820E+00	8.506E+02	25	M008	1 0 0 0 2	

456.   C$_5$H$_{10}$N$_2$O$_3$
D-Glutamine
D-2-Aminoglutaramic Acid
RN: 5959-95-5   **MP** (°C):
MW: 146.15   **BP** (°C):

Solubility (Moles/L)	Solubility (Grams/L)	Temp (°C)	Ref (#)	Evaluation (T P E A A)	Comments
2.910E-01	4.253E+01	ns	M025	0 2 0 1 2	

457.   C$_5$H$_{10}$N$_2$O$_3$
L-Glutamine
L(+)-Glutamin
L(+)-Glutamine
Glutamine
RN: 56-85-9   **MP** (°C):   185
MW: 146.15   **BP** (°C):

Solubility (Moles/L)	Solubility (Grams/L)	Temp (°C)	Ref (#)	Evaluation (T P E A A)	Comments
1.184E-01	1.730E+01	0	F300	1 0 0 0 2	
2.378E-01	3.475E+01	18	D041	1 0 0 0 1	
2.444E-01	3.572E+01	20	B032	1 2 2 1 2	
2.829E-01	4.135E+01	25	B032	1 2 2 1 2	
2.789E-01	4.077E+01	25	D041	1 0 0 0 2	
2.701E-01	3.948E+01	25	G315	1 0 2 2 2	
5.891E-02	8.610E+00	25	J303	2 0 2 2 2	
2.997E-01	4.380E+01	25.1	N024	2 0 2 2 2	
2.840E-01	4.150E+01	25.1	N025	2 0 2 2 2	
2.840E-01	4.150E+01	25.1	N026	2 0 2 2 2	
2.821E-01	4.123E+01	25.1	N027	1 1 2 2 2	
2.737E-01	4.000E+01	27	D036	2 1 2 2 2	
3.285E-01	4.801E+01	29.80	B032	1 2 2 1 2	
3.154E-01	4.610E+01	30	F300	1 0 0 0 2	
1.002E-01	1.464E+01	40	J303	2 0 2 2 2	
2.135E-01	3.120E+01	60	J303	2 0 2 2 2	

458. $C_5H_{10}N_2S_2$
Dazomet
3,5-Dimethyl-1,2,3,5-tetrahydro-1,3,5-thiadiazinethione-2
Thiazone
Thiazon
RN: 533-74-4   **MP** (°C):   106.5
MW: 162.28   **BP** (°C):

Solubility (Moles/L)	Solubility (Grams/L)	Temp (°C)	Ref (#)	Evaluation (T P E A A)	Comments
7.386E-03	1.199E+00	25	M061	1 0 0 0 1	
1.169E-02	1.896E+00	30	B185	1 0 0 0 1	
7.395E-03	1.200E+00	30	M161	1 0 0 0 1	

459. $C_5H_{10}N_6O_2$
Dinitrosopentamethylenetetramine
3,7-Dinitroso-1,3,5,7-tetraazabicyclo[3.3.1]nonane
RN: 101-25-7   **MP** (°C):   207
MW: 186.17   **BP** (°C):

Solubility (Moles/L)	Solubility (Grams/L)	Temp (°C)	Ref (#)	Evaluation (T P E A A)	Comments
5.318E-02	9.901E+00	ns	I313	0 0 0 0 0	

460. $C_5H_{10}O$
Tetrahydropyran
Pentamethylene Oxide
RN: 142-68-7   **MP** (°C):   -49.2
MW: 86.13   **BP** (°C):   88

Solubility (Moles/L)	Solubility (Grams/L)	Temp (°C)	Ref (#)	Evaluation (T P E A A)	Comments
1.372E+00	1.182E+02	0	B001	2 0 1 0 0	
1.122E+00	9.666E+01	10	B001	2 0 1 0 0	
1.021E+00	8.792E+01	15	B001	2 0 1 0 0	
9.351E-01	8.054E+01	20	B001	2 0 1 0 0	
8.620E-01	7.425E+01	25	B001	2 0 1 0 0	

461. $C_5H_{10}O$
Diethyl Ketone
3-Pentanone
RN: 96-22-0   **MP** (°C):   -42
MW: 86.13   **BP** (°C):   101.5

Solubility (Moles/L)	Solubility (Grams/L)	Temp (°C)	Ref (#)	Evaluation (T P E A A)	Comments
7.810E-01	6.727E+01	10	G032	1 2 1 1 2	
4.786E-01	4.123E+01	20	D052	1 1 0 0 1	
5.613E-01	4.834E+01	20	G030	1 2 0 0 2	
6.052E-01	5.213E+01	25	B019	1 0 1 2 0	

3.818E-01	3.288E+01	25	B060	2 0 1 1 1
5.328E-01	4.589E+01	25	G030	1 2 0 0 2
5.900E-01	5.082E+01	25	K012	1 0 0 0 1
4.999E-01	4.306E+01	30	G030	1 2 0 0 1
5.760E-01	4.961E+01	30	G032	1 2 1 1 2
4.560E-01	3.928E+01	50	G032	1 2 1 1 2

462.   C$_5$H$_{10}$O
1-Penten-3-ol
Penten-1-ol-3
RN: 616-25-1   **MP** (°C):
MW: 86.13   **BP** (°C):

Solubility (Moles/L)	Solubility (Grams/L)	Temp (°C)	Ref (#)	Evaluation (T P E A A)	Comments
9.312E-01	8.021E+01	20	G031	1 0 0 0 2	
8.798E-01	7.579E+01	25	G031	1 0 0 0 2	
8.340E-01	7.184E+01	30	G031	1 0 0 0 2	

463.   C$_5$H$_{10}$O
4-Penten-1-ol
Penten-4-ol-1
RN: 821-09-0   **MP** (°C):
MW: 86.13   **BP** (°C):   135.5

Solubility (Moles/L)	Solubility (Grams/L)	Temp (°C)	Ref (#)	Evaluation (T P E A A)	Comments
6.458E-01	5.562E+01	20	G031	1 0 0 0 2	
6.261E-01	5.393E+01	25	G031	1 0 0 0 2	
6.115E-01	5.267E+01	30	G031	1 0 0 0 2	

464.   C$_5$H$_{10}$O
3-Penten-2-ol
Penten-3-ol-2
RN: 1569-50-2   **MP** (°C):
MW: 86.13   **BP** (°C):   120

Solubility (Moles/L)	Solubility (Grams/L)	Temp (°C)	Ref (#)	Evaluation (T P E A A)	Comments
1.003E+00	8.642E+01	20	G031	1 0 0 0 2	
9.508E-01	8.189E+01	25	G031	1 0 0 0 2	
9.075E-01	7.817E+01	30	G031	1 0 0 0 2	

465.   C₅H₁₀O

3-Methyl-2-butanone
3-Methylbutanone-2
RN: 563-80-4   **MP** (°C):   -92
MW: 86.13   **BP** (°C):   94.5

Solubility (Moles/L)	Solubility (Grams/L)	Temp (°C)	Ref (#)	Evaluation (T P E A A)	Comments
8.130E-01	7.003E+01	10	G032	1 2 1 1 2	
7.116E-01	6.130E+01	20	G030	1 2 0 0 2	
6.654E-01	5.732E+01	25	G030	1 2 0 0 2	
6.240E-01	5.375E+01	30	G030	1 2 0 0 2	
6.080E-01	5.237E+01	30	G032	1 2 1 1 2	
5.940E-01	5.116E+01	50	G032	1 2 1 1 2	

466.   C₅H₁₀O

2-Methyl Tetrahydrofuran
2-Methyl Oxolane
β-Methyl Tetramethylene Oxide
RN: 96-47-9   **MP** (°C):   -136
MW: 86.13   **BP** (°C):   83

Solubility (Moles/L)	Solubility (Grams/L)	Temp (°C)	Ref (#)	Evaluation (T P E A A)	Comments
1.174E+00	1.011E+02	10	B001	2 0 1 0 0	

467.   C₅H₁₀O

1-Methyl Tetrahydrofuran
Methyl Oxolane
α-Methyl Tetramethylene Oxide
RN: 45376-90-7   **MP** (°C):
MW: 86.13   **BP** (°C):   80

Solubility (Moles/L)	Solubility (Grams/L)	Temp (°C)	Ref (#)	Evaluation (T P E A A)	Comments
2.101E+00	1.810E+02	0	B001	2 0 1 0 0	
1.788E+00	1.540E+02	10	B001	2 0 1 0 0	
1.646E+00	1.418E+02	15	B001	2 0 1 0 0	
1.519E+00	1.308E+02	20	B001	2 0 1 0 0	
1.414E+00	1.218E+02	25	B001	2 0 1 0 0	

468.   C$_5$H$_{10}$O
Cypreth Ether
Cyclopropane, Ethoxy-
Ethoxycyclopropane
Ethyl Cyclopropyl Ether
RN: 5614-38-0   **MP** (°C):
MW: 86.13   **BP** (°C):

Solubility (Moles/L)	Solubility (Grams/L)	Temp (°C)	Ref (#)	Evaluation (T P E A A)	Comments
3.162E-01	2.724E+01	25	K061	1 0 1 1 1	
2.500E-01	2.153E+01	25	K061	1 0 1 1 1	

469.   C$_5$H$_{10}$O
Valeraldehyde
n-Valeraldehyde
Valeral
n-Pentanal
RN: 110-62-3   **MP** (°C):   -92
MW: 86.13   **BP** (°C):   103

Solubility (Moles/L)	Solubility (Grams/L)	Temp (°C)	Ref (#)	Evaluation (T P E A A)	Comments
1.358E-01	1.170E+01	25	A049	1 0 0 0 2	
2.100E-01	1.809E+01	25	K012	1 0 0 0 1	

470.   C$_5$H$_{10}$O
Methy Propyl Ketone
Methyl Propyl Ketone
2-Pentanone
Pentan-2-one
RN: 107-87-9   **MP** (°C):   -78
MW: 86.13   **BP** (°C):   100.5

Solubility (Moles/L)	Solubility (Grams/L)	Temp (°C)	Ref (#)	Evaluation (T P E A A)	Comments
8.870E-01	7.640E+01	10	G032	1 2 1 1 2	
6.520E-01	5.616E+01	20	G030	1 2 0 0 2	
5.000E-01	4.307E+01	20	M312	1 0 0 0 1	
6.799E-01	5.857E+01	25	A356	2 1 2 2 1	
4.786E-01	4.123E+01	25	B060	2 0 1 1 1	
7.775E-01	6.697E+01	25	C333	2 2 2 2 2	
7.000E-01	6.029E+01	25	F044	1 0 0 0 1	
6.063E-01	5.222E+01	25	G030	1 2 0 0 2	
6.572E-01	5.660E+01	25	P055	1 0 0 0 2	
5.718E-01	4.925E+01	30	G030	1 2 0 0 2	
6.300E-01	5.426E+01	30	G032	1 2 1 1 2	
5.806E-01	5.001E+01	35	A356	2 1 2 2 1	

6.799E-01	5.857E+01	35	C333	2 2 2 2 2
5.302E-01	4.567E+01	45	A356	2 1 2 2 1
6.799E-01	5.857E+01	45	C333	2 2 2 2 2
5.150E-01	4.436E+01	50	G032	1 2 1 1 2
5.302E-01	4.567E+01	55	A356	2 1 2 2 1
5.302E-01	4.567E+01	55	A356	2 1 2 2 1
5.806E-01	5.001E+01	55	C333	2 2 2 2 2

471. $C_5H_{10}OS_2$
Butylxanthogenic Acid
RN: **MP** (°C):
MW: 150.26   **BP** (°C):

Solubility (Moles/L)	Solubility (Grams/L)	Temp (°C)	Ref (#)	Evaluation (T P E A A)	Comments
8.000E-04	1.202E-01	25	K012	1 0 0 0 0	

472. $C_5H_{10}O_2$
Butyl Formate
Formic Acid Butyl Ester
RN: 592-84-7   **MP** (°C):
MW: 102.13   **BP** (°C):   106.5

Solubility (Moles/L)	Solubility (Grams/L)	Temp (°C)	Ref (#)	Evaluation (T P E A A)	Comments
9.800E-02	1.001E+01	22	S006	1 0 0 0 1	
6.400E-02	6.537E+00	25	K012	1 0 0 0 1	
7.400E-02	7.558E+00	27	B052	1 0 1 1 2	
7.500E-02	7.660E+00	30.5	N014	1 1 1 0 2	
8.100E-02	8.273E+00	40.0	N014	1 1 1 0 2	

473. $C_5H_{10}O_2$
Ethyl Propionate
Propanoic Acid Ethyl Ester
RN: 105-37-3   **MP** (°C):   -73
MW: 102.13   **BP** (°C):   99

Solubility (Moles/L)	Solubility (Grams/L)	Temp (°C)	Ref (#)	Evaluation (T P E A A)	Comments
1.844E-01	1.884E+01	20	D052	1 1 0 0 2	
2.200E-01	2.247E+01	20	S006	1 0 0 0 1	
2.154E-01	2.200E+01	25	F300	1 0 0 0 1	
1.700E-01	1.736E+01	25	K012	1 0 0 0 1	
2.108E-01	2.153E+01	30	R318	1 1 0 1 1	

474. $C_5H_{10}O_2$
Isopropyl Acetate
Essigsaeureisopropyl Ester
Iso-propylacetat

RN: 108-21-4   **MP** (°C):   -73
MW: 102.13   **BP** (°C):   89

Solubility (Moles/L)	Solubility (Grams/L)	Temp (°C)	Ref (#)	Evaluation (T P E A A)	Comments
2.556E-01	2.610E+01	20	D052	1 1 0 0 2	average of 2
3.030E-01	3.095E+01	20	F001	1 0 1 2 2	
2.937E-01	3.000E+01	20	F300	1 0 0 0 2	
2.108E-01	2.153E+01	24.6	H121	2 0 0 0 1	
2.759E-01	2.818E+01	25	B060	2 0 1 1 1	
1.930E-01	1.971E+01	37	E028	1 0 1 1 2	

475. $C_5H_{10}O_2$
Isovaleric Acid
Isovaleriansaeure

RN: 503-74-2   **MP** (°C):   -29.3
MW: 102.13   **BP** (°C):   176.5

Solubility (Moles/L)	Solubility (Grams/L)	Temp (°C)	Ref (#)	Evaluation (T P E A A)	Comments
3.946E-01	4.031E+01	20	D041	1 0 0 0 1	
3.985E-01	4.070E+01	20	F300	1 0 0 0 2	

476. $C_5H_{10}O_2$
Methyl Butyrate
Buttersaeure-methyl Ester
n-Methyl n-Butyrate

RN: 623-42-7   **MP** (°C):   -95
MW: 102.13   **BP** (°C):   102

Solubility (Moles/L)	Solubility (Grams/L)	Temp (°C)	Ref (#)	Evaluation (T P E A A)	Comments
1.528E-01	1.561E+01	21	F001	1 0 1 2 2	
1.506E-01	1.538E+01	21	F300	1 0 0 0 2	
1.600E-01	1.634E+01	21	S006	1 0 0 0 2	
1.469E-01	1.500E+01	25	A049	1 0 0 0 2	

477. $C_5H_{10}O_2$
Propyl Acetate
Essigsaeurepropyl Ester
RN: 109-60-4   **MP** (°C):   -92
MW: 102.13   **BP** (°C):   101.6

Solubility (Moles/L)	Solubility (Grams/L)	Temp (°C)	Ref (#)	Evaluation (T P E A A)	Comments
2.222E-01	2.270E+01	20	E002	1 0 0 0 2	
1.850E-01	1.889E+01	20	F001	1 0 1 0 2	
1.821E-01	1.860E+01	20	F300	1 0 0 0 2	
1.800E-01	1.838E+01	20	M171	1 0 0 0 1	
2.220E-01	2.267E+01	21	S006	1 0 0 0 2	
1.920E-01	1.961E+01	25	B060	2 0 1 1 1	
1.731E-01	1.768E+01	30	R318	1 2 0 1 1	
1.960E-01	2.002E+01	37	E028	1 0 1 1 2	

478. $C_5H_{10}O_2$
Pivalic Acid
Trimethylacetic Acid
Trimethylessigsaeure
RN: 75-98-9   **MP** (°C):   35.5
MW: 102.13   **BP** (°C):   163.8

Solubility (Moles/L)	Solubility (Grams/L)	Temp (°C)	Ref (#)	Evaluation (T P E A A)	Comments
2.125E-01	2.170E+01	20	F300	1 0 0 0 2	

479. $C_5H_{10}O_2$
3-Hydroxy-2-methyltetrahydrofuran
3-Furanol, Tetrahydro-2-methyl-
RN: 29848-44-0   **MP** (°C):
MW: 102.13   **BP** (°C):

Solubility (Moles/L)	Solubility (Grams/L)	Temp (°C)	Ref (#)	Evaluation (T P E A A)	Comments
1.632E+00	1.667E+02	rt	B066	0 2 0 0 1	
4.896E+00	5.000E+02	rt	B066	0 2 0 0 2	

480. $C_5H_{10}O_2$
Valeric Acid
Valeric Acid, Normal
RN: 109-52-4   **MP** (°C):   -34.5
MW: 102.13   **BP** (°C):   185

Solubility (Moles/L)	Solubility (Grams/L)	Temp (°C)	Ref (#)	Evaluation (T P E A A)	Comments
2.295E-01	2.344E+01	25	B060	2 0 1 1 1	
4.636E-01	4.735E+01	25	H028	2 0 2 0 2	
3.697E-01	3.776E+01	25	H122	1 0 0 0 2	
4.055E-01	4.141E+01	25	H338	2 2 1 2 2	
3.750E-01	3.830E+01	25	K012	1 0 0 0 2	
4.893E-01	4.997E+01	35	H338	2 2 1 2 2	
2.936E-03	2.999E-01	c	L055	0 0 0 0 1	

481. $C_5H_{10}O_3$
Ethyl Carbonate
Diethyl Carbonate
RN: 105-58-8   **MP** (°C):   -43
MW: 118.13   **BP** (°C):   126

Solubility (Moles/L)	Solubility (Grams/L)	Temp (°C)	Ref (#)	Evaluation (T P E A A)	Comments
1.562E-01	1.845E+01	20	D052	1 1 0 0 2	

482. $C_5H_{10}O_3$
Methyl β-Methoxypropionate
Propionic Acid, 3-Methoxy-, Methyl Ester
Methyl 3-Methoxypropanoate
Methyl 3-Methoxypropionate
RN: 3852-09-3   **MP** (°C):
MW: 118.13   **BP** (°C):

Solubility (Moles/L)	Solubility (Grams/L)	Temp (°C)	Ref (#)	Evaluation (T P E A A)	Comments
3.628E+00	4.286E+02	25	R034	1 0 0 0 1	

483. $C_5H_{10}O_5$
L-Arabinose
L-Arabinopyranose
RN: 87-72-9   **MP** (°C):   158
MW: 150.13   **BP** (°C):

Solubility (Moles/L)	Solubility (Grams/L)	Temp (°C)	Ref (#)	Evaluation (T P E A A)	Comments
2.482E+00	3.726E+02	10	F300	1 0 0 0 2	

484. $C_5H_{10}O_5$
D-Xylose
α-Xylose
Wood sugar
RN: 58-86-6   **MP** (°C):   144.5
MW: 150.13   **BP** (°C):

Solubility (Moles/L)	Solubility (Grams/L)	Temp (°C)	Ref (#)	Evaluation (T P E A A)	Comments
2.879E+00	4.322E+02	25	G317	2 1 2 2 2	

485. $C_5H_{11}Br$
Isoamyl Bromide
1-Bromo-3-methylbutane
RN: 107-82-4   **MP** (°C):   -112
MW: 151.05   **BP** (°C):   120

Solubility (Moles/L)	Solubility (Grams/L)	Temp (°C)	Ref (#)	Evaluation (T P E A A)	Comments
1.324E-03	2.000E-01	16	F300	1 0 0 0 1	
1.300E-03	1.964E-01	16.5	F001	1 0 1 0 2	

486. $C_5H_{11}Br$
n-Amyl Bromide
1-Bromopentane
Pentyl Bromide
Amylene Bromide
RN: 110-53-2   **MP** (°C):   -87.9
MW: 151.05   **BP** (°C):   129.6

Solubility (Moles/L)	Solubility (Grams/L)	Temp (°C)	Ref (#)	Evaluation (T P E A A)	Comments
8.380E-04	1.266E-01	25	M342	1 0 1 1 2	
1.800E-02	2.719E+00	ns	H307	1 0 1 1 2	

487. $C_5H_{11}NO$
Pentanamide
Valeramide
RN: 626-97-1   **MP** (°C):   102-104
MW: 101.15   **BP** (°C):

Solubility (Moles/L)	Solubility (Grams/L)	Temp (°C)	Ref (#)	Evaluation (T P E A A)	Comments
5.530E-01	5.594E+01	6	H059	1 2 2 0 2	
6.360E-01	6.433E+01	16	H059	1 2 2 0 2	
7.880E-01	7.971E+01	25	H059	1 2 2 0 2	
1.108E+00	1.121E+02	37	H059	1 2 2 0 2	

488.   C₅H₁₁NO₂
DL-Norvaline
DL-2-Aminovaleric Acid
RN: 760-78-1   **MP** (°C):   303.0
MW: 117.15   **BP** (°C):

Solubility (Moles/L)	Solubility (Grams/L)	Temp (°C)	Ref (#)	Evaluation (T P E A A)	Comments
8.251E-01	9.666E+01	15	D041	1 0 0 0 2	
7.768E-01	9.100E+01	18	F300	1 0 0 0 1	
6.616E-01	7.751E+01	25	K031	2 1 2 1 2	

489.   C₅H₁₁NO₂
L-Norvaline
L-(+)-2-Aminovaleric Acid
RN: 6600-40-4   **MP** (°C):   >300
MW: 117.15   **BP** (°C):

Solubility (Moles/L)	Solubility (Grams/L)	Temp (°C)	Ref (#)	Evaluation (T P E A A)	Comments
8.286E-01	9.707E+01	15	D041	1 0 0 0 2	

490.   C₅H₁₁NO₂
3-Nitropentane
Pentane, 3-Nitro-
RN: 551-88-2   **MP** (°C):
MW: 117.15   **BP** (°C):   153

Solubility (Moles/L)	Solubility (Grams/L)	Temp (°C)	Ref (#)	Evaluation (T P E A A)	Comments
1.110E-02	1.300E+00	25	A049	1 0 0 0 1	

491.   C₅H₁₁NO₂
Betaine
Betain
RN: 107-43-7   **MP** (°C):   296
MW: 117.15   **BP** (°C):

Solubility (Moles/L)	Solubility (Grams/L)	Temp (°C)	Ref (#)	Evaluation (T P E A A)	Comments
5.216E+00	6.110E+02	19.30	F300	1 0 0 0 2	

492. $C_5H_{11}NO_2$
DL-Isovaline
DL-Isovalin
RN: 595-39-1   **MP** (°C):   315
MW: 117.15   **BP** (°C):

Solubility (Moles/L)	Solubility (Grams/L)	Temp (°C)	Ref (#)	Evaluation (T P E A A)	Comments
2.398E+00	2.809E+02	20	F300	1 0 0 0 2	

493. $C_5H_{11}NO_2$
DL-Valine
DL-Valin
RN: 516-06-3   **MP** (°C):   296
MW: 117.15   **BP** (°C):

Solubility (Moles/L)	Solubility (Grams/L)	Temp (°C)	Ref (#)	Evaluation (T P E A A)	Comments
5.593E-01	6.552E+01	0	D018	2 2 2 1 2	
5.711E-01	6.690E+01	25	C018	1 0 2 2 2	
6.035E-01	7.070E+01	25	D016	1 0 0 0 2	
5.912E-01	6.926E+01	25	D018	2 2 2 1 2	
5.614E-01	6.577E+01	25	D041	1 0 0 0 2	
5.975E-01	7.000E+01	25	F300	1 0 0 0 0	
7.352E-01	8.612E+01	50	D018	2 2 2 1 2	
7.170E-01	8.400E+01	50	F300	1 0 0 0 1	
1.003E+00	1.175E+02	75	D018	2 2 2 1 2	
9.559E-01	1.120E+02	75	D041	1 0 0 0 2	
9.560E-01	1.120E+02	75	F300	1 0 0 0 2	
1.351E+00	1.583E+02	99.99	P349	1 0 0 2 2	
1.349E+00	1.580E+02	100	F300	1 0 0 0 2	

494. $C_5H_{11}NO_2$
Isobutyl Carbamate
iso-Butyl Carbamate
RN: 543-28-2   **MP** (°C):   67
MW: 117.15   **BP** (°C):   206

Solubility (Moles/L)	Solubility (Grams/L)	Temp (°C)	Ref (#)	Evaluation (T P E A A)	Comments
5.000E-01	5.857E+01	37	H006	1 2 2 1 0	

495.   C$_5$H$_{11}$NO$_2$
L-Valine
Valine
L-(+)-valine
L-2-Amino-3-methylbutyric Acid
2-Amino-3-methylbutyric Acid
RN: 72-18-4   **MP** (°C):   315
MW: 117.15   **BP** (°C):

Solubility (Moles/L)	Solubility (Grams/L)	Temp (°C)	Ref (#)	Evaluation (T P E A A)	Comments
7.180E-01	8.411E+01	15	D349	2 1 1 2 2	
4.866E-01	5.701E+01	20	B032	1 2 2 1 2	
7.360E-01	8.622E+01	20	D349	2 1 1 2 2	
4.992E-01	5.848E+01	25	B032	1 2 2 1 2	
6.940E-01	8.130E+01	25	D041	1 0 0 0 2	
7.550E-01	8.845E+01	25	D349	2 1 1 2 2	
4.710E-01	5.518E+01	25	G092	2 1 1 1 1	
4.710E-01	5.518E+01	25	G315	1 0 2 2 2	
5.900E-01	6.912E+01	25	N001	2 0 2 1 0	EFG
4.740E-01	5.553E+01	25	N012	2 0 2 1 2	
5.019E-01	5.880E+01	27	D036	2 1 2 2 2	
5.114E-01	5.991E+01	29.80	B032	1 2 2 1 2	
7.929E-01	9.289E+01	65	D041	1 0 0 0 2	

496.   C$_5$H$_{11}$NO$_2$
n-Butyl Carbamate
Butyl Carbamate
RN: 592-35-8   **MP** (°C):   51
MW: 117.15   **BP** (°C):

Solubility (Moles/L)	Solubility (Grams/L)	Temp (°C)	Ref (#)	Evaluation (T P E A A)	Comments
2.200E-01	2.577E+01	37	H006	1 2 2 1 1	

497.   C$_5$H$_{11}$NO$_2$
tert-Butyl Carbamate
O-t-Butyl Carbamate
RN: 4248-19-5   **MP** (°C):   105
MW: 117.15   **BP** (°C):

Solubility (Moles/L)	Solubility (Grams/L)	Temp (°C)	Ref (#)	Evaluation (T P E A A)	Comments
1.250E+00	1.464E+02	37	H006	1 2 2 1 2	

498. $C_5H_{11}NO_2$
D-Valine
β-Amino-isovalerian-saeure
β-Aminoisovaleric Acid
RN: 640-68-6 **MP** (°C): >295
MW: 117.15 **BP** (°C):

Solubility (Moles/L)	Solubility (Grams/L)	Temp (°C)	Ref (#)	Evaluation (T P E A A)	Comments
1.291E-02	1.512E+00	10	D038	1 0 1 0 0	EFG, unit assumed, *sic*
4.296E-01	5.033E+01	20	D041	1 0 0 0 1	
7.053E-01	8.263E+01	25	C018	1 0 2 2 2	
1.343E-02	1.574E+00	25	D038	1 0 1 0 0	EFG, unit assumed, *sic*
1.384E-02	1.622E+00	33	D038	1 0 1 0 0	EFG, unit assumed, *sic*
1.426E-02	1.671E+00	40	D038	1 0 1 0 0	EFG, unit assumed, *sic*
1.455E-02	1.705E+00	49	D038	1 0 1 0 0	EFG, unit assumed, *sic*
1.500E-02	1.757E+00	57	D038	1 0 1 0 0	EFG, unit assumed, *sic*
1.592E-02	1.865E+00	65	D038	1 0 1 0 0	EFG, unit assumed, *sic*

499. $C_5H_{11}NO_2S$
Methionine
L-(-)-Methionine
2-Amino-4-(methylthio)butanoic Acid
RN: 63-68-3 **MP** (°C): -279
MW: 149.21 **BP** (°C):

Solubility (Moles/L)	Solubility (Grams/L)	Temp (°C)	Ref (#)	Evaluation (T P E A A)	Comments
3.504E-01	5.228E+01	20	B032	1 2 2 1 2	
3.791E-01	5.656E+01	25	B032	1 2 2 1 2	
3.566E-01	5.321E+01	25	G315	1 0 2 2 2	
3.753E-01	5.600E+01	25.1	N024	2 0 2 2 2	
3.746E-01	5.590E+01	25.1	N026	2 0 2 2 2	
3.548E-01	5.294E+01	25.1	N027	1 1 2 2 2	
3.498E-01	5.220E+01	27	D036	2 1 2 2 2	
4.093E-01	6.107E+01	29.80	B032	1 2 2 1 2	

500.   C₅H₁₁NO₂S
DL-Methionine
DL-Methionin
DL-2-Amino-4-(methylthio)butyric Acid
Acimetion

RN: 59-51-8   **MP** (°C):   281
MW: 149.21   **BP** (°C):

Solubility (Moles/L)	Solubility (Grams/L)	Temp (°C)	Ref (#)	Evaluation (T P E A A)	Comments
1.200E-01	1.790E+01	0	F300	1 0 0 0 2	
2.191E-01	3.269E+01	25	D041	1 0 0 0 2	
2.191E-01	3.270E+01	25	F300	1 0 0 0 2	
3.833E-01	5.720E+01	50	F300	1 0 0 0 2	
6.379E-01	9.519E+01	75	D041	1 0 0 0 2	
6.380E-01	9.520E+01	75	F300	1 0 0 0 2	
1.003E+00	1.497E+02	100	F300	1 0 0 0 2	

501.   C₅H₁₁NO₂S
Penicillamine
3,3-Dimethyl-D-(-)-cysteine
D-3-Mercaptovaline
D-Penicillamine

RN: 52-67-5   **MP** (°C):   198.0
MW: 149.21   **BP** (°C):

Solubility (Moles/L)	Solubility (Grams/L)	Temp (°C)	Ref (#)	Evaluation (T P E A A)	Comments
6.702E-01	1.000E+02	20	C120	0 0 0 0 0	

502.   C₅H₁₁NO₂.H₂O
Betaine (Monohydrate)
Trimethylammonioacetate (Monohydrate)

RN: 590-47-6   **MP** (°C):
MW: 135.16   **BP** (°C):

Solubility (Moles/L)	Solubility (Grams/L)	Temp (°C)	Ref (#)	Evaluation (T P E A A)	Comments
4.520E+00	6.109E+02	19	D041	1 0 0 0 2	

503. C_5H_{12}
2-Methylbutane
Isopentane
Izopentan
RN: 78-78-4   **MP** (°C):   -160
MW: 72.15   **BP** (°C):   30

Solubility (Moles/L)	Solubility (Grams/L)	Temp (°C)	Ref (#)	Evaluation (T P E A A)	Comments
1.003E-03	7.240E-02	0	P003	2 2 2 2 2	
6.653E-04	4.800E-02	25	K119	1 0 0 0 2	
6.625E-04	4.780E-02	25	M001	2 1 2 2 2	
6.625E-04	4.780E-02	25	M002	2 1 2 2 2	
6.874E-04	4.960E-02	25	P003	2 2 2 2 2	
6.653E-04	4.800E-02	25	P007	2 1 2 2 2	
6.653E-04	4.800E-02	25	P051	2 1 1 2 2	

504. C_5H_{12}
Neopentane
2,2-Dimethylpropane
RN: 463-82-1   **MP** (°C):
MW: 72.15   **BP** (°C):   9.5

Solubility (Moles/L)	Solubility (Grams/L)	Temp (°C)	Ref (#)	Evaluation (T P E A A)	Comments
2.220E-04	1.602E-02	25	D346	1 1 2 2 2	
4.601E-04	3.320E-02	25	M001	2 1 2 2 2	
5.611E-04	4.048E-02	25	S212	2 1 2 2 2	
3.833E-04	2.766E-02	40	S212	2 1 2 2 1	
2.667E-04	1.924E-02	60	S212	2 1 2 2 1	
2.389E-04	1.724E-02	80	S212	2 1 2 2 1	

505. C_5H_{12}
Pentane
n-Pentane
RN: 109-66-0   **MP** (°C):   -130
MW: 72.15   **BP** (°C):

Solubility (Moles/L)	Solubility (Grams/L)	Temp (°C)	Ref (#)	Evaluation (T P E A A)	Comments
9.106E-04	6.570E-02	0	P003	2 2 2 2 2	
5.666E-04	4.088E-02	4.0	N004	1 1 2 2 2	
1.516E-04	1.094E-02	4.8	L007	2 1 1 2 2	
1.516E-04	1.094E-02	5.1	L007	2 0 1 1 2	
5.944E-04	4.289E-02	10.0	N004	1 1 2 2 2	
1.635E-04	1.180E-02	14.8	L007	2 1 1 2 2	
2.425E-04	1.750E-02	20	M337	2 1 2 2 2	
5.444E-04	3.928E-02	20.0	N004	1 1 2 2 2	
1.563E-04	1.128E-02	24.8	L007	2 1 1 2 2	
5.267E-04	3.800E-02	25	A049	1 0 0 0 1	

5.475E-04	3.950E-02	25	K119	1 0 0 0 2
5.336E-04	3.850E-02	25	M001	2 1 2 2 2
5.336E-04	3.850E-02	25	M002	2 1 2 2 2
5.650E-04	4.077E-02	25	M342	1 0 1 1 2
6.597E-04	4.760E-02	25	P003	2 2 2 2 2
5.611E-04	4.048E-02	25.0	N004	1 1 2 2 2
5.475E-04	3.950E-02	25.0	P051	2 1 1 2 2
5.475E-04	3.950E-02	25.00	P007	2 1 2 2 2
5.611E-04	4.048E-02	30.0	N004	1 1 2 2 2
1.509E-04	1.089E-02	34.8	L007	2 1 1 2 2
5.516E-04	3.980E-02	40.1	P051	2 1 1 2 2
5.516E-04	3.980E-02	40.10	P007	2 1 2 2 2
5.793E-04	4.180E-02	55.7	P051	2 1 1 2 2
5.793E-04	4.180E-02	55.70	P007	2 1 2 2 2
9.619E-04	6.940E-02	99.1	P051	2 1 1 2 2
9.619E-04	6.940E-02	99.10	P007	2 1 2 2 2
1.525E-03	1.100E-01	121.3	P051	2 1 1 2 2
1.525E-03	1.100E-01	121.30	P007	2 1 2 2 2
2.786E-03	2.010E-01	137.3	P051	2 1 1 2 2
2.786E-03	2.010E-01	137.30	P007	2 1 2 2 2
4.130E-03	2.980E-01	149.5	P051	2 1 1 2 2
4.130E-03	2.980E-01	149.50	P007	2 1 2 2 2
1.010E-04	7.287E-03	ns	D348	0 0 2 2 2

506. $C_5H_{12}ClO_2PS_2$
Chlormephos
Dotan
Diethyl S-(Chloromethyl) Dithiophosphate
RN: 24934-91-6   **MP** (°C):
MW: 234.70   **BP** (°C):   83

Solubility (Moles/L)	Solubility (Grams/L)	Temp (°C)	Ref (#)	Evaluation (T P E A A)	Comments
2.556E-04	6.000E-02	20	L303	1 0 0 0 1	
2.556E-04	6.000E-02	20	M161	1 0 0 0 1	

507. $C_5H_{12}NO_3PS_2$
Dimethoate
O,O-Dimethyl S-(N-Methylcarbamoylmethyl) Dithiophosphate
RN: 60-51-5   **MP** (°C):   52.25
MW: 229.26   **BP** (°C):

Solubility (Moles/L)	Solubility (Grams/L)	Temp (°C)	Ref (#)	Evaluation (T P E A A)	Comments
1.096E-01	2.514E+01	20	B179	2 0 0 0 2	
1.309E-01	3.000E+01	20	G319	1 0 0 0 2	
1.090E-01	2.500E+01	21	M161	1 0 0 0 1	
1.701E-01	3.900E+01	ns	M061	0 0 0 0 1	

508. $C_5H_{12}N_2$
2-Methylpiperazine
2-Methyl-piperazin

RN: 109-07-9   **MP** (°C):   66
MW: 100.16   **BP** (°C):   155

Solubility (Moles/L)	Solubility (Grams/L)	Temp (°C)	Ref (#)	Evaluation (T P E A A)	Comments
4.343E+00	4.350E+02	20	F300	1 0 0 0 2	

509. $C_5H_{12}N_2O$
Methyl-n-butylnitrosamine
MBN

RN: 7068-83-9   **MP** (°C):
MW: 116.16   **BP** (°C):

Solubility (Moles/L)	Solubility (Grams/L)	Temp (°C)	Ref (#)	Evaluation (T P E A A)	Comments
2.000E-01	2.323E+01	24	M031	1 1 1 1 1	

510. $C_5H_{12}O$
2-Pentanol
iso-Amyl Alcohol
sec-Amyl Alcohol
Methyl Propyl Carbinol

RN: 6032-29-7   **MP** (°C):   -50
MW: 88.15   **BP** (°C):   119.3

Solubility (Moles/L)	Solubility (Grams/L)	Temp (°C)	Ref (#)	Evaluation (T P E A A)	Comments
7.708E-01	6.795E+01	0	S307	1 1 0 2 2	
6.189E-01	5.455E+01	10.1	S307	1 1 0 2 2	
5.030E-01	4.434E+01	19.5	S307	1 1 0 2 2	
4.573E-01	4.031E+01	20	C042	1 0 0 0 1	
1.473E-02	1.298E+00	20	D052	1 1 0 0 0	*sic*
4.538E-01	4.000E+01	20	F300	1 0 0 0 1	
5.258E-01	4.635E+01	20	G004	2 2 2 2 2	
3.836E-01	3.382E+01	25	B019	1 0 1 2 0	
4.843E-01	4.270E+01	25	G004	2 2 2 2 2	
4.499E-01	3.966E+01	30	G004	2 2 2 2 2	
4.300E-01	3.791E+01	30.6	S307	1 1 0 2 2	
3.900E-01	3.438E+01	40.0	S307	1 1 0 2 2	
3.645E-01	3.213E+01	50.0	S307	1 1 0 2 2	
3.432E-01	3.026E+01	60.0	S307	1 1 0 2 2	
3.379E-01	2.979E+01	70.1	S307	1 1 0 2 2	
3.443E-01	3.035E+01	79.9	S307	1 1 0 2 2	
3.368E-01	2.969E+01	90.3	S307	1 1 0 2 2	
5.149E-01	4.539E+01	ns	L003	0 0 2 1 2	

511.   C₅H₁₂O
Neopentyl Alcohol
t-Butyl Carbinol
RN: 75-84-3   **MP** (°C):   53
MW: 88.15   **BP** (°C):   114

Solubility (Moles/L)	Solubility (Grams/L)	Temp (°C)	Ref (#)	Evaluation (T P E A A)	Comments
4.048E-01	3.568E+01	12.0	S307	1 1 0 2 2	
3.826E-01	3.372E+01	18.8	S307	1 1 0 2 2	
4.090E-01	3.605E+01	20	G004	2 2 2 2 2	
3.836E-01	3.382E+01	25	G004	2 2 2 2 2	
3.603E-01	3.176E+01	30	G004	2 2 2 2 2	
3.229E-01	2.847E+01	30.0	S307	1 1 0 2 2	
2.982E-01	2.629E+01	40.0	S307	1 1 0 2 2	
2.616E-01	2.306E+01	50.0	S307	1 1 0 2 2	
2.778E-01	2.449E+01	60.0	S307	1 1 0 2 2	
2.399E-01	2.114E+01	70.2	S307	1 1 0 2 2	
2.864E-01	2.525E+01	80.0	S307	1 1 0 2 2	
2.637E-01	2.325E+01	90.0	S307	1 1 0 2 2	

512.   C₅H₁₂O
Methyl tert-Butyl Ether
tert-Butyl Methyl Ether
RN: 1634-04-4   **MP** (°C):   -109
MW: 88.15   **BP** (°C):   54.5

Solubility (Moles/L)	Solubility (Grams/L)	Temp (°C)	Ref (#)	Evaluation (T P E A A)	Comments
5.196E-01	4.580E+01	20	E019	1 0 1 1 1	
5.815E-01	5.126E+01	25	K072	1 0 1 1 1	
5.815E-01	5.126E+01	25	M087	1 1 2 1 2	

513.   C₅H₁₂O
3-Pentanol
Pentan-3-ol
Diethyl Carbinol
RN: 584-02-1   **MP** (°C):   <25
MW: 88.15   **BP** (°C):   115.6

Solubility (Moles/L)	Solubility (Grams/L)	Temp (°C)	Ref (#)	Evaluation (T P E A A)	Comments
8.704E-01	7.672E+01	0	S307	1 1 0 2 2	
7.382E-01	6.507E+01	10.2	S307	1 1 0 2 2	
6.026E-01	5.312E+01	20	G004	2 2 2 2 2	
6.280E-01	5.536E+01	20.0	S307	1 1 0 2 2	
5.505E-01	4.853E+01	25	C093	2 1 1 1 1	
5.556E-01	4.898E+01	25	G004	2 2 2 2 2	

5.144E-01	4.535E+01	30	G004	2 2 2 2 2	
5.730E-01	5.051E+01	30.0	S307	1 1 0 2 2	
4.510E-01	3.975E+01	40.0	S307	1 1 0 2 2	
4.604E-01	4.058E+01	50.0	S307	1 1 0 2 2	
3.889E-01	3.428E+01	60.0	S307	1 1 0 2 2	
3.783E-01	3.335E+01	70.0	S307	1 1 0 2 2	
3.635E-01	3.204E+01	80.0	S307	1 1 0 2 2	
3.773E-01	3.326E+01	90.0	S307	1 1 0 2 2	
1.392E+00	1.227E+02	ns	L003	0 0 2 1 1	
5.196E-01	4.580E+01	rt	H111	0 0 0 0 1	

514. $C_5H_{12}O$
3-Methyl-2-butanol
Methylisopropylcarbinol

RN: 598-75-4   **MP** (°C):   <25
MW: 88.15   **BP** (°C):   113

Solubility (Moles/L)	Solubility (Grams/L)	Temp (°C)	Ref (#)	Evaluation (T P E A A)	Comments
8.771E-01	7.732E+01	0	S307	1 1 0 2 2	
7.609E-01	6.708E+01	10.1	S307	1 1 0 2 2	
6.492E-01	5.723E+01	20	G004	2 2 2 2 2	
6.381E-01	5.625E+01	20.0	S307	1 1 0 2 2	
5.505E-01	4.853E+01	30	G004	2 2 2 2 2	
5.536E-01	4.880E+01	30.0	S307	1 1 0 2 2	
4.833E-01	4.260E+01	40.0	S307	1 1 0 2 2	
4.416E-01	3.892E+01	50.0	S307	1 1 0 2 2	
3.720E-01	3.279E+01	60.0	S307	1 1 0 2 2	
4.005E-01	3.531E+01	70.0	S307	1 1 0 2 2	
3.942E-01	3.475E+01	79.5	S307	1 1 0 2 2	
3.942E-01	3.475E+01	90.0	S307	1 1 0 2 2	

515. $C_5H_{12}O$
tert-Pentyl Alcohol
Dimethylethylcarbinol
tert-Amylalkohol

RN: 75-85-4   **MP** (°C):
MW: 88.15   **BP** (°C):   102.5

Solubility (Moles/L)	Solubility (Grams/L)	Temp (°C)	Ref (#)	Evaluation (T P E A A)	Comments
1.548E+00	1.364E+02	0.5	S307	1 1 0 2 2	
1.462E+00	1.289E+02	9.8	S307	1 1 0 2 2	
1.259E+00	1.110E+02	20	F300	1 0 0 0 2	
1.229E+00	1.083E+02	20	G004	2 2 2 2 2	
1.170E+00	1.031E+02	20.8	S307	1 1 0 2 2	
1.124E+00	9.910E+01	25	G004	2 2 2 2 2	
5.965E-01	5.258E+01	25	G004	2 2 2 2 2	
1.026E+00	9.041E+01	29.5	S307	1 1 0 2 2	
1.041E+00	9.173E+01	30	G004	2 2 2 2 2	

8.549E-01	7.536E+01	39.5	S307	1 1 0 2 2
7.649E-01	6.743E+01	49.0	S307	1 1 0 2 2
6.673E-01	5.882E+01	60.0	S307	1 1 0 2 2
6.391E-01	5.634E+01	70.2	S307	1 1 0 2 2
6.117E-01	5.393E+01	80.1	S307	1 1 0 2 2
5.883E-01	5.186E+01	90.2	S307	1 1 0 2 2
1.124E+00	9.910E+01	rt	H111	0 0 0 0 2

516. $C_5H_{12}O$
1-Pentanol
Amyl Alcohol
Pentanol
Pentyl Alcohol
n-Amyl Alcohol
RN: 71-41-0   **MP** (°C):   -79
MW: 88.15   **BP** (°C):   138

Solubility (Moles/L)	Solubility (Grams/L)	Temp (°C)	Ref (#)	Evaluation (T P E A A)	Comments
4.321E-01	3.809E+01	-0.5	F051	2 1 0 1 2	
3.358E-01	2.960E+01	0	E029	1 2 0 1 2	
3.635E-01	3.204E+01	0	S307	1 1 0 2 2	
3.709E-01	3.269E+01	7	F051	2 1 0 1 2	
2.982E-01	2.629E+01	10	E029	1 2 0 1 2	
2.864E-01	2.525E+01	10.2	S307	1 1 0 2 2	
3.068E-01	2.705E+01	14	F051	2 1 0 1 2	
3.004E-01	2.648E+01	15	F051	2 1 0 1 2	
5.395E+00	4.756E+02	15.5	F051	2 1 0 1 2	
2.875E-01	2.534E+01	16.5	F051	2 1 0 1 2	
2.821E-01	2.487E+01	18	F051	2 1 0 1 2	
2.453E-01	2.162E+01	20	A015	1 2 1 1 2	
1.020E-02	8.992E-01	20	D052	1 1 0 0 0	*sic*
2.605E-01	2.296E+01	20	E029	1 2 0 1 2	
2.616E-01	2.306E+01	20	G004	2 2 2 2 2	
1.676E-01	1.478E+01	20	L049	1 1 2 1 1	
3.070E-01	2.706E+01	20	M312	1 0 0 0 1	
2.496E-01	2.200E+01	20.2	S307	1 1 0 2 2	
3.607E-01	3.180E+01	22	H072	1 0 1 1 2	
2.691E-01	2.372E+01	23	F051	2 1 0 1 2	
3.730E-01	3.288E+01	25	B019	1 0 1 2 0	
2.451E-01	2.160E+01	25	B038	1 0 1 1 2	
1.896E-01	1.672E+01	25	B060	2 0 1 1 1	
2.442E-01	2.153E+01	25	C093	2 1 1 1 1	
1.000E+00	8.815E+01	25	F044	1 0 0 0 0	EFG
2.137E-01	1.884E+01	25	F317	2 1 1 1 2	
2.431E-01	2.143E+01	25	G004	2 2 2 2 2	
2.300E-01	2.027E+01	25	G075	1 0 1 0 1	
2.810E-01	2.477E+01	25	H028	2 0 2 0 2	

2.817E-01	2.483E+01	25	H104	1 0 0 0 1
2.500E-01	2.204E+01	25	K025	2 2 1 1 1
2.561E-01	2.258E+01	29	F051	2 1 0 1 2
2.333E-01	2.057E+01	30	E029	1 2 0 1 2
2.257E-01	1.990E+01	30	G004	2 2 2 2 2
2.246E-01	1.980E+01	30.6	S307	1 1 0 2 2
5.368E+00	4.732E+02	34.0	F051	2 1 0 1 2
2.475E-01	2.181E+01	36	F051	2 1 0 1 2
2.130E-01	1.878E+01	37	E028	1 0 1 1 2
2.115E-01	1.865E+01	40	E029	1 2 0 1 2
2.082E-01	1.836E+01	40.2	S307	1 1 0 2 2
2.006E-01	1.768E+01	50	E029	1 2 0 1 2
2.039E-01	1.797E+01	50.0	S307	1 1 0 2 2
2.475E-01	2.181E+01	58	F051	2 1 0 1 2
2.006E-01	1.768E+01	60	E029	1 2 0 1 2
2.039E-01	1.797E+01	60.3	S307	1 1 0 2 2
5.290E+00	4.664E+02	69.5	F051	2 1 0 1 2
2.061E-01	1.816E+01	70	E029	1 2 0 1 2
2.170E-01	1.913E+01	70.0	S307	1 1 0 2 2
2.561E-01	2.258E+01	72.0	F051	2 1 0 1 2
2.115E-01	1.865E+01	80	E029	1 2 0 1 2
2.213E-01	1.951E+01	80.0	S307	1 1 0 2 2
2.691E-01	2.372E+01	81	F051	2 1 0 1 2
2.821E-01	2.487E+01	87	F051	2 1 0 1 2
2.224E-01	1.961E+01	90	E029	1 2 0 1 2
2.453E-01	2.162E+01	90.7	S307	1 1 0 2 2
2.875E-01	2.534E+01	91	F051	2 1 0 1 2
3.004E-01	2.648E+01	95	F051	2 1 0 1 2
5.180E+00	4.566E+02	97.3	F051	2 1 0 1 2
3.068E-01	2.705E+01	98	F051	2 1 0 1 2
2.496E-01	2.200E+01	100	E029	1 2 0 1 2
2.875E-01	2.534E+01	110	E029	1 2 0 1 2
3.709E-01	3.269E+01	112	F051	2 1 0 1 2
3.304E-01	2.913E+01	120	E029	1 2 0 1 2
5.048E+00	4.450E+02	122.3	F051	2 1 0 1 2
4.321E-01	3.809E+01	126	F051	2 1 0 1 2
3.889E-01	3.428E+01	130	E029	1 2 0 1 2
4.677E-01	4.123E+01	140	E029	1 2 0 1 2
5.351E-01	4.717E+01	140	F051	2 1 0 1 2
4.896E+00	4.316E+02	141.6	F051	2 1 0 1 2
5.853E-01	5.159E+01	145	F051	2 1 0 1 2
6.290E-01	5.545E+01	148.5	F051	2 1 0 1 2
5.761E-01	5.078E+01	150	E029	1 2 0 1 2
4.707E+00	4.149E+02	157.3	F051	2 1 0 1 2
7.322E-01	6.455E+01	160	E029	1 2 0 1 2
9.060E-01	7.987E+01	167.0	F051	2 1 0 1 2
9.889E-01	8.717E+01	170	E029	1 2 0 1 2
1.001E+00	8.826E+01	171.2	F051	2 1 0 1 2
4.374E+00	3.856E+02	174.0	F051	2 1 0 1 2
1.690E+00	1.489E+02	180	E029	1 2 0 1 2

4.089E+00	3.605E+02	181.3	F051	2 1 0 1 2
1.435E+00	1.265E+02	182.5	F051	2 1 0 1 2
3.774E+00	3.327E+02	185.2	F051	2 1 0 1 2
1.833E+00	1.616E+02	186.0	F051	2 1 0 1 2
2.270E+00	2.001E+02	186.5	F051	2 1 0 1 2
3.472E+00	3.061E+02	186.5	F051	2 1 0 1 2
3.237E+00	2.854E+02	187.4	F051	2 1 0 1 2
3.040E+00	2.680E+02	187.5	F051	2 1 0 1 2
2.538E-01	2.237E+01	ns	L003	0 0 2 1 2
2.224E-01	1.961E+01	rt	H111	0 0 0 0 1

517. $C_5H_{12}O$
Ethylisopropyl Ether
Propane, 2-Ethoxy-
RN: 625-54-7   **MP** (°C):
MW: 88.15   **BP** (°C):

Solubility (Moles/L)	Solubility (Grams/L)	Temp (°C)	Ref (#)	Evaluation (T P E A A)	Comments
2.733E-01	2.409E+01	ns	J300	0 0 0 0 1	

518. $C_5H_{12}O$
2-Methyl-1-butanol
DL-2-Methyl-1-butanol
2-Methylbutan-1-ol
RN: 137-32-6   **MP** (°C):   -70
MW: 88.15   **BP** (°C):   128.0

Solubility (Moles/L)	Solubility (Grams/L)	Temp (°C)	Ref (#)	Evaluation (T P E A A)	Comments
4.269E-01	3.763E+01	0.5	S307	1 1 0 2 2	
3.720E-01	3.279E+01	9.7	S307	1 1 0 2 2	
3.122E-01	2.752E+01	19.6	S307	1 1 0 2 2	
3.496E-01	3.082E+01	20	G004	2 2 2 2 2	
3.304E-01	2.913E+01	25	C093	2 1 1 1 1	
3.272E-01	2.884E+01	25	G004	2 2 2 2 2	
2.778E-01	2.449E+01	29.6	S307	1 1 0 2 2	
3.122E-01	2.752E+01	30	G004	2 2 2 2 2	
2.616E-01	2.306E+01	39.3	S307	1 1 0 2 2	
2.453E-01	2.162E+01	49.6	S307	1 1 0 2 2	
2.301E-01	2.028E+01	59.3	S307	1 1 0 2 2	
2.485E-01	2.191E+01	69.5	S307	1 1 0 2 2	
2.551E-01	2.248E+01	79.7	S307	1 1 0 2 2	
2.724E-01	2.401E+01	90.8	S307	1 1 0 2 2	

519.   C$_5$H$_{12}$O
tert-Isoamyl Alcohol
3-Methyl-1-butanol
Isopentyl Alcohol
Isoamyl Alcohol
RN: 123-51-3   **MP** (°C):   -117
MW: 88.15   **BP** (°C):   130

Solubility (Moles/L)	Solubility (Grams/L)	Temp (°C)	Ref (#)	Evaluation (T P E A A)	Comments
4.079E-01	3.596E+01	0	S307	1 1 0 2 2	
3.090E-01	2.724E+01	10	A328	1 2 2 1 1	
3.454E-01	3.044E+01	10.1	S307	1 1 0 2 2	
3.347E-01	2.950E+01	15	K002	1 2 1 1 2	
3.130E-01	2.759E+01	18	F001	1 0 1 2 2	
2.918E-01	2.572E+01	19.8	S307	1 1 0 2 2	
3.120E-01	2.750E+01	20	F300	1 0 0 0 2	
3.144E-01	2.771E+01	20	G004	2 2 2 2 2	
3.111E-01	2.743E+01	20	K002	1 2 1 1 2	
9.586E-01	8.450E+01	20	K085	1 0 0 0 2	
2.659E-01	2.344E+01	25	A328	1 2 2 1 1	
3.411E-01	3.007E+01	25	C068	2 2 2 1 2	
2.982E-01	2.629E+01	25	C093	2 1 1 1 1	
3.251E-01	2.865E+01	25	F317	2 1 1 1 2	
2.950E-01	2.601E+01	25	G004	2 2 2 2 2	
2.950E-01	2.601E+01	25	K002	1 2 1 1 2	
2.799E-01	2.468E+01	30	G004	2 2 2 2 2	
2.832E-01	2.496E+01	30	K002	1 2 1 1 2	
2.842E-01	2.506E+01	30.1	H043	2 2 2 2 2	average of 3
2.540E-01	2.239E+01	30.2	S307	1 1 0 2 2	
2.442E-01	2.153E+01	40	A328	1 2 2 1 1	
2.420E-01	2.133E+01	40.0	S307	1 1 0 2 2	
2.257E-01	1.990E+01	49.9	S307	1 1 0 2 2	
2.431E-01	2.143E+01	59.8	S307	1 1 0 2 2	
2.344E-01	2.066E+01	70.0	S307	1 1 0 2 2	
2.442E-01	2.153E+01	80.0	S307	1 1 0 2 2	
2.518E-01	2.220E+01	90.0	S307	1 1 0 2 2	
2.836E-01	2.500E+01	ns	L003	0 0 2 1 2	
2.767E-01	2.439E+01	rt	H111	0 0 0 0 1	

520.   C$_5$H$_{12}$O$_2$
Formaldehyde Diethyl Acetal
Diethoxymethane
Diethylacetalformaldehyde
Formaldehyd-diaethyl-acetal
RN: 462-95-3   **MP** (°C):
MW: 104.15   **BP** (°C):   87.5

Solubility (Moles/L)	Solubility (Grams/L)	Temp (°C)	Ref (#)	Evaluation (T P E A A)	Comments
6.721E-01	7.000E+01	18	F300	1 0 0 0 1	
6.721E-01	7.000E+01	18	F300	1 0 0 0 1	

521.   C$_5$H$_{12}$O$_4$
Pentaerythritol
2,2-bis(Hydroxymethyl)-1,3-Propanediol
PE 200
Tetramethylolmethane
RN: 115-77-5   **MP** (°C):   260
MW: 136.15   **BP** (°C):

Solubility (Moles/L)	Solubility (Grams/L)	Temp (°C)	Ref (#)	Evaluation (T P E A A)	Comments
2.825E-01	3.846E+01	0	M043	1 0 0 0 0	
3.498E-01	4.762E+01	10	M043	1 0 0 0 0	
3.863E-01	5.260E+01	15	F300	1 0 0 0 2	
4.157E-01	5.660E+01	20	M043	1 0 0 0 0	
5.441E-01	7.407E+01	30	M043	1 0 0 0 0	
8.450E-01	1.150E+02	40	M043	1 0 0 0 1	
1.324E+00	1.803E+02	60	M043	1 0 0 0 1	
2.099E+00	2.857E+02	80	M043	1 0 0 0 1	
3.672E+00	5.000E+02	100	M043	1 0 0 0 2	

522.   C$_5$H$_{12}$O$_5$
DL-Arabinitol
(±)-Arabitol
RN: 2152-56-9   **MP** (°C):   103
MW: 152.15   **BP** (°C):

Solubility (Moles/L)	Solubility (Grams/L)	Temp (°C)	Ref (#)	Evaluation (T P E A A)	Comments
4.459E+00	6.785E+02	25	C346	2 0 2 1 2	

523. $C_5H_{12}O_5$
Adonitol
Adonit
Adonite
RN: 488-81-3   **MP** (°C):   104
MW: 152.15   **BP** (°C):

Solubility (Moles/L)	Solubility (Grams/L)	Temp (°C)	Ref (#)	Evaluation (T P E A A)	Comments
3.954E+00	6.016E+02	25	C346	2 0 2 1 2	

524. $C_5H_{13}N$
N-Methyldiethylamine
N,N-Diethylmethylamine
RN: 616-39-7   **MP** (°C):
MW: 87.17   **BP** (°C):   63

Solubility (Moles/L)	Solubility (Grams/L)	Temp (°C)	Ref (#)	Evaluation (T P E A A)	Comments
3.562E+00	3.105E+02	49.40	C086	2 2 2 2 2	average of 5
4.453E+00	3.881E+02	49.50	C086	2 2 2 2 2	
2.236E+00	1.949E+02	49.80	C086	2 2 2 2 2	
5.715E+00	4.982E+02	50.50	C086	2 2 2 2 2	
1.581E+00	1.378E+02	51.20	C086	2 2 2 2 2	
1.413E+00	1.231E+02	52.00	C086	2 2 2 2 2	
6.981E+00	6.085E+02	53.10	C086	2 2 2 2 2	
7.246E+00	6.316E+02	54.00	C086	2 2 2 2 2	

525. $C_5H_{13}O_3PS_2$
Demephion
O,O-Dimethyl 2-Methylmercaptoethyl Thiophosphate
Thiolo-Tinox
RN: 8065-62-1   **MP** (°C):
MW: 216.26   **BP** (°C):   109

Solubility (Moles/L)	Solubility (Grams/L)	Temp (°C)	Ref (#)	Evaluation (T P E A A)	Comments
2.312E-03	5.000E-01	20	M061	1 0 0 0 2	form II
9.248E-03	2.000E+00	ns	M061	0 0 0 0 2	form I
1.387E-02	3.000E+00	rt	M161	0 0 0 0 0	form II
1.387E-03	3.000E-01	rt	M161	0 0 0 0 2	form I

526.   C$_5$Cl$_6$
Hexachlorocyclopentadiene
1,2,3,4,5,5-Hexachloro-1,3-Cyclopentadiene
Hexachloro-1,3-cyclopentadiene
1,2,3,4,5,5-Hexachlorocyclopentadiene
RN: 77-47-4   **MP** (°C):   -9.9
MW: 272.77   **BP** (°C):   239

Solubility (Moles/L)	Solubility (Grams/L)	Temp (°C)	Ref (#)	Evaluation (T P E A A)	Comments
2.951E-06	8.050E-04	22.5	G301	2 1 0 1 2	

527.   C$_6$HCl$_3$N$_2$S
4,5,7-Trichloro-2,1,3-benzothiadiazole
PH 40-21
TH 052 H
RN: 1982-55-4   **MP** (°C):   131.5
MW: 239.51   **BP** (°C):

Solubility (Moles/L)	Solubility (Grams/L)	Temp (°C)	Ref (#)	Evaluation (T P E A A)	Comments
6.263E-06	1.500E-03	10	B200	1 0 0 0 1	
1.044E-05	2.500E-03	20	B200	1 0 0 0 1	
1.044E-05	2.500E-03	20	M061	1 0 0 0 1	
1.795E-05	4.300E-03	30	B200	1 0 0 0 1	

528.   C$_6$HCl$_4$NO$_2$
2,3,4,5-Tetrachloronitrobenzene
1,2,3,4-Tetrachloro-5-nitrobenzene
2,3,4,5-Tetrachloro-1-nitrobenzene
1-Nitro-2,3,4,5-tetrachlorobenzene
RN: 879-39-0   **MP** (°C):   66.0
MW: 260.89   **BP** (°C):

Solubility (Moles/L)	Solubility (Grams/L)	Temp (°C)	Ref (#)	Evaluation (T P E A A)	Comments
2.800E-05	7.305E-03	20	E308	1 2 2 1 1	

529.   C$_6$HCl$_4$NO$_2$
2,3,5,6-Tetrachloronitrobenzene
Tecnazene
RN: 117-18-0   **MP** (°C):   99.5
MW: 260.89   **BP** (°C):   304.0

Solubility (Moles/L)	Solubility (Grams/L)	Temp (°C)	Ref (#)	Evaluation (T P E A A)	Comments
8.000E-06	2.087E-03	20	E308	1 2 2 1 0	

530.   C₆HCl₄NO₂

2,3,4,6-Tetrachloronitrobenzene
Benzene, 1,2,3,5-Tetrachloro-4-nitro-
RN: 3714-62-3   **MP** (°C):
MW: 260.89   **BP** (°C):

Solubility (Moles/L)	Solubility (Grams/L)	Temp (°C)	Ref (#)	Evaluation (T P E A A)	Comments
2.900E-05	7.566E-03	20	E308	1 2 2 1 1	

531.   C₆HCl₅

Pentachlorobenzene
Penta-chlorobenzene
RN: 608-93-5   **MP** (°C):   82
MW: 250.34   **BP** (°C):   275

Solubility (Moles/L)	Solubility (Grams/L)	Temp (°C)	Ref (#)	Evaluation (T P E A A)	Comments
1.000E-06	2.503E-04	20	K337	1 0 0 0 2	
9.550E-07	2.391E-04	22	K305	1 0 1 1 0	
1.538E-06	3.850E-04	23	C305	1 1 2 2 2	
5.320E-06	1.332E-03	25	B173	2 0 2 2 2	
2.600E-06	6.509E-04	25	B317	1 0 0 0 2	
3.320E-06	8.311E-04	25	M342	1 0 1 1 2	
3.320E-06	8.311E-04	ns	M308	0 0 1 1 2	

532.   C₆HCl₅O

Pentachlorophenol
PCP
2,3,4,5,6-Pentachloro-phenol-
Phenol, 2,3,4,5,6-Pentachloro-
Dowicide 7
Fungifen
RN: 87-86-5   **MP** (°C):   174
MW: 266.34   **BP** (°C):

Solubility (Moles/L)	Solubility (Grams/L)	Temp (°C)	Ref (#)	Evaluation (T P E A A)	Comments
1.877E-05	5.000E-03	0	C310	1 0 0 0 1	
1.877E-05	5.000E-03	0	G310	1 0 0 0 0	
1.877E-05	5.000E-03	0	M061	1 0 0 0 0	
5.256E-05	1.400E-02	20	B185	1 0 0 0 1	
5.256E-05	1.400E-02	22.5	G301	2 1 0 1 2	
6.195E-05	1.650E-02	25	B183	0 0 0 0 1	
8.260E-05	2.200E-02	25	B185	1 0 0 0 1	
3.600E-05	9.588E-03	25	B316	1 0 2 1 1	
6.908E-05	1.840E-02	25	M373	1 0 2 1 2	
5.256E-05	1.400E-02	25	O320	1 0 1 1 1	
5.256E-05	1.400E-02	26.70	L095	2 2 1 1 2	
6.758E-05	1.800E-02	27	C310	1 0 0 0 1	

6.758E-05	1.800E-02	27	G310	1 0 0 0 1		
6.758E-05	1.800E-02	27	M061	1 0 0 0 1		
7.509E-05	2.000E-02	30	M161	1 0 0 0 1		
1.126E-04	3.000E-02	50	B200	1 0 0 0 0		
1.314E-04	3.500E-02	50	C310	1 0 0 0 1		
1.314E-04	3.500E-02	50	G310	1 0 0 0 1		
1.314E-04	3.500E-02	50	M061	1 0 0 0 1		
2.178E-04	5.800E-02	62	C310	1 0 0 0 1		
2.178E-04	5.800E-02	62	G310	1 0 0 0 1		
3.191E-04	8.499E-02	70	C310	1 0 0 0 1		
3.191E-04	8.499E-02	70	G310	1 0 0 0 1		
7.509E-05	2.000E-02	ns	L311	0 0 0 0 1		
7.134E-05	1.900E-02	ns	M110	0 0 0 0 0	EFG	
6.007E-06	1.600E-03	ns	N013	0 0 0 0 1		

533. C_6HF_5O
Pentafluorophenol
PFP

RN:	771-61-9	**MP** (°C):	34-36	
MW:	184.07	**BP** (°C):	143	

Solubility (Moles/L)	Solubility (Grams/L)	Temp (°C)	Ref (#)	Evaluation (T P E A A)	Comments
3.000E-01	5.522E+01	25	P031	1 1 2 2 2	

534. $C_6H_2Br_2ClNO_2$
2,6-Dibromoquinone-3-chlorimide
2,6-Dibromoquinonechloroimide

RN:		**MP** (°C):	
MW:	315.36	**BP** (°C):	

Solubility (Moles/L)	Solubility (Grams/L)	Temp (°C)	Ref (#)	Evaluation (T P E A A)	Comments
2.000E-04	6.307E-02	20	G043	1 0 1 1 0	

535. $C_6H_2ClN_3O_6$

2,4,6-Trinitro-1-chlorobenzene
Picryl Chloride
2-Chlor-1,3,5-trinitrobenzol
Chlorure de Picryle

RN: 88-88-0   **MP** (°C):
MW: 247.55   **BP** (°C):

Solubility (Moles/L)	Solubility (Grams/L)	Temp (°C)	Ref (#)	Evaluation (T P E A A)	Comments
7.190E-04	1.780E-01	15	D066	1 2 0 0 2	
7.189E-04	1.780E-01	15	D071	1 2 0 0 2	
7.271E-04	1.800E-01	15	F300	1 0 0 0 1	
2.141E-03	5.300E-01	16	D066	1 2 0 0 2	
2.140E-03	5.297E-01	50	D071	1 2 0 0 1	
1.398E-02	3.460E+00	100	D066	1 2 0 0 2	
1.393E-02	3.448E+00	100	D071	1 2 0 0 2	
1.454E-02	3.600E+00	100	F300	1 0 0 0 1	

536. $C_6H_2Cl_2O_4$

Chloranilic Acid
Chloranilsaeure

RN: 87-88-7   **MP** (°C):   283
MW: 208.99   **BP** (°C):

Solubility (Moles/L)	Solubility (Grams/L)	Temp (°C)	Ref (#)	Evaluation (T P E A A)	Comments
9.091E-03	1.900E+00	14	F300	1 0 0 0 1	
6.699E-02	1.400E+01	99	F300	1 0 0 0 1	

537. $C_6H_2Cl_3NO_2$

2,4,5-Trichloronitrobenzene
1,2,4-Trichloro-5-nitrobenzene
2,4,5-Trichloro-1-nitrobenzene
1,4,5-Trichloro-2-nitrobenzene
3,4,6-Trichloronitrobenzene

RN: 89-69-0   **MP** (°C):   57
MW: 226.45   **BP** (°C):   288

Solubility (Moles/L)	Solubility (Grams/L)	Temp (°C)	Ref (#)	Evaluation (T P E A A)	Comments
1.300E-04	2.944E-02	20	E308	1 2 2 1 2	

538. $C_6H_2Cl_3NO_2$
2,3,4-Trichloronitrobenzene
1,2,3-Trichloro-4-nitrobenzene
2,3,4-Trichloro-1-nitrobenzene
RN: 17700-09-3   **MP** (°C):   55.5
MW: 226.45   **BP** (°C):

Solubility (Moles/L)	Solubility (Grams/L)	Temp (°C)	Ref (#)	Evaluation (T P E A A)	Comments
1.150E-04	2.604E-02	20	E308	1 2 2 1 2	

539. $C_6H_2Cl_4$
1,2,3,4-Tetrachlorobenzene
Benzene, 1,2,3,4-Tetrachloro-
RN: 634-66-2   **MP** (°C):   48
MW: 215.89   **BP** (°C):   254

Solubility (Moles/L)	Solubility (Grams/L)	Temp (°C)	Ref (#)	Evaluation (T P E A A)	Comments
1.585E-05	3.422E-03	20	K337	1 0 0 0 2	
3.326E-05	7.180E-03	23	C305	1 1 2 2 2	
2.742E-05	5.920E-03	25	B304	2 0 2 2 2	
3.600E-05	7.772E-03	25	B317	1 0 0 0 2	
5.650E-05	1.220E-02	25	M342	1 0 1 1 2	
5.650E-05	1.220E-02	ns	M308	0 0 1 1 2	

540. $C_6H_2Cl_4$
1,2,3,5-Tetrachlorobenzene
1,2,4,6-Tetrachlorobenzene
RN: 634-90-2   **MP** (°C):   50
MW: 215.89   **BP** (°C):   246

Solubility (Moles/L)	Solubility (Grams/L)	Temp (°C)	Ref (#)	Evaluation (T P E A A)	Comments
1.000E-05	2.159E-03	20	K337	1 0 0 0 2	
1.148E-05	2.479E-03	22	K305	1 0 1 1 2	
1.496E-05	3.230E-03	23	C305	1 1 2 2 2	
1.860E-05	4.016E-03	25	B173	2 0 2 2 2	
2.362E-05	5.100E-03	25	B304	2 0 2 2 2	
1.660E-05	3.584E-03	25	B317	1 0 0 0 2	
1.340E-05	2.893E-03	25	M342	1 0 1 1 2	
1.654E-05	3.570E-03	ns	H123	0 0 0 0 2	
1.340E-05	2.893E-03	ns	M308	0 0 1 1 2	

541.   C$_6$H$_2$Cl$_4$
Trichlorobenzyl Chloride
TCBC
RN: 1344-32-7   **MP** (°C):
MW: 215.89   **BP** (°C):   93

Solubility (Moles/L)	Solubility (Grams/L)	Temp (°C)	Ref (#)	Evaluation (T P E A A)	Comments
9.264E-06	2.000E-03	25	B200	1 0 0 0 0	

542.   C$_6$H$_2$Cl$_4$
1,2,4,5-Tetrachlorobenzene
s-Tetrachlorobenzene
RN: 95-94-3   **MP** (°C):   139
MW: 215.89   **BP** (°C):   243

Solubility (Moles/L)	Solubility (Grams/L)	Temp (°C)	Ref (#)	Evaluation (T P E A A)	Comments
1.445E-06	3.121E-04	20	K337	1 0 0 0 2	
1.349E-06	2.912E-04	22	K305	1 0 1 1 1	
2.154E-06	4.650E-04	25	B304	2 0 2 2 2	
5.900E-06	1.274E-03	25	B317	1 0 0 0 2	
1.090E-05	2.353E-03	25	M342	1 0 1 1 2	
1.806E-06	3.900E-04	ns	B393	0 0 2 1 0	
1.090E-05	2.353E-03	ns	M308	0 0 1 1 2	

543.   C$_6$H$_2$Cl$_4$O
2,3,4,6-Tetrachlorophenol
Phenol, 2,3,4,6-Tetrachloro-
1-Hydroxy-2,3,4,6-tetrachlorobenzene
TCP
RN: 58-90-2   **MP** (°C):
MW: 231.89   **BP** (°C):

Solubility (Moles/L)	Solubility (Grams/L)	Temp (°C)	Ref (#)	Evaluation (T P E A A)	Comments
7.900E-04	1.832E-01	25	B316	1 0 2 1 1	

544.   C$_6$H$_2$Cl$_4$O
2,3,4,5-Tetrachlorophenol
Phenol, 2,3,4,5-Tetrachloro-
RN: 4901-51-3   **MP** (°C):   116
MW: 231.89   **BP** (°C):

Solubility (Moles/L)	Solubility (Grams/L)	Temp (°C)	Ref (#)	Evaluation (T P E A A)	Comments
7.158E-04	1.660E-01	25	M373	1 0 2 1 2	

545.   C$_6$H$_2$Cl$_4$O
2,3,5,6-Tetrachlorophenol
Phenol, 2,3,5,6-Tetrachloro-
RN: 935-95-5   **MP** (°C):   115
MW: 231.89   **BP** (°C):

Solubility (Moles/L)	Solubility (Grams/L)	Temp (°C)	Ref (#)	Evaluation (T P E A A)	Comments
4.312E-04	1.000E-01	25	M373	1 0 2 1 2	

546.   C$_6$H$_2$Cl$_4$O$_2$
Tetrachlorohydroquinone
2,3,5,6-Tetrachlorohydroquinone
RN: 87-87-6   **MP** (°C):
MW: 247.89   **BP** (°C):

Solubility (Moles/L)	Solubility (Grams/L)	Temp (°C)	Ref (#)	Evaluation (T P E A A)	Comments
8.673E-05	2.150E-02	ns	L311	0 0 0 0 1	

547.   C$_6$H$_2$F$_4$
1,2,3,5-Tetrafluorobenzene
1,2,4,6-Tetrafluorobenzene
m-Tetrafluorobenzene
1,3,4,5-Tetrafluorobenzene
RN: 2367-82-0   **MP** (°C):   -48
MW: 150.08   **BP** (°C):   83

Solubility (Moles/L)	Solubility (Grams/L)	Temp (°C)	Ref (#)	Evaluation (T P E A A)	Comments
4.952E-03	7.431E-01	25	B349	2 0 2 0 2	

548.   C$_6$H$_2$F$_4$
1,2,4,5-Tetrafluorobenzene
2,3,5,6-Tetrafluorobenzene
p-Tetrafluorobenzene
RN: 327-54-8   **MP** (°C):   4.5
MW: 150.08   **BP** (°C):   89.5

Solubility (Moles/L)	Solubility (Grams/L)	Temp (°C)	Ref (#)	Evaluation (T P E A A)	Comments
4.215E-03	6.326E-01	25	B349	2 0 2 0 2	

549. $C_6H_2F_4O$
2,3,5,6-Tetrafluorophenol
1,2,4,5-Tetrafluoro-3-hydroxybenzene
RN: 769-39-1   **MP** (°C):   38
MW: 166.08   **BP** (°C):   140

Solubility (Moles/L)	Solubility (Grams/L)	Temp (°C)	Ref (#)	Evaluation (T P E A A)	Comments
3.700E-01	6.145E+01	25	P031	1 1 2 2 2	

550. $C_6H_3Br_2NO_2$
2,6-Dibromoquinone Oxime
RN: **MP** (°C):
MW: 280.91   **BP** (°C):

Solubility (Moles/L)	Solubility (Grams/L)	Temp (°C)	Ref (#)	Evaluation (T P E A A)	Comments
8.500E-04	2.388E-01	20	G066	1 0 0 0 1	

551. $C_6H_3Br_3O$
2,4,6-Tribromobiphenyl
1,1'-Biphenyl, 2,4,6-Tribromo-
RN: 59080-33-0   **MP** (°C):   66
MW: 330.82   **BP** (°C):

Solubility (Moles/L)	Solubility (Grams/L)	Temp (°C)	Ref (#)	Evaluation (T P E A A)	Comments
4.111E-02	1.360E+01	26.5	G312	2 0 0 1 2	

552. $C_6H_3Br_3O$
2,4,6-Tribromophenol
2,4,6-Tribrom-phenol
Tribromophenol
Bromol
RN: 118-79-6   **MP** (°C):   95
MW: 330.82   **BP** (°C):   244

Solubility (Moles/L)	Solubility (Grams/L)	Temp (°C)	Ref (#)	Evaluation (T P E A A)	Comments
2.116E-04	7.000E-02	15	F300	1 0 0 0 1	
2.300E-04	7.609E-02	ns	O310	0 0 0 0 1	

553.   C₆H₃ClN₂O₄

$C_6H_3ClN_2O_4$

1-Chloro-2,4-dinitrobenzene
2,4-Dinitro-1-chlorobenzene
4-Chlor-1,3-dinitrobenzol
4-Chloro-1,3-dinitrobenzene

RN: 97-00-7   **MP** (°C):   53
MW: 202.55   **BP** (°C):   315

Solubility (Moles/L)	Solubility (Grams/L)	Temp (°C)	Ref (#)	Evaluation (T P E A A)	Comments
3.950E-05	8.000E-03	15	D071	1 2 0 0 0	
3.950E-05	8.000E-03	15	F300	1 0 0 0 0	
4.560E-05	9.236E-03	25	G090	2 2 1 1 1	
2.023E-03	4.098E-01	50	D071	1 2 0 0 1	
7.837E-03	1.587E+00	100	D071	1 2 0 0 2	
8.393E-03	1.700E+00	100	F300	1 0 0 0 1	

554.   C₆H₃ClN₄

$C_6H_3ClN_4$

7-Chloropteridine
Pteridine, 7-Chloro-

RN: 1125-84-4   **MP** (°C):   95
MW: 166.57   **BP** (°C):

Solubility (Moles/L)	Solubility (Grams/L)	Temp (°C)	Ref (#)	Evaluation (T P E A A)	Comments
1.305E-01	2.174E+01	20	A083	1 2 0 0 0	

555.   C₆H₃Cl₂NO₂

$C_6H_3Cl_2NO_2$

3,6-Dichloropicolinic Acid
3,6-Dichloro-2-pyridinecarboxylic Acid
Clopyralid
Lontrel
Stinger

RN: 1702-17-6   **MP** (°C):   151.5
MW: 192.00   **BP** (°C):

Solubility (Moles/L)	Solubility (Grams/L)	Temp (°C)	Ref (#)	Evaluation (T P E A A)	Comments
5.208E-03	1.000E+00	20	M161	1 0 0 0 0	
5.208E-03	1.000E+00	ns	K138	0 0 0 0 1	

556. $C_6H_3Cl_2NO_2$
3,4-Dichloronitrobenzene
1,2-Dichloro-4-nitrobenzene
RN: 99-54-7   **MP** (°C):   41.25
MW: 192.00   **BP** (°C):   255.5

Solubility (Moles/L)	Solubility (Grams/L)	Temp (°C)	Ref (#)	Evaluation (T P E A A)	Comments
6.290E-04	1.208E-01	20	E308	1 2 2 1 2	

557. $C_6H_3Cl_2NO_2$
2,5-Dichloronitrobenzene
1,4-Dichloro-2-nitrobenzene
RN: 89-61-2   **MP** (°C):   55.5
MW: 192.00   **BP** (°C):   267.5

Solubility (Moles/L)	Solubility (Grams/L)	Temp (°C)	Ref (#)	Evaluation (T P E A A)	Comments
4.800E-04	9.216E-02	20	E308	1 2 2 1 2	

558. $C_6H_3Cl_2NO_2$
2,3-Dichloronitrobenzene
1,2-Dichloro-3-nitrobenzene
RN: 3209-22-1   **MP** (°C):   61.5
MW: 192.00   **BP** (°C):   257.5

Solubility (Moles/L)	Solubility (Grams/L)	Temp (°C)	Ref (#)	Evaluation (T P E A A)	Comments
3.250E-04	6.240E-02	20	E308	1 2 2 1 2	

559. $C_6H_3Cl_3$
1,2,3-Trichlorobenzene
Benzene, 1,2,3-Trichloro-
vic-Trichlorobenzene
RN: 87-61-6   **MP** (°C):   51
MW: 181.45   **BP** (°C):   219

Solubility (Moles/L)	Solubility (Grams/L)	Temp (°C)	Ref (#)	Evaluation (T P E A A)	Comments
7.762E-05	1.408E-02	20	K337	1 0 0 0 2	
6.607E-05	1.199E-02	22	K305	1 0 1 1 2	
8.983E-05	1.630E-02	23	C305	1 1 2 2 2	
9.920E-05	1.800E-02	25	B304	2 0 2 2 2	
1.170E-04	2.123E-02	25	B317	1 0 0 0 2	
9.920E-05	1.800E-02	25	C313	1 0 2 2 2	
6.760E-05	1.227E-02	25	M342	1 0 1 1 2	
9.149E-05	1.660E-02	ns	H123	0 0 0 0 2	
6.760E-05	1.227E-02	ns	M308	0 0 1 1 2	

560. $C_6H_3Cl_3$
1,2,4-Trichlorobenzene
Benzene, 1,2,4-Trichloro-
RN: 120-82-1   **MP** (°C):   17
MW: 181.45   **BP** (°C):   213

Solubility (Moles/L)	Solubility (Grams/L)	Temp (°C)	Ref (#)	Evaluation (T P E A A)	Comments
1.653E-04	3.000E-02	19	M172	1 0 0 0 0	
1.950E-04	3.538E-02	20	K337	1 0 0 0 2	
1.072E-04	1.944E-02	22	K305	1 0 1 1 2	
1.725E-04	3.130E-02	25	B304	2 0 2 2 2	
2.200E-04	3.992E-02	25	B317	1 0 0 0 2	
2.692E-04	4.884E-02	25	C113	1 0 2 2 2	
2.540E-04	4.609E-02	25	M342	1 0 1 1 2	
3.555E-04	6.451E-02	30	M300	1 1 2 2 2	
3.555E-04	6.450E-02	30	M311	1 1 2 2 2	
2.540E-04	4.609E-02	ns	M308	0 0 1 1 2	

561. $C_6H_3Cl_3$
1,3,5-Trichlorobenzene
Benzene, 1,3,5-Trichloro-
RN: 108-70-3   **MP** (°C):   64
MW: 181.45   **BP** (°C):   208

Solubility (Moles/L)	Solubility (Grams/L)	Temp (°C)	Ref (#)	Evaluation (T P E A A)	Comments
2.399E-05	4.353E-03	20	K337	1 0 0 0 2	
3.236E-05	5.872E-03	22	K305	1 0 1 1 2	
5.842E-05	1.060E-02	23	C305	1 1 2 2 2	
3.312E-05	6.010E-03	25	B304	2 0 2 2 2	
2.900E-05	5.262E-03	25	B317	1 0 0 0 2	
2.270E-05	4.119E-03	25	M342	1 0 1 1 2	
2.270E-05	4.119E-03	ns	M308	0 0 1 1 2	

562. $C_6H_3Cl_3N_2O_2$
Picloram
4-Amino-3,5,6-trichloropicolinic Acid
RN: 1918-02-1   **MP** (°C):   241
MW: 241.46   **BP** (°C):

Solubility (Moles/L)	Solubility (Grams/L)	Temp (°C)	Ref (#)	Evaluation (T P E A A)	Comments
1.967E-03	4.750E-01	10	C031	2 0 2 2 2	pH 2.8
2.260E-03	5.457E-01	20	C031	2 0 2 2 2	pH 2.8
1.781E-03	4.300E-01	25	B185	1 0 0 0 2	
1.781E-03	4.300E-01	25	B200	1 0 0 0 1	
1.781E-03	4.300E-01	25	M161	1 0 0 0 2	

Solubility (Moles/L)	Solubility (Grams/L)	Temp (°C)	Ref (#)	Evaluation (T P E A A)	Comments
2.830E-03	6.833E-01	30	C031	2 0 2 2 2	pH 2.8
3.290E-03	7.944E-01	40	C031	2 0 2 2 2	pH 2.8
1.781E-03	4.300E-01	ns	K138	0 0 0 0 1	
1.780E-03	4.298E-01	ns	M061	0 0 0 0 1	
3.500E-04	8.451E-02	ns	O025	2 2 2 2 1	intrinsic

563. $C_6H_3Cl_3O$
2,4,6-Trichlorophenol
2,4,6-Trichlorphenol
Dowicide 25
RN: 88-06-2   **MP** (°C):   69
MW: 197.45   **BP** (°C):   246

Solubility (Moles/L)	Solubility (Grams/L)	Temp (°C)	Ref (#)	Evaluation (T P E A A)	Comments
2.532E-03	5.000E-01	11.20	F300	1 0 0 0 0	
4.558E-03	9.000E-01	22.5	G301	2 1 0 1 2	
2.200E-03	4.344E-01	25	B316	1 0 2 1 1	
3.586E-03	7.080E-01	25	M373	1 0 2 1 2	
4.554E-03	8.992E-01	25	R041	1 0 2 1 1	
4.558E-03	9.000E-01	25.40	F300	1 0 0 0 0	
1.266E-02	2.500E+00	96	F300	1 0 0 0 1	
<5.06E-03	<9.99E-01	ns	N034	0 0 0 0 0	

564. $C_6H_3Cl_3O$
2,3,4-Trichlorophenol
2,3,4-Trichlorphenol
RN: 15950-66-0   **MP** (°C):   80
MW: 197.45   **BP** (°C):

Solubility (Moles/L)	Solubility (Grams/L)	Temp (°C)	Ref (#)	Evaluation (T P E A A)	Comments
4.634E-03	9.150E-01	25	M373	1 0 2 1 2	

565. $C_6H_3Cl_3O$
2,3,5-Trichlorophenol
2,3,5-Trichlorphenol
RN: 933-78-8   **MP** (°C):   62
MW: 197.45   **BP** (°C):

Solubility (Moles/L)	Solubility (Grams/L)	Temp (°C)	Ref (#)	Evaluation (T P E A A)	Comments
3.905E-03	7.710E-01	25	M373	1 0 2 1 2	

566.   C$_6$H$_3$Cl$_3$O
2,3,6-Trichlorophenol
2,3,6-Trichlorphenol
RN: 933-75-5   **MP** (°C):   58
MW: 197.45   **BP** (°C):

Solubility (Moles/L)	Solubility (Grams/L)	Temp (°C)	Ref (#)	Evaluation (T P E A A)	Comments
2.993E-03	5.910E-01	25	M373	1 0 2 1 2	

567.   C$_6$H$_3$Cl$_3$O
2,4,5-Trichloro-phenol
Phenol, 2,4,5-Trichloro-
Dowicide 2
Preventol I
2,4,5-Trichlorophenol
Collunosol
RN: 95-95-4   **MP** (°C):   69
MW: 197.45   **BP** (°C):

Solubility (Moles/L)	Solubility (Grams/L)	Temp (°C)	Ref (#)	Evaluation (T P E A A)	Comments
4.800E-03	9.478E-01	25	B316	1 0 2 1 1	
3.287E-03	6.490E-01	25	M373	1 0 2 1 2	

568.   C$_6$H$_3$Cl$_4$N
Nitrapyrin
2-Chloro-6-(trichloromethyl)pyridine
Donco-163
N-Serve(R)
RN: 1929-82-4   **MP** (°C):   62.5
MW: 230.91   **BP** (°C):

Solubility (Moles/L)	Solubility (Grams/L)	Temp (°C)	Ref (#)	Evaluation (T P E A A)	Comments
1.738E-04	4.013E-02	20	B179	2 0 0 0 2	
1.732E-04	4.000E-02	20	G079	1 1 0 0 2	

569.   C$_6$H$_3$FN$_2$O$_4$
1-Fluoro-2,4-dinitrobenzene
FDNB
RN: 70-34-8   **MP** (°C):   26
MW: 186.10   **BP** (°C):

Solubility (Moles/L)	Solubility (Grams/L)	Temp (°C)	Ref (#)	Evaluation (T P E A A)	Comments
2.149E-03	4.000E-01	ns	B160	0 0 0 0 2	

570. $C_6H_3F_3O$
Trifluorophenol
2,3,4-Trifluorophenol
RN: 2822-41-5   **MP** (°C):
MW: 148.09   **BP** (°C):

Solubility (Moles/L)	Solubility (Grams/L)	Temp (°C)	Ref (#)	Evaluation (T P E A A)	Comments
4.200E-01	6.220E+01	25	P031	1 1 2 2 2	

571. $C_6H_3N_3O_6$
sym-Trinitrobenzene
1,3,5-Trinitro-benzol
1,3,5-Trinitrobenzene
RN: 99-35-4   **MP** (°C):   122.5
MW: 213.11   **BP** (°C):

Solubility (Moles/L)	Solubility (Grams/L)	Temp (°C)	Ref (#)	Evaluation (T P E A A)	Comments
1.305E-03	2.780E-01	15	D066	1 2 0 0 2	
1.304E-03	2.779E-01	15	D070	1 2 0 0 2	
1.314E-03	2.800E-01	15	F300	1 0 0 0 1	
4.786E-03	1.020E+00	50	D066	1 2 0 0 2	
4.781E-03	1.019E+00	50	D070	1 2 0 0 2	
2.337E-02	4.980E+00	100	D066	1 2 0 0 2	
2.325E-02	4.955E+00	100	D070	1 2 0 0 2	
2.393E-02	5.100E+00	100	F300	1 0 0 0 1	

572. $C_6H_3N_3O_7$
Picric Acid
2,4,6-Trinitrophenol
Picronitric Acid
Pikrinsaeure
RN: 88-89-1   **MP** (°C):   122.5
MW: 229.11   **BP** (°C):

Solubility (Moles/L)	Solubility (Grams/L)	Temp (°C)	Ref (#)	Evaluation (T P E A A)	Comments
2.948E-02	6.754E+00	0	D077	1 0 0 1 1	
4.322E-02	9.901E+00	0	M043	1 0 0 0 1	
4.364E-02	9.999E+00	7.10	E032	1 2 1 2 2	
4.232E-02	9.695E+00	9	D080	1 2 0 0 2	unit assumed
3.507E-02	8.035E+00	10	D077	1 0 0 1 1	
4.749E-02	1.088E+01	10	M043	1 0 0 0 1	
4.407E-02	1.010E+01	18.90	E032	1 2 1 2 2	
4.792E-02	1.098E+01	20	D077	1 0 0 1 2	
5.151E-02	1.180E+01	20	H048	1 0 0 0 2	unit assumed
4.300E-02	9.852E+00	20	K310	1 0 0 1 1	
5.176E-02	1.186E+01	20	M043	1 0 0 0 1	
4.932E-02	1.130E+01	23.50	F300	0 0 0 0 2	

5.327E-02	1.220E+01	25	D058	1 0 1 1 2	
5.520E-02	1.265E+01	25	F030	1 0 2 1 2	
5.684E-02	1.302E+01	25	H048	1 0 0 0 2	unit assumed
5.780E-02	1.324E+01	25	K040	1 0 2 1 2	
5.474E-02	1.254E+01	25	M094	1 0 0 1 2	
6.026E-02	1.381E+01	30	D077	1 0 0 1 2	
6.450E-02	1.478E+01	30	M043	1 0 0 0 1	
7.465E-02	1.710E+01	33.30	E032	1 2 1 2 2	
7.633E-02	1.749E+01	40	D077	1 0 0 1 2	
8.138E-02	1.865E+01	40	M043	1 0 0 0 1	
9.396E-02	2.153E+01	44.30	E032	1 2 1 2 2	
9.354E-02	2.143E+01	50	D077	1 0 0 1 2	
9.930E-02	2.275E+01	50	D080	1 2 0 0 2	unit assumed
1.193E-01	2.733E+01	60	D077	1 0 0 1 2	
1.312E-01	3.007E+01	60	M043	1 0 0 0 1	
1.398E-01	3.204E+01	62.90	E032	1 2 1 2 2	
1.464E-01	3.354E+01	70	D077	1 0 0 1 2	
1.703E-01	3.902E+01	72.60	E032	1 2 1 2 2	
1.844E-01	4.224E+01	80	D077	1 0 0 1 2	
1.920E-01	4.398E+01	80	M043	1 0 0 0 1	
1.956E-01	4.481E+01	82	D080	1 2 0 0 2	unit assumed
2.007E-01	4.598E+01	83.90	E032	1 2 1 2 2	
2.362E-01	5.411E+01	90	D077	1 0 0 1 2	
2.160E-01	4.949E+01	90	K310	1 0 0 1 2	
2.244E-01	5.141E+01	90.10	E032	1 2 1 2 2	
2.326E-01	5.330E+01	92.40	E032	1 2 1 2 2	
2.517E-01	5.767E+01	94.80	E032	1 2 1 2 2	
2.947E-01	6.751E+01	100	D077	1 0 0 1 2	
3.083E-01	7.063E+01	100	D080	1 2 0 0 2	unit assumed
3.055E-01	7.000E+01	100	F300	1 0 0 0 1	
2.932E-01	6.716E+01	100	M043	1 0 0 0 1	

573. $C_6H_3N_3O_8$
Styphnic Acid
Styphninsaeure
RN: 82-71-3   **MP** (°C):   176
MW: 245.11   **BP** (°C):

Solubility (Moles/L)	Solubility (Grams/L)	Temp (°C)	Ref (#)	Evaluation (T P E A A)	Comments
2.393E-02	5.865E+00	6.10	E032	1 2 1 2 2	
2.167E-02	5.312E+00	16.60	E032	1 2 1 2 2	
2.203E-02	5.400E+00	25	F300	1 0 0 0 1	
2.179E-02	5.341E+00	25	K040	1 0 2 1 2	
2.997E-02	7.346E+00	35.70	E032	1 2 1 2 2	
3.471E-02	8.507E+00	47.10	E032	1 2 1 2 2	
4.119E-02	1.010E+01	56.90	E032	1 2 1 2 2	
4.692E-02	1.150E+01	62	F300	1 0 0 0 2	

4.758E-02	1.166E+01	63.00	E032	1 2 1 2 2
6.109E-02	1.497E+01	71.20	E032	1 2 1 2 2
7.135E-02	1.749E+01	76.20	E032	1 2 1 2 2
8.000E-02	1.961E+01	80.30	E032	1 2 1 2 2
9.562E-02	2.344E+01	85.00	E032	1 2 1 2 2
1.096E-01	2.686E+01	89.80	E032	1 2 1 2 2
1.357E-01	3.326E+01	95.90	E032	1 2 1 2 2

574. C$_6$H$_4$BrF
1-Bromo-3-fluorobenzene
3-Bromofluorobenzene
RN: 1073-06-9 **MP** (°C):
MW: 175.01 **BP** (°C): 150

Solubility (Moles/L)	Solubility (Grams/L)	Temp (°C)	Ref (#)	Evaluation (T P E A A)	Comments
2.162E-03	3.784E-01	25	B349	2 0 2 0 2	

575. C$_6$H$_4$BrF
1-Bromo-2-fluorobenzene
2-Bromofluorobenzene
RN: 1072-85-1 **MP** (°C):
MW: 175.01 **BP** (°C): 151.5

Solubility (Moles/L)	Solubility (Grams/L)	Temp (°C)	Ref (#)	Evaluation (T P E A A)	Comments
2.018E-03	3.532E-01	25	B349	2 0 2 0 2	

576. C$_6$H$_4$BrNO$_3$
2-Bromo-4-nitrophenol
2-Brom-4-nitro-phenol
RN: 5847-59-6 **MP** (°C): 114
MW: 218.01 **BP** (°C):

Solubility (Moles/L)	Solubility (Grams/L)	Temp (°C)	Ref (#)	Evaluation (T P E A A)	Comments
1.009E-01	2.200E+01	100	F300	1 0 0 0 1	

577. C$_6$H$_4$Br$_2$
p-Dibromobenzene
1,4-Dibromobenzene
RN: 106-37-6 **MP** (°C): 87.3
MW: 235.92 **BP** (°C): 220.4

Solubility (Moles/L)	Solubility (Grams/L)	Temp (°C)	Ref (#)	Evaluation (T P E A A)	Comments
8.478E-05	2.000E-02	25	A003	1 0 1 2 1	
5.900E-03	1.392E+00	25	C316	1 0 2 2 2	0.1M NaCl
1.120E-04	2.642E-02	35	H077	2 2 2 2 2	

578. $C_6H_4Br_2$
m-Dibromobenzene
1,3-Dibromobenzene
RN: 108-36-1   **MP** (°C):   -7
MW: 235.92   **BP** (°C):   218

Solubility (Moles/L)	Solubility (Grams/L)	Temp (°C)	Ref (#)	Evaluation (T P E A A)	Comments
2.860E-04	6.747E-02	35	H077	2 2 2 2 2	

579. C_6H_4ClF
1-Chloro-3-fluorobenzene
3-Chlorofluorobenzene
RN: 625-98-9   **MP** (°C):
MW: 130.55   **BP** (°C):   127.6

Solubility (Moles/L)	Solubility (Grams/L)	Temp (°C)	Ref (#)	Evaluation (T P E A A)	Comments
4.517E-03	5.897E-01	25	B349	2 0 2 0 2	

580. C_6H_4ClF
1-Chloro-2-fluorobenzene
2-Chlorofluorobenzene
RN: 348-51-6   **MP** (°C):   -43
MW: 130.55   **BP** (°C):   137.6

Solubility (Moles/L)	Solubility (Grams/L)	Temp (°C)	Ref (#)	Evaluation (T P E A A)	Comments
3.845E-03	5.019E-01	25	B349	2 0 2 0 2	

581. $C_6H_4ClIO_2S$
Pipsyl Chloride
p-Iodobenzenesulfonyl Chloride
RN: 98-61-3   **MP** (°C):   81
MW: 302.52   **BP** (°C):

Solubility (Moles/L)	Solubility (Grams/L)	Temp (°C)	Ref (#)	Evaluation (T P E A A)	Comments
5.388E-05	1.630E-02	25	B048	1 0 2 2 2	
8.793E-05	2.660E-02	35	B048	1 0 2 2 2	
1.646E-04	4.980E-02	50	B048	1 0 2 2 2	

582. $C_6H_4ClNO_2$
6-Chloropicolinic Acid
Pyridinecarboxylic Acid, 6-Chloro-
RN: 4684-94-0   **MP** (°C):
MW: 157.56   **BP** (°C):

Solubility (Moles/L)	Solubility (Grams/L)	Temp (°C)	Ref (#)	Evaluation (T P E A A)	Comments
2.158E-02	3.400E+00	ns	K138	0 0 0 0 1	

583. $C_6H_4ClNO_2$
m-Chloronitrobenzene
1-Chloro-3-nitrobenzene
3-Chloronitrobenzene
m-Nitrochlorobenzene
RN: 121-73-3   **MP** (°C):   46.0
MW: 157.56   **BP** (°C):   236.0

Solubility (Moles/L)	Solubility (Grams/L)	Temp (°C)	Ref (#)	Evaluation (T P E A A)	Comments
1.732E-03	2.729E-01	20	E308	1 2 2 1 2	

584. $C_6H_4ClNO_2$
p-Chloronitrobenzene
4-Nitrochlorobenzene
4-CNB
RN: 100-00-5   **MP** (°C):   82
MW: 157.56   **BP** (°C):   242

Solubility (Moles/L)	Solubility (Grams/L)	Temp (°C)	Ref (#)	Evaluation (T P E A A)	Comments
1.777E-04	2.800E-02	17	D071	1 2 0 0 1	
1.777E-04	2.800E-02	17	F300	1 0 0 0 1	
2.877E-03	4.533E-01	20	E308	1 2 2 1 2	
1.429E-03	2.251E-01	20	H118	1 1 1 1 2	
1.429E-03	2.251E-01	20	H301	2 0 2 2 2	
<1.27E-03	<2.00E-01	25	B019	1 0 1 2 0	
1.600E-03	2.521E-01	25	G090	2 2 1 1 1	
7.933E-04	1.250E-01	50	D071	1 2 0 0 2	
9.709E-04	1.530E-01	100	D071	1 2 0 0 2	
1.016E-03	1.600E-01	100	F300	1 0 0 0 2	

585.   C₆H₄ClNO₂

o-Chloronitrobenzene

2-Nitrochlorobenzene

2-CNB

RN: 88-73-3   **MP** (°C):   32

MW: 157.56   **BP** (°C):   245

Solubility (Moles/L)	Solubility (Grams/L)	Temp (°C)	Ref (#)	Evaluation (T P E A A)	Comments
2.800E-03	4.412E-01	20	E308	1 2 2 1 2	
<1.27E-03	<2.00E-01	25	B019	1 0 1 2 0	
3.470E-03	5.467E-01	25	G090	2 2 1 1 1	

586.   C₆H₄Cl₂

1,2-Dichlorobenzene

o-Dichlorobenzene

RN: 95-50-1   **MP** (°C):   -17

MW: 147.00   **BP** (°C):   180

Solubility (Moles/L)	Solubility (Grams/L)	Temp (°C)	Ref (#)	Evaluation (T P E A A)	Comments
9.047E-04	1.330E-01	3.5	C094	1 0 0 0 2	
1.007E-03	1.480E-01	20	C094	1 0 0 0 2	
9.114E-04	1.340E-01	20	K056	1 0 2 2 2	
9.550E-04	1.404E-01	20	K337	1 0 0 0 2	
6.607E-04	9.713E-02	22	K305	1 0 1 1 2	
<1.36E-03	<2.00E-01	25	B019	1 0 1 2 0	
1.060E-03	1.558E-01	25	B173	2 0 2 2 2	
9.864E-04	1.450E-01	25	B185	1 0 0 0 2	
9.319E-04	1.370E-01	25	B304	2 0 2 2 2	
8.000E-04	1.176E-01	25	B317	1 0 0 0 2	
1.047E-03	1.539E-01	25	C113	1 0 2 2 2	
9.864E-04	1.450E-01	25	K056	1 0 2 2 2	
1.156E-03	1.700E-01	25	L319	1 0 2 1 1	
6.280E-04	9.232E-02	25	M342	1 0 1 1 2	
1.163E-03	1.710E-01	30	K056	1 0 2 2 2	
1.016E-03	1.494E-01	30	M300	1 1 2 2 2	
9.680E-04	1.423E-01	30	M311	1 1 2 2 2	
1.245E-03	1.830E-01	35	K056	1 0 2 2 2	
1.320E-03	1.940E-01	40	K056	1 0 2 2 2	
1.381E-03	2.030E-01	45	K056	1 0 2 2 2	
1.517E-03	2.230E-01	55	K056	1 0 2 2 2	
1.578E-03	2.320E-01	60	K056	1 0 2 2 2	
1.060E+03	1.558E+05	ns	A096	0 0 0 0 2	*sic*
6.280E-04	9.232E-02	ns	M308	0 0 1 1 2	

587.   C₆H₄Cl₂

1,4-Dichlorobenzene

p-Dichlorobenzene

RN: 106-46-7   **MP** (°C):   53.1

MW: 147.00   **BP** (°C):   173.4

Solubility (Moles/L)	Solubility (Grams/L)	Temp (°C)	Ref (#)	Evaluation (T P E A A)	Comments
4.680E-04	6.880E-02	20	K056	1 2 2 1 2	average of 4
3.020E-04	4.439E-02	20	K337	1 0 0 0 2	
2.252E-04	3.310E-02	20	T301	1 2 2 2 2	
3.311E-04	4.868E-02	22	K305	1 0 1 1 2	
5.292E-04	7.780E-02	22.20	W003	2 2 2 2 2	average of 2
5.673E-04	8.340E-02	24.60	W003	2 2 2 2 2	average of 3
5.170E-04	7.600E-02	25	A003	1 0 1 2 1	
5.928E-04	8.715E-02	25	A058	1 1 1 1 2	
<3.40E-03	<5.00E-01	25	B019	1 0 1 2 0	
5.020E-04	7.380E-02	25	B173	2 0 2 2 2	
4.442E-04	6.530E-02	25	B304	2 0 2 2 2	
5.270E-04	7.747E-02	25	B317	1 0 0 0 2	
3.990E-04	5.865E-02	25	C316	1 0 2 2 2	0.1M NaCl
5.374E-04	7.900E-02	25	F071	1 1 2 1 1	
5.374E-04	7.900E-02	25	H080	1 0 0 0 1	
5.381E-04	7.910E-02	25	K056	1 2 2 2 2	average of 2
5.646E-04	8.300E-02	25	M040	1 0 0 1 1	
5.442E-04	8.000E-02	25	M161	1 0 0 0 1	
2.100E-04	3.087E-02	25	M342	1 0 1 1 2	
6.932E-05	1.019E-02	25	N311	1 0 1 1 2	
5.898E-04	8.670E-02	25.50	W003	2 2 2 2 2	average of 2
5.238E-04	7.699E-02	30	G029	1 0 2 2 1	
6.347E-04	9.330E-02	30	K056	1 2 2 2 2	
6.267E-04	9.213E-02	30	M300	1 1 2 2 2	
6.422E-04	9.440E-02	30	M311	1 1 2 2 2	
6.299E-04	9.260E-02	30.00	W003	2 2 2 2 2	average of 2
6.939E-04	1.020E-01	34.50	W003	2 2 2 2 2	average of 3
5.646E-04	8.300E-02	35	K056	1 2 2 2 2	
8.231E-04	1.210E-01	38.40	W003	2 2 2 2 2	
6.857E-04	1.008E-01	40	K056	1 2 2 2 2	average of 2
8.292E-04	1.219E-01	45	K056	1 2 2 2 2	average of 2
1.082E-03	1.590E-01	47.50	W003	2 2 2 2 2	
1.184E-03	1.740E-01	50.10	W003	2 2 2 2 2	average of 2
1.061E-03	1.560E-01	55	K056	1 2 2 2 2	
1.429E-03	2.100E-01	59.20	W003	2 2 2 2 2	
1.109E-03	1.630E-01	60	K056	1 2 2 2 2	
1.483E-03	2.180E-01	60.70	W003	2 2 2 2 2	average of 2
1.565E-03	2.300E-01	65.10	W003	2 2 2 2 2	average of 3
1.612E-03	2.370E-01	65.20	W003	2 2 2 2 2	average of 3

1.912E-03	2.810E-01	73.40	W003	2 2 2 2 2
2.100E-04	3.087E-02	ns	M308	0 0 1 1 2
5.374E-04	7.900E-02	ns	M344	0 0 0 0 1
5.034E-04	7.400E-02	rt	S314	0 0 2 1 1

588.   C$_6$H$_4$Cl$_2$
1,3-Dichlorobenzene
m-Dichlorobenzene

RN: 541-73-1   **MP** (°C):   -24
MW: 147.00   **BP** (°C):   172-173

Solubility (Moles/L)	Solubility (Grams/L)	Temp (°C)	Ref (#)	Evaluation (T P E A A)	Comments
7.551E-04	1.110E-01	20	K056	1 0 2 2 2	
7.943E-04	1.168E-01	20	K337	1 0 0 0 2	
4.677E-04	6.876E-02	22	K305	1 0 1 1 2	
9.080E-04	1.335E-01	25	B173	2 0 2 2 2	
9.728E-04	1.430E-01	25	B304	2 1 2 1 2	
8.300E-04	1.220E-01	25	B317	1 0 0 0 2	
9.120E-04	1.341E-01	25	C113	1 0 2 2 2	
8.367E-04	1.230E-01	25	K056	1 0 2 2 2	
8.470E-04	1.245E-01	25	M342	1 0 1 1 2	
9.523E-04	1.400E-01	30	K056	1 0 2 2 2	
8.537E-04	1.255E-01	30	M300	1 1 2 2 2	
8.537E-04	1.255E-01	30	M311	1 1 2 2 2	
1.020E-03	1.500E-01	35	K056	1 0 2 2 2	
1.136E-03	1.670E-01	40	K056	1 0 2 2 2	
1.204E-03	1.770E-01	45	K056	1 0 2 2 2	
1.333E-03	1.960E-01	55	K056	1 0 2 2 2	
1.367E-03	2.010E-01	60	K056	1 0 2 2 2	
9.080E+02	1.335E+05	ns	A096	0 0 0 0 2	*sic*
8.470E-04	1.245E-01	ns	M308	0 0 1 1 2	

589.   C$_6$H$_4$Cl$_2$O
2,4-Dichlorophenol
2,4-Dichlor-phenol

RN: 120-83-2   **MP** (°C):   45
MW: 163.00   **BP** (°C):

Solubility (Moles/L)	Solubility (Grams/L)	Temp (°C)	Ref (#)	Evaluation (T P E A A)	Comments
2.748E-02	4.480E+00	19	D041	1 0 0 0 1	
~2.76E-02	~4.50E+00	20	F300	1 0 0 0 0	
2.748E-02	4.480E+00	20	N034	1 0 0 0 1	
3.403E-02	5.547E+00	25	M373	1 0 2 1 2	
3.052E-02	4.975E+00	25	R041	1 0 2 1 1	

590. C_6H_4Cl_2O

2,3-Dichlorophenol
Phenol, 2,3-Dichloro-
RN: 576-24-9 **MP** (°C): 59
MW: 163.00 **BP** (°C):

Solubility (Moles/L)	Solubility (Grams/L)	Temp (°C)	Ref (#)	Evaluation (T P E A A)	Comments
5.040E-02	8.215E+00	25	M373	1 0 2 1 2	

591. C_6H_4Cl_2O

3,5-Dichlorophenol
3,5-DCP
RN: 591-35-5 **MP** (°C): 68
MW: 163.00 **BP** (°C):

Solubility (Moles/L)	Solubility (Grams/L)	Temp (°C)	Ref (#)	Evaluation (T P E A A)	Comments
4.536E-02	7.394E+00	25	M373	1 0 2 1 2	

592. C_6H_4Cl_2O

3,4-Dichlorophenol
4,5-Dichlorophenol
3,4-DCP
RN: 95-77-2 **MP** (°C): 67
MW: 163.00 **BP** (°C):

Solubility (Moles/L)	Solubility (Grams/L)	Temp (°C)	Ref (#)	Evaluation (T P E A A)	Comments
5.678E-02	9.256E+00	25	M373	1 0 2 1 2	

593. C_6H_4Cl_2O

2,6-Dichlorophenol
2,6-DCP
RN: 87-65-0 **MP** (°C): 66.5
MW: 163.00 **BP** (°C):

Solubility (Moles/L)	Solubility (Grams/L)	Temp (°C)	Ref (#)	Evaluation (T P E A A)	Comments
1.610E-02	2.625E+00	25	M373	1 0 2 1 2	

594. C_6H_4Cl_2O

2,5-Dichlorophenol
2,5-Dichlor-phenol
RN: 583-78-8 **MP** (°C):
MW: 163.00 **BP** (°C):

Solubility (Moles/L)	Solubility (Grams/L)	Temp (°C)	Ref (#)	Evaluation (T P E A A)	Comments
3.800E-02	6.194E+00	25	B316	1 0 2 1 1	

595.   C$_6$H$_4$FI
1-Fluoro-4-iodobenzene
4-Fluoro-1-iodobenzene
p-Iodofluorobenzene
p-Fluoroiodobenzene
p-Fluorophenyl Iodide
RN: 352-34-1   **MP** (°C):   -27
MW: 222.00   **BP** (°C):   183

Solubility (Moles/L)	Solubility (Grams/L)	Temp (°C)	Ref (#)	Evaluation (T P E A A)	Comments
7.499E-04	1.665E-01	25	B349	2 0 2 0 2	

596.   C$_6$H$_4$I$_2$
1,4-Diiodobenzene
p-Diiodobenzene
4-Iodophenyl Iodide
RN: 624-38-4   **MP** (°C):   131
MW: 329.91   **BP** (°C):   285

Solubility (Moles/L)	Solubility (Grams/L)	Temp (°C)	Ref (#)	Evaluation (T P E A A)	Comments
4.244E-06	1.400E-03	25	A003	1 2 1 2 1	*sic*
3.100E-02	1.023E+01	25	C316	1 0 2 2 2	0.1M NaCl

597.   C$_6$H$_4$N$_2$O$_4$
o-Dinitrobenzene
1,2-Dinitrobenzene
RN: 528-29-0   **MP** (°C):   118
MW: 168.11   **BP** (°C):

Solubility (Moles/L)	Solubility (Grams/L)	Temp (°C)	Ref (#)	Evaluation (T P E A A)	Comments
8.328E-04	1.400E-01	20	F300	1 0 0 0 1	
7.910E-04	1.330E-01	25	I334	2 2 2 1 2	
7.418E-04	1.247E-01	25	L008	2 2 2 1 2	average of 3

598. $C_6H_4N_2O_4$
m-Dinitrobenzene
1,3-Dinitrobenzene

RN: 99-65-0   **MP** (°C):   89.5
MW: 168.11   **BP** (°C):   301.5

Solubility (Moles/L)	Solubility (Grams/L)	Temp (°C)	Ref (#)	Evaluation (T P E A A)	Comments
4.045E-04	6.800E-02	13	D070	1 2 0 0 1	
4.164E-04	7.000E-02	13	F300	1 0 0 0 0	
3.420E-03	5.749E-01	25	I334	2 2 2 1 2	
3.169E-03	5.328E-01	25	L008	2 2 2 1 2	average of 2
5.116E-03	8.600E-01	25.04	V013	2 2 2 2 2	
3.867E-03	6.500E-01	30	F300	1 0 0 0 1	
3.888E-03	6.536E-01	30	G029	1 0 2 2 2	
4.670E-03	7.851E-01	35	H077	2 2 2 2 2	
2.789E-03	4.688E-01	50	D070	1 2 0 0 2	
1.134E-02	1.906E+00	100	D070	1 2 0 0 2	
1.547E-02	2.600E+00	100	F300	1 0 0 0 1	
2.973E-03	4.998E-01	rt	D021	0 0 1 1 0	

599. $C_6H_4N_2O_4$
p-Dinitrobenzene
1,4-Dinitrobenzene

RN: 100-25-4   **MP** (°C):   173
MW: 168.11   **BP** (°C):

Solubility (Moles/L)	Solubility (Grams/L)	Temp (°C)	Ref (#)	Evaluation (T P E A A)	Comments
4.759E-04	8.000E-02	20	F300	1 0 0 0 0	
2.350E-04	3.951E-02	25	C316	1 0 2 2 2	0.1M NaCl
4.090E-04	6.876E-02	25	I334	2 2 2 1 2	
3.676E-04	6.180E-02	25	L008	2 2 2 1 2	average of 2
6.170E-04	1.037E-01	35	H077	2 2 2 2 2	
1.130E-02	1.900E+00	100	F300	1 0 0 0 1	

600. $C_6H_4N_2O_5$
3,5-Dinitrophenol
Phenol, θ-Dinitro-

RN: 586-11-8   **MP** (°C):
MW: 184.11   **BP** (°C):

Solubility (Moles/L)	Solubility (Grams/L)	Temp (°C)	Ref (#)	Evaluation (T P E A A)	Comments
7.288E-02	1.342E+01	51.6	S117	1 2 1 1 2	solid hydrate
2.373E+00	4.370E+02	54.1	S117	1 2 1 1 2	anhydrate
2.407E+00	4.431E+02	54.5	S117	1 2 1 1 2	anhydrate
2.442E+00	4.496E+02	55.5	S117	1 2 1 1 2	anhydrate
2.474E+00	4.555E+02	57.9	S117	1 2 1 1 2	anhydrate

2.516E+00	4.633E+02	61.9	S117	1 2 1 1 2	anhydrate
2.583E+00	4.756E+02	69.9	S117	1 2 1 1 2	anhydrate
2.617E+00	4.819E+02	81.3	S117	1 2 1 1 2	anhydrate
5.308E-01	9.772E+01	109.3	S117	1 0 1 1 2	
1.253E+00	2.307E+02	124.6	S117	1 0 1 1 2	

601.   C$_6$H$_4$N$_2$O$_5$
2,4-Dinitrophenol
α-Dinitrophenol
Aldifen
Fenoxyl Carbon N
RN: 51-28-5   **MP** (°C):   107.5
MW: 184.11   **BP** (°C):

Solubility (Moles/L)	Solubility (Grams/L)	Temp (°C)	Ref (#)	Evaluation (T P E A A)	Comments
1.097E-03	2.020E-01	12.5	D069	1 2 0 0 2	
1.086E-03	2.000E-01	12.50	F300	1 0 0 0 0	
1.629E-03	2.999E-01	15	D079	1 2 0 0 1	
3.025E-02	5.569E+00	18	D041	1 0 0 0 1	
2.800E-02	5.155E+00	20	K301	2 2 1 1 1	
2.524E-03	4.647E-01	25	H085	2 0 2 1 2	
1.467E-03	2.700E-01	25	P037	2 0 1 1 2	
4.356E-03	8.020E-01	50	D069	1 2 0 0 2	
9.504E-04	1.750E-01	50	D079	1 2 0 0 2	
7.431E-03	1.368E+00	54.50	E032	1 2 1 2 2	
1.192E-02	2.195E+00	67.60	E032	1 2 1 2 2	
1.630E-02	3.001E+00	75.80	E032	1 2 1 2 2	
3.414E-02	6.286E+00	85	D069	1 2 0 0 2	
3.170E-02	5.836E+00	87.40	E032	1 2 1 2 2	
4.845E-02	8.920E+00	92.40	E032	1 2 1 2 2	
6.547E-02	1.205E+01	96.20	E032	1 2 1 2 2	
7.163E-02	1.319E+01	100	D069	1 2 0 0 2	
8.964E-02	1.650E+01	100	D079	1 2 0 0 2	
7.061E-02	1.300E+01	100	F300	1 0 0 0 1	
2.444E-01	4.500E+01	h	F300	0 0 0 0 1	
2.702E-02	4.975E+00	ns	M061	0 0 0 0 0	

602. $C_6H_4N_2O_5$
2,6-Dinitrophenol
β-Dinitrophenol
RN: 573-56-8   **MP** (°C):
MW: 184.11   **BP** (°C):

Solubility (Moles/L)	Solubility (Grams/L)	Temp (°C)	Ref (#)	Evaluation (T P E A A)	Comments
1.710E-03	3.149E-01	15	D080	1 2 0 0 2	unit assumed
1.629E-03	3.000E-01	15	F300	1 0 0 0 0	
2.805E-02	5.164E+00	50	D080	1 2 0 0 2	unit assumed
6.547E-02	1.205E+01	100	D080	1 2 0 0 2	unit assumed
6.518E-02	1.200E+01	100	F300	1 0 0 0 1	

603. $C_6H_4N_2O_6$
2,4-Dinitroresorcinol
2,4-Dinitro-1,3-benzenediol
RN: 519-44-8   **MP** (°C):
MW: 200.11   **BP** (°C):

Solubility (Moles/L)	Solubility (Grams/L)	Temp (°C)	Ref (#)	Evaluation (T P E A A)	Comments
3.129E-02	6.261E+00	57.70	E032	1 2 1 2 2	
4.801E-02	9.607E+00	66.60	E032	1 2 1 2 2	
7.434E-02	1.488E+01	69.50	E032	1 2 1 2 2	
9.895E-02	1.980E+01	76.50	E032	1 2 1 2 2	
1.690E-01	3.382E+01	84.70	E032	1 2 1 2 2	
2.380E-01	4.762E+01	90.00	E032	1 2 1 2 2	
3.495E-01	6.994E+01	93.00	E032	1 2 1 2 2	

604. $C_6H_4N_2O_6$
4,6-Dinitroresorcinol
4,6-Dinitro-1,3-benzenediol
RN: 616-74-0   **MP** (°C):
MW: 200.11   **BP** (°C):

Solubility (Moles/L)	Solubility (Grams/L)	Temp (°C)	Ref (#)	Evaluation (T P E A A)	Comments
1.998E-03	3.998E-01	77.00	E032	1 2 1 2 2	
3.995E-03	7.994E-01	90.50	E032	1 2 1 2 2	
4.992E-03	9.990E-01	96.30	E032	1 2 1 2 2	

605.   C$_6$H$_4$N$_4$
Pteridine
1,3,5,8-Tetraazanaphthalene
Azinepurine
Pyrimido[4,5-b]pyrazine
Pyrazino[2,3-d]pyrimidine
RN: 91-18-9   **MP** (°C):   138
MW: 132.13   **BP** (°C):

Solubility (Moles/L)	Solubility (Grams/L)	Temp (°C)	Ref (#)	Evaluation (T P E A A)	Comments
9.461E-01	1.250E+02	20	A020	1 2 0 0 1	
9.461E-01	1.250E+02	20	B050	1 0 0 0 0	
9.230E-01	1.220E+02	22.5	A085	1 2 0 0 0	
3.784E+00	5.000E+02	100	B050	1 0 0 0 0	

606.   C$_6$H$_4$N$_4$O
7-Hydroxypteridine
7-Pteridinol
RN: 2432-27-1   **MP** (°C):
MW: 148.12   **BP** (°C):

Solubility (Moles/L)	Solubility (Grams/L)	Temp (°C)	Ref (#)	Evaluation (T P E A A)	Comments
7.493E-03	1.110E+00	20	B050	1 0 0 0 0	
8.768E-02	1.299E+01	100	B050	1 0 0 0 0	

607.   C$_6$H$_4$N$_4$O
6-Hydroxypteridine
6-Pteridinol
RN: 2432-26-0   **MP** (°C):
MW: 148.12   **BP** (°C):

Solubility (Moles/L)	Solubility (Grams/L)	Temp (°C)	Ref (#)	Evaluation (T P E A A)	Comments
1.928E-03	2.856E-01	20	A020	1 2 0 0 1	
1.928E-03	2.856E-01	20	B050	1 0 0 0 0	
2.923E-02	4.329E+00	100	B050	1 0 0 0 0	

608.   C₆H₄N₄O

$C_6H_4N_4O$

2-Hydroxypteridine
2-Pteridinol

RN: 25911-76-6   **MP** (°C):   240
MW: 148.12   **BP** (°C):

Solubility (Moles/L)	Solubility (Grams/L)	Temp (°C)	Ref (#)	Evaluation (T P E A A)	Comments
1.123E-02	1.664E+00	20	A020	1 2 0 0 1	
1.123E-02	1.664E+00	20	B050	1 0 0 0 0	
1.123E-02	1.664E+00	22.5	A085	1 2 0 0 0	
1.324E-01	1.961E+01	100	B050	1 0 0 0 0	

609.   C₆H₄N₄O

$C_6H_4N_4O$

4-Hydroxypteridine
4-Pteridinol

RN: 700-47-0   **MP** (°C):
MW: 148.12   **BP** (°C):

Solubility (Moles/L)	Solubility (Grams/L)	Temp (°C)	Ref (#)	Evaluation (T P E A A)	Comments
3.359E-02	4.975E+00	20	A020	1 2 0 0 1	
3.359E-02	4.975E+00	20	B050	1 0 0 0 0	
3.359E-02	4.975E+00	22.5	A085	1 2 0 0 0	
2.250E-01	3.333E+01	100	B050	1 0 0 0 0	

610.   C₆H₄N₄O₂

$C_6H_4N_4O_2$

2,4-Dihydroxypteridine
2:4-Dihydroxypteridine
Lumazine

RN: 487-21-8   **MP** (°C):   348.5
MW: 164.12   **BP** (°C):

Solubility (Moles/L)	Solubility (Grams/L)	Temp (°C)	Ref (#)	Evaluation (T P E A A)	Comments
7.607E-03	1.248E+00	20	B050	1 0 0 0 0	
7.607E-03	1.248E+00	22.5	A085	1 2 0 0 0	
5.035E-02	8.264E+00	100	B050	1 0 0 0 0	

611.   C₆H₄N₄O₂

$C_6H_4N_4O_2$

2,6-Dihydroxypteridine
2:6-Dihydroxypteridine

RN: 89324-38-9   **MP** (°C):
MW: 164.12   **BP** (°C):

Solubility (Moles/L)	Solubility (Grams/L)	Temp (°C)	Ref (#)	Evaluation (T P E A A)	Comments
1.354E-03	2.222E-01	100	A020	1 2 0 0 1	

612. $C_6H_4N_4O_2$
2,7-Dihydroxypteridine
2:7-Dihydroxypteridine
RN: 65882-62-4   **MP** (°C):
MW: 164.12   **BP** (°C):

Solubility (Moles/L)	Solubility (Grams/L)	Temp (°C)	Ref (#)	Evaluation (T P E A A)	Comments
6.033E-02	9.901E+00	100	A020	1 2 0 0 0	

613. $C_6H_4N_4O_2$
4,6-Dihydroxypteridine
4:6-Dihydroxypteridine
RN: 16310-36-4   **MP** (°C):
MW: 164.12   **BP** (°C):

Solubility (Moles/L)	Solubility (Grams/L)	Temp (°C)	Ref (#)	Evaluation (T P E A A)	Comments
1.108E-03	1.818E-01	20	A020	1 2 0 0 1	
1.218E-03	2.000E-01	20	B050	1 0 0 0 0	
2.024E-02	3.322E+00	100	B050	1 0 0 0 0	

614. $C_6H_4N_4O_2$
4,7-Dihydroxypteridine
4:7-Dihydroxypteridine
6,7-Dihydroxypteridine
6:7-Dihydroxypteridine
RN: 33669-70-4   **MP** (°C):
MW: 164.12   **BP** (°C):

Solubility (Moles/L)	Solubility (Grams/L)	Temp (°C)	Ref (#)	Evaluation (T P E A A)	Comments
2.030E-03	3.332E-01	20	A020	1 2 0 0 1	
1.523E-03	2.499E-01	20	A020	1 2 0 0 1	
2.030E-03	3.332E-01	20	B050	1 0 0 0 0	
1.523E-03	2.499E-01	20	B050	1 0 0 0 0	
2.094E-02	3.436E+00	100	B050	1 0 0 0 0	
1.014E-02	1.664E+00	100	B050	1 0 0 0 0	

615. $C_6H_4N_4O_3$
4,6,7-Trihydroxypteridine
4:6:7-Trihydroxypteridine
RN: 58947-88-9   **MP** (°C):
MW: 180.12   **BP** (°C):

Solubility (Moles/L)	Solubility (Grams/L)	Temp (°C)	Ref (#)	Evaluation (T P E A A)	Comments
2.056E-04	3.704E-02	20	A020	1 2 0 0 1	
2.056E-04	3.704E-02	20	B050	1 0 0 0 0	
7.930E-04	1.428E-01	100	B050	1 0 0 0 0	

616. $C_6H_4N_4O_3$
2,4,7-Trihydroxypteridine
2:4:7-Trihydroxypteridine
RN: 2577-38-0   **MP** (°C):
MW: 180.12   **BP** (°C):

Solubility (Moles/L)	Solubility (Grams/L)	Temp (°C)	Ref (#)	Evaluation (T P E A A)	Comments
4.626E-04	8.333E-02	20	A020	1 2 0 1 1	
4.626E-04	8.333E-02	20	B050	1 0 0 0 0	
3.963E-03	7.138E-01	100	A020	1 2 0 0 1	
3.963E-03	7.138E-01	100	B050	1 0 0 0 0	

617. $C_6H_4N_4O_4$
2,4,6,7-Tetrahydroxypteridine
2,4,6-Trihydroxypteridine
2:4:6-Trihydroxypteridine
RN: 2817-14-3   **MP** (°C):
MW: 196.12   **BP** (°C):

Solubility (Moles/L)	Solubility (Grams/L)	Temp (°C)	Ref (#)	Evaluation (T P E A A)	Comments
8.791E-05	1.724E-02	20	A020	1 2 0 1 1	
6.889E-04	1.351E-01	20	B050	1 0 0 0 0	
8.791E-05	1.724E-02	20	B050	1 0 0 0 0	
1.272E-02	2.494E+00	100	A020	1 2 0 0 0	
7.283E-04	1.428E-01	100	A020	1 2 0 0 0	
1.272E-02	2.494E+00	100	B050	1 0 0 0 0	

618.   C₆H₄N₄O₆

Picramine

2,4,6-Trinitroaniline

1-Amino-2,4,6-trinitrobenzene

MATB

RN: 489-98-5   **MP** (°C):   192

MW: 228.12   **BP** (°C):

Solubility (Moles/L)	Solubility (Grams/L)	Temp (°C)	Ref (#)	Evaluation (T P E A A)	Comments
8.710E-05	1.987E-02	25	B335	1 2 0 0 1	

619.   C₆H₄N₄S

7-Mercaptopteridine

7-Pteridinethiol

7(1H)-Pteridinethione

RN: 36653-71-1   **MP** (°C):

MW: 164.19   **BP** (°C):

Solubility (Moles/L)	Solubility (Grams/L)	Temp (°C)	Ref (#)	Evaluation (T P E A A)	Comments
1.964E-03	3.225E-01	20	A083	1 2 0 0 0	
6.760E-03	1.110E+00	100	A083	1 2 0 0 0	

620.   C₆H₄N₄S

4-Mercaptopteridine

4-Pteridinethiol

4(1H)-Pteridinethione

RN: 65882-61-3   **MP** (°C):   176dec

MW: 164.19   **BP** (°C):

Solubility (Moles/L)	Solubility (Grams/L)	Temp (°C)	Ref (#)	Evaluation (T P E A A)	Comments
1.691E-03	2.777E-01	22.5	A085	1 2 0 0 0	

621.   C₆H₄N₄S

2-Mercaptopteridine

2-Pteridinethiol

2(1H)-Pteridinethione

RN: 16878-76-5   **MP** (°C):   205

MW: 164.19   **BP** (°C):

Solubility (Moles/L)	Solubility (Grams/L)	Temp (°C)	Ref (#)	Evaluation (T P E A A)	Comments
4.347E-03	7.138E-01	22.5	A085	1 2 0 0 0	

622. $C_6H_4O_2$
Quinone
1,4-Benzoquinone
Benzochinhydrone
p-Quinone
RN: 106-51-4   **MP** (°C):   115.7
MW: 108.10   **BP** (°C):

Solubility (Moles/L)	Solubility (Grams/L)	Temp (°C)	Ref (#)	Evaluation (T P E A A)	Comments
8.630E-02	9.329E+00	11.85	L064	2 2 2 1 2	0.01N HCl
1.013E-01	1.095E+01	17.70	L065	1 0 0 0 2	0.01N HCl
1.021E-01	1.104E+01	17.90	L065	1 0 0 0 2	0.01N HCl
1.030E-01	1.113E+01	17.95	L065	1 0 0 0 2	0.01N HCl
1.030E-01	1.113E+01	18	L064	2 2 2 1 2	0.01N HCl
1.580E-02	1.708E+00	20	B113	1 2 2 1 2	
1.233E-01	1.333E+01	23.85	L064	2 2 2 1 2	0.01N HCl
1.295E-01	1.400E+01	24	F300	1 0 0 0 1	
1.266E-01	1.369E+01	25	G033	1 0 1 1 2	
1.397E-01	1.510E+01	25	K033	1 0 0 1 2	

623. $C_6H_4O_5$
2,5-Dicarboxyfuran
Furan-dicarbon-saeure-(2,5)
RN: 3238-40-2   **MP** (°C):
MW: 156.10   **BP** (°C):

Solubility (Moles/L)	Solubility (Grams/L)	Temp (°C)	Ref (#)	Evaluation (T P E A A)	Comments
6.406E-03	1.000E+00	18	F300	1 0 0 0 0	

624. $C_6H_4O_5$
2-Carboxy-5-hydroxy-4-pyrone
Komensaeure
Komenic Acid
RN: 499-78-5   **MP** (°C):
MW: 156.10   **BP** (°C):

Solubility (Moles/L)	Solubility (Grams/L)	Temp (°C)	Ref (#)	Evaluation (T P E A A)	Comments
3.267E-02	5.100E+00	25	F300	1 0 0 0 1	
3.921E-01	6.120E+01	100	F300	1 0 0 0 2	

625.   C₆H₅Br
Bromobenzene
Phenyl Bromide
Monobromobenzene
RN: 108-86-1   **MP** (°C):   -30
MW: 157.02   **BP** (°C):   156.2

Solubility (Moles/L)	Solubility (Grams/L)	Temp (°C)	Ref (#)	Evaluation (T P E A A)	Comments
2.611E-03	4.100E-01	25	A003	1 2 1 2 1	
2.620E-03	4.114E-01	25	W300	2 2 2 2 2	
2.840E-03	4.460E-01	30	F071	1 1 2 1 2	
2.966E-03	4.658E-01	30	G029	1 0 2 2 2	
2.840E-03	4.460E-01	30	H080	1 0 0 0 2	
2.102E-03	3.300E-01	30	M311	1 1 2 2 2	
2.799E-03	4.395E-01	30	V009	1 0 0 0 1	
2.920E-03	4.585E-01	35	H077	2 2 2 2 2	
5.110E-04	8.024E-02	ns	D348	0 0 2 2 2	
2.615E-03	4.106E-01	ns	M344	0 0 0 0 2	

626.   C₆H₅BrO
p-Bromophenol
4-Bromophenol
RN: 106-41-2   **MP** (°C):   66
MW: 173.02   **BP** (°C):   236

Solubility (Moles/L)	Solubility (Grams/L)	Temp (°C)	Ref (#)	Evaluation (T P E A A)	Comments
8.053E-02	1.393E+01	20	R087	1 1 2 2 2	0.15M NaCl
8.542E-02	1.478E+01	25	R041	1 0 2 1 1	

627.   C₆H₅BrO₃S
p-Bromobenzenesulfonic Acid
4-Bromobenzenesulfonic Acid
RN: 138-36-3   **MP** (°C):
MW: 237.08   **BP** (°C):

Solubility (Moles/L)	Solubility (Grams/L)	Temp (°C)	Ref (#)	Evaluation (T P E A A)	Comments
2.079E+00	4.929E+02	82.3	T023	1 2 2 1 2	
2.088E+00	4.949E+02	89.6	T023	1 2 2 1 2	
2.093E+00	4.961E+02	93.1	T023	1 2 2 1 2	
2.097E+00	4.972E+02	97.6	T023	1 2 2 1 2	

Solutions

628. C$_6$H$_5$BrO$_3$S.H$_2$O
p-Bromobenzenesulfonic Acid (Monohydrate)
RN: 138-36-3   MP (°C):
MW: 255.09   BP (°C):

Solubility (Moles/L)	Solubility (Grams/L)	Temp (°C)	Ref (#)	Evaluation (T P E A A)	Comments
1.799E+00	4.588E+02	43.8	T023	1 2 2 1 2	
1.821E+00	4.644E+02	60.2	T023	1 2 2 1 2	
1.586E+00	4.045E+02	71.2	T023	1 2 2 1 2	
1.924E+00	4.909E+02	76.6	T023	1 2 2 1 2	
1.922E+00	4.903E+02	78.5	T023	1 2 2 1 2	
1.855E+00	4.731E+02	80.3	T023	1 2 2 1 2	
1.868E+00	4.766E+02	86.2	T023	1 2 2 1 2	
1.907E+00	4.865E+02	87.2	T023	1 2 2 1 2	
1.889E+00	4.818E+02	90.2	T023	1 2 2 1 2	

629. C$_6$H$_5$BrO$_3$S.2.5H$_2$O
p-Bromobenzenesulfonic Acid (2.5 Hydrate)
RN: 138-36-3   MP (°C):
MW: 282.12   BP (°C):

Solubility (Moles/L)	Solubility (Grams/L)	Temp (°C)	Ref (#)	Evaluation (T P E A A)	Comments
1.375E+00	3.880E+02	-21.0	T023	1 2 2 1 2	
1.409E+00	3.975E+02	-10.5	T023	1 2 2 1 2	
1.447E+00	4.081E+02	0.0	T023	1 2 2 1 2	
1.495E+00	4.219E+02	12.5	T023	1 2 2 1 2	
1.522E+00	4.294E+02	19.9	T023	1 2 2 1 2	
1.566E+00	4.418E+02	27.6	T023	1 2 2 1 2	
1.613E+00	4.550E+02	34.6	T023	1 2 2 1 2	

630. C$_6$H$_5$Cl
Chlorobenzene
IP Carrier T 40
Phenyl Chloride
Tetrosin SP
Monochlorobenzene
MCB
RN: 108-90-7   MP (°C):   -45
MW: 112.56   BP (°C):   131

Solubility (Moles/L)	Solubility (Grams/L)	Temp (°C)	Ref (#)	Evaluation (T P E A A)	Comments
4.266E-03	4.802E-01	20	K337	1 0 0 0 2	
4.440E-03	4.998E-01	20	M312	1 0 0 0 2	
4.742E-03	5.337E-01	21	C024	2 1 1 2 2	
4.442E-03	5.000E-01	25	A003	1 2 1 2 1	
4.191E-03	4.717E-01	25	A058	1 1 1 1 2	
<1.78E-03	<2.00E-01	25	B019	1 0 1 2 0	

4.460E-03	5.020E-01	25	B304	2 0 2 2 2	
4.300E-03	4.840E-01	25	B317	1 0 0 0 2	
3.108E-03	3.499E-01	25	L319	1 0 2 1 1	
2.620E-03	2.949E-01	25	M342	1 0 1 1 2	
3.540E-02	3.984E+00	25	N309	1 0 0 0 1	*sic*
3.780E-03	4.255E-01	25	S359	2 1 2 2 2	
4.430E-03	4.986E-01	25	W300	2 2 2 2 2	
9.762E-03	1.099E+00	25.50	O005	2 0 2 2 1	*sic*
8.884E-04	1.000E-01	26.70	L095	2 2 1 1 2	
3.980E-03	4.480E-01	30	F071	1 1 2 1 2	
4.353E-03	4.900E-01	30	F300	1 0 0 0 1	
4.333E-03	4.878E-01	30	G029	1 0 2 2 2	
3.980E-03	4.480E-01	30	H080	1 0 0 0 2	
4.000E-03	4.502E-01	30	H332	2 2 2 2 0	
4.351E-03	4.898E-01	30	K065	2 0 2 1 2	
4.211E-03	4.740E-01	30	M300	1 1 2 2 2	
4.211E-03	4.740E-01	30	M311	1 1 2 2 2	
4.298E-03	4.838E-01	30	V009	1 0 0 0 1	
6.259E-03	7.045E-01	40	K065	2 0 2 1 2	
3.560E-03	4.007E-01	45	N043	1 0 2 2 2	
8.521E-03	9.591E-01	50	K065	2 0 2 1 2	
9.762E-03	1.099E+00	60	K065	2 0 2 1 2	
1.424E-02	1.602E+00	70	K065	2 0 2 1 2	
1.601E-02	1.802E+00	80	K065	2 0 2 1 2	
2.216E-02	2.494E+00	90	K065	2 0 2 1 2	
4.185E-03	4.711E-01	ns	H123	0 0 0 0 2	
2.620E-03	2.949E-01	ns	M308	0 0 1 1 2	
4.193E-03	4.720E-01	ns	M344	0 0 0 0 2	

631.   C₆H₅ClN₂O₄S

$C_6H_5ClN_2O_4S$

4-Chloro-3-nitro-benzenesulfonamide
Benzenesulfonamide, 4-Chloro-3-nitro-
RN: 97-09-6   **MP** (°C):
MW: 236.63   **BP** (°C):

Solubility (Moles/L)	Solubility (Grams/L)	Temp (°C)	Ref (#)	Evaluation (T P E A A)	Comments
9.500E-04	2.248E-01	15	K024	1 2 1 1 2	

632. C_6H_5ClO
m-Chlorophenol
3-Chlorophenol
Chlorophenate
3-Hydroxychlorobenzene
RN: 108-43-0   **MP** (°C):   33
MW: 128.56   **BP** (°C):   214

Solubility (Moles/L)	Solubility (Grams/L)	Temp (°C)	Ref (#)	Evaluation (T P E A A)	Comments
1.945E-01	2.500E+01	20	F300	1 0 0 0 1	
1.919E-01	2.468E+01	20	N034	1 0 0 0 2	
1.726E-01	2.219E+01	25	M373	1 0 2 1 2	

633. C_6H_5ClO
o-Chlorophenol
2-Chlorophenol
RN: 95-57-8   **MP** (°C):   9.3
MW: 128.56   **BP** (°C):   175

Solubility (Moles/L)	Solubility (Grams/L)	Temp (°C)	Ref (#)	Evaluation (T P E A A)	Comments
8.830E-02	1.135E+01	25	B173	2 0 2 2 2	
1.809E-01	2.326E+01	25	M373	1 0 2 1 2	
1.674E-01	2.153E+01	25	R041	1 0 2 1 1	
2.097E-01	2.695E+01	ns	N034	0 0 0 0 2	

634. C_6H_5ClO
p-Chlorophenol
4-Chloro-phenol-
Parachlorophenol
4-Hydroxychlorobenze
4-Chlorophenol
4-Hydroxychlorobenzene
RN: 106-48-9   **MP** (°C):   43.2
MW: 128.56   **BP** (°C):   220

Solubility (Moles/L)	Solubility (Grams/L)	Temp (°C)	Ref (#)	Evaluation (T P E A A)	Comments
2.022E-01	2.600E+01	20	F300	1 0 0 0 1	
1.022E-01	1.314E+01	20	H301	2 0 2 2 2	
1.993E-01	2.563E+01	20	N034	1 0 0 0 2	
1.839E-01	2.364E+01	20	R087	1 1 2 2 2	0.15M NaCl
2.100E-01	2.700E+01	25	B316	1 0 2 1 1	
2.053E-01	2.639E+01	25	M373	1 0 2 1 2	
1.823E-01	2.344E+01	25	R041	1 0 2 1 1	

635.   C₆H₅ClO₃S
p-Chlorobenzenesulfonic Acid
4-Chlor-benzolsulfosaeure
RN: 98-66-8   **MP** (°C):   67
MW: 192.62   **BP** (°C):   148

Solubility (Moles/L)	Solubility (Grams/L)	Temp (°C)	Ref (#)	Evaluation (T P E A A)	Comments
2.583E+00	4.975E+02	59.0	T023	1 2 2 1 2	
2.590E+00	4.988E+02	62.4	T023	1 2 2 1 2	

636.   C₆H₅ClO₃S.2.5H₂O
p-Chlorobenzenesulfonic Acid (2.5 Hydrate)
RN: 98-66-8   **MP** (°C):
MW: 237.66   **BP** (°C):

Solubility (Moles/L)	Solubility (Grams/L)	Temp (°C)	Ref (#)	Evaluation (T P E A A)	Comments
1.519E+00	3.609E+02	-26.0	T023	1 2 2 1 2	
1.553E+00	3.690E+02	-20.0	T023	1 2 2 1 2	
1.606E+00	3.816E+02	-11.0	T023	1 2 2 1 2	
1.653E+00	3.929E+02	-2.2	T023	1 2 2 1 2	
1.723E+00	4.095E+02	10.6	T023	1 2 2 1 2	
1.784E+00	4.240E+02	22.9	T023	1 2 2 1 2	
1.817E+00	4.318E+02	27.6	T023	1 2 2 1 2	
1.854E+00	4.406E+02	30.8	T023	1 2 2 1 2	

637.   C₆H₅Cl₂NO₂S
3,4-Dichloro-benzenesulfonamide
Benzenesulfonamide, 3,4-Dichloro-
RN: 23815-28-3   **MP** (°C):
MW: 226.08   **BP** (°C):

Solubility (Moles/L)	Solubility (Grams/L)	Temp (°C)	Ref (#)	Evaluation (T P E A A)	Comments
3.500E-03	7.913E-01	15	K024	1 2 1 1 2	

638.   C$_6$H$_5$F
Fluorobenzene
Fluorbenzol

RN: 462-06-6   **MP** (°C):   -42
MW: 96.11   **BP** (°C):   85

Solubility (Moles/L)	Solubility (Grams/L)	Temp (°C)	Ref (#)	Evaluation (T P E A A)	Comments
1.613E-02	1.550E+00	25	A003	1 2 1 2 2	
1.602E-02	1.540E+00	30	F071	1 1 2 1 2	
1.561E-02	1.500E+00	30	F300	1 0 0 0 1	
1.602E-02	1.540E+00	30	H080	1 0 0 0 2	
1.600E-02	1.538E+00	30	J036	1 2 0 0 2	
1.598E-02	1.535E+00	30	V009	1 0 0 0 2	
1.616E-02	1.553E+00	ns	M344	0 0 0 0 2	

639.   C$_6$H$_5$FN$_2$O$_3$
3-Acetyl-5-fluoro-2,4(1H,3H)-pyrimidinedi-one
3-Acetyl-5-fluorouracil

RN: 75410-15-0   **MP** (°C):   115-116
MW: 172.12   **BP** (°C):

Solubility (Moles/L)	Solubility (Grams/L)	Temp (°C)	Ref (#)	Evaluation (T P E A A)	Comments
2.487E-01	4.280E+01	22	B321	1 0 2 2 2	pH 4.0

640.   C$_6$H$_5$FN$_2$O$_4$
1-Methoxycarbonyl-5-fluorouracil
1(2H)-Pyrimidinecarboxylic Acid, 5-Fluoro-3,4-dihydro-2,4-dioxo-, Methyl Ester

RN: 71759-43-8   **MP** (°C):
MW: 188.12   **BP** (°C):

Solubility (Moles/L)	Solubility (Grams/L)	Temp (°C)	Ref (#)	Evaluation (T P E A A)	Comments
1.239E-01	2.330E+01	22	B332	1 1 0 0 1	pH 4.0

641.   C$_6$H$_5$FO
2-Fluorophenol
2-Fluor-phenol
o-Fluorophenol

RN: 367-12-4   **MP** (°C):   16.1
MW: 112.10   **BP** (°C):   171.5

Solubility (Moles/L)	Solubility (Grams/L)	Temp (°C)	Ref (#)	Evaluation (T P E A A)	Comments
7.200E-01	8.072E+01	25	P031	1 1 2 2 2	

642.   C₆H₅FO

p-Fluorophenol
4-Fluorophenol

RN: 371-41-5   **MP** (°C):   46-48
MW: 112.10   **BP** (°C):   185-188

Solubility (Moles/L)	Solubility (Grams/L)	Temp (°C)	Ref (#)	Evaluation (T P E A A)	Comments
5.671E-01	6.357E+01	20	R087	1 1 2 2 2	0.15M NaCl
7.200E-01	8.072E+01	25	P031	1 1 2 2 2	

643.   C₆H₅FO

m-Fluorophenol
3-Fluorophenol

RN: 372-20-3   **MP** (°C):   13.7
MW: 112.10   **BP** (°C):   178

Solubility (Moles/L)	Solubility (Grams/L)	Temp (°C)	Ref (#)	Evaluation (T P E A A)	Comments
6.900E-01	7.735E+01	25	P031	1 1 2 2 2	

644.   C₆H₅FO₃S.H₂O

p-Fluorobenzenesulfonic Acid (Monohydrate)

RN: 368-88-7   **MP** (°C):
MW: 194.18   **BP** (°C):

Solubility (Moles/L)	Solubility (Grams/L)	Temp (°C)	Ref (#)	Evaluation (T P E A A)	Comments
2.243E+00	4.355E+02	22.1	T023	1 2 2 1 2	
2.263E+00	4.394E+02	35.4	T023	1 2 2 1 2	
2.549E+00	4.950E+02	41.4	T023	1 2 2 1 2	
2.306E+00	4.477E+02	54.2	T023	1 2 2 1 2	
2.539E+00	4.930E+02	54.3	T023	1 2 2 1 2	
2.356E+00	4.575E+02	71.2	T023	1 2 2 1 2	
2.509E+00	4.872E+02	74.5	T023	1 2 2 1 2	
2.392E+00	4.644E+02	80.0	T023	1 2 2 1 2	
2.496E+00	4.847E+02	81.0	T023	1 2 2 1 2	
2.463E+00	4.782E+02	85.2	T023	1 2 2 1 2	
2.440E+00	4.739E+02	85.5	T023	1 2 2 1 2	

645. C₆H₅FO₃S.2.5H₂O

p-Fluorobenzenesulfonic Acid (2.5 Hydrate)

RN: 368-88-7   **MP** (°C):
MW: 221.21   **BP** (°C):

Solubility (Moles/L)	Solubility (Grams/L)	Temp (°C)	Ref (#)	Evaluation (T P E A A)	Comments
1.848E+00	4.088E+02	-15.5	T023	1 2 2 1 2	
1.880E+00	4.160E+02	-3.9	T023	1 2 2 1 2	
1.893E+00	4.187E+02	1.0	T023	1 2 2 1 2	
1.923E+00	4.254E+02	10.1	T023	1 2 2 1 2	
1.966E+00	4.349E+02	21.3	T023	1 2 2 1 2	

646. C₆H₅FO₃S.3H₂O

p-Fluorobenzenesulfonic Acid (Trihydrate)

RN: 368-88-7   **MP** (°C):
MW: 230.21   **BP** (°C):

Solubility (Moles/L)	Solubility (Grams/L)	Temp (°C)	Ref (#)	Evaluation (T P E A A)	Comments
1.731E+00	3.985E+02	-22.5	T023	1 2 2 1 2	
1.704E+00	3.922E+02	-21.4	T023	1 2 2 1 2	
1.751E+00	4.032E+02	-19.5	T023	1 2 2 1 2	
1.715E+00	3.949E+02	-18.5	T023	1 2 2 1 2	
1.760E+00	4.052E+02	-17.9	T023	1 2 2 1 2	
1.751E+00	4.032E+02	-13.0	T023	1 2 2 1 2	
1.784E+00	4.108E+02	-7.4	T023	1 2 2 1 2	

647. C₆H₅FO₃S.4H₂O

p-Fluorobenzenesulfonic Acid (Tetrahydrate)

RN: 368-88-7   **MP** (°C):
MW: 248.23   **BP** (°C):

Solubility (Moles/L)	Solubility (Grams/L)	Temp (°C)	Ref (#)	Evaluation (T P E A A)	Comments
1.469E+00	3.648E+02	-38.0	T023	1 2 2 1 2	
1.484E+00	3.684E+02	-35.4	T023	1 2 2 1 2	
1.498E+00	3.719E+02	-34.4	T023	1 2 2 1 2	
1.519E+00	3.771E+02	-32.5	T023	1 2 2 1 2	
1.532E+00	3.803E+02	-30.5	T023	1 2 2 1 2	
1.580E+00	3.922E+02	-26.4	T023	1 2 2 1 2	
1.605E+00	3.985E+02	-24.0	T023	1 2 2 1 2	

648.   C$_6$H$_5$I
Iodobenzene
RN: 591-50-4   **MP** (°C):   -30
MW: 204.01   **BP** (°C):   188

Solubility (Moles/L)	Solubility (Grams/L)	Temp (°C)	Ref (#)	Evaluation (T P E A A)	Comments
8.823E-04	1.800E-01	25	A003	1 2 1 2 1	
9.840E-04	2.007E-01	25	M342	1 0 1 1 2	
1.667E-03	3.400E-01	30	F071	1 1 2 1 2	
1.667E-03	3.400E-01	30	F300	1 0 0 0 2	
1.667E-03	3.400E-01	30	H080	1 0 0 0 2	
1.667E-03	3.400E-01	30	M344	1 0 0 0 2	
1.699E-03	3.467E-01	30	V009	1 0 0 0 1	

649.   C$_6$H$_5$IO
p-Iodophenol
4-Iodophenol
RN: 540-38-5   **MP** (°C):   94
MW: 220.01   **BP** (°C):   138 at 5 mm Hg

Solubility (Moles/L)	Solubility (Grams/L)	Temp (°C)	Ref (#)	Evaluation (T P E A A)	Comments
1.285E-02	2.828E+00	20	R087	1 1 2 2 2	0.15M NaCl

650.   C$_6$H$_5$NO$_2$
Nicotinic Acid
Niacin
RN: 59-67-6   **MP** (°C):   236
MW: 123.11   **BP** (°C):

Solubility (Moles/L)	Solubility (Grams/L)	Temp (°C)	Ref (#)	Evaluation (T P E A A)	Comments
1.208E-01	1.488E+01	1	H083	1 2 2 1 2	
2.679E-01	3.298E+01	16	C033	1 0 2 1 2	
1.358E-01	1.672E+01	20	D041	1 0 0 0 1	
1.436E-01	1.768E+01	20	H083	1 2 2 1 2	
1.381E-01	1.700E+01	20	M054	1 0 0 0 1	
3.652E-01	4.496E+01	28	C033	1 0 2 1 2	
2.595E-01	3.195E+01	42	H083	1 2 2 1 2	
3.735E-01	4.598E+01	60	H083	1 2 2 1 2	
5.604E-01	6.899E+01	80	H083	1 2 2 1 2	
6.809E-01	8.383E+01	88	H083	1 2 2 1 2	

651.   C₆H₅NO₂

Nitrobenzene
Nitrobenzol
Benzene, Nitro-
RN: 98-95-3   **MP** (°C):   6
MW: 123.11   **BP** (°C):   210

Solubility (Moles/L)	Solubility (Grams/L)	Temp (°C)	Ref (#)	Evaluation (T P E A A)	Comments
1.381E-02	1.700E+00	6	V004	1 0 1 2 2	
1.443E-02	1.777E+00	15	G029	1 0 2 2 2	
1.549E-02	1.907E+00	20	B179	2 0 0 0 2	
1.543E-02	1.900E+00	20	F300	1 0 0 0 1	
1.600E-02	1.970E+00	20	P073	1 0 0 1 2	
1.543E-02	1.900E+00	22.5	G301	2 1 0 1 2	
1.568E-02	1.930E+00	25	A003	1 2 1 2 2	
1.700E-02	2.093E+00	25	B173	2 0 2 2 2	
1.580E-02	1.945E+00	25	H071	2 2 2 1 2	
1.600E-02	1.970E+00	25	H332	2 2 2 2 1	
1.560E-02	1.921E+00	25	I334	2 2 2 1 2	
1.560E-02	1.921E+00	25	I335	2 2 2 2 2	
1.543E-02	1.900E+00	25	M087	1 1 2 1 2	
1.457E-02	1.794E+00	25.04	V013	2 2 2 2 2	
1.446E-02	1.780E+00	26.70	L095	2 2 1 1 2	
1.662E-02	2.046E+00	30	G029	1 0 2 2 2	
1.673E-02	2.060E+00	30	V004	1 0 1 2 2	
1.667E-02	2.052E+00	30	V009	1 0 0 0 2	
1.835E-02	2.259E+00	35	H077	2 2 2 2 2	
2.144E-02	2.640E+00	50	V004	1 0 1 2 2	
2.193E-02	2.700E+00	55	F300	1 0 0 0 1	
2.534E-02	3.120E+00	60	V004	1 0 1 2 2	
2.700E-03	3.324E-01	ns	D348	0 0 2 2 2	

652.   C₆H₅NO₃

o-Nitrophenol
2-Nitrophenol
RN: 88-75-5   **MP** (°C):   44
MW: 139.11   **BP** (°C):   214

Solubility (Moles/L)	Solubility (Grams/L)	Temp (°C)	Ref (#)	Evaluation (T P E A A)	Comments
1.000E-02	1.391E+00	20	H306	1 0 1 2 1	
9.906E-03	1.378E+00	23.10	E032	1 2 1 2 2	
1.793E-02	2.494E+00	25	D006	1 2 0 1 2	
1.797E-02	2.500E+00	25	D059	1 2 1 1 1	
1.163E-02	1.617E+00	30.40	E032	1 2 1 2 2	
1.456E-02	2.026E+00	36.20	E032	1 2 1 2 2	
2.300E-02	3.200E+00	38.40	F300	1 0 0 0 1	
1.936E-02	2.693E+00	39.80	E032	1 2 1 2 2	
2.157E-02	3.000E+00	40	D059	1 2 1 1 0	

2.864E-02	3.984E+00	54.60	E032	1 2 1 2 1
3.598E-02	5.005E+00	67.20	E032	1 2 1 2 2
4.429E-02	6.162E+00	72.10	E032	1 2 1 2 2
5.174E-02	7.198E+00	86.90	E032	1 2 1 2 2
6.560E-02	9.126E+00	93.80	E032	1 2 1 2 2
7.979E-02	1.110E+01	100	F300	1 0 0 0 2

653. $C_6H_5NO_3$
p-Nitrophenol
4-Nitrophenol
RN: 100-02-7   **MP** (°C):   113
MW: 139.11   **BP** (°C):

Solubility (Moles/L)	Solubility (Grams/L)	Temp (°C)	Ref (#)	Evaluation (T P E A A)	Comments
3.576E-02	4.975E+00	0	D006	1 2 0 1 1	
7.821E-02	1.088E+01	12.5	D006	1 2 0 1 1	
7.610E-02	1.059E+01	12.60	E032	1 2 1 2 2	
5.780E-02	8.040E+00	15	D069	1 2 0 0 2	
1.139E-01	1.584E+01	17.30	E032	1 2 1 2 2	
9.700E-02	1.349E+01	20	H306	1 0 1 2 1	
7.188E-02	9.999E+00	20	T301	1 2 2 2 2	
1.078E-01	1.500E+01	22.5	G301	2 1 0 1 2	
1.132E-01	1.575E+01	25	D006	1 2 0 1 1	
1.797E-01	2.500E+01	25	D059	1 2 1 1 1	
8.411E-02	1.170E+01	25	F300	1 0 0 0 2	
9.925E-02	1.381E+01	25	R041	1 0 2 1 1	
1.430E-01	1.990E+01	26.60	E032	1 2 1 2 2	
1.794E-01	2.496E+01	27.70	E032	1 2 1 2 2	
2.101E-01	2.922E+01	29.60	E032	1 2 1 2 2	
2.026E-01	2.818E+01	40	D006	1 2 0 1 1	
2.085E-01	2.900E+01	40	D059	1 2 1 1 1	
3.021E+00	4.203E+02	40.60	E032	1 2 1 2 2	
2.678E-01	3.726E+01	40.70	E032	1 2 1 2 2	
3.081E+00	4.286E+02	42.50	E032	1 2 1 2 2	
2.961E+00	4.120E+02	42.70	E032	1 2 1 2 2	
3.196E+00	4.447E+02	49.70	E032	1 2 1 2 2	
4.350E-01	6.052E+01	50	D069	1 2 0 0 2	
4.148E-01	5.770E+01	50	F300	1 0 0 0 2	
3.096E-01	4.306E+01	53.30	E032	1 2 1 2 2	
2.900E+00	4.034E+02	54.90	E032	1 2 1 2 2	
3.423E-01	4.762E+01	55.10	E032	1 2 1 2 2	
3.305E+00	4.598E+02	60.70	E032	1 2 1 2 2	
2.834E+00	3.942E+02	65.00	E032	1 2 1 2 2	
3.986E-01	5.545E+01	67.80	E032	1 2 1 2 2	
5.021E-01	6.985E+01	69.40	E032	1 2 1 2 2	
2.768E+00	3.850E+02	73.30	E032	1 2 1 2 2	
3.406E+00	4.739E+02	75.70	E032	1 2 1 2 2	

Solubility (Moles/L)	Solubility (Grams/L)	Temp (°C)	Ref (#)	Evaluation (T P E A A)
6.553E-01	9.116E+01	78.30	E032	1 2 1 2 2
6.837E-01	9.510E+01	79.80	E032	1 2 1 2 2
2.699E+00	3.754E+02	80.30	E032	1 2 1 2 2
7.124E-01	9.910E+01	80.70	E032	1 2 1 2 2
7.987E-01	1.111E+02	82.30	E032	1 2 1 2 2
9.431E-01	1.312E+02	85.70	E032	1 2 1 2 2
2.555E+00	3.554E+02	86.00	E032	1 2 1 2 2
1.076E+00	1.497E+02	88.50	E032	1 2 1 2 2
2.398E+00	3.336E+02	89.70	E032	1 2 1 2 2
1.320E+00	1.837E+02	90.70	E032	1 2 1 2 2
1.438E+00	2.000E+02	91.30	E032	1 2 1 2 2
2.234E+00	3.107E+02	91.30	E032	1 2 1 2 2
1.664E+00	2.315E+02	92.10	E032	1 2 1 2 2
2.056E+00	2.861E+02	92.70	E032	1 2 1 2 2
1.763E+00	2.453E+02	92.80	E032	1 2 1 2 2
1.865E+00	2.595E+02	92.90	E032	1 2 1 2 2
3.503E+00	4.873E+02	93.50	E032	1 2 1 2 2
5.100E-02	7.095E+00	ns	B157	0 0 0 0 1

654. $C_6H_5NO_3$
m-Nitrophenol
3-Nitrophenol

RN: 554-84-7   **MP** (°C):   97
MW: 139.11   **BP** (°C):   194

Solubility (Moles/L)	Solubility (Grams/L)	Temp (°C)	Ref (#)	Evaluation (T P E A A)	Comments
6.412E-02	8.920E+00	0	D006	1 2 0 1 1	
8.524E-02	1.186E+01	12.5	D006	1 2 0 1 1	
1.243E-01	1.730E+01	15.90	E032	1 2 1 2 2	
8.300E-02	1.155E+01	20	H306	1 0 1 2 1	
1.368E-01	1.903E+01	20.20	E032	1 2 1 2 2	
1.458E-01	2.028E+01	23.40	E032	1 2 1 2 2	
9.575E-02	1.332E+01	25	D006	1 2 0 1 2	
9.740E-02	1.355E+01	25	K040	1 0 2 1 2	
9.225E-02	1.283E+01	25	R041	1 0 2 1 1	
1.685E-01	2.344E+01	29.50	E032	1 2 1 2 2	
1.944E-01	2.705E+01	35.80	E032	1 2 1 2 2	
2.113E-01	2.940E+01	40	F300	1 0 0 0 2	
2.148E-01	2.988E+01	40.90	E032	1 2 1 2 2	
3.196E+00	4.445E+02	47.10	E032	1 2 1 2 2	
3.046E+00	4.237E+02	49.60	E032	1 2 1 2 2	
3.240E+00	4.507E+02	49.70	E032	1 2 1 2 2	
3.313E+00	4.609E+02	56.50	E032	1 2 1 2 2	
2.979E+00	4.145E+02	58.70	E032	1 2 1 2 2	
2.911E-01	4.049E+01	58.80	E032	1 2 1 2 2	
3.475E-01	4.834E+01	62.70	E032	1 2 1 2 2	
3.387E+00	4.712E+02	62.80	E032	1 2 1 2 2	
2.914E+00	4.054E+02	71.50	E032	1 2 1 2 2	
3.484E+00	4.846E+02	75.10	E032	1 2 1 2 2	

Solubility (Moles/L)	Solubility (Grams/L)	Temp (°C)	Ref (#)	Evaluation (T P E A A)	Comments
4.703E-01	6.542E+01	77.10	E032	1 2 1 2 2	
2.828E+00	3.935E+02	80.60	E032	1 2 1 2 2	
6.326E-01	8.801E+01	85.30	E032	1 2 1 2 2	
3.549E+00	4.937E+02	85.80	E032	1 2 1 2 2	
2.705E+00	3.762E+02	89.40	E032	1 2 1 2 2	
3.569E+00	4.965E+02	89.80	E032	1 2 1 2 2	
2.649E+00	3.684E+02	92.20	E032	1 2 1 2 2	
9.501E-01	1.322E+02	93.60	E032	1 2 1 2 2	
2.581E+00	3.591E+02	94.20	E032	1 2 1 2 2	
2.475E+00	3.443E+02	95.60	E032	1 2 1 2 2	
1.210E+00	1.683E+02	96.20	E032	1 2 1 2 2	
2.396E+00	3.333E+02	96.60	E032	1 2 1 2 2	
1.440E+00	2.004E+02	97.50	E032	1 2 1 2 2	
2.286E+00	3.181E+02	97.70	E032	1 2 1 2 2	
1.604E+00	2.232E+02	98.10	E032	1 2 1 2 2	
2.341E+00	3.256E+02	98.10	E032	1 2 1 2 2	
1.763E+00	2.453E+02	98.40	E032	1 2 1 2 2	
2.049E+00	2.851E+02	98.50	E032	1 2 1 2 2	
1.965E+00	2.734E+02	98.60	E032	1 2 1 2 2	
3.008E+00	4.184E+02	98.70	F300	1 0 0 0 2	

655. $C_6H_5NO_4$
3-Nitrocatechol
3-Nitro-1,2-benzenediol
RN: 6665-98-1   **MP** (°C):
MW: 155.11   **BP** (°C):

Solubility (Moles/L)	Solubility (Grams/L)	Temp (°C)	Ref (#)	Evaluation (T P E A A)	Comments
5.377E-02	8.340E+00	14.40	E032	1 2 1 2 2	
6.573E-02	1.019E+01	20.90	E032	1 2 1 2 2	
9.590E-02	1.488E+01	29.50	E032	1 2 1 2 2	
1.277E-01	1.980E+01	35.10	E032	1 2 1 2 2	
1.474E-01	2.286E+01	37.90	E032	1 2 1 2 2	
1.738E-01	2.695E+01	41.00	E032	1 2 1 2 2	
2.372E-01	3.679E+01	45.80	E032	1 2 1 2 2	
2.646E-01	4.104E+01	47.60	E032	1 2 1 2 2	
3.216E-01	4.988E+01	54.50	E032	1 2 1 2 2	
3.615E-01	5.607E+01	61.30	E032	1 2 1 2 2	
4.548E-01	7.055E+01	75.90	E032	1 2 1 2 2	
5.743E-01	8.909E+01	86.80	E032	1 2 1 2 2	
8.164E-01	1.266E+02	96.80	E032	1 2 1 2 2	

656. $C_6H_5NO_4$
4-Nitrocatechol
4-Nitro-1,2-benzenediol
RN: 3316-09-4   **MP** (°C):
MW: 155.11   **BP** (°C):

Solubility (Moles/L)	Solubility (Grams/L)	Temp (°C)	Ref (#)	Evaluation (T P E A A)	Comments
1.211E+00	1.878E+02	24.60	E032	1 2 1 2 2	
1.423E+00	2.208E+02	37.70	E032	1 2 1 2 2	
1.488E+00	2.308E+02	41.30	E032	1 2 1 2 2	
1.664E+00	2.582E+02	51.90	E032	1 2 1 2 2	
1.829E+00	2.837E+02	58.50	E032	1 2 1 2 2	
2.004E+00	3.109E+02	66.50	E032	1 2 1 2 2	
2.049E+00	3.179E+02	67.80	E032	1 2 1 2 2	
2.149E+00	3.334E+02	71.20	E032	1 2 1 2 2	

657. $C_6H_5NO_4$
Nitrohydroquinone
2-Nitroquinol
4-Hydroxy-2-nitrophenol
RN: 16090-33-8   **MP** (°C):
MW: 155.11   **BP** (°C):

Solubility (Moles/L)	Solubility (Grams/L)	Temp (°C)	Ref (#)	Evaluation (T P E A A)	Comments
6.933E-02	1.068E+01	30.20	E032	1 2 1 2 2	
1.022E-01	1.575E+01	34.60	E032	1 2 1 2 2	
1.583E-01	2.439E+01	44.60	E032	1 2 1 2 2	
2.012E-01	3.101E+01	49.60	E032	1 2 1 2 2	
3.149E-01	4.853E+01	54.50	E032	1 2 1 2 2	
4.527E-01	6.977E+01	59.10	E032	1 2 1 2 2	
6.446E-01	9.934E+01	61.70	E032	1 2 1 2 2	
7.210E-01	1.111E+02	64.20	E032	1 2 1 2 2	
8.464E-01	1.304E+02	65.00	E032	1 2 1 2 2	
1.082E+00	1.667E+02	93.80	E032	1 2 1 2 2	

658. $C_6H_5NO_4$
4-Nitroresorcinol
4-Nitro-1,3-benzenediol
RN: 3163-07-3   **MP** (°C):
MW: 155.11   **BP** (°C):

Solubility (Moles/L)	Solubility (Grams/L)	Temp (°C)	Ref (#)	Evaluation (T P E A A)	Comments
4.354E-02	6.754E+00	18.30	E032	1 2 1 2 2	
5.244E-02	8.133E+00	24.70	E032	1 2 1 2 2	
6.510E-02	1.010E+01	30.80	E032	1 2 1 2 2	
7.959E-02	1.235E+01	36.90	E032	1 2 1 2 2	
1.034E-01	1.604E+01	43.50	E032	1 2 1 2 2	

Solubility (Moles/L)	Solubility (Grams/L)	Temp (°C)	Ref (#)	Evaluation (T P E A A)
1.462E-01	2.267E+01	47.50	E032	1 2 1 2 2
1.817E-01	2.818E+01	49.10	E032	1 2 1 2 2
2.168E-01	3.363E+01	50.70	E032	1 2 1 2 2
2.497E-01	3.874E+01	51.20	E032	1 2 1 2 2
2.776E-01	4.306E+01	52.30	E032	1 2 1 2 2
3.286E-01	5.096E+01	53.90	E032	1 2 1 2 2
4.487E-01	6.959E+01	57.80	E032	1 2 1 2 2
5.951E-01	9.231E+01	62.70	E032	1 2 1 2 2
8.468E-01	1.313E+02	68.40	E032	1 2 1 2 2
1.075E+00	1.667E+02	71.90	E032	1 2 1 2 2
1.209E+00	1.875E+02	72.90	E032	1 2 1 2 2
1.325E+00	2.055E+02	73.30	E032	1 2 1 2 2
1.487E+00	2.307E+02	73.40	E032	1 2 1 2 2

659. $C_6H_5NO_4$
2-Nitroresorcinol
2-Nitro-1,3-benzenediol
RN: 601-89-8 MP (°C): 81
MW: 155.11 BP (°C):

Solubility (Moles/L)	Solubility (Grams/L)	Temp (°C)	Ref (#)	Evaluation (T P E A A)	Comments
8.435E-03	1.308E+00	28.40	E032	1 2 1 2 2	
1.306E-02	2.026E+00	36.70	E032	1 2 1 2 2	
2.319E-02	3.597E+00	47.60	E032	1 2 1 2 2	
3.635E-02	5.638E+00	54.90	E032	1 2 1 2 2	
6.276E-02	9.734E+00	67.20	E032	1 2 1 2 2	
8.399E-02	1.303E+01	74.40	E032	1 2 1 2 2	
1.208E-01	1.874E+01	82.90	E032	1 2 1 2 2	
1.529E-01	2.372E+01	92.30	E032	1 2 1 2 2	

660. $C_6H_5NO_5S$
p-Nitrobenzenesulfonic Acid
4-Nitrobenzenesulfonic Acid
RN: 138-42-1 MP (°C):
MW: 203.17 BP (°C):

Solubility (Moles/L)	Solubility (Grams/L)	Temp (°C)	Ref (#)	Evaluation (T P E A A)	Comments
2.343E+00	4.760E+02	100.5	T023	1 2 2 1 2	
2.412E+00	4.901E+02	105.0	T023	1 2 2 1 2	
2.461E+00	5.000E+02	110.0	T023	1 2 2 1 2	

661.   C₆H₅NO₅S.2H₂O

p-Nitrobenzenesulfonic Acid (Dihydrate)

RN: 15481-55-7   **MP** (°C):
MW: 239.21   **BP** (°C):

Solubility (Moles/L)	Solubility (Grams/L)	Temp (°C)	Ref (#)	Evaluation (T P E A A)	Comments
1.667E+00	3.987E+02	36.6	T023	1 2 2 1 2	
1.720E+00	4.113E+02	56.6	T023	1 2 2 1 2	
1.771E+00	4.235E+02	75.5	T023	1 2 2 1 2	
1.822E+00	4.359E+02	90.2	T023	1 2 2 1 2	
1.939E+00	4.638E+02	106.8	T023	1 2 2 1 2	
1.920E+00	4.592E+02	110.2	T023	1 2 2 1 2	

662.   C₆H₅NO₅S.4H₂O

p-Nitrobenzenesulfonic Acid (Tetrahydrate)

RN: 15481-55-7   **MP** (°C):
MW: 275.24   **BP** (°C):

Solubility (Moles/L)	Solubility (Grams/L)	Temp (°C)	Ref (#)	Evaluation (T P E A A)	Comments
1.060E+00	2.919E+02	-8.3	T023	1 2 2 1 2	
1.146E+00	3.153E+02	-1.0	T023	1 2 2 1 2	
1.273E+00	3.504E+02	10.8	T023	1 2 2 1 2	
1.318E+00	3.627E+02	16.0	T023	1 2 2 1 2	
1.409E+00	3.877E+02	26.3	T023	1 2 2 1 2	

663.   C₆H₅N₂OS

Methyl Acetylthiodiazole
Thiodiazolique Methyle Acetyle

RN: **MP** (°C):
MW: 153.18   **BP** (°C):

Solubility (Moles/L)	Solubility (Grams/L)	Temp (°C)	Ref (#)	Evaluation (T P E A A)	Comments
6.528E-04	1.000E-01	37	D084	1 0 1 0 1	

664.   C₆H₅N₃O₄

2,6-Dinitroaniline
2,6-Dinitrobenzenamine

RN: 606-22-4   **MP** (°C):   133
MW: 183.12   **BP** (°C):

Solubility (Moles/L)	Solubility (Grams/L)	Temp (°C)	Ref (#)	Evaluation (T P E A A)	Comments
4.365E-04	7.994E-02	25	B335	1 2 0 0 1	

665. $C_6H_5N_3O_4$
2,4-Dinitroaniline
2,4-Dinitrobenzenamine
2,4-Dinitroaminobenzene
1-Amino-2,4-dinitrobenzene
RN: 97-02-9   **MP** (°C):   176
MW: 183.12   **BP** (°C):

Solubility (Moles/L)	Solubility (Grams/L)	Temp (°C)	Ref (#)	Evaluation (T P E A A)	Comments
4.266E-04	7.812E-02	25	B335	1 2 0 0 1	

666. $C_6H_5N_3O_5$
Picramic Acid
2-Amino-4,6-dinitro-phenol
RN: 96-91-3   **MP** (°C):   169
MW: 199.12   **BP** (°C):

Solubility (Moles/L)	Solubility (Grams/L)	Temp (°C)	Ref (#)	Evaluation (T P E A A)	Comments
7.031E-03	1.400E+00	22	F300	1 0 0 0 1	

667. $C_6H_5N_5$
7-Aminopteridine
7-Pteridinamine
RN: 769-66-4   **MP** (°C):
MW: 147.14   **BP** (°C):

Solubility (Moles/L)	Solubility (Grams/L)	Temp (°C)	Ref (#)	Evaluation (T P E A A)	Comments
4.851E-03	7.138E-01	20	A083	1 2 0 0 0	
3.974E-02	5.848E+00	100	A083	1 2 0 0 0	

668. $C_6H_5N_5$
2-Aminopteridine
2-Pteridinamine
RN: 700-81-2   **MP** (°C):
MW: 147.14   **BP** (°C):

Solubility (Moles/L)	Solubility (Grams/L)	Temp (°C)	Ref (#)	Evaluation (T P E A A)	Comments
5.031E-03	7.402E-01	22.5	A085	1 2 0 0 0	

669.   C₆H₅N₅

4-Aminopteridine

4-Pteridinamine

RN: 6973-01-9   **MP** (°C):   305

MW: 147.14   **BP** (°C):

Solubility (Moles/L)	Solubility (Grams/L)	Temp (°C)	Ref (#)	Evaluation (T P E A A)	Comments
4.851E-03	7.138E-01	22.5	A085	1 2 0 0 0	

670.   C₆H₅N₅O

4-Amino-2-hydroxypteridine

4-Amino-2-oxopteridine

4-Aminopteridin-2-one

4-Amino-2-pteridone

RN: 22005-65-8   **MP** (°C):   >350

MW: 163.14   **BP** (°C):

Solubility (Moles/L)	Solubility (Grams/L)	Temp (°C)	Ref (#)	Evaluation (T P E A A)	Comments
4.378E-04	7.142E-02	20	A019	2 2 1 1 2	
5.104E-03	8.326E-01	100	A019	1 2 1 1 2	

671.   C₆H₅N₅O

7-Amino-6-hydroxypteridine

7-Amino-6-oxopteridine

7-Aminopteridin-6-one

7-Amino-6-pteridone

RN: 1008-85-1   **MP** (°C):

MW: 163.14   **BP** (°C):

Solubility (Moles/L)	Solubility (Grams/L)	Temp (°C)	Ref (#)	Evaluation (T P E A A)	Comments
1.226E-03	2.000E-01	100	A082	1 2 0 0 0	

672.   C₆H₅N₅O

2-Amino-4-hydroxypteridine

2-Amino-4(1H)-pteridinone

2-Amino-4(3H)-pteridinone

2-Amino-4-pteridone

2-Amino-4-oxopteridine

2-Aminopteridin-4-one

RN: 2236-60-4   **MP** (°C):

MW: 163.14   **BP** (°C):

Solubility (Moles/L)	Solubility (Grams/L)	Temp (°C)	Ref (#)	Evaluation (T P E A A)	Comments
1.075E-04	1.754E-02	22.5	A085	1 2 0 0 0	

673.   C₆H₅N₅O₂
Xanthopterin
2-Amino-4:6-dihydroxypteridine
RN: 119-44-8   **MP** (°C):
MW: 179.14   **BP** (°C):

Solubility (Moles/L)	Solubility (Grams/L)	Temp (°C)	Ref (#)	Evaluation (T P E A A)	Comments
1.396E-04	2.500E-02	22.5	A085	1 2 0 0 0	

674.   C₆H₅N₅O₃
Leucopterin
2-Amino-4:6:7-trihydroxypteridine
RN: 492-11-5   **MP** (°C):
MW: 195.14   **BP** (°C):

Solubility (Moles/L)	Solubility (Grams/L)	Temp (°C)	Ref (#)	Evaluation (T P E A A)	Comments
6.833E-06	1.333E-03	22.5	A085	1 2 0 0 0	

675.   C₆H₅N₅O₄S
3'-Nitrosoniridazole
2-Imidazolidinone, 1-Nitroso-3-(5-nitro-2-thiazolyl)-
RN: 34968-90-6   **MP** (°C):   202-203
MW: 243.20   **BP** (°C):

Solubility (Moles/L)	Solubility (Grams/L)	Temp (°C)	Ref (#)	Evaluation (T P E A A)	Comments
3.084E-04	7.500E-02	25	G051	1 0 1 1 0	

676.   C₆H₆
Benzene
Benzol
Phenyl Hydride
Cyclohexatriene
Benzolene
Phene
RN: 71-43-2   **MP** (°C):   5
MW: 78.11   **BP** (°C):   80

Solubility (Moles/L)	Solubility (Grams/L)	Temp (°C)	Ref (#)	Evaluation (T P E A A)	Comments
1.959E-02	1.530E+00	0	F300	1 0 0 0 2	
2.148E-02	1.678E+00	0	P003	2 2 2 2 2	
2.350E-02	1.836E+00	0.2	M151	2 1 2 2 2	
2.347E-02	1.833E+00	0.24	M183	1 2 1 1 2	
2.356E-02	1.840E+00	0.8	A004	1 2 2 1 2	
2.351E-02	1.837E+00	4.50	B086	2 1 2 2 2	

1.881E-02	1.469E+00	4.62	U013	1 0 0 0 0	EFG
2.646E-02	2.067E+00	4.8	L007	2 1 1 2 2	
1.178E-02	9.200E-01	5	S119	0 0 0 0 1	
2.646E-02	2.067E+00	5.0	L007	2 1 1 1 2	
1.838E-02	1.436E+00	5.39	U010	1 0 0 1 1	EFG
2.310E-02	1.804E+00	6.20	M151	2 1 2 2 2	
2.306E-02	1.802E+00	6.24	M183	1 2 1 1 2	
2.364E-02	1.847E+00	6.30	B086	2 1 2 2 2	
2.313E-02	1.807E+00	7.10	B086	2 1 2 2 2	
2.313E-02	1.807E+00	9	B086	2 1 2 2 2	
2.292E-02	1.790E+00	9.40	A004	1 2 2 1 2	
2.080E-02	1.625E+00	10	B149	2 1 1 2 2	
2.110E-02	1.648E+00	10	J302	2 1 2 2 2	
2.240E-02	1.750E+00	10	M130	1 0 0 0 2	
2.300E-02	1.797E+00	11.00	M151	2 1 2 2 2	
2.300E-02	1.796E+00	11.04	M183	1 2 1 1 2	
2.262E-02	1.767E+00	11.80	B086	2 1 2 2 2	
2.262E-02	1.767E+00	12.10	B086	2 1 2 2 2	
2.270E-02	1.773E+00	14.00	M151	2 1 2 2 2	
2.263E-02	1.767E+00	14.04	M183	1 2 1 1 2	
1.838E-02	1.436E+00	14.20	U013	1 0 0 0 0	EFG
2.655E-02	2.074E+00	14.8	L007	2 1 1 2 2	
2.655E-02	2.074E+00	14.9	L007	2 1 1 1 2	
2.290E-02	1.789E+00	15	I333	1 2 1 1 2	
2.150E-02	1.679E+00	15	S006	1 0 0 0 2	
1.971E-02	1.540E+00	15	S203	1 1 2 1 2	
1.797E-02	1.403E+00	15.02	U010	1 0 0 1 1	EFG
2.287E-02	1.787E+00	15.10	B086	2 1 2 2 2	
2.112E-02	1.650E+00	16	D047	1 0 0 1 2	
2.266E-02	1.770E+00	16.80	A004	1 2 2 1 2	
2.260E-02	1.765E+00	16.90	M151	2 1 2 2 2	
2.253E-02	1.760E+00	16.94	M183	1 2 1 1 2	
2.191E-02	1.711E+00	17	F002	2 2 2 2 2	
2.287E-02	1.787E+00	17.90	B086	2 1 2 2 2	
2.260E-02	1.765E+00	18.60	M151	2 1 2 2 2	
2.259E-02	1.764E+00	18.64	M183	1 2 1 1 2	
2.664E-02	2.081E+00	19.8	L007	2 1 1 2 2	
2.664E-02	2.081E+00	19.9	L007	2 1 1 1 2	
2.220E-02	1.734E+00	20	B149	2 1 1 2 2	
2.180E-02	1.703E+00	20	C006	1 2 1 1 2	
1.023E-02	7.994E-01	20	C121	0 0 0 0 0	unit assumed, *sic*
2.428E-02	1.896E+00	20	D052	1 1 0 0 1	
1.600E-02	1.250E+00	20	E009	1 0 0 0 1	
1.680E-02	1.312E+00	20	E025	1 0 2 2 2	
2.189E-02	1.710E+00	20	F071	1 1 2 1 2	
2.317E-02	1.810E+00	20	F300	1 0 0 0 2	
1.023E-02	7.994E-01	20	I310	0 0 0 0 0	
2.310E-02	1.804E+00	20	I333	1 2 1 1 2	
2.042E-02	1.595E+00	20	K337	1 0 0 0 2	
2.280E-02	1.781E+00	20	M312	1 0 0 0 1	

1.366E-02	1.067E+00	20	M337	2 1 2 2 2	
2.650E-02	2.070E+00	20	P073	1 0 0 1 2	
1.751E-02	1.368E+00	20.0	H043	2 2 2 2 2	
2.249E-02	1.757E+00	20.10	B086	2 1 2 2 2	
2.224E-02	1.737E+00	21	C024	2 1 1 2 2	
2.202E-02	1.720E+00	22	F002	2 2 2 2 2	
2.320E-02	1.812E+00	22.5	I333	1 2 1 1 2	
2.304E-02	1.800E+00	24	A004	1 2 2 1 2	
2.667E-02	2.084E+00	24.8	L007	2 1 1 2 2	
2.227E-02	1.740E+00	25	A001	1 2 2 2 2	
1.917E-02	1.498E+00	25	A037	2 2 2 2 2	
2.292E-02	1.790E+00	25	B003	2 2 2 2 2	
2.045E-02	1.597E+00	25	B019	1 0 1 2 0	
2.279E-02	1.780E+00	25	B060	2 0 1 1 1	
2.292E-02	1.790E+00	25	B090	2 2 2 1 2	
2.292E-02	1.790E+00	25	B151	1 2 2 1 2	
2.330E-02	1.820E+00	25	B153	2 1 1 1 2	
2.240E-02	1.750E+00	25	B173	2 0 2 2 2	
2.300E-02	1.797E+00	25	G323	2 2 2 2 2	
2.300E-02	1.797E+00	25	H332	2 2 2 2 1	
2.330E-02	1.820E+00	25	I333	1 2 1 1 2	
2.310E-02	1.804E+00	25	J302	2 1 2 2 2	
2.390E-02	1.867E+00	25	K001	2 2 2 2 2	
8.961E-03	7.000E-01	25	K072	1 0 1 1 1	
1.300E-02	1.015E+00	25	K123	1 0 2 2 1	
2.170E-02	1.695E+00	25	K316	2 2 2 2 2	
2.259E-02	1.765E+00	25	L002	2 2 2 2 2	
2.313E-02	1.807E+00	25	L319	1 0 2 1 1	
2.166E-02	1.692E+00	25	L322	1 1 2 2 1	
1.770E+00	1.383E+02	25	M021	2 2 2 1 2	sic
2.279E-02	1.780E+00	25	M131	1 0 0 0 2	
2.278E-02	1.780E+00	25	M132	2 2 2 1 2	
2.310E-02	1.804E+00	25	M151	2 1 2 2 2	average of 2
2.293E-02	1.791E+00	25	M151	2 1 1 2 2	
2.290E-02	1.789E+00	25	M342	1 0 1 1 2	
1.917E-02	1.498E+00	25	O015	0 0 0 0 0	
2.247E-02	1.755E+00	25	P003	2 2 2 2 2	
2.227E-02	1.740E+00	25	P051	2 1 1 2 2	
2.607E-02	2.036E+00	25	S010	2 1 2 1 2	
2.377E-02	1.857E+00	25	S012	2 0 2 2 2	
2.061E-02	1.610E+00	25	S203	1 1 2 1 2	
2.070E-02	1.617E+00	25	S359	2 1 2 2 2	
2.778E-02	2.170E+00	25	W057	2 0 2 2 2	
2.290E-02	1.789E+00	25	W300	2 2 2 2 2	
2.300E-02	1.797E+00	25.0	H043	2 2 2 2 2	
2.667E-02	2.084E+00	25.0	L007	2 1 1 1 2	
2.227E-02	1.740E+00	25.00	P007	2 1 2 2 2	
2.290E-02	1.789E+00	25.04	M183	1 2 1 1 2	

1.838E-02	1.436E+00	25.35	U010	1 0 0 1 1	EFG
1.881E-02	1.469E+00	25.35	U013	1 0 0 0 0	EFG
2.325E-02	1.816E+00	25.84	M183	1 2 1 1 2	
2.213E-02	1.729E+00	26	F002	2 2 2 2 2	
2.229E-02	1.742E+00	29	F002	2 2 2 2 2	
2.351E-02	1.837E+00	29.99	C349	2 1 2 2 2	
2.368E-02	1.850E+00	30	F300	1 0 0 0 2	
2.364E-02	1.847E+00	30	G029	1 0 2 2 2	
2.350E-02	1.836E+00	30	I333	1 2 1 1 2	
2.343E-02	1.830E+00	31	A004	1 2 2 1 2	
2.285E-02	1.785E+00	32	F002	2 2 2 2 2	
1.970E-02	1.539E+00	34.53	U013	1 0 0 0 0	EFG
2.685E-02	2.098E+00	34.8	L007	2 1 1 2 2	
2.329E-02	1.819E+00	35	F002	2 2 2 2 2	
2.253E-02	1.760E+00	35	S203	1 1 2 1 2	
2.685E-02	2.098E+00	35.1	L007	2 1 1 1 2	
1.925E-02	1.504E+00	35.48	U010	1 0 0 1 1	EFG
2.458E-02	1.920E+00	38	A004	1 2 2 1 2	
2.573E-02	2.010E+00	39.99	C349	2 1 2 2 2	
2.592E-02	2.025E+00	40	B151	1 2 1 1 2	
2.434E-02	1.902E+00	41	F002	2 2 2 2 2	
2.440E-02	1.906E+00	42	F002	2 2 2 2 2	
2.467E-02	1.927E+00	44	F002	2 2 2 2 2	
2.016E-02	1.574E+00	44.30	U010	1 0 0 1 1	EFG
2.062E-02	1.611E+00	44.30	U013	1 0 0 0 0	EFG
2.599E-02	2.030E+00	44.70	A004	1 2 2 1 2	
2.368E-02	1.850E+00	45	S203	1 1 2 1 2	
2.938E-02	2.295E+00	45.7	L007	2 1 1 1 2	
2.938E-02	2.295E+00	45.8	L007	2 1 1 2 2	
2.534E-02	1.979E+00	46	F002	2 2 2 2 2	
2.827E-02	2.208E+00	49.99	C349	2 1 2 2 2	
2.810E-02	2.195E+00	50	G323	2 2 2 2 1	
2.650E-02	2.070E+00	51	F002	2 2 2 2 2	
2.740E-02	2.140E+00	51.50	A004	1 2 2 1 2	
2.159E-02	1.687E+00	53.64	U010	1 0 0 1 1	EFG
2.210E-02	1.726E+00	54.71	U013	1 0 0 0 0	EFG
5.095E-02	3.980E+00	55.3	P051	2 1 1 2 2	
5.095E-02	3.980E+00	55.30	P007	2 1 2 2 2	
2.788E-02	2.178E+00	56	F002	2 2 2 2 2	
3.162E-02	2.470E+00	57	B124	2 2 2 1 2	
3.776E-02	2.950E+00	57.70	B124	1 2 2 1 2	
2.996E-02	2.340E+00	58.80	A004	1 2 2 1 2	
3.131E-02	2.446E+00	59.99	C349	2 1 2 2 2	
2.938E-02	2.295E+00	60	B126	1 0 1 1 1	
3.101E-02	2.422E+00	60	B151	1 2 1 1 2	
2.943E-02	2.299E+00	61	F002	2 2 2 2 2	
3.004E-02	2.347E+00	63	F002	2 2 2 2 2	
3.290E-02	2.570E+00	65.40	A004	1 2 2 1 2	
2.479E-02	1.936E+00	65.82	U013	1 0 0 0 0	EFG
3.597E-02	2.810E+00	69.20	B124	1 2 2 1 2	

3.587E-02	2.802E+00	69.30	B124	1 0 2 2 2	
3.463E-02	2.705E+00	69.99	C349	2 1 2 2 2	
8.280E-02	6.468E+00	74.7	P051	2 1 1 2 2	
8.280E-02	6.468E+00	74.70	P007	2 1 2 2 2	
3.872E-02	3.024E+00	79.99	C349	2 1 2 2 2	
4.429E-02	3.460E+00	89.99	C349	2 1 2 2 2	
5.256E-02	4.106E+00	99.99	C349	2 1 2 2 2	
2.560E-02	2.000E+00	100	J023	1 1 2 2 0	
7.681E-02	6.000E+00	150	J023	1 1 2 2 0	
2.688E-01	2.100E+01	200	J023	1 1 2 2 1	
9.345E-01	7.300E+01	250	J023	1 1 2 2 1	
1.357E+00	1.060E+02	285	J023	1 1 2 2 2	
1.869E+00	1.460E+02	300	J023	1 1 2 2 2	
2.200E-02	1.719E+00	ns	B059	0 0 1 1 2	
4.000E-03	3.125E-01	ns	D348	0 0 2 2 2	
2.279E-02	1.780E+00	ns	H123	0 0 0 0 2	
3.020E-01	2.359E+01	ns	H307	1 0 1 1 2	
4.500E-02	3.515E+00	ns	H333	0 1 0 1 0	EFG
2.330E-02	1.820E+00	ns	I332	0 0 0 0 2	
2.292E-02	1.790E+00	ns	K304	0 0 0 0 2	
1.933E-02	1.510E+00	ns	M010	0 0 0 0 2	
2.265E-02	1.769E+00	ns	M175	0 0 2 1 2	
2.279E-02	1.780E+00	ns	M344	0 0 0 0 2	

677. $C_6H_6BrNO_2S$
4-Bromobenzenesulfonamide
(4-Bromophenyl)sulfonamide
p-Bromobenzenesulfonamide
4-Aminosulfonyl-1-bromobenzene
RN: 701-34-8   **MP** (°C):
MW: 236.09   **BP** (°C):

Solubility (Moles/L)	Solubility (Grams/L)	Temp (°C)	Ref (#)	Evaluation (T P E A A)	Comments
4.200E-03	9.916E-01	15	K024	1 2 1 1 2	

678. $C_6H_6BrNO_3S$
p-Bromoaniline-m-sulfonic Acid
5-Amino-2-bromobenzenesulfonic Acid
RN: 150454-14-1   **MP** (°C):
MW: 252.09   **BP** (°C):

Solubility (Moles/L)	Solubility (Grams/L)	Temp (°C)	Ref (#)	Evaluation (T P E A A)	Comments
1.884E-02	4.750E+00	0.0	P038	1 2 2 2 2	anhydrous monoclinic
2.880E-02	7.260E+00	0.0	P038	1 2 2 2 2	anhydrous rhombic
3.511E-02	8.850E+00	9.8	P038	1 2 2 2 2	anhydrous rhombic
2.559E-02	6.450E+00	12.55	P038	1 2 2 2 2	anhydrous monoclinic

4.284E-02	1.080E+01	20.0	P038	1 2 2 2 2	anhydrous rhombic
3.419E-02	8.620E+00	25.0	P038	1 2 2 2 2	anhydrous monoclinic
4.740E-02	1.195E+01	25.0	P038	1 2 2 2 2	anhydrous rhombic
5.177E-02	1.305E+01	29.6	P038	1 2 2 2 2	anhydrous rhombic
5.732E-02	1.445E+01	34.7	P038	1 2 2 2 2	anhydrous rhombic
4.820E-02	1.215E+01	40.0	P038	1 2 2 2 2	anhydrous monoclinic
6.387E-02	1.610E+01	40.1	P038	1 2 2 2 2	anhydrous rhombic
6.922E-02	1.745E+01	44.5	P038	1 2 2 2 2	anhydrous rhombic
7.577E-02	1.910E+01	49.7	P038	1 2 2 2 2	anhydrous rhombic
8.330E-02	2.100E+01	54.8	P038	1 2 2 2 2	anhydrous rhombic
7.101E-02	1.790E+01	56.3	P038	1 2 2 2 2	anhydrous monoclinic
9.600E-02	2.420E+01	62.3	P038	1 2 2 2 2	anhydrous rhombic
9.679E-02	2.440E+01	70.0	P038	1 2 2 2 2	anhydrous monoclinic
1.115E-01	2.810E+01	70.4	P038	1 2 2 2 2	anhydrous rhombic
1.329E-01	3.350E+01	85.0	P038	1 2 2 2 2	anhydrous monoclinic
1.452E-01	3.660E+01	85.0	P038	1 2 2 2 2	anhydrous rhombic

679. $C_6H_6BrNO_3S$
p-Bromoaniline-o-sulfonic Acid
2-Amino-5-bromophenylsulfonic Acid
RN: 1576-59-6   **MP** (°C):
MW: 252.09   **BP** (°C):

Solubility (Moles/L)	Solubility (Grams/L)	Temp (°C)	Ref (#)	Evaluation (T P E A A)	Comments
8.846E-03	2.230E+00	0.0	P038	1 0 1 0 2	anhydrate
1.107E-02	2.790E+00	8.35	P038	1 0 1 0 2	anhydrate
1.424E-02	3.590E+00	16.75	P038	1 0 1 0 2	anhydrate
1.769E-02	4.460E+00	25.0	P038	1 0 1 0 2	anhydrate
2.578E-02	6.500E+00	40.0	P038	1 0 1 0 2	anhydrate
3.828E-02	9.650E+00	55.0	P038	1 0 1 0 2	anhydrate
5.454E-02	1.375E+01	70.0	P038	1 0 1 0 2	anhydrate
8.013E-02	2.020E+01	85.0	P038	1 0 1 0 2	anhydrate

680. $C_6H_6BrNO_3S.H_2O$
p-Bromoaniline-o-sulfonic Acid (Monohydrate)
2-Amino-5-bromophenylsulfonic Acid (Monohydrate)
RN: 1576-59-6   **MP** (°C):
MW: 270.11   **BP** (°C):

Solubility (Moles/L)	Solubility (Grams/L)	Temp (°C)	Ref (#)	Evaluation (T P E A A)	Comments
9.589E-03	2.590E+00	0.0	P038	1 0 1 0 2	monohydrate
1.303E-02	3.520E+00	8.35	P038	1 0 1 0 2	monohydrate
1.751E-02	4.730E+00	16.8	P038	1 0 1 0 2	monohydrate
2.244E-02	6.060E+00	25.0	P038	1 0 1 0 2	monohydrate

681.   C$_6$H$_6$ClN
m-Chloroaniline
3-Chloroaniline
RN: 108-42-9   **MP** (°C):   -10
MW: 127.57   **BP** (°C):   230.0

Solubility (Moles/L)	Solubility (Grams/L)	Temp (°C)	Ref (#)	Evaluation (T P E A A)	Comments
4.266E-02	5.442E+00	20	C113	1 0 2 1 2	

682.   C$_6$H$_6$ClN
p-Chloroaniline
4-Chloroaniline
RN: 106-47-8   **MP** (°C):   72.5
MW: 127.57   **BP** (°C):   232

Solubility (Moles/L)	Solubility (Grams/L)	Temp (°C)	Ref (#)	Evaluation (T P E A A)	Comments
2.157E-02	2.752E+00	20	H118	1 1 1 1 2	
2.157E-02	2.752E+00	20	H301	2 0 2 2 2	
3.057E-02	3.900E+00	22.5	G301	2 1 0 1 2	

683.   C$_6$H$_6$ClN
o-Chloroaniline
2-Chloroaniline
RN: 95-51-2   **MP** (°C):   -1
MW: 127.57   **BP** (°C):   208.8

Solubility (Moles/L)	Solubility (Grams/L)	Temp (°C)	Ref (#)	Evaluation (T P E A A)	Comments
2.951E-02	3.765E+00	20	C113	1 0 2 1 2	

684.   C$_6$H$_6$ClNO$_2$S
4-Chlorobenzenesulfonamide
p-Chlorobenzenesulfonamide
RN: 98-64-6   **MP** (°C):
MW: 191.64   **BP** (°C):

Solubility (Moles/L)	Solubility (Grams/L)	Temp (°C)	Ref (#)	Evaluation (T P E A A)	Comments
6.900E-03	1.322E+00	15	K024	1 2 1 1 2	

685.   C$_6$H$_6$ClNO$_2$S
m-Chlorobenzenesulfonamide
MON 5783
RN: 17260-71-8   **MP** (°C):
MW: 191.64   **BP** (°C):

Solubility (Moles/L)	Solubility (Grams/L)	Temp (°C)	Ref (#)	Evaluation (T P E A A)	Comments
3.500E-03	6.707E-01	15	K024	1 2 1 1 2	

686.   C$_6$H$_6$ClNO$_2$S
o-Chlorobenzenesulfonamide
2-Chlorobenzenesulfonamide
RN: 6961-82-6   **MP** (°C):
MW: 191.64   **BP** (°C):

Solubility (Moles/L)	Solubility (Grams/L)	Temp (°C)	Ref (#)	Evaluation (T P E A A)	Comments
2.600E-03	4.983E-01	15	K024	1 2 1 1 2	

687.   C$_6$H$_6$ClNO$_3$S
p-Chloroaniline-m-sulfonic Acid
1-Amino-4-chlorobenzene-3-sulfonic Acid
4-Chloro-3-sulfoaniline
3-Amino-6-chlorobenzenesulfonic Acid
RN: 88-43-7   **MP** (°C):
MW: 207.64   **BP** (°C):

Solubility (Moles/L)	Solubility (Grams/L)	Temp (°C)	Ref (#)	Evaluation (T P E A A)	Comments
5.447E-02	1.131E+01	0	P038	1 0 1 1 2	anhydrate

688.   C$_6$H$_6$ClNO$_3$S.H$_2$O
p-Chloroaniline-m-sulfonic Acid (Monohydrate)
1-Amino-4-chlorobenzene-3-sulfonic Acid (Monohydrate)
RN: 88-43-7   **MP** (°C):
MW: 225.65   **BP** (°C):

Solubility (Moles/L)	Solubility (Grams/L)	Temp (°C)	Ref (#)	Evaluation (T P E A A)	Comments
5.141E-02	1.160E+01	0	P038	1 0 1 1 2	metastable monohydrate

689. $C_6H_6ClNO_3S.H_2O$

p-Chloroaniline-o-sulfonic Acid (Monohydrate)

1-Amino-4-chloro-2-benzenesulfonic Acid (Monohydrate)

RN: 133-74-4   **MP** (°C):

MW: 225.65   **BP** (°C):

Solubility (Moles/L)	Solubility (Grams/L)	Temp (°C)	Ref (#)	Evaluation (T P E A A)	Comments
1.387E-02	3.130E+00	0	P038	1 2 2 1 2	monohydrate

690. $C_6H_6Cl_6$

δ-1,2,3,4,5,6-Hexachlorocyclohexane

δ-Benzene Hexachloride

RN: 608-73-1   **MP** (°C):

MW: 290.83   **BP** (°C):

Solubility (Moles/L)	Solubility (Grams/L)	Temp (°C)	Ref (#)	Evaluation (T P E A A)	Comments
3.438E-05	1.000E-02	20	C099	1 2 0 0 1	
1.080E-04	3.140E-02	25	W025	1 0 2 2 2	
4.009E-05	1.166E-02	28	K120	1 2 2 2 2	average of 4

691. $C_6H_6Cl_6$

Lindane

γ-BHC

Benzene Hexachloride

RN: 58-89-9   **MP** (°C):   112.5

MW: 290.83   **BP** (°C):   0

Solubility (Moles/L)	Solubility (Grams/L)	Temp (°C)	Ref (#)	Evaluation (T P E A A)	Comments
7.393E-06	2.150E-03	15	B083	2 2 1 2 2	
7.393E-06	2.150E-03	15	B162	1 0 0 0 2	
2.816E-05	8.190E-03	19	I018	1 0 0 0 2	
3.438E-05	1.000E-02	20	C099	1 2 0 0 1	
2.709E-05	7.880E-03	22	K137	1 1 2 1 0	
2.706E-05	7.870E-03	24	C313	1 0 2 2 2	
5.845E-05	1.700E-02	24	H116	2 1 0 0 2	
2.338E-05	6.800E-03	25	B083	2 2 1 2 2	
2.338E-05	6.800E-03	25	B162	1 0 0 0 2	
2.586E-05	7.520E-03	25	M060	2 2 1 2 2	
2.510E-05	7.300E-03	25	M130	1 0 0 0 1	
2.682E-05	7.800E-03	25	W025	1 0 2 2 2	
4.126E-05	1.200E-02	27	B161	2 1 2 2 0	EFG
2.235E-05	6.500E-03	28	K120	1 2 2 2 2	average of 4
3.920E-05	1.140E-02	35	B083	2 2 1 2 2	particle size ≤ 5 μm
7.221E-05	2.100E-02	35	B161	2 1 2 2 0	EFG
3.920E-05	1.140E-02	35	B162	1 0 0 0 2	

Solubility (Moles/L)	Solubility (Grams/L)	Temp (°C)	Ref (#)	Evaluation (T P E A A)	Comments
5.226E-05	1.520E-02	45	B083	2 2 1 2 2	particle size ≤ 5 µm
9.284E-05	2.700E-02	45	B161	2 1 2 2 0	EFG
1.135E-04	3.300E-02	50	B161	2 1 2 2 0	EFG
1.547E-04	4.500E-02	60	B161	2 1 2 2 0	EFG
2.400E-05	6.980E-03	ns	C318	0 2 2 1 2	
~3.44E-05	~1.00E-02	ns	I308	0 0 0 0 0	
5.158E-07	1.500E-04	ns	K138	0 0 0 0 2	*sic*
3.438E-06	1.000E-03	ns	M061	0 0 0 0 0	
2.407E-05	7.000E-03	ns	M110	0 0 0 0 0	EFG
3.438E-05	1.000E-02	rt	M161	0 0 0 0 1	

692. $C_6H_6Cl_6$

α-1,2,3,4,5,6-Hexachlorocyclohexane

α-Benzene Hexachloride

α-HCH

α-BHC

α-Hexachlorocyclohexane

RN: 319-84-6   **MP** (°C):   158
MW: 290.83   **BP** (°C):   288

Solubility (Moles/L)	Solubility (Grams/L)	Temp (°C)	Ref (#)	Evaluation (T P E A A)	Comments
3.438E-05	1.000E-02	20	C099	1 2 0 0 1	
6.877E-06	2.000E-03	25	W025	1 0 2 2 2	
5.570E-06	1.620E-03	28	K120	1 2 2 2 2	average of 4
3.438E-06	1.000E-03	ns	M061	0 0 0 0 0	

693. $C_6H_6Cl_6$

β-1,2,3,4,5,6-Hexachlorocyclohexane

β-Benzene Hexachloride

β-BHC

β-Hexachlorocyclohexane

RN: 319-85-7   **MP** (°C):   312
MW: 290.83   **BP** (°C):

Solubility (Moles/L)	Solubility (Grams/L)	Temp (°C)	Ref (#)	Evaluation (T P E A A)	Comments
1.719E-05	5.000E-03	20	C099	1 2 0 0 0	
8.252E-07	2.400E-04	25	W025	1 0 2 2 2	
5.501E-07	1.600E-04	28	K120	1 2 2 2 1	average of 2
1.719E-06	5.000E-04	ns	M061	0 0 0 0 0	

694. $C_6H_6FN_3O_3$
1-Methylcarbamoyl-5-fluorouracil
5-Fluoro-3,4-dihydro-N-methyl-2,4-dioxo-pyrimidinecarboxamide
1-Methylcarbamoyl-5-fluoro-2,4(1H,3H)-pyrimidinedi-one
RN: 56563-18-9   **MP** (°C):   225-228
MW: 187.13   **BP** (°C):

Solubility (Moles/L)	Solubility (Grams/L)	Temp (°C)	Ref (#)	Evaluation (T P E A A)	Comments
3.313E-03	6.200E-01	22	B321	1 0 2 2 2	pH 4.0
3.313E-03	6.200E-01	22	B388	1 0 2 2 2	

695. $C_6H_6INO_3S$
6-Iodoaniline-3-sulphonic Acid
Benzenesulfonic Acid, 3-Amino-6-iodo-
RN: **MP** (°C):
MW: 299.09   **BP** (°C):

Solubility (Moles/L)	Solubility (Grams/L)	Temp (°C)	Ref (#)	Evaluation (T P E A A)	Comments
1.597E-02	4.777E+00	25	B107	1 2 1 1 1	

696. $C_6H_6INO_3S$
5-Iodoaniline-3-sulphonic Acid
Benzenesulfonic Acid, 3-Amino-5-iodo-
RN: **MP** (°C):
MW: 299.09   **BP** (°C):

Solubility (Moles/L)	Solubility (Grams/L)	Temp (°C)	Ref (#)	Evaluation (T P E A A)	Comments
4.323E-02	1.293E+01	25	B107	1 2 1 1 2	

697. $C_6H_6INO_3S$
5-Iodoaniline-2-sulphonic Acid
Benzenesulfonic Acid, 2-Amino-5-iodo-
RN: **MP** (°C):
MW: 299.09   **BP** (°C):

Solubility (Moles/L)	Solubility (Grams/L)	Temp (°C)	Ref (#)	Evaluation (T P E A A)	Comments
8.671E-03	2.593E+00	25	B107	1 2 1 1 1	

698.   C$_6$H$_6$INO$_3$S
4-Iodoaniline-3-sulphonic Acid
Benzenesulfonic Acid, 3-Amino-4-iodo-
RN: **MP** (°C):
MW: 299.09   **BP** (°C):

Solubility (Moles/L)	Solubility (Grams/L)	Temp (°C)	Ref (#)	Evaluation (T P E A A)	Comments
4.486E-02	1.342E+01	25	B107	1 2 1 1 2	

699.   C$_6$H$_6$INO$_3$S
4-Iodoaniline-2-sulphonic Acid
Benzenesulfonic Acid, 2-Amino-4-iodo-
RN: 171664-62-3   **MP** (°C):
MW: 299.09   **BP** (°C):

Solubility (Moles/L)	Solubility (Grams/L)	Temp (°C)	Ref (#)	Evaluation (T P E A A)	Comments
1.697E-02	5.074E+00	25	B107	1 2 1 1 1	

700.   C$_6$H$_6$INO$_3$S
2-Iodoaniline-4-sulphonic Acid
Benzenesulfonic Acid, 4-Amino-2-iodo-
RN: 67877-88-7   **MP** (°C):
MW: 299.09   **BP** (°C):

Solubility (Moles/L)	Solubility (Grams/L)	Temp (°C)	Ref (#)	Evaluation (T P E A A)	Comments
6.781E-02	2.028E+01	25	B107	1 2 1 1 2	

701.   C$_6$H$_6$INO$_3$S
3-Iodoaniline-4-sulphonic Acid
Benzenesulfonic Acid, 4-Amino-3-iodo-
RN: 25210-30-4   **MP** (°C):
MW: 299.09   **BP** (°C):

Solubility (Moles/L)	Solubility (Grams/L)	Temp (°C)	Ref (#)	Evaluation (T P E A A)	Comments
6.474E-03	1.936E+00	25	B107	1 2 1 1 2	

702. $C_6H_6N_2O$
Nicotiamide
Niacinamide
Nicotinamide
RN: 98-92-0   **MP** (°C):   131
MW: 122.13   **BP** (°C):

Solubility (Moles/L)	Solubility (Grams/L)	Temp (°C)	Ref (#)	Evaluation (T P E A A)	Comments
4.094E+00	5.000E+02	20	D041	1 0 0 0 2	
8.188E+00	1.000E+03	20	M054	1 0 0 0 2	
2.900E-03	3.542E-01	25	A350	2 0 2 1 2	
8.188E+00	1.000E+03	25	D315	1 0 1 1 2	

703. $C_6H_6N_2O_2$
Urocanic Acid
Urocaninsaeure
RN: 104-98-3   **MP** (°C):   225
MW: 138.13   **BP** (°C):

Solubility (Moles/L)	Solubility (Grams/L)	Temp (°C)	Ref (#)	Evaluation (T P E A A)	Comments
1.086E-02	1.500E+00	17.40	F300	1 0 0 0 1	
4.318E-02	5.964E+00	37	D041	1 0 0 0 0	
5.575E-02	7.700E+00	50	F300	1 0 0 0 1	
4.098E-01	5.660E+01	100	D041	1 0 0 0 0	

704. $C_6H_6N_2O_2$
2-Nitroaniline
o-Nitroaniline
1-Amino-2-nitrobenzene
2-Nitro-aniline
RN: 88-74-4   **MP** (°C):   71.5
MW: 138.13   **BP** (°C):   284

Solubility (Moles/L)	Solubility (Grams/L)	Temp (°C)	Ref (#)	Evaluation (T P E A A)	Comments
6.467E-03	8.932E-01	20	T301	1 2 2 2 2	
8.764E-03	1.211E+00	25.0	C026	2 1 1 2 2	
1.750E-02	2.417E+00	40.1	C026	2 1 1 2 2	
6.134E-03	8.473E-01	50	T301	1 2 2 2 2	average of 4
6.799E-03	9.391E-01	80	T301	1 2 2 2 2	average of 4

705. $C_6H_6N_2O_2$
3-Nitroaniline
1-Amino-3-nitrobenzene
3-Nitrobenzenamine
m-Nitroaminobenzene
m-Nitroaniline
3-Nitro-anilin

RN: 99-09-2   **MP** (°C):   114
MW: 138.13   **BP** (°C):   306

Solubility (Moles/L)	Solubility (Grams/L)	Temp (°C)	Ref (#)	Evaluation (T P E A A)	Comments
8.710E-03	1.203E+00	20	B179	2 0 0 0 2	
5.370E-03	7.418E-01	25	B335	1 2 0 0 1	
6.516E-03	9.000E-01	25	F300	1 0 0 0 2	
3.020E-03	4.171E-01	25	L016	1 0 0 0 2	unit assumed
6.582E-03	9.092E-01	25.0	C026	2 1 1 2 2	
1.290E-02	1.782E+00	40.1	C026	2 1 1 2 2	

706. $C_6H_6N_2O_2$
p-Nitroaniline
4-Amino-nitrobenzene
Benzenamine
4-Nitroaniline
p-Aminonitrobenzene
4-Nitrobenzenamine

RN: 100-01-6   **MP** (°C):   146
MW: 138.13   **BP** (°C):   332

Solubility (Moles/L)	Solubility (Grams/L)	Temp (°C)	Ref (#)	Evaluation (T P E A A)	Comments
5.754E-03	7.948E-01	20	B179	2 0 0 0 2	
2.823E-03	3.900E-01	20	H300	1 2 2 2 1	sic
2.815E-03	3.888E-01	20	T301	1 2 2 2 2	
3.020E-03	4.171E-01	25	B335	1 2 0 0 1	
4.344E-03	6.000E-01	25	F300	1 0 0 0 2	sic
5.370E-03	7.418E-01	25	L016	1 0 0 0 2	unit assumed
4.110E-03	5.677E-01	25.0	C026	2 1 1 2 2	
5.267E-03	7.275E-01	30	G029	1 0 2 2 2	
8.367E-03	1.156E+00	40.1	C026	2 1 1 2 2	

707.   C$_6$H$_6$N$_2$O$_3$
5,5-Ethylenebarbituric Acid
Spirocyclopropane-1',5-barbituric Acid
5,7-Diazaspiro[2.5]octane-4,6,8-trione
RN: 6947-77-9   **MP** (°C):
MW: 154.13   **BP** (°C):

Solubility (Moles/L)	Solubility (Grams/L)	Temp (°C)	Ref (#)	Evaluation (T P E A A)	Comments
1.300E-02	2.004E+00	25	P350	2 1 1 1 2	intrinsic

708.   C$_6$H$_6$N$_2$O$_4$
1-Methylorotic Acid
4-Pyrimidinecarboxylic Acid, 1,2,3,6-Tetrahydro-1-methyl-2,6-dioxo-
RN: 705-36-2   **MP** (°C):
MW: 170.13   **BP** (°C):

Solubility (Moles/L)	Solubility (Grams/L)	Temp (°C)	Ref (#)	Evaluation (T P E A A)	Comments
1.200E-01	2.042E+01	20	N019	2 2 1 2 2	

709.   C$_6$H$_6$N$_2$O$_4$S
2-Nitrobenzenesulfonamide
o-Nitrobenzenesulfonamide
RN: 5455-59-4   **MP** (°C):
MW: 202.19   **BP** (°C):

Solubility (Moles/L)	Solubility (Grams/L)	Temp (°C)	Ref (#)	Evaluation (T P E A A)	Comments
1.600E-03	3.235E-01	15	K024	1 2 1 1 2	

710.   C$_6$H$_6$N$_2$O$_4$S
4-Nitrobenzenesulfonamide
p-Nitrobenzenesulfonamide
RN: 6325-93-5   **MP** (°C):
MW: 202.19   **BP** (°C):

Solubility (Moles/L)	Solubility (Grams/L)	Temp (°C)	Ref (#)	Evaluation (T P E A A)	Comments
3.000E-03	6.066E-01	15	K024	1 2 1 1 2	

711.   C$_6$H$_6$N$_2$O$_4$S
m-Nitrobenzenesulfonamide
3-Nitrobenzenesulfonamide
RN: 121-52-8   **MP** (°C):
MW: 202.19   **BP** (°C):

Solubility (Moles/L)	Solubility (Grams/L)	Temp (°C)	Ref (#)	Evaluation (T P E A A)	Comments
2.200E-03	4.448E-01	15	K024	1 2 1 1 2	

712.   C$_6$H$_6$N$_4$
8-Methylpurine
1H-Purine, 8-Methyl-
RN: 934-33-8   **MP** (°C):
MW: 134.14   **BP** (°C):

Solubility (Moles/L)	Solubility (Grams/L)	Temp (°C)	Ref (#)	Evaluation (T P E A A)	Comments
3.924E-01	5.263E+01	20	A022	1 0 0 0 0	

713.   C$_6$H$_6$N$_4$O
8-Hydroxymethylpurine
Purine-8-methanol
RN: 6642-26-8   **MP** (°C):
MW: 150.14   **BP** (°C):

Solubility (Moles/L)	Solubility (Grams/L)	Temp (°C)	Ref (#)	Evaluation (T P E A A)	Comments
3.014E-02	4.525E+00	20	A022	1 2 0 0 0	
4.440E-01	6.667E+01	100	A082	1 2 0 0 0	

714.   C$_6$H$_6$N$_4$O$_3$
9-Methyluric Acid
1H-Purine-2,6,8(3H)-trione, 7,9-Dihydro-9-methyl-
N9-Methyluric Acid
RN: 55441-71-9   **MP** (°C):
MW: 182.14   **BP** (°C):

Solubility (Moles/L)	Solubility (Grams/L)	Temp (°C)	Ref (#)	Evaluation (T P E A A)	Comments
2.999E-03	5.461E-01	ns	B115	0 0 1 1 0	

715.   C$_6$H$_6$N$_4$O$_3$
1-Methyluric Acid
α-Methyluric Acid
RN: 708-79-2   **MP** (°C):   400
MW: 182.14   **BP** (°C):

Solubility (Moles/L)	Solubility (Grams/L)	Temp (°C)	Ref (#)	Evaluation (T P E A A)	Comments
1.153E-02	2.101E+00	ns	B115	0 0 1 1 0	ζ form
8.701E-03	1.585E+00	ns	B115	0 0 1 1 0	γ form
2.731E-02	4.975E+00	ns	B115	0 0 1 1 0	

716.   C$_6$H$_6$N$_4$O$_3$S
Niridazole
Nirodazole
RN: 61-57-4   **MP** (°C):   261
MW: 214.20   **BP** (°C):

Solubility (Moles/L)	Solubility (Grams/L)	Temp (°C)	Ref (#)	Evaluation (T P E A A)	Comments
6.068E-04	1.300E-01	25	A081	1 0 1 1 0	EFG
1.634E-04	3.500E-02	25	G051	1 0 1 1 0	pH 2

717.   C$_6$H$_6$N$_4$O$_4$
5-Nitro-2-Furaldehyde Semicarbazone
Nitrofurazone
RN: 59-87-0   **MP** (°C):   236
MW: 198.14   **BP** (°C):

Solubility (Moles/L)	Solubility (Grams/L)	Temp (°C)	Ref (#)	Evaluation (T P E A A)	Comments
1.201E-03	2.380E-01	ns	I310	0 0 0 0 2	

718.   C$_6$H$_6$N$_6$
2,4-Diaminopteridine
2:4-Diaminopteridine
RN: 1127-93-1   **MP** (°C):
MW: 162.15   **BP** (°C):

Solubility (Moles/L)	Solubility (Grams/L)	Temp (°C)	Ref (#)	Evaluation (T P E A A)	Comments
2.055E-03	3.332E-01	20	A019	2 2 1 1 2	
4.708E-02	7.634E+00	100	A019	1 2 1 1 1	

719.   C_6H_6N_6 $C_6H_6N_6$
4-Hydrazinopteridine
4(1H)-Pteridinone, Hydrazone
RN: 77632-11-2   **MP** (°C):
MW: 162.15   **BP** (°C):

Solubility (Moles/L)	Solubility (Grams/L)	Temp (°C)	Ref (#)	Evaluation (T P E A A)	Comments
1.367E-02	2.217E+00	20	A083	1 2 0 0 0	
8.686E-02	1.408E+01	100	A083	1 2 0 0 0	

720.   C_6H_6N_6
4,7-Diaminopteridine
4:7-Diaminopteridine
RN: 771-41-5   **MP** (°C):
MW: 162.15   **BP** (°C):

Solubility (Moles/L)	Solubility (Grams/L)	Temp (°C)	Ref (#)	Evaluation (T P E A A)	Comments
1.233E-03	2.000E-01	20	A020	1 2 0 0 1	
2.049E-02	3.322E+00	100	A020	1 2 0 0 0	

721.   C_6H_6N_6
4,6-Diaminopteridine
4:6-Diaminopteridine
RN: 19167-60-3   **MP** (°C):
MW: 162.15   **BP** (°C):

Solubility (Moles/L)	Solubility (Grams/L)	Temp (°C)	Ref (#)	Evaluation (T P E A A)	Comments
2.569E-04	4.166E-02	20	A020	1 2 0 1 1	
6.554E-03	1.063E+00	100	A020	1 2 0 0 0	

722.   C_6H_6O C_6H_6O
Phenol
Carbolic Acid
Hydroxybenzene
RN: 108-95-2   **MP** (°C):   40.85
MW: 94.11   **BP** (°C):   182

Solubility (Moles/L)	Solubility (Grams/L)	Temp (°C)	Ref (#)	Evaluation (T P E A A)	Comments
7.164E-01	6.743E+01	0	A056	1 0 1 1 2	
7.136E-01	6.716E+01	0	B031	1 2 2 2 1	
7.164E-01	6.743E+01	0	L059	1 0 1 1 2	
6.858E-01	6.455E+01	8.60	C058	2 0 2 1 1	
7.321E-01	6.890E+01	10	A056	1 0 1 1 2	
7.321E-01	6.890E+01	10	L059	1 0 1 1 2	
6.672E-01	6.279E+01	16	D041	1 0 0 0 1	
7.779E-01	7.322E+01	20	B031	1 2 2 2 1	

8.710E-01	8.197E+01	20	B179	2 0 0 0 2	
4.866E+00	4.580E+02	20	C052	1 2 1 1 2	*sic*
8.235E-01	7.750E+01	20	F300	1 0 0 0 2	
8.198E-01	7.715E+01	20	H003	1 2 2 1 2	
1.600E+00	1.506E+02	20	H306	1 0 1 2 1	
8.500E-01	8.000E+01	20	K119	1 0 0 0 2	
7.130E-01	6.710E+01	20	K301	2 2 1 1 2	
6.175E-01	5.811E+01	20	R087	1 1 2 2 2	0.15M NaCl
9.490E-01	8.931E+01	22.70	M135	1 2 1 1 2	
1.000E+00	9.411E+01	25	A021	1 2 1 1 0	
9.882E-01	9.300E+01	25	B060	2 0 1 1 1	
9.400E-01	8.847E+01	25	B316	1 0 2 1 1	
9.000E-01	8.470E+01	25	F044	1 0 0 0 1	
8.468E-01	7.970E+01	25	H003	1 2 2 1 2	
8.245E-01	7.759E+01	25	H028	2 0 2 0 2	
1.527E-01	1.437E+01	25	K129	2 1 2 2 2	
8.854E-01	8.333E+01	25	L022	1 0 0 0 0	
9.000E-01	8.470E+01	25	L088	1 0 0 0 1	
7.413E-01	6.977E+01	25	M041	1 1 0 0 1	
9.300E-01	8.753E+01	25	P031	1 1 2 2 2	
7.688E-01	7.236E+01	25	R041	1 0 2 1 1	
9.900E-01	9.317E+01	26.90	M135	1 2 1 1 2	
8.970E-01	8.442E+01	30	H003	1 2 2 1 2	
8.297E-01	7.809E+01	30	V009	1 0 0 0 1	
1.048E+00	9.863E+01	32.20	M135	1 2 1 1 2	
9.598E-01	9.033E+01	34	B063	1 2 2 1 2	
9.580E-01	9.016E+01	35	H003	1 2 2 1 2	
1.107E+00	1.042E+02	36.00	M135	1 2 1 1 2	
9.130E-01	8.592E+01	40	B031	1 2 2 2 1	
1.158E+00	1.090E+02	43.70	M135	1 2 1 1 2	
1.369E+00	1.288E+02	47.70	M135	1 2 1 1 2	
1.172E+00	1.103E+02	48.00	C058	2 0 2 1 2	
1.138E+00	1.071E+02	50	M041	1 1 0 0 2	
1.476E+00	1.389E+02	50.50	M135	1 2 1 1 2	
1.183E+00	1.113E+02	51.90	B063	1 2 2 1 2	
1.592E+00	1.498E+02	53.50	M135	1 2 1 1 2	
1.725E+00	1.623E+02	55.80	M135	1 2 1 1 2	
1.388E+00	1.306E+02	55.90	B063	1 2 2 1 2	
1.375E+00	1.295E+02	57.30	H003	1 2 2 1 2	
1.856E+00	1.747E+02	57.80	M135	1 2 1 1 2	
1.590E+00	1.497E+02	60	B031	1 2 2 2 2	
2.163E+00	2.036E+02	60.90	M135	1 2 1 1 2	
1.612E+00	1.518E+02	61.70	B063	1 2 2 1 2	
1.723E+00	1.621E+02	62.74	H003	1 2 2 1 2	
1.771E+00	1.667E+02	63.20	B063	1 2 2 1 2	
2.109E+00	1.985E+02	65.40	B063	1 2 2 1 2	
3.064E+00	2.884E+02	65.50	B063	1 2 2 1 2	
2.567E+00	2.416E+02	65.55	B063	1 2 2 1 2	

Solubility (Moles/L)	Solubility (Grams/L)	Temp (°C)	Ref (#)	Evaluation (T P E A A)	Comments
2.767E+00	2.604E+02	65.60	B063	1 2 2 1 2	
2.388E+00	2.247E+02	65.79	H003	1 2 2 1 2	average of 2
2.590E+00	2.437E+02	65.84	H003	1 2 2 1 2	
2.624E+00	2.469E+02	65.86	H003	1 2 2 1 2	
2.536E+00	2.387E+02	65.90	H003	1 2 2 1 2	
2.818E+00	2.652E+02	66.0	H068	2 0 0 0 2	
2.397E+00	2.256E+02	66.01	H003	1 2 2 1 2	
1.734E+00	1.632E+02	66.30	C058	2 0 2 1 2	
8.594E-01	8.088E+01	ns	N330	2 2 2 1 2	
8.043E-01	7.570E+01	rt	N051	0 0 2 1 2	average of 3

723. $C_6H_6O_2$
Resorcinol
Resorcin
RN: 108-46-3   **MP** (°C): 110.0
MW: 110.11   **BP** (°C):

Solubility (Moles/L)	Solubility (Grams/L)	Temp (°C)	Ref (#)	Evaluation (T P E A A)	Comments
3.404E+00	3.748E+02	0	M022	1 0 0 0 2	
3.617E+00	3.983E+02	0	M043	1 0 0 0 2	
2.784E+00	3.066E+02	3.70	L090	1 0 0 1 2	
4.173E+00	4.595E+02	10	M043	1 0 0 0 1	
5.413E+00	5.960E+02	12.50	F300	1 0 0 0 2	
3.186E+00	3.508E+02	14.20	L090	1 0 0 1 2	
3.359E+00	3.699E+02	19.50	L090	1 0 0 1 2	
4.576E+00	5.038E+02	20	M022	1 0 0 0 2	
5.009E+00	5.516E+02	20	M043	1 0 0 0 2	
6.515E+00	7.174E+02	25	K040	1 0 2 1 2	
6.330E+00	6.970E+02	30	F300	1 0 0 0 2	
5.718E+00	6.296E+02	30	M043	1 0 0 0 2	
3.679E+00	4.051E+02	32.50	L090	1 0 0 1 2	
1.464E+01	1.612E+03	33.61	W038	2 2 2 1 2	
5.641E+00	6.211E+02	40	M022	1 0 0 0 2	
6.287E+00	6.923E+02	40	M043	1 0 0 0 2	
1.843E+01	2.030E+03	44.5	W038	2 2 2 1 2	
2.042E+01	2.249E+03	49.3	W038	2 2 2 1 2	
2.100E+01	2.312E+03	50.4	W038	2 2 2 1 2	
6.465E+00	7.119E+02	60	M022	1 0 0 0 2	
7.228E+00	7.959E+02	60	M043	1 0 0 0 2	
2.701E+01	2.974E+03	64.4	W038	2 2 2 1 2	
2.997E+01	3.300E+03	70.7	W038	2 2 2 1 2	
7.106E+00	7.825E+02	80	M022	1 0 0 0 2	
7.844E+00	8.638E+02	80	M043	1 0 0 0 2	
3.516E+01	3.871E+03	80.5	W038	2 2 2 1 2	
4.008E+01	4.414E+03	88.5	W038	2 2 2 1 2	
7.592E+00	8.360E+02	100	M022	1 0 0 0 2	
8.299E+00	9.138E+02	100	M043	1 0 0 0 2	
5.556E+01	6.117E+03	109.4	W038	2 2 2 1 2	
4.608E+00	5.074E+02	rt	D021	0 0 1 1 2	

724.   C_6H_6O_2

$C_6H_6O_2$

Pyrocatechol
Brenzkatechin
Catechol

RN: 120-80-9   **MP** (°C):   105
MW: 110.11   **BP** (°C):   245.5

Solubility (Moles/L)	Solubility (Grams/L)	Temp (°C)	Ref (#)	Evaluation (T P E A A)	Comments
2.824E+00	3.110E+02	20	F300	1 0 0 0 2	
2.823E+00	3.108E+02	20	M043	1 0 0 0 2	
4.190E+00	4.614E+02	25	K040	1 0 2 1 2	
5.743E+00	6.324E+02	40	M043	1 0 0 0 2	
1.278E+01	1.408E+03	41.2	W038	2 2 2 1 2	
2.061E+01	2.270E+03	56.7	W038	2 2 2 1 2	
2.068E+01	2.278E+03	57.1	W038	2 2 2 1 2	
7.308E+00	8.047E+02	60	M043	1 0 0 0 2	
2.617E+01	2.882E+03	66.2	W038	2 2 2 1 2	
8.337E+00	9.180E+02	80	M043	1 0 0 0 2	
8.974E+00	9.882E+02	100	M043	1 0 0 0 2	
5.556E+01	6.117E+03	104.5	W038	2 2 2 1 2	
2.823E+00	3.108E+02	rt	D021	0 0 1 1 2	

725.   C_6H_6O_2

$C_6H_6O_2$

Hydroquinone
Hydrochinon
Hydroquinol

RN: 123-31-9   **MP** (°C):   173.5
MW: 110.11   **BP** (°C):   286

Solubility (Moles/L)	Solubility (Grams/L)	Temp (°C)	Ref (#)	Evaluation (T P E A A)	Comments
3.493E-01	3.846E+01	0	M043	1 0 0 0 1	
4.653E-01	5.123E+01	10	M043	1 0 0 0 1	
4.904E-01	5.400E+01	15	F300	1 0 0 0 1	
5.077E-01	5.590E+01	17.70	L065	1 0 0 0 2	0.01N HCl
5.087E-01	5.601E+01	17.90	L065	1 0 0 0 2	0.01N HCl
5.101E-01	5.617E+01	17.95	L065	1 0 0 0 2	0.01N HCl
5.103E-01	5.619E+01	18	L064	2 2 2 1 2	0.01N HCl
6.100E-01	6.716E+01	20	M043	1 0 0 0 1	
6.357E-01	7.000E+01	22.5	G301	2 1 0 1 2	
6.180E-01	6.805E+01	23.75	L064	2 2 2 1 2	0.01N HCl
6.450E-01	7.102E+01	25	G033	1 0 1 1 2	
7.283E-01	8.020E+01	25	K033	1 0 0 1 2	
6.660E-01	7.334E+01	25	K040	1 0 2 1 2	
7.955E-01	8.759E+01	30	M043	1 0 0 0 1	
1.045E+00	1.150E+02	40	M043	1 0 0 0 1	
2.354E+00	2.593E+02	60	M043	1 0 0 0 1	

5.694E+00	6.270E+02	75.3	W038	2 2 2 1 2
4.251E+00	4.681E+02	80	M043	1 0 0 0 1
7.528E+00	8.289E+02	81.9	W038	2 2 2 1 2
6.034E+00	6.644E+02	100	M043	1 0 0 0 2
1.961E+01	2.159E+03	114.6	W038	2 2 2 1 2
2.180E+01	2.400E+03	120.3	W038	2 2 2 1 2
2.728E+01	3.004E+03	131.7	W038	2 2 2 1 2
2.942E+01	3.239E+03	136.0	W038	2 2 2 1 2
3.353E+01	3.692E+03	141.8	W038	2 2 2 1 2
3.621E+01	3.987E+03	147.2	W038	2 2 2 1 2
6.084E-01	6.699E+01	rt	D021	0 0 1 1 2

726. $C_6H_6O_3$

Phloroglucinol
1,3,5-Benzenetriol
1,3,5-Trihydroxybenzene
1,3,5-THB

RN: 108-73-6   **MP** (°C):   218.0
MW: 126.11   **BP** (°C):

Solubility (Moles/L)	Solubility (Grams/L)	Temp (°C)	Ref (#)	Evaluation (T P E A A)	Comments
8.405E-02	1.060E+01	20	F300	1 0 0 0 2	
8.860E-02	1.117E+01	rt	D021	0 0 1 1 2	

727. $C_6H_6O_3$

Methyl Furoate
5-Methyl-brenzschleimsaeure
5-Methylfuroic Acid

RN: 611-13-2   **MP** (°C):
MW: 126.11   **BP** (°C):   181

Solubility (Moles/L)	Solubility (Grams/L)	Temp (°C)	Ref (#)	Evaluation (T P E A A)	Comments
1.475E-01	1.860E+01	20	F300	1 0 0 0 2	

728. $C_6H_6O_3$

Maltol
3-Hydroxy-2-methyl-4-pyrone
Hydroxymethylpyrone
Palatone

RN: 118-71-8   **MP** (°C):   161.5
MW: 126.11   **BP** (°C):

Solubility (Moles/L)	Solubility (Grams/L)	Temp (°C)	Ref (#)	Evaluation (T P E A A)	Comments
8.643E-02	1.090E+01	15	F300	1 0 0 0 2	

729.   C$_6$H$_6$O$_3$
Pyrogallol
1,2,3-Trihydroxybenzene
1,2,3-Benzenetriol
Brown AP
Fourrine 85
RN: 87-66-1   **MP** (°C):   131
MW: 126.11   **BP** (°C):   309

Solubility (Moles/L)	Solubility (Grams/L)	Temp (°C)	Ref (#)	Evaluation (T P E A A)	Comments
2.379E+00	3.000E+02	13	F300	1 0 0 0 0	average
3.013E+00	3.800E+02	25	F300	1 0 0 0 1	
4.020E+00	5.070E+02	25	K040	1 0 2 1 2	

730.   C$_6$H$_6$O$_3$S
Benzenesulfonic Acid
Benzolsulfosaeure
RN: 98-11-3   **MP** (°C):   43
MW: 158.18   **BP** (°C):

Solubility (Moles/L)	Solubility (Grams/L)	Temp (°C)	Ref (#)	Evaluation (T P E A A)	Comments
3.088E+00	4.885E+02	31.4	T023	1 2 2 1 2	
3.109E+00	4.917E+02	42.6	T023	1 2 2 1 2	
3.136E+00	4.960E+02	56.0	T023	1 2 2 1 2	
3.154E+00	4.989E+02	61.3	T023	1 2 2 1 2	

731.   C$_6$H$_6$O$_3$S.H$_2$O
Benzenesulfonic Acid (Monohydrate)
RN: 98-11-3   **MP** (°C):
MW: 176.19   **BP** (°C):

Solubility (Moles/L)	Solubility (Grams/L)	Temp (°C)	Ref (#)	Evaluation (T P E A A)	Comments
2.542E+00	4.478E+02	21.3	T023	1 2 2 1 2	
2.568E+00	4.525E+02	31.0	T023	1 2 2 1 2	
2.770E+00	4.881E+02	32.6	T023	1 2 2 1 2	
2.598E+00	4.577E+02	39.5	T023	1 2 2 1 2	
2.751E+00	4.846E+02	39.8	T023	1 2 2 1 2	
2.722E+00	4.796E+02	49.0	T023	1 2 2 1 2	
2.641E+00	4.654E+02	49.0	T023	1 2 2 1 2	
2.682E+00	4.726E+02	52.4	T023	1 2 2 1 2	

732. $C_6H_6O_3S.2.5H_2O$
Benzenesulfonic Acid (2.5 Hydrate)
RN: 98-11-3   MP (°C):
MW: 203.22   BP (°C):

Solubility (Moles/L)	Solubility (Grams/L)	Temp (°C)	Ref (#)	Evaluation (T P E A A)	Comments
2.107E+00	4.281E+02	-4.0	T023	1 2 2 1 2	
2.122E+00	4.312E+02	-3.3	T023	1 2 2 1 2	
2.131E+00	4.331E+02	-2.5	T023	1 2 2 1 2	
2.150E+00	4.370E+02	-2.3	T023	1 2 2 1 2	

733. $C_6H_6O_3S.2H_2O$
Benzenesulfonic Acid (Dihydrate)
RN: 98-11-3   MP (°C):
MW: 194.21   BP (°C):

Solubility (Moles/L)	Solubility (Grams/L)	Temp (°C)	Ref (#)	Evaluation (T P E A A)	Comments
2.250E+00	4.370E+02	2.2	T023	1 2 2 1 2	
2.265E+00	4.399E+02	7.5	T023	1 2 2 1 2	
2.289E+00	4.446E+02	13.7	T023	1 2 2 1 2	
2.297E+00	4.460E+02	15.1	T023	1 2 2 1 2	

734. $C_6H_6O_3S.3H_2O$
Benzenesulfonic Acid (Trihydrate)
RN: 98-11-3   MP (°C):
MW: 212.22   BP (°C):

Solubility (Moles/L)	Solubility (Grams/L)	Temp (°C)	Ref (#)	Evaluation (T P E A A)	Comments
1.690E+00	3.586E+02	-40.8	T023	1 2 2 1 2	
1.766E+00	3.748E+02	-29.0	T023	1 2 2 1 2	
1.842E+00	3.909E+02	-18.5	T023	1 2 2 1 2	
1.922E+00	4.078E+02	-10.0	T023	1 2 2 1 2	
1.975E+00	4.191E+02	-5.9	T023	1 2 2 1 2	
2.011E+00	4.267E+02	-4.7	T023	1 2 2 1 2	

735. $C_6H_6O_4$
Muconic Acid
Muconsaeure
RN: 505-70-4   MP (°C):
MW: 142.11   BP (°C):

Solubility (Moles/L)	Solubility (Grams/L)	Temp (°C)	Ref (#)	Evaluation (T P E A A)	Comments
1.407E-03	2.000E-01	20	F300	1 0 0 0 2	

736.   C₆H₇F₃N₄OS

Thiazafluron

Urea, N,N'-Dimethyl-N-[5-(trifluoromethyl)-1,3,4-thiadiazol-2-yl]-

RN: 25366-23-8   **MP** (°C):   136.5
MW: 240.21   **BP** (°C):

Solubility (Moles/L)	Solubility (Grams/L)	Temp (°C)	Ref (#)	Evaluation (T P E A A)	Comments
8.724E-03	2.096E+00	20	E048	1 2 1 1 2	
8.742E-03	2.100E+00	20	M161	1 0 0 0 1	

737.   C₆H₇N

Aniline

Aminobenzene

C.I. Oxidation base 1

Aminophen

Kyanol

RN: 62-53-3   **MP** (°C):   -6.3
MW: 93.13   **BP** (°C):   184

Solubility (Moles/L)	Solubility (Grams/L)	Temp (°C)	Ref (#)	Evaluation (T P E A A)	Comments
3.531E-01	3.288E+01	8.60	C058	2 0 2 1 1	
3.877E-01	3.611E+01	13.8	K119	1 0 0 0 2	
3.747E-01	3.490E+01	18	F300	1 0 0 0 2	
3.818E-01	3.556E+01	18.15	P057	2 2 2 2 2	
3.612E-01	3.364E+01	22	H072	1 0 1 1 2	
3.930E-01	3.660E+01	22.5	G301	2 1 0 1 2	
3.931E-01	3.661E+01	25	B019	1 0 1 2 0	
3.931E-01	3.661E+01	25	B092	2 1 1 1 2	
4.000E-01	3.725E+01	25	F044	1 0 0 0 1	
3.791E-01	3.531E+01	25	G323	2 2 2 2 2	
3.800E-01	3.539E+01	25	H028	2 0 2 0 2	
3.791E-01	3.531E+01	25	H078	1 2 1 0 2	
3.650E-01	3.399E+01	25	M116	2 1 1 1 2	
3.731E-01	3.475E+01	25.40	C058	2 0 2 1 1	
3.930E-01	3.660E+01	26.70	L095	2 2 1 1 2	
4.229E-01	3.939E+01	48.00	C058	2 0 2 1 1	
4.328E-01	4.031E+01	50	G323	2 2 2 2 2	
5.016E-01	4.671E+01	60	B092	2 1 1 1 2	
5.016E-01	4.671E+01	66.30	C058	2 0 2 1 1	
7.025E-01	6.542E+01	96.70	C058	2 0 2 1 1	

738. C_6H_7NO
Phenylhydroxylamine
Phenylhydroxylamin
RN: 100-65-2   **MP** (°C):   82
MW: 109.13   **BP** (°C):

Solubility (Moles/L)	Solubility (Grams/L)	Temp (°C)	Ref (#)	Evaluation (T P E A A)	Comments
1.833E-01	2.000E+01	5	F300	1 0 0 0 0	
8.247E-01	9.000E+01	100	F300	1 0 0 0 0	

739. C_6H_7NO
p-Aminophenol
4-Aminophenol
RN: 123-30-8   **MP** (°C):   190
MW: 109.13   **BP** (°C):

Solubility (Moles/L)	Solubility (Grams/L)	Temp (°C)	Ref (#)	Evaluation (T P E A A)	Comments
1.008E-01	1.100E+01	0	F300	1 0 0 0 1	
9.970E-02	1.088E+01	0	M043	1 0 0 0 1	
1.176E-01	1.283E+01	10	M043	1 0 0 0 1	
1.443E-01	1.575E+01	20	M043	1 0 0 0 1	
1.709E-01	1.865E+01	30	M043	1 0 0 0 1	
2.060E-01	2.248E+01	40	M043	1 0 0 0 1	
2.678E-01	2.922E+01	59.0	S120	1 2 1 1 1	
3.184E-01	3.475E+01	60	M043	1 0 0 0 1	
5.544E-01	6.050E+01	77.0	S120	1 2 1 1 1	
6.709E-01	7.322E+01	80	M043	1 0 0 0 1	
8.399E-01	9.165E+01	86.7	S120	1 2 1 1 1	
1.497E+00	1.634E+02	96.6	S120	1 2 1 1 1	
2.475E+00	2.701E+02	100	M043	1 0 0 0 1	

740. C_6H_7NO
o-Aminophenol
2-Amino-phenol
RN: 95-55-6   **MP** (°C):   172
MW: 109.13   **BP** (°C):

Solubility (Moles/L)	Solubility (Grams/L)	Temp (°C)	Ref (#)	Evaluation (T P E A A)	Comments
1.558E-01	1.700E+01	0	F300	1 0 0 0 1	
1.532E-01	1.672E+01	0	M043	1 0 0 0 1	
1.709E-01	1.865E+01	10	M043	1 0 0 0 1	
1.797E-01	1.961E+01	20	M043	1 0 0 0 1	
1.973E-01	2.153E+01	30	M043	1 0 0 0 1	
2.148E-01	2.344E+01	40	M043	1 0 0 0 1	
2.409E-01	2.629E+01	60	M043	1 0 0 0 1	

Solubility (Moles/L)	Solubility (Grams/L)	Temp (°C)	Ref (#)	Evaluation (T P E A A)
2.669E-01	2.913E+01	80	M043	1 0 0 0 1
2.686E-01	2.931E+01	80.8	S120	1 2 1 1 1
3.558E-01	3.883E+01	88.0	S120	1 2 1 1 1
5.995E-01	6.542E+01	100	M043	1 0 0 0 1

741.   C₆H₇NO
m-Aminophenol
3-Aminophenol

RN: 591-27-5   **MP** (°C):   125
MW: 109.13   **BP** (°C):   164

Solubility (Moles/L)	Solubility (Grams/L)	Temp (°C)	Ref (#)	Evaluation (T P E A A)	Comments
1.797E-01	1.961E+01	10	M043	1 0 0 0 1	
2.291E-01	2.500E+01	20	F300	1 0 0 0 1	
2.409E-01	2.629E+01	20	M043	1 0 0 0 1	
3.355E-01	3.661E+01	30	M043	1 0 0 0 1	
3.261E-01	3.559E+01	32.6	S120	1 2 1 1 2	
4.859E-01	5.303E+01	40	M043	1 0 0 0 1	
6.788E-01	7.407E+01	47.9	S120	1 2 1 1 2	
8.850E-01	9.658E+01	53.0	S120	1 2 1 1 2	
1.590E+00	1.736E+02	60	M043	1 0 0 0 1	
1.406E+00	1.535E+02	60.4	S120	1 2 1 1 2	
2.148E+00	2.344E+02	66.4	S120	1 2 1 1 2	
2.627E+00	2.866E+02	68.9	S120	1 2 1 1 2	
2.927E+00	3.194E+02	70.2	S120	1 2 1 1 2	
3.161E+00	3.450E+02	71.5	S120	1 2 1 1 2	
3.410E+00	3.721E+02	73.2	S120	1 2 1 1 2	
3.737E+00	4.078E+02	77.2	S120	1 2 1 1 2	
6.752E+00	7.368E+02	80	M043	1 0 0 0 2	
4.098E+00	4.472E+02	85.2	S120	1 2 1 1 2	
4.311E+00	4.705E+02	96.0	S120	1 2 1 1 2	
8.291E+00	9.048E+02	100	M043	1 0 0 0 2	

742.   C₆H₇NO₂S
Benzenesulfonamide
Benzolsulfosaeure-amid

RN: 98-10-2   **MP** (°C):   151
MW: 157.19   **BP** (°C):

Solubility (Moles/L)	Solubility (Grams/L)	Temp (°C)	Ref (#)	Evaluation (T P E A A)	Comments
1.600E-02	2.515E+00	15	K024	1 2 1 1 2	
2.736E-02	4.300E+00	16	F300	1 0 0 0 1	

743.   C₆H₇NO₃S

Orthanilic Acid

Orthanilsaeure

RN: 88-21-1   **MP** (°C):   325

MW: 173.19   **BP** (°C):

Solubility (Moles/L)	Solubility (Grams/L)	Temp (°C)	Ref (#)	Evaluation (T P E A A)	Comments
4.585E-02	7.940E+00	0.0	P038	1 1 2 1 2	monohydrate
6.525E-02	1.130E+01	8.25	P038	1 1 2 1 2	monohydrate
7.535E-02	1.305E+01	12.3	P038	1 1 2 1 2	monohydrate
8.459E-02	1.465E+01	15.55	P038	1 1 2 1 2	anhydrate
8.776E-02	1.520E+01	16.75	P038	1 1 2 1 2	anhydrate
1.114E-01	1.930E+01	25	P038	1 1 2 1 2	anhydrate
1.738E-01	3.010E+01	41.3	P038	1 1 2 1 2	anhydrate
2.477E-01	4.290E+01	55.0	P038	1 1 2 1 2	anhydrate
3.672E-01	6.360E+01	70.0	P038	1 1 2 1 2	anhydrate
5.185E-01	8.980E+01	85.0	P038	1 1 2 1 2	anhydrate

744.   C₆H₇NO₃S

Sulfanilic Acid

4-Aminobenzenesulfonic Acid

Sulfanilsaeure

RN: 121-57-3   **MP** (°C):   122

MW: 173.19   **BP** (°C):

Solubility (Moles/L)	Solubility (Grams/L)	Temp (°C)	Ref (#)	Evaluation (T P E A A)	Comments
3.672E-02	6.359E+00	0	D077	1 0 0 1 1	
2.587E-02	4.480E+00	0	M043	1 0 0 0 1	
4.810E-02	8.330E+00	10	D077	1 0 0 1 1	
4.850E-02	8.400E+00	10	F300	1 0 0 0 1	
4.583E-02	7.937E+00	10	M043	1 0 0 0 1	
6.169E-02	1.068E+01	20	D077	1 0 0 1 2	
5.774E-02	1.000E+01	20	F300	1 0 0 0 1	
6.395E-02	1.108E+01	20	M043	1 0 0 0 2	
8.477E-02	1.468E+01	30	D077	1 0 0 1 2	
1.115E-01	1.932E+01	40	D077	1 0 0 1 2	
1.109E-01	1.920E+01	40	F300	1 0 0 0 2	
1.149E-01	1.990E+01	40	M043	1 0 0 0 2	
1.414E-01	2.449E+01	50	D077	1 0 0 1 2	
1.736E-01	3.007E+01	60	D077	1 0 0 1 2	
1.687E-01	2.922E+01	60	M043	1 0 0 0 2	
2.159E-01	3.740E+01	69.9	P038	1 0 2 1 2	anhydrate
2.103E-01	3.642E+01	70	D077	1 0 0 1 2	
2.492E-01	4.315E+01	80	D077	1 0 0 1 2	
2.492E-01	4.315E+01	80	M043	1 0 0 0 2	
2.737E-01	4.740E+01	85.0	P038	1 0 2 1 2	anhydrate
3.031E-01	5.249E+01	90	D077	1 0 0 1 2	

3.610E-01	6.253E+01	100	D077	1 0 0 1 2	
3.851E-01	6.670E+01	100	F300	1 0 0 0 2	
3.610E-01	6.253E+01	100	M043	1 0 0 0 2	
6.075E-02	1.052E+01	ns	K076	0 0 0 0 2	

745. $C_6H_7NO_3S$
Metanilic Acid
3-Aminobenzenesulfonic Acid
m-Sulfanilic Acid
RN: 121-47-1   **MP** (°C):   >300
MW: 173.19   **BP** (°C):

Solubility (Moles/L)	Solubility (Grams/L)	Temp (°C)	Ref (#)	Evaluation (T P E A A)	Comments
4.561E-02	7.900E+00	0.0	P038	1 2 2 1 2	anhydrate
5.901E-02	1.022E+01	7.75	P038	1 2 2 1 2	anhydrate
7.622E-02	1.320E+01	16.75	P038	1 2 2 1 2	anhydrate
9.440E-02	1.635E+01	24.95	P038	1 2 2 1 2	anhydrate
1.383E-01	2.395E+01	40.0	P038	1 2 2 1 2	anhydrate
1.975E-01	3.420E+01	55.0	P038	1 2 2 1 2	anhydrate
2.714E-01	4.700E+01	70.0	P038	1 2 2 1 2	anhydrate

746. $C_6H_7NO_3S.1.5H_2O$
Metanilic Acid (Sesquihydrate)
3-Aminobenzenesulfonic Acid (Sesquihydrate)
RN: 121-47-1   **MP** (°C):
MW: 200.21   **BP** (°C):

Solubility (Moles/L)	Solubility (Grams/L)	Temp (°C)	Ref (#)	Evaluation (T P E A A)	Comments
5.344E-02	1.070E+01	0.0	P038	1 2 2 1 2	
8.041E-02	1.610E+01	8.35	P038	1 2 2 1 2	
1.119E-01	2.240E+01	15.55	P038	1 2 2 1 2	
1.184E-01	2.370E+01	16.8	P038	1 2 2 1 2	
3.247E-01	6.500E+01	85.0	P038	1 2 2 1 2	

747. $C_6H_7NO_4S$
2-Aminophenol-4-sulfonic Acid
2-Amino-phenol-sulfosaeure-(4)
RN: 98-37-3   **MP** (°C):   >300
MW: 189.19   **BP** (°C):

Solubility (Moles/L)	Solubility (Grams/L)	Temp (°C)	Ref (#)	Evaluation (T P E A A)	Comments
5.286E-02	1.000E+01	14	F300	1 0 0 0 0	

748.   C₆H₇NO₄S

4-Aminophenol-2-sulfonic Acid
4-Amino-phenol-sulfosaeure-(2)
RN: 2835-04-3   **MP** (°C):
MW: 189.19   **BP** (°C):

Solubility (Moles/L)	Solubility (Grams/L)	Temp (°C)	Ref (#)	Evaluation (T P E A A)	Comments
3.700E-03	7.000E-01	14	F300	1 0 0 0 0	

749.   C₆H₇N₃O

Isoniazid
Isonicotinic Acid Hydrazide
Ianiazid
RN: 54-85-3   **MP** (°C):   171
MW: 137.14   **BP** (°C):

Solubility (Moles/L)	Solubility (Grams/L)	Temp (°C)	Ref (#)	Evaluation (T P E A A)	Comments
7.813E-01	1.071E+02	20	I307	0 0 0 0 1	
8.955E-01	1.228E+02	25	B187	1 0 0 0 1	
1.458E+00	2.000E+02	37	I307	0 0 0 0 1	
1.505E+00	2.063E+02	40	B187	1 0 0 0 1	

750.   C₆H₇N₃O₃

Orotic Acid Methylamide
Orotamide, N-Methyl-
RN: 1009-04-7   **MP** (°C):   284-286
MW: 169.14   **BP** (°C):

Solubility (Moles/L)	Solubility (Grams/L)	Temp (°C)	Ref (#)	Evaluation (T P E A A)	Comments
3.420E-01	5.785E+01	-4	N018	2 2 1 2 2	
6.840E-01	1.157E+02	16	N018	2 2 1 2 2	
8.340E-01	1.411E+02	25	N018	2 2 1 2 2	

751.   C₆H₇N₇

4,6,7-Triaminopteridine
4:6:7-Triaminopteridine
RN: 19167-62-5   **MP** (°C):
MW: 177.17   **BP** (°C):

Solubility (Moles/L)	Solubility (Grams/L)	Temp (°C)	Ref (#)	Evaluation (T P E A A)	Comments
4.515E-04	7.999E-02	20	A020	1 2 0 1 1	
1.252E-02	2.217E+00	100	A020	1 2 0 0 1	

752.   C₆H₇N₇
2,4,7-Triaminopteridine
2:4:7-Triaminopteridine
RN: 14439-13-5   **MP** (°C):
MW: 177.17   **BP** (°C):

Solubility (Moles/L)	Solubility (Grams/L)	Temp (°C)	Ref (#)	Evaluation (T P E A A)	Comments
1.254E-03	2.222E-01	20	A020	1 2 0 0 1	
2.808E-02	4.975E+00	100	A020	1 2 0 0 0	

753.   C₆H₇O₂P
Phenylphosphinic Acid
Phenyl-phosphinigsaeure
RN: 1779-48-2   **MP** (°C):   84
MW: 142.10   **BP** (°C):

Solubility (Moles/L)	Solubility (Grams/L)	Temp (°C)	Ref (#)	Evaluation (T P E A A)	Comments
4.757E-01	6.760E+01	14	F300	1 0 0 0 2	
4.843E+00	6.881E+02	100	F300	1 0 0 0 2	

754.   C₆H₇O₃P
Phenylphosphonic Acid
Phenylphosphonsaeure
RN: 1571-33-1   **MP** (°C):   164.5
MW: 158.09   **BP** (°C):

Solubility (Moles/L)	Solubility (Grams/L)	Temp (°C)	Ref (#)	Evaluation (T P E A A)	Comments
1.202E+00	1.900E+02	15	F300	1 0 0 0 2	

755.   C₆H₇O₃As
Benzenearsonic Acid
Phenylarsonsaeure
RN: 98-05-5   **MP** (°C):   160
MW: 202.04   **BP** (°C):

Solubility (Moles/L)	Solubility (Grams/L)	Temp (°C)	Ref (#)	Evaluation (T P E A A)	Comments
1.564E-01	3.160E+01	28	F300	1 0 0 0 2	
9.899E-01	2.000E+02	84	F300	1 0 0 0 1	

756. C₆H₈
1,4-Cyclohexadiene
1,4-Dihydrobenzene
RN: 628-41-1   **MP** (°C):   -49.2
MW: 80.13   **BP** (°C):   81

Solubility (Moles/L)	Solubility (Grams/L)	Temp (°C)	Ref (#)	Evaluation (T P E A A)	Comments
1.062E-02	8.512E-01	4.8	L007	2 2 1 2 2	
1.062E-02	8.512E-01	5.1	L007	2 1 1 1 2	
1.195E-02	9.576E-01	14.8	L007	2 2 1 2 2	
1.195E-02	9.576E-01	15.2	L007	2 1 1 1 2	
8.002E-03	6.412E-01	20	M337	2 1 2 2 2	
1.167E-02	9.353E-01	24.8	L007	2 2 1 2 2	
8.736E-03	7.000E-01	25	M001	2 1 2 2 2	
1.167E-02	9.353E-01	25.1	L007	2 1 1 1 2	
1.201E-02	9.625E-01	34.8	L007	2 2 1 2 2	
1.201E-02	9.625E-01	35.2	L007	2 1 1 1 2	
1.259E-02	1.009E+00	44.8	L007	2 2 1 2 2	
1.259E-02	1.009E+00	45.2	L007	2 1 1 1 2	

757. C₆H₈N₂
o-Phenylenediamine
o-Phenylendiamin
RN: 95-54-5   **MP** (°C):   102-103
MW: 108.14   **BP** (°C):   257

Solubility (Moles/L)	Solubility (Grams/L)	Temp (°C)	Ref (#)	Evaluation (T P E A A)	Comments
2.876E-01	3.110E+01	20	T301	1 2 2 2 2	
3.763E-01	4.070E+01	35	F300	1 0 0 0 2	
3.599E-01	3.892E+01	35.1	S115	1 2 1 1 2	
5.110E-01	5.527E+01	45.8	S115	1 2 1 1 2	
9.804E-01	1.060E+02	56.3	S115	1 2 1 1 2	
1.458E+00	1.577E+02	61.3	S115	1 2 1 1 2	
1.755E+00	1.898E+02	62.8	S115	1 2 1 1 2	
2.218E+00	2.398E+02	64.2	S115	1 2 1 1 2	
2.948E+00	3.188E+02	66.1	S115	1 2 1 1 2	
3.558E+00	3.847E+02	67.7	S115	1 2 1 1 2	
3.955E+00	4.277E+02	71.3	S115	1 2 1 1 2	
4.338E+00	4.691E+02	80.8	S115	1 2 1 1 2	
4.476E+00	4.841E+02	88.1	S115	1 2 1 1 2	
4.533E+00	4.902E+02	91.7	S115	1 2 1 1 2	
4.570E+00	4.942E+02	95.5	S115	1 2 1 1 2	

758. $C_6H_8N_2$
m-Phenylenediamine
m-Phenylendiamin
RN: 108-45-2   **MP** (°C):   63
MW: 108.14   **BP** (°C):   283

Solubility (Moles/L)	Solubility (Grams/L)	Temp (°C)	Ref (#)	Evaluation (T P E A A)	Comments
7.409E-01	8.012E+01	0.3	S115	1 2 1 1 2	α form
2.928E-01	3.166E+01	0.3	S115	1 2 1 1 2	β form
1.038E+00	1.122E+02	4.6	S115	1 2 1 1 2	α form
1.354E+00	1.465E+02	9.3	S115	1 2 1 1 2	α form
1.618E+00	1.750E+02	11.7	S115	1 2 1 1 2	α form
7.806E-01	8.442E+01	14.3	S115	1 2 1 1 2	β form
2.285E+00	2.472E+02	16.1	S115	1 2 1 1 2	α form
2.671E+00	2.889E+02	17.3	S115	1 2 1 1 2	α form
1.038E+00	1.122E+02	18.3	S115	1 2 1 1 2	β form
3.075E+00	3.326E+02	18.7	S115	1 2 1 1 2	α form
3.339E+00	3.611E+02	19.9	S115	1 2 1 1 2	α form
3.537E+00	3.825E+02	20.8	S115	1 2 1 1 2	α form
1.354E+00	1.465E+02	22.0	S115	1 2 1 1 2	β form
3.796E+00	4.105E+02	22.7	S115	1 2 1 1 2	α form
1.480E+00	1.600E+02	23.1	S115	1 2 1 1 2	β form
1.618E+00	1.750E+02	24.1	S115	1 2 1 1 2	β form
1.918E+00	2.074E+02	25.1	S115	1 2 1 1 2	β form
3.979E+00	4.303E+02	26.0	S115	1 2 1 1 2	α form
2.285E+00	2.472E+02	26.3	S115	1 2 1 1 2	β form
2.671E+00	2.889E+02	27.1	S115	1 2 1 1 2	β form
2.815E+00	3.044E+02	27.1	S115	1 2 1 1 2	β form
3.075E+00	3.326E+02	27.9	S115	1 2 1 1 2	β form
4.085E+00	4.418E+02	28.7	S115	1 2 1 1 2	α form
3.339E+00	3.611E+02	29.0	S115	1 2 1 1 2	β form
3.537E+00	3.825E+02	29.1	S115	1 2 1 1 2	β form
3.796E+00	4.105E+02	30.2	S115	1 2 1 1 2	β form
3.979E+00	4.303E+02	31.5	S115	1 2 1 1 2	β form
4.217E+00	4.560E+02	32.6	S115	1 2 1 1 2	α form
4.085E+00	4.418E+02	32.8	S115	1 2 1 1 2	β form
4.217E+00	4.560E+02	34.4	S115	1 2 1 1 2	β form
4.439E+00	4.800E+02	43.5	S115	1 2 1 1 2	α form
4.549E+00	4.919E+02	53.6	S115	1 2 1 1 2	α form
4.586E+00	4.960E+02	57.6	S115	1 2 1 1 2	α form
4.623E+00	5.000E+02	62.8	S115	1 2 1 1 2	α form

759.　$C_6H_8N_2$

p-Phenylenediamine

1,4-Phenylenediamine

RN: 106-50-3　**MP** (°C):　141

MW: 108.14　**BP** (°C):　267

Solubility (Moles/L)	Solubility (Grams/L)	Temp (°C)	Ref (#)	Evaluation (T P E A A)	Comments
9.880E-02	1.068E+01	3.6	S115	1 2 1 1 2	
3.299E-01	3.568E+01	23.7	S115	1 2 1 1 2	
4.180E-01	4.520E+01	25	F300	1 0 0 0 2	
8.292E-01	8.967E+01	37.8	S115	1 2 1 1 2	
1.460E+00	1.579E+02	49.9	S115	1 2 1 1 2	
1.978E+00	2.140E+02	59.2	S115	1 2 1 1 2	
2.368E+00	2.561E+02	64.6	S115	1 2 1 1 2	
2.724E+00	2.945E+02	69.2	S115	1 2 1 1 2	
3.155E+00	3.412E+02	75.5	S115	1 2 1 1 2	
3.432E+00	3.711E+02	80.3	S115	1 2 1 1 2	
3.809E+00	4.119E+02	88.5	S115	1 2 1 1 2	
4.055E+00	4.385E+02	95.9	S115	1 2 1 1 2	
1.500E-05	1.622E-03	98.59	M180	0 0 2 2 0	EFG
2.500E-05	2.704E-03	111.46	M180	0 0 2 2 0	EFG
4.000E-05	4.326E-03	117.47	M180	0 0 2 2 0	EFG
4.500E-05	4.866E-03	122.10	M180	0 0 2 2 0	EFG
5.000E-05	5.407E-03	126.84	M180	0 0 2 2 0	EFG
7.000E-05	7.570E-03	133.34	M180	0 0 2 2 0	EFG

760.　$C_6H_8N_2OS$

5,6-Dimethyl-2-thiouracil

4(1H)-Pyrimidinone, 2,3-Dihydro-5,6-dimethyl-2-thioxo-

5,6-Dimethylthiouracil

RN: 28456-54-4　**MP** (°C):

MW: 156.21　**BP** (°C):

Solubility (Moles/L)	Solubility (Grams/L)	Temp (°C)	Ref (#)	Evaluation (T P E A A)	Comments
8.790E-03	1.373E+00	25	G016	1 2 1 2 2	intrinsic

761.　$C_6H_8N_2O_2$

N,N-1,3-Dimethyluracil

1,3-Dimethyl-2,4-pyrimidinedione

N1,N3-Dimethyluracil

N,N'-Dimethyluracil

1,3-Dimethyluracil

RN: 874-14-6　**MP** (°C):

MW: 140.14　**BP** (°C):

Solubility (Moles/L)	Solubility (Grams/L)	Temp (°C)	Ref (#)	Evaluation (T P E A A)	Comments
3.568E+00	5.000E+02	ns	B177	0 0 0 0 2	

762.   C$_6$H$_8$N$_2$O$_2$S
m-Aminobenzenesulfonamide
Metanilamide
m-Amidobenzenesulfonamide
RN: 98-18-0   **MP** (°C):
MW: 172.21   **BP** (°C):

Solubility (Moles/L)	Solubility (Grams/L)	Temp (°C)	Ref (#)	Evaluation (T P E A A)	Comments
6.545E-02	1.127E+01	23	K034	2 2 2 2 2	
6.942E-02	1.196E+01	24	K034	2 2 2 2 2	
7.678E-02	1.322E+01	26	K034	2 2 2 2 2	
8.469E-02	1.458E+01	28	K034	2 2 2 2 2	
1.077E-01	1.855E+01	33	K034	2 2 2 2 2	
1.244E-01	2.143E+01	35.5	K034	2 2 2 2 2	
1.339E-01	2.306E+01	37	K034	2 2 2 2 2	
1.461E-01	2.515E+01	39	K034	2 2 2 2 2	
1.697E-01	2.922E+01	42	K034	2 2 2 2 2	
2.072E-01	3.568E+01	46	K034	2 2 2 2 2	
2.543E-01	4.379E+01	50	K034	2 2 2 2 2	

763.   C$_6$H$_8$N$_2$O$_2$S
o-Aminobenzenesulfonamide
Orthanilamide
RN: 3306-62-5   **MP** (°C):
MW: 172.21   **BP** (°C):

Solubility (Moles/L)	Solubility (Grams/L)	Temp (°C)	Ref (#)	Evaluation (T P E A A)	Comments
3.750E-02	6.458E+00	23	K034	2 2 2 2 1	
3.865E-02	6.655E+00	24	K034	2 2 2 2 1	
4.323E-02	7.444E+00	26	K034	2 2 2 2 1	
4.723E-02	8.133E+00	28	K034	2 2 2 2 1	
5.237E-02	9.018E+00	30.5	K034	2 2 2 2 1	
5.806E-02	9.999E+00	33	K034	2 2 2 2 2	
6.034E-02	1.039E+01	34	K034	2 2 2 2 2	
6.375E-02	1.098E+01	35.5	K034	2 2 2 2 2	
6.886E-02	1.186E+01	37	K034	2 2 2 2 2	
6.829E-02	1.176E+01	37	K034	2 2 2 2 2	
8.356E-02	1.439E+01	42	K034	2 2 2 2 2	
9.707E-02	1.672E+01	46	K034	2 2 2 2 2	
1.139E-01	1.961E+01	50	K034	2 2 2 2 2	

Solutions 273

764. $C_6H_8N_2O_2S$
Benzenesulfamide
Sulfanilamide
Sulfanilsaeure-amid
p-Aminobenzenesulphonamide

RN: 63-74-1 **MP** (°C): 165
MW: 172.21 **BP** (°C):

Solubility (Moles/L)	Solubility (Grams/L)	Temp (°C)	Ref (#)	Evaluation (T P E A A)	Comments
1.159E-02	1.996E+00	1	A047	1 0 0 0 0	EFG
1.057E-02	1.820E+00	4.40	B147	1 2 1 1 2	
1.458E-02	2.510E+00	10.20	B147	1 2 1 1 2	
1.957E-02	3.370E+00	15	B147	1 2 1 1 2	
2.323E-02	4.000E+00	15	F300	1 0 0 0 0	
2.660E-02	4.581E+00	15	K024	1 2 1 1 2	
2.241E-02	3.860E+00	15	S147	1 2 2 2 2	hydrate
2.889E-02	4.975E+00	16	A047	1 0 0 0 0	EFG
2.439E-02	4.200E+00	16	H114	1 0 0 0 2	
2.700E-02	4.650E+00	20	B147	1 2 1 1 2	
3.463E-02	5.964E+00	20	D041	1 0 0 0 0	
4.149E-02	7.145E+00	20	F073	1 2 2 2 2	
2.903E-02	5.000E+00	20	F300	1 0 0 0 0	
3.020E-02	5.200E+00	20	S147	1 2 2 2 2	hydrate
3.693E-02	6.359E+00	23	K034	2 2 2 2 1	
3.979E-02	6.853E+00	24	K034	2 2 2 2 1	
3.484E-02	6.000E+00	25	B147	1 2 1 1 2	
4.855E-02	8.360E+00	25	C102	2 0 2 2 2	
4.550E-02	7.835E+00	25	M116	2 1 1 1 2	
4.820E-02	8.300E+00	25	P015	2 2 2 2 1	
4.216E-02	7.260E+00	25	S147	1 2 2 2 2	hydrate
4.437E-02	7.641E+00	26	K034	2 2 2 2 1	
4.723E-02	8.133E+00	27	K034	2 2 2 2 1	
5.008E-02	8.625E+00	28	K034	2 2 2 2 1	
4.762E-02	8.200E+00	30	B147	1 2 1 1 2	
5.633E-02	9.700E+00	30	S147	1 2 2 2 2	hydrate
5.806E-02	9.999E+00	30.5	K034	2 2 2 2 2	
6.318E-02	1.088E+01	31	A047	1 0 0 0 0	EFG
6.205E-02	1.068E+01	31.7	K034	2 2 2 2 2	
6.829E-02	1.176E+01	33	K034	2 2 2 2 2	
7.282E-02	1.254E+01	34	K034	2 2 2 2 2	
6.388E-02	1.100E+01	35	B147	1 2 1 1 2	
7.543E-02	1.299E+01	35	S147	1 2 2 2 2	β form
7.848E-02	1.351E+01	35.5	K034	2 2 2 2 2	
1.259E-01	2.168E+01	37	A028	1 0 2 1 2	intrinsic
7.375E-02	1.270E+01	37	B147	1 2 1 1 2	
8.478E-02	1.460E+01	37	C102	2 0 2 2 2	
8.594E-02	1.480E+01	37	D084	1 0 1 0 2	
8.018E-02	1.381E+01	37	F072	1 0 0 0 2	
8.710E-02	1.500E+01	37	F300	1 0 0 0 1	

9.120E-02	1.571E+01	37	G028	2 2 1 1 2	α form, recrystallized
9.240E-02	1.591E+01	37	G028	2 2 1 1 2	γ form
9.070E-02	1.562E+01	37	G028	2 2 1 1 2	β form, recrystallized
8.920E-02	1.536E+01	37	G028	2 2 1 1 2	δ form, recrystallized
8.413E-02	1.449E+01	37	K034	2 2 2 2 2	
8.652E-02	1.490E+01	37	K086	1 0 0 0 2	
8.210E-02	1.414E+01	37	K095	2 0 0 0 2	intrinsic
8.710E-02	1.500E+01	37	L091	1 0 0 0 2	pH 5.5
8.469E-02	1.458E+01	37.50	M142	1 0 0 0 2	
9.201E-02	1.584E+01	39	K034	2 2 2 2 2	
8.362E-02	1.440E+01	40	B147	1 2 1 1 2	form II
9.750E-02	1.679E+01	40	G028	2 2 1 1 2	α form, recrystallized
9.680E-02	1.667E+01	40	G028	2 2 1 1 2	β form, recrystallized
9.640E-02	1.660E+01	40	G028	2 2 1 1 2	δ form, recrystallized
9.640E-02	1.660E+01	40	G028	2 2 1 1 2	γ form
9.518E-02	1.639E+01	40	S147	1 2 2 2 2	β form
1.049E-01	1.807E+01	42	K034	2 2 2 2 2	
1.086E-01	1.870E+01	45	B147	1 2 1 1 2	form II
1.201E-01	2.069E+01	45	S147	1 2 2 2 2	β form
1.256E-01	2.162E+01	46	K034	2 2 2 2 2	
1.527E-01	2.629E+01	50	A047	1 0 0 0 0	EFG
1.388E-01	2.390E+01	50	B147	1 2 1 1 2	form II
1.433E-01	2.468E+01	50	G028	2 2 1 1 2	δ form, recrystallized
1.435E-01	2.471E+01	50	G028	2 2 1 1 2	α form, recrystallized
1.419E-01	2.444E+01	50	G028	2 2 1 1 2	β form, recrystallized
1.430E-01	2.463E+01	50	G028	2 2 1 1 2	γ form
1.516E-01	2.610E+01	50	K034	2 2 2 2 2	
1.488E-01	2.562E+01	50	S147	1 2 2 2 2	β form
1.789E-01	3.080E+01	55	B147	1 2 1 1 2	form II
2.294E-01	3.950E+01	60	B147	1 2 1 1 2	form II
2.923E-01	5.033E+01	65	A047	1 0 0 0 0	EFG
2.962E-01	5.100E+01	65	B147	1 2 1 1 2	form II
3.833E-01	6.600E+01	70	B147	1 2 1 1 2	form II
4.599E-01	7.919E+01	75	A047	1 0 0 0 0	EFG
5.168E-01	8.900E+01	75	B147	1 2 1 1 2	form II
5.660E-01	9.747E+01	79	A047	1 0 0 0 0	EFG
6.272E-02	1.080E+01	ns	D035	0 0 0 0 2	
3.050E-02	5.252E+00	ns	L044	0 0 0 0 2	

765. $C_6H_8N_2O_2S \cdot H_2O$
Sulfanilamide (Monohydrate)
4-Aminobenzenesulfonamide (Monohydrate)
p-Anilinesulfonamide (Monohydrate)
Bacteramid (Monohydrate)

RN: 20203-81-0   **MP** (°C):
MW: 190.22   **BP** (°C):

Solubility (Moles/L)	Solubility (Grams/L)	Temp (°C)	Ref (#)	Evaluation (T P E A A)	Comments
2.200E-02	4.185E+00	15	G028	2 2 1 1 2	
4.320E-02	8.218E+00	26	G028	2 2 1 1 2	
5.600E-02	1.065E+01	30	G028	2 2 1 1 2	
8.420E-02	1.602E+01	37	G028	2 2 1 1 2	

766. $C_6H_8N_2O_3$
5,5-Dimethylbarbituric Acid
5,5-Dimethylbarbitursaeure
Barbituric Acid, 5,5-Dimethyl
2,4,6(1H,3H,5H)-Pyrimidinetrione, 5,5-Dimethyl
5,5-Dimethyl Barbituric Acid

RN: 24448-94-0   **MP** (°C):   278
MW: 156.14   **BP** (°C):

Solubility (Moles/L)	Solubility (Grams/L)	Temp (°C)	Ref (#)	Evaluation (T P E A A)	Comments
1.812E-02	2.829E+00	25	P350	2 1 1 1 2	intrinsic
1.549E-02	2.419E+00	ns	T003	0 0 0 0 2	

767. $C_6H_8N_2O_3S$
4-Phenylhydrazine Sulfonic Acid
Phenylhydrazin-sulfosaeure-(4)

RN: 98-71-5   **MP** (°C):
MW: 188.21   **BP** (°C):

Solubility (Moles/L)	Solubility (Grams/L)	Temp (°C)	Ref (#)	Evaluation (T P E A A)	Comments
3.029E-02	5.700E+00	11.50	F300	1 0 0 0 1	
1.860E-01	3.500E+01	100	F300	1 0 0 0 1	

768.   C₆H₈N₂O₈

Isosorbide Dinitrate
1,4:3,6-Dianhydro-D-glucitol dinitrate
Sorbidin
Isogen
Imdur
RN: 87-33-2   **MP** (°C):   70
MW: 236.14   **BP** (°C):

Solubility (Moles/L)	Solubility (Grams/L)	Temp (°C)	Ref (#)	Evaluation (T P E A A)	Comments
2.328E-03	5.497E-01	25	L033	1 0 2 1 2	

769.   C₆H₈N₄O

5-Amino-4-carboxymethylaminopyrimidine
RN: **MP** (°C):
MW: 152.16   **BP** (°C):

Solubility (Moles/L)	Solubility (Grams/L)	Temp (°C)	Ref (#)	Evaluation (T P E A A)	Comments
2.120E-01	3.226E+01	100	A082	1 2 0 0 0	

770.   C₆H₈N₈

2,4,6,7-Tetraminopteridine
2:4:6:7-Tetraminopteridine
RN: 19167-63-6   **MP** (°C):
MW: 192.18   **BP** (°C):

Solubility (Moles/L)	Solubility (Grams/L)	Temp (°C)	Ref (#)	Evaluation (T P E A A)	Comments
4.002E-04	7.692E-02	20	A020	1 2 0 1 1	

771.   C₆H₈O₂

Sorbic Acid
2,4-Hexadienoic Acid
2-Propenylacrylic Acid
Preservastat
Hexadienoic Acid
Sorbistat
RN: 110-44-1   **MP** (°C):   134.5
MW: 112.13   **BP** (°C):   228

Solubility (Moles/L)	Solubility (Grams/L)	Temp (°C)	Ref (#)	Evaluation (T P E A A)	Comments
1.700E-02	1.906E+00	30	L069	1 0 1 1 0	EFG

772. $C_6H_8O_6$
Ascorbic Acid
L-Ascorbic Acid
L-Ascorbinsaeure

RN: 50-81-7   **MP** (°C):   193
MW: 176.13   **BP** (°C):

Solubility (Moles/L)	Solubility (Grams/L)	Temp (°C)	Ref (#)	Evaluation (T P E A A)	Comments
9.269E-01	1.633E+02	6.99	A341	2 0 2 2 2	
9.509E-01	1.675E+02	7.99	A341	2 0 2 2 2	
9.880E-01	1.740E+02	9.99	A341	2 0 2 2 2	
1.026E+00	1.807E+02	11.99	A341	2 0 2 2 2	
1.142E+00	2.011E+02	15.99	A341	2 0 2 2 2	
1.418E+00	2.498E+02	20	D041	1 0 0 0 2	
1.283E+00	2.260E+02	20.99	A341	2 0 2 2 2	
1.397E+00	2.460E+02	24.99	A341	2 0 2 2 2	
1.891E+00	3.330E+02	25	D315	1 0 1 1 2	
9.757E-01	1.718E+02	25	N003	1 0 2 2 2	
1.551E+00	2.731E+02	28.99	A341	2 0 2 2 2	
1.718E+00	3.025E+02	33.99	A341	2 0 2 2 2	
1.758E+00	3.096E+02	35.99	A341	2 0 2 2 2	
1.856E+00	3.270E+02	38.99	A341	2 0 2 2 2	
1.028E+00	1.810E+02	40	N003	1 0 2 2 2	
2.009E+00	3.539E+02	42.99	A341	2 0 2 2 2	
2.021E+00	3.560E+02	43.99	A341	2 0 2 2 2	
2.066E+00	3.638E+02	44.99	A341	2 0 2 2 2	
2.132E+00	3.755E+02	47.69	A341	2 0 2 2 2	
2.184E+00	3.847E+02	48.49	A341	2 0 2 2 2	
2.235E+00	3.937E+02	49.99	A341	2 0 2 2 2	
2.255E+00	3.972E+02	50.39	A341	2 0 2 2 2	
2.275E+00	4.007E+02	50.99	A341	2 0 2 2 2	
2.373E+00	4.180E+02	52.49	A341	2 0 2 2 2	
2.383E+00	4.197E+02	53.99	A341	2 0 2 2 2	
2.413E+00	4.249E+02	54.09	A341	2 0 2 2 2	
2.449E+00	4.314E+02	54.99	A341	2 0 2 2 2	
2.520E+00	4.439E+02	60.02	A341	2 0 2 2 2	
2.551E+00	4.492E+02	61.99	A341	2 0 2 2 2	
2.635E+00	4.641E+02	64.99	A341	2 0 2 2 2	
1.891E+00	3.330E+02	ns	M054	0 0 0 0 2	

773.   C₆H₈O₆
Tricarballylic Acid
Tricarballylsaeure
1,2,3-Propanetricarboxylic Acid
RN: 99-14-9   **MP** (°C):   166
MW: 176.13   **BP** (°C):

Solubility (Moles/L)	Solubility (Grams/L)	Temp (°C)	Ref (#)	Evaluation (T P E A A)	Comments
1.885E+00	3.320E+02	18	F300	1 0 0 0 2	

774.   C₆H₈O₇
Citric Acid Anhydrous
2-Hydroxytricarballylic Acid
Citronensaeure
1,2,3-Propanetricarboxylic Acid
Citro
Citralite
RN: 77-92-9   **MP** (°C):   153
MW: 192.13   **BP** (°C):

Solubility (Moles/L)	Solubility (Grams/L)	Temp (°C)	Ref (#)	Evaluation (T P E A A)	Comments
2.549E+00	4.898E+02	0	M043	1 0 0 0 1	
1.885E+00	3.621E+02	0.0	K084	1 0 1 0 2	
1.881E+00	3.613E+02	1.2	K084	1 0 1 0 2	
1.875E+00	3.602E+02	1.6	K084	1 0 1 0 2	
2.562E+00	4.923E+02	4.99	A339	2 0 2 2 2	
2.684E+00	5.157E+02	9.99	A339	2 0 2 2 2	
1.825E+00	3.506E+02	10	D020	1 2 1 1 2	
2.571E+00	4.940E+02	10	F300	1 0 0 0 2	
1.825E+00	3.506E+02	10	F302	1 0 0 0 1	
2.817E+00	5.413E+02	10	M043	1 0 0 0 2	
1.938E+00	3.723E+02	10.0	K084	1 0 1 0 2	
1.927E+00	3.702E+02	10.8	K084	1 0 1 0 2	
2.811E+00	5.400E+02	14.99	A339	2 0 2 2 2	
1.933E+00	3.713E+02	15.0	K084	1 0 1 0 2	
2.918E+00	5.605E+02	19.99	A339	2 0 2 2 2	
3.089E+00	5.935E+02	20	D041	1 0 0 0 2	
2.816E+00	5.410E+02	20	F300	1 0 0 0 2	
1.935E+00	3.719E+02	20	F302	1 0 0 0 2	
3.089E+00	5.935E+02	20	M043	1 0 0 0 2	
3.045E+00	5.851E+02	24.99	A339	2 0 2 2 2	
1.994E+00	3.831E+02	25	D020	1 2 1 1 2	
1.254E+01	2.409E+03	25	K040	1 0 2 1 2	
3.201E+00	6.149E+02	29.99	A339	2 0 2 2 2	
2.037E+00	3.914E+02	30	F302	1 0 0 0 2	
3.366E+00	6.466E+02	30	M043	1 0 0 0 2	

3.296E+00	6.332E+02	34.99	A339	2 0 2 2 2	
2.100E+00	4.034E+02	35.8	D039	2 2 1 2 2	EFG
2.094E+00	4.023E+02	36.6	F302	1 0 0 0 2	
3.201E+00	6.150E+02	36.60	F300	1 0 0 0 2	
3.346E+00	6.429E+02	39.99	A339	2 0 2 2 2	
2.118E+00	4.069E+02	40	D020	1 2 1 1 2	
2.116E+00	4.065E+02	40	D039	2 2 1 2 0	EFG
2.118E+00	4.069E+02	40	F302	1 0 0 0 2	
3.553E+00	6.825E+02	40	M043	1 0 0 0 2	
3.438E+00	6.605E+02	44.99	A339	2 0 2 2 2	
3.488E+00	6.702E+02	49.99	A339	2 0 2 2 2	
2.161E+00	4.152E+02	50	D039	2 2 1 2 0	EFG
2.159E+00	4.149E+02	50	F302	1 0 0 0 2	
3.539E+00	6.800E+02	54.99	A339	2 0 2 2 2	
3.601E+00	6.918E+02	59.99	A339	2 0 2 2 2	
2.214E+00	4.253E+02	60	D039	2 2 1 2 0	EFG .
2.205E+00	4.236E+02	60	F302	1 0 0 0 2	
3.824E+00	7.347E+02	60	M043	1 0 0 0 2	
3.669E+00	7.050E+02	64.99	A339	2 0 2 2 2	
2.261E+00	4.344E+02	70	D039	2 2 1 2 0	EFG
2.251E+00	4.325E+02	70	F302	1 0 0 0 2	
2.300E+00	4.420E+02	80	D039	2 2 1 2 0	EFG
2.294E+00	4.407E+02	80	F302	1 0 0 0 2	
4.102E+00	7.881E+02	80	M043	1 0 0 0 2	
2.350E+00	4.515E+02	90	D039	2 2 1 2 0	EFG
2.336E+00	4.487E+02	90	F302	1 0 0 0 2	
2.391E+00	4.595E+02	100	D039	2 2 1 2 0	EFG
4.372E+00	8.400E+02	100	D041	1 0 0 0 2	
3.997E+00	7.680E+02	100	F300	1 0 0 0 2	
2.376E+00	4.565E+02	100	F302	1 0 0 0 1	
4.373E+00	8.403E+02	100	M043	1 0 0 0 2	

775. $C_6H_8O_7 \cdot H_2O$
Citric Acid (Monohydrate)
2-Hydroxytricarballylic Acid (Monohydrate)
RN: 5949-29-1   **MP** ($^\circ$C):
MW: 210.14   **BP** ($^\circ$C):

Solubility (Moles/L)	Solubility (Grams/L)	Temp ($^\circ$C)	Ref (#)	Evaluation (T P E A A)	Comments
1.554E+00	3.266E+02	0	D039	2 2 1 2 0	EFG
1.667E+00	3.502E+02	10	D039	2 2 1 2 0	EFG
3.005E+00	6.314E+02	17.20	L031	1 1 2 1 2	average of 2
3.077E+00	6.466E+02	19.80	L031	1 1 2 1 2	
1.771E+00	3.723E+02	20	D039	2 2 1 2 0	EFG
3.080E+00	6.473E+02	20.20	L031	1 1 2 1 2	
3.146E+00	6.610E+02	22.50	L031	1 1 2 1 2	
3.154E+00	6.627E+02	22.90	L031	1 1 2 1 2	
1.822E+00	3.830E+02	25	D039	2 2 1 2 2	EFG
3.214E+00	6.753E+02	25.10	L031	1 1 2 1 2	

3.216E+00	6.759E+02	25.30	L031	1 1 2 1 2	
3.272E+00	6.875E+02	27.00	L031	1 1 2 1 2	
3.276E+00	6.885E+02	27.60	L031	1 1 2 1 2	
3.303E+00	6.942E+02	28.60	L031	1 1 2 1 2	
1.864E+00	3.917E+02	30	D039	2 2 1 2 0	EFG
3.359E+00	7.059E+02	30.50	L031	1 1 2 1 2	
3.357E+00	7.054E+02	30.70	L031	1 1 2 1 2	
3.389E+00	7.122E+02	31.80	L031	1 1 2 1 2	
3.440E+00	7.230E+02	33.70	L031	1 1 2 1 2	
3.478E+00	7.308E+02	34.40	L031	1 1 2 1 2	
3.518E+00	7.392E+02	35.40	L031	1 1 2 1 2	

776. C_6H_8S
2-Ethylthiophene
Thiophene, 2-Ethyl-
RN: 872-55-9   **MP** (°C):   <25
MW: 112.19   **BP** (°C):   132

Solubility (Moles/L)	Solubility (Grams/L)	Temp (°C)	Ref (#)	Evaluation (T P E A A)	Comments
2.603E-03	2.920E-01	25	K119	1 0 0 0 2	
2.603E-03	2.920E-01	25	P051	2 1 1 2 2	
2.603E-03	2.920E-01	25.00	P007	2 1 2 2 2	

777. $C_6H_9NO_3$
4,6,10-Trioxa-1-azatricyclo[3.3.1.13,7]decane
Trimorpholin
Trimorpholine
RN: 281-36-7   **MP** (°C):
MW: 143.14   **BP** (°C):

Solubility (Moles/L)	Solubility (Grams/L)	Temp (°C)	Ref (#)	Evaluation (T P E A A)	Comments
1.167E+00	1.670E+02	0	F300	1 0 0 0 2	
2.375E+00	3.400E+02	80	F300	1 0 0 0 2	

778. $C_6H_9NO_3$
Trimethadione
3,5,5-Trimethyl-2,4-diketooxazolidine
3,5,5-Trimethyl-2,4-oxazolidinedione
Tridione
RN: 127-48-0   **MP** (°C):   46
MW: 143.14   **BP** (°C):

Solubility (Moles/L)	Solubility (Grams/L)	Temp (°C)	Ref (#)	Evaluation (T P E A A)	Comments
3.327E-01	4.762E+01	20	D041	1 0 0 0 0	

779.   C$_6$H$_9$NO$_6$

Triglycine
Complexon I
N,N-bis(Carboxymethyl)glycine
α,α',α''-Trimethylaminetricarboxylic Acid

RN: 139-13-9   **MP** (°C):   241.5
MW: 191.14   **BP** (°C):

Solubility (Moles/L)	Solubility (Grams/L)	Temp (°C)	Ref (#)	Evaluation (T P E A A)	Comments
3.090E-01	5.906E+01	25	M024	1 2 0 1 2	
3.395E-01	6.490E+01	25.1	N024	2 0 2 2 2	
3.374E-01	6.450E+01	25.1	N025	2 0 2 2 2	
3.348E-01	6.400E+01	25.1	N026	2 0 2 2 2	
3.101E-01	5.927E+01	25.1	N027	1 2 2 2 2	

780.   C$_6$H$_9$N$_3$

Kyanmethin
6-Amino-2,4-dimethyl-pyrimidin
6-Amino-2,4-dimethylpyrimidine

RN: 461-98-3   **MP** (°C):   182
MW: 123.16   **BP** (°C):

Solubility (Moles/L)	Solubility (Grams/L)	Temp (°C)	Ref (#)	Evaluation (T P E A A)	Comments
5.197E-02	6.400E+00	18	F300	1 0 0 0 1	

781.   C$_6$H$_9$N$_3$O$_2$

2-Isopropyl-4(5)-nitroimidazole
1H-Imidazole, 2-(1-Methylethyl)-4-nitro-
2-(1-Methylethyl)-4-nitro-1H-imidazole
2-Isopropyl-5-nitroimidazole
2-Isopropyl-4-nitroimidazole

RN: 13373-32-5   **MP** (°C):   182-183
MW: 155.16   **BP** (°C):

Solubility (Moles/L)	Solubility (Grams/L)	Temp (°C)	Ref (#)	Evaluation (T P E A A)	Comments
7.025E-02	1.090E+01	20	D344	1 1 2 2 2	
7.025E-02	1.090E+01	20	D344	1 1 2 2 2	
6.886E-02	1.068E+01	20	D344	1 1 2 2 2	
7.030E-02	1.091E+01	20	D344	1 1 2 2 2	

782. $C_6H_9N_3O_2$

L-Histidine
L-Histidin
Histidine
RN: 71-00-1   **MP** (°C):   287
MW: 155.16   **BP** (°C):

Solubility (Moles/L)	Solubility (Grams/L)	Temp (°C)	Ref (#)	Evaluation (T P E A A)	Comments
2.580E-01	4.003E+01	15	D349	2 1 1 2 2	
2.646E-01	4.106E+01	20	B032	1 2 2 1 2	
2.640E-01	4.096E+01	20	D349	2 1 1 2 2	
2.930E-01	4.546E+01	25	B032	1 2 2 1 2	
2.574E-01	3.994E+01	25	D041	1 0 0 0 2	
2.720E-01	4.220E+01	25	D349	2 1 1 2 2	
2.481E-01	3.850E+01	25	F300	1 0 0 0 2	
2.651E-01	4.114E+01	25	G315	1 0 2 2 2	
2.771E-01	4.300E+01	25.1	N024	2 0 2 2 2	
2.771E-01	4.300E+01	25.1	N025	2 0 2 2 2	
2.771E-01	4.300E+01	25.1	N026	2 0 2 2 2	
2.675E-01	4.150E+01	25.1	N027	1 1 2 2 2	
2.791E-01	4.330E+01	27	D036	2 1 2 2 2	
3.207E-01	4.976E+01	29.80	B032	1 2 2 1 2	
2.834E-01	4.398E+01	30	H062	2 2 2 0 1	EFG
5.213E-01	8.088E+01	50	H062	2 2 2 0 0	EFG
7.915E-01	1.228E+02	70	H062	2 2 2 0 0	EFG

783. $C_6H_9N_3O_3$

Metronidazole
Flagyl
2-Methyl-5-nitroimidazole-1-ethanol
Metrozine
Rozex
2-Methyl-5-nitro-1-imidazoleethanol
RN: 443-48-1   **MP** (°C):   158
MW: 171.16   **BP** (°C):

Solubility (Moles/L)	Solubility (Grams/L)	Temp (°C)	Ref (#)	Evaluation (T P E A A)	Comments
5.545E-02	9.490E+00	20	D344	1 1 2 2 2	
5.545E-02	9.490E+00	20	D344	1 1 2 2 2	
5.441E-02	9.312E+00	20	D344	1 1 2 2 2	
5.540E-02	9.482E+00	20	D344	1 1 2 2 2	
4.809E-02	8.232E+00	20	H324	1 0 2 2 1	
5.785E-02	9.901E+00	20	I315	0 0 0 0 0	
6.427E-02	1.100E+01	25	C062	1 1 2 1 2	

5.550E-02	9.500E+00	25	C124	2 0 1 1 2
5.727E-02	9.803E+00	26	H324	1 0 2 2 1
6.585E-02	1.127E+01	30	H324	1 0 2 2 1
5.843E-02	1.000E+01	ns	C324	0 0 2 2 0

784. C₆H₁₀
Cyclohexene
1,2,3,4-Tetrahydrobenzene
RN: 110-83-8   **MP** (°C):   -104
MW: 82.15   **BP** (°C):   83

Solubility (Moles/L)	Solubility (Grams/L)	Temp (°C)	Ref (#)	Evaluation (T P E A A)	Comments
3.408E-03	2.799E-01	4.8	L007	2 2 1 2 2	
3.408E-03	2.799E-01	5.1	L007	2 0 1 1 2	
3.633E-03	2.984E-01	14.8	L007	2 2 1 2 2	
3.633E-03	2.984E-01	15.2	L007	2 0 1 1 2	
1.583E-03	1.300E-01	20	C008	1 2 2 0 1	
2.769E-03	2.274E-01	20	M337	2 1 2 2 2	
3.450E-03	2.834E-01	23.5	S171	2 1 2 2 2	
3.639E-03	2.989E-01	24.8	L007	2 2 1 2 2	
2.593E-03	2.130E-01	25	M001	2 1 2 2 2	
3.639E-03	2.989E-01	25.1	L007	2 0 1 1 2	
3.681E-03	3.024E-01	34.8	L007	2 2 1 2 2	
3.681E-03	3.024E-01	35.2	L007	2 0 1 1 2	
6.000E-03	4.929E-01	40	P335	1 1 1 2 2	
3.779E-03	3.104E-01	44.8	L007	2 2 1 2 2	
3.779E-03	3.104E-01	45.2	L007	2 0 1 1 2	
1.800E-02	1.479E+00	140	P335	1 1 1 2 2	
1.583E-03	1.300E-01	ns	M010	0 0 0 0 1	

785. C₆H₁₀
3-Hexyne
Diethylacetylene
RN: 928-49-4   **MP** (°C):   -103
MW: 82.15   **BP** (°C):

Solubility (Moles/L)	Solubility (Grams/L)	Temp (°C)	Ref (#)	Evaluation (T P E A A)	Comments
6.800E-03	5.586E-01	25	H039	1 2 2 2 1	
6.400E-03	5.257E-01	35	H039	1 2 2 2 1	

786.   C₆H₁₀
1-Hexyne
Butylacetylene
n-Butylacetylene
RN: 693-02-7   **MP** (°C):   -132
MW: 82.15   **BP** (°C):   71

Solubility (Moles/L)	Solubility (Grams/L)	Temp (°C)	Ref (#)	Evaluation (T P E A A)	Comments
4.382E-03	3.600E-01	25	M001	2 1 2 2 2	
8.370E-03	6.876E-01	25	M342	1 0 1 1 2	

787.   C₆H₁₀
1,5-Hexadiene
Biallyl
Diallyl
RN: 592-42-7   **MP** (°C):   -141
MW: 82.15   **BP** (°C):   60

Solubility (Moles/L)	Solubility (Grams/L)	Temp (°C)	Ref (#)	Evaluation (T P E A A)	Comments
2.057E-03	1.690E-01	25	M001	2 1 2 2 2	

788.   C₆H₁₀BrNO₄
5-Bromo-2,2-dimethyl-5-nitro-1,3-dioxane
2,2-Dimethyl-5-bromo-5-nitro-1,3-dioxane
m-Dioxane, 5-Bromo-2,2-dimethyl-5-nitro-
RN: 60766-57-6   **MP** (°C):   79-81
MW: 240.06   **BP** (°C):

Solubility (Moles/L)	Solubility (Grams/L)	Temp (°C)	Ref (#)	Evaluation (T P E A A)	Comments
4.369E-03	1.049E+00	25	L013	1 0 2 1 2	

789.   C₆H₁₀BrNO₄
5-Bromo-2-ethyl-5-nitro-1,3-dioxane
2-Ethyl-5-bromo-5-nitro-1,3-dioxane
RN: 54010-85-4   **MP** (°C):   58-59
MW: 240.06   **BP** (°C):

Solubility (Moles/L)	Solubility (Grams/L)	Temp (°C)	Ref (#)	Evaluation (T P E A A)	Comments
3.205E-03	7.694E-01	25	L013	1 0 2 1 2	

790.   C$_6$H$_{10}$ClN$_5$
Deethylatrazine
2-Amino-4-isopropylamino-6-chloro-s-triazine
6-Chloro-N-(1-methylethyl)-1,3,5-triazine-2,4-diamine
RN: 6190-65-4   **MP** (°C):
MW: 187.63   **BP** (°C):

Solubility (Moles/L)	Solubility (Grams/L)	Temp (°C)	Ref (#)	Evaluation (T P E A A)	Comments
2.000E-03	3.692E-01	2	B193	1 1 0 0 1	

791.   C$_6$H$_{10}$O
Cyclohexanone
Cyclohexanon
RN: 108-94-1   **MP** (°C):   -47
MW: 98.15   **BP** (°C):

Solubility (Moles/L)	Solubility (Grams/L)	Temp (°C)	Ref (#)	Evaluation (T P E A A)	Comments
1.323E-02	1.298E+00	20	D052	1 1 0 0 1	*sic*
2.485E-01	2.439E+01	25	B060	2 0 1 1 1	
8.975E-01	8.809E+01	25	M323	2 2 1 1 2	

792.   C$_6$H$_{10}$O
Mesityl Oxide
Mesityloxid
RN: 141-79-7   **MP** (°C):   -57
MW: 98.15   **BP** (°C):   130

Solubility (Moles/L)	Solubility (Grams/L)	Temp (°C)	Ref (#)	Evaluation (T P E A A)	Comments
2.862E-01	2.809E+01	20	D052	1 1 0 0 0	
2.975E-01	2.920E+01	ns	F300	0 0 0 0 2	

793.   C$_6$H$_{10}$OS$_2$
Allicin
2-Propene-1-sulfinothioic Acid S-2-propenyl Ester
RN: 539-86-6   **MP** (°C):   <25
MW: 162.27   **BP** (°C):

Solubility (Moles/L)	Solubility (Grams/L)	Temp (°C)	Ref (#)	Evaluation (T P E A A)	Comments
1.479E-01	2.400E+01	10	F300	1 0 0 0 1	

794.   C$_6$H$_{10}$O$_2$
Methyl Vinyl Carbinol Acetate
1-Methylallyl Acetate
3-Buten-2-yl Acetate
RN: 6737-11-7   **MP** (°C):
MW: 114.15   **BP** (°C):

Solubility (Moles/L)	Solubility (Grams/L)	Temp (°C)	Ref (#)	Evaluation (T P E A A)	Comments
1.141E-01	1.303E+01	26	O012	1 2 1 1 2	
6.953E-02	7.937E+00	50	O012	1 2 1 1 2	
1.718E-01	1.961E+01	75	O012	1 2 1 1 2	

795.   C$_6$H$_{10}$O$_2$
3-Methyl-1,3-pentadione
1,2-Dimethyl-1,3-butadiene
3,4-Dimethylbutadiene
RN: 4549-74-0   **MP** (°C):   -5
MW: 114.15   **BP** (°C):   191

Solubility (Moles/L)	Solubility (Grams/L)	Temp (°C)	Ref (#)	Evaluation (T P E A A)	Comments
9.780E-01	1.116E+02	25	M078	2 0 1 0 2	

796.   C$_6$H$_{10}$O$_2$S$_4$
Dixanthogen
Ethyl Dixanthogen
RN: 502-55-6   **MP** (°C):   28
MW: 242.40   **BP** (°C):

Solubility (Moles/L)	Solubility (Grams/L)	Temp (°C)	Ref (#)	Evaluation (T P E A A)	Comments
1.300E-05	3.151E-03	22	P076	1 2 1 1 1	
1.140E-05	2.763E-03	25	H102	1 2 1 2 2	
<2.06E-06	<5.00E-04	25	M161	1 0 0 0 0	
1.250E-05	3.030E-03	ns	L083	0 0 0 0 0	EFG, pH 3-9

797.   C$_6$H$_{10}$O$_3$
Ethyl Acetoacetate
Acetessigsaeure-aethyl Ester
Acetoacetic Acid Ethyl Ester
RN: 141-97-9   **MP** (°C):   -45
MW: 130.14   **BP** (°C):   180.8

Solubility (Moles/L)	Solubility (Grams/L)	Temp (°C)	Ref (#)	Evaluation (T P E A A)	Comments
9.613E-01	1.251E+02	10.5	D041	1 0 0 0 2	
8.529E-01	1.110E+02	16.50	F300	1 0 0 0 2	

798. $C_6H_{10}O_4$
sym-Dimethylsuccinic Acid
Acide Dimethylsuccinique-sym
RN: 608-40-2   **MP** (°C):
MW: 146.14   **BP** (°C):

Solubility (Moles/L)	Solubility (Grams/L)	Temp (°C)	Ref (#)	Evaluation (T P E A A)	Comments
2.053E+00	3.000E+02	15	M051	1 0 0 0 2	

799. $C_6H_{10}O_4$
Adipic Acid
Adipinsaeure
RN: 124-04-9   **MP** (°C):   152
MW: 146.14   **BP** (°C):   337.5

Solubility (Moles/L)	Solubility (Grams/L)	Temp (°C)	Ref (#)	Evaluation (T P E A A)	Comments
5.431E-02	7.937E+00	0	M043	1 0 0 0 0	
6.766E-02	9.888E+00	4.99	A339	2 0 2 2 2	
7.853E-02	1.148E+01	9.99	A339	2 0 2 2 2	
6.775E-02	9.901E+00	10	M043	1 0 0 0 1	
1.061E-01	1.551E+01	14.99	A339	2 0 2 2 2	
9.580E-02	1.400E+01	15	F300	1 0 0 0 1	
9.580E-02	1.400E+01	15	L041	1 0 0 1 1	
9.580E-02	1.400E+01	15	M051	1 0 0 0 1	
1.303E-01	1.904E+01	19.99	A339	2 0 2 2 2	
1.011E-01	1.478E+01	20	D041	1 0 0 0 1	
1.276E-01	1.865E+01	20	M043	1 0 0 0 1	
9.856E-02	1.440E+01	20	M171	1 0 0 0 1	
9.000E-02	1.315E+01	20	S006	1 0 0 0 1	
4.824E-01	7.050E+01	21	B040	1 0 1 1 2	*sic*
1.664E-01	2.432E+01	24.99	A339	2 0 2 2 2	
2.216E-03	3.239E-01	25	K035	2 0 0 0 2	*sic*
2.053E-01	3.001E+01	29.99	A339	2 0 2 2 2	
1.993E-01	2.913E+01	30	M043	1 0 0 0 1	
2.045E-01	2.988E+01	34.10	A031	1 2 2 2 2	
2.546E-01	3.721E+01	34.99	A339	2 0 2 2 2	
2.933E-01	4.287E+01	39.3	G302	2 2 2 2 0	EFG
3.274E-01	4.785E+01	39.99	A339	2 0 2 2 2	
3.333E-01	4.871E+01	40	A031	1 2 2 2 2	
3.382E-01	4.943E+01	40	B088	1 0 0 0 1	
3.258E-01	4.762E+01	40	M043	1 0 0 0 1	
4.383E-01	6.406E+01	44.99	A339	2 0 2 2 2	
5.516E-01	8.062E+01	49.99	A339	2 0 2 2 2	
5.788E-01	8.458E+01	50	A031	1 2 2 2 2	
7.508E-01	1.097E+02	54.99	A339	2 0 2 2 2	
1.011E+00	1.477E+02	59.99	A339	2 0 2 2 2	
1.024E+00	1.497E+02	60	A031	1 2 2 2 2	
1.044E+00	1.525E+02	60	M043	1 0 0 0 1	

1.130E+00	1.652E+02	64.99	A339	2 0 2 2 2
1.740E+00	2.543E+02	70	A031	1 2 2 2 2
2.818E+00	4.118E+02	80	M043	1 0 0 0 1
3.330E+00	4.867E+02	87.10	A031	1 2 2 2 2
4.277E+00	6.250E+02	100	F300	1 0 0 0 2
4.211E+00	6.154E+02	100	M043	1 0 0 0 2

800.   C$_6$H$_{10}$O$_4$
2,2-Dimethylsuccinic Acid
α,α-Dimethylbernsteinsaeure
RN: 597-43-3   **MP** (°C):   140.5
MW: 146.14   **BP** (°C):

Solubility (Moles/L)	Solubility (Grams/L)	Temp (°C)	Ref (#)	Evaluation (T P E A A)	Comments
4.790E-01	7.000E+01	14	F300	1 0 0 0 2	

801.   C$_6$H$_{10}$O$_4$
DL-2,3-Dimethylsuccinic Acid
DL-α,α'-Dimethylbernsteinsaeure
RN: 13545-04-5   **MP** (°C):   120
MW: 146.14   **BP** (°C):

Solubility (Moles/L)	Solubility (Grams/L)	Temp (°C)	Ref (#)	Evaluation (T P E A A)	Comments
2.053E-01	3.000E+01	14	F300	1 0 0 0 0	

802.   C$_6$H$_{10}$O$_4$
Ethylene Glycol Diacetate
Glycol Diacetate
RN: 111-55-7   **MP** (°C):   -31
MW: 146.14   **BP** (°C):   190

Solubility (Moles/L)	Solubility (Grams/L)	Temp (°C)	Ref (#)	Evaluation (T P E A A)	Comments
1.202E+00	1.756E+02	20	D052	1 1 0 0 2	
9.661E-01	1.412E+02	20	M062	1 0 0 0 2	
8.526E-01	1.246E+02	22	F300	1 0 0 0 2	
1.034E+00	1.511E+02	24.50	O005	2 0 2 2 2	
1.070E+00	1.564E+02	25	F064	1 0 0 0 2	
1.220E-01	1.783E+01	ns	F014	0 0 0 0 2	

803. $C_6H_{10}O_4$
n-Propylmalonic Acid
Acide n-Propylmalonique
RN: 616-62-6   **MP** (°C):
MW: 146.14   **BP** (°C):

Solubility (Moles/L)	Solubility (Grams/L)	Temp (°C)	Ref (#)	Evaluation (T P E A A)	Comments
3.120E+00	4.560E+02	0	M051	1 0 0 0 2	
4.112E+00	6.010E+02	15	M051	1 0 0 0 2	
4.790E+00	7.000E+02	25	M051	1 0 0 0 2	
6.459E+00	9.440E+02	50	M051	1 0 0 0 2	

804. $C_6H_{10}O_4$
Methyl α-Acetoxypropionate
Methyl 2-Acetoxypropionate
Methyl O-Acetyllactate
Methyl 2-Acetyloxypropanoate
RN: 6284-75-9   **MP** (°C):
MW: 146.14   **BP** (°C):

Solubility (Moles/L)	Solubility (Grams/L)	Temp (°C)	Ref (#)	Evaluation (T P E A A)	Comments
5.556E-01	8.120E+01	25	R006	2 2 0 1 2	

805. $C_6H_{10}O_5$
Propanoic Acid, 2-[(Methoxycarbonyl)oxy]-, Methyl Ester
Carbonic Acid, Methyl Ester, Ester with Methyl Lactate
RN: 6288-11-5   **MP** (°C):
MW: 162.14   **BP** (°C):

Solubility (Moles/L)	Solubility (Grams/L)	Temp (°C)	Ref (#)	Evaluation (T P E A A)	Comments
2.412E-01	3.911E+01	25	R007	1 0 0 0 2	

806. $C_6H_{10}O_8$
D-Talogalactaric Acid
D-Taloschleimsaeure
D-Galactaric Acid
Galactaric Acid
Schleimsaeure
RN: 526-99-8   **MP** (°C):   >230
MW: 210.14   **BP** (°C):

Solubility (Moles/L)	Solubility (Grams/L)	Temp (°C)	Ref (#)	Evaluation (T P E A A)	Comments
1.565E-02	3.289E+00	14	D041	1 0 0 0 1	
1.570E-02	3.300E+00	14	F300	1 0 0 0 1	
8.090E-02	1.700E+01	100	F300	1 0 0 0 1	

807.   C$_6$H$_{11}$BrN$_2$O$_2$
3-Bromo-2-methyl-butanoic Ureide
Urea, (2-Bromo-2-methylbutyryl)-
DL-N-(2-Bromo-2-methylbutanoyl)urea
RN: 14368-76-4   **MP** (°C):
MW: 223.08   **BP** (°C):

Solubility (Moles/L)	Solubility (Grams/L)	Temp (°C)	Ref (#)	Evaluation (T P E A A)	Comments
1.390E-01	3.101E+01	ns	F056	0 2 2 2 1	

808.   C$_6$H$_{11}$BrN$_2$O$_2$
γ-Bromo-valeric Acid Ureide
RN: **MP** (°C):
MW: 223.08   **BP** (°C):

Solubility (Moles/L)	Solubility (Grams/L)	Temp (°C)	Ref (#)	Evaluation (T P E A A)	Comments
4.307E-02	9.607E+00	ns	F056	0 2 2 2 1	

809.   C$_6$H$_{11}$BrN$_2$O$_2$
α-Methyl-γ-bromo-butanoic Ureide
RN: **MP** (°C):
MW: 223.08   **BP** (°C):

Solubility (Moles/L)	Solubility (Grams/L)	Temp (°C)	Ref (#)	Evaluation (T P E A A)	Comments
4.658E-02	1.039E+01	ns	F056	0 2 2 2 1	

810.   C$_6$H$_{11}$BrN$_2$O$_2$
α-Bromo-valeric Acid Ureide
Pentanamide, N-(Aminocarbonyl)-2-bromo-
RN: 66947-87-3   **MP** (°C):
MW: 223.08   **BP** (°C):

Solubility (Moles/L)	Solubility (Grams/L)	Temp (°C)	Ref (#)	Evaluation (T P E A A)	Comments
3.690E-02	8.232E+00	ns	F056	0 2 2 2 1	
3.703E-02	8.261E+00	ns	F057	0 2 2 2 2	

811. $C_6H_{11}BrN_2O_2$
α-Bromo-isovaleric Ureide
Butanamide, N-(Aminocarbonyl)-2-bromo-3-methyl-
Dormigene
Pivadorn
Pivadorm
Isobromyl
RN: 496-67-3   **MP** (°C):
MW: 223.08   **BP** (°C):

Solubility (Moles/L)	Solubility (Grams/L)	Temp (°C)	Ref (#)	Evaluation (T P E A A)	Comments
8.531E-02	1.903E+01	ns	F057	0 2 2 2 2	

812. $C_6H_{11}BrN_2O_2$
β-Bromo-valeric Acid Ureide
RN: **MP** (°C):
MW: 223.08   **BP** (°C):

Solubility (Moles/L)	Solubility (Grams/L)	Temp (°C)	Ref (#)	Evaluation (T P E A A)	Comments
3.470E-02	7.740E+00	ns	F056	0 2 2 2 1	

813. $C_6H_{11}NO$
Caprolactam
ε-Caprolactam
RN: 105-60-2   **MP** (°C):   70
MW: 113.16   **BP** (°C):   180

Solubility (Moles/L)	Solubility (Grams/L)	Temp (°C)	Ref (#)	Evaluation (T P E A A)	Comments
3.776E+00	4.273E+02	5.70	B201	2 2 2 1 2	
3.850E+00	4.357E+02	10.30	B201	2 2 2 1 2	

814. $C_6H_{11}NO$
Cyclohexanone Oxime
Antioxidant D
(Hydroxyimino)Cyclohexane
RN: 100-64-1   **MP** (°C):   90
MW: 113.16   **BP** (°C):   208

Solubility (Moles/L)	Solubility (Grams/L)	Temp (°C)	Ref (#)	Evaluation (T P E A A)	Comments
1.409E-01	1.594E+01	25.5	K087	1 0 0 0 2	
1.580E-01	1.787E+01	32.0	K087	1 0 0 0 2	
1.648E-01	1.865E+01	36.8	K087	1 0 0 0 2	
1.936E-01	2.191E+01	44.0	K087	1 0 0 0 2	
2.155E-01	2.439E+01	48.8	K087	1 0 0 0 2	
2.715E-01	3.073E+01	60.4	K087	1 0 0 0 2	

2.922E-01	3.307E+01	63.7	K087	1 0 0 0 2
3.194E-01	3.614E+01	76.2	K087	1 0 0 0 2
3.456E-01	3.911E+01	83.1	K087	1 0 0 0 2
4.039E-01	4.571E+01	95.2	K087	1 0 0 0 2
4.939E-01	5.589E+01	110.7	K087	1 0 0 0 2
5.743E-01	6.498E+01	120	K087	1 0 0 0 2
7.386E-01	8.358E+01	131	K087	1 0 0 0 2

815. $C_6H_{11}NO_4$
α-Aminoadipic Acid
2-Aminohexanedioic Acid
α-Amino-adipinsaeure
RN: 542-32-5   **MP** (°C):
MW: 161.16   **BP** (°C):

Solubility (Moles/L)	Solubility (Grams/L)	Temp (°C)	Ref (#)	Evaluation (T P E A A)	Comments
1.365E-02	2.200E+00	20	F300	1 0 0 0 1	

816. $C_6H_{11}NO_4$
Glycine, N-(Carboxymethyl)-, 1-Ethyl Ester
AcGlyOEt
Acetic Acid, Iminodi-, Monoethyl Ester
RN: 21885-31-4   **MP** (°C):
MW: 161.16   **BP** (°C):

Solubility (Moles/L)	Solubility (Grams/L)	Temp (°C)	Ref (#)	Evaluation (T P E A A)	Comments
7.074E-03	1.140E+00	27	D036	2 1 2 2 2	

817. $C_6H_{11}N_2O_4PS_3$
Methidathion
Supracide
S-((5-Methoxy-2-oxo-1,3,4-thiadiazol-3(2H)-yl)methyl) O,O-Dimethyl Phosphorodithioate
Ultracide
Somanil
S-2,3-Dihydro-5-methoxy-2-oxo-1,3,4-thiadiazol-3-ylmethyl O,O-
dimethylphosphorodithioate
RN: 950-37-8   **MP** (°C):
MW: 302.33   **BP** (°C):

Solubility (Moles/L)	Solubility (Grams/L)	Temp (°C)	Ref (#)	Evaluation (T P E A A)	Comments
6.186E-04	1.870E-01	20	B300	2 2 1 1 2	
8.269E-04	2.500E-01	20	F311	1 2 2 2 1	
7.938E-04	2.400E-01	25	M161	1 0 0 0 2	

818. $C_6H_{11}N_3O_6$
Glycine Tripeptide
RN: **MP** ($^\circ$C):
MW: 221.17   **BP** ($^\circ$C):

Solubility (Moles/L)	Solubility (Grams/L)	Temp ($^\circ$C)	Ref (#)	Evaluation (T P E A A)	Comments
2.127E-01	4.705E+01	20	B032	1 2 2 1 2	
2.907E-01	6.430E+01	25	B032	1 2 2 1 2	
3.565E-01	7.884E+01	29.80	B032	1 2 2 1 2	

819. C_6H_{12}
2-Methyl-1-pentene
4-Methyl-4-pentene
RN: 763-29-1   **MP** ($^\circ$C):   -136
MW: 84.16   **BP** ($^\circ$C):   62

Solubility (Moles/L)	Solubility (Grams/L)	Temp ($^\circ$C)	Ref (#)	Evaluation (T P E A A)	Comments
9.268E-04	7.800E-02	25	M001	2 1 2 2 2	

820. C_6H_{12}
Methylcyclopentane
MCP
RN: 96-37-7   **MP** ($^\circ$C):   -142
MW: 84.16   **BP** ($^\circ$C):   72

Solubility (Moles/L)	Solubility (Grams/L)	Temp ($^\circ$C)	Ref (#)	Evaluation (T P E A A)	Comments
4.967E-04	4.180E-02	25	K119	1 0 0 0 2	
4.990E-04	4.200E-02	25	M001	2 1 2 2 2	
5.062E-04	4.260E-02	25	M002	2 1 2 2 2	
4.967E-04	4.180E-02	25	P051	2 1 1 2 2	
4.967E-04	4.180E-02	25.00	P007	2 1 2 2 2	
4.990E-04	4.200E-02	ns	H123	0 0 0 0 2	

821. C_6H_{12}
4-Methyl-1-pentene
4-Methylpentene
Isohexene
RN: 691-37-2   **MP** ($^\circ$C):   -154
MW: 84.16   **BP** ($^\circ$C):   53

Solubility (Moles/L)	Solubility (Grams/L)	Temp ($^\circ$C)	Ref (#)	Evaluation (T P E A A)	Comments
5.703E-04	4.800E-02	25	M001	2 1 2 2 1	

822. C$_6$H$_{12}$
1-Hexene
1-n-Hexene
Hexene
Dialen 6
RN: 592-41-6   **MP** (°C):   -140
MW: 84.16   **BP** (°C):   64

Solubility (Moles/L)	Solubility (Grams/L)	Temp (°C)	Ref (#)	Evaluation (T P E A A)	Comments
5.822E-04	4.900E-02	23	C332	2 0 2 2 1	
6.583E-04	5.540E-02	25	L002	2 2 2 2 2	
5.941E-04	5.000E-02	25	M001	2 1 2 2 2	
5.941E-04	5.000E-02	25	M040	1 0 0 1 1	
8.280E-04	6.969E-02	25	M342	1 0 1 1 2	

823. C$_6$H$_{12}$
Cyclohexane
Cyclohexan
RN: 110-82-7   **MP** (°C):   7
MW: 84.16   **BP** (°C):

Solubility (Moles/L)	Solubility (Grams/L)	Temp (°C)	Ref (#)	Evaluation (T P E A A)	Comments
9.734E-04	8.192E-02	4.8	L007	2 1 1 2 2	
9.734E-04	8.192E-02	5.1	L007	2 0 1 1 2	
1.054E-03	8.869E-02	14.8	L007	2 1 1 2 2	
1.054E-03	8.869E-02	15.2	L007	2 0 1 1 2	
9.505E-04	8.000E-02	16	D047	1 0 0 1 1	
<5.94E-04	<5.00E-02	17	F300	1 0 0 0 0	
4.396E-04	3.700E-02	20	M337	2 1 2 2 2	
6.178E-04	5.200E-02	23.5	S171	2 1 2 2 2	
1.055E-03	8.883E-02	24.8	L007	2 1 1 2 2	
9.505E-04	7.999E-02	25	G068	1 0 1 0 0	
6.939E-04	5.840E-02	25	G313	2 1 1 2 2	
1.426E-03	1.200E-01	25	K112	1 0 2 1 1	
7.901E-04	6.650E-02	25	K119	1 0 0 0 2	
6.737E-04	5.670E-02	25	L002	2 2 2 2 2	
6.535E-04	5.500E-02	25	M001	2 1 2 2 2	
6.535E-04	5.500E-02	25	M002	2 1 2 2 2	
6.535E-04	5.500E-02	25	M040	1 0 0 1 1	
6.832E-04	5.750E-02	25	M132	2 2 2 1 2	
7.901E-04	6.650E-02	25	P051	2 1 1 2 2	
6.270E-04	5.277E-02	25	S359	2 1 2 2 2	
7.901E-04	6.650E-02	25.00	P007	2 1 2 2 2	
1.055E-03	8.883E-02	34.8	L007	2 1 1 2 2	
1.055E-03	8.883E-02	35.2	L007	2 0 1 1 2	
5.389E-04	4.535E-02	38	K055	1 2 0 1 1	

1.085E-03	9.131E-02	44.8	L007	2 1 1 2 2	
1.085E-03	9.131E-02	45.2	L007	2 0 1 1 2	
1.426E-03	1.200E-01	50	L097	1 1 1 1 1	
2.020E-03	1.700E-01	56	G068	1 0 1 0 1	
3.222E-04	2.712E-02	71	K055	1 2 0 1 1	
3.326E-03	2.799E-01	94	G068	1 0 1 0 1	
1.200E-04	1.010E-02	ns	D348	0 0 2 2 2	
6.535E-04	5.500E-02	ns	H123	0 0 0 0 2	
5.000E-03	4.208E-01	ns	H333	0 1 0 1 0	EFG
9.505E-04	8.000E-02	ns	M010	0 0 0 0 0	
6.642E-04	5.590E-02	ns	M175	0 0 2 1 2	

824. $C_6H_{12}ClNO$
Acetamide, 2-chloro-N,N-diethyl-
CDEA

RN: 2315-36-8   **MP** (°C):
MW: 149.62   **BP** (°C):

Solubility (Moles/L)	Solubility (Grams/L)	Temp (°C)	Ref (#)	Evaluation (T P E A A)	Comments
5.264E-01	7.877E+01	25	B185	1 0 0 0 2	

825. $C_6H_{12}Cl_2O$
Dichloroisopropyl Ether
bis(2-Chloro-1-methylethyl) Ether
DCIP
β,β'-Dichlorodiisopropyl Ether
2,2'-Oxybis[1-chloropropane]
Pichloram

RN: 63283-80-7   **MP** (°C):
MW: 171.07   **BP** (°C):   187.3

Solubility (Moles/L)	Solubility (Grams/L)	Temp (°C)	Ref (#)	Evaluation (T P E A A)	Comments
9.921E-03	1.697E+00	20	M062	1 0 0 0 1	

826. $C_6H_{12}Cl_2O_2$
1,2-bis(2-Chloroethoxy)ethane
Triglycol Dichloride

RN: 112-26-5   **MP** (°C):   121
MW: 187.07   **BP** (°C):   235

Solubility (Moles/L)	Solubility (Grams/L)	Temp (°C)	Ref (#)	Evaluation (T P E A A)	Comments
9.916E-02	1.855E+01	20	M062	1 0 0 0 2	

827. $C_6H_{12}Cl_3O_4P$

Tris-(2-chloroethyl) Phosphate
Tri-β-chloroethyl Phosphate
RN: 115-96-8   **MP** (°C):
MW: 285.49   **BP** (°C):

Solubility (Moles/L)	Solubility (Grams/L)	Temp (°C)	Ref (#)	Evaluation (T P E A A)	Comments
<7.01E-04	<2.00E-01	25	B070	1 2 0 1 0	

828. $C_6H_{12}NO_3PS_2$

Diethyl 1,3-Dithietan-2-ylidenephosphoramidate
Nematak
AC 64475
Geofos
Fosthietan
CL 64475
RN: 21548-32-3   **MP** (°C):
MW: 241.27   **BP** (°C):

Solubility (Moles/L)	Solubility (Grams/L)	Temp (°C)	Ref (#)	Evaluation (T P E A A)	Comments
2.072E-01	5.000E+01	25	M161	1 0 0 0 1	

829. $C_6H_{12}NO_4PS_2$

Formothion
O,O-Dimethyl S-(N-Methyl-N-formylcarbamoylmethyl) Dithiophosphate
RN: 2540-82-1   **MP** (°C):
MW: 257.27   **BP** (°C):

Solubility (Moles/L)	Solubility (Grams/L)	Temp (°C)	Ref (#)	Evaluation (T P E A A)	Comments
1.011E-02	2.600E+00	24	M161	1 0 0 0 1	

830. $C_6H_{12}N_2O$

N-Nitrosohexamethyleneimine
NHMI
RN: 932-83-2   **MP** (°C):
MW: 128.18   **BP** (°C):

Solubility (Moles/L)	Solubility (Grams/L)	Temp (°C)	Ref (#)	Evaluation (T P E A A)	Comments
1.000E-01	1.282E+01	24	M031	1 1 1 1 1	

831. $C_6H_{12}N_2O_2$
Adipamide
Adipinsaeurediamid
RN: 628-94-4   **MP** (°C):
MW: 144.17   **BP** (°C):

Solubility (Moles/L)	Solubility (Grams/L)	Temp (°C)	Ref (#)	Evaluation (T P E A A)	Comments
3.052E-02	4.400E+00	12.20	F300	1 0 0 0 1	

832. $C_6H_{12}N_2O_2$
2,6-Dimethylnitrosomorpholine
DMNM
RN: 1456-28-6   **MP** (°C):
MW: 144.17   **BP** (°C):

Solubility (Moles/L)	Solubility (Grams/L)	Temp (°C)	Ref (#)	Evaluation (T P E A A)	Comments
8.600E-01	1.240E+02	24	M031	1 1 1 1 1	

833. $C_6H_{12}N_2O_3$
Daminozide
N-Dimethylamino-β-carbamyl Propionic Acid
Succinic Acid 2,2-Dimethylhydrazide
Alar
DMASA
RN: 1596-84-5   **MP** (°C):   155
MW: 160.17   **BP** (°C):

Solubility (Moles/L)	Solubility (Grams/L)	Temp (°C)	Ref (#)	Evaluation (T P E A A)	Comments
6.243E-01	1.000E+02	25	M161	1 0 0 0 2	

834. $C_6H_{12}N_2O_3$
δ-Aminovaleric Hydantoic Acid
δ-Uramidovaleric Acid
RN: **MP** (°C):   179
MW: 160.17   **BP** (°C):

Solubility (Moles/L)	Solubility (Grams/L)	Temp (°C)	Ref (#)	Evaluation (T P E A A)	Comments
1.740E-02	2.787E+00	25	M024	1 2 0 1 2	

835. $C_6H_{12}N_2O_4S_2$
L-Cystine
3,3'-Dithiobis(2-aminopropanoic Acid)
RN: 56-89-3   **MP** (°C):
MW: 240.30   **BP** (°C):

Solubility (Moles/L)	Solubility (Grams/L)	Temp (°C)	Ref (#)	Evaluation (T P E A A)	Comments
2.021E-03	4.858E-01	20	H082	1 2 1 1 2	isomeric
7.905E-04	1.900E-01	20	H082	1 2 1 1 2	plate cystine
4.536E-04	1.090E-01	25	D017	1 0 0 0 2	
4.577E-04	1.100E-01	25	D041	1 0 0 0 1	
4.661E-04	1.120E-01	25	L001	1 0 1 1 2	pH 6.0
4.910E-04	1.180E-01	27	D036	2 1 2 2 2	
2.163E-03	5.197E-01	75	D041	1 0 0 0 1	
4.536E-04	1.090E-01	rt	B103	0 0 0 0 2	

836. $C_6H_{12}N_2O_4S$
DL-Lanthionine
L-Cysteine, S-[(2R)-2-Amino-2-carboxyethyl]-
RN: 922-55-4   **MP** (°C):   280
MW: 208.24   **BP** (°C):

Solubility (Moles/L)	Solubility (Grams/L)	Temp (°C)	Ref (#)	Evaluation (T P E A A)	Comments
7.193E-03	1.498E+00	25	D041	1 0 0 0 1	

837. $C_6H_{12}N_2O_4S_2$
D-Cystine
D-(+)-3,3'-Dithiobis(2-aminopropanoic Acid)
RN: 349-46-2   **MP** (°C):   227
MW: 240.30   **BP** (°C):

Solubility (Moles/L)	Solubility (Grams/L)	Temp (°C)	Ref (#)	Evaluation (T P E A A)	Comments
4.577E-04	1.100E-01	25	D041	1 0 0 0 1	
4.702E-04	1.130E-01	25	L001	1 0 1 1 2	pH 6.0

838. $C_6H_{12}N_2O_4S_2$
DL-Cystine
Cystine
RN: 923-32-0   **MP** (°C):
MW: 240.30   **BP** (°C):

Solubility (Moles/L)	Solubility (Grams/L)	Temp (°C)	Ref (#)	Evaluation (T P E A A)	Comments
2.039E-04	4.900E-02	25	D041	1 0 0 0 1	
2.372E-04	5.700E-02	25	L001	1 0 1 1 1	pH 6.0

839.   C$_6$H$_{12}$N$_2$O$_4$S$_2$
Mesocystine
meso-Cystine
RN: 6020-39-9   **MP** (°C):
MW: 240.30   **BP** (°C):

Solubility (Moles/L)	Solubility (Grams/L)	Temp (°C)	Ref (#)	Evaluation (T P E A A)	Comments
2.330E-04	5.600E-02	25	L001	1 0 1 1 1	pH 6.0

840.   C$_6$H$_{12}$N$_2$S$_4$Zn
Ziram
Zinc bis Dimethyldithiocarbamate
Corozate
Karbam White
Fuklasin
Fuclasin
RN: 137-30-4   **MP** (°C):   240
MW: 305.81   **BP** (°C):

Solubility (Moles/L)	Solubility (Grams/L)	Temp (°C)	Ref (#)	Evaluation (T P E A A)	Comments
2.125E-04	6.500E-02	20	F300	1 0 0 0 1	
1.308E-05	4.000E-03	20	F311	1 2 2 2 1	sic
2.125E-04	6.500E-02	25	M161	1 0 0 0 1	

841.   C$_6$H$_{12}$N$_2$S$_4$
Thiram
Tetramethylthioperoxydicarbonothioic Diamine
Tetramethylthiuram Disulfide
N,N'-(Dithiodicarbonothioyl)bis(N-methylmethanamine)
Arasan
Nomersan
RN: 137-26-8   **MP** (°C):   155.5
MW: 240.43   **BP** (°C):

Solubility (Moles/L)	Solubility (Grams/L)	Temp (°C)	Ref (#)	Evaluation (T P E A A)	Comments
1.248E-04	3.000E-02	rt	M161	0 0 0 0 1	

842.   C$_6$H$_{12}$N$_4$
Methenamine
Hexamethylen-tetramin
RN: 100-97-0   **MP** (°C):
MW: 140.19   **BP** (°C):

Solubility (Moles/L)	Solubility (Grams/L)	Temp (°C)	Ref (#)	Evaluation (T P E A A)	Comments
3.200E+00	4.486E+02	12	F300	1 0 0 0 2	

843. $C_6H_{12}N_4O_2$
2,6-Dimethyldinitrosopiperazine
DMDNP
RN: 55380-34-2   **MP** (°C):
MW: 172.19   **BP** (°C):

Solubility (Moles/L)	Solubility (Grams/L)	Temp (°C)	Ref (#)	Evaluation (T P E A A)	Comments
1.200E-01	2.042E+01	24	M031	1 1 1 1 1	

844. $C_6H_{12}N_5O_2PS_2$
Menazon
O,O-Dimethyl S-(4,6-Diamino-1,3,5-triazinyl-2-methyl) Dithiophosphate
RN: 78-57-9   **MP** (°C):
MW: 281.30   **BP** (°C):

Solubility (Moles/L)	Solubility (Grams/L)	Temp (°C)	Ref (#)	Evaluation (T P E A A)	Comments
8.532E-04	2.400E-01	20	M161	1 0 0 0 1	
3.551E-03	9.990E-01	ns	M061	0 0 0 0 0	

845. $C_6H_{12}O$
Caproic Aldehyde
Hexaldehyde
n-Hexanal
RN: 66-25-1   **MP** (°C):
MW: 100.16   **BP** (°C):   131

Solubility (Moles/L)	Solubility (Grams/L)	Temp (°C)	Ref (#)	Evaluation (T P E A A)	Comments
4.992E-02	5.000E+00	25	A049	1 0 1 0 1	

846. $C_6H_{12}O$
1-Hexen-3-ol
Hexen-1-ol-3
RN: 4798-44-1   **MP** (°C):
MW: 100.16   **BP** (°C):   134

Solubility (Moles/L)	Solubility (Grams/L)	Temp (°C)	Ref (#)	Evaluation (T P E A A)	Comments
2.644E-01	2.648E+01	20	G031	1 0 0 0 2	
2.454E-01	2.458E+01	25	G031	1 0 0 0 2	
2.302E-01	2.306E+01	30	G031	1 0 0 0 2	

847.   C₆H₁₂O
Pinacolone
3,3-Dimethyl-2-butanone
3,3-Dimethylbutanone-2
RN: 75-97-8   **MP** (°C):   -52.5
MW: 100.16   **BP** (°C):   106.2

Solubility (Moles/L)	Solubility (Grams/L)	Temp (°C)	Ref (#)	Evaluation (T P E A A)	Comments
2.376E-01	2.380E+01	15	F300	1 0 0 0 2	
1.996E-01	1.999E+01	20	G030	1 2 0 0 2	
1.862E-01	1.865E+01	25	G030	1 2 0 0 2	
1.817E-01	1.820E+01	25	K072	1 0 1 1 1	
1.736E-01	1.739E+01	30	G030	1 2 0 0 2	

848.   C₆H₁₂O
Methyl Butyl Ketone
2-Hexanone
Methyl n-Butyl Ketone
RN: 591-78-6   **MP** (°C):   -57
MW: 100.16   **BP** (°C):   127

Solubility (Moles/L)	Solubility (Grams/L)	Temp (°C)	Ref (#)	Evaluation (T P E A A)	Comments
2.040E-01	2.043E+01	10	G032	1 2 1 1 2	
2.192E-02	2.195E+00	20	D052	1 1 0 0 1	*slc*
1.717E-01	1.720E+01	20	G030	1 2 0 0 2	
1.611E-01	1.614E+01	25	G030	1 2 0 0 2	
3.320E-01	3.326E+01	25	P055	1 0 0 0 2	
1.505E-01	1.507E+01	30	G030	1 2 0 0 2	
1.450E-01	1.452E+01	30	G032	1 2 1 1 2	
1.475E-01	1.478E+01	38	J020	2 1 2 1 1	
1.240E-01	1.242E+01	50	G032	1 2 1 1 2	

849.   C₆H₁₂O
Isopropylacetone
4-Methyl-2-pentanone
Methyl Isobutyl Ketone
RN: 108-10-1   **MP** (°C):   -80
MW: 100.16   **BP** (°C):   117

Solubility (Moles/L)	Solubility (Grams/L)	Temp (°C)	Ref (#)	Evaluation (T P E A A)	Comments
3.070E-01	3.075E+01	0	G032	1 2 1 1 2	
2.310E-01	2.314E+01	10	G032	1 2 1 1 2	
1.871E-01	1.874E+01	20	D052	1 1 0 0 2	
1.996E-01	1.999E+01	20	G030	1 2 0 0 2	
1.958E-01	1.961E+01	22.00	O005	2 0 2 2 0	
1.862E-01	1.865E+01	24.6	H121	2 0 0 0 1	
1.862E-01	1.865E+01	25	B060	2 0 1 1 1	

1.717E-01	1.720E+01	25	C329	1 1 1 1 1	average
1.871E-01	1.874E+01	25	G030	1 2 0 0 2	
2.340E-01	2.344E+01	25	K103	1 2 2 2 1	
1.862E-01	1.865E+01	25	L082	1 1 2 1 1	
1.736E-01	1.739E+01	25	L319	1 0 2 1 2	
1.817E-01	1.820E+01	25	M087	1 1 2 1 2	
1.669E-01	1.672E+01	25	R320	1 0 1 1 1	
1.746E-01	1.749E+01	30	G030	1 2 0 0 2	
1.660E-01	1.663E+01	30	G032	1 2 1 1 2	
1.410E-01	1.412E+01	50	G032	1 2 1 1 2	
4.720E+01	4.728E+03	53.0	R308	2 2 1 1 2	
1.669E-01	1.672E+01	70	L082	1 1 2 1 1	
1.370E-01	1.372E+01	75	G032	1 2 1 1 2	
4.300E+01	4.307E+03	97.0	R308	2 2 1 1 2	
4.088E+01	4.094E+03	108.0	R308	2 2 1 1 2	
3.902E+01	3.909E+03	120.0	R308	2 2 1 1 2	
3.333E-01	3.339E+01	125.0	R308	2 2 1 1 1	
5.278E-01	5.286E+01	151.0	R308	2 2 1 1 1	
3.425E+01	3.431E+03	153.0	R308	2 2 1 2 2	

850. $C_6H_{12}O$
Cyclohexanol
1-Cyclohexanol
Naxol
Cyclohexyl Alcoho
Adrona
Hydrophenol
RN: 108-93-0   **MP** (°C):   23
MW: 100.16   **BP** (°C):

Solubility (Moles/L)	Solubility (Grams/L)	Temp (°C)	Ref (#)	Evaluation (T P E A A)	Comments
5.357E-01	5.366E+01	11	F052	1 1 1 0 2	
5.391E-01	5.400E+01	11	F300	1 0 0 0 1	
1.296E-02	1.298E+00	20	D052	1 1 0 0 1	*sic*
3.283E-01	3.288E+01	25	B019	1 0 1 2 0	
3.283E-01	3.288E+01	25	B092	2 1 1 1 2	
3.469E-01	3.475E+01	25	C108	2 2 2 2 2	
3.800E-01	3.806E+01	25	F044	1 0 0 0 1	
3.766E-01	3.772E+01	25	H028	2 0 2 0 2	
3.655E-01	3.661E+01	35	C108	2 2 2 2 2	
3.264E-01	3.269E+01	60	B092	2 1 1 1 2	

851. C$_6$H$_{12}$O
4-Hexen-3-ol
Hexen-4-ol-3
RN: 4798-58-7   **MP** (°C):
MW: 100.16   **BP** (°C):

Solubility (Moles/L)	Solubility (Grams/L)	Temp (°C)	Ref (#)	Evaluation (T P E A A)	Comments
3.895E-01	3.902E+01	20	G031	1 0 0 0 2	
3.664E-01	3.670E+01	25	G031	1 0 0 0 2	
3.451E-01	3.456E+01	30	G031	1 0 0 0 2	

852. C$_6$H$_{12}$O
3-Methyl-2-pentanone
3-Methylpentanone-2
RN: 565-61-7   **MP** (°C):   <25
MW: 100.16   **BP** (°C):   118

Solubility (Moles/L)	Solubility (Grams/L)	Temp (°C)	Ref (#)	Evaluation (T P E A A)	Comments
2.206E-01	2.210E+01	20	G030	1 2 0 0 2	
2.044E-01	2.047E+01	25	G030	1 2 0 0 2	
1.890E-01	1.893E+01	30	G030	1 2 0 0 2	

853. C$_6$H$_{12}$O
2-Methyl-4-penten-3-ol
2-Methylpenten-4-ol-3
RN: 4798-45-2   **MP** (°C):
MW: 100.16   **BP** (°C):

Solubility (Moles/L)	Solubility (Grams/L)	Temp (°C)	Ref (#)	Evaluation (T P E A A)	Comments
3.180E-01	3.185E+01	20	G031	1 0 0 0 2	
2.964E-01	2.969E+01	25	G031	1 0 0 0 2	
2.804E-01	2.809E+01	30	G031	1 0 0 0 2	

854. C$_6$H$_{12}$O
3-Hexanone
Hexanone-3
RN: 589-38-8   **MP** (°C):   -55.5
MW: 100.16   **BP** (°C):   123

Solubility (Moles/L)	Solubility (Grams/L)	Temp (°C)	Ref (#)	Evaluation (T P E A A)	Comments
1.543E-01	1.546E+01	20	G030	1 2 0 0 2	
1.446E-01	1.449E+01	25	G030	1 2 0 0 2	
1.359E-01	1.361E+01	30	G030	1 2 0 0 2	

855.   C$_6$H$_{12}$O
4-Methyl-3-pentanone
4-Methylpentanone-3
RN: 565-69-5   **MP** (°C):
MW: 100.16   **BP** (°C):

Solubility (Moles/L)	Solubility (Grams/L)	Temp (°C)	Ref (#)	Evaluation (T P E A A)	Comments
1.601E-01	1.604E+01	20	G030	1 2 0 0 2	
1.495E-01	1.497E+01	25	G030	1 2 0 0 2	
1.398E-01	1.400E+01	30	G030	1 2 0 0 2	

856.   C$_6$H$_{12}$O$_2$
Ethyl Butyrate
Butanoic Acid Ethyl Ester
Ethyl Butanoate
Butyric Ether
RN: 105-54-4   **MP** (°C):   -135.4
MW: 116.16   **BP** (°C):   120

Solubility (Moles/L)	Solubility (Grams/L)	Temp (°C)	Ref (#)	Evaluation (T P E A A)	Comments
4.198E-02	4.876E+00	20	D052	1 1 0 0 1	
5.310E-02	6.168E+00	22	F001	1 0 1 2 2	
4.300E-02	4.995E+00	22	S006	1 0 0 0 1	
6.832E-02	7.937E+00	30	R318	1 1 0 1 0	

857.   C$_6$H$_{12}$O$_2$
Diethylacetic Acid
2-Ethylbutyric Acid
2-Ethyl-butanoic Acid
Ethylbutyric Acid
RN: 88-09-5   **MP** (°C):   -15
MW: 116.16   **BP** (°C):   194.5

Solubility (Moles/L)	Solubility (Grams/L)	Temp (°C)	Ref (#)	Evaluation (T P E A A)	Comments
2.147E-02	2.494E+00	25	O011	1 0 1 1 1	

858.   C$_6$H$_{12}$O$_2$
sec-Butyl Acetate
DL-sec-Butyl Acetate
RN: 105-46-4   **MP** (°C):
MW: 116.16   **BP** (°C):   114

Solubility (Moles/L)	Solubility (Grams/L)	Temp (°C)	Ref (#)	Evaluation (T P E A A)	Comments
5.305E-02	6.162E+00	20	D052	1 1 0 0 0	

859. $C_6H_{12}O_2$
n-Caproic Acid
n-Capronsaeure

RN: 142-62-1   **MP** (°C):   -3.4
MW: 116.16   **BP** (°C):   205

Solubility (Moles/L)	Solubility (Grams/L)	Temp (°C)	Ref (#)	Evaluation (T P E A A)	Comments
7.438E-02	8.640E+00	0	B136	1 0 2 1 2	
7.374E-02	8.566E+00	0.0	R001	1 1 1 1 2	
7.610E-02	8.840E+00	15	F300	1 0 0 0 2	
8.333E-02	9.680E+00	20	B136	1 0 2 1 2	
8.270E-02	9.607E+00	20	D041	1 0 0 0 1	
8.253E-02	9.587E+00	20	R001	1 1 1 1 2	
8.675E-02	1.008E+01	25	H028	2 0 2 0 2	
8.760E-02	1.018E+01	25	H122	1 0 0 0 2	
8.608E-02	9.999E+00	25	H339	2 2 1 2 2	
9.367E-02	1.088E+01	25	O011	1 0 1 1 1	
8.772E-02	1.019E+01	30	B136	1 0 2 1 2	
8.684E-02	1.009E+01	30	R001	1 1 1 1 2	
9.282E-02	1.078E+01	35	H339	2 2 1 2 2	
9.427E-02	1.095E+01	45	B136	1 0 2 1 2	
9.324E-02	1.083E+01	45	R001	1 1 1 1 2	
1.008E-01	1.171E+01	60	B136	1 0 2 1 2	
9.956E-02	1.156E+01	60	D041	1 0 0 0 2	
9.964E-02	1.157E+01	60	R001	1 1 1 1 2	

860. $C_6H_{12}O_2$
Isobutyl Acetate
Acetic Acid Isobutyl Ester
Essigsaeureisobutyl Ester

RN: 110-19-0   **MP** (°C):   -99
MW: 116.16   **BP** (°C):   118

Solubility (Moles/L)	Solubility (Grams/L)	Temp (°C)	Ref (#)	Evaluation (T P E A A)	Comments
6.502E-02	7.553E+00	14.60	L310	2 2 1 1 2	
5.729E-02	6.655E+00	20	D052	1 1 0 0 1	
5.800E-02	6.737E+00	20	F001	1 0 1 2 1	
5.768E-02	6.700E+00	20	F300	1 0 0 0 1	
6.154E-02	7.149E+00	24.90	L310	2 2 1 1 2	
5.390E-02	6.261E+00	25	B060	2 0 1 1 1	
5.967E-02	6.932E+00	47.90	L310	2 2 1 1 2	
6.154E-02	7.149E+00	67.60	L310	2 2 1 1 2	
6.493E-02	7.543E+00	74.90	L310	2 2 1 1 2	
6.502E-02	7.553E+00	75.20	L310	2 2 1 1 2	
6.875E-02	7.986E+00	84.80	L310	2 2 1 1 2	

7.205E-02	8.369E+00	93.20	L310	2 2 1 1 2
8.253E-02	9.587E+00	111.50	L310	2 2 1 1 2
8.540E-02	9.921E+00	115.70	L310	2 2 1 1 2
1.026E-01	1.192E+01	147.10	L310	2 2 1 1 2

861. $C_6H_{12}O_2$
3-Hydroxy-2,2-dimethyltetrahydrofuran
3-Furanol, Tetrahydro-2,2-dimethyl-
2,2-Dimethyltetrahydrofuran-3-ol
RN: 101398-19-0   **MP** (°C):
MW: 116.16   **BP** (°C):

Solubility (Moles/L)	Solubility (Grams/L)	Temp (°C)	Ref (#)	Evaluation (T P E A A)	Comments
7.826E-01	9.091E+01	rt	B066	0 2 0 0 1	

862. $C_6H_{12}O_2$
3-Hydroxy-2,5-dimethyltetrahydrofuran
3-Furanol, Tetrahydro-2,5-dimethyl-
RN: 30003-26-0   **MP** (°C):
MW: 116.16   **BP** (°C):

Solubility (Moles/L)	Solubility (Grams/L)	Temp (°C)	Ref (#)	Evaluation (T P E A A)	Comments
1.435E+00	1.667E+02	rt	B066	0 2 0 0 1	

863. $C_6H_{12}O_2$
n-Butyl Acetate
Essigsaeure-n-butyl Ester
n-Butylacetat
Butyl Acetate
1-Butyl Acetate
RN: 123-86-4   **MP** (°C):   -90
MW: 116.16   **BP** (°C):   117.5

Solubility (Moles/L)	Solubility (Grams/L)	Temp (°C)	Ref (#)	Evaluation (T P E A A)	Comments
3.686E-02	4.282E+00	20	D052	1 1 0 0 0	
8.609E-02	1.000E+01	22	F300	1 0 0 0 0	
5.814E-02	6.754E+00	25	B060	2 0 1 1 1	
7.171E-02	8.330E+00	25	L319	1 0 2 1 2	
1.935E-01	2.248E+01	25	P055	1 0 0 0 1	
2.489E-02	2.892E+00	30	N330	2 2 2 1 2	
7.679E-02	8.920E+00	30	R318	1 1 0 1 0	
5.020E-02	5.831E+00	37	E028	1 0 1 1 2	
5.899E-02	6.853E+00	50	O012	1 2 1 1 2	

864. C$_6$H$_{12}$O$_2$
Pentyl Formate
n-Amyl Formate
RN: 638-49-3   **MP** (°C):
MW: 116.16   **BP** (°C):

Solubility (Moles/L)	Solubility (Grams/L)	Temp (°C)	Ref (#)	Evaluation (T P E A A)	Comments
2.500E-02	2.904E+00	22	S006	1 0 0 0 1	

865. C$_6$H$_{12}$O$_2$
Propyl Propionate
Propionic Acid N-Propyl Ester
n-Propyl Propionate
RN: 106-36-5   **MP** (°C):
MW: 116.16   **BP** (°C):   123

Solubility (Moles/L)	Solubility (Grams/L)	Temp (°C)	Ref (#)	Evaluation (T P E A A)	Comments
5.000E-02	5.808E+00	22	S006	1 0 0 0 0	

866. C$_6$H$_{12}$O$_3$
2-Ethoxyethyl Acetate
Cellosolve Acetate
RN: 111-15-9   **MP** (°C):   -61
MW: 132.16   **BP** (°C):   156

Solubility (Moles/L)	Solubility (Grams/L)	Temp (°C)	Ref (#)	Evaluation (T P E A A)	Comments
1.499E+00	1.981E+02	20	D052	1 1 0 0 2	
1.415E+00	1.870E+02	20	M062	1 0 0 0 2	

867. C$_6$H$_{12}$O$_3$
Methyl β-Ethoxypropionate
Methyl 3-Ethoxyproplonate
3-Ethoxypropionic Acid Methyl Ester
RN: 14144-33-3   **MP** (°C):
MW: 132.16   **BP** (°C):

Solubility (Moles/L)	Solubility (Grams/L)	Temp (°C)	Ref (#)	Evaluation (T P E A A)	Comments
7.621E-01	1.007E+02	25	D002	1 2 1 1 2	
7.621E-01	1.007E+02	25	R034	0 0 0 0 2	

868.   C$_6$H$_{12}$O$_3$
Paraldehyde
Paraldehyd
RN: 123-63-7   **MP** (°C):   12.6
MW: 132.16   **BP** (°C):   128

Solubility (Moles/L)	Solubility (Grams/L)	Temp (°C)	Ref (#)	Evaluation (T P E A A)	Comments
8.853E-01	1.170E+02	8.5	P059	1 1 1 0 1	
8.377E-01	1.107E+02	11.5	P059	1 1 1 0 1	
8.287E-01	1.095E+02	12.0	P059	1 1 1 0 1	
8.323E-01	1.100E+02	13	F300	1 0 0 0 1	
8.047E-01	1.063E+02	13.5	P059	1 1 1 0 1	
7.621E-01	1.007E+02	17.0	P059	1 1 1 0 1	
6.311E-01	8.341E+01	27.0	P059	1 1 1 0 1	
8.475E-01	1.120E+02	30	F300	1 0 0 0 2	
5.377E-01	7.106E+01	40.0	P059	1 1 1 0 1	
5.246E-01	6.933E+01	42.5	P059	1 1 1 0 1	
4.283E-01	5.660E+01	68.0	P059	1 1 1 0 1	
4.148E-01	5.482E+01	75.0	P059	1 1 1 0 1	
4.540E-01	6.000E+01	100	F300	1 0 0 0 0	

869.   C$_6$H$_{12}$O$_5$
Rhamnose
α-L-Rhamnose
6-Deoxy-L-mannose
L-Mannomethylose
L-Rhamnose
RN: 3615-41-6   **MP** (°C):   82
MW: 164.16   **BP** (°C):

Solubility (Moles/L)	Solubility (Grams/L)	Temp (°C)	Ref (#)	Evaluation (T P E A A)	Comments
2.212E+00	3.631E+02	18	D041	1 0 0 0 1	
3.177E+00	5.215E+02	40	D041	1 0 0 0 1	

870.   C$_6$H$_{12}$O$_5$
D-Quercitol
D-Quercit
RN: 488-73-3   **MP** (°C):   234
MW: 164.16   **BP** (°C):

Solubility (Moles/L)	Solubility (Grams/L)	Temp (°C)	Ref (#)	Evaluation (T P E A A)	Comments
6.701E-01	1.100E+02	20	F300	1 0 0 0 2	

871. $C_6H_{12}O_6$

L-Sorbose
Sorbose
L-1,3,4,5,6-Pentahydroxyhexan-2-one
L-Xylo-2-Hexulose
RN: 87-79-6   **MP** (°C):   165
MW: 180.16   **BP** (°C):

Solubility (Moles/L)	Solubility (Grams/L)	Temp (°C)	Ref (#)	Evaluation (T P E A A)	Comments
1.970E+00	3.548E+02	17	D041	1 0 0 0 1	
1.998E+00	3.600E+02	17	F300	1 0 0 0 1	

872. $C_6H_{12}O_6$

D-Mannose
D-(+)-Mannose
Seminose
Carubinose
RN: 3458-28-4   **MP** (°C):   132
MW: 180.16   **BP** (°C):

Solubility (Moles/L)	Solubility (Grams/L)	Temp (°C)	Ref (#)	Evaluation (T P E A A)	Comments
3.956E+00	7.126E+02	17	D041	1 0 0 0 2	
3.957E+00	7.128E+02	17	F300	1 0 0 0 2	
2.399E+00	4.322E+02	25	G317	2 1 2 2 2	

873. $C_6H_{12}O_6$

Fructose
D-Fructose
D-(-)-Fructose
D-(-)-Levulose
Krystar 300
Nevulose
RN: 57-48-7   **MP** (°C):   129
MW: 180.16   **BP** (°C):

Solubility (Moles/L)	Solubility (Grams/L)	Temp (°C)	Ref (#)	Evaluation (T P E A A)	Comments
2.379E+00	4.286E+02	0	M043	1 0 0 0 1	
4.318E+00	7.780E+02	20	F300	1 0 0 0 2	
2.467E+00	4.444E+02	20	M043	1 0 0 0 1	
4.524E+00	8.150E+02	30	K122	1 1 1 1 2	
4.524E+00	8.150E+02	30	K135	1 1 1 1 2	
2.448E+01	4.410E+03	30	K136	1 1 1 1 2	
2.550E+00	4.595E+02	40	M043	1 0 0 0 1	
2.629E+00	4.737E+02	60	M043	1 0 0 0 1	

874.   C$_6$H$_{12}$O$_6$
Glucose
D-Glucose
D(+)-Glucose
Staleydex 111
Staleydex 333
RN: 50-99-7   **MP** (°C):   146
MW: 180.16   **BP** (°C):

Solubility (Moles/L)	Solubility (Grams/L)	Temp (°C)	Ref (#)	Evaluation (T P E A A)	Comments
1.749E+00	3.151E+02	0	M043	1 0 0 0 1	
2.227E+00	4.012E+02	0.0	Y020	1 1 2 1 2	
1.954E+00	3.520E+02	0.5	J019	1 0 1 2 2	
2.286E+00	4.118E+02	10	M043	1 0 0 0 1	
2.271E+00	4.091E+02	10.0	Y020	1 1 2 1 2	
3.365E+00	6.063E+02	15	D041	1 0 0 0 2	
2.660E+00	4.792E+02	20	M043	1 0 0 0 1	
2.314E+00	4.168E+02	20.0	Y020	1 1 2 1 2	
3.033E+00	5.464E+02	30	J019	1 0 1 2 2	
3.031E+00	5.460E+02	30	K122	1 1 1 1 2	
3.028E+00	5.455E+02	30	M043	1 0 0 0 2	
2.355E+00	4.244E+02	30.0	Y020	1 1 2 1 2	
1.901E+00	3.425E+02	30.50	M137	2 1 2 2 2	
2.042E+00	3.678E+02	35	B354	1 0 1 1 2	
3.416E+00	6.154E+02	40	M043	1 0 0 0 2	
2.396E+00	4.317E+02	40.0	Y020	1 1 2 1 2	
3.936E+00	7.091E+02	50	J019	1 0 1 2 2	
2.436E+00	4.388E+02	50.0	Y020	1 1 2 1 2	
4.090E+00	7.368E+02	60	M043	1 0 0 0 2	
4.523E+00	8.148E+02	80	M043	1 0 0 0 2	
2.501E+00	4.505E+02	rt	D021	0 0 1 1 2	

875. $C_6H_{12}O_6$
Inositol
Mesoinosit
cis-1,2,3,5-trans-4,6-Cyclohexanehexol
Dambose
Nucite
Phaseomannite

RN: 87-89-8   **MP** (°C):   226
MW: 180.16   **BP** (°C):

Solubility (Moles/L)	Solubility (Grams/L)	Temp (°C)	Ref (#)	Evaluation (T P E A A)	Comments
7.788E-01	1.403E+02	19	F300	1 0 0 0 2	
8.267E-01	1.489E+02	20	D041	1 0 0 0 2	
7.771E-01	1.400E+02	25	M054	1 0 0 0 1	
7.771E-01	1.400E+02	ns	L335	0 0 0 0 2	

876. $C_6H_{12}O_6$
Tagatose
Lyxo-2-Hexulose
DL-Tagatose

RN: 17598-81-1   **MP** (°C):
MW: 180.16   **BP** (°C):

Solubility (Moles/L)	Solubility (Grams/L)	Temp (°C)	Ref (#)	Evaluation (T P E A A)	Comments
2.084E+00	3.755E+02	22	F300	1 0 0 0 2	

877. $C_6H_{12}O_6$
D-Galactose
Galactose
(+)-Galactose
D(+)-Galactose

RN: 59-23-4   **MP** (°C):   169
MW: 180.16   **BP** (°C):

Solubility (Moles/L)	Solubility (Grams/L)	Temp (°C)	Ref (#)	Evaluation (T P E A A)	Comments
5.046E-01	9.091E+01	0	D041	1 0 0 0 1	
2.247E+00	4.048E+02	25	D041	1 0 0 0 1	
2.253E+00	4.058E+02	rt	D021	0 0 1 1 2	

878. $C_6H_{12}O_6$

D-Inositol
D(+)-Inositol
D-Chiro-Inositol
(+)-Chiro-Inositol
RN: 643-12-9   **MP** (°C):   249.5
MW: 180.16   **BP** (°C):

Solubility (Moles/L)	Solubility (Grams/L)	Temp (°C)	Ref (#)	Evaluation (T P E A A)	Comments
2.239E+00	4.034E+02	11	F300	1 0 0 0 2	

879. $C_6H_{12}O_6$

α-Glucose
α-D-Glucose
D-α-Glucose
Dextrose
RN: 492-62-6   **MP** (°C):   154.5
MW: 180.16   **BP** (°C):

Solubility (Moles/L)	Solubility (Grams/L)	Temp (°C)	Ref (#)	Evaluation (T P E A A)	Comments
1.355E+00	2.441E+02	0	D041	1 0 0 0 2	
1.942E+00	3.498E+02	0.0	Y020	1 1 2 1 2	
2.019E+00	3.638E+02	10.0	Y020	1 1 2 1 2	
2.775E+00	5.000E+02	20	F300	1 0 0 0 0	
2.096E+00	3.775E+02	20.0	Y020	1 1 2 1 2	
2.501E+00	4.505E+02	25	D041	1 0 0 0 2	
2.170E+00	3.909E+02	30.0	Y020	1 1 2 1 2	
2.242E+00	4.040E+02	40.0	Y020	1 1 2 1 2	
2.313E+00	4.168E+02	50.0	Y020	1 1 2 1 2	
2.346E+00	4.227E+02	54.7	Y020	1 1 2 1 2	

880. $C_6H_{12}O_6 \cdot H_2O$

Glucose (Monohydrate)
RN: 50-99-7   **MP** (°C):   83
MW: 198.17   **BP** (°C):

Solubility (Moles/L)	Solubility (Grams/L)	Temp (°C)	Ref (#)	Evaluation (T P E A A)	Comments
1.274E+00	2.525E+02	0.0	Y020	1 1 2 1 2	
1.449E+00	2.871E+02	10.0	Y020	1 1 2 1 2	
1.619E+00	3.209E+02	20.0	Y020	1 1 2 1 2	
1.781E+00	3.530E+02	30.0	Y020	1 1 2 1 2	
1.933E+00	3.831E+02	40.0	Y020	1 1 2 1 2	
2.072E+00	4.106E+02	50.0	Y020	1 1 2 1 2	
1.784E+00	3.536E+02	73.2	Y020	1 1 2 1 2	

881. $C_6H_{12}O_7$
Scyllitol
Scyllit
Quercinitol
Cocositol
RN: 488-59-5   **MP** (°C):   253
MW: 196.16   **BP** (°C):

Solubility (Moles/L)	Solubility (Grams/L)	Temp (°C)	Ref (#)	Evaluation (T P E A A)	Comments
5.149E-02	1.010E+01	18	F300	1 0 0 0 2	

882. $C_6H_{13}Br$
1-Bromohexane
Hexyl Bromide
RN: 111-25-1   **MP** (°C):   -84.7
MW: 165.08   **BP** (°C):   155.3

Solubility (Moles/L)	Solubility (Grams/L)	Temp (°C)	Ref (#)	Evaluation (T P E A A)	Comments
1.560E-04	2.575E-02	25	M342	1 0 1 1 2	

883. $C_6H_{13}NO$
Caproamide
n-Capronsaeure-amid
Hexanamide
Hexanoic Acid, Amide
RN: 628-02-4   **MP** (°C):   99
MW: 115.18   **BP** (°C):   255

Solubility (Moles/L)	Solubility (Grams/L)	Temp (°C)	Ref (#)	Evaluation (T P E A A)	Comments
1.610E-01	1.854E+01	6	H059	1 2 2 0 2	
2.030E-01	2.338E+01	16	H059	1 2 2 0 2	
2.580E-01	2.972E+01	25	H059	1 2 2 0 2	
2.750E-01	3.167E+01	29	H059	1 2 2 0 2	
3.150E-01	3.628E+01	33	H059	1 2 2 0 2	
3.250E-01	3.743E+01	35	H059	1 2 2 0 2	
3.390E-01	3.904E+01	37	H059	1 2 2 0 2	
3.890E-01	4.480E+01	41	H059	1 2 2 0 2	

884. $C_6H_{13}NO_2$
n-Amyl Carbamate
n-Pentyl Carbamate
O-Pentyl Carbamate
RN: 638-42-6   **MP** (°C):
MW: 131.18   **BP** (°C):

Solubility (Moles/L)	Solubility (Grams/L)	Temp (°C)	Ref (#)	Evaluation (T P E A A)	Comments
3.400E-02	4.460E+00	37	H006	1 2 2 1 1	

885. $C_6H_{13}NO_2$
Isopentyl Urethane
Isoamylurethan
Isoamylurethane
RN: 543-86-2   **MP** (°C):
MW: 131.18   **BP** (°C):

Solubility (Moles/L)	Solubility (Grams/L)	Temp (°C)	Ref (#)	Evaluation (T P E A A)	Comments
3.660E-02	4.801E+00	15.5	F001	1 0 1 2 2	

886. $C_6H_{13}NO_2$
D-Norleucine
D-2-Amino-n-caproic Acid
D-2-Aminohexanoic Acid
RN: 327-56-0   **MP** (°C):   >300
MW: 131.18   **BP** (°C):

Solubility (Moles/L)	Solubility (Grams/L)	Temp (°C)	Ref (#)	Evaluation (T P E A A)	Comments
1.201E-01	1.575E+01	19	D041	1 0 0 0 1	

887. $C_6H_{13}NO_2$
D-Leucine
D-2-Amino-4-methylvaleric Acid
D-2-Amino-4-methylpentanoic Acid
RN: 328-38-1   **MP** (°C):   >300
MW: 131.18   **BP** (°C):

Solubility (Moles/L)	Solubility (Grams/L)	Temp (°C)	Ref (#)	Evaluation (T P E A A)	Comments
1.641E-01	2.153E+01	25	D041	1 0 0 0 2	
1.975E-01	2.591E+01	50	D041	1 0 0 0 2	

888.   C$_6$H$_{13}$NO$_2$
tert-Amyl Carbamate
tert-Pentyl Carbamate
RN: 590-60-3   **MP** (°C):   85
MW: 131.18   **BP** (°C):

Solubility (Moles/L)	Solubility (Grams/L)	Temp (°C)	Ref (#)	Evaluation (T P E A A)	Comments
1.600E-01	2.099E+01	37	H006	1 2 2 1 1	

889.   C$_6$H$_{13}$NO$_2$
N-Propylurethane
Propylurethan
n-Propyl Urethane
RN: 623-85-8   **MP** (°C):
MW: 131.18   **BP** (°C):

Solubility (Moles/L)	Solubility (Grams/L)	Temp (°C)	Ref (#)	Evaluation (T P E A A)	Comments
7.475E-01	9.805E+01	15.5	F001	1 0 1 2 2	

890.   C$_6$H$_{13}$NO$_2$
L-Norleucine
Norleucine
α-Aminocaproic Acid
RN: 327-57-1   **MP** (°C):   327dec
MW: 131.18   **BP** (°C):

Solubility (Moles/L)	Solubility (Grams/L)	Temp (°C)	Ref (#)	Evaluation (T P E A A)	Comments
1.304E-01	1.710E+01	23	K060	1 2 0 0 2	
1.127E-01	1.478E+01	25	D041	1 0 0 0 1	
8.700E-02	1.141E+01	25	E015	1 2 1 1 1	
1.232E-01	1.616E+01	25	K031	2 1 2 1 2	

891. $C_6H_{13}NO_2$
L-Leucine
L(-)-Leucine
Leucine
2-Amino-4-methylpentanoic Acid
L-2-Amino-4-methylpentanoic Acid
(2S)-α-Leucine

RN: 61-90-5   **MP** (°C):   286-288
MW: 131.18   **BP** (°C):

Solubility (Moles/L)	Solubility (Grams/L)	Temp (°C)	Ref (#)	Evaluation (T P E A A)	Comments
1.692E-01	2.220E+01	0	F300	1 0 0 0 2	
1.740E-01	2.282E+01	15	D349	2 1 1 2 2	
1.601E-01	2.100E+01	20	B032	1 2 2 1 2	
1.800E-01	2.361E+01	20	D349	2 1 1 2 2	
1.695E-01	2.224E+01	21	P045	1 0 2 1 2	
1.640E-01	2.151E+01	25	B032	1 2 2 1 2	
1.851E-01	2.428E+01	25	C018	1 0 2 2 2	
1.712E-01	2.246E+01	25	C018	1 0 2 2 2	
1.883E-01	2.470E+01	25	D016	1 0 0 0 2	
1.634E-01	2.143E+01	25	D041	1 0 0 0 2	
1.860E-01	2.440E+01	25	D349	2 1 1 2 2	
1.807E-01	2.370E+01	25	F300	1 0 0 0 2	
1.626E-01	2.133E+01	25	G092	2 1 1 1 1	
1.626E-01	2.133E+01	25	G315	1 0 2 2 2	
1.647E-01	2.160E+01	25.1	N024	2 0 2 2 2	
1.654E-01	2.170E+01	25.1	N025	2 0 2 2 2	
1.647E-01	2.160E+01	25.1	N026	2 0 2 2 2	
1.612E-01	2.114E+01	25.1	N027	1 1 2 2 2	
1.765E-01	2.315E+01	27	D036	2 1 2 2 2	
1.601E-01	2.100E+01	27	D036	2 1 2 2 2	
1.682E-01	2.206E+01	29.80	B032	1 2 2 1 2	
2.142E-01	2.810E+01	50	F300	1 0 0 0 2	
2.805E-01	3.679E+01	75	D041	1 0 0 0 2	
2.805E-01	3.680E+01	75	F300	1 0 0 0 2	
2.886E-01	3.786E+01	92	M160	2 1 1 1 0	
4.069E-01	5.337E+01	99.99	P349	1 0 0 2 2	
4.071E-01	5.340E+01	100	F300	1 0 0 0 2	
1.830E-01	2.400E+01	ns	D072	0 0 0 0 1	

892. $C_6H_{13}NO_2$
L-Isoleucine
L(+)-Isoleucin
Isoleucine
RN: 73-32-5   **MP** (°C):   288
MW: 131.18   **BP** (°C):

Solubility (Moles/L)	Solubility (Grams/L)	Temp (°C)	Ref (#)	Evaluation (T P E A A)	Comments
2.844E-01	3.730E+01	15.50	F300	1 0 0 0 2	
2.533E-01	3.323E+01	20	B032	1 2 2 1 2	
2.619E-01	3.435E+01	25	B032	1 2 2 1 2	
3.017E-01	3.957E+01	25	D041	1 0 0 0 2	
2.364E-01	3.101E+01	25	O316	1 0 1 2 2	
2.358E-01	3.093E+01	25	O316	1 0 1 2 2	
2.714E-01	3.560E+01	27	D036	2 1 2 2 2	
2.690E-01	3.528E+01	29.80	B032	1 2 2 1 2	
4.369E-01	5.732E+01	75	D041	1 0 0 0 2	
3.801E-01	4.985E+01	84	M160	2 1 1 1 0	

893. $C_6H_{13}NO_2$
DL-Norleucine
DL-2-Amino-n-caproic Acid
2-Aminohexanoic Acid
DL-2-Aminohexanoic Acid
RN: 616-06-8   **MP** (°C):   >300
MW: 131.18   **BP** (°C):

Solubility (Moles/L)	Solubility (Grams/L)	Temp (°C)	Ref (#)	Evaluation (T P E A A)	Comments
6.863E-02	9.003E+00	0	D018	2 2 2 1 2	
8.660E-02	1.136E+01	25	C018	1 0 2 2 2	
8.767E-02	1.150E+01	25	D016	1 0 0 0 2	
8.906E-02	1.168E+01	25	D018	2 2 2 1 2	
8.891E-02	1.166E+01	25	D041	1 0 0 0 2	
8.118E-02	1.065E+01	25	K031	2 1 2 1 2	
8.660E-02	1.136E+01	25	M024	1 2 0 1 2	
1.348E-01	1.768E+01	50	D018	2 2 2 1 2	
2.135E-01	2.800E+01	75	D018	2 2 2 1 2	
2.134E-01	2.799E+01	75	D041	1 0 0 0 2	
3.788E-01	4.969E+01	99.99	P349	1 0 0 2 2	

894.   C$_6$H$_{13}$NO$_2$
DL-Leucine
DL-2-Amino-4-methylvaleric Acid
DL-2-Amino-4-methylpentanoic Acid
RN: 328-39-2   **MP** (°C):   295
MW: 131.18   **BP** (°C):

Solubility (Moles/L)	Solubility (Grams/L)	Temp (°C)	Ref (#)	Evaluation (T P E A A)	Comments
6.659E-02	8.735E+00	0	D018	2 2 2 1 2	
6.022E-02	7.900E+00	0	F300	1 0 0 0 1	
7.433E-02	9.750E+00	25	C018	1 0 2 2 2	
7.517E-02	9.860E+00	25	D016	1 0 0 0 2	
8.898E-02	1.167E+01	25	D018	2 2 2 1 2	
7.481E-02	9.813E+00	25	D041	1 0 0 0 2	
7.471E-02	9.800E+00	25	F300	1 0 0 0 1	
1.321E-01	1.733E+01	50	D018	2 2 2 1 2	
1.060E-01	1.390E+01	50	F300	1 0 0 0 2	
2.105E-01	2.762E+01	75	D018	2 2 2 1 2	
1.696E-01	2.225E+01	75	D041	1 0 0 0 2	
1.700E-01	2.230E+01	75	F300	1 0 0 0 2	
3.077E-01	4.036E+01	99.99	P349	1 0 0 2 2	
3.080E-01	4.040E+01	100	F300	1 0 0 0 2	

895.   C$_6$H$_{13}$NO$_2$
DL-Isoleucine
DL-2-Amino-3-methylpentanoic Acid
RN: 443-79-8   **MP** (°C):
MW: 131.18   **BP** (°C):

Solubility (Moles/L)	Solubility (Grams/L)	Temp (°C)	Ref (#)	Evaluation (T P E A A)	Comments
1.311E-01	1.720E+01	0	D018	2 2 2 1 2	
1.632E-01	2.141E+01	25	D018	2 2 2 1 2	
1.662E-01	2.180E+01	25	D041	1 0 0 0 2	
2.235E-01	2.931E+01	50	D018	2 2 2 1 2	
3.510E-01	4.605E+01	75	D018	2 2 2 1 2	
3.357E-01	4.404E+01	75	D041	1 0 0 0 2	
5.517E-01	7.237E+01	99.99	P349	1 0 0 2 2	

896. C₆H₁₃NO₂

α-Hydroxycaproamide
Hexanamide, 2-Hydroxy-
2-Hydroxyhexanamide

RN: 66461-73-2   **MP** (°C):
MW: 131.18   **BP** (°C):

Solubility (Moles/L)	Solubility (Grams/L)	Temp (°C)	Ref (#)	Evaluation (T P E A A)	Comments
8.300E-02	1.089E+01	25	M008	1 0 0 0 2	

897. C₆H₁₃NO₂

ε-Aminocaproic Acid
6-Aminocaproic Acid
ε-Amino-capronsaeure

RN: 60-32-2   **MP** (°C): 205
MW: 131.18   **BP** (°C):

Solubility (Moles/L)	Solubility (Grams/L)	Temp (°C)	Ref (#)	Evaluation (T P E A A)	Comments
3.848E+00	5.048E+02	25	M024	1 2 0 1 2	

898. C₆H₁₃NO₂

L-allo-Isoleucine
Alloisoleucine

RN: 1509-34-8   **MP** (°C): >280
MW: 131.18   **BP** (°C):

Solubility (Moles/L)	Solubility (Grams/L)	Temp (°C)	Ref (#)	Evaluation (T P E A A)	Comments
2.148E-01	2.818E+01	20	D041	1 0 0 0 1	

899. C₆H₁₄

Hexane
Normal Hexane
n-Hexane
Skellysolve B

RN: 110-54-3   **MP** (°C): -95
MW: 86.18   **BP** (°C): 65

Solubility (Moles/L)	Solubility (Grams/L)	Temp (°C)	Ref (#)	Evaluation (T P E A A)	Comments
1.915E-04	1.650E-02	0	P003	2 2 2 2 2	
1.900E-04	1.637E-02	4.0	N004	1 1 2 2 2	
1.761E-04	1.518E-02	14.0	N004	1 1 2 2 2	
1.600E-03	1.379E-01	15.5	F001	1 0 1 0 2	
6.382E-04	5.500E-02	16	D047	1 0 0 1 1	
1.427E-04	1.230E-02	25	A058	1 1 1 1 2	
1.624E-03	1.400E-01	25	A094	1 0 0 0 1	

1.625E-03	1.400E-01	25	K072	1 0 1 1 1	
1.857E-03	1.600E-01	25	K112	1 0 2 1 1	
1.860E-03	1.603E-01	25	K112	1 0 2 2 2	
1.099E-04	9.470E-03	25	K119	1 0 0 0 2	
1.427E-04	1.230E-02	25	L002	2 2 2 2 2	
1.102E-04	9.500E-03	25	M001	2 1 2 2 2	
1.102E-04	9.500E-03	25	M002	2 1 2 2 2	
1.102E-04	9.500E-03	25	M040	1 0 0 1 1	
1.625E-03	1.400E-01	25	M087	1 1 2 1 1	
1.430E-04	1.232E-02	25	M342	1 0 1 1 2	
1.439E-04	1.240E-02	25	P003	2 2 2 2 2	
1.624E-03	1.400E-01	25	S012	2 0 2 2 1	
2.128E-04	1.834E-02	25.0	N004	1 1 2 2 2	
1.099E-04	9.470E-03	25.0	P051	2 1 1 2 2	
1.099E-04	9.470E-03	25.00	P007	2 1 2 2 2	
1.494E-04	1.288E-02	35.0	N004	1 1 2 2 2	
4.623E-02	3.984E+00	38	J020	2 0 2 1 0	*sic*
1.172E-04	1.010E-02	40.1	P051	2 1 1 2 2	
1.172E-04	1.010E-02	40.10	P007	2 1 2 2 2	
2.578E-04	2.221E-02	45.0	N004	1 1 2 2 2	
2.553E-03	2.200E-01	50	L097	1 1 1 1 1	
2.456E-04	2.116E-02	55.0	N004	1 1 2 2 2	
1.532E-04	1.320E-02	55.7	P051	2 1 1 2 2	
1.532E-04	1.320E-02	55.70	P007	2 1 2 2 2	
1.775E-04	1.530E-02	69.7	P051	2 1 1 2 2	average of 2
1.764E-04	1.520E-02	69.70	P007	2 1 2 2 2	
1.787E-04	1.540E-02	69.70	P007	2 1 2 2 2	
2.599E-04	2.240E-02	99.1	P051	2 1 1 2 2	
2.599E-04	2.240E-02	99.10	P007	2 1 2 2 2	
3.388E-04	2.920E-02	114.4	P051	2 1 1 2 2	
3.388E-04	2.920E-02	114.40	P007	2 1 2 2 2	
4.363E-04	3.760E-02	121.3	P051	2 1 1 2 2	
4.363E-04	3.760E-02	121.30	P007	2 1 2 2 2	
6.603E-04	5.690E-02	137.3	P051	2 1 1 2 2	
6.603E-04	5.690E-02	137.30	P007	2 1 2 2 2	
1.230E-03	1.060E-01	151.8	P051	2 1 1 2 2	
1.230E-03	1.060E-01	151.80	P007	2 1 2 2 2	
1.102E-04	9.500E-03	ns	H123	0 0 0 0 2	
1.392E-03	1.200E-01	ns	M010	0 0 0 0 1	
1.880E-04	1.620E-02	ns	M175	0 0 2 1 2	

900.   C₆H₁₄
2,2-Dimethylbutane
Neohexane
RN: 75-83-2   **MP** (°C):   -100
MW: 86.18   **BP** (°C):   50

Solubility (Moles/L)	Solubility (Grams/L)	Temp (°C)	Ref (#)	Evaluation (T P E A A)	Comments
4.572E-04	3.940E-02	0	P003	2 2 2 2 2	
2.460E-04	2.120E-02	25	K119	1 0 0 0 2	
2.135E-04	1.840E-02	25	M001	2 1 2 2 2	
2.135E-04	1.840E-02	25	M002	2 1 2 2 2	
2.762E-04	2.380E-02	25	P003	2 2 2 2 2	
2.460E-04	2.120E-02	25	P051	2 1 1 2 2	
2.460E-04	2.120E-02	25.00	P007	2 1 2 2 2	
6.600E-04	5.687E-02	ns	J300	0 0 0 0 1	

901.   C₆H₁₄
3-Methylpentane
3-Metylopentan
RN: 96-14-0   **MP** (°C):   -118
MW: 86.18   **BP** (°C):   64

Solubility (Moles/L)	Solubility (Grams/L)	Temp (°C)	Ref (#)	Evaluation (T P E A A)	Comments
2.495E-04	2.150E-02	0	P003	2 2 2 2 2	
1.520E-04	1.310E-02	25	K119	1 0 0 0 2	
1.485E-04	1.280E-02	25	M001	2 1 2 2 2	
2.077E-04	1.790E-02	25	P003	2 2 2 2 2	
1.520E-04	1.310E-02	25	P051	2 1 1 2 2	
1.520E-04	1.310E-02	25.00	P007	2 1 2 2 2	
1.485E-04	1.280E-02	ns	H123	0 0 0 0 2	

902.   C₆H₁₄
2,3-Dimethylbutane
Diisopropyl
1,1,2,2-Tetramethylethane
RN: 79-29-8   **MP** (°C):   -129
MW: 86.18   **BP** (°C):   58

Solubility (Moles/L)	Solubility (Grams/L)	Temp (°C)	Ref (#)	Evaluation (T P E A A)	Comments
3.818E-04	3.290E-02	0	P003	2 2 2 2 2	
2.216E-04	1.910E-02	25	K119	1 0 0 0 2	
2.611E-04	2.250E-02	25	P003	2 2 2 2 2	
2.216E-04	1.910E-02	25.0	P051	2 1 1 2 2	
2.216E-04	1.910E-02	25.00	P007	2 1 2 2 2	
2.228E-04	1.920E-02	40.1	P051	2 1 1 2 2	
2.228E-04	1.920E-02	40.10	P007	2 1 2 2 2	
2.750E-04	2.370E-02	55.1	P051	2 1 1 2 2	

2.750E-04	2.370E-02	55.10	P007	2 1 2 2 2
4.653E-04	4.010E-02	99.1	P051	2 1 1 2 2
4.653E-04	4.010E-02	99.10	P007	2 1 2 2 2
6.591E-04	5.680E-02	121.3	P051	2 1 1 2 2
6.591E-04	5.680E-02	121.30	P007	2 1 2 2 2
1.136E-03	9.790E-02	137.3	P051	2 1 1 2 2
1.136E-03	9.790E-02	137.30	P007	2 1 2 2 2
1.984E-03	1.710E-01	149.5	P051	2 1 1 2 2
1.984E-03	1.710E-01	149.50	P007	2 1 2 2 2

903. C_6H_{14}
2-Methylpentane
2-Metylopentan
RN: 107-83-5   **MP** (°C):   -154
MW: 86.18   **BP** (°C):   62

Solubility (Moles/L)	Solubility (Grams/L)	Temp (°C)	Ref (#)	Evaluation (T P E A A)	Comments
2.257E-04	1.945E-02	0	P003	2 2 2 2 2	
5.976E-04	5.150E-02	23	C332	2 0 2 2 1	
1.508E-04	1.300E-02	25	K119	1 0 0 0 2	
1.648E-04	1.420E-02	25	L002	2 2 2 2 2	
1.601E-04	1.380E-02	25	M001	2 1 2 2 2	
1.601E-04	1.380E-02	25	M002	2 1 2 2 2	
1.822E-04	1.570E-02	25	P003	2 2 2 2 2	
1.508E-04	1.300E-02	25.0	P051	2 1 1 2 2	
1.508E-04	1.300E-02	25.00	P007	2 1 2 2 2	
1.601E-04	1.380E-02	40.1	P051	2 1 1 2 2	
1.601E-04	1.380E-02	40.10	P007	2 1 2 2 2	
1.822E-04	1.570E-02	55.7	P051	2 1 1 2 2	
1.822E-04	1.570E-02	55.70	P007	2 1 2 2 2	
3.145E-04	2.710E-02	99.1	P051	2 1 1 2 2	
3.145E-04	2.710E-02	99.10	P007	2 1 2 2 2	
5.210E-04	4.490E-02	118.0	P051	2 1 1 2 2	
5.210E-04	4.490E-02	118.00	P007	2 1 2 2 2	
1.007E-03	8.680E-02	137.3	P051	2 1 1 2 2	
1.007E-03	8.680E-02	137.30	P007	2 1 2 2 2	
1.311E-03	1.130E-01	149.50	P007	2 1 2 2 2	

904.   C$_6$H$_{14}$FO$_3$P
Isofluorphate
Diisopropylfluorophosphate
Phosphorofluoridic Acid bis(1-Methylethyl) Ester
Difluorophate
PF-3
T-1703
RN: 55-91-4   **MP** (°C):   -82
MW: 184.15   **BP** (°C):   183

Solubility (Moles/L)	Solubility (Grams/L)	Temp (°C)	Ref (#)	Evaluation (T P E A A)	Comments
8.236E-02	1.517E+01	25	D041	1 0 0 0 2	

905.   C$_6$H$_{14}$NO$_3$PS$_2$
Ethoate-methyl
O,O-Dimethyl S-(N-Ethylcarbamoylmethyl) Dithiophosphate
Fitios
RN: 116-01-8   **MP** (°C):   66.1
MW: 243.29   **BP** (°C):

Solubility (Moles/L)	Solubility (Grams/L)	Temp (°C)	Ref (#)	Evaluation (T P E A A)	Comments
3.494E-02	8.500E+00	25	M061	1 0 0 0 1	
3.494E-02	8.500E+00	25	M161	1 0 0 0 1	

906.   C$_6$H$_{14}$N$_2$
trans-2,5-Dimethylpiperazine
trans-2,5-Dimethyl-piperazin
RN: 2815-34-1   **MP** (°C):
MW: 114.19   **BP** (°C):

Solubility (Moles/L)	Solubility (Grams/L)	Temp (°C)	Ref (#)	Evaluation (T P E A A)	Comments
3.065E+00	3.500E+02	20	F300	1 0 0 0 1	

907.   C$_6$H$_{14}$N$_2$O
Methyl-n-amylnitrosamine
N-Nitroso(methyl)pentylamine
RN: 13256-07-0   **MP** (°C):
MW: 130.19   **BP** (°C):

Solubility (Moles/L)	Solubility (Grams/L)	Temp (°C)	Ref (#)	Evaluation (T P E A A)	Comments
8.400E-02	1.094E+01	24	D083	2 0 0 0 1	

908.   C₆H₁₄N₂O

Di-n-propylnitrosamine
N-Nitroso-N-propyl-1-propanamine
Dipropylnitrosamine
NDPA
DPNA
Nitrosodipropylamine

RN: 621-64-7   **MP** (°C):
MW: 130.19   **BP** (°C):

Solubility (Moles/L)	Solubility (Grams/L)	Temp (°C)	Ref (#)	Evaluation (T P E A A)	Comments
7.600E-02	9.895E+00	24	D083	2 0 0 0 1	

909.   C₆H₁₄N₂O

Di-isopropylnitrosamine
2-Propanamine, N-(1-Methylethyl)-N-nitroso-
N-Nitrosodiisopropylamine
NdiPA

RN: 601-77-4   **MP** (°C):
MW: 130.19   **BP** (°C):

Solubility (Moles/L)	Solubility (Grams/L)	Temp (°C)	Ref (#)	Evaluation (T P E A A)	Comments
1.000E-01	1.302E+01	24	D083	2 0 0 0 1	

910.   C₆H₁₄N₂O

Ethyl-n-butylnitrosamine
Nitroso-N-ethyl-n-butylamine
N-Nitroso-N-butylethylamine
N-Nitroso(ethyl)-n-butylamine
NEBA
Butanamine, N-Ethyl-N-nitroso-

RN: 4549-44-4   **MP** (°C):
MW: 130.19   **BP** (°C):

Solubility (Moles/L)	Solubility (Grams/L)	Temp (°C)	Ref (#)	Evaluation (T P E A A)	Comments
9.200E-02	1.198E+01	24	D083	2 0 0 0 1	

911. $C_6H_{14}N_2O_2$
L(+)-Lysine
L(+)-Lysin
Lysine
RN: 56-87-1   **MP** (°C):   224
MW: 146.19   **BP** (°C):

Solubility (Moles/L)	Solubility (Grams/L)	Temp (°C)	Ref (#)	Evaluation (T P E A A)	Comments
3.995E+00	5.840E+02	27	D036	2 1 2 2 2	

912. $C_6H_{14}N_4O_2$
DL-Arginine
(±)-Arginine
RN: 7200-25-1   **MP** (°C):
MW: 174.20   **BP** (°C):

Solubility (Moles/L)	Solubility (Grams/L)	Temp (°C)	Ref (#)	Evaluation (T P E A A)	Comments
1.382E+00	2.407E+02	20	J303	2 0 2 2 2	
1.978E+00	3.445E+02	40	J303	2 0 2 2 2	
2.781E+00	4.844E+02	50	J303	2 0 2 2 2	
3.851E+00	6.709E+02	60	J303	2 0 2 2 2	

913. $C_6H_{14}N_4O_2$
L-Arginine
L(+)-Arginin
Arginine
RN: 74-79-3   **MP** (°C):   244
MW: 174.20   **BP** (°C):

Solubility (Moles/L)	Solubility (Grams/L)	Temp (°C)	Ref (#)	Evaluation (T P E A A)	Comments
6.559E-01	1.143E+02	10	H062	1 2 2 0 0	EFG
8.588E-01	1.496E+02	20	B032	1 2 ? 1 2	
7.487E-01	1.304E+02	21	D041	1 0 0 0 1	
8.037E-01	1.400E+02	21	F300	1 0 0 0 0	average
1.044E+00	1.818E+02	25	B032	1 2 2 1 2	
9.230E-01	1.608E+02	25	G315	1 0 2 2 2	
3.060E+00	5.330E+02	27	D036	2 1 2 2 2	
1.241E+00	2.162E+02	29.80	B032	1 2 2 1 2	
1.111E+00	1.935E+02	30	H062	1 2 2 0 0	EFG
1.771E+00	3.084E+02	50	H062	1 2 2 0 0	EFG

914.   C_6H_14O
1-Hexanol
n-Hexanol
Amyl Carbinol
Caproic Alcohol
n-Hexyl Alcohol

RN: 111-27-3   **MP** (°C):
MW: 102.18   **BP** (°C):

Solubility (Moles/L)	Solubility (Grams/L)	Temp (°C)	Ref (#)	Evaluation (T P E A A)	Comments
7.864E-02	8.035E+00	0	E029	1 2 0 1 1	
9.344E-02	9.548E+00	0	S307	1 1 0 2 2	
7.706E-02	7.873E+00	5.54	H110	2 2 2 2 2	
7.487E-02	7.650E+00	6.84	H110	2 2 2 2 2	
7.213E-02	7.370E+00	8.64	H110	2 2 2 2 2	
6.803E-02	6.951E+00	10	E029	1 2 0 1 1	
7.372E-02	7.533E+00	10.2	S307	1 1 0 2 2	
6.906E-02	7.057E+00	11.04	H110	2 2 2 2 2	
6.671E-02	6.816E+00	12.94	H110	2 2 2 2 2	
6.506E-02	6.648E+00	14.64	H110	2 2 2 2 2	
6.287E-02	6.424E+00	17.04	H110	2 2 2 2 2	
6.861E-02	7.011E+00	20	A015	1 2 1 1 2	
6.224E-02	6.359E+00	20	E029	1 2 0 1 1	
6.070E-02	6.202E+00	20	H330	2 0 2 2 2	
4.869E-02	4.975E+00	20	L049	1 1 2 1 0	
5.150E-02	5.262E+00	20	P073	1 0 0 1 2	
6.475E-02	6.616E+00	20.0	S307	1 1 0 2 2	
5.991E-02	6.121E+00	20.74	H110	2 2 2 2 2	
5.854E-02	5.981E+00	22.94	H110	2 2 2 2 2	
6.250E-02	6.386E+00	24	H345	2 0 2 2 2	
6.069E-02	6.201E+00	25	B038	1 2 1 1 2	
5.644E-02	5.767E+00	25	B060	2 0 1 1 1	
5.837E-02	5.964E+00	25	C093	2 1 1 1 1	
1.000E+00	1.022E+02	25	F044	1 0 0 0 0	EFG
8.000E-02	8.174E+00	25	G075	1 0 1 0 0	
5.900E-02	6.028E+00	25	K025	2 2 1 1 2	
8.922E-02	9.116E+00	25	M323	2 2 1 1 2	
5.711E-02	5.835E+00	25.04	H110	2 2 2 2 2	
5.640E-02	5.762E+00	26.94	H110	2 2 2 2 2	
5.579E-02	5.701E+00	28.94	H110	2 2 2 2 2	
5.431E-02	5.549E+00	29.7	S307	1 1 0 2 2	
6.320E-02	6.458E+00	30	C091	1 2 1 1 1	
5.740E-02	5.865E+00	30	E029	1 2 0 1 1	
5.517E-02	5.637E+00	30.94	H110	2 2 2 2 2	
5.440E-02	5.558E+00	33.04	H110	2 2 2 2 2	
5.005E-02	5.114E+00	39.8	S307	1 1 0 2 2	
5.257E-02	5.371E+00	40	E029	1 2 0 1 1	

4.869E-02	4.975E+00	50	E029	1 2 0 1 1
4.840E-02	4.945E+00	50.0	S307	1 1 0 2 2
5.063E-02	5.173E+00	60	E029	1 2 0 1 1
5.043E-02	5.153E+00	60.0	S307	1 1 0 2 2
5.450E-02	5.569E+00	70	E029	1 2 0 1 1
5.540E-02	5.661E+00	70	F001	1 0 1 0 2
5.615E-02	5.737E+00	70.3	S307	1 1 0 2 2
5.934E-02	6.063E+00	80	E029	1 2 0 1 1
6.080E-02	6.212E+00	80	F001	1 0 1 0 2
6.079E-02	6.211E+00	80.3	S307	1 1 0 2 2
6.707E-02	6.853E+00	90	E029	1 2 0 1 1
6.660E-02	6.805E+00	90	F001	1 0 1 0 2
6.204E-02	6.340E+00	90.3	S307	1 1 0 2 2
7.767E-02	7.937E+00	100	E029	1 2 0 1 1
7.690E-02	7.857E+00	100	F001	1 0 1 0 2
8.826E-02	9.018E+00	110	E029	1 2 0 1 1
8.720E-02	8.910E+00	110	F001	1 0 1 0 2
1.007E-01	1.029E+01	120	E029	1 2 0 1 2
1.151E-01	1.176E+01	130	E029	1 2 0 1 2
1.323E-01	1.351E+01	140	E029	1 2 0 1 2
1.570E-01	1.604E+01	150	E029	1 2 0 1 2
1.966E-01	2.009E+01	160	E029	1 2 0 1 2
2.573E-01	2.629E+01	170	E029	1 2 0 1 2
3.410E-01	3.484E+01	180	E029	1 2 0 1 2
4.545E-01	4.644E+01	190	E029	1 2 0 1 2
6.188E-01	6.323E+01	200	E029	1 2 0 1 2
8.654E-01	8.842E+01	210	E029	1 2 0 1 2
1.372E+00	1.402E+02	220	E029	1 2 0 1 2
6.114E-02	6.247E+00	ns	L003	0 0 2 1 2

915. $C_6H_{14}O$
2,2-Dimethyl-1-butanol
t-Pentylcarbinol

RN: 1185-33-7   **MP** (°C):   -35
MW: 102.18   **BP** (°C):   136

Solubility (Moles/L)	Solubility (Grams/L)	Temp (°C)	Ref (#)	Evaluation (T P E A A)	Comments
7.960E-02	8.133E+00	20	G005	1 2 1 1 1	
7.382E-02	7.543E+00	25	G005	1 2 1 1 1	
6.900E-02	7.050E+00	30	G005	1 2 1 1 1	

916.   C₆H₁₄O
2,2-Dimethyl-3-butanol
t-Butylmethylcarbinol
RN: 464-07-3   **MP** (°C):
MW: 102.18   **BP** (°C):

Solubility (Moles/L)	Solubility (Grams/L)	Temp (°C)	Ref (#)	Evaluation (T P E A A)	Comments
2.517E-01	2.572E+01	20	G005	1 2 1 1 2	
2.322E-01	2.372E+01	25	G005	1 2 1 1 2	
2.163E-01	2.210E+01	30	G005	1 2 1 1 2	

917.   C₆H₁₄O
2,3-Dimethyl-1-butanol
Dimethyl-i-propylcarbinol
Dimethyl-isopropylcarbinol
RN: 594-60-5   **MP** (°C):   -14
MW: 102.18   **BP** (°C):

Solubility (Moles/L)	Solubility (Grams/L)	Temp (°C)	Ref (#)	Evaluation (T P E A A)	Comments
4.349E-01	4.443E+01	20	G005	1 2 1 1 2	
3.927E-01	4.012E+01	25	G005	1 2 1 1 2	
3.547E-01	3.624E+01	30	G005	1 2 1 1 2	

918.   C₆H₁₄O
2-Ethyl-1-butanol
2-Ethylbutanol
RN: 97-95-0   **MP** (°C):   -15
MW: 102.18   **BP** (°C):   146

Solubility (Moles/L)	Solubility (Grams/L)	Temp (°C)	Ref (#)	Evaluation (T P E A A)	Comments
6.127E-02	6.261E+00	20	D052	1 1 0 0 1	
3.899E-02	3.984E+00	25	C093	2 1 1 1 0	

919.   C₆H₁₄O
2-Hexanol
n-Butylmethylcarbinol
1-Methyl Pentanol
RN: 626-93-7   **MP** (°C):   <25
MW: 102.18   **BP** (°C):   136

Solubility (Moles/L)	Solubility (Grams/L)	Temp (°C)	Ref (#)	Evaluation (T P E A A)	Comments
1.975E-01	2.018E+01	0	S307	1 1 0 2 2	
1.617E-01	1.652E+01	10.1	S307	1 1 0 2 2	
1.246E-01	1.274E+01	19.8	S307	1 1 0 2 2	
1.456E-01	1.488E+01	20	G005	1 2 1 1 2	
1.690E-01	1.727E+01	20	H330	2 0 2 2 2	
1.323E-01	1.351E+01	25	G005	1 2 1 1 2	
1.141E-01	1.166E+01	29.9	S307	1 1 0 2 2	
1.237E-01	1.264E+01	30	G005	1 2 1 1 2	
1.055E-01	1.078E+01	40.0	S307	1 1 0 2 2	
9.306E-02	9.509E+00	50.0	S307	1 1 0 2 2	
8.826E-02	9.018E+00	60.2	S307	1 1 0 2 2	
9.498E-02	9.705E+00	70.0	S307	1 1 0 2 2	
1.094E-01	1.117E+01	80.1	S307	1 1 0 2 2	
9.114E-02	9.312E+00	90.2	S307	1 1 0 2 2	

920.   C₆H₁₄O
3-Methyl-1-pentanol
3-Methylpentanol
2-Ethyl-4-butanol
RN: 589-35-5   **MP** (°C):
MW: 102.18   **BP** (°C):   151

Solubility (Moles/L)	Solubility (Grams/L)	Temp (°C)	Ref (#)	Evaluation (T P E A A)	Comments
4.190E-02	4.282E+00	25	B060	2 0 1 1 1	

921.   C₆H₁₄O
Dipropyl Ether
Propyl Ether
Dipropylaether
Dipropylether
RN: 111-43-3   **MP** (°C):   -123
MW: 102.18   **BP** (°C):   89

Solubility (Moles/L)	Solubility (Grams/L)	Temp (°C)	Ref (#)	Evaluation (T P E A A)	Comments
5.644E-02	5.767E+00	0	B002	2 1 1 2 2	
3.996E-02	4.083E+00	10	B002	2 1 1 2 2	
3.705E-02	3.786E+00	15	B002	2 1 1 2 2	
2.927E-02	2.991E+00	20	B002	2 1 1 2 2	

2.936E-02	3.000E+00	20	F300	1 0 0 0 0
6.700E-02	6.846E+00	20	S006	1 0 0 0 1
2.441E-02	2.494E+00	25	B002	2 1 1 2 2
1.070E-01	1.093E+01	37	E028	1 0 1 1 2

922. $C_6H_{14}O$
2-Methyl-2-pentanol
Dimethyl-n-propylcarbinol
1,1-Dimethyl-1-butanol
RN: 590-36-3   **MP** (°C):   -107
MW: 102.18   **BP** (°C):   122

Solubility (Moles/L)	Solubility (Grams/L)	Temp (°C)	Ref (#)	Evaluation (T P E A A)	Comments
3.428E-01	3.503E+01	20	G005	1 2 1 1 2	
3.640E-01	3.719E+01	20	H330	2 0 2 2 2	
3.071E-01	3.138E+01	25	G005	1 2 1 1 2	
2.814E-01	2.875E+01	30	G005	1 2 1 1 2	

923. $C_6H_{14}O$
Isohexyl Alcohol
4-Methyl-1-pentanol
RN: 626-89-1   **MP** (°C):   <25
MW: 102.18   **BP** (°C):

Solubility (Moles/L)	Solubility (Grams/L)	Temp (°C)	Ref (#)	Evaluation (T P E A A)	Comments
1.020E-01	1.042E+01	20	H330	2 0 2 2 2	

924. $C_6H_{14}O$
3-Methyl-2-pentanol
3-Methyl-2-pentyl Alcohol
RN: 565-60-6   **MP** (°C):   <25
MW: 102.18   **BP** (°C):

Solubility (Moles/L)	Solubility (Grams/L)	Temp (°C)	Ref (#)	Evaluation (T P E A A)	Comments
2.004E-01	2.047E+01	20	G005	1 2 1 1 2	
1.863E-01	1.903E+01	25	G005	1 2 1 1 2	
1.721E-01	1.759E+01	30	G005	1 2 1 1 2	

925.   C$_6$H$_{14}$O
3-Methyl-3-pentanol
Diethylmethylcarbinol
RN: 77-74-7   **MP** (°C):   -24
MW: 102.18   **BP** (°C):

Solubility (Moles/L)	Solubility (Grams/L)	Temp (°C)	Ref (#)	Evaluation (T P E A A)	Comments
4.286E-01	4.379E+01	9.8	S307	1 1 0 2 2	
3.346E-01	3.419E+01	19.5	S307	1 1 0 2 2	
4.500E-01	4.598E+01	20	G005	1 2 1 1 2	
3.999E-01	4.086E+01	25	G005	1 2 1 1 2	
3.264E-01	3.335E+01	29.8	S307	1 1 0 2 2	
3.592E-01	3.670E+01	30	G005	1 2 1 1 2	
2.647E-01	2.705E+01	39.8	S307	1 1 0 2 2	
2.331E-01	2.382E+01	49.7	S307	1 1 0 2 2	
1.938E-01	1.980E+01	59.5	S307	1 1 0 2 2	
1.834E-01	1.874E+01	70.1	S307	1 1 0 2 2	
1.787E-01	1.826E+01	80.1	S307	1 1 0 2 2	
1.617E-01	1.652E+01	90.4	S307	1 1 0 2 2	

926.   C$_6$H$_{14}$O
tert-Amyl Methyl Ether
Methyl tert-Amyl Ether
RN: 994-05-8   **MP** (°C):
MW: 102.18   **BP** (°C):   85

Solubility (Moles/L)	Solubility (Grams/L)	Temp (°C)	Ref (#)	Evaluation (T P E A A)	Comments
1.208E-01	1.235E+01	20	E019	1 0 1 1 2	

927.   C$_6$H$_{14}$O
Propyl Isopropyl Ether
Propyl-isopropyl-aether
RN: 627-08-7   **MP** (°C):   <25
MW: 102.18   **BP** (°C):   83

Solubility (Moles/L)	Solubility (Grams/L)	Temp (°C)	Ref (#)	Evaluation (T P E A A)	Comments
7.285E-02	7.444E+00	10	B002	2 1 1 2 2	
7.242E-02	7.400E+00	10	F300	1 0 0 0 1	
5.837E-02	5.964E+00	15	B002	2 1 1 2 2	
5.872E-02	6.000E+00	15	F300	1 0 0 0 1	
4.966E-02	5.074E+00	20	B002	2 1 1 2 2	
4.578E-02	4.678E+00	25	B002	2 1 1 2 2	
4.600E-02	4.700E+00	25	F300	1 0 0 0 1	

928.   C$_6$H$_{14}$O
Isopropyl Ether
Diisopropyl Ether
RN: 108-20-3   **MP** (°C):   -60
MW: 102.18   **BP** (°C):   68.5

Solubility (Moles/L)	Solubility (Grams/L)	Temp (°C)	Ref (#)	Evaluation (T P E A A)	Comments
1.351E-01	1.381E+01	24.6	H121	2 0 0 0 1	
8.730E-02	8.920E+00	25	F048	2 0 0 0 0	
7.920E-02	8.092E+00	37	E028	1 0 1 1 2	

929.   C$_6$H$_{14}$O
3-Hexanol
n-Propylethylcarbinol
tert-Hexyl Alcohol
RN: 623-37-0   **MP** (°C):   <25
MW: 102.18   **BP** (°C):   134.5

Solubility (Moles/L)	Solubility (Grams/L)	Temp (°C)	Ref (#)	Evaluation (T P E A A)	Comments
2.619E-01	2.676E+01	0	S307	1 1 0 2 2	
1.881E-01	1.922E+01	10.1	S307	1 1 0 2 2	
3.062E-01	3.129E+01	20	A015	1 2 1 1 2	
1.683E-01	1.720E+01	20	G005	1 2 1 1 2	
1.608E-01	1.643E+01	20.0	S307	1 1 0 2 2	
1.551E-01	1.584E+01	25	G005	1 2 1 1 2	
1.437E-01	1.468E+01	30	G005	1 2 1 1 2	
1.342E-01	1.371E+01	30.0	S307	1 1 0 2 2	
1.189E-01	1.215E+01	39.8	S307	1 1 0 2 2	
1.065E-01	1.088E+01	50.0	S307	1 1 0 2 2	
9.882E-02	1.010E+01	60.1	S307	1 1 0 2 2	
9.882E-02	1.010E+01	70.2	S307	1 1 0 2 2	
1.036E-01	1.059E+01	80.2	S307	1 1 0 2 2	
1.065E-01	1.088E+01	90.3	S307	1 1 0 2 2	

930.   C₆H₁₄O

4-Methyl-2-pentanol
i-Butylmethylcarbinol
Methyl Amyl Alcohol

RN:	108-11-2	**MP** (°C):	-90	
MW:	102.18	**BP** (°C):	130	

Solubility (Moles/L)	Solubility (Grams/L)	Temp (°C)	Ref (#)	Evaluation (T P E A A)	Comments
2.684E-01	2.743E+01	0	S307	1 1 0 2 2	
2.004E-01	2.047E+01	9.7	S307	1 1 0 2 2	
1.664E-01	1.701E+01	20	D052	1 1 0 0 2	
1.721E-01	1.759E+01	20	G005	1 2 1 1 2	
1.570E-01	1.604E+01	20.0	S307	1 1 0 2 2	
1.636E-01	1.672E+01	25	C093	2 1 1 1 1	
1.579E-01	1.614E+01	25	G005	1 2 1 1 2	
1.465E-01	1.497E+01	30	G005	1 2 1 1 2	
1.475E-01	1.507E+01	30.0	S307	1 1 0 2 2	
1.246E-01	1.274E+01	40.3	S307	1 1 0 2 2	
1.151E-01	1.176E+01	50.0	S307	1 1 0 2 2	
1.074E-01	1.098E+01	60.1	S307	1 1 0 2 2	
1.094E-01	1.117E+01	70.2	S307	1 1 0 2 2	
1.199E-01	1.225E+01	80.2	S307	1 1 0 2 2	
1.132E-01	1.156E+01	90.2	S307	1 1 0 2 2	

931.   C₆H₁₄O

2-Methyl-3-pentanol
i-Propylethylcarbinol

RN:	565-67-3	**MP** (°C):	<25	
MW:	102.18	**BP** (°C):		

Solubility (Moles/L)	Solubility (Grams/L)	Temp (°C)	Ref (#)	Evaluation (T P E A A)	Comments
2.144E-01	2.191E+01	20	G005	1 2 0 0 2	
1.928E-01	1.970E+01	25	G005	1 2 1 1 2	
1.749E-01	1.787E+01	30	G005	1 2 1 1 2	

932.   C₆H₁₄O

2-Ethyl-4-butanol
3-Methylpentanol

RN:	105-30-6	**MP** (°C):	<25	
MW:	102.18	**BP** (°C):	148	

Solubility (Moles/L)	Solubility (Grams/L)	Temp (°C)	Ref (#)	Evaluation (T P E A A)	Comments
1.257E-01	1.284E+01	0	S307	1 1 0 2 2	
1.004E-01	1.025E+01	10.0	S307	1 1 0 2 2	
8.518E-02	8.704E+00	19.6	S307	1 1 0 2 2	
5.837E-02	5.964E+00	25	C093	2 1 1 1 1	
7.681E-02	7.848E+00	30.8	S307	1 1 0 2 2	

7.498E-02	7.661E+00	40.3	S307	1 1 0 2 2
7.295E-02	7.454E+00	50.0	S307	1 1 0 2 2
7.363E-02	7.523E+00	60.3	S307	1 1 0 2 2
7.478E-02	7.641E+00	70.1	S307	1 1 0 2 2
8.133E-02	8.310E+00	80.3	S307	1 1 0 2 2
8.931E-02	9.126E+00	90.7	S307	1 1 0 2 2

933. $C_6H_{14}O_2$
Diethyl Cellosolve
Ethylene Glycol Diethyl Ether
1,2-Diethoxyethane
3,6-Dioxaoctane
Ethyl Glyme
Diethoxyethane
RN: 629-14-1 **MP** (°C):
MW: 118.18 **BP** (°C): 119

Solubility (Moles/L)	Solubility (Grams/L)	Temp (°C)	Ref (#)	Evaluation (T P E A A)	Comments
2.273E-01	2.686E+01	20	D052	1 1 0 0 2	
1.469E+00	1.736E+02	20	M062	1 0 0 0 2	

934. $C_6H_{14}O_2$
Acetal
AcetaldehyD-diethylacetal
Acetaldehyde Diethyl Acetal
RN: 105-57-7 **MP** (°C):
MW: 118.18 **BP** (°C): 102.7

Solubility (Moles/L)	Solubility (Grams/L)	Temp (°C)	Ref (#)	Evaluation (T P E A A)	Comments
3.723E-01	4.400E+01	25	F300	1 0 0 0 1	

935. $C_6H_{14}O_3$
Carbitol
2-(2-Ethoxyethoxy)ethanol
RN: 111-90-0 **MP** (°C):
MW: 134.18 **BP** (°C): 196.0

Solubility (Moles/L)	Solubility (Grams/L)	Temp (°C)	Ref (#)	Evaluation (T P E A A)	Comments
3.610E+00	4.843E+02	4.50	C022	1 2 0 0 2	

936.   C$_6$H$_{14}$O$_6$
Galactitol
Dulcit
Dulcitol
RN: 608-66-2   **MP** (°C):   189.5
MW: 182.17   **BP** (°C):   277.5

Solubility (Moles/L)	Solubility (Grams/L)	Temp (°C)	Ref (#)	Evaluation (T P E A A)	Comments
1.599E-01	2.913E+01	14	D041	1 0 0 0 1	
1.702E-01	3.100E+01	15	F300	1 0 0 0 1	
2.086E+00	3.800E+02	100	F300	1 0 0 0 1	

937.   C$_6$H$_{14}$O$_6$
Mannitol
D-Mannit
D-Mannitol
RN: 87-78-5   **MP** (°C):   167
MW: 182.17   **BP** (°C):   292

Solubility (Moles/L)	Solubility (Grams/L)	Temp (°C)	Ref (#)	Evaluation (T P E A A)	Comments
5.081E-01	9.256E+01	0	C073	1 2 2 1 2	
5.171E-01	9.420E+01	0	M043	1 0 0 0 2	
6.614E-01	1.205E+02	10	M043	1 0 0 0 2	
7.734E-01	1.409E+02	15	C073	1 2 2 1 2	
7.740E-01	1.410E+02	15	F300	1 0 0 0 2	
7.408E-01	1.349E+02	18	D041	1 0 0 0 2	
7.936E-01	1.446E+02	19	N051	1 0 2 2 2	
8.609E-01	1.568E+02	20	M043	1 0 0 0 2	
9.732E-01	1.773E+02	25	B106	1 2 2 2 2	
9.739E-01	1.774E+02	25	B106	1 2 2 2 2	
9.762E-01	1.778E+02	25	B106	1 2 2 2 2	
9.639E-01	1.756E+02	25	C073	1 2 2 1 2	
8.255E-01	1.504E+02	25	H087	1 0 2 1 2	
1.000E+00	1.822E+02	30	D011	1 0 1 0 1	
1.105E+00	2.013E+02	30	M043	1 0 0 0 2	
1.254E+00	2.284E+02	35	C073	1 2 2 1 2	
1.411E+00	2.571E+02	40	M043	1 0 0 0 2	
1.760E+00	3.207E+02	50	C073	1 2 2 1 2	
1.827E+00	3.329E+02	51.50	B106	1 2 2 2 2	
2.083E+00	3.794E+02	60	C073	1 2 2 1 2	
2.104E+00	3.833E+02	60	F300	1 0 0 0 2	
2.150E+00	3.917E+02	60	M043	1 0 0 0 2	
2.416E+00	4.401E+02	67.40	B106	1 2 2 2 2	
2.504E+00	4.562E+02	70.50	B106	1 2 2 2 2	
2.936E+00	5.349E+02	80	M043	1 0 0 0 2	
3.015E+00	5.493E+02	82.90	B106	1 2 2 2 2	
3.253E+00	5.927E+02	88.10	B106	1 2 2 2 2	
3.299E+00	6.010E+02	90.10	B106	1 2 2 2 2	

3.590E+00	6.540E+02	98	B106	1 2 2 2 2
3.628E+00	6.610E+02	99.30	B106	1 2 2 2 2
3.641E+00	6.633E+02	100	M043	1 0 0 0 2
8.757E-01	1.595E+02	rt	D021	0 0 1 1 2

938. $C_6H_{14}O_6$
Sorbitol
D-Sorbitol
RN: 50-70-4   **MP** (°C):   110
MW: 182.17   **BP** (°C):

Solubility (Moles/L)	Solubility (Grams/L)	Temp (°C)	Ref (#)	Evaluation (T P E A A)	Comments
3.522E+00	6.416E+02	10	M043	1 0 0 0 2	
3.785E+00	6.894E+02	20	M043	1 0 0 0 2	
4.025E+00	7.333E+02	30	M043	1 0 0 0 2	
4.283E+00	7.802E+02	40	M043	1 0 0 0 2	

939. $C_6H_{15}N$
N-Ethyl-n-butylamine
Ethylbutylamine
N-Ethylbutan-1-amine
N-Ethylbutylamine
RN: 13360-63-9   **MP** (°C):   -78
MW: 101.19   **BP** (°C):   108

Solubility (Moles/L)	Solubility (Grams/L)	Temp (°C)	Ref (#)	Evaluation (T P E A A)	Comments
1.003E+00	1.015E+02	10	D332	2 2 1 1 2	
5.310E-01	5.373E+01	20	D332	2 2 1 1 2	
3.793E-01	3.838E+01	30	D332	2 2 1 1 2	
2.859E-01	2.893E+01	40	D332	2 2 1 1 2	

940. $C_6H_{15}N$
Triethylamine
Triaethylamin
RN: 121-44-8   **MP** (°C):   -115
MW: 101.19   **BP** (°C):   89

Solubility (Moles/L)	Solubility (Grams/L)	Temp (°C)	Ref (#)	Evaluation (T P E A A)	Comments
1.778E+00	1.799E+02	17.48	K142	1 0 0 0 2	
2.754E+00	2.787E+02	17.59	K142	1 0 0 0 2	
2.754E+00	2.787E+02	17.64	K142	1 0 0 0 2	
1.156E+00	1.170E+02	17.82	K142	1 0 0 0 2	
1.156E+00	1.170E+02	17.85	K142	1 0 0 0 2	
2.791E+00	2.824E+02	18	C088	2 2 2 2 1	
3.434E+00	3.475E+02	18.11	K142	1 0 0 0 2	

3.434E+00	3.475E+02	18.12	K142	1 0 0 0 2	
4.014E+00	4.062E+02	19.12	K142	1 0 0 0 2	
4.014E+00	4.062E+02	19.13	K142	1 0 0 0 2	
8.951E-01	9.058E+01	19.38	K142	1 0 0 0 2	
8.951E-01	9.058E+01	19.43	K142	1 0 0 0 2	
1.403E+00	1.420E+02	20	F300	1 0 0 0 2	
6.780E-01	6.861E+01	25.04	V013	2 2 2 2 2	
1.976E-01	2.000E+01	65	F300	1 0 0 0 1	

941. $C_6H_{15}N$
N-Ethyl-sec-butylamine
sec-Butylethylamine
2-Butanamine, N-Ethyl-
2-(Ethylamino)butane
RN: 21035-44-9   **MP** (°C):
MW: 101.19   **BP** (°C):

Solubility (Moles/L)	Solubility (Grams/L)	Temp (°C)	Ref (#)	Evaluation (T P E A A)	Comments
8.155E-01	8.253E+01	25	D332	2 2 1 1 2	
6.099E-01	6.172E+01	30	D332	2 2 1 1 2	
4.202E-01	4.252E+01	40	D332	2 2 1 1 2	

942. $C_6H_{15}N$
n-Dipropylamine
Dipropylamine
RN: 142-84-7   **MP** (°C):   -63
MW: 101.19   **BP** (°C):   110

Solubility (Moles/L)	Solubility (Grams/L)	Temp (°C)	Ref (#)	Evaluation (T P E A A)	Comments
5.470E-01	5.536E+01	12.2	H038	1 2 1 1 2	
2.794E-01	2.828E+01	36.1	H038	1 2 1 1 2	
2.335E-01	2.363E+01	44.1	H038	1 2 1 1 2	
1.900E-01	1.922E+01	52.6	H038	1 2 1 1 2	

943. $C_6H_{15}O_2PS_3$
Thiometon
O,O-Dimethyl S-(2-Ethylmercaptoethyl) Dithiophosphate
RN: 640-15-3   **MP** (°C):
MW: 246.35   **BP** (°C):   104

Solubility (Moles/L)	Solubility (Grams/L)	Temp (°C)	Ref (#)	Evaluation (T P E A A)	Comments
8.118E-04	2.000E-01	20	M061	1 0 0 0 2	
8.118E-04	2.000E-01	25	M161	1 0 0 0 2	

944.   C$_6$H$_{15}$O$_3$PS$_2$
Thiolo-Methylmercaptophos
Thiolo-Methyl Demeton
RN: **MP** (°C):
MW: 230.29   **BP** (°C):   89

Solubility (Moles/L)	Solubility (Grams/L)	Temp (°C)	Ref (#)	Evaluation (T P E A A)	Comments
1.433E-02	3.300E+00	20	M061	1 0 0 0 2	

945.   C$_6$H$_{15}$O$_3$PS$_2$
Thiono-Methylmercaptophos
Thiono-Methyl Demeton
RN: **MP** (°C):
MW: 230.29   **BP** (°C):   74

Solubility (Moles/L)	Solubility (Grams/L)	Temp (°C)	Ref (#)	Evaluation (T P E A A)	Comments
1.433E-03	3.300E-01	20	M061	1 0 0 0 2	

946.   C$_6$H$_{15}$O$_4$P
Triethyl Phosphate
Ethyl Phosphate
Phosphoric Acid, Triethyl Ester
TEP
RN: 78-40-0   **MP** (°C):   -56.4
MW: 182.16   **BP** (°C):   215

Solubility (Moles/L)	Solubility (Grams/L)	Temp (°C)	Ref (#)	Evaluation (T P E A A)	Comments
2.815E+00	5.128E+02	4.50	C022	1 2 0 0 2	
2.745E+00	5.000E+02	25	F300	1 0 0 0 1	

947.   C$_6$H$_{16}$FN$_2$OP
Mipafox
N,N'-Diisopropylphosphorodiamidic Fluoride
RN: 371-86-8   **MP** (°C):   65
MW: 182.18   **BP** (°C):

Solubility (Moles/L)	Solubility (Grams/L)	Temp (°C)	Ref (#)	Evaluation (T P E A A)	Comments
4.066E-01	7.407E+01	ns	M061	0 0 0 0 0	

948.   C$_6$H$_{16}$N$_2$
1,6-Hexanediamine
Hexamethylenediamine
RN: 124-09-4   **MP** (°C):   42
MW: 116.21   **BP** (°C):   205

Solubility (Moles/L)	Solubility (Grams/L)	Temp (°C)	Ref (#)	Evaluation (T P E A A)	Comments
6.123E+00	7.115E+02	4.50	C022	1 2 0 0 2	

949.   C$_6$H$_{17}$N$_3$O$_{10}$S
Glycine Sulfate
Triglycine Sulfate
RN: 513-29-1   **MP** (°C):
MW: 323.28   **BP** (°C):

Solubility (Moles/L)	Solubility (Grams/L)	Temp (°C)	Ref (#)	Evaluation (T P E A A)	Comments
3.314E-01	1.071E+02	0	M043	1 0 0 0 1	
5.155E-01	1.667E+02	10	M043	1 0 0 0 1	
6.576E-01	2.126E+02	20	M043	1 0 0 0 1	
8.188E-01	2.647E+02	30	M043	1 0 0 0 1	
9.600E-01	3.103E+02	40	M043	1 0 0 0 1	
1.326E+00	4.286E+02	60	M043	1 0 0 0 1	

950.   C$_6$H$_{18}$N$_4$
Triethylenetetramine
N,N'-bis(2-Aminoethyl)-ethylenediamine
1,8-Diamino-3,6-diazaoctane
1,4,7,10-Tetraazadecane
3,6-Diazaoctane-1,8-diamine
Trientine
RN: 112-24-3   **MP** (°C):   12
MW: 146.24   **BP** (°C):   266

Solubility (Moles/L)	Solubility (Grams/L)	Temp (°C)	Ref (#)	Evaluation (T P E A A)	Comments
5.655E+00	8.269E+02	4.50	C022	1 2 0 0 2	

951. $C_6Cl_4O_2$
Chloranil
Tetrachloro-p-benzoquinone
2,3,5,6-Tetrachloro-p-benzoquinone
2,3,5,6-Tetrachloro-2,5-cyclohexadiene-1,4-dione
Vulklor
Coversan
RN: 118-75-2   **MP** (°C):   290
MW: 245.88   **BP** (°C):

Solubility (Moles/L)	Solubility (Grams/L)	Temp (°C)	Ref (#)	Evaluation (T P E A A)	Comments
1.017E-03	2.500E-01	rt	M161	0 0 0 0 2	

952. $C_6Cl_5NO_2$
Quintozene
Pentachloronitrobenzene
Avical
Eorthcicle
Quintobenzene
RN: 82-68-8   **MP** (°C):   >139
MW: 295.34   **BP** (°C):   328

Solubility (Moles/L)	Solubility (Grams/L)	Temp (°C)	Ref (#)	Evaluation (T P E A A)	Comments
1.500E-06	4.430E-04	20	E308	1 2 2 1 1	
1.862E-06	5.500E-04	22	K137	1 1 2 1 0	
1.490E-06	4.400E-04	22.5	G301	2 1 0 1 2	

953. C_6Cl_6
Hexachlorobenzene
Benzene Hexachloride
HCB
Hexa-chlorobenzene
RN: 118-74-1   **MP** (°C):   228
MW: 284.78   **BP** (°C):   324.5

Solubility (Moles/L)	Solubility (Grams/L)	Temp (°C)	Ref (#)	Evaluation (T P E A A)	Comments
1.259E-07	3.585E-05	20	B179	2 0 0 0 2	
1.721E-08	4.900E-06	20	C113	1 0 1 1 1	
2.598E-08	7.400E-06	20	H300	1 1 2 2 1	
1.896E-08	5.400E-06	20	H300	1 1 2 2 1	
2.042E-08	5.815E-06	20	K337	1 0 0 0 2	
1.380E-08	3.931E-06	22	K305	1 0 1 1 2	
1.756E-08	5.000E-06	22.5	G301	2 1 0 1 2	
1.700E-08	4.841E-06	25	B317	1 0 0 0 2	
1.650E-08	4.699E-06	25	M342	1 0 1 1 2	

2.107E-08	6.000E-06	26.70	L095	2 2 1 1 2	
<3.51E-06	<1.00E-03	30	M311	1 1 2 2 0	
7.023E-08	2.000E-05	ns	L072	0 0 0 0 1	
2.107E-08	6.000E-06	ns	L311	0 0 0 0 1	
1.650E-07	4.699E-05	ns	M308	0 0 1 1 2	
2.458E-05	7.000E-03	rt	H053	0 2 2 2 0	γ isomer

954.   C₇H₃Br₂NO

Bromoxynil
3,5-Dibromo-4-hydroxybenzonitrile
4-Cyano-2,6-dibromophenol
RN: 1689-84-5   **MP** (°C):   190
MW: 276.93   **BP** (°C):

Solubility (Moles/L)	Solubility (Grams/L)	Temp (°C)	Ref (#)	Evaluation (T P E A A)	Comments
4.694E-04	1.300E-01	25	M161	1 0 0 0 2	
4.694E-04	1.300E-01	ns	M061	0 0 0 0 2	

955.   C₇H₃Br₃O₂

2,4,6-Tribromobenzoic Acid
2,4,6-Tribrom-benzoesaeure
RN: 633-12-5   **MP** (°C):
MW: 358.83   **BP** (°C):

Solubility (Moles/L)	Solubility (Grams/L)	Temp (°C)	Ref (#)	Evaluation (T P E A A)	Comments
9.754E-03	3.500E+00	15	F300	1 0 0 0 1	
1.533E-02	5.500E+00	100	F300	1 0 0 0 1	

956.   C₇H₃Cl₂N

Dichlobenil
2,6-Dichlorobenzonitrile
Benzonitrile, 2,6-Dichloro-
RN: 1194-65-6   **MP** (°C):   145
MW: 172.01   **BP** (°C):   270

Solubility (Moles/L)	Solubility (Grams/L)	Temp (°C)	Ref (#)	Evaluation (T P E A A)	Comments
1.046E-04	1.800E-02	20	B185	1 0 0 0 1	
1.046E-04	1.800E-02	20	B200	1 0 0 1 1	
1.046E-04	1.800E-02	20	G319	1 0 0 0 2	
1.046E-04	1.800E-02	20	M161	1 0 0 0 1	
1.163E-04	2.000E-02	25	B185	1 0 0 0 1	
5.813E-05	1.000E-02	25	M061	1 0 0 0 1	
1.046E-04	1.800E-02	ns	V303	0 0 0 0 1	

957. $C_7H_3Cl_3O_2$
2,3,6-Trichlorobenzoic Acid
2,3,6-TBA
RN: 50-31-7   **MP** (°C):   125
MW: 225.46   **BP** (°C):

Solubility (Moles/L)	Solubility (Grams/L)	Temp (°C)	Ref (#)	Evaluation (T P E A A)	Comments
3.726E-02	8.400E+00	20	B200	1 0 0 0 1	
3.415E-02	7.700E+00	22	M161	1 0 0 0 1	

958. $C_7H_3Cl_5O$
Pentachlorbenzyl Alcohol
Blastin
PCBA
RN: 16022-69-8   **MP** (°C):
MW: 280.37   **BP** (°C):

Solubility (Moles/L)	Solubility (Grams/L)	Temp (°C)	Ref (#)	Evaluation (T P E A A)	Comments
7.134E-07	2.000E-04	25	M061	0 0 0 0 0	

959. $C_7H_3I_2NO$
Ioxynil
4-Cyano-2,6-diiodophenol
4-Hydroxy-3,5-diiodobenzonitrile
RN: 1689-83-4   **MP** (°C):   212
MW: 370.92   **BP** (°C):

Solubility (Moles/L)	Solubility (Grams/L)	Temp (°C)	Ref (#)	Evaluation (T P E A A)	Comments
1.348E-04	5.000E-02	20	F311	1 2 2 2 1	
3.505E-04	1.300E-01	25	B200	1 0 0 0 2	
1.348E-04	5.000E-02	25	M161	1 0 0 0 1	

960. $C_7H_3N_3O_8$
2,4,6-Trinitrobenzoic Acid
2,4,6-Trinitrobenzoesaeure
Acide 2,4,6-Trinitrobenzoique
RN: 129-66-8   **MP** (°C):   228.7
MW: 257.12   **BP** (°C):

Solubility (Moles/L)	Solubility (Grams/L)	Temp (°C)	Ref (#)	Evaluation (T P E A A)	Comments
7.817E-02	2.010E+01	23	F300	1 0 0 0 2	
7.824E-02	2.012E+01	23.5	D067	1 2 0 0 2	
1.560E-01	4.012E+01	50	D067	1 2 0 0 2	
1.560E-01	4.010E+01	50	F300	1 0 0 0 2	

961. $C_7H_4BrNO_4$
3-Bromo-2-nitrobenzoic Acid
Benzoic Acid, 3-Bromo-2-nitro-
RN: 116529-61-4   **MP** (°C):
MW: 246.02   **BP** (°C):

Solubility (Moles/L)	Solubility (Grams/L)	Temp (°C)	Ref (#)	Evaluation (T P E A A)	Comments
3.012E-02	7.410E+00	25	H089	1 2 0 0 2	
1.341E-03	3.300E-01	25	H089	1 2 0 0 1	

962. C_7H_4BrNS
3-Bromophenyl Isothiocyanate
1-Bromo-3-isothiocyanato-benzene
RN: 2131-59-1   **MP** (°C):
MW: 214.09   **BP** (°C):   256.0

Solubility (Moles/L)	Solubility (Grams/L)	Temp (°C)	Ref (#)	Evaluation (T P E A A)	Comments
1.140E-04	2.441E-02	25	D019	1 1 1 1 2	
8.200E-05	1.756E-02	25	K032	2 2 0 1 1	

963. C_7H_4BrNS
4-Bromophenyl Isothiocyanate
1-Bromo-4-isothiocyanato-benzene
RN: 1985-12-2   **MP** (°C):   60.5
MW: 214.09   **BP** (°C):

Solubility (Moles/L)	Solubility (Grams/L)	Temp (°C)	Ref (#)	Evaluation (T P E A A)	Comments
5.400E-05	1.156E-02	25	D019	1 1 1 1 1	

964. $C_7H_4ClNO_4$
3-Chloro-2-nitrobenzoic Acid
2-Nitro-3-chlorobenzoic Acid
RN: 4771-47-5   **MP** (°C):
MW: 201.57   **BP** (°C):

Solubility (Moles/L)	Solubility (Grams/L)	Temp (°C)	Ref (#)	Evaluation (T P E A A)	Comments
2.332E-03	4.700E-01	25	H089	1 2 0 0 1	

965. $C_7H_4ClNO_4$
4-Chloro-3-nitrobenzoic Acid
3-Nitro-4-chlorobenzoic Acid
RN: 96-99-1   **MP** (°C):   181
MW: 201.57   **BP** (°C):

Solubility (Moles/L)	Solubility (Grams/L)	Temp (°C)	Ref (#)	Evaluation (T P E A A)	Comments
1.700E-03	3.427E-01	ns	C014	0 0 0 1 1	

966. $C_7H_4ClNO_4$
5-Chloro-2-nitrobenzoic Acid
2-Nitro-5-chlorobenzoic Acid
RN: 2516-95-2   **MP** (°C):
MW: 201.57   **BP** (°C):

Solubility (Moles/L)	Solubility (Grams/L)	Temp (°C)	Ref (#)	Evaluation (T P E A A)	Comments
4.797E-02	9.670E+00	25	H089	1 2 0 0 2	

967. C_7H_4ClNS
3-Chlorophenyl Isothiocyanate
1-Chloro-3-isothiocyanato-benzene
RN: 2392-68-9   **MP** (°C):
MW: 169.63   **BP** (°C):   249.5

Solubility (Moles/L)	Solubility (Grams/L)	Temp (°C)	Ref (#)	Evaluation (T P E A A)	Comments
2.000E-04	3.393E-02	25	D019	1 1 1 1 0	
1.120E-04	1.900E-02	25	K032	2 2 0 1 2	

968. $C_7H_4Cl_2O_2$
2,6-Dichlorobenzoic Acid
2,6-Dichlor-benzoesaeure
RN: 50-30-6   **MP** (°C):
MW: 191.01   **BP** (°C):

Solubility (Moles/L)	Solubility (Grams/L)	Temp (°C)	Ref (#)	Evaluation (T P E A A)	Comments
7.400E-02	1.414E+01	ns	C014	0 0 0 1 1	

969.   C$_7$H$_4$Cl$_2$O$_2$
3,4-Dichlorobenzoic Acid
Benzoic Acid, 3,4-Dichloro-
RN: 51-44-5   **MP** (°C):   208
MW: 191.01   **BP** (°C):

Solubility (Moles/L)	Solubility (Grams/L)	Temp (°C)	Ref (#)	Evaluation (T P E A A)	Comments
3.200E-04	6.112E-02	ns	C014	0 0 0 1 1	

970.   C$_7$H$_4$Cl$_2$O$_2$
2,4-Dichlorobenzoic Acid
2,4-Dichlor-benzoesaeure
RN: 50-84-0   **MP** (°C):
MW: 191.01   **BP** (°C):

Solubility (Moles/L)	Solubility (Grams/L)	Temp (°C)	Ref (#)	Evaluation (T P E A A)	Comments
2.500E-03	4.775E-01	ns	C014	0 2 0 1 1	

971.   C$_7$H$_4$Cl$_2$O$_2$
3,5-Dichlorobenzoic Acid
Benzoic Acid, 3,5-Dichloro-
RN: 51-36-5   **MP** (°C):   186
MW: 191.01   **BP** (°C):

Solubility (Moles/L)	Solubility (Grams/L)	Temp (°C)	Ref (#)	Evaluation (T P E A A)	Comments
7.700E-04	1.471E-01	ns	C014	0 0 0 1 1	

972.   C$_7$H$_4$Cl$_3$NO$_3$
Triclopyr
Garlon
(3,5,6-Trichloro-2-pyridinyl)oxyacetic Acid
Crossbow Turflon
RN: 55335-06-3   **MP** (°C):   149
MW: 256.47   **BP** (°C):   290

Solubility (Moles/L)	Solubility (Grams/L)	Temp (°C)	Ref (#)	Evaluation (T P E A A)	Comments
1.677E-03	4.300E-01	ns	K138	0 0 0 0 1	

973.   C₇H₄Cl₄O

2,4,5,6-Tetrachloro-3-methyl-phenol
m-Cresol, 2,4,5,6-Tetrachloro-
Phenol, 2,3,4,6-Tetrachloro-5-methyl-

RN: 10460-33-0   **MP** (°C):
MW: 245.92   **BP** (°C):

Solubility (Moles/L)	Solubility (Grams/L)	Temp (°C)	Ref (#)	Evaluation (T P E A A)	Comments
2.500E-05	6.148E-03	25	B316	1 0 2 1 1	

974.   C₇H₄Cl₄O

2,3,4,5-Tetrachloroanisole
Benzene, 1,2,3,4-Tetrachloro-5-methoxy-
Anisole, 2,3,4,5-Tetrachloro-

RN: 938-86-3   **MP** (°C):   88
MW: 245.92   **BP** (°C):

Solubility (Moles/L)	Solubility (Grams/L)	Temp (°C)	Ref (#)	Evaluation (T P E A A)	Comments
5.490E-06	1.350E-03	25	L348	1 2 2 1 2	

975.   C₇H₄INS

3-Iodophenyl Isothiocyanate
m-Iodophenyl Isothiocyanate

RN: 3125-73-3   **MP** (°C):
MW: 261.09   **BP** (°C):

Solubility (Moles/L)	Solubility (Grams/L)	Temp (°C)	Ref (#)	Evaluation (T P E A A)	Comments
2.100E-05	5.483E-03	25	K032	2 2 0 1 0	

976.   C₇H₄INS

4-Iodophenyl Isothiocyanate
4-Iodophenylisothiocyanate

RN: 2059-76-9   **MP** (°C):
MW: 261.09   **BP** (°C):

Solubility (Moles/L)	Solubility (Grams/L)	Temp (°C)	Ref (#)	Evaluation (T P E A A)	Comments
9.000E-05	2.350E-02	25	D019	1 1 1 1 1	

977. $C_7H_4I_2O_3$
3,5-Diiodosalicylic Acid
2-Hydroxy-3,5-diiod-benzoesaeure
RN: 133-91-5   **MP** (°C):   235.5
MW: 389.92   **BP** (°C):

Solubility (Moles/L)	Solubility (Grams/L)	Temp (°C)	Ref (#)	Evaluation (T P E A A)	Comments
4.274E-04	1.666E-01	10	C072	1 2 1 1 2	
1.795E-03	7.000E-01	15	F300	1 0 0 0 1	
4.931E-04	1.923E-01	25	C072	1 2 1 1 2	
3.847E-03	1.500E+00	h	F300	1 0 0 0 1	

978. $C_7H_4N_2O_2S$
3-Nitrophenyl Isothiocyanate
m-Nitrophenylisothiocyanate
RN: 3529-82-6   **MP** (°C):
MW: 180.19   **BP** (°C):

Solubility (Moles/L)	Solubility (Grams/L)	Temp (°C)	Ref (#)	Evaluation (T P E A A)	Comments
2.800E-04	5.045E-02	25	K032	2 2 0 1 2	

979. $C_7H_4N_2O_6$
3,5-Dinitrobenzoic Acid
3,5-Dinitrobenzoesaeure
RN: 99-34-3   **MP** (°C):   205
MW: 212.12   **BP** (°C):

Solubility (Moles/L)	Solubility (Grams/L)	Temp (°C)	Ref (#)	Evaluation (T P E A A)	Comments
6.350E-03	1.347E+00	25	K040	1 0 2 1 2	
2.923E-03	6.200E-01	25	P037	2 0 1 1 1	

980. $C_7H_4N_2O_6$
3,4-Dinitrobenzoic Acid
3,4-Dinitrobenzoesaeure
RN: 528-45-0   **MP** (°C):   166
MW: 212.12   **BP** (°C):

Solubility (Moles/L)	Solubility (Grams/L)	Temp (°C)	Ref (#)	Evaluation (T P E A A)	Comments
3.159E-02	6.700E+00	25	F300	1 0 0 0 1	

981.   C₇H₄N₂O₆
2,4-Dinitrobenzoic Acid
2,4-Dinitrobenzoesaeure
RN: 610-30-0   **MP** (°C):
MW: 212.12   **BP** (°C):

Solubility (Moles/L)	Solubility (Grams/L)	Temp (°C)	Ref (#)	Evaluation (T P E A A)	Comments
8.580E-02	1.820E+01	25	F300	1 0 0 0 2	
4.900E-02	1.039E+01	ns	C014	0 0 0 1 1	

982.   C₇H₄N₂O₆
2,6-Dinitrobenzoic Acid
2,6-Dinitrobenzoesaeure
RN: 603-12-3   **MP** (°C):
MW: 212.12   **BP** (°C):

Solubility (Moles/L)	Solubility (Grams/L)	Temp (°C)	Ref (#)	Evaluation (T P E A A)	Comments
7.600E-02	1.612E+01	ns	C014	0 2 0 1 1	

983.   C₇H₄N₄O₉
2,3,5,6-Tetranitroanisol
RN: **MP** (°C):
MW: 288.13   **BP** (°C):

Solubility (Moles/L)	Solubility (Grams/L)	Temp (°C)	Ref (#)	Evaluation (T P E A A)	Comments
6.941E-04	2.000E-01	50	F300	1 0 0 0 0	
4.165E-03	1.200E+00	100	F300	1 0 0 0 1	

984.   C₇H₄O₆
Chelidonic Acid
Chelidonsaeure
RN: 99-32-1   **MP** (°C):
MW: 184.11   **BP** (°C):

Solubility (Moles/L)	Solubility (Grams/L)	Temp (°C)	Ref (#)	Evaluation (T P E A A)	Comments
7.767E-02	1.430E+01	25	F300	1 0 0 0 2	
2.064E-01	3.800E+01	100	F300	1 0 0 0 1	

Solutions

985. $C_7H_4O_7$
Meconic Acid
Mekonsaeure
RN: 497-59-6   **MP** (°C):
MW: 200.11   **BP** (°C):

Solubility (Moles/L)	Solubility (Grams/L)	Temp (°C)	Ref (#)	Evaluation (T P E A A)	Comments
4.198E-02	8.400E+00	25	F300	1 0 0 0 1	
1.034E+00	2.070E+02	100	F300	1 0 0 0 2	

986. $C_7H_5BrO_2$
p-Bromobenzoic Acid
4-Bromobenzoic Acid
RN: 586-76-5   **MP** (°C):   252.0
MW: 201.03   **BP** (°C):

Solubility (Moles/L)	Solubility (Grams/L)	Temp (°C)	Ref (#)	Evaluation (T P E A A)	Comments
2.786E-04	5.600E-02	22.5	G301	2 1 0 1 2	
2.985E-04	6.000E-02	ns	B150	0 0 2 2 1	
2.885E-04	5.800E-02	ns	B150	0 0 2 2 1	
2.800E-04	5.629E-02	ns	C014	0 0 0 1 1	

987. $C_7H_5BrO_2$
m-Bromobenzoic Acid
3-Bromobenzoic Acid
RN: 585-76-2   **MP** (°C):   155
MW: 201.03   **BP** (°C):

Solubility (Moles/L)	Solubility (Grams/L)	Temp (°C)	Ref (#)	Evaluation (T P E A A)	Comments
2.000E-03	4.021E-01	ns	C014	0 0 0 1 1	

988. $C_7H_5ClO_2$
o-Chlorobenzoic Acid
2-Chlor-benzoesaeure
2-Chlorobenzoic Acid
RN: 118-91-2   **MP** (°C):   142
MW: 156.57   **BP** (°C):

Solubility (Moles/L)	Solubility (Grams/L)	Temp (°C)	Ref (#)	Evaluation (T P E A A)	Comments
2.100E-02	3.288E+00	24.99	B391	2 0 0 1 2	
1.341E-02	2.100E+00	25	F300	1 0 0 0 1	
8.686E-03	1.360E+00	25	P037	2 0 1 1 2	
1.865E-02	2.920E+00	37	M360	1 2 1 1 2	
2.574E-01	4.030E+01	100	F300	1 0 0 0 2	
1.330E-02	2.082E+00	ns	C014	0 0 0 1 2	
1.362E-02	2.132E+00	ns	O004	0 2 1 1 2	

989.   C$_7$H$_5$ClO$_2$
p-Chlorobenzoic Acid
4-Chlorobenzoic Acid
Chloradracylic
4-Chlor-benzoesaeure
RN: 74-11-3   **MP** (°C):   235
MW: 156.57   **BP** (°C):

Solubility (Moles/L)	Solubility (Grams/L)	Temp (°C)	Ref (#)	Evaluation (T P E A A)	Comments
5.748E-04	9.000E-02	22.5	G301	2 1 0 1 2	
8.000E-04	1.253E-01	24.99	B391	2 0 0 1 2	
4.918E-04	7.700E-02	25	F300	1 0 0 0 1	
4.639E-04	7.263E-02	25	T066	1 0 0 0 2	
7.026E-04	1.100E-01	37	M360	1 2 1 1 2	
4.918E-04	7.700E-02	ns	B150	0 0 2 2 1	
4.918E-04	7.700E-02	ns	B150	0 0 2 2 1	
4.350E-04	6.811E-02	ns	O004	0 2 1 1 2	

990.   C$_7$H$_5$ClO$_2$
meta-Chlorobenzoic Acid
3-Chlorobenzoic Acid
m-Chlorobenzoic Acid
3-Chlor-benzoesaeure
RN: 535-80-8   **MP** (°C):   154
MW: 156.57   **BP** (°C):

Solubility (Moles/L)	Solubility (Grams/L)	Temp (°C)	Ref (#)	Evaluation (T P E A A)	Comments
2.555E-04	4.000E-02	0	F300	1 0 0 0 0	
4.080E-03	6.388E-01	24.99	B391	2 0 0 1 2	
2.555E-03	4.000E-01	25	F300	1 0 0 0 0	
2.543E-03	3.982E-01	25	T066	1 0 0 0 2	
2.555E-03	4.000E-01	37	M360	1 2 1 1 2	
2.460E-03	3.852E-01	ns	O004	0 2 1 1 2	

991.   C$_7$H$_5$Cl$_2$NO
2,6-Dichlorobenzamide
Dichlorobenzamide
BAM
RN: 2008-58-4   **MP** (°C):   198
MW: 190.03   **BP** (°C):

Solubility (Moles/L)	Solubility (Grams/L)	Temp (°C)	Ref (#)	Evaluation (T P E A A)	Comments
1.421E-02	2.700E+00	22.5	G301	2 1 0 1 2	

992.   C₇H₅Cl₂NO₂

$C_7H_5Cl_2NO_2$

Chloramben

3-Amino-2,5-dichlorobenzoic Acid

RN: 133-90-4   **MP** (°C):   201
MW: 206.03   **BP** (°C):

Solubility (Moles/L)	Solubility (Grams/L)	Temp (°C)	Ref (#)	Evaluation (T P E A A)	Comments
3.398E-03	7.000E-01	25	B200	1 0 0 0 2	
3.398E-03	7.000E-01	25	M161	1 0 0 0 2	
3.398E-03	7.000E-01	ns	B185	0 0 0 0 2	

993.   C₇H₅Cl₂NS

$C_7H_5Cl_2NS$

2,6-Dichlorothiobenzamide

Prefix

Chlorthiamid

RN: 1918-13-4   **MP** (°C):   151.5
MW: 206.09   **BP** (°C):   0

Solubility (Moles/L)	Solubility (Grams/L)	Temp (°C)	Ref (#)	Evaluation (T P E A A)	Comments
4.561E-03	9.400E-01	20	M061	1 0 0 0 2	
4.610E-03	9.500E-01	21	M161	1 0 0 0 2	

994.   C₇H₅Cl₃O

$C_7H_5Cl_3O$

2,4,6-Trichloro-3-methylphenol

m-Cresol, 2,4,6-Trichloro-

2,4,6-Trichloro-m-cresol

RN: 551-76-8   **MP** (°C):
MW: 211.48   **BP** (°C):

Solubility (Moles/L)	Solubility (Grams/L)	Temp (°C)	Ref (#)	Evaluation (T P E A A)	Comments
5.300E-04	1.121E-01	25	B316	1 0 2 1 1	

995.   C₇H₅Cl₃O

$C_7H_5Cl_3O$

2,3,4-Trichloroanisole

1,2,3-Trichloro-4-methoxy-benzene

RN: 54135-80-7   **MP** (°C):   70
MW: 211.48   **BP** (°C):

Solubility (Moles/L)	Solubility (Grams/L)	Temp (°C)	Ref (#)	Evaluation (T P E A A)	Comments
5.107E-05	1.080E-02	25	L348	1 2 2 1 2	

996. $C_7H_5Cl_3O$
2,4,6-Trichloroanisole
1-Methoxy-2,4,6-trichlorobenzene
Methyl 2,4,6-Trichlorophenyl Ether
Tyrene
RN: 87-40-1   **MP** (°C):   61
MW: 211.48   **BP** (°C):

Solubility (Moles/L)	Solubility (Grams/L)	Temp (°C)	Ref (#)	Evaluation (T P E A A)	Comments
6.242E-05	1.320E-02	25	L348	1 2 2 1 2	

997. $C_7H_5FO_2$
m-Fluorobenzoic Acid
3-Fluor-benzoesaeure
3-Fluorobenzoic Acid
RN: 455-38-9   **MP** (°C):   123
MW: 140.12   **BP** (°C):

Solubility (Moles/L)	Solubility (Grams/L)	Temp (°C)	Ref (#)	Evaluation (T P E A A)	Comments
1.071E-02	1.500E+00	25	F300	1 0 0 0 1	
1.071E-02	1.500E+00	25	F300	1 0 0 0 1	

998. $C_7H_5FO_2$
o-Fluorobenzoic Acid
2-Fluorobenzoic Acid
RN: 445-29-4   **MP** (°C):   123
MW: 140.12   **BP** (°C):

Solubility (Moles/L)	Solubility (Grams/L)	Temp (°C)	Ref (#)	Evaluation (T P E A A)	Comments
5.139E-02	7.200E+00	25	F300	1 0 0 0 1	
5.139E-02	7.200E+00	25	F300	1 0 0 0 1	

999. $C_7H_5FO_2$
p-Fluorobenzoic Acid
4-Fluor-benzoesaeure
4-Fluorobenzoic Acid
RN: 456-22-4   **MP** (°C):   182.6
MW: 140.12   **BP** (°C):

Solubility (Moles/L)	Solubility (Grams/L)	Temp (°C)	Ref (#)	Evaluation (T P E A A)	Comments
8.564E-03	1.200E+00	25	F300	1 0 0 0 1	

1000. $C_7H_5F_3N_2O_4S$
3-Trifluoromethyl-4-nitrobenzenesulfonamide
4-Nitro-3-(trifluoromethyl)benzenesulfonamide
RN: 21988-05-6   **MP** (°C):
MW: 270.19   **BP** (°C):

Solubility (Moles/L)	Solubility (Grams/L)	Temp (°C)	Ref (#)	Evaluation (T P E A A)	Comments
6.500E-04	1.756E-01	15	K024	1 2 1 1 2	

1001. $C_7H_5IO_2$
m-Iodobenzoic Acid
3-Iodobenzoic Acid
RN: 618-51-9   **MP** (°C):   187
MW: 248.02   **BP** (°C):

Solubility (Moles/L)	Solubility (Grams/L)	Temp (°C)	Ref (#)	Evaluation (T P E A A)	Comments
5.380E-04	1.334E-01	15	D008	1 0 1 1 2	0.002N HCl

1002. $C_7H_5IO_2$
o-Iodobenzoic Acid
2-Iodobenzoic Acid
RN: 88-67-5   **MP** (°C):   162
MW: 248.02   **BP** (°C):

Solubility (Moles/L)	Solubility (Grams/L)	Temp (°C)	Ref (#)	Evaluation (T P E A A)	Comments
1.860E-03	4.613E-01	15	D008	1 0 1 1 2	0.002N HCl

1003. $C_7H_5IO_2$
p-Iodobenzoic Acid
4-Iodobenzoic Acid
RN: 619-58-9   **MP** (°C):
MW: 248.02   **BP** (°C):

Solubility (Moles/L)	Solubility (Grams/L)	Temp (°C)	Ref (#)	Evaluation (T P E A A)	Comments
1.120E-04	2.778E-02	15	D008	1 0 1 1 2	intrinsic

1004.   C$_7$H$_5$I$_2$NO$_3$

3,5-Diiodo-4-pyridone-N-acetic Acid
3,5-Diiod-pyridon-(4)-N-essigsaeure
3,5-Diiodo-4-pyridone-1-acetic Acid
Diodon
1,4-Dihydro-3,5-diiodo-4-oxopyridine-1-acetic Acid

RN: 101-29-1   **MP** (°C):   244
MW: 404.93   **BP** (°C):

Solubility (Moles/L)	Solubility (Grams/L)	Temp (°C)	Ref (#)	Evaluation (T P E A A)	Comments
6.883E-03	2.787E+00	ns	H055	0 1 0 2 2	

1005.   C$_7$H$_5$N

Benzonitrile
Benzonitril
Benzenenitrile
Benzoic Acid Nitrile
Phenyl Cyanide
Cyanobenzene

RN: 100-47-0   **MP** (°C):   -13
MW: 103.12   **BP** (°C):   190.7

Solubility (Moles/L)	Solubility (Grams/L)	Temp (°C)	Ref (#)	Evaluation (T P E A A)	Comments
1.839E-02	1.896E+00	24.0	P321	2 0 0 2 1	
4.200E-02	4.331E+00	25	M327	1 0 0 1 2	
3.671E-02	3.786E+00	35.5	P321	2 0 0 2 1	
5.400E-02	5.569E+00	50.0	P321	2 0 0 2 1	
4.056E-02	4.182E+00	57.0	P321	2 0 0 2 1	
5.496E-02	5.668E+00	62.5	P321	2 0 0 2 1	
8.268E-02	8.527E+00	85.0	P321	2 0 0 2 1	
8.459E-02	8.723E+00	90.5	P321	2 0 0 2 1	
9.981E-02	1.029E+01	95.5	P321	2 0 0 2 1	
9.697E-02	1.000E+01	100	F300	1 0 0 0 0	
1.065E-01	1.098E+01	101.0	P321	2 0 0 2 1	
1.339E-01	1.381E+01	116.0	P321	2 0 0 2 1	
1.920E-01	1.980E+01	127.5	P321	2 0 0 2 1	
2.171E-01	2.239E+01	142.0	P321	2 0 0 2 1	
2.888E-01	2.979E+01	148.0	P321	2 0 0 2 1	
2.834E-01	2.922E+01	149.0	P321	2 0 0 2 1	
3.873E-01	3.994E+01	160.5	P321	2 0 0 2 1	
5.747E-01	5.927E+01	164.5	P321	2 0 0 2 1	
1.373E+00	1.416E+02	201.0	P321	2 0 0 2 1	
2.937E+00	3.029E+02	211.0	P321	2 0 0 2 1	
9.696E-04	9.999E-02	ns	L055	0 0 0 0 1	

1006. C₇H₅NOS

3-Hydroxyphenyl Isothiocyanate
m-Hydroxyphenyl Isothiocyanate

RN: 3125-63-1   **MP** (°C):
MW: 151.19   **BP** (°C):

Solubility (Moles/L)	Solubility (Grams/L)	Temp (°C)	Ref (#)	Evaluation (T P E A A)	Comments
1.020E-02	1.542E+00	25	K032	2 2 0 1 2	

1007. C₇H₅NOS

4-Hydroxyphenyl Isothiocyanate
4-Hydroxyphenylisothiocyanate

RN: 2131-60-4   **MP** (°C):
MW: 151.19   **BP** (°C):

Solubility (Moles/L)	Solubility (Grams/L)	Temp (°C)	Ref (#)	Evaluation (T P E A A)	Comments
2.150E-03	3.251E-01	25	D019	1 1 1 1 2	

1008. C₇H₅NO₃

m-Nitrobenzaldehyde
3-Nitrobenzaldehyde
3-Nitro-benzaldehyd

RN: 99-61-6   **MP** (°C):   58
MW: 151.12   **BP** (°C):

Solubility (Moles/L)	Solubility (Grams/L)	Temp (°C)	Ref (#)	Evaluation (T P E A A)	Comments
6.617E-05	1.000E-02	25	F300	1 0 0 0 1	
3.309E+00	5.000E+02	58.0	S118	1 2 0 1 0	
6.292E-02	9.509E+00	75.1	S118	1 2 0 1 1	
3.272E+00	4.945E+02	85.2	S118	1 2 0 1 2	
1.266E-01	1.913E+01	111.9	S118	1 2 0 1 2	
1.934E-01	2.922E+01	136.4	S118	1 2 0 1 2	
3.103E-01	4.689E+01	157.3	S118	1 2 0 1 2	
6.293E-01	9.510E+01	181.0	S118	1 2 0 1 2	
8.142E-01	1.230E+02	191.4	S118	1 2 0 1 2	
1.253E+00	1.893E+02	205.4	S118	1 2 0 1 2	
1.878E+00	2.838E+02	211.8	S118	1 2 0 1 2	

1009.   C$_7$H$_5$NO$_3$
p-Nitrobenzaldehyde
4-Nitrobenzaldehyde
RN: 555-16-8   **MP** (°C):   106.5
MW: 151.12   **BP** (°C):

Solubility (Moles/L)	Solubility (Grams/L)	Temp (°C)	Ref (#)	Evaluation (T P E A A)	Comments
1.871E-01	2.828E+01	132.4	S118	1 2 0 1 2	
5.341E-01	8.071E+01	176.5	S118	1 2 0 1 2	
1.133E+00	1.713E+02	205.4	S118	1 2 0 1 2	
1.814E+00	2.742E+02	215.5	S118	1 2 0 1 2	

1010.   C$_7$H$_5$NO$_3$
o-Nitrobenzaldehyde
2-Nitrobenzaldehyde
2-Nitro-benzaldehyd
RN: 552-89-6   **MP** (°C):   44
MW: 151.12   **BP** (°C):   153

Solubility (Moles/L)	Solubility (Grams/L)	Temp (°C)	Ref (#)	Evaluation (T P E A A)	Comments
1.323E-04	2.000E-02	25	F300	1 0 0 0 1	
4.600E-02	6.951E+00	66.9	S118	1 2 0 1 1	
9.972E-02	1.507E+01	103.1	S118	1 2 0 1 1	
3.001E-01	4.535E+01	166.0	S118	1 2 0 1 1	

1011.   C$_7$H$_5$NO$_3$S
Saccharin
1,1-Dioxide-1,2-Benzisothiazol-3-(2H)-one
3-Benzisothiazolinone 1,1-dioxide
1,2-Benzisothiazol-3(2H)-one-1,1-dioxide
Kandiset
Glucid
RN: 81-07-2   **MP** (°C):   228.8
MW: 183.19   **BP** (°C):

Solubility (Moles/L)	Solubility (Grams/L)	Temp (°C)	Ref (#)	Evaluation (T P E A A)	Comments
2.347E-02	4.300E+00	25	F300	1 0 0 0 1	
1.880E-01	3.444E+01	30	M015	1 0 2 1 0	EFG

1012. $C_7H_5NO_4$
3,5-Pyridinedicarboxylic Acid
Dinicotinic Acid
RN: 499-81-0   **MP** (°C):
MW: 167.12   **BP** (°C):

Solubility (Moles/L)	Solubility (Grams/L)	Temp (°C)	Ref (#)	Evaluation (T P E A A)	Comments
6.400E-03	1.070E+00	25	C104	2 2 1 1 2	

1013. $C_7H_5NO_4$
Quinolinic Acid
2,3-Pyridinedicarboxylic Acid
Pyridine-2,3-Dicarboxylic Acid
Pyridine-2,3-Dicarboxylate
RN: 89-00-9   **MP** (°C):   190
MW: 167.12   **BP** (°C):

Solubility (Moles/L)	Solubility (Grams/L)	Temp (°C)	Ref (#)	Evaluation (T P E A A)	Comments
3.291E-02	5.500E+00	7	F300	1 0 0 0 1	
6.600E-02	1.103E+01	25	C104	2 2 1 1 2	
6.400E-02	1.070E+01	25	C104	2 2 1 1 2	

1014. $C_7H_5NO_4$
p-Nitrobenzoic Acid
4-Nitrobenzoic Acid
RN: 62-23-7   **MP** (°C):   242.4
MW: 167.12   **BP** (°C):

Solubility (Moles/L)	Solubility (Grams/L)	Temp (°C)	Ref (#)	Evaluation (T P E A A)	Comments
1.197E-03	2.000E-01	15	F300	1 0 0 0 2	
2.525E-03	4.220E-01	24.99	B391	2 0 0 1 2	
1.660E-03	2.774E-01	25	H071	2 2 2 1 2	
3.471E-03	5.800E-01	37	B171	2 0 1 1 2	

1015. $C_7H_5NO_4$
o-Nitrobenzoic Acid
2-Nitrobenzoic Acid
RN: 552-16-9   **MP** (°C):   147.5
MW: 167.12   **BP** (°C):

Solubility (Moles/L)	Solubility (Grams/L)	Temp (°C)	Ref (#)	Evaluation (T P E A A)	Comments
3.920E-02	6.551E+00	18	D058	1 0 1 1 2	
3.340E-02	5.582E+00	24.99	B391	2 0 0 1 2	
4.325E-02	7.228E+00	25	D058	1 0 1 1 2	
4.488E-02	7.500E+00	25	F300	1 0 0 0 1	
4.350E-02	7.270E+00	25	H071	2 2 2 1 2	

4.700E-02	7.855E+00	25	K040	1 0 2 1 2
4.360E-02	7.287E+00	25	K053	2 2 2 2 2
4.430E-02	7.404E+00	25	L050	2 0 1 2 2
4.415E-02	7.378E+00	25	R016	1 0 1 1 2
4.700E-02	7.855E+00	26.4	P043	2 0 1 1 2

1016. $C_7H_5NO_4$
m-Nitrobenzoic Acid
3-Nitrobenzoic Acid

RN: 121-92-6   **MP** (°C):   142.0
MW: 167.12   **BP** (°C):

Solubility (Moles/L)	Solubility (Grams/L)	Temp (°C)	Ref (#)	Evaluation (T P E A A)	Comments
1.436E-02	2.400E+00	15	F300	1 0 0 0 1	
1.530E-02	2.557E+00	24.99	B391	2 0 0 1 2	
2.121E-02	3.545E+00	25	C076	2 0 0 0 2	
2.140E-02	3.576E+00	25	K040	1 0 2 1 2	
1.227E-02	2.050E+00	25	P037	2 0 1 1 2	
6.582E-02	1.100E+01	37	B171	2 0 1 1 2	
2.334E-02	3.900E+00	ns	B361	0 0 0 2 2	

1017. $C_7H_5NO_4$
Lutidinic Acid
2,4-Pyridinedicarboxylic Acid

RN: 499-80-9   **MP** (°C):   248
MW: 167.12   **BP** (°C):

Solubility (Moles/L)	Solubility (Grams/L)	Temp (°C)	Ref (#)	Evaluation (T P E A A)	Comments
1.490E-02	2.490E+00	25	C104	2 2 1 1 2	
1.480E-02	2.473E+00	25	C104	2 2 1 1 2	

1018. $C_7H_5NO_4$
Isocinchomeronic Acid
2,5-Pyridinedicarboxylic Acid
Pyridine-2,5-Dicarboxylic Acid

RN: 100-26-5   **MP** (°C):   254
MW: 167.12   **BP** (°C):

Solubility (Moles/L)	Solubility (Grams/L)	Temp (°C)	Ref (#)	Evaluation (T P E A A)	Comments
7.400E-03	1.237E+00	25	C104	2 2 1 1 2	
7.100E-03	1.187E+00	25	C104	2 2 1 1 2	

1019. $C_7H_5NO_4$
Cinchomeronic Acid
3,4-Pyridinedicarboxylic Acid
RN: 490-11-9 **MP** (°C): 256
MW: 167.12 **BP** (°C):

Solubility (Moles/L)	Solubility (Grams/L)	Temp (°C)	Ref (#)	Evaluation (T P E A A)	Comments
1.400E-02	2.340E+00	25	C104	2 2 1 1 2	
1.380E-02	2.306E+00	25	C104	2 2 1 1 2	

1020. $C_7H_5NO_5$
5-Nitrosalicylic Acid
5-Nitrosalicylsaeure
RN: 96-97-9 **MP** (°C): 229-230
MW: 183.12 **BP** (°C):

Solubility (Moles/L)	Solubility (Grams/L)	Temp (°C)	Ref (#)	Evaluation (T P E A A)	Comments
1.092E-02	2.000E+00	45	F300	1 0 0 0 0	

1021. $C_7H_5NO_5$
3-Nitrosalicylic Acid
3-Nitro-salicylsaeure
RN: 85-38-1 **MP** (°C): 128
MW: 183.12 **BP** (°C):

Solubility (Moles/L)	Solubility (Grams/L)	Temp (°C)	Ref (#)	Evaluation (T P E A A)	Comments
7.099E-03	1.300E+00	16	F300	1 0 0 0 1	

1022. C_7H_5NS
Phenyl Isothiocyanate
Isothiocyanatobenzene
Phenyl Mustard Oil
PITC
RN: 103-72-0 **MP** (°C): -21.0
MW: 135.19 **BP** (°C): 221.0

Solubility (Moles/L)	Solubility (Grams/L)	Temp (°C)	Ref (#)	Evaluation (T P E A A)	Comments
6.650E-04	8.990E-02	25	D019	1 1 1 1 2	

1023. C$_7$H$_5$N$_3$O$_6$
2,4,6-Trinitrotoluene
2,4,6-Tronitrotoluol
RN: 118-96-7   **MP** (°C):   80.1
MW: 227.13   **BP** (°C):

Solubility (Moles/L)	Solubility (Grams/L)	Temp (°C)	Ref (#)	Evaluation (T P E A A)	Comments
4.843E-04	1.100E-01	0.3	D065	1 2 2 1 2	
4.843E-04	1.100E-01	0.3	F300	1 0 0 0 1	
4.842E-04	1.100E-01	0.3	T020	1 2 2 2 2	
4.975E-04	1.130E-01	5.9	D065	1 2 2 1 2	
4.974E-04	1.130E-01	5.9	T020	1 2 2 2 2	
5.283E-04	1.200E-01	20	D065	1 2 2 1 2	
5.283E-04	1.200E-01	20.0	T020	1 2 2 2 2	
8.937E-04	2.030E-01	33.1	D065	1 2 2 1 2	
8.936E-04	2.030E-01	33.1	T020	1 2 2 2 2	
1.497E-03	3.400E-01	44.2	D065	1 2 2 1 2	
1.496E-03	3.399E-01	44.2	T020	1 2 2 2 2	
1.629E-03	3.700E-01	45	D065	1 2 2 1 2	
1.628E-03	3.699E-01	45.0	T020	1 2 2 2 2	
2.351E-03	5.340E-01	53	D065	1 2 2 1 2	
2.350E-03	5.337E-01	53.0	T020	1 2 2 2 2	
2.703E-03	6.140E-01	57.1	D065	1 2 2 1 2	
2.702E-03	6.136E-01	57.1	T020	1 2 2 2 2	
4.240E-03	9.630E-01	73.2	D065	1 2 2 1 2	
4.236E-03	9.621E-01	73.2	T020	1 2 2 2 2	
6.054E-03	1.375E+00	94.4	D065	1 2 2 1 2	
6.045E-03	1.373E+00	94.4	T020	1 2 2 2 2	
6.459E-03	1.467E+00	99.5	D065	1 2 2 1 2	
6.449E-03	1.465E+00	99.5	T020	1 2 2 2 2	
6.459E-03	1.467E+00	99.50	F300	1 0 0 0 2	

1024. C$_7$H$_5$N$_3$O$_7$
2,4,6-Trinitroanisole
2-Methoxy-1,3,5-trinitro-benzene
Methyl Picrate
RN: 606-35-9   **MP** (°C):   69
MW: 243.13   **BP** (°C):

Solubility (Moles/L)	Solubility (Grams/L)	Temp (°C)	Ref (#)	Evaluation (T P E A A)	Comments
8.224E-04	2.000E-01	15	D079	1 2 0 0 1	
5.627E-03	1.368E+00	50	D079	1 2 0 0 2	
1.594E-02	3.875E+00	100	D079	1 2 0 0 2	

1025.   C₇H₅N₃O₇

2,4,6-Trinitro-m-cresol
2,4,6-Trinitro-m-kresol

RN: 3238-38-8   **MP** (°C):
MW: 243.13   **BP** (°C):

Solubility (Moles/L)	Solubility (Grams/L)	Temp (°C)	Ref (#)	Evaluation (T P E A A)	Comments
8.226E-03	2.000E+00	15	F300	1 0 0 0 0	

1026.   C₇H₅N₃O₇

Methyl Picric Acid
2,4,6-Trinitro-3-methylphenol
3-Methyl-2,4,6-trinitrophenol
2,4,6-Trinitro-m-cresol

RN: 602-99-3   **MP** (°C):
MW: 243.13   **BP** (°C):

Solubility (Moles/L)	Solubility (Grams/L)	Temp (°C)	Ref (#)	Evaluation (T P E A A)	Comments
1.000E-02	2.431E+00	25	K053	2 2 2 2 2	

1027.   C₇H₅N₅O₈

Nitramine
Tetryl
N-Methyl-N,2,4,5-tetranitroaniline

RN: 479-45-8   **MP** (°C):   131
MW: 287.15   **BP** (°C):

Solubility (Moles/L)	Solubility (Grams/L)	Temp (°C)	Ref (#)	Evaluation (T P E A A)	Comments
1.776E-04	5.100E-02	0.5	D066	1 2 2 1 2	
1.741E-04	5.000E-02	0.5	F300	1 0 0 0 0	
1.776E-04	5.100E-02	0.5	T015	1 2 0 1 1	
2.403E-04	6.900E-02	9.6	D066	1 2 2 1 2	
2.403E-04	6.900E-02	9.6	T015	1 2 0 1 1	
2.473E-04	7.100E-02	14.8	D066	1 2 2 1 1	
2.472E-04	7.099E-02	14.8	T015	1 2 0 1 1	
2.577E-04	7.400E-02	20.5	D066	1 2 2 1 1	
2.577E-04	7.399E-02	20.5	T015	1 2 0 1 1	
2.925E-04	8.400E-02	30	D066	1 2 2 1 1	
2.925E-04	8.399E-02	30.0	T015	1 2 0 1 1	
3.274E-04	9.400E-02	35	D066	1 2 2 1 1	
3.273E-04	9.399E-02	35.0	T015	1 2 0 1 1	
3.726E-04	1.070E-01	40	D066	1 2 2 1 2	
3.726E-04	1.070E-01	40.0	T015	1 2 0 1 2	
4.701E-04	1.350E-01	45	D066	1 2 2 1 2	
4.701E-04	1.350E-01	45.0	T015	1 2 0 1 2	
6.965E-04	2.000E-01	50	D066	1 2 2 1 2	
6.964E-04	2.000E-01	50.0	T015	1 2 0 1 2	

1.219E-03	3.500E-01	60	D066	0 0 0 0 0
1.218E-03	3.499E-01	60.05	T015	1 2 0 1 2
1.543E-03	4.430E-01	65	D065	1 2 2 1 2
1.542E-03	4.428E-01	65.05	T015	1 2 0 1 2
1.849E-03	5.310E-01	69.5	D065	1 2 2 1 2
1.848E-03	5.307E-01	69.5	T015	1 2 0 1 2
3.315E-03	9.520E-01	84.2	D065	1 2 2 1 2
3.312E-03	9.511E-01	84.2	T015	1 2 0 1 2
5.638E-03	1.619E+00	96.7	D065	1 2 2 1 2
5.629E-03	1.616E+00	96.7	T015	1 2 0 1 2
6.112E-03	1.755E+00	98.5	D065	1 2 2 1 2
6.101E-03	1.752E+00	98.55	T015	1 2 0 1 2
6.129E-03	1.760E+00	99	F300	1 0 0 0 2

1028. C_7H_6ClF
2-Fluorobenzyl Chloride
o-Fluorobenzyl Chloride
RN: 345-35-7   **MP** (°C):
MW: 144.58   **BP** (°C):

Solubility (Moles/L)	Solubility (Grams/L)	Temp (°C)	Ref (#)	Evaluation (T P E A A)	Comments
2.880E-03	4.164E-01	25	M342	1 0 1 1 2	

1029. C_7H_6ClF
3-Fluorobenzyl Chloride
m-Fluorobenzyl Chloride
RN: 456-42-8   **MP** (°C):
MW: 144.58   **BP** (°C):

Solubility (Moles/L)	Solubility (Grams/L)	Temp (°C)	Ref (#)	Evaluation (T P E A A)	Comments
2.860E-03	4.135E-01	25	M342	1 0 1 1 2	

1030. $C_7H_6ClN_3O_4S_2$
Chlorothiazide
Diuresal
RN: 58-94-6   **MP** (°C):   342
MW: 295.72   **BP** (°C):

Solubility (Moles/L)	Solubility (Grams/L)	Temp (°C)	Ref (#)	Evaluation (T P E A A)	Comments
9.560E-04	2.827E-01	25	A076	1 0 1 1 2	
9.000E-04	2.662E-01	30	A089	2 0 1 1 0	EFG
9.000E-04	2.662E-01	30	A093	2 0 1 1 0	EFG
6.763E-04	2.000E-01	ns	C114	0 0 0 0 0	
7.439E-04	2.200E-01	rt	A095	0 0 2 2 1	
9.806E-04	2.900E-01	rt	B181	0 0 1 1 2	

1031. $C_7H_6ClN_4O_5S_2$
4-Nitroso-hydrochlorothiazide
RN: **MP** (°C):   155-156
MW: 325.73   **BP** (°C):

Solubility (Moles/L)	Solubility (Grams/L)	Temp (°C)	Ref (#)	Evaluation (T P E A A)	Comments
7.368E-04	2.400E-01	25	G051	1 0 1 1 0	

1032. $C_7H_6Cl_2N_2O$
Chlorambenamide
3,5-Dichloroanthranilamide
Benzamide, 2-Amino-3,5-dichloro-
RN: 36765-01-2   **MP** (°C):   162.5
MW: 205.04   **BP** (°C):

Solubility (Moles/L)	Solubility (Grams/L)	Temp (°C)	Ref (#)	Evaluation (T P E A A)	Comments
8.291E-03	1.700E+00	rt	M161	0 0 0 0 1	

1033. $C_7H_6Cl_2O$
2,3-Dichloroanisole
1,2-Dichloro-3-methoxybenzene
RN: 1984-59-4   **MP** (°C):   32
MW: 177.03   **BP** (°C):

Solubility (Moles/L)	Solubility (Grams/L)	Temp (°C)	Ref (#)	Evaluation (T P E A A)	Comments
4.909E-04	8.690E-02	25	L348	1 2 2 1 2	

1034. $C_7H_6Cl_2O$
2,6-Dichloro-4-methyl-phenol
2,4-Dichloro-6-methyl-phenol-
RN: 2432-12-4   **MP** (°C):
MW: 177.03   **BP** (°C):

Solubility (Moles/L)	Solubility (Grams/L)	Temp (°C)	Ref (#)	Evaluation (T P E A A)	Comments
1.600E-03	2.833E-01	25	B316	1 0 2 1 1	
3.800E-03	6.727E-01	25	B316	1 0 2 1 1	

1035. $C_7H_6Cl_2O$
2,6-Dichloroanisole
Benzene, 1,3-Dichloro-2-methoxy-
RN: 1984-65-2   **MP** (°C):   31
MW: 177.03   **BP** (°C):

Solubility (Moles/L)	Solubility (Grams/L)	Temp (°C)	Ref (#)	Evaluation (T P E A A)	Comments
7.908E-04	1.400E-01	25	L348	1 2 2 1 2	

1036.   C$_7$H$_6$N$_2$O$_2$S
p-Cyanobenzenesulfonamide
4-Cyanobenzenesulfonamide
RN: 3119-02-6   **MP** (°C):
MW: 182.20   **BP** (°C):

Solubility (Moles/L)	Solubility (Grams/L)	Temp (°C)	Ref (#)	Evaluation (T P E A A)	Comments
6.100E-03	1.111E+00	15	K024	1 2 1 1 2	

1037.   C$_7$H$_6$N$_2$O$_4$
2,4-Dinitrotoluene
2,4-Dinitro-toluol
RN: 121-14-2   **MP** (°C):   71
MW: 182.14   **BP** (°C):   300

Solubility (Moles/L)	Solubility (Grams/L)	Temp (°C)	Ref (#)	Evaluation (T P E A A)	Comments
1.487E-03	2.709E-01	20	T301	1 2 2 2 2	
1.482E-03	2.699E-01	22	D070	1 2 0 0 1	
1.482E-03	2.700E-01	22	F300	1 0 0 0 1	
1.482E-03	2.699E-01	22	L053	1 1 0 0 1	
2.031E-03	3.699E-01	50	D070	1 2 0 0 1	
2.031E-03	3.699E-01	50	L053	1 1 0 0 1	
1.391E-02	2.534E+00	100	D070	1 2 0 0 2	
1.449E-02	2.640E+00	100	F300	1 0 0 0 2	
1.391E-02	2.534E+00	100	L053	1 1 0 0 2	

1038.   C$_7$H$_6$N$_2$O$_5$
Dinitrocresol
DNOC
2,4-Dinitro-6-methylphenol
Dinitro-o-cresol
RN: 534-52-1   **MP** (°C):   86
MW: 198.14   **BP** (°C):

Solubility (Moles/L)	Solubility (Grams/L)	Temp (°C)	Ref (#)	Evaluation (T P E A A)	Comments
6.561E-04	1.300E-01	15	M161	1 0 0 0 2	
6.309E-04	1.250E-01	ns	B185	0 0 0 0 2	
6.459E-04	1.280E-01	ns	M061	0 0 0 0 2	
1.000E-03	1.981E-01	ns	M163	0 0 0 0 0	EFG
1.262E-03	2.500E-01	ns	N013	0 0 0 0 2	

1039.   C$_7$H$_6$N$_2$O$_5$
2,4-Dinitroanisole
Dinitroanisole
Benzene, 1-Methoxy-2,4-dinitro-
RN: 119-27-7   **MP** (°C):   88
MW: 198.14   **BP** (°C):

Solubility (Moles/L)	Solubility (Grams/L)	Temp (°C)	Ref (#)	Evaluation (T P E A A)	Comments
7.822E-04	1.550E-01	15	D079	1 2 0 0 2	
6.863E-04	1.360E-01	50	D079	1 2 0 0 2	
2.401E-02	4.757E+00	100	D079	1 2 0 0 2	

1040.   C$_7$H$_6$N$_2$S
4-Thiocyanoaniline
Rhodan
RN: 2987-46-4   **MP** (°C):   142
MW: 150.20   **BP** (°C):

Solubility (Moles/L)	Solubility (Grams/L)	Temp (°C)	Ref (#)	Evaluation (T P E A A)	Comments
1.332E-03	2.000E-01	ns	M061	0 0 0 0 0	

1041.   C$_7$H$_6$N$_4$
4-Methylpteridine
Pteridine, 4-Methyl-
RN: 2432-21-5   **MP** (°C):   151
MW: 146.15   **BP** (°C):

Solubility (Moles/L)	Solubility (Grams/L)	Temp (°C)	Ref (#)	Evaluation (T P E A A)	Comments
3.258E-01	4.762E+01	20	A083	1 2 0 0 0	

1042.   C$_7$H$_6$N$_4$
7-Methylpteridine
Pteridine, 7-Methyl-
RN: 936-40-3   **MP** (°C):   196.5
MW: 146.15   **BP** (°C):

Solubility (Moles/L)	Solubility (Grams/L)	Temp (°C)	Ref (#)	Evaluation (T P E A A)	Comments
9.775E-01	1.429E+02	20	A083	1 2 0 0 0	

1043. $C_7H_6N_4$
2-Methylpteridine
Pteridine, 2-Methyl-
RN: 2432-20-4   **MP** (°C):   140
MW: 146.15   **BP** (°C):

Solubility (Moles/L)	Solubility (Grams/L)	Temp (°C)	Ref (#)	Evaluation (T P E A A)	Comments
6.842E-01	1.000E+02	20	A083	1 2 0 0 0	

1044. $C_7H_6N_4O$
4-Methoxypteridine
Pteridine, 4-Methoxy-
RN: 30564-38-6   **MP** (°C):   195
MW: 162.15   **BP** (°C):

Solubility (Moles/L)	Solubility (Grams/L)	Temp (°C)	Ref (#)	Evaluation (T P E A A)	Comments
7.614E-02	1.235E+01	20	A019	2 2 1 1 0	
6.167E-01	1.000E+02	100	A019	1 2 1 1 0	

1045. $C_7H_6N_4O$
7-Methoxypteridine
Pteridine, 7-Methoxy-
RN: 204443-27-6   **MP** (°C):
MW: 162.15   **BP** (°C):

Solubility (Moles/L)	Solubility (Grams/L)	Temp (°C)	Ref (#)	Evaluation (T P E A A)	Comments
1.209E-01	1.961E+01	20	A083	1 2 0 0 0	
1.233E+00	2.000E+02	100	A083	1 2 0 0 0	

1046. $C_7H_6N_4O$
4-Hydroxy-7-methylpteridine
4-Pteridinol, 7-Methyl-
RN: 34244-80-9   **MP** (°C):
MW: 162.15   **BP** (°C):

Solubility (Moles/L)	Solubility (Grams/L)	Temp (°C)	Ref (#)	Evaluation (T P E A A)	Comments
2.729E-02	4.425E+00	20	A019	2 2 1 1 2	
1.713E-01	2.778E+01	100	A019	1 2 1 1 1	

1047.   C₇H₆N₄O

4-Hydroxy-6-methylpteridine
4-Pteridinol, 6-Methyl-
RN: 16041-24-0   **MP** (°C):
MW: 162.15   **BP** (°C):

Solubility (Moles/L)	Solubility (Grams/L)	Temp (°C)	Ref (#)	Evaluation (T P E A A)	Comments
2.234E-02	3.623E+00	20	A019	2 2 1 1 2	
1.341E-01	2.174E+01	100	A019	1 2 1 1 1	

1048.   C₇H₆N₄O

3,4-Dihydro-4-keto-3-methylpteridine
3:4-Dihydro-4-keto-3-methylpteridine
RN: 24851-65-8   **MP** (°C):   286
MW: 162.15   **BP** (°C):

Solubility (Moles/L)	Solubility (Grams/L)	Temp (°C)	Ref (#)	Evaluation (T P E A A)	Comments
8.686E-02	1.408E+01	20	A019	2 2 1 1 0	
6.167E-01	1.000E+02	100	A019	1 2 1 1 0	

1049.   C₇H₆N₄O

2-Methoxypteridine
Pteridine, 2-Methoxy-
RN: 102170-44-5   **MP** (°C):   150
MW: 162.15   **BP** (°C):

Solubility (Moles/L)	Solubility (Grams/L)	Temp (°C)	Ref (#)	Evaluation (T P E A A)	Comments
7.614E-02	1.235E+01	20	A019	2 2 1 1 0	
1.233E+00	2.000E+02	100	A019	1 2 1 1 0	

1050.   C₇H₆N₄S

4-Methylthiopteridine
Pteridine, 4-(Methylthio)-
RN: 6966-78-5   **MP** (°C):   191
MW: 178.22   **BP** (°C):

Solubility (Moles/L)	Solubility (Grams/L)	Temp (°C)	Ref (#)	Evaluation (T P E A A)	Comments
4.313E-03	7.686E-01	20	A083	1 2 0 0 0	
3.100E-02	5.525E+00	100	A083	1 2 0 0 0	

1051.   C$_7$H$_6$N$_4$S
7-Methylthiopteridine
Pteridine, 7-(Methylthio)-
RN: 204443-30-1   **MP** (°C):
MW: 178.22   **BP** (°C):

Solubility (Moles/L)	Solubility (Grams/L)	Temp (°C)	Ref (#)	Evaluation (T P E A A)	Comments
2.792E-02	4.975E+00	20	A083	1 2 0 0 0	
1.439E-01	2.564E+01	100	A083	1 2 0 0 0	

1052.   C$_7$H$_6$N$_4$S
4-Mercapto-7-methylpteridine
4-Pteridinethiol, 7-Methyl-
RN: 98550-33-5   **MP** (°C):
MW: 178.22   **BP** (°C):

Solubility (Moles/L)	Solubility (Grams/L)	Temp (°C)	Ref (#)	Evaluation (T P E A A)	Comments
3.738E-03	6.662E-01	100	A083	1 2 0 0 0	

1053.   C$_7$H$_6$N$_4$S
2-Methylthiopteridine
Pteridine, 2-(Methylthio)-
RN: 16878-77-6   **MP** (°C):   136
MW: 178.22   **BP** (°C):

Solubility (Moles/L)	Solubility (Grams/L)	Temp (°C)	Ref (#)	Evaluation (T P E A A)	Comments
1.748E-02	3.115E+00	20	A083	1 2 0 0 0	
1.369E-01	2.439E+01	100	A083	1 2 0 0 0	

1054.   C$_7$H$_6$O
Benzaldehyde
Benzaldehyd
RN: 100-52-7   **MP** (°C):   -55
MW: 106.13   **BP** (°C):   179

Solubility (Moles/L)	Solubility (Grams/L)	Temp (°C)	Ref (#)	Evaluation (T P E A A)	Comments
3.251E-02	3.450E+00	20	C008	1 2 2 0 2	
2.827E-02	3.000E+00	20	F300	1 0 0 0 0	
3.754E-02	3.984E+00	25	B019	1 0 1 2 0	
3.754E-02	3.984E+00	25	B092	2 1 1 1 1	
6.549E-02	6.950E+00	25	C005	2 2 2 2 2	average

Solubility (Moles/L)	Solubility (Grams/L)	Temp (°C)	Ref (#)	Evaluation (T P E A A)	Comments
3.289E-02	3.490E+00	25	C008	1 2 2 0 2	
6.170E-02	6.548E+00	25	M017	1 2 0 1 2	
3.741E-02	3.970E+00	30	C008	1 2 2 0 2	
2.110E-02	2.239E+00	37	E028	1 0 1 1 2	
8.960E-02	9.509E+00	60	B092	2 0 1 1 1	

1055. C₇H₆O₂
Benzoic Acid
Benzenecarboxylic Acid
Benzoesaeure
RN: 65-85-0 **MP** (°C): 122
MW: 122.12 **BP** (°C): 249

Solubility (Moles/L)	Solubility (Grams/L)	Temp (°C)	Ref (#)	Evaluation (T P E A A)	Comments
1.390E-02	1.697E+00	0	F302	1 0 0 0 2	
1.390E-02	1.697E+00	0	M043	1 0 0 0 1	
1.720E-02	2.100E+00	10	F300	1 0 0 0 1	
1.716E-02	2.096E+00	10	F302	1 0 0 0 2	
1.634E-02	1.996E+00	10	M043	1 0 0 0 1	
2.010E-02	2.455E+00	15	P329	2 1 1 2 2	
1.982E-02	2.421E+00	15.5	K062	2 0 1 1 2	
2.200E-02	2.687E+00	17	B109	1 0 0 0 2	unit assumed, *sic*
2.237E-02	2.732E+00	17.7	K062	2 0 1 1 2	
2.260E-02	2.760E+00	18	B109	1 0 0 0 2	unit assumed, *sic*
2.211E-02	2.700E+00	18	F071	1 1 2 1 2	
2.100E-02	2.565E+00	18	H009	2 1 2 2 0	EFG, 0.01N HCl
2.211E-02	2.700E+00	18	H080	1 0 0 0 2	
2.257E-02	2.756E+00	18	L050	2 0 1 2 2	
2.211E-02	2.700E+00	18	M344	1 0 0 0 2	
2.308E-02	2.819E+00	19.0	K062	2 0 1 1 2	average of 2
2.368E-02	2.892E+00	20	D041	1 0 0 0 1	
2.339E-02	2.857E+00	20	F069	2 2 2 2 2	
2.375E-02	2.900E+00	20	F300	1 0 0 0 1	
2.368E-02	2.892E+00	20	F302	1 0 0 0 2	
2.200E-02	2.686E+00	20	M038	2 2 1 1 2	
2.368E-02	2.892E+00	20	M043	1 0 0 0 1	
2.457E-02	3.000E+00	20	M049	1 0 0 0 1	
2.400E-02	2.931E+00	20	P329	2 1 1 2 2	
2.825E-02	3.450E+00	20	W026	1 0 1 1 1	average of 2
2.540E-02	3.102E+00	22	E045	2 0 1 1 2	
2.605E-02	3.181E+00	23	E045	2 0 1 1 2	
2.807E-02	3.428E+00	24.6	W029	1 2 1 1 2	
2.449E-02	2.991E+00	25	B019	1 0 1 2 0	
2.751E-02	3.359E+00	25	B085	2 1 1 1 2	
2.683E-02	3.277E+00	25	B097	2 2 1 1 2	0.01M sodium benzoate
2.800E-02	3.420E+00	25	B128	1 0 1 1 2	
2.768E-02	3.381E+00	25	B302	1 0 0 0 0	pH 2.0
2.805E-02	3.426E+00	25	D058	1 0 1 1 2	
2.746E-02	3.354E+00	25	E045	2 0 1 1 2	

2.810E-02	3.432E+00	25	F001	1 0 1 2 2	
2.784E-02	3.400E+00	25	F300	1 0 0 0 1	
2.800E-02	3.419E+00	25	H009	2 1 2 2 0	EFG, 0.01N HCl
2.784E-02	3.400E+00	25	H015	1 0 0 0 1	
2.251E-03	2.749E-01	25	H060	2 0 2 0 2	*sic*
2.760E-02	3.371E+00	25	H071	2 2 2 1 2	
2.800E-02	3.419E+00	25	H084	1 0 0 0 1	
2.760E-02	3.371E+00	25	K005	1 0 0 1 2	
2.727E-02	3.330E+00	25	K047	1 2 1 2 2	
2.760E-02	3.371E+00	25	K057	2 2 1 1 2	
2.775E-02	3.389E+00	25	K064	2 2 2 1 2	
2.781E-02	3.396E+00	25	L048	1 2 2 1 2	
2.780E-02	3.395E+00	25	L050	2 0 1 2 2	
2.596E-02	3.170E+00	25	L338	1 0 1 1 2	
2.619E-02	3.199E+00	25	M038	2 2 1 1 2	
2.702E-02	3.300E+00	25	M049	1 0 0 0 1	
2.790E-02	3.407E+00	25	M116	2 1 1 1 2	
2.160E-02	2.638E+00	25	M149	2 0 2 2 2	intrinsic
2.900E-02	3.542E+00	25	O007	1 0 2 1 2	
2.268E-02	2.770E+00	25	P037	2 0 1 1 2	
2.807E-02	3.428E+00	25	P314	2 2 1 2 2	
8.820E+00	1.077E+03	25	P329	2 1 1 2 2	
2.793E-02	3.411E+00	25	R016	1 0 1 1 2	
2.781E-02	3.396E+00	25.0	K062	2 0 1 1 2	average of 2
2.700E-02	3.297E+00	25.00	M135	1 2 1 1 2	0.01N sodium benzoate
2.781E-02	3.396E+00	25.2	C096	1 0 0 1 2	
2.833E-02	3.460E+00	26	E045	2 0 1 1 2	
2.890E-02	3.529E+00	26.4	P043	2 0 1 1 2	
3.439E-02	4.200E+00	26.70	L095	2 2 1 1 2	
2.936E-02	3.586E+00	27	E045	2 0 1 1 2	
3.146E-02	3.842E+00	28	D050	1 2 1 2 2	
3.147E-02	3.843E+00	30	B109	1 0 0 0 2	unit assumed, *sic*
3.204E-02	3.913E+00	30	B109	1 0 0 0 2	unit assumed, *sic*
3.306E-02	4.037E+00	30	B118	1 0 0 0 2	
3.000E-02	3.664E+00	30	B142	2 0 1 1 0	EFG, 0.1N H$_2$SO$_4$
3.000E-02	3.664E+00	30	C077	1 0 2 2 0	
3.319E-02	4.054E+00	30	D033	2 2 1 2 2	
3.302E-02	4.033E+00	30	D061	1 0 0 0 2	
2.915E-02	3.560E+00	30	F005	1 2 2 2 2	
3.425E-02	4.182E+00	30	F302	1 0 0 0 2	
3.110E-02	3.799E+00	30	M038	2 2 1 1 2	
3.262E-02	3.984E+00	30	M043	1 0 0 0 1	
3.302E-02	4.033E+00	30	S204	2 0 1 0 2	
3.439E-02	4.200E+00	30	W026	1 0 1 1 1	average of 2
3.216E-02	3.927E+00	30.0	K062	2 0 1 1 2	average of 2
3.400E-02	4.152E+00	31	H009	2 1 2 2 0	EFG, 0.01N HCl
3.873E-02	4.730E+00	35	G052	2 1 1 1 2	
3.711E-02	4.532E+00	35	M038	2 2 1 1 2	

4.010E-02	4.897E+00	35	O007	1 0 2 1 2	
3.772E-02	4.607E+00	35	S204	2 0 1 0 2	
3.960E-02	4.836E+00	35.0	K062	2 0 1 1 2	
3.800E-02	4.641E+00	35.00	M135	1 2 1 1 2	0.01N sodium benzoate
4.201E-02	5.131E+00	37	B171	2 0 1 1 2	
3.611E-02	4.410E+00	37	F005	1 2 2 2 2	
4.200E-02	5.129E+00	37	H009	2 1 2 2 0	EFG, 0.01N HCl
3.734E-02	4.560E+00	37	M360	1 2 1 1 2	
4.528E-02	5.529E+00	40	D033	2 2 1 2 2	
4.884E-02	5.964E+00	40	F302	1 0 0 0 1	
4.376E-02	5.345E+00	40	M038	2 2 1 1 2	
4.560E-02	5.569E+00	40	M043	1 0 0 0 1	
4.424E-02	5.403E+00	40	S204	2 0 1 0 2	
5.110E-02	6.241E+00	42.4	W029	1 2 1 1 2	
4.774E-02	5.830E+00	45	F005	1 2 2 2 2	
5.000E-02	6.106E+00	45	H009	2 1 2 2 0	EFG, 0.01N HCl
5.282E-02	6.451E+00	45	M038	2 2 1 1 2	
5.254E-02	6.417E+00	45	S204	2 0 1 0 2	
5.324E-02	6.502E+00	45.0	K062	2 0 1 1 2	
5.500E-02	6.717E+00	45.00	M135	1 2 1 1 2	0.01N sodium benzoate
5.463E-02	6.672E+00	45.3	S124	1 0 0 1 1	
6.878E-02	8.400E+00	50	F300	1 0 0 0 1	
6.901E-02	8.428E+00	50	F302	1 0 0 0 2	
2.107E-02	2.573E+00	50	L006	1 0 0 0 2	
6.237E-02	7.617E+00	50	S204	2 0 1 0 2	
8.032E-02	9.809E+00	53.8	S124	1 0 0 1 2	
7.048E-02	8.607E+00	55	S204	2 0 1 0 2	
8.300E-02	1.014E+01	55.40	M135	1 2 1 1 2	0.01N sodium benzoate
8.853E-02	1.081E+01	57.8	W029	1 2 1 1 2	
9.710E-02	1.186E+01	60	F302	1 0 0 0 2	
9.550E-02	1.166E+01	60	L047	1 1 2 1 2	
9.390E-02	1.147E+01	60	M043	1 0 0 0 2	
1.000E-01	1.221E+01	60.20	M135	1 2 1 1 2	0.01N sodium benzoate
1.129E-01	1.378E+01	62.5	S124	1 0 0 1 2	
1.190E-01	1.453E+01	64.60	M135	1 2 1 1 2	0.01N sodium benzoate
1.390E-01	1.698E+01	68.50	M135	1 2 1 1 2	0.01N sodium benzoate
1.527E-01	1.864E+01	69.4	S124	1 0 0 1 2	
1.424E-01	1.739E+01	70	F302	1 0 0 0 2	
1.658E-01	2.025E+01	74.1	W029	1 2 1 1 2	
1.870E-01	2.284E+01	75.10	M135	1 2 1 1 2	0.01N sodium benzoate
2.242E-01	2.739E+01	79.0	S124	1 0 0 1 2	
2.210E-01	2.699E+01	79.30	M135	1 2 1 1 2	0.01N sodium benzoate
2.192E-01	2.676E+01	80	F302	1 0 0 0 2	
2.168E-01	2.648E+01	80	M043	1 0 0 0 2	
2.540E-01	3.102E+01	82.10	M135	1 2 1 1 2	0.01N sodium benzoate
2.567E-01	3.135E+01	82.3	S124	1 0 0 1 2	
2.485E-01	3.035E+01	83.1	W029	1 2 1 1 2	
3.124E-01	3.815E+01	88.3	W029	1 2 1 1 2	
4.211E-01	5.142E+01	88.6	S124	1 0 0 1 2	
3.550E-01	4.335E+01	88.60	M135	1 2 1 1 2	0.01N sodium benzoate

3.564E-01	4.352E+01	90	F302	1 0 0 0 2	
4.342E-01	5.302E+01	91.5	W029	1 2 1 1 2	average of 3
5.214E-01	6.367E+01	95	D041	1 0 0 0 1	
5.208E-01	6.360E+01	95	F300	1 0 0 0 2	
5.214E-01	6.367E+01	95	F302	1 0 0 0 2	
4.977E-01	6.078E+01	95.3	W029	1 2 1 1 2	
5.493E-01	6.708E+01	98.6	W029	1 2 1 1 2	
4.547E-01	5.553E+01	100	M043	1 0 0 0 2	
8.241E-01	1.006E+02	109.4	W029	1 2 1 1 2	
1.399E+00	1.709E+02	116.1	W029	1 2 1 1 2	
2.594E+00	3.168E+02	116.3	W029	1 2 1 1 2	
2.001E+00	2.444E+02	117.2	W029	1 2 1 1 2	
9.000E-04	1.099E-01	ns	D037	1 1 1 1 0	pH 3.0, intrinsic

1056. $C_7H_6O_2$
Salicylaldehyde
Salicylaldehyd
RN: 90-02-8   **MP** (°C):   -7
MW: 122.12   **BP** (°C):   197

Solubility (Moles/L)	Solubility (Grams/L)	Temp (°C)	Ref (#)	Evaluation (T P E A A)	Comments
6.614E-04	8.077E-02	25	K129	2 1 2 2 2	
1.392E-01	1.700E+01	86	F300	1 0 0 0 1	

1057. $C_7H_6O_2$
m-Hydroxybenzaldehyde
3-Hydroxy-benzaldehyd
RN: 100-83-4   **MP** (°C):   104
MW: 122.12   **BP** (°C):

Solubility (Moles/L)	Solubility (Grams/L)	Temp (°C)	Ref (#)	Evaluation (T P E A A)	Comments
2.252E-01	2.750E+01	43	F300	1 0 0 0 2	

1058. $C_7H_6O_2$
p-Hydroxybenzaldehyde
4-Hydroxy-benzaldehyd
RN: 123-08-0   **MP** (°C):   213.5
MW: 122.12   **BP** (°C):

Solubility (Moles/L)	Solubility (Grams/L)	Temp (°C)	Ref (#)	Evaluation (T P E A A)	Comments
1.056E-01	1.290E+01	30	F300	1 0 0 0 2	

1059. $C_7H_6O_3$

β-2-Furyncrylic Acid
β-2-Furylacrylic Acid
β-Furyl-(2)-acrylsaeure

RN: 539-47-9   **MP** (°C):   143
MW: 138.12   **BP** (°C):   286

Solubility (Moles/L)	Solubility (Grams/L)	Temp (°C)	Ref (#)	Evaluation (T P E A A)	Comments
1.448E-02	2.000E+00	20	F300	1 0 0 0 0	

1060. $C_7H_6O_3$

Salicylic Acid
2-Hydroxybenzoic Acid
o-Hydroxybenzoic Acid

RN: 69-72-7   **MP** (°C):   158
MW: 138.12   **BP** (°C):   211

Solubility (Moles/L)	Solubility (Grams/L)	Temp (°C)	Ref (#)	Evaluation (T P E A A)	Comments
6.799E-03	9.391E-01	0	C083	1 2 1 1 2	
5.792E-03	8.000E-01	0	F300	1 0 0 0 0	
9.400E-03	1.298E+00	0	M043	1 0 0 0 0	
9.400E-03	1.298E+00	0	M043	1 0 0 0 1	
1.108E-02	1.531E+00	9.99	Λ341	2 0 2 2 2	
9.472E-03	1.308E+00	10	B074	1 2 1 2 2	
8.688E-03	1.200E+00	10	F300	1 0 0 0 1	
1.084E-02	1.498E+00	10	M043	1 0 0 0 1	
1.084E-02	1.498E+00	10	M043	1 0 0 0 0	
9.327E-03	1.288E+00	10	W044	1 0 1 0 2	
1.009E-02	1.393E+00	12.1	W044	1 0 1 0 2	
1.207E-02	1.667E+00	14.5	D061	1 0 0 0 2	
1.209E-02	1.670E+00	14.50	B118	1 0 0 0 2	unit assumed
1.028E-02	1.420E+00	15	H022	1 2 2 2 2	
1.258E-02	1.737E+00	17	K046	1 0 0 0 2	spray-dried product
1.330E-02	1.837E+00	20	B074	1 2 1 2 2	
1.303E-02	1.800E+00	20	F071	1 1 2 1 2	
1.303E-02	1.800E+00	20	F300	1 0 0 0 1	
1.303E-02	1.800E+00	20	H080	1 0 0 0 2	
1.296E-02	1.790E+00	20	K047	1 2 1 2 2	
1.445E-02	1.996E+00	20	M043	1 0 0 0 0	
1.445E-02	1.996E+00	20	M043	1 0 0 0 1	
1.445E-02	1.996E+00	20	M107	2 2 1 1 0	EFG
1.303E-02	1.800E+00	20	M344	1 0 0 0 2	
1.593E-02	2.200E+00	20	W026	1 0 1 1 1	average of 2
1.330E-02	1.837E+00	20	W044	1 0 1 0 2	
1.520E-02	2.100E+00	21	B331	1 2 2 1 0	pH 7.4
1.390E-02	1.920E+00	22	E045	2 0 1 1 2	
1.470E-02	2.030E+00	23	E045	2 0 1 1 2	

1.474E-02	2.036E+00	23.0	W044	1 0 1 0 2	
1.550E-02	2.141E+00	24	E045	2 0 1 1 2	
1.847E-02	2.551E+00	24.99	A341	2 0 2 2 2	
1.590E-02	2.196E+00	25	B090	1 1 1 1 2	
1.230E-02	1.699E+00	25	B090	1 1 1 1 2	intrinsic
1.633E-02	2.255E+00	25	C083	1 2 1 1 2	
1.630E-02	2.251E+00	25	E045	2 0 1 1 2	
1.593E-02	2.200E+00	25	H007	1 0 2 2 1	
1.620E-02	2.238E+00	25	H084	1 0 0 0 2	
1.084E-02	1.498E+00	25	H129	1 0 0 1 0	
1.613E-02	2.228E+00	25	K040	1 0 2 1 2	
1.634E-02	2.257E+00	25	K053	2 2 2 2 2	
1.620E-02	2.238E+00	25	K057	2 2 1 1 2	
1.601E-02	2.211E+00	25	L050	2 0 1 2 2	
1.680E-02	2.320E+00	25	O007	1 0 2 1 2	
1.621E-02	2.239E+00	25	P314	2 2 1 2 2	
1.491E-02	2.059E+00	25.50	A012	2 2 2 2 2	
1.700E-02	2.348E+00	26	E045	2 0 1 1 2	
1.780E-02	2.459E+00	27	E045	2 0 1 1 2	
1.746E-02	2.411E+00	27	K046	1 0 0 0 2	spray-dried product
1.728E-02	2.387E+00	28	D050	1 2 1 2 2	
1.784E-02	2.464E+00	28.1	W044	1 0 1 0 2	
1.360E-02	1.878E+00	30	A065	2 0 2 2 1	
1.885E-02	2.603E+00	30	B074	1 2 1 2 2	
1.987E-02	2.745E+00	30	B118	1 0 0 0 2	unit assumed
1.750E-02	2.417E+00	30	B142	2 0 1 1 0	EFG, 0.1N H_2SO_4
1.800E-02	2.486E+00	30	C077	1 0 2 2 0	
1.986E-02	2.743E+00	30	D061	1 0 0 0 2	
1.426E-02	1.970E+00	30	F005	1 2 2 2 2	
1.796E-02	2.481E+00	30	H022	1 2 2 2 2	
1.700E-02	2.348E+00	30	K020	1 0 1 1 0	EFG
1.868E-02	2.580E+00	30	K047	1 2 1 2 2	
2.022E-02	2.792E+00	30	M043	1 0 0 0 1	
2.022E-02	2.792E+00	30	M043	1 0 0 0 0	
2.165E-02	2.991E+00	30	M107	2 2 1 1 0	EFG
2.244E-02	3.100E+00	30	W026	1 0 1 1 2	average of 2
1.906E-02	2.633E+00	30	W044	1 0 1 0 2	
2.172E-02	3.000E+00	30.6	P014	2 1 2 2 0	
2.442E-02	3.373E+00	33.99	A341	2 0 2 2 2	
2.201E-02	3.041E+00	34.4	W044	1 0 1 0 2	
2.273E-02	3.140E+00	35	K047	1 2 1 2 2	
2.390E-02	3.301E+00	35	O007	1 0 2 1 2	
1.332E-02	1.840E+00	37	B171	2 0 1 1 2	
1.861E-02	2.570E+00	37	C079	1 0 0 0 2	
1.897E-02	2.620E+00	37	F005	1 2 2 2 2	
2.452E-02	3.386E+00	37	K046	1 0 0 0 2	spray-dried product
2.590E-02	3.577E+00	38.7	W044	1 0 1 0 2	
2.848E-02	3.934E+00	40	B074	1 2 1 2 2	

2.679E-02	3.700E+00	40	F300	1 0 0 0 1	
2.672E-02	3.690E+00	40	K047	1 2 1 2 2	
3.028E-02	4.182E+00	40	M043	1 0 0 0 1	
3.028E-02	4.182E+00	40	M043	1 0 0 0 0	
2.884E-02	3.984E+00	40	M107	2 2 1 1 0	EFG
2.719E-02	3.756E+00	40	W044	1 0 1 0 2	
3.167E-02	4.374E+00	43.99	A341	2 0 2 2 2	
3.743E-02	5.170E+00	44.99	A341	2 0 2 2 2	
2.462E-02	3.400E+00	45	F005	1 2 2 2 2	
3.714E-02	5.130E+00	46.99	A341	2 0 2 2 2	
3.562E-02	4.921E+00	47	K046	1 0 0 0 2	spray-dried product
3.681E-02	5.084E+00	48.6	W044	1 0 1 0 2	
4.102E-02	5.665E+00	49.99	A341	2 0 2 2 2	
4.261E-02	5.885E+00	50	B074	1 2 1 2 2	
3.889E-02	5.371E+00	50	W044	1 0 1 0 2	
4.337E-02	5.991E+00	50.99	A341	2 0 2 2 2	
4.677E-02	6.461E+00	51.99	A341	2 0 2 2 2	
5.151E-02	7.115E+00	53.99	A341	2 0 2 2 2	
5.319E-02	7.347E+00	54.99	A341	2 0 2 2 2	
4.947E-02	6.833E+00	56.0	W044	1 0 1 0 2	
6.104E-02	8.431E+00	57.49	A341	2 0 2 2 2	
6.202E-02	8.566E+00	60	B074	1 2 1 2 2	
6.009E-02	8.300E+00	60	F300	1 0 0 0 1	
6.529E-02	9.018E+00	60	M043	1 0 0 0 1	
6.529E-02	9.018E+00	60	M043	1 0 0 0 0	
5.888E-02	8.133E+00	60	W044	1 0 1 0 2	
7.184E-02	9.922E+00	61.49	A341	2 0 2 2 2	
7.140E-02	9.862E+00	64.0	W044	1 0 1 0 2	
8.184E-02	1.130E+01	65.99	A341	2 0 2 2 2	
8.373E-02	1.156E+01	66.0	W044	1 0 1 0 2	
1.252E-01	1.730E+01	75.0	W044	1 0 1 0 2	
1.499E-01	2.070E+01	80	F300	1 0 0 0 2	
1.600E-01	2.210E+01	80	M043	1 0 0 0 0	
1.600E-01	2.210E+01	80	M043	1 0 0 0 2	
5.437E-01	7.510E+01	100	M043	1 0 0 0 2	
5.437E-01	7.510E+01	100	M043	1 0 0 0 0	
1.598E-02	2.207E+00	ns	O003	0 2 1 1 2	

1061. $C_7H_6O_3$
m-Hydroxybenzoic Acid
3-Hydroxy-benzoesaeure
3-Hydroxybenzoic Acid
m-Hydroxybenzoicacid

RN: 99-06-9   **MP** (°C):   202
MW: 138.12   **BP** (°C):

Solubility (Moles/L)	Solubility (Grams/L)	Temp (°C)	Ref (#)	Evaluation (T P E A A)	Comments
2.525E-02	3.488E+00	0	M043	1 0 0 0 1	
3.960E-02	5.470E+00	10	M043	1 0 0 0 1	
4.804E-02	6.636E+00	13.3	W044	1 0 1 0 2	
5.068E-02	7.000E+00	15	F300	1 0 0 0 1	
4.477E-02	6.184E+00	15	H022	1 2 2 2 2	
6.052E-02	8.360E+00	18.8	W044	1 0 1 0 2	
6.173E-02	8.527E+00	20	M043	1 0 0 0 1	
4.318E-02	5.964E+00	20	M107	2 2 1 1 0	EFG
7.551E-02	1.043E+01	24.3	W044	1 0 1 0 2	
5.249E-02	7.250E+00	25.50	A012	2 2 2 2 2	
7.800E-03	1.077E+00	30	A065	2 0 2 2 1	
8.600E-02	1.188E+01	30	C077	1 0 2 2 0	
8.800E-02	1.215E+01	30	H019	1 0 2 0 0	
8.300E-02	1.146E+01	30	H021	1 2 1 1 0	EFG
9.291E-02	1.283E+01	30	M043	1 0 0 0 1	
6.813E-02	9.411E+00	30	M107	2 2 1 1 0	EFG
9.552E-02	1.319E+01	30	W044	1 0 1 0 2	
9.855E-02	1.361E+01	30.9	W044	1 0 1 0 2	
1.271E-01	1.756E+01	36.2	W044	1 0 1 0 2	
1.420E-01	1.961E+01	40	M043	1 0 0 0 1	
1.105E-01	1.526E+01	40	M107	2 2 1 1 0	EFG
2.809E-01	3.880E+01	50	F300	1 0 0 0 1	
2.222E-01	3.070E+01	51.0	W044	1 0 1 0 2	
3.118E-01	4.306E+01	60	M043	1 0 0 0 1	
7.987E-01	1.103E+02	80	M043	1 0 0 0 2	
2.678E+00	3.699E+02	100	M043	1 0 0 0 2	
1.810E-02	2.500E+00	ns	B361	0 0 0 2 2	

1062.   C₇H₆O₃ — $C_7H_6O_3$

p-Hydroxybenzoic Acid
4-Hydroxy-benzoesaeure
4-Hydroxybenzoic Acid
p-Hydroxybenzoicacid
4-Hydroxybenzenecarboxylic Acid

RN:	99-96-7	**MP** (°C):	214.5
MW:	138.12	**BP** (°C):	

Solubility (Moles/L)	Solubility (Grams/L)	Temp (°C)	Ref (#)	Evaluation (T P E A A)	Comments
1.805E-02	2.494E+00	0	M043	1 0 0 0 1	
2.525E-02	3.488E+00	10	M043	1 0 0 0 1	
2.216E-02	3.061E+00	12.7	W044	1 0 1 0 2	
5.746E-02	7.937E+00	15	D041	1 0 0 0 0	
3.186E-02	4.400E+00	15	F300	1 0 0 0 1	
2.624E-02	3.624E+00	15	H022	1 2 2 2 2	
3.470E-02	4.793E+00	20	C006	1 2 1 1 2	
3.817E-02	5.272E+00	20	M043	1 0 0 0 1	
3.602E-02	4.975E+00	20	M107	2 2 1 1 0	EFG
3.545E-02	4.896E+00	20.9	W044	1 0 1 0 2	
3.545E-02	4.896E+00	25	D081	1 1 2 1 2	
6.580E-02	9.089E+00	25	D339	1 0 1 1 2	
4.634E-02	6.400E+00	25	H007	1 0 2 2 1	
3.318E-02	4.583E+00	25	M334	1 2 1 1 2	
6.241E-02	8.620E+00	25	N023	1 2 2 1 2	anhydrate
4.322E-02	5.970E+00	25	N023	1 2 2 1 2	hydrate
3.873E-02	5.350E+00	25.50	A012	2 2 2 2 2	
5.400E-02	7.459E+00	30	A065	2 0 2 2 1	
4.800E-02	6.630E+00	30	C077	1 0 2 2 0	
5.500E-02	7.597E+00	30	H019	1 0 2 0 0	
5.421E-02	7.488E+00	30	H022	1 2 2 2 2	
5.500E-02	7.597E+00	30	K020	1 0 1 1 0	EFG
5.746E-02	7.937E+00	30	M043	1 0 0 0 1	
5.746E-02	7.937E+00	30	M107	2 2 1 1 0	EFG
7.790E-02	1.076E+01	30	N023	1 2 2 1 2	anhydrate
5.538E-02	7.650E+00	30	N023	1 2 2 1 2	hydrate
5.496E-02	7.592E+00	30	W044	1 0 1 0 2	
7.076E-02	9.774E+00	34.4	W044	1 0 1 0 2	
7.247E-02	1.001E+01	35	N023	1 2 2 1 2	hydrate
9.781E-02	1.351E+01	35	N023	1 2 2 1 2	anhydrate
1.231E-01	1.700E+01	37	B171	2 0 1 1 2	
8.663E-02	1.197E+01	39.4	W044	1 0 1 0 2	
8.938E-02	1.235E+01	40	M043	1 0 0 0 2	
9.996E-02	1.381E+01	40	M107	2 2 1 1 0	EFG
1.203E-01	1.662E+01	40	N023	1 2 2 1 2	anhydrate
9.339E-02	1.290E+01	40	N023	1 2 2 1 2	hydrate
1.291E-01	1.783E+01	46.0	W044	1 0 1 0 2	
1.931E-01	2.667E+01	54.6	W044	1 0 1 0 2	
2.978E-01	4.114E+01	60	M043	1 0 0 0 2	

1.835E-01	2.534E+01	75	D041	1 0 0 0 1	
8.723E-01	1.205E+02	80	M043	1 0 0 0 2	
1.875E+00	2.590E+02	100	F300	1 0 0 0 2	
2.410E+00	3.329E+02	100	M043	1 0 0 0 2	

1063. $C_7H_6O_3$
Protocatechualdehyde
3,4-Dihydroxy-benzaldehyd
RN: 139-85-5   **MP** (°C):
MW: 138.12   **BP** (°C):

Solubility (Moles/L)	Solubility (Grams/L)	Temp (°C)	Ref (#)	Evaluation (T P E A A)	Comments
3.620E-01	5.000E+01	20	F300	1 0 0 0 0	
~1.88E+00	~2.60E+02	100	F300	1 0 0 0 0	

1064. $C_7H_6O_4$
Gentisic Acid
2,5-Dihydroxy-benzoesaeure
2,5-Dihydroxybenzoic Acid
2,5-Dihydroxybenzoicacid
Hydroquinonecarboxylic Acid
RN: 490-79-9   **MP** (°C):   205
MW: 154.12   **BP** (°C):

Solubility (Moles/L)	Solubility (Grams/L)	Temp (°C)	Ref (#)	Evaluation (T P E A A)	Comments
1.427E-01	2.200E+01	25	H007	1 0 2 2 1	

1065. $C_7H_6O_4$
Protocatechuic Acid
3,4-Dihydroxy-benzoesaeure
3,4-Dihydroxybenzoic Acid
RN: 99-50-3   **MP** (°C):
MW: 154.12   **BP** (°C):

Solubility (Moles/L)	Solubility (Grams/L)	Temp (°C)	Ref (#)	Evaluation (T P E A A)	Comments
1.181E-01	1.820E+01	14	F300	1 0 0 0 2	
1.440E+00	2.220E+02	80	F300	1 0 0 0 2	

1066. C$_7$H$_6$O$_4$
β-Resorcyclic Acid
2,4-Dihydroxy-benzoesaeure
2,4-Dihydroxybenzoic Acid
2,4-Dihydroxybenzoicacid
β-Resorcylic Acid
4-Hydroxysalicylic Acid
RN: 89-86-1   **MP** (°C): 225
MW: 154.12   **BP** (°C):

Solubility (Moles/L)	Solubility (Grams/L)	Temp (°C)	Ref (#)	Evaluation (T P E A A)	Comments
3.893E-02	6.000E+00	25	H007	1 0 2 2 1	

1067. C$_7$H$_6$O$_4$
2,6-Dihydroxybenzoic Acid
2,6-Dihydroxy-benzoesaeure
γ-Resorcylic Acid
RN: 303-07-1   **MP** (°C):
MW: 154.12   **BP** (°C):

Solubility (Moles/L)	Solubility (Grams/L)	Temp (°C)	Ref (#)	Evaluation (T P E A A)	Comments
6.200E-02	9.556E+00	ns	C014	0 0 0 1 1	

1068. C$_7$H$_6$O$_5$
Gallic Acid
3,4,5-Trihydroxybenzoesaeure
Gallussaeure
RN: 149-91-7   **MP** (°C): 250
MW: 170.12   **BP** (°C):

Solubility (Moles/L)	Solubility (Grams/L)	Temp (°C)	Ref (#)	Evaluation (T P E A A)	Comments
6.995E-02	1.190E+01	20	F300	1 0 0 0 2	
1.505E+00	2.561E+02	100	F300	1 0 0 0 2	

1069. C$_7$H$_6$O$_5$
2,3,4-Trihydroxybenzoic Acid
2,3,4-Trihydroxybenzoesaeure
RN: 610-02-6   **MP** (°C):
MW: 170.12   **BP** (°C):

Solubility (Moles/L)	Solubility (Grams/L)	Temp (°C)	Ref (#)	Evaluation (T P E A A)	Comments
5.878E-03	1.000E+00	12.50	F300	1 0 0 0 0	

1070.   C₇H₇Br
m-Bromotoluene
3-Bromotoluene
3-Methyl-1-bromobenzene
1-Bromo-3-methylbenzene
3-Bromo-1-methylbenzene
3-Methylphenyl Bromide

RN: 591-17-3   **MP** (°C):   -39.8
MW: 171.04   **BP** (°C):   183.7

Solubility (Moles/L)	Solubility (Grams/L)	Temp (°C)	Ref (#)	Evaluation (T P E A A)	Comments
3.000E-04	5.131E-02	ns	O013	0 1 0 1 0	

1071.   C₇H₇Cl
m-Chlorotoluene
3-Chlorotoluene
1-Chloro-3-methylbenzene
m-Tolyl Chloride

RN: 108-41-8   **MP** (°C):   -48
MW: 126.59   **BP** (°C):   161.8

Solubility (Moles/L)	Solubility (Grams/L)	Temp (°C)	Ref (#)	Evaluation (T P E A A)	Comments
3.000E-04	3.798E-02	ns	O013	0 1 0 1 0	

1072.   C₇H₇Cl
p-Chlorotoluene
4-Chlorotoluene
p-Tolyl Chloride
4-Chloro-1-methyl-benzene
PCT
1-Chloro-4-methylbenzene

RN: 106-43-4   **MP** (°C):   8
MW: 126.59   **BP** (°C):   162.0

Solubility (Moles/L)	Solubility (Grams/L)	Temp (°C)	Ref (#)	Evaluation (T P E A A)	Comments
8.415E-04	1.065E-01	20	H118	1 1 1 1 2	
1.084E-03	1.372E-01	20	H301	2 0 2 2 2	
3.000E-04	3.798E-02	ns	O013	0 1 0 1 0	

1073. C$_7$H$_7$Cl
o-Chlorotoluene
2-Chlorotoluene
2-Chloro-1-methylbenzene
2-Methylchlorobenzene
1-Methyl-2-chlorobenzene
OCT

RN: 95-49-8   **MP** (°C): -36
MW: 126.59   **BP** (°C): 159.0

Solubility (Moles/L)	Solubility (Grams/L)	Temp (°C)	Ref (#)	Evaluation (T P E A A)	Comments
3.000E-04	3.798E-02	ns	O013	0 1 0 1 0	

1074. C$_7$H$_7$ClO
Chlorocresol
3-Methyl-4-chlorophenol
4-Chloro-3-cresol
6-Chloro-3-hydroxytoluene
3-Methyl-4-chloro-phenol-
Phenol, 4-Chloro-3-methyl-

RN: 59-50-7   **MP** (°C): 67
MW: 142.59   **BP** (°C):

Solubility (Moles/L)	Solubility (Grams/L)	Temp (°C)	Ref (#)	Evaluation (T P E A A)	Comments
2.800E-02	3.992E+00	25	B316	1 0 2 1 1	
3.489E-02	4.975E+00	25	R041	1 0 2 1 1	
3.647E-02	5.200E+00	ns	G024	0 0 0 0 2	

1075. C$_7$H$_7$ClO
4-Chloroanisole
p-Chloroanisole
1-Chloro-4-methoxybenzene

RN: 623-12-1   **MP** (°C): -18
MW: 142.59   **BP** (°C):

Solubility (Moles/L)	Solubility (Grams/L)	Temp (°C)	Ref (#)	Evaluation (T P E A A)	Comments
1.662E-03	2.370E-01	25	L348	1 2 2 1 2	

1076.   C₇H₇ClO

3-Chloroanisole

m-Chloroanisole

1-Chloro-3-methoxybenzene

RN: 2845-89-8   **MP** (°C):   <25

MW: 142.59   **BP** (°C):

Solubility (Moles/L)	Solubility (Grams/L)	Temp (°C)	Ref (#)	Evaluation (T P E A A)	Comments
1.648E-03	2.350E-01	25	L348	1 2 2 1 2	

1077.   C₇H₇ClO

2-Chloroanisole

o-Chloroanisole

RN: 766-51-8   **MP** (°C):   -27

MW: 142.59   **BP** (°C):   196

Solubility (Moles/L)	Solubility (Grams/L)	Temp (°C)	Ref (#)	Evaluation (T P E A A)	Comments
3.437E-03	4.900E-01	25	L348	1 2 2 1 2	

1078.   C₇H₇ClO

2-Methyl-6-chloro-phenol

2-Chloro-6-methylphenol

6-Chloro-o-cresol

3-Chloro-2-hydroxytoluene

6-Chloro-2-methylphenol

RN: 87-64-9   **MP** (°C):

MW: 142.59   **BP** (°C):

Solubility (Moles/L)	Solubility (Grams/L)	Temp (°C)	Ref (#)	Evaluation (T P E A A)	Comments
2.500E-02	3.565E+00	25	B316	1 0 2 1 1	

1079.   C₇H₇ClO

2-Methyl-4-chloro-phenol

4-Chloro-o-cresol

4-Chloro-2-methylphenol

5-Chloro-2-hydroxytoluene

RN: 1570-64-5   **MP** (°C):   45-48

MW: 142.59   **BP** (°C):   220-225

Solubility (Moles/L)	Solubility (Grams/L)	Temp (°C)	Ref (#)	Evaluation (T P E A A)	Comments
4.800E-02	6.844E+00	25	B316	1 0 2 1 1	

1080. C₇H₇Cl₂NO

1080. $C_7H_7Cl_2NO$

Clopidol
3,5-Dichloro-2,6-Dimethyl-4-Pyridinol
Coyden
Methylchloropindol
RN: 2971-90-6   **MP** (°C):
MW: 192.05   **BP** (°C):

Solubility (Moles/L)	Solubility (Grams/L)	Temp (°C)	Ref (#)	Evaluation (T P E A A)	Comments
2.083E-04	4.000E-02	ns	K138	0 0 0 0 1	

1081. $C_7H_7Cl_3NO_3PS$

Chlorpyrifos-methyl
Chlorpyrifos-methy
RN: 5598-13-0   **MP** (°C):
MW: 322.54   **BP** (°C):

Solubility (Moles/L)	Solubility (Grams/L)	Temp (°C)	Ref (#)	Evaluation (T P E A A)	Comments
5.581E-06	1.800E-03	10	B324	2 2 2 2 2	
5.581E-06	1.800E-03	10	B324	2 2 2 2 2	
9.922E-06	3.200E-03	20	B300	2 1 1 1 2	
9.922E-06	3.200E-03	20	B324	2 2 2 2 2	
9.921E-06	3.200E-03	20	B324	2 2 2 2 2	
1.476E-05	4.760E-03	20	C053	1 0 2 2 1	
1.240E-05	4.000E-03	24	K069	2 0 0 1 1	
1.240E-05	4.000E-03	25	M161	1 0 0 0 0	
2.139E-05	6.899E-03	30	B324	2 2 2 2 2	
2.139E-05	6.900E-03	30	B324	2 2 2 2 2	
1.476E-05	4.760E-03	ns	F071	0 1 2 1 2	
1.240E-05	4.000E-03	ns	K138	0 0 0 0 1	
1.643E-05	5.300E-03	ns	M110	0 0 0 0 0	EFG

1082. $C_7H_7Cl_3NO_4P$

Torelle
Dimethyl 3,5,6-Trichloro-2-pyridinyl Phosphate
DOWCO 217
Fospirate
Phosphoric Acid, Dimethyl 3,5,6-Trichloro-2-Pyridyl Ester
RN: 5598-52-7   **MP** (°C):
MW: 306.47   **BP** (°C):

Solubility (Moles/L)	Solubility (Grams/L)	Temp (°C)	Ref (#)	Evaluation (T P E A A)	Comments
9.789E-04	3.000E-01	24	K069	2 0 0 1 1	

1083. C₇H₇FN₂O₃

1083. C$_7$H$_7$FN$_2$O$_3$

3-Propionyl-5-fluoro-2,4(1H,3H)-pyrimidinedi-one
3-Propionyl-5-fluorouracil
RN: 75410-16-1   **MP** (°C):   113-114
MW: 186.14   **BP** (°C):

Solubility (Moles/L)	Solubility (Grams/L)	Temp (°C)	Ref (#)	Evaluation (T P E A A)	Comments
1.896E-01	3.530E+01	22	B321	1 0 2 2 2	pH 4.0
1.896E-01	3.530E+01	22	B332	1 1 0 0 1	pH 4.0

1084. C$_7$H$_7$FN$_2$O$_4$

1-Acetoxymethyl-5-fluoro-2,4(1H,3H)-pyrimidinedi-one
1-Acetoxymethyl-5-fluorouracil
RN: 62113-41-1   **MP** (°C):   122-123
MW: 202.14   **BP** (°C):

Solubility (Moles/L)	Solubility (Grams/L)	Temp (°C)	Ref (#)	Evaluation (T P E A A)	Comments
2.132E-01	4.310E+01	22	B321	1 0 2 2 2	pH 4.0

1085. C$_7$H$_7$FN$_2$O$_4$

3-Acetoxymethyl-5-fluoro-2,4(1H,3H)-pyrimidinedi-one
3-Acetoxymethyl-5-fluorouracil
RN: 73042-04-3   **MP** (°C):   158-159
MW: 202.14   **BP** (°C):

Solubility (Moles/L)	Solubility (Grams/L)	Temp (°C)	Ref (#)	Evaluation (T P E A A)	Comments
9.894E-02	2.000E+01	22	B321	1 0 2 2 2	pH 4.0

1086. C$_7$H$_7$FN$_2$O$_4$

3-Ethyloxycarbonyl-5-fluoro-2,4(1H,3H)-pyrimidinedi-one
3-Ethyloxycarbonyl-5-fluorouracil
1-Ethyloxycarbonyl-5-fluorouracil
RN: 75410-27-4   **MP** (°C):   126-128
MW: 202.14   **BP** (°C):

Solubility (Moles/L)	Solubility (Grams/L)	Temp (°C)	Ref (#)	Evaluation (T P E A A)	Comments
3.562E-01	7.200E+01	22	B321	1 0 2 2 2	pH 4.0
3.413E-02	6.900E+00	22	B332	1 1 0 0 1	pH 4.0

1087.   C$_7$H$_7$NO
Benzamide
Benzamid
Phenyl Carboxamide
Benzoic Acid Amide

RN: 55-21-0   **MP** (°C):   130
MW: 121.14   **BP** (°C):   288

Solubility (Moles/L)	Solubility (Grams/L)	Temp (°C)	Ref (#)	Evaluation (T P E A A)	Comments
4.923E-02	5.964E+00	10	M043	1 0 0 0 0	
4.750E-02	5.754E+00	12	O019	1 0 0 1 2	
1.000E-01	1.211E+01	20	B139	2 1 1 1 1	
8.173E-02	9.901E+00	20	M043	1 0 0 0 1	
1.100E-01	1.333E+01	22	J037	1 0 1 1 1	
1.106E-01	1.340E+01	25	F300	1 0 0 0 2	
1.059E-01	1.283E+01	30	M043	1 0 0 0 1	
1.300E-01	1.575E+01	40	M043	1 0 0 0 1	
1.651E-01	2.000E+01	50	P064	2 0 1 1 1	
3.931E-01	4.762E+01	60	M043	1 0 0 0 0	
6.191E-01	7.500E+01	70	P064	2 0 1 1 1	
5.503E+00	6.667E+02	80	M043	1 0 0 0 2	
6.686E+00	8.100E+02	90	P064	2 0 1 1 2	
7.338E+00	8.889E+02	100	M043	1 0 0 0 2	
7.842E+00	9.500E+02	110	P064	2 0 1 1 2	
1.100E-01	1.332E+01	rt	D021	0 0 1 1 2	

1088.   C$_7$H$_7$NO$_2$
p-Aminobenzoic Acid
4-Amino-benzoesaeure
4-Aminobenzoic Acid
p-Aminobenzoicacid
1-Amino-4-carboxybenzene

RN: 150-13-0   **MP** (°C):   187.0
MW: 137.14   **BP** (°C):

Solubility (Moles/L)	Solubility (Grams/L)	Temp (°C)	Ref (#)	Evaluation (T P E A A)	Comments
2.479E-02	3.400E+00	12.80	F300	1 0 0 0 1	
3.609E-02	4.950E+00	18	C033	1 0 2 1 2	
3.628E-02	4.975E+00	25	D041	1 0 0 0 0	
3.930E-02	5.390E+00	25	L338	1 0 1 1 2	
3.646E-02	5.000E+00	25	M054	1 0 0 0 0	
3.500E-02	4.800E+00	25	P015	2 2 2 2 1	
4.455E-02	6.110E+00	30	C033	1 0 2 1 2	
4.579E-02	6.280E+00	30	H018	1 2 2 2 2	
4.500E-02	6.171E+00	30	L069	1 0 1 1 0	EFG
6.125E-02	8.400E+00	37	B171	2 0 1 1 2	
6.040E-02	8.283E+00	37	F006	1 1 2 2 2	

1089.　C₇H₇NO₂
o-Aminobenzoic Acid
2-Aminobenzoic Acid
Anthranilsaeure
RN:　118-92-3　　**MP** (°C):　145
MW:　137.14　　**BP** (°C):

Solubility (Moles/L)	Solubility (Grams/L)	Temp (°C)	Ref (#)	Evaluation (T P E A A)	Comments
2.181E-02	2.991E+00	10	M043	1 0 0 0 0	
2.543E-02	3.488E+00	14	D041	1 0 0 0 1	
2.552E-02	3.500E+00	14	F300	1 0 0 0 1	
2.543E-02	3.488E+00	20	M043	1 0 0 0 1	
4.349E-02	5.964E+00	30	M043	1 0 0 0 0	
6.504E-02	8.920E+00	40	M043	1 0 0 0 0	
3.552E+00	4.872E+02	100	M043	1 0 0 0 1	

1090.　C₇H₇NO₂
p-Nitrotoluene
4-Nitrotoluene
RN:　99-99-0　　**MP** (°C):　55
MW:　137.14　　**BP** (°C):

Solubility (Moles/L)	Solubility (Grams/L)	Temp (°C)	Ref (#)	Evaluation (T P E A A)	Comments
2.917E-04	4.000E-02	14.5	D070	1 2 0 0 1	
2.917E-04	4.000E-02	14.50	F300	1 0 0 0 1	
2.100E-03	2.880E-01	20	H306	1 0 1 2 1	
2.150E-03	2.949E-01	20	T301	1 2 2 2 2	
5.687E-04	7.799E-02	50	D070	1 2 0 0 1	
8.458E-04	1.160E-01	100	D070	1 2 0 0 2	

1091.　C₇H₇NO₂
Salicylamide
2-Hydroxybenzoicacidamide
Algamon
Amid-Sal
Amidosal
Algiamida
RN:　65-45-2　　**MP** (°C):　140
MW:　137.14　　**BP** (°C):

Solubility (Moles/L)	Solubility (Grams/L)	Temp (°C)	Ref (#)	Evaluation (T P E A A)	Comments
1.060E-02	1.454E+00	15	D012	1 1 0 1 2	
1.100E-02	1.509E+00	16	D012	1 1 0 1 2	
1.531E-02	2.100E+00	20	E046	1 0 0 0 0	EFG
1.900E-02	2.606E+00	22	J031	1 0 0 0 1	

1.604E-02	2.200E+00	23	B328	1 2 2 1 1	pH 4.0
1.500E-02	2.057E+00	25	D012	1 1 0 1 2	
1.750E-02	2.400E+00	25	E046	1 0 0 0 0	EFG
1.831E-02	2.511E+00	25	P314	2 2 1 2 2	
2.115E-02	2.900E+00	30	E046	1 0 0 0 0	EFG
2.771E-02	3.800E+00	35	E046	1 0 0 0 0	EFG
2.900E-02	3.977E+00	37	D012	1 1 0 1 2	
3.427E-02	4.700E+00	40	E046	1 0 0 0 0	EFG
4.280E-02	5.870E+00	45	D012	1 1 0 1 2	
5.323E-02	7.300E+00	50	E046	1 0 0 0 0	EFG
1.677E-03	2.300E-01	ns	B361	0 0 0 2 2	

1092. $C_7H_7NO_2$
o-Nitrotoluene
2-Nitro-toluol

RN: 88-72-2   **MP** (°C):   -9.5
MW: 137.14   **BP** (°C):   221.7

Solubility (Moles/L)	Solubility (Grams/L)	Temp (°C)	Ref (#)	Evaluation (T P E A A)	Comments
4.740E-03	6.500E-01	30	F300	1 0 0 0 2	

1093. $C_7H_7NO_2$
Methyl Nicotinate
Nicotinsaeure-methyl Ester

RN: 93-60-7   **MP** (°C):   39
MW: 137.14   **BP** (°C):   209

Solubility (Moles/L)	Solubility (Grams/L)	Temp (°C)	Ref (#)	Evaluation (T P E A A)	Comments
3.471E-01	4.760E+01	20	F300	1 0 0 0 2	*sic*
8.065E+00	1.106E+03	32	L346	1 0 0 1 0	

1094. $C_7H_7NO_2$
m-Nitrotoluene
3-Nitro-toluol

RN: 99-08-1   **MP** (°C):   16
MW: 137.14   **BP** (°C):   232.6

Solubility (Moles/L)	Solubility (Grams/L)	Temp (°C)	Ref (#)	Evaluation (T P E A A)	Comments
3.646E-03	5.000E-01	30	F300	1 0 0 0 2	

1095. $C_7H_7NO_2$

m-Aminobenzoic Acid
3-Amino-benzoesaeure
3-Aminobenzoic Acid

RN: 99-05-8   **MP** (°C):   174
MW: 137.14   **BP** (°C):

Solubility (Moles/L)	Solubility (Grams/L)	Temp (°C)	Ref (#)	Evaluation (T P E A A)	Comments
4.302E-02	5.900E+00	14.90	F300	1 0 0 0 1	
5.830E-02	7.995E+00	30	W007	2 0 2 2 2	

1096. $C_7H_7NO_3$

3-Methyl-4-nitrophenol
3-Nitro-p-cresol
3-Nitro-p-kresol
4-Nitro-5-methylphenol

RN: 2581-34-2   **MP** (°C):   128
MW: 153.14   **BP** (°C):

Solubility (Moles/L)	Solubility (Grams/L)	Temp (°C)	Ref (#)	Evaluation (T P E A A)	Comments
7.769E-03	1.190E+00	25	B104	1 2 1 1 1	

1097. $C_7H_7NO_3$

p-Aminosalicylic Acid
4-Amino-salicylsaeure
4-Aminosalicylic Acid

RN: 65-49-6   **MP** (°C):   150
MW: 153.14   **BP** (°C):

Solubility (Moles/L)	Solubility (Grams/L)	Temp (°C)	Ref (#)	Evaluation (T P E A A)	Comments
1.303E-02	1.996E+00	20	D041	1 0 0 0 0	
1.100E-02	1.685E+00	23	M072	1 2 1 1 0	EFG
2.100E-02	3.216E+00	30	L069	1 0 1 1 0	EFG
1.087E-02	1.664E+00	ns	H125	0 0 0 0 0	

1098. $C_7H_7NO_3$
p-Nitroanisol
4-Nitro-anisol
4-Nitroanisol
RN: 100-17-4   **MP** (°C):   54
MW: 153.14   **BP** (°C):   260

Solubility (Moles/L)	Solubility (Grams/L)	Temp (°C)	Ref (#)	Evaluation (T P E A A)	Comments
4.571E-04	7.000E-02	15	F300	1 0 0 0 1	
3.853E-03	5.900E-01	30	F300	1 0 0 0 2	

1099. $C_7H_7N_2OS$
Ethyl Acetylthiodiazole
Ethyle Acetyle Thiodiazolique
RN: **MP** (°C):
MW: 167.21   **BP** (°C):

Solubility (Moles/L)	Solubility (Grams/L)	Temp (°C)	Ref (#)	Evaluation (T P E A A)	Comments
1.196E-03	2.000E-01	37	D084	1 0 1 0 1	

1100. $C_7H_7N_5$
2-Methylaminopteridine
Pteridine, 2-(Methylamino)-
RN: 19167-57-8   **MP** (°C):   219
MW: 161.17   **BP** (°C):

Solubility (Moles/L)	Solubility (Grams/L)	Temp (°C)	Ref (#)	Evaluation (T P E A A)	Comments
1.933E-02	3.115E+00	20	A019	2 2 1 1 1	
1.724E-01	2.778E+01	100	A019	1 2 1 1 1	

1101. C_7H_8
Cycloheptatriene
1,3,5-Cycloheptatriene
Tropilidene
CHT
RN: 544-25-2   **MP** (°C):   -80
MW: 92.14   **BP** (°C):   116.0

Solubility (Moles/L)	Solubility (Grams/L)	Temp (°C)	Ref (#)	Evaluation (T P E A A)	Comments
6.301E-03	5.806E-01	4.8	L007	2 2 1 2 2	
6.301E-03	5.806E-01	5.1	L007	2 1 1 1 2	
7.207E-03	6.641E-01	14.8	L007	2 2 1 2 2	
7.207E-03	6.641E-01	15.2	L007	2 1 1 1 2	
7.260E-03	6.690E-01	24.8	L007	2 2 1 2 2	
6.729E-03	6.200E-01	25	M001	2 1 2 2 2	
7.260E-03	6.690E-01	25.1	L007	2 1 1 1 2	

8.045E-03	7.413E-01	34.8	L007	2 2 1 2 2
8.045E-03	7.413E-01	35.2	L007	2 1 1 1 2
8.294E-03	7.642E-01	44.8	L007	2 2 1 2 2
8.294E-03	7.642E-01	45.2	L007	2 1 1 1 2

1102. C_7H_8
Toluene
Methylbenzene

RN: 108-88-3   **MP** (°C):   -94
MW: 92.14   **BP** (°C):   110.6

Solubility (Moles/L)	Solubility (Grams/L)	Temp (°C)	Ref (#)	Evaluation (T P E A A)	Comments
7.857E-03	7.240E-01	0	P003	2 2 2 2 2	
5.819E-03	5.362E-01	0.06	U010	1 0 0 1 1	EFG
6.638E-03	6.116E-01	4.50	B086	2 1 2 2 2	
5.557E-03	5.120E-01	4.62	U010	1 0 0 1 1	EFG
5.557E-03	5.120E-01	4.62	U013	1 0 0 0 0	EFG
6.519E-03	6.006E-01	6.30	B086	2 1 2 2 2	
6.356E-03	5.857E-01	7.10	B086	2 1 2 2 2	
6.367E-03	5.867E-01	9	B086	2 1 2 2 2	
6.210E-03	5.722E-01	10	B149	2 1 1 2 2	
6.215E-03	5.727E-01	11.80	B086	2 1 2 2 2	
6.237E-03	5.747E-01	12.10	B086	2 1 2 2 2	
5.307E-03	4.890E-01	14.20	U013	1 0 0 0 0	EFG
5.785E-03	5.330E-01	15	S203	1 1 2 1 2	
6.172E-03	5.687E-01	15.10	B086	2 1 2 2 2	
5.424E-03	4.998E-01	16	D052	1 1 0 0 0	
5.100E-03	4.699E-01	16	F001	1 0 1 2 1	
5.101E-03	4.700E-01	16	F071	1 1 2 1 2	
5.101E-03	4.700E-01	16	F300	1 0 0 0 2	
5.101E-03	4.700E-01	16	H080	1 0 0 0 2	
5.100E-03	4.699E-01	16	S006	1 0 0 0 1	
6.370E-03	5.869E-01	20	B149	2 1 1 2 2	
6.154E-03	5.670E-01	20	B356	1 0 0 0 2	
5.424E-03	4.998E-01	20	C121	1 0 0 0 0	unit assumed, *sic*
5.590E-03	5.151E-01	20	M312	1 0 0 0 2	
4.982E-03	4.591E-01	20	M337	2 1 2 2 2	
6.139E-03	5.657E-01	20.10	B086	2 1 2 2 2	
5.196E-03	4.788E-01	21	C024	2 1 1 2 2	
5.752E-03	5.300E-01	25	A001	1 2 2 2 1	
5.098E-03	4.698E-01	25	A094	1 0 0 0 1	
6.805E-03	6.270E-01	25	B003	2 1 2 2 2	
5.589E-03	5.150E-01	25	B060	2 0 1 1 1	
6.690E-03	6.164E-01	25	B153	2 1 1 1 2	
1.680E-02	1.548E+00	25	B173	2 0 2 2 2	*sic*
5.687E-03	5.240E-01	25	B304	2 0 2 2 2	
8.000E-03	7.371E-01	25	H092	1 1 1 1 0	

6.500E-03	5.989E-01	25	H313	2 1 2 2 1	
6.000E-03	5.529E-01	25	H332	2 2 2 2 0	
6.370E-02	5.869E+00	25	I334	2 2 2 1 2	*sic*
6.370E-03	5.869E-01	25	I335	2 2 2 2 2	
5.430E-03	5.003E-01	25	K001	1 0 2 1 2	
5.318E-03	4.900E-01	25	K072	1 0 1 1 1	
6.290E-03	5.796E-01	25	K316	2 2 2 2 2	
5.641E-03	5.197E-01	25	L319	1 0 2 1 2	
5.589E-03	5.150E-01	25	M130	1 0 0 0 2	
5.638E-03	5.195E-01	25	M132	2 2 2 1 2	
6.280E-03	5.787E-01	25	M342	1 0 1 1 2	
6.219E-03	5.730E-01	25	P003	2 2 2 2 2	
6.012E-03	5.540E-01	25	P051	2 1 1 2 2	
6.045E-03	5.570E-01	25	S203	1 1 2 1 2	
5.804E-03	5.348E-01	25	S358	2 1 2 2 2	
5.650E-03	5.206E-01	25	S359	2 1 2 2 2	
6.280E-03	5.787E-01	25	W300	2 2 2 2 2	
5.307E-03	4.890E-01	25.35	U010	1 0 0 1 1	EFG
5.307E-03	4.890E-01	25.35	U013	1 0 0 0 0	EFG
3.255E-03	2.999E-01	30	F053	1 0 2 0 2	
6.183E-03	5.697E-01	30	G029	1 0 2 2 1	
5.067E-03	4.669E-01	30	M311	1 1 2 2 2	
1.409E-02	1.298E+00	30	S207	1 0 0 1 1	*sic*
5.557E-03	5.120E-01	34.53	U010	1 0 0 1 1	EFG
5.557E-03	5.120E-01	34.53	U013	1 0 0 0 0	EFG
6.371E-03	5.870E-01	35	S203	1 1 2 1 2	
5.954E-03	5.486E-01	44.30	U010	1 0 0 1 1	EFG
5.819E-03	5.362E-01	44.30	U013	1 0 0 0 0	EFG
6.892E-03	6.350E-01	45	S203	1 1 2 1 2	
1.517E-02	1.398E+00	45	S207	1 0 0 1 1	*sic*
6.529E-03	6.015E-01	54.71	U013	1 0 0 0 0	EFG
1.500E-02	1.382E+00	55	H092	1 1 1 1 1	
6.380E-03	5.879E-01	55.79	U010	1 0 0 1 1	EFG
1.734E-02	1.597E+00	60	S207	1 0 0 1 1	*sic*
7.325E-03	6.749E-01	65.82	U013	1 0 0 0 0	EFG
2.171E-02	2.000E+00	150	J023	1 1 2 2 0	
7.597E-02	7.000E+00	200	J023	1 1 2 2 0	
3.039E-01	2.800E+01	250	J023	1 1 2 2 1	
1.411E+00	1.300E+02	300	J023	1 1 2 2 2	
5.589E-03	5.150E-01	ns	H123	0 0 0 0 2	
1.380E-01	1.272E+01	ns	H307	1 0 1 1 2	*sic*
5.611E-03	5.170E-01	ns	M175	0 0 2 1 2	
5.589E-03	5.150E-01	ns	M344	0 0 0 0 2	

1103. C_7H_8
1,6-Heptadiyne
RN:	2396-63-6	**MP** (°C):	-85
MW:	92.14	**BP** (°C):	112

Solubility (Moles/L)	Solubility (Grams/L)	Temp (°C)	Ref (#)	Evaluation (T P E A A)	Comments
1.791E-02	1.650E+00	25	M001	2 1 2 2 2	

1104. $C_7H_8CIN_3O_4S_2$
Hydrochlorothiazide
Chlorozide
RN:	58-93-5	**MP** (°C):	274
MW:	297.74	**BP** (°C):	

Solubility (Moles/L)	Solubility (Grams/L)	Temp (°C)	Ref (#)	Evaluation (T P E A A)	Comments
2.425E-03	7.220E-01	25	A076	1 0 1 1 2	
2.045E-03	6.090E-01	25	D091	1 0 0 0 2	pH 6.2
2.687E-03	8.000E-01	25	G051	1 0 1 1 0	
2.800E-03	8.337E-01	30	A089	2 0 1 1 0	EFG
2.800E-03	8.337E-01	30	A093	2 0 1 1 0	EFG
2.520E-03	7.503E-01	30	E049	2 0 2 2 2	
3.627E-03	1.080E+00	37	D091	1 0 0 0 2	pH 7.2
7.650E-03	2.278E+00	50	M335	1 0 2 1 2	pH 5
1.982E-03	5.900E-01	rt	A095	0 0 2 2 0	

1105. $C_7H_8FN_3O_3$
1-(N,N-Dimethylcarbamoyl)-5-fluorouracil
1-Dimethylcarbamoyl-5-fluoro-2,4(1H,3H)-pyrimidinedi-one
RN:	60908-29-4	**MP** (°C):	226-227
MW:	201.16	**BP** (°C):	

Solubility (Moles/L)	Solubility (Grams/L)	Temp (°C)	Ref (#)	Evaluation (T P E A A)	Comments
2.983E-02	6.000E+00	22	B321	1 0 2 2 2	pH 4.0
2.983E-02	6.000E+00	22	B388	1 0 2 2 2	

1106.   C₇H₈FN₃O₃

1-Ethylcarbamoyl-5-fluorouracil
1-Ethylcarbamoyl-5-fluoro-2,4(1H,3H)-pyrimidinedi-one
N-Ethyl-5-fluoro-3,4-dihydro-2,4-dioxo-1-pyrimidinecarboxamide
RN: 58471-47-9   **MP** (°C):   190-196
MW: 201.16   **BP** (°C):

Solubility (Moles/L)	Solubility (Grams/L)	Temp (°C)	Ref (#)	Evaluation (T P E A A)	Comments
7.457E-03	1.500E+00	22	B321	1 0 2 2 2	pH 4.0
7.457E-03	1.500E+00	22	B388	1 0 2 2 2	

1107.   C₇H₈N₂O₂

3-Nitro-o-toluidine
3-Nitro-o-toluidin
RN: 603-83-8   **MP** (°C):   92
MW: 152.15   **BP** (°C):   305

Solubility (Moles/L)	Solubility (Grams/L)	Temp (°C)	Ref (#)	Evaluation (T P E A A)	Comments
8.807E-02	1.340E+01	100	F300	1 0 0 0 2	

1108.   C₇H₈N₂O₃

1-Methoxy-2-amino-4-nitrobenzene
RN: 99-59-2   **MP** (°C):   118
MW: 168.15   **BP** (°C):

Solubility (Moles/L)	Solubility (Grams/L)	Temp (°C)	Ref (#)	Evaluation (T P E A A)	Comments
3.388E-03	5.697E-01	rt	N015	0 0 2 2 2	

1109.   C₇H₈N₂O₃

5,5-Trimethylenebarbituric Acid
6,8-Diazaspiro[3.5]nonane-5,7,9-trione
RN: 6128-03-6   **MP** (°C):
MW: 168.15   **BP** (°C):

Solubility (Moles/L)	Solubility (Grams/L)	Temp (°C)	Ref (#)	Evaluation (T P E A A)	Comments
2.213E-02	3.721E+00	25	P350	2 1 1 1 2	intrinsic

1110.   C₇H₈N₂O₃S

5-Carboethoxy-2-thiouracil
Ethyl 2-Thiouracil-5-carboxylate
RN: 38026-46-9   **MP** (°C):   252
MW: 200.22   **BP** (°C):

Solubility (Moles/L)	Solubility (Grams/L)	Temp (°C)	Ref (#)	Evaluation (T P E A A)	Comments
7.970E-03	1.596E+00	25	G016	1 2 1 2 2	intrinsic

1111.   C₇H₈N₂O₄

Ethyl Orotate
1,2,3,6-Tetrahydro-2,6-dioxo-4-pyrimidine-carboxylic Acid, Ethyl Ester
RN: 1747-53-1   **MP** (°C):
MW: 184.15   **BP** (°C):

Solubility (Moles/L)	Solubility (Grams/L)	Temp (°C)	Ref (#)	Evaluation (T P E A A)	Comments
2.100E-02	3.867E+00	20	N019	2 2 1 2 2	

1112.   C₇H₈N₂S

1-Phenyl-2-thiourea
Phenylthioharnstoff
RN: 103-85-5   **MP** (°C):   149
MW: 152.22   **BP** (°C):

Solubility (Moles/L)	Solubility (Grams/L)	Temp (°C)	Ref (#)	Evaluation (T P E A A)	Comments
1.708E-02	2.600E+00	18	F300	1 0 0 0 1	
3.830E-01	5.830E+01	100	F300	1 0 0 0 2	

1113.   C₇H₈N₄O₂

Theophylline
1,3-Dimethylxanthine
Aerolate
Bronkotabs
Bronchodid Duracap
Bronkodyl
RN: 58-55-9   **MP** (°C):   272
MW: 180.17   **BP** (°C):

Solubility (Moles/L)	Solubility (Grams/L)	Temp (°C)	Ref (#)	Evaluation (T P E A A)	Comments
3.310E-02	5.964E+00	16	A072	1 0 1 0 1	
2.866E-02	5.164E+00	20	K052	1 1 1 1 2	
3.420E-02	6.162E+00	25	F009	2 2 2 2 0	EFG
3.675E-02	6.621E+00	25	L338	1 0 1 1 2	
4.089E-02	7.366E+00	25	M128	2 0 1 2 2	
4.083E-02	7.356E+00	25	M158	2 0 2 2 2	
3.580E-02	6.450E+00	25	N312	2 1 1 1 1	
4.607E-02	8.300E+00	25	P010	1 0 1 1 1	
4.607E-02	8.300E+00	25	P011	1 0 1 1 1	
4.440E-02	8.000E+00	25	P018	1 0 2 2 1	
4.440E-02	8.000E+00	25	P020	2 0 1 1 1	
4.607E-02	8.300E+00	25	P312	1 2 2 2 2	
4.500E-02	8.108E+00	30	B042	1 2 1 1 1	
4.500E-02	8.108E+00	30	G021	1 0 0 0 2	
4.100E-02	7.387E+00	30	H016	2 2 2 2 0	EFG

Solubility (Moles/L)	Solubility (Grams/L)	Temp (°C)	Ref (#)	Evaluation (T P E A A)	Comments
4.500E-02	8.108E+00	30	H020	1 0 0 0 1	
5.550E-02	1.000E+01	37	F076	2 0 2 2 0	
2.761E-02	4.975E+00	ns	J025	0 0 0 0 2	
3.580E-02	6.450E+00	ns	N062	2 0 1 2 2	
2.054E-04	3.700E-02	rt	N015	0 0 2 2 1	*sic*

1114. $C_7H_8N_4O_2$
Theobromine
Theobromin

RN: 83-67-0   **MP** (°C):   357
MW: 180.17   **BP** (°C):

Solubility (Moles/L)	Solubility (Grams/L)	Temp (°C)	Ref (#)	Evaluation (T P E A A)	Comments
1.665E-03	3.000E-01	18	F300	1 0 0 0 0	
3.328E-03	5.996E-01	19	A072	1 0 1 0 0	
2.419E-03	4.358E-01	20	K052	1 1 1 1 2	
1.830E-03	3.297E-01	25	M158	2 0 2 2 2	
2.489E-02	4.484E+00	25	O302	1 0 0 1 0	EFG, *sic*
2.775E-03	5.000E-01	25	P010	1 0 1 1 1	
3.330E-03	6.000E-01	25	P011	1 0 1 1 1	
3.386E-03	6.100E-01	25	P018	1 0 2 2 1	
3.108E-03	5.600E-01	25	P020	2 0 1 1 1	
3.000E-03	5.405E-01	30	B042	1 2 1 1 0	
~3.00E-03	~5.41E-01	30	H020	1 0 0 0 0	
3.830E-02	6.900E+00	100	F300	1 0 0 0 1	
2.774E-03	4.998E-01	c	D004	1 0 0 0 0	
3.676E-02	6.623E+00	h	D004	1 0 0 0 0	

1115. C_7H_8O
m-Cresol
3-Cresol
m-Methylphenol

RN: 108-39-4   **MP** (°C):   11
MW: 108.14   **BP** (°C):   202

Solubility (Moles/L)	Solubility (Grams/L)	Temp (°C)	Ref (#)	Evaluation (T P E A A)	Comments
1.060E-01	1.147E+01	0	M041	1 1 0 0 2	
2.167E-01	2.344E+01	20	B031	1 2 2 2 1	
2.112E-01	2.284E+01	20	R087	1 1 2 2 2	0.15M NaCl
2.149E-01	2.324E+01	20.3	L339	2 0 2 2 2	
1.420E-01	1.536E+01	25	A021	1 2 1 1 0	
1.991E-01	2.153E+01	25	B019	1 0 1 2 0	
2.053E-01	2.220E+01	25	C060	1 2 1 1 2	
2.099E-01	2.270E+01	25	F300	1 0 0 0 2	
1.946E-01	2.105E+01	25	M041	1 1 0 0 2	
2.255E-01	2.439E+01	25	R041	1 0 2 1 1	
2.292E-01	2.478E+01	40.0	L339	2 0 2 2 2	
2.682E-01	2.900E+01	46.2	K119	1 0 0 0 2	

Solubility (Moles/L)	Solubility (Grams/L)	Temp (°C)	Ref (#)	Evaluation (T P E A A)
2.326E-01	2.515E+01	50	M041	1 1 0 0 2
2.431E-01	2.629E+01	50.80	M098	1 2 0 1 1
2.712E-01	2.933E+01	58.4	L339	2 0 2 2 2
2.693E-01	2.913E+01	60	B031	1 2 2 2 1
3.331E-01	3.602E+01	77.2	L339	2 0 2 2 2
3.213E-01	3.475E+01	78.70	M098	1 2 0 1 1
3.982E-01	4.306E+01	92.20	M098	1 2 0 1 1
4.387E-01	4.744E+01	98.1	L339	2 0 2 2 2

1116. C_7H_8O

p-Cresol
4-Cresol
p-Methylphenol

RN: 106-44-5   **MP** (°C):   35.5
MW: 108.14   **BP** (°C):   201.8

Solubility (Moles/L)	Solubility (Grams/L)	Temp (°C)	Ref (#)	Evaluation (T P E A A)	Comments
1.813E-01	1.961E+01	20	B031	1 0 2 2 1	
1.701E-01	1.840E+01	20	R087	1 1 2 2 2	0.15M NaCl
1.990E-01	2.152E+01	25	A021	1 2 1 1 0	
1.902E-01	2.057E+01	25	B019	1 0 1 2 0	
1.813E-01	1.961E+01	25	L022	1 0 0 0 0	
1.967E-01	2.127E+01	25	P004	2 1 1 1 2	
1.902E-01	2.057E+01	25	R041	1 0 2 1 1	
2.044E-01	2.210E+01	29.5	K119	1 0 0 0 2	
1.999E-01	2.162E+01	29.50	M098	1 2 0 1 2	
2.090E-01	2.260E+01	40	F300	1 0 0 0 2	
3.334E-01	3.605E+01	82.10	M098	1 2 0 1 2	

1117. C_7H_8O

Benzyl Alcohol
Benzylalkohol
Benzenemethanol
Phenylmethanol
Phenylcarbinol
α-Hydroxytoluene

RN: 100-51-6   **MP** (°C):   -15.2
MW: 108.14   **BP** (°C):   204.7

Solubility (Moles/L)	Solubility (Grams/L)	Temp (°C)	Ref (#)	Evaluation (T P E A A)	Comments
3.606E-01	3.900E+01	17	F300	1 0 0 0 1	
3.488E-01	3.772E+01	20	H044	1 0 2 1 2	
3.520E-01	3.807E+01	20	S006	1 0 0 0 2	
3.967E-01	4.290E+01	25	B304	2 0 2 2 2	
3.540E-01	3.828E+01	25	H044	1 0 2 1 2	
4.260E-01	4.607E+01	25	L322	1 1 2 2 1	

3.616E-01	3.911E+01	30	H044	1 0 2 1 2
3.646E-01	3.943E+01	35	H044	1 0 2 1 2
3.676E-01	3.975E+01	40	H044	1 0 2 1 2
3.724E-01	4.027E+01	45	H044	1 0 2 1 2
3.722E-01	4.025E+01	50	H044	1 0 2 1 2
3.868E-01	4.182E+01	55	H044	1 0 2 1 2

1118. C_7H_8O
Anisole
Methoxybenzene
Methyl Phenyl Ether
Phenyl Methyl Ether

RN: 100-66-3   **MP** (°C): -37.3
MW: 108.14   **BP** (°C): 155.5

Solubility (Moles/L)	Solubility (Grams/L)	Temp (°C)	Ref (#)	Evaluation (T P E A A)	Comments
1.295E-03	1.400E-01	25	A003	1 2 1 2 1	*sic*
9.609E-02	1.039E+01	25	B019	1 0 1 2 0	
1.400E-02	1.514E+00	25	M327	1 0 0 1 2	
1.418E-02	1.533E+00	25.04	V013	2 2 2 2 2	
9.617E-02	1.040E+01	26.70	L095	2 2 1 1 2	

1119. C_7H_8O
2-Cresol
2-Methyl-phenol-
Phenol, 2-Methyl-
o-Cresol
o-Methylphenol

RN: 95-48-7   **MP** (°C): 31
MW: 108.14   **BP** (°C):

Solubility (Moles/L)	Solubility (Grams/L)	Temp (°C)	Ref (#)	Evaluation (T P E A A)	Comments
2.519E-01	2.724E+01	20	B031	1 0 2 2 1	
2.276E-01	2.461E+01	20	R087	1 1 2 2 2	0.15M NaCl
2.312E-01	2.500E+01	23	P332	2 1 1 2 2	
2.400E-01	2.595E+01	25	A021	1 2 1 1 0	
1.991E-01	2.153E+01	25	B019	1 0 1 2 0	
2.127E-01	2.300E+01	25	B060	2 0 1 1 1	
2.400E-01	2.595E+01	25	B316	1 0 2 1 1	
2.300E-01	2.487E+01	25	F044	1 0 0 0 1	
2.423E-01	2.620E+01	25	F300	1 0 0 0 2	
2.569E-01	2.778E+01	25	L022	1 0 0 0 0	
2.999E-01	3.244E+01	25	P004	2 1 1 1 2	
2.255E-01	2.439E+01	25	R041	1 0 2 1 1	
1.991E-01	2.153E+01	31	B092	2 1 1 1 2	

2.606E-01	2.818E+01	46.20	M098	1 2 0 1 1
2.497E-01	2.700E+01	50	K119	1 0 0 0 2
2.763E-01	2.988E+01	60	B092	2 1 1 1 2
3.557E-01	3.846E+01	86.70	M098	1 2 0 1 1

1120. C$_7$H$_8$O$_2$
3-Methoxyphenol
Resorcinol Monomethylether
p-Methoxyphenol

RN: 150-19-6   **MP** (°C):
MW: 124.14   **BP** (°C):

Solubility (Moles/L)	Solubility (Grams/L)	Temp (°C)	Ref (#)	Evaluation (T P E A A)	Comments
3.110E-01	3.861E+01	25	B314	1 0 0 1 2	
3.110E-01	3.861E+01	30	B315	1 0 1 1 2	
4.000E-03	4.966E-01	37	E028	1 0 1 1 1	*sic*

1121. C$_7$H$_8$O$_2$
Salicyl Alcohol
Salicylalkohol

RN: 90-01-7   **MP** (°C):   86
MW: 124.14   **BP** (°C):

Solubility (Moles/L)	Solubility (Grams/L)	Temp (°C)	Ref (#)	Evaluation (T P E A A)	Comments
5.075E-01	6.300E+01	22	F300	1 0 0 0 1	

1122. C$_7$H$_8$O$_2$
Guaiacol
o-Methoxyphenol

RN: 90-05-1   **MP** (°C):   28
MW: 124.14   **BP** (°C):   205

Solubility (Moles/L)	Solubility (Grams/L)	Temp (°C)	Ref (#)	Evaluation (T P E A A)	Comments
1.506E-01	1.870E+01	15	F300	1 0 0 0 2	
1.880E-01	2.334E+01	24.99	B353	2 1 1 1 2	
1.060E-02	1.316E+00	37	E028	1 0 1 1 2	*sic*

1123.   C₇H₈O₂
4,6-Dimethyl-1,2-pyrone
4,6-Dimethyl-α-pyrone
2,4-Dimethyl-α-pyrone
Mesitene Lactone
4,6-Dimethyl-2-pyranone
4,6-Dimethyl-2H-pyran-2-one

RN: 675-09-2   **MP** (°C):   49
MW: 124.14   **BP** (°C):

Solubility (Moles/L)	Solubility (Grams/L)	Temp (°C)	Ref (#)	Evaluation (T P E A A)	Comments
1.953E+00	2.424E+02	59.7	W022	2 2 1 1 0	EFG
2.088E+00	2.593E+02	86.3	W022	2 2 1 1 0	EFG

1124.   C₇H₈O₂
p-Methoxyphenol
p-Hydroxyanisole
Hydroquinone Monomethyl Ether
4-Methoxyphenol

RN: 150-76-5   **MP** (°C):   52.5
MW: 124.14   **BP** (°C):   243

Solubility (Moles/L)	Solubility (Grams/L)	Temp (°C)	Ref (#)	Evaluation (T P E A A)	Comments
2.073E-01	2.573E+01	20	R087	1 1 2 2 2	0.15M NaCl

1125.   C₇H₈O₃S
p-Toluenesulfonic Acid
4-Methylbenzenesulfonic Acid
Methylbenzenesulfonic Acid
Tosic Acid
PTSA
Toluene-4-sulfonic Acid

RN: 104-15-4   **MP** (°C):   106.5
MW: 172.20   **BP** (°C):

Solubility (Moles/L)	Solubility (Grams/L)	Temp (°C)	Ref (#)	Evaluation (T P E A A)	Comments
2.900E+00	4.993E+02	36.5	T023	1 2 2 1 2	
2.902E+00	4.997E+02	40.5	T023	1 2 2 1 2	
2.903E+00	4.999E+02	42.5	T023	1 2 2 1 2	

1126. $C_7H_8O_3S.H_2O$
p-Toluenesulfonic Acid (Monohydrate)
RN: 6192-52-5   **MP** (°C):   104.5
MW: 190.22   **BP** (°C):

Solubility (Moles/L)	Solubility (Grams/L)	Temp (°C)	Ref (#)	Evaluation (T P E A A)	Comments
2.107E+00	4.008E+02	-6.5	T023	1 2 2 1 2	
2.120E+00	4.033E+02	-1.5	T023	1 2 2 1 2	
2.129E+00	4.050E+02	1.5	T023	1 2 2 1 2	
2.168E+00	4.125E+02	20.1	T023	1 2 2 1 2	
2.210E+00	4.203E+02	38.8	T023	1 2 2 1 2	
2.616E+00	4.975E+02	45.3	T023	1 2 2 1 2	
2.257E+00	4.293E+02	55.2	T023	1 2 2 1 2	
2.593E+00	4.933E+02	73.9	T023	1 2 2 1 2	
2.329E+00	4.431E+02	78.4	T023	1 2 2 1 2	
2.566E+00	4.882E+02	89.1	T023	1 2 2 1 2	
2.375E+00	4.517E+02	89.9	T023	1 2 2 1 2	
2.446E+00	4.652E+02	101.1	T023	1 2 2 1 2	
2.525E+00	4.802E+02	102.9	T023	1 2 2 1 2	
2.498E+00	4.751E+02	104.8	T023	1 2 2 1 2	

1127. $C_7H_8O_3S.2H_2O$
o-Toluenesulfonic Acid (Dihydrate)
RN: 68066-37-5   **MP** (°C):
MW: 208.23   **BP** (°C):

Solubility (Moles/L)	Solubility (Grams/L)	Temp (°C)	Ref (#)	Evaluation (T P E A A)	Comments
1.718E+00	3.577E+02	-25.0	T023	1 2 2 1 2	
1.773E+00	3.691E+02	-13.0	T023	1 2 2 1 2	
1.823E+00	3.795E+02	0.8	T023	1 2 2 1 2	
1.891E+00	3.938E+02	16.8	T023	1 2 2 1 2	
1.954E+00	4.068E+02	31.2	T023	1 2 2 1 2	
2.264E+00	4.715E+02	48.2	T023	1 2 2 1 2	
2.055E+00	4.279E+02	50.0	T023	1 2 2 1 2	
2.243E+00	4.671E+02	54.0	T023	1 2 2 1 2	
2.090E+00	4.353E+02	56.0	T023	1 2 2 1 2	
2.207E+00	4.597E+02	60.4	T023	1 2 2 1 2	
2.148E+00	4.472E+02	61.2	T023	1 2 2 1 2	
2.179E+00	4.538E+02	62.0	T023	1 2 2 1 2	

1128.   C₇H₈O₃S.4H₂O

p-Toluenesulfonic Acid (Tetrahydrate)
RN: 104-15-4   **MP** (°C):
MW: 244.27   **BP** (°C):

Solubility (Moles/L)	Solubility (Grams/L)	Temp (°C)	Ref (#)	Evaluation (T P E A A)	Comments
1.422E+00	3.473E+02	-27.0	T023	1 2 2 1 2	
1.437E+00	3.510E+02	-26.0	T023	1 2 2 1 2	
1.450E+00	3.543E+02	-18.5	T023	1 2 2 1 2	
1.527E+00	3.730E+02	-16.5	T023	1 2 2 1 2	
1.592E+00	3.888E+02	-10.5	T023	1 2 2 1 2	
1.613E+00	3.939E+02	-8.5	T023	1 2 2 1 2	
1.640E+00	4.005E+02	-7.0	T023	1 2 2 1 2	
1.576E+00	3.848E+02	-5.9	T023	1 2 2 1 2	
1.605E+00	3.921E+02	-3.4	T023	1 2 2 1 2	
1.622E+00	3.961E+02	-2.2	T023	1 2 2 1 2	
1.641E+00	4.008E+02	-1.0	T023	1 2 2 1 2	

1129.   C₇H₈O₇

Methylenecitric Acid
Methylen-citronensaeure
RN: 144-16-1   **MP** (°C):
MW: 204.14   **BP** (°C):

Solubility (Moles/L)	Solubility (Grams/L)	Temp (°C)	Ref (#)	Evaluation (T P E A A)	Comments
2.337E-01	4.770E+01	20	F300	1 0 0 0 2	

1130.   C₇H₉ClN₂OS

TO-2
5-Chloro-4-methyl-2-propionamide-thiazole
CMPT
RN: 13915-79-2   **MP** (°C):   159
MW: 204.68   **BP** (°C):

Solubility (Moles/L)	Solubility (Grams/L)	Temp (°C)	Ref (#)	Evaluation (T P E A A)	Comments
8.794E-04	1.800E-01	ns	M061	0 0 0 0 2	

1131.   C₇H₉N
2,4-Lutidine
2,4-Dimethyl-pyridin
2,4-Dimethylpyridine
RN: 108-47-4   **MP** (°C):   -60
MW: 107.16   **BP** (°C):   159

Solubility (Moles/L)	Solubility (Grams/L)	Temp (°C)	Ref (#)	Evaluation (T P E A A)	Comments
3.961E+00	4.245E+02	23	J007	1 2 0 1 2	average of 2
1.896E+00	2.032E+02	23.4	C047	2 2 0 0 2	
1.896E+00	2.032E+02	23.40	A009	1 2 1 1 2	LCST
1.287E+00	1.379E+02	24.40	A009	1 2 1 1 2	EFG, LCST
2.419E+00	2.593E+02	25	A009	1 2 1 1 2	EFG, LCST
3.316E+00	3.553E+02	27.2	J007	1 2 0 1 2	
8.484E-01	9.091E+01	30	A009	1 2 1 1 2	EFG, LCST
3.111E+00	3.333E+02	32.50	A009	1 2 1 1 2	EFG, LCST
4.497E+00	4.819E+02	35.0	J007	1 2 0 1 2	
2.902E+00	3.110E+02	39.0	J007	1 2 0 1 2	
6.105E-01	6.542E+01	40	A009	1 2 1 1 2	EFG, LCST
3.500E+00	3.750E+02	50	A009	1 2 1 1 2	EFG, LCST
2.545E+00	2.727E+02	53	J007	1 2 0 1 2	
4.548E+00	4.873E+02	54.3	J007	1 2 0 1 2	
3.777E+00	4.048E+02	62.50	A009	1 2 1 1 2	EFG, LCST
2.204E+00	2.362E+02	68.5	J007	1 2 0 1 2	
6.105E-01	6.542E+01	149	A009	1 2 1 1 2	EFG, UCST
3.794E+00	4.065E+02	165	A009	1 2 1 1 2	EFG, UCST
1.287E+00	1.379E+02	180	A009	1 2 1 1 2	EFG, UCST
3.500E+00	3.750E+02	180	A009	1 2 1 1 2	EFG, UCST
3.111E+00	3.333E+02	186	A009	1 2 1 1 2	EFG, UCST
1.896E+00	2.032E+02	187	A009	1 2 1 1 2	EFG, UCST
2.419E+00	2.593E+02	187	A009	1 2 1 1 2	EFG, UCST
2.520E+00	2.701E+02	189	A009	1 2 1 1 2	UCST
2.520E+00	2.701E+02	189	C047	2 2 0 0 1	

1132.   C₇H₉N
o-Toluidine
2-Toluidine
RN: 95-53-4   **MP** (°C):   -15
MW: 107.16   **BP** (°C):   200

Solubility (Moles/L)	Solubility (Grams/L)	Temp (°C)	Ref (#)	Evaluation (T P E A A)	Comments
1.524E-01	1.633E+01	20	C113	1 0 2 1 2	
1.577E-01	1.690E+01	20	K119	1 0 0 0 2	
1.381E-01	1.480E+01	25	F300	1 0 0 0 2	

1133. C₇H₉N
Methylaniline
N-Methylaniline

RN:	100-61-8	**MP** (°C):	-57
MW:	107.16	**BP** (°C):	194

Solubility (Moles/L)	Solubility (Grams/L)	Temp (°C)	Ref (#)	Evaluation (T P E A A)	Comments
5.248E-02	5.624E+00	25	C113	1 0 2 1 2	

1134. C₇H₉N
m-Toluidine
3-Toluidine
4-Methylaniline
p-Toluidine
p-Toluidin

RN:	106-49-0	**MP** (°C):	43
MW:	107.16	**BP** (°C):	203

Solubility (Moles/L)	Solubility (Grams/L)	Temp (°C)	Ref (#)	Evaluation (T P E A A)	Comments
6.066E-02	6.500E+00	15	F300	1 0 0 0 1	
6.026E-02	6.457E+00	20	B179	2 0 0 0 2	
3.890E-01	4.169E+01	20	B179	2 0 0 0 2	
1.403E-01	1.503E+01	20	C113	1 0 2 1 2	
6.200E-02	6.644E+00	20	H306	1 0 1 2 1	
6.119E-02	6.557E+00	20	T301	1 2 2 2 2	

1135. C₇H₉N
4-Ethylpyridine
4-Aethyl-pyridin

RN:	536-75-4	**MP** (°C):	-90.5
MW:	107.16	**BP** (°C):	168.3

Solubility (Moles/L)	Solubility (Grams/L)	Temp (°C)	Ref (#)	Evaluation (T P E A A)	Comments
3.906E+00	4.186E+02	-19	C047	2 2 0 0 1	
2.495E+00	2.674E+02	182	C047	2 2 0 0 2	

1136. C₇H₉N
3-Ethylpyridine
3-Aethyl-pyridin
β-Lutidine

RN:	536-78-7	**MP** (°C):	
MW:	107.16	**BP** (°C):	163

Solubility (Moles/L)	Solubility (Grams/L)	Temp (°C)	Ref (#)	Evaluation (T P E A A)	Comments
2.520E+00	2.701E+02	196	C047	2 2 0 0 1	

1137.   C₇H₉N
3,5-Lutidine
3,5-Dimethylpyridine
RN: 591-22-0   **MP** (°C):   -9
MW: 107.16   **BP** (°C):   169

Solubility (Moles/L)	Solubility (Grams/L)	Temp (°C)	Ref (#)	Evaluation (T P E A A)	Comments
1.896E+00	2.032E+02	-12	C047	2 2 0 0 2	
2.520E+00	2.701E+02	192	C047	2 2 0 0 1	

1138.   C₇H₉N
3,4-Lutidine
3,4-Dimethylpyridine
RN: 583-58-4   **MP** (°C):   -12
MW: 107.16   **BP** (°C):   163

Solubility (Moles/L)	Solubility (Grams/L)	Temp (°C)	Ref (#)	Evaluation (T P E A A)	Comments
1.836E+00	1.968E+02	-3.6	C047	2 2 0 0 2	
2.470E+00	2.647E+02	163	C047	2 2 0 0 1	

1139.   C₇H₉N
2-Ethylpyridine
α-Lutidine
RN: 100-71-0   **MP** (°C):
MW: 107.16   **BP** (°C):   149

Solubility (Moles/L)	Solubility (Grams/L)	Temp (°C)	Ref (#)	Evaluation (T P E A A)	Comments
2.368E+00	2.537E+02	-5	C047	2 2 0 0 1	
2.760E+00	2.958E+02	231	C047	2 2 0 0 1	

1140.   C₇H₉N
2,3-Lutidine
2,3-Dimethylpyridine
RN: 583-61-9   **MP** (°C):   -15
MW: 107.16   **BP** (°C):   162

Solubility (Moles/L)	Solubility (Grams/L)	Temp (°C)	Ref (#)	Evaluation (T P E A A)	Comments
1.926E+00	2.063E+02	16.5	C047	2 2 0 0 1	
2.594E+00	2.780E+02	193	C047	2 2 0 0 2	

1141. C₇H₉N

C_7H_9N

2,5-Lutidine
2,5-Dimethyl-pyridin
2,5-Dimethylpyridine

RN: 589-93-5 **MP** (°C): -15
MW: 107.16 **BP** (°C): 157

Solubility (Moles/L)	Solubility (Grams/L)	Temp (°C)	Ref (#)	Evaluation (T P E A A)	Comments
1.984E+00	2.126E+02	13.1	C047	2 2 0 0 1	
7.186E-01	7.700E+01	23	F300	1 0 0 0 1	
2.570E+00	2.754E+02	207	C047	2 2 0 0 1	

1142. C₇H₉N

C_7H_9N

2,6-Lutidine
2,6-Dimethyl-pyridin
2,6-Dimethylpyridine

RN: 108-48-5 **MP** (°C): -6
MW: 107.16 **BP** (°C): 144

Solubility (Moles/L)	Solubility (Grams/L)	Temp (°C)	Ref (#)	Evaluation (T P E A A)	Comments
2.154E+00	2.308E+02	34	C047	2 2 0 0 1	
2.714E+00	2.908E+02	231	C047	2 2 0 0 1	

1143. C₇H₉NO

C_7H_9NO

p-Tolylhydroxylamine
p-Tolylhydroxylamin

RN: 623-10-9 **MP** (°C):
MW: 123.16 **BP** (°C):

Solubility (Moles/L)	Solubility (Grams/L)	Temp (°C)	Ref (#)	Evaluation (T P E A A)	Comments
8.120E-02	1.000E+01	5	F300	1 0 0 0 1	
4.027E-01	4.960E+01	100	F300	1 0 0 0 2	

1144. C₇H₉NO

C_7H_9NO

o-Anisidine
2-Anisidine
2-Methoxybenzenamine
o-Methoxyaniline
2-Methoxy-1-aminobenzene
o-Methoxyphenylamine

RN: 90-04-0 **MP** (°C): 5
MW: 123.16 **BP** (°C): 225

Solubility (Moles/L)	Solubility (Grams/L)	Temp (°C)	Ref (#)	Evaluation (T P E A A)	Comments
1.026E-01	1.264E+01	25	B019	1 0 1 2 0	

1145.   C₇H₉NO
p-Anisidine
4-Methoxybenzenamine
p-Methoxyaniline
4-Methoxy-1-aminobenzene
p-Methoxyphenylamine

RN: 104-94-9   **MP** (°C):   57
MW: 123.16   **BP** (°C):   246

Solubility (Moles/L)	Solubility (Grams/L)	Temp (°C)	Ref (#)	Evaluation (T P E A A)	Comments
9.311E-02	1.147E+01	20	T301	1 2 2 2 2	

1146.   C₇H₉NO₂
1,2-Dimethyl-3-hydroxy-4-pyridone
DMHP

RN: 30652-11-0   **MP** (°C):   271-273
MW: 139.16   **BP** (°C):

Solubility (Moles/L)	Solubility (Grams/L)	Temp (°C)	Ref (#)	Evaluation (T P E A A)	Comments
1.130E-01	1.572E+01	25	C340	1 0 2 1 2	pH 9.4

1147.   C₇H₉NO₂S
p-Toluenesulfonamide
p-Methylbenzenesulfonamide
4-Methylbenzenesulfonamide

RN: 70-55-3   **MP** (°C):   138
MW: 171.22   **BP** (°C):

Solubility (Moles/L)	Solubility (Grams/L)	Temp (°C)	Ref (#)	Evaluation (T P E A A)	Comments
1.110E-02	1.900E+00	9	F300	1 0 0 0 1	
1.180E-02	2.020E+00	15	K024	1 2 1 1 2	
1.843E-02	3.156E+00	25	H105	1 1 0 1 2	

1148.   C₇H₉NO₂S
o-Toluenesulfonamide
o-Methylbenzenesulfonamide

RN: 88-19-7   **MP** (°C):   156
MW: 171.22   **BP** (°C):

Solubility (Moles/L)	Solubility (Grams/L)	Temp (°C)	Ref (#)	Evaluation (T P E A A)	Comments
5.840E-03	1.000E+00	9	F300	1 0 0 0 0	
1.860E-02	3.185E+00	15	K024	1 2 1 1 2	
9.485E-03	1.624E+00	25	H105	1 1 0 1 2	

1149. C₇H₉NO₂S

m-Toluenesulfonamide
m-Methylbenzenesulfonamide
RN: 1899-94-1 **MP** (°C):
MW: 171.22 **BP** (°C):

Solubility (Moles/L)	Solubility (Grams/L)	Temp (°C)	Ref (#)	Evaluation (T P E A A)	Comments
1.750E-02	2.996E+00	15	K024	1 2 1 1 2	
4.563E-02	7.812E+00	25	H105	1 1 0 1 2	

1150. C₇H₉NO₃S

4-Amino-3-methylbenzene Sulfonic Acid
4-Amino-toluol-sulfosaeure-(3)
RN: 98-33-9 **MP** (°C):
MW: 187.22 **BP** (°C):

Solubility (Moles/L)	Solubility (Grams/L)	Temp (°C)	Ref (#)	Evaluation (T P E A A)	Comments
2.671E-02	5.000E+00	20	F300	1 0 0 0 0	

1151. C₇H₉NO₃S

4-Amino-2-methylbenzene Sulfonic Acid
4-Amino-toluol-sulfosaeure-(2)
RN: 133-78-8 **MP** (°C):
MW: 187.22 **BP** (°C):

Solubility (Moles/L)	Solubility (Grams/L)	Temp (°C)	Ref (#)	Evaluation (T P E A A)	Comments
2.404E-02	4.500E+00	20	F300	1 0 0 0 1	

1152. C₇H₉NO₃S

2-Amino-5-methylbenzene Sulfonic Acid
2-Amino-toluol-sulfosaeure-(5)
RN: 88-44-8 **MP** (°C): >300
MW: 187.22 **BP** (°C):

Solubility (Moles/L)	Solubility (Grams/L)	Temp (°C)	Ref (#)	Evaluation (T P E A A)	Comments
1.709E-01	3.200E+01	19	F300	1 0 0 0 1	

1153. C₇H₉NO₃S

p-Methoxybenzenesulfonamide
4-Methoxybenzenesulfonamide
RN: 1129-26-6 **MP** (°C):
MW: 187.22 **BP** (°C):

Solubility (Moles/L)	Solubility (Grams/L)	Temp (°C)	Ref (#)	Evaluation (T P E A A)	Comments
1.560E-02	2.921E+00	15	K024	1 2 1 1 2	

1154. C₇H₉N₃O

4-Phenylsemicarbazide
Phenylsemicarbazide

RN: 537-47-3   **MP** (°C):   123.5
MW: 151.17   **BP** (°C):

Solubility (Moles/L)	Solubility (Grams/L)	Temp (°C)	Ref (#)	Evaluation (T P E A A)	Comments
4.627E-03	6.995E-01	15	D068	1 2 0 0 0	

1155. C₇H₉N₃O₂S₂

Sulfathiourea
p-Aminobenzenesulfonylthiourea
p-Aminophenylsulfonylthiourea
Badional
Baldinol
Fontamide

RN: 515-49-1   **MP** (°C):   171.5
MW: 231.30   **BP** (°C):

Solubility (Moles/L)	Solubility (Grams/L)	Temp (°C)	Ref (#)	Evaluation (T P E A A)	Comments
2.365E-03	5.470E-01	20	F073	1 2 2 2 2	

1156. C₇H₉N₃O₃

Orotic Acid Ethylamide

RN: 1011-82-1   **MP** (°C):   263-265
MW: 183.17   **BP** (°C):

Solubility (Moles/L)	Solubility (Grams/L)	Temp (°C)	Ref (#)	Evaluation (T P E A A)	Comments
1.940E-01	3.553E+01	-4	N018	2 2 1 2 2	
3.240E-01	5.935E+01	16	N018	2 2 1 2 2	
3.980E-01	7.290E+01	25	N018	2 2 1 2 2	

1157. C₇H₉N₃O₃S

Sulfanilylurea
Sulfanilylharnstoff

RN: 547-44-4   **MP** (°C):   146
MW: 215.23   **BP** (°C):

Solubility (Moles/L)	Solubility (Grams/L)	Temp (°C)	Ref (#)	Evaluation (T P E A A)	Comments
1.084E-02	2.333E+00	20	F073	1 2 2 2 2	
5.575E-03	1.200E+00	37	F300	1 0 0 0 1	

1158. $C_7H_9N_3O_4$
Orotic Acid Ethanol Amide
RN: **MP** (°C):   217-218
MW: 199.17   **BP** (°C):

Solubility (Moles/L)	Solubility (Grams/L)	Temp (°C)	Ref (#)	Evaluation (T P E A A)	Comments
1.800E-01	3.585E+01	-4	N018	2 2 1 2 2	
3.460E-01	6.891E+01	16	N018	2 2 1 2 2	
4.470E-01	8.903E+01	25	N018	2 2 1 2 2	

1159. $C_7H_{10}N_2OS$
Propylthiouracil
6-Propyl-2-thiouracil
Propycil
RN: 51-52-5   **MP** (°C):   220.0
MW: 170.23   **BP** (°C):

Solubility (Moles/L)	Solubility (Grams/L)	Temp (°C)	Ref (#)	Evaluation (T P E A A)	Comments
6.520E-03	1.110E+00	20	A091	1 0 0 0 0	
6.455E-03	1.099E+00	20	I310	0 0 0 0 1	
7.070E-03	1.204E+00	25	G016	1 2 1 2 2	intrinsic
5.816E-02	9.901E+00	100	I310	0 0 0 0 1	

1160. $C_7H_{10}N_2O_2S$
N1-Methylsulfanilamide
4-Amino-N-methylbenzenesulfonamide
N-Methyl-p-aminobenzenesulfonamide
N-Methyl-4-aminobenzenesulfonamide
RN: 1709-52-0   **MP** (°C):
MW: 186.23   **BP** (°C):

Solubility (Moles/L)	Solubility (Grams/L)	Temp (°C)	Ref (#)	Evaluation (T P E A A)	Comments
9.450E-02	1.760E+01	37	K095	2 0 0 0 2	intrinsic

1161. $C_7H_{10}N_2O_2S$
p-Methylaminobenzenesulfonamide
4-Methylaminobenzenesulfonamide
RN: 16891-79-5   **MP** (°C):
MW: 186.23   **BP** (°C):

Solubility (Moles/L)	Solubility (Grams/L)	Temp (°C)	Ref (#)	Evaluation (T P E A A)	Comments
5.000E-03	9.312E-01	15	K024	1 2 1 1 2	

1162.   C$_7$H$_{10}$N$_2$O$_2$S
Toluenesulfamide
Sulfamide, (4-Methylphenyl)-
p-Tolylsulfamide
RN: 15853-38-0   **MP** (°C):
MW: 186.23   **BP** (°C):

Solubility (Moles/L)	Solubility (Grams/L)	Temp (°C)	Ref (#)	Evaluation (T P E A A)	Comments
3.020E-02	5.624E+00	37	A028	1 0 2 1 2	intrinsic

1163.   C$_7$H$_{10}$N$_2$O$_3$
5-Ethyl-5-methylbarbituric Acid
5-Methyl-5-ethylbarbituric Acid
RN: 27653-63-0   **MP** (°C):
MW: 170.17   **BP** (°C):

Solubility (Moles/L)	Solubility (Grams/L)	Temp (°C)	Ref (#)	Evaluation (T P E A A)	Comments
8.010E-02	1.363E+01	25	M310	2 2 2 2 2	
5.912E-02	1.006E+01	25	P350	2 1 1 1 2	intrinsic

1164.   C$_7$H$_{10}$N$_2$O$_3$
Isopropylbarbituric Acid
2,4,6(1H,3H,5H)-Pyrimidinetrione, 5-(1-Methylethyl)-
RN: 7391-69-7   **MP** (°C):
MW: 170.17   **BP** (°C):

Solubility (Moles/L)	Solubility (Grams/L)	Temp (°C)	Ref (#)	Evaluation (T P E A A)	Comments
3.482E-02	5.925E+00	20	J030	1 2 2 2 2	
5.905E-02	1.005E+01	37	J030	1 2 2 2 2	

1165.   C$_7$H$_{10}$N$_4$O$_2$S
Sulfanilylguanidine
Sulfaguanidine
Sulfaguanidin
Sulfanilguanidin
RN: 57-67-0   **MP** (°C):   190
MW: 214.25   **BP** (°C):

Solubility (Moles/L)	Solubility (Grams/L)	Temp (°C)	Ref (#)	Evaluation (T P E A A)	Comments
4.131E-03	8.850E-01	20	F073	1 2 2 2 2	
4.663E-03	9.990E-01	25	D041	1 0 0 0 0	
8.868E-03	1.900E+00	37	R045	1 2 1 1 2	
1.025E-02	2.195E+00	37.50	M142	1 2 0 0 2	
4.201E-01	9.000E+01	h	F300	0 0 0 0 0	

1166. $C_7H_{10}N_4O_3 \cdot H_2O$
Theopylline (Monohydrate)
1H-Purine-2,6-dione, 3,7-Dihydro-1,3-dimethyl-, Monohydrate
RN: 5967-84-0   **MP** (°C):   269-272
MW: 216.20   **BP** (°C):

Solubility (Moles/L)	Solubility (Grams/L)	Temp (°C)	Ref (#)	Evaluation (T P E A A)	Comments
3.823E-02	8.264E+00	c	D004	1 0 0 0 0	

1167. $C_7H_{10}O_4S \cdot H_2O$
o-Toluenesulfonic Acid (Monohydrate)
2-Methyl-benzenesulfonic Acid (Monohydrate)
RN: 88-20-0   **MP** (°C):
MW: 208.23   **BP** (°C):

Solubility (Moles/L)	Solubility (Grams/L)	Temp (°C)	Ref (#)	Evaluation (T P E A A)	Comments
2.348E+00	4.889E+02	32.5	T023	1 2 2 1 2	
2.335E+00	4.863E+02	38.6	T023	1 2 2 1 2	
2.318E+00	4.827E+02	45.7	T023	1 2 2 1 2	
2.266E+00	4.718E+02	48.5	T023	1 2 2 1 2	
2.302E+00	4.793E+02	48.6	T023	1 2 2 1 2	
2.273E+00	4.733E+02	49.0	T023	1 2 2 1 2	
2.289E+00	4.767E+02	49.6	T023	1 2 2 1 2	

1168. $C_7H_{10}O_5$
Shikimic Acid
Shikimisaeure
RN: 138-59-0   **MP** (°C):   190
MW: 174.15   **BP** (°C):

Solubility (Moles/L)	Solubility (Grams/L)	Temp (°C)	Ref (#)	Evaluation (T P E A A)	Comments
8.613E-01	1.500E+02	21	F300	1 0 0 0 1	

1169. $C_7H_{10}O_5$
Mesoxalic Acid Diethyl Ester
Mesooxalsaeure-diethyl Ester
RN: 609-09-6   **MP** (°C):   -30
MW: 174.15   **BP** (°C):   208

Solubility (Moles/L)	Solubility (Grams/L)	Temp (°C)	Ref (#)	Evaluation (T P E A A)	Comments
3.249E+00	5.658E+02	22	F300	1 0 0 0 2	

1170.   C$_7$H$_{11}$NO$_2$
Ethosuximide
Zarontin
2-Ethyl-2-methylsuccinimide
RN: 77-67-8   **MP** (°C):
MW: 141.17   **BP** (°C):

Solubility (Moles/L)	Solubility (Grams/L)	Temp (°C)	Ref (#)	Evaluation (T P E A A)	Comments
1.346E+00	1.900E+02	25	P061	1 0 0 0 2	pH 3-7.9

1171.   C$_7$H$_{11}$N$_3$O$_2$
1-Methyl-L-histidine
L-1-Methylhistidine
RN: 15507-76-3   **MP** (°C):   >254
MW: 169.18   **BP** (°C):

Solubility (Moles/L)	Solubility (Grams/L)	Temp (°C)	Ref (#)	Evaluation (T P E A A)	Comments
9.851E-01	1.667E+02	25	D041	1 0 0 0 0	

1172.   C$_7$H$_{11}$N$_3$O$_2$
Ipronidazole
1-Methyl-2-isopropyl-5-nitro-imidazole
RN: 14885-29-1   **MP** (°C):   58-60
MW: 169.18   **BP** (°C):

Solubility (Moles/L)	Solubility (Grams/L)	Temp (°C)	Ref (#)	Evaluation (T P E A A)	Comments
5.556E-02	9.400E+00	20	D344	1 1 2 2 2	
5.550E-02	9.390E+00	20	D344	1 1 2 2 2	
5.446E-02	9.214E+00	20	D344	1 1 2 2 2	
5.560E-02	9.407E+00	20	D344	1 1 2 2 2	

1173.   C$_7$H$_{11}$N$_7$S
Aziprotryne
2-Azido-4-isopropylamino-6-methylmercapto-s-triazine
C-7019
RN: 4658-28-0   **MP** (°C):   95
MW: 225.28   **BP** (°C):

Solubility (Moles/L)	Solubility (Grams/L)	Temp (°C)	Ref (#)	Evaluation (T P E A A)	Comments
2.441E-04	5.500E-02	20	M161	1 0 0 0 1	
3.329E-04	7.500E-02	ns	M061	0 0 0 0 1	

1174. C₇H₁₂
Cycloheptene
(1Z)-Cycloheptene
cis-Cycloheptene
RN: 628-92-2   **MP** (°C):   -56
MW: 96.17   **BP** (°C):   114.7

Solubility (Moles/L)	Solubility (Grams/L)	Temp (°C)	Ref (#)	Evaluation (T P E A A)	Comments
6.863E-04	6.600E-02	25	M001	2 1 2 2 1	

1175. C₇H₁₂
1-Heptyne
1-n-Heptyne
Pentylacetylene
Amylacetylene
RN: 628-71-7   **MP** (°C):   -81
MW: 96.17   **BP** (°C):   99

Solubility (Moles/L)	Solubility (Grams/L)	Temp (°C)	Ref (#)	Evaluation (T P E A A)	Comments
9.774E-04	9.400E-02	25	M001	2 1 2 2 2	

1176. C₇H₁₂
1-Methyl-1-cyclohexene
1-Methylcyclohexene
RN: 591-49-1   **MP** (°C):   -120
MW: 96.17   **BP** (°C):   110

Solubility (Moles/L)	Solubility (Grams/L)	Temp (°C)	Ref (#)	Evaluation (T P E A A)	Comments
5.407E-04	5.200E-02	25	M001	2 1 2 2 2	

1177. C₇H₁₂
1,6-Heptadiene
RN: 3070-53-9   **MP** (°C):   -129.0
MW: 96.17   **BP** (°C):   89

Solubility (Moles/L)	Solubility (Grams/L)	Temp (°C)	Ref (#)	Evaluation (T P E A A)	Comments
4.575E-04	4.400E-02	25	M001	2 1 2 2 1	

1178.   C₇H₁₂

2-Heptyne
1-Methyl-2-butylacetylene
Butyl(methyl)acetylene
RN: 1119-65-9   **MP** (°C):
MW: 96.17   **BP** (°C):

Solubility (Moles/L)	Solubility (Grams/L)	Temp (°C)	Ref (#)	Evaluation (T P E A A)	Comments
1.700E-03	1.635E-01	25	H039	1 2 2 2 2	

1179.   C₇H₁₂

2-Methyl-3-hexyne
1-Ethyl-2-isopropylacetylene
RN: 36566-80-0   **MP** (°C):
MW: 96.17   **BP** (°C):

Solubility (Moles/L)	Solubility (Grams/L)	Temp (°C)	Ref (#)	Evaluation (T P E A A)	Comments
1.800E-03	1.731E-01	25	H039	1 2 2 2 2	

1180.   C₇H₁₂BrNO₄

5-Bromo-2-propyl-5-nitro-1,3-dioxane
2-Propyl-5-bromo-5-nitro-1,3-dioxane
RN: 53983-01-0   **MP** (°C):   73-75
MW: 254.09   **BP** (°C):

Solubility (Moles/L)	Solubility (Grams/L)	Temp (°C)	Ref (#)	Evaluation (T P E A A)	Comments
1.102E-03	2.799E-01	25	L013	1 0 2 1 2	

1181.   C₇H₁₂ClN₅

2-Chloro-4-methyl Amino-6-propyl Amino-s-triazines
1,3,5-Triazine-2,4-diamine, 6-Chloro-N-methyl-N'-propyl-
s-Triazine, 2-Chloro-4-methylamino-6-propylamino-
RN: 73383-40-1   **MP** (°C):
MW: 201.66   **BP** (°C):

Solubility (Moles/L)	Solubility (Grams/L)	Temp (°C)	Ref (#)	Evaluation (T P E A A)	Comments
1.289E-03	2.600E-01	21	G099	2 0 0 1 0	

1182.   C$_7$H$_{12}$ClN$_5$
Simazine
2-Chloro-4-ethylamino-6-ethylamino-s-triazine
2-Chloro-4,6-bis(ethylamino)-s-triazine
Primatol S
RN: 122-34-9   **MP** (°C):   224
MW: 201.66   **BP** (°C):

Solubility (Moles/L)	Solubility (Grams/L)	Temp (°C)	Ref (#)	Evaluation (T P E A A)	Comments
9.918E-06	2.000E-03	10	B185	1 0 0 0 0	
2.512E-05	5.065E-03	20	B179	2 0 0 0 2	
2.479E-05	5.000E-03	20	B185	1 0 0 0 0	
2.827E-05	5.700E-03	20	C048	2 2 2 2 1	
1.736E-05	3.500E-03	20	F311	1 2 2 2 1	
2.479E-05	5.000E-03	21	B192	0 0 0 0 0	
2.479E-05	5.000E-03	21	G099	2 0 0 1 0	
2.479E-05	5.000E-03	22	M061	1 0 0 0 0	
7.500E-05	1.512E-02	26	G001	1 0 1 1 1	
1.310E-04	2.642E-02	50	G001	1 0 1 1 2	
4.165E-04	8.400E-02	85	B185	1 0 0 0 1	
4.110E-04	8.288E-02	85	B200	1 0 0 0 2	
1.736E-05	3.500E-03	ns	C101	0 0 0 0 1	
2.479E-05	5.000E-03	ns	G041	0 0 0 0 0	
2.479E-05	5.000E-03	ns	H112	0 0 0 0 0	
2.479E-05	5.000E-03	ns	J033	0 0 0 0 0	
2.479E-05	5.000E-03	rt	M161	0 0 0 0 0	

1183.   C$_7$H$_{12}$ClN$_5$
Norazine
2-Chloro-4-methylamino-6-isopropylamino-s-triazine
RN: 3004-71-5   **MP** (°C):   157-159
MW: 201.66   **BP** (°C):

Solubility (Moles/L)	Solubility (Grams/L)	Temp (°C)	Ref (#)	Evaluation (T P E A A)	Comments
1.289E-03	2.600E-01	20	J033	1 0 0 0 2	
1.289E-03	2.600E-01	21	B192	0 0 0 0 2	

1184.   C$_7$H$_{12}$N$_2$O$_2$
5-Isobutylhydantoin
Hydantoin of DL-Leucine
RN: 67337-73-9   **MP** (°C):   208
MW: 156.19   **BP** (°C):

Solubility (Moles/L)	Solubility (Grams/L)	Temp (°C)	Ref (#)	Evaluation (T P E A A)	Comments
1.240E-02	1.937E+00	ns	M025	0 2 0 1 2	

1185.   C$_7$H$_{12}$N$_4$O$_5$
Carbamidodiglycylglycine
Triglycine Hydantoin Acid
RN: **MP** (°C):   204
MW: 232.20   **BP** (°C):

Solubility (Moles/L)	Solubility (Grams/L)	Temp (°C)	Ref (#)	Evaluation (T P E A A)	Comments
4.460E-02	1.036E+01	25	M024	1 2 0 1 2	

1186.   C$_7$H$_{12}$N$_4$O$_5$
Diglycine Hydantoic Acid
Carbamidoglycylglycine
RN: **MP** (°C):   194
MW: 232.20   **BP** (°C):

Solubility (Moles/L)	Solubility (Grams/L)	Temp (°C)	Ref (#)	Evaluation (T P E A A)	Comments
1.260E-01	2.926E+01	25	M024	1 2 0 1 2	

1187.   C$_7$H$_{12}$O
2-Methylcyclohexanone
Methyl Anone
o-Methylcyohexanone
Methyl Cyclohexanone
RN: 583-60-8   **MP** (°C):
MW: 112.17   **BP** (°C):

Solubility (Moles/L)	Solubility (Grams/L)	Temp (°C)	Ref (#)	Evaluation (T P E A A)	Comments
1.135E-01	1.274E+01	23.50	O005	2 0 2 2 2	

1188.   C$_7$H$_{12}$O
3-Methylcyclohexanone
m-Methylcyclohexanone
RN: 591-24-2   **MP** (°C):   -75
MW: 112.17   **BP** (°C):   162

Solubility (Moles/L)	Solubility (Grams/L)	Temp (°C)	Ref (#)	Evaluation (T P E A A)	Comments
1.335E-02	1.498E+00	20	D052	1 1 0 0 0	

1189. C$_7$H$_{12}$O$_2$
Hexahydrobenzoic Acid
Cyclohexanecarboxylic Acid
Cyclohexan-carbonsaeure
RN: 98-89-5   **MP** (°C):   31
MW: 128.17   **BP** (°C):   232.5

Solubility (Moles/L)	Solubility (Grams/L)	Temp (°C)	Ref (#)	Evaluation (T P E A A)	Comments
1.565E-02	2.006E+00	15	L006	1 0 0 0 2	
1.560E-02	2.000E+00	21	F300	1 0 0 0 0	

1190. C$_7$H$_{12}$O$_4$
Pimelic Acid
Heptanedioc Acid
RN: 111-16-0   **MP** (°C):   105.7
MW: 160.17   **BP** (°C):   272

Solubility (Moles/L)	Solubility (Grams/L)	Temp (°C)	Ref (#)	Evaluation (T P E A A)	Comments
1.115E-01	1.786E+01	5.99	A341	2 0 2 2 2	
1.151E-01	1.844E+01	7.99	A341	2 0 2 2 2	
1.334E-01	2.137E+01	10.99	A341	2 0 2 2 2	
1.523E-01	2.439E+01	13	D041	1 0 0 0 1	
1.498E-01	2.400E+01	13.50	F300	1 0 0 0 1	
3.122E-01	5.000E+01	15	M051	1 0 0 0 1	
2.236E-01	3.582E+01	15.99	A341	2 0 2 2 2	
2.527E-01	4.048E+01	17.99	A341	2 0 2 2 2	
3.006E-01	4.815E+01	19.99	A341	2 0 2 2 2	
2.973E-01	4.762E+01	20	D041	1 0 0 0 0	
3.122E-01	5.000E+01	20	L041	1 0 0 1 1	
2.953E-01	4.730E+01	20	M171	1 0 0 0 1	
3.000E-02	4.805E+00	20	S006	1 0 0 0 1	
3.332E+00	5.337E+02	21	B040	1 0 1 1 2	*sic*
3.846E-01	6.160E+01	23.99	A341	2 0 2 2 2	
3.938E-01	6.307E+01	24.99	A341	2 0 2 2 2	
4.660E-01	7.464E+01	28.99	A341	2 0 2 2 2	
5.072E-01	8.124E+01	30.99	A341	2 0 2 2 2	
5.690E-01	9.114E+01	33.99	A341	2 0 2 2 2	
6.545E-01	1.048E+02	36.99	A341	2 0 2 2 2	
8.886E-01	1.423E+02	39.99	A341	2 0 2 2 2	
1.527E+00	2.446E+02	42.99	A341	2 0 2 2 2	
1.824E+00	2.922E+02	44.99	A341	2 0 2 2 2	
2.135E+00	3.420E+02	47.49	A341	2 0 2 2 2	
2.551E+00	4.086E+02	49.99	A341	2 0 2 2 2	
3.460E+00	5.542E+02	54.82	A341	2 0 2 2 2	
3.915E+00	6.270E+02	59.99	A341	2 0 2 2 2	
4.365E+00	6.991E+02	64.49	A341	2 0 2 2 2	
4.649E+00	7.446E+02	68.99	A341	2 0 2 2 2	

1191.   C$_7$H$_{12}$O$_4$
3-Methyladipic Acid
3-Methylhexanedioic Acid
RN: 3058-01-3   **MP** (°C):   101
MW: 160.17   **BP** (°C):   230

Solubility (Moles/L)	Solubility (Grams/L)	Temp (°C)	Ref (#)	Evaluation (T P E A A)	Comments
3.986E-01	6.385E+01	9.50	A031	1 2 2 2 2	
4.732E-01	7.579E+01	12.80	A031	1 2 2 2 2	
1.241E+00	1.987E+02	25.90	A031	1 2 2 2 2	
1.865E+00	2.987E+02	29.80	A031	1 2 2 2 2	
2.531E+00	4.055E+02	33.20	A031	1 2 2 2 2	
3.707E+00	5.938E+02	41.10	A031	1 2 2 2 2	
4.663E+00	7.468E+02	52.30	A031	1 2 2 2 2	
5.340E+00	8.553E+02	64.30	A031	1 2 2 2 2	

1192.   C$_7$H$_{12}$O$_4$
Diethyl Malonate
Malonic
Malonic Ester
Propanedioic Acid Diethyl Ester
Ethyl Propanedioate
Ethyl Methane Dicarboxylate
RN: 105-53-3   **MP** (°C):   -50
MW: 160.17   **BP** (°C):

Solubility (Moles/L)	Solubility (Grams/L)	Temp (°C)	Ref (#)	Evaluation (T P E A A)	Comments
1.450E-01	2.322E+01	37	E028	1 0 1 1 2	

1193.   C$_7$H$_{12}$O$_4$
n-Butylmalonic Acid
Acide n-Butylmalonique
RN: 534-59-8   **MP** (°C):   102
MW: 160.17   **BP** (°C):

Solubility (Moles/L)	Solubility (Grams/L)	Temp (°C)	Ref (#)	Evaluation (T P E A A)	Comments
7.242E-01	1.160E+02	0	M051	1 0 0 0 2	
1.898E+00	3.040E+02	15	M051	1 0 0 0 2	
2.735E+00	4.380E+02	25	M051	1 0 0 0 2	
4.951E+00	7.930E+02	50	M051	1 0 0 0 2	

1194. $C_7H_{12}O_4$
Ethyl α-Acetoxypropionate
Ethyl 2-(Acetyloxy)propanoate
Ethyl 2-Acetoxypropionate
RN: 2985-28-6   **MP** (°C):
MW: 160.17   **BP** (°C):

Solubility (Moles/L)	Solubility (Grams/L)	Temp (°C)	Ref (#)	Evaluation (T P E A A)	Comments
2.104E-01	3.370E+01	25	R006	2 2 0 1 2	

1195. $C_7H_{12}O_5$
Propanoic Acid, 2-[(Ethoxycarbonyl)oxy]-, Methyl Ester
RN: **MP** (°C):
MW: 176.17   **BP** (°C):

Solubility (Moles/L)	Solubility (Grams/L)	Temp (°C)	Ref (#)	Evaluation (T P E A A)	Comments
9.214E-02	1.623E+01	25	R007	1 0 0 0 2	

1196. $C_7H_{12}O_6$
Quinic Acid
Chinasaeure
D-(-)-Quinic Acid
1,3,4,5-Tetrahydroxycyclohexanecarboxylic Acid
RN: 77-95-2   **MP** (°C):   162
MW: 192.17   **BP** (°C):

Solubility (Moles/L)	Solubility (Grams/L)	Temp (°C)	Ref (#)	Evaluation (T P E A A)	Comments
1.509E+00	2.900E+02	9	F300	1 0 0 0 1	

1197. $C_7H_{13}BrN_2O_2$
Bromo-pivalate Ureide
RN: **MP** (°C):
MW: 237.10   **BP** (°C):

Solubility (Moles/L)	Solubility (Grams/L)	Temp (°C)	Ref (#)	Evaluation (T P E A A)	Comments
2.161E-01	5.123E+01	ns	F057	0 2 2 2 1	

1198.   C$_7$H$_{13}$BrN$_2$O$_2$
Carbromal
Adalin
Bromodiethylacetylurea
N-(Aminocarbonyl)-2-bromo-2-ethylbutanamide
1-Bromo-ethyl-butyryl-urea
Bromodiethylacetylcarbamide
RN: 77-65-6   **MP** (°C):   117
MW: 237.10   **BP** (°C):

Solubility (Moles/L)	Solubility (Grams/L)	Temp (°C)	Ref (#)	Evaluation (T P E A A)	Comments
2.109E-03	5.000E-01	20	F300	1 0 0 0 0	

1199.   C$_7$H$_{13}$NO$_2$S$_2$
2,2-(Dimethyl)-4-(methoxycarbamyl)-1,3-dithiolane
1,3-Dithiolane-4-methanol, 2,2-Dimethyl-, Carbamate
RN: 35801-62-8   **MP** (°C):
MW: 207.32   **BP** (°C):

Solubility (Moles/L)	Solubility (Grams/L)	Temp (°C)	Ref (#)	Evaluation (T P E A A)	Comments
6.000E-03	1.244E+00	rt	B174	0 0 1 0 0	

1200.   C$_7$H$_{13}$NO$_3$
N-Formylleucine
N-Formyl-DL-leucine
RN: 6113-61-7   **MP** (°C):
MW: 159.19   **BP** (°C):

Solubility (Moles/L)	Solubility (Grams/L)	Temp (°C)	Ref (#)	Evaluation (T P E A A)	Comments
1.850E-01	2.945E+01	ns	M025	0 2 0 1 2	

1201.   C$_7$H$_{13}$NO$_3$S
2,2-(Dimethyl)-4-(methoxycarbamyl)-1,3-oxathiolane
1,3-Oxathiolane-5-methanol, 2,2-Dimethyl-, Carbamate
RN: 78002-88-7   **MP** (°C):
MW: 191.25   **BP** (°C):

Solubility (Moles/L)	Solubility (Grams/L)	Temp (°C)	Ref (#)	Evaluation (T P E A A)	Comments
3.000E-02	5.738E+00	rt	B174	0 0 1 0 0	

1202.   C$_7$H$_{13}$N$_3$O$_3$S
Oxamyl
Vydate
Thioxamyl
N',N'-Dimethyl-N-[(methylcarbamoyl)oxy]-1-thiooxamimidic Acid Methyl Ester
N,N-Dimethyl-α-methylcarbamoyloxyimino-α-(methylthio)acetamide
DPX 1410
RN: 23135-22-0   **MP** (°C):   109
MW: 219.26   **BP** (°C):

Solubility (Moles/L)	Solubility (Grams/L)	Temp (°C)	Ref (#)	Evaluation (T P E A A)	Comments
1.288E+00	2.825E+02	20	B179	2 0 0 0 2	
1.277E+00	2.800E+02	25	M161	1 0 0 0 2	
9.977E-01	2.188E+02	ns	H308	0 0 0 0 1	

1203.   C$_7$H$_{13}$N$_5$O
Hydroxysimazine
1,3,5-Triazin-2(1H)-one, 4,6-bis(ethylamino)-
2-Hydroxysimazine
4,6-bis(Ethylamino)-s-triazin-2-ol
G 30414
RN: 2599-11-3   **MP** (°C):
MW: 183.21   **BP** (°C):

Solubility (Moles/L)	Solubility (Grams/L)	Temp (°C)	Ref (#)	Evaluation (T P E A A)	Comments
1.500E-04	3.280E-02	2	B193	1 1 0 0 1	

1204.   C$_7$H$_{14}$
1-Heptene
1-n-Heptene
n-Hept-1-ene
RN: 592-76-7   **MP** (°C):   -119
MW: 98.19   **BP** (°C):   93.6

Solubility (Moles/L)	Solubility (Grams/L)	Temp (°C)	Ref (#)	Evaluation (T P E A A)	Comments
1.850E-04	1.817E-02	25	M342	1 0 1 1 2	

1205. C₇H₁₄

Methylcyclohexane
Hexahydrotoluene
Methyl Cyclohexane

RN: 108-87-2   **MP** (°C):   -126
MW: 98.19   **BP** (°C):   101

Solubility (Moles/L)	Solubility (Grams/L)	Temp (°C)	Ref (#)	Evaluation (T P E A A)	Comments
1.711E-04	1.680E-02	20	B318	1 2 1 2 0	EFG
1.691E-04	1.660E-02	20	B356	1 0 0 0 2	
1.324E-04	1.300E-02	20	M337	2 1 2 2 2	
1.701E-04	1.670E-02	25	G313	2 1 1 2 2	
1.629E-04	1.600E-02	25	K119	1 0 0 0 2	
1.426E-04	1.400E-02	25	M001	2 1 2 2 2	
1.426E-04	1.400E-02	25	M002	2 1 2 2 2	
1.629E-04	1.600E-02	25.0	P051	2 1 1 2 2	
1.629E-04	1.600E-02	25.00	P007	2 1 2 2 2	
1.375E-04	1.350E-02	28	B348	2 1 2 2 2	
1.833E-04	1.800E-02	40.1	P051	2 1 1 2 2	
1.833E-04	1.800E-02	40.10	P007	2 1 2 2 2	
1.925E-04	1.890E-02	55.7	P051	2 1 1 2 2	
1.925E-04	1.890E-02	55.70	P007	2 1 2 2 2	
3.442E-04	3.380E-02	99.1	P051	2 1 1 2 2	
3.442E-04	3.380E-02	99.10	P007	2 1 2 2 2	
8.097E-04	7.950E-02	120.0	P051	2 1 1 2 2	
8.097E-04	7.950E-02	120.00	P007	2 1 2 2 2	
1.416E-03	1.390E-01	137.3	P051	2 1 1 2 2	
1.416E-03	1.390E-01	137.30	P007	2 1 2 2 2	
2.485E-03	2.440E-01	149.5	P051	2 1 1 2 2	
2.485E-03	2.440E-01	149.50	P007	2 1 2 2 2	
1.426E-04	1.400E-02	ns	H123	0 0 0 0 2	

1206. C₇H₁₄

2-Heptene

RN: 592-77-8   **MP** (°C):
MW: 98.19   **BP** (°C):

Solubility (Moles/L)	Solubility (Grams/L)	Temp (°C)	Ref (#)	Evaluation (T P E A A)	Comments
1.528E-04	1.500E-02	23.5	S171	2 1 2 2 2	
1.528E-04	1.500E-02	25	M001	2 1 2 2 1	

1207. C₇H₁₄

Cycloheptane
RN: 291-64-5   **MP** (°C): -12
MW: 98.19   **BP** (°C): 118.5

Solubility (Moles/L)	Solubility (Grams/L)	Temp (°C)	Ref (#)	Evaluation (T P E A A)	Comments
1.854E-04	1.820E-02	20	M337	2 1 2 2 2	
3.055E-04	3.000E-02	25	M001	2 1 2 2 2	
2.760E-04	2.710E-02	30	G313	2 1 1 2 2	

1208. C₇H₁₄N₂O₂S

Aldicarb
Temik
2-Methyl-2-(methylthio)propanal O-[(Methylamino)carbonyl]oxime
UC21149
N-Methylcarbamoyloxime, 2-Methyl-2-methylsulfenylpropionaldehyde
Methylcarbamic Acid
RN: 116-06-3   **MP** (°C): 99
MW: 190.27   **BP** (°C):

Solubility (Moles/L)	Solubility (Grams/L)	Temp (°C)	Ref (#)	Evaluation (T P E A A)	Comments
3.162E-02	6.017E+00	20	B179	2 0 0 0 2	
3.153E-02	6.000E+00	ns	H042	0 0 0 0 2	
3.135E-02	5.964E+00	ns	M061	0 0 0 0 0	
3.153E-02	6.000E+00	rt	M161	0 0 0 0 0	

1209. C₇H₁₄N₂O₃

ε-Aminocaproic Hydantoic Acid
ε-Uramidocaproic Acid
RN: **MP** (°C):
MW: 174.20   **BP** (°C):

Solubility (Moles/L)	Solubility (Grams/L)	Temp (°C)	Ref (#)	Evaluation (T P E A A)	Comments
6.900E-03	1.202E+00	25	M024	1 2 0 1 2	

1210. C₇H₁₄N₂O₃

α-Aminocaproic Hydantoic Acid
α-Uramidocaproic Acid
RN: **MP** (°C): 169
MW: 174.20   **BP** (°C):

Solubility (Moles/L)	Solubility (Grams/L)	Temp (°C)	Ref (#)	Evaluation (T P E A A)	Comments
6.900E-03	1.202E+00	25	M024	1 2 0 1 2	

1211.   C₇H₁₄N₂O₄S₂
Djenkoic Acid
Djenkolsaeure
RN: 498-59-9   **MP** (°C):
MW: 254.33   **BP** (°C):

Solubility (Moles/L)	Solubility (Grams/L)	Temp (°C)	Ref (#)	Evaluation (T P E A A)	Comments
1.966E-02	5.000E+00	100	F300	1 0 0 0 0	

1212.   C₇H₁₄N₆
N2,N2,N4,N4-Tetramethylmelamine
Tetramethylmelamine
RN: 2827-47-6   **MP** (°C):   227.0
MW: 182.23   **BP** (°C):

Solubility (Moles/L)	Solubility (Grams/L)	Temp (°C)	Ref (#)	Evaluation (T P E A A)	Comments
2.052E-03	3.740E-01	25	C051	1 2 1 1 2	pH 7

1213.   C₇H₁₄O
2-Heptanone
Heptan-2-one
RN: 110-43-0   **MP** (°C):   -31
MW: 114.19   **BP** (°C):   151.5

Solubility (Moles/L)	Solubility (Grams/L)	Temp (°C)	Ref (#)	Evaluation (T P E A A)	Comments
3.489E-02	3.984E+00	20	D052	1 1 0 0 0	
3.836E-02	4.381E+00	20	G030	1 2 0 0 1	
3.800E-02	4.339E+00	20	M312	1 0 0 0 1	
3.750E-02	4.282E+00	25	G030	1 2 0 0 1	
1.675E-01	1.913E+01	25	P055	1 0 0 0 1	
3.570E-02	4.077E+00	25	W300	2 2 2 2 2	
3.489E-02	3.984E+00	30	G030	1 2 0 0 1	

1214.   C₇H₁₄O
2,4-Dimethyl-3-pentanone
2,4-Dimethylpentanone-3
RN: 565-80-0   **MP** (°C):   -80
MW: 114.19   **BP** (°C):   124

Solubility (Moles/L)	Solubility (Grams/L)	Temp (°C)	Ref (#)	Evaluation (T P E A A)	Comments
5.137E-02	5.865E+00	20	G030	1 2 0 0 1	
4.963E-02	5.668E+00	25	G030	1 2 0 0 1	
4.877E-02	5.569E+00	30	G030	1 2 0 0 1	
4.972E-02	5.677E+00	ns	J300	0 0 0 0 1	

1215. C₇H₁₄O

Dipropyl Ketone
4-Heptanone
RN: 123-19-3   **MP** (°C): -32.6
MW: 114.19   **BP** (°C): 144

Solubility (Moles/L)	Solubility (Grams/L)	Temp (°C)	Ref (#)	Evaluation (T P E A A)	Comments
6.430E-02	7.342E+00	0	G032	1 2 1 1 2	
4.660E-02	5.321E+00	10	G032	1 2 1 1 2	
3.750E-02	4.282E+00	20	D052	1 1 0 0 1	
2.793E-02	3.190E+00	25.50	O005	2 0 2 2 1	
3.350E-02	3.825E+00	30	G032	1 2 1 1 2	
2.880E-02	3.289E+00	50	G032	1 2 1 1 2	
2.720E-02	3.106E+00	75	G032	1 2 1 1 2	

1216. C₇H₁₄O

Heptyl Aldehyde
Heptanal
Oenanthaldehyd
RN: 111-71-7   **MP** (°C): -43.3
MW: 114.19   **BP** (°C): 152.8

Solubility (Moles/L)	Solubility (Grams/L)	Temp (°C)	Ref (#)	Evaluation (T P E A A)	Comments
2.715E-02	3.100E+00	0	F300	1 0 0 0 1	
1.576E-02	1.800E+00	40	F300	1 0 0 0 1	

1217. C₇H₁₄O₂

Isoamyl Acetate
Acetic Acid Isoamyl Ester
Essigsaeureisoamyl Ester
RN: 123-92-2   **MP** (°C): -79
MW: 130.19   **BP** (°C): 142

Solubility (Moles/L)	Solubility (Grams/L)	Temp (°C)	Ref (#)	Evaluation (T P E A A)	Comments
1.920E-02	2.500E+00	15	F300	1 0 0 0 1	
1.222E-02	1.591E+00	20	E002	1 0 0 0 1	
1.227E-02	1.597E+00	23.50	O005	2 0 2 2 1	
1.533E-02	1.996E+00	25	L062	2 2 0 1 0	

1218. C$_7$H$_{14}$O$_2$
sec-Amyl Acetate
2-Pentyl Acetate
1-Methylbutyl Acetate
RN: 53496-15-4   **MP** (°C):
MW: 130.19   **BP** (°C):

Solubility (Moles/L)	Solubility (Grams/L)	Temp (°C)	Ref (#)	Evaluation (T P E A A)	Comments
1.457E-02	1.896E+00	20	D052	1 1 0 0 0	

1219. C$_7$H$_{14}$O$_2$
Propyl Butyrate
Buttersaeure-propyl Ester
n-Propyl n-Butyrate
RN: 105-66-8   **MP** (°C):   -95
MW: 130.19   **BP** (°C):   143

Solubility (Moles/L)	Solubility (Grams/L)	Temp (°C)	Ref (#)	Evaluation (T P E A A)	Comments
1.240E-02	1.614E+00	17	F001	1 0 1 0 2	
1.244E-02	1.620E+00	17	F300	1 0 0 0 2	
1.200E-02	1.562E+00	17	S006	1 0 0 0 1	

1220. C$_7$H$_{14}$O$_2$
n-Butyl Propionate
Butyl Propionate
RN: 590-01-2   **MP** (°C):   -89
MW: 130.19   **BP** (°C):   146.8

Solubility (Moles/L)	Solubility (Grams/L)	Temp (°C)	Ref (#)	Evaluation (T P E A A)	Comments
1.150E-02	1.498E+00	20	D052	1 1 0 0 0	
9.500E-03	1.237E+00	25	K012	1 0 0 0 1	

1221. C$_7$H$_{14}$O$_2$
3-Hydroxy-5-methyl-5-ethyltetrahydrofuran
3-Furanol, 5-Ethyltetrahydro-5-methyl-
RN: 30010-08-3   **MP** (°C):
MW: 130.19   **BP** (°C):

Solubility (Moles/L)	Solubility (Grams/L)	Temp (°C)	Ref (#)	Evaluation (T P E A A)	Comments
6.983E-01	9.091E+01	rt	B066	0 2 0 0 1	

1222. $C_7H_{14}O_2$
Methyl Hexanoate
Methyl Caproate

RN:	106-70-7	**MP** (°C):	-71.0	
MW:	130.19	**BP** (°C):	151.0	

Solubility (Moles/L)	Solubility (Grams/L)	Temp (°C)	Ref (#)	Evaluation (T P E A A)	Comments
1.018E-02	1.325E+00	20	M337	2 1 2 2 2	

1223. $C_7H_{14}O_2$
Isopropyl N-Butyrate
Isopropyl Butyrate
N-Butyric Acid Isopropyl Ester

RN:	638-11-9	**MP** (°C):		
MW:	130.19	**BP** (°C):		

Solubility (Moles/L)	Solubility (Grams/L)	Temp (°C)	Ref (#)	Evaluation (T P E A A)	Comments
1.198E-02	1.560E+00	ns	J300	0 0 0 0 1	

1224. $C_7H_{14}O_2$
Heptoic Acid
Heptanoic Acid
n-Heptanoic Acid

RN:	111-14-8	**MP** (°C):		
MW:	130.19	**BP** (°C):		

Solubility (Moles/L)	Solubility (Grams/L)	Temp (°C)	Ref (#)	Evaluation (T P E A A)	Comments
1.459E-02	1.900E+00	0	B136	1 0 2 1 2	
1.457E-02	1.896E+00	0.0	R001	1 1 1 1 2	
1.843E-02	2.400E+00	15	F300	1 0 0 0 1	
1.847E-02	2.404E+00	15	L006	1 0 0 0 2	
1.721E-02	2.240E+00	20	B136	1 0 2 1 2	
1.870E-02	2.434E+00	20.0	R001	1 1 1 1 2	
2.161E-02	2.813E+00	25	H122	1 0 0 0 2	
2.082E-02	2.710E+00	30	B136	1 0 2 1 2	
2.076E-02	2.703E+00	30.0	R001	1 1 1 1 2	
2.389E-02	3.110E+00	45	B136	1 0 2 1 2	
2.381E-02	3.100E+00	45.0	R001	1 1 1 1 2	
2.711E-02	3.530E+00	60	B136	1 0 2 1 2	
2.702E-02	3.518E+00	60.0	R001	1 1 1 1 2	

1225.   C$_7$H$_{14}$O$_2$
Pentyl Acetate
Amyl Acetate
RN: 628-63-7   **MP** (°C):   -100
MW: 130.19   **BP** (°C):   142

Solubility (Moles/L)	Solubility (Grams/L)	Temp (°C)	Ref (#)	Evaluation (T P E A A)	Comments
1.304E-02	1.697E+00	20	D052	1 1 0 0 1	
1.290E-02	1.679E+00	20	S006	1 0 0 0 2	
1.329E-02	1.730E+00	25	K072	1 0 1 1 1	
1.329E-02	1.730E+00	25	M087	1 1 2 1 2	
3.060E-02	3.984E+00	30	R318	1 1 0 1 0	

1226.   C$_7$H$_{14}$O$_3$
3-Methoxy Butyl Acetate
3-Methoxy-1-Butanol Acetate
Methyl-1,3-Butylene Glycol Acetate
3-Methoxybutyl Acetate
Butoxyl
Butoxyl (3-Methoxy-N-butyl Acetate)
RN: 4435-53-4   **MP** (°C):
MW: 146.19   **BP** (°C):

Solubility (Moles/L)	Solubility (Grams/L)	Temp (°C)	Ref (#)	Evaluation (T P E A A)	Comments
4.151E-01	6.068E+01	20	D052	1 1 0 0 2	

1227.   C$_7$H$_{14}$O$_3$
Butyl Lactate
Butyl α-Hydroxypropionate
2-Propanoic Acid
Lactic Acid Butyl Ester
Butyl 2-Hydroxypropanoate
RN: 138-22-7   **MP** (°C):   -28
MW: 146.19   **BP** (°C):   185

Solubility (Moles/L)	Solubility (Grams/L)	Temp (°C)	Ref (#)	Evaluation (T P E A A)	Comments
2.631E-01	3.846E+01	20	D052	1 1 0 0 1	
2.982E-01	4.360E+01	25	R006	2 2 0 1 2	

1228. $C_7H_{14}O_3$
n-Ethyl β-Ethoxypropionate
Ethyl β-Ethoxypropionate
RN: 763-69-9   **MP** (°C):
MW: 146.19   **BP** (°C):   166

Solubility (Moles/L)	Solubility (Grams/L)	Temp (°C)	Ref (#)	Evaluation (T P E A A)	Comments
3.597E-01	5.258E+01	25	D002	1 2 1 1 2	
3.566E-01	5.213E+01	25	R034	0 0 0 0 1	

1229. $C_7H_{14}O_3$
Methyl β-n-Propoxypropionate
Propanoic Acid, 3-Propoxy-, Methyl Ester
RN: 14144-39-9   **MP** (°C):
MW: 146.19   **BP** (°C):

Solubility (Moles/L)	Solubility (Grams/L)	Temp (°C)	Ref (#)	Evaluation (T P E A A)	Comments
2.249E-01	3.288E+01	25	R034	0 0 0 0 1	

1230. $C_7H_{14}O_3$
n-Propyl β-Methoxypropionate
Propionic Acid, 3-Methoxy-, Propyl Ester
RN: 5349-56-4   **MP** (°C):
MW: 146.19   **BP** (°C):

Solubility (Moles/L)	Solubility (Grams/L)	Temp (°C)	Ref (#)	Evaluation (T P E A A)	Comments
2.121E-01	3.101E+01	25	R034	0 0 0 0 1	

1231. $C_7H_{14}O_6$
β-Methyl-D-glucoside
β-Methyl-D-glucosid
RN: 709-50-2   **MP** (°C):
MW: 194.19   **BP** (°C):

Solubility (Moles/L)	Solubility (Grams/L)	Temp (°C)	Ref (#)	Evaluation (T P E A A)	Comments
1.892E+00	3.674E+02	17	F300	1 0 0 0 2	

1232. $C_7H_{14}O_6$
α-Methyl-D-mannoside
α-Methyl-D-mannosid

RN: 617-04-9   **MP** (°C):
MW: 194.19   **BP** (°C):

Solubility (Moles/L)	Solubility (Grams/L)	Temp (°C)	Ref (#)	Evaluation (T P E A A)	Comments
1.018E+00	1.976E+02	17	F300	1 0 0 0 2	

1233. $C_7H_{14}O_6$
α-D-Methylglucoside
α-Methyl-D-glucoside
α-Methyl-D-glucosid

RN: 97-30-3   **MP** (°C):   168
MW: 194.19   **BP** (°C):

Solubility (Moles/L)	Solubility (Grams/L)	Temp (°C)	Ref (#)	Evaluation (T P E A A)	Comments
1.992E+00	3.868E+02	17	F300	1 0 0 0 2	
2.543E+00	4.938E+02	17.8	W013	1 2 1 1 2	
2.637E+00	5.120E+02	22.5	W013	1 2 1 1 2	
2.657E+00	5.159E+02	25.5	W013	1 2 1 1 2	
2.696E+00	5.236E+02	26.6	W013	1 2 1 1 2	
2.699E+00	5.241E+02	27.3	W013	1 2 1 1 2	
2.751E+00	5.342E+02	31.8	W013	1 2 1 1 2	
2.806E+00	5.448E+02	33.9	W013	1 2 1 1 2	
2.849E+00	5.533E+02	37.2	W013	1 2 1 1 2	
2.951E+00	5.731E+02	43.2	W013	1 2 1 1 2	
3.060E+00	5.942E+02	49.0	W013	1 2 1 1 2	
3.078E+00	5.978E+02	49.6	W013	1 2 1 1 2	
3.131E+00	6.079E+02	51.8	W013	1 2 1 1 2	
3.166E+00	6.148E+02	54.4	W013	1 2 1 1 2	
3.213E+00	6.240E+02	57.3	W013	1 2 1 1 2	
3.297E+00	6.402E+02	60.6	W013	1 2 1 1 2	
3.332E+00	6.471E+02	62.7	W013	1 2 1 1 2	
3.360E+00	6.525E+02	64.2	W013	1 2 1 1 2	
3.403E+00	6.608E+02	66.2	W013	1 2 1 1 2	
3.435E+00	6.670E+02	67.8	W013	1 2 1 1 2	
3.542E+00	6.878E+02	73.2	W013	1 2 1 1 2	
3.651E+00	7.090E+02	78.0	W013	1 2 1 1 2	

1234. C$_7$H$_{14}$O$_7$
D-Mannoheptose
D-Sedoheptose
RN: 7634-39-1   **MP** (°C):
MW: 210.19   **BP** (°C):

Solubility (Moles/L)	Solubility (Grams/L)	Temp (°C)	Ref (#)	Evaluation (T P E A A)	Comments
>4.76E-01	>1.00E+02	20	F300	1 0 0 0 0	

1235. C$_7$H$_{14}$O$_7$
D-α-Glucoheptose
Gluco-Heptose
RN: 62475-58-5   **MP** (°C):
MW: 210.19   **BP** (°C):

Solubility (Moles/L)	Solubility (Grams/L)	Temp (°C)	Ref (#)	Evaluation (T P E A A)	Comments
4.128E-01	8.676E+01	20	D041	1 0 0 0 1	

1236. C$_7$H$_{15}$Br
1-Bromoheptane
Heptyl Bromide
RN: 629-04-9   **MP** (°C):   -56.1
MW: 179.11   **BP** (°C):   178.5

Solubility (Moles/L)	Solubility (Grams/L)	Temp (°C)	Ref (#)	Evaluation (T P E A A)	Comments
3.710E-05	6.645E-03	25	M342	1 0 1 1 2	

1237. C$_7$H$_{15}$Cl
1-Chloroheptane
Heptyl Chloride
RN: 629-06-1   **MP** (°C):   -69.5
MW: 134.65   **BP** (°C):   159

Solubility (Moles/L)	Solubility (Grams/L)	Temp (°C)	Ref (#)	Evaluation (T P E A A)	Comments
1.010E-04	1.360E-02	25	M342	1 0 1 1 2	

1238. C$_7$H$_{15}$I
1-Iodoheptane
Heptyl Iodide
RN: 4282-40-0   **MP** (°C):   -48.2
MW: 226.10   **BP** (°C):   204

Solubility (Moles/L)	Solubility (Grams/L)	Temp (°C)	Ref (#)	Evaluation (T P E A A)	Comments
1.550E-05	3.505E-03	25	M342	1 0 1 1 2	

1239. C$_7$H$_{15}$NO$_2$
n-Hexyl Carbamate
Hexyl Carbamate
RN: 2114-20-7   **MP** (°C):   62
MW: 145.20   **BP** (°C):

Solubility (Moles/L)	Solubility (Grams/L)	Temp (°C)	Ref (#)	Evaluation (T P E A A)	Comments
1.200E-02	1.742E+00	37	H006	1 2 2 1 1	

1240. C$_7$H$_{15}$NO$_2$
Isobutyl Urethane
Isobutylurethan
RN: 539-89-9   **MP** (°C):
MW: 145.20   **BP** (°C):

Solubility (Moles/L)	Solubility (Grams/L)	Temp (°C)	Ref (#)	Evaluation (T P E A A)	Comments
1.709E-01	2.482E+01	15.5	F001	1 0 1 2 2	

1241. C$_7$H$_{15}$NO$_2$
tert-Hexyl Carbamate
3,3-Dimethyl-1-butanol Carbamate
RN: 3124-38-7   **MP** (°C):
MW: 145.20   **BP** (°C):

Solubility (Moles/L)	Solubility (Grams/L)	Temp (°C)	Ref (#)	Evaluation (T P E A A)	Comments
3.400E-02	4.937E+00	37	H006	1 2 2 1 1	

1242. C$_7$H$_{16}$
2,2-Dimethylpentane
2,2-Dwumetylopentan
RN: 590-35-2   **MP** (°C):   -123
MW: 100.21   **BP** (°C):   79.2

Solubility (Moles/L)	Solubility (Grams/L)	Temp (°C)	Ref (#)	Evaluation (T P E A A)	Comments
4.391E-05	4.400E-03	25	K119	1 0 0 0 2	
4.391E-05	4.400E-03	25	P051	2 1 1 2 2	
4.391E-05	4.400E-03	25.00	P007	2 1 2 2 2	
4.100E-05	4.108E-03	ns	J300	0 0 0 0 1	

Solutions

1243.   C$_7$H$_{16}$
3,3-Dimethylpentane
3,3-Dwumetylopentan
RN: 562-49-2   **MP** (°C):   -135
MW: 100.21   **BP** (°C):   86

Solubility (Moles/L)	Solubility (Grams/L)	Temp (°C)	Ref (#)	Evaluation (T P E A A)	Comments
5.928E-05	5.940E-03	25	K119	1 0 0 0 2	
5.908E-05	5.920E-03	25.0	P051	2 1 1 2 2	
5.908E-05	5.920E-03	25.00	P007	2 1 2 2 2	
6.766E-05	6.780E-03	40.1	P051	2 1 1 2 2	
6.766E-05	6.780E-03	40.10	P007	2 1 2 2 2	
8.153E-05	8.170E-03	55.7	P051	2 1 1 2 2	
8.153E-05	8.170E-03	55.70	P007	2 1 2 2 2	
1.028E-04	1.030E-02	69.7	P051	2 1 1 2 2	
1.028E-04	1.030E-02	69.70	P007	2 1 2 2 2	
1.577E-04	1.580E-02	99.1	P051	2 1 1 2 2	
1.577E-04	1.580E-02	99.10	P007	2 1 2 2 2	
2.724E-04	2.730E-02	118.0	P051	2 1 1 2 2	
2.724E-04	2.730E-02	118.00	P007	2 1 2 2 2	
6.716E-04	6.730E-02	120.4	P051	2 1 1 2 2	
6.716E-04	6.730E-02	120.40	P007	2 1 2 2 2	
8.592E-04	8.610E-02	150.4	P051	2 1 1 2 2	
8.592E-04	8.610E-02	150.40	P007	2 1 2 2 2	

1244.   C$_7$H$_{16}$
2-Methylhexane
2-Metyloheksan
RN: 591-76-4   **MP** (°C):   -118
MW: 100.21   **BP** (°C):   90

Solubility (Moles/L)	Solubility (Grams/L)	Temp (°C)	Ref (#)	Evaluation (T P E A A)	Comments
1.397E-04	1.400E-02	23	C332	2 0 2 2 1	
2.535E-05	2.540E-03	25	K119	1 0 0 0 2	
2.535E-05	2.540E-03	25	P051	2 1 1 2 2	
2.535E-05	2.540E-03	25.00	P007	2 1 2 2 2	

1245.   C$_7$H$_{16}$
2,3-Dimethylpentane
2,3-Dwumetylopentan
RN: 565-59-3   **MP** (°C):   <25
MW: 100.21   **BP** (°C):   89

Solubility (Moles/L)	Solubility (Grams/L)	Temp (°C)	Ref (#)	Evaluation (T P E A A)	Comments
5.239E-05	5.250E-03	25	K119	1 0 0 0 2	
5.239E-05	5.250E-03	25	P051	2 1 1 2 2	
5.239E-05	5.250E-03	25.00	P007	2 1 2 2 2	

1246.   C₇H₁₆
3-Methylhexane
3-Metyloheksan
RN: 589-34-4   **MP** (°C):   -119
MW: 100.21   **BP** (°C):   91

Solubility (Moles/L)	Solubility (Grams/L)	Temp (°C)	Ref (#)	Evaluation (T P E A A)	Comments
5.229E-05	5.240E-03	0	P003	2 2 2 2 2	
1.048E-04	1.050E-02	23	C332	2 0 2 2 1	
2.635E-05	2.640E-03	25	K119	1 0 0 0 2	
4.940E-05	4.950E-03	25	P003	2 2 2 2 2	
2.635E-05	2.640E-03	25	P051	2 1 1 2 2	
2.635E-05	2.640E-03	25.00	P007	2 1 2 2 2	

1247.   C₇H₁₆
Heptane
n-Heptane
RN: 142-82-5   **MP** (°C):   -90.7
MW: 100.21   **BP** (°C):   98.4

Solubility (Moles/L)	Solubility (Grams/L)	Temp (°C)	Ref (#)	Evaluation (T P E A A)	Comments
4.381E-05	4.390E-03	0	P003	2 2 2 2 2	
1.950E-05	1.954E-03	4.3	N004	1 1 2 2 2	
2.017E-05	2.021E-03	13.5	N004	1 1 2 2 2	
4.990E-04	5.000E-02	15	F300	1 0 0 0 1	
5.200E-04	5.211E-02	15.50	F001	1 0 1 0 2	
1.497E-04	1.500E-02	16	D047	1 0 0 1 0	
2.694E-05	2.700E-03	20	M337	2 1 2 2 1	
3.990E-03	3.998E-01	25	G323	2 2 2 2 0	
4.990E-04	5.000E-02	25	K072	1 0 1 1 1	
2.235E-05	2.240E-03	25	K119	1 0 0 0 2	
2.924E-05	2.930E-03	25	M001	2 1 2 2 2	
2.924E-05	2.930E-03	25	M002	2 1 2 2 2	
4.990E-04	5.000E-02	25	M087	1 1 2 1 0	
3.050E-05	3.056E-03	25	M342	1 0 1 1 2	
3.363E-05	3.370E-03	25	P003	2 2 2 2 2	
4.989E-04	5.000E-02	25	S012	2 0 2 2 0	
2.656E-05	2.661E-03	25.0	N004	1 1 2 2 2	
2.235E-05	2.240E-03	25.0	P051	2 1 1 2 2	
2.235E-05	2.240E-03	25.00	P007	2 1 2 2 2	
2.261E-05	2.266E-03	35.0	N004	1 1 2 2 2	
2.625E-05	2.630E-03	40.1	P051	2 1 1 2 2	
2.400E-05	2.405E-03	45.0	N004	1 1 2 2 2	
8.973E-03	8.992E-01	50	G323	2 2 2 2 0	
3.104E-05	3.110E-03	55.7	P051	2 1 1 2 2	
3.104E-05	3.110E-03	55.70	P007	2 1 2 2 2	

5.589E-05	5.600E-03	99.1	P051	2 1 1 2 2
5.589E-05	5.600E-03	99.10	P007	2 1 2 2 2
1.138E-04	1.140E-02	118	P007	2 1 2 2 2
1.138E-04	1.140E-02	118.0	P051	2 1 1 2 2
2.724E-04	2.730E-02	136.6	P051	2 1 1 2 2
2.724E-04	2.730E-02	136.60	P007	2 1 2 2 2
4.361E-04	4.370E-02	150.4	P051	2 1 1 2 2
4.361E-04	4.370E-02	150.40	P007	2 1 2 2 2
3.692E-05	3.700E-03	ns	B151	0 2 1 1 1
7.000E-04	7.014E-02	ns	H012	0 2 2 0 0

1248. C_7H_{16}
2,4-Dimethylpentane
2,4-Dwumetylopentan

RN: 108-08-7 **MP** (°C): -123
MW: 100.21 **BP** (°C): 80

Solubility (Moles/L)	Solubility (Grams/L)	Temp (°C)	Ref (#)	Evaluation (T P E A A)	Comments
6.487E-05	6.500E-03	0	P003	2 2 2 2 2	
4.401E-05	4.410E-03	25	K119	1 0 0 0 2	
4.052E-05	4.060E-03	25	M001	2 1 2 2 2	
3.613E-05	3.620E-03	25	M002	2 1 2 2 2	
5.489E-05	5.500E-03	25	P003	2 2 2 2 2	
4.401E-05	4.410E-03	25	P051	2 1 1 2 2	
4.401E-05	4.410E-03	25.00	P007	2 1 2 2 2	
4.100E-05	4.108E-03	ns	J300	0 0 0 0 1	

1249. $C_7H_{16}O$
3-Ethyl-3-pentanol
3-Ethyl-pentanol-3
Triethyl Carbinol

RN: 597-49-9 **MP** (°C): -12
MW: 116.20 **BP** (°C): 141.0

Solubility (Moles/L)	Solubility (Grams/L)	Temp (°C)	Ref (#)	Evaluation (T P E A A)	Comments
1.613E-01	1.874E+01	20	G006	1 2 1 1 2	
1.422E-01	1.652E+01	25	G006	1 2 1 1 2	
1.272E-01	1.478E+01	30	G006	1 2 1 1 2	
1.071E-01	1.244E+01	40	G006	1 2 1 1 2	

1250.   C₇H₁₆O

3-Heptanol
(±)-3-Heptanol
3-Hydroxyheptane
1-Ethyl-1-pentanol

RN: 589-82-2   **MP** (°C):   -70
MW: 116.20   **BP** (°C):   156.0

Solubility (Moles/L)	Solubility (Grams/L)	Temp (°C)	Ref (#)	Evaluation (T P E A A)	Comments
4.100E-02	4.764E+00	20	H330	2 0 2 2 2	
3.428E-02	3.984E+00	25	C093	2 1 1 1 0	

1251.   C₇H₁₆O

2,3-Dimethyl-3-pentanol
2,3-Dimethylpentanol-3

RN: 595-41-5   **MP** (°C):   <25
MW: 116.20   **BP** (°C):   140

Solubility (Moles/L)	Solubility (Grams/L)	Temp (°C)	Ref (#)	Evaluation (T P E A A)	Comments
1.580E-01	1.836E+01	20	G006	1 2 1 1 2	
1.389E-01	1.614E+01	25	G006	1 2 1 1 2	
1.213E-01	1.410E+01	30	G006	1 2 1 1 2	

1252.   C₇H₁₆O

3-Methyl-3-hexanol
3-Methylhexanol-3

RN: 597-96-6   **MP** (°C):   <25
MW: 116.20   **BP** (°C):

Solubility (Moles/L)	Solubility (Grams/L)	Temp (°C)	Ref (#)	Evaluation (T P E A A)	Comments
1.146E-01	1.332E+01	20	G006	1 2 1 1 2	
1.012E-01	1.176E+01	25	G006	1 2 1 1 2	
9.110E-02	1.059E+01	30	G006	1 2 1 1 2	

1253.   C₇H₁₆O

Isopropyl tert-Butyl Ether
2-Methyl-2-(1-methylethoxy)-propane
t-Butyl Isopropyl Ether

RN: 17348-59-3   **MP** (°C):   -88
MW: 116.20   **BP** (°C):   87.6

Solubility (Moles/L)	Solubility (Grams/L)	Temp (°C)	Ref (#)	Evaluation (T P E A A)	Comments
4.303E-03	5.000E-01	25	K072	1 0 1 1 1	
4.303E-03	5.000E-01	25	M087	1 1 2 1 1 ·	

1254. C$_7$H$_{16}$O
2-Methyl-2-hexanol
2-Methylhexanol-2

RN: 625-23-0   **MP** (°C): <25
MW: 116.20   **BP** (°C): 141

Solubility (Moles/L)	Solubility (Grams/L)	Temp (°C)	Ref (#)	Evaluation (T P E A A)	Comments
9.195E-02	1.068E+01	20	G006	1 2 1 1 2	
8.267E-02	9.607E+00	25	G006	1 2 1 1 1	
7.422E-02	8.625E+00	30	G006	1 2 1 1 1	

1255. C$_7$H$_{16}$O
2,4-Dimethyl-3-pentanol
2,4-Dimethylpentanol-3
Diisopropyl Carbinol

RN: 600-36-2   **MP** (°C): -70
MW: 116.20   **BP** (°C):

Solubility (Moles/L)	Solubility (Grams/L)	Temp (°C)	Ref (#)	Evaluation (T P E A A)	Comments
1.009E-01	1.172E+01	0	S307	1 1 0 2 2	
8.942E-02	1.039E+01	10.0	S307	1 1 0 2 2	
6.660E-02	7.740E+00	20	G006	1 2 1 1 1	
6.067E-02	7.050E+00	20.2	S307	1 1 0 2 2	
1.935E-01	2.248E+01	24.50	O005	2 0 2 2 1	
5.982E-02	6.951E+00	25	G006	1 2 1 1 1	
5.727E-02	6.655E+00	30	G006	1 2 1 1 1	
5.489E-02	6.379E+00	30.6	S307	1 1 0 2 2	
4.562E-02	5.302E+00	39.5	S307	1 1 0 2 2	
4.332E-02	5.035E+00	49.7	S307	1 1 0 2 2	
3.992E-02	4.638E+00	60.3	S307	1 1 0 2 2	
3.778E-02	4.391E+00	70.2	S307	1 1 0 2 2	
3.667E-02	4.262E+00	80.2	S307	1 1 0 2 2	
3.855E-02	4.480E+00	90.6	S307	1 1 0 2 2	

1256. C$_7$H$_{16}$O
2,4-Dimethyl-2-pentanol
2,4-Dimethylpentanol-2

RN: 625-06-9   **MP** (°C): <-20
MW: 116.20   **BP** (°C):

Solubility (Moles/L)	Solubility (Grams/L)	Temp (°C)	Ref (#)	Evaluation (T P E A A)	Comments
1.272E-01	1.478E+01	20	G006	1 2 1 1 2	
1.138E-01	1.322E+01	25	G006	1 2 1 1 2	
1.037E-01	1.205E+01	30	G006	1 2 1 1 2	

1257.   C₇H₁₆O

2,3-Dimethyl-2-pentanol
2,3-Dimethylpentanol-2
RN: 4911-70-0   **MP** (°C):   <25
MW: 116.20   **BP** (°C):

Solubility (Moles/L)	Solubility (Grams/L)	Temp (°C)	Ref (#)	Evaluation (T P E A A)	Comments
1.430E-01	1.662E+01	20	G006	1 2 1 1 2	
1.305E-01	1.517E+01	25	G006	1 2 1 1 2	
1.188E-01	1.381E+01	30	G006	1 2 1 1 2	

1258.   C₇H₁₆O

2,3,3-Trimethyl-2-butanol
Dimethyl-tert-butylcarbinol
1,1,2,2-Tetramethylpropanol
1,1,2,2-Tetramethylpropyl alcohol
RN: 594-83-2   **MP** (°C):   17
MW: 116.20   **BP** (°C):   131

Solubility (Moles/L)	Solubility (Grams/L)	Temp (°C)	Ref (#)	Evaluation (T P E A A)	Comments
1.852E-01	2.153E+01	40	G006	1 2 1 1 2	

1259.   C₇H₁₆O

1-Heptanol
1-Hydroxyheptane
Heptan-1-ol
Heptanol-(1)
n-Heptyl Alcohol
RN: 111-70-6   **MP** (°C):   -34.6
MW: 116.20   **BP** (°C):   175.8

Solubility (Moles/L)	Solubility (Grams/L)	Temp (°C)	Ref (#)	Evaluation (T P E A A)	Comments
2.916E-02	3.388E+00	0	E029	1 2 0 1 1	
2.026E-02	2.354E+00	0	S307	1 1 0 2 2	
1.897E-02	2.205E+00	6.04	H110	2 2 2 2 2	
2.232E-02	2.593E+00	10	E029	1 2 0 1 1	
1.739E-02	2.020E+00	10.24	H110	2 2 2 2 2	
2.172E-02	2.524E+00	10.5	S307	1 1 0 2 2	
1.720E-02	1.999E+00	10.54	H110	2 2 2 2 2	
1.067E-02	1.240E+00	11.4	N042	1 0 2 1 1	
1.608E-02	1.869E+00	15.04	H110	2 2 2 2 2	
1.544E-02	1.795E+00	17.94	H110	2 2 2 2 2	
8.000E-03	9.296E-01	18	F001	1 0 1 0 2	
8.605E-03	1.000E+00	18	F300	1 0 0 0 1	
1.478E-02	1.717E+00	20	A015	1 2 1 1 2	

1.718E-02	1.996E+00	20	E029	1 2 0 1 1	
1.450E-02	1.685E+00	20	H330	2 0 2 2 2	
1.507E-02	1.751E+00	20.04	H110	2 2 2 2 2	
1.581E-02	1.837E+00	20.2	S307	1 1 0 2 2	
1.476E-02	1.716E+00	21.94	H110	2 2 2 2 2	
1.450E-02	1.685E+00	23.94	H110	2 2 2 2 2	
1.443E-02	1.677E+00	24.94	H110	2 2 2 2 2	
1.546E-02	1.797E+00	25	B038	1 2 1 1 2	
1.000E+00	1.162E+02	25	F044	1 0 0 0 0	EFG
1.460E-02	1.697E+00	25	K025	2 1 1 1 1	
1.434E-02	1.666E+00	25.04	H110	2 2 2 2 2	
1.423E-02	1.653E+00	26.04	H110	2 2 2 2 2	
1.411E-02	1.640E+00	28.04	H110	2 2 2 2 2	
1.375E-02	1.597E+00	30	E029	1 2 0 1 1	
1.397E-02	1.624E+00	30.14	H110	2 2 2 2 2	
1.399E-02	1.626E+00	30.14	H110	2 2 2 2 2	
1.323E-02	1.538E+00	30.6	S307	1 1 0 2 2	
1.386E-02	1.611E+00	32.94	H110	2 2 2 2 2	
1.426E-02	1.657E+00	39.8	S307	1 1 0 2 2	
1.117E-02	1.298E+00	40	E029	1 2 0 1 1	
9.456E-03	1.099E+00	50	E029	1 2 0 1 1	
1.392E-02	1.617E+00	50.1	S307	1 1 0 2 2	
9.456E-03	1.099E+00	60	E029	1 2 0 1 1	
1.529E-02	1.777E+00	60.0	S307	1 1 0 2 2	
1.289E-02	1.498E+00	70	E029	1 2 0 1 1	
1.080E-02	1.255E+00	70	F001	1 0 1 0 2	
1.752E-02	2.036E+00	70.1	S307	1 1 0 2 2	
1.632E-02	1.896E+00	80	E029	1 2 0 1 1	
1.460E-02	1.697E+00	80	F001	1 0 1 0 2	
1.863E-02	2.165E+00	80.1	S307	1 1 0 2 2	
1.975E-02	2.295E+00	90	E029	1 2 0 1 1	
1.940E-02	2.254E+00	90	F001	1 0 1 0 2	
2.086E-02	2.424E+00	90.5	S307	1 1 0 2 2	
2.488E-02	2.892E+00	100	E029	1 2 0 1 1	
2.460E-02	2.859E+00	100	F001	1 0 1 0 2	
2.582E-02	3.000E+00	100	F300	1 0 0 0 1	
3.001E-02	3.488E+00	110	E029	1 2 0 1 1	
3.060E-02	3.556E+00	110	F001	1 0 1 0 2	
3.685E-02	4.282E+00	120	E029	1 2 0 1 1	
4.537E-02	5.272E+00	130	E029	1 2 0 1 1	
5.557E-02	6.458E+00	140	E029	1 2 0 1 1	
6.830E-02	7.937E+00	150	E029	1 2 0 1 1	
8.352E-02	9.705E+00	160	E029	1 2 0 1 1	
1.046E-01	1.215E+01	170	E029	1 2 0 1 2	
1.355E-01	1.575E+01	180	E029	1 2 0 1 2	
1.753E-01	2.038E+01	190	E029	1 2 0 1 2	
2.213E-01	2.572E+01	200	E029	1 2 0 1 2	
2.894E-01	3.363E+01	210	E029	1 2 0 1 2	
3.847E-01	4.471E+01	220	E029	1 1 0 1 2	
5.404E-01	6.279E+01	230	E029	1 2 0 1 2	

7.894E-01	9.173E+01	240	E029	1 2 0 1 2
1.054E+00	1.225E+02	245	E029	1 2 0 1 2
1.029E-02	1.195E+00	ns	H012	0 2 2 0 2
1.558E-02	1.810E+00	ns	L003	0 0 2 1 2

1260. $C_7H_{16}O$
2-Heptanol
2-Hydroxyheptane
Amylmethylcarbinol

RN: 543-49-7   **MP** (°C):   <25
MW: 116.20   **BP** (°C):   159.00

Solubility (Moles/L)	Solubility (Grams/L)	Temp (°C)	Ref (#)	Evaluation (T P E A A)	Comments
5.532E-02	6.428E+00	0	S307	1 1 0 2 2	
3.966E-02	4.609E+00	10.2	S307	1 1 0 2 2	
3.633E-02	4.222E+00	19.5	S307	1 1 0 2 2	
3.001E-02	3.488E+00	30.7	S307	1 1 0 2 2	
2.813E-02	3.269E+00	40.0	S307	1 1 0 2 2	
2.514E-02	2.921E+00	50.0	S307	1 1 0 2 2	
2.471E-02	2.872E+00	60.3	S307	1 1 0 2 2	
2.754E-02	3.200E+00	70.3	S307	1 1 0 2 2	
2.754E-02	3.200E+00	80.0	S307	1 1 0 2 2	
2.942E-02	3.418E+00	90.2	S307	1 1 0 2 2	

1261. $C_7H_{16}O$
Heptanol
RN: 53535-33-4   **MP** (°C):   -36
MW: 116.20   **BP** (°C):   176

Solubility (Moles/L)	Solubility (Grams/L)	Temp (°C)	Ref (#)	Evaluation (T P E A A)	Comments
1.009E-01	1.173E+01	20	S006	1 0 0 0 2	
1.240E-02	1.441E+00	24	H345	2 0 2 2 2	

1262. $C_7H_{16}O$
4-Heptanol
Dipropyl Carbinol
RN: 589-55-9   **MP** (°C):   -42
MW: 116.20   **BP** (°C):

Solubility (Moles/L)	Solubility (Grams/L)	Temp (°C)	Ref (#)	Evaluation (T P E A A)	Comments
4.090E-02	4.753E+00	20	H330	2 0 2 2 2	

1263. $C_7H_{16}O$
2,2-Dimethyl-3-pentanol
2,2-Dimethylpentanol-3
RN: 3970-62-5 **MP** (°C): -5
MW: 116.20 **BP** (°C): 132

Solubility (Moles/L)	Solubility (Grams/L)	Temp (°C)	Ref (#)	Evaluation (T P E A A)	Comments
7.507E-02	8.723E+00	20	G006	1 2 1 1 1	
6.999E-02	8.133E+00	25	G006	1 2 1 1 1	
6.745E-02	7.838E+00	30	G006	1 2 1 1 1	

1264. $C_7H_{16}O_4S_2$
Sulfonmethane
Sulfonal
RN: 115-24-2 **MP** (°C): 125
MW: 228.33 **BP** (°C): 300

Solubility (Moles/L)	Solubility (Grams/L)	Temp (°C)	Ref (#)	Evaluation (T P E A A)	Comments
5.962E-02	1.361E+01	16	A072	1 0 1 0 2	
5.956E-02	1.360E+01	16	F300	1 0 0 0 2	
1.027E-02	2.345E+00	18	F062	1 0 2 2 2	
2.847E-01	6.500E+01	100	F300	1 0 0 0 1	

1265. $C_7H_{16}O_7$
(+)-Perseitol
D-Manno-α-heptit
RN: 527-06-0 **MP** (°C): 188
MW: 212.20 **BP** (°C):

Solubility (Moles/L)	Solubility (Grams/L)	Temp (°C)	Ref (#)	Evaluation (T P E A A)	Comments
3.044E-01	6.460E+01	18	F300	1 0 0 0 2	
1.466E+00	3.110E+02	74	F300	1 0 0 0 1	

1266.   C₇H₁₇O₂PS₃

Phorate

Thimet

Rampart

Phosphorodithioic Acid O,O-Diethyl S-[(Ethylthio)methyl] Ester

American Cyanamid 3911

CL 35,024

RN: 298-02-2   **MP** (°C):   -43

MW: 260.38   **BP** (°C):

Solubility (Moles/L)	Solubility (Grams/L)	Temp (°C)	Ref (#)	Evaluation (T P E A A)	Comments
6.874E-05	1.790E-02	20	B169	2 1 1 1 1	
1.905E-04	4.961E-02	20	B179	2 0 0 0 2	
7.681E-05	2.000E-02	24	F179	2 2 2 2 2	
2.688E-04	7.000E-02	ns	M061	0 0 0 0 1	
1.920E-04	5.000E-02	rt	M161	0 0 0 0 1	

1267.   C₇H₁₇O₂PS₃

S-2-Isopropylthioethyl O,O-Dimethyl Phosphorodithioate

Isothioate

O,O-Dimethyls-isopropylthioethyl Phosphoroditjioate

RN: 36614-38-7   **MP** (°C):

MW: 260.38   **BP** (°C):   55

Solubility (Moles/L)	Solubility (Grams/L)	Temp (°C)	Ref (#)	Evaluation (T P E A A)	Comments
3.725E-04	9.700E-02	25	M161	1 0 0 0 1	
3.725E-04	9.700E-02	25	N304	1 0 0 0 1	

1268.   C₇H₁₇O₄PS₃

Phorate Sulfone

O,O'-Diethyl S-Ethylsulfonylmethyl-phosphorodithioate

Thimet Sulfone

CL 18,161

Phosphorodithioic Acid O,O-Diethyl S-[(Ethylsulfonyl)methyl] Ester

RN: 2588-04-7   **MP** (°C):

MW: 292.38   **BP** (°C):

Solubility (Moles/L)	Solubility (Grams/L)	Temp (°C)	Ref (#)	Evaluation (T P E A A)	Comments
2.939E-03	8.593E-01	19	B169	2 0 1 1 2	

1269. $C_8H_2Cl_4N_2$
Chlorquinox
5,6,7,8-Tetrachloroquinoxaline
Lucel
Tetrachloroquinoxaline
RN: 3495-42-9   **MP** (°C):   190
MW: 267.93   **BP** (°C):

Solubility (Moles/L)	Solubility (Grams/L)	Temp (°C)	Ref (#)	Evaluation (T P E A A)	Comments
3.732E-06	1.000E-03	25	M161	1 0 0 0 0	

1270. $C_8H_2Cl_4O_4$
Tetrachlorophthalic Acid
Tetrachlorphthalsaeure
Tetrachloro-1,2-Benzenedicarboxylic Acid
RN: 632-58-6   **MP** (°C):
MW: 303.91   **BP** (°C):

Solubility (Moles/L)	Solubility (Grams/L)	Temp (°C)	Ref (#)	Evaluation (T P E A A)	Comments
1.876E-02	5.700E+00	14	F300	1 0 0 0 1	
1.007E-01	3.060E+01	99	F300	1 0 0 0 2	

1271. $C_8H_3Cl_2F_3N_2$
Chlorflurazole
4,5-Dichloro-2-(trifluoromethyl)-benzimidazole
Dichloro-2-(trifluoromethyl)benzimidazole
2-Trifluoromethyl-4,5-Dichlorobenzimidazole
RN: 3615-21-2   **MP** (°C):
MW: 255.03   **BP** (°C):

Solubility (Moles/L)	Solubility (Grams/L)	Temp (°C)	Ref (#)	Evaluation (T P E A A)	Comments
2.353E-04	6.000E-02	ns	B100	0 0 0 0 0	
2.353E-04	6.000E-02	ns	M061	0 0 0 0 1	

1272. $C_8H_3Cl_5O_2$
Pentachlorophenyl Acetate
Pentachlorophenol Acetate
Rabcon
RN: 1441-02-7   **MP** (°C):
MW: 308.38   **BP** (°C):

Solubility (Moles/L)	Solubility (Grams/L)	Temp (°C)	Ref (#)	Evaluation (T P E A A)	Comments
6.486E-05	2.000E-02	ns	L311	0 0 0 0 1	

1273. $C_8H_3Cl_5O_3$
2,3,4,5,6-Pentachlorophenoxyacetic Acid
Pentachlorophenoxyacetic Acid
RN: 2877-14-7   **MP** (°C):
MW: 324.38   **BP** (°C):

Solubility (Moles/L)	Solubility (Grams/L)	Temp (°C)	Ref (#)	Evaluation (T P E A A)	Comments
1.800E-04	5.839E-02	25	L030	1 0 2 1 1	

1274. $C_8H_4Cl_4O_3$
2,3,4,6-Tetrachlorophenoxyacetic Acid
Acetic Acid, (2,3,4,6-Tetrachlorophenoxy)-
RN: 10587-37-8   **MP** (°C):
MW: 289.93   **BP** (°C):

Solubility (Moles/L)	Solubility (Grams/L)	Temp (°C)	Ref (#)	Evaluation (T P E A A)	Comments
3.900E-04	1.131E-01	25	L030	1 0 2 1 1	

1275. $C_8H_4N_2$
1,4-Benzenedicarbonitrile
Terephthalonitrile
1,4-Dicyanobenzene
RN: 623-26-7   **MP** (°C):
MW: 128.13   **BP** (°C):

Solubility (Moles/L)	Solubility (Grams/L)	Temp (°C)	Ref (#)	Evaluation (T P E A A)	Comments
6.970E-04	8.931E-02	25	C316	1 0 2 2 2	0.1M NaCl

1276. $C_8H_4N_2S$
m-Cyanophenyl Isothiocyanate
3-Isothiocyanato-benzonitrile
3-Cyanophenyl Isothiocyanate
RN: 3125-78-8   **MP** (°C):
MW: 160.20   **BP** (°C):

Solubility (Moles/L)	Solubility (Grams/L)	Temp (°C)	Ref (#)	Evaluation (T P E A A)	Comments
6.410E-04	1.027E-01	25	K032	2 2 0 1 2	

1277. $C_8H_4N_2S_2$
m-Isothiocyanophenyl Isothiocyanate
3-Isothiocyanophenyl Isothiocyanate

RN: 3125-77-7   **MP** (°C):
MW: 192.26   **BP** (°C):

Solubility (Moles/L)	Solubility (Grams/L)	Temp (°C)	Ref (#)	Evaluation (T P E A A)	Comments
2.000E-05	3.845E-03	25	K032	2 2 0 1 1	

1278. $C_8H_4O_3$
Phthalic Anhydride
1,2-Benzenedicarboxylic Acid Anhydride
1,3-Isobenzofurandione
Phthalic Acid Anhydride
1,3-Dioxophthalan
1,3 Phthalandione

RN: 85-44-9   **MP** (°C):   130.8
MW: 148.12   **BP** (°C):   295.0

Solubility (Moles/L)	Solubility (Grams/L)	Temp (°C)	Ref (#)	Evaluation (T P E A A)	Comments
4.186E-02	6.200E+00	26.70	L095	2 2 1 1 2	
4.027E-02	5.964E+00	rt	D021	0 0 1 1 2	

1279. $C_8H_5ClO_4$
3-Chlorophthalic Acid
3-Chlor-phthalsaeure

RN: 27563-65-1   **MP** (°C):
MW: 200.58   **BP** (°C):

Solubility (Moles/L)	Solubility (Grams/L)	Temp (°C)	Ref (#)	Evaluation (T P E A A)	Comments
1.057E-01	2.120E+01	14	F300	1 0 0 0 2	

1280. $C_8H_5Cl_3O_2$
Chlorfenac
2,3,6-Trichlorophenylacetic Acid
Fenac

RN: 85-34-7   **MP** (°C):   161
MW: 239.49   **BP** (°C):

Solubility (Moles/L)	Solubility (Grams/L)	Temp (°C)	Ref (#)	Evaluation (T P E A A)	Comments
8.351E-04	2.000E-01	28	M161	1 0 0 0 2	
8.351E-04	2.000E-01	30	M061	1 0 0 0 2	

1281.   C$_8$H$_5$Cl$_3$O$_3$
2,3,4-Trichlorophenoxyacetic Acid
Acetic Acid, (2,3,4-Trichlorophenoxy)-
2,3,4-T
RN: 25141-27-9   **MP** (°C):
MW: 255.49   **BP** (°C):

Solubility (Moles/L)	Solubility (Grams/L)	Temp (°C)	Ref (#)	Evaluation (T P E A A)	Comments
8.000E-04	2.044E-01	25	L030	1 0 2 1 1	

1282.   C$_8$H$_5$Cl$_3$O$_3$
3,4,5-Trichlorophenoxyacetic Acid
Acetic Acid, (3,4,5-Trichlorophenoxy)-
3,4,5-T
RN: 80496-87-3   **MP** (°C):
MW: 255.49   **BP** (°C):

Solubility (Moles/L)	Solubility (Grams/L)	Temp (°C)	Ref (#)	Evaluation (T P E A A)	Comments
1.150E-03	2.938E-01	25	L030	1 0 2 1 2	

1283.   C$_8$H$_5$Cl$_3$O$_3$
2,4,6-Trichlorophenoxyacetic Acid
Acetic Acid, (2,4,6-Trichlorophenoxy)-
2,4,6-T
RN: 575-89-3   **MP** (°C):   45
MW: 255.49   **BP** (°C):

Solubility (Moles/L)	Solubility (Grams/L)	Temp (°C)	Ref (#)	Evaluation (T P E A A)	Comments
9.700E-04	2.478E-01	25	L030	1 0 2 1 1	

1284.   C$_8$H$_5$Cl$_3$O$_3$
2,3,5-Trichlorophenoxyacetic Acid
Acetic Acid, (2,3,5-Trichlorophenoxy)-
2,3,5-T
RN: 33433-95-3   **MP** (°C):
MW: 255.49   **BP** (°C):

Solubility (Moles/L)	Solubility (Grams/L)	Temp (°C)	Ref (#)	Evaluation (T P E A A)	Comments
1.000E-03	2.555E-01	25	L030	1 0 2 1 2	

1285. C$_8$H$_5$Cl$_3$O$_3$
2,4,5-Trichlorophenoxyacetic Acid
Acetic Acid, (2,4,5-Trichlorophenoxy)-
(2,4,5-Trichlorophenoxy)acetic Acid
2,4,5-T

RN: 93-76-5   **MP** (°C):   156
MW: 255.49   **BP** (°C):

Solubility (Moles/L)	Solubility (Grams/L)	Temp (°C)	Ref (#)	Evaluation (T P E A A)	Comments
9.316E-04	2.380E-01	20	B185	1 0 0 0 2	
7.398E-04	1.890E-01	20	M061	1 0 0 0 2	
1.100E-03	2.810E-01	25	B164	1 0 1 1 2	
1.096E-03	2.800E-01	25	B185	1 0 0 0 2	
1.050E-03	2.683E-01	25	L030	1 0 2 1 2	
1.088E-03	2.780E-01	25	M161	1 0 0 0 2	
9.316E-04	2.380E-01	30	B200	1 0 0 0 2	
9.783E-04	2.499E-01	ns	B100	0 0 0 0 1	
7.828E-04	2.000E-01	ns	B185	1 0 0 0 2	
8.000E-04	2.044E-01	ns	F184	0 0 0 0 1	
9.316E-04	2.380E-01	ns	K138	0 0 0 0 1	
9.824E-04	2.510E-01	ns	L024	0 0 0 0 2	
2.512E-04	6.418E-02	ns	M163	0 0 0 0 0	EFG
7.828E-04	2.000E-01	ns	N013	0 0 0 0 2	

1286. C$_8$H$_5$Cl$_3$O$_3$
2,3,6-Trichlorophenoxyacetic Acid
Acetic Acid, (2,3,6-Trichlorophenoxy)-
2,3,6-T

RN: 4007-00-5   **MP** (°C):   148
MW: 255.49   **BP** (°C):

Solubility (Moles/L)	Solubility (Grams/L)	Temp (°C)	Ref (#)	Evaluation (T P E A A)	Comments
2.400E-03	6.132E-01	25	L030	1 0 2 1 2	

1287. C$_8$H$_5$F$_3$O$_2$
α, α, α-Trifluoro-o-toluic Acid
Trifluoro-o-toluic Acid
Acide Orthotrifluortoluique

RN: 433-97-6   **MP** (°C):   111
MW: 190.12   **BP** (°C):   247

Solubility (Moles/L)	Solubility (Grams/L)	Temp (°C)	Ref (#)	Evaluation (T P E A A)	Comments
2.525E-02	4.800E+00	25	D064	1 2 1 1 2	

1288.   C$_8$H$_5$NO$_2$
Phthalimide
Phthalimid

RN: 85-41-6   **MP** (°C):   238.0
MW: 147.13   **BP** (°C):

Solubility (Moles/L)	Solubility (Grams/L)	Temp (°C)	Ref (#)	Evaluation (T P E A A)	Comments
2.447E-03	3.600E-01	25	F300	1 0 0 0 1	
2.719E-02	4.000E+00	100	F300	1 0 0 0 0	
4.075E-03	5.996E-01	rt	D021	0 0 1 1 0	

1289.   C$_8$H$_5$NO$_2$S
3-Carboxyphenylisothiocyanate
m-Isothiocyanobenzoic Acid

RN: 2131-63-7   **MP** (°C):
MW: 179.20   **BP** (°C):

Solubility (Moles/L)	Solubility (Grams/L)	Temp (°C)	Ref (#)	Evaluation (T P E A A)	Comments
5.600E-04	1.004E-01	25	D019	1 1 1 1 2	
8.000E-04	1.434E-01	25	K032	2 2 0 1 1	

1290.   C$_8$H$_5$NO$_2$S
4-Carboxyphenylisothiocyanate
p-Carboxyphenylisothiocyanate

RN: 2131-62-6   **MP** (°C):
MW: 179.20   **BP** (°C):

Solubility (Moles/L)	Solubility (Grams/L)	Temp (°C)	Ref (#)	Evaluation (T P E A A)	Comments
1.060E-04	1.900E-02	25	D019	1 1 1 1 2	

1291.   C$_8$H$_5$NO$_4$
6-Nitrophthalide
6-Nitro-phthalid

RN: 610-93-5   **MP** (°C):   145
MW: 179.13   **BP** (°C):

Solubility (Moles/L)	Solubility (Grams/L)	Temp (°C)	Ref (#)	Evaluation (T P E A A)	Comments
2.233E-03	4.000E-01	25	F300	1 0 0 0 2	

1292.   C$_8$H$_5$NO$_6$
2,3,4-Pyridinetricarboxylic Acid
Pyridin-tricarbonsaeure-(2,3,4)
RN: 632-95-1   **MP** (°C):   250
MW: 211.13   **BP** (°C):

Solubility (Moles/L)	Solubility (Grams/L)	Temp (°C)	Ref (#)	Evaluation (T P E A A)	Comments
5.684E-02	1.200E+01	15	F300	1 0 0 0 1	

1293.   C$_8$H$_5$NO$_6$
3-Nitrophthalic Acid
3-Nitro-phthalsaeure
RN: 603-11-2   **MP** (°C):   218
MW: 211.13   **BP** (°C):

Solubility (Moles/L)	Solubility (Grams/L)	Temp (°C)	Ref (#)	Evaluation (T P E A A)	Comments
9.520E-02	2.010E+01	25	F300	1 0 0 0 2	

1294.   C$_8$H$_6$
Ethynylbenzene
Phenylacetylene
RN: 536-74-3   **MP** (°C):   -44.8
MW: 102.14   **BP** (°C):   142.4

Solubility (Moles/L)	Solubility (Grams/L)	Temp (°C)	Ref (#)	Evaluation (T P E A A)	Comments
4.467E-03	4.562E-01	ns	D001	0 0 0 0 2	

1295.   C$_8$H$_6$BrNS
4-Bromobenzyl Isothiocyanate
p-Bromobenzyl Isothiocyanate
RN: 2076-56-4   **MP** (°C):
MW: 228.12   **BP** (°C):

Solubility (Moles/L)	Solubility (Grams/L)	Temp (°C)	Ref (#)	Evaluation (T P E A A)	Comments
6.500E-05	1.483E-02	25	D014	1 0 0 0 1	
1.500E-04	3.422E-02	25	D019	1 1 1 1 2	

1296.   C$_8$H$_6$BrNS
3-Bromobenzyl Isothiocyanate
m-Bromobenzyl Isothiocyanate
RN: 3845-33-8   **MP** (°C):
MW: 228.12   **BP** (°C):

Solubility (Moles/L)	Solubility (Grams/L)	Temp (°C)	Ref (#)	Evaluation (T P E A A)	Comments
1.070E-04	2.441E-02	25	D014	1 0 0 0 1	

1297.   C$_8$H$_6$ClNS
3-Chlorobenzyl Isothiocyanate
m-Chlorobenzyl Isothiocyanate
RN: 3694-58-4   **MP** (°C):
MW: 183.66   **BP** (°C):

Solubility (Moles/L)	Solubility (Grams/L)	Temp (°C)	Ref (#)	Evaluation (T P E A A)	Comments
1.370E-04	2.516E-02	25	D014	1 0 0 0 1	

1298.   C$_8$H$_6$ClNS
4-Chlorobenzyl Isothiocyanate
p-Chlorobenzyl Isothiocyanate
RN: 3694-45-9   **MP** (°C):
MW: 183.66   **BP** (°C):

Solubility (Moles/L)	Solubility (Grams/L)	Temp (°C)	Ref (#)	Evaluation (T P E A A)	Comments
1.480E-04	2.718E-02	25	D014	1 0 0 0 1	

1299.   C$_8$H$_6$Cl$_2$O$_3$
Dicamba
2-Methoxy-3,6-dichlorobenzoic Acid
RN: 1918-00-9   **MP** (°C):   98
MW: 221.04   **BP** (°C):

Solubility (Moles/L)	Solubility (Grams/L)	Temp (°C)	Ref (#)	Evaluation (T P E A A)	Comments
2.036E-02	4.500E+00	25	B200	1 0 0 0 1	
2.036E-02	4.500E+00	25	M161	1 0 0 0 1	
3.591E-02	7.937E+00	ns	B100	0 0 0 0 0	

1300.   C$_8$H$_6$Cl$_2$O$_3$
2,4-Dichlorophenoxyacetic Acid
2,4-D
(2,4-Dichlorophenoxy)acetic Acid
RN: 94-75-7   **MP** (°C):   138
MW: 221.04   **BP** (°C):

Solubility (Moles/L)	Solubility (Grams/L)	Temp (°C)	Ref (#)	Evaluation (T P E A A)	Comments
2.805E-03	6.200E-01	20	F311	1 2 2 2 1	
2.443E-03	5.400E-01	20	M061	1 0 0 0 2	
2.939E-03	6.496E-01	21.50	B200	1 0 0 0 0	
4.072E-03	9.000E-01	22.5	G301	2 1 0 1 2	
3.085E-03	6.820E-01	25	B164	1 0 1 1 2	
3.280E-03	7.250E-01	25	B185	1 0 0 0 2	
4.026E-03	8.900E-01	25	F071	1 1 2 1 2	

2.360E-03	5.217E-01	25	L030	1 0 2 1 2	
2.805E-03	6.200E-01	25	M161	1 0 0 0 2	
2.713E-03	5.996E-01	ns	B100	0 0 0 0 0	
4.072E-03	9.000E-01	ns	B185	0 0 0 0 2	
1.810E-03	4.000E-01	ns	B185	0 0 0 0 2	
2.500E-03	5.526E-01	ns	F184	0 0 0 0 1	
4.072E-03	9.000E-01	ns	K138	0 0 0 0 1	
2.805E-03	6.200E-01	ns	L024	0 0 0 0 2	
4.298E-03	9.500E-01	ns	M110	0 0 0 0 0	EFG
1.259E-03	2.783E-01	ns	M163	0 0 0 0 0	EFG
4.026E-03	8.900E-01	ns	M344	0 0 0 0 2	
2.488E-03	5.500E-01	ns	N013	0 0 0 0 2	

1301.   C₈H₆Cl₂O₃

2,3-Dichlorophenoxyacetic Acid

2,3-D

RN: 2976-74-1   **MP** (°C):   173
MW: 221.04   **BP** (°C):

Solubility (Moles/L)	Solubility (Grams/L)	Temp (°C)	Ref (#)	Evaluation (T P E A A)	Comments
1.550E-03	3.426E-01	25	L030	1 0 2 1 2	

1302.   C₈H₆Cl₂O₃

2,5-Dichlorophenoxyacetic Acid

2,5-D

RN: 582-54-7   **MP** (°C):
MW: 221.04   **BP** (°C):

Solubility (Moles/L)	Solubility (Grams/L)	Temp (°C)	Ref (#)	Evaluation (T P E A A)	Comments
2.420E-03	5.349E-01	25	L030	1 0 2 1 2	

1303.   C₈H₆Cl₂O₃

2,6-Dichlorophenoxyacetic Acid

2,6-D

RN: 575-90-6   **MP** (°C):
MW: 221.04   **BP** (°C):

Solubility (Moles/L)	Solubility (Grams/L)	Temp (°C)	Ref (#)	Evaluation (T P E A A)	Comments
7.050E-03	1.558E+00	25	L030	1 0 2 1 2	

1304.   C$_8$H$_6$Cl$_2$O$_3$
3,4-Dichlorophenoxyacetic Acid
3,4-D
RN: 588-22-7   **MP** (°C):   138
MW: 221.04   **BP** (°C):

Solubility (Moles/L)	Solubility (Grams/L)	Temp (°C)	Ref (#)	Evaluation (T P E A A)	Comments
2.070E-03	4.576E-01	25	L030	1 0 2 1 2	
2.090E-03	4.620E-01	ns	B185	0 0 0 0 2	

1305.   C$_8$H$_6$Cl$_2$O$_3$
3,5-Dichlorophenoxyacetic Acid
3,5-D
RN: 587-64-4   **MP** (°C):
MW: 221.04   **BP** (°C):

Solubility (Moles/L)	Solubility (Grams/L)	Temp (°C)	Ref (#)	Evaluation (T P E A A)	Comments
4.350E-03	9.615E-01	25	L030	1 0 2 1 2	

1306.   C$_8$H$_6$Cl$_4$O$_2$
Tetrachloroveratrole
3,4,5,6-Tetrachloro-1,2-dimethoxybenzene
RN: 944-61-6   **MP** (°C):
MW: 275.95   **BP** (°C):

Solubility (Moles/L)	Solubility (Grams/L)	Temp (°C)	Ref (#)	Evaluation (T P E A A)	Comments
5.762E-06	1.590E-03	25	L348	1 2 2 1 2	

1307.   C$_8$H$_6$Cl$_5$NO$_2$
Penclomedine
Pyridine
3,5-Dichloro-2,4-dimethoxy-6-(trichloromethyl)
NSC 338720
RN: 108030-77-9   **MP** (°C):
MW: 325.41   **BP** (°C):

Solubility (Moles/L)	Solubility (Grams/L)	Temp (°C)	Ref (#)	Evaluation (T P E A A)	Comments
1.229E-06	4.000E-04	25	P325	1 1 2 2 2	
1.229E-06	4.000E-04	25	P336	1 2 1 2 2	

1308. C₈H₆F₃N₃O₄S₂

Flumethiazide
6-(Trifluoromethyl)-2H-1,2,4-benzothiadiazine-7-sulfonamide 1,1-dioxide
6-Trifluoromethyl-7-sulfamoyl-4H-1,2,4-benzothiadiazine 1,1-dioxide
Trifluoromethylthiazide

RN: 148-56-1   **MP** (°C):
MW: 329.28   **BP** (°C):

Solubility (Moles/L)	Solubility (Grams/L)	Temp (°C)	Ref (#)	Evaluation (T P E A A)	Comments
3.189E-03	1.050E+00	rt	A095	0 0 2 2 2	

1309. C₈H₆INS

4-Iodobenzyl Isothiocyanate
p-Iodobenzyl Isothiocyanate

RN: 3694-49-3   **MP** (°C):
MW: 275.11   **BP** (°C):

Solubility (Moles/L)	Solubility (Grams/L)	Temp (°C)	Ref (#)	Evaluation (T P E A A)	Comments
5.100E-05	1.403E-02	25	D014	1 0 0 0 1	

1310. C₈H₆INS

3-Iodobenzyl Isothiocyanate
m-Iodobenzyl Isothiocyanate

RN: 3696-68-2   **MP** (°C):
MW: 275.11   **BP** (°C):

Solubility (Moles/L)	Solubility (Grams/L)	Temp (°C)	Ref (#)	Evaluation (T P E A A)	Comments
5.500E-05	1.513E-02	25	D014	1 0 0 0 1	

1311. C₈H₆N₂O₂S

3-Nitrobenzyl Isothiocyanate
m-Nitrobenzyl Isothiocyanate

RN: 3696-69-3   **MP** (°C):
MW: 194.21   **BP** (°C):

Solubility (Moles/L)	Solubility (Grams/L)	Temp (°C)	Ref (#)	Evaluation (T P E A A)	Comments
8.200E-05	1.593E-02	25	D014	1 0 0 0 1	

1312. $C_8H_6N_2O_2S$

4-Nitrobenzyl Isothiocyanate
p-Nitrobenzyl Isothiocyanate
RN: 3694-47-1   **MP** (°C):
MW: 194.21   **BP** (°C):

Solubility (Moles/L)	Solubility (Grams/L)	Temp (°C)	Ref (#)	Evaluation (T P E A A)	Comments
2.330E-04	4.525E-02	25	D014	1 0 0 0 1	

1313. $C_8H_6N_4O_5$

Nitrofurantoin
1-[(5-Nitrofurfurylidene)amino]hydantoin
Furatoin
Macrodantin
Macrobid
Welfurin
RN: 67-20-9   **MP** (°C):   268
MW: 238.16   **BP** (°C):

Solubility (Moles/L)	Solubility (Grams/L)	Temp (°C)	Ref (#)	Evaluation (T P E A A)	Comments
4.619E-04	1.100E-01	22	B154	1 1 1 1 1	pH 3.5
3.338E-04	7.950E-02	24	C034	2 0 2 2 2	
3.338E-04	7.950E-02	24	C118	1 0 0 0 2	
4.753E-04	1.132E-01	30	C011	2 0 2 1 0	EFG
4.761E-04	1.134E-01	30	C034	2 0 2 2 2	
4.761E-04	1.134E-01	30	C118	1 0 0 0 2	
8.264E-04	1.968E-01	37	A330	1 0 2 2 2	
1.142E-03	2.720E-01	37	B044	2 2 2 1 2	pH 7.2
7.310E-04	1.741E-01	37	C011	2 0 2 1 0	EFG
7.310E-04	1.741E-01	37	C034	2 0 2 2 2	
7.310E-04	1.741E-01	37	C118	1 0 0 0 2	
5.878E-04	1.400E-01	37	E044	1 0 1 1 2	
6.508E-04	1.550E-01	37	P034	1 0 0 0 2	pH 5
1.055E-03	2.512E-01	45	C034	2 0 2 2 2	
1.055E-03	2.512E-01	45	C118	1 0 0 0 2	
5.249E-04	1.250E-01	ns	P033	0 0 0 0 2	

1314.   C$_8$H$_6$N$_4$O$_8$
Alloxantin
Uroxine
Alloxantin Hydrate
RN: 76-24-4   **MP** (°C):   254dec
MW: 286.16   **BP** (°C):

Solubility (Moles/L)	Solubility (Grams/L)	Temp (°C)	Ref (#)	Evaluation (T P E A A)	Comments
1.753E-03	5.017E-01	25	B119	1 0 2 2 0	EFG
1.013E-02	2.900E+00	25	F300	1 0 0 0 1	
2.097E-01	6.000E+01	100	F300	1 0 0 0 0	

1315.   C$_8$H$_6$N$_4$S$_2$
Methylthiobenzothiazole
Benzothiazole
RN: 76006-86-5   **MP** (°C):
MW: 222.29   **BP** (°C):

Solubility (Moles/L)	Solubility (Grams/L)	Temp (°C)	Ref (#)	Evaluation (T P E A A)	Comments
4.948E-04	1.100E-01	22	P323	1 2 1 2 1	

1316.   C$_8$H$_6$O$_2$
Phthalic Dicarboxaldehyde
o-Phthalaldehyd
RN: 643-79-8   **MP** (°C):   56.5
MW: 134.14   **BP** (°C):

Solubility (Moles/L)	Solubility (Grams/L)	Temp (°C)	Ref (#)	Evaluation (T P E A A)	Comments
1.044E-01	1.400E+01	h	F300	0 0 0 0 1	

1317.   C$_8$H$_6$O$_2$
Terephthaldicarboxaldehyde
Terephthalaldehyd
RN: 623-27-8   **MP** (°C):   115
MW: 134.14   **BP** (°C):   246.5

Solubility (Moles/L)	Solubility (Grams/L)	Temp (°C)	Ref (#)	Evaluation (T P E A A)	Comments
1.491E-03	2.000E-01	20	F300	1 0 0 0 0	
1.297E-01	1.740E+01	100	F300	1 0 0 0 1	

1318. C$_8$H$_6$O$_3$

Piperonal
Heliotropine
3,4-Dihydroxybenzaldehyde Methylene Ketal
Methylenedioxy Procatechuic Aldehyde
Protocatechuic Aldehyde Methylene Ether
Piperonyl Aldehyde

RN: 120-57-0   **MP** (°C):   37
MW: 150.14   **BP** (°C):   263

Solubility (Moles/L)	Solubility (Grams/L)	Temp (°C)	Ref (#)	Evaluation (T P E A A)	Comments
2.331E-02	3.500E+00	20	F300	1 0 0 0 1	
4.463E-02	6.700E+00	78	F300	1 0 0 0 1	

1319. C$_8$H$_6$O$_3$

Benzoylformic Acid
Phenyglyoxilic Acid

RN: 611-73-4   **MP** (°C):   67
MW: 150.14   **BP** (°C):

Solubility (Moles/L)	Solubility (Grams/L)	Temp (°C)	Ref (#)	Evaluation (T P E A A)	Comments
6.128E+00	9.200E+02	0	C020	1 2 1 1 1	

1320. C$_8$H$_6$O$_4$

1,2-Benzenedicarboxylic Acid
o-Phthalic Acid
Phthalic Acid
Phthalsaeure
Benzene-1,2-dicarboxylic Acid

RN: 88-99-3   **MP** (°C):   230
MW: 166.13   **BP** (°C):

Solubility (Moles/L)	Solubility (Grams/L)	Temp (°C)	Ref (#)	Evaluation (T P E A A)	Comments
1.381E-02	2.295E+00	0	M043	1 0 0 0 1	
2.219E-02	3.686E+00	2	A027	1 0 0 0 1	
2.159E-02	3.587E+00	10	M043	1 0 0 0 1	
7.935E-03	1.318E+00	10	S198	2 1 2 2 2	
1.571E-02	2.611E+00	10.49	A341	2 0 2 2 2	
3.471E-02	5.767E+00	20	A027	1 0 0 0 1	
3.435E-02	5.707E+00	20	F069	2 2 2 2 2	
3.431E-02	5.700E+00	20	F300	1 0 0 0 1	
3.352E-02	5.569E+00	20	M043	1 0 0 0 1	
7.214E-03	1.199E+00	20	S198	2 1 2 2 2	
3.915E-02	6.504E+00	22.99	A341	2 0 2 2 2	
4.200E-02	6.978E+00	24.99	A341	2 0 2 2 2	

8.600E-02	1.429E+01	25	H084	1 0 0 0 1
8.520E-02	1.415E+01	25	K040	1 0 2 1 2
4.192E-02	6.965E+00	25	M030	2 1 0 1 2
4.279E-02	7.109E+00	25.8	W029	1 2 1 1 2
4.808E-02	7.988E+00	28	D050	1 2 1 2 2
5.152E-02	8.560E+00	29.49	A341	2 0 2 2 2
4.900E-02	8.141E+00	30	H019	1 0 2 0 0
4.777E-02	7.937E+00	30	M043	1 0 0 0 0
8.235E-03	1.368E+00	30	S198	2 1 2 2 2
5.865E-02	9.743E+00	33.99	A341	2 0 2 2 2
6.033E-02	1.002E+01	35	M030	2 1 0 1 2
6.561E-02	1.090E+01	35.99	A341	2 0 2 2 2
6.925E-02	1.150E+01	37.99	A341	2 0 2 2 2
7.137E-02	1.186E+01	40	M043	1 0 0 0 1
8.274E-02	1.375E+01	41.99	A341	2 0 2 2 2
7.865E-02	1.307E+01	43.7	W029	1 2 1 1 2
8.981E-02	1.492E+01	43.99	A341	2 0 2 2 2
8.991E-02	1.494E+01	44.99	A341	2 0 2 2 2
8.580E-02	1.425E+01	45	M030	2 1 0 1 2
9.890E-02	1.643E+01	45.99	A341	2 0 2 2 2
9.753E-02	1.620E+01	48.9	W029	1 2 1 1 2
1.116E-01	1.854E+01	49.99	A341	2 0 2 2 2
1.212E-01	2.014E+01	49.99	A341	2 0 2 2 2
1.349E-01	2.241E+01	53.99	A341	2 0 2 2 2
1.277E-01	2.122E+01	55	M030	2 1 0 1 2
1.339E-01	2.225E+01	58.0	W029	1 2 1 1 2
1.639E-01	2.724E+01	60	M043	1 0 0 0 1
1.741E-01	2.892E+01	60.99	A341	2 0 2 2 2
1.695E-01	2.815E+01	63.7	W029	1 2 1 1 2
2.145E-01	3.564E+01	64.99	A341	2 0 2 2 2
1.892E-01	3.144E+01	65	M030	2 1 0 1 2
2.826E-01	4.695E+01	75	M030	2 1 0 1 2
3.042E-01	5.053E+01	77.8	W029	1 2 1 1 2
3.567E-01	5.927E+01	80	M043	1 0 0 0 1
4.334E-01	7.200E+01	85	F300	1 0 0 0 0
4.297E-01	7.138E+01	85	M030	2 1 0 1 2
4.248E-01	7.058E+01	85.7	W029	1 2 1 1 2
6.377E-01	1.059E+02	94.8	W029	1 2 1 1 2
9.182E-01	1.525E+02	100	M043	1 0 0 0 2
8.208E-01	1.364E+02	101.1	W029	1 2 1 1 2
1.370E+00	2.276E+02	113.8	W029	1 2 1 1 2
9.015E-03	1.498E+00	ns	F014	0 0 0 0 2

1321. C$_8$H$_6$O$_4$
Isophthalic Acid
1,3-Benzenedicarboxylic Acid
m-Phthalic Acid
RN: 121-91-5   **MP** (°C):   345
MW: 166.13   **BP** (°C):

Solubility (Moles/L)	Solubility (Grams/L)	Temp (°C)	Ref (#)	Evaluation (T P E A A)	Comments
3.611E-04	6.000E-02	2	A027	1 0 0 0 0	
6.019E-04	9.999E-02	20	A027	1 0 0 0 0	
6.013E-03	9.990E-01	80	A027	1 0 0 0 0	

1322. C$_8$H$_6$O$_4$
1,4-Bezenedicarboxylic Acid
Terephthalic Acid
p-Phthalic Acid
RN: 100-21-0   **MP** (°C):
MW: 166.13   **BP** (°C):

Solubility (Moles/L)	Solubility (Grams/L)	Temp (°C)	Ref (#)	Evaluation (T P E A A)	Comments
9.029E-05	1.500E-02	20	F300	1 0 0 0 1	
1.920E-03	3.190E-01	25	C316	1 0 2 2 2	0.1M HCl
6.019E-04	9.999E-02	80	A027	1 0 0 0 0	

1323. C$_8$H$_6$O$_5$
2-Hydroxyisophthalic Acid
2-Hydroxy-iso-phthalsaeure
RN: 606-19-9   **MP** (°C):   244
MW: 182.13   **BP** (°C):

Solubility (Moles/L)	Solubility (Grams/L)	Temp (°C)	Ref (#)	Evaluation (T P E A A)	Comments
1.449E-01	2.640E+01	100	F300	1 0 0 0 2	

1324. C$_8$H$_6$O$_5$
5-Hydroxyisophthalic Acid
5-Hydroxy-iso-phthalsaeure
RN: 618-83-7   **MP** (°C):   293
MW: 182.13   **BP** (°C):

Solubility (Moles/L)	Solubility (Grams/L)	Temp (°C)	Ref (#)	Evaluation (T P E A A)	Comments
3.294E-03	6.000E-01	15	F300	1 0 0 0 1	
8.889E-01	1.619E+02	99	F300	1 0 0 0 2	

1325. $C_8H_6O_5$
4-Hydroxyisophthalic Acid
4-Hydroxy-iso-phthasaeure

RN: 636-46-4   **MP** (°C):   310
MW: 182.13   **BP** (°C):

Solubility (Moles/L)	Solubility (Grams/L)	Temp (°C)	Ref (#)	Evaluation (T P E A A)	Comments
1.647E-03	3.000E-01	24	F300	1 0 0 0 1	

1326. C_8H_6S
Thianaphthene
Benzo[b]thiophene
Benzothiofuran
1-Benzothiophene

RN: 95-15-8   **MP** (°C):   29-32
MW: 134.20   **BP** (°C):   221-222

Solubility (Moles/L)	Solubility (Grams/L)	Temp (°C)	Ref (#)	Evaluation (T P E A A)	Comments
1.611E-03	2.162E-01	59.0	L339	2 0 2 2 2	
2.610E-03	3.503E-01	78.5	L339	2 0 2 2 2	
4.386E-03	5.886E-01	99.0	L339	2 0 2 2 2	

1327. $C_8H_7BrN_2O_3$
o-Nitro-o-bromacetanilide
2-Bromo-5-nitroacetanilide

RN: 245115-83-7   **MP** (°C):
MW: 259.07   **BP** (°C):

Solubility (Moles/L)	Solubility (Grams/L)	Temp (°C)	Ref (#)	Evaluation (T P E A A)	Comments
7.720E-02	2.000E+01	rt	F043	0 0 2 1 1	

1328. $C_8H_7BrN_2O_3$
p-Nitro-o-bromacetanilide
2-Bromo-4-nitroacetanilide

RN: 57045-86-0   **MP** (°C):
MW: 259.07   **BP** (°C):

Solubility (Moles/L)	Solubility (Grams/L)	Temp (°C)	Ref (#)	Evaluation (T P E A A)	Comments
6.832E-02	1.770E+01	rt	F043	0 0 2 1 2	

1329.   C$_8$H$_7$ClN$_2$O$_3$
o-Nitro-o-chloracetanilide
2-Chloro-5-nitroacetanilide
RN: 72487-80-0   **MP** (°C):
MW: 214.61   **BP** (°C):

Solubility (Moles/L)	Solubility (Grams/L)	Temp (°C)	Ref (#)	Evaluation (T P E A A)	Comments
5.172E-02	1.110E+01	rt	F043	0 0 2 1 2	

1330.   C$_8$H$_7$ClN$_2$O$_3$
p-Nitro-o-chloracetanilide
2-Chloro-4-nitroacetanilide
RN: 881-87-8   **MP** (°C):
MW: 214.61   **BP** (°C):

Solubility (Moles/L)	Solubility (Grams/L)	Temp (°C)	Ref (#)	Evaluation (T P E A A)	Comments
5.172E-02	1.110E+01	rt	F043	0 0 2 1 2	

1331.   C$_8$H$_7$ClO$_3$
2-Chlorophenoxyacetic Acid
o-Chlorophenoxyacetic Acid
RN: 614-61-9   **MP** (°C):   146
MW: 186.60   **BP** (°C):

Solubility (Moles/L)	Solubility (Grams/L)	Temp (°C)	Ref (#)	Evaluation (T P E A A)	Comments
6.850E-03	1.278E+00	25	L030	1 0 2 1 2	

1332.   C$_8$H$_7$ClO$_3$
3-Chlorophenoxyacetic Acid
m-Chlorophenoxyacetic Acid
RN: 588-32-9   **MP** (°C):
MW: 186.60   **BP** (°C):

Solubility (Moles/L)	Solubility (Grams/L)	Temp (°C)	Ref (#)	Evaluation (T P E A A)	Comments
1.265E-02	2.360E+00	25	L030	1 0 2 1 2	

1333.   C$_8$H$_7$ClO$_3$
4-Chlorophenoxyacetic Acid
4-CPA
p-Chlorophenoxyacetic Acid

RN: 122-88-3   **MP** (°C):   157
MW: 186.60   **BP** (°C):

Solubility (Moles/L)	Solubility (Grams/L)	Temp (°C)	Ref (#)	Evaluation (T P E A A)	Comments
4.545E-03	8.480E-01	25	B164	1 0 1 1 2	
2.042E-03	3.810E-01	25	B185	1 0 0 0 2	
5.130E-03	9.572E-01	25	L030	1 0 2 1 2	

1334.   C$_8$H$_7$Cl$_2$NO$_2$
Chloramben Methyl Ester
Vegiben 2E
Methyl 3-amino-2,5-dichlorobenzoate
Amchem 65-81-B
Methyl Chloramben
Chloramben Methyl

RN: 7286-84-2   **MP** (°C):   63.5
MW: 220.06   **BP** (°C):

Solubility (Moles/L)	Solubility (Grams/L)	Temp (°C)	Ref (#)	Evaluation (T P E A A)	Comments
5.453E-04	1.200E-01	20	M161	1 0 0 0 2	

1335.   C$_8$H$_7$Cl$_3$O
2,4,6-Trichloro-3,5-dimethyl-phenol
3,5-Xylenol, 2,4,6-trichloro-

RN: 6972-47-0   **MP** (°C):
MW: 225.50   **BP** (°C):

Solubility (Moles/L)	Solubility (Grams/L)	Temp (°C)	Ref (#)	Evaluation (T P E A A)	Comments
2.200E-05	4.961E-03	25	B316	1 0 2 1 1	

1336.   C$_8$H$_7$Cl$_3$O$_2$
3,4,5-Trichloroveratrole
4,5,6-Trichloroveratrole

RN: 16766-29-3   **MP** (°C):   66
MW: 241.50   **BP** (°C):

Solubility (Moles/L)	Solubility (Grams/L)	Temp (°C)	Ref (#)	Evaluation (T P E A A)	Comments
4.265E-05	1.030E-02	25	L348	1 2 2 1 2	

1337.   C₈H₇N
Indole
2,3-Benzopyrrole
Benzopyrrole
1-Benzazole
1-Benzol β Pyrrol

RN: 120-72-9   **MP** (°C):   52
MW: 117.15   **BP** (°C):   253

Solubility (Moles/L)	Solubility (Grams/L)	Temp (°C)	Ref (#)	Evaluation (T P E A A)	Comments
9.219E-02	1.080E+01	25	K119	1 0 0 0 2	
3.037E-02	3.558E+00	25	P051	2 1 1 2 2	
3.037E-02	3.558E+00	25.00	P007	2 1 2 2 2	

1338.   C₈H₇N
p-Toluonitrile
p-Cyanotoluene
p-Methylbenzonitrile
4-Methylbenzenecarbonitrile

RN: 104-85-8   **MP** (°C):
MW: 117.15   **BP** (°C):

Solubility (Moles/L)	Solubility (Grams/L)	Temp (°C)	Ref (#)	Evaluation (T P E A A)	Comments
1.300E-02	1.523E+00	25	M327	1 0 0 1 2	

1339.   C₈H₇NOS
p-Methoxyphenyl Isothiocyanate
4-Methoxyphenylisothiocyanate

RN: 2284-20-0   **MP** (°C):   18.0
MW: 165.22   **BP** (°C):   280.5

Solubility (Moles/L)	Solubility (Grams/L)	Temp (°C)	Ref (#)	Evaluation (T P E A A)	Comments
2.500E-04	4.130E-02	25	D019	1 1 1 1 2	

1340.   C₈H₇NOS
m-Methoxyphenyl Isothiocyanate
3-Methoxyphenyl Isothiocyanate

RN: 3125-64-2   **MP** (°C):
MW: 165.22   **BP** (°C):

Solubility (Moles/L)	Solubility (Grams/L)	Temp (°C)	Ref (#)	Evaluation (T P E A A)	Comments
2.700E-04	4.461E-02	25	K032	2 2 0 1 2	

1341.   C$_8$H$_7$NO$_3$
Oxanilic Acid
N-Phenyloxalic Acid Monoamide
Oxanilsaure

RN: 500-72-1   **MP** (°C):   150
MW: 165.15   **BP** (°C):

Solubility (Moles/L)	Solubility (Grams/L)	Temp (°C)	Ref (#)	Evaluation (T P E A A)	Comments
4.990E-02	8.241E+00	25	D058	1 0 1 1 2	

1342.   C$_8$H$_7$NO$_4$
2-Nitro-3-methylbenzoic Acid
2-Nitro-m-toluic Acid
3-Methyl-2-nitrobenzoic Acid

RN: 5437-38-7   **MP** (°C):
MW: 181.15   **BP** (°C):

Solubility (Moles/L)	Solubility (Grams/L)	Temp (°C)	Ref (#)	Evaluation (T P E A A)	Comments
2.208E-03	4.000E-01	20	G063	1 0 0 0 1	
8.832E-03	1.600E+00	40	G063	1 0 0 0 1	
3.202E-02	5.800E+00	80	G063	1 0 0 0 1	
3.312E-02	6.000E+00	100	G063	1 0 0 0 0	

1343.   C$_8$H$_7$NO$_4$
6-Nitro-3-methylbenzoic Acid
2-Nitro-5-methylbenzoic Acid
5-Methyl-2-nitrobenzoic Acid
3-Methyl-6-nitrobenzoic Acid

RN: 3113-72-2   **MP** (°C):
MW: 181.15   **BP** (°C):

Solubility (Moles/L)	Solubility (Grams/L)	Temp (°C)	Ref (#)	Evaluation (T P E A A)	Comments
2.043E-02	3.700E+00	10	G063	1 0 0 0 1	
2.595E-02	4.700E+00	20	G063	1 0 0 0 1	
9.385E-02	1.700E+01	40	G063	1 0 0 0 1	
9.937E-02	1.800E+01	50	G063	1 0 0 0 1	
1.490E-01	2.700E+01	60	G063	1 0 0 0 1	
1.932E-01	3.500E+01	65	G063	1 0 0 0 1	
2.484E-01	4.500E+01	70	G063	1 0 0 0 1	
3.643E-01	6.600E+01	80	G063	1 0 0 0 1	
3.699E-01	6.700E+01	100	G063	1 0 0 0 1	

1344. C₈H₇NS
p-Tolyl Isothiocyanate
4-Tolylisothiocyanate
RN: 622-59-3   **MP** (°C):   25
MW: 149.22   **BP** (°C):   237

Solubility (Moles/L)	Solubility (Grams/L)	Temp (°C)	Ref (#)	Evaluation (T P E A A)	Comments
1.900E-05	2.835E-03	25	D019	1 1 1 1 1	

1345. C₈H₇NS
m-Methylphenyl Isothiocyanate
3-Methylphenyl Isothiocyanate
RN: 614-69-7   **MP** (°C):
MW: 149.22   **BP** (°C):

Solubility (Moles/L)	Solubility (Grams/L)	Temp (°C)	Ref (#)	Evaluation (T P E A A)	Comments
1.420E-04	2.119E-02	25	K032	2 2 0 1 2	

1346. C₈H₇NS
Benzyl Isothiocyanate
Benzylisothiocyanate
Isothiocyanatomethylbenzene
RN: 622-78-6   **MP** (°C):   112
MW: 149.22   **BP** (°C):   242

Solubility (Moles/L)	Solubility (Grams/L)	Temp (°C)	Ref (#)	Evaluation (T P E A A)	Comments
7.300E-04	1.089E-01	25	D014	1 0 0 0 2	

1347. C₈H₇N₅O
7-Acetamidopteridine
RN: **MP** (°C):
MW: 189.18   **BP** (°C):

Solubility (Moles/L)	Solubility (Grams/L)	Temp (°C)	Ref (#)	Evaluation (T P E A A)	Comments
4.035E-02	7.634E+00	100	A083	1 2 0 0 0	

1348. C₈H₇N₅O
4-Acetamidopteridine
RN: **MP** (°C):
MW: 189.18   **BP** (°C):

Solubility (Moles/L)	Solubility (Grams/L)	Temp (°C)	Ref (#)	Evaluation (T P E A A)	Comments
1.762E+00	3.333E+02	100	A083	1 2 0 0 0	

1349.   C$_8$H$_7$N$_5$O
2-Acetamidopteridine
RN: **MP** (°C):
MW: 189.18   **BP** (°C):

Solubility (Moles/L)	Solubility (Grams/L)	Temp (°C)	Ref (#)	Evaluation (T P E A A)	Comments
1.705E-01	3.226E+01	100	A083	1 2 0 0 0	

1350.   C$_8$H$_7$N$_5$O$_8$
2,4,6-Trinitrophenylethylnitramine
Tetrethyl
Trinitrophenylethylnitramine
Ethyl Tetryl
RN: 6052-13-7   **MP** (°C):
MW: 301.17   **BP** (°C):

Solubility (Moles/L)	Solubility (Grams/L)	Temp (°C)	Ref (#)	Evaluation (T P E A A)	Comments
1.992E-04	6.000E-02	22	D067	1 2 0 0 0	
8.633E-04	2.600E-01	50	D067	1 2 0 0 1	
8.998E-03	2.710E+00	100	D067	1 2 0 0 2	

1351.   C$_8$H$_8$
Styrene
Phenylethylene
Styrolene
Styrol
Ethenylbenzene
Annamene
RN: 100-42-5   **MP** (°C):   -30
MW: 104.15   **BP** (°C):   145

Solubility (Moles/L)	Solubility (Grams/L)	Temp (°C)	Ref (#)	Evaluation (T P E A A)	Comments
2.784E-03	2.899E-01	7	L028	1 0 1 1 1	
2.400E-03	2.499E-01	15	L028	1 0 1 1 1	
1.152E-03	1.200E-01	20	L096	1 2 0 2 2	
3.167E-03	3.299E-01	24	L028	1 0 1 1 1	
2.880E-03	3.000E-01	25	A002	1 2 1 1 1	
1.540E-03	1.604E-01	25	B173	2 0 2 2 2	
2.975E-03	3.099E-01	25	L028	1 0 1 1 1	
3.455E-03	3.599E-01	32	L028	1 0 1 1 1	
3.839E-03	3.998E-01	40	L028	1 0 1 1 1	
3.839E-03	3.998E-01	44	L028	1 0 1 1 1	
4.319E-03	4.498E-01	49	L028	1 0 1 1 1	
4.319E-03	4.498E-01	51	L028	1 0 1 1 1	
4.798E-03	4.998E-01	56	L028	1 0 1 1 1	
8.658E-02	9.018E+00	65	A324	2 2 2 1 1	
5.566E-03	5.797E-01	65	L028	1 0 1 1 1	

1352. C$_8$H$_8$BrCl$_2$O$_3$PS
Bromophos
O-(4-Bromo-2,5-dichlorophenyl) O,O-Dimethyl phosphorothioate
Nexion
Brofene
Brophene
Omexan

RN: 2104-96-3   **MP** (°C):   51
MW: 366.00   **BP** (°C):

Solubility (Moles/L)	Solubility (Grams/L)	Temp (°C)	Ref (#)	Evaluation (T P E A A)	Comments
6.557E-07	2.400E-04	10	B324	2 2 2 2 2	
6.558E-07	2.400E-04	10	B324	2 2 2 2 2	
8.197E-07	3.000E-04	20	B169	2 1 1 1 1	*sic*
9.290E-07	3.400E-04	20	B324	2 2 2 2 2	
9.290E-07	3.400E-04	20	B324	2 2 2 2 2	
2.732E-06	1.000E-03	20	F311	1 2 2 2 1	*sic*
1.093E-04	4.000E-02	20	M061	1 0 0 0 1	
1.093E-04	4.000E-02	20	W311	1 0 0 0 1	
2.634E-06	9.641E-04	30	B324	2 2 2 2 2	
2.623E-06	9.600E-04	30	B324	2 2 2 2 2	
1.093E-04	4.000E-02	ns	E050	0 0 0 0 1	
1.093E-04	4.000E-02	rt	M161	0 0 0 0 1	

1353. C$_8$H$_8$BrNO
4'-Bromoacetanilide
Acetamide, N-(4-Bromophenyl)-
Acetanilide, 4'-Bromo-
Bromoantifebrin

RN: 103-88-8   **MP** (°C):
MW: 214.07   **BP** (°C):

Solubility (Moles/L)	Solubility (Grams/L)	Temp (°C)	Ref (#)	Evaluation (T P E A A)	Comments
7.000E-04	1.498E-01	25	D044	1 1 1 1 2	

1354. C$_8$H$_8$ClNO
p-Chloroacetanilide
Acetamide, N-(4-Chlorophenyl)-
Acetanilide, 4'-Chloro-

RN: 539-03-7   **MP** (°C):
MW: 169.61   **BP** (°C):

Solubility (Moles/L)	Solubility (Grams/L)	Temp (°C)	Ref (#)	Evaluation (T P E A A)	Comments
1.000E-03	1.696E-01	25	D044	1 1 1 1 2	

1355. $C_8H_8Cl_2IO_3PS$

Iodofenphos
O-(2,5-Dichloro-4-iodophenyl) O,O-Dimethyl Phosphorothioate
Nuvanol-N
Dimethyl O-2,5-Dichloro-4-iodophenyl Thiophosphate
Alfacron
Jodfenphos

RN: 18181-70-9   **MP** (°C):   72
MW: 413.00   **BP** (°C):

Solubility (Moles/L)	Solubility (Grams/L)	Temp (°C)	Ref (#)	Evaluation (T P E A A)	Comments
2.421E-07	1.000E-04	20	B169	2 1 1 1 1	
4.843E-06	2.000E-03	20	M161	1 0 0 0 0	

1356. $C_8H_8Cl_2O$

2,4-Dichloro-6-ethyl-phenol
Phenol, 2,4-Dichloro-6-ethyl-

RN: 24539-94-4   **MP** (°C):
MW: 191.06   **BP** (°C):

Solubility (Moles/L)	Solubility (Grams/L)	Temp (°C)	Ref (#)	Evaluation (T P E A A)	Comments
1.300E-03	2.484E-01	25	B316	1 0 2 1 1	

1357. $C_8H_8Cl_2O_2$

Chloroneb
Demosan
Terraneb
Terraneb SP
1,4-Dichloro-2,5-dimethoxybenzene
Terraneb B

RN: 2675-77-6   **MP** (°C):   134.5
MW: 207.06   **BP** (°C):   268

Solubility (Moles/L)	Solubility (Grams/L)	Temp (°C)	Ref (#)	Evaluation (T P E A A)	Comments
3.864E-05	8.000E-03	25	M161	1 0 0 0 0	

1358. $C_8H_8Cl_2O_2$

4,5-Dichloroveratrole
Benzene, 1,2-Dichloro-4,5-dimethoxy-

RN: 2772-46-5   **MP** (°C):   83
MW: 207.06   **BP** (°C):

Solubility (Moles/L)	Solubility (Grams/L)	Temp (°C)	Ref (#)	Evaluation (T P E A A)	Comments
3.492E-04	7.230E-02	25	L348	1 2 2 1 2	average of 2

1359. $C_8H_8Cl_3O_3PS$
Ronnel
Fenchlorphos
Dermafos
Dimethyl Trichlorophenylthiophosphate
RN: 299-84-3   **MP** (°C):   35
MW: 321.55   **BP** (°C):

Solubility (Moles/L)	Solubility (Grams/L)	Temp (°C)	Ref (#)	Evaluation (T P E A A)	Comments
1.866E-06	6.000E-04	20	B169	2 2 1 1 1	
3.359E-06	1.080E-03	20	C053	1 0 2 2 1	
3.110E-06	1.000E-03	20	E048	1 2 1 1 0	
7.775E-06	2.500E-03	20	F311	1 2 2 2 1	
5.287E-06	1.700E-03	ns	F040	1 2 2 2 1	
3.359E-06	1.080E-03	ns	F071	0 1 2 1 2	
1.866E-05	6.000E-03	ns	K138	0 0 0 0 1	
1.368E-04	4.400E-02	ns	M061	0 0 0 0 1	
1.244E-04	4.000E-02	rt	M161	0 0 0 0 1	

1360. C_8H_8FNO
4'-Fluoroacetanilide
Acetamide, N-(4-Fluorophenyl)-
4-Fluoroacetanilide
RN: 351-83-7   **MP** (°C):
MW: 153.16   **BP** (°C):

Solubility (Moles/L)	Solubility (Grams/L)	Temp (°C)	Ref (#)	Evaluation (T P E A A)	Comments
1.630E-02	2.496E+00	25	D044	1 1 1 1 2	

1361. $C_8H_8F_3N_3O_4S_2$
Hydroflumethiazide
Diucardin
Saluron
RN: 135-09-1   **MP** (°C):   272
MW: 331.29   **BP** (°C):

Solubility (Moles/L)	Solubility (Grams/L)	Temp (°C)	Ref (#)	Evaluation (T P E A A)	Comments
1.449E-03	4.800E-01	37	C087	0 0 0 0 1	
2.048E-03	6.785E-01	37	C315	2 2 2 2 2	0.1N HCl, average of 4
9.958E-04	3.299E-01	rt	K144	0 0 0 0 1	

1362.   C₈H₈INO

p-Iodoaniline-N-acetate
4-Iodanilin-N-acetat
4-Iodoacetanilide
Acetanilide, 4'-Iodo-
4-Acetamidophenyl Iodide
p-Iodoacetanilide

RN: 622-50-4   **MP** (°C):
MW: 261.06   **BP** (°C):

Solubility (Moles/L)	Solubility (Grams/L)	Temp (°C)	Ref (#)	Evaluation (T P E A A)	Comments
7.000E-04	1.827E-01	25	D044	1 1 1 1 2	

1363.   C₈H₈N₂O₂

Phthalamide
1,2-Benzenedicarboxamide

RN: 88-96-0   **MP** (°C):   228
MW: 164.17   **BP** (°C):

Solubility (Moles/L)	Solubility (Grams/L)	Temp (°C)	Ref (#)	Evaluation (T P E A A)	Comments
1.218E-03	2.000E-01	20	A027	1 0 0 0 0	*sic*
3.594E-02	5.900E+00	30	K004	1 0 0 0 1	

1364.   C₈H₈N₂O₂

Ricinine
Ricinin

RN: 524-40-3   **MP** (°C):   201.5
MW: 164.17   **BP** (°C):

Solubility (Moles/L)	Solubility (Grams/L)	Temp (°C)	Ref (#)	Evaluation (T P E A A)	Comments
1.645E-02	2.700E+00	10	F300	1 0 0 0 1	

1365.   C₈H₈N₂O₃

2-Nitroaniline-N-acetate
2-Nitro-anilin-N-acetat
o-Nitroacetanilide

RN: 552-32-9   **MP** (°C):
MW: 180.16   **BP** (°C):

Solubility (Moles/L)	Solubility (Grams/L)	Temp (°C)	Ref (#)	Evaluation (T P E A A)	Comments
1.221E-02	2.200E+00	20	F300	1 0 0 0 1	
1.221E-02	2.200E+00	rt	F043	0 0 2 1 1	

1366. C$_8$H$_8$N$_2$O$_3$
4-Nitroaniline-N-acetate
4-Nitro-anilin-N-acetat
p-Nitroacetanilide
1-Nitro-4-acetylaminobenzene
RN: 104-04-1 **MP** (°C): 216
MW: 180.16 **BP** (°C):

Solubility (Moles/L)	Solubility (Grams/L)	Temp (°C)	Ref (#)	Evaluation (T P E A A)	Comments
1.221E-02	2.200E+00	20	F300	1 0 0 0 1	
6.000E-04	1.081E-01	25	D044	1 1 1 1 2	
1.221E-02	2.200E+00	rt	F043	0 0 2 1 1	

1367. C$_8$H$_8$N$_2$O$_6$S
MB 8882
Methyl N-(4-Nitrobenzenesulphonyl)carbamate
RN: 3337-70-0 **MP** (°C): 151
MW: 260.23 **BP** (°C):

Solubility (Moles/L)	Solubility (Grams/L)	Temp (°C)	Ref (#)	Evaluation (T P E A A)	Comments
3.839E-03	9.990E-01	ns	M061	0 0 0 0 0	

1368. C$_8$H$_8$N$_4$
6,7-Dimethylpteridine
6:7-Dimethylpteridine
RN: 704-61-0 **MP** (°C):
MW: 160.18 **BP** (°C):

Solubility (Moles/L)	Solubility (Grams/L)	Temp (°C)	Ref (#)	Evaluation (T P E A A)	Comments
3.468E-01	5.556E+01	20	A083	1 2 0 0 0	

1369. C$_8$H$_8$N$_4$O
4-Hydroxy-6,7-dimethylpteridine
4-Hydroxy-6:7-dimethylpteridine
RN: 14684-54-9 **MP** (°C):
MW: 176.18 **BP** (°C):

Solubility (Moles/L)	Solubility (Grams/L)	Temp (°C)	Ref (#)	Evaluation (T P E A A)	Comments
5.155E-03	9.083E-01	22.5	A085	1 2 0 0 0	

1370. C$_8$H$_8$N$_4$O$_2$S$_2$
2-Sulfanilamido-1,3,4-thiadiazole
Sulfathiadiazole
Sulfanilamide, N1-1,3,4-Thiadiazol-2-yl-
RN: 16806-29-4   **MP** (°C):
MW: 256.31   **BP** (°C):

Solubility (Moles/L)	Solubility (Grams/L)	Temp (°C)	Ref (#)	Evaluation (T P E A A)	Comments
2.848E-03	7.300E-01	37	R045	1 2 1 1 1	

1371. C$_8$H$_8$N$_4$O$_3$
1-Acetoxymethyl Allopurinol
4H-Pyrazolo[3,4-d]pyrimidin-4-one, 1-[(Acetyloxy)methyl]-1,5-dihydro-
RN: 98846-64-1   **MP** (°C):   257-258
MW: 208.18   **BP** (°C):

Solubility (Moles/L)	Solubility (Grams/L)	Temp (°C)	Ref (#)	Evaluation (T P E A A)	Comments
2.786E-03	5.800E-01	22	B322	1 0 2 2 2	

1372. C$_8$H$_8$N$_4$O$_4$
Nifuradene
1-[5-Nitrofurfuryllidene)Amino]-2-Imidazolidinone
RN: 555-84-0   **MP** (°C):   261.5
MW: 224.18   **BP** (°C):

Solubility (Moles/L)	Solubility (Grams/L)	Temp (°C)	Ref (#)	Evaluation (T P E A A)	Comments
3.925E-04	8.800E-02	ns	I310	0 0 0 0 1	

1373. C$_8$H$_8$N$_4$O$_4$S$_3$
CL 11,366
RN: **MP** (°C):
MW: 320.37   **BP** (°C):

Solubility (Moles/L)	Solubility (Grams/L)	Temp (°C)	Ref (#)	Evaluation (T P E A A)	Comments
1.405E-03	4.500E-01	ns	M032	0 0 0 0 1	

1374.   C_8H_8N_4O_6
2,4,6-Trinitroethylaniline
2-4-6-Trinitromonoethylaniline
RN: 7449-27-6   **MP** (°C):
MW: 256.18   **BP** (°C):

Solubility (Moles/L)	Solubility (Grams/L)	Temp (°C)	Ref (#)	Evaluation (T P E A A)	Comments
3.904E-04	1.000E-01	19	D067	1 2 0 0 2	
1.210E-03	3.100E-01	50	D067	1 2 0 0 2	
5.699E-03	1.460E+00	100	D067	1 2 0 0 2	

1375.   C_8H_8O
Styrene Oxide
1,2-Epoxyethylbenzene
RN: 96-09-3   **MP** (°C):   -36.8
MW: 120.15   **BP** (°C):   194.1

Solubility (Moles/L)	Solubility (Grams/L)	Temp (°C)	Ref (#)	Evaluation (T P E A A)	Comments
2.324E-02	2.792E+00	25	I313	0 0 0 0 1	

1376.   C_8H_8O
2,2,3-Trimethyl-3-pentanol
2,2,3-Trimethylpentanol-3
RN: 7294-05-5   **MP** (°C):   -6
MW: 120.15   **BP** (°C):

Solubility (Moles/L)	Solubility (Grams/L)	Temp (°C)	Ref (#)	Evaluation (T P E A A)	Comments
4.120E+00	4.950E+02	20	G007	1 2 0 1 2	
4.119E+00	4.949E+02	25	G007	1 2 0 1 2	
4.119E+00	4.949E+02	30	G007	1 2 0 1 2	

1377.   C_8H_8O
4-Methylbenzaldehyde
p-Methylbenzaldehyde
RN: 104-87-0   **MP** (°C):
MW: 120.15   **BP** (°C):   204

Solubility (Moles/L)	Solubility (Grams/L)	Temp (°C)	Ref (#)	Evaluation (T P E A A)	Comments
1.890E-02	2.271E+00	25	M017	1 2 0 1 2	

1378. C_8H_8O
Acetophenone
Acetophenon
Methyl Phenyl Ketone
RN: 98-86-2   **MP** (°C): 20.05
MW: 120.15   **BP** (°C): 202

Solubility (Moles/L)	Solubility (Grams/L)	Temp (°C)	Ref (#)	Evaluation (T P E A A)	Comments
4.503E-02	5.411E+00	24	H106	1 0 2 2 2	
4.611E-02	5.540E+00	24	M303	1 0 1 1 2	
5.243E-02	6.300E+00	25	A003	1 2 1 2 2	
4.470E-02	5.371E+00	25	B019	1 0 1 2 0	
4.470E-02	5.371E+00	25	B092	2 1 1 1 1	
5.600E-03	6.729E-01	25	F063	1 1 0 0 1	
6.605E-02	7.937E+00	60	B092	2 1 1 1 1	

1379. $C_8H_8O_2$
p-Toluic Acid
4-Methylbenzoic Acid
Toluenecarboxylic Acid
RN: 99-94-5   **MP** (°C): 180
MW: 136.15   **BP** (°C): 274

Solubility (Moles/L)	Solubility (Grams/L)	Temp (°C)	Ref (#)	Evaluation (T P E A A)	Comments
2.500E-03	3.404E-01	25	F001	1 0 1 0 2	
2.938E-03	4.000E-01	25	F300	1 0 0 0 2	
2.277E-03	3.100E-01	37	M360	1 2 1 1 2	
2.780E-03	3.785E-01	ns	C014	0 0 0 1 2	

1380. $C_8H_8O_2$
4-Hydroxyacetophenone
4'-Hydroxy-acetophenon
RN: 99-93-4   **MP** (°C): 110
MW: 136.15   **BP** (°C):

Solubility (Moles/L)	Solubility (Grams/L)	Temp (°C)	Ref (#)	Evaluation (T P E A A)	Comments
7.271E-02	9.900E+00	22	F300	1 0 0 0 1	

1381.   C$_8$H$_8$O$_2$
Phenylacetic Acid
Phenylessigsaeure
RN: 103-82-2   **MP** (°C):   76.5
MW: 136.15   **BP** (°C):   266

Solubility (Moles/L)	Solubility (Grams/L)	Temp (°C)	Ref (#)	Evaluation (T P E A A)	Comments
1.175E-01	1.600E+01	20	F071	1 1 2 1 2	
1.219E-01	1.660E+01	20	H080	1 0 0 0 2	
1.219E-01	1.660E+01	20	M344	1 0 0 0 2	
1.300E-01	1.770E+01	25	F300	1 0 0 0 2	
1.267E-01	1.725E+01	25	H071	2 2 2 1 2	
1.310E-01	1.784E+01	25	K040	1 0 2 1 2	
1.300E-01	1.770E+01	25.00	M135	1 2 1 1 2	0.01N sodium phenylacetate
1.451E-01	1.975E+01	30	D033	2 2 1 2 2	
1.910E-01	2.600E+01	35.00	M135	1 2 1 1 2	
2.113E-01	2.877E+01	40	D033	2 2 1 2 2	
2.880E-01	3.921E+01	41.50	M135	1 2 1 1 2	
2.900E-01	3.948E+01	45.00	M135	1 2 1 1 2	
3.650E-01	4.970E+01	58.40	M135	1 2 1 1 2	
4.350E-01	5.923E+01	68.80	M135	1 2 1 1 2	
5.130E-01	6.985E+01	76.50	M135	1 2 1 1 2	
6.110E-01	8.319E+01	83.00	M135	1 2 1 1 2	
6.860E-01	9.340E+01	86.70	M135	1 2 1 1 2	
7.712E-01	1.050E+02	100	F300	1 0 0 0 2	

1382.   C$_8$H$_8$O$_2$
p-Anisaldehyde
Anisaldehyd
p-Methoxybenzaldehyde
RN: 123-11-5   **MP** (°C):   0
MW: 136.15   **BP** (°C):   249.5

Solubility (Moles/L)	Solubility (Grams/L)	Temp (°C)	Ref (#)	Evaluation (T P E A A)	Comments
1.469E-02	2.000E+00	20	F300	1 0 0 0 0	
3.150E-02	4.289E+00	25	I019	1 0 1 2 2	

1383. $C_8H_8O_2$
o-Toluic Acid
o-Tolylsaeure
o-Toluylic Acid
2-Methylbenzoic Acid
RN: 118-90-1   **MP** (°C):   107
MW: 136.15   **BP** (°C):   258

Solubility (Moles/L)	Solubility (Grams/L)	Temp (°C)	Ref (#)	Evaluation (T P E A A)	Comments
8.700E-03	1.185E+00	25	F001	1 0 1 0 2	
8.780E-03	1.195E+00	25	R016	1 0 1 1 2	
1.014E-02	1.380E+00	37	M360	1 2 1 1 2	

1384. $C_8H_8O_2$
m-Toluic Acid
3-Methylbenzoic Acid
m-Methylbenzoic Acid
β-Methylbenzoic Acid
RN: 99-04-7   **MP** (°C):   112
MW: 136.15   **BP** (°C):   263

Solubility (Moles/L)	Solubility (Grams/L)	Temp (°C)	Ref (#)	Evaluation (T P E A A)	Comments
7.200E-03	9.803E-01	25	F001	1 0 1 0 2	
7.198E-03	9.800E-01	25	F300	1 0 0 0 2	
7.785E-03	1.060E+00	37	M360	1 2 1 1 2	

1385. $C_8H_8O_2$
Methyl Benzoate
Methyl p-Hydroxybenzoate
RN: 93-58-3   **MP** (°C):   -12
MW: 136.15   **BP** (°C):   198

Solubility (Moles/L)	Solubility (Grams/L)	Temp (°C)	Ref (#)	Evaluation (T P E A A)	Comments
7.337E-03	9.990E-01	15	G040	1 0 2 0 0	
3.085E-02	4.200E+00	22	N317	1 1 2 1 2	
2.926E-02	3.984E+00	25	G040	1 0 2 0 0	
1.447E-02	1.970E+00	25	L086	1 0 1 1 2	
1.497E-02	2.038E+00	25	M334	1 0 1 1 2	
1.777E-02	2.420E+00	30	L012	2 0 2 2 2	
1.796E-02	2.445E+00	30	L086	1 0 1 1 2	
3.654E-02	4.975E+00	35	G040	1 0 2 0 0	
2.221E-02	3.024E+00	35	L086	1 0 1 1 2	
2.723E-02	3.708E+00	40	L086	1 0 1 1 2	

1386. C₈H₈O₂Hg

Phenylmercuric Acetate
Ceresan
PMAC
Acetate, Phenylmercuric
PMA

RN: 62-38-4   **MP** (°C):   149
MW: 336.74   **BP** (°C):

Solubility (Moles/L)	Solubility (Grams/L)	Temp (°C)	Ref (#)	Evaluation (T P E A A)	Comments
7.335E-02	2.470E+01	20	M061	1 0 0 0 2	
1.389E-02	4.678E+00	ns	B185	0 0 0 0 1	
1.396E-02	4.700E+00	ns	N013	0 0 0 0 2	
1.298E-02	4.370E+00	rt	M161	0 0 0 0 2	

1387. C₈H₈O₃

4-Hydroxy-m-toluic Acid
4-Hydroxy-m-tolylsaeure-(1)
o-Cresotic Acid
2-Hydroxy-m-toluic Acid
2-Hydroxy-m-tolylsaeure-(1)

RN: 83-40-9   **MP** (°C):   165.5
MW: 152.15   **BP** (°C):

Solubility (Moles/L)	Solubility (Grams/L)	Temp (°C)	Ref (#)	Evaluation (T P E A A)	Comments
7.624E-02	1.160E+01	100	F300	1 0 0 0 2	
3.411E-01	5.190E+01	100	F300	1 0 0 0 2	

1388. C₈H₈O₃

p-Cresotic Acid
6-Hydroxy-m-toluic Acid
6-Hydroxy-m-tolylsaeure-(1)

RN: 89-56-5   **MP** (°C):   151
MW: 152.15   **BP** (°C):

Solubility (Moles/L)	Solubility (Grams/L)	Temp (°C)	Ref (#)	Evaluation (T P E A A)	Comments
1.439E-01	2.190E+01	100	F300	1 0 0 0 2	

1389.   C₈H₈O₃
o-Anisic Acid
2-Methoxybenzoic Acid
Salicylic Acid Methyl Ether
Salicylsaeure-methylaether
o-Methoxybenzoic Acid

RN:	579-75-9	**MP** (°C):	101
MW:	152.15	**BP** (°C):	200

Solubility (Moles/L)	Solubility (Grams/L)	Temp (°C)	Ref (#)	Evaluation (T P E A A)	Comments
2.760E-02	4.200E+00	25	H007	1 0 2 2 1	
3.286E-02	5.000E+00	30	F300	1 0 0 0 0	
3.503E-02	5.330E+00	37	M360	1 2 1 1 2	

1390.   C₈H₈O₃
Methylparaben
Me-paraben
Methyl p-Hydroxybenzoic Acid
Methyl 4-Hydroxybenzoate
Methyl Paraben

RN:	99-76-3	**MP** (°C):	131
MW:	152.15	**BP** (°C):	275

Solubility (Moles/L)	Solubility (Grams/L)	Temp (°C)	Ref (#)	Evaluation (T P E A A)	Comments
8.310E-03	1.264E+00	15	B355	1 1 1 1 2	
1.026E-02	1.561E+00	15	M352	1 1 1 1 2	
9.970E-03	1.517E+00	20	B355	1 1 1 1 2	
1.334E-02	2.030E+00	20	H056	1 0 2 1 2	
1.441E-02	2.193E+00	25	A059	1 0 1 1 2	
1.140E-02	1.735E+00	25	B355	1 1 1 1 2	
1.639E-02	2.494E+00	25	D081	1 2 2 1 2	
1.600E-02	2.434E+00	25	D339	1 0 1 1 2	
3.162E-02	4.811E+00	25	F322	2 0 1 1 0	EFG
1.364E-02	2.075E+00	25	L075	1 0 1 1 2	
1.393E-02	2.120E+00	25	L338	1 0 1 1 2	
1.460E-02	2.221E+00	25	M014	2 0 1 1 2	
1.585E-02	2.412E+00	25	M352	1 1 1 1 2	
1.643E-02	2.500E+00	25	O027	1 0 1 0 1	
1.485E-02	2.260E+00	25	P013	2 0 2 1 2	
1.446E-02	2.200E+00	25	P053	1 0 1 1 2	
1.600E-02	2.434E+00	27	B129	2 2 2 2 2	
1.500E-02	2.282E+00	27	G078	2 1 0 1 0	EFG
1.600E-02	2.434E+00	27	P019	1 2 1 1 0	EFG
1.450E-02	2.206E+00	27.0	G067	2 0 1 1 2	
1.828E-02	2.782E+00	30	A059	1 0 1 1 2	
1.564E-02	2.380E+00	30	M325	1 0 0 0 1	
2.275E-02	3.462E+00	35	A059	1 0 1 1 2	
2.550E-02	3.880E+00	37	B171	2 0 1 1 2	

2.268E-02	3.451E+00	39.3	G302	2 2 2 2 0	EFG
2.551E-02	3.882E+00	40	A059	1 0 1 1 2	
3.773E-02	5.740E+00	40	M352	1 1 1 1 2	
4.168E-02	6.341E+00	50	M352	1 1 1 1 2	

1391.   C$_8$H$_8$O$_3$
Methyl Salicylate
Salicylsaeure-methyl Ester
Methyl Hydroxybenzoate
Betula Oil
Panalgesic
Betula

RN: 119-36-8   **MP** (°C):   -8
MW: 152.15   **BP** (°C):   222

Solubility (Moles/L)	Solubility (Grams/L)	Temp (°C)	Ref (#)	Evaluation (T P E A A)	Comments
4.206E-03	6.400E-01	21	B331	1 2 2 1 1	
1.312E-02	1.996E+00	25	R041	1 0 2 1 1	
4.601E-03	7.000E-01	30	F300	1 0 0 0 0	
6.244E-03	9.500E-01	30	L012	2 0 2 2 1	

1392.   C$_8$H$_8$O$_3$
Mandelic Acid
Amygdalic Acid
α-Hydroxyphenylacetic Acid
Uromaline
α-Hydroxy-benzeneacetic Acid

RN: 90-64-2   **MP** (°C):   119.0
MW: 152.15   **BP** (°C):

Solubility (Moles/L)	Solubility (Grams/L)	Temp (°C)	Ref (#)	Evaluation (T P E A A)	Comments
1.191E+00	1.812E+02	25	K040	1 0 2 1 2	*sic*
8.795E-03	1.338E+00	25	R049	1 0 0 1 2	

1393.   C$_8$H$_8$O$_3$
DL-Mandelic Acid
DL-Mandelsaeure

RN: 611-72-3   **MP** (°C):   122
MW: 152.15   **BP** (°C):

Solubility (Moles/L)	Solubility (Grams/L)	Temp (°C)	Ref (#)	Evaluation (T P E A A)	Comments
9.050E-01	1.377E+02	20	F300	1 0 0 0 2	
1.134E+00	1.725E+02	24	F300	1 0 0 0 2	

1394. $C_8H_8O_3$
3-Methoxybenzoic Acid
3-Methoxy-benzoesaeure
m-Anisic Acid
m-Methoxybenzoic Acid

RN: 586-38-9   **MP** (°C): 110
MW: 152.15   **BP** (°C): 170

Solubility (Moles/L)	Solubility (Grams/L)	Temp (°C)	Ref (#)	Evaluation (T P E A A)	Comments
1.282E-02	1.950E+00	37	M360	1 2 1 1 2	
1.183E-03	1.800E-01	ns	B361	0 0 0 2 2	

1395. $C_8H_8O_3$
3-Hydroxy-p-toluic Acid
3-Hydroxy-p-tolylsaeure-(1)

RN: 586-30-1   **MP** (°C):
MW: 152.15   **BP** (°C):

Solubility (Moles/L)	Solubility (Grams/L)	Temp (°C)	Ref (#)	Evaluation (T P E A A)	Comments
2.859E-01	4.350E+01	100	F300	1 0 0 0 2	

1396. $C_8H_8O_3$
Vanillin
4-Hydroxy-3-methoxybenzaldehyde
3-Methoxy-4-hydroxybenzaldehyde
Methylprotocatechuic Aldehyde
Vanillic Aldehyde
Vanillaldehyde

RN: 121-33-5   **MP** (°C): 82
MW: 152.15   **BP** (°C): 285

Solubility (Moles/L)	Solubility (Grams/L)	Temp (°C)	Ref (#)	Evaluation (T P E A A)	Comments
4.439E-02	6.754E+00	.2	D073	1 1 2 1 1	
1.972E-02	3.000E+00	4.40	M096	1 1 2 1 1	
3.418E-02	5.200E+00	15.60	M096	1 1 2 1 2	
8.114E-02	1.235E+01	20	D073	1 1 2 1 2	
6.572E-02	1.000E+01	20	F300	1 0 0 0 0	
5.915E-02	9.000E+00	23.90	M096	1 1 2 1 2	
7.240E-02	1.102E+01	25	I019	1 0 1 2 2	
9.713E-02	1.478E+01	30	D073	1 1 2 1 2	
8.500E-02	1.293E+01	30	L069	1 0 1 1 0	EFG
1.697E-01	2.582E+01	40	D073	1 1 2 1 2	
3.010E-01	4.580E+01	50	D073	1 1 2 1 2	
3.160E-01	4.807E+01	60	D073	1 1 2 1 2	
3.286E-01	5.000E+01	80	F300	1 0 0 0 0	

1397. C$_8$H$_8$O$_3$
p-Methoxybenzoic Acid
4-Methoxybenzoic Acid
p-Anisic Acid
Anissaeure
RN: 100-09-4   **MP** (°C):   184
MW: 152.15   **BP** (°C):   275

Solubility (Moles/L)	Solubility (Grams/L)	Temp (°C)	Ref (#)	Evaluation (T P E A A)	Comments
1.775E-02	2.700E+00	19	F300	1 0 0 0 1	
3.483E-03	5.300E-01	37	B171	2 0 1 1 2	
1.380E-03	2.100E-01	37	M360	1 2 1 1 2	

1398. C$_8$H$_8$O$_3$
D-Mandelic Acid
(R)(-)Mandelic Acid
(S)-α-Hydroxybenzeneacetic Acid
L-Mandelic Acid
(S)-(+)-Mandelic Acid
RN: 17199-29-0   **MP** (°C):   132
MW: 152.15   **BP** (°C):

Solubility (Moles/L)	Solubility (Grams/L)	Temp (°C)	Ref (#)	Evaluation (T P E A A)	Comments
5.310E-01	8.080E+01	0	A043	1 2 1 1 2	
5.310E-01	8.080E+01	0	L035	1 2 2 1 2	
6.874E-01	1.046E+02	10	A043	1 2 1 1 2	
6.874E-01	1.046E+02	10	L035	1 2 2 1 2	
7.766E-01	1.182E+02	15	A043	1 2 1 1 2	
7.766E-01	1.182E+02	15	L035	1 2 2 1 2	
9.158E-01	1.393E+02	20	A043	1 2 1 1 2	
9.158E-01	1.393E+02	20	L035	1 2 2 1 2	
5.371E-01	8.173E+01	24.5	L035	1 2 2 1 1	
5.371E-01	8.173E+01	24.50	A043	1 2 1 1 1	
1.183E+00	1.800E+02	25	A043	1 2 1 1 2	
6.503E-01	9.894E+01	25	C045	2 2 0 1 2	
6.705E-01	1.020E+02	25	C045	2 2 0 1 2	
1.183E+00	1.800E+02	25	L035	1 2 2 1 2	
6.460E-01	9.829E+01	27.5	L035	1 2 2 1 2	
6.460E-01	9.829E+01	27.50	A043	1 2 1 1 2	
1.791E+00	2.725E+02	30	A043	1 2 1 1 2	
1.791E+00	2.725E+02	30	L035	1 2 2 1 2	
8.223E-01	1.251E+02	31.5	L035	1 2 2 1 2	
8.223E-01	1.251E+02	31.50	A043	1 2 1 1 2	
2.957E+00	4.499E+02	35	A043	1 2 1 1 2	
2.957E+00	4.499E+02	35	L035	1 2 2 1 2	
3.434E+00	5.224E+02	37	A043	1 2 1 1 2	

1.132E+00	1.722E+02	37	A043	1 2 1 1 2	
3.434E+00	5.224E+02	37	L035	1 2 2 1 2	
1.132E+00	1.722E+02	37	L035	1 2 2 1 2	
4.075E+00	6.201E+02	40	A043	1 2 1 1 2	
4.075E+00	6.201E+02	40	L035	1 2 2 1 2	
1.517E+00	2.308E+02	41.5	L035	1 2 2 1 2	
1.517E+00	2.308E+02	41.50	A043	1 2 1 1 2	
4.325E+00	6.580E+02	42.5	L035	1 2 2 1 2	
4.325E+00	6.580E+02	42.50	A043	1 2 1 1 2	
1.871E+00	2.847E+02	44	A043	1 2 1 1 2	
1.871E+00	2.847E+02	44	L035	1 2 2 1 2	
4.678E+00	7.118E+02	45	L035	1 2 2 1 2	
4.678E+00	7.118E+02	45.50	A043	1 2 1 1 2	
2.351E+00	3.577E+02	46.5	L035	1 2 2 1 2	
2.351E+00	3.577E+02	46.50	A043	1 2 1 1 2	
4.816E+00	7.328E+02	47	L035	1 2 2 1 2	
4.816E+00	7.328E+02	47.50	A043	1 2 1 1 2	
2.795E+00	4.253E+02	48.5	L035	1 2 2 1 2	
2.795E+00	4.253E+02	48.50	A043	1 2 1 1 2	
5.183E+00	7.886E+02	50	A043	1 2 1 1 2	
5.183E+00	7.886E+02	50	L035	1 2 2 1 2	
3.192E+00	4.856E+02	50.5	L035	1 2 2 1 2	
3.192E+00	4.856E+02	50.50	A043	1 2 1 1 2	
3.484E+00	5.301E+02	52.5	L035	1 2 2 1 2	
3.484E+00	5.301E+02	52.50	A043	1 2 1 1 2	
3.704E+00	5.635E+02	54.50	A043	1 2 1 1 2	
3.704E+00	5.635E+02	54.50	L035	1 2 2 1 2	
3.996E+00	6.080E+02	57	A043	1 2 1 1 2	
3.996E+00	6.080E+02	57	L035	1 2 2 1 2	
4.337E+00	6.599E+02	60.5	L035	1 2 2 1 2	
4.337E+00	6.599E+02	60.50	A043	1 2 1 1 2	
4.884E+00	7.431E+02	68	A043	1 2 1 1 2	
4.884E+00	7.431E+02	68	L035	1 2 2 1 2	

1399. $C_8H_8O_3$
m-Cresotic Acid
2-Hydroxy-p-tolylsaeure-(1)
m-Kresotinsaeure

RN: 50-85-1   **MP** (°C):   177
MW: 152.15   **BP** (°C):

Solubility (Moles/L)	Solubility (Grams/L)	Temp (°C)	Ref (#)	Evaluation (T P E A A)	Comments
6.638E-02	1.010E+01	100	F300	1 0 0 0 2	

1400. $C_8H_8O_3$
Phenoxyacetic Acid
Glycolic Acid Phenyl Ether
O-Phenylglycolic Acid
RN: 122-59-8   **MP** (°C):   98
MW: 152.15   **BP** (°C):   285

Solubility (Moles/L)	Solubility (Grams/L)	Temp (°C)	Ref (#)	Evaluation (T P E A A)	Comments
7.887E-02	1.200E+01	10	F071	1 1 2 1 2	
8.084E-03	1.230E+00	10	F300	1 0 0 0 2	
7.887E-02	1.200E+01	10	H080	1 0 0 0 2	
7.887E-02	1.200E+01	10	M344	1 0 0 0 2	
1.100E-04	1.674E-02	25	L030	1 0 2 1 2	

1401. $C_8H_8O_4$
Homogentisic Acid
2,5-Dihydroxyphenylacetic Acid
2,5-Dihydroxy-benzeneacetic Acid
RN: 451-13-8   **MP** (°C):   151
MW: 168.15   **BP** (°C):

Solubility (Moles/L)	Solubility (Grams/L)	Temp (°C)	Ref (#)	Evaluation (T P E A A)	Comments
2.732E+00	4.595E+02	25	D041	1 0 0 0 1	

1402. $C_8H_8O_4$
Vanillic Acid
Vanillinsaeure
RN: 121-34-6   **MP** (°C):   214
MW: 168.15   **BP** (°C):

Solubility (Moles/L)	Solubility (Grams/L)	Temp (°C)	Ref (#)	Evaluation (T P E A A)	Comments
8.921E-03	1.500E+00	14	F300	1 0 0 0 1	
1.546E-01	2.600E+01	100	F300	1 0 0 0 2	

1403. $C_8H_8O_5$
Methyl Gallate
Gallussaeuremethyl Ester
Methyl-3,4,5-trihydroxybenzoate
RN: 99-24-1   **MP** (°C):   201.5
MW: 184.15   **BP** (°C):

Solubility (Moles/L)	Solubility (Grams/L)	Temp (°C)	Ref (#)	Evaluation (T P E A A)	Comments
5.756E-02	1.060E+01	ns	F300	0 0 0 0 2	

1404. C$_8$H$_9$ClNO$_5$PS
Chlorthion
O,O-Dimethyl O-4-Nitro-3-Chlorophenyl Thiophosphate
RN: 500-28-7   **MP** (°C): 21
MW: 297.66   **BP** (°C):

Solubility (Moles/L)	Solubility (Grams/L)	Temp (°C)	Ref (#)	Evaluation (T P E A A)	Comments
1.344E-04	4.000E-02	20	M061	1 0 0 0 1	

1405. C$_8$H$_9$ClNO$_5$PS
Dicapthon
O-(2-Chloro-4-nitrophenyl) O,O-Dimethyl phosphorothioate
Dicaptan
Isochlorthion
RN: 2463-84-5   **MP** (°C):
MW: 297.66   **BP** (°C):

Solubility (Moles/L)	Solubility (Grams/L)	Temp (°C)	Ref (#)	Evaluation (T P E A A)	Comments
4.233E-05	1.260E-02	10	B324	2 2 2 2 2	
4.233E-05	1.260E-02	10	B324	2 2 2 2 2	
4.939E-05	1.470E-02	20	B300	2 1 1 1 2	
4.939E-05	1.470E-02	20	B324	2 2 2 2 2	
4.939E-05	1.470E-02	20	B324	2 2 2 2 2	
2.100E-05	6.250E-03	20	C053	1 0 2 2 1	
1.485E-04	4.420E-02	30	B324	2 2 2 2 2	
1.485E-04	4.420E-02	30	B324	2 2 2 2 2	
2.100E-05	6.250E-03	ns	F071	0 1 2 1 2	
1.176E-04	3.500E-02	ns	M061	0 0 0 0 1	
2.620E-05	7.800E-03	rt	F040	1 2 2 2 1	

1406. C$_8$H$_9$ClO
2,5-Dimethyl-4-chloro-phenol
4-Chloro-2,5-xylenol
4-Chloro-2,5-dimethylphenol
RN: 1124-06-7   **MP** (°C): 114-116
MW: 156.61   **BP** (°C):

Solubility (Moles/L)	Solubility (Grams/L)	Temp (°C)	Ref (#)	Evaluation (T P E A A)	Comments
5.700E-02	8.927E+00	25	B316	1 0 2 1 1	

1407. C₈H₉ClO

Chloroxylenol
3,5-Dimethyl-4-chloro-phenol-
RN: 88-04-0   **MP** (°C):   115.5
MW: 156.61   **BP** (°C):   246

Solubility (Moles/L)	Solubility (Grams/L)	Temp (°C)	Ref (#)	Evaluation (T P E A A)	Comments
1.596E-03	2.500E-01	20	M018	1 2 2 1 0	EFG
1.979E-03	3.099E-01	20	M093	1 0 0 1 1	
2.200E-02	3.445E+00	25	B316	1 0 2 1 1	*sic*
1.915E-03	2.999E-01	25	R041	1 0 2 1 1	

1408. C₈H₉ClO

2,6-Dimethyl-4-chloro-phenol
4-Chloro-2,6-xylenol
RN: 1123-63-3   **MP** (°C):
MW: 156.61   **BP** (°C):

Solubility (Moles/L)	Solubility (Grams/L)	Temp (°C)	Ref (#)	Evaluation (T P E A A)	Comments
3.300E-03	5.168E-01	25	B316	1 0 2 1 1	

1409. C₈H₉FN₂O₃

Ftorafur
THFFU
1-(2-Tetrahydrofuryl)-5-fluorouracil
RN: 37076-68-9   **MP** (°C):   167
MW: 200.17   **BP** (°C):

Solubility (Moles/L)	Solubility (Grams/L)	Temp (°C)	Ref (#)	Evaluation (T P E A A)	Comments
1.400E-01	2.802E+01	37	N017	1 0 2 2 1	

1410. C₈H₉FN₂O₄

1-Propionyloxymethyl-5-fluorouracil
1-Propionyloxymethyl-5-fluoro-2,4(1H,3H)-pyrimidinedi-one
RN: 66542-36-7   **MP** (°C):   100-102
MW: 216.17   **BP** (°C):

Solubility (Moles/L)	Solubility (Grams/L)	Temp (°C)	Ref (#)	Evaluation (T P E A A)	Comments
1.554E-01	3.360E+01	22	B321	1 0 2 2 2	pH 4.0

1411.   C$_8$H$_9$FN$_2$O$_4$
1-Isopropyloxyarbonyl-5-fluorouracil
1(2H)-Pyrimidinecarboxylic Acid, 5-Fluoro-3,4-dihydro-2,4-dioxo-, 1-Methylethyl Ester
RN: 109232-73-7   **MP** (°C):   180
MW: 216.17   **BP** (°C):

Solubility (Moles/L)	Solubility (Grams/L)	Temp (°C)	Ref (#)	Evaluation (T P E A A)	Comments
2.174E-02	4.700E+00	22	B332	1 1 0 0 1	pH 4.0

1412.   C$_8$H$_9$N
Indoline
2,3-Dihydro-1H-indole
2,3-Dihydroindole
RN: 496-15-1   **MP** (°C):   <25
MW: 119.17   **BP** (°C):   220.5

Solubility (Moles/L)	Solubility (Grams/L)	Temp (°C)	Ref (#)	Evaluation (T P E A A)	Comments
2.934E-02	3.497E+00	20.3	L339	2 0 2 2 2	
9.063E-02	1.080E+01	25	P051	2 1 1 2 2	
9.063E-02	1.080E+01	25.00	P007	2 1 2 2 1	
3.651E-02	4.350E+00	40.0	L339	2 0 2 2 2	
4.586E-02	5.465E+00	59.4	L339	2 0 2 2 2	
5.738E-02	6.838E+00	79.0	L339	2 0 2 2 2	
8.142E-02	9.703E+00	100.0	L339	2 0 2 2 2	

1413.   C$_8$H$_9$NO
p-Aminoacetophenone
4'-Aminoacetophenone
RN: 99-92-3   **MP** (°C):   106
MW: 135.17   **BP** (°C):   294

Solubility (Moles/L)	Solubility (Grams/L)	Temp (°C)	Ref (#)	Evaluation (T P E A A)	Comments
2.480E-02	3.352E+00	37.5	G002	1 1 1 1 2	

1414.   C$_8$H$_9$NO
Acetanilide
Acetanilid
RN: 103-84-4   **MP** (°C):   114
MW: 135.17   **BP** (°C):   304

Solubility (Moles/L)	Solubility (Grams/L)	Temp (°C)	Ref (#)	Evaluation (T P E A A)	Comments
2.652E-02	3.585E+00	0	L029	2 2 2 2 2	
3.534E-02	4.777E+00	10	M043	1 0 0 0 1	
3.251E-02	4.395E+00	10.1	L029	2 2 2 2 2	
2.970E-02	4.014E+00	14	O016	1 0 0 0 2	
3.688E-02	4.985E+00	15	L038	1 0 1 0 2	

3.710E-02	5.015E+00	20	B101	1 0 2 2 2	
3.666E-02	4.955E+00	20	K078	1 0 2 1 2	
4.129E-02	5.581E+00	20	L029	2 2 2 2 2	
3.827E-02	5.173E+00	20	M043	1 0 0 0 1	
3.330E-02	4.501E+00	20	O019	1 0 0 1 2	
3.884E-02	5.250E+00	20	W026	1 0 1 1 1	average of 2
4.142E-02	5.598E+00	25	B101	1 0 2 2 2	
4.160E-02	5.623E+00	25	D044	1 1 1 1 2	
4.143E-02	5.600E+00	25	F300	1 0 0 0 1	
4.697E-02	6.349E+00	25	L029	2 2 2 2 2	
4.486E-02	6.063E+00	25	M094	1 0 0 1 1	
3.699E-02	5.000E+00	25	P016	1 0 0 1 0	
4.887E-02	6.606E+00	30	B101	1 0 2 2 2	
5.351E-02	7.232E+00	30	L029	2 2 2 2 2	
4.632E-02	6.261E+00	30	M043	1 0 0 0 1	
5.253E-02	7.100E+00	30	W026	1 0 1 1 1	average of 2
5.792E-02	7.828E+00	32.6	L038	1 0 1 0 2	
5.930E-02	8.015E+00	35	B101	1 0 2 2 2	
7.134E-02	9.643E+00	40	L029	2 2 2 2 2	
6.381E-02	8.625E+00	40	M043	1 0 0 0 1	
9.682E-02	1.309E+01	50	L029	2 2 2 2 2	
1.349E-01	1.823E+01	60	L029	2 2 2 2 2	
1.522E-01	2.057E+01	60	M043	1 0 0 0 1	
1.928E-01	2.606E+01	70	L029	2 2 2 2 2	
3.321E-01	4.489E+01	80	M043	1 0 0 0 1	
4.047E-02	5.470E+00	rt	D021	0 0 1 1 1	

1415. C_8H_9NO
m-Aminoacetophenone
3'-Aminoacetophenone

RN: 99-03-6   **MP** (°C):   97
MW: 135.17   **BP** (°C):

Solubility (Moles/L)	Solubility (Grams/L)	Temp (°C)	Ref (#)	Evaluation (T P E A A)	Comments
5.220E-02	7.056E+00	37.5	G002	1 1 1 1 2	pH 6.8

1416.   C$_8$H$_9$NO$_2$

Acetaminophen
4-Acetamidophenol
4-Amino-phenol-N-acetat
p-Acetaminophen
p-Hydroxyacetanilide

RN: 103-90-2   **MP** (°C):   167
MW: 151.17   **BP** (°C):

Solubility (Moles/L)	Solubility (Grams/L)	Temp (°C)	Ref (#)	Evaluation (T P E A A)	Comments
7.307E-02	1.105E+01	15	M352	1 1 1 1 2	
1.323E-01	2.000E+01	25	B010	1 1 1 1 0	
9.500E-02	1.436E+01	25	C032	2 2 1 2 0	EFG
7.710E-02	1.165E+01	25	D044	1 1 1 1 2	
9.133E-02	1.381E+01	25	D078	1 2 1 1 2	
1.000E-01	1.512E+01	25	K041	1 0 0 0 0	
9.851E-02	1.489E+01	25	M352	1 1 1 1 2	
9.923E-02	1.500E+01	25	P016	1 0 0 1 1	
7.277E-02	1.100E+01	25	P312	1 2 2 2 2	
9.326E-02	1.410E+01	25	W019	1 0 1 1 2	
1.120E-01	1.693E+01	30	L069	1 0 1 1 0	EFG
1.323E-01	2.000E+01	37	F076	2 0 2 2 0	
1.442E-01	2.180E+01	37	K086	1 0 0 0 2	
1.349E-01	2.039E+01	39.3	G302	2 0 2 2 0	EFG
1.440E-01	2.177E+01	40	M352	1 1 1 1 2	
1.800E-01	2.720E+01	50	M352	1 1 1 1 2	

1417.   C$_8$H$_9$NO$_2$

Benzyl Carbamate
O-Benzyl Carbamate
Benzyloxycarbonyl Amine

RN: 621-84-1   **MP** (°C):   87
MW: 151.17   **BP** (°C):

Solubility (Moles/L)	Solubility (Grams/L)	Temp (°C)	Ref (#)	Evaluation (T P E A A)	Comments
4.500E-01	6.802E+01	37	H006	1 2 2 1 1	

1418.   C$_8$H$_9$NO$_2$

N-Methylanthranilic Acid
N-Methyl-anthranilsaeure

RN: 119-68-6   **MP** (°C):   171
MW: 151.17   **BP** (°C):

Solubility (Moles/L)	Solubility (Grams/L)	Temp (°C)	Ref (#)	Evaluation (T P E A A)	Comments
1.323E-03	2.000E-01	20	F300	1 0 0 0 2	
2.646E-03	4.000E-01	100	F300	1 0 0 0 2	

1419.   C$_8$H$_9$NO$_2$
Methyl-p-aminobenzoate
Methyl p-Aminobenzoate
4-Aminobenzoic Acid Methyl Ester
RN: 619-45-4   **MP** (°C):
MW: 151.17   **BP** (°C):

Solubility (Moles/L)	Solubility (Grams/L)	Temp (°C)	Ref (#)	Evaluation (T P E A A)	Comments
5.884E-03	8.894E-01	15	M352	1 1 1 1 2	
9.542E-03	1.442E+00	25	M352	1 1 1 1 2	
1.070E-02	1.618E+00	25	P303	2 0 2 2 2	
1.397E-02	2.112E+00	33	P303	2 0 2 2 2	
2.530E-02	3.825E+00	37	F006	1 1 2 2 2	
1.646E-02	2.488E+00	40	M352	1 1 1 1 2	
1.839E-02	2.780E+00	40	P303	2 0 2 2 2	
7.940E-03	1.200E+00	ns	M066	0 0 0 0 2	
7.940E-03	1.200E+00	rt	B016	0 0 1 1 2	pH 7.4

1420.   C$_8$H$_9$NO$_2$
DL-2-Phenylglycine
2-Amino-phenyl-essigsaeure
2-Aminophenylacetic Acid
RN: 2835-06-5   **MP** (°C):   255
MW: 151.17   **BP** (°C):

Solubility (Moles/L)	Solubility (Grams/L)	Temp (°C)	Ref (#)	Evaluation (T P E A A)	Comments
7.608E-01	1.150E+02	100	F300	1 0 0 0 2	

1421.   C$_8$H$_9$NO$_3$S
p-Acetylbenzenesulfonamide
4-Acetylbenzenesulfonamide
RN: 1565-17-9   **MP** (°C):
MW: 199.23   **BP** (°C):

Solubility (Moles/L)	Solubility (Grams/L)	Temp (°C)	Ref (#)	Evaluation (T P E A A)	Comments
2.300E-03	4.582E-01	15	K024	1 2 1 1 2	

1422.   C₈H₉NO₄

Biliverdic Acid
Biliverdinsaeure

RN:	487-65-0	**MP** (°C):	
MW:	183.17	**BP** (°C):	

Solubility (Moles/L)	Solubility (Grams/L)	Temp (°C)	Ref (#)	Evaluation (T P E A A)	Comments
2.129E-01	3.900E+01	20	F300	1 0 0 0 1	

1423.   C₈H₉N₃O₃

Orotic Acid Allylamide
4-Pyrimidinecarboxamide, 1,2,3,6-Tetrahydro-2,6-dioxo-N-2-propenyl-

RN:	292870-71-4	**MP** (°C):	259-262
MW:	195.18	**BP** (°C):	

Solubility (Moles/L)	Solubility (Grams/L)	Temp (°C)	Ref (#)	Evaluation (T P E A A)	Comments
1.780E-01	3.474E+01	-4	N018	2 2 1 2 2	
3.000E-01	5.855E+01	16	N018	2 2 1 2 2	
3.710E-01	7.241E+01	25	N018	2 2 1 2 2	

1424.   C₈H₉N₅

7-Dimethylaminopteridine
7-Pteridinamine, N,N-Dimethyl-

RN:	204443-26-5	**MP** (°C):	
MW:	175.19	**BP** (°C):	

Solubility (Moles/L)	Solubility (Grams/L)	Temp (°C)	Ref (#)	Evaluation (T P E A A)	Comments
8.154E-01	1.429E+02	20	A083	1 2 0 0 0	
1.903E+00	3.333E+02	100	A083	1 2 0 0 0	

1425.   C₈H₉N₅

2-Dimethylaminoptcridine
2-Pteridinamine, N,N-Dimethyl-

RN:	41047-52-3	**MP** (°C):	
MW:	175.19	**BP** (°C):	

Solubility (Moles/L)	Solubility (Grams/L)	Temp (°C)	Ref (#)	Evaluation (T P E A A)	Comments
1.631E+00	2.857E+02	22.5	A085	1 2 0 0 0	

1426.   C$_8$H$_9$N$_5$
4-Dimethylaminopteridine
4-Pteridinamine, N,N-Dimethyl-
RN: 14131-04-5   **MP** (°C):   165
MW: 175.19   **BP** (°C):

Solubility (Moles/L)	Solubility (Grams/L)	Temp (°C)	Ref (#)	Evaluation (T P E A A)	Comments
9.357E-02	1.639E+01	20	A019	2 2 1 1 0	
1.392E-01	2.439E+01	100	A019	1 2 1 1 0	

1427.   C$_8$H$_9$O$_3$PS
2-Methoxy-4H-benzo-1,3,2-dioxaphosphorin-2-thione
Dioxabenzofos
Salithion
Fenfosphorin
Dioxabenzophos
RN: 3811-49-2   **MP** (°C):   55.5
MW: 216.20   **BP** (°C):

Solubility (Moles/L)	Solubility (Grams/L)	Temp (°C)	Ref (#)	Evaluation (T P E A A)	Comments
2.683E-04	5.800E-02	30	M161	1 0 0 0 1	

1428.   C$_8$H$_{10}$
Xylene
Dimethylbenzene
Xylol
RN: 1330-20-7   **MP** (°C):
MW: 106.17   **BP** (°C):   137

Solubility (Moles/L)	Solubility (Grams/L)	Temp (°C)	Ref (#)	Evaluation (T P E A A)	Comments
8.469E-03	8.992E-01	20	C121	1 0 0 0 0	unit assumed, *sic*
1.000E-03	1.062E-01	25	H332	2 2 2 2 0	
<9.41E-03	<9.99E-01	25.50	O005	2 0 2 2 0	
9.419E-03	1.000E+00	150	J023	1 1 2 2 0	
3.297E-02	3.500E+00	200	J023	1 1 2 2 1	
1.036E-01	1.100E+01	250	J023	1 1 2 2 1	

1429. C_8H_{10}
m-Xylene
1,3-Xylene
RN: 108-38-3   **MP** (°C):   -47.4
MW: 106.17   **BP** (°C):   139.3

Solubility (Moles/L)	Solubility (Grams/L)	Temp (°C)	Ref (#)	Evaluation (T P E A A)	Comments
1.846E-03	1.960E-01	0	P003	2 2 2 2 2	
1.463E-03	1.554E-01	20	M337	2 1 2 2 2	
1.629E-03	1.730E-01	25	A001	1 2 2 2 2	
1.846E-03	1.960E-01	25	B003	2 2 2 2 2	
1.262E-03	1.340E-01	25	K119	1 0 0 0 2	
1.510E-03	1.603E-01	25	M342	1 0 1 1 2	
1.526E-03	1.620E-01	25	P003	2 2 2 2 2	
1.262E-03	1.340E-01	25	P051	2 1 1 2 2	
1.375E-03	1.460E-01	25	S005	2 2 2 2 2	
1.375E-03	1.460E-01	25	S191	1 2 2 2 2	
1.375E-03	1.460E-01	25	S358	2 1 2 2 2	
1.330E-03	1.412E-01	25	S359	2 1 2 2 2	
1.510E-03	1.603E-01	25	W300	2 2 2 2 2	
1.262E-03	1.340E-01	25.00	P007	2 1 2 2 2	
1.940E-03	2.059E-01	25.04	V013	2 2 2 2 2	
3.277E-03	3.479E-01	67.7	P005	1 1 2 1 2	
6.257E-03	6.643E-01	107.3	P005	1 1 2 1 2	
9.707E-03	1.031E+00	124.2	P005	1 1 2 1 2	
2.363E-02	2.509E+00	164.2	P005	1 1 2 1 2	
4.327E-02	4.594E+00	186.4	P005	1 1 2 1 2	
4.293E-02	4.557E+00	189.9	P005	1 1 2 1 2	
2.675E-01	2.840E+01	266.6	P005	1 1 2 1 2	
2.698E-01	2.865E+01	270.6	P005	1 1 2 1 2	

1430. C_8H_{10}
p-Xylene
1,4-Dimethylbenzene
1,4-Xylene
RN: 106-42-3   **MP** (°C):   13
MW: 106.17   **BP** (°C):   137

Solubility (Moles/L)	Solubility (Grams/L)	Temp (°C)	Ref (#)	Evaluation (T P E A A)	Comments
1.545E-03	1.640E-01	0	P003	2 2 2 2 2	
1.780E-03	1.890E-01	10	B149	2 1 1 2 2	
1.800E-03	1.911E-01	20	B149	2 1 1 2 2	
1.552E-03	1.648E-01	20	M337	2 1 2 2 2	
1.884E-03	2.000E-01	25	A001	1 2 2 2 2	
1.865E-03	1.980E-01	25	B003	2 2 2 2 2	
1.224E-03	1.300E-01	25	K072	1 0 1 1 1	
1.479E-03	1.570E-01	25	K119	1 0 0 0 2	
1.789E-03	1.900E-01	25	L319	1 0 2 1 1	

1.224E-03	1.300E-01	25	M087	1 1 2 1 1
2.020E-03	2.145E-01	25	M342	1 0 1 1 2
1.743E-03	1.850E-01	25	P003	2 2 2 2 2
1.479E-03	1.570E-01	25	P051	2 1 1 2 2
1.469E-03	1.560E-01	25	S005	2 2 2 2 2
1.469E-03	1.560E-01	25	S191	1 2 2 2 2
1.469E-03	1.560E-01	25	S358	2 1 2 2 2
1.510E-03	1.603E-01	25	S359	2 1 2 2 2
2.020E-03	2.145E-01	25	W300	2 2 2 2 2
1.479E-03	1.570E-01	25.00	P007	2 1 2 2 2
1.589E-03	1.687E-01	29.99	C350	2 1 2 2 2
1.766E-03	1.875E-01	39.99	C350	2 1 2 2 2
2.410E-03	2.559E-01	43.0	P005	1 1 2 1 2
1.911E-03	2.029E-01	49.99	C350	2 1 2 2 2
2.832E-03	3.007E-01	56.4	P005	1 1 2 1 2
2.244E-03	2.382E-01	59.99	C350	2 1 2 2 2
3.199E-03	3.396E-01	65.0	P005	1 1 2 1 2
2.683E-03	2.848E-01	69.99	C350	2 1 2 2 2
3.643E-03	3.868E-01	75.3	P005	1 1 2 1 2
3.171E-03	3.367E-01	79.99	C350	2 1 2 2 2
4.326E-03	4.593E-01	87.2	P005	1 1 2 1 2
3.721E-03	3.950E-01	89.99	C350	2 1 2 2 2
4.853E-03	5.152E-01	99.99	C350	2 1 2 2 2
2.363E-02	2.509E+00	162.5	P005	1 1 2 1 2
4.251E-02	4.513E+00	188.1	P005	1 1 2 1 2
1.614E-01	1.713E+01	243.2	P005	1 1 2 1 2
4.053E-01	4.303E+01	282.5	P005	1 1 2 1 2
4.011E-01	4.258E+01	294.9	P005	1 1 2 1 2
1.743E-03	1.850E-01	ns	H123	0 0 0 0 2

1431. C_8H_{10}

o-Xylene
1,2-Dimethylbenzene
1,2-Xylene

RN: 95-47-6   **MP** (°C):   -25
MW: 106.17   **BP** (°C):   144

Solubility (Moles/L)	Solubility (Grams/L)	Temp (°C)	Ref (#)	Evaluation (T P E A A)	Comments
1.337E-03	1.420E-01	0	P003	2 2 2 2 2	
2.000E-03	2.123E-01	10	B149	2 1 1 2 2	
2.260E-03	2.399E-01	20	B149	2 1 1 2 2	
1.605E-03	1.704E-01	20	M337	2 1 2 2 2	
1.921E-03	2.040E-01	25	A001	1 2 2 2 2	
1.648E-03	1.750E-01	25	B060	2 0 1 1 1	
1.573E-03	1.670E-01	25	K119	1 0 0 0 2	
1.648E-03	1.750E-01	25	M001	2 1 2 2 2	
1.648E-03	1.750E-01	25	M002	2 1 2 2 2	

1.648E-03	1.750E-01	25	M040	1 0 0 1 2
1.648E-03	1.750E-01	25	M130	1 0 0 0 2
2.080E-03	2.208E-01	25	M342	1 0 1 1 2
2.006E-03	2.130E-01	25	P003	2 2 2 2 2
1.573E-03	1.670E-01	25	P051	2 1 1 2 2
1.606E-03	1.705E-01	25	S005	2 2 2 2 2
1.606E-03	1.705E-01	25	S191	1 2 2 2 2
1.606E-03	1.705E-01	25	S358	2 1 2 2 2
1.680E-03	1.784E-01	25	S359	2 1 2 2 2
2.080E-03	2.208E-01	25	W300	2 2 2 2 2
1.573E-03	1.670E-01	25.00	P007	2 1 2 2 2
1.272E-03	1.350E-01	ns	B150	0 0 2 2 2
1.648E-03	1.750E-01	ns	M344	0 0 0 0 2

1432. C_8H_{10}
Ethylbenzene
Phenylethane
Ethylenzene
Ethylbenzol
EB

RN: 100-41-4   **MP** (°C):   -95
MW: 106.17   **BP** (°C):   136.3

Solubility (Moles/L)	Solubility (Grams/L)	Temp (°C)	Ref (#)	Evaluation (T P E A A)	Comments
1.856E-03	1.970E-01	0	P003	2 2 2 2 2	
1.846E-03	1.960E-01	4.50	B086	2 1 2 2 2	
1.808E-03	1.920E-01	6.30	B086	2 1 2 2 2	
1.752E-03	1.860E-01	7.10	B086	2 1 2 2 2	
1.761E-03	1.870E-01	9	B086	2 1 2 2 2	
1.910E-03	2.028E-01	10	B149	2 1 1 2 2	
1.850E-03	1.964E-01	10	O312	2 2 0 2 2	
1.705E-03	1.810E-01	11.80	B086	2 1 2 2 2	
1.723E-03	1.830E-01	12.10	B086	2 1 2 2 2	
1.812E-03	1.924E-01	14	O312	2 2 0 2 2	
1.300E-03	1.380E-01	15	F001	1 0 1 2 1	
1.300E-03	1.380E-01	15	S006	1 0 0 0 1	
1.658E-03	1.760E-01	15	S203	1 1 2 1 2	
1.695E-03	1.800E-01	15.10	B086	2 1 2 2 2	
1.776E-03	1.886E-01	17	O312	2 2 0 2 2	
1.733E-03	1.840E-01	17.90	B086	2 1 2 2 2	
2.901E-03	3.080E-01	18	F185	1 0 0 0 2	
2.788E-03	2.960E-01	18	F185	1 0 0 0 2	
1.725E-03	1.831E-01	18	O312	2 2 0 2 2	
3.080E-03	3.270E-01	19	F185	1 0 0 0 2	
1.676E-03	1.779E-01	19	O312	2 2 0 2 2	
2.000E-03	2.123E-01	20	B149	2 1 1 2 2	
1.695E-03	1.800E-01	20	B356	1 0 0 0 2	
1.770E-03	1.879E-01	20	O312	2 2 0 2 2	
1.695E-03	1.800E-01	20.10	B086	2 1 2 2 1	

1.724E-03	1.830E-01	21	O312	2 2 0 2 2
3.297E-03	3.500E-01	22	F185	1 0 0 0 2
1.713E-03	1.819E-01	22	O312	2 2 0 2 2
3.391E-03	3.600E-01	23	F185	1 0 0 0 2
1.751E-03	1.859E-01	23.5	O312	2 2 0 2 2
3.655E-03	3.880E-01	24	F185	1 0 0 0 2
1.582E-03	1.680E-01	25	A002	1 2 1 1 2
1.883E-03	2.000E-01	25	A094	1 0 0 0 0
1.959E-03	2.080E-01	25	B003	2 2 2 2 2
1.432E-03	1.520E-01	25	B060	2 0 1 1 1
2.000E-03	2.123E-01	25	B153	2 1 1 1 2
1.640E-03	1.741E-01	25	K001	1 0 2 1 2
1.319E-03	1.400E-01	25	K072	1 0 1 1 1
1.760E-03	1.869E-01	25	M342	1 0 1 1 2
1.811E-03	1.923E-01	25	O312	2 2 0 2 2
1.667E-03	1.770E-01	25	P003	2 2 2 2 2
1.234E-03	1.310E-01	25	P051	2 1 1 2 2
1.705E-03	1.810E-01	25	S203	1 1 2 1 2
1.518E-03	1.612E-01	25	S358	2 1 2 2 2
1.370E-03	1.455E-01	25	S359	2 1 2 2 2
1.760E-03	1.869E-01	25	W300	2 2 2 2 2
1.959E-03	2.080E-01	25.0	G035	1 0 0 0 2
1.753E-03	1.861E-01	25.8	O312	2 2 0 2 2
4.653E-03	4.940E-01	27	F185	1 0 0 0 2
1.677E-03	1.780E-01	28	B348	2 1 2 2 2
1.747E-03	1.855E-01	28	O312	2 2 0 2 2
5.604E-03	5.950E-01	29	F185	1 0 0 0 2
1.600E-03	1.698E-01	29.99	C350	2 1 2 2 2
1.391E-03	1.477E-01	30	M311	1 1 2 2 2
1.777E-03	1.887E-01	30	O312	2 2 0 2 2
6.103E-03	6.480E-01	31	F185	1 0 0 0 2
6.395E-03	6.790E-01	32	F185	1 0 0 0 2
7.017E-03	7.450E-01	34	F185	1 0 0 0 2
7.319E-03	7.770E-01	35	F185	1 0 0 0 2
1.818E-03	1.930E-01	35	O312	2 2 0 2 2
1.827E-03	1.940E-01	35	S203	1 1 2 1 2
7.865E-03	8.350E-01	36	F185	1 0 0 0 2
7.865E-03	8.350E-01	36	F185	1 0 0 0 2
8.637E-03	9.170E-01	38	F185	1 0 0 0 2
1.622E-03	1.722E-01	39.99	C350	2 1 2 2 2
1.928E-03	2.047E-01	40	O312	2 2 0 2 2
9.466E-03	1.005E+00	41	F185	1 0 0 0 2
1.991E-03	2.114E-01	45	O312	2 2 0 2 2
2.025E-03	2.150E-01	45	S203	1 1 2 1 2
1.154E-02	1.225E+00	47	F185	1 0 0 0 2
1.224E-02	1.300E+00	49	F185	1 0 0 0 2
1.861E-03	1.976E-01	49.99	C350	2 1 2 2 2
2.261E-03	2.400E-01	59.99	C350	2 1 2 2 2

2.738E-03	2.907E-01	69.99	C350	2 1 2 2 2	
3.327E-03	3.532E-01	79.99	C350	2 1 2 2 2	
3.860E-03	4.098E-01	89.99	C350	2 1 2 2 2	
4.742E-03	5.035E-01	99.99	C350	2 1 2 2 2	
4.829E-03	5.127E-01	115.0	G035	1 0 0 0 2	
1.120E-02	1.189E+00	140.5	G035	1 0 0 0 2	
3.332E-02	3.537E+00	170.5	G035	1 0 0 0 2	
6.185E-02	6.567E+00	210.0	G035	1 0 0 0 2	
1.052E-01	1.116E+01	233.5	G035	1 0 0 0 2	
1.432E-03	1.520E-01	ns	H123	0 0 0 0 2	
6.300E-02	6.689E+00	ns	H307	1 0 1 1 2	
1.432E-03	1.520E-01	ns	M344	0 0 0 0 2	

1433. $C_8H_{10}NO_5PS$

Methyl Parathion
Parathion-methyl
Methylparathion

RN: 298-00-0   **MP** (°C):   36
MW: 263.21   **BP** (°C):

Solubility (Moles/L)	Solubility (Grams/L)	Temp (°C)	Ref (#)	Evaluation (T P E A A)	Comments
8.282E-05	2.180E-02	10	B324	2 2 2 2 2	
8.283E-05	2.180E-02	10	B324	2 2 2 2 2	
1.432E-04	3.770E-02	19.50	B169	2 2 1 1 2	
1.444E-04	3.801E-02	20	B324	2 2 2 2 2	
1.444E-04	3.800E-02	20	B324	2 2 2 2 2	
9.498E-05	2.500E-02	20	M040	1 0 0 1 1	
2.090E-04	5.500E-02	25	M061	1 0 0 0 1	
2.185E-04	5.750E-02	25	M161	1 0 0 0 0	
2.223E-04	5.851E-02	30	B324	2 2 2 2 2	
2.222E-04	5.850E-02	30	B324	2 2 2 2 2	
1.900E-04	5.000E-02	ns	C117	0 0 0 0 0	

1434. $C_8H_{10}N_2O$

1-Methyl-3-phenylurea
Desfenuron
N-Phenyl-N'-methylurea
Desphenuron
N-Methyl-N'-phenylurea
IPO 4328

RN: 1007-36-9   **MP** (°C):
MW: 150.18   **BP** (°C):

Solubility (Moles/L)	Solubility (Grams/L)	Temp (°C)	Ref (#)	Evaluation (T P E A A)	Comments
4.927E+00	7.400E+02	45	W044	1 0 1 0 2	

1435.   C$_8$H$_{10}$N$_2$O
1-(4-Tolyl)urea
p-Tolylurea
RN: 622-51-5   **MP** (°C):
MW: 150.18   **BP** (°C):

Solubility (Moles/L)	Solubility (Grams/L)	Temp (°C)	Ref (#)	Evaluation (T P E A A)	Comments
2.044E-02	3.070E+00	45	W044	1 0 1 0 2	

1436.   C$_8$H$_{10}$N$_2$O
1-(2-Tolyl)urea
o-Tolylurea
RN: 614-77-7   **MP** (°C):
MW: 150.18   **BP** (°C):

Solubility (Moles/L)	Solubility (Grams/L)	Temp (°C)	Ref (#)	Evaluation (T P E A A)	Comments
1.667E-02	2.504E+00	45	W044	1 0 1 0 2	

1437.   C$_8$H$_{10}$N$_2$O
p-Phenylenediaminemono-N-acetate
p-Phenylendiamin-mono-N-acetat
RN: 589-29-7   **MP** (°C):
MW: 150.18   **BP** (°C):

Solubility (Moles/L)	Solubility (Grams/L)	Temp (°C)	Ref (#)	Evaluation (T P E A A)	Comments
4.128E-01	6.200E+01	57	F300	1 0 0 0 1	

1438.   C$_8$H$_{10}$N$_2$O
Methylbenzylnitrosamine
N-Nitroso(methyl)benzylamine
N-Nitroso-N-methylbenzylamine
N-Nitroso(benzyl)methylamine
N-Nitroso-N-methylbenzenemethanamine
RN: 937-40-6   **MP** (°C):
MW: 150.18   **BP** (°C):

Solubility (Moles/L)	Solubility (Grams/L)	Temp (°C)	Ref (#)	Evaluation (T P E A A)	Comments
3.000E-02	4.505E+00	24	D083	2 0 0 0 1	

1439. $C_8H_{10}N_2O$
Benzylurea
Benzyl-harnstoff
RN: 538-32-9 **MP** (°C): 147
MW: 150.18 **BP** (°C):

Solubility (Moles/L)	Solubility (Grams/L)	Temp (°C)	Ref (#)	Evaluation (T P E A A)	Comments
1.132E-01	1.700E+01	45	F300	1 0 0 0 2	
1.139E-01	1.710E+01	45	W044	1 0 1 0 2	

1440. $C_8H_{10}N_2O$
p-Aminoacetanilide
4-Aminoacetanilide
RN: 122-80-5 **MP** (°C): 164.5
MW: 150.18 **BP** (°C):

Solubility (Moles/L)	Solubility (Grams/L)	Temp (°C)	Ref (#)	Evaluation (T P E A A)	Comments
1.061E-01	1.593E+01	25	D044	1 1 1 1 2	
4.064E-01	6.103E+01	56.8	S115	1 2 1 1 2	
1.046E+00	1.570E+02	86.3	S115	1 2 1 1 2	
1.441E+00	2.165E+02	92.1	S115	1 2 1 1 2	
1.699E+00	2.552E+02	93.7	S115	1 2 1 1 2	
1.996E+00	2.998E+02	96.5	S115	1 2 1 1 2	
2.193E+00	3.293E+02	98.6	S115	1 2 1 1 2	

1441. $C_8H_{10}N_2O$
m-Aminoacetanilide
3-Aminoacetanilide
RN: 102-28-3 **MP** (°C):
MW: 150.18 **BP** (°C):

Solubility (Moles/L)	Solubility (Grams/L)	Temp (°C)	Ref (#)	Evaluation (T P E A A)	Comments
5.526E-01	8.299E+01	48.7	S115	1 2 1 1 2	
1.021E+00	1.534E+02	82.9	S115	1 2 1 1 2	

1442. $C_8H_{10}N_2O$
o-Aminoacetanilide
2-Aminoacetanilide
RN: 34801-09-7 **MP** (°C):
MW: 150.18 **BP** (°C):

Solubility (Moles/L)	Solubility (Grams/L)	Temp (°C)	Ref (#)	Evaluation (T P E A A)	Comments
2.189E-01	3.288E+01	7.2	S115	1 2 1 1 2	
7.161E-01	1.075E+02	22.0	S115	1 2 1 1 2	
1.215E+00	1.825E+02	33.5	S115	1 2 1 1 2	
1.612E+00	2.421E+02	42.1	S115	1 2 1 1 2	

1.958E+00	2.940E+02	50.4	S115	1 2 1 1 2
2.270E+00	3.409E+02	59.1	S115	1 2 1 1 2
2.601E+00	3.906E+02	69.9	S115	1 2 1 1 2
2.781E+00	4.177E+02	78.2	S115	1 2 1 1 2
2.943E+00	4.420E+02	88.1	S115	1 2 1 1 2
3.075E+00	4.618E+02	99.0	S115	1 2 1 1 2
3.213E+00	4.825E+02	115.4	S115	1 2 1 1 2

1443. $C_8H_{10}N_2O_3$

5,5-Tetramethylenebarbituric Acid
7,9-Diazaspiro[4.5]decane-6,8,10-trione
Spirocyclopentabarbituric Acid
RN: 56209-30-4   **MP** (°C):
MW: 182.18   **BP** (°C):

Solubility (Moles/L)	Solubility (Grams/L)	Temp (°C)	Ref (#)	Evaluation (T P E A A)	Comments
4.476E-03	8.154E-01	25	P350	2 1 1 1 2	intrinsic

1444. $C_8H_{10}N_2O_3$

5-Methyl-5-allylbarbituric Acid
2,4,6(1H,3H,5H)-Pyrimidinetrione, 5-Methyl-5-(2-propenyl)
RN: 143585-01-7   **MP** (°C):
MW: 182.18   **BP** (°C):

Solubility (Moles/L)	Solubility (Grams/L)	Temp (°C)	Ref (#)	Evaluation (T P E A A)	Comments
6.920E-02	1.261E+01	25	P350	2 1 1 1 2	intrinsic

1445. $C_8H_{10}N_2O_3S$

N4-Acetylsulfanilamide
N4-Acetylsulphanilamide
RN: 121-61-9   **MP** (°C):   216
MW: 214.24   **BP** (°C):

Solubility (Moles/L)	Solubility (Grams/L)	Temp (°C)	Ref (#)	Evaluation (T P E A A)	Comments
2.474E-02	5.300E+00	37	L091	1 0 0 0 2	pH 5.5
2.479E-02	5.312E+00	37.50	M142	1 0 0 0 2	

1446. $C_8H_{10}N_2O_3S$
N1-Acetylsulfanilamide
Sulfacetamide
Acetyl Sulfacetamide
RN: 144-80-9   **MP** (°C):   183
MW: 214.24   **BP** (°C):

Solubility (Moles/L)	Solubility (Grams/L)	Temp (°C)	Ref (#)	Evaluation (T P E A A)	Comments
5.881E-02	1.260E+01	20	F073	1 2 2 2 2	
5.834E-03	1.250E+00	37	B046	1 0 2 2 2	pH 4.5
5.834E-02	1.250E+01	37	B046	1 0 2 2 2	pH 5
6.908E-02	1.480E+01	37	D084	1 0 1 0 2	
5.601E-02	1.200E+01	37	K086	1 0 0 0 2	
5.134E-02	1.100E+01	37	L091	1 0 0 0 2	pH 5.5
2.327E-02	4.985E+00	ns	L044	0 0 0 0 2	

1447. $C_8H_{10}N_2O_3S$
Tosylurea
Tosyluree
RN: 1694-06-0   **MP** (°C):
MW: 214.24   **BP** (°C):

Solubility (Moles/L)	Solubility (Grams/L)	Temp (°C)	Ref (#)	Evaluation (T P E A A)	Comments
3.631E-03	7.779E-01	37	A028	1 0 2 1 2	intrinsic

1448. $C_8H_{10}N_2O_4S$
Asulam
Methyl N-(4-Aminobenzenesulphonyl)carbamate
RN: 3337-71-1   **MP** (°C):   144
MW: 230.24   **BP** (°C):

Solubility (Moles/L)	Solubility (Grams/L)	Temp (°C)	Ref (#)	Evaluation (T P E A A)	Comments
2.161E-02	4.975E+00	ns	M061	0 0 0 0 0	
2.172E-02	5.000E+00	rt	M161	0 0 0 0 0	

1449. $C_8H_{10}N_4O_2$
Caffeine
Coffein
RN: 58-08-2   **MP** (°C):   238
MW: 194.19   **BP** (°C):

Solubility (Moles/L)	Solubility (Grams/L)	Temp (°C)	Ref (#)	Evaluation (T P E A A)	Comments
3.887E-02	7.548E+00	0	H023	1 0 2 1 2	
3.800E-02	7.379E+00	1	M116	2 1 1 1 1	
3.757E-02	7.296E+00	2	C074	1 0 0 1 2	
6.603E-02	1.282E+01	15	H023	1 0 2 1 2	

Solubility (Moles/L)	Solubility (Grams/L)	Temp (°C)	Ref (#)	Evaluation (T P E A A)	Comments
5.800E-02	1.126E+01	15	O017	1 0 1 1 1	
5.770E-02	1.121E+01	15	O018	1 2 1 1 2	
5.770E-02	1.121E+01	15	O019	1 0 0 1 2	
6.859E-02	1.332E+01	16	A072	1 0 1 0 2	
7.415E-02	1.440E+01	20	F300	1 0 0 0 2	
6.779E-02	1.316E+01	20	J009	2 0 2 2 2	
1.242E-01	2.411E+01	25	A068	2 0 0 0 2	
1.066E-01	2.071E+01	25	E016	1 1 1 1 2	
1.081E-01	2.100E+01	25	F300	1 0 0 0 1	
1.080E-01	2.097E+01	25	L329	2 2 1 2 2	
1.110E-01	2.156E+01	25	M116	2 1 1 1 2	
1.244E-01	2.415E+01	25	M158	2 0 2 2 2	
1.000E-01	1.942E+01	25	O017	1 0 1 1 2	
1.002E-01	1.946E+01	25	O018	1 2 1 1 2	
1.098E-02	2.132E+00	25	O019	1 0 0 1 2	
4.615E+00	8.962E+02	25	O302	1 0 0 1 0	EFG, sic
1.107E-01	2.150E+01	25	P010	1 0 1 1 2	
1.123E-01	2.180E+01	25	P011	1 0 1 1 1	
1.195E-01	2.320E+01	25	P018	1 0 2 2 2	
1.081E-01	2.100E+01	25	P020	2 0 1 1 1	
1.330E-01	2.583E+01	30	B042	1 2 1 1 2	
1.330E-01	2.583E+01	30	G021	1 0 0 0 2	
1.330E-01	2.583E+01	30	H020	1 0 0 0 2	
1.333E-01	2.589E+01	30	H023	1 0 2 1 2	
1.330E-01	2.583E+01	30.60	M116	2 1 1 1 2	
1.670E-01	3.243E+01	35	O017	1 0 1 1 2	
1.909E-01	3.707E+01	37	C074	1 0 0 1 2	
1.930E-01	3.748E+01	37	M116	2 1 1 1 2	
2.266E-01	4.400E+01	40	F300	1 0 0 0 1	
5.211E-01	1.012E+02	57	C074	1 0 0 1 2	
1.408E+00	2.735E+02	83	C065	1 0 0 1 2	
1.407E+00	2.733E+02	85	C074	1 0 0 1 2	
1.739E+00	3.377E+02	87	C065	1 0 0 1 2	
2.343E+00	4.550E+02	90	C074	1 0 0 1 2	
1.287E-01	2.500E+01	ns	D035	0 0 0 0 2	
1.104E-01	2.143E+01	rt	D021	0 0 1 1 2	
1.596E-04	3.100E-02	rt	N015	0 0 2 2 1	sic

1450. $C_8H_{10}N_4O_2 \cdot H_2O$

Caffeine (Monohydrate)

1H-Purine-2,6-dione, 3,7-Dihydro-1,3,7-trimethyl-, Monohydrate

RN: 5743-12-4   **MP** (°C):   178
MW: 212.21   **BP** (°C):

Solubility (Moles/L)	Solubility (Grams/L)	Temp (°C)	Ref (#)	Evaluation (T P E A A)	Comments
1.011E-01	2.146E+01	25	D004	1 0 0 0 0	

1451. $C_8H_{10}N_4O_3$
1,3,7-Trimethyluric Acid
8-Oxy-caffeine
RN: 5415-44-1 **MP** (°C): 374
MW: 210.19 **BP** (°C):

Solubility (Moles/L)	Solubility (Grams/L)	Temp (°C)	Ref (#)	Evaluation (T P E A A)	Comments
1.142E-04	2.400E-02	rt	N015	0 0 2 2 1	

1452. $C_8H_{10}O$
2,4-Xylenol
2,4-Dimethylphenol
m-Xylenol
2,4-Dimethyl-phenol-
Phenol, 2,4-Dimethyl-
1-Hydroxy-2,4-dimethylbenzene
RN: 105-67-9 **MP** (°C): 26
MW: 122.17 **BP** (°C): 211.5

Solubility (Moles/L)	Solubility (Grams/L)	Temp (°C)	Ref (#)	Evaluation (T P E A A)	Comments
4.400E-02	5.375E+00	20	K132	1 0 1 1 1	
4.300E-02	5.253E+00	20	K309	1 0 0 1 1	
5.271E-02	6.440E+00	20	R087	1 1 2 2 2	0.15M NaCl
5.100E-02	6.231E+00	25	A021	1 2 1 1 2	
6.440E-02	7.868E+00	25	B173	2 0 2 2 2	
7.200E-02	8.796E+00	25	B316	1 0 2 1 1	
6.499E-02	7.940E+00	25	M127	1 0 0 0 2	
2.190E-01	2.675E+01	80	K309	1 0 0 1 2	

1453. $C_8H_{10}O$
Phenetole
Ethoxybenzene
RN: 103-73-1 **MP** (°C): -30
MW: 122.17 **BP** (°C): 171

Solubility (Moles/L)	Solubility (Grams/L)	Temp (°C)	Ref (#)	Evaluation (T P E A A)	Comments
4.500E-03	5.498E-01	25	M327	1 0 0 1 2	
4.657E-03	5.690E-01	25.04	V013	2 2 2 2 2	

1454. C$_8$H$_{10}$O
3,5-Xylenol
3,5-Dimethylphenol
RN: 108-68-9   **MP** (°C):   64
MW: 122.17   **BP** (°C):   219.5

Solubility (Moles/L)	Solubility (Grams/L)	Temp (°C)	Ref (#)	Evaluation (T P E A A)	Comments
3.300E-02	4.032E+00	20	K132	1 0 1 1 1	
2.961E-02	3.618E+00	20	R087	1 1 2 2 2	0.15M NaCl
4.000E-02	4.887E+00	25	A021	1 2 1 1 2	
4.000E-02	4.887E+00	25	B316	1 0 2 1 1	

1455. C$_8$H$_{10}$O
3,4-Xylenol
3,4-Dimethylphenol
As-o-xylenol
RN: 95-65-8   **MP** (°C):   62.5
MW: 122.17   **BP** (°C):   225

Solubility (Moles/L)	Solubility (Grams/L)	Temp (°C)	Ref (#)	Evaluation (T P E A A)	Comments
3.100E-02	3.787E+00	20	K132	1 0 1 1 1	
3.900E-02	4.765E+00	25	A021	1 2 1 1 2	
4.072E-02	4.975E+00	25	R041	1 0 2 1 1	
2.530E-02	3.091E+00	37	E028	1 0 1 1 2	

1456. C$_8$H$_{10}$O
2,6-Xylenol
1,3,2-Xylenol
2,6-Dimethylphenol
Vic-m-Xylenol
RN: 576-26-1   **MP** (°C):   49
MW: 122.17   **BP** (°C):   203

Solubility (Moles/L)	Solubility (Grams/L)	Temp (°C)	Ref (#)	Evaluation (T P E A A)	Comments
3.595E-02	4.392E+00	20	R087	1 1 2 2 2	0.15M NaCl
4.950E-02	6.047E+00	25	A021	1 2 1 1 2	
5.100E-02	6.231E+00	25	B316	1 0 2 1 1	

1457. C$_8$H$_{10}$O
2,5-Xylenol
2,5-Dimethylphenol
p-Xylenol
2,5-Dimethyl-phenol-
Phenol, 2,5-Dimethyl-

RN: 95-87-4   **MP** (°C):   75
MW: 122.17   **BP** (°C):   212

Solubility (Moles/L)	Solubility (Grams/L)	Temp (°C)	Ref (#)	Evaluation (T P E A A)	Comments
2.900E-02	3.543E+00	25	A021	1 2 1 1 2	
2.600E-02	3.176E+00	25	B316	1 0 2 1 1	

1458. C$_8$H$_{10}$O
2,3-Xylenol
2,3-Dimethylphenol

RN: 526-75-0   **MP** (°C):   75
MW: 122.17   **BP** (°C):   218

Solubility (Moles/L)	Solubility (Grams/L)	Temp (°C)	Ref (#)	Evaluation (T P E A A)	Comments
3.740E-02	4.569E+00	25	A021	1 2 1 1 2	

1459. C$_8$H$_{10}$O
4-Ethylphenol
p-Ethylphenol

RN: 123-07-9   **MP** (°C):   43.5
MW: 122.17   **BP** (°C):

Solubility (Moles/L)	Solubility (Grams/L)	Temp (°C)	Ref (#)	Evaluation (T P E A A)	Comments
4.854E-02	5.931E+00	20	R087	1 1 2 2 2	0.15M NaCl
2.332E-02	2.849E+00	25	L022	1 0 0 0 0	
4.011E-02	4.900E+00	25	M127	1 0 0 0 1	
4.072E-02	4.975E+00	25	R041	1 0 2 1 1	

1460. C$_8$H$_{10}$O
Phenylethylalcohol
Phenyl Ethyl Alcohol

RN: 60-12-8   **MP** (°C):
MW: 122.17   **BP** (°C):

Solubility (Moles/L)	Solubility (Grams/L)	Temp (°C)	Ref (#)	Evaluation (T P E A A)	Comments
1.470E-01	1.796E+01	20	S006	1 0 0 0 2	
1.432E-01	1.749E+01	25	H044	1 0 2 1 2	
1.455E-01	1.778E+01	30	H044	1 0 2 1 2	
1.487E-01	1.816E+01	35	H044	1 0 2 1 2	

1.518E-01	1.855E+01	40	H044	1 0 2 1 2
1.542E-01	1.884E+01	45	H044	1 0 2 1 2
1.562E-01	1.908E+01	50	H044	1 0 2 1 2
1.597E-01	1.951E+01	55	H044	1 0 2 1 2

1461. $C_8H_{10}O$
Phloral
RN: **MP** (°C):
MW: 122.17   **BP** (°C):   204.52

Solubility (Moles/L)	Solubility (Grams/L)	Temp (°C)	Ref (#)	Evaluation (T P E A A)	Comments
4.072E-02	4.975E+00	25	L022	1 0 0 0 0	

1462. $C_8H_{10}O_2$
1,3-Dimethoxybenzene
m-Dimethoxybenzene
Dimethylresorcinol
RN: 151-10-0   **MP** (°C):
MW: 138.17   **BP** (°C):   86

Solubility (Moles/L)	Solubility (Grams/L)	Temp (°C)	Ref (#)	Evaluation (T P E A A)	Comments
8.800E-03	1.216E+00	25	M327	1 0 0 1 2	

1463. $C_8H_{10}O_2$
Veratrole
o-Dimethoxybenzene
RN: 91-16-7   **MP** (°C):   15
MW: 138.17   **BP** (°C):   207

Solubility (Moles/L)	Solubility (Grams/L)	Temp (°C)	Ref (#)	Evaluation (T P E A A)	Comments
4.842E-02	6.690E+00	25	L348	1 2 2 1 2	

1464. $C_8H_{10}O_2$
p-Dimethoxybenzene
4-Dimethoxybenzene
RN: 150-78-7   **MP** (°C):
MW: 138.17   **BP** (°C):

Solubility (Moles/L)	Solubility (Grams/L)	Temp (°C)	Ref (#)	Evaluation (T P E A A)	Comments
5.530E-05	7.641E-03	25	C316	1 0 2 2 2	0.1M NaCl

1465. $C_8H_{10}O_2$
o-Ethoxyphenol
2-Ethoxyphenol
RN: 94-71-3   **MP** (°C):
MW: 138.17   **BP** (°C):

Solubility (Moles/L)	Solubility (Grams/L)	Temp (°C)	Ref (#)	Evaluation (T P E A A)	Comments
6.090E-02	8.414E+00	24.99	B353	2 1 1 1 2	

1466. $C_8H_{10}O_2$
3-Ethoxyphenol
m-Ethoxy Phenol
Resorcinol Monoethyl Ether
RN: 621-34-1   **MP** (°C):
MW: 138.17   **BP** (°C):

Solubility (Moles/L)	Solubility (Grams/L)	Temp (°C)	Ref (#)	Evaluation (T P E A A)	Comments
1.000E-01	1.382E+01	25	B314	1 0 0 1 2	
1.003E-01	1.386E+01	30	B315	1 0 1 1 2	

1467. $C_8H_{10}O_2$
2-Phenoxyethanol
Phenoxyethyl Alcohol
Ethylene Glycol Phenyl Ether
Arosol
1-Hydroxy-2-phenoxyethane
Phenoxethol
RN: 122-99-6   **MP** (°C):   12
MW: 138.17   **BP** (°C):   237

Solubility (Moles/L)	Solubility (Grams/L)	Temp (°C)	Ref (#)	Evaluation (T P E A A)	Comments
1.882E-01	2.601E+01	20	M062	1 0 0 0 2	
2.610E-01	3.606E+01	37	E028	1 0 1 1 2	

1468. $C_8H_{10}O_2$
p-Ethoxyphenol
Hydroquinone Monoethyl Ether
RN: 622-62-8   **MP** (°C):   64.5-67.5
MW: 138.17   **BP** (°C):   131 at 9 mm Hg

Solubility (Moles/L)	Solubility (Grams/L)	Temp (°C)	Ref (#)	Evaluation (T P E A A)	Comments
5.097E-02	7.043E+00	20	R087	1 1 2 2 2	0.15M NaCl

1469.   C$_8$H$_{10}$O$_3$
1,3-Dimethyl Ether Pyrogallol
Pyrogallol-1,3-dimethylaether
2,6-Dimethoxyphenol
RN: 91-10-1   **MP** (°C):   56
MW: 154.17   **BP** (°C):   262

Solubility (Moles/L)	Solubility (Grams/L)	Temp (°C)	Ref (#)	Evaluation (T P E A A)	Comments
1.116E-01	1.720E+01	13	F300	1 0 0 0 2	

1470.   C$_8$H$_{10}$O$_3$S
Benzene Sulfonic Acid Ethyl Ester
Ethyl Benzenesulfonate
Ethyl Phenylsulfonate
RN: 515-46-8   **MP** (°C):
MW: 186.23   **BP** (°C):

Solubility (Moles/L)	Solubility (Grams/L)	Temp (°C)	Ref (#)	Evaluation (T P E A A)	Comments
7.390E-03	1.376E+00	25	K097	2 0 2 2 2	

1471.   C$_8$H$_{10}$O$_4$
2-Cyclohexene-1,2-dicarboxylic Acid
Cyclohexen-(2)-dicarbonsaeure-(1,2)
RN: 38765-78-5   **MP** (°C):
MW: 170.17   **BP** (°C):

Solubility (Moles/L)	Solubility (Grams/L)	Temp (°C)	Ref (#)	Evaluation (T P E A A)	Comments
5.113E-02	8.700E+00	10	F300	1 0 0 0 1	

1472.   C$_8$H$_{10}$O$_4$
Cyclohexene-1,4-dicarboxylic Acid
Cyclohexen-(1)-dicarbonsaeure-(1,4)
RN: 2205-27-8   **MP** (°C):   312
MW: 170.17   **BP** (°C):

Solubility (Moles/L)	Solubility (Grams/L)	Temp (°C)	Ref (#)	Evaluation (T P E A A)	Comments
1.175E-03	2.000E-01	20	F300	1 0 0 0 0	

1473. C₈H₁₀O₅

Endothall
Endothal

RN: 145-73-3   **MP** (°C):   144
MW: 186.17   **BP** (°C):

Solubility (Moles/L)	Solubility (Grams/L)	Temp (°C)	Ref (#)	Evaluation (T P E A A)	Comments
4.883E-01	9.091E+01	20	B200	1 0 0 0 2	
5.372E-01	1.000E+02	20	M161	1 0 0 0 2	
4.883E-01	9.091E+01	ns	B100	0 0 0 0 0	
4.883E-01	9.091E+01	ns	C307	0 0 0 0 1	

1474. C₈H₁₀O₈

meso-1,2,3,4-Butanetetracarboxylic Acid
1,2,3,4-Butanetetracarboxylic Acid
Butanetetracarboxylic Acid
1,2,3,4,-Butane Tetracarboxylic Acid

RN: 1703-58-8   **MP** (°C):   196
MW: 234.16   **BP** (°C):

Solubility (Moles/L)	Solubility (Grams/L)	Temp (°C)	Ref (#)	Evaluation (T P E A A)	Comments
6.606E-01	1.547E+02	25	M370	1 2 2 1 2	

1475. C₈H₁₁BrN₂O₂

Isocil
Uracil, 5-Bromo-3-isopropyl-6-methyl-

RN: 314-42-1   **MP** (°C):   158-159
MW: 247.10   **BP** (°C):

Solubility (Moles/L)	Solubility (Grams/L)	Temp (°C)	Ref (#)	Evaluation (T P E A A)	Comments
8.701E-03	2.150E+00	25	B185	1 0 0 0 2	

1476. C₈H₁₁Cl₂NO

N,N-Diallyldichloroacetamide
Dichlormid
N, N-Diallyl Dichloroacetamide
2,2-Dichloro-N,N-di-2-propenylacetamide
R25788

RN: 37764-25-3   **MP** (°C):   5
MW: 208.09   **BP** (°C):

Solubility (Moles/L)	Solubility (Grams/L)	Temp (°C)	Ref (#)	Evaluation (T P E A A)	Comments
2.403E-02	5.000E+00	20	M161	1 0 0 0 0	

1477.   C$_8$H$_{11}$Cl$_3$O$_6$
Chloralose
1,2-O-(2,2,2-Trichloroethylidene)-α-D-glucofuranose
Anhydroglucochloral
Alfamat
Aphosal
Murex
RN: 15879-93-3   **MP** (°C):   187
MW: 309.53   **BP** (°C):

Solubility (Moles/L)	Solubility (Grams/L)	Temp (°C)	Ref (#)	Evaluation (T P E A A)	Comments
1.434E-02	4.440E+00	15	M161	1 0 0 0 2	

1478.   C$_8$H$_{11}$N
Xylidine
N,N-Dimethylaniline
Dimethylaminobenzene
Benzenamine
Aminodimethylbenzene
RN: 121-69-7   **MP** (°C):   2
MW: 121.18   **BP** (°C):   194

Solubility (Moles/L)	Solubility (Grams/L)	Temp (°C)	Ref (#)	Evaluation (T P E A A)	Comments
9.120E-03	1.105E+00	25	C113	1 0 2 1 2	

1479.   C$_8$H$_{11}$NO
Tyramine
Tyramin
4-Hydroxyphenylethylamine
4-(2-Aminoethyl)phenol
2-(p-Hydroxyphenyl)ethylamine
RN: 51-67-2   **MP** (°C):   164.5
MW: 137.18   **BP** (°C):   206

Solubility (Moles/L)	Solubility (Grams/L)	Temp (°C)	Ref (#)	Evaluation (T P E A A)	Comments
7.574E-02	1.039E+01	15	D041	1 0 0 0 2	
7.581E-02	1.040E+01	15	F300	1 0 0 0 2	

1480.   C$_8$H$_{11}$NO
Phenylethanolamine
Phenyl Ethanolamine
2-Anilinoethanol
β-Hydroxyethyl Aniline
N-Phenylethanolamine
PEA

RN: 7568-93-6   **MP** (°C):   56.5
MW: 137.18   **BP** (°C):   286.0

Solubility (Moles/L)	Solubility (Grams/L)	Temp (°C)	Ref (#)	Evaluation (T P E A A)	Comments
3.192E-01	4.379E+01	20	M062	1 0 0 0 2	

1481.   C$_8$H$_{11}$N$_2$O$_5$PS
Parathion-amino
Aminoparathion

RN: **MP** (°C):
MW: 278.23   **BP** (°C):

Solubility (Moles/L)	Solubility (Grams/L)	Temp (°C)	Ref (#)	Evaluation (T P E A A)	Comments
1.419E-03	3.948E-01	19.50	B169	2 2 1 1 2	

1482.   C$_8$H$_{12}$
4-Vinylcyclohexene
4-Vinyl-1-cyclohexene

RN: 100-40-3   **MP** (°C):   -101
MW: 108.18   **BP** (°C):   145

Solubility (Moles/L)	Solubility (Grams/L)	Temp (°C)	Ref (#)	Evaluation (T P E A A)	Comments
4.622E-04	5.000E-02	25	M001	2 1 2 2 1	

1483.   C₈H₁₂CINO
Allidochlor
CDAA
N,N-Diallyl-2-chloroacetamide
Randox
2-Chloro-N,N-diallylacetamide
CP 6,343
RN: 93-71-0   **MP** (°C):
MW: 173.64   **BP** (°C):

Solubility (Moles/L)	Solubility (Grams/L)	Temp (°C)	Ref (#)	Evaluation (T P E A A)	Comments
1.113E-01	1.932E+01	22	J008	1 0 0 0 2	
1.113E-01	1.932E+01	25	B185	1 0 0 0 2	
1.135E-01	1.970E+01	25	G319	1 0 0 0 2	
1.135E-01	1.970E+01	25	M161	1 0 0 0 2	
1.129E-01	1.961E+01	ns	B100	0 0 0 0 0	
1.130E-01	1.962E+01	ns	F184	0 0 0 0 2	
1.129E-01	1.961E+01	ns	M061	0 0 0 0 0	
3.162E-01	5.491E+01	ns	M163	0 0 0 0 0	EFG

1484.   C₈H₁₂N₂O₂S
N1-Dimethylsulfanilamide
p-Amino-N,N-dimethylbenzenesulfonamide
[(4-Aminophenyl)sulfonyl]dimethylamine
p-(Dimethylsulfamoyl)aniline
RN: 1709-59-7   **MP** (°C):
MW: 200.26   **BP** (°C):

Solubility (Moles/L)	Solubility (Grams/L)	Temp (°C)	Ref (#)	Evaluation (T P E A A)	Comments
3.130E-03	6.268E-01	37	K095	2 0 0 0 2	intrinsic

1485.   C₈H₁₂N₂O₂S
5,5-Diethyl-2-thiobarbituric Acid
4,6(1H,5H)-Pyrimidinedione, 5,5-Diethyldihydro-2-thioxo
Barbituric Acid, 5,5-Diethyl-2-thio
Certodorm
RN: 77-32-7   **MP** (°C):
MW: 200.26   **BP** (°C):

Solubility (Moles/L)	Solubility (Grams/L)	Temp (°C)	Ref (#)	Evaluation (T P E A A)	Comments
6.810E-03	1.364E+00	25	P350	2 1 1 1 2	intrinsic

1486. $C_8H_{12}N_2O_3$
Barbital
5,5-Diethylbarbituric Acid
Diethylmalonylurea
RN: 57-44-3 **MP** (°C): 190
MW: 184.20 **BP** (°C):

Solubility (Moles/L)	Solubility (Grams/L)	Temp (°C)	Ref (#)	Evaluation (T P E A A)	Comments
1.700E-02	3.131E+00	0	M143	1 2 1 1 0	
1.900E-02	3.500E+00	0	M143	1 2 1 1 2	
2.562E-02	4.720E+00	10	N007	1 2 2 2 2	form I
1.900E-02	3.500E+00	10	N007	1 2 2 2 2	form III
3.100E-02	5.710E+00	14	I006	1 0 0 0 1	
3.187E-02	5.870E+00	15	H018	1 2 2 2 2	
3.500E-02	6.447E+00	19	I006	1 0 0 0 1	
4.522E-02	8.330E+00	20	D041	1 0 0 0 1	
3.637E-02	6.700E+00	20	F300	1 0 0 0 1	
3.415E-02	6.290E+00	20	J030	1 2 2 2 2	
2.839E-02	5.230E+00	20	N007	1 2 2 2 2	form III
3.409E-02	6.280E+00	20	N007	1 2 2 2 2	form I
3.806E-02	7.011E+00	20	S146	2 2 2 1 2	form I
3.752E-02	6.912E+00	20	S146	2 2 2 1 2	form II
3.881E-02	7.149E+00	25	A023	1 0 0 1 2	
3.963E-02	7.300E+00	25	B011	2 0 0 1 0	
3.971E-02	7.314E+00	25	B065	1 1 1 1 1	
3.746E-02	6.900E+00	25	B167	1 1 0 0 1	pH 5.7
3.860E-02	7.110E+00	25	G003	1 1 1 1 2	pH 4.7
2.800E-02	5.158E+00	25	M143	1 2 1 1 2	
4.050E-02	7.460E+00	25	M310	2 2 2 2 2	
4.018E-02	7.401E+00	25	P350	2 1 1 1 2	intrinsic
4.239E-02	7.809E+00	25	S146	2 2 2 1 2	form II
4.010E-03	7.386E-01	25	V033	2 0 1 1 2	
4.010E-02	7.386E+00	25.00	T303	1 0 0 0 2	
4.300E-02	7.920E+00	27	I006	1 0 0 0 1	
4.300E-02	7.920E+00	30	G014	1 1 1 1 0	EFG, 0.003N H_2SO_4
2.704E-02	4.980E+00	30	H005	1 0 1 2 2	average of 4
4.408E-02	8.119E+00	30	H018	1 2 2 2 2	
4.400E-02	8.105E+00	30	I001	2 0 2 1 0	EFG, 0.003N H_2SO_4
4.260E-02	7.847E+00	30	K108	1 2 2 0 2	
4.425E-02	8.150E+00	30	N007	1 2 2 2 2	form I
4.207E-02	7.750E+00	30	N007	1 2 2 2 2	form III
4.720E-02	8.694E+00	30	S146	2 2 2 1 2	form I
4.618E-02	8.507E+00	30	S146	2 2 2 1 2	form II
5.162E-02	9.509E+00	35	S146	2 2 2 1 2	form I
5.184E-02	9.548E+00	35	S146	2 2 2 1 2	form II
5.150E-02	9.486E+00	35.00	T303	1 0 0 0 2	
4.843E-02	8.920E+00	36	A023	1 0 0 1 2	
5.152E-02	9.490E+00	37	J030	1 2 2 2 2	
5.300E-02	9.762E+00	37	K121	1 2 1 2 1	0.1N HCl

5.538E-02	1.020E+01	37	N007	1 2 2 2 2	form III
5.277E-02	9.720E+00	37	N007	1 2 2 2 2	form I
5.668E-02	1.044E+01	37	S146	2 2 2 1 2	form II
5.588E-02	1.029E+01	40	A023	1 0 0 1 1	
6.100E-01	1.124E+02	40	N008	1 0 1 1 2	*sic*
6.967E-02	1.283E+01	45	S146	2 2 2 1 2	form II
6.800E-02	1.253E+01	45.00	T303	1 0 0 0 2	
4.343E-01	8.000E+01	100	F300	1 0 0 0 1	
3.257E-02	6.000E+00	ns	T003	0 0 0 0 2	

1487. $C_8H_{12}O_2$
1-Epoxyethyl-3,4-Epoxycyclohexane
Vinylcyclohexene Dioxide
RN: 106-87-6   **MP** (°C): <-55
MW: 140.18   **BP** (°C): 227

Solubility (Moles/L)	Solubility (Grams/L)	Temp (°C)	Ref (#)	Evaluation (T P E A A)	Comments
1.103E+00	1.547E+02	20	I313	0 0 0 0 1	

1488. $C_8H_{12}O_4$
trans-Cyclohexane-1,4-dicarboxylic Acid
trans-Cyclohexan-dicarbonsaeure-(1,4)
RN: 619-82-9   **MP** (°C):
MW: 172.18   **BP** (°C):

Solubility (Moles/L)	Solubility (Grams/L)	Temp (°C)	Ref (#)	Evaluation (T P E A A)	Comments
4.646E-03	8.000E-01	17	F300	1 0 0 0 0	
7.550E-02	1.300E+01	100	F300	1 0 0 0 1	

1489. $C_8H_{12}O_4$
cis-Cyclohexane-1,2-dicarboxylic Acid
cis-Cyclohexan-dicarbonsaeure-(1,2)
RN: 610-09-3   **MP** (°C): 193
MW: 172.18   **BP** (°C):

Solubility (Moles/L)	Solubility (Grams/L)	Temp (°C)	Ref (#)	Evaluation (T P E A A)	Comments
>1.16E-02	>2.00E+00	20	F300	1 0 0 0 0	

1490. C$_8$H$_{12}$O$_4$
trans-Cyclohexane-1,2-dicarboxylic Acid
trans-Cyclohexan-dicarbonsaeure-(1,2)
RN: 2305-32-0   **MP** (°C):
MW: 172.18   **BP** (°C):

Solubility (Moles/L)	Solubility (Grams/L)	Temp (°C)	Ref (#)	Evaluation (T P E A A)	Comments
1.162E-02	2.000E+00	20	F300	1 0 0 0 0	

1491. C$_8$H$_{13}$BrN$_2$O$_2$
α-Bromethylpropylaceturea
RN: **MP** (°C):
MW: 249.11   **BP** (°C):

Solubility (Moles/L)	Solubility (Grams/L)	Temp (°C)	Ref (#)	Evaluation (T P E A A)	Comments
1.645E-03	4.098E-01	20	O021	1 0 0 0 0	

1492. C$_8$H$_{13}$NO
Diaalylacetamide
α,α-Diallylacetamide
2-(2-Propenyl)4-Pentenamide
RN: 60730-94-1   **MP** (°C):
MW: 139.20   **BP** (°C):

Solubility (Moles/L)	Solubility (Grams/L)	Temp (°C)	Ref (#)	Evaluation (T P E A A)	Comments
1.257E-01	1.750E+01	ns	H348	0 0 1 1 2	

1493. C$_8$H$_{13}$N$_2$O$_3$PS
Thionazin
O,O-Diethyl O-Pyrazinyl Thiophosphate
RN: 297-97-2   **MP** (°C):   -1.7
MW: 248.24   **BP** (°C):   80

Solubility (Moles/L)	Solubility (Grams/L)	Temp (°C)	Ref (#)	Evaluation (T P E A A)	Comments
4.592E-03	1.140E+00	25	M061	1 0 0 0 2	
4.592E-03	1.140E+00	27	M161	1 0 0 0 2	

1494.   C$_8$H$_{14}$
1-Octyne
Hexylacetylene
n-Hexylacetylene
RN: 629-05-0   **MP** (°C):   -80
MW: 110.20   **BP** (°C):   127

Solubility (Moles/L)	Solubility (Grams/L)	Temp (°C)	Ref (#)	Evaluation (T P E A A)	Comments
2.178E-04	2.400E-02	25	M001	2 1 2 2 2	

1495.   C$_8$H$_{14}$
2,2-Dimethyl-3-hexyne
1-Ethyl-2-tertbutylacetylene
RN: 4911-60-8   **MP** (°C):
MW: 110.20   **BP** (°C):

Solubility (Moles/L)	Solubility (Grams/L)	Temp (°C)	Ref (#)	Evaluation (T P E A A)	Comments
7.200E-04	7.934E-02	25	H039	1 2 2 2 1	

1496.   C$_8$H$_{14}$ClNS$_2$
Carbamic Acid, Diethyldithio-2chloroallyl Ester
2-Chloroallyl Diethyldithiocarbamate
CDEC
RN: 95-06-7   **MP** (°C):   <25
MW: 223.79   **BP** (°C):   128

Solubility (Moles/L)	Solubility (Grams/L)	Temp (°C)	Ref (#)	Evaluation (T P E A A)	Comments
4.469E-04	1.000E-01	25	B185	1 0 0 0 2	
4.111E-04	9.200E-02	25	B200	1 0 0 0 1	
4.469E-04	1.000E-01	25	F019	1 0 0 0 2	
4.111E-04	9.200E-02	25	G319	1 0 0 0 2	
4.111E-04	9.200E-02	25	M161	1 0 0 0 1	
4.468E-04	9.999E-02	ns	M061	0 0 0 0 0	approximate

1497.   C$_8$H$_{14}$ClN$_5$
Atrazine
2-Chloro-4-ethylamino-6-isopropylamino-s-triazine
RN: 1912-24-9   **MP** (°C):   172
MW: 215.69   **BP** (°C):

Solubility (Moles/L)	Solubility (Grams/L)	Temp (°C)	Ref (#)	Evaluation (T P E A A)	Comments
1.020E-04	2.200E-02	0	B185	1 0 0 0 1	
1.390E-04	2.998E-02	1	G091	1 0 1 2 2	pH 6.0
5.000E-04	1.078E-01	2	B193	1 2 0 0 0	
1.410E-04	3.041E-02	8	G091	1 0 1 2 2	pH 6.0
1.530E-04	3.300E-02	20	A314	1 1 0 0 1	
1.345E-04	2.900E-02	20	C048	2 2 2 2 1	
1.391E-04	3.000E-02	20	E048	1 2 1 1 1	
1.391E-04	3.000E-02	20	F311	1 2 2 2 1	
1.580E-04	3.408E-02	20	G091	1 0 1 2 2	pH 6.0
1.298E-04	2.800E-02	20	M161	1 0 0 0 1	
1.391E-04	3.000E-02	20	N333	1 0 0 0 1	
3.245E-04	7.000E-02	21	B192	0 0 0 0 1	
3.245E-04	7.000E-02	21	G099	2 0 0 1 0	
3.245E-04	7.000E-02	22	M061	1 0 0 0 1	
1.530E-04	3.300E-02	25	H024	2 2 2 2 2	
1.386E-04	2.990E-02	25	H073	2 1 1 2 2	
3.245E-04	7.000E-02	27	B185	1 0 0 0 1	
1.530E-04	3.300E-02	27	B200	1 0 0 0 1	
1.970E-04	4.249E-02	29	G091	1 0 1 2 2	pH 6.0
4.530E-04	9.771E-02	50	G001	1 0 0 1 2	
1.484E-03	3.200E-01	85	B185	1 0 0 0 2	
3.245E-04	7.000E-02	ns	C101	0 0 0 0 1	
3.245E-04	7.000E-02	ns	G041	0 0 0 0 1	
3.245E-04	7.000E-02	ns	H112	0 0 0 0 1	
1.530E-04	3.300E-02	ns	J033	0 0 0 0 1	
3.941E-04	8.500E-02	ns	M110	0 0 0 0 0	EFG

1498.   C$_8$H$_{14}$N$_2$O$_2$
cis-N,N,N',N'-Tetramethylfumaramide
2-Butenediamide, N,N,N',N'-Tetramethyl-, (Z)-
RN: 35075-35-5   **MP** (°C):
MW: 170.21   **BP** (°C):

Solubility (Moles/L)	Solubility (Grams/L)	Temp (°C)	Ref (#)	Evaluation (T P E A A)	Comments
1.730E+00	2.945E+02	30	K019	1 0 0 0 2	

1499.   C$_8$H$_{14}$N$_4$OS
Metribuzin
4-Amino-6-tert-butyl-3-(methylthio)-as-triazin-5(4H)-one
Bayer 6159H
Lexone
Sencor
Sencorex

RN: 21087-64-9   **MP** (°C):   125.8
MW: 214.29   **BP** (°C):

Solubility (Moles/L)	Solubility (Grams/L)	Temp (°C)	Ref (#)	Evaluation (T P E A A)	Comments
5.600E-03	1.200E+00	20	M161	1 0 0 0 1	
5.693E-03	1.220E+00	22.5	G301	2 1 0 1 2	
4.662E-03	9.990E-01	ns	B100	0 0 0 0 0	
7.000E-03	1.500E+00	ns	M110	0 0 0 0 0	EFG

1500.   C$_8$H$_{14}$O
Bicyclo[2.2.1]heptylcarbinol
2-Norcamphanemethanol

RN: 5240-72-2   **MP** (°C):
MW: 126.20   **BP** (°C):

Solubility (Moles/L)	Solubility (Grams/L)	Temp (°C)	Ref (#)	Evaluation (T P E A A)	Comments
7.916E-03	9.990E-01	ns	M061	0 0 0 0 0	

1501.   C$_8$H$_{14}$O$_2$
Cyclohexanol Acetate
Hexalin Acetate
Cyclohexyl Acetate

RN: 622-45-7   **MP** (°C):   <25
MW: 142.20   **BP** (°C):

Solubility (Moles/L)	Solubility (Grams/L)	Temp (°C)	Ref (#)	Evaluation (T P E A A)	Comments
1.123E-02	1.597E+00	20	D052	1 1 0 0 1	
2.033E-02	2.892E+00	23.50	O005	2 0 2 2 1	

1502.   C$_8$H$_{14}$O$_2$
3-Propyl-2,4-pentadione
3-Acetyl-2-hexanone

RN: 1540-35-8   **MP** (°C):   <25
MW: 142.20   **BP** (°C):

Solubility (Moles/L)	Solubility (Grams/L)	Temp (°C)	Ref (#)	Evaluation (T P E A A)	Comments
1.330E-01	1.891E+01	25	M078	2 0 1 0 2	

1503. C$_8$H$_{14}$O$_2$
5,5-Dimethyl-2,4-hexadione
Pivaloylacetone
Pivaloylacetylmethane
RN: 7307-04-2 **MP** (°C):
MW: 142.20 **BP** (°C):

Solubility (Moles/L)	Solubility (Grams/L)	Temp (°C)	Ref (#)	Evaluation (T P E A A)	Comments
2.340E-02	3.327E+00	25	M078	2 0 1 0 2	

1504. C$_8$H$_{14}$O$_2$
6-Methyl-2,4-heptadione
2-Methyl-4,6-heptanedione
Isovalerylacetone
RN: 3002-23-1 **MP** (°C): <25
MW: 142.20 **BP** (°C):

Solubility (Moles/L)	Solubility (Grams/L)	Temp (°C)	Ref (#)	Evaluation (T P E A A)	Comments
2.490E-02	3.541E+00	25	M078	2 0 1 0 2	

1505. C$_8$H$_{14}$O$_2$
2,4-Octadione
Valerylacetone
RN: 14090-87-0 **MP** (°C):
MW: 142.20 **BP** (°C):

Solubility (Moles/L)	Solubility (Grams/L)	Temp (°C)	Ref (#)	Evaluation (T P E A A)	Comments
2.760E-02	3.925E+00	25	M078	2 0 1 0 2	

1506. C$_8$H$_{14}$O$_2$S$_4$
Propyl Dixanthogen
bis(1-Propyl) Dixanthogen
Propyl Xanthogen Disulfide
Dipropyl Dixanthogen
Dipropyl Thioperoxydicarbonate
Dipropyl Xanthogen Disulfide
RN: 3750-28-5 **MP** (°C):
MW: 270.46 **BP** (°C):

Solubility (Moles/L)	Solubility (Grams/L)	Temp (°C)	Ref (#)	Evaluation (T P E A A)	Comments
1.500E-06	4.057E-04	25	H102	1 2 1 2 1	

1507. C₈H₁₄O₄
Tetramethyl Succinic Acid
Tetramethyl-bernsteinsaeure
RN: 630-51-3   **MP** (°C):
MW: 174.20   **BP** (°C):

Solubility (Moles/L)	Solubility (Grams/L)	Temp (°C)	Ref (#)	Evaluation (T P E A A)	Comments
2.755E-02	4.800E+00	13.5	F300	1 0 0 0 1	

1508. C₈H₁₄O₄
Suberic Acid
Korksaeure
RN: 505-48-6   **MP** (°C):   142
MW: 174.20   **BP** (°C):   279

Solubility (Moles/L)	Solubility (Grams/L)	Temp (°C)	Ref (#)	Evaluation (T P E A A)	Comments
4.592E-03	8.000E-01	0	L041	1 0 0 1 0	
5.301E-03	9.234E-01	6.99	A340	2 0 2 2 2	
7.097E-03	1.236E+00	12.69	A340	2 0 2 2 2	
8.037E-03	1.400E+00	15	F300	1 0 0 0 1	
7.463E-03	1.300E+00	15	L041	1 0 0 1 1	
7.463E-03	1.300E+00	15	M051	1 0 0 0 1	
9.789E-03	1.705E+00	18.69	A340	2 0 2 2 2	
9.185E-03	1.600E+00	20	L041	1 0 0 1 1	
8.986E-03	1.565E+00	20	M171	1 0 0 0 0	
1.206E-01	2.100E+01	21	B040	1 0 1 1 2	*sic*
1.388E-02	2.417E+00	24.99	A340	2 0 2 2 2	
3.387E-02	5.900E+00	25	F300	1 0 0 0 1	
6.800E-02	1.185E+01	25	K040	1 0 2 1 2	*sic*
1.700E-02	2.961E+00	30	H021	1 0 1 1 0	EFG
1.890E-02	3.293E+00	32.49	A340	2 0 2 2 2	
2.045E-02	3.563E+00	34.49	A340	2 0 2 2 2	
2.583E-02	4.500E+00	35	L041	1 0 0 1 1	
2.326E-02	4.051E+00	39.99	A340	2 0 2 2 2	
2.682E-02	4.673E+00	44.49	A340	2 0 2 2 2	
5.626E-02	9.800E+00	50	L041	1 0 0 1 1	
3.198E-02	5.571E+00	50.19	A340	2 0 2 2 2	
3.534E-02	6.156E+00	52.69	A340	2 0 2 2 2	
5.551E-02	9.670E+00	61.49	A340	2 0 2 2 2	
6.422E-02	1.119E+01	63.99	A340	2 0 2 2 2	
1.274E-01	2.220E+01	65	L041	1 0 0 1 2	
8.182E-02	1.425E+01	70.09	A340	2 0 2 2 2	
1.156E-01	2.013E+01	76.49	A340	2 0 2 2 2	

1509. $C_8H_{14}O_4$
Isoamylmalonic Acid
Acide Isoamylmalonique
RN: 616-87-5   **MP** (°C):
MW: 174.20   **BP** (°C):

Solubility (Moles/L)	Solubility (Grams/L)	Temp (°C)	Ref (#)	Evaluation (T P E A A)	Comments
2.210E+00	3.850E+02	0	M051	1 0 0 0 2	
2.974E+00	5.180E+02	15	M051	1 0 0 0 2	
3.490E+00	6.080E+02	25	M051	1 0 0 0 2	
4.788E+00	8.340E+02	50	M051	1 0 0 0 2	

1510. $C_8H_{14}O_4$
Butylene Glycol Diacetate
1,4-Diacetoxybutane
Tetramethylene Acetate
RN: 628-67-1   **MP** (°C):
MW: 174.20   **BP** (°C):

Solubility (Moles/L)	Solubility (Grams/L)	Temp (°C)	Ref (#)	Evaluation (T P E A A)	Comments
2.005E-01	3.494E+01	26	O012	1 2 1 1 2	
1.602E-01	2.790E+01	50	O012	1 2 1 1 2	
2.048E-01	3.568E+01	75	O012	1 2 1 1 2	

1511. $C_8H_{14}O_4$
Ethylene Glycol Dipropionate
1,2-Ethanediol, Dipropanoate
1,2-bis(Propionyloxy)ethane
RN: 123-80-8   **MP** (°C):
MW: 174.20   **BP** (°C):

Solubility (Moles/L)	Solubility (Grams/L)	Temp (°C)	Ref (#)	Evaluation (T P E A A)	Comments
9.480E-02	1.651E+01	25	F064	1 0 0 0 2	
9.170E-03	1.597E+00	ns	F014	0 0 0 0 2	

1512. $C_8H_{14}O_4$
Diethyl Succinate
Butanedioic Acid, Diethyl Ester
RN: 123-25-1   **MP** (°C):   -20
MW: 174.20   **BP** (°C):   217

Solubility (Moles/L)	Solubility (Grams/L)	Temp (°C)	Ref (#)	Evaluation (T P E A A)	Comments
1.089E-02	1.896E+00	ns	F014	0 0 0 0 2	

1513.   C$_8$H$_{14}$O$_4$
Propyl α-Acetoxypropionate
Hydracrylic Acid, Propyl Ester, Acetate
RN: 20473-73-8   **MP** (°C):
MW: 174.20   **BP** (°C):

Solubility (Moles/L)	Solubility (Grams/L)	Temp (°C)	Ref (#)	Evaluation (T P E A A)	Comments
5.683E-02	9.900E+00	25	R006	2 2 0 1 1	

1514.   C$_8$H$_{14}$O$_5$
Propanoic Acid, 2-[(Propoxycarbonyl)oxy]-, Methyl Ester
RN: **MP** (°C):
MW: 190.20   **BP** (°C):

Solubility (Moles/L)	Solubility (Grams/L)	Temp (°C)	Ref (#)	Evaluation (T P E A A)	Comments
2.720E-02	5.173E+00	25	R007	1 0 0 0 1	

1515.   C$_8$H$_{15}$ClN$_5$O
Hydroxyatrazine
4-(Ethylamino)-6-[(1-methylethyl)amino]-1,3,5-triazin-2(1H)-one
2-Hydroxy atrazine
RN: 2163-68-0   **MP** (°C):
MW: 232.69   **BP** (°C):

Solubility (Moles/L)	Solubility (Grams/L)	Temp (°C)	Ref (#)	Evaluation (T P E A A)	Comments
2.400E-04	5.585E-02	2	B193	1 2 0 0 1	

1516.   C$_8$H$_{15}$NO
Pelletierine
Pelletierin
RN: 2858-66-4   **MP** (°C):   <25
MW: 141.21   **BP** (°C):   195

Solubility (Moles/L)	Solubility (Grams/L)	Temp (°C)	Ref (#)	Evaluation (T P E A A)	Comments
3.541E-01	5.000E+01	20	F300	1 0 0 0 0	
3.372E-01	4.762E+01	25	D004	1 0 0 0 0	

1517. C₈H₁₅NO

Propylallylacetamide
2-Propyl-4-Pentenamide
PAD
RN: 90204-40-3 **MP** (°C):
MW: 141.21 **BP** (°C):

Solubility (Moles/L)	Solubility (Grams/L)	Temp (°C)	Ref (#)	Evaluation (T P E A A)	Comments
6.727E-02	9.500E+00	37	H347	1 1 2 2 1	

1518. C₈H₁₅N₃O₂

Isocarbamid
N-(2-Methylpropyl)-2-oxo-1-imidazolidinecarboxamide
RN: 30979-48-7 **MP** (°C): 95.5
MW: 185.23 **BP** (°C):

Solubility (Moles/L)	Solubility (Grams/L)	Temp (°C)	Ref (#)	Evaluation (T P E A A)	Comments
7.018E-03	1.300E+00	20	M161	1 0 0 0 1	

1519. C₈H₁₅N₃O₇

Streptozotocin
Streptozocin
D-2-Deoxy-2-(3-methyl-3-nitrosoureido)glucopyranose
RN: 18883-66-4 **MP** (°C): 115
MW: 265.22 **BP** (°C):

Solubility (Moles/L)	Solubility (Grams/L)	Temp (°C)	Ref (#)	Evaluation (T P E A A)	Comments
1.910E-02	5.066E+00	25	I307	0 0 0 0 2	

1520. C₈H₁₅N₅O

2-Methoxy-4-methylamino-6-isopropylamino-s-triazine
Noratone
RN: 3035-45-8 **MP** (°C):
MW: 197.24 **BP** (°C):

Solubility (Moles/L)	Solubility (Grams/L)	Temp (°C)	Ref (#)	Evaluation (T P E A A)	Comments
1.774E-02	3.500E+00	20	J033	1 0 0 0 2	
1.774E-02	3.500E+00	21	B192	0 0 0 0 2	

1521.   C$_8$H$_{15}$N$_5$O
Simetone
2-Methoxy-4,6-bis(ethylamino)-s-triazine
s-Triazole, 2,4-bis(Ethylamine)-6-methoxy-
RN: 673-04-1   **MP** (°C):   118-120
MW: 197.24   **BP** (°C):

Solubility (Moles/L)	Solubility (Grams/L)	Temp (°C)	Ref (#)	Evaluation (T P E A A)	Comments
1.622E-02	3.200E+00	21	B185	1 0 0 0 2	
1.622E-02	3.200E+00	21	B192	0 0 0 0 2	
1.622E-02	3.200E+00	21	G099	2 0 0 1 0	
3.550E-02	7.002E+00	50	G001	1 0 1 1 2	
1.622E-02	3.200E+00	ns	C101	0 0 0 0 1	

1522.   C$_8$H$_{15}$N$_5$S
Desmetryne
N-Methyl-N'-(1-methylethyl)-6-(methylthio)-1,3,5-triazine-2,4-diamine
Semeron
Methylamino-4-methylthio-6-isopropylamino-1,3,5-triazine
Topusyn
Methylthio-4-isopropylamino-6-methylamino-s-triazine
RN: 1014-69-3   **MP** (°C):
MW: 213.31   **BP** (°C):

Solubility (Moles/L)	Solubility (Grams/L)	Temp (°C)	Ref (#)	Evaluation (T P E A A)	Comments
2.813E-03	6.000E-01	20	F311	1 2 2 2 1	
2.719E-03	5.800E-01	20	M161	1 0 0 0 2	
2.811E-03	5.996E-01	ns	B100	0 0 0 0 0	
2.719E-03	5.800E-01	ns	J033	0 0 0 0 2	
2.719E-03	5.800E-01	ns	M061	0 0 0 0 2	

1523. $C_8H_{15}N_5S$
Simetryne
N,N'-Diethyl-6-(methylthio)-1,3,5-Triazine-2,4-diamine
G-32911
bis(Ethylamino)-6-(methylthio)-s-triazine
Methylthio-4,6-bis(ethylamino)-s-triazine
Cymetrin

RN: 1014-70-6   **MP** (°C): 82
MW: 213.31   **BP** (°C):

Solubility (Moles/L)	Solubility (Grams/L)	Temp (°C)	Ref (#)	Evaluation (T P E A A)	Comments
4.700E-03	1.003E+00	50	G001	1 0 1 1 1	
2.110E-03	4.500E-01	ns	C101	0 0 0 0 1	
2.110E-03	4.500E-01	ns	J033	0 0 0 0 2	
2.110E-03	4.500E-01	rt	M161	0 0 0 0 2	

1524. C_8H_{16}
trans-1,4-Dimethylcyclohexane
1,4-Transdimethylcyclohexane

RN: 2207-04-7   **MP** (°C): -37
MW: 112.22   **BP** (°C): 115

Solubility (Moles/L)	Solubility (Grams/L)	Temp (°C)	Ref (#)	Evaluation (T P E A A)	Comments
3.422E-05	3.840E-03	25	P051	2 1 1 2 2	
3.422E-05	3.840E-03	25.00	P007	2 1 2 2 2	

1525. C_8H_{16}
1,1,3-Trimethylcyclopentane
Cyclopentane, 1,1,3-Trimethyl-

RN: 4516-69-2   **MP** (°C): -142.4
MW: 112.22   **BP** (°C): 104.9

Solubility (Moles/L)	Solubility (Grams/L)	Temp (°C)	Ref (#)	Evaluation (T P E A A)	Comments
3.324E-05	3.730E-03	25	K119	1 0 0 0 2	
3.324E-05	3.730E-03	25	P051	2 1 1 2 2	
3.324E-05	3.730E-03	25.00	P007	2 1 2 2 2	

1526. C_8H_{16}
1,4-Dimethylcyclohexane
p-Dimethylcyclohexane

RN: 589-90-2   **MP** (°C): -87
MW: 112.22   **BP** (°C): 120

Solubility (Moles/L)	Solubility (Grams/L)	Temp (°C)	Ref (#)	Evaluation (T P E A A)	Comments
3.422E-05	3.840E-03	25	K119	1 0 0 0 2	

1527. C₈H₁₆
Caprylene
1-Octene
RN: 111-66-0   **MP** (°C):   -102
MW: 112.22   **BP** (°C):   121.0

Solubility (Moles/L)	Solubility (Grams/L)	Temp (°C)	Ref (#)	Evaluation (T P E A A)	Comments
3.208E-05	3.600E-03	23	C332	2 0 2 2 1	
2.406E-05	2.700E-03	25	M001	2 1 2 2 1	
3.650E-05	4.096E-03	25	M342	1 0 1 1 2	

1528. C₈H₁₆
cis-1,2-Dimethylcyclohexane
1-cis-2-Dimethylcyclohexane
RN: 2207-01-4   **MP** (°C):   -50
MW: 112.22   **BP** (°C):   129

Solubility (Moles/L)	Solubility (Grams/L)	Temp (°C)	Ref (#)	Evaluation (T P E A A)	Comments
6.773E-05	7.600E-03	20	M337	2 1 2 2 1	
5.347E-05	6.000E-03	25	M001	2 1 2 2 1	

1529. C₈H₁₆
Cyclooctane
RN: 292-64-8   **MP** (°C):   10
MW: 112.22   **BP** (°C):   151

Solubility (Moles/L)	Solubility (Grams/L)	Temp (°C)	Ref (#)	Evaluation (T P E A A)	Comments
1.619E-03	1.817E-01	20	M337	2 1 2 2 2	*sic*
7.040E-05	7.900E-03	25	M001	2 1 2 2 1	
7.040E-05	7.900E-03	ns	H123	0 0 0 0 2	

1530. C₈H₁₆
Ethyl Cyclohexane
Cyclohexane, Ethyl-
RN: 1678-91-7   **MP** (°C):
MW: 112.22   **BP** (°C):   131.8

Solubility (Moles/L)	Solubility (Grams/L)	Temp (°C)	Ref (#)	Evaluation (T P E A A)	Comments
5.614E-05	6.300E-03	20	M337	2 1 2 2 1	

1531. C_8H_{16}

n-Propylcyclopentane
1-Propylcyclopentane
RN: 2040-96-2   **MP** (°C):   -117
MW: 112.22   **BP** (°C):

Solubility (Moles/L)	Solubility (Grams/L)	Temp (°C)	Ref (#)	Evaluation (T P E A A)	Comments
1.818E-05	2.040E-03	25	K119	1 0 0 0 2	
1.818E-05	2.040E-03	25	P051	2 1 1 2 2	
1.818E-05	2.040E-03	25.00	P007	2 1 2 2 2	

1532. C_8H_{16}

trans-1,2-Dimethylcyclohexane
1,2-trans-Dimethylcyclohexane
RN: 6876-23-9   **MP** (°C):   -89
MW: 112.22   **BP** (°C):   123

Solubility (Moles/L)	Solubility (Grams/L)	Temp (°C)	Ref (#)	Evaluation (T P E A A)	Comments
4.634E-05	5.200E-03	20	M337	2 1 2 2 1	

1533. $C_8H_{16}N_2O_2$

N,N,N',N'-Tetramethylsuccinamide
N,N,N',N'-Tetramethylbutanediamide
RN: 7334-51-2   **MP** (°C):
MW: 172.23   **BP** (°C):

Solubility (Moles/L)	Solubility (Grams/L)	Temp (°C)	Ref (#)	Evaluation (T P E A A)	Comments
3.188E+00	5.490E+02	30	K004	1 0 0 0 2	

1534. $C_8H_{16}N_2O_4S_2$

DL-Homocystine
DL-meso-Homocystine
Oxidized DL-Homocysteine
RN: 870-93-9   **MP** (°C):   264
MW: 268.36   **BP** (°C):

Solubility (Moles/L)	Solubility (Grams/L)	Temp (°C)	Ref (#)	Evaluation (T P E A A)	Comments
7.451E-04	2.000E-01	25	D041	1 0 0 0 0	

1535. C_8H_16N_6
Pentamethylmelamine
1-(Methylamino)-3,5-bis(dimethylamino)-s-triazine
RN: 16268-62-5   **MP** (°C):   107.0
MW: 196.26   **BP** (°C):

Solubility (Moles/L)	Solubility (Grams/L)	Temp (°C)	Ref (#)	Evaluation (T P E A A)	Comments
1.679E-04	3.295E-02	25	B386	2 2 2 2 2	
1.010E-02	1.982E+00	25	B386	2 2 2 2 2	
1.101E-02	2.160E+00	25	C051	1 2 1 1 2	pH 7

1536. C_8H_16N_6O
N2-Hydroxy-N2,N4,N4,N6,N6-pentamethylmelamine
1-(Hydroxylamino)-3,5-bis(dimethylamino)-s-triazine
RN: 64124-14-7   **MP** (°C):   110.0
MW: 212.26   **BP** (°C):

Solubility (Moles/L)	Solubility (Grams/L)	Temp (°C)	Ref (#)	Evaluation (T P E A A)	Comments
4.412E-03	9.365E-01	25	B386	2 2 2 2 2	
4.259E-03	9.040E-01	25	C051	1 2 1 1 2	pH 7

1537. C_8H_16O
Caprylic Aldehyde
Octaldehyde
n-Octanal
RN: 124-13-0   **MP** (°C):
MW: 128.22   **BP** (°C):   163.4

Solubility (Moles/L)	Solubility (Grams/L)	Temp (°C)	Ref (#)	Evaluation (T P E A A)	Comments
4.368E-03	5.600E-01	25	A049	1 0 0 0 1	

1538. C_8H_16O
Hexyl Methyl Ketone
2-Octanone
Octan-2-one
RN: 111-13-7   **MP** (°C):   -16.0
MW: 128.22   **BP** (°C):   172.5

Solubility (Moles/L)	Solubility (Grams/L)	Temp (°C)	Ref (#)	Evaluation (T P E A A)	Comments
7.013E-03	8.992E-01	20	D052	1 1 0 0 0	

1539. C$_8$H$_{16}$O$_2$
3-Hydroxy-5-methyl-5-propyltetrahydrofuran
3-Furanol, 2-Ethyltetrahydro-5-methy-5-propyl-
RN: 29839-52-9   **MP** (°C):
MW: 144.22   **BP** (°C):

Solubility (Moles/L)	Solubility (Grams/L)	Temp (°C)	Ref (#)	Evaluation (T P E A A)	Comments
1.360E-01	1.961E+01	rt	B066	0 2 0 0 0	

1540. C$_8$H$_{16}$O$_2$
3-Hydroxy-2-ethyl-5,5-dimethyltetrahydrofuran
3-Furanol, 2-Ethyltetrahydro-5,5-dimethyl-
RN: 29839-59-6   **MP** (°C):
MW: 144.22   **BP** (°C):

Solubility (Moles/L)	Solubility (Grams/L)	Temp (°C)	Ref (#)	Evaluation (T P E A A)	Comments
3.302E-01	4.762E+01	rt	B066	0 2 0 0 0	

1541. C$_8$H$_{16}$O$_2$
3-Hydroxy-5-ethyl-2,5-dimethyltetrahydrofuran
3-Furanol, 2-Ethyltetrahydro-2,5-dimethyl-
RN: 29839-60-9   **MP** (°C):
MW: 144.22   **BP** (°C):

Solubility (Moles/L)	Solubility (Grams/L)	Temp (°C)	Ref (#)	Evaluation (T P E A A)	Comments
3.467E+00	5.000E+02	rt	B066	0 2 0 0 2	

1542. C$_8$H$_{16}$O$_2$
2-Ethylhexoic Acid
2-Ethyl-1-hexanoic Acid
3-Heptanecarboxylic Acid
Butylethylacetic Acid
RN: 149-57-5   **MP** (°C):
MW: 144.22   **BP** (°C):   228

Solubility (Moles/L)	Solubility (Grams/L)	Temp (°C)	Ref (#)	Evaluation (T P E A A)	Comments
1.039E-02	1.498E+00	25	O011	1 0 1 1 1	

1543.   C$_8$H$_{16}$O$_2$
Caprylic Acid
Caprylsaure
RN: 124-07-2   **MP** (°C):   16.7
MW: 144.22   **BP** (°C):   239.7

Solubility (Moles/L)	Solubility (Grams/L)	Temp (°C)	Ref (#)	Evaluation (T P E A A)	Comments
3.051E-03	4.400E-01	0	B136	1 0 2 1 1	
3.050E-03	4.398E-01	0.0	R001	1 1 1 1 1	
4.993E-03	7.200E-01	15	F300	1 0 0 0 1	
4.715E-03	6.800E-01	20	B136	1 0 2 1 1	
4.712E-03	6.795E-01	20	D041	1 0 0 0 1	
4.712E-03	6.795E-01	20.0	R001	1 1 1 1 1	
5.478E-03	7.900E-01	30	B136	1 0 2 1 1	
5.471E-03	7.890E-01	30	E005	2 1 1 2 2	
5.474E-03	7.894E-01	30.0	R001	1 1 1 1 1	
5.845E-03	8.430E-01	40	E005	2 1 1 2 2	
6.587E-03	9.500E-01	45	B136	1 0 2 1 1	
6.581E-03	9.491E-01	45.0	R001	1 1 1 1 1	
6.539E-03	9.430E-01	50	E005	2 1 1 2 2	
7.835E-03	1.130E+00	60	B136	1 0 2 1 2	
7.426E-03	1.071E+00	60	E005	2 1 1 2 2	
7.827E-03	1.129E+00	60.0	R001	1 1 1 1 2	
1.803E-02	2.600E+00	100	F300	1 0 0 0 1	

1544.   C$_8$H$_{16}$O$_2$
Isobutyl Isobutyrate
Isobutyl 2-Methylpropanoate
2-Methylpropyl 2-Methylpropanoate
IBIB
RN: 97-85-8   **MP** (°C):   -81
MW: 144.22   **BP** (°C):   147

Solubility (Moles/L)	Solubility (Grams/L)	Temp (°C)	Ref (#)	Evaluation (T P E A A)	Comments
3.952E-03	5.700E-01	25	A049	1 0 0 0 1	

1545.   C$_8$H$_{16}$O$_2$
n-Butyl n-Butyrate
Butyl Butyrate
RN: 109-21-7   **MP** (°C):
MW: 144.22   **BP** (°C):   165

Solubility (Moles/L)	Solubility (Grams/L)	Temp (°C)	Ref (#)	Evaluation (T P E A A)	Comments
3.465E-03	4.998E-01	20	D052	1 1 0 0 0	

1546. $C_8H_{16}O_2$
sec-Hexyl Acetate
Methyl Amyl Acetate

RN: 108-84-9   **MP** (°C):   -64
MW: 144.22   **BP** (°C):   140

Solubility (Moles/L)	Solubility (Grams/L)	Temp (°C)	Ref (#)	Evaluation (T P E A A)	Comments
5.543E-03	7.994E-01	20	D052	1 1 0 0 0	

1547. $C_8H_{16}O_2$
3-Hydroxy-2,2-diethyltetrahydrofuran

RN: **MP** (°C):
MW: 144.22   **BP** (°C):

Solubility (Moles/L)	Solubility (Grams/L)	Temp (°C)	Ref (#)	Evaluation (T P E A A)	Comments
1.360E-01	1.961E+01	rt	B066	0 2 0 0 0	

1548. $C_8H_{16}O_2$
3-Hydroxy-2,2,5,5-tetramethyltetrahydrofuran
3-Furanol, 2-Ethyltetrahydro-2,2,5,5-tetramethyl-

RN: 29839-74-5   **MP** (°C):
MW: 144.22   **BP** (°C):

Solubility (Moles/L)	Solubility (Grams/L)	Temp (°C)	Ref (#)	Evaluation (T P E A A)	Comments
6.304E-01	9.091E+01	rt	B066	0 2 0 0 1	

1549. $C_8H_{16}O_2$
Hexyl Acetate
2-Ethyl Butyl Acetate

RN: 142-92-7   **MP** (°C):   -80
MW: 144.22   **BP** (°C):   168

Solubility (Moles/L)	Solubility (Grams/L)	Temp (°C)	Ref (#)	Evaluation (T P E A A)	Comments
4.158E-03	5.996E-01	20	D052	1 1 0 0 0	
3.540E-03	5.105E-01	25	M124	2 1 2 2 2	

1550.   C$_8$H$_{16}$O$_2$
Pentyl Propionate
Propanoic Acid Pentyl Ester
Amyl n-Propanoate
n-Pentyl Propionate
RN: 624-54-4   **MP** (°C):
MW: 144.22   **BP** (°C):

Solubility (Moles/L)	Solubility (Grams/L)	Temp (°C)	Ref (#)	Evaluation (T P E A A)	Comments
4.900E-03	7.067E-01	20	S006	1 0 0 0 1	

1551.   C$_8$H$_{16}$O$_3$
Amyl Lactate
n-Pentyl Lactate
RN: 6382-06-5   **MP** (°C):
MW: 160.21   **BP** (°C):

Solubility (Moles/L)	Solubility (Grams/L)	Temp (°C)	Ref (#)	Evaluation (T P E A A)	Comments
6.242E-02	1.000E+01	25	R006	2 2 0 1 2	

1552.   C$_8$H$_{16}$O$_3$
Butylcellosolve Acetate
Ethylene Glycol Monobutyl Ether Acetate
Ektasolve EB Acetate
n-Butyl Cellosolve Acetate
Ethylene Glycol Mono-n-butyl Ether Acetate
RN: 112-07-2   **MP** (°C):
MW: 160.21   **BP** (°C):

Solubility (Moles/L)	Solubility (Grams/L)	Temp (°C)	Ref (#)	Evaluation (T P E A A)	Comments
5.567E-02	8.920E+00	20	D052	1 1 0 0 0	

1553.   C$_8$H$_{16}$O$_3$
2,2,5,5-Tetramethyltetrahydrofuran-3,4-diol
3,4-Furandiol, Tetrahydro-2,2,5,5-tetramethyl-
RN: 29839-67-6   **MP** (°C):
MW: 160.21   **BP** (°C):

Solubility (Moles/L)	Solubility (Grams/L)	Temp (°C)	Ref (#)	Evaluation (T P E A A)	Comments
5.674E-01	9.091E+01	rt	B066	0 2 0 0 1	

1554. $C_8H_{16}O_3$
n-Propyl β-Ethoxypropionate
Propionic Acid, 3-Ethoxy-, Propyl Ester
RN: 14144-34-4   **MP** (°C):
MW: 160.21   **BP** (°C):

Solubility (Moles/L)	Solubility (Grams/L)	Temp (°C)	Ref (#)	Evaluation (T P E A A)	Comments
9.466E-02	1.517E+01	25	D002	1 2 1 1 2	

1555. $C_8H_{16}O_3$
Methyl β-n-Butoxypropionate
Butanoic Acid, 3-Methoxy-3-oxopropyl Ester
RN: 40326-33-8   **MP** (°C):
MW: 160.21   **BP** (°C):

Solubility (Moles/L)	Solubility (Grams/L)	Temp (°C)	Ref (#)	Evaluation (T P E A A)	Comments
5.076E-02	8.133E+00	25	R034	0 0 0 0 1	

1556. $C_8H_{16}O_3$
n-Butyl β-Methoxypropionate
Propanoic Acid, 3-Methoxy-, Butyl Ester
RN: 4195-88-4   **MP** (°C):
MW: 160.21   **BP** (°C):

Solubility (Moles/L)	Solubility (Grams/L)	Temp (°C)	Ref (#)	Evaluation (T P E A A)	Comments
6.117E-02	9.800E+00	25	R034	0 0 0 0 1	

1557. $C_8H_{16}O_3S$
1,2-Oxathiolane, 5-pentyl-, 2,2-dioxide
1-Octanesulfonic Acid, 3-Hydroxy-, γ-Sultone
RN: 5633-87-4   **MP** (°C):
MW: 192.28   **BP** (°C):

Solubility (Moles/L)	Solubility (Grams/L)	Temp (°C)	Ref (#)	Evaluation (T P E A A)	Comments
1.300E-03	2.499E-01	20	B058	1 2 0 0 1	
7.938E-02	1.526E+01	100	B058	1 2 0 0 2	

1558. C$_8$H$_{16}$O$_4$
Metaldehyde
Acetaldehyde Homopolymer
Acetaldehyde Tetramer
RN: 9002-91-9   **MP** (°C):   112
MW: 176.21   **BP** (°C):

Solubility (Moles/L)	Solubility (Grams/L)	Temp (°C)	Ref (#)	Evaluation (T P E A A)	Comments
1.135E-03	2.000E-01	17	M161	1 0 0 0 2	

1559. C$_8$H$_{17}$Br
n-Octyl Bromide
1-Bromooctane
RN: 111-83-1   **MP** (°C):   -55
MW: 193.13   **BP** (°C):   200.8

Solubility (Moles/L)	Solubility (Grams/L)	Temp (°C)	Ref (#)	Evaluation (T P E A A)	Comments
8.650E-06	1.671E-03	25	M342	1 0 1 1 2	

1560. C$_8$H$_{17}$N
D-Coniine
α-Propylpiperidine
D-Coniin
Coniine
RN: 458-88-8   **MP** (°C):   -2
MW: 127.23   **BP** (°C):   166-167

Solubility (Moles/L)	Solubility (Grams/L)	Temp (°C)	Ref (#)	Evaluation (T P E A A)	Comments
1.415E-01	1.800E+01	19.5	F300	1 0 0 0 1	
7.782E-02	9.901E+00	25	D004	1 0 0 0 0	

1561. C$_8$H$_{17}$NO
Caprylylamide
Caprylsaeure-amid
RN: 629-01-6   **MP** (°C):
MW: 143.23   **BP** (°C):

Solubility (Moles/L)	Solubility (Grams/L)	Temp (°C)	Ref (#)	Evaluation (T P E A A)	Comments
3.288E-02	4.710E+00	100	F300	1 0 0 0 2	

1562.   C₈H₁₇NO
2-Isopropyl-3-methyl-butyramide
3-Methyl-2-(1-methylethyl)Butanamide
Diisopropylacetamide
RN: 5440-65-3   **MP** (°C):
MW: 143.23   **BP** (°C):

Solubility (Moles/L)	Solubility (Grams/L)	Temp (°C)	Ref (#)	Evaluation (T P E A A)	Comments
3.002E-02	4.300E+00	ns	H348	0 0 1 1 2	

1563.   C₈H₁₇NO
Propylisopropylacetamide
2-Isopropyl-2-propylacetamide
2-Isopropylvaleramide
PID
RN: 6098-19-7   **MP** (°C):
MW: 143.23   **BP** (°C):

Solubility (Moles/L)	Solubility (Grams/L)	Temp (°C)	Ref (#)	Evaluation (T P E A A)	Comments
2.444E-02	3.500E+00	37	H347	1 1 2 2 1	

1564.   C₈H₁₇NO
Methylpentylacetamide
2-Methyl-Heptanamide
MPD
RN: 4164-91-4   **MP** (°C):
MW: 143.23   **BP** (°C):

Solubility (Moles/L)	Solubility (Grams/L)	Temp (°C)	Ref (#)	Evaluation (T P E A A)	Comments
4.957E-02	7.100E+00	37	H347	1 1 2 2 1	

1565.   C₈H₁₇NO
Ethylbutylacetamide
2-Ethylhexanamide
EBD
RN: 4164-92-5   **MP** (°C):
MW: 143.23   **BP** (°C):

Solubility (Moles/L)	Solubility (Grams/L)	Temp (°C)	Ref (#)	Evaluation (T P E A A)	Comments
3.072E-02	4.400E+00	37	H347	1 1 2 2 1	

1566. C$_8$H$_{17}$NO
Valnoctamide
VCD
Valmethamide
2-Ethyl-3-methyl-Pentanamide
RN: 4171-13-5   **MP** (°C):
MW: 143.23   **BP** (°C):

Solubility (Moles/L)	Solubility (Grams/L)	Temp (°C)	Ref (#)	Evaluation (T P E A A)	Comments
6.074E-02	8.700E+00	ns	H348	0 0 1 1 2	

1567. C$_8$H$_{17}$NO
Dimethylbutylacetamide
2,2-DimethylHexanamide
DBD
RN: 20923-67-5   **MP** (°C):
MW: 143.23   **BP** (°C):

Solubility (Moles/L)	Solubility (Grams/L)	Temp (°C)	Ref (#)	Evaluation (T P E A A)	Comments
2.374E-02	3.400E+00	ns	H348	0 0 1 1 2	

1568. C$_8$H$_{17}$NO
Ethylisobutylacetamide
2-Ethyl-4-methylPentanamide
EID
RN: 130482-28-9   **MP** (°C):
MW: 143.23   **BP** (°C):

Solubility (Moles/L)	Solubility (Grams/L)	Temp (°C)	Ref (#)	Evaluation (T P E A A)	Comments
3.002E-02	4.300E+00	ns	H348	0 0 1 1 2	

1569. C$_8$H$_{17}$NO$_2$
n-Heptyl Carbamate
Heptyl Carbamate
RN: 4248-20-8   **MP** (°C):   66
MW: 159.23   **BP** (°C):

Solubility (Moles/L)	Solubility (Grams/L)	Temp (°C)	Ref (#)	Evaluation (T P E A A)	Comments
2.400E-03	3.822E-01	37	H006	1 2 2 1 1	

1570. C$_8$H$_{17}$NO$_3$
N-Isoamylurethane

RN: **MP** (°C):
MW: 175.23   **BP** (°C):

Solubility (Moles/L)	Solubility (Grams/L)	Temp (°C)	Ref (#)	Evaluation (T P E A A)	Comments
2.329E-02	4.082E+00	20	O021	1 0 0 0 0	

1571. C$_8$H$_{18}$
2,3,4-Trimethylpentane
2,3,4-Trojmetylopentan

RN: 565-75-3   **MP** (°C):   -110
MW: 114.23   **BP** (°C):   113

Solubility (Moles/L)	Solubility (Grams/L)	Temp (°C)	Ref (#)	Evaluation (T P E A A)	Comments
2.048E-05	2.340E-03	0	P003	2 2 2 2 2	
1.191E-05	1.360E-03	25	K119	1 0 0 0 2	
2.013E-05	2.300E-03	25	P003	2 2 2 2 2	
1.191E-05	1.360E-03	25	P051	2 1 1 2 2	
1.191E-05	1.360E-03	25.00	P007	2 1 2 2 2	

1572. C$_8$H$_{18}$
2,3-Dimethylhexane
2:3-Dimethylhexane

RN: 590-73-8   **MP** (°C):
MW: 114.23   **BP** (°C):   115

Solubility (Moles/L)	Solubility (Grams/L)	Temp (°C)	Ref (#)	Evaluation (T P E A A)	Comments
1.751E-06	2.000E-04	ns	B170	0 0 0 0 2	

1573. C$_8$H$_{18}$
3-Methylheptane
3-Metyloheptan

RN: 589-81-1   **MP** (°C):   -121
MW: 114.23   **BP** (°C):

Solubility (Moles/L)	Solubility (Grams/L)	Temp (°C)	Ref (#)	Evaluation (T P E A A)	Comments
2.539E-05	2.900E-03	23	C332	2 0 2 2 1	
6.933E-06	7.920E-04	25	K119	1 0 0 0 2	
6.933E-06	7.920E-04	25	P051	2 1 1 2 2	
6.933E-06	7.920E-04	25.00	P007	2 1 2 2 2	

1574.   C$_8$H$_{18}$
Isooctane
2:2:4-Trimethylpentane
RN: 540-84-1   **MP** (°C):
MW: 114.23   **BP** (°C):

Solubility (Moles/L)	Solubility (Grams/L)	Temp (°C)	Ref (#)	Evaluation (T P E A A)	Comments
2.153E-05	2.460E-03	0	P003	2 2 2 2 2	
1.226E-05	1.400E-03	20	M337	2 1 2 2 1	
9.980E-06	1.140E-03	25	K119	1 0 0 0 2	
2.136E-05	2.440E-03	25	M001	2 1 2 2 2	
2.136E-05	2.440E-03	25	M002	2 1 2 2 2	
2.136E-05	2.440E-03	25	M130	1 0 0 0 2	
1.795E-05	2.050E-03	25	P003	2 2 2 2 2	
9.980E-06	1.140E-03	25	P051	2 1 1 2 2	
9.980E-06	1.140E-03	25.00	P007	2 1 2 2 2	
7.879E-06	9.000E-04	ns	B170	0 0 0 0 2	
7.500E-05	8.567E-03	ns	J300	0 0 0 0 1	

1575.   C$_8$H$_{18}$
2-Methylheptane
RN: 592-27-8   **MP** (°C):   -109
MW: 114.23   **BP** (°C):

Solubility (Moles/L)	Solubility (Grams/L)	Temp (°C)	Ref (#)	Evaluation (T P E A A)	Comments
3.327E-05	3.800E-03	23	C332	2 0 2 2 1	

1576.   C$_8$H$_{18}$
n-Octane
Octane
RN: 111-65-9   **MP** (°C):   -56
MW: 114.23   **BP** (°C):

Solubility (Moles/L)	Solubility (Grams/L)	Temp (°C)	Ref (#)	Evaluation (T P E A A)	Comments
1.182E-05	1.350E-03	0	P003	2 2 2 2 2	
1.444E-05	1.650E-03	5.0	N004	1 1 2 2 1	
1.300E-04	1.485E-02	16	F001	1 0 1 2 1	
8.754E-05	1.000E-02	16	F300	1 0 0 0 1	
7.791E-06	8.900E-04	20	B318	1 2 1 2 0	EFG
7.739E-06	8.840E-04	20	B356	1 0 0 0 2	
1.420E-04	1.622E-02	20	S006	1 0 0 0 2	
1.800E-04	2.056E-02	25	H313	2 1 2 2 1	
1.751E-04	2.000E-02	25	K072	1 0 1 1 1	
3.773E-06	4.310E-04	25	K119	1 0 0 0 2	
5.778E-06	6.600E-04	25	M001	2 1 2 2 2	

5.778E-06	6.600E-04	25	M002	2 1 2 2 2
5.778E-06	6.600E-04	25	M040	1 0 0 1 1
1.751E-04	2.000E-02	25	M087	1 1 2 1 0
5.778E-06	6.600E-04	25	M130	1 0 0 0 1
5.970E-06	6.820E-04	25	M342	1 0 1 1 2
7.441E-06	8.500E-04	25	P003	2 2 2 2 1
1.751E-04	2.000E-02	25	S012	2 0 2 2 0
7.778E-06	8.885E-04	25.0	N004	1 1 2 2 1
3.773E-06	4.310E-04	25.0	P051	2 1 1 2 2
3.773E-06	4.310E-04	25.00	P007	2 1 2 2 2
4.587E-06	5.240E-04	40.1	P051	2 1 1 2 2
4.587E-06	5.240E-04	40.10	P007	2 1 2 2 2
1.611E-05	1.840E-03	45.0	N004	1 1 2 2 1
7.940E-06	9.070E-04	69.7	P051	2 1 1 2 2
7.940E-06	9.070E-04	69.70	P007	2 1 2 2 2
9.805E-06	1.120E-03	99.1	P051	2 1 1 2 2
9.805E-06	1.120E-03	99.10	P007	2 1 2 2 2
4.044E-05	4.620E-03	121.3	P051	2 1 1 2 2
4.044E-05	4.620E-03	121.30	P007	2 1 2 2 2
7.458E-05	8.520E-03	136.6	P051	2 1 1 2 2
7.458E-05	8.520E-03	136.60	P007	2 1 2 2 2
1.033E-04	1.180E-02	149.5	P051	2 1 1 2 2
1.033E-04	1.180E-02	149.50	P007	2 1 2 2 2
7.879E-07	9.000E-05	ns	B170	0 0 0 0 1
5.778E-06	6.600E-04	ns	H123	0 0 0 0 2

1577. $C_8H_{18}NO_4PS_2$
Vamidothion
O,O-Dimethyl S-2-(1-N-Methylcarbamoylethylmercapto)ethyl Thiophosphate
RN: 2275-23-2   **MP** (°C):   35.5
MW: 287.34   **BP** (°C):

Solubility (Moles/L)	Solubility (Grams/L)	Temp (°C)	Ref (#)	Evaluation (T P E A A)	Comments
1.392E+01	4.000E+03	20	M161	1 0 0 0 0	
1.392E+01	4.000E+03	ns	M061	0 0 0 0 2	

1578. $C_8H_{18}N_2O$
Di-n-butylnitrosamine
N-Nitroso-di-n-Butylamine
Dibutylnitrosamine
RN: 924-16-3   **MP** (°C):
MW: 158.25   **BP** (°C):   234

Solubility (Moles/L)	Solubility (Grams/L)	Temp (°C)	Ref (#)	Evaluation (T P E A A)	Comments
8.000E-03	1.266E+00	24	D083	2 0 0 0 0	
7.574E-03	1.199E+00	rt	I307	0 0 0 0 1	

1579. C$_8$H$_{18}$O
n-Butyl Ether
Butyl Ether
Dibutyl Ether
RN: 142-96-1   **MP** (°C):   -98
MW: 130.23   **BP** (°C):   142.5

Solubility (Moles/L)	Solubility (Grams/L)	Temp (°C)	Ref (#)	Evaluation (T P E A A)	Comments
1.418E-02	1.847E+00	24.80	O005	2 0 2 2 2	
2.700E-03	3.516E-01	25	K012	1 0 0 0 1	
6.138E-03	7.994E-01	25.50	O005	2 0 2 2 0	
1.720E-02	2.240E+00	37	E028	1 0 1 1 2	

1580. C$_8$H$_{18}$O
bis(2-Methyl Propyl) Ether
iso-Butyl Ether
Di-isobutyl Ether
RN: 628-55-7   **MP** (°C):
MW: 130.23   **BP** (°C):

Solubility (Moles/L)	Solubility (Grams/L)	Temp (°C)	Ref (#)	Evaluation (T P E A A)	Comments
1.059E+00	1.379E+02	25	M375	2 2 2 1 1	
1.227E-02	1.597E+00	51	M375	2 2 2 1 1	
1.002E+00	1.304E+02	60	M375	2 2 2 1 1	

1581. C$_8$H$_{18}$O
1-Octanol
Caprylic Alcohol
n-Octyl Alcohol
n-Octanol
RN: 111-87-5   **MP** (°C):   -16
MW: 130.23   **BP** (°C):   194

Solubility (Moles/L)	Solubility (Grams/L)	Temp (°C)	Ref (#)	Evaluation (T P E A A)	Comments
3.224E-03	4.198E-01	20	A015	1 2 1 1 2	
3.680E-03	4.793E-01	20	H330	2 0 2 2 2	
3.761E-03	4.898E-01	20.5	S307	1 1 0 2 1	
3.236E-03	4.214E-01	20.96	B178	1 1 0 1 2	EFG
3.162E-03	4.118E-01	23.58	B178	1 1 0 1 2	EFG
2.700E-03	3.516E-01	24	H345	2 0 2 2 2	
4.497E-03	5.857E-01	25	B038	1 2 1 1 2	
3.820E-02	4.975E+00	25	C093	2 1 1 1 0	*sic*
1.000E+00	1.302E+02	25	F044	1 0 0 0 0	EFG
1.060E-03	1.380E-01	25	J035	1 0 2 1 0	
3.830E-03	4.988E-01	25	J302	2 1 2 2 2	

Solubility (Moles/L)	Solubility (Grams/L)	Temp (°C)	Ref (#)	Evaluation (T P E A A)	Comments
3.800E-03	4.949E-01	25	K025	2 2 1 1 2	
4.530E-03	5.900E-01	25	K072	1 0 1 1 1	
3.970E-03	5.170E-01	25	L322	1 1 2 2 1	
4.530E-03	5.900E-01	25	M087	1 1 2 1 1	
4.110E-03	5.353E-01	25	S359	2 1 2 2 2	
7.671E-03	9.990E-01	30	R067	1 0 0 0 0	
4.911E-03	6.396E-01	30.6	S307	1 1 0 2 1	
3.236E-03	4.214E-01	34.53	B178	1 1 0 1 2	EFG
1.075E-03	1.400E-01	40	J035	1 0 2 1 0	
4.988E-03	6.496E-01	40.1	S307	1 1 0 2 1	
8.054E-03	1.049E+00	50.0	S307	1 1 0 2 2	
3.548E-03	4.621E-01	60	B178	1 1 0 1 2	EFG
6.751E-03	8.792E-01	60.3	S307	1 1 0 2 1	
3.548E-03	4.621E-01	69.31	B178	1 1 0 1 2	EFG
5.908E-03	7.694E-01	70.3	S307	1 1 0 2 1	
6.675E-03	8.692E-01	80.1	S307	1 1 0 2 1	
6.598E-03	8.593E-01	90.3	S307	1 1 0 2 1	
4.514E-03	5.879E-01	ns	L003	0 0 2 1 2	

1582. $C_8H_{18}O$

2-Ethyl-1-hexanol
Octyl Alcohol
Octyl-(2-ethyl Hexyl) Alcohol
2-Ethyl Hexanol
2-Ethylhexanol
2-Ethylhexan-1-ol

RN: 104-76-7   **MP** (°C):   -76
MW: 130.23   **BP** (°C):

Solubility (Moles/L)	Solubility (Grams/L)	Temp (°C)	Ref (#)	Evaluation (T P E A A)	Comments
1.012E-02	1.318E+00	10.2	S307	1 1 0 2 2	
9.586E-03	1.248E+00	19.8	S307	1 1 0 2 2	
4.604E-03	5.996E-01	20	D052	1 1 0 0 0	
6.760E-03	8.804E-01	20	H330	2 0 2 2 2	
9.982E-04	1.300E-01	25	K072	1 0 1 1 1	
7.441E-03	9.691E-01	29.6	S307	1 1 0 2 1	
8.437E-03	1.099E+00	40.1	S307	1 1 0 2 2	
5.678E-03	7.395E-01	50.2	S307	1 1 0 2 1	
6.598E-03	8.593E-01	60.3	S307	1 1 0 2 1	
7.594E-03	9.890E-01	70.1	S307	1 1 0 2 1	
8.284E-03	1.079E+00	80.1	S307	1 1 0 2 2	
8.973E-03	1.169E+00	90.3	S307	1 1 0 2 2	

1583. $C_8H_{18}O$
2-Octanol
sec-Caprylic Alcohol
sec-Octyl Alcohol
Methyl Hexyl Carbinol
RN: 123-96-6   **MP** (°C):   -38.6
MW: 130.23   **BP** (°C):   178.5

Solubility (Moles/L)	Solubility (Grams/L)	Temp (°C)	Ref (#)	Evaluation (T P E A A)	Comments
1.158E-02	1.508E+00	15	M073	1 0 2 2 2	
8.131E-03	1.059E+00	20	A015	1 2 1 1 2	
8.600E-03	1.120E+00	20	H330	2 0 2 2 2	
3.059E-02	3.984E+00	25	C093	2 1 1 1 0	
9.829E-03	1.280E+00	25	M073	1 0 2 2 2	
7.892E-03	1.028E+00	ns	J300	1 0 1 1 1	

1584. $C_8H_{18}O$
DL-2-Octanol
DL-Octanol-(2)
RN: 4128-31-8   **MP** (°C):   -31.6
MW: 130.23   **BP** (°C):   180

Solubility (Moles/L)	Solubility (Grams/L)	Temp (°C)	Ref (#)	Evaluation (T P E A A)	Comments
1.152E-02	1.500E+00	15	F300	1 0 0 0 1	
9.214E-03	1.200E+00	25	F300	1 0 0 0 1	

1585. $C_8H_{18}O_2$
Ethohexadiol
2-Ethyl-1,3-hexanediol
RN: 94-96-2   **MP** (°C):   -40
MW: 146.23   **BP** (°C):   244.2

Solubility (Moles/L)	Solubility (Grams/L)	Temp (°C)	Ref (#)	Evaluation (T P E A A)	Comments
4.103E-02	6.000E+00	20	M161	1 0 0 0 0	
2.756E-01	4.031E+01	25	C093	2 1 1 1 1	
2.756E-01	4.031E+01	ns	M061	0 0 0 0 1	

1586. $C_8H_{18}O_4S_2$
Sulfonethylmethane
Trional
RN: 76-20-0 **MP** (°C): 75
MW: 242.36 **BP** (°C):

Solubility (Moles/L)	Solubility (Grams/L)	Temp (°C)	Ref (#)	Evaluation (T P E A A)	Comments
2.053E-02	4.975E+00	16	A072	1 0 1 0 1	
2.063E-02	5.000E+00	16	F300	1 0 0 0 0	

1587. $C_8H_{19}N$
Octylamine
1-Aminooctane
1-Octanamine
Monoctylamine
n-Octylamine
RN: 111-86-4 **MP** (°C): -5
MW: 129.25 **BP** (°C): 175

Solubility (Moles/L)	Solubility (Grams/L)	Temp (°C)	Ref (#)	Evaluation (T P E A A)	Comments
1.547E-03	2.000E-01	25	K072	1 0 1 1 1	
1.547E-03	2.000E-01	25	M087	1 1 2 1 1	

1588. $C_8H_{19}N$
n-Dibutylamine
Di-n-butylamine
N,N-Dibutylamine
N-Butyl-1-butanamine
RN: 111-92-2 **MP** (°C): -62
MW: 129.25 **BP** (°C): 159

Solubility (Moles/L)	Solubility (Grams/L)	Temp (°C)	Ref (#)	Evaluation (T P E A A)	Comments
2.500E-02	3.231E+00	25	K012	1 0 0 0 1	

1589. $C_8H_{19}O_2PS_2$
Ethoprop
Ethoprophos
O-Ethyl-S,S-dipropylphosphorodithioate
Holdem
Rovokil
Ethyl S,S-Dipropyl phosphorodithioate
RN: 13194-48-4 **MP** (°C):
MW: 242.34 **BP** (°C): 88.5

Solubility (Moles/L)	Solubility (Grams/L)	Temp (°C)	Ref (#)	Evaluation (T P E A A)	Comments
3.095E-03	7.500E-01	ns	M161	0 0 0 0 2	

1590. C$_8$H$_{19}$O$_2$PS$_3$

Disulfoton
Phosphorodithioic Acid O,O-Diethyl S-[2-(Ethylthio)ethyl] Ester
Solvirex
Disyston
Thiodemeton
Ethylthiometon

RN:	298-04-4	MP (°C):	108
MW:	274.41	BP (°C):	

Solubility (Moles/L)	Solubility (Grams/L)	Temp (°C)	Ref (#)	Evaluation (T P E A A)	Comments
5.940E-05	1.630E-02	19.50	B169	2 1 1 1 2	
9.111E-05	2.500E-02	20	M061	1 0 0 0 1	
9.111E-05	2.500E-02	rt	M161	0 0 0 0 1	

1591. C$_8$H$_{19}$O$_3$P

Dibutyl Hydrogen Phosphonate
Di-n-Butyl Phosphite
Dibutoxyphosphine Oxide

RN:	1809-19-4	MP (°C):	
MW:	194.21	BP (°C):	

Solubility (Moles/L)	Solubility (Grams/L)	Temp (°C)	Ref (#)	Evaluation (T P E A A)	Comments
3.759E-02	7.300E+00	25	B070	1 2 0 1 1	

1592. C$_8$H$_{19}$O$_3$PS$_2$

Demetonthiol
Thiophosphorsaeure-O,O-diaethyl-S-[2-(aethylthio)-aethyl]-ester
O,O-Diethyl-S-(2-(ethylthio)-ethyl)ester Thiophosphoric Acid

RN:	126-75-0	MP (°C):	
MW:	258.34	BP (°C):	

Solubility (Moles/L)	Solubility (Grams/L)	Temp (°C)	Ref (#)	Evaluation (T P E A A)	Comments
7.742E-03	2.000E+00	20	F300	1 0 0 0 0	

1593.   C$_8$H$_{19}$O$_3$PS$_2$
Demetonthione
Thiophosphorsaeure-O,O-diaethyl-O-[2-(aethylthio)-aethyl]-ester
O,O-Diethyl-O-(2-(ethylthio)-ethyl)ester Thiophosphoric Acid
O,O-Diethyl 2-Ethylmercaptoethyl Thiophosphate
Systox
Thiolo-Demeton

RN: 298-03-3   **MP** (°C):
MW: 258.34   **BP** (°C):   134

Solubility (Moles/L)	Solubility (Grams/L)	Temp (°C)	Ref (#)	Evaluation (T P E A A)	Comments
2.323E-04	6.000E-02	20	M061	1 0 0 0 1	
7.742E-03	2.000E+00	rt	M161	0 0 0 0 0	form II
2.323E-04	6.000E-02	rt	M161	0 0 0 0 1	form I
1.277E-02	3.300E+00	rt	M161	0 0 0 0 1	

1594.   C$_8$H$_{19}$O$_4$P
Diethyl Isobutyl Phosphate
Ethyl Isobutyl Phosphate
Phosphoric Acid, Diethyl 2-Methylpropyl Ester

RN: 26628-97-7   **MP** (°C):
MW: 210.21   **BP** (°C):

Solubility (Moles/L)	Solubility (Grams/L)	Temp (°C)	Ref (#)	Evaluation (T P E A A)	Comments
6.660E-02	1.400E+01	25	B070	1 2 0 1 1	

1595.   C$_8$H$_{19}$O$_4$P
Diethyl Butyl Phosphate
Butyl Diethyl Phosphate

RN: 2737-00-0   **MP** (°C):
MW: 210.21   **BP** (°C):

Solubility (Moles/L)	Solubility (Grams/L)	Temp (°C)	Ref (#)	Evaluation (T P E A A)	Comments
7.136E-02	1.500E+01	25	B070	1 2 0 1 1	

1596.   C$_8$H$_{19}$O$_4$PS$_3$

Disulfoton Sulfone
Phosphorodithioic Acid O,O-Diethyl S-[2-(Ethylsulfonyl)ethyl] Ester
Disulfoton Dioxide
Diethyl S-(2-Ethylsulfonylethyl) Phosphorodithioate
Disyston Sulfone
Thiodemeton Sulfone

RN: 2497-06-5   **MP** (°C):
MW: 306.40   **BP** (°C):

Solubility (Moles/L)	Solubility (Grams/L)	Temp (°C)	Ref (#)	Evaluation (T P E A A)	Comments
2.716E-03	8.323E-01	20	B169	2 2 1 1 1	

1597.   C$_8$H$_{20}$Si

Tetraethylsilicane
Tetraethylsilane
Tetraethylsilicon

RN: 631-36-7   **MP** (°C):
MW: 144.33   **BP** (°C):

Solubility (Moles/L)	Solubility (Grams/L)	Temp (°C)	Ref (#)	Evaluation (T P E A A)	Comments
2.250E-06	3.248E-04	25	D346	1 1 2 2 2	

1598.   C$_8$H$_{20}$Sn

Tetraethyltin
Tetraethylstannane

RN: 597-64-8   **MP** (°C):   -112
MW: 234.94   **BP** (°C):   181

Solubility (Moles/L)	Solubility (Grams/L)	Temp (°C)	Ref (#)	Evaluation (T P E A A)	Comments
1.140E-06	2.678E-04	25	D346	1 1 2 2 2	

1599.   C$_8$H$_{20}$O$_5$P$_2$S$_2$

Sulfotepp
Pirofos
Tetraethyl Dithiopyrophosphate

RN: 3689-24-5   **MP** (°C):
MW: 322.32   **BP** (°C):   137.5

Solubility (Moles/L)	Solubility (Grams/L)	Temp (°C)	Ref (#)	Evaluation (T P E A A)	Comments
9.307E-05	3.000E-02	20	F300	1 0 0 0 0	
7.756E-05	2.500E-02	20	M061	1 0 0 0 1	
7.756E-05	2.500E-02	rt	M161	0 0 0 0 1	

1600. $C_8H_{23}N_5$
Tetraethylenepentamine
1,4,7,10,13-Pentaazatridecane
N-(2-Aminoethyl)-N'-(2-((2-aminoethyl)amino)ethyl)-1,2-ethanediamine
1,11-Diamino-3,6,9-triazaundecane
3,6,9-Triaza-1,11-undecanediamine
3,6,9-Triazaundecane-1,11-diamine

RN: 112-57-2   **MP** (°C):   -40
MW: 189.31   **BP** (°C):   340

Solubility (Moles/L)	Solubility (Grams/L)	Temp (°C)	Ref (#)	Evaluation (T P E A A)	Comments
4.582E+00	8.674E+02	4.50	C022	1 2 0 0 2	

1601. $C_8Cl_4N_2$
Chlorothalonil
2,4,5,6-Tetrachloro-1,3-benzenedicarbonitrile
Forturf
Exotherm
Bravo

RN: 1897-45-6   **MP** (°C):   250.5
MW: 265.91   **BP** (°C):

Solubility (Moles/L)	Solubility (Grams/L)	Temp (°C)	Ref (#)	Evaluation (T P E A A)	Comments
2.256E-06	6.000E-04	25	M161	1 0 0 0 0	

1602. $C_9H_4Cl_3NO_2S$
Folpet
N-(Trichloromethylthio)phthalimide
Folpan
Folpel
Phaltan
Phalton

RN: 133-07-3   **MP** (°C):   177
MW: 296.56   **BP** (°C):

Solubility (Moles/L)	Solubility (Grams/L)	Temp (°C)	Ref (#)	Evaluation (T P E A A)	Comments
3.388E-06	1.005E-03	20	B179	2 0 0 0 2	
3.372E-06	1.000E-03	20	F311	1 2 2 2 1	

1603. $C_9H_5Cl_3N_4$

Anilazine
4,6-Dichloro-N-(2-chlorophenyl)-1,3,5-triazin-2-amine
Triasyn
Direx
Dyrene
Kemate

RN: 101-05-3   **MP** (°C):   159.5
MW: 275.53   **BP** (°C):

Solubility (Moles/L)	Solubility (Grams/L)	Temp (°C)	Ref (#)	Evaluation (T P E A A)	Comments
3.629E-05	1.000E-02	ns	B160	0 0 0 0 1	

1604. $C_9H_6ClNO_3S$

Benazolin
7-Chloro-2-oxo-3(2H)-benzothiazolacetic Acid
Galipan
Herbazolin
Leymin
Metizolin

RN: 3813-05-6   **MP** (°C):   193
MW: 243.67   **BP** (°C):

Solubility (Moles/L)	Solubility (Grams/L)	Temp (°C)	Ref (#)	Evaluation (T P E A A)	Comments
2.462E-03	6.000E-01	20	M161	1 0 0 0 2	

1605. $C_9H_6Cl_2N_2O_3$

Methazole
2-(3,4-Dichlorophenyl)-4-methyl-1,2,4-oxadiazolidine-3,5-dione
Tunic
Paxilon
Chlormethazole
Mezopur

RN: 20354-26-1   **MP** (°C):   123
MW: 261.07   **BP** (°C):

Solubility (Moles/L)	Solubility (Grams/L)	Temp (°C)	Ref (#)	Evaluation (T P E A A)	Comments
5.746E-06	1.500E-03	24	C105	2 1 2 2 2	
5.746E-06	1.500E-03	25	M161	1 0 0 0 1	
5.746E-06	1.500E-03	25	W314	1 0 0 0 1	

1606. $C_9H_6Cl_6O_3S$

α-Endosulfan
5-Norbornene-2,3-dimethanol, 1,4,5,6,7,7-Hexachloro-, Cyclic Sulfite, endo-
Endosulfan I
Endosulfan A
Hexachloro-5-norbornene-2,3-dimethanol, Cyclic Sulfite, endo-
Thiodan I

RN: 959-98-8   **MP** (°C):   109
MW: 406.93   **BP** (°C):

Solubility (Moles/L)	Solubility (Grams/L)	Temp (°C)	Ref (#)	Evaluation (T P E A A)	Comments
1.253E-06	5.099E-04	20	B300	2 0 1 1 2	
1.302E-06	5.300E-04	25	W025	1 0 2 2 2	
4.030E-07	1.640E-04	ns	A069	0 0 0 0 2	

1607. $C_9H_6Cl_6O_3S$

β-Endosulfan
5-Norbornene-2,3-dimethanol, 1,4,5,6,7,7-Hexachloro-, Cyclic Sulfite, exo-
Endosulfan II
Hexachloro-5-norbornene-2,3-dimethanol, Cyclic Sulfite, exo-
Thiodan II

RN: 33213-65-9   **MP** (°C):   209
MW: 406.93   **BP** (°C):

Solubility (Moles/L)	Solubility (Grams/L)	Temp (°C)	Ref (#)	Evaluation (T P E A A)	Comments
1.106E-06	4.501E-04	20	B300	2 0 1 1 2	
6.881E-07	2.800E-04	25	W025	1 0 2 2 2	
1.720E-07	7.000E-05	ns	A069	0 0 0 0 1	

1608. $C_9H_6I_3NO_3$

2,4,6-Triiodo-3-acetaminobenzoic Acid
Acetrizoic Acid

RN: 85-36-9   **MP** (°C):
MW: 556.87   **BP** (°C):

Solubility (Moles/L)	Solubility (Grams/L)	Temp (°C)	Ref (#)	Evaluation (T P E A A)	Comments
2.299E-03	1.280E+00	25	L025	1 0 0 0 2	
3.232E-03	1.800E+00	50	L025	1 0 0 0 2	
5.387E-03	3.000E+00	100	L025	1 0 0 0 2	
2.442E-03	1.360E+00	ns	H055	0 1 0 2 2	

1609.   C$_9$H$_6$N$_2$S
4-Cyanobenzyl Isothiocyanate
p-Cyanobenzyl Isothiocyanate
Isothiocyanic Acid, p-Cyanobenzyl Ester
RN: 3694-48-2   **MP** (°C):
MW: 174.23   **BP** (°C):

Solubility (Moles/L)	Solubility (Grams/L)	Temp (°C)	Ref (#)	Evaluation (T P E A A)	Comments
3.200E-04	5.575E-02	25	D014	1 0 0 0 1	

1610.   C$_9$H$_6$O$_2$
Coumarin
Cumarin
1,2-Benzopyrone
2H-1-Benzopyran-2-one
Benzopyran-2-one
Benzopyrone
RN: 91-64-5   **MP** (°C):   70
MW: 146.15   **BP** (°C):

Solubility (Moles/L)	Solubility (Grams/L)	Temp (°C)	Ref (#)	Evaluation (T P E A A)	Comments
8.211E-03	1.200E+00	0	F300	1 0 0 0 1	
6.153E-03	8.992E-01	0.2	D073	1 1 2 1 0	
1.298E-02	1.896E+00	20	D073	1 1 2 1 1	
1.368E-02	2.000E+00	22.5	G301	2 1 0 1 2	
1.706E-02	2.494E+00	25	I312	0 0 0 0 1	
1.774E-02	2.593E+00	30	D073	1 1 2 1 1	
1.847E-02	2.700E+00	30	F300	1 0 0 0 1	
3.065E-02	4.480E+00	40	D073	1 1 2 1 1	
4.419E-02	6.458E+00	50	D073	1 1 2 1 1	
4.756E-02	6.951E+00	60	D073	1 1 2 1 1	
1.342E-01	1.961E+01	100	I312	0 0 0 0 0	
1.507E-02	2.203E+00	ns	R082	0 0 2 2 2	
6.842E-04	9.999E-02	rt	D021	0 0 1 1 0	*sic*

1611.   C$_9$H$_6$O$_3$
7-Hydroxycoumarin
Umbelliferone
RN: 93-35-6   **MP** (°C):   230
MW: 162.15   **BP** (°C):

Solubility (Moles/L)	Solubility (Grams/L)	Temp (°C)	Ref (#)	Evaluation (T P E A A)	Comments
1.918E-03	3.110E-01	ns	R082	0 0 2 2 2	

1612. C$_9$H$_6$O$_5$
Phthalonic Acid
Phthalonsaeure
RN: 528-46-1   **MP** (°C):
MW: 194.15   **BP** (°C):

Solubility (Moles/L)	Solubility (Grams/L)	Temp (°C)	Ref (#)	Evaluation (T P E A A)	Comments
2.756E+00	5.350E+02	15	F300	1 0 0 0 2	

1613. C$_9$H$_6$O$_6$
1,2,3-Benzenetricarboxylic Acid
Benzol-tricarbonsaeure-(1,2,3)
Hemimellitic Acid
RN: 569-51-7   **MP** (°C):   223
MW: 210.14   **BP** (°C):

Solubility (Moles/L)	Solubility (Grams/L)	Temp (°C)	Ref (#)	Evaluation (T P E A A)	Comments
1.456E-01	3.060E+01	19	F300	1 0 0 0 2	

1614. C$_9$H$_6$O$_6$
Hydrastic Acid
Hydrastsaeure
RN: 490-26-6   **MP** (°C):
MW: 210.14   **BP** (°C):

Solubility (Moles/L)	Solubility (Grams/L)	Temp (°C)	Ref (#)	Evaluation (T P E A A)	Comments
2.855E-02	6.000E+00	15	F300	1 0 0 0 1	

1615. C$_9$H$_6$O$_6$
Trimesic Acid
1,3,5-Benzenetricarboxylic Acid
Benzol-tricarbonsaeure-(1,3,5)
RN: 554-95-0   **MP** (°C):
MW: 210.14   **BP** (°C):

Solubility (Moles/L)	Solubility (Grams/L)	Temp (°C)	Ref (#)	Evaluation (T P E A A)	Comments
1.808E-02	3.800E+00	16	F300	1 0 0 0 1	
1.252E-01	2.630E+01	23	F300	1 0 0 0 2	

1616.   C₉H₇Cl₃O₃

Silvex
2-(2,4,5-Trichlorophenoxy)propionic Acid
Fenoprop
Propionic Acid, 2(2,4,5-Trichlorophenoxy)-

RN: 93-72-1   **MP** (°C):   181.6
MW: 269.51   **BP** (°C):   200

Solubility (Moles/L)	Solubility (Grams/L)	Temp (°C)	Ref (#)	Evaluation (T P E A A)	Comments
2.634E-04	7.100E-02	25	B164	1 0 1 1 1	
5.195E-04	1.400E-01	25	B185	1 0 0 0 2	
6.678E-04	1.800E-01	25	B200	1 0 0 0 1	
5.195E-04	1.400E-01	25	L024	1 0 0 0 2	
5.194E-04	1.400E-01	25	M061	1 0 0 0 1	
5.195E-04	1.400E-01	25	M161	1 0 0 0 2	
5.194E-04	1.400E-01	ns	B100	0 0 0 0 1	
5.195E-04	1.400E-01	ns	K138	0 0 0 0 1	

1617.   C₉H₇Cl₃O₃

Trichloroethyl Salicylate
Benzoic Acid, 2-Hydroxy-, 2,2,2-Trichloroethyl Ester

RN: 56529-85-2   **MP** (°C):
MW: 269.51   **BP** (°C):

Solubility (Moles/L)	Solubility (Grams/L)	Temp (°C)	Ref (#)	Evaluation (T P E A A)	Comments
4.081E-03	1.100E+00	37	D009	1 2 1 1 1	0.1N HCl

1618.   C₉H₇N

Quinoline
Chinolin
1-Azanaphthalene
Benzopyridine
1-Benzazine
Benzo[b]pyridine

RN: 91-22-5   **MP** (°C):   -15
MW: 129.16   **BP** (°C):   237.7

Solubility (Moles/L)	Solubility (Grams/L)	Temp (°C)	Ref (#)	Evaluation (T P E A A)	Comments
4.730E-02	6.110E+00	20	A050	1 0 1 1 2	
4.913E-02	6.346E+00	20.3	L339	2 0 2 2 2	
4.968E-02	6.417E+00	40.0	L339	2 0 2 2 2	
6.337E-02	8.185E+00	64.8	L339	2 0 2 2 2	
8.136E-02	1.051E+01	80.2	L339	2 0 2 2 2	
1.063E-01	1.373E+01	100.0	L339	2 0 2 2 2	

1619.   C$_9$H$_7$NO
Carbostyril
2-Hydroxyquinoline
2-Quinolinol

RN: 59-31-4   **MP** (°C):   199.0
MW: 145.16   **BP** (°C):

Solubility (Moles/L)	Solubility (Grams/L)	Temp (°C)	Ref (#)	Evaluation (T P E A A)	Comments
7.244E-03	1.052E+00	20	C035	1 0 2 2 1	

1620.   C$_9$H$_7$NO
3-Hydroxyquinoline
3-Quinolinol

RN: 580-18-7   **MP** (°C):
MW: 145.16   **BP** (°C):

Solubility (Moles/L)	Solubility (Grams/L)	Temp (°C)	Ref (#)	Evaluation (T P E A A)	Comments
4.050E-03	5.879E-01	20	A035	1 0 2 2 1	

1621.   C$_9$H$_7$NO
4-Hydroxyquinoline
4-Hydroxy-chinolin
4-Quinolinol

RN: 611-36-9   **MP** (°C):   201
MW: 145.16   **BP** (°C):

Solubility (Moles/L)	Solubility (Grams/L)	Temp (°C)	Ref (#)	Evaluation (T P E A A)	Comments
3.307E-02	4.800E+00	15	F300	1 0 0 0 1	

1622.   C$_9$H$_7$NO
5-Hydroxyquinoline
5-Quinolinol

RN: 578-67-6   **MP** (°C):   223
MW: 145.16   **BP** (°C):

Solubility (Moles/L)	Solubility (Grams/L)	Temp (°C)	Ref (#)	Evaluation (T P E A A)	Comments
2.869E-03	4.165E-01	20	A035	1 0 2 2 1	

1623. C$_9$H$_7$NO
6-Hydroxyquinoline
6-Quinolinol
RN: 580-16-5   **MP** (°C):   192
MW: 145.16   **BP** (°C):

Solubility (Moles/L)	Solubility (Grams/L)	Temp (°C)	Ref (#)	Evaluation (T P E A A)	Comments
6.882E-03	9.990E-01	20	A035	1 0 2 2 1	

1624. C$_9$H$_7$NO
8-Hydroxyquinoline
8-Quinolinol
Hydroxybenzopuridine
RN: 148-24-3   **MP** (°C):   76
MW: 145.16   **BP** (°C):   267

Solubility (Moles/L)	Solubility (Grams/L)	Temp (°C)	Ref (#)	Evaluation (T P E A A)	Comments
3.825E-03	5.552E-01	20	A035	1 0 2 2 1	
4.470E-03	6.489E-01	25.2	P024	2 1 1 1 2	
5.380E-03	7.810E-01	30.3	P024	2 1 1 1 2	

1625. C$_9$H$_7$NO
7-Hydroxyquinoline
7-Quinolinol
RN: 580-20-1   **MP** (°C):
MW: 145.16   **BP** (°C):

Solubility (Moles/L)	Solubility (Grams/L)	Temp (°C)	Ref (#)	Evaluation (T P E A A)	Comments
3.130E-03	4.543E-01	20	A035	1 0 2 2 1	

1626. C$_9$H$_7$NOS
m-Acetylphenyl Isothiocyanate
3-Acetylphenyl Isothiocyanate
RN: 3125-71-1   **MP** (°C):
MW: 177.23   **BP** (°C):

Solubility (Moles/L)	Solubility (Grams/L)	Temp (°C)	Ref (#)	Evaluation (T P E A A)	Comments
4.700E-05	8.330E-03	25	K032	2 2 0 1 1	

1627.   C$_9$H$_7$NOS
p-Acetylphenyl Isothiocyanate
4-Acetylphenyl Isothiocyanate
RN: 2131-57-9   **MP** (°C):
MW: 177.23   **BP** (°C):

Solubility (Moles/L)	Solubility (Grams/L)	Temp (°C)	Ref (#)	Evaluation (T P E A A)	Comments
9.500E-05	1.684E-02	25	D019	1 1 1 1 1	

1628.   C$_9$H$_7$NO$_2$S
m-Acetoxyphenyl Isothiocyanate
Methyl m-Isothiocyanobenzoate
RN: 3530-01-6   **MP** (°C):
MW: 193.23   **BP** (°C):

Solubility (Moles/L)	Solubility (Grams/L)	Temp (°C)	Ref (#)	Evaluation (T P E A A)	Comments
2.720E-04	5.256E-02	25	K032	2 2 0 1 2	
7.700E-04	1.488E-01	25	K032	2 2 0 1 2	

1629.   C$_9$H$_7$NO$_5$
2-(Oxalylamino)benzoic Acid
Oxanil-carbonsaeure-(2)
Oxanil-o-carboxylic Acid
RN: 5651-01-4   **MP** (°C):
MW: 209.16   **BP** (°C):

Solubility (Moles/L)	Solubility (Grams/L)	Temp (°C)	Ref (#)	Evaluation (T P E A A)	Comments
5.259E-03	1.100E+00	10	F300	1 0 0 0 1	

1630.   C$_9$H$_7$N$_3$S
Tricyclazole
Methyl-1,2,4-triazolo(3,4-b)benzothiazole
5-Methyl-1,2,4-triazolo[3,4-b]benzothiazole
RN: 41814-78-2   **MP** (°C):   187.5
MW: 189.24   **BP** (°C):

Solubility (Moles/L)	Solubility (Grams/L)	Temp (°C)	Ref (#)	Evaluation (T P E A A)	Comments
8.455E-03	1.600E+00	25	M161	1 0 0 0 1	

1631.   C₉H₇N₇O₂S
Azathioprine
Cytostatics
Imuran
Azatioprin
6-(1-Methyl-p-nitro-5-imidazolyl)-thiopurine
Ccucol
RN: 446-86-6   **MP** (°C):   243.5
MW: 277.27   **BP** (°C):

Solubility (Moles/L)	Solubility (Grams/L)	Temp (°C)	Ref (#)	Evaluation (T P E A A)	Comments
4.689E-04	1.300E-01	24	N016	2 0 1 1 1	
4.472E-04	1.240E-01	25	N063	1 1 1 1 2	intrinsic
4.689E-04	1.300E-01	25	N063	1 1 1 1 2	

1632.   C₉H₈Cl₂O₃
Dichlorprop
Dichloroprop
α-(2,4-Dichlorophenoxy)propionic Acid
RN: 120-36-5   **MP** (°C):   117.5
MW: 235.07   **BP** (°C):

Solubility (Moles/L)	Solubility (Grams/L)	Temp (°C)	Ref (#)	Evaluation (T P E A A)	Comments
1.489E-03	3.500E-01	20	L024	1 0 0 0 2	
1.489E-03	3.500E-01	20	M161	1 0 0 0 2	
3.527E-03	8.290E-01	25	B164	1 0 1 1 2	
3.020E-03	7.100E-01	28	B200	1 0 0 0 1	
1.484E-02	3.488E+00	ns	B100	0 0 0 0 1	

1633.   C₉H₈Cl₂O₃
Methyl (2,4-Dichlorophenoxy)acetate
2,4-Dichorophenoxyacetic Acid Methyl Ester
RN: 5335-03-5   **MP** (°C):
MW: 235.07   **BP** (°C):

Solubility (Moles/L)	Solubility (Grams/L)	Temp (°C)	Ref (#)	Evaluation (T P E A A)	Comments
7.657E-04	1.800E-01	ns	B185	0 0 0 0 2	
5.333E-04	1.254E-01	ns	M120	0 0 1 1 2	

1634.   C₉H₈Cl₃NO₂S

Captan
N-Trichloromethylthio-4-cyclohexene-1,2-dicarboximide
Vancide 89
Merpan 90
Orthocid-83
Pillarcap

RN: 133-06-2   **MP** (°C):   178
MW: 300.59   **BP** (°C):

Solubility (Moles/L)	Solubility (Grams/L)	Temp (°C)	Ref (#)	Evaluation (T P E A A)	Comments
1.660E-06	4.989E-04	20	B179	2 0 0 0 2	
<1.66E-06	<5.00E-04	20	F311	1 2 2 2 1	
1.544E-05	4.642E-03	ns	H322	0 0 0 2 2	
1.663E-06	5.000E-04	rt	M161	0 0 0 0 0	

1635.   C₉H₈N₂OS

m-Acetamidophenyl Isothiocyanate
3-Acetamidophenyl Isothiocyanate

RN: 3137-83-5   **MP** (°C):
MW: 192.24   **BP** (°C):

Solubility (Moles/L)	Solubility (Grams/L)	Temp (°C)	Ref (#)	Evaluation (T P E A A)	Comments
2.950E-04	5.671E-02	25	K032	2 2 0 1 2	

1636.   C₉H₈N₄O₆

Nifurtoinol
3-(Hydroxymethyl)nitrofurantoin

RN: 1088-92-2   **MP** (°C):
MW: 268.19   **BP** (°C):

Solubility (Moles/L)	Solubility (Grams/L)	Temp (°C)	Ref (#)	Evaluation (T P E A A)	Comments
1.230E-03	3.300E-01	22	B154	1 1 1 1 1	0.1M HCl

1637.   C₉H₈O

Cinnamaldehyde
3-Phenyl-2-propenal
Phenylacrolein
3-Phenyl-2-propenaldehyde
Zimtaldehyde

RN: 104-55-2   **MP** (°C):
MW: 132.16   **BP** (°C):   246

Solubility (Moles/L)	Solubility (Grams/L)	Temp (°C)	Ref (#)	Evaluation (T P E A A)	Comments
1.020E-02	1.348E+00	25	I019	1 0 1 2 2	
9.100E-03	1.203E+00	37	E028	1 0 1 1 1	

1638.   C$_9$H$_8$O$_2$
Cinnamic Acid
Phenylacrylic Acid
3-Phenylpropenoic Acid
2-Propenoic Acid, 3-Phenyl-
RN: 621-82-9   **MP** (°C):   133
MW: 148.16   **BP** (°C):   261.9

Solubility (Moles/L)	Solubility (Grams/L)	Temp (°C)	Ref (#)	Evaluation (T P E A A)	Comments
2.024E-03	2.999E-01	10	M043	1 0 0 0 0	
3.390E-03	5.023E-01	14.3	D061	1 0 0 0 2	
2.642E-03	3.914E-01	16.3	D061	1 0 0 0 2	
2.643E-03	3.916E-01	16.30	B118	1 0 0 0 2	unit assumed
1.515E-02	2.245E+00	20	C092	2 1 0 1 1	*sic*
2.699E-03	3.998E-01	20	M043	1 0 0 0 0	
3.170E-03	4.697E-01	22	E045	2 0 1 1 2	
3.260E-03	4.830E-01	23	E045	2 0 1 1 2	
3.360E-03	4.978E-01	24	E045	2 0 1 1 2	
3.450E-03	5.112E-01	25	E045	2 0 1 1 2	
3.850E-03	5.704E-01	25	K040	1 0 2 1 2	
3.340E-03	4.949E-01	25	L048	1 2 2 1 2	
3.340E-03	4.949E-01	25	L050	2 0 1 2 2	
3.540E-03	5.245E-01	26	E045	2 0 1 1 2	
3.800E-03	5.630E-01	26.4	P043	2 0 1 1 2	
3.630E-03	5.378E-01	27	E045	2 0 1 1 2	
4.963E-03	7.353E-01	28	D050	1 2 1 2 2	
4.688E-03	6.946E-01	30	B118	1 0 0 0 2	unit assumed
4.682E-03	6.937E-01	30	D061	1 0 0 0 2	
4.047E-03	5.996E-01	30	M043	1 0 0 0 0	
3.959E-02	5.865E+00	100	M043	1 0 0 0 1	

1639.   C$_9$H$_8$O$_2$
cis-Cinnamic Acid
cis-Zimtsaeure
RN: 102-94-3   **MP** (°C):
MW: 148.16   **BP** (°C):

Solubility (Moles/L)	Solubility (Grams/L)	Temp (°C)	Ref (#)	Evaluation (T P E A A)	Comments
4.657E-02	6.900E+00	18	F300	1 0 0 0 1	
4.644E-02	6.880E+00	18	M077	1 2 1 1 2	form III, mp 68 °C
5.143E-02	7.620E+00	18	M077	1 2 1 1 2	form II, mp 58 °C
6.041E-02	8.950E+00	18	M077	1 2 1 1 2	form I, mp 42 °C
5.703E-02	8.450E+00	25	M077	1 2 1 1 2	form III, mp 68 °C
6.324E-02	9.370E+00	25	M077	1 2 1 1 2	form II, mp 58 °C
7.445E-02	1.103E+01	25	M077	1 2 1 1 2	form I, mp 42 °C
7.519E-02	1.114E+01	35	M077	1 2 1 1 2	form III, mp 68 °C

8.362E-02	1.239E+01	35	M077	1 2 1 1 2	form II, mp 58 °C
9.861E-02	1.461E+01	35	M077	1 2 1 1 2	form I, mp 42 °C
9.760E-02	1.446E+01	45	M077	1 2 1 1 2	form III, mp 68 °C
1.086E-01	1.609E+01	45	M077	1 2 1 1 2	form II, mp 58 °C
1.245E-01	1.845E+01	55	M077	1 2 1 1 2	form III, mp 68 °C

1640. $C_9H_8O_2$

trans-Cinnamic Acid
trans-3-Phenyl-2-propenoic Acid
trans-β-Phenylacrylic Acid
(E)-3-Phenyl-2-Propenoic Acid

RN: 140-10-3   **MP** (°C):   133
MW: 148.16   **BP** (°C):   300

Solubility (Moles/L)	Solubility (Grams/L)	Temp (°C)	Ref (#)	Evaluation (T P E A A)	Comments
2.700E-03	4.000E-01	18	F300	1 0 0 0 0	
2.835E-03	4.200E-01	18	M077	1 2 1 1 2	
3.010E-03	4.460E-01	25	C090	1 2 2 2 2	
3.685E-03	5.460E-01	25	M077	1 2 1 1 2	
5.264E-03	7.800E-01	35	M077	1 2 1 1 2	
7.364E-03	1.091E+00	45	M077	1 2 1 1 2	

1641. $C_9H_8O_2$

Atropic Acid
Atropasaeure

RN: 492-38-6   **MP** (°C):   106
MW: 148.16   **BP** (°C):

Solubility (Moles/L)	Solubility (Grams/L)	Temp (°C)	Ref (#)	Evaluation (T P E A A)	Comments
8.774E-03	1.300E+00	20	F300	1 0 0 0 1	

1642. $C_9H_8O_4$

Aspirin
Acetyl-salicylsaeure
Acetylsalicylic Acid

RN: 50-78-2   **MP** (°C):   135
MW: 180.16   **BP** (°C):

Solubility (Moles/L)	Solubility (Grams/L)	Temp (°C)	Ref (#)	Evaluation (T P E A A)	Comments
3.121E-02	5.623E+00	4.62	M053	1 0 1 1 0	EFG, 0.1N HCl
1.107E-02	1.995E+00	12.55	M053	1 0 1 1 0	EFG, 0.1N HCl
3.200E-02	5.765E+00	14	O019	1 0 0 1 2	
1.998E-02	3.600E+00	15	E017	1 0 0 0 0	EFG
1.388E-02	2.500E+00	15	F300	1 0 0 0 1	
1.716E-02	3.091E+00	15	H022	1 2 2 2 2	
2.109E-02	3.800E+00	20	E017	1 0 0 0 0	EFG

1.460E-02	2.630E+00	20.96	M053	1 0 1 1 0	EFG, 0.1N HCl
2.553E-02	4.600E+00	25	E017	1 0 0 0 0	EFG
2.442E-02	4.400E+00	25	S304	1 2 1 2 2	form II
2.775E-02	5.000E+00	25	S304	1 2 1 2 2	form IV
2.131E-02	3.840E+00	25	S304	1 2 1 2 2	form I
1.890E-02	3.405E+00	25.6	G015	1 0 1 1 2	pH 1.00, pka 3.62, intrinsic
2.500E-02	4.504E+00	30	A065	2 0 2 2 1	
2.831E-02	5.100E+00	30	E017	1 0 0 0 0	EFG
2.387E-02	4.300E+00	30	G042	1 1 1 1 1	0.1N HCl
2.851E-02	5.137E+00	30	H022	1 2 2 2 2	
2.000E-02	3.603E+00	30	L069	1 0 1 1 0	EFG
3.108E-02	5.600E+00	30	S304	1 2 1 2 2	form II
3.275E-02	5.900E+00	30	S304	1 2 1 2 2	form IV
2.637E-02	4.750E+00	30	S304	1 2 1 2 2	form I
3.275E-02	5.900E+00	35	E017	1 0 0 0 0	EFG
2.942E-02	5.300E+00	37	D009	1 2 1 1 1	0.1N HCl
3.219E-02	5.800E+00	37	G042	1 1 1 1 1	0.1N HCl
3.569E-02	6.430E+00	37	K086	1 0 0 0 2	
3.031E-02	5.460E+00	37	M115	2 2 1 1 2	
4.052E-02	7.300E+00	37	S304	1 2 1 2 2	form II
3.830E-02	6.900E+00	37	S304	1 2 1 2 2	form I
4.218E-02	7.600E+00	37	S304	1 2 1 2 2	form IV
3.830E-02	6.900E+00	40	E017	1 0 0 0 0	EFG
4.607E-02	8.300E+00	40	S304	1 2 1 2 2	form IV
4.218E-02	7.600E+00	40	S304	1 2 1 2 2	form I
4.385E-02	7.900E+00	40	S304	1 2 1 2 2	form II
4.662E-02	8.400E+00	45	E017	1 0 0 0 0	EFG
4.274E-02	7.700E+00	45	G042	1 1 1 1 1	0.1N HCl
5.551E-02	1.000E+01	49.42	M053	1 0 1 1 0	EFG, 0.1N HCl
4.940E-02	8.900E+00	50	G042	1 1 1 1 1	0.1N HCl
6.829E-02	1.230E+01	60.17	M053	1 0 1 1 0	EFG, 0.1N HCl

1643.   C_9H_9ClO_3

$C_9H_9ClO_3$

DL-2-(2-Chlorophenoxy)propionic Acid
2-(o-Chlorophenoxy)propionic Acid
3-CP

RN: 76466-16-5   **MP** (°C):   113
MW: 200.62   **BP** (°C):

Solubility (Moles/L)	Solubility (Grams/L)	Temp (°C)	Ref (#)	Evaluation (T P E A A)	Comments
5.974E-03	1.199E+00	22	B200	1 0 0 0 1	
9.726E-02	1.951E+01	100	B200	1 0 0 0 2	

1644.   C$_9$H$_9$ClO$_3$
DL-2-(4-Chlorophenoxy)propionic Acid
RN: 3307-39-9   **MP** (°C):
MW: 200.62   **BP** (°C):

Solubility (Moles/L)	Solubility (Grams/L)	Temp (°C)	Ref (#)	Evaluation (T P E A A)	Comments
7.352E-03	1.475E+00	25	B164	1 0 1 1 2	
7.352E-03	1.475E+00	25	B185	1 0 0 0 2	

1645.   C$_9$H$_9$ClO$_3$
(4-Chloro-2-methylphenoxy)acetic Acid
MCPA
RN: 94-74-6   **MP** (°C):   120.0
MW: 200.62   **BP** (°C):

Solubility (Moles/L)	Solubility (Grams/L)	Temp (°C)	Ref (#)	Evaluation (T P E A A)	Comments
3.138E-03	6.296E-01	20	M061	1 0 0 0 1	
5.852E-03	1.174E+00	25	B164	1 0 1 1 2	
5.852E-03	1.174E+00	25	B185	1 0 0 0 2	
7.975E-03	1.600E+00	25	B185	1 0 0 0 2	
4.979E-03	9.990E-01	ns	B100	0 0 0 0 0	
3.190E-03	6.400E-01	ns	B185	0 0 0 0 2	
4.112E-03	8.250E-01	ns	L024	0 0 0 0 2	
4.112E-03	8.250E-01	rt	M161	0 0 0 0 2	

1646.   C$_9$H$_9$Cl$_2$NO
Propanil
3',4'-Dichloropropionanilide
DPA
RN: 709-98-8   **MP** (°C):   85
MW: 218.08   **BP** (°C):

Solubility (Moles/L)	Solubility (Grams/L)	Temp (°C)	Ref (#)	Evaluation (T P E A A)	Comments
5.961E-04	1.300E-01	20	F311	1 2 2 2 1	
2.293E-03	5.000E-01	ns	B185	0 0 0 0 2	
2.292E-03	4.998E-01	ns	B200	0 0 0 0 0	
2.293E-03	5.000E-01	ns	H042	0 0 0 0 2	
1.032E-03	2.250E-01	rt	M161	0 0 0 0 2	

1647. C$_9$H$_9$Cl$_2$NO$_2$
Dichlormate
3,4-Dichlorobenzyl N-Methylcarbamate
Romate
RN: 1966-58-1   **MP** (°C):   52
MW: 234.08   **BP** (°C):

Solubility (Moles/L)	Solubility (Grams/L)	Temp (°C)	Ref (#)	Evaluation (T P E A A)	Comments
7.262E-04	1.700E-01	25	B200	1 0 0 0 2	

1648. C$_9$H$_9$Cl$_2$NO$_2$
UC 22463
Sirmate 4E
Rowmate
Sirmate
RN: 62046-37-1   **MP** (°C):   52
MW: 234.08   **BP** (°C):

Solubility (Moles/L)	Solubility (Grams/L)	Temp (°C)	Ref (#)	Evaluation (T P E A A)	Comments
7.262E-04	1.700E-01	ns	H042	0 0 0 0 2	

1649. C$_9$H$_9$I$_2$NO$_3$
L-3,5-Diiodotyrosine
3,5-Diiodo-L-tyrosine
DIT
RN: 300-39-0   **MP** (°C):   213
MW: 432.99   **BP** (°C):

Solubility (Moles/L)	Solubility (Grams/L)	Temp (°C)	Ref (#)	Evaluation (T P E A A)	Comments
1.431E-03	6.196E-01	25	D041	1 0 0 0 1	

1650. C$_9$H$_9$I$_2$NO$_3$
3,5-Diiodotyrosine
3,5-Diiod-DL-tyrosin
DL-Thyronin
RN: 66-02-4   **MP** (°C):   204
MW: 432.99   **BP** (°C):

Solubility (Moles/L)	Solubility (Grams/L)	Temp (°C)	Ref (#)	Evaluation (T P E A A)	Comments
1.039E-03	4.500E-01	15	F300	1 0 0 0 1	
7.850E-04	3.399E-01	25	D041	1 0 0 0 1	
1.386E-03	6.000E-01	25	F300	1 0 0 0 0	
1.316E-02	5.700E+00	75	F300	1 0 0 0 1	

1651. C₉H₉N

Skatole
3-Methyl-indol
3-Methylindole

RN: 83-34-1   **MP** (°C): 95
MW: 131.18   **BP** (°C): 265.5

Solubility (Moles/L)	Solubility (Grams/L)	Temp (°C)	Ref (#)	Evaluation (T P E A A)	Comments
3.430E-03	4.500E-01	16	F300	1 0 0 0 0	

1652. C₉H₉NOS

p-Ethoxyphenyl Isothiocyanate
4-Ethoxyphenyl Isothiocyanate

RN: 25687-50-7   **MP** (°C):
MW: 179.24   **BP** (°C):

Solubility (Moles/L)	Solubility (Grams/L)	Temp (°C)	Ref (#)	Evaluation (T P E A A)	Comments
5.500E-05	9.858E-03	25	D019	1 1 1 1 1	

1653. C₉H₉NOS

m-Ethoxyphenyl Isothiocyanate
3-Ethoxyphenyl Isothiocyanate

RN: 3701-44-8   **MP** (°C):
MW: 179.24   **BP** (°C):

Solubility (Moles/L)	Solubility (Grams/L)	Temp (°C)	Ref (#)	Evaluation (T P E A A)	Comments
3.800E-04	6.811E-02	25	K032	2 2 0 1 2	

1654. C₉H₉NO₂

p-Acetamidobenzaldehyde
Acetamide, N-(4-Formylphenyl)-
Acetanilide, 4'-Formyl-
Micotiazone

RN: 122-85-0   **MP** (°C):
MW: 163.18   **BP** (°C):

Solubility (Moles/L)	Solubility (Grams/L)	Temp (°C)	Ref (#)	Evaluation (T P E A A)	Comments
1.990E-02	3.247E+00	25	D044	1 1 1 1 2	

1655.   C₉H₉NO₃

Hippuric Acid
Hippursaeure
N-Benzoylglycine
Benzoylaminoacetic Acid
RN: 495-69-2   **MP** (°C):   187
MW: 179.18   **BP** (°C):

Solubility (Moles/L)	Solubility (Grams/L)	Temp (°C)	Ref (#)	Evaluation (T P E A A)	Comments
1.836E-02	3.289E+00	20	D041	1 0 0 0 1	
2.177E-02	3.900E+00	20	F300	1 0 0 0 1	
2.050E-02	3.673E+00	25	B028	1 0 0 0 2	
2.048E-02	3.670E+00	25	K053	2 2 2 2 2	
2.095E-02	3.754E+00	25	L048	1 2 2 1 2	
2.095E-02	3.754E+00	25	L050	2 0 1 2 2	
2.048E-02	3.670E+00	25.1	N026	2 0 2 2 2	
3.320E-02	5.949E+00	38	B028	1 0 0 0 2	
2.334E-02	4.182E+00	rt	D021	0 0 1 1 1	

1656.   C₉H₉NO₃

Acetamide, 2-(Benzoyloxy)-
Glycolamide, Benzoate
RN: 64649-43-0   **MP** (°C):   121
MW: 179.18   **BP** (°C):

Solubility (Moles/L)	Solubility (Grams/L)	Temp (°C)	Ref (#)	Evaluation (T P E A A)	Comments
2.288E-02	4.100E+00	22	N317	1 1 2 1 2	

1657.   C₉H₉NO₄

Benzadox
((Benzoylamino)oxy)acetic Acid
Topcide
RN: 5251-93-4   **MP** (°C):
MW: 195.18   **BP** (°C):

Solubility (Moles/L)	Solubility (Grams/L)	Temp (°C)	Ref (#)	Evaluation (T P E A A)	Comments
8.069E-02	1.575E+01	ns	B100	0 0 0 0 1	

1658.   C$_9$H$_9$NS
p-Methylbenzyl Isothiocyanate
4-Methylbenzyl Isothiocyanate
RN: 3694-46-0   **MP** (°C):
MW: 163.24   **BP** (°C):

Solubility (Moles/L)	Solubility (Grams/L)	Temp (°C)	Ref (#)	Evaluation (T P E A A)	Comments
1.600E-04	2.612E-02	25	D014	1 0 0 0 1	

1659.   C$_9$H$_9$N$_3$OS
Benzthiazuron
Benzothiazol-2-yl-3-methylurea
N-2-Benzothiazolyl-N'-methylurea
Gatnon
RN: 1929-88-0   **MP** (°C):
MW: 207.26   **BP** (°C):

Solubility (Moles/L)	Solubility (Grams/L)	Temp (°C)	Ref (#)	Evaluation (T P E A A)	Comments
5.790E-05	1.200E-02	20	M161	1 0 0 0 1	

1660.   C$_9$H$_9$N$_3$O$_2$
Carbendazim
1H-Benzimidazol-2-ylcarbamic Acid Methyl Ester
RN: 10605-21-7   **MP** (°C):   302
MW: 191.19   **BP** (°C):

Solubility (Moles/L)	Solubility (Grams/L)	Temp (°C)	Ref (#)	Evaluation (T P E A A)	Comments
3.034E-05	5.800E-03	20	A064	1 0 1 1 1	
3.034E-05	5.800E-03	20	M161	1 0 0 0 1	pH 7

1661.   C$_9$H$_9$N$_3$O$_2$S$_2$
Sulfathiazole
Sulphathiazole
N1-2-Thiazolyl-
4-Amino-N-2-thiazolyl-
RN: 72-14-0   **MP** (°C):   202
MW: 255.32   **BP** (°C):

Solubility (Moles/L)	Solubility (Grams/L)	Temp (°C)	Ref (#)	Evaluation (T P E A A)	Comments
1.410E-03	3.600E-01	16	H114	1 0 0 0 1	
1.743E-03	4.450E-01	20	F073	1 2 2 2 2	
1.958E-03	5.000E-01	20	F074	1 0 0 0 2	
2.483E-03	6.340E-01	20	K028	2 1 2 1 2	pH 3.8, form I
4.426E-03	1.130E+00	20	K028	2 1 2 1 2	pH 7.3, form I
1.414E-03	3.610E-01	20	K028	2 1 2 1 2	pH 3.8, form II
2.460E-03	6.280E-01	20	K028	2 1 2 1 2	pH 7.3, form II

1.347E-03	3.439E-01	20	L058	1 0 1 1 1	
2.482E-03	6.336E-01	20	M042	1 0 0 0 2	pH 3.8, form I, mp 200-202 °C
1.413E-03	3.609E-01	20	M042	1 0 0 0 2	pH 3.8, form II, mp 175 °C
1.461E-03	3.730E-01	25	H005	1 0 1 2 2	average of 4
1.821E-03	4.650E-01	25	K096	1 2 2 2 2	α form
3.290E-03	8.400E-01	25	K096	1 2 2 2 2	β form
1.966E-03	5.020E-01	26	C102	2 0 2 2 2	
2.350E-03	6.000E-01	26	L052	1 0 0 0 0	
2.270E-03	5.796E-01	30	H018	1 2 2 2 2	
2.327E-03	5.940E-01	30	K096	1 2 2 2 2	α form
4.308E-03	1.100E+00	30	K096	1 2 2 2 2	β form
2.544E-03	6.496E-01	30	M046	1 0 0 0 1	
4.460E-03	1.139E+00	30.0	H010	2 2 1 1 2	
3.564E-03	9.100E-01	35	H114	1 0 0 0 1	
3.094E-03	7.900E-01	35	K096	1 2 2 2 2	α form
5.354E-03	1.367E+00	35	K096	1 2 2 2 2	β form
3.760E-03	9.600E-01	37	C102	2 0 2 2 2	
3.564E-03	9.100E-01	37	D084	1 0 1 0 1	
3.678E-03	9.391E-01	37	F072	1 0 0 0 2	
3.686E-03	9.411E-01	37	F075	1 0 2 2 2	
3.443E-03	8.790E-01	37	K091	1 0 0 0 2	
2.560E-03	6.536E-01	37	K095	2 0 0 0 2	intrinsic
3.838E-03	9.800E-01	37	L091	1 0 0 0 1	pH 5.5
3.721E-03	9.500E-01	37	M057	1 0 0 0 2	pH 5.5
3.799E-03	9.700E-01	37	R044	1 0 1 1 0	
3.756E-03	9.591E-01	37.50	M142	1 0 0 0 1	
3.603E-03	9.200E-01	38	K006	1 0 0 0 2	
6.619E-03	1.690E+00	40	K096	1 2 2 2 2	β form
4.073E-03	1.040E+00	40	K096	1 2 2 2 2	α form
8.284E-03	2.115E+00	45	K096	1 2 2 2 2	β form
5.288E-03	1.350E+00	45	K096	1 2 2 2 2	α form
9.964E-03	2.544E+00	49	K096	1 2 2 2 2	β form
6.592E-03	1.683E+00	49	K096	1 2 2 2 2	α form
1.683E-03	4.298E-01	ns	L044	0 0 0 0 2	
1.918E-03	4.898E-01	rt	N015	0 0 2 2 2	

1662. C$_9$H$_{10}$
Indan
2,3-Dihydroindene
Hydrindane
1H-Indene, 2,3-Dihydro-
Hydrindene
RN: 496-11-7   **MP** (°C):   -51.4
MW: 118.18   **BP** (°C):   176.5

Solubility (Moles/L)	Solubility (Grams/L)	Temp (°C)	Ref (#)	Evaluation (T P E A A)	Comments
9.232E-04	1.091E-01	25	M064	1 1 2 2 2	
7.522E-04	8.890E-02	25	P051	2 1 1 2 2	
9.232E-04	1.091E-01	ns	M344	0 0 0 0 2	

1663. C$_9$H$_{10}$
α-Methylstyrene
2-Phenyl-1-propene
Isopropenylbenzene
2-Phenylpropene
β-Phenylpropene
RN: 98-83-9   **MP** (°C):   -24.0
MW: 118.18   **BP** (°C):   167.0

Solubility (Moles/L)	Solubility (Grams/L)	Temp (°C)	Ref (#)	Evaluation (T P E A A)	Comments
9.772E-04	1.155E-01	ns	D001	0 0 0 0 2	

1664. C$_9$H$_{10}$BrClN$_2$O$_2$
Chlorbromuron
3-(4-Bromo-3-chlorophenyl)-1-methoxy-1-methylurea
N'-(4-Bromo-3-chlorophenyl)-N-methoxy-N-methylurea
Maloran
RN: 13360-45-7   **MP** (°C):
MW: 293.55   **BP** (°C):

Solubility (Moles/L)	Solubility (Grams/L)	Temp (°C)	Ref (#)	Evaluation (T P E A A)	Comments
1.202E-04	3.529E-02	20	B179	2 0 0 0 2	
1.192E-04	3.500E-02	20	M161	1 0 0 0 1	
1.703E-04	5.000E-02	ns	B200	0 0 0 0 1	
1.703E-04	5.000E-02	ns	G036	0 0 0 0 1	

1665. $C_9H_{10}Cl_2N_2O$
Diuron
1,1-Dimethyl-3-(3,4-dichlorophenyl)urea
3-(3,4-Dichlorophenyl)-1,1-dimethylurea
RN: 330-54-1   **MP** (°C):   158
MW: 233.10   **BP** (°C):

Solubility (Moles/L)	Solubility (Grams/L)	Temp (°C)	Ref (#)	Evaluation (T P E A A)	Comments
1.820E-04	4.242E-02	20	B179	2 0 0 0 2	
9.438E-05	2.200E-02	20	E048	1 2 1 1 1	
1.716E-04	4.000E-02	25	A039	1 1 0 0 2	
1.802E-04	4.200E-02	25	B185	1 0 0 0 1	
1.802E-04	4.200E-02	25	B200	1 0 0 0 1	
1.802E-04	4.200E-02	25	G036	1 0 0 0 1	
1.802E-04	4.200E-02	25	G099	1 0 0 1 0	
1.600E-04	3.730E-02	25	H073	2 1 1 2 2	
1.802E-04	4.200E-02	25	M061	1 0 0 0 1	
1.802E-04	4.200E-02	25	M161	1 0 0 0 1	
1.802E-04	4.200E-02	25	N333	1 0 0 0 1	
1.716E-04	4.000E-02	ns	B160	0 0 0 0 1	
1.802E-04	4.200E-02	ns	H042	0 0 0 0 1	
1.000E+02	2.331E+04	ns	H342	0 0 2 2 0	EFG, *sic*
1.802E-04	4.200E-02	ns	K007	0 0 0 0 1	
1.995E-04	4.651E-02	ns	M163	0 0 0 0 0	EFG

1666. $C_9H_{10}Cl_2N_2O_2$
Linuron
3-(3,4-Dichlorophenyl)-1-methoxy-1-methylurea
RN: 330-55-2   **MP** (°C):   93
MW: 249.10   **BP** (°C):

Solubility (Moles/L)	Solubility (Grams/L)	Temp (°C)	Ref (#)	Evaluation (T P E A A)	Comments
3.020E-04	7.523E-02	20	B179	2 0 0 0 2	
3.011E-04	7.500E-02	25	B185	1 0 0 0 1	
3.011E-04	7.500E-02	25	B200	1 0 0 0 1	
3.011E-04	7.500E-02	25	M061	1 0 0 0 1	
3.011E-04	7.500E-02	25	M161	1 0 0 0 1	
3.252E-04	8.100E-02	25	M162	1 1 0 0 1	
3.011E-04	7.500E-02	ns	K007	0 0 0 0 1	

1667. C₉H₁₀Cl₂O
2,4-Dichloro-6-propyl-phenol
RN: 91399-12-1   **MP** (°C):
MW: 205.09   **BP** (°C):

Solubility (Moles/L)	Solubility (Grams/L)	Temp (°C)	Ref (#)	Evaluation (T P E A A)	Comments
4.900E-04	1.005E-01	25	B316	1 0 2 1 1	

1668. C₉H₁₀Cl₃O₃PS
Trichlormetafos-3
O-Methyl O-Ethyl O-2,4,5-Trichlorophenyl Thiophosphate
RN: 2633-54-7   **MP** (°C):
MW: 335.58   **BP** (°C):   127

Solubility (Moles/L)	Solubility (Grams/L)	Temp (°C)	Ref (#)	Evaluation (T P E A A)	Comments
<1.19E-04	<4.00E-02	ns	M061	0 0 0 0 0	

1669. C₉H₁₀NO₃
2-Oxo-5-Indolinyl Acetate
5-Acetoxy-2-oxindole
RN: 74973-14-1   **MP** (°C):
MW: 180.18   **BP** (°C):

Solubility (Moles/L)	Solubility (Grams/L)	Temp (°C)	Ref (#)	Evaluation (T P E A A)	Comments
2.900E-02	5.225E+00	25	A066	1 0 1 1 1	

1670. C₉H₁₀NO₃PS
Cyanophos
Dimethyl O-(p-Cyanophenyl) Phosphorothioate
Ciafos
CYAP
RN: 2636-26-2   **MP** (°C):   14.5
MW: 243.22   **BP** (°C):

Solubility (Moles/L)	Solubility (Grams/L)	Temp (°C)	Ref (#)	Evaluation (T P E A A)	Comments
1.891E-04	4.600E-02	30	M161	1 0 0 0 1	

1671. C₉H₁₀N₂O₃
p-Ureidophenyl Acetate
4-Ureidophenyl Acetate
RN: 59746-11-1   **MP** (°C):
MW: 194.19   **BP** (°C):

Solubility (Moles/L)	Solubility (Grams/L)	Temp (°C)	Ref (#)	Evaluation (T P E A A)	Comments
3.200E-03	6.214E-01	25	A066	1 0 1 1 1	

1672. C$_9$H$_{10}$N$_2$O$_3$
o-Nitroacetotoluide
2-Nitroacetotoluide
RN: 612-45-3   **MP** (°C):
MW: 194.19   **BP** (°C):

Solubility (Moles/L)	Solubility (Grams/L)	Temp (°C)	Ref (#)	Evaluation (T P E A A)	Comments
1.133E-02	2.200E+00	rt	F043	0 0 2 1 1	

1673. C$_9$H$_{10}$N$_2$O$_3$
p-Nitroacetotoluide
4-Nitroacetotoluide
RN: **MP** (°C):
MW: 194.19   **BP** (°C):

Solubility (Moles/L)	Solubility (Grams/L)	Temp (°C)	Ref (#)	Evaluation (T P E A A)	Comments
1.133E-02	2.200E+00	rt	F043	0 0 2 1 1	

1674. C$_9$H$_{10}$N$_2$O$_3$S$_2$
Ethoxzolamide
6-Ethoxy-2-benzothiazolesulfonamide
Diuretic C
Cardrase
RN: 452-35-7   **MP** (°C):   188
MW: 258.32   **BP** (°C):

Solubility (Moles/L)	Solubility (Grams/L)	Temp (°C)	Ref (#)	Evaluation (T P E A A)	Comments
1.548E-04	4.000E-02	ns	M032	0 0 0 0 0	

1675. C$_9$H$_{10}$N$_2$S
4-Dimethylaminophenyl Isothiocyanate
4-Isothiocyanato-N,N-dimethyl-benzenamine
RN: 2131-64-8   **MP** (°C):
MW: 178.26   **BP** (°C):

Solubility (Moles/L)	Solubility (Grams/L)	Temp (°C)	Ref (#)	Evaluation (T P E A A)	Comments
7.500E-05	1.337E-02	25	D019	1 1 1 1 1	

1676. C₉H₁₀N₂S

1676. $C_9H_{10}N_2S$

3-Dimethylaminophenyl Isothiocyanate
N',N'-Dimethyl-m-aminophenyl Isothiocyanate
RN: 2392-67-8   **MP** (°C):
MW: 178.26   **BP** (°C):

Solubility (Moles/L)	Solubility (Grams/L)	Temp (°C)	Ref (#)	Evaluation (T P E A A)	Comments
4.200E-04	7.487E-02	25	D019	1 1 1 1 2	
1.950E-04	3.476E-02	25	K032	2 2 0 1 2	

1677. $C_9H_{10}N_4$

2,6,7-Trimethylpteridine
2:6:7-Trimethylpteridine
RN: 23767-00-2   **MP** (°C):
MW: 174.21   **BP** (°C):

Solubility (Moles/L)	Solubility (Grams/L)	Temp (°C)	Ref (#)	Evaluation (T P E A A)	Comments
7.087E-02	1.235E+01	20	A083	1 2 0 0 0	

1678. $C_9H_{10}N_4O_2S_2$

Sulfamethizole
Sulfamethylthiadiazole
RN: 144-82-1   **MP** (°C):   208
MW: 270.33   **BP** (°C):

Solubility (Moles/L)	Solubility (Grams/L)	Temp (°C)	Ref (#)	Evaluation (T P E A A)	Comments
1.957E-03	5.290E-01	20	F073	1 2 2 2 2	
3.320E-03	8.975E-01	37	A046	2 0 1 1 2	
3.884E-03	1.050E+00	37	B046	1 0 2 2 2	pH 4.5
3.270E-03	8.840E-01	37	K091	1 0 0 0 2	
3.270E-03	8.840E-01	37	W016	2 0 1 1 2	
2.938E-03	7.943E-01	ns	N057	1 0 2 2 0	EFG, intrinsic

1679. $C_9H_{10}O_2$

2,5-Dimethylbenzoic Acid
2-Carboxy-1,4-dimethylbenzene
Isoxylic Acid
RN: 610-72-0   **MP** (°C):   132.5-134.5
MW: 150.18   **BP** (°C):   268

Solubility (Moles/L)	Solubility (Grams/L)	Temp (°C)	Ref (#)	Evaluation (T P E A A)	Comments
1.199E-03	1.800E-01	25	H007	1 0 2 2 1	

1680.   C$_9$H$_{10}$O$_2$
Hydrocinnamic Acid
Hydrozimtsaeure
RN: 501-52-0   **MP** (°C):   48
MW: 150.18   **BP** (°C):   280

Solubility (Moles/L)	Solubility (Grams/L)	Temp (°C)	Ref (#)	Evaluation (T P E A A)	Comments
3.929E-02	5.900E+00	20	F300	1 0 0 0 2	
6.162E-02	9.254E+00	30	D033	2 2 1 2 2	
7.668E-02	1.152E+01	40	D033	2 2 1 2 2	

1681.   C$_9$H$_{10}$O$_2$
3,4-Dimethylbenzoic Acid
1-Carboxy-3,4-dimethylbenzene
RN: 619-04-5   **MP** (°C):   165
MW: 150.18   **BP** (°C):

Solubility (Moles/L)	Solubility (Grams/L)	Temp (°C)	Ref (#)	Evaluation (T P E A A)	Comments
8.600E-04	1.292E-01	ns	C014	0 0 0 1 1	

1682.   C$_9$H$_{10}$O$_2$
2,4-Dimethylbenzoic Acid
4-Carboxy-1,3-dimethylbenzene
RN: 611-01-8   **MP** (°C):   124-126
MW: 150.18   **BP** (°C):   267

Solubility (Moles/L)	Solubility (Grams/L)	Temp (°C)	Ref (#)	Evaluation (T P E A A)	Comments
1.065E-03	1.600E-01	25	H007	1 0 2 2 1	

1683.   C$_9$H$_{10}$O$_2$
Ethyl Benzoate
Ethyl p-Benzoate
Benzoesaeure-aethyl Ester
RN: 93-89-0   **MP** (°C):   -34
MW: 150.18   **BP** (°C):   212

Solubility (Moles/L)	Solubility (Grams/L)	Temp (°C)	Ref (#)	Evaluation (T P E A A)	Comments
7.990E-03	1.200E+00	22	N317	1 1 2 1 2	
4.794E-03	7.200E-01	25	A003	1 2 1 2 1	
6.659E-03	1.000E+00	60	F300	1 0 0 0 0	

1684. $C_9H_{10}O_2$
Benzyl Acetate
Phenylmethyl Acetate
Acetic Acid Phenylmethyl Ester
α-Acetoxytoluene

RN: 140-11-4   **MP** (°C):   -51.3
MW: 150.18   **BP** (°C):   206

Solubility (Moles/L)	Solubility (Grams/L)	Temp (°C)	Ref (#)	Evaluation (T P E A A)	Comments
9.973E-03	1.498E+00	25	M350	1 0 1 1 1	

1685. $C_9H_{10}O_3$
DL-Tropic Acid
DL-Tropasaeure

RN: 529-64-6   **MP** (°C):   118.5
MW: 166.18   **BP** (°C):

Solubility (Moles/L)	Solubility (Grams/L)	Temp (°C)	Ref (#)	Evaluation (T P E A A)	Comments
1.173E-01	1.950E+01	20	F300	1 0 0 0 2	

1686. $C_9H_{10}O_3$
Methyl-4-methoxybenzoate
Methyl Anisate

RN: 121-98-2   **MP** (°C):   49
MW: 166.18   **BP** (°C):   244

Solubility (Moles/L)	Solubility (Grams/L)	Temp (°C)	Ref (#)	Evaluation (T P E A A)	Comments
3.870E-03	6.431E-01	20	C006	1 0 1 1 2	

1687. $C_9H_{10}O_3$
Ethylparaben
4-Hydroxybenzoic Acid Ethyl Ester
Ethyl p-Hydroxybenzoate
Ethyl 4-hydroxybenzoate

RN: 120-47-8   **MP** (°C):   116
MW: 166.18   **BP** (°C):   297

Solubility (Moles/L)	Solubility (Grams/L)	Temp (°C)	Ref (#)	Evaluation (T P E A A)	Comments
2.750E-03	4.570E-01	15	B355	1 1 1 1 2	
3.370E-03	5.600E-01	20	B355	1 1 1 1 2	
4.910E-03	8.159E-01	20	C006	1 2 1 1 2	
5.329E-03	8.855E-01	25	A059	1 0 1 1 1	
4.090E-03	6.797E-01	25	B355	1 1 1 1 2	
4.510E-03	7.494E-01	25	D081	1 2 2 1 2	
5.300E-03	8.807E-01	25	D339	1 0 1 1 2	

6.310E-03	1.049E+00	25	F322	2 0 1 1 0	EFG
9.628E-03	1.600E+00	25	O027	1 0 1 0 1	
6.379E-03	1.060E+00	25	P013	2 0 2 1 2	
9.500E-03	1.579E+00	27	B129	2 2 2 2 1	
5.200E-03	8.641E-01	27	G078	2 1 0 1 0	EFG
5.400E-03	8.974E-01	27.0	G067	2 0 1 1 1	
6.770E-03	1.125E+00	30	A059	1 0 1 1 2	
8.266E-03	1.374E+00	35	A059	1 0 1 1 2	
7.568E-03	1.258E+00	39.3	G302	2 2 2 2 0	EFG
9.540E-03	1.585E+00	40	A059	1 0 1 1 2	

1688. $C_9H_{10}O_3$

Ethyl Salicylate
Ethyl o-Hydroxybenzoate
RN: 118-61-6   **MP** (°C):   1-3
MW: 166.18   **BP** (°C):

Solubility (Moles/L)	Solubility (Grams/L)	Temp (°C)	Ref (#)	Evaluation (T P E A A)	Comments
4.032E-02	6.700E+00	37	D009	1 2 1 1 1	0.1N HCl

1689. $C_9H_{10}O_4$

3,4-Methoxybenzoic Acid
Veratrumsaeure
RN: 93-07-2   **MP** (°C):
MW: 182.18   **BP** (°C):

Solubility (Moles/L)	Solubility (Grams/L)	Temp (°C)	Ref (#)	Evaluation (T P E A A)	Comments
2.745E-03	5.000E-01	14	F300	1 0 0 0 0	
3.293E-02	6.000E+00	100	F300	1 0 0 0 0	

1690.   C$_9$H$_{11}$BrN$_2$O$_2$
Metobromuron
3-(p-Bromophenyl)-1-methoxy-1-methylurea
Patoran
N'-(4-Bromophenyl)-N-methoxy-N-methylurea
Pattonex

RN: 3060-89-7   **MP** (°C):
MW: 259.11   **BP** (°C):

Solubility (Moles/L)	Solubility (Grams/L)	Temp (°C)	Ref (#)	Evaluation (T P E A A)	Comments
1.288E-03	3.338E-01	20	B179	2 0 0 0 2	
1.274E-03	3.300E-01	20	B200	1 0 0 0 2	
1.274E-03	3.300E-01	20	G036	1 0 0 0 2	
1.274E-03	3.300E-01	20	M061	1 0 0 0 1	
1.274E-03	3.300E-01	20	M161	1 0 0 0 2	
1.157E-03	2.999E-01	ns	B100	0 0 0 0 0	

1691.   C$_9$H$_{11}$ClN$_2$O
Monuron
N'-(4-Chlorophenyl)-N,N-dimethyl-urea
1,1-Dimethyl-3-(p-chlorophenyl)urea

RN: 150-68-5   **MP** (°C):   170.5
MW: 198.65   **BP** (°C):

Solubility (Moles/L)	Solubility (Grams/L)	Temp (°C)	Ref (#)	Evaluation (T P E A A)	Comments
1.007E-03	2.000E-01	18	F035	1 0 0 0 0	
1.175E-03	2.334E-01	20	B179	2 0 0 0 2	
1.007E-03	2.000E-01	20	E048	1 2 1 1 2	
1.007E-03	2.000E-01	20	F311	1 2 2 2 1	
1.158E-03	2.300E-01	25	A039	1 1 0 0 2	
1.158E-03	2.300E-01	25	B185	1 0 0 0 2	
1.158E-03	2.300E-01	25	B200	1 0 0 0 2	
1.158E-03	2.300E-01	25	G036	1 0 0 0 2	
1.158E-03	2.300E-01	25	G099	1 0 0 1 0	
1.319E-03	2.620E-01	25	H073	2 1 1 2 2	
1.158E-03	2.300E-01	25	M061	1 0 0 0 2	
1.158E-03	2.300E-01	25	M161	1 0 0 0 2	
1.007E-03	2.000E-01	ns	B100	0 0 0 0 0	
1.158E-03	2.300E-01	ns	B160	0 0 0 0 2	
9.000E-04	1.788E-01	ns	F184	0 0 0 0 0	
1.158E-03	2.300E-01	ns	H112	0 0 0 0 2	
1.158E-03	2.300E-01	ns	K007	0 0 0 0 2	
1.158E-03	2.300E-01	ns	N013	0 0 0 0 2	

1692.   C$_9$H$_{11}$ClN$_2$O$_2$
Monolinuron
3-(4-Chlorophenyl)-1-methoxy-1-methylurea
Arresin
Afesin
Aresin
RN: 1746-81-2   **MP** (°C):   80
MW: 214.65   **BP** (°C):

Solubility (Moles/L)	Solubility (Grams/L)	Temp (°C)	Ref (#)	Evaluation (T P E A A)	Comments
2.692E-03	5.777E-01	20	B179	2 0 0 0 2	
4.333E-03	9.300E-01	20	G036	1 0 0 0 2	
2.702E-03	5.800E-01	20	M061	1 0 0 0 2	
2.702E-03	5.800E-01	22.5	G301	2 1 0 1 2	
3.424E-03	7.350E-01	25	M162	1 1 0 0 2	
2.794E-03	5.996E-01	ns	B100	0 0 0 0 0	
2.702E-03	5.800E-01	rt	M161	0 0 0 0 2	

1693.   C$_9$H$_{11}$ClO
3-Methyl-5-ethyl-4-chloro-phenol
m-Cresol, 4-Chloro-5-ethyl-
RN: 1125-66-2   **MP** (°C):
MW: 170.64   **BP** (°C):

Solubility (Moles/L)	Solubility (Grams/L)	Temp (°C)	Ref (#)	Evaluation (T P E A A)	Comments
2.200E-03	3.754E-01	25	B316	1 0 2 1 1	

1694.   C$_9$H$_{11}$Cl$_2$N$_3$O$_4$S$_2$
Methylchlothiazide
2H-1,2,4-Benzothiadiazine -7-sulfonamide, 6-Chloro-3-(chloromethyl)-3,4-dihydro-2-methyl-1,1-dioxide
6-Chloro-3-(chloromethyl)-3,4-dihydro-2-methyl-2H-1,2,4-benzothiadiazine -7-sulfonamide 1,1-dioxide
RN: 135-07-9   **MP** (°C):
MW: 360.24   **BP** (°C):

Solubility (Moles/L)	Solubility (Grams/L)	Temp (°C)	Ref (#)	Evaluation (T P E A A)	Comments
1.388E-04	5.000E-02	rt	A095	0 0 2 2 0	

1695.   C$_9$H$_{11}$Cl$_3$NO$_3$PS
Chlorpyrifos
O,O-Diethyl O-3,5,6-Trichloro-2-pyridyl Phosphorothioate
DOWCO 179

RN: 2921-88-2   **MP** (°C):   41.5
MW: 350.59   **BP** (°C):

Solubility (Moles/L)	Solubility (Grams/L)	Temp (°C)	Ref (#)	Evaluation (T P E A A)	Comments
1.284E-06	4.500E-04	10	B324	2 2 2 2 2	
1.284E-06	4.502E-04	10	B324	2 2 2 2 2	
1.997E-06	7.000E-04	19	B169	2 1 1 1 1	
2.082E-06	7.299E-04	20	B300	2 1 1 1 2	
2.082E-06	7.299E-04	20	B324	2 2 2 2 2	
2.082E-06	7.300E-04	20	B324	2 2 2 2 2	
1.141E-06	4.000E-04	23	B096	1 2 0 0 0	
3.195E-06	1.120E-03	24	F179	2 2 2 2 2	
1.141E-06	4.000E-04	24	K069	2 0 0 1 1	
3.708E-06	1.300E-03	30	B324	2 2 2 2 2	
3.708E-06	1.300E-03	30	B324	2 2 2 2 2	
5.705E-06	2.000E-03	35	M161	1 0 0 0 0	
1.141E-06	4.000E-04	ns	F071	0 1 2 1 0	
8.557E-07	3.000E-04	ns	K138	0 0 0 0 1	
5.705E-06	2.000E-03	ns	M110	0 0 0 0 0	EFG

1696.   C$_9$H$_{11}$Cl$_3$NO$_4$P
Chlorpyrifos Oxon
Chlorpyrifos Oxygen Analog
Dursban Oxygen Analog
DOWCO 180
3,5,6-Trichloro-2-Pyridyl Diethyl Phosphate

RN: 5598-15-2   **MP** (°C):
MW: 334.53   **BP** (°C):

Solubility (Moles/L)	Solubility (Grams/L)	Temp (°C)	Ref (#)	Evaluation (T P E A A)	Comments
1.554E-03	5.200E-01	24	K069	2 0 0 1 1	

1697.   C$_9$H$_{11}$FN$_2$O$_4$
1-Butyloxycarbonyl-5-fluorouracil
5-Fluoro-1-(butoxycarbonyl)uracil

RN: 85326-32-5   **MP** (°C):
MW: 230.20   **BP** (°C):

Solubility (Moles/L)	Solubility (Grams/L)	Temp (°C)	Ref (#)	Evaluation (T P E A A)	Comments
2.563E-02	5.900E+00	22	B332	1 1 0 0 1	pH 4.0

1698.   C₉H₁₁FN₂O₄

1-Isobutyloxycarbonyl-5-fluorouracil

1(2H)-Pyrimidinecarboxylic Acid, 5-Fluoro-3,4-dihydro-2,4-dioxo-, 2-Methylpropyl Ester

RN: 71759-45-0   **MP** (°C):

MW: 230.20   **BP** (°C):

Solubility (Moles/L)	Solubility (Grams/L)	Temp (°C)	Ref (#)	Evaluation (T P E A A)	Comments
1.303E-02	3.000E+00	22	B332	1 1 0 0 1	pH 4.0

1699.   C₉H₁₁FN₂O₄

1-Butyryloxymethyl-5-fluorouracil

Butanoic Acid, (5-Fluoro-3,4-dihydro-2,4-dioxo-1(2H)-pyrimidinyl)methyl Ester

RN: 66542-37-8   **MP** (°C):

MW: 230.20   **BP** (°C):

Solubility (Moles/L)	Solubility (Grams/L)	Temp (°C)	Ref (#)	Evaluation (T P E A A)	Comments
4.170E-02	9.600E+00	22	B321	1 0 2 2 2	pH 4.0
4.170E-02	9.600E+00	22	B332	1 1 0 0 1	pH 4.0
4.952E-02	1.140E+01	22	M317	1 1 1 1 1	

1700.   C₉H₁₁IN₂O₅

2'-Deoxy-5-iodouridine

Idoxuridine

(+)-5-Iodo-2'-deoxyuridine

Herplex

RN: 54-42-2   **MP** (°C):   165

MW: 354.10   **BP** (°C):

Solubility (Moles/L)	Solubility (Grams/L)	Temp (°C)	Ref (#)	Evaluation (T P E A A)	Comments
5.650E+03	2.001E+06	25	N332	1 0 2 2 2	pH 7.4

1701.   C₉H₁₁N

1,2,3,4-Tetrahydroquinoline

Kusol

THQ

RN: 635-46-1   **MP** (°C):   15-17

MW: 133.19   **BP** (°C):   249

Solubility (Moles/L)	Solubility (Grams/L)	Temp (°C)	Ref (#)	Evaluation (T P E A A)	Comments
1.054E-02	1.404E+00	20.3	L339	2 0 2 2 2	
1.386E-02	1.847E+00	40.0	L339	2 0 2 2 2	
1.774E-02	2.362E+00	59.8	L339	2 0 2 2 2	
2.326E-02	3.098E+00	79.6	L339	2 0 2 2 2	
2.988E-02	3.980E+00	100.4	L339	2 0 2 2 2	

1702. C$_9$H$_{11}$NO
N-Methylacetanilide
Acetamide, N-Methyl-N-phenyl-
RN: 579-10-2 **MP** (°C): 102
MW: 149.19 **BP** (°C):

Solubility (Moles/L)	Solubility (Grams/L)	Temp (°C)	Ref (#)	Evaluation (T P E A A)	Comments
1.475E-01	2.200E+01	20	B101	1 0 2 2 2	
1.673E-01	2.496E+01	25	B101	1 0 2 2 2	
1.908E-01	2.847E+01	30	B101	1 0 2 2 2	
2.166E-01	3.232E+01	35	B101	1 0 2 2 2	
2.166E-01	3.232E+01	35	B101	1 0 2 2 2	

1703. C$_9$H$_{11}$NO
Methyl, [3-(Acetylamino)phenyl]-
m-Toluidin-N-acetat
m-Toluidine-N-acetate
RN: 113321-22-5 **MP** (°C):
MW: 149.19 **BP** (°C):

Solubility (Moles/L)	Solubility (Grams/L)	Temp (°C)	Ref (#)	Evaluation (T P E A A)	Comments
2.949E-02	4.400E+00	13	F300	1 0 0 0 1	

1704. C$_9$H$_{11}$NO
p-Aminopropiophenone
4'-Aminopropiophenone
RN: 70-69-9 **MP** (°C): 140
MW: 149.19 **BP** (°C):

Solubility (Moles/L)	Solubility (Grams/L)	Temp (°C)	Ref (#)	Evaluation (T P E A A)	Comments
2.360E-03	3.521E-01	37.5	G002	1 1 1 1 2	pH 6.8

1705. C$_9$H$_{11}$NO
Propionanilide
Propionsaeure-anilid
Propanilide
RN: 620-71-3 **MP** (°C): 106
MW: 149.19 **BP** (°C):

Solubility (Moles/L)	Solubility (Grams/L)	Temp (°C)	Ref (#)	Evaluation (T P E A A)	Comments
1.206E-02	1.800E+00	18	F300	1 0 0 0 1	
1.204E-02	1.797E+00	20	B101	1 0 2 2 2	

1706.   C₉H₁₁NO₂

Phe
(S)-(-)-Phenylalanine
(S)-Phenylalanine
2-Amino-3-phenylpropanoic Acid
Phenylalanine

RN: 63-91-2   **MP** (°C):   283
MW: 165.19   **BP** (°C):   295

Solubility (Moles/L)	Solubility (Grams/L)	Temp (°C)	Ref (#)	Evaluation (T P E A A)	Comments
6.047E-02	9.989E+00	0	D018	2 2 2 1 2	
1.174E-01	1.940E+01	0	F300	1 0 0 0 2	
1.740E-01	2.874E+01	15	D349	2 1 1 2 2	
1.515E-01	2.502E+01	20	B032	1 2 2 1 2	
1.770E-01	2.924E+01	20	D349	2 1 1 2 2	
1.637E-01	2.705E+01	25	B032	1 2 2 1 2	
1.740E-01	2.875E+01	25	D041	1 0 0 0 2	
1.800E-01	2.973E+01	25	D349	2 1 1 2 2	
1.816E-01	3.000E+01	25	F300	1 0 0 0 1	
1.649E-01	2.724E+01	25	G092	2 1 1 1 1	
1.649E-01	2.724E+01	25	G315	1 0 2 2 2	
1.589E-01	2.625E+01	25	K031	2 1 2 1 2	
1.200E-01	1.982E+01	25	M097	2 2 2 2 2	
1.494E-01	2.468E+01	25	M374	1 0 2 1 2	
2.100E-01	3.469E+01	25	N001	2 0 2 1 0	EFG
1.720E-01	2.841E+01	25	N012	2 0 2 1 2	
1.575E-01	2.601E+01	25	O316	1 0 1 2 2	
1.574E-01	2.601E+01	25	O316	1 0 1 2 2	
1.689E-01	2.790E+01	25.1	N024	2 0 2 2 2	
1.689E-01	2.790E+01	25.1	N025	2 0 2 2 2	
1.689E-01	2.790E+01	25.1	N026	2 0 2 2 2	
1.649E-01	2.724E+01	25.1	N027	1 1 2 2 2	
1.717E-01	2.837E+01	27	D036	2 1 2 2 2	
1.683E-01	2.780E+01	27	D036	2 1 2 2 2	
1.834E-01	3.030E+01	28	L081	2 1 2 2 2	
1.790E-01	2.957E+01	29.80	B032	1 2 2 1 2	
2.567E-01	4.240E+01	50	F300	1 0 0 0 2	
3.761E-01	6.212E+01	75	D041	1 0 0 0 2	
3.759E-01	6.210E+01	75	F300	1 0 0 0 2	
4.619E-01	7.630E+01	98	M160	2 1 1 1 0	
5.454E-01	9.010E+01	100	F300	1 0 0 0 2	

1707. $C_9H_{11}NO_2$

p-Methoxyacetanilide
p-Acetanisidine
N-(4-Methoxyphenyl)acetamide
N-(4-Methoxyphenyl)acetic Acid Amide
p-Acetanisidide
Acetamide, N-(4-methoxyphenyl)-

RN: 51-66-1   **MP** (°C): 400.3
MW: 165.19   **BP** (°C):

Solubility (Moles/L)	Solubility (Grams/L)	Temp (°C)	Ref (#)	Evaluation (T P E A A)	Comments
1.029E-02	1.700E+00	15	F300	1 0 0 0 1	
8.820E-03	1.457E+00	15	M352	1 1 1 1 2	
7.090E-02	1.171E+01	25	D044	1 1 1 1 2	
1.353E-02	2.234E+00	25	M352	1 1 1 1 2	
2.131E-02	3.521E+00	40	M352	1 1 1 1 2	
3.249E-02	5.367E+00	50	M352	1 1 1 1 2	

1708. $C_9H_{11}NO_2$

Ethyl p-Aminobenzoate
4-Aminobenzoic Acid Ethyl Ester
Ethyl p-Aminobenzoic Acid
Benzocaine

RN: 94-09-7   **MP** (°C): 89.0
MW: 165.19   **BP** (°C):

Solubility (Moles/L)	Solubility (Grams/L)	Temp (°C)	Ref (#)	Evaluation (T P E A A)	Comments
4.308E-03	7.117E-01	15	M352	1 1 1 1 2	
1.513E-02	2.500E+00	20	F300	1 0 0 0 1	
4.840E-03	7.995E-01	25	H008	1 2 2 2 2	
6.493E-03	1.073E+00	25	M352	1 1 1 1 2	
6.216E-03	1.027E+00	25	P303	2 0 2 2 2	
7.930E-03	1.310E+00	30	B071	1 2 1 1 2	
5.150E-03	8.507E-01	30	H018	1 2 2 2 2	
7.500E-03	1.239E+00	30	J018	1 2 0 1 1	0.05N NaOH
7.000E-03	1.156E+00	30	L069	1 0 1 1 0	EFG
7.680E-03	1.269E+00	30	R003	1 1 2 2 2	
8.156E-03	1.347E+00	33	P303	2 0 2 2 2	
1.020E-02	1.685E+00	37	F006	1 1 2 2 2	
1.164E-02	1.924E+00	40	M352	1 1 1 1 2	
1.032E-02	1.704E+00	40	P303	2 0 2 2 2	
1.701E-02	2.810E+00	50	M352	1 1 1 1 2	
4.810E-03	7.946E-01	ns	M066	0 0 0 0 2	
4.810E-03	7.946E-01	rt	B016	0 0 1 1 2	pH 7.4

1709. C$_9$H$_{11}$NO$_2$
D-Phenylalanine
D-α-Aminohydrocinnamic Acid
D-α-Amino-β-Phenylpropionic Acid
D-β-Phenyl-α-Aminopropionic Acid
D-PHE
RN: 673-06-3   **MP** (°C):   273
MW: 165.19   **BP** (°C):

Solubility (Moles/L)	Solubility (Grams/L)	Temp (°C)	Ref (#)	Evaluation (T P E A A)	Comments
1.763E-01	2.913E+01	25	D041	1 0 0 0 0	

1710. C$_9$H$_{11}$NO$_2$
2-Methyl-4-acetaminophenol
3-Methyl-4-hydroxyacetanilide
3-Methylparacetamol
RN: 16375-90-9   **MP** (°C):
MW: 165.19   **BP** (°C):

Solubility (Moles/L)	Solubility (Grams/L)	Temp (°C)	Ref (#)	Evaluation (T P E A A)	Comments
2.536E-02	4.189E+00	25	D078	1 2 1 1 2	

1711. C$_9$H$_{11}$NO$_2$
DL-Phenylalanine
DL-Phenylalanin
RN: 150-30-1   **MP** (°C):   166.5
MW: 165.19   **BP** (°C):

Solubility (Moles/L)	Solubility (Grams/L)	Temp (°C)	Ref (#)	Evaluation (T P E A A)	Comments
5.993E-02	9.900E+00	0	F300	1 0 0 0 1	
9.080E-02	1.500E+01	21	F300	1 0 0 0 1	
9.008E-02	1.488E+01	21	P045	1 0 2 1 2	
8.464E-02	1.398E+01	25	D018	2 2 2 1 2	
8.476E-02	1.400E+01	25	D041	1 0 0 0 2	
1.304E-01	2.154E+01	50	D018	2 2 2 1 2	
1.295E-01	2.140E+01	50	F300	1 0 0 0 2	
2.158E-01	3.564E+01	75	D018	2 2 2 1 2	
2.164E-01	3.575E+01	75	D041	1 0 0 0 2	
2.167E-01	3.580E+01	75	F300	1 0 0 0 2	
3.898E-01	6.440E+01	100	F300	1 0 0 0 2	

1712. $C_9H_{11}NO_2$
m-Tolyl Methylcarbamate
3-Tolyl Methylcarbamate
RN: 1129-41-5   **MP** (°C):   76.5
MW: 165.19   **BP** (°C):

Solubility (Moles/L)	Solubility (Grams/L)	Temp (°C)	Ref (#)	Evaluation (T P E A A)	Comments
1.574E-02	2.600E+00	30	M161	1 0 0 0 1	

1713. $C_9H_{11}NO_2$
4-(Dimethylamino)benzoic Acid
4-Dimethylaminobenzoic Acid
RN: 619-84-1   **MP** (°C):   242.5
MW: 165.19   **BP** (°C):

Solubility (Moles/L)	Solubility (Grams/L)	Temp (°C)	Ref (#)	Evaluation (T P E A A)	Comments
4.000E-04	6.608E-02	ns	C014	0 0 0 1 1	

1714. $C_9H_{11}NO_3$
L-Tyrosine
3-(4-Hydroxyphenyl)-L-alanine
Tyrosine
(S)-(-)-Tyrosine
p-Tyrosine
L-Tyrosin
RN: 60-18-4   **MP** (°C):   342dec
MW: 181.19   **BP** (°C):

Solubility (Moles/L)	Solubility (Grams/L)	Temp (°C)	Ref (#)	Evaluation (T P E A A)	Comments
1.241E-03	2.249E-01	0	D018	2 2 2 1 2	
1.104E-03	2.000E-01	0	F300	1 0 0 0 0	
2.042E-03	3.700E-01	20	B032	1 2 2 1 2	
2.495E-03	4.520E-01	21	P045	1 0 2 1 2	
2.285E-03	4.140E-01	22	A045	2 0 2 2 2	
2.642E-03	4.788E-01	25	D018	2 2 2 1 2	
2.482E-03	4.498E-01	25	D041	1 0 0 0 1	
2.759E-03	5.000E-01	25	F300	1 0 0 0 0	
2.620E-03	4.747E-01	25	H097	2 2 2 2 2	
2.622E-03	4.750E-01	25.1	N024	2 0 2 2 2	
2.495E-03	4.520E-01	25.1	N025	2 0 2 2 2	
2.489E-03	4.510E-01	25.1	N026	2 0 2 2 2	
2.488E-03	4.508E-01	25.1	N027	1 1 2 2 2	
2.753E-03	4.988E-01	27	D036	2 1 2 2 2	
2.677E-03	4.850E-01	27	D036	2 1 2 2 2	
3.195E-03	5.790E-01	28	L081	2 1 2 2 2	
6.064E-03	1.099E+00	50	D018	2 2 2 1 2	
6.071E-03	1.100E+00	50	F300	1 0 0 0 1	

1.309E-02	2.372E+00	75	D018	2 2 2 1 2
1.343E-02	2.434E+00	75	D041	1 0 0 0 2
1.325E-02	2.400E+00	75	F300	1 0 0 0 1
3.091E-02	5.600E+00	100	F300	1 0 0 0 1

1715. $C_9H_{11}NO_3$
D-Tyrosine
3-(4-Hydroxyphenyl)-D-alanine
RN: 556-02-5   **MP** (°C):   >300
MW: 181.19   **BP** (°C):

Solubility (Moles/L)	Solubility (Grams/L)	Temp (°C)	Ref (#)	Evaluation (T P E A A)	Comments
2.482E-03	4.498E-01	25	D041	1 0 0 0 1	
5.789E-03	1.049E+00	50	D041	1 0 0 0 2	

1716. $C_9H_{11}NO_3$
DL-Tyrosine
DL-Tyrosin
3-(4-Hydroxyphenyl)-DL-alanine
DL-2-Amino-3-(4-hydroxyphenyl)-propanoic Acid
RN: 556-03-6   **MP** (°C):   325
MW: 181.19   **BP** (°C):

Solubility (Moles/L)	Solubility (Grams/L)	Temp (°C)	Ref (#)	Evaluation (T P E A A)	Comments
5.519E-04	1.000E-01	0	F300	1 0 0 0 0	
2.208E-03	4.000E-01	20	F300	1 0 0 0 0	
1.936E-03	3.509E-01	25	D041	1 0 0 0 2	
4.610E-03	8.353E-01	50	D041	1 0 0 0 2	
4.415E-03	8.000E-01	50	F300	1 0 0 0 0	
3.753E-02	6.800E+00	100	F300	1 0 0 0 1	

1717. $C_9H_{11}NO_4$
Dopa
DL-3-(3,4-Dihydroxyphenyl)alanine
DL-Dopa
RN: 63-84-3   **MP** (°C):   >270
MW: 197.19   **BP** (°C):

Solubility (Moles/L)	Solubility (Grams/L)	Temp (°C)	Ref (#)	Evaluation (T P E A A)	Comments
2.523E-02	4.975E+00	20	D041	1 0 0 0 0	
1.237E-01	2.439E+01	100	D041	1 0 0 0 1	

1718.   C₉H₁₁NO₄

Levodopa
L-3,4-Dihydroxyphenylalanin
RN: 59-92-7   **MP** (°C):   277
MW: 197.19   **BP** (°C):

Solubility (Moles/L)	Solubility (Grams/L)	Temp (°C)	Ref (#)	Evaluation (T P E A A)	Comments
2.536E-02	5.000E+00	20	F300	1 0 0 0 0	
1.917E-02	3.780E+00	25	H015	1 0 0 0 2	
1.927E-02	3.800E+00	25.1	N025	2 0 2 2 2	
1.268E-01	2.500E+01	100	F300	1 0 0 0 1	

1719.   C₉H₁₁NS₂Hg

Phenylmercury Dimethyldithiocarbamate
Chipman Merbam
Merfenl 51
Phelam DP
RN: 32407-99-1   **MP** (°C):   175
MW: 397.91   **BP** (°C):

Solubility (Moles/L)	Solubility (Grams/L)	Temp (°C)	Ref (#)	Evaluation (T P E A A)	Comments
1.508E-05	6.000E-03	20	M161	1 0 0 0 0	

1720.   C₉H₁₁N₃O

Biacetyl Mono(2-pyridyl)-hydrazone
BPH
Biacetyl Mono(2-pyridyl)hydrazone
RN: 74158-10-4   **MP** (°C):   95
MW: 177.21   **BP** (°C):

Solubility (Moles/L)	Solubility (Grams/L)	Temp (°C)	Ref (#)	Evaluation (T P E A A)	Comments
5.643E-04	9.999E-02	ns	R080	2 0 2 2 0	

1721.   C₉H₁₁N₃O₂S₂

Sulfathiazoline
Benzenesulfonamide, 4-Amino-N-(4,5-dihydro-2-thiazolyl)-
RN: 32365-02-9   **MP** (°C):
MW: 257.33   **BP** (°C):

Solubility (Moles/L)	Solubility (Grams/L)	Temp (°C)	Ref (#)	Evaluation (T P E A A)	Comments
5.790E-04	1.490E-01	20	F073	1 2 2 2 2	

1722.   C$_9$H$_{11}$N$_3$O$_4$
Orotic Acid Morpholine
RN: **MP** (°C):   289-291
MW: 225.21   **BP** (°C):

Solubility (Moles/L)	Solubility (Grams/L)	Temp (°C)	Ref (#)	Evaluation (T P E A A)	Comments
4.400E-01	9.909E+01	-4	N018	2 2 1 2 2	
6.500E-01	1.464E+02	16	N018	2 2 1 2 2	
7.450E-01	1.678E+02	25	N018	2 2 1 2 2	

1723.   C$_9$H$_{12}$
1,2,4-Trimethylbenzene
Pseudocumene
RN: 95-63-6   **MP** (°C):   -44
MW: 120.20   **BP** (°C):   169

Solubility (Moles/L)	Solubility (Grams/L)	Temp (°C)	Ref (#)	Evaluation (T P E A A)	Comments
4.318E-04	5.190E-02	25	K119	1 0 0 0 2	
4.742E-04	5.700E-02	25	M001	2 1 2 2 2	
4.318E-04	5.190E-02	25	P051	2 1 1 2 2	
4.909E-04	5.900E-02	25	S005	2 2 2 2 2	
4.909E-04	5.900E-02	25	S191	1 2 2 2 2	
4.909E-04	5.900E-02	25	S358	2 1 2 2 2	
4.318E-04	5.190E-02	25.00	P007	2 1 2 2 2	
4.742E-04	5.700E-02	ns	M344	0 0 0 0 1	

1724.   C$_9$H$_{12}$
p-Ethyltoluene
4-Ethyltoluene
1-Ethyl-4-methylbenzene
RN: 622-96-8   **MP** (°C):   -62
MW: 120.20   **BP** (°C):   162

Solubility (Moles/L)	Solubility (Grams/L)	Temp (°C)	Ref (#)	Evaluation (T P E A A)	Comments
7.891E-04	9.485E-02	ns	H123	0 0 0 0 2	

1725. C_9H_{12}
1,2,3-Trimethylbenzene
Hemimellitene
Hemellitol

RN:	526-73-8	MP (°C):	-25
MW:	120.20	BP (°C):	175

Solubility (Moles/L)	Solubility (Grams/L)	Temp (°C)	Ref (#)	Evaluation (T P E A A)	Comments
5.450E-04	6.551E-02	25	M342	1 0 1 1 2	
6.256E-04	7.520E-02	25	S005	2 2 2 2 2	
6.256E-04	7.520E-02	25	S191	1 2 2 2 2	
6.256E-04	7.520E-02	25	S358	2 2 2 2 2	

1726. C_9H_{12}
1,8-Nonadiyne

RN:	2396-65-8	MP (°C):	-21
MW:	120.20	BP (°C):	55

Solubility (Moles/L)	Solubility (Grams/L)	Temp (°C)	Ref (#)	Evaluation (T P E A A)	Comments
1.040E-03	1.250E-01	25	M001	2 1 2 2 2	

1727. C_9H_{12}
1-Ethyl-2-methylbenzene
2-Ethyltoluene
o-Ethyltoluene
1-Methyl-2-ethylbenzene

RN:	611-14-3	MP (°C):	-80.8
MW:	120.20	BP (°C):	165.2

Solubility (Moles/L)	Solubility (Grams/L)	Temp (°C)	Ref (#)	Evaluation (T P E A A)	Comments
6.210E-04	7.464E-02	25	M342	1 0 1 1 2	
7.742E-04	9.305E-02	ns	H123	0 0 0 0 2	

1728. C_9H_{12}
Cumene
Isopropylbenzene
Cumol
2-Phenylpropane

RN:	98-82-8	MP (°C):	-96
MW:	120.20	BP (°C):	152

Solubility (Moles/L)	Solubility (Grams/L)	Temp (°C)	Ref (#)	Evaluation (T P E A A)	Comments
6.694E-04	8.046E-02	24.94	G034	1 2 2 2 2	
6.073E-04	7.300E-02	25	A002	1 2 1 1 1	
4.018E-04	4.830E-02	25	K119	1 0 0 0 2	
4.160E-04	5.000E-02	25	M001	2 1 2 2 2	

4.409E-04	5.300E-02	25	M002	2 2 1 2 1
4.160E-04	5.000E-02	25	M130	1 0 0 0 1
4.018E-04	4.830E-02	25	P051	2 1 1 2 2
5.433E-04	6.530E-02	25	S005	2 2 2 2 2
5.433E-04	6.530E-02	25	S191	1 2 2 2 2
5.433E-04	6.530E-02	25	S358	2 1 2 2 2
4.018E-04	4.830E-02	25.00	P007	2 1 2 2 2
6.897E-04	8.290E-02	29.94	G034	1 2 2 2 2
7.124E-04	8.563E-02	34.94	G034	1 2 2 2 2
7.469E-04	8.978E-02	39.94	G034	1 2 2 2 2
7.867E-04	9.456E-02	44.94	G034	1 2 2 2 2
8.353E-04	1.004E-01	49.94	G034	1 2 2 2 2
8.894E-04	1.069E-01	54.94	G034	1 2 2 2 2
9.566E-04	1.150E-01	59.94	G034	1 2 2 2 2
1.035E-03	1.243E-01	65.14	G034	1 2 2 2 2
1.128E-03	1.355E-01	70.34	G034	1 2 2 2 2
1.226E-03	1.473E-01	75.04	G034	1 2 2 2 2
1.345E-03	1.617E-01	80.24	G034	1 2 2 2 2
4.160E-04	5.000E-02	ns	H123	0 0 0 0 2
4.160E-04	5.000E-02	ns	M344	0 0 0 0 1

1729. C_9H_{12}

Mesitylene
1,3,5-Trimethylbenzene
Mesitelene

RN:	108-67-8	**MP** (°C):	-44.8
MW:	120.20	**BP** (°C):	164.7

Solubility (Moles/L)	Solubility (Grams/L)	Temp (°C)	Ref (#)	Evaluation (T P E A A)	Comments
3.794E-04	4.560E-02	15	S203	1 1 2 1 2	
3.111E-04	3.740E-02	20	M337	2 1 2 2 2	
8.070E-04	9.700E-02	25	A002	1 2 1 1 1	
4.010E-04	4.820E-02	25	S005	2 2 2 2 2	
4.010E-04	4.820E-02	25	S191	1 2 2 2 2	
4.118E-04	4.950E-02	25	S203	1 1 2 1 2	
4.010E-04	4.820E-02	25	S358	2 1 2 2 2	
3.280E-04	3.942E-02	25.04	V013	2 2 2 2 2	
5.322E-04	6.397E-02	29.99	C350	2 1 2 2 2	
4.509E-04	5.420E-02	35	S203	1 1 2 1 2	
5.555E-04	6.677E-02	39.99	C350	2 1 2 2 2	
4.701E-04	5.650E-02	45	S203	1 1 2 1 2	
6.166E-04	7.412E-02	49.99	C350	2 1 2 2 2	
7.555E-04	9.081E-02	59.99	C350	2 1 2 2 2	
9.221E-04	1.108E-01	69.99	C350	2 1 2 2 2	
1.161E-03	1.395E-01	79.99	C350	2 1 2 2 2	
1.361E-03	1.636E-01	89.99	C350	2 1 2 2 2	
1.616E-03	1.943E-01	99.99	C350	2 1 2 2 2	

1730.   C₉H₁₂

n-Propylbenzene
1-Phenylpropane
Propylbenzene
Isocomene

RN: 103-65-1   **MP** (°C):   -99.2
MW: 120.20   **BP** (°C):   159.2

Solubility (Moles/L)	Solubility (Grams/L)	Temp (°C)	Ref (#)	Evaluation (T P E A A)	Comments
4.470E-04	5.373E-02	10	O312	2 2 0 2 2	
5.000E-04	6.010E-02	15	F001	1 0 1 2 0	
4.350E-04	5.229E-02	15	O312	2 2 0 2 2	
4.520E-04	5.433E-02	20	O312	2 2 0 2 2	
4.576E-04	5.500E-02	25	A002	1 2 1 1 1	
1.000E-03	1.202E-01	25	K001	1 0 2 1 2	
4.340E-04	5.217E-02	25	M342	1 0 1 1 2	
4.430E-04	5.325E-02	25	O312	2 2 0 2 2	
8.319E-04	9.999E-02	25	S012	2 0 2 2 1	
4.150E-04	4.988E-02	25	S359	2 1 2 2 2	
3.920E-04	4.712E-02	25	T067	2 1 2 1 2	
4.340E-04	5.217E-02	25	W300	2 2 2 2 2	
4.370E-04	5.253E-02	30	O312	2 2 0 2 2	
4.710E-04	5.661E-02	35	O312	2 2 0 2 2	
5.320E-04	6.394E-02	40	O312	2 2 0 2 2	
5.540E-04	6.659E-02	45	O312	2 2 0 2 2	
1.098E-03	1.320E-01	85.8	G035	1 0 0 0 2	
1.381E-03	1.660E-01	114.5	G035	1 0 0 0 2	
2.670E-03	3.209E-01	140.5	G035	1 0 0 0 2	
7.232E-03	8.692E-01	188.0	G035	1 0 0 0 1	
2.033E-02	2.444E+00	222.0	G035	1 0 0 0 2	
4.576E-04	5.500E-02	ns	H123	0 0 0 0 2	
2.700E-02	3.245E+00	ns	H307	1 0 1 1 2	
4.576E-04	5.500E-02	ns	M344	0 0 0 0 1	

1731.   C₉H₁₂ClO₂PS₃

Carbophenothion-methyl
S-p-Chlorophenylthiomethyl O,O-Dimethyl Phosphorodithioate

RN: 953-17-3   **MP** (°C):
MW: 314.81   **BP** (°C):

Solubility (Moles/L)	Solubility (Grams/L)	Temp (°C)	Ref (#)	Evaluation (T P E A A)	Comments
4.669E-06	1.470E-03	10	B324	2 2 2 2 2	
4.670E-06	1.470E-03	10	B324	2 2 2 2 2	
5.178E-06	1.630E-03	20	B300	2 1 1 1 2	
5.083E-06	1.600E-03	20	B324	2 2 2 2 2	

5.082E-06	1.600E-03	20	B324	2 2 2 2 2
8.958E-06	2.820E-03	30	B324	2 2 2 2 2
8.958E-06	2.820E-03	30	B324	2 2 2 2 2
3.176E-06	1.000E-03	rt	M161	0 0 0 0 0

1732. $C_9H_{12}ClO_4P$

Heptenophos
7-Chlorobicyclo[3.2.0]hepta-2,6-dien-6-yl Dimethyl Phosphate
Ragadan
Hostaquick

RN: 23560-59-0   **MP** (°C):
MW: 250.62   **BP** (°C):   64

Solubility (Moles/L)	Solubility (Grams/L)	Temp (°C)	Ref (#)	Evaluation (T P E A A)	Comments
9.975E-03	2.500E+00	23	M161	1 0 0 0 1	

1733. $C_9H_{12}Cl_2N_4$

2,4-Dichloro-6-cyclohexylamino-1,3,5-triazine
2,4-Dichloro-6-(cyclohexylamino)triazine
1,3,5-Triazin-2-amine, 4,6-dichloro-N-cyclohexyl-

RN: 27282-86-6   **MP** (°C):
MW: 247.13   **BP** (°C):

Solubility (Moles/L)	Solubility (Grams/L)	Temp (°C)	Ref (#)	Evaluation (T P E A A)	Comments
4.046E-04	1.000E-01	ns	B160	0 0 0 0 2	

1734. $C_9H_{12}FN_3O_3$

1-Butylcarbamoyl-5-fluorouracil
N-Butyl-5-fluoro-2,4-dioxo-pyrimidinecarboxamide

RN: 64098-82-4   **MP** (°C):   136
MW: 229.21   **BP** (°C):

Solubility (Moles/L)	Solubility (Grams/L)	Temp (°C)	Ref (#)	Evaluation (T P E A A)	Comments
3.577E-03	8.200E-01	22	B321	1 0 2 2 2	pH 4.0
3.577E-03	8.200E-01	22	B388	1 0 2 2 2	

1735. $C_9H_{12}NO_5PS$
O-Methyl O-Ethyl O-4-Nitrophenyl Thiophosphate
Ethylmethylthiophos
Methylethylthiophos
Methylethylthiofos

RN: 2591-57-3   **MP** (°C):
MW: 277.24   **BP** (°C):   116

Solubility (Moles/L)	Solubility (Grams/L)	Temp (°C)	Ref (#)	Evaluation (T P E A A)	Comments
1.443E-04	4.000E-02	ns	M061	0 0 0 0 1	

1736. $C_9H_{12}NO_5PS$
Fenitrothion
Dimethyl O-(4-Nitro-m-tolyl) Phosphorothioate
Nuvanol
Novathion
Dybar
Metathionine

RN: 122-14-5   **MP** (°C):   3.4
MW: 277.24   **BP** (°C):

Solubility (Moles/L)	Solubility (Grams/L)	Temp (°C)	Ref (#)	Evaluation (T P E A A)	Comments
9.089E-05	2.520E-02	20	B169	2 0 1 1 2	
1.396E-04	3.870E-02	22	K137	1 1 2 1 0	*sic*
1.082E-04	3.000E-02	ns	F071	0 1 2 1 1	
1.082E-04	3.000E-02	ns	M061	0 0 0 0 1	
1.082E-04	3.000E-02	ns	M110	0 0 0 0 0	EFG

1737. $C_9H_{12}N_2O$
Fenuron
3-Phenyl-1,1-dimethylurea
N,N-Dimethyl-N-phenylurea
Beet-Klean

RN: 101-42-8   **MP** (°C):   133-134
MW: 164.21   **BP** (°C):

Solubility (Moles/L)	Solubility (Grams/L)	Temp (°C)	Ref (#)	Evaluation (T P E A A)	Comments
2.344E-02	3.849E+00	20	B179	2 0 0 0 2	
2.245E-02	3.686E+00	20	E048	1 2 1 1 2	
2.253E-02	3.700E+00	20	F311	1 2 2 2 1	
1.766E-02	2.900E+00	24	B185	1 0 0 0 2	
1.761E-02	2.892E+00	24	M061	1 0 0 0 1	
1.462E-02	2.400E+00	25	A039	1 1 0 0 2	
2.345E-02	3.850E+00	25	B200	1 0 0 0 0	
2.345E-02	3.850E+00	25	G036	1 0 0 0 2	
1.462E-02	2.400E+00	25	G099	1 0 0 1 0	
2.452E-02	4.027E+00	25	H073	2 1 1 2 2	

2.345E-02	3.850E+00	25	M161	1 0 0 0 2
2.426E-02	3.984E+00	ns	B100	0 0 0 0 0
1.462E-02	2.400E+00	ns	B160	0 0 0 0 2
2.345E-02	3.850E+00	ns	B185	1 0 0 0 2
1.761E-02	2.892E+00	ns	N013	0 0 0 0 1

1738. $C_9H_{12}N_2O_2$

Dulcin
(4-Ethoxyphenyl)urea
4-Aethoxy-phenylharnstoff

RN: 150-69-6   **MP** (°C):   173
MW: 180.21   **BP** (°C):

Solubility (Moles/L)	Solubility (Grams/L)	Temp (°C)	Ref (#)	Evaluation (T P E A A)	Comments
6.714E-03	1.210E+00	21	F300	1 0 0 0 2	
7.214E-03	1.300E+00	45	F300	1 0 0 0 1	
1.110E-01	2.000E+01	100	F300	1 0 0 0 0	
6.928E-03	1.248E+00	c	I314	0 0 0 0 2	
1.088E-01	1.961E+01	h	I314	0 0 0 0 0	

1739. $C_9H_{12}N_2O_2S$

3-Thio-2,4-diazaspiro[5.5]undecane-1,3,5-trione
2,4-Diazaspiro[5.5]undecane-1,5-dione, 3-Thioxo-
2,4-Diazaspiro[5.5]undecane-1,3,5-trione, 3-Thio

RN: 52-45-9   **MP** (°C):
MW: 212.27   **BP** (°C):

Solubility (Moles/L)	Solubility (Grams/L)	Temp (°C)	Ref (#)	Evaluation (T P E A A)	Comments
3.450E-04	7.323E-02	25	P350	2 1 1 1 2	intrinsic

1740. $C_9H_{12}N_2O_3$

2,4-Diazaspiro[5.5]undecane-1,3,5-trione
Spiro[barbituric Acid-5,1'-cyclohexane]

RN: 52-44-8   **MP** (°C):
MW: 196.21   **BP** (°C):

Solubility (Moles/L)	Solubility (Grams/L)	Temp (°C)	Ref (#)	Evaluation (T P E A A)	Comments
8.700E-04	1.707E-01	25	P350	2 1 1 1 2	intrinsic

1741. C₉H₁₂N₂O₃

1741. $C_9H_{12}N_2O_3$
5-Allyl-5-ethylbarbituric Acid
Barbituric Acid, 5-Allyl-5-ethyl
5-Ethyl-5-allylbarbituric Acid
Dormitiv
RN: 2373-84-4   **MP** (°C):
MW: 196.21   **BP** (°C):

Solubility (Moles/L)	Solubility (Grams/L)	Temp (°C)	Ref (#)	Evaluation (T P E A A)	Comments
2.433E-02	4.774E+00	25	P350	2 1 1 1 2	intrinsic

1742. $C_9H_{12}N_4O_2$
7-Ethyl Theophylline
7-Ethyl-1,3-dimethylxanthine
1H-Purine-2,6-dione, 7-Ethyl-3,7-dihydro-1,3-dimethyl-
RN: 23043-88-1   **MP** (°C):
MW: 208.22   **BP** (°C):

Solubility (Moles/L)	Solubility (Grams/L)	Temp (°C)	Ref (#)	Evaluation (T P E A A)	Comments
1.760E-01	3.665E+01	30	B042	1 2 1 1 2	
1.760E-01	3.665E+01	30	G021	1 0 0 0 2	

1743. $C_9H_{12}N_4O_2$
1H-Pyrazolo[3,4-d]pyrimidine, 4-(1-Ethoxyethoxy)-
1-Ethoxyethyl-4-allopurinyl Ether
RN: 52717-51-8   **MP** (°C):
MW: 208.22   **BP** (°C):

Solubility (Moles/L)	Solubility (Grams/L)	Temp (°C)	Ref (#)	Evaluation (T P E A A)	Comments
9.173E-03	1.910E+00	ns	H067	0 2 0 0 2	

1744. $C_9H_{12}N_4O_2$
1-Ethyl Theobromine
1-Ethyl-3,7-dimethylxanthine
1H-Purine-2,6-dione, 1-Ethyl-3,7-dihydro-3,7-dimethyl-
RN: 39832-36-5   **MP** (°C): 156
MW: 208.22   **BP** (°C):

Solubility (Moles/L)	Solubility (Grams/L)	Temp (°C)	Ref (#)	Evaluation (T P E A A)	Comments
1.910E-01	3.977E+01	30	B042	1 2 1 1 2	
1.910E-01	3.977E+01	30	G021	1 0 0 0 2	

1745. C$_9$H$_{12}$N$_4$O$_2$
8-Methyl Caffeine
1,3,7,8-Tetramethylxanthine
RN: 832-66-6   **MP** (°C):
MW: 208.22   **BP** (°C):

Solubility (Moles/L)	Solubility (Grams/L)	Temp (°C)	Ref (#)	Evaluation (T P E A A)	Comments
1.045E-02	2.175E+00	20	J009	1 0 2 2 2	

1746. C$_9$H$_{12}$N$_4$O$_3$
1,3,7,9-Tetramethyluric Acid
1H-Purine-2,6,8(3H)-trione, 7,9-Dihydro-1,3,7,9-tetramethyl-
Temorine
Temurin
Ba 2750
RN: 2309-49-1   **MP** (°C):
MW: 224.22   **BP** (°C):

Solubility (Moles/L)	Solubility (Grams/L)	Temp (°C)	Ref (#)	Evaluation (T P E A A)	Comments
1.472E-04	3.300E-02	rt	N015	0 0 2 2 1	

1747. C$_9$H$_{12}$N$_4$O$_3$
7-β-Hydroxyethyltheophylline
1H-Purine-2,6-dione, 3,7-Dihydro-7-(2-hydroxyethyl)-1,3-dimethyl-
Dilaphyllin
Etofylline
Corophyllin-N
RN: 519-37-9   **MP** (°C):
MW: 224.22   **BP** (°C):

Solubility (Moles/L)	Solubility (Grams/L)	Temp (°C)	Ref (#)	Evaluation (T P E A A)	Comments
1.439E-01	3.226E+01	ns	J025	0 0 0 0 1	

1748. C$_9$H$_{12}$N$_4$O$_3$
8-Methoxycaffeine
1H-Purine-2,6-dione, 3,7-Dihydro-8-methoxy-1,3,7-trimethyl-
RN: 569-34-6   **MP** (°C):
MW: 224.22   **BP** (°C):

Solubility (Moles/L)	Solubility (Grams/L)	Temp (°C)	Ref (#)	Evaluation (T P E A A)	Comments
1.140E-02	2.556E+00	25	K008	1 1 0 1 0	EFG
1.115E-04	2.500E-02	rt	N015	0 0 2 2 1	

1749. $C_9H_{12}N_4O_3S$
N4-Acetylsulfanilylguanidine
Acetamide, N-[4-[[(Aminoiminomethyl)amino]sulfonyl]phenyl]-
p-(Guanidinosulfonyl)acetanilide
Sulgin ASG

RN: 19077-97-5   **MP** (°C):
MW: 256.28   **BP** (°C):

Solubility (Moles/L)	Solubility (Grams/L)	Temp (°C)	Ref (#)	Evaluation (T P E A A)	Comments
1.560E-03	3.998E-01	37.50	M142	1 2 0 0 1	
5.766E-02	1.478E+01	h	M142	0 0 0 0 1	

1750. $C_9H_{12}O$
3-Methyl-5-ethyl-phenol
Phenol, 3-Ethyl-5-methyl-
m-Cresol, 5-Ethyl-

RN: 698-71-5   **MP** (°C):
MW: 136.20   **BP** (°C):

Solubility (Moles/L)	Solubility (Grams/L)	Temp (°C)	Ref (#)	Evaluation (T P E A A)	Comments
1.700E-02	2.315E+00	25	B316	1 0 2 1 1	

1751. $C_9H_{12}O$
4-Ethyl-3-methylphenol
3-Methyl-4-ethylphenol
4-Ethyl-m-cresol

RN: 1123-94-0   **MP** (°C):
MW: 136.20   **BP** (°C):

Solubility (Moles/L)	Solubility (Grams/L)	Temp (°C)	Ref (#)	Evaluation (T P E A A)	Comments
7.335E-03	9.990E-01	25	L020	1 0 0 0 0	

1752. $C_9H_{12}O$
2,3,5-Trimethyl-phenol
Isopseudocumenol
1-Hydroxy-2,3,5-trimethylbenzene

RN: 697-82-5   **MP** (°C):
MW: 136.20   **BP** (°C):

Solubility (Moles/L)	Solubility (Grams/L)	Temp (°C)	Ref (#)	Evaluation (T P E A A)	Comments
5.600E-03	7.627E-01	25	B316	1 0 2 1 1	

1753. C$_9$H$_{12}$O
2-Propylphenol
2-n-Propylphenol
2-Propyphenol
RN: 644-35-9   **MP** (°C):
MW: 136.20   **BP** (°C):   225

Solubility (Moles/L)	Solubility (Grams/L)	Temp (°C)	Ref (#)	Evaluation (T P E A A)	Comments
1.222E-02	1.664E+00	25	L022	1 0 0 0 0	

1754. C$_9$H$_{12}$O
4-Propylphenol
4-Propyphenol
p-n-Propylphenol
RN: 645-56-7   **MP** (°C):
MW: 136.20   **BP** (°C):   232

Solubility (Moles/L)	Solubility (Grams/L)	Temp (°C)	Ref (#)	Evaluation (T P E A A)	Comments
1.047E-02	1.427E+00	25	L022	1 0 0 0 0	

1755. C$_9$H$_{12}$O
2,4,6-Trimethylphenol
2-Hydroxymesitylene
1-Hydroxy-2,4,6-trimethylbenzene
Mesityl Alcohol
Hydroxymesitylene
RN: 527-60-6   **MP** (°C):   72
MW: 136.20   **BP** (°C):   220

Solubility (Moles/L)	Solubility (Grams/L)	Temp (°C)	Ref (#)	Evaluation (T P E A A)	Comments
7.400E-03	1.008E+00	25	B316	1 0 2 1 1	
4.892E-03	6.662E-01	25	L020	1 0 0 0 0	

1756. C$_9$H$_{12}$O$_2$
3-Propoxyphenol
m-Propoxy Phenol
Phenol, 3-Propoxy-
RN: 16533-50-9   **MP** (°C):
MW: 152.19   **BP** (°C):

Solubility (Moles/L)	Solubility (Grams/L)	Temp (°C)	Ref (#)	Evaluation (T P E A A)	Comments
2.590E-02	3.942E+00	30	B315	1 0 1 1 2	

1757. C$_9$H$_{12}$O$_2$
1-O-Benzylethanediol
Benzylcellosolve
Benzyl Cellosolve
RN: 622-08-2   **MP** (°C):
MW: 152.19   **BP** (°C):

Solubility (Moles/L)	Solubility (Grams/L)	Temp (°C)	Ref (#)	Evaluation (T P E A A)	Comments
2.618E-02	3.984E+00	20	D052	1 1 0 0 0	
2.813E-02	4.282E+00	23	M062	1 0 0 0 1	

1758. C$_9$H$_{12}$O$_2$
Cumene Hydroperoxide
CHP
RN: 80-15-9   **MP** (°C):
MW: 152.19   **BP** (°C):   100

Solubility (Moles/L)	Solubility (Grams/L)	Temp (°C)	Ref (#)	Evaluation (T P E A A)	Comments
9.140E-02	1.391E+01	25	K051	1 2 2 1 2	

1759. C$_9$H$_{12}$O$_2$
o-Propoxyphenol
2-Propoxyphenol
RN: 6280-96-2   **MP** (°C):
MW: 152.19   **BP** (°C):

Solubility (Moles/L)	Solubility (Grams/L)	Temp (°C)	Ref (#)	Evaluation (T P E A A)	Comments
1.550E-02	2.359E+00	24.99	B353	2 1 1 1 2	

1760. C$_9$H$_{13}$BrN$_2$O$_2$
5-Bromo-3-tert-butyl-6-methyluracil
Compound 733
RN: 7286-76-2   **MP** (°C):   188
MW: 261.13   **BP** (°C):

Solubility (Moles/L)	Solubility (Grams/L)	Temp (°C)	Ref (#)	Evaluation (T P E A A)	Comments
1.570E-03	4.100E-01	25	M061	1 0 0 0 0	
3.121E-03	8.150E-01	ns	B185	0 0 0 0 2	

1761.   C$_9$H$_{13}$BrN$_2$O$_2$
Bromacil
5-Bromo-6-methyl-3,5-butyluracil
RN: 314-40-9   **MP** (°C):   158.3
MW: 261.13   **BP** (°C):

Solubility (Moles/L)	Solubility (Grams/L)	Temp (°C)	Ref (#)	Evaluation (T P E A A)	Comments
2.719E-03	7.100E-01	25	B200	1 0 0 0 2	
3.119E-03	8.143E-01	25	B200	1 0 0 0 2	
3.121E-03	8.150E-01	25	M061	1 0 0 0 2	
3.121E-03	8.150E-01	25	M161	1 0 0 0 2	
3.061E-03	7.994E-01	ns	B100	0 0 0 0 0	

1762.   C$_9$H$_{13}$ClN$_2$O$_2$
Terbacil
3-tert-Butyl-5-chloro-6-methyluracil
5-Chloro-3-(1,1-dimethylethyl)-6-methyl-2,4(1H,3H)-pyrimidinedione
Sinbar 80W
Geonter
DPX-D732
RN: 5902-51-2   **MP** (°C):   176.0
MW: 216.67   **BP** (°C):

Solubility (Moles/L)	Solubility (Grams/L)	Temp (°C)	Ref (#)	Evaluation (T P E A A)	Comments
3.277E-03	7.100E-01	25	M061	1 0 0 0 2	
3.277E-03	7.100E-01	25	M161	1 0 0 0 2	
3.277E-03	7.100E-01	25	P307	1 0 0 0 1	
3.228E-03	6.995E-01	ns	B100	0 0 0 0 0	

1763. $C_9H_{13}ClN_6$

Cyanazine
Bladex
2-[[4-Chloro-6-(ethylamino)-1,3,5-triazin-2-yl]amino]-2-methylpropanenitrile
Fortrol
Payze
SD 45418

RN:	21725-46-2	**MP** (°C):	166.5
MW:	240.70	**BP** (°C):	

Solubility (Moles/L)	Solubility (Grams/L)	Temp (°C)	Ref (#)	Evaluation (T P E A A)	Comments
6.647E-04	1.600E-01	23	B200	1 0 0 0 2	
7.104E-04	1.710E-01	25	B200	1 0 0 0 2	
7.104E-04	1.710E-01	25	M061	1 0 0 0 2	
7.104E-04	1.710E-01	25	M161	1 0 0 0 2	
6.647E-04	1.600E-01	25	S309	1 0 0 0 2	
8.309E-04	2.000E-01	ns	M110	0 0 0 0 0	EFG

1764. $C_9H_{13}N$

2,4,5-Trimethylaniline
2,4,5-Trimethylanilin

RN:	137-17-7	**MP** (°C):	
MW:	135.21	**BP** (°C):	

Solubility (Moles/L)	Solubility (Grams/L)	Temp (°C)	Ref (#)	Evaluation (T P E A A)	Comments
8.875E-03	1.200E+00	19.40	F300	1 0 0 0 1	
1.109E-02	1.500E+00	28.70	F300	1 0 0 0 1	

1765. $C_9H_{13}NO_3$

Adrenaline
Adrenalin
Epinephrine
L-1-(3,4-Dihydroxyphenyl)-2-methylaminoethanol
Primatene
Epipen

RN:	51-43-4	**MP** (°C):	
MW:	183.21	**BP** (°C):	

Solubility (Moles/L)	Solubility (Grams/L)	Temp (°C)	Ref (#)	Evaluation (T P E A A)	Comments
9.825E-04	1.800E-01	20	F300	1 0 0 0 1	

1766. $C_9H_{13}N_3O_3$
Orotic Acid Diethylamine
Orotamide, N,N-Diethyl-
RN: 883-81-8   **MP** (°C):   192-194
MW: 211.22   **BP** (°C):

Solubility (Moles/L)	Solubility (Grams/L)	Temp (°C)	Ref (#)	Evaluation (T P E A A)	Comments
2.939E+00	6.208E+02	25	N018	2 2 1 2 2	

1767. $C_9H_{13}N_3O_3$
Orotic Acid n-Butylamide
Orotamide, N-Butyl-
RN: 13156-38-2   **MP** (°C):   276-277
MW: 211.22   **BP** (°C):

Solubility (Moles/L)	Solubility (Grams/L)	Temp (°C)	Ref (#)	Evaluation (T P E A A)	Comments
5.700E-02	1.204E+01	-4	N018	2 2 1 2 2	
9.600E-02	2.028E+01	16	N018	2 2 1 2 2	
1.180E-01	2.492E+01	25	N018	2 2 1 2 2	

1768. $C_9H_{13}N_3O_4$
Orotic Acid Isobutanolamine
RN: **MP** (°C):   247-249
MW: 227.22   **BP** (°C):

Solubility (Moles/L)	Solubility (Grams/L)	Temp (°C)	Ref (#)	Evaluation (T P E A A)	Comments
4.200E-01	9.543E+01	-4	N018	2 2 1 2 2	
7.060E-01	1.604E+02	16	N018	2 2 1 2 2	
8.410E-01	1.911E+02	25	N018	2 2 1 2 2	

1769. $C_9H_{13}N_3O_5$
Orotic Acid 2-Amide-2-methyl-1,3-propanediol
RN: **MP** (°C):   214-215
MW: 243.22   **BP** (°C):

Solubility (Moles/L)	Solubility (Grams/L)	Temp (°C)	Ref (#)	Evaluation (T P E A A)	Comments
3.450E-01	8.391E+01	-4	N018	2 2 1 2 2	
5.860E-01	1.425E+02	16	N018	2 2 1 2 2	
6.970E-01	1.695E+02	25	N018	2 2 1 2 2	

1770.   C$_9$H$_{13}$N$_5$O$_4$
Ganciclovir
2-Amino-1,9-dihydro-9-((2-hydroxy-1-(hydroxymethyl)ethoxy)methyl)-6H-purin-6-one
DHPG
RN: 82410-32-0   **MP** (°C):   250
MW: 255.24   **BP** (°C):

Solubility (Moles/L)	Solubility (Grams/L)	Temp (°C)	Ref (#)	Evaluation (T P E A A)	Comments
1.410E-02	3.600E+00	25	B360	1 0 2 2 2	

1771.   C$_9$H$_{13}$O$_2$P
Mesitylene Phosphinous Acid
Phosphinic Acid, (2,4,6-Trimethylphenyl)-
RN: 6781-97-1   **MP** (°C):   147.0
MW: 184.18   **BP** (°C):

Solubility (Moles/L)	Solubility (Grams/L)	Temp (°C)	Ref (#)	Evaluation (T P E A A)	Comments
1.565E-02	2.882E+00	1	C061	2 2 2 1 2	
1.619E-02	2.981E+00	25	C061	2 2 2 1 2	
1.754E-02	3.230E+00	35	C061	2 2 2 1 2	
2.082E-02	3.835E+00	45	C061	2 2 2 1 2	
2.836E-02	5.223E+00	65	C061	2 2 2 1 2	
3.774E-02	6.951E+00	85	C061	2 2 2 1 2	

1772.   C$_9$H$_{13}$O$_6$PS
Endothion
O,O-Dimethyl S-(5-Methoxypyronyl-2-methyl) Thiophosphate
RN: 2778-04-3   **MP** (°C):   90.5
MW: 280.24   **BP** (°C):

Solubility (Moles/L)	Solubility (Grams/L)	Temp (°C)	Ref (#)	Evaluation (T P E A A)	Comments
2.141E+00	6.000E+02	ns	M061	0 0 0 0 2	
5.353E+00	1.500E+03	ns	M161	0 0 0 0 1	

1773.   C$_9$H$_{14}$ClN$_5$
Cyprozine
2-Chloro-4-cyclopropylamino-6-isopropylamino-1,3,5-triazine
RN: 22936-86-3   **MP** (°C):   167
MW: 227.70   **BP** (°C):

Solubility (Moles/L)	Solubility (Grams/L)	Temp (°C)	Ref (#)	Evaluation (T P E A A)	Comments
3.030E-05	6.900E-03	25	B200	1 0 0 0 1	
8.582E-04	1.954E-01	40	B200	1 0 0 0 2	

1774.   C$_9$H$_{14}$N$_2$O$_3$
5-Ethyl-5-n-propylbarbituric Acid
2,4,6(1H,3H,5H)-Pyrimidinetrione, 5-Ethyl-5-propyl-
RN: 33376-25-9   **MP** (°C):   146.5
MW: 198.22   **BP** (°C):

Solubility (Moles/L)	Solubility (Grams/L)	Temp (°C)	Ref (#)	Evaluation (T P E A A)	Comments
2.872E-02	5.694E+00	25	B065	1 2 1 1 1	
3.610E-02	7.156E+00	25	M310	2 2 2 2 2	

1775.   C$_9$H$_{14}$N$_2$O$_3$
Metharbital
5,5'-Diethyl-1-methylbarbituric Acid
RN: 50-11-3   **MP** (°C):   155
MW: 198.22   **BP** (°C):

Solubility (Moles/L)	Solubility (Grams/L)	Temp (°C)	Ref (#)	Evaluation (T P E A A)	Comments
1.009E-02	2.000E+00	25	B011	2 0 0 1 0	
9.980E-03	1.978E+00	25	B065	1 1 1 1 1	
1.150E-02	2.280E+00	25	G003	1 1 1 1 2	pH 4.7
6.054E-03	1.200E+00	25	P061	1 0 0 0 2	
4.979E-03	9.870E-01	rt	M161	0 0 0 0 2	

1776.   C$_9$H$_{14}$N$_2$O$_3$
Probarbital
5-Ethyl-5-isopropylbarbituric Acid
2,4,6(1H,3H,5H)-Pyrimidinetrione, 5-Ethyl-5-(1-methylethyl)
RN: 76-76-6   **MP** (°C):   197.5
MW: 198.22   **BP** (°C):

Solubility (Moles/L)	Solubility (Grams/L)	Temp (°C)	Ref (#)	Evaluation (T P E A A)	Comments
6.104E-03	1.210E+00	25	B065	1 1 1 1 1	
7.111E-03	1.410E+00	25	P350	2 1 1 1 2	intrinsic
1.210E-01	2.399E+01	40	N008	1 0 1 1 2	sic

1777.   C$_9$H$_{14}$N$_6$
6-Amino-4-(diallylamino)-1,2-dihydro-1-hydroxy-2-imino-s-triazine
RN: **MP** (°C):
MW: 206.25   **BP** (°C):

Solubility (Moles/L)	Solubility (Grams/L)	Temp (°C)	Ref (#)	Evaluation (T P E A A)	Comments
1.459E-01	3.010E+01	37	H004	1 0 2 2 2	

1778.   C₉H₁₄O₆
L-Camphoronic Acid
L-Camphoronsaeure
RN: 2385-74-2   **MP** (°C):
MW: 218.21   **BP** (°C):

Solubility (Moles/L)	Solubility (Grams/L)	Temp (°C)	Ref (#)	Evaluation (T P E A A)	Comments
5.087E-01	1.110E+02	16	F300	1 0 0 0 2	

1779.   C₉H₁₄O₆
Triacetin
Propane-1,2,3-triyl Triacetate
Enzactin
Vanay
Triacetylglycerol
Glycerol Triacetate
RN: 102-76-1   **MP** (°C):   -78
MW: 218.21   **BP** (°C):   258

Solubility (Moles/L)	Solubility (Grams/L)	Temp (°C)	Ref (#)	Evaluation (T P E A A)	Comments
3.290E-01	7.180E+01	15	F300	1 0 0 0 2	
2.389E-01	5.213E+01	24.50	O005	1 0 2 2 1	
3.118E-02	6.803E+00	ns	F014	0 0 0 0 2	

1780.   C₉H₁₅Br₆O₄P
Tris-BP
Tris(2,3-dibromopropyl)Phosphate
2,3-Dibromo-1-propanol Phosphate (3:1)
2,3-Dibromopropyl Phosphate
Flamex t 23p
Anfram 3pb
RN: 126-72-7   **MP** (°C):   5.5
MW: 697.65   **BP** (°C):

Solubility (Moles/L)	Solubility (Grams/L)	Temp (°C)	Ref (#)	Evaluation (T P E A A)	Comments
1.147E-05	8.000E-03	24	H116	2 1 0 0 2	

1781. C$_9$H$_{15}$Cl$_6$O$_4$P
Fyrol FR-2
Tris(1,3-dichloroisopropyl) Phosphate
TCPP
Emulsion 212
TDCPP
PF 38
RN: 13674-87-8 **MP** (°C):
MW: 430.91 **BP** (°C):

Solubility (Moles/L)	Solubility (Grams/L)	Temp (°C)	Ref (#)	Evaluation (T P E A A)	Comments
1.624E-05	7.000E-03	24	H116	2 1 0 0 2	

1782. C$_9$H$_{15}$NO$_3$
Ecgonine
L-Ekgonin
3-Hydroxy-2-tropane Carboxylic Acid
RN: 481-37-8 **MP** (°C): 198
MW: 185.22 **BP** (°C):

Solubility (Moles/L)	Solubility (Grams/L)	Temp (°C)	Ref (#)	Evaluation (T P E A A)	Comments
9.610E-01	1.780E+02	ns	F300	0 0 0 0 2	

1783. C$_9$H$_{16}$
1-Nonyne
n-Heptylacetylene
Heptylacetylene
RN: 3452-09-3 **MP** (°C): -50
MW: 124.23 **BP** (°C): 150

Solubility (Moles/L)	Solubility (Grams/L)	Temp (°C)	Ref (#)	Evaluation (T P E A A)	Comments
5.796E-05	7.200E-03	25	M001	2 1 2 2 1	

1784. C$_9$H$_{16}$
2,2,5-Trimethyl-3-hexyne
3-Hexyne, 2,2,5-Trimethyl-
RN: 17530-23-3 **MP** (°C):
MW: 124.23 **BP** (°C):

Solubility (Moles/L)	Solubility (Grams/L)	Temp (°C)	Ref (#)	Evaluation (T P E A A)	Comments
2.410E-04	2.994E-02	25	H039	1 2 2 2 2	

1785. C$_9$H$_{16}$ClN$_4$

G 30451
2-Chloro-4-propylamino-6-isopropylamino-s-triazine
RN: 3567-85-9   **MP** (°C):
MW: 215.71   **BP** (°C):

Solubility (Moles/L)	Solubility (Grams/L)	Temp (°C)	Ref (#)	Evaluation (T P E A A)	Comments
1.947E-04	4.200E-02	21	B192	0 0 0 0 1	

1786. C$_9$H$_{16}$ClN$_5$

Trietazine
2-Chloro-4-diethylamino-6-ethylamino-s-triazine
2-Chloro-4-ethylamino-6-diethylamino-s-triazines
RN: 1912-26-1   **MP** (°C):   101
MW: 229.71   **BP** (°C):

Solubility (Moles/L)	Solubility (Grams/L)	Temp (°C)	Ref (#)	Evaluation (T P E A A)	Comments
8.706E-05	2.000E-02	20	B185	1 0 0 0 1	
8.706E-05	2.000E-02	21	B192	0 0 0 0 1	
8.706E-05	2.000E-02	21	G099	2 0 0 1 0	
8.706E-05	2.000E-02	25	M161	1 0 0 0 1	
8.706E-05	2.000E-02	ns	J033	0 0 0 0 1	

1787. C$_9$H$_{16}$ClN$_5$

Propazine
2-Chloro-4-isopropylamino-6-isopropylamino-s-triazine
RN: 139-40-2   **MP** (°C):   213
MW: 229.71   **BP** (°C):

Solubility (Moles/L)	Solubility (Grams/L)	Temp (°C)	Ref (#)	Evaluation (T P E A A)	Comments
3.744E-05	8.600E-03	20	B185	1 0 0 0 1	
4.000E-05	9.189E-03	20	B200	1 0 0 0 0	
2.307E-05	5.300E-03	20	C048	2 2 2 2 1	
2.177E-05	5.000E-03	20	F311	1 2 2 2 1	
3.744E-05	8.600E-03	20	M161	1 0 0 0 1	
3.744E-05	8.600E-03	21	B192	0 0 0 0 1	
3.744E-05	8.600E-03	21	G099	2 0 0 1 0	
3.744E-05	8.600E-03	22	M061	1 0 0 0 1	
7.700E-05	1.769E-02	50	G001	1 0 1 1 1	
3.744E-05	8.600E-03	ns	C101	0 0 0 0 1	
4.353E-05	1.000E-02	ns	G041	0 0 0 0 1	
3.744E-05	8.600E-03	ns	J033	0 0 0 0 1	

1788.   C₉H₁₆ClN₅

Terbuthylazine
Terbutylazine
2-Chloro-4-ethylamino-6-tert-butylamino-s-triazine
Primatol M

RN: 5915-41-3   **MP** (°C):   178
MW: 229.71   **BP** (°C):

Solubility (Moles/L)	Solubility (Grams/L)	Temp (°C)	Ref (#)	Evaluation (T P E A A)	Comments
2.177E-05	5.000E-03	20	F311	1 2 2 2 1	
3.700E-05	8.500E-03	20	M161	1 0 0 0 1	
3.700E-05	8.500E-03	ns	J033	0 0 0 0 1	

1789.   C₉H₁₆N₂O₄

Methyl-2,2-diethylmalonurate
Methyl 2,2-Diethylmalonurate

RN: 69577-07-7   **MP** (°C):   112
MW: 216.24   **BP** (°C):

Solubility (Moles/L)	Solubility (Grams/L)	Temp (°C)	Ref (#)	Evaluation (T P E A A)	Comments
1.100E-02	2.379E+00	23	B152	1 2 1 1 1	pH 3.5

1790.   C₉H₁₆N₄OS

Tebuthiuron
1-(5-tert-Butyl-1,3,4-thiadiazol-2-yl)-1,3-dimethylurea
Graslan
Spike
Spike 20P
Perflan

RN: 34014-18-1   **MP** (°C):   162.2
MW: 228.32   **BP** (°C):

Solubility (Moles/L)	Solubility (Grams/L)	Temp (°C)	Ref (#)	Evaluation (T P E A A)	Comments
1.007E-02	2.300E+00	ns	M161	0 0 0 0 1	

1791.   C₉H₁₆N₈

2-Azido-4-ethylamino-4-t-butylamino-s-triazine
WL 9385

RN: 2854-70-8   **MP** (°C):   102.5
MW: 236.28   **BP** (°C):

Solubility (Moles/L)	Solubility (Grams/L)	Temp (°C)	Ref (#)	Evaluation (T P E A A)	Comments
3.047E-04	7.200E-02	20	M061	1 0 0 0 1	

1792.   C$_9$H$_{16}$O$_2$
3-Hydroxy-5-spirocyclohexyltetrahydrofuran
1-Oxaspiro[4.5]decan-3-ol
RN: 29839-61-0   **MP** (°C):
MW: 156.23   **BP** (°C):

Solubility (Moles/L)	Solubility (Grams/L)	Temp (°C)	Ref (#)	Evaluation (T P E A A)	Comments
1.255E-01	1.961E+01	rt	B066	0 2 0 0 0	contains impurity

1793.   C$_9$H$_{16}$O$_2$
3-Hydroxy-2-methyl-5-spirocyclopentyltetrahydrofuran
1-Oxaspiro[4.4]nonan-3-ol, 2-Methyl-
RN: 29839-62-1   **MP** (°C):
MW: 156.23   **BP** (°C):

Solubility (Moles/L)	Solubility (Grams/L)	Temp (°C)	Ref (#)	Evaluation (T P E A A)	Comments
1.067E+00	1.667E+02	rt	B066	0 2 0 0 1	

1794.   C$_9$H$_{16}$O$_4$
Azelaic Acid
Azelainsaeure
Nonanedioic Acid
RN: 123-99-9   **MP** (°C):   106.5
MW: 188.23   **BP** (°C):   287

Solubility (Moles/L)	Solubility (Grams/L)	Temp (°C)	Ref (#)	Evaluation (T P E A A)	Comments
5.313E-03	1.000E+00	0	L041	1 0 0 1 1	
3.298E-03	6.208E-01	6.99	A340	2 0 2 2 2	
4.513E-03	8.494E-01	12.69	A340	2 0 2 2 2	
7.969F-03	1.500E+00	15	L041	1 0 0 1 1	
6.475E-03	1.219E+00	18.69	A340	2 0 2 2 2	
1.275E-02	2.400E+00	20	F300	1 0 0 0 1	
1.275E-02	2.400E+00	20	L041	1 0 0 1 1	
1.297E-02	2.441E+00	20	M171	1 0 0 0 1	
2.667E-01	5.020E+01	21	B040	1 0 1 1 2	*sic*
9.461E-03	1.781E+00	24.99	A340	2 0 2 2 2	
1.589E-02	2.990E+00	34.69	A340	2 0 2 2 2	
2.391E-02	4.500E+00	35	L041	1 0 0 1 1	
1.858E-02	3.498E+00	42.99	A340	2 0 2 2 2	
4.356E-02	8.200E+00	50	L041	1 0 0 1 1	
2.662E-02	5.010E+00	52.59	A340	2 0 2 2 2	
3.858E-02	7.263E+00	56.99	A340	2 0 2 2 2	
5.124E-02	9.645E+00	61.49	A340	2 0 2 2 2	
7.023E-02	1.322E+01	64.99	A340	2 0 2 2 2	
1.169E-01	2.200E+01	65	F300	1 0 0 0 1	
1.169E-01	2.200E+01	65	L041	1 0 0 1 1	

7.255E-02	1.366E+01	70.99	A340	2 0 2 2 2
8.355E-02	1.573E+01	74.49	A340	2 0 2 2 2
1.048E-01	1.972E+01	79.89	A340	2 0 2 2 2
9.430E-02	1.775E+01	84.49	A340	2 0 2 2 2

1795.   C$_9$H$_{16}$O$_4$
Butyl α-Acetoxypropionate
Hydracrylic Acid, Butyl Ester, Acetate
RN: 5422-69-5   **MP** (°C):
MW: 188.23   **BP** (°C):

Solubility (Moles/L)	Solubility (Grams/L)	Temp (°C)	Ref (#)	Evaluation (T P E A A)	Comments
1.700E-02	3.200E+00	25	R006	2 2 0 1 1	

1796.   C$_9$H$_{16}$O$_5$
Propanoic Acid, 2-[(Butoxycarbonyl)oxy]-, Methyl Ester
Propanoic Acid, 2-[(Methoxycarbonyl)oxy]-, Butyl Ester
RN: **MP** (°C):
MW: 204.22   **BP** (°C):

Solubility (Moles/L)	Solubility (Grams/L)	Temp (°C)	Ref (#)	Evaluation (T P E A A)	Comments
8.798E-03	1.797E+00	25	R007	1 0 0 0 1	
8.798E-03	1.797E+00	25	R007	1 0 0 0 1	

1797.   C$_9$H$_{17}$ClN$_3$O$_3$PS
Isazophos
Diethyl O-(5-Chloro-1-(1-methylethyl)-1H-1,2,4-triazol-3-yl) phosphorothioate
Miral
Triumph
CGA-12223
RN: 42509-80-8   **MP** (°C):
MW: 313.74   **BP** (°C):   100

Solubility (Moles/L)	Solubility (Grams/L)	Temp (°C)	Ref (#)	Evaluation (T P E A A)	Comments
4.780E-04	1.500E-01	20	E048	1 2 1 1 2	
4.781E-04	1.500E-01	20	M161	1 0 0 0 1	

1798.   C$_9$H$_{17}$NOS
Molinate
S-Ethyl Hexahydro-1H-azepine-1-carbothioate
Hydram
Carbothialate, Ethyl-1-hexa-methylene Imine-
Poperidinecarbothioic Acid, S-Ethyl Ester
RN: 2212-67-1   **MP** (°C):
MW: 187.31   **BP** (°C):

Solubility (Moles/L)	Solubility (Grams/L)	Temp (°C)	Ref (#)	Evaluation (T P E A A)	Comments
4.271E-03	8.000E-01	20	B200	1 0 0 0 2	
4.271E-03	8.000E-01	21	M161	1 0 0 0 2	
4.698E-03	8.800E-01	22	K137	1 1 2 1 0	
<5.33E-03	<9.99E-01	ns	B185	0 0 0 0 0	
4.869E-03	9.120E-01	ns	F019	0 0 0 0 2	
5.334E-03	9.990E-01	ns	M061	0 0 0 0 0	

1799.   C$_9$H$_{17}$NO$_3$
Diethylaceturethane
Detonal
RN: **MP** (°C):
MW: 187.24   **BP** (°C):

Solubility (Moles/L)	Solubility (Grams/L)	Temp (°C)	Ref (#)	Evaluation (T P E A A)	Comments
2.796E-02	5.236E+00	ns	O021	0 2 0 0 0	

1800.   C$_9$H$_{17}$NO$_4$
3,3-Dihydroxy-2,2,5,5-tetramethyl-4-carbamyltetrahydrofuran
3-Furamide, Tetrahydro-4,4-dihydroxy-2,2,5,5-tetramethyl-
RN: 29839-68-7   **MP** (°C):
MW: 203.24   **BP** (°C):

Solubility (Moles/L)	Solubility (Grams/L)	Temp (°C)	Ref (#)	Evaluation (T P E A A)	Comments
4.473E-01	9.091E+01	rt	B066	0 2 0 0 1	

1801.   C$_9$H$_{17}$N$_5$O
Atratone
2-Methoxy-4-ethylamino-6-isopropylamino-s-triazine
2-Methoxy-4-ethylamino-6-isopropylamino-s-triazines
RN: 1610-17-9   **MP** (°C):
MW: 211.27   **BP** (°C):

Solubility (Moles/L)	Solubility (Grams/L)	Temp (°C)	Ref (#)	Evaluation (T P E A A)	Comments
8.520E-03	1.800E+00	20	B185	1 0 0 0 2	
8.520E-03	1.800E+00	20	M061	1 0 0 0 2	
8.520E-03	1.800E+00	21	B192	0 0 0 0 2	
8.520E-03	1.800E+00	21	G099	2 0 0 1 0	
7.905E-03	1.670E+00	25	H073	2 1 1 2 2	
1.240E-02	2.620E+00	50	G001	1 0 1 1 2	
9.448E-03	1.996E+00	ns	B100	0 0 0 0 0	
8.520E-03	1.800E+00	ns	C101	0 0 0 0 1	
7.829E-03	1.654E+00	ns	J033	0 0 0 0 2	

1802.   C$_9$H$_{17}$N$_5$S
Ametryn
(2-Methylthio-4-ethylamino-6-isopropylamino-s-triazine
Ametryne
N-Ethyl-N'-(1-methylethyl)-6-(methylthio)-1,3,5-Triazine-2,4-diamine
Ametrex
RN: 834-12-8   **MP** (°C):   84
MW: 227.33   **BP** (°C):

Solubility (Moles/L)	Solubility (Grams/L)	Temp (°C)	Ref (#)	Evaluation (T P E A A)	Comments
8.100E-04	1.841E-01	20	B200	1 0 0 0 1	
8.358E-04	1.900E-01	20	F311	1 2 2 2 1	
8.138E-04	1.850E-01	20	M161	1 0 0 0 2	
9.194E-04	2.090E-01	25	H073	2 1 1 2 2	
1.660E-03	3.774E-01	50	G001	1 0 1 1 2	
8.138E-04	1.850E-01	ns	C101	0 0 0 0 1	
8.490E-04	1.930E-01	ns	J033	0 0 0 0 2	

1803. C$_9$H$_{18}$
1,1,3-Trimethylcyclohexane
Cyclogeraniolane
RN: 3073-66-3 **MP** (°C): -65.7
MW: 126.24 **BP** (°C): 136.6

Solubility (Moles/L)	Solubility (Grams/L)	Temp (°C)	Ref (#)	Evaluation (T P E A A)	Comments
1.402E-05	1.770E-03	25	K119	1 0 0 0 2	
1.402E-05	1.770E-03	25	P051	2 1 1 2 2	
1.402E-05	1.770E-03	25.00	P007	2 1 2 2 2	

1804. C$_9$H$_{18}$
1-Nonene
α-Nonene
1-n-Nonene
n-Non-1-ene
RN: 124-11-8 **MP** (°C): -81
MW: 126.24 **BP** (°C): 146.9

Solubility (Moles/L)	Solubility (Grams/L)	Temp (°C)	Ref (#)	Evaluation (T P E A A)	Comments
8.850E-06	1.117E-03	25	M342	1 0 1 1 2	

1805. C$_9$H$_{18}$N$_2$O$_2$S
Thiofanox
3,3-Dimethyl-1-(methylthio)-2-butanone O-((methylamino)carbonyl)oxime
Thiophanox
DS-15647
Dacamox
RN: 39196-18-4 **MP** (°C): 57
MW: 218.32 **BP** (°C):

Solublllty (Moles/L)	Solubility (Grams/L)	Temp (°C)	Ref (#)	Evaluation (T P E A A)	Comments
2.382E-02	5.200E+00	22	M161	1 0 0 0 1	

1806. C$_9$H$_{18}$N$_2$O$_4$
Meprobamate
2-Methyl-2-propyl-1,3-propanediol Dicarbamate
Deprol
Meprospan
Miltown
Pathibamate
RN: 57-53-4   **MP** (°C):   104
MW: 218.25   **BP** (°C):

Solubility (Moles/L)	Solubility (Grams/L)	Temp (°C)	Ref (#)	Evaluation (T P E A A)	Comments
2.841E-02	6.200E+00	25	C039	1 2 2 1 1	form II
1.512E-02	3.300E+00	25	C039	1 2 2 1 1	form I
1.512E-02	3.300E+00	25	D082	1 0 1 0 1	
3.757E-02	8.200E+00	30	C039	1 2 2 1 1	form II
1.970E-02	4.300E+00	30	C039	1 2 2 1 1	form I
2.612E-02	5.700E+00	35	C039	1 2 2 1 1	form I
4.857E-02	1.060E+01	35	C039	1 2 2 1 2	form II
3.391E-02	7.400E+00	40	C039	1 2 2 1 1	form I
5.865E-02	1.280E+01	40	C039	1 2 2 1 2	form II

1807. C$_9$H$_{18}$N$_3$S$_6$Fe
Ferbam
Tris(dimethyldithiocarbamate)iron
Knockmate
Ferbeck
Hexaferb
Trifungol
RN: 14484-64-1   **MP** (°C):
MW: 416.49   **BP** (°C):

Solubility (Moles/L)	Solubility (Grams/L)	Temp (°C)	Ref (#)	Evaluation (T P E A A)	Comments
2.881E-04	1.200E-01	rt	I314	0 0 0 0 2	
3.121E-04	1.300E-01	rt	M161	0 0 0 0 2	

1808.   C$_9$H$_{18}$N$_6$
1,3,5-Triazine-2,4,6-triamine, N,N',N''-Triethyl-
N2,N4,N6-Triethylmelamine
Tris(ethylamino)-1,3,5-triazine
2,4,6-Tris(ethylamino)-1,3,5-triazine
2,4,6-Tris(ethylamino)-s-triazine
N,N',N''-Triethyl-1,3,5-triazine-2,4,6-triamine
RN: 16268-92-1   **MP** (°C):
MW: 210.28   **BP** (°C):

Solubility (Moles/L)	Solubility (Grams/L)	Temp (°C)	Ref (#)	Evaluation (T P E A A)	Comments
7.318E-03	1.539E+00	25	B386	2 2 2 2 2	

1809.   C$_9$H$_{18}$N$_6$
Altretamine
Hexamethylmelamine
2,4,6-Tris(dimethylamino)-1,3,5-triazine
HMM
Hexastat
Hemel
RN: 645-05-6   **MP** (°C):   172.0
MW: 210.28   **BP** (°C):

Solubility (Moles/L)	Solubility (Grams/L)	Temp (°C)	Ref (#)	Evaluation (T P E A A)	Comments
3.846E-04	8.088E-02	25	B386	2 2 2 2 2	
4.327E-04	9.100E-02	25	C051	1 2 1 1 1	pH 7
4.150E-04	8.727E-02	25	K043	2 0 0 0 0	extrapolated

1810.   C$_9$H$_{18}$N$_6$O
Ethanol, 2-[[4,6-bis(Dimethylamino)-s-triazin-2-yl]amino]-
Ethanol, 2-[[4,6-bis(Dimethylamino)-1,3,5-triazin-2-yl]amino]-
RN: 31482-09-4   **MP** (°C):
MW: 226.28   **BP** (°C):

Solubility (Moles/L)	Solubility (Grams/L)	Temp (°C)	Ref (#)	Evaluation (T P E A A)	Comments
1.132E-02	2.562E+00	25	B386	2 2 2 2 2	

1811.   C$_9$H$_{18}$N$_6$O
N-Methylolpentamethylmelamine
N-(Hydroxymethyl)pentamethylmelamine
(Hydroxymethyl)pentamethylmelamine
RN: 16269-01-5   **MP** (°C):   121.0
MW: 226.28   **BP** (°C):

Solubility (Moles/L)	Solubility (Grams/L)	Temp (°C)	Ref (#)	Evaluation (T P E A A)	Comments
3.977E-03	9.000E-01	25	C051	1 2 1 1 0	pH 7, unstable in water

1812.   C$_9$H$_{18}$N$_6$O$_3$
N2,N4,N6-Trimethyl-N2,N4,N6-trimethylolmelamine
N,N',N''-Trimethyl-N,N',N''-trimethylolmelamine
Trimelamol
CB 10-375
RN: 64124-21-6   **MP** (°C):   129
MW: 258.28   **BP** (°C):

Solubility (Moles/L)	Solubility (Grams/L)	Temp (°C)	Ref (#)	Evaluation (T P E A A)	Comments
3.500E-02	9.040E+00	25	C051	1 2 1 1 2	pH 7

1813.   C$_9$H$_{18}$O
2,6-Dimethyl-4-heptanone
Diisobutyl Ketone
RN: 108-83-8   **MP** (°C):
MW: 142.24   **BP** (°C):   169

Solubility (Moles/L)	Solubility (Grams/L)	Temp (°C)	Ref (#)	Evaluation (T P E A A)	Comments
1.851E-02	2.633E+00	23.50	O005	2 0 2 2 2	

1814.   C$_9$H$_{18}$O
3-Hydroxy-2,3,4,5,5-pentamethyltetrahydrofuran
RN: **MP** (°C):
MW: 142.24   **BP** (°C):

Solubility (Moles/L)	Solubility (Grams/L)	Temp (°C)	Ref (#)	Evaluation (T P E A A)	Comments
6.391E-01	9.091E+01	rt	B066	0 2 0 0 1	

1815.   C$_9$H$_{18}$O
5-Nonanone
Dibutyl Ketone
RN: 502-56-7   **MP** (°C):   -50
MW: 142.24   **BP** (°C):   186.5

Solubility (Moles/L)	Solubility (Grams/L)	Temp (°C)	Ref (#)	Evaluation (T P E A A)	Comments
3.570E-03	5.078E-01	10	G032	1 2 1 1 2	
1.800E-03	2.560E-01	25	K012	1 0 0 0 1	
2.550E-03	3.627E-01	30	G032	1 2 1 1 2	
2.430E-03	3.457E-01	50	G032	1 2 1 1 2	

1816. C$_9$H$_{18}$O
Nonyl Aldehyde
n-Nonanal
RN: 124-19-6 **MP** (°C):
MW: 142.24 **BP** (°C): 93

Solubility (Moles/L)	Solubility (Grams/L)	Temp (°C)	Ref (#)	Evaluation (T P E A A)	Comments
6.749E-04	9.600E-02	25	A049	1 0 0 0 1	

1817. C$_9$H$_{18}$O$_2$
3-Hydroxy-2-isopropyl-5,5-dimethyltetrahydrofuran
3-Furanol, Tetrahydro-2-isopropyl-5,5-dimethyl-
RN: 29839-66-5 **MP** (°C):
MW: 158.24 **BP** (°C):

Solubility (Moles/L)	Solubility (Grams/L)	Temp (°C)	Ref (#)	Evaluation (T P E A A)	Comments
3.009E-01	4.762E+01	rt	B066	0 2 0 0 0	

1818. C$_9$H$_{18}$O$_2$
3-Hydroxy-5-propyl-2,5-dimethyltetrahydrofuran
3-Furanol, 2,5-Dimethyltetrahydro-5-propyl-
RN: **MP** (°C):
MW: 158.24 **BP** (°C):

Solubility (Moles/L)	Solubility (Grams/L)	Temp (°C)	Ref (#)	Evaluation (T P E A A)	Comments
1.841E-01	2.913E+01	rt	B066	0 2 0 0 0	

1819. C$_9$H$_{18}$O$_2$
3-Hydroxy-5-methyl-5-isobutyltetrahydrofuran
3-Furanol, 5-Isobutyltetrahydro-5-methyl-
RN: **MP** (°C):
MW: 158.24 **BP** (°C):

Solubility (Moles/L)	Solubility (Grams/L)	Temp (°C)	Ref (#)	Evaluation (T P E A A)	Comments
6.257E-02	9.901E+00	rt	B066	0 2 0 0 0	

1820. C$_9$H$_{18}$O$_2$
3-Hydroxy-5-methyl-5-butyltetrahydrofuran
3-Furanol, 5-Butyltetrahydro-5-methyl-
RN: **MP** (°C):
MW: 158.24 **BP** (°C):

Solubility (Moles/L)	Solubility (Grams/L)	Temp (°C)	Ref (#)	Evaluation (T P E A A)	Comments
2.518E-02	3.984E+00	rt	B066	0 2 0 0 0	

1821. $C_9H_{18}O_2$

3-Hydroxy-3-ethyl-2,2,5-trimethyltetrahydrofuranol
3-Furanol, 3-Ethyltetrahydro-2,2,5-trimethyl-

RN: 29839-58-5   **MP** (°C):
MW: 158.24   **BP** (°C):

Solubility (Moles/L)	Solubility (Grams/L)	Temp (°C)	Ref (#)	Evaluation (T P E A A)	Comments
4.134E-01	6.542E+01	rt	B066	0 2 0 0 0	

1822. $C_9H_{18}O_2$

Methyl Octanoate
Methyl Caprylate
Methyl Octylate

RN: 111-11-5   **MP** (°C):   -37
MW: 158.24   **BP** (°C):   194.5

Solubility (Moles/L)	Solubility (Grams/L)	Temp (°C)	Ref (#)	Evaluation (T P E A A)	Comments
4.069E-04	6.440E-02	20	M337	2 1 2 2 2	

1823. $C_9H_{18}O_2$

3-Hydroxy-2-methyl-2,5-diethyltetrahydrofuran
3-Furanol, 2,5-Diethyltetrahydro-2-methyl-

RN: 29839-64-3   **MP** (°C):
MW: 158.24   **BP** (°C):

Solubility (Moles/L)	Solubility (Grams/L)	Temp (°C)	Ref (#)	Evaluation (T P E A A)	Comments
1.239E-01	1.961E+01	rt	B066	0 2 0 0 0	

1824. $C_9H_{18}O_2$

Pelargonic Acid
1-Octanecarboxylic Acid
Nonylic Acid
n-Nonanoic Acid

RN: 112-05-0   **MP** (°C):   12
MW: 158.24   **BP** (°C):

Solubility (Moles/L)	Solubility (Grams/L)	Temp (°C)	Ref (#)	Evaluation (T P E A A)	Comments
8.847E-04	1.400E-01	0	B136	1 0 2 1 1	
8.846E-04	1.400E-01	0.0	R001	1 1 1 1 1	
1.795E-03	2.840E-01	20	B136	1 0 2 1 2	
1.643E-03	2.599E-01	20.0	R001	1 1 1 1 1	
2.003E-03	3.170E-01	30	B136	1 0 2 1 2	
1.340E-03	2.120E-01	30	E005	2 1 1 2 2	
2.022E-03	3.199E-01	30.0	R001	1 1 1 1 1	

2.496E-03	3.950E-01	40	B136	1 0 2 1 2
1.403E-03	2.220E-01	40	E005	2 1 1 2 2
2.591E-03	4.100E-01	45	B136	1 0 2 1 1
2.590E-03	4.098E-01	45.0	R001	1 1 1 1 1
1.668E-03	2.640E-01	50	E005	2 1 1 2 2
3.223E-03	5.100E-01	60	B136	1 0 2 1 1
1.890E-03	2.990E-01	60	E005	2 1 1 2 2
3.221E-03	5.097E-01	60.0	R001	1 1 1 1 1

1825. $C_9H_{18}O_2$
3-Hydroxy-2,2,4,5,5-pentamethyltetrahydrofuran
3-Furanol, Tetrahydro-2,2,4,5,5-pentamethyl-
RN: 29839-76-7 **MP** (°C):
MW: 158.24 **BP** (°C):

Solubility (Moles/L)	Solubility (Grams/L)	Temp (°C)	Ref (#)	Evaluation (T P E A A)	Comments
6.257E-02	9.901E+00	rt	B066	0 2 0 0 0	

1826. $C_9H_{18}O_2$
Butyl Valerate
n-Butyl Pentanoate
Butyl Valerianate
RN: 591-68-4 **MP** (°C):
MW: 158.24 **BP** (°C): 186-187

Solubility (Moles/L)	Solubility (Grams/L)	Temp (°C)	Ref (#)	Evaluation (T P E A A)	Comments
5.300E-04	8.387E-02	25	K012	1 0 0 0 1	

1827. $C_9H_{18}O_2$
3-Hydroxy-2-methyl-5,5-diethyltetrahydrofuran
3-Furanol, 5,5-Diethyltetrahydro-2-methyl-
RN: 6744-54-3 **MP** (°C):
MW: 158.24 **BP** (°C):

Solubility (Moles/L)	Solubility (Grams/L)	Temp (°C)	Ref (#)	Evaluation (T P E A A)	Comments
3.144E-02	4.975E+00	rt	B066	0 2 0 0 0	

1828. $C_9H_{18}O_2$
Pentyl Butyrate
n-Amyl n-Butyrate
Pentyl n-Butanoate
RN: 540-18-1 **MP** (°C):
MW: 158.24 **BP** (°C):

Solubility (Moles/L)	Solubility (Grams/L)	Temp (°C)	Ref (#)	Evaluation (T P E A A)	Comments
1.100E-03	1.741E-01	20	S006	1 0 0 0 1	

1829.   C₉H₁₈O₃
Hexyl Lactate
Propanoic Acid, 2-Hydroxy-, Hexyl Ester
RN: 20279-51-0   **MP** (°C):
MW: 174.24   **BP** (°C):

Solubility (Moles/L)	Solubility (Grams/L)	Temp (°C)	Ref (#)	Evaluation (T P E A A)	Comments
1.550E-02	2.700E+00	25	R006	2 2 0 1 1	

1830.   C₉H₁₈O₃
n-Butyl β-Ethoxypropionate
Propionic Acid, 3-Ethoxy-, Butyl Ester
RN: 14144-35-5   **MP** (°C):
MW: 174.24   **BP** (°C):

Solubility (Moles/L)	Solubility (Grams/L)	Temp (°C)	Ref (#)	Evaluation (T P E A A)	Comments
2.287E-02	3.984E+00	25	D002	1 2 1 1 1	

1831.   C₉H₁₈O₃
2,2-Diethyl-5-methyl-tetrahydrofuran-3,4-diol
3,4-Furandiol, 2,2-Diethyltetrahydro-5-methyl-
RN: 31889-35-7   **MP** (°C):
MW: 174.24   **BP** (°C):

Solubility (Moles/L)	Solubility (Grams/L)	Temp (°C)	Ref (#)	Evaluation (T P E A A)	Comments
9.565E-01	1.667E+02	rt	B066	0 2 0 0 1	

1832.   C₉H₁₈O₃
n-Propyl β-n-Propoxypropionate
Propanoic Acid, 3-Propoxy-, Propyl Ester
RN: 14144-41-3   **MP** (°C):
MW: 174.24   **BP** (°C):

Solubility (Moles/L)	Solubility (Grams/L)	Temp (°C)	Ref (#)	Evaluation (T P E A A)	Comments
2.059E-02	3.587E+00	25	R034	0 0 0 0 1	

1833.   C₉H₁₈O₃
1,3-Dioxolane-4-methanol, 2-Butyl-2-methyl
RN: 5694-76-8   **MP** (°C):
MW: 174.24   **BP** (°C):

Solubility (Moles/L)	Solubility (Grams/L)	Temp (°C)	Ref (#)	Evaluation (T P E A A)	Comments
1.940E-01	3.380E+01	25	P342	1 2 2 2 2	0.0001M Na₂CO₃

1834. $C_9H_{18}O_3$
n-Amyl β-Methoxypropionate
Pentyl 3-Methoxypropionate
RN: 10500-16-0   **MP** (°C):
MW: 174.24   **BP** (°C):

Solubility (Moles/L)	Solubility (Grams/L)	Temp (°C)	Ref (#)	Evaluation (T P E A A)	Comments
1.660E-02	2.892E+00	25	R034	0 0 0 0 1	

1835. $C_9H_{19}NOS$
Eptam
EPTC
Ethyl N,N'-di-n-Propylthiolcarbamate
S-Ethyl Dipropylthiocarbamate
S-Ethyl N,N-di-n-Propylthiocarbamate
RN: 759-94-4   **MP** (°C):   <25
MW: 189.32   **BP** (°C):   235

Solubility (Moles/L)	Solubility (Grams/L)	Temp (°C)	Ref (#)	Evaluation (T P E A A)	Comments
3.359E-03	6.360E-01	3	G319	1 0 0 0 2	
1.954E-03	3.700E-01	20	B200	1 0 0 0 2	
1.981E+01	3.750E+03	20	F019	1 0 0 0 2	*sic*
1.981E-03	3.750E-01	20	M061	1 0 0 0 2	
1.928E-03	3.650E-01	20	M161	1 0 0 0 2	
4.170E+00	7.895E+02	25	B185	1 0 0 0 2	*sic*
1.981E-03	3.750E-01	25	G319	1 0 0 0 2	
1.981E-03	3.750E-01	25	M131	0 0 0 0 2	
2.123E-03	4.020E-01	28	H109	1 0 0 0 2	

1836. $C_9H_{19}NO_2$
n-Octyl Carbamate
Carbamic Acid, Octyl Ester
RN: 2029-64-3   **MP** (°C):   67
MW: 173.26   **BP** (°C):

Solubility (Moles/L)	Solubility (Grams/L)	Temp (°C)	Ref (#)	Evaluation (T P E A A)	Comments
5.000E-04	8.663E-02	37	H006	1 2 2 1 0	

1837. $C_9H_{19}O_3$
3-Hydroxy-4-methylol-2,2,5,5-tetramethyltetrahydrofuran
RN: **MP** (°C):
MW: 175.25   **BP** (°C):

Solubility (Moles/L)	Solubility (Grams/L)	Temp (°C)	Ref (#)	Evaluation (T P E A A)	Comments
1.119E-01	1.961E+01	rt	B066	0 2 0 0 0	

1838. C$_9$H$_{20}$
3,3-Diethylpentane
Tetraethylmethane
RN: 1067-20-5   **MP** (°C):
MW: 128.26   **BP** (°C):

Solubility (Moles/L)	Solubility (Grams/L)	Temp (°C)	Ref (#)	Evaluation (T P E A A)	Comments
9.450E-06	1.212E-03	25	D346	1 1 2 2 2	

1839. C$_9$H$_{20}$
Nonane
n-Nonan
RN: 111-84-2   **MP** (°C):   -53
MW: 128.26   **BP** (°C):   151

Solubility (Moles/L)	Solubility (Grams/L)	Temp (°C)	Ref (#)	Evaluation (T P E A A)	Comments
<1.72E-05	<2.20E-03	20	M337	2 1 2 2 1	
9.512E-07	1.220E-04	25	K119	1 0 0 0 2	
1.715E-06	2.200E-04	25	M003	1 0 2 2 2	
9.512E-07	1.220E-04	25.0	P051	2 1 1 2 2	
9.512E-07	1.220E-04	25.00	P007	2 1 2 2 2	
2.409E-06	3.090E-04	69.7	P051	2 1 1 2 2	
3.275E-06	4.200E-04	99.1	P051	2 1 1 2 2	
3.275E-06	4.200E-04	99.10	P007	2 1 2 2 2	
1.325E-05	1.700E-03	121.3	P051	2 1 1 2 2	
1.325E-05	1.700E-03	121.30	P007	2 1 2 2 2	
3.953E-05	5.070E-03	136.6	P051	2 1 1 2 2	
3.953E-05	5.070E-03	136.60	P007	2 1 2 2 2	

1840. C$_9$H$_{20}$
4-Methyloctane
4-Metylooktan
RN: 2216-34-4   **MP** (°C):   -113
MW: 128.26   **BP** (°C):

Solubility (Moles/L)	Solubility (Grams/L)	Temp (°C)	Ref (#)	Evaluation (T P E A A)	Comments
8.966E-07	1.150E-04	25	K119	1 0 0 0 2	
8.966E-07	1.150E-04	25	P051	2 1 1 2 2	
8.966E-07	1.150E-04	25.00	P007	2 1 2 2 2	

1841. C_9H_20
2,2,5-Trimethylhexane
Hexane, 2,2,5-Trimethyl-
RN: 3522-94-9   **MP** (°C):   -120
MW: 128.26   **BP** (°C):   124.1

Solubility (Moles/L)	Solubility (Grams/L)	Temp (°C)	Ref (#)	Evaluation (T P E A A)	Comments
6.159E-06	7.900E-04	0	P003	2 2 2 2 1	
8.966E-06	1.150E-03	25	M001	2 1 2 2 2	
4.210E-06	5.400E-04	25	P003	2 2 2 2 1	

1842. C_9H_20
3-Methyloctane
Octane, 3-Methyl-
RN: 2216-33-3   **MP** (°C):
MW: 128.26   **BP** (°C):

Solubility (Moles/L)	Solubility (Grams/L)	Temp (°C)	Ref (#)	Evaluation (T P E A A)	Comments
6.237E-06	8.000E-04	23	C332	2 0 2 2 1	

1843. C_9H_20NO_3PS_2
Fostion
FAC 20
O,O-Diethyl S-(N-Isopropylcarbamylmethyl) Dithiophosphate
Prothoate
RN: 2275-18-5   **MP** (°C):   24.5
MW: 285.37   **BP** (°C):

Solubility (Moles/L)	Solubility (Grams/L)	Temp (°C)	Ref (#)	Evaluation (T P E A A)	Comments
8.761E-03	2.500E+00	20	M161	1 0 0 0 1	

1844. C_9H_20O
2,6-Dimethyl-4-heptanol
Diisobutylcarbinol
RN: 108-82-7   **MP** (°C):
MW: 144.26   **BP** (°C):

Solubility (Moles/L)	Solubility (Grams/L)	Temp (°C)	Ref (#)	Evaluation (T P E A A)	Comments
6.925E-03	9.990E-01	25	C093	2 1 1 1 1	

1845.   C$_9$H$_{20}$O
3,5,5-Trimethylhexanol
3.,5,5-Trimethyl Hexanol
Nonylol
3,5,5-Trimethyl-1-hexanol
RN: 3452-97-9   **MP** (°C):
MW: 144.26   **BP** (°C):

Solubility (Moles/L)	Solubility (Grams/L)	Temp (°C)	Ref (#)	Evaluation (T P E A A)	Comments
3.120E-03	4.501E-01	20	H330	2 0 2 2 2	
3.099E-03	4.470E-01	ns	J300	0 0 0 0 1	

1846.   C$_9$H$_{20}$O
n-Nonyl Alcohol
Nonanol
RN: 143-08-8   **MP** (°C):
MW: 144.26   **BP** (°C):   215

Solubility (Moles/L)	Solubility (Grams/L)	Temp (°C)	Ref (#)	Evaluation (T P E A A)	Comments
9.340E-04	1.347E-01	20	H330	2 0 2 2 2	
9.700E-04	1.399E-01	25	K025	2 2 1 1 2	

1847.   C$_9$H$_{20}$O
3-Nonanol
Hexyl Ethyl Carbinol
Ethyl n-Hexyl Carbinol
RN: 624-51-1   **MP** (°C):
MW: 144.26   **BP** (°C):

Solubility (Moles/L)	Solubility (Grams/L)	Temp (°C)	Ref (#)	Evaluation (T P E A A)	Comments
1.999E-03	2.884E-01	ns	J300	0 0 0 0 1	

1848.   C$_9$H$_{20}$O
Methyl-octyl-alcohol
2-Nonanol
Heptylmethylcarbinol
Methyl n-Heptyl Carbinol
RN: 628-99-9   **MP** (°C):
MW: 144.26   **BP** (°C):

Solubility (Moles/L)	Solubility (Grams/L)	Temp (°C)	Ref (#)	Evaluation (T P E A A)	Comments
≤4.00E-03	<5.77E-01	25	F044	1 0 0 0 0	

1849. C$_9$H$_{21}$N
Tripropylamine
Tri-n-propylamine
N,N-Dipropylpropanamine
N,N-Dipropyl-1-propanamine
RN: 102-69-2   **MP** (°C):   -93.5
MW: 143.27   **BP** (°C):   155

Solubility (Moles/L)	Solubility (Grams/L)	Temp (°C)	Ref (#)	Evaluation (T P E A A)	Comments
5.216E-03	7.473E-01	25.04	V013	2 2 2 2 2	

1850. C$_9$H$_{21}$O$_2$PS$_3$
Terbufos
O,O-Diethyl S-(((1,1-Dimethylethyl)thio)methyl) Phosphorodithoic Acid
Counter 15G
Contraven
ST 100
RN: 13071-79-9   **MP** (°C):
MW: 288.43   **BP** (°C):

Solubility (Moles/L)	Solubility (Grams/L)	Temp (°C)	Ref (#)	Evaluation (T P E A A)	Comments
1.907E-05	5.500E-03	19	B169	2 1 1 1 1	
1.758E-05	5.070E-03	24	F179	2 2 2 2 2	
1.907E-05	5.500E-03	ns	B325	0 1 0 0 1	
3.467E-05	1.000E-02	ns	M110	0 0 0 0 0	EFG
4.334E-05	1.250E-02	ns	M161	0 0 0 0 0	

1851. C$_9$H$_{21}$O$_3$P
Dibutyl Methyl Phosphonate
Di-n-butyl Methanephosphonate
RN: 2404-73-1   **MP** (°C):
MW: 208.24   **BP** (°C):

Solubility (Moles/L)	Solubility (Grams/L)	Temp (°C)	Ref (#)	Evaluation (T P E A A)	Comments
3.842E-02	8.000E+00	25	B070	1 2 0 1 0	

1852. $C_9H_{21}O_3PS_3$

S-Ethylsulphinylmethyl O,O-Di-isopropyl Phosphorodithioate
O,O-Diisopropyl S-[(Ethylsulfinyl)methyl] Dithiophosphate
Aphidan
PSP 204
IPSP

RN: 5827-05-4   **MP** (°C):
MW: 304.43   **BP** (°C):

Solubility (Moles/L)	Solubility (Grams/L)	Temp (°C)	Ref (#)	Evaluation (T P E A A)	Comments
4.927E-03	1.500E+00	15	M161	1 0 0 0 1	

1853. $C_9H_{21}O_3PS_3$

Terbufos Sulfoxide
Phosphorodithioic Acid, S-[[(1,1-Dimethylethyl)sulfinyl]methyl] O,O-Diethyl Ester
RN: 10548-10-4   **MP** (°C):
MW: 304.43   **BP** (°C):

Solubility (Moles/L)	Solubility (Grams/L)	Temp (°C)	Ref (#)	Evaluation (T P E A A)	Comments
>3.61E-03	>1.10E+00	ns	B325	0 1 0 0 1	

1854. $C_9H_{21}O_4P$

Dibutyl Methyl Phosphate
Methyl Dibutyl Phosphate
RN: 7242-59-3   **MP** (°C):
MW: 224.24   **BP** (°C):

Solubility (Moles/L)	Solubility (Grams/L)	Temp (°C)	Ref (#)	Evaluation (T P E A A)	Comments
3.166E-02	7.100E+00	25	B070	1 2 2 1 1	

1855. $C_9H_{21}O_4P$

Diethyl Amyl Phosphate
O,O-Diethyl O-Pentyl Phosphate
Diethyl Pentyl Phosphate
RN: 20195-08-8   **MP** (°C):
MW: 224.24   **BP** (°C):

Solubility (Moles/L)	Solubility (Grams/L)	Temp (°C)	Ref (#)	Evaluation (T P E A A)	Comments
3.345E-02	7.500E+00	25	B070	1 2 0 1 1	

1856.   C$_9$H$_{21}$O$_4$P
Tripropyl Phosphate
Tri-n-propyl Phosphate
RN: 513-08-6   **MP** (°C):
MW: 224.24   **BP** (°C):

Solubility (Moles/L)	Solubility (Grams/L)	Temp (°C)	Ref (#)	Evaluation (T P E A A)	Comments
3.100E-02	6.951E+00	30	V300	2 2 0 1 0	

1857.   C$_9$H$_{21}$O$_4$PS$_3$
Terbufos Sulfone
Phosphorodithioic Acid, S-[[(1,1-Dimethylethyl)sulfonyl]methyl] O,O-Diethyl Ester
Counter Sulfone
AC 94320
RN: 56070-16-7   **MP** (°C):
MW: 320.43   **BP** (°C):

Solubility (Moles/L)	Solubility (Grams/L)	Temp (°C)	Ref (#)	Evaluation (T P E A A)	Comments
1.273E-03	4.078E-01	18.50	B169	2 0 1 1 2	
1.273E-03	4.078E-01	ns	B325	0 1 0 0 1	

1858.   C$_9$H$_{22}$O$_4$P$_2$S$_4$
Ethion
O,O,O,O-Tetraethyl S,S-Methylene bisPhosphorodithioate
Nialate
Ethanox
Diethion
Hylemox
RN: 563-12-2   **MP** (°C):   -25
MW: 384.48   **BP** (°C):

Solubility (Moles/L)	Solubility (Grams/L)	Temp (°C)	Ref (#)	Evaluation (T P E A A)	Comments
1.483E-06	5.700E-04	10	B324	2 2 2 2 2	
1.483E-06	5.702E-04	10	B324	2 2 2 2 2	
2.861E-06	1.100E-03	19.50	B169	2 2 1 1 1	
1.769E-06	6.801E-04	20	B324	2 2 2 2 2	
1.769E-06	6.800E-04	20	B324	2 2 2 2 2	
1.977E-06	7.601E-04	30	B324	2 2 2 2 2	
1.977E-06	7.600E-04	30	B324	2 2 2 2 2	

1859. $C_{10}H_4Cl_2O_2$
Dichlone
2,3-Dichloro-1,4-naphthalenedione
Phygon XL
Phygon
Phygon Paste
USR 604
RN: 117-80-6   **MP** (°C):
MW: 227.05   **BP** (°C):

Solubility (Moles/L)	Solubility (Grams/L)	Temp (°C)	Ref (#)	Evaluation (T P E A A)	Comments
4.404E-07	1.000E-04	25	M161	1 0 0 0 0	
3.083E-05	7.000E-03	ns	B160	0 0 0 0 0	
4.404E-06	1.000E-03	ns	B185	0 0 0 0 0	

1860. $C_{10}H_5ClN_2O_4$
1-Chloro-2,4-dinitronaphthalene
2,4-Dinitro-1-naphthyl Chloride
2,4-Dinitrochloronaphthalene
2,4-Dinitro-1-chloronaphthalene
RN: 2401-85-6   **MP** (°C):   148
MW: 252.62   **BP** (°C):

Solubility (Moles/L)	Solubility (Grams/L)	Temp (°C)	Ref (#)	Evaluation (T P E A A)	Comments
3.959E-06	1.000E-03	25	M061	1 0 0 0 0	

1861. $C_{10}H_5Cl_7$
Heptachlor
1,4,5,6,7,8,8-Heptachloro-3α,4,7,7α-tetrahydro-4,7-methano-1H-indene
3-Chlorochlordene
Tetrahydro
Rhodiachlor
3,4,5,6,7,8,8α-Heptachlorodicyclopentadiene
RN: 76-44-8   **MP** (°C):   95.5
MW: 373.32   **BP** (°C):

Solubility (Moles/L)	Solubility (Grams/L)	Temp (°C)	Ref (#)	Evaluation (T P E A A)	Comments
2.679E-07	1.000E-04	15	B083	2 2 1 2 2	particle size ≤ 5 μm
4.822E-07	1.800E-04	25	B083	2 2 1 2 2	particle size ≤ 5 μm
1.500E-07	5.600E-05	25	I308	0 0 0 0 1	
1.500E-07	5.600E-05	26.5	P027	1 1 2 2 1	
1.500E-07	5.600E-05	27	M161	0 0 0 0 1	

8.438E-07	3.150E-04	35	B083	2 2 1 2 2	particle size ≤ 5 µm
1.313E-06	4.900E-04	45	B083	2 2 1 2 2	particle size ≤ 5 µm
8.036E-08	3.000E-05	ns	K138	0 0 0 0 2	
1.875E-07	7.000E-05	ns	M110	0 0 0 0 0	EFG

1862. $C_{10}H_5Cl_7O$

Heptachlor Epoxide

1,4,5,6,7,8,8-Heptachloro-2,3-epoxy-3α,4,7,7α-tetrahydro-4,7-methanoindan

Hepachlor Epoxide

RN: 1024-57-3   **MP** (°C):   160
MW: 389.32   **BP** (°C):

Solubility (Moles/L)	Solubility (Grams/L)	Temp (°C)	Ref (#)	Evaluation (T P E A A)	Comments
2.825E-07	1.100E-04	15	B083	2 2 1 2 2	particle size ≤ 5 µm
5.137E-07	2.000E-04	25	B083	2 2 1 2 2	particle size ≤ 5 µm
5.137E-07	2.000E-04	25	I308	0 0 0 0 1	
8.990E-07	3.500E-04	25	W025	1 0 2 2 2	
8.990E-07	3.500E-04	26.5	P027	1 1 2 2 1	
8.990E-07	3.500E-04	35	B083	2 2 1 2 2	particle size ≤ 5 µm
1.541E-06	6.000E-04	45	B083	2 2 1 2 2	particle size ≤ 5 µm
1.798E-06	7.000E-04	ns	M110	0 0 0 0 0	EFG

1863. $C_{10}H_5N_3O_6$

1,3,8-Trinitronaphthalene

1,3,8-Trinitronaphthalin

RN: 2364-46-7   **MP** (°C):
MW: 263.17   **BP** (°C):

Solubility (Moles/L)	Solubility (Grams/L)	Temp (°C)	Ref (#)	Evaluation (T P E A A)	Comments
6.840E-05	1.800E-02	15	F300	1 0 0 0 1	

1864. $C_{10}H_5N_3O_6$

1,4,5-Trinitronaphthalene

1,4,5-Trinitronaphthalin

RN: 2243-95-0   **MP** (°C):
MW: 263.17   **BP** (°C):

Solubility (Moles/L)	Solubility (Grams/L)	Temp (°C)	Ref (#)	Evaluation (T P E A A)	Comments
1.520E-04	4.000E-02	15	F300	1 0 0 0 1	

1865.   C$_{10}$H$_6$Br$_2$
1,4-Dibromonaphthalene
Naphthalene, 1,4-Dibromo-
RN: 83-53-4   **MP** (°C):   80-82
MW: 285.98   **BP** (°C):

Solubility (Moles/L)	Solubility (Grams/L)	Temp (°C)	Ref (#)	Evaluation (T P E A A)	Comments
4.333E-07	1.239E-04	4	D351	1 2 1 1 2	
1.217E-06	3.479E-04	25	D351	1 2 1 1 2	
3.006E-06	8.595E-04	40	D351	1 2 1 1 2	

1866.   C$_{10}$H$_6$Br$_2$
2,3-Dibromonaphthalene
Naphthalene, 2,3-Dibromo-
RN: 13214-70-5   **MP** (°C):
MW: 285.98   **BP** (°C):

Solubility (Moles/L)	Solubility (Grams/L)	Temp (°C)	Ref (#)	Evaluation (T P E A A)	Comments
1.922E-07	5.497E-05	4	D351	1 2 1 1 2	
4.778E-07	1.366E-04	25	D351	1 2 1 1 2	
1.222E-06	3.495E-04	40	D351	1 2 1 1 2	

1867.   C$_{10}$H$_6$Cl$_2$
1,4-Dichloronaphthalene
Naphthalene, 1,4-Dichloro-
RN: 1825-31-6   **MP** (°C):
MW: 197.07   **BP** (°C):

Solubility (Moles/L)	Solubility (Grams/L)	Temp (°C)	Ref (#)	Evaluation (T P E A A)	Comments
1.333E-06	2.628E-04	4	D351	1 2 1 1 2	
4.389E-06	8.649E-04	25	D351	1 2 1 1 2	
1.122E-05	2.212E-03	40	D351	1 2 1 1 2	

1868.   C$_{10}$H$_6$Cl$_4$O$_3$S
Glenbar
O,S-Dimethyl Tetrachlorothioterephthalate
RN: 3765-57-9   **MP** (°C):   161
MW: 348.03   **BP** (°C):

Solubility (Moles/L)	Solubility (Grams/L)	Temp (°C)	Ref (#)	Evaluation (T P E A A)	Comments
1.437E-06	5.000E-04	22	B200	1 0 0 0 0	
1.034E-06	3.600E-04	ns	M061	0 0 0 0 1	

1869. C₁₀H₆Cl₄O₄

$C_{10}H_6Cl_4O_4$

Dimethyl Tetrachloroterephthalate
DCPA

RN: 1861-32-1 **MP** (°C): 156
MW: 331.97 **BP** (°C):

Solubility (Moles/L)	Solubility (Grams/L)	Temp (°C)	Ref (#)	Evaluation (T P E A A)	Comments
1.506E-06	5.000E-04	25	B200	1 0 0 0 0	
<1.51E-06	<5.00E-04	25	M161	1 0 0 0 0	
≤1.51E-06	<5.00E-04	ns	B185	0 0 0 0 0	

1870. C₁₀H₆Cl₆

$C_{10}H_6Cl_6$

Chlordene
4,5,6,7,8,8-Hexachloro-3α,4,7,7α-tetrahydro-4,7-methanoindene

RN: 3734-48-3 **MP** (°C): -62
MW: 338.88 **BP** (°C):

Solubility (Moles/L)	Solubility (Grams/L)	Temp (°C)	Ref (#)	Evaluation (T P E A A)	Comments
2.281E-06	7.730E-04	26.70	L071	1 2 0 1 2	

1871. C₁₀H₆Cl₆O

$C_{10}H_6Cl_6O$

Chlordene Epoxide
2,3-Epoxy-4,5,6,7,8,8-hexachloro-3α,4,7,7α-tetrahydro-4,7-methanoindene
Chlordene Hydroxide
4,7-Methano-1H-inden-1-ol, 4,5,6,7,8,8-hexachloro-3α,4,7,7α-tetrahydro-

RN: 6058-23-7 **MP** (°C): 215
MW: 354.88 **BP** (°C):

Solubility (Moles/L)	Solubility (Grams/L)	Temp (°C)	Ref (#)	Evaluation (T P E A A)	Comments
3.829E-06	1.359E-03	26.70	L071	1 2 0 1 2	

1872. C₁₀H₆Cl₆O

$C_{10}H_6Cl_6O$

1-Hydroxychlordene
1-Hydroxy-4,5,6,7,8,8-hexachloro-3α,4,7,7α-tetra-hydro-4,7-methanoindene

RN: 2597-11-7 **MP** (°C): 194
MW: 354.88 **BP** (°C):

Solubility (Moles/L)	Solubility (Grams/L)	Temp (°C)	Ref (#)	Evaluation (T P E A A)	Comments
3.469E-06	1.231E-03	26.70	L071	1 2 0 1 2	

1873.   C$_{10}$H$_6$Cl$_6$O$_2$
1-Hydroxychlordene Epoxide
1-Hydroxy-2,3-epoxy-4,5,6,7,8,8-hexachloro-3α,4,7,7α-tetrahydro-4,7-methanoindene
RN: 24009-06-1   **MP** (°C):
MW: 370.88   **BP** (°C):

Solubility (Moles/L)	Solubility (Grams/L)	Temp (°C)	Ref (#)	Evaluation (T P E A A)	Comments
7.391E-06	2.741E-03	26.70	L071	1 1 1 1 2	

1874.   C$_{10}$H$_6$Cl$_8$
Chlordane
1,2,4,5,6,7,8,8-Octachloro-4,7-methano-3α,4,7,7α-Tetrahydroindane
Octachlor
Velsicol 1068
Toxichlor
Ortho-Klor
RN: 57-74-9   **MP** (°C):   105
MW: 409.78   **BP** (°C):

Solubility (Moles/L)	Solubility (Grams/L)	Temp (°C)	Ref (#)	Evaluation (T P E A A)	Comments
4.515E-06	1.850E-03	25	W025	1 0 2 2 2	
1.367E-07	5.600E-05	ns	K138	0 0 0 0 2	
1.708E-07	7.000E-05	ns	M110	0 0 0 0 0	EFG
1.367E-07	5.600E-05	ns	S187	0 2 2 1 1	

1875.   C$_{10}$H$_6$FN$_3$O$_3$
3-Nicotinoyl-5-fluorouracil
RN: **MP** (°C):
MW: 235.18   **BP** (°C):

Solubility (Moles/L)	Solubility (Grams/L)	Temp (°C)	Ref (#)	Evaluation (T P E A A)	Comments
1.148E-02	2.700E+00	22	B332	1 1 0 0 1	pH 4.0

1876.   C$_{10}$H$_6$N$_2$O$_4$
1,8-Dinitronaphthalene
1,8-Dinitronaphthalin
RN: 602-38-0   **MP** (°C):   107
MW: 218.17   **BP** (°C):

Solubility (Moles/L)	Solubility (Grams/L)	Temp (°C)	Ref (#)	Evaluation (T P E A A)	Comments
1.558E-04	3.400E-02	15	F300	1 0 0 0 1	

1877. $C_{10}H_6N_2O_4$
1,5-Dinitronaphthalene
1,5-Dinitronaphthalin

RN: 605-71-0   **MP** (°C): 216.5
MW: 218.17   **BP** (°C):

Solubility (Moles/L)	Solubility (Grams/L)	Temp (°C)	Ref (#)	Evaluation (T P E A A)	Comments
2.658E-04	5.800E-02	12	F300	1 0 0 0 1	

1878. $C_{10}H_6O_8$
Pyromellitic Acid
1,2,4,5-Benzenetetracarboxylic Acid
Benzol-tetracarbonsaeure-(1,2,4,5)

RN: 89-05-4   **MP** (°C):
MW: 254.15   **BP** (°C):

Solubility (Moles/L)	Solubility (Grams/L)	Temp (°C)	Ref (#)	Evaluation (T P E A A)	Comments
5.508E-02	1.400E+01	16	F300	1 0 0 0 2	

1879. $C_{10}H_7Br$
2-Bromonaphthalene
Naphthalene, 2-Bromo-

RN: 580-13-2   **MP** (°C): 53.5
MW: 207.08   **BP** (°C): 281.1

Solubility (Moles/L)	Solubility (Grams/L)	Temp (°C)	Ref (#)	Evaluation (T P E A A)	Comments
1.850E-05	3.831E-03	4	D351	1 2 1 1 2	
3.883E-05	8.041E-03	25	D351	1 2 1 1 2	
7.611E-05	1.576E-02	40	D351	1 2 1 1 2	
4.000E-05	8.283E-03	ns	L060	0 0 0 0 0	

1880. $C_{10}H_7Br$
1-Bromonaphthalene
Naphthalene, 1-Bromo-

RN: 90-11-9   **MP** (°C): 6.2
MW: 207.08   **BP** (°C): 281.1

Solubility (Moles/L)	Solubility (Grams/L)	Temp (°C)	Ref (#)	Evaluation (T P E A A)	Comments
4.383E-05	9.077E-03	4	D351	1 2 1 1 2	
4.733E-05	9.802E-03	10	D351	1 2 1 1 2	
4.500E-05	9.318E-03	21	A057	2 1 2 2 1	
6.444E-05	1.334E-02	25	D351	1 2 1 1 2	
9.166E-05	1.898E-02	40	D351	1 2 1 1 2	
6.000E-05	1.242E-02	ns	L060	0 0 0 0 0	

1881.   C₁₀H₇Cl
1-Chloronaphthalene
α-Chloronaphthalene
1-Naphthyl Chloride
RN: 90-13-1   **MP** (°C):   -20
MW: 162.62   **BP** (°C):   259.3

Solubility (Moles/L)	Solubility (Grams/L)	Temp (°C)	Ref (#)	Evaluation (T P E A A)	Comments
<1.23E-04	<2.00E-02	ns	L060	0 0 0 0 2	

1882.   C₁₀H₇Cl
β-Chloronaphthalene
2-Chloronaphthalene
RN: 91-58-7   **MP** (°C):   59.5
MW: 162.62   **BP** (°C):   256

Solubility (Moles/L)	Solubility (Grams/L)	Temp (°C)	Ref (#)	Evaluation (T P E A A)	Comments
<6.15E-06	<1.00E-03	30	M311	1 1 2 2 0	
8.000E-05	1.301E-02	ns	L060	0 0 0 0 0	

1883.   C₁₀H₇I
α-Iodonaphthalene
1-Iodonaphthalene
RN: 90-14-2   **MP** (°C):
MW: 254.07   **BP** (°C):

Solubility (Moles/L)	Solubility (Grams/L)	Temp (°C)	Ref (#)	Evaluation (T P E A A)	Comments
2.800E-05	7.114E-03	ns	L060	0 0 0 0 1	average

1884.   C₁₀H₇NO₂
1-Nitronaphthalene
1-Nitro-naphthalin
RN: 86-57-7   **MP** (°C):   59.5
MW: 173.17   **BP** (°C):   304

Solubility (Moles/L)	Solubility (Grams/L)	Temp (°C)	Ref (#)	Evaluation (T P E A A)	Comments
2.887E-04	5.000E-02	18	F300	1 0 0 0 1	

1885.   C$_{10}$H$_7$NO$_3$
Kynurenic Acid
4-Hydroxy-chinolin-carbonsaeure-(2)
Kynurensaeure

RN: 492-27-3   **MP** (°C):   282.5
MW: 189.17   **BP** (°C):

Solubility (Moles/L)	Solubility (Grams/L)	Temp (°C)	Ref (#)	Evaluation (T P E A A)	Comments
4.715E-02	8.920E+00	100	D041	1 0 0 0 0	
4.969E-03	9.400E-01	100	F300	1 0 0 0 1	

1886.   C$_{10}$H$_7$NO$_3$
1-Nitro-2-naphthol
1-Nitro-naphthol-(2)

RN: 550-60-7   **MP** (°C):   104
MW: 189.17   **BP** (°C):   115

Solubility (Moles/L)	Solubility (Grams/L)	Temp (°C)	Ref (#)	Evaluation (T P E A A)	Comments
1.057E-03	2.000E-01	20	F300	1 0 0 0 2	

1887.   C$_{10}$H$_7$N$_3$O$_3$
Orotic Acid Pyridine

RN: **MP** (°C):
MW: 217.19   **BP** (°C):

Solubility (Moles/L)	Solubility (Grams/L)	Temp (°C)	Ref (#)	Evaluation (T P E A A)	Comments
1.200E-01	2.606E+01	16	N018	2 2 1 2 2	

1888.   C$_{10}$H$_7$N$_3$S
Thiabendazole
2-(Thiazol-4-yl)benzimidazole
Mintezol
Apl-Luster
Mertect
Tecto

RN: 148-79-8   **MP** (°C):   304.5
MW: 201.25   **BP** (°C):

Solubility (Moles/L)	Solubility (Grams/L)	Temp (°C)	Ref (#)	Evaluation (T P E A A)	Comments
2.484E-04	5.000E-02	25	M161	1 0 0 0 1	intrinsic

1889.   C$_{10}$H$_8$
Naphthalene
Napthalene
Mothballs
Camphor Tar
RN: 91-20-3   **MP** (°C):   80.2
MW: 128.18   **BP** (°C):   217.9

Solubility (Moles/L)	Solubility (Grams/L)	Temp (°C)	Ref (#)	Evaluation (T P E A A)	Comments
1.350E-04	1.730E-02	4.99	P331	2 2 1 2 2	
1.320E-04	1.692E-02	8.20	M082	1 1 1 2 2	
1.320E-04	1.692E-02	8.20	M151	2 1 2 2 1	
1.320E-04	1.692E-02	8.24	M183	1 2 1 1 2	
1.580E-04	2.025E-02	9.99	P331	2 2 1 2 2	
1.390E-04	1.782E-02	10	J302	2 1 2 2 2	
1.500E-04	1.923E-02	11.50	M082	1 1 1 2 2	
1.500E-04	1.923E-02	11.50	M151	2 1 2 2 2	
1.502E-04	1.925E-02	11.54	M183	1 2 1 1 2	
1.570E-04	2.012E-02	12	S076	2 2 2 2 2	
1.590E-04	2.038E-02	13.40	M082	1 1 1 2 2	
1.590E-04	2.038E-02	13.40	M151	2 1 2 2 2	
1.591E-04	2.039E-02	13.44	M183	1 2 1 1 2	
1.900E-04	2.435E-02	14.99	P331	2 2 1 2 2	
1.716E-03	2.200E-01	15	F300	1 0 0 0 2	*sic*
1.716E-04	2.200E-02	15	M073	1 0 2 2 1	
1.680E-04	2.153E-02	15.10	M082	1 1 1 2 2	
1.680E-04	2.153E-02	15.10	M151	2 1 2 2 2	
1.677E-04	2.150E-02	15.14	M183	1 2 1 1 2	
1.900E-04	2.435E-02	18	S076	2 2 2 2 2	
2.010E-04	2.576E-02	19.30	M082	1 1 1 2 2	
2.010E-04	2.576E-02	19.30	M151	2 1 2 2 2	
2.013E-04	2.581E-02	19.34	M183	1 2 1 1 2	
2.240E-04	2.871E-02	19.99	P331	2 2 1 2 2	
1.748E-04	2.240E-02	20	A050	1 0 1 1 2	
7.412E-04	9.500E-02	20	B318	1 2 1 2 0	EFG
3.000E-04	3.845E-02	20	E009	1 0 0 0 1	
3.000E-04	3.845E-02	20	E025	1 0 2 2 1	
1.900E-04	2.435E-02	20	H306	1 0 1 2 1	
1.272E-04	1.630E-02	20	T301	1 2 2 2 2	
1.638E-04	2.100E-02	22	N311	1 0 1 1 2	
2.255E-04	2.890E-02	22.20	W003	2 2 2 2 2	average of 3
2.341E-04	3.000E-02	23	P332	2 1 1 2 2	
2.341E-04	3.000E-02	23	P339	2 0 1 2 2	
2.300E-04	2.948E-02	23.40	M082	1 1 1 2 2	
2.300E-04	2.948E-02	23.40	M151	2 1 2 2 2	
2.301E-04	2.949E-02	23.44	M183	1 2 1 1 2	
2.380E-04	3.050E-02	24.50	W003	2 2 2 2 2	average of 5

2.630E-04	3.371E-02	24.99	P331	2 2 1 2 2	
2.458E-04	3.150E-02	25	A001	1 2 2 2 2	
2.350E-04	3.012E-02	25	A325	2 1 2 2 2	
2.684E-04	3.440E-02	25	B003	2 2 2 2 2	
2.465E-04	3.160E-02	25	B319	2 0 1 2 2	average of 2
2.442E-04	3.130E-02	25	D337	2 1 2 2 2	
2.442E-04	3.130E-02	25	E004	2 1 2 2 2	
2.620E-04	3.358E-02	25	G047	2 2 2 2 2	
2.520E-04	3.230E-02	25	J302	2 1 2 2 2	
9.750E-05	1.250E-02	25	K001	2 2 2 2 2	
2.300E-04	2.948E-02	25	K123	1 0 2 2 1	
2.497E-04	3.200E-02	25	L332	1 1 1 1 0	
2.653E-04	3.400E-02	25	M040	1 0 0 1 1	
2.550E-04	3.268E-02	25	M058	2 2 2 2 2	
2.473E-04	3.170E-02	25	M064	1 1 2 2 2	
2.472E-04	3.169E-02	25	M071	2 2 2 2 2	
3.121E-04	4.000E-02	25	M073	1 0 2 2 1	
2.620E-04	3.358E-02	25	M123	1 0 0 0 2	
2.575E-04	3.300E-02	25	M130	1 0 0 0 1	
2.390E-04	3.063E-02	25	M342	1 0 1 1 2	
2.497E-04	3.200E-02	25	O320	1 0 1 1 1	
2.575E-05	3.300E-03	25	P340	1 1 2 2 1	
2.356E-04	3.020E-02	25	R042	1 2 2 2 2	
2.340E-04	2.999E-02	25	S076	2 2 2 2 2	
1.716E-04	2.200E-02	25	S227	1 2 1 1 1	
2.390E-04	3.063E-02	25	W300	2 2 2 2 2	
2.490E-04	3.192E-02	25.00	M082	1 1 1 2 2	
2.472E-04	3.169E-02	25.00	M151	2 1 1 2 2	
2.490E-04	3.192E-02	25.00	M151	2 1 2 2 2	
6.936E-04	8.890E-02	25.00	P007	2 1 2 2 2	
2.492E-04	3.194E-02	25.04	M183	1 2 1 1 2	
2.510E-04	3.217E-02	25.04	V013	2 2 2 2 2	
2.660E-04	3.409E-02	27.00	M082	1 1 1 2 2	
2.660E-04	3.409E-02	27.00	M151	2 1 2 2 2	
2.666E-04	3.417E-02	27.04	M183	1 2 1 1 2	
2.980E-04	3.820E-02	29.90	W003	2 2 2 2 2	average of 3
3.240E-04	4.153E-02	29.99	P331	2 2 1 2 2	
2.949E-04	3.780E-02	30.30	W003	2 2 2 2 2	average of 3
3.448E-04	4.420E-02	34.50	W003	2 2 2 2 2	average of 2
3.710E-04	4.755E-02	34.99	P331	2 2 1 2 2	
4.112E-04	5.270E-02	39.30	W003	2 2 2 2 2	average of 2
4.360E-04	5.588E-02	39.99	P331	2 2 1 2 2	
4.275E-04	5.480E-02	40.10	W003	2 2 2 2 2	
5.118E-04	6.560E-02	44.70	W003	2 2 2 2 2	average of 3
6.132E-04	7.860E-02	50.20	W003	2 2 2 2 2	
8.270E-04	1.060E-01	55.60	W003	2 2 2 2 2	
1.233E-03	1.580E-01	64.50	W003	2 2 2 2 2	average of 3
1.904E-03	2.440E-01	73.40	W003	2 2 2 2 2	average of 3
2.341E-04	3.000E-02	ns	F071	0 1 2 1 1	
2.341E-04	3.000E-02	ns	H080	0 0 0 0 1	

2.473E-04	3.170E-02	ns	H123	0 0 0 0 2	
2.473E-04	3.170E-02	ns	K304	0 0 0 0 2	
2.340E-04	2.999E-02	ns	L060	0 0 0 0 2	average
2.473E-04	3.170E-02	ns	M344	0 0 0 0 2	
2.341E-04	3.000E-02	ns	O009	0 0 0 0 0	
8.129E-04	1.042E-01	ns	R042	1 2 2 2 2	
2.341E-04	3.000E-02	rt	M161	0 0 0 0 1	
2.848E-04	3.650E-02	rt	S314	0 0 2 1 2	

1890.   C₁₀H₈BrN₃O
Bropirimine
2-Amino-5-bromo-6-phenyl-py-rimidin-4(3H)-one
ABPP
RN: 56741-95-8   **MP** (°C):
MW: 266.10   **BP** (°C):

Solubility (Moles/L)	Solubility (Grams/L)	Temp (°C)	Ref (#)	Evaluation (T P E A A)	Comments
2.931E-05	7.800E-03	37	A346	1 0 2 2 0	EFG

1891.   C₁₀H₈BrN₃O
Brompyrazone
Amino-4-bromo-2-phenyl-3(2H)-pyridazinone
1-Phenyl-4-amino-5-bromo-6-pyridazone
Pyridazinone, 5-Amino-4-bromo-2-phenyl-
RN: 3042-84-0   **MP** (°C):   223.5
MW: 266.10   **BP** (°C):

Solubility (Moles/L)	Solubility (Grams/L)	Temp (°C)	Ref (#)	Evaluation (T P E A A)	Comments
7.516E-04	2.000E-01	20	M161	1 0 0 0 2	

1892.   C₁₀H₈ClN₃O
Pyrazon
5-Amino-4-chloro-2-phenyl-3(2H)-pyridazinone
RN: 1698-60-8   **MP** (°C):   207
MW: 221.65   **BP** (°C):

Solubility (Moles/L)	Solubility (Grams/L)	Temp (°C)	Ref (#)	Evaluation (T P E A A)	Comments
1.353E-03	3.000E-01	20	B185	1 0 0 0 2	
1.353E-03	2.999E-01	20	B200	1 0 0 0 0	
1.353E-03	2.999E-01	20	M061	1 0 0 0 0	
1.805E-03	4.000E-01	20	M161	1 0 0 0 2	

1893. C$_{10}$H$_8$N$_2$
γ,γ'-Dipyridyl
4,4'-Bipyridyl
RN: 553-26-4   **MP** (°C):   69
MW: 156.19   **BP** (°C):

Solubility (Moles/L)	Solubility (Grams/L)	Temp (°C)	Ref (#)	Evaluation (T P E A A)	Comments
2.887E-02	4.509E+00	25	B095	2 0 1 1 2	

1894. C$_{10}$H$_8$N$_2$
α,α'-Dipyridyl
2,2'-Dipyridyl
α,α'-Bipyridyl
2,2'-Bipyridine
2,2'-Bipyridyl
RN: 366-18-7   **MP** (°C):   71.5
MW: 156.19   **BP** (°C):   273

Solubility (Moles/L)	Solubility (Grams/L)	Temp (°C)	Ref (#)	Evaluation (T P E A A)	Comments
3.201E-02	5.000E+00	20	F300	1 0 0 0 0	
3.778E-02	5.900E+00	25	B095	2 0 1 1 2	
4.094E-02	6.394E+00	25	K063	2 2 0 1 2	

1895. C$_{10}$H$_8$N$_2$O$_2$
4-Phenyluracil
4-Phenyl-uracil
RN: 21321-07-3   **MP** (°C):
MW: 188.19   **BP** (°C):

Solubility (Moles/L)	Solubility (Grams/L)	Temp (°C)	Ref (#)	Evaluation (T P E A A)	Comments
5.314E-02	1.000E+01	100	F300	1 0 0 0 0	

1896. C$_{10}$H$_8$O
1-Naphthol
α-Naphthol
RN: 90-15-3   **MP** (°C):   96
MW: 144.17   **BP** (°C):   288

Solubility (Moles/L)	Solubility (Grams/L)	Temp (°C)	Ref (#)	Evaluation (T P E A A)	Comments
6.030E-03	8.694E-01	11	K307	2 0 1 2 2	
7.700E-03	1.110E+00	20	K130	2 1 1 1 2	
7.700E-03	1.110E+00	20	K301	2 2 1 1 1	
7.700E-03	1.110E+00	20	K307	2 0 1 2 2	
6.001E-03	8.653E-01	24	H106	1 0 2 2 2	
6.007E-03	8.660E-01	24	M303	1 0 1 1 2	

3.029E-03	4.367E-01	25	L085	1 2 0 1 2
9.430E-03	1.360E+00	30	K307	2 0 1 2 2
1.490E-02	2.148E+00	40	K307	2 0 1 2 2
2.150E-02	3.100E+00	50	K307	2 0 1 2 2

1897.   C$_{10}$H$_8$O
2-Naphthol

β-Naphthol

RN: 135-19-3   **MP** (°C):   121

MW: 144.17   **BP** (°C):   285

Solubility (Moles/L)	Solubility (Grams/L)	Temp (°C)	Ref (#)	Evaluation (T P E A A)	Comments
2.462E-03	3.550E-01	6.90	M026	2 0 1 2 2	
3.378E-03	4.870E-01	13.45	M026	2 0 1 2 2	
3.473E-03	5.007E-01	15.60	M027	1 0 0 2 2	
3.646E-03	5.257E-01	16.20	M027	1 0 0 2 2	
3.891E-03	5.610E-01	17.70	M026	2 0 1 2 2	
4.450E-03	6.416E-01	20	K130	2 1 1 1 2	
4.500E-03	6.488E-01	20	K301	2 2 1 1 1	
4.450E-03	6.416E-01	20	K308	1 0 0 1 2	
5.800E-03	8.362E-01	20	M122	2 0 2 2 2	
4.945E-03	7.130E-01	21.50	M026	2 0 1 2 2	
4.713E-03	6.795E-01	23.20	M027	1 0 0 2 2	
3.954E-03	5.700E-01	25	F300	1 0 0 0 2	
5.240E-03	7.555E-01	25	K040	1 0 2 1 2	
5.356E-03	7.722E-01	25	L085	1 2 0 1 2	
6.929E-03	9.990E-01	25	R041	1 0 2 1 1	
6.076E-03	8.760E-01	29.50	M026	2 0 1 2 2	
6.431E-03	9.271E-01	31.30	M027	1 0 0 2 2	
6.832E-03	9.850E-01	33.30	M026	2 0 1 2 2	
9.045E-03	1.304E+00	38.70	M026	2 0 1 2 2	
1.116E-02	1.609E+00	44.50	M026	2 0 1 2 2	
1.388E-02	2.001E+00	49.50	M026	2 0 1 2 2	
1.706E-02	2.460E+00	55.20	M026	2 0 1 2 2	
2.104E-02	3.034E+00	60.00	M026	2 0 1 2 2	
2.928E-02	4.222E+00	68.10	M026	2 0 1 2 2	
3.810E-02	5.493E+00	75.00	M026	2 0 1 2 2	
4.670E-02	6.733E+00	80	K308	1 0 0 1 2	

1898. C₁₀H₈O₂

2,6-Dihydroxynaphthalene
2,6-Dihydroxy-naphthalin
RN: 581-43-1 **MP** (°C):
MW: 160.17 **BP** (°C):

Solubility (Moles/L)	Solubility (Grams/L)	Temp (°C)	Ref (#)	Evaluation (T P E A A)	Comments
6.243E-03	1.000E+00	14	F300	1 0 0 0 0	

1899. C₁₀H₈O₂

2,3-Dihydroxynaphthalene
2,3-Dihydroxy-naphthalin
RN: 92-44-4 **MP** (°C): 162
MW: 160.17 **BP** (°C):

Solubility (Moles/L)	Solubility (Grams/L)	Temp (°C)	Ref (#)	Evaluation (T P E A A)	Comments
1.830E-03	2.931E-01	20	M122	2 0 2 2 2	

1900. C₁₀H₉ClN₄O₂S

2-Sulfanilamido-5-chloropyrimidine
Benzenesulfonamide, 4-Amino-N-(5-chloro-2-pyrimidinyl)-
RN: 4482-46-6 **MP** (°C):
MW: 284.73 **BP** (°C):

Solubility (Moles/L)	Solubility (Grams/L)	Temp (°C)	Ref (#)	Evaluation (T P E A A)	Comments
6.322E-05	1.800E-02	37	R046	1 2 1 1 1	

1901. C₁₀H₉ClN₄O₂S

5-Sulfanilamido-2-chloropyrimidine
Benzenesulfonamide, 4-Amino-N-(2-chloro-5-pyrimidinyl)-
RN: 17103-49-0 **MP** (°C):
MW: 284.73 **BP** (°C):

Solubility (Moles/L)	Solubility (Grams/L)	Temp (°C)	Ref (#)	Evaluation (T P E A A)	Comments
1.127E-03	3.210E-01	37	R046	1 2 1 1 1	

1902. C₁₀H₉Cl₂NO

Acrylanilide, 3',4'-Dichloro-2-methyl-
Dicryl
RN: 2164-09-2 **MP** (°C): 127-128
MW: 230.10 **BP** (°C):

Solubility (Moles/L)	Solubility (Grams/L)	Temp (°C)	Ref (#)	Evaluation (T P E A A)	Comments
3.477E-05	8.000E-03	ns	B185	0 0 0 0 1	

1903. $C_{10}H_9Cl_3O_3$

2,4,5-Trichlorophenoxy-γ-butyric Acid
2,4,5-TB
4-(2,4,5-Trichlorophenoxy)butyric Acid
4-(2,4,5-TB)

RN: 93-80-1   **MP** (°C):   114.5
MW: 283.54   **BP** (°C):

Solubility (Moles/L)	Solubility (Grams/L)	Temp (°C)	Ref (#)	Evaluation (T P E A A)	Comments
1.481E-04	4.200E-02	25	B164	1 0 1 1 1	
1.481E-04	4.200E-02	ns	B185	1 0 0 0 1	

1904. $C_{10}H_9Cl_3O_3$

2,4-Dichlorophenoxyacetic Acid β-Monochloroethyl Ester
Ethanol, 2-Chloro-, (2,4-Dichlorophenoxy)acetate

RN: 19810-30-1   **MP** (°C):
MW: 283.54   **BP** (°C):

Solubility (Moles/L)	Solubility (Grams/L)	Temp (°C)	Ref (#)	Evaluation (T P E A A)	Comments
1.910E-04	5.415E-02	ns	M120	0 0 1 1 2	

1905. $C_{10}H_9Cl_4NO_2S$

Captafol
cis-3α,4,7,7α-Tetrahydro-2-(1,1,2,2-tetrachloroethyl)thio-1H-Isoindole-1,3(2H)-dione
Crisfolatan
Difolatan
Folcid

RN: 2939-80-2   **MP** (°C):   160.5
MW: 349.06   **BP** (°C):

Solubility (Moles/L)	Solubility (Grams/L)	Temp (°C)	Ref (#)	Evaluation (T P E A A)	Comments
4.074E-06	1.422E-03	20	B179	2 0 0 0 2	
4.011E-06	1.400E-03	ns	M161	0 0 0 0 1	

1906. C$_{10}$H$_9$Cl$_4$O$_4$P
Tetrachlorvinphos
2-Chloro-1-(2,4,5-trichlorophenyl)vinyl Dimethyl Phosphate
Rabon
Gardona
SD 8447
Stirofos
RN: 961-11-5 **MP** (°C): 96
MW: 365.97 **BP** (°C):

Solubility (Moles/L)	Solubility (Grams/L)	Temp (°C)	Ref (#)	Evaluation (T P E A A)	Comments
3.006E-05	1.100E-02	20	M161	1 0 0 0 1	

1907. C$_{10}$H$_9$Cl$_4$O$_4$P
Gardona
2-Chloro-1-(2,4,5-trichlorophenyl)vinyldimethylphosphate
RN: 22248-79-9 **MP** (°C): 97.5
MW: 365.97 **BP** (°C):

Solubility (Moles/L)	Solubility (Grams/L)	Temp (°C)	Ref (#)	Evaluation (T P E A A)	Comments
3.006E-05	1.100E-02	20	M061	1 0 0 0 1	

1908. C$_{10}$H$_9$N
1-Naphthylamine
1-Aminonaphthalene
α-Naphthoylamine
α-Naphthylamin
α-Naphthylamine
RN: 134-32-7 **MP** (°C): 50
MW: 143.19 **BP** (°C): 300.8

Solubility (Moles/L)	Solubility (Grams/L)	Temp (°C)	Ref (#)	Evaluation (T P E A A)	Comments
1.187E-02	1.700E+00	20	F300	1 0 0 0 1	
3.600E-04	5.155E-02	ns	L060	0 0 0 0 1	average

1909. C$_{10}$H$_9$N
2-Naphthylamine
Naphthylamine-(2)
β-Naphthylamin
β-Naphthylamine
RN: 91-59-8 **MP** (°C): 113
MW: 143.19 **BP** (°C): 306.1

Solubility (Moles/L)	Solubility (Grams/L)	Temp (°C)	Ref (#)	Evaluation (T P E A A)	Comments
1.320E-03	1.890E-01	rt	N015	0 0 2 2 2	

1910. C$_{10}$H$_9$N
3-Methyl-isoquinoline
Isoquinoline, 3-Methyl-
RN: 1125-80-0   **MP** (°C):
MW: 143.19   **BP** (°C):   519.2

Solubility (Moles/L)	Solubility (Grams/L)	Temp (°C)	Ref (#)	Evaluation (T P E A A)	Comments
6.418E-03	9.190E-01	20	A050	1 0 1 1 2	

1911. C$_{10}$H$_9$NO
8-Hydroxyquinaldine
2-Methyl 8-Quinolinol
RN: 826-81-3   **MP** (°C):   72.5
MW: 159.19   **BP** (°C):

Solubility (Moles/L)	Solubility (Grams/L)	Temp (°C)	Ref (#)	Evaluation (T P E A A)	Comments
2.460E+03	3.916E+05	25.2	P024	2 2 1 1 2	
2.670E+03	4.250E+05	30.3	P024	2 2 1 1 2	

1912. C$_{10}$H$_9$NO
4-Hydroxy-2-methylquinoline
4-Hydroxy-2-methyl-chinolin
RN: 607-67-0   **MP** (°C):   234
MW: 159.19   **BP** (°C):

Solubility (Moles/L)	Solubility (Grams/L)	Temp (°C)	Ref (#)	Evaluation (T P E A A)	Comments
6.282E-02	1.000E+01	20	F300	1 0 0 0 1	
5.936E-01	9.450E+01	100	F300	1 0 0 0 2	

1913. C$_{10}$H$_9$NO$_2$S
Ethyl m-Isothiocyanobenzoate
Ethyl 3-Isothiocyanobenzoate
RN: 3137-84-6   **MP** (°C):
MW: 207.25   **BP** (°C):

Solubility (Moles/L)	Solubility (Grams/L)	Temp (°C)	Ref (#)	Evaluation (T P E A A)	Comments
2.500E-04	5.181E-02	25	K032	2 2 0 1 2	

1914.   C₁₀H₉NO₂S

1914. $C_{10}H_9NO_2S$

Ethyl 4-Isothiocyanatobenzoate
4-Carbethoxyphenylisothiocyanate
Ethyl p-Isothiocyanatobenzoate
RN: 1205-06-7   **MP** (°C):
MW: 207.25   **BP** (°C):

Solubility (Moles/L)	Solubility (Grams/L)	Temp (°C)	Ref (#)	Evaluation (T P E A A)	Comments
9.000E-05	1.865E-02	25	D019	1 1 1 1 1	

1915. $C_{10}H_9NO_3S$.

2-Naphthylamine-5-sulfonic Acid
Dahl's Acid
Naphthylamin-(2)-sulfosaeure-(5)
RN: 81-05-0   **MP** (°C):
MW: 223.25   **BP** (°C):

Solubility (Moles/L)	Solubility (Grams/L)	Temp (°C)	Ref (#)	Evaluation (T P E A A)	Comments
1.478E-03	3.300E-01	20	F300	1 0 0 0 2	

1916. $C_{10}H_9NO_3S$

1-Naphthylamine-5-sulfonic Acid
Laurent's Acid
Naphthylamin-(1)-sulfosaeure-(5)
RN: 84-89-9   **MP** (°C):
MW: 223.25   **BP** (°C):

Solubility (Moles/L)	Solubility (Grams/L)	Temp (°C)	Ref (#)	Evaluation (T P E A A)	Comments
4.479E-03	1.000E+00	20	F300	1 0 0 0 2	

1917. $C_{10}H_9NO_3S$

Cassella's Acid F
2-Naphthylamine-7-sulfonic Acid
Naphthylamin-(2)-sulfosaeure-(7)
RN: 494-44-0   **MP** (°C):
MW: 223.25   **BP** (°C):

Solubility (Moles/L)	Solubility (Grams/L)	Temp (°C)	Ref (#)	Evaluation (T P E A A)	Comments
8.958E-04	2.000E-01	20	F300	1 0 0 0 1	
1.389E-02	3.100E+00	100	F300	1 0 0 0 1	

1918.   C$_{10}$H$_9$NO$_3$S
Badische Acid
2-Naphthylamine-8-sulfonic Acid
Naphthylamin-(2)-sulfosaeure-(8)
RN: 86-60-2   **MP** (°C):
MW: 223.25   **BP** (°C):

Solubility (Moles/L)	Solubility (Grams/L)	Temp (°C)	Ref (#)	Evaluation (T P E A A)	Comments
2.688E-03	6.000E-01	20	F300	1 0 0 0 2	

1919.   C$_{10}$H$_9$NO$_3$S
1-Naphthylamine-8-sulfonic Acid
Naphthylamin-(1)-sulfosaeure-(8)
Peri Acid
RN: 82-75-7   **MP** (°C):
MW: 223.25   **BP** (°C):

Solubility (Moles/L)	Solubility (Grams/L)	Temp (°C)	Ref (#)	Evaluation (T P E A A)	Comments
8.958E-04	2.000E-01	21	F300	1 0 0 0 0	
1.971E-02	4.400E+00	100	F300	1 0 0 0 1	

1920.   C$_{10}$H$_9$NO$_3$S
1-Naphthylamine-4-sulfonic Acid
4-Amino-1-naphthalenesulfonic Acid
Naphthionic Acid
Naphthylamin-(1)-sulfosaeure-(4)
Pirias Acid
RN: 84-86-6   **MP** (°C):   000
MW: 223.25   **BP** (°C):

Solubility (Moles/L)	Solubility (Grams/L)	Temp (°C)	Ref (#)	Evaluation (T P E A A)	Comments
1.209E-03	2.699E-01	0	D077	1 0 0 1 1	
1.299E-03	2.899E-01	10	D077	1 0 0 1 1	
1.388E-03	3.099E-01	20	D077	1 0 0 1 1	
1.344E-03	3.000E-01	20	F300	1 0 0 0 0	
1.657E-03	3.699E-01	30	D077	1 0 0 1 1	
2.149E-03	4.798E-01	40	D077	1 0 0 1 1	
2.641E-03	5.897E-01	50	D077	1 0 0 1 1	
3.357E-03	7.494E-01	60	D077	1 0 0 1 1	
4.341E-03	9.691E-01	70	D077	1 0 0 1 1	
5.815E-03	1.298E+00	80	D077	1 0 0 1 2	
7.825E-03	1.747E+00	90	D077	1 0 0 1 2	
1.021E-03	2.279E-01	100	D077	1 0 0 1 2	
1.075E-02	2.400E+00	100	F300	1 0 0 0 1	

1921. C$_{10}$H$_9$NO$_3$S
1-Naphthylamine-2-sulfonic Acid
Naphthylamin-(1)-sulfosaeure-(2)
RN: 81-06-1 **MP** (°C):
MW: 223.25 **BP** (°C):

Solubility (Moles/L)	Solubility (Grams/L)	Temp (°C)	Ref (#)	Evaluation (T P E A A)	Comments
1.836E-02	4.100E+00	20	F300	1 0 0 0 1	
1.402E-01	3.130E+01	100	F300	1 0 0 0 2	

1922. C$_{10}$H$_9$NO$_3$S
Bronner's Acid
2-Naphthylamine-6-sulfonic Acid
Naphthylamin-(2)-sulfosaeure-(6)
RN: 93-00-5 **MP** (°C):
MW: 223.25 **BP** (°C):

Solubility (Moles/L)	Solubility (Grams/L)	Temp (°C)	Ref (#)	Evaluation (T P E A A)	Comments
5.375E-04	1.200E-01	20	F300	1 0 0 0 1	
7.615E-03	1.700E+00	100	F300	1 0 0 0 1	

1923. C$_{10}$H$_9$NO$_3$S
1,6-Cleve's Acid
1-Naphthylamine-6-sulfonic Acid
Naphthylamin-(1)-sulfosaeure-(6)
RN: 119-79-9 **MP** (°C):
MW: 223.25 **BP** (°C):

Solubility (Moles/L)	Solubility (Grams/L)	Temp (°C)	Ref (#)	Evaluation (T P E A A)	Comments
4.479E-03	1.000E+00	16	F300	1 0 0 0 2	

1924. C$_{10}$H$_9$NO$_3$S
2-Naphthylamine-1-sulfonic Acid
α-Naphthylamine-o-monosulfonic Acid
RN: 81-16-3 **MP** (°C):
MW: 223.25 **BP** (°C):

Solubility (Moles/L)	Solubility (Grams/L)	Temp (°C)	Ref (#)	Evaluation (T P E A A)	Comments
1.072E-02	2.394E+00	0	D077	1 0 0 1 1	
1.429E-02	3.190E+00	10	D077	1 0 0 1 1	
1.829E-02	4.083E+00	20	D077	1 0 0 1 1	
2.317E-02	5.173E+00	30	D077	1 0 0 1 1	
2.893E-02	6.458E+00	40	D077	1 0 0 1 1	
3.555E-02	7.937E+00	50	D077	1 0 0 1 1	
4.435E-02	9.901E+00	60	D077	1 0 0 1 2	

6.010E-02	1.342E+01	70	D077	1 0 0 1 2
7.834E-02	1.749E+01	80	D077	1 0 0 1 2
1.028E-01	2.296E+01	90	D077	1 0 0 1 2
1.347E-01	3.007E+01	100	D077	1 0 0 1 2

1925. $C_{10}H_9NO_4S$
7-Amino-1-naphthol-3-sulfonic Acid
7-Amino-naphtol-(1)-sulfosaeure-(3)
RN: 90-51-7   **MP** (°C):
MW: 239.25   **BP** (°C):

Solubility (Moles/L)	Solubility (Grams/L)	Temp (°C)	Ref (#)	Evaluation (T P E A A)	Comments
1.881E-02	4.500E+00	h	F300	0 0 0 0 1	

1926. $C_{10}H_9NO_9S_3$
1-Naphthylamine-2,4,7-trisulfonic Acid
1,3,6-Naphthalenetrisulfonic Acid, 4-Amino-
RN: 61986-93-4   **MP** (°C):
MW: 383.38   **BP** (°C):

Solubility (Moles/L)	Solubility (Grams/L)	Temp (°C)	Ref (#)	Evaluation (T P E A A)	Comments
4.799E-01	1.840E+02	20	F054	1 2 1 1 2	
8.216E-01	3.150E+02	80	F054	1 2 1 1 2	

1927. $C_{10}H_9N_3O_3S$
1-Sulfanilyl-3-methyl-5-pyrazolone
RN: **MP** (°C):
MW: 251.27   **BP** (°C):

Solubility (Moles/L)	Solubility (Grams/L)	Temp (°C)	Ref (#)	Evaluation (T P E A A)	Comments
1.827E-03	4.590E-01	37	R045	1 2 1 1 2	

1928. $C_{10}H_9N_4O_5$
Picrolonic Acid
Pikrolonsaeure
RN: 550-74-3   **MP** (°C):   116
MW: 265.21   **BP** (°C):

Solubility (Moles/L)	Solubility (Grams/L)	Temp (°C)	Ref (#)	Evaluation (T P E A A)	Comments
3.394E-02	9.000E+00	17	F300	1 0 0 0 0	
3.582E-02	9.500E+00	100	F300	1 0 0 0 1	

1929.   C₁₀H₁₀Fe

Ferrocene
bis-Cyclopentadienyliron
Ferrotsen
Iron bis(Cyclopentadiene)
RN: 102-54-5   **MP** (°C):
MW: 186.04   **BP** (°C):

Solubility (Moles/L)	Solubility (Grams/L)	Temp (°C)	Ref (#)	Evaluation (T P E A A)	Comments
3.388E-05	6.304E-03	25	B335	1 2 0 0 1	

1930.   C₁₀H₁₀BrNO₄

5-Bromo-2-p-phenyl-5-nitro-1,3-dioxane
m-Dioxane, 5-Bromo-5-nitro-2-phenyl-
1,3-Dioxane, 5-Bromo-5-nitro-2-phenyl-
RN: 58522-87-5   **MP** (°C):   82-84
MW: 288.10   **BP** (°C):

Solubility (Moles/L)	Solubility (Grams/L)	Temp (°C)	Ref (#)	Evaluation (T P E A A)	Comments
1.596E-03	4.598E-01	25	L013	1 0 2 1 2	

1931.   C₁₀H₁₀BrNO₅

5-Bromo-2-p-phenol-5-nitro-1,3-dioxane
m-Dioxane, 5-Bromo-5-nitro-2-phenol-
RN: 60766-61-2   **MP** (°C):   142-144
MW: 304.10   **BP** (°C):

Solubility (Moles/L)	Solubility (Grams/L)	Temp (°C)	Ref (#)	Evaluation (T P E A A)	Comments
1.413E-03	4.298E-01	25	L013	1 0 2 1 2	

1932.   C₁₀H₁₀ClNO₃

Chloroacetyl Acetaminophen
Acetic Acid, Chloro-, 4-(Acetylamino)phenyl Ester
Acetanilide, 4'-Hydroxy-, Chloroacetate (Ester)
RN: 17321-63-0   **MP** (°C):   184.5-185
MW: 227.65   **BP** (°C):

Solubility (Moles/L)	Solubility (Grams/L)	Temp (°C)	Ref (#)	Evaluation (T P E A A)	Comments
1.230E-03	2.800E-01	37	D029	1 0 1 1 1	

1933.   C₁₀H₁₀Cl₂F₂N₂OS

3-[3-Chloro-4-(chlorodifluoromethylthio)phenyl]-1,1-dimethylurea
N-[3-Chloro-4-(chlorodifluoromethylthiol)phenyl]-N',N'-dimethylurea
N-(3-Chloro-4-difluorochloromethylthiophenyl)-N',N'-dimethylurea
Thiochlormethyl
N-[3-Chloro-4-(chlorodifluoromethylthio)phenyl]-N',N'-dimethylurea

RN: 33439-45-1   **MP** (°C):   113.5
MW: 315.17   **BP** (°C):

Solubility (Moles/L)	Solubility (Grams/L)	Temp (°C)	Ref (#)	Evaluation (T P E A A)	Comments
2.159E-01	6.803E+01	20	M161	1 0 0 0 1	

1934.   C₁₀H₁₀Cl₂O₂

Chlorfenprop-methyl
Methyl 2-chloro-3-(p-chlorophenyl)propionate
Methyl α-p-Dichlorohydrocinnamate
Bidisin
Fatex

RN: 14437-17-3   **MP** (°C):
MW: 233.10   **BP** (°C):   111.5

Solubility (Moles/L)	Solubility (Grams/L)	Temp (°C)	Ref (#)	Evaluation (T P E A A)	Comments
1.716E-04	4.000E-02	20	M161	1 0 0 0 1	

1935.   C₁₀H₁₀Cl₂O₃

4-(2,4-Dichlorophenoxy)propionic Acid
2,4-DB

RN: 94-82-6   **MP** (°C):   118
MW: 249.10   **BP** (°C):

Solubility (Moles/L)	Solubility (Grams/L)	Temp (°C)	Ref (#)	Evaluation (T P E A A)	Comments
2.690E-04	6.700E-02	25	B164	1 0 1 1 1	
1.847E-04	4.600E-02	25	M161	1 0 0 0 1	
2.128E-04	5.300E-02	ns	B185	1 0 0 0 1	
1.847E-04	4.600E-02	ns	L024	1 0 0 0 1	
2.128E-04	5.300E-02	rt	M061	0 0 0 0 1	

1936. C$_{10}$H$_{10}$Cl$_2$O$_3$

Ethyl (2,4-Dichlorophenoxy)acetate
2,4-Dichlorophenoxyacetic Acid Ethyl Ester

RN: 533-23-3   **MP** (°C):
MW: 249.10   **BP** (°C):

Solubility (Moles/L)	Solubility (Grams/L)	Temp (°C)	Ref (#)	Evaluation (T P E A A)	Comments
2.529E-04	6.300E-02	ns	M120	0 0 1 1 2	

1937. C$_{10}$H$_{10}$Cl$_8$

Toxaphene
Camphechlor
Campheclor
PhenAcide
Toxakil
Chlorinated Champhene

RN: 8001-35-2   **MP** (°C):   65
MW: 413.82   **BP** (°C):

Solubility (Moles/L)	Solubility (Grams/L)	Temp (°C)	Ref (#)	Evaluation (T P E A A)	Comments
1.329E-06	5.500E-04	20	M336	2 0 2 2 2	
9.666E-07	4.000E-04	25	C100	1 0 2 1 0	
1.208E-06	5.000E-04	25	P085	1 0 1 1 2	
1.788E-06	7.400E-04	25	W025	1 0 2 2 2	
1.450E-06	6.000E-04	ns	M110	0 0 0 0 0	EFG
7.250E-06	3.000E-03	rt	M161	0 0 0 0 0	

1938. C$_{10}$H$_{10}$N$_4$O

Metamitron
3-Methyl-4-amino-6-phenyl-1,2,4-triazin-5(4H)-one
4-Amino-3-methyl-6-phenyl-1,2,4-triazin-5-one
Goltix

RN: 41394-05-2   **MP** (°C):   166.6
MW: 202.22   **BP** (°C):

Solubility (Moles/L)	Solubility (Grams/L)	Temp (°C)	Ref (#)	Evaluation (T P E A A)	Comments
8.901E-03	1.800E+00	20	M161	1 0 0 0 1	

1939.   C$_{10}$H$_{10}$N$_4$O$_2$S
Sulfadiazine
Sulphadiazine
N1-(2-Pyrimidinyl)-sulfanilamide
Debenal
RN: 68-35-9   **MP** (°C):   254
MW: 250.28   **BP** (°C):

Solubility (Moles/L)	Solubility (Grams/L)	Temp (°C)	Ref (#)	Evaluation (T P E A A)	Comments
2.360E-04	5.907E-02	20	C006	1 2 1 1 2	
1.814E-04	4.540E-02	20	E003	2 2 1 1 2	
5.993E-04	1.500E-01	20	F073	1 2 2 2 2	
2.917E-04	7.299E-02	20	L058	1 0 1 1 1	
3.077E-04	7.700E-02	25	C102	2 0 2 2 2	
2.637E-03	6.600E-01	25	K048	1 2 2 1 1	pH 1.26
3.036E-04	7.599E-02	30	E003	2 2 1 1 2	
3.640E-04	9.110E-02	30	H018	1 2 2 2 2	
3.200E-04	8.009E-02	30	L069	1 0 1 1 0	EFG
7.192E-04	1.800E-01	35	H114	1 0 0 0 1	
5.074E-04	1.270E-01	37	C102	2 0 2 2 2	
4.914E-04	1.230E-01	37	F072	1 0 0 0 2	
4.794E-04	1.200E-01	37	F075	1 0 2 2 2	
5.114E-04	1.280E-01	37	K091	1 0 0 0 2	
5.194E-04	1.300E-01	37	L091	1 0 0 0 1	pH 5.5
7.192E-04	1.800E-01	37	M057	1 0 0 0 2	pH 5.5
8.790E-04	2.200E-01	37	R044	1 0 1 1 0	EFG, intrinsic
4.914E-04	1.230E-01	37	R045	1 2 1 1 1	
6.712E-04	1.680E-01	37	S192	1 0 1 1 2	pH 6.0
5.074E-04	1.270E-01	37	W016	2 0 1 1 2	
4.914E-04	1.230E-01	37	W053	1 0 0 0 2	
3.956E-04	9.900E-02	38	K006	1 0 0 0 1	
5.154E-04	1.290E-01	40	E003	2 2 1 1 2	
5.194E-04	1.300E-01	ns	G083	0 0 0 0 1	pH 5.5

1940.   C$_{10}$H$_{10}$N$_4$O$_2$S
Sulfapyrazine
Sulphapyrazine
RN: 116-44-9   **MP** (°C):   255
MW: 250.28   **BP** (°C):

Solubility (Moles/L)	Solubility (Grams/L)	Temp (°C)	Ref (#)	Evaluation (T P E A A)	Comments
1.998E-04	5.000E-02	37	L091	1 0 0 0 0	pH 5.5

1941. $C_{10}H_{10}N_4O_2S$
4-Sulfanilamidopyrimidine
4-Sulfapyrimidine
Sulfanilamide, N1-4-Pyrimidinyl-
RN: 599-82-6   **MP** (°C):
MW: 250.28   **BP** (°C):

Solubility (Moles/L)	Solubility (Grams/L)	Temp (°C)	Ref (#)	Evaluation (T P E A A)	Comments
1.414E-02	3.540E+00	37	R045	1 2 1 1 2	

1942. $C_{10}H_{10}N_4O_2S$
5-Sulfanilamidopyrimidine
5-Sulfapyrimidine
Sulfanilamide, N1-5-Pyrimidinyl-
RN: 17103-48-9   **MP** (°C):
MW: 250.28   **BP** (°C):

Solubility (Moles/L)	Solubility (Grams/L)	Temp (°C)	Ref (#)	Evaluation (T P E A A)	Comments
3.916E-04	9.800E-02	37	R046	1 2 1 1 1	

1943. $C_{10}H_{10}N_4O_4S$
5-Sulfanilamidouracil
Benzenesulfonamide, 4-Amino-N-(1,2,3,4-tetrahydro-2,4-dioxo-5-pyrimidinyl)-
RN: 6912-98-7   **MP** (°C):
MW: 282.28   **BP** (°C):

Solubility (Moles/L)	Solubility (Grams/L)	Temp (°C)	Ref (#)	Evaluation (T P E A A)	Comments
1.722E-03	4.860E-01	37	R045	1 2 1 1 0	

1944. $C_{10}H_{10}O$
Benzalacetone
4-Phenyl-3-buten-2-one
Methyl Styryl Ketone
RN: 122-57-6   **MP** (°C):
MW: 146.19   **BP** (°C):

Solubility (Moles/L)	Solubility (Grams/L)	Temp (°C)	Ref (#)	Evaluation (T P E A A)	Comments
9.560E-03	1.398E+00	25	R070	1 2 2 2 2	

1945.   C₁₀H₁₀O₂

p-Acetylacetophenone
Ethanone, 1,1'-(1,4-Phenylene)bis-
RN: 1009-61-6   **MP** (°C):
MW: 162.19   **BP** (°C):

Solubility (Moles/L)	Solubility (Grams/L)	Temp (°C)	Ref (#)	Evaluation (T P E A A)	Comments
3.890E-05	6.309E-03	25	C316	1 0 2 2 2	0.1M NaCl

1946.   C₁₀H₁₀O₂

Methyl Cinnamate
2-Propenoic Acid
3-Phenyl-, Methyl Ester
RN: 103-26-4   **MP** (°C):
MW: 162.19   **BP** (°C):

Solubility (Moles/L)	Solubility (Grams/L)	Temp (°C)	Ref (#)	Evaluation (T P E A A)	Comments
2.500E-03	4.055E-01	25	R070	1 2 2 2 2	

1947.   C₁₀H₁₀O₂

trans-α-Methyl-cinnamic Acid
α-Methyl-trans-zimtsaeure
RN: 1895-97-2   **MP** (°C):
MW: 162.19   **BP** (°C):

Solubility (Moles/L)	Solubility (Grams/L)	Temp (°C)	Ref (#)	Evaluation (T P E A A)	Comments
7.399E-03	1.200E+00	h	F300	0 0 0 0 1	

1948.   C₁₀H₁₀O₄

Acetyl-r-mandelic Acid
(R)(-)O-Acetylmandelic Acid
[R]-[-]-α-(Acetoxy)phenylacetic Acid
O-Acetylmandelic Acid
RN: 5438-68-6   **MP** (°C):
MW: 194.19   **BP** (°C):

Solubility (Moles/L)	Solubility (Grams/L)	Temp (°C)	Ref (#)	Evaluation (T P E A A)	Comments
2.919E-02	5.668E+00	0	A043	1 2 1 1 1	
2.919E-02	5.668E+00	0	L035	1 2 2 1 1	
3.478E-02	6.754E+00	10	A043	1 2 1 1 1	
3.478E-02	6.754E+00	10	L035	1 2 2 1 1	
3.884E-02	7.543E+00	15	A043	1 2 1 1 1	
3.884E-02	7.543E+00	15	L035	1 2 2 1 1	
4.897E-02	9.509E+00	20	A043	1 2 1 1 1	

4.897E-02	9.509E+00	20	L035	1 2 2 1 1	
5.804E-02	1.127E+01	25	A043	1 2 1 1 2	
5.804E-02	1.127E+01	25	L035	1 2 2 1 2	
7.060E-02	1.371E+01	30	A043	1 2 1 1 2	
7.060E-02	1.371E+01	30	L035	1 2 2 1 2	
1.005E-01	1.951E+01	35	A043	1 2 1 1 2	
1.587E-01	3.082E+01	40	A043	1 2 1 1 2	
2.795E-01	5.428E+01	45	A043	1 2 1 1 2	
2.795E-01	5.428E+01	45	L035	1 2 2 1 2	
6.125E-01	1.189E+02	50	A043	1 2 1 1 2	
6.125E-01	1.189E+02	50	L035	1 2 2 1 2	

1949. $C_{10}H_{10}O_4$

Terephthalate Acid Dimethyl Ester
Terephthalsaeure-dimethyl Ester
1,4-Benzenedicarboxylic Acid Dimethyl Ester
Terephthalic Acid
Dimethyl TerePhthalate
Dimethyl 1,4-Benzenedicarboxylate

RN: 120-61-6   **MP** (°C):   140
MW: 194.19   **BP** (°C):

Solubility (Moles/L)	Solubility (Grams/L)	Temp (°C)	Ref (#)	Evaluation (T P E A A)	Comments
1.690E-04	3.282E-02	25	C316	1 0 2 2 2	0.1M NaCl
1.540E-02	2.991E+00	h	F070	1 0 0 0 1	

1950. $C_{10}H_{10}O_4$

Meconin
Mekonin

RN: 569-31-3   **MP** (°C):   102
MW: 194.19   **BP** (°C):

Solubility (Moles/L)	Solubility (Grams/L)	Temp (°C)	Ref (#)	Evaluation (T P E A A)	Comments
1.287E-02	2.500E+00	25	F300	1 0 0 0 0	
2.420E-02	4.700E+00	100	F300	1 0 0 0 1	

1951. $C_{10}H_{10}O_4$

Acetylsalicylic Acid, Methyl Ester
Methyl 2-Acetoxybenzoate
Benzoic Acid, 2-(Acetyloxy)-, Methyl Ester

RN: 580-02-9   **MP** (°C):   48
MW: 194.19   **BP** (°C):

Solubility (Moles/L)	Solubility (Grams/L)	Temp (°C)	Ref (#)	Evaluation (T P E A A)	Comments
1.447E-02	2.810E+00	21	N335	1 2 1 1 2	

1952. $C_{10}H_{10}O_4$
Dimethyl Phthalate
1,2-Benzenedicarboxylic Acid, Dimethyl Ester
Fermine
Unimoll DM
Mipax
Palatinol M
RN: 131-11-3   **MP** (°C):   5.5
MW: 194.19   **BP** (°C):   283.7

Solubility (Moles/L)	Solubility (Grams/L)	Temp (°C)	Ref (#)	Evaluation (T P E A A)	Comments
2.210E-02	4.292E+00	20	L300	2 1 0 2 2	
4.087E-02	7.937E+00	20.00	D343	1 0 1 1 0	
2.317E-01	4.500E+01	25	F067	1 0 2 2 2	*sic*
2.307E-02	4.480E+00	c	F070	1 0 0 0 0	
1.566E-02	3.041E+00	ns	F014	0 0 0 0 2	
2.052E-02	3.984E+00	ns	H069	0 0 1 1 1	
2.214E-02	4.300E+00	rt	M161	0 0 0 0 1	

1953. $C_{10}H_{10}O_5$
Opianic Acid
Opiansaeure
RN: 519-05-1   **MP** (°C):   150
MW: 210.19   **BP** (°C):

Solubility (Moles/L)	Solubility (Grams/L)	Temp (°C)	Ref (#)	Evaluation (T P E A A)	Comments
1.189E-02	2.500E+00	20	F300	1 0 0 0 1	
8.088E-02	1.700E+01	h	F300	0 0 0 0 1	

1954. $C_{10}H_{11}ClO_3$
4-(4-Chlorophenoxy)butyric Acid
4-(4-CPB)
RN: 3547-07-7   **MP** (°C):
MW: 214.65   **BP** (°C):

Solubility (Moles/L)	Solubility (Grams/L)	Temp (°C)	Ref (#)	Evaluation (T P E A A)	Comments
5.125E-04	1.100E-01	25	B164	1 0 1 1 2	

1955. C$_{10}$H$_{11}$ClO$_3$

Mecoprop
2-(4-Chloro-2-methylphenoxy)propionic Acid
2-(2-Methyl-4-chlorophenoxy)propionic Acid
2-(MCPP)

RN: 93-65-2   **MP** (°C): 93
MW: 214.65   **BP** (°C):

Solubility (Moles/L)	Solubility (Grams/L)	Temp (°C)	Ref (#)	Evaluation (T P E A A)	Comments
2.888E-03	6.200E-01	20	B185	1 0 0 0 2	
2.795E-03	6.000E-01	20	B200	1 0 0 0 2	
2.887E-03	6.196E-01	20	M061	1 0 0 0 1	
2.888E-03	6.200E-01	20	M161	1 0 0 0 2	
4.170E-03	8.950E-01	25	B164	1 0 1 1 2	
4.170E-03	8.950E-01	25	B185	1 0 0 0 2	
2.794E-03	5.996E-01	ns	B100	0 0 0 0 0	
2.050E-04	4.400E-02	ns	B185	1 0 0 0 1	
2.888E-03	6.200E-01	ns	L024	1 0 0 0 2	

1956. C$_{10}$H$_{11}$Cl$_3$O$_2$

2,3,6-Trichlorobenzyloxypropanol
1-Propanol, 3-[(2,3,6-Trichlorobenzyl)oxy]-

RN: 1591-82-8   **MP** (°C):
MW: 269.56   **BP** (°C):

Solubility (Moles/L)	Solubility (Grams/L)	Temp (°C)	Ref (#)	Evaluation (T P E A A)	Comments
2.708E-04	7.300E-02	25	B185	1 0 0 0 1	
2.708E-04	7.300E-02	25	B200	1 0 0 0 1	

1957. C$_{10}$H$_{11}$FN$_2$O$_6$

1,3-bis(Acetoxymethyl)-5-fluoro-2,4(1H,3H)-pyrimidinedi-one
1,3-bis(Acetoxymethyl)-5-fluorouracil

RN: 66542-48-1   **MP** (°C): 105-106
MW: 274.21   **BP** (°C):

Solubility (Moles/L)	Solubility (Grams/L)	Temp (°C)	Ref (#)	Evaluation (T P E A A)	Comments
1.568E-02	4.300E+00	22	B321	1 0 2 2 2	pH 4.0

1958.   C₁₀H₁₁F₃N₂O
Fluometuron
1,1-Dimethyl-3-(α,α,α-trifluoro-m-tolyl)urea
RN: 2164-17-2   **MP** (°C):   163
MW: 232.21   **BP** (°C):

Solubility (Moles/L)	Solubility (Grams/L)	Temp (°C)	Ref (#)	Evaluation (T P E A A)	Comments
4.571E-04	1.061E-01	20	B179	2 0 0 0 2	
4.522E-04	1.050E-01	20	M161	1 0 0 0 2	
3.661E-04	8.500E-02	24	C105	2 1 2 2 2	
3.876E-04	9.000E-02	25	B200	1 0 0 0 1	
3.876E-04	9.000E-02	25	G036	1 0 0 0 1	
3.876E-04	9.000E-02	25	M061	1 0 0 0 1	

1959.   C₁₀H₁₁F₃N₂O₃S
Fluoridamid
Acetamide, N-{4-Methyl-3-{{(trifluoromethyl)sulfonyl}amino}phenyl}-
Sustar
MBR6033
RN: 47000-92-0   **MP** (°C):   182-184
MW: 296.27   **BP** (°C):

Solubility (Moles/L)	Solubility (Grams/L)	Temp (°C)	Ref (#)	Evaluation (T P E A A)	Comments
4.388E-04	1.300E-01	22	G307	1 0 0 0 1	

1960.   C₁₀H₁₁NO
N-Methylcinnamide
2-Propenamide, N-Methyl-3-phenyl-
RN: 2757-10-0   **MP** (°C):
MW: 161.21   **BP** (°C):

Solubility (Moles/L)	Solubility (Grams/L)	Temp (°C)	Ref (#)	Evaluation (T P E A A)	Comments
1.310E-02	2.112E+00	ns	H350	0 0 0 0 2	

1961.   C₁₀H₁₁NOS
m-Isopropoxyphenyl Isothiocyanate
3-Isopropoxyphenyl Isothiocyanate
RN: 3528-90-3   **MP** (°C):
MW: 193.27   **BP** (°C):

Solubility (Moles/L)	Solubility (Grams/L)	Temp (°C)	Ref (#)	Evaluation (T P E A A)	Comments
4.700E-04	9.084E-02	25	K032	2 2 0 1 2	

1962. C₁₀H₁₁NO₃

Acetamide, 2-(Benzoyloxy)-N-methyl-

RN: 106231-50-9   **MP** (°C):   111
MW: 193.20   **BP** (°C):

Solubility (Moles/L)	Solubility (Grams/L)	Temp (°C)	Ref (#)	Evaluation (T P E A A)	Comments
1.915E-02	3.700E+00	22	N317	1 1 2 1 2	

1963. C₁₀H₁₁NO₃

p-Acetoxy-acetanilide
p-Acetoxyacetanilide
Acetaminophen Acetate
Acetyl Acetaminophen

RN: 2623-33-8   **MP** (°C):   153
MW: 193.20   **BP** (°C):

Solubility (Moles/L)	Solubility (Grams/L)	Temp (°C)	Ref (#)	Evaluation (T P E A A)	Comments
1.656E-03	3.200E-01	25	B010	1 1 1 1 0	
1.237E-02	2.390E+00	25	E016	1 1 1 1 2	
1.139E-02	2.200E+00	25	M333	1 1 0 0 2	
1.760E-02	3.400E+00	37	D029	1 0 1 1 1	

1964. C₁₀H₁₁NO₄

Carbobenzoxyglycine
N-Carbobenzyloxyglycine
N-CBZ-Glycine
Benzyloxycarbonyl Glycine

RN: 1138-80-3   **MP** (°C):
MW: 209.20   **BP** (°C):

Solubility (Moles/L)	Solubility (Grams/L)	Temp (°C)	Ref (#)	Evaluation (T P E A A)	Comments
2.180E-02	4.560E+00	25.1	N026	2 0 2 2 2	
2.170E-02	4.539E+00	25.1	N027	1 1 2 2 2	

1965. C₁₀H₁₁NO₄

O-(acetoxymethyl) Salicylamide
2-[(Acetyloxy)methoxy]-benzamide
Benzamide, 2-[(Acetyloxy)methoxy]-
O-Acetoxymethyl Methyl Salicylamide

RN: 102273-25-6   **MP** (°C):   92.5
MW: 209.20   **BP** (°C):

Solubility (Moles/L)	Solubility (Grams/L)	Temp (°C)	Ref (#)	Evaluation (T P E A A)	Comments
>2.39E-02	>5.00E+00	23	B328	1 2 2 1 1	pH 4
2.390E-02	5.000E+00	23	B328	1 2 2 1 1	

1966.   C₁₀H₁₁NO₄
Methyl Acetaminophen
Carbonic Acid, 4-(Acetylamino)phenyl Methyl Ester
Acetanilide, 4'-Hydroxy-, Methyl Carbonate (Ester)
RN: 17321-62-9   **MP** (°C):   115.5-116.5
MW: 209.20   **BP** (°C):

Solubility (Moles/L)	Solubility (Grams/L)	Temp (°C)	Ref (#)	Evaluation (T P E A A)	Comments
2.868E-02	6.000E+00	37	D029	1 0 1 1 1	

1967.   C₁₀H₁₁NO₅
Acido D-Feniltartrammico Tartranilico
RN: **MP** (°C):   194
MW: 225.20   **BP** (°C):

Solubility (Moles/L)	Solubility (Grams/L)	Temp (°C)	Ref (#)	Evaluation (T P E A A)	Comments
1.232E-01	2.774E+01	17.40	C070	1 2 2 1 2	

1968.   C₁₀H₁₁NO₆
Acido p-Ossifeniltartrammico
RN: **MP** (°C):   218
MW: 241.20   **BP** (°C):

Solubility (Moles/L)	Solubility (Grams/L)	Temp (°C)	Ref (#)	Evaluation (T P E A A)	Comments
1.677E-01	4.045E+01	14	C071	1 2 0 1 2	

1969.   C₁₀H₁₁N₃OS
Methabenzthiazuron
N-2-Benzothiazolyl-N,N'-dimethylurea
1,3-Dimethyl-3-(2-benzothiazolyl)urea
Methyl-N'-methyl-N'-(2-benzothiazolyl)urea
Tribunil
Preparation 5633
RN: 18691-97-9   **MP** (°C):   119.5
MW: 221.28   **BP** (°C):

Solubility (Moles/L)	Solubility (Grams/L)	Temp (°C)	Ref (#)	Evaluation (T P E A A)	Comments
2.666E-04	5.900E-02	20	M161	1 0 0 0 1	

1970. $C_{10}H_{11}N_3O_2S_2$
Methyl Sulfathiazole
Sulfathiazol Methyle

RN: 15251-46-4   **MP** (°C):
MW: 269.35   **BP** (°C):

Solubility (Moles/L)	Solubility (Grams/L)	Temp (°C)	Ref (#)	Evaluation (T P E A A)	Comments
9.653E-04	2.600E-01	37	D084	1 0 1 0 1	

1971. $C_{10}H_{11}N_3O_2S$
Sulfapyrrole

RN: **MP** (°C):
MW: 237.28   **BP** (°C):

Solubility (Moles/L)	Solubility (Grams/L)	Temp (°C)	Ref (#)	Evaluation (T P E A A)	Comments
2.023E-02	4.800E+00	20	F073	1 2 2 2 2	

1972. $C_{10}H_{11}N_3O_2S_2$
N1-Methyl-N1-2-thiazolyl-sulfanilamide
N1-Methylsulfathiazole

RN: 51203-19-1   **MP** (°C):
MW: 269.35   **BP** (°C):

Solubility (Moles/L)	Solubility (Grams/L)	Temp (°C)	Ref (#)	Evaluation (T P E A A)	Comments
1.150E-03	3.097E-01	37	K095	2 0 0 0 2	intrinsic

1973. $C_{10}H_{11}N_3O_2S_2$
Sulfamethylthiazole
4-Methyl-2-sulfanilamidothiazole
2-(p-Aminobenzenesulfonamido)-4-methylthiazole
2-Sulfanilamido-4-methylthiazole
Aseptil 2
Ciba 3753

RN: 515-59-3   **MP** (°C):   239
MW: 269.35   **BP** (°C):

Solubility (Moles/L)	Solubility (Grams/L)	Temp (°C)	Ref (#)	Evaluation (T P E A A)	Comments
4.084E-04	1.100E-01	20	F073	1 2 2 2 2	
4.084E-04	1.100E-01	20	F074	1 0 0 0 2	

1974.   C₁₀H₁₁N₃O₃

α-Semicarbazono-p-tolyl Acetate

RN: **MP** (°C):
MW: 221.22   **BP** (°C):

Solubility (Moles/L)	Solubility (Grams/L)	Temp (°C)	Ref (#)	Evaluation (T P E A A)	Comments
1.400E-03	3.097E-01	25	A066	1 0 1 1 1	

1975.   C₁₀H₁₁N₃O₃S

Sulfamethoxazole

4-Amino-N-(5-methyl-3-isoxazolyl)benzenesulfonamide

Cotrimoxazole

Septra

Bactrim

Cotrim

RN: 723-46-6   **MP** (°C):   167
MW: 253.28   **BP** (°C):

Solubility (Moles/L)	Solubility (Grams/L)	Temp (°C)	Ref (#)	Evaluation (T P E A A)	Comments
1.109E-03	2.810E-01	25	D308	1 0 2 2 2	pH 3.22
1.974E-03	5.000E-01	25	R025	1 0 0 0 0	
1.488E-03	3.770E-01	32	D308	1 0 2 2 2	pH 4.0
1.824E-03	4.620E-01	37	D308	1 0 2 2 2	pH 3.43
2.408E-03	6.100E-01	37	H120	1 1 1 1 1	normal saline
2.480E-03	6.281E-01	37	K095	2 0 0 0 2	intrinsic
5.527E-03	1.400E+00	37	M321	1 0 0 0 2	intrinsic

1976.   C₁₀H₁₁N₅O₂S

5-Sulfanilamido-2-aminopyrimidine

Benzenesulfonamide, 4-Amino-N-(2-amino-5-pyrimidinyl)-

RN: 71119-38-5   **MP** (°C):
MW: 265.30   **BP** (°C):

Solubility (Moles/L)	Solubility (Grams/L)	Temp (°C)	Ref (#)	Evaluation (T P E A A)	Comments
3.129E-04	8.300E-02	37	R046	1 2 1 1 1	

1977. C$_{10}$H$_{12}$

Tetralin
1,2,3,4-Tetrahydronaphthalene
RN: 119-64-2   **MP** (°C):   -31.0
MW: 132.21   **BP** (°C):   207.2

Solubility (Moles/L)	Solubility (Grams/L)	Temp (°C)	Ref (#)	Evaluation (T P E A A)	Comments
3.404E-04	4.500E-02	20	B356	1 0 0 0 2	
3.532E-04	4.670E-02	28	B348	2 1 2 2 2	
1.513E-03	2.000E-01	150	J023	1 1 2 2 0	
3.026E-03	4.000E-01	200	J023	1 1 2 2 0	
3.026E-02	4.000E+00	250	J023	1 1 2 2 0	
3.236E-04	4.278E-02	ns	D001	0 0 0 0 2	

1978. C$_{10}$H$_{12}$BrCl$_2$O$_3$PS

Bromophos-ethyl
O-(4-Bromo-2,5-dichlorophenyl) O,O-Diethyl Phosphorothioate
Nexagan
Filariol
RN: 4824-78-6   **MP** (°C):
MW: 394.06   **BP** (°C):

Solubility (Moles/L)	Solubility (Grams/L)	Temp (°C)	Ref (#)	Evaluation (T P E A A)	Comments
5.329E-07	2.100E-04	10	B324	2 2 2 2 2	
5.329E-07	2.100E-04	10	B324	2 2 2 2 2	
8.629E-07	3.400E-04	20	B324	2 2 2 2 2	
8.628E-07	3.400E-04	20	B324	2 2 2 2 2	
7.613E-06	3.000E-03	20	F311	1 2 2 2 1	
5.075E-06	2.000E-03	20	W312	1 0 0 0 0	
1.269E-06	5.001E-04	30	B324	2 2 2 2 2	
1.269E-06	5.000E-04	30	B324	2 2 2 2 2	
5.075E-06	2.000E-03	ns	E050	0 0 0 0 0	
5.075E-06	2.000E-03	rt	M161	0 0 0 0 0	

1979. C$_{10}$H$_{12}$ClNO$_2$

Baclofen
Lioresal
β-(Aminomethyl)-p-chlorohydrocinnamic Acid
RN: 1134-47-0   **MP** (°C):
MW: 213.67   **BP** (°C):

Solubility (Moles/L)	Solubility (Grams/L)	Temp (°C)	Ref (#)	Evaluation (T P E A A)	Comments
2.129E-02	4.549E+00	25	M374	1 0 2 1 2	

1980.   C₁₀H₁₂ClNO₂

Chloro-IPC
Furloe
Taterpex
Chlorpropham
Isopropyl m-Chlorocarbanilate

RN: 101-21-3   **MP** (°C):   38
MW: 213.67   **BP** (°C):

Solubility (Moles/L)	Solubility (Grams/L)	Temp (°C)	Ref (#)	Evaluation (T P E A A)	Comments
5.055E-04	1.080E-01	20	B185	1 0 0 0 2	
3.744E-04	8.000E-02	25	G099	1 0 0 1 0	
3.744E-04	8.000E-02	25	G319	1 0 0 0 2	
4.165E-04	8.900E-02	25	M161	1 0 0 0 1	
3.744E-04	8.000E-02	ns	B185	0 0 0 0 1	
4.119E-04	8.800E-02	ns	B200	0 0 0 0 1	
3.744E-04	8.000E-02	ns	F035	0 0 0 0 0	
4.119E-04	8.800E-02	ns	H042	0 0 0 0 1	
3.744E-04	8.000E-02	ns	M061	0 0 0 0 1	
3.548E-04	7.581E-02	ns	M163	0 0 0 0 0	EFG
5.055E-04	1.080E-01	ns	N013	0 0 0 0 2	

1981.   C₁₀H₁₂ClN₃O₂

Tranid
3-Chloro-6-cyanonorbornanone-2-oxime-O,N-methylcarbamate

RN: 15271-41-7   **MP** (°C):   143.5
MW: 241.68   **BP** (°C):

Solubility (Moles/L)	Solubility (Grams/L)	Temp (°C)	Ref (#)	Evaluation (T P E A A)	Comments
8.259E-03	1.996E+00	ns	M061	0 0 0 0 0	

1982.   C₁₀H₁₂ClN₃O₃S

Quinethazone
7-Chloro-2-ethyl-1,2,3,4-tetrahydro-4-oxo-6-quinazolinesulfonamide
Hydromox
CL 36010
Aquamox

RN: 73-49-4   **MP** (°C):   251
MW: 289.74   **BP** (°C):

Solubility (Moles/L)	Solubility (Grams/L)	Temp (°C)	Ref (#)	Evaluation (T P E A A)	Comments
5.176E-04	1.500E-01	25	A081	1 0 1 1 0	EFG

1983.   C₁₀H₁₂ClN₅O₂

$C_{10}H_{12}ClN_5O_2$

2-Chloro-2',3'-dideoxyadenosine
2-ClDDA

RN: 114849-58-0   **MP** (°C):
MW: 269.69   **BP** (°C):

Solubility (Moles/L)	Solubility (Grams/L)	Temp (°C)	Ref (#)	Evaluation (T P E A A)	Comments
3.745E-03	1.010E+00	25	A336	2 2 2 2 2	

1984.   C₁₀H₁₂Cl₂O

$C_{10}H_{12}Cl_2O$

2,4-Dichloro-6-butyl-phenol
Phenol, 2-Butyl-4,6-dichloro-

RN: 91399-13-2   **MP** (°C):
MW: 219.11   **BP** (°C):

Solubility (Moles/L)	Solubility (Grams/L)	Temp (°C)	Ref (#)	Evaluation (T P E A A)	Comments
2.400E-04	5.259E-02	25	B316	1 0 2 1 1	

1985.   C₁₀H₁₂Cl₃O₂PS

$C_{10}H_{12}Cl_3O_2PS$

Trichloronate
Trichloronat
Ethyl O-(2,4,5-Trichlorophenyl) Ethylphosphonothioate
Agritox
BAY 37289

RN: 327-98-0   **MP** (°C):
MW: 333.60   **BP** (°C):   108

Solubility (Moles/L)	Solubility (Grams/L)	Temp (°C)	Ref (#)	Evaluation (T P E A A)	Comments
2.458E-06	8.200E-04	10	B324	2 2 2 2 2	
2.458E-06	8.200E-04	10	B324	2 2 2 2 2	
1.769E-06	5.901E-04	20	B300	2 1 1 1 2	
2.638E-06	8.800E-04	20	B324	2 2 2 2 2	
2.638E-06	8.800E-04	20	B324	2 2 2 2 2	
1.499E-04	5.000E-02	20	M161	1 0 0 0 1	*sic*
3.208E-06	1.070E-03	30	B324	2 2 2 2 2	
3.207E-06	1.070E-03	30	B324	2 2 2 2 2	

1986.   C₁₀H₁₂N₂O₂

$C_{10}H_{12}N_2O_2$

Acetone N-(Phenylcarbamoyl)oxime
Acetone Oxime N-Phenylcarbamate
Proxypham

RN: **MP** (°C):   109.5
MW: 192.22   **BP** (°C):

Solubility (Moles/L)	Solubility (Grams/L)	Temp (°C)	Ref (#)	Evaluation (T P E A A)	Comments
2.601E-03	5.000E-01	ns	M061	0 0 0 0 2	approximate

1987. $C_{10}H_{12}N_2O_3$
Allobarbital
5,5-Diallylbarbituric Acid
RN: 52-43-7   **MP** (°C):   171
MW: 208.22   **BP** (°C):

Solubility (Moles/L)	Solubility (Grams/L)	Temp (°C)	Ref (#)	Evaluation (T P E A A)	Comments
6.003E-03	1.250E+00	20	J030	1 2 2 2 2	
7.193E-03	1.498E+00	25	A023	1 0 0 1 2	
8.500E-03	1.770E+00	25	G003	1 1 1 1 1	pH 4.7
8.650E-03	1.801E+00	25	V033	2 0 1 1 2	
8.700E-03	1.812E+00	25.00	T303	1 0 0 0 1	
9.250E-03	1.926E+00	30	G014	1 1 1 1 0	EFG
9.200E-03	1.916E+00	30	I001	2 0 2 1 0	EFG, 0.003N H_2SO_4
9.200E-03	1.916E+00	30	K108	1 2 2 0 1	
1.150E-02	2.394E+00	35	A023	1 0 0 1 2	
1.110E-02	2.311E+00	35.00	T303	1 0 0 0 2	
1.215E-02	2.530E+00	37	J030	1 2 2 2 2	
1.200E-02	2.499E+00	37	K121	1 2 1 2 1	0.1N HCl
1.675E-02	3.488E+00	40	A023	1 0 0 1 2	
1.370E-01	2.853E+01	40	N008	1 0 1 1 2	*sic*
1.690E-02	3.519E+00	45.00	T303	1 0 0 0 2	
7.036E-03	1.465E+00	ns	T003	0 0 0 0 2	

1988. $C_{10}H_{12}N_2O_3$
Barbituric-2-14C Acid, 5,5-Diallyl
RN: 112599-90-3   **MP** (°C):
MW: 208.22   **BP** (°C):

Solubility (Moles/L)	Solubility (Grams/L)	Temp (°C)	Ref (#)	Evaluation (T P E A A)	Comments
8.381E-03	1.745E+00	25	P350	2 1 1 1 2	intrinsic

1989. C₁₀H₁₂N₂O₃S

1989. $C_{10}H_{12}N_2O_3S$
Bentazon
2,1,3-Benzothiadiazin-4(3H)-one
Thiadiazinol
Basagran 4E
Adagio
BAS 351H

RN: 25057-89-0 **MP** (°C): 138.0
MW: 240.28 **BP** (°C):

Solubility (Moles/L)	Solubility (Grams/L)	Temp (°C)	Ref (#)	Evaluation (T P E A A)	Comments
2.081E-03	5.000E-01	20	M161	1 0 0 0 2	
2.080E-03	4.998E-01	ns	B100	0 0 0 0 0	
3.329E-03	8.000E-01	ns	M110	0 0 0 0 0	EFG

1990. $C_{10}H_{12}N_2O_4S$
N1,N4-Diacetylsulfanilamide
N4-Acetylsulphacetamide

RN: 5626-90-4 **MP** (°C):
MW: 256.28 **BP** (°C):

Solubility (Moles/L)	Solubility (Grams/L)	Temp (°C)	Ref (#)	Evaluation (T P E A A)	Comments
8.389E-03	2.150E+00	37	L091	1 0 0 0 2	pH 5.5

1991. $C_{10}H_{12}N_2O_5$
D-Monofeniltartramide Tartranilamide

RN: **MP** (°C): 226
MW: 240.22 **BP** (°C):

Solubility (Moles/L)	Solubility (Grams/L)	Temp (°C)	Ref (#)	Evaluation (T P E A A)	Comments
1.958E-02	4.704E+00	21.50	C070	1 2 2 1 2	

1992.   C$_{10}$H$_{12}$N$_2$O$_5$
2,4-Dinitro-6-sec-butylphenol
Dinoseb
4,6-Dinitro-2-S-butylphenol
Phenol, 4,6-Dinitro-2-sec-butyl-
RN: 88-85-7   **MP** (°C):   38
MW: 240.22   **BP** (°C):

Solubility (Moles/L)	Solubility (Grams/L)	Temp (°C)	Ref (#)	Evaluation (T P E A A)	Comments
2.165E-04	5.200E-02	25	B200	1 0 0 0 1	
2.165E-04	5.200E-02	25	G319	1 0 0 0 2	
3.053E-03	7.335E-01	25	M061	1 0 0 0 2	
4.159E-03	9.990E-01	ns	B100	0 0 0 0 0	
2.081E-04	5.000E-02	ns	B185	0 0 0 0 1	
1.413E-03	3.393E-01	ns	M163	0 0 0 0 0	EFG
4.163E-04	1.000E-01	rt	M161	0 0 0 0 2	

1993.   C$_{10}$H$_{12}$N$_3$O$_3$PS$_2$
Azinphos-methyl
Guthion
S-(3,4-Dihydro-4-oxobenzo[d][1,2,3]triazin-3-ylmethyl) O,O-Dimethyl Phosphorodithioate
Methyl Gusathion
RN: 86-50-0   **MP** (°C):   74
MW: 317.33   **BP** (°C):

Solubility (Moles/L)	Solubility (Grams/L)	Temp (°C)	Ref (#)	Evaluation (T P E A A)	Comments
2.994E-05	9.501E-03	10	B324	2 2 2 2 2	
2.994E-05	9.500E-03	10	B324	2 2 2 2 2	
4.412E-05	1.400E-02	15	A087	1 0 0 1 0	
6.587E-05	2.090E-02	20	B300	2 1 1 1 2	
6.587E-05	2.090E-02	20	B324	2 2 2 2 2	
6.586E-05	2.090E-02	20	B324	2 2 2 2 2	
9.454E-05	3.000E-02	20	M061	1 0 0 0 1	
9.139E-05	2.900E-02	25	A087	1 0 0 1 0	
1.374E-04	4.360E-02	30	B324	2 2 2 2 2	
1.374E-04	4.360E-02	30	B324	2 2 2 2 2	
1.481E-04	4.700E-02	35	A087	1 0 0 1 0	
1.040E-04	3.300E-02	rt	M161	0 0 0 0 1	

1994. C$_{10}$H$_{12}$N$_4$
6,7-Diethylpteridine
RN: **MP** (°C): 52
MW: 188.23 **BP** (°C):

Solubility (Moles/L)	Solubility (Grams/L)	Temp (°C)	Ref (#)	Evaluation (T P E A A)	Comments
6.641E-01	1.250E+02	20	A019	2 2 1 1 0	

1995. C$_{10}$H$_{12}$N$_4$O
2-Hydroxy-6,7-diethylpteridine
2-Hydroxy-6:7-diethylpteridine
4-Hydroxy-6,7-diethylpteridine
4-Hydroxy-6:7-diethylpteridine
RN: 90870-76-1 **MP** (°C):
MW: 204.23 **BP** (°C):

Solubility (Moles/L)	Solubility (Grams/L)	Temp (°C)	Ref (#)	Evaluation (T P E A A)	Comments
1.221E-02	2.494E+00	20	A019	2 2 1 1 2	
5.434E-03	1.110E+00	20	A019	2 2 1 1 2	

1996. C$_{10}$H$_{12}$N$_4$O$_2$
2,4-Dihydroxy-6,7-diethylpteridine
2,4-Dihydroxy-6:7-diethylpteridine
RN: 113222-29-0 **MP** (°C): 218
MW: 220.23 **BP** (°C):

Solubility (Moles/L)	Solubility (Grams/L)	Temp (°C)	Ref (#)	Evaluation (T P E A A)	Comments
4.124E-03	9.083E-01	20	A019	2 2 1 1 2	

1997. C$_{10}$H$_{12}$N$_4$O$_2$
1H-Pyrazolo[3,4-d]pyrimidine, 4-[(Tetrahydro-2H-pyran-2-yl)oxy]-
2-Tetrahydropuran-4-allopurinyl Ether
RN: 52717-52-9 **MP** (°C):
MW: 220.23 **BP** (°C):

Solubility (Moles/L)	Solubility (Grams/L)	Temp (°C)	Ref (#)	Evaluation (T P E A A)	Comments
1.653E-02	3.640E+00	ns	H067	0 2 0 0 2	

1998. $C_{10}H_{12}N_4O_2S$
Sulfaethidole
Ethyl Thiodiazole
Sulfaethylthiadiazole
Thiodiazolique Ethyle
RN: 94-19-9   **MP** (°C):   188
MW: 252.30   **BP** (°C):

Solubility (Moles/L)	Solubility (Grams/L)	Temp (°C)	Ref (#)	Evaluation (T P E A A)	Comments
8.522E-04	2.150E-01	20	F073	1 2 2 2 2	
1.288E-02	3.250E+00	37	B046	1 0 2 2 2	pH 5
1.585E-03	4.000E-01	37	D084	1 0 1 0 1	

1999. $C_{10}H_{12}N_4O_3$
2',3'-Dideoxyinosine
Videx
Didanosine
CCRIS 805
CCRIS 805Didanosine
RN: 69655-05-6   **MP** (°C):   175
MW: 236.23   **BP** (°C):

Solubility (Moles/L)	Solubility (Grams/L)	Temp (°C)	Ref (#)	Evaluation (T P E A A)	Comments
4.614E-02	1.090E+01	4	A337	1 0 2 2 2	
1.156E-01	2.730E+01	25	A337	1 0 2 2 2	

2000. $C_{10}H_{12}N_4O_3$
1-Butyryloxymethyl Allopurinol
Butanoic Acid, (4,5-Dihydro-4-oxo-1H-pyrazolo[3,4-d]pyrimidin-1-yl)methyl Ester
RN: 98827-21-5   **MP** (°C):   224-226
MW: 236.23   **BP** (°C):

Solubility (Moles/L)	Solubility (Grams/L)	Temp (°C)	Ref (#)	Evaluation (T P E A A)	Comments
1.482E-03	3.500E-01	22	B322	1 0 2 2 2	

2001. $C_{10}H_{12}N_4O_3$
2-Butyryloxymethyl Allopurinol
Butanoic Acid, (4,5-Dihydro-4-oxo-2H-pyrazolo[3,4-d]pyrimidin-1-yl)methyl Ester
RN: 98827-22-6   **MP** (°C):   182-183
MW: 236.23   **BP** (°C):

Solubility (Moles/L)	Solubility (Grams/L)	Temp (°C)	Ref (#)	Evaluation (T P E A A)	Comments
6.350E-03	1.500E+00	22	B322	1 0 2 2 2	

2002. C₁₀H₁₂N₄O₅

Inosine

Inosin

Hypoxanthine Ribonucleoside

RN: 58-63-9 **MP** (°C): 212dec
MW: 268.23 **BP** (°C):

Solubility (Moles/L)	Solubility (Grams/L)	Temp (°C)	Ref (#)	Evaluation (T P E A A)	Comments
5.871E-02	1.575E+01	20	D041	1 0 0 0 1	
5.890E-02	1.580E+01	20	F300	1 0 0 0 2	

2003. C₁₀H₁₂N₄O₆

2,4,6-Trinitrodiethylaniline

2-4-6-Trinitrodiethylaniline

RN: 106415-21-8 **MP** (°C):
MW: 284.23 **BP** (°C):

Solubility (Moles/L)	Solubility (Grams/L)	Temp (°C)	Ref (#)	Evaluation (T P E A A)	Comments
1.759E-04	5.000E-02	50	D067	1 2 0 0 0	
7.037E-04	2.000E-01	100	D067	1 2 0 0 1	

2004. C₁₀H₁₂N₅O₆P

Adenosine 3':5'-Monophosphate

Adenosine, Cyclic 3',5'-(hydrogen phosphate)

4H-Furo[3,2-d]-1,3,2-dioxaphosphorin, Adenosine Deriv

RN: 60-92-4 **MP** (°C):
MW: 329.21 **BP** (°C):

Solubility (Moles/L)	Solubility (Grams/L)	Temp (°C)	Ref (#)	Evaluation (T P E A A)	Comments
2.360E-02	7.769E+00	20	D034	1 1 2 1 2	pH 7.0

2005. C₁₀H₁₂N₆O₂S

2-S-Cysteinyl-4,6-bis-(dimethylamino)-s-triazine

RN: **MP** (°C): 173
MW: 280.31 **BP** (°C):

Solubility (Moles/L)	Solubility (Grams/L)	Temp (°C)	Ref (#)	Evaluation (T P E A A)	Comments
7.991E-03	2.240E+00	25	C051	1 2 1 1 2	pH 7

2006.   C$_{10}$H$_{12}$O
5,6,7,8-Tetrahydro-2-naphthol
5,6,7,8-Tetrahydro-naphthol-(2)
RN: 1125-78-6   **MP** (°C):   56.5
MW: 148.21   **BP** (°C):   275.5

Solubility (Moles/L)	Solubility (Grams/L)	Temp (°C)	Ref (#)	Evaluation (T P E A A)	Comments
1.012E-02	1.500E+00	20	F300	1 0 0 0 1	

2007.   C$_{10}$H$_{12}$O
Anethole
Methoxy-4-propenylbenzene
Propenylanisole
p-Propenylanisole
Anise Camphor
Isoestragole
RN: 104-46-1   **MP** (°C):   21.4
MW: 148.21   **BP** (°C):

Solubility (Moles/L)	Solubility (Grams/L)	Temp (°C)	Ref (#)	Evaluation (T P E A A)	Comments
7.490E-04	1.110E-01	25	I019	1 0 1 2 2	

2008.   C$_{10}$H$_{12}$O
Estragole
1-Methoxy-4-(2-propen-1-yl)benzene
Chavicyl Methyl Ether
4-Allylanisole
Tarragon
RN: 140-67-0   **MP** (°C):   <25
MW: 148.21   **BP** (°C):   216

Solubility (Moles/L)	Solubility (Grams/L)	Temp (°C)	Ref (#)	Evaluation (T P E A A)	Comments
1.200E-03	1.778E-01	25	I019	1 0 1 2 2	

2009.   C$_{10}$H$_{12}$O$_2$
n-Propyl Benzoate
Propyl Benzoate
Benzoicacidpropyl Ester
RN: 2315-68-6   **MP** (°C):   -51
MW: 164.21   **BP** (°C):

Solubility (Moles/L)	Solubility (Grams/L)	Temp (°C)	Ref (#)	Evaluation (T P E A A)	Comments
1.531E-03	2.514E-01	20	H301	2 0 2 2 2	

2010.   C$_{10}$H$_{12}$O$_2$
2,4,6-Trimethylbenzoic Acid
Mesitylenecarboxylic Acid
RN: 480-63-7   **MP** (°C):   154
MW: 164.21   **BP** (°C):

Solubility (Moles/L)	Solubility (Grams/L)	Temp (°C)	Ref (#)	Evaluation (T P E A A)	Comments
4.400E-03	7.225E-01	ns	C014	0 2 0 1 1	

2011.   C$_{10}$H$_{12}$O$_2$
Eugenol
1-Allyl-3-methoxy-4-hydroxybenzene
2-Methoxy-4-allylphenol
2-Methoxy-4-(2-propenyl)phenol
4-Allylguaiacol
Allylguaiacol
RN: 97-53-0   **MP** (°C):   15
MW: 164.21   **BP** (°C):

Solubility (Moles/L)	Solubility (Grams/L)	Temp (°C)	Ref (#)	Evaluation (T P E A A)	Comments
1.500E-02	2.463E+00	25	I019	1 0 1 2 2	
4.020E-02	6.601E+00	37	E028	1 0 1 1 2	

2012.   C$_{10}$H$_{12}$O$_2$
β-Phenylbutyric Acid
3-Phenyl-n-butyric Acid
RN: 4593-90-2   **MP** (°C):   38
MW: 164.21   **BP** (°C):   171

Solubility (Moles/L)	Solubility (Grams/L)	Temp (°C)	Ref (#)	Evaluation (T P E A A)	Comments
5.635E-02	9.254E+00	30	D033	2 2 1 2 2	
7.013E-02	1.152E+01	40	D033	2 2 1 2 2	

2013.   C$_{10}$H$_{12}$O$_3$
Propylparaben
Pr-paraben
Propyl p-Hydroxybenzoic Acid
Propyl 4-Hydroxybenzoate
Propyl Paraben
RN: 94-13-3   **MP** (°C):   96.5
MW: 180.21   **BP** (°C):

Solubility (Moles/L)	Solubility (Grams/L)	Temp (°C)	Ref (#)	Evaluation (T P E A A)	Comments
2.050E-03	3.694E-01	15	B355	1 1 1 1 2	
1.172E-03	2.112E-01	15	M352	1 1 1 1 2	
2.410E-03	4.343E-01	20	B355	1 1 1 1 2	
2.055E-03	3.703E-01	25	A059	1 0 1 1 1	
2.570E-03	4.631E-01	25	B355	1 1 1 1 2	
2.773E-03	4.998E-01	25	D081	1 2 2 1 2	
1.990E-03	3.586E-01	25	D339	1 0 1 1 2	
1.778E-03	3.205E-01	25	F322	2 0 1 1 0	EFG
1.844E-03	3.323E-01	25	M352	1 1 1 1 2	
2.775E-03	5.000E-01	25	O027	1 0 1 0 0	
2.863E-03	5.160E-01	25	P013	2 0 2 1 2	
2.300E-03	4.145E-01	27	B129	2 2 2 2 1	
2.443E-03	4.403E-01	30	A059	1 0 1 1 1	
2.053E-03	3.700E-01	30	M325	1 0 0 0 1	
3.054E-03	5.503E-01	35	A059	1 0 1 1 1	
3.403E-03	6.132E-01	39.3	G302	2 2 2 2 0	EFG
4.053E-03	7.303E-01	40	A059	1 0 1 1 1	
3.925E-03	7.073E-01	40	M352	1 1 1 1 2	
6.492E-03	1.170E+00	50	M352	1 1 1 1 2	

2014.   C$_{10}$H$_{12}$O$_4$
Cantharidin
Dimethyl-3,6-epoxyperhydrophthalic Anhydride
Cantharides
Hexahydro-3α,7α-dimethyl-4β,7β-epoxyisobenzofuran-1,3-dione
Spanish Fly
RN: 56-25-7   **MP** (°C):
MW: 196.20   **BP** (°C):

Solubility (Moles/L)	Solubility (Grams/L)	Temp (°C)	Ref (#)	Evaluation (T P E A A)	Comments
1.529E-04	3.000E-02	20	F300	1 0 0 0 0	
3.058E-01	6.000E+01	100	F300	1 0 0 0 0	

2015. C$_{10}$H$_{12}$O$_8$

Dilactone

α-Oxo-β-methylol-γ-butyrolactone Betrachten

RN:		MP (°C):	140
MW:	260.20	BP (°C):	

Solubility (Moles/L)	Solubility (Grams/L)	Temp (°C)	Ref (#)	Evaluation (T P E A A)	Comments
9.374E-02	2.439E+01	0	F023	1 1 0 0 1	unit assumed
1.900E-01	4.943E+01	25	F023	1 1 0 0 1	unit assumed
5.972E-01	1.554E+02	50	F023	1 1 0 0 2	unit assumed
1.788E+00	4.652E+02	75	F023	1 1 0 0 2	unit assumed
2.451E+00	6.377E+02	100	F023	1 1 0 0 2	unit assumed

2016. C$_{10}$H$_{13}$ClN$_2$

Chlordimeform

N'-(4-Chloro-2-methylphenyl)-N,N-dimethylmethanimidamide

Bermat

Fundex

Galecon

Chlorophenamidine

RN:	6164-98-3	MP (°C):	32
MW:	196.68	BP (°C):	

Solubility (Moles/L)	Solubility (Grams/L)	Temp (°C)	Ref (#)	Evaluation (T P E A A)	Comments
1.032E-03	2.030E-01	10	B324	2 2 2 2 2	
1.032E-03	2.030E-01	10	B324	2 2 2 2 2	
1.373E-03	2.700E-01	20	B300	2 0 1 1 2	
1.373E-03	2.700E-01	20	B324	2 2 2 2 2	
1.372E-03	2.699E-01	20	B324	2 2 2 2 2	
1.271E-03	2.500E-01	20	M161	1 0 0 0 2	

2017. C$_{10}$H$_{13}$ClN$_2$O

Chlortoluron

N'-(3-Chloro-4-methylphenyl)-N,N-dimethylurea

Dicuran

Chlortokem

Tolurex

RN:	15545-48-9	MP (°C):	147.5
MW:	212.68	BP (°C):	

Solubility (Moles/L)	Solubility (Grams/L)	Temp (°C)	Ref (#)	Evaluation (T P E A A)	Comments
3.311E-04	7.043E-02	20	B179	2 0 0 0 2	
3.291E-04	7.000E-02	20	F311	1 2 2 2 1	
3.291E-04	7.000E-02	20	M161	1 0 0 0 1	

2018. C$_{10}$H$_{13}$ClN$_2$O
Trimeturon
N'-4-Chlorophenyl-O,N,N-trimethylisourea
RN: 3050-27-9   **MP** (°C):   147.5
MW: 212.68   **BP** (°C):

Solubility (Moles/L)	Solubility (Grams/L)	Temp (°C)	Ref (#)	Evaluation (T P E A A)	Comments
3.289E-03	6.995E-01	ns	M061	0 0 0 0 1	

2019. C$_{10}$H$_{13}$ClN$_2$O$_2$
Metoxuron
N'-(3-Chloro-4-methoxyphenyl)-N,N-dimethylurea
Purivel
Sulerex
Dosanex
Dosaflo
RN: 19937-59-8   **MP** (°C):   125
MW: 228.68   **BP** (°C):

Solubility (Moles/L)	Solubility (Grams/L)	Temp (°C)	Ref (#)	Evaluation (T P E A A)	Comments
3.020E-03	6.906E-01	20	B179	2 0 0 0 2	
2.622E-03	5.996E-01	20	E048	1 2 1 1 2	
2.965E-03	6.780E-01	23	M161	0 0 0 0 2	
3.059E-03	6.995E-01	ns	B100	0 0 0 0 0	

2020. C$_{10}$H$_{13}$ClN$_2$O$_3$S
Chlorpropamide
N3-Butyl-N1-p-chlorobenzenesulfonylurea
Diabinese
Glucamide
Catanil
Diabaril
RN: 94-20-2   **MP** (°C):   128
MW: 276.74   **BP** (°C):

Solubility (Moles/L)	Solubility (Grams/L)	Temp (°C)	Ref (#)	Evaluation (T P E A A)	Comments
9.311E-04	2.577E-01	37	A028	1 0 2 1 2	intrinsic
9.250E-04	2.560E-01	37	A046	2 0 1 1 2	
~1.26E-03	~3.50E-01	37	B140	2 2 1 2 0	pH 1.5, form V
1.203E-03	3.330E-01	37	B140	2 2 1 2 2	pH 1.5, form I
1.384E-03	3.830E-01	37	B140	2 2 1 2 2	pH 1.5, form II
8.925E-04	2.470E-01	37	B140	2 2 1 2 2	pH 1.5, form III
1.153E-03	3.190E-01	37	B140	2 2 1 2 2	pH 1.5, form IV

2021. $C_{10}H_{13}Cl_2FN_2O_2S_2$

Tolylfluanid
1,1-Dichloro-N-((dimethylamino)sulfonyl)-1-fluoro-N-(4-methylphenyl)methanesulfenamide
Dichlofluanid-methyl
Euparen M
BAY 5712α
BAY 49854

RN: 731-27-1	**MP** (°C):	96
MW: 347.26	**BP** (°C):	

Solubility (Moles/L)	Solubility (Grams/L)	Temp (°C)	Ref (#)	Evaluation (T P E A A)	Comments
1.152E-02	4.000E+00	rt	M161	0 0 0 0 0	

2022. $C_{10}H_{13}Cl_2O_3PS$

Dichlofenthion
Diethyl O-Dichlorophenyl phosphorothioate
Hexanema
Diclophenthion
Nemacide
TRI-VC13

RN: 97-17-6	**MP** (°C):	
MW: 315.16	**BP** (°C):	164

Solubility (Moles/L)	Solubility (Grams/L)	Temp (°C)	Ref (#)	Evaluation (T P E A A)	Comments
7.774E-07	2.450E-04	25	M161	1 0 0 0 2	
7.774E-07	2.450E-04	ns	F071	0 1 2 1 2	
7.774E-04	2.450E-01	ns	M061	0 0 0 0 2	sic

2023. $C_{10}H_{13}FN_2O_3$

1-Pivaloyloxymethyl-5-fluorouracil

RN:	**MP** (°C):	
MW: 228.23	**BP** (°C):	

Solubility (Moles/L)	Solubility (Grams/L)	Temp (°C)	Ref (#)	Evaluation (T P E A A)	Comments
1.095E-02	2.500E+00	22	M317	1 1 1 1 1	

2024. $C_{10}H_{13}FN_2O_4$

1-Pivaloyloxymethyl-5-fluoro-2,4(1H,3H)-pyrimidinedi-one
1-Pivaloyloxymethyl-5-fluorouracil

RN: 62113-42-2	**MP** (°C):	158-160
MW: 244.22	**BP** (°C):	

Solubility (Moles/L)	Solubility (Grams/L)	Temp (°C)	Ref (#)	Evaluation (T P E A A)	Comments
9.418E-03	2.300E+00	22	B321	1 0 2 2 2	pH 4.0

2025.   C$_{10}$H$_{13}$NO$_2$
Phenacetin
p-Ethoxyacetanilide
p-Acetophenetidide
RN: 62-44-2   **MP** (°C):   134.5
MW: 179.22   **BP** (°C):

Solubility (Moles/L)	Solubility (Grams/L)	Temp (°C)	Ref (#)	Evaluation (T P E A A)	Comments
3.010E-04	5.395E-02	14	O019	1 0 0 1 2	
2.010E-03	3.603E-01	15	M352	1 1 1 1 2	
3.903E-03	6.995E-01	20	M043	1 0 0 0 0	
4.300E-02	7.706E+00	25	D044	1 1 1 1 2	
4.464E-03	8.000E-01	25	F300	1 0 0 0 0	
2.801E-03	5.020E-01	25	M333	1 1 0 0 2	
2.799E-03	5.016E-01	25	M352	1 1 1 1 2	
5.483E-03	9.828E-01	40	M352	1 1 1 1 2	
7.878E-03	1.412E+00	50	M352	1 1 1 1 2	
6.616E-02	1.186E+01	100	I315	0 0 0 0 1	
7.867E-02	1.410E+01	100	M043	1 0 0 0 2	
4.237E-03	7.594E-01	c	I315	0 0 0 0 1	
6.584E-03	1.180E+00	ns	F059	1 0 2 2 2	0.1N HCl
5.574E-03	9.990E-01	rt	D021	0 0 1 1 1	

2026.   C$_{10}$H$_{13}$NO$_2$
Propyl-p-aminobenzoate
Risocaine
4-Aminobenzoic Acid Propyl Ester
RN: 94-12-2   **MP** (°C):   75.5
MW: 179.22   **BP** (°C):

Solubility (Moles/L)	Solubility (Grams/L)	Temp (°C)	Ref (#)	Evaluation (T P E A A)	Comments
1.655E-03	2.966E-01	15	M352	1 1 1 1 2	
2.220E-03	3.979E-01	25	H008	1 2 2 2 2	
2.860E-03	5.125E-01	25	M352	1 1 1 1 2	
3.553E-03	6.368E-01	25	P303	2 0 2 2 2	
4.219E-03	7.561E-01	33	P303	2 0 2 2 2	
4.700E-03	8.423E-01	37	F006	1 1 2 2 2	
4.629E-03	8.297E-01	40	M352	1 1 1 1 2	
5.217E-03	9.351E-01	40	P303	2 0 2 2 2	
7.047E-03	1.263E+00	50	M352	1 1 1 1 2	
1.890E-03	3.387E-01	ns	M066	0 0 0 0 2	
1.890E-03	3.387E-01	rt	B016	0 0 1 1 2	pH 7.4

2027.　$C_{10}H_{13}NO_2$
3,4-Xylyl Methylcarbamate
3,4-Dimethylphenyl Methylcarbamate
3,4-Dimethylphenyl N-Methylcarbamate
MPMC
Meobal

RN:	2425-10-7	**MP** (°C):	79.5
MW:	179.22	**BP** (°C):	126.5

Solubility (Moles/L)	Solubility (Grams/L)	Temp (°C)	Ref (#)	Evaluation (T P E A A)	Comments
7.254E-03	1.300E+00	30	M161	1 0 0 0 1	

2028.　$C_{10}H_{13}NO_2$
2,6-Dimethyl-4-acetaminophenol
4-Acetamido-2,6-dimethylphenol

RN:	22900-79-4	**MP** (°C):	
MW:	179.22	**BP** (°C):	

Solubility (Moles/L)	Solubility (Grams/L)	Temp (°C)	Ref (#)	Evaluation (T P E A A)	Comments
1.227E-02	2.200E+00	25	D078	1 2 1 1 2	

2029.　$C_{10}H_{13}NO_2$
Propham
Isopropyl Carbanilate
Isopropyl-N-phenyl Carbamate
IPC

RN:	122-42-9	**MP** (°C):	87
MW:	179.22	**BP** (°C):	

Solubility (Moles/L)	Solubility (Grams/L)	Temp (°C)	Ref (#)	Evaluation (T P E A A)	Comments
5.580E-04	1.000E-01	25	G099	1 0 0 1 0	
1.116E-04	2.000E-02	ns	B185	0 0 0 0 1	
1.786E-04	3.200E-02	ns	B185	0 0 0 0 1	
1.395E-03	2.500E-01	ns	B200	0 0 0 0 2	
5.580E-04	1.000E-01	ns	F035	0 0 0 0 0	
1.395E-03	2.500E-01	ns	H042	0 0 0 0 2	
1.000E-03	1.792E-01	ns	M163	0 0 0 0 0	EFG
1.395E-03	2.500E-01	ns	N013	0 0 0 0 2	

2030.   C$_{10}$H$_{13}$NO$_2$
Butyl Nicotinate
n-Butyl Nicotinate
RN: 6938-06-3   **MP** (°C):
MW: 179.22   **BP** (°C):

Solubility (Moles/L)	Solubility (Grams/L)	Temp (°C)	Ref (#)	Evaluation (T P E A A)	Comments
1.367E-02	2.450E+00	32	L346	1 0 0 1 2	

2031.   C$_{10}$H$_{13}$NO$_2$
2,5-Dimethyl-4-acetaminophenol
4-Acetamido-2,5-dimethylphenol
RN: 69477-71-0   **MP** (°C):
MW: 179.22   **BP** (°C):

Solubility (Moles/L)	Solubility (Grams/L)	Temp (°C)	Ref (#)	Evaluation (T P E A A)	Comments
9.694E-03	1.737E+00	25	D078	1 2 1 1 2	

2032.   C$_{10}$H$_{13}$NO$_2$
Methyl p-Dimethylaminobenzoic Acid
Methyl 4-Dimethylaminobenzoate
RN: 1202-25-1   **MP** (°C):   371.7
MW: 179.22   **BP** (°C):

Solubility (Moles/L)	Solubility (Grams/L)	Temp (°C)	Ref (#)	Evaluation (T P E A A)	Comments
3.400E-04	6.093E-02	15	M352	1 1 1 1 2	
4.988E-04	8.940E-02	25	M352	1 1 1 1 2	
8.277E-04	1.483E-01	40	M352	1 1 1 1 2	
1.111E-03	1.991E-01	50	M352	1 1 1 1 2	

2033.   C$_{10}$H$_{13}$NO$_3$
m-Ethoxyphenyl N-Methylcarbamate
1,3-Ethoxyphenyl N-Methylcarbamate
RN: 7225-96-9   **MP** (°C):   57
MW: 195.22   **BP** (°C):

Solubility (Moles/L)	Solubility (Grams/L)	Temp (°C)	Ref (#)	Evaluation (T P E A A)	Comments
6.403E-03	1.250E+00	30	D089	2 2 0 0 0	

2034. $C_{10}H_{13}NO_3$
o-Ethoxyphenyl N-Methylcarbamate
1,2-Ethoxyphenyl N-Methylcarbamate
RN: 23409-17-8 **MP** (°C): 79.5
MW: 195.22 **BP** (°C):

Solubility (Moles/L)	Solubility (Grams/L)	Temp (°C)	Ref (#)	Evaluation (T P E A A)	Comments
1.178E-02	2.300E+00	30	D089	2 2 0 0 0	

2035. $C_{10}H_{13}N_3O_2S_2$
3-Methyl-2-sulfanilamide-2,3-dihydrothiazole
Benzenesulfonamide, 4-Amino-N-(2,3-dihydro-3-methyl-2-thiazolyl)-
RN: 51203-20-4 **MP** (°C):
MW: 271.36 **BP** (°C):

Solubility (Moles/L)	Solubility (Grams/L)	Temp (°C)	Ref (#)	Evaluation (T P E A A)	Comments
5.690E-04	1.544E-01	37	K095	2 0 0 0 2	intrinsic

2036. $C_{10}H_{13}N_4O_3$
Spasmolysin
β-Hydroxypropyltheophylline
RN: 603-00-9 **MP** (°C):
MW: 237.24 **BP** (°C):

Solubility (Moles/L)	Solubility (Grams/L)	Temp (°C)	Ref (#)	Evaluation (T P E A A)	Comments
1.204E+00	2.857E+02	ns	J025	0 0 0 0 1	

2037. $C_{10}H_{13}N_5$
4-Amino-6,7-diethylpteridine
RN: **MP** (°C):
MW: 203.25 **BP** (°C):

Solubility (Moles/L)	Solubility (Grams/L)	Temp (°C)	Ref (#)	Evaluation (T P E A A)	Comments
1.171E-03	2.380E-01	20	A019	2 2 1 1 2	

2038. $C_{10}H_{13}N_5$
2-Amino-6,7-diethylpteridine
RN: **MP** (°C):
MW: 203.25 **BP** (°C):

Solubility (Moles/L)	Solubility (Grams/L)	Temp (°C)	Ref (#)	Evaluation (T P E A A)	Comments
9.110E-04	1.852E-01	20	A019	2 2 1 1 2	

2039. C$_{10}$H$_{13}$N$_5$O
2-Amino-4-hydroxy-6,7-diethylpteridine
2-Amino-4-hydroxy-6:7-diethylpteridine
RN: **MP** (°C):   >350
MW: 219.25   **BP** (°C):

Solubility (Moles/L)	Solubility (Grams/L)	Temp (°C)	Ref (#)	Evaluation (T P E A A)	Comments
5.303E-05	1.163E-02	20	A019	2 2 1 1 2	

2040. C$_{10}$H$_{13}$N$_5$O
4-Amino-2-hydroxy-6,7-diethylpteridine
4-Amino-2-hydroxy-6:7-diethylpteridine
RN: **MP** (°C):
MW: 219.25   **BP** (°C):

Solubility (Moles/L)	Solubility (Grams/L)	Temp (°C)	Ref (#)	Evaluation (T P E A A)	Comments
2.850E-04	6.250E-02	20	A019	2 2 1 1 2	

2041. C$_{10}$H$_{13}$N$_5$O$_2$
2',3'-Dideoxyadenosine
DDA
RN: 4097-22-7   **MP** (°C):   181-184
MW: 235.25   **BP** (°C):

Solubility (Moles/L)	Solubility (Grams/L)	Temp (°C)	Ref (#)	Evaluation (T P E A A)	Comments
1.228E-01	2.890E+01	4	A337	1 0 2 2 2	
1.836E-01	4.320E+01	25	A337	1 0 2 2 2	

2042. C$_{10}$H$_{13}$N$_5$O$_3$
Deoxyadenosine
2'-Deoxyadenosine
dA
RN: 958-09-8   **MP** (°C):
MW: 251.25   **BP** (°C):

Solubility (Moles/L)	Solubility (Grams/L)	Temp (°C)	Ref (#)	Evaluation (T P E A A)	Comments
2.690E-02	6.759E+00	25	H061	1 2 2 0 2	

2043. $C_{10}H_{13}N_5O_4$
Adenosine
Adenosin
9-B-D-Ribofuranosyl-9H-purin-6-amine Adenine Riboside
Adenocard
9-β-D-Ribofuranosyladenine
RN: 58-61-7   **MP** (°C):   234
MW: 267.25   **BP** (°C):

Solubility (Moles/L)	Solubility (Grams/L)	Temp (°C)	Ref (#)	Evaluation (T P E A A)	Comments
1.920E-02	5.131E+00	25	H061	1 2 2 0 2	
2.000E-02	5.345E+00	ns	R030	0 0 0 0 0	
8.232E-05	2.200E-02	rt	N015	0 0 2 2 1	*sic*

2044. $C_{10}H_{13}N_5O_5$
Guanosine
Guanosin
2-Amino-9-β-D-ribofuranosyl-9H-purine-6-(1H)-one
Guanine Riboside
rG
RN: 118-00-3   **MP** (°C):
MW: 283.25   **BP** (°C):

Solubility (Moles/L)	Solubility (Grams/L)	Temp (°C)	Ref (#)	Evaluation (T P E A A)	Comments
2.471E-03	7.000E-01	18	F300	1 0 0 0 1	
1.820E-03	5.155E-01	25	H061	1 2 2 0 2	
1.073E-01	3.040E+01	100	F300	1 0 0 0 1	

2045. $C_{10}H_{14}$
sec-Butylbenzene
1-Methylpropylbenzene
RN: 135-98-8   **MP** (°C):   -82.7
MW: 134.22   **BP** (°C):   173.5

Solubility (Moles/L)	Solubility (Grams/L)	Temp (°C)	Ref (#)	Evaluation (T P E A A)	Comments
2.302E-03	3.090E-01	25	A002	1 2 1 1 2	*sic*
7.525E-05	1.010E-02	25	K119	1 0 0 0 2	
1.311E-04	1.760E-02	25	S005	2 2 2 2 2	
1.311E-04	1.760E-02	25	S191	1 2 2 2 2	
1.311E-04	1.760E-02	25	S358	2 1 2 2 2	

2046. C$_{10}$H$_{14}$
Butylbenzene
1-Phenylbutane
n-Butylbenzene
RN: 68411-44-9   **MP** (°C):   -88
MW: 134.22   **BP** (°C):

Solubility (Moles/L)	Solubility (Grams/L)	Temp (°C)	Ref (#)	Evaluation (T P E A A)	Comments
1.300E-02	1.745E+00	ns	H307	1 0 1 1 2	

2047. C$_{10}$H$_{14}$
1,2-Diethylbenzene
o-Diethylbenzene
RN: 135-01-3   **MP** (°C):   -31
MW: 134.22   **BP** (°C):   183

Solubility (Moles/L)	Solubility (Grams/L)	Temp (°C)	Ref (#)	Evaluation (T P E A A)	Comments
5.300E-04	7.114E-02	10	B149	2 1 1 2 1	
5.300E-04	7.114E-02	20	B149	2 1 1 2 1	

2048. C$_{10}$H$_{14}$
1,4-Diethylbenzene
p-Diethylbenzene
RN: 105-05-5   **MP** (°C):   -43
MW: 134.22   **BP** (°C):   184

Solubility (Moles/L)	Solubility (Grams/L)	Temp (°C)	Ref (#)	Evaluation (T P E A A)	Comments
1.850E-04	2.483E-02	10	B149	2 1 1 2 2	
1.850E-04	2.483E-02	20	B149	2 1 1 2 2	

2049. C$_{10}$H$_{14}$
Durene
1,2,4,5-Tetramethylbenzene
Durol
RN: 95-93-2   **MP** (°C):   80.0
MW: 134.22   **BP** (°C):   192.0

Solubility (Moles/L)	Solubility (Grams/L)	Temp (°C)	Ref (#)	Evaluation (T P E A A)	Comments
2.593E-05	3.480E-03	25	K119	1 0 0 0 2	
2.593E-05	3.480E-03	25	P051	2 1 1 2 2	
2.593E-05	3.480E-03	25.00	P007	2 1 2 2 2	
1.445E-04	1.940E-02	ns	D001	0 0 0 0 2	
7.152E-05	9.600E-03	ns	H123	0 0 0 0 2	

2050.   C$_{10}$H$_{14}$
tert-Butylbenzene
1,1-Dimethylethylbenzene
t-Butylbenzene
RN: 98-06-6   **MP** (°C):   -58
MW: 134.22   **BP** (°C):   168.5

Solubility (Moles/L)	Solubility (Grams/L)	Temp (°C)	Ref (#)	Evaluation (T P E A A)	Comments
2.533E-04	3.400E-02	25	A002	1 2 1 1 1	
2.198E-04	2.950E-02	25	S005	2 2 2 2 2	
1.311E-04	1.760E-02	25	S191	1 2 2 2 2	
2.198E-04	2.950E-02	25	S358	2 1 2 2 2	

2051.   C$_{10}$H$_{14}$
Isobutylbenzene
2-Methyl-1-phenylpropane
(2-Methylpropyl)-benzene
RN: 538-93-2   **MP** (°C):   -51
MW: 134.22   **BP** (°C):   170.5

Solubility (Moles/L)	Solubility (Grams/L)	Temp (°C)	Ref (#)	Evaluation (T P E A A)	Comments
7.525E-05	1.010E-02	25	P051	2 1 1 2 2	
7.525E-05	1.010E-02	25.00	P007	2 1 2 2 2	
7.525E-05	1.010E-02	ns	H123	0 0 0 0 2	

2052.   C$_{10}$H$_{14}$
n-Butylbenzene
1-Phenylbutane
Butylbenzene
RN: 104-51-8   **MP** (°C):   -88.5
MW: 134.22   **BP** (°C):   183.1

Solubility (Moles/L)	Solubility (Grams/L)	Temp (°C)	Ref (#)	Evaluation (T P E A A)	Comments
9.940E-05	1.334E-02	7	O312	2 2 0 2 2	
9.670E-05	1.298E-02	10	O312	2 2 0 2 2	
9.790E-05	1.314E-02	12.5	O312	2 2 0 2 2	
9.660E-05	1.297E-02	15	O312	2 2 0 2 2	
9.790E-05	1.314E-02	17.5	O312	2 2 0 2 2	
9.909E-05	1.330E-02	20	B356	1 0 0 0 2	
1.018E-04	1.366E-02	20	O312	2 2 0 2 2	
9.387E-06	1.260E-03	25	A002	1 2 1 1 2	*sic*
3.700E-04	4.966E-02	25	K001	1 0 2 1 2	
1.320E-04	1.772E-02	25	M124	2 1 2 2 2	
1.030E-04	1.382E-02	25	M342	1 0 1 1 2	
1.025E-04	1.376E-02	25	O312	2 2 0 2 2	
8.791E-05	1.180E-02	25	S005	2 2 2 2 2	
3.725E-04	5.000E-02	25	S012	2 0 2 2 0	

8.791E-05	1.180E-02	25	S191	1 2 2 2 2
8.791E-05	1.180E-02	25	S358	2 1 2 2 2
1.030E-04	1.382E-02	25	W300	2 2 2 2 2
1.244E-04	1.670E-02	29.99	C350	2 1 2 2 2
1.086E-04	1.458E-02	30	O312	2 2 0 2 2
1.147E-04	1.540E-02	35	O312	2 2 0 2 2
1.328E-04	1.782E-02	39.99	C350	2 1 2 2 2
1.234E-04	1.656E-02	40	O312	2 2 0 2 2
1.411E-04	1.894E-02	45	O312	2 2 0 2 2
1.517E-04	2.036E-02	49.99	C350	2 1 2 2 2
2.006E-04	2.692E-02	59.99	C350	2 1 2 2 2
2.389E-04	3.206E-02	69.99	C350	2 1 2 2 2
3.555E-04	4.772E-02	79.99	C350	2 1 2 2 2
4.555E-04	6.114E-02	89.99	C350	2 1 2 2 2
6.222E-04	8.351E-02	99.99	C350	2 1 2 2 2
9.387E-05	1.260E-02	ns	H123	0 0 0 0 2

2053. $C_{10}H_{14}$
p-Cymene
1-Methyl-4-isopropylbenzene
4-Cymene
Dolcymine
RN: 99-87-6   **MP** (°C):   -68
MW: 134.22   **BP** (°C):   177

Solubility (Moles/L)	Solubility (Grams/L)	Temp (°C)	Ref (#)	Evaluation (T P E A A)	Comments
2.979E-03	3.998E-01	25	B019	1 0 1 2 0	*sic*
1.740E-04	2.335E-02	25	B173	2 0 2 2 2	*sic*

2054. $C_{10}H_{14}Cl_2NO_2PS$
DMPA
Isopropylphosphoramidothioate
O-(2,4-Dichlorophenyl)-O-methyl
Phosphoramidothioic Acid, Isopropyl-o-(2,4-dichlorophenyl)-o-methyl Ester
RN: 299-85-4   **MP** (°C):   51.4
MW: 314.17   **BP** (°C):

Solubility (Moles/L)	Solubility (Grams/L)	Temp (°C)	Ref (#)	Evaluation (T P E A A)	Comments
1.595E-05	5.010E-03	25	B185	1 0 0 0 2	
1.591E-05	5.000E-03	25	B200	1 0 0 0 0	
1.591E-05	5.000E-03	ns	M061	0 0 0 0 0	

2055. $C_{10}H_{14}Cl_6N_4O_2$

Triforine
N,N'-[1,4-Piperazinediylbis(2,2,2-trichloroethylidene)] Bisformamide
Funginex
Denarin
Biformylchlorazin
Saprol

RN: 26644-46-2   **MP** (°C):   155
MW: 434.97   **BP** (°C):

Solubility (Moles/L)	Solubility (Grams/L)	Temp (°C)	Ref (#)	Evaluation (T P E A A)	Comments
~1.38E-05	~6.00E-03	rt	D303	0 0 0 0 0	
6.437E-05	2.800E-02	rt	M161	0 0 0 0 0	

2056. $C_{10}H_{14}NO_5PS$

Parathion
O,O-Diethyl O-p-Nitrophenyl Phosphorothioate
Foliclal
Rhodiatox
Alkron
Fosferno

RN: 56-38-2   **MP** (°C):   6
MW: 291.26   **BP** (°C):

Solubility (Moles/L)	Solubility (Grams/L)	Temp (°C)	Ref (#)	Evaluation (T P E A A)	Comments
3.536E-05	1.030E-02	10	B324	2 2 2 2 2	
3.536E-05	1.030E-02	10	B324	2 2 2 2 2	
4.257E-05	1.240E-02	20	B169	2 1 1 1 1	
8.318E-05	2.423E-02	20	B179	2 0 0 0 2	
4.429E-05	1.290E-02	20	B324	2 2 2 2 2	
4.429E-05	1.290E-02	20	B324	2 2 2 2 2	
2.245E-05	6.540E-03	24	F179	2 2 2 2 2	
8.240E-05	2.400E-02	25	M161	1 0 0 0 1	
5.219E-05	1.520E-02	30	B324	2 2 2 2 2	
5.219E-05	1.520E-02	30	B324	2 2 2 2 2	
4.086E-05	1.190E-02	ns	F071	0 1 2 1 2	
8.240E-05	2.400E-02	ns	M061	0 0 0 0 1	
6.867E-05	2.000E-02	ns	M110	0 0 0 0 0	EFG
8.240E-05	2.400E-02	ns	M344	0 0 0 0 1	

2057.   C$_{10}$H$_{14}$NO$_6$P
Paraoxon
Diethyl p-Nitrophenyl Phosphate
Fosfacol
Eticol
Ethyl Paraoxon
Miotisal

RN:	311-45-5	**MP** (°C):
MW:	275.20	**BP** (°C): 169-170

Solubility (Moles/L)	Solubility (Grams/L)	Temp (°C)	Ref (#)	Evaluation (T P E A A)	Comments
1.318E-02	3.627E+00	20	B169	2 0 1 1 2	
3.634E-03	1.000E+00	20	F300	1 0 0 0 0	

2058.   C$_{10}$H$_{14}$N$_2$O
N-(Dimethylaminomethyl)benzamide
Benzamide, N-[(Dimethylamino)methyl]-

RN:	59917-58-7	**MP** (°C):
MW:	178.24	**BP** (°C):

Solubility (Moles/L)	Solubility (Grams/L)	Temp (°C)	Ref (#)	Evaluation (T P E A A)	Comments
2.600E+00	4.634E+02	22	J037	1 0 1 1 1	

2059.   C$_{10}$H$_{14}$N$_2$O
N-(Ethylaminomethyl)benzamide
Benzamide, N-[(Ethylamino)methyl]-

RN:	73239-20-0	**MP** (°C):
MW:	178.24	**BP** (°C):

Solubility (Moles/L)	Solubility (Grams/L)	Temp (°C)	Ref (#)	Evaluation (T P E A A)	Comments
7.300E-02	1.301E+01	22	J037	1 0 1 1 1	

2060.   C$_{10}$H$_{14}$N$_2$O$_2$
m-N,N-Dimethylaminophenyl N-Methylcarbamate
1,3-N,N-Dimethylaminophenyl N-Methylcarbamate

RN:	2631-39-2	**MP** (°C): 86
MW:	194.24	**BP** (°C):

Solubility (Moles/L)	Solubility (Grams/L)	Temp (°C)	Ref (#)	Evaluation (T P E A A)	Comments
3.604E-03	7.000E-01	30	D089	2 2 0 0 0	

2061. C₁₀H₁₄N₂O₃

5-Isopropyl-5-allylbarbituric Acid
Aprobarbital
5-(1-Methylethyl)-5-(2-propenyl)-2,4,6(1H,3H,5H)-pyrimidinetrione
5-Allyl-5-isopropylbarbituric Acid
Aprobarbitone

RN:	77-02-1	MP (°C):	141
MW:	210.23	BP (°C):	

Solubility (Moles/L)	Solubility (Grams/L)	Temp (°C)	Ref (#)	Evaluation (T P E A A)	Comments
1.617E-02	3.400E+00	20	J030	1 2 2 2 2	
1.960E-02	4.121E+00	25	P350	2 1 1 1 2	intrinsic
1.940E-02	4.079E+00	25	V033	2 0 1 1 2	
1.940E-02	4.079E+00	25.00	T303	1 0 0 0 2	
2.600E-02	5.466E+00	35.00	T303	1 0 0 0 2	
2.664E-02	5.600E+00	37	J030	1 2 2 2 2	
3.340E-02	7.022E+00	45.00	T303	1 0 0 0 2	
1.912E-02	4.020E+00	ns	T003	0 0 0 0 2	

2062. C₁₀H₁₄N₂O₃

5-Methyl-5-(3-methylbut-2-enyl)barbituric Acid
2,4,6(1H,3H,5H)-Pyrimidinetrione, 5-Methyl-5-(3-methyl-2-butenyl)

RN:	66843-01-4	MP (°C):	
MW:	210.23	BP (°C):	

Solubility (Moles/L)	Solubility (Grams/L)	Temp (°C)	Ref (#)	Evaluation (T P E A A)	Comments
2.503E-03	5.262E-01	25	P350	2 1 1 1 2	intrinsic

2063. C₁₀H₁₄N₂O₃

2,4-Diazaspiro[5.6]dodecane-1,3,5-trione

RN:	143288-61-3	MP (°C):	
MW:	210.23	BP (°C):	

Solubility (Moles/L)	Solubility (Grams/L)	Temp (°C)	Ref (#)	Evaluation (T P E A A)	Comments
6.790E-04	1.427E-01	25	P350	2 1 1 1 2	intrinsic

2064. C₁₀H₁₄N₂S

Methiuron
N,N-Dimethyl-N'-3-methylphenylthiourea

RN:	21540-35-2	MP (°C):	145
MW:	194.30	BP (°C):	

Solubility (Moles/L)	Solubility (Grams/L)	Temp (°C)	Ref (#)	Evaluation (T P E A A)	Comments
2.059E-03	4.000E-01	ns	M061	0 0 0 0 2	

2065. C$_{10}$H$_{14}$N$_4$O$_2$
1-Propyl Theobromine
3,7-Dimethyl-1-propyl-xanthine
RN: 204443-29-8   **MP** (°C):   99
MW: 222.25   **BP** (°C):

Solubility (Moles/L)	Solubility (Grams/L)	Temp (°C)	Ref (#)	Evaluation (T P E A A)	Comments
6.190E-02	1.376E+01	30	B042	1 2 1 1 2	

2066. C$_{10}$H$_{14}$N$_4$O$_2$
7-Propyl Theophylline
3,7-Dimethyl-7-propyl-xanthine
RN: 27760-74-3   **MP** (°C):
MW: 222.25   **BP** (°C):

Solubility (Moles/L)	Solubility (Grams/L)	Temp (°C)	Ref (#)	Evaluation (T P E A A)	Comments
1.044E+00	2.320E+02	30	B042	1 2 1 1 2	
1.040E+00	2.311E+02	30	G021	1 0 0 0 2	

2067. C$_{10}$H$_{14}$N$_4$O$_3$
Ethoxycaffeine
1,3,7-Trimethyl-2,6-dioxo-8-ethoxypurine
RN: 577-66-2   **MP** (°C):   143
MW: 238.25   **BP** (°C):

Solubility (Moles/L)	Solubility (Grams/L)	Temp (°C)	Ref (#)	Evaluation (T P E A A)	Comments
1.255E-02	2.991E+00	19	A072	1 2 1 0 1	

2068. C$_{10}$H$_{14}$N$_4$O$_4$
Dyphylline
7-(2,3-Dihydroxypropyl)theophylline
Lufyllin-EPG
Neothylline
Airet
RN: 479-18-5   **MP** (°C):   158
MW: 254.25   **BP** (°C):

Solubility (Moles/L)	Solubility (Grams/L)	Temp (°C)	Ref (#)	Evaluation (T P E A A)	Comments
6.686E-01	1.700E+02	37	F076	2 0 2 2 1	

2069. C₁₀H₁₄N₅O₇P

2069. $C_{10}H_{14}N_5O_7P$
2'-Adenylic Acid
2'-Adenylsaeure
RN: 130-49-4   **MP** (°C):
MW: 347.23   **BP** (°C):

Solubility (Moles/L)	Solubility (Grams/L)	Temp (°C)	Ref (#)	Evaluation (T P E A A)	Comments
1.440E-03	5.000E-01	15	F300	1 0 0 0 0	

2070. $C_{10}H_{14}N_5O_7P$
3'-Adenylic Acid
3'-Adenylsaeure
RN: 84-21-9   **MP** (°C):   197
MW: 347.23   **BP** (°C):

Solubility (Moles/L)	Solubility (Grams/L)	Temp (°C)	Ref (#)	Evaluation (T P E A A)	Comments
1.440E-03	5.000E-01	15	F300	1 0 0 0 0	

2071. $C_{10}H_{14}O$
L-Carvone
r-(-)-p-Mentha-6,8-dien-2-one
1-Methyl-4-isopropenyl-6-cyclohexen-2-one
p-Mentha-6,8-dien-2-one
RN: 6485-40-1   **MP** (°C):   <25
MW: 150.22   **BP** (°C):   230

Solubility (Moles/L)	Solubility (Grams/L)	Temp (°C)	Ref (#)	Evaluation (T P E A A)	Comments
8.654E-03	1.300E+00	18	F300	1 0 0 0 1	
8.654E-03	1.300E+00	25	A049	1 0 0 0 1	
1.100E-02	1.652E+00	37	E028	1 0 1 1 2	

2072. $C_{10}H_{14}O$
Carvacrol
2-Methyl-5-isopropylphenol
RN: 499-75-2   **MP** (°C):   3
MW: 150.22   **BP** (°C):

Solubility (Moles/L)	Solubility (Grams/L)	Temp (°C)	Ref (#)	Evaluation (T P E A A)	Comments
6.650E-03	9.990E-01	25	L021	1 0 0 0 0	
8.321E-03	1.250E+00	25	M127	1 0 0 0 2	

2073.   C$_{10}$H$_{14}$O
Thymol
6-Isopropyl-m-cresol
3-Hydroxy-p-cymene
5-Methyl-2-isopropyl-1-phenol
2-Isopropyl-5-methyl Phenol
5-Methyl-2-(1-methylethyl)phenol

RN: 89-83-8   **MP** (°C):   48-51
MW: 150.22   **BP** (°C):

Solubility (Moles/L)	Solubility (Grams/L)	Temp (°C)	Ref (#)	Evaluation (T P E A A)	Comments
5.991E-03	9.000E-01	20	F300	1 0 0 0 0	
5.700E-03	8.563E-01	25	F044	1 0 0 0 1	
6.046E-03	9.083E-01	25	L021	1 0 0 0 0	
6.650E-03	9.990E-01	25	R041	1 0 2 1 1	
5.990E-02	8.998E+00	37	E028	1 0 1 1 2	*sic*
8.654E-03	1.300E+00	37	F300	1 0 0 0 1	

2074.   C$_{10}$H$_{14}$O
4-sec-Butylphenol
p-sec-Butylphenol

RN: 99-71-8   **MP** (°C):
MW: 150.22   **BP** (°C):

Solubility (Moles/L)	Solubility (Grams/L)	Temp (°C)	Ref (#)	Evaluation (T P E A A)	Comments
6.391E-03	9.600E-01	25	M127	1 0 0 0 1	

2075.   C$_{10}$H$_{14}$O
o-n-Butylphenol
2-n-Butylphenol

RN: 28805-86-9   **MP** (°C):
MW: 150.22   **BP** (°C):

Solubility (Moles/L)	Solubility (Grams/L)	Temp (°C)	Ref (#)	Evaluation (T P E A A)	Comments
2.662E-03	3.998E-01	25	L022	1 0 0 0 0	

2076.   C₁₀H₁₄O
p-n-Butylphenol
4-n-Butylphenol
RN: 1638-22-8   **MP** (°C):
MW: 150.22   **BP** (°C):

Solubility (Moles/L)	Solubility (Grams/L)	Temp (°C)	Ref (#)	Evaluation (T P E A A)	Comments
3.038E-03	4.563E-01	20	R087	1 1 2 2 2	0.15M NaCl
2.662E-03	3.998E-01	25	L022	1 0 0 0 0	

2077.   C₁₀H₁₄O
p-tert-Butylphenol
4-t-Butylphenol
RN: 98-54-4   **MP** (°C):   99.5
MW: 150.22   **BP** (°C):   237

Solubility (Moles/L)	Solubility (Grams/L)	Temp (°C)	Ref (#)	Evaluation (T P E A A)	Comments
4.327E-03	6.500E-01	22.5	G301	2 1 0 1 2	
3.327E-03	4.998E-01	25	L021	1 0 0 0 0	
3.861E-03	5.800E-01	25	M127	1 0 0 0 1	
4.427E-03	6.650E-01	25	P004	2 1 1 1 2	
5.076E-03	7.625E-01	30	P004	2 1 1 1 2	
5.785E-03	8.690E-01	35	P004	2 1 1 1 2	
6.534E-03	9.815E-01	40	P004	2 1 1 1 2	

2078.   C₁₀H₁₄O₂
3-Butoxyphenol
m-Butoxy Phenol
Phenol, 3-Butoxy-
RN: 18979-72-1   **MP** (°C):
MW: 166.22   **BP** (°C):

Solubility (Moles/L)	Solubility (Grams/L)	Temp (°C)	Ref (#)	Evaluation (T P E A A)	Comments
8.240E-03	1.370E+00	30	B315	1 0 1 1 2	

2079.   C₁₀H₁₄O₂
p-Diethoxybenzene
4-Diethoxybenzene
RN: 122-95-2   **MP** (°C):
MW: 166.22   **BP** (°C):

Solubility (Moles/L)	Solubility (Grams/L)	Temp (°C)	Ref (#)	Evaluation (T P E A A)	Comments
4.560E-04	7.580E-02	25	C316	1 0 2 2 2	0.1M NaCl

2080.   C$_{10}$H$_{14}$O$_2$
o-Butoxyphenol
2-Butoxyphenol
RN: 39075-90-6   **MP** (°C):
MW: 166.22   **BP** (°C):

Solubility (Moles/L)	Solubility (Grams/L)	Temp (°C)	Ref (#)	Evaluation (T P E A A)	Comments
3.920E-03	6.516E-01	24.99	B353	2 1 1 1 2	

2081.   C$_{10}$H$_{14}$O$_8$
1,1,2,2-Ethanetetrol, Tetraacetate
Glyoxal-tetraacetat
Glyoxal Tetraacetate
RN: 59602-16-3   **MP** (°C):
MW: 262.22   **BP** (°C):

Solubility (Moles/L)	Solubility (Grams/L)	Temp (°C)	Ref (#)	Evaluation (T P E A A)	Comments
3.051E-05	8.000E-03	25	F300	1 0 0 0 1	

2082.   C$_{10}$H$_{15}$N
Diethylaniline
2,6-Diethylaniline
RN: 579-66-8   **MP** (°C):   -38
MW: 149.24   **BP** (°C):   215

Solubility (Moles/L)	Solubility (Grams/L)	Temp (°C)	Ref (#)	Evaluation (T P E A A)	Comments
4.489E-03	6.700E-01	26.70	L095	2 2 1 1 2	

2083.   C$_{10}$H$_{15}$NO
Ephedrine
L-Erythro-2-(methylamino)-1-phenylpropan-1-ol
(1R,2S)-(-)-Ephedrine
L-α-(1-Methylaminoethyl)benzyl Alcohol
RN: 299-42-3   **MP** (°C):   38-39
MW: 165.24   **BP** (°C):

Solubility (Moles/L)	Solubility (Grams/L)	Temp (°C)	Ref (#)	Evaluation (T P E A A)	Comments
2.882E-01	4.762E+01	25	D004	1 0 0 0 0	
3.442E-01	5.688E+01	25	L338	1 0 1 1 2	
3.850E-01	6.362E+01	30	L069	1 0 1 1 0	EFG
1.160E+00	1.917E+02	ns	F007	0 0 0 0 2	

2084.   C₁₀H₁₅NO

2084. $C_{10}H_{15}NO$
Ethyl Phenyl Ethanolamine
2-(N-Ethylanilino)ethanol
N-Phenyl-N-ethylethanolamine
RN: 92-50-2   **MP** (°C):
MW: 165.24   **BP** (°C):   268

Solubility (Moles/L)	Solubility (Grams/L)	Temp (°C)	Ref (#)	Evaluation (T P E A A)	Comments
3.011E-02	4.975E+00	20	M062	1 0 0 0 1	

2085. $C_{10}H_{15}NO_2$
N-Phenyldiethanolamine
Phenyl Diethanolamine
N,N-di(Hydroxyethyl)aniline
2,2'-(Phenylimino)diethanol
PDEA
RN: 120-07-0   **MP** (°C):   57
MW: 181.24   **BP** (°C):

Solubility (Moles/L)	Solubility (Grams/L)	Temp (°C)	Ref (#)	Evaluation (T P E A A)	Comments
1.783E-01	3.232E+01	20	M062	1 0 0 0 2	

2086. $C_{10}H_{15}N_5O_5$
Arabinosyladenine
9-β-D-Arabino Furanosyl Adenine
Vidarabine
β-D-Arabinosyladenine
Spongoadenosine
RN: 24356-66-9   **MP** (°C):   208
MW: 285.26   **BP** (°C):

Solubility (Moles/L)	Solubility (Grams/L)	Temp (°C)	Ref (#)	Evaluation (T P E A A)	Comments
1.800E-03	5.135E-01	ns	R030	0 0 0 0 1	

2087.   C₁₀H₁₅OPS₂

Fonofos
Ethyl S-Phenyl Ethylphosphonothiolthionate
Diphonate
Dyfonate®
Stauffer N-2790

RN: 944-22-9   **MP** (°C):
MW: 246.33   **BP** (°C):

Solubility (Moles/L)	Solubility (Grams/L)	Temp (°C)	Ref (#)	Evaluation (T P E A A)	Comments
6.373E-05	1.570E-02	20	B169	2 1 1 1 2	
6.089E-05	1.500E-02	ns	M110	0 0 0 0 0	EFG

2088.   C₁₀H₁₅O₃PS₂

Fenthion
4-Methylmercapto-3-methylphenyl Dimethyl Thiophosphate
Mercaptofos
Thiophos
Baycid
Entex

RN: 55-38-9   **MP** (°C):   7.5
MW: 278.33   **BP** (°C):

Solubility (Moles/L)	Solubility (Grams/L)	Temp (°C)	Ref (#)	Evaluation (T P E A A)	Comments
2.299E-05	6.400E-03	10	B324	2 2 2 2 2	
2.300E-05	6.402E-03	10	B324	2 2 2 2 2	
2.698E-05	7.509E-03	20	B300	2 1 1 1 2	
3.244E-05	9.029E-03	20	B324	2 2 2 2 2	
3.341E-05	9.300E-03	20	B324	2 2 2 2 2	
1.940E-04	5.400E-02	20	M061	1 0 0 0 1	
4.074E-05	1.134E-02	30	B324	2 2 2 2 2	
4.060E-05	1.130E-02	30	B324	2 2 2 2 2	
1.976E-04	5.500E-02	rt	M161	0 0 0 0 0	

2089.   C₁₀H₁₆

Limonene
p-Mentha-1,8-diene
Cyclil Decene
Acintene DP Dipentene

RN: 138-86-3   **MP** (°C):   73.97
MW: 136.24   **BP** (°C):   175.5

Solubility (Moles/L)	Solubility (Grams/L)	Temp (°C)	Ref (#)	Evaluation (T P E A A)	Comments
6.390E-05	8.706E-03	25	I019	1 0 1 2 2	
2.202E-04	3.000E-02	25	M350	1 0 1 1 1	

2090.   C$_{10}$H$_{16}$
D-Limonene
D-1,8-p-Menthadiene
(R)-1-Methyl-4-(1-methylethenyl)cyclohexene
(R)-(+)-Limonene
Hemo-sol
RN: 5989-27-5   **MP** (°C):   95
MW: 136.24   **BP** (°C):   176

Solubility (Moles/L)	Solubility (Grams/L)	Temp (°C)	Ref (#)	Evaluation (T P E A A)	Comments
7.080E-01	9.646E+01	0	M124	2 1 2 2 1	
7.670E-01	1.045E+02	5	M124	2 1 2 2 2	
1.011E-04	1.377E-02	25	M124	2 1 2 2 1	

2091.   C$_{10}$H$_{16}$Cl$_3$NOS
Triallate
S-(2,3,3-Trichloroallyl)diisopropylthiocarbamate
RN: 2303-17-5   **MP** (°C):   29
MW: 304.67   **BP** (°C):

Solubility (Moles/L)	Solubility (Grams/L)	Temp (°C)	Ref (#)	Evaluation (T P E A A)	Comments
1.313E-05	4.000E-03	25	B200	1 0 0 1 0	
1.313E-05	4.000E-03	25	M161	1 0 0 0 0	
1.313E-05	4.000E-03	ns	F019	0 0 0 0 0	

2092.   C$_{10}$H$_{16}$NO$_2$S$_2$
2-Cyclopentamethylene-4-methoxycarbamyl-1,3-dithiolane
2-Cyclopentyl-4-methoxycarbamyl-1,3-dithiolane
RN: **MP** (°C):
MW: 246.37   **BP** (°C):

Solubility (Moles/L)	Solubility (Grams/L)	Temp (°C)	Ref (#)	Evaluation (T P E A A)	Comments
3.000E-04	7.391E-02	rt	B174	0 0 1 0 0	

2093.   C$_{10}$H$_{16}$NO$_3$S
2-Cyclopentamethylene-4-methoxycarbamyl-1,3-oxathiolane
RN: **MP** (°C):
MW: 230.31   **BP** (°C):

Solubility (Moles/L)	Solubility (Grams/L)	Temp (°C)	Ref (#)	Evaluation (T P E A A)	Comments
1.500E-03	3.455E-01	rt	B174	0 0 1 0 1	

2094.   C₁₀H₁₆NO₄

2-Cyclopentamethylene-4-methoxycarbamyl-1,3-dioxolane

RN:　　　　　　　　**MP** (°C):
MW:　　214.24　　**BP** (°C):

Solubility (Moles/L)	Solubility (Grams/L)	Temp (°C)	Ref (#)	Evaluation (T P E A A)	Comments
1.300E-02	2.785E+00	rt	B174	0 0 1 0 1	

2095.   C₁₀H₁₆N₂O₃

5-Ethyl-5-(2-methylpropyl)barbituric Acid

RN:　　125-40-6　　**MP** (°C):　　174.5
MW:　　212.25　　**BP** (°C):

Solubility (Moles/L)	Solubility (Grams/L)	Temp (°C)	Ref (#)	Evaluation (T P E A A)	Comments
3.997E-03	8.483E-01	25	B065	1 2 1 1 1	

2096.   C₁₀H₁₆N₂O₃

Butabarbital
Butethal
5-Ethyl-5-n-butylbarbituric Acid
5-Butyl-5-ethylbarbituric Acid

RN:　　77-28-1　　**MP** (°C):　　127
MW:　　212.25　　**BP** (°C):

Solubility (Moles/L)	Solubility (Grams/L)	Temp (°C)	Ref (#)	Evaluation (T P E A A)	Comments
1.602E-02	3.400E+00	0	D089	0 0 0 0 2	form I
1.484E-02	3.150E+00	20	J030	1 2 2 2 2	
1.044E-02	2.215E+00	20	K078	1 0 2 1 2	
4.052E-03	8.600E-01	25	B011	2 0 0 1 0	
4.218E-03	8.954E-01	25	B065	1 1 1 1 1	
1.936E-02	4.110E+00	25	B065	1 1 1 1 1	
8.000E-03	1.698E+00	25	G003	1 1 1 1 1	pH 4.7
2.300E-02	4.882E+00	25	M310	2 2 2 2 2	
2.130E-02	4.521E+00	25	V033	2 0 1 1 2	
4.070E-03	8.639E-01	25	V033	2 0 1 1 2	
2.130E-02	4.521E+00	25.00	T303	1 0 0 0 2	
7.400E-03	1.571E+00	25.00	T303	1 0 0 0 1	
1.950E-02	4.139E+00	30	I001	2 0 2 1 0	EFG, 0.003N H₂SO₄
9.900E-03	2.101E+00	35.00	T303	1 0 0 0 1	
2.430E-02	5.158E+00	35.00	T303	1 0 0 0 2	
2.299E-02	4.880E+00	37	J030	1 2 2 2 2	
3.090E-02	6.559E+00	45.00	T303	1 0 0 0 2	
1.370E-02	2.908E+00	45.00	T303	1 0 0 0 2	
1.743E-02	3.700E+00	amb	D092	0 2 2 1 2	form II

1.602E-02	3.400E+00	amb	D092	0 2 2 1 2	0.1N HCl, form III, mp 124 °C
1.743E-02	3.700E+00	amb	D092	0 2 2 1 2	form I
9.362E-03	1.987E+00	ns	T003	0 0 0 0 2	
8.952E-03	1.900E+00	ns	T003	0 0 0 0 2	

2097. $C_{10}H_{16}N_2O_3$
5,5-Dipropylbarbituric Acid
5,5-Dipropylbarbitursaeure
Proponal
RN: 2217-08-5 **MP** (°C): 146
MW: 212.25 **BP** (°C):

Solubility (Moles/L)	Solubility (Grams/L)	Temp (°C)	Ref (#)	Evaluation (T P E A A)	Comments
2.827E-03	6.000E-01	20	F300	1 0 0 0 0	
2.968E-03	6.300E-01	20	J030	1 2 2 2 1	
5.088E-03	1.080E+00	37	J030	1 2 2 2 2	
6.926E-02	1.470E+01	100	F300	1 0 0 0 2	

2098. $C_{10}H_{16}N_2O_3$
5,5-Diisopropylbarbituric Acid
Barbituric Acid, 5,5-Diisopropyl
2,4,6(1H,3H,5H)-Pyrimidinetrione, 5,5-bis(1-Methylethyl)
RN: 99167-69-8 **MP** (°C):
MW: 212.25 **BP** (°C):

Solubility (Moles/L)	Solubility (Grams/L)	Temp (°C)	Ref (#)	Evaluation (T P E A A)	Comments
1.715E-03	3.640E-01	25	P350	2 1 1 1 2	intrinsic

2099. $C_{10}H_{16}N_2O_3S$
Biotin d
D-Biotin
Biotin
RN: 58-85-5 **MP** (°C): 232
MW: 244.31 **BP** (°C):

Solubility (Moles/L)	Solubility (Grams/L)	Temp (°C)	Ref (#)	Evaluation (T P E A A)	Comments
9.003E-04	2.200E-01	25	D041	1 0 0 0 1	
1.433E-03	3.500E-01	25	D315	1 0 1 1 2	
8.186E-04	2.000E-01	25	M054	1 0 0 0 0	

2100.   C$_{10}$H$_{16}$N$_2$O$_4$
Methyl-2,2-diallylmalonurate
Methyl 2,2-Diallylmalonurate
RN: 73632-82-3   **MP** (°C):   84
MW: 228.25   **BP** (°C):

Solubility (Moles/L)	Solubility (Grams/L)	Temp (°C)	Ref (#)	Evaluation (T P E A A)	Comments
6.800E-03	1.552E+00	23	B152	1 2 1 1 1	pH 3.5

2101.   C$_{10}$H$_{16}$N$_4$O$_2$
7-Butyl Theophylline
1H-Purine-2,6-dione, 7-Butyl-3,7-dihydro-1,3-dimethyl-
7-Butyl-1,3-dimethylxanthine
RN: 1021-65-4   **MP** (°C):
MW: 224.26   **BP** (°C):

Solubility (Moles/L)	Solubility (Grams/L)	Temp (°C)	Ref (#)	Evaluation (T P E A A)	Comments
1.560E-02	3.499E+00	30	B042	1 2 1 1 2	
1.560E-02	3.499E+00	30	G021	1 0 0 0 2	

2102.   C$_{10}$H$_{16}$N$_4$O$_2$S
3-(5-tert-Butyl-1,3,4-thiadiazol-2-yl)-4-hydroxy-1
2-Imidazolidinone, 3-[5-(1,1-Dimethylethyl)-1,3,4-thiadiazol-2-yl]-4-hydroxy-1-methyl-
Buthidazole
Ravage
VEL 5026
RN: 55511-98-3   **MP** (°C):   133.5
MW: 256.33   **BP** (°C):

Solubility (Moles/L)	Solubility (Grams/L)	Temp (°C)	Ref (#)	Evaluation (T P E A A)	Comments
1.322E-02	3.388E+00	25	M161	1 0 0 0 1	

2103. C$_{10}$H$_{16}$O
D-Camphor
D-Campher
Camphor
RN: 76-22-2   **MP** (°C):   179.7
MW: 152.24   **BP** (°C):

Solubility (Moles/L)	Solubility (Grams/L)	Temp (°C)	Ref (#)	Evaluation (T P E A A)	Comments
1.095E-02	1.667E+00	15.50	L073	1 2 2 1 2	
6.569E-03	1.000E+00	20	F300	1 0 0 0 0	
1.363E-02	2.076E+00	20	K078	1 0 2 1 2	
1.030E-02	1.568E+00	25	I019	1 0 1 2 2	
1.340E-02	2.040E+00	25	L338	1 0 1 1 2	
1.630E-02	2.481E+00	37	E028	1 0 1 1 2	
1.115E-02	1.697E+00	ns	F014	0 0 0 0 2	

2104. C$_{10}$H$_{16}$O
D-Fenchone
D-1,3,3-Trimethyl-2-norbornanone
Bicyclo[2.2.1]heptan-2-one, 1,3,3-Trimethyl-, (1S)-
α-Fenchone
(+)-Fenchone
RN: 4695-62-9   **MP** (°C):   6.1
MW: 152.24   **BP** (°C):   193.5

Solubility (Moles/L)	Solubility (Grams/L)	Temp (°C)	Ref (#)	Evaluation (T P E A A)	Comments
1.311E-02	1.996E+00	20	D052	1 1 0 0 0	
1.410E-02	2.147E+00	25	I019	1 0 1 2 2	

2105. C$_{10}$H$_{16}$O
L-Dihydrocarvone
L-Dihydro-carvon
RN: 619-02-3   **MP** (°C):
MW: 152.24   **BP** (°C):   221

Solubility (Moles/L)	Solubility (Grams/L)	Temp (°C)	Ref (#)	Evaluation (T P E A A)	Comments
6.569E-03	1.000E+00	20	F300	1 0 0 0 0	

2106.   C$_{10}$H$_{16}$O
Citral
trans-3,7-Dimethyl-2,6-octadienal
Geranialdehyde
Neral
Geranial
Citral A
RN: 5392-40-5   **MP** (°C):   <10
MW: 152.24   **BP** (°C):   92.5

Solubility (Moles/L)	Solubility (Grams/L)	Temp (°C)	Ref (#)	Evaluation (T P E A A)	Comments
1.970E-03	2.999E-01	25	M350	1 0 1 1 1	
8.800E-03	1.340E+00	37	E028	1 0 1 1 1	

2107.   C$_{10}$H$_{16}$O
Carvotan Acetone
Carvotan-aceton
RN: 499-71-8   **MP** (°C):
MW: 152.24   **BP** (°C):

Solubility (Moles/L)	Solubility (Grams/L)	Temp (°C)	Ref (#)	Evaluation (T P E A A)	Comments
5.912E-03	9.000E-01	20	F300	1 0 0 0 0	

2108.   C$_{10}$H$_{16}$O$_2$
3-Hydroxy-3-ethynyl-2,2,5,5-tetramethyltetrahydrofuran
3-Furanol, 3-Ethynyltetrahydro-2,2,5,5-tetramethyl-
RN: 24270-82-4   **MP** (°C):
MW: 168.24   **BP** (°C):

Solubility (Moles/L)	Solubility (Grams/L)	Temp (°C)	Ref (#)	Evaluation (T P E A A)	Comments
1.165E-01	1.961E+01	rt	B066	0 2 0 0 0	

2109.   C$_{10}$H$_{16}$O$_4$
L-Isocamphoric Acid
L-Isocamphersaeure
RN: 5394-83-2   **MP** (°C):   173
MW: 200.24   **BP** (°C):

Solubility (Moles/L)	Solubility (Grams/L)	Temp (°C)	Ref (#)	Evaluation (T P E A A)	Comments
1.698E-02	3.400E+00	20	F300	1 0 0 0 1	

2110. C$_{10}$H$_{16}$O$_4$
D-Camphoric Acid
D-Camphersaeure
RN: 124-83-4 **MP** (°C):
MW: 200.24 **BP** (°C):

Solubility (Moles/L)	Solubility (Grams/L)	Temp (°C)	Ref (#)	Evaluation (T P E A A)	Comments
3.796E-02	7.600E+00	25	F300	1 0 0 0 1	

2111. C$_{10}$H$_{16}$O$_5$
DL-Cineolic Acid
DL-Cineolsaeure
RN: 473-18-7 **MP** (°C): 208
MW: 216.24 **BP** (°C):

Solubility (Moles/L)	Solubility (Grams/L)	Temp (°C)	Ref (#)	Evaluation (T P E A A)	Comments
6.474E-02	1.400E+01	15	F300	1 0 0 0 1	
3.006E-01	6.500E+01	100	F300	1 0 0 0 1	

2112. C$_{10}$H$_{17}$Cl$_2$NOS
Diallate
DATC
S-(2,3-Dichloroallyl-N,N-diisopropylthiocarbamate
RN: 2303-16-4 **MP** (°C): -10
MW: 270.22 **BP** (°C):

Solubility (Moles/L)	Solubility (Grams/L)	Temp (°C)	Ref (#)	Evaluation (T P E A A)	Comments
1.480E-04	4.000E-02	25	B185	1 0 0 0 1	
5.181E-05	1.400E-02	25	B200	1 0 0 1 1	
1.480E-04	4.000E-02	25	M061	1 0 0 0 1	
5.181E-05	1.400E-02	25	M161	1 0 0 0 1	
1.480E-04	4.000E-02	ns	F019	0 0 0 0 1	
1.480E-04	4.000E-02	rt	I314	0 0 0 0 1	

2113. C$_{10}$H$_{17}$NO$_2$
Methyprylon
Dimerin
3,3-Diethyl-5-methyl-2,4-piperidinedione
RN: 125-64-4 **MP** (°C):
MW: 183.25 **BP** (°C):

Solubility (Moles/L)	Solubility (Grams/L)	Temp (°C)	Ref (#)	Evaluation (T P E A A)	Comments
4.147E-01	7.600E+01	25	R027	1 0 0 0 1	

2114. $C_{10}H_{17}N_2O_4PS$
Etrimfos
Dimethyl O-(2-Ethyl-4-ethoxy-pyrimidin-6-yl)thionophosphate
Ekamet G
Ekamet ULV
Etrimphos
RN: 38260-54-7   **MP** (°C):
MW: 292.30   **BP** (°C):

Solubility (Moles/L)	Solubility (Grams/L)	Temp (°C)	Ref (#)	Evaluation (T P E A A)	Comments
3.421E-02	1.000E+01	20	M161	1 0 0 0 1	

2115. $C_{10}H_{17}N_3O_5$
Orotic Acid Choline
RN: **MP** (°C):   102-104
MW: 259.26   **BP** (°C):

Solubility (Moles/L)	Solubility (Grams/L)	Temp (°C)	Ref (#)	Evaluation (T P E A A)	Comments
2.697E+00	6.992E+02	25	N018	2 2 1 2 2	

2116. $C_{10}H_{17}N_3O_6S$
Glutathione
Glutathion
RN: 70-18-8   **MP** (°C):   193.4
MW: 307.33   **BP** (°C):

Solubility (Moles/L)	Solubility (Grams/L)	Temp (°C)	Ref (#)	Evaluation (T P E A A)	Comments
2.958E-01	9.090E+01	0	F300	1 0 0 0 2	

2117. $C_{10}H_{17}O_3P$
Diethyl Phenyl Phosphonate
Diethyl Benzenephosphonate
Diethyl Phenylphosphonate
RN: 1754-49-0   **MP** (°C):
MW: 216.22   **BP** (°C):   110

Solubility (Moles/L)	Solubility (Grams/L)	Temp (°C)	Ref (#)	Evaluation (T P E A A)	Comments
<9.25E-04	<2.00E-01	25	B070	1 2 0 1 0	

2118.   C$_{10}$H$_{18}$
2,2,5,5-Tetramethyl-3-hexyne
Di-tert-butylacetylene
Di-tert-butylethyne
RN: 17530-24-4   **MP** (°C):
MW: 138.25   **BP** (°C):

Solubility (Moles/L)	Solubility (Grams/L)	Temp (°C)	Ref (#)	Evaluation (T P E A A)	Comments
1.470E-04	2.032E-02	25	H039	1 2 2 2 2	
7.700E-05	1.065E-02	35	H039	1 2 2 2 1	

2119.   C$_{10}$H$_{18}$
cis-Decalin
cis-Decahydronaphthalene
cis-Bicyclo[4.4.0]decane
RN: 493-01-6   **MP** (°C):   -43.2
MW: 138.25   **BP** (°C):   195.7

Solubility (Moles/L)	Solubility (Grams/L)	Temp (°C)	Ref (#)	Evaluation (T P E A A)	Comments
6.452E-02	8.920E+00	300	S355	1 1 1 2 0	EFG

2120.   C$_{10}$H$_{18}$
Decalin
Decahydronaphthalene
RN: 91-17-8   **MP** (°C):   -31
MW: 138.25   **BP** (°C):

Solubility (Moles/L)	Solubility (Grams/L)	Temp (°C)	Ref (#)	Evaluation (T P E A A)	Comments
<1.45E-03	<2.00E-01	25	B019	1 0 1 2 0	
6.430E-06	8.890E-04	25	P051	2 1 1 2 2	
6.430E-06	8.890E-04	25.00	P007	2 1 2 2 2	
4.492E-05	6.210E-03	ns	H123	0 0 0 0 2	

2121.   C$_{10}$H$_{18}$ClN$_5$
Ipazine
2-Chloro-4-diethylamino-6-isopropylamino-s-triazine
2-Chloro-4-isopropylamino-6-biethylamino-s-triazines
RN: 1912-25-0   **MP** (°C):
MW: 243.74   **BP** (°C):

Solubility (Moles/L)	Solubility (Grams/L)	Temp (°C)	Ref (#)	Evaluation (T P E A A)	Comments
1.641E-04	4.000E-02	21	B192	0 0 0 0 1	
1.641E-04	4.000E-02	21	G099	2 0 0 1 0	
1.641E-04	4.000E-02	ns	B185	1 0 0 0 1	

2122.   C₁₀H₁₈N₂O₄

Ethyl-2,2-diethylmalonurate
Ethyl 2,2-Diethylmalnurate
RN: 73632-76-5   **MP** (°C):   84.5
MW: 230.27   **BP** (°C):

Solubility (Moles/L)	Solubility (Grams/L)	Temp (°C)	Ref (#)	Evaluation (T P E A A)	Comments
8.400E-03	1.934E+00	23	B152	1 2 1 1 1	pH 3.5

2123.   C₁₀H₁₈N₂O₅

Methoxymethyl-2,2-diethylmalonurate
Methoxymethyl 2,2-Diethylmalonurate
RN: 73632-79-8   **MP** (°C):   113
MW: 246.27   **BP** (°C):

Solubility (Moles/L)	Solubility (Grams/L)	Temp (°C)	Ref (#)	Evaluation (T P E A A)	Comments
6.800E-03	1.675E+00	23	B152	1 2 1 1 1	pH 3.5

2124.   C₁₀H₁₈N₆O₂

1-(Sarcosino)-3,5-bis(dimethylamino)-s-triazine
N2-Carboxymethyl-N2,N4,N4,N6,N6-pentamethylmelamine
RN: 64124-17-0   **MP** (°C):
MW: 254.29   **BP** (°C):

Solubility (Moles/L)	Solubility (Grams/L)	Temp (°C)	Ref (#)	Evaluation (T P E A A)	Comments
7.360E-02	1.872E+01	25	B386	2 2 2 2 2	

2125.   C₁₀H₁₈O

L-Menthone
trans-p-Menthan-3-one
p-Menthan-3-one
(-)-5-Methyl-2-(1-methylethyl)cyclohexanone
(-)-Menthone
RN: 14073-97-3   **MP** (°C):   -6
MW: 154.25   **BP** (°C):   207

Solubility (Moles/L)	Solubility (Grams/L)	Temp (°C)	Ref (#)	Evaluation (T P E A A)	Comments
3.220E-03	4.967E-01	25	I019	1 0 1 2 2	

2126.   C$_{10}$H$_{18}$O
D-Borneol
Borneocamphor
Sumatra Camphor
endo-2-Bornanol
RN: 464-43-7   **MP** (°C):   208
MW: 154.25   **BP** (°C):   212

Solubility (Moles/L)	Solubility (Grams/L)	Temp (°C)	Ref (#)	Evaluation (T P E A A)	Comments
4.797E-03	7.400E-01	25	F300	1 0 0 0 1	

2127.   C$_{10}$H$_{18}$O
1,8-Cineole
Eucalyptol
Cineol
Cineole
RN: 470-82-6   **MP** (°C):   36.5
MW: 154.25   **BP** (°C):

Solubility (Moles/L)	Solubility (Grams/L)	Temp (°C)	Ref (#)	Evaluation (T P E A A)	Comments
4.123E-02	6.359E+00	1.5	E036	1 0 1 1 1	
4.187E-02	6.458E+00	4.0	B352	1 1 0 0 1	
3.674E-02	5.668E+00	7.5	E036	1 0 1 1 1	
3.482E-02	5.371E+00	10	E036	1 0 1 1 1	
3.610E-02	5.569E+00	10.0	B352	1 1 0 0 1	
1.297E-02	2.000E+00	15	F300	1 0 0 0 1	
3.097E-02	4.777E+00	15.0	B352	1 1 0 0 1	
2.261E-02	3.488E+00	21	E036	1 0 1 1 1	
2.454E-02	3.786E+00	21.0	B352	1 1 0 0 1	
2.010E-02	3.100E+00	25	A049	1 0 0 0 1	
1.746E-02	2.693E+00	30.0	B352	1 1 0 0 1	
1.552E-02	2.394E+00	35.0	B352	1 1 0 0 1	
9.100E-03	1.404E+00	37	E028	1 0 1 1 1	
1.359E-02	2.096E+00	40	E036	1 0 1 1 1	
1.423E-02	2.195E+00	40.0	B352	1 1 0 0 1	
1.294E-02	1.996E+00	45.0	B352	1 1 0 0 1	
1.229E-02	1.896E+00	50	E036	1 0 1 1 1	
1.100E-02	1.697E+00	50.0	B352	1 1 0 0 1	

2128.   C$_{10}$H$_{18}$O
Linalool
3,7-Dimethylocta-1,6-dien-3-ol
2,6-Dimethylocta-2,7-dien-6-ol
Linalol
3,7-Dimethyl-1,6-octadien-3-ol
RN: 78-70-6   **MP** (°C):   <25
MW: 154.25   **BP** (°C):   195.5

Solubility (Moles/L)	Solubility (Grams/L)	Temp (°C)	Ref (#)	Evaluation (T P E A A)	Comments
1.030E-02	1.589E+00	25	I019	1 0 1 2 2	
9.710E-03	1.498E+00	25	M350	1 0 1 1 1	
3.800E-02	5.862E+00	37	E028	1 0 1 1 2	

2129.   C$_{10}$H$_{18}$O
Borneol
endo-1,7,7-Trimethyl-bicyclo[2.2.1]heptan-2-ol
L-Borneol
RN: 507-70-0   **MP** (°C):   206
MW: 154.25   **BP** (°C):   210

Solubility (Moles/L)	Solubility (Grams/L)	Temp (°C)	Ref (#)	Evaluation (T P E A A)	Comments
4.512E-03	6.960E-01	15	M073	1 0 2 2 2	
4.784E-03	7.380E-01	25	M073	1 0 2 2 2	

2130.   C$_{10}$H$_{18}$O$_2$
2,4-Decadione
Acetylmethyl Hexyl Ketone
RN: 13329-78-7   **MP** (°C):
MW: 170.25   **BP** (°C):

Solubility (Moles/L)	Solubility (Grams/L)	Temp (°C)	Ref (#)	Evaluation (T P E A A)	Comments
2.600E-03	4.427E-01	25	M078	2 0 1 0 1	

2131.   C$_{10}$H$_{18}$O$_2$
3-Pentyl-2,4-pentadione
3-Amyl-2,4-pentanedione
RN: 27970-50-9   **MP** (°C):
MW: 170.25   **BP** (°C):

Solubility (Moles/L)	Solubility (Grams/L)	Temp (°C)	Ref (#)	Evaluation (T P E A A)	Comments
1.410E-02	2.401E+00	25	M078	2 0 1 0 2	

2132. $C_{10}H_{18}O_2$
D-Campholic Acid
D-Campholsaeure
RN: 464-88-0   **MP** (°C):
MW: 170.25   **BP** (°C):

Solubility (Moles/L)	Solubility (Grams/L)	Temp (°C)	Ref (#)	Evaluation (T P E A A)	Comments
9.398E-04	1.600E-01	19	F300	1 0 0 0 1	

2133. $C_{10}H_{18}O_2$
Sobrerol
Pinolhydrat
RN: 498-71-5   **MP** (°C):   130
MW: 170.25   **BP** (°C):   170

Solubility (Moles/L)	Solubility (Grams/L)	Temp (°C)	Ref (#)	Evaluation (T P E A A)	Comments
1.880E-01	3.200E+01	15	F300	1 0 0 0 1	
1.938E-01	3.300E+01	ns	L335	0 0 0 0 2	

2134. $C_{10}H_{18}O_3$
2,2,5,5-Tetramethyl-tetrahydro-3-hydroxy-3-furanyl Methyl Ketone
Ketone, Methyl Tetrahydro-3-hydroxy-2,2,5,5-tetramethyl-3-furyl
RN: 24282-51-7   **MP** (°C):
MW: 186.25   **BP** (°C):

Solubility (Moles/L)	Solubility (Grams/L)	Temp (°C)	Ref (#)	Evaluation (T P E A A)	Comments
1.053E-01	1.961E+01	rt	B066	0 2 0 0 0	

2135. $C_{10}H_{18}O_4$
Amyl α-Acetoxypropionate
Hydracrylic Acid, Pentyl Ester, Acetate
RN: 20473-77-2   **MP** (°C):
MW: 202.25   **BP** (°C):

Solubility (Moles/L)	Solubility (Grams/L)	Temp (°C)	Ref (#)	Evaluation (T P E A A)	Comments
3.461E-03	7.000E-01	25	R006	2 2 0 1 1	

2136. $C_{10}H_{18}O_4$
Dimethyl Cyclohexyl Oxalate
RN: **MP** (°C):
MW: 202.25   **BP** (°C):

Solubility (Moles/L)	Solubility (Grams/L)	Temp (°C)	Ref (#)	Evaluation (T P E A A)	Comments
≤9.89E-06	<2.00E-03	15	H069	1 0 1 1 0	

2137. C₁₀H₁₈O₄

Ethylene Glycol Dibutyrate
Ethylene Glycol Di-N-Butyrate
RN: 105-72-6   **MP** (°C):
MW: 202.25   **BP** (°C):

Solubility (Moles/L)	Solubility (Grams/L)	Temp (°C)	Ref (#)	Evaluation (T P E A A)	Comments
8.220E-03	1.663E+00	25	F064	1 0 0 0 2	
2.471E-03	4.998E-01	ns	F014	0 0 0 0 1	

2138. C₁₀H₁₈O₄

Diethoxyethyl Adipate
Diethyl Adipate
RN: 141-28-6   **MP** (°C):   -18
MW: 202.25   **BP** (°C):   251

Solubility (Moles/L)	Solubility (Grams/L)	Temp (°C)	Ref (#)	Evaluation (T P E A A)	Comments
2.965E-03	5.996E-01	ns	F014	0 0 0 0 1	
1.223E-02	2.474E+00	ns	F014	0 0 0 0 2	

2139. C₁₀H₁₈O₄

Sebacic Acid
Sebacinsaeure
RN: 111-20-6   **MP** (°C):   134.5
MW: 202.25   **BP** (°C):   294.5

Solubility (Moles/L)	Solubility (Grams/L)	Temp (°C)	Ref (#)	Evaluation (T P E A A)	Comments
1.978E-04	4.000E-02	0	F300	1 0 0 0 0	
1.978E-04	4.000E-02	0	L041	1 0 0 1 0	
4.944E-03	1.000E+00	20	F300	1 0 0 0 1	
4.944E-03	1.000E+00	20	L041	1 0 0 1 1	
9.889E-03	2.000E+00	21	B040	1 0 1 1 1	sic
7.911E-03	1.600E+00	35	L041	1 0 0 1 1	
1.088E-02	2.200E+00	50	L041	1 0 0 1 1	
2.077E-02	4.200E+00	65	F300	1 0 0 0 1	
2.077E-02	4.200E+00	65	L041	1 0 0 1 1	
8.898E-04	1.800E-01	ns	F014	0 0 0 0 1	

2140. $C_{10}H_{18}O_5$
Propanoic Acid, 2-[(Ethoxycarbonyl)oxy]-, Butyl Ester
Propanoic Acid, 2-[(Amoxycarbonyl)oxy]-, Methyl Ester
RN: **MP** (°C):
MW: 218.25   **BP** (°C):

Solubility (Moles/L)	Solubility (Grams/L)	Temp (°C)	Ref (#)	Evaluation (T P E A A)	Comments
2.290E-03	4.998E-01	25	R007	1 0 0 0 0	
3.205E-03	6.995E-01	25	R007	1 0 0 0 0	

2141. $C_{10}H_{18}O_5$
Diethylene Glycol Dipropionate
Ethanol, 2,2'-Oxybis-, Dipropanoate
RN: 6942-59-2   **MP** (°C):
MW: 218.25   **BP** (°C):

Solubility (Moles/L)	Solubility (Grams/L)	Temp (°C)	Ref (#)	Evaluation (T P E A A)	Comments
1.592E-01	3.475E+01	ns	F014	0 0 0 0 2	

2142. $C_{10}H_{19}NO_3$
Ethylpropylaceturethane
RN: **MP** (°C):
MW: 201.27   **BP** (°C):

Solubility (Moles/L)	Solubility (Grams/L)	Temp (°C)	Ref (#)	Evaluation (T P E A A)	Comments
7.088E-03	1.427E+00	c	O021	0 2 0 0 0	

2143. $C_{10}H_{19}NO_3$
Oenanthylylurethane
RN: **MP** (°C):
MW: 201.27   **BP** (°C):

Solubility (Moles/L)	Solubility (Grams/L)	Temp (°C)	Ref (#)	Evaluation (T P E A A)	Comments
1.043E-03	2.100E-01	ns	O021	0 0 0 0 0	

2144. $C_{10}H_{19}NO_4S$
2-Amino-5-naphthol-1-sulfonic Acid
RN: **MP** (°C):
MW: 249.33   **BP** (°C):

Solubility (Moles/L)	Solubility (Grams/L)	Temp (°C)	Ref (#)	Evaluation (T P E A A)	Comments
8.503E-03	2.120E+00	c	B125	1 2 0 0 2	

2145.   C₁₀H₁₉N₂O₄PS
Cyanthoate
Phosphorothioic Acid, S-(2-((1-Cyano-1-methylethyl)amino)-2-oxoethyl) O,O-Diethyl
Ester
Tartran
RN: 3734-95-0   **MP** (°C):
MW: 294.31   **BP** (°C):

Solubility (Moles/L)	Solubility (Grams/L)	Temp (°C)	Ref (#)	Evaluation (T P E A A)	Comments
2.378E-01	7.000E+01	20	M161	1 0 0 0 1	

2146.   C₁₀H₁₉N₅O
Terebumeton
1,3,5-Triazine-2,4-diamine, N-(1,1-Dimethylethyl)-N'-ethyl-6-methoxy-
2-Methoxy-4-ethylamino-6-tert-butylamino-s-triazine
Karagard
4-(Ethylamino)-2-methoxy-6-(tert-butylamino)-s-triazine
Caragard
RN: 33693-04-8   **MP** (°C):   123.5
MW: 225.30   **BP** (°C):

Solubility (Moles/L)	Solubility (Grams/L)	Temp (°C)	Ref (#)	Evaluation (T P E A A)	Comments
5.770E-04	1.300E-01	20	M161	1 0 0 0 2	

2147.   C₁₀H₁₉N₅O
Secbumeton
2-sec-Butylamino-4-ethylamino-6-methoxy-s-triazine
GS-14254
RN: 26259-45-0   **MP** (°C):   86
MW: 225.30   **BP** (°C):

Solubility (Moles/L)	Solubility (Grams/L)	Temp (°C)	Ref (#)	Evaluation (T P E A A)	Comments
2.930E-03	6.601E-01	1	G091	1 0 1 2 2	pH 6.0
3.250E-03	7.322E-01	8	G091	1 0 1 2 2	pH 6.0
2.750E-03	6.196E-01	20	B200	1 0 0 0 2	
2.663E-03	6.000E-01	20	F311	1 2 2 2 1	
3.070E-03	6.917E-01	20	G091	1 0 1 2 2	pH 6.0
2.752E-03	6.200E-01	20	M161	1 0 0 0 2	
3.300E-03	7.435E-01	29	G091	1 0 1 2 2	pH 6.0

2148.   C₁₀H₁₉N₅O

Prometone
2-Methoxy-4,6-bis-isopropylamino-s-triazine
Pramitol
Primatol O
Prometon
2-Methoxy-4,6-bis-(isopropyl-amino)-s-triazine

RN: 1610-18-0   **MP** (°C):   91.5
MW: 225.30   **BP** (°C):   91-92

Solubility (Moles/L)	Solubility (Grams/L)	Temp (°C)	Ref (#)	Evaluation (T P E A A)	Comments
3.330E-03	7.502E-01	20	B200	1 0 0 0 2	
2.752E-03	6.200E-01	20	F311	1 2 2 2 1	
3.329E-03	7.500E-01	20	M161	1 0 0 0 2	
3.329E-03	7.500E-01	21	B192	0 0 0 0 2	
1.554E-02	3.500E+00	21	G099	2 0 0 1 0	
3.329E-03	7.500E-01	21	G099	2 0 0 1 0	
4.680E-03	1.054E+00	50	G001	1 0 1 1 2	
3.548E-03	7.994E-01	ns	B100	0 0 0 0 0	
3.329E-03	7.500E-01	ns	B185	0 0 0 0 2	
3.329E-03	7.500E-01	ns	C101	0 0 0 0 1	
3.329E-03	7.500E-01	ns	G041	0 0 0 0 2	
3.329E-03	7.500E-01	ns	H112	0 0 0 0 2	
3.329E-03	7.500E-01	ns	J033	0 0 0 0 2	

2149.   C₁₀H₁₉N₅O

2-Methoxy-4-ethylamino-6-diethylamino-s-triazine
G31432

RN: 13532-26-8   **MP** (°C):
MW: 225.30   **BP** (°C):

Solubility (Moles/L)	Solubility (Grams/L)	Temp (°C)	Ref (#)	Evaluation (T P E A A)	Comments
1.775E-04	4.000E-02	20	J033	1 0 0 0 1	

2150.   C₁₀H₁₉N₅OS

Hydroxyprometryne
1,3,5-Triazin-2(1H)-one, 4,6-bis[(1-methylethyl)amino]-
bis(Isopropylamino)hydroxy-s-triazine
GS 11526

RN: 7374-53-0   **MP** (°C):
MW: 257.36   **BP** (°C):

Solubility (Moles/L)	Solubility (Grams/L)	Temp (°C)	Ref (#)	Evaluation (T P E A A)	Comments
4.000E-04	1.029E-01	2	B193	1 2 0 0 0	

2151. C$_{10}$H$_{19}$N$_5$S
Terbutryn
2-Methylthio-4-ethylamino-6-tert-butylamino-s-triazine
Terbutryne
N-(1,1-Dimethylethyl)-N'-ethyl-6-(methylthio)-1,3,5-Triazine-2,4-diamine
Terbutrex

RN:	886-50-0	**MP** (°C):	104		
MW:	241.36	**BP** (°C):	157		

Solubility (Moles/L)	Solubility (Grams/L)	Temp (°C)	Ref (#)	Evaluation (T P E A A)	Comments
1.090E-04	2.631E-02	1	G091	1 0 1 2 2	pH 6.0
1.100E-04	2.655E-02	8	G091	1 0 1 2 2	pH 6.0
2.400E-04	5.793E-02	20	B200	1 0 0 0 1	
1.036E-04	2.500E-02	20	E048	1 2 1 1 1	
1.036E-04	2.500E-02	20	F311	1 2 2 2 1	
1.460E-04	3.524E-02	20	G091	1 0 1 2 2	pH 6.0
2.403E-04	5.800E-02	20	M161	1 0 0 0 1	
1.660E-04	4.007E-02	29	G091	1 0 1 2 2	pH 6.0
2.403E-04	5.800E-02	ns	J033	0 0 0 0 1	

2152. C$_{10}$H$_{19}$N$_5$S
s-Triazole, 2,4-bis(isopropylamine)-6-methylmercapto-

RN:		**MP** (°C):	
MW:	241.36	**BP** (°C):	

Solubility (Moles/L)	Solubility (Grams/L)	Temp (°C)	Ref (#)	Evaluation (T P E A A)	Comments
1.989E-04	4.800E-02	20	B185	1 0 0 0 1	

2153. C$_{10}$H$_{19}$N$_5$S
Prometryne
N,N'-bis(1-Methylethyl)-6-methylthio-1,3,5-triazine-2,4-diamine
Caparol
Primatol Q
Gesagard
Caparol 80W

RN:	7287-19-6	**MP** (°C):	118	
MW:	241.36	**BP** (°C):		

Solubility (Moles/L)	Solubility (Grams/L)	Temp (°C)	Ref (#)	Evaluation (T P E A A)	Comments
2.400E-04	5.793E-02	2	B193	1 2 0 0 0	
2.000E-04	4.827E-02	20	B200	1 0 0 0 0	
1.657E-04	4.000E-02	20	F311	1 2 2 2 1	
1.988E-03	4.798E-01	20	M061	1 0 0 0 1	
1.989E-04	4.800E-02	20	M161	1 0 0 0 1	
1.989E-04	4.800E-02	24	C105	2 1 2 2 2	

4.200E-04	1.014E-01	50	G001	1 0 1 1 2	
1.989E-04	4.800E-02	ns	C101	0 0 0 0 1	
1.989E-04	4.800E-02	ns	H112	0 0 0 0 1	
1.989E-04	4.800E-02	ns	J033	0 0 0 0 1	

2154. $C_{10}H_{19}O_6PS_2$

Malathion
Dicarboethoxyethyl O,O-Dimethyl Phosphorodithioate
Carbofos
Cythion
Mercaptothion
Phosphothion

RN: 121-75-5   **MP** (°C):   3
MW: 330.36   **BP** (°C):

Solubility (Moles/L)	Solubility (Grams/L)	Temp (°C)	Ref (#)	Evaluation (T P E A A)	Comments
4.267E-04	1.410E-01	10	B324	2 2 2 2 2	
4.268E-04	1.410E-01	10	B324	2 2 2 2 2	
4.329E-04	1.430E-01	20	B300	2 1 1 1 2	
4.389E-04	1.450E-01	20	B324	2 2 2 2 2	
4.388E-04	1.450E-01	20	B324	2 2 2 2 2	
4.389E-04	1.450E-01	20	F311	1 2 2 2 1	
4.389E-04	1.450E-01	20	M061	1 0 0 0 2	
4.389E-04	1.450E-01	20	M344	1 0 0 0 2	
4.964E-04	1.640E-01	30	B324	2 2 2 2 2	
4.963E-04	1.640E-01	30	B324	2 2 2 2 2	
4.389E-04	1.450E-01	rt	M161	0 0 0 0 2	

2155. $C_{10}H_{20}$

n-Pentylcyclopentane
1-Pentylcyclopentane

RN: 3741-00-2   **MP** (°C):
MW: 140.27   **BP** (°C):

Solubility (Moles/L)	Solubility (Grams/L)	Temp (°C)	Ref (#)	Evaluation (T P E A A)	Comments
8.198E-07	1.150E-04	25	K119	1 0 0 0 2	
8.198E-07	1.150E-04	25	P051	2 1 1 2 2	
8.198E-07	1.150E-04	25.00	P007	2 1 2 2 2	

2156. $C_{10}H_{20}NO_4PS$
Propetamphos
Methylethyl (E)-3-((((Ethylamino)methoxyphosphinothioyl)oxy)-2-butenoate
Safrotin
Seraphos
Zoecon
RN: 31218-83-4   **MP** (°C):
MW: 281.31   **BP** (°C):   88

Solubility (Moles/L)	Solubility (Grams/L)	Temp (°C)	Ref (#)	Evaluation (T P E A A)	Comments
3.910E-04	1.100E-01	24	M161	1 0 0 0 2	

2157. $C_{10}H_{20}NO_5PS_2$
Mecarbam
O,O-Diethyl S-(N-Methyl-N-carboethoxycarbamoylmethyl) Dithiophosphate
RN: 2595-54-2   **MP** (°C):
MW: 329.38   **BP** (°C):   144

Solubility (Moles/L)	Solubility (Grams/L)	Temp (°C)	Ref (#)	Evaluation (T P E A A)	Comments
3.033E-03	9.990E-01	rt	M061	0 0 0 0 0	
<3.04E-03	<1.00E+00	rt	M161	0 0 0 0 0	

2158. $C_{10}H_{20}N_2S_4$
Disulfiram
Tetraethylthioperoxydicarbonothioic Diamide
Tetraethylthiuram Disulfide
Antadix
Antabuse
Esperal
RN: 97-77-8   **MP** (°C):   70
MW: 296.54   **BP** (°C):

Solubility (Moles/L)	Solubility (Grams/L)	Temp (°C)	Ref (#)	Evaluation (T P E A A)	Comments
6.744E-04	2.000E-01	25	I314	0 0 0 0 0	
1.379E-05	4.090E-03	25	L033	1 0 2 1 2	*sic*
1.012E-03	3.000E-01	ns	N061	0 0 0 0 0	

2159. $C_{10}H_{20}N_6O$
N-(Methoxymethyl)pentamethylmelamine
N-Methylolpentamethylmelamine Methyl Ether
RN: 64124-15-8 **MP** (°C): 39
MW: 240.31 **BP** (°C):

Solubility (Moles/L)	Solubility (Grams/L)	Temp (°C)	Ref (#)	Evaluation (T P E A A)	Comments
6.242E-03	1.500E+00	25	C051	1 2 1 1 1	pH 7, unstable in water

2160. $C_{10}H_{20}O$
Menthol
Cyclohexanol, 5-Methyl-2-(1-methylethyl)-, (1α,2β,5α)-
3-p-Menthanol
RN: 89-78-1 **MP** (°C): 42
MW: 156.27 **BP** (°C): 212

Solubility (Moles/L)	Solubility (Grams/L)	Temp (°C)	Ref (#)	Evaluation (T P E A A)	Comments
2.560E-03	4.000E-01	20	F300	1 0 0 0 2	
2.920E-03	4.563E-01	25	I019	1 0 1 2 2	
8.600E-03	1.344E+00	37	E028	1 0 1 1 1	

2161. $C_{10}H_{20}O$
Citronellol
3,7-Dimethyl-6-octen-1-ol
Levo-citronellol
β-Citronellol
RN: 106-22-9 **MP** (°C):
MW: 156.27 **BP** (°C): 222

Solubility (Moles/L)	Solubility (Grams/L)	Temp (°C)	Ref (#)	Evaluation (T P E A A)	Comments
1.280E-03	2.000E-01	25	M350	1 0 1 1 1	

2162. $C_{10}H_{20}O_2$
3-Hydroxy-2,5,5-triethyltetrahydrofuran
3-Furanol, 2,5,5-Triethyltetrahydro-
RN: 29839-70-1 **MP** (°C):
MW: 172.27 **BP** (°C):

Solubility (Moles/L)	Solubility (Grams/L)	Temp (°C)	Ref (#)	Evaluation (T P E A A)	Comments
5.747E-02	9.901E+00	rt	B066	0 2 0 0 0	

2163. $C_{10}H_{20}O_2$
3-Hydroxy-2-pentyl-5-methyltetrahydrofuran
3-Furanol, 5-Methyltetrahydro-2-pentyl-
RN: 29848-45-1   **MP** (°C):
MW: 172.27   **BP** (°C):

Solubility (Moles/L)	Solubility (Grams/L)	Temp (°C)	Ref (#)	Evaluation (T P E A A)	Comments
1.159E-02	1.996E+00	rt	B066	0 2 0 0 0	

2164. $C_{10}H_{20}O_2$
3-Hydroxy-2,5-dimethyl-2,5-diethyltetrahydrofuran
3-Furanol, 2,5-Diethyltetrahydro-2,5-dimethyl-
RN: 30010-09-4   **MP** (°C):
MW: 172.27   **BP** (°C):

Solubility (Moles/L)	Solubility (Grams/L)	Temp (°C)	Ref (#)	Evaluation (T P E A A)	Comments
1.138E-01	1.961E+01	rt	B066	0 2 0 0 0	

2165. $C_{10}H_{20}O_2$
3-Hydroxy-2,5-dipropyltetrahydrofuran
3-Furanol, 2,5-Dipropyltetrahydro-
RN: 30003-27-1   **MP** (°C):
MW: 172.27   **BP** (°C):

Solubility (Moles/L)	Solubility (Grams/L)	Temp (°C)	Ref (#)	Evaluation (T P E A A)	Comments
1.159E-02	1.996E+00	rt	B066	0 2 0 0 0	

2166. $C_{10}H_{20}O_2$
3-Hydroxy-2-butyl-5,5-methyltetrahydrofuran
3-Furanol, 2-Butyltetrahydro-5,5-dimethyl-
RN: 29839-71-2   **MP** (°C):
MW: 172.27   **BP** (°C):

Solubility (Moles/L)	Solubility (Grams/L)	Temp (°C)	Ref (#)	Evaluation (T P E A A)	Comments
2.888E-02	4.975E+00	rt	B066	0 2 0 0 0	

2167. $C_{10}H_{20}O_2$
n-Capric Acid
Caprinsaeure
Decanoic Acid
Nonanecarboxylic Acid

RN: 334-48-5   **MP** (°C):   31.4
MW: 172.27   **BP** (°C):   270

Solubility (Moles/L)	Solubility (Grams/L)	Temp (°C)	Ref (#)	Evaluation (T P E A A)	Comments
5.515E-04	9.500E-02	0	B136	1 0 2 1 1	
5.514E-04	9.499E-02	0.0	R001	1 1 1 1 1	
1.509E-04	2.600E-02	15	F300	1 0 0 0 1	
2.902E-04	5.000E-02	20	A011	1 2 1 1 1	
9.462E-04	1.630E-01	20	B136	1 0 2 1 2	
8.706E-04	1.500E-01	20	D041	1 0 0 0 1	
8.706E-04	1.500E-01	20.0	R001	1 1 1 1 1	
3.590E-04	6.184E-02	25	J001	1 0 2 1 2	
1.115E-03	1.920E-01	30	B136	1 0 2 1 2	
3.715E-04	6.400E-02	30	E005	2 1 1 2 1	
1.045E-03	1.800E-01	30.0	R001	1 1 1 1 1	
1.294E-03	2.230E-01	40	B136	1 0 2 1 2	
4.179E-04	7.200E-02	40	E005	2 1 1 2 1	
1.335E-03	2.300E-01	45	B136	1 0 2 1 1	
1.335E-03	2.299E-01	45.0	R001	1 1 1 1 1	
4.702E-04	8.100E-02	50	E005	2 1 1 2 1	
5.000E-04	8.613E-02	50	J001	1 0 2 1 2	
1.567E-03	2.700E-01	60	B136	1 0 2 1 1	
5.805E-04	1.000E-01	60	E005	2 1 1 2 2	
1.567E-03	2.699E-01	60.0	R001	1 1 1 1 1	

2168. $C_{10}H_{20}O_2$
3-Hydroxy-5,5-dipropyltetrahydrofuran
3-Furanol, 5,5-Dipropyltetrahydro-

RN: 29839-54-1   **MP** (°C):
MW: 172.27   **BP** (°C):

Solubility (Moles/L)	Solubility (Grams/L)	Temp (°C)	Ref (#)	Evaluation (T P E A A)	Comments
5.747E-02	9.901E+00	rt	B066	0 2 0 0 0	

2169. $C_{10}H_{20}O_2$
3-Hydroxy-5,5-diisopropyltetrahydrofuran
3-Furanol, 5,5-Diisopropyltetrahydro-

RN: 29839-55-2   **MP** (°C):
MW: 172.27   **BP** (°C):

Solubility (Moles/L)	Solubility (Grams/L)	Temp (°C)	Ref (#)	Evaluation (T P E A A)	Comments
5.747E-02	9.901E+00	rt	B066	0 2 0 0 0	

2170.   C₁₀H₂₀O₂
3-Hydroxy-2-propyl-5-methyl-5-ethyltetrahydrofuran
3-Furanol, 5-Ethyltetrahydro-5-methyl-2-propyl-
RN: 29839-72-3   **MP** (°C):
MW: 172.27   **BP** (°C):

Solubility (Moles/L)	Solubility (Grams/L)	Temp (°C)	Ref (#)	Evaluation (T P E A A)	Comments
5.747E-02	9.901E+00	rt	B066	0 2 0 0 0	

2171.   C₁₀H₂₀O₂
3-Hydroxy-2-ethyl-5-propyl-5-methyltetrahydrofuran
3-Furanol, 2-Ethyltetrahydro-5-methyl-5-propyl-
RN: 29839-73-4   **MP** (°C):
MW: 172.27   **BP** (°C):

Solubility (Moles/L)	Solubility (Grams/L)	Temp (°C)	Ref (#)	Evaluation (T P E A A)	Comments
5.747E-02	9.901E+00	rt	B066	0 2 0 0 0	

2172.   C₁₀H₂₀O₂
3-Hydroxy-2,2-dimethyl-5,5-diethyltetrahydrofuran
3-Furanol, 5,5-Diethyltetrahydro-2,2-dimethyl-
RN: 29839-77-8   **MP** (°C):
MW: 172.27   **BP** (°C):

Solubility (Moles/L)	Solubility (Grams/L)	Temp (°C)	Ref (#)	Evaluation (T P E A A)	Comments
2.888E-02	4.975E+00	rt	B066	0 2 0 0 0	

2173.   C₁₀H₂₀O₂.H₂O
Terpin (Monohydrate)
Terpin-hydrat
RN: 2451-01-6   **MP** (°C):   116
MW: 190.29   **BP** (°C):

Solubility (Moles/L)	Solubility (Grams/L)	Temp (°C)	Ref (#)	Evaluation (T P E A A)	Comments
2.102E-02	4.000E+00	15	F300	1 0 0 0 0	
1.799E-02	3.424E+00	25	M012	1 0 2 1 2	
1.661E-01	3.160E+01	100	F300	1 0 0 0 2	

2174. C$_{10}$H$_{20}$O$_3$
n-Amyl β-Ethoxypropionate
Propionic Acid, 3-Ethoxy-, Pentyl Ester

RN: 14144-36-6 **MP** (°C):
MW: 188.27 **BP** (°C):

Solubility (Moles/L)	Solubility (Grams/L)	Temp (°C)	Ref (#)	Evaluation (T P E A A)	Comments
6.366E-03	1.199E+00	25	D002	1 2 1 1 1	

2175. C$_{10}$H$_{20}$O$_3$
1,3-Dioxolane-4-methanol, 2-Methyl-2-pentyl
2-Heptanone, Cyclic (hydroxymethyl)ethylene Acetal
2-Methyl-2-n-amyl-4-hydroxymethyl-1,3-dioxolane
2-Methyl-2-pentyl-1,3-dioxolane-4-methanol

RN: 4361-59-5 **MP** (°C):
MW: 188.27 **BP** (°C):

Solubility (Moles/L)	Solubility (Grams/L)	Temp (°C)	Ref (#)	Evaluation (T P E A A)	Comments
5.090E-02	9.583E+00	25	P342	1 2 2 2 2	0.0001M Na$_2$CO$_3$

2176. C$_{10}$H$_{20}$O$_4$
Butyl Carbitol Acetate
Diethylene Glycol Acetate Butyl Ether
Diethylene Glycol Butyl Ether Acetate
Diglykol-monobutylaether-acetat

RN: 124-17-4 **MP** (°C): -32
MW: 204.27 **BP** (°C): 245

Solubility (Moles/L)	Solubility (Grams/L)	Temp (°C)	Ref (#)	Evaluation (T P E A A)	Comments
7.709E-02	1.575E+01	20	D052	1 1 0 0 1	
1.792E-01	3.661E+01	20	M062	1 0 0 0 1	

2177.   C$_{10}$H$_{21}$NOS
Vernolate
S-Propyl Dipropylthiocarbamate
Carbamic Acid, Dipropylthio-, S-propyl Ester
Carbamate, n-Propyl-di-n-propylthio-
Vernam
RN: 1929-77-7   **MP** (°C):   <25
MW: 203.35   **BP** (°C):

Solubility (Moles/L)	Solubility (Grams/L)	Temp (°C)	Ref (#)	Evaluation (T P E A A)	Comments
4.426E-04	9.000E-02	20	B200	1 0 0 0 1	
5.262E-04	1.070E-01	21	F019	1 0 0 0 2	
5.262E-04	1.070E-01	21	M161	1 0 0 0 2	
<4.92E-04	<1.00E-01	ns	B185	1 0 0 0 1	
4.917E-04	9.999E-02	ns	M061	0 0 0 0 0	

2178.   C$_{10}$H$_{21}$NOS
Pebulate
S-Propyl Butylethylthiocarbamate
RN: 1114-71-2   **MP** (°C):   <25
MW: 203.35   **BP** (°C):

Solubility (Moles/L)	Solubility (Grams/L)	Temp (°C)	Ref (#)	Evaluation (T P E A A)	Comments
2.951E-04	6.000E-02	20	M161	1 0 0 0 1	
4.524E-04	9.200E-02	21	F019	1 0 0 0 1	
4.524E-04	9.200E-02	21	M061	1 0 0 0 1	
2.951E-04	6.000E-02	ns	B200	0 0 0 0 1	

2179.   C$_{10}$H$_{22}$
n-Decane
Decane
Decyl Hydride
RN: 124-18-5   **MP** (°C):   -30.0
MW: 142.29   **BP** (°C):   174.0

Solubility (Moles/L)	Solubility (Grams/L)	Temp (°C)	Ref (#)	Evaluation (T P E A A)	Comments
1.389E-07	1.976E-05	20	B165	1 0 1 1 1	
1.124E-07	1.600E-05	25	B069	1 0 1 1 1	
1.389E-07	1.976E-05	25	F004	1 2 2 2 1	
3.655E-07	5.200E-05	25	M003	1 0 2 2 1	
3.655E-07	5.200E-05	25	M040	1 0 0 1 1	
1.546E-07	2.200E-05	ns	B033	0 0 0 0 2	
1.546E-07	2.200E-05	ns	B033	0 2 2 2 1	
3.655E-07	5.200E-05	ns	H123	0 0 0 0 2	

2180. C$_{10}$H$_{22}$
4,4-Dimethyloctane
RN: 15869-95-1   **MP** (°C):
MW: 142.29   **BP** (°C):   157.5

Solubility (Moles/L)	Solubility (Grams/L)	Temp (°C)	Ref (#)	Evaluation (T P E A A)	Comments
1.546E-05	2.200E-03	20	M337	2 1 2 2 1	

2181. C$_{10}$H$_{22}$O
n-Decyl Alcohol
Alcohol C-10
Nonyl Acarbinol
Capric Alcohol
RN: 36729-58-5   **MP** (°C):
MW: 158.29   **BP** (°C):

Solubility (Moles/L)	Solubility (Grams/L)	Temp (°C)	Ref (#)	Evaluation (T P E A A)	Comments
2.690E-04	4.258E-02	20	H330	2 0 2 2 2	
2.000E-04	3.166E-02	24	H345	2 0 2 2 2	
2.340E-04	3.704E-02	25	K025	2 2 1 1 2	
2.527E-05	4.000E-03	40	W305	1 0 0 1 0	EFG
3.000E-04	4.748E-02	ns	H012	0 2 2 0 0	

2182. C$_{10}$H$_{23}$O$_3$P
Ethyl Dibutyl Phosphonate
Dibutyl Ethyl Phosphonate
RN: 2404-58-2   **MP** (°C):
MW: 222.27   **BP** (°C):

Solubility (Moles/L)	Solubility (Grams/L)	Temp (°C)	Ref (#)	Evaluation (T P E A A)	Comments
2.699E-02	6.000E+00	25	B070	1 2 0 1 0	
5.849E-02	1.300E+01	25	B070	1 2 0 1 1	

2183. C$_{10}$H$_{23}$O$_4$P
Dibutyl Ethyl Phosphate
RN: 7242-58-2   **MP** (°C):
MW: 238.27   **BP** (°C):

Solubility (Moles/L)	Solubility (Grams/L)	Temp (°C)	Ref (#)	Evaluation (T P E A A)	Comments
1.427E-02	3.400E+00	25	B070	1 2 2 1 1	

2184.   C$_{10}$Cl$_{10}$O
Chlordecone
Kepone
1,2,3,5,6,7,8,9,10,10-Decachloropentacyclo[5.2.1.0(2,6).0(3,9).0(5,8)]decano-4-one
Merex
Decachloroketone
RN: 143-50-0   **MP** (°C):
MW: 490.64   **BP** (°C):

Solubility (Moles/L)	Solubility (Grams/L)	Temp (°C)	Ref (#)	Evaluation (T P E A A)	Comments
8.153E-03	4.000E+00	100	M161	1 0 0 0 0	

2185.   C$_{10}$Cl$_{12}$
Mirex
1,2,3,4,5,5-Hexachloro-1,3-cyclopentadiene dimer
Bichlorendo
Ferriamicide
Dechlorane 4070
RN: 2385-85-5   **MP** (°C):
MW: 545.55   **BP** (°C):

Solubility (Moles/L)	Solubility (Grams/L)	Temp (°C)	Ref (#)	Evaluation (T P E A A)	Comments
1.558E-07	8.500E-05	25	M134	1 2 1 1 1	
1.741E-07	9.500E-05	ns	M110	0 0 0 0 0	EFG

2186.   C$_{11}$H$_6$BrNS
1-Bromo-2-naphthylisothiocyanate
RN: 2392-80-5   **MP** (°C):
MW: 264.15   **BP** (°C):

Solubility (Moles/L)	Solubility (Grams/L)	Temp (°C)	Ref (#)	Evaluation (T P E A A)	Comments
4.800E-05	1.268E-02	25	D019	1 1 1 1 1	

2187.   C$_{11}$H$_6$O$_3$
Psoralen
7H-Furo[3,2-g][1]benzopyran-7-one
RN: 66-97-7   **MP** (°C):   158-161
MW: 186.17   **BP** (°C):

Solubility (Moles/L)	Solubility (Grams/L)	Temp (°C)	Ref (#)	Evaluation (T P E A A)	Comments
3.500E-04	6.516E-02	25	A355	1 0 2 2 1	

2188.   C$_{11}$H$_7$FN$_2$O$_3$
3-Benzoyl-5-fluorouracil
RN: 61251-77-2   **MP** (°C):   169-170
MW: 234.19   **BP** (°C):

Solubility (Moles/L)	Solubility (Grams/L)	Temp (°C)	Ref (#)	Evaluation (T P E A A)	Comments
5.551E-03	1.300E+00	22	B321	1 0 2 2 2	pH 4.0
5.551E-03	1.300E+00	22	B332	1 1 0 0 1	pH 4.0

2189.   C$_{11}$H$_7$FN$_2$O$_4$
1-Phenyloxycarbonyl-5-fluorouracil
RN: 75410-28-5   **MP** (°C):
MW: 250.19   **BP** (°C):

Solubility (Moles/L)	Solubility (Grams/L)	Temp (°C)	Ref (#)	Evaluation (T P E A A)	Comments
3.597E-03	9.000E-01	22	B332	1 1 0 0 1	pH 4.0

2190.   C$_{11}$H$_7$FN$_2$O$_4$
3-Phenyloxycarbonyl-5-fluoro-2,4(1H,3H)-pyrimidinedi-one
3-Phenyloxycarbonyl-5-fluorouracil
RN: 66999-97-1   **MP** (°C):   169-170
MW: 250.19   **BP** (°C):

Solubility (Moles/L)	Solubility (Grams/L)	Temp (°C)	Ref (#)	Evaluation (T P E A A)	Comments
5.995E-04	1.500E-01	22	B321	1 0 2 2 2	pH 4.0

2191.   C$_{11}$H$_7$NS
2-Naphthyl Isothiocyanate
2-Isothiocyanatonaphthalene
β-Naphthyl Mustard Oil
RN: 1636-33-5   **MP** (°C):
MW: 185.25   **BP** (°C):

Solubility (Moles/L)	Solubility (Grams/L)	Temp (°C)	Ref (#)	Evaluation (T P E A A)	Comments
3.600E-05	6.669E-03	25	D019	1 1 1 1 1	

2192.   C₁₁H₇NS

1-Naphthyl Isothiocyanate
1-Isothiocyanatonaphthalene
α-Naphthyl Mustard Oil
Kesscocide
ANI
ANIT

RN:	551-06-4	**MP** (°C):	58.0
MW:	185.25	**BP** (°C):	

Solubility (Moles/L)	Solubility (Grams/L)	Temp (°C)	Ref (#)	Evaluation (T P E A A)	Comments
2.500E-05	4.631E-03	25	D019	1 1 1 1 1	

2193.   C₁₁H₈N₄O₄

Orotic Acid Nicotinmide

RN:		**MP** (°C):	252-253
MW:	260.21	**BP** (°C):	

Solubility (Moles/L)	Solubility (Grams/L)	Temp (°C)	Ref (#)	Evaluation (T P E A A)	Comments
6.800E-02	1.769E+01	25	N018	2 2 1 2 2	

2194.   C₁₁H₈O₂

Menadione
2-Methyl-1,4-naphthoquinone
Vitamin K3
Kativ-G
Panosine
Menaphthone

RN:	58-27-5	**MP** (°C):	106
MW:	172.19	**BP** (°C):	

Solubility (Moles/L)	Solubility (Grams/L)	Temp (°C)	Ref (#)	Evaluation (T P E A A)	Comments
9.291E-04	1.600E-01	25	P096	1 0 2 2 2	
6.969E-04	1.200E-01	30	K090	1 2 2 2 0	EFG
8.700E-04	1.498E-01	30	O321	2 2 2 2 1	
8.710E-04	1.500E-01	30	O321	2 2 2 2 1	
9.291E-04	1.600E-01	30.00	E033	1 0 2 1 0	EFG
1.161E-03	2.000E-01	37.00	E033	1 0 2 1 0	EFG

2195. C₁₁H₈O₂
2-Naphthoic Acid
β-Naphthoic Acid
2-Naphthalenecarboxylic Acid
RN: 93-09-4   **MP** (°C):
MW: 172.19   **BP** (°C):

Solubility (Moles/L)	Solubility (Grams/L)	Temp (°C)	Ref (#)	Evaluation (T P E A A)	Comments
1.300E-04	2.238E-02	25	M149	2 2 2 2 1	intrinsic, *sic*
1.617E-06	2.785E-04	30	K148	1 1 0 0 2	
2.323E-06	4.000E-04	40	K148	1 1 0 0 1	
3.165E-06	5.450E-04	50	K148	1 1 0 0 2	
3.949E-06	6.800E-04	60	K148	1 1 0 0 2	
4.652E-06	8.010E-04	70	K148	1 1 0 0 2	
5.459E-06	9.400E-04	80	K148	1 1 0 0 2	
6.261E-06	1.078E-03	90	K148	1 1 0 0 2	

2196. C₁₁H₈O₃
8-Hydroxypsoralon
RN: **MP** (°C):
MW: 188.18   **BP** (°C):

Solubility (Moles/L)	Solubility (Grams/L)	Temp (°C)	Ref (#)	Evaluation (T P E A A)	Comments
6.100E-04	1.148E-01	25	A355	1 0 2 2 1	

2197. C₁₁H₉ClO₂S
Tianafac
RN: 51527-19-6   **MP** (°C):
MW: 240.71   **BP** (°C):

Solubility (Moles/L)	Solubility (Grams/L)	Temp (°C)	Ref (#)	Evaluation (T P E A A)	Comments
1.444E-04	3.476E-02	25	C314	1 1 2 2 2	
1.442E-04	3.470E-02	25	C314	1 1 2 2 2	

2198.   C₁₁H₉Cl₂NO₂
Barban
4-Chloro-2-butynyl-N-(3-chlorophenyl)carbamate
4-Chloro-2-butynyl-m-chlorocarbanilate
RN: 101-27-9   **MP** (°C):   75
MW: 258.11   **BP** (°C):

Solubility (Moles/L)	Solubility (Grams/L)	Temp (°C)	Ref (#)	Evaluation (T P E A A)	Comments
4.262E-05	1.100E-02	25	B200	1 0 0 0 2	
4.262E-05	1.100E-02	25	M161	1 0 0 0 1	
3.874E-05	1.000E-02	ns	H042	0 0 0 0 1	
4.262E-04	1.100E-01	ns	M061	0 0 0 0 2	

2199.   C₁₁H₉Cl₄NO₄
OCS-21693
TMMT
Methyl-2,3,5,6-tetrachloro-N-methoxy-N-methylterephthalamate
RN: 14419-01-3   **MP** (°C):   96
MW: 361.01   **BP** (°C):

Solubility (Moles/L)	Solubility (Grams/L)	Temp (°C)	Ref (#)	Evaluation (T P E A A)	Comments
1.385E-05	5.000E-03	25	B200	1 0 0 0 0	

2200.   C₁₁H₉I₃N₂O₄
3,5-Diacetylamino-2,4,6-triiodobenzoic Acid
Iothalamic Acid
Diatrazoic Acid
RN: 117-96-4   **MP** (°C):
MW: 613.92   **BP** (°C):

Solubility (Moles/L)	Solubility (Grams/L)	Temp (°C)	Ref (#)	Evaluation (T P E A A)	Comments
8.144E-01	5.000E+02	25	L100	1 0 0 0 2	
9.773E-01	6.000E+02	50	L100	1 0 0 0 2	
1.189E+00	7.297E+02	90	L100	1 0 0 0 2	
2.557E-03	1.570E+00	ns	H055	0 1 0 2 2	

2201. C₁₁H₁₀
2-Methylnaphthalene
2-Methyl Naphthalene
β-Methyl Naphthalenes

RN: 91-57-6   **MP** (°C): 35
MW: 142.20   **BP** (°C): 241.5

Solubility (Moles/L)	Solubility (Grams/L)	Temp (°C)	Ref (#)	Evaluation (T P E A A)	Comments
1.730E-04	2.460E-02	25	E004	2 1 2 2 2	
1.828E-04	2.600E-02	25	L332	1 1 1 1 0	
1.786E-04	2.540E-02	25	M064	1 1 2 2 2	
1.800E-04	2.560E-02	25	M342	1 0 1 1 1	
1.758E-04	2.500E-02	25	O320	1 0 1 1 1	
1.786E-04	2.540E-02	ns	H123	0 0 0 0 2	
8.000E-05	1.138E-02	ns	L060	0 0 0 0 0	
1.786E-04	2.540E-02	ns	M344	0 0 0 0 2	

2202. C₁₁H₁₀
1-Methylnaphthalene
1-Methyl Naphthalene
1-Methyl-napthalene
α-Methyl Naphthalenes
α-Methylnaphthalene

RN: 90-12-0   **MP** (°C): -22
MW: 142.20   **BP** (°C): 244

Solubility (Moles/L)	Solubility (Grams/L)	Temp (°C)	Ref (#)	Evaluation (T P E A A)	Comments
1.739E-04	2.473E-02	4	D351	1 2 1 1 2	
1.600E-04	2.275E-02	10	S076	2 2 2 2 1	
2.000E-04	2.844E-02	14	S076	2 2 2 2 1	
1.195E-04	1.700E-02	20	A050	1 0 1 1 2	
2.145E-04	3.050E-02	20	B318	1 2 1 2 0	EFG
2.124E-04	3.020E-02	20	B356	1 0 0 0 2	
2.000E-04	2.844E-02	20	S076	2 2 2 2 1	
2.100E-04	2.986E-02	21	A057	2 1 2 2 1	
2.489E-04	3.539E-02	25	D351	1 2 1 1 2	
1.814E-04	2.580E-02	25	E004	2 1 2 2 2	
1.899E-04	2.700E-02	25	L332	1 1 1 1 0	
2.004E-04	2.850E-02	25	M064	1 1 2 2 2	
2.000E-04	2.844E-02	25	M342	1 0 1 1 2	
2.100E-04	2.986E-02	25	S076	2 2 2 2 1	
2.440E-04	3.470E-02	28	B348	2 2 2 2 2	
2.955E-04	4.203E-02	40	D351	1 2 1 1 2	
2.004E-04	2.850E-02	ns	H123	0 0 0 0 2	
1.600E-04	2.275E-02	ns	L060	0 0 0 0 1	
2.004E-04	2.850E-02	ns	M344	0 0 0 0 2	

2203. C₁₁H₁₀BrN₃O₂S

2-Sulfanilamido-5-bromopyridine
Benzenesulfonamide, 4-Amino-N-(5-bromo-2-pyridinyl)-
RN: 16805-99-5 **MP** (°C):
MW: 328.19 **BP** (°C):

Solubility (Moles/L)	Solubility (Grams/L)	Temp (°C)	Ref (#)	Evaluation (T P E A A)	Comments
1.158E-04	3.800E-02	37	R058	1 2 1 1 1	

2204. C₁₁H₁₀BrN₃O₂S

5-Sulfanilamido-2-bromopyridine
Benzenesulfonamide, 4-Amino-N-(2-bromo-5-pyridinyl)-
RN: 17103-43-4 **MP** (°C):
MW: 328.19 **BP** (°C):

Solubility (Moles/L)	Solubility (Grams/L)	Temp (°C)	Ref (#)	Evaluation (T P E A A)	Comments
3.717E-04	1.220E-01	37	R058	1 2 1 1 2	

2205. C₁₁H₁₀ClNO₂

Chlorbupham
1-Methylpropyn-2-yl N-(m-Chlorophenyl)carbamate
Chlorbufam
Bi-PC
RN: 1967-16-4 **MP** (°C): 45.5
MW: 223.66 **BP** (°C):

Solubility (Moles/L)	Solubility (Grams/L)	Temp (°C)	Ref (#)	Evaluation (T P E A A)	Comments
2.414E-03	5.400E-01	20	B185	1 0 0 0 2	
2.414E-03	5.400E-01	20	M161	1 0 0 0 2	

2206. C₁₁H₁₀ClN₃O₂S

5-Sulfanilamido-2-chloropyridine
N1-(6-Chloro-3-pyridyl)sulfanilamide
RN: 34392-82-0 **MP** (°C):
MW: 283.74 **BP** (°C):

Solubility (Moles/L)	Solubility (Grams/L)	Temp (°C)	Ref (#)	Evaluation (T P E A A)	Comments
6.344E-04	1.800E-01	37	R058	1 2 1 1 1	

2207. C$_{11}$H$_{10}$Cl$_2$O$_3$
2,4-Dichlorophenoxyacetic Acid Allyl Ester
Allyl 2,4-Dichlorophenoxyacetate
RN: 58965-05-2   **MP** (°C):
MW: 261.11   **BP** (°C):

Solubility (Moles/L)	Solubility (Grams/L)	Temp (°C)	Ref (#)	Evaluation (T P E A A)	Comments
1.426E-04	3.722E-02	ns	M120	0 0 1 1 2	

2208. C$_{11}$H$_{10}$IN$_3$O$_2$S
2-Sulfanilamido-5-iodopyridine
Benzenesulfonamide, 4-Amino-N-(5-iodo-2-pyridinyl)-
RN: 71119-21-6   **MP** (°C):
MW: 375.19   **BP** (°C):

Solubility (Moles/L)	Solubility (Grams/L)	Temp (°C)	Ref (#)	Evaluation (T P E A A)	Comments
3.465E-05	1.300E-02	37	R058	1 2 1 1 1	

2209. C$_{11}$H$_{10}$N$_2$O
Vasicinone
Pyrrolo[2,1-b]quinazolin-9(1H)-one, 2,3-Dihydro-3-hydroxy-, (3S)-
(-)-Vasicinone
L-Vasicinone
RN: 486-64-6   **MP** (°C):   204
MW: 186.22   **BP** (°C):

Solubility (Moles/L)	Solubility (Grams/L)	Temp (°C)	Ref (#)	Evaluation (T P E A A)	Comments
8.578E-03	1.597E+00	25	B194	2 2 2 2 1	

2210. C$_{11}$H$_{10}$N$_2$O
3-o-Toluoxypyridazine
Credazine
3-(2-Methylphenoxy)-pyridazine
RN: 14491-59-9   **MP** (°C):   78
MW: 186.22   **BP** (°C):

Solubility (Moles/L)	Solubility (Grams/L)	Temp (°C)	Ref (#)	Evaluation (T P E A A)	Comments
1.072E-02	1.996E+00	ns	B100	0 0 0 0 0	
1.074E-02	2.000E+00	rt	M161	0 0 0 0 0	

2211.   C₁₁H₁₀N₂O₃

Phenylmethylbarbituric Acid
Barbituric Acid, 5-Methyl-5-phenyl
2,4,6(1H,3H,5H)-Pyrimidinetrione, 5-Methyl-5-phenyl
2,4,6-Trioxo-5-methyl-5-phenylhexahydropyrimidine
Heptobarbital

RN: 76-94-8   **MP** (°C):   226
MW: 218.21   **BP** (°C):

Solubility (Moles/L)	Solubility (Grams/L)	Temp (°C)	Ref (#)	Evaluation (T P E A A)	Comments
3.480E-03	7.594E-01	20	J030	1 2 2 2 1	
4.170E-03	9.100E-01	25	P350	2 1 1 1 2	intrinsic
6.133E-03	1.338E+00	37	J030	1 2 2 2 2	

2212.   C₁₁H₁₀N₂S

1-Naphthylthiourea
ANTU

RN: 86-88-4   **MP** (°C):   198
MW: 202.28   **BP** (°C):

Solubility (Moles/L)	Solubility (Grams/L)	Temp (°C)	Ref (#)	Evaluation (T P E A A)	Comments
2.966E-03	6.000E-01	rt	M161	0 0 0 0 2	

2213.   C₁₁H₁₀N₄O₄S

2-Sulfanilamido-5-nitropyridine
Benzenesulfonamide, 4-Amino-N-(5-nitro-2-pyridinyl)-

RN: 39588-36-8   **MP** (°C):
MW: 294.29   **BP** (°C):

Solubility (Moles/L)	Solubility (Grams/L)	Temp (°C)	Ref (#)	Evaluation (T P E A A)	Comments
1.257E-04	3.700E-02	37	R058	1 2 1 1 1	

2214. $C_{11}H_{11}ClO_3$

Alclofenac
(4-Allyloxy-3-chlorophenyl)acetic Acid
(3-Chloro-4-allyloxyphenyl)acetic Acid

RN: 22131-79-9 **MP** (°C):
MW: 226.66 **BP** (°C):

Solubility (Moles/L)	Solubility (Grams/L)	Temp (°C)	Ref (#)	Evaluation (T P E A A)	Comments
4.850E-05	1.099E-02	5	F306	1 0 1 2 2	intrinsic
5.780E-05	1.310E-02	25	C314	1 1 2 2 2	
5.780E-05	1.310E-02	25	C314	1 1 2 2 2	
6.200E-05	1.405E-02	25	F306	1 0 1 2 2	intrinsic
8.000E-05	1.813E-02	37	F306	1 0 1 2 2	intrinsic

2215. $C_{11}H_{11}N$

2,7-Dimethylquinoline
Quinoline, 2,7-Dimethyl-

RN: 93-37-8 **MP** (°C): 58
MW: 157.22 **BP** (°C):

Solubility (Moles/L)	Solubility (Grams/L)	Temp (°C)	Ref (#)	Evaluation (T P E A A)	Comments
1.142E-02	1.795E+00	25	P051	2 1 1 2 2	
1.142E-02	1.795E+00	25.00	P007	2 1 2 2 2	

2216. $C_{11}H_{11}N$

2,4-Dimethylquinoline
Quinoline, 2,4-Dimethyl-

RN: 1198-37-4 **MP** (°C): 264
MW: 157.22 **BP** (°C):

Solubility (Moles/L)	Solubility (Grams/L)	Temp (°C)	Ref (#)	Evaluation (T P E A A)	Comments
1.142E-02	1.795E+00	25	K119	1 0 0 0 2	

2217. $C_{11}H_{11}NO$

Aziridine, 1-(1-Oxo-3-phenyl-2-propenyl)-
N-Cyclopropylcinnamamide

RN: 53162-40-6 **MP** (°C):
MW: 173.22 **BP** (°C):

Solubility (Moles/L)	Solubility (Grams/L)	Temp (°C)	Ref (#)	Evaluation (T P E A A)	Comments
3.150E-03	5.456E-01	ns	H350	0 0 0 0 2	

2218.   C₁₁H₁₁NO₂

Phensuximide
Milontin
N-Methyl-2-phenyl-succinimide
RN: 86-34-0   **MP** (°C):   71-73
MW: 189.22   **BP** (°C):

Solubility (Moles/L)	Solubility (Grams/L)	Temp (°C)	Ref (#)	Evaluation (T P E A A)	Comments
2.220E-02	4.200E+00	25	P061	1 0 0 0 2	

2219.   C₁₁H₁₁NO₂S

Butyric Acid, p-Isothiocyanatophenyl Ester
RN: 96933-13-0   **MP** (°C):
MW: 221.28   **BP** (°C):

Solubility (Moles/L)	Solubility (Grams/L)	Temp (°C)	Ref (#)	Evaluation (T P E A A)	Comments
8.200E-05	1.814E-02	25	K032	2 2 0 1 1	

2220.   C₁₁H₁₁NO₄

Acetamide, N-Acetyl-2-(benzoyloxy)-
RN: 68659-48-3   **MP** (°C):   104.5
MW: 221.21   **BP** (°C):

Solubility (Moles/L)	Solubility (Grams/L)	Temp (°C)	Ref (#)	Evaluation (T P E A A)	Comments
3.978E-03	8.800E-01	22	N317	1 1 2 1 2	

2221.   C₁₁H₁₁NO₅

Benzoxydiglycine
RN: **MP** (°C):
MW: 237.21   **BP** (°C):

Solubility (Moles/L)	Solubility (Grams/L)	Temp (°C)	Ref (#)	Evaluation (T P E A A)	Comments
1.391E-02	3.300E+00	25.1	N026	2 0 2 2 2	

2222.   C₁₁H₁₁NO₅

Benzoic Acid, 2-(Acetyloxy)-, 2-Amino-2-oxoethyl Ester
(O-Acetylsalicyloyloxy)acetamide
RN: 50785-22-3   **MP** (°C):   128.5
MW: 237.21   **BP** (°C):

Solubility (Moles/L)	Solubility (Grams/L)	Temp (°C)	Ref (#)	Evaluation (T P E A A)	Comments
1.619E-02	3.840E+00	21	N335	1 2 1 1 2	

2223. $C_{11}H_{11}N_3OS$
Seedvax
2-Amino-4-methyl-5-carboxanilidothiazole
RN: 21452-14-2   **MP** (°C):   221
MW: 233.29   **BP** (°C):

Solubility (Moles/L)	Solubility (Grams/L)	Temp (°C)	Ref (#)	Evaluation (T P E A A)	Comments
4.282E-03	9.990E-01	ns	M061	0 0 0 0 0	

2224. $C_{11}H_{11}N_3O_2S$
Sulfapyridine
2-(Aminobenzene-4'-sulfamido)-pyridine
2-[Aminobenzol-4'-sulfamid]-pyridin
Sulphapyridine
2-Sulfapyridine
N-(2-Pyridyl)sulfanilamide
RN: 144-83-2   **MP** (°C):   192
MW: 249.29   **BP** (°C):

Solubility (Moles/L)	Solubility (Grams/L)	Temp (°C)	Ref (#)	Evaluation (T P E A A)	Comments
6.819E-04	1.700E-01	16	H114	1 0 0 0 2	
2.006E-03	5.000E-01	20	C103	1 2 0 0 2	
1.323E-03	3.299E-01	20	D041	1 0 0 0 1	
8.023E-04	2.000E-01	20	F073	1 2 2 2 2	
8.023E-04	2.000E-01	20	F073	1 2 2 2 2	
1.075E-03	2.680E-01	25	C102	2 0 2 2 2	
1.645E-03	4.100E-01	35	H114	1 0 0 0 1	
1.950E-03	4.860E-01	37	C102	2 0 2 2 2	
1.805E-03	4.500E-01	37	D084	1 0 1 0 1	
1.985E-03	4.948E-01	37	F072	1 0 0 0 2	
1.985E-03	4.948E-01	37	F075	1 0 2 2 2	
2.006E-03	5.000E-01	37	F300	1 0 0 0 0	
4.047E-03	1.009E+00	37	G037	2 2 2 1 0	EFG, form V
3.807E-03	9.491E-01	37	G073	2 2 2 1 0	EFG, form I
6.128E-03	1.528E+00	37	G073	2 2 2 1 0	EFG, amorphous
3.807E-03	9.491E-01	37	G073	2 2 2 1 0	EFG, form II
2.090E-03	5.210E-01	37	K095	2 0 0 0 2	intrinsic
2.447E-03	6.100E-01	37	M057	1 0 0 0 2	pH 5.5
2.607E-03	6.500E-01	37	R044	1 0 1 1 0	
2.165E-03	5.397E-01	37.50	M142	1 0 0 0 1	
6.417E-04	1.600E-01	37.50	M142	1 0 0 0 1	
2.006E-03	5.000E-01	38	K006	1 0 0 0 2	
4.412E-03	1.100E+00	40	C103	1 2 0 0 2	
4.212E-02	1.050E+01	100	C103	1 2 0 0 2	
3.972E-02	9.901E+00	100	D041	1 0 0 0 0	
1.484E-03	3.699E-01	rt	N015	0 0 2 2 2	

2225. C$_{11}$H$_{11}$N$_3$O$_3$S$_2$
Acetyl Sulfathiazole
Sulfathiazol Acetyle
N4-Acetylsulfathiazole
N4-Acetylsulphathiazole
RN: 127-76-4   **MP** (°C):
MW: 297.36   **BP** (°C):

Solubility (Moles/L)	Solubility (Grams/L)	Temp (°C)	Ref (#)	Evaluation (T P E A A)	Comments
3.363E-04	1.000E-01	37	D084	1 0 1 0 1	
2.186E-04	6.500E-02	37	F075	1 0 2 2 1	
2.354E-04	7.000E-02	37	L091	1 0 0 0 0	pH 5.5
1.951E-04	5.800E-02	37	M057	1 0 0 0 1	pH 5.5
2.018E-04	6.000E-02	37.50	M142	1 0 0 0 0	
2.388E-04	7.100E-02	38	K006	1 0 0 0 1	

2226. C$_{11}$H$_{11}$N$_3$O$_3$S
5-Sulfanilamido-2-hydroxypyridine
RN: 71119-20-5   **MP** (°C):
MW: 265.29   **BP** (°C):

Solubility (Moles/L)	Solubility (Grams/L)	Temp (°C)	Ref (#)	Evaluation (T P E A A)	Comments
9.725E-03	2.580E+00	37	R058	1 2 1 1 1	

2227. C$_{11}$H$_{12}$ClNO$_4$
Chloroethyl Acetaminophen
Carbonic Acid, 4-(Acetylamino)phenyl 2-chloroethyl Ester
Acetanilide, 4'-Hydroxy-, 2-Chloroethyl Carbonate (Ester)
RN: 17243-29-7   **MP** (°C):   122.5-123
MW: 257.68   **BP** (°C):

Solubility (Moles/L)	Solubility (Grams/L)	Temp (°C)	Ref (#)	Evaluation (T P E A A)	Comments
1.514E-03	3.900E-01	37	D029	1 0 1 1 1	

Solutions

2228.   C$_{11}$H$_{12}$Cl$_2$N$_2$O$_5$
Chloramphenicol
D-(-)-Threo-1-(p-nitrophenyl)-2-dichloroacetamido-1,3-propanediol
Amphicol
Leukomycin
Cloramical
Intramyctin
RN: 56-75-7   **MP** (°C):   150.5
MW: 323.13   **BP** (°C):

Solubility (Moles/L)	Solubility (Grams/L)	Temp (°C)	Ref (#)	Evaluation (T P E A A)	Comments
7.717E-03	2.494E+00	20	D041	1 0 0 0 1	
5.570E-03	1.800E+00	23	M168	2 0 0 0 0	EFG
1.200E-02	3.878E+00	25	A352	2 0 1 1 1	
7.717E-03	2.494E+00	25	I312	0 0 0 0 1	
1.156E-02	3.736E+00	25.5	J011	1 0 2 1 2	pH 4.7
1.370E-02	4.427E+00	30	K020	1 0 1 1 0	EFG
1.238E-02	4.000E+00	37	G010	1 0 1 1 0	EFG

2229.   C$_{11}$H$_{12}$Cl$_2$O$_3$
2,4-D Isopropyl Ester
2,4-D-isopropyl Ester
2,4-Dichlorophenoxyacetic Acid Isopropyl Ester
2,4-Dichlorophenoxyacetic Acid Iso-Propyl Ester
RN: 94-11-1   **MP** (°C):
MW: 263.12   **BP** (°C):   139

Solubility (Moles/L)	Solubility (Grams/L)	Temp (°C)	Ref (#)	Evaluation (T P E A A)	Comments
1.040E-04	2.736E-02	ns	M120	0 0 1 1 2	
1.419E-04	3.734E-02	ns	M120	0 0 1 1 2	

2230.   C$_{11}$H$_{12}$I$_3$NO$_2$
Iopanoic Acid
β-(3-Amino-2,4,6-triiodophenyl)-α-ethylpropionic Acid
Bilijodon
Cholevid
Choladine
Colepax
RN: 96-83-3   **MP** (°C):   155.2
MW: 570.94   **BP** (°C):

Solubility (Moles/L)	Solubility (Grams/L)	Temp (°C)	Ref (#)	Evaluation (T P E A A)	Comments
6.100E-04	3.483E-01	37	J016	1 0 0 0 1	pH 7.4
2.627E-05	1.500E-02	ns	H055	0 1 0 2 2	

2231.   C$_{11}$H$_{12}$NO$_4$PS$_2$
Phosmet
Phosphorodithioic Scid S-[(1,3-Dihydro-1,3-dioxo-2H-isoindol-2-yl)methyl] O,O-Dimethyl
Ester
Decemthion
Smidan
Appa
Imidan
RN: 732-11-6   **MP** (°C):
MW: 317.32   **BP** (°C):

Solubility (Moles/L)	Solubility (Grams/L)	Temp (°C)	Ref (#)	Evaluation (T P E A A)	Comments
7.690E-05	2.440E-02	20	B300	2 1 1 1 2	
7.878E-05	2.500E-02	25	M061	1 0 0 0 1	
7.878E-05	2.500E-02	25	M161	1 0 0 0 1	
7.878E-05	2.500E-02	ns	F071	0 1 2 1 1	

2232.   C$_{11}$H$_{12}$N$_2$O
Antipyrine
Antipyrin
2,3-Dimethyl-1-phenyl-3-pyrazolin-5-one
1,2-Dihydro-1,5-dimethyl-2-phenyl-3H-pyrazol-3-one
Phenazone
RN: 60-80-0   **MP** (°C):   114
MW: 188.23   **BP** (°C):   319

Solubility (Moles/L)	Solubility (Grams/L)	Temp (°C)	Ref (#)	Evaluation (T P E A A)	Comments
1.493E+00	2.811E+02	0.0	K075	1 0 0 0 2	
1.550E+00	2.918E+02	2.5	K075	1 0 0 0 2	
1.968E+00	3.705E+02	4.62	M109	2 1 1 1 0	EFG
1.472E-01	2.771E+01	5	L089	1 0 0 0 2	sic
1.613E+00	3.036E+02	6.1	K075	1 0 0 0 2	
1.777E-01	3.344E+01	10	L089	1 0 0 0 2	sic
2.084E+00	3.922E+02	11.74	M109	2 1 1 1 0	EFG
2.261E+00	4.256E+02	14.20	M109	2 1 1 1 0	EFG
1.771E+00	3.333E+02	20	D041	1 0 0 0 0	
2.205E-01	4.150E+01	20	L089	1 0 0 0 2	sic
2.472E+00	4.654E+02	20.96	M109	2 1 1 1 0	EFG
2.621E-01	4.934E+01	25	L089	1 0 0 0 2	sic
3.294E+00	6.200E+02	25	P012	1 0 1 2 0	
3.294E+00	6.200E+02	25	P016	1 0 0 1 2	
3.559E+00	6.700E+02	25	P020	2 0 1 1 2	
2.717E+00	5.114E+02	25.35	M109	2 1 1 1 0	EFG
3.020E+00	5.685E+02	29.87	M109	2 1 1 1 0	EFG
2.621E-01	4.934E+01	30	L089	1 0 0 0 2	sic
2.983E-01	5.616E+01	35	L089	1 0 0 0 2	sic

3.968E+00	7.468E+02	39.34	M109	2 1 1 1 0	*EFG*
3.359E-01	6.323E+01	40	L089	1 0 0 0 2	*sic*
5.637E-01	1.061E+02	50	L089	1 0 0 0 2	*sic*
2.656E+00	5.000E+02	rt	D021	0 0 1 1 2	

2233. $C_{11}H_{12}N_2O_2$

Tryptophan
2-Amino-3-(lH-indol-3-yl)-propanoic Acid
3-Indol-3-ylalanine
L-β-3-Indolylalanine
Trp
(S)-(-)-Tryptophan

RN: 73-22-3   **MP** (°C):
MW: 204.23   **BP** (°C):

Solubility (Moles/L)	Solubility (Grams/L)	Temp (°C)	Ref (#)	Evaluation (T P E A A)	Comments
4.015E-02	8.200E+00	0	F300	1 0 0 0 1	
6.042E-02	1.234E+01	20	B032	1 2 2 1 2	
6.395E-02	1.306E+01	22.5	P045	0 0 2 1 2	
6.551E-02	1.338E+01	25	B032	1 2 2 1 2	
5.519E-02	1.127E+01	25	D041	1 0 0 0 2	
5.337E-02	1.090E+01	25	F300	1 0 0 0 2	
6.665E-02	1.361E+01	25	G092	2 1 1 1 1	
6.665E-02	1.361E+01	25	G315	1 0 2 2 2	
5.519E-02	1.127E+01	25	H070	1 0 0 0 2	
6.267E-02	1.280E+01	25.1	N024	2 0 2 2 2	
6.757E-02	1.380E+01	25.1	N025	2 0 2 2 2	
6.757E-02	1.380E+01	25.1	N026	2 0 2 2 2	
6.665E-02	1.361E+01	25.1	N027	1 1 2 2 2	
1.787E-01	3.650E+01	27	D036	2 1 2 2 2	
5.386E-02	1.100E+01	28	L081	2 1 2 2 2	
7.056E-02	1.441E+01	29.80	B032	1 2 2 1 2	
8.100E-02	1.654E+01	30	N009	1 0 2 2 2	
9.480E-02	1.936E+01	40	N009	1 0 2 2 2	
8.226E-02	1.680E+01	50	F300	1 0 0 0 2	
1.122E-01	2.291E+01	50	N009	1 0 2 2 2	
1.200E-01	2.450E+01	70	F300	1 0 0 0 2	
1.334E-01	2.724E+01	75	D041	1 0 0 0 2	
2.448E-01	5.000E+01	100	F300	1 0 0 0 1	

2234.   C₁₁H₁₂N₂O₂

DL-Tryptophan

1H-Indole-3-alanine

DL-α-Amino-3-indolepropionic Acid

RN: 54-12-6   **MP** (°C):   289

MW: 204.23   **BP** (°C):

Solubility (Moles/L)	Solubility (Grams/L)	Temp (°C)	Ref (#)	Evaluation (T P E A A)	Comments
1.020E-02	2.083E+00	20	N006	1 0 2 2 2	
1.140E-02	2.328E+00	25	N006	1 0 2 2 2	
1.221E-02	2.494E+00	30	D041	1 0 0 0 1	
1.250E-02	2.553E+00	30	N006	1 0 2 2 2	
1.200E-02	2.451E+00	30	N009	1 0 2 2 2	
1.640E-02	3.349E+00	40	N006	1 0 2 2 2	
1.570E-02	3.206E+00	40	N009	1 0 2 2 2	
2.150E-02	4.391E+00	50	N006	1 0 2 2 2	

2235.   C₁₁H₁₂N₂O₂

5-Ethyl-5-phenylhydantoin

2,4-Imidazolidinedione, 5-Ethyl-5-phenyl-

Nirvanol

5-Phenyl-5-ethylhydantoin

Normephenytoin

RN: 631-07-2   **MP** (°C):

MW: 204.23   **BP** (°C):

Solubility (Moles/L)	Solubility (Grams/L)	Temp (°C)	Ref (#)	Evaluation (T P E A A)	Comments
3.938E-03	8.044E-01	37	F183	1 0 1 1 1	intrinsic

2236.   C₁₁H₁₂N₂O₄

Acetamide, N-(2-Amino-2-oxoethyl)-2-(benzoyloxy)-

RN: 106231-53-2   **MP** (°C):   151.5

MW: 236.23   **BP** (°C):

Solubility (Moles/L)	Solubility (Grams/L)	Temp (°C)	Ref (#)	Evaluation (T P E A A)	Comments
3.175E-02	7.500E+00	22	N317	1 1 2 1 2	

2237. $C_{11}H_{12}N_4O_2S$
4-Sulfanilamido-2-methylpyrimidine
Benzenesulfonamide, 4-Amino-N-(2-methyl-4-pyrimidinyl)-

RN: 599-84-8   **MP** (°C):
MW: 264.31   **BP** (°C):

Solubility (Moles/L)	Solubility (Grams/L)	Temp (°C)	Ref (#)	Evaluation (T P E A A)	Comments
2.357E-02	6.230E+00	37	R046	1 2 1 1 2	

2238. $C_{11}H_{12}N_4O_2S$
2-Sulfanilamido-5-aminopyridine
Benzenesulfonamide, 4-Amino-N-(5-amino-2-pyridinyl)-

RN: 16840-28-1   **MP** (°C):
MW: 264.31   **BP** (°C):

Solubility (Moles/L)	Solubility (Grams/L)	Temp (°C)	Ref (#)	Evaluation (T P E A A)	Comments
1.581E-02	4.180E+00	37	R058	1 2 1 1 2	

2239. $C_{11}H_{12}N_4O_2S$
Sulfamethylpyrimidine
Ulfamerazine
Sulfamerazine

RN: 127-79-7   **MP** (°C):   234
MW: 264.31   **BP** (°C):

Solubility (Moles/L)	Solubility (Grams/L)	Temp (°C)	Ref (#)	Evaluation (T P E A A)	Comments
8.967E-04	2.370E-01	20	F073	1 2 2 2 2	
7.641E-04	2.020E-01	20	L058	1 0 1 1 2	
1.400E-03	3.700E-01	37	L091	1 0 0 0 1	pH 5.5
1.203E-03	3.180E-01	37	R045	1 2 1 1 2	
1.381E-03	3.650E-01	37	S192	1 0 1 1 2	pH 6.0
1.551E-03	4.100E-01	38	K006	1 0 0 0 1	

2240.   C₁₁H₁₂N₄O₃S

Sulfameter
Sulphamethoxydiazine
RN: 651-06-9   **MP** (°C):   213
MW: 280.31   **BP** (°C):

Solubility (Moles/L)	Solubility (Grams/L)	Temp (°C)	Ref (#)	Evaluation (T P E A A)	Comments
1.677E-03	4.700E-01	30	M113	2 2 2 2 0	form III, EFG, 0.1N HCl
2.604E-03	7.300E-01	30	M113	2 2 2 2 0	form II, EFG, 0.1N HCl
1.891E-03	5.300E-01	30	M113	2 2 2 2 0	form I, EFG, 0.1N HCl
2.462E-03	6.900E-01	30	M113	2 2 2 2 0	EFG, 0.1N HCl, amorphous
3.211E-04	9.000E-02	37.5	C081	1 0 1 0 0	EFG, form III
6.243E-04	1.750E-01	37.5	C081	1 0 1 0 0	EFG, form II
4.281E-04	1.200E-01	37.5	C081	1 0 1 0 0	EFG, form I

2241.   C₁₁H₁₂N₄O₃S

Sulfamethoxypyridazine
Sulphamethoxypyridazine
4-Amino-N-(6-methoxy-3-pyridazinyl)-benzenesulfonamide
RN: 80-35-3   **MP** (°C):   182.5
MW: 280.31   **BP** (°C):

Solubility (Moles/L)	Solubility (Grams/L)	Temp (°C)	Ref (#)	Evaluation (T P E A A)	Comments
2.067E-03	5.795E-01	25	E314	2 0 2 2 2	intrinsic
2.569E-02	7.200E+00	37	B046	1 0 2 2 2	pH 4.5

2242.   C₁₁H₁₂N₄O₃S₂

N4-Acetyl Sulfamethizole
Acetyl Sulfamethylthiazole
RN: 39719-87-4   **MP** (°C):
MW: 312.37   **BP** (°C):

Solubility (Moles/L)	Solubility (Grams/L)	Temp (°C)	Ref (#)	Evaluation (T P E A A)	Comments
1.313E-03	4.100E-01	37	B046	1 0 2 2 1	pH 4.5

2243.   C₁₁H₁₂N₄O₃S

2-Sulfanilamido-4-methoxypyrimidine
Benzenesulfonamide, 4-Amino-N-(4-methoxy-2-pyrimidinyl)-
RN: 3213-22-7   **MP** (°C):
MW: 280.31   **BP** (°C):

Solubility (Moles/L)	Solubility (Grams/L)	Temp (°C)	Ref (#)	Evaluation (T P E A A)	Comments
6.493E-04	1.820E-01	37	R046	1 2 1 1 2	

2244. C₁₁H₁₂N₄O₃S

$C_{11}H_{12}N_4O_3S$

5-Sulfanilamido-2-methoxypyrimidine
Benzenesulfonamide, 4-Amino-N-(2-methoxy-5-pyrimidinyl)-

RN: 71119-37-4 **MP** (°C):
MW: 280.31 **BP** (°C):

Solubility (Moles/L)	Solubility (Grams/L)	Temp (°C)	Ref (#)	Evaluation (T P E A A)	Comments
3.282E-04	9.200E-02	37	R046	1 2 1 1 1	

2245. C₁₁H₁₂N₄O₅

$C_{11}H_{12}N_4O_5$

2,5-Diacetoxymethyl Allopurinol
4H-Pyrazolo[3,4-d]pyrimidin-4-one, 2,5-bis[(Acetyloxy)methyl]-2,5-dihydro-

RN: 98827-24-8 **MP** (°C): 153-154
MW: 280.24 **BP** (°C):

Solubility (Moles/L)	Solubility (Grams/L)	Temp (°C)	Ref (#)	Evaluation (T P E A A)	Comments
1.035E-02	2.900E+00	22	B322	1 0 2 2 2	

2246. C₁₁H₁₂N₆O₂S

$C_{11}H_{12}N_6O_2S$

6-Sulfapurine

RN: **MP** (°C):
MW: 292.32 **BP** (°C):

Solubility (Moles/L)	Solubility (Grams/L)	Temp (°C)	Ref (#)	Evaluation (T P E A A)	Comments
4.447E-05	1.300E-02	20	F073	1 2 2 2 1	

2247. C₁₁H₁₂O₂

$C_{11}H_{12}O_2$

Ethyl Cinnamate
Ethyl (E)-Cinnamate
Ethyl 3-Phenyl Propenoate
Ethyl Phenylacrylate

RN: 103-36-6 **MP** (°C): 6
MW: 176.22 **BP** (°C): 271

Solubility (Moles/L)	Solubility (Grams/L)	Temp (°C)	Ref (#)	Evaluation (T P E A A)	Comments
1.010E-03	1.780E-01	25	A002	1 2 1 1 2	

2248. $C_{11}H_{12}O_4$
Propionyl-r-mandelic Acid
RN: **MP** (°C): 126
MW: 208.22 **BP** (°C):

Solubility (Moles/L)	Solubility (Grams/L)	Temp (°C)	Ref (#)	Evaluation (T P E A A)	Comments
1.389E-02	2.892E+00	0	A043	1 2 1 1 1	
1.389E-02	2.892E+00	0	L035	1 2 2 1 1	
1.675E-02	3.488E+00	10	A043	1 2 1 1 1	
1.675E-02	3.488E+00	10	L035	1 2 2 1 1	
1.770E-02	3.686E+00	15	A043	1 2 1 1 1	
1.770E-02	3.686E+00	15	L035	1 2 2 1 1	
1.818E-02	3.786E+00	20	A043	1 2 1 1 1	
1.818E-02	3.786E+00	20	L035	1 2 2 1 1	
2.484E-02	5.173E+00	25	A043	1 2 1 1 1	
2.484E-02	5.173E+00	25	L035	1 2 2 1 1	
2.817E-02	5.865E+00	30	A043	1 2 1 1 1	
2.817E-02	5.865E+00	30	L035	1 2 2 1 1	
3.528E-02	7.346E+00	35	A043	1 2 1 1 1	
3.528E-02	7.346E+00	35	L035	1 2 2 1 1	
5.789E-02	1.205E+01	40	A043	1 2 1 1 2	
5.789E-02	1.205E+01	40	L035	1 2 2 1 2	
8.724E-02	1.816E+01	45	A043	1 2 1 1 2	
8.724E-02	1.816E+01	45	L035	1 2 2 1 2	
1.606E-01	3.344E+01	50	A043	1 2 1 1 2	
1.606E-01	3.344E+01	50	L035	1 2 2 1 2	

2249. $C_{11}H_{12}O_4$
3,5-Dimethoxycinnamic Acid
Predominantly Trans Isomer
RN: 16909-11-8 **MP** (°C): 174.5
MW: 208.22 **BP** (°C):

Solubility (Moles/L)	Solubility (Grams/L)	Temp (°C)	Ref (#)	Evaluation (T P E A A)	Comments
1.510E-04	3.144E-02	25	R070	1 2 2 2 2	

2250. $C_{11}H_{12}O_4S$
Benzoic Acid, 2-(Acetyloxy)-, (Methylthio)methyl Ester
RN: 76432-30-9 **MP** (°C):
MW: 240.28 **BP** (°C):

Solubility (Moles/L)	Solubility (Grams/L)	Temp (°C)	Ref (#)	Evaluation (T P E A A)	Comments
2.289E-03	5.500E-01	21	N335	1 2 1 1 2	

2251. $C_{11}H_{12}O_5S$
2-(Acetoxy)-benzoic Acid, (Methylsulfinyl)methyl Ester
RN: 76432-33-2 **MP** (°C): 80.5
MW: 256.28 **BP** (°C):

Solubility (Moles/L)	Solubility (Grams/L)	Temp (°C)	Ref (#)	Evaluation (T P E A A)	Comments
1.651E-02	4.230E+00	21	N335	1 2 1 1 2	

2252. $C_{11}H_{12}O_6S$
2-(Acetoxy)-benzoic Acid, (Methylsulfonyl)methyl Ester
RN: 76432-35-4 **MP** (°C): 150
MW: 272.28 **BP** (°C):

Solubility (Moles/L)	Solubility (Grams/L)	Temp (°C)	Ref (#)	Evaluation (T P E A A)	Comments
4.040E-04	1.100E-01	21	N335	1 2 1 1 2	

2253. $C_{11}H_{13}ClO_3$
Bexone
4-(2-Methyl-4-chlorophenoxy)butyric Acid
4-(MCPB)
MCPB
RN: 94-81-5 **MP** (°C):
MW: 228.68 **BP** (°C):

Solubility (Moles/L)	Solubility (Grams/L)	Temp (°C)	Ref (#)	Evaluation (T P E A A)	Comments
2.099E-04	4.800E-02	25	B164	1 0 1 1 1	
1.924E-04	4.400E-02	ns	L024	0 0 0 0 1	
1.924E-04	4.400E-02	ns	M061	0 0 0 0 1	
1.924E-04	4.400E-02	rt	M161	0 0 0 0 1	

2254. $C_{11}H_{13}FN_2O_4$
1-Cyclohexyloxycarbonyl-5-fluorouracil
1(2H)-Pyrimidinecarboxylic Acid, 5-Fluoro-3,4-dihydro-2,4-dioxo-, Cyclohexyl Ester
RN: 109232-74-8 **MP** (°C):
MW: 256.24 **BP** (°C):

Solubility (Moles/L)	Solubility (Grams/L)	Temp (°C)	Ref (#)	Evaluation (T P E A A)	Comments
3.590E-03	9.200E-01	22	B332	1 1 0 0 1	pH 4.0

2255.   C₁₁H₁₃F₃N₂O₃S

Mefluidide

N-(2,4-Dimethyl-5-(((trifluoromethyl)sulfonyl)amino)phenyl)acetamide

Vistar

Embark

MBR 12325

Methafluoridamid

RN: 53780-34-0   **MP** (°C):   184

MW: 310.30   **BP** (°C):

Solubility (Moles/L)	Solubility (Grams/L)	Temp (°C)	Ref (#)	Evaluation (T P E A A)	Comments
5.801E-04	1.800E-01	23	M161	1 0 0 0 2	

2256.   C₁₁H₁₃F₃N₄O₄

Dinitramine

1,3-Benzenediamine, N1,N1-Diethyl-2,6-dinitro-4-(trifluoromethyl)-

N3,N3-Diethyl-2,4-dinitro-6-(trifluoromethyl)-m-phenylenediamine

N3,N3-Diethyl-2,4-dinitro-6-(trifluoromethyl)-1,3-phenylenediamine

USB 3584

RN: 29091-05-2   **MP** (°C):   98.5

MW: 322.25   **BP** (°C):

Solubility (Moles/L)	Solubility (Grams/L)	Temp (°C)	Ref (#)	Evaluation (T P E A A)	Comments
3.414E-06	1.100E-03	25	M161	1 0 0 0 1	

2257.   C₁₁H₁₃NO

N-Ethylcinnamamide

N-Ethyl-3-phenyl-2-propenamide

RN: 23784-45-4   **MP** (°C):

MW: 175.23   **BP** (°C):

Solubility (Moles/L)	Solubility (Grams/L)	Temp (°C)	Ref (#)	Evaluation (T P E A A)	Comments
6.390E-03	1.120E+00	ns	H350	0 0 0 0 2	

2258.   C₁₁H₁₃NO

N,N-Dimethylcinnamide

Cinnamic Acid Dimethylamide

N,N-Dimethyl-3-phenyl-2-propenamide

RN: 13156-74-6   **MP** (°C):

MW: 175.23   **BP** (°C):

Solubility (Moles/L)	Solubility (Grams/L)	Temp (°C)	Ref (#)	Evaluation (T P E A A)	Comments
1.670E-02	2.926E+00	ns	H350	0 0 0 0 2	

2259. C₁₁H₁₃NO₃

Acetamide, 2-(Benzoyloxy)-N-ethyl-
2-(Benzoyloxy)-N-ethylacetamide

RN: 64649-57-6 **MP** (°C): 106
MW: 207.23 **BP** (°C):

Solubility (Moles/L)	Solubility (Grams/L)	Temp (°C)	Ref (#)	Evaluation (T P E A A)	Comments
5.791E-03	1.200E+00	22	N317	1 1 2 1 2	

2260. C₁₁H₁₃NO₃

Acetaminophen Propionate
Propionic Acid, p-Acetamidophenyl Ester

RN: 54942-42-6 **MP** (°C): 130
MW: 207.23 **BP** (°C):

Solubility (Moles/L)	Solubility (Grams/L)	Temp (°C)	Ref (#)	Evaluation (T P E A A)	Comments
1.544E-03	3.200E-01	25	B010	1 1 1 1 0	

2261. C₁₁H₁₃NO₃

Acetamide, 2-(Benzoyloxy)-N,N-dimethyl-
2-(Benzoyloxy)-N,N-dimethylacetamide

RN: 106231-54-3 **MP** (°C): 81.5
MW: 207.23 **BP** (°C):

Solubility (Moles/L)	Solubility (Grams/L)	Temp (°C)	Ref (#)	Evaluation (T P E A A)	Comments
4.246E-02	8.800E+00	22	N317	1 1 2 1 2	

2262. C₁₁H₁₃NO₄

Dioxacarb
2-(1,3-Dioxolan-2-yl)phenyl Methylcarbamate
2-(1,3-Dioxolan-2-yl)-phenyl N-methylcarbamate
Elocron
Famid

RN: 6988-21-2 **MP** (°C): 114.5
MW: 223.23 **BP** (°C):

Solubility (Moles/L)	Solubility (Grams/L)	Temp (°C)	Ref (#)	Evaluation (T P E A A)	Comments
2.688E-02	6.000E+00	20	M161	1 0 0 0 0	

2263. C$_{11}$H$_{13}$NO$_4$
N,N-Dimethyl Glycolamide Salicylate
2-Hydroxybenzoic Acid, 2-(dimethylamino)-2-oxoethyl Ester
RN: 114665-08-6   **MP** (°C):   68
MW: 223.23   **BP** (°C):

Solubility (Moles/L)	Solubility (Grams/L)	Temp (°C)	Ref (#)	Evaluation (T P E A A)	Comments
1.971E-02	4.400E+00	21	B331	1 2 2 1 0	pH 7.4
1.971E-02	4.400E+00	21	B331	1 2 2 1 1	

2264. C$_{11}$H$_{13}$NO$_4$
Bendiocarb
2,2-Dimethyl-1,3-benzodioxol-4-ol methylcarbamate
Fuam
Multimet
Garvox
RN: 22781-23-3   **MP** (°C):   129.5
MW: 223.23   **BP** (°C):

Solubility (Moles/L)	Solubility (Grams/L)	Temp (°C)	Ref (#)	Evaluation (T P E A A)	Comments
1.792E-04	4.000E-02	25	M161	1 0 0 0 1	
1.792E-04	4.000E-02	25	W310	1 0 0 0 0	

2265. C$_{11}$H$_{13}$NO$_4$
Ethyl Acetaminophen
Carbonic Acid, 4-(Acetylamino)phenyl Ethyl Ester
Acetanilide, 4'-Hydroxy-, Ethyl Carbonate (Ester)
RN: 17243-26-4   **MP** (°C):   121-122
MW: 223.23   **BP** (°C):

Solubility (Moles/L)	Solubility (Grams/L)	Temp (°C)	Ref (#)	Evaluation (T P E A A)	Comments
4.928E-03	1.100E+00	37	D029	1 0 1 1 1	

2266. C$_{11}$H$_{13}$N$_3$O

Ampyrone
4-Aminoantipyrine
Aminophenazone

RN: 83-07-8	**MP** (°C): 109
MW: 203.25	**BP** (°C):

Solubility (Moles/L)	Solubility (Grams/L)	Temp (°C)	Ref (#)	Evaluation (T P E A A)	Comments
9.053E-01	1.840E+02	5.39	M109	2 1 1 1 0	EFG
1.088E+00	2.211E+02	10.93	M109	2 1 1 1 0	EFG
1.252E+00	2.544E+02	14.20	M109	2 1 1 1 0	EFG
1.527E+00	3.103E+02	20.96	M109	2 1 1 1 0	EFG
2.076E+00	4.218E+02	25.35	M109	2 1 1 1 0	EFG
2.384E+00	4.845E+02	29.87	M109	2 1 1 1 0	EFG
2.400E-01	4.878E+01	30	I010	2 1 2 2 1	EFG, *sic*
2.862E+00	5.816E+02	39.34	M109	2 1 1 1 0	EFG

2267. C$_{11}$H$_{13}$N$_3$O$_3$S

Sulfisoxazole
4-Amino-N-(3,4-dimethyl-5-isoxazolyl)benzenesulfonamide
3,4-Dimethyl-5-sulfanilamidoisoxazole
Gantrisin
Urogan
Urisoxin

RN: 127-69-5	**MP** (°C): 194
MW: 267.31	**BP** (°C):

Solubility (Moles/L)	Solubility (Grams/L)	Temp (°C)	Ref (#)	Evaluation (T P E A A)	Comments
1.235E-03	3.300E-01	37	B046	1 0 2 2 1	pH 4.5
3.142E-04	8.400E-02	37	K022	1 0 1 1 0	intrinsic
1.092E-03	2.920E-01	37	K091	1 0 0 0 2	

2268. C$_{11}$H$_{13}$N$_3$O$_3$S

Sulfamoxole
Sulfuno
N-(4,5-Dimethyloxazol-2-yl)sulfanilamide

RN: 729-99-7	**MP** (°C): 193
MW: 267.31	**BP** (°C):

Solubility (Moles/L)	Solubility (Grams/L)	Temp (°C)	Ref (#)	Evaluation (T P E A A)	Comments
3.595E-03	9.610E-01	20	K028	2 1 2 1 2	pH 6.0, form I
3.430E-03	9.170E-01	20	K028	2 1 2 1 2	pH 3.8, form I
3.277E-03	8.760E-01	20	K028	2 1 2 1 2	pH 6.0, form II
3.165E-03	8.460E-01	20	K028	2 1 2 1 2	pH 3.8, form II

6.274E-03	1.677E+00	20	K028	2 1 2 1 2	pH 7.3, form I
5.447E-03	1.456E+00	20	K028	2 1 2 1 2	pH 7.3, form II
3.427E-03	9.162E-01	20	M042	1 0 0 0 2	pH 3.8, form I, mp 205-211 °C
3.162E-03	8.453E-01	20	M042	1 0 0 0 2	pH 3.8, form II, mp 188-195 °C

2269. $C_{11}H_{13}N_3O_3S$
N1-Methyl-N1-(5-methyl-3-isoxazolyl)sulfanilamide
N1-Methylsulfamethoxazole

RN: 51543-31-8   **MP** (°C):
MW: 267.31   **BP** (°C):

Solubility (Moles/L)	Solubility (Grams/L)	Temp (°C)	Ref (#)	Evaluation (T P E A A)	Comments
6.280E-04	1.679E-01	37	K095	2 0 0 0 2	intrinsic

2270. $C_{11}H_{13}N_5O_2$
Carbovir
9-[4α-(Hydroxymethyl)-cyclopent-2-ene-1α-yl]guanine

RN: 118353-05-2   **MP** (°C):
MW: 247.26   **BP** (°C):

Solubility (Moles/L)	Solubility (Grams/L)	Temp (°C)	Ref (#)	Evaluation (T P E A A)	Comments
5.015E-03	1.240E+00	25	A338	1 0 2 2 2	

2271. $C_{11}H_{13}N_5O_5$
Arabinosyladenine 5'-Formate
Arabinosyladenine 5'-O-Formate Ester
NSC 171240

RN: 55648-40-3   **MP** (°C):   168-170
MW: 295.26   **BP** (°C):

Solubility (Moles/L)	Solubility (Grams/L)	Temp (°C)	Ref (#)	Evaluation (T P E A A)	Comments
1.152E-01	3.400E+01	ns	R030	0 0 0 0 1	

2272. C₁₁H₁₄ClNO

Propachlor
2-Chloro-N-isopropylacetanilide
N-Isopropyl-2-chloroacetanilide
N-Isopropyl-α-chloroacetanilide

RN: 1918-16-7 **MP** (°C): 67
MW: 211.69 **BP** (°C): 110

Solubility (Moles/L)	Solubility (Grams/L)	Temp (°C)	Ref (#)	Evaluation (T P E A A)	Comments
3.307E-03	7.000E-01	20	B200	1 0 0 0 2	
3.307E-03	7.000E-01	20	M161	1 0 0 0 2	
3.304E-03	6.995E-01	ns	J008	0 0 0 0 0	
3.304E-03	6.995E-01	ns	M061	0 0 0 0 0	
2.362E-03	5.000E-01	ns	M110	0 0 0 0 0	EFG

2273. C₁₁H₁₄N₂O

Cytisine
Cytisin

RN: 485-35-8 **MP** (°C): 155
MW: 190.25 **BP** (°C):

Solubility (Moles/L)	Solubility (Grams/L)	Temp (°C)	Ref (#)	Evaluation (T P E A A)	Comments
2.308E+00	4.390E+02	16	F300	1 0 0 0 2	

2274. C₁₁H₁₄N₂O₃S

Sulfadicramide
2-Butenamide, N-[(4-Aminophenyl)sulfonyl]-3-methyl-
N-Sulfanilyl-β,β-dimethylacrylamide
Sulfirgamid
Irgamide
Sulfirgamide

RN: 115-68-4 **MP** (°C): 184.5
MW: 254.31 **BP** (°C):

Solubility (Moles/L)	Solubility (Grams/L)	Temp (°C)	Ref (#)	Evaluation (T P E A A)	Comments
1.026E-03	2.610E-01	20	F073	1 2 2 2 2	

2275. C₁₁H₁₄N₄O₂S₂

4-Amino-N-(5-isopropyl-1,3,4-thiadiazol-2-yl)benzenesulfonamide
N1-(5-Isopropyl-1,3,4-thiadiazol-2-yl)sulfanilamide
Sulfaisopropylthiadiazole
Glyprothiazole
PASIT
RP 2254

RN: 80-34-2   **MP** (°C):
MW: 298.39   **BP** (°C):

Solubility (Moles/L)	Solubility (Grams/L)	Temp (°C)	Ref (#)	Evaluation (T P E A A)	Comments
7.330E-04	2.187E-01	37	A046	2 0 1 1 2	

2276. C₁₁H₁₄N₄O₂S₂

4-Amino-N-(5-propyl-1,3,4-thiadiazol-2-yl)benzenesulfonamide
N1-(5-Propyl-1,3,4-thiadiazol-2-yl)sulfanilamide

RN: 71119-32-9   **MP** (°C):
MW: 298.39   **BP** (°C):

Solubility (Moles/L)	Solubility (Grams/L)	Temp (°C)	Ref (#)	Evaluation (T P E A A)	Comments
8.980E-04	2.680E-01	37	A046	2 0 1 1 2	

2277. C₁₁H₁₄N₄O₃

1-Pivaloyloxymethyl Allopurinol
Propanoic Acid, 2,2-Dimethyl-, (4,5-Dihydro-4-oxo-2H-pyrazolo[3,4-d]pyrimidin-1-yl)-methyl Ester

RN: 98827-18-0   **MP** (°C):   185-187
MW: 250.26   **BP** (°C):

Solubility (Moles/L)	Solubility (Grams/L)	Temp (°C)	Ref (#)	Evaluation (T P E A A)	Comments
2.078E-03	5.200E-01	22	B322	1 0 2 2 2	

2278. C₁₁H₁₄N₄O₃

2-Pivaloyloxymethyl Allopurinol
Propanoic Acid, 2,2-Dimethyl-, (4,5-Dihydro-4-oxo-2H-pyrazolo[3,4-d]pyrimidin-2-yl)-methyl Ester

RN: 98827-15-7   **MP** (°C):   180-181
MW: 250.26   **BP** (°C):

Solubility (Moles/L)	Solubility (Grams/L)	Temp (°C)	Ref (#)	Evaluation (T P E A A)	Comments
6.793E-03	1.700E+00	22	B322	1 0 2 2 2	

2279.   C$_{11}$H$_{14}$N$_4$O$_5$
6-Methoxypurine Arabinoside
9H-Purine, 9-β-D-Arabinofuranosyl-6-methoxy-
RN: 91969-06-1   **MP** (°C):
MW: 282.26   **BP** (°C):

Solubility (Moles/L)	Solubility (Grams/L)	Temp (°C)	Ref (#)	Evaluation (T P E A A)	Comments
4.980E-02	1.406E+01	37	C348	1 2 2 2 2	pH 7.00

2280.   C$_{11}$H$_{14}$O
o-2-Pentenylphenol
Phenol, 2-(2-Pentenyl)-
RN: 62536-86-1   **MP** (°C):
MW: 162.23   **BP** (°C):

Solubility (Moles/L)	Solubility (Grams/L)	Temp (°C)	Ref (#)	Evaluation (T P E A A)	Comments
2.054E-03	3.332E-01	25	L021	1 0 0 0 0	

2281.   C$_{11}$H$_{14}$O$_2$
Ethyl Hydrocinnamate
Ethyl 3-Phenylpropionate
Benzenepropanoic Acid, Ethyl Ester
RN: 2021-28-5   **MP** (°C):
MW: 178.23   **BP** (°C):   122

Solubility (Moles/L)	Solubility (Grams/L)	Temp (°C)	Ref (#)	Evaluation (T P E A A)	Comments
1.234E-03	2.200E-01	25	A002	1 2 1 1 1	

2282.   C$_{11}$H$_{14}$O$_2$
δ-Phenylvaleric Acid
Benzenepentanoic Acid
5-Phenylvaleric Acid
RN: 2270-20-4   **MP** (°C):   59
MW: 178.23   **BP** (°C):

Solubility (Moles/L)	Solubility (Grams/L)	Temp (°C)	Ref (#)	Evaluation (T P E A A)	Comments
9.969E-03	1.777E+00	30	D033	2 2 1 2 2	
1.159E-02	2.066E+00	40	D033	2 2 1 2 2	

2283.   C$_{11}$H$_{14}$O$_3$
Butylparaben
Bu-paraben
Butyl 4-Hydroxybenzoate
RN: 94-26-8   **MP** (°C):   68.5
MW: 194.23   **BP** (°C):

Solubility (Moles/L)	Solubility (Grams/L)	Temp (°C)	Ref (#)	Evaluation (T P E A A)	Comments
7.040E-04	1.367E-01	15	B355	1 1 1 1 2	
8.350E-04	1.622E-01	20	B355	1 1 1 1 2	
1.065E-03	2.069E-01	20	C006	1 2 1 1 2	
1.277E-03	2.481E-01	25	A059	1 0 1 1 1	
1.050E-03	2.039E-01	25	B355	1 1 1 1 2	
8.751E-04	1.700E-01	25	D081	1 2 2 1 2	
1.130E-03	2.195E-01	25	D339	1 0 1 1 2	
5.623E-04	1.092E-01	25	F322	2 0 1 1 0	EFG
1.030E-03	2.000E-01	25	O027	1 0 1 0 0	
7.465E-04	1.450E-01	25	P013	2 0 2 1 2	
1.200E-03	2.331E-01	27	B129	2 2 2 2 1	
1.200E-03	2.331E-01	27	G078	2 1 0 1 0	EFG
1.777E-03	3.452E-01	30	A059	1 0 1 1 1	
2.221E-03	4.314E-01	35	A059	1 0 1 1 1	
2.064E-03	4.009E-01	39.3	G302	2 2 2 2 0	EFG
2.610E-03	5.069E-01	40	A059	1 0 1 1 1	
1.100E-03	2.137E-01	ns	G067	2 0 1 1 1	

2284.   C$_{11}$H$_{14}$O$_3$
n-Butyl Salicylate
2-Hydroxy-benzoic Acid, Butyl Ester
Salicylic Acid n-Butyl Ester
Butyl Salicylate
Benzoic Acid, 2-Hydroxy-, Butyl Ester
RN: 2052-14-4   **MP** (°C):
MW: 194.23   **BP** (°C):

Solubility (Moles/L)	Solubility (Grams/L)	Temp (°C)	Ref (#)	Evaluation (T P E A A)	Comments
1.442E-02	2.800E+00	37	D009	1 2 1 1 1	0.1N HCl

2285. C$_{11}$H$_{14}$O$_4$
Dimethyl Carbate
Dimelone
RN: 5826-73-3   **MP** (°C):   38
MW: 210.23   **BP** (°C):

Solubility (Moles/L)	Solubility (Grams/L)	Temp (°C)	Ref (#)	Evaluation (T P E A A)	Comments
6.197E-02	1.303E+01	35	M061	1 0 0 0 2	

2286. C$_{11}$H$_{15}$BrClO$_3$PS
Profenofos
O-(4-Bromo-2-chlorophenyl)-O-ethyl-S-propyl phosphorothioate
Selecron
Curacron
Polycron
RN: 41198-08-7   **MP** (°C):
MW: 373.64   **BP** (°C):   110

Solubility (Moles/L)	Solubility (Grams/L)	Temp (°C)	Ref (#)	Evaluation (T P E A A)	Comments
5.353E-05	2.000E-02	20	E048	1 2 1 1 1	
5.353E-05	2.000E-02	20	M161	1 0 0 0 1	

2287. C$_{11}$H$_{15}$BrN$_2$O
Butallylonal
5-(2-Bromoallyl)-5-sec-butylbarbituric Acid
Dial
RN: 1142-70-7   **MP** (°C):   131.5
MW: 271.16   **BP** (°C):

Solubility (Moles/L)	Solubility (Grams/L)	Temp (°C)	Ref (#)	Evaluation (T P E A A)	Comments
2.522E-03	6.840E-01	ns	T003	0 0 0 0 2	

2288. C$_{11}$H$_{15}$FN$_2$O$_4$
1-Hexyloxycarbonyl-5-fluorouracil
1(2H)-Pyrimidinecarboxylic Acid, 5-Fluoro-3,4-dihydro-2,4-dioxo-, Hexyl Ester
RN: 66999-99-3   **MP** (°C):   68
MW: 258.25   **BP** (°C):

Solubility (Moles/L)	Solubility (Grams/L)	Temp (°C)	Ref (#)	Evaluation (T P E A A)	Comments
5.808E-03	1.500E+00	22	B332	1 1 0 0 1	pH 4.0

2289.   C$_{11}$H$_{15}$NO$_2$
Butamben
4-Aminobenzoic Acid Butyl Ester
Butyl p-Aminobenzoate
RN:　94-25-7　　　**MP** (°C):　　58.0
MW:　193.25　　　**BP** (°C):

Solubility (Moles/L)	Solubility (Grams/L)	Temp (°C)	Ref (#)	Evaluation (T P E A A)	Comments
1.030E-03	1.990E-01	25	H008	1 2 2 2 2	
8.332E-04	1.610E-01	25	P303	2 0 2 2 2	
1.200E-03	2.319E-01	30	J018	1 2 0 1 1	0.05N NaOH
1.200E-03	2.319E-01	30	J022	1 0 2 1 1	
1.200E-03	2.319E-01	30	N045	1 2 2 2 0	EFG
1.389E-03	2.683E-01	33	P303	2 0 2 2 2	
1.720E-03	3.324E-01	37	F006	1 1 2 2 2	
1.700E-03	3.285E-01	37	J026	2 2 2 1 1	
2.221E-03	4.293E-01	40	P303	2 0 2 2 2	
7.140E-04	1.380E-01	ns	M066	0 0 0 0 2	
7.140E-04	1.380E-01	rt	B016	0 0 1 1 2	pH 7.4

2290.   C$_{11}$H$_{15}$NO$_2$
m-Isopropylphenyl N-Methylcarbamate
3-Isopropylphenyl N-Methylcarbamate
UC-10854
RN:　64-00-6　　　**MP** (°C):　　53
MW:　193.25　　　**BP** (°C):

Solubility (Moles/L)	Solubility (Grams/L)	Temp (°C)	Ref (#)	Evaluation (T P E A A)	Comments
4.398E-04	8.500E-02	30	D089	2 2 0 0 0	
4.398E-04	8.500E-02	30	M061	1 0 0 0 1	

2291.   C$_{11}$H$_{15}$NO$_2$S
Ethiofencarb
2-((Ethylthio)methyl)phenyl Methylcarbamate
Ethylmercaptomethylphenyl-N-methylcarbamate
Ethiophencarp
Croneton
HOX 1901
RN:　29973-13-5　　　**MP** (°C):　　<25
MW:　225.31　　　**BP** (°C):

Solubility (Moles/L)	Solubility (Grams/L)	Temp (°C)	Ref (#)	Evaluation (T P E A A)	Comments
8.078E-03	1.820E+00	20	M161	1 0 0 0 2	

2292. C₁₁H₁₅NO₃
α, 3-o-Isopropylidene Pyriridoxine
RN: **MP** (°C):
MW: 209.25   **BP** (°C):

Solubility (Moles/L)	Solubility (Grams/L)	Temp (°C)	Ref (#)	Evaluation (T P E A A)	Comments
1.196E-02	2.503E+00	37	M067	2 0 1 1 2	

2293. C₁₁H₁₅NO₃
Propoxur
o-Isopropoxyphenyl Methylcarbamate
Baygon
Blattanex
Blattosep
Suncide
RN: 114-26-1   **MP** (°C):   91
MW: 209.25   **BP** (°C):

Solubility (Moles/L)	Solubility (Grams/L)	Temp (°C)	Ref (#)	Evaluation (T P E A A)	Comments
8.301E-03	1.737E+00	10	B324	2 2 2 2 2	
8.316E-03	1.740E+00	10	B324	2 2 2 2 2	
8.885E-03	1.859E+00	20	B300	2 2 1 1 2	
9.244E-03	1.934E+00	20	B324	2 2 2 2 2	
9.206E-03	1.926E+00	20	B324	2 2 2 2 2	
9.558E-03	2.000E+00	20	M161	1 0 0 0 0	
1.166E-02	2.440E+00	30	B324	2 2 2 2 2	
1.163E-02	2.434E+00	30	B324	2 2 2 2 2	
4.732E-02	9.901E+00	ns	M061	0 0 0 0 0	approximate
4.301E-04	9.000E-02	ns	M110	0 0 0 0 0	EFG

2294. C₁₁H₁₅NO₄
n-Ethyl-6-hydroxynorbornane-2-carboxamide-3,5-lactone
RN: **MP** (°C):
MW: 225.25   **BP** (°C):

Solubility (Moles/L)	Solubility (Grams/L)	Temp (°C)	Ref (#)	Evaluation (T P E A A)	Comments
2.908E-01	6.550E+01	20	K050	1 1 1 1 2	

2295.   C$_{11}$H$_{15}$N$_3$O$_2$

Formetanate

Methylcarbamic Acid, Ester with N'-(m-hydroxyphenyl)-N,N-dimethylformamidine

RN: 22259-30-9   **MP** (°C):   102.5

MW: 221.26   **BP** (°C):

Solubility (Moles/L)	Solubility (Grams/L)	Temp (°C)	Ref (#)	Evaluation (T P E A A)	Comments
4.520E-03	1.000E+00	rt	M161	0 0 0 0 0	

2296.   C$_{11}$H$_{15}$N$_3$O$_3$

Orotic Acid Cyclohexylamide

Orotamide, N-Cyclohexyl-

RN: 4558-58-1   **MP** (°C):   284-285

MW: 237.26   **BP** (°C):

Solubility (Moles/L)	Solubility (Grams/L)	Temp (°C)	Ref (#)	Evaluation (T P E A A)	Comments
7.500E-02	1.779E+01	-4	N018	2 2 1 2 2	
1.100E-01	2.610E+01	16	N018	2 2 1 2 2	
1.330E-01	3.156E+01	25	N018	2 2 1 2 2	

2297.   C$_{11}$H$_{15}$N$_3$O$_5$

Triglycidylurazol

Anaxirone

RN: 77658-97-0   **MP** (°C):   91

MW: 269.26   **BP** (°C):

Solubility (Moles/L)	Solubility (Grams/L)	Temp (°C)	Ref (#)	Evaluation (T P E A A)	Comments
7.426E-04	2.000E-01	ns	D319	0 0 0 0 0	

2298.   C$_{11}$H$_{15}$O$_3$P

Diethyl Benzoyl Phosphonate

Methylene, (Diethoxyphosphinyl)phenyl-

RN: 105394-75-0   **MP** (°C):

MW: 226.21   **BP** (°C):

Solubility (Moles/L)	Solubility (Grams/L)	Temp (°C)	Ref (#)	Evaluation (T P E A A)	Comments
<8.84E-04	<2.00E-01	25	B070	1 2 0 1 0	

2299. C$_{11}$H$_{16}$
tert-Amylbenzene
t-Amylbenzene
RN: 2049-95-8   **MP** (°C):   -57.8
MW: 148.25   **BP** (°C):

Solubility (Moles/L)	Solubility (Grams/L)	Temp (°C)	Ref (#)	Evaluation (T P E A A)	Comments
7.083E-05	1.050E-02	25	A002	1 2 1 1 2	

2300. C$_{11}$H$_{16}$
Pentamethylbenzene
1,2,3,4,5-Pentamethyl Benzene
RN: 700-12-9   **MP** (°C):   50.8
MW: 148.25   **BP** (°C):   231.0

Solubility (Moles/L)	Solubility (Grams/L)	Temp (°C)	Ref (#)	Evaluation (T P E A A)	Comments
1.047E-04	1.552E-02	ns	D001	0 0 0 0 2	

2301. C$_{11}$H$_{16}$
Amylbenzene
n-Pentylbenzene
Pentylbenzene
n-Amylbenzene
n-Pentylbenzene1-Phenylpentane
RN: 538-68-1   **MP** (°C):   -75
MW: 148.25   **BP** (°C):   205.4

Solubility (Moles/L)	Solubility (Grams/L)	Temp (°C)	Ref (#)	Evaluation (T P E A A)	Comments
2.348E-05	3.481E-03	7	O312	2 2 0 2 2	
2.144E-05	3.178E-03	10	O312	2 2 0 2 2	
2.323E-05	3.444E-03	12.5	O312	2 2 0 2 2	
2.153E-05	3.192E-03	15	O312	2 2 0 2 2	
2.311E-05	3.426E-03	17.5	O312	2 2 0 2 2	
2.142E-05	3.176E-03	20	O312	2 2 0 2 2	
2.590E-05	3.840E-03	25	M342	1 0 1 1 2	
2.276E-05	3.374E-03	25	O312	2 2 0 2 2	
2.433E-05	3.607E-03	30	O312	2 2 0 2 2	
2.642E-05	3.917E-03	35	O312	2 2 0 2 2	
2.868E-05	4.252E-03	40	O312	2 2 0 2 2	
3.163E-05	4.689E-03	45	O312	2 2 0 2 2	
6.000E-03	8.895E-01	ns	H307	1 0 1 1 2	

2302. C₁₁H₁₆ClO₂PS₃

Carbophenothion
O,O-Diethyl S-(4-Chlorophenylthiomethyl) Dithiophosphate
Trithion
Garrathion
Nephocarp
Lethox

RN: 786-19-6   **MP** (°C): <25
MW: 342.87   **BP** (°C):

Solubility (Moles/L)	Solubility (Grams/L)	Temp (°C)	Ref (#)	Evaluation (T P E A A)	Comments
1.779E-06	6.100E-04	10	B324	2 2 2 2 2	
1.779E-06	6.100E-04	10	B324	2 2 2 2 2	
1.838E-06	6.302E-04	20	B300	2 1 1 1 2	
1.837E-06	6.300E-04	20	B324	2 2 2 2 2	
1.838E-06	6.302E-04	20	B324	2 2 2 2 2	
2.129E-06	7.300E-04	30	B324	2 2 2 2 2	
2.129E-06	7.300E-04	30	B324	2 2 2 2 2	
<1.17E-04	<4.00E-02	ns	M161	0 0 0 0 0	

2303. C₁₁H₁₆N₂O₂

Aminocarb
Phenol, 4-(Dimethylamino)-3-methyl, Methylcarbamate (Ester)
Carbamic Acid, Methyl-, 4-(Dimethylamino)-m-tolyl Ester

RN: 2032-59-9   **MP** (°C): 93
MW: 208.26   **BP** (°C):

Solubility (Moles/L)	Solubility (Grams/L)	Temp (°C)	Ref (#)	Evaluation (T P E A A)	Comments
4.187E-03	8.720E-01	10	B324	2 2 2 2 2	
4.183E-03	8.712E-01	10	B324	2 2 2 2 2	
4.394E-03	9.151E-01	20	B300	2 2 1 1 2	
4.389E-03	9.142E-01	20	B324	2 2 2 2 2	
4.394E-03	9.151E-01	20	B324	2 2 2 2 2	
4.393E-03	9.150E-01	20	G300	1 0 0 0 2	
6.521E-03	1.358E+00	30	B324	2 2 2 2 2	
6.540E-03	1.362E+00	30	B324	2 2 2 2 2	

2304. C₁₁H₁₆N₂O₂

4-Aminobenzoic Acid-2-(ethyl-amino)ethyl Ester
2-(Ethylamino)ethyl 4-Aminobenzoate

RN: **MP** (°C):
MW: 208.26   **BP** (°C):

Solubility (Moles/L)	Solubility (Grams/L)	Temp (°C)	Ref (#)	Evaluation (T P E A A)	Comments
2.700E-02	5.623E+00	ns	M066	0 0 0 0 1	

2305.   C₁₁H₁₆N₂O₃

$C_{11}H_{16}N_2O_3$

Butalbital
Itobarbital
5-Allyl-5-isobutylbarbituric Acid
Fioricet
Phrenilin
Medigesic

RN: 77-26-9   **MP** (°C):   138
MW: 224.26   **BP** (°C):

Solubility (Moles/L)	Solubility (Grams/L)	Temp (°C)	Ref (#)	Evaluation (T P E A A)	Comments
7.590E-03	1.702E+00	25	V033	2 0 1 1 2	
7.600E-03	1.704E+00	25.00	T303	1 0 0 0 1	
1.030E-02	2.310E+00	35.00	T303	1 0 0 0 2	
1.410E-02	3.162E+00	45.00	T303	1 0 0 0 2	

2306.   C₁₁H₁₆N₂O₃

$C_{11}H_{16}N_2O_3$

Vinbarbital
5-Ethyl-5-(1-methyl-1-butenyl)barbituric Acid

RN: 125-42-8   **MP** (°C):   161
MW: 224.26   **BP** (°C):

Solubility (Moles/L)	Solubility (Grams/L)	Temp (°C)	Ref (#)	Evaluation (T P E A A)	Comments
3.121E-03	7.000E-01	25	B011	2 0 0 1 0	
3.164E-03	7.097E-01	25	B065	1 1 1 1 1	
4.870E-03	1.092E+00	25	V033	2 0 1 1 2	
4.900E-03	1.099E+00	25.00	T303	1 0 0 0 1	
7.000E-03	1.570E+00	35.00	T303	1 0 0 0 1	
8.000E-03	1.794E+00	45.00	T303	1 0 0 0 1	

2307.   C₁₁H₁₆N₂O₃

$C_{11}H_{16}N_2O_3$

5-Allyl-5-butylbarbituric Acid
n-Butylallylbarbitone
n-Butylallylbarbituric Acid
Allylbutylbarbituric Acid
Idobutal

RN: 3146-66-5   **MP** (°C):
MW: 224.26   **BP** (°C):

Solubility (Moles/L)	Solubility (Grams/L)	Temp (°C)	Ref (#)	Evaluation (T P E A A)	Comments
6.723E-03	1.508E+00	20	J030	1 2 2 2 2	
8.945E-03	2.006E+00	37	J030	1 2 2 2 2	

2308.   C₁₁H₁₆N₂O₃
2,4-Diazaspiro[5.7]tridecane-1,3,5-trione
RN: 143288-62-4   **MP** (°C):
MW: 224.26   **BP** (°C):

Solubility (Moles/L)	Solubility (Grams/L)	Temp (°C)	Ref (#)	Evaluation (T P E A A)	Comments
1.042E-03	2.337E-01	25	P350	2 1 1 1 2	intrinsic

2309.   C₁₁H₁₆N₂O₃
Barbituric Acid, 5-Ethyl-5-(3-methyl-2-butenyl)
5-Ethyl-5-(3'-methylbut-2'-enyl)barbituric Acid
2,4,6(1H,3H,5H)-Pyrimidinetrione, 5-Ethyl-5-(3-methyl-2-butenyl)-
2,4,6(1H,3H,5H)-Pyrimidinetrione, 5-Ethyl-5-(3-methyl-2-butenyl)
RN: 21149-88-2   **MP** (°C):
MW: 224.26   **BP** (°C):

Solubility (Moles/L)	Solubility (Grams/L)	Temp (°C)	Ref (#)	Evaluation (T P E A A)	Comments
5.583E-03	1.252E+00	25	P350	2 1 1 1 2	intrinsic

2310.   C₁₁H₁₆N₂O₃
Talbutal
Allyl-sec-butyl-barbituric Acid
5-Allyl-5-sec-butylbarbituric Acid
RN: 115-44-6   **MP** (°C):   109
MW: 224.26   **BP** (°C):

Solubility (Moles/L)	Solubility (Grams/L)	Temp (°C)	Ref (#)	Evaluation (T P E A A)	Comments
9.632E-03	2.160E+00	ns	T003	0 0 0 0 2	

2311.   C₁₁H₁₆N₂O₃S
Phenbutamide
N-(Phenylsulfonyl)-N'-butylurea
N-Benzenesulfonyl-N'-n-butylurea
RN: 3149-00-6   **MP** (°C):   131
MW: 256.33   **BP** (°C):

Solubility (Moles/L)	Solubility (Grams/L)	Temp (°C)	Ref (#)	Evaluation (T P E A A)	Comments
8.995E-04	2.306E-01	37	A028	1 0 2 1 2	intrinsic
9.000E-04	2.307E-01	37	A046	2 0 1 1 2	

2312. C$_{11}$H$_{16}$N$_2$O$_4$
Methyl-2-ethyl-2-allylmalonurate
Methyl 2-Ethyl-2-allylmalonurate
RN: 73632-83-4   **MP** (°C):   78.5
MW: 240.26   **BP** (°C):

Solubility (Moles/L)	Solubility (Grams/L)	Temp (°C)	Ref (#)	Evaluation (T P E A A)	Comments
1.200E-02	2.883E+00	23	B152	1 2 1 1 1	pH 3.5

2313. C$_{11}$H$_{16}$N$_2$O$_5$
Methoxycarbonylmethyl-2,2-diethylmalonurate
Methoxycarbonylmethyl 2,2-Diethylmalonurate
RN: **MP** (°C):   89
MW: 256.26   **BP** (°C):

Solubility (Moles/L)	Solubility (Grams/L)	Temp (°C)	Ref (#)	Evaluation (T P E A A)	Comments
9.700E-03	2.486E+00	23	B152	1 2 1 1 1	pH 3.5

2314. C$_{11}$H$_{16}$N$_4$O$_2$
1-Butyl Theobromine
1-Butyl-3,7-dimethylxanthine
1-n-Butyl-3,7-dimethylxanthine
RN: 1143-30-2   **MP** (°C):   108
MW: 236.28   **BP** (°C):

Solubility (Moles/L)	Solubility (Grams/L)	Temp (°C)	Ref (#)	Evaluation (T P E A A)	Comments
2.370E-02	5.600E+00	30	B042	1 2 1 1 2	

2315. C$_{11}$H$_{16}$N$_4$O$_4$
2,6-Piperazinedione, 4,4'-(1-Methyl-1,2-ethanediyl)bis-
1,2-Di(4-piperazine-2,6-dione)propane
2,6-Piperazinedione, 4,4'-(1-Methyl-1,2-ethanediyl)bis-, (±)-, Polymer with 1,3-dibromopropane
Propane, 1,3-Dibromo-, Polymer with (±)-4,4'-(1-methyl-1,2-ethanediyl)bis[2,6-piperazinedione]
RN: 21416-67-1   **MP** (°C):   192 dec
MW: 268.27   **BP** (°C):   233 dec

Solubility (Moles/L)	Solubility (Grams/L)	Temp (°C)	Ref (#)	Evaluation (T P E A A)	Comments
1.118E-02	3.000E+00	25	P326	0 1 0 2 0	
~5.59E-02	~1.50E+01	25	R017	1 2 2 2 1	enantimer (R)
~1.12E-02	~3.00E+00	25	R017	1 2 2 2 0	

2316.   C$_{11}$H$_{16}$O
p-tert-Pentylphenol
p-(α,α-Dimethylpropyl)phenol
p-(1,1-Dimethylpropyl)phenol
1-Hydroxy-4(2-methyl-2-butyl)benzene
PTAP
RN: 80-46-6   **MP** (°C):
MW: 164.25   **·BP** (°C):

Solubility (Moles/L)	Solubility (Grams/L)	Temp (°C)	Ref (#)	Evaluation (T P E A A)	Comments
7.162E-04	1.176E-01	25	L021	1 0 0 0 0	
1.023E-03	1.680E-01	25	M127	1 0 0 0 2	

2317.   C$_{11}$H$_{16}$O
2-Methyl-5-t-butylphenol
5-tert-Butyl-2-methylphenol
5-tert-Butyl-o-cresol
o-Cresol, 5-tert-Butyl-
RN: 5781-02-2   **MP** (°C):
MW: 164.25   **BP** (°C):

Solubility (Moles/L)	Solubility (Grams/L)	Temp (°C)	Ref (#)	Evaluation (T P E A A)	Comments
2.533E-03	4.160E-01	25	M127	1 0 0 0 2	

2318.   C$_{11}$H$_{16}$O
o-n-Amylphenol
2-n-Amylphenol
RN: 87-26-3   **MP** (°C):
MW: 164.25   **BP** (°C):

Solubility (Moles/L)	Solubility (Grams/L)	Temp (°C)	Ref (#)	Evaluation (T P E A A)	Comments
9.365E-04	1.538E-01	25	L022	1 0 0 0 0	

2319.   C$_{11}$H$_{16}$O
p-n-Amylphenol
4-n-Pentylphenol
RN: 14938-35-3   **MP** (°C):
MW: 164.25   **BP** (°C):

Solubility (Moles/L)	Solubility (Grams/L)	Temp (°C)	Ref (#)	Evaluation (T P E A A)	Comments
6.088E-04	9.999E-02	25	L022	1 0 0 0 0	

2320. C₁₁H₁₆O

o-2-Hexenylphenol
2-2-Hexenylphenol
RN: 75121-79-8   **MP** (°C):
MW: 164.25   **BP** (°C):

Solubility (Moles/L)	Solubility (Grams/L)	Temp (°C)	Ref (#)	Evaluation (T P E A A)	Comments
7.162E-04	1.176E-01	25	L021	1 0 0 0 0	

2321. C₁₁H₁₆O

p-sec-Amylphenol
4-sec-Amylphenol
RN: 25735-67-5   **MP** (°C):
MW: 164.25   **BP** (°C):

Solubility (Moles/L)	Solubility (Grams/L)	Temp (°C)	Ref (#)	Evaluation (T P E A A)	Comments
6.408E-04	1.053E-01	25	L021	1 0 0 0 0	

2322. C₁₁H₁₆O₂

4-n-Amyl Resorcinol
4-n-Amyl-resorcin
RN: 533-24-4   **MP** (°C):
MW: 180.25   **BP** (°C):

Solubility (Moles/L)	Solubility (Grams/L)	Temp (°C)	Ref (#)	Evaluation (T P E A A)	Comments
1.110E-02	2.000E+00	20	F300	1 0 0 0 0	

2323. C₁₁H₁₆O₂

3-Pentoxyphenol
m-Pentoxy Phenol
Phenol, 3-Pentoxy-
RN: 18979-73-2   **MP** (°C):
MW: 180.25   **BP** (°C):

Solubility (Moles/L)	Solubility (Grams/L)	Temp (°C)	Ref (#)	Evaluation (T P E A A)	Comments
2.130E-03	3.839E-01	30	B315	1 0 1 1 2	

2324. C₁₁H₁₇NO₃

Dimetan
5,5-Dimethyldihydroresorcinyl N,N-Dimethylcarbamate
RN: 122-15-6   **MP** (°C): 45.5
MW: 211.26   **BP** (°C):

Solubility (Moles/L)	Solubility (Grams/L)	Temp (°C)	Ref (#)	Evaluation (T P E A A)	Comments
1.379E-01	2.913E+01	ns	M061	0 0 0 0 0	approximate

2325.   C$_{11}$H$_{17}$N$_3$O$_3$
Orotic Acid Triethylamide
RN: **MP** (°C):   200-202
MW: 239.28   **BP** (°C):

Solubility (Moles/L)	Solubility (Grams/L)	Temp (°C)	Ref (#)	Evaluation (T P E A A)	Comments
2.261E+00	5.410E+02	25	N018	2 2 1 2 2	

2326.   C$_{11}$H$_{17}$N$_3$O$_3$S
Carbutamide
4-Amino-N-[(butylamino)carbonyl]-benzenesulfonamide
1-Butyl-3-sulfanilyl Urea
RN: 339-43-5   **MP** (°C):   144.5
MW: 271.34   **BP** (°C):

Solubility (Moles/L)	Solubility (Grams/L)	Temp (°C)	Ref (#)	Evaluation (T P E A A)	Comments
1.972E-03	5.352E-01	37	A028	1 0 2 1 2	intrinsic
1.950E-03	5.291E-01	37	A046	2 0 1 1 2	
6.634E-03	1.800E+00	37	C054	2 0 2 1 2	0.1N HCl

2327.   C$_{11}$H$_{17}$N$_3$O$_6$
Orotic Acid Triethanolamide
RN: **MP** (°C):   104-108
MW: 287.27   **BP** (°C):

Solubility (Moles/L)	Solubility (Grams/L)	Temp (°C)	Ref (#)	Evaluation (T P E A A)	Comments
1.315E+00	3.778E+02	-4	N018	2 2 1 2 2	
1.882E+00	5.407E+02	16	N018	2 2 1 2 2	
2.187E+00	6.283E+02	25	N018	2 2 1 2 2	

2328.   C$_{11}$H$_{17}$O$_3$PS
Kitazin
O,O-Diethyl S-Benzyl Thiophosphate
IBP
S-Benzyl O,O-di-isopropyl Phosphorothioate
RN: 13286-32-3   **MP** (°C):
MW: 260.29   **BP** (°C):   115

Solubility (Moles/L)	Solubility (Grams/L)	Temp (°C)	Ref (#)	Evaluation (T P E A A)	Comments
1.537E-03	4.000E-01	22	K137	1 1 2 1 0	

2329.   C₁₁H₁₇O₃PS₂

$C_{11}H_{17}O_3PS_2$

Fensulfothion Sulfide
O,O-Diethyl O-[p-(Methylthio)phenyl] Phosphorothioate
Phosphorothioic Acid, O,O-Diethyl O-[4-(Methylthio)phenyl] Ester

RN: 3070-15-3   **MP** (°C):
MW: 292.36   **BP** (°C):

Solubility (Moles/L)	Solubility (Grams/L)	Temp (°C)	Ref (#)	Evaluation (T P E A A)	Comments
1.266E-05	3.700E-03	20	M318	2 2 0 0 2	

2330.   C₁₁H₁₇O₄PS₂

$C_{11}H_{17}O_4PS_2$

Fensulfothion
O,O-Diethyl O-(4-(Methylsulfinyl)phenyl) Phosphorothioate
Dasanit
Bay 25141
Agricur
Chemagro 25141

RN: 115-90-2   **MP** (°C):   <25
MW: 308.36   **BP** (°C):

Solubility (Moles/L)	Solubility (Grams/L)	Temp (°C)	Ref (#)	Evaluation (T P E A A)	Comments
6.473E-03	1.996E+00	20	B169	2 2 1 1 2	
6.473E-03	1.996E+00	20	F318	2 2 0 0 2	
4.994E-03	1.540E+00	25	M161	1 0 0 0 2	

2331.   C₁₁H₁₇O₅PS₂

$C_{11}H_{17}O_5PS_2$

Fensulfothion Sulfone
Phosphorothioic Acid, O,O-Diethyl O-[p-(Methylsulfonyl)phenyl] Ester
Dasanit Sulfone
Dasanit Sulphone

RN: 14255-72-2   **MP** (°C):
MW: 324.36   **BP** (°C):

Solubility (Moles/L)	Solubility (Grams/L)	Temp (°C)	Ref (#)	Evaluation (T P E A A)	Comments
1.242E-04	4.030E-02	10	B324	2 2 2 2 2	
1.243E-04	4.032E-02	10	B324	2 2 2 2 2	
2.300E-04	7.459E-02	20	B169	2 2 1 1 2	
2.633E-04	8.540E-02	20	B324	2 2 2 2 2	
2.633E-04	8.539E-02	20	B324	2 2 2 2 2	
2.300E-04	7.459E-02	20	M318	2 2 0 0 2	
3.576E-04	1.160E-01	30	B324	2 2 2 2 2	
3.576E-04	1.160E-01	30	B324	2 2 2 2 2	

2332. C₁₁H₁₈N₂O₂S
Thiopental
5-Ethyl-5-(1-methyl-butyl)-2-thiobarbituric Acid
5-Ethyl-5-(1-methylbutyl)-2-thiobarbituric Acid
Barbituric Acid, 5-Ethyl-5-(1-methylbutyl)-2-thio
4,6(1H,5H)-Pyrimidinedione, 5-Ethyldihydro-5-(1-methylbutyl)-2-thioxo
Pentothiobarbital
RN: 76-75-5   **MP** (°C):   158
MW: 242.34   **BP** (°C):

Solubility (Moles/L)	Solubility (Grams/L)	Temp (°C)	Ref (#)	Evaluation (T P E A A)	Comments
2.063E-04	5.000E-02	25	A023	1 0 0 1 1	
3.301E-04	8.000E-02	25	B011	2 0 0 1 0	
3.333E-04	8.077E-02	25	B065	1 1 1 1 1	
8.200E-04	1.987E-01	25	G003	1 1 1 1 1	pH 4.7
2.094E-04	5.075E-02	25	P350	2 1 1 1 2	intrinsic
3.000E-04	7.270E-02	30	K108	1 2 2 0 0	
3.301E-04	7.999E-02	35	A023	1 0 0 1 1	
4.126E-04	9.999E-02	40	A023	1 0 0 1 1	

2333. C₁₁H₁₈N₂O₃
Pilocarpic Acid
1,2-Secopilocarpin-2-oic Acid
RN: 28406-15-7   **MP** (°C):
MW: 226.28   **BP** (°C):

Solubility (Moles/L)	Solubility (Grams/L)	Temp (°C)	Ref (#)	Evaluation (T P E A A)	Comments
5.303E-04	1.200E-01	23	B340	1 1 2 1 1	pH 9

2334. C₁₁H₁₈N₂O₃
Amobarbital
5-Ethyl-5-isoamylbarbituric Acid
Amylobarbitone
RN: 57-43-2   **MP** (°C):   157
MW: 226.28   **BP** (°C):

Solubility (Moles/L)	Solubility (Grams/L)	Temp (°C)	Ref (#)	Evaluation (T P E A A)	Comments
2.828E-03	6.400E-01	20	J030	1 2 2 2 1	
3.533E-03	7.994E-01	25	A023	1 0 0 1 1	
2.475E-03	5.600E-01	25	B011	2 0 0 1 0	
2.665E-03	6.030E-01	25	B065	1 1 1 1 1	
3.900E-03	8.825E-01	25	G003	1 1 1 1 1	pH 4.7
2.170E-03	4.910E-01	25	V033	2 0 1 1 2	
2.200E-03	4.978E-01	25.00	T303	1 0 0 0 1	
3.000E-03	6.788E-01	30	G014	1 1 1 1 0	EFG

3.100E-03	7.015E-01	30	I001	2 0 2 1 0	EFG, 0.003N H_2SO_4
2.846E-03	6.440E-01	30	I015	1 2 2 1 2	pH 6.0, 3 forms
3.200E-03	7.241E-01	30	K108	1 2 2 0 1	
3.300E-03	7.467E-01	35.00	T303	1 0 0 0 1	
4.375E-03	9.900E-01	37	J030	1 2 2 2 1	
4.000E-03	9.051E-01	37	K121	1 2 1 2 0	0.1N HCl
5.517E-03	1.248E+00	40	A023	1 0 0 1 1	
3.820E-02	8.644E+00	40	N008	1 0 1 1 2	*sic*
4.300E-03	9.730E-01	45.00	T303	1 0 0 0 1	
2.342E-03	5.300E-01	ns	T003	0 0 0 0 2	

2335. $C_{11}H_{18}N_2O_3$
Pentobarbital
5-Ethyl-5-(1-methyl-butyl)-barbituric Acid
RN: 76-74-4   **MP** (°C):   130
MW: 226.28   **BP** (°C):

Solubility (Moles/L)	Solubility (Grams/L)	Temp (°C)	Ref (#)	Evaluation (T P E A A)	Comments
4.415E-03	9.990E-01	25	A023	1 0 0 1 1	
2.210E-03	5.000E-01	25	B011	2 0 0 1 0	
2.221E-03	5.026E-01	25	B065	1 1 1 1 1	
3.000E-03	6.788E-01	25	G003	1 1 1 1 1	pH 4.7
4.070E-03	9.210E-01	25	V033	2 0 1 1 2	
4.100E-03	9.277E-01	25.00	T303	1 0 0 0 1	
6.000E-03	1.358E+00	30	K108	1 2 2 0 1	
6.178E-03	1.398E+00	35	A023	1 0 0 1 1	
5.700E-03	1.290E+00	35.00	T303	1 0 0 0 1	
7.000E-03	1.584E+00	37	K121	1 2 1 2 0	0.1N HCl
7.060E-03	1.597E+00	40	A023	1 0 0 1 1	
7.640E-02	1.729E+01	40	N008	1 0 1 1 2	*sic*
6.900E-03	1.561E+00	45.00	T303	1 0 0 0 1	

2336. $C_{11}H_{18}N_2O_3$
5-n-Pentyl-5-ethylbarbituric Acid
5-Ethyl-5-pentylbarbituric Acid
RN: 115-58-2   **MP** (°C):   135.5
MW: 226.28   **BP** (°C):

Solubility (Moles/L)	Solubility (Grams/L)	Temp (°C)	Ref (#)	Evaluation (T P E A A)	Comments
6.657E-03	1.506E+00	25	B065	1 2 1 1 1	
2.448E-03	5.540E-01	ns	T003	0 0 0 0 2	

2337.   C$_{11}$H$_{18}$N$_4$O$_2$

Pirimicarb
2-(Dimethylamino)-5,6-dimethyl-4-pyrimidinyl Dimethylcarbamate
Abol
Rapid
Fernos
Aphox

RN: 23103-98-2   **MP** (°C):   90.5
MW: 238.29   **BP** (°C):

Solubility (Moles/L)	Solubility (Grams/L)	Temp (°C)	Ref (#)	Evaluation (T P E A A)	Comments
1.133E-02	2.700E+00	25	M161	1 0 0 0 1	

2338.   C$_{11}$H$_{19}$N$_3$O

Ethirimol
5-Butyl-2-(ethylamino)-4-hydroxy-6-methylpyrimidine
Milgo
Milcurb Super
Milstem

RN: 23947-60-6   **MP** (°C):   159.5
MW: 209.29   **BP** (°C):

Solubility (Moles/L)	Solubility (Grams/L)	Temp (°C)	Ref (#)	Evaluation (T P E A A)	Comments
9.556E-04	2.000E-01	rt	M161	0 0 0 0 0	

2339.   C$_{11}$H$_{19}$N$_3$O

Dimethirimol
2-Dimethylamino-4-hyroxy-5-n-butyl-6-methylpyrimidine

RN: 5221-53-4   **MP** (°C):   102
MW: 209.29   **BP** (°C):

Solubility (Moles/L)	Solubility (Grams/L)	Temp (°C)	Ref (#)	Evaluation (T P E A A)	Comments
5.734E-03	1.200E+00	25	M161	1 0 0 0 1	
5.727E-03	1.199E+00	ns	M061	0 0 0 0 1	

2340.   C$_{11}$H$_{20}$

2-Methyldecalin
Decahydro-2-methylnaphthalene

RN: 2958-76-1   **MP** (°C):
MW: 152.28   **BP** (°C):

Solubility (Moles/L)	Solubility (Grams/L)	Temp (°C)	Ref (#)	Evaluation (T P E A A)	Comments
2.666E-07	4.060E-05	25	B069	1 0 1 1 2	

2341.   C$_{11}$H$_{20}$ClN$_5$
Chlorazine
2-Chloro-4-diethylamino-6-diethylamino-s-triazine
2-Chloro-4,6-bis-(diethylamino)-s-triazine Chlorazine
1,3,5-Triazine
1,3,5-Triazine-2,4-diamine
RN: 580-48-3   **MP** (°C):
MW: 257.77   **BP** (°C):

Solubility (Moles/L)	Solubility (Grams/L)	Temp (°C)	Ref (#)	Evaluation (T P E A A)	Comments
3.492E-05	9.000E-03	20	J033	1 0 0 0 0	
3.879E-05	1.000E-02	21	B192	0 0 0 0 1	
3.492E-05	9.000E-03	21	G099	2 0 0 1 0	

2342.   C$_{11}$H$_{20}$N$_2$O$_4$
Isopropyl-2,2-diethylmalonurate
Isopropyl 2,2-Diethylmalonurate
RN: 73632-77-6   **MP** (°C):   99.5
MW: 244.29   **BP** (°C):

Solubility (Moles/L)	Solubility (Grams/L)	Temp (°C)	Ref (#)	Evaluation (T P E A A)	Comments
1.700E-03	4.153E-01	23	B152	1 2 1 1 1	pH 3.5

2343.   C$_{11}$H$_{20}$N$_3$O$_3$PS
Pirimiphos-methyl
Pirimiphosmethyl
RN: 29232-93-7   **MP** (°C):   15
MW: 305.34   **BP** (°C):

Solubility (Moles/L)	Solubility (Grams/L)	Temp (°C)	Ref (#)	Evaluation (T P E A A)	Comments
7.139E-05	2.180E-02	10	B324	2 2 2 2 2	
7.946E-05	2.426E-02	10	B324	2 2 2 2 2	
7.363E-05	2.248E-02	20	B300	2 1 1 1 2	
1.119E-04	3.417E-02	20	B324	2 2 2 2 2	
1.005E-04	3.070E-02	20	B324	2 2 2 2 2	
1.640E-04	5.008E-02	30	B324	2 2 2 2 2	
1.474E-04	4.500E-02	30	B324	2 2 2 2 2	
1.638E-05	5.000E-03	30	M161	1 0 0 0 0	sic

2344.   C₁₁H₂₀N₆
1-(Pyrrolidinyl)-3,5-bis(dimethylamino)-s-triazine
1-Pyrrolidino-3,5-bis(Dimethylamino)-s-triazine
RN: 13452-85-2   **MP** (°C):
MW: 236.32   **BP** (°C):

Solubility (Moles/L)	Solubility (Grams/L)	Temp (°C)	Ref (#)	Evaluation (T P E A A)	Comments
1.641E-04	3.878E-02	25	B386	2 2 2 2 2	

2345.   C₁₁H₂₀N₆O
1-(Morpholinyl)-3,5-bis(dimethylamino)-s-triazine
s-Triazine, 2,4-bis(Dimethylamino)-6-morpholino-
RN: 16269-02-6   **MP** (°C):
MW: 252.32   **BP** (°C):

Solubility (Moles/L)	Solubility (Grams/L)	Temp (°C)	Ref (#)	Evaluation (T P E A A)	Comments
1.303E-03	3.288E-01	25	B386	2 2 2 2 2	

2346.   C₁₁H₂₀N₆S
1-(Thiomorpholinyl)-3,5-bis(dimethylamino)-s-triazine
1,3,5-Triazine-2,4-diamine, N,N,N',N'-Tetramethyl-6-(4-thiomorpholinyl)-
RN: 41492-69-7   **MP** (°C):
MW: 268.39   **BP** (°C):

Solubility (Moles/L)	Solubility (Grams/L)	Temp (°C)	Ref (#)	Evaluation (T P E A A)	Comments
5.689E-05	1.527E-02	25	B386	2 2 2 2 2	

2347.   C₁₁H₂₀O₂
Undecylenic Acid
10-Undecylenic Acid
Hendecenoic Acid
RN: 112-38-9   **MP** (°C):   25
MW: 184.28   **BP** (°C):

Solubility (Moles/L)	Solubility (Grams/L)	Temp (°C)	Ref (#)	Evaluation (T P E A A)	Comments
4.000E-04	7.371E-02	30	D051	2 0 0 1 2	
1.074E-04	1.980E-02	30	E005	2 1 1 2 2	
1.248E-04	2.300E-02	40	E005	2 1 1 2 1	
1.411E-04	2.600E-02	50	E005	2 1 1 2 1	
1.000E-03	1.843E-01	60	D051	2 0 0 1 2	
1.736E-04	3.200E-02	60	E005	2 1 1 2 1	

2348. C$_{11}$H$_{20}$O$_4$
Undecanedioic Acid
1,9-Nonanedicarboxylic Acid
Nonan-dicarbonsaeure-(1,9)
RN: 1852-04-6   **MP** (°C):
MW: 216.28   **BP** (°C):

Solubility (Moles/L)	Solubility (Grams/L)	Temp (°C)	Ref (#)	Evaluation (T P E A A)	Comments
2.358E-02	5.100E+00	21	B040	1 0 1 1 1	*sic*
6.473E-04	1.400E-01	ns	F300	0 0 0 0 2	

2349. C$_{11}$H$_{20}$O$_4$
Hexyl α-Acetoxypropionate
Propanoic Acid, 2-(Acetyloxy)-, Hexyl Ester
RN: 96884-73-0   **MP** (°C):
MW: 216.28   **BP** (°C):

Solubility (Moles/L)	Solubility (Grams/L)	Temp (°C)	Ref (#)	Evaluation (T P E A A)	Comments
9.247E-04	2.000E-01	25	R006	2 2 0 1 1	

2350. C$_{11}$H$_{20}$O$_5$
Propanoic Acid, 2-[(Hexthoxycarbonyl)oxy]-, Methyl Ester
RN: **MP** (°C):
MW: 232.28   **BP** (°C):

Solubility (Moles/L)	Solubility (Grams/L)	Temp (°C)	Ref (#)	Evaluation (T P E A A)	Comments
4.305E-04	9.999E-02	25	R007	1 0 0 0 0	

2351. C$_{11}$H$_{21}$BrO$_2$
11-Bromoundecanoic Acid
Bromo-11-undecanoique Acide
RN: 2834-05-1   **MP** (°C):   49.5
MW: 265.20   **BP** (°C):   173.5

Solubility (Moles/L)	Solubility (Grams/L)	Temp (°C)	Ref (#)	Evaluation (T P E A A)	Comments
2.000E-04	5.304E-02	30	D051	2 0 0 1 2	
7.500E-04	1.989E-01	60	D051	2 0 0 1 2	

2352.   C$_{11}$H$_{21}$NOS
Cycloate
S-Ethyl N-Ethylthiocyclohexanecarbamate
RO-Neet
S-Ethyl N,N-Ethylcyclohexylthiocarbamate
RN: 1134-23-2   **MP** (°C):   12
MW: 215.36   **BP** (°C):

Solubility (Moles/L)	Solubility (Grams/L)	Temp (°C)	Ref (#)	Evaluation (T P E A A)	Comments
3.947E-04	8.500E-02	22	B200	1 0 0 0 1	
3.947E-04	8.500E-02	22	F019	1 0 0 0 1	
3.947E-04	8.500E-02	22	M161	1 0 0 0 1	

2353.   C$_{11}$H$_{21}$NO$_3$
Dipropylaceturethane
RN: **MP** (°C):
MW: 215.29   **BP** (°C):

Solubility (Moles/L)	Solubility (Grams/L)	Temp (°C)	Ref (#)	Evaluation (T P E A A)	Comments
1.857E-03	3.998E-01	20	O021	1 2 0 0 0	

2354.   C$_{11}$H$_{21}$N$_5$O
Ipatone
1,3,5-Triazine, 2-(Diethylamino)-4-(isopropylamino)-6-methoxy
1,3,5-Triazine-2,4-diamine, N,N-Diethyl-6-methoxy-N'-(1-methylethyl)
RN: 3004-70-4   **MP** (°C):
MW: 239.32   **BP** (°C):

Solubility (Moles/L)	Solubility (Grams/L)	Temp (°C)	Ref (#)	Evaluation (T P E A A)	Comments
4.178E-04	1.000E-01	20	J033	1 0 0 0 2	

2355.   C$_{11}$H$_{21}$N$_5$OS
Gesaran
2-Methylthio-4-isopropylamino-6-(3-methoxypropylamino)-s-triazine
Methoprotryne
RN: 841-06-5   **MP** (°C):   69
MW: 271.39   **BP** (°C):

Solubility (Moles/L)	Solubility (Grams/L)	Temp (°C)	Ref (#)	Evaluation (T P E A A)	Comments
1.179E-03	3.200E-01	20	F311	1 2 2 2 1	
1.179E-03	3.200E-01	20	M161	1 0 0 0 2	
1.179E-03	3.200E-01	ns	J033	0 0 0 0 2	
3.681E-03	9.990E-01	ns	M061	0 0 0 0 0	

2356.   C₁₁H₂₁N₅S

Dipropetryn

2-(Ethylthio)-4,6-bis(isopropylamino)-s-triazine

Cotofor

Sancap

Sancap 80W

RN: 4147-51-7   **MP** (°C):   105

MW: 255.39   **BP** (°C):

Solubility	Solubility	Temp	Ref	Evaluation	
(Moles/L)	(Grams/L)	(°C)	(#)	(T P E A A)	Comments
6.265E-05	1.600E-02	rt	M161	0 0 0 0 1	

2357.   C₁₁H₂₁N₅S

Dimethametryn

N-(1,2-Dimethylpropyl)-N'-ethyl-6-(methylthio)-1,3,5-triazine-2,4-diamine

Belclene 310

RN: 22936-75-0   **MP** (°C):

MW: 255.39   **BP** (°C):   152

Solubility	Solubility	Temp	Ref	Evaluation	
(Moles/L)	(Grams/L)	(°C)	(#)	(T P E A A)	Comments
1.958E-04	5.000E-02	20	M161	1 0 0 0 1	

2358.   C₁₁H₂₁N₅S

Ipatryne

2-Methylmercapto-4-isopropylamino-6-diethylamino-s-triazine

RN: **MP** (°C):

MW: 255.39   **BP** (°C):

Solubility	Solubility	Temp	Ref	Evaluation	
(Moles/L)	(Grams/L)	(°C)	(#)	(T P E A A)	Comments
2.100E-05	5.363E-03	26	G001	1 0 1 1 1	

2359.   C₁₁H₂₁N₇

1-(1-Piperizinyl)-3,5-bis(dimethylamino)-s-triazine

1,3,5-Triazine-2,4-diamine, N,N,N',N'-Tetramethyl-6-(1-piperazinyl)-

RN: 125867-94-9   **MP** (°C):

MW: 251.34   **BP** (°C):

Solubility	Solubility	Temp	Ref	Evaluation	
(Moles/L)	(Grams/L)	(°C)	(#)	(T P E A A)	Comments
1.081E-02	2.717E+00	25	B386	2 2 2 2 2	

2360.   C$_{11}$H$_{21}$O$_5$

Propanoic Acid, 2-[(Proxycarbonyl)oxy]-, Butyl Ester

RN: **MP** (°C):
MW: 233.29   **BP** (°C):

Solubility (Moles/L)	Solubility (Grams/L)	Temp (°C)	Ref (#)	Evaluation (T P E A A)	Comments
4.286E-04	9.999E-02	25	R007	1 0 0 0 0	

2361.   C$_{11}$H$_{22}$N$_2$O

Cycluron
N'-Cyclooctyl-N,N-dimethylurea
Cyclooctyl-1,1-dimethylurea
OMU

RN: 2163-69-1   **MP** (°C):   138
MW: 198.31   **BP** (°C):

Solubility (Moles/L)	Solubility (Grams/L)	Temp (°C)	Ref (#)	Evaluation (T P E A A)	Comments
7.564E-04	1.500E-01	20	B185	1 0 0 0 2	
6.051E-03	1.200E+00	20	G036	1 0 0 0 2	
5.541E-03	1.099E+00	20	M061	1 0 0 0 1	
5.547E-03	1.100E+00	20	M161	1 0 0 0 1	
6.310E-04	1.251E-01	ns	M163	0 0 0 0 0	EFG

2362.   C$_{11}$H$_{22}$N$_6$

N6,N6-Diethyl-N2,N2,N4,N4-tetramethylmelamine
1,3,5-Triazine-2,4,6-triamine, N,N-Diethyl-N',N',N'',N''-Tetramethyl-

RN: 16268-75-0   **MP** (°C):   42.0
MW: 238.34   **BP** (°C):

Solubility (Moles/L)	Solubility (Grams/L)	Temp (°C)	Ref (#)	Evaluation (T P E A A)	Comments
2.979E-04	7.100E-02	25	C051	1 2 1 1 1	pH 7

2363.   C$_{11}$H$_{22}$O$_2$

3-Hydroxy-2-propyl-5,5-diethyltetrahydrofuran

RN: **MP** (°C):
MW: 186.30   **BP** (°C):

Solubility (Moles/L)	Solubility (Grams/L)	Temp (°C)	Ref (#)	Evaluation (T P E A A)	Comments
1.053E-01	1.961E+01	rt	B066	0 2 0 0 0	

2364.   C₁₁H₂₂O₂
Undecanoic Acid
Undecanoique Acide

RN:	112-37-8	**MP** (°C):	28.5
MW:	186.30	**BP** (°C):	228

Solubility (Moles/L)	Solubility (Grams/L)	Temp (°C)	Ref (#)	Evaluation (T P E A A)	Comments
3.382E-03	6.300E-01	0	B136	1 0 2 1 1	
3.381E-04	6.300E-02	0.0	R001	1 1 1 1 1	
5.744E-04	1.070E-01	20	B136	1 0 2 1 2	
4.992E-04	9.299E-02	20.0	R001	1 1 1 1 1	
6.978E-04	1.300E-01	30	B136	1 0 2 1 2	
2.800E-04	5.216E-02	30	D051	2 0 0 1 2	
5.904E-04	1.100E-01	30.0	R001	1 1 1 1 1	
7.730E-04	1.440E-01	40	B136	1 0 2 1 2	
6.978E-04	1.300E-01	45	B136	1 0 2 1 1	
6.977E-04	1.300E-01	45.0	R001	1 1 1 1 1	
8.052E-04	1.500E-01	60	B136	1 0 2 1 1	
6.000E-04	1.118E-01	60	D051	2 0 0 1 2	
8.050E-04	1.500E-01	60.0	R001	1 1 1 1 1	

2365.   C₁₁H₂₂O₂
Methyl Caprate
Capric Acid Methyl Ester
Methyl Decanoate

RN:	110-42-9	**MP** (°C):	-13
MW:	186.30	**BP** (°C):	223

Solubility (Moles/L)	Solubility (Grams/L)	Temp (°C)	Re 2368 (#)	aluation (T P E A A)	Comments
<2.36E-05	<4.40E-03	20	M337	2 1 2 2 1	

2366.   C₁₁H₂₂O₃
n-Hexyl β-Ethoxypropionate
Propionic Acid, 3-Ethoxy-, Hexyl Ester

RN:	14144-37-7	**MP** (°C):	
MW:	202.30	**BP** (°C):	

Solubility (Moles/L)	Solubility (Grams/L)	Temp (°C)	Ref (#)	Evaluation (T P E A A)	Comments
1.483E-03	2.999E-01	25	D002	1 2 1 1 0	

2367.   C₁₁H₂₂O₃
Octyl Lactate
Propanoic Acid, 2-Hydroxy-, Octyl Ester
RN: 5464-71-1   **MP** (°C):
MW: 202.30   **BP** (°C):

Solubility (Moles/L)	Solubility (Grams/L)	Temp (°C)	Ref (#)	Evaluation (T P E A A)	Comments
3.955E-03	8.000E-01	25	R006	2 2 0 1 0	

2368.   C₁₁H₂₂O₃
1,3-Dioxolane-4-methanol, 2-Hexyl-2-methyl
2-Octanone, Cyclic (hydroxymethyl)ethylene Acetal
RN: 5660-52-6   **MP** (°C):
MW: 202.30   **BP** (°C):

Solubility (Moles/L)	Solubility (Grams/L)	Temp (°C)	Ref (#)	Evaluation (T P E A A)	Comments
1.360E-02	2.751E+00	25	P342	1 2 2 2 2	0.0001M Na₂CO₃

2369.   C₁₁H₂₂O₃
n-Butyl β-n-Butoxypropionate
Butyl 3-Butoxypropionate
Propanoic Acid, 3-Butoxy-, Butyl Ester
RN: 14144-48-0   **MP** (°C):
MW: 202.30   **BP** (°C):

Solubility (Moles/L)	Solubility (Grams/L)	Temp (°C)	Ref (#)	Evaluation (T P E A A)	Comments
3.951E-03	7.994E-01	25	R034	0 0 0 0 0	

2370.   C₁₁H₂₂O₄
1,3-Dioxolane-4-methanol, 2-(2-Butoxyethyl)-2-methyl
RN: 143458-55-3   **MP** (°C):
MW: 218.30   **BP** (°C):

Solubility (Moles/L)	Solubility (Grams/L)	Temp (°C)	Ref (#)	Evaluation (T P E A A)	Comments
2.640E-01	5.763E+01	25	P342	1 2 2 2 2	0.0001M Na₂CO₃

2371. $C_{11}H_{23}NOS$
Butylate
S-Ethyl Diisobutylthiocarbamate
RN: 2008-41-5   **MP** (°C):   <25
MW: 217.38   **BP** (°C):

Solubility (Moles/L)	Solubility (Grams/L)	Temp (°C)	Ref (#)	Evaluation (T P E A A)	Comments
2.070E-04	4.500E-02	22	B200	1 0 0 0 1	
2.070E-04	4.500E-02	22	F019	1 0 0 0 1	
2.070E-04	4.500E-02	rt	M161	0 0 0 0 1	

2372. $C_{11}H_{23}NO_2$
11-Aminoundecanoic Acid
Amino-11-undecanoique Acide
RN: 2432-99-7   **MP** (°C):   191
MW: 201.31   **BP** (°C):

Solubility (Moles/L)	Solubility (Grams/L)	Temp (°C)	Ref (#)	Evaluation (T P E A A)	Comments
1.986E-03	3.998E-01	20	E039	2 0 1 1 1	smoothed
1.600E-03	3.221E-01	30	D051	2 0 0 1 2	
4.962E-03	9.990E-01	30	E039	2 0 1 1 2	smoothed
8.925E-03	1.797E+00	40	E039	2 0 1 1 2	smoothed
1.486E-02	2.991E+00	50	E039	2 0 1 1 2	smoothed
1.000E-02	2.013E+00	60	D051	2 0 0 1 2	
2.471E-02	4.975E+00	60	E039	2 0 1 1 2	smoothed
3.453E-02	6.951E+00	65	E039	2 0 1 1 2	smoothed
4.431E-02	8.920E+00	70	E039	2 0 1 1 2	smoothed
5.405E-02	1.088E+01	75	E039	2 0 1 1 2	smoothed
6.858E-02	1.381E+01	80	E039	2 0 1 1 2	smoothed
8.183E-02	1.647E+01	85	E039	2 0 1 1 2	smoothed
9.740E-02	1.961E+01	90	E039	2 0 1 1 2	smoothed
1.145E-01	2.306E+01	95	E039	2 0 1 1 2	smoothed
1.259E-01	2.534E+01	100	E039	2 0 1 1 2	smoothed

2373. $C_{11}H_{24}$
Undecane
n-Undecane
n-Hendecane
RN: 1120-21-4   **MP** (°C):   -26
MW: 156.31   **BP** (°C):   196

Solubility (Moles/L)	Solubility (Grams/L)	Temp (°C)	Ref (#)	Evaluation (T P E A A)	Comments
<9.60E-06	<1.50E-03	20	M337	2 1 2 2 1	
2.815E-08	4.400E-06	25	M003	1 0 2 2 1	

2374.   C$_{12}$HCl$_7$O
1,2,3,4,6,7,8-Heptachlorodibenzofuran
1,2,3,4,6,7,8-HpCDF
PCDF 131
F 131
RN: 67562-39-4   **MP** (°C):   236
MW: 409.31   **BP** (°C):

Solubility (Moles/L)	Solubility (Grams/L)	Temp (°C)	Ref (#)	Evaluation (T P E A A)	Comments
3.310E-12	1.355E-09	22.5	F314	1 1 0 2 2	

2375.   C$_{12}$HCl$_7$O$_2$
1,2,3,4,6,7,8-Heptachlorodibenzo-p-dioxin
1,2,3,4,6,7,8-HpCDD
PCDD 73
D 73
Heptachlorodibenzo-p-dioxin
RN: 35822-46-9   **MP** (°C):   265
MW: 425.31   **BP** (°C):

Solubility (Moles/L)	Solubility (Grams/L)	Temp (°C)	Ref (#)	Evaluation (T P E A A)	Comments
2.200E-12	9.357E-10	7.0	F315	1 2 0 2 2	
2.690E-12	1.144E-09	11.5	F315	1 2 0 2 2	
3.040E-12	1.293E-09	17.0	F315	1 2 0 2 2	
5.400E-12	2.297E-09	21.0	F315	1 2 0 2 2	
6.030E-12	2.565E-09	26.0	F315	1 2 0 2 2	
1.481E-11	6.300E-09	40	F303	1 2 1 2 1	
1.490E-11	6.337E-09	41.0	F315	1 2 0 2 2	

2376.   C$_{12}$HCl$_9$
2,2',3,3',4,4',5,5',6-Nonachlorobiphenyl
2,3,4,5,6,2',3',4',5'-Nonachlorbiphenyl
RN: 40186-72-9   **MP** (°C):   204.5
MW: 464.22   **BP** (°C):

Solubility (Moles/L)	Solubility (Grams/L)	Temp (°C)	Ref (#)	Evaluation (T P E A A)	Comments
1.680E-10	7.800E-08	22	O311	2 2 1 2 1	
5.490E-11	2.549E-08	25	D331	2 1 2 2 2	
5.493E-11	2.550E-08	25	D335	1 0 0 0 2	
2.413E-10	1.120E-07	25	W025	1 0 2 2 2	
5.490E-11	2.549E-08	25.0	M324	1 2 1 1 2	
1.100E-10	5.106E-08	32	D331	2 1 2 2 2	
1.100E-10	5.106E-08	32.0	M324	1 2 1 1 2	

1.420E-10	6.592E-08	40	D331	2 1 2 2 2	
1.420E-10	6.592E-08	40.0	M324	1 2 1 1 2	
2.840E-10	1.318E-07	50	D331	2 1 2 2 2	
2.840E-10	1.318E-07	50.0	M324	1 2 1 1 2	

2377. $C_{12}HCl_9$
2,2',3,3',4,5,5',6,6'-Nonachlorobiphenyl
1,1'-Biphenyl, 2,2',3,3',4,5,5',6,6'-Nonachloro-
PCB 208
RN: 52663-77-1   **MP** (°C):   182
MW: 464.22   **BP** (°C):

Solubility (Moles/L)	Solubility (Grams/L)	Temp (°C)	Ref (#)	Evaluation (T P E A A)	Comments
3.880E-11	1.801E-08	25	M342	1 0 1 1 2	

2378. $C_{12}H_2Br_8$
Octabromobiphenyl
OBBP
Bromkal 80
RN: 27858-07-7   **MP** (°C):   225.0
MW: 785.42   **BP** (°C):

Solubility (Moles/L)	Solubility (Grams/L)	Temp (°C)	Ref (#)	Evaluation (T P E A A)	Comments
3.183E-08	2.500E-05	25	N326	1 0 0 0 1	average

2379. $C_{12}H_2Cl_6O$
1,2,3,4,7,8-Hexachlorodibenzofuran
1,2,3,4,7,8-HxCDF
F 118
PCDF 118
RN: 70648-26-9   **MP** (°C):   226
MW: 374.87   **BP** (°C):

Solubility (Moles/L)	Solubility (Grams/L)	Temp (°C)	Ref (#)	Evaluation (T P E A A)	Comments
2.200E-11	8.247E-09	22.5	F314	1 1 0 2 2	

2380. C$_{12}$H$_2$Cl$_6$O
1,2,3,6,7,8-Hexachlorodibenzofuran
1,2,3,6,7,8-HxCDF
F 121
PCDF 121
2,3,4,7,8,9-Hexachlorodibenzofuran
RN: 57117-44-9   **MP** (°C):   233
MW: 374.87   **BP** (°C):

Solubility (Moles/L)	Solubility (Grams/L)	Temp (°C)	Ref (#)	Evaluation (T P E A A)	Comments
4.720E-11	1.769E-08	22.5	F314	1 1 0 2 2	

2381. C$_{12}$H$_2$Cl$_6$O$_2$
1,2,3,4,7,8-Hexachlorodibenzo-p-dioxin
1,2,3,4,7,8-Hexachlorodibenzo[b,e][1,4]dioxin
1,2,3,4,7,8-Hexachlorodibenzo[1,4]dioxin
1,2,3,4,7,8-HxCDD
D 66
PCDD 66
RN: 39227-28-6   **MP** (°C):   273
MW: 390.87   **BP** (°C):

Solubility (Moles/L)	Solubility (Grams/L)	Temp (°C)	Ref (#)	Evaluation (T P E A A)	Comments
5.910E-12	2.310E-09	7.0	F315	1 2 0 2 2	
7.980E-12	3.119E-09	11.5	F315	1 2 0 2 2	
1.070E-11	4.182E-09	17.0	F315	1 2 0 2 2	
1.126E-11	4.400E-09	20	F303	1 2 1 2 1	
1.250E-11	4.886E-09	21.0	F315	1 2 0 2 2	
2.020E-11	7.896E-09	26.0	F315	1 2 0 2 2	
4.861E-11	1.900E-08	40	F303	1 2 1 2 2	
4.860E-11	1.900E-08	41.0	F315	1 2 0 2 2	

2382. C$_{12}$H$_2$Cl$_8$
2,2',3,3',4,4',5,5'-Octachlorobiphenyl
2,3,4,5,2',3',4',5'-Octachlorbiphenyl
PCB 194
RN: 35694-08-7   **MP** (°C):   156
MW: 429.77   **BP** (°C):

Solubility (Moles/L)	Solubility (Grams/L)	Temp (°C)	Ref (#)	Evaluation (T P E A A)	Comments
2.885E-10	1.240E-07	22	O311	2 2 1 2 2	
6.329E-10	2.720E-07	25	W025	1 0 2 2 2	

2383.   C$_{12}$H$_2$Cl$_8$
2,2',3,3',5,5',6,6'-Octachlorobiphenyl
2,3,5,6,2',3',5',6'-Octachlorbiphenyl
RN: 2136-99-4   **MP** (°C):   161
MW: 429.77   **BP** (°C):

Solubility (Moles/L)	Solubility (Grams/L)	Temp (°C)	Ref (#)	Evaluation (T P E A A)	Comments
2.650E-10	1.139E-07	20	D331	2 1 2 2 2	
2.650E-10	1.139E-07	20.0	M324	1 2 1 1 2	
3.420E-10	1.470E-07	25	D331	2 1 2 2 2	
3.420E-10	1.470E-07	25	D335	1 0 0 0 2	
9.150E-10	3.932E-07	25	M342	1 0 1 1 2	
4.188E-10	1.800E-07	25	W025	1 0 2 2 1	
3.420E-10	1.470E-07	25.0	M324	1 2 1 1 2	
4.930E-10	2.119E-07	32	D331	2 1 2 2 2	
4.930E-10	2.119E-07	32.0	M324	1 2 1 1 2	
1.780E-09	7.650E-07	50	D331	2 1 2 2 2	
1.780E-09	7.650E-07	50.0	M324	1 2 1 1 2	

2384.   C$_{12}$H$_3$Cl$_5$O
2,3,4,7,8-Pentachlorodibenzofuran
2,3,4,7,8-P5CDF
PeCDF, 2,3,4,7,8-
RN: 57117-31-4   **MP** (°C):   195.5
MW: 340.42   **BP** (°C):

Solubility (Moles/L)	Solubility (Grams/L)	Temp (°C)	Ref (#)	Evaluation (T P E A A)	Comments
6.920E-10	2.356E-07	22.5	F314	1 1 0 2 2	

2385.   C$_{12}$H$_3$Cl$_5$O$_2$
1,2,3,4,7-Pentachlorodibenzo-p-dioxin
Dibenzo[b,e][1,4]dioxin, 1,2,3,4,7-Pentachloro-
PCDD 50
RN: 39227-61-7   **MP** (°C):   195
MW: 356.42   **BP** (°C):

Solubility (Moles/L)	Solubility (Grams/L)	Temp (°C)	Ref (#)	Evaluation (T P E A A)	Comments
1.420E-10	5.061E-08	7.0	F315	1 2 0 2 2	
1.880E-10	6.701E-08	11.5	F315	1 2 0 2 2	
2.440E-10	8.697E-08	17.0	F315	1 2 0 2 2	
3.367E-10	1.200E-07	20	F303	1 2 1 2 1	
3.450E-10	1.230E-07	21.0	F315	1 2 0 2 2	
4.630E-10	1.650E-07	26.0	F315	1 2 0 2 2	
1.291E-09	4.600E-07	40	F303	1 2 1 2 1	
1.280E-09	4.562E-07	41.0	F315	1 2 0 2 2	

2386.   C$_{12}$H$_3$Cl$_7$
2,2',3,3',4,4',6-Heptachlorobiphenyl
1,1'-Biphenyl, 2,2',3,3',4,4',6-Heptachloro-
PCB 171
RN: 52663-71-5   **MP** (°C):   117
MW: 395.33   **BP** (°C):

Solubility (Moles/L)	Solubility (Grams/L)	Temp (°C)	Ref (#)	Evaluation (T P E A A)	Comments
1.042E-08	4.120E-06	20	M336	2 0 2 2 2	
5.490E-09	2.170E-06	25	M342	1 0 1 1 2	
5.490E-09	2.170E-06	ns	M308	0 0 1 1 2	

2387.   C$_{12}$H$_3$Cl$_7$
2,2',3,4,4',5,5'-Heptachlorobiphenyl
1,1'-Biphenyl, 2,2',3,4,4',5,5'-Heptachloro-
PCB 180
RN: 35065-29-3   **MP** (°C):   112
MW: 395.33   **BP** (°C):

Solubility (Moles/L)	Solubility (Grams/L)	Temp (°C)	Ref (#)	Evaluation (T P E A A)	Comments
9.739E-09	3.850E-06	20	M336	2 0 2 2 2	

2388.   C$_{12}$H$_3$Cl$_7$
2,2',3,3',4,5,6-Heptachlorobiphenyl
1,1'-Biphenyl, 2,2',3,3',4,5',6'-Heptachloro-
PCB 177
RN: 52663-70-4   **MP** (°C):
MW: 395.33   **BP** (°C):

Solubility (Moles/L)	Solubility (Grams/L)	Temp (°C)	Ref (#)	Evaluation (T P E A A)	Comments
1.219E-08	4.820E-06	20	M336	2 0 2 2 2	

2389.   C$_{12}$H$_3$Cl$_7$
2,2',3,3',4,4',5-Heptachlorobiphenyl
1,1'-Biphenyl, 2,2',3,3',4,4',5-Heptachloro-
PCB 170
CB 170
RN: 35065-30-6   **MP** (°C):   134.5
MW: 395.33   **BP** (°C):

Solubility (Moles/L)	Solubility (Grams/L)	Temp (°C)	Ref (#)	Evaluation (T P E A A)	Comments
8.778E-09	3.470E-06	20	M336	2 0 2 2 2	

2390. C$_{12}$H$_3$Cl$_7$
2,2',3,3',4,5',6-Heptachlorobiphenyl
1,1'-Biphenyl, 2,2',3,3',4,5',6-Heptachloro-
PCB 175
RN: 40186-70-7   **MP** (°C):
MW: 395.33   **BP** (°C):

Solubility (Moles/L)	Solubility (Grams/L)	Temp (°C)	Ref (#)	Evaluation (T P E A A)	Comments
2.261E-08	8.940E-06	20	M336	2 0 2 2 2	

2391. C$_{12}$H$_3$Cl$_7$
2,2',3,3',4,5,6'-Heptachlorobiphenyl
1,1'-Biphenyl, 2,2',3,3',4,5,6'-Heptachloro-
PCB 174
RN: 38411-25-5   **MP** (°C):
MW: 395.33   **BP** (°C):

Solubility (Moles/L)	Solubility (Grams/L)	Temp (°C)	Ref (#)	Evaluation (T P E A A)	Comments
1.328E-08	5.250E-06	20	M336	2 0 2 2 2	

2392. C$_{12}$H$_3$Cl$_7$
2,2',3,3',4,5,6-Heptachlorobiphenyl
1,1'-Biphenyl, 2,2',3,3',4,5,6-Heptachloro-
PCB 173
RN: 68194-16-1   **MP** (°C):
MW: 395.33   **BP** (°C):

Solubility (Moles/L)	Solubility (Grams/L)	Temp (°C)	Ref (#)	Evaluation (T P E A A)	Comments
1.052E-08	4.160E-06	20	M336	2 0 2 2 2	

2393. C$_{12}$H$_3$Cl$_7$
2,2',3,3',4,6,6'-Heptachlorobiphenyl
1,1'-Biphenyl, 2,2',3,3',4,6,6'-Heptachloro-
PCB 176
RN: 52663-65-7   **MP** (°C):
MW: 395.33   **BP** (°C):

Solubility (Moles/L)	Solubility (Grams/L)	Temp (°C)	Ref (#)	Evaluation (T P E A A)	Comments
1.480E-08	5.850E-06	20	M336	2 0 2 2 2	

2394.   C$_{12}$H$_3$Cl$_7$
2,2',3,3',5,5',6-Heptachlorobiphenyl
1,1'-Biphenyl, 2,2',3,3',5,5',6-Heptachloro-
PCB 178
RN: 52663-67-9   **MP** (°C):
MW: 395.33   **BP** (°C):

Solubility (Moles/L)	Solubility (Grams/L)	Temp (°C)	Ref (#)	Evaluation (T P E A A)	Comments
2.236E-08	8.840E-06	20	M336	2 0 2 2 2	

2395.   C$_{12}$H$_3$Cl$_7$
2,2',3,3',4,5,5'-Heptachlorobiphenyl
1,1'-Biphenyl, 2,2',3,3',4,5,5'-Heptachloro-
PCB 172
RN: 52663-74-8   **MP** (°C):
MW: 395.33   **BP** (°C):

Solubility (Moles/L)	Solubility (Grams/L)	Temp (°C)	Ref (#)	Evaluation (T P E A A)	Comments
1.088E-08	4.300E-06	20	M336	2 0 2 2 2	

2396.   C$_{12}$H$_3$Cl$_7$
2,2',3,4,4',5',6-Heptachlorobiphenyl
1,1'-Biphenyl, 2,2',3,4,4',5',6-Heptachloro-
PCB 183
RN: 52663-69-1   **MP** (°C):   83
MW: 395.33   **BP** (°C):

Solubility (Moles/L)	Solubility (Grams/L)	Temp (°C)	Ref (#)	Evaluation (T P E A A)	Comments
1.239E-08	4.900E-06	20	M336	2 0 2 2 2	

2397.   C$_{12}$H$_3$Cl$_7$
2,2',3,4,5,5',6-Heptachlorobiphenyl
2,3,4,5,6,2',5'-Heptachlorbiphenyl
PCB 185
RN: 52712-05-7   **MP** (°C):   147
MW: 395.33   **BP** (°C):

Solubility (Moles/L)	Solubility (Grams/L)	Temp (°C)	Ref (#)	Evaluation (T P E A A)	Comments
1.381E-08	5.460E-06	20	M336	2 0 2 2 2	*sic*
1.189E-09	4.700E-07	25	W025	1 0 2 2 1	

2398. C₁₂H₃Cl₇
Heptachlorobiphenyl
1,1'-Biphenyl, Heptachloro-
Heptachlorodiphenyl
RN: 28655-71-2   **MP** (°C):
MW: 395.33   **BP** (°C):

Solubility (Moles/L)	Solubility (Grams/L)	Temp (°C)	Ref (#)	Evaluation (T P E A A)	Comments
1.581E-08	6.250E-06	11.5	D085	2 0 2 2 2	mixed isomers

2399. C₁₂H₃Cl₇
2,2',3,4',5,5',6-Heptachlorobiphenyl
1,1'-Biphenyl, 2,2',3,4',5,5',6-Heptachloro-
PCB 187
RN: 52663-68-0   **MP** (°C):   104
MW: 395.33   **BP** (°C):

Solubility (Moles/L)	Solubility (Grams/L)	Temp (°C)	Ref (#)	Evaluation (T P E A A)	Comments
1.141E-08	4.510E-06	20	M336	2 0 2 2 2	

2400. C₁₂H₄Br₆
FireMaster FF-1 (Hexabromobiphenyl mixture)
RN: **MP** (°C):
MW: 627.62   **BP** (°C):

Solubility (Moles/L)	Solubility (Grams/L)	Temp (°C)	Ref (#)	Evaluation (T P E A A)	Comments
1.753E-08	1.100E-05	25	H303	1 0 0 0 1	

2401. C₁₂H₄Br₆
Fire Master BP-6 (Hexabromophenyl mixture)
RN: 59536-65-1   **MP** (°C):
MW: 627.62   **BP** (°C):

Solubility (Moles/L)	Solubility (Grams/L)	Temp (°C)	Ref (#)	Evaluation (T P E A A)	Comments
1.753E-08	1.100E-05	25	H303	1 0 0 0 1	

2402. C₁₂H₄Br₆
2,2',4,4',6,6'-Hexabromobiphenyl
Hexabromobiphenyl
Polybromilated Biphenyl
RN: 36355-01-8   **MP** (°C):   72
MW: 627.62   **BP** (°C):

Solubility (Moles/L)	Solubility (Grams/L)	Temp (°C)	Ref (#)	Evaluation (T P E A A)	Comments
9.954E-04	6.247E-01	26.5	G312	2 0 0 1 2	

2403.   C$_{12}$H$_4$Cl$_4$O
2,3,7,8-Tetrachlorodibenzofuran
2,3,7,8-T4CDF
RN: 51207-31-9   **MP** (°C):   227
MW: 305.98   **BP** (°C):

Solubility (Moles/L)	Solubility (Grams/L)	Temp (°C)	Ref (#)	Evaluation (T P E A A)	Comments
1.370E-09	4.192E-07	22.5	F314	1 1 0 2 2	

2404.   C$_{12}$H$_4$Cl$_4$O$_2$
1,2,3,4-Tetrachlorodibenzo-p-dioxin
1,2,3,4-TCDD
1,2,3,4-Tetrachlorodibenzo[b,e][1,4]dioxin
RN: 30746-58-8   **MP** (°C):   184-186
MW: 321.98   **BP** (°C):

Solubility (Moles/L)	Solubility (Grams/L)	Temp (°C)	Ref (#)	Evaluation (T P E A A)	Comments
3.510E-10	1.130E-07	4.0	D330	2 2 1 2 2	
4.007E-11	1.290E-08	4.3	L321	2 1 2 2 2	
1.065E-09	3.430E-07	5	S352	2 2 0 2 2	
1.401E-09	4.510E-07	15	S352	2 2 0 2 2	
1.500E-09	4.830E-07	17.3	L321	2 1 2 2 2	
1.708E-09	5.500E-07	25	S352	2 2 0 2 1	average of 2
1.957E-09	6.300E-07	25	S352	2 2 0 2 1	
1.460E-09	4.701E-07	25.0	D330	2 2 1 2 2	
3.541E-09	1.140E-06	35	S352	2 2 0 2 2	
3.630E-09	1.169E-06	40.0	D330	2 2 1 2 2	
6.476E-09	2.085E-06	45	S352	2 2 0 2 2	

2405.   C$_{12}$H$_4$Cl$_4$O$_2$
1,2,3,7-Tetrachlorodibenzo-p-dioxin
PCDD 29
RN: 67028-18-6   **MP** (°C):   175
MW: 321.98   **BP** (°C):

Solubility (Moles/L)	Solubility (Grams/L)	Temp (°C)	Ref (#)	Evaluation (T P E A A)	Comments
7.560E-10	2.434E-07	7.0	F315	1 2 0 2 2	
8.120E-10	2.614E-07	11.5	F315	1 2 0 2 2	
1.250E-09	4.025E-07	17.0	F315	1 2 0 2 2	
1.336E-09	4.300E-07	20	F303	1 2 1 2 1	
1.490E-09	4.797E-07	21.0	F315	1 2 0 2 2	
2.260E-09	7.277E-07	26.0	F315	1 2 0 2 2	
3.944E-09	1.270E-06	40	F303	1 2 1 2 1	
4.330E-09	1.394E-06	41.0	F315	1 2 0 2 2	

2406. $C_{12}H_4Cl_4O_2$
1,3,6,8-Tetrachlorodibenzo-p-dioxin
PCDD 42
1,3,6,8-Tetrachlorodibenzo[1,4]dioxin
RN: 33423-92-6   **MP** (°C):   219
MW: 321.98   **BP** (°C):

Solubility (Moles/L)	Solubility (Grams/L)	Temp (°C)	Ref (#)	Evaluation (T P E A A)	Comments
9.939E-10	3.200E-07	20	F303	1 2 1 2 1	
9.939E-10	3.200E-07	20	W319	1 2 1 2 1	
1.211E-09	3.900E-07	40	F303	1 2 1 2 1	
1.211E-09	3.900E-07	40	W319	1 2 1 2 1	
9.845E-10	3.170E-07	ns	W332	0 1 0 2 2	

2407. $C_{12}H_4Cl_4O_2$
2,3,7,8-Tetrachlorodibenzo-p-dioxin
TCDD
2,3,7,8-Tetrachlorodibenzodioxin
RN: 1746-01-6   **MP** (°C):   310
MW: 321.98   **BP** (°C):

Solubility (Moles/L)	Solubility (Grams/L)	Temp (°C)	Ref (#)	Evaluation (T P E A A)	Comments
5.994E-11	1.930E-08	22	M340	1 2 2 1 2	
6.212E-10	2.000E-07	ns	C098	0 0 0 0 0	
6.212E-10	2.000E-07	ns	K138	0 0 0 0 2	
6.212E-10	2.000E-07	ns	N320	0 0 0 0 2	
2.457E-11	7.910E-09	rt	A323	0 2 2 1 2	

2408. $C_{12}H_4Cl_6$
Hexachlorobiphenyl
1,1'-Biphenyl, Hexachloro-
RN: 26601-64-9   **MP** (°C):
MW: 360.88   **BP** (°C):

Solubility (Moles/L)	Solubility (Grams/L)	Temp (°C)	Ref (#)	Evaluation (T P E A A)	Comments
2.754E-08	9.940E-06	11.5	D085	2 0 2 2 2	mixed isomers

2409.   C₁₂H₄Cl₆
2,2',4,4',5,5'-Hexachlorobiphenyl
2,4,5,2',4',5'-PCB
2,4,5,2',4',5'-Hexachlorobiphenyl
RN: 35065-27-1   **MP** (°C):   103
MW: 360.88   **BP** (°C):

Solubility (Moles/L)	Solubility (Grams/L)	Temp (°C)	Ref (#)	Evaluation (T P E A A)	Comments
1.280E-08	4.619E-06	4.0	D330	2 2 1 2 2	
2.533E-08	9.140E-06	20	M336	2 0 2 2 2	*sic*
3.187E-09	1.150E-06	22	O311	2 2 1 2 2	
2.632E-09	9.500E-07	24	C053	1 0 2 2 1	
2.632E-09	9.500E-07	24	F071	1 1 2 1 1	
2.632E-09	9.500E-07	24	M344	1 0 0 0 1	
2.390E-09	8.625E-07	25	D306	2 1 2 2 2	
3.325E-09	1.200E-06	25	W025	1 0 2 2 1	
2.340E-08	8.445E-06	25.0	D330	2 2 1 2 2	
3.540E-08	1.278E-05	40	D330	2 2 1 2 2	
2.641E-09	9.530E-07	ns	H058	0 1 2 1 2	

2410.   C₁₂H₄Cl₆
2,2',3,3',4,5-Hexachlorobiphenyl
2,3,4,5,2',3'-Hexachlorbiphenyl
2,2',3,3',4,5'-Hexachlorobiphenyl
PCB 129
RN: 55215-18-4   **MP** (°C):   101
MW: 360.88   **BP** (°C):

Solubility (Moles/L)	Solubility (Grams/L)	Temp (°C)	Ref (#)	Evaluation (T P E A A)	Comments
1.577E-08	5.690E-06	20	M336	2 0 2 2 2	
1.610E-08	5.810E-06	25	D306	2 1 2 2 2	
2.355E-09	8.500E-07	25	W025	1 0 2 2 1	

2411.   C₁₂H₄Cl₆
2,2',3,4,4',6-Hexachlorobiphenyl
1,1'-Biphenyl, 2,2',3,4,4',6-Hexachloro-
PCB 139
RN: 56030-56-9   **MP** (°C):   73
MW: 360.88   **BP** (°C):

Solubility (Moles/L)	Solubility (Grams/L)	Temp (°C)	Ref (#)	Evaluation (T P E A A)	Comments
3.372E-08	1.217E-05	20	M336	2 0 2 2 2	

2412. C₁₂H₄Cl₆
2,3,3',4,4',5'-Hexachlorobiphenyl
2,3,3',4,4',5-Hexachlorobiphenyl
RN: 38380-08-4 **MP** (°C): 127
MW: 360.88 **BP** (°C):

Solubility (Moles/L)	Solubility (Grams/L)	Temp (°C)	Ref (#)	Evaluation (T P E A A)	Comments
1.477E-08	5.330E-06	20	M336	2 0 2 2 2	

2413. C₁₂H₄Cl₆
2,2',3,3',6,6'-Hexachlorobiphenyl
1,1'-Biphenyl, 2,2',3,3',6,6'-Hexachloro-, (+)-
(+)-PCB 136
RN: 207004-30-6 **MP** (°C): 114
MW: 360.88 **BP** (°C):

Solubility (Moles/L)	Solubility (Grams/L)	Temp (°C)	Ref (#)	Evaluation (T P E A A)	Comments
3.050E-09	1.101E-06	4	D331	2 1 2 2 2	
3.050E-09	1.101E-06	4.0	M324	1 2 1 1 2	
9.010E-09	3.252E-06	20	D331	2 1 2 2 2	
5.586E-08	2.016E-05	20	M336	2 0 2 2 2	
9.010E-09	3.252E-06	20.0	M324	1 2 1 1 2	
1.250E-08	4.511E-06	25	D331	2 1 2 2 2	
1.250E-08	4.510E-06	25	D335	1 0 0 0 2	
1.670E-08	6.027E-06	25	M342	1 0 1 1 2	
1.250E-08	4.511E-06	25.0	M324	1 2 1 1 2	
1.850E-08	6.676E-06	32	D331	2 1 2 2 2	
1.850E-08	6.676E-06	32.0	M324	1 2 1 1 2	
1.670E-08	6.027E-06	ns	M308	0 0 1 1 2	

2414. C₁₂H₄Cl₆
2,2',3,4',5,5'-Hexachlorobiphenyl
1,1'-Biphenyl, 2,2',3,4',5,5'-Hexachloro-
PCB 146
RN: 51908-16-8 **MP** (°C):
MW: 360.88 **BP** (°C):

Solubility (Moles/L)	Solubility (Grams/L)	Temp (°C)	Ref (#)	Evaluation (T P E A A)	Comments
2.103E-08	7.590E-06	20	M336	2 0 2 2 2	

2415. C₁₂H₄Cl₆

2,2',3,4,4',5'-Hexachlorobiphenyl
1,1'-Biphenyl, 2,2',3,4,4',5'-Hexachloro-
PCB 138
CB 138
K 138
RN: 35065-28-2   **MP** (°C):   80.5
MW: 360.88   **BP** (°C):

Solubility (Moles/L)	Solubility (Grams/L)	Temp (°C)	Ref (#)	Evaluation (T P E A A)	Comments
2.020E-08	7.290E-06	20	M336	2 0 2 2 2	

2416. C₁₂H₄Cl₆

2,2',3,4,4',5-Hexachlorobiphenyl
1,1'-Biphenyl, 2,2',3,4,4',5-Hexachloro-
PCB 137
RN: 35694-06-5   **MP** (°C):   77
MW: 360.88   **BP** (°C):

Solubility (Moles/L)	Solubility (Grams/L)	Temp (°C)	Ref (#)	Evaluation (T P E A A)	Comments
2.328E-08	8.400E-06	20	M336	2 0 2 2 2	

2417. C₁₂H₄Cl₆

2,3,3',4,4',6-Hexachlorobiphenyl
1,1'-Biphenyl, 2,3,3',4,4',6-Hexachloro-
PCB 158
RN: 74472-42-7   **MP** (°C):   107
MW: 360.88   **BP** (°C):

Solubility (Moles/L)	Solubility (Grams/L)	Temp (°C)	Ref (#)	Evaluation (T P E A A)	Comments
2.236E-08	8.070E-06	20	M336	2 0 2 2 2	

2418.   C$_{12}$H$_4$Cl$_6$
2,2',3,3',4,4'-Hexachlorobiphenyl
2,3,4,2',3',4'-Hexachlorbiphenyl
PCB 128
1,1'-Biphenyl, 2,2',3,3',4,4'-Hexachloro-
RN: 38380-07-3   **MP** (°C):   150
MW: 360.88   **BP** (°C):

Solubility (Moles/L)	Solubility (Grams/L)	Temp (°C)	Ref (#)	Evaluation (T P E A A)	Comments
1.857E-08	6.700E-06	20	M336	2 0 2 2 2	*sic*
9.690E-10	3.497E-07	25	D306	2 1 2 2 2	
7.840E-10	2.829E-07	25	M342	1 0 1 1 2	
1.219E-09	4.400E-07	25	W025	1 0 2 2 1	

2419.   C$_{12}$H$_4$Cl$_6$
Aroclor 1260
Arochlor 1260
RN: 11096-82-5   **MP** (°C):
MW: 360.88   **BP** (°C):   402.5

Solubility (Moles/L)	Solubility (Grams/L)	Temp (°C)	Ref (#)	Evaluation (T P E A A)	Comments
3.879E-08	1.400E-05	4	M336	2 0 2 2 1	
3.990E-08	1.440E-05	20	M336	2 0 2 2 2	
6.927E-08	2.500E-05	20	N326	1 0 0 0 1	

2420.   C$_{12}$H$_4$Cl$_6$
2,2',3,3',4,6-Hexachlorobiphenyl
2,2',3,4',5',6'-Hexachlorobiphenyl
PCB 131
RN: 61798-70-7   **MP** (°C):
MW: 360.88   **BP** (°C):

Solubility (Moles/L)	Solubility (Grams/L)	Temp (°C)	Ref (#)	Evaluation (T P E A A)	Comments
3.358E-08	1.212E-05	20	M336	2 0 2 2 2	

2421.   C$_{12}$H$_4$Cl$_6$
2,2',3,3',5,6'-Hexachlorobiphenyl
1,1'-Biphenyl, 2,2',3,3',5,6'-Hexachloro-
PCB 135
RN: 52744-13-5   **MP** (°C):
MW: 360.88   **BP** (°C):

Solubility (Moles/L)	Solubility (Grams/L)	Temp (°C)	Ref (#)	Evaluation (T P E A A)	Comments
3.586E-08	1.294E-05	20	M336	2 0 2 2 2	

2422.   C$_{12}$H$_4$Cl$_6$
2,2',3,3',5,6-Hexachlorobiphenyl
2,3,5,6,2',3'-Hexachlorbiphenyl
RN: 52704-70-8   **MP** (°C):   132
MW: 360.88   **BP** (°C):

Solubility (Moles/L)	Solubility (Grams/L)	Temp (°C)	Ref (#)	Evaluation (T P E A A)	Comments
3.588E-08	1.295E-05	20	M336	2 0 2 2 2	*sic*
2.522E-09	9.100E-07	25	W025	1 0 2 2 1	

2423.   C$_{12}$H$_4$Cl$_6$
2,3,3',4',5,6-Hexachlorobiphenyl
1,1'-Biphenyl, 2,3,3',4',5,6-Hexachloro-
PCB 163
RN: 74472-44-9   **MP** (°C):   122
MW: 360.88   **BP** (°C):

Solubility (Moles/L)	Solubility (Grams/L)	Temp (°C)	Ref (#)	Evaluation (T P E A A)	Comments
1.469E-08	5.300E-06	25	B319	2 0 1 2 1	
1.471E-08	5.310E-06	25	H341	1 0 0 0 2	

2424.   C$_{12}$H$_4$Cl$_6$
2,2',3,5,5',6-Hexachlorobiphenyl
1,1'-Biphenyl, 2,2',3,5,5',6-Hexachloro-
PCB 151
RN: 52663-63-5   **MP** (°C):   100
MW: 360.88   **BP** (°C):

Solubility (Moles/L)	Solubility (Grams/L)	Temp (°C)	Ref (#)	Evaluation (T P E A A)	Comments
3.755E-08	1.355E-05	20	M336	2 0 2 2 2	

2425.   C$_{12}$H$_4$Cl$_6$
2,2',3,4,5,5'-Hexachlorobiphenyl
1,1'-Biphenyl, 2,2',3,4,5,5'-Hexachloro-
PCB 141
RN: 52712-04-6   **MP** (°C):   85
MW: 360.88   **BP** (°C):

Solubility (Moles/L)	Solubility (Grams/L)	Temp (°C)	Ref (#)	Evaluation (T P E A A)	Comments
2.092E-08	7.550E-06	20	M336	2 0 2 2 2	

2426. $C_{12}H_4Cl_6$

2,2',3,4,5',6-Hexachlorobiphenyl
1,1'-Biphenyl, 2,2',3,4,5',6-Hexachloro-
PCB 144

RN: 68194-14-9   **MP** (°C):
MW: 360.88   **BP** (°C):

Solubility (Moles/L)	Solubility (Grams/L)	Temp (°C)	Ref (#)	Evaluation (T P E A A)	Comments
3.586E-08	1.294E-05	20	M336	2 0 2 2 2	

2427. $C_{12}H_4Cl_6$

2,2',4,4',6,6'-Hexachlorobiphenyl
1,1'-Biphenyl, 2,2',4,4',6,6'-Hexachloro-
PCB 155

RN: 33979-03-2   **MP** (°C):   112.5
MW: 360.88   **BP** (°C):

Solubility (Moles/L)	Solubility (Grams/L)	Temp (°C)	Ref (#)	Evaluation (T P E A A)	Comments
3.020E-09	1.090E-06	22	O311	2 2 1 2 2	
6.280E-09	2.266E-06	25	D306	2 1 2 2 2	
9.120E-09	3.291E-06	25	L322	1 1 2 2 2	
1.130E-09	4.078E-07	25	M342	1 0 1 1 2	
2.494E-09	9.000E-07	25	W025	1 0 2 2 1	
1.130E-09	4.078E-07	ns	M308	0 0 1 1 2	

2428. $C_{12}H_5Br_5$

2,2',4,5,5'-Pentabromobiphenyl
1,1'-Biphenyl, 2,2',4,5,5'-Pentabromo-
PBB 101

RN: 67888-96-4   **MP** (°C):
MW: 548.72   **BP** (°C):

Solubility (Moles/L)	Solubility (Grams/L)	Temp (°C)	Ref (#)	Evaluation (T P E A A)	Comments
1.880E-10	1.032E-07	4.0	D330	2 2 1 2 2	
8.060E-10	4.423E-07	25	D330	2 2 1 2 2	
1.790E-09	9.822E-07	40.0	D330	2 2 1 2 2	

2429. C₁₂H₅Cl₃O₂

1,2,4-Trichlorodibenzo-p-dioxin
Dibenzo[b,e][1,4]dioxin, 1,2,4-trichloro-
PCDD 14
RN: 39227-58-2   **MP** (°C):   129
MW: 287.53   **BP** (°C):

Solubility (Moles/L)	Solubility (Grams/L)	Temp (°C)	Ref (#)	Evaluation (T P E A A)	Comments
7.617E-09	2.190E-06	5	S352	2 2 0 2 2	
1.659E-08	4.770E-06	15	S352	2 2 0 2 2	
2.925E-08	8.410E-06	25	S352	2 2 0 2 2	
2.925E-08	8.410E-06	25	S352	2 2 0 2 2	
5.801E-08	1.668E-05	35	S352	2 2 0 2 2	
9.815E-08	2.822E-05	45	S352	2 2 0 2 2	

2430. C₁₂H₅Cl₅

2,2',4,5,5'-Pentachlorobiphenyl
2,4,5,2',5'-PCB
2,2',4,5,5'-PCB
RN: 37680-73-2   **MP** (°C):   77
MW: 326.44   **BP** (°C):

Solubility (Moles/L)	Solubility (Grams/L)	Temp (°C)	Ref (#)	Evaluation (T P E A A)	Comments
1.880E-08	6.137E-06	4	D331	2 1 2 2 2	
1.880E-08	6.137E-06	4.0	M324	1 2 1 1 2	
3.710E-08	1.211E-05	20	D331	2 1 2 2 2	
8.044E-08	2.626E-05	20	M336	2 0 2 2 2	
3.710E-08	1.211E-05	20.0	M324	1 2 1 1 2	
3.063E-08	1.000E-05	24	C053	1 0 2 2 1	
3.370E-08	1.100E-05	24	C311	1 0 2 2 1	EFG
3.063E-08	1.000E-05	24	F071	1 1 2 1 1	
3.063E-08	1.000E-05	24	M344	1 0 0 0 1	
3.370E-08	1.100E-05	25	C313	1 0 2 2 2	
2.070E-08	6.757E-06	25	D306	2 1 2 2 2	
4.720E-08	1.541E-05	25	D331	2 1 2 2 2	
4.718E-08	1.540E-05	25	D335	1 0 0 0 2	
5.920E-08	1.933E-05	25	M342	1 0 1 1 2	
1.287E-08	4.200E-06	25	W025	1 0 2 2 1	
4.720E-08	1.541E-05	25.0	M324	1 2 1 1 2	
6.830E-08	2.230E-05	32	D331	2 1 2 2 2	
6.830E-08	2.230E-05	32.0	M324	1 2 1 1 2	
3.155E-08	1.030E-05	ns	H058	0 1 2 1 2	
5.820E-08	1.900E-05	ns	M118	0 1 1 1 1	
5.920E-08	1.933E-05	ns	M308	0 0 1 1 2	

2431. C$_{12}$H$_5$Cl$_5$
2,2',3,3',4-Pentachlorobiphenyl
1,1'-Biphenyl, 2,2',3,3',4-Pentachloro-
PCB 82
RN: 52663-62-4   **MP** (°C):   119
MW: 326.44   **BP** (°C):

Solubility (Moles/L)	Solubility (Grams/L)	Temp (°C)	Ref (#)	Evaluation (T P E A A)	Comments
8.908E-08	2.908E-05	20	M336	2 0 2 2 2	

2432. C$_{12}$H$_5$Cl$_5$
Pentachlorobiphenyl
2,2',4,4',6-Pentachlorobiphenyl
Kanekrol 500
RN: 25429-29-2   **MP** (°C):
MW: 326.44   **BP** (°C):

Solubility (Moles/L)	Solubility (Grams/L)	Temp (°C)	Ref (#)	Evaluation (T P E A A)	Comments
6.341E-08	2.070E-05	11.5	D085	2 0 2 2 2	mixed isomers
9.496E-08	3.100E-05	22.5	G301	2 1 0 1 2	

2433. C$_{12}$H$_5$Cl$_5$
2,2',3,3',5-Pentachlorobiphenyl
1,1'-Biphenyl, 2,2',3,3',5-Pentachloro-
PCB 83
RN: 60145-20-2   **MP** (°C):
MW: 326.44   **BP** (°C):

Solubility (Moles/L)	Solubility (Grams/L)	Temp (°C)	Ref (#)	Evaluation (T P E A A)	Comments
8.648E-08	2.823E-05	20	M336	2 0 2 2 2	

2434. C$_{12}$H$_5$Cl$_5$
2,2',3,3',6-Pentachlorobiphenyl
1,1'-Biphenyl, 2,2',3,3',6-Pentachloro-
PCB 84
RN: 52663-60-2   **MP** (°C):
MW: 326.44   **BP** (°C):

Solubility (Moles/L)	Solubility (Grams/L)	Temp (°C)	Ref (#)	Evaluation (T P E A A)	Comments
1.440E-07	4.702E-05	20	M336	2 0 2 2 2	

2435.   C$_{12}$H$_5$Cl$_5$
2',3,4,5,5'-Pentachlorobiphenyl
1,1'-Biphenyl, 2,3',4',5,5'-Pentachloro-
PCB 124
RN: 70424-70-3   **MP** (°C):   105
MW: 326.44   **BP** (°C):

Solubility (Moles/L)	Solubility (Grams/L)	Temp (°C)	Ref (#)	Evaluation (T P E A A)	Comments
4.843E-08	1.581E-05	20	M336	2 0 2 2 2	

2436.   C$_{12}$H$_5$Cl$_5$
2,2',3,4',6-Pentachlorobiphenyl
2,2',4,6,6'-Pentachlorobiphenyl
PCB 104
RN: 56558-16-8   **MP** (°C):   85
MW: 326.44   **BP** (°C):

Solubility (Moles/L)	Solubility (Grams/L)	Temp (°C)	Ref (#)	Evaluation (T P E A A)	Comments
1.208E-07	3.945E-05	20	M336	2 0 2 2 2	
4.770E-08	1.557E-05	25	D306	2 1 2 2 2	

2437.   C$_{12}$H$_5$Cl$_5$
2,2',3',4,5-Pentachlorobiphenyl
1,1'-Biphenyl, 2,2',3,4',5-Pentachloro-
PCB 87
RN: 41464-51-1   **MP** (°C):   81
MW: 326.44   **BP** (°C):

Solubility (Moles/L)	Solubility (Grams/L)	Temp (°C)	Ref (#)	Evaluation (T P E A A)	Comments
8.703E-08	2.841E-05	20	M336	2 0 2 2 2	

2438.   C$_{12}$H$_5$Cl$_5$
2,2',3,4,4'-Pentachlorobiphenyl
1,1'-Biphenyl, 2,2',3,4,4'-Pentachloro-
PCB 85
RN: 65510-45-4   **MP** (°C):
MW: 326.44   **BP** (°C):

Solubility (Moles/L)	Solubility (Grams/L)	Temp (°C)	Ref (#)	Evaluation (T P E A A)	Comments
6.712E-08	2.191E-05	20	M336	2 0 2 2 2	

2439. $C_{12}H_5Cl_5$
2,2',3,4,5'-Pentachlorobiphenyl
2,3,4,2',5'-Pentachlorbiphenyl
PCB 87
RN: 38380-02-8   **MP** (°C):   112
MW: 326.44   **BP** (°C):

Solubility (Moles/L)	Solubility (Grams/L)	Temp (°C)	Ref (#)	Evaluation (T P E A A)	Comments
9.009E-08	2.941E-05	20	M336	2 0 2 2 2	
1.379E-08	4.500E-06	25	W025	1 0 2 2 1	

2440. $C_{12}H_5Cl_5$
2,2',3,4,5-Pentachlorobiphenyl
2,3,4,5,2'-Pentachlorbiphenyl
PCB 86
RN: 55312-69-1   **MP** (°C):   112
MW: 326.44   **BP** (°C):

Solubility (Moles/L)	Solubility (Grams/L)	Temp (°C)	Ref (#)	Evaluation (T P E A A)	Comments
7.046E-08	2.300E-05	23	W024	0 0 0 0 0	
7.046E-08	2.300E-05	23	W024	0 0 0 0 0	
1.042E-07	3.400E-05	25	B319	2 0 1 2 1	
1.069E-07	3.490E-05	25	H341	1 0 0 0 2	
3.002E-08	9.800E-06	25	W025	1 0 2 2 2	

2441. $C_{12}H_5Cl_5$
2,2',3,4,6-Pentachlorobiphenyl
2,3,4,6,2'-Pentachlorbiphenyl
PCB 88
RN: 55215-17-3   **MP** (°C):   63
MW: 326.44   **BP** (°C):

Solubility (Moles/L)	Solubility (Grams/L)	Temp (°C)	Ref (#)	Evaluation (T P E A A)	Comments
3.676E-08	1.200E-05	25	W025	1 0 2 2 2	

2442. $C_{12}H_5Cl_5$
2,2',3,5',6-Pentachlorobiphenyl
1,1'-Biphenyl, 2,2',3,5',6-Pentachloro-
PCB 95
RN: 38379-99-6   **MP** (°C):   94
MW: 326.44   **BP** (°C):

Solubility (Moles/L)	Solubility (Grams/L)	Temp (°C)	Ref (#)	Evaluation (T P E A A)	Comments
1.658E-07	5.413E-05	20	M336	2 0 2 2 2	

2443. C$_{12}$H$_5$Cl$_5$
2',3,3',4,5-Pentachlorobiphenyl
1,1'-Biphenyl, 2,3,3',4',5'-Pentachloro-
PCB 122
RN: 76842-07-4   **MP** (°C):
MW: 326.44   **BP** (°C):

Solubility (Moles/L)	Solubility (Grams/L)	Temp (°C)	Ref (#)	Evaluation (T P E A A)	Comments
3.933E-08	1.284E-05	20	M336	2 0 2 2 2	

2444. C$_{12}$H$_5$Cl$_5$
2,3,4,5,6-Pentachlorobiphenyl
1,1'-Biphenyl, 2,3,4,5,6-Pentachloro-
PCB 116
RN: 18259-05-7   **MP** (°C):   123
MW: 326.44   **BP** (°C):

Solubility (Moles/L)	Solubility (Grams/L)	Temp (°C)	Ref (#)	Evaluation (T P E A A)	Comments
4.166E-08	1.360E-05	22	O311	2 2 1 2 2	
1.230E-08	4.015E-06	25	D306	2 1 2 2 2	
1.680E-08	5.484E-06	25	M342	1 0 1 1 2	
2.083E-08	6.800E-06	25	W025	1 0 2 2 1	
1.680E-08	5.484E-06	ns	M308	0 0 1 1 2	

2445. C$_{12}$H$_5$Cl$_5$
2,2',4,4',5-Pentachlorobiphenyl
1,1'-Biphenyl, 2,2',4,4',5-Pentachloro-
PCB 99
RN: 38380-01-7   **MP** (°C):
MW: 326.44   **BP** (°C):

Solubility (Moles/L)	Solubility (Grams/L)	Temp (°C)	Ref (#)	Evaluation (T P E A A)	Comments
6.798E-08	2.219E-05	20	M336	2 0 2 2 2	

2446. C$_{12}$H$_5$Cl$_5$
2,3',4,4',5-Pentachlorobiphenyl
1,1'-Biphenyl, 2,3',4,4',5-Pentachloro-
PCB 118
CB 118
RN: 31508-00-6   **MP** (°C):   109
MW: 326.44   **BP** (°C):

Solubility (Moles/L)	Solubility (Grams/L)	Temp (°C)	Ref (#)	Evaluation (T P E A A)	Comments
4.117E-08	1.344E-05	20	M336	2 0 2 2 2	

2447. $C_{12}H_5Cl_5$
2,3,3',4',5-Pentachlorobiphenyl
1,1'-Biphenyl, 2,3,3',4',5-Pentachloro-
PCB 107
RN: 70424-68-9   **MP** (°C):
MW: 326.44   **BP** (°C):

Solubility (Moles/L)	Solubility (Grams/L)	Temp (°C)	Ref (#)	Evaluation (T P E A A)	Comments
4.546E-08	1.484E-05	20	M336	2 0 2 2 2	

2448. $C_{12}H_5Cl_5$
2,3,3',4',6-Pentachlorobiphenyl
1,1'-Biphenyl, 2,3,3',4',6-Pentachloro-
PCB 110
RN: 38380-03-9   **MP** (°C):
MW: 326.44   **BP** (°C):

Solubility (Moles/L)	Solubility (Grams/L)	Temp (°C)	Ref (#)	Evaluation (T P E A A)	Comments
8.829E-08	2.882E-05	20	M336	2 0 2 2 2	

2449. $C_{12}H_5Cl_5$
2,3,4,4',5-Pentachlorobiphenyl
1,1'-Biphenyl, 2,3,4,4',5-Pentachloro-
PCB 114
RN: 74472-37-0   **MP** (°C): 98
MW: 326.44   **BP** (°C):

Solubility (Moles/L)	Solubility (Grams/L)	Temp (°C)	Ref (#)	Evaluation (T P E A A)	Comments
4.895E-08	1.598E-05	20	M336	2 0 2 2 2	

2450. $C_{12}H_5N_5O_{11}$
Pentanitrophenylether
Benzene, 2-(2,4-Dinitrophenoxy)-1,3,5-trinitro-
RN: 5950-87-8   **MP** (°C):
MW: 395.20   **BP** (°C):

Solubility (Moles/L)	Solubility (Grams/L)	Temp (°C)	Ref (#)	Evaluation (T P E A A)	Comments
1.771E-04	7.000E-02	27	D067	1 2 0 0 0	
4.302E-04	1.700E-01	50	D067	1 2 0 0 1	
2.404E-03	9.500E-01	100	D067	1 2 0 0 1	

2451. $C_{12}H_5N_7O_{12}$
Hexanitrodiphenylamine
Benzenamine, 2,4,6-Trinitro-N-(2,4,6-trinitrophenyl)-
RN: 131-73-7   **MP** (°C):
MW: 439.21   **BP** (°C):

Solubility (Moles/L)	Solubility (Grams/L)	Temp (°C)	Ref (#)	Evaluation (T P E A A)	Comments
1.366E-04	6.000E-02	17	D070	1 2 0 0 0	
4.325E-04	1.900E-01	50	D070	1 2 0 0 1	
7.738E-04	3.399E-01	100	D070	1 2 0 0 1	

2452. $C_{12}H_6Br_4$
2,2',5,5'-Tetrabromobiphenyl
Tetrabromobiphenyl
RN: 59080-37-4   **MP** (°C):   143
MW: 469.82   **BP** (°C):

Solubility (Moles/L)	Solubility (Grams/L)	Temp (°C)	Ref (#)	Evaluation (T P E A A)	Comments
8.630E-03	4.054E+00	26.5	G312	2 0 0 1 2	

2453. $C_{12}H_6Cl_2O$
2,8-Dichlorodibenzofuran
2,8-DCDF
DCDF
RN: 5409-83-6   **MP** (°C):   184
MW: 237.09   **BP** (°C):

Solubility (Moles/L)	Solubility (Grams/L)	Temp (°C)	Ref (#)	Evaluation (T P E A A)	Comments
1.620E-08	3.841E-06	4.5	D330	2 2 1 2 2	
6.110E-08	1.449E-05	25	D330	2 2 1 2 2	
1.462E-17	3.467E-15	25	O320	0 0 1 1 1	
1.430E-07	3.390E-05	39.5	D330	2 2 1 2 2	

2454.   C$_{12}$H$_6$Cl$_2$O$_2$
2,7-Dichlorodibenzo-p-dioxin
2,7-DCDD
2,8-Dichlorodibenzodioxin
RN: 33857-26-0   **MP** (°C):   201
MW: 253.09   **BP** (°C):

Solubility (Moles/L)	Solubility (Grams/L)	Temp (°C)	Ref (#)	Evaluation (T P E A A)	Comments
4.307E-09	1.090E-06	5	S352	2 2 0 2 2	
7.942E-09	2.010E-06	15	S352	2 2 0 2 2	
1.482E-08	3.750E-06	25	S352	2 2 0 2 2	
1.482E-08	3.750E-06	25	S352	2 2 0 2 2	
2.873E-08	7.270E-06	35	S352	2 2 0 2 2	
5.295E-08	1.340E-05	45	S352	2 2 0 2 2	

2455.   C$_{12}$H$_6$Cl$_2$O$_2$
2,3-Dichlorodibenzo-p-dioxin
2,3-Dichlorodibenzodioxin
PCDD 10
RN: 29446-15-9   **MP** (°C):   160
MW: 253.09   **BP** (°C):

Solubility (Moles/L)	Solubility (Grams/L)	Temp (°C)	Ref (#)	Evaluation (T P E A A)	Comments
1.454E-08	3.680E-06	5	S352	2 2 0 2 2	
2.829E-08	7.160E-06	15	S352	2 2 0 2 2	
5.887E-08	1.490E-05	25	S352	2 2 0 2 2	
5.887E-08	1.490E-05	25	S352	2 2 0 2 2	
1.201E-07	3.040E-05	35	S352	2 2 0 2 2	
2.315E-07	5.860E-05	45	S352	2 2 0 2 2	

2456.   C$_{12}$H$_6$Cl$_2$O$_2$
2,8-Dichlorodibenzo-p-dioxin
2,8-Dichlorodibenzodioxin
PCDD 12
3,6-Dichloro-9,10-dioxaanthracene
RN: 38964-22-6   **MP** (°C):   151
MW: 253.09   **BP** (°C):

Solubility (Moles/L)	Solubility (Grams/L)	Temp (°C)	Ref (#)	Evaluation (T P E A A)	Comments
1.746E-08	4.420E-06	5	S352	2 2 0 2 2	
3.394E-08	8.590E-06	15	S352	2 2 0 2 2	
6.599E-08	1.670E-05	25	S352	2 2 0 2 2	
6.614E-08	1.674E-05	25	S352	2 2 0 2 2	
1.088E-07	2.753E-05	35	S352	2 2 0 2 2	
2.035E-07	5.150E-05	45	S352	2 2 0 2 2	

2457.   C$_{12}$H$_6$Cl$_3$NO$_3$
Quinonamid
2-(Dichloroacetamido)-3-chloro-1,4-naphthoquinone
HOE 13465OH
Chinonamid
2-[(Dichloroacetyl)amino]-3-chloro-1,4-naphthoquinone
RN: 27541-88-4   **MP** (°C):   212.5
MW: 318.55   **BP** (°C):

Solubility (Moles/L)	Solubility (Grams/L)	Temp (°C)	Ref (#)	Evaluation (T P E A A)	Comments
9.418E-06	3.000E-03	23	M161	1 0 0 0 0	pH 4.6

2458.   C$_{12}$H$_6$Cl$_3$NO$_3$
Chlornitrofen
4-Nitrophenyl 2,4,6-trichlorophenyl Ether
1,3,5-Trichloro-2-(4-nitrophenoxy)benzene
1',3',5'-Trichlorophenyl-4-nitrophenyl Ether
RN: 1836-77-7   **MP** (°C):
MW: 318.55   **BP** (°C):

Solubility (Moles/L)	Solubility (Grams/L)	Temp (°C)	Ref (#)	Evaluation (T P E A A)	Comments
2.398E-06	7.640E-04	22	K137	1 1 2 1 0	

2459.   C$_{12}$H$_6$Cl$_4$
Tetrachlorobiphenyl
1,1'-Biphenyl, Tetrachloro-
Pyralene 1498
RN: 26914-33-0   **MP** (°C):
MW: 291.99   **BP** (°C):

Solubility (Moles/L)	Solubility (Grams/L)	Temp (°C)	Ref (#)	Evaluation (T P E A A)	Comments
1.825E-07	5.330E-05	11.5	D085	2 0 2 2 2	mixed isomers

2460.   C$_{12}$H$_6$Cl$_4$
Aroclor 1248
Arochlor 1248
RN: 12672-29-6   **MP** (°C):
MW: 291.99   **BP** (°C):   357.5

Solubility (Moles/L)	Solubility (Grams/L)	Temp (°C)	Ref (#)	Evaluation (T P E A A)	Comments
3.425E-07	1.000E-04	20	N326	1 0 0 0 2	

2461. $C_{12}H_6Cl_4$

Aroclor 1254
Arochlor 1254
RN: 11097-69-1   **MP** (°C):
MW: 291.99   **BP** (°C):

Solubility (Moles/L)	Solubility (Grams/L)	Temp (°C)	Ref (#)	Evaluation (T P E A A)	Comments
1.336E-07	3.900E-05	4	M336	2 0 2 2 1	
8.288E-08	2.420E-05	11.5	D085	2 0 2 2 2	
9.623E-08	2.810E-05	16.50	W033	1 0 2 2 2	
8.459E-08	2.470E-05	16.50	W033	1 0 2 2 2	
1.473E-07	4.300E-05	20	M336	2 0 2 2 1	
1.712E-07	5.000E-05	20	N326	1 0 0 0 1	
~1.92E-07	~5.60E-05	ns	H117	0 2 2 2 0	
1.541E-07	4.500E-05	ns	L106	0 0 2 1 1	
1.370E-07	4.000E-05	ns	M184	0 0 0 0 0	

2462. $C_{12}H_6Cl_4$

3,3',5,5'-Tetrachlorobiphenyl
1,1'-Biphenyl, 3,3',5,5'-Tetrachloro-
PCB 80
RN: 33284-52-5   **MP** (°C):   164
MW: 291.99   **BP** (°C):

Solubility (Moles/L)	Solubility (Grams/L)	Temp (°C)	Ref (#)	Evaluation (T P E A A)	Comments
4.220E-09	1.232E-06	25	D306	2 1 2 2 2	

2463. $C_{12}H_6Cl_4$

3,3',4,4'-Tetrachlorobiphenyl
3,4,3',4'-Tetrachlorbiphenyl
RN: 32598-13-3   **MP** (°C):   183
MW: 291.99   **BP** (°C):

Solubility (Moles/L)	Solubility (Grams/L)	Temp (°C)	Ref (#)	Evaluation (T P E A A)	Comments
5.000E-10	1.460E-07	4	D331	2 1 2 2 2	
5.000E-10	1.460E-07	4.0	M324	1 2 1 1 2	
1.490E-09	4.351E-07	20	D331	2 1 2 2 2	
1.490E-09	4.351E-07	20.0	M324	1 2 1 1 2	
6.165E-09	1.800E-06	22	O311	2 2 1 2 1	
1.404E-07	4.100E-05	23	W024	0 0 0 0 0	*sic*
1.880E-09	5.489E-07	25	D306	2 1 2 2 2	
1.950E-09	5.694E-07	25	D331	2 1 2 2 2	
1.949E-09	5.690E-07	25	D335	1 0 0 0 2	
2.569E-09	7.500E-07	25	W025	1 0 2 2 1	
1.950E-09	5.694E-07	25.0	M324	1 2 1 1 2	
4.040E-09	1.180E-06	32	D331	2 1 2 2 2	
4.040E-09	1.180E-06	32.0	M324	1 2 1 1 2	

2464.   C₁₂H₆Cl₄
2,4,4',6-Tetrachlorobiphenyl
1,1'-Biphenyl, 2,4,4',6-Tetrachloro-
PCB 75
RN: 32598-12-2   **MP** (°C):   65
MW: 291.99   **BP** (°C):

Solubility (Moles/L)	Solubility (Grams/L)	Temp (°C)	Ref (#)	Evaluation (T P E A A)	Comments
3.120E-07	9.110E-05	25	D306	2 1 2 2 2	

2465.   C₁₂H₆Cl₄
2,3,4,5-Tetrachlorobiphenyl
1,1'-Biphenyl, 2,3,4,5-Tetrachloro-
PCB 61
RN: 33284-53-6   **MP** (°C):   92
MW: 291.99   **BP** (°C):

Solubility (Moles/L)	Solubility (Grams/L)	Temp (°C)	Ref (#)	Evaluation (T P E A A)	Comments
3.390E-08	9.900E-06	25	B319	2 0 1 2 1	
4.780E-08	1.396E-05	25	D306	2 1 2 2 2	
4.677E-08	1.366E-05	25	L322	1 1 2 2 2	
7.170E-08	2.094E-05	25	M342	1 0 1 1 2	
6.575E-08	1.920E-05	25	W025	1 0 2 2 2	
7.170E-08	2.094E-05	ns	M308	0 0 1 1 2	

2466.   C₁₂H₆Cl₄
2,4,4',5-Tetrachlorobiphenyl
1,1'-Biphenyl, 2,4,4',5-Tetrachloro-
PCB 74
RN: 32690-93-0   **MP** (°C):
MW: 291.99   **BP** (°C):

Solubility (Moles/L)	Solubility (Grams/L)	Temp (°C)	Ref (#)	Evaluation (T P E A A)	Comments
1.049E-07	3.064E-05	20	M336	2 0 2 2 2	

2467. C$_{12}$H$_6$Cl$_4$
2,2',4,5-Tetrachlorobiphenyl
1,1'-Biphenyl, 2,2',4,5-Tetrachloro-
PCB 48

RN: 70362-47-9   **MP** (°C):   63.9
MW: 291.99   **BP** (°C):

Solubility (Moles/L)	Solubility (Grams/L)	Temp (°C)	Ref (#)	Evaluation (T P E A A)	Comments
1.026E-07	2.995E-05	20	M336	2 0 2 2 2	
5.630E-08	1.644E-05	25	M342	1 0 1 1 2	

2468. C$_{12}$H$_6$Cl$_4$
2,2',4,5'-Tetrachlorobiphenyl
2,2',4',5-Tetrachlorobiphenyl
PCB 49

RN: 41464-40-8   **MP** (°C):   67
MW: 291.99   **BP** (°C):

Solubility (Moles/L)	Solubility (Grams/L)	Temp (°C)	Ref (#)	Evaluation (T P E A A)	Comments
2.676E-07	7.814E-05	20	M336	2 0 2 2 2	
5.630E-08	1.644E-05	ns	M308	0 0 1 1 2	

2469. C$_{12}$H$_6$Cl$_4$
2',3,4,5-Tetrachlorobiphenyl
1,1'-Biphenyl, 2,3',4',5'-Tetrachloro-
PCB 76

RN: 70362-48-0   **MP** (°C):   92.0
MW: 291.99   **BP** (°C):

Solubility (Moles/L)	Solubility (Grams/L)	Temp (°C)	Ref (#)	Evaluation (T P E A A)	Comments
1.888E-07	5.513E-05	20	M336	2 0 2 2 2	

2470. C$_{12}$H$_6$Cl$_4$
2,2',3,3'-Tetrachlorobiphenyl
1,1'-Biphenyl, 2,2',3,3'-Tetrachloro-
PCB 40

RN: 38444-93-8   **MP** (°C):   121.0
MW: 291.99   **BP** (°C):

Solubility (Moles/L)	Solubility (Grams/L)	Temp (°C)	Ref (#)	Evaluation (T P E A A)	Comments
2.764E-07	8.070E-05	20	M336	2 0 2 2 2	
5.822E-07	1.700E-04	23	W024	0 0 0 0 0	
5.340E-08	1.559E-05	25	D306	2 1 2 2 2	

2471. C$_{12}$H$_6$Cl$_4$
2,2',3,4'-Tetrachlorobiphenyl
1,1'-Biphenyl, 2,2',3,4'-Tetrachloro-
PCB 42
RN: 36559-22-5   **MP** (°C):   68
MW: 291.99   **BP** (°C):

Solubility (Moles/L)	Solubility (Grams/L)	Temp (°C)	Ref (#)	Evaluation (T P E A A)	Comments
2.083E-07	6.083E-05	20	M336	2 0 2 2 2	

2472. C$_{12}$H$_6$Cl$_4$
2,2',3,4-Tetrachlorobiphenyl
1,1'-Biphenyl, 2,2',3,4-Tetrachloro-
PCB 41
RN: 52663-59-9   **MP** (°C):
MW: 291.99   **BP** (°C):

Solubility (Moles/L)	Solubility (Grams/L)	Temp (°C)	Ref (#)	Evaluation (T P E A A)	Comments
2.219E-07	6.480E-05	20	M336	2 0 2 2 2	

2473. C$_{12}$H$_6$Cl$_4$
2,2',3,5'-Tetrachlorobiphenyl
1,1'-Biphenyl, 2,2',3,5'-Tetrachloro-
PCB 44
RN: 41464-39-5   **MP** (°C):   47
MW: 291.99   **BP** (°C):

Solubility (Moles/L)	Solubility (Grams/L)	Temp (°C)	Ref (#)	Evaluation (T P E A A)	Comments
3.426E-07	1.001E-04	20	M336	2 0 2 2 2	
2.226E-07	6.500E-05	23	W024	0 0 0 0 0	
2.740E-07	8.000E-05	25	B319	2 0 1 2 0	

2474. C$_{12}$H$_6$Cl$_4$
2,2',4,4'-Tetrachlorobiphenyl
1,1'-Biphenyl, 2,2',4,4'-Tetrachloro-
PCB 47
RN: 2437-79-8   **MP** (°C):   42.0
MW: 291.99   **BP** (°C):

Solubility (Moles/L)	Solubility (Grams/L)	Temp (°C)	Ref (#)	Evaluation (T P E A A)	Comments
1.853E-07	5.410E-05	22	O311	2 2 1 2 2	
5.993E-07	1.750E-04	23	W024	0 0 0 0 0	
7.534E-07	2.200E-04	25	B351	1 0 0 1 1	

2475. C$_{12}$H$_6$Cl$_4$
2,3,4,4'-Tetrachlorobiphenyl
1,1'-Biphenyl, 2,3,4,4'-Tetrachloro-
PCB 60
RN: 33025-41-1   **MP** (°C):   142
MW: 291.99   **BP** (°C):

Solubility (Moles/L)	Solubility (Grams/L)	Temp (°C)	Ref (#)	Evaluation (T P E A A)	Comments
1.333E-07	3.893E-05	20	M336	2 0 2 2 2	

2476. C$_{12}$H$_6$Cl$_4$
2,2',5,5'-Tetrachlorobiphenyl
1,1'-Biphenyl, 2,2',5,5'-Tetrachloro-
PCB 52
RN: 35693-99-3   **MP** (°C):   87
MW: 291.99   **BP** (°C):

Solubility (Moles/L)	Solubility (Grams/L)	Temp (°C)	Ref (#)	Evaluation (T P E A A)	Comments
3.855E-07	1.126E-04	20	M336	2 0 2 2 2	
5.240E-08	1.530E-05	22	O311	2 2 1 2 2	
1.575E-07	4.600E-05	23	W024	0 0 0 0 0	
5.822E-07	1.700E-04	25	B319	2 0 1 2 2	
3.750E-07	1.095E-04	25	D306	2 1 2 2 2	
1.250E-07	3.650E-05	25	H341	1 0 0 0 2	
1.884E-07	5.500E-05	ns	B301	0 2 1 1 1	
9.076E-08	2.650E-05	ns	H058	0 1 2 1 2	
5.480E-08	1.600E-05	ns	M118	0 1 1 1 1	

2477. C$_{12}$H$_6$Cl$_4$
2,2',5,6'-Tetrachlorobiphenyl
1,1'-Biphenyl, 2,2',5,6'-Tetrachloro-
PCB 53
RN: 41464-41-9   **MP** (°C):   103
MW: 291.99   **BP** (°C):

Solubility (Moles/L)	Solubility (Grams/L)	Temp (°C)	Ref (#)	Evaluation (T P E A A)	Comments
3.717E-07	1.085E-04	20	M336	2 0 2 2 2	
1.630E-07	4.759E-05	25	D306	2 1 2 2 2	

2478. C_{12}H_{6}Cl_{4}
2,2',6,6'-Tetrachlorobiphenyl
1,1'-Biphenyl, 2,2',6,6'-Tetrachloro-
PCB 54
RN: 15968-05-5   **MP** (°C):   198.0
MW: 291.99   **BP** (°C):

Solubility (Moles/L)	Solubility (Grams/L)	Temp (°C)	Ref (#)	Evaluation (T P E A A)	Comments
9.247E-09	2.700E-06	22	O311	2 2 1 2 1	
4.070E-08	1.188E-05	25	D306	2 1 2 2 2	

2479. C_{12}H_{6}Cl_{4}
2,3',4',5-Tetrachlorobiphenyl
1,1'-Biphenyl, 2,3',4',5-Tetrachloro-
PCB 70
RN: 32598-11-1   **MP** (°C):   106
MW: 291.99   **BP** (°C):

Solubility (Moles/L)	Solubility (Grams/L)	Temp (°C)	Ref (#)	Evaluation (T P E A A)	Comments
1.239E-07	3.618E-05	20	M336	2 0 2 2 2	
2.055E-07	6.000E-05	23	W024	0 0 0 0 0	
7.534E-08	2.200E-05	ns	B301	0 2 1 1 1	

2480. C_{12}H_{6}Cl_{4}
2,3',4,4'-Tetrachlorobiphenyl
1,1'-Biphenyl, 2,3',4,4'-Tetrachloro-
PCB 66
RN: 32598-10-0   **MP** (°C):   128.0
MW: 291.99   **BP** (°C):

Solubility (Moles/L)	Solubility (Grams/L)	Temp (°C)	Ref (#)	Evaluation (T P E A A)	Comments
1.259E-07	3.676E-05	20	M336	2 0 2 2 2	

2481. C_{12}H_{6}Cl_{4}
2,3',4,6-Tetrachlorobiphenyl
1,1'-Biphenyl, 2,3',4,6-Tetrachloro-
PCB 69
RN: 60233-24-1   **MP** (°C):   46
MW: 291.99   **BP** (°C):

Solubility (Moles/L)	Solubility (Grams/L)	Temp (°C)	Ref (#)	Evaluation (T P E A A)	Comments
7.004E-08	2.045E-05	20	M336	2 0 2 2 2	

2482. C$_{12}$H$_6$Cl$_4$
2,3,3',4'-Tetrachlorobiphenyl
1,1'-Biphenyl, 2,3,3',4'-Tetrachloro-
PCB 56
RN: 41464-43-1   **MP** (°C):
MW: 291.99   **BP** (°C):

Solubility (Moles/L)	Solubility (Grams/L)	Temp (°C)	Ref (#)	Evaluation (T P E A A)	Comments
1.334E-07	3.894E-05	20	M336	2 0 2 2 2	

2483. C$_{12}$H$_6$Cl$_4$
2,3,4',5-Tetrachlorobiphenyl
1,1'-Biphenyl, 2,3,4',5-Tetrachloro-
PCB 63
RN: 74472-34-7   **MP** (°C):
MW: 291.99   **BP** (°C):

Solubility (Moles/L)	Solubility (Grams/L)	Temp (°C)	Ref (#)	Evaluation (T P E A A)	Comments
8.997E-08	2.627E-05	20	M336	2 0 2 2 2	

2484. C$_{12}$H$_6$Cl$_4$
2,3,4',6-Tetrachlorobiphenyl
1,1'-Biphenyl, 2,3,4',6-Tetrachloro-
PCB 64
RN: 52663-58-8   **MP** (°C):
MW: 291.99   **BP** (°C):

Solubility (Moles/L)	Solubility (Grams/L)	Temp (°C)	Ref (#)	Evaluation (T P E A A)	Comments
3.207E-07	9.365E-05	20	M336	2 0 2 2 2	

2485. C$_{12}$H$_6$Cl$_4$
2,2',3,6'-Tetrachlorobiphenyl
1,1'-Biphenyl, 2,2',3,6'-Tetrachloro-
PCB 46
RN: 41464-47-5   **MP** (°C):
MW: 291.99   **BP** (°C):

Solubility (Moles/L)	Solubility (Grams/L)	Temp (°C)	Ref (#)	Evaluation (T P E A A)	Comments
3.628E-07	1.059E-04	20	M336	2 0 2 2 2	

2486.   C$_{12}$H$_6$Cl$_4$O$_2$S
Tetradifon
2,4,5,4'-Tetrachlorodiphenyl Sulfone
Tedion
Aracnol K
Akaritox
Rotetra
RN: 116-29-0   **MP** (°C):   148.5
MW: 356.06   **BP** (°C):

Solubility (Moles/L)	Solubility (Grams/L)	Temp (°C)	Ref (#)	Evaluation (T P E A A)	Comments
1.404E-07	5.000E-05	10	V301	1 0 0 0 0	
5.617E-04	2.000E-01	50	M161	1 0 0 0 0	
9.549E-07	3.400E-04	50	V301	1 0 0 0 1	

2487.   C$_{12}$H$_7$BrClNO$_2$
Halacrinate
7-Bromo-5-chloro-8-quinolinyl 2-propenoate
Halocrinate
RN: 34462-96-9   **MP** (°C):   100.5
MW: 312.56   **BP** (°C):

Solubility (Moles/L)	Solubility (Grams/L)	Temp (°C)	Ref (#)	Evaluation (T P E A A)	Comments
1.920E-05	6.000E-03	20	M161	1 0 0 0 0	

2488.   C$_{12}$H$_7$ClO$_2$
2-Chlorodibenzo-p-dioxin
2-Monochlorodibenzo-p-dioxin
PCDD 2
RN: 39227-54-8   **MP** (°C):   89
MW: 218.64   **BP** (°C):

Solubility (Moles/L)	Solubility (Grams/L)	Temp (°C)	Ref (#)	Evaluation (T P E A A)	Comments
6.100E-07	1.334E-04	3.90	D330	2 2 1 2 2	
2.904E-07	6.350E-05	5	S352	2 2 0 2 2	
6.266E-07	1.370E-04	15	S352	2 2 0 2 2	
1.363E-06	2.980E-04	25	S352	2 2 0 2 2	average of 2
1.271E-06	2.780E-04	25	S352	2 2 0 2 2	
1.460E-06	3.192E-04	25.0	D330	2 2 1 2 2	
2.987E-06	6.530E-04	35	S352	2 2 0 2 2	
3.430E-06	7.499E-04	39.0	D330	2 2 1 2 2	
5.072E-06	1.109E-03	45	S352	2 2 0 2 2	

2489. C$_{12}$H$_7$ClO$_2$
1-Chlorodibenzo-p-dioxin
1-Monochlorodibenzodioxin
PCDD 1
RN: 39227-53-7   **MP** (°C):   98
MW: 218.64   **BP** (°C):

Solubility (Moles/L)	Solubility (Grams/L)	Temp (°C)	Ref (#)	Evaluation (T P E A A)	Comments
6.220E-07	1.360E-04	5	S352	2 2 0 2 2	
1.066E-06	2.330E-04	15	S352	2 2 0 2 2	
1.907E-06	4.170E-04	25	S352	2 2 0 2 2	
1.907E-06	4.170E-04	25	S352	2 2 0 2 2	
3.316E-06	7.250E-04	35	S352	2 2 0 2 2	
5.671E-06	1.240E-03	45	S352	2 2 0 2 2	

2490. C$_{12}$H$_7$Cl$_2$NO$_3$
Nitrofen
2,4-Dichlorophenyl-4-nitrophenyl Ether
RN: 1836-75-5   **MP** (°C):   70.5
MW: 284.10   **BP** (°C):

Solubility (Moles/L)	Solubility (Grams/L)	Temp (°C)	Ref (#)	Evaluation (T P E A A)	Comments
3.520E-06	1.000E-03	22	M061	1 0 0 0 0	
3.344E-05	9.500E-03	22	M161	1 0 0 0 0	
3.520E-06	1.000E-03	ns	B100	0 0 0 0 0	
2.144E-06	6.090E-04	ns	H322	0 0 0 2 2	

2491. C$_{12}$H$_7$Cl$_3$
2,3',6-Trichlorobiphenyl
1,1'-Biphenyl, 2,3',6-Trichloro-
RN: 38444-76-7   **MP** (°C):
MW: 257.55   **BP** (°C):

Solubility (Moles/L)	Solubility (Grams/L)	Temp (°C)	Ref (#)	Evaluation (T P E A A)	Comments
1.498E-07	3.858E-05	20	M336	2 0 2 2 2	

2492. C$_{12}$H$_7$Cl$_3$
3,4,4'-Trichlorobiphenyl
3,4,4'-Trichlorbiphenyl
RN: 38444-90-5   **MP** (°C):   88
MW: 257.55   **BP** (°C):

Solubility (Moles/L)	Solubility (Grams/L)	Temp (°C)	Ref (#)	Evaluation (T P E A A)	Comments
2.791E-07	7.189E-05	20	M336	2 0 2 2 2	
3.106E-07	8.000E-05	23	W024	0 0 0 0 0	
5.902E-08	1.520E-05	25	W025	1 0 2 2 2	

2493.   C₁₂H₇Cl₃

$\text{2493.} \quad C_{12}H_7Cl_3$

2,4,6-Trichlorobiphenyl
1,1'-Biphenyl, 2,4,6-Trichloro-
RN: 35693-92-6   **MP** (°C):   62.5
MW: 257.55   **BP** (°C):

Solubility (Moles/L)	Solubility (Grams/L)	Temp (°C)	Ref (#)	Evaluation (T P E A A)	Comments
3.120E-07	8.036E-05	4.0	D330	2 2 1 2 2	
9.800E-07	2.524E-04	25	D306	2 1 2 2 2	
9.333E-07	2.404E-04	25	L322	1 1 2 2 2	
8.760E-07	2.256E-04	25	M342	1 0 1 1 2	
7.250E-07	1.867E-04	25.0	D330	2 2 1 2 2	
1.690E-06	4.353E-04	40.0	D330	2 2 1 2 2	
8.760E-07	2.256E-04	ns	M308	0 0 1 1 2	

2494.   C₁₂H₇Cl₃

$\text{2494.} \quad C_{12}H_7Cl_3$

2,4,5-Trichlorobiphenyl
1,1'-Biphenyl, 2,4,5-Trichloro-
RN: 15862-07-4   **MP** (°C):   77
MW: 257.55   **BP** (°C):

Solubility (Moles/L)	Solubility (Grams/L)	Temp (°C)	Ref (#)	Evaluation (T P E A A)	Comments
3.300E-07	8.500E-05	23	W024	0 0 0 0 0	
5.436E-07	1.400E-04	25	B319	2 0 1 2 1	
5.514E-07	1.420E-04	25	H341	1 0 0 0 2	
6.320E-07	1.628E-04	25	M342	1 0 1 1 2	
3.572E-07	9.200E-05	25	W025	1 0 2 2 1	
6.320E-07	1.628E-04	ns	M308	0 0 1 1 2	

2495.   C₁₂H₇Cl₃

$\text{2495.} \quad C_{12}H_7Cl_3$

2,4,4'-Trichlorobiphenyl
2,4,4'-PCB
RN: 7012-37-5   **MP** (°C):   57
MW: 257.55   **BP** (°C):

Solubility (Moles/L)	Solubility (Grams/L)	Temp (°C)	Ref (#)	Evaluation (T P E A A)	Comments
4.465E-07	1.150E-04	20	C302	1 1 2 2 2	
5.559E-07	1.432E-04	20	M336	2 0 2 2 2	
2.601E-07	6.700E-05	22	O311	2 2 1 2 1	
4.271E-07	1.100E-04	24	C311	1 0 2 2 1	EFG
4.504E-07	1.160E-04	25	C313	1 0 2 2 2	
4.530E-07	1.167E-04	25	D306	2 1 2 2 2	
1.010E-06	2.600E-04	25	W025	1 0 2 2 2	

2496.   C₁₂H₇Cl₃

2',3,4-Trichlorobiphenyl
1,1'-Biphenyl, 2',3,4-Trichloro-
RN: 38444-86-9   **MP** (°C):   60.0
MW: 257.55   **BP** (°C):

Solubility (Moles/L)	Solubility (Grams/L)	Temp (°C)	Ref (#)	Evaluation (T P E A A)	Comments
5.147E-07	1.326E-04	20	M336	2 0 2 2 2	
1.165E-07	3.000E-05	23	W024	0 0 0 0 0	

2497.   C₁₂H₇Cl₃

Trichlorobiphenyl
Apirolio 1431C
Pyranol 1499
Pyralene 3011
RN: 25323-68-6   **MP** (°C):
MW: 257.55   **BP** (°C):

Solubility (Moles/L)	Solubility (Grams/L)	Temp (°C)	Ref (#)	Evaluation (T P E A A)	Comments
4.620E-07	1.190E-04	11.5	D085	2 0 2 2 2	mixed isomers

2498.   C₁₂H₇Cl₃

2,3,6-Trichlorobiphenyl
1,1'-Biphenyl, 2,3,6-Trichloro-
RN: 55702-45-9   **MP** (°C):   49
MW: 257.55   **BP** (°C):

Solubility (Moles/L)	Solubility (Grams/L)	Temp (°C)	Ref (#)	Evaluation (T P E A A)	Comments
5.126E-07	1.320E-04	20	M336	2 0 2 2 2	

2499.   C₁₂H₇Cl₃

Aroclor 1242
Arochlor 1242
RN: 53469-21-9   **MP** (°C):
MW: 257.55   **BP** (°C):

Solubility (Moles/L)	Solubility (Grams/L)	Temp (°C)	Ref (#)	Evaluation (T P E A A)	Comments
7.377E-07	1.900E-04	4	M336	2 0 2 2 2	
5.160E-07	1.329E-04	11.5	D085	2 0 2 2 2	
1.076E-06	2.770E-04	20	M336	2 0 2 2 2	
7.766E-07	2.000E-04	20	N326	1 0 0 0 2	
1.747E-07	4.500E-05	ns	L106	0 0 2 1 1	
7.766E-07	2.000E-04	ns	M184	0 0 0 0 0	

2500. C$_{12}$H$_7$Cl$_3$
2,2',3-Trichlorobiphenyl
1,1'-Biphenyl, 2,2',3-Trichloro-
RN: 38444-78-9   **MP** (°C):   28.1
MW: 257.55   **BP** (°C):

Solubility (Moles/L)	Solubility (Grams/L)	Temp (°C)	Ref (#)	Evaluation (T P E A A)	Comments
1.138E-06	2.930E-04	20	M336	2 0 2 2 2	

2501. C$_{12}$H$_7$Cl$_3$
2,2',4-Trichlorobiphenyl
1,1'-Biphenyl, 2,2',4-Trichloro-
RN: 37680-66-3   **MP** (°C):
MW: 257.55   **BP** (°C):

Solubility (Moles/L)	Solubility (Grams/L)	Temp (°C)	Ref (#)	Evaluation (T P E A A)	Comments
1.006E-06	2.592E-04	20	M336	2 0 2 2 2	

2502. C$_{12}$H$_7$Cl$_3$
2,2',5-Trichlorobiphenyl
1,1'-Biphenyl, 2,2',5-Trichloro-
PCB 18
RN: 37680-65-2   **MP** (°C):   44
MW: 257.55   **BP** (°C):

Solubility (Moles/L)	Solubility (Grams/L)	Temp (°C)	Ref (#)	Evaluation (T P E A A)	Comments
1.160E-06	2.986E-04	20	M336	2 0 2 2 2	
1.980E-06	5.099E-04	25	D306	2 1 2 2 2	
2.485E-06	6.400E-04	25	W025	1 0 2 2 2	
4.271E-07	1.100E-04	ns	B301	0 2 1 1 2	
9.629E-07	2.480E-04	ns	H058	0 1 2 1 2	
6.212E-08	1.600E-05	ns	M118	0 1 1 1 1	

2503. C$_{12}$H$_7$Cl$_3$
2,2',6-Trichlorobiphenyl
1,1'-Biphenyl, 2,2',6-Trichloro-
RN: 38444-73-4   **MP** (°C):
MW: 257.55   **BP** (°C):

Solubility (Moles/L)	Solubility (Grams/L)	Temp (°C)	Ref (#)	Evaluation (T P E A A)	Comments
1.741E-06	4.483E-04	20	M336	2 0 2 2 2	

2504. $C_{12}H_7Cl_3$
2,3',5-Trichlorobiphenyl
1,1'-Biphenyl, 2,3',5-Trichloro-
RN: 38444-81-4 **MP** (°C): 40
MW: 257.55 **BP** (°C):

Solubility (Moles/L)	Solubility (Grams/L)	Temp (°C)	Ref (#)	Evaluation (T P E A A)	Comments
5.374E-07	1.384E-04	20	M336	2 0 2 2 2	
9.810E-07	2.527E-04	25	D306	2 1 2 2 2	

2505. $C_{12}H_7Cl_3$
2,3,4'-Trichlorobiphenyl
1,1'-Biphenyl, 2,3,4'-Trichloro-
RN: 38444-85-8 **MP** (°C): 69
MW: 257.55 **BP** (°C):

Solubility (Moles/L)	Solubility (Grams/L)	Temp (°C)	Ref (#)	Evaluation (T P E A A)	Comments
5.500E-07	1.417E-04	20	M336	2 0 2 2 2	

2506. $C_{12}H_7Cl_3$
2,4',5-Trichlorobiphenyl
2,5,4'-Trichlorobiphenyl
PCB 31
RN: 16606-02-3 **MP** (°C): 67
MW: 257.55 **BP** (°C):

Solubility (Moles/L)	Solubility (Grams/L)	Temp (°C)	Ref (#)	Evaluation (T P E A A)	Comments
5.559E-07	1.432E-04	20	M336	2 0 2 2 2	
3.494E-07	9.000E-05	22	O311	2 2 1 2 1	
4.271E-07	1.100E-04	22.5	G301	2 1 0 1 2	
2.912E-07	7.500E-05	ns	B301	0 2 1 1 1	

2507. $C_{12}H_7Cl_3O_2$
Triclosan
5-Chloro-2-(2,4-dichlorophenoxy)-phenol
RN: 3380-34-5 **MP** (°C): 55.2
MW: 289.55 **BP** (°C):

Solubility (Moles/L)	Solubility (Grams/L)	Temp (°C)	Ref (#)	Evaluation (T P E A A)	Comments
3.454E-05	1.000E-02	20	A067	1 0 0 0 0	

2508. C$_{12}$H$_7$N$_3$O$_2$
5-Nitro-1,10-phenanthroline
5-Nitro-o-phenanthroline
RN: 4199-88-6   **MP** (°C):
MW: 225.21   **BP** (°C):

Solubility (Moles/L)	Solubility (Grams/L)	Temp (°C)	Ref (#)	Evaluation (T P E A A)	Comments
1.210E-04	2.725E-02	25.04	B094	1 2 1 2 2	

2509. C$_{12}$H$_7$N$_5$O$_8$
2,4,2',4'-Tetranitrodiphenylamine
2,4,2',4-Tetranitro-diphenylamin
RN: 2908-76-1   **MP** (°C):
MW: 349.22   **BP** (°C):

Solubility (Moles/L)	Solubility (Grams/L)	Temp (°C)	Ref (#)	Evaluation (T P E A A)	Comments
5.727E-04	2.000E-01	100	F300	1 0 0 0 2	

2510. C$_{12}$H$_7$N$_5$O$_8$
2,4,5,6-Tetranitrodiphenylamine
RN: **MP** (°C):
MW: 349.22   **BP** (°C):

Solubility (Moles/L)	Solubility (Grams/L)	Temp (°C)	Ref (#)	Evaluation (T P E A A)	Comments
2.348E-04	8.199E-02	13.5	D070	1 2 0 0 1	
2.949E-04	1.030E-01	50	D070	1 2 0 0 2	
5.783E-04	2.020E-01	100	D070	1 2 0 0 2	

2511. C$_{12}$H$_8$
Acenaphthylene
1,2-Dehydroacenaphthalene
Acenaphthalene
RN: 208-96-8   **MP** (°C):   93.5-94.5
MW: 152.20   **BP** (°C):

Solubility (Moles/L)	Solubility (Grams/L)	Temp (°C)	Ref (#)	Evaluation (T P E A A)	Comments
2.582E-05	3.930E-03	25	L332	1 1 1 1 2	

2512. $C_{12}H_8Br_2$
4,4'-Dibromobiphenyl
p,p'-Dibromobiphenyl

RN: 92-86-4   **MP** (°C):   170
MW: 312.02   **BP** (°C):   357

Solubility (Moles/L)	Solubility (Grams/L)	Temp (°C)	Ref (#)	Evaluation (T P E A A)	Comments
1.841E-02	5.743E+00	26.5	G312	2 0 0 1 2	

2513. $C_{12}H_8Cl_2$
4,4'-Dichlorobiphenyl
4,4'-PCB
Dichlorobiphenyl

RN: 2050-68-2   **MP** (°C):   149
MW: 223.10   **BP** (°C):   317

Solubility (Moles/L)	Solubility (Grams/L)	Temp (°C)	Ref (#)	Evaluation (T P E A A)	Comments
1.488E-06	3.320E-04	11.5	D085	2 0 2 2 2	mixed isomers
2.779E-07	6.200E-05	20	C053	1 0 2 2 1	
2.779E-07	6.200E-05	20	F071	1 1 1 1 1	
2.779E-07	6.200E-05	20	M344	1 0 0 0 1	
2.689E-07	6.000E-05	24	H100	2 0 2 2 0	
2.376E-07	5.300E-05	25	B319	2 0 1 2 2	average of 2
2.062E-07	4.600E-05	25	B350	1 0 0 0 1	
1.630E-07	3.637E-05	25	D306	2 1 2 2 2	
2.913E-07	6.500E-05	25	H341	1 0 0 0 1	
2.510E-07	5.600E-05	25	W025	1 0 2 2 1	

2514. $C_{12}H_8Cl_2$
2,6-Dichlorobiphenyl
1,1'-Biphenyl, 2,6-Dichloro-
PCB 10

RN: 33146-45-1   **MP** (°C):   35
MW: 223.10   **BP** (°C):

Solubility (Moles/L)	Solubility (Grams/L)	Temp (°C)	Ref (#)	Evaluation (T P E A A)	Comments
2.420E-06	5.400E-04	22	O311	2 2 1 2 2	
1.080E-05	2.410E-03	25	D306	2 1 2 2 2	
6.230E-06	1.390E-03	25	M342	1 0 1 1 2	
6.230E-06	1.390E-03	ns	M308	0 0 1 1 2	

2515. $C_{12}H_8Cl_2$
2,3'-Dichlorobiphenyl
1,1'-Biphenyl, 2,3'-Dichloro-
RN: 25569-80-6   **MP** (°C):
MW: 223.10   **BP** (°C):

Solubility (Moles/L)	Solubility (Grams/L)	Temp (°C)	Ref (#)	Evaluation (T P E A A)	Comments
2.599E-06	5.798E-04	20	M336	2 0 2 2 2	

2516. $C_{12}H_8Cl_2$
3,4-Dichlorobiphenyl
1,1'-Biphenyl, 3,4-Dichloro-
RN: 2974-92-7   **MP** (°C):   49.5
MW: 223.10   **BP** (°C):   197.5

Solubility (Moles/L)	Solubility (Grams/L)	Temp (°C)	Ref (#)	Evaluation (T P E A A)	Comments
3.550E-08	7.920E-06	25	D306	2 1 2 2 2	

2517. $C_{12}H_8Cl_2$
3,3'-Dichlorobiphenyl
1,1'-Biphenyl, 3,3'-Dichloro-
RN: 2050-67-1   **MP** (°C):   29
MW: 223.10   **BP** (°C):   323.0

Solubility (Moles/L)	Solubility (Grams/L)	Temp (°C)	Ref (#)	Evaluation (T P E A A)	Comments
1.590E-06	3.547E-04	25	D306	2 1 2 2 2	

2518. $C_{12}H_8Cl_2$
2,4'-Dichlorobiphenyl
2,4'-PCB
RN: 34883-43-7   **MP** (°C):   43
MW: 223.10   **BP** (°C):

Solubility (Moles/L)	Solubility (Grams/L)	Temp (°C)	Ref (#)	Evaluation (T P E A A)	Comments
2.855E-06	6.370E-04	20	C302	1 1 2 2 2	
2.413E-06	5.383E-04	20	M336	2 0 2 2 2	
2.241E-06	5.000E-04	24	H100	2 0 2 2 0	
2.779E-06	6.200E-04	25	W025	1 0 2 2 2	
2.855E-06	6.370E-04	ns	H058	0 1 2 1 2	

2519. $C_{12}H_8Cl_2$
2,4-Dichlorobiphenyl
1,1'-Biphenyl, 2,4-Dichloro-
RN: 33284-50-3   **MP** (°C):   25.0
MW: 223.10   **BP** (°C):

Solubility (Moles/L)	Solubility (Grams/L)	Temp (°C)	Ref (#)	Evaluation (T P E A A)	Comments
2.747E-06	6.129E-04	20	M336	2 0 2 2 2	
3.138E-07	7.000E-05	23	W024	0 0 0 0 0	*sic*
5.065E-06	1.130E-03	25	B319	2 0 1 2 2	
5.065E-06	1.130E-03	25	B350	1 0 0 0 2	
5.150E-06	1.149E-03	25	D306	2 1 2 2 2	

2520. $C_{12}H_8Cl_2$
2,5-Dichlorobiphenyl
1,1'-Biphenyl, 2,5-Dichloro-
RN: 34883-39-1   **MP** (°C):   23
MW: 223.10   **BP** (°C):

Solubility (Moles/L)	Solubility (Grams/L)	Temp (°C)	Ref (#)	Evaluation (T P E A A)	Comments
6.454E-06	1.440E-03	23	W024	0 0 0 0 0	
5.000E-06	1.116E-03	25	D306	2 1 2 2 2	
8.700E-06	1.941E-03	25	M342	1 0 1 1 1	
2.600E-06	5.800E-04	25	W025	1 0 2 2 2	
8.516E-07	1.900E-04	ns	B301	0 2 1 1 2	
2.680E-05	5.979E-03	ns	M308	0 0 1 1 2	

2521. $C_{12}H_8Cl_2$
2,2'-Dichlorobiphenyl
2,2'-PCB
RN: 13029-08-8   **MP** (°C):   61
MW: 223.10   **BP** (°C):

Solubility (Moles/L)	Solubility (Grams/L)	Temp (°C)	Ref (#)	Evaluation (T P E A A)	Comments
3.214E-06	7.170E-04	20	C302	1 1 2 2 2	
5.038E-06	1.124E-03	20	M336	2 0 2 2 2	
3.541E-06	7.900E-04	22.5	G301	2 1 0 1 2	
6.275E-06	1.400E-03	23	W024	0 0 0 0 0	
4.034E-06	9.000E-04	24	H100	2 0 2 2 0	
5.410E-06	1.207E-03	25	D306	2 1 2 2 2	
3.541E-06	7.900E-04	25	W025	1 0 2 2 2	

2522.   C$_{12}$H$_8$Cl$_6$

Aldrin

1,2,3,4,10,10-Hexachloro-1,4,4α,5,8,8α-hexahydro-1,4:5,8-dimethanonaphthalene

Aldrite

Seedrin

Aldrosol

HHDN

RN: 309-00-2   **MP** (°C):   104.3

MW: 364.92   **BP** (°C):

Solubility (Moles/L)	Solubility (Grams/L)	Temp (°C)	Ref (#)	Evaluation (T P E A A)	Comments
2.877E-07	1.050E-04	15	B083	2 2 1 2 2	particle size ≤ 5 µm
7.413E-08	2.705E-05	20	B179	2 0 0 0 2	
4.659E-08	1.700E-05	22.5	G301	2 1 0 1 2	
4.933E-07	1.800E-04	25	B083	2 2 1 2 2	particle size ≤ 5 µm
5.481E-07	2.000E-04	25	M130	1 0 0 0 0	
4.659E-08	1.700E-05	25	W025	1 0 2 2 2	
7.399E-08	2.700E-05	26.5	P027	1 1 2 2 1	
5.481E-07	2.000E-04	26.70	L095	2 2 1 1 2	
7.399E-08	2.700E-05	27	M161	0 0 0 0 1	
9.591E-07	3.500E-04	35	B083	2 2 1 2 2	particle size ≤ 5 µm
1.644E-06	6.000E-04	45	B083	2 2 1 2 2	particle size ≤ 5 µm
7.399E-08	2.700E-05	ns	I308	0 0 0 0 1	
3.562E-08	1.300E-05	ns	K138	0 0 0 0 2	
1.096E-07	4.000E-05	ns	M110	0 0 0 0 0	EFG

2523.   C$_{12}$H$_8$Cl$_6$O

Endrin

1,2,3,4,10,10-Hexachloro-6,7-epoxy-1,4,4α,5,6,7,8,8α-octahydro-1,4-endo-endo-5,8-dimethano-naphthalene

Mendrin

Nendrin

RN: 72-20-8   **MP** (°C):   228.0

MW: 380.91   **BP** (°C):

Solubility (Moles/L)	Solubility (Grams/L)	Temp (°C)	Ref (#)	Evaluation (T P E A A)	Comments
3.413E-07	1.300E-04	15	B083	2 2 1 2 2	particle size ≤ 5 µm
6.563E-07	2.500E-04	25	B083	2 2 1 2 2	particle size ≤ 5 µm
6.826E-07	2.600E-04	25	W025	1 0 2 2 2	
1.103E-06	4.200E-04	35	B083	2 2 1 2 2	particle size ≤ 5 µm
1.641E-06	6.250E-04	45	B083	2 2 1 2 2	particle size ≤ 5 µm
6.301E-08	2.400E-05	ns	K138	0 0 0 0 2	
1.050E-06	4.000E-04	ns	M110	0 0 0 0 0	EFG
<2.63E-07	<1.00E-04	ns	N034	0 0 0 0 0	

2524. $C_{12}H_8Cl_6O$
Dieldrin
3,4,5,6,9,9-Hexachloro-1α,2,2α,3,6,6α,7,7α-octahydro-2,7:3,6-Dimethanonaphth[2,3-b]-oxirene
Alvit
Quintox
Oxralox

RN:	60-57-1	**MP** (°C):	175.5	
MW:	380.91	**BP** (°C):		

Solubility (Moles/L)	Solubility (Grams/L)	Temp (°C)	Ref (#)	Evaluation (T P E A A)	Comments
2.100E-07	7.999E-05	10	B324	2 2 2 2 2	
2.100E-07	8.000E-05	10	B324	2 2 2 2 2	
2.363E-07	9.000E-05	15	B083	2 2 1 2 1	particle size ≤ 5 μm
4.898E-07	1.866E-04	20	B179	2 0 0 0 2	
3.675E-07	1.400E-04	20	B324	2 2 2 2 2	
3.676E-07	1.400E-04	20	B324	2 2 2 2 2	
1.229E-06	4.680E-04	22	K137	1 1 2 1 0	
5.119E-07	1.950E-04	25	B083	2 2 1 2 2	particle size ≤ 5 μm
4.883E-07	1.860E-04	25	I308	0 0 0 0 2	
6.563E-07	2.500E-04	25	M130	1 0 0 0 1	
5.251E-07	2.000E-04	25	W025	1 0 2 2 2	
1.313E-07	5.000E-05	26	M061	1 0 0 0 0	
4.883E-07	1.860E-04	26.5	P027	1 1 2 2 2	
5.251E-07	2.000E-04	27	B161	2 1 2 2 0	EFG
4.883E-07	1.860E-04	27	M161	0 0 0 0 2	
5.251E-07	2.000E-04	30	B324	2 2 2 2 2	
5.251E-07	2.000E-04	30	B324	2 2 2 2 2	
1.050E-06	4.000E-04	35	B083	2 2 1 2 2	particle size ≤ 5 μm
1.313E-06	5.000E-04	40	B161	2 1 2 2 0	EFG
1.706E-06	6.500E-04	45	B083	2 2 1 2 2	particle size ≤ 5 μm
2.363E-06	9.000E-04	50	B161	2 1 2 2 0	EFG
3.544E-06	1.350E-03	60	B161	2 1 2 2 0	EFG
6.511E-06	2.480E-03	70	B161	2 1 2 2 0	EFG
6.563E-07	2.500E-04	ns	H322	0 0 0 2 2	
5.776E-08	2.200E-05	ns	K138	0 0 0 0 2	
7.876E-07	3.000E-04	ns	M110	0 0 0 0 0	EFG
≤2.63E-07	<1.00E-04	ns	N034	0 0 0 0 0	

2525. $C_{12}H_8N_2$
m-Phenanthroline
m-Phenanthrolin

RN:	230-46-6	**MP** (°C):		
MW:	180.21	**BP** (°C):		

Solubility (Moles/L)	Solubility (Grams/L)	Temp (°C)	Ref (#)	Evaluation (T P E A A)	Comments
4.000E-03	7.208E-01	ns	K114	0 0 0 0 0	

2526.   C$_{12}$H$_8$N$_2$
o-Phenanthroline
1,10-Phenanthroline
o-Phenanthrolin
RN: 66-71-7   **MP** (°C):   115
MW: 180.21   **BP** (°C):   >300

Solubility (Moles/L)	Solubility (Grams/L)	Temp (°C)	Ref (#)	Evaluation (T P E A A)	Comments
1.526E-02	2.750E+00	25	M155	1 0 1 1 0	EFG
1.490E-02	2.685E+00	25.04	B094	1 2 1 2 2	
1.850E-02	3.334E+00	31	B094	1 2 1 2 2	
2.090E-02	3.766E+00	35	B094	1 2 1 2 2	
2.550E-02	4.595E+00	40.04	B094	1 2 1 2 2	
2.880E-02	5.190E+00	45.44	B094	1 2 1 2 2	
3.410E-02	6.145E+00	50.04	B094	1 2 1 2 2	

2527.   C$_{12}$H$_8$N$_2$
p-Phenanthroline
p-Phenanthrolin
RN: 230-07-9   **MP** (°C):
MW: 180.21   **BP** (°C):

Solubility (Moles/L)	Solubility (Grams/L)	Temp (°C)	Ref (#)	Evaluation (T P E A A)	Comments
8.000E-03	1.442E+00	ns	K114	0 0 0 0 0	

2528.   C$_{12}$H$_8$N$_2$
Phenazine
Dibenzopyrazine
RN: 92-82-0   **MP** (°C):   175.5
MW: 180.21   **BP** (°C):

Solubility (Moles/L)	Solubility (Grams/L)	Temp (°C)	Ref (#)	Evaluation (T P E A A)	Comments
1.400E-04	2.523E-02	25	K009	1 2 1 1 0	EFG

2529.   C$_{12}$H$_8$N$_4$O$_6$
Picrylaniline
2,4,6-Trinitrodiphenyllamine
RN: 2919-12-2   **MP** (°C):
MW: 304.22   **BP** (°C):

Solubility (Moles/L)	Solubility (Grams/L)	Temp (°C)	Ref (#)	Evaluation (T P E A A)	Comments
5.888E-05	1.791E-02	25	B335	1 2 0 0 1	

2530.   C₁₂H₈O

Dibenzofuran
Diphenylene Oxide
DBF
RN: 132-64-9   **MP** (°C):   83
MW: 168.20   **BP** (°C):   154

Solubility (Moles/L)	Solubility (Grams/L)	Temp (°C)	Ref (#)	Evaluation (T P E A A)	Comments
9.820E-06	1.652E-03	4.0	D330	2 2 1 2 2	
5.960E-05	1.002E-02	25	B173	2 0 2 2 2	
1.850E-05	3.112E-03	25	L301	1 1 2 2 2	
2.510E-05	4.222E-03	25.0	D330	2 2 1 2 2	
4.140E-05	6.963E-03	39.8	D330	2 2 1 2 2	

2531.   C₁₂H₈O₂

Dibenzo-p-dioxin
Dibenzo[1,4]dioxin
Oxanthrene
Phenodioxin
RN: 262-12-4   **MP** (°C):   119
MW: 184.20   **BP** (°C):

Solubility (Moles/L)	Solubility (Grams/L)	Temp (°C)	Ref (#)	Evaluation (T P E A A)	Comments
1.150E-06	2.118E-04	4.10	D330	2 2 1 2 2	
1.113E-06	2.050E-04	5	S352	2 2 0 2 2	
2.497E-06	4.600E-04	15	S352	2 2 0 2 2	
4.729E-06	8.710E-04	25	S352	2 2 0 2 2	average of 2
4.571E-06	8.420E-04	25	S352	2 2 0 2 2	
4.890E-06	9.007E-04	25.0	D330	2 2 1 2 2	
9.566E-06	1.762E-03	35	S352	2 2 0 2 2	
1.300E-05	2.395E-03	40.0	D330	2 2 1 2 2	
1.771E-05	3.262E-03	45	S352	2 2 0 2 2	

2532.   C₁₂H₈O₄

Methoxsalen
Ammoidin
8-Methoxy-2',3',6,7-furocoumarin
Methoxalen
8-Methoxyfuranocoumarin
Oxypsoralen
RN: 298-81-7   **MP** (°C):   148
MW: 216.20   **BP** (°C):

Solubility (Moles/L)	Solubility (Grams/L)	Temp (°C)	Ref (#)	Evaluation (T P E A A)	Comments
2.200E-04	4.756E-02	30	E012	1 2 1 1 0	

2533.   C$_{12}$H$_8$S
Dibenzothiophene
Diphenylene Sulfide
RN: 132-65-0   **MP** (°C):   97
MW: 184.26   **BP** (°C):   332

Solubility (Moles/L)	Solubility (Grams/L)	Temp (°C)	Ref (#)	Evaluation (T P E A A)	Comments
7.978E-06	1.470E-03	24	H106	1 0 2 2 2	
7.978E-06	1.470E-03	24	M303	1 0 1 1 2	
2.871E-06	5.291E-04	25	L301	1 1 2 2 2	
7.978E-06	1.470E-03	ns	H107	0 0 0 0 2	

2534.   C$_{12}$H$_9$Br
4-Bromobiphenyl
1,1'-Biphenyl, 4-Bromo-
Bromodiphenyl
RN: 92-66-0   **MP** (°C):   91.5
MW: 233.11   **BP** (°C):   310.0

Solubility (Moles/L)	Solubility (Grams/L)	Temp (°C)	Ref (#)	Evaluation (T P E A A)	Comments
1.010E-06	2.354E-04	4.0	D330	2 2 1 2 2	
2.800E-06	6.527E-04	25.0	D330	2 2 1 2 2	
3.740E-06	8.718E-04	40.0	D330	2 2 1 2 2	

2535.   C$_{12}$H$_9$Cl
2-Chlorobiphenyl
2-PCB
RN: 2051-60-7   **MP** (°C):   32
MW: 188.66   **BP** (°C):   274

Solubility (Moles/L)	Solubility (Grams/L)	Temp (°C)	Ref (#)	Evaluation (T P E A A)	Comments
1.993E-05	3.760E-03	20	C302	1 1 2 2 2	
3.074E-05	5.800E-03	23	W024	0 0 0 0 0	
4.771E-06	9.000E-04	24	H100	2 0 2 2 0	
4.134E-05	7.800E-03	25	B351	1 0 0 1 1	
2.680E-05	5.056E-03	25	M342	1 0 1 1 2	
2.189E-05	4.130E-03	25	W025	1 0 2 2 2	
2.680E-05	5.056E-03	ns	M308	0 0 1 1 2	

2536. C₁₂H₉Cl

Aroclor 1221
Arochlor 1221
RN: 11104-28-2   **MP** (°C):
MW: 188.66   **BP** (°C):

Solubility (Moles/L)	Solubility (Grams/L)	Temp (°C)	Ref (#)	Evaluation (T P E A A)	Comments
>1.06E-06	>2.00E-04	ns	M184	0 0 0 0 0	

2537. C₁₂H₉Cl

4-Chlorobiphenyl
1-Chloro-4-phenyl benzene
4-Monochloro-biphenyl
RN: 2051-62-9   **MP** (°C):   77
MW: 188.66   **BP** (°C):   291

Solubility (Moles/L)	Solubility (Grams/L)	Temp (°C)	Ref (#)	Evaluation (T P E A A)	Comments
6.202E-06	1.170E-03	23	W024	0 0 0 0 0	
2.120E-06	4.000E-04	24	H100	2 0 2 2 0	
7.103E-06	1.340E-03	25	B319	2 0 1 2 2	average of 2
6.891E-06	1.300E-03	25	B350	1 0 0 0 2	
6.361E-06	1.200E-03	25	B351	1 0 0 1 1	
6.361E-06	1.200E-03	25	H341	1 0 0 0 2	
7.087E-06	1.337E-03	25	L322	1 1 2 2 2	average of 2
7.079E-06	1.336E-03	25	L322	1 1 2 2 2	average of 2
4.771E-06	9.000E-04	25	W025	1 0 2 2 2	

2538. C₁₂H₉Cl

3-Chlorobiphenyl
3-Chlorbiphenyl
RN: 2051-61-8   **MP** (°C):   16
MW: 188.66   **BP** (°C):   285

Solubility (Moles/L)	Solubility (Grams/L)	Temp (°C)	Ref (#)	Evaluation (T P E A A)	Comments
1.908E-05	3.600E-03	23	W024	0 0 0 0 0	
9.806E-06	1.850E-03	23	W024	0 0 0 0 0	
1.924E-05	3.630E-03	25	B319	2 0 1 2 2	
6.891E-06	1.300E-03	25	W025	1 0 2 2 2	

2539. C$_{12}$H$_9$ClF$_3$N$_3$O
Norflurazon
4-Chloro-5-(methylamino)-2-(α,α,α-trifluoro-m-tolyl)-3(2H)-pyridazinone
Zorial
RN: 27314-13-2   **MP** (°C):   177
MW: 303.67   **BP** (°C):

Solubility (Moles/L)	Solubility (Grams/L)	Temp (°C)	Ref (#)	Evaluation (T P E A A)	Comments
9.220E-05	2.800E-02	23	M161	1 0 0 0 1	
9.220E-05	2.800E-02	24	C105	2 1 2 2 2	
9.220E-05	2.800E-02	25	B310	1 1 0 0 1	

2540. C$_{12}$H$_9$ClN$_2$
4-Chloroazobenzene
Diazene, (4-Chlorophenyl)phenyl-, (E)-
RN: 4340-77-6   **MP** (°C):   88
MW: 216.67   **BP** (°C):

Solubility (Moles/L)	Solubility (Grams/L)	Temp (°C)	Ref (#)	Evaluation (T P E A A)	Comments
2.000E-06	4.333E-04	25	B333	1 0 0 0 1	

2541. C$_{12}$H$_9$ClO
4-Chlorophenyl Phenyl Ether
1-Chloro-4-phenoxybenzene
p-Chlorodiphenyl Oxide
RN: 7005-72-3   **MP** (°C):
MW: 204.66   **BP** (°C):

Solubility (Moles/L)	Solubility (Grams/L)	Temp (°C)	Ref (#)	Evaluation (T P E A A)	Comments
1.612E-05	3.300E-03	25	B131	1 0 0 0 1	

2542. C$_{12}$H$_9$Cl$_2$NO$_3$
Vinclozolin
3-(3,5-Dichlorophenyl)-5-ethenyl-5-methyl-2,4-oxazolidinedione
Ornalin
Vinclozalin
Ronilan
RN: 50471-44-8   **MP** (°C):   108
MW: 286.12   **BP** (°C):

Solubility (Moles/L)	Solubility (Grams/L)	Temp (°C)	Ref (#)	Evaluation (T P E A A)	Comments
3.495E-03	1.000E+00	20	M161	1 0 0 0 0	

2543. C$_{12}$H$_9$Cl$_3$NO$_2$S
Reserptyl
4'-[Chlorophenyl}-3,4-dichlorophenylbenzene-sulphonamide
RN: **MP** (°C): 127-129
MW: 337.63 **BP** (°C):

Solubility (Moles/L)	Solubility (Grams/L)	Temp (°C)	Ref (#)	Evaluation (T P E A A)	Comments
1.066E-04	3.600E-02	25	L014	1 0 1 1 1	

2544. C$_{12}$H$_9$FN$_2$O$_4$
1-Benzyloxycarbonyl-5-fluorouracil
1(2H)-Pyrimidinecarboxylic Acid, 5-Fluoro-3,4-dihydro-2,4-dioxo-, Phenylmethyl Ester
RN: 66999-98-2 **MP** (°C):
MW: 264.21 **BP** (°C):

Solubility (Moles/L)	Solubility (Grams/L)	Temp (°C)	Ref (#)	Evaluation (T P E A A)	Comments
3.028E-04	8.000E-02	22	B332	1 1 0 0 1	pH 4.0

2545. C$_{12}$H$_9$N
Carbazole
9-Azafluorene
Dibenzo[b,d]pyrrole
Diphenylenimine
9H-Carbazole
Dibenzopyrrole
RN: 86-74-8 **MP** (°C): 245
MW: 167.21 **BP** (°C): 355

Solubility (Moles/L)	Solubility (Grams/L)	Temp (°C)	Ref (#)	Evaluation (T P E A A)	Comments
7.177E-06	1.200E-03	20	H300	1 1 2 2 1	
5.427E-06	9.075E-04	25	L301	1 1 2 2 2	

2546. C$_{12}$H$_9$NO$_3$
Furo[3,4-b]quinolin-3(1H)-one, 9-Hydroxy-1-methyl-
RN: 74103-11-0 **MP** (°C):
MW: 215.21 **BP** (°C):

Solubility (Moles/L)	Solubility (Grams/L)	Temp (°C)	Ref (#)	Evaluation (T P E A A)	Comments
3.253E-07	7.000E-05	25	P089	2 1 2 2 2	
4.321E-07	9.300E-05	37	P089	2 1 2 2 2	
5.529E-07	1.190E-04	51	P089	2 1 2 2 2	

2547.   C$_{12}$H$_9$NS
Phenothiazine
Dibenzo-1,4-thiazine
Thiodiphenylamine
RN: 92-84-2   **MP** (°C):   185.1
MW: 199.28   **BP** (°C):

Solubility (Moles/L)	Solubility (Grams/L)	Temp (°C)	Ref (#)	Evaluation (T P E A A)	Comments
6.000E-06	1.196E-03	20	M177	2 2 2 2 0	EFG
8.000E-06	1.594E-03	25	M177	2 2 2 2 0	EFG
1.000E-05	1.993E-03	30	M177	2 2 2 2 0	EFG

2548.   C$_{12}$H$_9$N$_3$O$_2$
4-Nitroazobenzene
Diazene, (p-Nitrophenyl)phenyl-, (E)-
RN: 2491-52-3   **MP** (°C):
MW: 227.22   **BP** (°C):

Solubility (Moles/L)	Solubility (Grams/L)	Temp (°C)	Ref (#)	Evaluation (T P E A A)	Comments
2.800E-06	6.362E-04	25	B333	1 0 0 0 1	

2549.   C$_{12}$H$_9$N$_3$O$_3$
Dis. A. 3
4-[(4-Nitrophenyl)azo]phenol
p-Nitrophenylazophenol
p-Hydroxy-p'-nitroazobenzene
RN: 1435-60-5   **MP** (°C):   216
MW: 243.22   **BP** (°C):

Solubility (Moles/L)	Solubility (Grams/L)	Temp (°C)	Ref (#)	Evaluation (T P E A A)	Comments
1.600E-05	3.892E-03	25	B333	1 0 0 0 1	

2550. C$_{12}$H$_9$N$_3$O$_4$
2,4-Dinitrodiphenylamine
2,4-Dinitrodiphenylamin
C.I. Disperse Yellow 14
RN: 961-68-2   **MP** (°C):   160
MW: 259.22   **BP** (°C):

Solubility (Moles/L)	Solubility (Grams/L)	Temp (°C)	Ref (#)	Evaluation (T P E A A)	Comments
1.466E-04	3.800E-02	15	D070	1 2 0 0 1	
1.543E-04	4.000E-02	15	F300	1 0 0 0 0	
5.100E-06	1.322E-03	25	B333	1 0 0 0 1	*sic*
3.240E-04	8.399E-02	50	D070	1 2 0 0 1	
5.516E-04	1.430E-01	100	D070	1 2 0 0 2	

2551. C$_{12}$H$_9$N$_3$O$_5$
C.I. Disperse Yellow 1
C.I.Disperse Yellow 1
p-(2,4-Dinitroanilino)
2,4-Dinitro-4'-hydroxydiphenylamine
4-Hydroxy-2',4'-dinitrodiphenylamine
RN: 119-15-3   **MP** (°C):   194
MW: 275.22   **BP** (°C):

Solubility (Moles/L)	Solubility (Grams/L)	Temp (°C)	Ref (#)	Evaluation (T P E A A)	Comments
9.000E-06	2.477E-03	25	B333	1 0 0 0 1	
6.195E-05	1.705E-02	60	P313	1 2 1 2 2	average of 2
1.546E-04	4.255E-02	70	P313	1 2 1 2 2	average of 2
2.954E-04	8.130E-02	80	P313	1 2 1 2 2	average of 2
5.559E-04	1.530E-01	90	P313	1 2 1 2 2	average of 2
1.163E-03	3.200E-01	100	P313	1 2 1 2 2	

2552. C$_{12}$H$_9$N$_5$O$_3$
1-Nicotinoyloxymethyl Allopurinol
3-Pyridinecarboxylic Acid, (4,5-Dihydro-4-oxo-1H-pyrazolo[3,4-d]pyrimidin-1-yl)methyl Ester
RN: 98846-66-3   **MP** (°C):   242-243
MW: 271.24   **BP** (°C):

Solubility (Moles/L)	Solubility (Grams/L)	Temp (°C)	Ref (#)	Evaluation (T P E A A)	Comments
3.429E-04	9.300E-02	22	B322	1 0 2 2 2	

2553. $C_{12}H_{10}$
Acenaphthene
1,2-Dihydroacenaphthene
1,8-Ethylenenaphthalene
peri-Ethylenenaphthalene
RN: 83-32-9   **MP** (°C):   95
MW: 154.21   **BP** (°C):   279

Solubility (Moles/L)	Solubility (Grams/L)	Temp (°C)	Ref (#)	Evaluation (T P E A A)	Comments
2.315E-05	3.570E-03	22.20	W003	2 2 2 2 2	
4.780E-05	7.371E-03	25	B173	2 0 2 2 2	
2.250E-05	3.470E-03	25	E004	2 1 2 2 2	
2.218E-05	3.420E-03	25	L332	1 1 1 1 2	
2.548E-05	3.930E-03	25	M064	1 1 2 2 2	
2.550E-05	3.932E-03	25	M342	1 0 1 1 2	
8.889E-07	1.371E-04	25	R084	2 2 2 2 1	*sic*
2.330E-05	3.593E-03	25.04	V013	2 2 2 2 2	
3.041E-05	4.690E-03	30.00	W003	2 2 2 2 2	average of 3
3.761E-05	5.800E-03	34.50	W003	2 2 2 2 2	average of 3
4.520E-05	6.970E-03	39.30	W003	2 2 2 2 1	average of 3
6.076E-05	9.370E-03	44.70	W003	2 2 2 2 1	average of 3
8.060E-05	1.243E-02	50.10	W003	2 2 2 2 2	average of 3
1.038E-04	1.600E-02	55.60	W003	2 2 2 2 2	average of 3
1.741E-04	2.685E-02	64.50	W003	2 2 2 2 2	average of 3
1.511E-04	2.330E-02	65.20	W003	2 2 2 2 2	average of 3
2.118E-04	3.267E-02	69.80	W003	2 2 2 2 2	average of 3
2.283E-04	3.520E-02	71.90	W003	2 2 2 2 2	
2.568E-04	3.960E-02	73.40	W003	2 2 2 2 2	average of 2
2.597E-04	4.005E-02	74.70	W003	2 2 2 2 2	average of 2
3.981E-05	6.139E-03	ns	D001	0 0 0 0 2	
2.248E-05	3.467E-03	ns	I332	0 0 0 0 1	
2.000E-05	3.084E-03	ns	L060	0 0 0 0 0	average
2.548E-05	3.930E-03	ns	M344	0 0 0 0 2	

2554. C$_{12}$H$_{10}$
Diphenyl
Biphenyl
Phenylbenzene
1,1'-Biphenyl
Lemonene

RN: 92-52-4   **MP** (°C):   69.1
MW: 154.21   **BP** (°C):   254

Solubility (Moles/L)	Solubility (Grams/L)	Temp (°C)	Ref (#)	Evaluation (T P E A A)	Comments
1.718E-05	2.650E-03	-0.7	N053	1 0 0 1 0	EFG
1.973E-05	3.042E-03	4.62	N053	1 0 0 1 0	EFG
2.670E-05	4.118E-03	10	J302	2 1 2 2 2	
2.372E-05	3.658E-03	10.13	N053	1 0 0 1 0	EFG
2.918E-05	4.500E-03	14.20	N053	1 0 0 1 0	EFG
3.800E-05	5.860E-03	20	H306	1 0 1 2 1	
4.182E-05	6.450E-03	20	T301	1 2 2 2 2	
3.590E-05	5.536E-03	20.10	N053	1 0 0 1 0	EFG
4.100E-05	6.323E-03	21	A057	2 1 2 2 1	
4.850E-05	7.480E-03	22.5	G301	2 1 0 1 2	
1.187E-04	1.830E-02	23.5	S171	2 1 2 2 2	
2.983E-05	4.600E-03	24	H100	2 0 2 2 1	
5.512E-05	8.500E-03	24	H116	2 1 0 0 2	
4.708E-05	7.260E-03	24.60	W003	2 2 2 2 2	average of 3
3.852E-05	5.940E-03	25	A001	1 0 2 2 2	
4.570E-05	7.048E-03	25	A325	2 1 2 2 2	
4.850E-05	7.480E-03	25	B003	2 2 2 2 2	
3.910E-05	6.030E-03	25	B173	2 0 2 2 2	
4.799E-05	7.400E-03	25	B319	2 0 1 2 1	average of 2
4.409E-05	6.800E-03	25	B351	1 0 0 1 1	
4.831E-05	7.450E-03	25	E004	2 1 2 2 2	
4.850E-05	7.479E-03	25	J302	2 1 2 2 2	
4.863E-05	7.500E-03	25	M040	1 0 0 1 1	
4.539E-05	7.000E-03	25	M064	1 1 2 2 1	
4.850E-05	7.480E-03	25	M130	1 0 0 0 2	
4.350E-05	6.708E-03	25	M342	1 0 1 1 2	
4.540E-05	7.001E-03	25	M342	1 0 1 1 2	
4.234E-04	6.530E-02	25	S005	2 2 2 2 2	
4.910E-05	7.572E-03	25.04	V013	2 2 2 2 2	
4.416E-05	6.811E-03	25.35	N053	1 0 0 1 0	EFG
5.689E-05	8.774E-03	28.95	N053	1 0 0 1 0	EFG
5.700E-05	8.790E-03	29.90	W003	2 2 2 2 2	average of 3
5.525E-05	8.520E-03	30.30	W003	2 2 2 2 2	average of 3
8.624E-05	1.330E-02	38.40	W003	2 2 2 2 2	average of 3
8.624E-05	1.330E-02	40.10	W003	2 2 2 2 2	average of 3
1.219E-04	1.880E-02	47.50	W003	2 2 2 2 2	average of 3
1.381E-04	2.130E-02	50.10	W003	2 2 2 2 2	average of 3
1.381E-04	2.130E-02	50.20	W003	2 2 2 2 2	average of 2
1.855E-04	2.860E-02	54.70	W003	2 2 2 2 2	average of 3

Solubility (Moles/L)	Solubility (Grams/L)	Temp (°C)	Ref (#)	Evaluation (T P E A A)	Comments
2.347E-04	3.620E-02	59.20	W003	2 2 2 2 2	average of 3
2.620E-04	4.040E-02	60.50	W003	2 2 2 2 2	
2.918E-04	4.500E-02	64.50	W003	2 2 2 2 2	average of 3
4.539E-05	7.000E-03	ns	H123	0 0 0 0 2	
4.350E-05	6.708E-03	ns	M308	0 0 1 1 2	
4.539E-05	7.000E-03	ns	M344	0 0 0 0 1	

2555. C$_{12}$H$_{10}$CIN
4-Amino-4'-chlorodiphenyl
4-Chloro-4'-aminobiphenyl
p-Amino-p'-chlorobiphenyl
p'-Chloro-p-phenylaniline
RN: 135-68-2   **MP** (°C):
MW: 203.67   **BP** (°C):

Solubility (Moles/L)	Solubility (Grams/L)	Temp (°C)	Ref (#)	Evaluation (T P E A A)	Comments
2.300E-05	4.684E-03	ns	B305	0 2 0 0 1	

2556. C$_{12}$H$_{10}$Cl$_2$N$_2$
3,3'-Dichlorobenzidine
3,3'-Dichloro-4,4'-biphenyldiamine
o,o'-Dichlorobenzidine
4,4'-Diamino-3,3'-dichlorobiphenyl
RN: 91-94-1   **MP** (°C):   132
MW: 253.13   **BP** (°C):

Solubility (Moles/L)	Solubility (Grams/L)	Temp (°C)	Ref (#)	Evaluation (T P E A A)	Comments
1.230E-05	3.114E-03	25	B173	2 0 2 2 2	
<3.95E-06	<1.00E-03	30	M311	1 1 2 2 0	

2557. C$_{12}$H$_{10}$N$_2$
Azobenzene
Diphenyl Diimide
Benzeneazobenzene
Diphenyldiazene
Azobenzide
Azobenzol
RN: 103-33-3   **MP** (°C):   68
MW: 182.23   **BP** (°C):   293

Solubility (Moles/L)	Solubility (Grams/L)	Temp (°C)	Ref (#)	Evaluation (T P E A A)	Comments
1.660E-03	3.024E-01	20	B179	2 0 0 0 2	*sic*
1.921E-05	3.500E-03	20	J009	1 0 2 2 1	
4.610E-05	8.400E-03	20	J027	1 0 0 0 1	
3.500E-05	6.378E-03	25	B333	1 0 0 0 1	

2.415E-05	4.400E-03	25	H050	1 2 2 1 1	
2.579E-05	4.700E-03	25	P096	1 0 2 2 2	
5.202E+00	9.480E+02	37	H052	1 1 2 2 1	EFG, *sic* , pH 6.3
1.646E-03	2.999E-01	rt	D021	0 0 1 1 1	*sic*

2558. $C_{12}H_{10}N_2O$
Diphenylnitrosamine
Redax
N-Nitroso-N-Phenylaniline
RN: 86-30-6 **MP** (°C): 67
MW: 198.23 **BP** (°C):

Solubility (Moles/L)	Solubility (Grams/L)	Temp (°C)	Ref (#)	Evaluation (T P E A A)	Comments
1.770E-04	3.509E-02	25	B173	2 0 2 2 2	

2559. $C_{12}H_{10}N_2O$
4-Phenylazophenol
4-Hydroxyazobenzene
p-Hydroxyazobenzene
C.I. Solvent Yellow 7
RN: 1689-82-3 **MP** (°C): 150
MW: 198.23 **BP** (°C): 220

Solubility (Moles/L)	Solubility (Grams/L)	Temp (°C)	Ref (#)	Evaluation (T P E A A)	Comments
4.540E-04	9.000E-02	20	F300	1 0 0 0 1	
1.100E-04	2.180E-02	25	B333	1 0 0 0 1	
1.715E-04	3.400E-02	37	H120	1 1 1 1 1	normal saline
4.036E-03	8.000E-01	100	F300	1 0 0 0 1	

2560. $C_{12}H_{10}N_2O_2$
2,4-Dihydroxyazobenzene
2,4-Dihydroxy-azobenzol
RN: 2051-85-6 **MP** (°C): 170
MW: 214.23 **BP** (°C):

Solubility (Moles/L)	Solubility (Grams/L)	Temp (°C)	Ref (#)	Evaluation (T P E A A)	Comments
9.336E-04	2.000E-01	20	F300	1 0 0 0 0	

2561. $C_{12}H_{10}N_2O_3$
3-Hydroxyazobenzene
3-Hydroxy-azobenzol
RN: 40038-46-8 **MP** (°C):
MW: 230.23 **BP** (°C):

Solubility (Moles/L)	Solubility (Grams/L)	Temp (°C)	Ref (#)	Evaluation (T P E A A)	Comments
3.475E-03	8.000E-01	100	F300	1 0 0 0 1	

2562.   C$_{12}$H$_{10}$N$_4$O$_2$
C.I. Disperse Orange 3
4'-Nitro-4-aminoazobenzene
4-Amino-4'-nitroazobenzene
4-(4-Nitrophenylazo)aniline
RN: 730-40-5   **MP** (°C):   211
MW: 242.24   **BP** (°C):

Solubility (Moles/L)	Solubility (Grams/L)	Temp (°C)	Ref (#)	Evaluation (T P E A A)	Comments
1.200E-06	2.907E-04	25	B333	1 0 0 0 1	

2563.   C$_{12}$H$_{10}$N$_4$O$_4$
C.I. Disperse Yellow 9
2,4-Dinitro-4'-aminodiphenylamine
4-Amino-2',4'-dinitrodiphenylamine
C.I. 10375
RN: 6373-73-5   **MP** (°C):   188
MW: 274.24   **BP** (°C):

Solubility (Moles/L)	Solubility (Grams/L)	Temp (°C)	Ref (#)	Evaluation (T P E A A)	Comments
6.000E-06	1.645E-03	25	B333	1 0 0 0 1	

2564.   C$_{12}$H$_{10}$O
Phenyl Ether
Diphenyl Ether
RN: 101-84-8   **MP** (°C):   28
MW: 170.21   **BP** (°C):   259

Solubility (Moles/L)	Solubility (Grams/L)	Temp (°C)	Ref (#)	Evaluation (T P E A A)	Comments
2.341E-02	3.984E+00	25	B019	1 0 1 2 0	*sic*
1.060E-04	1.804E-02	25	B173	2 0 2 2 2	
1.234E-04	2.100E-02	25	F071	1 1 2 1 1	
1.100E-04	1.872E-02	25.04	V013	2 2 2 2 2	

2565. C$_{12}$H$_{10}$O
o-Phenylphenol
2-Phenylphenol
RN: 90-43-7   **MP** (°C):   56.5
MW: 170.21   **BP** (°C):   282

Solubility (Moles/L)	Solubility (Grams/L)	Temp (°C)	Ref (#)	Evaluation (T P E A A)	Comments
9.790E-04	1.666E-01	25	L021	1 0 0 0 0	
4.110E-03	6.995E-01	25	M061	0 0 0 0 0	
4.112E-03	7.000E-01	25	M161	1 0 0 0 0	
3.162E-04	5.383E-02	rt	D056	0 1 1 1 0	EFG, pH 6-8, *sic*

2566. C$_{12}$H$_{10}$O
p-Phenylphenol
p-Hydroxybiphenyl
RN: 92-69-3   **MP** (°C):   164.5
MW: 170.21   **BP** (°C):   306.5

Solubility (Moles/L)	Solubility (Grams/L)	Temp (°C)	Ref (#)	Evaluation (T P E A A)	Comments
3.300E-04	5.617E-02	25	E014	2 2 2 1 2	pH 7.2
5.875E-05	1.000E-02	25	L021	1 0 0 0 0	

2567. C$_{12}$H$_{10}$O$_2$
2-Hydroxydiphenyl Ether
2-Hydroxy-diphenyl-aether
RN: 2417-10-9   **MP** (°C):
MW: 186.21   **BP** (°C):

Solubility (Moles/L)	Solubility (Grams/L)	Temp (°C)	Ref (#)	Evaluation (T P E A A)	Comments
5.907E-04	1.100E-01	20	F300	1 0 0 0 1	

2568. C$_{12}$H$_{10}$O$_2$
1-Naphthaleneacetic Acid
NAA
RN: 86-87-3   **MP** (°C):   134
MW: 186.21   **BP** (°C):

Solubility (Moles/L)	Solubility (Grams/L)	Temp (°C)	Ref (#)	Evaluation (T P E A A)	Comments
2.040E-03	3.799E-01	17	B200	1 0 0 0 1	
2.255E-03	4.198E-01	20	B200	1 0 0 0 1	
1.179E-02	2.195E+00	20	C092	2 2 0 1 2	
2.228E-03	4.148E-01	25	M061	1 0 0 0 2	average of 2

2569. C$_{12}$H$_{10}$O$_3$
β-Naphthoxyacetic Acid
(2-Naphthoxy)acetic Acid
Phyomone
BNOA
RN: 120-23-0   **MP** (°C):   155-157
MW: 202.21   **BP** (°C):

Solubility (Moles/L)	Solubility (Grams/L)	Temp (°C)	Ref (#)	Evaluation (T P E A A)	Comments
4.330E-04	8.756E-02	25	D088	1 2 2 2 2	
8.100E-04	1.638E-01	35	D088	1 2 2 2 2	
1.100E-05	2.224E-03	45	D088	1 2 2 2 2	

2570. C$_{12}$H$_{10}$O$_4$
Quinhydrone
Chinhydron
RN: 106-34-3   **MP** (°C):   171
MW: 218.21   **BP** (°C):

Solubility (Moles/L)	Solubility (Grams/L)	Temp (°C)	Ref (#)	Evaluation (T P E A A)	Comments
1.861E-02	4.061E+00	25	B121	1 2 2 1 2	average of 4

2571. C$_{12}$H$_{11}$ClN$_2$O$_5$S
Furosemide
Frusemide
RN: 54-31-9   **MP** (°C):   206
MW: 330.75   **BP** (°C):

Solubility (Moles/L)	Solubility (Grams/L)	Temp (°C)	Ref (#)	Evaluation (T P E A A)	Comments
2.210E-04	7.310E-02	30	E049	2 0 2 2 2	

2572. C$_{12}$H$_{11}$Cl$_2$NO
Propyzamide
3,5-Dichloro-N-(1,1-dimethyl-2-propynyl)benzamide
Pronamide
Kerb 50W
RH-315
RN: 23950-58-5   **MP** (°C):   155.5
MW: 256.13   **BP** (°C):

Solubility (Moles/L)	Solubility (Grams/L)	Temp (°C)	Ref (#)	Evaluation (T P E A A)	Comments
5.856E-05	1.500E-02	25	M161	1 0 0 0 1	

2573. C₁₂H₁₁I₃N₂O₄

2573. $C_{12}H_{11}I_3N_2O_4$

Iodamide
3-Acetamido-5-acetamidomethyl-2,4,6-triiodobenzoic Acid
3-Acetylamino-5-acetylaminomethyl-2,4,6-triiodobenzoic Acid
Jodomiron 380
Uromiro
Uromiron

RN: 440-58-4   **MP** (°C):
MW: 627.95   **BP** (°C):

Solubility (Moles/L)	Solubility (Grams/L)	Temp (°C)	Ref (#)	Evaluation (T P E A A)	Comments
4.777E-03	3.000E+00	20	F045	1 2 2 2 1	
5.096E-03	3.200E+00	40	F045	1 2 2 2 1	
6.211E-03	3.900E+00	60	F045	1 2 2 2 1	

2574. $C_{12}H_{11}N$

Diphenylamine
4-Aminobiphenyl

RN: 122-39-4   **MP** (°C):   53.5
MW: 169.23   **BP** (°C):   302.0

Solubility (Moles/L)	Solubility (Grams/L)	Temp (°C)	Ref (#)	Evaluation (T P E A A)	Comments
1.820E-03	3.079E-01	20	B179	2 0 0 0 2	
3.132E-04	5.300E-02	20	H300	1 2 2 2 1	
3.274E-04	5.540E-02	20	T301	1 2 2 2 2	
2.765E-04	4.680E-02	25	F029	1 0 0 0 2	
3.415E-04	5.780E-02	50	T301	1 2 2 2 2	average of 5
3.557E-04	6.020E-02	80	T301	1 2 2 2 2	average of 5
1.772E-03	2.999E-01	rt	D021	0 0 1 1 0	

2575. $C_{12}H_{11}NO_2$

Carbaryl
1-Naphthyl N-Methylcarbamate
Devicarb
Hexavin
Karbaspray
Murvin

RN: 63-25-2   **MP** (°C):   142
MW: 201.23   **BP** (°C):

Solubility (Moles/L)	Solubility (Grams/L)	Temp (°C)	Ref (#)	Evaluation (T P E A A)	Comments
2.710E-04	5.453E-02	5	H343	1 0 2 2 2	
3.598E-04	7.239E-02	10	B324	2 2 2 2 2	
3.444E-04	6.930E-02	10	B324	2 2 2 2 2	
3.150E-04	6.339E-02	10	H343	1 0 2 2 2	
3.740E-04	7.526E-02	15	H343	1 0 2 2 2	
1.995E-04	4.015E-02	20	B179	2 0 0 0 2	

5.164E-04	1.039E-01	20	B300	2 1 1 1 2	
4.947E-04	9.955E-02	20	B324	2 2 2 2 2	
5.168E-04	1.040E-01	20	B324	2 2 2 2 2	
2.485E-04	5.000E-02	20	F311	1 2 2 2 1	
4.450E-04	8.955E-02	20	H343	1 0 2 2 2	
1.690E-04	3.400E-02	22	K137	1 1 2 1 0	
1.988E-04	4.000E-02	22.5	G301	2 1 0 1 2	
5.210E-04	1.048E-01	25	H343	1 0 2 2 2	
6.184E-04	1.244E-01	30	B324	2 2 2 2 2	
6.460E-04	1.300E-01	30	B324	2 2 2 2 2	
1.988E-04	4.000E-02	30	D089	2 2 0 0 0	
6.520E-04	1.312E-01	30	H343	1 0 2 2 2	
1.988E-04	4.000E-02	30	M161	1 0 0 0 1	
7.860E-04	1.582E-01	35	H343	1 0 2 2 2	
8.990E-04	1.809E-01	40	H343	1 0 2 2 2	
1.006E-03	2.024E-01	45	H343	1 0 2 2 2	
1.988E-04	4.000E-02	ns	H042	0 0 0 0 1	
2.783E-04	5.600E-02	ns	M110	0 0 0 0 0	EFG

2576. $C_{12}H_{11}NO_2$
Fenfuram
2-Methyl-N-phenyl-3-furancarboxamide
Pano-ram
RN: 24691-80-3   **MP** (°C):   109.5
MW: 201.23   **BP** (°C):

Solubility (Moles/L)	Solubility (Grams/L)	Temp (°C)	Ref (#)	Evaluation (T P E A A)	Comments
4.970E-04	1.000E-01	20	M161	1 0 0 0 0	

2577. $C_{12}H_{11}N_3$
Diazoaminobenzene
1,3-Diphenyltriazene
Anilinoazobenzene
N-(Phenylazo)aniline
RN: 136-35-6   **MP** (°C):   98.0
MW: 197.24   **BP** (°C):

Solubility (Moles/L)	Solubility (Grams/L)	Temp (°C)	Ref (#)	Evaluation (T P E A A)	Comments
2.534E-03	4.998E-01	rt	D021	0 0 1 1 0	

2578. $C_{12}H_{11}N_3$
C.I. Solvent Yellow 1
p-Aminoazobenzene
4-Aminoazobenzene
4-Amino-azobenzol

RN: 60-09-3   **MP** (°C):   125
MW: 197.24   **BP** (°C):   >360

Solubility (Moles/L)	Solubility (Grams/L)	Temp (°C)	Ref (#)	Evaluation (T P E A A)	Comments
6.591E-04	1.300E-01	18	F300	1 0 0 0 1	
1.500E-04	2.959E-02	25	B333	1 0 0 0 1	
2.484E-04	4.900E-02	37	H120	1 1 1 1 1	normal saline
5.510E-04	1.087E-01	60	B198	1 2 1 1 2	
1.041E-03	2.053E-01	71.80	B198	1 2 1 1 2	
1.907E-03	3.761E-01	84.10	B198	1 2 1 1 2	
3.431E-03	6.767E-01	97.40	B198	1 2 1 1 2	

2579. $C_{12}H_{11}N_3O_3$
Orotic Acid Benzylamide
Orotamide, N-Benzyl-

RN: 13156-36-0   **MP** (°C):   260-263
MW: 245.24   **BP** (°C):

Solubility (Moles/L)	Solubility (Grams/L)	Temp (°C)	Ref (#)	Evaluation (T P E A A)	Comments
4.600E-02	1.128E+01	-4	N018	2 2 1 2 2	
8.700E-02	2.134E+01	16	N018	2 2 1 2 2	
1.180E-01	2.894E+01	25	N018	2 2 1 2 2	

2580. $C_{12}H_{11}O_4P$
Diphenyl Phosphate
Phosphoric Acid, Diphenyl Ester

RN: 838-85-7   **MP** (°C):   63
MW: 250.19   **BP** (°C):

Solubility (Moles/L)	Solubility (Grams/L)	Temp (°C)	Ref (#)	Evaluation (T P E A A)	Comments
>1.08E-03	>2.70E-01	24	H116	2 1 0 0 0	

2581. $C_{12}H_{12}$
2,6-Dimethylnaphthalene

RN: 581-42-0   **MP** (°C):   109
MW: 156.23   **BP** (°C):

Solubility (Moles/L)	Solubility (Grams/L)	Temp (°C)	Ref (#)	Evaluation (T P E A A)	Comments
1.280E-05	2.000E-03	25	M064	1 1 2 2 1	
1.280E-05	2.000E-03	25	M342	1 0 1 1 2	
1.280E-05	2.000E-03	ns	M344	0 0 0 0 1	

2582. C$_{12}$H$_{12}$
2-Ethylnaphthalene

RN: 939-27-5   **MP** (°C):   -7.4
MW: 156.23   **BP** (°C):   251.5

Solubility (Moles/L)	Solubility (Grams/L)	Temp (°C)	Ref (#)	Evaluation (T P E A A)	Comments
5.895E-05	9.210E-03	20	B356	1 0 0 0 2	
5.121E-05	8.000E-03	25	E004	2 1 2 2 2	

2583. C$_{12}$H$_{12}$
2,3-Dimethylnaphthalene

RN: 581-40-8   **MP** (°C):   103
MW: 156.23   **BP** (°C):   269

Solubility (Moles/L)	Solubility (Grams/L)	Temp (°C)	Ref (#)	Evaluation (T P E A A)	Comments
1.274E-05	1.990E-03	25	E004	2 1 2 2 2	
1.920E-05	3.000E-03	25	M064	1 1 2 2 1	
1.920E-05	3.000E-03	25	M342	1 0 1 1 2	
1.920E-05	3.000E-03	ns	M344	0 0 0 0 1	

2584. C$_{12}$H$_{12}$
1-Ethylnaphthalene

RN: 1127-76-0   **MP** (°C):   -15
MW: 156.23   **BP** (°C):   258

Solubility (Moles/L)	Solubility (Grams/L)	Temp (°C)	Ref (#)	Evaluation (T P E A A)	Comments
5.200E-05	8.124E-03	10	S076	2 2 2 2 1	
5.200E-05	8.124E-03	14	S076	2 2 2 2 1	
6.400E-05	9.999E-03	20	S076	2 2 2 2 1	
6.849E-05	1.070E-02	25	M064	1 1 2 2 2	
6.850E-05	1.070E-02	25	M342	1 0 1 1 2	
6.400E-05	9.999E-03	25	S076	2 2 2 2 1	
6.849E-05	1.070E-02	ns	M344	0 0 0 0 2	

2585. C$_{12}$H$_{12}$
1,5-Dimethylnaphthalene

RN: 571-61-9   **MP** (°C):   81
MW: 156.23   **BP** (°C):   265.5

Solubility (Moles/L)	Solubility (Grams/L)	Temp (°C)	Ref (#)	Evaluation (T P E A A)	Comments
1.754E-05	2.740E-03	25	E004	2 1 2 2 2	
2.163E-05	3.380E-03	25	M064	1 1 2 2 2	
2.160E-05	3.375E-03	25	M342	1 0 1 1 2	
2.163E-05	3.380E-03	ns	M344	0 0 0 0 2	

2586. $C_{12}H_{12}$
1,4-Dimethylnaphthalene
RN: 571-58-4   **MP** (°C):   7.6
MW: 156.23   **BP** (°C):   262

Solubility (Moles/L)	Solubility (Grams/L)	Temp (°C)	Ref (#)	Evaluation (T P E A A)	Comments
4.544E-05	7.100E-03	4	D351	1 2 1 1 2	
4.744E-05	7.412E-03	10	D351	1 2 1 1 2	
6.081E-05	9.500E-03	20	B318	1 2 1 2 0	EFG
6.062E-05	9.470E-03	20	B356	1 0 0 0 2	
6.167E-05	9.634E-03	25	D351	1 2 1 1 2	
7.297E-05	1.140E-02	25	M064	1 1 2 2 2	
7.300E-05	1.140E-02	25	M342	1 0 1 1 1	
7.944E-05	1.241E-02	40	D351	1 2 1 1 2	
7.297E-05	1.140E-02	ns	M344	0 0 0 0 2	

2587. $C_{12}H_{12}$
1,3-Dimethylnaphthalene
RN: 575-41-7   **MP** (°C):   -5
MW: 156.23   **BP** (°C):   263

Solubility (Moles/L)	Solubility (Grams/L)	Temp (°C)	Ref (#)	Evaluation (T P E A A)	Comments
5.121E-05	8.000E-03	25	M064	1 1 2 2 1	
5.120E-05	7.999E-03	25	M342	1 0 1 1 2	
5.121E-05	8.000E-03	ns	M344	0 0 0 0 1	

2588. $C_{12}H_{12}ClNO$
2-Chloro-N-(1-methyl-2-propynyl)acetanilide
Basamaize
RN: 35846-47-0   **MP** (°C):   40
MW: 221.69   **BP** (°C):

Solubility (Moles/L)	Solubility (Grams/L)	Temp (°C)	Ref (#)	Evaluation (T P E A A)	Comments
2.255E-03	5.000E-01	20	B200	1 0 0 0 0	

2589.   C$_{12}$H$_{12}$N$_2$
Benzidine
Benzidin
p-Diaminobiphenyl
RN: 92-87-5   **MP** (°C):   117
MW: 184.24   **BP** (°C):   400

Solubility (Moles/L)	Solubility (Grams/L)	Temp (°C)	Ref (#)	Evaluation (T P E A A)	Comments
1.953E-03	3.599E-01	24	H106	1 0 2 2 2	
1.954E-03	3.600E-01	24	M303	1 0 1 1 2	pH 5.9
2.712E-03	4.998E-01	25	B019	1 0 1 2 0	
2.822E-03	5.200E-01	25	B068	2 0 1 1 1	
2.700E-04	4.975E-02	25	H091	1 2 2 2 1	*sic*
1.465E-03	2.699E-01	rt	N015	0 0 2 2 2	

2590.   C$_{12}$H$_{12}$N$_2$
m-Benzidine
3-Benzidine
RN: 2050-89-7   **MP** (°C):   117
MW: 184.24   **BP** (°C):   400

Solubility (Moles/L)	Solubility (Grams/L)	Temp (°C)	Ref (#)	Evaluation (T P E A A)	Comments
5.970E-02	1.100E+01	100	F300	1 0 0 0 1	

2591.   C$_{12}$H$_{12}$N$_2$OS
2,4-Dimethyl-5-carboxanilidothiazole
G-696
RN: 21452-18-6   **MP** (°C):   141
MW: 232.31   **BP** (°C):

Solubility (Moles/L)	Solubility (Grams/L)	Temp (°C)	Ref (#)	Evaluation (T P E A A)	Comments
1.056E-02	2.454E+00	25	M061	1 0 0 0 2	

2592.   C$_{12}$H$_{12}$N$_2$O$_2$S
Dapsone
4,4'-Diaminodiphenyl Sulphone
RN: 80-08-0   **MP** (°C):   175
MW: 248.31   **BP** (°C):

Solubility (Moles/L)	Solubility (Grams/L)	Temp (°C)	Ref (#)	Evaluation (T P E A A)	Comments
5.638E-04	1.400E-01	25	P351	2 2 1 2 1	pH 7.4
6.444E-04	1.600E-01	25	P351	2 2 1 2 1	
1.530E-03	3.800E-01	37	L037	1 2 2 1 1	

2593. $C_{12}H_{12}N_2O_2S$
Sulfabenz
Sulfanilid
RN: 127-77-5 **MP** (°C):
MW: 248.31 **BP** (°C):

Solubility (Moles/L)	Solubility (Grams/L)	Temp (°C)	Ref (#)	Evaluation (T P E A A)	Comments
2.819E-02	7.000E+00	100	F300	1 0 0 0 0	

2594. $C_{12}H_{12}N_2O_3$
Nalidixic Acid
NegGRAM
1-Ethyl-1,4-dihydro-7-methyl-4-oxo-1,8-naphthyridine-3-carboxylic Acid
Nalidic Acid
RN: 389-08-2 **MP** (°C): 228
MW: 232.24 **BP** (°C):

Solubility (Moles/L)	Solubility (Grams/L)	Temp (°C)	Ref (#)	Evaluation (T P E A A)	Comments
4.306E-04	1.000E-01	23	G098	1 0 0 0 0	
7.079E-01	1.644E+02	37	O307	1 0 1 2 1	pH 2, EFG

2595. $C_{12}H_{12}N_2O_3$
Phenobarbital
5-Ethyl-5-Phenylbarbituric Acid
Phenylethylmalonylurea
RN: 50-06-6 **MP** (°C): 176
MW: 232.24 **BP** (°C):

Solubility (Moles/L)	Solubility (Grams/L)	Temp (°C)	Ref (#)	Evaluation (T P E A A)	Comments
3.980E-03	9.243E-01	15	H018	1 2 2 2 2	
3.180E-03	7.385E-01	15	S149	1 2 2 1 2	hydrate
3.680E-03	8.546E-01	15	S149	1 2 2 1 2	anhydrate
4.736E-03	1.100E+00	20	I009	1 2 2 1 1	EFG, 0.005M HCl
3.789E-03	8.800E-01	20	J030	1 2 2 2 1	
5.081E-03	1.180E+00	20	K143	1 2 2 2 2	form III
4.521E-03	1.050E+00	20	K143	1 2 2 2 2	form II
3.143E-03	7.300E-01	20	N023	1 2 2 1 1	hydrate
4.866E-03	1.130E+00	20	N023	1 2 2 1 2	anhydrate
4.510E-03	1.047E+00	20	S149	1 2 2 1 2	anhydrate
3.920E-03	9.104E-01	20	S149	1 2 2 1 2	hydrate
4.731E-03	1.099E+00	25	A023	1 0 0 1 1	
5.167E-03	1.200E+00	25	B011	2 0 0 1 0	
4.994E-03	1.160E+00	25	B065	1 1 1 1 0	
5.590E-03	1.298E+00	25	E011	2 1 1 2 1	
7.737E-03	1.797E+00	25	E011	2 1 1 2 1	pH 7.0
3.078E-02	7.149E+00	25	E011	2 1 1 2 1	pH 8.0
4.731E-03	1.099E+00	25	F009	2 2 2 2 0	EFG

4.600E-03	1.068E+00	25	G003	1 1 1 1 1	pH 4.7
2.734E-03	6.350E-01	25	H005	1 0 1 2 2	
5.161E-03	1.199E+00	25	K010	2 0 0 1 1	
6.114E-03	1.420E+00	25	K143	1 2 2 2 2	form III
5.512E-03	1.280E+00	25	K143	1 2 2 2 2	form II
4.650E-03	1.080E+00	25	L032	2 1 2 0 2	
4.790E-03	1.112E+00	25	M056	2 2 2 2 2	
5.684E-03	1.320E+00	25	N023	1,2 2 1 2	anhydrate
4.995E-03	1.160E+00	25	N023	1 2 2 1 2	hydrate
6.020E-03	1.398E+00	25	P006	2 0 2 2 1	
4.306E-03	1.000E+00	25	P015	2 2 2 2 1	
4.761E-03	1.106E+00	25	P350	2 1 1 1 2	intrinsic
5.320E-03	1.236E+00	25	S149	1 2 2 1 2	anhydrate
4.830E-03	1.122E+00	25	S149	1 2 2 1 2	hydrate
5.170E-03	1.201E+00	25	V033	2 0 1 1 2	
5.200E-03	1.208E+00	25.00	T303	1 0 0 0 1	
6.700E-03	1.556E+00	30	A065	2 0 2 2 1	
6.310E-03	1.465E+00	30	H018	1 2 2 2 2	
6.000E-03	1.393E+00	30	I001	2 0 2 1 0	EFG, 0.003N H_2SO_4
6.100E-03	1.417E+00	30	K108	1 2 2 0 1	
7.148E-03	1.660E+00	30	K143	1 2 2 2 2	form III
6.502E-03	1.510E+00	30	K143	1 2 2 2 2	form II
6.502E-03	1.510E+00	30	N023	1 2 2 1 2	anhydrate
6.071E-03	1.410E+00	30	N023	1 2 2 1 2	hydrate
6.020E-03	1.398E+00	30	O321	2 2 2 2 1	
6.000E-03	1.393E+00	30	O321	2 2 2 2 1	
8.612E-03	2.000E+00	32	M157	2 0 1 1 0	EFG
7.737E-03	1.797E+00	35	A023	1 0 0 1 2	
7.750E-03	1.800E+00	35	S149	1 2 2 1 2	anhydrate
7.700E-03	1.788E+00	35	S149	1 2 2 1 2	hydrate
8.500E-03	1.974E+00	35.00	T303	1 0 0 0 1	
7.923E-03	1.840E+00	37	J030	1 2 2 2 2	
8.000E-03	1.858E+00	37	K121	1 2 1 2 0	0.1N HCl
9.023E-03	2.096E+00	40	A023	1 0 0 1 2	
9.000E-02	2.090E+01	40	N008	1 0 1 1 2	*sic*
1.055E-02	2.450E+00	45	S149	1 2 2 1 2	anhydrate
1.108E-02	2.573E+00	45	S149	1 2 2 1 2	hydrate
1.130E-02	2.624E+00	45.00	T303	1 0 0 0 2	
1.506E-02	3.498E+00	50	S149	1 2 2 1 2	hydrate
1.266E-02	2.940E+00	50	S149	1 2 2 1 2	anhydrate
1.698E-02	3.943E+00	55	S149	1 2 2 1 2	hydrate
1.499E-02	3.481E+00	55	S149	1 2 2 1 2	anhydrate
1.033E-02	2.400E+00	60	I009	1 2 2 1 1	EFG, 0.005M HCl
4.177E-03	9.700E-01	ns	T003	0 0 0 0 2	

2596. C$_{12}$H$_{12}$N$_2$O$_6$S$_2$
Benzidine-2,2'-disulfonic Acid
Benzidin-disulfosaeure-(2,2')
RN: 117-61-3 **MP** (°C):
MW: 344.37 **BP** (°C):

Solubility (Moles/L)	Solubility (Grams/L)	Temp (°C)	Ref (#)	Evaluation (T P E A A)	Comments
2.323E-03	8.000E-01	25	F300	1 0 0 0 0	

2597. C$_{12}$H$_{12}$N$_2$S
Thiopyrine
1-Phenyl-2,3-dimethyl-3-pyrazoline-5-thione
RN: 5702-69-2 **MP** (°C): 166
MW: 216.31 **BP** (°C):

Solubility (Moles/L)	Solubility (Grams/L)	Temp (°C)	Ref (#)	Evaluation (T P E A A)	Comments
6.600E-02	1.428E+01	ns	D087	0 2 0 0 2	

2598. C$_{12}$H$_{12}$N$_4$O$_3$S
N4-Acetyl Sulfadiazine
N4-Acetylsulfadiazine
Acetyl Sulfadiazine
2-N4-Acetylsulfanilamidopyrimidine
RN: 127-74-2 **MP** (°C):
MW: 292.32 **BP** (°C):

Solubility (Moles/L)	Solubility (Grams/L)	Temp (°C)	Ref (#)	Evaluation (T P E A A)	Comments
5.131E-04	1.500E-01	37	F075	1 0 2 2 2	
7.200E-04	2.105E-01	37	G026	1 0 1 1 0	EFG, pH 5.4
6.842E-04	2.000E-01	37	L091	1 0 0 0 1	pH 5.5
8.723E-04	2.550E-01	37	M057	1 0 0 0 2	pH 5.5
5.131E-04	1.500E-01	37	R045	1 2 1 1 1	
5.131E-04	1.500E-01	37	R045	1 2 1 1 1	

2599. C$_{12}$H$_{12}$N$_4$O$_3$S
N4-Acetylsulfapyrazine
N4-Acetylsulphapyrazine
RN: 5433-91-0 **MP** (°C):
MW: 292.32 **BP** (°C):

Solubility (Moles/L)	Solubility (Grams/L)	Temp (°C)	Ref (#)	Evaluation (T P E A A)	Comments
1.710E-04	5.000E-02	37	L091	1 0 0 0 0	pH 5.5

2600. $C_{12}H_{12}N_6O_6$
TMPPT
1,3,7,9-Tetramethylpyrimido(5,4-γ) Pteridine-2,4,6,8(1H,3H,7H,9H)-tetrone
RN: **MP** (°C):
MW: 336.27   **BP** (°C):

Solubility (Moles/L)	Solubility (Grams/L)	Temp (°C)	Ref (#)	Evaluation (T P E A A)	Comments
3.860E-04	1.298E-01	25	K008	1 1 0 1 0	EFG
3.900E-04	1.311E-01	25	K009	1 2 1 1 0	EFG

2601. $C_{12}H_{12}O_6$
Benzoic Acid, 2-(Acetyloxy)-, (Acetyloxy)methyl Ester
Salicylic Acid Acetate, Hydroxymethyl Ester Acetate
RN: 32620-68-1   **MP** (°C):   oil
MW: 252.23   **BP** (°C):

Solubility (Moles/L)	Solubility (Grams/L)	Temp (°C)	Ref (#)	Evaluation (T P E A A)	Comments
9.634E-03	2.430E+00	21	N335	1 2 1 1 2	

2602. $C_{12}H_{13}ClN_2O$
Buturon
3-(para-Chlorophenyl)-1-methyl-1-(1-methyl-2-propynyl) Urea
Urea, N'-(4-Chlorophenyl)-N-methyl-N-(1-methyl-2-propynyl)
Eptapur
RN: 3766-60-7   **MP** (°C):   145.5
MW: 236.70   **BP** (°C):

Solubility (Moles/L)	Solubility (Grams/L)	Temp (°C)	Ref (#)	Evaluation (T P E A A)	Comments
1.267E-04	3.000E-02	20	G036	1 0 0 0 1	
1.267E-04	3.000E-02	20	M161	1 0 0 0 1	

2603. $C_{12}H_{13}I_3N_2O_2$
Iopodic Acid
Ipodic Acid
RN: 5587-89-3   **MP** (°C):
MW: 597.96   **BP** (°C):

Solubility (Moles/L)	Solubility (Grams/L)	Temp (°C)	Ref (#)	Evaluation (T P E A A)	Comments
3.027E-03	1.810E+00	ns	H055	0 1 0 2 2	

2604. $C_{12}H_{13}I_3N_2O_3$
Iocetamic Acid
N-(3-Amino-2,4,6-triiodophenyl)-3-acetamido-2-methylpropionic Acid
Cholebrine
MP 620
DRC 1201
RN: 16034-77-8 **MP** (°C): 224
MW: 613.96 **BP** (°C):

Solubility (Moles/L)	Solubility (Grams/L)	Temp (°C)	Ref (#)	Evaluation (T P E A A)	Comments
8.610E-03	5.286E+00	37	J016	1 0 0 0 2	pH 7.4

2605. $C_{12}H_{13}NO_2$
Methsuximide
Celontin
N-Methyl-α-methyl-α-phenylsuccinimide
RN: 77-41-8 **MP** (°C): 52-53
MW: 203.24 **BP** (°C):

Solubility (Moles/L)	Solubility (Grams/L)	Temp (°C)	Ref (#)	Evaluation (T P E A A)	Comments
1.378E-02	2.800E+00	25	P061	1 0 0 0 2	

2606. $C_{12}H_{13}NO_2S$
Carboxin
2,3-Dihydro-5-carboxanilido-6-methyl-1,4-oxathiin
Vitavax
RN: 5234-68-4 **MP** (°C): 94
MW: 235.31 **BP** (°C):

Solubility (Moles/L)	Solubility (Grams/L)	Temp (°C)	Ref (#)	Evaluation (T P E A A)	Comments
7.225E-04	1.700E-01	25	M061	1 0 0 0 2	
7.225E-04	1.700E-01	25	M161	1 0 0 0 2	

2607. $C_{12}H_{13}NO_3$
Azetidine, 1-[(Benzoyloxy)acetyl]-
RN: 115178-66-0 **MP** (°C): 74.5
MW: 219.24 **BP** (°C):

Solubility (Moles/L)	Solubility (Grams/L)	Temp (°C)	Ref (#)	Evaluation (T P E A A)	Comments
2.463E-02	5.400E+00	22	N317	1 1 2 1 2	

2608.   C₁₂H₁₃NO₃
Crotonyl Acetaminophen
Crotonic Acid, Ester with 4'-hydroxyacetanilide
Acetanilide, 4'-Hydroxy-, Crotonate (Ester)
RN: 20675-24-5   **MP** (°C):   146-147
MW: 219.24   **BP** (°C):

Solubility (Moles/L)	Solubility (Grams/L)	Temp (°C)	Ref (#)	Evaluation (T P E A A)	Comments
1.961E-03	4.300E-01	37	D029	1 0 1 1 1	

2609.   C₁₂H₁₃NO₄
Acetamide, N-Acetyl-2-(benzoyloxy)-N-methyl-
RN: 115178-80-8   **MP** (°C):
MW: 235.24   **BP** (°C):

Solubility (Moles/L)	Solubility (Grams/L)	Temp (°C)	Ref (#)	Evaluation (T P E A A)	Comments
1.360E-03	3.200E-01	22	N317	1 1 2 1 2	

2610.   C₁₂H₁₃NO₄S
Plantvax
2,3-Dihydro-5-carboxanilido-6-methyl-1,4-oxathiin-4,4-dioxide
Oxycarboxin
RN: 5259-88-1   **MP** (°C):   128.7
MW: 267.31   **BP** (°C):

Solubility (Moles/L)	Solubility (Grams/L)	Temp (°C)	Ref (#)	Evaluation (T P E A A)	Comments
3.741E-03	1.000E+00	25	M161	1 0 0 0 0	
3.741E-03	1.000E+00	ns	M061	0 0 0 0 2	

2611.   C₁₂H₁₃NO₄S₂
4-Ethylsulfonylnaphthalene-1-sulfonamide
ENS
4-ENS
RN: 842-00-2   **MP** (°C):
MW: 299.37   **BP** (°C):

Solubility (Moles/L)	Solubility (Grams/L)	Temp (°C)	Ref (#)	Evaluation (T P E A A)	Comments
3.775E-04	1.130E-01	c	K042	2 2 2 2 2	

2612. $C_{12}H_{13}NO_5$
Glycine, N-[(Benzoyloxy)acetyl]-N-methyl-
RN: 106231-64-5 **MP** (°C): 160.5
MW: 251.24 **BP** (°C):

Solubility (Moles/L)	Solubility (Grams/L)	Temp (°C)	Ref (#)	Evaluation (T P E A A)	Comments
5.572E-03	1.400E+00	22	N317	1 1 2 1 2	

2613. $C_{12}H_{13}NO_5$
Acid succinyl Acetaminophen
Butanedioic Acid, Mono[4-(acetylamino)phenyl] Ester
Acetanilide, 4'-Hydroxy-, Hydrogen Succinate (Ester)
RN: 20675-25-6 **MP** (°C): 145.5-146.5
MW: 251.24 **BP** (°C):

Solubility (Moles/L)	Solubility (Grams/L)	Temp (°C)	Ref (#)	Evaluation (T P E A A)	Comments
2.587E-02	6.500E+00	37	D029	1 0 1 1 1	

2614. $C_{12}H_{13}NO_6$
Carbobenzoxydiglycine
RN: **MP** (°C):
MW: 267.24 **BP** (°C):

Solubility (Moles/L)	Solubility (Grams/L)	Temp (°C)	Ref (#)	Evaluation (T P E A A)	Comments
2.432E-03	6.500E-01	25.1	N026	2 0 2 2 2	
2.804E-03	7.494E-01	25.1	N027	1 1 2 2 2	

2615. $C_{12}H_{13}N_3O_2$
Isocarboxazid
Marplan
RN: 59-63-2 **MP** (°C):
MW: 231.26 **BP** (°C):

Solubility (Moles/L)	Solubility (Grams/L)	Temp (°C)	Ref (#)	Evaluation (T P E A A)	Comments
3.459E-03	8.000E-01	25	R024	1 0 0 0 0	

2616. $C_{12}H_{13}N_3O_2S$
N1-Methyl-N1-2-pyridyl-sulfanilamide
N1-Methyl-N1-(2-pyridyl)sulfanilamide
RN: 51543-29-4 **MP** (°C):
MW: 263.32 **BP** (°C):

Solubility (Moles/L)	Solubility (Grams/L)	Temp (°C)	Ref (#)	Evaluation (T P E A A)	Comments
4.740E-03	1.248E+00	37	K095	2 0 0 0 2	intrinsic

2617.   C$_{12}$H$_{13}$N$_3$O$_3$S$_2$
Methyl Acetyl Sulfathiazole
Sulfathiazol Methyle Acetyle
RN: **MP** (°C):
MW: 311.38   **BP** (°C):

Solubility (Moles/L)	Solubility (Grams/L)	Temp (°C)	Ref (#)	Evaluation (T P E A A)	Comments
2.248E-04	7.000E-02	37	D084	1 0 1 0 0	

2618.   C$_{12}$H$_{13}$N$_3$O$_4$S
Acetylsulfamethoxazole
Acetanilide, 4'-[(5-Methyl-3-isoxazolyl)sulfamoyl]-
4'-Acetyl-3-sulfa-5-methylisoxazole
RN: 21312-10-7   **MP** (°C):
MW: 295.32   **BP** (°C):

Solubility (Moles/L)	Solubility (Grams/L)	Temp (°C)	Ref (#)	Evaluation (T P E A A)	Comments
2.573E-04	7.600E-02	37	H120	1 1 1 1 1	normal saline

2619.   C$_{12}$H$_{14}$Cl$_2$O$_3$
2,4-Dichlorophenoxyacetic Acid sec-Butyl Ester
RN: 94-79-1   **MP** (°C):
MW: 277.15   **BP** (°C):

Solubility (Moles/L)	Solubility (Grams/L)	Temp (°C)	Ref (#)	Evaluation (T P E A A)	Comments
6.252E-05	1.733E-02	ns	M120	0 0 1 1 2	

2620.   C$_{12}$H$_{14}$Cl$_2$O$_3$
2,4-Dichlorophenoxyacetic Acid n-Butyl Ester
2,4-Dichlorophenoxyacetic Acid Butyl Ester
RN: 94-80-4   **MP** (°C):
MW: 277.15   **BP** (°C):

Solubility (Moles/L)	Solubility (Grams/L)	Temp (°C)	Ref (#)	Evaluation (T P E A A)	Comments
5.495E-05	1.523E-02	ns	M120	0 0 1 1 2	

2621.   C$_{12}$H$_{14}$Cl$_3$O$_4$P
Chlorfenvinphos
2-Chloro-1-(2,4-dichlorophenyl)Ethenyl Phosphoric Acid, Diethyl Ester
Dermaton
Birlanex
Birlane
Steladone
RN: 470-90-6   **MP** (°C):
MW: 359.58   **BP** (°C):

Solubility (Moles/L)	Solubility (Grams/L)	Temp (°C)	Ref (#)	Evaluation (T P E A A)	Comments
3.476E-04	1.250E-01	10	B324	2 2 2 2 2	
3.476E-04	1.250E-01	10	B324	2 2 2 2 2	
4.074E-04	1.465E-01	20	B179	2 0 0 0 2	
3.449E-04	1.240E-01	20	B300	2 1 1 1 2	
3.449E-04	1.240E-01	20	B324	2 2 2 2 2	
3.448E-04	1.240E-01	20	B324	2 2 2 2 2	
3.893E-04	1.400E-01	20	F311	1 2 2 2 1	
4.033E-04	1.450E-01	20	M061	1 0 0 0 2	
4.033E-04	1.450E-01	23	M161	1 0 0 0 2	
2.976E-04	1.070E-01	30	B324	2 2 2 2 2	
2.975E-04	1.070E-01	30	B324	2 2 2 2 2	

2622.   C$_{12}$H$_{14}$NO$_4$PS
Ditalimfos
O,O-Diethyl (1,3-Dihydro-1,3-dioxo-2H-isoindol-2-yl) Phosphonothioate
Laptran
Plondrel
RN: 5131-24-8   **MP** (°C):   83.5
MW: 299.29   **BP** (°C):

Solubility (Moles/L)	Solubility (Grams/L)	Temp (°C)	Ref (#)	Evaluation (T P E A A)	Comments
4.444E-04	1.330E-01	rt	M161	0 0 0 0 2	

2623.   C$_{12}$H$_{14}$N$_2$O$_2$
Primidone
5-Ethyldihydro-5-phenyl-4,6(1H,5H)-pyrimidinedione
Desoxyphenobarbitone
2-Deoxyphenobarbital
RN: 125-33-7   **MP** (°C):   281.5
MW: 218.26   **BP** (°C):

Solubility (Moles/L)	Solubility (Grams/L)	Temp (°C)	Ref (#)	Evaluation (T P E A A)	Comments
2.200E-03	4.802E-01	30	K108	1 2 2 0 1	
2.747E-03	5.996E-01	37	P061	1 0 0 0 2	
2.291E-03	5.000E-01	rt	D025	0 0 0 0 0	

2624. $C_{12}H_{14}N_2O_4$
Propanamide, 2-[[(Benzoyloxy)acetyl]amino]-
RN: 115193-30-1   **MP** (°C):   201.5
MW: 250.26   **BP** (°C):

Solubility (Moles/L)	Solubility (Grams/L)	Temp (°C)	Ref (#)	Evaluation (T P E A A)	Comments
1.918E-03	4.800E-01	22	N317	1 1 2 1 2	

2625. $C_{12}H_{14}N_2O_4$
Acetamide, N-(2-Amino-2-oxoethyl)-2-(benzoyloxy)-N-methyl-
RN: 106231-62-3   **MP** (°C):   101.5
MW: 250.26   **BP** (°C):

Solubility (Moles/L)	Solubility (Grams/L)	Temp (°C)	Ref (#)	Evaluation (T P E A A)	Comments
1.207E-01	3.020E+01	22	N317	1 1 2 1 2	

2626. $C_{12}H_{14}N_2O_5$
2-Cyclohexyl-4,6-dinitrophenol
Dinex
4,6-Dinitro-2-cyclohexylphenol
2,4-Dinitro-6-cyclohexylphenol
RN: 131-89-5   **MP** (°C):   106
MW: 266.26   **BP** (°C):

Solubility (Moles/L)	Solubility (Grams/L)	Temp (°C)	Ref (#)	Evaluation (T P E A A)	Comments
5.634E-05	1.500E-02	25	M061	1 0 0 0 1	pH 6.5
6.760E-06	1.800E-03	25	M061	1 0 0 0 1	pH 1

2627. $C_{12}H_{14}N_2O_6$
Dinoseb Acetate
Aretit
RN: 2813-95-8   **MP** (°C):   26.5
MW: 282.26   **BP** (°C):

Solubility (Moles/L)	Solubility (Grams/L)	Temp (°C)	Ref (#)	Evaluation (T P E A A)	Comments
7.794E-03	2.200E+00	rt	M161	0 0 0 0 1	

2628. $C_{12}H_{14}N_4O_2S$
6-Sulfanilamido-2,4-dimethylpyrimidine
6-Sulfanilamido-2,4-dimethylpyrimidin
Sulfisomidine
Sulphasomidine
RN: 515-64-0   **MP** (°C):   243.0
MW: 278.33   **BP** (°C):

Solubility (Moles/L)	Solubility (Grams/L)	Temp (°C)	Ref (#)	Evaluation (T P E A A)	Comments
4.965E-03	1.382E+00	25	M319	2 1 1 1 2	
6.862E-03	1.910E+00	37	K086	1 0 0 0 2	
5.802E-03	1.615E+00	ns	B133	0 2 0 1 2	pH 7.4
1.075E-02	2.991E+00	ns	M141	0 0 0 0 0	

2629. $C_{12}H_{14}N_4O_2S$
Sulfamethazine
Sulfadimezine
2-Sulfanilamido-4,6,-dimethylpyrimidine
RN: 57-68-1   **MP** (°C):   176
MW: 278.33   **BP** (°C):

Solubility (Moles/L)	Solubility (Grams/L)	Temp (°C)	Ref (#)	Evaluation (T P E A A)	Comments
5.317E-03	1.480E+00	20	F073	1 2 2 2 2	
1.544E-03	4.298E-01	20	L058	1 0 1 1 2	
1.893E-03	5.269E-01	20	O032	1 0 0 0 2	
1.424E-03	3.963E-01	24	N021	2 0 1 2 2	pH 5.6
5.389E-03	1.500E+00	29	C049	1 2 0 0 1	
2.695E-03	7.500E-01	37	L091	1 0 0 0 1	pH 5.5
6.862E-03	1.910E+00	37	M057	1 0 0 0 2	pH 5.5
2.414E-03	6.720E-01	37	S192	1 0 1 1 2	pH 6.0
2.299E-03	6.400E-01	38	K006	1 0 0 0 1	
1.185E-03	3.299E-01	ns	L044	0 0 0 0 2	

2630. $C_{12}H_{14}N_4O_2S.0.5H_2O$
Sulphamethazine (Hemihydrate)
Sulfamethazine Hemihydrate
RN: 57-68-1   **MP** (°C):
MW: 287.34   **BP** (°C):

Solubility (Moles/L)	Solubility (Grams/L)	Temp (°C)	Ref (#)	Evaluation (T P E A A)	Comments
6.786E-03	1.950E+00	37	R044	1 0 1 1 0	

2631. C$_{12}$H$_{14}$N$_4$O$_2$S
2-Sulfanilylamino-4-ethylpyrimidine
RN: 2276-96-2   **MP** (°C):
MW: 278.33   **BP** (°C):

Solubility (Moles/L)	Solubility (Grams/L)	Temp (°C)	Ref (#)	Evaluation (T P E A A)	Comments
6.180E-04	1.720E-01	37	R076	1 2 0 0 2	

2632. C$_{12}$H$_{14}$N$_4$O$_2$S
2-Sulfanilamido-4,5-dimethylpyrimidine
RN: 4462-43-5   **MP** (°C):   225.7
MW: 278.33   **BP** (°C):

Solubility (Moles/L)	Solubility (Grams/L)	Temp (°C)	Ref (#)	Evaluation (T P E A A)	Comments
7.186E-04	2.000E-01	29	C049	1 2 0 0 1	

2633. C$_{12}$H$_{14}$N$_4$O$_3$S$_2$
Acetyl Sulfaethylthiadiazole
Acetamide, N-[4-[[(5-Ethyl-1,3,4-thiadiazol-2-yl)amino]sulfonyl]phenyl]-
RN: 1037-51-0   **MP** (°C):
MW: 326.40   **BP** (°C):

Solubility (Moles/L)	Solubility (Grams/L)	Temp (°C)	Ref (#)	Evaluation (T P E A A)	Comments
4.963E-03	1.620E+00	37	B046	1 0 2 2 2	pH 4.6

2634. C$_{12}$H$_{14}$N$_4$O$_3$S
Sulfamethomidine
Sulphamethomidine
RN: 3772-76-7   **MP** (°C):   146.0
MW: 294.33   **BP** (°C):

Solubility (Moles/L)	Solubility (Grams/L)	Temp (°C)	Ref (#)	Evaluation (T P E A A)	Comments
2.864E-03	8.430E-01	ns	B133	0 2 0 1 2	pH 7.4

2635. C$_{12}$H$_{14}$N$_4$O$_3$S
2-Sulfanilamido-4-ethoxypyrimidine
RN: 71138-72-2   **MP** (°C):
MW: 294.33   **BP** (°C):

Solubility (Moles/L)	Solubility (Grams/L)	Temp (°C)	Ref (#)	Evaluation (T P E A A)	Comments
1.801E-04	5.300E-02	37	R046	1 2 1 1 2	

2636. C$_{12}$H$_{14}$N$_4$O$_4$S
Sulfadimethoxine
Sulphadimethoxine
RN: 122-11-2   **MP** (°C):   202.0
MW: 310.33   **BP** (°C):

Solubility (Moles/L)	Solubility (Grams/L)	Temp (°C)	Ref (#)	Evaluation (T P E A A)	Comments
1.492E-04	4.630E-02	37	W055	1 2 0 1 2	
1.105E-03	3.430E-01	ns	B133	0 2 0 1 2	pH 7.4

2637. C$_{12}$H$_{14}$O$_4$
Trimethylacetyl Salicylate
Salicylic Acid, Pivalate
2-Carboxyphenyl Pivalate
RN: 2704-58-7   **MP** (°C):
MW: 222.24   **BP** (°C):

Solubility (Moles/L)	Solubility (Grams/L)	Temp (°C)	Ref (#)	Evaluation (T P E A A)	Comments
9.730E-04	2.162E-01	25.6	G015	1 0 1 1 2	pH 1.00, pka 3.74, intrinsic

2638. C$_{12}$H$_{14}$O$_4$
Diethyl Phthalate
Ethyl Phthalate
Di-ethyl Phthalate
Phthalic Acid Ethyl Ester
Phthalsaeure-diaethyl Ester
RN: 84-66-2   **MP** (°C):   -40.5
MW: 222.24   **BP** (°C):   296.1

Solubility (Moles/L)	Solubility (Grams/L)	Temp (°C)	Ref (#)	Evaluation (T P E A A)	Comments
4.495E-03	9.990E-01	20	F070	1 0 0 0 0	
4.180E-03	9.290E-01	20	L300	2 1 0 2 2	
1.793E-02	3.984E+00	20.00	D343	1 0 1 1 0	
5.399E-03	1.200E+00	25	F067	1 0 2 2 2	
4.500E-03	1.000E+00	25	F300	1 0 0 0 0	

2639.   C₁₂H₁₅ClNO₄PS₂
Phosalone
Diethyl S-((6-chloro-2-oxobenzoxazolin-3-yl)methyl) phosphorodithioate
Rubitox
Benzophosphate
RN: 2310-17-0   **MP** (°C):
MW: 367.81   **BP** (°C):

Solubility (Moles/L)	Solubility (Grams/L)	Temp (°C)	Ref (#)	Evaluation (T P E A A)	Comments
3.263E-06	1.200E-03	10	B324	2 2 2 2 2	
3.263E-06	1.200E-03	10	B324	2 2 2 2 2	
7.069E-06	2.600E-03	20	B300	2 2 1 1 2	
7.069E-06	2.600E-03	20	B324	2 2 2 2 2	
7.069E-06	2.600E-03	20	B324	2 2 2 2 2	
5.845E-06	2.150E-03	20	C053	1 0 2 2 1	
1.006E-05	3.700E-03	30	B324	2 2 2 2 2	
1.006E-05	3.700E-03	30	B324	2 2 2 2 2	
5.845E-06	2.150E-03	ns	F071	0 1 2 1 2	
2.719E-05	1.000E-02	rt	M161	0 0 0 0 1	

2640.   C₁₂H₁₅ClO₃
Clofibrate
2-(p-Chlorophenoxy)-2-methylpropionic Acid Ethyl Ester
Abitrate
Atromid S
RN: 637-07-0   **MP** (°C):
MW: 242.70   **BP** (°C):

Solubility (Moles/L)	Solubility (Grams/L)	Temp (°C)	Ref (#)	Evaluation (T P E A A)	Comments
4.000E-04	9.708E-02	rt	G093	0 1 1 1 2	

2641.   C₁₂H₁₅IN₂O₆
Uridine, 2'-Deoxy-5-iodo-, 5'-Propanoate
5'-Propionyl 5-iodo-2'-deoxyuridine
5-Iodo-2'-deoxyuridine 5'-propionate
RN: 84043-25-4   **MP** (°C):   167.5
MW: 410.17   **BP** (°C):

Solubility (Moles/L)	Solubility (Grams/L)	Temp (°C)	Ref (#)	Evaluation (T P E A A)	Comments
3.480E+03	1.427E+06	25	N332	1 0 2 2 2	pH 7.4

2642.   C$_{12}$H$_{15}$NO
n-propylcinnamamide
Cinnamamide, N-Propyl-
2-Propenamide, 3-Phenyl-N-propyl-
RN: 6329-15-3   **MP** (°C):
MW: 189.26   **BP** (°C):

Solubility (Moles/L)	Solubility (Grams/L)	Temp (°C)	Ref (#)	Evaluation (T P E A A)	Comments
2.300E-03	4.353E-01	ns	H350	0 0 0 0 2	

2643.   C$_{12}$H$_{15}$NO$_3$
Carbofuran
2,3-Dihydro-2,2-dimethyl-7-benzofuranol methylcarbamate
Crisfuran
Furadanx
Curaterr
RN: 1563-66-2   **MP** (°C):   152
MW: 221.26   **BP** (°C):

Solubility (Moles/L)	Solubility (Grams/L)	Temp (°C)	Ref (#)	Evaluation (T P E A A)	Comments
1.315E-03	2.909E-01	10	B324	2 2 2 2 2	
1.315E-03	2.910E-01	10	B324	2 2 2 2 2	
1.446E-03	3.199E-01	19	B169	2 1 1 1 1	
1.446E-03	3.199E-01	20	B324	2 2 2 2 2	
1.446E-03	3.199E-01	20	B324	2 2 2 2 2	
3.164E-03	7.000E-01	25	M161	1 0 0 0 2	
1.695E-03	3.750E-01	30	B324	2 2 2 2 2	
1.694E-03	3.749E-01	30	B324	2 2 2 2 2	

2644.   C$_{12}$H$_{15}$NO$_3$
Propanamide, 3-(Benzoyloxy)-N,N-dimethyl-
RN: 115178-77-3   **MP** (°C):
MW: 221.26   **BP** (°C):

Solubility (Moles/L)	Solubility (Grams/L)	Temp (°C)	Ref (#)	Evaluation (T P E A A)	Comments
7.955E-02	1.760E+01	22	N317	1 1 2 1 2	

2645.   C$_{12}$H$_{15}$NO$_3$
Acetamide, 2-(Benzoyloxy)-N-propyl-
RN: 106231-51-0   **MP** (°C):   89.5
MW: 221.26   **BP** (°C):

Solubility (Moles/L)	Solubility (Grams/L)	Temp (°C)	Ref (#)	Evaluation (T P E A A)	Comments
2.893E-03	6.400E-01	22	N317	1 1 2 1 2	

2646.   C$_{12}$H$_{15}$NO$_3$
Acetamide, 2-(Benzoyloxy)-N-(1-methylethyl)-
RN: 115193-27-6   **MP** (°C):   129.5
MW: 221.26   **BP** (°C):

Solubility (Moles/L)	Solubility (Grams/L)	Temp (°C)	Ref (#)	Evaluation (T P E A A)	Comments
1.853E-03	4.100E-01	22	N317	1 1 2 1 2	

2647.   C$_{12}$H$_{15}$NO$_3$
Acetaminophen Butyrate
Butyryl Acetaminophen
Butanoic Acid, 4-(Acetylamino)phenyl Ester
Acetanilide, 4'-Hydroxy-, Butyrate
RN: 14771-98-3   **MP** (°C):   140
MW: 221.26   **BP** (°C):

Solubility (Moles/L)	Solubility (Grams/L)	Temp (°C)	Ref (#)	Evaluation (T P E A A)	Comments
1.491E-03	3.300E-01	25	B010	1 1 1 1 0	
2.441E-03	5.400E-01	37	D029	1 0 1 1 1	

2648.   C$_{12}$H$_{15}$NO$_3$
Acetamide, N-[2-(Benzoyloxy)ethyl]-N-methyl-
RN: 57440-16-1   **MP** (°C):
MW: 221.26   **BP** (°C):

Solubility (Moles/L)	Solubility (Grams/L)	Temp (°C)	Ref (#)	Evaluation (T P E A A)	Comments
1.415E-01	3.130E+01	22	N317	1 1 2 1 2	

2649.   C$_{12}$H$_{15}$NO$_4$
O-(Butyryloxymethyl) Salicylamide
O-Butyryloxymethyl Salicylamide
Butanoic Acid, [2-(Aminocarbonyl)phenoxy]methyl Ester
RN: 103951-39-9   **MP** (°C):   57
MW: 237.26   **BP** (°C):

Solubility (Moles/L)	Solubility (Grams/L)	Temp (°C)	Ref (#)	Evaluation (T P E A A)	Comments
1.054E-02	2.500E+00	23	B328	1 2 2 1 1	pH 4.0
1.054E-02	2.500E+00	23	B328	1 2 2 1 1	

2650. C₁₂H₁₅NO₄

Acetamide, 2-(Benzoyloxy)-N-(2-hydroxyethyl)-N-methyl-
RN: 106231-59-8 **MP** (°C): 79
MW: 237.26 **BP** (°C):

Solubility (Moles/L)	Solubility (Grams/L)	Temp (°C)	Ref (#)	Evaluation (T P E A A)	Comments
8.135E-02	1.930E+01	22	N317	1 1 2 1 2	

2651. C₁₂H₁₅NO₄

Isopropyl Acetaminophen
Carbonic Acid, 4-(Acetylamino)phenyl 1-Methylethyl Ester
Acetanilide, 4'-Hydroxy-, Isopropyl Carbonate
RN: 17239-27-9 **MP** (°C): 131.5-132
MW: 237.26 **BP** (°C):

Solubility (Moles/L)	Solubility (Grams/L)	Temp (°C)	Ref (#)	Evaluation (T P E A A)	Comments
4.636E-03	1.100E+00	37	D029	1 0 1 1 1	

2652. C₁₂H₁₅NO₅

Benzoic Acid, 2-Hydroxy-, 2-[(2-Hydroxyethyl)methylamino]-2-oxoethyl Ester
N-Methyl-N-carbamoylmethyl Glycolamide Salicylate
RN: 114665-09-7 **MP** (°C): 92.5
MW: 253.26 **BP** (°C):

Solubility (Moles/L)	Solubility (Grams/L)	Temp (°C)	Ref (#)	Evaluation (T P E A A)	Comments
2.488E-02	6.300E+00	21	B331	1 2 2 1 0	pH 7.4
2.488E-02	6.300E+00	21	B331	1 2 2 1 1	

2653. C₁₂H₁₅NO₆

Ethonyphenyl Tartramic Acid
RN: **MP** (°C): 201
MW: 269.26 **BP** (°C):

Solubility (Moles/L)	Solubility (Grams/L)	Temp (°C)	Ref (#)	Evaluation (T P E A A)	Comments
1.481E-02	3.989E+00	14	C069	1 2 0 1 2	

2654. $C_{12}H_{15}N_2O_3PS$
Quinalphos
Diethyl O-(2-Quinoxalyl) phosphorothioate
Diethquinalphion
Bayrusil
Ekalux

RN: 13593-03-8 **MP** (°C): 33.5
MW: 298.30 **BP** (°C):

Solubility (Moles/L)	Solubility (Grams/L)	Temp (°C)	Ref (#)	Evaluation (T P E A A)	Comments
7.375E-05	2.200E-02	24	M161	1 0 0 0 1	

2655. $C_{12}H_{15}N_2O_3PS$
Phoxim
4-Ethoxy-7-phenyl-3,5-dioxa-6-aza-4-phosphaoct-6-ene-8-nitrile 4-sulfide
Baythion
Sebacil
Volation

RN: 14816-18-3 **MP** (°C):
MW: 298.30 **BP** (°C):

Solubility (Moles/L)	Solubility (Grams/L)	Temp (°C)	Ref (#)	Evaluation (T P E A A)	Comments
1.106E-05	3.300E-03	10	B324	2 2 2 2 2	
1.106E-05	3.299E-03	10	B324	2 2 2 2 2	
1.374E-05	4.099E-03	20	B300	2 1 1 1 2	
1.374E-05	4.099E-03	20	B324	2 2 2 2 2	
1.374E-05	4.100E-03	20	B324	2 2 2 2 2	
2.347E-05	7.000E-03	20	M161	1 0 0 0 0	
1.643E-05	4.901E-03	30	B324	2 2 2 2 2	
1.643E-05	4.900E-03	30	B324	2 2 2 2 2	

2656. $C_{12}H_{15}N_3O_2S$
1-Methyl-2-sulfanilamide-1,2-dihydropyridine
Benzenesulfonamide, 4-Amino-N-(1,2-dihydro-1-methyl-2-pyridinyl)-

RN: 51543-30-7 **MP** (°C):
MW: 265.34 **BP** (°C):

Solubility (Moles/L)	Solubility (Grams/L)	Temp (°C)	Ref (#)	Evaluation (T P E A A)	Comments
3.690E-03	9.791E-01	37	K095	2 0 0 0 2	intrinsic

2657.   C$_{12}$H$_{15}$N$_3$O$_3$
Triallyl Cyanurate
Cyanursaeure-triallylaether
RN: 101-37-1   **MP** (°C):   26-28
MW: 249.27   **BP** (°C):   119-120

Solubility (Moles/L)	Solubility (Grams/L)	Temp (°C)	Ref (#)	Evaluation (T P E A A)	Comments
2.407E-02	6.000E+00	20	F300	1 0 0 0 0	

2658.   C$_{12}$H$_{15}$N$_3$O$_6$
1,3,5-Triglycidyl-S-triazinetrione
α-TGT
RN: 2451-62-9   **MP** (°C):
MW: 297.27   **BP** (°C):

Solubility (Moles/L)	Solubility (Grams/L)	Temp (°C)	Ref (#)	Evaluation (T P E A A)	Comments
4.373E-02	1.300E+01	0	A088	0 0 1 1 1	

2659.   C$_{12}$H$_{15}$N$_5$O$_5$
Pivaloyl Salicylate
9-(2-O-Acetyl-β-D-arabinofuranosyl)adenine
RN: 87970-03-4   **MP** (°C):   195
MW: 309.28   **BP** (°C):

Solubility (Moles/L)	Solubility (Grams/L)	Temp (°C)	Ref (#)	Evaluation (T P E A A)	Comments
3.026E-01	9.360E+01	37	B306	1 2 0 1 2	pH 7.3

2660.   C$_{12}$H$_{15}$N$_5$O$_5$
9-[5'-(O-Acetyl)-β-D-arabinofuranosyl]adenine Ester
Vidarabine 5'-Acetate
RN: 65926-28-5   **MP** (°C):   198.0
MW: 309.28   **BP** (°C):

Solubility (Moles/L)	Solubility (Grams/L)	Temp (°C)	Ref (#)	Evaluation (T P E A A)	Comments
2.134E-02	6.600E+00	ns	B134	0 1 1 1 1	

2661.   C$_{12}$H$_{15}$O$_3$P
Diallyl Phenyl Phosphonate
Phosphonic Acid, Phenyl-, Di-2-Propenyl Ester
RN: 2948-89-2   **MP** (°C):
MW: 238.23   **BP** (°C):

Solubility (Moles/L)	Solubility (Grams/L)	Temp (°C)	Ref (#)	Evaluation (T P E A A)	Comments
1.259E-03	3.000E-01	25	B070	1 2 0 1 0	

2662.   C₁₂H₁₆ClNOS

Thiobencarb
S-4-Chlorobenzyl Diethylthiocarbamate
Diethylcarbamothioic Acid S-[(4-Chlorophenyl)methyl] Ester
4-Chlorobenzyl N,N-diethylthiocarbamate
RN: 28249-77-6   **MP** (°C):
MW: 257.78   **BP** (°C):   127.5

Solubility (Moles/L)	Solubility (Grams/L)	Temp (°C)	Ref (#)	Evaluation (T P E A A)	Comments
1.164E-04	3.000E-02	22	K137	1 1 2 1 0	

2663.   C₁₂H₁₆Cl₂N₂O

Neburon
1-Butyl-3-(3,4-dichlorophenyl)-1-methylurea
RN: 555-37-3   **MP** (°C):   101.5
MW: 275.18   **BP** (°C):

Solubility (Moles/L)	Solubility (Grams/L)	Temp (°C)	Ref (#)	Evaluation (T P E A A)	Comments
1.744E-05	4.800E-03	20	F311	1 2 2 2 1	
1.744E-05	4.800E-03	24	B185	1 0 0 0 1	
1.744E-05	4.800E-03	24	G036	1 0 0 0 1	
1.744E-05	4.800E-03	24	M061	1 0 0 0 1	
1.744E-05	4.800E-03	24	M161	1 0 0 0 1	
1.744E-05	4.800E-03	25	A039	1 1 0 0 1	
1.744E-05	4.800E-03	25	G099	1 0 0 1 0	
1.744E-05	4.800E-03	ns	K007	0 0 0 0 1	

2664.   C₁₂H₁₆N₂

Etryptamine
α-Ethyltryptamine
RN: 2235-90-7   **MP** (°C):   97
MW: 188.27   **BP** (°C):

Solubility (Moles/L)	Solubility (Grams/L)	Temp (°C)	Ref (#)	Evaluation (T P E A A)	Comments
2.709E-03	5.100E-01	rt	M011	0 0 2 1 1	intrinsic

2665.   C₁₂H₁₆N₂O

N-(Piperidinomethyl)benzamide
Benzamide, N-(1-Pyrrolidinylmethyl)-
RN: 92788-60-8   **MP** (°C):
MW: 204.27   **BP** (°C):

Solubility (Moles/L)	Solubility (Grams/L)	Temp (°C)	Ref (#)	Evaluation (T P E A A)	Comments
7.100E-03	1.450E+00	22	J037	1 0 1 1 1	

2666.   C$_{12}$H$_{16}$N$_2$O$_2$
N,N,N',N'-Tetramethylphthalamide
1,2-Benzenedicarboxamide, N,N,N',N'-Tetramethyl-
RN: 6329-16-4   **MP** (°C):
MW: 220.27   **BP** (°C):

Solubility (Moles/L)	Solubility (Grams/L)	Temp (°C)	Ref (#)	Evaluation (T P E A A)	Comments
3.223E+00	7.100E+02	30	K004	1 0 0 0 2	

2667.   C$_{12}$H$_{16}$N$_2$O$_2$
N,N,N',N'-Tetramethylisophthalamide
1,3-Benzenedicarboxamide, N,N,N',N'-Tetramethyl-
RN: 14334-36-2   **MP** (°C):
MW: 220.27   **BP** (°C):

Solubility (Moles/L)	Solubility (Grams/L)	Temp (°C)	Ref (#)	Evaluation (T P E A A)	Comments
3.069E+00	6.760E+02	30	K004	1 0 0 0 2	
3.070E+00	6.762E+02	30	K019	1 0 0 0 2	

2668.   C$_{12}$H$_{16}$N$_2$O$_2$
N,N,N',N'-Tetramethylterephthalamide
1,4-Benzenedicarboxamide, N,N,N',N'-Tetramethyl-
RN: 13158-31-1   **MP** (°C):
MW: 220.27   **BP** (°C):

Solubility (Moles/L)	Solubility (Grams/L)	Temp (°C)	Ref (#)	Evaluation (T P E A A)	Comments
1.843E+00	4.060E+02	30	K004	1 0 0 0 2	
1.840E+00	4.053E+02	30	K019	1 0 0 0 2	

2669.   C$_{12}$H$_{16}$N$_2$O$_3$
Hexobarbital
5-(1-Cyclohexen-1-yl)-1,5-dimethylbarbituric Acid
5-(1-Cyclohexenyl)-1,5-dimethylbarbituric Acid
Hexabarital
RN: 56-29-1   **MP** (°C):   146
MW: 236.27   **BP** (°C):

Solubility (Moles/L)	Solubility (Grams/L)	Temp (°C)	Ref (#)	Evaluation (T P E A A)	Comments
1.227E-03	2.900E-01	20	J030	1 2 2 2 1	
1.840E-03	4.347E-01	25	M056	2 2 2 2 2	
2.000E-03	4.725E-01	30	K108	1 2 2 0 1	
2.709E-03	6.400E-01	37	J030	1 2 2 2 1	

2670. C₁₂H₁₆N₂O₃
Cyclobarbital
Phanodorm
RN: 52-31-3   **MP** (°C):   173
MW: 236.27   **BP** (°C):

Solubility (Moles/L)	Solubility (Grams/L)	Temp (°C)	Ref (#)	Evaluation (T P E A A)	Comments
6.772E-03	1.600E+00	20	F300	1 0 0 0 1	
6.941E-03	1.640E+00	20	J030	1 2 2 2 2	
3.500E-02	8.270E+00	25	G003	1 1 1 1 1	pH 4.7
8.000E-03	1.890E+00	30	G014	1 1 1 1 0	EFG
7.800E-03	1.843E+00	30	I001	2 0 2 1 0	EFG, 0.003N H₂SO₄
8.000E-03	1.890E+00	30	K108	1 2 2 0 1	
9.735E-03	2.300E+00	37	F300	1 0 0 0 1	
9.523E-03	2.250E+00 ·	37	J030	1 2 2 2 2	
9.140E-02	2.160E+01	40	N008	1 2 1 1 2	*sic*

2671. C₁₂H₁₆N₂O₃
Carbetamide
N-Ethyl-2-((((phenylamino)carbonyl)oxy)propanamide
Leguarme
RN: 16118-49-3   **MP** (°C):   >110
MW: 236.27   **BP** (°C):

Solubility (Moles/L)	Solubility (Grams/L)	Temp (°C)	Ref (#)	Evaluation (T P E A A)	Comments
1.481E-02	3.500E+00	20	M161	1 0 0 0 1	

2672. C₁₂H₁₆N₃O₃PS
Triazophos
O,O-Diethyl O-(1-Phenyl-1H-1,2,4-triazol-3-yl) Phosphorothioate
Hostathion
RN: 24017-47-8   **MP** (°C):
MW: 313.32   **BP** (°C):

Solubility (Moles/L)	Solubility (Grams/L)	Temp (°C)	Ref (#)	Evaluation (T P E A A)	Comments
7.884E-05	2.470E-02	20	B300	2 1 1 1 2	
1.245E-04	3.900E-02	23	M161	1 0 0 0 1	
1.245E-04	3.900E-02	23	T305	1 0 0 0 1	

2673. C$_{12}$H$_{16}$N$_3$O$_3$PS$_2$
Azinphos-ethyl
O,O-Diethyl S-[(4-Oxo-1,2,3-benzotriazin-3(4H)-yl)methyl] Phosphorodithioate
Azinos
Ethyl Guthion
RN: 2642-71-9   **MP** (°C):
MW: 345.38   **BP** (°C):

Solubility (Moles/L)	Solubility (Grams/L)	Temp (°C)	Ref (#)	Evaluation (T P E A A)	Comments
1.940E-05	6.700E-03	10	B324	2 2 2 2 2	
1.940E-05	6.700E-03	10	B324	2 2 2 2 2	
3.040E-05	1.050E-02	20	B300	2 2 1 1 2	
3.040E-05	1.050E-02	20	B324	2 2 2 2 2	
3.040E-05	1.050E-02	20	B324	2 2 2 2 2	
7.152E-05	2.470E-02	30	B324	2 2 2 2 2	
7.151E-05	2.470E-02	30	B324	2 2 2 2 2	

2674. C$_{12}$H$_{16}$N$_4$O$_2$
2,5-Diaziridinyl-3,6-bis(methylamino)-1,4-benzoquinone
Benzoquinone-2,5-bisaziridinyl-3,6-bismethyl Amino
RN: 59886-52-1   **MP** (°C): 220
MW: 248.29   **BP** (°C):

Solubility (Moles/L)	Solubility (Grams/L)	Temp (°C)	Ref (#)	Evaluation (T P E A A)	Comments
<4.03E-04	<1.00E-01	rt	C317	0 2 0 0 0	

2675. C$_{12}$H$_{16}$N$_4$O$_2$S$_2$
4-Amino-N-(5-butyl-1,3,4-thiadiazol-2-yl)benzenesulfonamide
Sulfanilamide, N1-(5-Butyl-1,3,4-thiadiazol-2-yl)-
RN: 71119-31-8   **MP** (°C):
MW: 312.41   **BP** (°C):

Solubility (Moles/L)	Solubility (Grams/L)	Temp (°C)	Ref (#)	Evaluation (T P E A A)	Comments
2.710E-04	8.466E-02	37	A046	2 0 1 1 2	

2676. C$_{12}$H$_{16}$N$_4$O$_2$S$_2$
Glybuthiazole
p-Aminobenzenesulfamido-tert-butylthiodiazole
Glipasol
Glypasol
RN: 535-65-9   **MP** (°C): 222
MW: 312.41   **BP** (°C):

Solubility (Moles/L)	Solubility (Grams/L)	Temp (°C)	Ref (#)	Evaluation (T P E A A)	Comments
1.820E-04	5.686E-02	37	A046	2 0 1 1 2	

2677.   C₁₂H₁₆N₄O₇S

2'-Methylsulfonyl-6-methoxypurine Arabinoside

9H-Purine, 6-Methoxy-9-[2-O-(methylsulfonyl)-β-D-arabinofuranosyl]-

RN: 145913-48-0   **MP** (°C):   188-190
MW: 360.35   **BP** (°C):

Solubility (Moles/L)	Solubility (Grams/L)	Temp (°C)	Ref (#)	Evaluation (T P E A A)	Comments
1.720E-02	6.198E+00	37	C348	1 2 2 2 2	pH 7.00

2678.   C₁₂H₁₆N₅O₃PS₂

Azinphos-ethyl O-Analog

RN: **MP** (°C):
MW: 373.39   **BP** (°C):

Solubility (Moles/L)	Solubility (Grams/L)	Temp (°C)	Ref (#)	Evaluation (T P E A A)	Comments
1.017E-02	3.797E+00	10	B300	2 2 1 1 2	

2679.   C₁₂H₁₆O

p-Cyclohexylphenol

4-Cyclohexylphenol

RN: 1131-60-8   **MP** (°C):
MW: 176.26   **BP** (°C):

Solubility (Moles/L)	Solubility (Grams/L)	Temp (°C)	Ref (#)	Evaluation (T P E A A)	Comments
3.782E-04	6.666E-02	25	L021	1 0 0 0 0	

2680.   C₁₂H₁₆O

o-Cyclohexylphenol

2-Cyclohexylphenol

RN: 119-42-6   **MP** (°C):
MW: 176.26   **BP** (°C):

Solubility (Moles/L)	Solubility (Grams/L)	Temp (°C)	Ref (#)	Evaluation (T P E A A)	Comments
4.727E-04	8.333E-02	25	L021	1 0 0 0 0	

2681.   C₁₂H₁₆O₂

4-Cyclohexylresorcinol

p-Cyclohexylresorcinol

RN: 2138-20-7   **MP** (°C):
MW: 192.26   **BP** (°C):

Solubility (Moles/L)	Solubility (Grams/L)	Temp (°C)	Ref (#)	Evaluation (T P E A A)	Comments
2.599E-03	4.998E-01	25	L021	1 0 0 0 0	

2682. C₁₂H₁₆O₂

$C_{12}H_{16}O_2$

ε-Phenylcaproic Acid
6-Phenylcaproic Acid
6-Phenylhexanoic Acid
RN: 5581-75-9   **MP** (°C):
MW: 192.26   **BP** (°C):

Solubility (Moles/L)	Solubility (Grams/L)	Temp (°C)	Ref (#)	Evaluation (T P E A A)	Comments
2.495E-03	4.798E-01	30	D033	2 2 1 2 2	
4.002E-03	7.694E-01	40	D033	2 2 1 2 2	

2683. C₁₂H₁₆O₃

$C_{12}H_{16}O_3$

Isoamyl Salicylate
Isoamyl o-Hydroxybenzoate
3-Methylbutyl Salicylate
3-Methylbutyl o-Hydroxybenzoate
RN: 87-20-7   **MP** (°C):
MW: 208.26   **BP** (°C):

Solubility (Moles/L)	Solubility (Grams/L)	Temp (°C)	Ref (#)	Evaluation (T P E A A)	Comments
6.961E-04	1.450E-01	25	D081	1 2 2 1 2	

2684. C₁₂H₁₆O₇.H₂O

$C_{12}H_{16}O_7 \cdot H_2O$

Arbutin (Monohydrate)
Hydroquinone-β-D-glucopyranoside Monohydrate
RN: 6058-77-1   **MP** (°C): 195-200
MW: 290.27   **BP** (°C):

Solubility (Moles/L)	Solubility (Grams/L)	Temp (°C)	Ref (#)	Evaluation (T P E A A)	Comments
3.828E-01	1.111E+02	c	D004	1 0 0 0 0	
1.723E+00	5.000E+02	h	D004	1 0 0 0 0	

2685. C₁₂H₁₇NO₂

$C_{12}H_{17}NO_2$

2-sec-Butylphenyl Methylcarbamate
BPMC
2-(1-Methylpropyl)phenol Methylcarbamate
N-Methyl O-sec-butylPhenylcarbamate
RN: 3766-81-2   **MP** (°C): 32
MW: 207.27   **BP** (°C): 112.5

Solubility (Moles/L)	Solubility (Grams/L)	Temp (°C)	Ref (#)	Evaluation (T P E A A)	Comments
4.294E-04	8.900E-02	22	K137	1 1 2 1 0	
3.184E-03	6.600E-01	30	M161	1 0 0 0 2	

2686.   C₁₂H₁₇NO₂
m-tert-Butylphenyl N-Methylcarbamate
3-tert-Butylphenyl N-Methylcarbamate
RN: 780-11-0   **MP** (°C):   144.0
MW: 207.27   **BP** (°C):

Solubility (Moles/L)	Solubility (Grams/L)	Temp (°C)	Ref (#)	Evaluation (T P E A A)	Comments
<2.41E-06	<5.00E-04	30	D089	2 2 0 0 0	

2687.   C₁₂H₁₇NO₂
Pentyl p-Aminobenzoate
4-Aminobenzoic Acid Pentyl Ester
RN: 13110-37-7   **MP** (°C):
MW: 207.27   **BP** (°C):

Solubility (Moles/L)	Solubility (Grams/L)	Temp (°C)	Ref (#)	Evaluation (T P E A A)	Comments
3.900E-04	8.084E-02	37	F006	1 1 2 2 1	
1.890E-04	3.917E-02	ns	M066	0 0 0 0 2	
1.890E-04	3.917E-02	rt	B016	0 0 1 1 2	pH 7.4

2688.   C₁₂H₁₇NO₂
Promecarb
5-Isopropyl-m-tolyl Methylcarbamate
Carbamult
RN: 2631-37-0   **MP** (°C):   87.5
MW: 207.27   **BP** (°C):   117

Solubility (Moles/L)	Solubility (Grams/L)	Temp (°C)	Ref (#)	Evaluation (T P E A A)	Comments
4.439E-04	9.200E-02	rt	M161	0 0 0 0 1	

2689.   C₁₂H₁₇NO₂
Hexyl Nicotinate
n-Hexyl Nicotinoate
Nicotinic Acid n-Hexyl Ester
RN: 23597-82-2   **MP** (°C):
MW: 207.27   **BP** (°C):

Solubility (Moles/L)	Solubility (Grams/L)	Temp (°C)	Ref (#)	Evaluation (T P E A A)	Comments
8.202E-04	1.700E-01	32	L346	1 0 0 1 2	

2690.   C$_{12}$H$_{17}$NO$_2$
2,6-Diethyl-4-acetaminophenol
3,5-Diethylparacetamol
4-Acetamido-2,6-diethylphenol
RN: 55205-89-5   **MP** (°C):
MW: 207.27   **BP** (°C):

Solubility (Moles/L)	Solubility (Grams/L)	Temp (°C)	Ref (#)	Evaluation (T P E A A)	Comments
2.943E-03	6.101E-01	25	D078	1 2 1 1 2	

2691.   C$_{12}$H$_{17}$NO$_3$
Acetamide, N-[4-(1-ethoxyethoxy)phenyl]-
1-(p-Acetaminophenoxy)-1-ethoxyethane
RN: 51736-24-4   **MP** (°C):
MW: 223.27   **BP** (°C):

Solubility (Moles/L)	Solubility (Grams/L)	Temp (°C)	Ref (#)	Evaluation (T P E A A)	Comments
3.000E-03	6.698E-01	ns	H076	0 0 0 0 0	

2692.   C$_{12}$H$_{17}$NO$_3$
m-n-Butoxyphenyl N-Methylcarbamate
3-n-Butoxyphenyl N-Methylcarbamate
RN: 3978-68-5   **MP** (°C):   54.5
MW: 223.27   **BP** (°C):

Solubility (Moles/L)	Solubility (Grams/L)	Temp (°C)	Ref (#)	Evaluation (T P E A A)	Comments
4.031E-04	9.000E-02	30	D089	2 2 0 0 0	

2693.   C$_{12}$H$_{17}$NO$_3$
m-sec-Butoxyphenyl N-Methylcarbamate
3-sec-Butoxyphenyl N-Methylcarbamate
RN: 13538-22-2   **MP** (°C):   53
MW: 223.27   **BP** (°C):

Solubility (Moles/L)	Solubility (Grams/L)	Temp (°C)	Ref (#)	Evaluation (T P E A A)	Comments
3.583E-04	8.000E-02	30	D089	2 2 0 0 0	

2694.   C$_{12}$H$_{17}$NO$_4$
3,5-Dimethoxy-acetophenetide
RN: **MP** (°C):
MW: 239.27   **BP** (°C):

Solubility (Moles/L)	Solubility (Grams/L)	Temp (°C)	Ref (#)	Evaluation (T P E A A)	Comments
4.904E-01	1.173E+02	21.80	B102	2 0 1 1 1	solid hydrate
3.344E+00	8.000E+02	35.60	B102	2 0 1 1 2	liquid hydrate
8.778E-01	2.100E+02	39.40	B102	2 0 1 1 1	solid hydrate
3.233E+00	7.736E+02	45.60	B102	2 0 1 1 2	liquid hydrate
1.586E+00	3.795E+02	57	B102	2 0 1 1 1	solid hydrate
3.172E+00	7.591E+02	58.10	B102	2 0 1 1 2	liquid hydrate
3.172E+00	7.591E+02	68.50	B102	2 0 1 1 2	liquid hydrate
2.100E+00	5.026E+02	69.50	B102	2 0 1 1 1	solid hydrate
2.288E+00	5.474E+02	72.80	B102	2 0 1 1 1	solid hydrate
2.569E+00	6.147E+02	77.10	B102	2 0 1 1 2	solid hydrate
2.790E+00	6.675E+02	80.20	B102	2 0 1 1 2	solid hydrate
2.947E+00	7.053E+02	82.60	B102	2 0 1 1 2	solid hydrate
3.049E+00	7.296E+02	84.20	B102	2 0 1 1 2	solid hydrate
3.233E+00	7.736E+02	84.30	B102	2 0 1 1 2	liquid hydrate
3.172E+00	7.591E+02	86	B102	2 0 1 1 2	solid hydrate
3.233E+00	7.736E+02	86.90	B102	2 0 1 1 2	solid hydrate
3.348E+00	8.011E+02	99.80	B102	2 0 1 1 2	liquid hydrate
3.459E+00	8.275E+02	111.10	B102	2 0 1 1 2	liquid hydrate
3.527E+00	8.440E+02	118.40	B102	2 0 1 1 2	liquid hydrate
3.632E+00	8.690E+02	129.20	B102	2 0 1 1 2	liquid hydrate
4.031E+00	9.645E+02	173.60	B102	2 0 1 1 2	liquid hydrate

2695.   C$_{12}$H$_{17}$N$_2$O$_2$
4-Aminobenzoic Acid-2-(propyl-amino)ethyl Ester
2-(Propylamino)ethyl 4-Aminobenzoate
4-Aminobenzoic Acid 2-(Propyl-amino)ethyl Ester
RN: **MP** (°C):
MW: 221.28   **BP** (°C):

Solubility (Moles/L)	Solubility (Grams/L)	Temp (°C)	Ref (#)	Evaluation (T P E A A)	Comments
3.000E-04	6.638E-02	ns	M066	0 0 0 0 0	

2696.   C$_{12}$H$_{17}$N$_3$O$_4$S
3'-Nitroso-tolbutamide
RN: **MP** (°C):
MW: 299.35   **BP** (°C):

Solubility (Moles/L)	Solubility (Grams/L)	Temp (°C)	Ref (#)	Evaluation (T P E A A)	Comments
3.341E-04	1.000E-01	25	G051	1 0 1 1 0	

2697. $C_{12}H_{17}N_5O_3$

N,N-Diethylglycyloxymethyl-1-allopurinol

Glycine, N,N-Diethyl-, (4,5-Dihydro-4-oxo-1H-pyrazolo[3,4-d]pyrimidin-1-yl)Methyl Ester

RN: 98204-08-1   **MP** (°C):

MW: 279.30   **BP** (°C):

Solubility (Moles/L)	Solubility (Grams/L)	Temp (°C)	Ref (#)	Evaluation (T P E A A)	Comments
1.611E-02	4.500E+00	22	B323	1 0 2 2 2	

2698. $C_{12}H_{17}O_4PS_2$

Phenthoate

Dimethyl-S-(α-ethoxycarbonylbenzyl) Phosphorodithioate

Elsan

Fenthoate

Phent

Cidial

RN: 2597-03-7   **MP** (°C):

MW: 320.37   **BP** (°C):

Solubility (Moles/L)	Solubility (Grams/L)	Temp (°C)	Ref (#)	Evaluation (T P E A A)	Comments
6.243E-04	2.000E-01	20	M161	1 0 0 0 2	
3.434E-05	1.100E-02	22	K137	1 1 2 1 0	

2699. $C_{12}H_{18}$

1-Phenylhexane

Hexylbenzene

n-Hexylbenzene

RN: 1077-16-3   **MP** (°C):   -61

MW: 162.28   **BP** (°C):   226

Solubility (Moles/L)	Solubility (Grams/L)	Temp (°C)	Ref (#)	Evaluation (T P E A A)	Comments
5.678E-06	9.214E-04	5.04	M183	1 2 1 1 2	
5.678E-06	9.214E-04	6.04	M183	1 2 1 1 2	
5.140E-06	8.341E-04	7	O312	2 2 0 2 2	
5.667E-06	9.196E-04	8.04	M183	1 2 1 1 2	
5.583E-06	9.060E-04	9.04	M183	1 2 1 1 2	
5.150E-06	8.357E-04	10	O312	2 2 0 2 2	
5.572E-06	9.042E-04	10.04	M183	1 2 1 1 2	
5.717E-06	9.277E-04	11.04	M183	1 2 1 1 2	
5.733E-06	9.304E-04	12.04	M183	1 2 1 1 2	
5.667E-06	9.196E-04	13.04	M183	1 2 1 1 2	
5.700E-06	9.250E-04	14.04	M183	1 2 1 1 2	
5.090E-06	8.260E-04	15	O312	2 2 0 2 2	
5.594E-06	9.079E-04	15.04	M183	1 2 1 1 2	
5.661E-06	9.187E-04	16.04	M183	1 2 1 1 2	
5.606E-06	9.097E-04	17.04	M183	1 2 1 1 2	

5.678E-06	9.214E-04	18.04	M183	1 2 1 1 2
5.811E-06	9.430E-04	19.04	M183	1 2 1 1 2
5.860E-06	9.509E-04	20	O312	2 2 0 2 2
5.850E-06	9.493E-04	20.04	M183	1 2 1 1 2
5.889E-06	9.556E-04	21.04	M183	1 2 1 1 2
5.872E-06	9.529E-04	22.04	M183	1 2 1 1 2
6.056E-06	9.827E-04	23.04	M183	1 2 1 1 2
6.133E-06	9.953E-04	24.04	M183	1 2 1 1 2
6.270E-06	1.017E-03	25	M342	1 0 1 1 2
5.560E-06	9.023E-04	25	O312	2 2 0 2 2
6.156E-06	9.989E-04	25.04	M183	1 2 1 1 2
6.156E-06	9.989E-04	26.04	M183	1 2 1 1 2
6.239E-06	1.012E-03	27.04	M183	1 2 1 1 2
6.261E-06	1.016E-03	29.04	M183	1 2 1 1 2
6.140E-06	9.964E-04	30	O312	2 2 0 2 2
6.590E-06	1.069E-03	35	O312	2 2 0 2 2
6.590E-06	1.069E-03	40	O312	2 2 0 2 2
8.000E-06	1.298E-03	45	O312	2 2 0 2 2
2.000E-03	3.246E-01	ns	H307	1 0 1 1 2

2700. $C_{12}H_{18}N_2O$

Isoproturon
N,N-Dimethyl-N'-(4-(1-methylethyl)phenyl)urea
3-(4-Isopropylphenyl)-1,1-dimethylurea
Tolkan
DPX 6774

RN: 34123-59-6   **MP** (°C):   158.5
MW: 206.29   **BP** (°C):

Solubility (Moles/L)	Solubility (Grams/L)	Temp (°C)	Ref (#)	Evaluation (T P E A A)	Comments
2.909E-04	6.000E-02	20	M161	1 0 0 0 1	

2701. $C_{12}H_{18}N_2O_2$

Zectran
4-Dimethylamino-3,5-Dimethylphenol Methylcarbamate Ester
Mexacarbole
Mexacarbate

RN: 315-18-4   **MP** (°C):   85
MW: 222.29   **BP** (°C):

Solubility (Moles/L)	Solubility (Grams/L)	Temp (°C)	Ref (#)	Evaluation (T P E A A)	Comments
4.498E-04	9.999E-02	25	I314	0 0 0 0 0	

2702. $C_{12}H_{18}N_2O_2S$
Thiamylal
5-Allyl-5-(1-methyl-butyl)-barbituric Acid
5-Allyl-5-(1-methylbutyl)-2-thiobarbituric Acid
RN: 77-27-0   **MP** (°C):   132
MW: 254.35   **BP** (°C):

Solubility (Moles/L)	Solubility (Grams/L)	Temp (°C)	Ref (#)	Evaluation (T P E A A)	Comments
5.104E-03	1.298E+00	25	A023	1 0 0 1 1	
1.966E-04	5.000E-02	25	B011	2 0 0 1 0	
1.944E-04	4.946E-02	25	B065	1 1 1 1 2	
3.480E-04	8.852E-02	25	G003	1 1 1 1 1	pH 4.7
7.500E-03	1.908E+00	30	G014	1 1 1 1 0	EFG
6.600E-03	1.679E+00	30	I001	2 0 2 1 0	EFG, 0.003N H_2SO_4
8.630E-03	2.195E+00	40	A023	1 0 0 1 1	
3.750E-03	9.538E-01	40	N008	1 2 1 1 2	sic
8.792E-03	2.236E+00	ns	G039	0 0 0 0 0	EFG

2703. $C_{12}H_{18}N_2O_3$
Secobarbital
5-Allyl-5-(1-methylbutyl)barbituric Acid
Seconal
RN: 76-73-3   **MP** (°C):   98
MW: 238.29   **BP** (°C):

Solubility (Moles/L)	Solubility (Grams/L)	Temp (°C)	Ref (#)	Evaluation (T P E A A)	Comments
7.250E-03	1.728E+00	25	G003	1 1 1 1 2	pH 7
4.410E-03	1.051E+00	25	V033	2 0 1 1 2	
4.400E-03	1.048E+00	25.00	T303	1 0 0 0 1	
6.300E-03	1.501E+00	35.00	T303	1 0 0 0 1	
7.900E-02	1.882E+01	40	N008	1 0 1 1 2	sic
9.400E-03	2.240E+00	45.00	T303	1 0 0 0 1	

2704. $C_{12}H_{18}N_2O_3$
5-Isopropyl-5-(3-methylbut-2-enyl)barbituric Acid
2,4,6(1H,3H,5H)-Pyrimidinetrione, 5-(3-Methyl-2-butenyl)-5-(1-methylethyl)
RN: 67051-26-7   **MP** (°C):
MW: 238.29   **BP** (°C):

Solubility (Moles/L)	Solubility (Grams/L)	Temp (°C)	Ref (#)	Evaluation (T P E A A)	Comments
2.555E-03	6.088E-01	25	P350	2 1 1 1 2	intrinsic

2705. $C_{12}H_{18}N_2O_3S$
Tolbutamide
1-Butyl-3-(para-tolylsulfonyl) Urea
Oramide
Orinase
RN: 64-77-7   **MP** (°C):   129
MW: 270.35   **BP** (°C):

Solubility (Moles/L)	Solubility (Grams/L)	Temp (°C)	Ref (#)	Evaluation (T P E A A)	Comments
5.178E-04	1.400E-01	25	G051	1 0 1 1 0	
4.068E-04	1.100E-01	25	P096	1 0 2 2 2	
3.900E-04	1.054E-01	30	G318	2 0 1 2 1	EFG
4.027E-04	1.089E-01	37	A028	1 0 2 1 2	intrinsic
4.030E-04	1.090E-01	37	A046	2 0 1 1 2	
5.659E-04	1.530E-01	37	B138	1 2 0 0 2	pH 1.5, form II
5.289E-04	1.430E-01	37	B138	1 2 0 0 2	pH 1.5, form III
5.067E-04	1.370E-01	37	B138	1 2 0 0 2	pH 1.5, form I
3.699E-04	1.000E-01	37.0	H033	1 0 2 1 0	pH 1.4, intrinsic
3.031E-03	8.193E-01	37.5	F015	1 0 2 2 1	pH 6.0, pKa 5.32
2.535E-02	6.853E+00	37.5	F015	1 0 2 2 2	pH 7.0, pKa 5.32

2706. $C_{12}H_{18}N_2O_4S$
Anisylbutamide
Methoxyphenylbutazolamide
Methoxytolbutamide
RN: 24535-67-9   **MP** (°C):
MW: 286.35   **BP** (°C):

Solubility (Moles/L)	Solubility (Grams/L)	Temp (°C)	Ref (#)	Evaluation (T P E A A)	Comments
4.236E-04	1.213E-01	37	A028	1 0 2 1 2	intrinsic
4.260E-04	1.220E-01	37	A046	2 0 1 1 2	

2707. $C_{12}H_{18}N_2O_5$
D-Mannosephenylhydrazone
D-Mannose-phenylhydrazon
RN: 6147-14-4   **MP** (°C):   195.5
MW: 270.29   **BP** (°C):

Solubility (Moles/L)	Solubility (Grams/L)	Temp (°C)	Ref (#)	Evaluation (T P E A A)	Comments
3.811E-02	1.030E+01	100	F300	1 0 0 0 2	

2708. $C_{12}H_{18}N_4O_6S$

Oryzalin
3,5-Dinitro-N4,N4-dipropylsulfanilamide
RN: 19044-88-3 **MP** (°C): 137
MW: 346.36 **BP** (°C):

Solubility (Moles/L)	Solubility (Grams/L)	Temp (°C)	Ref (#)	Evaluation (T P E A A)	Comments
2.454E-04	8.500E-02	25	B200	1 0 0 0 1	
6.929E-06	2.400E-03	25	M161	1 0 0 0 1	

2709. $C_{12}H_{18}O$

4-Butyl-2,5-dimethylphenol
2,5-Xylenol, 4-Butyl-
RN: 91763-77-8 **MP** (°C):
MW: 178.28 **BP** (°C):

Solubility (Moles/L)	Solubility (Grams/L)	Temp (°C)	Ref (#)	Evaluation (T P E A A)	Comments
2.244E-04	4.000E-02	25	L020	1 0 0 0 0	

2710. $C_{12}H_{18}O$

2,4-Dipropylphenol
Phenol, 2,4-Dipropyl-
RN: 23167-99-9 **MP** (°C):
MW: 178.28 **BP** (°C):

Solubility (Moles/L)	Solubility (Grams/L)	Temp (°C)	Ref (#)	Evaluation (T P E A A)	Comments
1.402E-04	2.500E-02	25	L020	1 0 0 0 0	

2711. $C_{12}H_{18}O$

4-Butyl-2,6-dimethylphenol
Phenol, 4-Butyl-2,6-dimethyl-
2,6-Xylenol, 4-Butyl-
RN: 6676-26-2 **MP** (°C):
MW: 178.28 **BP** (°C):

Solubility (Moles/L)	Solubility (Grams/L)	Temp (°C)	Ref (#)	Evaluation (T P E A A)	Comments
2.244E-04	4.000E-02	25	L020	1 0 0 0 0	

2712.   C$_{12}$H$_{18}$O
2-Butyl-6-ethylphenol
Phenol, 2-Butyl-6-ethyl-
RN: 22496-45-3   **MP** (°C):
MW: 178.28   **BP** (°C):

Solubility (Moles/L)	Solubility (Grams/L)	Temp (°C)	Ref (#)	Evaluation (T P E A A)	Comments
1.870E-04	3.333E-02	25	L020	1 0 0 0 0	

2713.   C$_{12}$H$_{18}$O
2-Butyl-4-ethylphenol
Phenol, 2-Butyl-4-ethyl-
RN: 3781-74-6   **MP** (°C):
MW: 178.28   **BP** (°C):

Solubility (Moles/L)	Solubility (Grams/L)	Temp (°C)	Ref (#)	Evaluation (T P E A A)	Comments
1.402E-04	2.500E-02	25	L020	1 0 0 0 0	

2714.   C$_{12}$H$_{18}$O
2-Butyl-4,6-dimethylphenol
2,6-Xylenol, 2-Butyl-
RN: 6483-60-9   **MP** (°C):
MW: 178.28   **BP** (°C):

Solubility (Moles/L)	Solubility (Grams/L)	Temp (°C)	Ref (#)	Evaluation (T P E A A)	Comments
1.603E-04	2.857E-02	25	L020	1 0 0 0 0	

2715.   C$_{12}$H$_{18}$O
2-Butyl-4,5-dimethylphenol
Phenol, 2-Butyl-4,5-dimethyl-
RN: **MP** (°C):
MW: 178.28   **BP** (°C):

Solubility (Moles/L)	Solubility (Grams/L)	Temp (°C)	Ref (#)	Evaluation (T P E A A)	Comments
1.870E-04	3.333E-02	25	L020	1 0 0 0 0	

2716. C₁₂H₁₈O

2,6-Dipropylphenol
Phenol, 2,6-Dipropyl-

RN: 6626-32-0   **MP** (°C):
MW: 178.28   **BP** (°C):

Solubility (Moles/L)	Solubility (Grams/L)	Temp (°C)	Ref (#)	Evaluation (T P E A A)	Comments
1.402E-04	2.500E-02	25	L020	1 0 0 0 0	

2717. C₁₂H₁₈O

o-n-Hexylphenol
2-n-Hexylphenol

RN: 3226-32-2   **MP** (°C):
MW: 178.28   **BP** (°C):

Solubility (Moles/L)	Solubility (Grams/L)	Temp (°C)	Ref (#)	Evaluation (T P E A A)	Comments
2.244E-04	4.000E-02	25	L022	1 0 0 0 0	

2718. C₁₂H₁₈O

p-n-Hexylphenol
4-n-Hexylphenol

RN: 2446-69-7   **MP** (°C):
MW: 178.28   **BP** (°C):

Solubility (Moles/L)	Solubility (Grams/L)	Temp (°C)	Ref (#)	Evaluation (T P E A A)	Comments
1.603E-04	2.857E-02	25	L022	1 0 0 0 0	

2719. C₁₂H₁₈O₂

4-Hexylresorcinol
4-n-Hexylresorcin

RN: 136-77-6   **MP** (°C): 68
MW: 194.28   **BP** (°C): 334

Solubility (Moles/L)	Solubility (Grams/L)	Temp (°C)	Ref (#)	Evaluation (T P E A A)	Comments
2.574E-03	5.000E-01	18	F300	1 0 0 0 1	

2720.   C$_{12}$H$_{18}$O$_4$S$_2$
Di-isopropyl 1,3-Dithiolan-2-ylidinemalonate
Isoprothiolane
Fuji-one
bis(1-Methylethyl) 1,3-Dithiolan-2-ylidenepropanedioate
RN: 50512-35-1   **MP** (°C):   52.25
MW: 290.40   **BP** (°C):   168

Solubility (Moles/L)	Solubility (Grams/L)	Temp (°C)	Ref (#)	Evaluation (T P E A A)	Comments
1.653E-04	4.800E-02	20	H309	1 0 0 0 1	
1.653E-04	4.800E-02	20	M161	1 0 0 0 1	

2721.   C$_{12}$H$_{19}$ClNO$_3$P
Crufomate
O-Methyl O-2-Chloro-4-tert-butyphenyl N-Methylamidophosphate
RN: 299-86-5   **MP** (°C):   60.25
MW: 291.72   **BP** (°C):   117.5

Solubility (Moles/L)	Solubility (Grams/L)	Temp (°C)	Ref (#)	Evaluation (T P E A A)	Comments
1.705E-02	4.975E+00	ns	M061	0 0 0 0 0	

2722.   C$_{12}$H$_{19}$N$_3$O$_8$
Orotic Acid Methylglucamide
RN: **MP** (°C):   184-186
MW: 333.30   **BP** (°C):

Solubility (Moles/L)	Solubility (Grams/L)	Temp (°C)	Ref (#)	Evaluation (T P E A A)	Comments
4.470E-01	1.490E+02	-4	N018	2 2 1 2 2	
7.090E-01	2.363E+02	16	N018	2 2 1 2 2	
8.150E-01	2.716E+02	25	N018	2 2 1 2 2	

2723.   C$_{12}$H$_{19}$N$_6$OP
Triamiphos
5-Amino-1-(bis(dimethylamino)phosphoryl)-3-phenyl-1,2,4-triazole
Triamifos
Wepsyn 155
Wepsyn
bis(Dimethylamino)-(3-amino-5-phenyl-1,2,4-triazol-1-yl)-phosphine Oxide
RN: 1031-47-6   **MP** (°C):   167.5
MW: 294.30   **BP** (°C):

Solubility (Moles/L)	Solubility (Grams/L)	Temp (°C)	Ref (#)	Evaluation (T P E A A)	Comments
8.495E-04	2.500E-01	20	M161	1 0 0 0 2	

2724. C₁₂H₂₀

Triisobutene
1,8-Nonadiene, 2,8-Dimethyl-5-methylene-
RN: 36370-80-6 **MP** (°C):
MW: 164.29 **BP** (°C):

Solubility (Moles/L)	Solubility (Grams/L)	Temp (°C)	Ref (#)	Evaluation (T P E A A)	Comments
4.944E-08	8.123E-06	20	B165	1 0 1 1 1	
5.838E-03	9.591E-01	97.30	B165	1 0 1 1 1	

2725. C₁₂H₂₀N₂O₃

5-Ethyl-5-n-hexylbarbituric Acid
2,4,6(1H,3H,5H)-Pyrimidinetrione, 5-Ethyl-5-hexyl-
Hexethal
Ortal
Ortol
RN: 77-30-5 **MP** (°C):
MW: 240.30 **BP** (°C):

Solubility (Moles/L)	Solubility (Grams/L)	Temp (°C)	Ref (#)	Evaluation (T P E A A)	Comments
8.930E-04	2.146E-01	25	M310	2 2 2 2 2	

2726. C₁₂H₂₀N₄O₂

3-Cyclohexyl-6-dimethylamino-1-methyl-1,3,5-triazine-2,4-dione
1,3,5-Triazine-2,4(1H,3H)-dione, 3-Cyclohexyl-6-(dimethylamino)-1-methyl-
Hexazinone
Pronone
DPX 3674
RN: 51235-04-2 **MP** (°C): 116
MW: 252.32 **BP** (°C):

Solubility (Moles/L)	Solubility (Grams/L)	Temp (°C)	Ref (#)	Evaluation (T P E A A)	Comments
1.308E-01	3.300E+01	25	M161	1 0 0 0 1	

2727. C₁₂H₂₀N₄O₆

Acetyltetraglycine Ethyl Ester
Glycine, N-Acetylglycylglycylglycyl-, Ethyl Ester
RN: 637-83-2 **MP** (°C): 264
MW: 316.32 **BP** (°C):

Solubility (Moles/L)	Solubility (Grams/L)	Temp (°C)	Ref (#)	Evaluation (T P E A A)	Comments
8.220E-04	2.600E-01	0	R036	2 1 2 2 1	
2.466E-03	7.800E-01	25	R036	2 1 2 2 1	
5.216E-03	1.650E+00	40	R036	2 1 2 2 2	

2728.   C$_{12}$H$_{20}$O$_2$
Linalyl Acetate
Bergamol
3,7-Dimethyl-1,6-octadien-3-yl Acetate
Linalyl

RN:	115-95-7	**MP** (°C):	
MW:	196.29	**BP** (°C):	220

Solubility (Moles/L)	Solubility (Grams/L)	Temp (°C)	Ref (#)	Evaluation (T P E A A)	Comments
2.546E-03	4.998E-01	25	M350	1 0 1 1 1	

2729.   C$_{12}$H$_{20}$O$_4$
Dibutyl Maleate
Di-n-butyl Maleate

RN:	105-76-0	**MP** (°C):	
MW:	228.29	**BP** (°C):	

Solubility (Moles/L)	Solubility (Grams/L)	Temp (°C)	Ref (#)	Evaluation (T P E A A)	Comments
1.073E-03	2.450E-01	25	F067	1 0 2 2 2	

2730.   C$_{12}$H$_{20}$O$_6$
Tripropionin
1,2,3-Propanetriol, Tripropanoate
1,2,3-Propanetriyl Tripropionate
Tripropionylglycerol
Tripropanoylglycerol

RN:	139-45-7	**MP** (°C):	
MW:	260.29	**BP** (°C):	

Solubility (Moles/L)	Solubility (Grams/L)	Temp (°C)	Ref (#)	Evaluation (T P E A A)	Comments
1.199E-02	3.120E+00	ns	F014	0 0 0 0 2	

2731. C₁₂H₂₁N₂O₃PS

Diazinon
O,O-Diethyl O-(2-Isopropyl-6-methyl-4-pyrimidinyl), Phosphorothioate
Dimpylate
Basudin
Spectracide
Fezudin

RN: 333-41-5 **MP** (°C): >120
MW: 304.35 **BP** (°C):

Solubility (Moles/L)	Solubility (Grams/L)	Temp (°C)	Ref (#)	Evaluation (T P E A A)	Comments
2.336E-04	7.109E-02	10	B324	2 2 2 2 2	
2.336E-04	7.110E-02	10	B324	2 2 2 2 2	
1.318E-04	4.012E-02	20	B179	2 0 0 0 2	
2.261E-04	6.881E-02	20	B300	2 1 1 1 2	
1.758E-04	5.350E-02	20	B324	2 2 2 2 2	
1.758E-04	5.350E-02	20	B324	2 2 2 2 2	
1.314E-04	4.000E-02	20	M061	1 0 0 0 1	
2.260E-04	6.880E-02	22	B169	2 1 1 1 2	
1.331E-04	4.050E-02	22	K137	1 1 2 1 0	
1.436E-04	4.370E-02	30	B324	2 2 2 2 2	
1.436E-04	4.370E-02	30	B324	2 2 2 2 2	
1.314E-04	4.000E-02	rt	M161	0 0 0 0 1	

2732. C₁₂H₂₁N₇O

1-(4'-Formyl-1-piperizinyl)-3,5-bis(dimethylamino)-s-triazine
1-Piperazinecarboxaldehyde, 4-[4,6-bis(Dimethylamino)-1,3,5-triazin-2-yl]-

RN: 126974-79-6 **MP** (°C):
MW: 279.35 **BP** (°C):

Solubility (Moles/L)	Solubility (Grams/L)	Temp (°C)	Ref (#)	Evaluation (T P E A A)	Comments
3.670E-03	1.025E+00	25	B386	2 2 2 2 2	

2733. C₁₂H₂₂N₂O₂

N,N,N',N'-Tetraethylfumaramide
2-Butenediamide, N,N,N',N'-Tetraethyl-

RN: 111328-65-5 **MP** (°C):
MW: 226.32 **BP** (°C):

Solubility (Moles/L)	Solubility (Grams/L)	Temp (°C)	Ref (#)	Evaluation (T P E A A)	Comments
6.900E-01	1.562E+02	30	K019	1 0 0 0 1	

2734.　C₁₂H₂₂N₆

1-(Piperidinyl)-3,5-bis(dimethylamino)-s-triazine
s-Triazine, 2,4-bis(Dimethylamino)-6-piperidino-
RN:　　16268-79-4　　**MP** (°C):
MW:　　250.35　　　　**BP** (°C):

Solubility (Moles/L)	Solubility (Grams/L)	Temp (°C)	Ref (#)	Evaluation (T P E A A)	Comments
1.758E-04	4.402E-02	25	B386	2 2 2 2 2	

2735.　C₁₂H₂₂O₄

Dibutyl Succinate
Succinic Acid di-n-Butyl Ester
Tabutrex
RN:　　141-03-7　　**MP** (°C):　　-29
MW:　　230.31　　　**BP** (°C):　　108

Solubility (Moles/L)	Solubility (Grams/L)	Temp (°C)	Ref (#)	Evaluation (T P E A A)	Comments
9.984E-04	2.299E-01	ns	F014	0 0 0 0 1	

2736.　C₁₂H₂₂O₄

1,10-Decanedicarboxylic Acid
Decan-dicarbonsaeure-(1,10)
Dodecanedioc Acid
RN:　　693-23-2　　**MP** (°C):　　128
MW:　　230.31　　　**BP** (°C):

Solubility (Moles/L)	Solubility (Grams/L)	Temp (°C)	Ref (#)	Evaluation (T P E A A)	Comments
1.737E-04	4.000E-02	20	F300	1 0 0 0 0	
3.039E-03	7.000E-01	21	B040	1 0 1 1 0	*sic*
5.124E-03	1.180E+00	100	F300	1 0 0 0 2	

2737.　C₁₂H₂₂O₄

Ethylene Glycol Divalerate
RN:　　　　　　　　　**MP** (°C):
MW:　　230.31　　　**BP** (°C):

Solubility (Moles/L)	Solubility (Grams/L)	Temp (°C)	Ref (#)	Evaluation (T P E A A)	Comments
6.460E-04	1.488E-01	25	F064	1 0 0 0 2	

2738.   C$_{12}$H$_{22}$O$_6$
Dimethoxyethyl Adipate
RN: **MP** (°C):
MW: 262.31   **BP** (°C):

Solubility (Moles/L)	Solubility (Grams/L)	Temp (°C)	Ref (#)	Evaluation (T P E A A)	Comments
5.338E-02	1.400E+01	ns	F014	0 0 0 0 2	

2739.   C$_{12}$H$_{22}$O$_6$
Dibutyl Tartrate
(2R,3R)-Di-n-Butyl Tartrate
ENT 396
RN: 87-92-3   **MP** (°C):   21
MW: 262.31   **BP** (°C):

Solubility (Moles/L)	Solubility (Grams/L)	Temp (°C)	Ref (#)	Evaluation (T P E A A)	Comments
1.840E-02	4.827E+00	ns	F014	0 0 0 0 2	

2740.   C$_{12}$H$_{22}$O$_6$
Triethylene Glycol Dipropionate
Ethanol, 2,2'-[1,2-Ethanediylbis(oxy)]bis-, Dipropanoate
RN: 141-34-4   **MP** (°C):
MW: 262.31   **BP** (°C):

Solubility (Moles/L)	Solubility (Grams/L)	Temp (°C)	Ref (#)	Evaluation (T P E A A)	Comments
2.394E-01	6.279E+01	ns	F014	0 0 0 0 2	

2741.   C$_{12}$H$_{22}$O$_{11}$
Lactose
4-O-B-D-Galactopyranosyl-D-glucose
Milk Sugar
RN: 63-42-3   **MP** (°C):   201
MW: 342.30   **BP** (°C):

Solubility (Moles/L)	Solubility (Grams/L)	Temp (°C)	Ref (#)	Evaluation (T P E A A)	Comments
3.177E-01	1.087E+02	0	M043	1 0 0 0 2	
3.116E-01	1.067E+02	0	P052	1 0 2 2 2	
4.701E-01	1.609E+02	1	P049	1 0 1 1 1	
3.811E-01	1.304E+02	10	M043	1 0 0 0 2	
4.767E-01	1.632E+02	20	M043	1 0 0 0 2	
5.189E-01	1.776E+02	25	D041	1 0 0 0 2	
5.470E-01	1.873E+02	25	P049	1 0 1 1 1	
6.000E-01	2.054E+02	30	D011	1 0 1 0 1	
5.880E-01	2.013E+02	30	M043	1 0 0 0 2	
7.298E-01	2.498E+02	40	M043	1 0 0 0 2	
1.067E+00	3.651E+02	60	M043	1 0 0 0 2	

1.475E+00	5.050E+02	80	M043	1 0 0 0 2
1.699E+00	5.816E+02	89	D041	1 0 0 0 2
1.767E+00	6.047E+02	100	M043	1 0 0 0 2
4.775E-01	1.635E+02	rt	D021	0 0 1 1 2

2742. $C_{12}H_{22}O_{11}$
β-Lactose
B-Lactose
Milchzucker
4-O-β-D-Galactopyranosyl-D-glucose
RN: 5965-66-2   **MP** (°C):   253
MW: 342.30   **BP** (°C):

Solubility (Moles/L)	Solubility (Grams/L)	Temp (°C)	Ref (#)	Evaluation (T P E A A)	Comments
1.525E-01	5.220E+01	20	F300	1 0 0 0 2	
7.303E-02	2.500E+01	h	F300	0 0 0 0 1	

2743. $C_{12}H_{22}O_{11}$
Cellobiose
4-O-β-D-Glucopyranosyl-D-glucose
4-β-D-Glucopyransoyl-D-glucopyranose
D-(+)-Cellobiose
RN: 528-50-7   **MP** (°C):
MW: 342.30   **BP** (°C):

Solubility (Moles/L)	Solubility (Grams/L)	Temp (°C)	Ref (#)	Evaluation (T P E A A)	Comments
3.243E-01	1.110E+02	15	F300	1 0 0 0 2	
3.475E-01	1.189E+02	30.50	M137	2 1 2 2 2	
1.198E+00	4.100E+02	h	F300	0 0 0 0 1	

2744. $C_{12}H_{22}O_{11}$
Sucrose
Saccharose
β-D-Fructofuranosyl-α-D-glucopyranoside
α-D-Glucopyranosyl β-D-fructofuranoside
Beet Sugar
Cane Sugar
RN: 57-50-1   **MP** (°C):   191
MW: 342.30   **BP** (°C):

Solubility (Moles/L)	Solubility (Grams/L)	Temp (°C)	Ref (#)	Evaluation (T P E A A)	Comments
1.878E+00	6.429E+02	0	D041	1 0 0 0 2	
1.876E+00	6.421E+02	0	G046	1 0 1 1 2	
1.142E+00	3.909E+02	0	H094	1 0 0 0 2	
1.874E+00	6.416E+02	0	M043	1 0 0 0 2	

1.884E+00	6.450E+02	0	P052	1 0 2 2 2	
1.880E+00	6.435E+02	0.9	M074	1 0 0 0 2	average of 3
1.157E+00	3.961E+02	10	H094	1 0 0 0 2	
1.914E+00	6.552E+02	10	M043	1 0 0 0 2	
1.943E+00	6.650E+02	12.5	F300	1 0 0 0 2	
1.938E+00	6.633E+02	15	D041	1 0 0 0 2	
1.934E+00	6.622E+02	15.80	M074	1 0 0 0 2	average of 3
1.931E+00	6.609E+02	18.5	W013	1 2 1 1 2	
1.946E+00	6.660E+02	20	F300	1 0 0 0 2	
1.170E+00	4.005E+02	20	G060	1 0 0 0 2	
1.173E+00	4.015E+02	20	H094	1 0 0 0 2	
1.960E+00	6.711E+02	20	M043	1 0 0 0 2	
1.956E+00	6.697E+02	23.9	W013	1 2 1 1 2	
1.954E+00	6.689E+02	24.4	W013	1 2 1 1 2	
1.964E+00	6.723E+02	24.9	W013	1 2 1 1 2	
1.986E+00	6.798E+02	25	G046	1 0 1 1 2	
1.179E+00	4.036E+02	25	G060	1 0 0 0 2	
1.981E+00	6.779E+02	25.60	M074	1 0 0 0 2	average of 3
1.963E+00	6.721E+02	25.9	W013	1 2 1 1 2	
1.188E+00	4.067E+02	30	G060	1 0 0 0 2	
1.190E+00	4.072E+02	30	H094	1 0 0 0 2	
2.006E+00	6.865E+02	30	M043	1 0 0 0 2	
1.997E+00	6.836E+02	30.0	W013	1 2 1 1 2	
1.996E+00	6.831E+02	30.5	W013	1 2 1 1 2	
2.003E+00	6.855E+02	30.50	M074	1 0 0 0 2	average of 3
2.008E+00	6.873E+02	31.5	W013	1 2 1 1 2	
2.005E+00	6.862E+02	33.1	W013	1 2 1 1 2	
2.025E+00	6.932E+02	34.5	W013	1 2 1 1 2	
2.030E+00	6.950E+02	35	G046	1 0 1 1 2	
1.198E+00	4.100E+02	35	G060	1 0 0 0 2	
2.028E+00	6.942E+02	36.0	W013	1 2 1 1 2	
2.028E+00	6.941E+02	36.4	W013	1 2 1 1 2	
1.207E+00	4.133E+02	40	G060	1 0 0 0 2	
1.207E+00	4.132E+02	40	H094	1 0 0 0 2	
2.057E+00	7.041E+02	40	M043	1 0 0 0 2	
2.050E+00	7.017E+02	40.2	W013	1 2 1 1 2	
2.052E+00	7.023E+02	40.7	W013	1 2 1 1 2	
2.055E+00	7.035E+02	41.0	W013	1 2 1 1 2	
2.061E+00	7.055E+02	42.2	W013	1 2 1 1 2	
2.067E+00	7.074E+02	42.3	W013	1 2 1 1 2	
2.080E+00	7.120E+02	45	F300	1 0 0 0 2	
1.217E+00	4.167E+02	45	G060	1 0 0 0 2	
2.093E+00	7.163E+02	46.1	W013	1 2 1 1 2	
2.107E+00	7.212E+02	49.6	W013	1 2 1 1 2	
2.111E+00	7.225E+02	50	G046	1 0 1 1 2	
1.228E+00	4.202E+02	50	G060	1 0 0 0 2	
7.596E+00	2.600E+03	50	H063	1 0 0 0 2	
1.225E+00	4.194E+02	50	H094	1 0 0 0 2	
2.101E+00	7.191E+02	50.2	W013	1 2 1 1 2	
2.118E+00	7.251E+02	51.1	W013	1 2 1 1 2	

2.124E+00	7.272E+02	52.2	W013	1 2 1 1 2
2.126E+00	7.276E+02	52.6	W013	1 2 1 1 2
2.134E+00	7.304E+02	53.6	W013	1 2 1 1 2
2.134E+00	7.305E+02	53.8	W013	1 2 1 1 2
2.126E+00	7.278E+02	54.1	W013	1 2 1 1 2
1.237E+00	4.235E+02	55	G060	1 0 0 0 2
2.137E+00	7.316E+02	55.8	W013	1 2 1 1 2
2.147E+00	7.350E+02	56.1	W013	1 2 1 1 2
2.154E+00	7.372E+02	56.4	W013	1 2 1 1 2
2.151E+00	7.364E+02	57.5	W013	1 2 1 1 2
2.154E+00	7.374E+02	57.8	W013	1 2 1 1 2
2.152E+00	7.368E+02	58.4	W013	1 2 1 1 2
2.165E+00	7.410E+02	58.6	W013	1 2 1 1 2
2.166E+00	7.415E+02	59.7	W013	1 2 1 1 2
1.248E+00	4.273E+02	60	G060	1 0 0 0 2
1.244E+00	4.259E+02	60	H094	1 0 0 0 2
2.167E+00	7.416E+02	60	M043	1 0 0 0 2
2.176E+00	7.448E+02	61.1	W013	1 2 1 1 2
2.176E+00	7.447E+02	61.4	W013	1 2 1 1 2
2.182E+00	7.469E+02	62.6	W013	1 2 1 1 2
2.189E+00	7.493E+02	62.9	W013	1 2 1 1 2
2.193E+00	7.505E+02	64.6	W013	1 2 1 1 2
1.258E+00	4.307E+02	65	G060	1 0 0 0 2
2.204E+00	7.543E+02	65.5	W013	1 2 1 1 2
2.214E+00	7.580E+02	66.4	W013	1 2 1 1 2
2.219E+00	7.595E+02	66.5	W013	1 2 1 1 2
2.222E+00	7.607E+02	68.2	W013	1 2 1 1 2
2.221E+00	7.603E+02	69.0	W013	1 2 1 1 2
1.269E+00	4.344E+02	70	G060	1 0 0 0 2
2.230E+00	7.632E+02	70.1	W013	1 2 1 1 2
2.233E+00	7.645E+02	70.4	W013	1 2 1 1 2
2.251E+00	7.706E+02	72.8	W013	1 2 1 1 2
2.249E+00	7.698E+02	73.8	W013	1 2 1 1 2
2.267E+00	7.760E+02	74.5	W013	1 2 1 1 2
2.265E+00	7.752E+02	74.6	W013	1 2 1 1 2
2.256E+00	7.724E+02	75	G046	1 0 1 1 2
1.280E+00	4.380E+02	75	G060	1 0 0 0 2
2.266E+00	7.758E+02	75.1	W013	1 2 1 1 2
2.290E+00	7.840E+02	79.5	W013	1 2 1 1 2
1.291E+00	4.417E+02	80	G060	1 0 0 0 2
1.090E+01	3.730E+03	80	H063	1 0 0 0 2
2.289E+00	7.835E+02	80	M043	1 0 0 0 2
2.304E+00	7.886E+02	82.3	W013	1 2 1 1 2
2.333E+00	7.985E+02	85.1	W013	1 2 1 1 2
2.335E+00	7.994E+02	85.3	W013	1 2 1 1 2
2.337E+00	7.999E+02	85.5	W013	1 2 1 1 2
2.344E+00	8.022E+02	86.6	W013	1 2 1 1 2
2.346E+00	8.032E+02	88.0	W013	1 2 1 1 2

2.355E+00	8.061E+02	90	G046	1 0 1 1 2	
2.363E+00	8.087E+02	90.2	W013	1 2 1 1 2	
2.388E+00	8.176E+02	95	G046	1 0 1 1 2	
2.409E+00	8.247E+02	98	G046	1 0 1 1 2	
2.424E+00	8.296E+02	100	D041	1 0 0 0 2	
2.424E+00	8.296E+02	100	G046	1 0 1 1 2	
2.424E+00	8.296E+02	100	M043	1 0 0 0 2	

2745. $C_{12}H_{22}O_{11}$
Maltose
D-Glucose, 4-O-α-D-glucopyranosyl-
α-Maltose
Malt Sugar
RN: 69-79-4   **MP** (°C):   102.5
MW: 342.30   **BP** (°C):

Solubility (Moles/L)	Solubility (Grams/L)	Temp (°C)	Ref (#)	Evaluation (T P E A A)	Comments
1.061E+00	3.631E+02	0	M043	1 0 0 0 1	
1.151E+00	3.939E+02	10	M043	1 0 0 0 1	
1.517E+00	5.192E+02	20	D041	1 0 0 0 2	
1.280E+00	4.382E+02	20	M043	1 0 0 0 1	
1.408E+00	4.819E+02	30	M043	1 0 0 0 1	
1.530E+00	5.238E+02	40	M043	1 0 0 0 2	
1.859E+00	6.364E+02	60	M043	1 0 0 0 2	
2.191E+00	7.500E+02	80	M043	1 0 0 0 2	
1.517E+00	5.192E+02	rt	D021	0 0 1 1 2	

2746. $C_{12}H_{23}NO_3$
Propylbutylaceturethane
RN: **MP** (°C):
MW: 229.32   **BP** (°C):

Solubility (Moles/L)	Solubility (Grams/L)	Temp (°C)	Ref (#)	Evaluation (T P E A A)	Comments
1.395E-03	3.199E-01	20	O021	1 2 0 0 0	

2747. $C_{12}H_{23}N_7$
1-(4'-Methyl-1-piperizinyl)-3,5-bis(dimethylamino)-s-triazine
2-(4-Methyl-1-piperazinyl)-4,6-bis(dimethylamino)-s-triazine
RN: 5512-05-0   **MP** (°C):
MW: 265.36   **BP** (°C):

Solubility (Moles/L)	Solubility (Grams/L)	Temp (°C)	Ref (#)	Evaluation (T P E A A)	Comments
4.514E-03	1.198E+00	25	B386	2 2 2 2 2	

2748. C₁₂H₂₄N₂O₂

N,N,N',N'-Tetramethylsuberamide
Octanediamide, N,N,N',N'-Tetramethyl-
RN: 27397-05-3   **MP** (°C):
MW: 228.34   **BP** (°C):

Solubility (Moles/L)	Solubility (Grams/L)	Temp (°C)	Ref (#)	Evaluation (T P E A A)	Comments
2.520E+00	5.754E+02	30	D010	1 2 1 1 2	

2749. C₁₂H₂₄N₃O₃PS

Thiophosphoryl Trimorpholide
Morpholine, 4,4',4''-Phosphinothioylidynetris-
Phosphine Sulfide, Trimorpholino-
RN: 14129-98-7   **MP** (°C):
MW: 321.38   **BP** (°C):

Solubility (Moles/L)	Solubility (Grams/L)	Temp (°C)	Ref (#)	Evaluation (T P E A A)	Comments
9.987E-03	3.210E+00	25	A040	1 0 0 0 2	

2750. C₁₂H₂₄N₃O₄P

Phosphoryl Trimorpholide
Morpholine, 4,4',4''-Phosphinylidynetris-
Phosphine Oxide, Trimorpholino-
RN: 4441-12-7   **MP** (°C):
MW: 305.32   **BP** (°C):

Solubility (Moles/L)	Solubility (Grams/L)	Temp (°C)	Ref (#)	Evaluation (T P E A A)	Comments
1.989E+00	6.072E+02	25	A040	1 0 0 0 2	

2751. C₁₂H₂₄N₆

N2,N4,N6-Triethyl-N2,N4,N6-trimethylmelamine
1,3,5-Triazine-2,4,6-triamine, N,N',N''-Triethyl-N,N',N''-trimethyl-
RN: 64124-20-5   **MP** (°C):
MW: 252.37   **BP** (°C):

Solubility (Moles/L)	Solubility (Grams/L)	Temp (°C)	Ref (#)	Evaluation (T P E A A)	Comments
1.981E-04	5.000E-02	25	C051	1 2 1 1 0	pH 7

2752. $C_{12}H_{24}N_9P_3$

Hexaziridinocyclotriphosphazene
2,2,4,4,6,6-Hexahydro-2,2,4,4,6,6-hexakis(1-aziridinyl)-1,3,5,2,4,6-triazatriphosphorine
2,2,4,4,6,6-Hexakis(1-aziridinyl)cyclotriphosphaza-1,3,5-triene
Apholate
APN
ENT 26316

RN: 52-46-0   **MP** (°C):
MW: 387.31   **BP** (°C):

Solubility (Moles/L)	Solubility (Grams/L)	Temp (°C)	Ref (#)	Evaluation (T P E A A)	Comments
3.172E-01	1.000E+02	ns	L076	0 1 0 0 0	approximate

2753. $C_{12}H_{24}O_2$

Lauric Acid
Dodecanoic Acid
Laurostearic Acid

RN: 143-07-7   **MP** (°C):   44
MW: 200.32   **BP** (°C):

Solubility (Moles/L)	Solubility (Grams/L)	Temp (°C)	Ref (#)	Evaluation (T P E A A)	Comments
1.847E-04	3.700E-02	0	B136	1 0 2 1 1	
1.847E-04	3.700E-02	0.0	R001	1 1 1 1 1	
2.895E-04	5.800E-02	20	B136	1 0 2 1 1	
2.745E-04	5.500E-02	20	D041	1 0 0 0 1	
2.745E-04	5.500E-02	20.0	R001	1 1 1 1 1	
2.400E-05	4.808E-03	25	J001	1 0 2 1 2	
8.486E-06	1.700E-03	25	M083	1 0 0 1 1	
1.150E-05	2.304E-03	25	R002	1 2 2 2 2	intrinsic
2.080E-05	4.167E-03	25	R002	1 2 2 2 2	
3.345E-04	6.700E-02	30	B136	1 0 2 1 1	
3.145E-04	6.300E-02	30.0	R001	1 1 1 1 1	
3.494E-04	7.000E-02	40	B136	1 0 2 1 1	
3.844E-05	7.700E-03	40	E005	2 1 1 2 1	
3.744E-04	7.500E-02	45	B136	1 0 2 1 1	
3.744E-04	7.499E-02	45.0	R001	1 1 1 1 1	
4.593E-05	9.200E-03	50	E005	2 1 1 2 1	
5.470E-05	1.096E-02	50	J001	1 0 2 1 2	
4.343E-04	8.700E-02	60	B136	1 0 2 1 1	
5.791E-05	1.160E-02	60	E005	2 1 1 2 2	
4.343E-04	8.699E-02	60.0	R001	1 1 1 1 1	

2754.   C₁₂H₂₄O₂

3-Hydroxy-2,2,5,5-tetraethyltetrahydrofuran
3-Furanol, 2,2,5,5-Tetraethyltetrahydro-
RN: 29839-78-9   **MP** (°C):
MW: 200.32   **BP** (°C):

Solubility (Moles/L)	Solubility (Grams/L)	Temp (°C)	Ref (#)	Evaluation (T P E A A)	Comments
1.493E-02	2.991E+00	rt	B066	0 2 0 0 0	

2755.   C₁₂H₂₄O₃

1,3-Dioxolane-4-methanol, 2-Heptyl-2-methyl
2-Heptyl-4-hydroxymethyl-2-methyl-1,3-dioxolane
RN: 5660-50-4   **MP** (°C):
MW: 216.32   **BP** (°C):

Solubility (Moles/L)	Solubility (Grams/L)	Temp (°C)	Ref (#)	Evaluation (T P E A A)	Comments
3.560E-03	7.701E-01	25	P342	1 2 2 2 2	0.0001M Na₂CO₃

2756.   C₁₂H₂₄O₄

1,3-Dioxolane-4-methanol, 2-Methyl-2-[2-(pentyloxy)ethyl]
RN: 143458-56-4   **MP** (°C):
MW: 232.32   **BP** (°C):

Solubility (Moles/L)	Solubility (Grams/L)	Temp (°C)	Ref (#)	Evaluation (T P E A A)	Comments
6.250E-02	1.452E+01	25	P342	1 2 2 2 2	0.0001M Na₂CO₃

2757.   C₁₂H₂₆

Dodecane
N-Dodecane
Alkane C(12)
Duodecane
Bihexyl
Adakane 12
RN: 112-40-3   **MP** (°C):   -9.6
MW: 170.34   **BP** (°C):   216.3

Solubility (Moles/L)	Solubility (Grams/L)	Temp (°C)	Ref (#)	Evaluation (T P E A A)	Comments
4.931E-08	8.400E-06	22.5	G301	2 1 0 1 2	
2.055E-08	3.500E-06	23	C332	2 0 2 2 1	
1.068E-08	1.820E-06	25	B156	1 0 2 2 2	
4.944E-08	8.422E-06	25	F004	1 2 2 2 1	
3.900E-09	6.643E-07	ns	D348	0 0 2 2 2	
2.231E-08	3.800E-06	ns	H123	0 0 0 0 2	

2758.   C$_{12}$H$_{26}$O
Dodecanol
Dodecyl Alcohol
Lauryl Alcohol
Undecyl Carbinol
RN: 112-53-8   **MP** (°C):   24
MW: 186.34   **BP** (°C):   261

Solubility (Moles/L)	Solubility (Grams/L)	Temp (°C)	Ref (#)	Evaluation (T P E A A)	Comments
9.100E-06	1.696E-03	16	K011	1 2 1 1 2	
2.300E-05	4.286E-03	25	R002	1 2 2 2 2	
1.560E-05	2.907E-03	34	K011	1 2 1 1 2	
1.930E-05	3.596E-03	49	K011	1 2 1 1 2	

2759.   C$_{12}$H$_{27}$N
Tributylamine
Tris-n-butylamine
N,N-Dibutyl-1-butanamine
RN: 102-82-9   **MP** (°C):   -70
MW: 185.36   **BP** (°C):   216

Solubility (Moles/L)	Solubility (Grams/L)	Temp (°C)	Ref (#)	Evaluation (T P E A A)	Comments
7.649E-04	1.418E-01	25.04	V013	2 2 2 2 2	

2760.   C$_{12}$H$_{27}$N.4H$_2$O
Dodecylamine (Tetrahydrate)
RN: 124-22-1   **MP** (°C):
MW: 257.42   **BP** (°C):

Solubility (Moles/L)	Solubility (Grams/L)	Temp (°C)	Ref (#)	Evaluation (T P E A A)	Comments
2.776E-03	7.145E-01	ns	R037	0 2 2 1 0	

2761.   C$_{12}$H$_{27}$OP
Tributyl Phosphine Oxide
Tributylphosphine Oxide
TBPO
RN: 814-29-9   **MP** (°C):   64
MW: 218.32   **BP** (°C):

Solubility (Moles/L)	Solubility (Grams/L)	Temp (°C)	Ref (#)	Evaluation (T P E A A)	Comments
1.035E+00	2.260E+02	13.20	H031	1 2 2 2 2	
8.794E-01	1.920E+02	13.40	H031	1 2 2 2 2	
4.718E-01	1.030E+02	16.30	H031	1 2 2 2 2	
1.832E-01	4.000E+01	25	B070	1 2 0 1 1	
2.551E-01	5.570E+01	25.00	H031	1 2 2 2 2	
2.299E-01	5.020E+01	27.00	H032	1 1 2 1 2	

2.244E-01	4.900E+01	27.8	H032	1 1 2 1 2
2.125E-01	4.640E+01	29.0	H032	1 1 2 1 2
2.020E-01	4.410E+01	30.2	H032	1 1 2 1 2
1.974E-01	4.310E+01	31.1	H032	1 1 2 1 2
1.892E-01	4.130E+01	32.0	H032	1 1 2 1 2
1.818E-01	3.970E+01	32.5	H032	1 1 2 1 2
1.626E-01	3.550E+01	34.50	H031	1 2 2 2 2
1.530E-01	3.340E+01	36.0	H032	1 1 2 1 2
1.205E-01	2.630E+01	42.6	H032	1 1 2 1 2
1.063E-01	2.320E+01	46.0	H032	1 1 2 1 2
1.035E-01	2.260E+01	46.70	H031	1 2 2 2 2
8.932E-02	1.950E+01	50.4	H032	1 1 2 1 2
7.466E-02	1.630E+01	56.00	H031	1 2 2 2 2
5.176E-02	1.130E+01	76.50	H031	1 2 2 2 2
4.306E-02	9.400E+00	99.00	H031	1 2 2 2 2

2762. $C_{12}H_{27}O_2P$

Butyl Dibutyl Phosphinate
Butoxydibutylphosphine Oxide
Dibutylbutoxyphosphine Oxide
Butyl Dibutylphosphinate

RN: 2950-47-2   **MP** (°C):
MW: 234.32   **BP** (°C):

Solubility (Moles/L)	Solubility (Grams/L)	Temp (°C)	Ref (#)	Evaluation (T P E A A)	Comments
1.920E-02	4.500E+00	25	B070	1 2 0 1 1	

2763. $C_{12}H_{27}O_3P$

Diethyl Octyl Phosphonate
Diethyl Octanephosphonate

RN: 1068-07-1   **MP** (°C):
MW: 250.32   **BP** (°C):

Solubility (Moles/L)	Solubility (Grams/L)	Temp (°C)	Ref (#)	Evaluation (T P E A A)	Comments
<7.99E-04	<2.00E-01	25	B070	1 2 0 1 0	

2764. C$_{12}$H$_{27}$O$_3$P
Dibutyl Butyl Phosphonate
Dibutoxybutylphosphine Oxide
Dibutyl Butanephosphonate
Dibutyl Butylphosphonate
TC 44
RN: 78-46-6   **MP** (°C):
MW: 250.32   **BP** (°C):

Solubility (Moles/L)	Solubility (Grams/L)	Temp (°C)	Ref (#)	Evaluation (T P E A A)	Comments
1.997E-03	5.000E-01	25	B070	1 2 0 1 0	

2765. C$_{12}$H$_{27}$O$_4$P
Tributyl Phosphate
Tri-n-butyl Phosphate
RN: 126-73-8   **MP** (°C):
MW: 266.32   **BP** (°C):   289.0

Solubility (Moles/L)	Solubility (Grams/L)	Temp (°C)	Ref (#)	Evaluation (T P E A A)	Comments
4.036E-03	1.075E+00	3.4	H027	2 1 2 2 2	
3.800E-03	1.012E+00	4.0	H027	2 1 2 2 2	
3.593E-03	9.570E-01	5.0	H027	2 1 2 2 2	
2.403E-03	6.400E-01	13.0	H027	2 1 2 2 2	
1.500E-03	3.995E-01	25	B070	1 2 0 1 2	
1.464E-03	3.900E-01	25	B070	1 2 0 1 1	
2.253E-02	6.000E+00	25	F300	1 0 0 0 0	
1.585E-03	4.220E-01	25.0	H027	2 1 2 2 2	
1.570E-03	4.180E-01	25.0	H032	2 2 2 1 1	EFG
1.070E-03	2.850E-01	50.0	H027	2 1 2 2 2	
1.239E-03	3.299E-01	ns	F014	0 0 0 0 1	

2766. C$_{12}$H$_{28}$Ge
Tetrapropylgermanium
Tetra-n-propylgermane
RN: 994-65-0   **MP** (°C):
MW: 244.96   **BP** (°C):

Solubility (Moles/L)	Solubility (Grams/L)	Temp (°C)	Ref (#)	Evaluation (T P E A A)	Comments
3.320E-08	8.133E-06	25	D346	1 1 2 2 2	

2767.   C$_{12}$Br$_{10}$O
Decabromodiphenyl Ether
DBDPO
Decabromodiphenyl Oxide
RN: 1163-19-5   **MP** (°C):   298.0
MW: 959.22   **BP** (°C):

Solubility (Moles/L)	Solubility (Grams/L)	Temp (°C)	Ref (#)	Evaluation (T P E A A)	Comments
2.606E-08	2.500E-05	25	N326	1 0 0 0 1	average

2768.   C$_{12}$Cl$_8$O
Octachlorodibenzofuran
1,2,3,4,6,7,8,9-Octachlorodibenzofuran
OCDF
O8CDF
RN: 39001-02-0   **MP** (°C):   258
MW: 443.76   **BP** (°C):

Solubility (Moles/L)	Solubility (Grams/L)	Temp (°C)	Ref (#)	Evaluation (T P E A A)	Comments
2.610E-12	1.158E-09	25.0	D330	2 2 1 2 2	
8.680E-12	3.852E-09	39.50	D330	2 2 1 2 2	
3.150E-11	1.398E-08	58.6	D330	2 2 1 2 2	
1.370E-11	6.079E-09	80.0	D330	2 2 1 2 2	

2769.   C$_{12}$Cl$_8$O$_2$
Octachlorodibenzo-p-dioxin
OCDD
1,2,3,4,6,7,8,9-Octachlorodibenzodioxin
O8CDD
Octachlorodibenzo[b,e][1,4]dioxin
RN: 3268-87-9   **MP** (°C):   330
MW: 459.76   **BP** (°C):

Solubility (Moles/L)	Solubility (Grams/L)	Temp (°C)	Ref (#)	Evaluation (T P E A A)	Comments
8.700E-13	4.000E-10	20	F303	1 2 1 2 0	
8.700E-13	4.000E-10	20	W319	1 2 1 2 1	
1.610E-13	7.400E-11	25	S352	2 2 0 2 1	
1.610E-13	7.402E-11	25.0	D330	2 2 1 2 2	
4.350E-12	2.000E-09	40	F303	1 2 1 2 1	
4.350E-12	2.000E-09	40	W319	1 2 1 2 1	
6.750E-13	3.103E-10	40.0	D330	2 2 1 2 2	
3.960E-12	1.821E-09	60.0	D330	2 2 1 2 2	
1.710E-12	7.862E-10	80.0	D330	2 2 1 2 2	
8.374E-13	3.850E-10	ns	W332	0 1 0 2 2	

2770. $C_{12}Cl_{10}$
Decachlorobiphenyl
Decachlorbiphenyl
2,2',3,3',4,4',5,5',6,6'-Decachlorobiphenyl
RN: 2051-24-3 **MP** (°C): 305
MW: 498.66 **BP** (°C):

Solubility (Moles/L)	Solubility (Grams/L)	Temp (°C)	Ref (#)	Evaluation (T P E A A)	Comments
4.211E-11	2.100E-08	22	O311	2 2 1 2 1	
1.300E-12	6.483E-10	25	D331	2 1 2 2 2	
1.303E-11	6.500E-09	25	D335	1 0 0 0 1	
1.490E-11	7.430E-09	25	M342	1 0 1 1 2	
3.209E-11	1.600E-08	25	W025	1 0 2 2 1	
1.300E-12	6.483E-10	25.0	M324	1 2 1 1 2	
1.680E-11	8.378E-09	60	D331	2 1 2 2 2	
1.680E-11	8.378E-09	60.0	M324	1 2 1 1 2	
3.530E-11	1.760E-08	70	D331	2 1 2 2 2	
3.530E-11	1.760E-08	70.0	M324	1 2 1 1 2	
9.930E-11	4.952E-08	80	D331	2 1 2 2 2	
9.930E-11	4.952E-08	80.0	M324	1 2 1 1 2	

2771. $C_{13}H_6Cl_5NO_3$
Oxyclozanide
3,5,6,3',5'-Pentachloro-2,2'-dihydroxybenzanilide
Zanilox
Diplin
ICI 46638
Zanil
RN: 2277-92-1 **MP** (°C):
MW: 401.46 **BP** (°C):

Solubility (Moles/L)	Solubility (Grams/L)	Temp (°C)	Ref (#)	Evaluation (T P E A A)	Comments
7.224E-05	2.900E-02	25	P036	2 2 2 2 1	average of 3, form III
2.665E-06	1.070E-03	25	P036	2 2 2 2 1	average of 3, form II
6.227E-07	2.500E-04	25	P036	2 2 2 2 1	average of 3, form I

2772.   C₁₃H₆Cl₆O₂

Hexachlorophene

2,2'-Methylenebis[3,4,6-trichlorophenol]

Bilevon

AT-7

Dermadex

Exofene

RN: 70-30-4   **MP** (°C):   164.5

MW: 406.91   **BP** (°C):

Solubility (Moles/L)	Solubility (Grams/L)	Temp (°C)	Ref (#)	Evaluation (T P E A A)	Comments
6.142E-04	2.499E-01	22	M048	1 0 1 1 0	EFG
4.669E-05	1.900E-02	25	A008	1 0 0 0 0	EFG
3.441E-04	1.400E-01	25	A010	2 2 2 1 1	0.003N HCl
7.373E-07	3.000E-04	ns	V302	0 0 0 0 0	*sic*

2773.   C₁₃H₇Br₂N₃O₆

Bromofenoxim

3,5-Dibromo-4-hydroxybenzaldehyde-2,4-dinitrophenyloxime

Faneron

Bromfenim

RN: 13181-17-4   **MP** (°C):   196.5

MW: 461.04   **BP** (°C):

Solubility (Moles/L)	Solubility (Grams/L)	Temp (°C)	Ref (#)	Evaluation (T P E A A)	Comments
2.169E-07	1.000E-04	20	M161	1 0 0 0 0	

2774.   C₁₃H₇F₃N₂O₅

Fluorodifen

p-Nitrophenyl α,α,α-Trifluoro-2-nitro-p-tolyl Ether

RN: 15457-05-3   **MP** (°C):   90

MW: 328.21   **BP** (°C):

Solubility (Moles/L)	Solubility (Grams/L)	Temp (°C)	Ref (#)	Evaluation (T P E A A)	Comments
6.094E-06	2.000E-03	20	E048	1 2 1 1 0	
6.094E-06	2.000E-03	20	M161	1 0 0 0 0	
<6.09E-06	<2.00E-03	ns	B200	0 0 0 0 0	
6.094E-06	2.000E-03	ns	M061	0 0 0 0 0	

2775.   C$_{13}$H$_8$ClFO$_2$
4'-Chloro-5-fluoro-2-hydroxy Benzophenone
SL79.182
RN: 62433-26-5   **MP** (°C):
MW: 250.66   **BP** (°C):

Solubility (Moles/L)	Solubility (Grams/L)	Temp (°C)	Ref (#)	Evaluation (T P E A A)	Comments
3.590E-05	8.999E-03	37	F309	1 0 2 2 2	

2776.   C$_{13}$H$_8$ClNO
CP 31675
2-Chloro-N-(2-methyl-6-t-butylphenyl)acetamide
RN: 3785-20-4   **MP** (°C):   115
MW: 229.67   **BP** (°C):

Solubility (Moles/L)	Solubility (Grams/L)	Temp (°C)	Ref (#)	Evaluation (T P E A A)	Comments
1.306E-03	3.000E-01	ns	M061	0 0 0 0 2	

2777.   C$_{13}$H$_8$ClN$_3$O
RJ-64
3,4-Pyridyl-(5)-2-chlorophenyl-1,2,4-oxadiazole
RN: 27199-40-2   **MP** (°C):
MW: 257.68   **BP** (°C):

Solubility (Moles/L)	Solubility (Grams/L)	Temp (°C)	Ref (#)	Evaluation (T P E A A)	Comments
5.045E-03	1.300E+00	37	C054	2 2 2 1 2	0.1N HCl

2778.   C$_{13}$H$_8$Cl$_2$N$_2$O$_4$
Niclosamide
2',5-Dichloro-4'-nitrosalicylanilide
2-Chloro-4-nitrophenylamide-6-chlorosalicylic Acid
Cestocid
Devermine
Bayluscid
RN: 50-65-7   **MP** (°C):   230
MW: 327.13   **BP** (°C):

Solubility (Moles/L)	Solubility (Grams/L)	Temp (°C)	Ref (#)	Evaluation (T P E A A)	Comments
1.987E-05	6.500E-03	rt	M161	0 0 0 0 0	

2779. $C_{13}H_8N_2O_2S$

m-Pyridine Carboxyphenylisothiocyanate
Picolinic Acid, m-Isothiocyanatophenyl Ester

RN: 5174-37-8   **MP** (°C):
MW: 256.28   **BP** (°C):

Solubility (Moles/L)	Solubility (Grams/L)	Temp (°C)	Ref (#)	Evaluation (T P E A A)	Comments
5.000E-05	1.281E-02	25	K032	2 2 0 1 1	

2780. $C_{13}H_9Cl_2NO_4$

2,4-Dichlorophenyl 3-Methoxy-4-nitrophenyl Ether
Chlomethoxyfen
Chlomethoxynil

RN: 32861-85-1   **MP** (°C):   113.5
MW: 314.13   **BP** (°C):

Solubility (Moles/L)	Solubility (Grams/L)	Temp (°C)	Ref (#)	Evaluation (T P E A A)	Comments
9.550E-07	3.000E-04	15	M161	1 0 0 0 0	

2781. $C_{13}H_9F_3N_2O_2$

Niflumic Acid
2-[3-(Trifluoromethyl)anilino]nicotinic Acid
Actol
Flogovital
Donalgin
Landruma

RN: 4394-00-7   **MP** (°C):   204
MW: 282.22   **BP** (°C):

Solubility (Moles/L)	Solubility (Grams/L)	Temp (°C)	Ref (#)	Evaluation (T P E A A)	Comments
6.732E-05	1.900E-02	rt	H302	0 0 2 1 1	intrinsic

2782. $C_{13}H_9N$

Phenanthridine
Phenanthridin
9-Azaphenanthrene
3,4-Benzoisoquinoline
5-Azaphenanthrene

RN: 229-87-8   **MP** (°C):   106.5
MW: 179.22   **BP** (°C):   349

Solubility (Moles/L)	Solubility (Grams/L)	Temp (°C)	Ref (#)	Evaluation (T P E A A)	Comments
1.674E-03	3.000E-01	20	F300	1 0 0 0 1	

2783. C₁₃H₉N

$C_{13}H_9N$

Acridine

2,3,5,6-Dibenzopyridine

Acridin

RN: 260-94-6 **MP** (°C): 107

MW: 179.22 **BP** (°C): 346

Solubility (Moles/L)	Solubility (Grams/L)	Temp (°C)	Ref (#)	Evaluation (T P E A A)	Comments
3.200E-04	5.735E-02	24	A029	2 0 0 0 1	0.01N KOH
2.142E-04	3.840E-02	24	H106	1 0 2 2 2	
2.143E-04	3.840E-02	24	M303	1 0 1 1 2	
3.348E-04	6.000E-02	30	K090	1 2 2 2 0	EFG
3.348E-04	6.000E-02	30	K090	1 2 2 2 0	

2784. C₁₃H₉NO

$C_{13}H_9NO$

2-Hydroxyacridine

o-Hydroxyacridine

RN: 22817-17-0 **MP** (°C):

MW: 195.22 **BP** (°C):

Solubility (Moles/L)	Solubility (Grams/L)	Temp (°C)	Ref (#)	Evaluation (T P E A A)	Comments
2.000E-05	3.904E-03	20	A029	1 0 0 0 0	

2785. C₁₃H₉NS

$C_{13}H_9NS$

p-Biphenyl Isothiocyanate

4-Biphenyl Isothiocyanate

RN: 25687-48-3 **MP** (°C):

MW: 211.29 **BP** (°C):

Solubility (Moles/L)	Solubility (Grams/L)	Temp (°C)	Ref (#)	Evaluation (T P E A A)	Comments
1.400E-05	2.958E-03	25	D019	1 1 1 1 1	

2786. C₁₃H₉NS

$C_{13}H_9NS$

m-Biphenyl Isothiocyanate

3-Biphenyl Isothiocyanate

RN: 1510-25-4 **MP** (°C):

MW: 211.29 **BP** (°C):

Solubility (Moles/L)	Solubility (Grams/L)	Temp (°C)	Ref (#)	Evaluation (T P E A A)	Comments
3.000E-05	6.339E-03	25	K032	2 2 0 1 1	

2787. $C_{13}H_{10}$
Fluorene
o-Biphenylmethane
2,3-Benzindene
o-Biphenylenemethane
Diphenylenemethane
2,2'-Methylenebiphenyl
RN: 86-73-7   **MP** (°C):   116
MW: 166.22   **BP** (°C):   295

Solubility (Moles/L)	Solubility (Grams/L)	Temp (°C)	Ref (#)	Evaluation (T P E A A)	Comments
4.320E-06	7.181E-04	6.60	M082	1 1 1 2 2	
4.320E-06	7.181E-04	6.60	M151	2 1 2 2 2	
4.326E-06	7.190E-04	6.64	M183	1 2 1 1 2	
5.820E-06	9.674E-04	13.20	M082	1 1 1 2 2	
5.820E-06	9.674E-04	13.20	M151	2 1 2 2 2	
5.822E-06	9.678E-04	13.24	M183	1 2 1 1 2	
7.240E-06	1.203E-03	18.00	M082	1 1 1 2 2	
7.240E-06	1.203E-03	18.00	M151	2 1 2 2 2	
7.244E-06	1.204E-03	18.04	M183	1 2 1 1 2	
9.720E-06	1.616E-03	24.00	M082	1 1 1 2 2	
9.720E-06	1.616E-03	24.00	M151	2 1 2 2 2	
9.728E-06	1.617E-03	24.04	M183	1 2 1 1 2	
1.137E-05	1.890E-03	24.60	W003	2 2 2 2 2	average of 3
1.179E-05	1.960E-03	25	B319	2 0 1 2 2	
2.790E-05	4.638E-03	25	L301	1 1 2 2 2	
1.143E-05	1.900E-03	25	L332	1 1 1 1 1	
1.191E-05	1.980E-03	25	M064	1 1 2 2 2	
1.014E-05	1.685E-03	25	M071	2 2 2 2 2	
1.190E-05	1.978E-03	25	M342	1 0 1 1 2	
1.010E-05	1.679E-03	25	W300	2 2 2 2 2	
1.014E-05	1.685E-03	25.00	M151	2 1 1 2 2	
1.110E-05	1.845E-03	27.00	M082	1 1 1 2 2	
1.110E-05	1.845E-03	27.00	M151	2 1 2 2 2	
1.111E-05	1.847E-03	27.04	M183	1 2 1 1 2	
1.420E-05	2.360E-03	29.90	W003	2 2 2 2 2	average of 3
1.317E-05	2.190E-03	30.30	W003	2 2 2 2 2	average of 3
1.350E-05	2.244E-03	31.10	M082	1 1 1 2 2	
1.350E-05	2.244E-03	31.10	M151	2 1 2 2 2	
1.353E-05	2.250E-03	31.14	M183	1 2 1 1 2	
2.244E-05	3.730E-03	38.40	W003	2 2 2 2 2	average of 2
2.322E-05	3.860E-03	40.10	W003	2 2 2 2 2	average of 3
3.387E-05	5.630E-03	47.50	W003	2 2 2 2 2	average of 3
3.862E-05	6.420E-03	50.10	W003	2 2 2 2 2	average of 3
3.772E-05	6.270E-03	50.20	W003	2 2 2 2 2	
5.071E-05	8.430E-03	54.70	W003	2 2 2 2 2	average of 3
6.317E-05	1.050E-02	59.20	W003	2 2 2 2 2	

6.678E-05	1.110E-02	60.50	W003	2 2 2 2 2	average of 3
8.543E-05	1.420E-02	65.10	W003	2 2 2 2 2	average of 3
1.119E-04	1.860E-02	70.70	W003	2 2 2 2 2	average of 3
1.131E-04	1.880E-02	71.90	W003	2 2 2 2 2	
1.293E-04	2.150E-02	73.40	W003	2 2 2 2 2	
1.191E-05	1.980E-03	ns	M344	0 0 0 0 2	

2788. $C_{13}H_{10}BrCl_2O_2PS$

Leptophos
Phenylphosphonothioic Acid O-(4-Bromo-2,5-dichlorophenyl) O-Methyl Ester
Phosvel
NK 711
Velsicol 506
Oleophosvel

RN: 21609-90-5   **MP** (°C):   60
MW: 412.08   **BP** (°C):

Solubility (Moles/L)	Solubility (Grams/L)	Temp (°C)	Ref (#)	Evaluation (T P E A A)	Comments
7.280E-09	3.000E-06	10	B324	2 2 2 2 2	
8.707E-09	3.588E-06	10	B324	2 2 2 2 2	
1.699E-07	7.000E-05	20	B169	2 2 1 1 0	
6.095E-08	2.512E-05	20	B300	2 2 1 1 2	
6.095E-08	2.512E-05	20	B324	2 2 2 2 2	
5.096E-08	2.100E-05	20	B324	2 2 2 2 2	
1.141E-08	4.700E-06	20	C053	1 0 2 2 1	
1.213E-08	5.000E-06	22	K137	1 1 2 1 0	
7.280E-08	3.000E-05	24	C105	2 1 2 2 2	
5.824E-06	2.400E-03	25	M161	1 0 0 0 1	*sic*
1.306E-07	5.382E-05	30	B324	2 2 2 2 2	
1.092E-07	4.500E-05	30	B324	2 2 2 2 2	
2.184E-08	9.000E-06	ns	F040	1 2 2 2 0	
1.141E-08	4.700E-06	ns	F071	0 1 2 1 1	
1.699E-07	7.000E-05	ns	M110	0 0 0 0 0	EFG

2789. $C_{13}H_{10}BrCl_2O_3P$

Leptophos Oxon
O-(4-Bromo-2,5-dichlorophenyl) O-Methyl phenylphosphonate
Phosvel Oxon

RN: 25006-32-0   **MP** (°C):
MW: 396.01   **BP** (°C):

Solubility (Moles/L)	Solubility (Grams/L)	Temp (°C)	Ref (#)	Evaluation (T P E A A)	Comments
8.586E-06	3.400E-03	20.50	B169	2 2 1 1 2	

2790. C$_{13}$H$_{10}$ClNO$_2$
4'-Chloro Salicylanilide
N-(p-Chlorophenyl)-o-hydroxybenzamide
N-(p-Chlorophenyl)salicylamide
RN: 3679-63-8   **MP** (°C):
MW: 247.68   **BP** (°C):

Solubility (Moles/L)	Solubility (Grams/L)	Temp (°C)	Ref (#)	Evaluation (T P E A A)	Comments
4.885E-08	1.210E-05	ns	N336	0 0 2 2 2	intrinsic

2791. C$_{13}$H$_{10}$Cl$_2$O
2,4,-Dichloro-6-benzyl-phenol
o-Cresol, 4,6-Dichloro-α-phenyl-
2-Benzyl-4,6-dichlorophenol
RN: 19578-81-5   **MP** (°C):
MW: 253.13   **BP** (°C):

Solubility (Moles/L)	Solubility (Grams/L)	Temp (°C)	Ref (#)	Evaluation (T P E A A)	Comments
2.300E-05	5.822E-03	25	B316	1 0 2 1 1	

2792. C$_{13}$H$_{10}$Cl$_2$O$_2$
Dichlorophen
2,2'-Dihydroxy-5,5'-dichlorodiphenylmethane
G-4
RN: 97-23-4   **MP** (°C):   177-178
MW: 269.13   **BP** (°C):

Solubility (Moles/L)	Solubility (Grams/L)	Temp (°C)	Ref (#)	Evaluation (T P E A A)	Comments
1.115E-04	3.000E-02	25	M061	1 0 0 0 0	
1.115E-04	3.000E-02	25	M161	1 0 0 0 1	

2793. C$_{13}$H$_{10}$INO
Benodanil
2-Iodo-N-phenylbenzamide
Iodobenzanilide
Calirus
RN: 15310-01-7   **MP** (°C):   137
MW: 323.14   **BP** (°C):

Solubility (Moles/L)	Solubility (Grams/L)	Temp (°C)	Ref (#)	Evaluation (T P E A A)	Comments
6.189E-05	2.000E-02	20	M161	1 0 0 0 1	

2794. C$_{13}$H$_{10}$N$_2$
4-Aminoacridine
4-Acridinamine
RN: 578-07-4   **MP** (°C):   108.5
MW: 194.24   **BP** (°C):   346

Solubility (Moles/L)	Solubility (Grams/L)	Temp (°C)	Ref (#)	Evaluation (T P E A A)	Comments
7.000E-05	1.360E-02	24	A029	2 0 0 1 0	0.01N KOH

2795. C$_{13}$H$_{10}$N$_2$
9-Aminoacridine
10-Amino-5-azaanthracene
Monacrin
Izoacridina
Aminacrine
9AA
RN: 90-45-9   **MP** (°C):   241
MW: 194.24   **BP** (°C):

Solubility (Moles/L)	Solubility (Grams/L)	Temp (°C)	Ref (#)	Evaluation (T P E A A)	Comments
6.000E-05	1.165E-02	24	A029	2 0 0 1 0	0.01N KOH

2796. C$_{13}$H$_{10}$N$_2$
1-Aminoacridine
1-Acridinamine
RN: 578-06-3   **MP** (°C):   183
MW: 194.24   **BP** (°C):   346

Solubility (Moles/L)	Solubility (Grams/L)	Temp (°C)	Ref (#)	Evaluation (T P E A A)	Comments
6.000E-05	1.165E-02	24	A029	2 0 0 0 1	intrinsic

2797. C$_{13}$H$_{10}$N$_2$
2-Aminoacridine
2-Acridinamine
RN: 581-28-2   **MP** (°C):   108.5
MW: 194.24   **BP** (°C):   346

Solubility (Moles/L)	Solubility (Grams/L)	Temp (°C)	Ref (#)	Evaluation (T P E A A)	Comments
5.000E-05	9.712E-03	24	A029	2 0 0 1 0	0.01N KOH

2798. C_{13}H_{10}N_2
3-Aminoacridine
3-Acridinamine
RN: 581-29-3   **MP** (°C):   108.5
MW: 194.24   **BP** (°C):   346

Solubility (Moles/L)	Solubility (Grams/L)	Temp (°C)	Ref (#)	Evaluation (T P E A A)	Comments
1.500E-04	2.914E-02	24	A029	2 0 0 1 1	0.01N KOH

2799. C_{13}H_{10}N_4O_3
1-Benzoyloxymethyl Allopurinol
4H-Pyrazolo[3,4-d]pyrimidin-4-one, 1-[(Benzoyloxy)methyl]-1,5-dihydro-
RN: 98846-65-2   **MP** (°C):   217-219
MW: 270.25   **BP** (°C):

Solubility (Moles/L)	Solubility (Grams/L)	Temp (°C)	Ref (#)	Evaluation (T P E A A)	Comments
8.881E-05	2.400E-02	22	B322	1 0 2 2 2	

2800. C_{13}H_{10}O
Benzophenone
α-Oxodiphenylmethane
Diphenylmethanone
Benzoylbenzene
α-Oxoditane
Oxoditane
RN: 119-61-9   **MP** (°C):   48.5
MW: 182.22   **BP** (°C):   305.4

Solubility (Moles/L)	Solubility (Grams/L)	Temp (°C)	Ref (#)	Evaluation (T P E A A)	Comments
4.121E-04	7.510E-02	20	H301	2 0 2 2 2	
7.500E-04	1.367E-01	25	F063	1 1 0 0 1	
3.292E-04	6.000E-02	ns	F014	0 0 0 0 0	

2801. C_{13}H_{10}O_3
Phenyl Salicylate
Salol
2-Hydroxybenzoic Acid Phenyl Ester
RN: 118-55-8   **MP** (°C):   42.0
MW: 214.22   **BP** (°C):   173.0

Solubility (Moles/L)	Solubility (Grams/L)	Temp (°C)	Ref (#)	Evaluation (T P E A A)	Comments
7.002E-04	1.500E-01	25	F300	1 0 0 0 1	
1.866E-03	3.998E-01	rt	D021	0 0 1 1 0	

2802.   C₁₃H₁₀O₄

2,4,6-Trihydroxybenzophenone

2,4,6-Trihydroxy-benzophenon

RN: 3555-86-0   **MP** (°C):

MW: 230.22   **BP** (°C):

Solubility (Moles/L)	Solubility (Grams/L)	Temp (°C)	Ref (#)	Evaluation (T P E A A)	Comments
1.347E-02	3.100E+00	22	F300	1 0 0 0 1	

2803.   C₁₃H₁₀O₆

Maclurin

2,4,6,3',4'-Penta-hydroxy-benzophenol

2,4,6,3',4'-Pentahydroxybenzophenon

RN: 519-34-6   **MP** (°C):   222.5

MW: 262.22   **BP** (°C):

Solubility (Moles/L)	Solubility (Grams/L)	Temp (°C)	Ref (#)	Evaluation (T P E A A)	Comments
1.907E-02	5.000E+00	14	F300	1 0 0 0 0	

2804.   C₁₃H₁₁ClF₃N₃O

San 6706

4-Chloro-5-(dimethylamino)-2-(α,α,α-trifluoro-m-tolyl)-3(2H)-pyridazinone

RN: 23576-23-0   **MP** (°C):   151

MW: 317.70   **BP** (°C):

Solubility (Moles/L)	Solubility (Grams/L)	Temp (°C)	Ref (#)	Evaluation (T P E A A)	Comments
3.305E-05	1.050E-02	23.50	B200	2 0 0 0 2	

2805.   C₁₃H₁₁ClN₄O

6H-Dipyrido[3,2-b:2',3'-e][1,4]diazepin-6-one, 2-Chloro-11-ethyl-5,11-dihydro-

RN: 134698-40-1   **MP** (°C):

MW: 274.71   **BP** (°C):

Solubility (Moles/L)	Solubility (Grams/L)	Temp (°C)	Ref (#)	Evaluation (T P E A A)	Comments
4.365E-06	1.199E-03	ns	M381	0 1 1 1 2	pH 7.0

2806.   C$_{13}$H$_{11}$ClO
Chlorophene
5-Chloro-2-hydroxydiphenylmethane
Benzylchlorophenol
RN: 120-32-1   **MP** (°C):   48.5
MW: 218.69   **BP** (°C):

Solubility (Moles/L)	Solubility (Grams/L)	Temp (°C)	Ref (#)	Evaluation (T P E A A)	Comments
1.900E-02	4.155E+00	20	A008	1 0 0 0 0	EFG
1.100E-01	2.406E+01	ns	B047	0 0 0 0 0	EFG

2807.   C$_{13}$H$_{11}$N
2-Aminofluorene
9H-Fluoren-2-amine
2-Fluorenamine
RN: 153-78-6   **MP** (°C):   129
MW: 181.24   **BP** (°C):

Solubility (Moles/L)	Solubility (Grams/L)	Temp (°C)	Ref (#)	Evaluation (T P E A A)	Comments
1.710E-04	3.100E-02	rt	N015	0 0 2 2 1	

2808.   C$_{13}$H$_{11}$NO$_2$
Salicylanilide
2-Hydroxy-N-phenylbenzamide
2-Hydroxybenzanilide
RN: 87-17-2   **MP** (°C):   136
MW: 213.24   **BP** (°C):

Solubility (Moles/L)	Solubility (Grams/L)	Temp (°C)	Ref (#)	Evaluation (T P E A A)	Comments
2.579E-04	5.500E-02	23	M061	1 0 0 0 1	
2.579E-04	5.500E-02	25	M161	1 0 0 0 1	

2809.   C$_{13}$H$_{11}$NO$_3$
Furo[3,4-b]quinolin-3(1H)-one, 9-Hydroxy-1,7-dimethyl-
RN: 74103-12-1   **MP** (°C):
MW: 229.24   **BP** (°C):

Solubility (Moles/L)	Solubility (Grams/L)	Temp (°C)	Ref (#)	Evaluation (T P E A A)	Comments
9.597E-08	2.200E-05	25	P089	2 1 2 2 2	
1.527E-07	3.500E-05	37	P089	2 1 2 2 2	
2.116E-07	4.850E-05	51	P089	2 1 2 2 2	

2810. C₁₃H₁₁NO₃

Furo[3,4-b]quinolin-3(1H)-one, 9-Hydroxy-1,6-dimethyl-

RN: **MP** (°C):
MW: 229.24 **BP** (°C):

Solubility (Moles/L)	Solubility (Grams/L)	Temp (°C)	Ref (#)	Evaluation (T P E A A)	Comments
2.530E-07	5.800E-05	25	P089	2 1 2 2 2	
3.054E-07	7.000E-05	37	P089	2 1 2 2 2	
3.817E-07	8.750E-05	51	P089	2 1 2 2 2	

2811. C₁₃H₁₁N₃O₂

Benquinox
Cerenox
Seredon
Benzoylhydrazone of Quinone Oxime

RN: 495-73-8 **MP** (°C):
MW: 241.25 **BP** (°C):

Solubility (Moles/L)	Solubility (Grams/L)	Temp (°C)	Ref (#)	Evaluation (T P E A A)	Comments
2.073E-05	5.000E-03	ns	M061	0 0 0 0 0	

2812. C₁₃H₁₁N₃O₂S₂

2-Sulfanilamidobenzothiazole

RN: **MP** (°C):
MW: 305.38 **BP** (°C):

Solubility (Moles/L)	Solubility (Grams/L)	Temp (°C)	Ref (#)	Evaluation (T P E A A)	Comments
3.275E-06	1.000E-03	37	R045	1 2 1 1 1	

2813. C₁₃H₁₁N₇O₄S

5-p-Nitrobenzenesulfonamidotetrazole

RN: **MP** (°C):
MW: 361.34 **BP** (°C):

Solubility (Moles/L)	Solubility (Grams/L)	Temp (°C)	Ref (#)	Evaluation (T P E A A)	Comments
2.214E-05	8.000E-03	37	R045	1 2 1 1 0	

2814. C_{13}H_{12}
4-Methylbiphenyl
4-Phenyltoluene
RN: 644-08-6   **MP** (°C):   49.5
MW: 168.24   **BP** (°C):   267.5

Solubility (Moles/L)	Solubility (Grams/L)	Temp (°C)	Ref (#)	Evaluation (T P E A A)	Comments
1.090E-05	1.834E-03	4.9	D330	2 2 1 2 2	
2.410E-05	4.055E-03	25.0	D330	2 2 1 2 2	
4.180E-05	7.032E-03	40.0	D330	2 2 1 2 2	

2815. C_{13}H_{12}
Diphenylmethane
1,1'-Methylenebis-benzene
Phenylbenzyl
Benzylbenzene
RN: 101-81-5   **MP** (°C):   25.9
MW: 168.24   **BP** (°C):   264.5

Solubility (Moles/L)	Solubility (Grams/L)	Temp (°C)	Ref (#)	Evaluation (T P E A A)	Comments
1.783E-05	3.000E-03	24	H116	2 1 0 0 2	
8.381E-05	1.410E-02	25	A001	1 2 2 2 2	
8.381E-05	1.410E-02	25	A017	1 0 0 0 2	
8.710E-05	1.465E-02	25	D001	1 0 0 0 2	

2816. C_{13}H_{12}N_2O
Carbanilide
Diphenylurea
N,N'-Diphenylurea
RN: 102-07-8   **MP** (°C):   238.0
MW: 212.25   **BP** (°C):

Solubility (Moles/L)	Solubility (Grams/L)	Temp (°C)	Ref (#)	Evaluation (T P E A A)	Comments
7.066E-04	1.500E-01	rt	D021	0 0 1 1 1	

2817.   C$_{13}$H$_{12}$N$_2$O$_3$
Phenallymal
5-Allyl-5-phenylbarbituric Acid
2,4,6(1H,3H,5H)-Pyrimidinetrione, 5-Phenyl-5-(2-propenyl)
Barbituric Acid, 5-Allyl-5-phenyl
RN: 115-43-5   **MP** (°C):   156.5
MW: 244.25   **BP** (°C):

Solubility (Moles/L)	Solubility (Grams/L)	Temp (°C)	Ref (#)	Evaluation (T P E A A)	Comments
4.499E-03	1.099E+00	20	J030	1 2 2 2 2	
4.272E-03	1.043E+00	25	P350	2 1 1 1 2	intrinsic
7.764E-03	1.896E+00	37	J030	1 2 2 2 2	

2818.   C$_{13}$H$_{12}$O
Benzhydrol
Diphenylmethanol
RN: 91-01-0   **MP** (°C):   69
MW: 184.24   **BP** (°C):   298

Solubility (Moles/L)	Solubility (Grams/L)	Temp (°C)	Ref (#)	Evaluation (T P E A A)	Comments
2.714E-03	5.000E-01	20	F300	1 0 0 0 0	
2.800E-03	5.159E-01	25	D007	2 0 1 1 1	

2819.   C$_{13}$H$_{12}$O
o-Benzylphenol
2-Benzylphenol
RN: 28994-41-4   **MP** (°C):   53.5
MW: 184.24   **BP** (°C):   312

Solubility (Moles/L)	Solubility (Grams/L)	Temp (°C)	Ref (#)	Evaluation (T P E A A)	Comments
1.085E-03	2.000E-01	25	L021	1 0 0 0 0	

2820.   C$_{13}$H$_{12}$O
p-Benzylphenol
4-Benzylphenol
RN: 101-53-1   **MP** (°C):   81.5
MW: 184.24   **BP** (°C):   322

Solubility (Moles/L)	Solubility (Grams/L)	Temp (°C)	Ref (#)	Evaluation (T P E A A)	Comments
5.427E-04	9.999E-02	25	L021	1 0 0 0 0	

2821. C₁₃H₁₃Cl₂N₃O₃

Glycophen
1-Imidazolidinecarboxamide, 3-(3,5-Dichlorophenyl)-N-(1-methylethyl)-2,4-dioxo-
Iprodial
LFA 2043
Iprodione
RN: 36734-19-7   **MP** (°C):   136
MW: 330.17   **BP** (°C):

Solubility (Moles/L)	Solubility (Grams/L)	Temp (°C)	Ref (#)	Evaluation (T P E A A)	Comments
3.937E-05	1.300E-02	20	M161	1 0 0 0 1	

2822. C₁₃H₁₃NO₂

β-(α-Naphthyl)-β-alanine
Alanine, 3-(1(4H)-Naphthylidene)-
RN: 13913-40-1   **MP** (°C):
MW: 215.25   **BP** (°C):

Solubility (Moles/L)	Solubility (Grams/L)	Temp (°C)	Ref (#)	Evaluation (T P E A A)	Comments
2.260E-03	4.865E-01	25	M097	2 2 2 2 2	

2823. C₁₃H₁₃NO₅

2-Azetidinecarboxylic Acid, 1-[(Benzoyloxy)acetyl]-
RN: 115178-74-0   **MP** (°C):   149.5
MW: 263.25   **BP** (°C):

Solubility (Moles/L)	Solubility (Grams/L)	Temp (°C)	Ref (#)	Evaluation (T P E A A)	Comments
7.217E-03	1.900E+00	22	N317	1 1 2 1 2	

2824. C₁₃H₁₃N₃O₃S

N4-Acetyl Sulfapyridine
Acetylsulfapyridine
Sulfapyridine Acetylee
RN: 19077-98-6   **MP** (°C):
MW: 291.33   **BP** (°C):

Solubility (Moles/L)	Solubility (Grams/L)	Temp (°C)	Ref (#)	Evaluation (T P E A A)	Comments
1.098E-03	3.200E-01	37	D084	1 0 1 0 1	
7.207E-04	2.100E-01	37	F075	1 0 2 2 2	
1.119E-03	3.260E-01	37	M057	1 0 0 0 2	pH 5.5

2825. C$_{13}$H$_{13}$N$_3$O$_5$S$_2$
Succinylsulfathiazole
2-(N(4)-Succinylsulfanilamido)thiazole
p-2-Thiazolylsulfamoylsuccinanilic Acid
Kaoxidin
Colistatin
Cremosuxidine
RN: 116-43-8   **MP** (°C):
MW: 355.39   **BP** (°C):

Solubility (Moles/L)	Solubility (Grams/L)	Temp (°C)	Ref (#)	Evaluation (T P E A A)	Comments
1.379E-03	4.900E-01	38	K006	1 0 0 0 1	

2826. C$_{13}$H$_{13}$O$_4$P
Diphenyl Methyl Phosphate
Methyl Diphenyl Phosphate
RN: 115-89-9   **MP** (°C):
MW: 264.22   **BP** (°C):

Solubility (Moles/L)	Solubility (Grams/L)	Temp (°C)	Ref (#)	Evaluation (T P E A A)	Comments
3.633E-06	9.600E-04	24	H116	2 1 0 0 2	*sic*
7.569E-03	2.000E+00	25	A044	1 0 0 0 0	*sic*

2827. C$_{13}$H$_{14}$
1,4,5-Trimethylnaphthalene
Naphthalene, 1,4,5-Trimethyl-
RN: 2131-41-1   **MP** (°C):   58
MW: 170.26   **BP** (°C):

Solubility (Moles/L)	Solubility (Grams/L)	Temp (°C)	Ref (#)	Evaluation (T P E A A)	Comments
1.233E-05	2.100E-03	25	M064	1 1 2 2 1	
1.190E-05	2.026E-03	25	M342	1 0 1 1 2	
1.233E-05	2.100E-03	ns	M344	0 0 0 0 1	

2828. C$_{13}$H$_{14}$F$_3$N$_3$O$_4$
Ethalfluralin
N-Ethyl-N-(2-methyl-2-propenyl)-2,6-dinitro-4-(trifluoromethyl)benzenamine
Buvilan
Solanan
RN: 55283-68-6   **MP** (°C):   55.5
MW: 333.27   **BP** (°C):

Solubility (Moles/L)	Solubility (Grams/L)	Temp (°C)	Ref (#)	Evaluation (T P E A A)	Comments
6.001E-07	2.000E-04	25	M161	1 0 0 0 0	pH 7
9.002E-07	3.000E-04	ns	D304	1 0 0 0 0	

2829. C₁₃H₁₄N₂
4,4'-Methylenedianiline
4,4'-Methylenebisbenzeneamine
Tonox
HT 972
RN: 101-77-9   **MP** (°C):   93
MW: 198.27   **BP** (°C):   398

Solubility (Moles/L)	Solubility (Grams/L)	Temp (°C)	Ref (#)	Evaluation (T P E A A)	Comments
5.044E-03	1.000E+00	19	I307	0 0 0 0 0	

2830. C₁₃H₁₄N₂O₃
Mephobarbital
5-Ethyl-1-methyl-5-phenylbarbituric Acid
5-Ethyl-N-methyl-5-phenylbarbituric Acid
Mebaral
Prominal
Methylphenobarbital
RN: 115-38-8   **MP** (°C):   176
MW: 246.27   **BP** (°C):

Solubility (Moles/L)	Solubility (Grams/L)	Temp (°C)	Ref (#)	Evaluation (T P E A A)	Comments
6.090E-04	1.500E-01	20	J030	1 2 2 2 1	
4.872E-04	1.200E-01	37	J030	1 2 2 2 1	

2831. C₁₃H₁₄N₂O₆
Benzoic Acid, 2-(Acetyloxy)-, 2-[(2-Amino-2-oxoethyl)amino]-2-oxoethyl Ester
RN: 118247-02-2 **MP** (°C):   186
MW: 294.27   **BP** (°C):

Solubility (Moles/L)	Solubility (Grams/L)	Temp (°C)	Ref (#)	Evaluation (T P E A A)	Comments
2.990E-03	8.800E-01	21	N335	1 2 1 1 2	

2832. $C_{13}H_{14}N_4O_3S$

N4-Acetylsulfamerazine
N4-Acetylsulphamerazine
2-N4-Acetylsulfanilamido-4-methylpyrimidine
RN: 127-73-1   **MP** (°C):
MW: 306.35   **BP** (°C):

Solubility (Moles/L)	Solubility (Grams/L)	Temp (°C)	Ref (#)	Evaluation (T P E A A)	Comments
1.200E-03	3.676E-01	37	G026	1 0 1 1 0	EFG, pH 5.4
2.579E-03	7.900E-01	37	L091	1 0 0 0 1	pH 5.5
9.140E-04	2.800E-01	37	R045	1 2 1 1 2	
9.140E-04	2.800E-01	37	R045	1 2 1 1 1	
1.234E-03	3.780E-01	37	S192	1 0 1 1 2	pH 6.0
2.611E-03	8.000E-01	38	K006	1 0 0 0 1	

2833. $C_{13}H_{14}N_4O_4S$

Acetyl Sulfamethoxypyridazine
3-(N1-Acetylsulfanilamido)-6-methoxypyridazine
Acetylmidicel
RN: 127-75-3   **MP** (°C):
MW: 322.34   **BP** (°C):

Solubility (Moles/L)	Solubility (Grams/L)	Temp (°C)	Ref (#)	Evaluation (T P E A A)	Comments
6.825E-04	2.200E-01	37	B046	1 0 2 2 1	pH 4.5

2834. $C_{13}H_{14}O_6$

Salicylic Acid Acetate, Hydroxymethyl Ester Propionate
RN: 32620-70-5   **MP** (°C):   51.5
MW: 266.25   **BP** (°C):

Solubility (Moles/L)	Solubility (Grams/L)	Temp (°C)	Ref (#)	Evaluation (T P E A A)	Comments
2.629E-03	7.000E-01	21	N335	1 2 1 1 2	

2835. $C_{13}H_{14}O_6$

Methylphthalyl Ethyl Glycolate
2-Ethoxy-2-oxoethyl Methyl Ester
RN: 85-71-2   **MP** (°C):   <-35
MW: 266.25   **BP** (°C):   189

Solubility (Moles/L)	Solubility (Grams/L)	Temp (°C)	Ref (#)	Evaluation (T P E A A)	Comments
1.990E-03	5.297E-01	20	F070	1 0 0 0 2	

2836.   C₁₃H₁₅NO₂
Glutethimide
Doriden
Noxyron
RN: 77-21-4   **MP** (°C):   84
MW: 217.27   **BP** (°C):

Solubility (Moles/L)	Solubility (Grams/L)	Temp (°C)	Ref (#)	Evaluation (T P E A A)	Comments
4.372E-03	9.500E-01	27	B043	1 0 1 2 0	EFG
4.600E-03	9.994E-01	30	D010	1 2 1 1 2	
4.603E-03	1.000E+00	32	B043	1 0 1 2 0	EFG
5.753E-03	1.250E+00	37	B043	1 0 1 2 0	EFG
5.523E-05	1.200E-02	37	B045	1 0 1 1 2	
4.603E-03	1.000E+00	ns	A090	0 0 0 0 1	sic
4.600E-03	9.994E-01	ns	R010	0 1 0 0 2	

2837.   C₁₃H₁₅NO₂
Pyracarbolid
3,4-Dihydro-6-methyl-N-phenyl-2H-pyran-5-carboxamide
Sicarol
RN: 24691-76-7   **MP** (°C):   110.5
MW: 217.27   **BP** (°C):

Solubility (Moles/L)	Solubility (Grams/L)	Temp (°C)	Ref (#)	Evaluation (T P E A A)	Comments
2.762E-03	6.000E-01	40	M161	1 0 0 0 0	

2838.   C₁₃H₁₅NO₂S
m-Carboxylpentylphenylisothiocyanate
RN: **MP** (°C):
MW: 249.33   **BP** (°C):

Solubility (Moles/L)	Solubility (Grams/L)	Temp (°C)	Ref (#)	Evaluation (T P E A A)	Comments
7.300E-05	1.820E-02	25	K032	2 2 0 1 1	

2839.   C₁₃H₁₅NO₃
Pyrrolidine, 1-[(Benzoyloxy)acetyl]-
RN: 115178-67-1   **MP** (°C):   58
MW: 233.27   **BP** (°C):

Solubility (Moles/L)	Solubility (Grams/L)	Temp (°C)	Ref (#)	Evaluation (T P E A A)	Comments
2.701E-02	6.300E+00	22	N317	1 1 2 1 2	

2840. C₁₃H₁₅NO₄

$C_{13}H_{15}NO_4$

Morpholine, 4-[(Benzoyloxy)acetyl]-

RN: 106231-68-9   **MP** (°C): 103.5
MW: 249.27   **BP** (°C):

Solubility (Moles/L)	Solubility (Grams/L)	Temp (°C)	Ref (#)	Evaluation (T P E A A)	Comments
1.685E-02	4.200E+00	22	N317	1 1 2 1 2	

2841. C₁₃H₁₅NO₅

$C_{13}H_{15}NO_5$

Benzoic Acid, 2-(Acetyloxy)-, 2-(Ethylamino)-2-oxoethyl Ester

RN: 118247-01-1   **MP** (°C): 80.5
MW: 265.27   **BP** (°C):

Solubility (Moles/L)	Solubility (Grams/L)	Temp (°C)	Ref (#)	Evaluation (T P E A A)	Comments
2.081E-02	5.520E+00	21	N335	1 2 1 1 2	

2842. C₁₃H₁₅NO₅

$C_{13}H_{15}NO_5$

Benzoic Acid, 2-(Acetyloxy)-, 2-(Dimethylamino)-2-oxoethyl Ester

RN: 118247-04-4   **MP** (°C): 75.5
MW: 265.27   **BP** (°C):

Solubility (Moles/L)	Solubility (Grams/L)	Temp (°C)	Ref (#)	Evaluation (T P E A A)	Comments
2.827E-02	7.500E+00	21	N335	1 2 1 1 2	

2843. C₁₃H₁₅N₃O₂

$C_{13}H_{15}N_3O_2$

Pyrolan
1-Phenyl-3-methylpyrazolyl-5-dimethylcarbamate

RN: 87-47-8   **MP** (°C): 50
MW: 245.28   **BP** (°C): 161

Solubility (Moles/L)	Solubility (Grams/L)	Temp (°C)	Ref (#)	Evaluation (T P E A A)	Comments
8.138E-03	1.996E+00	ns	M061	0 0 0 0 0	

2844. C₁₃H₁₅N₃O₃S

$C_{13}H_{15}N_3O_3S$

2-Sulfanilamido-3-ethoxypyridine
Benzenesulfonamide, 4-Amino-N-(3-ethoxy-2-pyridinyl)-

RN: 71119-19-2   **MP** (°C):
MW: 293.35   **BP** (°C):

Solubility (Moles/L)	Solubility (Grams/L)	Temp (°C)	Ref (#)	Evaluation (T P E A A)	Comments
8.011E-04	2.350E-01	37	R058	1 2 1 1 2	

2845.   C$_{13}$H$_{15}$N$_3$O$_3$S
5-Sulfanilamido-2-ethoxypyridine
Benzenesulfonamide, 4-Amino-N-(6-ethoxy-3-pyridinyl)-
RN: 71720-65-5   **MP** (°C):
MW: 293.35   **BP** (°C):

Solubility (Moles/L)	Solubility (Grams/L)	Temp (°C)	Ref (#)	Evaluation (T P E A A)	Comments
1.227E-04	3.600E-02	37	R058	1 2 1 1 1	

2846.   C$_{13}$H$_{15}$N$_3$O$_4$S
N1-(3,4-Dimethyl-5-isoxazolyl)-N4-acetylsulfanilamide
Acetylsulfadimethylisoxazole
N4-Acetylsulfisoxazole
4-N-Acetylsulfisoxazole
N-Acetylsulfisoxazole
RN: 4206-74-0   **MP** (°C):
MW: 309.35   **BP** (°C):

Solubility (Moles/L)	Solubility (Grams/L)	Temp (°C)	Ref (#)	Evaluation (T P E A A)	Comments
2.450E-02	7.579E+00	37	B110	1 0 2 2 2	pH 6.7

2847.   C$_{13}$H$_{15}$N$_3$O$_4$S
Acetyl Sulfisoxazole
N1-Acetyl-sulfaisoxazole
RN: 80-74-0   **MP** (°C):   193.5
MW: 309.35   **BP** (°C):

Solubility (Moles/L)	Solubility (Grams/L)	Temp (°C)	Ref (#)	Evaluation (T P E A A)	Comments
2.586E-04	8.000E-02	37	B046	1 0 2 2 0	pH 4.5
1.199E-04	3.710E-02	37	M117	2 1 1 1 2	pH 6.0

2848.   C$_{13}$H$_{16}$Cl$_2$O$_3$
2,4-Dichlorophenoxyacetic Acid n-Pentyl Ester
2,4-D Pentyl Ester
Pentyl 2,4-Dichlorophenoxyacetate
Amyl 2,4-Dichlorophenoxyacetate
RN: 1917-92-6   **MP** (°C):
MW: 291.18   **BP** (°C):

Solubility (Moles/L)	Solubility (Grams/L)	Temp (°C)	Ref (#)	Evaluation (T P E A A)	Comments
2.897E-05	8.436E-03	ns	M120	0 0 1 1 2	

2849. C₁₃H₁₆Cl₂O₃

2,4-Dichlorophenoxyacetic Acid 1-Ethylpropyl Ester

RN: 65267-94-9 **MP** (°C):
MW: 291.18 **BP** (°C):

Solubility (Moles/L)	Solubility (Grams/L)	Temp (°C)	Ref (#)	Evaluation (T P E A A)	Comments
1.667E-05	4.855E-03	ns	M120	0 0 1 1 2	

2850. C₁₃H₁₆Cl₂O₃

2,4-Dichlorophenoxyacetic Acid 2-Methylbutyl Ester

RN: **MP** (°C):
MW: 291.18 **BP** (°C):

Solubility (Moles/L)	Solubility (Grams/L)	Temp (°C)	Ref (#)	Evaluation (T P E A A)	Comments
1.291E-05	3.760E-03	ns	M120	0 0 1 1 2	

2851. C₁₃H₁₆F₃N₃O₄

Trifluralin

α,α,α-Trifluoro-2,6-dinitro-N,N-dipropyl-p-toluidine

RN: 1582-09-8 **MP** (°C): 48.5
MW: 335.29 **BP** (°C):

Solubility (Moles/L)	Solubility (Grams/L)	Temp (°C)	Ref (#)	Evaluation (T P E A A)	Comments
1.193E-05	4.000E-03	20	F311	1 2 2 2 1	
2.419E-05	8.110E-03	22	K137	1 1 2 1 0	
1.730E-06	5.800E-04	25	G319	1 0 0 0 2	
<2.98E-06	<1.00E-03	27	B200	1 0 0 0 0	
<2.98E-06	<1.00E-03	27	M161	1 0 0 0 0	
<2.98E-06	<1.00E-03	27	P028	1 0 0 0 0	
7.158E-05	2.400E-02	ns	B185	0 0 0 0 1	
1.193E-04	4.000E-02	ns	M061	0 0 0 0 1	
2.088E-06	7.000E-04	ns	M110	0 0 0 0 0	EFG

2852. C₁₃H₁₆F₃N₃O₄

Benefin

Benfluralin

RN: 1861-40-1 **MP** (°C): 65
MW: 335.29 **BP** (°C):

Solubility (Moles/L)	Solubility (Grams/L)	Temp (°C)	Ref (#)	Evaluation (T P E A A)	Comments
<2.98E-06	<1.00E-03	25	B200	1 0 0 0 0	
<2.98E-06	<1.00E-03	25	M161	1 0 0 0 0	
<2.98E-06	<1.00E-03	25	P028	1 0 0 0 0	
2.088E-04	7.000E-02	ns	M061	0 0 0 0 1	

2853. C₁₃H₁₆NO₄PS

Isoxathion
O,O-Diethyl O-5-Phenylisoxazol-3-yl Phosphorothioate
E-48
Karphos
SI-6711
RN: 18854-01-8   **MP** (°C):
MW: 313.31   **BP** (°C):   160

Solubility (Moles/L)	Solubility (Grams/L)	Temp (°C)	Ref (#)	Evaluation (T P E A A)	Comments
6.064E-06	1.900E-03	25	N305	1 0 0 0 1	'

2854. C₁₃H₁₆N₂

3-(1-Methyl-2-pyrrolidinyl)-indole
RN: **MP** (°C):
MW: 200.29   **BP** (°C):

Solubility (Moles/L)	Solubility (Grams/L)	Temp (°C)	Ref (#)	Evaluation (T P E A A)	Comments
3.510E-03	7.030E-01	37	H004	1 0 2 2 2	
3.510E-03	7.030E-01	37	H011	1 2 2 2 2	

2855. C₁₃H₁₆N₂O₄

N-Acetyl-L-tyrosinamide Acetate
RN: **MP** (°C):
MW: 264.28   **BP** (°C):

Solubility (Moles/L)	Solubility (Grams/L)	Temp (°C)	Ref (#)	Evaluation (T P E A A)	Comments
1.300E-02	3.436E+00	25	A066	1 0 1 1 1	

2856. C₁₃H₁₆N₂O₄

Methyl-2-ethyl-2-phenylmalonurate
Methyl 2-Ethyl-2-phenylmalonurate
RN: 73632-81-2   **MP** (°C):   105
MW: 264.28   **BP** (°C):

Solubility (Moles/L)	Solubility (Grams/L)	Temp (°C)	Ref (#)	Evaluation (T P E A A)	Comments
1.800E-03	4.757E-01	23	B152	1 2 1 1 1	pH 3.5

2857.   C$_{13}$H$_{16}$N$_2$O$_6$
Medinoterb Acetate
m-Cresol, 6-Tert-Butyl-2,4-dinitro-, Acetate
MC 1488
RN: 2487-01-6   **MP** (°C):   86.5
MW: 296.28   **BP** (°C):

Solubility (Moles/L)	Solubility (Grams/L)	Temp (°C)	Ref (#)	Evaluation (T P E A A)	Comments
3.375E-05	1.000E-02	rt	M161	0 0 0 0 1	

2858.   C$_{13}$H$_{16}$N$_4$O$_2$S
2-p-Aminobenzenesulphonamido-4,5,6-trimethylpyrimidine
Sulfanilamide, N1-(4,5,6-Trimethyl-2-pyrimidinyl)-
RN: 5433-64-7   **MP** (°C):
MW: 292.36   **BP** (°C):

Solubility (Moles/L)	Solubility (Grams/L)	Temp (°C)	Ref (#)	Evaluation (T P E A A)	Comments
5.131E-04	1.500E-01	37	R075	1 0 0 0 1	

2859.   C$_{13}$H$_{16}$N$_4$O$_2$S
2-Sulfanilylamino-4-ethyl-5-methylpyrimidine
RN: **MP** (°C):
MW: 292.36   **BP** (°C):

Solubility (Moles/L)	Solubility (Grams/L)	Temp (°C)	Ref (#)	Evaluation (T P E A A)	Comments
8.551E-04	2.500E-01	37	R076	1 2 0 0 1	

2860.   C$_{13}$H$_{16}$N$_4$O$_6$.0.5H$_2$O
9-[5-O-(Acetate-β-D-arabinofuranosyl)]-6-methoxy-9H-purine (Hemihydrate)
2'-Acetyl-6-methoxypurine Arabinoside (Hemihydrate)
RN: 121032-43-7   **MP** (°C):   174-176
MW: 333.30   **BP** (°C):

Solubility (Moles/L)	Solubility (Grams/L)	Temp (°C)	Ref (#)	Evaluation (T P E A A)	Comments
3.250E-02	1.083E+01	37	C348	1 2 2 2 2	pH 7.00
5.310E-02	1.770E+01	37	M378	1 2 1 1 2	pH 7.2

2861.   C₁₃H₁₆O₄
Diethylacetyl Salicylate
Salicylic Acid, 2-Ethylbutyrate
RN: 100613-21-6   **MP** (°C):
MW: 236.27   **BP** (°C):

Solubility (Moles/L)	Solubility (Grams/L)	Temp (°C)	Ref (#)	Evaluation (T P E A A)	Comments
2.800E-03	6.616E-01	25.6	G015	1 0 1 1 2	pH 1.00, pka 4.00, intrinsic

2862.   C₁₃H₁₆O₆
Methyl Phthalyl Ethyl Glycollate
RN: **MP** (°C):
MW: 268.27   **BP** (°C):

Solubility (Moles/L)	Solubility (Grams/L)	Temp (°C)	Ref (#)	Evaluation (T P E A A)	Comments
4.096E-03	1.099E+00	15	H069	1 0 1 1 1	
1.975E-03	5.297E-01	ns	F014	0 0 0 0 1	

2863.   C₁₃H₁₆O₇.0.75H₂O
Helicin (0.75 Hydrate)
Salicylaldehyde β-D-Glucoside
Benzaldehyde, 2-(β-D-Glucopyranosyloxy)-, Hydrate (4:3)
RN: 618-65-5   **MP** (°C):
MW: 297.78   **BP** (°C):

Solubility (Moles/L)	Solubility (Grams/L)	Temp (°C)	Ref (#)	Evaluation (T P E A A)	Comments
5.505E-02	1.639E+01	c	D004	1 0 0 0 0	

2864.   C₁₃H₁₇IN₂O₆
Uridine, 2'-Deoxy-5-iodo-, 5'-(2-Methylpropanoate)
5'-Isobutyryl 5-iodo-2'-deoxyuridine
5-Iodo-2'-deoxyuridine 5'-isobutyrate
RN: 84043-27-6   **MP** (°C):   144.5
MW: 424.19   **BP** (°C):

Solubility (Moles/L)	Solubility (Grams/L)	Temp (°C)	Ref (#)	Evaluation (T P E A A)	Comments
1.750E+03	7.423E+05	25	N332	1 0 2 2 2	pH 7.4

2865. $C_{13}H_{17}IN_2O_6$
Uridine, 2'-Deoxy-5-iodo-, 5'-Butanoate
5'-Butyryl 5-iodo-2'-deoxyuridine
5-Iodo-2'-deoxyuridine 5'-butyrate
RN: 84043-26-5   **MP** (°C):   145.5
MW: 424.19   **BP** (°C):

Solubility (Moles/L)	Solubility (Grams/L)	Temp (°C)	Ref (#)	Evaluation (T P E A A)	Comments
1.450E+03	6.151E+05	25	N332	1 0 2 2 2	pH 7.4

2866. $C_{13}H_{17}NO$
N-Butylcinnamamide
N-Butyl-3-phenyl-2-propenamide
RN: 6299-56-5   **MP** (°C):
MW: 203.29   **BP** (°C):

Solubility (Moles/L)	Solubility (Grams/L)	Temp (°C)	Ref (#)	Evaluation (T P E A A)	Comments
9.700E-04	1.972E-01	ns	H350	0 0 0 0 2	

2867. $C_{13}H_{17}NO$
N,N-Diethylcinnamamide
N,N-Diethyl-3-phenyl-2-propenamide
RN: 3680-04-4   **MP** (°C):
MW: 203.29   **BP** (°C):

Solubility (Moles/L)	Solubility (Grams/L)	Temp (°C)	Ref (#)	Evaluation (T P E A A)	Comments
7.450E-03	1.514E+00	ns	H350	0 0 0 0 2	

2868. $C_{13}H_{17}NO_3$
Acetamide, 2-(Benzoyloxy)-N-butyl-
RN: 115193-28-7   **MP** (°C):   69.5
MW: 235.29   **BP** (°C):

Solubility (Moles/L)	Solubility (Grams/L)	Temp (°C)	Ref (#)	Evaluation (T P E A A)	Comments
1.743E-03	4.100E-01	22	N317	1 1 2 1 2	

2869. $C_{13}H_{17}NO_3$
Butanamide, 4-(Benzoyloxy)-N,N-dimethyl-
RN: 115178-78-4   **MP** (°C):   40.5
MW: 235.29   **BP** (°C):

Solubility (Moles/L)	Solubility (Grams/L)	Temp (°C)	Ref (#)	Evaluation (T P E A A)	Comments
5.908E-02	1.390E+01	22	N317	1 1 2 1 2	

2870. C₁₃H₁₇NO₃
2-(p-Acetaminophenoxy)tetrahydropyran
RN: 51453-65-7   **MP** (°C):   60
MW: 235.29   **BP** (°C):

Solubility (Moles/L)	Solubility (Grams/L)	Temp (°C)	Ref (#)	Evaluation (T P E A A)	Comments
3.000E-03	7.059E-01	ns	H076	0 0 0 0 0	

2871. C₁₃H₁₇NO₃
N-Acetyl-L-phenylalanine Ethyl Ester
RN: 2361-96-8   **MP** (°C):
MW: 235.29   **BP** (°C):

Solubility (Moles/L)	Solubility (Grams/L)	Temp (°C)	Ref (#)	Evaluation (T P E A A)	Comments
1.084E-02	2.550E+00	5	L081	2 1 2 2 2	
1.755E-02	4.130E+00	28	L081	2 1 2 2 2	
2.814E-02	6.620E+00	40	L081	2 1 2 2 2	
3.417E-02	8.040E+00	55	L081	2 1 2 2 2	
7.268E-02	1.710E+01	65	L081	2 1 2 2 2	

2872. C₁₃H₁₇NO₃
Acetamide, 2-(Benzoyloxy)-N,N-diethyl-
RN: 64649-63-4   **MP** (°C):   72.5
MW: 235.29   **BP** (°C):

Solubility (Moles/L)	Solubility (Grams/L)	Temp (°C)	Ref (#)	Evaluation (T P E A A)	Comments
8.500E-03	2.000E+00	22	N317	1 1 2 1 2	

2873. C₁₃H₁₇NO₃
Pivalyl Acetaminophen
Propanoic Acid, 2,2-Dimethyl-, 4-(Acetylamino)phenyl Ester
Acetanilide, 4'-Hydroxy-, Pivalate (Ester)
RN: 20675-23-4   **MP** (°C):   162.5-163
MW: 235.29   **BP** (°C):

Solubility (Moles/L)	Solubility (Grams/L)	Temp (°C)	Ref (#)	Evaluation (T P E A A)	Comments
4.675E-04	1.100E-01	37	D029	1 0 1 1 1	

2874.   C$_{13}$H$_{17}$NO$_4$
Propanoic Acid, 2,2-Dimethyl-, [2-(Aminocarbonyl)phenoxy]methyl Ester
O-Pivaloyloxymethyl Salicylamide

RN: 103951-40-2   **MP** (°C):   94-96
MW: 251.28   **BP** (°C):

Solubility (Moles/L)	Solubility (Grams/L)	Temp (°C)	Ref (#)	Evaluation (T P E A A)	Comments
2.428E-03	6.100E-01	23	B328	1 2 2 1 1	

2875.   C$_{13}$H$_{17}$NO$_4$
Acetamide, 2-(Benzoyloxy)-N-ethyl-N-(2-hydroxyethyl)-

RN: 106231-60-1   **MP** (°C):   79.5
MW: 251.28   **BP** (°C):

Solubility (Moles/L)	Solubility (Grams/L)	Temp (°C)	Ref (#)	Evaluation (T P E A A)	Comments
4.298E-02	1.080E+01	22	N317	1 1 2 1 2	

2876.   C$_{13}$H$_{17}$NO$_4$
O-(Pivaloyloxymethyl) Salicylamide

RN: **MP** (°C):   95
MW: 251.28   **BP** (°C):

Solubility (Moles/L)	Solubility (Grams/L)	Temp (°C)	Ref (#)	Evaluation (T P E A A)	Comments
2.428E-03	6.100E-01	23	B328	1 2 2 1 1	pH 4

2877.   C$_{13}$H$_{17}$NO$_4$
Benzoic Acid, 2-Hydroxy-, 2-(Diethylamino)-2-oxoethyl Ester
N,N-Diethylglycolamide Salicylate
N,N-Diethyl Glycolamide Salicylate

RN: 65783-69-9   **MP** (°C):   74-75
MW: 251.28   **BP** (°C):

Solubility (Moles/L)	Solubility (Grams/L)	Temp (°C)	Ref (#)	Evaluation (T P E A A)	Comments
2.786E-03	7.000E-01	21	B331	1 2 2 1 1	pH 7.4
2.786E-03	7.000E-01	21	B331	1 2 2 1 1	

2878.   C$_{13}$H$_{17}$NO$_4$
N-Acetyl-L-tyrosine Ethyl Ester
Ethyl N-Acetyl-L-tyrosinate

RN: 840-97-1   **MP** (°C):
MW: 251.28   **BP** (°C):

Solubility (Moles/L)	Solubility (Grams/L)	Temp (°C)	Ref (#)	Evaluation (T P E A A)	Comments
5.571E-03	1.400E+00	5	L081	2 1 2 2 2	
1.385E-02	3.480E+00	28	L081	2 1 2 2 2	

2879.   C₁₃H₁₇NO₄

Isobutyl Acetaminophen
Carbonic Acid, Isobutyl Ester, Ester with 4'-hydroxyacetanilide
Acetanilide, 4'-Hydroxy-, Isobutyl Carbonate (Ester)

RN: 20460-96-2   **MP** (°C):   119-121
MW: 251.28   **BP** (°C):

Solubility (Moles/L)	Solubility (Grams/L)	Temp (°C)	Ref (#)	Evaluation (T P E A A)	Comments
1.512E-03	3.800E-01	37	D029	1 0 1 1 1	

2880.   C₁₃H₁₇NO₄

Butyl Acetaminophen
Carbonic Acid, Butyl Ester, Ester with 4'-Hydroxyacetanilide
Acetanilide, 4'-Hydroxy-, Butyl Carbonate (Ester)

RN: 19872-68-5   **MP** (°C):   119.5-120
MW: 251.28   **BP** (°C):

Solubility (Moles/L)	Solubility (Grams/L)	Temp (°C)	Ref (#)	Evaluation (T P E A A)	Comments
6.367E-04	1.600E-01	37	D029	1 0 1 1 1	

2881.   C₁₃H₁₇NO₅

Acetamide, 2-(Benzoyloxy)-N,N-bis(2-hydroxyethyl)-

RN: 106231-61-2   **MP** (°C):   81
MW: 267.28   **BP** (°C):

Solubility (Moles/L)	Solubility (Grams/L)	Temp (°C)	Ref (#)	Evaluation (T P E A A)	Comments
2.694E+00	7.200E+02	22	N317	1 1 2 1 2	

2882.   C₁₃H₁₇NO₆

Acetamide, 2-(Benzoyloxy)-N-[2-hydroxy-1,1-bis(hydroxymethyl)ethyl]-

RN: 115193-31-2   **MP** (°C):   126.5
MW: 283.28   **BP** (°C):

Solubility (Moles/L)	Solubility (Grams/L)	Temp (°C)	Ref (#)	Evaluation (T P E A A)	Comments
5.401E-02	1.530E+01	22	N317	1 1 2 1 2	

2883. $C_{13}H_{17}N_3O$
Aminopyrine
Amidopyrine
4-Dimethylaminoantipyrine
Febrinina
Febron
Itamidone
RN: 58-15-1   **MP** (°C):   108
MW: 231.30   **BP** (°C):

Solubility (Moles/L)	Solubility (Grams/L)	Temp (°C)	Ref (#)	Evaluation (T P E A A)	Comments
2.827E-01	6.540E+01	0	C025	0 0 0 0 2	form A
5.607E-01	1.297E+02	4.62	M109	2 1 1 1 0	EFG
5.463E-01	1.264E+02	10.93	M109	2 1 1 1 0	EFG
5.430E-01	1.256E+02	15.02	M109	2 1 1 1 0	EFG
2.291E-01	5.300E+01	20	C025	0 0 0 0 2	form A
5.452E-01	1.261E+02	20.96	M109	2 1 1 1 0	EFG
2.291E-01	5.300E+01	25	P012	1 0 1 2 0	
2.162E-01	5.000E+01	25	P016	1 0 0 1 1	
2.075E-01	4.800E+01	25	P020	2 0 1 1 1	
1.773E+00	4.100E+02	25	P020	2 0 1 1 2	
5.618E-01	1.300E+02	25.35	M109	2 1 1 1 0	EFG
5.965E-01	1.380E+02	29.87	M109	2 1 1 1 0	EFG
2.350E-01	5.436E+01	30	A078	2 1 2 1 0	FFG
2.291E-01	5.300E+01	37	C025	0 0 0 0 2	form A
6.329E-01	1.464E+02	38.37	M109	2 1 1 1 0	EFG
6.646E-01	1.537E+02	49.42	M109	2 1 1 1 0	EFG
3.415E-01	7.900E+01	55	C025	0 0 0 0 2	form A
5.638E-01	1.304E+02	65	C025	0 0 0 0 2	form A
2.162E+00	5.000E+02	69.50	C025	0 0 0 0 2	form A
1.729E+00	4.000E+02	70	C025	0 0 0 0 2	form B
1.167E+00	2.700E+02	70.50	C025	0 0 0 0 2	form B
2.879E+00	6.660E+02	74.40	C025	0 0 0 0 2	form B
8.647E-01	2.000E+02	77.50	C025	0 0 0 0 2	torm B
6.485E-01	1.500E+02	81	C025	0 0 0 0 2	form B
3.243E+00	7.500E+02	84	C025	0 0 0 0 2	form B
3.359E+00	7.770E+02	92	C025	0 0 0 0 2	form B

2884. $C_{13}H_{17}N_5O_5$
9-(2-O-Propionyl-β-D-arabinofuranosyl)adenine
RN: 65174-99-4   **MP** (°C):
MW: 323.31   **BP** (°C):

Solubility (Moles/L)	Solubility (Grams/L)	Temp (°C)	Ref (#)	Evaluation (T P E A A)	Comments
3.618E-04	1.170E-01	37	B306	1 2 0 1 2	pH 7.3

2885.   C$_{13}$H$_{17}$N$_5$O$_5$
9-[5'-(O-Propionyl)-β-D-arabinofuranosyl]adenine Ester
RN: 14000-32-9   **MP** (°C):   202.0
MW: 323.31   **BP** (°C):

Solubility (Moles/L)	Solubility (Grams/L)	Temp (°C)	Ref (#)	Evaluation (T P E A A)	Comments
2.846E-02	9.200E+00	ns	B134	0 1 1 1 1	

2886.   C$_{13}$H$_{17}$N$_5$O$_6$
9-(1,3-Diacetate-2-propoxymethyl)guanine
RN: 86357-19-9   **MP** (°C):   238
MW: 339.31   **BP** (°C):

Solubility (Moles/L)	Solubility (Grams/L)	Temp (°C)	Ref (#)	Evaluation (T P E A A)	Comments
1.709E-03	5.800E-01	25	B360	1 0 2 2 2	

2887.   C$_{13}$H$_{17}$N$_5$O$_8$
9-(1,3-Dimethoxycarbonyl-2-propoxymethyl)guanine
RN: 91625-66-0   **MP** (°C):   178
MW: 371.31   **BP** (°C):

Solubility (Moles/L)	Solubility (Grams/L)	Temp (°C)	Ref (#)	Evaluation (T P E A A)	Comments
3.851E-04	1.430E-01	25	B360	1 0 2 2 2	

2888.   C$_{13}$H$_{18}$ClNO
Pentanochlor
Solan
Pentamide, N-(3-Chloro-4-methylphenyl)-2-methyl-
RN: 2307-68-8   **MP** (°C):   84
MW: 239.75   **BP** (°C):

Solubility (Moles/L)	Solubility (Grams/L)	Temp (°C)	Ref (#)	Evaluation (T P E A A)	Comments
3.337E-05	8.000E-03	ns	B185	1 0 0 0 1	
3.545E-05	8.500E-03	rt	M161	0 0 0 0 0	

2889.   C₁₃H₁₈ClNO

Monalide
N-(4-Chlorophenyl)-2,2-dimethylvaleramide
RN: 7287-36-7   **MP** (°C): 87.5
MW: 239.75   **BP** (°C):

Solubility (Moles/L)	Solubility (Grams/L)	Temp (°C)	Ref (#)	Evaluation (T P E A A)	Comments
9.510E-05	2.280E-02	23	M161	1 0 0 0 2	
9.510E-05	2.280E-02	ns	M061	0 0 0 0 2	

2890.   C₁₃H₁₈ClN₃O₄S₂

Cyclopenthiazide
6-Chloro-3-cyclopentylmethyl-3,4-dihydro-2H-1,2,4-benzothiadiazine-7-sulphonamide 1,1-dioxide
RN: 742-20-1   **MP** (°C): 235
MW: 379.89   **BP** (°C):

Solubility (Moles/L)	Solubility (Grams/L)	Temp (°C)	Ref (#)	Evaluation (T P E A A)	Comments
1.316E-04	5.000E-02	rt	A095	0 0 2 2 0	

2891.   C₁₃H₁₈Cl₂N₂O₂

Melphalan
4-[bis(2-Chloroethyl)amino]-L-Phenylalanine
RN: 148-82-3   **MP** (°C):
MW: 305.21   **BP** (°C):

Solubility (Moles/L)	Solubility (Grams/L)	Temp (°C)	Ref (#)	Evaluation (T P E A A)	Comments
1.442E-02	4.400E+00	30	L343	2 1 1 1 0	EFG

2892.   C₁₃H₁₈N₂O₂

Lenacil
3-Cyclohexyl-5,6-trimethyleneuracil
RN: 2164-08-1   **MP** (°C): 290
MW: 234.30   **BP** (°C):

Solubility (Moles/L)	Solubility (Grams/L)	Temp (°C)	Ref (#)	Evaluation (T P E A A)	Comments
2.561E-05	6.000E-03	25	M061	1 0 0 0 0	
2.561E-05	6.000E-03	25	M161	1 0 0 0 0	

2893. $C_{13}H_{18}N_2O_3$

Heptabarbital
5-(1-Cyclohepten-1-yl)-5-ethyl-2,4,6(1H,3H,5H)-pyrimidinetrione
5-(1-Cyclohepten-1-yl)-5-ethylbarbituric Acid
Heptabarbitone

RN: 509-86-4   **MP** (°C):   174
MW: 250.30   **BP** (°C):

Solubility (Moles/L)	Solubility (Grams/L)	Temp (°C)	Ref (#)	Evaluation (T P E A A)	Comments
1.000E-03	2.503E-01	25	V033	2 0 1 1 2	
1.000E-03	2.503E-01	25.00	T303	1 0 0 0 1	
1.400E-03	3.504E-01	35.00	T303	1 0 0 0 1	
1.170E-02	2.929E+00	40	N008	1 0 1 1 2	*sic*
1.800E-03	4.505E-01	45.00	T303	1 0 0 0 1	

2894. $C_{13}H_{18}N_2O_3S$

Tosylcyclopentylurea
Tosylcyclopentyluree

RN: 1027-87-8   **MP** (°C):
MW: 282.36   **BP** (°C):

Solubility (Moles/L)	Solubility (Grams/L)	Temp (°C)	Ref (#)	Evaluation (T P E A A)	Comments
2.649E-04	7.478E-02	37	A028	1 0 2 1 2	intrinsic
2.650E-04	7.483E-02	37	A046	2 0 1 1 2	

2895. $C_{13}H_{18}N_2O_4$

Methyl-2-methyl-2-cyclohexenyl-6-methylmalonurate
Methyl 2-Methyl-2-cyclohexenyl-6-methylmalonurate

RN: **MP** (°C):   94
MW: 266.30   **BP** (°C):

Solubility (Moles/L)	Solubility (Grams/L)	Temp (°C)	Ref (#)	Evaluation (T P E A A)	Comments
2.100E-03	5.592E-01	23	B152	1 2 1 1 1	pH 3.5

2896. $C_{13}H_{18}N_4O_2S_2$

4-Amino-N-(5-isopentyl-1,3,4-thiadiazol-2-yl)benzenesulfonamide
Benzenesulfonamide, 4-Amino-N-[5-(3-methylbutyl)-1,3,4-thiadiazol-2-yl]-

RN: 71119-29-4   **MP** (°C):
MW: 326.44   **BP** (°C):

Solubility (Moles/L)	Solubility (Grams/L)	Temp (°C)	Ref (#)	Evaluation (T P E A A)	Comments
9.000E-05	2.938E-02	37	A046	2 0 1 1 2	

2897. $C_{13}H_{18}N_4O_2S_2$
4-Amino-N-(5-pentyl-1,3,4-thiadiazol-2-yl)benzenesulfonamide
Benzenesulfonamide, 4-Amino-N-(5-pentyl-1,3,4-thiadiazol-2-yl)-
RN: 71119-30-7   **MP** (°C):
MW: 326.44   **BP** (°C):

Solubility (Moles/L)	Solubility (Grams/L)	Temp (°C)	Ref (#)	Evaluation (T P E A A)	Comments
1.120E-04	3.656E-02	37	A046	2 0 1 1 2	

2898. $C_{13}H_{18}O_2$
Ibuprofen
2-(4-Isobutylphenyl)propionic Acid
Advil
Ebufac
Rufen
RN: 15687-27-1   **MP** (°C):   75
MW: 206.29   **BP** (°C):

Solubility (Moles/L)	Solubility (Grams/L)	Temp (°C)	Ref (#)	Evaluation (T P E A A)	Comments
3.340E-05	6.890E-03	5	F306	1 0 1 2 2	intrinsic
7.271E-05	1.500E-02	20	N316	1 0 1 1 0	EFG
3.102E-04	6.400E-02	21	B331	1 2 2 1 2	pH 7.4
5.478E-05	1.130E-02	25	C314	1 1 2 2 2	
5.560E-05	1.147E-02	25	C314	1 1 2 2 2	
9.430E-04	1.945E-01	25	D345	1 1 2 2 2	
4.300E-05	8.870E-03	25	F301	1 1 0 0 1	pH 2.0, *sic*
4.300E-05	8.870E-03	25	F306	1 0 1 2 2	intrinsic
1.212E-04	2.500E-02	30	N316	1 0 1 1 0	EFG
5.210E-05	1.075E-02	37	F306	1 0 1 2 2	intrinsic
1.551E-04	3.200E-02	37	N316	1 0 1 1 0	EFG
1.600E-04	3.301E-02	50	M335	1 0 2 1 2	pH 5
2.036E-04	4.200E-02	50	N316	1 0 1 1 0	EFG
2.327E-04	4.800E-02	60	N316	1 0 1 1 0	EFG
2.600E-04	5.363E-02	ns	F327	0 0 1 2 2	
1.018E-04	2.100E-02	rt	H302	0 0 2 1 2	intrinsic

2899. $C_{13}H_{18}O_2$
r-Ibuprofen
(R)-2-(4-Isobutylphenyl)propanoic Acid
r-(-)-p-Isobutylhydratropic Acid
l-Ibuprofen
RN: 51146-57-7   **MP** (°C):
MW: 206.29   **BP** (°C):

Solubility (Moles/L)	Solubility (Grams/L)	Temp (°C)	Ref (#)	Evaluation (T P E A A)	Comments
1.790E-03	3.693E-01	25	D345	1 1 2 2 2	

2900.   C$_{13}$H$_{18}$O$_2$
S-Ibuprofen
(S)-(+)-2-(4-Isobutylphenyl)propionic Acid
D-Ibuprofen
Seractil
Dexibuprofen
RN: 51146-56-6   **MP** (°C):
MW: 206.29   **BP** (°C):

Solubility (Moles/L)	Solubility (Grams/L)	Temp (°C)	Ref (#)	Evaluation (T P E A A)	Comments
1.790E-03	3.693E-01	25	D345	1 1 2 2 2	

2901.   C$_{13}$H$_{18}$O$_3$
Hexyl p-Hydroxybenzoate
4-Hydroxybenzoic Acid N-Hexyl Ester
RN: 1083-27-8   **MP** (°C):
MW: 222.29   **BP** (°C):

Solubility (Moles/L)	Solubility (Grams/L)	Temp (°C)	Ref (#)	Evaluation (T P E A A)	Comments
3.680E-04	8.180E-02	15	B355	1 1 1 1 2	
3.810E-04	8.469E-02	20	B355	1 1 1 1 2	
6.190E-04	1.376E-01	25	B355	1 1 1 1 2	
1.704E-03	3.789E-01	25	D081	1 2 2 1 2	
3.162E-04	7.029E-02	25	F322	2 0 1 1 0	EFG

2902.   C$_{13}$H$_{18}$O$_3$
n-Hexyl Salicylate
n-Hexyl 2-hydroxybenzoate
RN: 6259-76-3   **MP** (°C):
MW: 222.29   **BP** (°C):

Solubility (Moles/L)	Solubility (Grams/L)	Temp (°C)	Ref (#)	Evaluation (T P E A A)	Comments
1.260E-03	2.800E-01	37	D009	1 2 1 1 1	0.1N HCl

2903. $C_{13}H_{18}O_5S$
Ethofumesate
2-Ethoxy-2,3-dihydro-3,3-dimethyl-5-benzofuranyl Methanesulfonate
Nortran
Tramat
RN: 26225-79-6   **MP** (°C): 71
MW: 286.35   **BP** (°C):

Solubility (Moles/L)	Solubility (Grams/L)	Temp (°C)	Ref (#)	Evaluation (T P E A A)	Comments
3.841E-04	1.100E-01	25	M161	1 0 0 0 2	
3.841E-04	1.100E-01	25	W313	1 0 0 0 1	

2904. $C_{13}H_{18}O_7$
Salicin
2-(Hydroxymethyl)phenyl-β-D-glucopyranoside
Salicoside
RN: 138-52-3   **MP** (°C): 199
MW: 286.28   **BP** (°C):

Solubility (Moles/L)	Solubility (Grams/L)	Temp (°C)	Ref (#)	Evaluation (T P E A A)	Comments
1.397E-01	4.000E+01	25	F300	1 0 0 0 0	
9.082E-01	2.600E+02	100	F300	1 0 0 0 1	
1.455E-01	4.167E+01	c	D004	1 0 0 0 0	
8.733E-01	2.500E+02	h	D004	1 0 0 0 0	

2905. $C_{13}H_{19}NO_2$
Hexyl p-Aminobenzoate
4-Aminobenzoic Acid Hexyl Ester
RN: 55791-76-9   **MP** (°C):
MW: 221.30   **BP** (°C):

Solubility (Moles/L)	Solubility (Grams/L)	Temp (°C)	Ref (#)	Evaluation (T P E A A)	Comments
1.040E-04	2.302E-02	37	F006	1 1 2 2 2	
4.500E-05	9.959E-03	ns	M066	0 0 0 0 1	
4.300E-05	9.516E-03	rt	B016	0 0 1 1 1	pH 7.4

2906. $C_{13}H_{19}NO_2$
Ibuproxam
2-(4-Isobutylphenyl)propionohydroxamic Acid
Ibudros
RN: 53648-05-8   **MP** (°C): 123
MW: 221.30   **BP** (°C):

Solubility (Moles/L)	Solubility (Grams/L)	Temp (°C)	Ref (#)	Evaluation (T P E A A)	Comments
9.037E-04	2.000E-01	ns	M148	0 2 0 0 0	

2907.   C$_{13}$H$_{19}$NO$_4$
N,N-Diethyl-6-hydroxynorbornane-2-carboxamide-3,5-lactone
RN: **MP** (°C):
MW: 253.30   **BP** (°C):

Solubility (Moles/L)	Solubility (Grams/L)	Temp (°C)	Ref (#)	Evaluation (T P E A A)	Comments
1.153E-01	2.920E+01	20	K050	1 1 1 1 2	

2908.   C$_{13}$H$_{19}$N$_3$O$_4$
N-(1-Ethylpropyl)-2,6-dinitro-3,4-xylidine
Pendimethalin
RN: 40487-42-1   **MP** (°C):   56.5
MW: 281.31   **BP** (°C):

Solubility (Moles/L)	Solubility (Grams/L)	Temp (°C)	Ref (#)	Evaluation (T P E A A)	Comments
1.066E-06	3.000E-04	20	M161	1 0 0 0 0	
1.081E-03	3.040E-01	ns	B185	0 0 0 0 2	

2909.   C$_{13}$H$_{19}$N$_3$O$_6$S
Nitralin
4-(Methylsulfonyl)-2,6-dinitro-N,N-dipropylaniline
RN: 4726-14-1   **MP** (°C):   151
MW: 345.38   **BP** (°C):

Solubility (Moles/L)	Solubility (Grams/L)	Temp (°C)	Ref (#)	Evaluation (T P E A A)	Comments
1.737E-06	6.000E-04	22	M161	1 0 0 0 0	
1.737E-06	6.000E-04	25	B200	1 0 0 0 0	
1.737E-07	6.000E-05	25	P028	1 0 0 0 1	
1.737E-06	6.000E-04	ns	M061	0 0 0 0 0	

2910.   C$_{13}$H$_{20}$N$_2$O$_2$
Procaine
Novacaine
Novokain
RN: 59-46-1   **MP** (°C):   60
MW: 236.32   **BP** (°C):

Solubility (Moles/L)	Solubility (Grams/L)	Temp (°C)	Ref (#)	Evaluation (T P E A A)	Comments
4.000E-02	9.453E+00	30	L068	1 0 0 1 0	EFG
4.200E-02	9.925E+00	37.5	L034	2 2 0 1 2	pH 7.4
5.494E-03	1.298E+00	ns	E031	0 0 2 1 2	
2.700E-02	6.381E+00	ns	M066	0 0 0 0 1	

2911.   C$_{13}$H$_{20}$N$_2$O$_2$
N,N'-Diethyl-bicyclo(2.2.1)hept-5-ene-2,3-trans-dicarboxamide
RN: **MP** (°C):
MW: 236.32   **BP** (°C):

Solubility (Moles/L)	Solubility (Grams/L)	Temp (°C)	Ref (#)	Evaluation (T P E A A)	Comments
3.216E-02	7.600E+00	20	K050	1 1 1 1 2	

2912.   C$_{13}$H$_{20}$N$_2$O$_2$
4-Aminobenzoic Acid-2-(butyl-amino)ethyl Ester
2-(Butylamino)ethyl 4-Aminobenzoate
RN: **MP** (°C):
MW: 236.32   **BP** (°C):

Solubility (Moles/L)	Solubility (Grams/L)	Temp (°C)	Ref (#)	Evaluation (T P E A A)	Comments
1.700E-04	4.017E-02	ns	M066	0 0 0 0 1	

2913.   C$_{13}$H$_{20}$N$_2$O$_3$
5-Allyl-5-ethylbutylbarbituric Acid
RN: **MP** (°C):
MW: 252.32   **BP** (°C):

Solubility (Moles/L)	Solubility (Grams/L)	Temp (°C)	Ref (#)	Evaluation (T P E A A)	Comments
1.587E-02	4.004E+00	20	J030	1 2 2 2 2	
2.579E-02	6.507E+00	37	J030	1 2 2 2 2	

2914.   C$_{13}$H$_{20}$N$_2$O$_3$
2,4,6(1H,3H,5H)-Pyrimidinetrione, 5-(1,1-Dimethylethyl)-5-(3-methyl-2-butenyl)
RN: 143585-02-8   **MP** (°C):
MW: 252.32   **BP** (°C):

Solubility (Moles/L)	Solubility (Grams/L)	Temp (°C)	Ref (#)	Evaluation (T P E A A)	Comments
2.810E-04	7.090E-02	25	P350	2 1 1 1 2	intrinsic

2915.   C$_{13}$H$_{20}$O
2-Hexyl-6-methylphenol
o-Cresol, 6-Hexyl-
RN: 106593-25-3   **MP** (°C):
MW: 192.30   **BP** (°C):

Solubility (Moles/L)	Solubility (Grams/L)	Temp (°C)	Ref (#)	Evaluation (T P E A A)	Comments
2.600E-05	5.000E-03	25	L020	1 0 0 0 0	

2916.   C$_{13}$H$_{20}$O
2-Hexyl-4-methylphenol
2-Hexyl-p-cresol
RN: 54612-53-2   **MP** (°C):
MW: 192.30   **BP** (°C):

Solubility (Moles/L)	Solubility (Grams/L)	Temp (°C)	Ref (#)	Evaluation (T P E A A)	Comments
3.467E-05	6.667E-03	25	L020	1 0 0 0 0	

2917.   C$_{13}$H$_{20}$O
4-Hexyl-2-methylphenol
o-Cresol, 4-Hexyl-
RN: 3280-61-3   **MP** (°C):
MW: 192.30   **BP** (°C):

Solubility (Moles/L)	Solubility (Grams/L)	Temp (°C)	Ref (#)	Evaluation (T P E A A)	Comments
2.600E-05	5.000E-03	25	L020	1 0 0 0 0	

2918.   C$_{13}$H$_{20}$O
p-n-Heptylphenol
4-n-Heptylphenol
RN: 1987-50-4   **MP** (°C):
MW: 192.30   **BP** (°C):

Solubility (Moles/L)	Solubility (Grams/L)	Temp (°C)	Ref (#)	Evaluation (T P E A A)	Comments
5.778E-05	1.111E-02	25	L022	1 0 0 0 0	

2919.   C$_{13}$H$_{20}$O
o-n-Heptylphenol
2-n-Heptylphenol
RN: 5284-22-0   **MP** (°C):
MW: 192.30   **BP** (°C):

Solubility (Moles/L)	Solubility (Grams/L)	Temp (°C)	Ref (#)	Evaluation (T P E A A)	Comments
6.118E-05	1.176E-02	25	L022	1 0 0 0 0	

2920.   C₁₃H₂₁NO₃

Salbutamol
Albuterol
Ventolin
RN: 18559-94-9   **MP** (°C):   151
MW: 239.32   **BP** (°C):

Solubility (Moles/L)	Solubility (Grams/L)	Temp (°C)	Ref (#)	Evaluation (T P E A A)	Comments
7.400E-02	1.771E+01	20	M380	1 0 2 1 0	EFG
7.500E-02	1.795E+01	25	M380	1 0 2 1 0	EFG
7.400E-02	1.771E+01	37	M380	1 0 2 1 0	EFG
5.885E-02	1.408E+01	ns	A092	0 0 0 0 0	

2921.   C₁₃H₂₁O₃PS

S-Benzyl O,O-Di-isopropyl Phosphorothioate
Isokitazine
Kitazin P
IBP
Iprobenfos
Kitazin L
RN: 26087-47-8   **MP** (°C):
MW: 288.35   **BP** (°C):   126

Solubility (Moles/L)	Solubility (Grams/L)	Temp (°C)	Ref (#)	Evaluation (T P E A A)	Comments
3.468E-03	1.000E+00	18	M161	1 0 0 0 0	

2922.   C₁₃H₂₁O₄PS

4-(Methylthio)phenyl Dipropyl Phosphate
O,O-Dipropyl O-4-Methylthiophenyl Phosphate
Propaphos
Kayaphos
Kayphosnac
RN: 7292-16-2   **MP** (°C):
MW: 304.35   **BP** (°C):   176

Solubility (Moles/L)	Solubility (Grams/L)	Temp (°C)	Ref (#)	Evaluation (T P E A A)	Comments
4.107E-04	1.250E-01	25	M161	1 0 0 0 2	

2923.   C₁₃H₂₂NO₃PS

Fenamiphos
1-(Methylethyl)-O-ethyl-O-(3-methyl-4-(methylthio)phenyl)phosphoramidate
Nemacur
Bay 68138

RN: 22224-92-6   **MP** (°C):
MW: 303.36   **BP** (°C):

Solubility (Moles/L)	Solubility (Grams/L)	Temp (°C)	Ref (#)	Evaluation (T P E A A)	Comments
1.008E-03	3.059E-01	10	B324	2 2 2 2 2	
1.009E-03	3.061E-01	10	B324	2 2 2 2 2	
2.291E-03	6.950E-01	20	B179	2 0 0 0 2	
1.084E-03	3.288E-01	20	B300	2 1 1 1 2	
1.085E-03	3.291E-01	20	B324	2 2 2 2 2	
1.084E-03	3.289E-01	20	B324	2 2 2 2 2	
1.381E-03	4.189E-01	30	B324	2 2 2 2 2	
1.381E-03	4.188E-01	30	B324	2 2 2 2 2	
2.307E-03	7.000E-01	rt	M161	0 0 0 0 2	

2924.   C₁₃H₂₂N₂O

Noruron
3-(Hexahydro-4,7-methanoindan-5-yl)-1,1-dimethylurea
Norea

RN: 18530-56-8   **MP** (°C):   171
MW: 222.33   **BP** (°C):

Solubility (Moles/L)	Solubility (Grams/L)	Temp (°C)	Ref (#)	Evaluation (T P E A A)	Comments
6.747E-04	1.500E-01	20	M061	1 0 0 0 2	
6.747E-04	1.500E-01	25	B200	1 0 0 0 2	
6.747E-04	1.500E-01	ns	G036	0 0 0 0 2	

2925.   C₁₃H₂₂N₂O

Isonoruron
Urea, 3-[Hexahydro-4,7-methanoindan-1(or 2)-yl]-1,1-dimethyl-
Tricuron
BAS 2103H

RN: 28346-65-8   **MP** (°C):   165
MW: 222.33   **BP** (°C):

Solubility (Moles/L)	Solubility (Grams/L)	Temp (°C)	Ref (#)	Evaluation (T P E A A)	Comments
9.895E-04	2.200E-01	20	M161	1 0 0 0 2	

2926. C₁₃H₂₂N₂O₃

2926. $C_{13}H_{22}N_2O_3$

5-Ethyl-5-n-heptylbarbituric Acid
5-Ethyl-5-heptylbarbituric Acid
RN: 60784-70-5   **MP** (°C):
MW: 254.33   **BP** (°C):

Solubility (Moles/L)	Solubility (Grams/L)	Temp (°C)	Ref (#)	Evaluation (T P E A A)	Comments
6.050E-04	1.539E-01	25	M310	2 2 2 2 2	

2927. $C_{13}H_{22}O_3$

Methyl Dihydrojasmonate
Hedione
Methyl 3-Oxo-2-pentylcyclopentaneacetate
Claigeon
RN: 24851-98-7   **MP** (°C):
MW: 226.32   **BP** (°C):

Solubility (Moles/L)	Solubility (Grams/L)	Temp (°C)	Ref (#)	Evaluation (T P E A A)	Comments
1.767E-03	3.998E-01	25	M350	1 0 1 1 1	

2928. $C_{13}H_{24}N_3O_3PS$

Pirimiphos-ethyl
Diethyl O-(2-(Diethylamino)-6-methyl-4-pyrimidinyl) phosphorothioate
Fernex
Primotec
Solgard
RN: 23505-41-1   **MP** (°C):
MW: 333.39   **BP** (°C):

Solubility (Moles/L)	Solubility (Grams/L)	Temp (°C)	Ref (#)	Evaluation (T P E A A)	Comments
1.190E-05	3.967E-03	20	B300	2 1 1 1 2	
<3.00E-06	<1.00E-03	30	M161	1 0 0 0 0	

2929. $C_{13}H_{24}N_4O_3S$

Bupirimate
5-Butyl-2-(ethylamino)-6-methyl-4-pyrimidinyl dimethylsulfamate
Nimrod
RN: 41483-43-6   **MP** (°C):   50.5
MW: 316.43   **BP** (°C):

Solubility (Moles/L)	Solubility (Grams/L)	Temp (°C)	Ref (#)	Evaluation (T P E A A)	Comments
6.953E-05	2.200E-02	rt	M161	0 0 0 0 1	

2930.   C$_{13}$H$_{24}$N$_6$
1-(Hexamethyleneiminel)-3,5-bis(dimethylamino)-s-triazine
1,3,5-Triazine-2,4-diamine, 6-(Hexahydro-1H-azepin-1-yl)-N,N,N',N'-tetramethyl-
RN: 125867-92-7   **MP** (°C):
MW: 264.38   **BP** (°C):

Solubility (Moles/L)	Solubility (Grams/L)	Temp (°C)	Ref (#)	Evaluation (T P E A A)	Comments
2.265E-05	5.988E-03	25	B386	2 2 2 2 2	

2931.   C$_{13}$H$_{24}$O$_4$
1,11-Undecanedicarboxylic Acid
1,13-Tridecanedioic Acid
Brassylic Acid
RN: 505-52-2   **MP** (°C):   111
MW: 244.33   **BP** (°C):

Solubility (Moles/L)	Solubility (Grams/L)	Temp (°C)	Ref (#)	Evaluation (T P E A A)	Comments
6.139E-03	1.500E+00	21	B040	1 0 1 1 1	*sic*
1.637E-04	4.000E-02	24	F300	1 0 0 0 0	*sic*

2932.   C$_{13}$H$_{24}$O$_4$
Octyl α-Acetoxypropionate
Propanoic Acid, 2-(Acetyloxy)-, Octyl Ester
RN: 6283-90-5   **MP** (°C):
MW: 244.33   **BP** (°C):

Solubility (Moles/L)	Solubility (Grams/L)	Temp (°C)	Ref (#)	Evaluation (T P E A A)	Comments
4.093E-04	1.000E-01	25	R006	2 2 0 1 1	

2933.   C$_{13}$H$_{25}$NO$_3$
Dibutylaceturethane
RN: **MP** (°C):
MW: 243.35   **BP** (°C):

Solubility (Moles/L)	Solubility (Grams/L)	Temp (°C)	Ref (#)	Evaluation (T P E A A)	Comments
3.287E-04	7.999E-02	44	O021	1 2 0 0 0	

2934. C₁₃H₂₆N₂O₂

N,N,N',N'-Tetramethylazelamide
Nonanediamide, N,N,N',N'-Tetramethyl-
RN: 13424-87-8 **MP** (°C):
MW: 242.36 **BP** (°C):

Solubility (Moles/L)	Solubility (Grams/L)	Temp (°C)	Ref (#)	Evaluation (T P E A A)	Comments
3.900E+00	9.452E+02	30	D010	1 2 1 1 2	

2935. C₁₃H₂₆O₂

Methyl Laurate
Dodecanoic Acid Methyl Ester
Methyl Dodecanoate
RN: 111-82-0 **MP** (°C): 41
MW: 214.35 **BP** (°C): 261

Solubility (Moles/L)	Solubility (Grams/L)	Temp (°C)	Ref (#)	Evaluation (T P E A A)	Comments
<2.05E-05	<4.40E-03	20	M337	2 1 2 2 1	

2936. C₁₃H₂₆O₂

n-Tridecanoic Acid
Tridecanoic Acid
RN: 638-53-9 **MP** (°C): 41.5
MW: 214.35 **BP** (°C): 236

Solubility (Moles/L)	Solubility (Grams/L)	Temp (°C)	Ref (#)	Evaluation (T P E A A)	Comments
9.797E-05	2.100E-02	0	B136	1 0 2 1 1	
9.797E-05	2.100E-02	0.0	R001	1 1 1 1 1	
1.540E-04	3.300E-02	20	B136	1 0 2 1 1	
1.539E-04	3.300E-02	20.0	R001	1 1 1 1 1	
1.773E-04	3.800E-02	30	B136	1 0 2 1 1	
1.773E-04	3.800E-02	30.0	R001	1 1 1 1 1	
2.053E-04	4.400E-02	45	B136	1 0 2 1 1	
2.053E-04	4.400E-02	45.0	R001	1 1 1 1 1	
2.519E-04	5.400E-02	60	B136	1 0 2 1 1	
2.519E-04	5.400E-02	60.0	R001	1 1 1 1 1	

2937. C₁₃H₂₆O₃

Decyl Lactate
2-Hydroxypropionic Acid Decyl Ester
RN: 42175-34-8 **MP** (°C):
MW: 230.35 **BP** (°C):

Solubility (Moles/L)	Solubility (Grams/L)	Temp (°C)	Ref (#)	Evaluation (T P E A A)	Comments
8.682E-04	2.000E-01	25	R006	2 2 0 1 0	

2938.   C$_{13}$H$_{26}$O$_3$
n-Octyl β-Ethoxypropionate
RN: **MP** (°C):
MW: 230.35   **BP** (°C):

Solubility (Moles/L)	Solubility (Grams/L)	Temp (°C)	Ref (#)	Evaluation (T P E A A)	Comments
≤4.34E-04	<10.00E-02	25	D002	1 2 1 1 0	

2939.   C$_{13}$H$_{26}$O$_4$
1,3-Dioxolane-4-methanol, 2-[2-(Hexyloxy)ethyl]-2-methyl
RN: 124485-63-8   **MP** (°C):
MW: 246.35   **BP** (°C):

Solubility (Moles/L)	Solubility (Grams/L)	Temp (°C)	Ref (#)	Evaluation (T P E A A)	Comments
1.600E-02	3.942E+00	25	P342	1 2 2 2 2	0.0001M Na$_2$CO$_3$

2940.   C$_{14}$H$_4$N$_2$O$_2$S$_2$
Dithianon
1,4-Dithiaanthraquinone-2,3-dinitrile
2,3-Dicyano-1,4-dithiaantraquinone
RN: 3347-22-6   **MP** (°C):   225
MW: 296.33   **BP** (°C):

Solubility (Moles/L)	Solubility (Grams/L)	Temp (°C)	Ref (#)	Evaluation (T P E A A)	Comments
1.687E-06	5.000E-04	ns	A305	0 0 0 0 0	

2941.   C$_{14}$H$_7$ClO$_5$S
1,5-Chloroanthraquinone Sulfonic Acid
1-Anthracenesulfonic Acid, 5-Chloro-9,10-dihydro-9,10-dioxo-
RN: **MP** (°C):
MW: 322.73   **BP** (°C):

Solubility (Moles/L)	Solubility (Grams/L)	Temp (°C)	Ref (#)	Evaluation (T P E A A)	Comments
1.033E+00	3.333E+02	18	F047	1 2 1 1 0	

2942.   C$_{14}$H$_7$ClO$_5$S
1,6-Chloroanthraquinone Sulfonic Acid
2-Anthracenesulfonic Acid, 5-Chloro-9,10-dihydro-9,10-dioxo-
RN: 300360-23-0   **MP** (°C):
MW: 322.73   **BP** (°C):

Solubility (Moles/L)	Solubility (Grams/L)	Temp (°C)	Ref (#)	Evaluation (T P E A A)	Comments
6.197E-01	2.000E+02	18	F047	1 2 1 1 0	

2943. C$_{14}$H$_7$ClO$_5$S
1,7-Chloroanthraquinone Sulfonic Acid
RN: **MP** (°C):
MW: 322.73   **BP** (°C):

Solubility (Moles/L)	Solubility (Grams/L)	Temp (°C)	Ref (#)	Evaluation (T P E A A)	Comments
6.197E-01	2.000E+02	18	F047	1 2 1 1 0	

2944. C$_{14}$H$_8$Cl$_4$
p,p'-Dichlorodiphenyldichloroethylene
2,2-bis(4-Chlorophenyl)-1,1-dichloroethylene
p,p'-DDE
RN: 72-55-9   **MP** (°C):   89.0
MW: 318.03   **BP** (°C):

Solubility (Moles/L)	Solubility (Grams/L)	Temp (°C)	Ref (#)	Evaluation (T P E A A)	Comments
1.729E-07	5.500E-05	15	B083	2 2 1 2 1	particle size ≤ 5 µm
1.258E-07	4.000E-05	20	C053	1 0 2 2 1	
1.258E-07	4.000E-05	20	F071	1 1 2 1 1	
3.773E-07	1.200E-04	25	B083	2 2 1 2 2	particle size < 5 µm
3.773E-07	1.200E-04	25	I308	0 0 0 0 1	
4.088E-09	1.300E-06	25	M134	1 2 1 1 1	
4.402E-08	1.400E-05	25	W025	1 0 1 1 1	
7.389E-07	2.350E-04	35	B083	2 2 1 2 2	particle size ≤ 5 µm
1.415E-06	4.500E-04	45	B083	2 2 1 2 2	particle size ≤ 5 µm
4.717E-09	1.500E-06	ns	M110	0 0 0 0 0	EFG
4.088E-09	1.300E-06	ns	M118	0 1 1 1 1	

2945. C$_{14}$H$_8$Cl$_4$
2,4'-Dichlorodiphenyldichloroethylene
1-(2-Chlorophenyl)-1-(4-chlorophenyl)-2,2-dichloroethylene
o,p'-DDE
RN: 3424-82-6   **MP** (°C):   76.5
MW: 318.03   **BP** (°C):

Solubility (Moles/L)	Solubility (Grams/L)	Temp (°C)	Ref (#)	Evaluation (T P E A A)	Comments
4.402E-07	1.400E-04	25	B083	2 2 1 2 2	particle size ≤ 5 µm

2946.   C$_{14}$H$_8$O$_2$
Anthraquinone
9,10-Anthraquinone
9,10-Dioxoanthracene
Corbit
Morkit
Hoelite

RN:	84-65-1	**MP** (°C):	286	
MW:	208.22	**BP** (°C):	377	

Solubility (Moles/L)	Solubility (Grams/L)	Temp (°C)	Ref (#)	Evaluation (T P E A A)	Comments
6.500E-06	1.353E-03	25	E014	2 2 2 1 1	pH 7.3
3.000E-06	6.247E-04	ns	G077	0 0 0 0 1	

2947.   C$_{14}$H$_8$O$_4$
Alizarin
Alizarine
C.I. Mordant Red 11

RN:	72-48-0	**MP** (°C):	290	
MW:	240.22	**BP** (°C):	430	

Solubility (Moles/L)	Solubility (Grams/L)	Temp (°C)	Ref (#)	Evaluation (T P E A A)	Comments
1.300E-05	3.123E-03	25	B333	1 0 0 0 1	*sic*
1.664E-03	3.998E-01	rt	D021	0 0 1 1 1	*sic*

2948.   C$_{14}$H$_8$O$_4$
Quinizarin
1,4-Dihydroxyanthraquinone
C.I. Pigment Violet 12

RN:	81-64-1	**MP** (°C):	192	
MW:	240.22	**BP** (°C):		

Solubility (Moles/L)	Solubility (Grams/L)	Temp (°C)	Ref (#)	Evaluation (T P E A A)	Comments
4.000E-07	9.609E-05	25	B333	1 0 0 0 1	
6.000E-05	1.441E-02	98.59	M180	0 0 2 2 0	EFG
9.200E-05	2.210E-02	111.46	M180	0 0 2 2 0	EFG
1.100E-04	2.642E-02	117.47	M180	0 0 2 2 0	EFG
1.800E-04	4.324E-02	123.67	M180	0 0 2 2 0	EFG
2.000E-04	4.804E-02	126.84	M180	0 0 2 2 0	EFG
2.100E-04	5.045E-02	135.00	M180	0 0 2 2 0	EFG
4.900E-04	1.177E-01	141.78	M180	0 0 2 2 0	EFG
7.500E-04	1.802E-01	152.37	M180	0 0 2 2 0	EFG

2949. $C_{14}H_8O_5$
Purpurin
1,2,4-Trihydroxy-anthrachinon
RN: 81-54-9   **MP** (°C):
MW: 256.22   **BP** (°C):

Solubility (Moles/L)	Solubility (Grams/L)	Temp (°C)	Ref (#)	Evaluation (T P E A A)	Comments
2.500E-05	6.405E-03	25	B333	1 0 0 0 1	

2950. $C_{14}H_8O_6$
Quinalizarin
1,2,5,8-Tetrahydroxyanthraquinone
9,10-Anthracenedione
Alizarine Bordeaux B
Mordant Violet 26
RN: 81-61-8   **MP** (°C):
MW: 272.22   **BP** (°C):

Solubility (Moles/L)	Solubility (Grams/L)	Temp (°C)	Ref (#)	Evaluation (T P E A A)	Comments
9.500E-06	2.586E-03	25	B333	1 0 0 0 1	

2951. $C_{14}H_8O_8S_2$
Anthraquinone-1,8-disulfonic Acid
1,8-Disulfonic Acid Anthraquinone
Anthrachinon-disulfosaeure-(1,8)
1,8-Anthraquinone Disulfonic Acid
RN: 82-48-4   **MP** (°C):   293
MW: 368.34   **BP** (°C):

Solubility (Moles/L)	Solubility (Grams/L)	Temp (°C)	Ref (#)	Evaluation (T P E A A)	Comments
1.086E+00	4.000E+02	18	F047	1 2 1 1 1	

2952. $C_{14}H_8O_8S_2$
1,6-Anthraquinone Disulfonic Acid
Anthraquinone-1,6-Disulfonic Acid
RN: 14486-58-9   **MP** (°C):   216
MW: 368.34   **BP** (°C):

Solubility (Moles/L)	Solubility (Grams/L)	Temp (°C)	Ref (#)	Evaluation (T P E A A)	Comments
1.357E+00	5.000E+02	18	F047	1 2 1 1 0	

2953. $C_{14}H_8O_8S_2$
1,5-Anthraquinone Disulfonic Acid
Anthraquinone-1,5-Disulfonic Acid
RN: 252967-17-2   **MP** (°C):   310.0
MW: 368.34   **BP** (°C):

Solubility (Moles/L)	Solubility (Grams/L)	Temp (°C)	Ref (#)	Evaluation (T P E A A)	Comments
1.086E+00	4.000E+02	18	F047	1 2 1 1 1	

2954. $C_{14}H_9ClF_2N_2O_2$
Difluron
Diflubenzuron
TH 6040
RN: 35367-38-5   **MP** (°C):   239
MW: 310.69   **BP** (°C):

Solubility (Moles/L)	Solubility (Grams/L)	Temp (°C)	Ref (#)	Evaluation (T P E A A)	Comments
6.437E-07	2.000E-04	20	M161	1 0 0 0 0	
6.437E-07	2.000E-04	20	R303	1 0 0 0 0	
9.656E-07	3.000E-04	24	C105	2 1 2 2 2	
1.609E-06	5.000E-04	ns	M110	0 0 0 0 0	EFG

2955. $C_{14}H_9Cl_2NO_5$
Bifenox
5-(2,4-Dichlorphenoxy)-2-nitro-benzoic Acid Methyl Ester
Modown 4 Flowable
Modown
RN: 42576-02-3   **MP** (°C):   85
MW: 342.14   **BP** (°C):

Solubility (Moles/L)	Solubility (Grams/L)	Temp (°C)	Ref (#)	Evaluation (T P E A A)	Comments
1.461E-06	5.000E-04	ns	M110	0 0 0 0 0	EFG
1.023E-06	3.500E-04	ns	M161	0 0 0 0 1	

2956. $C_{14}H_9Cl_5$

o,p'-DDT
1-(2-Chlorophenyl)-1-(4-chlorophenyl)-2,2,2-trichloroethane
2,4'-DDT
2-(2-Chlorophenyl)-2-(4-chlorophenyl)-1,1,1-trichloroethane

RN: 789-02-6   **MP** (°C):   74.0
MW: 354.49   **BP** (°C):

Solubility (Moles/L)	Solubility (Grams/L)	Temp (°C)	Ref (#)	Evaluation (T P E A A)	Comments
1.410E-07	5.000E-05	15	B083	2 2 1 2 1	particle size ≤ 5 μm
2.398E-07	8.500E-05	25	B083	2 2 1 2 1	particle size ≤ 5 μm
2.398E-07	8.500E-05	25	I308	0 0 0 0 1	
7.334E-08	2.600E-05	25	W025	1 0 2 2 1	
3.808E-07	1.350E-04	35	B083	2 2 1 2 2	particle size ≤ 5 μm
5.642E-07	2.000E-04	45	B083	2 2 1 2 2	particle size ≤ 5 μm

2957. $C_{14}H_9Cl_5$

p,p'-DDT
2,2-bis(p-Chlorophenyl)-1,1,1-trichloroethane
p,p'-TDEE

RN: 50-29-3   **MP** (°C):   108.5
MW: 354.49   **BP** (°C):   260

Solubility (Moles/L)	Solubility (Grams/L)	Temp (°C)	Ref (#)	Evaluation (T P E A A)	Comments
3.385E-09	1.200E-06	0	G319	1 0 0 0 2	
1.664E-08	5.900E-06	2	B186	2 0 2 2 2	
4.796E-08	1.700E-05	15	B083	2 2 1 2 1	particle size ≤ 5 μm
1.834E-07	6.500E-05	15	B083	2 2 1 2 1	particle size ≤ 5 μm
2.800E-07	9.926E-05	18	G054	1 0 1 0 1	
1.410E-08	5.000E-06	20	C111	1 0 0 0 0	
1.410E-08	5.000E-06	20	C113	1 0 2 1 1	
1.128E-07	4.000E-05	20	E048	1 2 1 1 0	
2.172E-08	7.700E-06	20	F303	1 2 1 2 1	
2.172E-08	7.700E-06	20	W319	1 2 1 2 1	
1.552E-08	5.500E-06	24	C311	1 0 2 2 1	EFG
1.523E-08	5.400E-06	24	C313	1 0 2 2 2	
2.821E-09	1.000E-06	24	K069	2 0 0 1 1	
3.385E-09	1.200E-06	25	B036	1 1 0 1 1	
3.949E-07	1.400E-04	25	B083	2 2 1 2 2	particle size ≤ 5 μm
7.052E-08	2.500E-05	25	B083	2 2 1 2 1	particle size ≤ 5 μm
4.796E-09	1.700E-06	25	B093	2 2 2 2 1	
1.055E-07	3.740E-05	25	B186	2 0 2 2 2	
9.168E-09	3.250E-06	25	F071	1 1 2 1 1	
3.385E-09	1.200E-06	25	M040	1 0 0 1 1	
3.385E-09	1.200E-06	25	M130	1 0 0 0 1	
2.821E-09	1.000E-06	25	P085	1 0 1 1 1	
1.552E-08	5.500E-06	25	W025	1 0 2 2 1	
3.385E-09	1.200E-06	26.70	L095	2 2 1 1 2	

1.044E-07	3.700E-05	35	B083	2 2 1 2 1	particle size ≤ 5 μm
7.334E-07	2.600E-04	35	B083	2 2 1 2 2	particle size ≤ 5 μm
1.269E-07	4.500E-05	37.50	B186	2 0 2 2 2	
1.269E-07	4.500E-05	45	B083	2 2 1 2 1	particle size ≤ 5 μm
1.439E-06	5.100E-04	45	B083	2 2 1 2 2	particle size ≤ 5 μm
1.552E-08	5.500E-06	ns	C318	0 2 2 1 2	
3.385E-09	1.200E-06	ns	I300	0 0 0 0 1	
4.796E-09	1.700E-06	ns	K138	0 0 0 0 2	
2.821E-09	1.000E-06	ns	M061	0 0 0 0 0	
3.103E-09	1.100E-06	ns	M110	0 0 0 0 0	EFG
5.642E-09	2.000E-06	ns	M138	0 0 0 0 0	
8.745E-09	3.100E-06	ns	M344	0 0 0 0 1	

2958. $C_{14}H_9Cl_5O$

Dicofol
4-Chloro-α-(4-chlorophenyl)-α-(trichloromethyl)benzenemethanol
4,4'-Dichloro-α-(trichloromethyl)benzhydrol
Acarin
Carbox
Cekudifol

RN: 115-32-2   **MP** (°C):   79
MW: 370.49   **BP** (°C):

Solubility (Moles/L)	Solubility (Grams/L)	Temp (°C)	Ref (#)	Evaluation (T P E A A)	Comments
3.563E-06	1.320E-03	25	W025	1 0 2 2 2	

2959. $C_{14}H_9F$

1-Fluoroanthracene

RN: 7651-80-1   **MP** (°C):
MW: 196.23   **BP** (°C):

Solubility (Moles/L)	Solubility (Grams/L)	Temp (°C)	Ref (#)	Evaluation (T P E A A)	Comments
1.325E-06	2.600E-04	ns	M344	0 0 0 0 2	

2960. $C_{14}H_9NO_2$

2-Phenyl-3,1-benzoxazin-4-one
Bentranil
Linarotox
Linurotox

RN: 1022-46-4   **MP** (°C):   123.5
MW: 223.23   **BP** (°C):

Solubility (Moles/L)	Solubility (Grams/L)	Temp (°C)	Ref (#)	Evaluation (T P E A A)	Comments
2.464E-05	5.500E-03	20	M161	1 0 0 0 0	

2961. C$_{14}$H$_9$NO$_2$
1-Aminoanthraquinone
1-Amino-9,10-anthracenedione
1-Amino-9,10-anthraquinone

RN: 82-45-1   **MP** (°C):   254
MW: 223.23   **BP** (°C):

Solubility (Moles/L)	Solubility (Grams/L)	Temp (°C)	Ref (#)	Evaluation (T P E A A)	Comments
1.400E-06	3.125E-04	25	B333	1 0 0 0 1	

2962. C$_{14}$H$_9$NO$_2$
2-Aminoanthraquinone
2-Amino-9,10-Anthracenedione
2-Amino-9,10-anthraquinone
Aminoanthraquinone
AAQ

RN: 117-79-3   **MP** (°C):   310
MW: 223.23   **BP** (°C):

Solubility (Moles/L)	Solubility (Grams/L)	Temp (°C)	Ref (#)	Evaluation (T P E A A)	Comments
7.300E-07	1.630E-04	25	B333	1 0 0 0 1	

2963. C$_{14}$H$_9$NO$_2$S
4-Benzoyl Phenylisothiocyanate
4-Isothiocyanatobenzophenone

RN: 26328-59-6   **MP** (°C):
MW: 255.30   **BP** (°C):

Solubility (Moles/L)	Solubility (Grams/L)	Temp (°C)	Ref (#)	Evaluation (T P E A A)	Comments
1.400E-05	3.574E-03	25	K032	2 2 0 1 1	

2964. C$_{14}$H$_9$NO$_3$
1-Amino-4-hydroxyanthraquinone
C.I. Disperse Red 15
Disperse Red 15
Celliton Fast Pink B

RN: 116-85-8   **MP** (°C):   208
MW: 239.23   **BP** (°C):

Solubility (Moles/L)	Solubility (Grams/L)	Temp (°C)	Ref (#)	Evaluation (T P E A A)	Comments
1.200E-06	2.871E-04	25	B333	1 0 0 0 1	
1.129E-05	2.700E-03	60	P313	1 2 1 2 2	average of 2
1.797E-05	4.300E-03	70	P313	1 2 1 2 2	average of 2
2.320E-05	5.550E-03	80	P313	1 2 1 2 2	average of 2
4.828E-05	1.155E-02	90	P313	1 2 1 2 2	average of 2
1.500E-04	3.588E-02	98.59	M180	0 0 2 2 0	EFG

2.500E-04	5.981E-02	111.46	M180	0 0 2 2 0	EFG
3.000E-04	7.177E-02	114.44	M180	0 0 2 2 0	EFG
4.500E-04	1.077E-01	122.10	M180	0 0 2 2 0	EFG
6.000E-04	1.435E-01	126.84	M180	0 0 2 2 0	EFG
6.500E-04	1.555E-01	130.07	M180	0 0 2 2 0	EFG
1.500E-03	3.588E-01	152.37	M180	0 0 2 2 0	EFG

2965. $C_{14}H_{10}$
Anthracene
Paranaphthalene
Anthracin
Green Oil
Anthraxcene

RN: 120-12-7 **MP** (°C): 218
MW: 178.24 **BP** (°C): 342

Solubility (Moles/L)	Solubility (Grams/L)	Temp (°C)	Ref (#)	Evaluation (T P E A A)	Comments
7.125E-08	1.270E-05	5.20	M063	2 1 2 2 2	
7.100E-08	1.265E-05	5.20	M082	1 1 1 2 1	
7.100E-08	1.265E-05	5.20	M151	2 1 2 2 1	
7.133E-08	1.271E-05	5.24	M183	1 2 1 1 2	
9.094E-08	1.621E-05	9.74	M183	1 2 1 1 2	
9.818E-08	1.750E-05	10.00	M063	2 1 2 2 2	
9.800E-08	1.747E-05	10.00	M082	1 1 1 2 1	
9.800E-08	1.747E-05	10.00	M151	2 1 2 2 1	
9.828E-08	1.752E-05	10.04	M183	1 2 1 1 2	
1.246E-07	2.220E-05	14.10	M063	2 1 2 2 2	
1.250E-07	2.228E-05	14.10	M082	1 1 1 2 2	
1.250E-07	2.228E-05	14.10	M151	2 1 2 2 2	
1.247E-07	2.223E-05	14.14	M183	1 2 1 1 2	
1.212E-07	2.160E-05	15	B385	2 0 2 2 2	
1.409E-07	2.512E-05	16.64	M183	1 2 1 1 2	
1.633E-07	2.910E-05	18.30	M063	2 1 2 2 2	
1.630E-07	2.905E-05	18.30	M082	1 1 1 2 2	
1.630E-07	2.905E-05	18.30	M151	2 1 2 2 2	
1.634E-07	2.912E-05	18.34	M183	1 2 1 1 2	
2.400E-07	4.278E-05	20	E009	1 0 0 0 1	
2.240E-07	3.992E-05	20	E025	1 0 2 2 2	
1.851E-07	3.300E-05	20	H300	1 1 2 2 1	
2.087E-07	3.720E-05	22.40	M063	2 1 2 2 2	
2.090E-07	3.725E-05	22.40	M082	1 1 1 2 2	
2.090E-07	3.725E-05	22.40	M151	2 1 2 2 2	
2.089E-07	3.723E-05	22.44	M183	1 2 1 1 2	
2.974E-07	5.300E-05	22.5	G301	2 1 0 1 2	
3.927E-07	7.000E-05	23	P332	2 1 1 2 2	
3.927E-07	7.000E-05	23	P339	2 0 1 2 2	
2.123E-07	3.784E-05	23.24	M183	1 2 1 1 2	

2.435E-07	4.340E-05	24.60	M063	2 1 2 2 2	
2.440E-07	4.349E-05	24.60	M082	1 1 1 2 2	
2.440E-07	4.349E-05	24.60	M151	2 1 2 2 2	
2.437E-07	4.344E-05	24.64	M183	1 2 1 1 2	
2.500E-07	4.456E-05	25	A325	2 1 2 2 1	
2.188E-07	3.900E-05	25	B319	2 0 1 2 1	average of 2
2.174E-07	3.875E-05	25	B385	2 0 2 2 2	
4.470E-07	7.967E-05	25	K001	2 2 2 2 2	
3.800E-07	6.773E-05	25	K123	1 0 2 2 1	
4.152E-07	7.400E-05	25	L301	1 1 2 2 2	
3.927E-07	7.000E-05	25	L332	1 1 1 1 2	
4.096E-07	7.300E-05	25	M064	1 1 2 2 1	
4.100E-06	7.308E-04	25	M342	1 0 1 1 2	
1.683E-07	3.000E-05	25	S227	1 2 1 1 1	
4.211E-07	7.506E-05	25	T066	1 0 0 0 2	
2.500E-07	4.456E-05	25	W300	2 2 2 2 2	
2.502E-07	4.460E-05	25.00	M151	2 1 1 2 2	
4.208E-07	7.500E-05	27	D003	1 0 0 1 1	
3.125E-07	5.570E-05	28.70	M063	2 1 2 2 2	
3.130E-07	5.579E-05	28.70	M082	1 1 1 2 2	
3.130E-07	5.579E-05	28.70	M151	2 1 2 2 2	
3.128E-07	5.575E-05	28.74	M183	1 2 1 1 2	
3.198E-07	5.700E-05	29	M071	2 2 2 2 2	
3.198E-07	5.700E-05	29.00	M151	2 1 1 2 2	
3.212E-07	5.724E-05	29.34	M183	1 2 1 1 2	
3.512E-07	6.260E-05	35	B385	2 0 2 2 2	
6.845E-07	1.220E-04	35.40	W003	2 2 2 2 2	average of 3
8.416E-07	1.500E-04	39.30	W003	2 2 2 2 2	average of 3
1.167E-06	2.080E-04	44.70	W003	2 2 2 2 2	average of 3
1.565E-06	2.790E-04	47.50	W003	2 2 2 2 2	
1.683E-06	3.000E-04	50.10	W003	2 2 2 2 2	average of 3
2.211E-06	3.940E-04	54.70	W003	2 2 2 2 2	average of 3
2.794E-06	4.980E-04	59.20	W003	2 2 2 2 2	average of 3
3.703E-06	6.600E-04	64.50	W003	2 2 2 2 1	average of 3
3.703E-06	6.600E-04	65.10	W003	2 2 2 2 1	average of 3
5.162E-06	9.200E-04	69.80	W003	2 2 2 2 1	
5.274E-06	9.400E-04	70.70	W003	2 2 2 2 1	average of 3
5.106E-06	9.100E-04	71.90	W003	2 2 2 2 2	
6.677E-06	1.190E-03	74.70	W003	2 2 2 2 2	average of 3
2.356E-07	4.200E-05	ns	H123	0 0 0 0 2	
1.800E-07	3.208E-05	ns	H306	1 0 1 2 1	
4.096E-07	7.300E-05	ns	K304	0 0 0 0 1	
4.096E-07	7.300E-05	ns	M344	0 0 0 0 2	
5.000E-07	8.912E-05	ns	W005	0 0 1 2 0	

2966. C_{14}H_{10}
Phenanthrene
Phenanthracene
RN: 85-01-8   **MP** (°C):   100
MW: 178.24   **BP** (°C):   340

Solubility (Moles/L)	Solubility (Grams/L)	Temp (°C)	Ref (#)	Evaluation (T P E A A)	Comments
1.462E-06	2.607E-04	-0.7	N053	1 0 0 1 0	EFG
1.970E-06	3.511E-04	4.00	M082	1 1 1 2 2	
1.970E-06	3.511E-04	4.00	M151	2 1 2 2 2	
2.027E-06	3.613E-04	4.04	M183	1 2 1 1 2	
2.265E-06	4.037E-04	4.62	N053	1 0 0 1 0	EFG
2.373E-06	4.230E-04	8.50	M063	2 1 2 2 2	
2.370E-06	4.224E-04	8.50	M082	1 1 1 2 2	
2.370E-06	4.224E-04	8.50	M151	2 1 2 2 2	
2.375E-06	4.233E-04	8.54	M183	1 2 1 1 2	
2.626E-06	4.680E-04	10.00	M063	2 1 2 2 2	
2.630E-06	4.688E-04	10.00	M082	1 1 1 2 2	
2.630E-06	4.688E-04	10.00	M151	2 1 2 2 2	
2.628E-06	4.684E-04	10.04	M183	1 2 1 1 2	
3.055E-06	5.446E-04	10.13	N053	1 0 0 1 0	EFG
2.873E-06	5.120E-04	12.50	M063	2 1 2 2 2	
2.870E-06	5.115E-04	12.50	M082	1 1 1 2 2	
2.870E-06	5.115E-04	12.50	M151	2 1 2 2 2	
2.875E-06	5.124E-04	12.54	M183	1 2 1 1 2	
3.759E-06	6.700E-04	14.20	N053	1 0 0 1 0	EFG
3.372E-06	6.010E-04	15.00	M063	2 1 2 2 2	
3.370E-06	6.007E-04	15.00	M082	1 1 1 2 2	
3.370E-06	6.007E-04	15.00	M151	2 1 2 2 2	
3.375E-06	6.015E-04	15.04	M183	1 2 1 1 2	
1.500E-05	2.674E-03	20	E025	1 0 2 2 2	
6.200E-06	1.105E-03	20	H306	1 0 1 2 1	
4.420E-06	7.878E-04	20.00	M082	1 1 1 2 2	
4.420E-06	7.878E-04	20.00	M151	2 1 2 2 2	
4.419E-06	7.877E-04	20.04	M183	1 2 1 1 2	
4.578E-06	8.160E-04	21.00	M063	2 1 2 2 2	
4.580E-06	8.163E-04	21.00	M082	1 1 1 2 2	
4.580E-06	8.163E-04	21.00	M151	2 1 2 2 2	
4.582E-06	8.167E-04	21.04	M183	1 2 1 1 2	
5.582E-06	9.950E-04	24.30	M063	2 1 2 2 2	
5.360E-06	9.553E-04	24.30	M082	1 1 1 2 2	
5.360E-06	9.553E-04	24.30	M151	2 1 2 2 2	
5.363E-06	9.558E-04	24.34	M183	1 2 1 1 2	
6.284E-06	1.120E-03	24.60	W003	2 2 2 2 2	average of 2
5.577E-06	9.940E-04	25	A001	1 2 2 2 2	
6.059E-06	1.080E-03	25	B319	2 0 1 2 1	
6.003E-06	1.070E-03	25	E004	2 1 2 2 2	

9.000E-06	1.604E-03	25	K001	2 2 2 2 0	
5.611E-06	1.000E-03	25	L332	1 1 1 1 1	
7.238E-06	1.290E-03	25	M064	1 1 2 2 2	
6.620E-06	1.180E-03	25	M342	1 0 1 1 2	
3.815E-06	6.800E-04	25	P340	1 1 2 2 1	
7.278E-06	1.297E-03	25	T066	1 0 0 0 2	
5.610E-06	9.999E-04	25	W300	2 2 2 2 2	
5.622E-06	1.002E-03	25.00	M151	2 1 1 2 2	
6.800E-06	1.212E-03	25.04	V013	2 2 2 2 2	
5.690E-06	1.014E-03	25.35	N053	1 0 0 1 0	EFG
8.977E-06	1.600E-03	27	D003	1 0 0 1 1	
9.257E-06	1.650E-03	27	D043	2 0 0 0 2	average of 2
7.854E-06	1.400E-03	28.95	N053	1 0 0 1 0	EFG
6.845E-06	1.220E-03	29	M071	2 2 2 2 2	
6.845E-06	1.220E-03	29.00	M151	2 1 1 2 2	
7.165E-06	1.277E-03	29.90	M063	2 1 2 2 2	
7.160E-06	1.276E-03	29.90	M082	1 1 1 2 2	
7.160E-06	1.276E-03	29.90	M151	2 1 2 2 2	
8.360E-06	1.490E-03	29.90	W003	2 2 2 2 2	
6.867E-06	1.224E-03	29.94	M183	1 2 1 1 2	
8.304E-06	1.480E-03	30.30	W003	2 2 2 2 2	average of 2
1.035E-05	1.845E-03	34.53	N053	1 0 0 1 0	EFG
1.375E-05	2.450E-03	38.40	W003	2 2 2 2 2	average of 2
1.274E-05	2.270E-03	40.10	W003	2 2 2 2 2	average of 3
2.171E-05	3.870E-03	47.50	W003	2 2 2 2 2	average of 3
2.429E-05	4.330E-03	50.10	W003	2 2 2 2 2	average of 3
2.289E-05	4.080E-03	50.20	W003	2 2 2 2 2	average of 3
3.164E-05	5.640E-03	54.70	W003	2 2 2 2 2	average of 3
4.034E-05	7.190E-03	59.20	W003	2 2 2 2 2	average of 3
4.096E-05	7.300E-03	60.50	W003	2 2 2 2 1	average of 3
5.498E-05	9.800E-03	65.10	W003	2 2 2 2 1	average of 3
7.013E-05	1.250E-02	70.70	W003	2 2 2 2 2	average of 3
7.238E-05	1.290E-02	71.90	W003	2 2 2 2 2	
8.528E-05	1.520E-02	73.40	W003	2 2 2 2 2	
7.238E-06	1.290E-03	ns	H123	0 0 0 0 2	
7.238E-06	1.290E-03	ns	K304	0 0 0 0 2	
7.238E-06	1.290E-03	ns	M344	0 0 0 0 2	
1.500E-05	2.674E-03	ns	W005	0 0 1 2 1	

2967. $C_{14}H_{10}Cl_2O_3$
Fenclofenac
Benzeneacetic Acid, 2-(2,4-Dichlorophenoxy)-
RX 67408
RN: 34645-84-6   **MP** (°C):   136
MW: 297.14   **BP** (°C):

Solubility (Moles/L)	Solubility (Grams/L)	Temp (°C)	Ref (#)	Evaluation (T P E A A)	Comments
2.840E-05	8.439E-03	25	C314	1 1 2 2 2	
2.827E-05	8.400E-03	25	C314	1 1 2 2 2	

2968.   C₁₄H₁₀Cl₄
1-(2-Chlorophenyl)-1-(4-chlorophenyl)-2,2-dichloroethane
o,p'-DDD
RN: 53-19-0   **MP** (°C):   76
MW: 320.05   **BP** (°C):

Solubility (Moles/L)	Solubility (Grams/L)	Temp (°C)	Ref (#)	Evaluation (T P E A A)	Comments
1.875E-07	6.000E-05	15	B083	2 2 1 2 1	particle size ≤ 5 µm
3.125E-07	1.000E-04	25	B083	2 2 1 2 2	particle size ≤ 5 µm
8.749E-07	2.800E-04	35	B083	2 2 1 2 2	particle size ≤ 5 µm
9.842E-07	3.150E-04	45	B083	2 2 1 2 2	particle size ≤ 5 µm

2969.   C₁₄H₁₀Cl₄
DDD
1,1-Dichloro-2,2-bis(p-chlorophenyl)ethane
p,p'-TDE
Dichlorodiphenyldichloroethane
RN: 72-54-8   **MP** (°C):   109.5
MW: 320.05   **BP** (°C):   193

Solubility (Moles/L)	Solubility (Grams/L)	Temp (°C)	Ref (#)	Evaluation (T P E A A)	Comments
1.562E-07	5.000E-05	15	B083	2 2 1 2 1	particle size ≤ 5 µm
2.812E-07	9.000E-05	25	B083	2 2 1 2 1	particle size ≤ 5 µm
6.249E-08	2.000E-05	25	W025	1 0 2 2 1	
4.687E-07	1.500E-04	35	B083	2 2 1 2 2	particle size ≤ 5 µm
7.499E-07	2.400E-04	45	B083	2 2 1 2 2	particle size ≤ 5 µm
9.374E-09	3.000E-06	ns	M110	0 0 0 0 0	EFG

2970.   C₁₄H₁₀F₃NO₂
Flufenamic Acid
N-(α,α,α-Trifluoro-m-tolyl)anthranilic Acid
N-(3-Trifluoromethylphenyl)anthranilic Acid
RN: 530-78-9   **MP** (°C):   132-135
MW: 281.24   **BP** (°C):

Solubility (Moles/L)	Solubility (Grams/L)	Temp (°C)	Ref (#)	Evaluation (T P E A A)	Comments
3.890E-06	1.094E-03	25	G085	1 0 0 0 1	EFG
4.000E-05	1.125E-02	25	I007	1 2 2 2 0	EFG
1.031E-04	2.900E-02	30	D015	2 0 1 1 0	EFG
6.670E-06	1.876E-03	35	G085	2 0 0 0 0	EFG
6.200E-04	1.744E-01	35	H091	1 2 2 2 1	sic
2.133E-04	6.000E-02	37	D015	2 0 1 1 0	EFG
3.556E-05	1.000E-02	rt	H302	0 0 2 1 2	intrinsic

2971.   C$_{14}$H$_{10}$N$_2$O$_2$
C.I. Disperse Violet 1
1,4-Diamino-9,10-anthraquinone
Acetate Red Violet R
Acetoquinone Light Heliotrope NL
Supracet Brilliant Violet 3R
Violet 14447
RN: 128-95-0   **MP** (°C):   275
MW: 238.25   **BP** (°C):

Solubility (Moles/L)	Solubility (Grams/L)	Temp (°C)	Ref (#)	Evaluation (T P E A A)	Comments
9.600E-07	2.287E-04	25	B333	1 0 0 0 1	

2972.   C$_{14}$H$_{10}$N$_2$O$_6$
Dipentum
Olsalazine
RN: 15722-48-2   **MP** (°C):
MW: 302.25   **BP** (°C):

Solubility (Moles/L)	Solubility (Grams/L)	Temp (°C)	Ref (#)	Evaluation (T P E A A)	Comments
3.800E-08	1.149E-05	25	D311	1 0 2 2 2	0.1M NaCl

2973.   C$_{14}$H$_{10}$O
1-Anthranol
1-Anthrol
Anthranol
RN: 529-86-2   **MP** (°C):   152
MW: 194.24   **BP** (°C):

Solubility (Moles/L)	Solubility (Grams/L)	Temp (°C)	Ref (#)	Evaluation (T P E A A)	Comments
1.850E-04	3.593E-02	25	L085	1 2 0 1 2	

2974.   C$_{14}$H$_{10}$O
2-Anthranol
2-Anthrol
RN: 613-14-9   **MP** (°C):
MW: 194.24   **BP** (°C):

Solubility (Moles/L)	Solubility (Grams/L)	Temp (°C)	Ref (#)	Evaluation (T P E A A)	Comments
4.720E-04	9.167E-02	25	L085	1 2 0 1 2	

2975.   C$_{14}$H$_{10}$O$_3$
Diphenyleneglycollic Acid
RN: **MP** (°C):
MW: 226.23   **BP** (°C):

Solubility (Moles/L)	Solubility (Grams/L)	Temp (°C)	Ref (#)	Evaluation (T P E A A)	Comments
1.082E-02	2.448E+00	25	K040	1 0 2 1 2	

2976.   C$_{14}$H$_{10}$O$_4$
Benzoyl Peroxide
Benzoyl-peroxid
RN: 94-36-0   **MP** (°C):   105
MW: 242.23   **BP** (°C):

Solubility (Moles/L)	Solubility (Grams/L)	Temp (°C)	Ref (#)	Evaluation (T P E A A)	Comments
6.399E-07	1.550E-04	rt	C342	1 0 2 1 2	

2977.   C$_{14}$H$_{10}$O$_4$
Diphenic Acid
1,1'-Biphenyl-2,2'-dicarboxylic Acid
2,2'-Biphenyldicarboxylic Acid
RN: 482-05-3   **MP** (°C):   228
MW: 242.23   **BP** (°C):

Solubility (Moles/L)	Solubility (Grams/L)	Temp (°C)	Ref (#)	Evaluation (T P E A A)	Comments
5.200E-03	1.260E+00	25	K040	1 0 2 1 2	

2978.   C$_{14}$H$_{10}$O$_5$
Gentisin
9H-Xanthen-9-one, 1,7-Dihydroxy-3-methoxy-
Gentianic Acid
Gentianin
RN: 437-50-3   **MP** (°C):   266.5
MW: 258.23   **BP** (°C):

Solubility (Moles/L)	Solubility (Grams/L)	Temp (°C)	Ref (#)	Evaluation (T P E A A)	Comments
1.162E-03	3.000E-01	16	F300	1 0 0 0 2	

2979. $C_{14}H_{10}O_9$
Digallic Acid
m-Digallic Acid
m-Digallussaeure
RN: 536-08-3   **MP** (°C):
MW: 322.23   **BP** (°C):

Solubility (Moles/L)	Solubility (Grams/L)	Temp (°C)	Ref (#)	Evaluation (T P E A A)	Comments
1.552E-03	5.000E-01	25	F300	1 0 0 0 0	
5.896E-02	1.900E+01	100	F300	1 0 0 0 1	

2980. $C_{14}H_{11}CINO_2$
7-Chloro-5,11-dihydrodibenz[b,e][1,4]oxazepine-5-carboxamide
RN: **MP** (°C):
MW: 260.70   **BP** (°C):

Solubility (Moles/L)	Solubility (Grams/L)	Temp (°C)	Ref (#)	Evaluation (T P E A A)	Comments
1.534E-04	4.000E-02	37	G020	1 0 0 0 1	

2981. $C_{14}H_{11}CIN_2O_4S$
Chlorthalidone
2-Chloro-5-(2,3-dihydro-1-hydroxy-3-oxo-1H-isoindol-1-yl)benzenesulfonamide
Hygroton
Thalitone
Chlortalidone
RN: 77-36-1   **MP** (°C):
MW: 338.77   **BP** (°C):

Solubility (Moles/L)	Solubility (Grams/L)	Temp (°C)	Ref (#)	Evaluation (T P E A A)	Comments
3.542E-04	1.200E-01	25	P312	1 2 2 2 2	
4.510E-04	1.528E-01	ns	I304	0 0 2 2 2	

2982. $C_{14}H_{11}Cl_3O_2$
2,2-bis(-p-Hydroxyphenyl)-1,1,1-trichloroethylene
Hydroxychlor
p,p'-Hydroxy-DDT
RN: 2971-36-0   **MP** (°C):   194
MW: 317.60   **BP** (°C):

Solubility (Moles/L)	Solubility (Grams/L)	Temp (°C)	Ref (#)	Evaluation (T P E A A)	Comments
2.393E-04	7.600E-02	ns	K117	0 1 2 1 1	

2983. C$_{14}$H$_{11}$FN$_2$O$_5$
1-Acetoxymethyl-3-benzoyl-5-fluoro-2,4(1H,3H)-pyrimidinedi-one
1-Acetoxymethyl-3-benzoyl-5-fluorouracil
RN: 97096-67-8   **MP** (°C):   127-128
MW: 306.25   **BP** (°C):

Solubility (Moles/L)	Solubility (Grams/L)	Temp (°C)	Ref (#)	Evaluation (T P E A A)	Comments
4.571E-04	1.400E-01	22	B321	1 0 2 2 2	pH 4.0

2984. C$_{14}$H$_{11}$N
Acetonitrile, Diphenyl-
Diphenatrile
RN: 86-29-3   **MP** (°C):   74
MW: 193.25   **BP** (°C):

Solubility (Moles/L)	Solubility (Grams/L)	Temp (°C)	Ref (#)	Evaluation (T P E A A)	Comments
1.138E-03	2.200E-01	ns	B185	1 0 0 0 2	

2985. C$_{14}$H$_{11}$N
2-Aminoanthracene
2-Anthrylamine
β-Aminoanthracene
2-Anthracenamine
2-Anthramine
Anthracene Amine
RN: 613-13-8   **MP** (°C):   238
MW: 193.25   **BP** (°C):

Solubility (Moles/L)	Solubility (Grams/L)	Temp (°C)	Ref (#)	Evaluation (T P E A A)	Comments
6.727E-06	1.300E-03	24	H106	1 0 2 2 2	
6.727E-09	1.300E-06	ns	M349	0 2 1 1 2	

2986. C$_{14}$H$_{11}$NO$_2$
N-Benzoylbenzamide
Dibenzamid
RN: 614-28-8   **MP** (°C):   152
MW: 225.25   **BP** (°C):

Solubility (Moles/L)	Solubility (Grams/L)	Temp (°C)	Ref (#)	Evaluation (T P E A A)	Comments
5.327E-03	1.200E+00	15	F300	1 0 0 0 1	

2987.   C$_{14}$H$_{11}$N$_3$O$_2$
Salicylolhydrazone of Picolinealdehyde
RN: **MP** (°C):
MW: 253.26   **BP** (°C):

Solubility (Moles/L)	Solubility (Grams/L)	Temp (°C)	Ref (#)	Evaluation (T P E A A)	Comments
7.897E-04	2.000E-01	ns	G089	0 1 2 0 1	

2988.   C$_{14}$H$_{12}$
1-Methylfluorene
1-Methyl-9H-fluorene
RN: 1730-37-6   **MP** (°C):   87
MW: 180.25   **BP** (°C):

Solubility (Moles/L)	Solubility (Grams/L)	Temp (°C)	Ref (#)	Evaluation (T P E A A)	Comments
6.047E-06	1.090E-03	25	B319	2 0 1 2 2	
6.060E-06	1.092E-03	25	M342	1 0 1 1 2	

2989.   C$_{14}$H$_{12}$
1,1-Diphenylethene
1,1-Diphenylethylene
RN: 530-48-3   **MP** (°C):   8.2
MW: 180.25   **BP** (°C):   277

Solubility (Moles/L)	Solubility (Grams/L)	Temp (°C)	Ref (#)	Evaluation (T P E A A)	Comments
3.662E-05	6.600E-03	25	A002	1 0 1 1 1	

2990.   C$_{14}$H$_{12}$
trans-Stilbene
trans-Diphenylethylene
1,2-Diphenylethene
trans-1,2-Diphenylethylene
trans-α, β-Diphenylethylene
Toluylene
RN: 103-30-0   **MP** (°C):   124
MW: 180.25   **BP** (°C):   306

Solubility (Moles/L)	Solubility (Grams/L)	Temp (°C)	Ref (#)	Evaluation (T P E A A)	Comments
1.609E-06	2.900E-04	25	A002	1 0 1 1 1	

2991.   C₁₄H₁₂F₃NO₄S₂
Perfluidone
Methyl-4-(phenylsulfonyl)trifluoromethanesulfonanilide
1,1,1-Trifluoro-N-(2-methyl-4-(phenylsulfonyl)phenyl)methanesulfonamide
Destun
MBR 8251
Trifluoro-N-(2-methyl-4-(phenylsulfonyl)phenyl)methanesulfonamide
RN: 37924-13-3   **MP** (°C):   143
MW: 379.38   **BP** (°C):

Solubility (Moles/L)	Solubility (Grams/L)	Temp (°C)	Ref (#)	Evaluation (T P E A A)	Comments
1.582E-04	6.000E-02	22	G306	1 0 0 0 1	
1.582E-04	6.000E-02	22	M161	1 0 0 0 1	

2992.   C₁₄H₁₂N₂S
2-(4-Aminophenyl)-6-methyl-benzothiazole
Dehydrothio-N-toluidin
Dehydrothio-N-toluidine
RN: 92-36-4   **MP** (°C):   194.8
MW: 240.33   **BP** (°C):   434

Solubility (Moles/L)	Solubility (Grams/L)	Temp (°C)	Ref (#)	Evaluation (T P E A A)	Comments
2.080E-04	5.000E-02	100	F300	1 0 0 0 0	

2993.   C₁₄H₁₂N₄O₂
C.I. Disperse Blue 1
9,10-Anthracenedione, 1,4,5,8-Tetraamino-
RN: 2475-45-8   **MP** (°C):   332
MW: 268.28   **BP** (°C):

Solubility (Moles/L)	Solubility (Grams/L)	Temp (°C)	Ref (#)	Evaluation (T P E A A)	Comments
1.000E-07	2.683E-05	25	B333	1 0 0 0 1	

2994. $C_{14}H_{12}O_2$
Benzoin
2-Hydroxy-1,2-diphenylethanone
Benzoylphenylcarbinol
2-Hydroxy-2-phenylacetophenone
Hydroxy-2-phenyl Acetophenone
RN: 579-44-2   **MP** (°C):   137
MW: 212.25   **BP** (°C):   344

Solubility (Moles/L)	Solubility (Grams/L)	Temp (°C)	Ref (#)	Evaluation (T P E A A)	Comments
1.413E-03	3.000E-01	25	F300	1 0 0 0 0	
1.413E-03	2.999E-01	rt	D021	0 0 1 1 0	

2995. $C_{14}H_{12}O_2$
Diphenylacetic Acid
Diphenyl-essigsaeure
RN: 117-34-0   **MP** (°C):   148
MW: 212.25   **BP** (°C):

Solubility (Moles/L)	Solubility (Grams/L)	Temp (°C)	Ref (#)	Evaluation (T P E A A)	Comments
6.000E-04	1.274E-01	25	K040	1 0 2 1 2	

2996. $C_{14}H_{12}O_2$
Benzyl Benzoate
Ascabin
Scabagen
Benzoic Acid Phenylmethyl Ester
Benylate
Phenylmethyl Benzoate
RN: 120-51-4   **MP** (°C):   19
MW: 212.25   **BP** (°C):   323

Solubility (Moles/L)	Solubility (Grams/L)	Temp (°C)	Ref (#)	Evaluation (T P E A A)	Comments
1.225E-04	2.600E-02	15	H069	1 0 1 1 1	

2997. $C_{14}H_{12}O_2$
4-Biphenylacetic Acid
Felbinac
RN: 5728-52-9   **MP** (°C):
MW: 212.25   **BP** (°C):

Solubility (Moles/L)	Solubility (Grams/L)	Temp (°C)	Ref (#)	Evaluation (T P E A A)	Comments
1.850E-04	3.927E-02	25	P344	1 1 2 2 0	EFG

2998.   C$_{14}$H$_{12}$O$_3$
Benzilic Acid
2,2-Diphenyl-2-hydroxyacetic Acid
Diphenylglycolic Acid
Benzeneacetic Acid, α-Hydroxy-α-phenyl-
2-Hydroxy-2,2-diphenylethanoic Acid
RN: 76-93-7   **MP** (°C):   150
MW: 228.25   **BP** (°C):

Solubility (Moles/L)	Solubility (Grams/L)	Temp (°C)	Ref (#)	Evaluation (T P E A A)	Comments
7.690E-03	1.755E+00	25	K040	1 0 2 1 2	
6.190E-03	1.413E+00	25	L050	2 0 1 2 2	

2999.   C$_{14}$H$_{12}$O$_3$
Benzylparaben
Benzyl 4-hydroxybenzoate
Phenylmethyl Ester
RN: 94-18-8   **MP** (°C):
MW: 228.25   **BP** (°C):

Solubility (Moles/L)	Solubility (Grams/L)	Temp (°C)	Ref (#)	Evaluation (T P E A A)	Comments
4.031E-04	9.200E-02	25	P013	2 0 2 1 2	

3000.   C$_{14}$H$_{12}$O$_5$
Khellin
Amicardine
RN: 82-02-0   **MP** (°C):   154.5
MW: 260.25   **BP** (°C):

Solubility (Moles/L)	Solubility (Grams/L)	Temp (°C)	Ref (#)	Evaluation (T P E A A)	Comments
9.500E-01	2.472E+02	25	E312	2 0 2 2 0	EFG, *sic*
1.153E-04	3.000E-02	25	J028	1 2 0 2 0	
7.000E-04	1.822E-01	30	E012	1 2 1 1 0	
1.300E-03	3.383E-01	42	E012	1 2 1 1 0	

3001.   C$_{14}$H$_{13}$ClN$_4$O
6H-Dipyrido[3,2-b:2',3'-e][1,4]diazepin-6-one, 2-Chloro-11-ethyl-5,11-dihydro-5-methyl-
RN: 133627-12-0   **MP** (°C):
MW: 288.74   **BP** (°C):

Solubility (Moles/L)	Solubility (Grams/L)	Temp (°C)	Ref (#)	Evaluation (T P E A A)	Comments
7.691E-05	2.221E-02	ns	M381	0 1 1 1 2	pH 7.0

3002. C$_{14}$H$_{13}$NO$_6$
Benzoic Acid, 2-(Acetyloxy)-, (2,5-Dioxo-1-pyrrolidinyl)methyl Ester
Salicylic Acid Acetate, Ester with N-(hydroxymethyl)succinimide
RN: 32620-72-7   **MP** (°C):   117.5
MW: 291.26   **BP** (°C):

Solubility (Moles/L)	Solubility (Grams/L)	Temp (°C)	Ref (#)	Evaluation (T P E A A)	Comments
1.717E-03	5.000E-01	21	N335	1 2 1 1 2	

3003. C$_{14}$H$_{13}$N$_2$
4,7-Dimethyl-1,10-phenanthroline
4,7-Dimethyl-o-phenanthroline
RN: 3248-05-3   **MP** (°C):   193
MW: 209.27   **BP** (°C):

Solubility (Moles/L)	Solubility (Grams/L)	Temp (°C)	Ref (#)	Evaluation (T P E A A)	Comments
1.070E-04	2.239E-02	25.04	B094	1 2 1 2 2	

3004. C$_{14}$H$_{13}$N$_3$O$_2$
Pyrido[2,3-b][1,5]benzoxazepin-5(6H)-one, 3-Amino-6,9-dimethyl-
RN: 134894-45-4   **MP** (°C):
MW: 255.28   **BP** (°C):

Solubility (Moles/L)	Solubility (Grams/L)	Temp (°C)	Ref (#)	Evaluation (T P E A A)	Comments
9.057E-04	2.312E-01	ns	M381	0 1 1 1 2	pH 7.0

3005. C$_{14}$H$_{14}$
4,4'-Dimethylbiphenyl
4,4'-Dimethyl-1,1'-biphenyl
p,p'-Bitoluene
RN: 613-33-2   **MP** (°C):   125.0
MW: 182.27   **BP** (°C):   295.0

Solubility (Moles/L)	Solubility (Grams/L)	Temp (°C)	Ref (#)	Evaluation (T P E A A)	Comments
3.770E-07	6.871E-05	4.0	D330	2 2 1 2 2	
9.590E-07	1.748E-04	25.0	D330	2 2 1 2 2	
2.420E-06	4.411E-04	40.0	D330	2 2 1 2 2	

3006. $C_{14}H_{14}$

Bibenzyl
1,2-Diphenylethane
Benzene, 1,1'-(1,2-Ethanediyl)bis-
RN: 103-29-7   **MP** (°C):   52.0
MW: 182.27   **BP** (°C):   284

Solubility (Moles/L)	Solubility (Grams/L)	Temp (°C)	Ref (#)	Evaluation (T P E A A)	Comments
2.359E-05	4.300E-03	25	A002	1 0 1 1 1	

3007. $C_{14}H_{14}NO_4PS$

EPN
Ethyl O-(p-Nitrophenyl) Phenylphosphonothionate
O-Ethyl O-p-Nitrophenyl Benzenephosphonothioate
Ethyl O-(p-nitrophenyl) benzenethiophosphonate
RN: 2104-64-5   **MP** (°C):   36
MW: 323.31   **BP** (°C):

Solubility (Moles/L)	Solubility (Grams/L)	Temp (°C)	Ref (#)	Evaluation (T P E A A)	Comments
9.629E-06	3.113E-03	22	K137	1 1 2 1 0	

3008. $C_{14}H_{14}N_4O$

6H-Dipyrido[3,2-b:2',3'-e][1,4]diazepin-6-one, 11-Ethyl-5,11-dihydro-5-methyl
RN: 132312-85-7   **MP** (°C):
MW: 254.29   **BP** (°C):

Solubility (Moles/L)	Solubility (Grams/L)	Temp (°C)	Ref (#)	Evaluation (T P E A A)	Comments
2.399E-03	6.100E-01	ns	M381	0 1 1 1 2	pH 7.0

3009. $C_{14}H_{14}N_4O_2$

Dye II
4-[[(4-Dimethylamino)phenyl]azo]nitrobenzene
RN: **MP** (°C):
MW: 270.29   **BP** (°C):

Solubility (Moles/L)	Solubility (Grams/L)	Temp (°C)	Ref (#)	Evaluation (T P E A A)	Comments
7.800E-07	2.108E-04	84.10	B198	1 2 1 1 1	
2.040E-06	5.514E-04	97.40	B198	1 2 1 1 2	

3010. $C_{14}H_{14}N_4O_2$
Dis. A. 7
RN: 2491-74-9 **MP** (°C): 236
MW: 270.29 **BP** (°C):

Solubility (Moles/L)	Solubility (Grams/L)	Temp (°C)	Ref (#)	Evaluation (T P E A A)	Comments
2.000E-09	5.406E-07	25	B333	1 0 0 0 1	

3011. $C_{14}H_{14}N_4O_4$
β,γ-Dihydroxypropyltheophylline
RN: 180262-60-6 **MP** (°C):
MW: 302.29 **BP** (°C):

Solubility (Moles/L)	Solubility (Grams/L)	Temp (°C)	Ref (#)	Evaluation (T P E A A)	Comments
3.007E-01	9.091E+01	ns	J025	0 0 0 0 1	

3012. $C_{14}H_{14}N_4S$
6H-Dipyrido[3,2-b:2',3'-e][1,4]diazepine-6-thione, 11-Ethyl-5,11-dihydro-5-methyl
RN: 134698-27-4 **MP** (°C):
MW: 270.36 **BP** (°C):

Solubility (Moles/L)	Solubility (Grams/L)	Temp (°C)	Ref (#)	Evaluation (T P E A A)	Comments
2.323E-05	6.280E-03	ns	M381	0 1 1 1 2	pH 7.0

3013. $C_{14}H_{14}O$
6-Benzyl-m-cresol
Phenol, 5-Methyl-2-(phenylmethyl)-
RN: 30091-04-4 **MP** (°C):
MW: 198.27 **BP** (°C):

Solubility (Moles/L)	Solubility (Grams/L)	Temp (°C)	Ref (#)	Evaluation (T P E A A)	Comments
1.441E-04	2.857E-02	25	L021	1 0 0 0 0	

3014. $C_{14}H_{14}O$
DL-1,2-Diphenylethanol
DL-1,2-Diphenyl-aethanol
RN: 614-29-9 **MP** (°C): 67
MW: 198.27 **BP** (°C):

Solubility (Moles/L)	Solubility (Grams/L)	Temp (°C)	Ref (#)	Evaluation (T P E A A)	Comments
3.026E-03	6.000E-01	100	F300	1 0 0 0 0	

3015. $C_{14}H_{14}O_2$
DL-Hydrobenzoin
Hydrobenzoin
RN: 27134-24-3   **MP** (°C):   139
MW: 214.27   **BP** (°C):

Solubility (Moles/L)	Solubility (Grams/L)	Temp (°C)	Ref (#)	Evaluation (T P E A A)	Comments
1.167E-02	2.500E+00	15	F300	1 0 0 0 1	
8.867E-03	1.900E+00	15	F300	1 0 0 0 1	
6.021E-02	1.290E+01	100	F300	1 0 0 0 2	
6.021E-02	1.290E+01	100	F300	1 0 0 0 2	

3016. $C_{14}H_{14}O_3$
Pindone
2-Pivaloylindandione-1,3
RN: 83-26-1   **MP** (°C):   109
MW: 230.27   **BP** (°C):

Solubility (Moles/L)	Solubility (Grams/L)	Temp (°C)	Ref (#)	Evaluation (T P E A A)	Comments
7.817E-05	1.800E-02	25	M061	1 0 0 0 1	
7.817E-05	1.800E-02	25	M161	1 0 0 0 1	

3017. $C_{14}H_{14}O_3$
Naproxen
6-Methoxy-α-methyl-2-naphthaleneacetic Acid
(S)-6-Methoxy-α-methyl-2-naphthaleneacetic Acid
Laraflex
RN: 22204-53-1   **MP** (°C):   155.3
MW: 230.27   **BP** (°C):

Solubility (Moles/L)	Solubility (Grams/L)	Temp (°C)	Ref (#)	Evaluation (T P E A A)	Comments
4.310E-05	9.924E-03	5	F306	1 0 1 2 2	intrinsic
6.948E-05	1.600E-02	21	B331	1 2 2 1 2	pH 7.4
6.905E-05	1.590E-02	25	C059	1 2 1 1 2	
6.900E-05	1.589E-02	25	F306	1 0 1 2 2	intrinsic
1.146E-04	2.639E-02	37	F306	1 0 1 2 2	intrinsic
5.646E-05	1.300E-02	rt	H302	0 0 2 1 2	intrinsic

3018.   C$_{14}$H$_{14}$O$_3$S
o-Cresyl-p-toluene Sulfonate
2-Methylphenyl Tosylate
o-Tolyl Tosylate
2-Tolyl Tosylate

RN: 599-75-7   **MP** (°C):
MW: 262.33   **BP** (°C):

Solubility (Moles/L)	Solubility (Grams/L)	Temp (°C)	Ref (#)	Evaluation (T P E A A)	Comments
1.144E-04	3.000E-02	ns	F014	0 0 0 0 0	

3019.   C$_{14}$H$_{14}$O$_4$
Diallyl Phthalate
Di-2-propenyl Phthalate

RN: 131-17-9   **MP** (°C):   -70
MW: 246.27   **BP** (°C):   165

Solubility (Moles/L)	Solubility (Grams/L)	Temp (°C)	Ref (#)	Evaluation (T P E A A)	Comments
<4.06E-04	<10.00E-02	20	F070	1 0 0 0 1	
7.390E-04	1.820E-01	20	L300	2 1 0 2 2	

3020.   C$_{14}$H$_{15}$N
p-Aminostilbene
4-Aminostilbene

RN: 834-24-2   **MP** (°C):
MW: 197.28   **BP** (°C):

Solubility (Moles/L)	Solubility (Grams/L)	Temp (°C)	Ref (#)	Evaluation (T P E A A)	Comments
2.534E-05	5.000E-03	rt	N015	0 0 2 2 0	

3021.   C$_{14}$H$_{15}$NO$_5$
L-Proline, 1-[(Benzoyloxy)acetyl]-

RN: 115178-75-1   **MP** (°C):   72.5
MW: 277.28   **BP** (°C):

Solubility (Moles/L)	Solubility (Grams/L)	Temp (°C)	Ref (#)	Evaluation (T P E A A)	Comments
2.561E-02	7.100E+00	22	N317	1 1 2 1 2	

3022. C₁₄H₁₅N₃
o-Aminoazotoluene
2-Amino-5-azotoluene
RN: 97-56-3   **MP** (°C):   101
MW: 225.30   **BP** (°C):

Solubility (Moles/L)	Solubility (Grams/L)	Temp (°C)	Ref (#)	Evaluation (T P E A A)	Comments
3.107E-05	7.000E-03	37	H120	1 1 1 1 1	normal saline

3023. C₁₄H₁₅N₃
p-Dimethylaminoazobenzene
4-Dimethylaminoazobenzol
Dimethylaminoazobenzene
Methylgelb
C.I. Solvent Yellow 2
RN: 60-11-7   **MP** (°C):   116
MW: 225.30   **BP** (°C):

Solubility (Moles/L)	Solubility (Grams/L)	Temp (°C)	Ref (#)	Evaluation (T P E A A)	Comments
8.877E-04	2.000E-01	20	F300	1 0 0 0 0	
6.214E-06	1.400E-03	20	J027	1 0 0 0 1	
1.700E-06	3.830E-04	25	B333	1 0 0 0 1	sic
7.101E-04	1.600E-01	rt	D021	0 0 1 1 1	sic

3024. C₁₄H₁₅N₃O₃S
Gly-Dapsone
Acetamide, 2-Amino-N-[4-[(4-aminophenyl)sulfonyl]phenyl]
RN: 160349-02-0   **MP** (°C):
MW: 305.36   **BP** (°C):

Solubility (Moles/L)	Solubility (Grams/L)	Temp (°C)	Ref (#)	Evaluation (T P E A A)	Comments
2.849E-03	8.700E-01	25	P351	2 2 1 2 1	pH 7.4
>4.91E-02	>1.50E+01	25	P351	2 2 1 2 1	

3025. C₁₄H₁₅N₅O₅
9-(2-O-Butyryl-β-D-arabinofuranosyl)adenine
9H-Purin-6-amine, 9-[3,5-bis-O-[(1,1-Dimethylethyl)dimethylsilyl]-2-O-(1-oxobutyl)-β-D-arabinofuranosyl]-
RN: 87970-05-6   **MP** (°C):
MW: 333.31   **BP** (°C):

Solubility (Moles/L)	Solubility (Grams/L)	Temp (°C)	Ref (#)	Evaluation (T P E A A)	Comments
1.023E-04	3.410E-02	37	B306	1 2 0 1 2	pH 7.3

3026. C$_{14}$H$_{16}$ClN$_3$O$_2$
Triadimefon
1-(4-Chlorophenoxy)-3,3-dimethyl-1-(1H-1,2,4-triazol-1-yl)-2-butanone
Triamefon
Bayleton

RN:	43121-43-3	**MP** (°C):	82.3
MW:	293.76	**BP** (°C):	

Solubility (Moles/L)	Solubility (Grams/L)	Temp (°C)	Ref (#)	Evaluation (T P E A A)	Comments
8.851E-04	2.600E-01	20	M161	1 0 0 0 2	

3027. C$_{14}$H$_{16}$ClO$_5$PS
Coumaphos
O,O-Diethyl O-(3-Chloro-4-methylcoumarinyl-7) Thiophosphate

RN:	56-72-4	**MP** (°C):	91
MW:	362.77	**BP** (°C):	

Solubility (Moles/L)	Solubility (Grams/L)	Temp (°C)	Ref (#)	Evaluation (T P E A A)	Comments
4.135E-06	1.500E-03	20	M061	1 0 0 0 1	

3028. C$_{14}$H$_{16}$Cl$_2$O$_3$
2,4-Dichlorophenoxyacetic Acid Cyclohexyl Ester
Cyclohexyl 2,4-Dichlorophenoxyacetate

RN:	65267-97-2	**MP** (°C):	
MW:	303.19	**BP** (°C):	

Solubility (Moles/L)	Solubility (Grams/L)	Temp (°C)	Ref (#)	Evaluation (T P E A A)	Comments
1.811E-05	5.492E-03	ns	M120	0 0 1 1 2	

3029. C$_{14}$H$_{16}$FN$_3$O$_3$
2,5-Diaziridinyl-3-floro-6-morpholino-1,4-benzoquinone
2,5-Cyclohexadiene-1,4-dione, 2,5-bis(1-Aziridinyl)-3-fluoro-6-(4-morpholinyl)-

RN:	59886-45-2	**MP** (°C):	157
MW:	293.30	**BP** (°C):	

Solubility (Moles/L)	Solubility (Grams/L)	Temp (°C)	Ref (#)	Evaluation (T P E A A)	Comments
6.819E-03	2.000E+00	rt	C317	0 2 0 0 0	

3030.   C$_{14}$H$_{16}$F$_3$N$_3$O$_4$
Profluralin
N-(Cyclopropylmethyl)-2,6-dinitro-N-propyl-4-(trifluoromethyl)benzenamine
Pregard
Tolban
ER-5461
RN: 26399-36-0   **MP** (°C):   32
MW: 347.30   **BP** (°C):

Solubility (Moles/L)	Solubility (Grams/L)	Temp (°C)	Ref (#)	Evaluation (T P E A A)	Comments
2.879E-07	1.000E-04	20	E048	1 2 1 1 0	
2.879E-07	1.000E-04	20	M161	1 0 0 0 0	
2.879E-07	1.000E-04	27	K315	1 0 0 0 1	

3031.   C$_{14}$H$_{16}$N$_2$
o-Tolidine
3,3'-Dimethylbenzidine
RN: 119-93-7   **MP** (°C):   130.0
MW: 212.30   **BP** (°C):

Solubility (Moles/L)	Solubility (Grams/L)	Temp (°C)	Ref (#)	Evaluation (T P E A A)	Comments
6.123E-03	1.300E+00	25	B068	2 0 1 1 1	

3032.   C$_{14}$H$_{16}$N$_2$O$_2$
3,3'-Dimethoxybenzidine
o-Dianisidine
Dianisidine
RN: 119-90-4   **MP** (°C):   137
MW: 244.30   **BP** (°C):

Solubility (Moles/L)	Solubility (Grams/L)	Temp (°C)	Ref (#)	Evaluation (T P E A A)	Comments
2.456E-04	6.000E-02	25	B068	2 0 1 1 0	

3033.   C$_{14}$H$_{16}$N$_2$O$_4$
2-Pyrrolidinecarboxamide, 1-[(Benzoyloxy)acetyl]-
RN: 116482-82-7   **MP** (°C):   194.5
MW: 276.29   **BP** (°C):

Solubility (Moles/L)	Solubility (Grams/L)	Temp (°C)	Ref (#)	Evaluation (T P E A A)	Comments
5.429E-03	1.500E+00	22	N317	1 1 2 1 2	

3034. C$_{14}$H$_{16}$N$_4$
Disperse Black 3
N,N-Dimethyl-4,4'-azodian
4-Amino-4'-(dimethylamino)azobenzene
C.I. 11025
RN: 539-17-3   **MP** (°C):
MW: 240.31   **BP** (°C):

Solubility (Moles/L)	Solubility (Grams/L)	Temp (°C)	Ref (#)	Evaluation (T P E A A)	Comments
5.000E-07	1.202E-04	25	B333	1 0 0 0 1	

3035. C$_{14}$H$_{16}$N$_4$O$_2$S
2-Sulfanilamido-5,6,7,8-tetrahydroquinazoline
2-Sulfanilamido-5,6,7,8,-tetrahydroquinazoline
RN: 71119-34-1   **MP** (°C):   255
MW: 304.37   **BP** (°C):

Solubility (Moles/L)	Solubility (Grams/L)	Temp (°C)	Ref (#)	Evaluation (T P E A A)	Comments
2.234E-04	6.800E-02	29	C049	1 2 0 0 1	

3036. C$_{14}$H$_{16}$N$_4$O$_3$S
N4-Acetylsulfamethazine
N4-Acetylsulfamezathine
N4-Acetylsulphamethazine
Acetylsulfamethazine
2-p-Acetamidobenzenesulphonamido-4:6-dimethylpyri-
RN: 100-90-3   **MP** (°C):
MW: 320.37   **BP** (°C):

Solubility (Moles/L)	Solubility (Grams/L)	Temp (°C)	Ref (#)	Evaluation (T P E A A)	Comments
2.900E-03	9.291E-01	37	G026	1 0 1 1 0	EFG, pH 5.4
3.590E-03	1.150E+00	37	L091	1 0 0 0 2	pH 5.5
3.590E-03	1.150E+00	37	M057	1 0 0 0 2	pH 5.5
3.590E-03	1.150E+00	37	R075	1 2 0 0 2	
2.197E-03	7.040E-01	37	S192	1 0 1 1 2	pH 6.0
2.622E-03	8.400E-01	38	K006	1 0 0 0 1	

3037. C$_{14}$H$_{16}$N$_4$O$_3$S
N4-Acetylsulphasomidine
Acetamide, N-[4-[[(2,6-Dimethyl-4-pyrimidinyl)amino]sulfonyl]phenyl]-
RN: 3163-31-3   **MP** (°C):
MW: 320.37   **BP** (°C):

Solubility (Moles/L)	Solubility (Grams/L)	Temp (°C)	Ref (#)	Evaluation (T P E A A)	Comments
1.373E-04	4.400E-02	ns	B133	0 2 0 0 1	pH 7.4

3038. C₁₄H₁₆N₄O₃S

2-(N4-Acetylsulfanilylamino)-4-ethylpyrimidine

RN: **MP** (°C):
MW: 320.37 **BP** (°C):

Solubility (Moles/L)	Solubility (Grams/L)	Temp (°C)	Ref (#)	Evaluation (T P E A A)	Comments
2.435E-05	7.800E-03	37	R076	1 2 0 0 2	

3039. C₁₄H₁₆N₄O₄S

N4-Acetylsulphamethomidine

RN: **MP** (°C):
MW: 336.37 **BP** (°C):

Solubility (Moles/L)	Solubility (Grams/L)	Temp (°C)	Ref (#)	Evaluation (T P E A A)	Comments
7.730E-04	2.600E-01	ns	B133	0 2 0 0 2	pH 7.4

3040. C₁₄H₁₆N₄O₅S

N4-Acetylsulphadimethoxine
N4-Acetyl-2,4-dimethoxy-6-sulfanilamidopyrimidine
N4-Acetylsulfadimethoxypyrimidine
Sulfadimethoxine N4-Acetate

RN: 555-25-9 **MP** (°C):
MW: 352.37 **BP** (°C):

Solubility (Moles/L)	Solubility (Grams/L)	Temp (°C)	Ref (#)	Evaluation (T P E A A)	Comments
5.392E-04	1.900E-01	ns	B133	0 2 0 0 2	pH 7.4

3041. C₁₄H₁₆O₆

Ethylphthalyl Ethyl Glycolate
Ethoxycarbonylmethyl Ethyl Phthalate
Ethylphthalyl Ethylglycolate

RN: 84-72-0 **MP** (°C): 20
MW: 280.28 **BP** (°C): 320

Solubility (Moles/L)	Solubility (Grams/L)	Temp (°C)	Ref (#)	Evaluation (T P E A A)	Comments
<2.85E-03	<7.99E-01	20	F070	1 0 0 0 1	

3042. $C_{14}H_{16}O_6$

Benzoic Acid, 2-(Acetyloxy)-, (1-Oxobutoxy)methyl Ester

RN: 118247-07-7 **MP** (°C): Oil
MW: 280.28 **BP** (°C):

Solubility (Moles/L)	Solubility (Grams/L)	Temp (°C)	Ref (#)	Evaluation (T P E A A)	Comments
1.249E-03	3.500E-01	21	N335	1 2 1 1 2	

3043. $C_{14}H_{17}CINO_4PS_2$

Dialifos
Dialifor
Diethyl S-(2-Chloro-1-phthalimidoethyl) Phosphorodithioate
Torak
Hercules 14503

RN: 10311-84-9 **MP** (°C): 67
MW: 393.85 **BP** (°C):

Solubility (Moles/L)	Solubility (Grams/L)	Temp (°C)	Ref (#)	Evaluation (T P E A A)	Comments
4.570E-07	1.800E-04	ns	F071	0 1 2 1 1	

3044. $C_{14}H_{17}NO$

N-Cyclopentylcinnamamide
2-Propenamide, N-Cyclopentyl-3-phenyl-

RN: 59831-97-9 **MP** (°C):
MW: 215.30 **BP** (°C):

Solubility (Moles/L)	Solubility (Grams/L)	Temp (°C)	Ref (#)	Evaluation (T P E A A)	Comments
2.280E-04	4.909E-02	ns	H350	0 0 0 0 2	

3045. $C_{14}H_{17}NO$

1-Cinnamoylpiperidine
N,N-Pentamethylenecinnamamide
1-(1-Oxo-3-phenyl-2-propenyl)-Piperidine

RN: 5422-81-1 **MP** (°C):
MW: 215.30 **BP** (°C):

Solubility (Moles/L)	Solubility (Grams/L)	Temp (°C)	Ref (#)	Evaluation (T P E A A)	Comments
9.600E-04	2.067E-01	ns	H350	0 0 0 0 2	

3046. $C_{14}H_{17}NO_2S$
m-Carboxylhexylphenylisothiocyanate
3-Carboxylhexylphenylisothiocyanate
RN: **MP** (°C):
MW: 263.36   **BP** (°C):

Solubility (Moles/L)	Solubility (Grams/L)	Temp (°C)	Ref (#)	Evaluation (T P E A A)	Comments
7.000E-05	1.844E-02	25	K032	2 2 0 1 1	

3047. $C_{14}H_{17}NO_3$
Piperidine, 1-[(Benzoyloxy)acetyl]-
RN: 106231-67-8   **MP** (°C):   88
MW: 247.30   **BP** (°C):

Solubility (Moles/L)	Solubility (Grams/L)	Temp (°C)	Ref (#)	Evaluation (T P E A A)	Comments
3.154E-03	7.800E-01	22	N317	1 1 2 1 2	

3048. $C_{14}H_{17}NO_4$
4-Piperidinol, 1-[(Benzoyloxy)acetyl]-
RN: 115178-71-7   **MP** (°C):   121.5
MW: 263.30   **BP** (°C):

Solubility (Moles/L)	Solubility (Grams/L)	Temp (°C)	Ref (#)	Evaluation (T P E A A)	Comments
4.482E-02	1.180E+01	22	N317	1 1 2 1 2	

3049. $C_{14}H_{17}NO_5$
Glycine, N-[(Benzoyloxy)acetyl]-N-methyl-, Ethyl Ester
RN: 106231-63-4   **MP** (°C):   39.5
MW: 279.30   **BP** (°C):

Solubility (Moles/L)	Solubility (Grams/L)	Temp (°C)	Ref (#)	Evaluation (T P E A A)	Comments
2.148E-02	6.000E+00	22	N317	1 1 2 1 2	

3050. $C_{14}H_{17}N_5O_3$
Pipemidic Acid
Pipemidique Acide
RN: 51940-44-4   **MP** (°C):   253
MW: 303.32   **BP** (°C):

Solubility (Moles/L)	Solubility (Grams/L)	Temp (°C)	Ref (#)	Evaluation (T P E A A)	Comments
1.060E-03	3.215E-01	25	D051	2 0 0 1 2	0.05N NaCl
1.160E-03	3.519E-01	37	D051	2 0 0 1 2	0.05N NaCl

3051. C$_{14}$H$_{18}$ClN$_3$S

Chlorothen
N,N-Dimethyl-N'-(2-pyridyl)-N'-(5-chloro-2-thenyl)ethylenediamine
Chloromethapyrilene
5-Chloro-N-(2-(dimethylamino)ethyl)-N-(2-pyridyl)-2-thenylamine
Chloropyrilene

RN: 148-65-2   **MP** (°C):
MW: 295.84   **BP** (°C): 155.5

Solubility (Moles/L)	Solubility (Grams/L)	Temp (°C)	Ref (#)	Evaluation (T P E A A)	Comments
6.800E-03	2.012E+00	37.5	L034	2 2 0 1 2	pH 7.4

3052. C$_{14}$H$_{18}$Cl$_2$O$_3$

2,4-Dichlorophenoxyacetic Acid n-Hexyl Ester
Chloroxone
Agrotect
Amoxone
BH 2,4-D

RN: 1917-95-9   **MP** (°C):
MW: 305.20   **BP** (°C):

Solubility (Moles/L)	Solubility (Grams/L)	Temp (°C)	Ref (#)	Evaluation (T P E A A)	Comments
1.941E-05	5.924E-03	ns	M120	0 0 1 1 2	

3053. C$_{14}$H$_{18}$N$_2$O

Propyphenazone
Isopropylantipyrine
1,2-Dihydro-1,5-dimethyl-4-(isopropyl)-2-phenyl-pyrazol-3-one
4-Isopropyl-2,3-dimethyl-5-oxo-1-phenyl-3-pyrazoline

RN: 479-92-5   **MP** (°C): 103
MW: 230.31   **BP** (°C):

Solubility (Moles/L)	Solubility (Grams/L)	Temp (°C)	Ref (#)	Evaluation (T P E A A)	Comments
3.383E+00	7.791E+02	4.62	M109	2 1 1 1 0	EFG
3.330E+00	7.670E+02	10.93	M109	2 1 1 1 0	EFG
3.257E+00	7.501E+02	15.02	M109	2 1 1 1 0	EFG
3.238E+00	7.458E+02	20.96	M109	2 1 1 1 0	EFG
3.229E+00	7.436E+02	25.35	M109	2 1 1 1 0	EFG
3.238E+00	7.458E+02	29.87	M109	2 1 1 1 0	EFG
3.257E+00	7.501E+02	38.37	M109	2 1 1 1 0	EFG
3.348E+00	7.711E+02	40.32	M109	2 1 1 1 0	EFG

3054. C$_{14}$H$_{18}$N$_2$O$_3$
Reposal
5-Bicyclo[3.2.1]oct-2-en-3-yl-5-ethyl-2,4,6(1H,3H,5H)-pyrimidinetri-one
5-Bicyclo[3.2.1]oct-2-en-3-yl-5-ethylbarbituric Acid
RN: 3625-25-0   **MP** (°C):   213
MW: 262.31   **BP** (°C):

Solubility (Moles/L)	Solubility (Grams/L)	Temp (°C)	Ref (#)	Evaluation (T P E A A)	Comments
1.680E-03	4.407E-01	25	V033	2 0 1 1 2	
1.700E-03	4.459E-01	25.00	T303	1 0 0 0 1	
2.300E-03	6.033E-01	35.00	T303	1 0 0 0 1	
2.500E-03	6.558E-01	45.00	T303	1 0 0 0 1	

3055. C$_{14}$H$_{18}$N$_4$O$_2$S
2-Sulfanilylamino-4-isobutylpyrimidine
RN: 106596-34-3   **MP** (°C):
MW: 306.39   **BP** (°C):

Solubility (Moles/L)	Solubility (Grams/L)	Temp (°C)	Ref (#)	Evaluation (T P E A A)	Comments
3.264E-04	1.000E-01	37	R076	1 2 0 0 1	

3056. C$_{14}$H$_{18}$N$_4$O$_3$
Benomyl
(1-(Butylamino)carbonyl)-1H-benzimidazol-2-yl)carbamic Acid Methyl Ester
RN: 17804-35-2   **MP** (°C):
MW: 290.32   **BP** (°C):

Solubility (Moles/L)	Solubility (Grams/L)	Temp (°C)	Ref (#)	Evaluation (T P E A A)	Comments
1.309E-05	3.800E-03	20	A064	1 0 1 1 1	
1.309E-05	3.800E-03	20	M161	1 0 0 0 1	pH 7
~6.89E-06	~2.00E-03	ns	B309	0 0 0 0 0	

3057. $C_{14}H_{18}N_4O_3$
Trimethoprim
5-(3,4,5-Trimethoxybenzyl)-2,4-diaminopyrimidine
Monotrim
Syraprim
Proloprim
Trimpex

RN: 738-70-5   **MP** (°C):   201
MW: 290.32   **BP** (°C):

Solubility (Moles/L)	Solubility (Grams/L)	Temp (°C)	Ref (#)	Evaluation (T P E A A)	Comments
1.378E-03	4.000E-01	25	M167	1 0 0 0 0	
1.722E-03	5.000E-01	32	D308	1 0 2 2 2	pH 8.54
2.711E-03	7.870E-01	37	G086	1 0 0 0 1	
1.378E-03	4.000E-01	37	M321	1 0 0 0 2	intrinsic

3058. $C_{14}H_{18}N_4O_6.0.5H_2O$
2'-Propionyl-6-methoxypurine Arabinoside (Hemihydrate)

RN: 145913-38-8   **MP** (°C):   60-65
MW: 347.33   **BP** (°C):

Solubility (Moles/L)	Solubility (Grams/L)	Temp (°C)	Ref (#)	Evaluation (T P E A A)	Comments
1.100E-01	3.821E+01	37	C348	1 2 2 2 2	pH 7.00

3059. $C_{14}H_{18}N_4O_7.0.5H_2O$
9-[5-O-(Methoxyacetate-β-D-arabinofuranosyl)]-6-methoxy-9H-purine (Hemihydrate)

RN: 121032-38-0   **MP** (°C):   137-139
MW: 363.33   **BP** (°C):

Solubility (Moles/L)	Solubility (Grams/L)	Temp (°C)	Ref (#)	Evaluation (T P E A A)	Comments
7.810E-02	2.838E+01	37	M378	1 2 1 1 2	pH 7.2

3060. $C_{14}H_{18}N_4O_7.0.9H_2O$
2'-Methoxyacetyl-6-methoxypurine Arabinoside (0.9 Hydrate)

RN: 145913-47-9   **MP** (°C):
MW: 370.54   **BP** (°C):

Solubility (Moles/L)	Solubility (Grams/L)	Temp (°C)	Ref (#)	Evaluation (T P E A A)	Comments
9.090E-02	3.368E+01	37	C348	1 2 2 2 2	pH 7.00

3061.   C₁₄H₁₈N₆O₄

2,5-Diaziridinyl-3,6-bis(glycinamide)-1,4-benzoquinone

RN: 59886-49-6   **MP** (°C):   200
MW: 334.34   **BP** (°C):

Solubility (Moles/L)	Solubility (Grams/L)	Temp (°C)	Ref (#)	Evaluation (T P E A A)	Comments
1.495E-03	5.000E-01	rt	C317	0 2 0 0 0	

3062.   C₁₄H₁₈O₄

Di-n-propyl Phthalate

Dipropyl Phthalate

RN: 131-16-8   **MP** (°C):
MW: 250.30   **BP** (°C):

Solubility (Moles/L)	Solubility (Grams/L)	Temp (°C)	Ref (#)	Evaluation (T P E A A)	Comments
4.320E-04	1.081E-01	20	L300	2 1 0 2 2	

3063.   C₁₄H₁₈O₄

Diisopropyl Phthalate

bis(1-Methyl-ethyl) Phthalate

RN: 605-45-8   **MP** (°C):
MW: 250.30   **BP** (°C):

Solubility (Moles/L)	Solubility (Grams/L)	Temp (°C)	Ref (#)	Evaluation (T P E A A)	Comments
1.330E-03	3.329E-01	20	L300	2 1 0 2 2	

3064.   C₁₄H₁₈O₆

Ethyl Phthalyl Ethyl Glycollate

RN: **MP** (°C):
MW: 282.30   **BP** (°C):

Solubility (Moles/L)	Solubility (Grams/L)	Temp (°C)	Ref (#)	Evaluation (T P E A A)	Comments
1.770E-03	4.998E-01	15	H069	1 0 1 1 0	
1.770E-03	4.998E-01	ns	F014	0 0 0 0 1	

3065. $C_{14}H_{18}O_6$

Dimethoxyethyl Phthalate
1,2-Benzenedicarboxylic Acid, di(2-Methoxyethyl) Ester
RN: 34006-76-3 **MP** (°C):
MW: 282.30 **BP** (°C):

Solubility (Moles/L)	Solubility (Grams/L)	Temp (°C)	Ref (#)	Evaluation (T P E A A)	Comments
2.986E-02	8.428E+00	20	F070	1 0 0 0 1	
2.944E-02	8.310E+00	ns	F014	0 0 0 0 2	

3066. $C_{14}H_{18}O_6$

Methyl Glycol Phthalate
bis(2-Methoxyethyl) Phthalate
RN: 117-82-8 **MP** (°C):
MW: 282.30 **BP** (°C):

Solubility (Moles/L)	Solubility (Grams/L)	Temp (°C)	Ref (#)	Evaluation (T P E A A)	Comments
3.090E-02	8.723E+00	15	H069	1 0 1 1 1	

3067. $C_{14}H_{19}Cl_2NO_2$

Chlorambucil
N,N-di-(2-Chloroethyl)-γ-(p-aminophenyl)butyric Acid
Linfolysin
Elcoril
Linfolizin
Leukersan
RN: 305-03-3 **MP** (°C): 64
MW: 304.22 **BP** (°C):

Solubility (Moles/L)	Solubility (Grams/L)	Temp (°C)	Ref (#)	Evaluation (T P E A A)	Comments
<3.29E-03	<1.00E+00	30	L343	2 1 1 1 0	EFG

3068. $C_{14}H_{19}IN_2O_6$

Uridine, 2'-Deoxy-5-iodo-, 5'-Pentanoate
5'-Valeryl 5-iodo-2'-deoxyuridine
RN: 84052-69-7 **MP** (°C): 142.5
MW: 438.22 **BP** (°C):

Solubility (Moles/L)	Solubility (Grams/L)	Temp (°C)	Ref (#)	Evaluation (T P E A A)	Comments
4.000E+02	1.753E+05	25	N332	1 0 2 2 2	pH 7.4

3069. $C_{14}H_{19}IN_2O_6$
Uridine, 2'-Deoxy-5-iodo-, 5'-(2,2-Dimethylpropanoate)
5'-Pivaloyl 5-iodo-2'-deoxyuridine
5-Iodo-2'-deoxyuridine 5'-pivalate
RN: 84043-28-7   **MP** (°C):   106.5
MW: 438.22   **BP** (°C):

Solubility (Moles/L)	Solubility (Grams/L)	Temp (°C)	Ref (#)	Evaluation (T P E A A)	Comments
4.400E+02	1.928E+05	25	N332	1 0 2 2 2	pH 7.4

3070. $C_{14}H_{19}NO$
n-Pentylcinnamamide
2-Propenamide, N-Pentyl-3-phenyl-
RN: 23784-51-2   **MP** (°C):
MW: 217.31   **BP** (°C):

Solubility (Moles/L)	Solubility (Grams/L)	Temp (°C)	Ref (#)	Evaluation (T P E A A)	Comments
8.200E-05	1.782E-02	ns	H350	0 0 0 0 2	

3071. $C_{14}H_{19}NO_3$
Propanamide, 2-(Benzoyloxy)-N,N-diethyl-
RN: 115178-79-5   **MP** (°C):   53.5
MW: 249.31   **BP** (°C):

Solubility (Moles/L)	Solubility (Grams/L)	Temp (°C)	Ref (#)	Evaluation (T P E A A)	Comments
5.214E-03	1.300E+00	22	N317	1 1 2 1 2	

3072. $C_{14}H_{19}NO_3$
Acetaminophen Hexanoate
Hexanyl Acetaminophen
Hexanoic Acid, 4-(Acetylamino)phenyl Ester
4'-Hydroxyacetanilide Hexanoate
RN: 20675-21-2   **MP** (°C):   107
MW: 249.31   **BP** (°C):

Solubility (Moles/L)	Solubility (Grams/L)	Temp (°C)	Ref (#)	Evaluation (T P E A A)	Comments
7.220E-05	1.800E-02	25	B010	1 1 1 1 0	
2.286E-04	5.700E-02	37	D029	1 0 1 1 1	

3073.   C$_{14}$H$_{19}$NO$_4$
Anisomycin
(2R,3R,4R)-2-(4-Methoxybenzyl)-3,4-pyrrolidinediol-3-acetate
RN: 22862-76-6   **MP** (°C):   140.5
MW: 265.31   **BP** (°C):

Solubility (Moles/L)	Solubility (Grams/L)	Temp (°C)	Ref (#)	Evaluation (T P E A A)	Comments
2.469E-02	6.550E+00	28	A038	2 0 1 1 2	

3074.   C$_{14}$H$_{19}$N$_3$S
Methapyrilene
N,N-Dimethyl-N',2-pyridinyl-N'-(2-thienylmethyl)-1,2-ethanediamine
Cope
A3322
AH-42
Semiken
RN: 91-80-5   **MP** (°C):   <25
MW: 261.39   **BP** (°C):

Solubility (Moles/L)	Solubility (Grams/L)	Temp (°C)	Ref (#)	Evaluation (T P E A A)	Comments
2.300E-03	6.012E-01	30	L068	1 0 0 1 0	EFG
1.700E-02	4.444E+00	37.5	L034	2 2 0 1 2	pH 7.4

3075.   C$_{14}$H$_{19}$N$_3$S
Thenyldiamine
1,2-Ethanediamine, N,N-Dimethyl-N'-2-pyridinyl-N'-(3-thienylmethyl)-
N-(2-Dimethylaminoethyl)-N-2-pyridyl-3-thenylamine
Thefanil
Thenfadil
Tenfidil
RN: 91-79-2   **MP** (°C):
MW: 261.39   **BP** (°C):

Solubility (Moles/L)	Solubility (Grams/L)	Temp (°C)	Ref (#)	Evaluation (T P E A A)	Comments
1.700E-02	4.444E+00	37.5	L034	2 2 0 1 2	pH 7.4

3076.   C$_{14}$H$_{19}$N$_5$O$_4$
N,N-Diethylsuccinamyloxymethyl-1-allopurinol
Butanoic Acid, 4-(Diethylamino)-4-oxo-, (4,5-Dihydro-4-oxo-1H-pyrazolo[3,4-d]pyrimidin-1-yl)methyl Ester
RN: 98827-27-1   **MP** (°C):   138-140
MW: 321.34   **BP** (°C):

Solubility (Moles/L)	Solubility (Grams/L)	Temp (°C)	Ref (#)	Evaluation (T P E A A)	Comments
1.027E-01	3.300E+01	22	B322	1 0 2 2 2	

3077.   C₁₄H₁₉N₅O₅

9-[5'-(O-Butyryl)-β-D-arabinofuranosyl]adenine Ester
Vidarabine 5'-Butyrate
RN: 65926-30-9   **MP** (°C):
MW: 337.34   **BP** (°C):

Solubility (Moles/L)	Solubility (Grams/L)	Temp (°C)	Ref (#)	Evaluation (T P E A A)	Comments
4.773E-02	1.610E+01	ns	B134	0 1 1 1 2	

3078.   C₁₄H₁₉O₆P

Crotoxyphos
Dimethylphosphate of α-Methylbenzyl-3-hydroxy-cis-crotonate
RN: 7700-17-6   **MP** (°C):
MW: 314.28   **BP** (°C):   135

Solubility (Moles/L)	Solubility (Grams/L)	Temp (°C)	Ref (#)	Evaluation (T P E A A)	Comments
3.179E-03	9.990E-01	ns	M061	0 0 0 0 0	
3.182E-03	1.000E+00	rt	M161	0 0 0 0 0	

3079.   C₁₄H₂₀ClNO₂

Alachlor
2-Chloro-2',6'-diethyl-N-(methoxymethyl)acetanilide
RN: 15972-60-8   **MP** (°C):   39.5
MW: 269.77   **BP** (°C):

Solubility (Moles/L)	Solubility (Grams/L)	Temp (°C)	Ref (#)	Evaluation (T P E A A)	Comments
8.896E-04	2.400E-01	23	M161	1 0 0 0 2	
5.486E-04	1.480E-01	25	B200	1 0 0 0 2	
5.486E-04	1.480E-01	ns	M061	0 0 0 0 2	
5.560E-04	1.500E-01	ns	M110	0 0 0 0 0	EFG

3080.   C₁₄H₂₀N₂O

Siduron
1-(2-Methylcyclohexyl)-3-phenylurea
RN: 1982-49-6   **MP** (°C):   133
MW: 232.33   **BP** (°C):

Solubility (Moles/L)	Solubility (Grams/L)	Temp (°C)	Ref (#)	Evaluation (T P E A A)	Comments
7.748E-05	1.800E-02	25	B200	1 0 0 0 1	
7.748E-05	1.800E-02	25	G036	1 0 0 0 1	
7.748E-05	1.800E-02	25	M161	1 0 0 0 1	

3081. C$_{14}$H$_{20}$N$_2$O$_3$S
Tolcyclamide
1-Cyclohexyl-3-para-tolylsulfonylurea
Glycyclamide
RN: 664-95-9 **MP** (°C): 175
MW: 296.39 **BP** (°C):

Solubility (Moles/L)	Solubility (Grams/L)	Temp (°C)	Ref (#)	Evaluation (T P E A A)	Comments
6.194E-05	1.836E-02	37	A028	1 0 2 1 2	intrinsic
6.200E-05	1.838E-02	37	A046	2 0 1 1 2	

3082. C$_{14}$H$_{20}$N$_3$O$_5$PS
Pyrazophos
2-[(Diethoxyphosphinothioyl)oxy]-5-methylpyrazolo[1,5-a]pyrimidine-6-carboxylic Acid
Ethyl Ester
Afugan
Curamil
RN: 13457-18-6 **MP** (°C): 50.5
MW: 373.37 **BP** (°C):

Solubility (Moles/L)	Solubility (Grams/L)	Temp (°C)	Ref (#)	Evaluation (T P E A A)	Comments
1.125E-05	4.200E-03	20	A306	0 0 0 0 1	
1.125E-05	4.200E-03	20	M161	1 0 0 0 1	

3083. C$_{14}$H$_{20}$N$_4$O$_2$
2,5-Diaziridinyl-3,6-bis(dimethylamino)-1,4-benzoquinone
RN: 59886-50-9 **MP** (°C): 112
MW: 276.34 **BP** (°C):

Solubility (Moles/L)	Solubility (Grams/L)	Temp (°C)	Ref (#)	Evaluation (T P E A A)	Comments
3.619E-02	1.000E+01	rt	C317	0 2 0 0 0	

3084. C$_{14}$H$_{20}$N$_4$O$_2$
2,5-Diaziridinyl-3,6-bis(ethylamino)-1,4-benzoquinone
RN: 59886-53-2 **MP** (°C): 157
MW: 276.34 **BP** (°C):

Solubility (Moles/L)	Solubility (Grams/L)	Temp (°C)	Ref (#)	Evaluation (T P E A A)	Comments
1.809E-03	5.000E-01	rt	C317	0 2 0 0 0	

3085.   C₁₄H₂₀N₄O₂
2,5-bis(Methylaziridinyl)-3,6-bis(methylamino)-1,4-benzoquinone
RN: 64947-06-4   **MP** (°C):   179
MW: 276.34   **BP** (°C):

Solubility (Moles/L)	Solubility (Grams/L)	Temp (°C)	Ref (#)	Evaluation (T P E A A)	Comments
<3.62E-04	<1.00E-01	rt	C317	0 2 0 0 0	

3086.   C₁₄H₂₀N₄O₄
2,5-Diaziridinyl-3,6-bis(hydroxyethylamino)-1,4-benzoquinone
RN: 59886-54-3   **MP** (°C):   188
MW: 308.34   **BP** (°C):

Solubility (Moles/L)	Solubility (Grams/L)	Temp (°C)	Ref (#)	Evaluation (T P E A A)	Comments
6.486E-03	2.000E+00	rt	C317	0 2 0 0 0	

3087.   C₁₄H₂₀O₃
Heptyl p-Hydroxybenzoate
n-Heptyl 4-hydroxybenzoate
RN: 1085-12-7   **MP** (°C):   48
MW: 236.31   **BP** (°C):

Solubility (Moles/L)	Solubility (Grams/L)	Temp (°C)	Ref (#)	Evaluation (T P E A A)	Comments
2.630E-04	6.215E-02	-244	B355	1 1 1 1 2	
2.010E-04	4.750E-02	15	B355	1 1 1 1 2	
2.520E-04	5.955E-02	20	B355	1 1 1 1 2	
5.827E-03	1.377E+00	25	D081	1 2 2 1 2	*sic*
1.259E-04	2.975E-02	25	F322	2 0 1 1 0	EFG

3088.   C₁₄H₂₁NO₂
Benzenepropanamide, N-Hydroxy-α2,4,6-pentamethyl
RN: 60631-10-9   **MP** (°C):
MW: 235.33   **BP** (°C):

Solubility (Moles/L)	Solubility (Grams/L)	Temp (°C)	Ref (#)	Evaluation (T P E A A)	Comments
3.000E-04	7.060E-02	26	G076	1 0 0 0 1	

3089. C$_{14}$H$_{21}$NO$_2$
Benzeneacetamide, N-hydroxy-α-dipropyl
RN: 60631-09-6   **MP** (°C):
MW: 235.33   **BP** (°C):

Solubility (Moles/L)	Solubility (Grams/L)	Temp (°C)	Ref (#)	Evaluation (T P E A A)	Comments
1.300E-03	3.059E-01	26	G076	1 0 0 0 1	

3090. C$_{14}$H$_{21}$NO$_2$
Octyl Nicotinate
Nicotinic Acid n-Octyl Ester
RN: 70136-02-6   **MP** (°C):
MW: 235.33   **BP** (°C):

Solubility (Moles/L)	Solubility (Grams/L)	Temp (°C)	Ref (#)	Evaluation (T P E A A)	Comments
4.249E-05	1.000E-02	32	L346	1 0 0 1 2	

3091. C$_{14}$H$_{21}$NO$_2$
Heptyl p-Aminobenzoate
Heptyl 4-Aminobenzoate
RN: 14309-40-1   **MP** (°C):
MW: 235.33   **BP** (°C):

Solubility (Moles/L)	Solubility (Grams/L)	Temp (°C)	Ref (#)	Evaluation (T P E A A)	Comments
2.000E-05	4.707E-03	37	F006	1 1 2 2 1	
3.300E-05	7.766E-03	ns	M066	0 0 0 0 1	

3092. C$_{14}$H$_{21}$NO$_2$
2,6-Diisopropyl-4-acetaminophenol
3,5-Diisopropylparacetamol
4-Acetamido-2,6-diisopropylphenol
RN: 1988-14-3   **MP** (°C):
MW: 235.33   **BP** (°C):

Solubility (Moles/L)	Solubility (Grams/L)	Temp (°C)	Ref (#)	Evaluation (T P E A A)	Comments
5.844E-04	1.375E-01	25	D078	1 2 2 1 2	

3093. $C_{14}H_{21}NO_3$
4-Methoxybenzoic Acid-2-(diethylamino)ethyl Ester
Diethylaminoethyl p-Anisate
RN: 10367-84-7   **MP** (°C):
MW: 251.33   **BP** (°C):

Solubility (Moles/L)	Solubility (Grams/L)	Temp (°C)	Ref (#)	Evaluation (T P E A A)	Comments
5.300E-03	1.332E+00	ns	M066	0 0 0 0 1	

3094. $C_{14}H_{21}NO_4P$
Phenyl(di-morpholido)-phosphate
RN: **MP** (°C):
MW: 298.30   **BP** (°C):

Solubility (Moles/L)	Solubility (Grams/L)	Temp (°C)	Ref (#)	Evaluation (T P E A A)	Comments
2.583E+00	7.706E+02	25	A040	1 0 0 0 2	

3095. $C_{14}H_{21}N_3O_3$
Karbutilate
m-(3,3-Dimethylureido)phenyl-tert-butylcarbamate
Tandex
RN: 4849-32-5   **MP** (°C):   176.3
MW: 279.34   **BP** (°C):

Solubility (Moles/L)	Solubility (Grams/L)	Temp (°C)	Ref (#)	Evaluation (T P E A A)	Comments
1.163E-03	3.250E-01	20	B200	1 0 0 0 2	
1.163E-03	3.250E-01	rt	M161	0 0 0 0 2	

3096. $C_{14}H_{21}N_3O_3S$
Tolazamide
N-(((Hexahydro-1H-azepin-1-yl)amino)carbonyl)-4-methylbenzenesulfonamide
Tolinase
N-(p-Toluenesulfonyl)-N'-hexamethyleniminourea
U 17835
RN: 1156-19-0   **MP** (°C):   170
MW: 311.41   **BP** (°C):

Solubility (Moles/L)	Solubility (Grams/L)	Temp (°C)	Ref (#)	Evaluation (T P E A A)	Comments
2.100E-04	6.540E-02	30	H025	1 0 2 1 1	intrinsic

3097. C₁₄H₂₂

2-Octylbenzene
(1-Methylheptyl)benzene

RN: 777-22-0	**MP** (°C):
MW: 190.33	**BP** (°C):

Solubility (Moles/L)	Solubility (Grams/L)	Temp (°C)	Ref (#)	Evaluation (T P E A A)	Comments
1.585E-06	3.017E-04	ns	D001	0 0 0 0 2	

3098. C₁₄H₂₂N₂O

Lidocaine
2-(Diethylamino)-N-(2,6-dimethylphenyl)acetamide
2-Diethylamino-2',6'-acetoxylidide
Lignocaine
Leostesin
Xylocaine

RN: 137-58-6	**MP** (°C): 68
MW: 234.34	**BP** (°C):

Solubility (Moles/L)	Solubility (Grams/L)	Temp (°C)	Ref (#)	Evaluation (T P E A A)	Comments
1.850E-02	4.335E+00	14.5	N046	2 0 1 2 2	intrinsic
1.643E-02	3.850E+00	25	L338	1 0 1 1 2	
1.630E-02	3.820E+00	25	N046	2 0 1 2 2	intrinsic
1.750E-02	4.101E+00	30	L068	1 0 0 1 0	EFG
1.460E-02	3.421E+00	34.5	N046	2 0 1 2 2	intrinsic
1.440E-02	3.375E+00	37	N044	2 1 1 2 2	intrinsic

3099. C₁₄H₂₂N₂O₂

4-Methylaminobenzoic Acid-2-(diethyl-amino)ethyl Ester
Benzoic Acid, 4-(Methylamino)-, 2-(Diethylamino)ethyl Ester
Benzoic Acid, p-(Methylamino)-, 2-(Diethylamino)ethyl Ester

RN: 16488-52-1	**MP** (°C):
MW: 250.34	**BP** (°C):

Solubility (Moles/L)	Solubility (Grams/L)	Temp (°C)	Ref (#)	Evaluation (T P E A A)	Comments
7.750E-03	1.940E+00	ns	M066	0 0 0 0 2	

3100. C₁₄H₂₂N₂O₂

4-Aminobenzoic Acid-2-(diethyl-amino)propyl Ester
2-Diethylamino)propyl 4-Aminobenzoate

RN: 5878-13-7	**MP** (°C):
MW: 250.34	**BP** (°C):

Solubility (Moles/L)	Solubility (Grams/L)	Temp (°C)	Ref (#)	Evaluation (T P E A A)	Comments
1.290E-02	3.229E+00	ns	M066	0 0 0 0 2	

3101. $C_{14}H_{22}N_2O_3$
2,4-Diazaspiro[5.10]hexadecane-1,3,5-trione
RN: 143288-63-5   **MP** (°C):
MW: 266.34   **BP** (°C):

Solubility (Moles/L)	Solubility (Grams/L)	Temp (°C)	Ref (#)	Evaluation (T P E A A)	Comments
2.600E-05	6.925E-03	25	P350	2 1 1 1 2	intrinsic

3102. $C_{14}H_{22}N_2O_4$
Ethyl-2-methyl-2-cyclohexenyl-6-methylmalonurate
Ethyl 2-Methyl-2-cyclohexenyl-6-methylmalonurate
RN: **MP** (°C):   97.5
MW: 282.34   **BP** (°C):

Solubility (Moles/L)	Solubility (Grams/L)	Temp (°C)	Ref (#)	Evaluation (T P E A A)	Comments
1.000E-03	2.823E-01	23	B152	1 2 1 1 1	pH 3.5

3103. $C_{14}H_{22}N_2O_5$
Methoxymethyl-2-methyl-2-cyclohexenyl-6-methylmalonurate
RN: **MP** (°C):   73
MW: 298.34   **BP** (°C):

Solubility (Moles/L)	Solubility (Grams/L)	Temp (°C)	Ref (#)	Evaluation (T P E A A)	Comments
3.800E-03	1.134E+00	23	B152	1 2 1 1 1	pH 3.5

3104. $C_{14}H_{22}O$
Methyl Ionone
6-Methylionone
RN: 1335-46-2   **MP** (°C):
MW: 206.33   **BP** (°C):

Solubility (Moles/L)	Solubility (Grams/L)	Temp (°C)	Ref (#)	Evaluation (T P E A A)	Comments
9.693E-05	2.000E-02	25	M350	1 0 1 1 1	

3105. $C_{14}H_{22}O$
o-n-Octylphenol
2-n-Octylphenol
RN: 949-13-3   **MP** (°C):
MW: 206.33   **BP** (°C):

Solubility (Moles/L)	Solubility (Grams/L)	Temp (°C)	Ref (#)	Evaluation (T P E A A)	Comments
1.385E-05	2.857E-03	25	L022	1 0 0 0 0	

3106. $C_{14}H_{22}O$
p-n-Octylphenol
4-Octylphenol
RN: 1806-26-4 **MP** (°C): 44.5
MW: 206.33 **BP** (°C):

Solubility (Moles/L)	Solubility (Grams/L)	Temp (°C)	Ref (#)	Evaluation (T P E A A)	Comments
6.107E-05	1.260E-02	20.5	A335	1 0 2 2 2	
6.120E-05	1.263E-02	20.5	A335	1 0 2 2 2	
8.812E-06	1.818E-03	25	L022	1 0 0 0 0	

3107. $C_{14}H_{23}O_3P$
Dibutyl Phenyl Phosphonate
Dibutoxyphenylphosphine Oxide
Dibutyl Phenylphosphonate
RN: 1024-34-6 **MP** (°C):
MW: 270.31 **BP** (°C):

Solubility (Moles/L)	Solubility (Grams/L)	Temp (°C)	Ref (#)	Evaluation (T P E A A)	Comments
<7.40E-04	<2.00E-01	25	B070	1 2 0 1 0	

3108. $C_{14}H_{24}NO_4PS_3$
Bensulide
O,O-bis(1-Methylethyl) S-(2-((Phenylsulfonyl)amino)ethyl) Phosphorodithioate
Betasan
Betamec
Exporsan
Benzulfide
RN: 741-58-2 **MP** (°C): 34.4
MW: 397.52 **BP** (°C):

Solubility (Moles/L)	Solubility (Grams/L)	Temp (°C)	Ref (#)	Evaluation (T P E A A)	Comments
6.289E-05	2.500E-02	20	B200	1 2 0 0 1	
6.289E-05	2.500E-02	rt	M161	0 0 0 0 1	

3109. $C_{14}H_{24}N_2O_3$
p-5-Ethyl-5-methylhexylcarbinylbarbituric Acid
RN: **MP** (°C):
MW: 268.36 **BP** (°C):

Solubility (Moles/L)	Solubility (Grams/L)	Temp (°C)	Ref (#)	Evaluation (T P E A A)	Comments
1.543E-03	4.140E-01	ns	T003	0 0 0 0 2	

3110.   C$_{14}$H$_{24}$N$_2$O$_3$
5-Ethyl-5-n-octylbarbituric Acid
2,4,6(1H,3H,5H)-Pyrimidinetrione, 5-Ethyl-5-octyl-
RN: 64810-90-8   **MP** (°C):
MW: 268.36   **BP** (°C):

Solubility (Moles/L)	Solubility (Grams/L)	Temp (°C)	Ref (#)	Evaluation (T P E A A)	Comments
1.140E-04	3.059E-02	25	M310	2 2 2 2 2	

3111.   C$_{14}$H$_{24}$O$_2$
3-Hydroxy-2,5-dispirocyclohexyltetrahydrofuran
7-Oxadispiro[5.1.5.2]pentadecan-14-ol
RN: 29839-63-2   **MP** (°C):
MW: 224.35   **BP** (°C):

Solubility (Moles/L)	Solubility (Grams/L)	Temp (°C)	Ref (#)	Evaluation (T P E A A)	Comments
3.098E-02	6.951E+00	rt	B066	0 2 0 0 0	contains impurity

3112.   C$_{14}$H$_{26}$O$_4$
1,12-Dodecanedicarboxylic Acid
Tetradecanedioic Acid
RN: 821-38-5   **MP** (°C):   127
MW: 258.36   **BP** (°C):

Solubility (Moles/L)	Solubility (Grams/L)	Temp (°C)	Ref (#)	Evaluation (T P E A A)	Comments
7.741E-04	2.000E-01	21	B040	1 0 1 1 0	sic

3113.   C$_{14}$H$_{27}$NO$_2$
Pentanamide, N-hydroxy-α,α-dipropyl
RN: **MP** (°C):
MW: 241.38   **BP** (°C):

Solubility (Moles/L)	Solubility (Grams/L)	Temp (°C)	Ref (#)	Evaluation (T P E A A)	Comments
5.000E-04	1.207E-01	26	G076	1 0 0 0 1	

3114.   C$_{14}$H$_{28}$NO$_3$PS$_2$
Piperophos
S-(2-(2-Methyl-1-piperidinyl)-2-oxoethyl) O,O-dipropyl Phosphorodithioate
RN: 24151-93-7   **MP** (°C):
MW: 353.49   **BP** (°C):

Solubility (Moles/L)	Solubility (Grams/L)	Temp (°C)	Ref (#)	Evaluation (T P E A A)	Comments
7.072E-05	2.500E-02	20	M161	1 0 0 0 1	

3115. $C_{14}H_{28}N_2O_2$
N,N,N',N'-Tetramethylsebacamide
Decanediamide, N,N,N',N'-Tetramethyl-
RN: 13424-83-4   **MP** (°C):
MW: 256.39   **BP** (°C):

Solubility (Moles/L)	Solubility (Grams/L)	Temp (°C)	Ref (#)	Evaluation (T P E A A)	Comments
5.270E-01	1.351E+02	30	D010	1 2 1 1 2	

3116. $C_{14}H_{28}O_2$
Myristic Acid
Tetradecanoic Acid
Crodacid
1-Tridecanecarboxylic Acid
RN: 544-63-8   **MP** (°C):   54
MW: 228.38   **BP** (°C):

Solubility (Moles/L)	Solubility (Grams/L)	Temp (°C)	Ref (#)	Evaluation (T P E A A)	Comments
5.692E-05	1.300E-02	0	B136	1 0 2 1 1	
5.692E-05	1.300E-02	0.0	R001	1 1 1 1 1	
8.757E-05	2.000E-02	20	B136	1 0 2 1 1	
8.757E-05	2.000E-02	20	D041	1 0 0 0 0	
8.757E-05	2.000E-02	20	R001	1 1 1 1 1	
4.700E-06	1.073E-03	25	J001	1 0 2 1 1	average of 2
8.000E-07	1.827E-04	25	R002	1 2 2 2 2	intrinsic
3.710E-06	8.473E-04	25	R002	1 2 2 2 2	
9.633E-05	2.200E-02	30	B136	1 0 2 1 1	
1.051E-04	2.400E-02	30	R001	1 1 1 1 1	
1.270E-04	2.900E-02	40	B136	1 0 2 1 1	
1.270E-04	2.900E-02	45	B136	1 0 2 1 1	
1.270E-04	2.900E-02	45	R001	1 1 1 1 1	
1.839E-05	4.200E-03	50	E005	2 1 1 2 1	
9.700E-06	2.215E-03	50	J001	1 0 2 1 1	
1.489E-04	3.400E-02	60	B136	1 0 2 1 1	
2.452E-05	5.600E-03	60	E005	2 1 1 2 1	
1.489E-04	3.400E-02	60	R001	1 1 1 1 1	

3117. $C_{14}H_{28}O_4$
1,3-Dioxolane-4-methanol, 2-[2-(Heptyloxy)ethyl]-2-methyl
RN: 143458-57-5   **MP** (°C):
MW: 260.38   **BP** (°C):

Solubility (Moles/L)	Solubility (Grams/L)	Temp (°C)	Ref (#)	Evaluation (T P E A A)	Comments
4.440E-03	1.156E+00	25	P342	1 2 2 2 2	0.0001M Na_2CO_3

3118. C$_{14}$H$_{29}$NO$_2$
Benzenepropanamide, N-Hydroxy- α2,3-pentamethyl
Octanamide, N-Hydroxy-2,2-dipropyl
RN: 60631-08-5   **MP** (°C):
MW: 243.39   **BP** (°C):

Solubility (Moles/L)	Solubility (Grams/L)	Temp (°C)	Ref (#)	Evaluation (T P E A A)	Comments
4.500E-04	1.095E-01	26	G076	1 0 0 0 1	
1.500E-03	3.651E-01	26	G076	1 0 0 0 1	

3119. C$_{14}$H$_{29}$NO$_2$
Octanamide, 2,2,4-triethyl-N-hydroxy
RN: 60631-07-4   **MP** (°C):
MW: 243.39   **BP** (°C):

Solubility (Moles/L)	Solubility (Grams/L)	Temp (°C)	Ref (#)	Evaluation (T P E A A)	Comments
4.500E-04	1.095E-01	26	G076	1 0 0 0 1	

3120. C$_{14}$H$_{29}$NO$_2$
Tetradecanamide, N-hydroxy
Myristohydroxamic Acid
N-Hydroxytetradecanamide
RN: 17698-03-2   **MP** (°C):
MW: 243.39   **BP** (°C):

Solubility (Moles/L)	Solubility (Grams/L)	Temp (°C)	Ref (#)	Evaluation (T P E A A)	Comments
1.000E-04	2.434E-02	26	G076	1 0 0 0 1	

3121. C$_{14}$H$_{29}$NO$_2$
Decanamide, 2,2-Diethyl-N-hydroxy
RN: 60631-06-3   **MP** (°C):
MW: 243.39   **BP** (°C):

Solubility (Moles/L)	Solubility (Grams/L)	Temp (°C)	Ref (#)	Evaluation (T P E A A)	Comments
6.000E-06	1.460E-03	26	G076	1 0 0 0 1	

3122.   C$_{14}$H$_{29}$NO$_2$
Hexanamide, 2,2-Dibutyl-N-hydroxy
2,2-Dibutyl-N-hydroxyhexanamide
Tri-n-butylacetohydroxamic Acid
RN: 52061-82-2   **MP** (°C):
MW: 243.39   **BP** (°C):

Solubility (Moles/L)	Solubility (Grams/L)	Temp (°C)	Ref (#)	Evaluation (T P E A A)	Comments
7.000E-05	1.704E-02	26	G076	1 0 0 0 1	

3123.   C$_{14}$H$_{29}$NO$_2$
Dodecanamide, N-Hydroxy-2,2-dimethyl
RN: 60631-05-2   **MP** (°C):
MW: 243.39   **BP** (°C):

Solubility (Moles/L)	Solubility (Grams/L)	Temp (°C)	Ref (#)	Evaluation (T P E A A)	Comments
1.600E-05	3.894E-03	26	G076	1 0 0 0 1	

3124.   C$_{14}$H$_{29}$NO$_2$
Pentanamide, N-hydroxy-4-methyl-2,2-bis(2-methylpropyl)
RN: 60469-53-6   **MP** (°C):
MW: 243.39   **BP** (°C):

Solubility (Moles/L)	Solubility (Grams/L)	Temp (°C)	Ref (#)	Evaluation (T P E A A)	Comments
1.000E+01	2.434E+03	26	G076	1 0 0 0 1	

3125.   C$_{14}$H$_{30}$
n-Tetradecane
Tetradecane
RN: 629-59-4   **MP** (°C):   5.89
MW: 198.40   **BP** (°C):   253.7

Solubility (Moles/L)	Solubility (Grams/L)	Temp (°C)	Ref (#)	Evaluation (T P E A A)	Comments
1.663E-09	3.300E-07	23	C332	2 0 2 2 1	
3.500E-08	6.944E-06	25	F004	1 2 2 2 1	
1.159E-08	2.300E-06	ns	H123	0 0 0 0 2	

3126.   C$_{14}$H$_{30}$O
Tetradecanol
RN: 27196-00-5   **MP** (°C):
MW: 214.39   **BP** (°C):

Solubility (Moles/L)	Solubility (Grams/L)	Temp (°C)	Ref (#)	Evaluation (T P E A A)	Comments
1.460E-06	3.130E-04	25	R002	1 2 2 2 2	

3127.   C₁₄H₃₀O

Myristyl Alcohol
Tetradecanol

RN: 112-72-1   **MP** (°C):   38
MW: 214.39   **BP** (°C):   289

Solubility (Moles/L)	Solubility (Grams/L)	Temp (°C)	Ref (#)	Evaluation (T P E A A)	Comments
9.049E-08	1.940E-05	4	H030	2 2 2 2 2	
9.049E-08	1.940E-05	4	H103	1 2 2 2 2	
8.909E-07	1.910E-04	25	H103	1 2 2 2 2	
5.737E-07	1.230E-04	32	H030	2 2 2 2 2	
5.737E-07	1.230E-04	32	H103	1 2 2 2 2	
1.105E-06	2.370E-04	45	H030	2 2 2 2 2	
1.105E-06	2.370E-04	45	H103	1 2 2 2 2	
2.094E-06	4.490E-04	61	H030	2 2 2 2 2	
2.094E-06	4.490E-04	61	H103	1 2 2 2 2	

3128.   C₁₄H₃₁O₂P

Ethyl Dihexyl Phosphinate
Phosphinic Acid, Dihexyl-, Ethyl Ester

RN: 113977-19-8   **MP** (°C):
MW: 262.38   **BP** (°C):

Solubility (Moles/L)	Solubility (Grams/L)	Temp (°C)	Ref (#)	Evaluation (T P E A A)	Comments
<3.81E-04	<1.00E-01	25	B070	1 2 0 1 0	

3129.   C₁₄H₃₁O₃P

Dibutyl Hexyl Phosphonate
Phosphinic Acid, Hexyl-, Dibutyl Ester

RN: 5929-66-8   **MP** (°C):
MW: 278.38   **BP** (°C):

Solubility (Moles/L)	Solubility (Grams/L)	Temp (°C)	Ref (#)	Evaluation (T P E A A)	Comments
<7.18E-04	<2.00E-01	25	B070	1 2 0 1 0	

3130.   C₁₄H₃₁O₃P

Diethyl Hexyl Phosphonate
Phosphinic Acid, Hexyl-, Diethyl Ester

RN: 16165-66-5   **MP** (°C):
MW: 278.38   **BP** (°C):

Solubility (Moles/L)	Solubility (Grams/L)	Temp (°C)	Ref (#)	Evaluation (T P E A A)	Comments
2.155E-03	6.000E-01	25	B070	1 2 0 1 0	

3131. $C_{14}H_{31}O_4P$
Diethyl Decyl Phosphate
Phosphoric Acid, Decyl Ester
RN: 20195-16-8   **MP** (°C):
MW: 294.37   **BP** (°C):

Solubility (Moles/L)	Solubility (Grams/L)	Temp (°C)	Ref (#)	Evaluation (T P E A A)	Comments
<3.40E-04	<1.00E-01	25	B070	1 2 0 1 0	

3132. $C_{14}H_{31}O_4P$
Dibutyl Hexyl Phosphate
Phosphoric Acid, Dibutyl Hexyl Ester
RN: 80421-90-5   **MP** (°C):
MW: 294.37   **BP** (°C):

Solubility (Moles/L)	Solubility (Grams/L)	Temp (°C)	Ref (#)	Evaluation (T P E A A)	Comments
<3.40E-04	<1.00E-01	25	B070	1 2 0 1 0	

3133. $C_{14}H_{31}O_5P$
Dibutyl Ethoxybutyl Phosphate
RN: 100888-67-3   **MP** (°C):
MW: 310.37   **BP** (°C):

Solubility (Moles/L)	Solubility (Grams/L)	Temp (°C)	Ref (#)	Evaluation (T P E A A)	Comments
2.255E-03	7.000E-01	25	B070	1 2 0 1 0	

3134. $C_{15}H_{10}$
4,5-Methylenephenanthrene
4H-Cyclopenta[def]phenanthrene
RN: 203-64-5   **MP** (°C):   76
MW: 190.25   **BP** (°C):

Solubility (Moles/L)	Solubility (Grams/L)	Temp (°C)	Ref (#)	Evaluation (T P E A A)	Comments
5.782E-06	1.100E-03	27	D003	1 0 0 1 1	

3135.   C$_{15}$H$_{10}$Cl$_2$N$_2$O$_2$

Lorazepam
Alzapam
Ativan
Apo-Lorazepam
7-Chloro-5-(o-chlorophenyl)-1,3-dihydro-3-hydroxy-2H-1,4-benzodiazepin-2-one
RN: 846-49-1   **MP** (°C):   167
MW: 321.17   **BP** (°C):

Solubility (Moles/L)	Solubility (Grams/L)	Temp (°C)	Ref (#)	Evaluation (T P E A A)	Comments
1.681E-04	5.400E-02	ns	N315	0 2 2 1 2	pH 7.09

3136.   C$_{15}$H$_{10}$O$_2$

9-Anthracenecarboxylic Acid
Anthracene-9-carboxylic Acid
RN: 723-62-6   **MP** (°C):   214
MW: 222.25   **BP** (°C):

Solubility (Moles/L)	Solubility (Grams/L)	Temp (°C)	Ref (#)	Evaluation (T P E A A)	Comments
3.824E-04	8.499E-02	24	H106	1 0 2 2 2	
3.825E-07	8.500E-05	ns	M349	0 2 1 1 2	

3137.   C$_{15}$H$_{10}$O$_4$S

7-Methylthio-2-xanthonecarboxylic Acid
RN: 40363-76-6   **MP** (°C):
MW: 286.31   **BP** (°C):

Solubility (Moles/L)	Solubility (Grams/L)	Temp (°C)	Ref (#)	Evaluation (T P E A A)	Comments
9.081E-07	2.600E-04	25	C059	1 2 1 1 1	

3138.   C$_{15}$H$_{10}$O$_5$S

7-Methylsulfinyl-2-xanthonecarboxylic Acid
RN: 40691-50-7   **MP** (°C):
MW: 302.31   **BP** (°C):

Solubility (Moles/L)	Solubility (Grams/L)	Temp (°C)	Ref (#)	Evaluation (T P E A A)	Comments
9.064E-06	2.740E-03	25	C059	1 2 1 1 2	

3139. $C_{15}H_{10}O_6$
Eriodictyol
5,7,3',4'-Tetra-hydroxyflavon
RN: 552-58-9 **MP** (°C): 257dec
MW: 286.24 **BP** (°C):

Solubility (Moles/L)	Solubility (Grams/L)	Temp (°C)	Ref (#)	Evaluation (T P E A A)	Comments
2.445E-04	7.000E-02	20	F300	1 0 0 0 1	
6.987E-04	2.000E-01	100	F300	1 0 0 0 2	

3140. $C_{15}H_{10}O_7$
Morin
3,5,7,2',4',-Penta-hydroxyflavon
RN: 480-16-0 **MP** (°C): 299.5
MW: 302.24 **BP** (°C):

Solubility (Moles/L)	Solubility (Grams/L)	Temp (°C)	Ref (#)	Evaluation (T P E A A)	Comments
8.271E-04	2.500E-01	20	F300	1 0 0 0 1	
2.978E-03	9.000E-01	100	F300	1 0 0 0 0	

3141. $C_{15}H_{11}ClN_2O_2$
Oxazepam
Serax
7-Chloro-1,3-dihydro-3-hydroxy-5-phenyl-2H-1,4-benzodiazepin-2-one
Apo-Oxazepam
Abboxampam
RN: 604-75-1 **MP** (°C): 205.5
MW: 286.72 **BP** (°C):

Solubility (Moles/L)	Solubility (Grams/L)	Temp (°C)	Ref (#)	Evaluation (T P E A A)	Comments
6.975E-05	2.000E 02	22	N319	1 0 2 2 0	
7.673E-05	2.200E-02	c	B362	1 0 2 2 2	

3142. $C_{15}H_{11}ClO_3$
Chlorflurecol-methyl
Chlorflurenol
Methyl-2-chloro-9-hydroxyfluorene-9-carboxylate
RN: 2536-31-4 **MP** (°C): 152
MW: 274.71 **BP** (°C):

Solubility (Moles/L)	Solubility (Grams/L)	Temp (°C)	Ref (#)	Evaluation (T P E A A)	Comments
6.552E-05	1.800E-02	20	A308	1 0 0 0 1	
7.936E-05	2.180E-02	20	B200	1 0 0 0 2	
6.552E-05	1.800E-02	20	M161	1 0 0 0 1	

3143. C$_{15}$H$_{11}$NO$_2$
C.I. Disperse Red 9
1-(Methylamino)-9,10-anthraquinone
Serilene Fast Pink BT
Smoke Red M
RN: 82-38-2   **MP** (°C):   161
MW: 237.26   **BP** (°C):

Solubility (Moles/L)	Solubility (Grams/L)	Temp (°C)	Ref (#)	Evaluation (T P E A A)	Comments
3.100E-07	7.355E-05	25	B333	1 0 0 0 1	

3144. C$_{15}$H$_{11}$NO$_2$
C.I. Disperse Orange 11
1-Amino-2-methylanthraquinone
2-Methyl-1-anthraquinonylamine
Acetate Fast Orange R
RN: 82-28-0   **MP** (°C):   208
MW: 237.26   **BP** (°C):

Solubility (Moles/L)	Solubility (Grams/L)	Temp (°C)	Ref (#)	Evaluation (T P E A A)	Comments
1.400E-06	3.322E-04	25	B333	1 0 0 0 1	

3145. C$_{15}$H$_{11}$N$_3$O$_3$
Nitrazepam
1,3-Dihydro-7-nitro-5-phenyl-2H-1,4-benzodiazepin-2-one
Mogadon
Unisomnia
RN: 146-22-5   **MP** (°C):   224
MW: 281.27   **BP** (°C):

Solubility (Moles/L)	Solubility (Grams/L)	Temp (°C)	Ref (#)	Evaluation (T P E A A)	Comments
1.529E-04	4.300E-02	30	O321	2 2 2 2 1	

3146. C$_{15}$H$_{12}$
9-Methylanthracene
RN: 779-02-2   **MP** (°C):   79
MW: 192.26   **BP** (°C):   196

Solubility (Moles/L)	Solubility (Grams/L)	Temp (°C)	Ref (#)	Evaluation (T P E A A)	Comments
1.358E-06	2.610E-04	25	M064	1 1 2 2 2	
1.330E-06	2.557E-04	25	M342	1 0 1 1 2	
1.358E-06	2.610E-04	ns	M344	0 0 0 0 2	

3147.   C₁₅H₁₂

2-Methylanthracene

RN: 613-12-7   **MP** (°C):   204
MW: 192.26   **BP** (°C):

Solubility (Moles/L)	Solubility (Grams/L)	Temp (°C)	Ref (#)	Evaluation (T P E A A)	Comments
3.672E-08	7.060E-06	6.30	M063	2 1 2 2 2	
3.670E-08	7.056E-06	6.30	M082	1 1 1 2 2	
3.670E-08	7.056E-06	6.30	M151	2 1 2 2 2	
3.675E-08	7.066E-06	6.34	M183	1 2 1 1 2	
4.411E-08	8.480E-06	9.10	M063	2 1 2 2 2	
4.410E-08	8.479E-06	9.10	M082	1 1 1 2 2	
4.410E-08	8.479E-06	9.10	M151	2 1 2 2 2	
4.414E-08	8.487E-06	9.14	M183	1 2 1 1 2	
4.905E-08	9.430E-06	10.80	M063	2 1 2 2 2	
4.900E-08	9.421E-06	10.80	M082	1 1 1 2 2	
4.900E-08	9.421E-06	10.80	M151	2 1 2 2 2	
4.909E-08	9.438E-06	10.84	M183	1 2 1 1 2	
5.773E-08	1.110E-05	13.90	M063	2 1 2 2 2	
5.750E-08	1.106E-05	13.90	M082	1 1 1 2 2	
5.750E-08	1.106E-05	13.90	M151	2 1 2 2 2	
5.778E-08	1.111E-05	13.94	M183	1 2 1 1 2	
7.542E-08	1.450E-05	18.30	M063	2 1 2 2 2	
7.540E-08	1.450E-05	18.30	M082	1 1 1 2 2	
7.540E-08	1.450E-05	18.30	M151	2 1 2 2 2	
7.550E-08	1.452E-05	18.34	M183	1 2 1 1 2	
9.934E-08	1.910E-05	23.10	M063	2 1 2 2 2	
9.940E-08	1.911E-05	23.10	M082	1 1 1 2 2	
9.940E-08	1.911E-05	23.10	M151	2 1 2 2 2	
9.944E-08	1.912E-05	23.14	M183	1 2 1 1 2	
2.028E-07	3.900E-05	25	M064	1 1 2 2 1	
1.108E-07	2.130E-05	25.00	M151	2 1 1 2 2	
1.259E-07	2.420E-05	27.00	M063	2 1 2 2 2	
1.260E-07	2.423E-05	27.00	M082	1 1 1 2 2	
1.260E-07	2.423E-05	27.00	M151	2 1 2 2 2	
1.260E-07	2.423E-05	27.04	M183	1 2 1 1 2	
1.670E-07	3.210E-05	31.10	M063	2 1 2 2 2	
1.670E-07	3.211E-05	31.10	M082	1 1 1 2 2	
1.670E-07	3.211E-05	31.10	M151	2 1 2 2 2	
1.671E-07	3.213E-05	31.14	M183	1 2 1 1 2	

3148. C₁₅H₁₂

1-Methylphenanthrene

RN: 832-69-9   **MP** (°C):   118
MW: 192.26   **BP** (°C):   358

Solubility (Moles/L)	Solubility (Grams/L)	Temp (°C)	Ref (#)	Evaluation (T P E A A)	Comments
4.952E-07	9.520E-05	6.60	M063	2 1 2 2 2	
4.950E-07	9.517E-05	6.60	M082	1 1 1 2 2	
4.950E-07	9.517E-05	6.60	M151	2 1 2 2 2	
4.956E-06	9.529E-04	6.64	M183	1 2 1 1 2	
5.929E-07	1.140E-04	8.90	M063	2 1 2 2 2	
5.940E-07	1.142E-04	8.90	M082	1 1 1 2 2	
5.940E-07	1.142E-04	8.90	M151	2 1 2 2 2	
5.933E-07	1.141E-04	8.94	M183	1 2 1 1 2	
7.646E-07	1.470E-04	14.00	M063	2 1 2 2 2	
7.650E-07	1.471E-04	14.00	M082	1 1 1 2 2	
7.650E-07	1.471E-04	14.00	M151	2 1 2 2 2	
7.650E-07	1.471E-04	14.04	M183	1 2 1 1 2	
1.004E-06	1.930E-04	19.20	M063	2 1 2 2 2	
1.010E-06	1.942E-04	19.20	M082	1 1 1 2 2	
1.010E-06	1.942E-04	19.20	M151	2 1 2 2 2	
1.004E-06	1.931E-04	19.24	M183	1 2 1 1 2	
1.326E-06	2.550E-04	24.10	M063	2 1 2 2 2	
1.320E-06	2.538E-04	24.10	M082	1 1 1 2 2	
1.320E-06	2.538E-04	24.10	M151	2 1 2 2 2	
1.327E-06	2.552E-04	24.14	M183	1 2 1 1 2	
1.399E-06	2.690E-04	25.00	M151	2 1 1 2 2	
1.581E-06	3.040E-04	26.90	M063	2 1 2 2 2	
1.580E-06	3.038E-04	26.90	M082	1 1 1 2 2	
1.580E-06	3.038E-04	26.90	M151	2 1 2 2 2	
1.583E-06	3.043E-04	26.94	M183	1 2 1 1 2	
1.846E-06	3.550E-04	29.90	M063	2 1 2 2 2	
1.850E-06	3.557E-04	29.90	M082	1 1 1 2 2	
1.850E-06	3.557E-04	29.90	M151	2 1 2 2 2	
1.848E-06	3.553E-04	29.94	M183	1 2 1 1 2	

3149. C₁₅H₁₂Cl₂O₃

Ethanol, 2-(2,4-Dicholrophenoxy)-, Benzoate
Benzoate, 2-(2,4-Dichlorophenoxy)ethyl-
2,4-DEB

RN: 94-83-7   **MP** (°C):   74
MW: 311.17   **BP** (°C):

Solubility (Moles/L)	Solubility (Grams/L)	Temp (°C)	Ref (#)	Evaluation (T P E A A)	Comments
1.543E-04	4.800E-02	ns	B185	1 0 0 0 1	

3150. $C_{15}H_{12}Cl_2O_3$
2,4-Dichlorophenoxyacetic Acid Benzyl Ester
Benzyl 2,4-Dichlorophenoxyacetate
2,4-DBE
RN: 13246-97-4 **MP** (°C):
MW: 311.17 **BP** (°C):

Solubility (Moles/L)	Solubility (Grams/L)	Temp (°C)	Ref (#)	Evaluation (T P E A A)	Comments
4.955E-05	1.542E-02	ns	M120	0 0 1 1 2	

3151. $C_{15}H_{12}I_3NO_4$
Liothyronine
3,3',5-Triiodothyronine
RN: 6893-02-3 **MP** (°C): 236dec
MW: 650.98 **BP** (°C):

Solubility (Moles/L)	Solubility (Grams/L)	Temp (°C)	Ref (#)	Evaluation (T P E A A)	Comments
6.080E-06	3.958E-03	37	L094	2 0 0 1 2	pH 4-5, zwitterion

3152. $C_{15}H_{12}N_2O$
5H-Dibenz[b,f]azepine-5-carboxamide
Carbazepine
5-Carbamoyl-5H-Dibenz[b,f]azepine
Iminostilbene
Carbamazepine
Epitol
RN: 298-46-4 **MP** (°C): 190-193
MW: 236.28 **BP** (°C):

Solubility (Moles/L)	Solubility (Grams/L)	Temp (°C)	Ref (#)	Evaluation (T P E A A)	Comments
4.655E-04	1.100E-01	20	B196	1 0 0 0 1	
4.700E-04	1.110E-01	20	B196	1 0 0 0 1	
4.000E-03	9.451E-01	rt	B397	1 0 2 2 2	EFG

3153. $C_{15}H_{12}N_2O_2$
Phenytoin
5,5-Diphenyl-2,4-imidazolidinedione
Dilantin
5,5-Diphenylhydantoin
Ekko
Zentropil
RN: 57-41-0   **MP** (°C):   296.5
MW: 252.28   **BP** (°C):

Solubility (Moles/L)	Solubility (Grams/L)	Temp (°C)	Ref (#)	Evaluation (T P E A A)	Comments
3.765E-04	9.499E-02	0	B114	1 1 1 2 1	pH 6-7
1.268E-04	3.200E-02	22	B154	1 1 1 1 1	0.1M HCl
5.549E-05	1.400E-02	25	P061	1 0 0 0 2	pH 1-7
1.526E-04	3.850E-02	37	F183	1 0 1 1 2	intrinsic
2.600E-04	6.559E-02	50	M335	1 0 2 1 2	pH 5

3154. $C_{15}H_{12}N_2O_2$
Disperse Violet 4
1-Amino-4-(N-methylamino)anthraquinone
Interchem Acetate Violet 6B
RN: 1220-94-6   **MP** (°C):   193
MW: 252.28   **BP** (°C):

Solubility (Moles/L)	Solubility (Grams/L)	Temp (°C)	Ref (#)	Evaluation (T P E A A)	Comments
2.300E-06	5.802E-04	25	B333	1 0 0 0 1	

3155. $C_{15}H_{12}N_2O_3$
5-Phenyl-5-(p-hydroxy)phenyl-hydantoin
DL-5-(p-Hydroxyphenyl-5-phenylhydantoin
p-Hydroxyphenytoin
Hydroxydiphenylhydantoin
p-Hydroxydiphenylhydantoin
RN: 2784-27-2   **MP** (°C):
MW: 268.27   **BP** (°C):

Solubility (Moles/L)	Solubility (Grams/L)	Temp (°C)	Ref (#)	Evaluation (T P E A A)	Comments
1.342E-04	3.600E-02	37	F183	1 0 1 1 2	intrinsic

3156. $C_{15}H_{12}N_2O_3$

Furfurin
1H-Imidazole, 2,4,5-tri-2-furanyl-4,5-dihydro-
2-Imidazoline, 2,4,5-Tri-2-furyl-
RN: 550-23-2   **MP** (°C):
MW: 268.27   **BP** (°C):

Solubility (Moles/L)	Solubility (Grams/L)	Temp (°C)	Ref (#)	Evaluation (T P E A A)	Comments
7.455E-04	2.000E-01	8	F300	1 0 0 0 0	
2.870E-02	7.700E+00	100	F300	1 0 0 0 1	

3157. $C_{15}H_{12}O_4$

Benzoyl-r-mandelic Acid
p-Benzoylmandelic Acid
RN: 100915-04-6   **MP** (°C): 177
MW: 256.26   **BP** (°C):

Solubility (Moles/L)	Solubility (Grams/L)	Temp (°C)	Ref (#)	Evaluation (T P E A A)	Comments
1.980E-02	5.074E+00	0	A043	1 2 1 1 1	
1.980E-02	5.074E+00	0	L035	1 2 2 1 1	
2.327E-02	5.964E+00	10	A043	1 2 1 1 1	
2.327E-02	5.964E+00	10	L035	1 2 2 1 1	
2.520E-02	6.458E+00	15	A043	1 2 1 1 1	
2.520E-02	6.458E+00	15	L035	1 2 2 1 1	
2.828E-02	7.247E+00	20	A043	1 2 1 1 1	
2.828E-02	7.247E+00	20	L035	1 2 2 1 1	
3.059E-02	7.838E+00	25	A043	1 2 1 1 1	
3.059E-02	7.838E+00	25	L035	1 2 2 1 1	
3.557E-02	9.116E+00	30	A043	1 2 1 1 1	
3.557E-02	9.116E+00	30	L035	1 2 2 1 1	
4.017E-02	1.029E+01	35	A043	1 2 1 1 2	
4.017E-02	1.029E+01	35	L035	1 2 2 1 2	
4.894E-02	1.254E+01	40	A043	1 2 1 1 2	
4.894E-02	1.254E+01	40	L035	1 2 2 1 2	
6.032E-02	1.546E+01	45	A043	1 2 1 1 2	
6.032E-02	1.546E+01	45	L035	1 2 2 1 2	
7.201E-02	1.845E+01	50	A043	1 2 1 1 2	
7.201E-02	1.845E+01	50	L035	1 2 2 1 2	

3158.   C$_{15}$H$_{12}$O$_4$

Benzoic Acid, 2-(Acetyloxy)-, Phenyl Ester
Phennin
Phenyl 2-Acetoxybenzoate
Vesipyrin
Spiroform
Phenyl Acetylsalicylate
RN: 134-55-4   **MP** (°C):   97.5
MW: 256.26   **BP** (°C):

Solubility (Moles/L)	Solubility (Grams/L)	Temp (°C)	Ref (#)	Evaluation (T P E A A)	Comments
7.805E-05	2.000E-02	21	N335	1 2 1 1 2	

3159.   C$_{15}$H$_{13}$Cl$_3$O$_2$

2-p-Methoxyphenyl-2-p-hydroxyphenyl-1,1,1-trichloro-ethane
Phenol, 4-[2,2,2-Trichloro-1-(4-methoxyphenyl)ethyl]-
RN: 28463-03-8   **MP** (°C):   112-114
MW: 331.63   **BP** (°C):

Solubility (Moles/L)	Solubility (Grams/L)	Temp (°C)	Ref (#)	Evaluation (T P E A A)	Comments
2.412E-06	8.000E-04	ns	K117	0 1 2 1 1	

3160.   C$_{15}$H$_{13}$FO$_2$

Flurbiprofen
3-Fluoro-4-phenylhydratropic Acid
Froben
Ansaid
RN: 5104-49-4   **MP** (°C):   110
MW: 244.27   **BP** (°C):

Solubility (Moles/L)	Solubility (Grams/L)	Temp (°C)	Ref (#)	Evaluation (T P E A A)	Comments
2.530E-05	6.180E-03	5	F306	1 0 1 2 2	intrinsic
1.332E-04	3.254E-02	25	C314	1 1 2 2 2	
1.331E-04	3.250E-02	25	C314	1 1 2 2 2	
3.870E-05	9.453E-03	25	F306	1 0 1 2 2	intrinsic
1.940E-04	4.739E-02	25	O303	1 0 0 1 0	EFG
4.600E-05	1.124E-02	37	F306	1 0 1 2 2	intrinsic
2.700E-04	6.595E-02	ns	O304	0 0 1 2 2	
3.275E-05	8.000E-03	rt	H302	0 0 2 1 2	intrinsic

3161. $C_{15}H_{13}F_3N_4O$
6H-Dipyrido[3,2-b:2',3'-e][1,4]diazepin-6-one, 11-Ethyl-5,11-dihydro-2-methyl-4-
(trifluoromethyl)-
RN: 135794-72-8   **MP** (°C):
MW: 322.29   **BP** (°C):

Solubility (Moles/L)	Solubility (Grams/L)	Temp (°C)	Ref (#)	Evaluation (T P E A A)	Comments
6.209E-05	2.001E-02	ns	M381	0 1 1 1 2	pH 7.0

3162. $C_{15}H_{13}NO$
7-Benzoylindoline
U-26,952
RN: 33244-57-4   **MP** (°C):   124
MW: 223.28   **BP** (°C):

Solubility (Moles/L)	Solubility (Grams/L)	Temp (°C)	Ref (#)	Evaluation (T P E A A)	Comments
1.026E-05	2.290E-03	25	C046	1 0 1 1 2	

3163. $C_{15}H_{13}NO_2$
Dibenz[b,f][1,4]oxazepin-11(10H)-one, 10-Ethyl-
RN: 17296-50-3   **MP** (°C):
MW: 239.28   **BP** (°C):

Solubility (Moles/L)	Solubility (Grams/L)	Temp (°C)	Ref (#)	Evaluation (T P E A A)	Comments
2.089E-04	4.999E-02	ns	M381	0 1 1 1 2	pH 7.0

3164. $C_{15}H_{13}NO_2S$
Metiazinic Acid
Methiazinic Acid
RN: 13993-65-2   **MP** (°C):   146
MW: 271.34   **BP** (°C):

Solubility (Moles/L)	Solubility (Grams/L)	Temp (°C)	Ref (#)	Evaluation (T P E A A)	Comments
1.142E-04	3.100E-02	30	D015	2 0 1 1 0	EFG
2.211E-04	6.000E-02	37	D015	2 0 1 1 0	EFG

3165. $C_{15}H_{13}NO_3$
Benzoyl Acetaminophen
Acetamide, N-[4-(Benzoyloxy)phenyl]-
Acetanilide, 4'-Hydroxy-, Benzoate (Ester)
RN: 537-52-0   **MP** (°C):   170.5-171.5
MW: 255.28   **BP** (°C):

Solubility (Moles/L)	Solubility (Grams/L)	Temp (°C)	Ref (#)	Evaluation (T P E A A)	Comments
6.659E-05	1.700E-02	37	D029	1 0 1 1 1	

3166.   C$_{15}$H$_{13}$NO$_4$
Phenyl Acetaminophen
Carbonic Acid, 4-(Acetylamino)phenyl Phenyl Ester
Acetanilide, 4'-Hydroxy-, Phenyl Carbonate (Ester)
RN: 17239-23-5   **MP** (°C):   139-140.5
MW: 271.28   **BP** (°C):

Solubility (Moles/L)	Solubility (Grams/L)	Temp (°C)	Ref (#)	Evaluation (T P E A A)	Comments
2.322E-04	6.300E-02	37	D029	1 0 1 1 1	

3167.   C$_{15}$H$_{13}$N$_3$O$_4$S
Piroxicam
2H-1,2-Benzothiazine-3-carboxamide, 4-Hydroxy-2-methyl-N-2-pyridinyl-, 1,1-dioxide
Fensaid
Feldene
Candyl
Mobilis
RN: 36322-90-4   **MP** (°C):   198
MW: 331.35   **BP** (°C):

Solubility (Moles/L)	Solubility (Grams/L)	Temp (°C)	Ref (#)	Evaluation (T P E A A)	Comments
6.941E-05	2.300E-02	rt	H302	0 0 2 1 2	intrinsic

3168.   C$_{15}$H$_{14}$ClN$_3$O$_4$S
Cefaclor
5-Thia-1-azabicyclo[4.2.0]oct-2-ene-2-carboxylic Acid, 7-[[(2R)-
Aminophenylacetyl]amino]-3-chloro-8-oxo-, (6R,7R)-
Ceclor
Alfacet
Cephaclor
RN: 53994-73-3   **MP** (°C):
MW: 367.81   **BP** (°C):

Solubility (Moles/L)	Solubility (Grams/L)	Temp (°C)	Ref (#)	Evaluation (T P E A A)	Comments
2.592E-02	1.000E+01	ns	L099	0 0 0 0 0	

3169. $C_{15}H_{14}Cl_2N_4O_3$
C.I. Disperse Orange 5
Ethanol, 2-[[4-[(2,6-Dichloro-4-nitrophenyl)azo]phenyl]methylamino]
Amacel Fast Brown 3R
Celliton Fast Brown 3R

RN: 6232-56-0   **MP** (°C): 127
MW: 369.21   **BP** (°C):

Solubility (Moles/L)	Solubility (Grams/L)	Temp (°C)	Ref (#)	Evaluation (T P E A A)	Comments
4.300E-07	1.588E-04	25	B333	1 0 0 0 1	
8.938E-06	3.300E-03	60	P313	1 2 1 2 2	average of 2
1.530E-05	5.650E-03	70	P313	1 2 1 2 2	average of 2
2.939E-05	1.085E-02	80	P313	1 2 1 2 2	average of 2
6.378E-05	2.355E-02	90	P313	1 2 1 2 2	average of 2
1.354E-04	5.000E-02	100	P313	1 2 1 2 2	

3170. $C_{15}H_{14}F_3N_3O_4S_2$
Bendroflumethiazide
Corzide
Rauzide
Naturetin

RN: 73-48-3   **MP** (°C): 222
MW: 421.42   **BP** (°C):

Solubility (Moles/L)	Solubility (Grams/L)	Temp (°C)	Ref (#)	Evaluation (T P E A A)	Comments
1.200E-04	5.057E-02	20	A080	1 0 2 1 2	
2.570E-04	1.083E-01	25	A076	1 0 1 1 2	
9.492E-05	4.000E-02	rt	A095	0 0 2 2 0	

3171. $C_{15}H_{14}NO_2PS$
Cyanofenphos
O-(4-Cyanophenyl) O-Ethyl Phenylphosphonothioate
Surecide

RN: 13067-93-1   **MP** (°C): 83
MW: 303.32   **BP** (°C):

Solubility (Moles/L)	Solubility (Grams/L)	Temp (°C)	Ref (#)	Evaluation (T P E A A)	Comments
1.978E-06	6.000E-04	30	M161	1 0 0 0 0	

3172. $C_{15}H_{14}N_2O_2$
Dibenz[b,f][1,4]oxazepin-11(10H)-one, 8-Amino-2-methyl-

RN: 155206-47-6   **MP** (°C):
MW: 254.29   **BP** (°C):

Solubility (Moles/L)	Solubility (Grams/L)	Temp (°C)	Ref (#)	Evaluation (T P E A A)	Comments
1.180E-04	3.001E-02	ns	M381	0 1 1 1 2	pH 7.0

3173.   C₁₅H₁₄N₂O₃
p-(3-Phenylureido)phenyl Acetate
Benzeneacetic Acid, 4-[[(Phenylamino)carbonyl]amino]-
RN: 181518-40-1   **MP** (°C):
MW: 270.29   **BP** (°C):

Solubility (Moles/L)	Solubility (Grams/L)	Temp (°C)	Ref (#)	Evaluation (T P E A A)	Comments
3.600E-05	9.730E-03	25	A066	1 0 1 1 1	

3174.   C₁₅H₁₄N₄O
Nevarapine
6H-Dipyrido[3,2-b:2',3'-e][1,4]diazepin-6-one, 11-cyclopropyl-5,11-dihydro-4-methyl
Nevirapine
BI-RG 587
RN: 129618-40-2   **MP** (°C):   248
MW: 266.31   **BP** (°C):

Solubility (Moles/L)	Solubility (Grams/L)	Temp (°C)	Ref (#)	Evaluation (T P E A A)	Comments
6.412E-04	1.708E-01	ns	M381	0 1 1 1 2	pH 7.0

3175.   C₁₅H₁₄O₃
Methyl Benzoyl Benzoate
Benzoic Acid, 4-Hydroxy-, (4-Methylphenyl)methyl Ester
RN: 84833-58-9   **MP** (°C):
MW: 242.28   **BP** (°C):

Solubility (Moles/L)	Solubility (Grams/L)	Temp (°C)	Ref (#)	Evaluation (T P E A A)	Comments
2.064E-04	5.000E-02	ns	F014	0 0 0 0 0	

3176.   C₁₅H₁₅ClN₂O₂
Chlorooxuron
(N'-4-(4-Chlorophenoxy)phenyl-N,N-dimethylurea)
3-[p-(p'-Chlorophenoxy)phenyl]-1,1-dimethylurea
N-4-(4'-Chlorophenoxy)phenyl-N',N'-dimethylurea
Tenoran
RN: 1982-47-4   **MP** (°C):   151
MW: 290.75   **BP** (°C):

Solubility (Moles/L)	Solubility (Grams/L)	Temp (°C)	Ref (#)	Evaluation (T P E A A)	Comments
1.273E-05	3.700E-03	20	B185	1 0 0 0 1	
1.273E-05	3.700E-03	20	G036	1 0 0 0 1	
1.273E-05	3.700E-03	20	M161	1 0 0 0 1	pH 7
9.286E-06	2.700E-03	ns	B200	0 0 0 0 1	
1.273E-04	3.700E-02	ns	M061	0 0 0 0 1	

3177.   C₁₅H₁₅CIN₂O₄S

Xipamide

2',6'-Salicyloxylidide, 4-Chloro-5-sulfamoyl-

Aquaphor

Aquaphor (Diuretic)

BEI 1293

Diurex

RN: 14293-44-8   **MP** (°C):   256

MW: 354.81   **BP** (°C):

Solubility (Moles/L)	Solubility (Grams/L)	Temp (°C)	Ref (#)	Evaluation (T P E A A)	Comments
1.635E-04	5.800E-02	25	H074	1 2 2 1 1	

3178.   C₁₅H₁₅CIO

2-Benzyl-3,5-dimethyl-4-chloro-phenol

RN: 1867-85-2   **MP** (°C):

MW: 246.74   **BP** (°C):

Solubility (Moles/L)	Solubility (Grams/L)	Temp (°C)	Ref (#)	Evaluation (T P E A A)	Comments
5.000E-05	1.234E-02	25	B316	1 0 2 1 1	

3179.   C₁₅H₁₅NO₂

Mefenamic Acid

2',3'-Dimethyl-N-phenyl-anthranilic Acid

Forte Mefenamic Acid

N-(2,3-Xylyl)anthranilic Acid

Ponstel

Ponstan

RN: 61-68-7   **MP** (°C):   230.5

MW: 241.29   **BP** (°C):

Solubility (Moles/L)	Solubility (Grams/L)	Temp (°C)	Ref (#)	Evaluation (T P E A A)	Comments
8.289E-05	2.000E-02	30	D015	2 0 1 1 0	EFG
2.800E-05	6.756E-03	35	H091	1 2 2 2 1	*sic*
1.658E-04	4.000E-02	37	D015	2 0 1 1 0	EFG
1.100E-04	2.654E-02	ns	O304	0 0 1 2 2	

3180.   C₁₅H₁₅N₃O

5H-Pyrido[2,3-b][1,5]benzodiazepine-5-one, 11-Ethyl-6,11-dihydro-6-methyl-

RN: 132686-75-0   **MP** (°C):

MW: 253.31   **BP** (°C):

Solubility (Moles/L)	Solubility (Grams/L)	Temp (°C)	Ref (#)	Evaluation (T P E A A)	Comments
1.782E-05	4.515E-03	ns	M381	0 1 1 1 2	pH 7.0
4.742E-04	1.201E-01	ns	M381	0 1 1 1 2	pH 7.0

3181. C₁₅H₁₅N₃O₂
Pyrido[2,3-b][1,5]benzoxazepin-5(6H)-one, 3-Amino-6,7,9-trimethyl-
RN: **MP** (°C):
MW: 269.31   **BP** (°C):

Solubility (Moles/L)	Solubility (Grams/L)	Temp (°C)	Ref (#)	Evaluation (T P E A A)	Comments
1.730E-04	4.658E-02	ns	M381	0 1 1 1 2	pH 7.0

3182. C₁₅H₁₅N₃O₂
C.I. Disperse Yellow 3
Acetamide, N-[4-[(2-Hydroxy-5-methylphenyl)azo]phenyl]-
RN: 2832-40-8   **MP** (°C):   195
MW: 269.31   **BP** (°C):

Solubility (Moles/L)	Solubility (Grams/L)	Temp (°C)	Ref (#)	Evaluation (T P E A A)	Comments
1.200E-07	3.232E-05	25	B333	1 0 0 0 1	

3183. C₁₅H₁₅N₃S
5H-Pyrido[2,3-b][1,5]benzodiazepine-5-thione, 11-Ethyl-6,11-dihydro-6-methyl-
RN: 132686-95-4   **MP** (°C):
MW: 269.37   **BP** (°C):

Solubility (Moles/L)	Solubility (Grams/L)	Temp (°C)	Ref (#)	Evaluation (T P E A A)	Comments
1.968E-05	5.301E-03	ns	M381	0 1 1 1 2	pH 7.0

3184. C₁₅H₁₆N₂O₂
Ancymidol
α-Cyclopropyl-α-(4-methoxyphenyl)-5-pyrimidinemethanol
A-Rest
RN: 12771-68-5   **MP** (°C):   110.5
MW: 256.31   **BP** (°C):

Solubility (Moles/L)	Solubility (Grams/L)	Temp (°C)	Ref (#)	Evaluation (T P E A A)	Comments
2.536E-03	6.500E-01	25	M161	1 0 0 0 2	

3185. C₁₅H₁₆N₄O
6H-Dipyrido[3,2-b:2',3'-e][1,4]diazepin-6-one, 5,11-Dihydro-5-methyl-11-propyl-
RN: 132312-81-3   **MP** (°C):
MW: 268.32   **BP** (°C):

Solubility (Moles/L)	Solubility (Grams/L)	Temp (°C)	Ref (#)	Evaluation (T P E A A)	Comments
1.327E-03	3.562E-01	ns	M381	0 1 1 1 2	pH 7.0

3186. $C_{15}H_{16}N_4O$
6H-Dipyrido[3,2-b:2',3'-e][1,4]diazepin-6-one, 11-Ethyl-5,11-dihydro-2,4-dimethyl-
RN: 134698-31-0   **MP** (°C):
MW: 268.32   **BP** (°C):

Solubility (Moles/L)	Solubility (Grams/L)	Temp (°C)	Ref (#)	Evaluation (T P E A A)	Comments
2.793E-05	7.493E-03	ns	M381	0 1 1 1 2	pH 7.0

3187. $C_{15}H_{16}N_4O$
6H-Dipyrido[3,2-b:2',3'-e][1,4]diazepin-6-one, 5,11-Diethyl-5,11-dihydro-
RN: 132312-82-4   **MP** (°C):
MW: 268.32   **BP** (°C):

Solubility (Moles/L)	Solubility (Grams/L)	Temp (°C)	Ref (#)	Evaluation (T P E A A)	Comments
1.380E-03	3.704E-01	ns	M381	0 1 1 1 2	pH 7.0

3188. $C_{15}H_{16}N_4O_2$
6H-Dipyrido[3,2-b:2',3'-e][1,4]diazepin-6-one, 11-Ethyl-5,11-dihydro-2-methoxy-4-methyl-
RN: 135794-75-1   **MP** (°C):
MW: 284.32   **BP** (°C):

Solubility (Moles/L)	Solubility (Grams/L)	Temp (°C)	Ref (#)	Evaluation (T P E A A)	Comments
7.031E-06	1.999E-03	ns	M381	0 1 1 1 2	pH 7.0

3189. $C_{15}H_{16}N_4O_2$
1H-Purine-2,6-dione, 1,3-Diethyl-3,7-dihydro-8-phenyl-
1,3-Diethyl-8-phenylxanthine
8-Phenyl-1,3-diethylxanthine
RN: 75922-48-4   **MP** (°C):
MW: 284.32   **BP** (°C):

Solubility (Moles/L)	Solubility (Grams/L)	Temp (°C)	Ref (#)	Evaluation (T P E A A)	Comments
3.517E-06	1.000E-03	ns	H316	0 2 1 1 0	0.1N HCl
2.110E-05	6.000E-03	ns	H316	0 2 1 1 0	pH 7.4

3190. $C_{15}H_{16}N_4O_5S$
Benzenesulfonic Acid, 4-(1,3-Diethyl-2,3,6,7-tetrahydro-2,6-dioxo-1H-purin-8-yl)-
RN: 89073-47-2   **MP** (°C):   >360
MW: 364.38   **BP** (°C):

Solubility (Moles/L)	Solubility (Grams/L)	Temp (°C)	Ref (#)	Evaluation (T P E A A)	Comments
>1.56E-01	>5.70E+01	ns	H316	0 2 1 1 0	pH 7.4
≥2.20E-02	>8.00E+00	ns	H316	0 2 1 1 0	0.1N HCl

3191.   C₁₅H₁₆O₂

Bisphenol A

2,2-bis-[4-Hydroxyphenyl]-propan

2,2-bis-(4-Hydroxypheny)-propane

RN: 80-05-7   **MP** (°C):

MW: 228.29   **BP** (°C):

Solubility (Moles/L)	Solubility (Grams/L)	Temp (°C)	Ref (#)	Evaluation (T P E A A)	Comments
1.533E-03	3.500E-01	20	F300	1 0 0 0 1	

3192.   C₁₅H₁₆O₃

Osthole

2H-1-Benzopyran-2-one, 7-Methoxy-8-(3-methyl-2-butenyl)-

RN: 484-12-8   **MP** (°C):   83.5

MW: 244.29   **BP** (°C):

Solubility (Moles/L)	Solubility (Grams/L)	Temp (°C)	Ref (#)	Evaluation (T P E A A)	Comments
4.912E-05	1.200E-02	30	B144	1 0 1 0 1	

3193.   C₁₅H₁₆O₉.2H₂O

Aesculin (Dihydrate)

Esculin

6,7-Dihydroxycoumarin 6-glucoside

2H-1-Benzopyran-2-one, 6-(β-D-Glucopyranosyloxy)-7-hydroxy-

RN: 531-75-9   **MP** (°C):   205dec

MW: 376.32   **BP** (°C):

Solubility (Moles/L)	Solubility (Grams/L)	Temp (°C)	Ref (#)	Evaluation (T P E A A)	Comments
4.605E-03	1.733E+00	c	D004	1 0 0 0 0	

3194.   C₁₅H₁₇FN₄O₂

Flupirtine

Carbamic Acid, [2-Amino-6-[[(4-fluorophenyl)methyl]amino]-3-pyridinyl]-, Ethyl Ester

RN: 56995-20-1   **MP** (°C):   175.8-177.7

MW: 304.33   **BP** (°C):

Solubility (Moles/L)	Solubility (Grams/L)	Temp (°C)	Ref (#)	Evaluation (T P E A A)	Comments
3.286E-03	1.000E+00	ns	D321	0 0 0 0 0	

3195.   C$_{15}$H$_{17}$NO$_3$

Acetamide, 2-(Benzoyloxy)-N,N-di-Acetamide, 2-(benzoyloxy)-N,N-di-2-propenyl-

RN: 106231-58-7   **MP** (°C):   42.5
MW: 259.31   **BP** (°C):

Solubility (Moles/L)	Solubility (Grams/L)	Temp (°C)	Ref (#)	Evaluation (T P E A A)	Comments
2.738E-03	7.100E-01	22	N317	1 1 2 1 2	

3196.   C$_{15}$H$_{17}$NO$_5$

L-Proline, 1-[(Benzoyloxy)acetyl]-, Methyl Ester

RN: 115178-76-2   **MP** (°C):   72.5
MW: 291.31   **BP** (°C):

Solubility (Moles/L)	Solubility (Grams/L)	Temp (°C)	Ref (#)	Evaluation (T P E A A)	Comments
8.239E-03	2.400E+00	22	N317	1 1 2 1 2	

3197.   C$_{15}$H$_{17}$NO$_7$

Glycine, N-[[[2-(Acetyloxy)benzoyl]oxy]acetyl]-, Ethyl Ester

RN: 118247-03-3   **MP** (°C):   68.5
MW: 323.31   **BP** (°C):

Solubility (Moles/L)	Solubility (Grams/L)	Temp (°C)	Ref (#)	Evaluation (T P E A A)	Comments
1.336E-02	4.320E+00	21	N335	1 2 1 1 2	

3198.   C$_{15}$H$_{17}$N$_3$O$_3$S

L-Ala-Dapsone
2-Amino-N-[4-[(4-aminophenyl)sulfonyl]phenyl]-, (S)-
Propanamide

RN: 160348-99-2   **MP** (°C):
MW: 319.39   **BP** (°C):

Solubility (Moles/L)	Solubility (Grams/L)	Temp (°C)	Ref (#)	Evaluation (T P E A A)	Comments
2.066E-02	6.600E+00	25	P351	2 2 1 2 1	pH 7.4
≥9.39E-02	>3.00E+01	25	P351	2 2 1 2 1	

3199.   C$_{15}$H$_{18}$Cl$_2$N$_2$O$_3$
Oxadiazon
3-[2,4-Dichloro-5-(1-methylethoxy)phenyl]-5-(1,1-dimethylethyl)-1,3,4-oxadiazol-2(3H)-
one
Ronstar
Scotts OH I
RP-17623
RN: 19666-30-9   **MP** (°C):   88
MW: 345.23   **BP** (°C):

Solubility (Moles/L)	Solubility (Grams/L)	Temp (°C)	Ref (#)	Evaluation (T P E A A)	Comments
2.028E-06	7.000E-04	20	M161	1 0 0 0 0	
2.028E-06	7.000E-04	24	C105	2 1 2 2 2	

3200.   C$_{15}$H$_{18}$I$_3$NO$_5$
Iopronic Acid
Butanoic Acid, 2-[[2-[3-(Acetylamino)-2,4,6-triiodophenoxy]ethoxy]methyl]-
RN: 37723-78-7   **MP** (°C):   130
MW: 673.03   **BP** (°C):

Solubility (Moles/L)	Solubility (Grams/L)	Temp (°C)	Ref (#)	Evaluation (T P E A A)	Comments
2.984E-02	2.008E+01	37	J016	1 0 0 0 2	pH 7.4
1.456E-04	9.799E-02	50	F013	1 0 1 1 1	

3201.   C$_{15}$H$_{18}$N$_2$O$_3$
N-Acetyl-L-tryptophan Ethyl Ester
RN: 2382-80-1   **MP** (°C):   106
MW: 274.32   **BP** (°C):

Solubility (Moles/L)	Solubility (Grams/L)	Temp (°C)	Ref (#)	Evaluation (T P E A A)	Comments
1.896E-03	5.200E-01	5	L081	2 2 2 2 1	
5.359E-03	1.470E+00	28	L081	2 1 2 2 2	

3202.   C$_{15}$H$_{18}$N$_4$O$_3$S
2-(N4-Acetylsulfanilylamino)-4-n-propylpyrimidine
RN: **MP** (°C):
MW: 334.40   **BP** (°C):

Solubility (Moles/L)	Solubility (Grams/L)	Temp (°C)	Ref (#)	Evaluation (T P E A A)	Comments
1.914E-05	6.400E-03	37	R076	1 2 0 0 2	

3203. $C_{15}H_{18}N_4O_3S$
2-(N4-Acetylsulfanilylamino)-4-ethyl-5-methylpyrimidine
RN: **MP** (°C):
MW: 334.40   **BP** (°C):

Solubility (Moles/L)	Solubility (Grams/L)	Temp (°C)	Ref (#)	Evaluation (T P E A A)	Comments
1.077E-05	3.600E-03	37	R076	1 2 0 0 1	

3204. $C_{15}H_{18}N_4O_5$
Mitomycin C
MMC
6-Amino-8-[[(aminocarbonyl)oxy]methyl]-1,1α,2,8,8α,8β-hexahydro-8α-methoxy-5-methyl, Mitomycinum
RN: 50-07-7   **MP** (°C):   >360
MW: 334.33   **BP** (°C):

Solubility (Moles/L)	Solubility (Grams/L)	Temp (°C)	Ref (#)	Evaluation (T P E A A)	Comments
2.730E-03	9.127E-01	25	M316	1 1 1 1 2	

3205. $C_{15}H_{18}O_3$
Santonin
Naphtho[1,2-b]furan-2,8(3H,4H)-dione, 3α,5,5α,9β-Tetrahydro-3,5α,9-trimethyl-, (3S,3αS,5αS,9βS)-
RN: 481-06-1   **MP** (°C):   170
MW: 246.31   **BP** (°C):

Solubility (Moles/L)	Solubility (Grams/L)	Temp (°C)	Ref (#)	Evaluation (T P E A A)	Comments
8.120E-04	2.000E-01	17.5	F300	1 0 0 0 0	
1.624E-02	4.000E+00	100	F300	1 0 0 0 0	

3206. $C_{15}H_{18}O_4$
β-Cyclopentylpropionyl Salicylate
RN: **MP** (°C):
MW: 262.31   **BP** (°C):

Solubility (Moles/L)	Solubility (Grams/L)	Temp (°C)	Ref (#)	Evaluation (T P E A A)	Comments
1.060E-04	2.780E-02	25.6	G015	1 0 1 1 2	pH 1.00, pka 3.91, intrinsic

3207.   C₁₅H₁₉ClO₂
1,1-Drichloro-1-methyl-2,2-bis(p-methoxylphenyl)ethane
RN: 56288-27-8   **MP** (°C):
MW: 266.77   **BP** (°C):

Solubility (Moles/L)	Solubility (Grams/L)	Temp (°C)	Ref (#)	Evaluation (T P E A A)	Comments
6.373E-06	1.700E-03	rt	C122	0 2 2 2 2	

3208.   C₁₅H₁₉NO
N-Cyclohexylcinnamamide
2-Propenamide, N-Cyclohexyl-3-phenyl-
RN: 6750-98-7   **MP** (°C):
MW: 229.32   **BP** (°C):

Solubility (Moles/L)	Solubility (Grams/L)	Temp (°C)	Ref (#)	Evaluation (T P E A A)	Comments
4.040E-05	9.265E-03	ns	H350	0 0 0 0 2	

3209.   C₁₅H₁₉NO
N,N-Hexamethylenecinnamamide
Hexahydro-1-(1-oxo-3-phenyl-2-propenyl)1H-Azepine
RN: 59832-05-2   **MP** (°C):
MW: 229.32   **BP** (°C):

Solubility (Moles/L)	Solubility (Grams/L)	Temp (°C)	Ref (#)	Evaluation (T P E A A)	Comments
2.460E-04	5.641E-02	ns	H350	0 0 0 0 2	

3210.   C₁₅H₁₉NO₂
Tropacocaine
RN: 537-26-8   **MP** (°C):   49
MW: 245.32   **BP** (°C):

Solubility (Moles/L)	Solubility (Grams/L)	Temp (°C)	Ref (#)	Evaluation (T P E A A)	Comments
4.300E-03	1.055E+00	15	K059	2 2 2 0 1	

3211.   C₁₅H₁₉NO₃
1H-Azepine, 1-[(Benzoyloxy)acetyl]hexahydro-
RN: 115178-68-2   **MP** (°C):   107.5
MW: 261.32   **BP** (°C):

Solubility (Moles/L)	Solubility (Grams/L)	Temp (°C)	Ref (#)	Evaluation (T P E A A)	Comments
2.870E-03	7.500E-01	22	N317	1 1 2 1 2	

3212. C$_{15}$H$_{19}$NO$_5$
Benzoic Acid, 2-(Acetyloxy)-, 2-(Diethylamino)-2-oxoethyl Ester
RN: 116482-56-5   **MP** (°C):   76.5
MW: 293.32   **BP** (°C):

Solubility (Moles/L)	Solubility (Grams/L)	Temp (°C)	Ref (#)	Evaluation (T P E A A)	Comments
7.773E-03	2.280E+00	21	N335	1 2 1 1 2	

3213. C$_{15}$H$_{20}$N$_2$O$_4$
Benzyl-2,2-diethylmalonurate
Benzyl 2,2-Diethylmalonurate
RN: 73632-78-7   **MP** (°C):   107
MW: 292.34   **BP** (°C):

Solubility (Moles/L)	Solubility (Grams/L)	Temp (°C)	Ref (#)	Evaluation (T P E A A)	Comments
2.200E-04	6.431E-02	23	B152	1 2 1 1 1	pH 3.5

3214. C$_{15}$H$_{20}$N$_2$O$_4$S
Acetohexamide
Acetohexamid
1-(p-Acetylbenzenesulfonyl)-3-cyclohexylurea
Dymelor
Dimelin
RN: 968-81-0   **MP** (°C):   189
MW: 324.40   **BP** (°C):

Solubility (Moles/L)	Solubility (Grams/L)	Temp (°C)	Ref (#)	Evaluation (T P E A A)	Comments
7.706E-04	2.500E-01	25	K023	1 0 2 2 1	EFG, pH 6.5, average of 2
3.483E-05	1.130E-02	37	B130	1 2 1 1 2	pH 1.5, form II
4.963E-05	1.610E-02	37	B130	1 2 1 1 2	pH 1.5, form III
8.015E-05	2.600E-02	37	K106	1 2 2 2 0	EFG, form I
9.556E-05	3.100E-02	37	K106	1 2 2 2 0	EFG, form II

3215. C$_{15}$H$_{20}$N$_4$O$_2$S
2-Sulfanilylamino-4-amylpyrimidine
RN: 107203-72-5   **MP** (°C):
MW: 320.42   **BP** (°C):

Solubility (Moles/L)	Solubility (Grams/L)	Temp (°C)	Ref (#)	Evaluation (T P E A A)	Comments
6.242E-04	2.000E-01	37	R076	1 2 0 0 1	

3216.   C$_{15}$H$_{20}$N$_4$O$_5$
1,5-Dibutyryloxymethyl Allopurinol
RN: 98827-19-1   **MP** (°C):   122-123
MW: 336.35   **BP** (°C):

Solubility (Moles/L)	Solubility (Grams/L)	Temp (°C)	Ref (#)	Evaluation (T P E A A)	Comments
1.487E-04	5.000E-02	22	B322	1 0 2 2 2	

3217.   C$_{15}$H$_{20}$N$_4$O$_5$
2,5-Dibutyryloxymethyl Allopurinol
RN: 98827-20-4   **MP** (°C):   133-135
MW: 336.35   **BP** (°C):

Solubility (Moles/L)	Solubility (Grams/L)	Temp (°C)	Ref (#)	Evaluation (T P E A A)	Comments
2.795E-04	9.400E-02	22	B322	1 0 2 2 2	

3218.   C$_{15}$H$_{20}$N$_4$O$_6$.0.25H$_2$O
9-[5-O-(Isobutyrate-β-D-arabinofuranosyl)]-6-methoxy-9H-purine (0.25 Hydrate)
RN: 121032-44-8   **MP** (°C):   glass
MW: 356.85   **BP** (°C):

Solubility (Moles/L)	Solubility (Grams/L)	Temp (°C)	Ref (#)	Evaluation (T P E A A)	Comments
3.830E-02	1.367E+01	37	M378	1 2 1 1 2	pH 7.2

3219.   C$_{15}$H$_{20}$N$_4$O$_6$
9-[5-O-(Butyrate-β-D-arabinofuranosyl)]-6-methoxy-9H-purine
RN: 121032-41-5   **MP** (°C):   108-110
MW: 352.35   **BP** (°C):

Solubility (Moles/L)	Solubility (Grams/L)	Temp (°C)	Ref (#)	Evaluation (T P E A A)	Comments
9.680E-03	3.411E+00	37	M378	1 2 1 1 2	pH 7.2

3220.   C$_{15}$H$_{20}$N$_4$O$_6$
2'-Isobutyryl-6-methoxypurine Arabinoside
RN: 121032-44-8   **MP** (°C):
MW: 352.35   **BP** (°C):

Solubility (Moles/L)	Solubility (Grams/L)	Temp (°C)	Ref (#)	Evaluation (T P E A A)	Comments
6.700E-01	2.361E+02	37	C348	1 2 2 2 2	pH 7.00

3221. C$_{15}$H$_{20}$N$_4$O$_6$.0.3H$_2$O
2'-Butyryl-6-methoxypurine Arabinoside (0.3 Hydrate)
RN: 121032-41-5 **MP** (°C):
MW: 357.75 **BP** (°C):

Solubility (Moles/L)	Solubility (Grams/L)	Temp (°C)	Ref (#)	Evaluation (T P E A A)	Comments
2.310E-01	8.264E+01	37	C348	1 2 2 2 2	pH 7.00

3222. C$_{15}$H$_{21}$NO
N,N-Dipropylcinnamamide
Cinnamamide, N,N-Dipropyl-
RN: 23784-56-7 **MP** (°C):
MW: 231.34 **BP** (°C):

Solubility (Moles/L)	Solubility (Grams/L)	Temp (°C)	Ref (#)	Evaluation (T P E A A)	Comments
2.890E-03	6.686E-01	ns	H350	0 0 0 0 2	

3223. C$_{15}$H$_{21}$NO$_2$
Meperidine
Ethyl 1-Methyl-4-phenylpiperidine-4-carboxylate
Demerol
Dolantin
Pethidine
RN: 57-42-1 **MP** (°C): 30
MW: 247.34 **BP** (°C):

Solubility (Moles/L)	Solubility (Grams/L)	Temp (°C)	Ref (#)	Evaluation (T P E A A)	Comments
1.300E-02	3.215E+00	30	L068	1 0 0 1 0	EFG

3224. C$_{15}$H$_{21}$NO$_2$S$_2$
2-(p-Isopropylphenyl)-2-methyl-4-(methoxycarbamyl)-1,3-dithiolane
RN: 35801-67-3 **MP** (°C):
MW: 311.47 **BP** (°C):

Solubility (Moles/L)	Solubility (Grams/L)	Temp (°C)	Ref (#)	Evaluation (T P E A A)	Comments
2.500E-05	7.787E-03	rt	B174	0 0 1 0 1	

3225. C$_{15}$H$_{21}$NO$_3$
Acetamide, 2-(Benzoyloxy)-N,N-bis(1-methylethyl)-
RN: 106231-56-5 **MP** (°C): 105.5
MW: 263.34 **BP** (°C):

Solubility (Moles/L)	Solubility (Grams/L)	Temp (°C)	Ref (#)	Evaluation (T P E A A)	Comments
4.557E-04	1.200E-01	22	N317	1 1 2 1 2	

3226. C₁₅H₂₁NO₃

Acetamide, 2-(Benzoyloxy)-N,N-dipropyl-

RN: 106231-55-4 **MP** (°C): 20
MW: 263.34 **BP** (°C):

Solubility (Moles/L)	Solubility (Grams/L)	Temp (°C)	Ref (#)	Evaluation (T P E A A)	Comments
4.177E-03	1.100E+00	22	N317	1 1 2 1 2	

3227. C₁₅H₂₁NO₃

Acetamide, 2-(Benzoyloxy)-N-hexyl-

RN: 115193-29-8 **MP** (°C): 130.5
MW: 263.34 **BP** (°C):

Solubility (Moles/L)	Solubility (Grams/L)	Temp (°C)	Ref (#)	Evaluation (T P E A A)	Comments
1.253E-04	3.300E-02	22	N317	1 1 2 1 2	

3228. C₁₅H₂₁NO₃S

2-(p-Isopropylphenyl)-2-methyl-4-(methoxycarbamyl)-1,3-oxathiolane

RN: 24606-94-8 **MP** (°C):
MW: 295.40 **BP** (°C):

Solubility (Moles/L)	Solubility (Grams/L)	Temp (°C)	Ref (#)	Evaluation (T P E A A)	Comments
6.000E-05	1.772E-02	rt	B174	0 0 1 0 0	

3229. C₁₅H₂₁NO₄

Metalaxyl
Methyl N-(2,6-Dimethyl-phenyl)-N-(2'-methoxyacetyl)-DL-alaninate
Apron
Ridomil
Subdue
Fubol

RN: 57837-19-1 **MP** (°C): 72
MW: 279.34 **BP** (°C):

Solubility (Moles/L)	Solubility (Grams/L)	Temp (°C)	Ref (#)	Evaluation (T P E A A)	Comments
2.488E-02	6.951E+00	20	E048	1 2 1 1 2	

3230. C$_{15}$H$_{21}$NO$_4$
2-(p-Isopropylphenyl)-2-methyl-4-(methoxycarbamyl)-1,3-dioxolane
RN: 35858-24-3   **MP** (°C):
MW: 279.34   **BP** (°C):

Solubility (Moles/L)	Solubility (Grams/L)	Temp (°C)	Ref (#)	Evaluation (T P E A A)	Comments
9.000E-04	2.514E-01	rt	B174	0 0 1 0 0	

3231. C$_{15}$H$_{21}$NO$_4$
Hexyl Acetaminophen
Carbonic Acid, 4-(Acetylamino)phenyl Hexyl Ester
Acetanilide, 4'-Hydroxy-, Hexyl Carbonate (Ester)
RN: 17239-22-4   **MP** (°C):   112.5-113.5
MW: 279.34   **BP** (°C):

Solubility (Moles/L)	Solubility (Grams/L)	Temp (°C)	Ref (#)	Evaluation (T P E A A)	Comments
1.325E-04	3.700E-02	37	D029	1 0 1 1 1	

3232. C$_{15}$H$_{21}$NO$_5$
Acetamide, 2-(Benzoyloxy)-N,N-bis(2-hydroxypropyl)-
RN: 115178-63-7   **MP** (°C):   105.5
MW: 295.34   **BP** (°C):

Solubility (Moles/L)	Solubility (Grams/L)	Temp (°C)	Ref (#)	Evaluation (T P E A A)	Comments
6.636E-02	1.960E+01	22	N317	1 1 2 1 2	

3233. C$_{15}$H$_{21}$NO$_5$
Acetamide, 2-(Benzoyloxy)-N,N-bis(2-methoxyethyl)-
RN: 115178-64-8   **MP** (°C):   57.5
MW: 295.34   **BP** (°C):

Solubility (Moles/L)	Solubility (Grams/L)	Temp (°C)	Ref (#)	Evaluation (T P E A A)	Comments
2.672E-02	7.890E+00	22	N317	1 1 2 1 2	

3234. C$_{15}$H$_{21}$N$_2$O$_3$
C.I. Disperse Red 11
RN: 2872-48-2   **MP** (°C):   242
MW: 277.35   **BP** (°C):

Solubility (Moles/L)	Solubility (Grams/L)	Temp (°C)	Ref (#)	Evaluation (T P E A A)	Comments
2.500E-06	6.934E-04	25	B333	1 0 0 0 1	

3235. $C_{15}H_{21}N_5O_5$
9-[5'-(O-Isovaleryl)-β-D-arabinofuranosyl]adenine Ester
RN: 65926-32-1 **MP** (°C):
MW: 351.37 **BP** (°C):

Solubility (Moles/L)	Solubility (Grams/L)	Temp (°C)	Ref (#)	Evaluation (T P E A A)	Comments
5.635E-02	1.980E+01	ns	B134	0 1 1 1 2	

3236. $C_{15}H_{21}N_5O_5$
9-[5'-(O-Pivaloyl)-β-D-arabinofuranosyl]adenine Ester
RN: 65926-33-2 **MP** (°C):
MW: 351.37 **BP** (°C):

Solubility (Moles/L)	Solubility (Grams/L)	Temp (°C)	Ref (#)	Evaluation (T P E A A)	Comments
1.992E-02	7.000E+00	ns	B134	0 1 1 1 1	

3237. $C_{15}H_{21}N_5O_5$
9-(2-O-Valeryl-β-D-arabinofuranosyl)adenine
RN: 87984-85-8 **MP** (°C):
MW: 351.37 **BP** (°C):

Solubility (Moles/L)	Solubility (Grams/L)	Temp (°C)	Ref (#)	Evaluation (T P E A A)	Comments
2.960E-04	1.040E-01	37	B306	1 2 0 1 2	pH 7.3

3238. $C_{15}H_{21}N_5O_5$
9-[5'-(O-Valeryl)-β-D-arabinofuranosyl]adenine Ester
RN: 65926-31-0 **MP** (°C):
MW: 351.37 **BP** (°C):

Solubility (Moles/L)	Solubility (Grams/L)	Temp (°C)	Ref (#)	Evaluation (T P E A A)	Comments
2.391E-02	8.400E+00	ns	B134	0 1 1 1 1	

3239. $C_{15}H_{21}N_5O_6$
9-(1,3-Dipropionate-2-propoxymethyl)guanine
RN: 86357-20-2 **MP** (°C): 192
MW: 367.36 **BP** (°C):

Solubility (Moles/L)	Solubility (Grams/L)	Temp (°C)	Ref (#)	Evaluation (T P E A A)	Comments
7.622E-03	2.800E+00	25	B360	1 0 2 2 2	

3240. C₁₅H₂₂ClNO₂

Metolachlor
2-Chloro-N-(2-ethyl-6-methylphenyl)-N-(2-methoxy-1-methylethyl)acetamide
Dual
Cotoran Multi
Ontrack 8E
Bicep 6L
RN: 51218-45-2   **MP** (°C):   <25
MW: 283.80   **BP** (°C):   100

Solubility (Moles/L)	Solubility (Grams/L)	Temp (°C)	Ref (#)	Evaluation (T P E A A)	Comments
1.867E-03	5.297E-01	20	E048	1 2 1 1 2	
1.868E-03	5.300E-01	20	M161	1 0 0 0 2	

3241. C₁₅H₂₂ClNO₂

CP 52223
2-Chloro-N-(2,6-dimethyl)phenyl-N-isopropoxymethylacetamide
RN: 24353-58-0   **MP** (°C):
MW: 283.80   **BP** (°C):   137.5

Solubility (Moles/L)	Solubility (Grams/L)	Temp (°C)	Ref (#)	Evaluation (T P E A A)	Comments
2.079E-04	5.900E-02	ns	M061	0 0 0 0 1	

3242. C₁₅H₂₂N₂O

DL-Mepivacaine
Carbocaine
1-Methyl-2',6'-pipecoloxylidide
Carbocain
RN: 96-88-8   **MP** (°C):   150
MW: 246.36   **BP** (°C):

Solubility (Moles/L)	Solubility (Grams/L)	Temp (°C)	Ref (#)	Evaluation (T P E A A)	Comments
1.360E-02	3.350E+00	14.9	N046	2 0 1 2 2	intrinsic
3.653E-02	9.000E+00	23	F176	2 0 0 2 0	EFG, pH 7.4, intrinsic
2.841E-02	7.000E+00	23	F176	2 0 0 2 0	EFG, pH 7.4, intrinsic
1.020E-02	2.513E+00	25	N046	2 0 1 2 2	intrinsic
9.910E-03	2.441E+00	34.5	N046	2 0 1 2 2	intrinsic
7.970E-03	1.963E+00	37	N044	2 1 1 2 2	intrinsic

3243. $C_{15}H_{22}O_3$
Octyl p-Hydroxybenzoate
n-Octyl 4-Hydroxybenzoate
RN: 1219-38-1   **MP** (°C):   54
MW: 250.34   **BP** (°C):

Solubility (Moles/L)	Solubility (Grams/L)	Temp (°C)	Ref (#)	Evaluation (T P E A A)	Comments
1.470E-05	3.680E-03	15	B355	1 1 1 1 2	
2.300E-04	5.758E-02	20	B355	1 1 1 1 2	
4.650E-04	1.164E-01	25	B355	1 1 1 1 2	
3.273E-03	8.193E-01	25	D081	1 2 2 1 2	
3.162E-04	7.916E-02	25	F322	2 0 1 1 0	EFG

3244. $C_{15}H_{23}NO_2$
Octyl p-Aminobenzoate
4-Aminobenzoic Acid Octyl Ester
RN: 14309-41-2   **MP** (°C):
MW: 249.36   **BP** (°C):

Solubility (Moles/L)	Solubility (Grams/L)	Temp (°C)	Ref (#)	Evaluation (T P E A A)	Comments
3.200E-06	7.979E-04	37	F006	1 1 2 2 1	

3245. $C_{15}H_{23}NO_2$
Octyl m-Aminobenzoate
Octyl 3-Aminobenzoate
RN: 52222-35-2   **MP** (°C):
MW: 249.36   **BP** (°C):

Solubility (Moles/L)	Solubility (Grams/L)	Temp (°C)	Ref (#)	Evaluation (T P E A A)	Comments
3.000E-05	7.481E-03	ns	M066	0 0 0 0 0	

3246. $C_{15}H_{23}NO_3$
Parethoxycaine
4-Ethoxybenzoic Acid-2-(diethylamino)ethyl Ester
RN: 94-23-5   **MP** (°C):   173.0
MW: 265.36   **BP** (°C):

Solubility (Moles/L)	Solubility (Grams/L)	Temp (°C)	Ref (#)	Evaluation (T P E A A)	Comments
1.930E-03	5.121E-01	ns	M066	0 0 0 0 2	

3247. C₁₅H₂₃NO₃

Oxprenolol
Corbeton
1-[o-(Allyloxy)phenoxy]-3-(isopropylamino)-2-propanol

RN: 6452-71-7   **MP** (°C):
MW: 265.36   **BP** (°C):

Solubility (Moles/L)	Solubility (Grams/L)	Temp (°C)	Ref (#)	Evaluation (T P E A A)	Comments
4.786E-01	1.270E+02	25	P312	1 2 2 2 2	

3248. C₁₅H₂₃NO₄

Cycloheximide
3-((R)-2-((1S,3S,5S)-3,5-Dimethyl-2-oxocyclohexyl)-2-hydroxyethyl)glutarimide
Actidione
Actispray
Naramycin
Kaken

RN: 66-81-9   **MP** (°C):   116.3
MW: 281.35   **BP** (°C):

Solubility (Moles/L)	Solubility (Grams/L)	Temp (°C)	Ref (#)	Evaluation (T P E A A)	Comments
7.464E-02	2.100E+01	2	M161	1 0 0 0 1	

3249. C₁₅H₂₃N₃O₄

Isopropalin
2,6-Dinitro-N,N-dipropylcumidene
4-Isopropyl-2,6-dinitro-N,N-dipropylaniline
2,6-Dinitro-N,N-dipropylcumidine
Paarlan
Paarlan EC

RN: 33820-53-0   **MP** (°C):
MW: 309.37   **BP** (°C):

Solubility (Moles/L)	Solubility (Grams/L)	Temp (°C)	Ref (#)	Evaluation (T P E A A)	Comments
3.232E-07	1.000E-04	25	M161	1 0 0 0 0	

3250. C₁₅H₂₃N₃O₄S
Cyclacilllin
Anhydrous 6-(1-Aminocyclohexanecarboxamido)penicillanic Acid
RN: 3485-14-1   **MP** (°C):
MW: 341.43   **BP** (°C):

Solubility (Moles/L)	Solubility (Grams/L)	Temp (°C)	Ref (#)	Evaluation (T P E A A)	Comments
1.611E-01	5.500E+01	7	P035	1 1 1 1 0	EFG
1.054E-01	3.600E+01	20	P035	1 1 1 1 0	EFG
9.372E-02	3.200E+01	25	P035	1 1 1 1 0	EFG
7.908E-02	2.700E+01	30	P035	1 1 1 1 0	EFG
6.736E-02	2.300E+01	40	P035	1 1 1 1 0	EFG
6.151E-02	2.100E+01	50	P035	1 1 1 1 0	EFG
5.858E-02	2.000E+01	60	P035	1 1 1 1 0	EFG

3251. C₁₅H₂₃N₃O₄S
Sulpiride
N-[(1-Ethyl-2-pyrrolidinyl)methyl]-2-methoxy-5-sulfamoylbenzamide
RN: 15676-16-1   **MP** (°C):
MW: 341.43   **BP** (°C):

Solubility (Moles/L)	Solubility (Grams/L)	Temp (°C)	Ref (#)	Evaluation (T P E A A)	Comments
<6.15E-04	<2.10E-01	25	P312	1 2 2 2 2	

3252. C₁₅H₂₃N₃O₄S.2H₂O
Cyclacillin (Dihydrate)
Dihydrate 6-(1-Aminocyclohexanecarboxamido)penicillanic Acid
RN: 3485-14-1   **MP** (°C):
MW: 377.46   **BP** (°C):

Solubility (Moles/L)	Solubility (Grams/L)	Temp (°C)	Ref (#)	Evaluation (T P E A A)	Comments
3.709E-02	1.400E+01	10	P035	1 1 1 1 0	EFG
3.709E-02	1.400E+01	20	P035	1 1 1 1 0	EFG
3.656E-02	1.380E+01	25	P035	1 1 1 1 0	EFG
3.656E-02	1.380E+01	30	P035	1 1 1 1 0	EFG
3.682E-02	1.390E+01	40	P035	1 1 1 1 0	EFG
3.762E-02	1.420E+01	50	P035	1 1 1 1 0	EFG
4.504E-02	1.700E+01	60	P035	1 1 1 1 0	EFG

3253. C$_{15}$H$_{24}$NO$_4$PS
Isofenphos
Methylethyl 2-((Ethoxy((1-methylethyl)amino)phosphinothioyl)oxy)benzoate
Amaze
Oftanol
Pryfon
RN: 25311-71-1   **MP** (°C):
MW: 345.40   **BP** (°C):

Solubility (Moles/L)	Solubility (Grams/L)	Temp (°C)	Ref (#)	Evaluation (T P E A A)	Comments
6.399E-05	2.210E-02	20	B300	2 1 1 1 2	*sic*
6.891E-02	2.380E+01	20	M161	1 0 0 0 2	*sic*

3254. C$_{15}$H$_{24}$N$_2$O$_2$
Tetracaine
Pantocaine
Cetacaine
RN: 94-24-6   **MP** (°C):
MW: 264.37   **BP** (°C):

Solubility (Moles/L)	Solubility (Grams/L)	Temp (°C)	Ref (#)	Evaluation (T P E A A)	Comments
5.900E-04	1.560E-01	ns	E031	0 0 2 1 2	

3255. C$_{15}$H$_{24}$N$_2$O$_2$
N,N,N'-Triethyl-bicyclo(2.2.1)hept-5-ene-2,3-trans-dicarboxamide
RN: 62249-37-0   **MP** (°C):
MW: 264.37   **BP** (°C):

Solubility (Moles/L)	Solubility (Grams/L)	Temp (°C)	Ref (#)	Evaluation (T P E A A)	Comments
2.232E-01	5.900E+01	20	K050	1 1 1 1 2	

3256. C$_{15}$H$_{24}$N$_2$O$_2$
4-Aminobenzoic Acid-2-(diethyl-amino)butyl Ester
2-(Diethyl(amino)butyl 4-Aminobenzoate
RN: 5878-14-8   **MP** (°C):
MW: 264.37   **BP** (°C):

Solubility (Moles/L)	Solubility (Grams/L)	Temp (°C)	Ref (#)	Evaluation (T P E A A)	Comments
4.300E-03	1.137E+00	ns	M066	0 0 0 0 1	

3257. C₁₅H₂₄N₂O₂

3257. $C_{15}H_{24}N_2O_2$

4-Ethylaminobenzoic Acid-2-(diethyl-amino)ethyl Ester

RN: 16488-53-2   **MP** (°C):
MW: 264.37   **BP** (°C):

Solubility (Moles/L)	Solubility (Grams/L)	Temp (°C)	Ref (#)	Evaluation (T P E A A)	Comments
4.600E-03	1.216E+00	ns	M066	0 0 0 0 1	

3258. $C_{15}H_{24}N_2O_3$

2,4-Diazaspiro[5.11]heptadecane-1,3,5-trione

RN: 143288-64-6   **MP** (°C):
MW: 280.37   **BP** (°C):

Solubility (Moles/L)	Solubility (Grams/L)	Temp (°C)	Ref (#)	Evaluation (T P E A A)	Comments
1.600E-06	4.486E-04	25	P350	2 1 1 1 2	intrinsic

3259. $C_{15}H_{24}O$

4-Nonylphenol
4-t-Nonylphenol

RN: 104-40-5   **MP** (°C):
MW: 220.36   **BP** (°C):

Solubility (Moles/L)	Solubility (Grams/L)	Temp (°C)	Ref (#)	Evaluation (T P E A A)	Comments
2.090E-05	4.605E-03	2	A335	1 0 2 2 2	
2.088E-05	4.600E-03	2	A335	1 0 2 2 2	
2.230E-05	4.914E-03	10	A335	1 0 2 2 2	
2.233E-05	4.920E-03	10	A335	1 0 2 2 2	
2.380E-05	5.245E-03	14	A335	1 0 2 2 2	
2.378E-05	5.240E-03	14	A335	1 0 2 2 2	
2.470E-05	5.443E-03	20.5	A335	1 0 2 2 2	
2.464E-05	5.430E-03	20.5	A335	1 0 2 2 2	
2.882E-05	6.350E-03	25	A335	1 0 2 2 2	
2.890E-05	6.368E-03	25	A335	1 0 2 2 2	
3.177E-05	7.000E-03	25	M127	1 0 0 0 0	

3260. $C_{15}H_{24}O$

Butylated Hydroxytoluene
2,6-Di-tert-Butyl-p-Cresol
2,6-Di-tert-Butyl-1-Hydroxy-4-Methylbenzene
4-Hydroxy-3,5-Di-tert-Butyltoluene

RN: 128-37-0   **MP** (°C):   71
MW: 220.36   **BP** (°C):   265

Solubility (Moles/L)	Solubility (Grams/L)	Temp (°C)	Ref (#)	Evaluation (T P E A A)	Comments
≤4.54E-05	<1.00E-02	25	P312	1 2 2 2 2	

3261. $C_{15}H_{26}N_2$

Sparteine
(-)-Spartein

RN: 90-39-1   **MP** (°C):   30
MW: 234.39   **BP** (°C):

Solubility (Moles/L)	Solubility (Grams/L)	Temp (°C)	Ref (#)	Evaluation (T P E A A)	Comments
1.297E-02	3.040E+00	22	F300	1 0 0 0 2	
1.297E-02	3.040E+00	25	D004	1 0 0 0 0	

3262. $C_{15}H_{26}N_2O_3$

5-Ethyl-5-n-nonylbarbituric Acid

RN: 64810-91-9   **MP** (°C):
MW: 282.39   **BP** (°C):

Solubility (Moles/L)	Solubility (Grams/L)	Temp (°C)	Ref (#)	Evaluation (T P E A A)	Comments
3.450E-04	9.742E-02	25	M310	2 2 2 2 2	

3263. $C_{15}H_{26}N_2O_3$

5-Allyl-5-methylhexylcarbinylbarbituric Acid

RN: **MP** (°C):
MW: 282.39   **BP** (°C):

Solubility (Moles/L)	Solubility (Grams/L)	Temp (°C)	Ref (#)	Evaluation (T P E A A)	Comments
1.084E-02	3.060E+00	ns	T003	0 0 0 0 2	

3264. C₁₅H₂₆O₆

Tributyrin
Glyceryl Tributyrate
Tributanoylglycerol
1,2,3-Propanetriyl Tributyrate
RN: 60-01-5   **MP** (°C):   173
MW: 302.37   **BP** (°C):   287.5

Solubility (Moles/L)	Solubility (Grams/L)	Temp (°C)	Ref (#)	Evaluation (T P E A A)	Comments
3.307E-04	9.999E-02	ns	F014	0 0 0 0 1	

3265. C₁₅H₂₈O₄

1,13-Tridecanedicarboxylic Acid
1,15-Pentadecandioic Acid
RN: 1460-18-0   **MP** (°C):
MW: 272.39   **BP** (°C):

Solubility (Moles/L)	Solubility (Grams/L)	Temp (°C)	Ref (#)	Evaluation (T P E A A)	Comments
1.285E-03	3.500E-01	21	B040	1 0 1 1 1	*sic*

3266. C₁₅H₃₀

1-Pentadecene
RN: 13360-61-7   **MP** (°C):
MW: 210.41   **BP** (°C):

Solubility (Moles/L)	Solubility (Grams/L)	Temp (°C)	Ref (#)	Evaluation (T P E A A)	Comments
1.778E-09	3.740E-07	23	C332	2 0 2 2 1	

3267. C₁₅H₃₀O₂

Pentadecylic Acid
Pentadecanoic Acid
RN: 1002-84-2   **MP** (°C):   52
MW: 242.41   **BP** (°C):

Solubility (Moles/L)	Solubility (Grams/L)	Temp (°C)	Ref (#)	Evaluation (T P E A A)	Comments
3.135E-05	7.600E-03	0	B136	1 0 2 1 1	
3.135E-05	7.600E-03	0.0	R001	1 1 1 1 1	
4.950E-05	1.200E-02	20	B136	1 0 2 1 1	
4.950E-05	1.200E-02	20.0	R001	1 1 1 1 1	
5.775E-05	1.400E-02	30	B136	1 0 2 1 1	
5.775E-05	1.400E-02	30.0	R001	1 1 1 1 1	

7.013E-05	1.700E-02	45	B136	1 0 2 1 1
7.013E-05	1.700E-02	45.0	R001	1 1 1 1 1
8.251E-05	2.000E-02	60	B136	1 0 2 1 1
8.250E-05	2.000E-02	60.0	R001	1 1 1 1 1

3268. C₁₅H₃₀O₃
Dodecyl Lactate
Propanoic Acid, 2-Hydroxy-, Dodecyl Ester
RN: 6283-92-7   **MP** (°C):
MW: 258.40   **BP** (°C):

Solubility (Moles/L)	Solubility (Grams/L)	Temp (°C)	Ref (#)	Evaluation (T P E A A)	Comments
3.870E-04	1.000E-01	25	R006	2 2 0 1 0	

3269. C₁₅H₃₂O
Pentadecanol
Pentadecan-1-ol
1-Pentadecanol
RN: 629-76-5   **MP** (°C):   46
MW: 228.42   **BP** (°C):

Solubility (Moles/L)	Solubility (Grams/L)	Temp (°C)	Ref (#)	Evaluation (T P E A A)	Comments
4.500E-07	1.028E-04	25	R002	1 2 2 2 2	

3270. C₁₆H₁₀
Fluoranthene
1,2-Benzacenaphthene
1,2-(1,8-Naphthalenediyl)benzene
Benzo[j,k]fluorene
Idryl
FA
RN: 206-44-0   **MP** (°C):   107
MW: 202.26   **BP** (°C):   384

Solubility (Moles/L)	Solubility (Grams/L)	Temp (°C)	Ref (#)	Evaluation (T P E A A)	Comments
4.050E-07	8.191E-05	8.10	M082	1 1 1 2 2	
4.050E-07	8.191E-05	8.10	M151	2 1 2 2 2	
4.058E-07	8.207E-05	8.14	M183	1 1 1 1 2	
5.290E-07	1.070E-04	13.20	M082	1 1 1 2 2	
5.290E-07	1.070E-04	13.20	M151	2 1 2 2 2	
5.295E-07	1.071E-04	13.24	M183	1 1 1 1 2	
7.330E-07	1.483E-04	19.70	M082	1 1 1 2 2	
7.330E-07	1.483E-04	19.70	M151	2 1 2 2 2	
7.339E-07	1.484E-04	19.74	M183	1 2 1 1 2	
1.190E-06	2.407E-04	20	E009	1 0 0 1 2	
9.394E-07	1.900E-04	20	H300	1 1 2 2 1	
9.394E-07	1.900E-04	20	H300	1 1 2 2 1	

5.933E-07	1.200E-04	24	H116	2 1 0 0 2
1.000E-06	2.023E-04	24.60	M082	1 1 1 2 2
1.000E-06	2.023E-04	24.60	M151	2 1 2 2 2
1.003E-06	2.028E-04	24.64	M183	1 2 1 1 2
1.400E-06	2.832E-04	25	A325	2 1 2 2 1
1.320E-06	2.670E-04	25	K001	2 2 2 2 2
1.335E-06	2.700E-04	25	L332	1 1 1 1 2
1.285E-06	2.600E-04	25	M064	1 1 2 2 1
1.019E-06	2.060E-04	25	M071	2 2 2 2 2
1.300E-06	2.629E-04	25	M342	1 0 1 1 1
1.167E-06	2.360E-04	25	S227	1 2 1 1 2
1.019E-06	2.060E-04	25.00	M151	2 1 1 2 2
1.187E-06	2.400E-04	27	D003	1 0 0 1 1
1.305E-06	2.640E-04	29	M071	2 2 2 2 2
1.305E-06	2.640E-04	29.00	M151	2 1 1 2 2
1.380E-06	2.791E-04	29.90	M082	1 1 1 2 2
1.380E-06	2.791E-04	29.90	M151	2 1 2 2 2
1.382E-06	2.796E-04	29.94	M183	1 2 1 1 2
1.300E-06	2.630E-04	ns	I332	0 0 0 0 1

3271. $C_{16}H_{10}$

Pyrene

Benzo[def]phenanthrene

RN: 129-00-0   **MP** (°C):   156

MW: 202.26   **BP** (°C):   404

Solubility (Moles/L)	Solubility (Grams/L)	Temp (°C)	Ref (#)	Evaluation (T P E A A)	Comments
<1.00E-07	<2.02E-05	4	K049	1 2 1 1 0	
2.430E-07	4.915E-05	4.70	M082	1 1 1 2 2	
2.430E-07	4.915E-05	4.70	M151	2 1 2 2 2	
2.434E-07	4.924E-05	4.74	M183	1 2 1 1 2	
2.890E-07	5.845E-05	9.50	M082	1 1 1 2 2	
2.890E-07	5.845E-05	9.50	M151	2 1 2 2 2	
2.895E-07	5.855E-05	9.54	M183	1 2 1 1 2	
3.560E-07	7.200E-05	14.30	M082	1 1 1 2 2	
3.560E-07	7.200E-05	14.30	M151	2 1 2 2 2	
3.563E-07	7.206E-05	14.34	M183	1 2 1 1 2	
3.588E-07	7.258E-05	15	B385	2 0 2 2 2	
4.610E-07	9.324E-05	18.70	M082	1 1 1 2 2	
4.610E-07	9.324E-05	18.70	M151	2 1 2 2 2	
4.617E-07	9.338E-05	18.74	M183	1 2 1 1 2	
5.200E-07	1.052E-04	20	E009	1 0 0 0 1	
5.200E-07	1.052E-04	20	E025	1 0 1 2 1	
4.700E-07	9.506E-05	20	H306	1 0 1 2 1	
5.370E-07	1.086E-04	21.20	M082	1 1 1 2 2	
5.370E-07	1.086E-04	21.20	M151	2 1 2 2 2	
5.394E-07	1.091E-04	21.24	M183	1 2 1 1 2	

6.279E-07	1.270E-04	22.20	W003	2 1 2 2 2	average of 3
6.675E-07	1.350E-04	24	H106	1 0 2 2 2	
1.582E-07	3.200E-05	24	H116	2 1 0 0 1	
6.675E-07	1.350E-04	24	M129	1 2 1 1 2	
5.834E-07	1.180E-04	25	B319	2 0 1 2 2	
6.490E-07	1.313E-04	25	B385	2 0 2 2 2	
7.700E-07	1.557E-04	25	K001	1 0 2 1 2	
4.700E-07	9.506E-05	25	K123	1 0 2 2 1	
7.911E-07	1.600E-04	25	L332	1 1 1 1 2	
6.675E-07	1.350E-04	25	M064	1 1 2 2 2	
6.526E-07	1.320E-04	25	M071	2 2 2 2 2	
6.675E-07	1.350E-04	25	M156	1 2 1 1 2	
6.670E-07	1.349E-04	25	M342	1 0 1 1 2	
3.955E-07	8.000E-05	25	P340	1 1 2 2 1	
3.556E-08	7.191E-06	25	R084	2 2 2 2 1	*sic*
7.400E-07	1.497E-04	25	R302	1 2 1 2 1	
8.455E-07	1.710E-04	25	S227	1 2 1 1 2	
6.526E-07	1.320E-04	25.00	M151	2 1 1 2 2	
6.730E-07	1.361E-04	25.50	M082	1 1 1 2 2	
6.730E-07	1.361E-04	25.50	M151	2 1 2 2 2	
6.728E-07	1.361E-04	25.54	M183	1 2 1 1 2	
8.158E-07	1.650E-04	27	D003	1 0 0 1 1	
8.010E-07	1.620E-04	29	M071	2 2 2 2 2	
8.010E-07	1.620E-04	29.00	M151	2 1 1 2 2	
8.390E-07	1.697E-04	29.90	M082	1 1 1 2 2	
8.390E-07	1.697E-04	29.90	M151	2 1 2 2 2	
8.411E-07	1.701E-04	29.94	M183	1 2 1 1 2	
1.147E-06	2.320E-04	34.50	W003	2 1 2 2 2	average of 2
9.888E-07	2.000E-04	35	B385	2 0 2 2 2	
1.973E-06	3.990E-04	44.70	W003	2 1 2 2 2	average of 3
2.784E-06	5.630E-04	50.10	W003	2 1 2 2 2	average of 3
3.758E-06	7.600E-04	55.60	W003	2 1 2 2 1	average of 3
3.659E-06	7.400E-04	56.00	W003	2 1 2 2 1	
4.648E-06	9.400E-04	60.70	W003	2 1 2 2 1	average of 3
6.329E-06	1.280E-03	65.20	W003	2 1 2 2 2	average of 2
9.196E-06	1.860E-03	71.90	W003	2 1 2 2 2	average of 3
1.093E-05	2.210E-03	74.70	W003	2 1 2 2 2	
6.675E-07	1.350E-04	ns	H123	0 0 0 0 2	
6.675E-07	1.350E-04	ns	K304	0 0 0 0 2	
6.675E-07	1.350E-04	ns	M344	0 0 0 0 2	
5.000E-07	1.011E-04	ns	M383	0 2 1 1 0	
1.000E-06	2.023E-04	ns	W005	0 0 1 2 0	

3272.   C$_{16}$H$_{10}$N$_2$O$_8$S$_2$
C.I. Acid Blue 74(Free Acid)
Indigo-disulfosaeure-(5,5')
Indigotinsulfonic Acid

RN:	860-22-0	MP (°C):
MW:	422.39	BP (°C):

Solubility (Moles/L)	Solubility (Grams/L)	Temp (°C)	Ref (#)	Evaluation (T P E A A)	Comments
~2.37E-02	~1.00E+01	25	F300	1 0 0 0 0	

3273.   C$_{16}$H$_{11}$NO$_2$
Cinchophen
2-Phenyl-4-quinolinecarboxylic Acid
2-Phenylcinchoninic Acid

RN:	132-60-5	MP (°C):	213
MW:	249.27	BP (°C):	

Solubility (Moles/L)	Solubility (Grams/L)	Temp (°C)	Ref (#)	Evaluation (T P E A A)	Comments
6.418E-04	1.600E-01	25	L074	2 2 1 1 2	

3274.   C$_{16}$H$_{12}$F$_3$NO
6H-Dibenz[b,e]azepin-6-one, 5,11-Dihydro-5-(2,2,2-trifluoroethyl)-

RN:	155206-49-8	MP (°C):
MW:	291.28	BP (°C):

Solubility (Moles/L)	Solubility (Grams/L)	Temp (°C)	Ref (#)	Evaluation (T P E A A)	Comments
1.589E-05	4.627E-03	ns	M381	0 1 1 1 2	pH 7.0

3275.   C$_{16}$H$_{12}$N$_2$O$_3$
5,5-Diphenylbarbituric Acid
2,4,6(1H,3H,5H)-Pyrimidinetrione, 5,5-Diphenyl
Barbituric Acid, 5,5-Diphenyl

RN:	21914-07-8	MP (°C):
MW:	280.29	BP (°C):

Solubility (Moles/L)	Solubility (Grams/L)	Temp (°C)	Ref (#)	Evaluation (T P E A A)	Comments
6.370E-05	1.785E-02	25	P350	2 1 1 1 2	intrinsic

3276.   C₁₆H₁₂N₂O₄S

Sulfanaphthoquinone

RN: **MP** (°C):
MW: 328.35   **BP** (°C):

Solubility (Moles/L)	Solubility (Grams/L)	Temp (°C)	Ref (#)	Evaluation (T P E A A)	Comments
1.370E-04	4.500E-02	20	F073	1 2 2 2 1	

3277.   C₁₆H₁₂O₆

Hematein

Haematein

Benz[b]indeno[1,2-d]pyran-9(6H)-one, 6α,7-Dihydro-3,4,6α,10-Tetrahydroxy-

RN: 475-25-2   **MP** (°C): >200
MW: 300.27   **BP** (°C):

Solubility (Moles/L)	Solubility (Grams/L)	Temp (°C)	Ref (#)	Evaluation (T P E A A)	Comments
1.998E-03	6.000E-01	20	F300	1 0 0 0 1	

3278.   C₁₆H₁₂O₆

Benzoic Acid, 2-(Acetyloxy)-, 2-Carboxyphenyl Ester

RN: 530-75-6   **MP** (°C):   166.5
MW: 300.27   **BP** (°C):

Solubility (Moles/L)	Solubility (Grams/L)	Temp (°C)	Ref (#)	Evaluation (T P E A A)	Comments
6.661E-05	2.000E-02	21	N335	1 2 1 1 2	

3279.   C₁₆H₁₃ClN₂O

Diazepam

7-Chloro-1-methyl-5-phenyl-2H-1,4-benzodiazepin-2-one

Valium

Valrelease

Vazepam

Diazemuls

RN: 439-14-5   **MP** (°C):   125
MW: 284.75   **BP** (°C):

Solubility (Moles/L)	Solubility (Grams/L)	Temp (°C)	Ref (#)	Evaluation (T P E A A)	Comments
1.475E-04	4.200E-02	20	N059	2 0 2 2 2	average of 2
1.756E-04	5.000E-02	25	G084	2 0 2 2 1	
1.756E-04	5.000E-02	25	G095	2 1 2 2 1	
1.756E-04	5.000E-02	25	M159	1 0 2 2 0	EFG, pH 7.0
2.320E-04	6.606E-02	25	M320	2 2 1 1 2	
1.510E-04	4.300E-02	25	N055	2 0 2 2 1	
1.580E-04	4.500E-02	25	N055	2 0 2 1 2	
1.721E-04	4.900E-02	25	N055	2 0 2 0 2	

1.405E-04	4.000E-02	30	R081	1 2 2 2 0	
2.900E-04	8.258E-02	50	M335	1 0 2 1 2	pH 6.0
1.200E-04	3.417E-02	ns	F327	0 0 1 2 2	
1.756E-04	5.000E-02	ns	M036	0 0 0 0 0	

3280. $C_{16}H_{13}I_3N_2O_3$

Iobenzamic Acid
N-(3-Amino-2,4,6-triiodobenzoyl)-N-phenyl-β-alanine
Orbil
Osbiland
Razebil
Osbil
RN: 3115-05-7   **MP** (°C):
MW: 662.01   **BP** (°C):

Solubility (Moles/L)	Solubility (Grams/L)	Temp (°C)	Ref (#)	Evaluation (T P E A A)	Comments
1.737E-04	1.150E-01	ns	H055	0 1 0 2 2	

3281. $C_{16}H_{13}NO_3$

C.I. Disperse Red 3
N-(2-Hydroxyethyl)-1-aminoanthraquinone
Disperse Red 3
Disperse Red 66
RN: 4465-58-1   **MP** (°C):   168
MW: 267.29   **BP** (°C):

Solubility (Moles/L)	Solubility (Grams/L)	Temp (°C)	Ref (#)	Evaluation (T P E A A)	Comments
1.600E-05	4.277E-03	25	B333	1 0 0 0 1	

3282. $C_{16}H_{13}N_3$

Yellow AB
1-Phenylazo-2-naphthylamine
RN: 85-84-7   **MP** (°C):   102
MW: 247.30   **BP** (°C):

Solubility (Moles/L)	Solubility (Grams/L)	Temp (°C)	Ref (#)	Evaluation (T P E A A)	Comments
1.213E-06	3.000E-04	37	H120	1 1 1 1 0	normal saline

3283. C$_{16}$H$_{13}$N$_3$O$_3$
Mebendazole
Methyl 5-Benzoyl Benzimidazole-2-carbamate
Pantelmin
Methyl 5-Benzoyl-2-benzimidazolecarbamate
RN: 31431-39-7   **MP** (°C):   288.5
MW: 295.30   **BP** (°C):

Solubility (Moles/L)	Solubility (Grams/L)	Temp (°C)	Ref (#)	Evaluation (T P E A A)	Comments
1.693E-06	5.000E-04	21	N337	1 0 2 2 0	pH 5
1.700E-06	5.020E-04	21	N337	1 0 2 2 0	pH 5
1.199E-04	3.540E-02	25	H075	1 0 2 1 2	polymorph C
2.414E-04	7.130E-02	25	H075	1 0 2 1 2	polymorph B
3.332E-05	9.840E-03	25	H075	1 0 2 1 2	polymorph A

3284. C$_{16}$H$_{14}$
9,10-Dimethylanthracene
RN: 781-43-1   **MP** (°C):   182
MW: 206.29   **BP** (°C):

Solubility (Moles/L)	Solubility (Grams/L)	Temp (°C)	Ref (#)	Evaluation (T P E A A)	Comments
2.715E-07	5.600E-05	25	M064	1 1 2 2 1	
2.700E-07	5.570E-05	25	M342	1 0 1 1 1	
2.715E-07	5.600E-05	ns	M344	0 0 0 0 2	

3285. C$_{16}$H$_{14}$ClN$_3$O
Chlordiazepoxide
7-Chloro-2-(methylamino)-5-phenyl-3H-1,4-benzodiazepine-4-oxide
Librium
Menrium
Tropium
SK-Lygen
RN: 58-25-3   **MP** (°C):   236
MW: 299.76   **BP** (°C):

Solubility (Moles/L)	Solubility (Grams/L)	Temp (°C)	Ref (#)	Evaluation (T P E A A)	Comments
6.672E-03	2.000E+00	rt	M035	0 0 0 0 0	

3286.   C$_{16}$H$_{14}$Cl$_2$N$_2$O$_2$
Phenobenzuron
Benzoyl-1-(3,4-dichlorophenyl)-3,3-dimethylurea
Benzomarc
Urea, N-Benzoyl-N-(3,4-dichlorophenyl)-N',N'-dimethyl-
RN: 3134-12-1   **MP** (°C):   119
MW: 337.21   **BP** (°C):

Solubility (Moles/L)	Solubility (Grams/L)	Temp (°C)	Ref (#)	Evaluation (T P E A A)	Comments
4.745E-05	1.600E-02	22	M161	1 0 0 0 1	

3287.   C$_{16}$H$_{14}$Cl$_2$O$_3$
Chlorobenzilate
Ethyl 4,4'-Dichlorobenzilate
Acaraben
Benzilen
Folbex
Kopmite
RN: 510-15-6   **MP** (°C):   36
MW: 325.19   **BP** (°C):   157

Solubility (Moles/L)	Solubility (Grams/L)	Temp (°C)	Ref (#)	Evaluation (T P E A A)	Comments
3.998E-05	1.300E-02	20	F311	1 2 2 2 1	

3288.   C$_{16}$H$_{14}$Cl$_2$O$_4$
Diclotop-methyl
Methyl (+/-)-2-[4-(2,4-Dichlorophenoxy)phenoxy]propionate
RN: 51338-27-3   **MP** (°C):   40
MW: 341.19   **BP** (°C):

Solubility (Moles/L)	Solubility (Grams/L)	Temp (°C)	Ref (#)	Evaluation (T P E A A)	Comments
1.465E-04	5.000E-02	22	M161	1 0 0 0 1	

3289.   C$_{16}$H$_{14}$FNO
6H-Dibenz[b,e]azepin-6-one, 5-(2-Fluoroethyl)-5,11-dihydro-
RN: 155206-48-7   **MP** (°C):
MW: 255.29   **BP** (°C):

Solubility (Moles/L)	Solubility (Grams/L)	Temp (°C)	Ref (#)	Evaluation (T P E A A)	Comments
2.917E-04	7.448E-02	ns	M381	0 1 1 1 2	pH 7.0

3290. C$_{16}$H$_{14}$N$_2$O
Methaqualone
Quaalude
Mandrax
Somnafac
RN: 72-44-6   **MP** (°C):   114-117
MW: 250.30   **BP** (°C):

Solubility (Moles/L)	Solubility (Grams/L)	Temp (°C)	Ref (#)	Evaluation (T P E A A)	Comments
1.198E-03	2.999E-01	23	P094	1 0 0 0 2	

3291. C$_{16}$H$_{14}$N$_2$O$_2$
C.I. Disperse Blue 14
9,10-Anthracenedione, 1,4-bis(Methylamino)-
RN: 2475-44-7   **MP** (°C):   226
MW: 266.30   **BP** (°C):

Solubility (Moles/L)	Solubility (Grams/L)	Temp (°C)	Ref (#)	Evaluation (T P E A A)	Comments
1.400E-07	3.728E-05	25	B333	1 0 0 0 1	

3292. C$_{16}$H$_{14}$N$_2$O$_3$
3-(Hydroxymethyl)phenytoin
3-(Hydroxymethyl)-5,5-diphenyl-2,4-imidazolidinedione
RN: 21616-46-6   **MP** (°C):
MW: 282.30   **BP** (°C):

Solubility (Moles/L)	Solubility (Grams/L)	Temp (°C)	Ref (#)	Evaluation (T P E A A)	Comments
4.959E-04	1.400E-01	22	B154	1 1 1 1 1	0.1M HCl

3293. C$_{16}$H$_{14}$N$_2$O$_4$
C.I. Disperse Blue 26
9,10-Anthracenedione, 1,5-Dihydroxy-4,8-bis(methylamino)-
Resiren Blue TG
Navilene Blue GL
PTB 31
RN: 3860-63-7   **MP** (°C):   217
MW: 298.30   **BP** (°C):

Solubility (Moles/L)	Solubility (Grams/L)	Temp (°C)	Ref (#)	Evaluation (T P E A A)	Comments
6.800E-08	2.028E-05	25	B333	1 0 0 0 1	

3294.   C$_{16}$H$_{14}$O$_3$
Fenbufen
3-(4-Biphenylylcarbonyl) Propionic Acid
Lederfen

RN:	36330-85-5	**MP** (°C):	185
MW:	254.29	**BP** (°C):	

Solubility (Moles/L)	Solubility (Grams/L)	Temp (°C)	Ref (#)	Evaluation (T P E A A)	Comments
3.700E-06	9.409E-04	5	F306	1 0 1 2 2	intrinsic
6.430E-05	1.635E-02	25	C314	1 1 2 2 2	
6.410E-05	1.630E-02	25	C314	1 1 2 2 2	
8.700E-06	2.212E-03	25	F301	1 1 0 0 1	pH 2.0, *sic*
8.700E-06	2.212E-03	25	F306	1 0 1 2 2	intrinsic
1.800E-05	4.577E-03	37	F306	1 0 1 2 2	intrinsic
7.865E-06	2.000E-03	rt	H302	0 0 2 1 1	intrinsic

3295.   C$_{16}$H$_{14}$O$_3$
Ketoprofen
2-(meta-Benzoylphenyl) Propionic Acid
Orudis
Alrheumat
Oruvail

RN:	22071-15-4	**MP** (°C):	94
MW:	254.29	**BP** (°C):	

Solubility (Moles/L)	Solubility (Grams/L)	Temp (°C)	Ref (#)	Evaluation (T P E A A)	Comments
2.509E-04	6.380E-02	5	F306	1 0 1 2 2	intrinsic
9.045E-04	2.300E-01	21	B331	1 2 2 1 1	pH 7.4
5.646E-04	1.436E-01	25	F306	1 0 1 2 2	intrinsic
8.066E-04	2.051E-01	37	F306	1 0 1 2 2	intrinsic
2.006E-04	5.100E-02	rt	H302	0 0 2 1 2	intrinsic

3296.   C$_{16}$H$_{15}$ClN$_2$
Medazepam
7-Chloro-2,3-dihydro-1-methyl-5-phenyl-1H-1,4-benzodiazepine
Nobrium

RN:	2898-12-6	**MP** (°C):	
MW:	270.76	**BP** (°C):	

Solubility (Moles/L)	Solubility (Grams/L)	Temp (°C)	Ref (#)	Evaluation (T P E A A)	Comments
4.000E-05	1.083E-02	37	L011	1 0 2 1 1	

3297.   C₁₆H₁₅Cl₃OS₂

3297. $C_{16}H_{15}Cl_3OS_2$

2-(p-Methylthiophenyl)-2-(p-methylsulfinylphenyl)-1,1,1-trichloroethane

RN: 28463-05-0   **MP** (°C):   133-136
MW: 393.78   **BP** (°C):

Solubility (Moles/L)	Solubility (Grams/L)	Temp (°C)	Ref (#)	Evaluation (T P E A A)	Comments
3.174E-06	1.250E-03	ns	K117	0 1 2 1 1	

3298.   C₁₆H₁₅Cl₃O₂

3298. $C_{16}H_{15}Cl_3O_2$

Methoxychlor
1,1'-(2,2,2-Trichloroethylidene)-bis[4-methoxybenzene]
Maralate
Methoxy DDT
Marlate
Chemform

RN: 72-43-5   **MP** (°C):   82.5
MW: 345.66   **BP** (°C):

Solubility (Moles/L)	Solubility (Grams/L)	Temp (°C)	Ref (#)	Evaluation (T P E A A)	Comments
5.786E-08	2.000E-05	15	B083	2 2 1 2 1	particle size ≤ 5 µm
1.302E-07	4.500E-05	25	B083	2 2 1 2 1	particle size ≤ 5 µm
1.447E-07	5.000E-05	25	P085	1 0 1 1 2	
2.893E-07	1.000E-04	25	W025	1 0 2 2 2	
2.748E-07	9.500E-05	35	B083	2 2 1 2 1	particle size ≤ 5 µm
5.352E-07	1.850E-04	45	B083	2 2 1 2 2	particle size ≤ 5 µm
1.794E-06	6.200E-04	ns	K117	0 1 2 1 1	
8.679E-09	3.000E-06	ns	K138	0 0 0 0 2	
2.314E-06	8.000E-04	ns	M110	0 0 0 0 0	EFG
1.794E-06	6.200E-04	ns	M138	0 1 0 0 1	
3.472E-07	1.200E-04	ns	M344	0 0 0 0 1	

3299.   C₁₆H₁₅Cl₃O₂S₂

3299. $C_{16}H_{15}Cl_3O_2S_2$

2,2-bis(p-Methylsulfinylphenyl)-1,1,1-trichloroethane
2-(p-Methylsulfoxidephenyl)-1,1,1-trichloroethane

RN: 28396-87-4   **MP** (°C):   150-153
MW: 409.78   **BP** (°C):

Solubility (Moles/L)	Solubility (Grams/L)	Temp (°C)	Ref (#)	Evaluation (T P E A A)	Comments
7.077E-05	2.900E-02	ns	K117	0 1 2 1 1	

3300.   C$_{16}$H$_{15}$Cl$_3$O$_4$S$_2$
2,2-bis(p-Methylsulfonylphenyl)-1,1,1-trichloroethane
RN: 30665-94-2   **MP** (°C):   236.0
MW: 441.78   **BP** (°C):

Solubility (Moles/L)	Solubility (Grams/L)	Temp (°C)	Ref (#)	Evaluation (T P E A A)	Comments
3.395E-06	1.500E-03	ns	K117	0 1 2 1 1	

3301.   C$_{16}$H$_{15}$Cl$_3$S$_2$
2,2-bis-(p-Methylthiophenyl)-1,1,1-trichloroethane
RN: 19679-38-0   **MP** (°C):   115-117
MW: 377.78   **BP** (°C):

Solubility (Moles/L)	Solubility (Grams/L)	Temp (°C)	Ref (#)	Evaluation (T P E A A)	Comments
1.509E-06	5.700E-04	ns	K117	0 1 2 1 1	

3302.   C$_{16}$H$_{15}$FN$_2$O$_5$
1-Butyryloxymethyl-3-benzoyl-5-fluoro-2,4(1H,3H)-pyrimidinedi-one
1-Butyryloxymethyl-3-benzoyl-5-fluorouracil
RN: 97108-48-0   **MP** (°C):   81-82
MW: 334.31   **BP** (°C):

Solubility (Moles/L)	Solubility (Grams/L)	Temp (°C)	Ref (#)	Evaluation (T P E A A)	Comments
1.855E-04	6.200E-02	22	B321	1 0 2 2 2	pH 4.0

3303.   C$_{16}$H$_{15}$NO
4-Cyano-4'-propyloxybiphenyl
3 COB
RN: 52709-86-1   **MP** (°C):
MW: 237.30   **BP** (°C):

Solubility (Moles/L)	Solubility (Grams/L)	Temp (°C)	Ref (#)	Evaluation (T P E A A)	Comments
9.000E-07	2.136E-04	21	D300	2 2 1 1 2	

3304. C₁₆H₁₅NO₂

Cinnamyl Anthranilate
2-Propen-1-ol, 3-Phenyl-, 2-Aminobenzoate
2-Aminobenzoic Acid 3-Phenyl-2-propenyl Ester
3-Phenyl-2-propen-1-yl Anthranilate
3-Phenyl-2-propenyl 2-Aminobenzoate
Cinnamyl Alcohol

RN: 87-29-6   **MP** (°C): 60
MW: 253.30   **BP** (°C): 332

Solubility (Moles/L)	Solubility (Grams/L)	Temp (°C)	Ref (#)	Evaluation (T P E A A)	Comments
9.080E-07	2.300E-04	ns	B338	0 0 0 0 1	
9.080E-07	2.300E-04	ns	B338	0 0 0 0 1	

3305. C₁₆H₁₅NO₃

Benzoylphenylalanine
N-Benzoyl-DL-phenylalanine

RN: 2901-76-0   **MP** (°C):
MW: 269.30   **BP** (°C):

Solubility (Moles/L)	Solubility (Grams/L)	Temp (°C)	Ref (#)	Evaluation (T P E A A)	Comments
3.156E-03	8.500E-01	25.1	N026	2 0 2 2 2	

3306. C₁₆H₁₅NO₄

Benzoyltyrosine
N-benzoyl-L-tyrosine

RN: 2566-23-6   **MP** (°C):
MW: 285.30   **BP** (°C):

Solubility (Moles/L)	Solubility (Grams/L)	Temp (°C)	Ref (#)	Evaluation (T P E A A)	Comments
1.290E-02	3.680E+00	25.1	N026	2 0 2 2 2	

3307. C₁₆H₁₅N₅

6H-Dipyrido[3,2-b:2',3'-e][1,4]diazepin-6-nitrile, 11-Cyclopropyl-5,11-dihydro-4-methyl
RN: **MP** (°C):
MW: 277.33   **BP** (°C):

Solubility (Moles/L)	Solubility (Grams/L)	Temp (°C)	Ref (#)	Evaluation (T P E A A)	Comments
1.816E-05	5.035E-03	ns	M381	0 1 1 1 2	pH 7.0

3308.   C$_{16}$H$_{15}$N$_5$O$_4$S
2,5-Disulfanilamidopyridine
RN: **MP** (°C):
MW: 373.39   **BP** (°C):

Solubility (Moles/L)	Solubility (Grams/L)	Temp (°C)	Ref (#)	Evaluation (T P E A A)	Comments
1.326E-03	4.950E-01	37	R058	1 2 1 1 2	

3309.   C$_{16}$H$_{16}$
1,2,3,6,7,8-Hexahydropyrene
RN: 1732-13-4   **MP** (°C):   133
MW: 208.31   **BP** (°C):

Solubility (Moles/L)	Solubility (Grams/L)	Temp (°C)	Ref (#)	Evaluation (T P E A A)	Comments
1.100E-06	2.291E-04	4	K049	1 0 2 1 1	

3310.   C$_{16}$H$_{16}$ClN$_3$O$_3$S
Metolazone
2-Methyl-3-(o-tolyl)-6-sulfamyl-7-chloro-1,2,3,4-tetrahydro-4-quinazolinone
Zaroxolyn
Mykrox
Diulo
RN: 17560-51-9   **MP** (°C):   256.0
MW: 365.84   **BP** (°C):

Solubility (Moles/L)	Solubility (Grams/L)	Temp (°C)	Ref (#)	Evaluation (T P E A A)	Comments
9.321E-05	3.410E-02	10	B030	1 0 1 1 2	
1.339E-04	4.900E-02	20	B030	1 0 1 1 2	
1.648E-04	6.030E-02	25	B030	1 0 1 1 2	
1.971E-04	7.210E-02	30	B030	1 0 1 1 2	
2.236E-04	8.180E-02	35	B030	1 0 1 1 2	
2.733E-04	1.000E-01	36	B030	1 0 1 1 2	
1.640E-04	6.000E-02	37	H013	1 0 0 0 0	
2.952E-04	1.080E-01	40	B030	1 0 1 1 2	
3.799E-04	1.390E-01	45	B030	1 0 1 1 2	
4.155E-04	1.520E-01	50	B030	1 0 1 1 2	

3311.   C$_{16}$H$_{16}$N$_2$
3,4,7,8-Tetramethyl-1,10-phenanthroline
RN: 1660-93-1   **MP** (°C):   278.5
MW: 236.32   **BP** (°C):

Solubility (Moles/L)	Solubility (Grams/L)	Temp (°C)	Ref (#)	Evaluation (T P E A A)	Comments
6.400E-06	1.512E-03	25.04	B094	1 2 1 2 1	

3312. C$_{16}$H$_{16}$N$_2$O$_4$
Desmedipham
Ethyl m-Hydroxycarbanilate Carbanilate
Carbamic Acid, N-Phenyl-, 3-((Ethoxycarbonyl)amino)phenyl Ester
Betanex
Betanal-475
Betamix 70 WP
RN: 13684-56-5   **MP** (°C):   120
MW: 300.32   **BP** (°C):

Solubility (Moles/L)	Solubility (Grams/L)	Temp (°C)	Ref (#)	Evaluation (T P E A A)	Comments
2.331E-05	7.000E-03	rt	M161	0 0 0 0 0	
2.331E-05	7.000E-03	rt	R304	0 0 0 0 0	

3313. C$_{16}$H$_{16}$N$_2$O$_4$
Phenmedipham
Methyl m-Hydroxycarbanilate m-Methylcarbanilate
RN: 13684-63-4   **MP** (°C):   143
MW: 300.32   **BP** (°C):

Solubility (Moles/L)	Solubility (Grams/L)	Temp (°C)	Ref (#)	Evaluation (T P E A A)	Comments
<3.33E-05	<1.00E-02	20	B200	1 0 0 0 0	
3.330E-06	1.000E-03	20	F311	1 2 2 2 1	
3.330E-05	1.000E-02	ns	M061	0 0 0 0 1	
9.989E-06	3.000E-03	rt	M161	0 0 0 0 0	

3314. C$_{16}$H$_{16}$N$_4$
Disperse Black 1
RN: 6054-48-4   **MP** (°C):
MW: 264.33   **BP** (°C):

Solubility (Moles/L)	Solubility (Grams/L)	Temp (°C)	Ref (#)	Evaluation (T P E A A)	Comments
3.000E-07	7.930E-05	25	B333	1 0 0 0 1	

3315. C$_{16}$H$_{16}$N$_4$O
6H-Dipyrido[3,2-b:2',3'-e][1,4]diazepin-6-one, 11-Cyclobutyl-5,11-dihydro-5-methyl-
RN: 135794-88-6   **MP** (°C):
MW: 280.33   **BP** (°C):

Solubility (Moles/L)	Solubility (Grams/L)	Temp (°C)	Ref (#)	Evaluation (T P E A A)	Comments
2.911E-04	8.160E-02	ns	M381	0 1 1 1 2	pH 7.0

3316. C$_{16}$H$_{16}$N$_4$O
6H-Dipyrido[3,2-b:2'3'-e][1,4]diazepin-6-one, 11-Cyclopropyl-5,11-dihydro-2,4-dimethyl-
RN: 135794-77-3   **MP** (°C):
MW: 280.33   **BP** (°C):

Solubility (Moles/L)	Solubility (Grams/L)	Temp (°C)	Ref (#)	Evaluation (T P E A A)	Comments
5.346E-05	1.499E-02	ns	M381	0 1 1 1 2	pH 7.0

3317. C$_{16}$H$_{16}$N$_6$O$_4$S
2,5-Disulfanilamidopyrimidine
RN: **MP** (°C):
MW: 388.41   **BP** (°C):

Solubility (Moles/L)	Solubility (Grams/L)	Temp (°C)	Ref (#)	Evaluation (T P E A A)	Comments
5.664E-05	2.200E-02	37	R046	1 2 1 1 1	

3318. C$_{16}$H$_{16}$O$_2$
4-Methoxy-3,3'-dimethylbenzophenone
RN: 41295-28-7   **MP** (°C):   62.25
MW: 240.30   **BP** (°C):

Solubility (Moles/L)	Solubility (Grams/L)	Temp (°C)	Ref (#)	Evaluation (T P E A A)	Comments
8.323E-06	2.000E-03	20	M161	1 0 0 0 0	

3319. C$_{16}$H$_{16}$O$_3$
Ethyl Benzoyl Benzoate
RN: 106396-19-4   **MP** (°C):
MW: 256.30   **BP** (°C):

Solubility (Moles/L)	Solubility (Grams/L)	Temp (°C)	Ref (#)	Evaluation (T P E A A)	Comments
3.901E-04	9.999E-02	ns	F014	0 0 0 0 1	

3320. C$_{16}$H$_{17}$ClN$_2$S
Chlorphenethazine
2-Chloro-N,N-dimethyl-10H-phenothiazine-10-ethanamide
RN: 2095-24-1   **MP** (°C):
MW: 304.84   **BP** (°C):

Solubility (Moles/L)	Solubility (Grams/L)	Temp (°C)	Ref (#)	Evaluation (T P E A A)	Comments
1.500E-05	4.573E-03	ns	G023	0 0 1 1 1	

3321. C₁₆H₁₇ClN₄O₃

C.I. Disperse Red 13
4-Nitro-2-chloro-4'-[ethyl(2-hydroxyethyl)amino]azobenzene
Acetoquinone Light Rubine BLZ
Acetamine Rubine B
Acetate Fast Rubine B
RN: 3180-81-2 **MP** (°C): 133
MW: 348.79 **BP** (°C):

Solubility (Moles/L)	Solubility (Grams/L)	Temp (°C)	Ref (#)	Evaluation (T P E A A)	Comments
3.300E-08	1.151E-05	25	B333	1 0 0 0 1	

3322. C₁₆H₁₇ClN₄O₄

C.I. Disperse Red 7
Ethanol, 2,2'-[[3-Chloro-4-[(4-nitrophenyl)azo]phenyl]imino]bis-
RN: 4540-00-5 **MP** (°C): 190
MW: 364.79 **BP** (°C):

Solubility (Moles/L)	Solubility (Grams/L)	Temp (°C)	Ref (#)	Evaluation (T P E A A)	Comments
1.100E-06	4.013E-04	25	B333	1 0 0 0 1	

3323. C₁₆H₁₇NO

Diphenamid
Dyamid
Enide
N,N-Dimethyl-α-phenylbenzeneacetamide
N,N-Dimethyldiphenylacetamide
Diherbid
RN: 957-51-7 **MP** (°C): 132
MW: 239.32 **BP** (°C):

Solubility (Moles/L)	Solubility (Grams/L)	Temp (°C)	Ref (#)	Evaluation (T P E A A)	Comments
1.003E-03	2.399E-01	25	M061	1 0 0 0 1	
1.086E-03	2.600E-01	25	M161	1 0 0 0 2	
1.090E-03	2.609E-01	27	B200	1 0 0 0 2	
1.086E-03	2.600E-01	ns	B185	0 0 0 0 2	
2.079E-02	4.975E+00	ns	B200	0 0 0 0 0	
1.086E-03	2.600E-01	ns	H042	0 0 0 0 2	

3324.   C₁₆H₁₇NO₄
2-Naphthaleneacetic Acid, 6-Methoxy-α-methyl-, 2-Amino-2-oxoethyl Ester, (S)
Naproxen, N,N-Glycolamide Ester
2-Naphthaleneacetic Acid, 6-Methoxy-α-methyl-, 2-Amino-2-oxoethyl Ester
Naproxen N,N-Glycolamide Ester
RN: 114665-17-7   **MP** (°C):   139.5
MW: 287.32   **BP** (°C):

Solubility (Moles/L)	Solubility (Grams/L)	Temp (°C)	Ref (#)	Evaluation (T P E A A)	Comments
1.183E-04	3.400E-02	21	B331	1 2 2 1 2	pH 7.4
1.183E-04	3.400E-02	21	B331	1 2 2 1 1	

3325.   C₁₆H₁₇N₃O₄S
Cephalexin
Cefanex
C-Lexin
Keflex
Cefalexin
RN: 15686-71-2   **MP** (°C):
MW: 347.40   **BP** (°C):

Solubility (Moles/L)	Solubility (Grams/L)	Temp (°C)	Ref (#)	Evaluation (T P E A A)	Comments
1.724E-02	5.990E+00	10	O305	2 2 1 2 2	noncrystalline
1.569E-01	5.450E+01	15	O305	2 2 1 2 2	noncrystalline
1.416E-01	4.920E+01	20	O305	2 2 1 2 2	noncrystalline
3.598E-02	1.250E+01	25	P311	1 2 2 2 0	EFG
3.500E-03	1.216E+00	35	E311	1 0 2 2 2	

3326.   C₁₆H₁₇N₃O₄S.H₂O
Cephalexin (Monohydrate)
RN: 23325-78-2   **MP** (°C):
MW: 365.41   **BP** (°C):

Solubility (Moles/L)	Solubility (Grams/L)	Temp (°C)	Ref (#)	Evaluation (T P E A A)	Comments
3.694E-02	1.350E+01	25	M165	1 0 0 0 2	

3327.   C₁₆H₁₇N₅O₅
Dis. A. 12
Ethanol, 2-[[4-[(2,4-Dinitrophenyl)azo]phenyl]ethylamino]-
RN: 62570-20-1   **MP** (°C):
MW: 359.34   **BP** (°C):

Solubility (Moles/L)	Solubility (Grams/L)	Temp (°C)	Ref (#)	Evaluation (T P E A A)	Comments
2.000E-06	7.187E-04	25	B333	1 0 0 0 1	

3328.   C$_{16}$H$_{17}$N$_5$O$_6$
Dis. A. 14
4-[bis(2-Hydroxyethyl)amino]-2',4'-dinitroazobenzene
RN: 60129-67-1   **MP** (°C):
MW: 375.34   **BP** (°C):

Solubility (Moles/L)	Solubility (Grams/L)	Temp (°C)	Ref (#)	Evaluation (T P E A A)	Comments
6.000E-06	2.252E-03	25	B333	1 0 0 0 1	

3329.   C$_{16}$H$_{18}$ClNO$_4$S
Oxathiin Carboxanilide
Benzoic Acid, 2-Chlloro-5-[[(5,6-dihydro-2-methyl-1,4-oxathiin-3-yl)-
carnonyl]amino]isopropyl Ester
RN: 135812-04-3   **MP** (°C):   130
MW: 355.84   **BP** (°C):

Solubility (Moles/L)	Solubility (Grams/L)	Temp (°C)	Ref (#)	Evaluation (T P E A A)	Comments
3.653E-06	1.300E-03	25	O319	0 0 0 0 1	

3330.   C$_{16}$H$_{18}$NO$_5$P
Diphenylmorpholidophosphate
RN: **MP** (°C):
MW: 335.30   **BP** (°C):

Solubility (Moles/L)	Solubility (Grams/L)	Temp (°C)	Ref (#)	Evaluation (T P E A A)	Comments
6.844E-03	2.295E+00	25	A040	1 0 0 0 2	

3331.   C$_{16}$H$_{18}$N$_2$O$_3$
Difenoxuron
N-4-(4'-Methoxyphenoxy)phenyl-N',N'-dimethylurea
C-3470
RN: 14214-32-5   **MP** (°C):   138.5
MW: 286.33   **BP** (°C):

Solubility (Moles/L)	Solubility (Grams/L)	Temp (°C)	Ref (#)	Evaluation (T P E A A)	Comments
6.985E-05	2.000E-02	20	M161	1 0 0 0 1	
6.985E-05	2.000E-02	ns	M061	0 0 0 0 1	

3332. C$_{16}$H$_{18}$N$_4$O

6H-Dipyrido[3,2-b:2',3'-e][1,4]diazepin-6-one, 11-(1,1-Dimethylethyl)-5,11-dihydro-5-
methyl-

RN: 135794-80-8   **MP** (°C):
MW: 282.35   **BP** (°C):

Solubility (Moles/L)	Solubility (Grams/L)	Temp (°C)	Ref (#)	Evaluation (T P E A A)	Comments
1.416E-05	3.997E-03	ns	M381	0 1 1 1 2	pH 7.0

3333. C$_{16}$H$_{18}$N$_4$O$_2$

Dis. A. 5
4-Nitro-4'-diethylaminoazobenzene
4-Nitro-4'-N,N-diethylaminoazobenzene
DEANAB

RN: 3025-52-3   **MP** (°C):   152
MW: 298.35   **BP** (°C):

Solubility (Moles/L)	Solubility (Grams/L)	Temp (°C)	Ref (#)	Evaluation (T P E A A)	Comments
4.000E-11	1.193E-08	25	B333	1 0 0 0 1	

3334. C$_{16}$H$_{18}$N$_4$O$_2$

Dye III
4[[(4-Diethylamino)phenyl]azo]nitrobenzene

RN: **MP** (°C):
MW: 298.35   **BP** (°C):

Solubility (Moles/L)	Solubility (Grams/L)	Temp (°C)	Ref (#)	Evaluation (T P E A A)	Comments
9.100E-07	2.715E-04	97.40	B198	1 2 1 1 1	

3335. C$_{16}$H$_{18}$N$_4$O$_3$

Disperse Red 1
Dye IV
C.I. Disperse Red 1
1-[N-Ethyl-N-(2-hydroxyethyl)amino]-4-(4-nitrophenylazo)benzene
4-Nitro-4'-[ethyl(2-hydroxyethyl)amino]azobenzene

RN: 2872-52-8   **MP** (°C):   161
MW: 314.35   **BP** (°C):

Solubility (Moles/L)	Solubility (Grams/L)	Temp (°C)	Ref (#)	Evaluation (T P E A A)	Comments
5.400E-07	1.697E-04	25	B333	1 0 0 0 1	
5.400E-06	1.697E-03	60	B198	1 2 1 1 1	
6.521E-06	2.050E-03	60	P313	1 2 1 2 2	average of 2
1.082E-05	3.400E-03	70	P313	1 2 1 2 2	average of 2
1.310E-05	4.118E-03	71.80	B198	1 2 1 1 2	

1.797E-05	5.650E-03	80	P313	1 2 1 2 2	average of 2
3.120E-05	9.808E-03	84.10	B198	1 2 1 1 2	
3.388E-05	1.065E-02	90	P313	1 2 1 2 2	average of 2
7.130E-05	2.241E-02	97.40	B198	1 2 1 1 2	

3336.   C$_{16}$H$_{18}$N$_4$O$_4$

Disperse Red 19
Dye V
C.I. Disperse Red 19
2-[(2-Hydroxyethyl)[4-(4-nitrophenylazo)phenyl]amino]ethanol
4'-[(N,N-Dihydroxyethyl)amino]-4-nitroazobenzene

RN: 2734-52-3   **MP** (°C):   209
MW: 330.35   **BP** (°C):

Solubility (Moles/L)	Solubility (Grams/L)	Temp (°C)	Ref (#)	Evaluation (T P E A A)	Comments
7.100E-07	2.345E-04	25	B333	1 0 0 0 1	
1.170E-05	3.865E-03	60	B198	1 2 1 1 2	
3.030E-05	1.001E-02	71.80	B198	1 2 1 1 2	
8.330E-05	2.752E-02	84.10	B198	1 2 1 1 2	
2.100E-04	6.937E-02	97.40	B198	1 2 1 1 2	

3337.   C$_{16}$H$_{18}$O$_3$

Naproxen Ethyl Ester
2-Naphthaleneacetic Acid, 6-Methoxy-α-methyl-, Ethyl Ester, (alphaS)-

RN: 31220-35-6   **MP** (°C):
MW: 258.32   **BP** (°C):

Solubility (Moles/L)	Solubility (Grams/L)	Temp (°C)	Ref (#)	Evaluation (T P E A A)	Comments
4.645E-06	1.200E-03	21	B331	1 2 2 1 2	pH 7.4
4.645E-06	1.200E-03	21	B331	1 2 2 1 1	

3338.   C$_{16}$H$_{19}$ClN$_2$

Chlorpheniramine
1-(p-Chlorophenyl)-1-(2-pyridyl)-3-dimethylaminopropane

RN: 132-22-9   **MP** (°C):   <25
MW: 274.80   **BP** (°C):   142

Solubility (Moles/L)	Solubility (Grams/L)	Temp (°C)	Ref (#)	Evaluation (T P E A A)	Comments
2.000E-02	5.496E+00	37.5	L034	2 2 0 1 2	pH 7.4

3339.   C₁₆H₁₉NO₇
Benzoic Acid, 2-(Acetyloxy)-, 2-[(2-Ethoxy-2-oxoethyl)methylamino]-2-oxoethyl Ester
RN: 116482-77-0   **MP** (°C):   47.5
MW: 337.33   **BP** (°C):

Solubility (Moles/L)	Solubility (Grams/L)	Temp (°C)	Ref (#)	Evaluation (T P E A A)	Comments
2.846E-03	9.600E-01	21	N335	1 2 1 1 2	

3340.   C₁₆H₁₉N₃O₂
C.I. Solvent Yellow 58
p-[bis(2-Hydroxyethyl)amino]azobenzene
4-[bis(2-Hydroxyethyl)amino]azobenzene
RN: 2452-84-8   **MP** (°C):   134
MW: 285.35   **BP** (°C):

Solubility (Moles/L)	Solubility (Grams/L)	Temp (°C)	Ref (#)	Evaluation (T P E A A)	Comments
1.100E-04	3.139E-02	25	B333	1 0 0 0 1	

3341.   C₁₆H₁₉N₃O₄S
Ampicillin
(2S,5R,6R)-6-[(R)-2-Amino-2-phenylacetamido]-3,3-dimethyl-7-oxo-4-thia-1-
azabicyclo[3.2.0]heptane-2-carboxylic Acid
Aminobenzylpenicillin
Unasyn
Wymox
Totacillin
RN: 69-53-4   **MP** (°C):
MW: 349.41   **BP** (°C):

Solubility (Moles/L)	Solubility (Grams/L)	Temp (°C)	Ref (#)	Evaluation (T P E A A)	Comments
4.293E-02	1.500E+01	7.5	P009	1 0 2 1 0	EFG
3.721E-02	1.300E+01	20	P009	1 0 2 1 0	EFG
2.890E-02	1.010E+01	21	M044	2 0 2 2 2	
3.978E-02	1.390E+01	25	H051	1 2 2 2 2	
3.434E-02	1.200E+01	30	P009	1 0 2 1 0	EFG
3.291E-02	1.150E+01	40	P009	1 0 2 1 0	EFG

3342. C₁₆H₁₉N₃O₄S

$C_{16}H_{19}N_3O_4S$

Cephradine
Anspor
Velosef

RN: 38821-53-3   **MP** (°C): 140
MW: 349.41   **BP** (°C):

Solubility (Moles/L)	Solubility (Grams/L)	Temp (°C)	Ref (#)	Evaluation (T P E A A)	Comments
6.096E-02	2.130E+01	ns	F181	0 0 0 0 2	

3343. C₁₆H₁₉N₃O₄S.3H₂O

$C_{16}H_{19}N_3O_4S \cdot 3H_2O$

Ampicillin (Trihydrate)

RN: 7177-48-2   **MP** (°C): 198
MW: 403.46   **BP** (°C):

Solubility (Moles/L)	Solubility (Grams/L)	Temp (°C)	Ref (#)	Evaluation (T P E A A)	Comments
1.413E-02	5.700E+00	7.5	P009	1 0 2 1 0	EFG
1.487E-02	6.000E+00	20	P009	1 0 2 1 0	EFG
1.873E-02	7.558E+00	21	M044	2 0 2 2 2	
1.983E-02	8.000E+00	30	P009	1 0 2 1 0	EFG
2.479E-02	1.000E+01	40	P009	1 0 2 1 0	EFG

3344. C₁₆H₁₉N₃O₅S.3H₂O

$C_{16}H_{19}N_3O_5S \cdot 3H_2O$

Amoxicillin (Trihydrate)
4-Thia-1-azabicyclo(3,2,0)heptane-2-carboxylic Acid (Trihydrate)

RN: 61336-70-7   **MP** (°C):
MW: 419.46   **BP** (°C):

Solubility (Moles/L)	Solubility (Grams/L)	Temp (°C)	Ref (#)	Evaluation (T P E A A)	Comments
~9.54E-03	~4.00E+00	ns	B188	0 0 0 0 0	

3345. C₁₆H₁₉N₅O

$C_{16}H_{19}N_5O$

6H-Dipyrido[3,2-b:2',3'-e][1,4]diazepin-6-one, 2-Dimethylamino)-11-ethyl-5,11-dihydro-4-methyl-

RN: 135795-08-3   **MP** (°C):
MW: 297.36   **BP** (°C):

Solubility (Moles/L)	Solubility (Grams/L)	Temp (°C)	Ref (#)	Evaluation (T P E A A)	Comments
1.346E-05	4.002E-03	ns	M381	0 1 1 1 2	pH 7.0

3346.   C₁₆H₁₉N₅O₂

6H-Dipyrido[3,2-b:2',3'-e][1,4]diazepin-6-one, 11-Ethyl-5,11-dihydro-2-[(2-hydroxyethyl)methylamino

RN: 155206-46-5   **MP** (°C):
MW: 313.36   **BP** (°C):

Solubility (Moles/L)	Solubility (Grams/L)	Temp (°C)	Ref (#)	Evaluation (T P E A A)	Comments
4.365E-04	1.368E-01	ns	M381	0 1 1 1 2	pH 7.0

3347.   C₁₆H₁₉O₄P

Butyl Diphenyl Phosphate

RN: 2752-95-6   **MP** (°C):
MW: 306.30   **BP** (°C):

Solubility (Moles/L)	Solubility (Grams/L)	Temp (°C)	Ref (#)	Evaluation (T P E A A)	Comments
<6.53E-04	<2.00E-01	25	B070	1 2 0 1 0	

3348.   C₁₆H₂₀I₃N₃O₇

1,3-Benzenedicarboxamide, N-(2,3-Dihydroxypropyl)-N'-(2-hydroxyethyl)-5-[(2-hydroxy-1-oxopropyl)amino]-2,4,6-triiodo-(RS)

RN: 77868-43-0   **MP** (°C):
MW: 747.07   **BP** (°C):

Solubility (Moles/L)	Solubility (Grams/L)	Temp (°C)	Ref (#)	Evaluation (T P E A A)	Comments
6.374E-02	4.762E+01	25	P091	1 0 0 0 1	

3349.   C₁₆H₂₀I₃N₃O₇

1,3-Benzenedicarboxamide, N-(2-Hydroxyethyl)-N'-[2-hydroxy-1-(hydroxymethyl)ethyl]-5-[(2-hydroxy-1-oxopropyl)amino]-2,4,6-triiodo-, (S)-

RN: 77868-44-1   **MP** (°C):
MW: 747.07   **BP** (°C):

Solubility (Moles/L)	Solubility (Grams/L)	Temp (°C)	Ref (#)	Evaluation (T P E A A)	Comments
2.625E-02	1.961E+01	25	P091	1 0 0 0 1	

3350.   C₁₆H₂₀I₃N₃O₈

1,3-Benzenedicarboxamide, N,N'-bis(2,3-Dihydroxypropyl)-5S-[(hydroxyacetyl)amino]-2,4,6-triiodo- [RS-(RS*,RS*)]-

RN: 77868-40-7   **MP** (°C):
MW: 763.07   **BP** (°C):

Solubility (Moles/L)	Solubility (Grams/L)	Temp (°C)	Ref (#)	Evaluation (T P E A A)	Comments
2.317E-02	1.768E+01	25	P091	1 0 0 0 1	

3351. $C_{16}H_{20}I_3N_3O_8$
1,3-Benzenedicarboxamide, 5-[(Hydroxyacetyl)amino]-N,N'-bis[2-hydroxy-1-(hydroxymethyl)ethyl]-2,4,6-triiodo-

RN: 77868-41-8   **MP** (°C):
MW: 763.07   **BP** (°C):

Solubility (Moles/L)	Solubility (Grams/L)	Temp (°C)	Ref (#)	Evaluation (T P E A A)	Comments
5.282E-02	4.031E+01	25	P091	1 0 0 0 1	

3352. $C_{16}H_{20}N_4O_2$
Apazone
APZ
Azapropazone

RN: 13539-59-8   **MP** (°C):   247
MW: 300.36   **BP** (°C):

Solubility (Moles/L)	Solubility (Grams/L)	Temp (°C)	Ref (#)	Evaluation (T P E A A)	Comments
4.900E-04	1.472E-01	35	H091	1 2 2 2 1	*sic*
2.896E-04	8.700E-02	rt	H302	0 0 2 1 1	intrinsic

3353. $C_{16}H_{20}N_4O_3S$
2-(N4-Acetylsulfanilylamino)-4-isobutylpyrimidine

RN: **MP** (°C):
MW: 348.43   **BP** (°C):

Solubility (Moles/L)	Solubility (Grams/L)	Temp (°C)	Ref (#)	Evaluation (T P E A A)	Comments
1.091E-05	3.800E-03	37	R076	1 2 0 0 1	

3354. $C_{16}H_{20}N_8O_2S$
6-[D-2-amino-2-(4-aminophenyl)-acetamido]-3,3-dimethyl-7-oxo-4-thia-1-azabicyclo[3,2,0]hept-2-yl-5-t

RN: **MP** (°C):
MW: 388.45   **BP** (°C):

Solubility (Moles/L)	Solubility (Grams/L)	Temp (°C)	Ref (#)	Evaluation (T P E A A)	Comments
5.277E-03	2.050E+00	25	B148	2 2 2 1 2	

3355.   C₁₆H₂₀O₆P₂S₃

Temephos
O,O'-(Thiodi-4,1-phenylene)bis(O,O'-dimethylphosphorothioate)
Abate
Tetramethyl O,O'-Thiodi-p-phenylene Phosphorothioate
Abaphos
Tetrafenphos

RN: 3383-96-8   **MP** (°C):
MW: 466.47   **BP** (°C):

Solubility (Moles/L)	Solubility (Grams/L)	Temp (°C)	Ref (#)	Evaluation (T P E A A)	Comments
1.929E-08	9.000E-06	10	B324	2 2 2 2 2	
1.929E-08	8.998E-06	10	B324	2 2 2 2 2	
5.788E-07	2.700E-04	20	B300	2 1 1 1 2	
5.788E-07	2.700E-04	20	B324	2 2 2 2 2	
5.788E-07	2.700E-04	20	B324	2 2 2 2 2	
1.501E-06	7.002E-04	30	B324	2 2 2 2 2	
1.501E-06	7.000E-04	30	B324	2 2 2 2 2	

3356.   C₁₆H₂₁ClN₃S

Methylene Blue
Methylenblau
C.I. 52015

RN: 61-73-4   **MP** (°C):
MW: 322.88   **BP** (°C):

Solubility (Moles/L)	Solubility (Grams/L)	Temp (°C)	Ref (#)	Evaluation (T P E A A)	Comments
~1.02E-01	~3.30E+01	20	F300	1 0 0 0 0	

3357.   C₁₆H₂₁NO

N,N-Heptamethylenecinnamamide
Octahydro-1-(1-oxo-3-phenyl-2-propenyl) Azocine,

RN: 59832-06-3   **MP** (°C):
MW: 243.35   **BP** (°C):

Solubility (Moles/L)	Solubility (Grams/L)	Temp (°C)	Ref (#)	Evaluation (T P E A A)	Comments
2.560E-04	6.230E-02	ns	H350	0 0 0 0 2	

3358. C₁₆H₂₁NO

3358. $C_{16}H_{21}NO$

N-Cycloheptylcinnamamide
N-Cycloheptyl-3-phenyl- 2-Propenamide,
RN: 59831-98-0 **MP** (°C):
MW: 243.35 **BP** (°C):

Solubility (Moles/L)	Solubility (Grams/L)	Temp (°C)	Ref (#)	Evaluation (T P E A A)	Comments
3.570E-06	8.688E-04	ns	H350	0 0 0 0 2	

3359. $C_{16}H_{21}NO_2S$

m-Carboxyloctylphenylisothiocyanate
3-Carboxyloctylphenylisothiocyanate
RN: **MP** (°C):
MW: 291.42 **BP** (°C):

Solubility (Moles/L)	Solubility (Grams/L)	Temp (°C)	Ref (#)	Evaluation (T P E A A)	Comments
6.000E-05	1.748E-02	25	K032	2 2 0 1 1	

3360. $C_{16}H_{21}NO_3$

Piperidine, 1-[(Benzoyloxy)acetyl]-2-ethyl-
RN: 115178-69-3 **MP** (°C): 54.5
MW: 275.35 **BP** (°C):

Solubility (Moles/L)	Solubility (Grams/L)	Temp (°C)	Ref (#)	Evaluation (T P E A A)	Comments
1.889E-03	5.200E-01	22	N317	1 1 2 1 2	

3361. $C_{16}H_{21}NO_3$

Piperidine, 1-[(Benzoyloxy)acetyl]-2,6-dimethyl-
RN: 115178-70-6 **MP** (°C): 118
MW: 275.35 **BP** (°C):

Solubility (Moles/L)	Solubility (Grams/L)	Temp (°C)	Ref (#)	Evaluation (T P E A A)	Comments
5.448E-04	1.500E-01	22	N317	1 1 2 1 2	

3362. $C_{16}H_{21}NO_5$

Benzoic Acid, 2-(Acetyloxy)-, 2-(Diethylamino)-1-methyl-2-oxoethyl Ester
RN: 118247-09-9 **MP** (°C): 40.5
MW: 307.35 **BP** (°C):

Solubility (Moles/L)	Solubility (Grams/L)	Temp (°C)	Ref (#)	Evaluation (T P E A A)	Comments
2.499E-02	7.680E+00	21	N335	1 2 1 1 2	

3363.　C$_{16}$H$_{21}$N$_3$

Tripelennamine
N-Benzyl-N',N'-dimethyl-N-2-pyridylethylenediamine
PBZ
Pelamine

RN:　91-81-6　　　　**MP** (°C):　<25
MW:　255.37　　　　**BP** (°C):

Solubility (Moles/L)	Solubility (Grams/L)	Temp (°C)	Ref (#)	Evaluation (T P E A A)	Comments
2.300E-03	5.873E-01	30	L068	1 0 0 1 0	EFG
1.500E-02	3.830E+00	37.5	L034	2 2 0 1 2	pH 7.4

3364.　C$_{16}$H$_{22}$Cl$_2$O$_3$

2,4-Dichlorophenoxyacetic Acid n-Octyl Ester
2,4-Dichlorophenoxyacetic Acid Capryl Ester

RN:　1928-44-5　　**MP** (°C):
MW:　333.26　　　　**BP** (°C):

Solubility (Moles/L)	Solubility (Grams/L)	Temp (°C)	Ref (#)	Evaluation (T P E A A)	Comments
2.128E-05	7.092E-03	ns	M120	0 0 1 1 2	

3365.　C$_{16}$H$_{22}$N$_4$O

Neohetramine
N,N-Dimethyl-N'-(p-methoxybenzyl)-N'-(2-pyrimidyl)ethylenediamine
Tonzilamine

RN:　91-85-0　　　　**MP** (°C):
MW:　286.38　　　　**BP** (°C):

Solubility (Moles/L)	Solubility (Grams/L)	Temp (°C)	Ref (#)	Evaluation (T P E A A)	Comments
1.900E-02	5.441E+00	37.5	L034	2 2 0 1 2	pH 7.4

3366.　C$_{16}$H$_{22}$N$_4$O$_2$S

2-Sulfanilamido-4-methyl-5-n-amylpyrimidine

RN:　71119-35-2　**MP** (°C):　188-190
MW:　334.44　　　　**BP** (°C):

Solubility (Moles/L)	Solubility (Grams/L)	Temp (°C)	Ref (#)	Evaluation (T P E A A)	Comments
8.372E-05	2.800E-02	29	C049	1 2 0 0 1	

3367.   C$_{16}$H$_{22}$N$_4$O$_6$.0.5H$_2$O
6-Methoxy-9-(5-O-pivalate-β-D-arabinofuranosyl)]-9H-purine (Hemihydrate)
RN: 121032-42-6   **MP** (°C):   glass
MW: 375.38   **BP** (°C):

Solubility (Moles/L)	Solubility (Grams/L)	Temp (°C)	Ref (#)	Evaluation (T P E A A)	Comments
3.560E-02	1.336E+01	37	M378	1 2 1 1 2	pH 7.2

3368.   C$_{16}$H$_{22}$N$_4$O$_6$.0.5H$_2$O
6-Methoxy-9-(5-O-valerate-β-D-arabinofuranosyl)]-6-methoxy-9H-purine (Hemihydrate)
RN: 142963-77-7   **MP** (°C):   foam
MW: 375.38   **BP** (°C):

Solubility (Moles/L)	Solubility (Grams/L)	Temp (°C)	Ref (#)	Evaluation (T P E A A)	Comments
1.720E-03	6.457E-01	37	M378	1 2 1 1 2	pH 7.2

3369.   C$_{16}$H$_{22}$N$_4$O$_6$
2'-Valeryl-6-methoxypurine Arabinoside
2'-Trimethylacetyl-6-methoxypurine Arabinoside
RN: 121032-22-2   **MP** (°C):   118-120
MW: 366.38   **BP** (°C):

Solubility (Moles/L)	Solubility (Grams/L)	Temp (°C)	Ref (#)	Evaluation (T P E A A)	Comments
2.400E-01	8.793E+01	37	C348	1 2 2 2 2	pH 7.00
1.070E-01	3.920E+01	37	C348	1 2 2 2 2	pH 7.00

3370.   C$_{16}$H$_{22}$O$_4$
tere-Butyl Phthalate
RN: 30448-43-2   **MP** (°C):
MW: 278.35   **BP** (°C):

Solubility (Moles/L)	Solubility (Grams/L)	Temp (°C)	Ref (#)	Evaluation (T P E A A)	Comments
3.952E-06	1.100E-03	25	D336	2 1 2 2 2	

3371. C₁₆H₂₂O₄
Dibutyl Phthalate
n-Butyl Phthalate

RN: 84-74-2   **MP** (°C): -35
MW: 278.35   **BP** (°C): 430

Solubility (Moles/L)	Solubility (Grams/L)	Temp (°C)	Ref (#)	Evaluation (T P E A A)	Comments
4.455E-05	1.240E-02	10	S198	2 1 2 2 2	
3.952E-05	1.100E-02	15	H069	1 0 1 1 1	
3.630E-05	1.010E-02	20	L300	2 1 0 2 2	
3.880E-05	1.080E-02	20	S198	2 1 2 2 2	
3.593E-04	1.000E-01	22	N311	1 0 1 1 2	
6.574E-05	1.830E-02	23.5	S171	2 1 2 2 2	
3.126E-05	8.700E-03	25	D336	2 1 2 2 2	
3.449E-05	9.600E-03	25	D336	2 1 2 2 2	
4.670E-05	1.300E-02	25	F067	1 0 2 2 2	
1.609E-02	4.480E+00	25	F070	1 0 0 0 2	*sic*
4.095E-05	1.140E-02	30	S198	2 1 2 2 2	
1.437E-03	4.000E-01	rt	M161	0 0 0 0 2	

3372. C₁₆H₂₂O₄
Diisobutyl Phthalate
1,2-Benzenedicarboxylic Acid, bis(2-Methylpropyl) Esterpalatinol
Phthalic Acid Diisobutyl Ester
Palatinolic

RN: 84-69-5   **MP** (°C):
MW: 278.35   **BP** (°C):

Solubility (Moles/L)	Solubility (Grams/L)	Temp (°C)	Ref (#)	Evaluation (T P E A A)	Comments
3.592E-04	9.999E-02	20	F070	1 0 0 0 2	
7.300E-05	2.032E-02	20	L300	2 1 0 2 2	
2.227E-05	6.200E-03	24	H116	2 1 0 0 2	
5.030E-06	1.400E-03	25	D336	2 1 2 2 2	

3373. C₁₆H₂₂O₆
Diethoxyethyl Phthalate
bis(2-Ethoxyethyl) Phthalate

RN: 605-54-9   **MP** (°C):
MW: 310.35   **BP** (°C):

Solubility (Moles/L)	Solubility (Grams/L)	Temp (°C)	Ref (#)	Evaluation (T P E A A)	Comments
6.271E-03	1.946E+00	ns	F014	0 0 0 0 2	

3374. C₁₆H₂₂O₈·2H₂O

3374. $C_{16}H_{22}O_8 \cdot 2H_2O$

Coniferin (Dihydrate)
4-Hydroxy-3-methoxy-1-(γ-hydroxypropenyl)benzene-4-D-glucoside (Dihydrate)
Abietin(Dihydrate)
Coniferosi(Dihydrate)

RN: 531-29-3   **MP** (°C): 185
MW: 378.38   **BP** (°C):

Solubility (Moles/L)	Solubility (Grams/L)	Temp (°C)	Ref (#)	Evaluation (T P E A A)	Comments
1.315E-02	4.975E+00	c	D004	1 0 0 0 0	

3375. $C_{16}H_{22}O_{11}$

β-D-Glucose Pentaacetate
β-Glucose-penta-acetat

RN: 604-69-3   **MP** (°C): 131
MW: 390.35   **BP** (°C):

Solubility (Moles/L)	Solubility (Grams/L)	Temp (°C)	Ref (#)	Evaluation (T P E A A)	Comments
2.306E-03	9.000E-01	18	F300	1 0 0 0 0	

3376. $C_{16}H_{22}O_{11}$

α-Glucose Pentaacetate
α-Glucose-penta-acetat

RN: 3891-59-6   **MP** (°C): 110
MW: 390.35   **BP** (°C):

Solubility (Moles/L)	Solubility (Grams/L)	Temp (°C)	Ref (#)	Evaluation (T P E A A)	Comments
3.843E-03	1.500E+00	18	F300	1 0 0 0 1	

3377. $C_{16}H_{23}FN_2O_6$

1,3-bis(Pivaloyloxymethyl)-5-fluoro-2,4(1H,3H)-pyrimidinedi-one
1,3-bis(Pivaloyloxymethyl)-5-fluorouracil

RN: 66542-50-5   **MP** (°C): 102-104
MW: 358.37   **BP** (°C):

Solubility (Moles/L)	Solubility (Grams/L)	Temp (°C)	Ref (#)	Evaluation (T P E A A)	Comments
1.256E-04	4.500E-02	22	B321	1 0 2 2 2	pH 4.0

3378.   C$_{16}$H$_{23}$NO
n-Heptylcinnamamide
2-Propenamide, N-Heptyl-3-phenyl-
RN: 59831-99-1   **MP** (°C):
MW: 245.37   **BP** (°C):

Solubility (Moles/L)	Solubility (Grams/L)	Temp (°C)	Ref (#)	Evaluation (T P E A A)	Comments
7.600E-06	1.865E-03	ns	H350	0 0 0 0 2	

3379.   C$_{16}$H$_{23}$NO$_2$
Etoxadrol
(+)-2-(2-Ethyl-2-phenyl-1,3-dioxolan-4-yl)piperidine
RN: 28189-85-7   **MP** (°C):
MW: 261.37   **BP** (°C):

Solubility (Moles/L)	Solubility (Grams/L)	Temp (°C)	Ref (#)	Evaluation (T P E A A)	Comments
2.487E-03	6.500E-01	20	K017	1 2 2 2 2	pH 10, intrinsic
1.098E-02	2.870E+00	30	K017	1 2 2 2 2	pH 10, intrinsic
4.668E-02	1.220E+01	40	K017	1 2 2 2 2	pH 10, intrinsic

3380.   C$_{16}$H$_{23}$NO$_3$
Acetaminophen Octanoate
Octanoic Acid, 4-(Acetylamino)phenyl Ester
RN: 54942-41-5   **MP** (°C):   103
MW: 277.37   **BP** (°C):

Solubility (Moles/L)	Solubility (Grams/L)	Temp (°C)	Ref (#)	Evaluation (T P E A A)	Comments
3.605E-05	1.000E-02	25	B010	1 1 1 1 0	

3381.   C$_{16}$H$_{23}$NO$_6$
Monocrotaline
(-)-Monocrotaline
RN: 315-22-0   **MP** (°C):   202
MW: 325.36   **BP** (°C):

Solubility (Moles/L)	Solubility (Grams/L)	Temp (°C)	Ref (#)	Evaluation (T P E A A)	Comments
3.644E-02	1.186E+01	ns	I312	0 0 0 0 1	

3382.   C$_{16}$H$_{23}$N$_5$O$_5$
9-[5'-(O-Caproyl)-β-D-arabinofuranosyl]adenine Ester
RN: 65926-34-3   **MP** (°C):
MW: 365.39   **BP** (°C):

Solubility (Moles/L)	Solubility (Grams/L)	Temp (°C)	Ref (#)	Evaluation (T P E A A)	Comments
6.842E-03	2.500E+00	ns	B134	0 1 1 1 1	

3383.   C$_{16}$H$_{23}$N$_5$O$_5$
9-[5'-(O-tert-Butylacetyl)-β-D-arabinofuranosyl]adenine Ester
RN: 68325-42-8   **MP** (°C):
MW: 365.39   **BP** (°C):

Solubility (Moles/L)	Solubility (Grams/L)	Temp (°C)	Ref (#)	Evaluation (T P E A A)	Comments
2.135E-02	7.800E+00	ns	B134	0 1 1 1 1	

3384.   C$_{16}$H$_{24}$N$_2$O$_2$
N,N,N',N'-Tetraethylisophthalamide
RN: 13698-87-8   **MP** (°C):
MW: 276.38   **BP** (°C):

Solubility (Moles/L)	Solubility (Grams/L)	Temp (°C)	Ref (#)	Evaluation (T P E A A)	Comments
7.200E-01	1.990E+02	30	K019	1 0 0 0 2	

3385.   C$_{16}$H$_{24}$N$_2$O$_2$
N,N,N',N'-Tetraethylterephthalamide
RN: 15394-30-6   **MP** (°C):
MW: 276.38   **BP** (°C):

Solubility (Moles/L)	Solubility (Grams/L)	Temp (°C)	Ref (#)	Evaluation (T P E A A)	Comments
2.000E-02	5.528E+00	30	K019	1 0 0 0 1	

3386.   C$_{16}$H$_{24}$N$_4$O$_2$
2,5-Diaziridinyl-3,6-bis(propylamino)-1,4-benzoquinone
RN: 59886-47-4   **MP** (°C):   140
MW: 304.40   **BP** (°C):

Solubility (Moles/L)	Solubility (Grams/L)	Temp (°C)	Ref (#)	Evaluation (T P E A A)	Comments
≤3.29E-04	<1.00E-01	rt	C317	0 2 0 0 0	

3387. C$_{16}$H$_{24}$N$_4$O$_6$
2,5-Diaziridinyl-3,6-bis(2'-hydroxyl-3'-hydroxylpropylamino)-1,4-benzoquinone
2,5-Diaziridinyl-3,6-bis(hydroxylethylmethylamino)-1,4-benzoquinone
RN: 59886-55-4   **MP** (°C):   273
MW: 368.39   **BP** (°C):

Solubility (Moles/L)	Solubility (Grams/L)	Temp (°C)	Ref (#)	Evaluation (T P E A A)	Comments
1.629E-01	6.000E+01	rt	C317	0 2 0 0 0	
8.143E-02	3.000E+01	rt	C317	0 2 0 0 0	

3388. C$_{16}$H$_{24}$N$_6$
1-(Methylphenethylamino)-3,5-bis(dimethylamino)-s-triazine
RN: 125867-93-8   **MP** (°C):
MW: 300.41   **BP** (°C):

Solubility (Moles/L)	Solubility (Grams/L)	Temp (°C)	Ref (#)	Evaluation (T P E A A)	Comments
2.427E-05	7.291E-03	25	B386	2 2 2 2 2	

3389. C$_{16}$H$_{24}$O$_3$
Nonyl p-Hydroxybenzoate
Nonyl 4-Hydroxybenzoate
RN: 38713-56-3   **MP** (°C):
MW: 264.37   **BP** (°C):

Solubility (Moles/L)	Solubility (Grams/L)	Temp (°C)	Ref (#)	Evaluation (T P E A A)	Comments
4.824E-03	1.275E+00	25	D081	1 2 2 1 2	

3390. C$_{16}$H$_{24}$O$_4$
3,4-Epoxy-6-Methylcyclohexylmethyl-3,4-Epoxy-6-Methylcyclohexane Carboxylate
EP 201
RN: 141-37-7   **MP** (°C):
MW: 280.37   **BP** (°C):

Solubility (Moles/L)	Solubility (Grams/L)	Temp (°C)	Ref (#)	Evaluation (T P E A A)	Comments
1.067E-02	2.991E+00	ns	I313	0 0 0 0 0	

3391. C₁₆H₂₅NOS

3391. $C_{16}H_{25}NOS$

S-Benzyl Di-sec-butylthiocarbamate
Thiocarbazil
Tiocarbazil

RN: 36756-79-3   **MP** (°C):
MW: 279.45   **BP** (°C):

Solubility (Moles/L)	Solubility (Grams/L)	Temp (°C)	Ref (#)	Evaluation (T P E A A)	Comments
8.946E-06	2.500E-03	30	M161	1 0 0 0 1	

3392. $C_{16}H_{25}NO_2$

Butacarb
Carbamic Acid, N-Methyl-, 3,5-di-tert-Butylphenyl Ester
3,5-di-tert-Butylphenyl Methylcarbamate

RN: 2655-19-8   **MP** (°C):   102.9
MW: 263.38   **BP** (°C):

Solubility (Moles/L)	Solubility (Grams/L)	Temp (°C)	Ref (#)	Evaluation (T P E A A)	Comments
5.695E-05	1.500E-02	20	M161	1 0 0 0 1	

3393. $C_{16}H_{25}NO_2$

Nonyl p-Aminobenzoate
Nonyl 4-Aminobenzoate

RN: 37139-21-2   **MP** (°C):
MW: 263.38   **BP** (°C):

Solubility (Moles/L)	Solubility (Grams/L)	Temp (°C)	Ref (#)	Evaluation (T P E A A)	Comments
1.020E-06	2.687E-04	37	F006	1 1 2 2 2	

3394. $C_{16}H_{25}NO_3$

4-Propoxybenzoic Acid-2-(diethyl-amino)ethyl Ester

RN: 15788-85-9   **MP** (°C):
MW: 279.38   **BP** (°C):

Solubility (Moles/L)	Solubility (Grams/L)	Temp (°C)	Ref (#)	Evaluation (T P E A A)	Comments
4.500E-04	1.257E-01	ns	M066	0 0 0 0 1	

3395. $C_{16}H_{26}N_2O_2$

4-Propylaminobenzoic Acid-2-(diethyl-amino)ethyl Ester

RN: 16488-54-3   **MP** (°C):
MW: 278.40   **BP** (°C):

Solubility (Moles/L)	Solubility (Grams/L)	Temp (°C)	Ref (#)	Evaluation (T P E A A)	Comments
1.030E-03	2.867E-01	ns	M066	0 0 0 0 2	

3396. C$_{16}$H$_{26}$O$_2$
4-Octylphenol Monoethoxylate
RN: 51437-89-9   **MP** (°C):
MW: 250.38   **BP** (°C):

Solubility (Moles/L)	Solubility (Grams/L)	Temp (°C)	Ref (#)	Evaluation (T P E A A)	Comments
3.195E-05	8.000E-03	20.5	A335	1 0 2 2 2	
3.200E-05	8.012E-03	20.5	A335	1 0 2 2 2	

3397. C$_{16}$H$_{26}$O$_6$
Triethylene Glycol Dibutyrate
RN: 26962-26-5   **MP** (°C):
MW: 314.38   **BP** (°C):

Solubility (Moles/L)	Solubility (Grams/L)	Temp (°C)	Ref (#)	Evaluation (T P E A A)	Comments
2.524E-02	7.937E+00	ns	F014	0 0 0 0 2	

3398. C$_{16}$H$_{28}$N$_3$O$_2$
Dioxyethylaminoazobenzene
RN: **MP** (°C):
MW: 294.42   **BP** (°C):

Solubility (Moles/L)	Solubility (Grams/L)	Temp (°C)	Ref (#)	Evaluation (T P E A A)	Comments
2.945E-04	8.670E-02	0	K036	1 0 0 0 2	
4.212E-04	1.240E-01	25	K036	1 0 0 0 2	
2.819E-03	8.300E-01	90	K036	1 0 0 0 2	

3399. C$_{16}$H$_{32}$O$_2$
Palmitic Acid
hexadecanoic Acid
RN: 57-10-3   **MP** (°C):   56
MW: 256.43   **BP** (°C):

Solubility (Moles/L)	Solubility (Grams/L)	Temp (°C)	Ref (#)	Evaluation (T P E A A)	Comments
1.794E-05	4.600E-03	0	B136	1 0 2 1 1	
1.794E-05	4.600E-03	0.0	R001	1 1 1 1 1	
2.808E-05	7.200E-03	20	B136	1 0 2 1 1	
2.808E-05	7.200E-03	20.0	R001	1 1 1 1 1	
3.200E-06	8.206E-04	25	J001	1 0 2 1 1	
1.200E-07	3.077E-05	25	R002	1 2 2 2 2	intrinsic
2.680E-06	6.872E-04	25	R002	1 2 2 2 2	
3.237E-05	8.300E-03	30	B136	1 0 2 1 1	
3.237E-05	8.300E-03	30.0	R001	1 1 1 1 1	
3.900E-05	1.000E-02	45	B136	1 0 2 1 1	

3.900E-05	1.000E-02	45.0	R001	1 1 1 1 1
4.000E-06	1.026E-03	50	J001	1 0 2 1 1
4.680E-05	1.200E-02	60	B136	1 0 2 1 1
4.680E-05	1.200E-02	60.0	R001	1 1 1 1 1

3400. C₁₆H₃₄

3-Methylpentadecane

RN: 2882-96-4 **MP** (°C): -22
MW: 226.45 **BP** (°C): 282

Solubility (Moles/L)	Solubility (Grams/L)	Temp (°C)	Ref (#)	Evaluation (T P E A A)	Comments
4.328E-10	9.800E-08	23	C332	2 0 2 2 1	

3401. C₁₆H₃₄

Hexadecane
n-Hexadecane
Cetane

RN: 544-76-3 **MP** (°C): 18.17
MW: 226.45 **BP** (°C):

Solubility (Moles/L)	Solubility (Grams/L)	Temp (°C)	Ref (#)	Evaluation (T P E A A)	Comments
2.778E-08	6.290E-06	25	F004	1 2 2 2 1	

3402. C₁₆H₃₄

2-Methylpentadecane

RN: 1560-93-6 **MP** (°C): -7
MW: 226.45 **BP** (°C): 282

Solubility (Moles/L)	Solubility (Grams/L)	Temp (°C)	Ref (#)	Evaluation (T P E A A)	Comments
4.681E-10	1.060E-07	23	C332	2 0 2 2 1	

3403. C₁₆H₃₄O

Hexadecanol
Cetyl Alcohol

RN: 36653-82-4 **MP** (°C): 49
MW: 242.45 **BP** (°C): 344

Solubility (Moles/L)	Solubility (Grams/L)	Temp (°C)	Ref (#)	Evaluation (T P E A A)	Comments
1.699E-07	4.120E-05	22.5	G301	2 1 0 1 2	
1.700E-07	4.122E-05	25	R002	1 2 2 2 2	
3.300E-08	8.001E-06	34	K011	1 2 1 1 2	
6.393E-08	1.550E-05	43	H030	2 2 2 2 2	
6.393E-08	1.550E-05	43	H103	1 2 2 2 2	
1.270E-07	3.079E-05	55	K011	1 2 1 1 2	
1.675E-07	4.060E-05	61	H030	2 2 2 2 2	
1.675E-07	4.060E-05	61	H103	1 2 2 2 2	

3404. C₁₆H₃₅O₃P
Dibutyl Isooctyl Phosphonate
RN: 108979-58-4   **MP** (°C):
MW: 306.43   **BP** (°C):

Solubility (Moles/L)	Solubility (Grams/L)	Temp (°C)	Ref (#)	Evaluation (T P E A A)	Comments
<6.53E-04	<2.00E-01	25	B070	1 2 0 1 0	

3405. C₁₆H₃₅O₄P
Dibutyl Octyl Phosphate
RN: 25786-28-1   **MP** (°C):
MW: 322.43   **BP** (°C):

Solubility (Moles/L)	Solubility (Grams/L)	Temp (°C)	Ref (#)	Evaluation (T P E A A)	Comments
<3.10E-04	<1.00E-01	25	B070	1 2 0 1 0	

3406. C₁₇H₁₁NO₃
Furo[3,4-b]quinolin-3(1H)-one, 9-Hydroxy-1-phenyl-
RN: 74103-09-6   **MP** (°C):
MW: 277.28   **BP** (°C):

Solubility (Moles/L)	Solubility (Grams/L)	Temp (°C)	Ref (#)	Evaluation (T P E A A)	Comments
1.190E-07	3.300E-05	25	P089	2 1 2 2 2	
1.388E-07	3.850E-05	37	P089	2 1 2 2 2	
1.677E-07	4.650E-05	51	P089	2 1 2 2 2	

3407. C₁₇H₁₂
1,2-Benzofluorene
Benzo[a]fluorene
11H-Benzo[a]fluorene
RN: 238-84-6   **MP** (°C):   187
MW: 216.29   **BP** (°C):   407

Solubility (Moles/L)	Solubility (Grams/L)	Temp (°C)	Ref (#)	Evaluation (T P E A A)	Comments
2.081E-07	4.500E-05	25	M064	1 1 2 2 1	
2.100E-07	4.542E-05	25	M342	1 0 1 1 1	
2.081E-07	4.500E-05	ns	M344	0 0 0 0 2	

3408.   C$_{17}$H$_{12}$
2,3-Benzofluorene
Benzo[b]fluorene
11H-Benzo[b]fluorene
RN: 243-17-4   **MP** (°C):   209
MW: 216.29   **BP** (°C):

Solubility (Moles/L)	Solubility (Grams/L)	Temp (°C)	Ref (#)	Evaluation (T P E A A)	Comments
1.849E-08	4.000E-06	25	B319	2 0 1 2 0	
9.247E-09	2.000E-06	25	M064	1 1 2 2 1	
9.250E-09	2.001E-06	25	M342	1 0 1 1 2	

3409.   C$_{17}$H$_{12}$ClFN$_3$O$_2$
α-(4-Chlorophenyl)-α-(1-2-(2-chloro)phenylethenyl)-1H-1,2,4-triazole-1-ethanol
X-7801
DuP 860
RN: **MP** (°C):
MW: 344.76   **BP** (°C):

Solubility (Moles/L)	Solubility (Grams/L)	Temp (°C)	Ref (#)	Evaluation (T P E A A)	Comments
4.612E-06	1.590E-03	22	M362	1 1 2 1 1	

3410.   C$_{17}$H$_{12}$ClNO$_2$S
Fentiazac
4-(p-Chlorophenyl)-2-phenyl-5-thiazoleacetic Acid
RN: 18046-21-4   **MP** (°C):   161.1
MW: 329.81   **BP** (°C):

Solubility (Moles/L)	Solubility (Grams/L)	Temp (°C)	Ref (#)	Evaluation (T P E A A)	Comments
9.400E-06	3.100E-03	5	F306	1 0 1 2 2	intrinsic
9.600E-05	3.166E-02	25	C314	1 1 2 2 2	
9.612E-05	3.170E-02	25	C314	1 1 2 2 2	
1.080E-05	3.562E-03	25	F306	1 0 1 2 2	intrinsic
1.310E-05	4.320E-03	37	F306	1 0 1 2 2	intrinsic

3411.   C$_{17}$H$_{12}$Cl$_2$N$_2$O
Fenarimol
2,4'-Dichloro-α-(5-pyrimidinyl)benzhydryl Alcohol
α-(2-Chlorophenyl)-α-(4-chlorophenyl)-5-Pyrimidinemethanol
Tebulan
Rubigan 4AS
Rimidin
RN: 60168-88-9   **MP** (°C):   118
MW: 331.20   **BP** (°C):

Solubility (Moles/L)	Solubility (Grams/L)	Temp (°C)	Ref (#)	Evaluation (T P E A A)	Comments
4.136E-05	1.370E-02	25	M161	1 0 0 0 2	pH 7

3412.   C$_{17}$H$_{12}$Cl$_{10}$O$_3$
Kelevan
Allied GC 9160
Despirol
RN: 4234-79-1   **MP** (°C):   91
MW: 618.81   **BP** (°C):

Solubility (Moles/L)	Solubility (Grams/L)	Temp (°C)	Ref (#)	Evaluation (T P E A A)	Comments
8.888E-06	5.500E-03	20	M164	1 0 0 0 1	

3413.   C$_{17}$H$_{12}$I$_2$O$_3$
Benziodarone
Algocor
Amplivix
Dilafurane
RN: 68-90-6   **MP** (°C):
MW: 518.09   **BP** (°C):

Solubility (Moles/L)	Solubility (Grams/L)	Temp (°C)	Ref (#)	Evaluation (T P E A A)	Comments
1.135E-05	5.881E-03	20	H301	2 0 2 2 2	

3414.   C$_{17}$H$_{12}$O$_6$
Aflatoxin B1
AFB1
RN: 1162-65-8   **MP** (°C):   268
MW: 312.28   **BP** (°C):

Solubility (Moles/L)	Solubility (Grams/L)	Temp (°C)	Ref (#)	Evaluation (T P E A A)	Comments
4.803E-05	1.500E-02	ns	I306	0 0 0 0 1	

3415.   C$_{17}$H$_{12}$O$_7$
Aflatoxin G1
RN: 1165-39-5   **MP** (°C):   244
MW: 328.28   **BP** (°C):

Solubility (Moles/L)	Solubility (Grams/L)	Temp (°C)	Ref (#)	Evaluation (T P E A A)	Comments
4.569E-05	1.500E-02	ns	I306	0 0 0 0 1	

3416.   C$_{17}$H$_{13}$ClO$_3$
Itanoxone
2'-Chloro-α-methylene-γ-oxo[1,1'-biphenyl]-4-butanoic Acid
F 1379
RN: 58182-63-1   **MP** (°C):   212
MW: 300.74   **BP** (°C):

Solubility (Moles/L)	Solubility (Grams/L)	Temp (°C)	Ref (#)	Evaluation (T P E A A)	Comments
6.318E-04	1.900E-01	20	C112	2 0 1 1 2	

3417.   C$_{17}$H$_{13}$Cl$_2$N$_3$O$_2$
α-(2,4-Difluorophenyl)-α-(1-2-(2-chloro)phenylethenyl)-1H-1,2,4-triazole-1-ethanol
A-9991
DuP 991
RN: **MP** (°C):
MW: 362.22   **BP** (°C):

Solubility (Moles/L)	Solubility (Grams/L)	Temp (°C)	Ref (#)	Evaluation (T P E A A)	Comments
1.933E-05	7.000E-03	22	M362	1 1 2 1 1	

3418.   C$_{17}$H$_{14}$N$_2$O
1-o-Tolylazo-2-naphthol
Orange OT
Oil Orange SS
1-(o-Tolylazo)-2-naphthol
RN: 2646-17-5   **MP** (°C):   131
MW: 262.31   **BP** (°C):

Solubility (Moles/L)	Solubility (Grams/L)	Temp (°C)	Ref (#)	Evaluation (T P E A A)	Comments
1.000E-07	2.623E-05	rt	M163	0 0 0 0 1	

3419.   C₁₇H₁₄O₆
Aflatoxin B2
RN: 7220-81-7   **MP** (°C):   286
MW: 314.30   **BP** (°C):

Solubility (Moles/L)	Solubility (Grams/L)	Temp (°C)	Ref (#)	Evaluation (T P E A A)	Comments
4.773E-05	1.500E-02	ns	I306	0 0 0 0 1	

3420.   C₁₇H₁₄O₇
Aflatoxin G2
RN: 7241-98-7   **MP** (°C):   237
MW: 330.30   **BP** (°C):

Solubility (Moles/L)	Solubility (Grams/L)	Temp (°C)	Ref (#)	Evaluation (T P E A A)	Comments
4.541E-05	1.500E-02	ns	I306	0 0 0 0 1	

3421.   C₁₇H₁₅NO₃
Cinnamyl Acetaminophen
Cinnamic Acid, Ester with 4'-Hydroxyacetanilide
Acetanilide, 4'-Hydroxy-, Cinnamate (Ester)
RN: 20682-28-4   **MP** (°C):   200-201
MW: 281.31   **BP** (°C):

Solubility (Moles/L)	Solubility (Grams/L)	Temp (°C)	Ref (#)	Evaluation (T P E A A)	Comments
4.977E-06	1.400E-03	37	D029	1 0 1 1 1	

3422.   C₁₇H₁₅NO₅
Benzoic Acid, 2-(Acetyloxy)-, 4-(Acetylamino)phenyl Ester
RN: 5003-48-5   **MP** (°C):   174.5
MW: 313.31   **BP** (°C):

Solubility (Moles/L)	Solubility (Grams/L)	Temp (°C)	Ref (#)	Evaluation (T P E A A)	Comments
6.383E-05	2.000E-02	21	N335	1 2 1 1 2	

3423. $C_{17}H_{16}Br_2O_3$

Bromopropylate
1-Methylethyl-4-bromo-α-(4-bromophenyl)-α-hydroxybenzeneacetate
Neoron
GS-19851
Phenisobromolate

RN: 18181-80-1   **MP** (°C):   77
MW: 428.13   **BP** (°C):

Solubility (Moles/L)	Solubility (Grams/L)	Temp (°C)	Ref (#)	Evaluation (T P E A A)	Comments
<1.17E-06	<5.00E-04	20	F311	1 2 2 2 1	
1.168E-05	5.000E-03	20	M161	1 0 0 0 0	

3424. $C_{17}H_{16}ClFN_2O_2$

Progabide
Butanamide, 4-[[(4-Chlorophenyl)(5-fluoro-2-hydroxyphenyl)methylene]amino]-
Gabrene
SL 76-002

RN: 62666-20-0   **MP** (°C):
MW: 334.78   **BP** (°C):

Solubility (Moles/L)	Solubility (Grams/L)	Temp (°C)	Ref (#)	Evaluation (T P E A A)	Comments
1.110E-04	3.716E-02	37	F309	1 0 2 2 2	
1.110E-04	3.716E-02	37	F318	2 2 0 0 2	

3425. $C_{17}H_{16}Cl_2O_3$

Chloropropylate
1-Methylethyl-4-chloro-α-(4-chlorophenyl)-α-hydroxybenzenacetate
Chlormite
Acaralate
G-24163
Rospin

RN: 5836-10-2   **MP** (°C):   74
MW: 339.22   **BP** (°C):

Solubility (Moles/L)	Solubility (Grams/L)	Temp (°C)	Ref (#)	Evaluation (T P E A A)	Comments
4.422E-06	1.500E-03	20	F311	1 2 2 2 1	
2.948E-05	1.000E-02	rt	M161	0 0 0 0 1	

3426.   C$_{17}$H$_{16}$N$_2$O$_2$S
1-Sulfamethylnaphthalene
RN: MP (°C):
MW: 312.39   BP (°C):

Solubility (Moles/L)	Solubility (Grams/L)	Temp (°C)	Ref (#)	Evaluation (T P E A A)	Comments
3.201E-05	1.000E-02	20	F073	1 2 2 2 1	

3427.   C$_{17}$H$_{16}$N$_2$O$_3$
C.I. Disperse Blue 3
1-[(2-Hydroxyethyl)amino]-4-(methylamino)-9,10-anthracenedione
C.I. 61505
RN: 2475-46-9   MP (°C):   187
MW: 296.33   BP (°C):

Solubility (Moles/L)	Solubility (Grams/L)	Temp (°C)	Ref (#)	Evaluation (T P E A A)	Comments
1.200E-07	3.556E-05	25	B333	1 0 0 0 1	

3428.   C$_{17}$H$_{16}$N$_2$O$_3$S
4-Sulfahydroxymethylnaphthalene
RN: MP (°C):
MW: 328.39   BP (°C):

Solubility (Moles/L)	Solubility (Grams/L)	Temp (°C)	Ref (#)	Evaluation (T P E A A)	Comments
1.675E-04	5.500E-02	20	F073	1 2 2 2 1	

3429.   C$_{17}$H$_{16}$N$_2$O$_4$
p-(p-Acetamidobenzamido)phenyl Acetate
RN: 74973-19-6   MP (°C):
MW: 312.33   BP (°C):

Solubility (Moles/L)	Solubility (Grams/L)	Temp (°C)	Ref (#)	Evaluation (T P E A A)	Comments
3.900E-05	1.218E-02	25	A066	1 0 1 1 1	

3430.   C$_{17}$H$_{16}$N$_2$O$_4$S
1-Benzenesulfonyl-5-ethyl-5-phenyl-hydantoin
5-Ethyl-5phenyl-1(phenylsulfonyl)-2,4-imidazolidinedione
RN: 21413-25-2   MP (°C):
MW: 344.39   BP (°C):

Solubility (Moles/L)	Solubility (Grams/L)	Temp (°C)	Ref (#)	Evaluation (T P E A A)	Comments
9.782E-04	3.369E-01	37	F183	1 0 1 1 1	intrinsic

3431. $C_{17}H_{16}N_2O_5$

p-4-Acetaminophenyl Acetaminophen
Acetamide, N,N'-[Carbonylbis(oxy-4,1-phenylene)]bis-
Acetanilide, 4'-Hydroxy-, Carbonate (2:1) (Ester)

RN: 19872-72-1   **MP** (°C):   219.5-220
MW: 328.33   **BP** (°C):

Solubility (Moles/L)	Solubility (Grams/L)	Temp (°C)	Ref (#)	Evaluation (T P E A A)	Comments
1.827E-04	6.000E-02	37	D029	1 0 1 1 1	

3432. $C_{17}H_{17}ClO_6$

Griseofulvin
(2S-trans)-7-Chloro-2',4,6-trimethoxy-6'-methylspiro[benzofuran-2(3H),1'-[2]cyclohexene]-
3,4'-dione
Fulvicin
Grisactin
Grifulvin
Griseostatin

RN: 126-07-8   **MP** (°C):   220.0
MW: 352.77   **BP** (°C):

Solubility (Moles/L)	Solubility (Grams/L)	Temp (°C)	Ref (#)	Evaluation (T P E A A)	Comments
1.830E-05	6.456E-03	15	E010	2 2 2 2 2	
2.466E-05	8.700E-03	20	N322	1 0 2 2 1	
3.260E-05	1.150E-02	21	E316	1 1 2 2 1	
4.025E-04	1.420E-01	21	M044	2 0 2 2 2	microsize, *sic*
3.175E-04	1.120E-01	21	M044	2 0 2 2 2	*sic*
2.126E-05	7.500E-03	22	C040	2 0 2 2 0	EFG
2.076E-05	7.325E-03	22	M382	2 1 1 1 1	average of 2
2.523E-05	8.900E-03	23	B362	1 0 2 2 2	
2.268E-05	8.000E-03	25	C037	2 1 2 2 2	
2.450E-05	8.643E-03	25	E010	2 2 2 2 2	
3.685E-05	1.300E-02	25	H015	1 0 0 0 1	
2.835E-05	1.000E-02	25	L033	1 0 2 1 1	
2.750E-05	9.700E-03	25	P096	1 0 2 2 2	
2.551E-05	9.000E-03	27	B043	1 0 1 2 0	EFG
2.835E-05	1.000E-02	30	M045	2 0 0 0 0	
4.000E-05	1.411E-02	30	O321	2 2 2 2 1	
4.252E-05	1.500E-02	30	O321	2 2 2 2 1	
3.510E-05	1.238E-02	35	E010	2 2 2 2 2	
3.969E-05	1.400E-02	37	B039	2 1 1 1 0	EFG
4.252E-05	1.500E-02	37	B043	1 0 1 2 0	EFG
3.969E-05	1.400E-02	37	B045	1 0 1 1 1	
4.054E-05	1.430E-02	37	F033	2 0 2 0 2	
3.968E-05	1.400E-02	37	G011	1 0 1 1 0	EFG
4.252E-05	1.500E-02	37	K018	1 0 0 0 1	

5.669E-05	2.000E-02	45	B043	1 0 1 2 0	EFG
6.140E-05	2.166E-02	45	E010	2 2 2 2 2	
3.798E-05	1.340E-02	ns	D340	0 0 1 1 2	
2.466E-05	8.700E-03	ns	N323	0 0 2 2 1	

3433.   C₁₇H₁₇NO₂

$C_{17}H_{17}NO_2$

Apomorphine
Apomorphin
RN: 58-00-4   **MP** (°C):
MW: 267.33   **BP** (°C):

Solubility (Moles/L)	Solubility (Grams/L)	Temp (°C)	Ref (#)	Evaluation (T P E A A)	Comments
4.000E-04	1.069E-01	15	K059	2 2 2 0 0	
7.481E-02	2.000E+01	25	P312	1 2 2 2 2	

3434.   C₁₇H₁₇NO₅

$C_{17}H_{17}NO_5$

N-Benzyloxycarbonyl-L-tyrosine
Carbobenzoxytyrosine
RN: 1164-16-5   **MP** (°C):
MW: 315.33   **BP** (°C):

Solubility (Moles/L)	Solubility (Grams/L)	Temp (°C)	Ref (#)	Evaluation (T P E A A)	Comments
4.852E-03	1.530E+00	25.1	N026	2 0 2 2 2	

3435.   C₁₇H₁₇N₅O₅

$C_{17}H_{17}N_5O_5$

9-[5'-(O-Benzoyl)-β-D-arabinofuranosyl]adenine Ester
RN: 42782-57-0   **MP** (°C):   223.0
MW: 371.36   **BP** (°C):

Solubility (Moles/L)	Solubility (Grams/L)	Temp (°C)	Ref (#)	Evaluation (T P E A A)	Comments
2.154E-04	8.000E-02	ns	B134	0 1 1 1 0	

3436.   C₁₇H₁₈ClNO₆

$C_{17}H_{18}ClNO_6$

Griseofulvin-4'-oxime
Spiro[benzofuran-2(3H),1'-[2]cyclohexene]-3,4'-dione, 7-Chloro-2',4,6-trimethoxy-6'-methyl-, 4'-oxime
RN: 13215-54-8   **MP** (°C):
MW: 367.79   **BP** (°C):

Solubility (Moles/L)	Solubility (Grams/L)	Temp (°C)	Ref (#)	Evaluation (T P E A A)	Comments
3.589E-04	1.320E-01	37	F033	2 0 2 0 2	

3437.   C₁₇H₁₈ClN₅O₆
Dis. A. 8
Ethanol, 2,2'-[[4-[(2-Chloro-4,6-dinitrophenyl)azo]-3-methylphenyl]imino]bis-
RN: 65125-87-3   **MP** (°C):
MW: 423.82   **BP** (°C):

Solubility (Moles/L)	Solubility (Grams/L)	Temp (°C)	Ref (#)	Evaluation (T P E A A)	Comments
5.000E-07	2.119E-04	25	B333	1 0 0 0 1	

3438.   C₁₇H₁₈Cl₂N₄O₄
Dis. A. 10
Ethanol, 2,2'-[4-(2,6-Dichloro-4-nitrophenylazo)-m-tolylimino]di-
RN: 58528-60-2   **MP** (°C):
MW: 413.26   **BP** (°C):

Solubility (Moles/L)	Solubility (Grams/L)	Temp (°C)	Ref (#)	Evaluation (T P E A A)	Comments
1.100E-06	4.546E-04	25	B333	1 0 0 0 1	

3439.   C₁₇H₁₈N₂O₆
Nifedipine
3,5-Pyridinedicarboxylicacid
RN: 21829-25-4   **MP** (°C):   172-174
MW: 346.34   **BP** (°C):

Solubility (Moles/L)	Solubility (Grams/L)	Temp (°C)	Ref (#)	Evaluation (T P E A A)	Comments
1.675E-05	5.800E-03	25	B387	2 0 2 2 2	

3440.   C₁₇H₁₈N₄O₃S
4-Sulfanilamido-1-phenyl-2,3-dimethyl-5-pyrazolone
RN: 71119-16-9   **MP** (°C):
MW: 358.42   **BP** (°C):

Solubility (Moles/L)	Solubility (Grams/L)	Temp (°C)	Ref (#)	Evaluation (T P E A A)	Comments
4.352E-04	1.560E-01	37	R045	1 2 1 1 2	

3441.   C₁₇H₁₉ClN₂S
1-Chloropromazine
1-Chloro-N,N-dimethyl-10H-phenothiazine-10-propanamide
RN: 13100-13-5   **MP** (°C):
MW: 318.87   **BP** (°C):

Solubility (Moles/L)	Solubility (Grams/L)	Temp (°C)	Ref (#)	Evaluation (T P E A A)	Comments
1.200E-05	3.826E-03	0	G023	0 0 0 0 1	

3442.   C₁₇H₁₉ClN₂S
3-Chloropromazine
3-Chloro-N,N-dimethyl-10H-phenothiazine-10-propanamide
RN: 484-19-5   **MP** (°C):
MW: 318.87   **BP** (°C):

Solubility (Moles/L)	Solubility (Grams/L)	Temp (°C)	Ref (#)	Evaluation (T P E A A)	Comments
1.000E-05	3.189E-03	ns	G023	0 0 1 1 1	

3443.   C₁₇H₁₉ClN₂S
4-Chloropromazine
4-Chloro-N,N-dimethyl-10H-phenothiazine-10-propanamide
RN: 13094-24-1   **MP** (°C):
MW: 318.87   **BP** (°C):

Solubility (Moles/L)	Solubility (Grams/L)	Temp (°C)	Ref (#)	Evaluation (T P E A A)	Comments
1.100E-05	3.508E-03	ns	G023	0 0 1 1 1	

3444.   C₁₇H₁₉ClN₂S
Chlorpromazine
2-Chloro-N,N-dimethyl-10H-phenothiazine-10-propanamine
Thorazine
Thor-prom
Ormazine
Largactil
RN: 50-53-3   **MP** (°C):   56.5
MW: 318.87   **BP** (°C):

Solubility (Moles/L)	Solubility (Grams/L)	Temp (°C)	Ref (#)	Evaluation (T P E A A)	Comments
8.000E-06	2.551E-03	24	G023	2 0 1 1 1	
2.195E-06	7.000E-04	30	P044	1 0 1 0 1	
8.000E-06	2.551E-03	ns	G023	0 0 1 1 0	

3445.   C₁₇H₁₉ClN₄O₄
C.I. Disperse Red 5
Ethanol, 2,2'-[[4-[(2-Chloro-4-nitrophenyl)azo]-3-methylphenyl]imino]bis-
RN: 3769-57-1   **MP** (°C):   192
MW: 378.82   **BP** (°C):

Solubility (Moles/L)	Solubility (Grams/L)	Temp (°C)	Ref (#)	Evaluation (T P E A A)	Comments
3.800E-07	1.440E-04	25	B333	1 0 0 0 1	

3446. C₁₇H₁₉ClO₆

Griseofulvin-4'-ol

Spiro[benzofuran-2(3H),1'-[2]cyclohexen]-3-one, 7-Chloro-4'-hydroxy-2',4,6-trimethoxy-6'-methyl-

RN: 13215-53-7   **MP** (°C):
MW: 354.79   **BP** (°C):

Solubility (Moles/L)	Solubility (Grams/L)	Temp (°C)	Ref (#)	Evaluation (T P E A A)	Comments
7.129E-04	2.529E-01	37	F033	2 0 2 0 2	average of 2

3447. C₁₇H₁₉NO₃

1-Methyl-1-nitro-2-(p-methylphenyl)-2-p-Ethoxylphenyl)ethane

RN: 53982-07-3   **MP** (°C):
MW: 285.35   **BP** (°C):

Solubility (Moles/L)	Solubility (Grams/L)	Temp (°C)	Ref (#)	Evaluation (T P E A A)	Comments
8.060E-06	2.300E-03	rt	C122	0 2 2 2 2	

3448. C₁₇H₁₉NO₃

Piperine

Piperidine, 1-[5-(1,3-Benzodioxol-5-yl)-1-oxo-2,4-pentadienyl]-, (E,E)-
N-[(E,E)-Piperoyl]piperidine

RN: 94-62-2   **MP** (°C):   130.0
MW: 285.35   **BP** (°C):

Solubility (Moles/L)	Solubility (Grams/L)	Temp (°C)	Ref (#)	Evaluation (T P E A A)	Comments
1.400E-04	3.995E-02	15	K059	2 2 2 0 1	
1.402E-04	4.000E-02	18	F300	1 0 0 0 0	
3.504E-04	9.999E-02	rt	D021	0 0 1 1 0	

3449. C₁₇H₁₉NO₃

Morphine

Morphin

7,8-Didehydro-4,5-epoxy-17-methylmorphinan-3,6-diol

RN: 57-27-2   **MP** (°C):   254dec
MW: 285.35   **BP** (°C):

Solubility (Moles/L)	Solubility (Grams/L)	Temp (°C)	Ref (#)	Evaluation (T P E A A)	Comments
5.000E-04	1.427E-01	15	K059	2 2 2 0 0	
5.222E-04	1.490E-01	20	B061	1 0 1 1 2	
5.257E-04	1.500E-01	20	F300	1 0 0 0 0	
7.200E-04	2.054E-01	30	L068	1 0 0 1 0	EFG
1.000E-03	2.853E-01	30	L069	1 0 1 1 0	EFG
1.051E-03	2.999E-01	rt	D021	0 0 1 1 0	

3450.   C$_{17}$H$_{19}$NO$_3$.H$_2$O
Morphine (Monohydrate)
Morphinan-3,6-diol, 7,8-Didehydro-4,5-epoxy-17-methyl- (5α,6α)-, Monohydrate
RN: 6009-81-0   **MP** (°C):   254dec
MW: 303.36   **BP** (°C):

Solubility (Moles/L)	Solubility (Grams/L)	Temp (°C)	Ref (#)	Evaluation (T P E A A)	Comments
9.328E-04	2.830E-01	c	D004	1 0 0 0 0	
3.064E-03	9.294E-01	h	D004	1 0 0 0 0	

3451.   C$_{17}$H$_{19}$NO$_4$
1-Methyl-1-nitro-2,2-bis(p-methoxylphenyl)ethane
RN: 34197-26-7   **MP** (°C):
MW: 301.35   **BP** (°C):

Solubility (Moles/L)	Solubility (Grams/L)	Temp (°C)	Ref (#)	Evaluation (T P E A A)	Comments
2.854E-05	8.600E-03	rt	C122	0 2 2 2 2	

3452.   C$_{17}$H$_{19}$N$_3$
Antazoline
Albalon-A
RN: 91-75-8   **MP** (°C):   120
MW: 265.36   **BP** (°C):

Solubility (Moles/L)	Solubility (Grams/L)	Temp (°C)	Ref (#)	Evaluation (T P E A A)	Comments
2.500E-03	6.634E-01	30	L068	1 0 0 1 0	EFG
1.900E-02	5.042E+00	37.5	L034	2 2 0 1 2	pH 7.4

3453.   C$_{17}$H$_{19}$N$_5$O$_6$
Dis. A. 1
Ethanol, 2,2'-[4-(2,4-Dinitrophenylazo)-m-tolylimino]di-
Disperse Violet 4K
Terasil Violet P 4RT
RN: 41541-13-3   **MP** (°C):   190
MW: 389.37   **BP** (°C):

Solubility (Moles/L)	Solubility (Grams/L)	Temp (°C)	Ref (#)	Evaluation (T P E A A)	Comments
7.000E-07	2.726E-04	25	B333	1 0 0 0 1	

3454. C₁₇H₂₀ClN₅O₂

$C_{17}H_{20}ClN_5O_2$

1H-Purine-2,6-dione, 8-(2-Amino-4-chlorophenyl)-3,7-dihydro-1,3-dipropyl-
1,3-Dipropyl-8-(2-amino-4-chlorophenyl)xanthine
PACPX

RN: 85872-51-1   **MP** (°C):
MW: 361.83   **BP** (°C):

Solubility (Moles/L)	Solubility (Grams/L)	Temp (°C)	Ref (#)	Evaluation (T P E A A)	Comments
<2.76E-07	<1.00E-04	ns	H316	0 2 1 1 0	pH 7.4
1.105E-06	4.000E-04	ns	H316	0 2 1 1 0	0.1N HCl

3455. C₁₇H₂₀N₂O

$C_{17}H_{20}N_2O$

Michler's Ketone
Tetramethyldiaminobenzophenone
bis[4-(Dimethylamino)phenyl]-methanone
p,p'-bis(N,N-Dimethylamino)benzophenone
4,4[-bis(Dimethylamino)benzophenone

RN: 90-94-8   **MP** (°C): 172.0
MW: 268.36   **BP** (°C):

Solubility (Moles/L)	Solubility (Grams/L)	Temp (°C)	Ref (#)	Evaluation (T P E A A)	Comments
1.490E-03	3.998E-01	rt	D021	0 0 1 1 0	

3456. C₁₇H₂₀N₂S

$C_{17}H_{20}N_2S$

Promazine
Primazine
Sparine
Prozine

RN: 58-40-2   **MP** (°C): 32
MW: 284.43   **BP** (°C):

Solubility (Moles/L)	Solubility (Grams/L)	Temp (°C)	Ref (#)	Evaluation (T P E A A)	Comments
5.000E-05	1.422E-02	24	G023	2 0 1 1 1	
5.000E-05	1.422E-02	ns	G023	0 0 0 0 1	

3457. C$_{17}$H$_{20}$N$_2$S
Promethazine
10-(2-Dimethylaminopropyl)phenothiazine
10-(2-Dimethylamino-2-methylethyl)phenothiazine
Fenergan
Protazine
Thiergan
RN: 60-87-7   **MP** (°C):   60
MW: 284.43   **BP** (°C):   191

Solubility (Moles/L)	Solubility (Grams/L)	Temp (°C)	Ref (#)	Evaluation (T P E A A)	Comments
5.500E-05	1.564E-02	24	G023	2 0 1 1 1	

3458. C$_{17}$H$_{20}$N$_4$O$_4$
C.I. Disperse Red 17
Ethanol, 2,2'-[[3-Methyl-4-[(4-nitrophenyl)azo]phenyl]imino]bis-
RN: 3179-89-3   **MP** (°C):   160
MW: 344.37   **BP** (°C):

Solubility (Moles/L)	Solubility (Grams/L)	Temp (°C)	Ref (#)	Evaluation (T P E A A)	Comments
1.800E-06	6.199E-04	25	B333	1 0 0 0 1	

3459. C$_{17}$H$_{20}$N$_4$O$_5$
Dis. A. 13
4-Nitro-2-methoxy-4'-di(β-hydroxyethyl)-aminoazobenzene
Ethanol, 2,2'-[[4-[(2-Methoxy-4-nitrophenyl)azo]phenyl]imino]bis
Ethanol, 2,2'-[p-(2-Methoxy-4-nitrophenylazo)phenylimino]di-
RN: 41541-14-4   **MP** (°C):
MW: 360.37   **BP** (°C):

Solubility (Moles/L)	Solubility (Grams/L)	Temp (°C)	Ref (#)	Evaluation (T P E A A)	Comments
2.000E-05	7.207E-03	25	B333	1 0 0 0 1	
6.826E-04	2.460E-01	100	P313	1 2 1 2 2	

3460. C$_{17}$H$_{20}$N$_4$O$_5$S
Benzenesulfonic Acid, 4-(2,3,6,7-Tetrahydro-2,6-dioxo-1,3-dipropyl-1H-purin-8-yl)-
RN: 89073-57-4   **MP** (°C):
MW: 392.44   **BP** (°C):

Solubility (Moles/L)	Solubility (Grams/L)	Temp (°C)	Ref (#)	Evaluation (T P E A A)	Comments
3.313E-03	1.300E+00	ns	H316	0 2 1 1 1	0.1N HCl
>6.12E-02	>2.40E+01	ns	H316	0 2 1 1 1	pH 7.4

3461. $C_{17}H_{20}N_4O_6$

Riboflavine
Riboflavin
Robiflavine
7,8-Dimethyl-10-ribitylisoalloxazine
Zinvit-G
E-101

RN: 83-88-5 **MP** (°C): 290
MW: 376.37 **BP** (°C):

Solubility (Moles/L)	Solubility (Grams/L)	Temp (°C)	Ref (#)	Evaluation (T P E A A)	Comments
2.657E-04	9.999E-02	20	A022	1 0 0 0 0	
2.250E-04	8.468E-02	25	A079	1 0 1 1 2	
2.657E-04	9.999E-02	25	D041	1 0 0 0 0	
1.754E-04	6.600E-02	25	D315	1 0 1 1 2	
3.959E-04	1.490E-01	37	E018	1 0 2 1 2	

3462. $C_{17}H_{20}O_6$

Mycophenolic Acid
6-(1,3-Dihydro-7-hydroxy-5-methoxy-4-methyl-1-oxoisobenzofuran-6-yl)-4-methyl-4-Hexanoic Acid

RN: 24280-93-1 **MP** (°C):
MW: 320.35 **BP** (°C):

Solubility (Moles/L)	Solubility (Grams/L)	Temp (°C)	Ref (#)	Evaluation (T P E A A)	Comments
4.058E-05	1.300E-02	25	L333	1 1 1 1 0	

3463. $C_{17}H_{21}NO_2$

Napropamide
N,N-Diethyl-2-(1-naphthyloxy)propanamide
Devrinol 50W
Devrinol
Devrinol 10G
Devrinol 2E

RN: 15299-99-7 **MP** (°C): 75.1
MW: 271.36 **BP** (°C):

Solubility (Moles/L)	Solubility (Grams/L)	Temp (°C)	Ref (#)	Evaluation (T P E A A)	Comments
2.690E-04	7.300E-02	20	M161	1 0 0 0 1	

3464. C$_{17}$H$_{21}$NO$_4$
Cocaine
L-Cocaine
L-Cocain
RN: 50-36-2   **MP** (°C):   98
MW: 303.36   **BP** (°C):

Solubility (Moles/L)	Solubility (Grams/L)	Temp (°C)	Ref (#)	Evaluation (T P E A A)	Comments
4.000E-03	1.213E+00	15	K059	2 2 2 0 0	
5.934E-03	1.800E+00	22	F300	1 0 0 0 1	
5.485E-03	1.664E+00	25	D004	1 0 0 0 0	
5.266E-03	1.597E+00	25	D041	1 0 0 0 1	
1.248E-02	3.786E+00	80	D041	1 0 0 0 1	

3465. C$_{17}$H$_{21}$NO$_4$
Scopolamine
Scopolamin
Hyoscine
Murocoll
Plexonal
Transderm-SCOP
RN: 51-34-3   **MP** (°C):   59
MW: 303.36   **BP** (°C):

Solubility (Moles/L)	Solubility (Grams/L)	Temp (°C)	Ref (#)	Evaluation (T P E A A)	Comments
3.132E-01	9.500E+01	15	F300	1 0 0 0 1	
3.296E-01	1.000E+02	ns	C109	0 0 0 0 1	

3466. C$_{17}$H$_{21}$N$_3$O$_2$
Dis. A. 2.
Ethanol, 2,2'-[[3-Methyl-4-(phenylazo)phenyl]imino]bis-
4-[bis(2-Hydroxyethyl)amino]-2-methylazobenzene
RN: 3771-38-8   **MP** (°C):   111
MW: 299.38   **BP** (°C):

Solubility (Moles/L)	Solubility (Grams/L)	Temp (°C)	Ref (#)	Evaluation (T P E A A)	Comments
7.600E-05	2.275E-02	25	B333	1 0 0 0 1	

3467.   C₁₇H₂₁N₅O₂

1H-Purine-2,6-dione, 8-(2-Aminophenyl)-3,7-dihydro-1,3-dipropyl-

RN: 96445-34-0   **MP** (°C):   276dec
MW: 327.39   **BP** (°C):

Solubility (Moles/L)	Solubility (Grams/L)	Temp (°C)	Ref (#)	Evaluation (T P E A A)	Comments
<3.05E-06	<1.00E-03	ns	H316	0 2 1 1 0	pH 7.4
1.222E-05	4.000E-03	ns	H316	0 2 1 1 0	0.1N HCl

3468.   C₁₇H₂₁N₅O₁₀

9-(1,3-Dihemisuccinate-2-propoxymethyl)guanine

RN: 88110-76-3   **MP** (°C):   167
MW: 455.38   **BP** (°C):

Solubility (Moles/L)	Solubility (Grams/L)	Temp (°C)	Ref (#)	Evaluation (T P E A A)	Comments
1.039E-01	4.730E+01	25	B360	1 0 2 2 2	

3469.   C₁₇H₂₂I₃N₃O₈

1,3-Benzenedicarboxamide, N,N'-bis(2,3-Dihydroxypropyl)-5S-[(2-hydroxy-1-oxopropyl)amino]-2,4,6-triiodo-[S-(S*,S*)]-

RN: **MP** (°C):
MW: 777.09   **BP** (°C):

Solubility (Moles/L)	Solubility (Grams/L)	Temp (°C)	Ref (#)	Evaluation (T P E A A)	Comments
1.379E-01	1.071E+02	25	P091	1 0 0 0 1	

3470.   C₁₇H₂₂I₃N₃O₈

1,3-Benzenedicarboxamide, N,N'-bis(2,3-Dihydroxypropyl)-5S-[(2-hydroxy-1-oxopropyl)amino]-2,4,6-triiodo-(RS,S)-

RN: **MP** (°C):
MW: 777.09   **BP** (°C):

Solubility (Moles/L)	Solubility (Grams/L)	Temp (°C)	Ref (#)	Evaluation (T P E A A)	Comments
1.379E-01	1.071E+02	25	P091	1 0 0 0 1	

3471.   C₁₇H₂₂I₃N₃O₈

1,3-Benzenedicarboxamide, N,N'-bis(2,3-Dihydroxypropyl)-5RS-[(2-hydroxy-1-oxopropyl)amino]-2,4,6-triiodo-[RS-(RS*,RS*)]-

RN: 60166-94-1   **MP** (°C):
MW: 777.09   **BP** (°C):

Solubility (Moles/L)	Solubility (Grams/L)	Temp (°C)	Ref (#)	Evaluation (T P E A A)	Comments
1.379E-01	1.071E+02	25	P091	1 0 0 0 1	

3472. $C_{17}H_{22}I_3N_3O_8$

1,3-Benzenedicarboxamide, N,N'-bis(2,3-Dihydroxypropyl)-5RS-[(2-hydroxy-1-oxopropyl)amino]-2,4,6-triiodo-[RS-(RS*,S*)]-

RN: **MP** (°C):
MW: 777.09   **BP** (°C):

Solubility (Moles/L)	Solubility (Grams/L)	Temp (°C)	Ref (#)	Evaluation (T P E A A)	Comments
1.379E-01	1.071E+02	25	P091	1 0 0 0 1	

3473. $C_{17}H_{22}I_3N_3O_8$

1,3-Benzenedicarboxamide, N,N'-bis[2-Hydroxy-1-(hydroxymethyl)ethyl]-5-[(2-hydroxy-1-oxopropyl)amino]-2,4,6-triiodo-, (RS)-

RN: 60208-45-9   **MP** (°C):
MW: 777.09   **BP** (°C):

Solubility (Moles/L)	Solubility (Grams/L)	Temp (°C)	Ref (#)	Evaluation (T P E A A)	Comments
1.775E-01	1.379E+02	25	P091	1 0 0 0 1	

3474. $C_{17}H_{22}I_3N_3O_8$

1,3-Benzenedicarboxamide, N-(2,3-Dihydroxypropyl)-N'-[2-hydroxy-1-(hydroxymethyl)ethyl]-5-[(2-hydroxy-1-oxopropyl)amino]-2,4,6-triiodo-

RN: 77868-45-2   **MP** (°C):
MW: 777.09   **BP** (°C):

Solubility (Moles/L)	Solubility (Grams/L)	Temp (°C)	Ref (#)	Evaluation (T P E A A)	Comments
1.379E-01	1.071E+02	25	P091	1 0 0 0 1	

3475. $C_{17}H_{22}I_3N_3O_8$

DL-Iopamidol
1,3-Benzenedicarboxamide, N,N'-bis[2-Hydroxy-1-(hydroxymethyl)ethyl]-5-[(2-hydroxy-1-oxopropyl)amino]-2,4,6-triiodo-, (S)-
L-Iopamidol
1,3-Benzenedicarboxamide, N,N'-bis(2,3-Dihydroxypropyl)-5RS-[(2-hydroxy-1-oxopropyl)amino]-2,4,6-triiodo-[RS-(S*,S*)]-

RN: 60166-93-0   **MP** (°C):
MW: 777.09   **BP** (°C):

Solubility (Moles/L)	Solubility (Grams/L)	Temp (°C)	Ref (#)	Evaluation (T P E A A)	Comments
6.096E-01	4.737E+02	20	F178	1 0 0 0 1	EFG
1.580E-01	1.228E+02	20	F178	1 0 0 0 1	EFG
6.096E-01	4.737E+02	25	P091	1 0 0 0 1	
1.580E-01	1.228E+02	25	P091	1 0 0 0 1	
5.798E-01	4.505E+02	40	F178	1 0 0 0 1	EFG
1.963E-01	1.525E+02	40	F178	1 0 0 0 1	EFG

5.679E-01	4.413E+02	60	F178	1 0 0 0 1	EFG
3.120E-01	2.424E+02	60	F178	1 0 0 0 1	EFG
6.235E-01	4.845E+02	80	F178	1 0 0 0 1	EFG
5.209E-01	4.048E+02	80	F178	1 0 0 0 1	EFG
6.911E-01	5.370E+02	100	F178	1 0 0 0 1	EFG
7.098E-01	5.516E+02	100	F178	1 0 0 0 1	EFG

3476. $C_{17}H_{22}I_3N_3O_8$
1,3-Benzenedicarboxamide, N,N'-bis(2,3-Dihydroxypropyl)-5S-[(2-hydroxy-1-oxopropyl)amino]-2,4,6-triiodo-[RS-(RS*,RS*)]-
RN: 77942-93-9   **MP** (°C):
MW: 777.09   **BP** (°C):

Solubility (Moles/L)	Solubility (Grams/L)	Temp (°C)	Ref (#)	Evaluation (T P E A A)	Comments
1.480E-01	1.150E+02	25	P091	1 0 0 0 1	

3477. $C_{17}H_{22}I_3N_3O_9$
1,3-Benzenedicarboxamide, 5-[(2,3-Dihydroxy-1-oxopropyl)amino]-N,N'-bis[2-hydroxy-1-(hydroxymethyl)ethyl]-2,4,6-triiodo-
RN: 69698-47-1   **MP** (°C):
MW: 793.09   **BP** (°C):

Solubility (Moles/L)	Solubility (Grams/L)	Temp (°C)	Ref (#)	Evaluation (T P E A A)	Comments
6.573E-02	5.213E+01	25	P091	1 0 0 0 1	

3478. $C_{17}H_{22}I_3N_3O_9$
1,3-Benzenedicarboxamide, 5-[(2,3-Dihydroxy-1-oxobutyl)amino]-N,N'-bis[2-hydroxy-1-(hydroxymethyl)ethyl]-2,4,6-triiodo-
RN: 129968-26-9   **MP** (°C):
MW: 793.09   **BP** (°C):

Solubility (Moles/L)	Solubility (Grams/L)	Temp (°C)	Ref (#)	Evaluation (T P E A A)	Comments
5.430E-02	4.306E+01	25	P091	1 0 0 0 1	

3479. $C_{17}H_{22}N_4O_3S$
2-(N4-Acetylsulfanilylamino)-4-n-amylpyrimidine
RN: **MP** (°C):
MW: 362.45   **BP** (°C):

Solubility (Moles/L)	Solubility (Grams/L)	Temp (°C)	Ref (#)	Evaluation (T P E A A)	Comments
1.214E-05	4.400E-03	37	R076	1 2 0 0 1	

3480.　C$_{17}$H$_{22}$N$_4$O$_7$.0.75H$_2$O

2'-(2-Methyl-3-one-pentanyl)-6-methoxypurine Arabinoside (0.75 Hydrate)

RN:　　145913-50-4　　**MP** (°C):　　55-60
MW:　　407.90　　　　　**BP** (°C):

Solubility (Moles/L)	Solubility (Grams/L)	Temp (°C)	Ref (#)	Evaluation (T P E A A)	Comments
8.770E-02	3.577E+01	37	C348	1 2 2 2 2	pH 7.00

3481.　C$_{17}$H$_{23}$NO

N-Cyclooctylcinnamamide
2-Propenamide, N-Cyclooctyl-3-phenyl-

RN:　　59832-00-7　　**MP** (°C):
MW:　　257.38　　　　　**BP** (°C):

Solubility (Moles/L)	Solubility (Grams/L)	Temp (°C)	Ref (#)	Evaluation (T P E A A)	Comments
2.660E-06	6.846E-04	ns	H350	0 0 0 0 2	

3482.　C$_{17}$H$_{23}$NO

N,N-Octamethylenecinnamamide
Octahydro-1-(1-oxo-3-phenyl-2-propenyl)1H-Azonine

RN:　　59832-07-4　　**MP** (°C):
MW:　　257.38　　　　　**BP** (°C):

Solubility (Moles/L)	Solubility (Grams/L)	Temp (°C)	Ref (#)	Evaluation (T P E A A)	Comments
2.460E-04	6.332E-02	ns	H350	0 0 0 0 2	

3483.　C$_{17}$H$_{23}$NO$_3$

Atropine
Atropin
8-Methyl-8-azabicyclo[3.2.1]octan-3-yl 3-hydroxy-2-phenylpropionate
Neo-Diophen
Minims

RN:　　51-55-8　　　　**MP** (°C):　　115
MW:　　289.38　　　　　**BP** (°C):

Solubility (Moles/L)	Solubility (Grams/L)	Temp (°C)	Ref (#)	Evaluation (T P E A A)	Comments
5.500E-03	1.592E+00	15	K059	2 2 2 0 1	
5.529E-03	1.600E+00	18	F300	1 0 0 0 1	
6.898E-03	1.996E+00	20	D041	1 0 0 0 0	
1.032E-02	2.987E+00	20	K052	1 1 1 1 2	
1.148E-02	3.322E+00	25	D004	1 0 0 0 0	
7.586E-03	2.195E+00	rt	D021	0 0 1 1 1	

3484. C₁₇H₂₃NO₃

Hyoscyamine

Hyoscyamin

Benzeneacetic Acid, α-(Hydroxymethyl)-, 8-Methyl-8-azabicyclo[3.2.1]oct-3-yl Ester, [3(S)-Endo]-

Daturine

Duboisine

L-Hyoscyamine

RN: 101-31-5   **MP** (°C):   108.5

MW: 289.38   **BP** (°C):

Solubility (Moles/L)	Solubility (Grams/L)	Temp (°C)	Ref (#)	Evaluation (T P E A A)	Comments
1.244E-02	3.600E+00	20	F300	1 0 0 0 2	
1.225E-02	3.546E+00	c	D004	1 0 0 0 0	

3485. C₁₇H₂₃NO₅

Benzoic Acid, 2-(Acetyloxy)-, 2-[bis(1-Methylethyl)amino]-2-oxoethyl Ester

RN: 116482-76-9   **MP** (°C):   108.9

MW: 321.38   **BP** (°C):

Solubility (Moles/L)	Solubility (Grams/L)	Temp (°C)	Ref (#)	Evaluation (T P E A A)	Comments
5.601E-04	1.800E-01	21	N335	1 2 1 1 2	

3486. C₁₇H₂₃NO₅

Benzoic Acid, 2-(Acetyloxy)-, 2-(Dipropylamino)-2-oxoethyl Ester

RN: 116482-75-8   **MP** (°C):   50.5

MW: 321.38   **BP** (°C):

Solubility (Moles/L)	Solubility (Grams/L)	Temp (°C)	Ref (#)	Evaluation (T P E A A)	Comments
2.240E-03	7.200E-01	21	N335	1 2 1 1 2	

3487. C₁₇H₂₃N₃O

Aeo-antergan

1,2-Ethanediamine, N-[(4-Methoxyphenyl)methyl]-N',N'-dimethyl-N-2-pyridinyl-

Dorantamin

Anthisan

Dipane

Copsamine

RN: 91-84-9   **MP** (°C):

MW: 285.39   **BP** (°C):

Solubility (Moles/L)	Solubility (Grams/L)	Temp (°C)	Ref (#)	Evaluation (T P E A A)	Comments
1.200E-02	3.425E+00	37.5	L034	2 2 0 1 2	pH 7.4

3488.   C₁₇H₂₃N₃O₂

2-Methoxy-N-[2-(diethyl-amino)ethyl]-4-quinoline Carboxamide
N-[2-(Diethylamino)ethyl]-2-methoxyquinoline-4-carboxamide
RN: 2716-98-5   **MP** (°C):
MW: 301.39   **BP** (°C):

Solubility (Moles/L)	Solubility (Grams/L)	Temp (°C)	Ref (#)	Evaluation (T P E A A)	Comments
3.000E-03	9.042E-01	ns	B018	0 0 0 0 1	
3.000E-03	9.042E-01	ns	M066	0 0 0 0 0	

3489.   C₁₇H₂₄N₄O₅

2,5-Dipivaloyloxymethyl Allopurinol
RN: 98827-17-9   **MP** (°C):   145-146
MW: 364.40   **BP** (°C):

Solubility (Moles/L)	Solubility (Grams/L)	Temp (°C)	Ref (#)	Evaluation (T P E A A)	Comments
1.235E-04	4.500E-02	22	B322	1 0 2 2 2	

3490.   C₁₇H₂₄N₄O₅

1,5-Dipivaloyloxymethyl Allopurinol
RN: 98827-16-8   **MP** (°C):   136-137
MW: 364.40   **BP** (°C):

Solubility (Moles/L)	Solubility (Grams/L)	Temp (°C)	Ref (#)	Evaluation (T P E A A)	Comments
5.488E-05	2.000E-02	22	B322	1 0 2 2 2	

3491.   C₁₇H₂₄N₄O₆

2'-Hexanyl-6-methoxypurine Arabinoside
RN: 145913-39-9   **MP** (°C):
MW: 380.40   **BP** (°C):

Solubility (Moles/L)	Solubility (Grams/L)	Temp (°C)	Ref (#)	Evaluation (T P E A A)	Comments
1.890E-02	7.190E+00	37	C348	1 2 2 2 2	pH 7.00

3492.   C₁₇H₂₅NO

N-Octylcinnamamide
2-Propenamide, N-Octyl-3-phenyl-
RN: 55030-48-3   **MP** (°C):
MW: 259.39   **BP** (°C):

Solubility (Moles/L)	Solubility (Grams/L)	Temp (°C)	Ref (#)	Evaluation (T P E A A)	Comments
1.390E-06	3.606E-04	ns	H350	0 0 0 0 2	

3493. C₁₇H₂₅NO₃

Acetamide, 2-(Benzoyloxy)-N,N-bis(2-methylpropyl)-

RN: 115193-33-4   **MP** (°C):   44.5
MW: 291.39   **BP** (°C):

Solubility (Moles/L)	Solubility (Grams/L)	Temp (°C)	Ref (#)	Evaluation (T P E A A)	Comments
2.745E-04	8.000E-02	22	N317	1 1 2 1 2	

3494. C₁₇H₂₅NO₃

Acetamide, 2-(Benzoyloxy)-N,N-Acetamide, 2-(benzoyloxy)-N,N-dibutyl-

RN: 106231-57-6   **MP** (°C):   25
MW: 291.39   **BP** (°C):

Solubility (Moles/L)	Solubility (Grams/L)	Temp (°C)	Ref (#)	Evaluation (T P E A A)	Comments
2.745E-04	8.000E-02	22	N317	1 1 2 1 2	

3495. C₁₇H₂₅NO₄

Octyl Acetaminophen
Carbonic Acid, Octyl Ester, Ester with 4'-hydroxyacetanilide
Acetanilide, 4'-Hydroxy-, Octyl Carbonate (Ester)

RN: 19872-70-9   **MP** (°C):   82.5-83
MW: 307.39   **BP** (°C):

Solubility (Moles/L)	Solubility (Grams/L)	Temp (°C)	Ref (#)	Evaluation (T P E A A)	Comments
1.431E-05	4.400E-03	37	D029	1 0 1 1 1	

3496. C₁₇H₂₅N₅O₆

9-(1,3-Dibutyrate-2-propoxymethyl)guanine

RN: 88110-71-8   **MP** (°C):   200
MW: 395.42   **BP** (°C):

Solubility (Moles/L)	Solubility (Grams/L)	Temp (°C)	Ref (#)	Evaluation (T P E A A)	Comments
3.541E-04	1.400E-01	25	B360	1 0 2 2 2	

3497. C$_{17}$H$_{26}$ClNO$_2$
Butachlor
N-(Butoxymethyl)-2-chloro-N-(2,6-diethylphenyl)acetamide
N-(Butoxymethyl)-2-chloro-2',6'-diethylacetanilide
Machete
Butanex
Hiltachlor
RN: 23184-66-9   **MP** (°C):   <-5
MW: 311.86   **BP** (°C):   196

Solubility (Moles/L)	Solubility (Grams/L)	Temp (°C)	Ref (#)	Evaluation (T P E A A)	Comments
6.413E-05	2.000E-02	20	M161	1 0 0 0 1	

3498. C$_{17}$H$_{26}$O$_3$
Decyl-p-hydroxybenzoate
Decyl p-hydroxybenzoate
n-Decyl p-hydroxybenzoate
RN: 69679-30-7   **MP** (°C):   58
MW: 278.39   **BP** (°C):

Solubility (Moles/L)	Solubility (Grams/L)	Temp (°C)	Ref (#)	Evaluation (T P E A A)	Comments
3.200E-05	8.909E-03	15	B355	1 1 1 1 2	
3.710E-05	1.033E-02	20	B355	1 1 1 1 2	
8.800E-05	2.450E-02	25	B355	1 1 1 1 2	
1.303E-03	3.629E-01	25	D081	1 2 2 1 2	*sic*
7.943E-05	2.211E-02	25	F322	2 0 1 1 0	EFG

3499. C$_{17}$H$_{27}$NO$_2$
Terbutol
2,6-Di-tert-butyl-p-tolyl Methylcarbamate
RN: 1918-11-2   **MP** (°C):   185
MW: 277.41   **BP** (°C):

Solubility (Moles/L)	Solubility (Grams/L)	Temp (°C)	Ref (#)	Evaluation (T P E A A)	Comments
2.343E-05	6.500E-03	25	B200	1 0 0 0 0	
2.523E-05	7.000E-03	ns	H042	0 0 0 0 0	

3500. $C_{17}H_{27}NO_3$
Stadacain
4-Butoxybenzoic Acid 2-(Diethyl-amino)ethyl Ester
RN: 2350-32-5 **MP** (°C): 146
MW: 293.41 **BP** (°C):

Solubility (Moles/L)	Solubility (Grams/L)	Temp (°C)	Ref (#)	Evaluation (T P E A A)	Comments
1.300E-04	3.814E-02	ns	M066	0 0 0 0 1	

3501. $C_{17}H_{28}N_2O_2$
4-Butylaminobenzoic Acid 2-(Diethyl-amino)ethyl Ester
RN: 3772-42-7 **MP** (°C):
MW: 292.42 **BP** (°C):

Solubility (Moles/L)	Solubility (Grams/L)	Temp (°C)	Ref (#)	Evaluation (T P E A A)	Comments
4.100E-04	1.199E-01	ns	M066	0 0 0 0 1	

3502. $C_{17}H_{28}N_2O_2$
Endomid
N,N,N',N'-Tetraethyl-bicyclo(2.2.1)hept-5-ene-2,3-dicarboxamide
RN: 4582-18-7 **MP** (°C):
MW: 292.42 **BP** (°C):

Solubility (Moles/L)	Solubility (Grams/L)	Temp (°C)	Ref (#)	Evaluation (T P E A A)	Comments
5.916E-02	1.730E+01	20	K050	1 1 1 1 2	

3503. $C_{17}H_{28}O_2$
4-Nonylphenol Monoethoxylate
Ethanol, 2-(4-Nonylphenoxy)-
RN: 104-35-8 **MP** (°C):
MW: 264.41 **BP** (°C):

Solubility (Moles/L)	Solubility (Grams/L)	Temp (°C)	Ref (#)	Evaluation (T P E A A)	Comments
1.048E-05	2.770E-03	2	A335	1 0 2 2 2	
1.050E-05	2.776E-03	2	A335	1 0 2 2 2	
1.063E-05	2.810E-03	10	A335	1 0 2 2 2	
1.060E-05	2.803E-03	10	A335	1 0 2 2 2	
1.074E-05	2.840E-03	14	A335	1 0 2 2 2	
1.080E-05	2.856E-03	14	A335	1 0 2 2 2	
1.140E-05	3.014E-03	20.5	A335	1 0 2 2 2	
1.142E-05	3.020E-03	20.5	A335	1 0 2 2 2	
1.280E-05	3.384E-03	25	A335	1 0 2 2 2	
1.275E-05	3.370E-03	25	A335	1 0 2 2 2	

3504.   C$_{17}$H$_{34}$O$_2$
Margaric Acid
Heptadecanoic Acid
RN: 506-12-7   **MP** (°C):
MW: 270.46   **BP** (°C):

Solubility (Moles/L)	Solubility (Grams/L)	Temp (°C)	Ref (#)	Evaluation (T P E A A)	Comments
1.035E-05	2.800E-03	0	B136	1 0 2 1 1	
1.035E-05	2.800E-03	0.0	R001	1 1 1 1 1	
1.553E-05	4.200E-03	20	B136	1 0 2 1 1	
1.553E-05	4.200E-03	20.0	R001	1 1 1 1 1	
1.997E-05	5.400E-03	30	B136	1 0 2 1 1	
2.034E-05	5.500E-03	30.0	R001	1 1 1 1 1	
2.551E-05	6.900E-03	45	B136	1 0 2 1 1	
2.551E-05	6.900E-03	45.0	R001	1 1 1 1 1	
2.995E-05	8.100E-03	60	B136	1 0 2 1 1	
2.995E-05	8.100E-03	60.0	R001	1 1 1 1 1	

3505.   C$_{17}$H$_{36}$O
Heptadecanol
1-Heptadecanol
RN: 1454-85-9   **MP** (°C):   58
MW: 256.48   **BP** (°C):

Solubility (Moles/L)	Solubility (Grams/L)	Temp (°C)	Ref (#)	Evaluation (T P E A A)	Comments
<=1E-7	<=2.56E-5	25	R002	1 2 2 2 2	

3506.   C$_{18}$H$_{10}$Cl$_4$
2,4,4'',6-Tetrachloro-p-terphenyl
2,4,4'',6-Tetrachloro-1,1':4',1''-terphenyl
RN: 61576-97-4   **MP** (°C):
MW: 368.09   **BP** (°C):

Solubility (Moles/L)	Solubility (Grams/L)	Temp (°C)	Ref (#)	Evaluation (T P E A A)	Comments
1.606E-10	5.910E-08	4	D351	1 2 1 1 2	
4.728E-10	1.740E-07	25	D351	1 2 1 1 2	
1.106E-09	4.069E-07	40	D351	1 2 1 1 2	

3507.   C$_{18}$H$_{10}$I$_6$N$_2$O$_7$
Ioglycamic Acid
N,N'-bis(3-Carboxy-2,4,6-triiodophenyl)-diglycolamide
BE 419
RN: 2618-25-9   **MP** (°C):
MW: 1127.72   **BP** (°C):

Solubility (Moles/L)	Solubility (Grams/L)	Temp (°C)	Ref (#)	Evaluation (T P E A A)	Comments
1.773E-04	2.000E-01	ns	H055	0 1 0 2 2	

3508.   C$_{18}$H$_{10}$N$_2$O$_2$S
Disperse Brightner
2,2'-(2,5-Thiophenediyl)bisbenzoxazole
Unitex OB
Uvitex EBF
RN: 2866-43-5   **MP** (°C):   219
MW: 318.36   **BP** (°C):

Solubility (Moles/L)	Solubility (Grams/L)	Temp (°C)	Ref (#)	Evaluation (T P E A A)	Comments
3.000E-08	9.551E-06	25	B333	1 0 0 0 1	

3509.   C$_{18}$H$_{11}$Cl$_3$
2,4'',5-Trichloro-p-terphenyl
2,4'',5-Trichloro-1,1':4',1''-terphenyl
RN: 61576-93-0   **MP** (°C):
MW: 333.65   **BP** (°C):

Solubility (Moles/L)	Solubility (Grams/L)	Temp (°C)	Ref (#)	Evaluation (T P E A A)	Comments
3.028E-10	1.010E-07	4	D351	1 2 1 1 2	
1.233E-09	4.115E-07	25	D351	1 2 1 1 2	
2.567E-09	8.564E-07	39	D351	1 2 1 1 2	

3510.   C$_{18}$H$_{11}$NO$_3$
Samaron Yellow
Supra Light Yellow GGL(IG)
RN: 1326-08-5   **MP** (°C):
MW: 289.29   **BP** (°C):

Solubility (Moles/L)	Solubility (Grams/L)	Temp (°C)	Ref (#)	Evaluation (T P E A A)	Comments
4.000E-06	1.157E-03	98.59	M180	0 0 2 2 0	EFG
8.000E-06	2.314E-03	111.46	M180	0 0 2 2 0	EFG
1.000E-05	2.893E-03	112.94	M180	0 0 2 2 0	EFG
1.100E-05	3.182E-03	119.00	M180	0 0 2 2 0	EFG
1.300E-05	3.761E-03	125.25	M180	0 0 2 2 0	EFG
1.400E-05	4.050E-03	128.45	M180	0 0 2 2 0	EFG
2.200E-05	6.364E-03	152.37	M180	0 0 2 2 0	EFG

3511.   C₁₈H₁₁NO₃
Disperse Yellow 54
C.I. Disperse Yellow 54
RN: 7576-65-0   **MP** (°C):
MW: 289.29   **BP** (°C):

Solubility (Moles/L)	Solubility (Grams/L)	Temp (°C)	Ref (#)	Evaluation (T P E A A)	Comments
1.000E-07	2.893E-05	25	B333	1 0 0 0 1	
2.400E-07	6.943E-05	60.0	D093	1 2 1 2 0	EFG
6.500E-07	1.880E-04	71.8	D093	1 2 1 2 0	EFG
1.600E-06	4.629E-04	84.1	D093	1 2 1 2 0	EFG
4.000E-06	1.157E-03	97.4	D093	1 2 1 2 0	EFG

3512.   C₁₈H₁₂
1,2-Benzanthracene
Benzanthracene
1,2-Benzoanthracene
RN: 56-55-3   **MP** (°C):   155
MW: 228.30   **BP** (°C):

Solubility (Moles/L)	Solubility (Grams/L)	Temp (°C)	Ref (#)	Evaluation (T P E A A)	Comments
1.310E-08	2.991E-06	6.90	M082	1 1 1 2 2	
1.310E-08	2.991E-06	6.90	M151	2 1 2 2 2	
1.311E-08	2.992E-06	6.94	M183	1 2 1 1 2	
1.660E-08	3.790E-06	10.70	M082	1 1 1 2 2	
1.660E-08	3.790E-06	10.70	M151	2 1 2 2 2	
1.657E-08	3.783E-06	11.14	M183	1 2 1 1 2	
2.100E-08	4.794E-06	14.24	M183	1 2 1 1 2	
2.100E-08	4.794E-06	14.30	M082	1 1 1 2 2	
2.100E-08	4.794E-06	14.30	M151	2 1 2 2 2	
1.583E-08	3.613E-06	14.34	M183	1 2 1 1 2	
2.365E-08	5.400E-06	15	B385	2 0 2 2 2	
2.446E-08	5.584E-06	18.14	M183	1 2 1 1 2	
2.770E-08	6.324E-06	19.30	M082	1 1 1 2 2	
2.770E-08	6.324E-06	19.30	M151	2 1 2 2 2	
2.775E-08	6.335E-06	19.34	M183	1 2 1 1 2	
3.670E-08	8.378E-06	23.10	M082	1 1 1 2 2	
3.670E-08	8.378E-06	23.10	M151	2 1 2 2 2	
3.669E-08	8.377E-06	23.14	M183	1 2 1 1 2	
3.507E-08	8.007E-06	23.64	M183	1 2 1 1 2	
1.927E-07	4.400E-05	24	H116	2 1 0 0 1	
4.117E-08	9.400E-06	25	B319	2 0 1 2 1	
4.056E-08	9.260E-06	25	B385	2 0 2 2 2	
4.310E-08	9.840E-06	25	K001	2 2 2 2 2	
3.900E-09	8.904E-07	25	K123	1 0 2 2 1	sic
2.497E-08	5.700E-06	25	L332	1 1 1 1 2	

6.132E-08	1.400E-05	25	M064	1 1 2 2 1
4.117E-08	9.400E-06	25	M071	2 2 2 2 2
6.130E-08	1.399E-05	25	M342	1 0 1 1 2
4.117E-08	9.400E-06	25.00	M151	2 1 1 2 1
3.774E-08	8.617E-06	25.04	M183	1 2 1 1 2
4.818E-08	1.100E-05	27	D003	1 0 0 1 1
5.344E-08	1.220E-05	29	M071	2 2 2 2 2
5.344E-08	1.220E-05	29.00	M151	2 1 1 2 2
5.436E-08	1.241E-05	29.54	M183	1 2 1 1 2
5.580E-08	1.274E-05	29.70	M082	1 1 1 2 2
5.580E-08	1.274E-05	29.70	M151	2 1 2 2 2
5.567E-08	1.271E-05	29.74	M183	1 2 1 1 2
7.635E-08	1.743E-05	35	B385	2 0 2 2 2
6.132E-08	1.400E-05	ns	M344	0 0 0 0 2

3513. $C_{18}H_{12}$
Chrysene
1,2-Benzphenanthrene
RN: 218-01-9   **MP** (°C):   254
MW: 228.30   **BP** (°C):   448

Solubility (Moles/L)	Solubility (Grams/L)	Temp (°C)	Ref (#)	Evaluation (T P E A A)	Comments
3.100E-09	7.077E-07	6.50	M082	1 1 1 2 2	
3.100E-09	7.077E-07	6.50	M151	2 1 2 2 2	
3.500E-09	7.990E-07	11.00	M082	1 1 1 2 2	
3.500E-09	7.990E-07	11.00	M151	2 1 2 2 2	
6.130E-09	1.399E-06	20.40	M082	1 1 1 2 2	
6.130E-09	1.399E-06	20.40	M151	2 1 2 2 2	
6.139E-09	1.401E-06	20.44	M183	1 2 1 1 2	
9.199E-09	2.100E-06	23	P339	2 0 1 2 2	
7.446E-08	1.700E-05	24	H116	2 1 0 0 1	
7.360E-09	1.680E-06	24.00	M082	1 1 1 2 2	
7.360E-09	1.680E-06	24.00	M151	2 1 2 2 2	
7.367E-09	1.682E-06	24.04	M183	1 2 1 1 2	
4.818E-09	1.100E-06	25	B319	2 0 1 2 1	average of 2
2.760E-08	6.301E-06	25	K001	2 2 2 2 2	
2.628E-08	6.000E-06	25	L332	1 1 1 1 2	
8.761E-09	2.000E-06	25	M064	1 1 2 2 1	
7.884E-09	1.800E-06	25	M071	2 2 2 2 2	
8.760E-09	2.000E-06	25	M342	1 0 1 1 2	
7.884E-09	1.800E-06	25.00	M151	2 1 1 2 1	
8.280E-09	1.890E-06	25.30	M082	1 1 1 2 2	
8.280E-09	1.890E-06	25.30	M151	2 1 2 2 2	
8.283E-09	1.891E-06	25.34	M183	1 2 1 1 2	
6.570E-09	1.500E-06	27	D003	1 0 0 1 1	
9.680E-09	2.210E-06	28.70	M082	1 1 1 2 2	
9.680E-09	2.210E-06	28.70	M151	2 1 2 2 2	
9.689E-09	2.212E-06	28.74	M183	1 2 1 1 2	

9.637E-09	2.200E-06	29	M071	2 2 2 2 2	
9.637E-09	2.200E-06	29.00	M151	2 1 1 2 1	
8.761E-09	2.000E-06	ns	M344	0 0 0 0 2	
3.400E-06	7.762E-04	ns	W005	0 0 1 2 1	*sic*

3514. $C_{18}H_{12}$
Triphenylene
9,10-Benzphenanthrene
Isochrysene

RN: 217-59-4   **MP** (°C):   199
MW: 228.30   **BP** (°C):   425

Solubility (Moles/L)	Solubility (Grams/L)	Temp (°C)	Ref (#)	Evaluation (T P E A A)	Comments
1.180E-08	2.694E-06	8	M082	1 1 1 2 2	
1.180E-08	2.694E-06	8	M151	2 1 2 2 2	
1.311E-08	2.992E-06	8.04	M183	1 2 1 1 2	
1.330E-08	3.036E-06	12.00	M082	1 1 1 2 2	
1.330E-08	3.036E-06	12.00	M151	2 1 2 2 2	
1.328E-08	3.033E-06	12.04	M183	1 2 1 1 2	
1.490E-08	3.402E-06	14.80	M082	1 1 1 2 2	
1.490E-08	3.402E-06	14.80	M151	2 1 2 2 2	
2.500E-07	5.707E-05	20	E009	1 0 0 1 1	
2.140E-08	4.886E-06	20.50	M082	1 1 1 2 2	
2.140E-08	4.886E-06	20.50	M151	2 1 2 2 2	
2.144E-08	4.894E-06	20.54	M183	1 2 1 1 2	
1.800E-07	4.109E-05	25	A325	2 1 2 2 1	
1.880E-07	4.292E-05	25	K001	1 0 2 1 2	
1.884E-07	4.300E-05	25	M064	1 1 2 2 1	
2.891E-08	6.600E-06	25.00	M151	2 1 1 2 1	
1.665E-07	3.800E-05	27	D003	1 0 0 1 1	
3.350E-08	7.648E-06	27.30	M082	1 1 1 2 2	
3.350E-08	7.648E-06	27.30	M151	2 1 2 2 2	
3.354E-08	7.657E-06	27.34	M183	1 2 1 1 2	
3.550E-08	8.105E-06	28.20	M082	1 1 1 2 2	
3.550E-08	8.105E-06	28.20	M151	2 1 2 2 2	
3.556E-08	8.117E-06	28.24	M183	1 2 1 1 2	
1.486E-08	3.393E-06	114.84	M183	1 2 1 1 2	
1.884E-07	4.300E-05	ns	M344	0 0 0 0 2	

3515. C$_{18}$H$_{12}$
Tetracene
Naphthacene
2,3-Benzanthracene
RN: 92-24-0   **MP** (°C):   341
MW: 228.30   **BP** (°C):

Solubility (Moles/L)	Solubility (Grams/L)	Temp (°C)	Ref (#)	Evaluation (T P E A A)	Comments
1.580E-08	3.607E-06	20	E009	1 0 0 1 2	
6.600E-09	1.507E-06	25	K001	2 2 2 2 1	
2.497E-09	5.700E-07	25	M064	1 1 2 2 1	
2.500E-09	5.707E-07	25	M342	1 0 1 1 1	
4.380E-09	1.000E-06	27	D003	1 0 0 1 1	
2.497E-09	5.700E-07	ns	M344	0 0 0 0 2	

3516. C$_{18}$H$_{12}$N$_2$
2,2'-Biquinoline
2,2'-Biquinolyl
RN: 119-91-5   **MP** (°C):   193
MW: 256.31   **BP** (°C):

Solubility (Moles/L)	Solubility (Grams/L)	Temp (°C)	Ref (#)	Evaluation (T P E A A)	Comments
3.980E-06	1.020E-03	24	H106	1 0 2 2 2	
3.980E-06	1.020E-03	24	M303	1 0 1 1 2	

3517. C$_{18}$H$_{12}$N$_4$O
4-Hydroxy-6,7-diphenylpteridine
4-Hydroxy-6:7-diphenylpteridine
RN: 102943-71-5   **MP** (°C):
MW: 300.32   **BP** (°C):

Solubility (Moles/L)	Solubility (Grams/L)	Temp (°C)	Ref (#)	Evaluation (T P E A A)	Comments
6.658E-04	2.000E-01	20	A019	2 2 1 1 2	

3518. C$_{18}$H$_{13}$N
6-Aminochrysene
6-Chrysenamine
RN: 2642-98-0   **MP** (°C):   210
MW: 243.31   **BP** (°C):

Solubility (Moles/L)	Solubility (Grams/L)	Temp (°C)	Ref (#)	Evaluation (T P E A A)	Comments
6.370E-07	1.550E-04	24	H106	1 0 2 2 2	
6.370E-10	1.550E-07	ns	M349	0 2 1 1 2	

3519. C₁₈H₁₃NO₃

Furo[3,4-b]quinolin-3(1H)-one, 9-Hydroxy-6-methyl-1-phenyl-
RN: 74103-08-5   **MP** (°C):
MW: 291.31   **BP** (°C):

Solubility (Moles/L)	Solubility (Grams/L)	Temp (°C)	Ref (#)	Evaluation (T P E A A)	Comments
4.463E-08	1.300E-05	25	P089	2 1 2 2 2	
1.270E-07	3.700E-05	37	P089	2 1 2 2 2	
2.163E-07	6.300E-05	51	P089	2 1 2 2 2	

3520. C₁₈H₁₃NO₃

N-1-Naphthylphthalamic Acid
Naptalam
2-((1-Naphthylamino)carbonyl)benzoic Acid
Naphthylphthalamic Acid
ALANAP-1
NPA
RN: 132-66-1   **MP** (°C):   185
MW: 291.31   **BP** (°C):

Solubility (Moles/L)	Solubility (Grams/L)	Temp (°C)	Ref (#)	Evaluation (T P E A A)	Comments
6.866E-04	2.000E-01	25	B200	1 0 0 0 2	
6.866E-04	2.000E-01	ns	B185	0 0 0 0 2	
6.866E-04	2.000E-01	ns	N013	0 0 0 0 2	
6.866E-04	2.000E-01	rt	M161	0 0 0 0 2	

3521. C₁₈H₁₄

o-Terphenyl
1,2-Diphenyl Benzene
RN: 84-15-1   **MP** (°C):   58
MW: 230.31   **BP** (°C):   332

Solubility (Moles/L)	Solubility (Grams/L)	Temp (°C)	Ref (#)	Evaluation (T P E A A)	Comments
5.380E-06	1.239E-03	25	A325	2 1 2 2 2	

3522. C₁₈H₁₄

m-Terphenyl
1,3-Diphenyl Benzene
RN: 92-06-8   **MP** (°C):   89
MW: 230.31   **BP** (°C):   365

Solubility (Moles/L)	Solubility (Grams/L)	Temp (°C)	Ref (#)	Evaluation (T P E A A)	Comments
6.560E-06	1.511E-03	25	A325	2 1 2 2 2	

3523.   C$_{18}$H$_{14}$
p-Terphenyl
1,4-Diphenyl Benzene
RN: 92-94-4   **MP** (°C):   213
MW: 230.31   **BP** (°C):

Solubility (Moles/L)	Solubility (Grams/L)	Temp (°C)	Ref (#)	Evaluation (T P E A A)	Comments
7.800E-08	1.796E-05	25	A325	2 1 2 2 1	

3524.   C$_{18}$H$_{14}$Cl$_4$N$_2$O
Miconazole
1-[2-(2,4-Dichlorophenyl)-2-[(2,4-dichlorophenyl)methoxy]ethyl]-1H-imidazole
1-[2,4-Dichloro-β-[(2,4-dichlorobenzyl)oxy]phenethyl]imidazole
Conoderm
RN: 22916-47-8   **MP** (°C):
MW: 416.14   **BP** (°C):

Solubility (Moles/L)	Solubility (Grams/L)	Temp (°C)	Ref (#)	Evaluation (T P E A A)	Comments
<4.80E-09	<2.00E-06	25	P348	0 1 1 2 2	

3525.   C$_{18}$H$_{14}$N$_4$O
Disperse Yellow 23
Phenol, 4-[[4-(Phenylazo)phenyl]azo]-
p-Hydroxy-p-bis(azobenzene)
RN: 6250-23-3   **MP** (°C):
MW: 302.34   **BP** (°C):

Solubility (Moles/L)	Solubility (Grams/L)	Temp (°C)	Ref (#)	Evaluation (T P E A A)	Comments
2.000E-10	6.047E-08	25	B333	1 0 0 0 1	
1.300E-07	3.930E-05	71.8	D093	1 2 1 2 0	EFG
5.500E-07	1.663E-04	84.1	D093	1 2 1 2 0	EFG
2.300E-06	6.954E-04	97.4	D093	1 2 1 2 0	EFG

3526. $C_{18}H_{14}N_4O_2$
Disperse Orange 1
Dye VI
C.I. Disperse Organe 1
4-(p-Nitrophenylazo)diphenylamine
4-Anilino-4'-nitroazobenzene
4-(4-Nitrophenylazo)diphenylamine
RN: 2581-69-3   **MP** (°C): 157
MW: 318.34   **BP** (°C):

Solubility (Moles/L)	Solubility (Grams/L)	Temp (°C)	Ref (#)	Evaluation (T P E A A)	Comments
1.500E-09	4.775E-07	25	B333	1 0 0 0 1	
3.000E-07	9.550E-05	84.10	B198	1 2 1 1 0	
1.420E-06	4.520E-04	97.40	B198	1 2 1 1 2	
4.900E-06	1.560E-03	111.60	B198	1 2 1 1 1	
1.950E-05	6.208E-03	127	B198	1 2 1 1 2	

3527. $C_{18}H_{15}Cl_4N_3O_4$
Miconazole Nitrate-β Cyclidextrin Complexant
RN: 22832-87-7   **MP** (°C):
MW: 479.15   **BP** (°C):

Solubility (Moles/L)	Solubility (Grams/L)	Temp (°C)	Ref (#)	Evaluation (T P E A A)	Comments
3.700E-04	1.773E-01	25	P348	0 1 1 2 2	

3528. $C_{18}H_{15}N_3O_5$
1H-Benzimidazole-1-carboxylic Acid, 6-Benzoyl-2-[(methoxycarbonyl)amino]-, Methyl Ester
RN: 104663-14-1   **MP** (°C): 156.5
MW: 353.34   **BP** (°C):

Solubility (Moles/L)	Solubility (Grams/L)	Temp (°C)	Ref (#)	Evaluation (T P E A A)	Comments
1.981E-05	7.000E-03	21	N337	1 0 2 2 0	pH 5
1.900E-05	6.713E-03	21	N337	1 0 2 2 0	pH 5

3529.   C₁₈H₁₅O₄P

Triphenyl Phosphate
Phosphoric Acid Triphenyl Ester
Triphenyl Phosphoric Acid Ester
Phenyl Phosphate
TPP

RN: 115-86-6   **MP** (°C):   49
MW: 326.29   **BP** (°C):   245

Solubility (Moles/L)	Solubility (Grams/L)	Temp (°C)	Ref (#)	Evaluation (T P E A A)	Comments
2.237E-06	7.300E-04	24	H116	2 1 0 0 2	
6.129E-05	2.000E-02	ns	F014	0 0 0 0 0	

3530.   C₁₈H₁₆Cl₃N₃O₄

Econazole Nitrate
Pevaryl
Spectazole
R 14827

RN: 68797-31-9   **MP** (°C):
MW: 444.70   **BP** (°C):

Solubility (Moles/L)	Solubility (Grams/L)	Temp (°C)	Ref (#)	Evaluation (T P E A A)	Comments
1.600E-03	7.115E-01	25	P348	0 1 1 2 2	

3531.   C₁₈H₁₆N₂O₃

Benzoyltryptophan
N-Benzoyl-DL-tryptophan

RN: 2901-79-3   **MP** (°C):
MW: 308.34   **BP** (°C):

Solubility (Moles/L)	Solubility (Grams/L)	Temp (°C)	Ref (#)	Evaluation (T P E A A)	Comments
1.816E-03	5.600E-01	25.1	N026	2 0 2 2 2	

3532.   C₁₈H₁₆N₄O₃S

2-(N4-Acetylsulfanilylamino)-4-phenylpyrimidine

RN: **MP** (°C):
MW: 368.42   **BP** (°C):

Solubility (Moles/L)	Solubility (Grams/L)	Temp (°C)	Ref (#)	Evaluation (T P E A A)	Comments
9.772E-06	3.600E-03	37	R076	1 2 0 0 1	

3533. $C_{18}H_{17}ClN_4O_6 \cdot 0.5H_2O$

9-[5-O-(4-Chlorobenzoyl-β-D-arabinofuranosyl)]-6-methoxy-9H-purine (Hemihydrate)

RN: 121032-34-6   **MP** (°C):   122-124
MW: 429.82   **BP** (°C):

Solubility (Moles/L)	Solubility (Grams/L)	Temp (°C)	Ref (#)	Evaluation (T P E A A)	Comments
1.880E-04	8.081E-02	37	M378	1 2 1 1 2	pH 7.2

3534. $C_{18}H_{17}Cl_2NO_3$

Benzoylprop-ethyl
Ethyl N-Benzoyl-N-(3,4-dichlorophenyl)-2-aminopropionate
FX 2182
N-Benzoyl-N-(3,4-dichlorophenyl)-DL-alanine Ethyl Ester
Enaven
Suffix

RN: 22212-55-1   **MP** (°C):   70.5
MW: 366.25   **BP** (°C):

Solubility (Moles/L)	Solubility (Grams/L)	Temp (°C)	Ref (#)	Evaluation (T P E A A)	Comments
5.461E-05	2.000E-02	25	M161	1 0 0 0 1	

3535. $C_{18}H_{17}N_5O_8$

6-Methoxy-9-(5-O-[4-nitrobenzoyl]-β-D-arabinofuranosyl)-9H-purine

RN: 121032-21-1   **MP** (°C):   202-203
MW: 431.36   **BP** (°C):

Solubility (Moles/L)	Solubility (Grams/L)	Temp (°C)	Ref (#)	Evaluation (T P E A A)	Comments
3.400E-05	1.467E-02	37	M378	1 2 1 1 2	pH 7.2

3536. $C_{18}H_{18}ClNO_4$

Clanobutin
Butanoic Acid, 4-[(4-Chlorobenzoyl)(4-methoxyphenyl)amino]-
Bykahepar

RN: 30544-61-7   **MP** (°C):
MW: 347.80   **BP** (°C):

Solubility (Moles/L)	Solubility (Grams/L)	Temp (°C)	Ref (#)	Evaluation (T P E A A)	Comments
1.270E-04	4.417E-02	37	K093	1 2 1 1 2	pH 3.0

3537. C$_{18}$H$_{18}$CINO$_5$

Etofibrate
3-Pyridinecarboxylic Acid, 2-[2-(4-Chlorophenoxy)-2-methyl-1-oxopropoxy]ethyl Ester
Tricerol
Lipo-Merz

RN: 31637-97-5   **MP** (°C):
MW: 363.80   **BP** (°C):

Solubility (Moles/L)	Solubility (Grams/L)	Temp (°C)	Ref (#)	Evaluation (T P E A A)	Comments
2.000E-05	7.276E-03	rt	G093	0 1 1 1 2	pH≥4

3538. C$_{18}$H$_{18}$CINS

Chlorprothixene
Taractan
1-Propanamine, 3-(2-Chloro-9H-thioxanthen-9-ylidene)-N,N-dimethyl-, (3Z)-
Rentovet

RN: 113-59-7   **MP** (°C):
MW: 315.87   **BP** (°C):

Solubility (Moles/L)	Solubility (Grams/L)	Temp (°C)	Ref (#)	Evaluation (T P E A A)	Comments
3.936E-05	1.243E-02	20	H301	2 0 2 2 2	

3539. C$_{18}$H$_{18}$N$_2$O$_4$

C.I. Disperse Blue 23
1,4-bis[(2-Hydroxyethyl)amino]anthraquinone
Acetoquinone Blue BF

RN: 4471-41-4   **MP** (°C):   248
MW: 326.36   **BP** (°C):

Solubility (Moles/L)	Solubility (Grams/L)	Temp (°C)	Ref (#)	Evaluation (T P E A A)	Comments
2.400F-06	7.833F-04	25	B333	1 0 0 0 1	

3540. C$_{18}$H$_{18}$N$_4$O$_6$

9-[5-O-(Benzoyl-β-D-arabinofuranosyl)]-6-methoxy-9H-purine

RN: 121032-31-3   **MP** (°C):   202-204
MW: 386.37   **BP** (°C):

Solubility (Moles/L)	Solubility (Grams/L)	Temp (°C)	Ref (#)	Evaluation (T P E A A)	Comments
7.400E-05	2.859E-02	37	M378	1 2 1 1 2	pH 7.2

3541.   C$_{18}$H$_{18}$N$_4$O$_6$.0.75H$_2$O
2'-Benzoyl-6-methoxypurine Arabinoside (0.75 Hydrate)
RN: 145913-44-6   **MP** (°C):   84-86
MW: 399.88   **BP** (°C):

Solubility (Moles/L)	Solubility (Grams/L)	Temp (°C)	Ref (#)	Evaluation (T P E A A)	Comments
1.780E-02	7.118E+00	37	C348	1 2 2 2 2	pH 7.00

3542.   C$_{18}$H$_{18}$N$_8$O$_6$
7,7'-Succinylditheophylline
RN: 58447-18-0   **MP** (°C):
MW: 442.39   **BP** (°C):

Solubility (Moles/L)	Solubility (Grams/L)	Temp (°C)	Ref (#)	Evaluation (T P E A A)	Comments
1.630E-03	7.211E-01	25	L067	1 0 1 1 2	

3543.   C$_{18}$H$_{18}$O$_2$
Dienestrol
3,4-bis(4-Hydroxyphenyl)-2,4-hexadiene
Dehydrostilbestrol
RN: 84-17-3   **MP** (°C):   227.5
MW: 266.34   **BP** (°C):

Solubility (Moles/L)	Solubility (Grams/L)	Temp (°C)	Ref (#)	Evaluation (T P E A A)	Comments
1.126E-05	3.000E-03	37	B039	2 1 1 1 0	EFG

3544.   C$_{18}$H$_{18}$O$_2$
Equilenin
3-Hydroxy-17-keto-δ(1,3,5-10,6,8)estrapentaene
1,3,5-10,6,8-Estrapentaen-3-ol-17-one
RN: 517-09-9   **MP** (°C):   258
MW: 266.34   **BP** (°C):

Solubility (Moles/L)	Solubility (Grams/L)	Temp (°C)	Ref (#)	Evaluation (T P E A A)	Comments
5.707E-06	1.520E-03	25	L033	1 0 2 1 2	

3545. C$_{18}$H$_{18}$O$_3$
Flurecol-butyl
Flurenol-n-butyl Ester
n-Butyl-9-hydroxyfluorene-(9)-carboxylate
RN: 2314-09-2   **MP** (°C): 70
MW: 282.34   **BP** (°C):

Solubility (Moles/L)	Solubility (Grams/L)	Temp (°C)	Ref (#)	Evaluation (T P E A A)	Comments
1.293E-02	3.650E+00	20	B200	1 0 0 0 2	*sic*
1.293E-04	3.650E-02	20	M161	1 0 0 0 2	*sic*

3546. C$_{18}$H$_{19}$Cl$_2$NO$_4$
Felodipine
3,5-Pyridinedicarboxylic Acid, 4-(2,3-Dichlorophenyl)-1,4-dihydro-2,6-dimethyl-, Ethyl Methyl Ester
Plendil
RN: 72509-76-3   **MP** (°C):
MW: 384.26   **BP** (°C):

Solubility (Moles/L)	Solubility (Grams/L)	Temp (°C)	Ref (#)	Evaluation (T P E A A)	Comments
1.301E-06	5.000E-04	20	N322	1 0 2 2 1	
1.179E-05	4.530E-03	22	M382	2 1 1 1 1	

3547. C$_{18}$H$_{19}$F$_3$N$_2$S
4-Trifluoromethyl-N,N-dimethyl-10H-phenothiazine-10-propanamide
RN: 3852-94-6   **MP** (°C):
MW: 352.42   **BP** (°C):

Solubility (Moles/L)	Solubility (Grams/L)	Temp (°C)	Ref (#)	Evaluation (T P E A A)	Comments
7.000E-06	2.467E-03	ns	G023	0 0 1 1 0	

3548. C$_{18}$H$_{19}$F$_3$N$_2$S
Fluopromazine
Triflupromazine
RN: 146-54-3   **MP** (°C): <25
MW: 352.42   **BP** (°C):

Solubility (Moles/L)	Solubility (Grams/L)	Temp (°C)	Ref (#)	Evaluation (T P E A A)	Comments
5.000E-06	1.762E-03	24	G022	2 0 1 1 1	
5.000E-06	1.762E-03	ns	F027	0 0 0 0 0	

3549. C$_{18}$H$_{19}$NO
Desmethyldoxepin
1-Propanamine, 3-Dibenz[b,e]oxepin-11(6H)-ylidene-N-methyl-
RN: 1225-56-5   **MP** (°C):
MW: 265.36   **BP** (°C):

Solubility (Moles/L)	Solubility (Grams/L)	Temp (°C)	Ref (#)	Evaluation (T P E A A)	Comments
3.950E-04	1.048E-01	25	E051	1 0 2 1 2	

3550. C$_{18}$H$_{19}$N$_2$O$_4$
N-Benzoyl-L-tyrosinamide Acetate
RN: **MP** (°C):
MW: 327.36   **BP** (°C):

Solubility (Moles/L)	Solubility (Grams/L)	Temp (°C)	Ref (#)	Evaluation (T P E A A)	Comments
1.300E-04	4.256E-02	25	A066	1 0 1 1 1	

3551. C$_{18}$H$_{19}$N$_3$O$_6$S
Cephaloglycin
5-Thia-1-azabicyclo[4.2.0]oct-2-ene-2-carboxylic Acid
RN: 3577-01-3   **MP** (°C):
MW: 405.43   **BP** (°C):

Solubility (Moles/L)	Solubility (Grams/L)	Temp (°C)	Ref (#)	Evaluation (T P E A A)	Comments
2.590E-02	1.050E+01	25	P311	1 2 2 2 0	EFG

3552. C$_{18}$H$_{19}$N$_5$O$_3$
C.I. Disperse Dye
Propanenitrile, 3-[(2-Hydroxyethyl)[3-methyl-4-[(4-nitrophenyl)azo]phenyl]amino]-
Celliton Discharge Scarlet RNL
Celliton Fast Scarlet RN
RN: 6054-58-6   **MP** (°C):   156
MW: 353.38   **BP** (°C):

Solubility (Moles/L)	Solubility (Grams/L)	Temp (°C)	Ref (#)	Evaluation (T P E A A)	Comments
1.900E-07	6.714E-05	25	B333	1 0 0 0 1	

3553. C$_{18}$H$_{19}$N$_5$O$_6$
2'-(o-Aminobenzoyl)-6-methoxypurine Arabinoside
RN: 121032-55-1   **MP** (°C):
MW: 401.38   **BP** (°C):

Solubility (Moles/L)	Solubility (Grams/L)	Temp (°C)	Ref (#)	Evaluation (T P E A A)	Comments
2.060E-02	8.268E+00	37	C348	1 2 2 2 2	pH 7.00

3554. C$_{18}$H$_{19}$N$_5$O$_6$.0.3H$_2$O
9-[5-O-(4-Aminobenzoyl-β-D-arabinofuranosyl)]-6-methoxy-9H-purine (0.3 Hydrate)
RN: 121032-39-1   **MP** (°C):   198-200
MW: 406.79   **BP** (°C):

Solubility (Moles/L)	Solubility (Grams/L)	Temp (°C)	Ref (#)	Evaluation (T P E A A)	Comments
3.400E-05	1.383E-02	37	M378	1 2 1 1 2	pH 7.2

3555. C$_{18}$H$_{20}$
2,4-Diphenyl-4-methyl-2-pentene
RN: 6362-80-7   **MP** (°C):
MW: 236.36   **BP** (°C):

Solubility (Moles/L)	Solubility (Grams/L)	Temp (°C)	Ref (#)	Evaluation (T P E A A)	Comments
1.047E-07	2.475E-05	ns	D001	0 0 0 0 2	

3556. C$_{18}$H$_{20}$Cl$_2$O$_2$
1-Dichloro-2,2-bis(p-ethoxylphenyl)ethane
RN: 7388-32-1   **MP** (°C):
MW: 339.26   **BP** (°C):

Solubility (Moles/L)	Solubility (Grams/L)	Temp (°C)	Ref (#)	Evaluation (T P E A A)	Comments
7.664E-08	2.600E-05	rt	C122	0 2 2 2 2	

3557. C$_{18}$H$_{20}$N$_4$O$_7$S
2'-(p-Methylbenzenesulfonyl)-6-methoxypurine Arabinoside
RN: 145913-49-1   **MP** (°C):   214-215
MW: 436.45   **BP** (°C):

Solubility (Moles/L)	Solubility (Grams/L)	Temp (°C)	Ref (#)	Evaluation (T P E A A)	Comments
1.240E-04	5.412E-02	37	C348	1 2 2 2 2	pH 7.00

3558.   C$_{18}$H$_{20}$O$_2$
Equilin
3-Hydroxy-17-keto-δ(1,3,5-10,7)estratetraene
1,3,5(10),7-Estratetraen-3-ol-17-one
RN: 474-86-2   **MP** (°C):   238
MW: 268.36   **BP** (°C):

Solubility (Moles/L)	Solubility (Grams/L)	Temp (°C)	Ref (#)	Evaluation (T P E A A)	Comments
5.217E-06	1.400E-03	25	H049	2 2 2 2 1	
5.254E-06	1.410E-03	25	L033	1 0 2 1 2	

3559.   C$_{18}$H$_{20}$O$_2$
Diethylstilbestrol
Diethylstilboestrol
Destrol
4,4'-(1,2-Diethyl-1,2-ethenediyl)bisphenol
Tylosterone
Vagestrol
RN: 56-53-1   **MP** (°C):   169
MW: 268.36   **BP** (°C):

Solubility (Moles/L)	Solubility (Grams/L)	Temp (°C)	Ref (#)	Evaluation (T P E A A)	Comments
4.472E-05	1.200E-02	25	G009	1 0 1 1 1	
9.316E-05	2.500E-02	30	M007	2 2 1 2 2	average of 6

3560.   C$_{18}$H$_{21}$ClN$_2$
Chlorocyclizine
Chlorcyclizine
RN: 82-93-9   **MP** (°C):
MW: 300.83   **BP** (°C):

Solubility (Moles/L)	Solubility (Grams/L)	Temp (°C)	Ref (#)	Evaluation (T P E A A)	Comments
1.000E-03	3.008E-01	37.5	L034	2 2 0 1 2	pH 7.4

3561.   C$_{18}$H$_{21}$ClN$_2$S
2-Chloro-N,N-dimethyl-10H-phenothiazine-10-butanamine
RN: 13094-23-0   **MP** (°C):
MW: 332.90   **BP** (°C):

Solubility (Moles/L)	Solubility (Grams/L)	Temp (°C)	Ref (#)	Evaluation (T P E A A)	Comments
5.000E-06	1.664E-03	ns	G023	0 0 1 1 0	

3562.   C$_{18}$H$_{21}$ClO
1-Chloro-1-methyl-2-(p-methylphenyl)-2-p-Ethoxylphenyl)ethane
RN: 56265-27-1   **MP** (°C):
MW: 288.82   **BP** (°C):

Solubility (Moles/L)	Solubility (Grams/L)	Temp (°C)	Ref (#)	Evaluation (T P E A A)	Comments
5.540E-06	1.600E-03	rt	C122	0 2 2 2 2	

3563.   C$_{18}$H$_{21}$NO$_3$
Thebainone A
Morphinan-6-one, 7,8-Didehydro-4-hydroxy-3-methoxy-17-methyl-
Thebainon
RN: 467-98-1   **MP** (°C):   146
MW: 299.37   **BP** (°C):

Solubility (Moles/L)	Solubility (Grams/L)	Temp (°C)	Ref (#)	Evaluation (T P E A A)	Comments
1.336E-02	4.000E+00	20	F300	1 0 0 0 0	
2.839E-02	8.500E+00	100	F300	1 0 0 0 1	

3564.   C$_{18}$H$_{21}$NO$_3$
Codeine
Codein
Methylmorphin
7,8-Didehydro-4,5-α-epoxy-3-methoxy-17-methylmorphinan-6-α-ol
Nucofed
Robitussin AC
RN: 76-57-3   **MP** (°C):   155
MW: 299.37   **BP** (°C):

Solubility (Moles/L)	Solubility (Grams/L)	Temp (°C)	Ref (#)	Evaluation (T P E A A)	Comments
3.006E-02	9.000E+00	20	Λ073	1 1 1 1 0	
2.672E-02	8.000E+00	20	F300	1 0 0 0 0	
2.760E-02	8.264E+00	20	K052	1 1 1 1 2	
1.591E-01	4.762E+01	25	E041	2 2 2 2 0	EFG, form III, recrystallized
3.242E-02	9.705E+00	25	E041	2 2 2 2 0	EFG, form II, recrystallized
3.176E-02	9.509E+00	25	E041	2 2 2 2 0	EFG, form I, recrystallized
3.340E-02	1.000E+01	30	A073	1 1 1 1 1	
3.674E-02	1.100E+01	40	A073	1 1 1 1 1	
4.342E-02	1.300E+01	50	A073	1 1 1 1 1	
5.010E-02	1.500E+01	60	A073	1 1 1 1 1	
6.013E-02	1.800E+01	70	A073	1 1 1 1 1	
6.347E-02	1.900E+01	80	A073	1 1 1 1 1	
5.578E-02	1.670E+01	80	F300	1 0 0 0 2	
8.017E-02	2.400E+01	90	A073	1 1 1 1 1	
1.069E-01	3.200E+01	100	A073	1 1 1 1 1	

3565. C₁₈H₂₁NO₃.H₂O

Codeine (Monohydrate)

Morphinan-6-ol, 7,8-Didehydro-4,5-epoxy-3-methoxy-17-methyl-, Monohydrate, (5α,6α)

RN: 6059-47-8   **MP** (°C):   155
MW: 317.39   **BP** (°C):

Solubility (Moles/L)	Solubility (Grams/L)	Temp (°C)	Ref (#)	Evaluation (T P E A A)	Comments
2.604E-02	8.264E+00	c	D004	1 0 0 0 0	

3566. C₁₈H₂₁NO₄

2-Naphthaleneacetic Acid, 6-Methoxy-α-methyl-, 2-(Dimethylamino)-2-oxoethyl Ester, (S)
Naproxen, N,N-Dimethyl Glycolamide Ester
2-Naphthaleneacetic Acid, 6-Methoxy-α-methyl-, 2-(Dimethylamino)-2-oxoethyl Ester
Naproxen N,N-Dimethyl Glycolamide Ester

RN: 114665-18-8   **MP** (°C):   150.5
MW: 315.37   **BP** (°C):

Solubility (Moles/L)	Solubility (Grams/L)	Temp (°C)	Ref (#)	Evaluation (T P E A A)	Comments
1.268E-05	4.000E-03	21	B331	1 2 2 1 2	pH 7.4
1.268E-05	4.000E-03	21	B331	1 2 2 1 1	

3567. C₁₈H₂₂N₂

Desipramine
Norpramine
Pertofran

RN: 50-47-5   **MP** (°C):   212
MW: 266.39   **BP** (°C):

Solubility (Moles/L)	Solubility (Grams/L)	Temp (°C)	Ref (#)	Evaluation (T P E A A)	Comments
2.200E-04	5.861E-02	24	G022	2 0 1 1 1	

3568. C₁₈H₂₂N₄O₅

Dis. A. 9
Ethanol, 2,2'-[[4-[(2-Methoxy-4-nitrophenyl)azo]-3-methylphenyl]imino]bis-
4-[bis(2-Hydroxyethyl)amino]-2'-methoxy-2-methyl-4'-nitroazobenzene

RN: 41541-11-1   **MP** (°C):
MW: 374.40   **BP** (°C):

Solubility (Moles/L)	Solubility (Grams/L)	Temp (°C)	Ref (#)	Evaluation (T P E A A)	Comments
4.500E-06	1.685E-03	25	B333	1 0 0 0 1	

3569. C$_{18}$H$_{22}$O$_2$
Hexestrol
4,4'-(1,2-Diethylethylene)diphenol
Dihydrodiethylstilbestrol
Esestrolo
RN: 5635-50-7   **MP** (°C):   186.5
MW: 270.37   **BP** (°C):

Solubility (Moles/L)	Solubility (Grams/L)	Temp (°C)	Ref (#)	Evaluation (T P E A A)	Comments
4.438E-05	1.200E-02	37	B039	2 1 1 1 0	EFG
3.699E-05	1.000E-02	37	B045	1 0 1 1 1	

3570. C$_{18}$H$_{22}$O$_2$
Estrone
Oestrone
Folliculin
1,3,5(10)-Estratrien-3-ol-17-one
Estra-1,3,5(10)-Trien-17-one, 3-Hydroxy-
Oestrin
RN: 53-16-7   **MP** (°C):   252.5
MW: 270.37   **BP** (°C):

Solubility (Moles/L)	Solubility (Grams/L)	Temp (°C)	Ref (#)	Evaluation (T P E A A)	Comments
2.959E-06	8.000E-04	25	H049	2 2 2 2 1	
1.110E-04	3.000E-02	25	I309	0 0 0 0 1	sic
2.959E-06	8.000E-04	25	L033	1 0 2 1 1	
1.109E-03	2.999E-01	25	P324	0 1 0 2 1	
8.200E-06	2.217E-03	37	H034	1 0 2 1 1	pH 7.4
1.184E-05	3.200E-03	37	L010	1 0 0 1 1	
3.162E-06	8.550E-04	ns	A074	0 0 0 0 0	EFG

3571. C$_{18}$H$_{23}$N$_3$O$_3$S
L-Leu-dapsone
2-Amino-N-[4-[(4-aminophenyl)sulfonyl]phenyl]-4-methyl-, (S)-
Pentanamide
RN: 160349-00-8   **MP** (°C):
MW: 361.47   **BP** (°C):

Solubility (Moles/L)	Solubility (Grams/L)	Temp (°C)	Ref (#)	Evaluation (T P E A A)	Comments
8.576E-04	3.100E-01	25	P351	2 2 1 2 1	pH 7.4
≥6.92E-02	>2.50E+01	25	P351	2 2 1 2 1	

3572.   C₁₈H₂₃N₃O₄S

3572. $C_{18}H_{23}N_3O_4S$

Phentolamine Methanesulfonate
Vasomax
Regitine Mesylate
Regitine Methanesulfonate

RN: 65-28-1   **MP** (°C):   177
MW: 377.47   **BP** (°C):

Solubility (Moles/L)	Solubility (Grams/L)	Temp (°C)	Ref (#)	Evaluation (T P E A A)	Comments
3.979E+00	1.502E+03	30	D011	1 0 1 0 2	

3573. $C_{18}H_{24}I_3N_3O_9$

1,3-Benzenedicarboxamide, 5RS-[(2,3-Dihydroxy-1-oxobutyl)amino]-N,N'-bis(2,3-dihydroxypropyl)-2,4,6-triiodo-[RS-(RS*,S*)]-

RN: 77868-48-5   **MP** (°C):
MW: 807.12   **BP** (°C):

Solubility (Moles/L)	Solubility (Grams/L)	Temp (°C)	Ref (#)	Evaluation (T P E A A)	Comments
1.327E-01	1.071E+02	25	P091	1 0 0 0 1	

3574. $C_{18}H_{24}N_4O_2$

2,5-Diaziridinyl-3,6-dipyrrolidino-1,4-benzoquinone

RN: 59886-43-0   **MP** (°C):   160
MW: 328.42   **BP** (°C):

Solubility (Moles/L)	Solubility (Grams/L)	Temp (°C)	Ref (#)	Evaluation (T P E A A)	Comments
1.522E-03	5.000E-01	rt	C317	0 2 0 0 0	

3575. $C_{18}H_{24}N_4O_2S$

2-Sulfanilamidobornylenepyrimidine

RN: **MP** (°C):   276
MW: 360.48   **BP** (°C):

Solubility (Moles/L)	Solubility (Grams/L)	Temp (°C)	Ref (#)	Evaluation (T P E A A)	Comments
8.322E-05	3.000E-02	29	C049	1 2 0 0 1	

3576. $C_{18}H_{24}N_4O_2S$

2-Sulfanilamido-5,6,7,8,-tetrahydro-8-isopropyl-5-methyl-quinazoline

RN: 71119-36-3   **MP** (°C):   185-187
MW: 360.48   **BP** (°C):

Solubility (Moles/L)	Solubility (Grams/L)	Temp (°C)	Ref (#)	Evaluation (T P E A A)	Comments
6.658E-05	2.400E-02	29	C049	1 2 0 0 1	

3577.   C$_{18}$H$_{24}$N$_4$O$_3$S
L-Lys-Dapsone
Hexanamide, 2,6-Diamino-N-[4-[(4-aminophenyl)sulfonyl]phenyl]-, (S)
RN: 160349-03-1   **MP** (°C):
MW: 376.48   **BP** (°C):

Solubility (Moles/L)	Solubility (Grams/L)	Temp (°C)	Ref (#)	Evaluation (T P E A A)	Comments
>1.73E-01	>6.50E+01	25	P351	2 2 1 2 1	pH 7.4
>1.73E-01	>6.50E+01	25	P351	2 2 1 2 1	

3578.   C$_{18}$H$_{24}$O$_2$
Estradiol
17-β-Estradiol
Estradiol-17β
RN: 50-28-2   **MP** (°C):   176
MW: 272.39   **BP** (°C):

Solubility (Moles/L)	Solubility (Grams/L)	Temp (°C)	Ref (#)	Evaluation (T P E A A)	Comments
1.652E-05	4.500E-03	20	G072	1 2 2 1 2	
6.200E-06	1.689E-03	20	L077	1 2 2 2 1	
2.566E-05	6.990E-03	23	B014	0 0 1 2 2	
7.413E-06	2.019E-03	25	B041	1 0 2 2 0	EFG
6.000E-07	1.634E-04	25	E014	2 2 2 1 1	pH 7.3
1.432E-05	3.900E-03	25	H049	2 2 2 2 1	
1.836E-05	5.000E-03	25	K003	2 1 1 1 1	
1.320E-05	3.596E-03	27.34	L077	1 2 2 2 2	
2.060E-05	5.611E-03	35	L077	1 2 2 2 2	
1.500E-05	4.086E-03	37	H034	1 0 2 1 2	pH 7.4
2.350E-05	6.401E-03	37	H035	1 1 1 1 2	pH 7.4
1.430E-05	3.895E-03	37	H054	1 2 2 2 2	
1.880E-05	5.120E-03	37	R069	1 0 2 2 2	pH 7.4
1.000E-05	2.724E-03	37.50	B041	1 0 2 2 0	EFG
2.830E-05	7.709E-03	42	L077	1 2 2 2 2	
3.560E-05	9.697E-03	50	L077	1 2 2 2 2	

3579.   C$_{18}$H$_{24}$O$_2$
α-Estradiol
17-α-Estradiol
RN: 57-91-0   **MP** (°C):   220
MW: 272.39   **BP** (°C):

Solubility (Moles/L)	Solubility (Grams/L)	Temp (°C)	Ref (#)	Evaluation (T P E A A)	Comments
1.432E-05	3.900E-03	25	L033	1 0 2 1 2	

3580.   C$_{18}$H$_{24}$O$_3$
Estriol
Oestriol
Drihydroxyestrin
RN: 50-27-1   **MP** (°C):   284.5
MW: 288.39   **BP** (°C):

Solubility (Moles/L)	Solubility (Grams/L)	Temp (°C)	Ref (#)	Evaluation (T P E A A)	Comments
1.110E-05	3.200E-03	25	H049	2 2 2 2 1	
1.000E-04	2.884E-02	30	O321	2 2 2 2 1	
1.006E-04	2.900E-02	30	O321	2 2 2 2 1	

3581.   C$_{18}$H$_{24}$O$_6$
Butylphthalyl Butyl Glycolate
1,2-Benzenedicarboxylic Acid 2-Butoxy-2-oxoethyl Butyl Ester
Butyl Carbobutoxymethyl Phthalate
RN: 85-70-1   **MP** (°C):   <-35
MW: 336.39   **BP** (°C):   219

Solubility (Moles/L)	Solubility (Grams/L)	Temp (°C)	Ref (#)	Evaluation (T P E A A)	Comments
3.567E-05	1.200E-02	20	F070	1 0 0 0 2	

3582.   C$_{18}$H$_{25}$I$_3$N$_3$O$_9$
3,5-Diacetylamino-2,4,6-triiodobenzoic Acid Methyl-glucamide
RN: **MP** (°C):   191
MW: 808.13   **BP** (°C):

Solubility (Moles/L)	Solubility (Grams/L)	Temp (°C)	Ref (#)	Evaluation (T P E A A)	Comments
1.101E+00	8.900E+02	20	L100	1 0 0 0 1	

3583.   C$_{18}$H$_{25}$NO
Racemethorphan
Dextromethorphan HBr
RN: 510-53-2   **MP** (°C):
MW: 271.41   **BP** (°C):

Solubility (Moles/L)	Solubility (Grams/L)	Temp (°C)	Ref (#)	Evaluation (T P E A A)	Comments
1.326E-01	3.600E+01	37	F008	1 1 2 2 2	0.1N HCl

3584.   C$_{18}$H$_{25}$N$_3$O$_2$

2-Ethoxy-N-[2-(diethyl-amino)ethyl]-4-quinoline Carboxamide
N-[2-(Diethylamino)ethyl]-2-ethoxyquinoline-4-carboxamide

RN: 2716-99-6   **MP** (°C):
MW: 315.42   **BP** (°C):

Solubility (Moles/L)	Solubility (Grams/L)	Temp (°C)	Ref (#)	Evaluation (T P E A A)	Comments
6.600E-04	2.082E-01	ns	M066	0 0 0 0 1	

3585.   C$_{18}$H$_{26}$NO$_4$

Ibuprofen N-Methyl-N-crabamoyl Methyl Glycolamide Ester

RN: **MP** (°C):   100.5
MW: 320.41   **BP** (°C):

Solubility (Moles/L)	Solubility (Grams/L)	Temp (°C)	Ref (#)	Evaluation (T P E A A)	Comments
4.057E-04	1.300E-01	0	B331	1 2 2 1 1	pH 7.4

3586.   C$_{18}$H$_{26}$N$_2$O$_4$

Benzeneacetic Acid, α-Methyl-4-(2-methylpropyl)-, 2-[(2-Amino-2-oxoethyl)methylamino]-2-oxoethyl Ester
Ibuprofen N-Methyl-N-carbamoyl Methyl Glycolamide Ester

RN: 114665-11-1   **MP** (°C):   100-101
MW: 334.42   **BP** (°C):

Solubility (Moles/L)	Solubility (Grams/L)	Temp (°C)	Ref (#)	Evaluation (T P E A A)	Comments
3.887E-04	1.300E-01	21	B331	1 2 2 1 1	

3587.   C$_{18}$H$_{26}$N$_4$O$_6$.0.5H$_2$O

2'-Heptanyl-6-methoxypurine Arabinoside (Hemihydrate)

RN: 145913-40-2   **MP** (°C):   83-85
MW: 403.44   **BP** (°C):

Solubility (Moles/L)	Solubility (Grams/L)	Temp (°C)	Ref (#)	Evaluation (T P E A A)	Comments
2.780E-03	1.122E+00	37	C348	1 2 2 2 2	pH 7.00

3588.   C$_{18}$H$_{26}$N$_4$O$_6$

9-[5-O-(Heptylate-β-D-arabinofuranosyl)]-6-methoxy-9H-purine

RN: 142963-79-9   **MP** (°C):   foam
MW: 394.43   **BP** (°C):

Solubility (Moles/L)	Solubility (Grams/L)	Temp (°C)	Ref (#)	Evaluation (T P E A A)	Comments
2.120E-04	8.362E-02	37	M378	1 2 1 1 2	pH 7.2

3589. C₁₈H₂₆O

Acetyl Ethyl Tetramethyl Tetralin
1-(3-Ethyl-5,6,7,8-tetrahydro-5,5,8,8-tetramethyl-2-naphthalenyl)ethanone
AETT
1,1,4,4-Tetramethyl-6-ethyl-7-acetyl-1,2,3,4-tetrahydronaphthalene
Ethanone, 1-(3-Ethyl-5,6,7,8-tetrahydro-5,5,8,8-tetramethyl-2-naphthyl)-

RN: 88-29-9   **MP** (°C):
MW: 258.41   **BP** (°C):

Solubility (Moles/L)	Solubility (Grams/L)	Temp (°C)	Ref (#)	Evaluation (T P E A A)	Comments
4.644E-08	1.200E-05	ns	B338	0 0 0 0 1	
4.644E-08	1.200E-05	ns	B338	0 0 0 0 1	

3590. C₁₈H₂₆O₂

Nortestosterone
Estr-4-en-3-one, 17-Hydroxy-, (17β)

RN: 434-22-0   **MP** (°C):
MW: 274.41   **BP** (°C):

Solubility (Moles/L)	Solubility (Grams/L)	Temp (°C)	Ref (#)	Evaluation (T P E A A)	Comments
1.126E-02	3.090E+00	25	P324	0 1 0 2 1	

3591. C₁₈H₂₆O₄

Dipentyl Phthalate
Diamyl Phthalate

RN: 131-18-0   **MP** (°C):   <-55
MW: 306.41   **BP** (°C):   342

Solubility (Moles/L)	Solubility (Grams/L)	Temp (°C)	Ref (#)	Evaluation (T P E A A)	Comments
1.450E-06	4.443E-04	20	L300	2 1 0 2 2	
9.791E-07	3.000E-04	25	F067	1 0 2 2 0	
3.263E-04	9.999E-02	ns	F014	0 0 0 0 0	

3592. C₁₈H₂₆O₆

Butyl Phthalyl Butyl Glycollate

RN: **MP** (°C):
MW: 338.40   **BP** (°C):

Solubility (Moles/L)	Solubility (Grams/L)	Temp (°C)	Ref (#)	Evaluation (T P E A A)	Comments
2.955E-05	1.000E-02	15	H069	1 0 1 1 0	
5.318E-04	1.800E-01	ns	F014	0 0 0 0 1	

3593. C₁₈H₂₇NO
N-Nonylcinnamamide
2-Propenamide, N-Nonyl-3-phenyl-
RN: 59832-01-8 **MP** (°C):
MW: 273.42 **BP** (°C):

Solubility (Moles/L)	Solubility (Grams/L)	Temp (°C)	Ref (#)	Evaluation (T P E A A)	Comments
2.220E-06	6.070E-04	ns	H350	0 0 0 0 2	

3594. C₁₈H₂₇NO₃
p-Acetamidophenyl Decanoate
Acetaminophen Decanoate
RN: 54942-37-9 **MP** (°C): 107
MW: 305.42 **BP** (°C):

Solubility (Moles/L)	Solubility (Grams/L)	Temp (°C)	Ref (#)	Evaluation (T P E A A)	Comments
2.947E-05	9.000E-03	25	B010	1 1 1 1 0	

3595. C₁₈H₂₇N₅O₅
9-[5'-(O-Caprylyl)-β-D-arabinofuranosyl]adenine Ester
RN: 66460-51-3 **MP** (°C):
MW: 393.45 **BP** (°C):

Solubility (Moles/L)	Solubility (Grams/L)	Temp (°C)	Ref (#)	Evaluation (T P E A A)	Comments
2.542E-04	1.000E-01	ns	B134	0 1 1 1 0	

3596. C₁₈H₂₈N₂O
DL-Bupivacaine
Bupivacaine
Marcaine
Bupivicaine
Marcaine (Hydrochloride Monohydrate)
RN: 2180-92-9 **MP** (°C): 107
MW: 288.44 **BP** (°C):

Solubility (Moles/L)	Solubility (Grams/L)	Temp (°C)	Ref (#)	Evaluation (T P E A A)	Comments
3.750E-04	1.082E-01	14.9	N046	2 0 1 2 2	intrinsic
1.733E-03	5.000E-01	23	F176	2 0 0 2 0	EFG, pH 7.4, intrinsic
3.180E-04	9.172E-02	25	N046	2 0 1 2 2	intrinsic
3.130E-04	9.028E-02	34.5	N046	2 0 1 2 2	intrinsic
4.170E-04	1.203E-01	37	N044	2 1 1 2 2	intrinsic

3597.   C$_{18}$H$_{28}$N$_4$O$_2$
2,5-Diaziridinyl-3,6-bis(butylamino)-1,4-benzoquinone
RN: 59886-48-5   **MP** (°C):   95
MW: 332.45   **BP** (°C):

Solubility (Moles/L)	Solubility (Grams/L)	Temp (°C)	Ref (#)	Evaluation (T P E A A)	Comments
<3.01E-04	<1.00E-01	rt	C317	0 2 0 0 0	

3598.   C$_{18}$H$_{28}$O$_3$
Undecyl p-Hydroxybenzoate
Undecyl 4-Hydroxybenzoate
RN: 69679-31-8   **MP** (°C):
MW: 292.42   **BP** (°C):

Solubility (Moles/L)	Solubility (Grams/L)	Temp (°C)	Ref (#)	Evaluation (T P E A A)	Comments
8.079E-03	2.362E+00	25	D081	1 2 2 1 2	

3599.   C$_{18}$H$_{29}$NO$_2$
Penbutolol
Levatol
2-Propanol, 1-(2-Cyclopentylphenoxy)-3-[(1,1-dimethylethyl)amino]-, (S)-
RN: 38363-40-5   **MP** (°C):   70
MW: 291.44   **BP** (°C):

Solubility (Moles/L)	Solubility (Grams/L)	Temp (°C)	Ref (#)	Evaluation (T P E A A)	Comments
2.402E-02	7.000E+00	rt	H096	1 0 0 0 0	

3600.   C$_{18}$H$_{29}$NO$_3$
4-Pentoxybenzoic Acid-2-(diethyl-amino)ethyl Ester
RN: 38973-73-8   **MP** (°C):
MW: 307.44   **BP** (°C):

Solubility (Moles/L)	Solubility (Grams/L)	Temp (°C)	Ref (#)	Evaluation (T P E A A)	Comments
6.000E-05	1.845E-02	ns	M066	0 0 0 0 1	

3601.   C$_{18}$H$_{30}$N$_2$O$_2$
4-Pentylaminobenzoic Acid-2-(diethylamino)ethyl Ester
RN: 16488-56-5   **MP** (°C):
MW: 306.45   **BP** (°C):

Solubility (Moles/L)	Solubility (Grams/L)	Temp (°C)	Ref (#)	Evaluation (T P E A A)	Comments
2.100E-04	6.435E-02	ns	M066	0 0 0 0 1	

3602. C$_{18}$H$_{30}$O$_3$
4-Octylphenol Diethoxylate
2-[2-(p-Octylphenoxy)ethoxy]ethanol
RN: 51437-90-2   **MP** (°C):
MW: 294.44   **BP** (°C):

Solubility (Moles/L)	Solubility (Grams/L)	Temp (°C)	Ref (#)	Evaluation (T P E A A)	Comments
4.483E-05	1.320E-02	20.5	A335	1 0 2 2 2	
4.490E-05	1.322E-02	20.5	A335	1 0 2 2 2	

3603. C$_{18}$H$_{30}$O$_{15}$.4H$_2$O
Triamylose (Tetrahydrate)
RN: **MP** (°C):
MW: 558.49   **BP** (°C):

Solubility (Moles/L)	Solubility (Grams/L)	Temp (°C)	Ref (#)	Evaluation (T P E A A)	Comments
2.298E-02	1.283E+01	20	P048	1 2 1 1 1	

3604. C$_{18}$H$_{31}$O$_4$P
Butyl Octyl Phenyl Phosphate
RN: 110459-55-7   **MP** (°C):
MW: 342.42   **BP** (°C):

Solubility (Moles/L)	Solubility (Grams/L)	Temp (°C)	Ref (#)	Evaluation (T P E A A)	Comments
<5.84E-04	<2.00E-01	25	B070	1 2 0 1 0	

3605. C$_{18}$H$_{32}$O$_7$
Tributyl Citrate
Tri-n-butyl Citrate
Butyl Citrate
RN: 77-94-1   **MP** (°C):   -20
MW: 360.45   **BP** (°C):

Solubility (Moles/L)	Solubility (Grams/L)	Temp (°C)	Ref (#)	Evaluation (T P E A A)	Comments
1.664E-04	6.000E-02	15	H069	1 0 1 1 0	
2.219E-04	7.999E-02	ns	F014	0 0 0 0 0	

3606.   C$_{18}$H$_{32}$O$_{16}$

Raffinose

6G-α-D-Galactosylsucrose

Melitose

Gossypose

Melitriose

RN: 512-69-6   **MP** (°C):   80.0

MW: 504.45   **BP** (°C):

Solubility (Moles/L)	Solubility (Grams/L)	Temp (°C)	Ref (#)	Evaluation (T P E A A)	Comments
6.518E-02	3.288E+01	0.0	H040	1 2 2 2 1	
6.556E-02	3.307E+01	0.02	H040	1 2 2 2 2	
1.227E-01	6.191E+01	10.00	H040	1 2 2 2 1	
1.879E-01	9.478E+01	16.38	H040	1 2 2 2 2	
1.937E-01	9.772E+01	16.90	H040	1 2 2 2 2	
2.480E-01	1.251E+02	20	D041	1 0 0 0 2	
2.373E-01	1.197E+02	20.00	H040	1 2 2 2 2	
3.192E-01	1.610E+02	24.80	H040	1 2 2 2 2	
4.555E-01	2.298E+02	25	P049	1 0 1 1 1	
3.228E-01	1.628E+02	25.05	H040	1 2 2 2 2	
3.340E-01	1.685E+02	25.50	H040	1 2 2 2 2	
4.227E-01	2.132E+02	30.00	H040	1 2 2 2 2	
6.398E-01	3.227E+02	39.38	H040	1 2 2 2 2	
6.599E-01	3.329E+02	40.00	H040	1 2 2 2 2	
9.217E-01	4.650E+02	50.00	H040	1 2 2 2 2	
1.016E+00	5.125E+02	53.20	H040	1 2 2 2 2	
1.201E+00	6.060E+02	60.00	H040	1 2 2 2 2	
1.239E+00	6.250E+02	61.60	H040	1 2 2 2 2	
1.473E+00	7.430E+02	70.00	H040	1 2 2 2 2	
1.682E+00	8.484E+02	78.00	H040	1 2 2 2 2	
2.480E-01	1.251E+02	rt	D021	0 0 1 1 2	

3607.   C$_{18}$H$_{32}$O$_{16}$.5H$_2$O

Raffinose (Pentahydrate)

6G-α-D-Galactosylsucrose (Pentahydrate)

RN: 17629-30-0   **MP** (°C):   80

MW: 594.52   **BP** (°C):

Solubility (Moles/L)	Solubility (Grams/L)	Temp (°C)	Ref (#)	Evaluation (T P E A A)	Comments
5.531E-02	3.288E+01	0	M043	1 0 0 0 1	
1.041E-01	6.191E+01	10	M043	1 0 0 0 1	
2.014E-01	1.197E+02	20	M043	1 0 0 0 2	
3.586E-01	2.132E+02	30	M043	1 0 0 0 2	
5.599E-01	3.329E+02	40	M043	1 0 0 0 2	
7.821E-01	4.650E+02	60	M043	1 0 0 0 2	
1.019E+00	6.060E+02	80	M043	1 0 0 0 2	

3608. C₁₈H₃₄OSn
Cyhexatin
Tricyclohexylhydroxystannane
Tricyclohexyltin Hydroxide
Plictran
Dowco 213
RN: 13121-70-5 **MP** (°C): 196.5
MW: 385.16 **BP** (°C):

Solubility (Moles/L)	Solubility (Grams/L)	Temp (°C)	Ref (#)	Evaluation (T P E A A)	Comments
<2.60E-06	<1.00E-03	25	M161	1 0 0 0 0	
<2.60E-06	<1.00E-03	ns	K138	0 0 0 0 1	

3609. C₁₈H₃₄O₄
Dibutyl Sebacate
Di-n-butyl Sebacate
Decanedioic Acid Dibutyl Ester
Dibutyl Decanedioate
RN: 109-43-3 **MP** (°C):
MW: 314.47 **BP** (°C):

Solubility (Moles/L)	Solubility (Grams/L)	Temp (°C)	Ref (#)	Evaluation (T P E A A)	Comments
1.590E-04	5.000E-02	ns	F014	0 0 0 0 0	

3610. C₁₈H₃₆O₂
Stearic Acid
Stearinsaeure
Octadecanoic Acid
RN: 57-11-4 **MP** (°C): 70
MW: 284.49 **BP** (°C):

Solubility (Moles/L)	Solubility (Grams/L)	Temp (°C)	Ref (#)	Evaluation (T P E A A)	Comments
6.327E-06	1.800E-03	0	B136	1 0 2 1 1	
6.327E-06	1.800E-03	0.0	R001	1 1 1 1 1	
9.842E-06	2.800E-03	20	B136	1 0 2 1 1	
1.055E-05	3.000E-03	20	F300	1 0 0 0 0	
1.019E-05	2.900E-03	20.0	R001	1 1 1 1 1	
2.100E-06	5.974E-04	25	J001	1 0 2 1 1	
1.970E-06	5.604E-04	25	R002	1 2 2 2 2	
1.195E-05	3.400E-03	30	B136	1 0 2 1 1	
1.195E-05	3.400E-03	30.0	R001	1 1 1 1 1	
1.700E-05	4.836E-03	35	M004	2 0 0 0 2	
1.476E-05	4.200E-03	45	B136	1 0 2 1 1	
1.476E-05	4.200E-03	45.0	R001	1 1 1 1 1	
2.700E-06	7.681E-04	50	J001	1 0 2 1 1	
5.770E-05	1.641E-02	50	M004	2 0 0 0 2	

1.758E-05	5.000E-03	60	B136	1 0 2 1 1
1.758E-05	5.000E-03	60	F300	1 0 0 0 0
1.758E-05	5.000E-03	60.0	R001	1 1 1 1 1
1.145E-05	3.257E-03	62.5	M004	1 0 0 0 2

3611. $C_{18}H_{38}$
n-Octadecane
Octadecane

RN: 593-45-3   **MP** (°C):   29.5
MW: 254.50   **BP** (°C):   317.0

Solubility (Moles/L)	Solubility (Grams/L)	Temp (°C)	Ref (#)	Evaluation (T P E A A)	Comments
4.715E-07	1.200E-04	10	C331	1 2 0 2 1	
2.358E-08	6.000E-06	25	B069	1 0 1 1 1	
2.240E-08	5.700E-06	25	B069	1 0 1 1 1	
5.894E-07	1.500E-04	30	C331	1 2 0 2 1	
6.680E-07	1.700E-04	60	C331	1 2 0 2 1	
3.045E-08	7.750E-06	ns	B003	0 2 2 2 2	
3.045E-08	7.750E-06	ns	B033	0 0 0 0 2	

3612. $C_{18}H_{38}O$
Octadecanol
Stearyl Alcohol
Octadecyl Alcohol
Steraffine

RN: 112-92-5   **MP** (°C):   61
MW: 270.50   **BP** (°C):   336

Solubility (Moles/L)	Solubility (Grams/L)	Temp (°C)	Ref (#)	Evaluation (T P E A A)	Comments
4.000E-09	1.082E-06	34	K011	1 2 1 1 1	
2.200E-08	5.951E-06	65	K011	1 2 1 1 1	

3613. $C_{18}H_{39}N.2H_2O$
Octadecylamine (Dihydrate)
1-Aminooctadecane (Dihydrate)

RN: 124-30-1   **MP** (°C):
MW: 305.55   **BP** (°C):

Solubility (Moles/L)	Solubility (Grams/L)	Temp (°C)	Ref (#)	Evaluation (T P E A A)	Comments
5.891E-09	1.800E-06	ns	R037	0 2 2 1 1	

3614. C$_{18}$H$_{39}$O$_3$P
Dibutyl Decyl Phosphonate
RN: 36378-71-9   **MP** (°C):
MW: 334.48   **BP** (°C):

Solubility (Moles/L)	Solubility (Grams/L)	Temp (°C)	Ref (#)	Evaluation (T P E A A)	Comments
<5.98E-04	<2.00E-01	25	B070	1 2 0 1 0	

3615. C$_{18}$H$_{39}$O$_4$P
Dibutyl Decyl Phosphate
RN: 111440-78-9   **MP** (°C):
MW: 350.48   **BP** (°C):

Solubility (Moles/L)	Solubility (Grams/L)	Temp (°C)	Ref (#)	Evaluation (T P E A A)	Comments
<2.85E-04	<1.00E-01	25	B070	1 2 0 1 0	

3616. C$_{18}$H$_{39}$O$_7$P
Tributoxyethyl Phosphate
RN: 78-51-3   **MP** (°C): -70
MW: 398.48   **BP** (°C):

Solubility (Moles/L)	Solubility (Grams/L)	Temp (°C)	Ref (#)	Evaluation (T P E A A)	Comments
2.760E-03	1.100E+00	25	B070	1 2 0 1 1	

3617. C$_{19}$H$_{12}$O$_6$
Dicumarol
3,3'-Methylene-bis(4-hydroxycoumarin)
Dicoumarol
RN: 66-76-2   **MP** (°C): 290
MW: 336.30   **BP** (°C):

Solubility (Moles/L)	Solubility (Grams/L)	Temp (°C)	Ref (#)	Evaluation (T P E A A)	Comments
<4.46E-04	<1.50E-01	25	P312	1 2 2 2 2	

3618. C$_{19}$H$_{13}$Cl
6-Chloro-10-methyl-1,2-benzanthracene
RN: 188124-97-2   **MP** (°C):
MW: 276.77   **BP** (°C):

Solubility (Moles/L)	Solubility (Grams/L)	Temp (°C)	Ref (#)	Evaluation (T P E A A)	Comments
3.613E-08	1.000E-05	27	D003	1 0 0 1 0	

3619.   C$_{19}$H$_{13}$Cl
4-Fluoro-10-methyl-1,2-benzanthracene
4-FMBA
RN: 2990-70-7   **MP** (°C):
MW: 276.77   **BP** (°C):

Solubility (Moles/L)	Solubility (Grams/L)	Temp (°C)	Ref (#)	Evaluation (T P E A A)	Comments
1.900E-08	5.259E-06	22	B062	1 0 2 1 2	

3620.   C$_{19}$H$_{13}$Cl
3-Fluoro-10-methyl-1,2-benzanthracene
3-FMBA
RN: 20629-50-9   **MP** (°C):
MW: 276.77   **BP** (°C):

Solubility (Moles/L)	Solubility (Grams/L)	Temp (°C)	Ref (#)	Evaluation (T P E A A)	Comments
1.900E-08	5.259E-06	22	B062	1 0 2 1 2	

3621.   C$_{19}$H$_{14}$
1'-Methyl-1,2-benzanthracene
RN: 2498-77-3   **MP** (°C):   138
MW: 242.32   **BP** (°C):

Solubility (Moles/L)	Solubility (Grams/L)	Temp (°C)	Ref (#)	Evaluation (T P E A A)	Comments
2.270E-07	5.500E-05	27	D003	1 0 0 1 2	

3622.   C$_{19}$H$_{14}$
10-Methyl-1,2-benzanthracene
RN: 2541-69-7   **MP** (°C):   141
MW: 242.32   **BP** (°C):

Solubility (Moles/L)	Solubility (Grams/L)	Temp (°C)	Ref (#)	Evaluation (T P E A A)	Comments
4.539E-08	1.100E-05	24	H116	2 1 0 0 1	

3623.   C$_{19}$H$_{14}$
9-Methyl-1,2-benzanthracene
RN: 2381-16-0   **MP** (°C):   138
MW: 242.32   **BP** (°C):

Solubility (Moles/L)	Solubility (Grams/L)	Temp (°C)	Ref (#)	Evaluation (T P E A A)	Comments
1.527E-07	3.700E-05	24	H116	2 1 0 0 1	

3624.   C₁₉H₁₄
5-Methylchrysene
RN: 3697-24-3   **MP** (°C):   117.1
MW: 242.32   **BP** (°C):

Solubility (Moles/L)	Solubility (Grams/L)	Temp (°C)	Ref (#)	Evaluation (T P E A A)	Comments
2.559E-07	6.200E-05	27	D003	1 0 0 1 1	

3625.   C₁₉H₁₄
6-Methylchrysene
RN: 1705-85-7   **MP** (°C):   149
MW: 242.32   **BP** (°C):

Solubility (Moles/L)	Solubility (Grams/L)	Temp (°C)	Ref (#)	Evaluation (T P E A A)	Comments
2.682E-07	6.500E-05	27	D003	1 0 0 1 1	

3626.   C₁₉H₁₄O₃
Aurin
Rosolic Acid
4-[bis-(p-Hydroxyphenyl)methylene]-2,5-cyclohexadien-1-one
RN: 603-45-2   **MP** (°C):
MW: 290.32   **BP** (°C):

Solubility (Moles/L)	Solubility (Grams/L)	Temp (°C)	Ref (#)	Evaluation (T P E A A)	Comments
4.128E-03	1.199E+00	rt	D021	0 0 1 1 1	

3627.   C₁₉H₁₄O₅S
Phenolsulfonaphthalein
Phenolrot
RN: 143-74-8   **MP** (°C):   >300
MW: 354.38   **BP** (°C):

Solubility (Moles/L)	Solubility (Grams/L)	Temp (°C)	Ref (#)	Evaluation (T P E A A)	Comments
8.748E-04	3.100E-01	100	F300	1 0 0 0 2	

3628.   C$_{19}$H$_{16}$CINO$_4$
Indomethacin
Indomethacine,Form IV
RN: 53-86-1   **MP** (°C):   134.0
MW: 357.80   **BP** (°C):

Solubility (Moles/L)	Solubility (Grams/L)	Temp (°C)	Ref (#)	Evaluation (T P E A A)	Comments
5.590E-06	2.000E-03	15	N314	1 1 1 1 0	EFG
2.376E-05	8.500E-03	25	B072	2 2 1 1 0	EFG
1.677E-05	6.000E-03	25	B072	2 2 1 1 0	EFG
1.118E-05	4.000E-03	25	B072	2 2 1 1 0	EFG
3.913E-05	1.400E-02	25	K026	2 0 2 2 1	
2.795E-05	1.000E-02	25	K027	2 0 2 2 0	
2.620E-06	9.374E-04	25	M149	2 2 2 2 2	intrinsic
1.397E-05	5.000E-03	25	N314	1 1 1 1 0	EFG
2.515E-05	9.000E-03	30	D015	2 2 1 1 0	EFG
3.500E-04	1.252E-01	35	H091	1 2 2 2 1	sic
4.472E-05	1.600E-02	37	D015	2 2 1 1 0	EFG
2.795E-05	1.000E-02	40	N314	1 1 1 1 0	EFG
6.652E-05	2.380E-02	50	N314	1 1 1 1 0	EFG
1.010E-03	3.614E-01	ns	O304	0 0 1 2 2	
2.515E-05	9.000E-03	rt	H302	0 0 2 1 1	intrinsic

3629.   C$_{19}$H$_{16}$O
Triphenylcarbinol
Triphenylmethanol
RN: 76-84-6   **MP** (°C):   164.2
MW: 260.34   **BP** (°C):

Solubility (Moles/L)	Solubility (Grams/L)	Temp (°C)	Ref (#)	Evaluation (T P E A A)	Comments
5.500E-03	1.432E+00	25	D007	2 0 1 1 2	

3630.   C$_{19}$H$_{16}$O$_4$
Warfarin
3-(1'-Phenyl-2'-acetylethyl)-4-hydroxycoumarin
Rosex
Kypfarin
RN: 81-81-2   **MP** (°C):   161
MW: 308.34   **BP** (°C):

Solubility (Moles/L)	Solubility (Grams/L)	Temp (°C)	Ref (#)	Evaluation (T P E A A)	Comments
1.297E-04	4.000E-02	ns	C036	0 0 0 0 0	

3631. $C_{19}H_{17}ClN_2O$
Prazepam
Centrax
7-Chloro-1-(cyclopropylmethyl)-1,3-dihydro-5-phenyl-2H-1,4-benzodiazepin-2-one
Demetrin
Verstran
RN: 2955-38-6   **MP** (°C):
MW: 324.81   **BP** (°C):

Solubility (Moles/L)	Solubility (Grams/L)	Temp (°C)	Ref (#)	Evaluation (T P E A A)	Comments
2.800E-05	9.095E-03	25	M320	2 2 1 1 2	

3632. $C_{19}H_{17}ClN_2O_4$
Glafenine
N-(7-Chloro-4-quinolyl)anthranilate
2,3-Dihydroxypropyl-N-(7-chloro-4-quinolinyl)anthranilate
RN: 3820-67-5   **MP** (°C):   169.5
MW: 372.81   **BP** (°C):

Solubility (Moles/L)	Solubility (Grams/L)	Temp (°C)	Ref (#)	Evaluation (T P E A A)	Comments
1.032E-01	3.846E+01	ns	M152	0 0 0 0 0	pH 1.0, intrinsic

3633. $C_{19}H_{17}N_3O_4S_2$
Cephaloridine
Glaxoridin
Keflodin
Loridine
RN: 50-59-9   **MP** (°C):   184
MW: 415.49   **BP** (°C):

Solubility (Moles/L)	Solubility (Grams/L)	Temp (°C)	Ref (#)	Evaluation (T P E A A)	Comments
>4.81E-02	>2.00E+01	21	M044	2 0 2 2 0	

3634. $C_{19}H_{17}N_3O_4S_2$
Sugordomycin
RN: 1405-50-1   **MP** (°C):
MW: 415.49   **BP** (°C):

Solubility (Moles/L)	Solubility (Grams/L)	Temp (°C)	Ref (#)	Evaluation (T P E A A)	Comments
2.304E-02	9.572E+00	21	M044	2 0 2 2 2	

3635.   C$_{19}$H$_{17}$N$_3$O$_5$

1H-Benzimidazole-1-carboxylic Acid, 6-Benzoyl-2-[(methoxycarbonyl)amino]-, Ethyl Ester

RN: 153474-30-7   **MP** (°C):   165.5
MW: 367.36   **BP** (°C):

Solubility (Moles/L)	Solubility (Grams/L)	Temp (°C)	Ref (#)	Evaluation (T P E A A)	Comments
2.722E-05	1.000E-02	21	N337	1 0 2 2 0	pH 5
2.700E-05	9.919E-03	21	N337	1 0 2 2 0	pH 5

3636.   C$_{19}$H$_{18}$

1,2,3,4-Tetrahydro-10-methyl-1,2-benzanthracene
10-Methyl-1,2-cyclohexane Anthracene

RN: 6366-18-3   **MP** (°C):   117
MW: 246.36   **BP** (°C):

Solubility (Moles/L)	Solubility (Grams/L)	Temp (°C)	Ref (#)	Evaluation (T P E A A)	Comments
1.786E-07	4.400E-05	27	D003	1 0 0 1 1	

3637.   C$_{19}$H$_{18}$Cl$_2$N$_2$O$_2$

G-20
p,p-Dichlorophenylbutazone

RN: 4047-57-8   **MP** (°C):
MW: 377.27   **BP** (°C):

Solubility (Moles/L)	Solubility (Grams/L)	Temp (°C)	Ref (#)	Evaluation (T P E A A)	Comments
2.386E-04	9.000E-02	ns	B158	0 0 0 0 1	pH 7.0

3638.   C$_{19}$H$_{18}$N$_2$O$_3$

Kebuzone
3,5-Pyrazolidinedione

RN: 853-34-9   **MP** (°C):   128
MW: 322.37   **BP** (°C):

Solubility (Moles/L)	Solubility (Grams/L)	Temp (°C)	Ref (#)	Evaluation (T P E A A)	Comments
5.402E-04	1.742E-01	20	M140	2 0 1 1 1	

3639. C$_{19}$H$_{18}$N$_2$O$_3$
G-23
1-Oxybutylphenylbutazone
3,5-Pyrazolidinedione, 4-Butyryl-1,2-diphenyl-
RN: 13167-98-1   **MP** (°C):
MW: 322.37   **BP** (°C):

Solubility (Moles/L)	Solubility (Grams/L)	Temp (°C)	Ref (#)	Evaluation (T P E A A)	Comments
3.722E-04	1.200E-01	ns	B158	0 0 0 0 1	pH 7.0

3640. C$_{19}$H$_{19}$ClFNO$_3$
Flamprop-isopropyl
Flufenprop-isopropyl
Isopropyl N-Benzoyl-N-(3-chloro-4-fluorophenyl)alanine
1-Methylethyl N-Benzoyl-N-(3-chloro-4-fluorophenyl)-DL-alanine
RN: 52756-22-6   **MP** (°C):   56.5
MW: 363.82   **BP** (°C):

Solubility (Moles/L)	Solubility (Grams/L)	Temp (°C)	Ref (#)	Evaluation (T P E A A)	Comments
4.948E-05	1.800E-02	20	M161	1 0 0 0 0	

3641. C$_{19}$H$_{19}$N$_7$O$_6$
Folic Acid
N-(p-(((2-Amino-4-hydroxy-6-pteridinyl)methyl)amino)benzoyl)-L-glutamic Acid
Vitamin M
Pteroylglutamic Acid
Folcysteine
Folacin
RN: 59-30-3   **MP** (°C):
MW: 441.41   **BP** (°C):

Solubility (Moles/L)	Solubility (Grams/L)	Temp (°C)	Ref (#)	Evaluation (T P E A A)	Comments
3.619E-03	1.597E+00	25	D041	1 0 0 0 1	sic
3.625E-06	1.600E-03	25	D315	1 0 1 1 2	
2.243E-02	9.901E+00	100	D041	1 0 0 0 0	sic

3642. C$_{19}$H$_{20}$ClNO$_9$
Griseofulvin-4-carboxy-methoxime
RN: **MP** (°C):
MW: 441.83   **BP** (°C):

Solubility (Moles/L)	Solubility (Grams/L)	Temp (°C)	Ref (#)	Evaluation (T P E A A)	Comments
1.704E-04	7.529E-02	37	F033	2 0 2 0 2	

3643.   C₁₉H₂₀N₂O

Cinchoninone
Cinchoninon
9-Deoxy-9-oxocinchonine
RN: 14509-68-3   **MP** (°C):
MW: 292.38   **BP** (°C):

Solubility (Moles/L)	Solubility (Grams/L)	Temp (°C)	Ref (#)	Evaluation (T P E A A)	Comments
6.498E-04	1.900E-01	20	F300	1 0 0 0 1	

3644.   C₁₉H₂₀N₂O₂

Phenylbutazone
1,2-Diphenyl-4-butyl-3,5-dioxopyrazolidine
Butazolidin
Equiphen
Butazone
RN: 50-33-9   **MP** (°C):   107
MW: 308.38   **BP** (°C):

Solubility (Moles/L)	Solubility (Grams/L)	Temp (°C)	Ref (#)	Evaluation (T P E A A)	Comments
8.415E-05	2.595E-02	20	H301	2 0 2 2 2	
4.864E-05	1.500E-02	20	P026	1 0 1 1 1	
1.102E-04	3.400E-02	25	P096	1 0 2 2 2	
1.540E-04	4.750E-02	30	D015	2 0 1 1 0	EFG
1.000E-03	3.084E-01	35	H091	1 2 2 2 1	*sic*
9.362E-03	2.887E+00	36	I002	2 2 1 1 2	pH 6.95, recrystallized
9.076E-03	2.799E+00	36	I002	2 2 1 1 2	pH 6.95, recrystallized
7.575E-03	2.336E+00	36	I002	2 2 1 1 2	pH 6.95, recrystallized
6.907E-03	2.130E+00	36	I002	2 2 1 1 2	pH 6.95, recrystallized
2.108E-04	6.500E-02	37	D015	2 0 1 1 0	EFG
1.816E-04	5.600E-02	37	E047	1 0 1 1 1	
7.134E-03	2.200E+00	ns	B158	0 0 0 0 1	pH 7.0
1.300E-04	4.009E-02	ns	O304	0 0 1 2 2	
2.594E-05	8.000E-03	rt	H302	0 0 2 1 2	intrinsic
1.310E-01	4.040E+01	rt	N056	0 0 1 1 2	average of 2

3645.   C₁₉H₂₀N₂O₂

G-21
p,p-Dimethylphenylbutazone
RN: 745-27-7   **MP** (°C):
MW: 308.38   **BP** (°C):

Solubility (Moles/L)	Solubility (Grams/L)	Temp (°C)	Ref (#)	Evaluation (T P E A A)	Comments
3.891E-04	1.200E-01	ns	B158	0 0 0 0 1	pH 7.0

3646.   C$_{19}$H$_{20}$N$_2$O$_3$
Oxyphenbutazone
p-Hydroxyphenylbutazone
RN: 129-20-4   **MP** (°C):   124
MW: 324.38   **BP** (°C):

Solubility (Moles/L)	Solubility (Grams/L)	Temp (°C)	Ref (#)	Evaluation (T P E A A)	Comments
1.850E-04	6.000E-02	30	D015	2 0 1 1 0	EFG
2.497E-04	8.100E-02	37	D015	2 0 1 1 0	EFG
3.083E-02	1.000E+01	ns	B158	0 0 0 0 1	pH 7.0, *sic*
6.166E-05	2.000E-02	rt	H302	0 0 2 1 2	intrinsic

3647.   C$_{19}$H$_{20}$N$_4$O$_6$.0.1H$_2$O
9-[5-O-(Benzyl Formyl-β-D-arabinofuranosyl)]-6-methoxy-9H-purine (0.1 Hydrate)
RN: 121032-36-8   **MP** (°C):   foam
MW: 402.20   **BP** (°C):

Solubility (Moles/L)	Solubility (Grams/L)	Temp (°C)	Ref (#)	Evaluation (T P E A A)	Comments
1.050E-02	4.223E+00	37	M378	1 2 1 1 2	pH 7.2

3648.   C$_{19}$H$_{20}$N$_4$O$_6$.0.5H$_2$O
6-Methoxy-9-(5-O-[4-methylbenzoyl]-β-D-arabinofuranosyl)-9H-purine (Hemihydrate)
RN: 121032-20-0   **MP** (°C):   127-128
MW: 409.40   **BP** (°C):

Solubility (Moles/L)	Solubility (Grams/L)	Temp (°C)	Ref (#)	Evaluation (T P E A A)	Comments
3.500E-05	1.433E-02	37	M378	1 2 1 1 2	pH 7.2

3649.   C$_{19}$H$_{20}$N$_4$O$_6$
2'-(p-Toluylyl)-6-methoxypurine Arabinoside
2'-Phenylacetyl-6-methoxypurine Arabinoside
RN: 121032-52-8   **MP** (°C):   69-73
MW: 400.39   **BP** (°C):

Solubility (Moles/L)	Solubility (Grams/L)	Temp (°C)	Ref (#)	Evaluation (T P E A A)	Comments
5.870E-02	2.350E+01	37	C348	1 2 2 2 2	pH 7.00
5.840E-03	2.338E+00	37	C348	1 2 2 2 2	pH 7.00

3650.   C₁₉H₂₀N₄O₇.0.25H₂O

9-[5-O-(4-Methoxybenzoyl-β-D-arabinofuranosyl)]-6-methoxy-9H-purine (0.25 Hydrate)

RN: 121032-35-7   **MP** (°C):   195-197
MW: 420.90   **BP** (°C):

Solubility (Moles/L)	Solubility (Grams/L)	Temp (°C)	Ref (#)	Evaluation (T P E A A)	Comments
1.960E-04	8.250E-02	37	M378	1 2 1 1 2	pH 7.2

3651.   C₁₉H₂₀N₄O₇.0.05H₂O

9-[5-O-(Benzyl Acetate-β-D-arabinofuranosyl)]-6-methoxy-9H-purine (0.05 Hydrate)

RN: 121032-37-9   **MP** (°C):   193-195
MW: 417.29   **BP** (°C):

Solubility (Moles/L)	Solubility (Grams/L)	Temp (°C)	Ref (#)	Evaluation (T P E A A)	Comments
3.930E-04	1.640E-01	37	M378	1 2 1 1 2	pH 7.2

3652.   C₁₉H₂₀N₄O₇.0.5H₂O

2'-Phenoxyacetyl-6-methoxypurine Arabinoside (Hemihydrate)

RN: 145913-46-8   **MP** (°C):   123-125
MW: 425.40   **BP** (°C):

Solubility (Moles/L)	Solubility (Grams/L)	Temp (°C)	Ref (#)	Evaluation (T P E A A)	Comments
>2.21E-02	>9.40E+00	37	C348	1 2 2 2 2	pH 7.00

3653.   C₁₉H₂₀N₄O₇

2'-(p-Methoxybenzoyl)-6-methoxypurine Arabinoside

RN: 121032-51-7   **MP** (°C):   71-75
MW: 416.39   **BP** (°C):

Solubility (Moles/L)	Solubility (Grams/L)	Temp (°C)	Ref (#)	Evaluation (T P E A A)	Comments
6.660E-03	2.773E+00	37	C348	1 2 2 2 2	pH 7.00

3654. $C_{19}H_{20}O_4$
Butylbenzyl Phthalate
Butyl Phenyl-methyl Phthalate
Benzylbutyl Phthalate
Phthalate Butyl Benzyl Ester
Butyl Benzyl Phthalate
1,2-Benzenedicarboxylic Acid Butyl Phenylmethyl Ester
RN: 85-68-7 **MP** (°C): <-35
MW: 312.37 **BP** (°C): 370

Solubility (Moles/L)	Solubility (Grams/L)	Temp (°C)	Ref (#)	Evaluation (T P E A A)	Comments
9.020E-06	2.818E-03	20	L300	2 1 0 2 2	
2.273E-06	7.100E-04	24	H116	2 1 0 0 2	
8.644E-06	2.700E-03	25	F067	1 0 2 2 1	

3655. $C_{19}H_{21}ClO_4$
Isobutyl (+/-)-2-[4-(4-Chlorophenoxy)phenoxy]propionate
RN: 51337-71-4 **MP** (°C): 39.5
MW: 348.83 **BP** (°C):

Solubility (Moles/L)	Solubility (Grams/L)	Temp (°C)	Ref (#)	Evaluation (T P E A A)	Comments
5.160E-04	1.800E-01	22	M161	1 0 0 0 2	

3656. $C_{19}H_{21}F_3N_2S$
2-Trifluoromethyl-N,N-dimethyl-10H-phenothiazine-10-propanamide
RN: 2340-66-1 **MP** (°C):
MW: 366.45 **BP** (°C):

Solubility (Moles/L)	Solubility (Grams/L)	Temp (°C)	Ref (#)	Evaluation (T P E A A)	Comments
5.000E-06	1.832E-03	ns	G023	0 0 1 1 0	

3657. $C_{19}H_{21}NO$
Doxepin
Adapin
Deptran
Sinequan
RN: 1668-19-5 **MP** (°C): 120
MW: 279.39 **BP** (°C):

Solubility (Moles/L)	Solubility (Grams/L)	Temp (°C)	Ref (#)	Evaluation (T P E A A)	Comments
1.130E-04	3.157E-02	25	E051	1 0 2 1 2	

3658.   C$_{19}$H$_{21}$NO$_3$
Thebaine
Paramorphine
Morphinan, 6,7,8,14-Tetradehydro-4,5α-epoxy-3,6-dimethoxy-17-methyl-
RN: 115-37-7   **MP** (°C):
MW: 311.38   **BP** (°C):

Solubility (Moles/L)	Solubility (Grams/L)	Temp (°C)	Ref (#)	Evaluation (T P E A A)	Comments
2.200E-03	6.850E-01	15	K059	2 2 2 0 1	

3659.   C$_{19}$H$_{21}$N$_5$O$_2$
Dye VII
4-[[(4-(N-Butyl-N-ethylnitrile)amino)phenyl]azo]nitrobenzene
RN: **MP** (°C):
MW: 351.41   **BP** (°C):

Solubility (Moles/L)	Solubility (Grams/L)	Temp (°C)	Ref (#)	Evaluation (T P E A A)	Comments
4.800E-07	1.687E-04	71.80	B198	1 2 1 1 1	
9.700E-07	3.409E-04	84.10	B198	1 2 1 1 1	
2.020E-06	7.099E-04	97.40	B198	1 2 1 1 2	

3660.   C$_{19}$H$_{21}$N$_5$O$_2$
Dis. A. 6
Propanenitrile, 3-[Butyl[4-[(4-nitrophenyl)azo]phenyl]amino]-
RN: 69472-19-1   **MP** (°C):   118
MW: 351.41   **BP** (°C):

Solubility (Moles/L)	Solubility (Grams/L)	Temp (°C)	Ref (#)	Evaluation (T P E A A)	Comments
2.000E-08	7.028E-06	25	B333	1 0 0 0 1	

3661.   C19H21N5O5
9-[5'-(O-Hydrocinnamoyl)-☐-D-arabinofuranosyl]adenine Ester
RN: 68325-41-7   MP (°C):
MW: 399.41   BP (°C):

Solubility (Moles/L)	Solubility (Grams/L)	Temp (°C)	Ref (#)	Evaluation (T P E A A)	Comments
3.756E-03	1.500E+00	ns	B134	0 1 1 1 1	

3662.   C19H22Cl2O2
1-Methyl-1,1-dichloro-2,2-bis(p-ethoxylphenyl)ethane

RN:	56265-23-7	**MP (°C):**	
MW:	353.29	**BP (°C):**	

Solubility (Moles/L)	Solubility (Grams/L)	Temp (°C)	Ref (#)	Evaluation (T P E A A)	Comments
1.415E-07	5.000E-05	rt	C122	0 2 2 2 2	

3663.   C19H22N2O
Cinchonine
Cinchonan-9-ol
(+)-Cinchonine
(9S)-Cinchonan-9-ol

RN:	118-10-5	**MP (°C):**	265
MW:	294.40	**BP (°C):**	

Solubility (Moles/L)	Solubility (Grams/L)	Temp (°C)	Ref (#)	Evaluation (T P E A A)	Comments
4.800E-06	1.413E-03	15	K059	2 2 2 0 1	
9.253E-04	2.724E-01	25	D004	1 0 0 0 0	
9.171E-04	2.700E-01	100	F300	1 0 0 0 1	
8.150E-04	2.399E-01	rt	D021	0 0 1 1 1	

3664. $C_{19}H_{22}N_2O$
Cinchonidine
Cinchonidin
(8α,9R)-Cinchonan-9-ol
L-Cinchonidine

RN:	485-71-2	**MP (°C):**	210
MW:	294.40	**BP (°C):**	

Solubility (Moles/L)	Solubility (Grams/L)	Temp (°C)	Ref (#)	Evaluation (T P E A A)	Comments
9.000E-04	2.650E-01	15	K059	2 2 2 0 0	
6.793E-04	2.000E-01	25	F300	1 0 0 0 0	
1.970E-03	5.800E-01	100	F300	1 0 0 0 1	
6.792E-04	2.000E-01	c	D004	1 0 0 0 0	
8.490E-04	2.499E-01	rt	D021	0 0 1 1 1	

3665.   C₁₉H₂₂N₂OS
Acetylpromazine
3-Acetyl-10-(3-dimethylaminopropyl)phenothiazine
Plegicil
Vetranquil
Notensil
Plivafen
RN: 61-00-7   **MP** (°C):
MW: 326.46   **BP** (°C):

Solubility (Moles/L)	Solubility (Grams/L)	Temp (°C)	Ref (#)	Evaluation (T P E A A)	Comments
4.901E-05	1.600E-02	25	L045	1 1 1 1 2	intrinsic

3666.   C₁₉H₂₂N₂O₅
2-Naphthaleneacetic Acid, 6-Methoxy-α-methyl-, 2-[(2-Amino-2-oxoethyl)methylamino]-2-oxoethyl Ester
Naproxen N-Methyl -N-Carbamoyl Methyl Glycolamide Ester
RN: 114681-69-5 **MP** (°C):   179
MW: 358.40   **BP** (°C):

Solubility (Moles/L)	Solubility (Grams/L)	Temp (°C)	Ref (#)	Evaluation (T P E A A)	Comments
1.646E-04	5.900E-02	21	B331	1 2 2 1 1	

3667.   C₁₉H₂₂N₂S
Mepazine
Pecazine
RN: 60-89-9   **MP** (°C):   80
MW: 310.46   **BP** (°C):   233

Solubility (Moles/L)	Solubility (Grams/L)	Temp (°C)	Ref (#)	Evaluation (T P E A A)	Comments
1.800E-05	5.588E-03	24	G022	2 0 1 1 1	

3668.   C₁₉H₂₃ClO₂
1-Chloro-1-methyl-2,2-bis(p-ethoxylphenyl)ethane
RN: 56265-22-6 **MP** (°C):
MW: 318.85   **BP** (°C):

Solubility (Moles/L)	Solubility (Grams/L)	Temp (°C)	Ref (#)	Evaluation (T P E A A)	Comments
2.760E-06	8.800E-04	rt	C122	0 2 2 2 2	

3669. C₁₉H₂₃NO₃

3669. $C_{19}H_{23}NO_3$

Ethylmorphine

7,8-Didehydro-4,5-epoxy-3-ethoxy-17-methylmorphinan-6-ol

RN: 76-58-4   **MP** (°C):

MW: 313.40   **BP** (°C):

Solubility (Moles/L)	Solubility (Grams/L)	Temp (°C)	Ref (#)	Evaluation (T P E A A)	Comments
8.916E-03	2.794E+00	20	K052	1 1 1 1 2	

3670. $C_{19}H_{23}NO_4$

1-Methyl-1-nitro-2,2-bis(p-ethoxylphenyl)ethane

RN: 26258-70-8   **MP** (°C):

MW: 329.40   **BP** (°C):

Solubility (Moles/L)	Solubility (Grams/L)	Temp (°C)	Ref (#)	Evaluation (T P E A A)	Comments
1.093E-06	3.600E-04	rt	C122	0 2 2 2 2	

3671. $C_{19}H_{23}NO_5$

2-Naphthaleneacetic Acid, 6-Methoxy-α-methyl-, 2-[(2-Hydroxyethyl)methylamino]-2-oxoethyl Ester

Naproxen N-Methyl-N-Ethanol Glycolamide Ester

RN: 114665-19-9   **MP** (°C):   110

MW: 345.40   **BP** (°C):

Solubility (Moles/L)	Solubility (Grams/L)	Temp (°C)	Ref (#)	Evaluation (T P E A A)	Comments
4.053E-04	1.400E-01	21	B331	1 2 2 1 1	

3672. $C_{19}H_{23}N_3$

Amitraz

1,5-Di(2,4-dimethylphenyl)-3-methyl-1,3,5-triazapenta-1,4-diene

Ovasyn

Mitac

Triazid

Baam

RN: 33089-61-1   **MP** (°C):   86.5

MW: 293.42   **BP** (°C):

Solubility (Moles/L)	Solubility (Grams/L)	Temp (°C)	Ref (#)	Evaluation (T P E A A)	Comments
3.408E-06	1.000E-03	rt	M161	0 0 0 0 0	

3673. C₁₉H₂₄N₂

Imipramine

10,11-Dihydro-N,N-dimethyl-5H-Dibenz[b,f]azepine-5-propanamine

5-[3-(Dimethylamino)propyl]-10,11-dihydro-5H-dibenz[b,f]azepine

RN: 50-49-7   **MP** (°C):   174

MW: 280.42   **BP** (°C):

Solubility (Moles/L)	Solubility (Grams/L)	Temp (°C)	Ref (#)	Evaluation (T P E A A)	Comments
6.500E-05	1.823E-02	24	G022	2 0 1 1 1	

3674. C₁₉H₂₄N₂O

Hydrocinchonine

Hydrocinchonin

Cinchotine

RN: 485-65-4   **MP** (°C):   268

MW: 296.42   **BP** (°C):

Solubility (Moles/L)	Solubility (Grams/L)	Temp (°C)	Ref (#)	Evaluation (T P E A A)	Comments
2.362E-03	7.000E-01	16	F300	1 0 0 0 1	
2.593E-03	7.686E-01	25	D004	1 0 0 0 0	

3675. C₁₉H₂₄N₂OS

Methotrimeprazine

Levomepromazine

RN: 60-99-1   **MP** (°C):   117

MW: 328.48   **BP** (°C):

Solubility (Moles/L)	Solubility (Grams/L)	Temp (°C)	Ref (#)	Evaluation (T P E A A)	Comments
6.089E-05	2.000E-02	25	A081	1 0 1 1 0	EFG

3676. C₁₉H₂₄N₂O₂S

Cyclohexyl-p-toluene Sulfonamide

Cyclohexyl-4-toluene Sulfonamide

RN: **MP** (°C):

MW: 344.48   **BP** (°C):

Solubility (Moles/L)	Solubility (Grams/L)	Temp (°C)	Ref (#)	Evaluation (T P E A A)	Comments
1.742E-04	6.000E-02	ns	F014	0 0 0 0 0	

3677. $C_{19}H_{24}N_4O_7$
Propyloxycarbonyl-mitomycin C
RN: **MP** (°C):
MW: 420.43   **BP** (°C):

Solubility (Moles/L)	Solubility (Grams/L)	Temp (°C)	Ref (#)	Evaluation (T P E A A)	Comments
3.300E-04	1.387E-01	25	M316	1 1 1 1 2	

3678. $C_{19}H_{24}O$
1,1-Dimethyl-2-(p-methylphenyl)-2-p-Ethoxylphenyl)ethane
RN: 56265-26-0   **MP** (°C):
MW: 268.40   **BP** (°C):

Solubility (Moles/L)	Solubility (Grams/L)	Temp (°C)	Ref (#)	Evaluation (T P E A A)	Comments
6.706E-07	1.800E-04	rt	C122	0 2 2 2 2	

3679. $C_{19}H_{24}O_2$
1,1,1-Trimethyl-2,2-bis(p-methyloxylphenyl)ethane
RN: 4741-74-6   **MP** (°C):
MW: 284.40   **BP** (°C):

Solubility (Moles/L)	Solubility (Grams/L)	Temp (°C)	Ref (#)	Evaluation (T P E A A)	Comments
2.426E-06	6.900E-04	rt	C122	0 2 2 2 2	

3680. $C_{19}H_{24}O_3$
Adrenosterone
Androstene-3,11,17-trione
RN: 382-45-6   **MP** (°C):   220
MW: 300.40   **BP** (°C):

Solubility (Moles/L)	Solubility (Grams/L)	Temp (°C)	Ref (#)	Evaluation (T P E A A)	Comments
3.279E-04	9.849E-02	23.5	J003	2 0 2 1 2	average of 2
2.610E-04	7.840E-02	37	H004	1 0 2 2 2	
5.059E-04	1.520E-01	37	J003	1 0 2 1 2	

3681. $C_{19}H_{25}NO$
N,N-Dicyclopentylcinnamamide
2-Propenamide, N,N-Dicyclopentyl-3-phenyl-
RN: 59832-08-5   **MP** (°C):
MW: 283.42   **BP** (°C):

Solubility (Moles/L)	Solubility (Grams/L)	Temp (°C)	Ref (#)	Evaluation (T P E A A)	Comments
7.750E-07	2.196E-04	ns	H350	0 0 0 0 2	

3682.   C$_{19}$H$_{26}$I$_3$N$_3$O$_{10}$

1,3-Benzenedicarboxamide, N,N'-bis[2-Hydroxy-1,1-bis(hydroxymethyl)ethyl]-5-[(2-hydroxy-1-oxopropyl)amino]-2,4,6-triiodo-, (S)-

RN: 77868-46-3   **MP** (°C):
MW: 837.15   **BP** (°C):

Solubility (Moles/L)	Solubility (Grams/L)	Temp (°C)	Ref (#)	Evaluation (T P E A A)	Comments
2.342E-02	1.961E+01	25	P091	1 0 0 0 1	

3683.   C$_{19}$H$_{26}$N$_6$O$_4$S

Benzenesulfonamide, 4-(1,3-Diethyl-2,3,6,7-tetrahydro-2,6-dioxo-1H-purin-8-yl)-N-[2-(dimethylamino)ethyl]-

RN: 89073-49-4   **MP** (°C):   264
MW: 434.52   **BP** (°C):

Solubility (Moles/L)	Solubility (Grams/L)	Temp (°C)	Ref (#)	Evaluation (T P E A A)	Comments
2.532E-04	1.100E-01	ns	H316	0 2 1 1 2	pH 7.4
2.647E-02	1.150E+01	ns	H316	0 2 1 1 2	0.1N HCl

3684.   C$_{19}$H$_{26}$O

δ-4-Androstene-3-one

RN: **MP** (°C):
MW: 270.42   **BP** (°C):

Solubility (Moles/L)	Solubility (Grams/L)	Temp (°C)	Ref (#)	Evaluation (T P E A A)	Comments
<1.00E-06	<2.70E-04	25	E014	2 2 2 1 0	pH 7.3

3685.   C$_{19}$H$_{26}$O$_2$

Androstenedione
4-Androstene-3,17-dione
Androst-4-en-3,17-dion

RN: 63-05-8   **MP** (°C):
MW: 286.42   **BP** (°C):

Solubility (Moles/L)	Solubility (Grams/L)	Temp (°C)	Ref (#)	Evaluation (T P E A A)	Comments
2.000E-04	5.728E-02	25	E014	2 2 2 1 2	pH 7.3
2.840E-02	8.133E+00	25	P324	0 1 0 2 1	
1.399E-04	4.007E-02	37	H034	1 0 2 1 2	pH 7.4
1.700E-04	4.870E-02	37	L010	1 0 0 1 1	

3686.   C$_{19}$H$_{27}$N$_3$O
Doxylamine Ethanamine
RN: **MP** (°C):
MW: 313.45   **BP** (°C):

Solubility (Moles/L)	Solubility (Grams/L)	Temp (°C)	Ref (#)	Evaluation (T P E A A)	Comments
3.000E-02	9.403E+00	37.5	L034	2 2 0 1 2	pH 7.4

3687.   C$_{19}$H$_{27}$N$_3$O$_2$
2-Propoxy-N-[2-(diethyl-amino)ethyl]-4-quinoline Carboxamide
N-[2-(Diethylamino)ethyl]-2-propoxyquinoline-4-carboxamide
RN: 2717-00-2   **MP** (°C):
MW: 329.45   **BP** (°C):

Solubility (Moles/L)	Solubility (Grams/L)	Temp (°C)	Ref (#)	Evaluation (T P E A A)	Comments
3.980E-04	1.311E-01	ns	B018	0 0 0 0 2	
3.980E-04	1.311E-01	ns	M066	0 0 0 0 2	

3688.   C$_{19}$H$_{28}$Cl$_2$O$_3$
2,4-Dichlorophenoxyacetic Acid n-Undecyl Ester
RN: 65267-95-0   **MP** (°C):
MW: 375.34   **BP** (°C):

Solubility (Moles/L)	Solubility (Grams/L)	Temp (°C)	Ref (#)	Evaluation (T P E A A)	Comments
1.977E-05	7.420E-03	ns	M120	0 0 1 1 2	

3689.   C$_{19}$H$_{28}$N$_4$O$_6$
2'-Octanyl-6-methoxypurine Arabinoside
RN: 145913-41-3   **MP** (°C):   75-77
MW: 408.46   **BP** (°C):

Solubility (Moles/L)	Solubility (Grams/L)	Temp (°C)	Ref (#)	Evaluation (T P E A A)	Comments
6.110E-04	2.496E-01	37	C348	1 2 2 2 2	pH 7.00

3690.   C$_{19}$H$_{28}$O
7α-Methyl-19-nortestosterone
Trestolone
19-Nor-7α-methyltestosterone
RN: 3764-87-2   **MP** (°C):
MW: 272.43   **BP** (°C):

Solubility (Moles/L)	Solubility (Grams/L)	Temp (°C)	Ref (#)	Evaluation (T P E A A)	Comments
3.377E-04	9.200E-02	37	H004	1 0 2 2 2	

3691.   C$_{19}$H$_{28}$O$_2$
5,6-Dehydroisoandrosterone
Prasterone
Dehydroepiandrosterone
Dehydroisoandrosterone
RN: 53-43-0   **MP** (°C):   140.5
MW: 288.43   **BP** (°C):

Solubility (Moles/L)	Solubility (Grams/L)	Temp (°C)	Ref (#)	Evaluation (T P E A A)	Comments
7.558E-05	2.180E-02	23.5	J003	2 0 2 1 2	average of 6
1.000E-04	2.884E-02	37	E014	2 2 2 1 2	pH 7.3
1.040E-04	3.000E-02	37	H034	1 0 2 1 2	pH 7.4
1.144E-04	3.300E-02	37	J003	1 0 2 1 2	average of 4
8.633E-05	2.490E-02	ns	B057	0 2 1 1 2	

3692.   C$_{19}$H$_{28}$O$_2$
Androstanedione
5α-Androstane-3,17-dione
RN: 846-46-8   **MP** (°C):   142
MW: 288.43   **BP** (°C):

Solubility (Moles/L)	Solubility (Grams/L)	Temp (°C)	Ref (#)	Evaluation (T P E A A)	Comments
1.141E-04	3.290E-02	23.5	J003	1 0 2 1 2	average of 2
2.200E-04	6.346E-02	25	E014	2 2 2 1 2	pH 7.3
1.685E-04	4.860E-02	37	J003	1 0 2 1 2	average of 2

3693.   C$_{19}$H$_{28}$O$_2$
Testosterone
17β-Hydroxyandrost-4-en-3-one
Halotensin
Virilon
Oreton
Testex
RN: 58-22-0   **MP** (°C):   155
MW: 288.43   **BP** (°C):

Solubility (Moles/L)	Solubility (Grams/L)	Temp (°C)	Ref (#)	Evaluation (T P E A A)	Comments
5.600E-05	1.615E-02	10	B012	2 0 1 1 0	
6.390E-05	1.843E-02	10	L017	2 2 2 2 2	
2.254E-04	6.500E-02	15	F042	2 2 2 2 1	
7.550E-05	2.178E-02	15	L017	2 2 2 2 2	
7.900E-05	2.279E-02	20	B012	2 0 1 1 0	
2.430E-04	7.009E-02	20	F012	1 0 1 1 1	
2.392E-04	6.900E-02	20	F042	2 2 2 2 1	
8.460E-05	2.440E-02	20	G072	1 2 2 1 2	

7.790E-05	2.247E-02	20	L017	2 2 2 2 2	
8.000E-05	2.307E-02	20	L070	1 2 0 2 0	EFG
6.870E-05	1.982E-02	20	L077	1 2 2 2 2	
8.000E-04	2.307E-01	20	L087	1 1 2 1 0	EFG
8.100E-05	2.336E-02	25	B012	2 0 1 1 0	
9.500E-05	2.740E-02	25	B041	1 0 2 2 1	
8.913E-05	2.571E-02	25	B041	1 0 2 2 0	EFG
2.531E-04	7.300E-02	25	F042	2 2 2 2 1	
8.321E-05	2.400E-02	25	K003	2 1 1 1 1	
1.664E-04	4.800E-02	25	L009	1 0 0 1 1	
8.480E-05	2.446E-02	25	L017	2 2 2 2 2	
6.934E-05	2.000E-02	25	L338	1 0 1 1 2	
1.040E-04	3.000E-02	27.34	L077	1 2 2 2 2	
1.060E-04	3.057E-02	30	B012	2 0 1 1 0	
2.670E-04	7.700E-02	30	F042	2 2 2 2 1	
9.790E-05	2.824E-02	30	L017	2 2 2 2 2	
1.100E-04	3.173E-02	30	L068	1 0 0 1 0	EFG
2.500E-04	7.211E-02	30	L344	2 0 1 1 0	
1.040E-04	3.000E-02	30	M007	2 2 1 2 2	average of 8
8.876E-05	2.560E-02	30	T005	2 0 2 2 2	
1.096E-04	3.163E-02	31	A025	2 2 2 2 0	EFG
1.300E-04	3.750E-02	35	L017	2 2 2 2 2	
1.397E-04	4.029E-02	35	L077	1 2 2 2 2	
1.950E-04	5.624E-02	37	B013	1 0 2 2 0	average
1.250E-04	3.605E-02	37	E014	2 2 2 1 2	pH 7.3
1.013E-04	2.922E-02	37	H034	1 0 2 1 2	pH 7.4
1.259E-04	3.631E-02	37.50	B041	1 0 2 2 0	EFG
1.260E-04	3.634E-02	37.50	B041	1 0 2 2 2	
1.400E-04	4.038E-02	40	B012	2 0 1 1 0	
1.570E-04	4.528E-02	40	L017	2 2 2 2 2	
3.000E-04	8.653E-02	40	L070	1 2 0 2 0	EFG
1.702E-04	4.909E-02	42.34	L077	1 2 2 2 2	
1.870E-04	5.394E-02	45	L017	2 2 2 2 2	
2.100E-04	6.057E-02	50	B012	2 0 1 1 0	
2.350E-04	6.778E-02	50	L017	2 2 2 2 2	
2.053E-04	5.922E-02	50	L077	1 2 2 2 2	
6.795E-05	1.960E-02	ns	B057	0 2 1 1 2	
3.814E-05	1.100E-02	ns	B338	0 0 0 0 1	

3694. $C_{19}H_{28}O_2 \cdot H_2O$
Testosterone (Monohydrate)
Testosterone Monohydrate -I
RN: 58-22-0   **MP** (°C):
MW: 306.45   **BP** (°C):

Solubility (Moles/L)	Solubility (Grams/L)	Temp (°C)	Ref (#)	Evaluation (T P E A A)	Comments
6.265E-05	1.920E-02	15	F042	2 2 2 2 2	crystal-II
5.352E-05	1.640E-02	15	F042	2 2 2 2 2	crystal-I
7.081E-05	2.170E-02	20	F042	2 2 2 2 2	crystal-II
6.265E-05	1.920E-02	20	F042	2 2 2 2 2	crystal-I
8.256E-05	2.530E-02	25	F042	2 2 2 2 2	crystal-II
7.310E-05	2.240E-02	25	F042	2 2 2 2 2	crystal-I
9.333E-05	2.860E-02	30	F042	2 2 2 2 2	crystal-II
8.484E-05	2.600E-02	30	F042	2 2 2 2 2	crystal-I

3695. $C_{19}H_{28}O_3$
11-Ketoetiocholanolone
3α-Hydroxy-5β-androstane-11,17-dione
Etiocholanol-11-one
Ba 2684
RN: 739-27-5   **MP** (°C):
MW: 304.43   **BP** (°C):

Solubility (Moles/L)	Solubility (Grams/L)	Temp (°C)	Ref (#)	Evaluation (T P E A A)	Comments
7.455E-04	2.269E-01	23	J003	2 0 2 1 2	average of 4
9.457E-04	2.879E-01	37	J003	1 0 2 1 2	average of 2

3696. $C_{19}H_{29}NO$
n-Decylcinnamamide
2-Propenamide, N-Decyl-3-phenyl-
RN: 59832-02-9   **MP** (°C):
MW: 287.45   **BP** (°C):

Solubility (Moles/L)	Solubility (Grams/L)	Temp (°C)	Ref (#)	Evaluation (T P E A A)	Comments
2.530E-06	7.272E-04	ns	H350	0 0 0 0 2	

3697. $C_{19}H_{29}N_5O_6$
9-(1,3-Dipivaloate-2-propoxymethyl)guanine
RN: 88110-72-9   **MP** (°C):   231
MW: 423.47   **BP** (°C):

Solubility (Moles/L)	Solubility (Grams/L)	Temp (°C)	Ref (#)	Evaluation (T P E A A)	Comments
1.653E-05	7.000E-03	25	B360	1 0 2 2 2	

3698. C$_{19}$H$_{30}$O
Androstane-17-one

RN: 36378-49-1 **MP** (°C): 119
MW: 274.45 **BP** (°C):

Solubility (Moles/L)	Solubility (Grams/L)	Temp (°C)	Ref (#)	Evaluation (T P E A A)	Comments
<2.00E-07	<5.49E-05	25	E014	2 2 2 1 0	pH 7.3

3699. C$_{19}$H$_{30}$OS
Epitiostanol

RN: 2363-58-8 **MP** (°C): 127
MW: 306.51 **BP** (°C):

Solubility (Moles/L)	Solubility (Grams/L)	Temp (°C)	Ref (#)	Evaluation (T P E A A)	Comments
3.915E-06	1.200E-03	37	H120	1 1 1 1 1	normal saline

3700. C$_{19}$H$_{30}$O$_2$
Etiocholanolone
3α-Hydroxy-5β-androstane-17-one
5-Isoandrosterone

RN: 53-42-9 **MP** (°C):
MW: 290.45 **BP** (°C):

Solubility (Moles/L)	Solubility (Grams/L)	Temp (°C)	Ref (#)	Evaluation (T P E A A)	Comments
1.002E-04	2.910E-02	23.5	J003	2 0 2 1 2	average of 2
7.000E-05	2.033E-02	25	E014	2 2 2 1 1	pH 7.3, pyrogen

3701. C$_{19}$H$_{30}$O$_2$
Stanolone
Androstanolone

RN: 521-18-6 **MP** (°C): 181.0
MW: 290.45 **BP** (°C):

Solubility (Moles/L)	Solubility (Grams/L)	Temp (°C)	Ref (#)	Evaluation (T P E A A)	Comments
1.185E+00	3.443E+02	ns	B057	0 2 1 1 2	

3702.   C$_{19}$H$_{30}$O$_2$
Epiandrosterone
Isoandrosterone
RN: 481-29-8   **MP** (°C):   161
MW: 290.45   **BP** (°C):

Solubility (Moles/L)	Solubility (Grams/L)	Temp (°C)	Ref (#)	Evaluation (T P E A A)	Comments
6.955E-05	2.020E-02	23.5	J003	2 0 2 1 2	average of 5
8.160E-05	2.370E-02	37	J003	1 0 2 1 2	average of 3

3703.   C$_{19}$H$_{30}$O$_2$
Androsterone
3α-Hydroxy-17-androstanone
3α-Hydroxy-5α-androstan-17-one
Hydroxy-5α-androstan-17-one
Epihydroxyetioallocholan-17-one
Hydroxy-17-androstanone
RN: 53-41-8   **MP** (°C):   185
MW: 290.45   **BP** (°C):

Solubility (Moles/L)	Solubility (Grams/L)	Temp (°C)	Ref (#)	Evaluation (T P E A A)	Comments
3.959E-05	1.150E-02	23.5	J003	2 0 2 1 2	average of 2
4.300E-05	1.249E-02	37	E014	2 2 2 1 1	pH 7.3
6.163E-05	1.790E-02	37	J003	1 0 2 1 2	average of 2

3704.   C$_{19}$H$_{30}$O$_3$
Androstane-3-β,11-β-diol-17-one
Hydroxyisoandrosterone
RN: 514-17-0   **MP** (°C):   235
MW: 306.45   **BP** (°C):

Solubility (Moles/L)	Solubility (Grams/L)	Temp (°C)	Ref (#)	Evaluation (T P E A A)	Comments
2.552E-04	7.819E-02	23.5	J003	1 0 2 1 2	average of 2

3705.   C$_{19}$H$_{30}$O$_3$
p-(Dodecyloxy)benzoic Acid
Dodecyl p-Hydroxybenzoate
RN: 2312-15-4   **MP** (°C):   95
MW: 306.45   **BP** (°C):

Solubility (Moles/L)	Solubility (Grams/L)	Temp (°C)	Ref (#)	Evaluation (T P E A A)	Comments
3.569E-03	1.094E+00	25	D081	1 2 2 1 2	

3706.   C$_{19}$H$_{30}$O$_3$
11-Hydroxyetiocholanolone
5β-Androstan-17-one, 3α,11-Dihydroxy-
RN: 3272-49-9   **MP** (°C):
MW: 306.45   **BP** (°C):

Solubility (Moles/L)	Solubility (Grams/L)	Temp (°C)	Ref (#)	Evaluation (T P E A A)	Comments
1.400E-04	4.290E-02	23.5	J003	2 0 1 1 2	average of 2

3707.   C$_{19}$H$_{31}$NO$_2$
Dodecyl p-Aminobenzoate
p-Aminobenzoic Acid Dodecyl Ester
RN: 20043-94-1   **MP** (°C):
MW: 305.46   **BP** (°C):

Solubility (Moles/L)	Solubility (Grams/L)	Temp (°C)	Ref (#)	Evaluation (T P E A A)	Comments
1.600E-08	4.887E-06	37	F006	1 1 2 2 1	

3708.   C$_{19}$H$_{31}$NO$_3$
4-Hexoxybenzoic Acid-2-(diethyl-amino)ethyl Ester
RN: 38973-74-9   **MP** (°C):
MW: 321.46   **BP** (°C):

Solubility (Moles/L)	Solubility (Grams/L)	Temp (°C)	Ref (#)	Evaluation (T P E A A)	Comments
4.000E-05	1.286E-02	ns	M066	0 0 0 0 1	

3709.   C$_{19}$H$_{32}$N$_2$O$_2$
4-Hexylaminobenzoic Acid-2-(diethyl-amino)ethyl Ester
RN: 16488-57-6   **MP** (°C):
MW: 320.48   **BP** (°C):

Solubility (Moles/L)	Solubility (Grams/L)	Temp (°C)	Ref (#)	Evaluation (T P E A A)	Comments
1.900E-04	6.089E-02	ns	M066	0 0 0 0 1	

3710.   C$_{19}$H$_{32}$O$_3$
4-Nonylphenol Diethoxylate
RN: 20427-84-3   **MP** (°C):
MW: 308.47   **BP** (°C):

Solubility (Moles/L)	Solubility (Grams/L)	Temp (°C)	Ref (#)	Evaluation (T P E A A)	Comments
1.180E-05	3.640E-03	2	A335	1 0 2 2 2	
1.080E-05	3.331E-03	10	A335	1 0 2 2 2	
1.096E-05	3.380E-03	10	A335	1 0 2 2 2	
9.700E-06	2.992E-03	14	A335	1 0 2 2 2	

9.726E-06	3.000E-03	14	A335	1 0 2 2 2
1.100E-05	3.393E-03	20.5	A335	1 0 2 2 2
1.096E-05	3.380E-03	20.5	A335	1 0 2 2 2
1.200E-05	3.702E-03	25	A335	1 0 2 2 2
1.196E-05	3.690E-03	25	A335	1 0 2 2 2

3711. $C_{19}H_{34}O_3$

Methoprene
Isopropyl (2E,4E)-11-Methoxy-3,7,11-trimethyl-2,4-dodecadienoate
Kabat
Precor
Dianex
Pharorid

| **RN:** | 40596-69-8 | **MP** (°C): | 164 |
| **MW:** | 310.48 | **BP** (°C): | 100 |

Solubility (Moles/L)	Solubility (Grams/L)	Temp (°C)	Ref (#)	Evaluation (T P E A A)	Comments
4.477E-06	1.390E-03	25	D302	1 0 0 0 2	
6.442E-06	2.000E-03	ns	M110	0 0 0 0 0	EFG

3712. $C_{20}H_{12}$

Benzo(a)pyrene
1,2-Benzopyrene
3,4-Benzpyrene
Benzo[a]pyrene
Benz[a]pyrene

| **RN:** | 50-32-8 | **MP** (°C): | 179 |
| **MW:** | 252.32 | **BP** (°C): | 310 |

Solubility (Moles/L)	Solubility (Grams/L)	Temp (°C)	Ref (#)	Evaluation (T P E A A)	Comments
3.309E-09	8.350E-07	15	B385	2 0 2 2 2	
2.000E-09	5.046E-07	20	E009	1 0 0 0 1	
2.972E-05	7.500E-03	23	T025	1 2 0 1 1	*sic*
6.341E-09	1.600E-06	25	B319	2 0 1 2 1	
5.667E-09	1.430E-06	25	B385	2 0 2 2 2	
4.400E-10	1.110E-07	25	K123	1 0 2 2 1	
1.506E-08	3.800E-06	25	L332	1 1 1 1 2	
1.506E-08	3.800E-06	25	M064	1 1 2 2 1	
1.500E-08	3.785E-06	25	M342	1 0 1 1 1	
6.428E-09	1.622E-06	25.04	M183	1 2 1 1 2	
1.585E-08	4.000E-06	27	D003	1 0 0 1 1	
9.083E-09	2.292E-06	30.04	M183	1 2 1 1 2	
1.098E-08	2.770E-06	35	B385	2 0 2 2 2	
1.506E-08	3.800E-06	ns	M344	0 0 0 0 2	
2.400E-08	6.056E-06	ns	W005	0 0 1 2 1	
4.756E-09	1.200E-06	ns	W302	0 0 0 0 1	

3713. C$_{20}$H$_{12}$
Benzo(b)fluoranthene
3,4-Benzofluoranthene
2,3-Benzofluoranthene
B[B]F
RN: 205-99-2   **MP** (°C):   108
MW: 252.32   **BP** (°C):

Solubility (Moles/L)	Solubility (Grams/L)	Temp (°C)	Ref (#)	Evaluation (T P E A A)	Comments
5.945E-09	1.500E-06	ns	W302	0 0 0 0 1	

3714. C$_{20}$H$_{12}$
Benzo(e)pyrene
4,5-Benzopyrene
B[E]P
RN: 192-97-2   **MP** (°C):   178.5
MW: 252.32   **BP** (°C):

Solubility (Moles/L)	Solubility (Grams/L)	Temp (°C)	Ref (#)	Evaluation (T P E A A)	Comments
3.900E-09	9.840E-07	25	K123	1 0 2 2 1	
~1.59E-08	~4.00E-06	25	S227	1 2 1 1 0	
6.625E-02	1.672E+01	318	S355	1 1 1 2 0	EFG
1.192E-01	3.007E+01	330	S355	1 1 1 2 0	EFG
1.524E-01	3.846E+01	335	S355	1 1 1 2 0	EFG
2.066E-01	5.213E+01	342	S355	1 1 1 2 0	EFG
4.246E-01	1.071E+02	361	S355	1 1 1 2 0	EFG
4.559E-01	1.150E+02	365	S355	1 1 1 2 0	EFG

3715. C$_{20}$H$_{12}$
Benzo(j)fluoranthene
Benzo[l]fluoranthene
Benzo-12,13-fluoranthene
10,11-Benzofluoranthene
RN: 205-82-3   **MP** (°C):   165
MW: 252.32   **BP** (°C):

Solubility (Moles/L)	Solubility (Grams/L)	Temp (°C)	Ref (#)	Evaluation (T P E A A)	Comments
9.908E-09	2.500E-06	ns	W302	0 0 0 0 1	

3716. $C_{20}H_{12}$
Benzo(k)fluoranthene
11,12-Benzo[k]fluoranthene
11,12-Benzofluoranthene
8,9-Benzofluoranthene
2,3,1',8'-Binaphthylene
B[K]F

RN: 207-08-9   **MP** (°C):   216
MW: 252.32   **BP** (°C):

Solubility (Moles/L)	Solubility (Grams/L)	Temp (°C)	Ref (#)	Evaluation (T P E A A)	Comments
3.171E-09	8.000E-07	ns	W302	0 0 0 0 0	

3717. $C_{20}H_{12}$
Perylene
Dibenz[de,kl]anthracene
peri-Dinaphthalene

RN: 198-55-0   **MP** (°C):   273
MW: 252.32   **BP** (°C):

Solubility (Moles/L)	Solubility (Grams/L)	Temp (°C)	Ref (#)	Evaluation (T P E A A)	Comments
4.200E-10	1.060E-07	20	E009	1 0 0 1 1	
1.585E-09	4.000E-07	25	M064	1 1 2 2 0	
1.600E-09	4.037E-07	25	M342	1 0 1 1 1	
<1.98E-09	<5.00E-07	27	D003	1 0 0 1 0	
1.585E-09	4.000E-07	ns	M344	0 0 0 0 1	

3718. $C_{20}H_{13}N$
3,4,5,6-Dibenzocarbazole
3:4,5:6-Dibenzocarbazole

RN: 194-59-2   **MP** (°C):   158
MW: 267.33   **BP** (°C):

Solubility (Moles/L)	Solubility (Grams/L)	Temp (°C)	Ref (#)	Evaluation (T P E A A)	Comments
2.000E-07	5.347E-05	22	B175	1 0 1 1 0	

3719. $C_{20}H_{13}N$
13H-Dibenzo(a,i)carbazole
1:2,7:8-Dibenzocarbazole
RN: 239-64-5 MP (°C): 220
MW: 267.33 BP (°C):

Solubility (Moles/L)	Solubility (Grams/L)	Temp (°C)	Ref (#)	Evaluation (T P E A A)	Comments
<5.00E-08	<1.34E-05	22	B175	1 0 1 1 0	*sic*
3.890E-08	1.040E-05	24	H106	1 0 2 2 2	
3.890E-08	1.040E-05	24	M303	1 0 1 1 2	

3720. $C_{20}H_{13}N$
1,2,5,6-Dibenzocarbazole
1:2,5:6-Dibenzocarbazole
RN: 207-84-1 MP (°C):
MW: 267.33 BP (°C):

Solubility (Moles/L)	Solubility (Grams/L)	Temp (°C)	Ref (#)	Evaluation (T P E A A)	Comments
5.000E-08	1.337E-05	22	B175	1 0 1 1 0	

3721. $C_{20}H_{14}$
Cholanthrene
1,2-Dihydroxybenz[j]aceanthrylene
RN: 479-23-2 MP (°C): 173
MW: 254.33 BP (°C):

Solubility (Moles/L)	Solubility (Grams/L)	Temp (°C)	Ref (#)	Evaluation (T P E A A)	Comments
1.376E-08	3.500E-06	27	D003	1 0 0 1 1	

3722. $C_{20}H_{14}$
3,4'-Ace-1,2-benzanthracene
Benz[k]acephenanthrene
RN: 5779-79-3 MP (°C):
MW: 254.33 BP (°C):

Solubility (Moles/L)	Solubility (Grams/L)	Temp (°C)	Ref (#)	Evaluation (T P E A A)	Comments
1.062E-08	2.700E-06	27	D003	1 0 0 1 1	

3723.   C₂₀H₁₄I₆N₂O₆

Di(3-carboxy-2,4,6-triiodoanilido)adipic Acid

Iodipamide

RN: 606-17-7   **MP** (°C):   306
MW: 1139.77   **BP** (°C):

Solubility (Moles/L)	Solubility (Grams/L)	Temp (°C)	Ref (#)	Evaluation (T P E A A)	Comments
4.036E-04	4.600E-01	20	N035	1 1 2 1 1	
1.404E-04	1.600E-01	ns	H055	0 1 0 2 2	

3724.   C₂₀H₁₄N₂O₂

Disperse Blue 19

C.I. Disperse Blue 19

RN: 4395-65-7   **MP** (°C):   194
MW: 314.35   **BP** (°C):

Solubility (Moles/L)	Solubility (Grams/L)	Temp (°C)	Ref (#)	Evaluation (T P E A A)	Comments
6.100E-10	1.918E-07	25	B333	1 0 0 0 1	
2.100E-07	6.601E-05	60.0	D093	1 2 1 2 0	EFG
5.000E-07	1.572E-04	71.8	D093	1 2 1 2 0	EFG
1.700E-06	5.344E-04	81.4	D093	1 2 1 2 0	EFG
4.200E-06	1.320E-03	97.4	D093	1 2 1 2 0	EFG

3725.   C₂₀H₁₄O₂

3,3-Diphenylphthalide

3,3-Diphenyl-phthalid

RN: 596-29-2   **MP** (°C):
MW: 286.33   **BP** (°C):

Solubility (Moles/L)	Solubility (Grams/L)	Temp (°C)	Ref (#)	Evaluation (T P E A A)	Comments
1.397E-04	4.000E-02	25	F300	1 0 0 0 0	

3726.   C₂₀H₁₄O₄

Phenolphthalein
2-[bis(4-Hydroxyphenyl)methyl]benzoic Acid
Espotabs
Alophen
Figsen
Laxettes

RN: 77-09-8   **MP** (°C):   260.0
MW: 318.33   **BP** (°C):

Solubility (Moles/L)	Solubility (Grams/L)	Temp (°C)	Ref (#)	Evaluation (T P E A A)	Comments
6.283E-06	2.000E-03	25	H064	1 2 2 0 2	
7.476E-04	2.380E-01	100	H064	1 2 2 0 2	
1.256E-03	3.998E-01	rt	D021	0 0 1 1 0	

3727.   C₂₀H₁₄O₄

Phenyl Phthalate
Diphenyl Phthalate

RN: 84-62-8   **MP** (°C):   71
MW: 318.33   **BP** (°C):

Solubility (Moles/L)	Solubility (Grams/L)	Temp (°C)	Ref (#)	Evaluation (T P E A A)	Comments
2.576E-07	8.200E-05	24	H116	2 1 0 0 1	

3728.   C₂₀H₁₆

10-Ethyl-1,2-benzanthracene
10-Ethylbenz[a]anthracene

RN: 14854-08-1   **MP** (°C):   114
MW: 256.35   **BP** (°C):

Solubility (Moles/L)	Solubility (Grams/L)	Temp (°C)	Ref (#)	Evaluation (T P E A A)	Comments
1.755E-07	4.500E-05	27	D003	1 0 0 1 1	
1.560E-07	4.000E-05	27	D043	2 0 0 0 0	average of 2

3729.   C₂₀H₁₆

3729. $C_{20}H_{16}$

9,10-Dimethyl-1,2-benzanthracene
7,12-Dimethyl-1,2-benzanthracene
7,12-Dimethylbenz[a]anthracene
9,10-Dimethyl-benz[a]anthracene

RN: 56-56-4   **MP** (°C):   122
MW: 256.35   **BP** (°C):

Solubility (Moles/L)	Solubility (Grams/L)	Temp (°C)	Ref (#)	Evaluation (T P E A A)	Comments
9.518E-08	2.440E-05	24	H106	1 0 2 2 2	
2.145E-07	5.500E-05	24	H116	2 1 0 0 1	
9.752E-08	2.500E-05	24	M129	1 2 1 1 1	
2.380E-07	6.100E-05	25	M064	1 1 2 2 1	
9.518E-08	2.440E-05	25	M156	1 2 1 1 2	
1.677E-07	4.300E-05	27	D003	1 0 0 1 1	

3730. $C_{20}H_{16}$

5,6-Dimethylchrysene
Chrysene, 5,6-Dimethyl-

RN: 3697-27-6   **MP** (°C):   127
MW: 256.35   **BP** (°C):   200

Solubility (Moles/L)	Solubility (Grams/L)	Temp (°C)	Ref (#)	Evaluation (T P E A A)	Comments
9.752E-08	2.500E-05	27	D003	1 0 0 1 1	

3731. $C_{20}H_{16}O_{4}$

Phenolphthalin
Benzoic Acid, 2-[bis(4-Hydroxyphenyl)methyl]-

RN: 81-90-3   **MP** (°C):   237
MW: 320.35   **BP** (°C):

Solubility (Moles/L)	Solubility (Grams/L)	Temp (°C)	Ref (#)	Evaluation (T P E A A)	Comments
5.463E-04	1.750E-01	20	F300	1 0 0 0 2	

3732. $C_{20}H_{18}O_{2}Sn$

Triphenyltin Hydroxide Acetate
Fentin Acetate

RN: 900-95-8   **MP** (°C):   120
MW: 409.06   **BP** (°C):

Solubility (Moles/L)	Solubility (Grams/L)	Temp (°C)	Ref (#)	Evaluation (T P E A A)	Comments
6.845E-05	2.800E-02	20	M161	1 0 0 0 1	

3733. C$_{20}$H$_{19}$NO$_3$
Acronine
3,12-Dihydro-6-methoxy-3,3,12-trimethyl-7H-pyrano(2,3-c)acridin-7-one
Acronycine
RN: 7008-42-6 **MP** (°C): 175-176
MW: 321.38 **BP** (°C):

Solubility (Moles/L)	Solubility (Grams/L)	Temp (°C)	Ref (#)	Evaluation (T P E A A)	Comments
7.779E-06	2.500E-03	22	B064	1 0 1 1 0	
8.401E-06	2.700E-03	25	R071	1 2 2 2 1	

3734. C$_{20}$H$_{19}$NO$_5$.6H$_2$O
Berberine (Hexahydrate)
Berberine
RN: 2086-83-1 **MP** (°C): 145dec
MW: 461.47 **BP** (°C):

Solubility (Moles/L)	Solubility (Grams/L)	Temp (°C)	Ref (#)	Evaluation (T P E A A)	Comments
9.422E-02	4.348E+01	25	D004	1 0 0 0 0	

3735. C$_{20}$H$_{19}$N$_3$
Rosaniline
Basic Violet 14
C.I. 42510
Calcozine Magenta xx
Cerise B
RN: 632-99-5 **MP** (°C):
MW: 301.39 **BP** (°C):

Solubility (Moles/L)	Solubility (Grams/L)	Temp (°C)	Ref (#)	Evaluation (T P E A A)	Comments
9.951E-04	2.999E-01	rt	D021	0 0 1 1 0	

3736. C$_{20}$H$_{19}$N$_3$O$_5$
1H-Benzimidazole-1-carboxylic Acid, 6-Benzoyl-2-[(methoxycarbonyl)amino]-, Propyl Ester
RN: 153474-31-8 **MP** (°C): 113.5
MW: 381.39 **BP** (°C):

Solubility (Moles/L)	Solubility (Grams/L)	Temp (°C)	Ref (#)	Evaluation (T P E A A)	Comments
1.311E-05	5.000E-03	21	N337	1 0 2 2 2	pH 5
1.311E-05	5.000E-03	21	N337	1 0 2 2 0	pH 5

3737. C$_{20}$H$_{20}$N$_2$O$_6$
Succinyl Acetaminophen
Butanedioic Acid, bis[4-(Acetylamino)phenyl] Ester
Acetanilide, 4'-Hydroxy-, Succinate
Acetanilide, 4'-Hydroxy-, Succinate (2:1) (Ester)
RN: 2725-63-5   **MP** (°C):   229-230
MW: 384.39   **BP** (°C):

Solubility (Moles/L)	Solubility (Grams/L)	Temp (°C)	Ref (#)	Evaluation (T P E A A)	Comments
1.769E-05	6.800E-03	37	D029	1 0 1 1 1	

3738. C$_{20}$H$_{20}$N$_6$O$_6$S$_2$
2,5-di-(N4-Acetylsulfanilylamino)pyrimidine
RN: **MP** (°C):
MW: 504.55   **BP** (°C):

Solubility (Moles/L)	Solubility (Grams/L)	Temp (°C)	Ref (#)	Evaluation (T P E A A)	Comments
9.910E-06	5.000E-03	37	R076	1 2 0 0 1	

3739. C$_{20}$H$_{21}$NO$_4$
Papaverine
Pantoyl Taurine
RN: 58-74-2   **MP** (°C):   147
MW: 339.39   **BP** (°C):

Solubility (Moles/L)	Solubility (Grams/L)	Temp (°C)	Ref (#)	Evaluation (T P E A A)	Comments
1.100E-04	3.733E-02	37.5	L034	2 2 0 1 2	pH 7.4

3740. C$_{20}$H$_{21}$NO$_5$
Aspirin Phenylalanine Ethyl Ester
L-Phenylalanine, N-[2-(Acetyloxy)benzoyl]-, Ethyl Ester
RN: 76748-72-6   **MP** (°C):
MW: 355.39   **BP** (°C):

Solubility (Moles/L)	Solubility (Grams/L)	Temp (°C)	Ref (#)	Evaluation (T P E A A)	Comments
4.700E-04	1.670E-01	25	B182	2 2 1 1 1	

3741. C$_{20}$H$_{22}$ClN
Pyrrobutamine
Pyrrolidine, 1-[4-(4-Chlorophenyl)-3-phenyl-2-butenyl]-
RN: 91-82-7   **MP** (°C):
MW: 311.86   **BP** (°C):

Solubility (Moles/L)	Solubility (Grams/L)	Temp (°C)	Ref (#)	Evaluation (T P E A A)	Comments
8.700E-04	2.713E-01	37.5	L034	2 2 0 1 2	pH 7.4

3742. C$_{20}$H$_{22}$N$_2$O$_2$
Quininone
Chininon
Cinchonan-9-one, 6'-Methoxy-, (8α)-
RN: 84-31-1   **MP** (°C):   212
MW: 322.41   **BP** (°C):

Solubility (Moles/L)	Solubility (Grams/L)	Temp (°C)	Ref (#)	Evaluation (T P E A A)	Comments
9.305E-06	3.000E-03	20	F300	1 0 0 0 0	

3743. C$_{20}$H$_{23}$N
Amitriptyline
5-(3-Dimethylaminopropylidene)-5H-dibenzo[a,d]-10,11-dihydrocycloheptene
Adepress
Adepril
RN: 50-48-6   **MP** (°C):   196
MW: 277.41   **BP** (°C):

Solubility (Moles/L)	Solubility (Grams/L)	Temp (°C)	Ref (#)	Evaluation (T P E A A)	Comments
3.500E-05	9.709E-03	24	G022	2 0 1 1 1	

3744. C$_{20}$H$_{23}$NO$_2$
Dexoxadrol
(+)-2-(2,2-Diphenyl-1,3-dioxolan-4-yl)piperidine
Relane
CL 911C
RN: 4741-41-7   **MP** (°C):
MW: 309.41   **BP** (°C):

Solubility (Moles/L)	Solubility (Grams/L)	Temp (°C)	Ref (#)	Evaluation (T P E A A)	Comments
2.262E-04	7.000E-02	rt	K017	0 2 2 2 2	intrinsic

3745.   C$_{20}$H$_{24}$ClN$_3$S
Prochlorperazine
Compazine
Ultrazine
Cotranzine
Compa-Z

RN: 58-38-8   **MP** (°C):   228
MW: 373.95   **BP** (°C):

Solubility (Moles/L)	Solubility (Grams/L)	Temp (°C)	Ref (#)	Evaluation (T P E A A)	Comments
4.000E-05	1.496E-02	24	G022	2 0 1 1 1	

3746.   C$_{20}$H$_{24}$N$_2$
Dimethindene
Dimetindene
Pyridine, 2-[1-[2-[2-(Dimethylamino)ethyl]inden-3-yl]ethyl]-

RN: 5636-83-9   **MP** (°C):
MW: 292.43   **BP** (°C):

Solubility (Moles/L)	Solubility (Grams/L)	Temp (°C)	Ref (#)	Evaluation (T P E A A)	Comments
8.160E-04	2.386E-01	37	L094	2 0 0 1 2	pH>10.03, intrinsic

3747.   C$_{20}$H$_{24}$N$_2$O$_2$
Quinidine
Chinidin
Cinchonan-9-ol, 6'-Methoxy-, (9S)-

RN: 56-54-2   **MP** (°C):   174
MW: 324.43   **BP** (°C):

Solubility (Moles/L)	Solubility (Grams/L)	Temp (°C)	Ref (#)	Evaluation (T P E A A)	Comments
7.200E-04	2.336E-01	15	K059	2 2 2 0 1	
4.315E-04	1.400E-01	25	F300	1 0 0 0 1	
1.540E-03	4.998E-01	c	D004	1 0 0 0 0	
3.848E-03	1.248E+00	h	D004	1 0 0 0 0	

3748. $C_{20}H_{24}N_2O_2$
Quinine
Chinin
Quinine Alkaloid
RN: 130-95-0 **MP** (°C): 177
MW: 324.43 **BP** (°C):

Solubility (Moles/L)	Solubility (Grams/L)	Temp (°C)	Ref (#)	Evaluation (T P E A A)	Comments
1.541E-03	5.000E-01	15	F300	1 0 0 0 0	
1.760E-03	5.711E-01	25	D004	1 0 0 0 0	
9.247E-04	3.000E-01	25	P015	2 2 2 2 1	
4.007E-03	1.300E+00	100	.F300	1 0 0 0 1	
1.756E-03	5.697E-01	rt	D021	0 0 1 1 1	

3749. $C_{20}H_{24}N_2O_2.3H_2O$
Quinine (Trihydrate)
Quinine, Compd. with Valeric Acid (1:1), Hydrate
Cinchonan-9-ol, 6'-Methoxy-, Trihydrate, (8α,9R)-
RN: 6151-51-5 **MP** (°C): 57
MW: 378.47 **BP** (°C):

Solubility (Moles/L)	Solubility (Grams/L)	Temp (°C)	Ref (#)	Evaluation (T P E A A)	Comments
1.693E-03	6.406E-01	c	D004	1 0 0 0 0	
3.299E-03	1.248E+00	h	D004	1 0 0 0 0	

3750. $C_{20}H_{24}N_2O_4$
Pheniramine Maleate
1-Phenyl-1-(2-pyridyl)-3-dimethylaminopropane Maleate
Prophenpyridamine Maleate
RN: 132-20-7 **MP** (°C):
MW: 356.43 **BP** (°C):

Solubility (Moles/L)	Solubility (Grams/L)	Temp (°C)	Ref (#)	Evaluation (T P E A A)	Comments
3.100E-02	1.105E+01	37.5	L034	2 2 0 1 2	pH 7.4

3751. $C_{20}H_{24}N_2O_5$
Naproxen, N-Methyl-N-carbamoyl Methyl-glycolamide Ester
RN: **MP** (°C): 179.5
MW: 372.42 **BP** (°C):

Solubility (Moles/L)	Solubility (Grams/L)	Temp (°C)	Ref (#)	Evaluation (T P E A A)	Comments
1.584E-04	5.900E-02	21	B331	1 2 2 1 1	pH 7.4

3752.   C$_{20}$H$_{24}$O$_2$
Ethinyl Estradiol
Ethinyloestradiol
Ethynyl Estradiol
Estone
RN: 57-63-6   **MP** (°C):   182
MW: 296.41   **BP** (°C):

Solubility (Moles/L)	Solubility (Grams/L)	Temp (°C)	Ref (#)	Evaluation (T P E A A)	Comments
3.441E-05	1.020E-02	20	G072	1 2 2 1 2	
1.000E-04	2.964E-02	20	L070	1 2 0 2 0	EFG
2.560E-05	7.588E-03	20	L077	1 2 2 2 2	
3.272E-05	9.700E-03	25	H049	2 2 2 2 1	
3.374E-05	1.000E-02	25	K003	2 1 1 1 1	
3.810E-05	1.129E-02	27.34	L077	1 2 2 2 2	
5.060E-05	1.500E-02	35	L077	1 2 2 2 2	
1.000E-04	2.964E-02	40	L070	1 2 0 2 0	EFG
6.240E-05	1.850E-02	42.34	L077	1 2 2 2 2	
7.420E-05	2.199E-02	50	L077	1 2 2 2 2	
1.349E-05	4.000E-03	ns	N302	0 2 1 2 1	

3753.   C$_{20}$H$_{24}$O$_3$
Methylsecodione
RN: 80702-24-5   **MP** (°C):
MW: 312.41   **BP** (°C):

Solubility (Moles/L)	Solubility (Grams/L)	Temp (°C)	Ref (#)	Evaluation (T P E A A)	Comments
1.919E-03	5.996E-01	25	P324	0 1 0 2 1	

3754.   C$_{20}$H$_{24}$O$_4$
3,11-Dioxo-4,17(20)-cis-pregnadien-21-oic Acid Methyl Ester
U-2726
RN: **MP** (°C):
MW: 328.41   **BP** (°C):

Solubility (Moles/L)	Solubility (Grams/L)	Temp (°C)	Ref (#)	Evaluation (T P E A A)	Comments
1.309E-05	4.300E-03	ns	K029	0 0 2 1 1	

3755.   C$_{20}$H$_{24}$O$_6$
Dibenzo-18-crown-6
DBC
RN: 14187-32-7   **MP** (°C):
MW: 360.41   **BP** (°C):

Solubility (Moles/L)	Solubility (Grams/L)	Temp (°C)	Ref (#)	Evaluation (T P E A A)	Comments
2.025E-05	7.300E-03	25	M127	1 2 1 1 1	
9.000E-05	3.244E-02	26	P029	2 0 0 0 1	

3756.   C$_{20}$H$_{25}$ClO$_2$
1-Chloro-1,1-dimethyl-2,2-bis(p-ethoxylphenyl)ethane
RN: 56265-24-8   **MP** (°C):
MW: 332.87   **BP** (°C):

Solubility (Moles/L)	Solubility (Grams/L)	Temp (°C)	Ref (#)	Evaluation (T P E A A)	Comments
5.708E-07	1.900E-04	rt	C122	0 2 2 2 2	

3757.   C$_{20}$H$_{25}$NO$_2$
Adiphenine
2-Diethylaminoethyl Diphenylacetate
Tranzetil
Patrovine
SKF 962A
RN: 64-95-9   **MP** (°C):   113.5
MW: 311.43   **BP** (°C):

Solubility (Moles/L)	Solubility (Grams/L)	Temp (°C)	Ref (#)	Evaluation (T P E A A)	Comments
1.000E-02	3.114E+00	30	L068	1 0 0 1 0	EFG

3758.   C$_{20}$H$_{26}$NO$_4$
2-Naphthaleneacetic Acid, 6-Methoxy-α-methyl-, 2-(Diethylamino)-2-oxoethyl Ester, (S)
Naproxen, N,N-Diethyl Glycolamide Ester
2-Naphthaleneacetic Acid, 6-Methoxy-α-methyl-, 2-(Diethylamino)-2-oxoethyl Ester
Naproxen N,N-Diethyl Glycolamide Ester
RN: 106231-74-7   **MP** (°C):   89
MW: 343.43   **BP** (°C):

Solubility (Moles/L)	Solubility (Grams/L)	Temp (°C)	Ref (#)	Evaluation (T P E A A)	Comments
3.494E-05	1.200E-02	21	B331	1 2 2 1 1	pH 7.4
3.494E-05	1.200E-02	21	B331	1 2 2 1 1	

3759. C₂₀H₂₅NO₄
3,11-Dioxo-4,17(20)-cis-pregnadien-20-oic Acid Methyl Ester 3-Oxime
RN: **MP** (°C):
MW: 343.43   **BP** (°C):

Solubility (Moles/L)	Solubility (Grams/L)	Temp (°C)	Ref (#)	Evaluation (T P E A A)	Comments
1.543E-05	5.300E-03	ns	K029	0 0 2 1 1	

3760. C₂₀H₂₅NO₅
Naproxen, N-Methyl-N-hydroxyethyl Glycolamide Ester
RN: **MP** (°C):   110
MW: 359.43   **BP** (°C):

Solubility (Moles/L)	Solubility (Grams/L)	Temp (°C)	Ref (#)	Evaluation (T P E A A)	Comments
3.895E-04	1.400E-01	21	B331	1 2 2 1 1	pH 7.4

3761. C₂₀H₂₅NO₆
2-Naphthaleneacetic Acid, 6-Methoxy-α-methyl-, 2-[bis(2-Hydroxyethyl)amino]-2-oxoethyl Ester
Naproxen N,N-Diethanol Glycolamide Ester
Naproxen,N,N-Dihydroxyethyl Glycolamide Ester
RN: 114665-20-2   **MP** (°C):   113
MW: 375.43   **BP** (°C):

Solubility (Moles/L)	Solubility (Grams/L)	Temp (°C)	Ref (#)	Evaluation (T P E A A)	Comments
1.092E-03	4.100E-01	21	B331	1 2 2 1 1	pH 7.4
1.092E-03	4.100E-01	21	B331	1 2 2 1 1	

3762. C₂₀H₂₆N₂O₂
Hydroquinine
Cinchonan-9-ol, 10,11-Dihydro-6'-methoxy-, (8α,9R)-
10,11-Dihydroquinine
RN: 522-66-7   **MP** (°C):   173.5
MW: 326.44   **BP** (°C):

Solubility (Moles/L)	Solubility (Grams/L)	Temp (°C)	Ref (#)	Evaluation (T P E A A)	Comments
3.063E-04	9.999E-02	20	K059	2 2 2 0 1	

3763.   C$_{20}$H$_{26}$N$_2$O$_2$
Ajmaline
Rauwolfine
Ajmalan-17,21-diol, (17R,21α)-
Merabitol
Raugalline
RN: 4360-12-7   **MP** (°C):   159
MW: 326.44   **BP** (°C):

Solubility (Moles/L)	Solubility (Grams/L)	Temp (°C)	Ref (#)	Evaluation (T P E A A)	Comments
1.100E-03	3.591E-01	0	M106	2 1 1 1 0	EFG
1.300E-03	4.244E-01	15	M106	2 1 1 1 0	EFG
1.500E-03	4.897E-01	30	M106	2 1 1 1 0	EFG

3764.   C$_{20}$H$_{26}$O$_2$
Norethindrone
Norethisterone
RN: 68-22-4   **MP** (°C):   203
MW: 298.43   **BP** (°C):

Solubility (Moles/L)	Solubility (Grams/L)	Temp (°C)	Ref (#)	Evaluation (T P E A A)	Comments
1.334E-05	3.981E-03	10	L078	1 0 1 2 0	EFG
1.679E-05	5.012E-03	20	L078	1 0 1 2 0	EFG
2.360E-05	7.043E-03	25	H099	1 0 2 2 2	
1.884E-05	5.623E-03	25	L078	1 0 1 2 2	
8.377E-03	2.500E+00	25	P312	1 2 2 2 2	
2.114E-05	6.310E-03	30	L078	1 0 1 2 0	EFG
3.610E-05	1.077E-02	37	C004	1 0 2 2 2	EFG
2.986E-05	8.912E-03	40	L078	1 0 1 2 0	EFG
4.218E-05	1.259E-02	50	L078	1 0 1 2 0	EFG

3765.   C$_{20}$H$_{26}$O$_2$
1,1-Dimethyl-2,2-bis(p-ethoxylphenyl)ethane
RN: 56265-21-5   **MP** (°C):
MW: 298.43   **BP** (°C):

Solubility (Moles/L)	Solubility (Grams/L)	Temp (°C)	Ref (#)	Evaluation (T P E A A)	Comments
1.441E-07	4.300E-05	rt	C122	0 2 2 2 2	

3766.   C$_{20}$H$_{26}$O$_4$
Dicyclohexyl Phthalate
1,2-Benzenedicarboxylic Acid, Dicyclohexyl Ester
RN: 84-61-7   **MP** (°C):   66
MW: 330.43   **BP** (°C):

Solubility (Moles/L)	Solubility (Grams/L)	Temp (°C)	Ref (#)	Evaluation (T P E A A)	Comments
1.211E-05	4.000E-03	24	H116	2 1 0 0 2	

3767.   C$_{20}$H$_{27}$NO$_{11}$
Amygdalin
(R)-Amygdalin
(R)-Laenitrile
(R)-Amygdaloside
RN: 29883-15-6   **MP** (°C):   223
MW: 457.44   **BP** (°C):

Solubility (Moles/L)	Solubility (Grams/L)	Temp (°C)	Ref (#)	Evaluation (T P E A A)	Comments
1.705E-01	7.800E+01	10	F300	1 0 0 0 1	

3768.   C$_{20}$H$_{27}$NO$_{11}$.3H$_2$O
Amygdalin (Trihydrate)
D-(-)-Amygdalin
(R)-Amygdalin
RN: 29883-15-6   **MP** (°C):   214-216
MW: 511.48   **BP** (°C):

Solubility (Moles/L)	Solubility (Grams/L)	Temp (°C)	Ref (#)	Evaluation (T P E A A)	Comments
1.504E-01	7.692E+01	c	D004	1 0 0 0 0	

3769.   C$_{20}$H$_{27}$O$_4$P
Octyldiphenyl Phosphate
Disflamoll DPO
RN: 115-88-8   **MP** (°C):
MW: 362.41   **BP** (°C):

Solubility (Moles/L)	Solubility (Grams/L)	Temp (°C)	Ref (#)	Evaluation (T P E A A)	Comments
3.863E-07	1.400E-04	24	H116	2 1 0 0 2	

3770. $C_{20}H_{28}O$
Vitamin A Aldehyde
Retinal
All-trans-Retinal
All-trans Vitamin A Aldehyde
Retinene

RN: 116-31-4   **MP** (°C):   63
MW: 284.45   **BP** (°C):

Solubility (Moles/L)	Solubility (Grams/L)	Temp (°C)	Ref (#)	Evaluation (T P E A A)	Comments
≤2.46E-04	<7.00E-02	25	P312	1 2 2 2 2	

3771. $C_{20}H_{28}O_2$
19-Norprogesterone
19-Norpregn-4-ene-3,20-dione

RN: 472-54-8   **MP** (°C):
MW: 300.44   **BP** (°C):

Solubility (Moles/L)	Solubility (Grams/L)	Temp (°C)	Ref (#)	Evaluation (T P E A A)	Comments
1.202E-04	3.610E-02	37	L010	2 0 2 1 1	

3772. $C_{20}H_{28}O_2$
Retinoic Acid
All-trans retinoic Acid
3,7-Dimethyl-9-(2,6,6-trimethyl-1-cyclohexen-1-yl)-2,4,6,8-nonatetraenoic Acid
β-All-trans-Retinoic Acid

RN: 302-79-4   **MP** (°C):   180-181
MW: 300.44   **BP** (°C):

Solubility (Moles/L)	Solubility (Grams/L)	Temp (°C)	Ref (#)	Evaluation (T P E A A)	Comments
≤2.33E-04	<7.00E-02	25	P312	1 2 2 2 2	

3773. $C_{20}H_{28}O_3$
5,6-Dehydroisoandrosterone Formate
Androst-5-en-17-one, 3α-Hydroxy-, Formate

RN: 4589-84-8   **MP** (°C):
MW: 316.44   **BP** (°C):

Solubility (Moles/L)	Solubility (Grams/L)	Temp (°C)	Ref (#)	Evaluation (T P E A A)	Comments
4.424E-05	1.400E-02	ns	B057	0 2 1 1 2	

3774.   C$_{20}$H$_{28}$O$_3$
Testosterone Formate
Androst-4-en-17β-ol-3-one Formate
Testosterone 17-Formate
RN: 3129-42-8   **MP** (°C):
MW: 316.44   **BP** (°C):

Solubility (Moles/L)	Solubility (Grams/L)	Temp (°C)	Ref (#)	Evaluation (T P E A A)	Comments
1.389E-05	4.395E-03	25	J004	1 0 1 1 2	
1.390E-05	4.400E-03	ns	B057	0 2 1 1 1	

3775.   C$_{20}$H$_{29}$N$_3$O$_2$
Dibucaine
Cinchocaine
RN: 85-79-0   **MP** (°C):   64
MW: 343.47   **BP** (°C):

Solubility (Moles/L)	Solubility (Grams/L)	Temp (°C)	Ref (#)	Evaluation (T P E A A)	Comments
1.980E-04	6.801E-02	ns	B018	0 0 0 0 2	
1.980E-04	6.801E-02	ns	M066	0 0 0 0 2	

3776.   C$_{20}$H$_{30}$N$_4$O$_6$
2'-Nonyl-6-methoxypurine Arabinoside
4-Quinolinecarboxamide, 2-Butoxy-N-[2-(diethylamino)ethyl]-
RN: 145913-42-4   **MP** (°C):   88-90
MW: 422.49   **BP** (°C):

Solubility (Moles/L)	Solubility (Grams/L)	Temp (°C)	Ref (#)	Evaluation (T P E A A)	Comments
1.030E-04	4.352E-02	37	C348	1 2 2 2 2	pH 7.00

3777.   C$_{20}$H$_{30}$O
D 263
4,6-Diisopropyl-1,1-dimethyl-7-propionylindan
RN: 290294-31-4   **MP** (°C):   117
MW: 286.46   **BP** (°C):

Solubility (Moles/L)	Solubility (Grams/L)	Temp (°C)	Ref (#)	Evaluation (T P E A A)	Comments
3.491E-06	1.000E-03	ns	M061	0 0 0 0 0	

3778. $C_{20}H_{30}O$

Vitamin A
Retinol
Afaxin
α-Sterol

RN: 68-26-8   **MP** (°C):   62
MW: 286.46   **BP** (°C):   137-138

Solubility (Moles/L)	Solubility (Grams/L)	Temp (°C)	Ref (#)	Evaluation (T P E A A)	Comments
≤3.49E-05	<1.00E-02	25	P312	1 2 2 2 2	

3779. $C_{20}H_{30}O_2$

17-Methyltestosterone
17-α-Methyltestosterone
Methyltestosterone
Methyl-testosterone

RN: 58-18-4   **MP** (°C):   161
MW: 302.46   **BP** (°C):

Solubility (Moles/L)	Solubility (Grams/L)	Temp (°C)	Ref (#)	Evaluation (T P E A A)	Comments
1.230E-04	3.720E-02	20	F012	1 0 1 1 1	
1.120E-04	3.388E-02	25	H099	1 0 2 2 2	
1.058E-04	3.200E-02	25	K003	2 1 1 1 1	
4.400E-02	1.331E+01	25	M379	1 0 1 1 0	EFG,sic
<5.62E-04	<1.70E-01	25	P312	1 2 2 2 2	
2.313E-03	6.995E-01	25	P324	0 1 0 2 1	
1.018E-04	3.080E-02	30	T005	2 0 2 2 2	
1.200E-04	3.630E-02	37	E014	2 2 2 1 2	pH 7.3
7.472E-05	2.260E-02	ns	B057	0 2 1 1 2	
9.918E-05	3.000E-02	rt	N302	0 2 1 2 1	

3780. $C_{20}H_{30}O_2$

Abietic Acid
13-Isopropylpodocarpa-7,13-dien-15-oic Acid
Sylvic Acid

RN: 514-10-3   **MP** (°C):   172
MW: 302.46   **BP** (°C):

Solubility (Moles/L)	Solubility (Grams/L)	Temp (°C)	Ref (#)	Evaluation (T P E A A)	Comments
1.600E-04	4.839E-02	20	B009	2 2 1 2 0	

3781. C$_{20}$H$_{30}$O$_3$

Androstanolone Formate

5α-Androstan-3-one, 17-Hydroxy-, Formate

RN: 4589-90-6   **MP** (°C):
MW: 318.46   **BP** (°C):

Solubility (Moles/L)	Solubility (Grams/L)	Temp (°C)	Ref (#)	Evaluation (T P E A A)	Comments
4.679E-06	1.490E-03	ns	B057	0 2 1 1 2	

3782. C$_{20}$H$_{30}$O$_6$

Butyl Glycol Phthalate

bis(2-Butoxyethyl) Phthalate

Dibutoxyethyl Phthalate

bis(2-N-Butoxyethyl) Phthalate

RN: 117-83-9   **MP** (°C):   230
MW: 366.46   **BP** (°C):   210

Solubility (Moles/L)	Solubility (Grams/L)	Temp (°C)	Ref (#)	Evaluation (T P E A A)	Comments
5.458E-05	2.000E-02	15	H069	1 0 1 1 0	
<8.18E-04	<3.00E-01	20	F070	1 0 0 0 1	

3783. C$_{20}$H$_{31}$NO$_3$

Acetaminophen Laurate

Acetaminophen Dodecanoate

RN: 54942-38-0   **MP** (°C):   111
MW: 333.47   **BP** (°C):

Solubility (Moles/L)	Solubility (Grams/L)	Temp (°C)	Ref (#)	Evaluation (T P E A A)	Comments
1.799E-05	6.000E-03	25	B010	1 1 1 1 0	

3784. C$_{20}$H$_{32}$O$_3$

Tridecyl p-Hydroxybenzoate

p-Hydroxybenzoic Acid Tridecyl Ester

RN: 69679-32-9   **MP** (°C):
MW: 320.48   **BP** (°C):

Solubility (Moles/L)	Solubility (Grams/L)	Temp (°C)	Ref (#)	Evaluation (T P E A A)	Comments
1.135E-03	3.639E-01	25	D081	1 2 2 1 2	

3785.   C$_{20}$H$_{32}$O$_5$
Dinoprostone
Prostaglandin E2
RN: 363-24-6   **MP** (°C):   66-68
MW: 352.48   **BP** (°C):

Solubility (Moles/L)	Solubility (Grams/L)	Temp (°C)	Ref (#)	Evaluation (T P E A A)	Comments
3.123E-03	1.101E+00	8.53	F068	0 0 2 2 0	
4.022E-03	1.418E+00	19.24	F068	0 0 2 2 0	
4.173E-03	1.471E+00	25.35	F068	0 0 2 2 0	
4.575E-03	1.613E+00	29.9	F068	0 0 2 2 0	

3786.   C$_{20}$H$_{33}$NO$_3$
4-Heptoxybenzoic Acid-2-(diethyl-amino)ethyl Ester
RN: 38973-75-0   **MP** (°C):
MW: 335.49   **BP** (°C):

Solubility (Moles/L)	Solubility (Grams/L)	Temp (°C)	Ref (#)	Evaluation (T P E A A)	Comments
5.000E-05	1.677E-02	ns	M066	0 0 0 0 1	

3787.   C$_{20}$H$_{34}$N$_2$O$_2$
4-Heptylaminobenzoic Acid-2-(diethyl-amino)ethyl Ester
RN: **MP** (°C):
MW: 334.51   **BP** (°C):

Solubility (Moles/L)	Solubility (Grams/L)	Temp (°C)	Ref (#)	Evaluation (T P E A A)	Comments
2.100E-04	7.025E-02	ns	M066	0 0 0 0 1	

3788.   C$_{20}$H$_{34}$O$_4$
4-Octylphenol Triethoxylate
RN: 51437-91-3   **MP** (°C):
MW: 338.49   **BP** (°C):

Solubility (Moles/L)	Solubility (Grams/L)	Temp (°C)	Ref (#)	Evaluation (T P E A A)	Comments
5.436E-05	1.840E-02	20.5	A335	1 0 2 2 2	
5.440E-05	1.841E-02	20.5	A335	1 0 2 2 2	

3789. $C_{20}H_{34}O_8$
Acetyl Tributyl Citrate
1,2,3-Propanetricarboxylic Acid
Tributyl Acetylcitrate
RN: 77-90-7   **MP** (°C):
MW: 402.49   **BP** (°C):

Solubility (Moles/L)	Solubility (Grams/L)	Temp (°C)	Ref (#)	Evaluation (T P E A A)	Comments
4.224E-06	1.700E-03	25	F067	1 0 2 2 1	

3790. $C_{20}H_{36}O_4$
Dioctyl Maleate
2-Butenedioic Acid (Z)-
Dioctyl Ester
RN: 2915-53-9   **MP** (°C):
MW: 340.51   **BP** (°C):

Solubility (Moles/L)	Solubility (Grams/L)	Temp (°C)	Ref (#)	Evaluation (T P E A A)	Comments
1.762E-06	6.000E-04	25	F067	1 0 2 2 2	

3791. $C_{20}H_{36}O_6$
Dicyclohexyl-18-crown-6
Dibenzo[b,k][1,4,7,10,13,16]hexaoxacyclooctadecin, Icosahydro-
Dicyclohexano-18-crown-6
cis-Dicyclohexano-18-crown-6
RN: 16069-36-6   **MP** (°C):
MW: 372.51   **BP** (°C):

Solubility (Moles/L)	Solubility (Grams/L)	Temp (°C)	Ref (#)	Evaluation (T P E A A)	Comments
3.600E-02	1.341E+01	26	P029	2 0 0 0 2	
2.200E-02	8.195E+00	53	P029	2 0 0 0 2	
1.000E-02	3.725E+00	82	P029	2 0 0 0 2	

3792. $C_{20}H_{40}$
1-Eicosene
n-Eicosene
RN: 3452-07-1   **MP** (°C):
MW: 280.54   **BP** (°C):

Solubility (Moles/L)	Solubility (Grams/L)	Temp (°C)	Ref (#)	Evaluation (T P E A A)	Comments
1.907E-12	5.350E-10	23	C332	2 0 2 2 1	

3793. C$_{21}$H$_{13}$N
1:2,6:7-Dibenzacridine
RN: 226-92-6   **MP** (°C):
MW: 279.34   **BP** (°C):

Solubility (Moles/L)	Solubility (Grams/L)	Temp (°C)	Ref (#)	Evaluation (T P E A A)	Comments
5.000E-08	1.397E-05	22	B175	1 0 1 1 0	

3794. C$_{21}$H$_{13}$N
1:2,8:9-Dibenzacridine
RN: 224-53-3   **MP** (°C):
MW: 279.34   **BP** (°C):

Solubility (Moles/L)	Solubility (Grams/L)	Temp (°C)	Ref (#)	Evaluation (T P E A A)	Comments
7.000E-08	1.955E-05	22	B175	1 0 1 1 0	

3795. C$_{21}$H$_{13}$N
3:4,6:7-Dibenzacridine
RN: 226-97-1   **MP** (°C):
MW: 279.34   **BP** (°C):

Solubility (Moles/L)	Solubility (Grams/L)	Temp (°C)	Ref (#)	Evaluation (T P E A A)	Comments
2.500E-07	6.984E-05	22	B175	1 0 1 1 1	

3796. C$_{21}$H$_{14}$
5-Methyl-3,4-benzpyrene
RN: 31647-36-6   **MP** (°C): 216
MW: 266.35   **BP** (°C):

Solubility (Moles/L)	Solubility (Grams/L)	Temp (°C)	Ref (#)	Evaluation (T P E A A)	Comments
3.004E-09	8.000E-07	27	D003	1 0 0 1 0	

3797. C$_{21}$H$_{15}$ClN$_2$O$_4$S
1-(p-Chlorobenzenesulfonyl)-5,5-diphenyl-hydantoin
RN: 24759-38-4   **MP** (°C):
MW: 426.88   **BP** (°C):

Solubility (Moles/L)	Solubility (Grams/L)	Temp (°C)	Ref (#)	Evaluation (T P E A A)	Comments
7.965E-07	3.400E-04	37	F183	1 0 1 1 2	intrinsic

3798.   C₂₁H₁₅N₃O₆S

1-(p-Nitrobenzenesulfonyl)-5,5-diphenyl-hydantoin

RN: 21413-53-6   **MP** (°C):
MW: 437.43   **BP** (°C):

Solubility (Moles/L)	Solubility (Grams/L)	Temp (°C)	Ref (#)	Evaluation (T P E A A)	Comments
1.486E-06	6.500E-04	37	F183	1 0 1 1 2	intrinsic

3799.   C₂₁H₁₆

3-Methylcholanthrene
1,2-Dihydro-3-methyl-benz[j]aceanthrylene
20-Methylcholanthrene

RN: 56-49-5   **MP** (°C):   179
MW: 268.36   **BP** (°C):   280

Solubility (Moles/L)	Solubility (Grams/L)	Temp (°C)	Ref (#)	Evaluation (T P E A A)	Comments
1.204E-08	3.230E-06	24	H106	1 0 2 2 2	
1.081E-08	2.900E-06	25	M064	1 1 2 2 1	
1.204E-08	3.230E-06	25	M156	1 2 1 1 2	
1.100E-08	2.952E-06	25	M342	1 0 1 1 1	
5.589E-09	1.500E-06	27	D003	1 0 0 1 1	
1.081E-08	2.900E-06	ns	M344	0 0 0 0 1	

3800.   C₂₁H₁₆N₂O₂

C.I. Disperse Blue 24
9,10-Anthracenedione, 1-Amino-4-hydroxy-2-phenoxy-
Serilene Red 2BL
Sumikaron Red E-FBL
Solvent Red 146

RN: 17418-58-5   **MP** (°C):   151
MW: 328.37   **BP** (°C):

Solubility (Moles/L)	Solubility (Grams/L)	Temp (°C)	Ref (#)	Evaluation (T P E A A)	Comments
5.000E-08	1.642E-05	25	B333	1 0 0 0 1	

3801.   C₂₁H₁₆N₂O₄S

1-Benzenesulfonyl-5,5-diphenyl-hydantoin

RN: 21413-28-5   **MP** (°C):
MW: 392.44   **BP** (°C):

Solubility (Moles/L)	Solubility (Grams/L)	Temp (°C)	Ref (#)	Evaluation (T P E A A)	Comments
4.587E-06	1.800E-03	37	F183	1 0 1 1 2	intrinsic

3802. $C_{21}H_{16}N_2O_5S$
1-(p-Hydroxylbenzenesulfonyl)-5,5-diphenyl-hydantoin
RN: 24759-35-1 **MP** (°C):
MW: 408.44 **BP** (°C):

Solubility (Moles/L)	Solubility (Grams/L)	Temp (°C)	Ref (#)	Evaluation (T P E A A)	Comments
8.080E-06	3.300E-03	37	F183	1 0 1 1 2	intrinsic

3803. $C_{21}H_{17}N_3O_2S_2$
2-Sulfanilamido-4-p-diphenylthiazole
RN: **MP** (°C):
MW: 407.52 **BP** (°C):

Solubility (Moles/L)	Solubility (Grams/L)	Temp (°C)	Ref (#)	Evaluation (T P E A A)	Comments
2.454E-06	1.000E-03	37	R045	1 2 1 1 0	

3804. $C_{21}H_{17}N_3O_4S$
1-(p-Aminobenzenesulfonyl)-5,5-diphenyl-hydantoin
RN: 24759-34-0 **MP** (°C):
MW: 407.45 **BP** (°C):

Solubility (Moles/L)	Solubility (Grams/L)	Temp (°C)	Ref (#)	Evaluation (T P E A A)	Comments
3.436E-06	1.400E-03	37	F183	1 0 1 1 2	intrinsic

3805. $C_{21}H_{19}NO_4$
Cinmetacin
1-Cinnamoyl-2-methyl-5-methoxyindolyl-3-acetic Acid
Indolacin
RN: 20168-99-4 **MP** (°C): 170
MW: 349.39 **BP** (°C):

Solubility (Moles/L)	Solubility (Grams/L)	Temp (°C)	Ref (#)	Evaluation (T P E A A)	Comments
≤2.86E-06	<1.00E-03	25	K027	2 0 2 2 0	

3806.   C$_{21}$H$_{20}$Cl$_2$O$_3$
Permethrin
3-(2,2-Dichloroethenyl)-2,2-dimethylcyclopropanecarboxylic Acid (3-Phenoxyphenyl)methyl Ester
Ambush
Pounce
Ectiban
RN: 52645-53-1   **MP** (°C):   36.5
MW: 391.30   **BP** (°C):   200

Solubility (Moles/L)	Solubility (Grams/L)	Temp (°C)	Ref (#)	Evaluation (T P E A A)	Comments
5.111E-07	2.000E-04	ns	M161	0 0 0 0 0	

3807.   C$_{21}$H$_{21}$ClN$_2$O$_8$
Demeclocycline
Declomycin
Methylchlorotetracycline
Demethylchlortetracycline
RN: 127-33-3   **MP** (°C):
MW: 464.86   **BP** (°C):

Solubility (Moles/L)	Solubility (Grams/L)	Temp (°C)	Ref (#)	Evaluation (T P E A A)	Comments
3.259E-03	1.515E+00	21	M044	2 0 2 2 2	
3.012E-03	1.400E+00	25	B191	1 0 0 0 1	neutral pH

3808.   C$_{21}$H$_{21}$NO$_6$
Hydrastine
Hydrastin
(1R,9S)-β-Hydrastine
RN: 118-08-1   **MP** (°C):   132
MW: 383.40   **BP** (°C):

Solubility (Moles/L)	Solubility (Grams/L)	Temp (°C)	Ref (#)	Evaluation (T P E A A)	Comments
8.200E-04	3.144E-01	15	K059	2 2 2 0 1	
7.825E-05	3.000E-02	20	F300	1 0 0 0 1	

3809.   C₂₁H₂₁NO₆

3809. $C_{21}H_{21}NO_6$

Rhoeadine

[1,3]Dioxolo[4,5-h]-1,3-dioxolo[7,8][2]benzopyrano[3,4-a][3]benzazepine,

5β,6,7,8,13β,15-Hexahydro-15-methoxy-6-methyl-, (5bR,13bR,15S)

8-Methoxy-16-methyl-2,3:10,11-bis[methylenebis(oxy)]-, (8β)-

RN: 2718-25-4   **MP** (°C):   245-247dec

MW: 383.40   **BP** (°C):

Solubility (Moles/L)	Solubility (Grams/L)	Temp (°C)	Ref (#)	Evaluation (T P E A A)	Comments
2.172E-03	8.326E-01	25	D004	1 0 0 0 0	

3810. $C_{21}H_{21}N_3O_3S$

L-Phe-Dapsone

Benzenepropanamide, α-Amino-N-[4-[(4-aminophenyl)sulfonyl]phenyl]-, (S)-

RN: 160349-01-9   **MP** (°C):

MW: 395.48   **BP** (°C):

Solubility (Moles/L)	Solubility (Grams/L)	Temp (°C)	Ref (#)	Evaluation (T P E A A)	Comments
5.057E-06	2.000E-03	25	P351	2 2 1 2 1	pH 7.4
3.287E-03	1.300E+00	25	P351	2 2 1 2 1	

3811. $C_{21}H_{21}O_4P$

Tricresyl Phosphate

Tritolyl Phosphate

Tri-p-cresyl Phosphate

RN: 1330-78-5   **MP** (°C):

MW: 368.37   **BP** (°C):   265

Solubility (Moles/L)	Solubility (Grams/L)	Temp (°C)	Ref (#)	Evaluation (T P E A A)	Comments
2.009E-07	7.400E-05	24	H116	2 1 0 0 1	
2.715E-07	1.000E-04	25	F067	1 0 2 2 1	
2.172E-04	7.999E-02	ns	F014	0 0 0 0 0	

3812.   C₂₁H₂₂N₂O₂
Strychnine
Strychnidin-10-one
Gopher Getter
L-Strychnine
Gopher Bait
RN: 57-24-9   **MP** (°C):   275
MW: 334.42   **BP** (°C):

Solubility (Moles/L)	Solubility (Grams/L)	Temp (°C)	Ref (#)	Evaluation (T P E A A)	Comments
2.700E-04	9.029E-02	15	K059	2 2 2 0 1	
4.186E-04	1.400E-01	20.0	N002	2 1 2 2 1	
5.980E-04	2.000E-01	30.0	N002	2 1 2 2 1	
1.017E-03	3.400E-01	40.0	N002	2 1 2 2 1	
1.196E-03	4.000E-01	50.0	N002	2 1 2 2 1	
1.346E-03	4.500E-01	60.0	N002	2 1 2 2 1	
1.794E-03	6.000E-01	75.0	N002	2 1 2 2 1	
4.672E-04	1.562E-01	c	D004	1 0 0 0 0	
9.643E-04	3.225E-01	h	D004	1 0 0 0 0	
4.276E-04	1.430E-01	rt	M161	0 0 0 0 2	

3813.   C₂₁H₂₂N₂O₅
Benzeneacetic Acid, 4-Benzoyl-α-methyl-, 2-[(2-Amino-2-oxoethyl)methylamino]-2-oxoethyl Ester
N-Methyl-N-Carbamoyl Methyl Glycolamide Salicylate
RN: 114665-16-6   **MP** (°C):   83
MW: 382.42   **BP** (°C):

Solubility (Moles/L)	Solubility (Grams/L)	Temp (°C)	Ref (#)	Evaluation (T P E A A)	Comments
3.792E-03	1.450E+00	21	B331	1 2 2 1 1	

3814.   C₂₁H₂₂N₂O₅
Ketoprofen, N-methyl-N-carbamoylmethyl Glycolamide Ester
Benzeneacetic Acid, 3-Benzoyl-α-methyl-, 2-[(2-Amino-2-oxoethyl)methylamino]-2-oxoethyl Ester
RN: 116482-84-9   **MP** (°C):   83.5
MW: 382.42   **BP** (°C):

Solubility (Moles/L)	Solubility (Grams/L)	Temp (°C)	Ref (#)	Evaluation (T P E A A)	Comments
3.792E-03	1.450E+00	21	B331	1 2 2 1 1	pH 7.4

3815. C$_{21}$H$_{23}$ClFNO$_2$
Haloperidol
Haldol
4-[4-(p-Chlorophenyl)-4-hydroxypiperidino]-4'-fluorobutyrophenone
Serenace

RN: 52-86-8   **MP** (°C):   148
MW: 375.87   **BP** (°C):

Solubility (Moles/L)	Solubility (Grams/L)	Temp (°C)	Ref (#)	Evaluation (T P E A A)	Comments
7.981E-06	3.000E-03	30	P044	1 0 1 0 1	

3816. C$_{21}$H$_{23}$N$_3$OS
Pericyazine
2-Cyano-10-[3'-(4"-hydroxypiperidino)propyl]phenothiazine
Periciazine

RN: 2622-26-6   **MP** (°C):   116
MW: 365.50   **BP** (°C):

Solubility (Moles/L)	Solubility (Grams/L)	Temp (°C)	Ref (#)	Evaluation (T P E A A)	Comments
1.040E-04	3.801E-02	37	F011	1 0 1 1 2	pH 7.4

3817. C$_{21}$H$_{24}$F$_3$N$_3$S
Trifluoperazine
Stelazine

RN: 117-89-5   **MP** (°C):   232
MW: 407.50   **BP** (°C):   206

Solubility (Moles/L)	Solubility (Grams/L)	Temp (°C)	Ref (#)	Evaluation (T P E A A)	Comments
3.000E-05	1.223E-02	24	G022	2 0 1 1 1	
3.600E-05	1.467E-02	37	F011	1 0 1 1 1	pH 7.4

3818. C$_{21}$H$_{24}$O$_{10}$·2H$_2$O
Phloridzin (Dihydrate)
1-Propanone, 1-[2-(β-D-Glucopyranosyloxy)-4,6-dihydroxyphenyl]-3-(4-hydroxyphenyl)-,
Dihydrate

RN: 7061-54-3   **MP** (°C):   108
MW: 472.45   **BP** (°C):

Solubility (Moles/L)	Solubility (Grams/L)	Temp (°C)	Ref (#)	Evaluation (T P E A A)	Comments
2.115E-03	9.990E-01	c	d004	1 0 0 0 0	

3819.   C$_{21}$H$_{25}$NO
4-Cyano-4'-octyloxybiphenyl
8 COB

RN: **MP** (°C):
MW: 307.44   **BP** (°C):

Solubility (Moles/L)	Solubility (Grams/L)	Temp (°C)	Ref (#)	Evaluation (T P E A A)	Comments
2.700E-07	8.301E-05	21	D300	2 2 1 1 2	

3820.   C$_{21}$H$_{26}$ClN$_3$OS
Perphenazine
4-(3-(2-Chlorophenothiazin-10-YL)propyl)-1-piperazineethanol
Etrafon
Trilafon

RN: 58-39-9   **MP** (°C):   97
MW: 403.98   **BP** (°C):   280

Solubility (Moles/L)	Solubility (Grams/L)	Temp (°C)	Ref (#)	Evaluation (T P E A A)	Comments
7.000E-05	2.828E-02	24	G022	2 0 1 1 1	

3821.   C$_{21}$H$_{26}$N$_2$O$_3$
1-(2,3-Dihydro-5-methoxybenzo[b]furan-2-ylmethyl)-4-(o-methoxyphenyl)piperazine

RN: **MP** (°C):
MW: 354.45   **BP** (°C):

Solubility (Moles/L)	Solubility (Grams/L)	Temp (°C)	Ref (#)	Evaluation (T P E A A)	Comments
5.642E-05	2.000E-02	37	L079	1 0 1 1 0	intrinsic

3822.   C$_{21}$H$_{26}$O$_4$
17-Hydroxy-6-methyl-16-methylenepregna-4,6-diene-3,20-dione Acetate

RN: **MP** (°C):
MW: 342.44   **BP** (°C):

Solubility (Moles/L)	Solubility (Grams/L)	Temp (°C)	Ref (#)	Evaluation (T P E A A)	Comments
8.469E-06	2.900E-03	37	H004	1 0 2 2 2	

3823. C$_{21}$H$_{26}$O$_5$
Prednisone
1,4-Pregnadiene-17α,21-diol-3,11,20-trione
1,4-Pregnadiene-17x,21-diol-3,11,20-trione
Delcortin
Metocorten
Panasol

RN: 53-03-2   **MP** (°C): 234
MW: 358.44   **BP** (°C):

Solubility (Moles/L)	Solubility (Grams/L)	Temp (°C)	Ref (#)	Evaluation (T P E A A)	Comments
3.208E-04	1.150E-01	25	K003	2 1 1 1 1	

3824. C$_{21}$H$_{27}$FO$_5$
Fluprednisolone
6α-Fluoro-11β,17,21-trihydroxypregna-1,4-diene-3,20-dione17,21-trihydroxypregna-1,4-diene-3,20-dione
Alphadrol

RN: 53-34-9   **MP** (°C):
MW: 378.44   **BP** (°C):

Solubility (Moles/L)	Solubility (Grams/L)	Temp (°C)	Ref (#)	Evaluation (T P E A A)	Comments
2.748E-03	1.040E+00	37	H004	1 0 2 2 2	

3825. C$_{21}$H$_{27}$FO$_5$.H$_2$O
Fluprednisolone (Monohydrate)

RN: 53-34-9   **MP** (°C):
MW: 396.46   **BP** (°C):

Solubility (Moles/L)	Solubility (Grams/L)	Temp (°C)	Ref (#)	Evaluation (T P E A A)	Comments
1.478E-03	5.860E-01	37	H004	1 0 2 2 2	

3826. C$_{21}$H$_{27}$FO$_6$
Triamcinolone
9α-Fluoro-11β,16α,17α,21-tetrahydroxy-1,4-pregnadiene-3,20-dione
9α-Fluoro-16α-hydroxyprednisolone
Aristocort

RN: 124-94-7   **MP** (°C): 269
MW: 394.44   **BP** (°C):

Solubility (Moles/L)	Solubility (Grams/L)	Temp (°C)	Ref (#)	Evaluation (T P E A A)	Comments
2.028E-04	7.999E-02	25	F024	1 0 0 0 0	

3827.   C$_{21}$H$_{28}$N$_4$O$_7$
Pentyloxycarbonyl-mitomycin C
RN: **MP** (°C):
MW: 448.48   **BP** (°C):

Solubility (Moles/L)	Solubility (Grams/L)	Temp (°C)	Ref (#)	Evaluation (T P E A A)	Comments
5.900E-04	2.646E-01	25	M316	1 1 1 1 2	

3828.   C$_{21}$H$_{28}$O$_2$
1,1,1-Trimethyl-2,2-bis(p-ethoxylphenyl)ethane
RN: 27955-87-9   **MP** (°C):
MW: 312.46   **BP** (°C):

Solubility (Moles/L)	Solubility (Grams/L)	Temp (°C)	Ref (#)	Evaluation (T P E A A)	Comments
4.481E-07	1.400E-04	rt	C122	0 2 2 2 2	

3829.   C$_{21}$H$_{28}$O$_2$
Ethisterone
17α-Ethynyl Testosterone
Ethynyl Testosterone
Gestoral
Pregneninolone
Anhydrohydroxyprogesterone
RN: 434-03-7   **MP** (°C):   269
MW: 312.46   **BP** (°C):

Solubility (Moles/L)	Solubility (Grams/L)	Temp (°C)	Ref (#)	Evaluation (T P E A A)	Comments
1.920E-06	5.999E-04	20	G072	1 2 2 1 2	
1.600E-06	4.999E-04	20	L077	1 2 2 2 1	
1.280E-06	4.000E-04	25	K003	2 1 1 1 1	
2.200E-06	6.874E-04	27.34	L077	1 2 2 2 1	
3.200E-06	9.999E-04	35	L077	1 2 2 2 1	
3.500E-06	1.094E-03	42.34	L077	1 2 2 2 1	
4.200E-06	1.312E-03	50	L077	1 2 2 2 1	

3830.   C$_{21}$H$_{28}$O$_5$
Cortisone
17-Hydroxy-11-dehydrocorticosterone
Cortate
RN: 53-06-5   **MP** (°C):   222
MW: 360.45   **BP** (°C):

Solubility (Moles/L)	Solubility (Grams/L)	Temp (°C)	Ref (#)	Evaluation (T P E A A)	Comments
7.766E-04	2.799E-01	20	D041	1 0 0 0 0	
6.379E-04	2.299E-01	25	K003	2 1 1 1 1	
7.768E-04	2.800E-01	25	M023	1 0 2 1 1	
7.500E-04	2.703E-01	30	L344	2 0 1 1 0	EFG
6.000E-04	2.163E-01	37	E014	2 2 2 1 2	pH 7.3
7.768E-04	2.800E-01	ns	B338	0 0 0 0 1	

3831.   C$_{21}$H$_{28}$O$_5$
Prednisolone
11β,17α,21-Trihydroxypregna-1,4-diene-3,20-dione
Ropredlone
Predonin
Hostacortin H
Nisolone
RN: 50-24-8   **MP** (°C):   240
MW: 360.45   **BP** (°C):

Solubility (Moles/L)	Solubility (Grams/L)	Temp (°C)	Ref (#)	Evaluation (T P E A A)	Comments
6.173E-03	2.225E+00	25	G008	1 2 1 1 2	*sic*
5.963E-04	2.150E-01	25	K003	2 1 1 1 1	
1.379E-03	4.970E-01	25	K021	1 2 2 2 1	
7.000E-04	2.523E-01	30	H016	2 2 2 2 0	EFG
1.268E-03	4.570E-01	30	T002	1 0 2 0 2	anhydrous, form A
1.398E-03	5.040E-01	30	T002	1 0 2 0 2	anhydrous, form B
6.658E-04	2.400E-01	30	T002	1 0 2 0 2	hydrate
6.658E-04	2.400E-01	30	W006	2 2 2 1 2	hydrate, form C
9.738E-04	3.510E-01	37	H004	1 0 2 2 2	
5.500E-04	1.982E-01	ns	F327	0 0 1 2 2	
1.398E-03	5.040E-01	ns	W006	2 2 2 1 2	anhydrous, form B

3832. C$_{21}$H$_{28}$O$_5$
Aldosterone
18-Oxocorticosterone
Aldocortin
Electrocortin
18-Oxo-11β,21-dihydroxy-4-pregnene-3,20-dione
RN: 52-39-1   **MP** (°C):   108
MW: 360.45   **BP** (°C):

Solubility (Moles/L)	Solubility (Grams/L)	Temp (°C)	Ref (#)	Evaluation (T P E A A)	Comments
1.420E-04	5.118E-02	37	H034	1 0 2 1 2	pH 7.4

3833. C$_{21}$H$_{29}$FO$_5$
Fludrocortisone
9α-Fluoro-17-hydroxycorticosterone
9α-Fluorohydrocortisone
Florinef
RN: 127-31-1   **MP** (°C):   260dec
MW: 380.46   **BP** (°C):

Solubility (Moles/L)	Solubility (Grams/L)	Temp (°C)	Ref (#)	Evaluation (T P E A A)	Comments
2.918E-04	1.110E-01	25	K021	1 2 2 2 1	
8.516E-04	3.240E-01	25	L009	1 0 0 1 1	

3834. C$_{21}$H$_{29}$NO
N,N-DicyclohexylCinnamamide
N,N-Dicyclohexyl-3-phenyl2-Propenamide
RN: 6631-21-6   **MP** (°C):
MW: 311.47   **BP** (°C):

Solubility (Moles/L)	Solubility (Grams/L)	Temp (°C)	Ref (#)	Evaluation (T P E A A)	Comments
5.680E-06	1.769E-03	ns	H350	0 0 0 0 2	

3835. C$_{21}$H$_{30}$N$_4$O$_{10}$
Methylol Riboflavine
Methylol-riboflavin
RN: **MP** (°C):
MW: 498.49   **BP** (°C):

Solubility (Moles/L)	Solubility (Grams/L)	Temp (°C)	Ref (#)	Evaluation (T P E A A)	Comments
2.387E-02	1.190E+01	20	F300	1 0 0 0 2	compound not stable

3836.   C$_{21}$H$_{30}$N$_6$O$_4$S

Benzenesulfonamide, N-[2-(Dimethylamino)ethyl]-4-(2,3,4,5,6,7-hexahydro-2,6-dioxo-1,3-dipropyl-1H-purin-8-yl)-

RN: 89073-58-5   **MP** (°C):   270dec
MW: 462.58   **BP** (°C):

Solubility (Moles/L)	Solubility (Grams/L)	Temp (°C)	Ref (#)	Evaluation (T P E A A)	Comments
4.302E-02	1.990E+01	ns	H316	0 2 1 1 2	0.1N HCl
1.081E-04	5.000E-02	ns	H316	0 2 1 1 2	pH 7.4

3837.   C$_{21}$H$_{30}$O$_2$

Tetrahydrocannabinol
THC
Dronabinol
δ9-Tetrahydrocannabinol

RN: 1972-08-3   **MP** (°C):
MW: 314.47   **BP** (°C):

Solubility (Moles/L)	Solubility (Grams/L)	Temp (°C)	Ref (#)	Evaluation (T P E A A)	Comments
8.904E-06	2.800E-03	23	G018	1 0 0 1 0	

3838.   C$_{21}$H$_{30}$O$_2$

Progesterone
δ4-Pregnene-3,20-dione
Corlutin
Corlutina
Lutein
Pregn-4-ene-3,20-dione

RN: 57-83-0   **MP** (°C):   121
MW: 314.47   **BP** (°C):

Solubility (Moles/L)	Solubility (Grams/L)	Temp (°C)	Ref (#)	Evaluation (T P E A A)	Comments
1.700E-05	5.346E-03	10	B012	2 0 1 1 0	
2.200E-05	6.918E-03	20	B012	2 0 1 1 0	
3.210E-05	1.009E-02	20	L077	1 2 2 2 2	
2.600E-05	8.176E-03	21.70	M108	1 2 1 1 2	form A
4.837E-05	1.521E-02	23	B014	0 0 1 2 2	
3.720E-05	1.170E-02	24.00	M108	1 2 1 1 2	form B
2.800E-05	8.805E-03	25	B012	2 0 1 1 0	
2.512E-05	7.899E-03	25	B041	1 0 2 2 0	EFG
3.802E-05	1.196E-02	25	F312	1 1 2 2 2	units assumed
2.862E-05	9.000E-03	25	K003	2 1 1 1 1	
6.359E-04	2.000E-01	25	P324	0 1 0 2 1	
2.810E-05	8.837E-03	25.30	M108	1 2 1 1 2	form A
3.690E-05	1.160E-02	27.34	L077	1 2 2 2 2	
3.600E-05	1.132E-02	30	B012	2 0 1 1 0	

Solubility (Moles/L)	Solubility (Grams/L)	Temp (°C)	Ref (#)	Evaluation (T P E A A)	Comments
3.498E-05	1.100E-02	30	M007	2 2 1 2 2	average of 8
3.800E-05	1.195E-02	30.20	M108	1 2 1 1 2	form A
4.520E-05	1.421E-02	30.50	M108	1 2 1 1 2	form B
4.230E-05	1.330E-02	35	L077	1 2 2 2 2	
5.390E-05	1.695E-02	35.50	M108	1 2 1 1 2	form B
4.690E-05	1.475E-02	36.40	M108	1 2 1 1 2	form A
3.816E-05	1.200E-02	37	A086	1 0 1 1 2	
4.800E-05	1.509E-02	37	H034	1 0 2 1 2	pH 7.4
4.260E-05	1.340E-02	37	H035	1 1 1 1 2	pH 7.4
4.007E-05	1.260E-02	37	L010	1 0 0 1 1	
4.260E-05	1.340E-02	37.50	B041	1 0 2 2 2	
3.981E-05	1.252E-02	37.50	B041	1 0 2 2 0	EFG
3.800E-05	1.195E-02	40	B012	2 0 1 1 0	
6.750E-05	2.123E-02	40.70	M108	1 2 1 1 2	form B
6.370E-05	2.003E-02	41.30	M108	1 2 1 1 2	form A
4.580E-05	1.440E-02	42.34	L077	1 2 2 2 2	
6.500E-05	2.044E-02	46.10	M108	1 2 1 1 2	form A
4.900E-05	1.541E-02	50	B012	2 0 1 1 0	
4.930E-05	1.550E-02	50	L077	1 2 2 2 2	

3839. $C_{21}H_{30}O_3$

17-α-Hydroxyprogesterone
Pregn-4-ene-3,20-dione, 17-Hydroxy-
Prodix
Prodox
U 3096

RN: 68-96-2   **MP** (°C):   222
MW: 330.47   **BP** (°C):

Solubility (Moles/L)	Solubility (Grams/L)	Temp (°C)	Ref (#)	Evaluation (T P E A A)	Comments
1.530E-05	5.056E-03	20	L077	1 2 2 2 2	
1.960E-05	6.477E-03	27.34	L077	1 2 2 2 2	
2.760E-05	9.121E-03	35	L077	1 2 2 2 2	
3.580E-05	1.183E-02	42.34	L077	1 2 2 2 2	
4.290E-05	1.418E-02	50	L077	1 2 2 2 2	

3840. $C_{21}H_{30}O_3$

5,6-Dehydroisoandrosterone Acetate
Androst-5-en-17-one, 3-(Acetyloxy)-, (3β)-

RN: 853-23-6   **MP** (°C):   166
MW: 330.47   **BP** (°C):

Solubility (Moles/L)	Solubility (Grams/L)	Temp (°C)	Ref (#)	Evaluation (T P E A A)	Comments
3.480E-05	1.150E-02	ns	B057	0 2 1 1 2	

3841. $C_{21}H_{30}O_3$

Deoxycorticosterone
21-Hydroxyprogesterone
4-Pregnen-21-ol-3,20-dione
11-Deoxycorticosterone
21-Hydroxypregn-4-ene-3,20-dione
RN: 64-85-7   **MP** (°C): 141.5
MW: 330.47   **BP** (°C):

Solubility (Moles/L)	Solubility (Grams/L)	Temp (°C)	Ref (#)	Evaluation (T P E A A)	Comments
4.387E-04	1.450E-01	25	K003	2 1 1 1 1	
1.800E-04	5.948E-02	37	E014	2 2 2 1 2	pH 7.3
1.070E-04	3.536E-02	37	H034	1 0 2 1 2	pH 7.4

3842. $C_{21}H_{30}O_3$

Testosterone Acetate
17-O-Acetyltestosterone
Androst-4-en-3-one, 17-(Acetyloxy)-, (17β)-
RN: 1045-69-8   **MP** (°C): 140
MW: 330.47   **BP** (°C):

Solubility (Moles/L)	Solubility (Grams/L)	Temp (°C)	Ref (#)	Evaluation (T P E A A)	Comments
7.111E-06	2.350E-03	25	J004	1 0 1 1 2	
7.111E-06	2.350E-03	ns	B057	0 2 1 1 2	

3843. $C_{21}H_{30}O_5$

Hydrocortisone
11β,17,21-Trihydroxypregn-4-ene-3,20-dione
Colifoam
Cortaid
Cortef
Bactine
RN: 50-23-7   **MP** (°C): 218.5
MW: 362.47   **BP** (°C):

Solubility (Moles/L)	Solubility (Grams/L)	Temp (°C)	Ref (#)	Evaluation (T P E A A)	Comments
4.780E-04	1.733E-01	10	B012	2 0 1 1 0	
7.725E-04	2.800E-01	20	A067	0 0 0 0 1	
7.430E-04	2.693E-01	20	B012	2 0 1 1 0	
8.820E-04	3.197E-01	25	B012	2 0 1 1 0	
7.725E-04	2.800E-01	25	H015	1 0 0 0 1	
8.194E-04	2.970E-01	25	H098	1 0 2 0 2	
8.190E-04	2.969E-01	25	H320	1 0 2 2 2	
8.194E-04	2.970E-01	25	H320	1 0 2 2 2	
7.860E-04	2.849E-01	25	K003	2 1 1 1 1	
1.614E-03	5.850E-01	25	K021	1 2 2 2 1	

7.725E-04	2.800E-01	25	M023	1 0 2 1 1	
9.896E-03	3.587E+00	25	P324	0 1 0 2 1	
1.034E-03	3.748E-01	30	B012	2 0 1 1 0	
1.000E-03	3.625E-01	30	L344	2 0 1 1 0	EFG
1.070E-03	3.878E-01	37	H036	1 0 2 2 2	EFG
1.265E-03	4.585E-01	40	B012	2 0 1 1 0	
1.519E-03	5.506E-01	50	B012	2 0 1 1 0	
7.725E-04	2.800E-01	298	F016	0 0 0 0 2	

3844. $C_{21}H_{30}O_6$
Cortisone Acetate
Pregn-4-ene-3,11,20-trione, 21-(Acetyloxy)-17-hydroxy-
RN: 50-04-4   **MP** (°C):   235
MW: 378.47   **BP** (°C):

Solubility (Moles/L)	Solubility (Grams/L)	Temp (°C)	Ref (#)	Evaluation (T P E A A)	Comments
5.284E-05	2.000E-02	22.5	G301	2 1 0 1 2	
5.020E-05	1.900E-02	25	K003	2 1 1 1 1	
5.284E-05	2.000E-02	25	M023	1 0 2 1 0	
7.398E-05	2.800E-02	25	P096	1 0 2 2 2	
1.000E-04	3.785E-02	30	L068	1 0 0 1 0	EFG

3845. $C_{21}H_{31}NO$
N-Cyclododecylcinnamamide
2-Propenamide, N-Cyclododecyl-3-phenyl
RN: 59832-03-0   **MP** (°C):
MW: 313.49   **BP** (°C):

Solubility (Moles/L)	Solubility (Grams/L)	Temp (°C)	Ref (#)	Evaluation (T P E A A)	Comments
3.910E-08	1.226E-05	ns	H350	0 0 0 0 2	

3846. $C_{21}H_{31}N_3O_2$
2-Pentoxy-N-[2-(diethyl-amino)ethyl]-4-quinoline Carboxamide
N-[2-(Diethylamino)ethyl]-2-pentoxyquinoline-4-carboxamide
RN: 2717-02-4   **MP** (°C):
MW: 357.50   **BP** (°C):

Solubility (Moles/L)	Solubility (Grams/L)	Temp (°C)	Ref (#)	Evaluation (T P E A A)	Comments
5.300E-05	1.895E-02	ns	B018	0 0 0 0 1	
5.300E-05	1.895E-02	ns	M066	0 0 0 0 1	

3847.   C$_{21}$H$_{32}$O$_2$
3,20-Pregnanedione
7α-17-Dimethyltestosterone
Bolasterone
RN: 128-23-4   **MP** (°C):
MW: 316.49   **BP** (°C):

Solubility (Moles/L)	Solubility (Grams/L)	Temp (°C)	Ref (#)	Evaluation (T P E A A)	Comments
1.833E-04	5.800E-02	37	H004	1 0 2 2 2	

3848.   C$_{21}$H$_{32}$O$_2$
7α,17-Dimethyl-19-nortestosterone
RN: **MP** (°C):
MW: 316.49   **BP** (°C):

Solubility (Moles/L)	Solubility (Grams/L)	Temp (°C)	Ref (#)	Evaluation (T P E A A)	Comments
1.434E-04	4.540E-02	37	H004	1 0 2 2 2	

3849.   C$_{21}$H$_{32}$O$_2$
Pregnenolone
3β-Hydroxy-5-pregnen-20-one
5-Pregnen-3β-ol-20-one
3β-Hydroxypregn-5-en-20-one
RN: 145-13-1   **MP** (°C):   193
MW: 316.49   **BP** (°C):

Solubility (Moles/L)	Solubility (Grams/L)	Temp (°C)	Ref (#)	Evaluation (T P E A A)	Comments
2.230E-05	7.058E-03	37	H034	1 0 2 1 2	pH 7.4

3850.   C$_{21}$H$_{32}$O$_3$
Androstanolone Acetate
Androstan-3-one, 17-(Acetyloxy)-, (5α,17β)-
Stanolone Acetate
RN: 1164-91-6   **MP** (°C):
MW: 332.49   **BP** (°C):

Solubility (Moles/L)	Solubility (Grams/L)	Temp (°C)	Ref (#)	Evaluation (T P E A A)	Comments
2.672E-01	8.884E+01	ns	B057	0 2 1 1 2	

3851.   C$_{21}$H$_{33}$NO
2-Propenamide, N-Dodecyl-3-phenyl-
RN: 55125-24-1   **MP** (°C):
MW: 315.50   **BP** (°C):

Solubility (Moles/L)	Solubility (Grams/L)	Temp (°C)	Ref (#)	Evaluation (T P E A A)	Comments
2.100E-06	6.626E-04	ns	H350	0 0 0 0 2	

3852.   C$_{21}$H$_{33}$NO$_7$
Lasiocarpine
(7α-Angelyloxy-5,6,7,8α-Tetrahydro-3H-pyrrolizin-1-yl)methyl-2,3-dihydroxy-2-(1'-methoxyethyl)-3-methylbutyrate
RN: 303-34-4   **MP** (°C):   97
MW: 411.50   **BP** (°C):

Solubility (Moles/L)	Solubility (Grams/L)	Temp (°C)	Ref (#)	Evaluation (T P E A A)	Comments
1.641E-02	6.754E+00	ns	I312	0 0 0 0 1	

3853.   C$_{21}$H$_{34}$O$_3$
Tetradecyl p-Hydroxybenzoate
Tetradecyl 4-Hydroxybenzoate
RN: 71177-53-2   **MP** (°C):
MW: 334.50   **BP** (°C):

Solubility (Moles/L)	Solubility (Grams/L)	Temp (°C)	Ref (#)	Evaluation (T P E A A)	Comments
1.088E-03	3.639E-01	25	D081	1 2 2 1 2	

3854.   C$_{21}$H$_{35}$NO$_3$
4-Octoxybenzoic Acid-2-(diethyl-amino)ethyl Ester
RN: 38973-76-1   **MP** (°C):
MW: 349.52   **BP** (°C):

Solubility (Moles/L)	Solubility (Grams/L)	Temp (°C)	Ref (#)	Evaluation (T P E A A)	Comments
4.000E-05	1.398E-02	ns	M066	0 0 0 0 1	

3855. C$_{21}$H$_{36}$O$_4$
4-Nonylphenol Triethoxylate
Ethanol, 2-[2-[2-(4-Nonylphenoxy)ethoxy]ethoxy]-
RN: 51437-95-7   **MP** (°C):
MW: 352.52   **BP** (°C):

Solubility (Moles/L)	Solubility (Grams/L)	Temp (°C)	Ref (#)	Evaluation (T P E A A)	Comments
1.668E-05	5.880E-03	20.5	A335	1 0 2 2 2	
1.670E-05	5.887E-03	20.5	A335	1 0 2 2 2	

3856. C$_{21}$H$_{40}$O$_4$
α-Monoolein
1-Monoolein
Glycerol Monooleate
9-Octadecenoic Acid (Z)-, Monoester with 1,2,3-propanetriol
1-Oleoyl-sn-glycerol
RN: 25496-72-4   **MP** (°C):
MW: 356.55   **BP** (°C):

Solubility (Moles/L)	Solubility (Grams/L)	Temp (°C)	Ref (#)	Evaluation (T P E A A)	Comments
<1.00E-05	<3.57E-03	30	O321	2 2 2 2 1	

3857. C$_{21}$H$_{44}$
2-Methyleicosane
19-Methyleicosane
RN: 1560-84-5   **MP** (°C):
MW: 296.58   **BP** (°C):

Solubility (Moles/L)	Solubility (Grams/L)	Temp (°C)	Ref (#)	Evaluation (T P E A A)	Comments
5.091E-13	1.510E-10	23	C332	2 0 2 2 1	

3858. C$_{21}$H$_{44}$
3-Methyleicosane
18-Methyleicosane
RN: 6418-46-8   **MP** (°C):
MW: 296.58   **BP** (°C):

Solubility (Moles/L)	Solubility (Grams/L)	Temp (°C)	Ref (#)	Evaluation (T P E A A)	Comments
5.294E-13	1.570E-10	23	C332	2 0 2 2 1	

3859.   C$_{22}$H$_{12}$
Benzo[g,h,i]perylene
Benz[g,h,i]perylene
RN: 191-24-2   **MP** (°C):   279
MW: 276.34   **BP** (°C):   >500

Solubility (Moles/L)	Solubility (Grams/L)	Temp (°C)	Ref (#)	Evaluation (T P E A A)	Comments
6.500E-10	1.796E-07	25	K123	1 0 2 2 1	
9.409E-10	2.600E-07	25	M064	1 1 2 2 1	
9.400E-10	2.598E-07	25	M342	1 0 1 1 1	
9.409E-10	2.600E-07	ns	M344	0 0 0 0 1	
2.533E-09	7.000E-07	ns	W302	0 0 0 0 0	

3860.   C$_{22}$H$_{12}$
Indeno(1,2,3-cd)pyrene
Indeno[1,2,3-cd]pyrene
o-Phenylenepyrene
RN: 193-39-5   **MP** (°C):   162.5
MW: 276.34   **BP** (°C):   536

Solubility (Moles/L)	Solubility (Grams/L)	Temp (°C)	Ref (#)	Evaluation (T P E A A)	Comments
6.876E-10	1.900E-07	ns	W302	0 0 0 0 1	

3861.   C$_{22}$H$_{14}$
Picene
1,2,7,8-Dibenzphenanthrene
3,4-Benzchrysene
RN: 213-46-7   **MP** (°C):   366
MW: 278.36   **BP** (°C):   518

Solubility (Moles/L)	Solubility (Grams/L)	Temp (°C)	Ref (#)	Evaluation (T P E A A)	Comments
1.550E-08	4.315E-06	20	E009	1 0 0 1 2	
8.981E-09	2.500E-06	27	D003	1 0 0 1 1	

3862.   C$_{22}$H$_{14}$
1,2:3,4-Dibenzanthracene
RN: 215-58-7   **MP** (°C):   205
MW: 278.36   **BP** (°C):   518

Solubility (Moles/L)	Solubility (Grams/L)	Temp (°C)	Ref (#)	Evaluation (T P E A A)	Comments
5.748E-09	1.600E-06	25	B319	2 0 1 2 1	
8.200E-08	2.283E-05	25	K123	1 0 2 2 1	

3863. C22H14
1,2:5,6-Dibenzanthracene
1,2,5,6-Dibenzanthracene
RN: 53-70-3   **MP** (°C): 266
MW: 278.36   **BP** (°C): 524

Solubility (Moles/L)	Solubility (Grams/L)	Temp (°C)	Ref (#)	Evaluation (T P E A A)	Comments
8.945E-09	2.490E-06	24	H106	1 0 2 2 2	
7.904E-09	2.200E-06	25	B319	2 0 1 2 2	
2.150E-09	5.985E-07	25	K001	2 2 2 2 2	
1.100E-07	3.062E-05	25	K123	1 0 2 2 1	*sic*
8.945E-09	2.490E-06	25	M156	1 2 1 1 2	
1.800E-09	5.010E-07	25	M342	1 0 1 1 2	
1.796E-09	5.000E-07	27	D003	1 0 0 1 1	

3864. C22H14
1,2:7,8-Dibenzanthracene
Dibenz[a,j]anthracene
Dinaphthanthracene
RN: 224-41-9   **MP** (°C): 196
MW: 278.36   **BP** (°C):

Solubility (Moles/L)	Solubility (Grams/L)	Temp (°C)	Ref (#)	Evaluation (T P E A A)	Comments
3.100E-08	8.629E-06	25	K123	1 0 2 2 1	
4.311E-08	1.200E-05	27	D003	1 0 0 1 1	

3865. C22H16F3N3
Fluotrimazole
1H-1,2,3-Triazole, 1-[Diphenyl[3-(trifluoromethyl)phenyl]methyl]-
RN: 57381-79-0   **MP** (°C): 132
MW: 379.39   **BP** (°C):

Solubility (Moles/L)	Solubility (Grams/L)	Temp (°C)	Ref (#)	Evaluation (T P E A A)	Comments
3.954E-09	1.500E-06	20	M161	1 0 0 0 1	

3866. C22H16O8
Ethyl Biscoumacetate
Tromexan
RN: 548-00-5   **MP** (°C): 154
MW: 408.37   **BP** (°C):

Solubility (Moles/L)	Solubility (Grams/L)	Temp (°C)	Ref (#)	Evaluation (T P E A A)	Comments
2.179E-04	8.900E-02	20	K028	2 1 2 1 2	pH 3.8, form I
3.747E-04	1.530E-01	20	K028	2 1 2 1 2	pH 3.8, form II
2.179E-04	8.899E-02	20	M042	1 0 0 0 1	pH 3.8, form I, mp 172-182 °C
3.761E-04	1.536E-01	20	M042	1 0 0 0 2	pH 3.8, form II, mp 153-160 °C

3867.   C$_{22}$H$_{17}$ClN$_2$
Clotrimazole
1-(o-Chloro-α,α-diphenylbenzyl)imidazole
1-[α-(2-Chlorophenyl)benzhydryl]imidazole
Lotrimin
RN: 23593-75-1   **MP** (°C):   147-149
MW: 344.85   **BP** (°C):

Solubility (Moles/L)	Solubility (Grams/L)	Temp (°C)	Ref (#)	Evaluation (T P E A A)	Comments
<2.90E-05	<1.00E-02	25	H328	1 0 0 1 0	

3868.   C$_{22}$H$_{18}$N$_2$O$_4$S
Hydantoin, 5,5-Diphenyl-1-(o-tolylsulfonyl)-
1-(o-Methylbenzenesulfonyl)-5,5-diphenyl-hydantoin
RN: 24759-41-9   **MP** (°C):
MW: 406.46   **BP** (°C):

Solubility (Moles/L)	Solubility (Grams/L)	Temp (°C)	Ref (#)	Evaluation (T P E A A)	Comments
1.870E-06	7.600E-04	37	F183	1 0 1 1 2	intrinsic

3869.   C$_{22}$H$_{18}$N$_2$O$_5$S
1-(p-Methoxylbenzenesulfonyl)-5,5-diphenyl-hydantoin
RN: 24759-37-3   **MP** (°C):
MW: 422.46   **BP** (°C):

Solubility (Moles/L)	Solubility (Grams/L)	Temp (°C)	Ref (#)	Evaluation (T P E A A)	Comments
1.207E-06	5.100E-04	37	F183	1 0 1 1 2	intrinsic

3870.   C$_{22}$H$_{19}$Br$_2$NO$_3$
Deltamethrin
3-(2,2-Dibromoethenyl)-2,2-dimethylcyclopropanecarboxylic Acid Cyano(3-
phenoxyphenyl)methylEster
RN: 52918-63-5   **MP** (°C):   98-101
MW: 505.22   **BP** (°C):   300

Solubility (Moles/L)	Solubility (Grams/L)	Temp (°C)	Ref (#)	Evaluation (T P E A A)	Comments
3.959E-09	2.000E-06	25	M364	1 0 0 0 1	

3871. C$_{22}$H$_{19}$F$_6$NOS
α-Piperidyl-3,6-bis(trifluoromethyl)-9-phenanthrenemethanol
RN: 31817-24-0   **MP** (°C):   215
MW: 459.46   **BP** (°C):

Solubility (Moles/L)	Solubility (Grams/L)	Temp (°C)	Ref (#)	Evaluation (T P E A A)	Comments
1.632E-05	7.500E-03	25	A013	1 0 2 2 0	average

3872. C$_{22}$H$_{20}$
10-Butyl-1,2-benzanthracene
RN: 188124-94-9   **MP** (°C):   97
MW: 284.40   **BP** (°C):

Solubility (Moles/L)	Solubility (Grams/L)	Temp (°C)	Ref (#)	Evaluation (T P E A A)	Comments
2.813E-08	8.000E-06	27	D003	1 0 0 1 1	

3873. C$_{22}$H$_{20}$O$_{13}$
Carminic Acid
Carmine
Carminsaeure
RN: 1260-17-9   **MP** (°C):
MW: 492.40   **BP** (°C):

Solubility (Moles/L)	Solubility (Grams/L)	Temp (°C)	Ref (#)	Evaluation (T P E A A)	Comments
2.637E-03	1.298E+00	rt	D021	0 0 1 1 1	

3874. C$_{22}$H$_{22}$FN$_3$O$_2$
Droperidol
2H-Benzimidazol-2-one, 1-[1-[4-(4-Fluorophenyl)-4-oxobutyl]-1,2,3,6-tetrahydro-4-pyridinyl]-1,3-dihydro-
Sintodril
Neurolidol
R 4749
RN: 548-73-2   **MP** (°C):
MW: 379.44   **BP** (°C):

Solubility (Moles/L)	Solubility (Grams/L)	Temp (°C)	Ref (#)	Evaluation (T P E A A)	Comments
1.081E-05	4.100E-03	30	P044	1 0 1 0 1	

3875.　C$_{22}$H$_{22}$N$_2$O$_8$
Methacycline Base
Oxytetracycline, 6-Methylene-
Tri-methacycline
Rondomycin
RN:　　914-00-1　　　**MP** (°C):
MW:　　442.43　　　　**BP** (°C):

Solubility (Moles/L)	Solubility (Grams/L)	Temp (°C)	Ref (#)	Evaluation (T P E A A)	Comments
1.706E-02	7.548E+00	21	M044	2 0 2 2 2	

3876.　C$_{22}$H$_{22}$N$_4$O$_6$
Benzoyl-mitomycin C
RN:　　　　　　　　　**MP** (°C):
MW:　　438.44　　　　**BP** (°C):

Solubility (Moles/L)	Solubility (Grams/L)	Temp (°C)	Ref (#)	Evaluation (T P E A A)	Comments
1.000E-05	4.384E-03	25	M316	1 1 1 1 2	

3877.　C$_{22}$H$_{23}$ClN$_2$O$_8$
Chlortetracycline
7-Chlortetracycline
Acronize PD
Acronize
RN:　　57-62-5　　　**MP** (°C):
MW:　　478.89　　　　**BP** (°C):

Solubility (Moles/L)	Solubility (Grams/L)	Temp (°C)	Ref (#)	Evaluation (T P E A A)	Comments
1.316E-03	6.300E-01	25	B191	1 0 0 0 1	
2.297E-03	1.100E+00	37	M104	1 2 1 1 0	form II, EFG, recrystallized
1.566E-03	7.500E-01	37	M104	1 2 1 1 0	form I, EFG, recrystallized
2.088E-04	1.000E-01	37	M105	1 2 1 1 0	EFG

3878.   C$_{22}$H$_{23}$NO$_7$

Noscapine
Narcotine
O-Methylnarcotoline
Opianin
Opian

RN: 128-62-1	**MP** (°C): 176	
MW: 413.43	**BP** (°C):	

Solubility (Moles/L)	Solubility (Grams/L)	Temp (°C)	Ref (#)	Evaluation (T P E A A)	Comments
4.000E-05	1.654E-02	15	K059	2 2 2 0 0	
7.327E-04	3.029E-01	25	D004	1 0 0 0 0	
7.256E-04	3.000E-01	30	A073	1 1 1 1 0	
1.693E-03	7.000E-01	40	A073	1 1 1 1 0	
2.419E-03	1.000E+00	50	A073	1 1 1 1 1	
2.419E-03	1.000E+00	60	A073	1 1 1 1 1	
2.419E-03	1.000E+00	70	A073	1 1 1 1 1	
2.419E-03	1.000E+00	80	A073	1 1 1 1 1	
3.628E-03	1.500E+00	90	A073	1 1 1 1 1	
4.838E-03	2.000E+00	100	A073	1 1 1 1 1	

3879.   C$_{22}$H$_{24}$N$_2$O$_8$

Tetracycline
Achromycin V
Sumycin
Robitet
Panmycin

RN: 60-54-8	**MP** (°C): 176dec	
MW: 444.45	**BP** (°C):	

Solubility (Moles/L)	Solubility (Grams/L)	Temp (°C)	Ref (#)	Evaluation (T P E A A)	Comments
9.900E-04	4.400E-01	25	B191	1 0 0 0 1	neutral pH
5.200E-04	2.311E-01	25	G012	2 0 2 1 0	EFG, pH 5.0
5.700E-04	2.533E-01	25	H017	1 2 2 2 0	EFG, pH 5.0
2.655E-03	1.180E+00	29	N031	1 2 2 2 0	EFG, pH 5.0
7.600E-04	3.378E-01	30	L069	1 0 1 1 0	EFG
1.777E-03	7.900E-01	35	N031	1 2 2 2 0	EFG, pH 5.0
7.875E-02	3.500E+01	37	M104	1 2 1 1 2	form II, recrystallized
6.232E-02	2.770E+01	37	M104	1 2 1 1 2	form I, recrystallized
6.478E-04	2.879E-01	ns	N302	0 2 1 2 2	

3880.　C$_{22}$H$_{24}$N$_2$O$_8$.H$_2$O
Doxycycline (Monohydrate)
Doxylin
Monodox
Vibra-tabs
Doxy-caps
Vibramycin
RN:　　564-25-0　　　**MP** (°C):　　201dec
MW:　　462.46　　　　**BP** (°C):

Solubility (Moles/L)	Solubility (Grams/L)	Temp (°C)	Ref (#)	Evaluation (T P E A A)	Comments
1.362E-03	6.300E-01	25	B132	2 1 1 1 0	EFG

3881.　C$_{22}$H$_{24}$N$_2$O$_9$
Oxytetracycline
Glomycin
Hydroxytetracycline
Riomitsin
Terrafungine
Stevacin
RN:　　79-57-2　　　**MP** (°C):　　184
MW:　　460.44　　　　**BP** (°C):

Solubility (Moles/L)	Solubility (Grams/L)	Temp (°C)	Ref (#)	Evaluation (T P E A A)	Comments
4.234E-04	1.950E-01	20	L051	1 0 0 0 2	
9.990E-04	4.600E-01	25	B191	1 0 0 0 1	neutral pH
4.800E-04	2.210E-01	25	G012	2 0 2 1 0	EFG, pH 5.0
6.798E-04	3.130E-01	25	H005	1 0 1 2 2	pH 5.8
5.000E-04	2.302E-01	25	H017	1 2 2 2 0	EFG, pH 5.0
6.515E-04	3.000E-01	29	N031	1 2 2 2 0	EFG, pH 5.0
8.687E-04	4.000E-01	37	M104	1 2 1 1 0	form II, EFG, recrystallized
6.515E-04	3.000E-01	37	M104	1 2 1 1 0	form I, EFG, recrystallized

3882.　C$_{22}$H$_{24}$N$_4$O$_5$
Benzyl-mitomycin C
RN:　　　　　　　　　**MP** (°C):
MW:　　424.46　　　　**BP** (°C):

Solubility (Moles/L)	Solubility (Grams/L)	Temp (°C)	Ref (#)	Evaluation (T P E A A)	Comments
1.490E-03	6.324E-01	25	M316	1 1 1 1 2	

3883.   C₂₂H₂₅NO₆
Colchicine
Colchicin
RN: 64-86-8   **MP** (°C):
MW: 399.45   **BP** (°C):

Solubility (Moles/L)	Solubility (Grams/L)	Temp (°C)	Ref (#)	Evaluation (T P E A A)	Comments
9.629E-02	3.846E+01	20	D041	1 0 0 0 0	
1.088E-01	4.348E+01	25	D004	1 0 0 0 0	

3884.   C₂₂H₂₆F₃N₃OS
Fluphenazine
Permitil
Modecate
Prolixin
RN: 69-23-8   **MP** (°C):   <25
MW: 437.53   **BP** (°C):   271

Solubility (Moles/L)	Solubility (Grams/L)	Temp (°C)	Ref (#)	Evaluation (T P E A A)	Comments
7.100E-05	3.106E-02	37	F011	1 0 1 1 1	pH 7.4

3885.   C₂₂H₂₈F₂O₅
Flumethasone
Flumethasonpivalate
RN: 2135-17-3   **MP** (°C):
MW: 410.46   **BP** (°C):

Solubility (Moles/L)	Solubility (Grams/L)	Temp (°C)	Ref (#)	Evaluation (T P E A A)	Comments
2.436E-06	1.000E-03	20	A067	0 0 0 0 0	

3886.   C₂₂H₂₈O₃
Norethindrone Acetate
Norethisterone Acetate
RN: 51-98-9   **MP** (°C):   161
MW: 340.47   **BP** (°C):

Solubility (Moles/L)	Solubility (Grams/L)	Temp (°C)	Ref (#)	Evaluation (T P E A A)	Comments
9.288E-06	3.162E-03	10	L078	1 0 1 2 0	EFG
1.312E-05	4.467E-03	20	L078	1 0 1 2 0	EFG
1.570E-05	5.345E-03	25	H099	1 0 2 2 2	
1.652E-05	5.623E-03	25	L078	1 0 1 2 2	
1.853E-05	6.310E-03	30	L078	1 0 1 2 0	EFG
2.937E-05	1.000E-02	40	L078	1 0 1 2 0	EFG

3887.   C$_{22}$H$_{28}$O$_3$
Canrenone
17-Hydroxy-3-oxo-17α-pregna-4,6-diene-21-carboxylic Acid Lactone
RN: 976-71-6   **MP** (°C):   149-151
MW: 340.47   **BP** (°C):

Solubility (Moles/L)	Solubility (Grams/L)	Temp (°C)	Ref (#)	Evaluation (T P E A A)	Comments
8.000E-07	2.724E-04	25	G017	1 0 1 0 0	EFG
8.100E-05	2.758E-02	37	C004	1 0 2 2 1	*sic*
8.958E-07	3.050E-04	37	O306	1 0 1 2 2	
6.374E-07	2.170E-04	rt	O306	0 0 1 2 2	

3888.   C$_{22}$H$_{29}$FO$_4$
Fluorometholone
9-Fluoro-11β,17-dihydroxy-6α-methylpregna-1,4-diene-3,20-dione
21-Desoxy-9α-fluoro-6α-methyl-prednisolone
RN: 426-13-1   **MP** (°C):
MW: 376.47   **BP** (°C):

Solubility (Moles/L)	Solubility (Grams/L)	Temp (°C)	Ref (#)	Evaluation (T P E A A)	Comments
7.968E-05	3.000E-02	25	G008	1 2 1 1 0	

3889.   C$_{22}$H$_{29}$FO$_5$
Betamethasone
Pregna-1,4-diene-3,20-dione, 9-Fluoro-11,17,21-trihydroxy-16-methyl-, (11β,16β)-
RN: 378-44-9   **MP** (°C):   230
MW: 392.47   **BP** (°C):

Solubility (Moles/L)	Solubility (Grams/L)	Temp (°C)	Ref (#)	Evaluation (T P E A A)	Comments
1.478E-04	5.800E-02	25	K003	2 1 1 1 1	
1.936E-04	7.599E-02	25	P096	1 0 2 2 2	
1.500E-04	5.887E-02	30	O321	2 2 2 2 1	
1.529E-04	6.000E-02	30	O321	2 2 2 2 1	

3890.   C$_{22}$H$_{29}$FO$_5$
Dexamethasone
Dexamethasone Alcohol
RN: 50-02-2   **MP** (°C):   262
MW: 392.47   **BP** (°C):

Solubility (Moles/L)	Solubility (Grams/L)	Temp (°C)	Ref (#)	Evaluation (T P E A A)	Comments
8.200E-05	3.218E-02	10	B012	2 0 1 1 0	
1.580E-04	6.201E-02	20	B012	2 0 1 1 0	
2.800E-04	1.099E-01	23	L345	1 0 1 1 2	
2.270E-04	8.909E-02	25	B012	2 0 1 1 0	
2.140E-04	8.399E-02	25	K003	2 1 1 1 1	
3.083E-04	1.210E-01	25	K021	1 2 2 2 1	
2.548E-04	1.000E-01	25	P312	1 2 2 2 2	
2.520E-04	9.890E-02	30	B012	2 0 1 1 0	
2.955E-04	1.160E-01	37	D026	1 2 1 2 2	
3.560E-04	1.397E-01	40	B012	2 0 1 1 0	
4.600E-04	1.805E-01	50	B012	2 0 1 1 0	
1.707E-04	6.700E-02	ns	N302	0 2 1 2 1	

3891.   C$_{22}$H$_{30}$Cl$_2$N$_{10}$
Chlorhexidin
Chlorhexidine
bis(5-(p-Chlorophenyl)biguanidinio)hexane
RN: 55-56-1   **MP** (°C):
MW: 505.46   **BP** (°C):

Solubility (Moles/L)	Solubility (Grams/L)	Temp (°C)	Ref (#)	Evaluation (T P E A A)	Comments
1.583E-04	7.999E-02	20	D341	1 0 1 1 0	
8.309E-05	4.200E-02	22.5	G301	2 1 0 1 2	

3892.   C$_{22}$H$_{30}$N$_2$O$_2$
Aspidospermine
Aspidospermidine, 1-Acetyl-17-methoxy-
RN: 466-49-9   **MP** (°C):   208
MW: 354.50   **BP** (°C):

Solubility (Moles/L)	Solubility (Grams/L)	Temp (°C)	Ref (#)	Evaluation (T P E A A)	Comments
4.701E-04	1.666E-01	c	D004	1 0 0 0 0	

3893.   C$_{22}$H$_{30}$O$_5$
Methylprednisolone
6α-Methylprednisolone
Medrol
Solumedrol
Metrisone
Promacortine

RN: 83-43-2   **MP** (°C):   232.5
MW: 374.48   **BP** (°C):

Solubility (Moles/L)	Solubility (Grams/L)	Temp (°C)	Ref (#)	Evaluation (T P E A A)	Comments
3.204E-04	1.200E-01	25	A014	1 0 1 1 0	EFG
2.403E-04	9.000E-02	25	A014	1 0 1 1 0	EFG, pH 5.0
2.534E-03	9.491E-01	25	G008	1 2 1 1 1	
3.445E-04	1.290E-01	25	K021	1 2 2 2 1	
1.335E-04	5.000E-02	27.14	H026	1 0 2 1 0	EFG, form I
1.923E-04	7.199E-02	30.0	H010	2 2 1 1 1	
4.273E-04	1.600E-01	31.72	H026	1 0 2 1 0	EFG, form II
3.124E-04	1.170E-01	37	H004	1 0 2 2 2	polymorph I
3.765E-04	1.410E-01	37	H004	1 0 2 2 2	polymorph II
5.341E-04	2.000E-01	40.32	H026	1 0 2 1 0	EFG, form II
2.937E-04	1.100E-01	40.32	H026	1 0 2 1 0	EFG, form I
4.273E-04	1.600E-01	51.52	H026	1 0 2 1 0	EFG, form I
1.362E-03	5.100E-01	81.45	H026	1 0 2 1 0	EFG, form II
1.068E-03	4.000E-01	81.45	H026	1 0 2 1 0	EFG, form I
2.670E-04	1.000E-01	ns	M169	0 0 0 0 1	

3894.   C$_{22}$H$_{30}$O$_6$
5,16-β-Dihydroxy-6-β-methyl-3,11-dioxo-5-α-pregn-17(20)-ene-cis-20-carboxylic Acid
Methyl Ester
U-20235

RN: **MP** (°C):
MW: 390.48   **BP** (°C):

Solubility (Moles/L)	Solubility (Grams/L)	Temp (°C)	Ref (#)	Evaluation (T P E A A)	Comments
6.402E-04	2.500E-01	ns	K029	0 0 2 1 1	

3895. C$_{22}$H$_{32}$O$_3$
Methyltestosterone Acetate
17-α-Methyltestosterone Acetate
RN: 1099-79-2   **MP** (°C):   164
MW: 344.50   **BP** (°C):

Solubility (Moles/L)	Solubility (Grams/L)	Temp (°C)	Ref (#)	Evaluation (T P E A A)	Comments
1.430E-05	4.926E-03	25	H099	1 0 2 2 2	
5.196E-06	1.790E-03	ns	B057	0 2 1 1 2	

3896. C$_{22}$H$_{32}$O$_3$
Testosterone Propionate
17-(1-Oxopropoxy)-(17β)-androst-4-en-3-one
Testosterone-17-Propionate
Agovirin
Androsan
Androgen
RN: 57-85-2   **MP** (°C):   120
MW: 344.50   **BP** (°C):

Solubility (Moles/L)	Solubility (Grams/L)	Temp (°C)	Ref (#)	Evaluation (T P E A A)	Comments
1.710E-04	5.891E-02	20	F012	1 0 1 1 1	
4.300E-06	1.481E-03	25	J004	1 0 1 1 2	
5.806E-06	2.000E-03	25	K003	2 1 1 1 1	
6.096E-06	2.100E-03	30	T005	2 0 2 2 1	
1.060E-05	3.652E-03	37.50	B054	1 0 1 1 2	
4.296E-06	1.480E-03	ns	B057	0 2 1 1 2	

3897. C$_{22}$H$_{32}$O$_3$
5,6-Dehydroisoandrosterone Propionate
RN: 1167-87-9   **MP** (°C):
MW: 344.50   **BP** (°C):

Solubility (Moles/L)	Solubility (Grams/L)	Temp (°C)	Ref (#)	Evaluation (T P E A A)	Comments
2.415E-05	8.320E-03	ns	B057	0 2 1 1 2	

3898. C$_{22}$H$_{32}$O$_3$
Nandrolone Butyrate
RN: **MP** (°C):
MW: 344.50   **BP** (°C):

Solubility (Moles/L)	Solubility (Grams/L)	Temp (°C)	Ref (#)	Evaluation (T P E A A)	Comments
1.460E-05	5.030E-03	37	C026	2 2 1 2 2	

3899.   C$_{22}$H$_{33}$N$_3$O$_2$

2-Hexoxy-N-[2-(diethyl-amino)ethyl]-4-quinoline Carboxamide

N-[2-(Diethylamino)ethyl]-2-hexoxyquinoline-4-carboxamide

RN: 2717-03-5   **MP** (°C):

MW: 371.53   **BP** (°C):

Solubility (Moles/L)	Solubility (Grams/L)	Temp (°C)	Ref (#)	Evaluation (T P E A A)	Comments
6.700E-06	2.489E-03	ns	B018	0 0 0 0 1	
6.700E-06	2.489E-03	ns	M066	0 0 0 0 1	

3900.   C$_{22}$H$_{34}$Cl$_2$O$_3$

2,4-Dichlorophenoxyacetic Acid n-Tetradecyl Ester

RN: 65267-96-1   **MP** (°C):

MW: 417.42   **BP** (°C):

Solubility (Moles/L)	Solubility (Grams/L)	Temp (°C)	Ref (#)	Evaluation (T P E A A)	Comments
1.161E-05	4.848E-03	ns	M120	0 0 1 1 2	

3901.   C$_{22}$H$_{34}$N$_6$O$_4$

2,5-Diaziridinyl-3,6-di(1'-piperazineethanol)-1,4-benzoquinone

RN: 59886-40-7   **MP** (°C):   170

MW: 446.55   **BP** (°C):

Solubility (Moles/L)	Solubility (Grams/L)	Temp (°C)	Ref (#)	Evaluation (T P E A A)	Comments
4.479E-02	2.000E+01	rt	C317	0 2 0 0 0	

3902.   C$_{22}$H$_{34}$O$_3$

Androstanolone Propionate

Androstan-3-one, 17-(1-Oxopropoxy)-, (5α,17β)-

RN: 855-22-1   **MP** (°C):

MW: 346.51   **BP** (°C):

Solubility (Moles/L)	Solubility (Grams/L)	Temp (°C)	Ref (#)	Evaluation (T P E A A)	Comments
1.789E-06	6.200E-04	ns	B057	0 2 1 1 2	

3903.   C$_{22}$H$_{35}$NO$_3$

Acetaminophen Myristate

Acetaminophen Tetradecanoate

RN: 54942-39-1   **MP** (°C):   114

MW: 361.53   **BP** (°C):

Solubility (Moles/L)	Solubility (Grams/L)	Temp (°C)	Ref (#)	Evaluation (T P E A A)	Comments
1.660E-05	6.000E-03	25	B010	1 1 1 1 0	

3904. C$_{22}$H$_{38}$O$_5$
4-Octylphenol Tetraethoxylate
Ethanol, 2-[2-[2-[2-(4-Octylphenoxy)ethoxy]ethoxy]ethoxy]-
RN: 51437-92-4 **MP** (°C):
MW: 382.55 **BP** (°C):

Solubility (Moles/L)	Solubility (Grams/L)	Temp (°C)	Ref (#)	Evaluation (T P E A A)	Comments
6.404E-05	2.450E-02	20.5	A335	1 0 2 2 2	
6.410E-05	2.452E-02	20.5	A335	1 0 2 2 2	

3905. C$_{22}$H$_{39}$O$_3$P
Dioctyl Phenyl Phosphonate
Di-n-Octyl Phenylphosphonate
DOPP
RN: 1754-47-8 **MP** (°C):
MW: 382.53 **BP** (°C):

Solubility (Moles/L)	Solubility (Grams/L)	Temp (°C)	Ref (#)	Evaluation (T P E A A)	Comments
≤5.23E-04	<2.00E-01	25	B070	1 2 0 1 0	

3906. C$_{22}$H$_{39}$O$_3$P
Diisooctyl Phenyl Phosphonate
RN: **MP** (°C):
MW: 382.53 **BP** (°C):

Solubility (Moles/L)	Solubility (Grams/L)	Temp (°C)	Ref (#)	Evaluation (T P E A A)	Comments
≤2.61E-04	<1.00E-01	25	B070	1 2 0 1 0	

3907. C$_{22}$H$_{42}$O$_4$
Dioctyl Adipate
bis(2-Ethylhexyl) Adipate
RN: 103-23-1 **MP** (°C):
MW: 370.58 **BP** (°C):

Solubility (Moles/L)	Solubility (Grams/L)	Temp (°C)	Ref (#)	Evaluation (T P E A A)	Comments
8.095E-06	3.000E-03	25	F067	1 0 2 2 1	

3908. C$_{22}$H$_{43}$N$_5$O$_{13}$
Amikacin
Antibiotic BB-K8
RN: 37517-28-5 **MP** (°C): 203
MW: 585.61 **BP** (°C):

Solubility (Moles/L)	Solubility (Grams/L)	Temp (°C)	Ref (#)	Evaluation (T P E A A)	Comments
3.159E-01	1.850E+02	25	K044	1 0 0 0 2	pH 10.4

3909.   C₂₃H₁₆O₆
Pamoic Acid
4,4'-Methylenebis[3-hydroxy-2-naphthalenecarboxylic Acid]
3,3'-Dihydroxy-4,4'-methylenedi-2-naphthoic Acid
Embonic Acid
RN: 130-85-8   **MP** (°C):
MW: 388.38   **BP** (°C):

Solubility (Moles/L)	Solubility (Grams/L)	Temp (°C)	Ref (#)	Evaluation (T P E A A)	Comments
2.800E-01	1.087E+02	ns	F007	0 0 0 0 1	

3910.   C₂₃H₁₈F₂N₄O
α-(2,4-Difluorophenyl)-α-(1-2-(2-pyridyl)phenylethenyl)-1H-1,2,4-triazole-1-ethanol
XD405
RN: 124669-93-8   **MP** (°C):
MW: 404.42   **BP** (°C):

Solubility (Moles/L)	Solubility (Grams/L)	Temp (°C)	Ref (#)	Evaluation (T P E A A)	Comments
7.418E-06	3.000E-03	22	M372	1 2 1 1 1	intrinsic

3911.   C₂₃H₂₀N₂O₂S
G-1
p-Phenylthioethylphenylbutazone
1,2-Diphenyl-4-(2-phenylthioethyl)-3,5-pyrazolidinedione
RN: 3736-92-3   **MP** (°C):
MW: 388.49   **BP** (°C):

Solubility (Moles/L)	Solubility (Grams/L)	Temp (°C)	Ref (#)	Evaluation (T P E A A)	Comments
4.118E-03	1.600E+00	ns	B158	0 0 0 0 1	pH 7.0

3912.   C₂₃H₂₂
10-Amyl-1,2-benzanthracene
RN: 188124-96-1   **MP** (°C):
MW: 298.43   **BP** (°C):

Solubility (Moles/L)	Solubility (Grams/L)	Temp (°C)	Ref (#)	Evaluation (T P E A A)	Comments
2.681E-09	8.000E-07	27	D003	1 0 0 1 0	

3913. C₂₃H₂₂O₆

Rotenone
Tubatoxin
Derris
1,2,12,12α-Tetrahydro-2α-isopropenyl-8,9-dimethoxy(1)benzopyrano(3,4-b)furo(2,3-h)(1)benzopyran-6(6α H)-one

RN: 83-79-4   **MP** (°C): 163
MW: 394.43   **BP** (°C):

Solubility (Moles/L)	Solubility (Grams/L)	Temp (°C)	Ref (#)	Evaluation (T P E A A)	Comments
4.310E-07	1.700E-04	25	C100	1 0 2 1 1	
3.803E-05	1.500E-02	100	M161	1 0 0 0 1	

3914. C₂₃H₂₃NO

Trifenmorph
Frescon
N-Tritylmorpholine

RN: 1420-06-0   **MP** (°C): 175
MW: 329.45   **BP** (°C):

Solubility (Moles/L)	Solubility (Grams/L)	Temp (°C)	Ref (#)	Evaluation (T P E A A)	Comments
6.071E-08	2.000E-05	20	M161	1 0 0 0 1	

3915. C₂₃H₂₄N₄O₂

Diantipyrylmethane
4,4'-Methylenediantipyrine
4,4'-Diantipyrylmethane

RN: 1251-85-0   **MP** (°C): 182
MW: 388.47   **BP** (°C):

Solubility (Moles/L)	Solubility (Grams/L)	Temp (°C)	Ref (#)	Evaluation (T P E A A)	Comments
1.130E-03	4.390E-01	20	P054	1 2 2 2 2	
1.132E-03	4.398E-01	20	P054	1 2 2 2 2	

3916. C₂₃H₂₄N₄O₆

Benzylcarbonyl-mitomycin C

RN: **MP** (°C):
MW: 452.47   **BP** (°C):

Solubility (Moles/L)	Solubility (Grams/L)	Temp (°C)	Ref (#)	Evaluation (T P E A A)	Comments
2.240E-03	1.014E+00	25	M316	1 1 1 1 2	

3917.   C$_{23}$H$_{24}$N$_{4}$O$_{7}$
Benzyloxycarbonyl-mitomycin C
RN: **MP** (°C):
MW: 468.47   **BP** (°C):

Solubility (Moles/L)	Solubility (Grams/L)	Temp (°C)	Ref (#)	Evaluation (T P E A A)	Comments
5.200E-04	2.436E-01	25	M316	1 1 1 1 2	

3918.   C$_{23}$H$_{24}$N$_{4}$S$_{2}$
Dithiodiantipyrinylmethane
3H-Pyrazole-3-thione, 4,4'-Methylenebis[1,2-dihydro-1,5-dimethyl-2-phenyl-
RN: 53799-78-3   **MP** (°C):   166
MW: 420.60   **BP** (°C):

Solubility (Moles/L)	Solubility (Grams/L)	Temp (°C)	Ref (#)	Evaluation (T P E A A)	Comments
5.000E-04	2.103E-01	ns	D087	0 2 0 0 1	

3919.   C$_{23}$H$_{26}$N$_{2}$O$_{4}$
Brucine
Brucin
RN: 357-57-3   **MP** (°C):   178
MW: 394.47   **BP** (°C):

Solubility (Moles/L)	Solubility (Grams/L)	Temp (°C)	Ref (#)	Evaluation (T P E A A)	Comments
8.112E-03	3.200E+00	15	F300	1 0 0 0 1	
1.330E-03	5.247E-01	15	K059	2 2 2 0 2	
1.698E-02	6.700E+00	100	F300	1 0 0 0 1	
1.267E-03	4.998E-01	rt	D021	0 0 1 1 1	

3920.   C$_{23}$H$_{26}$N$_{2}$O$_{4}$.4H$_{2}$O
Brucine (Tetrahydrate)
Strychnidin-10-one, 2,3-Dimethoxy-, Tetrahydrate
RN: 5892-11-5   **MP** (°C):   105
MW: 466.54   **BP** (°C):

Solubility (Moles/L)	Solubility (Grams/L)	Temp (°C)	Ref (#)	Evaluation (T P E A A)	Comments
6.677E-03	3.115E+00	c	D004	1 0 0 0 0	
1.420E-02	6.623E+00	h	D004	1 0 0 0 0	

3921. $C_{23}H_{26}O_3$
Phenothrin
(3-Phenoxylphenyl)methyl 2,2-dimethyl-3-(2-methyl-1-propenyl)cyclopropanecarboxylate
Sumithrin
3-Phenoxybenzyl D-cis and trans-2,2-dimethyl-3-(2-methylpropenyl)-
cyclopropanecarboxylate
RN: 26002-80-2   **MP** (°C):   <25
MW: 350.46   **BP** (°C):

Solubility (Moles/L)	Solubility (Grams/L)	Temp (°C)	Ref (#)	Evaluation (T P E A A)	Comments
5.707E-06	2.000E-03	30	M161	1 0 0 0 0	

3922. $C_{23}H_{27}ClO_4$
Delmadinone Acetate
Pregna-1,4,6-triene-3,20-dione, 17-(Acetyloxy)-6-chloro-
RN: 13698-49-2   **MP** (°C):   168
MW: 402.92   **BP** (°C):

Solubility (Moles/L)	Solubility (Grams/L)	Temp (°C)	Ref (#)	Evaluation (T P E A A)	Comments
1.506E-05	6.070E-03	37	K070	1 0 0 1 2	
1.134E-05	4.570E-03	ns	K070	1 0 0 1 2	

3923. $C_{23}H_{27}NO_8$
Narceine
o-Veratric Acid, 6-[[6-[2-(Dimethylamino)ethyl]-2-methoxy-3,4-(methylenedioxy)-
phenyl]acetyl]-
NIH 10760
RN: 131-28-2   **MP** (°C):   138
MW: 445.47   **BP** (°C):

Solubility (Moles/L)	Solubility (Grams/L)	Temp (°C)	Ref (#)	Evaluation (T P E A A)	Comments
1.300E-03	5.791E-01	15	K059	2 2 2 0 1	
2.915E-03	1.299E+00	c	D004	1 0 0 0 0	
1.016E-02	4.525E+00	h	D004	1 0 0 0 0	

3924. $C_{23}H_{27}N_3O_7$
Minocycline
Dynacin
Minocin
RN: 10118-90-8   **MP** (°C):
MW: 457.49   **BP** (°C):

Solubility (Moles/L)	Solubility (Grams/L)	Temp (°C)	Ref (#)	Evaluation (T P E A A)	Comments
1.137E-01	5.200E+01	25	B191	1 0 0 0 1	neutral pH

3925. C$_{23}$H$_{28}$ClN$_3$O$_2$S

Thiopropazate
1-(2-Acetoxyethyl)-4-[3-(2-chloro-10-phenothiazinyl)propyl]piperazine
RN: 84-06-0   **MP** (°C):
MW: 446.02   **BP** (°C):

Solubility (Moles/L)	Solubility (Grams/L)	Temp (°C)	Ref (#)	Evaluation (T P E A A)	Comments
2.000E-05	8.920E-03	24	G022	2 0 1 1 1	

3926. C$_{23}$H$_{28}$ClN$_3$O$_5$S

Glyburide
HB 419
Glibenclamide
Diabeta
1-((p-(2-(5-Chloro-o-anisamido)ethyl)phenyl)-sulfonyl)-3-cyclohexylurea
RN: 10238-21-8   **MP** (°C):   169
MW: 494.01   **BP** (°C):

Solubility (Moles/L)	Solubility (Grams/L)	Temp (°C)	Ref (#)	Evaluation (T P E A A)	Comments
1.137E-05	5.615E-03	22	M382	2 1 1 1 1	average of 2
6.275E-05	3.100E-02	25	G088	1 1 1 1 0	
8.097E-06	4.000E-03	27	H093	1 0 1 1 0	

3927. C$_{23}$H$_{28}$O$_7$

Prednisone Acetate
Pregna-1,4-diene-3,11,20-trione, 21-(Acetyloxy)-17-hydroxy-
RN: 125-10-0   **MP** (°C):
MW: 416.48   **BP** (°C):

Solubility (Moles/L)	Solubility (Grams/L)	Temp (°C)	Ref (#)	Evaluation (T P E A A)	Comments
5.522E-05	2.300E-02	25	K003	2 1 1 1 1	

3928. C$_{23}$H$_{31}$Cl$_2$NO$_3$

Estramustine
Estradiol 3-[bis(2-Chloroethyl)carbamate]
3-[bis(2-Chloroethyl)carbamate
RN: 2998-57-4   **MP** (°C):
MW: 440.41   **BP** (°C):

Solubility (Moles/L)	Solubility (Grams/L)	Temp (°C)	Ref (#)	Evaluation (T P E A A)	Comments
~2.27E-06	~1.00E-03	30	L334	1 0 1 1 0	

3929. C$_{23}$H$_{31}$FO$_6$

9α-Fluorohydrocortisone Acetate

Pregn-4-ene-3,20-dione, 21-(Acetyloxy)-9-fluoro-11,17-dihydroxy-, (11β)-

RN: 514-36-3   **MP** (°C):

MW: 422.50   **BP** (°C):

Solubility (Moles/L)	Solubility (Grams/L)	Temp (°C)	Ref (#)	Evaluation (T P E A A)	Comments
1.278E-04	5.400E-02	25	K021	1 2 2 2 1	

3930. C$_{23}$H$_{31}$O$_7$

Cortisone-21-hemi-succinate

RN: **MP** (°C):

MW: 419.50   **BP** (°C):

Solubility (Moles/L)	Solubility (Grams/L)	Temp (°C)	Ref (#)	Evaluation (T P E A A)	Comments
4.768E-04	2.000E-01	ns	E307	0 0 1 1 0	

3931. C$_{23}$H$_{32}$O$_2$

Medrogestone

Pregna-4,6-diene-3,20-dione, 6,17-Dimethyl-

RN: 977-79-7   **MP** (°C):   144

MW: 340.51   **BP** (°C):

Solubility (Moles/L)	Solubility (Grams/L)	Temp (°C)	Ref (#)	Evaluation (T P E A A)	Comments
5.345E-06	1.820E-03	25	L033	1 0 2 1 2	

3932. C$_{23}$H$_{32}$O$_4$

Deoxycorticosterone Acetate

Pregn-4-ene-3,20-dione, 21-(Acetyloxy)-

RN: 56-47-3   **MP** (°C):   156

MW: 372.51   **BP** (°C):

Solubility (Moles/L)	Solubility (Grams/L)	Temp (°C)	Ref (#)	Evaluation (T P E A A)	Comments
1.074E-05	4.000E-03	25	K003	2 1 1 1 1	

3933.   C$_{23}$H$_{32}$O$_6$
Hydrocortisone Acetate
Hydrocortisone-21-Acetate
Cortisol Acetate
Cortisol 21-acetate
RN: 50-03-3   **MP** (°C):   223dec
MW: 404.51   **BP** (°C):

Solubility (Moles/L)	Solubility (Grams/L)	Temp (°C)	Ref (#)	Evaluation (T P E A A)	Comments
3.486E-05	1.410E-02	25	C037	2 1 2 2 2	
1.555E-05	6.290E-03	25	H098	1 0 2 0 2	
1.555E-05	6.290E-03	25	H320	1 0 2 2 2	
1.550E-05	6.270E-03	25	H320	1 0 2 2 2	
2.472E-05	1.000E-02	25	K003	2 1 1 1 1	
3.461E-05	1.400E-02	25	K021	1 2 2 2 1	
2.472E-05	1.000E-02	25	M023	1 0 2 1 0	
2.472E-05	1.000E-02	ns	M169	0 0 0 0 1	
1.904E-05	7.700E-03	ns	N323	0 0 2 2 1	

3934.   C$_{23}$H$_{34}$O$_3$
Testosterone Butyrate
Androst-4-en-3-one, 17-(1-Oxobutoxy)-, (17bet)-
RN: 3410-54-6   **MP** (°C):
MW: 358.53   **BP** (°C):

Solubility (Moles/L)	Solubility (Grams/L)	Temp (°C)	Ref (#)	Evaluation (T P E A A)	Comments
1.406E-06	5.039E-04	25	J004	1 0 1 1 2	
1.403E-06	5.030E-04	ns	B057	0 2 1 1 2	

3935.   C$_{23}$H$_{34}$O$_3$
17-α-Methyltestosterone Propionate
RN: **MP** (°C):
MW: 358.53   **BP** (°C):

Solubility (Moles/L)	Solubility (Grams/L)	Temp (°C)	Ref (#)	Evaluation (T P E A A)	Comments
2.845E-06	1.020E-03	ns	B057	0 2 1 1 2	

3936. C$_{23}$H$_{34}$O$_3$
5,6-Dehydroisoandrosterone Butyrate
Androst-5-en-17-one, 3-(1-Oxobutoxy)-, (3β)-
RN: 15253-51-7   **MP** (°C):
MW: 358.53   **BP** (°C):

Solubility (Moles/L)	Solubility (Grams/L)	Temp (°C)	Ref (#)	Evaluation (T P E A A)	Comments
1.231E+00	4.413E+02	ns	B057	0 2 1 1 2	

3937. C$_{23}$H$_{34}$O$_4$
Digitoxigenin
Card-20(22)-enolide, 3,14-Dihydroxy-, (3β,5β)-
RN: 143-62-4   **MP** (°C):
MW: 374.53   **BP** (°C):

Solubility (Moles/L)	Solubility (Grams/L)	Temp (°C)	Ref (#)	Evaluation (T P E A A)	Comments
3.000E-05	1.124E-02	30	O321	2 2 2 2 1	
3.000E-05	1.124E-02	30	O321	2 2 2 2 1	

3938. C$_{23}$H$_{35}$NOS
5-Pregnene-20-one-3-spiro-2'-(1',2'-thiazolidine)
RN: **MP** (°C):   127-136
MW: 373.61   **BP** (°C):

Solubility (Moles/L)	Solubility (Grams/L)	Temp (°C)	Ref (#)	Evaluation (T P E A A)	Comments
~1.34E-05	~5.00E-03	ns	B199	0 0 0 0 0	

3939. C$_{23}$H$_{36}$O$_3$
Androstanolone Butyrate
Androstan-3-one, 17-(1-Oxobutoxy)-, (5α,17β)-
RN: 18069-66-4   **MP** (ⁿC):
MW: 360.54   **BP** (°C):

Solubility (Moles/L)	Solubility (Grams/L)	Temp (°C)	Ref (#)	Evaluation (T P E A A)	Comments
1.220E-06	4.400E-04	ns	B057	0 2 1 1 2	

3940. C$_{23}$H$_{38}$O$_3$
Hexadecyl p-Hydroxybenzoate
Hexadecyl 4-Hydroxybenzoate
RN: 71067-09-9   **MP** (°C):
MW: 362.56   **BP** (°C):

Solubility (Moles/L)	Solubility (Grams/L)	Temp (°C)	Ref (#)	Evaluation (T P E A A)	Comments
1.045E-03	3.789E-01	25	D081	1 2 2 1 2	

3941.   C₂₃H₄₀O₅

4-Nonylphenol Tetraethoxylate

p-Nonylphenol Tetraethoxylate

RN: 7311-27-5   **MP** (°C):

MW: 396.57   **BP** (°C):

Solubility (Moles/L)	Solubility (Grams/L)	Temp (°C)	Ref (#)	Evaluation (T P E A A)	Comments
1.929E-05	7.650E-03	20.5	A335	1 0 2 2 2	
1.930E-05	7.654E-03	20.5	A335	1 0 2 2 2	

3942.   C₂₄H₁₂

Coronene

Coronen

RN: 191-07-1   **MP** (°C):   438

MW: 300.36   **BP** (°C):   525

Solubility (Moles/L)	Solubility (Grams/L)	Temp (°C)	Ref (#)	Evaluation (T P E A A)	Comments
4.680E-09	1.406E-06	20	E009	1 0 0 1 2	
3.329E-10	1.000E-07	25	B319	2 0 1 2 1	
4.661E-10	1.400E-07	25	M064	1 1 2 2 1	
4.660E-10	1.400E-07	25	M342	1 0 1 1 2	

3943.   C₂₄H₂₀N₂

N,N'-Diphenylbenzidine

RN: 531-91-9   **MP** (°C):   247

MW: 336.44   **BP** (°C):

Solubility (Moles/L)	Solubility (Grams/L)	Temp (°C)	Ref (#)	Evaluation (T P E A A)	Comments
1.783E-07	6.000E-05	50	K068	1 0 2 2 0	buffer
1.783E-07	6.000E-05	rt	K068	0 0 2 2 0	buffer

3944.   C₂₄H₂₂N₂O₂

G-3

p-Phenylpropylphenylbutazone

3,5-Pyrazolidinedione, 1,2-Diphenyl-4-(3-phenylpropyl)-

RN: 32060-78-9   **MP** (°C):

MW: 370.46   **BP** (°C):

Solubility (Moles/L)	Solubility (Grams/L)	Temp (°C)	Ref (#)	Evaluation (T P E A A)	Comments
3.779E-04	1.400E-01	ns	B158	0 0 0 0 1	pH 7.0

3945.   C$_{24}$H$_{26}$N$_4$O$_2$
Methyldiantipyrylmethane
MDAM
RN: 1606-56-0   **MP** (°C):
MW: 402.50   **BP** (°C):

Solubility (Moles/L)	Solubility (Grams/L)	Temp (°C)	Ref (#)	Evaluation (T P E A A)	Comments
1.118E-03	4.498E-01	20	P054	1 2 2 2 2	

3946.   C$_{24}$H$_{26}$N$_4$S$_2$
Methyldithiopyrylmethane
3H-Pyrazole-3-thione, 4,4'-Ethylidenebis[1,2-dihydro-1,5-dimethyl-2-phenyl-
RN: 74713-70-5   **MP** (°C):   229
MW: 434.63   **BP** (°C):

Solubility (Moles/L)	Solubility (Grams/L)	Temp (°C)	Ref (#)	Evaluation (T P E A A)	Comments
5.000E-04	2.173E-01	ns	D087	0 2 0 0 1	

3947.   C$_{24}$H$_{27}$BrN$_6$O$_{10}$
C.I. Disperse Blue 79
2'-Acetylamino-4'-[bis(acetoxyethyl)amino]-6-bromo-2,4-dinitro-5'-ethoxyazobenzene
RN: 12239-34-8   **MP** (°C):   146
MW: 639.43   **BP** (°C):

Solubility (Moles/L)	Solubility (Grams/L)	Temp (°C)	Ref (#)	Evaluation (T P E A A)	Comments
1.000E-09	6.394E-07	25	B333	1 0 0 0 1	

3948.   C$_{24}$H$_{27}$N
Prenylamine
N-(3,3-Diphenylpropyl)-α-methylphenylethylamine
RN: 390-64-7   **MP** (°C):   36.5
MW: 329.49   **BP** (°C):

Solubility (Moles/L)	Solubility (Grams/L)	Temp (°C)	Ref (#)	Evaluation (T P E A A)	Comments
1.517E-04	5.000E-02	37	C054	2 0 2 1 0	

3949.   C₂₄H₃₀F₂O₆
Fluocinolone Acetonide
6α,9α-Difluoro-16α Hydroxyprednisolone-16,17-acetonide
6α,9α-Difluoro-16α,17α-isopropylidenedioxy-1,4-pregnadiene-3,20-dione
RN: 67-73-2   **MP** (°C):   260.5
MW: 452.50   **BP** (°C):

Solubility (Moles/L)	Solubility (Grams/L)	Temp (°C)	Ref (#)	Evaluation (T P E A A)	Comments
2.387E-04	1.080E-01	25	K021	1 2 2 2 1	
4.641E-05	2.100E-02	25	O001	2 0 2 2 2	
2.210E-04	1.000E-01	25	P008	1 0 1 1 1	EFG

3950.   C₂₄H₃₁FO₅S
Timobesone Acetate
17-β-Methythiocarbonyl-9α-fluoro-11β
RN: 79578-14-6   **MP** (°C):
MW: 450.57   **BP** (°C):

Solubility (Moles/L)	Solubility (Grams/L)	Temp (°C)	Ref (#)	Evaluation (T P E A A)	Comments
6.000E-03	2.703E+00	25	O318	1 0 1 2 2	

3951.   C₂₄H₃₁FO₆
Triamcinolone Acetonide
9α-Fluoro-16α-hydroxyprednisolone acetonide
Triamcinolone 16α,17-Acetonide
Aristoderm
Adcortyl-A
RN: 76-25-5   **MP** (°C):   293
MW: 434.51   **BP** (°C):

Solubility (Moles/L)	Solubility (Grams/L)	Temp (°C)	Ref (#)	Evaluation (T P E A A)	Comments
9.205E-05	4.000E-02	23	F025	1 0 0 0 0	
9.436E-05	4.100E-02	25	K021	1 2 2 2 1	
6.076E-04	2.640E-01	25	L009	1 0 0 1 1	
4.833E-05	2.100E-02	28	B055	2 0 2 2 2	
4.027E-05	1.750E-02	28	B056	1 2 1 1 2	
5.869E-05	2.550E-02	37	B055	2 0 2 2 2	
4.764E-05	2.070E-02	37	B056	1 2 1 1 2	
9.205E-05	4.000E-02	37	F025	1 0 0 0 0	
7.733E-05	3.360E-02	50	B055	2 0 2 2 2	
6.099E-05	2.650E-02	50	B056	1 2 1 1 2	

3952. C$_{24}$H$_{31}$FO$_6$
Betamethasone Acetate
Betamethasone-17-acetate
9α-Fluoro-16β-methylprednisolone-21-acetate
RN: 987-24-6   **MP** (°C):   200dec
MW: 434.51   **BP** (°C):

Solubility (Moles/L)	Solubility (Grams/L)	Temp (°C)	Ref (#)	Evaluation (T P E A A)	Comments
6.904E-05	3.000E-02	25	K003	2 1 1 1 1	

3953. C$_{24}$H$_{31}$FO$_6$
Dexamethasone Acetate
Dexamethasone-17-acetate
Dexamethasone Acetate
RN: 1177-87-3   **MP** (°C):   263
MW: 434.51   **BP** (°C):

Solubility (Moles/L)	Solubility (Grams/L)	Temp (°C)	Ref (#)	Evaluation (T P E A A)	Comments
2.992E-05	1.300E-02	25	K003	2 1 1 1 1	
6.214E-05	2.700E-02	37	D026	1 2 1 2 2	

3954. C$_{24}$H$_{31}$NO$_4$
Drotaverine
1-(3,4-Diethoxybenzylidene)-6,7-diethoxy-1,2,3,4-tetrahydroisoquinoline
RN: 14009-24-6   **MP** (°C):
MW: 397.52   **BP** (°C):

Solubility (Moles/L)	Solubility (Grams/L)	Temp (°C)	Ref (#)	Evaluation (T P E A A)	Comments
3.459E-02	1.375E+01	37	C054	2 0 2 1 2	

3955. C$_{24}$H$_{32}$O$_4$
Ethynodiol Diacetate
Ovulen-50
RN: 297-76-7   **MP** (°C):   126
MW: 384.52   **BP** (°C):

Solubility (Moles/L)	Solubility (Grams/L)	Temp (°C)	Ref (#)	Evaluation (T P E A A)	Comments
3.641E-06	1.400E-03	25	L027	1 0 0 0 2	

3956.　C$_{24}$H$_{32}$O$_4$S
Spironolactone
17-Hydroxy-7α-mercapto-3-oxo-17α-pregn-4-ene-21-carboxylic Acid γ-Lactone Acetate
Spiractin
RN:　52-01-7　　**MP** (°C):　134
MW:　416.58　　**BP** (°C):

Solubility (Moles/L)	Solubility (Grams/L)	Temp (°C)	Ref (#)	Evaluation (T P E A A)	Comments
7.200E-06	2.999E-03	25	A348	1 0 2 2 0	
5.281E-05	2.200E-02	25	C037	2 1 2 2 2	
5.281E-05	2.200E-02	25	G084	2 0 2 2 1	
4.801E-05	2.000E-02	25	G095	2 1 2 2 1	
6.649E-05	2.770E-02	37	K092	2 0 0 1 2	

3957.　C$_{24}$H$_{32}$O$_5$
7-Carboxylic Acid Methyl Ester　Canrenone
RN:　　　　　　**MP** (°C):
MW:　400.52　　**BP** (°C):

Solubility (Moles/L)	Solubility (Grams/L)	Temp (°C)	Ref (#)	Evaluation (T P E A A)	Comments
1.960E-04	7.850E-02	37	C004	1 0 2 2 2	EFG

3958.　C$_{24}$H$_{32}$O$_6$
Cortisone 17-Propionate
Pregn-4-ene-3,11,20-trione, 21-Hydroxy-17-(1-oxopropoxy)-
RN:　136370-32-6　**MP** (°C):
MW:　416.52　　**BP** (°C):

Solubility (Moles/L)	Solubility (Grams/L)	Temp (°C)	Ref (#)	Evaluation (T P E A A)	Comments
1.921E-05	8.000E-03	25	M023	1 0 2 1 0	

3959.　C$_{24}$H$_{33}$FO$_6$
Flurandrenolone
Fludroxycortide
6-Fluoro-16α-hydroxyhydrocortisone-16,17-acetonide
RN:　1524-88-5　　**MP** (°C):
MW:　436.53　　**BP** (°C):

Solubility (Moles/L)	Solubility (Grams/L)	Temp (°C)	Ref (#)	Evaluation (T P E A A)	Comments
6.758E-04	2.950E-01	25	K021	1 2 2 2 1	

3960. $C_{24}H_{34}N_2O$
Bepridil
1-Isobutoxy-2-pyrrolidino-3-N-benzylanilino-propane
Bepadin
RN: 64706-54-3   **MP** (°C):
MW: 366.55   **BP** (°C):

Solubility (Moles/L)	Solubility (Grams/L)	Temp (°C)	Ref (#)	Evaluation (T P E A A)	Comments
2.027E-02	7.430E+00	37	N032	1 0 1 1 2	

3961. $C_{24}H_{34}N_2O_3$
Lysine Estrone Ester
RN: **MP** (°C):
MW: 398.55   **BP** (°C):

Solubility (Moles/L)	Solubility (Grams/L)	Temp (°C)	Ref (#)	Evaluation (T P E A A)	Comments
3.162E-01	1.260E+02	ns	A074	0 0 0 0 0	EFG

3962. $C_{24}H_{34}O_5$
Dehydrocholic Acid
3,7,12-Trioxo-5β-cholanic Acid
RN: 81-23-2   **MP** (°C):   237
MW: 402.54   **BP** (°C):

Solubility (Moles/L)	Solubility (Grams/L)	Temp (°C)	Ref (#)	Evaluation (T P E A A)	Comments
4.472E-04	1.800E-01	15	G081	1 0 1 1 1	
1.615E-04	6.500E-02	30	O321	2 2 2 2 1	
1.600E-04	6.441E-02	30	O321	2 2 2 2 1	

3963. $C_{24}H_{34}O_6$
Hydrocortisone Propionate
Hydrocortisone-21-propionate
RN: 6677-98-1   **MP** (°C):
MW: 418.53   **BP** (°C):

Solubility (Moles/L)	Solubility (Grams/L)	Temp (°C)	Ref (#)	Evaluation (T P E A A)	Comments
2.772E-05	1.160E-02	25	H098	1 0 2 0 2	
2.772E-05	1.160E-02	25	H320	1 0 2 2 2	
2.770E-05	1.159E-02	25	H320	1 0 2 2 2	

3964.   C$_{24}$H$_{36}$O$_3$
Testosterone Valerate
Androst-4-en-3-one, 17-[(1-Oxopentyl)oxy]-, (17β)-
Testosterone 17-Valerate
RN: 3129-43-9   **MP** (°C):
MW: 372.55   **BP** (°C):

Solubility (Moles/L)	Solubility (Grams/L)	Temp (°C)	Ref (#)	Evaluation (T P E A A)	Comments
7.778E-07	2.898E-04	25	J004	1 0 1 1 2	
7.811E-07	2.910E-04	ns	B057	0 2 1 1 2	

3965.   C$_{24}$H$_{36}$O$_3$
5,6-Dehydroisoandrosterone Valerate
Androst-5-en-17-one, 3-[(1-Oxopentyl)oxy]-, (3β)-
RN: 7642-68-4   **MP** (°C):
MW: 372.55   **BP** (°C):

Solubility (Moles/L)	Solubility (Grams/L)	Temp (°C)	Ref (#)	Evaluation (T P E A A)	Comments
2.061E-05	7.680E-03	ns	B057	0 2 1 1 2	

3966.   C$_{24}$H$_{38}$O$_3$
Androstanolone Valerate
Androstan-3-one, 17-[(1-Oxopentyl)oxy]-, (5α,17β)-
RN: 26271-72-7   **MP** (°C):
MW: 374.57   **BP** (°C):

Solubility (Moles/L)	Solubility (Grams/L)	Temp (°C)	Ref (#)	Evaluation (T P E A A)	Comments
8.143E-07	3.050E-04	ns	B057	0 2 1 1 2	

3967. C$_{24}$H$_{38}$O$_4$
Octyl Phthalate
Di(2-ethylhexyl)phthalate
Di-(2-ethylhexyl)-phthalate
Di-sec-octyl Phthalate
bis(2-Ethylhexyl) Phthalate
bis-(2-Ethylhexyl) 1,2-Benzenedicarboxylate

RN: 117-81-7   **MP** (°C):   -50
MW: 390.57   **BP** (°C):   386.9

Solubility (Moles/L)	Solubility (Grams/L)	Temp (°C)	Ref (#)	Evaluation (T P E A A)	Comments
2.560E-04	9.999E-02	20	F070	1 0 0 0 1	*sic*
1.050E-07	4.101E-05	20	L300	2 1 0 2 2	
1.536E-06	6.000E-04	22.5	G301	2 1 0 1 2	
7.297E-07	2.850E-04	24	H116	2 1 0 0 2	
6.913E-07	2.700E-04	25	D336	2 1 2 2 2	
1.280E-06	5.000E-04	25	F067	1 0 2 2 0	

3968. C$_{24}$H$_{38}$O$_4$
bis(Tereoctyl) Phthalate

RN: **MP** (°C):
MW: 390.57   **BP** (°C):

Solubility (Moles/L)	Solubility (Grams/L)	Temp (°C)	Ref (#)	Evaluation (T P E A A)	Comments
5.633E-08	2.200E-05	25	D336	2 1 2 2 2	

3969. C$_{24}$H$_{38}$O$_4$
bis(Isooctyl) Phthalate
Diisooctyl Phthalate
1,2-Benzenedicarboxylic Acid Diisooctyl Ester

RN: 27554-26-3   **MP** (°C):   -4
MW: 390.57   **BP** (°C):   239

Solubility (Moles/L)	Solubility (Grams/L)	Temp (°C)	Ref (#)	Evaluation (T P E A A)	Comments
1.024E-07	4.000E-05	25	D336	2 1 2 2 2	

3970. C$_{24}$H$_{38}$O$_4$
Di-2-ethylhexyl Isophthalate
D-(2-Ethylhexyl) Isophthalate
Dioctyl Isophthalate

RN: 137-89-3   **MP** (°C):
MW: 390.57   **BP** (°C):   400

Solubility (Moles/L)	Solubility (Grams/L)	Temp (°C)	Ref (#)	Evaluation (T P E A A)	Comments
2.816E-08	1.100E-05	24	H116	2 1 0 0 2	

3971. C₂₄H₃₈O₄
bis(n-Octyl) Phthalate
Di-n-Octyl Phthalate
1,2-Benzenedicarboxylic Acid
RN: 117-84-0   **MP** (°C):   -25
MW: 390.57   **BP** (°C):   220

Solubility (Moles/L)	Solubility (Grams/L)	Temp (°C)	Ref (#)	Evaluation (T P E A A)	Comments
5.121E-08	2.000E-05	25	D336	2 1 2 2 2	

3972. C₂₄H₃₈O₄
Apocholic Acid
RN: 641-81-6   **MP** (°C):   175.5
MW: 390.57   **BP** (°C):

Solubility (Moles/L)	Solubility (Grams/L)	Temp (°C)	Ref (#)	Evaluation (T P E A A)	Comments
2.048E-03	8.000E-01	15	G081	1 0 1 1 0	

3973. C₂₄H₃₉NO₃
Acetaminophen Palmitate
Acetaminophen Hexadecanoate
RN: 54942-40-4   **MP** (°C):   117
MW: 389.58   **BP** (°C):

Solubility (Moles/L)	Solubility (Grams/L)	Temp (°C)	Ref (#)	Evaluation (T P E A A)	Comments
1.283E-05	5.000E-03	25	B010	1 1 1 1 0	

3974. C₂₄H₄₀O₃
Lithocholic Acid
3α-Hydroxy-5β-cholan-24-oic Acid
3α-Hydroxycholanic Acid
RN: 434-13-9   **MP** (°C):   184
MW: 376.58   **BP** (°C):

Solubility (Moles/L)	Solubility (Grams/L)	Temp (°C)	Ref (#)	Evaluation (T P E A A)	Comments
3.800E-08	1.431E-05	10	F307	1 2 2 2 2	pH 3.0
4.000E-08	1.506E-05	15	F307	1 2 2 2 2	pH 3.0
4.600E-08	1.732E-05	20	F307	1 2 2 2 2	pH 3.0
1.000E-06	3.766E-04	20	I012	1 2 2 1 0	pH 2.4
5.000E-08	1.883E-05	25	F307	1 2 2 2 2	pH 3.0
6.000E-08	2.260E-05	30	F307	1 2 2 2 2	pH 3.0
7.500E-08	2.824E-05	35	F307	1 2 2 2 2	pH 3.0

1.000E-06	3.766E-04	37	I012	1 2 2 1 0	pH 2.4
1.000E-07	3.766E-05	40	F307	1 2 2 2 2	pH 3.0
1.100E-07	4.142E-05	45	F307	1 2 2 2 2	pH 3.0
1.400E-07	5.272E-05	50	F307	1 2 2 2 2	pH 3.0

3975.   C$_{24}$H$_{40}$O$_{3}$
3β-Hydroxy-5β-cholanoic Acid
7α-Hydroxy-5β-cholanoic Acid
RN: **MP** (°C):
MW: 376.58   **BP** (°C):

Solubility (Moles/L)	Solubility (Grams/L)	Temp (°C)	Ref (#)	Evaluation (T P E A A)	Comments
1.800E-07	6.779E-05	10	F307	1 2 2 2 2	pH 3.0
4.400E-07	1.657E-04	10	F307	1 2 2 2 2	pH 3.0
2.200E-07	8.285E-05	15	F307	1 2 2 2 2	pH 3.0
5.200E-07	1.958E-04	15	F307	1 2 2 2 2	pH 3.0
2.400E-07	9.038E-05	20	F307	1 2 2 2 2	pH 3.0
6.500E-07	2.448E-04	20	F307	1 2 2 2 2	pH 3.0
2.800E-07	1.054E-04	25	F307	1 2 2 2 2	pH 3.0
7.900E-07	2.975E-04	25	F307	1 2 2 2 2	pH 3.0
3.500E-07	1.318E-04	30	F307	1 2 2 2 2	pH 3.0
9.700E-07	3.653E-04	30	F307	1 2 2 2 2	pH 3.0
5.300E-07	1.996E-04	35	F307	1 2 2 2 2	pH 3.0
1.190E-06	4.481E-04	35	F307	1 2 2 2 2	pH 3.0
8.200E-07	3.088E-04	40	F307	1 2 2 2 2	pH 3.0
1.490E-06	5.611E-04	40	F307	1 2 2 2 2	pH 3.0
1.770E-06	6.666E-04	45	F307	1 2 2 2 2	pH 3.0
1.280E-06	4.820E-04	45	F307	1 2 2 2 2	pH 3.0
2.150E-06	8.097E-04	50	F307	1 2 2 2 2	pH 3.0
1.500E-06	5.649E-04	50	F307	1 2 2 2 2	pH 3.0
2.150E-06	8.097E-04	50	F307	1 2 2 2 2	pH 3.0

3976.   C$_{24}$H$_{40}$O$_{4}$
Chenodeoxycholic Acid
CDCA
RN: 474-25-9   **MP** (°C):   119
MW: 392.58   **BP** (°C):

Solubility (Moles/L)	Solubility (Grams/L)	Temp (°C)	Ref (#)	Evaluation (T P E A A)	Comments
2.500E-05	9.815E-03	10	F307	1 2 2 2 2	pH 3.0
2.500E-05	9.815E-03	15	F307	1 2 2 2 2	pH 3.0
2.600E-05	1.021E-02	20	F307	1 2 2 2 2	pH 3.0
2.290E-04	8.990E-02	20	I012	1 2 2 1 2	pH 2.4
2.700E-05	1.060E-02	25	F307	1 2 2 2 2	pH 3.0
2.800E-05	1.099E-02	30	F307	1 2 2 2 2	pH 3.0
3.000E-05	1.178E-02	35	F307	1 2 2 2 2	pH 3.0
2.560E-04	1.005E-01	37	I008	1 0 0 1 2	

2.560E-04	1.005E-01	37	I012	1 2 2 1 2	pH 2.4
3.150E-05	1.237E-02	40	F307	1 2 2 2 2	pH 3.0
3.400E-05	1.335E-02	45	F307	1 2 2 2 2	pH 3.0
3.600E-05	1.413E-02	50	F307	1 2 2 2 2	pH 3.0

3977.   C$_{24}$H$_{40}$O$_4$
Ursodeoxycholic Acid
UDCA
RN: 128-13-2   **MP** (°C):   203
MW: 392.58   **BP** (°C):

Solubility (Moles/L)	Solubility (Grams/L)	Temp (°C)	Ref (#)	Evaluation (T P E A A)	Comments
7.000E-06	2.748E-03	10	F307	1 2 2 2 2	pH 3.0
7.500E-06	2.944E-03	15	F307	1 2 2 2 2	pH 3.0
8.000E-06	3.141E-03	20	F307	1 2 2 2 2	pH 3.0
5.100E-05	2.002E-02	20	I012	1 2 2 1 1	pH 2.4
9.000E-06	3.533E-03	25	F307	1 2 2 2 2	pH 3.0
1.000E-05	3.926E-03	30	F307	1 2 2 2 2	pH 3.0
1.150E-05	4.515E-03	35	F307	1 2 2 2 2	pH 3.0
5.300E-05	2.081E-02	37	I008	1 0 0 1 1	
5.300E-05	2.081E-02	37	I012	1 2 2 1 1	pH 2.4
1.200E-05	4.711E-03	40	F307	1 2 2 2 2	pH 3.0
1.300E-05	5.104E-03	45	F307	1 2 2 2 2	pH 3.0
1.400E-05	5.496E-03	50	F307	1 2 2 2 2	pH 3.0

3978.   C$_{24}$H$_{40}$O$_4$
Deoxycholic Acid
Cholan-24-oic Acid, 3,12-Dihydroxy-, (3α,5β,12α)-
3α,12α-Dihydroxy-5β-cholanoic Acid
RN: 83-44-3   **MP** (°C):   176
MW: 392.58   **BP** (°C):

Solubility (Moles/L)	Solubility (Grams/L)	Temp (°C)	Ref (#)	Evaluation (T P E A A)	Comments
2.400E-05	9.422E-03	10	F307	1 2 2 2 2	pH 3.0
2.600E-05	1.021E-02	15	F307	1 2 2 2 2	pH 3.0
6.113E-04	2.400E-01	15	G081	1 0 1 1 1	
5.093E-04	2.000E-01	20	D041	1 0 0 0 0	
2.700E-05	1.060E-02	20	F307	1 2 2 2 2	pH 3.0
1.110E-04	4.358E-02	20	I012	1 2 2 1 2	pH 2.4
2.800E-05	1.099E-02	25	F307	1 2 2 2 2	pH 3.0
2.800E-05	1.099E-02	30	F307	1 2 2 2 2	pH 3.0
2.900E-05	1.138E-02	35	F307	1 2 2 2 2	pH 3.0
1.140E-04	4.475E-02	37	I012	1 2 2 1 2	pH 2.4
2.900E-05	1.138E-02	40	F307	1 2 2 2 2	pH 3.0
3.000E-05	1.178E-02	45	F307	1 2 2 2 2	pH 3.0
3.200E-05	1.256E-02	50	F307	1 2 2 2 2	pH 3.0

3979.   C$_{24}$H$_{40}$O$_4$
Hyodeoxycholic Acid
3α,6α-Dihydroxy-5β-cholanoic Acid
RN: 83-49-8   **MP** (°C):   198
MW: 392.58   **BP** (°C):

Solubility (Moles/L)	Solubility (Grams/L)	Temp (°C)	Ref (#)	Evaluation (T P E A A)	Comments
1.000E-05	3.926E-03	10	F307	1 2 2 2 2	pH 3.0
1.200E-05	4.711E-03	15	F307	1 2 2 2 2	pH 3.0
1.300E-05	5.104E-03	20	F307	1 2 2 2 2	pH 3.0
1.500E-05	5.889E-03	25	F307	1 2 2 2 2	pH 3.0
1.700E-05	6.674E-03	30	F307	1 2 2 2 2	pH 3.0
1.800E-05	7.067E-03	35	F307	1 2 2 2 2	pH 3.0
2.000E-05	7.852E-03	40	F307	1 2 2 2 2	pH 3.0
2.200E-05	8.637E-03	45	F307	1 2 2 2 2	pH 3.0
2.600E-05	1.021E-02	50	F307	1 2 2 2 2	pH 3.0

3980.   C$_{24}$H$_{40}$O$_5$
Ursocholic Acid
3α,7β,12α-Trihydroxy-5β-cholanoic Acid
RN: 2955-27-3   **MP** (°C):
MW: 408.58   **BP** (°C):

Solubility (Moles/L)	Solubility (Grams/L)	Temp (°C)	Ref (#)	Evaluation (T P E A A)	Comments
1.590E-03	6.496E-01	10	F307	1 2 2 2 2	pH 3.0
1.610E-03	6.578E-01	15	F307	1 2 2 2 2	pH 3.0
1.640E-03	6.701E-01	20	F307	1 2 2 2 2	pH 3.0
1.670E-03	6.823E-01	25	F307	1 2 2 2 2	pH 3.0
1.710E-03	6.987E-01	30	F307	1 2 2 2 2	pH 3.0
1.762E-03	7.199E-01	35	F307	1 2 2 2 2	pH 3.0
1.828E-03	7.469E-01	40	F307	1 2 2 2 2	pH 3.0
1.872E-03	7.649E-01	45	F307	1 2 2 2 2	pH 3.0
2.000E-03	8.172E-01	50	F307	1 2 2 2 2	pH 3.0

3981.   C$_{24}$H$_{40}$O$_5$
Cholic Acid
Cholsaeure
RN: 81-25-4   **MP** (°C):   198
MW: 408.58   **BP** (°C):

Solubility (Moles/L)	Solubility (Grams/L)	Temp (°C)	Ref (#)	Evaluation (T P E A A)	Comments
2.210E-04	9.030E-02	10	F307	1 2 2 2 2	pH 3.0
6.486E-04	2.650E-01	15	F300	1 0 0 0 0	
2.140E-04	8.744E-02	15	F307	1 2 2 2 2	pH 3.0
6.853E-04	2.800E-01	15	G081	1 0 1 1 1	
6.851E-04	2.799E-01	20	D041	1 0 0 0 1	

2.247E-04	9.180E-02	20	E008	1 0 2 0 2	average of 3
2.200E-04	8.989E-02	20	F307	1 2 2 2 2	pH 3.0
4.280E-04	1.749E-01	20	I012	1 2 2 1 2	pH 2.4
2.350E-04	9.602E-02	25	F307	1 2 2 2 2	pH 3.0
2.670E-04	1.091E-01	30	F307	1 2 2 2 2	pH 3.0
3.240E-04	1.324E-01	35	F307	1 2 2 2 2	pH 3.0
4.600E-04	1.879E-01	37	I012	1 2 2 1 2	pH 2.4
3.830E-04	1.565E-01	40	F307	1 2 2 2 2	pH 3.0
4.830E-04	1.973E-01	45	F307	1 2 2 2 2	pH 3.0
6.390E-04	2.611E-01	50	F307	1 2 2 2 2	pH 3.0

3982. $C_{24}H_{40}O_5$
3α, 6α, 7α -Trihydroxy-5β-cholanate
RN: **MP** (°C):
MW: 408.58   **BP** (°C):

Solubility (Moles/L)	Solubility (Grams/L)	Temp (°C)	Ref (#)	Evaluation (T P E A A)	Comments
3.700E-05	1.512E-02	10	F307	1 2 2 2 2	pH 3.0
3.800E-05	1.553E-02	15	F307	1 2 2 2 2	pH 3.0
4.100E-05	1.675E-02	20	F307	1 2 2 2 2	pH 3.0
4.500E-05	1.839E-02	25	F307	1 2 2 2 2	pH 3.0
5.500E-05	2.247E-02	30	F307	1 2 2 2 2	pH 3.0
6.900E-05	2.819E-02	35	F307	1 2 2 2 2	pH 3.0
8.600E-05	3.514E-02	40	F307	1 2 2 2 2	pH 3.0
1.160E-04	4.740E-02	45	F307	1 2 2 2 2	pH 3.0
1.600E-04	6.537E-02	50	F307	1 2 2 2 2	pH 3.0

3983. $C_{24}H_{50}$
Tetracosane
n-Tetracosane
Alkane C(24)
RN: 646-31-1   **MP** (°C):   54
MW: 338.67   **BP** (°C):   391.3

Solubility (Moles/L)	Solubility (Grams/L)	Temp (°C)	Ref (#)	Evaluation (T P E A A)	Comments
1.264E-02	4.282E+00	321	S355	1 1 1 2 0	EFG
8.878E-02	3.007E+01	369	S355	1 1 1 2 0	EFG

3984. C$_{24}$H$_{51}$OP
tri-n-Octylphosphine Oxide
TOPO
Trioctylphosphine oxide
RN: 78-50-2   **MP** (°C):
MW: 386.65   **BP** (°C):

Solubility (Moles/L)	Solubility (Grams/L)	Temp (°C)	Ref (#)	Evaluation (T P E A A)	Comments
7.242E-06	2.800E-03	0	O002	2 0 2 2 1	
3.880E-06	1.500E-03	25	O002	2 0 2 2 1	

3985. C$_{24}$H$_{51}$O$_3$P
Dibutyl Hexadecyl Phosphonate
Phosphonic Acid, Hexadecyl-, Dibutyl Ester
RN: 84869-93-2   **MP** (°C):
MW: 418.65   **BP** (°C):

Solubility (Moles/L)	Solubility (Grams/L)	Temp (°C)	Ref (#)	Evaluation (T P E A A)	Comments
<4.78E-04	<2.00E-01	25	B070	1 2 0 1 0	

3986. C$_{24}$H$_{51}$O$_4$P
Tris-(2-ethylhexyl) Phosphate
Disflamoll TOF
TEHP
Flexol TOF
RN: 78-42-2   **MP** (°C):
MW: 434.65   **BP** (°C):

Solubility (Moles/L)	Solubility (Grams/L)	Temp (°C)	Ref (#)	Evaluation (T P E A A)	Comments
1.380E-06	6.000E-04	24	H116	2 1 0 0 2	

3987. C$_{24}$H$_{54}$OSn$_2$
bis(Tributyltin) Oxide
6-Oxa-5,7-distannaundecane, 5,5,7,7-Tetrabutyl-
RN: 56-35-9   **MP** (°C):
MW: 596.08   **BP** (°C): 180

Solubility (Moles/L)	Solubility (Grams/L)	Temp (°C)	Ref (#)	Evaluation (T P E A A)	Comments
1.678E-04	1.000E-01	rt	M161	0 0 0 0 2	

3988. C$_{25}$H$_{24}$N$_2$O$_2$S
G-8
o,p-Dimethylphenylthioethylphenylbutazone
3,5-Pyrazolidinedione, 1,2-Diphenyl-4-[2-(2,4-xylylthio)ethyl]-
RN: 102892-46-6   **MP** (°C):
MW: 416.55   **BP** (°C):

Solubility (Moles/L)	Solubility (Grams/L)	Temp (°C)	Ref (#)	Evaluation (T P E A A)	Comments
3.121E-04	1.300E-01	ns	B158	0 0 0 0 1	pH 7.0

3989. C$_{25}$H$_{28}$N$_4$O$_2$
Ethyldiantipyrylmethane
EDAM
RN: 61358-28-9   **MP** (°C):
MW: 416.53   **BP** (°C):

Solubility (Moles/L)	Solubility (Grams/L)	Temp (°C)	Ref (#)	Evaluation (T P E A A)	Comments
3.601E-04	1.500E-01	20	P054	1 2 2 2 2	

3990. C$_{25}$H$_{28}$O$_3$
Estradiol Benzoate
Estradiol Monobenzoate
7β-Estradiol-3-benzoate
RN: 50-50-0   **MP** (°C):   190
MW: 376.50   **BP** (°C):

Solubility (Moles/L)	Solubility (Grams/L)	Temp (°C)	Ref (#)	Evaluation (T P E A A)	Comments
1.062E-06	4.000E-04	25	K003	2 1 1 1 1	

3991. C$_{25}$H$_{29}$I$_2$NO$_3$
Amiodarone
Cordarone
Aratac
RN: 1951-25-3   **MP** (°C):
MW: 645.32   **BP** (°C):

Solubility (Moles/L)	Solubility (Grams/L)	Temp (°C)	Ref (#)	Evaluation (T P E A A)	Comments
1.110E-03	7.164E-01	25	B337	2 2 2 1 2	

3992.   C$_{25}$H$_{31}$FO$_8$
Triamcinolone 16, 21-Diacetate
Pregna-1,4-diene-3,20-dione, 16,21-bis(Acetyloxy)-9-fluoro-11,17-dihydroxy-,
(11β,16apha)-
RN: 67-78-7   **MP** (°C):   235
MW: 478.52   **BP** (°C):

Solubility (Moles/L)	Solubility (Grams/L)	Temp (°C)	Ref (#)	Evaluation (T P E A A)	Comments
1.003E-04	4.800E-02	25	F026	1 0 0 0 2	

3993.   C$_{25}$H$_{31}$NO$_2$
3-Hydroxy-17β-{[(1-methyl-1,4-dihydropyridin-3-yl)-carbonyl]oxy}-estra-1,3,5(10)-triene
RN: **MP** (°C):
MW: 377.53   **BP** (°C):

Solubility (Moles/L)	Solubility (Grams/L)	Temp (°C)	Ref (#)	Evaluation (T P E A A)	Comments
1.743E-07	6.580E-05	25	B366	1 2 2 2 2	

3994.   C$_{25}$H$_{34}$O$_3$
Norethindrone Dimethylpropionate
19-Norpregn-4-en-20-yn-3-one, 17-(2,2-Dimethyl-1-oxopropoxy)-, (17α)-
RN: 65445-09-2   **MP** (°C):
MW: 382.55   **BP** (°C):

Solubility (Moles/L)	Solubility (Grams/L)	Temp (°C)	Ref (#)	Evaluation (T P E A A)	Comments
7.894E-08	3.020E-05	25	L078	1 0 1 2 2	

3995.   C$_{25}$H$_{34}$O$_6$
Budesonide
16,17-Butylidenebis(oxy)-11-,21-dihydroxypregna-1,4-diene-3,20-dione
Rhinocort
RN: 51333-22-3   **MP** (°C):
MW: 430.55   **BP** (°C):

Solubility (Moles/L)	Solubility (Grams/L)	Temp (°C)	Ref (#)	Evaluation (T P E A A)	Comments
5.000E-05	2.153E-02	ns	F327	0 0 1 2 2	

3996. $C_{25}H_{34}O_9$

6-(1,3-Dihydro-7-acetate-5-methoxy-4-methyl-1-oxoisobenzofuran-6-yl)-4-methyl-4-hexanoic Solketal Ester

RN: **MP** (°C):
MW: 478.54 **BP** (°C):

Solubility (Moles/L)	Solubility (Grams/L)	Temp (°C)	Ref (#)	Evaluation (T P E A A)	Comments
1.881E-05	9.000E-03	25	L333	1 1 1 1 0	

3997. $C_{25}H_{36}N_4O_7$

Nonyloxycarbonyl-mitomycin C
2'-(2-Hexanoyl-2-pentanyl-acetyl)-6-methoxypurine Arabinoside

RN: **MP** (°C):
MW: 504.59 **BP** (°C):

Solubility (Moles/L)	Solubility (Grams/L)	Temp (°C)	Ref (#)	Evaluation (T P E A A)	Comments
2.500E-07	1.261E-04	25	M316	1 1 1 1 2	
2.020E-03	1.019E+00	37	C348	1 2 2 2 2	pH 7.00

3998. $C_{25}H_{36}O_6$

Hydrocortisone Butyrate
Hydrocortisone-21-butyrate
11,17-Dihydroxy-21-(1-oxobutoxy)-pregn-4-ene-3,20-dione

RN: 6677-99-2 **MP** (°C):
MW: 432.56 **BP** (°C):

Solubility (Moles/L)	Solubility (Grams/L)	Temp (°C)	Ref (#)	Evaluation (T P E A A)	Comments
1.787E-05	7.730E-03	25	H098	1 0 2 0 2	
1.787E-05	7.730E-03	25	H320	1 0 2 2 2	
1.780E-05	7.700E-03	25	H320	1 0 2 2 2	

3999. $C_{25}H_{36}O_7$

5,16-β-Dihydroxy-6-β-methyl-3,11-dioxo-5-α-pregn-17(20)-ene-cis-20-carboxylic Acid Methyl Ester Cycl

RN: **MP** (°C):
MW: 448.56 **BP** (°C):

Solubility (Moles/L)	Solubility (Grams/L)	Temp (°C)	Ref (#)	Evaluation (T P E A A)	Comments
1.672E-04	7.500E-02	ns	K029	0 0 2 1 1	

4000.   C$_{25}$H$_{40}$O$_3$Si$_2$
Norethindrone Pentamethyldisiloxyl Ether
RN: **MP** (°C):
MW: 444.77   **BP** (°C):

Solubility (Moles/L)	Solubility (Grams/L)	Temp (°C)	Ref (#)	Evaluation (T P E A A)	Comments
2.301E-07	1.023E-04	25	L078	1 0 1 2 2	

4001.   C$_{25}$H$_{42}$O$_3$
Octadecyl-p-hydroxybenzoate
RN: 71067-10-2   **MP** (°C):
MW: 390.61   **BP** (°C):

Solubility (Moles/L)	Solubility (Grams/L)	Temp (°C)	Ref (#)	Evaluation (T P E A A)	Comments
8.343E-04	3.259E-01	25	D081	1 2 2 1 2	

4002.   C$_{25}$H$_{44}$
Nonadecylbenzene
1-Phenylnonadecane
RN: 29136-19-4   **MP** (°C):
MW: 344.63   **BP** (°C):   419

Solubility (Moles/L)	Solubility (Grams/L)	Temp (°C)	Ref (#)	Evaluation (T P E A A)	Comments
1.530E-02	5.272E+00	328	S355	1 1 1 2 0	EFG
2.396E-01	8.257E+01	363	S355	1 1 1 2 0	EFG

4003.   C$_{25}$H$_{44}$O$_6$
4-Nonylphenol Pentaethoxylate
RN: 20636-48-0   **MP** (°C):
MW: 440.63   **BP** (°C):

Solubility (Moles/L)	Solubility (Grams/L)	Temp (°C)	Ref (#)	Evaluation (T P E A A)	Comments
2.151E-05	9.480E-03	20.5	A335	1 0 2 2 2	
2.150E-05	9.473E-03	20.5	A335	1 0 2 2 2	

4004.   C$_{25}$H$_{48}$O$_4$
Dioctyl Azelate
Di(2-ethylhexyl) Azelate
RN: 103-24-2   **MP** (°C):
MW: 412.66   **BP** (°C):

Solubility (Moles/L)	Solubility (Grams/L)	Temp (°C)	Ref (#)	Evaluation (T P E A A)	Comments
2.423E-07	1.000E-04	25	F067	1 0 2 2 0	

4005. $C_{25}H_{54}O_2P_2$
bis(Di-n-hexyl-phosphinyl)methane
HDPM
RN: 2785-33-3   **MP** (°C):
MW: 448.66   **BP** (°C):

Solubility (Moles/L)	Solubility (Grams/L)	Temp (°C)	Ref (#)	Evaluation (T P E A A)	Comments
1.426E-04	6.400E-02	0	O002	2 0 2 2 0	EFG
8.849E-05	3.970E-02	25	O002	2 0 2 2 1	average of 2
6.241E-05	2.800E-02	35	O002	2 0 2 2 0	EFG
4.458E-05	2.000E-02	40	O002	2 0 2 2 0	EFG
4.458E-05	2.000E-02	40	O002	2 0 2 2 0	EFG
3.377E-03	1.515E+00	45	O002	2 0 2 2 0	EFG

4006. $C_{26}H_{18}N_2O_4$
Samaron Violet
Mowilith Red 3B(IG)
RN: 6408-72-6   **MP** (°C):
MW: 422.44   **BP** (°C):

Solubility (Moles/L)	Solubility (Grams/L)	Temp (°C)	Ref (#)	Evaluation (T P E A A)	Comments
3.000E-06	1.267E-03	98.59	M180	0 0 2 2 0	EFG
4.000E-06	1.690E-03	109.98	M180	0 0 2 1 0	EFG
4.500E-06	1.901E-03	120.54	M180	0 0 2 2 0	EFG
6.000E-06	2.535E-03	133.34	M180	0 0 2 2 0	EFG
8.000E-06	3.380E-03	141.78	M180	0 0 2 2 0	EFG

4007. $C_{26}H_{20}N_2O_8S_2$
1,8-Anthraquinone Disulfonic Acid Anilide
RN: **MP** (°C):
MW: 552.59   **BP** (°C):

Solubility (Moles/L)	Solubility (Grams/L)	Temp (°C)	Ref (#)	Evaluation (T P E A A)	Comments
4.209E-02	2.326E+01	18	F047	1 2 1 1 1	

4008. $C_{26}H_{20}N_2O_8S_2$
1,5-Anthraquinone Disulfonic Acid Anilide
RN: **MP** (°C):
MW: 552.59   **BP** (°C):

Solubility (Moles/L)	Solubility (Grams/L)	Temp (°C)	Ref (#)	Evaluation (T P E A A)	Comments
7.210E-03	3.984E+00	18	F047	1 2 1 1 1	

4009. C$_{26}$H$_{28}$Cl$_2$N$_4$O$_4$
Ketoconazole
(±)-cis-1-Acetyl-4-(4-[(2-[2,4-dichlorophenyl]-2-[1H-imidazol-1-ylmethyl]-1,3-dioxolan-4-yl)-methoxy]phenyl)piperazine
RN: 65277-42-1 **MP** (°C):
MW: 531.44 **BP** (°C):

Solubility (Moles/L)	Solubility (Grams/L)	Temp (°C)	Ref (#)	Evaluation (T P E A A)	Comments
1.505E-04	8.000E-02	37	C323	1 0 1 1 0	EFG

4010. C$_{26}$H$_{28}$N$_2$
Cinnarizine
Stugeron
RN: 298-57-7 **MP** (°C):
MW: 368.53 **BP** (°C):

Solubility (Moles/L)	Solubility (Grams/L)	Temp (°C)	Ref (#)	Evaluation (T P E A A)	Comments
2.035E-03	7.500E-01	ns	B155	0 0 1 1 0	EFG, pH 3.0

4011. C$_{26}$H$_{28}$N$_4$O$_2$
Propyldiantipyrylmethane
PDAM
RN: 1461-17-2 **MP** (°C):
MW: 428.54 **BP** (°C):

Solubility (Moles/L)	Solubility (Grams/L)	Temp (°C)	Ref (#)	Evaluation (T P E A A)	Comments
1.400E-04	6.000E-02	20	P054	1 2 2 2 2	

4012. C$_{26}$H$_{30}$N$_4$O$_2$
Isopropyldiantipyrylmethane
IPDAM
RN: 15536-49-9 **MP** (°C):
MW: 430.55 **BP** (°C):

Solubility (Moles/L)	Solubility (Grams/L)	Temp (°C)	Ref (#)	Evaluation (T P E A A)	Comments
4.644E-04	2.000E-01	20	P054	1 2 2 2 2	

4013. C$_{26}$H$_{30}$N$_4$S$_2$
Propyldithiopyrylmethane
3H-Pyrazole-3-thione, 4,4'-Butylidenebis[1,2-dihydro-1,5-dimethyl-2-phenyl-
RN: 57094-83-4 **MP** (°C): 222
MW: 462.68 **BP** (°C):

Solubility (Moles/L)	Solubility (Grams/L)	Temp (°C)	Ref (#)	Evaluation (T P E A A)	Comments
2.400E-04	1.110E-01	ns	D087	0 2 0 0 1	

4014. C₂₆H₃₂F₂O₇

Diflorasone Diacetate

U-34865

RN: 33564-31-7   **MP** (°C):

MW: 494.54   **BP** (°C):

Solubility (Moles/L)	Solubility (Grams/L)	Temp (°C)	Ref (#)	Evaluation (T P E A A)	Comments
1.314E-05	6.500E-03	25	F003	2 2 1 0 1	
1.254E-05	6.200E-03	37	F003	2 2 1 0 1	
2.629E-05	1.300E-02	50	F003	2 2 1 0 1	

4015. C₂₆H₃₂F₂O₇

Fluocinolide

Fluocinonide

Fluocinolone Acetonide Acetate

RN: 356-12-7   **MP** (°C):

MW: 494.54   **BP** (°C):

Solubility (Moles/L)	Solubility (Grams/L)	Temp (°C)	Ref (#)	Evaluation (T P E A A)	Comments
1.072E-06	5.300E-04	25	O001	2 0 2 2 2	
2.022E-05	1.000E-02	25	P008	1 0 1 1 1	EFG

4016. C₂₆H₃₂O₃

Testosterone Benzoate

Androst-4-en-3-one, 17-(Benzoyloxy)-, (17β)-

RN: 2088-71-3   **MP** (°C):

MW: 392.54   **BP** (°C):

Solubility (Moles/L)	Solubility (Grams/L)	Temp (°C)	Ref (#)	Evaluation (T P E A A)	Comments
3.312E-05	1.300E-02	25	L342	1 0 1 1 2	

4017. C₂₆H₃₆O₃

Norethisterone Heptanoate

RN: **MP** (°C):

MW: 396.58   **BP** (°C):

Solubility (Moles/L)	Solubility (Grams/L)	Temp (°C)	Ref (#)	Evaluation (T P E A A)	Comments
1.521E-07	6.030E-05	25	E301	1 0 1 1 2	

4018. $C_{26}H_{36}O_6$
Prednisolone 21-Trimethylacetate
Prednisolone Acetate
RN: 52-21-1   **MP** (°C):   233
MW: 444.57   **BP** (°C):

Solubility (Moles/L)	Solubility (Grams/L)	Temp (°C)	Ref (#)	Evaluation (T P E A A)	Comments
2.609E-05	1.160E-02	25	C037	2 1 2 2 2	
6.298E-05	2.800E-02	25	K021	1 2 2 2 1	
2.699E-05	1.200E-02	ns	N302	0 2 1 2 1	

4019. $C_{26}H_{37}FO_5$
Dexamethasone TBA
RN: **MP** (°C):
MW: 448.58   **BP** (°C):

Solubility (Moles/L)	Solubility (Grams/L)	Temp (°C)	Ref (#)	Evaluation (T P E A A)	Comments
2.229E-05	1.000E-02	37	D026	1 2 1 2 2	

4020. $C_{26}H_{38}NO_8$
Glucosamine Testosterone
17-β-(4-Androsten-3-one)-N-2-(2-desoxyglucosyl)
RN: **MP** (°C):   185-190
MW: 492.59   **BP** (°C):

Solubility (Moles/L)	Solubility (Grams/L)	Temp (°C)	Ref (#)	Evaluation (T P E A A)	Comments
1.332E-03	6.560E-01	25	L009	1 0 0 1 1	

4021. $C_{26}H_{38}O_4$
Trimethylcyclohexyl Phthalate
bis(cis-3,3,5-Trimethylcyclohexyl) Phthalate
RN: 245652-81-7   **MP** (°C):   93
MW: 414.59   **BP** (°C):

Solubility (Moles/L)	Solubility (Grams/L)	Temp (°C)	Ref (#)	Evaluation (T P E A A)	Comments
2.894E-07	1.200E-04	24	H116	2 1 0 0 2	

4022.   C$_{26}$H$_{38}$O$_6$
Hydrocortisone Valerate
Hydrocortisone-21-valerate
RN: 6678-00-8   **MP** (°C):
MW: 446.59   **BP** (°C):

Solubility (Moles/L)	Solubility (Grams/L)	Temp (°C)	Ref (#)	Evaluation (T P E A A)	Comments
6.830E-06	3.050E-03	25	H098	1 0 2 0 2	
6.830E-06	3.050E-03	25	H320	1 0 2 2 2	
6.780E-06	3.028E-03	25	H320	1 0 2 2 2	

4023.   C$_{26}$H$_{39}$NO$_3$S
4-Pregnene-20-one-3-spiro-2'-(4'-ethoxycarbonyl-1',3'-thiazolidine)
RN: **MP** (°C):   131-135
MW: 445.67   **BP** (°C):

Solubility (Moles/L)	Solubility (Grams/L)	Temp (°C)	Ref (#)	Evaluation (T P E A A)	Comments
~3.81E-06	~1.70E-03	ns	B199	0 0 0 0 0	

4024.   C$_{26}$H$_{43}$NO$_3$
Acetaminophen Stearate
Acetaminophen Octadecanoate
Stearoyl Acetaminophen
Octadecanoic Acid, 4-(Acetylamino)phenyl Ester
Acetanilide, 4'-Hydroxy-, Stearate (Ester)
RN: 20675-22-3   **MP** (°C):   117
MW: 417.64   **BP** (°C):

Solubility (Moles/L)	Solubility (Grams/L)	Temp (°C)	Ref (#)	Evaluation (T P E A A)	Comments
1.197E-05	5.000E-03	25	B010	1 1 1 1 0	
3.592E-05	1.500E-02	37	D029	1 0 1 1 1	

4025.   C$_{26}$H$_{43}$NO$_6$
Glycocholic Acid
Glycine, N-[(3α,5β,7α,12α)-3,7,12-Trihydroxy-24-oxocholan-24-yl]-
RN: 475-31-0   **MP** (°C):   130
MW: 465.64   **BP** (°C):

Solubility (Moles/L)	Solubility (Grams/L)	Temp (°C)	Ref (#)	Evaluation (T P E A A)	Comments
7.085E-04	3.299E-01	20	E035	1 2 0 0 1	
2.188E-03	1.019E+00	60	E035	1 2 0 0 2	
5.035E-03	2.344E+00	80	E035	1 2 0 0 2	
1.810E-02	8.428E+00	100	E035	1 2 0 0 1	

4026.   C$_{26}$H$_{50}$O$_4$
Dioctyl Sebacate
Sebacic Acid bis(2-Ethylhexyl) Ester
RN: 122-62-3   **MP** (°C):   -67
MW: 426.69   **BP** (°C):   248

Solubility (Moles/L)	Solubility (Grams/L)	Temp (°C)	Ref (#)	Evaluation (T P E A A)	Comments
2.344E-07	1.000E-04	25	F067	1 0 2 2 0	

4027.   C$_{26}$H$_{56}$O$_2$P$_2$
bis(Di-n-hexyl-phosphinyl)ethane
HDPE
RN: 2785-34-4   **MP** (°C):
MW: 462.68   **BP** (°C):

Solubility (Moles/L)	Solubility (Grams/L)	Temp (°C)	Ref (#)	Evaluation (T P E A A)	Comments
2.810E-05	1.300E-02	0	O002	2 0 2 2 2	EFG
6.484E-06	3.000E-03	25	O002	2 0 2 2 2	
6.484E-06	3.000E-03	60	O002	2 0 2 2 2	EFG

4028.   C$_{27}$H$_{22}$Cl$_2$N$_4$
Clofazimine
Lamprene
N,5-bis(4-Chlorophenyl)-3,4-dihydro-3-((1-methylethyl)imino)-2-phenazinamine
3-(p-Chloroanilino)-10-(p-chlorophenyl)-2,10-dihydro-2-(isopropylimino)phenazine
RN: 2030-63-9   **MP** (°C):   211
MW: 473.41   **BP** (°C):

Solubility (Moles/L)	Solubility (Grams/L)	Temp (°C)	Ref (#)	Evaluation (T P E A A)	Comments
2.000E-04	9.468E-02	ns	O322	0 0 1 2 0	EFG

4029.   C$_{27}$H$_{29}$NO$_{11}$
Adriamycin
Adriblastin
RN: 23214-92-8   **MP** (°C):   205
MW: 543.53   **BP** (°C):

Solubility (Moles/L)	Solubility (Grams/L)	Temp (°C)	Ref (#)	Evaluation (T P E A A)	Comments
3.607E-02	1.961E+01	ns	I312	0 0 0 0 0	

4030.   C$_{27}$H$_{30}$O$_3$
Norethindrone Benzoate
RN: **MP** (°C):
MW: 402.54   **BP** (°C):

Solubility (Moles/L)	Solubility (Grams/L)	Temp (°C)	Ref (#)	Evaluation (T P E A A)	Comments
2.019E-08	8.128E-06	25	L078	1 0 1 2 2	

4031.   C$_{27}$H$_{32}$N$_4$O$_2$
Isobutyldiantipyrylmethane
IBDAM
RN: 16671-34-4   **MP** (°C):
MW: 444.58   **BP** (°C):

Solubility (Moles/L)	Solubility (Grams/L)	Temp (°C)	Ref (#)	Evaluation (T P E A A)	Comments
1.350E-04	6.000E-02	20	P054	1 2 2 2 2	

4032.   C$_{27}$H$_{32}$N$_4$O$_2$
Butyldiantipyrylmethane
BDAM
RN: 61358-30-3   **MP** (°C):
MW: 444.58   **BP** (°C):

Solubility (Moles/L)	Solubility (Grams/L)	Temp (°C)	Ref (#)	Evaluation (T P E A A)	Comments
6.748E-05	3.000E-02	20	P054	1 2 2 2 2	

4033.   C$_{27}$H$_{32}$N$_4$S$_2$
Isobutyldithiopyrylmethane
3H-Pyrazole-3-thione, 4,4'-(3-Methylbutylidene)bis[1,2-dihydro-1,5-dimethyl-2-phenyl-
RN: 73429-89-7   **MP** (°C):   209
MW: 476.71   **BP** (°C):

Solubility (Moles/L)	Solubility (Grams/L)	Temp (°C)	Ref (#)	Evaluation (T P E A A)	Comments
1.600E-04	7.627E-02	ns	D087	0 2 0 0 1	

4034. C$_{27}$H$_{32}$O$_{14}$
Naringin
4H-1-Benzopyran-4-one, 7-[[2-O-(6-Deoxy-α-L-mannopyranosyl)-β-D-glucopyranosyl]oxy]-2,3-dihydro-5-hydroxy-2-(4-hydroxyphenyl)-, (S)-
RN: 10236-47-2 **MP** (°C):
MW: 580.55 **BP** (°C):

Solubility (Moles/L)	Solubility (Grams/L)	Temp (°C)	Ref (#)	Evaluation (T P E A A)	Comments
2.928E-04	1.700E-01	6	P070	1 2 1 1 1	
8.613E-04	5.000E-01	20	P070	1 2 1 1 1	
1.361E-03	7.900E-01	35	P070	1 2 1 1 1	
3.376E-03	1.960E+00	45	P070	1 2 1 1 2	
1.233E-02	7.160E+00	55	P070	1 2 1 1 2	
7.271E-02	4.221E+01	65	P070	1 2 1 1 2	
1.864E-01	1.082E+02	75	P070	1 2 1 1 2	

4035. C$_{27}$H$_{33}$N$_3$O$_8$
Rolitetracycline
N-(1-Pyrrolidinylmethyl)tetracycline
Syntetrin
Tetraverin
Synotodecin
RN: 751-97-3 **MP** (°C): 162dec
MW: 527.58 **BP** (°C):

Solubility (Moles/L)	Solubility (Grams/L)	Temp (°C)	Ref (#)	Evaluation (T P E A A)	Comments
>3.79E-02	>2.00E+01	21	M044	2 0 2 2 0	

4036. C$_{27}$H$_{34}$O$_3$
Testosterone Phenylacetate
Androst-4-en-3-one, 17-[(Phenylacetyl)oxy]-, (17β)-
RN: 5704-03-0 **MP** (°C):
MW: 406.57 **BP** (°C):

Solubility (Moles/L)	Solubility (Grams/L)	Temp (°C)	Ref (#)	Evaluation (T P E A A)	Comments
2.206E-05	8.970E-03	25	L342	1 0 1 1 2	

4037. C$_{27}$H$_{34}$O$_{10}$
Cortisone Tricarballylate
RN: **MP** (°C):
MW: 518.57 **BP** (°C):

Solubility (Moles/L)	Solubility (Grams/L)	Temp (°C)	Ref (#)	Evaluation (T P E A A)	Comments
1.350E-04	7.000E-02	25	M023	1 0 2 1 0	

4038.   C$_{27}$H$_{38}$N$_2$O$_6$
p-Ureidophenyl Prostaglandin E2
RN: **MP** (°C):
MW: 486.61   **BP** (°C):

Solubility (Moles/L)	Solubility (Grams/L)	Temp (°C)	Ref (#)	Evaluation (T P E A A)	Comments
2.800E-05	1.363E-02	25	A066	1 0 1 1 1	

4039.   C$_{27}$H$_{38}$O$_3$
Norethindrone Heptanoate
RN: **MP** (°C):
MW: 410.60   **BP** (°C):

Solubility (Moles/L)	Solubility (Grams/L)	Temp (°C)	Ref (#)	Evaluation (T P E A A)	Comments
1.468E-07	6.026E-05	25	L078	1 0 1 2 2	

4040.   C$_{27}$H$_{40}$N$_2$O$_6$
p-Ureidophenyl Prostaglandin F2 α
RN: **MP** (°C):
MW: 488.63   **BP** (°C):

Solubility (Moles/L)	Solubility (Grams/L)	Temp (°C)	Ref (#)	Evaluation (T P E A A)	Comments
6.900E-05	3.372E-02	25	A066	1 0 1 1 1	

4041.   C$_{27}$H$_{40}$O$_6$
Hydrocortisone Tebutate
Hydrocortisone-21-hexanoate
Hydrocortisone-21-caproate
RN: 508-96-3   **MP** (°C):   168
MW: 460.62   **BP** (°C):

Solubility (Moles/L)	Solubility (Grams/L)	Temp (°C)	Ref (#)	Evaluation (T P E A A)	Comments
3.083E-06	1.420E-03	25	H098	1 0 2 0 2	
3.083E-06	1.420E-03	25	H320	1 0 2 2 2	
3.060E-06	1.409E-03	25	H320	1 0 2 2 2	

4042. C$_{27}$H$_{42}$Cl$_2$N$_2$O$_6$
α-Chloramphenicol Palmitate
β-Chloramphenicol Palmitate
Chloramphenicol Palmitate
RN: 530-43-8   **MP** (°C):   359
MW: 561.55   **BP** (°C):

Solubility (Moles/L)	Solubility (Grams/L)	Temp (°C)	Ref (#)	Evaluation (T P E A A)	Comments
1.100E-08	6.177E-06	20	M006	2 2 1 2 1	
8.500E-08	4.773E-05	20	M006	2 2 1 2 1	
1.500E-08	8.423E-06	25	M006	2 2 1 2 1	
9.600E-08	5.391E-05	25	M006	2 2 1 2 1	
7.123E-06	4.000E-03	28	R004	2 1 1 1 0	
1.800E-08	1.011E-05	29	M006	2 2 1 2 1	
1.440E-07	8.086E-05	29	M006	2 2 1 2 2	
2.700E-08	1.516E-05	32	M006	2 2 1 2 1	
2.600E-07	1.460E-04	32	M006	2 2 1 2 2	
3.100E-08	1.741E-05	35	M006	2 2 1 2 1	
3.800E-07	2.134E-04	35	M006	2 2 1 2 2	

4043. C$_{27}$H$_{42}$N$_4$O$_7$.0.3H$_2$O
2'-(2-Heptanoyl-2-hexanyl-acetyl)-6-methoxypurine Arabinoside (0.3 Hydrate)
RN: 145913-52-6   **MP** (°C):
MW: 540.06   **BP** (°C):

Solubility (Moles/L)	Solubility (Grams/L)	Temp (°C)	Ref (#)	Evaluation (T P E A A)	Comments
2.990E-04	1.615E-01	37	C348	1 2 2 2 2	pH 7.00

4044. C$_{27}$H$_{42}$O$_3$
Diosgenin
(25R)-Spirost-5-en-3β-ol
RN: 512-04-9   **MP** (°C):   204
MW: 414.63   **BP** (°C):

Solubility (Moles/L)	Solubility (Grams/L)	Temp (°C)	Ref (#)	Evaluation (T P E A A)	Comments
4.824E-08	2.000E-05	25	L033	1 0 2 1 0	

4045. C$_{27}$H$_{42}$O$_3$
Nandrolone Nonanoate
RN: **MP** (°C):
MW: 414.63   **BP** (°C):

Solubility (Moles/L)	Solubility (Grams/L)	Temp (°C)	Ref (#)	Evaluation (T P E A A)	Comments
2.233E-06	9.260E-04	37	C026	2 2 1 2 2	

4046. $C_{27}H_{43}NO_8$
N-Methylglucamine Testosterone
17-β-(4-Androsten-3-one)-N-methyl-N-1-(1-desoxyglucosyl) Carbamate
RN: **MP** (°C):   183-185
MW: 509.65   **BP** (°C):

Solubility (Moles/L)	Solubility (Grams/L)	Temp (°C)	Ref (#)	Evaluation (T P E A A)	Comments
8.633E-05	4.400E-02	25	L009	1 0 0 1 1	

4047. $C_{27}H_{44}N_4O_6$
2'-Hexadecyl-6-methoxypurine Arabinoside
RN: 145913-43-5   **MP** (°C):   97-99
MW: 520.67   **BP** (°C):

Solubility (Moles/L)	Solubility (Grams/L)	Temp (°C)	Ref (#)	Evaluation (T P E A A)	Comments
1.900E-05	9.893E-03	37	C348	1 2 2 2 2	pH 7.00

4048. $C_{27}H_{44}O$
Vitamin D3
Cholecalciferol
Activated 7-Dehydrocholesterol
Oleovitamin d3
RN: 67-97-0   **MP** (°C):   85
MW: 384.65   **BP** (°C):

Solubility (Moles/L)	Solubility (Grams/L)	Temp (°C)	Ref (#)	Evaluation (T P E A A)	Comments
<5.98E-04	<2.30E-01	25	P312	1 2 2 2 2	

4049. $C_{27}H_{46}O$
Cholesterol
(3β)-Cholest-5-en-3-ol
3β-Hydroxycholest-5-ene
Cholest-5-en-3-ol (3β)-
Cholest-5-ene-3β-ol
RN: 57-88-5   **MP** (°C):   148.0
MW: 386.67   **BP** (°C):

Solubility (Moles/L)	Solubility (Grams/L)	Temp (°C)	Ref (#)	Evaluation (T P E A A)	Comments
<1.00E-06	<3.87E-04	25	E014	2 2 2 1 0	pH 7.3
1.345E-04	5.200E-02	25	L009	1 0 0 1 1	sic
2.069E-07	8.000E-05	30	C107	2 2 2 2 0	
1.707E-07	6.600E-05	30	M007	2 2 1 2 1	average of 16
2.457E-07	9.500E-05	30	M052	2 0 2 2 0	EFG
1.300E-08	5.027E-06	37	C338	1 0 2 1 1	

5.172E-06	2.000E-03	ns	G027	0 0 0 0 0	
4.655E-06	1.800E-03	ns	H119	0 1 2 1 1	
5.172E-06	2.000E-03	ns	I312	0 0 0 0 0	
6.707E-03	2.593E+00	rt	D021	0 0 1 1 1	*sic*

4050. $C_{27}H_{58}O_2P_2$
bis(Di-n-hexyl-phosphinyl)propane
HDPP
RN: 2896-56-2   **MP** (°C):
MW: 476.71   **BP** (°C):

Solubility (Moles/L)	Solubility (Grams/L)	Temp (°C)	Ref (#)	Evaluation (T P E A A)	Comments
2.727E-04	1.300E-01	0	O002	2 0 2 2 0	EFG
1.154E-04	5.500E-02	15	O002	2 0 2 2 0	EFG
3.566E-05	1.700E-02	25	O002	2 0 2 2 0	

4051. $C_{28}H_{29}F_2N_3O$
Pimozide
2-Benzimidazolinone, 1-[1-[4,4-bis(p-Fluorophenyl)butyl]-4-piperidyl]-
Orap
RN: 2062-78-4   **MP** (°C):
MW: 461.56   **BP** (°C):

Solubility (Moles/L)	Solublilty (Grams/L)	Temp (°C)	Ref (#)	Evaluation (T P E A A)	Comments
6.283E-06	2.900E-03	30	P044	1 0 1 0 1	

4052. $C_{28}H_{36}O_3$
Testosterone Phenyl Propionate
Androst-4-en-3-one, 17-(1-Oxo-3-phenylpropoxy)-, (17β)-
RN: 1255-49-8   **MP** (°C):
MW: 420.60   **BP** (°C):

Solubility (Moles/L)	Solubility (Grams/L)	Temp (°C)	Ref (#)	Evaluation (T P E A A)	Comments
5.350E-06	2.250E-03	25	L342	1 0 1 1 2	

4053. $C_{28}H_{39}NO_6$
p-Acetamidophenyl Prostaglandin E2
RN: **MP** (°C):
MW: 485.63   **BP** (°C):

Solubility (Moles/L)	Solubility (Grams/L)	Temp (°C)	Ref (#)	Evaluation (T P E A A)	Comments
5.400E-05	2.622E-02	25	A066	1 0 1 1 1	

4054. $C_{28}H_{39}NO_6$

2-Oxo-5-Indolinyl Prostaglandin F2α

Prosta-5,13-dien-1-oic Acid, 9,11,15-Trihydroxy-, 2,3-Dihydro-2-oxo-1H-indol-5-yl Ester, (5Z,9α,11α,13E,15S)-

RN: 74973-22-1   **MP** (°C):

MW: 485.63   **BP** (°C):

Solubility (Moles/L)	Solubility (Grams/L)	Temp (°C)	Ref (#)	Evaluation (T P E A A)	Comments
6.000E-05	2.914E-02	25	A066	1 0 1 1 1	

4055. $C_{28}H_{39}N_3O_6$

α-Semicarbazono-p-tolyl Prostaglandin E2

RN: **MP** (°C):

MW: 513.64   **BP** (°C):

Solubility (Moles/L)	Solubility (Grams/L)	Temp (°C)	Ref (#)	Evaluation (T P E A A)	Comments
2.500E-06	1.284E-03	25	A066	1 0 1 1 1	

4056. $C_{28}H_{40}FNO_{11}.H_2O$

Glucosamine 9-α-Fluorohyfrocortisome (Monohydrate)

21-(9-α-Fluoro-11β, 17α-Dihydroxy-4-pregnen-3,20-dione)-N-2-(2-desoxyglucosyl) Carbamate

RN: **MP** (°C):   176-178

MW: 603.64   **BP** (°C):

Solubility (Moles/L)	Solubility (Grams/L)	Temp (°C)	Ref (#)	Evaluation (T P E A A)	Comments
5.964E-04	3.600E-01	25	L009	1 0 0 1 1	

4057. $C_{28}H_{41}N_3O_6$

α-Semicarbazono-p-tolyl Prostaglandin F2 α

RN: **MP** (°C):

MW: 515.66   **BP** (°C):

Solubility (Moles/L)	Solubility (Grams/L)	Temp (°C)	Ref (#)	Evaluation (T P E A A)	Comments
1.600E-05	8.250E-03	25	A066	1 0 1 1 1	

4058. $C_{28}H_{42}FNO_{11}.H_2O$
Glucamine 9-α-Fluorohyfrocortisome (Monohydrate)
RN: **MP** (°C):   105-110
MW: 605.66   **BP** (°C):

Solubility (Moles/L)	Solubility (Grams/L)	Temp (°C)	Ref (#)	Evaluation (T P E A A)	Comments
4.456E-03	2.699E+00	25	L009	1 0 0 1 1	

4059. $C_{28}H_{42}O_6$
Hydrocortisone Heptanoate
Hydrocortisone-21-heptanoate
RN: **MP** (°C):
MW: 474.64   **BP** (°C):

Solubility (Moles/L)	Solubility (Grams/L)	Temp (°C)	Ref (#)	Evaluation (T P E A A)	Comments
2.082E-06	9.880E-04	25	H098	1 0 2 0 2	
2.082E-06	9.880E-04	25	H320	1 0 2 2 2	
2.060E-06	9.778E-04	25	H320	1 0 2 2 2	

4060. $C_{28}H_{44}O_3$
Nandrolone Decanoate
Deca-Durabolln
Norandrostenolone Decanoate
RN: 360-70-3   **MP** (°C):
MW: 428.66   **BP** (°C):

Solubility (Moles/L)	Solubility (Grams/L)	Temp (°C)	Ref (#)	Evaluation (T P E A A)	Comments
1.549E-06	6.640E-04	37	C026	2 2 1 2 2	

4061. $C_{28}H_{46}O_4$
Di-n-decyl Phthalate
RN: 84-77-5   **MP** (°C):
MW: 446.68   **BP** (°C):

Solubility (Moles/L)	Solubility (Grams/L)	Temp (°C)	Ref (#)	Evaluation (T P E A A)	Comments
7.388E-07	3.300E-04	24	H116	2 1 0 0 2	

4062. $C_{28}H_{46}O_4$
Diisodecyl Phthalate
RN: 26761-40-0   **MP** (°C):
MW: 446.68   **BP** (°C):

Solubility (Moles/L)	Solubility (Grams/L)	Temp (°C)	Ref (#)	Evaluation (T P E A A)	Comments
6.269E-07	2.800E-04	24	H116	2 1 0 0 2	

4063.   C$_{28}$H$_{60}$O$_2$P$_2$
bis(Di-n-hexyl-phosphinyl)butane
HDPB
RN: 2785-35-5   **MP** (°C):
MW: 490.74   **BP** (°C):

Solubility (Moles/L)	Solubility (Grams/L)	Temp (°C)	Ref (#)	Evaluation (T P E A A)	Comments
3.627E-04	1.780E-01	0	O002	2 0 2 2 0	EFG
1.284E-04	6.300E-02	15	O002	2 0 2 2 0	EFG
4.076E-05	2.000E-02	25	O002	2 0 2 2 0	

4064.   C$_{29}$H$_{20}$N$_2$O$_4$
1,4-Dibenzoylaminoanthraquinone
Benzamide, N,N'-(9,10-Dihydro-3-methyl-9,10-dioxo-1,8-anthracenediyl)bis
RN: 4627-15-0   **MP** (°C):
MW: 460.49   **BP** (°C):

Solubility (Moles/L)	Solubility (Grams/L)	Temp (°C)	Ref (#)	Evaluation (T P E A A)	Comments
2.200E-05	1.013E-02	50	G077	1 0 0 0 1	

4065.   C$_{29}$H$_{27}$N$_5$O$_4$
m-Nitrophenyldiantipyrylmethane
m-NPhDAM
RN: 1606-53-7   **MP** (°C):
MW: 509.57   **BP** (°C):

Solubility (Moles/L)	Solubility (Grams/L)	Temp (°C)	Ref (#)	Evaluation (T P E A A)	Comments
5.887E-05	3.000E-02	20	P054	1 2 2 2 2	

4066.   C$_{29}$H$_{27}$N$_5$O$_4$
o-Nitrophenyldiantipyrylmethane
o-NPhDAM
RN: 14957-18-7   **MP** (°C):
MW: 509.57   **BP** (°C):

Solubility (Moles/L)	Solubility (Grams/L)	Temp (°C)	Ref (#)	Evaluation (T P E A A)	Comments
3.925E-05	2.000E-02	20	P054	1 2 2 2 2	

4067. C$_{29}$H$_{27}$N$_5$O$_4$
p-Nitrophenyldiantipyrylmethane
p-NPhDAM
RN: 55774-19-1   **MP** (°C):
MW: 509.57   **BP** (°C):

Solubility (Moles/L)	Solubility (Grams/L)	Temp (°C)	Ref (#)	Evaluation (T P E A A)	Comments
3.925E-05	2.000E-02	20	P054	1 2 2 2 2	

4068. C$_{29}$H$_{28}$N$_4$O$_2$
Phenyldiantipyrylmethane
PhDAM
RN: 1861-84-3   **MP** (°C):
MW: 464.57   **BP** (°C):

Solubility (Moles/L)	Solubility (Grams/L)	Temp (°C)	Ref (#)	Evaluation (T P E A A)	Comments
5.165E-04	2.399E-01	20	P054	1 2 2 2 2	

4069. C$_{29}$H$_{28}$N$_4$O$_3$
o-Hydroxylphenyldiantipyrylmethane
o-HPhDAM
RN: 1606-55-9   **MP** (°C):
MW: 480.57   **BP** (°C):

Solubility (Moles/L)	Solubility (Grams/L)	Temp (°C)	Ref (#)	Evaluation (T P E A A)	Comments
<2.08E-05	<1.00E-02	20	P054	1 2 2 2 0	

4070. C$_{29}$H$_{28}$N$_4$S$_2$
Phenyldithiopyrylmethane
3H-Pyrazole-3-thione, 4,4'-(Phenylmethylene)bis[1,2-dihydro-1,5-dimethyl-2-phenyl-
RN: 74713-68-1   **MP** (°C):   160
MW: 496.70   **BP** (°C):

Solubility (Moles/L)	Solubility (Grams/L)	Temp (°C)	Ref (#)	Evaluation (T P E A A)	Comments
4.200E-05	2.086E-02	ns	D087	0 2 0 0 1	

4071. C$_{29}$H$_{32}$O$_{13}$
Etoposide
4'-Demethylepipodophyllotoxin ethylidene-β-D-glucoside
Vepesid
VP-16
RN: 33419-42-0   **MP** (°C):   236-251
MW: 588.57   **BP** (°C):

Solubility (Moles/L)	Solubility (Grams/L)	Temp (°C)	Ref (#)	Evaluation (T P E A A)	Comments
3.398E-04	2.000E-01	ns	D347	0 0 2 2 0	

4072. C$_{29}$H$_{36}$N$_{4}$O$_{2}$
Hexyldiantipyrylmethane
HDAM
RN: 7660-44-8   **MP** (°C):
MW: 472.64   **BP** (°C):

Solubility (Moles/L)	Solubility (Grams/L)	Temp (°C)	Ref (#)	Evaluation (T P E A A)	Comments
4.230E-05	1.999E-02	20	P054	1 2 2 2 2	
4.232E-05	2.000E-02	20	P054	1 2 2 2 2	

4073. C$_{29}$H$_{36}$N$_{4}$S$_{2}$
Hexyldithiopyrylmethane
3H-Pyrazole-3-thione, 4,4'-Heptylidenebis[1,2-dihydro-1,5-dimethyl-2-phenyl-
RN: 74713-69-2   **MP** (°C):   169
MW: 504.77   **BP** (°C):

Solubility (Moles/L)	Solubility (Grams/L)	Temp (°C)	Ref (#)	Evaluation (T P E A A)	Comments
4.100E-05	2.070E-02	0	D087	0 2 0 0 1	

4074. C$_{29}$H$_{38}$Cl$_{2}$N$_{2}$O$_{3}$
3β-Hydroxy-13α-amino-13,17-seco-5α-androstan-17-oic-13,17-lactam-4-N,N-bis-
(chloroethyl)amino Phenyl-acetate
RN: **MP** (°C):
MW: 533.54   **BP** (°C):

Solubility (Moles/L)	Solubility (Grams/L)	Temp (°C)	Ref (#)	Evaluation (T P E A A)	Comments
3.186E-07	1.700E-04	25	P022	2 1 1 1 2	
3.599E-07	1.920E-04	30	P022	2 1 1 1 2	
4.517E-07	2.410E-04	44	P022	2 1 1 1 2	
6.110E-07	3.260E-04	73	P022	2 1 1 1 2	

4075.   C$_{29}$H$_{38}$O$_3$
Testosterone Phenylbutyrate
RN: **MP** (°C):
MW: 434.62   **BP** (°C):

Solubility (Moles/L)	Solubility (Grams/L)	Temp (°C)	Ref (#)	Evaluation (T P E A A)	Comments
3.681E-06	1.600E-03	25	L342	1 0 1 1 2	

4076.   C$_{29}$H$_{40}$N$_2$O$_4$
Emetine
Emetan, 6',7',10,11-Tetramethoxy-
NSC 33669
RN: 483-18-1   **MP** (°C):   74
MW: 480.65   **BP** (°C):

Solubility (Moles/L)	Solubility (Grams/L)	Temp (°C)	Ref (#)	Evaluation (T P E A A)	Comments
2.000E-03	9.613E-01	15	K059	2 2 2 0 0	
2.078E-03	9.990E-01	c	D004	1 0 0 0 0	

4077.   C$_{29}$H$_{42}$O$_6$
Cortisone Caprylate
RN: **MP** (°C):
MW: 486.65   **BP** (°C):

Solubility (Moles/L)	Solubility (Grams/L)	Temp (°C)	Ref (#)	Evaluation (T P E A A)	Comments
4.110E-06	2.000E-03	25	M023	1 0 2 1 0	

4078.   C$_{29}$H$_{44}$FNO$_{11}$.H$_2$O
N-Methylglucamine 9-α-Fluorohyfrocortisome (Monohydrate)
21-(9-α-Fluoro-11β, 17α-Dihydroxy-4-pregnen-3,20-dione)-N-methyl-N-1-(1-desoxyglucosyl)
Carbamate
RN: **MP** (°C):   120
MW: 619.69   **BP** (°C):

Solubility (Moles/L)	Solubility (Grams/L)	Temp (°C)	Ref (#)	Evaluation (T P E A A)	Comments
6.358E-03	3.940E+00	25	L009	1 0 0 1 1	

4079. C$_{29}$H$_{44}$O$_{12}$
Oubain
γ-Strophanthin
Ouabain
Quabain
RN: 630-60-4   **MP** (°C):   185
MW: 584.67   **BP** (°C):

Solubility (Moles/L)	Solubility (Grams/L)	Temp (°C)	Ref (#)	Evaluation (T P E A A)	Comments
2.223E-02	1.300E+01	25	P312	1 2 2 2 2	
1.693E-02	9.901E+00	c	D004	1 0 0 0 0	
2.851E-01	1.667E+02	h	D004	1 0 0 0 0	

4080. C$_{29}$H$_{46}$N$_4$O$_7$.0.4H$_2$O
2'-(2-Octanoyl-2-heptanyl-acetyl)-6-methoxypurine Arabinoside (0.4 Hydrate)
RN: 145913-53-7   **MP** (°C):
MW: 569.92   **BP** (°C):

Solubility (Moles/L)	Solubility (Grams/L)	Temp (°C)	Ref (#)	Evaluation (T P E A A)	Comments
2.810E-05	1.601E-02	37	C348	1 2 2 2 2	pH 7.00

4081. C$_{29}$H$_{46}$O$_3$
Nandrolone Undecanoate
RN: **MP** (°C):
MW: 442.69   **BP** (°C):

Solubility (Moles/L)	Solubility (Grams/L)	Temp (°C)	Ref (#)	Evaluation (T P E A A)	Comments
1.360E-06	6.020E-04	37	C026	2 2 1 2 2	

4082. C$_{30}$H$_{28}$N$_4$O$_3$
Benzoyldiantipyrylmethane
BenzDAM
RN: 55774-17-9   **MP** (°C):
MW: 492.58   **BP** (°C):

Solubility (Moles/L)	Solubility (Grams/L)	Temp (°C)	Ref (#)	Evaluation (T P E A A)	Comments
<2.03E-05	<1.00E-02	20	P054	1 2 2 2 0	

4083.   C$_{30}$H$_{34}$O$_{13}$
Picrotoxin
Picrotoxine
RN: 124-87-8   **MP** (°C):
MW: 602.60   **BP** (°C):

Solubility (Moles/L)	Solubility (Grams/L)	Temp (°C)	Ref (#)	Evaluation (T P E A A)	Comments
4.964E-03	2.991E+00	20	D041	1 0 0 0 0	
6.776E-03	4.083E+00	rt	D021	0 0 1 1 1	

4084.   C$_{30}$H$_{48}$O$_{12}$
Periplocin
Card-20(22)-enolide, 3-[(2,6-Dideoxy-4-O-β-D-glucopyranosyl-3-O-methyl-β-D-ribo-hexopyranosyl)oxy]-5,14-dihydroxy-, (3β,5β)-
Periplocoside
RN: 13137-64-9   **MP** (°C):   205
MW: 600.71   **BP** (°C):

Solubility (Moles/L)	Solubility (Grams/L)	Temp (°C)	Ref (#)	Evaluation (T P E A A)	Comments
1.321E-02	7.937E+00	c	D004	1 0 0 0 0	

4085.   C$_{31}$H$_{33}$N$_{5}$O$_{2}$
p-Dimethylaminophenyldiantipyrylmethane
p-DMAPhDAM
RN: 2088-76-8   **MP** (°C):
MW: 507.64   **BP** (°C):

Solubility (Moles/L)	Solubility (Grams/L)	Temp (°C)	Ref (#)	Evaluation (T P E A A)	Comments
1.576E-04	7.999E-02	20	P054	1 2 2 2 2	

4086.   C$_{31}$H$_{38}$N$_{2}$O$_{11}$
Dihydronovobiocin
Benzamide, N-[7-[[3-O-(Aminocarbonyl)-6-deoxy-5-C-methyl-4-O-methyl-β-L-lyxo-hexopyranosyl]oxy]-4-hydroxy-8-methyl-2-oxo-2H-1-benzopyran-3-yl]-4-hydroxy-3-(3-methylbutyl)-
RN: 29826-16-2   **MP** (°C):
MW: 614.66   **BP** (°C):

Solubility (Moles/L)	Solubility (Grams/L)	Temp (°C)	Ref (#)	Evaluation (T P E A A)	Comments
2.928E-04	1.800E-01	28	A038	2 0 1 1 2	

4087. $C_{31}H_{42}FNO_{12}.H_2O$

Glucosamine Triamcinolone Acetonide (Monohydrate)
21-(9-α-Fluoro-11β-hydroxy-16α, 17α-Isopropylidenedioxy-1,4-pregnadien-3,20-dione)-
N-2-(2-desoxyglucosyl) Carbamate

RN: **MP** (°C):   250-255
MW: 657.69   **BP** (°C):

Solubility (Moles/L)	Solubility (Grams/L)	Temp (°C)	Ref (#)	Evaluation (T P E A A)	Comments
5.717E-04	3.760E-01	25	L009	1 0 0 1 1	

4088. $C_{31}H_{44}FNO_{12}.H_2O$

GlucamineTriamcinolone Acetonide (Monohydrate)
21-(9-α-Fluoro-11β-hydroxy-16α, 17α-Isopropylidenedioxy-1,4-pregnadien-3,20-dione)-
N-1-(1-desoxyglucosyl) Carbamate

RN: **MP** (°C):   150
MW: 659.71   **BP** (°C):

Solubility (Moles/L)	Solubility (Grams/L)	Temp (°C)	Ref (#)	Evaluation (T P E A A)	Comments
5.366E-03	3.540E+00	25	L009	1 0 0 1 1	

4089. $C_{31}H_{44}N_2O_7$

N-Acetyl-L-tyrosinamide Prostaglandin E2

RN: **MP** (°C):
MW: 556.71   **BP** (°C):

Solubility (Moles/L)	Solubility (Grams/L)	Temp (°C)	Ref (#)	Evaluation (T P E A A)	Comments
1.700E-04	9.464E-02	25	A066	1 0 1 1 1	

4090. $C_{31}H_{46}N_2O_7$

N-Acetyl-L-tyrosinamide Prostaglandin F2 α

RN: **MP** (°C):
MW: 558.72   **BP** (°C):

Solubility (Moles/L)	Solubility (Grams/L)	Temp (°C)	Ref (#)	Evaluation (T P E A A)	Comments
1.400E-04	7.822E-02	25	A066	1 0 1 1 1	

4091. $C_{31}H_{48}O_{12}$
Strophanthin
k-Strophanthin

RN: 11005-63-3   **MP** (°C):   179
MW: 612.72   **BP** (°C):

Solubility (Moles/L)	Solubility (Grams/L)	Temp (°C)	Ref (#)	Evaluation (T P E A A)	Comments
3.709E-02	2.273E+01	25	D004	1 0 0 0 0	

4092. $C_{32}H_{32}O_{14}$
Chartreusin
Lambdamycin
NSC 5159
Antibiotic X 465A

RN: 6377-18-0   **MP** (°C):   246-249
MW: 640.60   **BP** (°C):

Solubility (Moles/L)	Solubility (Grams/L)	Temp (°C)	Ref (#)	Evaluation (T P E A A)	Comments
2.342E-05	1.500E-02	25	P067	2 0 2 2 2	

4093. $C_{32}H_{37}NO_5S$
Dextropropoxyphene Napsylate
Darvocet N-50
Darvocet N-100
Darvon-N

RN: 17140-78-2   **MP** (°C):
MW: 547.72   **BP** (°C):

Solubility (Moles/L)	Solubility (Grams/L)	Temp (°C)	Ref (#)	Evaluation (T P E A A)	Comments
2.556E-03	1.400E+00	22	N319	1 0 2 2 1	

4094. $C_{32}H_{46}FNO_{12}.H_2O$
N-Methylglucamine Triamcinolone Acetonide (Monohydrate)
21-(9-α-Fluoro-11β-hydroxy-16α, 17α-Isopropylidenedioxy-1,4-pregnadien-3,20-dione)-
N-methyl-N-1-(1-desoxyglucosyl) Carbamate

RN: **MP** (°C):   152
MW: 673.74   **BP** (°C):

Solubility (Moles/L)	Solubility (Grams/L)	Temp (°C)	Ref (#)	Evaluation (T P E A A)	Comments
4.744E-03	3.196E+00	25	L009	1 0 0 1 1	

4095. $C_{32}H_{49}NO_9$
Cevadine
Cevane-3,4,12,14,16,17,20-heptol, 4,9-epoxy-, 3-[(2Z)-2-methyl-2-butenoate],
(3β,4α,16β)-
Veratrine
RN: 62-59-9   **MP** (°C):   213.5
MW: 591.75   **BP** (°C):

Solubility (Moles/L)	Solubility (Grams/L)	Temp (°C)	Ref (#)	Evaluation (T P E A A)	Comments
8.000E-03	4.734E+00	15	K059	2 2 2 0 0	

4096. $C_{32}H_{54}O_4$
Didodecyl Phthalate
1,2-Benzenedicarboxylic Acid, Didodecyl Ester
RN: 2432-90-8   **MP** (°C):
MW: 502.78   **BP** (°C):

Solubility (Moles/L)	Solubility (Grams/L)	Temp (°C)	Ref (#)	Evaluation (T P E A A)	Comments
2.784E-07	1.400E-04	24	H116	2 1 0 0 2	

4097. $C_{33}H_{25}N_3O_3$
Norbormide
5-(α-Hydroxy-α-2-pyridylbenzyl)-7-(α-2-pyridylbenzylidene)-5-norbornene-2,3-
dicaboximide
Shoxin
RN: 991-42-4   **MP** (°C):   >160
MW: 511.59   **BP** (°C):

Solubility (Moles/L)	Solubility (Grams/L)	Temp (°C)	Ref (#)	Evaluation (T P E A A)	Comments
1.173E-04	6.000E-02	rt	M161	0 0 0 0 1	

4098. $C_{33}H_{34}O_3$
Norethindrone Biphenyl-4-carboxylate
RN: **MP** (°C):
MW: 478.64   **BP** (°C):

Solubility (Moles/L)	Solubility (Grams/L)	Temp (°C)	Ref (#)	Evaluation (T P E A A)	Comments
7.762E-09	3.715E-06	25	L078	1 0 1 2 2	

4099.   C₃₃H₃₄O₄

Norethindrone 4-Phenoxybenzoate
RN: **MP** (°C):
MW: 494.64   **BP** (°C):

Solubility (Moles/L)	Solubility (Grams/L)	Temp (°C)	Ref (#)	Evaluation (T P E A A)	Comments
1.431E-07	7.079E-05	25	L078	1 0 1 2 2	

4100.   C₃₃H₃₆N₄O₆

Bilirubin
21H-Biline-8,12-dipropanoic Acid, 2,17-Diethenyl-1,10,19,22,23,24-hexahydro-3,7,13,18-tetramethyl-1,19-dioxo-
RN: 635-65-4   **MP** (°C):
MW: 584.68   **BP** (°C):

Solubility (Moles/L)	Solubility (Grams/L)	Temp (°C)	Ref (#)	Evaluation (T P E A A)	Comments
7.000E-09	4.093E-06	18	K104	1 0 0 0 2	intrinsic

4101.   C₃₃H₄₀N₂O₉

Reserpine
3,4,5-Trimethoxybenzoyl Methyl Reserpate
Rauwilid
Rauwiloid
RN: 50-55-5   **MP** (°C):
MW: 608.69   **BP** (°C):

Solubility (Moles/L)	Solubility (Grams/L)	Temp (°C)	Ref (#)	Evaluation (T P E A A)	Comments
1.200E-04	7.304E-02	30	L068	1 0 0 1 0	EFG

4102.   C₃₃H₄₁N₅O₆S₂

Kynostatin
KNI-272
4-Thiazolidinecarboxamide, N-(1,1-Dimethylethyl)-3-[(2S,3S)-2-hydroxy-3-[[(2R)-2-[[(5-isoquinolinyloxy)acetyl]amino]-3-(methylthio)-1-oxopropyl]amino]-1-oxo-4-phenylbutyl]-, (4R)-
RN: 147318-81-8 **MP** (°C):
MW: 667.85   **BP** (°C):

Solubility (Moles/L)	Solubility (Grams/L)	Temp (°C)	Ref (#)	Evaluation (T P E A A)	Comments
6.289E-06	4.200E-03	25	J308	2 0 2 2 2	

4103.   C$_{33}$H$_{45}$NO$_9$
Delphinine
Indaconitine, N-Deethyl-3-deoxy-N-methyl-
RN: 561-07-9   **MP** (°C):   198-200
MW: 599.73   **BP** (°C):

Solubility (Moles/L)	Solubility (Grams/L)	Temp (°C)	Ref (#)	Evaluation (T P E A A)	Comments
3.335E-05	2.000E-02	25	D004	1 0 0 0 0	

4104.   C$_{33}$H$_{47}$NO$_{13}$
Natamycin
Pimafucin
RN: 7681-93-8   **MP** (°C):
MW: 665.74   **BP** (°C):

Solubility (Moles/L)	Solubility (Grams/L)	Temp (°C)	Ref (#)	Evaluation (T P E A A)	Comments
4.506E-05	3.000E-02	20	B190	1 2 1 1 0	
6.159E-04	4.100E-01	21	M044	2 0 2 2 2	*sic*

4105.   C$_{34}$H$_{34}$N$_4$O$_4$
Protoprophyrin IX
Protoporphyrin IX
RN: 553-12-8   **MP** (°C):
MW: 562.67   **BP** (°C):

Solubility (Moles/L)	Solubility (Grams/L)	Temp (°C)	Ref (#)	Evaluation (T P E A A)	Comments
1.900E-04	1.069E-01	25	C097	2 0 1 1 1	EFG

4106.   C$_{34}$H$_{47}$NO$_{11}$
Aconitine
Acetylbenzoylaconine
RN: 302-27-2   **MP** (°C):   204
MW: 645.75   **BP** (°C):

Solubility (Moles/L)	Solubility (Grams/L)	Temp (°C)	Ref (#)	Evaluation (T P E A A)	Comments
4.691E-04	3.029E-01	25	D004	1 0 0 0 0	

4107. $C_{34}H_{50}O_7$

Carbenoxolone

Olean-12-en-29-oic Acid, 3-(3-Carboxy-1-oxopropoxy)-11-oxo-, (3β,20β)-

RN: 5697-56-3   **MP** (°C):
MW: 570.77   **BP** (°C):

Solubility (Moles/L)	Solubility (Grams/L)	Temp (°C)	Ref (#)	Evaluation (T P E A A)	Comments
1.160E-05	6.621E-03	24	B363	1 2 2 2 2	
1.630E-05	9.304E-03	37	B363	1 2 2 2 2	

4108. $C_{34}H_{57}NO_7$

Glucosamine Cholesterol

3-β-(5-Cholestenyl)-N-2-(2-desoxyglucosyl) Carbamate

RN: **MP** (°C):   155-158
MW: 591.84   **BP** (°C):

Solubility (Moles/L)	Solubility (Grams/L)	Temp (°C)	Ref (#)	Evaluation (T P E A A)	Comments
9.530E-04	5.640E-01	25	L009	1 0 0 1 1	

4109. $C_{34}H_{58}O_4$

Ditridecyl Phthalate
Staflex DTDP
Truflex DTDP
Hexaplas DTDP
Jayflex DTDP
Polycizer 962BPA

RN: 119-06-2   **MP** (°C):
MW: 530.84   **BP** (°C):

Solubility (Moles/L)	Solubility (Grams/L)	Temp (°C)	Ref (#)	Evaluation (T P E A A)	Comments
6.405E-07	3.400E-04	24	H116	2 1 0 0 2	

4110. $C_{34}H_{68}N_3O_8S_2$

Lincomycin Hexadecylsulfamate

RN: **MP** (°C):
MW: 711.06   **BP** (°C):

Solubility (Moles/L)	Solubility (Grams/L)	Temp (°C)	Ref (#)	Evaluation (T P E A A)	Comments
5.738E-04	4.080E-01	21	M044	2 0 2 2 2	

4111. C$_{35}$H$_{44}$N$_2$O$_7$
p-(p-Acetamidobenzamido)phenyl Prostaglandin E2
RN: **MP** (°C):
MW: 604.75   **BP** (°C):

Solubility (Moles/L)	Solubility (Grams/L)	Temp (°C)	Ref (#)	Evaluation (T P E A A)	Comments
9.800E-08	5.927E-05	25	A066	1 0 1 1 1	

4112. C$_{35}$H$_{46}$N$_2$O$_7$
p-(p-Acetamidobenzamido)phenyl Prostaglandin F2 α
RN: **MP** (°C):
MW: 606.77   **BP** (°C):

Solubility (Moles/L)	Solubility (Grams/L)	Temp (°C)	Ref (#)	Evaluation (T P E A A)	Comments
2.800E-07	1.699E-04	25	A066	1 0 1 1 1	

4113. C$_{35}$H$_{47}$NO$_9$
Rhizoxin
RN: 90996-54-6   **MP** (°C):
MW: 625.77   **BP** (°C):

Solubility (Moles/L)	Solubility (Grams/L)	Temp (°C)	Ref (#)	Evaluation (T P E A A)	Comments
1.918E-05	1.200E-02	25	P336	1 2 1 2 2	

4114. C$_{35}$H$_{61}$NO$_7$
N-Methylglucamine Cholesterol
3-β-(5-Cholestenyl)-N-methyl-N-1-(1-desoxyglucosyl) Carbamate
RN: **MP** (°C):   131-133
MW: 607.88   **BP** (°C):

Solubility (Moles/L)	Solubility (Grams/L)	Temp (°C)	Ref (#)	Evaluation (T P E A A)	Comments
1.842E-04	1.120E-01	25	L009	1 0 0 1 1	

4115. C$_{36}$H$_{47}$N$_2$O$_7$
N-Benzoyl-L-tyrosinamide Prostaglandin E2
RN: **MP** (°C):
MW: 619.79   **BP** (°C):

Solubility (Moles/L)	Solubility (Grams/L)	Temp (°C)	Ref (#)	Evaluation (T P E A A)	Comments
4.700E-07	2.913E-04	25	A066	1 0 1 1 1	

4116.   C$_{36}$H$_{49}$N$_2$O$_7$
N-Benzoyl-L-tyrosinamide Prostaglandin F2 α
RN: **MP** (°C):
MW: 621.80   **BP** (°C):

Solubility (Moles/L)	Solubility (Grams/L)	Temp (°C)	Ref (#)	Evaluation (T P E A A)	Comments
1.800E-06	1.119E-03	25	A066	1 0 1 1 1	

4117.   C$_{36}$H$_{56}$O$_{14}$
Digitalin
Card-20(22)-enolide, 3-[(6-Deoxy-4-O-β-D-glucopyranosyl-3-O-methyl-β-D-galactopyranosyl)oxy]-14,16-dihydroxy-, (3β,5β,16β)-
Digitalinum Verum
RN: 752-61-4   **MP** (°C):   229
MW: 712.84   **BP** (°C):

Solubility (Moles/L)	Solubility (Grams/L)	Temp (°C)	Ref (#)	Evaluation (T P E A A)	Comments
1.401E-03	9.990E-01	25	D004	1 0 0 0 0	

4118.   C$_{36}$H$_{60}$O$_2$
Vitamin A Palmitate
Retinol, Hexadecanoate
Retinyl Palmitate
RN: 79-81-2   **MP** (°C):
MW: 524.88   **BP** (°C):

Solubility (Moles/L)	Solubility (Grams/L)	Temp (°C)	Ref (#)	Evaluation (T P E A A)	Comments
5.000E-07	2.624E-04	25	P343	0 1 1 1 2	

4119.   C$_{36}$H$_{60}$O$_{30}$
α-Cyclodextrin
β-Hexaamylose
(C6H10O5)6
α-Dextrin
RN: 10016-20-3   **MP** (°C):
MW: 972.86   **BP** (°C):

Solubility (Moles/L)	Solubility (Grams/L)	Temp (°C)	Ref (#)	Evaluation (T P E A A)	Comments
9.345E-02	9.091E+01	20	F186	1 2 1 1 1	
2.409E-02	2.344E+01	20	P048	1 0 1 1 1	*sic*
1.204E-01	1.171E+02	23.7	J305	2 0 2 2 2	
1.118E-01	1.088E+02	23.7	J305	2 0 2 2 2	
1.460E-01	1.420E+02	25	B396	1 0 2 2 2	
1.800E-01	1.751E+02	25	O321	2 2 2 2 1	
1.211E-01	1.178E+02	25.0	J305	2 0 2 2 2	

1.318E-01	1.282E+02	25.0	J305	2 0 2 2 2
1.678E-01	1.632E+02	30.0	J305	2 0 2 2 2
1.501E-01	1.460E+02	30.0	J305	2 0 2 2 2
1.696E-01	1.650E+02	33.0	J305	2 0 2 2 2
1.912E-01	1.860E+02	33.0	J305	2 0 2 2 2
2.161E-01	2.102E+02	35.0	J305	2 0 2 2 2
1.885E-01	1.834E+02	35.0	J305	2 0 2 2 2
2.331E-01	2.268E+02	38.0	J305	2 0 2 2 2
2.023E-01	1.968E+02	38.0	J305	2 0 2 2 2
2.100E-01	2.043E+02	40	O321	2 2 2 2 1
2.171E-01	2.112E+02	40.0	J305	2 0 2 2 2
2.532E-01	2.463E+02	40.0	J305	2 0 2 2 2
2.229E-01	2.169E+02	42.0	J305	2 0 2 2 2
2.616E-01	2.545E+02	42.0	J305	2 0 2 2 2
2.677E-01	2.604E+02	43.0	J305	2 0 2 2 2
2.283E-01	2.221E+02	43.0	J305	2 0 2 2 2
2.492E-01	2.424E+02	45.0	J305	2 0 2 2 2
2.982E-01	2.901E+02	45.0	J305	2 0 2 2 2
3.397E-01	3.305E+02	48.0	J305	2 0 2 2 2
2.773E-01	2.698E+02	48.0	J305	2 0 2 2 2
4.700E-01	4.572E+02	55	O321	2 2 2 2 1
1.302E-01	1.266E+02	ns	M335	0 0 2 0 1
1.490E-01	1.450E+02	rt	F041	0 2 2 0 2

4120. $C_{36}H_{72}N_3O_8S_2$
Lincomycin Octadecylsulfamate
RN: **MP** (°C):
MW: 739.12   **BP** (°C):

Solubility (Moles/L)	Solubility (Grams/L)	Temp (°C)	Ref (#)	Evaluation (T P E A A)	Comments
3.897E-04	2.880E-01	21	M044	2 0 2 2 2	

4121. $C_{36}H_{74}$
n-Hexatriacontane
Hexatriacontane
RN: 630-06-8   **MP** (°C):   75.0
MW: 506.99   **BP** (°C):

Solubility (Moles/L)	Solubility (Grams/L)	Temp (°C)	Ref (#)	Evaluation (T P E A A)	Comments
3.353E-09	1.700E-06	25	B069	1 0 1 1 1	
4.122E-09	2.090E-06	ns	B033	0 0 0 0 2	
4.122E-09	2.090E-06	ns	B033	0 2 2 2 2	

4122. $C_{37}H_{67}NO_{13}$
Erythromycin
E.E.S
E-Mycin
Erypar
Erythromycin Anhydrate

RN: 114-07-8   **MP** (°C):   139
MW: 733.95   **BP** (°C):

Solubility (Moles/L)	Solubility (Grams/L)	Temp (°C)	Ref (#)	Evaluation (T P E A A)	Comments
1.825E-03	1.339E+00	20	N334	1 0 1 1 0	EFG
1.570E-03	1.152E+00	30	F310	1 0 2 2 2	
1.107E-03	8.124E-01	37	N334	1 0 1 1 0	EFG
9.701E-04	7.120E-01	40	F310	1 0 2 2 2	
6.799E-04	4.990E-01	50	F310	1 0 2 2 2	
9.063E-04	6.651E-01	50	N334	1 0 1 1 0	EFG
5.341E-04	3.920E-01	60	F310	1 0 2 2 2	
5.137E-04	3.770E-01	70	F310	1 0 2 2 2	
5.055E-04	3.710E-01	80	F310	1 0 2 2 2	

4123. $C_{37}H_{67}NO_{13} \cdot 2H_2O$
Erythromycin (Dihydrate)

RN: 114-07-8   **MP** (°C):
MW: 769.98   **BP** (°C):

Solubility (Moles/L)	Solubility (Grams/L)	Temp (°C)	Ref (#)	Evaluation (T P E A A)	Comments
6.857E-04	5.280E-01	30	F310	1 0 2 2 2	
4.922E-04	3.790E-01	40	F310	1 0 2 2 2	
4.377E-04	3.370E-01	50	F310	1 0 2 2 2	
4.143E-04	3.190E-01	60	F310	1 0 2 2 2	
4.598E-04	3.540E-01	70	F310	1 0 2 2 2	
5.688E-04	4.380E-01	80	F310	1 0 2 2 2	

4124. $C_{38}H_{69}NO_{13}$
Clarithromycin
Biaxin
A-56268
TE-031

RN: 81103-11-9   **MP** (°C):   218.5
MW: 747.97   **BP** (°C):

Solubility (Moles/L)	Solubility (Grams/L)	Temp (°C)	Ref (#)	Evaluation (T P E A A)	Comments
1.330E-04	9.948E-02	20	N334	1 0 1 1 0	EFG
1.089E-04	8.145E-02	37	N334	1 0 1 1 0	EFG
4.893E-05	3.660E-02	50	N334	1 0 1 1 0	EFG

4125.   C$_{40}$H$_{51}$NO$_{14}$
Streptovaricin C
Streptovaricin
RN: 1404-74-6   **MP** (°C):   189
MW: 769.85   **BP** (°C):

Solubility (Moles/L)	Solubility (Grams/L)	Temp (°C)	Ref (#)	Evaluation (T P E A A)	Comments
1.604E-03	1.235E+00	21	M044	2 0 2 2 2	

4126.   C$_{41}$H$_{64}$O$_{13}$
Digitoxin
(3β,5β)-3-[(0-2,6-Dideoxy-β-D-ribo-hexopyranosyl-(1->4)-O-2,6-dideoxy-β-D-ribo-hexopyranosyl-(1->4)-2,6-dideoxy-β-D-ribo-hexopyranosyl)oxy]-14-hydroxycard-20(22)-enolide
Crystodigin
Digifortis
RN: 71-63-6   **MP** (°C):   256
MW: 764.96   **BP** (°C):

Solubility (Moles/L)	Solubility (Grams/L)	Temp (°C)	Ref (#)	Evaluation (T P E A A)	Comments
1.307E-05	1.000E-02	20	J010	1 0 0 0 0	
5.098E-06	3.900E-03	25	M301	1 1 2 2 1	anhydrate
2.000E-05	1.530E-02	30	O321	2 2 2 2 1	
2.222E-05	1.700E-02	30	O321	2 2 2 2 1	
1.447E-05	1.107E-02	37	C303	2 2 2 2 2	average of 3
3.255E-06	2.490E-03	37	M301	1 1 2 2 1	anhydrate
1.300E-05	9.944E-03	ns	M070	0 0 0 0 1	
9.151E-06	7.000E-03	ns	N302	0 2 1 2 0	

4127.   C$_{41}$H$_{64}$O$_{14}$
Gitoxin
Anhydrogitalin
Pseudodigitoxin
Bigitalin
RN: 4562-36-1   **MP** (°C):
MW: 780.96   **BP** (°C):

Solubility (Moles/L)	Solubility (Grams/L)	Temp (°C)	Ref (#)	Evaluation (T P E A A)	Comments
3.000E-06	2.343E-03	ns	M070	0 0 0 0 0	

4128. $C_{41}H_{64}O_{14}$

Digoxin

3β-((O-2,6-Dideoxy-β-D-Ribo-hexopyranosyl-(1rightarrow4)-O-2,6-dideoxy-β-D-Ribo-hexopyranosyl-(1rightarrow4)-2,6-dideoxy-β-D-Ribo-hexopyranosyl)oxy)-12β,14-dihydroxy-5β-card-20(22)-enolide

Lanoxicaps

Lanoxin

RN: 20830-75-5   **MP** (°C):   260
MW: 780.96   **BP** (°C):

Solubility (Moles/L)	Solubility (Grams/L)	Temp (°C)	Ref (#)	Evaluation (T P E A A)	Comments
1.253E-04	9.789E-02	25	F010	2 1 2 2 2	Swiss micron
6.786E-05	5.300E-02	25	F010	2 1 2 2 2	
7.375E-05	5.760E-02	25	F010	2 1 2 2 2	Swiss standard
8.297E-05	6.480E-02	25	F010	2 1 2 2 2	
1.000E-04	7.810E-02	25	H066	1 0 0 0 0	EFG
3.585E-05	2.800E-02	25	M301	1 1 2 2 1	
3.675E-05	2.870E-02	25	N301	2 0 2 2 2	
3.841E-05	3.000E-02	27	E052	2 0 2 2 0	EFG
3.585E-05	2.800E-02	30	O321	2 2 2 2 1	
4.000E-05	3.124E-02	30	O321	2 2 2 2 1	
6.312E-05	4.930E-02	37	C303	2 2 2 2 2	average of 6
3.457E-05	2.700E-02	37	M301	1 1 2 2 1	
3.483E-05	2.720E-02	37	N301	2 0 2 2 2	
4.443E-05	3.470E-02	37	R009	1 0 0 0 2	
2.817E-05	2.200E-02	100	D027	1 2 0 0 1	
8.963E-06	7.000E-03	ns	F037	0 0 2 0 2	mp 225.5 °C
5.570E-06	4.350E-03	ns	F037	0 0 2 0 2	mp 228.5 °C
6.915E-06	5.400E-03	ns	F037	0 0 2 0 2	mp 235.5 °C
7.363E-06	5.750E-03	ns	F037	0 0 2 0 2	mp 225.5 °C
4.097E-05	3.200E-02	ns	N302	0 2 1 2 1	
5.900E-05	4.608E-02	rt	J034	0 0 1 1 1	

4129. $C_{41}H_{67}NO_{15}$

Troleandomycin

Triacetyloleandomycin

RN: 2751-09-9   **MP** (°C):
MW: 813.99   **BP** (°C):

Solubility (Moles/L)	Solubility (Grams/L)	Temp (°C)	Ref (#)	Evaluation (T P E A A)	Comments
3.071E-04	2.500E-01	28	A038	2 0 1 1 1	

4130. C₄₂H₇₀O₃₅

6-O-α-D-Glucosyl-α-cyclodextrin

RN: **MP** (°C):

MW: 1135.01   **BP** (°C):

Solubility (Moles/L)	Solubility (Grams/L)	Temp (°C)	Ref (#)	Evaluation (T P E A A)	Comments
8.000E-01	9.080E+02	25	O321	2 2 2 2 1	
1.030E+00	1.169E+03	40	O321	2 2 2 2 1	
1.190E+00	1.351E+03	55	O321	2 2 2 2 1	

4131. C₄₂H₇₀O₃₅

β-Cyclodextrin

β-Cyclodextrin Hydrate

Cycloheptaamylose Hydrate

Cyclodextrin Hydrate

RN: 7585-39-9   **MP** (°C):   298-300

MW: 1135.01   **BP** (°C):

Solubility (Moles/L)	Solubility (Grams/L)	Temp (°C)	Ref (#)	Evaluation (T P E A A)	Comments
1.044E-02	1.185E+01	15	W317	2 2 1 0 2	
1.216E-02	1.381E+01	20	F186	1 2 1 1 1	
1.282E-02	1.455E+01	20	W317	2 2 1 0 2	
1.540E-02	1.748E+01	23.7	J305	2 0 2 2 2	
1.630E-02	1.850E+01	25	B396	1 0 2 2 2	
1.558E-02	1.768E+01	25	H319	1 0 2 2 0	
1.600E-02	1.816E+01	25	O304	1 2 2 2 2	
1.600E-02	1.816E+01	25	O321	2 2 2 2 1	
1.551E-02	1.760E+01	25	W317	2 2 1 0 2	
1.630E-02	1.850E+01	25.0	J305	2 0 2 2 2	
1.895E-02	2.151E+01	30	W317	2 2 1 0 2	
2.440E-02	2.769E+01	35.0	J305	2 0 2 2 2	
3.100E-02	3.519E+01	40	O321	2 2 2 2 1	
2.980E-02	3.382E+01	40.0	J305	2 0 2 2 2	
3.850E-02	4.370E+01	45.0	J305	2 0 2 2 2	
4.430E-02	5.028E+01	48.0	J305	2 0 2 2 2	
4.400E-02	4.994E+01	55	O321	2 2 2 2 1	
1.558E-02	1.768E+01	ns	M335	0 0 2 0 1	

4132.   C$_{43}$H$_{58}$N$_4$O$_{12}$
Rifampin
Rifampicin
RN: 13292-46-1   **MP** (°C):
MW: 822.96   **BP** (°C):

Solubility (Moles/L)	Solubility (Grams/L)	Temp (°C)	Ref (#)	Evaluation (T P E A A)	Comments
1.300E-01	1.070E+02	25	B073	2 1 2 2 2	pH 2.12,*sic*
4.374E-03	3.600E+00	25	B073	2 1 2 2 1	pH 2.5
1.701E-03	1.400E+00	25	B073	2 1 2 2 1	pH 5.33
1.215E-03	1.000E+00	25	B073	2 1 2 2 1	pH 3.99
1.215E-03	1.000E+00	25	B073	2 1 2 2 1	pH 3.03
1.580E-03	1.300E+00	25	G096	1 0 0 0 0	pH 4.3
3.393E-03	2.792E+00	rt	F182	0 0 0 0 1	pH 7.5

4133.   C$_{43}$H$_{75}$NO$_{16}$
Erythromycin Ethyl Succinate
RN: 1264-62-6   **MP** (°C):
MW: 862.07   **BP** (°C):

Solubility (Moles/L)	Solubility (Grams/L)	Temp (°C)	Ref (#)	Evaluation (T P E A A)	Comments
2.262E-04	1.950E-01	21	M044	2 0 2 2 2	

4134.   C$_{44}$H$_{74}$O$_{34}$
n-Ethyl-paba-β-cyclodextrin
RN: **MP** (°C):
MW: 1147.06   **BP** (°C):

Solubility (Moles/L)	Solubility (Grams/L)	Temp (°C)	Ref (#)	Evaluation (T P E A A)	Comments
5.100E-03	5.850E+00	ns	F327	0 0 1 2 2	

4135.   C$_{44}$H$_{74}$O$_{35}$
Hydroxyethyl-β-cyclodextrin
RN: **MP** (°C):
MW: 1163.06   **BP** (°C):

Solubility (Moles/L)	Solubility (Grams/L)	Temp (°C)	Ref (#)	Evaluation (T P E A A)	Comments
3.224E-01	3.750E+02	ns	M335	0 0 2 0 1	

4136.   C$_{45}$H$_{73}$NO$_{15}$
Solanine
β-D-Galactopyranoside, (3β)-Solanid-5-en-3-yl O-6-deoxy-α-L-mannopyranosyl-(1®2)-O-
[β-D-glucopyranosyl-(1-3)]-
Solanidane, β-D-Galactopyranoside Deriv
RN: 20562-02-1   **MP** (°C):
MW: 868.08   **BP** (°C):

Solubility (Moles/L)	Solubility (Grams/L)	Temp (°C)	Ref (#)	Evaluation (T P E A A)	Comments
3.000E-05	2.604E-02	15	K059	2 2 2 0 0	

4137.   C$_{45}$H$_{76}$O$_{35}$
n-Propyl-paba-β-cyclodextrin
RN: **MP** (°C):
MW: 1177.09   **BP** (°C):

Solubility (Moles/L)	Solubility (Grams/L)	Temp (°C)	Ref (#)	Evaluation (T P E A A)	Comments
2.100E-03	2.472E+00	ns	F327	0 0 1 2 2	

4138.   C$_{46}$H$_{77}$NO$_{17}$
Tylosin
Vubityl 200
Vetil(R)
RN: 1401-69-0   **MP** (°C):   128
MW: 916.12   **BP** (°C):

Solubility (Moles/L)	Solubility (Grams/L)	Temp (°C)	Ref (#)	Evaluation (T P E A A)	Comments
8.195E-03	7.508E+00	21	M044	2 0 2 2 2	

4139.   C$_{46}$H$_{78}$O$_{35}$
n-Butyl-paba-β-cyclodextrin
RN: **MP** (°C):
MW: 1191.11   **BP** (°C):

Solubility (Moles/L)	Solubility (Grams/L)	Temp (°C)	Ref (#)	Evaluation (T P E A A)	Comments
7.000E-04	8.338E-01	ns	F327	0 0 1 2 2	

4140. C₄₇H₇₃NO₁₇

Amphotericin B

RN: 1397-89-3   **MP** (°C):
MW: 924.10   **BP** (°C):

Solubility (Moles/L)	Solubility (Grams/L)	Temp (°C)	Ref (#)	Evaluation (T P E A A)	Comments
8.116E-04	7.500E-01	28	A038	2 0 1 1 1	
3.246E-06	3.000E-03	ns	K067	0 0 2 1 0	intrinsic

4141. C₄₇H₇₅NO₁₇

Nystatin
Mycostatin
Biofanal
Nystex
Fungicidin

RN: 1400-61-9   **MP** (°C):
MW: 926.12   **BP** (°C):

Solubility (Moles/L)	Solubility (Grams/L)	Temp (°C)	Ref (#)	Evaluation (T P E A A)	Comments
3.887E-04	3.600E-01	24	M166	2 0 0 0 1	

4142. C₄₈H₈₀O₄₀

γ-Cyclodextrin
Cyclooctaamylose
Ringdex C
Cyclomaltooctaose
Dexy Pearl γ-100

RN: 17465-86-0   **MP** (°C):
MW: 1297.15   **BP** (°C):

Solubility (Moles/L)	Solubility (Grams/L)	Temp (°C)	Ref (#)	Evaluation (T P E A A)	Comments
1.338E-01	1.736E+02	20	F186	1 2 1 1 1	
1.789E-01	2.320E+02	25	B396	1 0 2 2 2	
2.000E-01	2.594E+02	25	O321	2 2 2 2 1	
1.680E-01	2.179E+02	25.0	J305	2 0 2 2 2	
2.040E-01	2.646E+02	30.0	J305	2 0 2 2 2	
2.430E-01	3.152E+02	35.0	J305	2 0 2 2 2	
4.300E-01	5.578E+02	40	O321	2 2 2 2 1	
2.680E-01	3.476E+02	40.0	J305	2 0 2 2 2	
3.110E-01	4.034E+02	42.0	J305	2 0 2 2 2	
6.400E-01	8.302E+02	55	O321	2 2 2 2 1	
1.452E-01	1.883E+02	ns	M335	0 0 2 0 1	

4143. C$_{48}$H$_{80}$O$_{40}$
6-O-α-D-Maltosyl-α-cyclodextrin
6-O-α-Maltosyl-α-cyclodextrin
RN: **MP** (°C):
MW: 1297.15   **BP** (°C):

Solubility (Moles/L)	Solubility (Grams/L)	Temp (°C)	Ref (#)	Evaluation (T P E A A)	Comments
7.700E-01	9.988E+02	25	O321	2 2 2 2 1	
2.400E-01	3.113E+02	25	O321	2 2 2 2 1	
7.700E-01	9.988E+02	40	O321	2 2 2 2 1	
3.500E-01	4.540E+02	40	O321	2 2 2 2 1	
1.330E+00	1.725E+03	55	O321	2 2 2 2 1	
5.400E-01	7.005E+02	55	O321	2 2 2 2 1	

4144. C$_{49}$H$_{87}$NS
Erythromycin Lactobionate
RN: 3847-29-8   **MP** (°C):   145
MW: 722.31   **BP** (°C):

Solubility (Moles/L)	Solubility (Grams/L)	Temp (°C)	Ref (#)	Evaluation (T P E A A)	Comments
>2.77E-02	>2.00E+01	21	M044	2 0 2 2 0	

4145. C$_{50}$H$_{82}$N$_{10}$O$_{31}$S$_{10}$
Decane(S-(carboxymethyl)-L-cysteine))
RN: **MP** (°C):
MW: 1639.90   **BP** (°C):

Solubility (Moles/L)	Solubility (Grams/L)	Temp (°C)	Ref (#)	Evaluation (T P E A A)	Comments
5.820E-05	9.544E-02	15	N331	1 0 2 2 2	
5.730E-04	9.397E-01	25	N331	1 0 2 2 2	

4146. C$_{51}$H$_{70}$N$_{12}$O$_{11}$
His-pro-D-phe-his-leu-leu-thr-tyr
RN: **MP** (°C):
MW: 1027.20   **BP** (°C):

Solubility (Moles/L)	Solubility (Grams/L)	Temp (°C)	Ref (#)	Evaluation (T P E A A)	Comments
8.100E-05	8.320E-02	20	B141	1 2 0 0 1	pH 7.5

4147. $C_{51}H_{74}O_{19}$
Penta-acetyl-gitoxin
RN: 7242-04-8   **MP** (°C):
MW: 991.15   **BP** (°C):

Solubility (Moles/L)	Solubility (Grams/L)	Temp (°C)	Ref (#)	Evaluation (T P E A A)	Comments
1.200E-05	1.189E-02	ns	M070	0 0 0 0 1	

4148. $C_{52}H_{72}N_{12}O_{10}$
His-pro-phe-his-leu-leu-val-tyr
RN: **MP** (°C):
MW: 1025.23   **BP** (°C):

Solubility (Moles/L)	Solubility (Grams/L)	Temp (°C)	Ref (#)	Evaluation (T P E A A)	Comments
1.610E-04	1.651E-01	ns	B141	0 2 0 0 2	pH 7.5

4149. $C_{52}H_{72}N_{12}O_{10}$
His-pro-phe-his-leu-D-leu-val-tyr
RN: **MP** (°C):
MW: 1025.23   **BP** (°C):

Solubility (Moles/L)	Solubility (Grams/L)	Temp (°C)	Ref (#)	Evaluation (T P E A A)	Comments
1.370E-04	1.405E-01	ns	B141	0 2 0 0 2	pH 7.5

4150. $C_{52}H_{88}O_{39}$
n-Butyl-paba-γ-cyclodextrin
RN: **MP** (°C):
MW: 1337.26   **BP** (°C):

Solubility (Moles/L)	Solubility (Grams/L)	Temp (°C)	Ref (#)	Evaluation (T P E A A)	Comments
7.000E-04	9.361E-01	ns	F327	0 0 1 2 2	

4151. $C_{52}H_{97}NO_{18}S$
Erythromycin Estolate
RN: 3521-62-8   **MP** (°C):   135
MW: 1056.41   **BP** (°C):

Solubility (Moles/L)	Solubility (Grams/L)	Temp (°C)	Ref (#)	Evaluation (T P E A A)	Comments
1.515E-04	1.600E-01	21	M044	2 0 2 2 2	

4152. C$_{54}$H$_{90}$O$_{45}$
6-O-α-D-Maltosyl-β-cyclodextrin
6-O-α-Maltosyl-β-cyclodextrin
RN: **MP** (°C):
MW: 1459.29   **BP** (°C):

Solubility (Moles/L)	Solubility (Grams/L)	Temp (°C)	Ref (#)	Evaluation (T P E A A)	Comments
1.040E+00	1.518E+03	25	O321	2 2 2 2 1	
1.040E+00	1.518E+03	40	O321	2 2 2 2 1	
1.220E+00	1.780E+03	55	O321	2 2 2 2 1	

4153. C$_{54}$H$_{90}$O$_{45}$
6-O-α-D-Glucosyl-γ-cyclodextrin
RN: **MP** (°C):
MW: 1459.29   **BP** (°C):

Solubility (Moles/L)	Solubility (Grams/L)	Temp (°C)	Ref (#)	Evaluation (T P E A A)	Comments
9.800E-01	1.430E+03	25	O321	2 2 2 2 1	
1.010E+00	1.474E+03	40	O321	2 2 2 2 1	
1.180E+00	1.722E+03	55	O321	2 2 2 2 1	

4154. C$_{54}$H$_{90}$O$_{45}$
6-O-α-D-Maltotriosyl-α-cyclodextrin
6-O-α-Maltotriosyl-α-cyclodextrin
RN: **MP** (°C):
MW: 1459.29   **BP** (°C):

Solubility (Moles/L)	Solubility (Grams/L)	Temp (°C)	Ref (#)	Evaluation (T P E A A)	Comments
1.070E+00	1.561E+03	25	O321	2 2 2 2 1	
1.220E+00	1.780E+03	40	O321	2 2 2 2 1	
1.370E+00	1.999E+03	55	O321	2 2 2 2 1	

4155. C$_{55}$H$_{70}$N$_{12}$O$_{10}$
His-pro-phe-his-leu-phe-val-tyr
RN: **MP** (°C):
MW: 1059.25   **BP** (°C):

Solubility (Moles/L)	Solubility (Grams/L)	Temp (°C)	Ref (#)	Evaluation (T P E A A)	Comments
1.760E-04	1.864E-01	ns	B141	0 2 0 0 2	pH 7.5

4156. C₅₅H₇₉N₁₃O₁₁

His-pro-D-phe-his-leu-leu-val-tyr-serinol

RN: **MP** (°C):
MW: 1098.32   **BP** (°C):

Solubility (Moles/L)	Solubility (Grams/L)	Temp (°C)	Ref (#)	Evaluation (T P E A A)	Comments
3.000E-04	3.295E-01	20	B141	1 2 0 0 2	pH 7.5

4157. C₅₅H₉₀N₁₁O₃₄S₁₁

Undecane(S-(carboxymethyl)-L-cysteine))

RN: **MP** (°C):
MW: 1802.09   **BP** (°C):

Solubility (Moles/L)	Solubility (Grams/L)	Temp (°C)	Ref (#)	Evaluation (T P E A A)	Comments
9.200E-06	1.658E-02	15	N331	1 0 2 2 2	
1.340E-04	2.415E-01	25	N331	1 0 2 2 2	
2.900E-04	5.226E-01	35	N331	1 0 2 2 2	

4158. C₅₆H₉₈O₃₅

β-Cyclodextrin, Tetradeca-O-methyl-
Heptakis(2,6-di-O-methyl)-β-cyclodextrin

RN: 188367-19-3   **MP** (°C):
MW: 1331.38   **BP** (°C):

Solubility (Moles/L)	Solubility (Grams/L)	Temp (°C)	Ref (#)	Evaluation (T P E A A)	Comments
2.727E-01	3.631E+02	25	H319	1 0 2 2 0	

4159. C₅₇H₇₉N₁₃O₁₁

Pro-his-pro-phe-his-leu-D-leu-val-tyr

RN: **MP** (°C):
MW: 1122.35   **BP** (°C):

Solubility (Moles/L)	Solubility (Grams/L)	Temp (°C)	Ref (#)	Evaluation (T P E A A)	Comments
4.100E-05	4.602E-02	ns	B141	0 2 0 0 1	pH 7.5

4160. C₅₇H₇₉N₁₃O₁₁

Pro-his-pro-phe-his-leu-leu-val-tyr

RN: **MP** (°C):
MW: 1122.35   **BP** (°C):

Solubility (Moles/L)	Solubility (Grams/L)	Temp (°C)	Ref (#)	Evaluation (T P E A A)	Comments
3.240E-04	3.636E-01	ns	B141	0 2 0 0 2	pH 7.5

4161. $C_{60}H_{77}N_{13}O_{11}$
Pro-his-pro-phe-his-leu-phe-val-tyr
RN: **MP** (°C):
MW: 1156.36 **BP** (°C):

Solubility (Moles/L)	Solubility (Grams/L)	Temp (°C)	Ref (#)	Evaluation (T P E A A)	Comments
3.430E-04	3.966E-01	ns	B141	0 2 0 0 2	pH 7.5

4162. $C_{60}H_{92}N_{12}O_{10}$
Gramicidin S
Gramicidin
Cyclo(L-leucyl-D-phenylalanyl-L-prolyl-L-valyl-L-ornithyl-L-leucyl-D-phenylalanyl-L-prolyl-L-valyl-L-ornithyl)
Gramicidin S-A
RN: 113-73-5 **MP** (°C):
MW: 1141.48 **BP** (°C):

Solubility (Moles/L)	Solubility (Grams/L)	Temp (°C)	Ref (#)	Evaluation (T P E A A)	Comments
1.226E-04	1.400E-01	28	A038	2 0 1 1 2	

4163. $C_{60}H_{98}N_{12}O_{37}S_{12}$
Dodecane(S-(carboxymethyl)-L-cystein))
RN: **MP** (°C):
MW: 1964.28 **BP** (°C):

Solubility (Moles/L)	Solubility (Grams/L)	Temp (°C)	Ref (#)	Evaluation (T P E A A)	Comments
2.300E-06	4.518E-03	15	N331	1 0 2 2 2	
2.400E-05	4.714E-02	25	N331	1 0 2 2 2	
5.880E-05	1.155E-01	35	N331	1 0 2 2 2	

4164. $C_{60}H_{100}O_{50}$

6-O-α-D-Maltotriosyl-β-cyclodextrin
6-O-α-Maltotriosyl-β-cyclodextrin
6-O-α-D-Maltosyl-γ-cyclodextrin
6-O-α-Maltosyl-γ-cyclodextrin

RN: **MP** (°C):
MW: 1621.44   **BP** (°C):

Solubility (Moles/L)	Solubility (Grams/L)	Temp (°C)	Ref (#)	Evaluation (T P E A A)	Comments
9.400E-01	1.524E+03	25	O321	2 2 2 2 1	
9.400E-01	1.524E+03	25	O321	2 2 2 2 1	
9.400E-01	1.524E+03	40	O321	2 2 2 2 1	
9.400E-01	1.524E+03	40	O321	2 2 2 2 1	
1.140E+00	1.848E+03	55	O321	2 2 2 2 1	
1.100E+00	1.784E+03	55	O321	2 2 2 2 1	

4165. $C_{62}H_{86}N_{12}O_{16}$

Actinomycin D
Actactinomycin A IV
Actinomycin AIV
Actinomycin I1

RN: 50-76-0   **MP** (°C):
MW: 1255.45   **BP** (°C):

Solubility (Moles/L)	Solubility (Grams/L)	Temp (°C)	Ref (#)	Evaluation (T P E A A)	Comments
3.983E-04	5.000E-01	37	G025	1 0 0 0 1	
7.965E-04	1.000E+00	rt	G025	0 0 0 0 1	

4166. $C_{62}H_{111}N_{11}O_{12}$

Cyclosporin A
1,4,7,10,13,16,19,22,25,28,31-Undecaazacyclotritriacontane, Cyclic Peptide Deriv.
Sandimmun Neoral
Sandimmun
Sang-35
SDZ-OXL 400

RN: 59865-13-3   **MP** (°C):   148-151
MW: 1202.64   **BP** (°C):

Solubility (Moles/L)	Solubility (Grams/L)	Temp (°C)	Ref (#)	Evaluation (T P E A A)	Comments
3.326E-05	4.000E-02	25	B376	2 0 0 0 0	

4167. $C_{63}H_{85}N_{21}O_{19}$
Candicidin
Candeptin
Vanobid
RN: 1403-17-4 **MP** (°C):
MW: 1440.51 **BP** (°C):

Solubility (Moles/L)	Solubility (Grams/L)	Temp (°C)	Ref (#)	Evaluation (T P E A A)	Comments
9.349E-03	1.347E+01	21	M044	2 0 2 2 2	

4168. $C_{63}H_{88}N_{14}O_{14}PCo$
Vitamin B12
Cyanoject
Hydrobexan
Alphamine
Crystamine
Cyomin
RN: 68-19-9 **MP** (°C):
MW: 1355.40 **BP** (°C):

Solubility (Moles/L)	Solubility (Grams/L)	Temp (°C)	Ref (#)	Evaluation (T P E A A)	Comments
9.149E-03	1.240E+01	20	F300	1 0 0 0 2	

4169. $C_{64}H_{112}O_{40}$
Dimethyl-β-cyclodextrin
β-Cyclodextrin, 2A,2B,2C,2D,2E,2F,2G,6A,6B,6C,6D,6E,6F,6G-Tetradeca-O-methyl-
Heptakis(2,6-di-O-methyl)-β-cyclodextrin
Tetradeca-O-methyl-β-cyclodextrin
Tetradecakis-2,6-O-methylcycloheptaamylose
RN: 51166-71-3 **MP** (°C): 298-300
MW: 1521.58 **BP** (°C):

Solubility (Moles/L)	Solubility (Grams/L)	Temp (°C)	Ref (#)	Evaluation (T P E A A)	Comments
1.397E-01	2.126E+02	c	D316	0 0 0 0 1	

4170. $C_{65}H_{106}N_{13}O_{40}S_{13}$
Tridecane(S-(carboxymethyl)-L-cyateine))
RN: **MP** (°C):
MW: 2126.46 **BP** (°C):

Solubility (Moles/L)	Solubility (Grams/L)	Temp (°C)	Ref (#)	Evaluation (T P E A A)	Comments
6.200E-06	1.318E-02	25	N331	1 0 2 2 2	
1.600E-05	3.402E-02	35	N331	1 0 2 2 2	

4171. $C_{66}H_{110}O_{55}$
6-O-α-D-Maltotriosyl-γ-cyclodextrin
6-O-α-Maltotriosyl-γ-cyclodextrin
RN: **MP** (°C):
MW: 1783.58 **BP** (°C):

Solubility (Moles/L)	Solubility (Grams/L)	Temp (°C)	Ref (#)	Evaluation (T P E A A)	Comments
8.500E-01	1.516E+03	25	O321	2 2 2 2 1	
8.500E-01	1.516E+03	40	O321	2 2 2 2 1	
1.040E+00	1.855E+03	55	O321	2 2 2 2 1	

4172. $C_{67}H_{93}N_{15}O_{13}$
Pro-pro-pro-his-pro-phe-his-leu-D-leu-val-tyr
RN: **MP** (°C):
MW: 1316.58 **BP** (°C):

Solubility (Moles/L)	Solubility (Grams/L)	Temp (°C)	Ref (#)	Evaluation (T P E A A)	Comments
3.650E-04	4.806E-01	ns	B141	0 2 0 0 2	pH 7.5

4173. $C_{67}H_{93}N_{15}O_{13}$
Pro-pro-pro-his-pro-phe-his-leu-leu-val-tyr
RN: **MP** (°C):
MW: 1316.58 **BP** (°C):

Solubility (Moles/L)	Solubility (Grams/L)	Temp (°C)	Ref (#)	Evaluation (T P E A A)	Comments
3.750E-04	4.937E-01	ns	B141	0 2 0 0 2	pH 7.5

4174. $C_{70}H_{89}N_{15}O_{13}$
Pro-pro-pro-his-pro-phe-his-leu-phe-val-tyr
RN: **MP** (°C):
MW: 1348.58 **BP** (°C):

Solubility (Moles/L)	Solubility (Grams/L)	Temp (°C)	Ref (#)	Evaluation (T P E A A)	Comments
2.240E-04	3.021E-01	ns	B141	0 2 0 0 2	pH 7.5

4175. $C_{70}H_{126}O_{35}$
β-Cyclodextrin, Tetradeca-O-ethyl-
Heptakis(2,6-di-O-ethyl)-β-cyclodextrin
RN: 194715-43-0 **MP** (°C):
MW: 1527.76 **BP** (°C):

Solubility (Moles/L)	Solubility (Grams/L)	Temp (°C)	Ref (#)	Evaluation (T P E A A)	Comments
3.273E-05	5.000E-02	25	H319	1 0 2 2 0	

4176.   C$_{72}$H$_{85}$N$_{19}$O$_{18}$S$_5$
Thiostrepton
Bryamycin
RN: 1393-48-2   **MP** (°C):   210
MW: 1664.92   **BP** (°C):

Solubility (Moles/L)	Solubility (Grams/L)	Temp (°C)	Ref (#)	Evaluation (T P E A A)	Comments
5.286E-05	8.800E-02	21	M044	2 0 2 2 1	
1.442E-04	2.400E-01	28	A038	2 0 1 1 1	

4177.   C$_{72}$H$_{100}$N$_{18}$O$_{17}$PCo
Coenzyme B12
Cobamamide
Cobalamin, Co-(5'-deoxy-5'-adenosyl)-
Dibencozide
Funacomide
Deoxyadenosylcobalamin
RN: 13870-90-1   **MP** (°C):
MW: 1579.62   **BP** (°C):

Solubility (Moles/L)	Solubility (Grams/L)	Temp (°C)	Ref (#)	Evaluation (T P E A A)	Comments
1.646E-02	2.600E+01	24	M054	1 0 0 0 1	

4178.   C$_{74}$H$_{100}$ClN$_{15}$O$_{14}$
Antarelix
AcDNal-DCpa-Ser-Tyr-Dhai-Leu-Lys(iPr)-Pro-Dala-NH2
RN: 151272-78-5   **MP** (°C):
MW: 1459.17   **BP** (°C):

Solubility (Moles/L)	Solubility (Grams/L)	Temp (°C)	Ref (#)	Evaluation (T P E A A)	Comments
>6.85E-03	>1.00E+01	ns	D350	0 1 0 1 1	

4179.   C$_{75}$H$_{122}$N$_{15}$O$_{46}$S$_{15}$
Pendecane(S-(carboxymethyl)-L-cysteine))
RN: **MP** (°C):
MW: 2450.84   **BP** (°C):

Solubility (Moles/L)	Solubility (Grams/L)	Temp (°C)	Ref (#)	Evaluation (T P E A A)	Comments
3.400E-07	8.333E-04	25	N331	1 0 2 2 2	

4180. $C_{77}H_{107}N_{17}O_{15}$

Pro-pro-pro-pro-pro-his-pro-phe-his-leu-leu-val-tyr

RN: **MP** ($°$C):
MW: 1510.82   **BP** ($°$C):

Solubility (Moles/L)	Solubility (Grams/L)	Temp ($°$C)	Ref (#)	Evaluation (T P E A A)	Comments
1.328E-03	2.006E+00	ns	B141	0 2 0 0 2	pH 7.5

4181. $C_{80}H_{105}N_{17}O_{15}$

Pro-pro-pro-pro-pro-his-pro-phe-his-leu-phe-val-tyr

RN: **MP** ($°$C):
MW: 1544.83   **BP** ($°$C):

Solubility (Moles/L)	Solubility (Grams/L)	Temp ($°$C)	Ref (#)	Evaluation (T P E A A)	Comments
8.400E-04	1.298E+00	ns	B141	0 2 0 0 2	pH 7.5

4182. $C_{85}H_{117}N_{20}O_{18}$

Asp-arg-val-tyr-ile-his-pro-D-phe-his-leu-phe-val-tyr

RN: **MP** ($°$C):
MW: 1707.00   **BP** ($°$C):

Solubility (Moles/L)	Solubility (Grams/L)	Temp ($°$C)	Ref (#)	Evaluation (T P E A A)	Comments
6.200E-05	1.058E-01	20	B141	1 2 0 0 1	pH 7.5

References

A001   L.J. Andrews and R.M. Keefer (1949) Cation Complexes of Compounds Containing Carbon–Carbon Double Bonds. IV. The Argentation of Aromatic Hydrocarbons, *Journal of the American Chemical Society*, 71, 3644-3647.

A002   L.J. Andrews and R.M. Keefer (1950) Cation Complexes of Compounds Containing Carbon–Carbon Double Bonds. VII. Further Studies on the Argentation of Substituted Benzenes, *Journal of the American Chemical Society*, 72, 5034-5037.

A003   L.J. Andrews and R.M. Keefer (1950) Cation Complexes of Compounds Containing Carbon–Carbon Double Bonds. VI. The Argentation of Substituted Benzenes, *Journal of the American Chemical Society*, 72, 3113-3116.

A004   D.M. Alexander (1959) The Solubility of Benzene in Water, *Journal of Physical Chemistry*, 63, 1021-1022.

A008   N.A. Allawala and S. Riegelman (1953) The Release of Antimicrobial Agents from Solutions of Surface-Active Agents, *Journal of the American Pharmaceutical Association, Scientific Edition*, 42, 267-275.

A009   R.J.L. Andon and J.D. Cox (1952) Phase Relationships in the Pyridine Series. Part I. The Miscibility of some Pyridine Homologues with Water, *Journal of the Chemical Society (London)*, 4601-4606.

A010   R.A. Anderson and K.J. Morgan (1966) Effect of Solubilisation on the Antibacterial Activity of Hexachlorophane, *Journal of Pharmacy and Pharmacology*, 18, 449-456.

A011   C.C. Addison (1946) The Properties of Freshly Formed Surfaces. Part VI. The Influence of Temperature and Concentration on the Dynamic and Static Surface Tensions of Aqueous Decoic Acid Solutions, *Journal of the Chemical Society (London)*, 579-585.

A012   D.M. Altwein, J.N. Delgado, and F.P. Cosgrove (1965) Effect of Urea Concentrations on the Solubility of the Isomeric Monohydroxybenzoic Acids, *Journal of Pharmaceutical Sciences*, 54, 603-606.

A013   S. Agharkar, S. Lindenbaum, and T. Higuchi (1976) Enhancement of Solubility of Drug Salts by Hydrophilic Counterions: Properties of Organic Salts of an Antimalarial Drug, *Journal of Pharmaceutical Sciences*, 65, 747-749.

A014   M.I. Amin and J.T. Bryan (1973) Kinetics and Factors Affecting Stability of Methylprednisolone in Aqueous Formulation, *Journal of Pharmaceutical Sciences*, 62, 1768-1771.

A015   C.C. Addison (1945) The Properties of Freshly Formed Surfaces. Part IV. The Influence of Chain Length and Structure on the Static and the Dynamic Surface Tensions of Aqueous-Alcoholic Solutions, *Journal of the Chemical Society (London)*, 98-106.

A016   A.P. Altshuller and H.E. Everson (1953) The Solubility of Ethyl Acetate in Water, *Journal of the American Chemical Society*, 75, 1727-1727.

A017   L.J. Andrews and R.M. Keefer (1950) Cation Complexes of Compounds Containing Carbon-Carbon Double Bonds. IV. The Argentation of Aromatic Hydrocarbons, *Journal of the American Chemical Society*, 72, 5801-5801.

A018   A. Albert and D.J. Brown (1954) Purine Studies. Part I. Stability to Acid and Alkali. Solubility. Ionization. Comparison with Pteridines, *Journal of the Chemical Society (London)*, 2060-2071.

A019   A. Albert, D.J. Brown, and G. Cheeseman (1952) Pteridine Studies. Part III. The Solubility and the Stability to Hydrolysis of Pteridines, *Journal of the Chemical Society (London)*, 4219-4232.

A020   A. Albert, J.H. Lister, and C. Pedersen (1956) Pteridine Studies. Part X. Pteridines with more than One Hydroxy- or Amino-group, *Journal of the Chemical Society (London)*, 4621-4628.

A021   E. Azaz and M. Donbrow (1976) Solubilization of Phenolic Compounds in Nonionic Surface-Active Agents, *Journal of Colloid and Interface Science*, 57, 11-15.

A022   A. Albert (1955) The Solubility of 8-Hydroxymethylpurine, *Chemistry and Industry (London)*, 202-202.

A023   U. Avico, E. Ciranni Signoretti, R. Di Francesco..., and E. Cingolani (1976) Physical Parameters and Biological Activity of Organic Compounds, *Bollettino Chimico Farmaceutico*, 115, 242-253.

A025   D. Abelson, C. Depatie, and V. Craddock (1960) Interactions of Testosterone with Amino Acids, *Archives of Biochemistry and Biophysics*, 91, 71-74.

A027   Z.K. Abidova and G.K. Khodzhaev (1960) A Separation Method for a Mixture of Acids: Benzoic, Phthalic (o-, m-, p-), Trimellitic Trimesic, and Hemimellitic, *Uzbekskii Khimicheskii Zhurnal*, 1, 69-76.

A028   R. Alric and R. Puech (1972) Coefficient de Partage, Solubilite Intrinseque et Puissance Relative de Sulfonylurees Hypoglycemiantes, *Journal of Pharmacology*, 3, 435-447.

A029   A. Albert (1966) Surface Activity and Association. Ionization. Dipole Moments, 147-261.

A031   E.C. Attane and T.F. Doumani (1949) Solubilities of Aliphatic Dicarboxylic Acids in Water, *Industrial and Engineering Chemistry*, 41, 2015-2017.

A032   L.J. Andrews and R.M. Keefer (1951) The Argentation of Organic Iodides, *Journal of the American Chemical Society*, 73, 5733-5736.

A034   Y. Arakawa, M. Nakano, K. Juni, and T. Arita (1976) Physical Properties of Pyrimidine and Purine Antimetabolites. I. The Effects of Salts and Temperature on the Solubility of 5-Fluorouracil, 1-(2-Tetrahydrofuryl)-5-fluorouracil, 6-Mercaptopurine, and Thioinosine, *Chemical and Pharmaceutical Bulletin*, 24, 1654-1657.

A035   A. Albert (1956) The Solubility of Quinoline and the Hydroxyquinolines, *Chemistry and Industry (London)*, 252-252.

A037   R.A. Alberty and E.R. Washburn (1945) The Ternary System Isobutyl Alcohol-Benzene-Water at 25 C, *Journal of Physical Chemistry*, 49, 4-8.

A038   M.L. Andrew and P.J. Weiss (1959) Solubility of Antibiotics in Twenty-four Solvents. II, *Antibiotics and Chemotherapy (Washington, D.C.)*, 9, 277-279.

A039   A.L. Abel (1957) The Substituted Urea Herbicides, *Chemistry and Industry (London)*, 1106-1112.

A040   L.F. Audrieth and A.D.F. Toy (1942) The Aquo Ammono Phosphoric Acids. III. The N-Substituted Derivatives of Phosphoryl and Thiophosphoryl Triamide as Hydrogen Bonding Agents, *Journal of the American Chemical Society*, 64, 1553-1555.

A043   W.R. Angus and R.P. Owen (1943) Aqueous Solubilities of r- and l-Mandelic Acids and Three O-Acyl-r-mandelic Acids, *Journal of the Chemical Society (London)*, 231-232.

A044   A. Apelblat (1971) Extraction of Nitric Acid and Hydrochloric Acid by Methyl Diphenyl Phosphate, *Journal of the Chemical Society A: Inorganic, Physical, Theoretical*, 3459-3463.

A045   K. Ando (1926) Der EinfluB der Salze auf die Loslichkeit des Glykokolls und des Tyrosins, *Biochemische Zeitschrift (1948-1967)*, 173, 426-432.

A046   R. Alric and R. Puech (1971) Mesure de la Solubilite Intrinseque et de la Constante Apparente d'Ionisation Acide de Sulfamidothiodiazols et de Sulfonylurees Hypoglycemiants en Solution Aqueuse a 37 C, *Journal of Pharmacology*, 2, 141-154.

A047   N.L. Allport (1936) p-Aminobenzenesulphonamide-Research Paper, *Quarterly Journal of Pharmacy and Pharmcology*, 9, 560-566.

A048   M.H. Abraham, E. Ah-Sing, R.E. Marks, and R.A. Schulz (1977) Thermodynamics of Solution of Two Forms of DL-alpha-Amino-n-butyric Acid in Water, *Journal of the Chemical Society, Faraday Transactions* 1, 181-185.

A049   J.E. Amoore and R.G. Buttery (1978) Partition Coefficients and Comparative Olfactometry, *Chemical Senses and Flavor*, 3, 57-71.

A050   W. Albersmeyer (1958) Quantitative Determination of Aromatic Hydrocarbons in Aqueous Solution, Gas- u. Wasserfach, 99, 269.

A052   A. Azarnoosh and J.J. McKetta (1959) Solubility of Propylene in Water, *Journal of Chemical and Engineering Data*, 4, 211-212.

A055   L.F. Audrieth and A.W. Browne (1930) Azido-Carbondisulfide. IV. Preparation and Properties of the New Inter-halogenoid, Cyanogen Azido-dithiocarbonate, *Journal of the American Chemical Society*, 52, 2799-2805.

A056   E. Angelescu (1928) Uber Loslichkeit in Losungsmittelgemischen. II. Die Loslichkeit eines Stoffes, der in jedem Verhaltnis mit einem der Losungsmittel Mischbar ist, *Zeitschrift fuer Physikalische Chemie*, Stoechiometrie und Verwandschaftslehre, 138, 300-310.

A057   M. Almgren, F. Grieser, and J.K. Thomas (1979) Dynamic and Static Aspects of Solubilization of Neutral Arenes in Ionic Micellar Solutions, *Journal of the American Chemical Society*, 101, 279-291.

A058   M. Aquan-Yuen, D. MacKay, and W.Y. Shiu (1979) Solubility of Hexane, Phenanthrene, Chlorobenzene, and p-Dichlorobenzene in Aqueous Electrolyte Solutions, *Journal of Chemical and Engineering Data*, 24, 30-34.

A059   K.S. Alexander, B. Laprade, J.W. Mauger, and A.N. Paruta (1978) Thermodynamics of Aqueous Solutions of Parabens, *Journal of Pharmaceutical Sciences*, 67, 624-627.

A064   D.J. Austin, K.A. Lord, and I.H. Williams (1976) High Pressure Liquid Chromatography of Benzimidazoles, *Pesticide Science*, 7, 211-222.

A065   S.S. Ahsan and S.M. Blaug (1960) Interactions of Tweens with some Pharmaceuticals, *Drug Standards*, 28, 95-100.

A066   B.D. Anderson and R.A. Conradi (1980) Prostaglandin Prodrugs VI: Structure-Thermodynamic Activity and Structure-Aqueous Solubility Relationships, *Journal of Pharmaceutical Sciences*, 69, 424-430.

A067   H. Asche (1979) Wirkstofffreigabe aus Externa, *Fette, Seifen, Anstrichmittel*, 81, 370-373.

A068   A. Adjei, J. Newburger, and A. Martin (1980) Extended Hildebrand Approach: Solubility of Caffeine in Dioxane-Water Mixtures, *Journal of Pharmaceutical Sciences*, 69, 659-661.

A069   S. Ali (1978) Degradation and Environmental Fate of Endosulfan Isomers and Endosulfan Sulfate in Mouse, Insect and Laboratory Model Ecosystem, unpublished.

A070   A.I. Altsybeeva, V.P. Belousov, N.V. Ovtrakht, and A.G. Morachevskii (1964) Phase Equilibria in and Thermodynamic Properties of the s-Butanol Water System, *Russian Journal of Physical Chemistry*, 38, 676-679.

A072   G. Aiello (1921) Uber die Verteilungskoeffizienten der Diuretica und Narkotica und die Theorie der Narkose, *Biochemische Zeitschrift (1948-1967)*, 124, 192-205.

A073   J.N. Rakshit (1921) Morphine, Codeine, and Narcotine in Indian Opium, *Analyst (London)*, 46, 481-492.

A074   G.L. Amidon, G.D. Leesman, and R.L. Elliott (1980) Improving Intestinal Absorption of Water-Insoluble Compounds: A Membrane Metabolism Strategy, *Journal of Pharmaceutical Sciences*, 69, 1363-1367.

A075   A.I. Altsybeeva and A.G. Morachevskii (1964) Phase Equilibria in the Ternary System sec-Butanol-Methyl-ethylketone-Water, *Zhurnal Fizicheskoi Khimii (Moscow)*, 38, 1574-1579.

A076   H.O. Ammar and H.A. Salama (1980) Solubilization of Benzothiadiazide Diuretics by Cetomacrogol, *Pharmazeutische Industrie*, 42, 849-851.

A078   H.O. Ammar, S.A. Ibrahim, A.A. Kassem, and S.S. Abu-Zaid (1980) Interaction of Aromatic Monocarboxylic Acid Derivatives with Amidopyrine, *Pharmazeutische Industrie*, 42, 1312-1315.

A079   H.O. Ammar and H.A. Salama (1981) Effect of Sodium Salts of Toluic Acids on the Water-Solubility of Riboflavine, *Pharmazeutische Industrie*, 43, 194-197.

A080   H.O. Ammar and H.A. Salama (1981) Interaction Between Bendroflumethiazide and Caffeine, *Pharmazie (Berlin)*, 36, 265-266.

A081   A.E. Aboutaleb, A.A. Ali, and R.B. Salama (1981) Micellar Solubilization of Quinethazone, Levomepromazine, and Niridazole, *Pharmazie (Berlin)*, 36, 35 37.

A082   A. Albert (1955) Pteridine Studies. Part VII. The Degradation of 4-, 6-, and 7-Hydroxypteridine by Acid and Alkali, *Journal of the Chemical Society (London)*, 2690-2699.

A083   A. Albert, D.J. Brown, and H.C.S. Wood (1954) Pteridine Studies. Part V. The Monosubstituted Pteridines, *Journal of the Chemical Society (London)*, 3832-3839.

A085   A. Albert, D.J. Brown, and G. Cheeseman (1951) Pteridine Studies. Part I. Pteridine, and 2- and 4-Amino- and 2- and 4-Hydroxy-pteridines, *Journal of the Chemical Society (London)*, 474-485.

A086   G.E. Amidon, W.I. Higuchi, and N.F.H. Ho (1982) Theoretical and Experimental Studies of Transport of Micelle-Solubilized Solutes, *Journal of Pharmaceutical Sciences*, 71, 77-84.

A087   C.A. Anderson, J.C. Cavagnol, C.J. Cohen..., and J.W. Young (1974) Guthion (Azinphosmethyl): Organophosphorus Insecticide, *Residue Reviews*, 51, 123-130.

A088   M.N. Azmin, A. Setanoians, R.G.G. Blackie..., and J.F.B. Stuart (1982) Formulation of 1,3,5-Triglycidyl-S-triazinetrione (alpha-TGT) for Intravenous Injection, *International Journal of Pharmaceutics*, 10, 109-118.

A089   H.O. Ammar, S.A. Ibrahim, and T.H. El-Faham (1981) Interaction of Chlorothiazide and Hydrochlorothiazide with Certain Amides, Imides and Xanthines, *Pharmazeutische Industrie*, 43, 292-295.

A090   H.Y. Aboul-Enein (1976) Glutethimide, *Analytical Profiles of Drug Substances*, 5, 142-149.

A091   H.Y. Aboul-Enein (1977) Propylthiouracil, *Analytical Profiles of Drug Substances*, 6, 458-463.

A092   H.Y. Aboul-Enein, A.A. Al-Badr, and S.E. Irahim (1981) Salbutamol, *Analytical Profiles of Drug Substances*, 10, 665-669.

A093   H.O. Ammar, S.A. Ibrahim, and T.H. El-Faham (1982) Effect of Aromatic Hydrotropes on the Solubility of Some Benzothiadiazines, *Pharmazie (Berlin)*, 37, 36-40.

A094   W.L. Archer and V.L. Stevens (1977) Comparison of Chlorinated, Aliphatic, Aromatic, and Oxygenated Hydrocarbons as Solvents, *Industrial and Engineering Chemistry*, *Product Research and Development*, 16, 319-325.

A095   D.K. Agrawal and A.V. Deshpande (1982) Spectrophotometric Determination and Solubility Studies of Some Benzothiadiazine Derivatives, *Pharmazie (Berlin)*, 37, 150-150.

A096   W.B. Arbuckle (1983) Estimating Activity Coefficients for Use in Calculating Environmental Parameters, *Environmental Science and Technology*, 17, 537-542.

A305   E. Amadori and W. Heupt (1978) Dithianon, *Analytical Methods for Pesticides and Plant Growth Regulators*, 10, 181-187.

A306   J. Asshauer, K. Hommel, and T. Hoppe (1978) Pyrazophos, *Analytical Methods for Pesticides and Plant Growth Regulators*, 10, 237-241.

A308   E. Amadori and W. Heupt (1978) Chlorflurecol-methyl, *Analytical Methods for Pesticides and Plant Growth Regulators*, 10, 525-532.

A314   R. Anliker and P. Moser (1987) The Limits of Bioaccumulation of Organic Pigments in Fish: Their Relation to the Partition Coefficient and the Solubility in Water and Octanol, *Ecotoxicology and Environmental Safety*, 13, 43-52.

A323   W.J. Adams and K.M. Blaine (1986) A Water Solubility Determination of 2,3,7,8-TCDD, *Chemosphere*, 15, 1397-1400.

A324   M. Alonso and F. Recasens (1986) Liquid-Liquid Equilibrium for the System Acrylonitrile-Styrene-Water at 338 K, *Journal of Chemical and Engineering Data*, 31, 164-166.

A325   M. Akiyoshi, T. Deguchi, and I. Sanemasa (1987) The Vapor Saturation Method for Preparing Aqueous Solutions of Solid Aromatic Hydrocarbons, *Bulletin of the Chemical Society of Japan* (Nippon Kagakukai Bulletin), 60, 3935-3939.

A326   K. Akita and F. Yoshida (1963) Phase-Equilibria in Methanol-Ethyl Acetate-Water System, *Journal of Chemical and Engineering Data*, 8, 484-490.

A328   V.W. Arnold and E.R. Washburn (1958) Ternary System Isoamyl Alcohol-Isopropyl Alcohol-Water at 10, 25 and 40 degrees, *Journal of Physical Chemistry*, 62, 1088-1090.

A330   M.A. Akade, D.K. Agrawal, and J.A.K. Lauwo (1986) Influence of Polyethylene Glycol 6000 and Mannitol on the In-vitro Dissolution Properties of Nitrofurantoin by the Dispersion Technique, *Pharmazie (Berlin)*, 41, 849-851.

A335   M. Ahel and W. Giger (1993) Aqueous Solubility of Alkylphenols and Alkylphenol Polyethoxylates, *Chemosphere*, 26, 1461-1470.

A336   L.A. Al-Razzak and V.J. Stella (1990) Stability and Solubility of 2-chloro-2',3'-dideoxyadenosine, *International Journal of Pharmaceutics*, 60, 53-60.

A337   B.D. Anderson, M.B. Wygant, T.X. Xiang..., and V.J. Stella (1988) Preformulation Solubility and Kinetic Studies of 2',3'-dideoxypurine nucleosides: Potential Anti-AIDS Agents, *International Journal of Pharmaceutics*, 45, 27-37.

A338   B.D. Anderson and C.Y. Chiang (1990) Physicochemical Properties of Carbovir, a Potential Anti-HIV Agent, *Journal of Pharmaceutical Sciences*, 79, 787-790.

A339   A. Apelblat and E. Manzurola (1987) Solubility of Oxalic, Malonic, Succinic, Adipic, Maleic, Malic, Citric, and Tartaric Acids in Water from 278.15 to 338.15 K, *Journal of Chemical Thermodynamics*, 19, 317-320.

A340   A. Apelblat and E. Manzurola (1990) Solubility of Suberic, Azelaic, Levulinic, Glycolic, and Diglycolic Acids in Water from 278.25 to 361.35 K, *Journal of Chemical Thermodynamics*, 22, 289-292.

A341   A. Apelblat and E. Manzurola (1989) Solubility of Ascorbic, 2-Furancarboxylic, Glutaric, Pimelic, Salicylic, and o-Phthalic Acids in Water from 279.15 to 342.15 K, and Apparent Molar Volumes of Ascorbic, Glutaric, and Pimelic Acids in Water at 298.15 K, *Journal of Chemical Thermodynamics*, 21, 1003-1008.

A346   S.M. Ahmed, A.A. Abdel-Rahman, S.I. Saleh, and M.O. Ahmed (1993) Comparative Dissolution Characteristics of Bropirimine-beta-cyclodextrin Inclusion Complex and its Solid Dispersion with PEG 6000, *International Journal of Pharmaceutics*, 96, 5-11.

A348   A. Acarturk, A. Sencan, and N. Celebi (1993) Evaluation of the Effect of Low-molecular Chitosan on the Solubility and Dissolution Characteristics of Spironolactone, *Pharmazie (Berlin)*, 48, 605-607.

A350   H.O. Ammar and S.A. El-Nahhas (1993) Effect of Aromatic Hydrotropes on the Solubility of Allopurinol. Part 2. Effect of Nicotinamide and Sodium Salts of Benzoic, Naphthoic and Nicotinic Acids, *Pharmazie (Berlin)*, 48, 534-536.

A351   H.O. Ammar and S.A. El-Nahhas (1993) Effect of Aromatic Hydrotropes on the Solubility of Allopurinol. Part 3. Sodium Salts of Toluic Acids, *Pharmazie (Berlin)*, 48, 751-754.

A352   A.E. Aboutaleb, A.A.A. Rahman, and S. Ismail (1986) Studies of Cyclodextrin Inclusion Complexes. I. Inclusion Complexes between alpha-and beta-Cyclodextrins and Chloramphenicol in Aqueous Solutions, *Drug Development and Industrial Pharmacy*, 12, 2259-2265.

A355   H.A. Alschaibani and E-E.A. Abu-Gharib (1995) Transfer Chemical Potentials, Solubility and Reactivity of Psoralen and 8-Hydroxypsoralen in Binary Aqueous Methanol Mixtures, *Journal of the Chinese Society*, 42, 37-42.

A356   A. Arce, A. Blanco, P. Souza, and I. Vidal (1995) Liquid-Liquid Equilibria of the Ternary Mixture Water + Propanoic Acid + Methyl Ethyl Ketone and Water + Propanoic Acid + Methyl Propyl Ketone, *Journal of Chemical and Engineering Data*, 40, 225-229.

B001   G.M. Bennett and W.G. Philip (1928) The Influence of Structure on the Solubilities of Ethers. Part II. Some Cyclic Ethers, *Journal of the Chemical Society (London)*, 1937-1942.

B002   G.M. Bennett and W.G. Philip (1928) The Influence of Structure on the Solubilities of Ethers. Part I. Aliphatic Ethers, *Journal of the Chemical Society (London)*, 1930-1937.

B003   R.L. Bohon and W.F. Claussen (1951) The Solubility of Aromatic Hydrocarbons in Water, *Journal of the American Chemical Society*, 73, 1571-1578.

B004   J.A.V. Butler and C.N. Ramchandani (1935) The Solubility of Non-electrolytes. Part II. The Influence of the Polar Group on the Free Energy of Hydration of Aliphatic Compounds, *Journal of the Chemical Society (London)*, 952-955.

B009   E. Back and B. Steenberg (1950) Simultaneous Determination of Ionization Constant, Solubility Product and Solubility for Slightly Soluble Acids and Bases. Electrolytic Constants for Abietic Acid, *Acta Chemica Scandinavica*, 4, 810-815.

B010   C.T. Bauguess, F. Sadik, J.H. Fincher, and C.W. Hartman (1975) Hydrolysis of Fatty Acid Esters of Acetaminophen in Buffered Pancreatic Lipase Systems I, *Journal of Pharmaceutical Sciences*, 64, 117-120.

B011   T.L. Breon and A.N. Paruta (1970) Solubility Profiles for Several Barbiturates in Hydroalcoholic Mixtures, *Journal of Pharmaceutical Sciences*, 59, 1306-1313.

B012   B.W. Barry and D.I.D. El Eini (1976) Solubilization of Hydrocortisone, Dexamethasone, Testosterone and Progesterone by Long-Chain Polyoxyethylene Surfactants, *Journal of Pharmacy and Pharmacology*, 28, 210-218.

B013   F. Bischoff and R.D. Stauffer (1954) The Dispersion of Testosterone in Aqueous Bovine Serum Albumin Solution, *Journal of the American Chemical Society*, 76, 1962-1965.

B014   S. Batra (1975) Aqueous Solubility of Steroid Hormones: An Explanation for the Discrepancy in the Published Data, *Journal of Pharmacy and Pharmacology*, 27, 777-779.

B016   J. Buchi, X. Perlia, and A. Strassle (1966) Beziehungen Zwischen den Physikalisch-Chemischen Eigenschaften, der Chemischen Reaktivitat und der Lokalanasthetischen Wirkung in der Reihe der 4-Aminobenzoesaure-alkylester, *Arzneimittel-Forschung*, 16, 1657-1668.

B018   J. Buchi and X. Perlia (1960) Water-Solubility and Turbidity-pH of Local Anesthetic Bases in Homologous Series, *Arzneimittel-Forschung*, 10, 544-549.

B019   H.S. Booth and H.E. Everson (1948) Hydrotropic Solubilities—Solubilities in 40 Per Cent Sodium Xylenesulfonate, *Industrial and Engineering Chemistry*, 40, 1491-1493.

B028   H. Borsook and J.W. Dubnoff (1940) The Biological Synthesis of Hippuric Acid in Vitro, *Journal of Biological Chemistry*, 132, 307-324.

B030   A. Burger (1975) Dissolution and Polymorphism of Metolazone, *Arzneimittel-Forschung*, 25, 24-27.

B031   C.R. Bailey (1923) The Increased Solubility of Phenolic Substances in Water on Addition of a Third Substance, *Journal of the Chemical Society (London)*, 123, 2579-2589.

B032   H.B. Bull, K. Breese, and C.A. Swenson (1978) Solubilities of Amino Acids, *Biopolymers*, 17, 1091-1100.

B033   E.G. Baker (1959) Origin and Migration of Oil, *Science*, 129, 871-874.

B036   M.C. Bowman, F. Acree, Jr., and M.K. Corbett (1960) Solubility of Carbon-14 DDT in Water, *Journal of Agricultural and Food Chemistry*, 8, 406-408.

B038   J.A.V. Butler, D.W. Thomson, and W.H. Maclennan (1933) The Free Energy of the Normal Aliphatic Alcohols in Aqueous Solution. Part I. The Partial Vapour Pressures of Aqueous Solutions of Methyl, n-Propyl, and n-Butyl Alcohols. Part II. The Solubilities.... Part III. The Theory of..., *Journal of the Chemical Society (London)*, 674-686.

B039   T.R. Bates, S-L. Lin, and M. Gibaldi (1967) Solubilization and Rate of Dissolution of Drugs in the Presence of Physiologic Concentrations of Lysolecithin, *Journal of Pharmaceutical Sciences*, 56, 1492-1495.

B040   F.L. Breusch and E. Ulusoy (1964) Physikalische Eigenschaften Homologer Kristallisierter Reihen mit Alternierendem und Nicht Alternierendem Schmelzpunkt, *Fette, Seifen, Anstrichmittel*, 66, 739-742.

B041   F. Bischoff and H.R. Pilhorn (1948) The State and Distribution of Steroid Hormones in Biologic Systems. III. Solubilities of Testosterone, Progesterone, and alpha-Estradiol in Aqueous Salt and Protein Solution and in Serum, *Journal of Biological Chemistry*, 174, 663-682.

B042   S. Bolton, D. Guttman, and T. Higuchi (1957) Complexes Formed in Solution by Homologs of Caffeine, *Journal of the American Pharmaceutical Association, Scientific Edition*, 46, 38-41.

B043   T.R. Bates, M. Gibaldi, and J.L. Kanig (1966) Solubilizing Properties of Bile Salt Solutions I, *Journal of Pharmaceutical Sciences*, 55, 191-199.

B044   T.R. Bates, J.M. Young, C.M. Wu, and H.A. Rosenberg (1974) pH-Dependent Dissolution Rate of Nitrofurantoin from Commercial Suspensions, Tablets, and Capsules, *Journal of Pharmaceutical Sciences*, 63, 643-645.

B045   T.R. Bates, M. Gibaldi, and J.L. Kanig (1966) Solubilizing Properties of Bile Salt Solutions. II, *Journal of Pharmaceutical Sciences*, 55, 901-906.

B046   F.J. Bandelin and W. Malesh (1959) The Solubility of Various Sulfonamides Employed in Urinary Tract Infections, *Journal of the American Pharmaceutical Association, Scientific Edition*, 48, 177-181.

B047   H.S. Bean and H. Berry (1951) The Bactericidal Activity of Phenols in Aqueous Solutions of Soap, *Journal of Pharmacy and Pharmacology*, 3, 639-649.

B048   W.J. Blaedel and M.A. Evenson (1964) The Solubility of p-Iodobenzenesulfonyl Chloride, *Journal of Chemical and Engineering Data*, 9, 138-139.

B049   A. Bendich, G.B. Brown, F.S. Philips, and J.B. Thiersch (1950) The Direct Oxidation of Adenine In Vivo, *Journal of Biological Chemistry*, 183, 267-277.

B050   A. Albert (1955) Six-Membered Heteroaromatic Rings Containing Nitrogen: Correlation of Structure, 124-133.

B052   M.S. Bidner and M. de Santiago (1971) Solubilite de Liquides Non-electrolytes dans des Solution Aqueuses d'Electrolytes, *Chemical Engineering Science*, 26, 1484-1488.

B054   F. Bischoff, R.E. Katherman, Y.S. Yee, and J.J. Moran (1952) Solubilities of Testosterone and Estradiol Esters in Biologic Systems, *Federation Proceedings, Federation of American Societies for Experimental Biology*, 11, 189-189.

B055   L.H. Block and R.N. Patel (1973) Solubility and Dissolution of Triamcinolone Acetonide, *Journal of Pharmaceutical Sciences*, 62, 617-621.

B056   C.R. Behl, L.H. Block, and M.L. Borke (1976) Aqueous Solubility of 14C-Triamcinolone Acetonide, *Journal of Pharmaceutical Sciences*, 65, 429-430.

B057   D.B. Bowen, K.C. James, and M. Roberts (1970) An Investigation of the Distribution Coefficients of Some Androgen Esters Using Paper Chromatography, *Journal of Pharmacy and Pharmacology*, 22, 518-522.

B058   E.L. Baldeschwieler and H.A. Cassar (1929) A New Petroleum By-product: Octane-sultone, *Journal of the American Chemical Society*, 51, 2969-2975.

B059   C. Buffington and H. Turndorf (1976) Anesthetics Alter the Solubility of Nonpolar Compounds in Water, *Bulletin of New York Academy Medicine*, 52, 838-841.

B060   G. Belfort (1981) Selective Adsorption of Organic Homologs onto Activated Carbon from Dilute Aqueo, 2, 207-241.

B061   H. Baggesgaard-Rasmussen and F. Reimers (1935) Die Loslichkeit des Morphins in Verschiedenen Losungsmitteln, *Archiv der Pharmazie und Berichte der Deutschen Pharmazeutischen Gesellschaft*, 273, 129-139.

B062   E. Boyland and B. Green (1962) The Interaction of Polycyclic Hydrocarbons and Purines, *British Journal of Cancer*, 16, 347-360.

B063   S.T. Bowden and J.H. Purnell (1954) The Influence of Uranyl and Thorium Salts on the Miscibility of Phenol and Water, *Journal of the Chemical Society (London)*, 535-538.

B064   D.W.A. Bourne, T. Higuchi, and A.J. Repta (1977) Acetylacroninium Salts as Soluble Prodrugs of the Antineoplastic Agent Acronine, *Journal of Pharmaceutical Sciences*, 66, 628-631.

B065   T.L. Breon, J.W. Mauger, G.E. Osborne..., and A.N. Paruta (1976) The Aqueous Solubility of Variously Substituted Barbituric Acids. I. Chemical Effects, *Drug Development Communications*, 2, 521-535.

B066   D.G. Bamford, D.F. Biggs, M.F. Cuthbert..., and W.R. Wragg (1970) The Preparation and Intravenous Anaesthetic Activity of Tetrahydrofuran-3-ols, *Journal of Pharmacy and Pharmacology*, 22, 694-699.

B068   M.C. Bowman, J.R. King, and C.L. Holder (1976) Benzidine and Congeners: Analytical Chemical Properties and Trace Analysis in Five Substrates, *International Journal of Environmental Analytical Chemistry*, 4, 205-223.

B069   E.G. Baker (1958) Crude Oil Composition and Hydrocarbon Solubility, *American Chemical Society*, Division of Petroleum Chemistry, Preprints, 3, 61-69.

B070   L.L. Burger and R.M. Wagner (1958) Preparation and Properties of Some Organophosphorus Compounds, *Chemical & Engineering Data Series*, 3, 310-313.

B071   F. Bottari, G. Di Colo, E. Nannipieri..., and M.F. Serafini (1977) Release of Drugs from Ointment Bases II: In Vitro Release of Benzocaine from Suspension-Type Aqueous Gels, *Journal of Pharmaceutical Sciences*, 66, 926-928.

B072   L. Borka (1974) The Polymorphism of Indomethacine, *Acta Pharmaceutica Suecica*, 11, 295-303.

B073   G. Boman, P. Lundgren, and G. Stjernstrom (1975) Mechanism of the Inhibitory Effects of PAS Granules on the Absorption of Rifampicin: Adsorption of Rifampicin by an Excipient, Bentonite, *European Journal of Clinical Pharmacology*, 8, 293-299.

B074   C.R. Bailey (1925) The Condensed Ternary System Phenol-Water-Salicylic Acid, *Journal of the Chemical Society (London)*, 126, 1951-1965.

B075   H.J. Backer (1930) L'Acide Chloromethionique, *Recueil des Travaux Chimiques des Pays-Bas*, 49, 729-734.

B076   H.J. Backer (1929) Preparation Simple de l'Acide Methionique, *Recueil des Travaux Chimiques des Pays-Bas*, 48, 949-935.

B077   H.J. Backer (1929) Quelques Syntheses de l'Acide Bromomethionique, *Recueil des Travaux Chimiques des Pays-Bas*, 48, 616-621.

B078   C. Boulin and L-J. Simon (1920) Action de l'Eau sur le Sulfate Dimethylique, *Comptes Rendus Hebdomadaires des Seances de l'Academie des Sciences*, 170, 392-394.

B079   C. Boulin and L-J. Simon (1920) Action de l'Eau sur le Sulfure d'Ethyle Dichlore, *Comptes Rendus Hebdomadaires des Seances de l'Academie des Sciences*, 170, 845-848.

B080   L.A. Balykova, G.P. Verkholetova, N.S. Lebedeva..., and A.V. Starkov (1967) Solubility and Bactericidal Activity of 1,3-Dichlorohydantoin, 1,3-Dichloro-5-methylhydantoin and Trichloroisocyanuric Acid, *Zhurnal Mikrobiologii, Epidemiologii i Immunobiologii*, 44, 14-18.

B083   J.W. Biggar and R.L. Riggs (1974) Apparent Solubility of Organochlorine Insecticides in Water at Various Temperatures, *Hilgardia*, 42, 383-391.

B085   J.O. Bockris and H. Egan (1947) The Salting-Out Effect and Dielectric Constant, *Transactions of the Faraday Society*, 43, 151-159.

B086   R.L. Brown and S.P. Wasik (1974) A Method of Measuring the Solubilities of Hydrocarbons in Aqueous Solutions, *Journal of Research of the National Bureau of Standards*, 78, 453-460.

B088   W.D. Bancroft and F.J.C. Butler (1932) Solubility of Succinic Acid in Binary Mixtures, *Journal of Physical Chemistry*, 36, 2515-2520.

B090   R.L. Bergen, Jr. and F.A. Long (1956) The Salting in of Substituted Benzenes by Large Ion Salts, *Journal of Physical Chemistry*, 60, 1131-1135.

B092   H.S. Booth and H.E. Everson (1950) Hydrotropic Solubilities—Solubilities in Aqueous Sodium o-, m-, and p-Xylenesulfonate Solutions, *Industrial and Engineering Chemistry*, 42, 1536-1537.

B093   J.W. Biggar, G.R. Dutt, and R.L. Riggs (1967) Predicting and Measuring the Solubility of p,p'-DDT in Water, *Bulletin of Environmental Contamination and Toxicology*, 2, 90-100.

B094   J. Burgess and R.I. Haines (1978) Solubilities of 1,10-Phenanthroline and Substituted Derivatives in Water and in Aqueous Methanol, *Journal of Chemical and Engineering Data*, 23, 196-197.

B095   H.P. Bennetto and J.W. Letcher (1972) Solubility and Solvation of Bipyridyls and Biphenyl in Water, *Chemistry and Industry (London)*, 847-848.

B096   H.F. Brust (1966) A Summary of Chemical and Physical Properties of DURSBAN, *Down to Earth*, 22, 21-22.

B097   J.O. Bockris, J. Bowler-Reed, and J.A. Kitchener (1951) The Salting-In Effect, *Transactions of the Faraday Society*, 47, 184-192.

B099   F.F. Blicke and E.S. Blake (1930) Local Anesthetics in the Pyrrole Series, *Journal of the American Chemical Society*, 52, 235-240.

B100   R.C. Brian (1976) The History and Classification of Herbicides, *Herbicides: Physiology, Biochemistry, Ecology*, 1, 13-54, Academic Press.

B101   A.E. Bradfield and A.F. Williams (1929) The Solubility of Certain Anilides in Water-Acetic Acid Mixtures, *Journal of the Chemical Society (London)*, 2542-2544.

B102   M.T. Bogert and J. Ehrlich (1919) A Unique Case of a Liquid that Exhibits a Minimum Solubility in an Unstable Region, *Journal of the American Chemical Society*, 41, 741-745.

B103   G. Blix (1928) Uber die Loslichkeitsverhaltnisse von Cystin im Harn, *Hospodarsky Zpravodaj*, 178, 109-115.

B104   H. Buchowski, W. Jodzewicz, R. Milek..., and A. Maczynski (1975) Solubility and Hydrogen Bonding. Part I. Solubility of 4-Nitro-5-methylphenol in One-Component Solvents, *Roczniki Chemii*, 49, 1879-1887.

B106   J.M. Braham (1919) Some Physical Properties of Mannite and its Aqueous Solutions, *Journal of the American Chemical Society*, 41, 1707-1719.

B107   M. Boyle (1919) The Conductivities of Iodoanilinesulphonic Acids, *Journal of the Chemical Society (London)*, 115, 1505-1517.

B108   D.G. Beech and S. Glasstone (1938) Solubility Influences. Part V. The Influence of Aliphatic Alcohols on the Solubility of Ethyl Acetate in Water, *Journal of the Chemical Society (London)*, 67-70.

B109   W.V. Bhagwat and N.R. Dhar (1929) Dissociation Constants of Some Inorganic Acids from Solubility Measurements, *Journal of the Indian Chemical Society*, 6, 807-822.

B110   A.R. Biamonte and G.H. Schneller (1952) Observations on the Solubility of Certain Sulfonamides, *Journal of the American Pharmaceutical Association, Scientific Edition*, 41, 341-345.

B111   A. Boutaric and G. Corbet (1926) Sur la Temperature Critique de Dissolution de l'Acroleine et de l'Eau et sur la Masse Moleculaire de la Resine d'Acroleine Soluble, *Comptes Rendus Hebdomadaires des Seances de l'Academie des Sciences*, 183, 42-44.

B112   E.V. Bell and G.M. Bennett (1929) Stereoisomerism of Disulphoxides and Related Substances. Part IV. Di- and Tri-sulphoxides of Trimethylene Trisulphide, *Journal of the Chemical Society (London)*, 15-19.

B113   A.E. Brodsky and M.I. Alferow (1929) Uber die Loslichkeit des Benzochinhydrons in Wasserigem Alkohol, *Berichte der Deutschen Chemischen Gesellschaft*, 62, 2132-2133.

B114   S.M. Bastami and M.J. Groves (1978) Some Factors Influencing the In Vitro Release of Phenytoin from Formulations, *International Journal of Pharmaceutics*, 1, 151-153.

B115   H. Biltz and M. Heyn (1917) Alpha, zeta, und delta-Methylharnsaure, *Annalen der Chemie*, 98-162.

B116   H. Biltz and L. Herrmann (1923) Uber die Loslichkeit von Harnsaure in Wasser, *Annalen der Chemie*, 104-111.

B117   A. Bourgom (1924) Contribution a l'Etude du Methylal comme Solvant, *Bulletin des Societes Chimiques Belges*, 33, 101-115.

B118   W.V. Bhagwat and S.S. Doosaj (1933) Limitations of Solubility Method for Determining Dissociation Constant, *Journal of the Indian Chemical Society*, 10, 477-490.

B119   E. Biilmann and J. Bentzon (1918) Uber Alloxan und Alloxanthin, *Berichte der Deutschen Chemischen Gesellschaft*, 51, 522-532.

B121   A. Berthoud and S. Kunz (1938) Solubilite et Dissociation de la Quinhydrone, *Helvetica Chimica Acta*, 21, 17-21.

B123   W.B. Brooks and J.J. McKetta (1955) The Solubility of 1-Butene in Water, *Petroleum Refiner*, 34, 143-144.

B124   J. Barbaudy (1926) Contribution a l'Etude de la Distillation des Melanges Ternaires Heterogenes-Le Systeme Eau-Benzene-Toluene, *Journal de Chimie Physique et de Physico-Chimie Biologique*, 23, 290-298.

B125   H.T. Bucherer and R. Wahl (1921) Uber die 2,5,1-Aminonaphtolsulfonsaure (A-Saure) und ihre Derivate, *Journal fuer Praktische Chemie*, 103, 129-150.

B126   J. Barbaudy (1926) Systeme Alcool Ethylique-Benzene-Eau. I. Etude de la Surface de Trouble, *Recueil des Travaux Chimiques des Pays-Bas*, 45, 207-213.

B128   R.J. Braun and E.L. Parrott (1972) Influence of Viscosity and Solubilization on Dissolution Rate, *Journal of Pharmaceutical Sciences*, 61, 175-178.

B129   S.M. Blaug and S.S. Ahsan (1961) Interaction of Parabens with Nonionic Macromolecules, *Journal of Pharmaceutical Sciences*, 50, 441-443.

B130   A. Burger (1978) Zur Polymorphie Oraler Antidiabetika, *Scientia Pharmaceutica*, 46, 207-222.

B131   D.R. Branson (1977) A New Capacitor Fluid—A Case Study in Product Stewardship, *unpublished*, 44-61.

B132   J.B. Bogardus and R.K. Blackwood, Jr. (1979) Solubility of Doxycycline in Aqueous Solution, *Journal of Pharmaceutical Sciences*, 68, 188-194.

B133   J.W. Bridges, S.R. Walker, and R.T. Williams (1969) Species Differences in the Metabolism and Excretion of Sulphasomidine and Sulphamethomidine, *Biochemical Journal*, 111, 173-179.

B134   D.C. Baker, T.H. Haskell, and S.R. Putt (1978) Prodrugs of 9-beta-D-Arabinofuranosyladenine. 1. Synthesis and Evaluation of Some 5-(O-Acyl) Derivatives, *Journal of Medicinal Chemistry*, 21, 1218-1221.

B135   M. Bachstez (1930) Uber die Konstitution der Orotsaure, *Chemische Berichte*, 63, 1000-1007.

B136   S. Baykut (1956) The Solubility of the Higher Fatty Acids in Water, Chemie-Physique Serie C, 21, 36-45.

B138   A. Burger (1975) Zur Polymorphie Oraler Antidiabetika, *Scientia Pharmaceutica*, 43, 161-168.

B139   B.A. Beremzhanov, N.N. Nura, and R.S. Erkasov (1975) Solubility of Benzamide in Aqueous Solutions of Sulfuric, Selenic, and Phosphoric Acids at 20 C, *Journal of General Chemistry of the USSR*, 45, 1191-1194.

B140   A. Burger (1975) ZurPolymorphie Oraler Antidiabetika I. Mitteilung: Chlorpropamid, *Scientia Pharmaceutica*, 43, 152-161.

B141   J. Burton, K. Poulsen, and E. Haber (1978) Solubility and Lipophilicity Relationships in the Design of Renin Inhibitors, 219-237.

B142   S. Bolton (1963) Interaction of Urea and Thiourea with Benzoic and Salicylic Acids, *Journal of Pharmaceutical Sciences*, 52, 1071-1074.

B144   S.K. Baveja, V.S. Raju, M.P. Pakhetra, and S. Kaur (1978) Formulation of Intravenous Osthole Solution, *Indian Journal of Pharmacy*, 40, 230.

B147   A. Burger (1973) Das Auflosungsverhalten von Sulfanilamid in Wasser, *Pharmazeutische Industrie*, 35, 626-633.

B148   J.B. Bogardus and N.R. Palepu (1979) Ionization and Solubility of an Amphoteric beta-Lactam Antibiotic, *International Journal of Pharmaceutics*, 4, 159-170.

B149   A. Ben-Naim and J. Wilf (1979) A Direct Measurement of Intramolecular Hydrophobic Interactions, *Journal of Chemical Physics*, 70, 771-777.

B150   P.J. Brooker and M. Ellison (1974) The Determination of the Water Solubility of Organic Compounds by a Rapid Turbidimetric Method, *Chemistry and Industry (London)*, 5, 785-787.

B151   H-J. Bittrich, H. Gedan, and G. Feix (1979) Zur Loslichkeitsbeeinflussung von Kohlen-wasserstoffen in Wasser (Effects on the Solubility of Hydrocarbons in Water), *Zeitschrift fuer Physikalische Chemie (Leipzig)*, 260, 1009-1013.

B152   H. Bundgaard, A.B. Hansen, and C. Larsen (1979) Pro-Drugs as Drug Delivery Systems. III. Esters of Malonuric Acids as Novel Pro-Drug Types for Barbituric Acids, *International Journal of Pharmaceutics*, 3, 341-353.

B153   A. Ben-Naim and J. Wilf (1980) Solubilities and Hydrophobic Interactions in Aqueous Solutions of Monoalkylbenzene Molecules, *Journal of Physical Chemistry*, 84, 583-586.

B154   H. Bundgaard and M. Johansen (1980) Pro-Drugs as Drug Delivery Systems. VIII. Bioreversible Derivatization of Hydantoins by N-Hydroxymethylation, *International Journal of Pharmaceutics*, 5, 67-77.

B155   S.V. Bogdanova, N. Lambov, and E. Minkov (1981) Physicochemical Studies of Cinnarizine-Polyvinylpyrrolidone Solid Dispersion, *Pharmazie (Berlin)*, 36, 197-199.

B156   D.K. Button (1976) The Influence of Clay and Bacteria on the Concentration of Dissolved Hydrocarbon in Saline Solution, *Geochimica et Cosmochimica Acta*, 40, 435-440.

B157   A.A. Bugaevskii, N.R. Sumskaya, and V.O. Kruglov (1977) The Salting-Out of p-Nitrophenol in Aqueous Sodium Chloride Solutions, *Russian Journal of Physical Chemistry*, 51, 1072-1073.

B158   B.B. Brodie and C.A.M. Hogben (1957) Some Physico-Chemical Factors in Drug Action, *Journal of Pharmacy and Pharmacology*, 9, 345-380.

B160   H.P. Burchfield (1960) Performance of Fungicides on Plants and in Soil—Physical, Chemical, and Biological Considerations, *Plant Pathology*, 3, 447-520.

B161   H.M. Bhavnagary and M. Jayaram (1974) Determination of Water Solubilities of Lindane and Dieldrin at Different Temperatures, *Bulletin of Grain Technology*, 12, 95-99.

B162   J.W. Biggar, L.D. Donnen, and R.L. Riggs (1968) Soil Interaction with Organically Polluted Water. Summary Report Dept. of Water Science and Engineering, University of California (1966) cf. F.A. Gunther, W.E. Westlake, and P.S. Jaglan Eds., *Residue Reviews*, 20, 1.

B164   R. Behrens and H.L. Morton (1963) Some Factors Influencing Activity of 12 Phenoxy Acids on Mesquite Root Inhibition, *Plant Physiology*, 38, 165-170.

B165   H. Becke and G. Quitzsch (1977) Das Phasengleichgewichtsverhalten Ternarer Systeme der Art C4-Alkohol-Wasser-Kohlenwasserstoff, *Chemische Technik (Leipzig)*, 29, 49-51.

B166   B.E. Ballard (1961) The Physicochemical Properties of Drugs that Control their Absorption Rate after Implantation, Ph.D. thesis, 210-239.

B167   F.W. Barnes, Jr. and W.F. Seip (1960) Hollow Crystals from Buffer Solutions of Sodium Diethyl Barbiturate, *Science*, 131, 161-161.

B169   B.T. Bowman and W.W. Sans (1979) The Aqueous Solubility of Twenty-seven Insecticides and Related Compounds, *Journal of Environmental Science & Health, Series B*, B14, 625-634.

B170   E.G. Baker (1960) A Hypothesis Concerning the Accumulation of Sediment Hydrocarbons to Form Crude Oil, *Geochimica et Cosmochimica Acta*, 19, 309-317.

B171   D. Banerjee and B.K. Gupta (1980) The Estimation of Compound Hydrophobicities and their Relevance to Partition Coefficients, *Canadian Journal of Pharmaceutical Science*, 15, 61-63.

B173   S. Banerjee, S.H. Yalkowsky, and S.C. Valvani (1980) Water Solubility and Octanol/Water Partition Coefficients of Organics. Limitations of the Solubility-Partition Coefficient Correlation, *Environmental Science and Technology*, 14, 1227-1229.

B174   R. Bohm (1980) Physico-chemical Properties of the Cyclic Ketals and Thioketals, *Pharmazie (Berlin)*, 35, 802-803.

B175   J. Booth and E. Boyland (1953) The Reaction of the Carcinogenic Dibenzcarbazoles and Dibenzacridines with Purines and Nucleic Acid, *Biochimica et Biophysica Acta*, 12, 75-87.

B177   P.C. Bansal, I.H. Pitman, J.N.S. Tam..., and J.J. Kaminski (1981) N-Hydroxymethyl Derivatives of Nitrogen Heterocycles as Possible Prodrugs. I. N-Hydroxymethylation of Uracils, *Journal of Pharmaceutical Sciences*, 70, 850-854.

B178   A. Brodin, B. Sandin, and B. Faijerson (1976) Rates of Transfer of Organic Protolytic Solutes Between an Aqueous and an Organic Phase. V. The Thermodynamics of Mass Transfer, *Acta Pharmaceutica Suecica*, 13, 331-352.

B179   G.G. Briggs (1981) Theoretical and Experimental Relationships Between Soil Adsorption, Octanol-Water Partition Coefficients, Water Solubilities, Bioconcentration Factors, and the Parachor, *Journal of Agricultural and Food Chemistry*, 29, 1050-1059.

B181   J.E. Baer, H.L. Leidy, A.V. Brooks, and K.H. Beyer (1959) The Physiological Disposition of Chlorothiazide (Diuril) in the Dog, *Journal of Pharmacology and Experimental Therapeutics*, 125, 295-302.

B182   P.K. Banerjee and G.L. Amidon (1981) Physicochemical Property Modification Strategies Based on Enzyme Substrate Specificities. I. Rationale, Synthesis, and Pharmaceutical Properties of Aspirin Derivatives, *Journal of Pharmaceutical Sciences*, 70, 1299-1309.

B183   A. Bevenue and H. Beckman (1967) Pentachlorophenol: A Discussion of its Properties and its Occurrence as a Residue in Human and Animal Tissues, *Residue Reviews*, 19, 83-87.

B185   G.W. Bailey and J.L. White (1965) Herbicides: A Compilation of Their Physical, Chemical, and Biological Properties, *Residue Reviews*, 10, 97-120.

B186   F.H. Babers (1955) The Solubility of DDT in Water Determined Radiometrically, *Journal of the American Chemical Society*, 77, 4666-4666.

B187   G.A. Brewer (1977) Isoniazid, *Analytical Profiles of Drug Substances*, 6, 183-229.

B188   P.K. Bhattacharyya and W.M. Cort (1978) Amoxicillin, *Analytical Profiles of Drug Substances*, 7, 19-35.

B189   S.A. Benezra and T.R. Bennett (1978) Allopurinol, *Analytical Profiles of Drug Substances*, 7, 1-4.

B190   H. Brik (1981) Natamycin, *Analytical Profiles of Drug Substances*, 10, 513-541.

B191 W.C. Barringer, W. Shultz, G.M. Sieger, and R.A. Nash (1974) Minocycline Hydrochloride and its Relationship to Other Tetracycline Antibiotics, *American Journal of Pharmacy*, 146, 179-191.

B192 C.E. Bartley (1959) Triazine Compounds, *Farm Chemicals*, 122, 28-34.

B193 P. Beilstein, A.M. Cook, and R. Hutter (1981) Determination of Seventeen s-Triazine Herbicides and Derivatives by High-Pressure Liquid Chromatography, *Journal of Agricultural and Food Chemistry*, 29, 1132-1135.

B194 H.L. Bhalla (1981) Preformulation Studies on Vasicinone—A Bronchodilatory Alkaloid (Study of Some Physico-Chemical Aspects), *Drug Development and Industrial Pharmacy*, 7, 755-768.

B196 H. Bundgaard, M. Johansen, V. Stella, and M. Cortese (1982) Pro-drugs as Drug Delivery Systems. XXI. Preparation, Physicochemical Properties and Bioavailability of a Novel Water-soluble Pro-drug Type for Carbamazepine, *International Journal of Pharmaceutics*, 10, 181-192.

B197 M.K. Baranaev, I.S. Gilman, L.M. Kogan, and N.P. Rodinova (1954) Separating Dichloroethane From its Aqueous Solutions, *Journal of Applied Chemistry of the USSR*, 27, 1031-1036.

B198 W. Biedermann and A. Datyner (1981) The Interaction of Nonionic Dyestuffs with Sodium Dodecyl Sulfate and its Correlation with Lipophilic Parameters, *Journal of Colloid and Interface Science*, 82, 276-285.

B199 N. Bodor and K.B. Sloan (1982) Soft Drugs. V. Thiazolidine-type Derivatives of Progesterone and Testosterone, *Journal of Pharmaceutical Sciences*, 71, 514-520.

B200 G.E. Barrier, J.L. Hilton, R.E. Frans..., and D.E. Moreland (1970) *Herbicide Handbook of the Weed Science Society of America*, 1-353.

B201 A.J. Barduhn and M. Handley (1982) Low-temperature Solubility of Caprolactam in Water, *Journal of Chemical and Engineering Data*, 27, 306-308.

B300 B.T. Bowman and W.W. Sans (1983) Further Water Solubility Determinations of Insecticidal Compounds, *Journal of Environmental Science & Health, Series B*, B18, 221-227.

B301 W.A. Bruggeman, L.B.J. Martron, D. Kooiman, and O. Hutzinger (1981) Accumulation and Elimination Kinetics of di-, tri-, and tetra-Chlorobiphenyls by Goldfish after Dietary and Aqueous Exposure, *Chemosphere*, 10, 811-832.

B302 A. Beerbower, P.L. Wu, and A. Martin (1984) Expanded Solubility Parameter Approach I: Naphthalene and Benzoic Acid in Individual Solvents, *Journal of Pharmaceutical Sciences*, 73, 179-188.

B304 S. Banerjee (1984) Solubility of Organic Mixtures in Water, *Environmental Science and Technology*, 18, 587-591.

B305 T.A. Bengtsson (1958) 4-Amino-4'-chlorodiphenyl as Analytical Reagent for Sulphate, *Analytica Chimica Acta*, 18, 353-359.

B306 D.C. Baker, S.D. Kumar, W.J. Waites..., and W.J. Lambert (1984) Synthesis and Evaluation of a Series of 2'-O-Acyl Derivatives of 9-beta-D-Arabinofuranosyladenine as Antiherpes Agents, *Journal of Medicinal Chemistry*, 27, 270-274.

B309 W.E. Bleidner, R. Morales, and R.F. Holt (1978) Benomyl, *Analytical Methods for Pesticides and Plant Growth Regulators*, 10, 157-171.

B310 S.S. Brady, C. Van Hoek, and V.F. Boyd (1978) Norflurazon, *Analytical Methods for Pesticides and Plant Growth Regulators*, 10, 415-435.

B314 A.E. Beezer, P.L.O. Volpe, M.C.P. Lima, and W.H. Hunter (1986) Solution Thermodynamics for Alkoxy Phenols in Alcohol and in Water-Alcohol Systems, *Journal of Solution Chemistry*, 15, 341-363.

B315 A.E. Beezer, W.H. Hunter, and D.E. Storey (1983) Enthalpies of Solution of a Series of m-Alkoxy Phenols in Water, n-Octanol and Water-n-Octanol Mutually Saturated: Derivation of the Thermodynamic Parameters for Solute Transfer Between These Solvents, *Journal of Pharmacy and Pharmacology*, 35, 350-357.

B316 G.E. Blackmann, M.H. Parke, and G. Garton (1955) The Physiological Activity of Substituted Phenols. II. Relationships between Physical Properties and Physiological Activity, *Archives of Biochemistry and Biophysics*, 54, 55-70.

B317 A. Bobra, W.Y. Shiu, and D. Mackay (1985) Quantitative Structure-Activity Relationships for the Acute Toxicity of Chlorobenzenes to Daphnia Magna, *Environmental Toxicology and Chemistry*, 4, 297-305.

B318 D.R. Burris and W.G. MacIntyre (1985) Water Solubility Behavior of Binary Hydrocarbon Mixtures, *Environmental Toxicology and Chemistry*, 4, 371-377.

B319   J.W. Billington, G-L. Huang, F. Szeto..., and D. MacKay (1988) Preparation of Aqueous Solutions of Sparingly Soluble Organic Substances. I. Single Component Systems, *Environmental Toxicology and Chemistry*, 7, 117-124.

B321   A. Buur, H. Bundgaard, and E. Falch (1985) Prodrugs of 5-fluorouracil. IV. Hydrolysis Kinetics, Bioactivation and Physicochemical Properties of Various N-acyloxymethyl Derivatives of 5-fluorouracil, *International Journal of Pharmaceutics*, 24, 43-60.

B322   H. Bundgaard and E. Falch (1985) Allopurinol Prodrugs. II. Synthesis, Hydrolysis Kinetics and Physicochemical Properties of Various N-Acyloxymethyl Allopurinol Derivatives, *International Journal of Pharmaceutics*, 24, 307-325.

B323   H. Bundgaard and E. Falch (1985) Allopurinol Prodrugs. III. Water-soluble N-Acyloxymethyl Allopurinol Derivatives for Rectal or Parenteral Use, *International Journal of Pharmaceutics*, 25, 27-39.

B324   B.T. Bowman and W.W. Sans (1985) Effect of Temperature on the Water Solubility of Insecticides, *Journal of Environmental Science & Health, Series B*, 20, 625-631.

B325   B.T. Bowman and W.W. Sans (1982) Adsorption, Desorption, Soil Mobility, Aqueous Persistence and Octanol-Water Partitioning Coefficients of Terbufos, Terbufos Sulfoxide and Terbufos Sulfone, *Journal of Environmental Science & Health, Series B*, 17, 447-462.

B328   H. Bundgaard, U. Klixbull, and E. Falch (1986) Prodrugs as Drug Delivery Systems. 44. O-Acyloxymethyl, O-Acyl and N-Acyl Salicylamide Derivatives as Possible Prodrugs for Salicylamide, *International Journal of Pharmaceutics*, 30, 111-121.

B331   H. Bundgaard and N.M. Nielsen (1988) Glycolamide Esters as a Novel Biolabile Prodrug Type for Non-steroidal Anti-inflammatory Carboxylic Acid Drugs, *International Journal of Pharmaceutics*, 43, 101-110.

B332   A. Buur and H. Bundgaard (1987) Prodrugs of 5-Fluorouracil. VIII. Improved Rectal and Oral Delivery of 5-Fluorouracil via Various Prodrugs. Structure-Rectal Absorption Relationships, *International Journal of Pharmaceutics*, 36, 41-49.

B333   G.L. Baughman and T.A. Perenich (1988) Fate of Dyes in Aquatic Systems. I. Solubility and Partitioning of Some Hydrophobic Dyes and Related Compounds, *Environmental Toxicology and Chemistry*, 7, 183-199.

B335   J.L. Brisset (1985) Solubilities of Various Nitroanilines in Water-Pyridine, Water-Acetonitrile, and Water-Ethylene Glycol Solvents, *Journal of Chemical and Engineering Data*, 30, 381-383.

B337   M. Bonati, F. Gaspari, V. D'Aranno, and G. Tognoni (1984) Physicochemical and Analytical Characteristics of Amiodarone, *Journal of Pharmaceutical Sciences*, 73, 829-831.

B338   R.L. Bronaugh and R.F. Stewart (1984) Methods for In Vitro Percutaneous Absorption. Studies III. Hydrophobic Compounds, *Journal of Pharmaceutical Sciences*, 73, 1255-1258.

B340   H. Bundgaard, E. Falch, C. Larsen, and T.J. Mikkelson (1986) Pilocarpine Prodrugs. II. Synthesis, Stability, Bioconversion, and Physicochemical Properties of Sequentially Labile Pilocarpine Acid Diesters, *Journal of Pharmaceutical Sciences*, 75, 775-783.

B342   P.L. Bolden, J.C. Hoskins, and A.D. King, Jr. (1983) The Solubility of Gases in Solutions Containing Sodium Alkylsulfates of Various Chain Lengths, *Journal of Colloid and Interface Science*, 91, 454-463.

B348   D.R. Burris and W.G. MacIntyre (1986) Solution of Hydrocarbons in a Hydrocarbon-Water System with Changing Phase Composition Due to Evaporation, *Environmental Science and Technology*, 20, 296-299.

B349   R.J. Baker, B.J. Donelan, L.J. Peterson..., and C-C. Tsai (1987) Correlation and Estimation of Aqueous Solubilities of Halogenated Benzenes, *Physics and Chemistry of Liquids*, 16, 279-292.

B350   J.B. Billington (1986) Physical Chemical Properties of Polychlorinated Biphenyls, *Journal of Physical Chemical Reference Data*, 15, 7-9.

B351   R.E. Bailey, W.L. Rhinehart, S.J. Gonsior..., and W.B. Neely (1981) Hazard Assessment of Monochloro Biphenyl in the Aquatic Environment. A Case History, presentation at a Meeting.

B352   A.F.M. Barton and J. Tjandra (1988) Ternary Phase Equilibrium Studies of the Water-Ethanol-1,8-Cineole System, *Fluid Phase Equilibria*, 44, 117-123.

B353   A.E. Beezer, M.C.P. Lima, G.G. Fox..., and B.V. Smith (1987) Solution Thermodynamics for o-Alkoxyphenols in Water and in Water-Alcohol Systems, *Thermochimica Acta*, 116, 329-335.

B354   G.L. Bockstanz, M. Buffa, and C.T. Lira (1989) Solubilities of Alpha-Anhydrous Glucose in Ethanol/Water Mixtures, *Journal of Chemical and Engineering Data*, 34, 426-429.

B355   A.E. Beezer, S. Forster, W.B. Park, and G.J. Rimmer (1991) Solution Thermodynamics of 4-Hydroxybenzoates in Water, 95% Ethanol-Water, 1-Octanol and Hexane, *Thermochimica Acta*, 178, 59-65.

B356   D.R. Burris and W.G. MacIntyre (1987) Water Solubility Behavior of Hydrocarbon Mixtures—Implications for Petroleum Dis.

B360   E.J. Benjamin, B.A. Firestone, R. Bergstrom..., and Y.Y.T. Lin (1987) Selection of a Derivative of the Antiviral Agent 9-[(1,3-Dihydroxy-2-propoxy)-methyl]Guanine (DHPG) with Improved Oral Absorption, *Pharmaceutical Research*, 4, 120-123.

B361   B. Berner, D.R. Wilson, R.H. Guy..., and H.I. Maibach (1988) The Relationship of pKa and Acute Skin Irritation in Man, *Pharmaceutical Research*, 5, 660-663.

B362   M. Bisrat, E.K. Anderberg, M.I. Barnett, and C. Nystrom (1992) Physicochemical Aspects of Drug Release. XV. Investigation of Diffusional Transport in Dissolution of Suspended, Sparingly Soluble Drugs, *International Journal of Pharmaceutics*, 80, 191-201.

B363   J. Blanchard, J.O. Boyle, and S.V. Wagenen (1988) Determination of the Partition Coefficients, Acid Dissociation Constants, and Intrinsic Solubility of Carbenoxolone, *Journal of Pharmaceutical Sciences*, 77, 548-552.

B366   M.E. Brewster, K.S. Estes, T. Loftsson..., and N. Bodor (1988) Improved Delivery through Biological Membranes. XXXI. Solubilization and Stabilization of an Estradiol Chemical Delivery System by Modified B-Cyclodextrins, *Journal of Pharmaceutical Sciences*, 77, 981-985.

B376   J.F. Borel (1986) Formulation of Dosage Forms, *Ciclosporin*, Karger.

B384   F. Belaj, R. Tripolt, and E. Nachbaur (1990) Kristallstruktur und Thermisches Verhalten der Additionsverbindungen von Trithiocyanursaure mit Tetrahydrofuran und 1,4-Dioxan, *Monatshefte fuer Chemie*, 121, 99-108.

B385   L.A. Blyshak, K.Y. Dodson, G. Patonay..., and W.E. May (1989) Determination of Cyclodextrin Formation Constants Using Dynamic Coupled-Column Liquid Chromatography, *Analytical Chemistry*, 61, 955-960.

B386   B.K. Braxton and J.H. Rytting (1989) Solubilities and Solution Thermodynamics of Several Substituted Melamines, *Thermochimica Acta*, 154, 27-47.

B387   K.M. Boje, M. Sak, and H-L. Fung (1988) Complexation of Nifedipine with Substituted Phenolic Ligands, *Pharmaceutical Research*, 5, 655-659.

B388   A. Buur and H. Bundgaard (1985) Prodrugs of 5-Fluorouracil. III. Hydrolysis Kinetics in Aqueous Solution and Biological Media, Lipophilicity and Solubility of Various 1-Carbamoyl Derivatives of 5-Fluorouracil, *International Journal of Pharmaceutics*, 23, 209-222.

B390   K. Baba, Y. Takeichi, and Y. Nakai (1990) Molecular Behavior and Dissolution Characteristics of Uracil in Ground Mixtures, *Chemical and Pharmaceutical Bulletin*, 38, 2542-2546.

B391   P.K. Biswas, S.C. Lahiri, and B.P. Dey (1993) Solvational Behavior of Some Substituted Benzoic Acids in Ethanol-Water Mixtures at 298.15 K, *Bulletin of the Chemical Society of Japan (Nippon Kagakukai Bulletin)*, 66, 2785-2789.

B393   A. Bharath, C. Mallard, D. Orr..., and A. Smith (1984) Problems in Determining the Water Solubility of Organic Compounds, *Bulletin of Environmental Contamination and Toxicology*, 33, 133-137.

B394   S. Brandani, V. Brandani, and D. Flammini (1994) Solubility of Trioxane in Water, *Journal of Chemical and Engineering Data*, 39, 201-201.

B396   M.E. Brewster, J.W. Simpkins, M.S. Hora..., and N. Bodor (1989) The Potential Use of Cyclodextrins in Parenteral Formulations, *Journal of Parenteral Science and Technology*, 43, 231-240.

B397   M.E. Brewster, W.R. Anderson, K.S. Estes, and N. Bodor (1991) Development of Aqueous Parenteral Formulations for Carbamazepine through the Use of Modified Cyclodextrins, *Journal of Parenteral Science and Technology*, 80, 380-383.

C004   Y.W. Chien and H.J. Lambert (1975) Solubilization of Steroids by Multiple Co-solvent Systems, *Chemical and Pharmaceutical Bulletin*, 23, 1085-1090.

C005   J.E. Carless and J. Swarbrick (1964) The Solubility of Benzaldehyde in Water as Determined by Reactive Index Measurements, *Journal of Pharmacy and Pharmacology*, 16, 633-634.

C006   T.C. Corby and P.H. Elworthy (1971) The Solubility of Some Compounds in Hexadecylpoly-oxyethylene Monoethers, Polyethylene Glycols, Water and Hexane, *Journal of Pharmacy and Pharmacology*, 23, 39-48.

C008   J.E. Carless and J.R. Nixon (1960) The Oxidation of Solubilised and Emulsified Oils (Research Paper), *Journal of Pharmacy and Pharmacology*, 12, 340-347.

C011   D.E. Cadwallader, H.W. Jun, and L.K. Chen (1975) Nitrofurantoin Solubility in Aqueous Urea Solutions, *Journal of Pharmaceutical Sciences*, 64, 886-887.

C014   M.K. Chantooni, Jr. and I.M. Kolthoff (1974) Transfer Activity Coefficients of ortho-Substituted and Non-ortho-substituted Benzoates Between Water, Methanol, and Polar Aprotic Solvents, *Journal of Physical Chemistry*, 78, 839-846.

C017   L. Costantino and V. Vitagliano (1967) The Influence of Solvation of Purinic and Pyrimidinic Bases on the Conformational Stability of DNA Solutions, *Biochimica et Biophysica Acta*, 134, 204-206.

C018   E.J. Cohn, T.L. McMeekin, J.T. Edsall, and J.H. Weare (1934) Studies in the Physical Chemistry of Amino Acids, Peptides and Related Substances. II. The Solubility of alpha-Amino Acids in Water and in Alcohol-Water Mixtures, *Journal of the American Chemical Society*, 56, 2270-2282.

C020   B.B. Corson, N.E. Sanborn, and P.R. Van Ess (1930) Some Observations on Benzoylformic Acid, *Journal of the American Chemical Society*, 52, 1623-1626.

C022   M.J. Copley, E. Ginsberg, G.F. Zellhoefer, and C.S. Marvel (1941) Hydrogen Bonding and the Solubility of Alcohols and Amines in Organic Solvents. XIII, *Journal of the American Chemical Society*, 63, 254-256.

C023   R.P. Chapman, P.R. Averell, and R.R. Harris (1943) *Solubility of Melamine in Water*, *Industrial and Engineering Chemistry*, 35, 137-138.

C024   W. Chey and G.V. Calder (1972) Method for Determining Solubility of Slightly Soluble Organic Compounds, *Journal of Chemical and Engineering Data*, 17, 199-200.

C025   R. Charonnat (1927) La Solubilite de la 1-Phenyl-2,3-dimethyl-4-dimethylamino-5-pyrazolone dans L'eau, *Comptes Rendus Hebdomadaires des Seances de l'Academie des Sciences*, 185, 284-286.

C026   M.A.Q. Chaudry and K.C. James (1974) A Hansch Analysis of the Anabolic Activities of Some Nandrolone Esters, *Journal of Medicinal Chemistry*, 17, 157-161.

C031   M.W. Cheung and J.W. Biggar (1974) Solubility and Molecular Structure of 4-Amino-3,5,6-trichloropicolinic Acid in Relation to pH and Temperature, *Journal of Agricultural and Food Chemistry*, 22, 202-206.

C032   Y.P. Chow and A.J. Repta (1972) Complexation of Acetaminophen with Methyl Xanthines, *Journal of Pharmaceutical Sciences*, 61, 1454-1458.

C033   G. Caronna (1948) Antagonismo Batterico e Influenze di Solubilita, *Gazzetta Chimica Italiana*, 78, 827-835.

C034   L-K. Chen, D.E. Cadwallader, and H.W. Jun (1976) Nitrofurantoin Solubility in Aqueous Urea and Creatinine Solutions, *Journal of Pharmaceutical Sciences*, 65, 868-872.

C035   N.M. Cone, S.E. Forman, and J.C. Krantz, Jr. (1941) Relationship Between Anesthetic Potency and Physical Properties, *Proceedings of the Society for Experimental Biology and Medicine*, 48, 461-463.

C036   F.B. Coon, E.F. Richter, L.W. Hein, and C.H. Krieger (1954) Problems Encountered in Physicochemical Determination of Warfarin, *Journal of Agricultural and Food Chemistry*, 2, 739-741.

C037   W.L. Chiou (1975) Possibility of Errors in Using Filter Paper for Solubility Determination, *Canadian Journal of Pharmaceutical Science*, 10, 112-114.

C038   J.T. Carstensen and M. Patel (1975) Dissolution Patterns of Polydisperse Powders: Oxalic Acid Dihydrate, *Journal of Pharmaceutical Sciences*, 64, 1770-1776.

C039   J.A. Clements and S.D. Popli (1973) The Preparation and Properties of Crystal Modifications of Meprobamate, *Canadian Journal of Pharmaceutical Science*, 8, 88-92.

C040   W.L. Chiou (1977) Pharmaceutical Applications of Solid Dispersion Systems: X-Ray Diffraction and Aqueous Solubility Studies on Griseofulvin-Polyethylene Glycol 6000 Systems, *Journal of Pharmaceutical Sciences*, 66, 989-991.

C042   W.W. Clough and C.O. Johns (1923) Higher Alcohols from Petroleum Olefins, *Industrial and Engineering Chemisry*, 15, 1030-1032.

C045   A.N. Campbell and F.C. Garrow (1930) The Physical Identity of Enantiomers, *Transactions of the Faraday Society*, 26, 560-565.

C046   M.J. Cho and M.J. Peterman (1978) Pre-formulation Studies of 7-Benzoylindoline (U-26,952) and Possible Utilization of Molecular Interaction with beta-Cyclodextrin in Development of an Oral Dosage Form, *Pharmaceutical Research and Development*, 2-21.

C047   J.D. Cox (1954) Phase Relationships in the Pyridine Series. Part IV. The Miscibility of the Ethylpyridines and Dimethylpyridines with Water, *Journal of the Chemical Society (London)*, 3183-3187.

C048   R. Calvet, M. Terce, and J. Le Renard (1975) Cinetique de Dissolution dans l'Eau de l'Atrazine, de la Propazine et de la Simazine, *Weed Research*, 15, 387-392.

C049   W.T. Caldwell, E.C. Kornfeld, and C.K. Donnell (1941) Substituted 2-Sulfanilamidopyrimidines, *Journal of the American Chemical Society*, 63, 2188-2190.

C051   A.J. Cumber and W.C.J. Ross (1977) Analogues of Hexamethylmelamine. The Anti-neoplastic Activity of Derivatives with Enhanced Water Solubility, *Chemico-Biological Interactions*, 17, 349-357.

C052   A.N. Campbell and A.J.R. Campbell (1940) The Heats of Solution, Heats of Formation, Specific Heats and Equilibrium Diagrams of Certain Molecular Compounds, *Journal of the American Chemical Society*, 62, 291-297.

C053   C.T. Chiou, V.H. Freed, D.W. Schmedding, and R.L. Kohnert (1977) Partition Coefficient and Bioaccumulation of Selected Organic Chemicals, *Environmental Science and Technology*, 11, 475-478.

C054   A. Csontos, I. Racz, and L. Gyarmati (1977) A Contribution to the Kinetics of Dissolution of Some Modern Drugs, *Pharmazie (Berlin)*, 32, 498-500.

C055   J.J. Conti, D.F. Othmer, and R. Gilmont (1960) Composition of Vapors from Boiling Binary Solutions, *Journal of Chemical and Engineering Data*, 5, 301-307.

C056   J.B. Conway and J.J. Norton (1951) Ternary System Furfural-Ethylene Glycol-Water, *Industrial and Engineering Chemistry*, 43, 1433-1435.

C057   P.J. Carlisle and A.A. Levine (1932) Stability of Chlorohydrocarbons. I. Methylene Chloride, *Industrial and Engineering Chemisry*, 24, 146-147.

C058   A.N. Campbell (1945) The System Aniline-Phenol-Water, *Journal of the American Chemical Society*, 67, 981-987.

C059   Z.T. Chowhan (1978) pH-Solubility Profiles of Organic Carboxylic Acids and Their Salts, *Journal of Pharmaceutical Sciences*, 67, 1257-1260.

C060   J.S. Carter and R.K. Hardy (1928) The Salting-out Effect. Influence of Electrolytes on the Solubility of m-Cresol in Water, *Journal of the Chemical Society (London)*, 131, 127-129.

C061   H.J.M. Creighton (1926) Solubility and Electrolytic Conductance of Mesitylene Phosphinous Acid, *Journal of Physical Chemistry*, 30, 1207-1208.

C062   M.J. Cho and J.J. Biermacher (1976) Water-soluble Prodrug of Metronidazole: Synthesis and Serum Hydrolysis of Metronidazole Phosphate, Technical Report, 5-10.

C064   G.A. Clarke, T.R. Williams, and R.W. Taft (1962) A Manometric Determination of the Solvolysis Rate of Gaseous t-Butyl Chloride in Aqueous Solution, *Journal of the American Chemical Society*, 84, 2292-2295.

C065   M. Chambon, J. Bouvier, and P. Duron (1937) Etude Physico-Chimique du Phenomene de Solubilisation de la Cafeine par le Benzoate de Soude, *Journal de Pharmacie et de Chimie*, 26, 216-231.

C066   E.M. Chapin and J.M. Bell (1931) The Solubility of Oxalic Acid in Aqueous Solutions of Hydrochloric Acid, *Journal of the American Chemical Society*, 53, 3284-3287.

C068   J. Coull and H.B. Hope (1935) The Ternary System Isoamyl Alcohol-Propyl Alcohol-Water, *Journal of Physical Chemistry*, 39, 967-971.

C069   L. Casale (1917) Amidi ed Imidi Tartariche. Nota III, *Gazzetta Chimica Italiana*, 47, 63-68.

C070   L. Casale (1917) Amidi ed Imidi Tartariche. Nota I, *Gazzetta Chimica Italiana*, 47, 272-285.

C071   L. Casale (1918) Amidi ed Imidi Tartariche. Nota II, *Gazzetta Chimica Italiana*, 48, 114-120.

C072   V. Cofman (1920) Sulla Preparazione dell'Acido Diiodosalicilico e la sua Solubilita nell' Acqua, *Gazzetta Chimica Italiana*, 50, 296-299.

C073   H.J.M. Creighton and D.S. Klauder, Jr. (1923) Solubility of Mannite in Mixtures of Ethyl Alcohol and Water, *Journal of the Franklin Institute*, 195, 687-691.

C074   M. Chambon, J. Bouvier, and P. Duron (1937) Etude du Systeme Cafeine-Benzoate de Sodium-Eau, *Bulletin Society of Chemistry, French*, 4, 1401-1407.

C075   L. Czerski and A. Czaplinski (1962) Solubility of Ethane in Water and NaCl and CaCl Solutions at 0 and Pressures above 1 Atmosphere, *Roczniki Chemii*, 36, 1827-1834.

C076   A.R. Collett and C.L. Lazzell (1930) Solubility Relations of the Isomeric Nitro Benzoic Acids, *Journal of Physical Chemistry*, 34, 1838-1847.

C077   J. Cohen and J.L. Lach (1963) Interaction of Pharmaceuticals with Schardinger Dextrins. I. Interaction with Hydroxybenzoic Acids and p-Hydroxybenzoates, *Journal of Pharmaceutical Sciences*, 52, 132-136.

C079　J.H. Collett and B.L. Flood (1976) Some Effects of Urea on Drug Dissolution, *Journal of Pharmacy and Pharmacology*, 28, 206-209.

C081　C. Caramella, P. Colombo, U. Conte..., and A. La Manna (1975) On the Direct Compression of Sulfamethoxydiazine Polymorphic Forms. II, *Farmaco, Edizione Pratica (PAVIA)*, 30, 496-501.

C083　E. Cohen and H. Goedhart (1931) Die Metastabilitat der Materie und deren Bedeutung fur unsere Kalorimetrischen Standarde, *Proceedings of the Koninklijke Nederlandse Akadamie van Wetenschappen*, 34, 3-14.

C086　J.L. Copp (1955) Thermodynamics of Binary Systems Containing Amines. Part 2, *Transactions of the Faraday Society*, 51, 1056-1061.

C087　O.I. Corrigan, C.A. Murphy, and R.F. Timoney (1979) Dissolution Properties of Polyethylene Glycols and Polyethylene Glycol-Drug Systems, *International Journal of Pharmaceutics*, 4, 67-74.

C088　J.L. Copp and D.H. Everett (1953) Thermodynamics of Binary Mixtures Containing Amines, *Discussions of the Faraday Society*, 15, 174-188.

C090　K.A. Connors and T.W. Rosanske (1980) trans-Cinnamic Acid-alpha-Cyclodextrin System as Studied by Solubility, Spectral, and Potentiometric Techniques, *Journal of Pharmaceutical Sciences*, 69, 173-179.

C091　C.A. Chandy and M.R. Rao (1962) Ternary Liquid Equilibria: 1-Hexanol-Water-Fatty Acids, *Journal of Chemical and Engineering Data*, 7, 473-475.

C092　A.K. Charykov and T.V. Tal'nikova (1974) pH-Metric Method of Determining the Solubility and Distribution Ratios of Some Organic Compounds in Extraction Systems, *Journal of Analytical Chemistry of the USSR*, 29, 818-822.

C093　E.D. Crittenden, Jr. and A.N. Hixson (1954) Extraction of Hydrogen Chloride from Aqueous Solutions, *Industrial and Engineering, Process Design and Development*, 46, 265-274.

C094　C.T. Chiou, L.J. Peters, and V.H. Freed (1979) A Physical Concept of Soil-Water Equilibria for Nonionic Organic Compounds, *Science*, 206, 831-832.

C095　J.H. Collett and G. Kesteven (1978) The Solubility of Allopurinol in Aqueous Solutions of Polyvinylpyrrolidone, *Drug Development and Industrial Pharmacy*, 4, 555-568.

C096　E.F. Chase and M. Kilpatrick, Jr. (1931) The Classical Dissociation Constant of Benzoic Acid and the Activity Coefficient of Molecular Benzoic Acid in Potassium Chloride Solutions, *Journal of the American Chemical Society*, 53, 2589-2597.

C097　M.E. Carlotti, M. Trotta, and M.R. Gasco (1982) Behaviour of Hematoporphyrin and Protoporphyrin with Antidepressant Drugs, *Pharmazie (Berlin)*, 37, 194-196.

C098　W.B. Crummett and R.H. Stehl (1973) Determination of Chlorinated Dibenzo-p-dioxins and Dibenzofurans in Various Materials, *Environmental Health Perspectives*, 5, 15-25.

C099　G.R. Chamlin (1946) The Chemistry of Benzene Hexachloride and Its Insecticidal Properties, *Journal of Chemical Education*, 23, 283-284.

C100　J.M. Cohen, L.J. Kamphake, A.E. Lemke..., and R.L. Woodward (1960) Effect of Fish Poisons on Water Supplies. Part 1. Removal of Toxic Materials, *Journal of the American Water Works Association*, 52, 1151-1566.

C101　J.R. Cox (1962) Triazine Derivatives as Non-selective Herbicides, *Journal of the Science of Food and Agriculture*, 13, 99-103.

C102　W.G. Clark, E.A. Strakosch, and N.I. Levitan (1942) Solubility and pH Data of Some of the Commonly Used Sulfonamides, *Journal of Laboratory and Clinical Medicine*, 28, 188-189.

C103　M.L. Crossley, E.H. Northey, and M.E. Hultquist (1940) Sulfanilamide Derivatives. V. Constitution and Properties of 2-Sulfanilamidopyridine, *Journal of the American Chemical Society*, 62, 372-374.

C104　L. Campanella, T. Ferri, and P. Mazzoni (1979) Solubility of Pyridinedicarboxylic Acids, *Journal of Inorganic and Nuclear Chemistry*, 41, 1054-1055.

C105　R.D. Carringer, J.B. Weber, and T.J. Monaco (1975) Adsorption-Desorption of Selected Pesticides by Organic Matter and Montmorillonite, *Journal of Agricultural and Food Chemistry*, 23, 568-572.

C107　D.E. Cadwallader and D.K. Madan (1981) Effect of Macromolecules on Aqueous Solubility of Cholesterol and Hormone Drugs, *Journal of Pharmaceutical Sciences*, 70, 442-446.

C108　J.M. Castaneda, F.J. Lozano, and S. Trejo (1981) Ternary Equilibrium for the System Water/Cyclohexanol/2-Ethyl-2-(hydroxymethyl)-1,3-propanediol, *Journal of Chemical and Engineering Data*, 26, 133-135.

C109　S.K. Chandrasekaran, P.S. Campbell, and A.S. Michaels (1977) Effect of Dimethyl Sulfoxide on Drug Permeation Through Human Skin, *American Institute of Chemical Engineers Journal*, 23, 810-816.

C111   C.T. Chiou, D.W. Schmedding, and J.H. Block (1981) Correlation of Water Solubility with Octanol-Water Partition Coefficient, *Journal of Pharmaceutical Sciences*, 70, 1176-1177.

C112   H. Cousse, G. Mouzin, J-P. Ribet, and J-C. Vezin (1981) Physicochemical and Analytical Characteristics of Itanoxone, *Journal of Pharmaceutical Sciences*, 70, 1245-1248.

C113   C.T. Chiou, D.W. Schmedding, and M. Manes (1982) Partitioning of Organic Compounds in Octanol-Water Systems, *Environmental Science and Technology*, 16, 4-10.

C114   W.F. Charnicki, F.A. Bacher, S.A. Freeman, and D.H. DeCesare (1959) The Pharmacy of Chlorothiazide (6-Chloro-7-sulfamyl-1,2,4-benzothiadiazine-1,1-dioxide): A New Orally Effective Diuretic Agent, *Journal of the American Pharmaceutical Association, Scientific Edition*, 48, 656-659.

C115   B.A. Cosgrove and J. Walkley (1981) Solubilities of Gases in H_2O and $2H_2O$, *Journal of Chromatography*, 216, 161-167.

C116   H. Clever, E.R. Baker, and W.R. Hale (1970) Solubility of Ethylene in Aqueous Silver Nitrate and Potassium Nitrate Solutions, *Journal of Chemical and Engineering Data*, 15, 411-413.

C117   D.E. Coffin (1967) Residues of Parathion, Methyl Parathion, EPN, and Their Oxons in Canadian Fruits and Vegetables, *Residue Reviews*, 7, 61-63.

C118   D.E. Cadwallader and H.W. Jun (1976) Nitrofurantoin, *Analytical Profiles of Drug Substances*, 5, 348-369.

C119   J. Coca and R. Diaz (1980) Extraction of Furfural from Aqueous Solutions with Chlorinated Hydrocarbons, *Journal of Chemical and Engineering Data*, 25, 80-83.

C120   C.C. Chiu and L.T. Grady (1981) Penicillamine, *Analytical Profiles of Drug Substances*, 10, 602-613.

C121   B.G. Chitwood (1952) Nematocidal Action of Halogenated Hydrocarbons, *Advances in Chemistry Series*, 7, 91-99.

C122   J.R. Coats, R.L. Metcalf, I.P. Kapoor..., and P.A. Boyle (1979) Physical-chemical and Biological Degradation Studies on DDT Analogues with Altered Aliphatic Moieties, *Journal of Agricultural and Food Chemistry*, 27, 1016-1022.

C124   M.J. Cho, R.R. Kurtz, C. Lewis..., and D.J. Houser (1982) Metronidazole Phosphate—A Water-soluble Prodrug for Parenteral Solutions of Metronidazole, *Journal of Pharmaceutical Sciences*, 71, 410-414.

C302   C.T. Chiou, P.E. Porter, and D.W. Schmedding (1983) Partition Equilibria of Nonionic Organic Compounds between Soil Organic Matter and Water, *Environmental Science and Technology*, 17, 227-231.

C303   W.L. Chiou and L.E. Kyle (1979) Differential Thermal, Solubility, and Aging Studies on Various Sources of Digoxin and Digitoxin Powder: Biopharmaceutical Implications, *Journal of Pharmaceutical Sciences*, 68, 1224-1229.

C305   C.T. Chiou (1985) Partition Coefficients of Organic Compounds in Lipid-Water Systems and Correlations with Fish Bioconcentration Factors, *Environmental Science and Technology*, 19, 57-62.

C307   R. Carlson, R. Whitaker, and A. Landskov (1978) Endothall, *Analytical Methods for Pesticides and Plant Growth Regulators*, 10, 327-340.

C309   J.M. Correa, A. Arce, A. Blanco, and A. Correa (1987) Liquid-Liquid Equilibria of the System Water + Acetic Acid + Methyl Ethyl Ketone at Several Temperatures, *Fluid Phase Equilibria*, 32, 151-162.

C310   T.S. Carswell and H.K. Nason (1938) Properties and Uses of Pentachlorophenol, *Industrial and Engineering Chemistry*, 30, 622-626.

C311   C.T. Chiou, D.E. Kile, T.I. Brinton..., and J.A. Leenheer (1987) A Comparison of Water Solubility Enhancements of Organic Solutes by Aquatic Humic Materials and Commercial Humic Acids, *Environmental Science and Technology*, 21, 1231-1234.

C313   C.T. Chiou, R.L. Macolm, T.I. Brinton, and D.E. Kile (1986) Water Solubility Enhancement of Some Organic Pollutants and Pesticides by Dissolved Humic and Fulvic Acids, *Environmental Science and Technology*, 20, 502-508.

C314   A. Chiarini and A. Tartarini (1984) pH-Solubility Relationship and Partition Coefficients for Some Anti-inflammatory Arylaliphatic Acids, *Archiv der Pharmazie*, 317, 268-273.

C315   O.I. Corrigan and R.F. Timoney (1974) Anomalous Behaviour of Some Hydroflumethiazide Crystal Samples, *Journal of Pharmacy and Pharmacology*, 26, 838-840.

C316   K.A. Connors and D.D. Pendergast (1984) Microscopic Binding Constants in Cyclodextrin Systems: Complexation of alpha-Cyclodextrin with sym-1,4-Disubstituted Benzenes, *Journal of the American Chemical Society*, 106, 7607-7614.

C317   F.T. Chou, A.H. Khan, and J.S. Driscoll (1976) Potential Central Nervous System Antitumor Agents. Aziridinylbenzoquinones, *Journal of Medicinal Chemistry*, 19, 1302-1308.

C318   G. Caron, I.H. Suffet, and T. Belton (1985) Effect of Dissolved Organic Carbon on the Environmental Distribution of Nonpolar Organic Compounds, *Chemosphere*, 14, 993-1000.

C323   J.A. Carlson, H.J. Mann, and D.M. Canafax (1983) Effect of pH on Disintegration and Dissolution of Ketoconazole Tablets, *American Journal of Hospital Pharmacy*, 40, 1334-1336.

C324   Y.E.W. Chien (1984) Solubilization of Metronidazole by Water-Miscible Multi-Cosolvents and Water-Soluble Vitamins, *Journal of Parenteral Science and Technology*, 38, 32-36.

C329   J.B. Conway and J.B. Philip (1953) Ternary System: Furfural-Methyl Isobutyl Ketone-Water at 25 C, *Industrial and Engineering Chemistry*, 45, 1083-1085.

C331   S-S. Chang, J.R. Maurey, and W.J. Pummer (1983) Solubilities of Two n-Alkanes in Various Solvents, *Journal of Chemical and Engineering Data*, 28, 187-189.

C332   M. Coates, D.W. Connell, and D.M. Barron (1985) Aqueous Solubility and Octan-1-ol to Water Partition Coefficients of Aliphatic Hydrocarbons, *Environmental Science and Technology*, 19, 628-632.

C333   J.M. Correa, A. Blanco, and A. Arce (1989) Liquid-Liquid Equilibria of the System Water + Acetic Acid + Methyl Isopropyl Ketone Between 25 and 55 C, *Journal of Chemical and Engineering Data*, 34, 415-419.

C338   K. Chijiiwa and M. Nagai (1989) Bile Salt Micelle Can Sustain More Cholesterol in the Intermicellar Aqueous Phase than the Maximal Aqueous Solubility, *Archives of Biochemistry and Biophysics*, 270, 472-477.

C340   H.K. Chan, S. Venkataram, D.J.W. Grant, and Y.E. Rahman (1991) Solid State Properties of an Oral Iron Chelator, 1,2-Dimethyl-3-hydroxy-4-pyridone, and Its Acetic Acid Solvate. I. Physicochemical Characterization, Intrinsic Dissolution Rate, and Solution Thermodynamics, *Journal of Pharmaceutical Sciences*, 80, 677-685.

C342   E.M. Chellquist and W.G. Gorman (1992) Benzoyl Peroxide Solubility and Stability in Hydric Solvents, *Pharmaceutical Research*, 9, 1341-1345.

C346   S. Cohen, Y. Marcus, Y. Migron..., and A. Shafran (1993) Water Sorption, Binding and Solubility of Polyols, *Journal of the Chemical Society, Faraday Transactions* 1, 89, 3271-3275.

C347   C-C. Chen, Y. Zhu, and L.B. Evans (1989) Phase Partitioning of Biomolecules: Solubilities of Amino Acids, *Biotechnology Progress*, 5, 111-118.

C348   S.D. Chamberlain, A.R. Moorman, L.A. Jones..., and T.A. Krenitsky (1992) 2'-Ester Prodrugs of the Varicella-zoster Antiviral Agent, 6-methoxypurine arabinoside, *Antiviral Chemistry and Chemotherapy*, 3, 371-378.

C349   H. Chen and J. Wagner (1994) An Apparatus and Procedure for Measuring Mutual Solubilities of Hydrocarbons + Water: Benzene + Water from 303 to 373 K, *Journal of Chemical and Engineering Data*, 39, 470-474.

C350   H. Chen and J. Wagner (1994) Mutual Solubilities of Alkylbenzene + Water Systems at Temperatures from 303 to 373 K: Ethylbenzene, p-Xylene, 1,3,5-Trimethylbenzene, and Butyl-benzene, *Journal of Chemical and Engineering Data*, 39, 679-684.

D001   N.C. Deno and H.E. Berkheimer (1960) Phase Equilibria Molecular Transport Thermodynamics, *Journal of Chemical and Engineering Data*, 5, 1-5.

D002   M.R. Dixon, C.E. Rehberg, and C.H. Fisher (1948) Preparation and Physical Properties of n-Alkyl beta-Ethoxypropionates, *Journal of the American Chemical Society*, 70, 3733-3738.

D003   W.W. Davis, M.E. Krahl, and G.H.A. Clowes (1942) Solubility of Carcinogenic and Related Hydrocarbons in Water, *Journal of the American Chemical Society*, 64, 108-110.

D004   J.A. Dean (1973) Physical Constants of Alkaloids, 394-417.

D005   M.S. Dunn, M.P. Stoddard, L.B. Rubin, and R.C. Bovie (1943) Investigations of Amino Acids and Peptides, *Journal of Biological Chemistry*, 151, 241-258.

D006   J.C. Duff and E.J. Bills (1930) The Solubilities of Nitrophenols in Aqueous Ethyl-alcoholic Solutions, *Journal of the Chemical Society (London)*, 1331-1338.

D007   N.C. Deno and C. Perizzolo (1957) The Application of Activity Coefficient Data to the Relations Between Kinetics and Acidity Functions, *Journal of the American Chemical Society*, 79, 1345-1348.

D008   M. Donbrow and H. Ben-Shalom (1967) Molecular Interactions of Caffeine with o-, m-, and p-Iodobenzoic Acids and o-, m-, and p-Fluorobenzoic Acids, *Journal of Pharmacy and Pharmacology*, 19, 495-501.

D009   L.W. Dittert, H.C. Caldwell, T. Ellison..., and J.V. Swintosky (1968) Carbonate Ester Prodrugs of Salicylic Acid, *Journal of Pharmaceutical Sciences*, 57, 828-831.

D010   P.P. DeLuca, L. Lachman, and H.G. Schroeder (1973) Physical-chemical Properties of Substituted Amides in Aqueous Solution and Evaluation of their Potential Use as Solubilizing Agents, *Journal of Pharmaceutical Sciences*, 62, 1320-1327.

D011   P.P. DeLuca and L. Lachman (1965) Lyophilization of Pharmaceuticals. I. Effect of Certain Physical-Chemical Properties, *Journal of Pharmaceutical Sciences*, 54, 617-624.

D012   M. Donbrow, E. Touitou, and H. Ben-Shalom (1976) Stability of Salicylamide-Caffeine Complex at Different Temperatures and its Thermodynamic Parameters, *Journal of Pharmacy and Pharmacology*, 28, 766-769.

D013   I. Dunstan, J.V. Griffiths, and S.A. Harvey (1965) Nitric Esters. Part I. Characterisation of the Isomeric Glycerol Dinitrates, *Journal of the Chemical Society (London)*, 1319-1324.

D014   L. Drobnica, M. Zemanova, P. Nemec..., and P. Nemec, Jr. (1967) Antifungal Activity of Isothiocyanates and Related Compounds, *Applied Microbiology*, 15, 701-709.

D015   N.A. Daabis, S.A. Khalil, and V.F. Naggar (1976) The Effect of Urea, Amidopyrine, Phenazone and Paracetamol on the Solubility of Some Sparingly Soluble Antirheumatics, *Canadian Journal of Pharmaceutical Science*, 11, 114-117.

D016   J.B. Dalton and C.L.A. Schmidt (1933) The Solubilities of Certain Amino Acids in Water, the Densities of Their Solutions at Twenty-five Degrees, and the Calculated Heats of Solution and Partial Molal Volumes, *Journal of Biological Chemistry*, 103, 549-575.

D017   J.B. Dalton and C.L.A. Schmidt (1935) The Solubilities of Certain Amino Acids and Related Compounds in Water, the Densities of Their Solutions at Twenty-five Degrees, and the Calculated Heats of Solution and Partial Molal Volumes. II, *Journal of Biological Chemistry*, 109, 241-248.

D018   M.S. Dunn, F.J. Ross, and L.S. Read (1933) The Solubility of the Amino Acids in Water, *Journal of Biological Chemistry*, 103, 579-595.

D019   L. Drobnica and J. Augustin (1965) Reaction of Isothiocyanates with Amino Acids, Peptides and Proteins. I. Kinetics of the Reaction of Aromatic Isothiocyanates with Glycine, *Collection of Czechoslovak Chemical Communications*, 30, 99-105.

D020   L.H. Dalman (1937) Ternary Systems of Urea and Acids. IV. Urea, Citric Acid and Water. V. Urea, Acetic Acid and Water. VI. Urea, Tartaric Acid and Water, *Journal of the American Chemical Society*, 59, 775-779.

D021   W.M. Dehn (1917) Comparative Solubilities in Water, in Pyridine and in Aqueous Pyridine, *Journal of the American Chemical Society*, 39, 1399-1404.

D022   T.L. Davis, A.A. Ashdown, and H.R. Couch (1925) Two Forms of Nitroguanidine, *Journal of the American Chemical Society*, 47, 1063-1066.

D025   R.D. Daley (1973) Primidone, *Analytical Profiles of Drug Substances*, 2, 409-421.

D026   R.E. Dempski, J.B. Portnoff, and A.W. Wase (1969) In Vitro Release and In Vitro Penetration Studies of a Topical Steroid from Nonaqueous Vehicles, *Journal of Pharmaceutical Sciences*, 58, 579-582.

D027   L. Desvergnes (1929) Sur Quelques Proprietes Physiques de Certains Derives Nitres, *The Reviews of Chemical Industry*, 38, 265-266.

D029   L.W. Dittert, H.C. Caldwell, H.J. Adams..., and J.V. Swintosky (1968) Acetaminophen Prodrugs. I. Synthesis, Physicochemical Properties, and Analgesic Activity, *Journal of Pharmaceutical Sciences*, 57, 774-780.

D031   R.J. Daniel and W. Doran (1926) Some Chemical Constituents of the Mussel (Mytilus Edulis), *Biochemical Journal*, 20, 676-684.

D033   T.C. Daniels and R.E. Lyons (1931) Concerning the Physical Properties of Solutions of Certain Phenyl-substituted Acids in Relation to their Bactericidal Power, *Journal of Physical Chemistry*, 35, 2049-2060.

D034   M. Dworkin and K.H. Keller (1977) Solubility and Diffusion Coefficient of Adenosine 3':5'-Monophosphate, *Journal of Biological Chemistry*, 252, 864-865.

D035   S.J. Desai (1966) Quantitative Mechanistic Studies of Drug Release from Inert Matrices, Ph.D. thesis, 80-80.

D036   K.H. Dooley and F.J. Castellino (1972) Solubility of Amino Acids in Aqueous Guanidinium Thiocyanate Solutions, *Biochemistry*, 11, 1870-1874.

D037   M. Donbrow and C.T. Rhodes (1964) Potentiometric Studies on Solubilisation in Non-ionic Micellar Solutions. Part I. Interpretation of pH Changes and Mechanism of Solubilisation of Benzoic Acid, *Journal of the Chemical Society (London)*, 6166-6171.

D038   J.B. Dalton and C.L.A. Schmidt (1936) The Solubility of d-Valine in Water, *Journal of General Physiology*, 19, 767-771.

D039   L.H. Dalman (1937) The Solubility of Citric and Tartaric Acids in Water, *Journal of the American Chemical Society*, 59, 2547-2549.

D040   R. De Santis, L. Marrelli, and P.N. Muscetta (1976) Influence of Temperature on the Liquid-Liquid Equilibrium of the Water-n-Butyl Alcohol-Sodium Chloride System, *Journal of Chemical and Engineering Data*, 21, 324-327.

D041   R.M.C. Dawson, D.C. Elliott, W.H. Elliott, and K.M. Jones (1969) Data for Biochemical Research, 1, 1, Oxford.

D043   W.W. Davis and T.V. Parke, Jr. (1942) A Nephelometric Method for Determination of Solubilities of Extremely Low Order, *Journal of the American Chemical Society*, 64, 101-107.

D044   J.C. Dearden, J.H. Collett, and E. Tomlinson (1976) In Vitro Dissolution Rate as a Parameter in Structure-activity Studies, *Experientia*, 23, 37-40.

D046   H.S. Davis and O.F. Wiedeman (1945) Physical Properties of Acrylonitrile, *Industrial and Engineering Chemistry*, 37, 482-485.

D047   R. Durand (1948) Recherches sur l'Hydrotropie. Etude de la Solubilite de l'Heptane, de l'Hexane et du Cyclohexane dans les Solutions Aqueuses de Quelques sels d'Acides Gras, *Comptes Rendus Hebdomadaires des Seances de l'Academie des Sciences*, 226, 409-410.

D049   H. Druckrey, R. Preussmann, N. Nashed, and S. Ivankovic (1966) Carcinogene Alkylierende Substanzen. I. Dimethylsulfat, Carcinogene Wirkung an Ratten und Wahrscheinliche Ursache von Berufskrebs, *Zeitschrift fuer Krebsforschung*, 68, 103-111.

D050   P.G. Desai and A.M. Patel (1935) Effect of Polarity on the Solubilities of some Organic Acids, *Journal of the Indian Chemical Society*, 12, 131-136.

D051   J. Desbarres and H.O. El Sayed (1977) Determination des pK et des Solubilites d'Acides peu Solubles (Acide Pipemidique, Acide Undecanoique et Derives), *Comptes Rendus Hebdomadaires des Seances de l'Academie des Sciences, Serie C: Sciences Chimiques*, 285, 431-434.

D052   A.K. Doolittle (1935) Lacquer Solvents in Commercial Use, *Industrial and Engineering Chemistry*, 27, 1169-1179.

D055   R.A. Dawe (1965) Thesis, Ph.D. thesis.

D056   D.L. Dyer (1959) The Effect of pH on Solubilization of Weak Acids and Bases, *Journal of Colloid Science*, 14, 640-645.

D058   C. Drucker (1929) Experimentelle Beitrage zur Frage der Elektrolytischen Dissoziation, *Monatshefte fuer Chemie*, 53, 62-68.

D059   J.C. Duff (1929) The Solubilities of o- and p-Nitrophenols in Aqueous Methyl-alcoholic Solutions at 25 and 40 Degrees. Formation of beta-p-Nitrophenol, *Journal of the Chemical Society (London)*, 2789-2796.

D060   L.H. Dalman (1934) Ternary Systems of Urea and Acids. I. Urea, Nitric Acid and Water. II. Urea, Sulfuric Acid and Water. III. Urea, Oxalic Acid and Water, *Journal of the American Chemical Society*, 56, 549-553.

D061   S.S. Doosaj and W.V. Bhagwat (1933) Solubilities of Weak Acids in Salts of Weak Acids at Very High Concentrations, *Journal of the Indian Chemical Society*, 10, 225-232.

D062   H.R. Dittmar (1930) The Decomposition of Malic Acid by Sulfuric Acid, *Journal of the American Chemical Society*, 52, 2746-2754.

D063   F. Drouillon (1925) Etude du Melange Ternaire: Eau, Alcool Ethylique, Alcool Butylique Normal, *Journal de Chimie Physique et de Physico-Chimie Biologique*, 22, 149-160.

D064   S. De Brouwer (1930) Sur l'Acide Orthotrifluortoluique et le Nitrotrifluorcresol 1-3-6, *Bulletin des Societes Chimiques Belges*, 39, 298-308.

D065   L. Desvergnes (1924) Sur la Solubilite du 2-4-6-Trinitrotoluene du Tetryl et de la Tetranitraniline dans les Solvants Organiques, *Moniteur Scientifique*, 14, 121-130.

D066   L. Desvergnes (1931) Le 1-3-5-Trinitrobenzene ou Benzite, *Chimica e l'Industria (Milan)*, 25, 3-16.

D067   L. Desvergnes (1926) Sur Quelques Proprietes Physiques des Derives Nitres: l'acide 2-4-6-Trinitrobenzoique, *Moniteur Scientifique*, 16, 201-208.

D068   M.A. Doucet (1923) Travaux Originaux-De l'Action de l'Iode sur Quelques Semi-carbazides Substituees en (1); Application a leur Dosage, *Journal de Pharmacie et de Chimie*, 27, 361-365.

D069   L. Desvergnes (1927) Sur Quelques Proprietes Physiques des Nitrophenols-Orthonitrophenol, *The Reviews of Chemical Industry*, 36, 194-196.

D070   L. Desvergnes (1925) Sur Quelques Proprietes Physiques des Derives Nitres: 1-3-Dinitrobenzene, *Moniteur Scientifique*, 15, 149-158.

D071   L. Desvergnes (1925) Sur Quelques Proprietes Physiques des Derives Nitres: 4-Nitro-4-Chlorobenzene, *Moniteur Scientifique*, 15, 73-78.

D072   J. Duclaux and A. Durand-Gasselin (1938) Les Perchlorates et la Serie Lyotrope. II, *Journal de Chimie Physique et de Physico-Chimie Biologique*, 35, 189-192.

D073   M. De Groote (1920) The Solubility of Vanillin and Coumarin in Glycerine Solutions, *American Perfumer (APRFA)*, 15, 372-374.

D077   J.H. Dolinski (1905) Ueber die Loslichkeit einiger Organischer Verbindungen in Wasser bei Verschiedenen Temperaturen, *Berichte der Deutschen Chemischen Gesellschaft*, 38, 1835-1837.

D078   J.C. Dearden and N.C. Patel (1978) Dissolution Kinetics of some Alkyl Derivatives of Acetaminophen, *Drug Development and Industrial Pharmacy*, 4, 529-535.

D079   L. Desvergnes (1924) Sur Quelques Proprietes Physiques des Derives Nitres: 1-2-3 Dinitranisol, *Moniteur Scientifique*, 14, 249-257.

D080   L. Desvergnes (1927) Sur Quelques Proprietes Physiques des Nitrophenols-2.6-Dinitrophenol, *The Reviews of Chemical Industry*, 36, 224-226.

D081   M. Dymicky and C.N. Huhtanen (1979) Inhibition of Clostridium Botulinum by p-Hydroxybenzoic Acid n-Alkyl Esters, *Antimicrobial Agents & Chemotherapy*, 15, 798-801.

D082   M. Draguet-Brughmans, M. Azibi, and R. Bouche (1979) Solubilite et Vitesse de Dissolution du Meprobamate: Des cas Significatifs, *Journa de Pharmacie de Belgique*, 34, 267-271.

D083   H. Druckrey, R. Preussmann, and S. Ivankovic (1967) Organotrope Carcinogene Wirkungen bei 65 Verschiedenen N-Nitroso-verbindungen an BD-Ratten, *Zeitschrift fuer Krebsforschung*, 69, 103-201.

D084   P. Durel and M. Allinne (1941) Sur la Precipitation des Produits Sulfamides dans l'Urine, *Bulletins et Memoires de la Societe Medicale des Hopitaux de Paris*, 251-259.

D085   R.N. Dexter and S.P. Pavlou (1978) Mass Solubility and Aqueous Activity Coefficients of Stable Organic Chemicals in the Marine Environment: Polychlorinated Biphenyls, *Marine Chemistry*, 6, 41-53.

D086   P.T. DeLassus and D.D. Schmidt (1981) Solubilities of Vinyl Chloride and Vinylidene Chloride in Water, *Journal of Chemical and Engineering Data*, 26, 274-276.

D087   A.V. Dolgorev, Y.G. Lysak, Y.F. Zibarova, and A.P. Lukoyanov (1980) Dithiopyrylmethane and Its Analogs as Analytical Reagents. Synthesis and Properties, *Journal of Analytical Chemistry of the USSR*, 35, 560-567.

D088   M. Donbrow and P. Sax (1982) Thermodynamic Parameters of Molecular Complexes in Aqueous Solution: Enthalpy-entropy Compensation in a Series of Complexes of Caffeine with beta-Naphthoxyacetic Acid and Drug-related Aromatic Compounds, *Journal of Pharmacy and Pharmacology*, 34, 215-224.

D089   W.A.L. David, R.L. Metcalf, and M. Winton (1960) The Systemic Insecticidal Properties of Certain Carbamates, *Journal of Economic Entomology*, 53, 1021-1025.

D091   II.P. Deppeler (1981) Hydrochlorothiazide, *Analytical Profiles of Drug Substances*, 10, 406-423.

D092   M. Draguet-Brughmans, P. Draux, and R. Bouche (1981) Polymorphisme du Butobarbital, *Journa de Pharmacie de Belgique*, 36, 397-403.

D093   A. Datyner (1978) The Solubilization of Nonionic Dyestuffs at Elevated Temperatures in Aqueous Solutions of Sodium Dodecyl Sulfate, *Journal of Colloid and Interface Science*, 65, 527-532.

D300   C. David, E. Szalai, and D. Baeyens-Volant (1982) Photophysical Processes in Cyanobiphenyl Derivatives. II. Cyanobiphenyl Derivatives as Fluorescent Probes in Micellar Environment, *Berichte der Bunsengesellschaft fuer Physikalische Chemie*, 86, 710-716.

D302   L.L. Dunham and W.W. Miller (1978) Methoprene, *Analytical Methods for Pesticides and Plant Growth Regulators*, 10, 95-109.

D303   R. Darskus and D. Eichler (1978) Triforine, *Analytical Methods for Pesticides and Plant Growth Regulators*, 10, 243-253.

D304   E.W. Day (1978) Ethalfluralin, *Analytical Methods for Pesticides and Plant Growth Regulators*, 10, 341-352.

D305   E.A. Dietz, Jr. and L.O. Moore (1978) Monomethylarsonic Acid, Cacodylic Acid, and their Sodium Salts, *Analytical Methods for Pesticides and Plant Growth Regulators*, 10, 385-401.

D306   F.M. Dunnivant and A.W. Elzerman (1988) Aqueous Solubility and Henry's Law Constant Data for PCB Congeners for Evaluation of Quantitative Structure-property Relationships (QSPRs), *Chemosphere*, 17, 525-541.

D307   H. DeVoe and S.P. Wasik (1984) Aqueous Solubilities and Enthalpies of Solution of Adenine and Guanine, *Journal of Solution Chemistry*, 13, 51-61.

D308   R. Dahlan, C. McDonald, and V.B. Sunderland (1987) Solubilities and Intrinsic Dissolution Rates of Sulphamethoxazole and Trimethoprim, *Journal of Pharmacy and Pharmacology*, 39, 246-251.

D311   M. Dahlund and A. Olin (1987) Chemical Equilibria in Aqueous Solutions of Olsalazine, 3,3'-azo-Bis(6-hydroxybenzoic Acid), *Acta Pharmaceutica Suecica*, 24, 219-232.

D315   E. DeRitter (1982) Vitamins in Pharmaceutical Formulations, *Journal of Pharmaceutical Sciences*, 71, 1073-1075.

D316   (1984) Dimethyl-beta-cyclodextrin, *Drugs of the Future*, 9, 576-578.

D319   (1983) Triglycidulurazol, *Drugs of the Future*, 9, 209-210.

D321   (1983) Flupirtine, *Drugs of the Future*, 8, 773-775.

D330   W.J. Doucette and A.W. Andren (1988) Aqueous Solubility of Selected Biphenyl, Furan, and Dioxin Congeners, *Chemosphere*, 17, 243-252.

D331   R.M. Dickhut, A.W. Andren, and D.E. Armstrong (1986) Aqueous Solubilities of Six Polychlorinated Biphenyl Congeners at Four Temperatures, *Environmental Science and Technology*, 20, 807-810.

D332   R.R. Davison and W.H. Smith (1969) Vapor-liquid Equilibrium of N-Ethyl-n-butylamine-Water and N-Ethyl-sec-butylamine-Water, *Journal of Chemical and Engineering Data*, 14, 296-298.

D335   R.M. Dickhut (1985) Dissertation or Masters Thesis.

D336   D.L. DeFoe, G.W. Holcombe, D.E. Hammermeister, and K.E. Biesinger (1990) Solubility and Toxicity of Eight Phthalate Esters to Four Aquatic Organisms, *Environmental Toxicology and Chemistry*, 9, 623-636.

D337   R.M. Dickhut, A.W. Andren, and D.E. Armstrong (1989) Naphthalene Solubility in Selected Organic Solvent/Water Mixtures, *Journal of Chemical and Engineering Data*, 34, 438-443.

D339   G. Dempsey and P. Molyneux (1992) Quantitative Investigations of Amino Acids and Peptides. IV. The Solubilities of the Amino Acids in Water-Ethyl Alcohol Mixtures, *Journal of the Chemical Society*, *Faraday Transactions* 1, 88, 971-977.

D340   J.H. De Smidt, J.C.A. Offringa, and D.J.A. Crommelin (1991) Dissolution Rate of Griseofulvin in Bile Salt Solutions, *Journal of Pharmaceutical Sciences*, 80, 399-401.

D341   G.W. Denton (1991) Chlohexidine, 274-275.

D343   U. Dramur and B. Tatli (1993) Liquid-Liquid Equilibria of Water + Acetic Acid + Phthalic Esters (Dimethyl Phthalate and Diethyl Phthalate) Ternaries, *Journal of Chemical and Engineering Data*, 38, 23-25.

D344   D. Dumanovic, J. Jovanovic, S. Popovic, and D. Kosanovic (1986) The Solubility of some 4(5)- and 5-nitroimidazoles in Water and Twenty Common Organic Solvents, *Journal of the Serbian Chemical Society*, 51, 411-416.

D345   S.K. Dwivedi, S. Sattari, F. Jamali, and A.G. Mitchell (1992) Ibuprofen racemate and enantiomers: Phase diagram, solubility and thermodynamic studies, *International Journal of Pharmaceutics*, 87, 95-104.

D346   C.L. De Ligny and N.G. Van Der Veen (1971) Solubilities of Some Tetra-Alkyl-Carbon, -Silicon, -Germanium and -Tin Compounds in Mixtures of Water with Methanol, Ethanol, Dioxane, Acetone and Acetic Acid and Differences Between the Standard Chemical Potentials of these Solutes.., *Recueil des Travaux Chimiques des Pays-Bas*, 90, 984-1009.

D347   A. Darwish, A.T. Florence, and A.M. Saleh (1989) Effects of Hydrotropic Agents on the Solubility, Precipitation, and Protein Binding of Etoposide, *Journal of Pharmaceutical Sciences*, 78, 577-581.

D348   A.H. Demond and A.S. Lindner (1993) Estimation of Interfacial Tension between Organic Liquids and Water, *Environmental Science and Technology*, 27, 2318-2331.

D349   B.P. Dey and S.C. Lahiri (1988) Solubilities of Amino Acids in Methanol + Water Mixtures at Different Temperatures, *Indian Journal of Chemistry*, 27A, 297-302.

D350   R. Deghenghi, F. Boutignon, P. Wuthrich, and V. Lenaerts (1993) Antarelix (EP 24332) a Novel Water Soluble LHRH Antagonist, *Biomedical and Pharmacotherapy*, 47, 107-110.

D351   R.M. Dickhut, K.E. Miller, and A.W. Andren (1994) Evaluation of Total Molecular Surface Area for Predicting Air-Water Partitioning Properties of Hydrophobic Aromatic Chemicals, *Chemosphere*, 29, 283-297.

E002   L.V. Erichsen (1952) Das Loslichkeitsdekrement der Methylengruppe und die Funktionsloslichkeit in Homologen Reihen, *Naturwissenschaften*, 39, 189-189.

E003   P.H. Elworthy and H.E.C. Worthington (1968) The Solubility of Sulphadiazine in Water-Dimethyl-Formamide Mixtures, *Journal of Pharmacy and Pharmacology*, 20, 830-835.

E004   R.P. Eganhouse and J.A. Calder (1976) The Solubility of Medium Molecular Weight Aromatic Hydrocarbons and the Effects of Hydrocarbon Co-solutes and Salinity, *Geochimica et Cosmochimica Acta*, 40, 555-561.

E005   D.N. Eggenberger, F.K. Broome, A.W. Ralston, and H.J. Harwood (1949) The Solubilities of the Normal Saturated Fatty Acids in Water, *Journal of Organic Chemistry*, 14, 1108-1110.

E008   P. Ekwall, T. Rosendahl, and A. Sten (1958) Solubility of Bile Acids, *Acta Chemica Scandinavica*, 12, 1622.

E009   J. Eisenbrand and K. Baumann (1970) Uber die Bestimmung der Wasserloslichkeit von Coronen, Fluoranthen, Perylen, Picen, Tetracen, und Triphenylen und uber die Bildung Wasserloslicher Komplexe dieser Kohlenwasserstoffe mit Coffein, *Zeitschrift fuer Lebensmittel-Untersuchung und -Forschung*, 144, 312-317.

E010   P.H. Elworthy and F.J. Lipscomb (1968) A Note on the Solubility of Griseofulvin, *Journal of Pharmacy and Pharmacology*, 20, 790-792.

E011   T.D. Edmonson and J.E. Goyan (1958) The Effect of Hydrogen Ion and Alcohol Concentration on the Solubility of Phenobarbital, *Journal of the American Pharmaceutical Association, Scientific Edition*, 47, 810-812.

E012   H.A.M. El-Shibini, S. Abd-Elfattah, and M.M. Motawi (1972) Die Solubilisation des Khellins durch Coffein und Dihydroxypropyltheophyllin, *Pharmazie (Berlin)*, 27, 570-573.

E014   K. Eik-Nes, J.A. Schellman, R. Lumry, and L.T. Samuels (1954) The Binding of Steroids to Protein. I. Solubility Determinations, *Journal of Biological Chemistry*, 206, 411-419.

E015   A. England, Jr. and E.J. Cohn (1935) Studies in the Physical Chemistry of Amino Acids, Peptides and Related Substances. IV. The Distribution Coefficients of Amino Acids between Water and Butyl Alcohol, *Journal of the American Chemical Society*, 57, 634-637.

E016   W.O. Emery and C.D. Wright (1921) Distribution of Certain Drugs Between Immiscible Solvents, *Journal of the American Chemical Society*, 43, 2323-2335.

E017   L.J. Edwards (1951) The Dissolution and Diffusion of Aspirin in Aqueous Media, *Transactions of the Faraday Society*, 47, 1191-1210.

E018   N.A. El-Gindy and F. El-Khawas (1977) Solubility and Dissolution Enhancement of Riboflavine by Solid Dispersion Systems, *Pharmazeutische Industrie*, 39, 84-86.

E019   T.W. Evans (1936) The Hill Method for Solubility Determinations, *Industrial and Engineering Chemistry, Analytical Edition*, 8, 206-208.

E022   E.I. Eger II, R. Shargel, and G. Merkel (1963) Solubility of Diethyl Ether in Water, Blood and Oil, *Anesthesiology*, 24, 676-678.

E025   J. Eisenbrand and K. Baumann (1969) Uber die Bestimmung der Wasserloslichkeit von Benzol, Naphthalin, Anthracen und Pyren und uber die Bildung Wasserloslicher Komplexe dieser Kohlenwasserstoffe mit Coffein, *Zeitschrift fuer Lebensmittel-Untersuchung und -Forschung*, 140, 210-216.

E028   B.K. Evans, K.C. James, and D.K. Luscombe (1978) Quantitative Structure—Activity Relationships and Carminative Activity, *Journal of Pharmaceutical Sciences*, 67, 277-278.

E029   L.V. Erichsen (1952) Die Kritischen Losungstemperaturen in der Homologen Reihe der Primaren Normalen Alkohole, *Brennstoff-Chemie*, 33, 166-172.

E031   J. Eisenbrand and H. Picher (1938) Bestimmung der Dissoziationskonstanten, Loslichkeiten und Verteilungskoeffizienten von Pantokain- und Novokainbase, *Archiv der Pharmazie und Berichte der Deutschen Pharmazeutischen Gesellschaft*, 276, 1-17.

E032   N.N. Efremov (1940) On the Solubility in Water of the Nitro Derivatives of Phenol and of the Dihydroxybenzenes, *Bulletin Academic Science USSR Division of Chemical Science*, 1-29.

E033   S.A. El-Fattah and N.A. Daabis (1977) The Effect of Dihydroxypropyl Theophylline on the Solubility and Stability of Menadione (Vitamin K3), *Pharmazie (Berlin)*, 32, 232-234.

E035   Emich (1882) Loslichkeit der Glycocholsaure, *Monatshefte fuer Chemie*, 3, 336-340.

E036   J.C. Earle (1918) Notes on Cineol, *Journal of the Society of Chemical Industry*, London, 37, 274-274.

E037   W.V. Evans and M.B. Aylesworth (1926) Some Critical Constants of Furfural, *Industrial and Engineering Chemistry*, 18, 24-27.

E039   Y.V. Efremov and I.F. Golubev (1962) The Solubility of omega-Aminoundecanoic Acid in Aqueous Alcoholic Solutions, *Russian Journal of Physical Chemistry*, 36, 516-516.

E041   A.R. Ebian and N.A. El-Gindy (1978) Codeine Crystal Forms. I. Preparation, Identification, and Characterization, *Scientia Pharmaceutica*, 46, 1-7.

E044   S. El Gamal, N. Borie, and Y. Hammouda (1978) The Influence of Urea, Polyethylene Glycol 6000 and Polyvinyl Pyrrolidone on the Dissolution Properties of Nitrofurantoin, *Pharmazeutische Industrie*, 40, 1373-1376.

E045   M. Eisenberg, P. Chang, C.W. Tobias, and C.R. Wilke (1955) Physical Properties of Organic Acids, *American Institute of Chemical Engineers Journal*, 1, 558-558.

E046   L.J. Edwards (1953) Salicylamide: Thermodynamic Dissociation Constant. Solubility and Quantitative Estimation by U.V. Absorption Spectrophotometry, *Transactions of the Faraday Society*, 49, 234-236.

E047   H.M. El-Banna and O.Y. Abdallah (1980) Physicochemical and Dissolution Studies of Phenylbutazone Binary Systems, *Pharmaceutica Acta Helvetiae*, 55, 256-260.

E048   H. Ellgehausen, C. D'Hondt, and R. Fuerer (1981) Reversed-phase Chromatography as a General Method for Determining Octan-1-ol/Water Partition Coefficients, *Pesticide Science*, 12, 219-227.

E049   Z.A. El Gholmy (1979) Effect of Urea and Sodium Chloride on the Aqueous Solubility of Acetazolamide, Hydrochlorothiazide and Frusemide, *Journal of Drug Research*, 11, 181-189.

E050   D. Eichler (1972) Bromophos and Bromophos-Ethyl Residues, *Residue Reviews*, 41, 65-67.

E051   K. Embil and G. Torosian (1982) Solubility and Ionization Characteristics of Doxepin and Desmethyldoxepin, *Journal of Pharmaceutical Sciences*, 71, 191-193.

E052   A.E.M. El-Nimr, S.M. Omar, and M.A. Kassem (1980) Effect of Bile Salt-polyvinylpyrrolidone Complexes on the Solubilization Profiles of Diethylstilbestrol and Digoxin, *Pharmazeutische Industrie*, 42, 311-314.

E301   R.P. Enever, K. Fotherby, S. Naderi, and G.A. Lewis (1983) Long-acting Contraceptive Agents: The Influence of Physico-chemical Properties of Some Esters of Norethisterone upon the Plasma Levels of Free Norethisterone, *Steroids, An International Journal (San Francisco)*, 41, 381-396.

E305   A.A. El-Harakany and A.O. Barakat (1985) Solubility of Tris-(hydroxymethyl)-aminomethane in Water-2-Methoxyethanol Solvent Mixtures and the Solvent Effect on the Dissociation of the Protonated Base, *Journal of Solution Chemistry*, 14, 263-269.

E307   P. Ekwall, L. Sjoblom, and J. Olsen (1953) The Spectrophotometric Determination of Steroid Hormones Solubilized in Aqueous Solutions of Association Colloids, *Acta Chemica Scandinavica*, 7, 347-351.

E308   J.W. Eckert (1962) Fungistatic and Phytotoxic Properties of Some Derivatives of Nitrobenzene, *Phytopathology*, 52, 642-649.

E311   H. Egawa, S. Maeda, E. Yonemochi..., and Y. Nakai (1992) Solubility Parameter and Dissolution Behavior of Cefalexin Powders with Different Crystallinity, *Chemical and Pharmaceutical Bulletin*, 40, 819-820.

E312   G.M. El-Mahrouk, S.Y. Amin, and R.A. Shoukry (1991) Complexation of Khellin with Different Cyclodextrins, *Journal of Drug Research. Egypt*, 20, 91-101.

E314   J.B. Escalera, P. Bustamante, and A. Martin (1994) Predicting the Solubility of Drugs in Solvent Mixtures: Multiple Solubility Maxima and the Chameleonic Effect, *Journal of Pharmacy and Pharmacology*, 46, 172-176.

E316   A.A. Elamin, C. Ahlneck, G. Alderborn, and C. Nystrom (1994) Increased Metastable Solubility of Milled Griseofulvin, Depending on the Formation of a Disordered Surface Structure, *International Journal of Pharmaceutics*, 111, 159-170.

F001   H. Fuhner (1924) Die Wasserloslichkeit in Homologen Reihen, *Berichte der Deutschen Chemischen Gesellschaft*, 57, 510-515.

F002   F. Franks, M. Gent, and H.H. Johnson (1963) The Solubility of Benzene in Water, *Journal of the Chemical Society (London)*, 2716-2723.

F003   G.L. Flynn, R.W. Smith, and S.H. Yalkowsky (1972) Solubility of Hydrophobic Species in Aqueous Systems. I. Solubility of U-34,865 in Propylene Glycol-Water Mixtures as a Function of Solvent Composition and Temperature, Technical Report.

F004   F. Franks (1966) Solute-Water Interactions and the Solubility Behaviour of Long-Chain Paraffin Hydrocarbons, *Nature (London)*, 210, 87-88.

F005   S. Feldman and M. Gibaldi (1967) Effect of Urea on Solubility—Role of Water Structure, *Journal of Pharmaceutical Sciences*, 56, 370-375.

F006   G.L. Flynn and S.H. Yalkowsky (1972) Correlation and Prediction of Mass Transport Across Membranes. I. Influence of Alkyl Chain Length on Flux-determining Properties of Barrier and Diffusant, *Journal of Pharmaceutical Sciences*, 61, 838-852.

F007   A.T. Florence and A. Rahman (1975) Polyvinylpyrrolidones and Their Influence on the Dissolution Rates of Compounds of Varying Aqueous Solubilities, *Journal of Pharmacy and Pharmacology*, 27, 55-55.

F008   B. Farhadieh, S. Borodkin, and J.D. Buddenhagen (1971) Drug Release from Methyl Acrylate-Methyl Methacrylate Copolymer Matrix. I. Kinetics of Release, *Journal of Pharmaceutical Sciences*, 60, 209-212.

F009   A. Fritz, J.L. Lach, and L.D. Bighley (1971) Solubility Analysis of Multicomponent Systems Capable of Interacting in Solution, *Journal of Pharmaceutical Sciences*, 60, 1617-1619.

F010   A.T. Florence and E.G. Salole (1976) Changes in Crystallinity and Solubility on Comminution of Digoxin and Observations on Spironolactone and Oestradiol, *Journal of Pharmacy and Pharmacology*, 28, 637-642.

F011   A.T. Florence, A.W. Jenkins, and A.H. Loveless (1976) Effect of Formulation of Intramuscular Injections of Phenothiazines on Duration of Activity, *Journal of Pharmaceutical Sciences*, 65, 1665-1668.

F012   E.A. Fedorova, L.F. Shashkina, and V.K. Fedorov (1976) On the Method of Determination of the Solubility of Derivative Testosterone, *Khimiko-Farmatsevticheskii Zhurnal*, 10, 139-142.

F013   E. Felder, D. Pitre, and M. Grandi (1976) Radiopaque Contrast Media. XXXV. Physical Properties of Iopronic Acid, a New Oral Cholecystografic Agent, *Farmaco, Edizione Scientifica*, 31, 426-437.

F014   C.F. Fondyce and L.W.A. Meyer (1940) Industrial and Engineering Chemistry, 32, 1053-1053.

F015   A.A. Forist and T. Chulski (1956) pH-Solubility Relationships for 1-Butyl-3-p-tolylsulfonylurea (Orinase) and Its Metabolite, 1-Butyl-3-p-carboxy-phenylsulfonylurea, *Metabolism, Clinical and Experimental*, 5, 807-812.

F016   G.L. Flynn, N.F.H. Ho, S. Hwang, and J. Park (1976) Interfacing Matrix Release and Membrane Absorption—Analysis of Steroid Absorption from a Vaginal Device in the Rabbit Doe, *Controlled Release Polymeric Formulations*, 87-122.

F017   A. Findlay and A.N. Campbell (1928) The Influence of Constitution on the Stability of Racemates, *Journal of the Chemical Society (London)*, 1768-1775.

F018   C.L. Foy (1969) The Chorinated Aliphatic Acids, in *Degradation of Herbicides*, Marcel Dekker, 207-211.

F019   S.C. Fang (1969) Thiolcarbamates, in *Degradation of Herbicides*, 9, 147-149, Marcel Dekker.

F023   W.W. Feofilaktow (1926) Uber die Kondensation von Brenztraubensaure mit Paraformaldehyd unter Zusatz von Schwefelsaure, *Berichte der Deutschen Chemischen Gesellschaft*, 59, 2765-2777.

F024   K. Florey (1972) Triamcinolone, *Analytical Profiles of Drug Substances*, 1, 378-381.

F025   K. Florey (1972) Triamcinolone Acetonide, *Analytical Profiles of Drug Substances*, 1, 398-409.

F026   K. Florey (1972) Triamcinolone Diacetate, *Analytical Profiles of Drug Substances*, 1, 423-433.

F027   K. Florey (1973) Triflupromazine Hydrochloride, *Analytical Profiles of Drug Substances*, 2, 523-546.

F029   R.C. Farmer (1920) The Decomposition of Nitric Esters, *Journal of the Chemical Society (London)*, 117, 806-818.

F030   R. Flatt and A. Jordan (1933) Contribution au Probleme de la Solvatation: Determination des Rayons d'Ions Dissous, *Helvetica Chimica Acta*, 16, 37-53.

F033   L.J. Fischer and S. Riegelman (1967) Absorption and Activity of Some Derivatives of Griseofulvin, *Journal of Pharmaceutical Sciences*, 56, 469-476.

F035   V.H. Freed (1953) Mode of Action other than Aryl Oxyalkyl Acids, *Journal of Agricultural and Food Chemistry*, 1, 47-50.

F037   A.T. Florence, E.G. Salole, and J.B. Stenlake (1974) The Effect of Particle Size Reduction on Digoxin Crystal Properties, *Journal of Pharmacy and Pharmacology*, 26, 479-480.

F040   V.H. Freed, R. Haque, D. Schmedding, and R. Kohnert (1976) Physicochemical Properties of Some Organophosphates in Relation to Their Chronic Toxicity, *Environmental Health Perspectives*, 13, 77-81.

F041   D. French, M.L. Levine, J.H. Pazur, and E. Norberg (1949) Studies on the Schardinger Dextrins. The Preparation and Solubility Characteristics of alpha, beta and gamma Dextrins, *Journal of the American Chemical Society*, 71, 353-356.

F042   S. Frokjoer and V.S. Andersen (1974) Application of Differential Scanning Calorimetry to the Determination of the Solubility of a Metastable Drug, *Archive of Pharmacy Chemistry Science*, 2, 50-59.

F043   H. Franzen and E. Engel (1921) Mitteillung aus dem Chemischen Institut der Technischen Hochschule zu Karlsruhe, *Journal fuer Praktische Chemie*, 102, 156-186.

F044   J. Ferguson (1927) The Use of Chemical Potentials as Indices of Toxicity, *Proceedings of the Royal Society of London, Series B: Biological Sciences*, 127, 387-404.

F045   E. Felder, D. Pitre, and M. Grandi (1977) Radiopaque Contrast Media, *Farmaco, Edizione Scientifica*, 32, 755-766.

F047   H.E. Fierz-David (1927) Uber die Anthrachinon-sulfosauren, *Helvetica Chimica Acta*, 10, 197-227.

F048   F.J. Frere (1949) Ternary System Diisopropyl Ether-Isopropyl Alcohol-Water at 25 C, *Industrial and Engineering Chemistry*, 41, 2365-2367.

F049   A.R. Fowler and H. Hunt (1941) The System Nitromethane-n-Propanol-Water: Vapor-Liquid Equilibria in the Ternary and the Three Binary Systems, *Industrial and Engineering Chemistry*, 33, 90-95.

F050   R.H. Fritzsche and D.L. Stockton (1946) Systems Containing Isobutanol and Tetrachloroethane : Liquid-Vapor and Liquid-Liquid Equilibria, *Industrial and Engineering Chemistry*, 38, 737-740.

F051   F. Fontein (1910) Gleichgewichte in Ternaren und Quaternaren Systemen, Wobei Zwei Flussige Schichten Auftreten konnen, *Zeitschrift fuer Physikalische Chemie, Abteilung A: Chemische Thermodynamik, Kinetik, Electrochemie, Eigenschaftslehre*, 73, 212-251.

F052   M.D. Forcrand (1912) Sur quelques Constantes Physiques du Cyclohexanol, *Comptes Rendus Hebdomadaires des Seances de l'Academie des Sciences*, 154, 1327-1330.

F053   R.M. Fuoss (1943) The System Water-n-Butanol-Toluene at 30 Degrees, *Journal of the American Chemical Society*, 65, 78-81.

F054   F. Frisch (1930) Uber das Naphthalingrun V, *Helvetica Chimica Acta*, 13, 768-785.

F055   G.S. Forbes and A.S. Coolidge (1919) Relations Between Distribution Ratio, Temperature and Concentration in System: Water, Ether, Succinic Acid, *Journal of the American Chemical Society*, 41, 150-167.

F056   E. Fourneau and G. Florence (1928) Contribution a l'Etude des Ureides des Acides Bromo-valerianiques. III. Influences sur les Proprietes Physiologiques de la Migration de l'Halogene dans la Chaine de l'Acide, *Bulletin Society of Chemistry, French*, 43, 1027-1040.

F057   E. Fourneau and G. Florence (1928) Contribution a l'Etude des Ureides des Acides Bromo-valerianiques. II. Influence de la Ramification de la Chaine sur les Proprietes Physiologiques, *Bulletin Society of Chemistry, French*, 43, 211-216.

F059   P. Finholt and S. Solvang (1968) Dissolution Kinetics of Drugs in Human Gastric Juice—The Role of Surface Tension, *Journal of Pharmaceutical Sciences*, 57, 1322-1326.

F062   V.A. Falck (1919) Beitrag zur Kenntnis des Sulfonal, *Pharmazeutische Zentralhalle*, 60, 409-416.

F063   J.H. Fendler, E.J. Fendler, G.A. Infante..., and L.K. Patterson (1975) Absorption and Proton Magnetic Resonance Spectroscopic Investigation of the Environment of Acetophenone and Benzophenone in Aqueous Micellar Solutions, *Journal of the American Chemical Society*, 97, 89-95.

F064   N. Funasaki, S. Hada, and K. Kawamura (1976) The Surface Tension of Aqueous Solutions of Ethylene Glycol Diesters, *Nippon Kagaku Kaishi*, 12, 1944-1946.

F066   F. Flottmann (1928) Uber Loslichkeitsgleichgewichte, *Ztschrft fur Analytische Chemie*, 73, 1-39.

F067   I. Fukano and Y. Obata (1976) Solubility of Phthalates in Water, *Purosuchikkusu*, 27, 48-49.

F068   S.G. Frank and M.J. Cho (1978) Phase Solubility Analysis and PMR Study of Complexing Behavior of Dinoprostone with beta-Cyclodextrin in Water, *Journal of Pharmaceutical Sciences*, 67, 1665-1668.

F069   H. Freundlich and G.V. Slottman (1927) Uber das Gelten der Traubeschen Regel bei der Hydrotropie, *Biochemische Zeitschrift (1948-1967)*, 188, 101-111.

F070   L. Fishbein and P.W. Albro (1972) Chromatographic and Biological Aspects of the Phthalate Esters, *Journal of Chromatography*, 70, 365-412.

F071   V.H. Freed, C.T. Chiou, D. Schmedding, and R. Kohnert (1979) Some Physical Factors in Toxicological Assessment Tests, *Environmental Health Perspectives*, 30, 75-80.

F072   C.L. Fox, Jr. and H.M. Rose (1942) Ionization of Sulfonamides, *Proceedings of the Society for Experimental Biology and Medicine*, 50, 142-145.

F073   A.R. Frisk (1943) Sulfanilamide Derivatives—Chemotherapeutic Evaluation of N1-Substituted Sulfanilamides, *Acta Medical Scandanavia*, 142, 1-20.

F074   A.R. Frisk (1941) Blood Concentration, Acetylation and Urinary Excretion of Sulfapyridine and Sulfathiazole after Various Sulfapyridine and Sulfathiazole Derivatives Administered by Different Routes, *Acta Medica Scandinavica*, 106, 369-403.

F075   W.J. Feinstone, R.D. Williams, R.T. Wolff..., and M.L. Crossley (1940) The Toxicity, Absorption and Chemotherapeutic Activity of 2-Sulfanilamidopyrimidine (Sulfadiazine), *Bulletin of the John Hopkins Hospital*, 67, 427-456.

F076   H. Fessi, J-P. Marty, F. Puisieux, and J.T. Carstensen (1982) Square Root of Time Dependence of Matrix Formulations with Low Drug Content, *Journal of Pharmaceutical Sciences*, 71, 749-752.

F176   P. Friberger and G. Aberg (1971) Some Physicochemical Properties of the Racemates and the Optically Active Isomers of Two Local Anaesthetic Compounds, *Acta Pharmaceutica Suecica*, 8, 361-364.

F178   E. Felder, D. Pitre, and P. Tirone (1977) Radiopaque Contrast Media. XLIV. Preclinical Studies with a New Nonionic Contrast Agent, *Farmaco, Edizione Scientifica*, 32, 835-844.

F179   A. Felsot and P.A. Dahm (1979) Sorption of Organophosphorus and Carbamate Insecticides by Soil, *Journal of Agricultural and Food Chemistry*, 27, 557-563.

F181   K. Florey (1976) Cephradine, *Analytical Profiles of Drug Substances*, 5, 22-37.

F182   S. Furesz (1970) Chemical and Biological Properties of Rifampicin, *Antibiotica et Chemotherapia (1954-68)*, 16, 316-351.

F183   H. Fujioka and T. Tan (1981) Biopharmaceutical Studies on Hydantoin Derivatives. I. Physico-chemical Properties of Hydantoin Derivatives and Their Intestinal Absorption, *Journal of Pharmaceutics Dynamics*, 4, 759-770.

F184   V.H. Freed and P. Burschel (1957) The Relationship of Water Solubility to Dosage of Herbicides, *Zeitschrift fuer Pflanzenkrankheiten und Pflanzenschutz*, 64, 477-479.

F185   T.C. Filippov and A.A. Firman (1952) *Zhurnal Prikladnoi Khimii (Leningrad)*, 25, 895-897.

F186   K. Freudenberg, E. Plankenhorn, and H. Knauber (1947) Schardinger's Dextrins—Derived from Starch, *Chemistry and Industry (London)*, 731-735.

F300   R.K. Freier (1976) Aqueous Solutions. Volume 1. Data for Inorganic and Organic Compounds, 1.

F301   A. Fini, V. Zecchi, L. Rodriguez, and A. Tartarini (1984) Solubility-dissolution Relationship for Ibuprofen, Fenbufen and Their Sodium Salts in Acid Medium, *Pharmaceutica Acta Helvetiae*, 59, 106-108.

F302   R.K. Freier (1978) *Data for Inorganic and Organic Compounds*. Supplements 2, 2, 17-443, Walter deGruyter.

F303   K.J. Friesen, L.P. Sarna, and G.R.B. Webster (1985) Aqueous Solubility of Polychlorinated Dibenzo-p-dioxins Determined by High Pressure Liquid Chromatography, *Chemosphere*, 14, 1267-1274.

F306   A. Fini, M. Laus, I. Orienti, and V. Zecchi (1986) Dissolution and Partition Thermodynamic Functions of Some Nonsteroidal Anti-inflammatory Drugs, *Journal of Pharmaceutical Sciences*, 75, 23-25.

F307   A. Fini, A. Roda, R. Fugazza, and B. Grigolo (1985) Chemical Properties of Bile Acids. III. Bile Acid Structure and Solubility in Water, *Journal of Solution Chemistry*, 14, 595-603.

F309   N.F. Farraj, S.S. Davis, G.D. Parr, and H.N.E. Stevens (1988) The Stability and Solubility of Progabide and Its Related Metabolic Derivatives, *Pharmaceutical Research*, 5, 226-231.

F310   Y. Fukumori, T. Fukuda, Y. Yamamoto..., and N. Sato (1983) Physical Characterization of Erythromycin Dihydrate, Anhydrate and Amorphous Solid and Their Dissolution Properties, *Chemical and Pharmaceutical Bulletin*, 31, 4029-4039.

F311   R. Furer and M. Geiger (1977) A Simple Method of Determining the Aqueous Solubility of Organic Substances, *Pesticide Science*, 8, 337-344.

F312   M.D. Fulford, J.E. Slonek, and M.J. Groves (1986) A Note on the Solubility of Progesterone in Aqueous Polyethylene Gylcol 400, *Drug Development and Industrial Pharmacy*, 12, 631-635.

F314   K.J. Friesen, J. Vilk, and D.C.G. Muir (1990) Aqueous Solubilities of Selected 2,3,7,8-Substituted Polychlorinated Dibenzofurans (PCDFs), *Chemosphere*, 20, 27-32.

F315   K.J. Friesen and G.R.B. Webster (1990) Temperature Dependence of the Aqueous Solubilities of Highly Chlorinated Dibenzo-p-dioxins, *Environmental Science and Technology*, 24, 97-101.

F317   M. Faizal, F.J. Smagghe, G.H. Malmary..., and J.R. Molinier (1990) Equilibrium Diagrams at 25 degrees C of Water-Oxalic Acid-2-Methyl-1-Propanol, Water-Oxalic Acid-1-Pentanol, and Water-Oxalic Acid-3-Methyl-1-Butanol ternary Systems, *Journal of Chemical and Engineering Data*, 35, 352-354.

F318   N.F. Farraj, S.S. Davis, G.D. Parr, and H.N.E. Stevens (1989) Modification of the Aqueous Solubility and Stability of Progabide, *International Journal of Pharmaceutics*, 52, 11-18.

F322   S. Forster, G. Buckton, and A.E. Beezer (1991) The Importance of Chain Length on the Wettability and Solubility of Organic Homologs, *International Journal of Pharmaceutics*, 72, 29-34.

F325   A.F. Frolov, M.A. Loginova, and A.P. Karaseva (1968) Mutual Solubility in the System Butyl Alcohol-Ethyl Alcohol-Methyl Alcohol-Water, *Journal of General Chemistry of the USSR*, 38, 1164-1166.

F327   H.W. Frilink, A.C. Eissens, A.J.M. Schoonen, and C.F. Lerk (1991) The Effects of Cyclodextrins on Drug Release from Fatty Suppository Bases. I. In Vitro Observations, *European Journal of Pharmaceutics and Biopharmaceutics*, 37, 178-182.

G001   F.W. Getzen and T.M. Ward (1971) Influence of Water Structure on Aqueous Solubility, *Industrial and Engineering Chemistry, Product Research and Development*, 10, 122-132.

G002   E.R. Garrett and P.B. Chemburkar (1968) Evaluation, Control, and Prediction of Drug Diffusion Through Polymeric Membranes. II, *Journal of Pharmaceutical Sciences*, 57, 949-959.

G003   E.R. Garrett and P.B. Chemburkar (1968) Evaluation, Control, and Prediction of Drug Diffusion Through Polymeric Membranes. III, *Journal of Pharmaceutical Sciences*, 57, 1401-1409.

G004   P.M. Ginnings and R. Baum (1937) Aqueous Solubilities of the Isomeric Pentanols, *Journal of the American Chemical Society*, 59, 1111-1113.

G005   P.M. Ginnings and R. Webb (1938) Aqueous Solubilities of Some Isomeric Hexanols, *Journal of the American Chemical Society*, 60, 1388-1389.

G006   P.M. Ginnings and M. Hauser (1938) Aqueous Solubilities of Some Isomeric Heptanols, *Journal of the American Chemical Society*, 60, 2581-2582.

G007   P.M. Ginnings and D. Coltrane (1939) Aqueous Solubility of 2,2,3-Trimethylpentanol-3, *Journal of the American Chemical Society*, 61, 525-525.

G008   D.E. Guttman, W.E. Hamlin, J.W. Shell, and J.G. Wagner (1961) Solubilization of Anti-inflammatory Steroids by Aqueous Solutions of Triton WR-1339, *Journal of Pharmaceutical Sciences*, 50, 305-307.

G009   M. Gabaldon, J. Sanchez, and A. Llombart, Jr. (1968) In Vitro Utilization of Diethylstilbestrol by Rat Liver, *Journal of Pharmaceutical Sciences*, 57, 1744-1747.

G010   A.H. Goldberg, M. Gibaldi, J.L. Kanig, and M. Mayersohn (1966) Increasing Dissolution Rates and Gastrointestinal Absorption of Drugs via Solid Solutions and Eutectic Mixtures. IV, *Journal of Pharmaceutical Sciences*, 55, 581-583.

G011   A.H. Goldberg, M. Gibaldi, and J.L. Kanig (1966) Increasing Dissolution Rates and Gastrointestinal Absorption of Drugs via Solid Solutions and Eutectic Mixtures. III, *Journal of Pharmaceutical Sciences*, 55, 487-492.

G012   E.H. Gans and T. Higuchi (1957) The Solubility and Complexing Properties of Oxytetracycline and Tetracycline I, *Journal of the American Pharmaceutical Association, Scientific Edition*, 46, 458-466.

G014   M.W. Gouda, A.A. Ismail, and M.M. Motawi (1970) Micellar Solubilization of Barbiturates. II. Solubilities of Certain Barbiturates in Polyoxyethylene Stearates of Varying Hydrophilic Chain Length, *Journal of Pharmaceutical Sciences*, 59, 1402-1405.

G015   E.R. Garrett (1957) Prediction of Stability in Pharmaceutical Preparations. IV, *Journal of the American Pharmaceutical Association, Scientific Edition*, 46, 584-586.

G016   E.R. Garrett and D.J. Weber (1970) Metal Complexes of Thiouracils. I. Stability Constants by Potentiometric Titration Studies and Structures of Complexes, *Journal of Pharmaceutical Sciences*, 59, 1383-1391.

G017   E.R. Garrett and C.M. Won (1971) Prediction of Stability in Pharmaceutical Preparations. XVI. Kinetics of Hydrolysis of Canrenone and Lactonization of Canrenoic Acid, *Journal of Pharmaceutical Sciences*, 60, 1801-1809.

G018   E.R. Garrett and C.A. Hunt (1974) Physicochemical Properties, Solubility, and Protein Binding of delta-9-Tetrahydrocannabinol, *Journal of Pharmaceutical Sciences*, 63, 1056-1064.

G020   I.S. Gibbs, A. Heald, H. Jacobson..., and I. Weliky (1976) Physical Characterization and Activity In Vivo of Polymorphic Forms of 7-Chloro-5,11-dihydrodibenz[b,e][1,4]oxazepine-5-carboxamide, a Potential Tricyclic Antidepressant, *Journal of Pharmaceutical Sciences*, 65, 1380-1385.

G021   D. Guttman and T. Higuchi (1957) Reversible Association of Caffeine and of Some Caffeine Homologs in Aqueous Solution, *Journal of the American Pharmaceutical Association, Scientific Edition*, 46, 4-9.

G022   A.L. Green (1967) Ionization Constants and Water Solubilities of Some Aminoalkylphenothiazine Tranquillizers and Related Compounds, *Journal of Pharmacy and Pharmacology*, 19, 10-16.

G023   A.L. Green (1967) Activity Correlations and the Mode of Action of Aminoalkylphenothiazine Tranquillizers, *Journal of Pharmacy and Pharmacology*, 19, 207-208.

G024   M.A.F. Gadalla, A.M. Saleh, and M.M. Motawi (1974) Effect of Electrolytes on the Solubility and Solubilization of Chlorocresol, *Pharmazie (Berlin)*, 29, 105-107.

G025   S.N. Giri and L.R. Kartt (1975) Temperature Dependent Aqueous Solubility of Actinomycin D, *Specialia*, 31, 482-483.

G026   D.R. Gilligan and M.N. Plummer (1944) Comparative Solubilities of Sulfadiazine, Sulfamerizine and Sulfamethazine and Their N4-Acetyl Derivatives at Varying pH Levels, *Proceedings of the Society for Experimental Biology and Medicine*, 53, 142-145.

G027   A. Gemant (1962) Solubilization of Cholesterol, *Life Sciences*, 1, 233-238.

G028   J.K. Guillory and H.O. Lin (1976) Some Properties of Sulfanilamide Monohydrate, *Chemical and Pharmaceutical Bulletin*, 24, 1675-1678.

G029   P.M. Gross and J.H. Saylor (1931) The Solubilities of Certain Slightly Soluble Organic Compounds in Water, *Journal of the American Chemical Society*, 53, 1744-1751.

G030   P.M. Ginnings, D. Plonk, and E. Carter (1940) Aqueous Solubilities of Some Aliphatic Ketones, *Journal of the American Chemical Society*, 62, 1923-1924.

G031   P.M. Ginnings, E. Herring, and D. Coltrane (1939) Aqueous Solubilities of Some Unsaturated Alcohols, *Journal of the American Chemical Society*, 61, 807-808.

G032   P. Gross, J.C. Rintelen, and J.H. Saylor (1939) Energy and Volume Relations in the Solubilities of Some Ketones in Water, *Journal of Physical Chemistry*, 43, 197-205.

G033   F.S. Granger and J.M. Nelson (1921) Oxidation and Reduction of Hydroquinone and Quinone from the Standpoint of Electromotive-force Measurements, *Journal of the American Chemical Society*, 43, 1401-1415.

G034   D.N. Glew and R.E. Robertson (1956) The Spectrophotometric Determination of the Solubility of Cumene in Water by a Kinetic Method, *Journal of Physical Chemistry*, 60, 332-337.

G035   A.N. Guseva and E.I. Parnov (1964) Isothermal Sections of Monocyclic Arene-Water Binary Systems at 25, 100, and 200 Degrees [Ethylbenzene and Propylbenzene (Ed. of Translation)], *Russian Journal of Physical Chemistry*, 38, 439-440.

G036   H. Geissbuhler (1969) The Subsituted Ureas, Degradation of Herbicides, 79-83.

G037   P.M. Gross, J.H. Saylor, and M.A. Gorman (1933) Solubility Studies. IV. The Solubilities of Certain Slightly Soluble Organic Compounds in Water, *Journal of the American Chemical Society*, 55, 650-652.

G038   P. Gross (1929) The Determination of the Solubility of Slightly Soluble Liquids in Water and the Solubilities of the Dichloro-ethanes and -propanes, *Journal of the American Chemical Society*, 51, 2362-2366.

G039   W.G. Gorman and G.D. Hall (1964) Dielectric Constant Correlations with Solubility and Solubility Parameters, *Journal of Pharmaceutical Sciences*, 53, 1017-1020.

G040   E.C. Gilbert and B.E. Lauer (1927) A Study of the Ternary System Methyl Benzoate, Methanol, Water, *Journal of Physical Chemistry*, 31, 1050-1052.

G041   H. Gysin (1962) Triazine Herbicides Their Chemistry, Biological Properties and Mode of Action, *Chemistry and Industry (London)*, 31, 1393-1400.

G042   R.V. Griffiths and A.G. Mitchell (1971) Surface Transformation During Dissolution of Aspirin, *Journal of Pharmaceutical Sciences*, 60, 267-270.

G043   H.D. Gibbs (1927) Phenol Tests, *Journal of Biological Chemistry*, 72, 649-655.

G046   V.G. Grube and M. Nubaum (1928) Phasentheoretische Untersuchungen Uber die Entzuckerung der Melasse, *Zeitschrift fuer Elektrochemie*, 34, 91-98.

G047   J.E. Gordon and R.L. Thorne (1967) Salt Effects on the Activity Coefficient of Naphthalene in Mixed Aqueous Electrolyte Solutions. I. Mixtures of Two Salts, *Journal of Physical Chemistry*, 71, 4390-4392.

G050   J. Griswold, M.E. Klecka, and R.V. West, Jr. (1948) Conjugate Liquid Phase Equilibria — C4-Hydrocarbon-Furfural-Water Systems, *Chemical Engineering Progress Symposium Series – "Phase-Equilibria"*, 44, 839-846.

G051   B. Gold and S.S. Mirvish (1977) N-Nitroso Derivatives of Hydrochlorothiazide, Niridazole, and Tolbutamide, *Toxicology and Applied Pharmacology*, 40, 131-136.

G052   G.M. Goeller and A. Osol (1937) The Salting-out of Molecular Benzoic Acid in Aqueous Salt Solutions at 35 Degrees, *Journal of the American Chemical Society*, 59, 2132-2134.

G053   P. Gross (1929) Uber den Aussalzeffekt an Dichlorathanen und -Propanen, *Zeitschrift fur Physical Chemistry Abt B*, 6, 215-220.

G054   P. Gavaudan and H. Poussel (1947) Le Mecanisme de l'Action Insecticide du Dichlorodiphenyl-trichlorethane (DDT) et la Regle Thermodynamique des Narcotiques Indifferents, *Comptes Rendus Hebdomadaires des Seances de l'Academie des Sciences*, 224, 683-685.

G055   D.N. Glew (1960) The Gas Hydrate of Bromochlorodifluoromethane, *Canadian Journal of Chemistry*, 38, 208-221.

G056   G.P. Gladis (1960) Effects of Moisture on Corrosion in Petrochemical Environments, *Chemical Engineering Progress Symposium Series* – "Phase-Equilibria", 56, 43-51.

G058   W. Gerrard (1972) Significance of the Solubility of Hydrocarbon Gases in Liquids in Relation to the Intermolecular Structure of Liquids, *Chemistry and Industry (London)*, 21, 804-805.

G060   D. Grut (1937) The Solubility of Sucrose, *Zeitschrift fuer die Zuckerindustrie Czechoslov*, 61, 345.

G061   D.N. Glew and E.A. Moelwyn-Hughes (1953) Chemical Statics of the Methyl Halides in Water, *Discussions of the Faraday Society*, 15, 150-161.

G062   S. Glasstone and A. Pound (1925) Solubility Influences. Part I. The Effect of Some Salts, Sugars, and Temperature on the Solubility of Ethyl Acetate in Water, *Journal of the Chemical Society (London)*, 107, 2660-2667.

G063   A. Giacalone (1935) Solubilita dell'Acido 6-Nitro-3-metilbenzoico in Benzene, Toluene ed Acqua, *Gazzetta Chimica Italiana*, 65, 844-850.

G066   H.D. Gibbs (1927) Phenol Tests, *Journal of Physical Chemistry*, 31, 1057-1081.

G067   A. Goto, F. Endo, and K. Ito (1977) Gel Filtration of Solubilized Systems. I. On the Gel Filtration of Aqueous Sodium Lauryl Sulfate Solution Solubilizing Alkyl Paraben on Sephadex G-50, *Chemical and Pharmaceutical Bulletin*, 25, 1165-1173.

G068   A.N. Guseva and E.I. Parnov (1963) The Solubility of Cyclohexane in Water, *Russian Journal of Physical Chemistry*, 37, 1494-1494.

G072   M.M. Gale and L. Saunders (1971) The Solubilisation of Steroids by Lysophosphatidylcholine Testosterone, Estradiol and Their 17-alpha-Ethinyl Derivations, *Biochimica et Biophysica Acta*, 248, 466-470.

G073   M.W. Gouda, A.R. Ebian, M.A. Moustafa, and S.A. Khalil (1977) Sulphapyridine Crystal Forms, *Drug Development and Industrial Pharmacy*, 3, 273-290.

G075   J. Gettins, D. Hall, P.L. Jobling..., and E. Wyn-Jones (1978) Thermodynamic and Kinetic Parameters Associated with the Exchange Process Involving Alcohols and Micelles, *Journal of the Chemistry Society, Faraday Transactions II*, 71, 1957-1964.

G076   G.M. Gasparini (1979) The Preparation and Properties of Trialkylacetohydroxamic Acids: Effect of the Neoalkyl Structure with Regard to the Solubility, the Stability and Some Extractive Capacities, *Gazzetta Chimica Italiana*, 109, 357-363.

G077   A. Geake and J.T. Lemon (1938) Semiquinone Formation by Anthraquinone and Some Simple Derivatives, *Transactions of the Faraday Society*, 34, 1409-1427.

G078   A. Goto, R. Sakura, and F. Endo (1980) Gel Filtration of Solubilized Systems. V. Effects of Sodium Chloride on Micellar Sodium Lauryl Sulfate Solutions Solubilizing Alkylparabens, *Chemical and Pharmaceutical Bulletin*, 28, 14-22.

G079   C.A.I. Goring (1962) Control of Nitrification by 2-Chloro-6-(trichloromethyl) Pyridine, *Soil Science*, 93, 211-218.

G080   C.A.I. Goring (1962) Theory and Principles of Soil Fumigation, *Advances Pest Control Research*, 5, 47-84.

G081   E. Gillert (1926) Cholerese und Choleretica, ein Beitrag zur Physiologie der Galle, *Zfrieidie Gesamte Experimental Medizin*, 48, 255-275.

G083   D.R. Gilligan, S. Garb, C. Wheeler, and M.N. Plummer (1943) Adjuvant Alkali Therapy in the Prevention of Renal Complications from Sulfadiazine, *Journal of the American Medical Association*, 122, 1160-1165.

G084   A.S. Geneidi and H. Hamacher (1980) Enhancement of Dissolution Rates of Spironolactone and Diazepam via Polyols and PEG Solid Dispersion Systems, *Pharmazeutische Industrie*, 42, 401-404.

G085   A.H. Ghanem, H. El-Sabbagh, and H. Abdel-Alim (1980) Solubilization of Flufenamic Acid, *Pharmazeutische Industrie*, 42, 854-856.

G086   A. Ghanem, M. Meshali, and I. Ramadaan (1980) Dissolution Rate of Trimethoprim Polyvinylpyrrolidone Coprecipitate, *Pharmazie (Berlin)*, 35, 689-690.

G088   A.S. Geneidi, M.S. Adel, and E. Shehata (1980) Enhanced Dissolution of Gilbenclamide From Gilbenclamide-Poloxamer and Glibenclamide-PVP Coprecipitates, *Canadian Journal of Pharmaceutical Science*, 15, 81-84.

G089   M. Gallego, M. Garcia-Vargas, F. Pino, and M. Valcarcel (1978) Analytical Applications of Picolinealdehyde Salicyloylhydrazone, *Microchemical Journal*, 23, 353-359.

G090   Z.E. Grabovskaya and M.I. Vinnik (1966) Activity Coefficients of Certain Aromatic Compounds in Concentrated Sulphuric Acid Solutions, *Russian Journal of Physical Chemistry*, 40, 1221-1223.

G091   J.D. Gaynor and V. Van Volk (1981) s-Triazine Solubility in Chloride Salt Solutions, *Journal of Agricultural and Food Chemistry*, 29, 1143-1146.

G092   K. Gekko (1981) Mechanism of Poly-induced Protein Stabilization: Solubility of Amino Acids and Diglycine in Aqueous Polyol Solutions, *Journal of Biochemistry (Tokyo)*, 90, 1633-1641.

G093   E.R. Garrett and M.R. Gardner (1982) Prediction of Stability in Pharmaceutical Preparations. XIX. Stability Evaluation and Bioanalysis of Clofibric Acid Esters by High-pressure Liquid Chromatography, *Journal of Pharmaceutical Sciences*, 71, 14-25.

G095   A.S. Geneidi and H. Hamacher (1980) Physical Characterization and Dissolution Profiles of Spironolactone and Diazepam Coprecipitates, *Pharmazeutische Industrie*, 42, 315-319.

G096   G.G. Gallo and P. Radaelli (1976) Rifampin, *Analytical Profiles of Drug Substances*, 5, 468-509.

G098   P.E. Grubb (1979) Nalidixic Acid, *Analytical Profiles of Drug Substances*, 8, 371-381.

G099   H. Gysin and E. Knusli (1960) Chemistry and Herbicidal Properties of Triazine Derivatives, *Advances Pest Control Research*, 3, 289-355.

G101   J. Griswold, P.L. Chu, and W.O. Winsauer (1949) Phase Equilibria in Ethyl Alcohol-Ethyl Acetate-Water System, *Industrial and Engineering Chemistry*, 41, 2352-2358.

G300   H. Geyer, P. Sheehan, D. Kotzias..., and F. Korte (1982) Prediction of Ecotoxicological Behaviour of Chemicals: Relationship Between Physico-chemical Properties and Bioaccumulation of Organic Chemicals in the Mussel Mytilus Edulis, *Chemosphere*, 11, 1121-1134.

G301   H. Geyer, R. Viswanathan, D. Freitag, and F. Korte (1981) Relationship Between Water Solubility of Organic Chemicals and Their Bioaccumulation by the Alga Chlorella, *Chemosphere*, 10, 1307-1313.

G302   D.J.W. Grant, M. Mehdizadeh, A.H-L. Chow, and J.E. Fairbrother (1984) Non-linear van't Hoff Solubility—Temperature Plots and Their Pharmaceutical Interpretation, *International Journal of Pharmaceutics*, 18, 25-38.

G306   C.D. Green (1978) Perfluidone, *Analytical Methods for Pesticides and Plant Growth Regulators*, 10, 437-450.

G307   C.D. Green (1978) Fluoridamid, *Analytical Methods for Pesticides and Plant Growth Regulators*, 10, 533-543.

G310   W.E. Gledhill, R.G. Kaley, W.J. Adams..., and V.W. Saeger (1980) An Environmental Safety Assessment of Butyl Benzyl Phthalate, *Environmental Science and Technology*, 14, 301-305.

G312   F.A.P. Gobas, J.M. Lahittete, G. Garofalo..., and D. Mackay (1988) A Novel Method for Measuring Membrane-Water Partition Coefficients of Hydrophobic Organic Chemicals: Comparison with 1-Octanol-Water Partitioning, *Journal of Pharmaceutical Sciences*, 77, 265-272.

G313   F.R. Groves (1988) Solubility of Cycloparaffins in Distilled Water and Salt Water, *Journal of Chemical and Engineering Data*, 33, 136-138.

G315   K. Gekko and S. Koga (1984) The Stability of Protein Structure in Aqueous Propylene Glycol Amino Acid Solubility and Preferential Solvation of Protein, *Biochimica et Biophysica Acta*, 786, 151-160.

G317   N. Gabas, T. Carillon, and N. Hiquily (1988) Solubilities of D-Xylose and D-Mannose in Water-Ethanol Mixtures at 25 C, *Journal of Chemical and Engineering Data*, 33, 128-130.

G318   R.B. Gandhi and A.H. Karara (1988) Characterization, Dissolution and Diffusion Properties of Tolbutamide-beta-Cyclodextrin, *Drug Development and Industrial Pharmacy*, 14, 657-682.

G319   C.A.I. Goring and J.W. Hamaker (1972) Decomposition: Quantitative Aspects, *Organic Chemicals in the Soil Environment*, 1, 384-385.

G323   J. Griswold, J-N. Chew, and M.E. Klecka (1950) Pure Hydrocarbons from Petroleum: Recovery of Aniline Solvent from Distex Hydrocarbon Products by Water Extraction, *Industrial and Engineering Chemistry*, 42, 1246-1251.

H002   A.E. Hill (1923) The Mutual Solubility of Liquids. I. The Mutual Solubility of Ethyl Ether and Water. II. The Solubility of Water in Benzene, *Journal of the American Chemical Society*, 45, 1143-1155.

H003   A.E. Hill and W.M. Malisoff (1926) The Mutual Solubility of Liquids. III. The Mutual Solubility of Phenol and Water. IV. The Mutual Solubility of Normal Butyl Alcohol and Water, *Journal of the American Chemical Society*, 48, 918-927.

H004   W.E. Hamlin, J.I. Northam, and J.G. Wagner (1965) Relationship Between In Vitro Dissolution Rates and Solubilities of Numerous Compounds Representative of Various Chemical Species, *Journal of Pharmaceutical Sciences*, 54, 1651-1653.

H005   T. Higuchi, M. Gupta, and L.W. Busse (1953) Influence of Electrolytes, pH, and Alcohol Concentration on the Solubilities of Acidic Drugs, *Journal of the American Pharmaceutical Association, Scientific Edition*, 42, 157-161.

H006   J.B. Houston, D.G. Upshall, and J.W. Bridges (1974) A Re-evaluation of the Importance of Partition Coefficients in the Gastrointestinal Absorption of Anutrients, *Journal of Pharmacology and Experimental Therapeutics*, 189, 244-254.

H007   K.A. Herzog and J. Swarbrick (1971) Drug Permeation through Thin-model Membranes. III. Correlations between In Vitro Transfer, In Vivo Absorption, and Physicochemical Parameters of Substituted Benzoic Acids, *Journal of Pharmaceutical Sciences*, 60, 1666-1668.

H008   M.J. Hunt and L. Saunders (1975) The Solubilization of Some Local Anaesthetic Esters of p-Aminobenzoic Acid by Lysophosphatidylcholine, *Journal of Pharmacy and Pharmacology*, 27, 119-124.

H009   K.J. Humphreys and C.T. Rhodes (1968) Effect of Temperature Upon Solubilization by a Series of Nonionic Surfactants, *Journal of Pharmaceutical Sciences*, 57, 79-83.

H010   W.I. Higuchi, P.D. Bernardo, and S.C. Mehta (1967) Polymorphism and Drug Availability. II. Dissolution Rate Behavior of the Polymorphic Forms of Sulfathiazole and Methylpredisolone, *Journal of Pharmaceutical Sciences*, 56, 200-207.

H011   W.E. Hamlin and W.I. Higuchi (1966) Dissolution Rate-solubility Behavior of 3-(1-Methyl-2-pyrrolidinyl)-indole as a Function of Hydrogen-ion Concentration, *Journal of Pharmaceutical Sciences*, 55, 205-207.

H012   W.D. Harkins and H. Oppenheimer (1949) Solubilization of Polar-Non-polar Substances in Solutions of Long Chain Electrolytes, *Journal of the American Chemical Society*, 71, 808-811.

H013   O.N. Hinsvark, W. Zazulak, and A.I. Cohen (1972) Liquid Chromatography: Its Use in the Biological Characterization and Study of Metolazone—A New Diuretic, *Journal of Chromatographic Science*, 10, 379-382.

H015   A. Hussain (1972) Prediction of Dissolution Rates of Slightly Water-soluble Powders from Simple Mathematical Relationships, *Journal of Pharmaceutical Sciences*, 61, 811-813.

H016   T. Higuchi and A. Drubulis (1961) Complexation of Organic Substances in Aqueous Solution by Hydroxyaromatic Acids and Their Salts, *Journal of Pharmaceutical Sciences*, 50, 905-909.

H017   T. Higuchi and S. Bolton (1959) The Solubility and Complexing Properties of Oxytetracycline and Tetracycline. III, *Journal of the American Pharmaceutical Association, Scientific Edition*, 48, 557-564.

H018   T. Higuchi and J.L. Lach (1954) Investigation of Some Complexes Formed in Solution by Caffeine. IV. Interactions Between Caffeine and Sulfathiazole, Sulfadiazine, p-Aminobenzoic Acid, Benzocaine, Phenobarbital, and Barbital, *Journal of the American Pharmaceutical Association, Scientific Edition*, 43, 349-354.

H019   T. Higuchi and J.L. Lach (1954) Study of Possible Complex Formation Between Macromolecules and Certain Pharmaceuticals. III. Interaction of Polyethylene Glycols with Several Organic Acids, *Journal of the American Pharmaceutical Association, Scientific Edition*, 43, 465-470.

H020   T. Higuchi and J.L. Lach (1954) Investigation of Complexes Formed in Solution by Caffeine. VI. Comparison of Complexing Behaviors of Methylated Xanthines with p-Aminobenzoic Acid, Salicylic Acid, Acetylsalicylic Acid, and p-Hydroxybenzoic Acid, *Journal of the American Pharmaceutical Association, Scientific Edition*, 43, 527-530.

H021   T. Higuchi and J.L. Lach (1954) Investigation of Some Complexes Formed in Solution by Caffeine. V. Interactions Between Caffeine and p-Aminobenzoic Acid, m-Hydroxybenzoic Acid, Picric Acid, o-Phthalic Acid, Suberic Acid, and Valeric Acid, *Journal of the American Pharmaceutical Association, Scientific Edition*, 43, 524-527.

H022   T. Higuchi and D.A. Zuck (1953) Investigation of Some Complexes Formed in Solution by Caffeine. III. Interactions Between Caffeine and Aspirin, p-Hydroxybenzoic Acid, m-Hydroxybenzoic Acid, Salicylic Acid, Salicylate Ion, and Butyl Paraben, *Journal of the American Pharmaceutical Association, Scientific Edition*, 42, 138-145.

H023   T. Higuchi and D.A. Zuck (1953) Investigation of Some Complexes Formed in Solution by Caffeine. II. Benzoic Acid and Benzoate Ion, *Journal of the American Pharmaceutical Association, Scientific Edition*, 42, 132-138.

H024   W.D. Hormann and D.O. Eberle (1972) The Aqueous Solubility of 2-Chloro-4-ethylamino-6-isopropylamino-1,3,5-triazine (Atrazine) Obtained by an Improved Analytical Method, *Weed Research*, 12, 199-202.

H025   W.I. Higuchi, N.A. Mir, A.P. Parker, and W.E. Hamlin (1965) Dissolution Kinetics of a Weak Acid, 1,1-Hexamethylene p-Tolylsulfonylsemicarbazide, and Its Sodium Salt, *Journal of Pharmaceutical Sciences*, 54, 8-11.

H026   W.I. Higuchi, P.K. Lau, T. Higuchi, and J.W. Shell (1963) Polymorphism and Drug Availability: Solubility Relationships in the Methylprednisolone System, *Journal of Pharmaceutical Sciences*, 52, 150-153.

H027   C.E. Higgins, W.H. Balwin, and B.A. Soldano (1959) Effects of Electrolytes and Temperature on the Solubility of Tributyl Phosphate in Water, *Journal of Physical Chemistry*, 63, 113-118.

H028   R.S. Hansen, Y. Fu, and F.E. Bartell (1949) Multimolecular Adsorption from Binary Liquid Solutions, *Journal of Physical Chemistry*, 53, 769-785.

H030   C.S. Hoffman, Jr. (1967) Water Solubilities of Tetradecanol and Hexadecanol by Gas-liquid Chromatography, Ph.D. thesis.

H031   C.E. Higgins and W.H. Baldwin (1960) Refractometric Determination of Mutual Solubility as a Function of Temperature: Tributyl Phosphine Oxide and Water, *Analytical Chemistry*, 32, 233-236.

H032   C.E. Higgins and W.H. Baldwin (1960) Effect of Centrifugation on Solution Temperature and Solubility of Tributyl Phosphate and Tributyl Phosphine Oxide in Water, *Analytical Chemistry*, 32, 236-238.

H033   A. Haussler and P. Hajdu (1958) Mitteilung uber die Dissoziationskonstante und Loslichkeit von Rastinon 'Hoechst', *Archiv der Pharmazie*, 291, 531-536.

H034   R.B. Heap, A.M. Symons, and J.C. Watkins (1970) Steroids and Their Interactions with Phospholipids: Solubility, Distribution Coefficient and Effect on Potassium Permeability of Liposomes, *Biochimica et Biophysica Acta*, 218, 482-495.

II035   R.B. Heap, A.M. Symons, and J.C. Watkins (1971) An Interaction Between Oestradiol and Progesterone in Aqueous Solutions and in a Model Membrane System, *Biochimica et Biophysica Acta*, 233, 307-314.

II036   B.R. Hajratwala and H. Taylor (1976) Effect of Non-ionic Surfactants on the Dissolution and Solubility of Hydrocortisone, *Journal of Pharmacy and Pharmacology*, 28, 934-935.

H037   H.C. Hetherington and J.M. Braham (1923) Preparation of Dicyanodiamide from Calcium Cyanamide, *Industrial and Engineering Chemistry*, 15, 1060-1063.

H038   R.W. Hobson, R.J. Hartman, and E.W. Kanning (1941) A Solubility Study of Di-n-propylamine, *Journal of the American Chemical Society*, 63, 2094-2095.

H039   G.K. Helmkamp, F.L. Carter, and H.J. Lucas (1957) Coordination of Silver Ion with Unsaturated Compounds. VIII. Alkynes, *Journal of the American Chemical Society*, 79, 1306-1310.

H040   E.H. Hungerford and A.R. Nees (1934) Raffinose Preparation and Propertics, *Industrial and Engineering Chemistry*, 26, 462-464.

H041   A.L. Horvath (1976) Temperature and Pressure Effects on Solubility: Selection and Consistency, *Chemie-Ingenieur-Technik*, 48, 144-146.

H042   R.A. Herrett (1969) Methyl- and Phenylcarbamates, 113-141.

H043   M. Hayashi and T. Sasaki (1956) Measurement of Solubilities of Sparingly Soluble Liquids in Water and Aqueous Detergent Solutions Using Non-ionic Surfactant, *Bulletin of the Chemical Society of Japan (Nippon Kagakukai Bulletin)*, 29, 857-859.

H044   W. Huckel, M-T. Niesel, and L. Buchs (1944) Die Anomalitaten des Benzylalkohols und Seiner Losungen, *Chemische Berichte*, 77, 334-337.

H046   M.L. Huang and S. Niazi (1977) Polymorphic and Dissolution Properties of Mercaptopurine, *Journal of Pharmaceutical Sciences*, 66, 608-609.

H048   H.V. Halban and H. Kortschak (1938) Uber die Loslichkeit der Pikrinsaure in Wasser und Wasserigen Elektrolytlosungen, *Helvetica Chimica Acta*, 21, 392-401.

H049   A.R. Hurwitz and S.T. Liu (1977) Determination of Aqueous Solubility and pKa Values of Estrogens, *Journal of Pharmaceutical Sciences*, 66, 624-627.

H050   G.S. Hartley (1938) The Solvent Properties of Aqueous Solutions of Paraffin-chain Salts. Part I. The Solubility of trans-Azobenzene in Solutions of Cetylpyridinium Salts, *Journal of the Chemical Society (London)*, 1968-1975.

H051   J.P. Hou and J.W. Poole (1969) The Amino Acid Nature of Ampicillin and Related Penicillins, *Journal of Pharmaceutical Sciences*, 58, 1510-1515.

H052   A.F. Hofmann (1963) The Function of Bile Salts in Fat Absorption, *Biochemical Journal*, 89, 57-63.

H053   W. Hancock and E.Q. Laws (1955) The Determination of Traces of Benzene Hexachloride in Water and Sewage Effluents, *Analyst (London)*, 80, 665-674.

H054   R. Hahnel (1971) Interactions of Estradiol-17-beta with Amino Acids, *Journal of Steroid Biochemistry*, 2, 61-65.

H055   H. Hoevel-Kestermann and H. Muhlemann (1972) Analytische Untersuchungen einiger Jodhaltiger Rontgenkontrastmittel im Hinblick auf die Ph. Helv. VI, *Pharmaceutica Acta Helvetiae*, 47, 394-423.

H056   R. Huttenrauch and I. Keiner (1976) Molekulargalenik, *Pharmazie (Berlin)*, 31, 489-491.

H058   R. Haque and D. Schmedding (1975) A Method of Measuring the Water Solubility of Hydrophobic Chemicals: Solubility of Five Polychlorinated Biphenyls, *Bulletin of Environmental Contamination and Toxicology*, 14, 13-17.

H059   A. Hamabata, S. Chang, and P.H. von Hippel (1973) Model Studies on the Effects of Neutral Salts on the Conformational Stability of Biological Macromolecules. III. Solubility of Fatty Acid Amides in Ionic Solutions, *Biochemistry*, 12, 1271-1277.

H060   J.O. Halford (1933) Relative Strength of Benzoic and Salicylic Acids in Alcohol-water Solutions, *Journal of the American Chemical Society*, 55, 2272-2278.

H061   T.T. Herskovits and J.P. Harrington (1972) Solution Studies of the Nucleic Acid Bases and Related Model Compounds. Solubility in Aqueous Alcohol and Glycol Solutions, *Biochemistry*, 11, 4800-4810.

H062   K. Hayashi, T. Matsuda, T. Takeyama, and T. Hino (1966) Solubilities Studies of Basic Amino Acids, *Agricultural and Biological Chemistry*, 30, 378-384.

H063   R. Hruby and V. Kasjanov (1940) Solubility of Sucrose, *The International Sugar Journal*, 42, 21-24.

H064   M.H. Hubacher (1945) Solubility, Density and Melting Point of Phenolphthalein, *Journal of the American Pharmaceutical Association, Scientific Edition*, 34, 76-78.

H066   T. Higuchi and M. Ikeda (1974) Rapidly Dissolving Forms of Digoxin: Hydroquinone Complex, *Journal of Pharmaceutical Sciences*, 63, 809-811.

H067   A. Hussain and J.H. Rytting (1974) Prodrug Approach to Enhancement of Rate of Dissolution of Allopurinol, *Journal of Pharmaceutical Sciences*, 63, 798-799.

H068   J.E. Howard and W.H. Patterson (1926) Miscibility Tests of Dilute Solutions of Chromic Chloride Hexahydrates, *Journal of the Chemical Society (London)*, 129, 2791-2796.

H069   R.N. Haward (1943) Determination of the Solubility of Plasticisers in Water, *Analyst (London)*, 68, 303-305.

H070   R.A. Harte and J.L. Chen (1949) Tryptophan as Solubilizing Agent for Riboflavin, *Journal of the American Pharmaceutical Association, Scientific Edition*, 38, 568-570.

H071   L.P. Hammett and R.P. Chapman (1934) The Solubilities of Some Organic Oxygen Compounds in Sulfuric Acid-Water Mixtures, *Journal of the American Chemical Society*, 56, 1282-1285.

H072   W. Herz (1898) Ueber die Loslichkeit einiger mit Wasser Schwer Mischbarer Flussigkeiten, *Berichte der Deutschen Chemischen Gesellschaft*, 31, 2668-2673.

H073   K.B. Hurle and V.H. Freed (1972) Effect of Electrolytes on the Solubility of Some 1,3,5-Triazines and Substituted Ureas and Their Adsorption on Soil, *Weed Research*, 12, 1-10.

H074   F.W. Hempelmann (1977) Studies on Xipamid (4-Chlor-5-sulfamoyl-2',6'-salicyloxylidide), *Arzneimittel-Forschung*, 27, 2140-2143.

H075   M. Himmelreich, B.J. Rawson, and T.R. Watson (1977) Polymorphic Forms of Mebendazole, *Australian Journal of Pharmaceutical Sciences*, 6, 123-125.

H076   A. Hussain, P. Kulkarni, and D. Perrier (1978) Prodrug Approaches to Enhancement of Physicochemical Properties of Drugs. IX. Acetaminophen Prodrug, *Journal of Pharmaceutical Sciences*, 67, 545-546.

H077   J. Hine, H.W. Haworth, and O.B. Ramsay (1963) Polar Effects on Rates and Equilibria. VI. The Effect of Solvent on the Transmission of Polar Effects, *Journal of the American Chemical Society*, 85, 1473-1476.

H078   A.E. Hill and R. Macy (1924) Ternary Systems. II. Silver Perchlorate, Aniline and Water, *Journal of the American Chemical Society*, 46, 1132-1150.

H080   C.R. Hodgman (1952) Chemodynamics: Transport and Behavior of Chemicals in the Environment—A Problem in Environmental Health, 59-59.

H081   G.P. Haight, Jr. (1951) Solubility of Methyl Bromide in Water and in Some Fruit Juices, *Industrial and Engineering Chemistry*, 43, 1827-1828.

H082   W.F. Hoffman and R.A. Gortner (1922) Sulfur in Proteins. I. The Effect of Acid Hydrolysis Upon Cystine, *Journal of the American Chemical Society*, 44, 341-360.

H083   L. Hertelendi (1943) Zur Loslichkeit der Nicotinsaure (beta-Pyridincarbonsaure, Vitamin B Faktor), *Zeitschrift fuer Physikalische Chemie, Abteilung A: Chemische Thermodynamik, Kinetik, Electrochemie, Eigenschaftslehre*, 192, 379-380.

H084   W. Herz and F. Hiebenthal (1928) Uber Loslichkeitsbeeinflussungen, *Zeitschrift fuer Anorganische Chemie*, 177, 363-380.

H085   H.V. Halban, G. Kortum, and M. Seiler (1935) Die Dissoziationskonstanten Schwacher und Mittelstarker Elektrolyte, *Zeitschrift fuer Physikalische Chemie, Abteilung A: Chemische Thermodynamik, Kinetik, Electrochemie, Eigenschaftslehre*, 173, 449-463.

H087   P.H. Hermans (1925) Die Loslichkeitskurven der Systeme Mannit-Borsaure-Wasser und cis-Tetrahydronaphthalin 1,2-diol-Borsaure-Wasser bei 25, *Zeitschrift fuer Anorganische Chemie*, 142, 111-114.

H089   A.F. Holleman (1910) Sur l'Analyse Quantitative des Produits de la Nitration des Acides Metachloro- et Metabromo-benzoiques, *Recueil des Travaux Chimiques des Pays-Bas et de la Belgique*, 29, 394-402.

H090   B. Holmberg (1921) Stereokemiska Studier V. Diklorbarnstensyrornas Sterokemi, *Arkiv for Kemi, Mineralogi och Geologi*, 8, 1-35.

H091   Y. Hamada, N. Nambu, and T. Nagai (1975) Interactions of alpha- and beta-Cyclodextrin with Several Non-steroidal Antiinflammatory Drugs in Aqueous Solution, *Chemical and Pharmaceutical Bulletin*, 23, 1205-1211.

H092   P.C. Ho, C-H. Ho, and K.A. Kraus (1979) Solubility of Toluene in Aqueous Sodium Alkylbenzenesulfonate Solutions, *Journal of Chemical and Engineering Data*, 24, 115-118.

H093   P. Hajdu, K.F. Kohler, F.H. Schmidt, and H. Spingler (1969) Physico-chemical and Analytical Studies with HB 419, *Arzneimittel-Forschung*, 19, 1381-1386.

H094   S. Horiba (1917) Sucrose-Water, Memoirs of the College of Engineering, Kyoto Imperial University, 2, 519-519.

H096   P. Hajdu and D. Damm (1979) Physico-chemical and Analytical Studies of Penbutolol, *Arzneimittel-Forschung*, 29, 602-606.

H097   D.I. Hitchcock (1924) The Solubility of Tyrosine in Acid and in Alkali, *Journal of General Physiology*, 6, 747-756.

H098   T.A. Hagen (1979) Physicochemical Study of Hydrocortisone and Hydrocortison N-Alkyl-21-esters, Ph.D. thesis.

H099   T. Higuchi, F-M.L. Shih, T. Kimura, and J.H. Rytting (1979) Solubility Determination of Barely Aqueous-soluble Organic Solids, *Journal of Pharmaceutical Sciences*, 68, 1267-1272.

H100   T.B. Hoover (1971) Water Solubilities of PCB Isomers, *PCB Newsletter*, 3, 4-5.

H101   S. Horiba (1917) Studies of Solution. I. The Change of Molecular Solution Volumes in Solutions, Memoirs of the College of Science and Engineering, Kyoto Imperial University, 2, 1-43.

H102   I.C. Hamilton and R. Woods (1979) The Effect of Alkyl Chain Length on the Aqueous Solubility and Redox Properties of Symmetrical Dixanthogens, *Australian Journal of Chemistry*, 32, 2171-2179.

H103   C.S. Hoffman and E.W. Anacker (1967) Water Solubilities of Tetradecanol and Hexadecanol, *Journal of Chromatography*, 30, 390-396.

H104   M-R. Hakala and J.B. Rosenholm (1980) Thermodynamics of Micellization and Solubilization in the System Water+Sodium n-Octanoate+n-Pentanol at 25 C, *Journal of the Chemical Society, Faraday Transactions* 1, 76, 473-488.

H105   A.F. Holleman and P. Caland (1911) Quantitative Untersuchungen uber die Sulfonierung des Toluols, *Berichte der Deutschen Chemischen Gesellschaft*, 44, 162-163.

H106   J.J. Hassett, J.C. Means, W.L. Banwart, and S.G. Wood (1980) Sorption Properties of Sediments and Energy-related Pollutants, *Environmental Protection Agency*, 103-103.

H107   J.J. Hassett, J.C. Means, W.L. Banwart..., and A. Khan (1980) Sorption of Dibenzothiophene by Soils and Sediments, *Journal of Environmental Quality*, 9, 184-186.

H109   R.E. Hughes, Jr. and V.H. Freed (1961) The Determination of Ethyl N,N-Di-n-propy-lthiolcarbamate (EPTC) in Soil by Gas Chromatography, *Journal of Agricultural and Food Chemistry*, 9, 381-382.

H110   D.J.T. Hill and L.R. White (1974) The Enthalpies of Solution of Hexan-1-ol and Heptan-1-ol in Water, *Australian Journal of Chemistry*, 27, 1905-1916.

H111   A.J. Hyde, D.M. Langbridge, and A.S.C. Lawrence (1954) Soap + Water + Amphiphile Systems, *Discussions of the Faraday Society*, 18, 239-258.

H112   C.I. Harris (1966) Adsorption, Movement, and Phytotoxicity of Monuron and s-Triazine Herbicides in Soil, *Weed Science*, 14, 6-10.

H114   F. Hawking (1941) The Rate of Diffusion of Sulphonamide Compounds, *Quarterly Journal of Pharmacy and Pharmacology*, 14, 226-233.

H116   H.C. Hollifield (1979) Rapid Nephelometric Estimate of Water Solubility of Highly Insoluble Organic Chemicals of Environmental Interest, *Bulletin of Environmental Contamination and Toxicology*, 23, 579-586.

H117   R. Haque, D.W. Schmedding, and V.H. Freed (1974) Aqueous Solubility, Adsorption, and Vapor Behavior of Polychlorinated Biphenyl Aroclor 1254, *Environmental Science and Technology*, 8, 139-142.

H118   T.L. Hafkenscheid and E. Tomlinson (1981) Estimation of Aqueous Solubilities of Organic Non-electrolytes Using Liquid Chromatographic Retention Data, *Journal of Chromatography*, 218, 409-425.

H119   M.E. Haberland and J.A. Reynolds (1973) Self-association of Cholesterol in Aqueous Solution, *Proceedings of the National Academy of Science of the United States of America*, 70, 2313-2316.

H120   K. Hirano, T. Ichihashi, and H. Yamada (1981) Studies on the Absorption of Practically Water-insoluble Drugs Following Injection. II. Intramuscular Absorption from Aqueous Suspensions in Rats, *Chemical and Pharmaceutical Bulletin*, 29, 817-827.

H121   K. Hlavaty and J. Linek (1973) Liquid-Liquid Equilibria in Four Ternary Acetic Acid-organic Solvent-Water Systems at 24.6 C, *Collection of Czechoslovak Chemical Communications*, 38, 374-378.

H122   R.S. Hansen, F.A. Miller, and S.D. Christian (1955) Activity Coefficients of Components in the Systems Water-Acetic Acid, Water-Propionic Acid and Water-n-Butyric Acid at 25, *Journal of Physical Chemistry*, 59, 391-395.

H123   T.C. Hutchinson, J.A. Hellebust, D. Tam..., and W.Y. Shiu (1978) The Correlation of the Toxicity to Algae of Hydrocarbons and Halogenated Hydrocarbons with Their Physical-chemical Properties, *Environment Science Research*, 13, 577-586.

H124   W. Hayduk and V.K. Malik (1971) Density, Viscosity, and Carbon Dioxide Solubility and Diffusivity in Aqueous Ethylene Glycol Solutions, *Journal of Chemical and Engineering Data*, 16, 143-146.

H125   M.M.A. Hassan, A.I. Jado, and M.U. Zubair (1981) Aminosalicylic Acid, *Analytical Profiles of Drug Substances*, 10, 1-27.

H127   H. Harms (1939) Uber die Energieverhaltnisse der OH-OH-Bindung, *Zeitschrift fur Physical Chemistry Abt B*, 43, 257-270.

H129   R. Huttenrauch and S. Fricke (1982) Zur Beziehung zwischen Ordnungsgrad und Losungsvermogen des Wassers, *Pharmazie (Berlin)*, 37, 147-148.

H300   Y. Hashimoto, K. Tokura, K. Ozaki, and W.M.J. Strachan (1982) A Comparison of Water Solubilities by the Flask and Micro-column Methods, *Chemosphere*, 11, 991-1001.

H301   T.L. Hafkenscheid and E. Tomlinson (1983) Isocratic Chromatographic Retention Data for Estimating Aqueous Solubilities of Acidic, Basic and Neutral Drugs, *International Journal of Pharmaceutics*, 17, 1-21.

H302   C.D. Herzfeldt and R. Kummel (1983) Dissociation Constants, Solubilities and Dissolution Rates of Some Selected Nonsteroidal Antiinflammatories, *Drug Development and Industrial Pharmacy*, 9, 767-793.

H303   J.L. Hesse and R.A. Powers (1978) Polybrominated Biphenyl (PBB) Contamination of the Pine River, Gratiot, and Midland Counties, Michigan, *Environmental Health Perspectives*, 23, 19-25.

H306   Y. Hashimoto, K. Tokura, H. Kishi, and W.M.J. Strachan (1984) Prediction of Seawater Solubility of Aromatic Compounds, *Chemosphere*, 13, 881-888.

H307   K.O. Hiller, B. Masloch, and H.J. Mockel (1977) Zusammenhang Zwischen Loslichkeit und Kapazitatsfaktor bei der Reverse-Phase-Bonded-Phase-Chromatographie von Alkylbenzolen, Alkylbromiden und Alkyldisulfiden, *Ztschrft fur Analytische Chemie*, 283, 109-113.

H308   R.F. Holt and R.E. Leitch (1978) Oxamyl, *Analytical Methods for Pesticides and Plant Growth Regulators*, 10, 111-118.

H309   T. Hattori and M. Kanauchi (1978) Isoprothiolane, *Analytical Methods for Pesticides and Plant Growth Regulators*, 10, 229-236.

H313   P.C. Ho (1985) Solubilities of Toluene and n-Octane in Aqueous Protosurfactant and Surfactant Solutions, *Journal of Chemical and Engineering Data*, 30, 88-90.

H316   H.W. Hamilton, D.F. Ortwine, D.F. Worth..., and R.P. Steffen (1985) Synthesis of Xanthines as Adenosine Antagonists, a Practical Quantitative Structure-activity Relationship Application, *Journal of Medicinal Chemistry*, 28, 1071-1079.

H319   T. Hirayama, N. Hirashima, K. Abe, and M. Ueno (1988) Utilitization of Diethyl-beta-cyclodextrin as a Sustained-release Carrier for Isosorbide Dinitrate, *Journal of Pharmaceutical Sciences*, 77, 233-236.

H320   T.A. Hagen and G.L. Flynn (1987) Permeation of Hydrocortisone and Hydrocortisone 21-Alkyl Esters Through Silicone Rubber Membranes—Relationship to Regular Solution Solubility Behavior, *Journal of Membrane Science*, 30, 47-65.

H322   F. Herzel and A.S. Murty (1984) Do Carrier Solvents Enhance the Water Solubility of Hydrophobic Compunds?, *Bulletin of Environmental Contamination and Toxicology*, 32, 53-58.

H324   J. Hughes, P. Tenni, C. McDonald, and V.B. Sunderland (1982) Solubility of Metronidazole for Topical Application, *Australian Journal of Hospital Pharmacy*, 12, 58-58.

H328   J.G. Hoogerheide and B.E. Wyka (1982) Clotrimazole, *Analytical Profiles of Drug Substances*, 11, 225-229.

H330   J.R. Hommelen (1959) The Elimination of Errors Due to Evaporation of the Solute in the Determination of Surface Tensions, *Journal of the American Chemical Society*, 14, 385-400.

H332   E. Hogfeldt and B. Bolander (1963) On the Extraction of Water and Nitric Acid by Aromatic Hydrocarbons, *Arkiv foer Kemi*, 21, 161-186.

H333   U. Henriksson, T. Klason, L. Odberg, and J.C. Eriksson (1977) Solubilization of Benzene and Cyclohexane in Aqueous Solutions of Hexadecyltrimethylammonium Bromide: A Deuterium Magnetic Resonance Study, *Chemical Physics Letters*, 52, 554-558.

H337   H.W. Haggard (1923) An Accurate Method of Determining Small Amounts of Ethyl Ether in Air, Blood, and Other Fluids, Together with a Determination of the Coefficient of Distribution of Ether Between Air and Blood at Various Temperatures, *Journal of Biological Chemistry*, 55, 131-143.

H338   E.L. Heric and R.E. Langford (1972) System Furfural-Water-Valeric Acid at 25 and 35 C, *Journal of Chemical and Engineering Data*, 17, 209-211.

H339   E.L. Heric and R.E. Langford (1972) System Furfural-Water-Caproic Acid at 25 and 35 C, *Journal of Chemical and Engineering Data*, 17, 471-473.

H340   E.L. Heric, B.H. Blackwell, L.J. Gaissert..., and J.W. Pierce (1966) Distribution of Butyric Acid Between Furfural and Water at 25 and 35 C, *Journal of Chemical and Engineering Data*, 11, 38-40.

H341   G.L. Huang (1983) Dissertation or Masters Thesis.

H342   L.A. Hardaway and S.H. Yalkowsky (1991) Cosolvent Effects on Diuron Solubility, *Journal of Pharmaceutical Sciences*, 80, 197-198.

H343   M.A. Huerta-Diaz and S. Rodriguez (1992) Solubility Measurements and Determination of Setschenow Constants for the Pesticide Carbaryl in Seawater and Other Electrolyte Solutions, *Canadian Journal of Chemistry*, 70, 2864-2868.

H345   O. Harva (1956) The Effect of Long-Chain Alcohols on the Properties of Sodium Laurate Solutions, *Recueil des Travaux Chimiques des Pays-Bas*, 75, 101-111.

H347   A. Haj-Yehia and M. Bialer (1989) Structure-Pharmacokinetic Relationships in a Series of Valpromide Derivatives with Antiepileptic Activity, *Pharmaceutical Research*, 6, 683-689.

H348   A. Haj-Yehia and M. Bialer (1990) Structure-Pharmacokinetic Relationships in a Series of Short Fatty Acid Amides that Possess Anticonvulsant Activity, *Journal of Pharmaceutical Sciences*, 79, 719-724.

H350   H.L. Holmes (1975) *Structure-Activity Relationships for Some Conjugated Heteroenoid Compounds, Catechol Monoethers and Morphine Alkaloids*, Volume II. Defence Research Establishment, Suffield, 2.

I001   A.A. Ismail, M.W. Gouda, and M.M. Motawi (1970) Micellar Solubilization of Barbiturates. I. Solubilities of Certain Barbiturates in Polysorbates of Varying Hydrophobic Chain Length, *Journal of Pharmaceutical Sciences*, 59, 220-224.

I002   H.G. Ibrahim, F. Pisano, and A. Bruno (1977) Polymorphism of Phenylbutazone: Properties and Compressional Behavior of Crystals, *Journal of Pharmaceutical Sciences*, 66, 669-673.

I006   L. Irrera (1931) Influenze di Solubilita, *Gazzetta Chimica Italiana*, 61, 614-618.

I007   K. Ikeda, K. Uekama, and M. Otagiri (1975) Inclusion Complexes of beta-Cyclodextrin with Antiinflammatory Drugs Fenamates in Aqueous Solution, *Chemical and Pharmaceutical Bulletin*, 23, 201-208.

I008   H. Igimi and M.C. Carey (1979) Dissimilar pH-Solubility Relations of Chenodeoxycholic (CDCA) and Ursodeoxycholic (UDCA) Acids, *Gastroenterology*, 76, 1159-1159.

I009   K. Ikeda, K. Kato, and T. Tukamoto (1971) Solubilization of Barbiturates by Polyoxyethylene Lauryl Ether, *Chemical and Pharmaceutical Bulletin*, 19, 2510-2517.

I010   S.A. Ibrahim, H.O. Ammar, A.A. Kassem, and S.S. Abu-Zaid (1979) Effect of Some Hydrotropic Agents on the Water Solubility of Aminophenazone, *Pharmazie (Berlin)*, 34, 809-812.

I011   R.F. Inga and J.J. McKetta (1961) Solubility of Propyne in Water, *Journal of Chemical and Engineering Data*, 6, 337-338.

I012   H. Igimi and M.C. Carey (1980) pH-Solubility Relations of Chenodeoxycholic and Ursodeoxycholic Acids: Physical-chemical Basis for Dissimilar Solution and Membrane Phenomena, *Journal of Lipid Research*, 21, 72-90.

I015   A. Ikekawa and S. Hayakawa (1981) Mechanochemical Change in the Solid State and the Solubility of Amobarbital, *Bulletin of the Chemical Society of Japan (Nippon Kagakukai Bulletin)*, 54, 2587-2591.

I017   R.F. Inga and J.J. McKetta (1961) Solubility of Cyclopropane in Water, *Petroleum Refiner*, 40, 191-192.

I018   K.A. Ivanov (1956) Solubility of Lindane in H$_2$0, *Gigiena i Sanitariya*, 21, 82-83.

I019   Y. Ikeda, K. Matsumoto, K. Kunihiro..., and K. Uekama (1982) Inclusion Complexation of Essential Oils with alpha- and beta-Cyclodextrins, *Yakugaku Zasshi (Tokyo)*, 102, 83-88.

I300   F. Irmann (1965) Eine Einfache Korrelation zwischen Wasserloslichkeit und Struktur von Kohlenwasserstoffen und Halogenkohlenwasserstoffen, *Chemie-Ingenieur-Technik*, 37, 789-798.

I304   S.A. Ibrahim and S. Shawky (1983) Effect of Some Aliphatic Acids, Their Sodium and Potassium Salts on Aqueous Solubility of Certain Diuretics, *Pharmazeutische Industrie*, 45, 207-212.

I306   IARC Committee (1971) *Iarc Monographs on the Evaluation of the Carcinogenic Risk of Chemicals to Man*, 1, 1-181.

I307   IARC Committee (1973) Some Aromatic Amines, Hydrazine and Related Substances, N-Nitroso Compounds and Miscellaneous Alkylating Agents, in *Iarc Monographs on the Evaluation of the Carcinogenic Risk of Chemicals to Man*, 4, 1-259.

I308   IARC Committee (1973) Some Organochlorine Pesticides, in *Iarc Monographs on the Evaluation of the Carcinogenic Risk of Chemicals to Man*, 5, 1-211.

I309   IARC Committee (1974) Sex Hormones, in *Iarc Monographs on the Evaluation of the Carcinogenic Risk of Chemicals to Man*, 6, 1-210.

I310   IARC Committee (1974) Some Anti-thyroid and Related Substances, Nitrofurans and Industrial Chemicals,in *Iarc Monographs on the Evaluation of the Carcinogenic Risk of Chemicals to Man*, 7, 1-291.

I312   IARC Committee (1975) Some Naturally Occurring Substances, in *Iarc Monographs on the Evaluation of the Carcinogenic Risk of Chemicals to Man*, 10, 1-328.

I313   IARC Committee (1976) Cadmium, Nickel, Some Epoxides, Miscellaneous Industrial Chemicals and General Considerations on Volatile Anaesthetics, in *Iarc Monographs on the Evaluation of the Carcinogenic Risk of Chemicals to Man*, 11, 1-277.

I314   IARC Committee (1976) Some Carbamated, Thiocarbamates and Carbazides, in *Iarc Monographs on the Evaluation of the Carcinogenic Risk of Chemicals to Man*, 12, 1-260.

I315   IARC Committee (1976) Some Miscellaneous Pharmaceutical Substances, in *Iarc Monographs on the Evaluation of the Carcinogenic Risk of chemicals to Man*, 13, 1-233.

I316   IARC Committee (1977) Some Fumigants, the Herbicides 2,4-D and 2,4,5-T, Chlorinated Dibenzodioxins and Miscellaneous Industrial Chemicals, in *Iarc Monographs on the Evaluation of the Carcinogenic Risk of Chemicals to Man*, 15, 1-265.

I332   P. Isnard and S. Lambert (1988) Estimating Bioconcentration Factors from Octanol-Water Partition Coefficient and Aqueous Solubility, *Chemosphere*, 17, 21-34.

I333   E. Iwamoto, Y. Tanaka, H. Kimura, and Y. Yamamoto (1980) Solute-solvent Interactions with Metal Chelate Electrolytes. Part III. Salting in of Tris(acetylacetonato)cobalt (III) and Benzene by Aromatic and Aliphatic Ions, *Journal of Solution Chemistry*, 9, 841-856.

I334   E. Iwamoto, Y. Hiyama, and Y. Yamamoto (1977) Hydrophobic and Charge-dipole Interactions in Aqueous Solutions of Highly Charged Metal Chelate Cations and Nitrobenzene, Dinitrobenzenes, and Toluene at 25 C, *Journal of Solution Chemistry*, 6, 371-383.

I335   E. Iwamoto, M. Yamamoto, and Y. Yamamoto (1974) Salting-in of Nitrobenzene and Toluene by Metal Chelate Electrolytes, *Inorganic Nuclear Chemistry Letters*, 10, 1069-1076.

J001   L.M. John and J.W. McBain (1948) The Hydrolysis of Soap Solutions. II. The Solubilities of Higher Fatty Acids, *Journal of the American Oil Chemists' Society*, 25, 40-41.

J003   G.M. Jacobsohn and D. Levenberg (1964) Solubilities of 17-Ketosteroids in Water, *Steroids, An International Journal (San Francisco)*, 4, 849-853.

J004   K.C. James and M. Roberts (1968) The Solubilities of the Lower Testosterone Esters, *Journal of Pharmacy and Pharmacology*, 20, 709-714.

J005   G.V. Jeffreys (1963) Phase Equilibrium for System Methylethylketone, Cyclohexane, and Water, *Journal of Chemical and Engineering Data*, 8, 320-323.

J007   W.J. Jones and J.B. Speakman (1921) Some Physical Properties of Aqueous Solutions of Certain Pyridine Bases, *Journal of the American Chemical Society*, 43, 1867-1870.

J008   E.G. Jaworski (1969) Chloroacetamides, *In Degradation of Herbicides*, Marcel Dekker, 165-167.

J009   R.K. Joshi, L. Krasnec, and I. Lacko (1971) Studies on Solubilization. Part II, *Pharmaceutica Acta Helvetiae*, 46, 570-582.

J010   I.M. Jakovljevic (1974) Digitoxin, *Analytical Profiles of Drug Substances*, 3, 149-159.

J011   K.C. James and R.H. Leach (1970) A Borax-chloramphenicol Complex in Aqueous Solution, *Journal of Pharmacy and Pharmacology*, 22, 612-614.

J012   D.C. Jones (1929) The Systems n-Butyl Alcohol-Water and n-Butyl Alcohol-Acetone-Water, *Journal of the Chemical Society (London)*, 799-813.

J016   J.O. Janes, P.M. Loeb, R.N. Berk, and J.M. Dietschy (1977) Intestinal Absorption of Oral Cholecystographic Agents, *Clinical Research*, 25, 312-312.

J017   E. Janecke (1933) Uber das System Methylalkohol-Isobutylalkohol-Wasser, *Zeitschrift fuer Physikalische Chemie*, Abteilung A: Chemische Thermodynamik, Kinetik, Electrochemie, Eigenschaftslehre, 164, 401-416.

J018   K. Juni, M. Nakano, and T. Arita (1977) Controlled Drug Permeation. II. Comparative Permeability and Stability of Butamben and Benzocaine, *Chemical and Pharmaceutical Bulletin*, 25, 1098-1100.

J019   R.F. Jackson and C.G. Silsbee (1922) The Solubility of Dextrose in Water, *Scientific Papers, National Bureau of Standards*, 17, 715-724.

J020   J.H. Jones and J.F. McCants (1954) Ternary Solubility Data: 1-Butanol-Methyl 1 Butyl Ketone-Water, 1-Butyraldehyde-Ethyl Acetate-Water, 1-Hexane-Methyl Ethyl Ketone-Water, *Industrial and Engineering Chemistry*, 46, 1956-1958.

J021   E. Janecke-Heidelberg (1930) Einzelvortrage, *Zeitschrift fuer Elektrochemie*, 36, 645-654.

J022   K. Juni, T. Tomitsuka, M. Nakano, and T. Arita (1978) Analysis of Permeation Profiles of Drugs from Systems Containing Micelles, *Chemical and Pharmaceutical Bulletin*, 26, 837-841.

J023   A. Jaeger (1923) Uber die Loslichkeit von Flussigen Kohlenwasserstoffen in Uberhitztem Wasser, Brennstoff-Chemie, 4, 259-260.

J025   H. Jacobi, A. Lange, and K. Pfleger (1956) Vergleichende Untersuchungen Wasserloslicher Theophyllin-derivate, Arzneimittel-Forschung, 6, 41-43.

J026   K. Juni, K. Nomoto, M. Nakano, and T. Arita (1979) Drug Release Through a Silicone Capsular Membrane From Micellar Solution, Emulsion, and Cosolvent Systems and the Correlation of Release Data in Vivo with Release Profile in Vitro, *Journal of Membrane Science*, 5, 295-304.

J027   M. Janado, K. Takenaka, H. Nakamori, and Y. Yano (1980) Solubilities of Water-Insoluble Dyes in Internal Water of Swollen Sephadex Gels, *Journal of Biochemistry (Tokyo)*, 87, 57-62.

J028   R. Jachowicz, L. Krowczynski, and Z. Kubiak (1977) Increasing the Solubility of Khellia for the Preparation of Injection Solutions, *Farmagia Polska*, 33, 419-422.

J030   J.C. Jespersen and K.T. Larsen (1934) Identificering af Terapeutisk Vigtige Barbitursyrer, *Yakugaku Zasshi*, 8, 212-226.

J031   M. Johansen and H. Bundgaard (1980) Pro-drugs as Drug Delivery Systems. XIII. Kinetics of Decomposition of N-Mannich Bases of Salicylamide and Assessment of Their Suitability as Possible Pro-drugs for Amines, *International Journal of Pharmaceutics*, 7, 119-127.

J033   L.S. Jordan (1970) Residue Reviews, *Residue Reviews*, 32, 7-13.

J034   H. Jones (1981) Complex Formation Between Digoxin and beta-Cyclodextrin, *Journal of Pharmacy and Pharmacology*, 33, 27-27.

J035   M. Janado and T. Nishida (1981) Effect of Sugars on the Solubility of Hydrophobic Solutes in Water, *Journal of Solution Chemistry*, 10, 489-500.

J036   D.C. Jones, R.H. Ottewill, and A.P.J. Chater (1957) The Adsorption of Insoluble Vapours on Water Surfaces, 188-199.

J037   M. Johansen and H. Bundgaard (1980) Pro-drugs as Drug Delivery Systems. XII. Solubility, Dissolution and Partitioning Behaviour of N-Mannich Bases and N-Hydroxymethyl Derivatives, *Archive of Pharmacy Chemistry Science*, 8, 717-727.

J300   K.C. James (1984) Calculation of Molecular Surface Areas and Aqueous Solubilities at Ambient Temperatures, *International Journal of Pharmaceutics*, 21, 123-128.

J302   M. Janado and Y. Yano (1985) The Nature of the Cosolvent Effects of Sugars on the Aqueous Solubilities of Hydrocarbons, *Bulletin of the Chemical Society of Japan (Nippon Kagakukai Bulletin)*, 58, 1913-1917.

J303   X.Z. Jin and K.C. Chao (1992) Solubility of Four Amino Acids in Water and of Four Pairs of Amino Acids in Their Water Solutions, *Journal of Chemical and Engineering Data*, 37, 199-203.

J305   M.J. Jozwiakowski and K.A. Connors (1985) Aqueous Solubility Behavior of Three Cyclodextrins, *Carbohydrate Research*, 143, 51-59.

J308   M.D. Johnson, B.L. Hoesterey, and B.D. Anderson (1984) Solubilization of a Tripeptide HIV Protease Inhibitor Using a Combination of Ionization and Complexation with Chemically Modified Cyclodextrins, *Journal of Pharmaceutical Sciences*, 83, 1142-1146.

K001   H.B. Klevens (1950) Solubilization of Polycyclic Hydrocarbons, *Journal of Physical and Colloid Chemistry*, 54, 283-297.

K002   I.A. Kablukov and V.T. Malischeva (1925) The Volumetric Method of Measurement of the Mutual Solubility of Liquids. The Mutual Solubility of the Systems Ethyl Ether-Water and Iso-amyl Alcohol-Water, *Journal of the American Chemical Society*, 47, 1553-1561.

K003   P. Kabasakalian, E. Britt, and M.D. Yudis (1966) Solubility of Some Steroids in Water, *Journal of Pharmaceutical Sciences*, 55, 642-642.

K004   H.B. Kostenbauder and T. Higuchi (1957) A Note on the Water Solubility of some N,N-Dialkylamides, *Journal of the American Pharmaceutical Association, Scientific Edition*, 46, 205-206.

K005   I.M. Korenman (1971) Hydrotropic Dissolution, *Russian Journal of Physical Chemistry*, 45, 1011-1011.

K006   H.A. Krebs and J.C. Speakman (1946) The Solubility of Sulphonamides in Relation to Hydrogen-ion Concentration, *British Medical Journal*, 1, 47-50.

K007   J.J. Kirkland (1971) Columns for Modern Analytical Liquid Chromatography, *Analytical Chemistry*, 43, 36-48.

K008   K. Kakemi, H. Sezaki, M. Nakano, and K. Ohsuga (1969) Effect of Structure of Pyridinecarboxylic Acids and Hydroxypyridines on Molecular Interaction in Water, *Journal of Pharmaceutical Sciences*, 58, 699-702.

K009   K. Kakemi, H. Sezaki, T. Mitsunaga, and M. Nakano (1970) Effect of Structural Similarity on Molecular Interaction in Aqueous Solution: Interaction of Phenazine and Tetramethylpyrimido-pteridinetetrone with Alkylxanthines and Benzene Derivatives, *Journal of Pharmaceutical Sciences*, 59, 1597-1601.

K010   G.M. Krause and J.M. Cross (1951) Solubility of Phenobarbital in Alcohol-Glycerin-Water Systems, *Journal of the American Pharmaceutical Association, Scientific Edition*, 40, 137-139.

K011   F.P. Krause and W. Lange (1965) Aqueous Solubilities of n-Dodecanol, n-Hexadecanol, and n-Octadecanol by a New Method, *Journal of Physical Chemistry*, 69, 3171-3173.

K012   I.A. Kakovsky (1957) Physicochemical Properties of Some Flotation Reagents and Their Salts with Ions of Heavy Non-ferrous Metals, *Solubilization and Micelles*, 4, 225-237.

K013   H. Kakinuma (1941) The Solubility of Urea in Water, *Journal of Physical Chemistry*, 45, 1045-1046.

K017   S.F. Kramer and G.L. Flynn (1972) Solubility of Organic Hydrochlorides, *Journal of Pharmaceutical Sciences*, 61, 1896-1903.

K018   B. Katchen and S. Symchowicz (1967) Correlation of Dissolution Rate and Griseofulvin Absorption in Man, *Journal of Pharmaceutical Sciences*, 56, 1108-1111.

K019   H.B. Kostenbauder and T. Higuchi (1956) Formation of Molecular Complexes by Some Water-soluble Amides. II. Effect of Decreasing Water Solubility on Degree of Complex Formation, *Journal of the American Pharmaceutical Association, Scientific Edition*, 45, 810-813.

K020   H.B. Kostenbauder and T. Higuchi (1956) Formation of Molecular Complexes by Some Water-soluble Amides I. Interaction of Several Amides with p-Hydroxybenzoic Acid, Salicylic Acid, Chloramphenicol, and Phenol, *Journal of the American Pharmaceutical Association, Scientific Edition*, 45, 518-522.

K021   M. Katz and Z.I. Shaikh (1965) Percutaneous Corticosteroid Absorption Correlated to Partition Coefficient, *Journal of Pharmaceutical Sciences*, 54, 591-594.

K022   S.A. Kaplan, R.E. Weinfeld, C.W. Abruzzo, and M. Lewis (1972) Pharmacokinetic Profile of Sulfisoxazole Following Intravenous, Intramuscular, and Oral Administration to Man, *Journal of Pharmaceutical Sciences*, 61, 773-778.

K023   N. Khalafallah and Y. Hammouda (1973) The Solubility and Complexing Properties of Acetohexamide in the Presence of Hydrotropic Agents, *Pharmazie (Berlin)*, 28, 452-454.

K024   N. Kakeya, M. Aoki, A. Kamada, and N. Yata (1969) Biological Activities of Drugs. VI. Structure-Activity Relationship of Sulfonamide Carbonic Anhydrase Inhibitors, *Chemical and Pharmaceutical Bulletin*, 17, 1010-1018.

K025   K. Kinoshita, H. Ishikawa, and K. Shinoda (1958) Solubility of Alcohols in Water Determined by the Surface Tension Measurements, *Bulletin of the Chemical Society of Japan (Nippon Kagakukai Bulletin)*, 31, 1081-1083.

K026   H. Krasowska, L. Krowczynski, and E. Glab (1972) Solubility of Indomethacin in Organic Solvent and Solvent-Water Systems, *Dissertationes Pharmaceuticae et Pharmacologicae*, 24, 623-630.

K027   H. Krasowska (1976) Solubilization of Indomethacin and Cinmetacin by Non-ionic Surfactants of the Polyoxyethylene Type, *Farmaco, Edizione Pratica (PAVIA)*, 31, 463-472.

K028   M. Kuhnert-Brandstatter and A. Martinek (1965) Uber den Einfluss der Polymorphie auf die Loslichkeit von Arzneimitteln, *Mikrochimica et Ichnoanalytica Acta*, 909-919.

K029   D.G. Kaiser, W.C. Krueger, L.M. Pschigoda, and B.F. Zimmer (1969) Aqueous Solubilities and Distribution Coefficients of U-20,235, U-25,312, U-2726 and U-22,338, Technical Report, 1-7.

K031   G.C. Kresheck, H. Schneider, and H.A. Scheraga (1965) The Effect of D_2O on the Thermal Stability of Proteins. Thermodynamic Parameters for the Transfer of Model Compounds from H_2O to D_2O, *Journal of Physical Chemistry*, 69, 3132-3144.

K032   P. Kristian and L. Drobnica (1966) Reactions of Isothiocyanates with Amino Acids, Peptides and Proteins. IV. Kinetics of the Reaction of Substituted Phenylisothiocyanates with Glycine, *Collection of Czechoslovak Chemical Communications*, 31, 1333-1339.

K033   S. Korman and V.K. La Mer (1936) Deuterium Exchange Equilibria in Solution and the Quinhydrone Electrode, *Journal of the American Chemical Society*, 58, 1396-1403.

K034   R.H. Kienle and J.M. Sayward (1942) Solubilities of Orthanilamide, Metanilamide and Sulfanilamide, *Journal of the American Chemical Society*, 64, 2464-2468.

K035   N.P. Komar, V.V. Mel'nik, K.V. Zimina, and A.G. Kozachenko (1971) Solubility of Adipic Acid, *Vestnik Khar'Kovskogo Universiteta*, 67-71.

K036   A.K.N. Kuroki and K. Konishi (1964) Distribution of Disperse Dye Between Water and Benzene Phases, *Sen'i Kikai Gakkaishi*, 20, 256-261.

K040   J. Knox and M.B. Richards (1919) The Basic Properties of Oxygen in Organic Acids and Phenols; and the Quadrivalency of Oxygen, *Journal of the Chemical Society (London)*, 115, 508-531.

K041   I.M. Kovach, I.H. Pitman, and T. Higuchi (1975) Amino Acid Esters of Phenolic Drugs as Potentially Useful Prodrugs, *Journal of Pharmaceutical Sciences*, 64, 1070-1071.

K042   J.R. King and M.C. Bowman (1975) 4-Ethylsulfonylnaphthalene-1-sulfonamide (ENS): Analytical Chemical Behavior and Trace Analysis in Five Substrates, *Biochemical Medicine*, 12, 313-330.

K043   B. Kreilgard, T. Higuchi, and A.J. Repta (1975) Complexation in Formulation of Parenteral Solutions: Solubilization of the Cytotoxic Agent Hexamethylmelamine by Complexation with Gentisic Acid Species, *Journal of Pharmaceutical Sciences*, 64, 1850-1855.

K044   M.A. Kaplan, W.P. Coppola, B.C. Nunning, and A.P. Granatek (1976) Pharmaceutical Properties and Stability of Amikacin. Part I, *Current Therapeutic Research, Clinical and Experimental*, 20, 352-358.

K046   Y. Kawashima, M. Saito, and H. Takenaka (1975) Improvement of Solubility and Dissolution Rate of Poorly Water-soluble Salicylic Acid by a Spray-drying Technique, *Journal of Pharmacy and Pharmacology*, 27, 1-5.

K047   M. Kanke and K. Sekiguchi (1973) Dissolution Behavior of Solid Drugs. I. Improvement and Simplification of Dissolution Rate Measurement, and Its Application to Solubility Determinations, *Chemical and Pharmaceutical Bulletin*, 21, 871-877.

K048   H.A. Krebs and J.C. Speakman (1945) Dissociation Constant, Solubility, and the pH Value of the Solvent, *Journal of the Chemical Society (London)*, 593-595.

K049   M. Kodama, Y. Tagashira, A. Imamura, and C. Nagata (1966) Effect of Secondary Structure of DNA upon Solubility of Aromatic Hydrocarbons, *Journal of Biochemistry (Tokyo)*, 59, 257-264.

K050   H. Koch and R. Bodmann (1976) Akute Toxizitat von Endomid und seinen Metaboliten Korrelation zwischen Biologischer Aktivitat und Lipophilen Eigenschaften, *Archiv der Pharmazie*, 309, 812-822.

K051   I.M. Kolthoff and A.I. Medalia (1949) The Reaction between Ferrous Iron and Peroxides. III. Reaction with Cumene Hydroperoxide, in Aqueous Solution, *Journal of the American Chemical Society*, 71, 3789-3792.

K052   D. Kuttel (1968) Die Solubilisierungsmoglichkeit einiger Alkaloidbasen mit Tween 80, *Pharmazeutische Zentralhalle*, 107, 593-600.

K053   J. Kendall (1911) On the Ionic Solubility-product, *Proceedings Royal Society*, 85, 200-219.

K055   A.P. Kudchadker and J.J. McKetta (1961) Solubility of Cyclohexane in Water, *American Institute of Chemical Engineers Journal*, 7, 707-707.

K056   A. Klemenc and M. Low (1930) Die Loslichkeit in Wasser und ihr Zusammenhang der Drei Dichlorbenzole. Eine Methode zur Bestimmung der Loslichkeit sehr Wenig Loslicher und Zugleich sehr Fluchtiger Stoffe, Recueil des Travaux Chimiques des Pays-Bas, 49, 629-640.

K057   J. Kendall and J.C. Andrews (1921) The Solubilities of Acids in Aqueous Solutions of Other Acids, *Journal of the American Chemical Society*, 43, 1545-1560.

K058   J.C. Krantz, Jr., W.E. Evens, Jr., S.E. Forman, and H.L. Wollenweber (1942) Anesthesia. VI. The Anesthetic, Action of Cyclopropyl Vinyl Ether, *Journal of Pharmacology and Experimental Therapeutics*, 75, 30-37.

K059   J.M. Kolthoff (1925) Die Dissoziationskonstante, das Loslichkeitsprodukt und die Titrierbarkeit von Alkaloiden, Biochemische Zeitschrift (1948-1967), 162, 289-353.

K060   H. Kudielka (1908) Zur Kenntnis der alpha-Amino-n-capronsaure, *Monatshefte fuer Chemie*, 29, 351-358.

K061   J.C. Krantz, Jr., C.J. Carr, S.E. Forman..., and H. Wollenweber (1941) Anesthesia. IV. The Anesthetic Action of Cyclopropyl Ethyl Ether, *Journal of Pharmacology and Experimental Therapeutics*, 72, 233-244.

K062   S. Kumar, S.N. Upadhyay, and V.K. Mathur (1978) On the Solubility of Benzoic Acid in Aqueous Carboxymethylcellulose Solutions, *Journal of Chemical and Engineering Data*, 23, 139-141.

K063   N.P. Komar and G.S. Zaslavskaya (1973) The Solubility of alpha-alpha'-Bipyridyl in Aqueous Salt Solutions, *Russian Journal of Physical Chemistry*, 47, 1642-1643.

K064   I.M. Kolthoff and W. Bosch (1932) The Activity Coefficient of Benzoic Acid in Solutions of Neutral Salts and of Sodium Benzoate, *Journal of Physical Chemistry*, 36, 1685-1694.

K065   V.M. Kisarov (1962) Solubility of Chlorobenzene in Water, *Journal of Applied Chemistry of the USSR*, 35, 2252-2253.

K067   F. Kral and G. Strauss (1978) A Biologically Active Borate Derivative of Amphotericin B Soluble in Saline Solution, *The Journal of Antibiotics*, 31, 257-259.

K068   I.M. Kolthoff and L.A. Sarver (1930) Properties of Diphenylamine and Diphenylbenzidine as Oxidation-reduction Indicators, *Journal of the American Chemical Society*, 52, 4179-4191.

K069   E.E. Kenaga (1971) Some Physical, Chemical, and Insecticidal Properties of Some O,O-Dialkyl O-(3,5,6-Trichloro-2-pyridyl) Phosphates and Phosphorothioates, *Bulletin of the World Health Organization*, 44, 225-228.

K070   J.S. Kent (1976) Controlled Release of Delmadinone Acetate From Silicone Polymer Tubing: In Vitro-In Vivo Correlations, *American Chemical Society*, Division of Organic Coating and Plastics Chemistry, 36, 356-361.

K072   H.B. Klevens (1950) Solubilization, *Chemical Reviews*, 47, 1-73.

K075   R. Kremann and E. Janetzky (1923) Das Ternare System Antipyrin-Coffein-Wasser ein Beitrag zur Kenntnis des Migranins, *Monatshefte fuer Chemie*, 44, 49-63.

K076   I.M. Kolthoff (1926) The Hydration of Dissolved Saccharose and the Expression of the Concentration in Measuring the Activity of Ions, *Proceedings Royal Academy of Science of Amsterdam*, 29, 885-898.

K077   E.A. Klobbie (1894) Gleichgewichte in den Systemen Aether-Wasser und Aether-Wasser-Malonsaure, *Zeitschrift fuer Physikalische Chemie (Leipzig)*, 14, 615-632.

K078   D. Kuttel (1964) Die Solubilisationsfahigkeit des Tween 20, 60, 80 bei Einigen in Wasser Schlecht Loslichen Medikamenten, *Pharmazeutische Zentralhalle*, 103, 10-16.

K079   J. Kendall and L.E. Harrison (1928) Compound Formation in Ester-Water Systems, *Transactions of the Faraday Society*, 24, 588-596.

K084   R. Kremann and H. Eitel (1923) Das Ternare System Zucker-Zitronensaure-Wasser ein Beitrag zur Theorie der Speiseeise vom Standpunkt der Phasenlehre, Recueil des Travaux Chimiques des Pays-Bas, 42, 539-546.

K085   I.L. Krupatkin (1956) Ternary Systems with Layering without Formation of Chemical Compounds, *Journal of General Chemistry of the USSR*, 26, 1815-1819.

K086   N. Kaneniwa, N. Watari, and H. Iijima (1978) Dissolution of Slightly Soluble Drugs. V. Effect of Particle Size on Gastrointestinal Drug Absorption and Its Relation to Solubility, *Chemical and Pharmaceutical Bulletin*, 26, 2603-2614.

K087   N.E. Khazanova (1954) Liquid-Liquid Equilibrium in the System Cyclohexane-liquid Ammonia, *Trudy Gosudarstevennoy Institute Azotnoi Promyshlennosi*, 4, 5-12.

K090   N. Kaneniwa and A. Ikekawa (1975) Solubilization of Water-Insoluble Organic Powders by Ball-Milling in the Presence of Polyvinylpyrrolidone, *Chemical and Pharmaceutical Bulletin*, 23, 2973-2986.

K091   N. Kaneniwa and N. Watari (1978) Dissolution of Slightly Soluble Drugs. IV. Effect of Particle Size of Sulfonamides on In Vitro Dissolution Rate and In Vivo Absorption Rate, and Their Relation to Solubility, *Chemical and Pharmaceutical Bulletin*, 26, 813-826.

K092   M. Kata and L. Haragh (1981) Spray-embedding of Spironolactone with beta-Cyclodextrin, *Pharmazie (Berlin)*, 36, 784-785.

K093   K. Klemm, W. Krastinat, and U. Kruger (1979) Synthese und Physikalisch-chemische Eigenschaften von Clanobutin, *Arzneimittel-Forschung*, 29, 1-2.

K095   K. Kitao, K. Kubo, T. Morishita..., and A. Kamada (1973) Studies on Absorption of Drugs. VII. Absorption of Isomeric N-Heterocylic Sulfonamides from the Rat Small Intestines and Relations between Physicochemical Property and Absorption of Unionized Sulfonamides, *Chemical and Pharmaceutical Bulletin*, 21, 2417-2426.

K096   M. Kanke and K. Sekiguchi (1973) Dissolution Behavior of Solid Drugs. II. Determination of the Transition Temperature of Sulfathiazole Polymorphs by Measuring the Initial Dissolution Rates, *Chemical and Pharmaceutical Bulletin*, 21, 878-884.

K097   K.G. Karlsson (1925) Uber die Zersetzungsgeschwindigkcit einiger Ester in ihrer Abhangigkeit von der Wasserstoffionenkonzentration, *Zeitschrift Fur anorganische und Allgemeine Chemie*, 145, 1-57.

K103   H. Kanal, V. Inouye, R. Goo, and H. Wakatsuki (1979) Solubility of 4-Methyl-2-pentanone in Aqueous Phase of Various Salt Concentrations, *Analytical Chemistry*, 51, 1019-1021.

K104   I.V. Kolosov and E.P. Shapovalenko (1977) Study of Acid-base Equilibria in Aqueous Solutions of Bilirubin by the Solubility Method, *Journal of General Chemistry of the USSR*, 47, 1967-1967.

K105   I.L. Krupatkin, L.D. Vorob'eva, V.P. Maskhuliya, and M.E. Veselova (1975) Liquid-phase Equilibria in the Systems Water-2-Furaldehyde-Thiocyanates and Water-Butanone-Thiocyanates, *Journal of General Chemistry of the USSR*, 45, 973-977.

K106   K. Kuroda, T. Yokoyama, T. Umeda, and Y. Takagishi (1978) Studies on Drug Nonequivalence. VI. Physico-chemical Studies on Polymorphism of Acctohexamide, *Chemical and Pharmaceutical Bulletin*, 26, 2565-2568.

K108   K. Koizumi, K. Mitsui, and K. Higuchi (1974) Comparison Between Interactions of alpha- and beta-Cyclodextrin with Barbituric Acid Derivatives, *Yakugaku Zasshi (Tokyo)*, 94, 1515-1519.

K112   I.M. Korenman and R.P. Arefeva (1978) Determination of the Solubility of Hydrocarbons in Water, *Zhurnal Prikladnoi Khimii (Leningrad)*, 51, 957-958.

K114   P. Krumholz (1951) Structural Studies on Polynuclear Pyridine Compounds, *Journal of the American Chemical Society*, 73, 3487-3492.

K117   I.P. Kapoor, R.L. Metcalf, R.F. Nystrom, and G.K. Sangha (1970) Comparative Metabolism of Methoxychlor, Methiochlor, and DDT in Mouse, Insects, and in a Model Ecosystem, *Journal of Agricultural and Food Chemistry*, 18, 1145-1152.

K119   T. Krzyzanowska and J. Szeliga (1978) Determination of the Solubility of Individual Hydrocarbons, *Nafta (Katowice, Poland)*, 34, 413-417.

K120   N. Kurihara, M. Uchida, T. Fujita, and M. Nakajima (1973) Studies on BHC Isomers and Related Compounds. V. Some Physicochemical Properties of BHC Isomers, *Pesticide Biochemistry and Physiology*, 2, 383-390.

K121   K. Koizumi, H. Miki, and Y. Kubota (1980) Enhancement of the Hypnotic Potency of Barbiturates by Inclusion Complexation with beta-Cyclodextrin, *Chemical and Pharmaceutical Bulletin*, 28, 319-322.

K122   F.H.C. Kelly (1954) Phase Equilibria in Sugar Solutions. II. Ternary Systems of Water-Sucrose-Hexose, *Journal of Applied Chemistry*, 4, 405-406.

K123   R. Krasnoschekova and M. Gubergrits (1976) The Relationship Between Reactivity and Hydrophobicity of Polycyclic Aromatic Hydrocarbons, *Organic Reaction*, 13, 432-439.

K129   F. Kralj and D. Sincic (1980) Mutual Solubilities of Phenol, Salicylaldehyde, Phenol-salicylaldehyde Mixture, and Water with and without the Presence of Sodium Chloride or Sodium Chloride plus Sodium Sulfate, *Journal of Chemical and Engineering Data*, 25, 335-338.

K130   Y.I. Korenman, E.I. Polumestnaya, and E.V. Lyubeznykh (1979) The Extraction and Solubility of Naphthols in the Presence of Neutral Salts, *Russian Journal of Physical Chemistry*, 53, 1663-1665.

K132   Y.I. Korenman and V.S. Smirnov (1980) The Solubilities and Distribution Constants of Xylenols in Water-Organic Solvent Systems, *Russian Journal of Physical Chemistry*, 54, 1553-1554.

K135   F.H.C. Kelly (1954) Phase Equilibria in Sugar Solutions. III. Ternary Systems of Water-hexose-inorganic Salt, *Journal of Applied Chemistry*, 4, 407-408.

K136   F.H.C. Kelly (1954) Phase Equilibria in Sugar Solutions. IV. Ternary System of Water-Glucose-Fructose, *Journal of Applied Chemistry*, 4, 409-411.

K137   J. Kanazawa (1981) Measurement of the Bioconcentration Factors of Pesticides by Freshwater Fish and Their Correlation with Physicochemical Properties or Acute Toxicities, *Pesticide Science*, 12, 417-424.

K138   E.E. Kenaga (1980) Correlation of Bioconcentration Factors of Chemicals in Aquatic and Terrestrial Organisms with Their Physical and Chemical Properties, *Environmental Science and Technology*, 14, 553-556.

K142   V.V. Kotel'nikov and V.P. Skripov (1959) Isotope Effect in the Mutual Solubility of Water and Triethylamine, *Nauchnaya Doklady Vysshei Shkoly Khimiya i Khimicheskaya Teknologiya*, 53, 248-249.

K143   Y. Kato, Y. Okamoto, S. Nagasawa, and T. Ueki (1981) Solubility of a New Polymorph of Phenobarbital Obtained by Crystallization in the Presence of Phenytoin, *Chemical and Pharmaceutical Bulletin*, 29, 3410-3413.

K144   W. Kobinger and F.J. Lund (1959) Investigations Into a New Oral Diuretic, Rontyl (6-Trifluoromethyl-7-sulfamyl-3,4-dihydro-1,2,4-benzothiadiazine-1,1-dioxide), *Acta Pharmacology et Toxicology*, 15, 265-274.

K148   V.N. Kulakov, A.G. Artyukh, I.M. Nikiforov..., and T.V. Barinova (1981) Solubility of 2-Naphthoic Acid in Aqueous Solutions of N-Methylpyrrolidone, *Journal of Applied Chemistry of the USSR*, 54, 335-338.

K301   Y.I. Korenman (1983) Correlation Between the Partition Constants of Organic Substances and Their Solubilities in Water and Extractants, *Russian Journal of Physical Chemistry*, 57, 382-384.

K304   S.W. Karickhoff (1981) Semi-empirical Estimation of Sorption of Hydrophobic Pollutants on Natural Sediments and Soils, *Chemosphere*, 10, 833-846.

K305   H. Konemann (1981) Quantitative Structure-activity Relationships in Fish Toxicity Studies. Part I. Relationship for 50 Industrial Pollutants, *Toxicology*, 19, 209-221.

K307   Y.I. Korenman and E.I. Polumestnaya (1977) The Solubility of 1-Naphthol in Water at Different Temperatures, *Russian Journal of Physical Chemistry*, 51, 1392-1392.

K308   Y.I. Korenman, E.I. Polumestnaya, and L.I. Shestakova (1977) The Solubility of 2-Naphthol in Water at Different Temperatures, *Russian Journal of Physical Chemistry*, 51, 608-608.

K309   Y.I. Korenman, I.V. Karmaeva, and L.N. Sergeeva (1977) The Solubility of 2,4-Xylenol in Water at Different Temperatures, *Russian Journal of Physical Chemistry*, 51, 165-165.

K310   Y.I. Korenman, S.N. Taldykina, and T.N. Bogomolova (1976) The Solubility of Picric Acid in Water at Different Temperatures, *Russian Journal of Physical Chemistry*, 50, 1780-1781.

K315   R.A. Kahrs (1978) Profluralin, *Analytical Methods for Pesticides and Plant Growth Regulators*, 10, 451-459.

K316   D.F. Keeley, M.A. Hoffpauir, and J.R. Meriwether (1988) Solubility of Aromatic Hydrocarbons in Water and Sodium Chloride Solutions of Different Ionic Strengths: Benzene and Toluene, *Journal of Chemical and Engineering Data*, 33, 87-89.

K337   H. Kishii, M. Nakamura, and Y. Hashimoto (1987) Prediction of Solubility of Aromatic Compounds in Water by Using Total Molecular Surface Area, *Nippon Kagaku Kaishi*, 8, 1615-1622.

L001   H.S. Loring and V. Du Vigneaud (1934) The Solubility of the Stereoisomers of Cystine with a Note on the Identity of Stone and Hair Cystine, *Journal of Biological Chemistry*, 107, 267-274.

L002   P.J. Leinonen and D. MacKay (1973) The Multicomponent Solubility of Hydrocarbons in Water, *Canadian Journal of Chemical Engineering*, 51, 230-233.

L003   B.A. Lindenberg (1951) Sur la Solubilite des Substances Organiques Amphipatiques dans les Glycerides Neutres et Hydroxyles, *Journal de Chimie Physique et de Physico-Chimie Biologique*, 48, 350-355.

L006   J.S. Lumsden (1905) The Physical Properties of Heptoic, Hexahydrobenzoic, and Benzoic Acids and Their Derivatives, *Journal of the Chemical Society (London)*, 87, 90-99.

L007   A.A. Liabastre (1974) Experimental Determination of the Solubility of Small Organic Molecules in H_2O and D_2O and the Application of the Scaled Particle Theory to Aqueous and Nonaqueous Solutions, Ph.D. thesis, 48-167.

L008   A.G. Leiga and J.N. Sarmousakis (1966) The Effect of Certain Salts on the Aqueous Solubilities of o-, m-, and p-Dinitrobenzene, *Journal of Physical Chemistry*, 70, 3544-3549.

L009   W.E. Lange and M.E. Amundson (1962) Soluble Steroids I-Sugar Derivatives, *Journal of Pharmaceutical Sciences*, 51, 1102-1106.

L010   R.E. Lacey and D.R. Cowsar (1973) Factors Affecting the Release of Steroids from Silicones, in *Controlled Release of Biologically Active Agents*, Plenum Press, 117-136.

L011   G.F. Le Petit (1976) Medazepam pKa Determined by Spectrophotometric and Solubility Methods, *Journal of Pharmaceutical Sciences*, 65, 1094-1095.

L012   R.E. Lindstrom and A.R. Giaquinto (1970) Salt Effects in Aqueous Solutions of Urea, *Journal of Pharmaceutical Sciences*, 59, 1625-1630.

L013   L.C. Lappas, C.A. Hirsch, and C.L. Winely (1976) Substituted 5-Nitro-1,3-dioxanes: Correlation of Chemical Structure and Antimicrobial Activity, *Journal of Pharmaceutical Sciences*, 65, 1301-1305.

L014   B. Lang (1971) Solubility and Solubilization Studies on Reseptyl, *Pharmazie*, 26, 689-691.

L015   J.H. Lister and D.S. Caldbick (1976) An Investigation into the Factors Governing the Aqueous Solubility of Xanthine (Purine-2,6-dione), *Journal of Applied Chemistry Biotechnology*, 26, 351-354.

L016   P. Letellier (1973) Influence du Solvant sur les Solubilites de Composes Organiques. Correlations avec les Variations des Effects de Substituants, *Bulletin de la Societe Chimique de France*, 5, 1569-1575.

L017   G.F. Lata and L.K. Dac (1965) Steroid Solubility Studies with Aqueous Solutions of Urea and Ureides, *Archives of Biochemistry and Biophysics*, 109, 434-441.

L020   P.D. Lamson, H.W. Brown, R.W. Stoughton..., and A. Bass (1935) Anthelmintic Studies on Alkyl-hydroxybenzenes. IV. Isomerism in Polyalkylphenols, *Journal of Pharmacology and Experimental Therapeutics*, 53, 234-238.

L021   P.D. Lamson, H.W. Brown, R.W. Stoughton..., and A.D. Bass (1935) Anthelmintic Studies on Alkyl-hydroxybenzenes. V. Phenols with Other than Normal Alkyl Side Chains, *Journal of Pharmacology and Experimental Therapeutics*, 53, 239-241.

L022   P.D. Lamson, H.W. Brown, R.W. Stoughton..., and A. Bass (1935) Anthelmintic Studies on Alkyl-hydroxybenzenes. II. ortho and para-n-Alkylphenols, *Journal of Pharmacology and Experimental Therapeutics*, 53, 218-226.

L024   M.A. Loos (1969) Phenoxyalkanoic Acids, 1-5.

L025   H. Langecker, A. Harwart, and K. Junkmann (1953) 2,4,6-Trijod-3-acetaminobenzoesaure-abkommlinge als Kontrastmittel, *Archiv fuer Experimentelle Pathologie und Pharmakologie*, 220, 195-206.

L027   E.P.K. Lau and J.L. Sutter (1974) Ethynodiol Diacetate, *Analytical Profiles of Drug Substances*, 3, 254-277.

L028   W.H. Lane (1946) Determination of the Solubility of Styrene in Water and of Water in Styrene, *Industrial and Engineering Chemistry, Analytical Edition*, 18, 295-296.

L029   T.S. Logan (1945) The Aqueous Solubility of Acetanilide, *Journal of the American Chemical Society*, 67, 1182-1184.

L030   A.C. Leopold, P. Van Schaik, and M. Neal (1960) Molecular Structure and Herbicide Adsorption, *Weeds*, 8, 48-54.

L031   C. Laguerie, M. Aubry, and J-P. Couderc (1976) Some Physicochemical Data on Monohydrate Citric Acid Solutions in Water: Solubility, Density, Viscosity, Diffusivity, pH of Standard Solution, and Refractive Index, *Journal of Chemical and Engineering Data*, 21, 85-87.

L032   E.E. Leuallen (1949) Solubility of Phenobarbital in Ethanol-Water Systems, *Journal of the American Pharmaceutical Association, Practical Pharmacy Edition*, 10, 722-724.

L033   S-T. Liu, C.F. Carney, and A.R. Hurwitz (1977) Adsorption as a Possible Limitation in Solubility Determination, *Journal of Pharmacy and Pharmacology*, 29, 319-321.

L034   N.G. Lordi and J.E. Christian (1956) Physical Properties and Pharmacological Activity: Antihistaminics, *Journal of the American Pharmaceutical Association*, Scientific Edition, 45, 300-305.

L035   Lewkowitsch (1883) Aqueous Solubilities of r- and l-Mandelic Acids and Three O-Acyl-r-mandelic Acids, *Berichte der Deutschen Chemischen Gesellschaft*, 16, 1566.

L037   C.U. Linderstrom-Lang and R.F. Naylor (1962) 4,4'-Diaminodiphenyl Sulphone: Solubility and Distribution in Blood, *Biochemical Journal*, 83, 417-420.

L038   T.S. Logan (1946) The Effect of KCl, NaCl and Na$_2$SO$_4$ on the Aqueous Solubility of Acetanilide, *Journal of the American Chemical Society*, 68, 1660-1661.

L039   O. Lutz (1902) Ueber Einige Falle von Sauerstoffwanderung in der Molekel. I, *Berichte der Deutschen Chemischen Gesellschaft*, 35, 2460-2466.

L041   F. Lamouroux (1899) Sur la Solubilite dans l'Eau des Acides Normaux de la Serie Oxalique, *Comptes Rendus Hebdomadaires des Seances de l'Academie des Sciences*, 128, 998-1000.

L042   L. Levine, J.A. Gordon, and W.P. Jencks (1963) The Relationship of Structure to the Effectiveness of Denaturing Agents of Deoxyribonucleic Acid, *Biochemistry*, 2, 168-175.

L044   N.M. Lykhol'ot (1965) Studies of the Solubility of Sulfanilamide Preparations. V. The Solubility of Sulfanilamide in Phosphate-Citrate Buffer Mixtures, *Farmatsevty Zhurnal*, 20, 44-46.

L045   S. Liu and A. Hurwitz (1977) The Effect of Micelle Formation on Solubility and pKa Determination of Acetylpromazine Maleate, *Journal of Colloid and Interface Science*, 60, 410-413.

L047   W. Licht, Jr. and L.D. Wiener (1950) Hydrotropic Solvents for Benzoic Acid, *Industrial and Engineering Chemistry*, 42, 1538-1542.

L048   E. Larsson (1926) Zur Elektrolytischen Dissoziation der Zweibasischen Sauren. III. Bestimmung Zweiter Dissoziationskonstanten aus Loslichkeitsversuchen, *Zeitschrift fuer Anorganische Chemie*, 115, 247-254.

L049   G.S. Laddha and J.M. Smith (1948) The Systems: Glycol-n-amyl Alcohol-Water and Glycol-n-hexyl Alcohol-Water, *Industrial and Engineering Chemistry*, 40, 494-496.

L050   E. Larsson (1927) Die Loslichkeit von Sauren in Salzlosungen. I, *Zeitschrift fuer Physikalische Chemie (Leipzig)*, 127, 233-248.

L051   G.I. Linkov (1975) Studies on Oxytetracylcline Solubility in Aqueous Media, *Antibiotiki (Moscow)*, 20, 53-58.

L052   W.A. Lott and F.B. Bergeim (1939) 2-(p-Aminobenzenesulfonamido)-thiazole: A New Chemo-therapeutic Agent, *Journal of the American Chemical Society*, 61, 3593-3594.

L053   E. Leikola and I. Suihkonen (1960) Amino-, metyyli- ja Nitroryhmien Aseman Vaikutus Substituoidun Bentseenimolekyylin Liukoisuuteen, *Farmaseuttinen Aikakauslehti*, 69, 193-201.

L055   J.H. Lee (1962) Hydrophilics, *School Science Review*, 43, 391-393.

L058   D. Lehr (1945) Inhibition of Drug Precipitation in the Urinary Tract by the Use of Sulfonamide Mixtures. I. Sulfathiazole-sulfadiazine Mixture, *Proceedings of the Society for Experimental Biology and Medicine*, 58, 11-14.

L059   P. Leone and E. Angelescu (1922) Variazioni di Solubilita di un Corpo per la Presenza di Altri Corpi. I. Acqua-fenolo-difenoli, *Gazzetta Chimica Italiana*, 52, 61-74.

L060   A. Levan and G. Ostergren (1943) The Mechanism of C-Mitotic Action Observations on the Naphthalene Series, *Hereditas*, 29, 381-432.

L061   P. Leone and M. Benelli (1922) Variazioni di Solubilita di un Corpo per Presenza di Altri Corpi (II). Acqua-Epicloridrina-Acido Acetico, *Gazzetta Chimica Italiana*, 52, 75-86.

L062   B.A. Lloyd, S.O. Thompson, and J.B. Ferguson (1937) Equilibria in Liquid Systems Containing Furfural, *Canadian Journal of Research, Section B: Chemical Sciences*, 15, 98-102.

L063   W. Ledbury and C.W. Frost (1927) The Solubility of Nitroglycerol in Water, *Journal of the Society of Chemical Industry*, London, 46, 120-120.

L064   K. Linderstrom-Lang (1924) Solubility of Hydroquinone, *Comptes Rendus des Travaux du Laboratoire Carlsbreg*, 15, 4-28.

L065   K. Linderstrom-Lang (1929) On the Relation Between the Sizes of Ions and the Salting-out of Hydroquinone and Quinone, *Comptes Rendus des Travaux du Laboratoire Carlsbreg*, 17, 1-6.

L067   H.K. Lee, H. Lambert, V.J. Stella..., and T. Higuchi (1979) Hydrolysis and Dissolution Behavior of a Prolonged-release Prodrug of Theophylline: 7,7'-Succinylditheophylline, *Journal of Pharmaceutical Sciences*, 68, 288-293.

L068   J.L. Lach and W.A. Pauli (1966) Interaction of Pharmaceuticals with Schardinger Dextrins. VI. Interactions of beta-cyclodextrin, Sodium Deoxycholate, and Deoxycholic Acid with Amines and Pharmaceutical Agents, *Journal of Pharmaceutical Sciences*, 55, 32-38.

L069   J.L. Lach and J. Cohen (1963) Interaction of Pharmaceuticals with Schardinger Dextrins. II. Interaction with Selected Compounds, *Journal of Pharmaceutical Sciences*, 52, 137-142.

L070   B. Lundberg, T. Lovgren, and B. Heikius (1979) Simultaneous Solubilization of Steroid Hormones II: Androgens and Estrogens, *Journal of Pharmaceutical Sciences*, 68, 542-545.

L071   P-Y. Lu, R.L. Metcalf, A.S. Hirwe, and J.W. Williams (1975) Evaluation of Environmental Distribution and Fate of Hexachlorocyclopentadiene, Chlordene, Heptachlor, and Heptachlor Epoxide in a Laboratory Model Ecosystem, *Journal of Agricultural and Food Chemistry*, 23, 967-973.

L072   J.L. Laseter, C.K. Bartell, A.L. Laska..., and R.L. Evans (1976) An Ecological Study of Hexachlorobenzene, unpublished, 77-77.

L073   H. Leo and E. Rimbach (1919) Uber die Wasserloslichkeit des Camphers, *Biochemische Zeitschrift (1948-1967)*, 95, 306-312.

L074   P.N. Leech, W. Rabak, and A.H. Clark (1919) American-made Synthetic Drugs. II, *Journal of the American Medical Association*, 73, 754-759.

L075   R.E. Lindstrom (1979) Solubility in Amide-Water Cosolvent Systems. II. Cosolvent Excess at Solute Surface, *Journal of Pharmaceutical Sciences*, 68, 1141-1143.

L076   J-F. Labarre, J-P. Faucher, G. Levy..., and G. Francois (1979) Antitumour Activity of Some Cyclophosphazenes, *European Journal of Cancer*, 15, 637-643.

L077   B. Lundberg (1979) Temperature Effect on the Water Solubility and Water-Octanol Partition of Some Steroids, *Acta Pharmaceutica Suecica*, 16, 151-159.

L078   G.A. Lewis and R.P. Enever (1979) Solution Thermodynamics of Some Potentially Long-acting Norethinedrone Derivatives. III. Measurement of Aqueous Solubilities and the Use of Group Free Energy Contributions in Predicting Partition Coefficients, *International Journal of Pharmaceutics*, 3, 319-333.

L079   S-L. Lin, L. Lachman, C.J. Swartz, and C.F. Huebner (1972) Preformulation Investigation. I. Relation of Salt Forms and Biological Activity of an Experimental Antihypertensive, *Journal of Pharmaceutical Sciences*, 61, 1418-1422.

L080   T.S. Lakshmi and P.K. Nandi (1978) Interaction of Adenine and Thymine with Aqueous Sugar Solutions, *Journal of Solution Chemistry*, 7, 283-289.

L081   T.S. Lakshmi and P.K. Nandi (1976) Effects of Sugar Solutions on the Activity Coefficients of Aromatic Amino Acids and Their N-Acetyl Ethyl Esters, *Journal of Physical Chemistry*, 80, 249-252.

L082   F.J. Lozano (1981) Ternary Equilibrium for the System Water/Methyl Isobutyl Ketone/ 2-Ethyl-2-(hydroxymethyl)-1,3-propanediol, *Journal of Chemical and Engineering Data*, 26, 131-133.

L083   J. Leja (1973) Some Electrochemical and Chemical Studies Related to Froth Flotation with Xanthates, *Minerals Science and Engineering*, 5, 278-286.

L084   T.M. Lesteva, S.K. Orgorodnikov, and T.N. Tyvina (1968) Liquid-Liquid Phase Equilibria in the System 3-Methylbutanediol-1,3-n-butanol-Water, *Journal of Applied Chemistry of the USSR*, 41, 1103-1105.

L085   K. Lauer (1937) Der Einfluss des Losungsmittels auf den Ablauf Chemischer Reaktionen. XIV. Mitteil: Zur Kenntnis der Aromatischen Kohlenwasserstoffe, *Berichte der Deutschen Chemischen Gesellschaft*, 70, 1127-1133.

L086   R.E. Lindstrom and C.H. Lee (1980) Solubility in Amide-Water Cosolvent Systems: A Thermodynamic View, *Journal of Pharmacy and Pharmacology*, 32, 245-247.

L087   B. Lundberg, T. Lovgren, and C. Blomqvist (1979) The Effect of Salt on the Solubility of Steroids in Tetradecyltrimethylammonium Bromide, *Acta Pharmaceutica Suecica*, 16, 144-150.

L088   F.A. Long and W.F. McDevit (1952) Activity Coefficients of Nonelectrolyte Solutes in Aqueous Salt Solutions, *Chemical Reviews*, 51, 119-169.

L089   M.C. Lantsman (1973) *Izvestiya Tomskogo Politekhnicheskogo Instituta*, 257, 202-204.

L090   K. Loskit (1924) Uber Polymorphie, *Zeitschrift fuer Physikalische Chemie (Leipzig)*, 134, 156-159.

L091   D. Lehr (1950) Choice of Sulphonamides for Mixture Therapy, *British Medical Journal*, 2, 601-604.

L094   B.H. Lippold and J.F. Lichey (1981) Loslichkeits-pH-profile Mehrprotoniger Arzneistoffe am Beispiel des Assoziierenden Dimetindens und des Zwitterionischen Liothyronins, *Archiv der Pharmazie*, 314, 541-556.

L095   P-Y. Lu and R.L. Metcalf (1975) Environmental Fate and Biodegradability of Benzene Derivatives as Studied in a Model Aquatic Ecosystem, *Environmental Health Perspectives*, 10, 269-284.

L096   G.J.M. Ley, D.O. Hummel, and C. Schneider (1967) Gamma-radiation-induced Polymerization of Some Vinyl Monomers in Emulsion Systems, *Advances in Chemistry Series*, 66, 184-202.

L097   G. Lohr (1978) Estimation of the Uptake Rate of Solvents into Latex Particles, *Polymers as Colloid Systems*, 2, 71-81.

L099   L.J. Lorenz (1980) Cefaclor, *Analytical Profiles of Drug Substances*, 9, 107-115.

L100   H. Langecker, A. Harwart, and K. Junkmann (1954) 3,5-Diacetylamino-2,4,6-trijodbenzoesaure als Rontgenkontrastmittel, *Archiv fuer Experimentelle Pathologie und Pharmakologie*, 222, 584-590.

L103   M.C. Likhosherstov, S.V. Alekseev, and T.V. Shalaeva (1935) 2,3-Dichlorobutane Solubility, *Zhurnal Khimicheskoi Promyshlennosti*, 12, 705-709.

L106   J. Lawrence and H.M. Tosine (1976) Adsorption of Polychlorinated Biphenyls from Aqueous Solutions and Sewage, *Environmental Science and Technology*, 10, 381-383.

L300   F. Leyder and P. Boulanger (1983) Ultraviolet Absorption, Aqueous Solubility, and Octanol-Water Partition for Several Phthalates, *Bulletin of Environmental Contamination and Toxicology*, 30, 152-157.

L301   P-Y. Lu, R.L. Metcalf, and E.M. Carlson (1978) Environmental Fate of Five Radiolabeled Coal Conversion By-products Evaluated on a Laboratory Model Ecosystem, *Environmental Health Perspectives*, 24, 201-208.

L303   V.P. Lynch (1978) Chlormephos, *Analytical Methods for Pesticides and Plant Growth Regulators*, 10, 49-55.

L310   J. Linek (1976) Liquid-Liquid Equilibrium in the Isobutyl Acetate-Water System, *Collection of Czechoslovak Chemical Communications*, 41, 1714-1717.

L311   P.Y. Lu, R.L. Metcalf, and L.K. Cole (1977) The Environmental Fate of 14C-Pentachlorophenol in Laboratory Model Ecosystems, in *Pentachlorophenol: Chemistry, Pharmacology, and Environmental Toxicology*, Plenum Press, 53.

L319   J.M. Lo, C.L. Tseng, and J.Y. Yang (1986) Radiometric Method for Determining Solubility of Organic Solvents in Water, *Analytical Chemistry*, 58, 1596-1597.

L320   R.E. Langford and E.L. Heric (1972) Furfural-Water-Formic Acid System at 25 and 35 C, *Journal of Chemical and Engineering Data*, 17, 87-89.

L321   K.B. Lodge (1989) Solubility Studies Using a Generator Column for 2,3,7,8-Tetrachlorodibenzo-p-dioxin, *Chemosphere*, 18, 933-940.

L322   A. Li, W.J. Doucette, and A.W. Andren (1992) Solubility of Polychlorinated Biphenyls in Binary Water/Organic Solvent Systems, *Chemosphere*, 24, 1347-1360.

L329   T.H. Lilley, H. Linsdell, and A. Maestre (1992) Association of Caffeine in Water and in Aqueous Solutions of Sucrose, *Journal of Chemical Society*, Faraday Transactions 1, 88, 2865-2870.

L332   L.S. Lee, P.S.C. Rao, and I. Okuda (1992) Equilibrium Partitioning of Polycyclic Aromatic Hydrocarbons from Coal Tar into Water, *Environmental Science and Technology*, 26, 2110-2115.

L333   W.A. Lee, L. Gu, A.R. Miksztal, and P.H. Nelson (1990) Bioavailability Improvement of Mycophenolic Acid through Amino Ester Derivatization, *Pharmaceutical Research*, 7, 161-166.

L334   T. Loftsson, B.J. Olafsdottir, and J. Baldvinsdottir (1992) Estramustine: Hydrolysis, Solubilization, and Stabilization in Aqueous Solutions, *International Journal of Pharmaceutics*, 79, 107-112.

L335   R.T. Lipnick, D.E. Johnson, J.H. Gilford, and L.D. Newsome (1985) Comparison of Fish Toxicity Screening Data for 55 Alcohols with the Quantitative Structure-Activity Relationship Predictions of Minimum Toxicity for Nonreactive Nonelectrolyte Organic Compounds, *Environmental Toxicology and Chemistry*, 4, 281-296.

L338   H.M. Lin and R.A. Nash (1993) An Experimental Method for Determining the Hildebrand Solubility Parameter of Organic Nonelectrolytes, *Journal of Pharmaceutical Sciences*, 82, 1018-1026.

L339   W.A. Leet, H-M. Lin, and K-C. Chao (1987) Mutual Solubities in Six Binary Mixtures of Water + a Heavy Hydrocarbon or a Derivative, *Journal of Chemical and Engineering Data*, 32, 37-40.

L342   S.L. Leung, G. Becker, R. Karunanithy, and J.T. Fell (1989) Studies on Long-acting Aryl Carboxylic Acid Esters of Testosterone, *Pharmaceutica Acta Helvetiae*, 64, 121-124.

L343   T. Loftsson, S. Bjornsdottir, G. Palsdottir, and N. Bodor (1989) The Effects of 2-Hydroxypropyl-beta-Cyclodextrin on the Solubility and Stability of Chlorambucil and Melphalan in Aqueous Solution, *International Journal of Pharmaceutics*, 57, 63-72.

L344   F. Liu, D.O. Kildsig, and A.K. Mitra (1990) beta-Cyclodextrin/Steroid Complexation: Effect of Steroid Structure on Association Equilibria, *Pharmaceutical Research*, 7, 869-873.

L345   T. Loftson, H. Frioriksdottir, S. Thorisdottir, and E. Stefansson (1994) The Effect of Hydroxy-propyl Methylcellulose on the Release of Dexamethasone from Aqueous 2-Hydroxypropyl-beta-cyclodextrin Formulations, *International Journal of Pharmaceutics*, 104, 181-184.

L346   V.P. Le and B.C. Lippold (1995) Influence of Physicochemical Properties of Homologous Esters of Nicotinic Acid on Skin Permeability and Maximum Flux, *International Journal of Pharmaceutics*, 124, 285-292.

L348   R. Lunl, W. Shiu, and D. Mackay (1995) Aqueous Solubilities and Octanol-Water Partition Coeffients of Chloroveratroles and Chloroanisoles, *Journal of Chemical Engineering Data*, 40, 959-962.

M001   C. McAuliffe (1966) Solubility in Water of Paraffin, Cycloparaffin, Olefin, Acetylene, Cycloolefin, and Aromatic Hydrocarbons, *Journal of Physical Chemistry*, 70, 1267-1275.

M002   C. McAuliffe (1963) Solubility in Water of C1-C9 Hydrocarbons, *Nature (London)*, 200, 1092-1093.

M003   C. McAuliffe (1969) Solubility in Water of Normal C9 and C10 Alkane Hydrocarbons, *Science*, 163, 478-479.

M004   S. Mukherjee and N.P. Datta (1939) Electrochemical Properties of Stearic Acid Hydrosol. Part I, *Journal of the Indian Chemical Society*, 16, 563-582.

M006   M. Muramatsu, M. Iwahashi, and K. Masumoto (1975) Polymorphic Effects of Chloramphenicol Palmitate on Thermodynamic Stability in Crystals and Solubilities in Water and in Aqueous Urea Solution, *Journal of Chemical and Engineering Data*, 20, 6-9.

M007   D.K. Madan and D.E. Cadwallader (1973) Solubility of Cholesterol and Hormone Drugs in Water, *Journal of Pharmaceutical Sciences*, 62, 1567-1569.

M008   T.L. McMeekin (1943) Unpublished data cited in C007, 201-202.

M010   J.W. McBain and K.J. Lissant (1951) The Solubilization of Four Typical Hydrocarbons in Aqueous Solution by Three Typical Detergents, *Journal of Physical Chemistry*, 65, 655-658.

M011   W. Morozowich, T. Chulski, W.E. Hamlin..., and J.G. Wagner (1962) Relationship Between In Vitro Dissolution Rates, Solubilities, and LT50's in Mice of Some Salts of Benzphetamine and Etryptamine, *Journal of Pharmaceutical Sciences*, 51, 993-996.

M012   L.B. Mascardo and M. Barr (1953) The Solubility of Terpin Hydrate in Hydroalcoholic Solutions, *Journal of the American Pharmaceutical Association, Practical Pharmacy Edition*, 14, 772-773.

M013   E.J. Merrill (1965) Solubility of Pentaerythritol Tetranitrate-1,2-14C in Water and Saline, *Journal of Pharmaceutical Sciences*, 54, 1670-1671.

M014   C. McDonald and R.E. Lindstrom (1974) The Effect of Urea on the Solubility of Methyl p-Hydroxybenzoate in Aqueous Sodium Chloride Solution, *Journal of Pharmacy and Pharmacology*, 26, 39-45.

M015   J.R. Marvel and A.P. Lemberger (1960) Complexing Tendencies of Saccharin in Aqueous Solutions, *Journal of the American Pharmaceutical Association, Scientific Edition*, 49, 417-419.

M017   A.G. Mitchell and L.S.C. Wan (1964) Oxidation of Aldehydes Solubilized in Nonionic Surfactants. I. Solubility of Benzaldehyde and Methylbenzaldehyde in Aqueous Solutions of Polyoxyethylene Glycol Ethers, *Journal of Pharmaceutical Sciences*, 53, 1467-1470.

M018   B.A. Mulley and A.D. Metcalf (1956) Non-ionic Surface-active Agents. Part I. The Solubility of Chloroxylenol in Aqueous Solutions of Polyethylene Glycol 1000 Monocetyl Ether, *Journal of Pharmacy and Pharmacology*, 8, 774-780.

M020   H. Marshall and D. Bain (1910) Sodium Succinates, *Journal of the Chemical Society (London)*, 97, 1074-1085.

M021   W.F. McDevit and F.A. Long (1952) The Activity Coefficient of Benzene in Aqueous Salt Solutions, *Journal of the American Chemical Society*, 74, 1773-1773.

M022   F.S. Mortimer (1923) The Solubility Relations in Mixtures Containing Polar Components, *Journal of the American Chemical Society*, 45, 633-641.

M023   T.J. Macek, W.H. Baade, A. Bornn, and F.A. Bacher (1952) Observations on the Solubility of Some Cortical Hormones, *Science*, 116, 399-399.

M024   T.L. McMeekin, E.J. Cohn, and J.H. Weare (1936) Studies in the Physical Chemistry of Amino Acids, Peptides and Related Substances. A Comparison of the Solubility of Amino Acids, Peptides and Their Derivatives, *Journal of the American Chemical Society*, 58, 2173-2181.

M025   T.L. McMeekin, E.J. Cohn, and J.H. Weare (1935) Studies in the Physical Chemistry of Amino Acids, Peptides and Related Substances. III. The Solubility of Derivatives of the Amino Acids in Alcohol-Water Mixtures, *Journal of the American Chemical Society*, 57, 626-633.

M026   M.P. Moyle and M. Tyner (1953) Solubility and Diffusivity of 2-Naphthol in Water, *Industrial and Engineering Chemistry*, 45, 1794-1797.

M027   L.K. McCune and R.H. Wilhelm (1949) Mass and Momentum Transfer in Solid-liquid System, *Industrial and Engineering Chemistry*, 41, 1124-1127.

M028   W. McBride, R.A. Henry, J. Cohen, and S. Skolnik (1951) Solubility of Nitroguanidine in Water, *Journal of the American Chemical Society*, 73, 485-486.

M029   L.S. Mason (1947) The Solubilities of Four Amino Butyric Acids and the Densities of Aqueous Solutions of the Acids at 25 Degrees, *Journal of the American Chemical Society*, 69, 3000-3002.

M030   L. McMaster, E. Bender, and E. Weil (1921) The Solubility of Phthalic Acid in Water and Sodium Sulfate Solutions, *Journal of the American Chemical Society*, 43, 1205-1207.

M031   S.S. Mirvish, P. Issenberg, and H.C. Sornson (1976) Air-water and Ether-water Distribution of N-Nitroso Compounds: Implications for Laboratory Safety, Analytic Methodology, and Carcinogenicity for the Rat Esophagus, Nose, and Liver, *Journal of the National Cancer Institute*, 56, 1125-1129.

M032   T.H. Maren (1969) Renal Carbonic Anhydrase and the Pharmacology of Sulfonamide Inhibitors, 24, 225-228.

M035   A. MacDonald, A.F. Michaelis, and B.Z. Senkowski (1972) Chlordiazepoxide, *Analytical Profiles of Drug Substances*, 1, 15-25.

M036   A. MacDonald, A.F. Michaelis, and B.Z. Senkowski (1972) Diazepam, *Analytical Profiles of Drug Substances*, 1, 79-89.

M037   E.W. McGovern (1943) Chlorohydrocarbon Solvents, *Industrial and Engineering Chemistry*, 35, 1230-1239.

M038   J.W. Mauger and A.N. Paruta (1974) Entropy of Transfer of Molecular Benzoic Acid from a Pure Liquid to an Aqueous Solution, *Journal of Pharmaceutical Sciences*, 63, 576-579.

M040   D. Mackay (1981) Environmental and Laboratory Rates of Volatilization of Toxic Chemicals from Wat, 1, 303-319.

M041   N.J.L. Megson (1938) The Solubility of Phenols in Formalin, *Transactions of the Faraday Society*, 34, 525-532.

M042   K. Munzel (1970) Galenische Formgebung und Arzneimittelwirkung Neue Erkenntnisse und Feststellung, 14, 269-337.

M043   J.W. Mullin (1972) Crystallisation, 425-426.

M044   J.R. Marsh and P.J. Weiss (1967) Solubility of Antibiotics in Twenty-six Solvents. III, *Journal of the Association of Official Analytical Chemists*, 50, 457-462.

M045   J.R. Marvel, D.A. Schlichting, C. Denton..., and M.M. Cahn (1964) The Effect of a Surfactant and of Particle Size on Griseofulvin Plasma Levels, *Journal of Investigative Dermatology*, 42, 197-203.

M046   S.C. Mehta (1969) Mechanistic Studies of: I. Crystal Growth of Sulfathiazole and its Inhibition by PVP; II. Dissolution of High-energy Sulfathiazole-PVP Coprecipitates, Ph.D thesis, 19-19.

M047   W. McBride, R.A. Henry, and G.B.L. Smith (1949) Solubility of Nitroaminoguanidine, *Journal of the American Chemical Society*, 71, 2937-2938.

M048   E. Minkov, M. Zahariewa, B. Botev..., and T. Trandafilov (1969) Solubilisierung und Emulgierung von Hexachlorophen, *Pharmazie (Berlin)*, 24, 353-356.

M049   B.A. Mulley and A.J. Winfield (1970) Non-ionic Surface-active Agents. Part VII. Solubility of Benzoic Acid in Aqueous Solutions of Some Monododecyl Polyoxyalkanols, *Journal of the Chemical Society A: Inorganic, Physical, Theoretical*, 1459-1464.

M051   G. Massol and F. Lamouroux (1899) Sur la Solubilite dans l'Eau des Acides Maloniques Substitues, *Comptes Rendus Hebdomadaires des Seances de l'Academie des Sciences*, 128, 1000-1002.

M052   D.K. Madan and D.E. Cadwallader (1970) Effect of Macromolecules on Aqueous Solubility of Cholesterol, *Journal of Pharmaceutical Sciences*, 59, 1362-1363.

M053   A.G. Mitchell and D.J. Saville (1969) The Dissolution of Commercial Aspirin, *Journal of Pharmacy and Pharmacology*, 21, 28-34.

M054   J.G. Morris and E.R. Redfearn (1969) Vitamins and Coenzymes, 191-215.

M056   N. Maher and G. Sirois (1971) Solubilite du Phenobarbital et de l'Hexobarbital en Relation avec leur Activite Biologique (Solubility of Phenobarbital and Hexobarbital in Relation With Their Biological Activity), *Revue Canadienne de Biologie*, 30, 45-49.

M057   D.W. Macartney, R.W. Luxton, G.S. Smith..., and J. Goldman (1942) Sulphamethazine Trial of a New Sulphonamide, *Lancet*, 1, 639-642.

M058   P. Mukerjee and J.R. Cardinal (1976) Solubilization as a Method for Studying Self-Association: Solubility of Naphthalene in the Bile Salt Sodium Cholate and the Complex Pattern of Its Aggregation, *Journal of Pharmaceutical Sciences*, 65, 882-885.

M059   F.W. Miller, Jr. and H.R. Dittmar (1934) The Solubility of Urea in Water. The Heat of Fusion of Urea, *Journal of the American Chemical Society*, 56, 848-849.

M060   W.L. Masterton and T.P. Lee (1972) Effect of Dissolved Salts on Water Solubility of Lindane, *Environmental Science and Technology*, 6, 919-921.

M061   N.N. Melnikov, F.A. Gunther, and J.D. Gunther (1971) Residue Reviews, in *Chemistry of Pesticides*, vol. 36, Springer-Verlag, 36-439.

M062   H.B. McClure (1939) Industrial Applications of the Glycols, *Industrial and Engineering Chemistry, News Edition*, 17, 149-153.

M063   W.E. May, S.P. Wasik, and D.H. Freeman (1978) Determination of the Aqueous Solubility of Polynuclear Aromatic Hydrocarbons by a Coupled Column Liquid Chromatographic Technique, *Analytical Chemistry*, 50, 175-179.

M064   D. Mackay and W.Y. Shiu (1977) Aqueous Solubility of Polynuclear Aromatic Hydrocarbons, *Journal of Chemical and Engineering Data*, 22, 399-402.

M065   M. Mannheimer (1956) Mutual Solubility of Liquefied Gases and Water at Room Temperature, *Chemist-Analyst*, 45, 8-10.

M066   O. Meyer (1967) Dissertation or Masters Thesis.

M067   N. Mizuno, Y. Iwayama, H. Takagi, and A. Kamada (1973) Stability and Intestinal Absorption of alpha,3-o-Isopropylidene Pyridoxine, *Yakuzaigaku*, 33, 172-178.

M068   E.T. McBee and R.E. Hatton (1949) Production of Hexachlorobutadiene, *Industrial and Engineering Chemistry*, 41, 809-812.

M069   A. McKeown (1922) The Influence of Electrolytes on the Solubility of Non-electrolytes, *Journal of the American Chemical Society*, 44, 1203-1209.

M070   R. Mcggcs, H.J. Portius, and K.R.H. Repke (1977) Penta-acetyl-gitoxin: The Prototype of a Prodrug in the Cardiac Glycoside Series, *Pharmazie (Berlin)*, 32, 665-667.

M071   W.E. May, S.P. Wasik, and D.H. Freeman (1978) Determination of the Solubility Behavior of Some Polycyclic Aromatic Hydrocarbons in Water, *Analytical Chemistry*, 50, 997-1000.

M072   F. Muller and R. Suverkrup (1977) Die Zersetzung von p-Aminosalicylsaure in Gegenwart Begrenzter Wassermengen, *Pharmazeutische Industrie*, 39, 1115-1122.

M073   S. Mitchell (1926) A Method for Determining the Solubility of Sparingly Soluble Substances, *Journal of the Chemical Society (London)*, 1333-1336.

M074   P. Mondain-Monval (1925) Sur la Solubilite du Saccharose, *Comptes Rendus Hebdomadaires des Seances de l'Academie des Sciences*, 181, 37-40.

M075   K.H. Meyer and O. Klemm (1940) La Solubilite de l'Anhydride du Glycocolle, *Helvetica Chimica Acta*, 23, 25-27.

M077   J. Meyer (1911) Die Polymorphie der Allozimtsaure, *Zeitschrift fuer Elektrochemie*, 17, 976-984.

M078   L.M. Moiseeva and N.M. Kuznetsova (1971) Determination of the Solubility of Aliphatic beta-Diketones and Their Compounds with Beryllium in Water, *Zhurnal Analiticheskoi Khimii*, 26, 2094-2096.

M081   I.W. Mehl (1935) Ein Uebersichtsdiagramm Log p-1/T fuer das Stoffpaar Methylamin-Wasser, *Zeitschrift fuer die Gesamte Kaelte-Industrie*, 42, 13-14.

M082   W.E. May (1977) The Solubility Behavior of Polynuclear Aromatic Hydrocarbons in Aqueous Systems, Ph.D thesis, 113-135.

M083   J.W. McBain and M. Eaton (1928) Hydrolysis in Solutions of Potassium Laurate as Measured by Extraction with Benzene, *Journal of the Chemical Society (London)*, 131, 2166-2179.

M087   J.W. McBain and P.H. Richards (1946) Solubilization of Insoluble Organic Liquids by Detergents, *Industrial and Engineering Chemistry*, 38, 642-646.

M088   M. Mion and G. Urbain (1931) Contribution a l'Etude du Systeme Eau, Alcool Ethylique, Acide Acetique, Acetate d'Ethyle, *Comptes Rendus Hebdomadaires des Seances de l'Academie des Sciences*, 193, 1330-1333.

M091   W.L. Masterton (1954) Partial Molal Volumes of Hydrocarbons in Water Solution, *Journal of Chemical Physics*, 22, 1830-1833.

M093   A.G. Mitchell (1964) Bactericidal Activity of Chloroxylenol in Aqueous Solutions of Cetomacrogol, *Journal of Pharmacy and Pharmacology*, 16, 533-537.

M094   J. Mahieu (1936) La Solubilite dans les Melanges de Deux Solvants Miscibles, *Bulletin des Societes Chimiques Belges*, 45, 667-674.

M095   A.J. Mueller, L.I. Pugsley, and J.B. Ferguson (1931) The System Normal Butyl Alcohol-Methyl Alcohol-Water, *Journal of Physical Chemistry*, 35, 1314-1327.

M096   C.E. Mange and O. Ehler (1924) Solubilities of Vanillin, *Industrial and Engineering Chemistry*, 16, 1258-1260.

M097   V.V. Monblanova and V.M. Rodinov (1953) Solubilities of Some beta-Amino Acids, *Journal of General Chemistry of the USSR*, 23, 1899-1901.

M098   A. Michels and E.C.F. Ten Haaf (1927) The Three-phase-lines of the Systems: Water-ortho-Cresol, Water-Metacresol, and Water-Paracresol, *Proceedings of the Koninklijke Nederlandse Akadamie van Wetenschappen*, 30, 52-54.

M099   G.H. Mains (1922) The System Furfural—Water. I. A Study of Its Properties with Reference to Their Commercial Application in the Production of Furfural... *Chemical and Metallurgical Engineering*, 26, 779-784.

M101   W. Manchot, M. Jahrstorfer, and H. Zepter (1924) Untersuchungen uber Gasloslichkeit und Hydratation, *Zeitschrift fuer Anorganische Chemie*, 141, 45-81.

M102   I.B.C. Matheson and A.D. King, Jr. (1978) Solubility of Gases in Micellar Solutions, *Journal of Colloid and Interface Science*, 66, 464-469.

M104   S. Miyazaki, M. Nakano, and T. Arita (1975) Effect of Crystal Forms on the Dissolution Behavior and Bioavailability of Tetracycline, Chlortetracycline, and Oxytetracycline Bases, *Chemical and Pharmaceutical Bulletin*, 23, 552-558.

M105   S. Miyazaki, M. Nakano, and T. Arita (1975) A Comparison of Solubility Characteristics of Free Bases and Hydrochloride Salts of Tetracycline Antibiotics in Hydrochloric Acid Solutions, *Chemical and Pharmaceutical Bulletin*, 23, 1197-1204.

M106   H. Matsumaru, S. Tsuchiya, and T. Hosono (1977) Interaction and Dissolution Characteristics of Ajmaline-PVP Coprecipitate, *Chemical and Pharmaceutical Bulletin*, 25, 2504-2509.

M107   J.W. Mullin and T.P. Cook (1965) Diffusion and Dissolution of the Hydroxybenzoic Acids in Water, *Journal of Applied Chemistry*, 15, 145-151.

M108   M. Muramatsu, M. Iwahashi, and U. Takeuchi (1979) Thermodynamic Relationship between alpha- and beta-Forms of Crystalline Progesterone, *Journal of Pharmaceutical Sciences*, 68, 175-177.

M109   Y. Morimoto, R. Hori, and T. Arita (1974) Solubilities of Antipyrine Derivatives in Water and Non-polar Solvents, *Chemical and Pharmaceutical Bulletin*, 22, 2217-2222.

M110   R.L. Metcalf and J.R. Sanborn (1975) Pesticides and Environmental Quality in Illinois, *Illinois, Natural History Survey*, Bulletin, 31, 381-438.

M111   R.W. Merriman (1913) The Mutual Solubilities of Ethyl Acetate and Water and the Densities of Mixtures of Ethyl Acetate and Ethyl Alcohol, *Journal of the Chemical Society (London)*, 103, 1774-1789.

M112   A.G. Morachevskii and Z.P. Popovich (1965) Liquid-Vapor Equilibrium and Mutual Solubility of Components in the System Tertiary-Butyl Alcohol-sec-Butyl Alcohol-Water, *Journal of Applied Chemistry of the USSR*, 38, 2085-2088.

M113   M.A. Moustafa, A.R. Ebian, S.A. Khalil, and M.M. Motawi (1971) Sulphamethoxydiazine Crystal Forms, *Journal of Pharmacy and Pharmacology*, 23, 868-874.

M114   T. Moriyoshi, Y. Aoki, and H. Kamiyama (1977) Mutual Solubility of i-Butanol + Water Under High Pressure, *Journal of Chemical Thermodynamics*, 9, 495-502.

M115   A.G. Mitchell and J.F. Broadhead (1967) Hydrolysis of Solubilized Aspirin, *Journal of Pharmaceutical Sciences*, 56, 1261-1266.

M116   H. Matsuura and K. Sekiguchi (1960) Studies on the Effect of Inorganic Salts on the Solubility of Organic Pharmaceutical Compounds, *Yakuzaigaku*, 20, 213-218.

M117   S. Mizukami and K. Nagata (1956) On the Solubility of N1-Acetyl-Sulfaisoxazole and its Decomposed Rate in the Simulated Intestinal Fluid, *Shionogi Kenkyusho Nempo*, 6, 58-64.

M118   R.L. Metcalf, J.R. Sanborn, P-Y. Lu, and D. Nye (1975) Laboratory Model Ecosystem Studies of the Degradation and Fate of Radiolabeled Tri-, Tetra-, and Pentachlorobiphenyl Compared with DDE, *Archives of Environmental Contamination and Toxicology*, 3, 151-165.

M119   T. Moriyoshi, S. Kaneshina, K. Aihara, and K. Yabumoto (1975) Mutual Solubility of 2-Butanol + Water Under High Pressure, *Journal of Chemical Thermodynamics*, 7, 537-545.

M120   R. Mitzner and C-R. Kramer (1977) Zur Loslichkeit Einiger 2,4-Dichlorphenoxyessig-saurealkylester in Wasser, *Zeitschrift fuer Chemie*, 17, 379-380.

M122   W. Markowski, E. Soczewinski, and K. Czapinska (1978) Analogy of Solubility and Chromatographic Parameters in Reversed Phase Partition Chromatography, *Polish Journal of Chemistry*, 52, 1775-1780.

M123   Y.A. Mirgorod (1978) Assessing the Initiation of Submicelle Formation in Aqueous Solutions of Diphilic Molecules, *Kolloid Zhurnal*, 40, 483-488.

M124   H.A. Massaldi and C.J. King (1973) Simple Technique to Determine Solubilities of Sparingly Soluble Organics: Solubility and Activity Coefficients of d-Limonene, n-Butylbenzene, and n-Hexyl Acetate in Water and Sucrose Solutions, *Journal of Chemical and Engineering Data*, 18, 393-397.

M125   J.F. McCants, J.H. Jones, and W.H. Hopson (1953) Ternary Solubility Data for Systems Involving 1-Propanol and Water, *Industrial and Engineering Chemistry*, 45, 454-456.

M127   Y. Marcus, L.E. Asher, J. Hormadaly, and E. Pross (1981) Selective Extraction of Potassium Chloride by Crown Ethers in Substituted Phenol Solvents, *Hydrometallurgy*, 7, 27-39.

M128   A. Martin, J. Newburger, and A. Adjei (1980) Extended Hildebrand Solubility Approach: Solubility of Theophylline in Polar Binary Solvents, *Journal of Pharmaceutical Sciences*, 69, 487-491.

M129   J.C. Means, J.J. Hassett, S.G. Wood, and W.L. Banwart (1979) Sorption Properties of Energy-related Pollutants and Sediments, 327-339.

M130   D. Mackay and P.J. Leinonen (1975) Rate of Evaporation of Low-solubility Contaminants from Water Bodies to Atmosphere, *Environmental Science and Technology*, 9, 1178-1180.

M131   M.M. Mortland and W.F. Meggitt (1966) Interaction of Ethyl N,N-Di-n-propylthiolcarbamate (EPTC) with Montmorillonite, *Journal of Agricultural and Food Chemistry*, 14, 126-129.

M132   D. Mackay and W.Y. Shiu (1975) The Determination of the Solubility of Hydrocarbons in Aqueous Sodium Chloride Solutions, *Canadian Journal of Chemical Engineering*, 53, 239-242.

M133   G. McConnell, D.M. Ferguson, and C.R. Pearson (1975) Chlorinated Hydrocarbons and the Environment, *Endeavour*, 34, 13-18.

M134   R.L. Metcalf, I.P. Kapoor, P-Y. Lu..., and P. Sherman (1973) Model Ecosystem Studies of the Environmental Fate of Six Organochlorine Pesticides, *Environmental Health Perspectives*, 4, 35-44.

M135   T.J. Morrison (1944) The Salting-Out Effect, *Transactions of the Faraday Society*, 40, 43-48.

M136   M.J. McGuire and I.H. Suffet (1980) The Calculated Net Adsorption Energy Concept, 1, 91-115.

M137   F.R. Mistry and S.M. Barnett (1980) An Equilibrium Phase Diagram for the Glucose-Cellobiose-Water System at 30.5 C, *Journal of Chemical and Engineering Data*, 25, 223-226.

M138   R.L. Metcalf (1972) DDT Substitutes, *Critical Reviews in Environmental Control*, 3, 25-59.

M139   M.J. McGuire, I.H. Suffet, and J.V. Radziul (1978) Assessment of Unit Processes for the Removal of Trace Organic Compounds from Drinking Water, *Journal of the American Water Works Association*, 70, 565-572.

M140   L. Mitterhauszerova, K. Kralova, and L. Krasnec (1980) Wechselwirkung zwischen Kebuzon (Ketophenylbutazon) und Modifizierten Starken, *Pharmazie (Berlin)*, 35, 159-160.

M141   R. Meier, O. Allemann, and H.V. Meyenburg (1944) 6-Sulfanilamido-2,4-dimethylpyrimidin, *Schweizerische Medizinische Wochenschrift*, 74, 1091-1095.

M142   E.K. Marshall, Jr., A.C. Bratton, H.J. White, and J.T. Litchfield, Jr. (1940) Sulfanilylguanidine: A Chemotherapeutic Agent for Intestinal Infections, *Bulletin of the John Hopkins Hospital*, 67, 163-188.

M143   G.G. Manov, K.E. Schuette, and F.S. Kirk (1952) Ionization Constant of 5-5'-Diethylbarbituric Acid from to 60 C, *Journal of Research of the National Bureau of Standards*, 48, 84-91.

M145   R.A. McNabb and F.M.A. McNabb (1980) Physiological Chemistry of Uric Acid: Solubility, Colloid and Ion-binding Properties, *Comparative Biochemistry Physiology*, 67, 27-34.

M146   A.G. Morachevskii, N.A. Smirnova, and R.V. Lyzlova (1965) Phase Equilibria in the Ternary Systems Isobutyraldehyde-Isobutyl Alcohol-Water and Isovaleraldehyde-Isobutyl Alcohol-Water, *Journal of Applied Chemistry of the USSR*, 38, 1245-1248.

M147   N.B. Matin, E.N. Zil'berman, V.I. Trachenko..., and V.A. Afanas'ev (1980) Solubility of Acrylamide in Water and Some Organic Solvents, *Journal of Applied Chemistry of the USSR*, 52, 2228-2229.

M148   M. Mannelli, P. Gigli, G. Orzalesi..., and T. Bisagno (1980) Physico-chemical Properties of Ibuproxam, a New Non-Steroideal Anti-Inflammatory Agent, *Bollettino Chimico Farmaceutico*, 119, 203-208.

M149   K.G. Mooney, M.A. Mintun, K.J. Himmelstein, and V.J. Stella (1981) Dissolution Kinetics of Carboxylic Acids. I. Effect of pH under Unbuffered Conditions, *Journal of Pharmaceutical Sciences*, 70, 13-22.

M151   W.E. May (1980) The Solubility Behavior of Polycyclic Aromatic Hydrocarbons in Aqueous Systems, *Advances in Chemistry Series*, 185, 143-192.

M152   F. Moolenaar, J. Visser, and T. Huizinga (1980) Biopharmaceutics of Rectal Administration of Drugs in Man. Absorption Rate and Bioavailability of Glafenine After Oral and Rectal Administration, *International Journal of Pharmaceutics*, 4, 195-203.

M153   J.W. Mullin and M.J.L. Whiting (1980) Succinic Acid Crystal Growth Rates in Aqueous Solution, *Industrial and Engineering Chemistry, Fundamentals*, 19, 117-121.

M155   M. Montagu-Bourin, P. Levillain, R. Ceolin..., and C. Souleau (1981) Le Systeme Ternaire Eau-phenanthroline-1,10-acide Perchlorique. I. Etude de la Solubilite a 25 C, dans la Region Riche en Phenanthroline-1,10 et Cristallochimie du Perchlorate d'Hydrogenato-bis (Phenanthroline-1,10), *Bulletin de la Societe Chimique de France*, 1, 109-112.

M156   J.C. Means, S.G. Wood, J. Hassett, and W.L. Banwart (1980) Sorption of Polynuclear Aromatic Hydrocarbons by Sediments and Soils, *Environmental Science and Technology*, 14, 1524-1528.

M157   M.A. Moustafa, A.M. Molokhia, and M.W. Gouda (1981) Phenobarbital Solubility in Propylene Glycol-Glycerol-Water Systems, *Journal of Pharmaceutical Sciences*, 70, 1172-1174.

M158   A. Martin, A.M. Paruta, and A. Adjei (1981) Extended Hildebrand Solubility Approach: Methylxanthines in Mixed Solvents, *Journal of Pharmaceutical Sciences*, 70, 1115-1120.

M159   N.A. Mason, S. Cline, M.L. Hyneck..., and G.L. Flynn (1981) Factors Affecting Diazepam Infusion: Solubility, Administration-set Composition, and Flow Rate, *American Journal of Hospital Pharmacy*, 38, 1449-1454.

M160   C.E. Messer, G. Malakoff, J. Well, and S. Labib (1981) Phase Equilibrium Behavior of Certain Pairs of Amino Acids in Aqueous Solution, *Journal of Physical Chemistry*, 85, 3533-3540.

M161   H. Martin and C.R. Worthing (1977) Pesticide Manual: Basic Information on the Chemicals Used as Active Components of Pesticides.

M162   H. Maier-Bode and K. Hartel (1981) Linuron and Monolinuron, *Residue Reviews*, 77, 4-6.

M163   K.J. Mysels (1969) Contribution of Micelles to the Transport of a Water-Insoluble Substance through a Membrane, *Advances in Chemistry Series*, 86, 24-77.

M164   H. Maier-Bode (1976) The Insecticide "Kelevan," *In Residues of Pesticides and Other Contaminants in the Total Environment*, Volume 63, Springer-Verlag, 44-48.

M165   L.P. Marrelli (1975) Cephalexin, *Analytical Profiles of Drug Substances*, 4, 21.

M166   G. Michel (1977) Nystatin, *Analytical Profiles of Drug Substances*, 6, 341-371.

M167   G.J. Manius (1978) Trimethoprim, *Analytical Profiles of Drug Substances*, 7, 445.

M168   M. Moriyama, A. Inoue, M. Isoya..., and M. Hanano (1978) Dissolution Properties and Gastrointestinal Absorption of Chloramphenicol from Hydrophilic High Molecular Compound Coprecipitates, *Yakugaku Zasshi (Tokyo)*, 98, 1012-1018.

M169   J.W. Mauger, S.A. Howard, and E.Z. Damewood (1978) A Simulation Experiment for the Dissolution of Monosized and Multisized Drug Particles, *American Journal of Pharmaceutical Education*, 42, 60-63.

M171   J.H.C. Merckel (1937) Die Loslichkeit der Dicarbonsauren, *Recueil des Travaux Chimiques des Pays-Bas*, 56, 811-814.

M172   A.F. Meleschenko (1960) Gigiena i Sanitariya, 25, 54-57.

M175   D. Mackay, W.Y. Shiu, and A.W. Wolkoff (1975) Gas Chromatographic Determination of Low Concentrations of Hydrocarbons in Water by Vapor Phase Extraction, *American Society for Testing and Materials*, 573, 251-258.

M177   Y. Moroi, K. Sato, and R. Matuura (1982) Solubilization of Phenothiazine in Aqueous Surfactant Micelles, *Journal of Physical Chemistry*, 86, 2463-2468.

M180   W. McDowell and R. Weingarten (1969) New Experimental Evidence about the Dyeing of Polyester Materials with Disperse Dyes, *Journal of the Society of Dyers and Colourists*, 85, 589-597.

M183   W.E. May, S.P. Wasik, M.M. Miller..., and R.N. Goldberg (1983) Solution Thermodynamics of Some Slightly Soluble Hydrocarbons in Water, *Journal of Chemical and Engineering Data*, 28, 197-200.

M184   J.P. Mieure, O. Hicks, R.G. Kaley, and V.W. Saeger (1976) Characterization of Polychlorinated Biphenyls, *Proceedings of National Conference on Polychlorinated Biphenyls*, 84-87.

M300   M.E. McNally and R.L. Grob (1983) Determination of the Solubility Limits of Organic Priority Pollutants by Gas Chromatographic Headspace Analysis, *Journal of Chromatography*, 260, 23-32.

M301   L. Molin, G. Dahlstrom, M-I. Nilsson..., and L. Tekenbergs (1983) Solubility, Partition, and Adsorption of Digitalis Glycosides, *Acta Pharmaceutica Suecica*, 20, 129-144.

M303   J.C. Means, J.J. Hassett, S.G. Wood..., and A. Khan (1980) Sorption Properties of Polynuclear Aromatic Hydrocarbons and Sediments: Heterocy, 395-404.

M308   M.M. Miller, S. Ghodbane, S.P. Wasik..., and D.E. Martire (1984) Aqueous Solubilities, Octanol/Water Partition Coefficients, and Entropies of Melting of Chlorinated Benzenes and Biphenyls, *Journal of Chemical and Engineering Data*, 29, 184-190.

M310   J.M. Mayer and M. Rowland (1984) Determination of Aqueous Solubilities of a Series of 5-Ethyl-5-alkylbarbituric Acids and Their Correlation with Log P and Melting Points, *Drug Development and Industrial Pharmacy*, 10, 69-83.

M311   M.E. McNally and R.L. Grob (1984) Headspace Determination of Solubility Limits of the Base Neutral and Volatile Components From the Environmental Protection Agency's List of Priority Pollutants, *Journal of Chromatography*, 284, 105-116.

M312   D. Mackay and A.T.K. Yeun (1983) Mass Transfer Coefficient Correlations for Volatilization of Organic Solutes from Water, *Environmental Science and Technology*, 17, 211-217.

M314   E. Meeussen and P. Huyskens (1966) Etude de la Structure du Butanol-n en Solution par les Coefficients de Partage, *Journal de Chimie Physique et de Physico-Chimie Biologique*, 63, 845-854.

M315   E. Mentasti, C. Rinaudo, and R. Boistelle (1983) Solubility and Mechanism of Dissolution of Dihydrated and Anhydrous Uric Acid, *Journal of Chemical and Engineering Data*, 28, 247-251.

M316   E. Mukai, K. Arase, M. Hashida, and H. Sezaki (1985) Enhanced Delivery of Mitomycin C Prodrugs through the Skin, *International Journal of Pharmaceutics*, 25, 95-103.

M317   B. Mollgaard, A. Hoelgaard, and H. Bundgaard (1982) Prodrugs as Drug Delivery Systems. XXIII. Improved Dermal Delivery of 5-Fluorouracil through Human Skin via N-Acyloxymethyl Pro-drug Derivatives, *International Journal of Pharmaceutics*, 12, 153-162.

M318   J.R.W. Miles, B.T. Bowman, and C.R. Harris (1981) Adsorption, Desorption, Soil Mobility and Aqueous Persistence of Fensulfothion and Its Sulfide and Sulfone Metabolites, Journal of Environmental Science & Health, Series B, 16, 309-324.

M319   A. Martin, P.L. Wu, and T. Velasquez (1985) Extended Hildebrand Solubility Approach: Sulfonamides in Binary and Ternary Solvents, *Journal of Pharmaceutical Sciences*, 74, 277-282.

M320   M.E. Moro, M.M. Velazquez, J.M. Cachaza, and L.J. Rodriguez (1986) Solubility of Diazepam and Prazepam in Aqueous Non-ionic Surfactants, *Journal of Pharmacy and Pharmacology*, 38, 294-296.

M321   M. Meshali, H. El-Sabbagh, and I. Ramadan (1984) Simultaneous Solubility and Dissolution Rate of Sulfamethoxazole and Trimethoprim in Binary Mixture, *Pharmazie (Berlin)*, 39, 407-408.

M323   J.M. Marco, M.I. Galan, and J. Costa (1988) Liquid-Liquid Equilibria for the Quaternary System Water-Phosphoric Acid-1-Hexanol-Cyclohexanone at 25 C, *Journal of Chemical and Engineering Data*, 33, 211-214.

M324   D. Mackay, S. Paterson, and W.H. Schroeder (1986) Model Describing the Rates of Transfer Processes of Organic Chemicals between Atmosphere and Water, *Environmental Science and Technology*, 20, 807-810.

M325   C. McDonald and C. Richardson (1981) The Effect of Added Salts on Solubilization by a Non-ionic Surfactant, *Journal of Pharmacy and Pharmacology*, 33, 38-39.

M327   J.C. McGowan, P.N. Atkinson, and L.H. Ruddle (1966) The Physical Toxicity of Chemicals. V. Interaction Terms for Solubilities and Partition Coefficients, *Journal of Applied Chemistry*, 16, 99-102.

M333   R.H. Manzo, A.A. Ahumada, and E. Luna (1984) Effects of Solvent Medium on Solubility. IV. Comparison of the Hydrophilic-lipophilic Character Exhibited by Functional Groups in Ethanol-Water and Ethanol-Cyclohexane Mixtures, *Journal of Pharmaceutical Sciences*, 73, 1869-1871.

M334   A. Martin, P.L. Wu, and A. Beerbower (1984) Expanded Solubility Parameter Approach. II. p-Hydroxybenzoic Acid and Methyl p-Hydroxybenzoate in Individual Solvents, *Journal of Pharmaceutical Sciences*, 73, 188-194.

M335   F.A. Menard, M.G. Dedhiya, and C.T. Rhodes (1988) Potential Pharmaceutical Applications of a New beta-Cyclodextrin Derivative, *Drug Development and Industrial Pharmacy*, 14, 1529-1547.

M336   T.J. Murphy, M.D. Mullin, and J.A. Meyer (1987) Equilibration of Polychlorinated Biphenyls and Toxaphene with Air and Water, *Environmental Science and Technology*, 21, 155-162.

M337   A. Mohammadzadeh-K., R.E. Feeney, and L.M. Smith (1969) Hydrophobic Binding of Hydrocarbons by Proteins. I. Relationship of Hydrocarbon Structure, *Biochimica et Biophysica Acta*, 194, 246-255.

M339   C. Munz and P.V. Roberts (1986) Effects of Solute Concentration and Cosolvents on the Aqueous Activity Coefficient of Halogenated Hydrocarbons, *Environmental Science and Technology*, 20, 830-836.

M340   L. Marple, R. Brunck, and L. Throop (1986) Water Solubility of 2,3,7,8-Tetrachlorodibenzo-p-dioxin, *Environmental Science and Technology*, 20, 180-182.

M342   M.M. Miller, S.P. Wasik, G-L. Huang..., and D. Mackay (1985) Relationships between Octanol-Water Partition Coefficient and Aqueous Solubility, *Environmental Science and Technology*, 19, 522-529.

M344   D. Mackay, A. Bobra, W.Y. Shiu, and S.H. Yalkowsky (1980) Relationships Between Aqueous Solubility and Octanol-Water Partition Coefficients, *Chemosphere*, 9, 701-711.

M345   C.J. Major and O.J. Swenson (1946) Acetic Acid-Ethyl Ether-Water System. Mutual Solubility and Tie Line Data, *Industrial and Engineering Chemistry*, 38, 834-836.

M346   J.W. Malone and R.W. Vining (1967) Phase Equilibria Data for the System n-Propyl Alcohol-Water-Nitroethane, *Journal of Chemical and Engineering Data*, 12, 387-389.

M347   J. Matous, J.P. Novak, J. Sobr, and J. Pick (1972) Phase Equilibria in the System Tetra-hydrofuran(1)-Water(2), *Collection of Czechoslovak Chemical Communications*, 37, 2653-2663.

M348   I. Mertl (1972) Liquid-Vapour Equilibrium. II. Phase Equilibria in the Ternary System Ethyl Acetate-Ethanol-Water, *Collection of Czechoslovak Chemical Communications*, 37, 366-374.

M349   J.C. Means, S.G. Wood, J.J. Hassett, and W.L. Banwart (1982) Sorption of Amino- and Carboxy-Substituted Polynuclear Aromatic Hydrocarbons by Sediments and Soils, *Environmental Science and Technology*, 16, 93-98.

M350   H. Matsuda, K. Ito, M. Tanaka..., and H. Sumiyoshi (1991) Inclusion Complexes of Various Fragrance Materials with 2-Hydroxypropyl-beta-cyclodextrin, *STP Pharma Science*, 1, 211-215.

M352   R.H. Manzo and A.A. Ahumada (1990) Effects of Solvent Medium on Solubility. V. Enthalpic and Entropic Contributions to the Free Energy Changes of Di-substituted Benzene Derivatives in Ethanol:Water and Ethanol:Cyclohexane Mixtures, *Journal of Pharmaceutical Sciences*, 79, 1109-1115.

M360   M.B. Maurin, L.W. Dittert, and A.A. Hussain (1992) Mechanism of Diffusion of Monosubstituted Benzoic Acids through Ethylene-Vinyl Acetate Copolymers, *Journal of Pharmaceutical Sciences*, 81, 79-84.

M362   M.B. Maurin, W.J. Addicks, S.M. Rowe, and R. Hogan (1993) Physical Chemical Properties of Alpha Styryl Carbinol Antifungal Agents, *Pharmaceutical Research*, 10, 309-312.

M364   R. Mestres and G. Mestres (1992) Deltamethrin: Uses and Environmental Safety, *Reviews of Environmental Contamination and Toxicology*, 124, 1-3.

M368   K. Muraoka and T. Hirata (1988) Hydraulic Behaviour of Chlorinated Organic Compounds in Water, *Water Research*, 22, 485-489.

M370   C.E. Morris (1992) Solubility of meso-1,2,3,4-Butanetetracarboxylic Acid and Some of Its Salts in Water, *Journal of Chemical and Engineering Data*, 37, 330-331.

M372   M.B. Maurin, R.D. Vickery, C.A. Gerard, and·M. Hussain (1993) Solubility of Ionization Behavior of the Antifungal alpha-(2,4-difluorophenyl)-alpha-[(1-(2-(2-pyridly)phenylethenyl)]-1H-1,2,4-triazole-1-ethanol bismesylate (XD405), *International Journal of Pharmaceutics*, 94, 11-14.

M373   K-C. Ma, W-Y. Shiu, and D. Mackay (1993) Aqueous Solubilities of Chlorinated Phenol at 25 C, *Journal of Chemical and Engineering Data*, 38, 364-366.

M374   G.C. Mazzenga and B. Berner (1991) The Transdermal Delivery of Zwitterionic Drugs. I. The Solubility of Zwitterion Salts, *Journal of Controlled Release*, 16, 77-88.

M375   A.F. Marcilla, F. Ruiz, and M.C. Sabater (1994) Two-Phase and Three-Phase Liquid-Liquid Equilibrium for Bis(2-methylpropyl) Ester + Phosphoric Acid + Water, *Journal of Chemical and Engineering Data*, 39, 14-18.

M378   A.R. Moorman, S.D. Chamberlain, L.A. Jones..., and T.A. Krenitsky (1992) 5'-ester Prodrugs of the Varicella-Zoster Antiviral Agent, 6-methoxypurine arabinoside, *Antiviral Chemistry and Chemotherapy*, 3, 141-146.

M379   B.W. Muller and E. Albers (1991) Effect of Hydrotropic Substances on the Complexation of Sparingly Soluble Drugs with Cyclodextrin Derivatives and the Influence of Cyclodextrin Complexation on the Pharmacokinetics of the Drugs, *Journal of Pharmaceutical Sciences*, 80, 599-604.

M380   H.M.C. Marques, J. Hadgraft, and I.W. Kellaway (1990) Studies of Cyclodextrin Inclusion Complexes. I. The Salbutamol-cyclodextrin Complex as Studied by Phase Solubility and DSC, *International Journal of Pharmaceutics*, 63, 259-266.

M381   M.M. Morelock, L.L. Choi, G.L. Bell, and J.L. Wright (1994) Estimation and Correlation of Drug Water Solubility with Pharmacological Parameters Required for Biological Activity, *Journal of Pharmaceutical Sciences*, 83, 948-951.

M382   M. Mosharraf and C. Nystrom (1995) Solubility Characterization of Practically Insoluble Drugs Using the Coulter Counter Principle, *International Journal of Pharmaceutics*, 122, 57-67.

M383   A. Munoz de la Pena, T. Ndou, J.B. Zung, and I.M. Warner (1991) Stoichiometry and Formation Constants of Pyrene Inclusion Complexes with beta-and gamma-Cyclodextrin, *Journal of Physical Chemistry*, 95, 3330-3334.

N001   T.E. Needham, Jr., A.N. Paruta, and R.J. Gerraughty (1971) Solubility of Amino Acids in Mixed Solvent Systems, *Journal of Pharmaceutical Sciences*, 60, 258-260.

N002   W.F. Ng and C.F. Poe (1956) A Note on the Solubility of Strychnine in Alcohol and Water Mixtures, *Journal of the American Pharmaceutical Association, Scientific Edition*, 45, 351-353.

N003   J.R. Nixon and B.P.S. Chawla (1969) Solubilization and Rheology of the System Ascorbic Acid-Water-Polysorbate 80: Temperature Effects, *Journal of Pharmacy and Pharmacology*, 21, 79-84.

N004   H.D. Nelson and C.L. De Ligny (1968) The Determination of the Solubilities of Some n-Alkanes in Water at Different Temperatures, by Means of Gas Chromatography, *Recueil des Travaux Chimiques des Pays-Bas*, 87, 528-544.

N006   H. Nogami, T. Nagai, and H. Uchida (1968) Physico-chemical Approach to Biopharmaceutical Phenomena. II. Hydrophobic Hydration of Tryptophan in Aqueous Solution, *Chemical and Pharmaceutical Bulletin*, 16, 2257-2262.

N007   H. Nogami, T. Nagai, E. Fukuoka, and T. Yotsuyanagi (1969) Dissolution Kinetics of Barbital Polymorphs, *Chemical and Pharmaceutical Bulletin*, 17, 23-31.

N008   H. Nogami, T. Nagai, and H. Uchida (1969) Physico-chemical Approach to Biopharmaceutical Phenomena. IV. Adsorption of Barbituric Acid Derivatives by Carbon Black from Aqueous Solution, *Chemical and Pharmaceutical Bulletin*, 17, 168-175.

N009   H. Nogami, T. Nagai, and H. Umeyama (1970) Effect of Third Component on Water Structure around Tryptophan in Aqueous Solution, *Chemical and Pharmaceutical Bulletin*, 18, 328-334.

N012   T.E. Needham, Jr., A.N. Paruta, and R.J. Gerraughty (1971) Solubility of Amino Acids in Pure Solvent Systems, *Journal of Pharmaceutical Sciences*, 60, 565-567.

N013   R.W. Nex and A.W. Swezey (1954) Some Chemical and Physical Properties of Weed Killers, *Weeds*, 3, 241-253.

N014   A.K. Nanda and M.M. Sharma (1966) Effective Interfacial Area in Liquid-Liquid Extraction, *Chemical Engineering Science*, 21, 707-714.

N015   W.J.P. Neish (1948) On the Solubilisation of Aromatic Amines by Purines, *Recueil des Travaux Chimiques des Pays-Bas*, 67, 361-373.

N016   D.W. Newton, W.J. Murray, and S. Ratanamaneichatara (1982) Evaluation of Solubility Data Useful for Phase Solubility Determination of Azathioprine, *Analytica Chemica Acta*, 135, 343-346.

N017   M. Nakano, Y. Arakawa, K. Juni, and T. Arita (1976) Physical Properties of Pyrimidine and Purine Antimetabolites. II. Permeation of 5-Fluorouracil and 1-(2-Tetrahydrofuryl)-5-fluorouracil through Cellophane, Collagen, and Silicone Membranes, *Chemical and Pharmaceutical Bulletin*, 24, 2716-2722.

N018   H. Nakatani (1963) Studies on Pharmaceutical Preparations of Orotic Acid. II. Isolation of Reaction Products of Orotic Acid and Amines, and Their Solubility in Water, *Yakugaku Zasshi (Tokyo)*, 83, 6-9.

N019   H. Nakatani (1964) Studies on Pharmaceutical Preparations of Orotic Acid. VI. Water Soluble Properties of Orotic Acid Salts, *Yakugaku Zasshi*, 84, 1057-1061.

N021   R.N. Nasipuri and S.A.H. Khalil (1973) Adsorption-Dissolution Relationship in Sulfamethazine-Benzoic Acid System, *Journal of Pharmaceutical Sciences*, 62, 473-475.

N023   H. Nogami, T. Nagai, and T. Yotsuyanagi (1969) Dissolution Phenomena of Organic Medicinals Involving Simultaneous Phase Changes, *Chemical and Pharmaceutical Bulletin*, 17, 499-509.

N024   Y. Nozaki and C. Tanford (1970) The Solubility of Amino Acids, Diglycine, and Triglycine in Aqueous Guanidine Hydrochloride Solutions, *Journal of Biological Chemistry*, 245, 1648-1652.

N025   Y. Nozaki and C. Tanford (1971) The Solubility of Amino Acids and Two Glycine Peptides in Aqueous Ethanol and Dioxane Solutions, *Journal of Biological Chemistry*, 246, 2211-2217.

N026   Y. Nozaki and C. Tanford (1965) The Solubility of Amino Acids and Related Compounds in Aqueous Ethylene Glycol Solutions, *Journal of Biological Chemistry*, 240, 3568-3573.

N027   Y. Nozaki and C. Tanford (1963) The Solubility of Amino Acids and Related Compounds in Aqueous Urea Solutions, *Journal of Biological Chemistry*, 238, 4074-4081.

N028   P. Nylen (1926) Zur Kenntnis der Organischen Phosphorverbindungen. II. Uber beta-Phosphon-propionsaure und gamma-Phosphon-n-buttersaure, *Berichte der Deutschen Chemischen Gesellschaft*, 59, 1119-1123.

N031   V. Naggar, N.A. Daabis, and M.M. Motawi (1974) Solubilization of Tetracycline and Oxytetracycline, *Pharmazie (Berlin)*, 29, 122-125.

N032   L.S. Nang, D. Cosnier, G. Terrie, and J. Moleyre (1977) Consequence of Solubility Alteration by Salt Effect on Dissolution Enhancement and Biological Response of a Solid Dispersion of an Experimental Antianginal Drug, *Pharmacology*, 15, 545-550.

N034   M.F. Nathan (1978) Choosing a Process for Chloride Removal, *Chemical Engineering (New York)*, 85, 93-100.

N035   W. Neudert and H. Ropke (1954) Uber das Physikalischchemische Verhalten des Dinatriumsalzes des Adipinsaure-bis-[2.4.6-trijod-3-carboxy-anilids] und Anderer Trijodbenzolderivate, *Chemische Berichte*, 87, 659-667.

N038   A. Niini (1938) The Determination of the Reciprocal Solubilities of Water and Certain Organic Liquids by Means of a Pycnometer and Refractometer, *Suomen Kemistilehti A*, 11, 19-20.

N041   H. Negoro, T. Miki, and S. Ueda (1959) Interaction between Pyrazinamide and Sodium p-Aminosalicylate or Sodium Hydroxybenzoates in Aqueous Solution, *Chemical and Pharmaceutical Bulletin*, 7, 91-95.

N042   A.C. Noorduyn (1919) Sur des Hydrates d'Oenanthol, *Recueil des Travaux Chimiques des Pays-Bas et de la Belgique*, 38, 345-350.

N043   H.D. Nelson and J.H. Smit (1978) Gas Chromatographic Determination of the Water Solubility of the Halogenobenzenes, *Suid-Afrikaanse Tydskrf vir Chemie*, 31, 76-76.

N044   N.I. Nakano, N. Kawahara, T. Amiya..., and D. Furukawa (1978) 3-Hydroxy-2-naphthoates of Lidocaine, Mepivacaine, and Bupivacaine and Their Dissolution Characteristics, *Chemical and Pharmaceutical Bulletin*, 26, 936-941.

N045   M. Nakano, K. Juni, and T. Arita (1976) Controlled Drug Permeation. I. Controlled Release of Butamben through Silicone Membrane by Complexation, *Journal of Pharmaceutical Sciences*, 65, 709-712.

N046   N.I. Nakano (1979) Temperature-Dependent Aqueous Solubilities of Lidocaine, Mepivacaine, and Bupivacaine, *Journal of Pharmaceutical Sciences*, 68, 667-668.

N050   S. Niazi (1978) Thermodynamics of Mercaptopurine Dehydration, *Journal of Pharmaceutical Sciences*, 67, 488-491.

N051   F.C. Nachod (1938) Keto-enoltautomerien in Leichten und Schweren Losungsmitteln, *Zeitschrift fuer Physikalische Chemie, Abteilung A: Chemische Thermodynamik, Kinetik, Electrochemie, Eigenschaftslehre*, 182, 193-219.

N053   M. Nango, H. Yamamoto, K. Joukou..., and N. Kuroki (1980) Solubility of Aromatic Hydrocarbons in Water and Aqueous Solutions of Sugars, *Journal of the Chemical Society, Chemical Communications*, 104-105.

N055   D.W. Newton, D.F. Driscoll, J.L. Goudreau, and S. Ratanamaneichatara (1981) Solubility Characteristics of Diazepam in Aqueous Admixture Solutions: Theory and Practice, *American Journal of Hospital Pharmacy*, 38, 179-182.

N056   V.F.B. Naggar, S. El-Gamal, and M.A. Shams-Eldeen (1980) Physicochemical Studies of Phenylbutazone Recrystallized from Polysorbate 80, *Scientia Pharmaceutica*, 48, 335-343.

N057   M. Nicklasson, A. Brodin, and H. Nyqvist (1981) Studies on the Relationship Between Solubility and Intrinsic Rate of Dissolution as a Function of pH, *Acta Pharmaceutica Suecica*, 18, 119-128.

N059   M.C.O. Neira, J.M. Fernando, P. de Leon, and L. Fernanda (1980) Effect of Dielectric Constants on the Solubility of Diazepam, *Revista Colombiana de Ciencias Quimico-Farmaceuticas*, 3, 37-61.

N061   N.G. Nash and R.D. Daley (1975) Disulfiram, *Analytical Profiles of Drug Substances*, 4, 170-177.

N062   J. Nishijo, K. Ohno, K. Nishimura..., and I. Yonetani (1982) Soluble Complex Formation of Theophylline with Aliphatic Di- and Monoamines in Aqueous Solution, *Chemical and Pharmaceutical Bulletin*, 30, 771-776.

N063   D.W. Newton, S. Ratanamaneichatara, and W.J. Murray (1982) Dissociation, Solubility and Lipophilicity of Azathioprine, *International Journal of Pharmaceutics*, 11, 209-213.

N301   L. Nyberg, L. Bratt, A. Forsgren, and S. Hugosson (1974) Bioavailability of Digoxin from Tablets. I. In Vitro Characterization of Digoxin Tablets, *Acta Pharmaceutica Suecica*, 11, 447-458.

N302   E. Nurnberg (1976) Neuere Untersuchungsergebnisse von Pharmazeutischen Spruhtrocknungsprodukten, *Pharmazeutische Industrie*, 38, 228-232.

N304   T. Nakagawa and M. Kanauchi (1978) Isothioate, *Analytical Methods for Pesticides and Plant Growth Regulators*, 10, 75-82.

N305   T. Nakamura and K. Yamaoka (1978) Isoxathion, *Analytical Methods for Pesticides and Plant Growth Regulators*, 10, 83-94.

N306   T. Nakamura, K. Yamaoka, and M. Kotakemori (1978) Hymexazol, *Analytical Methods for Pesticides and Plant Growth Regulators*, 10, 215-228.

N309   M. Newman, C.B. Hayworth, and R.E. Treybal (1949) Dehydration of Aqueous Methyl Ethyl Ketone: Equilibrium Data for Extractive Distillation and Solvent Extraction, *Industrial and Engineering Chemistry*, 41, 2039-2043.

N311   G.A. Nyssen, E.T. Miller, T.F. Glass, and C.R. Quinn II (1987) Solubilities of Hydrophobic Compounds in Aqueous-Organic Solvent Mixtures, *Environmental Monitoring and Assessment*, 9, 1-11.

N312   J. Nishijo and I. Yonetani (1982) The Interaction of Theophylline with Benzylamine in Aqueous Solution, *Chemical and Pharmaceutical Bulletin*, 30, 4507-4511.

N314   N.M. Najib and M.S. Suleiman (1985) The Effect of Hydrophilic Polymers and Surface Active Agents on the Solubility of Indomethacin, *International Journal of Pharmaceutics*, 24, 165-171.

N315   D.W. Newton, W.A. Narducci, W.A. Leet, and C.T. Ueda (1983) Lorazepam Solubility in and Sorption from Intravenous Admixture Solutions, *American Journal of Hospital Pharmacy*, 40, 424-427.

N316   N.M. Najib and M.A.S. Salem (1987) Release of Ibuprofen from Polyethylene Glycol Solid Dispersions: Equilibrium Solubility Approach, *Drug Development and Industrial Pharmacy*, 13, 2263-2275.

N317   N.M. Nielsen and H. Bundgaard (1988) Gycolamide Esters as Biolabile Prodrugs of Carboxylic Acid Agents: Synthesis, Stability, Bioconversion, and Physicochemical Properties, *Journal of Pharmaceutical Sciences*, 77, 285-298.

N319   M. Nicklasson and A. Brodin (1984) The Relationship between Intrinsic Dissolution Rates and Solubilities in the Water-Ethanol Binary Solvent System, *International Journal of Pharmaceutics*, 18, 149-156.

N320   W.B. Neely (1979) Estimating Rate Constants for the Uptake and Clearance of Chemicals by Fish, *Environmental Science and Technology*, 13, 1506-1508.

N322   C. Nystrom and M. Bisrat (1986) Coulter Counter Measurements of Solubility and Dissolution Rate of Sparingly Soluble Compounds Using Micellar Solutions, *Journal of Pharmacy and Pharmacology*, 38, 420-425.

N323   C. Nystrom, J. Mazur, M.I. Barnett, and M. Glazer (1985) Dissolution Rate Measurements of Sparingly Soluble Compounds with the Coulter Counter Model TAII, *Journal of Pharmacy and Pharmacology*, 37, 217-221.

N326   J.M. Norris, J.W. Ehrmantraut, C.L. Gibbons..., and J.S. Brosier (1973) Toxicological and Environmental Factors Involved in the Selection of Decabromodiphenyl Oxide as a Fire Retardant Chemical, *Applied Polymer Symposia*, 22, 195-219.

N330   K.S. Narasimhan, C.C. Reddy, and K.S. Chari (1962) Solubility and Equilibrium Data of Phenol-Water-n-Butyl Acetate System at 30 C, *Journal of Chemical and Engineering Data*, 7, 340-343.

N331   A. Nakaishi, H. Maeda, T. Tomiyama..., and Y. Kyogoku (1988) Chain Length Dependence of Solubility of Monodisperse Polypeptides in Aqueous Solutions and the Stability of the Beta-Structure, *Journal of Physical Chemistry*, 92, 6161-6166.

N332   M.M. Narurkar and A.K. Mitra (1988) Synthesis, Physicochemical Properties, and Cytotoxicity of a Series of 5'-Ester Prodrugs of 5-Iodo-2'-deoxyuridine, *Pharmaceutical Research*, 5, 734-737.

N333   P. Nkedi-Kizza, M.L. Brusseau, P.S.C. Rao, and A.G. Hornsby (1989) Nonequilibrium Sorption during Displacement of Hydrophobic Organic Chemicals and 45-Ca through Soil Columns with Aqueous and Mixed Solvents, *Environmental Science and Technology*, 23, 814-820.

N334   Y. Nakagawa, S. Itai, T. Yoshida, and T. Nagai (1992) Physicochemical Properties and Stability in the Acidic Solution of a New Macrolide Antibiotic, Clarithromycin, in Comparison with Erythromycin, *Chemical and Pharmaceutical Bulletin*, 40, 725-728.

N335   N.M. Nielsen and H. Bundgaard (1989) Evaluation of Glycolamide Esters and Various Other Esters of Aspirin as True Aspirin Prodrugs, *Journal of Medicinal Chemistry*, 32, 727-734.

N336   A. Natarajan, V. Sapre, U.B. Hadkar, and P.Y. Shirodkar (1992) The Correlation of pKa with Biological Activity of Substituted Salicylanide Derivatives, *Indian Drugs*, 29, 545-552.

N337   L.S. Nielsen, F. Slok, and H. Bundgaard (1994) N-Alkoxycarbonyl Prodrugs of Mebendazole with Increased Water Solubility, *International Journal of Pharmaceutics*, 102, 231-239.

O001   J.A. Ostrenga and C. Steinmetz (1970) Estimation of Steroid Solubility: Use of Fractional Molar Attraction Constants, *Journal of Pharmaceutical Sciences*, 59, 414-416.

O002   J.W. O'Laughlin, F.W. Sealock, and C.V. Banks (1964) Determination of Solubility of Several Phosphine Oxides in Aqueous Solutions Using a New Spectrophotometric Procedure, *Analytical Chemistry*, 34, 224-226.

O003   A. Osol and M. Kilpatrick (1933) The "Salting-Out" and "Salting-In" of Weak Acids. II. The Activity Coefficients of the Molecules of Ortho, Meta, and Para-Hydroxybenzoic Acids in Aqueous Salt Solutions,, 55, 4440-4444.

O004   A. Osol and M. Kilpatrick (1933) The "Salting-Out" and "Salting-In" of Weak Acids. I. The Activity Coefficients of the Molecules of Ortho, Meta, and Para-Chlorobenzoic Acids in Aqueous Salt Solutions,, 55, 4430-4440.

O005   D.F. Othmer, R.E. White, and E. Trueger (1941) Liquid-Liquid Extraction Data, *Industrial and Engineering Chemistry*, 33, 1240-1248.

O006   W.L. O'Connell (1963) Properties of Heavy Liquids, *Transactions of Society of Mining Engineers*, 226, 126-132.

O007   T. Okano, K. Uekama, and K. Ikeda (1968) Electronic Properties of N-Heteroaromatics. XVI. Charge Transfer Properties of Pyrazolone Antipyretics. On the Complex Formation of Aminopyrine with Benzoic Acid and Salicylic Acid, *Chemical and Pharmaceutical Bulletin*, 16, 6-12.

O009   R.P. Ouellette and J.A. King (1977) Chemical Week Pesticides Register.

O011   D.F. Othmer and J. Serrano, Jr. (1949) Solubility Data for Ternary Liquid Systems: Systems of Acetic Acid, Higher Boiling Homologous Acids, and Water, *Industrial and Engineering Chemistry*, 41, 1030-1032.

O012   D.F. Othmer, W.S. Bergen, N. Shlechter, and P.F. Bruins (1945) Liquid-Liquid Extraction Data: Systems Used in Butadien Manufacture from Butylene Glycol, *Industrial and Engineering Chemistry*, 37, 890-894.

O013   G. Ostergren and A. Levan (1943) The Connection Between c-Mitotic Activity and Water Solubility in Some Monocyclic Compounds, *Hereditas*, 29, 496-498.

O015   A.L. Olsen and E.R. Washburn (1935) Study of Solutions of Isopropyl Alcohol in Benzene, in Water and in Benzene and Water, *Journal of the American Chemical Society*, 57, 303-305.

O016   E. Oliveri-Mandala and F. Forni (1925) Influenze di Solubilita. (Acetanilide-antipirina, Acetanilide-Piramidone). Nota IV, *Gazzetta Chimica Italiana*, 55, 783-788.

O017   E. Oliveri-Mandala and L. Irrera (1930) Influenze di Solubilitia (Coppie: Tiourea-Antipirina, Caffeina-Antipirina). Nota VII, *Gazzetta Chimica Italiana*, 60, 872-877.

O018   E. Oliveri-Mandala (1926) Influenze di Solubilita (Coppie: Cloralio-Caffeina, Urotropina-Antipirina, Urotropina-Cloralio).-Nota V, *Gazzetta Chimica Italiana*, 56, 889-896.

O019   E. Oliveri-Mandala (1926) Influenze di Solubilita (Constituzione Chimica e Solubilita). Nota VI, *Gazzetta Chimica Italiana*, 56, 896-901.

O021   I. Odaira (1915) Synthesis of the Different Acylurethanes and Some Allied Compounds, *The Influence of the Acid Radicals*, 1, 324-330.

O025   J. Osteryoung and J.W. Whittaker (1980) Picloram: Solubility and Acid-Base Equilibria Determined by Normal Pulse Polarography, *Journal of Agricultural and Food Chemistry*, 28, 95-97.

O026   D.F. Othmer and P.L. Ku (1960) Solubility Data for Ternary Liquid Systems Acetic Acid and Formic Acid Distributed between Chloroform and Water, *Journal of Chemical and Engineering Data*, 5, 42-44.

O027   J.J. O'Neill, P.L. Peelor, A.F. Peterson, and C.H. Strube (1979) Selection of Parabens as Preservatives for Cosmetics and Toiletries, *Journal of the Society of Cosmetic Chemistry*, 30, 25-38.

O028   D.F. Othmer, M.M. Chudgar, and S.L. Levy (1952) Binary and Ternary Systems of Acetone, Methyl Ethyl Ketone, and Water, *Industrial and Engineering Chemistry*, 44, 1872-1881.

O032   J.E. Ojile, C.B. Macfarlane, and A.B. Selkirk (1982) Drug Distribution During Massing and Its Effect on Dose Uniformity in Granules, *International Journal of Pharmaceutics*, 10, 99-107.

O300   M. Ottnad, N.A. Jenny, and C-H. Roder (1978) Methyl Isothiocyanate, *Analytical Methods for Pesticides and Plant Growth Regulators*, 10, 563-573.

O302   A.B. Ochsner, R.J. Belloto, Jr., and T.D. Sokoloski (1985) Prediction of Xanthine Solubilities Using Statistical Techniques, *Journal of Pharmaceutical Sciences*, 74, 132-135.

O303   M. Otagiri, T. Imai, F. Hirayama, and K. Uekama (1983) Inclusion Complex Formations of the Antiinflammatory Drug Flurbiprofen with Cyclodextrins in Aqueous Solution and in Solid State, *Acta Pharmaceutica Suecica*, 20, 11-20.

O304   Y. Okada, S. Horiyama, and K. Koizumi (1986) Studies on Inclusion Complexes of Non-steroidal Anti-inflammatory Agents with Cyclosophoraose-A, *Yakugaku Zasshi (Tokyo)*, 106, 240-247.

O305   M. Otsuka and N. Kaneniwa (1983) Hygroscopicity and Solubility of Noncrystalline Cephalexin, *Chemical and Pharmaceutical Bulletin*, 31, 230-236.

O306   A. Obikili, M. Deyme, D. Wouessidjewe, and D. Duchene (1988) Improvement of Aqueous Solubility and Dissolution Kinetics of Canrenone by Solid Dispersion in Sucroester, *Drug Development and Industrial Pharmacy*, 14, 791-803.

O307   S. Othman, H. Muti, O. Shaheen..., and W.A. Al-Turk (1988) Studies on the Adsorption and Solubility of Nalidixic Acid, *International Journal of Pharmaceutics*, 41, 197-203.

O310   A.G. Ogston (1936) Some Dissociation Constants, *Journal of the Chemical Society (London)*, 1713-1713.

O311   A. Opperhuizen, F.A.P. Gobas, J.M.D. Van der Steen, and O. Hutzinger (1988) Aqueous Solubility of Polychlorinated Biphenyls Related to Molecular Structure, *Environmental Science and Technology*, 22, 638-646.

O312   J.W. Owens, S.P. Wasik, and H. DeVoe (1986) Aqueous Solubilities and Enthalpies of Solution of n-Alkylbenzenes, *Journal of Chemical and Engineering Data*, 31, 47-51.

O316   C.J. Orella and D.J. Kirwan (1989) The Solubility of Amino Acids in Mixtures of Water and Aliphatic Alcohols, *Biotechnology Progress*, 5, 89-91.

O317   C.J. Orella and D.J. Kirwan (1991) Correlation of Amino Acid Solubilities in Aqueous Aliphatic Alcohol Solutions, *Industrial and Engineering Chemistry, Product Research and Development*, 30, 1040-1045.

O318   J.T.H. Ong and E. Manoukian (1988) Micellar Solubilization of Timobesone Acetate in Aqueous and Aqueous Propylene Glycol Solutions of Nonionic Surfactants, *Pharmaceutical Research*, 5, 704-708.

O319   I. Oh, S.C. Chi, B.R. Vishnuvajjala, and B.D. Anderson (1991) Stability and Solubilization of Oxathiin Carboxanilide, a Novel anti-HIV Agent, *International Journal of Pharmaceutics*, 73, 23-31.

O320   C.E. Orazio, S. Kapila, R.K. Puri, and A.F. Yanders (1992) Persistence of Chlorinated Dioxins and Furans in the Soil Environment, *Chemosphere*, 25, 1469-1474.

O321   Y. Okada, Y. Kubota, K. Koizumi..., and K. Ogata (1988) Some Properties and the Inclusion Behavior of Branched Cyclodextrins, *Chemical and Pharmaceutical Bulletin*, 36, 2176-2185.

O322   J.R. O'Reilly, O.I. Corrigan, and C.M. O'Driscoll (1994) The Effect of Mixed Micellar Systems, Bile Salt/Fatty Acids, on the Solubility and Intestinal Absorption of Clofazimine (B663) in the Anaesthetised Rat, *International Journal of Pharmaceutics*, 109, 147-154.

P003   J. Polak and B. Lu (1973) Mutual Solubilities of Hydrocarbons and Water at and 25 C, *Canadian Journal of Chemistry*, 51, 4018-4023.

P004   G.H. Parsons, C.H. Rochester, A. Rostron, and P.C. Sykes (1972) The Thermodynamics of Hydration of Phenols, *Journal of Chemical Social Perkin*, 2, 136-138.

P005   W.A. Pryor and R.E. Jentoft, Jr. (1961) Solubility of m- and p-Xylene in Water and in Aqueous Ammonia from to 300 C, *Journal of Chemical and Engineering Data*, 6, 36-37.

P006   C.F. Peterson and R.E. Hopponen (1953) Solubility of Phenobarbital in Propylene Glycol- Alcohol-Water Systems, *Journal of the American Pharmaceutical Association, Scientific Edition*, 42, 540-541.

P007   L. Price (1973) The Solubility of Hydrocarbons and Petroleum in Water as Applied to the Primary Migration of Petroleum, Ph.D. thesis, 60-261.

P008   B.J. Poulsen, E. Young, V. Coquilla, and M. Katz (1968) Effect of Topical Vehicle Composition on the In Vitro Release of Fluocinolone Acetonide and its Acetate Ester, *Journal of Pharmaceutical Sciences*, 57, 928-933.

P009   J.W. Poole and C.K. Bahal (1968) Dissolution Behavior and Solubility of Anhydrous and Trihydrate Forms of Ampicillin, *Journal of Pharmaceutical Sciences*, 57, 1945-1948.

P010   A.N. Paruta and S.A. Irani (1966) Solubility Profiles for the Xanthines in Aqueous Solutions of a Glycol Ether. II, *Journal of Pharmaceutical Sciences*, 55, 1060-1064.

P011   A.N. Paruta and S.A. Irani (1966) Solubility Profiles for the Xanthines in Aqueous Alcoholic Mixtures. I. Ethanol and Methanol, *Journal of Pharmaceutical Sciences*, 55, 1055-1059.

P012   A.N. Paruta (1967) Solubility Profiles for Antipyrine and Aminopyrine in Hydroalcoholic Solutions, *Journal of Pharmaceutical Sciences*, 56, 1565-1569.

P013   A.N. Paruta and B.B. Sheth (1966) Solubility of Parabens in Syrup Vehicles, *Journal of Pharmaceutical Sciences*, 55, 1208-1211.

P014   A.N. Paruta, B.J. Sciarrone, and N.G. Lordi (1964) Solubility of Salicylic Acid as a Function of Dielectric Constant, *Journal of Pharmaceutical Sciences*, 53, 1349-1353.

P015   A.N. Paruta (1964) Solubility of Several Solutes as a Function of the Dielectric Constant of Sugar Solutions, *Journal of Pharmaceutical Sciences*, 53, 1252-1254.

P016   A.N. Paruta and S.A. Irani (1965) Dielectric Solubility Profiles in Dioxane-Water Mixtures for Several Antipyretic Drugs, *Journal of Pharmaceutical Sciences*, 54, 1334-1338.

P018   A.N. Paruta, B.J. Sciarrone, and N.G. Lordi (1965) Solubility Profiles for the Xanthines in Dioxane-Water Mixtures, *Journal of Pharmaceutical Sciences*, 54, 838-841.

P019   N.K. Patel and H.B. Kostenbauder (1958) Interaction of Preservatives with Macromolecules. I. Binding of Parahydroxyethylene 80 Sorbitan Monooleate (Tween 80), *Journal of the American Pharmaceutical Association, Scientific Edition*, 47, 289-293.

P020   A.N. Paruta and B.B. Sheth (1966) Solubility of the Xanthines, Antipyrine, and Several Derivatives in Syrup Vehicles, *Journal of Pharmaceutical Sciences*, 55, 896-901.

P022   F.M. Plakogiannis and P. Catsoulakos (1976) Solubility Behavior of 3-beta-Hydroxy-13-alpha-amino-13,17-seco-5-alpha-androstan-17-oic-13,17-lactam-4-N,N-bis-(chloroethyl) Amino Phenyl-acetate, a New Anti-cancer Agent, *Pharmaceutica Acta Helvetiae*, 51, 249-252.

P023   L.A. Pinck and M.A. Kelly (1925) The Solubility of Urea in Water, *Journal of the American Chemical Society*, 47, 2170-2172.

P024   J.P. Phillips and H.P. Price (1951) Spectrophotometric Study of 8-Hydroxyquinaldine Chelates. Notes on 8-Quinolinol Chelates, *Journal of the American Chemical Society*, 73, 4414-4415.

P026   R. Pulver, B. Exer, and B. Herrmann (1956) Uber die Beeinflussung Enzymatischer Reaktionen durch Phenylbutazon und die Ubertragbarkeit Fermentchemischer Befunde auf die Stoffwechselprozesse der Zelle, *Schweizerische Medizinische Wochenschrift*, 86, 1080-1085.

P027   K.S. Park and W.N. Bruce (1968) The Determination of the Water Solubility of Aldrin, Dieldrin, Heptachlor, and Heptachlor Epoxide, *Journal of Economic Entomology*, 61, 770-774.

P028   G.W. Probst and J.B. Tepe (1969) Trifluralin and Related Compounds, 225-257.

P029   C.J. Pedersen (1967) Cyclic Polyethers and Their Complexes with Metal Salts, *Journal of the American Chemical Society*, 89, 7017-7036.

P031   R.J. Pinney and V. Walters (1969) The Relation Between the Bactericidal Activities and Certain Physico-chemical Properties of Some Fluorophenols, *Journal of Pharmacy and Pharmacology*, 21, 415-422.

P033   M.F. Paul, R.C. Bender, and E.G. Nohle (1959) Renal Excretion of Nitrofurantoin (Furadantin), *American Journal of Physiology*, 197, 580-584.

P034   M.F. Paul, C. Harrington, R.C. Bender..., and M.H. Bryson (1967) Effect of pH and of Urea on Nitrofurantoin Activity, *Proceedings of the Society for Experimental Biology and Medicine*, 125, 941-947.

P035   J.W. Poole and C.K. Bahal (1970) Dissolution Behavior and Solubility of Anhydrous and Dihydrate Forms of Wy-4508, and Aminoalicyclic Penicillin, *Journal of Pharmaceutical Sciences*, 59, 1265-1267.

P036   J.T. Pearson and G. Varney (1973) The Anomalous Behaviour of Some Oxyclozanide Polymorphs, *Journal of Pharmacy and Pharmacology*, 25, 62-70.

P037   J.C. Philip and F.B. Garner (1909) Influence of Various Sodium Salts on the Solubility of Sparingly Soluble Acids. Part II, *Journal of the Chemical Society (London)*, 95, 1466-1473.

P038   J.C. Philip and R.S. Colborne (1924) The Solubility of Anilinesulphonic Acids, *Journal of the Chemical Society (London)*, 125, 492-500.

P040   E.A. Pasquinelli (1956) Correlation of the Mutual Solubilities of Two Liquids with the Electric and Magnetic Properties of the Pure Components, *Transactions of the Faraday Society*, 53, 932-938.

P041   J.F. Powell (1947) The Solubility or Distribution Coefficient of Trichlorethylene in Water, Whole Blood, and Plasma, *British Journal of Industrial Medicine*, 4, 233-236.

P043   J.C. Philip (1905) Influence of Various Sodium Salts on the Solubility of Sparingly Soluble Acids, *Journal of the Chemical Society (London)*, 87, 987-1003.

P044   M.D. Parker (1974) The Influence of Anions on the Physical Properties of Butyrophenone-type Molecules, *Dissertation Abstracts of International*, 35, 34-68.

P045   P. Pfeiffer and O. Angern (1924) Das Aussalzen der Aminosaeuren, *Zeitschrift fuer Physiologische Chemie*, 133, 180-192.

P046   C.R. Pearson and G. McConnell (1975) Chlorinated C_1 and C_2 Hydrocarbons in the Marine Environment, *Proceedings of the Royal Society of London, Series B: Biological Sciences*, 189, 305-332.

P048   H. Pringsheim and D. Dernikos (1922) Weiteres uber die Polyamylosen (Beitrage zur Chemie der Starke. VI), *Berichte der Deutschen Chemischen Gesellschaft*, 55, 1433-1449.

P049   G. Pucher and W.M. Dehn (1921) Solubilities in Mixtures of Two Solvents, *Journal of the American Chemical Society*, 43, 1753-1758.

P051   L.C. Price (1976) Aqueous Solubility of Petroleum as Applied to Its Origin and Primary Migration, *The American Association of Petroleum Geologists Bulletin*, 60, 213-244.

P052   P.N. Peter (1928) Solubility Relationships of Lactose-sucrose Solutions. I. Lactose-Sucrose Solubilities at Low Temperatures, *Journal of Physical Chemistry*, 32, 1856-1864.

P053   M-C. Poelman, F. Puisieux, and J-C. Chaumeil (1975) Interactions between Antiseptics and Surfactants. II. Study the Solubility Method of the Interaction of Methyl-parahydroxy-benzoate with Polyoxyethylene Fatty Alcohol Ethers, *Annals Pharmaceutiques Francaises*, 33, 551-557.

P054   B.I. Petrov, V.P. Zhivopistsev, I.A. Kislitsyn, and M.A. Volkova (1977) Solubility of Diantipyryl-methanes in Aqueous Solution and in Organic Solvents, *Journal of Analytical Chemistry of the USSR*, 32, 1180-1186.

P055   J.G. Park and H.E. Hofmann (1932) Aliphatic Ketones as Solvents, *Industrial and Engineering Chemistry*, 24, 132-134.

P057   K.J. Pedersen (1934) Studies of Complex Formation between Aniline and Picrate Ion by Solubility Measurements, *Journal of the American Chemical Society*, 56, 2615-2619.

P059   P. Pascal (1920) Etude Physico-chimique des Melanges d'Eau, d'Aldehyde et de Paraldehyde, *Bulletin de la Societe Chimique de France*, 27, 353-362.

P060   W.H. Patterson (1938) Estimation of Deuterium Oxide-Water Mixtures. Part II. The Solubility Curves with n-Butyric Acid and with Isobutyric Acid, *Journal of the Chemical Society (London)*, 1559-1561.

P061   C.E. Pippenger, J.K. Penry, and H. Kutt (1978) Physiochemical and Pharmacological Properties of Antiepileptic Drugs, 321-333.

P064   G. Pleuger (1925) Beitrage zur Kenntnis der Loslichkeit in Flussigkeitsgemischen, *Pleuger, Zur Kenntnis der Loslichkeit in Flussigkeisgemischen*, 26, 167-170.

P065   T. Palitzsch (1929) Studien uber die Oberflachenspannung von Losungen. IV. Uber den Gegenseitigen Einfluss von Urethan und Salzen auf ihr Losungsvolum und ihre Loslichkeit in Wasser, *Zeitschrift fuer Physikalische Chemie, Abteilung A: Chemische Thermodynamik, Kinetik, Electrochemie, Eigenschaftslehre*, 145, 97-108.

P067   G.K. Poochikian and J.C. Cradock (1979) Enhanced Chartreusin Solubility by Hydroxybenzoate Hydrotropy, *Journal of Pharmaceutical Sciences*, 68, 728-732.

P068   C.C. Peck and L.Z. Benet (1978) General Method for Determining Macrodissociation Constants of Polyprotic, Amphoteric Compounds from Solubility Measurements, *Journal of Pharmaceutical Sciences*, 67, 12-16.

P070   G.N. Pulley (1936) Solubility of Naringin in Water, *Industrial and Engineering Chemistry*, Analytical Edition, 8, 360-360.

P073   E.M. Pavlovskaya, A.K. Charykov, and V.I. Tikhomirov (1977) pH-Potentiometric Determination of the Solubility of Sparingly Soluble Organic Extractants in Water and Aqueous Solutions of Neutral Salts, *Journal of General Chemistry of the USSR*, 47, 2230-2234.

P076   A. Pomianowski and J. Leja (1963) Spectrophotometric Study of Xanthate and Dixanthogen Solutions, *Canadian Journal of Chemistry*, 41, 2219-2228.

P077   V.E. Petritis and C.J. Geankoplis (1959) Phase Equilibria in 1-Butanol-Water-Lactic Acid System, *Journal of Chemical and Engineering Data*, 4, 197-198.

P081   J.E. Peachey (1963) Chemical Control of Plant Parasitic Nematodes in the United Kingdom, *Chemistry and Industry (London)*, 1736-1740.

P085   D.F. Paris, D.L. Lewis, and J.T. Barnett (1977) Bioconcentration of Toxaphene by Micro-organisms, *Bulletin of Environmental Contamination and Toxicology*, 17, 564-572.

P089   F.M. Plakogiannis, C. Iordanides, and C. Siakali-Kiolafa (1980) Solubility Studies of Certain 2,3 Quinolino-phthalides, *Drug Development and Industrial Pharmacy*, 6, 61-75.

P091   D. Pitre and E. Felder (1980) Development, Chemistry, and Physical Properties of Iopamidol and Its Analogues, *Investigative Radiology*, 15, 301-309.

P094   D.M. Patel, A.J. Visalli, J.J. Zalipsky, and N.H. Reavey-Cantwell (1975) Methaqualone, *Analytical Profiles of Drug Substances*, 4, 245-255.

P096   M.S. Patel, P.H. Elworthy, and A.K. Dewsnup (1981) Solubilisation of Drugs in Nonionic Surfactants, *Journal of Pharmacy and Pharmacology*, 63, 64-64.

P303   A.N. Paruta (1984) Thermodynamics of Aqueous Solutions of Alkyl p-Aminobenzoate, *Drug Development and Industrial Pharmacy*, 10, 453-465.

P307   H.L. Pease, R.E. Leitch, and O.R. Hunt (1978) Terbacil, *Analytical Methods for Pesticides and Plant Growth Regulators*, 10, 483-492.

P311   R.R. Pfeiffer, K.S. Yang, and M.A. Tucker (1970) Crystal Pseudopolymorphism of Cephaloglycin and Cephalexin, *Journal of Pharmaceutical Sciences*, 59, 1809-1814.

P312   J. Pitha, J. Milecki, H. Fales..., and K. Uekama (1986) Hydroxypropyl-beta-cyclodextrin: Preparation and Characterization; Effects on Solubility of Drugs, *International Journal of Pharmaceutics*, 29, 73-82.

P313   D. Patterson and R.P. Sheldon (1959) The Dyeing of Polyester Fibers with Disperse Dyes: Mechanism and Kinetics of the Process for Purified Dyes, *Transactions of the Faraday Society*, 55, 1254-1264.

P314   E.L. Parrott, M. Simpson, and D.R. Flanagan (1983) Dissolution Kinetics of a Three-component Solid. II. Benzoic Acid, Salicylic Acid, and Salicylamide, *Journal of Pharmaceutical Sciences*, 72, 765-766.

P315   P.F. Pascoe and W.A. Sherbrock-Cox (1963) The Reaction Between Anhydrous Ethyleneimine and Water, *Journal of Applied Chemistry*, 13, 564-572.

P321   R.O. Pryanikova and I.A. Markina (1985) Liquid-Liquid Equilibrium in the Benzonitrile-Water and Benzonitrile-Water-Ammonia Systems. Generalised Sechenov Equation, *Russian Journal of Physical Chemistry*, 59, 1306-1308.

P323   R.F. Platford (1983) The Octanol-Water Partitioning of Some Hydrophobic and Hydrophilic Compounds, *Chemosphere*, 12, 1107-1111.

P324   S.D. Bruck (1983) Molecular Encapsulation of Drugs by Cyclodextrins and Congeners, in *Controlled Drug Delivery*, vol. 1, CRC Press, Boca Raton, 125-148.

P325   R.J. Prankerd, S.G. Frank, and V.J. Stella (1988) Preliminary Development and Evaluation of a Parenteral Emulsion Formulation of Penclomedine (NSC-338720; 3,5-Dichloro-2,4-dimethoxy-6-trichloromethylpyridine): A Novel, Practically Water Insoluble Cytotoxic Agent, *Journal of Parenteral Science and Technology*, 42, 76-81.

P326   I.H. Pitman (1976) Three Chemical Approaches Towards the Solubilisation of Drugs: Control of Enantiomer Composition, Salt Selection, and Pro-drug Formation, *Australian Journal of Pharmaceutical Sciences*, 17-19.

P329   A. Pal and S.C. Lahiri (1989) Solubility and the Thermodynamics of Transfer of Benzoic Acid in Mixed Solvents, *Indian Journal of Chemistry*, 28A, 276-279.

P331   P. Perez-Tejeda, C. Yanes, and A. Maestre (1990) Solubility of Naphthalene in Water + Alcohol Solutions at Various Temperatures, *Journal of Chemical and Engineering Data*, 35, 244-246.

P332   R. Pinal, P.S.C. Rao, L.S. Lee, and P.V. Cline (1990) Cosolvency of Partially Miscible Organic Solvents on the Solubility of Hydrophobic Organic Chemicals, *Environmental Science and Technology*, 24, 639-647.

P335   H.J. Panneman and A.A.C. Beenackers (1992) Solvent Effects on the Hydration of Cyclohexene Catalyzed by a Strong Acid Ion Exchange Resin. I. Solubility of Cyclohexene in Aqueous Sulfolane Mixtures, *Industrial and Engineering Chemistry, Product Research and Development*, 31, 1227-1231.

P336 R.J. Prankerd and V.J. Stella (1990) The Use of Oil-in-Water Emulsions as a Vehicle for Parenteral Drug Administration, *Journal of Parenteral Science and Technology*, 44, 139-149.

P339 R. Pinal, L.S. Lee, and P.S.C. Rao (1991) Prediction of the Solubility of Hydrophobic Compounds in Nonideal Solvent Mixtures, *Chemosphere*, 22, 939-951.

P340 C.A. Peters and R.G. Luthy (1993) Coal Tar in Water-Miscible Solvents: Experimental Evaluation, *Environmental Science and Technology*, 27, 2831-2843.

P342 A. Piasecki (1989) Chemical Structure and Surface Activity. Part XXI. The Amphiphilic Properties of 2-(2-Alkoxyethyl)-2-Methyl-4-Hydroxymethyl-1-1,3-Dioxolanes, *Colloids and Surfaces*, 36, 383-390.

P343 G.F. Palmieri, P. Wehrle, G. Duportail, and A. Stamm (1992) Inclusion Complexation of Vitamin A Palmitate with B-Cyclodextrin in Aqueous Solution, *Drug Development and Industrial Pharmacy*, 18, 2117-2121.

P344 G. Puglisi, N.A. Santagati, R. Pignatello, and G. Mazzone (1990) Inclusion Complexation of 4-Biphenylacetic Acid with B-Cyclodextrin, *Drug Development and Industrial Pharmacy*, 16, 395-413.

P348 M. Pedersen, M. Edelsten, V.F. Nielsen..., and C. Slot (1993) Formation and Antimycotic Effect of Cyclodextrin Inclusion Complexes of Econazole and Miconazole, *International Journal of Pharmaceutics*, 90, 247-254.

P349 S.P. Pinho, C.M. Silva, and E.A. Macedo (1994) Solubility of Amino Acids: A Group-Contribution Model Involving Phase and Chemical Equilibria, *Industrial and Engineering Chemistry*, *Product Research and Development*, 33, 1341-1347.

P350 R.J. Prankerd and R.H. McKeown (1994) Physico-chemical Properties of Barbituric Acid Derivatives. IV. Solubilities of 5,5-Disubstituted Barbituric Acids in Water, *International Journal of Pharmaceutics*, 112, 1-15.

P351 N.L. Plchopin, W.N. Charman, and V.J. Stella (1995) Amino Acid Derivatives of Dapsone as Water-soluble Prodrugs, *International Journal of Pharmaceutics*, 121, 157-167.

R001 A.W. Ralston and C.W. Hoerr (1942) The Solubilities of the Normal Saturated Fatty Acids, *Journal of Organic Chemistry*, 7, 546-555.

R002 I.D. Robb (1966) Determination of the Aqueous Solubility of Fatty Acids and Alcohols, *Australian Journal of Chemistry*, 19, 2281-2284.

R003 N.E. Richardson and B.J. Meakin (1975) The Influence of Cosolvents and Substrate Substituents on the Sorption of Benzoic Acid Derivatives by Polyamides, *Journal of Pharmacy and Pharmacology*, 27, 145-151.

R004 J.A. Rogers and J.G. Nairn (1973) Solubility and Dielectric Constant Correlations of the Systems Chloramphenicol Palmitate-Propylene Glycol-Water, *Canadian Journal of Pharmaceutical Science*, 8, 75-77.

R006 C.E. Rehberg and M.B. Dixon (1950) n-Alkyl Lactates and Their Acetates, *Journal of the American Chemical Society*, 72, 1918-1922.

R007 C.E. Rehberg and M.B. Dixon (1950) Mixed Esters of Lactic and Carbonic Acids, n-Alkyl Carbonates of Methyl and Butyl Lactates, and Butyl Carbonates of n-Alkyl Lactates, *Journal of Organic Chemistry*, 15, 565-571.

R009 R.K. Reddy, S.A. Khalil, and M.W. Gouda (1976) Dissolution Characteristics and Oral Absorption of Digitoxin and Digoxin Coprecipitates, *Journal of Pharmaceutical Sciences*, 65, 1753-1758.

R010 T. Rebagay and P. DeLuca (1976) Correlation of Dielectric Constant and Solubilizing Properties of Tetramethyldicarboxamides, *Journal of Pharmaceutical Sciences*, 65, 1645-1647.

R016 M. Randall and C.F. Failey (1927) The Activity Coefficient of the Undissociated Part of Weak Electrolytes, *Chemical Reviews*, 4, 291-318.

R017 A.J. Repta, M.J. Baltezor, and P.C. Bansal (1976) Utilization of an Enantiomer as a Solution to a Pharmaceutical Problem: Application to Solubilization of 1,2-Di(4-piperazine-2,6-dione)propane, *Journal of Pharmaceutical Sciences*, 65, 238-242.

R023 B.C. Rudy and B.Z. Senkowski (1973) Fluorouracil, *Analytical Profiles of Drug Substances*, 2, 221-242.

R024 B.C. Rudy and B.Z. Senkowski (1973) Isocarboxazid, *Analytical Profiles of Drug Substances*, 2, 295-314.

R025 B.C. Rudy and B.Z. Senkowski (1973) Sulfamethoxazole, *Analytical Profiles of Drug Substances*, 2, 467-486.

R027 B.C. Rudy and B.Z. Senkowski (1973) Methyprylon: An Analytical Profile, *Analytical Profiles of Drug Substances*, 2, 363-382.

R028   J. Raventos (1956) The Action of Fluothane—A New Volatile Anaesthetic, *British Journal of Pharmacology*, 11, 394-410.

R030   A.J. Repta, B.J. Rawson, R.D. Shaffer..., and T. Higuchi (1975) Rational Development of a Soluble Prodrug of a Cytotoxic Nucleoside: Preparation and Properties of Arabinosyladenine 5'-Formate, *Journal of Pharmaceutical Sciences*, 64, 392-396.

R034   C.E. Rehberg, M.B. Dixon, and C.H. Fisher (1947) Preparation and Physical Properties of n-Alkyl beta-n-Alkoxypropionates, *Journal of the American Chemical Society*, 69, 2966-2970.

R036   D.R. Robinson and W.P. Jencks (1965) The Effect of Compounds of the Urea-Guanidinium Class on the Activity Coefficient of Acetyltetraglycine Ethyl Ester and Related Compounds, *Journal of the American Chemical Society*, 87, 2462-2470.

R037   A.W. Ralston, C.W. Hoerr, and E.J. Hoffman (1942) Studies on High Molecular Weight Aliphatic Amines and Their Salts. VII. The Systems Octylamine-, Dodecylamine- and Octadecylamine-Water, *Journal of the American Chemical Society*, 64, 1516-1523.

R039   D.R. Robinson and M.E. Grant (1966) The Effects of Aqueous Salt Solutions on the Activity Coefficients of Purine and Pyrimidine Bases and Their Relation to the Denaturation of Deoxyribonucleic Acid by Salts, *Journal of Biological Chemistry*, 241, 4030-4042.

R041   M.S. Roberts, R.A. Anderson, and J. Swarbrick (1977) Permeability of Human Epidermis to Phenolic Compounds, *Journal of Pharmacy and Pharmacology*, 29, 677-683.

R042   M. Roseman and W.P. Jencks (1975) Interactions of Urea and Other Polar Compounds in Water, *Journal of the American Chemical Society*, 97, 631-640.

R044   F.L. Rose, A.R. Martin, and H.G.L. Bevan (1943) Sulphamethazine (2-4'-Aminobenzenesulphonyl-amino-4:6-dimethylpyrimidine) A New Heterocyclic Derivative of Sulphanilamide, *Journal of Pharmacology and Experimental Therapeutics*, 77, 127-141.

R045   R.O. Roblin, Jr., J.H. Williams, P.S. Winnek, and J.P. English (1940) Chemotherapy. II. Some Sulfanilamido Heterocycles, *Journal of the American Chemical Society*, 62, 2002-2005.

R046   R.O. Roblin, Jr., P.S. Winnek, and J.P. English (1942) Studies in Chemotherapy. IV. Sulfanilamidopyrimidines, *Journal of the American Chemical Society*, 64, 567-570.

R047   W.L. Ruigh and A.E. Erickson (1941) The Variation of the Oil-Water Distribution Ratio of Divinyl Ether with Concentration, *Anesthesiology*, 2, 546-551.

R048   A.G. Rauws, M. Olling, and A.E. Wibowo (1973) The Determination of Fluorochlorocarbons in Air and Body Fluids, *Journal of Pharmacy and Pharmacology*, 25, 718-722.

R049   J.D.M. Ross and T.J. Morrison (1936) Acid Salts of Monobasic Organic Acids. Part II, *Journal of the Chemical Society (London)*, 867-872.

R058   R.O. Roblin, Jr. and P.S. Winnek (1940) Chemotherapy. I. Substituted Sulfanilamidopyridines, *Journal of the American Chemical Society*, 62, 1999-2001.

R060   B.H. Robbins (1936) Studies of Cyclopropane. I. The Quantitative Determination of Cyclopropane in Air, Water, and Blood by Means of Iodine Pentoxide, *Journal of Pharmacology and Experimental Therapeutics*, 58, 243-259.

R063   P.A. Rice, R.P. Gale, and A.J. Barduhn (1976) Solubility of Butane in Water and Salt Solutions at Low Temperatures, *Journal of Chemical and Engineering Data*, 21, 204-206.

R067   K.S. Rao, M.V.R. Rao, and C.V. Rao (1961) Ternary Liquid Equilibria: Acetone-water-n-heptanol & Acetone-water-n-octanol Systems, *Journal of Scientific and Industrial Research*, 20B, 283-286.

R069   D.E. Resetarits, K.C. Cheng, B.A. Bolton..., and T.R. Bates (1979) Dissolution Behavior of 17-beta-Estradiol (E2) from Povidone Coprecipitates, Comparison with Microcrystalline and Macrocrystalline E2, *International Journal of Pharmaceutics*, 2, 113-123.

R070   T.W. Rosanske and K.A. Connors (1980) Stoichiometric Model of alpha-Cyclodextrin Complex Formation, *Journal of Pharmaceutical Sciences*, 69, 564-567.

R071   A.J. Repta and A.A. Hincal (1980) Complexation and Solubilization of Acronine with Alkylgentisates, *International Journal of Pharmaceutics*, 5, 149-155.

R072   L.A. Reber, W.M. McNabb, and W.W. Lucasse (1942) The Effect of Salts on the Mutual Miscibility of Normal Butyl Alcohol and Water, *Journal of Physical Chemistry*, 46, 500-515.

R075   F.L. Rose and G. Swain (1945) 2-p-Aminobenzenesulphonamidopyrimidines. Preparation by a Novel Route, *Journal of the Chemical Society (London)*, 689-692.

R076   G.W. Raiziss and M. Freifelder (1942) N1-Sulfanilylamino-alkyl-pyrimidines, *Journal of the American Chemical Society*, 64, 2340-2342.

R078   H.H. Reamer, B.H. Sage, and W.N. Lacey (1952) Phase Equilibria in Hydrocarbon Systems n-Butane-Water System in the Two-phase Region, *Industrial and Engineering Chemistry*, 44, 609-615.

R080   M.M. Rodriguez and A.G. Asuero (1980) Studies on Pyridylhydrazones Derived from Biacetyl as Analytical Reagents, *Microchemical Journal*, 25, 309-322.

R081   M. Rosoff and A.T.M. Serajuddin (1980) Solubilization of Diazepam in Bile Salts and in Sodium Cholate-Lecithin-Water Phases, *International Journal of Pharmaceutics*, 6, 137-146.

R082   W.A. Ritschel, K.W. Grummich, S. Kaul, and T.J. Hardt (1981) Biopharmaceutical Parameters of Coumarin and 7-Hydroxycoumarin, *Pharmazeutische Industrie*, 43, 271-276.

R084   S.S. Rossi and W.H. Thomas (1981) Solubility Behavior of Three Aromatic Hydrocarbons in Distilled Water and Natural Seawater, *Environmental Science and Technology*, 15, 715-716.

R087   J.A. Rogers (1982) Solution Thermodynamics of Phenols, *International Journal of Pharmaceutics*, 10, 89-97.

R302   B. Roland, K. Kimura, and J. Smid (1984) Interaction of Neutral Arenes with Poly(vinylbenzo-18-crown-6) and Poly(vinylbenzoglyme) in Aqueous Media, *Journal of Colloid and Interface Science*, 97, 392-400.

R303   B. Rabenort, P.C. DeWilde, F.G. DeBoer..., and R.D. Cannizzaro (1978) Diflubenzuron, *Analytical Methods for Pesticides and Plant Growth Regulators*, 10, 57-72.

R304   C-H. Roder, N.A. Jenny, and M. Ottnad (1978) Desmedipham, *Analytical Methods for Pesticides and Plant Growth Regulators*, 10, 293-303.

R308   B.S. Rawat and S. Krishna (1984) Isobaric Vapor-liquid Equilibria for the Partially Miscible System of Water-Methyl Isobutyl Ketone, *Journal of Chemical and Engineering Data*, 29, 403-406.

R318   R.J. Rao and C.V. Rao (1959) Ternary Liquid Equilibria Systems: n-Propanol-Water-Esters, *Journal of Applied Chemistry*, 9, 69-73.

R319   M.R. Rao and C.V. Rao (1957) Ternary Liquid Equilibria. IV. Various Systems, *Journal of Applied Chemistry*, 7, 659-666.

R320   E.A. Regna and P.F. Bruins (1956) Recovery of Aconitic Acid from Molasses, *Industrial and Engineering Chemistry*, 48, 1268-1277.

R321   W. Reinders and C.H. De Minjer (1947) Vapour-Liquid Equilibria in Ternary Systems. VI. The System Water-Acetone-Chloroform, *Recueil des Travaux Chimiques des Pays-Bas*, 66, 564-604.

S005   C. Smith and J.A. Calder (1975) Solubility of Alkylbenzenes in Distilled Water and Seawater at 25.0 C, *Journal of Chemical and Engineering Data*, 20, 320-322.

S006   G. Saracco and E. Spaccamela-Marchetti (1958) Influenza della Catena Idrocarburica sulla Solubilita in Acqua di Serie Omologhe, *Annales de Chimica*, 48, 1357-1370.

S010   E. Sada, S. Kito, and Y. Ito (1975) Solubility of Toluene in Aqueous Salt Solutions, *Journal of Chemical and Engineering Data*, 20, 373-375.

S012   R.S. Stearns, H. Oppenheimer, E. Simon, and W.D. Harkins (1947) Solubilization by Solutions of Long-chain Colloidal Electrolytes, *Journal of Chemical Physics*, 15, 496-507.

S076   F.P. Schwarz and S.P. Wasik (1977) A Fluorescence Method for the Measurement of the Partition Coefficients of Naphthalene, 1-Methylnaphthalene, and 1-Ethylnaphthalene in Water, *Journal of Chemical and Engineering Data*, 22, 270-273.

S115   N.V. Sidgwick and J.A. Neill (1923) The Solubility of the Phenylenediamines and of their Monoacetyl Derivatives, *Journal of the Chemical Society (London)*, 123, 2813-2819.

S117   N.V. Sidgwick and T.W.J. Taylor (1922) The Solubility and Volatility of 3:5-Dinitrophenol, *Journal of the Chemical Society (London)*, 121, 1853-1859.

S118   N.V. Sidgwick and W.M. Dash (1922) The Solubility and Volatility of the Nitrobenzaldehydes, *Journal of the Chemical Society (London)*, 121, 2586-2592.

S119   N.V. Sidgwick and W.J. Spurrell (1920) The System Benzene-Ethyl Alcohol-Water between +25 and -5 Degrees, *Journal of the Chemical Society (London)*, 117, 1397-1404.

S120   N.V. Sidgwick and R.K. Callow (1924) The Solubility of the Aminophenols, *Journal of the Chemical Society (London)*, 125, 522-527.

S124   V.F. Sergeeva and M.I. Usanovich (1959) Effect of Some Electrolytes on Solubility of Benzoic Acid in Water, *Journal of General Chemistry of the USSR*, 29, 1369-1372.

S131   L. Shnidman and A.A. Sunier (1932) The Solubility of Urea in Water, *Journal of Physical Chemistry*, 30, 1232-1240.

S133   H.A. Smith and M. Berman (1937) The Solubility Curves of the Systems Carbon Tetrachloride-n-Alkyl Acids-Water at 25 Degrees, *Journal of the American Chemical Society*, 59, 2390-2391.

S146   K. Sekiguchi, M. Kanke, N. Nakamura, and Y. Tsuda (1975) Dissolution Behavior of Solid Drugs. V. Determination of the Transition Temperature and Heat of Transition between Barbital

Polymorphs by Initial Dissolution Rate Measurements, *Chemical and Pharmaceutical Bulletin*, 23, 1347-1352.

S147   K. Sekiguchi, Y. Tsuda, and M. Kanke (1975) Dissolution Behavior of Solid Drugs. VI. Determination of Transition Temperatures of Various Physical Forms of Sulfanilamide by Initial Dissolution Rate Measurements, *Chemical and Pharmaceutical Bulletin*, 23, 1353-1362.

S149   K. Sekiguchi, M. Kanke, Y. Tsuda..., and Y. Tsuda (1973) Dissolution Behavior of Solid Drugs. III. Determination of the Transition Temperature between the Hydrate and Anhydrous Forms of Phenobarbital by Measuring Their Dissolution Rates, *Chemical and Pharmaceutical Bulletin*, 21, 1592-1600.

S171   F.P. Schwarz (1980) Measurement of the Solubilities of Slightly Soluble Organic Liquids in Water by Elution Chromatography, *Analytical Chemistry*, 52, 10-15.

S187   J.R. Sanborn, R.L. Metcalf, W.N. Bruce, and P-Y. Lu (1976) The Fate of Chlordane and Toxaphene in a Terrestrial-Aquatic Model Ecosystem, *Environmental Entomology*, 5, 533-538.

S191   C. Sutton (1974) The Solubility of Aromatic Hydrocarbons and the Geochemistry of Hydrocarbons in the Eastern Gulf of Mexico, Ph.D. thesis, 1-198.

S192   L.H. Schmidt, H.B. Hughes, E.A. Badger, and I.G. Schmidt (1944) The Toxicity of Sulfamerazine and Sulfamethazine, *Journal of Pharmacology and Experimental Therapeutics*, 81, 17-42.

S198   F.P. Schwarz and J. Miller (1980) Determination of the Aqueous Solubilities of Organic Liquids at 10.0, 20.0, and 30.0 C by Elution Chromatography, *Analytical Chemistry*, 52, 2162-2164.

S200   Y.S. Shenkin and L.N. Zaikina (1978) The Temperature Dependence of the Solubility of Urea in Water, *Russian Journal of Physical Chemistry*, 52, 1017-1018.

S203   I. Sanemasa, M. Araki, T. Deguchi, and H. Nagai (1981) Solubilities of Benzene and the Alkylbenzenes in Water — Method for Obtaining Aqueous Solutions Saturated with Vapours in Equilibrium with Organic Liquids, *Chemistry Letters*, 225-228.

S204   H. Sahay, S. Kumar, S.N. Upadhyay, and Y.D. Upadhya (1981) Solubility of Benzoic Acid in Aqueous Polymeric Solutions, *Journal of Chemical and Engineering Data*, 26, 181-183.

S207   A-B. Salem (1979) Solubility Data of the System Acetic Acid-Toluene-Water at Different Temperatures, *Journal of Chemical Engineering of Japan*, 12, 236-238.

S212   S.K. Shoor, R.D. Walker, Jr., and K.E. Gubbins (1969) Salting Out of Nonpolar Gases in Aqueous Potassium Hydroxide Solutions, *Journal of Physical Chemistry*, 73, 312-317.

S227   F.P. Schwarz and S.P. Wasik (1976) Fluorescence Measurements of Benzene, Naphthalene, Anthracene, Pyrene, Fluoranthene, and Benzo[e]pyrene in Water, *Analytical Chemistry*, 48, 524-528.

S304   M.P. Summers, R.P. Enever, and J.E. Carless (1972) Studies of the Dissolution Characteristics of Three Crystal Forms of Aspirin, *Symposium on Particle Growth in Suspensions*, 247-259.

S306   J.N. Spencer and T.A. Judge (1983) Hydrophobic Hydration of Thymine, *Journal of Solution Chemistry*, 12, 847-853.

S307   R. Stephenson, J. Stuart, and M. Tabak (1984) Mutual Solubility of Water and Aliphatic Alcohols, *Journal of Chemical and Engineering Data*, 29, 287-290.

S309   Shell Development Company Analytical Department Biological Sci Res Center (1978) Cyanazine, *Analytical Methods for Pesticides and Plant Growth Regulators*, 10, 275-292.

S314   M. Sarker and D. Wilson (1987) Solubilities of p-Dichlorobenzene and Naphthalene in Several Aqueous-Organic Solvent Mixtures, *Journal of the Tennessee Academy of Science*, 69-74.

S352   W.Y. Shiu, W. Doucette, F.A.P. Gobas..., and D. Mackay (1988) Physical-chemical Properties of Chlorinated Dibenzo-p-dioxins, *Environmental Science and Technology*, 22, 651-658.

S355   N.D. Sanders (1986) Visual Observation of the Solubility of Heavy Hydrocarbons in Near-critical Water, *Industrial and Engineering Chemistry, Fundamentals*, 25, 169-171.

S357   V.R. Sohoni and U.R. Warhadpande (1952) System Ethyl Acetate-Acetic Acid-Water at 30 C, *Industrial and Engineering Chemistry*, 44, 1428-1429.

S358   C. Sutton and J.A. Calder (1975) Solubility of Alkylbenzenes in Distilled Water and Seawater at 25.0C, *Journal of Chemical and Engineering Data*, 20, 320-322.

S359   I. Sanemasa, Y. Miyazaki, S. Arakawa..., and T. Deguchi (1987) The Solubility of Benzene-hydrocarbon Binary Mixtures in Water, *Bulletin of the Chemical Society of Japan (Nippon Kagakukai Bulletin)*, 60, 517-523.

T002   P.W. Taylor, Jr. and D.E. Wurster (1965) Dissolution Kinetics of Certain Crystalline Forms of Prednisolone. II. Influence of Low Concentrations of Sodium Lauryl Sulfate, *Journal of Pharmaceutical Sciences*, 54, 1654-1658.

T003   D.L. Tabern and E.F. Shelberg (1933) Physico-chemical Properties and Hypnotic Action of Substituted Barbituric Acids, *Journal of the American Chemical Society*, 55, 328-332.

T005   A.L. Thakkar and P.B. Kuehn (1969) Solubilization of Some Steroids in Aqueous Solutions of a Steroidal Nonionic Surfactant, *Journal of Pharmaceutical Sciences*, 58, 850-852.

T008   P.O.P. Ts'o, I.S. Melvin, and A.C. Olson (1963) Interaction and Association of Bases and Nucleosides in Aqueous Solutions, *Journal of the American Chemical Society*, 85, 1289-1296.

T015   C.A. Taylor and W.H. Rinkenbach (1923) The Solubility of Trinitro-phenylmethyl-nitramine (Tetryl) in Organic Solvents, *Journal of the American Chemical Society*, 45, 104-107.

T020   C.A. Taylor and W.H. Rinkenbach (1923) The Solubility of Trinitrotoluene in Organic Solvents, *Journal of the American Chemical Society*, 45, 44-59.

T023   D. Taylor and G.C. Vincent (1952) Phase Equilibria in Sulphonic Acid-Water Systems, *Journal of the Chemical Society (London)*, 3218-3224.

T025   P.O.P. Ts'o and P. Lu (1964) Interaction of Nucleic Acids. II. Chemical Linkage of the Carcinogen 3,4-Benzpyrene to DNA Induced by Photoradiation, *Proceedings of the National Academy of Science of the United States of America*, 51, 272-280.

T033   P.C.L. Thorne (1921) The Solubility of Ethyl Ether in Solutions of Sodium Chloride, *Journal of the Chemical Society (London)*, 119, 262-268.

T066   C. Tsonopoulos and J.M. Prausnitz (1971) Activity Coefficients of Aromatic Solutes in Dilute Aqueous Solutions, *Industrial and Engineering Chemistry, Fundamentals*, 10, 593-599.

T067   Y.B. Tewari, M.M. Miller, S.P. Wasik, and D.E. Martire (1982) Aqueous Solubility and Octanol/Water Partition Coefficient of Organic Compounds at 25.0 C, *Journal of Chemical and Engineering Data*, 27, 451-454.

T301   J. Takano, T. Yauoka, and S. Mitsuzawa (1982) Solubility Measurements of Solid Organic Compounds in Water by TOC Method, *Nippon Kagaku Kaishi*, 1830-1834.

T303   C. Treiner, C. Vaution, and G.N. Cave (1982) Correlations Between Solubilities, Heats of Fusion and Partition Coefficients for Barbituric Acids in Octanol + Water and in Aqueous Micellar Solutions, *Journal of Pharmacy and Pharmacology*, 34, 539-540.

T305   W.G. Thier, K. Hommel, and T. Hoppe (1978) Triazophos, *Analytical Methods for Pesticides and Plant Growth Regulators*, 10, 127-137.

U010   M. Ueda, A. Katayama, N. Kuroki, and T. Urahata (1978) Effect of Urea on the Solubility of Benzene and Toluene in Water, *Progress in Colloid and Polymer Science*, 63, 116-119.

U013   M. Ueda, A. Katayama, T. Urahata, and N. Kuroki (1980) Effect of Alcohols on the Solubilities of Aromatic Hydrocarbons in Water, *Kagaku To Kogyo (Tokyo)*, 54, 252-258.

V004   H.E. Vermillion, B. Werbel, J.H. Saylor, and P.M. Gross (1941) Solubility Studies. VI. The Solubility of Nitrobenzene in Deuterium Water and in Ordinary Water, *Journal of the American Chemical Society*, 63, 1346-1347.

V009   A.E. Van Arkel and S.E. Vles (1936) Loslichkeit von Organischen Verbindungen in Wasser, *Recueil des Travaux Chimiques des Pays-Bas*, 55, 407-411.

V013   A. Vesala (1974) Thermodynamics of Transfer of Nonelectrolytes from Light to Heavy Water. I. Linear Free Energy Correlations of Free Energy of Transfer with Solubility and Heat of Melting of a Nonelectrolyte, *Acta Chemica Scandinavica, Series A: Physical and Inorganic Chemistry*, 28, 839-845.

V033   C. Vaution, C. Treiner, F. Puisieux, and J.T. Carstensen (1981) Solubility Behavior of Barbituric Acids in Aqueous Solution of Sodium Alkyl Sulfonate as a Function of Concentration and Temperature, *Journal of Pharmaceutical Sciences*, 70, 1238-1242.

V300   E.I. Valko and M.B. Epstein (1957) Comicellization, *Solubilization and Micelles*, 1, 334-339.

V301   A. Van Rossum, P.C. DeWilde, F.G. DeBoer, and P.K. Korver (1978) Tetradifon, *Analytical Methods for Pesticides and Plant Growth Regulators*, 10, 119-126.

V302   O.W. Van Auken and M. Hulse (1978) Hexachlorophene, *Analytical Methods for Pesticides and Plant Growth Regulators*, 10, 189-214.

V303   A. Van Rossum, P.C. DeWilde, F.G. DeBoer, and P.K. Korver (1978) Dichlobenil, *Analytical Methods for Pesticides and Plant Growth Regulators*, 10, 311-320.

W003   D. Wauchope and F.W. Getzen (1972) Temperature Dependence of Solubilities in Water and Heats of Fusion of Solid Aromatic Hydrocarbons, *Journal of Chemical and Engineering Data*, 17, 38-41.

W005   H. Weil-Malherbe (1946) The Solubilization of Polycyclic Aromatic Hydrocarbons by Purines, *Biochemical Journal*, 40, 351-363.

W006   D.E. Wurster and P.W. Taylor, Jr. (1965) Dissolution Kinetics of Certain Crystalline Forms of Prednisolone, *Journal of Pharmaceutical Sciences*, 54, 670-676.

W007   D.E. Wurster and D.O. Kildsig (1965) Effect of Complex Formation on Dissolution Kinetics of m-Aminobenzoic Acid, *Journal of Pharmaceutical Sciences*, 54, 1491-1494.

W011   J.M. Weiss and C.R. Downs (1923) The Physical Properties of Maleic, Fumaric and Malic Acids, *Journal of the American Chemical Society*, 45, 1003-1008.

W013   W.S. Wise and E.B. Nicholson (1955) The Solubilities and Heats of Crystallisation of Sucrose and Methyl alpha-D-Glucoside in Aqueous Solution, *Journal of the Chemical Society (London)*, 2714-2716.

W016   N. Watari and N. Kaneniwa (1976) Dissolution of Slightly Soluble Drugs. II. Effect of Particle Size on Dissolution Behavior in Sodium Lauryl Sulfate Solutions, *Chemical and Pharmaceutical Bulletin*, 24, 2577-2584.

W019   V. Walters (1968) The Dissolution of Paracetamol Tablets and the In Vitro Transfer of Paracetamol With and Without Sorbitol, *Journal of Pharmacy and Pharmacology*, 20, 228-231.

W022   R.H. Wiley and N.R. Smith (1951) Reciprocal Solubility of 4,6-Dimethyl-1,2-pyrone and Water, *Journal of the American Chemical Society*, 73, 1383-1384.

W024   P.R. Wallnofer, M. Koniger, and O. Hutzinger (1973) Anrnx, 13, 14.

W025   L. Weil, G. Dure, and K. Quentin (1974) Wasserloslichkeit von Insektiziden Chlorierten Kohlenwasserstoffen und Polychlorierten Biphenylen im Hinblick auf eine Gewasserbelastung mit diesen Stoffen, *Zeitschrift fuer Wasser und Abwasser Forschung*, 7, 169-175.

W026   R. Wright (1927) Selective Solvent Action. Part VI. The Effect of Temperature on the Solubilities of Semisolutes in Aqueous Alcohol, *Journal of the Chemical Society (London)*, 130, 1334-1337.

W029   H.L. Ward and S.S. Cooper (1930) The System, Benzoic Acid, Ortho Phthalic Acid, Water, *Journal of Physical Chemistry*, 34, 1484-1493.

W033   C.S. Wiese and D.A. Griffin (1978) The Solubility of Aroclor 1254 in Seawater, *Bulletin of Environmental Contamination and Toxicology*, 19, 403-411.

W038   W.H. Walker, A.R. Collett, and C.L. Lazzell (1931) The Solubility Relations of the Isomeric Dihydroxybenzenes, *Journal of Physical Chemistry*, 35, 3259-3271.

W044   J. Walker and J.K. Wood (1898) Solubility of Isomeric Substances, *Journal of the Chemical Society (London)*, 73, 618-627.

W053   A.S. Wilkerson (1942) Optical Properties of 2-Sulfanilamidopyrimidine (Sulfadiazine), *Journal of the American Chemical Society*, 64, 2230-2230.

W055   N. Watari, M. Hanano, and N. Kaneniwa (1980) Dissolution of Slightly Soluble Drugs. VI. Effect of Particle Size of Sulfadimethoxine on the Oral Bioavailability, *Chemical and Pharmaceutical Bulletin*, 28, 2221-2225.

W057   J.D. Worley (1967) Benzene as a Solute in Water, *Canadian Journal of Chemistry*, 45, 2465-2467.

W300   S.P. Wasik, M.M. Miller, Y.B. Tewari..., and W.H. Zoller (1983) Determination of the Vapor Pressure, Aqueous Solubility, and Octanol/Water Partition Coefficient of Hydrophobic Substances by Coupled Generator Column/Liquid Chromatographic Methods, *Residue Reviews*, 85, 29-42.

W302   S.A. Wise, W.J. Bonnett, F.R. Guenther, and W.E. May (1981) A Relationship Between Reversed-phase C18 Liquid Chromatographic Retention and the Shape of Polycyclic Aromatic Hydrocarbons, *Journal of Chromatographic Science*, 19, 457-465.

W305   A.F.H. Ward and A.G. Chitale (1957) A Study of Solubilization by Electrical Conductivity Measurements, *Solubilization and Micelles*, 1, 405-409.

W310   R.J. Whiteoak, J.B. Reary, and K.C. Overton (1978) Bendiocarb, *Analytical Methods for Pesticides and Plant Growth Regulators*, 10, 3-17.

W311   R.D. Weeren and D. Eichler (1978) Bromophos, *Analytical Methods for Pesticides and Plant Growth Regulators*, 10, 31-40.

W312   R.D. Weeren and D. Eichler (1978) Bromophos-Ethyl, *Analytical Methods for Pesticides and Plant Growth Regulators*, 10, 41-43.

W313   R.J. Whiteoak, M. Crofts, R.J. Harris, and K.C. Overton (1978) Ethofumesate, *Analytical Methods for Pesticides and Plant Growth Regulators*, 10, 353-366.

W314   D.M. Whitacre, Y.H. Atallah, J.E. Forrette, and H.K. Suzuki (1978) Methazole, *Analytical Methods for Pesticides and Plant Growth Regulators*, 10, 367-384.

W317   N. Wiedenhof and J.N.J. Lammers (1968) Properties of Cyclodextrins. Part II. Preparation of a Stable beta-Cyclodextrin Hydrate and Determination of Its Water Content and Enthalpy of Solution in Water from 15-30 C, *Carbohydrate Research*, 7, 1-6.

W319   G.R.B. Webster, K.J. Friesen, L.P. Sarna, and D.C.G. Muir (1985) Environmental Fate Modelling of Chlorodioxins: Determination of Physical Constants, *Chemosphere*, 14, 609-622.

W332   G.R.B. Webster, L.P. Sarna, and D.C. Muir (1985) Octanol-Water Partition Coefficient of 1,3,6,8-TCDD and OCDD by Reverse Phase HPLC, in *Chlorinated Dioxins and Dibenzofurans in the Total Environment* vol. II, Butterworth, 79-87.

Y020   F.E. Young (1957) D-Glucose-Water Phase Diagram, *Journal of Physical Chemistry*, 61, 616-619.

Z008   C.O. Zerpa, P.B. Dharmawardhana, W.R. Parrish, and E.D. Sloan (1979) Solubility of Cyclopropane in Aqueous Solutions of Potassium Chloride, *Journal of Chemical and Engineering Data*, 24, 26-28.

Index 1: Molecular Formula

CHBrCl$_2$, 1
CHBr$_2$Cl, 2
CHBr$_3$, 3
CHClF$_2$, 4
CHCl$_3$, 5
CHI$_3$, 6
CH$_2$BrCl, 7
CH$_2$Br$_2$, 8
CH$_2$Cl$_2$, 9
CH$_2$I$_2$, 10
CH$_2$N$_2$, 11
CH$_3$Br, 12
CH$_3$BrO$_6$S$_2$, 13
CH$_3$Cl, 14
CH$_3$ClO$_6$S$_2$, 15
CH$_3$F, 16
CH$_3$I, 17
CH$_3$NO, 18
CH$_3$NO$_2$, 19
CH$_3$N$_5$, 20
CH$_4$, 21
CH$_4$N$_2$O, 22
CH$_4$N$_2$S, 23
CH$_4$N$_4$O$_2$, 24
CH$_4$O, 25
CH$_4$O$_6$S$_2$, 26
CH$_4$O$_6$S$_2$.H$_2$O, 27
CH$_5$N, 28
CH$_5$N$_5$O$_2$, 29
CH$_5$O$_3$As, 30
CH$_5$As, 31
CBrClF$_2$, 32
CBr$_3$F, 33
CBr$_4$, 34
CClN, 35
CClN$_3$O$_6$, 36
CCl$_2$F$_2$, 37
CCl$_3$F, 38
CCl$_3$NO$_2$, 39
CCl$_4$, 40
CF$_4$, 41
COS, 42
CO$_2$, 43
CS$_2$, 44
C$_2$HBrClF$_3$, 45
C$_2$HCl$_3$, 46
C$_2$HCl$_3$O.H$_2$O, 47
C$_2$HCl$_3$O$_2$, 48
C$_2$HCl$_5$, 49
C$_2$H$_2$, 50
C$_2$H$_2$Br$_4$, 51
C$_2$H$_2$Cl$_2$, 52-54
C$_2$H$_2$Cl$_3$As, 55
C$_2$H$_2$Cl$_4$, 56, 57

C$_2$H$_2$O$_4$, 58
C$_2$H$_2$O$_4$.2H$_2$O, 59
C$_2$H$_3$Br$_3$O, 60
C$_2$H$_3$Cl, 61
C$_2$H$_3$ClNO$_2$, 62
C$_2$H$_3$Cl$_3$, 63, 64
C$_2$H$_3$FO$_2$, 65
C$_2$H$_3$N, 66, 67
C$_2$H$_3$NS, 68
C$_2$H$_4$, 69
C$_2$H$_4$BrCl, 70
C$_2$H$_4$Br$_2$, 71
C$_2$H$_4$ClNO, 72
C$_2$H$_4$ClNO$_2$, 73
C$_2$H$_4$Cl$_2$, 74, 75
C$_2$H$_4$F$_2$, 76
C$_2$H$_4$N$_2$O$_2$, 77
C$_2$H$_4$N$_4$, 78, 79
C$_2$H$_4$N$_2$O$_2$S$_2$, 80
C$_2$H$_4$O$_2$, 81
C$_2$H$_4$O$_3$, 82
C$_2$H$_5$Br, 83
C$_2$H$_5$Cl, 84
C$_2$H$_5$I, 85
C$_2$H$_5$N, 86
C$_2$H$_5$NO, 87
C$_2$H$_5$NO$_2$, 88-91
C$_2$H$_5$NS, 92
C$_2$H$_5$N.2H$_2$O, 93
C$_2$H$_5$N$_3$O$_2$, 94, 95
C$_2$H$_5$N$_5$O$_3$, 96
C$_2$H$_5$O$_5$P, 97
C$_2$H$_5$O$_5$As, 98
C$_2$H$_6$, 99
C$_2$H$_6$O, 100
C$_2$H$_6$O$_2$, 101
C$_2$H$_6$O$_3$S, 102
C$_2$H$_6$O$_4$S, 103
C$_2$H$_7$N, 104
C$_2$H$_7$NO$_3$S, 105
C$_2$H$_7$O$_2$As, 106
C$_2$H$_7$As, 107
C$_2$Cl$_2$F$_4$, 108
C$_2$Cl$_3$F$_3$, 109
C$_2$Cl$_4$, 110
C$_2$Cl$_6$, 111
C$_2$N$_2$, 112
C$_2$N$_4$S$_2$, 113
C$_2$N$_6$S$_4$, 114
C$_3$H$_2$Cl$_2$N$_2$O$_2$, 115
C$_3$H$_2$N$_2$, 116
C$_3$H$_2$N$_2$O$_3$, 117
C$_3$H$_3$Cl$_3$O$_3$, 118
C$_3$H$_3$N, 119

$C_3H_3NOS_2$, 120
$C_3H_3N_3O_3$, 121, 122
$C_3H_3N_3S_3$, 123
C_3H_4, 124
$C_3H_4ClN_5$, 125
$C_3H_4Cl_2$, 126-129
$C_3H_4Cl_2O_2$, 130
$C_3H_4N_2O$, 131
$C_3H_4N_2O_2$, 132
$C_3H_4N_2O_3S$, 133
$C_3H_4N_4O_2$, 134
C_3H_4O, 135
$C_3H_4O_4$, 136
C_3H_5Br, 137
$C_3H_5Br_2Cl$, 138
C_3H_5Cl, 139
C_3H_5ClO, 140, 141
$C_3H_5Cl_2NO_2$, 142
$C_3H_5Cl_3$, 143
$C_3H_5IO_2$, 144
C_3H_5N, 145, 146
C_3H_5NO, 147
$C_3H_5NO_3$, 148
$C_3H_5N_3O$, 149
$C_3H_5N_3O_9$, 150
$C_3H_5N_5O$, 151
C_3H_6, 152, 153
C_3H_6BrCl, 154
$C_3H_6BrNO_4$, 155
$C_3H_6Br_2$, 156
$C_3H_6ClNO_2$, 157, 158
$C_3H_6Cl_2$, 159, 160
$C_3H_6Cl_2O$, 161
$C_3H_6N_2O_2$, 162-165
$C_3H_6N_2O_3$, 166
$C_3H_6N_2O_7$, 167, 168
$C_3H_6N_2S$, 169
$C_3H_6N_4Hg$, 170
$C_3H_6N_6$, 171
$C_3H_6N_6O_6$, 172
C_3H_6O, 173, 174
$C_3H_6O_2$, 175-177
$C_3H_6O_2S_3$, 178, 179
$C_3H_6O_3$, 180-182
$C_3H_6O_3S_3$, 183, 184
$C_3H_6O_3S$, 185
C_3H_7Br, 186, 187
C_3H_7BrO, 188
C_3H_7Cl, 189, 190
C_3H_7ClO, 191
C_3H_7I, 192, 193
$C_3H_7NO_2$, 194-202
$C_3H_7NO_2S$, 203
$C_3H_7NO_3$, 204-207
$C_3H_7NO_5$, 208
$C_3H_7N_3O_2$, 209, 210
$C_3H_7O_5P$, 211
C_3H_8, 212
$C_3H_8NO_5P$, 213
C_3H_8O, 214, 215
$C_3H_8OS_2$, 216
$C_3H_8O_2$, 217
$C_3H_8O_3$, 218
C_3H_9N, 219, 220

$C_3H_9O_4P$, 221
$C_3H_{12}N_6O_3$, 222
$C_3Cl_3N_3O_3$, 223
C_4HI_4N, 224
C_4H_2, 225
$C_4H_2N_2O_4$, 226
$C_4H_3FN_2O_2$, 227
$C_4H_3N_2S$, 228
$C_4H_3N_3O_5$, 229
$C_4H_4Br_2O_4$, 230
$C_4H_4Cl_2N_2O_2$, 231
$C_4H_4Cl_2O_4$, 232
$C_4H_4N_2$, 233
$C_4H_4N_2O$, 234, 235
$C_4H_4N_2OS$, 236
$C_4H_4N_2O_2$, 237-239
$C_4H_4N_2O_3$, 240, 241
$C_4H_4O_4$, 242, 243
C_4H_4S, 244
$C_4H_5BrO_4$, 245
$C_4H_5ClO_2$, 246-249
$C_4H_5ClO_4$, 250
$C_4H_5F_3O$, 251
C_4H_5N, 252, 253
$C_4H_5NO_2$, 254, 255
C_4H_5NS, 256
$C_4H_5N_3O$, 257
$C_4H_5N_3OS$, 258
$C_4H_5N_3O_2$, 259, 260
C_4H_6, 261, 262
$C_4H_6BrNO_4$, 263
$C_4H_6Cl_2O_2S$, 264
$C_4H_6N_2O_2$, 265
$C_4H_6N_2S_4Zn$, 266
$C_4H_6N_4O_3$, 267
$C_4H_6N_4O_3S_2$, 268
C_4H_6O, 269-271
$C_4H_6O_2$, 272-277
$C_4H_6O_2S_4$, 278
$C_4H_6O_3$, 279
$C_4H_6O_4$, 280-282
$C_4H_6O_5$, 283-285
$C_4H_6O_6$, 286-289
C_4H_7Br, 290
$C_4H_7BrN_2O_2$, 291
$C_4H_7BrO_2$, 292
C_4H_7Cl, 293
$C_4H_7Cl_2O_4P$, 294
$C_4H_7Cl_3O$, 295
C_4H_7N, 296
$C_4H_7NO_3$, 297
$C_4H_7NO_4$, 298-302
$C_4H_7N_2O_4$, 303
$C_4H_7N_3O$, 304
C_4H_8, 305, 306
$C_4H_8Cl_2$, 307
$C_4H_8Cl_2O$, 308
$C_4H_8Cl_2OS$, 309
$C_4H_8Cl_2O_2S$, 310
$C_4H_8Cl_2S$, 311
$C_4H_8Cl_3O_4P$, 312
$C_4H_8N_2O_2$, 313, 314
$C_4H_8N_2O_3$, 315-319
$C_4H_8N_2O_3 \cdot H_2O$, 320

$C_4H_8N_4O_2$, 321
C_4H_8O, 322-326
$C_4H_8O_2$, 327-333
C_4H_9Br, 334, 335
C_4H_9Cl, 336-339
C_4H_9I, 340
C_4H_9NO, 341, 342
$C_4H_9NO_2$, 343-349
$C_4H_9NO_3$, 350-353
$C_4H_9N_3O_2$, 354
$C_4H_9O_5P$, 355
C_4H_{10}, 356, 357
$C_4H_{10}NO_3PS$, 358
$C_4H_{10}N_2O$, 359
$C_4H_{10}O$, 360-366
$C_4H_{10}O_2S$, 367
$C_4H_{10}O_4$, 368, 369
$C_4H_{10}S$, 370
$C_4H_{11}N$, 371, 372
$C_4H_{11}NO_3$, 373
$C_4H_{11}NO_8P_2$, 374
C_4Cl_6, 375
$C_5H_2Cl_3NO$, 376, 377
$C_5H_4ClN_5$, 378
$C_5H_4N_2O_4$, 379-381
$C_5H_4N_4$, 382
$C_5H_4N_4O$, 383-385
$C_5H_4N_4O_2$, 386
$C_5H_4N_4O_2.H_2O$, 387
$C_5H_4N_4O_3$, 388
$C_5H_4N_4O_3.2H_2O$, 389
$C_5H_4N_4S$, 390
$C_5H_4O_2$, 391
$C_5H_4O_2S$, 392
$C_5H_4O_3$, 393, 394
C_5H_5NO, 395-397
$C_5H_5NO_2$, 398
$C_5H_5N_3O$, 399
$C_5H_5N_5$, 400
$C_5H_5N_5O$, 401, 402
$C_5H_5N_5O_2$, 403
$C_5H_6Cl_2N_2$, 404
$C_5H_6Cl_3N_5O_2$, 405
$C_5H_6N_2OS$, 406, 407
$C_5H_6N_2O_2$, 408, 409
$C_5H_6N_2O_4$, 410
$C_5H_6O_2$, 411
$C_5H_6O_4$, 412-414
$C_5H_7NO_2$, 415
$C_5H_7N_2O_2$, 416
$C_5H_7N_3O$, 417
$C_5H_7N_3O_2$, 418
C_5H_8, 419-422
$C_5H_8BrNO_4$, 423
$C_5H_8N_2O_2$, 424, 425
$C_5H_8N_2O_3S_2$, 426
$C_5H_8N_4O_{12}$, 427
C_5H_8O, 428, 429
$C_5H_8O_2$, 430-432
$C_5H_8O_3$, 433, 434
$C_5H_8O_4$, 435-438
$C_5H_9BrO_2$, 439, 440
$C_5H_9NO_2$, 441, 442
$C_5H_9NO_3$, 443, 444

$C_5H_9NO_4$, 445-447
C_5H_{10}, 448-451
$C_5H_{10}Cl_3O_3P$, 452
$C_5H_{10}N_2O$, 453
$C_5H_{10}N_2O_2S$, 454
$C_5H_{10}N_2O_3$, 455-457
$C_5H_{10}N_2S_2$, 458
$C_5H_{10}N_6O_2$, 459
$C_5H_{10}O$, 460-470
$C_5H_{10}OS_2$, 471
$C_5H_{10}O_2$, 472-480
$C_5H_{10}O_3$, 481, 482
$C_5H_{10}O_5$, 483, 484
$C_5H_{11}Br$, 485, 486
$C_5H_{11}NO$, 487
$C_5H_{11}NO_2$, 488-498
$C_5H_{11}NO_2S$, 499-501
$C_5H_{11}NO_2.H_2O$, 502
C_5H_{12}, 503-505
$C_5H_{12}ClO_2PS_2$, 506
$C_5H_{12}NO_3PS_2$, 507
$C_5H_{12}N_2$, 508
$C_5H_{12}N_2O$, 509
$C_5H_{12}O$, 510-519
$C_5H_{12}O_2$, 520
$C_5H_{12}O_4$, 521
$C_5H_{12}O_5$, 522, 523
$C_5H_{13}N$, 524
$C_5H_{13}O_3PS_2$, 525
C_6Cl_6, 526
$C_6HCl_5N_2S$, 527
$C_6HCl_4NO_2$, 528-530
C_6HCl_5, 531
C_6HCl_5O, 532
C_6HF_5O, 533
$C_6H_2Br_2ClNO_2$, 534
$C_6H_2ClN_3O_6$, 535
$C_6H_2Cl_2O_4$, 536
$C_6H_2Cl_3NO_2$, 537, 538
$C_6H_2Cl_4$, 539-542
$C_6H_2Cl_4O$, 543-545
$C_6H_2Cl_4O_2$, 546
$C_6H_2F_4$, 547, 548
$C_6H_2F_4O$, 549
$C_6H_3Br_2NO_2$, 550
$C_6H_3Br_3O$, 551, 552
$C_6H_3ClN_2O_4$, 553
$C_6H_3ClN_4$, 554
$C_6H_3Cl_2NO_2$, 555-558
$C_6H_3Cl_3$, 559-561
$C_6H_3Cl_3N_2O_2$, 562
$C_6H_3Cl_3O$, 563-567
$C_6H_3Cl_4N$, 568
$C_6H_3FN_2O_4$, 569
$C_6H_3F_3O$, 570
$C_6H_3N_3O_6$, 571
$C_6H_3N_3O_7$, 572
$C_6H_3N_3O_8$, 573
C_6H_4BrF, 574, 575
$C_6H_4BrNO_3$, 576
$C_6H_4Br_2$, 577, 578
C_6H_4ClF, 579, 580
$C_6H_4ClIO_2S$, 581
$C_6H_4ClNO_2$, 582-585

$C_6H_4Cl_2$, 586-588
$C_6H_4Cl_2O$, 589-594
C_6H_4FI, 595
$C_6H_4I_2$, 596
$C_6H_4N_2O_4$, 597-599
$C_6H_4N_2O_5$, 600-602
$C_6H_4N_2O_6$, 603, 604
$C_6H_4N_4$, 605
$C_6H_4N_4O$, 606-609
$C_6H_4N_4O_2$, 610-614
$C_6H_4N_4O_3$, 615, 616
$C_6H_4N_4O_4$, 617
$C_6H_4N_4O_6$, 618
$C_6H_4N_4S$, 619-621
$C_6H_4O_2$, 622
$C_6H_4O_5$, 623, 624
C_6H_5Br, 625
C_6H_5BrO, 626
$C_6H_5BrO_3S$, 627
$C_6H_5BrO_3S.H_2O$, 628
$C_6H_5BrO_3S.2.5H_2O$, 629
C_6H_5Cl, 630
$C_6H_5ClN_2O_4S$, 631
C_6H_5ClO, 632-634
$C_6H_5ClO_3S$, 635
$C_6H_5ClO_3S.2.5H_2O$, 636
$C_6H_5Cl_2NO_2S$, 637
C_6H_5F, 638
$C_6H_5FN_2O_3$, 639
$C_6H_5FN_2O_4$, 640
C_6H_5FO, 641-643
$C_6H_5FO_3S.H_2O$, 644
$C_6H_5FO_3S.2.5H_2O$, 645
$C_6H_5FO_3S.3H_2O$, 646
$C_6H_5FO_3S.4H_2O$, 647
C_6H_5I, 648
C_6H_5IO, 649
$C_6H_5NO_2$, 650, 651
$C_6H_5NO_3$, 652-654
$C_6H_5NO_4$, 655-659
$C_6H_5NO_5S$, 660
$C_6H_5NO_5S.2H_2O$, 661
$C_6H_5NO_5S.4H_2O$, 662
$C_6H_5N_2OS$, 663
$C_6H_5N_3O_4$, 664, 665
$C_6H_5N_3O_5$, 666
$C_6H_5N_5$, 667-669
$C_6H_5N_5O$, 670-672
$C_6H_5N_5O_2$, 673
$C_6H_5N_5O_3$, 674
$C_6H_5N_5O_4S$, 675
C_6H_6, 676
$C_6H_6BrNO_2S$, 677
$C_6H_6BrNO_3S$, 678, 679
$C_6H_6BrNO_3S.H_2O$, 680
C_6H_6ClN, 681-683
$C_6H_6ClNO_2S$, 684-686
$C_6H_6ClNO_3S$, 687
$C_6H_6ClNO_3S.H_2O$, 688, 689
$C_6H_6Cl_6$, 690-693
$C_6H_6FN_3O_3$, 694
$C_6H_6INO_3S$, 695-701
$C_6H_6N_2O$, 702
$C_6H_6N_2O_2$, 703-706

$C_6H_6N_2O_3$, 707
$C_6H_6N_2O_4$, 708
$C_6H_6N_2O_4S$, 709-711
$C_6H_6N_4$, 712
$C_6H_6N_4O$, 713
$C_6H_6N_4O_3$, 714, 715
$C_6H_6N_4O_3S$, 716
$C_6H_6N_4O_4$, 717
$C_6H_6N_6$, 718-721
C_6H_6O, 722
$C_6H_6O_2$, 723-725
$C_6H_6O_3$, 726-729
$C_6H_6O_3S$, 730
$C_6H_6O_3S.H_2O$, 731
$C_6H_6O_3S.2.5H_2O$, 732
$C_6H_6O_3S.2H_2O$, 733
$C_6H_6O_3S.3H_2O$, 734
$C_6H_6O_4$, 735
$C_6H_6F_3N_4OS$, 736
C_6H_7N, 737
C_6H_7NO, 738-741
$C_6H_7NO_2S$, 742
$C_6H_7NO_3S$, 743-745
$C_6H_7NO_3S.1.5H_2O$, 746
$C_6H_7NO_4S$, 747, 748
$C_6H_7N_3O$, 749
$C_6H_7N_3O_3$, 750
$C_6H_7N_7$, 751, 752
$C_6H_7O_2P$, 753
$C_6H_7O_3P$, 754
$C_6H_7O_3As$, 755
C_6H_8, 756
$C_6H_8N_2$, 757-759
$C_6H_8N_2OS$, 760
$C_6H_8N_2O_2$, 761
$C_6H_8N_2O_2S$, 762-764
$C_6H_8N_2O_2S.H_2O$, 765
$C_6H_8N_2O_3$, 766
$C_6H_8N_2O_3S$, 767
$C_6H_8N_2O_8$, 768
$C_6H_8N_4O$, 769
$C_6H_8N_8$, 770
$C_6H_8O_2$, 771
$C_6H_8O_6$, 772, 773
$C_6H_8O_7$, 774
$C_6H_8O_7.H_2O$, 775
C_6H_8S, 776
$C_6H_9NO_3$, 777, 778
$C_6H_9NO_6$, 779
$C_6H_9N_3$, 780
$C_6H_9N_3O_2$, 781, 782
$C_6H_9N_3O_3$, 783
C_6H_{10}, 784-787
$C_6H_{10}BrNO_4$, 788, 789
$C_6H_{10}ClN_5$, 790
$C_6H_{10}O$, 791, 792
$C_6H_{10}OS_2$, 793
$C_6H_{10}O_2$, 794, 795
$C_6H_{10}O_2S_4$, 796
$C_6H_{10}O_3$, 797
$C_6H_{10}O_4$, 798-804
$C_6H_{10}O_5$, 805
$C_6H_{10}O_8$, 806
$C_6H_{11}BrN_2O_2$, 807-812

C$_6$H$_{11}$NO, 813, 814
C$_6$H$_{11}$NO$_4$, 815, 816
C$_6$H$_{11}$N$_3$O$_4$PS$_3$, 817
C$_6$H$_{11}$N$_3$O$_6$, 818
C$_6$H$_{12}$, 819-823
C$_6$H$_{12}$ClNO, 824
C$_6$H$_{12}$Cl$_2$O, 825
C$_6$H$_{12}$Cl$_2$O$_2$, 826
C$_6$H$_{12}$Cl$_3$O$_4$P, 827
C$_6$H$_{12}$NO$_3$PS$_2$, 828
C$_6$H$_{12}$NO$_4$PS$_2$, 829
C$_6$H$_{12}$N$_2$O, 830
C$_6$H$_{12}$N$_2$O$_2$, 831, 832
C$_6$H$_{12}$N$_2$O$_3$, 833, 834
C$_6$H$_{12}$N$_2$O$_4$S$_2$, 835
C$_6$H$_{12}$N$_2$O$_4$S, 836
C$_6$H$_{12}$N$_2$O$_4$S$_2$, 837-839
C$_6$H$_{12}$N$_2$S$_4$Zn, 840
C$_6$H$_{12}$N$_2$S$_4$, 841
C$_6$H$_{12}$N$_4$, 842
C$_6$H$_{12}$N$_4$O$_2$, 843
C$_6$H$_{12}$N$_5$O$_5$PS$_2$, 844
C$_6$H$_{12}$O, 845-855
C$_6$H$_{12}$O$_2$, 856-865
C$_6$H$_{12}$O$_3$, 866-868
C$_6$H$_{12}$O$_5$, 869, 870
C$_6$H$_{12}$O$_6$, 871-879
C$_6$H$_{12}$O$_6$.H$_2$O, 880
C$_6$H$_{12}$O$_7$, 881
C$_6$H$_{13}$Br, 882
C$_6$H$_{13}$NO, 883
C$_6$H$_{13}$NO$_2$, 884-898
C$_6$H$_{14}$, 899-903
C$_6$H$_{14}$FO$_3$P, 904
C$_6$H$_{14}$NO$_3$PS$_2$, 905
C$_6$H$_{14}$N$_2$, 906
C$_6$H$_{14}$N$_2$O, 907-910
C$_6$H$_{14}$N$_2$O$_2$, 911
C$_6$H$_{14}$N$_4$O$_2$, 912, 913
C$_6$H$_{14}$O, 914-932
C$_6$H$_{14}$O$_2$, 933, 934
C$_6$H$_{14}$O$_3$, 935
C$_6$H$_{14}$O$_6$, 936-938
C$_6$H$_{15}$N, 939-942
C$_6$H$_{15}$O$_2$PS$_3$, 943
C$_6$H$_{15}$O$_3$PS$_2$, 944, 945
C$_6$H$_{15}$O$_4$P, 946
C$_6$H$_{16}$FN$_2$OP, 947
C$_6$H$_{16}$N$_2$, 948
C$_6$H$_{17}$N$_3$O$_{10}$S, 949
C$_6$H$_{18}$N$_4$, 950
C$_6$Cl$_4$O$_2$, 951
C$_6$Cl$_5$NO$_2$, 952
C$_6$Cl$_6$, 953
C$_7$H$_3$Br$_2$NO, 954
C$_7$H$_3$Br$_3$O$_2$, 955
C$_7$H$_3$Cl$_2$N, 956
C$_7$H$_3$Cl$_3$O$_2$, 957
C$_7$H$_3$Cl$_3$O, 958
C$_7$H$_3$I$_2$NO, 959
C$_7$H$_3$N$_3$O$_8$, 960
C$_7$H$_4$BrNO$_4$, 961
C$_7$H$_4$BrNS, 962, 963
C$_7$H$_4$ClNO$_4$, 964-966

C$_7$H$_4$ClNS, 967
C$_7$H$_4$Cl$_2$O$_2$, 968-971
C$_7$H$_4$Cl$_3$NO$_3$, 972
C$_7$H$_4$Cl$_4$O, 973, 974
C$_7$H$_4$INS, 975, 976
C$_7$H$_4$I$_2$O$_3$, 977
C$_7$H$_4$N$_2$O$_2$S, 978
C$_7$H$_4$N$_2$O$_6$, 979-982
C$_7$H$_4$N$_4$O$_9$, 983
C$_7$H$_4$O$_6$, 984
C$_7$H$_4$O$_7$, 985
C$_7$H$_5$BrO$_2$, 986, 987
C$_7$H$_5$ClO$_2$, 988-990
C$_7$H$_5$Cl$_2$NO, 991
C$_7$H$_5$Cl$_2$NO$_2$, 992
C$_7$H$_5$Cl$_2$NS, 993
C$_7$H$_5$Cl$_3$O, 994-996
C$_7$H$_5$FO$_2$, 997-999
C$_7$H$_5$F$_3$N$_2$O$_4$S, 1000
C$_7$H$_5$IO$_2$, 1001-1003
C$_7$H$_5$I$_2$NO$_3$, 1004
C$_7$H$_5$N, 1005
C$_7$H$_5$NOS, 1006, 1007
C$_7$H$_5$NO$_3$, 1008-1010
C$_7$H$_5$NO$_3$S, 1011
C$_7$H$_5$NO$_4$, 1012-1019
C$_7$H$_5$NO$_5$, 1020, 1021
C$_7$H$_5$NS, 1022
C$_7$H$_5$N$_3$O$_6$, 1023
C$_7$H$_5$N$_3$O$_7$, 1024-1026
C$_7$H$_5$N$_5$O$_8$, 1027
C$_7$H$_6$ClF, 1028, 1029
C$_7$H$_6$ClN$_3$O$_4$S$_2$, 1030
C$_7$H$_6$ClN$_4$O$_5$S$_2$, 1031
C$_7$H$_6$Cl$_2$N$_2$O, 1032
C$_7$H$_6$Cl$_2$O, 1033-1035
C$_7$H$_6$N$_2$O$_2$S, 1036
C$_7$H$_6$N$_2$O$_4$, 1037
C$_7$H$_6$N$_2$O$_5$, 1038, 1039
C$_7$H$_6$N$_2$S, 1040
C$_7$H$_6$N$_4$, 1041-1043
C$_7$H$_6$N$_4$O, 1044-1049
C$_7$H$_6$N$_4$S, 1050-1053
C$_7$H$_6$O, 1054
C$_7$H$_6$O$_2$, 1055-1058
C$_7$H$_6$O$_3$, 1059-1063
C$_7$H$_6$O$_4$, 1064-1067
C$_7$H$_6$O$_5$, 1068, 1069
C$_7$H$_7$Br, 1070
C$_7$H$_7$Cl, 1071-1073
C$_7$H$_7$ClO, 1074-1079
C$_7$H$_7$Cl$_2$NO, 1080
C$_7$H$_7$Cl$_2$NO$_3$PS, 1081
C$_7$H$_7$Cl$_3$NO$_4$P, 1082
C$_7$H$_7$FN$_2$O$_3$, 1083
C$_7$H$_7$FN$_2$O$_4$, 1084-1086
C$_7$H$_7$NO, 1087
C$_7$H$_7$NO$_2$, 1088-1095
C$_7$H$_7$NO$_3$, 1096-1098
C$_7$H$_7$N$_2$OS, 1099
C$_7$H$_7$N$_5$, 1100
C$_7$H$_8$, 1101-1103
C$_7$H$_8$ClN$_3$O$_4$S$_2$, 1104
C$_7$H$_8$FN$_3$O$_3$, 1105, 1106

$C_7H_8N_2O_2$, 1107
$C_7H_8N_2O_3$, 1108, 1109
$C_7H_8N_2O_3S$, 1110
$C_7H_8N_2O_4$, 1111
$C_7H_8N_2S$, 1112
$C_7H_8N_4O_2$, 1113, 1114
C_7H_8O, 1115-1119
$C_7H_8O_2$, 1120-1124
$C_7H_8O_3S$, 1125
$C_7H_8O_3S.H_2O$, 1126
$C_7H_8O_3S.2H_2O$, 1127
$C_7H_8O_3S.4H_2O$, 1128
$C_7H_8O_7$, 1129
$C_7H_9ClN_2OS$, 1130
C_7H_9N, 1131-1142
C_7H_9NO, 1143-1145
$C_7H_9NO_2$, 1146
$C_7H_9NO_2S$, 1147-1149
$C_7H_9NO_3S$, 1150-1153
$C_7H_9N_3O$, 1154
$C_7H_9N_3O_2S_2$, 1155
$C_7H_9N_3O_3$, 1156
$C_7H_9N_3O_3S$, 1157
$C_7H_9N_3O_4$, 1158
$C_7H_{10}N_2OS$, 1159
$C_7H_{10}N_2O_2S$, 1160-1162
$C_7H_{10}N_2O_3$, 1163, 1164
$C_7H_{10}N_4O_2S$, 1165
$C_7H_{10}N_4O_3.H_2O$, 1166
$C_7H_{10}O_4S.H_2O$, 1167
$C_7H_{10}O_3$, 1168, 1169
$C_7H_{11}NO_2$, 1170
$C_7H_{11}N_3O_2$, 1171, 1172
$C_7H_{11}N_7S$, 1173
C_7H_{12}, 1174-1179
$C_7H_{12}BrNO_4$, 1180
$C_7H_{12}ClN_5$, 1181-1183
$C_7H_{12}N_2O_2$, 1184
$C_7H_{12}N_4O_5$, 1185, 1186
$C_7H_{12}O$, 1187, 1188
$C_7H_{12}O_2$, 1189
$C_7H_{12}O_4$, 1190-1194
$C_7H_{12}O_5$, 1195
$C_7H_{12}O_6$, 1196
$C_7H_{13}BrN_2O_2$, 1197, 1198
$C_7H_{13}NO_2S_2$, 1199
$C_7H_{13}NO_3$, 1200
$C_7H_{13}NO_3S$, 1201
$C_7H_{13}N_3O_3S$, 1202
$C_7H_{13}N_5O$, 1203
C_7H_{14}, 1204-1207
$C_7H_{14}N_2O_2S$, 1208
$C_7H_{14}N_2O_3$, 1209, 1210
$C_7H_{14}N_2O_4S_2$, 1211
$C_7H_{14}N_6$, 1212
$C_7H_{14}O$, 1213-1216
$C_7H_{14}O_2$, 1217-1225
$C_7H_{14}O_3$, 1226-1230
$C_7H_{14}O_6$, 1231-1233
$C_7H_{14}O_7$, 1234, 1235
$C_7H_{15}Br$, 1236
$C_7H_{15}Cl$, 1237
$C_7H_{15}I$, 1238
$C_7H_{15}NO_2$, 1239-1241

C_7H_{16}, 1242-1248
$C_7H_{16}O$, 1249-1263
$C_7H_{16}O_4S_2$, 1264
$C_7H_{16}O_7$, 1265
$C_7H_{17}O_2PS_3$, 1266, 1267
$C_7H_{17}O_4PS_3$, 1268
$C_8H_2Cl_4N_2$, 1269
$C_8H_2Cl_4O_4$, 1270
$C_8H_3Cl_2F_3N_2$, 1271
$C_8H_3Cl_5O_2$, 1272
$C_8H_3Cl_5O_3$, 1273
$C_8H_4Cl_4O_3$, 1274
$C_8H_4N_2$, 1275
$C_8H_4N_2S$, 1276
$C_8H_4N_2S_2$, 1277
$C_8H_4O_3$, 1278
$C_8H_5ClO_4$, 1279
$C_8H_5Cl_3O_2$, 1280
$C_8H_5Cl_3O_3$, 1281-1286
$C_8H_5F_3O_2$, 1287
$C_8H_5NO_2$, 1288
$C_8H_5NO_2S$, 1289, 1290
$C_8H_5NO_4$, 1291
$C_8H_5NO_6$, 1292, 1293
C_8H_6, 1294
C_8H_6BrNS, 1295, 1296
C_8H_6ClNS, 1297, 1298
$C_8H_6Cl_2O_3$, 1299-1305
$C_8H_6Cl_4O_2$, 1306
$C_8H_6Cl_5NO_2$, 1307
$C_8H_6F_3N_3O_4S_2$, 1308
C_8H_6INS, 1309, 1310
$C_8H_6N_2O_2S$, 1311, 1312
$C_8H_6N_4O_5$, 1313
$C_8H_6N_4O_8$, 1314
$C_8H_6N_2S$, 1315
$C_8H_6O_2$, 1316, 1317
$C_8H_6O_3$, 1318, 1319
$C_8H_6O_4$, 1320-1322
$C_8H_6O_5$, 1323-1325
C_8H_6S, 1326
$C_8H_7BrN_2O_3$, 1327, 1328
$C_8H_7ClN_2O_3$, 1329, 1330
$C_8H_7ClO_3$, 1331-1333
$C_8H_7Cl_2NO_2$, 1334
$C_8H_7Cl_3O$, 1335
$C_8H_7Cl_3O_2$, 1336
C_8H_7N, 1337, 1338
C_8H_7NOS, 1339, 1340
$C_8H_7NO_3$, 1341
$C_8H_7NO_4$, 1342, 1343
C_8H_7NS, 1344-1346
$C_8H_7N_5O$, 1347-1349
$C_8H_7N_5O_8$, 1350
C_8H_8, 1351
$C_8H_8BrCl_2O_3PS$, 1352
C_8H_8BrNO, 1353
C_8H_8ClNO, 1354
$C_8H_8Cl_2IO_3PS$, 1355
$C_8H_8Cl_2O$, 1356
$C_8H_8Cl_2O_2$, 1357, 1358
$C_8H_8Cl_2O_3PS$, 1359
C_8H_8FNO, 1360
$C_8H_8F_3N_3O_4S_2$, 1361

C_8H_8INO, 1362
$C_8H_8N_2O_2$, 1363, 1364
$C_8H_8N_2O_3$, 1365, 1366
$C_8H_8N_2O_6S$, 1367
$C_8H_8N_4$, 1368
$C_8H_8N_4O$, 1369
$C_8H_8N_4O_2S_2$, 1370
$C_8H_8N_4O_3$, 1371
$C_8H_8N_4O_4$, 1372
$C_8H_8N_4O_4S_3$, 1373
$C_8H_8N_4O_6$, 1374
C_8H_8O, 1375-1378
$C_8H_8O_2$, 1379-1385
$C_8H_8O_2Hg$, 1386
$C_8H_8O_3$, 1387-1400
$C_8H_8O_4$, 1401, 1402
$C_8H_8O_5$, 1403
$C_8H_9ClNO_5PS$, 1404, 1405
C_8H_9ClO, 1406-1408
$C_8H_9FN_2O_3$, 1409
$C_8H_9FN_2O_4$, 1410, 1411
C_8H_9N, 1412
C_8H_9NO, 1413-1415
$C_8H_9NO_2$, 1416-1420
$C_8H_9NO_3S$, 1421
$C_8H_9NO_4$, 1422
$C_8H_9N_3O_3$, 1423
$C_8H_9N_5$, 1424-1426
$C_8H_9O_3PS$, 1427
C_8H_{10}, 1428-1432
$C_8H_{10}NO_5PS$, 1433
$C_8H_{10}N_2O$, 1434-1442
$C_8H_{10}N_2O_3$, 1443, 1444
$C_8H_{10}N_2O_3S$, 1445-1447
$C_8H_{10}N_2O_4S$, 1448
$C_8H_{10}N_4O_2$, 1449
$C_8H_{10}N_4O_2 \cdot H_2O$, 1450
$C_8H_{10}N_4O_3$, 1451
$C_8H_{10}O$, 1452-1461
$C_8H_{10}O_2$, 1462-1468
$C_8H_{10}O_3$, 1469
$C_8H_{10}O_3S$, 1470
$C_8H_{10}O_4$, 1471, 1472
$C_8H_{10}O_5$, 1473
$C_8H_{10}O_8$, 1474
$C_8H_{11}BrN_2O_2$, 1475
$C_8H_{11}Cl_2NO$, 1476
$C_8H_{11}Cl_3O_6$, 1477
$C_8H_{11}N$, 1478
$C_8H_{11}NO$, 1479, 1480
$C_8H_{11}N_2O_5PS$, 1481
C_8H_{12}, 1482
$C_8H_{12}ClNO$, 1483
$C_8H_{12}N_2O_2S$, 1484, 1485
$C_8H_{12}N_2O_3$, 1486
$C_8H_{12}O_2$, 1487
$C_8H_{12}O_4$, 1488-1490
$C_8H_{13}BrN_2O_2$, 1491
$C_8H_{13}NO$, 1492
$C_8H_{13}N_2O_3PS$, 1493
C_8H_{14}, 1494, 1495
$C_8H_{14}ClNS_2$, 1496
$C_8H_{14}ClN_5$, 1497
$C_8H_{14}N_2O_2$, 1498

$C_8H_{14}N_4OS$, 1499
$C_8H_{14}O$, 1500
$C_8H_{14}O_2$, 1501-1505
$C_8H_{14}O_2S_4$, 1506
$C_8H_{14}O_4$, 1507-1513
$C_8H_{14}O_5$, 1514
$C_8H_{15}ClN_5O$, 1515
$C_8H_{15}NO$, 1516, 1517
$C_8H_{15}N_3O_2$, 1518
$C_8H_{15}N_3O_7$, 1519
$C_8H_{15}N_5O$, 1520, 1521
$C_8H_{15}N_5S$, 1522, 1523
C_8H_{16}, 1524-1532
$C_8H_{16}N_2O_2$, 1533
$C_8H_{16}N_2O_4S_2$, 1534
$C_8H_{16}N_6$, 1535
$C_8H_{16}N_6O$, 1536
$C_8H_{16}O$, 1537, 1538
$C_8H_{16}O_2$, 1539-1550
$C_8H_{16}O_3$, 1551-1556
$C_8H_{16}O_3S$, 1557
$C_8H_{16}O_4$, 1558
$C_8H_{17}Br$, 1559
$C_8H_{17}N$, 1560
$C_8H_{17}NO$, 1561-1568
$C_8H_{17}NO_2$, 1569
$C_8H_{17}NO_3$, 1570
C_8H_{18}, 1571-1576
$C_8H_{18}NO_4PS_2$, 1577
$C_8H_{18}N_2O$, 1578
$C_8H_{18}O$, 1579-1584
$C_8H_{18}O_2$, 1585
$C_8H_{18}O_4S_2$, 1586
$C_8H_{19}N$, 1587, 1588
$C_8H_{19}O_2PS_2$, 1589
$C_8H_{19}O_2PS_3$, 1590
$C_8H_{19}O_3P$, 1591
$C_8H_{19}O_3PS_2$, 1592, 1593
$C_8H_{19}O_4P$, 1594, 1595
$C_8H_{19}O_4PS_3$, 1596
$C_8H_{20}Si$, 1597
$C_8H_{20}Sn$, 1598
$C_8H_{20}O_5P_2S_2$, 1599
$C_8H_{23}N_5$, 1600
$C_9Cl_4N_2$, 1601
$C_9H_4Cl_3NO_2S$, 1602
$C_9H_5Cl_3N_4$, 1603
$C_9H_6ClNO_3S$, 1604
$C_9H_6Cl_2N_2O_3$, 1605
$C_9H_6Cl_6O_3S$, 1606, 1607
$C_9H_6INO_3$, 1608
$C_9H_6N_2S$, 1609
$C_9H_6O_2$, 1610
$C_9H_6O_3$, 1611
$C_9H_6O_5$, 1612
$C_9H_6O_6$, 1613-1615
$C_9H_7Cl_3O_3$, 1616, 1617
C_9H_7N, 1618
C_9H_7NO, 1619-1625
C_9H_7NOS, 1626, 1627
$C_9H_7NO_2S$, 1628
$C_9H_7NO_5$, 1629
$C_9H_7N_3S$, 1630
$C_9H_7N_3O_2S$, 1631

C$_9$H$_8$Cl$_2$O$_3$, 1632, 1633
C$_9$H$_8$Cl$_3$NO$_2$S, 1634
C$_9$H$_8$N$_2$OS, 1635
C$_9$H$_8$N$_4$O$_6$, 1636
C$_9$H$_8$O, 1637
C$_9$H$_8$O$_2$, 1638-1641
C$_9$H$_8$O$_4$, 1642
C$_9$H$_9$ClO$_3$, 1643-1645
C$_9$H$_9$Cl$_2$NO, 1646
C$_9$H$_9$Cl$_2$NO$_2$, 1647, 1648
C$_9$H$_9$I$_2$NO$_3$, 1649, 1650
C$_9$H$_9$N, 1651
C$_9$H$_9$NOS, 1652, 1653
C$_9$H$_9$NO$_2$, 1654
C$_9$H$_9$NO$_3$, 1655, 1656
C$_9$H$_9$NO$_4$, 1657
C$_9$H$_9$NS, 1658
C$_9$H$_9$N$_3$OS, 1659
C$_9$H$_9$N$_3$O$_2$, 1660
C$_9$H$_9$N$_3$O$_2$S$_2$, 1661
C$_9$H$_{10}$, 1662, 1663
C$_9$H$_{10}$BrClN$_2$O$_2$, 1664
C$_9$H$_{10}$Cl$_2$N$_2$O, 1665
C$_9$H$_{10}$Cl$_2$N$_2$O$_2$, 1666
C$_9$H$_{10}$Cl$_2$O, 1667
C$_9$H$_{10}$Cl$_3$O$_3$PS, 1668
C$_9$H$_{10}$NO$_3$, 1669
C$_9$H$_{10}$NO$_3$PS, 1670
C$_9$H$_{10}$N$_2$O$_3$, 1671-1673
C$_9$H$_{10}$N$_2$O$_3$S$_2$, 1674
C$_9$H$_{10}$N$_2$S, 1675, 1676
C$_9$H$_{10}$N$_4$, 1677
C$_9$H$_{10}$N$_4$O$_2$S$_2$, 1678
C$_9$H$_{10}$O$_2$, 1679-1684
C$_9$H$_{10}$O$_3$, 1685-1688
C$_9$H$_{10}$O$_4$, 1689
C$_9$H$_{11}$BrN$_2$O$_2$, 1690
C$_9$H$_{11}$ClN$_2$O, 1691
C$_9$H$_{11}$ClN$_2$O$_2$, 1692
C$_9$H$_{11}$ClO, 1693
C$_9$H$_{11}$Cl$_2$N$_3$O$_4$S$_2$, 1694
C$_9$H$_{11}$Cl$_3$NO$_3$PS, 1695
C$_9$H$_{11}$Cl$_3$NO$_4$P, 1696
C$_9$H$_{11}$FN$_2$O$_4$, 1697-1699
C$_9$H$_{11}$IN$_2$O$_5$, 1700
C$_9$H$_{11}$N, 1701
C$_9$H$_{11}$NO, 1702-1705
C$_9$H$_{11}$NO$_2$, 1706-1713
C$_9$H$_{11}$NO$_3$, 1714-1716
C$_9$H$_{11}$NO$_4$, 1717, 1718
C$_9$H$_{11}$NS$_2$Hg, 1719
C$_9$H$_{11}$N$_3$O, 1720
C$_9$H$_{11}$N$_3$O$_2$S$_2$, 1721
C$_9$H$_{11}$N$_3$O$_4$, 1722
C$_9$H$_{12}$, 1723-1730
C$_9$H$_{12}$ClO$_2$PS$_3$, 1731
C$_9$H$_{12}$ClO$_4$P, 1732
C$_9$H$_{12}$Cl$_3$N$_4$, 1733
C$_9$H$_{12}$FN$_3$O$_3$, 1734
C$_9$H$_{12}$NO$_5$PS, 1735, 1736
C$_9$H$_{12}$N$_2$O, 1737
C$_9$H$_{12}$N$_2$O$_2$, 1738
C$_9$H$_{12}$N$_2$O$_2$S, 1739
C$_9$H$_{12}$N$_2$O$_3$, 1740, 1741

C$_9$H$_{12}$N$_4$O$_2$, 1742-1745
C$_9$H$_{12}$N$_4$O$_3$, 1746-1748
C$_9$H$_{12}$N$_4$O$_3$S, 1749
C$_9$H$_{12}$O, 1750-1755
C$_9$H$_{12}$O$_2$, 1756-1759
C$_9$H$_{13}$BrN$_2$O$_2$, 1760, 1761
C$_9$H$_{13}$ClN$_2$O$_2$, 1762
C$_9$H$_{13}$ClN$_6$, 1763
C$_9$H$_{13}$N, 1764
C$_9$H$_{13}$NO$_3$, 1765
C$_9$H$_{13}$N$_3$O$_3$, 1766, 1767
C$_9$H$_{13}$N$_3$O$_4$, 1768
C$_9$H$_{13}$N$_3$O$_5$, 1769
C$_9$H$_{13}$N$_5$O$_4$, 1770
C$_9$H$_{13}$O$_2$P, 1771
C$_9$H$_{13}$O$_6$PS, 1772
C$_9$H$_{14}$ClN$_5$, 1773
C$_9$H$_{14}$N$_2$O$_3$, 1774-1776
C$_9$H$_{14}$N$_6$, 1777
C$_9$H$_{14}$O$_6$, 1778, 1779
C$_9$H$_{15}$Br$_6$O$_4$P, 1780
C$_9$H$_{15}$Cl$_6$O$_4$P, 1781
C$_9$H$_{15}$NO$_3$, 1782
C$_9$H$_{16}$, 1783, 1784
C$_9$H$_{16}$ClN$_4$, 1785
C$_9$H$_{16}$ClN$_5$, 1786-1788
C$_9$H$_{16}$N$_2$O$_4$, 1789
C$_9$H$_{16}$N$_4$OS, 1790
C$_9$H$_{16}$N$_8$, 1791
C$_9$H$_{16}$O$_2$, 1792, 1793
C$_9$H$_{16}$O$_4$, 1794, 1795
C$_9$H$_{16}$O$_5$, 1796
C$_9$H$_{17}$ClN$_3$O$_3$PS, 1797
C$_9$H$_{17}$NOS, 1798
C$_9$H$_{17}$NO$_3$, 1799
C$_9$H$_{17}$NO$_4$, 1800
C$_9$H$_{17}$N$_5$O, 1801
C$_9$H$_{17}$N$_5$S, 1802
C$_9$H$_{18}$, 1803, 1804
C$_9$H$_{18}$N$_2$O$_2$S, 1805
C$_9$H$_{18}$N$_2$O$_4$, 1806
C$_9$H$_{18}$N$_3$S$_6$Fe, 1807
C$_9$H$_{18}$N$_6$, 1808, 1809
C$_9$H$_{18}$N$_6$O, 1810, 1811
C$_9$H$_{18}$N$_6$O$_3$, 1812
C$_9$H$_{18}$O, 1813-1816
C$_9$H$_{18}$O$_2$, 1817-1828
C$_9$H$_{18}$O$_3$, 1829-1834
C$_9$H$_{19}$NOS, 1835
C$_9$H$_{19}$NO$_2$, 1836
C$_9$H$_{19}$O$_3$, 1837
C$_9$H$_{20}$, 1838-1842
C$_9$H$_{20}$NO$_3$PS$_2$, 1843
C$_9$H$_{20}$O, 1844-1848
C$_9$H$_{21}$N, 1849
C$_9$H$_{21}$O$_2$PS$_3$, 1850
C$_9$H$_{21}$O$_3$P, 1851
C$_9$H$_{21}$O$_3$PS$_3$, 1852, 1853
C$_9$H$_{21}$O$_4$P, 1854-1856
C$_9$H$_{21}$O$_4$PS$_3$, 1857
C$_9$H$_{22}$O$_4$P$_2$S$_4$, 1858
C$_{10}$H$_4$Cl$_2$O$_2$, 1859
C$_{10}$H$_5$ClN$_2$O$_4$, 1860
C$_{10}$H$_5$Cl$_7$, 1861

$C_{10}H_5Cl_3O$, 1862
$C_{10}H_5N_3O_6$, 1863, 1864
$C_{10}H_6Br_2$, 1865, 1866
$C_{10}H_6Cl_2$, 1867
$C_{10}H_6Cl_4O_3S$, 1868
$C_{10}H_6Cl_4O_4$, 1869
$C_{10}H_6Cl_6$, 1870
$C_{10}H_6Cl_6O$, 1871, 1872
$C_{10}H_6Cl_6O_2$, 1873
$C_{10}H_6Cl_8$, 1874
$C_{10}H_6FN_3O_3$, 1875
$C_{10}H_6N_2O_4$, 1876, 1877
$C_{10}H_6O_8$, 1878
$C_{10}H_7Br$, 1879, 1880
$C_{10}H_7Cl$, 1881, 1882
$C_{10}H_7I$, 1883
$C_{10}H_7NO_2$, 1884
$C_{10}H_7NO_3$, 1885, 1886
$C_{10}H_7N_3O_3$, 1887
$C_{10}H_7N_3S$, 1888
$C_{10}H_8$, 1889
$C_{10}H_8BrN_3O$, 1890, 1891
$C_{10}H_8ClN_3O$, 1892
$C_{10}H_8N_2$, 1893, 1894
$C_{10}H_8N_2O_2$, 1895
$C_{10}H_8O$, 1896, 1897
$C_{10}H_8O_2$, 1898, 1899
$C_{10}H_9ClN_4O_2S$, 1900, 1901
$C_{10}H_9Cl_2NO$, 1902
$C_{10}H_9Cl_3O_3$, 1903, 1904
$C_{10}H_9Cl_4NO_2S$, 1905
$C_{10}H_9Cl_4O_4P$, 1906, 1907
$C_{10}H_9N$, 1908-1910
$C_{10}H_9NO$, 1911, 1912
$C_{10}H_9NO_2S$, 1913, 1914
$C_{10}H_9NO_3S$, 1915-1924
$C_{10}H_9NO_4S$, 1925
$C_{10}H_9NO_9S_3$, 1926
$C_{10}H_9N_3O_3S$, 1927
$C_{10}H_9N_4O_5$, 1928
$C_{10}H_{10}Fe$, 1929
$C_{10}H_{10}BrNO_4$, 1930
$C_{10}H_{10}BrNO_5$, 1931
$C_{10}H_{10}ClNO_3$, 1932
$C_{10}H_{10}Cl_2F_3N_2OS$, 1933
$C_{10}H_{10}Cl_2O_2$, 1934
$C_{10}H_{10}Cl_2O_3$, 1935, 1936
$C_{10}H_{10}Cl_8$, 1937
$C_{10}H_{10}N_4O$, 1938
$C_{10}H_{10}N_4O_2S$, 1939-1942
$C_{10}H_{10}N_4O_4S$, 1943
$C_{10}H_{10}O$, 1944
$C_{10}H_{10}O_2$, 1945-1947
$C_{10}H_{10}O_4$, 1948-1952
$C_{10}H_{10}O_5$, 1953
$C_{10}H_{11}ClO_3$, 1954, 1955
$C_{10}H_{11}Cl_3O_2$, 1956
$C_{10}H_{11}FN_2O_6$, 1957
$C_{10}H_{11}F_3N_2O$, 1958
$C_{10}H_{11}F_3N_2O_3S$, 1959
$C_{10}H_{11}NO$, 1960
$C_{10}H_{11}NOS$, 1961
$C_{10}H_{11}NO_3$, 1962, 1963
$C_{10}H_{11}NO_4$, 1964-1966

$C_{10}H_{11}NO_5$, 1967
$C_{10}H_{11}NO_6$, 1968
$C_{10}H_{11}N_3OS$, 1969
$C_{10}H_{11}N_3O_2S_2$, 1970
$C_{10}H_{11}N_3O_2S$, 1971
$C_{10}H_{11}N_3O_2S_2$, 1972, 1973
$C_{10}H_{11}N_3O_3$, 1974
$C_{10}H_{11}N_3O_3S$, 1975
$C_{10}H_{11}N_5O_2S$, 1976
$C_{10}H_{12}$, 1977
$C_{10}H_{12}BrCl_2O_3PS$, 1978
$C_{10}H_{12}ClNO_2$, 1979, 1980
$C_{10}H_{12}ClN_3O_2$, 1981
$C_{10}H_{12}ClN_3O_3S$, 1982
$C_{10}H_{12}ClN_5O_2$, 1983
$C_{10}H_{12}Cl_2O$, 1984
$C_{10}H_{12}Cl_3O_2PS$, 1985
$C_{10}H_{12}N_2O_2$, 1986
$C_{10}H_{12}N_2O_3$, 1987, 1988
$C_{10}H_{12}N_2O_3S$, 1989
$C_{10}H_{12}N_2O_4S$, 1990
$C_{10}H_{12}N_2O_5$, 1991, 1992
$C_{10}H_{12}N_3O_3PS_2$, 1993
$C_{10}H_{12}N_4$, 1994
$C_{10}H_{12}N_4O$, 1995
$C_{10}H_{12}N_4O_2$, 1996, 1997
$C_{10}H_{12}N_4O_2S$, 1998
$C_{10}H_{12}N_4O_3$, 1999-2001
$C_{10}H_{12}N_4O_5$, 2002
$C_{10}H_{12}N_4O_6$, 2003
$C_{10}H_{12}N_5O_6P$, 2004
$C_{10}H_{12}N_6O_2S$, 2005
$C_{10}H_{12}O$, 2006-2008
$C_{10}H_{12}O_2$, 2009-2012
$C_{10}H_{12}O_3$, 2013
$C_{10}H_{12}O_4$, 2014
$C_{10}H_{12}O_8$, 2015
$C_{10}H_{13}ClN_2$, 2016
$C_{10}H_{13}ClN_2O$, 2017, 2018
$C_{10}H_{13}ClN_2O_2$, 2019
$C_{10}H_{13}ClN_2O_3S$, 2020
$C_{10}H_{13}Cl_2FN_2O_2S_2$, 2021
$C_{10}H_{13}Cl_2O_3PS$, 2022
$C_{10}H_{13}FN_2O_3$, 2023
$C_{10}H_{13}FN_2O_4$, 2024
$C_{10}H_{13}NO_2$, 2025-2032
$C_{10}H_{13}NO_3$, 2033, 2034
$C_{10}H_{13}N_3O_2S_2$, 2035
$C_{10}H_{13}N_4O_3$, 2036
$C_{10}H_{13}N_5$, 2037, 2038
$C_{10}H_{13}N_5O$, 2039, 2040
$C_{10}H_{13}N_5O_2$, 2041
$C_{10}H_{13}N_5O_3$, 2042
$C_{10}H_{13}N_5O_4$, 2043
$C_{10}H_{13}N_5O_5$, 2044
$C_{10}H_{14}$, 2045-2053
$C_{10}H_{14}Cl_2NO_2PS$, 2054
$C_{10}H_{14}Cl_2N_4O_2$, 2055
$C_{10}H_{14}NO_5PS$, 2056
$C_{10}H_{14}NO_5P$, 2057
$C_{10}H_{14}N_2O$, 2058, 2059
$C_{10}H_{14}N_2O_2$, 2060
$C_{10}H_{14}N_2O_3$, 2061-2063
$C_{10}H_{14}N_2S$, 2064

$C_{10}H_{14}N_4O_2$, 2065, 2066
$C_{10}H_{14}N_4O_3$, 2067
$C_{10}H_{14}N_4O_4$, 2068
$C_{10}H_{14}N_5O_7P$, 2069, 2070
$C_{10}H_{14}O$, 2071-2077
$C_{10}H_{14}O_2$, 2078-2080
$C_{10}H_{14}O_8$, 2081
$C_{10}H_{15}N$, 2082
$C_{10}H_{15}NO$, 2083, 2084
$C_{10}H_{15}NO_2$, 2085
$C_{10}H_{15}N_5O_5$, 2086
$C_{10}H_{15}OPS_2$, 2087
$C_{10}H_{15}O_3PS_2$, 2088
$C_{10}H_{16}$, 2089, 2090
$C_{10}H_{16}Cl_3NOS$, 2091
$C_{10}H_{16}NO_5S_2$, 2092
$C_{10}H_{16}NO_3S$, 2093
$C_{10}H_{16}NO_4$, 2094
$C_{10}H_{16}N_2O_3$, 2095-2098
$C_{10}H_{16}N_2O_3S$, 2099
$C_{10}H_{16}N_2O_4$, 2100
$C_{10}H_{16}N_4O_2$, 2101
$C_{10}H_{16}N_4O_2S$, 2102
$C_{10}H_{16}O$, 2103-2107
$C_{10}H_{16}O_2$, 2108
$C_{10}H_{16}O_4$, 2109, 2110
$C_{10}H_{16}O_5$, 2111
$C_{10}H_{17}Cl_2NOS$, 2112
$C_{10}H_{17}NO_2$, 2113
$C_{10}H_{17}N_2O_4PS$, 2114
$C_{10}H_{17}N_3O_5$, 2115
$C_{10}H_{17}N_3O_6S$, 2116
$C_{10}H_{17}O_3P$, 2117
$C_{10}H_{18}$, 2118-2120
$C_{10}H_{18}ClN_5$, 2121
$C_{10}H_{18}N_2O_4$, 2122
$C_{10}H_{18}N_2O_5$, 2123
$C_{10}H_{18}N_6O_2$, 2124
$C_{10}H_{18}O$, 2125-2129
$C_{10}H_{18}O_2$, 2130-2133
$C_{10}H_{18}O_3$, 2134
$C_{10}H_{18}O_4$, 2135-2139
$C_{10}H_{18}O_5$, 2140, 2141
$C_{10}H_{19}NO_3$, 2142, 2143
$C_{10}H_{19}NO_4S$, 2144
$C_{10}H_{19}N_2O_4PS$, 2145
$C_{10}H_{19}N_5O$, 2146-2149
$C_{10}H_{19}N_5OS$, 2150
$C_{10}H_{19}N_5S$, 2151-2153
$C_{10}H_{19}O_6PS_2$, 2154
$C_{10}H_{20}$, 2155
$C_{10}H_{20}NO_4PS$, 2156
$C_{10}H_{20}NO_5PS_2$, 2157
$C_{10}H_{20}N_3S_4$, 2158
$C_{10}H_{20}N_6O$, 2159
$C_{10}H_{20}O$, 2160, 2161
$C_{10}H_{20}O_2$, 2162-2172
$C_{10}H_{20}O_2 \cdot H_2O$, 2173
$C_{10}H_{20}O_3$, 2174, 2175
$C_{10}H_{20}O_4$, 2176
$C_{10}H_{21}NOS$, 2177, 2178
$C_{10}H_{22}$, 2179, 2180
$C_{10}H_{22}O$, 2181
$C_{10}H_{23}O_3P$, 2182

$C_{10}H_{23}O_4P$, 2183
$C_{10}Cl_{10}O$, 2184
$C_{10}Cl_{12}$, 2185
$C_{11}H_6BrNS$, 2186
$C_{11}H_6O_3$, 2187
$C_{11}H_7FN_2O_3$, 2188
$C_{11}H_7FN_2O_4$, 2189, 2190
$C_{11}H_7NS$, 2191, 2192
$C_{11}H_8N_4O_4$, 2193
$C_{11}H_8O_2$, 2194, 2195
$C_{11}H_8O_3$, 2196
$C_{11}H_9ClO_2S$, 2197
$C_{11}H_9Cl_2NO_2$, 2198
$C_{11}H_9Cl_4NO_4$, 2199
$C_{11}H_9I_3N_2O_4$, 2200
$C_{11}H_{10}$, 2201, 2202
$C_{11}H_{10}BrN_3O_2S$, 2203, 2204
$C_{11}H_{10}ClNO_2$, 2205
$C_{11}H_{10}ClN_3O_2S$, 2206
$C_{11}H_{10}Cl_2O_3$, 2207
$C_{11}H_{10}IN_3O_2S$, 2208
$C_{11}H_{10}N_2O$, 2209, 2210
$C_{11}H_{10}N_2O_3$, 2211
$C_{11}H_{10}N_2S$, 2212
$C_{11}H_{10}N_4O_4S$, 2213
$C_{11}H_{11}ClO_3$, 2214
$C_{11}H_{11}N$, 2215, 2216
$C_{11}H_{11}NO$, 2217
$C_{11}H_{11}NO_2$, 2218
$C_{11}H_{11}NO_3S$, 2219
$C_{11}H_{11}NO_4$, 2220
$C_{11}H_{11}NO_5$, 2221, 2222
$C_{11}H_{11}N_3OS$, 2223
$C_{11}H_{11}N_3O_2S$, 2224
$C_{11}H_{11}N_3O_3S_2$, 2225
$C_{11}H_{11}N_3O_3S$, 2226
$C_{11}H_{12}ClNO_4$, 2227
$C_{11}H_{12}Cl_2N_2O_5$, 2228
$C_{11}H_{12}Cl_2O_3$, 2229
$C_{11}H_{12}I_3NO_2$, 2230
$C_{11}H_{12}NO_4PS_2$, 2231
$C_{11}H_{12}N_2O$, 2232
$C_{11}H_{12}N_2O_2$, 2233-2235
$C_{11}H_{12}N_2O_4$, 2236
$C_{11}H_{12}N_4O_2S$, 2237-2239
$C_{11}H_{12}N_4O_3S$, 2240, 2241
$C_{11}H_{12}N_4O_3S_2$, 2242
$C_{11}H_{12}N_4O_5S$, 2243, 2244
$C_{11}H_{12}N_4O_5$, 2245
$C_{11}H_{12}N_6O_2S$, 2246
$C_{11}H_{12}O_2$, 2247
$C_{11}H_{12}O_4$, 2248, 2249
$C_{11}H_{12}O_4S$, 2250
$C_{11}H_{12}O_5S$, 2251
$C_{11}H_{12}O_6S$, 2252
$C_{11}H_{13}ClO_3$, 2253
$C_{11}H_{13}FN_2O_4$, 2254
$C_{11}H_{13}F_3N_2O_3S$, 2255
$C_{11}H_{13}F_3N_4O_4$, 2256
$C_{11}H_{13}NO$, 2257, 2258
$C_{11}H_{13}NO_3$, 2259-2261
$C_{11}H_{13}NO_4$, 2262-2265
$C_{11}H_{13}N_3O$, 2266
$C_{11}H_{13}N_3O_3S$, 2267-2269

$C_{11}H_{13}N_5O_2$, 2270
$C_{11}H_{13}N_5O_3$, 2271
$C_{11}H_{14}ClNO$, 2272
$C_{11}H_{14}N_2O$, 2273
$C_{11}H_{14}N_2O_3S$, 2274
$C_{11}H_{14}N_4O_2S_2$, 2275, 2276
$C_{11}H_{14}N_4O_3$, 2277, 2278
$C_{11}H_{14}N_4O_5$, 2279
$C_{11}H_{14}O$, 2280
$C_{11}H_{14}O_2$, 2281, 2282
$C_{11}H_{14}O_3$, 2283, 2284
$C_{11}H_{14}O_4$, 2285
$C_{11}H_{15}BrClO_5PS$, 2286
$C_{11}H_{15}BrN_2O$, 2287
$C_{11}H_{15}FN_2O_4$, 2288
$C_{11}H_{15}NO_2$, 2289, 2290
$C_{11}H_{15}NO_2S$, 2291
$C_{11}H_{15}NO_3$, 2292, 2293
$C_{11}H_{15}NO_4$, 2294
$C_{11}H_{15}N_3O_2$, 2295
$C_{11}H_{15}N_3O_3$, 2296
$C_{11}H_{15}N_3O_5$, 2297
$C_{11}H_{15}O_3P$, 2298
$C_{11}H_{16}$, 2299-2301
$C_{11}H_{16}ClO_2PS_3$, 2302
$C_{11}H_{16}N_2O_2$, 2303, 2304
$C_{11}H_{16}N_2O_3$, 2305-2310
$C_{11}H_{16}N_2O_3S$, 2311
$C_{11}H_{16}N_2O_4$, 2312
$C_{11}H_{16}N_2O_5$, 2313
$C_{11}H_{16}N_4O_2$, 2314
$C_{11}H_{16}N_4O_4$, 2315
$C_{11}H_{16}O$, 2316-2321
$C_{11}H_{16}O_2$, 2322, 2323
$C_{11}H_{17}NO_3$, 2324
$C_{11}H_{17}N_3O_3$, 2325
$C_{11}H_{17}N_3O_3S$, 2326
$C_{11}H_{17}N_3O_6$, 2327
$C_{11}H_{17}O_3PS$, 2328
$C_{11}H_{17}O_3PS_2$, 2329
$C_{11}H_{17}O_4PS_2$, 2330
$C_{11}H_{17}O_5PS_2$, 2331
$C_{11}H_{18}N_2O_2S$, 2332
$C_{11}H_{18}N_2O_3$, 2333-2336
$C_{11}H_{18}N_4O_2$, 2337
$C_{11}H_{19}N_3O$, 2338, 2339
$C_{11}H_{20}$, 2340
$C_{11}H_{20}ClN_5$, 2341
$C_{11}H_{20}N_2O_4$, 2342
$C_{11}H_{20}N_3O_3PS$, 2343
$C_{11}H_{20}N_6$, 2344
$C_{11}H_{20}N_6O$, 2345
$C_{11}H_{20}N_6S$, 2346
$C_{11}H_{20}O_2$, 2347
$C_{11}H_{20}O_4$, 2348, 2349
$C_{11}H_{20}O_5$, 2350
$C_{11}H_{21}BrO_2$, 2351
$C_{11}H_{21}NOS$, 2352
$C_{11}H_{21}NO_3$, 2353
$C_{11}H_{21}N_5O$, 2354
$C_{11}H_{21}N_5OS$, 2355
$C_{11}H_{21}N_5S$, 2356-2358
$C_{11}H_{21}N_7$, 2359
$C_{11}H_{21}O_5$, 2360

$C_{11}H_{22}N_2O$, 2361
$C_{11}H_{22}N_6$, 2362
$C_{11}H_{22}O_2$, 2363-2365
$C_{11}H_{22}O_3$, 2366-2369
$C_{11}H_{22}O_4$, 2370
$C_{11}H_{23}NOS$, 2371
$C_{11}H_{23}NO_2$, 2372
$C_{11}H_{24}$, 2373
$C_{12}HCl_7O$, 2374
$C_{12}HCl_7O_2$, 2375
$C_{12}HCl_9$, 2376, 2377
$C_{12}H_2Br_8$, 2378
$C_{12}H_2Cl_6O$, 2379, 2380
$C_{12}H_2Cl_6O_2$, 2381
$C_{12}H_2Cl_8$, 2382, 2383
$C_{12}H_3Cl_5O$, 2384
$C_{12}H_3Cl_5O_2$, 2385
$C_{12}H_3Cl_7$, 2386-2399
$C_{12}H_4Br_6$, 2400-2402
$C_{12}H_4Cl_4O$, 2403
$C_{12}H_4Cl_4O_2$, 2404-2407
$C_{12}H_4Cl_6$, 2408-2427
$C_{12}H_5Br_5$, 2428
$C_{12}H_5Cl_3O_2$, 2429
$C_{12}H_5Cl_5$, 2430-2449
$C_{12}H_5N_5O_{11}$, 2450
$C_{12}H_5N_5O_{12}$, 2451
$C_{12}H_6Br_4$, 2452
$C_{12}H_6Cl_2O$, 2453
$C_{12}H_6Cl_2O_2$, 2454-2456
$C_{12}H_6Cl_3NO_3$, 2457, 2458
$C_{12}H_6Cl_4$, 2459-2485
$C_{12}H_6Cl_4O_2S$, 2486
$C_{12}H_7BrClNO_2$, 2487
$C_{12}H_7ClO_2$, 2488, 2489
$C_{12}H_7Cl_2NO_3$, 2490
$C_{12}H_7Cl_3$, 2491-2506
$C_{12}H_7Cl_3O_2$, 2507
$C_{12}H_7N_3O_2$, 2508
$C_{12}H_7N_5O_8$, 2509, 2510
$C_{12}H_8$, 2511
$C_{12}H_8Br_2$, 2512
$C_{12}H_8Cl_2$, 2513-2521
$C_{12}H_8Cl_6$, 2522
$C_{12}H_8Cl_6O$, 2523, 2524
$C_{12}H_8N_2$, 2525-2528
$C_{12}H_8N_4O_6$, 2529
$C_{12}H_8O$, 2530
$C_{12}H_8O_2$, 2531
$C_{12}H_8O_4$, 2532
$C_{12}H_8S$, 2533
$C_{12}H_9Br$, 2534
$C_{12}H_9Cl$, 2535-2538
$C_{12}H_9ClF_3N_3O$, 2539
$C_{12}H_9ClN_2$, 2540
$C_{12}H_9ClO$, 2541
$C_{12}H_9Cl_2NO_3$, 2542
$C_{12}H_9Cl_3NO_2S$, 2543
$C_{12}H_9FN_2O_4$, 2544
$C_{12}H_9N$, 2545
$C_{12}H_9NO_3$, 2546
$C_{12}H_9NS$, 2547
$C_{12}H_9N_3O_2$, 2548
$C_{12}H_9N_3O_3$, 2549

$C_{12}H_9N_3O_4$, 2550
$C_{12}H_9N_3O_5$, 2551
$C_{12}H_9N_5O_3$, 2552
$C_{12}H_{10}$, 2553, 2554
$C_{12}H_{10}ClN$, 2555
$C_{12}H_{10}Cl_2N_2$, 2556
$C_{12}H_{10}N_2$, 2557
$C_{12}H_{10}N_2O$, 2558, 2559
$C_{12}H_{10}N_2O_2$, 2560
$C_{12}H_{10}N_2O_3$, 2561
$C_{12}H_{10}N_4O_2$, 2562
$C_{12}H_{10}N_4O_4$, 2563
$C_{12}H_{10}O$, 2564-2566
$C_{12}H_{10}O_2$, 2567, 2568
$C_{12}H_{10}O_3$, 2569
$C_{12}H_{10}O_4$, 2570
$C_{12}H_{11}ClN_2O_5S$, 2571
$C_{12}H_{11}Cl_2NO$, 2572
$C_{12}H_{11}I_3N_2O_4$, 2573
$C_{12}H_{11}N$, 2574
$C_{12}H_{11}NO_2$, 2575, 2576
$C_{12}H_{11}N_3$, 2577, 2578
$C_{12}H_{11}N_3O_3$, 2579
$C_{12}H_{11}O_4P$, 2580
$C_{12}H_{12}$, 2581-2587
$C_{12}H_{12}ClNO$, 2588
$C_{12}H_{12}N_2$, 2589, 2590
$C_{12}H_{12}N_2OS$, 2591
$C_{12}H_{12}N_2O_2S$, 2592, 2593
$C_{12}H_{12}N_2O_3$, 2594, 2595
$C_{12}H_{12}N_2O_6S_2$, 2596
$C_{12}H_{12}N_2S$, 2597
$C_{12}H_{12}N_4O_3S$, 2598, 2599
$C_{12}H_{12}N_6O_6$, 2600
$C_{12}H_{12}O_6$, 2601
$C_{12}H_{13}ClN_2O$, 2602
$C_{12}H_{13}I_3N_2O_2$, 2603
$C_{12}H_{13}I_3N_2O_3$, 2604
$C_{12}H_{13}NO_2$, 2605
$C_{12}H_{13}NO_2S$, 2606
$C_{12}H_{13}NO_3$, 2607, 2608
$C_{12}H_{13}NO_4$, 2609
$C_{12}H_{13}NO_4S$, 2610
$C_{12}H_{13}NO_4S_2$, 2611
$C_{12}H_{13}NO_5$, 2612, 2613
$C_{12}H_{13}NO_6$, 2614
$C_{12}H_{13}N_3O_2$, 2615
$C_{12}H_{13}N_3O_2S$, 2616
$C_{12}H_{13}N_3O_3S_2$, 2617
$C_{12}H_{13}N_3O_4S$, 2618
$C_{12}H_{14}Cl_2O_3$, 2619, 2620
$C_{12}H_{14}Cl_3O_4P$, 2621
$C_{12}H_{14}NO_4PS$, 2622
$C_{12}H_{14}N_2O_2$, 2623
$C_{12}H_{14}N_2O_4$, 2624, 2625
$C_{12}H_{14}N_2O_5$, 2626
$C_{12}H_{14}N_2O_6$, 2627
$C_{12}H_{14}N_4O_2S$, 2628, 2629
$C_{12}H_{14}N_4O_2S.0.5H_2O$, 2630
$C_{12}H_{14}N_4O_4S$, 2631, 2632
$C_{12}H_{14}N_4O_3S_2$, 2633
$C_{12}H_{14}N_4O_3S$, 2634, 2635
$C_{12}H_{14}N_4O_4S$, 2636
$C_{12}H_{14}O_4$, 2637, 2638

$C_{12}H_{15}ClNO_4PS_2$, 2639
$C_{12}H_{15}ClO_3$, 2640
$C_{12}H_{15}IN_2O_6$, 2641
$C_{12}H_{15}NO$, 2642
$C_{12}H_{15}NO_3$, 2643-2648
$C_{12}H_{15}NO_4$, 2649-2651
$C_{12}H_{15}NO_5$, 2652
$C_{12}H_{15}NO_6$, 2653
$C_{12}H_{15}N_2O_3PS$, 2654, 2655
$C_{12}H_{15}N_3O_2S$, 2656
$C_{12}H_{15}N_3O_3$, 2657
$C_{12}H_{15}N_3O_6$, 2658
$C_{12}H_{15}N_5O_5$, 2659, 2660
$C_{12}H_{15}O_3P$, 2661
$C_{12}H_{16}ClNOS$, 2662
$C_{12}H_{16}Cl_2N_2O$, 2663
$C_{12}H_{16}N_2$, 2664
$C_{12}H_{16}N_2O$, 2665
$C_{12}H_{16}N_2O_2$, 2666-2668
$C_{12}H_{16}N_2O_3$, 2669-2671
$C_{12}H_{16}N_3O_3PS$, 2672
$C_{12}H_{16}N_3O_5PS_2$, 2673
$C_{12}H_{16}N_4O_2$, 2674
$C_{12}H_{16}N_4O_4S_2$, 2675, 2676
$C_{12}H_{16}N_4O_7S$, 2677
$C_{12}H_{16}N_5O_5PS_2$, 2678
$C_{12}H_{16}O$, 2679, 2680
$C_{12}H_{16}O_2$, 2681, 2682
$C_{12}H_{16}O_3$, 2683
$C_{12}H_{16}O_7.H_2O$, 2684
$C_{12}H_{17}NO_2$, 2685-2690
$C_{12}H_{17}NO_3$, 2691-2693
$C_{12}H_{17}NO_4$, 2694
$C_{12}H_{17}N_2O_2$, 2695
$C_{12}H_{17}N_3O_4S$, 2696
$C_{12}H_{17}N_5O_3$, 2697
$C_{12}H_{17}O_4PS_2$, 2698
$C_{12}H_{18}$, 2699
$C_{12}H_{18}N_2O$, 2700
$C_{12}H_{18}N_2O_2$, 2701
$C_{12}H_{18}N_2O_2S$, 2702
$C_{12}H_{18}N_2O_3$, 2703, 2704
$C_{12}H_{18}N_2O_3S$, 2705
$C_{12}H_{18}N_2O_4S$, 2706
$C_{12}H_{18}N_2O_5$, 2707
$C_{12}H_{18}N_4O_6S$, 2708
$C_{12}H_{18}O$, 2709-2718
$C_{12}H_{18}O_2$, 2719
$C_{12}H_{18}O_4S_2$, 2720
$C_{12}H_{19}ClNO_3P$, 2721
$C_{12}H_{19}N_3O_8$, 2722
$C_{12}H_{19}N_6OP$, 2723
$C_{12}H_{20}$, 2724
$C_{12}H_{20}N_2O_3$, 2725
$C_{12}H_{20}N_4O_2$, 2726
$C_{12}H_{20}N_4O_6$, 2727
$C_{12}H_{20}O_2$, 2728
$C_{12}H_{20}O_4$, 2729
$C_{12}H_{20}O_6$, 2730
$C_{12}H_{21}N_2O_3PS$, 2731
$C_{12}H_{21}N_7O$, 2732
$C_{12}H_{22}N_2O_2$, 2733
$C_{12}H_{22}N_6$, 2734
$C_{12}H_{22}O_4$, 2735-2737

$C_{12}H_{22}O_6$, 2738-2740
$C_{12}H_{22}O_{11}$, 2741-2745
$C_{12}H_{23}NO_3$, 2746
$C_{12}H_{23}N_7$, 2747
$C_{12}H_{24}N_2O_2$, 2748
$C_{12}H_{24}N_6O_3PS$, 2749
$C_{12}H_{24}N_6O_4P$, 2750
$C_{12}H_{24}N_6$, 2751
$C_{12}H_{24}N_9P_3$, 2752
$C_{12}H_{24}O_2$, 2753, 2754
$C_{12}H_{24}O_3$, 2755
$C_{12}H_{24}O_4$, 2756
$C_{12}H_{26}$, 2757
$C_{12}H_{26}O$, 2758
$C_{12}H_{27}N$, 2759
$C_{12}H_{27}N.4H_2O$, 2760
$C_{12}H_{27}OP$, 2761
$C_{12}H_{27}O_2P$, 2762
$C_{12}H_{27}O_3P$, 2763, 2764
$C_{12}H_{27}O_4P$, 2765
$C_{12}H_{28}Ge$, 2766
$C_{12}Br_{10}O$, 2767
$C_{12}Cl_8O$, 2768
$C_{12}Cl_8O_2$, 2769
$C_{12}Cl_{10}$, 2770
$C_{13}H_6Cl_5NO_3$, 2771
$C_{13}H_6Cl_6O_7$, 2772
$C_{13}H_7Br_3N_3O_6$, 2773
$C_{13}H_7F_3N_2O_5$, 2774
$C_{13}H_8ClFO_2$, 2775
$C_{13}H_8ClNO$, 2776
$C_{13}H_8ClN_3O$, 2777
$C_{13}H_8Cl_2N_2O_4$, 2778
$C_{13}H_8N_2O_2S$, 2779
$C_{13}H_9Cl_3NO_4$, 2780
$C_{13}H_9F_3N_2O_2$, 2781
$C_{13}H_9N$, 2782, 2783
$C_{13}H_9NO$, 2784
$C_{13}H_9NS$, 2785, 2786
$C_{13}H_{10}$, 2787
$C_{13}H_{10}BrCl_2O_2PS$, 2788
$C_{13}H_{10}BrCl_2O_3P$, 2789
$C_{13}H_{10}ClNO_2$, 2790
$C_{13}H_{10}Cl_2O$, 2791
$C_{13}H_{10}Cl_2O_2$, 2792
$C_{13}H_{10}INO$, 2793
$C_{13}H_{10}N_2$, 2794-2798
$C_{13}H_{10}N_4O_3$, 2799
$C_{13}H_{10}O$, 2800
$C_{13}H_{10}O_3$, 2801
$C_{13}H_{10}O_4$, 2802
$C_{13}H_{10}O_6$, 2803
$C_{13}H_{11}ClF_3N_3O$, 2804
$C_{13}H_{11}ClN_4O$, 2805
$C_{13}H_{11}ClO$, 2806
$C_{13}H_{11}N$, 2807
$C_{13}H_{11}NO_2$, 2808
$C_{13}H_{11}NO_3$, 2809, 2810
$C_{13}H_{11}N_3O_2$, 2811
$C_{13}H_{11}N_3O_2S_2$, 2812
$C_{13}H_{11}N_3O_3S$, 2813
$C_{13}H_{12}$, 2814, 2815
$C_{13}H_{12}N_2O$, 2816
$C_{13}H_{12}N_2O_3$, 2817

$C_{13}H_{12}O$, 2818-2820
$C_{13}H_{13}Cl_2N_3O_3$, 2821
$C_{13}H_{13}NO_2$, 2822
$C_{13}H_{13}NO_5$, 2823
$C_{13}H_{13}N_3O_3S$, 2824
$C_{13}H_{13}N_3O_5S_2$, 2825
$C_{13}H_{13}O_4P$, 2826
$C_{13}H_{14}$, 2827
$C_{13}H_{14}F_3N_3O_4$, 2828
$C_{13}H_{14}N_2$, 2829
$C_{13}H_{14}N_2O_3$, 2830
$C_{13}H_{14}N_2O_6$, 2831
$C_{13}H_{14}N_4O_3S$, 2832
$C_{13}H_{14}N_4O_4S$, 2833
$C_{13}H_{14}O_6$, 2834, 2835
$C_{13}H_{15}NO_2$, 2836, 2837
$C_{13}H_{15}NO_2S$, 2838
$C_{13}H_{15}NO_3$, 2839
$C_{13}H_{15}NO_4$, 2840
$C_{13}H_{15}NO_5$, 2841, 2842
$C_{13}H_{15}N_3O_2$, 2843
$C_{13}H_{15}N_3O_3S$, 2844, 2845
$C_{13}H_{15}N_3O_4S$, 2846, 2847
$C_{13}H_{16}Cl_2O_3$, 2848-2850
$C_{13}H_{16}F_3N_3O_4$, 2851, 2852
$C_{13}H_{16}NO_4PS$, 2853
$C_{13}H_{16}N_2$, 2854
$C_{13}H_{16}N_2O_4$, 2855, 2856
$C_{13}H_{16}N_2O_6$, 2857
$C_{13}H_{16}N_4O_2S$, 2858, 2859
$C_{13}H_{16}N_4O_6.0.5H_2O$, 2860
$C_{13}H_{16}O_4$, 2861
$C_{13}H_{16}O_6$, 2862
$C_{13}H_{16}O_7.0.75H_2O$, 2863
$C_{13}H_{17}IN_2O_6$, 2864, 2865
$C_{13}H_{17}NO$, 2866, 2867
$C_{13}H_{17}NO_3$, 2868-2873
$C_{13}H_{17}NO_4$, 2874-2880
$C_{13}H_{17}NO_5$, 2881
$C_{13}H_{17}NO_6$, 2882
$C_{13}H_{17}N_3O$, 2883
$C_{13}H_{17}N_5O_5$, 2884, 2885
$C_{13}H_{17}N_5O_6$, 2886
$C_{13}H_{17}N_5O_8$, 2887
$C_{13}H_{18}ClNO$, 2888, 2889
$C_{13}H_{18}ClN_3O_4S_2$, 2890
$C_{13}H_{18}Cl_2N_2O_2$, 2891
$C_{13}H_{18}N_2O_2$, 2892
$C_{13}H_{18}N_2O_3$, 2893
$C_{13}H_{18}N_2O_3S$, 2894
$C_{13}H_{18}N_2O_4$, 2895
$C_{13}H_{18}N_2O_2S_2$, 2896, 2897
$C_{13}H_{18}O_2$, 2898-2900
$C_{13}H_{18}O_3$, 2901, 2902
$C_{13}H_{18}O_5S$, 2903
$C_{13}H_{18}O_7$, 2904
$C_{13}H_{19}NO_2$, 2905, 2906
$C_{13}H_{19}NO_4$, 2907
$C_{13}H_{19}N_3O_4$, 2908
$C_{13}H_{19}N_3O_6S$, 2909
$C_{13}H_{20}N_2O_2$, 2910-2912
$C_{13}H_{20}N_2O_3$, 2913, 2914
$C_{13}H_{20}O$, 2915-2919
$C_{13}H_{21}NO_3$, 2920

$C_{13}H_{21}O_3PS$, 2921
$C_{13}H_{21}O_4PS$, 2922
$C_{13}H_{22}NO_3PS$, 2923
$C_{13}H_{22}N_2O$, 2924, 2925
$C_{13}H_{22}N_2O_3$, 2926
$C_{13}H_{22}O_3$, 2927
$C_{13}H_{24}N_3O_3PS$, 2928
$C_{13}H_{24}N_4O_3S$, 2929
$C_{13}H_{24}N_6$, 2930
$C_{13}H_{24}O_4$, 2931, 2932
$C_{13}H_{25}NO_3$, 2933
$C_{13}H_{26}N_2O_2$, 2934
$C_{13}H_{26}O_2$, 2935, 2936
$C_{13}H_{26}O_3$, 2937, 2938
$C_{13}H_{26}O_4$, 2939
$C_{14}H_7N_2O_2S_2$, 2940
$C_{14}H_7ClO_5S$, 2941-2943
$C_{14}H_8Cl_4$, 2944, 2945
$C_{14}H_8O_2$, 2946
$C_{14}H_8O_4$, 2947, 2948
$C_{14}H_8O_5$, 2949
$C_{14}H_8O_6$, 2950
$C_{14}H_8O_8S_2$, 2951-2953
$C_{14}H_9ClF_2N_2O_2$, 2954
$C_{14}H_9Cl_2NO_5$, 2955
$C_{14}H_9Cl_5$, 2956, 2957
$C_{14}H_9Cl_5O$, 2958
$C_{14}H_9F$, 2959
$C_{14}H_9NO_2$, 2960-2962
$C_{14}H_9NO_2S$, 2963
$C_{14}H_9NO_3$, 2964
$C_{14}H_{10}$, 2965, 2966
$C_{14}H_{10}Cl_2O_3$, 2967
$C_{14}H_{10}Cl_4$, 2968, 2969
$C_{14}H_{10}F_3NO_2$, 2970
$C_{14}H_{10}N_2O_2$, 2971
$C_{14}H_{10}N_2O_6$, 2972
$C_{14}H_{10}O$, 2973, 2974
$C_{14}H_{10}O_3$, 2975
$C_{14}H_{10}O_4$, 2976, 2977
$C_{14}H_{10}O_5$, 2978
$C_{14}H_{10}O_9$, 2979
$C_{14}H_{11}ClNO_2$, 2980
$C_{14}H_{11}ClN_2O_4S$, 2981
$C_{14}H_{11}Cl_2O_2$, 2982
$C_{14}H_{11}FN_2O_5$, 2983
$C_{14}H_{11}N$, 2984, 2985
$C_{14}H_{11}NO_2$, 2986
$C_{14}H_{11}N_3O_2$, 2987
$C_{14}H_{12}$, 2988-2990
$C_{14}H_{12}F_3NO_4S_2$, 2991
$C_{14}H_{12}N_2S$, 2992
$C_{14}H_{12}N_4O_2$, 2993
$C_{14}H_{12}O_2$, 2994-2997
$C_{14}H_{12}O_3$, 2998, 2999
$C_{14}H_{12}O_5$, 3000
$C_{14}H_{13}ClN_4O$, 3001
$C_{14}H_{13}NO_6$, 3002
$C_{14}H_{13}N_3$, 3003
$C_{14}H_{13}N_3O_2$, 3004
$C_{14}H_{14}$, 3005, 3006
$C_{14}H_{14}NO_4PS$, 3007
$C_{14}H_{14}N_4O$, 3008
$C_{14}H_{14}N_4O_2$, 3009, 3010

$C_{14}H_{14}N_4O_4$, 3011
$C_{14}H_{14}N_4S$, 3012
$C_{14}H_{14}O$, 3013, 3014
$C_{14}H_{14}O_2$, 3015
$C_{14}H_{14}O_3$, 3016, 3017
$C_{14}H_{14}O_3S$, 3018
$C_{14}H_{14}O_4$, 3019
$C_{14}H_{15}N$, 3020
$C_{14}H_{15}NO_5$, 3021
$C_{14}H_{15}N_3$, 3022, 3023
$C_{14}H_{15}N_3O_3S$, 3024
$C_{14}H_{15}N_5O_5$, 3025
$C_{14}H_{16}ClN_3O_2$, 3026
$C_{14}H_{16}ClO_5PS$, 3027
$C_{14}H_{16}Cl_2O_3$, 3028
$C_{14}H_{16}FN_3O_3$, 3029
$C_{14}H_{16}F_3N_3O_4$, 3030
$C_{14}H_{16}N_2$, 3031
$C_{14}H_{16}N_2O_2$, 3032
$C_{14}H_{16}N_2O_4$, 3033
$C_{14}H_{16}N_4$, 3034
$C_{14}H_{16}N_4O_2S$, 3035
$C_{14}H_{16}N_4O_3S$, 3036-3038
$C_{14}H_{16}N_4O_4S$, 3039
$C_{14}H_{16}N_4O_5S$, 3040
$C_{14}H_{16}O_6$, 3041, 3042
$C_{14}H_{17}ClNO_4PS_2$, 3043
$C_{14}H_{17}NO$, 3044, 3045
$C_{14}H_{17}NO_2S$, 3046
$C_{14}H_{17}NO_3$, 3047
$C_{14}H_{17}NO_4$, 3048
$C_{14}H_{17}NO_5$, 3049
$C_{14}H_{17}N_5O_3$, 3050
$C_{14}H_{18}ClN_3S$, 3051
$C_{14}H_{18}Cl_2O_3$, 3052
$C_{14}H_{18}N_2O$, 3053
$C_{14}H_{18}N_2O_3$, 3054
$C_{14}H_{18}N_4O_2S$, 3055
$C_{14}H_{18}N_4O_3$, 3056, 3057
$C_{14}H_{18}N_4O_6 \cdot 0.5H_2O$, 3058
$C_{14}H_{18}N_4O_7 \cdot 0.5H_2O$, 3059
$C_{14}H_{18}N_4O_7 \cdot 0.9H_2O$, 3060
$C_{14}H_{18}N_6O_4$, 3061
$C_{14}H_{18}O_4$, 3062, 3063
$C_{14}H_{18}O_6$, 3064-3066
$C_{14}H_{19}Cl_2NO_2$, 3067
$C_{14}H_{19}IN_2O_6$, 3068, 3069
$C_{14}H_{19}NO$, 3070
$C_{14}H_{19}NO_3$, 3071, 3072
$C_{14}H_{19}NO_4$, 3073
$C_{14}H_{19}N_3S$, 3074, 3075
$C_{14}H_{19}N_5O_4$, 3076
$C_{14}H_{19}N_5O_5$, 3077
$C_{14}H_{19}O_6P$, 3078
$C_{14}H_{20}ClNO_2$, 3079
$C_{14}H_{20}N_2O$, 3080
$C_{14}H_{20}N_2O_3S$, 3081
$C_{14}H_{20}N_3O_5PS$, 3082
$C_{14}H_{20}N_4O_2$, 3083-3085
$C_{14}H_{20}N_4O_4$, 3086
$C_{14}H_{20}O_3$, 3087
$C_{14}H_{21}NO_2$, 3088-3092
$C_{14}H_{21}NO_3$, 3093
$C_{14}H_{21}NO_4P$, 3094

$C_{14}H_{21}N_3O_3$, 3095
$C_{14}H_{21}N_3O_3S$, 3096
$C_{14}H_{22}$, 3097
$C_{14}H_{22}N_2O$, 3098
$C_{14}H_{22}N_2O_2$, 3099, 3100
$C_{14}H_{22}N_2O_3$, 3101
$C_{14}H_{22}N_2O_4$, 3102
$C_{14}H_{22}N_2O_5$, 3103
$C_{14}H_{22}O$, 3104-3106
$C_{14}H_{23}O_3P$, 3107
$C_{14}H_{24}NO_4PS_3$, 3108
$C_{14}H_{24}N_2O_3$, 3109, 3110
$C_{14}H_{24}O_2$, 3111
$C_{14}H_{26}O_4$, 3112
$C_{14}H_{27}NO_2$, 3113
$C_{14}H_{28}NO_3PS_2$, 3114
$C_{14}H_{28}N_2O_2$, 3115
$C_{14}H_{28}O_2$, 3116
$C_{14}H_{28}O_4$, 3117
$C_{14}H_{29}NO_2$, 3118-3124
$C_{14}H_{30}$, 3125
$C_{14}H_{30}O$, 3126, 3127
$C_{14}H_{31}O_2P$, 3128
$C_{14}H_{31}O_3P$, 3129, 3130
$C_{14}H_{31}O_4P$, 3131, 3132
$C_{14}H_{31}O_5P$, 3133
$C_{15}H_{10}$, 3134
$C_{15}H_{10}Cl_2N_2O_2$, 3135
$C_{15}H_{10}O_2$, 3136
$C_{15}H_{10}O_4S$, 3137
$C_{15}H_{10}O_5S$, 3138
$C_{15}H_{10}O_6$, 3139
$C_{15}H_{10}O_7$, 3140
$C_{15}H_{11}ClN_2O_2$, 3141
$C_{15}H_{11}ClO_3$, 3142
$C_{15}H_{11}NO_2$, 3143, 3144
$C_{15}H_{11}N_3O_3$, 3145
$C_{15}H_{12}$, 3146-3148
$C_{15}H_{12}Cl_2O_3$, 3149, 3150
$C_{15}H_{12}I_3NO_4$, 3151
$C_{15}H_{12}N_2O$, 3152
$C_{15}H_{12}N_2O_2$, 3153, 3154
$C_{15}H_{12}N_2O_3$, 3155, 3156
$C_{15}H_{12}O_4$, 3157, 3158
$C_{15}H_{13}ClO_2$, 3159
$C_{15}H_{13}FO_2$, 3160
$C_{15}H_{13}F_3N_4O$, 3161
$C_{15}H_{13}NO$, 3162
$C_{15}H_{13}NO_2$, 3163
$C_{15}H_{13}NO_2S$, 3164
$C_{15}H_{13}NO_3$, 3165
$C_{15}H_{13}NO_4$, 3166
$C_{15}H_{13}N_3O_4S$, 3167
$C_{15}H_{14}ClN_3O_4S$, 3168
$C_{15}H_{14}Cl_2N_4O_3$, 3169
$C_{15}H_{14}F_3N_3O_4S_2$, 3170
$C_{15}H_{14}NO_2PS$, 3171
$C_{15}H_{14}N_2O_2$, 3172
$C_{15}H_{14}N_2O_3$, 3173
$C_{15}H_{14}N_4O$, 3174
$C_{15}H_{14}O_3$, 3175
$C_{15}H_{15}ClN_2O_2$, 3176
$C_{15}H_{15}ClN_2O_4S$, 3177
$C_{15}H_{15}ClO$, 3178

$C_{15}H_{15}NO_2$, 3179
$C_{15}H_{15}N_3O$, 3180
$C_{15}H_{15}N_3O_2$, 3181, 3182
$C_{15}H_{15}N_3S$, 3183
$C_{15}H_{16}N_2O_2$, 3184
$C_{15}H_{16}N_4O$, 3185-3187
$C_{15}H_{16}N_4O_2$, 3188, 3189
$C_{15}H_{16}N_4O_5S$, 3190
$C_{15}H_{16}O_2$, 3191
$C_{15}H_{16}O_3$, 3192
$C_{15}H_{16}O_9.2H_2O$, 3193
$C_{15}H_{17}FN_4O_2$, 3194
$C_{15}H_{17}NO_4$, 3195
$C_{15}H_{17}NO_5$, 3196
$C_{15}H_{17}NO_7$, 3197
$C_{15}H_{17}N_3O_3S$, 3198
$C_{15}H_{18}ClN_2O_3$, 3199
$C_{15}H_{18}I_3NO_5$, 3200
$C_{15}H_{18}N_2O_3$, 3201
$C_{15}H_{18}N_4O_3S$, 3202, 3203
$C_{15}H_{18}N_4O_5$, 3204
$C_{15}H_{18}O_3$, 3205
$C_{15}H_{18}O_4$, 3206
$C_{15}H_{19}ClO_2$, 3207
$C_{15}H_{19}NO$, 3208, 3209
$C_{15}H_{19}NO_2$, 3210
$C_{15}H_{19}NO_3$, 3211
$C_{15}H_{19}NO_5$, 3212
$C_{15}H_{20}N_2O_4$, 3213
$C_{15}H_{20}N_2O_4S$, 3214
$C_{15}H_{20}N_4O_2S$, 3215
$C_{15}H_{20}N_4O_5$, 3216, 3217
$C_{15}H_{20}N_4O_6.0.25H_2O$, 3218
$C_{15}H_{20}N_4O_6$, 3219, 3220
$C_{15}H_{20}N_4O_6.0.3H_2O$, 3221
$C_{15}H_{21}NO$, 3222
$C_{15}H_{21}NO_2$, 3223
$C_{15}H_{21}NO_2S_2$, 3224
$C_{15}H_{21}NO_3$, 3225-3227
$C_{15}H_{21}NO_3S$, 3228
$C_{15}H_{21}NO_4$, 3229-3231
$C_{15}H_{21}NO_5$, 3232, 3233
$C_{15}H_{21}N_3O_3$, 3234
$C_{15}H_{21}N_5O_5$, 3235-3238
$C_{15}H_{21}N_5O_6$, 3239
$C_{15}H_{22}ClNO_2$, 3240, 3241
$C_{15}H_{22}N_2O$, 3242
$C_{15}H_{22}O_3$, 3243
$C_{15}H_{23}NO_2$, 3244, 3245
$C_{15}H_{23}NO_3$, 3246, 3247
$C_{15}H_{23}NO_4$, 3248
$C_{15}H_{23}N_3O_4$, 3249
$C_{15}H_{23}N_3O_3S$, 3250, 3251
$C_{15}H_{23}N_3O_3S.2H_2O$, 3252
$C_{15}H_{24}NO_4PS$, 3253
$C_{15}H_{24}N_2O_2$, 3254-3257
$C_{15}H_{24}N_2O_3$, 3258
$C_{15}H_{24}O$, 3259, 3260
$C_{15}H_{26}N_2$, 3261
$C_{15}H_{26}N_2O_3$, 3262, 3263
$C_{15}H_{26}O_6$, 3264
$C_{15}H_{28}O_4$, 3265
$C_{15}H_{30}$, 3266
$C_{15}H_{30}O_2$, 3267

$C_{15}H_{30}O_3$, 3268
$C_{15}H_{32}O$, 3269
$C_{16}H_{10}$, 3270, 3271
$C_{16}H_{10}N_2O_8S_2$, 3272
$C_{16}H_{11}NO_2$, 3273
$C_{16}H_{12}F_3NO$, 3274
$C_{16}H_{12}N_2O_3$, 3275
$C_{16}H_{12}N_2O_4S$, 3276
$C_{16}H_{12}O_6$, 3277, 3278
$C_{16}H_{13}ClN_2O$, 3279
$C_{16}H_{13}IN_2O_3$, 3280
$C_{16}H_{13}NO_3$, 3281
$C_{16}H_{13}N_3$, 3282
$C_{16}H_{13}N_3O_3$, 3283
$C_{16}H_{14}$, 3284
$C_{16}H_{14}ClN_3O$, 3285
$C_{16}H_{14}Cl_2N_2O_2$, 3286
$C_{16}H_{14}Cl_2O_3$, 3287
$C_{16}H_{14}Cl_2O_4$, 3288
$C_{16}H_{14}FNO$, 3289
$C_{16}H_{14}N_2O$, 3290
$C_{16}H_{14}N_2O_2$, 3291
$C_{16}H_{14}N_2O_3$, 3292
$C_{16}H_{14}N_2O_4$, 3293
$C_{16}H_{14}O_3$, 3294, 3295
$C_{16}H_{15}ClN_2$, 3296
$C_{16}H_{15}Cl_3OS_2$, 3297
$C_{16}H_{15}Cl_3O_2$, 3298
$C_{16}H_{15}Cl_3O_2S_2$, 3299
$C_{16}H_{15}Cl_3O_4S_2$, 3300
$C_{16}H_{15}Cl_3S_2$, 3301
$C_{16}H_{15}FN_2O_5$, 3302
$C_{16}H_{15}NO$, 3303
$C_{16}H_{15}NO_2$, 3304
$C_{16}H_{15}NO_3$, 3305
$C_{16}H_{15}NO_4$, 3306
$C_{16}H_{15}N_5$, 3307
$C_{16}H_{15}N_5O_4S$, 3308
$C_{16}H_{16}$, 3309
$C_{16}H_{16}ClN_3O_3S$, 3310
$C_{16}H_{16}N_2$, 3311
$C_{16}H_{16}N_2O_4$, 3312, 3313
$C_{16}H_{16}N_4$, 3314
$C_{16}H_{16}N_4O$, 3315, 3316
$C_{16}H_{16}N_6O_4S$, 3317
$C_{16}H_{16}O_2$, 3318
$C_{16}H_{16}O_3$, 3319
$C_{16}H_{17}ClN_2S$, 3320
$C_{16}H_{17}ClN_4O_3$, 3321
$C_{16}H_{17}ClN_4O_4$, 3322
$C_{16}H_{17}NO$, 3323
$C_{16}H_{17}NO_4$, 3324
$C_{16}H_{17}N_3O_4S$, 3325
$C_{16}H_{17}N_3O_4S.H_2O$, 3326
$C_{16}H_{17}N_5O_5$, 3327
$C_{16}H_{17}N_2O_6$, 3328
$C_{16}H_{18}ClNO_4S$, 3329
$C_{16}H_{18}NO_3P$, 3330
$C_{16}H_{18}N_2O_3$, 3331
$C_{16}H_{18}N_4O$, 3332
$C_{16}H_{18}N_4O_2$, 3333, 3334
$C_{16}H_{18}N_4O_3$, 3335
$C_{16}H_{18}N_4O_4$, 3336
$C_{16}H_{18}O_3$, 3337

$C_{16}H_{19}ClN_2$, 3338
$C_{16}H_{19}NO_7$, 3339
$C_{16}H_{19}N_3O_2$, 3340
$C_{16}H_{19}N_3O_4S$, 3341, 3342
$C_{16}H_{19}N_3O_4S.3H_2O$, 3343
$C_{16}H_{19}N_3O_5S.3H_2O$, 3344
$C_{16}H_{19}N_5O$, 3345
$C_{16}H_{19}N_5O_2$, 3346
$C_{16}H_{19}O_4P$, 3347
$C_{16}H_{20}I_3N_3O_7$, 3348, 3349
$C_{16}H_{20}I_3N_3O_8$, 3350, 3351
$C_{16}H_{20}N_4O_2$, 3352
$C_{16}H_{20}N_4O_3S$, 3353
$C_{16}H_{20}N_8O_2S$, 3354
$C_{16}H_{20}O_6P_2S_3$, 3355
$C_{16}H_{21}ClN_3S$, 3356
$C_{16}H_{21}NO$, 3357, 3358
$C_{16}H_{21}NO_2S$, 3359
$C_{16}H_{21}NO_3$, 3360, 3361
$C_{16}H_{21}NO_5$, 3362
$C_{16}H_{21}N_3$, 3363
$C_{16}H_{22}Cl_2O_3$, 3364
$C_{16}H_{22}N_4O$, 3365
$C_{16}H_{22}N_2O_2S$, 3366
$C_{16}H_{22}N_4O_6.0.5H_2O$, 3367, 3368
$C_{16}H_{22}N_4O_6$, 3369
$C_{16}H_{22}O_4$, 3370-3372
$C_{16}H_{22}O_6$, 3373
$C_{16}H_{22}O_8.2H_2O$, 3374
$C_{16}H_{22}O_{11}$, 3375, 3376
$C_{16}H_{23}FN_2O_6$, 3377
$C_{16}H_{23}NO$, 3378
$C_{16}H_{23}NO_2$, 3379
$C_{16}H_{23}NO_3$, 3380
$C_{16}H_{23}NO_6$, 3381
$C_{16}H_{23}N_5O_5$, 3382, 3383
$C_{16}H_{24}N_2O_2$, 3384, 3385
$C_{16}H_{24}N_4O_2$, 3386
$C_{16}H_{24}N_4O_6$, 3387
$C_{16}H_{24}N_6$, 3388
$C_{16}H_{24}O_3$, 3389
$C_{16}H_{24}O_4$, 3390
$C_{16}H_{25}NOS$, 3391
$C_{16}H_{25}NO_2$, 3392, 3393
$C_{16}H_{25}NO_3$, 3394
$C_{16}H_{26}N_2O_2$, 3395
$C_{16}H_{26}O_2$, 3396
$C_{16}H_{26}O_6$, 3397
$C_{16}H_{28}N_3O_2$, 3398
$C_{16}H_{32}O_2$, 3399
$C_{16}H_{34}$, 3400-3402
$C_{16}H_{34}O$, 3403
$C_{16}H_{35}O_3P$, 3404
$C_{16}H_{35}O_4P$, 3405
$C_{17}H_{11}NO_3$, 3406
$C_{17}H_{12}$, 3407, 3408
$C_{17}H_{12}ClFN_3O_2$, 3409
$C_{17}H_{12}ClNO_2S$, 3410
$C_{17}H_{12}Cl_3N_2O$, 3411
$C_{17}H_{12}Cl_{10}O_3$, 3412
$C_{17}H_{12}I_2O_3$, 3413
$C_{17}H_{12}O_6$, 3414
$C_{17}H_{12}O_7$, 3415
$C_{17}H_{13}ClO_3$, 3416

$C_{17}H_{13}Cl_2N_3O_2$, 3417
$C_{17}H_{14}N_2O$, 3418
$C_{17}H_{14}O_6$, 3419
$C_{17}H_{14}O_7$, 3420
$C_{17}H_{15}NO_3$, 3421
$C_{17}H_{15}NO_5$, 3422
$C_{17}H_{16}Br_2O_3$, 3423
$C_{17}H_{16}ClFN_2O_2$, 3424
$C_{17}H_{16}Cl_2O_3$, 3425
$C_{17}H_{16}N_2O_2S$, 3426
$C_{17}H_{16}N_2O_3$, 3427
$C_{17}H_{16}N_2O_3S$, 3428
$C_{17}H_{16}N_2O_4$, 3429
$C_{17}H_{16}N_2O_4S$, 3430
$C_{17}H_{16}N_2O_5$, 3431
$C_{17}H_{17}ClO_6$, 3432
$C_{17}H_{17}NO_2$, 3433
$C_{17}H_{17}NO_5$, 3434
$C_{17}H_{17}N_5O_5$, 3435
$C_{17}H_{18}ClNO_6$, 3436
$C_{17}H_{18}ClN_5O_6$, 3437
$C_{17}H_{18}Cl_2N_4O_4$, 3438
$C_{17}H_{18}N_2O_6$, 3439
$C_{17}H_{18}N_4O_3S$, 3440
$C_{17}H_{19}ClN_2S$, 3441-3444
$C_{17}H_{19}ClN_4O_4$, 3445
$C_{17}H_{19}ClO_6$, 3446
$C_{17}H_{19}NO_3$, 3447-3449
$C_{17}H_{19}NO_3.H_2O$, 3450
$C_{17}H_{19}NO_4$, 3451
$C_{17}H_{19}N_3$, 3452
$C_{17}H_{19}N_5O_6$, 3453
$C_{17}H_{20}ClN_3O_2$, 3454
$C_{17}H_{20}N_2O$, 3455
$C_{17}H_{20}N_2S$, 3456, 3457
$C_{17}H_{20}N_4O_4$, 3458
$C_{17}H_{20}N_4O_5$, 3459
$C_{17}H_{20}N_4O_5S$, 3460
$C_{17}H_{20}N_4O_6$, 3461
$C_{17}H_{20}O_6$, 3462
$C_{17}H_{21}NO_2$, 3463
$C_{17}H_{21}NO_4$, 3464, 3465
$C_{17}H_{21}N_3O_2$, 3466
$C_{17}H_{21}N_5O_2$, 3467
$C_{17}H_{21}N_5O_{10}$, 3468
$C_{17}H_{22}I_2N_2O_8$, 3469-3476
$C_{17}H_{22}I_3N_3O_9$, 3477, 3478
$C_{17}H_{22}N_4O_3S$, 3479
$C_{17}H_{22}N_4O_7.0.75H_2O$, 3480
$C_{17}H_{23}NO$, 3481, 3482
$C_{17}H_{23}NO_3$, 3483, 3484
$C_{17}H_{23}NO_5$, 3485, 3486
$C_{17}H_{23}N_3O$, 3487
$C_{17}H_{23}N_3O_2$, 3488
$C_{17}H_{24}N_4O_5$, 3489, 3490
$C_{17}H_{24}N_4O_6$, 3491
$C_{17}H_{25}NO$, 3492
$C_{17}H_{25}NO_3$, 3493, 3494
$C_{17}H_{25}NO_4$, 3495
$C_{17}H_{25}N_5O_6$, 3496
$C_{17}H_{26}ClNO_2$, 3497
$C_{17}H_{26}O_3$, 3498
$C_{17}H_{27}NO_2$, 3499
$C_{17}H_{27}NO_3$, 3500

$C_{17}H_{28}N_2O_2$, 3501, 3502
$C_{17}H_{28}O_2$, 3503
$C_{17}H_{34}O_2$, 3504
$C_{17}H_{36}O$, 3505
$C_{18}H_{10}Cl_4$, 3506
$C_{18}H_{10}I_6N_2O_7$, 3507
$C_{18}H_{10}N_2O_2S$, 3508
$C_{18}H_{11}Cl_3$, 3509
$C_{18}H_{11}NO_3$, 3510, 3511
$C_{18}H_{12}$, 3512-3515
$C_{18}H_{12}N_2$, 3516
$C_{18}H_{12}N_4O$, 3517
$C_{18}H_{13}N$, 3518
$C_{18}H_{13}NO_3$, 3519, 3520
$C_{18}H_{14}$, 3521-3523
$C_{18}H_{14}Cl_4N_2O$, 3524
$C_{18}H_{14}N_4O$, 3525
$C_{18}H_{14}N_4O_2$, 3526
$C_{18}H_{15}Cl_4N_3O_4$, 3527
$C_{18}H_{15}N_3O_5$, 3528
$C_{18}H_{15}O_4P$, 3529
$C_{18}H_{16}Cl_3N_3O_4$, 3530
$C_{18}H_{16}N_2O_3$, 3531
$C_{18}H_{16}N_4O_3S$, 3532
$C_{18}H_{17}ClN_4O_6.0.5H_2O$, 3533
$C_{18}H_{17}Cl_3NO_3$, 3534
$C_{18}H_{17}N_5O_8$, 3535
$C_{18}H_{18}ClNO_4$, 3536
$C_{18}H_{18}ClNO_5$, 3537
$C_{18}H_{18}ClNS$, 3538
$C_{18}H_{18}N_2O_4$, 3539
$C_{18}H_{18}N_4O_6$, 3540
$C_{18}H_{18}N_4O_6.0.75H_2O$, 3541
$C_{18}H_{18}N_8O_6$, 3542
$C_{18}H_{18}O_2$, 3543, 3544
$C_{18}H_{18}O_3$, 3545
$C_{18}H_{19}Cl_2NO_4$, 3546
$C_{18}H_{19}F_3N_2S$, 3547, 3548
$C_{18}H_{19}NO$, 3549
$C_{18}H_{19}N_2O_4$, 3550
$C_{18}H_{19}N_3O_6S$, 3551
$C_{18}H_{19}N_5O_3$, 3552
$C_{18}H_{19}N_5O_6$, 3553
$C_{18}H_{19}N_5O_6.0.3H_2O$, 3554
$C_{18}H_{20}$, 3555
$C_{18}H_{20}Cl_2O_2$, 3556
$C_{18}H_{20}N_4O_5S$, 3557
$C_{18}H_{20}O_2$, 3558, 3559
$C_{18}H_{21}ClN_2$, 3560
$C_{18}H_{21}ClN_5S$, 3561
$C_{18}H_{21}ClO$, 3562
$C_{18}H_{21}NO_3$, 3563, 3564
$C_{18}H_{21}NO_3.H_2O$, 3565
$C_{18}H_{21}NO_4$, 3566
$C_{18}H_{22}N_2$, 3567
$C_{18}H_{22}N_4O_5$, 3568
$C_{18}H_{22}O_2$, 3569, 3570
$C_{18}H_{23}N_3O_3S$, 3571
$C_{18}H_{23}N_3O_4S$, 3572
$C_{18}H_{24}I_3N_3O_9$, 3573
$C_{18}H_{24}N_4O_2$, 3574
$C_{18}H_{24}N_4O_2S$, 3575, 3576
$C_{18}H_{24}N_4O_3S$, 3577
$C_{18}H_{24}O_2$, 3578, 3579

$C_{18}H_{24}O_3$, 3580
$C_{18}H_{24}O_6$, 3581
$C_{18}H_{25}I_3N_3O_9$, 3582
$C_{18}H_{25}NO$, 3583
$C_{18}H_{25}N_3O_2$, 3584
$C_{18}H_{26}NO_4$, 3585
$C_{18}H_{26}N_2O_4$, 3586
$C_{18}H_{26}N_4O_6 \cdot 0.5H_2O$, 3587
$C_{18}H_{26}N_4O_6$, 3588
$C_{18}H_{26}O$, 3589
$C_{18}H_{26}O_2$, 3590
$C_{18}H_{26}O_4$, 3591
$C_{18}H_{26}O_6$, 3592
$C_{18}H_{27}NO$, 3593
$C_{18}H_{27}NO_3$, 3594
$C_{18}H_{27}N_5O_5$, 3595
$C_{18}H_{28}N_2O$, 3596
$C_{18}H_{28}N_4O_2$, 3597
$C_{18}H_{28}O_3$, 3598
$C_{18}H_{29}NO_2$, 3599
$C_{18}H_{29}NO_3$, 3600
$C_{18}H_{30}N_2O_2$, 3601
$C_{18}H_{30}O_3$, 3602
$C_{18}H_{30}O_{15} \cdot 4H_2O$, 3603
$C_{18}H_{31}O_4P$, 3604
$C_{18}H_{32}O_7$, 3605
$C_{18}H_{32}O_{16}$, 3606
$C_{18}H_{32}O_{16} \cdot 5H_2O$, 3607
$C_{18}H_{34}OSn$, 3608
$C_{18}H_{34}O_4$, 3609
$C_{18}H_{36}O_2$, 3610
$C_{18}H_{38}$, 3611
$C_{18}H_{38}O$, 3612
$C_{18}H_{39}N \cdot 2H_2O$, 3613
$C_{18}H_{39}O_3P$, 3614
$C_{18}H_{39}O_4P$, 3615
$C_{18}H_{39}O_7P$, 3616
$C_{19}H_{12}O_6$, 3617
$C_{19}H_{13}Cl$, 3618-3620
$C_{19}H_{14}$, 3621-3625
$C_{19}H_{14}O_3$, 3626
$C_{19}H_{14}O_5S$, 3627
$C_{19}H_{16}ClNO_4$, 3628
$C_{19}H_{16}O$, 3629
$C_{19}H_{16}O_4$, 3630
$C_{19}H_{17}ClN_2O$, 3631
$C_{19}H_{17}ClN_2O_4$, 3632
$C_{19}H_{17}N_3O_4S_2$, 3633, 3634
$C_{19}H_{17}N_3O_5$, 3635
$C_{19}H_{18}$, 3636
$C_{19}H_{18}Cl_2N_2O_2$, 3637
$C_{19}H_{18}N_2O_3$, 3638, 3639
$C_{19}H_{19}ClFNO_3$, 3640
$C_{19}H_{19}N_7O_6$, 3641
$C_{19}H_{20}ClNO_9$, 3642
$C_{19}H_{20}N_2O$, 3643
$C_{19}H_{20}N_2O_2$, 3644, 3645
$C_{19}H_{20}N_2O_3$, 3646
$C_{19}H_{20}N_4O_6 \cdot 0.1H_2O$, 3647
$C_{19}H_{20}N_4O_6 \cdot 0.5H_2O$, 3648
$C_{19}H_{20}N_4O_6$, 3649
$C_{19}H_{20}N_4O_7 \cdot 0.25H_2O$, 3650
$C_{19}H_{20}N_4O_7 \cdot 0.05H_2O$, 3651
$C_{19}H_{20}N_4O_7 \cdot 0.5H_2O$, 3652

$C_{19}H_{20}N_4O_7$, 3653
$C_{19}H_{20}O_4$, 3654
$C_{19}H_{21}ClO_4$, 3655
$C_{19}H_{21}F_3N_2S$, 3656
$C_{19}H_{21}NO$, 3657
$C_{19}H_{21}NO_3$, 3658
$C_{19}H_{21}N_5O_2$, 3659, 3660
$C_{19}H_{21}N_5O_5$, 3661
$C_{19}H_{22}Cl_2O_2$, 3662
$C_{19}H_{22}N_2O$, 3663, 3664
$C_{19}H_{22}N_2OS$, 3665
$C_{19}H_{22}N_2O_5$, 3666
$C_{19}H_{22}N_2S$, 3667
$C_{19}H_{23}ClO_2$, 3668
$C_{19}H_{23}NO_3$, 3669
$C_{19}H_{23}NO_4$, 3670
$C_{19}H_{23}NO_5$, 3671
$C_{19}H_{23}N_3$, 3672
$C_{19}H_{24}N_2$, 3673
$C_{19}H_{24}N_2O$, 3674
$C_{19}H_{24}N_2OS$, 3675
$C_{19}H_{24}N_2O_2S$, 3676
$C_{19}H_{24}N_4O_7$, 3677
$C_{19}H_{24}O$, 3678
$C_{19}H_{24}O_2$, 3679
$C_{19}H_{24}O_3$, 3680
$C_{19}H_{25}NO$, 3681
$C_{19}H_{26}IN_3O_{10}$, 3682
$C_{19}H_{26}N_6O_4S$, 3683
$C_{19}H_{26}O$, 3684
$C_{19}H_{26}O_2$, 3685
$C_{19}H_{27}N_3O$, 3686
$C_{19}H_{27}N_3O_2$, 3687
$C_{19}H_{28}Cl_2O_3$, 3688
$C_{19}H_{28}N_4O_6$, 3689
$C_{19}H_{28}O$, 3690
$C_{19}H_{28}O_2$, 3691-3693
$C_{19}H_{28}O_2 \cdot H_2O$, 3694
$C_{19}H_{28}O_3$, 3695
$C_{19}H_{29}NO$, 3696
$C_{19}H_{29}N_5O_6$, 3697
$C_{19}H_{30}O$, 3698
$C_{19}H_{30}OS$, 3699
$C_{19}H_{30}O_2$, 3700-3703
$C_{19}H_{30}O_3$, 3704-3706
$C_{19}H_{31}NO_2$, 3707
$C_{19}H_{31}NO_3$, 3708
$C_{19}H_{32}N_2O_2$, 3709
$C_{19}H_{32}O_3$, 3710
$C_{19}H_{34}O_3$, 3711
$C_{20}H_{12}$, 3712-3717
$C_{20}H_{13}N$, 3718-3720
$C_{20}H_{14}$, 3721, 3722
$C_{20}H_{14}I_6N_2O_6$, 3723
$C_{20}H_{14}N_2O_2$, 3724
$C_{20}H_{14}O_2$, 3725
$C_{20}H_{14}O_4$, 3726, 3727
$C_{20}H_{16}$, 3728-3730
$C_{20}H_{16}O_4$, 3731
$C_{20}H_{18}O_2Sn$, 3732
$C_{20}H_{19}NO_3$, 3733
$C_{20}H_{19}NO_5 \cdot 6H_2O$, 3734
$C_{20}H_{19}N_3$, 3735
$C_{20}H_{19}N_3O_5$, 3736

$C_{20}H_{20}N_2O_6$,   3737
$C_{20}H_{20}N_6O_6S_2$,   3738
$C_{20}H_{21}NO_4$,   3739
$C_{20}H_{21}NO_5$,   3740
$C_{20}H_{22}ClN$,   3741
$C_{20}H_{22}N_2O_2$,   3742
$C_{20}H_{23}N$,   3743
$C_{20}H_{23}NO_2$,   3744
$C_{20}H_{24}ClN_3S$,   3745
$C_{20}H_{24}N_2$,   3746
$C_{20}H_{24}N_2O_2$,   3747, 3748
$C_{20}H_{24}N_2O_2 \cdot 3H_2O$,   3749
$C_{20}H_{24}N_2O_4$,   3750
$C_{20}H_{24}N_2O_5$,   3751
$C_{20}H_{24}O_2$,   3752
$C_{20}H_{24}O_3$,   3753
$C_{20}H_{24}O_4$,   3754
$C_{20}H_{24}O_6$,   3755
$C_{20}H_{25}ClO_2$,   3756
$C_{20}H_{25}NO_2$,   3757
$C_{20}H_{25}NO_4$,   3758, 3759
$C_{20}H_{25}NO_5$,   3760
$C_{20}H_{25}NO_6$,   3761
$C_{20}H_{26}N_2O_2$,   3762, 3763
$C_{20}H_{26}O_2$,   3764, 3765
$C_{20}H_{26}O_4$,   3766
$C_{20}H_{27}NO_{11}$,   3767
$C_{20}H_{27}NO_{11} \cdot 3H_2O$,   3768
$C_{20}H_{27}O_4P$,   3769
$C_{20}H_{28}O$,   3770
$C_{20}H_{28}O_2$,   3771, 3772
$C_{20}H_{28}O_3$,   3773, 3774
$C_{20}H_{29}N_3O_2$,   3775
$C_{20}H_{30}N_4O_6$,   3776
$C_{20}H_{30}O$,   3777, 3778
$C_{20}H_{30}O_2$,   3779, 3780
$C_{20}H_{30}O_3$,   3781
$C_{20}H_{30}O_6$,   3782
$C_{20}H_{31}NO_3$,   3783
$C_{20}H_{32}O_3$,   3784
$C_{20}H_{32}O_5$,   3785
$C_{20}H_{33}NO_3$,   3786
$C_{20}H_{34}N_2O_2$,   3787
$C_{20}H_{34}O_4$,   3788
$C_{20}H_{34}O_8$,   3789
$C_{20}H_{36}O_4$,   3790
$C_{20}H_{36}O_6$,   3791
$C_{20}H_{40}$,   3792
$C_{21}H_{13}N$,   3793-3795
$C_{21}H_{14}$,   3796
$C_{21}H_{15}ClN_2O_4S$,   3797
$C_{21}H_{15}N_3O_6S$,   3798
$C_{21}H_{16}$,   3799
$C_{21}H_{16}N_2O_2$,   3800
$C_{21}H_{16}N_2O_4S$,   3801
$C_{21}H_{16}N_2O_5S$,   3802
$C_{21}H_{17}N_3O_2S_2$,   3803
$C_{21}H_{17}N_3O_4S$,   3804
$C_{21}H_{19}NO_4$,   3805
$C_{21}H_{20}Cl_2O_3$,   3806
$C_{21}H_{21}ClN_2O_8$,   3807
$C_{21}H_{21}NO_6$,   3808, 3809
$C_{21}H_{21}N_3O_3S$,   3810
$C_{21}H_{21}O_4P$,   3811

$C_{21}H_{22}N_2O_2$,   3812
$C_{21}H_{22}N_2O_5$,   3813, 3814
$C_{21}H_{23}ClFNO_2$,   3815
$C_{21}H_{23}N_3OS$,   3816
$C_{21}H_{24}F_3N_5S$,   3817
$C_{21}H_{24}O_{10} \cdot 2H_2O$,   3818
$C_{21}H_{25}NO$,   3819
$C_{21}H_{26}ClN_3OS$,   3820
$C_{21}H_{26}N_2O_3$,   3821
$C_{21}H_{26}O_4$,   3822
$C_{21}H_{26}O_5$,   3823
$C_{21}H_{27}FO_5$,   3824
$C_{21}H_{27}FO_5 \cdot H_2O$,   3825
$C_{21}H_{27}FO_6$,   3826
$C_{21}H_{28}N_4O_7$,   3827
$C_{21}H_{28}O_2$,   3828, 3829
$C_{21}H_{28}O_5$,   3830-3832
$C_{21}H_{29}FO_5$,   3833
$C_{21}H_{29}NO$,   3834
$C_{21}H_{30}N_4O_{10}$,   3835
$C_{21}H_{30}N_6O_4S$,   3836
$C_{21}H_{30}O_2$,   3837, 3838
$C_{21}H_{30}O_3$,   3839-3842
$C_{21}H_{30}O_5$,   3843
$C_{21}H_{30}O_6$,   3844
$C_{21}H_{31}NO$,   3845
$C_{21}H_{31}N_3O_2$,   3846
$C_{21}H_{32}O_2$,   3847-3849
$C_{21}H_{32}O_3$,   3850
$C_{21}H_{33}NO$,   3851
$C_{21}H_{33}NO_7$,   3852
$C_{21}H_{34}O_3$,   3853
$C_{21}H_{35}NO_3$,   3854
$C_{21}H_{36}O_4$,   3855
$C_{21}H_{40}O_4$,   3856
$C_{21}H_{44}$,   3857, 3858
$C_{22}H_{12}$,   3859, 3860
$C_{22}H_{14}$,   3861-3864
$C_{22}H_{16}F_3N_3$,   3865
$C_{22}H_{16}O_8$,   3866
$C_{22}H_{17}ClN_2$,   3867
$C_{22}H_{18}N_2O_4S$,   3868
$C_{22}H_{18}N_2O_5S$,   3869
$C_{22}H_{19}Br_2NO_3$,   3870
$C_{22}H_{19}F_6NOS$,   3871
$C_{22}H_{20}$,   3872
$C_{22}H_{20}O_{13}$,   3873
$C_{22}H_{22}FN_3O_2$,   3874
$C_{22}H_{22}N_2O_8$,   3875
$C_{22}H_{22}N_4O_6$,   3876
$C_{22}H_{23}ClN_2O_8$,   3877
$C_{22}H_{23}NO_7$,   3878
$C_{22}H_{24}N_2O_8$,   3879
$C_{22}H_{24}N_2O_8 \cdot H_2O$,   3880
$C_{22}H_{24}N_2O_9$,   3881
$C_{22}H_{24}N_4O_5$,   3882
$C_{22}H_{25}NO_6$,   3883
$C_{22}H_{26}F_3N_3OS$,   3884
$C_{22}H_{28}F_2O_5$,   3885
$C_{22}H_{28}O_3$,   3886, 3887
$C_{22}H_{29}FO_4$,   3888
$C_{22}H_{29}FO_5$,   3889, 3890
$C_{22}H_{30}Cl_2N_{10}$,   3891
$C_{22}H_{30}N_2O_2$,   3892

$C_{22}H_{30}O_5$, 3893
$C_{22}H_{30}O_6$, 3894
$C_{22}H_{32}O_3$, 3895-3898
$C_{22}H_{33}N_3O_2$, 3899
$C_{22}H_{34}Cl_2O_3$, 3900
$C_{22}H_{34}N_6O_4$, 3901
$C_{22}H_{34}O_3$, 3902
$C_{22}H_{35}NO_3$, 3903
$C_{22}H_{38}O_5$, 3904
$C_{22}H_{39}O_3P$, 3905, 3906
$C_{22}H_{42}O_4$, 3907
$C_{22}H_{43}N_5O_{13}$, 3908
$C_{23}H_{16}O_6$, 3909
$C_{23}H_{18}F_2N_4O$, 3910
$C_{23}H_{20}N_2O_2S$, 3911
$C_{23}H_{22}$, 3912
$C_{23}H_{22}O_6$, 3913
$C_{23}H_{23}NO$, 3914
$C_{23}H_{24}N_4O_2$, 3915
$C_{23}H_{24}N_4O_6$, 3916
$C_{23}H_{24}N_4O_7$, 3917
$C_{23}H_{24}N_4S_2$, 3918
$C_{23}H_{26}N_2O_4$, 3919
$C_{23}H_{26}N_2O_4.4H_2O$, 3920
$C_{23}H_{26}O_3$, 3921
$C_{23}H_{27}ClO_4$, 3922
$C_{23}H_{27}NO_8$, 3923
$C_{23}H_{27}N_3O_7$, 3924
$C_{23}H_{28}ClN_3O_2S$, 3925
$C_{23}H_{28}ClN_3O_5S$, 3926
$C_{23}H_{28}O_7$, 3927
$C_{23}H_{31}Cl_2NO_3$, 3928
$C_{23}H_{31}FO_6$, 3929
$C_{23}H_{31}O_7$, 3930
$C_{23}H_{32}O_2$, 3931
$C_{23}H_{32}O_4$, 3932
$C_{23}H_{32}O_6$, 3933
$C_{23}H_{34}O_3$, 3934-3936
$C_{23}H_{34}O_4$, 3937
$C_{23}H_{35}NOS$, 3938
$C_{23}H_{36}O_3$, 3939
$C_{23}H_{38}O_3$, 3940
$C_{23}H_{40}O_5$, 3941
$C_{24}H_{12}$, 3942
$C_{24}H_{20}N_2$, 3943
$C_{24}H_{22}N_2O_2$, 3944
$C_{24}H_{26}N_4O_2$, 3945
$C_{24}H_{26}N_4S_2$, 3946
$C_{24}H_{27}BrN_6O_{10}$, 3947
$C_{24}H_{27}N$, 3948
$C_{24}H_{30}F_2O_6$, 3949
$C_{24}H_{31}FO_3S$, 3950
$C_{24}H_{31}FO_6$, 3951-3953
$C_{24}H_{31}NO_4$, 3954
$C_{24}H_{32}O_4$, 3955
$C_{24}H_{32}O_4S$, 3956
$C_{24}H_{32}O_5$, 3957
$C_{24}H_{32}O_6$, 3958
$C_{24}H_{33}FO_6$, 3959
$C_{24}H_{34}N_2O$, 3960
$C_{24}H_{34}N_2O_3$, 3961
$C_{24}H_{34}O_5$, 3962
$C_{24}H_{34}O_6$, 3963
$C_{24}H_{36}O_3$, 3964, 3965

$C_{24}H_{38}O_3$, 3966
$C_{24}H_{38}O_4$, 3967-3972
$C_{24}H_{39}NO_3$, 3973
$C_{24}H_{40}O_3$, 3974, 3975
$C_{24}H_{40}O_4$, 3976-3979
$C_{24}H_{40}O_5$, 3980-3982
$C_{24}H_{50}$, 3983
$C_{24}H_{51}OP$, 3984
$C_{24}H_{51}O_3P$, 3985
$C_{24}H_{51}O_4P$, 3986
$C_{24}H_{54}OSn_2$, 3987
$C_{25}H_{24}N_2O_2S$, 3988
$C_{25}H_{28}N_4O_2$, 3989
$C_{25}H_{28}O_3$, 3990
$C_{25}H_{29}I_2NO_3$, 3991
$C_{25}H_{31}FO_8$, 3992
$C_{25}H_{31}NO_2$, 3993
$C_{25}H_{34}O_3$, 3994
$C_{25}H_{34}O_6$, 3995
$C_{25}H_{34}O_9$, 3996
$C_{25}H_{36}N_4O_7$, 3997
$C_{25}H_{36}O_6$, 3998
$C_{25}H_{36}O_7$, 3999
$C_{25}H_{40}O_3Si_2$, 4000
$C_{25}H_{42}O_3$, 4001
$C_{25}H_{44}$, 4002
$C_{25}H_{44}O_6$, 4003
$C_{25}H_{48}O_4$, 4004
$C_{25}H_{54}O_2P_2$, 4005
$C_{26}H_{18}N_2O_4$, 4006
$C_{26}H_{20}N_2O_8S_2$, 4007, 4008
$C_{26}H_{28}Cl_2N_4O_4$, 4009
$C_{26}H_{28}N_2$, 4010
$C_{26}H_{28}N_4O_2$, 4011
$C_{26}H_{30}N_4O_2$, 4012
$C_{26}H_{30}N_4S_2$, 4013
$C_{26}H_{32}F_2O_7$, 4014, 4015
$C_{26}H_{32}O_3$, 4016
$C_{26}H_{36}O_3$, 4017
$C_{26}H_{36}O_6$, 4018
$C_{26}H_{37}FO_5$, 4019
$C_{26}H_{38}NO_8$, 4020
$C_{26}H_{38}O_4$, 4021
$C_{26}H_{38}O_6$, 4022
$C_{26}H_{39}NO_3S$, 4023
$C_{26}H_{43}NO_3$, 4024
$C_{26}H_{43}NO_6$, 4025
$C_{26}H_{50}O_4$, 4026
$C_{26}H_{56}O_2P_2$, 4027
$C_{27}H_{22}Cl_2N_4$, 4028
$C_{27}H_{29}NO_{11}$, 4029
$C_{27}H_{30}O_3$, 4030
$C_{27}H_{32}N_4O_2$, 4031, 4032
$C_{27}H_{32}N_4S_2$, 4033
$C_{27}H_{32}O_{14}$, 4034
$C_{27}H_{33}N_3O_8$, 4035
$C_{27}H_{34}O_3$, 4036
$C_{27}H_{34}O_{10}$, 4037
$C_{27}H_{38}N_2O_6$, 4038
$C_{27}H_{38}O_3$, 4039
$C_{27}H_{40}N_2O_6$, 4040
$C_{27}H_{40}O_6$, 4041
$C_{27}H_{42}Cl_2N_2O_6$, 4042
$C_{27}H_{42}N_4O_7.0.3H_2O$, 4043

$C_{27}H_{42}O_3$, 4044, 4045
$C_{27}H_{43}NO_8$, 4046
$C_{27}H_{44}N_4O_6$, 4047
$C_{27}H_{44}O$, 4048
$C_{27}H_{46}O$, 4049
$C_{27}H_{58}O_2P_2$, 4050
$C_{28}H_{29}F_2N_3O$, 4051
$C_{28}H_{36}O_3$, 4052
$C_{28}H_{39}NO_6$, 4053, 4054
$C_{28}H_{39}N_3O_6$, 4055
$C_{28}H_{40}FNO_{11}.H_2O$, 4056
$C_{28}H_{41}N_3O_6$, 4057
$C_{28}H_{42}FNO_{11}.H_2O$, 4058
$C_{28}H_{42}O_6$, 4059
$C_{28}H_{44}O_3$, 4060
$C_{28}H_{46}O_4$, 4061, 4062
$C_{28}H_{60}O_2P_2$, 4063
$C_{29}H_{20}N_2O_4$, 4064
$C_{29}H_{27}N_5O_4$, 4065-4067
$C_{29}H_{28}N_4O_2$, 4068
$C_{29}H_{28}N_4O_3$, 4069
$C_{29}H_{28}N_4S_2$, 4070
$C_{29}H_{32}O_{13}$, 4071
$C_{29}H_{36}N_4O_2$, 4072
$C_{29}H_{36}N_4S_2$, 4073
$C_{29}H_{38}Cl_2N_2O_3$, 4074
$C_{29}H_{38}O_3$, 4075
$C_{29}H_{40}N_2O_4$, 4076
$C_{29}H_{42}O_6$, 4077
$C_{29}H_{44}FNO_{11}.H_2O$, 4078
$C_{29}H_{44}O_{12}$, 4079
$C_{29}H_{46}N_4O_7.0.4H_2O$, 4080
$C_{29}H_{46}O_3$, 4081
$C_{30}H_{28}N_4O_3$, 4082
$C_{30}H_{34}O_{13}$, 4083
$C_{30}H_{48}O_{12}$, 4084
$C_{31}H_{33}N_5O_2$, 4085
$C_{31}H_{38}N_2O_{11}$, 4086
$C_{31}H_{42}FNO_{12}.H_2O$, 4087
$C_{31}H_{44}FNO_{12}.H_2O$, 4088
$C_{31}H_{44}N_2O_7$, 4089
$C_{31}H_{46}N_2O_7$, 4090
$C_{31}H_{48}O_{12}$, 4091
$C_{32}H_{32}O_{14}$, 4092
$C_{32}H_{37}NO_5S$, 4093
$C_{32}H_{46}FNO_{12}.H_2O$, 4094
$C_{32}H_{49}NO_9$, 4095
$C_{32}H_{54}O_4$, 4096
$C_{33}H_{25}N_3O_3$, 4097
$C_{33}H_{34}O_3$, 4098
$C_{33}H_{34}O_4$, 4099
$C_{33}H_{36}N_4O_6$, 4100
$C_{33}H_{40}N_2O_9$, 4101
$C_{33}H_{41}N_5O_6S_2$, 4102
$C_{33}H_{45}NO_9$, 4103
$C_{33}H_{47}NO_{13}$, 4104
$C_{34}H_{34}N_4O_4$, 4105
$C_{34}H_{47}NO_{11}$, 4106
$C_{34}H_{50}O_7$, 4107
$C_{34}H_{57}NO_7$, 4108
$C_{34}H_{58}O_4$, 4109
$C_{34}H_{68}N_3O_8S_2$, 4110
$C_{35}H_{44}N_2O_7$, 4111

$C_{35}H_{46}N_2O_7$, 4112
$C_{35}H_{47}NO_9$, 4113
$C_{35}H_{61}NO_7$, 4114
$C_{36}H_{47}N_2O_7$, 4115
$C_{36}H_{49}N_2O_7$, 4116
$C_{36}H_{56}O_{14}$, 4117
$C_{36}H_{60}O_2$, 4118
$C_{36}H_{60}O_{30}$, 4119
$C_{36}H_{72}N_3O_8S_2$, 4120
$C_{36}H_{74}$, 4121
$C_{37}H_{67}NO_{13}$, 4122
$C_{37}H_{67}NO_{13}.2H_2O$, 4123
$C_{38}H_{69}NO_{13}$, 4124
$C_{40}H_{51}NO_{14}$, 4125
$C_{41}H_{64}O_{13}$, 4126
$C_{41}H_{64}O_{14}$, 4127, 4128
$C_{41}H_{67}NO_{15}$, 4129
$C_{42}H_{70}O_{35}$, 4130, 4131
$C_{43}H_{58}N_4O_{12}$, 4132
$C_{43}H_{75}NO_{16}$, 4133
$C_{44}H_{74}O_{34}$, 4134
$C_{44}H_{74}O_{35}$, 4135
$C_{45}H_{73}NO_{15}$, 4136
$C_{45}H_{76}O_{35}$, 4137
$C_{46}H_{77}NO_{17}$, 4138
$C_{46}H_{78}O_{35}$, 4139
$C_{47}H_{73}NO_{17}$, 4140
$C_{47}H_{75}NO_{17}$, 4141
$C_{48}H_{80}O_{40}$, 4142, 4143
$C_{49}H_{87}NS$, 4144
$C_{50}H_{82}N_{10}O_{31}S_{10}$, 4145
$C_{51}H_{70}N_{12}O_{11}$, 4146
$C_{51}H_{74}O_{19}$, 4147
$C_{52}H_{72}N_{12}O_{10}$, 4148, 4149
$C_{52}H_{88}O_{39}$, 4150
$C_{52}H_{97}NO_{18}S$, 4151
$C_{54}H_{90}O_{45}$, 4152-4154
$C_{55}H_{70}N_{12}O_{10}$, 4155
$C_{55}H_{79}N_{13}O_{11}$, 4156
$C_{55}H_{90}N_{11}O_{34}S_{11}$, 4157
$C_{56}H_{98}O_{35}$, 4158
$C_{57}H_{79}N_{13}O_{11}$, 4159, 4160
$C_{60}H_{77}N_{13}O_{11}$, 4161
$C_{60}H_{92}N_{12}O_{10}$, 4162
$C_{66}H_{98}N_{12}O_{37}S_{12}$, 4163
$C_{60}H_{100}O_{50}$, 4164
$C_{62}H_{86}N_{12}O_{16}$, 4165
$C_{62}H_{111}N_{11}O_{12}$, 4166
$C_{63}H_{85}N_{21}O_{19}$, 4167
$C_{63}H_{88}N_{14}O_{14}PCo$, 4168
$C_{64}H_{112}O_{40}$, 4169
$C_{65}H_{106}N_{13}O_{40}S_{13}$, 4170
$C_{66}H_{110}O_{55}$, 4171
$C_{67}H_{93}N_{15}O_{13}$, 4172, 4173
$C_{70}H_{89}N_{15}O_{13}$, 4174
$C_{70}H_{126}O_{35}$, 4175
$C_{72}H_{85}N_9O_{18}S_5$, 4176
$C_{72}H_{100}N_{18}O_{17}PCo$, 4177
$C_{74}H_{100}ClN_{15}O_{14}$, 4178
$C_{75}H_{122}N_{15}O_{46}S_{15}$, 4179
$C_{77}H_{107}N_{17}O_{15}$, 4180
$C_{80}H_{105}N_{17}O_{15}$, 4181
$C_{85}H_{117}N_{20}O_{18}$, 4182

Index 2: Chemical Abstracts Service Registry Number (RN)

50-02-2, 3890
50-03-3, 3933
50-04-4, 3844
50-06-6, 2595
50-07-7, 3204
50-11-3, 1775
50-23-7, 3843
50-24-8, 3831
50-27-1, 3580
50-28-2, 3578
50-29-3, 2957
50-30-6, 968
50-31-7, 957
50-32-8, 3712
50-33-9, 3644
50-36-2, 3464
50-44-2, 390
50-47-5, 3567
50-48-6, 3743
50-49-7, 3673
50-50-0, 3990
50-53-3, 3444
50-55-5, 4101
50-59-9, 3633
50-65-7, 2778
50-70-4, 938
50-71-5, 226
50-76-0, 4165
50-78-2, 1642
50-81-7, 772
50-84-0, 970
50-85-1, 1399
50-99-7, 874, 880
51-21-8, 227
51-28-5, 601
51-34-3, 3465
51-35-4, 444
51-36-5, 971
51-43-4, 1765
51-44-5, 969
51-52-5, 1159
51-55-8, 3483
51-66-1, 1707
51-67-2, 1479
51-79-6, 202
51-98-9, 3886
52-01-7, 3956
52-21-1, 4018
52-31-3, 2670
52-39-1, 3832
52-43-7, 1987

52-44-8, 1740
52-45-9, 1739
52-46-0, 2752
52-51-7, 155
52-67-5, 501
52-68-6, 312
52-86-8, 3815
53-03-2, 3823
53-06-5, 3830
53-16-7, 3570
53-19-0, 2968
53-34-9, 3824, 3825
53-41-8, 3703
53-42-9, 3700
53-43-0, 3691
53-70-3, 3863
53-86-1, 3628
54-12-6, 2234
54-31-9, 2571
54-42-2, 1700
54-85-3, 749
55-18-5, 359
55-21-0, 1087
55-38-9, 2088
55-56-1, 3891
55-63-0, 150
55-91-4, 904
56-04-2, 407
56-12-2, 349
56-23-5, 40
56-25-7, 2014
56-29-1, 2669
56-35-9, 3987
56-38-2, 2056
56-40-6, 88
56-41-7, 198
56-45-1, 206
56-47-3, 3932
56-49-5, 3799
56-53-1, 3559
56-54-2, 3747
56-55-3, 3512
56-56-4, 3729
56-72-4, 3027
56-75-7, 2228
56-81-5, 218
56-82-6, 182
56-84-8, 300
56-85-9, 457
56-86-0, 445
56-87-1, 911

56-89-3, 835
57-00-1, 354
57-06-7, 256
57-10-3, 3399
57-11-4, 3610
57-13-6, 22
57-15-8, 295
57-24-9, 3812
57-27-2, 3449
57-41-0, 3153
57-42-1, 3223
57-43-2, 2334
57-44-3, 1486
57-48-7, 873
57-50-1, 2744
57-53-4, 1806
57-62-5, 3877
57-63-6, 3752
57-67-0, 1165
57-68-1, 2629, 2630
57-74-9, 1874
57-83-0, 3838
57-85-2, 3896
57-88-5, 4049
57-91-0, 3579
58-00-4, 3433
58-08-2, 1449
58-15-1, 2883
58-18-4, 3779
58-22-0, 3693, 3694
58-25-3, 3285
58-27-5, 2194
58-38-8, 3745
58-39-9, 3820
58-40-2, 3456
58-55-9, 1113
58-61-7, 2043
58-63-9, 2002
58-74-2, 3739
58-85-5, 2099
58-86-6, 484
58-89-9, 691
58-90-2, 543
58-93-5, 1104
58-94-6, 1030
59-23-4, 877
59-30-3, 3641
59-31-4, 1619
59-46-1, 2910
59-50-7, 1074
59-51-8, 500
59-52-9, 216
59-63-2, 2615
59-66-5, 268
59-67-6, 650
59-87-0, 717
59-92-7, 1718
60-01-5, 3264
60-09-3, 2578
60-11-7, 3023
60-12-8, 1460
60-18-4, 1714
60-27-5, 304
60-29-7, 362

60-32-2, 897
60-35-5, 87
60-51-5, 507
60-54-8, 3879
60-57-1, 2524
60-80-0, 2232
60-87-7, 3457
60-89-9, 3667
60-92-4, 2004
60-99-1, 3675
61-00-7, 3665
61-57-4, 716
61-68-7, 3179
61-73-4, 3356
61-82-5, 79
61-90-5, 891
62-23-7, 1014
62-38-4, 1386
62-44-2, 2025
62-53-3, 737
62-55-5, 92
62-56-6, 23
62-57-7, 347
62-59-9, 4095
62-73-7, 294
63-05-8, 3685
63-25-2, 2575
63-42-3, 2741
63-68-3, 499
63-74-1, 764
63-84-3, 1717
63-91-2, 1706
64-00-6, 2290
64-19-7, 81
64-77-7, 2705
64-85-7, 3841
64-86-8, 3883
64-95-9, 3757
65-28-1, 3572
65-45-2, 1091
65-49-6, 1097
65-71-4, 409
65-85-0, 1055
65-86-1, 380
66-02-4, 1650
66-22-8, 239
66-25-1, 845
66-27-3, 102
66-71-7, 2526
66-76-2, 3617
66-81-9, 3248
66-97-7, 2187
67-20-9, 1313
67-52-7, 241
67-56-1, 25
67-63-0, 215
67-66-3, 5
67-72-1, 111
67-73-2, 3949
67-78-7, 3992
67-97-0, 4048
68-19-9, 4168
68-22-4, 3764
68-26-8, 3778

68-35-9, 1939
68-90-6, 3413
68-94-0, 384
68-96-2, 3839
69-23-8, 3884
69-53-4, 3341
69-72-7, 1060
69-79-4, 2745
69-89-6, 386, 387
69-93-2, 388, 389
70-18-8, 2116
70-25-7, 96
70-30-4, 2772
70-34-8, 569
70-47-3, 316
70-55-3, 1147
70-69-9, 1704
71-00-1, 782
71-23-8, 214
71-30-7, 257
71-36-3, 360
71-41-0, 516
71-43-2, 676
71-55-6, 63
71-63-6, 4126
72-14-0, 1661
72-18-4, 495
72-19-5, 353
72-20-8, 2523
72-43-5, 3298
72-44-6, 3290
72-48-0, 2947
72-54-8, 2969
72-55-9, 2944
73-22-3, 2233
73-24-5, 400
73-32-5, 892
73-40-5, 401
73-48-3, 3170
73-49-4, 1982
74-11-3, 989
74-79-3, 913
74-82-8, 21
74-83-9, 12
74-84-0, 99
74-85-1, 69
74-86-2, 50
74-87-3, 14
74-88-4, 17
74-89-5, 28
74-95-3, 8
74-96-4, 83
74-97-5, 7
74-98-6, 212
74-99-7, 124
75-00-3, 84
75-01-4, 61
75-03-6, 85
75-04-7, 104
75-05-8, 66
75-09-2, 9
75-11-6, 10
75-15-0, 44
75-17-2, 18

75-19-4, 153
75-25-2, 3
75-26-3, 186
75-27-4, 1
75-28-5, 357
75-29-6, 190
75-30-9, 192
75-34-3, 75
75-35-4, 53
75-37-6, 76
75-45-6, 4
75-47-8, 6
75-50-3, 220
75-52-5, 19
75-56-9, 173
75-60-5, 106
75-65-0, 366
75-69-4, 38
75-71-8, 37
75-73-0, 41
75-80-9, 60
75-83-2, 900
75-84-3, 511
75-85-4, 515
75-97-8, 847
75-98-9, 478
75-99-0, 130
76-01-7, 49
76-03-9, 48
76-06-2, 39
76-13-1, 109
76-14-2, 108
76-20-0, 1586
76-22-2, 2103
76-24-4, 1314
76-25-5, 3951
76-44-8, 1861
76-57-3, 3564
76-58-4, 3669
76-73-3, 2703
76-74-4, 2335
76-75-5, 2332
76-76-6, 1776
76-84-6, 3629
76-93-7, 2998
76-94-8, 2211
77-02-1, 2061
77-09-8, 3726
77-21-4, 2836
77-26-9, 2305
77-27-0, 2702
77-28-1, 2096
77-30-5, 2725
77-32-7, 1485
77-36-1, 2981
77-41-8, 2605
77-47-4, 526
77-65-6, 1198
77-67-8, 1170
77-71-4, 424
77-74-7, 925
77-78-1, 103
77-86-1, 373
77-90-7, 3789

77-92-9,	774	
77-94-1,	3605	
77-95-2,	1196	
78-11-5,	427	
78-40-0,	946	
78-42-2,	3986	
78-46-6,	2764	
78-50-2,	3984	
78-51-3,	3616	
78-57-9,	844	
78-70-6,	2128	
78-77-3,	334	
78-78-4,	503	
78-79-5,	420	
78-83-1,	364	
78-84-2,	323	
78-85-3,	269	
78-86-4,	336	
78-87-5,	160	
78-92-2,	361	
78-93-3,	325	
78-95-5,	140	
79-00-5,	64	
79-01-6,	46	
79-06-1,	147	
79-07-2,	72	
79-09-4,	177	
79-14-1,	82	
79-20-9,	176	
79-24-3,	91	
79-27-6,	51	
79-29-8,	902	
79-31-2,	330	
79-34-5,	56	
79-46-9,	199	
79-57-2,	3881	
79-81-2,	4118	
80-05-7,	3191	
80-08-0,	2592	
80-15-9,	1758	
80-34-2,	2275	
80-35-3,	2241	
80-46-6,	2316	
80-58-0,	292	
80-60-4,	346	
80-62-6,	431	
80-68-2,	350	
80-74-0,	2847	
81-05-0,	1915	
81-06-1,	1921	
81-07-2,	1011	
81-16-3,	1924	
81-23-2,	3962	
81-25-4,	3981	
81-54-9,	2949	
81-61-8,	2950	
81-64-1,	2948	
81-81-2,	3630	
81-90-3,	3731	
82-02-0,	3000	
82-28-0,	3144	
82-38-2,	3143	
82-45-1,	2961	
82-48-4,	2951	

82-68-8,	952	
82-71-3,	573	
82-75-7,	1919	
82-93-9,	3560	
83-07-8,	2266	
83-26-1,	3016	
83-32-9,	2553	
83-34-1,	1651	
83-40-9,	1387	
83-43-2,	3893	
83-44-3,	3978	
83-49-8,	3979	
83-53-4,	1865	
83-67-0,	1114	
83-79-4,	3913	
83-88-5,	3461	
84-06-0,	3925	
84-15-1,	3521	
84-17-3,	3543	
84-21-9,	2070	
84-31-1,	3742	
84-61-7,	3766	
84-62-8,	3727	
84-65-1,	2946	
84-66-2,	2638	
84-69-5,	3372	
84-72-0,	3041	
84-74-2,	3371	
84-77-5,	4061	
84-86-6,	1920	
84-89-9,	1916	
85-01-8,	2966	
85-34-7,	1280	
85-36-9,	1608	
85-38-1,	1021	
85-41-6,	1288	
85-44-9,	1278	
85-68-7,	3654	
85-70-1,	3581	
85-71-2,	2835	
85-79-0,	3775	
85-84-7,	3282	
86-29-3,	2984	
86-30-6,	2558	
86-34-0,	2218	
86-50-0,	1993	
86-57-7,	1884	
86-60-2,	1918	
86-73-7,	2787	
86-74-8,	2545	
86-87-3,	2568	
86-88-4,	2212	
87-17-2,	2808	
87-20-7,	2683	
87-26-3,	2318	
87-29-6,	3304	
87-33-2,	768	
87-40-1,	996	
87-47-8,	2843	
87-58-1,	224	
87-61-6,	559	
87-64-9,	1078	
87-65-0,	593	
87-66-1,	729	

87-68-3,   375
87-69-4,   286
87-72-9,   483
87-78-5,   937
87-79-6,   871
87-86-5,   532
87-87-6,   546
87-88-7,   536
87-89-8,   875
87-90-1,   223
87-92-3,   2739
88-04-0,   1407
88-06-2,   563
88-09-5,   857
88-13-1,   392
88-14-2,   394
88-19-7,   1148
88-20-0,   1167
88-21-1,   743
88-29-9,   3589
88-43-7,   687, 688
88-44-8,   1152
88-67-5,   1002
88-72-2,   1092
88-73-3,   585
88-74-4,   704
88-75-5,   652
88-85-7,   1992
88-88-0,   535
88-89-1,   572
88-96-0,   1363
88-99-3,   1320
89-00-9,   1013
89-05-4,   1878
89-56-5,   1388
89-61-2,   557
89-69-0,   537
89-78-1,   2160
89-83-8,   2073
89-86-1,   1066
90-01-7,   1121
90-02-8,   1056
90-04-0,   1144
90-05-1,   1122
90-11-9,   1880
90-12-0,   2202
90-13-1,   1881
90-14-2,   1883
90-15-3,   1896
90-39-1,   3261
90-43-7,   2565
90-45-9,   2795
90-51-7,   1925
90-64-2,   1392
90-94-8,   3455
91-01-0,   2818
91-10-1,   1469
91-16-7,   1463
91-17-8,   2120
91-18-9,   605
91-20-3,   1889
91-22-5,   1618
91-57-6,   2201
91-58-7,   1882

91-59-8,   1909
91-64-5,   1610
91-75-8,   3452
91-79-2,   3075
91-80-5,   3074
91-81-6,   3363
91-82-7,   3741
91-84-9,   3487
91-85-0,   3365
91-94-1,   2556
92-06-8,   3522
92-24-0,   3515
92-36-4,   2992
92-44-4,   1899
92-50-2,   2084
92-52-4,   2554
92-66-0,   2534
92-69-3,   2566
92-82-0,   2528
92-84-2,   2547
92-86-4,   2512
92-87-5,   2589
92-94-4,   3523
93-00-5,   1922
93-07-2,   1689
93-09-4,   2195
93-35-6,   1611
93-37-8,   2215
93-58-3,   1385
93-60-7,   1093
93-65-2,   1955
93-71-0,   1483
93-72-1,   1616
93-76-5,   1285
93-80-1,   1903
93-89-0,   1683
94-09-7,   1708
94-11-1,   2229
94-12-2,   2026
94-13-3,   2013
94-18-8,   2999
94-19-9,   1998
94-20-2,   2020
94-23-5,   3246
94-24-6,   3254
94-25-7,   2289
94-26-8,   2283
94-36-0,   2976
94-62-2,   3448
94-71-3,   1465
94-74-6,   1645
94-75-7,   1300
94-79-1,   2619
94-80-4,   2620
94-81-5,   2253
94-82-6,   1935
94-83-7,   3149
94-96-2,   1585
95-06-7,   1496
95-15-8,   1326
95-45-4,   313
95-47-6,   1431
95-48-7,   1119
95-49-8,   1073

95-50-1, 586
95-51-2, 683
95-53-4, 1132
95-54-5, 757
95-55-6, 740
95-57-8, 633
95-63-6, 1723
95-65-8, 1455
95-77-2, 592
95-87-4, 1457
95-93-2, 2049
95-94-3, 542
95-95-4, 567
96-09-3, 1375
96-12-8, 138
96-14-0, 901
96-18-4, 143
96-22-0, 461
96-23-1, 161
96-33-3, 272
96-37-7, 820
96-45-7, 169
96-47-9, 466
96-83-3, 2230
96-88-8, 3242
96-91-3, 666
96-97-9, 1020
96-99-1, 965
97-00-7, 553
97-02-9, 665
97-09-6, 631
97-17-6, 2022
97-23-4, 2792
97-30-3, 1233
97-53-0, 2011
97-56-3, 3022
97-59-6, 267
97-65-4, 413
97-77-8, 2158
97-85-8, 1544
97-95-0, 918
98-01-1, 391
98-05-5, 755
98-06-6, 2050
98-10-2, 742
98-11-3, 730-734
98-18-0, 762
98-33-9, 1150
98-37-3, 747
98-54-4, 2077
98-61-3, 581
98-64-6, 684
98-66-8, 635, 636
98-71-5, 767
98-82-8, 1728
98-83-9, 1663
98-86-2, 1378
98-89-5, 1189
98-92-0, 702
98-95-3, 651
98-96-4, 399
99-03-6, 1415
99-04-7, 1384
99-05-8, 1095

99-06-9, 1061
99-08-1, 1094
99-09-2, 705
99-14-9, 773
99-24-1, 1403
99-32-1, 984
99-34-3, 979
99-35-4, 571
99-50-3, 1065
99-54-7, 556
99-59-2, 1108
99-61-6, 1008
99-65-0, 598
99-71-8, 2074
99-76-3, 1390
99-87-6, 2053
99-92-3, 1413
99-93-4, 1380
99-94-5, 1379
99-96-7, 1062
99-99-0, 1090
100-00-5, 584
100-01-6, 706
100-02-7, 653
100-09-4, 1397
100-17-4, 1098
100-21-0, 1322
100-25-4, 599
100-26-5, 1018
100-40-3, 1482
100-41-4, 1432
100-42-5, 1351
100-47-0, 1005
100-51-6, 1117
100-52-7, 1054
100-61-8, 1133
100-64-1, 814
100-65-2, 738
100-66-3, 1118
100-71-0, 1139
100-75-4, 453
100-83-4, 1057
100-90-3, 3036
100-97-0, 842
101-05-3, 1603
101-21-3, 1980
101-25-7, 459
101-27-9, 2198
101-29-1, 1004
101-31-5, 3484
101-37-1, 2657
101-42-8, 1737
101-53-1, 2820
101-77-9, 2829
101-81-5, 2815
101-84-8, 2564
102-07-8, 2816
102-28-3, 1441
102-54-5, 1929
102-69-2, 1849
102-76-1, 1779
102-82-9, 2759
102-94-3, 1639
103-23-1, 3907

103-24-2, 4004
103-26-4, 1946
103-29-7, 3006
103-30-0, 2990
103-33-3, 2557
103-36-6, 2247
103-65-1, 1730
103-72-0, 1022
103-73-1, 1453
103-82-2, 1381
103-84-4, 1414
103-85-5, 1112
103-88-8, 1353
103-90-2, 1416
104-04-1, 1366
104-15-4, 1125, 1128
104-35-8, 3503
104-40-5, 3259
104-46-1, 2007
104-51-8, 2052
104-55-2, 1637
104-76-7, 1582
104-85-8, 1338
104-87-0, 1377
104-94-9, 1145
104-98-3, 703
105-05-5, 2048
105-30-6, 932
105-37-3, 473
105-46-4, 858
105-53-3, 1192
105-54-4, 856
105-56-6, 415
105-57-7, 934
105-58-8, 481
105-60-2, 813
105-66-8, 1219
105-67-9, 1452
105-72-6, 2137
105-76-0, 2729
106-22-9, 2161
106-34-3, 2570
106-36-5, 865
106-37-6, 577
106-41-2, 626
106-42-3, 1430
106-43-4, 1072
106-44-5, 1116
106-46-7, 587
106-47-8, 682
106-48-9, 634
106-49-0, 1134
106-50-3, 759
106-51-4, 622
106-57-0, 265
106-70-7, 1222
106-87-6, 1487
106-89-8, 141
106-93-4, 71
106-94-5, 187
106-95-6, 137
106-97-8, 356
106-98-9, 306
106-99-0, 262

107-00-6, 261
107-02-8, 135
107-04-0, 70
107-05-1, 139
107-06-2, 74
107-08-4, 193
107-10-8, 219
107-12-0, 146
107-13-1, 119
107-21-1, 101
107-35-7, 105
107-38-0, 98
107-43-7, 491
107-82-4, 485
107-83-5, 903
107-87-9, 470
107-91-5, 131
107-92-6, 331
107-93-7, 277
107-95-9, 197
107-97-1, 196
108-03-2, 200
108-05-4, 274
108-08-7, 1248
108-10-1, 849
108-11-2, 930
108-13-4, 163
108-19-0, 94
108-20-3, 928
108-21-4, 474
108-24-7, 279
108-36-1, 578
108-38-3, 1429
108-39-4, 1115
108-41-8, 1071
108-42-9, 681
108-43-0, 632
108-45-2, 758
108-46-3, 723
108-47-4, 1131
108-48-5, 1142
108-67-8, 1729
108-68-9, 1454
108-70-3, 561
108-73-6, 726
108-78-1, 171
108-80-5, 122
108-82-7, 1844
108-83-8, 1813
108-84-9, 1546
108-86-1, 625
108-87-2, 1205
108-88-3, 1102
108-90-7, 630
108-93-0, 850
108-94-1, 791
108-95-2, 722
109-00-2, 397
109-07-9, 508
109-21-7, 1545
109-43-3, 3609
109-52-4, 480
109-60-4, 477
109-64-8, 156

109-65-9,	335	112-57-2,	1600
109-66-0,	505	112-72-1,	3127
109-67-1,	450	112-92-5,	3612
109-68-2,	449	113-59-7,	3538
109-69-3,	339	113-73-5,	4162
109-70-6,	154	114-07-8,	4122, 4123
109-73-9,	372	114-26-1,	2293
109-74-0,	296	115-07-1,	152
109-77-3,	116	115-10-6,	100
109-87-5,	217	115-11-7,	305
109-92-2,	322	115-24-2,	1264
109-93-3,	270	115-32-2,	2958
109-94-4,	175	115-37-7,	3658
109-97-7,	252	115-38-8,	2830
109-99-9,	326	115-43-5,	2817
110-02-1,	244	115-44-6,	2310
110-14-5,	314	115-58-2,	2336
110-15-6,	281	115-68-4,	2274
110-16-7,	242	115-77-5,	521
110-17-8,	243	115-86-6,	3529
110-19-0,	860	115-88-8,	3769
110-42-9,	2365	115-89-9,	2826
110-43-0,	1213	115-90-2,	2330
110-44-1,	771	115-95-7,	2728
110-53-2,	486	115-96-8,	827
110-54-3,	899	116-01-8,	905
110-61-2,	233	116-06-3,	1208
110-62-3,	469	116-29-0,	2486
110-74-7,	328	116-31-4,	3770
110-82-7,	823	116-43-8,	2825
110-83-8,	784	116-44-9,	1940
110-88-3,	181	116-85-8,	2964
110-94-1,	437	117-18-0,	529
110-99-6,	283	117-34-0,	2995
111-11-5,	1822	117-61-3,	2596
111-13-7,	1538	117-79-3,	2962
111-14-8,	1224	117-80-6,	1859
111-15-9,	866	117-81-7,	3967
111-16-0,	1190	117-82-8,	3066
111-20-6,	2139	117-83-9,	3782
111-25-1,	882	117-84-0,	3971
111-27-3,	914	117-89-5,	3817
111-43-3,	921	117-96-4,	2200
111-44-4,	308	118-00-3,	2044
111-55-7,	802	118-08-1,	3808
111-65-9,	1576	118-10-5,	3663
111-66-0,	1527	118-52-5,	405
111-70-6,	1259	118-55-8,	2801
111-71-7,	1216	118-61-6,	1688
111-82-0,	2935	118-71-8,	728
111-83-1,	1559	118-74-1,	953
111-84-2,	1839	118-75-2,	951
111-86-4,	1587	118-79-6,	552
111-87-5,	1581	118-90-1,	1383
111-90-0,	935	118-91-2,	988
111-92-2,	1588	118-92-3,	1089
112-05-0,	1824	118-96-7,	1023
112-07-2,	1552	119-06-2,	4109
112-24-3,	950	119-15-3,	2551
112-26-5,	826	119-27-7,	1039
112-37-8,	2364	119-36-8,	1391
112-38-9,	2347	119-42-6,	2680
112-40-3,	2757	119-44-8,	673
112-53-8,	2758	119-61-9,	2800

119-64-2, 1977
119-68-6, 1418
119-79-9, 1923
119-90-4, 3032
119-91-5, 3516
119-93-7, 3031
120-07-0, 2085
120-12-7, 2965
120-23-0, 2569
120-32-1, 2806
120-36-5, 1632
120-47-8, 1687
120-51-4, 2996
120-57-0, 1318
120-61-6, 1949
120-72-9, 1337
120-73-0, 382
120-80-9, 724
120-82-1, 560
120-83-2, 589
120-89-8, 117
121-14-2, 1037
121-33-5, 1396
121-34-6, 1402
121-44-8, 940
121-47-1, 745, 746
121-52-8, 711
121-57-3, 744
121-61-9, 1445
121-69-7, 1478
121-73-3, 583
121-75-5, 2154
121-82-4, 172
121-91-5, 1321
121-92-6, 1016
121-98-2, 1686
122-11-2, 2636
122-14-5, 1736
122-15-6, 2324
122-34-9, 1182
122-39-4, 2574
122-42-9, 2029
122-57-6, 1944
122-59-8, 1400
122-62-3, 4026
122-80-5, 1440
122-85-0, 1654
122-88-3, 1333
122-95-2, 2079
122-99-6, 1467
123-07-9, 1459
123-08-0, 1058
123-11-5, 1382
123-19-3, 1215
123-25-1, 1512
123-30-8, 739
123-31-9, 725
123-33-1, 238
123-38-6, 174
123-51-3, 519
123-54-6, 430
123-56-8, 254
123-63-7, 868
123-72-8, 324

123-73-9, 271
123-76-2, 433
123-80-8, 1511
123-86-4, 863
123-91-1, 332
123-92-2, 1217
123-96-6, 1583
123-99-9, 1794
124-04-9, 799
124-07-2, 1543
124-09-4, 948
124-11-8, 1804
124-13-0, 1537
124-17-4, 2176
124-18-5, 2179
124-19-6, 1816
124-22-1, 2760
124-30-1, 3613
124-38-9, 43
124-48-1, 2
124-58-3, 30
124-83-4, 2110
124-87-8, 4083
124-94-7, 3826
125-10-0, 3927
125-33-7, 2623
125-40-6, 2095
125-42-8, 2306
125-64-4, 2113
126-07-8, 3432
126-72-7, 1780
126-73-8, 2765
126-75-0, 1592
126-98-7, 253
127-18-4, 110
127-19-5, 342
127-31-1, 3833
127-33-3, 3807
127-48-0, 778
127-69-5, 2267
127-73-1, 2832
127-74-2, 2598
127-75-3, 2833
127-76-4, 2225
127-77-5, 2593
127-79-7, 2239
128-13-2, 3977
128-23-4, 3847
128-37-0, 3260
128-62-1, 3878
128-95-0, 2971
129-00-0, 3271
129-20-4, 3646
129-66-8, 960
130-49-4, 2069
130-85-8, 3909
130-95-0, 3748
131-11-3, 1952
131-16-8, 3062
131-17-9, 3019
131-18-0, 3591
131-28-2, 3923
131-73-7, 2451
131-89-5, 2626

132-20-7, 3750
132-22-9, 3338
132-60-5, 3273
132-64-9, 2530
132-65-0, 2533
132-66-1, 3520
133-06-2, 1634
133-07-3, 1602
133-37-9, 287
133-74-4, 689
133-78-8, 1151
133-90-4, 992
133-91-5, 977
134-32-7, 1908
134-55-4, 3158
135-01-3, 2047
135-07-9, 1694
135-09-1, 1361
135-19-3, 1897
135-68-2, 2555
135-98-8, 2045
136-35-6, 2577
136-77-6, 2719
137-17-7, 1764
137-26-8, 841
137-30-4, 840
137-32-6, 518
137-58-6, 3098
137-89-3, 3970
138-22-7, 1227
138-36-3, 627-629
138-42-1, 660
138-52-3, 2904
138-59-0, 1168
138-86-3, 2089
139-13-9, 779
139-40-2, 1787
139-45-7, 2730
139-85-5, 1063
140-10-3, 1640
140-11-4, 1684
140-67-0, 2008
140-79-4, 321
140-88-5, 432
141-03-7, 2735
141-28-6, 2138
141-34-4, 2740
141-37-7, 3390
141-76-4, 144
141-78-6, 333
141-79-7, 792
141-82-2, 136
141-84-4, 120
141-90-2, 236
141-97-9, 797
142-28-9, 159
142-29-0, 419
142-62-1, 859
142-68-7, 460
142-73-4, 299
142-82-5, 1247
142-84-7, 942
142-92-7, 1549
142-96-1, 1579

143-07-7, 2753
143-08-8, 1846
143-50-0, 2184
143-62-4, 3937
143-74-8, 3627
144-16-1, 1129
144-49-0, 65
144-62-7, 58
144-80-9, 1446
144-82-1, 1678
144-83-2, 2224
144-98-9, 351
145-13-1, 3849
145-73-3, 1473
146-22-5, 3145
146-54-3, 3548
147-71-7, 288
147-73-9, 289
147-85-3, 442
148-24-3, 1624
148-56-1, 1308
148-65-2, 3051
148-79-8, 1888
148-82-3, 2891
149-32-6, 369
149-57-5, 1542
149-91-7, 1068
150-13-0, 1088
150-19-6, 1120
150-30-1, 1711
150-68-5, 1691
150-69-6, 1738
150-76-5, 1124
150-78-7, 1464
151-10-0, 1462
151-56-4, 86, 93
151-67-7, 45
153-78-6, 2807
156-59-2, 52
156-60-5, 54
191-07-1, 3942
191-24-2, 3859
192-97-2, 3714
193-39-5, 3860
194-59-2, 3718
198-55-0, 3717
203-64-5, 3134
205-82-3, 3715
205-99-2, 3713
206-44-0, 3270
207-08-9, 3716
207-84-1, 3720
208-96-8, 2511
213-46-7, 3861
215-58-7, 3862
217-59-4, 3514
218-01-9, 3513
224-41-9, 3864
224-53-3, 3794
226-92-6, 3793
226-97-1, 3795
229-87-8, 2782
230-07-9, 2527
230-46-6, 2525

238-84-6, 3407
239-64-5, 3719
243-17-4, 3408
260-94-6, 2783
262-12-4, 2531
281-36-7, 777
287-92-3, 448
291-64-5, 1207
292-64-8, 1529
297-76-7, 3955
297-97-2, 1493
298-00-0, 1433
298-02-2, 1266
298-03-3, 1593
298-04-4, 1590
298-46-4, 3152
298-57-7, 4010
298-81-7, 2532
299-42-3, 2083
299-84-3, 1359
299-85-4, 2054
299-86-5, 2721
300-39-0, 1649
302-17-0, 47
302-27-2, 4106
302-72-7, 194
302-79-4, 3772
302-84-1, 205
303-07-1, 1067
303-34-4, 3852
305-03-3, 3067
309-00-2, 2522
311-45-5, 2057
312-84-5, 207
314-40-9, 1761
314-42-1, 1475
315-18-4, 2701
315-22-0, 3381
315-30-0, 385
319-84-6, 692
319-85-7, 693
327-54-8, 548
327-56-0, 886
327-57-1, 890
327-98-0, 1985
328-38-1, 887
328-39-2, 894
330-54-1, 1665
330-55-2, 1666
333-41-5, 2731
334-48-5, 2167
338-69-2, 201
339-43-5, 2326
345-35-7, 1028
348-51-6, 580
349-46-2, 837
351-83-7, 1360
352-34-1, 595
352-93-2, 370
352-97-6, 210
353-54-8, 33
353-59-3, 32
356-12-7, 4015
357-57-3, 3919

360-70-3, 4060
363-24-6, 3785
366-18-7, 1894
367-12-4, 641
368-88-7, 644-647
371-41-5, 642
371-86-8, 947
372-20-3, 643
378-44-9, 3889
382-45-6, 3680
389-08-2, 2594
390-64-7, 3948
406-90-6, 251
420-04-2, 11
426-13-1, 3888
431-03-8, 276
433-97-6, 1287
434-03-7, 3829
434-13-9, 3974
434-22-0, 3590
437-50-3, 2978
439-14-5, 3279
440-58-4, 2573
443-48-1, 783
443-79-8, 895
445-29-4, 998
446-86-6, 1631
451-13-8, 1401
452-35-7, 1674
453-20-3, 327
455-38-9, 997
456-22-4, 999
456-42-8, 1029
458-88-8, 1560
460-12-8, 225
460-19-5, 112
461-58-5, 78
461-72-3, 132
461-98-3, 780
462-02-2, 121
462-06-6, 638
462-60-2, 166
462-95-3, 520
463-58-1, 42
463-82-1, 504
464-07-3, 916
464-43-7, 2126
464-88-0, 2132
466-49-9, 3892
467-98-1, 3563
470-82-6, 2127
470-90-6, 2621
471-03-4, 310
471-46-5, 77
472-54-8, 3771
473-18-7, 2111
474-25-9, 3976
474-86-2, 3558
475-25-2, 3277
475-31-0, 4025
479-18-5, 2068
479-23-2, 3721
479-45-8, 1027
479-92-5, 3053

480-16-0, 3140
480-63-7, 2010
481-06-1, 3205
481-29-8, 3702
481-37-8, 1782
482-05-3, 2977
483-18-1, 4076
484-12-8, 3192
484-19-5, 3442
485-35-8, 2273
485-65-4, 3674
485-71-2, 3664
486-64-6, 2209
487-21-8, 610
487-65-0, 1422
488-59-5, 881
488-73-3, 870
488-81-3, 523
489-98-5, 618
490-11-9, 1019
490-26-6, 1614
490-79-9, 1064
492-11-5, 674
492-27-3, 1885
492-38-6, 1641
492-62-6, 879
493-01-6, 2119
494-44-0, 1917
495-69-2, 1655
495-73-8, 2811
496-11-7, 1662
496-15-1, 1412
496-64-0, 393
496-67-3, 811
497-59-6, 985
498-21-5, 435
498-23-7, 412
498-24-8, 414
498-59-9, 1211
498-71-5, 2133
499-71-8, 2107
499-75-2, 2072
499-78-5, 624
499-80-9, 1017
499-81-0, 1012
500-28-7, 1404
500-72-1, 1341
501-52-0, 1680
502-39-6, 170
502-55-6, 796
502-56-7, 1815
503-40-2, 26, 27
503-66-2, 180
503-74-2, 475
505-48-6, 1508
505-52-2, 2931
505-60-2, 311
505-70-4, 735
506-12-7, 3504
506-77-4, 35
507-20-0, 338
507-70-0, 2129
508-96-3, 4041
509-86-4, 2893

510-15-6, 3287
510-53-2, 3583
512-04-9, 4044
512-56-1, 221
512-69-6, 3606
513-08-6, 1856
513-29-1, 949
513-36-0, 337
514-10-3, 3780
514-17-0, 3704
514-36-3, 3929
515-46-8, 1470
515-49-1, 1155
515-59-3, 1973
515-64-0, 2628
516-05-2, 280
516-06-3, 493
517-09-9, 3544
519-05-1, 1953
519-34-6, 2803
519-37-9, 1747
519-44-8, 603
521-18-6, 3701
522-66-7, 3762
524-40-3, 1364
526-73-8, 1725
526-75-0, 1458
526-78-3, 230
526-99-8, 806
527-06-0, 1265
527-60-6, 1755
528-29-0, 597
528-45-0, 980
528-46-1, 1612
528-50-7, 2743
529-64-6, 1685
529-86-2, 2973
530-43-8, 4042
530-48-3, 2989
530-75-6, 3278
530-78-9, 2970
531-29-3, 3374
531-75-9, 3193
531-91-9, 3943
533-23-3, 1936
533-24-4, 2322
533-74-4, 458
534-52-1, 1038
534-59-8, 1193
535-11-5, 440
535-65-9, 2676
535-80-8, 990
536-08-3, 2979
536-74-3, 1294
536-75-4, 1135
536-78-7, 1136
537-26-8, 3210
537-47-3, 1154
537-52-0, 3165
538-32-9, 1439
538-68-1, 2301
538-93-2, 2051
539-03-7, 1354
539-17-3, 3034

539-47-9, 1059
539-86-6, 793
539-89-9, 1240
540-18-1, 1828
540-38-5, 649
540-54-5, 189
540-84-1, 1574
541-25-3, 55
541-35-5, 341
541-73-1, 588
542-32-5, 815
542-69-8, 340
542-75-6, 129
543-24-8, 297
543-28-2, 494
543-49-7, 1260
543-86-2, 885
544-25-2, 1101
544-63-8, 3116
544-76-3, 3401
547-44-4, 1157
548-00-5, 3866
548-73-2, 3874
550-23-2, 3156
550-60-7, 1886
550-74-3, 1928
551-06-4, 2192
551-76-8, 994
551-88-2, 490
551-92-8, 418
552-16-9, 1015
552-32-9, 1365
552-58-9, 3139
552-89-6, 1010
553-12-8, 4105
553-26-4, 1893
553-90-2, 282
554-01-8, 417
554-12-1, 329
554-57-4, 426
554-84-7, 654
554-95-0, 1615
555-16-8, 1009
555-25-9, 3040
555-37-3, 2663
555-84-0, 1372
556-02-5, 1715
556-03-6, 1716
556-50-3, 318
556-61-6, 68
556-88-7, 24
557-17-5, 363
558-13-4, 34
561-07-9, 4103
562-49-2, 1243
563-12-2, 1858
563-45-1, 451
563-80-4, 465
564-25-0, 3880
565-59-3, 1245
565-60-6, 924
565-61-7, 852
565-67-3, 931
565-69-5, 855

565-75-3, 1571
565-80-0, 1214
569-31-3, 1950
569-34-6, 1748
569-51-7, 1613
570-22-9, 379
571-58-4, 2586
571-61-9, 2585
573-56-8, 602
575-41-7, 2587
575-89-3, 1283
575-90-6, 1303
576-24-9, 590
576-26-1, 1456
577-66-2, 2067
578-06-3, 2796
578-07-4, 2794
578-67-6, 1622
579-10-2, 1702
579-44-2, 2994
579-66-8, 2082
579-75-9, 1389
580-02-9, 1951
580-13-2, 1879
580-16-5, 1623
580-18-7, 1620
580-20-1, 1625
580-48-3, 2341
581-28-2, 2797
581-29-3, 2798
581-40-8, 2583
581-42-0, 2581
581-43-1, 1898
582-54-7, 1302
583-58-4, 1138
583-60-8, 1187
583-61-9, 1140
583-78-8, 594
584-02-1, 513
585-76-2, 987
586-11-8, 600
586-30-1, 1395
586-38-9, 1394
586-76-5, 986
587-64-4, 1305
588-22-7, 1304
588-32-9, 1332
589-29-7, 1437
589-34-4, 1246
589-35-5, 920
589-38-8, 854
589-55-9, 1262
589-81-1, 1573
589-82-2, 1250
589-90-2, 1526
589-93-5, 1141
590-01-2, 1220
590-35-2, 1242
590-36-3, 922
590-47-6, 502
590-60-3, 888
590-73-8, 1572
591-07-1, 162
591-12-8, 411

591-17-3, 1070
591-22-0, 1137
591-24-2, 1188
591-27-5, 741
591-35-5, 591
591-49-1, 1176
591-50-4, 648
591-68-4, 1826
591-76-4, 1244
591-78-6, 848
591-93-5, 422
591-97-9, 293
592-27-8, 1575
592-35-8, 496
592-41-6, 822
592-42-7, 787
592-76-7, 1204
592-77-8, 1206
592-84-7, 472
593-45-3, 3611
593-52-2, 31
593-53-3, 16
593-59-9, 107
593-75-9, 67
594-60-5, 917
594-72-9, 62
594-83-2, 1258
595-39-1, 492
595-41-5, 1251
595-44-8, 142
595-46-0, 438
596-29-2, 3725
597-35-3, 367
597-43-3, 800
597-49-9, 1249
597-64-8, 1598
597-96-6, 1252
598-42-5, 89
598-53-8, 365
598-55-0, 90
598-75-4, 514
598-92-5, 73
599-01-9, 118
599-75-7, 3018
599-82-6, 1941
599-84-8, 2237
600-13-5, 247
600-25-9, 157
600-36-2, 1255
601-75-2, 436
601-77-4, 909
601-89-8, 659
602-38-0, 1876
602-99-3, 1026
603-00-9, 2036
603-11-2, 1293
603-12-3, 982
603-45-2, 3626
603-83-8, 1107
604-69-3, 3375
604-75-1, 3141
605-45-8, 3063
605-54-9, 3373
605-71-0, 1877

606-17-7, 3723
606-19-9, 1323
606-22-4, 664
606-35-9, 1024
607-67-0, 1912
608-34-4, 404
608-40-2, 798
608-66-2, 936
608-73-1, 690
608-93-5, 531
609-09-6, 1169
609-36-9, 441
610-02-6, 1069
610-09-3, 1489
610-30-0, 981
610-72-0, 1679
610-93-5, 1291
611-01-8, 1682
611-13-2, 727
611-14-3, 1727
611-36-9, 1621
611-72-3, 1393
611-73-4, 1319
612-45-3, 1672
613-12-7, 3147
613-13-8, 2985
613-14-9, 2974
613-33-2, 3005
614-28-8, 2986
614-29-9, 3014
614-61-9, 1331
614-69-7, 1345
614-77-7, 1436
615-53-2, 315
615-77-0, 408
616-06-8, 893
616-25-1, 462
616-39-7, 524
616-62-6, 803
616-74-0, 604
616-87-5, 1509
617-04-9, 1232
617-45-8, 302
617-65-2, 446
618-51-9, 1001
618-65-5, 2863
618-83-7, 1324
619-02-3, 2105
619-04-5, 1681
619-45-4, 1419
619-58-9, 1003
619-82-9, 1488
619-84-1, 1713
620-71-3, 1705
621-34-1, 1466
621-64-7, 908
621-82-9, 1638
621-84-1, 1417
622-08-2, 1757
622-45-7, 1501
622-50-4, 1362
622-51-5, 1435
622-59-3, 1344
622-62-8, 1468

622-78-6, 1346
622-96-8, 1724
623-10-9, 1143
623-12-1, 1075
623-26-7, 1275
623-27-8, 1317
623-36-9, 429
623-37-0, 929
623-42-7, 476
623-85-8, 889
623-87-0, 167
624-38-4, 596
624-51-1, 1847
624-54-4, 1550
624-79-3, 145
625-06-9, 1256
625-23-0, 1254
625-54-7, 517
625-98-9, 579
626-03-9, 398
626-48-2, 416
626-64-2, 396
626-89-1, 923
626-93-7, 919
626-97-1, 487
627-05-4, 343
627-08-7, 927
627-12-3, 344
627-18-9, 188
627-19-0, 421
627-30-5, 191
628-02-4, 883
628-41-1, 756
628-55-7, 1580
628-63-7, 1225
628-67-1, 1510
628-71-7, 1175
628-92-2, 1174
628-94-4, 831
628-99-9, 1848
629-01-6, 1561
629-04-9, 1236
629-05-0, 1494
629-06-1, 1237
629-14-1, 933
629-59-4, 3125
629-76-5, 3269
630-06-8, 4121
630-20-6, 57
630-51-3, 1507
630-60-4, 4079
631-07-2, 2235
631-36-7, 1597
632-12-2, 204
632-58-6, 1270
632-95-1, 1292
632-99-5, 3735
633-12-5, 955
634-66-2, 539
634-90-2, 540
635-46-1, 1701
635-65-4, 4100
636-26-0, 406
636-46-4, 1325

636-61-3, 285
637-07-0, 2640
637-83-2, 2727
638-11-9, 1223
638-16-4, 123
638-42-6, 884
638-49-3, 864
638-53-9, 2936
640-15-3, 943
640-68-6, 498
641-81-6, 3972
643-12-9, 878
643-79-8, 1316
644-08-6, 2814
644-35-9, 1753
645-05-6, 1809
645-56-7, 1754
645-92-1, 151
645-93-2, 134
646-31-1, 3983
651-06-9, 2240
664-95-9, 3081
673-04-1, 1521
673-06-3, 1709
675-09-2, 1123
684-93-5, 95
691-37-2, 821
693-02-7, 786
693-23-2, 2736
696-23-1, 259
697-82-5, 1752
698-71-5, 1750
700-12-9, 2300
700-47-0, 609
700-81-2, 668
701-34-8, 677
704-61-0, 1368
705-36-2, 708
708-79-2, 715
709-50-2, 1231
709-98-8, 1646
723-46-6, 1975
723-62-6, 3136
729-99-7, 2268
730-40-5, 2562
731-27-1, 2021
732-11-6, 2231
738-70-5, 3057
739-27-5, 3695
741-58-2, 3108
742-20-1, 2890
745-27-7, 3645
751-97-3, 4035
752-61-4, 4117
759-05-7, 434
759-73-9, 209
759-94-4, 1835
760-78-1, 488
763-29-1, 819
763-69-9, 1228
766-51-8, 1077
769-39-1, 549
769-66-4, 667
771-41-5, 720

771-61-9,　533
777-22-0,　3097
779-02-2,　3146
780-11-0,　2686
781-43-1,　3284
786-19-6,　2302
789-02-6,　2956
814-29-9,　2761
821-09-0,　463
821-38-5,　3112
826-81-3,　1911
832-66-6,　1745
832-69-9,　3148
834-12-8,　1802
834-24-2,　3020
838-85-7,　2580
840-97-1,　2878
841-06-5,　2355
842-00-2,　2611
846-46-8,　3692
846-49-1,　3135
853-23-6,　3840
853-34-9,　3638
855-22-1,　3902
860-22-0,　3272
866-23-9,　452
870-93-9,　1534
872-55-9,　776
874-14-6,　761
879-39-0,　528
881-87-8,　1330
883-81-8,　1766
886-50-0,　2151
900-95-8,　3732
914-00-1,　3875
922-55-4,　836
923-06-8,　245
923-32-0,　838
924-16-3,　1578
928-45-0,　352
928-49-4,　785
932-52-5,　260
932-83-2,　830
933-75-5,　566
933-78-8,　565
934-33-8,　712
935-95-5,　545
936-40-3,　1042
937-40-6,　1438
938-86-3,　974
939-27-5,　2582
944-22-9,　2087
944-61-6,　1306
949-13-3,　3105
950-37-8,　817
953-17-3,　1731
957-51-7,　3323
958-09-8,　2042
959-98-8,　1606
961-11-5,　1906
961-68-2,　2550
968-81-0,　3214
976-71-6,　3887
977-79-7,　3931

987-24-6,　3952
991-42-4,　4097
994-05-8,　926
994-65-0,　2766
1002-84-2,　3267
1004-40-6,　258
1007-28-9,　125
1007-36-9,　1434
1008-85-1,　671
1009-04-7,　750
1009-61-6,　1945
1011-82-1,　1156
1014-69-3,　1522
1014-70-6,　1523
1021-65-4,　2101
1022-46-4,　2960
1024-34-6,　3107
1024-57-3,　1862
1027-87-8,　2894
1031-47-6,　2723
1037-51-0,　2633
1045-69-8,　3842
1067-20-5,　1838
1068-07-1,　2763
1071-83-6,　213
1072-85-1,　575
1073-06-9,　574
1077-16-3,　2699
1083-27-8,　2901
1085-12-7,　3087
1088-92-2,　1636
1099-79-2,　3895
1114-71-2,　2178
1119-65-9,　1178
1120-21-4,　2373
1120-71-4,　185
1123-63-3,　1408
1123-94-0,　1751
1124-06-7,　1406
1125-66-2,　1693
1125-78-6,　2006
1125-80-0,　1910
1125-84-4,　554
1127-76-0,　2584
1127-93-1,　718
1129-26-6,　1153
1129-41-5,　1712
1131-60-8,　2679
1134-23-2,　2352
1134-47-0,　1979
1138-80-3,　1964
1142-70-7,　2287
1143-30-2,　2314
1156-19-0,　3096
1162-65-8,　3414
1163-19-5,　2767
1164-16-5,　3434
1164-91-6,　3850
1165-39-5,　3415
1167-87-9,　3897
1177-87-3,　3953
1185-33-7,　915
1193-24-4,　237
1194-65-6,　956

1198-37-4, 2216
1202-25-1, 2032
1205-06-7, 1914
1219-38-1, 3243
1220-94-6, 3154
1225-56-5, 3549
1251-85-0, 3915
1255-49-8, 4052
1260-17-9, 3873
1264-62-6, 4133
1326-08-5, 3510
1330-20-7, 1428
1330-78-5, 3811
1335-46-2, 3104
1344-32-7, 541
1393-48-2, 4176
1397-89-3, 4140
1400-61-9, 4141
1401-69-0, 4138
1403-17-4, 4167
1404-74-6, 4125
1405-50-1, 3634
1420-06-0, 3914
1435-60-5, 2549
1441-02-7, 1272
1454-85-9, 3505
1456-28-6, 832
1460-18-0, 3265
1461-17-2, 4011
1468-37-7, 278
1509-34-8, 898
1510-25-4, 2786
1524-88-5, 3959
1540-35-8, 1502
1560-84-5, 3857
1560-93-6, 3402
1563-66-2, 2643
1565-17-9, 1421
1569-50-2, 464
1570-64-5, 1079
1571-33-1, 754
1576-59-6, 679, 680
1582-09-8, 2851
1591-82-8, 1956
1596-84-5, 833
1606-53-7, 4065
1606-55-9, 4069
1606-56-0, 3945
1610-17-9, 1801
1610-18-0, 2148
1634-04-4, 512
1636-33-5, 2191
1638-22-8, 2076
1660-93-1, 3311
1668-19-5, 3657
1678-91-7, 1530
1689-82-3, 2559
1689-83-4, 959
1689-84-5, 954
1694-06-0, 1447
1698-60-8, 1892
1702-17-6, 555
1703-58-8, 1474
1705-85-7, 3625

1709-52-0, 1160
1709-59-7, 1484
1730-37-6, 2988
1732-13-4, 3309
1746-01-6, 2407
1746-81-2, 1692
1747-53-1, 1111
1754-47-8, 3905
1754-49-0, 2117
1779-48-2, 753
1804-15-5, 164
1806-26-4, 3106
1809-19-4, 1591
1825-31-6, 1867
1836-75-5, 2490
1836-77-7, 2458
1839-18-5, 378
1852-04-6, 2348
1861-32-1, 1869
1861-40-1, 2852
1861-84-3, 4068
1867-85-2, 3178
1895-97-2, 1947
1897-45-6, 1601
1899-94-1, 1149
1912-24-9, 1497
1912-25-0, 2121
1912-26-1, 1786
1917-92-6, 2848
1917-95-9, 3052
1918-00-9, 1299
1918-02-1, 562
1918-11-2, 3499
1918-13-4, 993
1918-16-7, 2272
1928-44-5, 3364
1929-77-7, 2177
1929-82-4, 568
1929-88-0, 1659
1943-16-4, 36
1951-25-3, 3991
1966-58-1, 1647
1967-16-4, 2205
1970-40-7, 377
1972-08-3, 3837
1982-47-4, 3176
1982-49-6, 3080
1982-55-4, 527
1984-59-4, 1033
1984-65-2, 1035
1985-12-2, 963
1987-50-4, 2918
1988-14-3, 3092
2008-41-5, 2371
2008-58-4, 991
2021-28-5, 2281
2029-64-3, 1836
2030-63-9, 4028
2032-59-9, 2303
2040-96-2, 1531
2043-43-8, 195
2049-95-8, 2299
2050-67-1, 2517
2050-68-2, 2513

2050-89-7, 2590
2051-24-3, 2770
2051-60-7, 2535
2051-61-8, 2538
2051-62-9, 2537
2051-85-6, 2560
2052-14-4, 2284
2059-76-9, 976
2062-78-4, 4051
2076-56-4, 1295
2086-83-1, 3734
2088-71-3, 4016
2088-76-8, 4085
2095-24-1, 3320
2104-64-5, 3007
2104-96-3, 1352
2114-20-7, 1239
2131-41-1, 2827
2131-57-9, 1627
2131-59-1, 962
2131-60-4, 1007
2131-62-6, 1290
2131-63-7, 1289
2131-64-8, 1675
2135-17-3, 3885
2136-99-4, 2383
2138-20-7, 2681
2152-56-9, 522
2163-68-0, 1515
2163-69-1, 2361
2164-08-1, 2892
2164-09-2, 1902
2164-17-2, 1958
2180-92-9, 3596
2205-27-8, 1472
2207-01-4, 1528
2207-04-7, 1524
2212-67-1, 1798
2216-33-3, 1842
2216-34-4, 1840
2217-08-5, 2097
2235-90-7, 2664
2236-60-4, 672
2243-95-0, 1864
2270-20-4, 2282
2275-18-5, 1843
2275-23-2, 1577
2276-96-2, 2631
2277-92-1, 2771
2284-20-0, 1339
2303-16-4, 2112
2303-17-5, 2091
2305-32-0, 1490
2307-68-8, 2888
2309-49-1, 1746
2310-17-0, 2639
2312-15-4, 3705
2314-09-2, 3545
2315-36-8, 824
2315-68-6, 2009
2340-66-1, 3656
2350-32-5, 3500
2361-96-8, 2871
2363-58-8, 3699

2364-46-7, 1863
2367-82-0, 547
2373-84-4, 1741
2381-16-0, 3623
2382-80-1, 3201
2385-74-2, 1778
2385-85-5, 2185
2392-67-8, 1676
2392-68-9, 967
2392-80-5, 2186
2396-63-6, 1103
2396-65-8, 1726
2401-85-6, 1860
2404-58-2, 2182
2404-73-1, 1851
2417-10-9, 2567
2425-10-7, 2027
2432-12-4, 1034
2432-20-4, 1043
2432-21-5, 1041
2432-26-0, 607
2432-27-1, 606
2432-90-8, 4096
2432-99-7, 2372
2437-79-8, 2474
2439-99-8, 374
2446-69-7, 2718
2451-01-6, 2173
2451-62-9, 2658
2452-84-8, 3340
2463-84-5, 1405
2475-44-7, 3291
2475-45-8, 2993
2475-46-9, 3427
2487-01-6, 2857
2491-15-8, 148
2491-52-3, 2548
2491-74-9, 3010
2497-06-5, 1596
2498-77-3, 3621
2516-95-2, 966
2536-31-4, 3142
2540-82-1, 829
2541-69-7, 3622
2566-23-6, 3306
2577-38-0, 616
2581-34-2, 1096
2581-69-3, 3526
2588-04-7, 1268
2591-57-3, 1735
2595-54-2, 2157
2597-03-7, 2698
2597-11-7, 1872
2599-11-3, 1203
2618-25-9, 3507
2622-26-6, 3816
2623-33-8, 1963
2631-37-0, 2688
2631-39-2, 2060
2633-54-7, 1668
2636-26-2, 1670
2642-71-9, 2673
2642-98-0, 3518
2646-17-5, 3418

2655-19-8, 3392
2675-77-6, 1357
2704-58-7, 2637
2716-98-5, 3488
2716-99-6, 3584
2717-00-2, 3687
2717-02-4, 3846
2717-03-5, 3899
2718-25-4, 3809
2725-63-5, 3737
2734-52-3, 3336
2737-00-0, 1595
2751-09-9, 4129
2752-95-6, 3347
2757-10-0, 1960
2772-46-5, 1358
2778-04-3, 1772
2784-27-2, 3155
2785-33-3, 4005
2785-34-4, 4027
2785-35-5, 4063
2813-95-8, 2627
2815-34-1, 906
2817-14-3, 617
2822-41-5, 570
2827-47-6, 1212
2832-40-8, 3182
2834-05-1, 2351
2835-04-3, 748
2835-06-5, 1420
2835-81-6, 345
2835-82-7, 348
2845-89-8, 1076
2854-70-8, 1791
2858-66-4, 1516
2866-43-5, 3508
2872-48-2, 3234
2872-52-8, 3335
2877-14-7, 1273
2882-96-4, 3400
2896-56-2, 4050
2898-12-6, 3296
2901-76-0, 3305
2901-79-3, 3531
2908-76-1, 2509
2915-53-9, 3790
2919-12-2, 2529
2921-88-2, 1695
2939-80-2, 1905
2948-89-2, 2661
2950-47-2, 2762
2955-27-3, 3980
2955-38-6, 3631
2958-76-1, 2340
2958-99-8, 115
2971-36-0, 2982
2971-90-6, 1080
2974-92-7, 2516
2976-74-1, 1301
2985-28-6, 1194
2987-46-4, 1040
2990-70-7, 3619
2998-57-4, 3928
3001-57-8, 264

3002-23-1, 1504
3004-70-4, 2354
3004-71-5, 1183
3025-52-3, 3333
3035-45-8, 1520
3042-84-0, 1891
3050-27-9, 2018
3058-01-3, 1191
3060-89-7, 1690
3068-88-0, 275
3070-15-3, 2329
3070-53-9, 1177
3073-66-3, 1803
3113-72-2, 1343
3115-05-7, 3280
3119-02-6, 1036
3124-38-7, 1241
3125-63-1, 1006
3125-64-2, 1340
3125-71-1, 1626
3125-73-3, 975
3125-77-7, 1277
3125-78-8, 1276
3129-42-8, 3774
3129-43-9, 3964
3134-12-1, 3286
3137-83-5, 1635
3137-84-6, 1913
3146-66-5, 2307
3149-00-6, 2311
3163-07-3, 658
3163-31-3, 3037
3179-89-3, 3458
3180-81-2, 3321
3209-22-1, 558
3213-22-7, 2243
3226-32-2, 2717
3238-38-8, 1025
3238-40-2, 623
3248-05-3, 3003
3268-87-9, 2769
3272-49-9, 3706
3280-61-3, 2917
3306-62-5, 763
3307-39-9, 1644
3316-09-4, 656
3337-70-0, 1367
3337-71-1, 1448
3347-22-6, 2940
3373-53-3, 402
3374-22-9, 203
3380-34-5, 2507
3383-96-8, 3355
3410-54-6, 3934
3424-82-6, 2945
3425-08-9, 222
3452-07-1, 3792
3452-09-3, 1783
3452-97-9, 1845
3458-28-4, 872
3485-14-1, 3250, 3252
3495-42-9, 1269
3521-62-8, 4151
3522-94-9, 1841

3528-90-3,	1961	4248-19-5,	497
3529-82-6,	978	4248-20-8,	1569
3530-01-6,	1628	4282-40-0,	1238
3547-07-7,	1954	4340-77-6,	2540
3555-86-0,	2802	4360-12-7,	3763
3567-85-9,	1785	4361-59-5,	2175
3577-01-3,	3551	4378-43-2,	355
3615-21-2,	1271	4394-00-7,	2781
3615-41-6,	869	4395-65-7,	3724
3625-25-0,	3054	4408-78-0,	97
3679-63-8,	2790	4418-61-5,	20
3680-04-4,	2867	4435-53-4,	1226
3689-24-5,	1599	4441-12-7,	2750
3694-45-9,	1298	4462-43-5,	2632
3694-46-0,	1658	4465-58-1,	3281
3694-47-1,	1312	4471-41-4,	3539
3694-48-2,	1609	4482-46-6,	1900
3694-49-3,	1309	4516-69-2,	1525
3694-58-4,	1297	4540-00-5,	3322
3696-68-2,	1310	4549-44-4,	910
3696-69-3,	1311	4549-74-0,	795
3697-24-3,	3624	4558-58-1,	2296
3697-27-6,	3730	4562-36-1,	4127
3701-44-8,	1653	4582-18-7,	3502
3714-62-3,	530	4589-84-8,	3773
3724-65-0,	273	4589-90-6,	3781
3734-48-3,	1870	4593-90-2,	2012
3734-95-0,	2145	4627-15-0,	4064
3736-92-3,	3911	4658-28-0,	1173
3741-00-2,	2155	4684-94-0,	582
3750-28-5,	1506	4695-62-9,	2104
3764-87-2,	3690	4726-14-1,	2909
3765-57-9,	1868	4741-41-7,	3744
3766-60-7,	2602	4741-74-6,	3679
3766-81-2,	2685	4771-47-5,	964
3769-57-1,	3445	4798-44-1,	846
3771-38-8,	3466	4798-45-2,	853
3772-42-7,	3501	4798-58-7,	851
3772-76-7,	2634	4824-78-6,	1978
3781-74-6,	2713	4849-32-5,	3095
3785-20-4,	2776	4901-51-3,	544
3811-49-2,	1427	4911-60-8,	1495
3813-05-6,	1604	4911-70-0,	1257
3820-67-5,	3632	5003-48-5,	3422
3845-33-8,	1296	5104-49-4,	3160
3847-29-8,	4144	5131-24-8,	2622
3852-09-3,	482	5162-44-7,	290
3852-94-6,	3547	5174-37-8,	2779
3860-63-7,	3293	5221-53-4,	2339
3891-59-6,	3376	5234-68-4,	2606
3970-62-5,	1263	5240-72-2,	1500
3978-68-5,	2692	5251-93-4,	1657
4007-00-5,	1286	5259-88-1,	2610
4047-57-8,	3637	5284-22-0,	2919
4097-22-7,	2041	5335-03-5,	1633
4128-31-8,	1584	5349-56-4,	1230
4147-51-7,	2356	5392-40-5,	2106
4164-91-4,	1564	5394-83-2,	2109
4164-92-5,	1565	5409-83-6,	2453
4171-13-5,	1566	5415-44-1,	1451
4195-88-4,	1556	5422-69-5,	1795
4199-88-6,	2508	5422-81-1,	3045
4206-74-0,	2846	5427-26-9,	410
4234-79-1,	3412	5433-64-7,	2858

5433-91-0, 2599
5437-38-7, 1342
5438-68-6, 1948
5440-65-3, 1562
5455-59-4, 709
5464-71-1, 2367
5512-05-0, 2747
5581-75-9, 2682
5587-89-3, 2603
5598-13-0, 1081
5598-15-2, 1696
5598-52-7, 1082
5614-38-0, 468
5626-90-4, 1990
5633-87-4, 1557
5635-50-7, 3569
5636-83-9, 3746
5651-01-4, 1629
5660-50-4, 2755
5660-52-6, 2368
5694-76-8, 1833
5697-56-3, 4107
5702-69-2, 2597
5704-03-0, 4036
5728-52-9, 2997
5743-12-4, 1450
5779-79-3, 3722
5781-02-2, 2317
5794-13-8, 320
5819-08-9, 309
5826-73-3, 2285
5827-05-4, 1852
5836-10-2, 3425
5847-59-6, 576
5878-13-7, 3100
5878-14-8, 3256
5892-11-5, 3920
5902-51-2, 1762
5915-41-3, 1788
5929-66-8, 3129
5949-29-1, 775
5950-87-8, 2450
5959-95-5, 456
5962 42-5, 211
5965-66-2, 2742
5967-84-0, 1166
5989-27-5, 2090
6009-81-0, 3450
6020-39-9, 839
6032-29-7, 510
6052-13-7, 1350
6054-48-4, 3314
6054-58-6, 3552
6058-23-7, 1871
6058-77-1, 2684
6059-47-8, 3565
6098-19-7, 1563
6113-61-7, 1200
6128-03-6, 1109
6147-14-4, 2707
6151-51-5, 3749
6153-56-6, 59
6164-98-3, 2016
6190-65-4, 790

6192-52-5, 1126
6214-28-4, 249
6232-56-0, 3169
6250-23-3, 3525
6259-76-3, 2902
6280-96-2, 1759
6283-90-5, 2932
6283-92-7, 3268
6284-75-9, 804
6288-11-5, 805
6299-56-5, 2866
6325-93-5, 710
6329-15-3, 2642
6329-16-4, 2666
6362-80-7, 3555
6366-18-3, 3636
6373-73-5, 2563
6377-18-0, 4092
6382-06-5, 1551
6408-72-6, 4006
6418-46-8, 3858
6452-71-7, 3247
6483-60-9, 2714
6485-40-1, 2071
6515-38-4, 376
6600-40-4, 489
6625-00-9, 246
6626-32-0, 2716
6631-21-6, 3834
6642-26-8, 713
6665-98-1, 655
6676-26-2, 2711
6677-98-1, 3963
6677-99-2, 3998
6678-00-8, 4022
6737-11-7, 794
6744-54-3, 1827
6750-98-7, 3208
6781-97-1, 1771
6876-23-9, 1532
6893-02-3, 3151
6893-26-1, 447
6912-98-7, 1943
6915-15-7, 284
6938-06-3, 2030
6942-59-2, 2141
6947-77-9, 707
6961-82-6, 686
6966-78-5, 1050
6968-16-7, 368
6972-47-0, 1335
6973-01-9, 669
6988-21-2, 2262
7005-72-3, 2541
7008-42-6, 3733
7012-37-5, 2495
7061-54-3, 3818
7068-83-9, 509
7177-48-2, 3343
7200-25-1, 912
7220-81-7, 3419
7225-96-9, 2033
7241-98-7, 3420
7242-04-8, 4147

7242-58-2,	2183	13167-98-1,	3639
7242-59-3,	1854	13181-17-4,	2773
7286-76-2,	1760	13194-48-4,	1589
7286-84-2,	1334	13214-70-5,	1866
7287-19-6,	2153	13215-53-7,	3446
7287-36-7,	2889	13215-54-8,	3436
7292-16-2,	2922	13246-97-4,	3150
7294-05-5,	1376	13256-07-0,	907
7307-04-2,	1503	13286-32-3,	2328
7311-27-5,	3941	13292-46-1,	4132
7334-51-2,	1533	13329-78-7,	2130
7374-53-0,	2150	13360-45-7,	1664
7388-32-1,	3556	13360-61-7,	3266
7391-69-7,	1164	13360-63-9,	939
7417-67-6,	165	13373-32-5,	781
7449-27-6,	1374	13424-83-4,	3115
7568-93-6,	1480	13424-87-8,	2934
7576-65-0,	3511	13452-85-2,	2344
7581-97-7,	307	13457-18-6,	3082
7585-39-9,	4131	13532-26-8,	2149
7634-39-1,	1234	13538-22-2,	2693
7642-68-4,	3965	13539-59-8,	3352
7651-80-1,	2959	13545-04-5,	801
7660-44-8,	4072	13593-03-8,	2654
7681-93-8,	4104	13674-87-8,	1781
7700-17-6,	3078	13684-56-5,	3312
8001-35-2,	1937	13684-63-4,	3313
8065-62-1,	525	13698-49-2,	3922
9002-91-9,	1558	13698-87-8,	3384
10004-44-1,	255	13870-90-1,	4177
10016-20-3,	4119	13913-40-1,	2822
10061-01-5,	127	13915-79-2,	1130
10061-02-6,	128	13952-84-6,	371
10118-90-8,	3924	13993-65-2,	3164
10236-47-2,	4034	14000-32-9,	2885
10238-21-8,	3926	14009-24-6,	3954
10311-84-9,	3043	14073-97-3,	2125
10367-84-7,	3093	14090-87-0,	1505
10460-33-0,	973	14129-98-7,	2749
10500-16-0,	1834	14131-04-5,	1426
10548-10-4,	1853	14144-33-3,	867
10587-37-8,	1274	14144-34-4,	1554
10605-21-7,	1660	14144-35-5,	1830
11005-63-3,	4091	14144-36-6,	2174
11096-82-5,	2419	14144-37-7,	2366
11097-69-1,	2461	14144-39-9,	1229
11104-28-2,	2536	14144-41-3,	1832
12122-67-7,	266	14144-48-0,	2369
12239-34-8,	3947	14187-32-7,	3755
12672-29-6,	2460	14214-32-5,	3331
12771-68-5,	3184	14255-72-2,	2331
13029-08-8,	2521	14293-44-8,	3177
13067-93-1,	3171	14299-55-9,	291
13071-79-9,	1850	14309-40-1,	3091
13094-23-0,	3561	14309-41-2,	3244
13094-24-1,	3443	14334-36-2,	2667
13100-13-5,	3441	14368-76-4,	807
13110-37-7,	2687	14419-01-3,	2199
13121-70-5,	3608	14437-17-3,	1934
13137-64-9,	4084	14439-13-5,	752
13156-36-0,	2579	14484-64-1,	1807
13156-38-2,	1767	14486-58-9,	2952
13156-74-6,	2258	14491-59-9,	2210
13158-31-1,	2668	14509-68-3,	3643

14684-54-9, 1369
14771-98-3, 2647
14816-18-3, 2655
14854-08-1, 3728
14885-29-1, 1172
14938-35-3, 2319
14949-00-9, 80
14957-18-7, 4066
15216-12-3, 231
15251-46-4, 1970
15253-51-7, 3936
15271-41-7, 1981
15299-99-7, 3463
15310-01-7, 2793
15394-30-6, 3385
15414-82-1, 425
15457-05-3, 2774
15481-55-7, 661, 662
15507-76-3, 1171
15536-49-9, 4012
15545-48-9, 2017
15676-16-1, 3251
15686-71-2, 3325
15687-27-1, 2898
15722-48-2, 2972
15788-85-9, 3394
15853-38-0, 1162
15862-07-4, 2494
15869-95-1, 2180
15879-93-3, 1477
15950-66-0, 564
15968-05-5, 2478
15972-60-8, 3079
16022-69-8, 958
16034-77-8, 2604
16041-24-0, 1047
16045-92-4, 250
16069-36-6, 3791
16090-33-8, 657
16118-49-3, 2671
16165-66-5, 3130
16268-62-5, 1535
16268-75-0, 2362
16268-79-4, 2734
16268-92-1, 1808
16269-01-5, 1811
16269-02-6, 2345
16310-36-4, 613
16375-90-9, 1710
16488-52-1, 3099
16488-53-2, 3257
16488-54-3, 3395
16488-56-5, 3601
16488-57-6, 3709
16533-50-9, 1756
16606-02-3, 2506
16671-34-4, 4031
16752-77-5, 454
16766-29-3, 1336
16805-99-5, 2203
16806-29-4, 1370
16840-28-1, 2238
16878-76-5, 621
16878-77-6, 1053

16891-79-5, 1161
16909-11-8, 2249
17103-43-4, 2204
17103-48-9, 1942
17103-49-0, 1901
17140-78-2, 4093
17199-29-0, 1398
17239-22-4, 3231
17239-23-5, 3166
17239-27-9, 2651
17243-26-4, 2265
17243-29-7, 2227
17260-71-8, 685
17296-50-3, 3163
17321-62-9, 1966
17321-63-0, 1932
17348-59-3, 1253
17418-58-5, 3800
17465-86-0, 4142
17530-23-3, 1784
17530-24-4, 2118
17560-51-9, 3310
17598-81-1, 876
17629-30-0, 3607
17698-03-2, 3120
17700-09-3, 538
17804-35-2, 3056
18046-21-4, 3410
18069-66-4, 3939
18181-70-9, 1355
18181-80-1, 3423
18259-05-7, 2444
18264-75-0, 29
18530-56-8, 2924
18559-94-9, 2920
18691-97-9, 1969
18854-01-8, 2853
18883-66-4, 1519
18979-72-1, 2078
18979-73-2, 2323
19044-88-3, 2708
19077-97-5, 1749
19077-98-6, 2824
19167-57-8, 1100
19167-60-3, 721
19167-62-5, 751
19167-63-6, 770
19578-81-5, 2791
19666-30-9, 3199
19679-38-0, 3301
19810-30-1, 1904
19872-68-5, 2880
19872-70-9, 3495
19872-72-1, 3431
19922-87-3, 232
19937-59-8, 2019
20043-94-1, 3707
20168-99-4, 3805
20195-08-8, 1855
20195-16-8, 3131
20203-81-0, 765
20279-51-0, 1829
20354-26-1, 1605
20427-84-3, 3710

20460-96-2, 2879
20473-73-8, 1513
20473-77-2, 2135
20562-02-1, 4136
20629-50-9, 3620
20636-48-0, 4003
20675-21-2, 3072
20675-22-3, 4024
20675-23-4, 2873
20675-24-5, 2608
20675-25-6, 2613
20682-28-4, 3421
20830-75-5, 4128
20923-67-5, 1567
21035-44-9, 941
21087-64-9, 1499
21149-88-2, 2309
21312-10-7, 2618
21321-07-3, 1895
21413-25-2, 3430
21413-28-5, 3801
21413-53-6, 3798
21416-67-1, 2315
21452-14-2, 2223
21452-18-6, 2591
21540-35-2, 2064
21548-32-3, 828
21609-90-5, 2788
21616-46-6, 3292
21725-46-2, 1763
21829-25-4, 3439
21885-31-4, 816
21914-07-8, 3275
21988-05-6, 1000
22005-65-8, 670
22071-15-4, 3295
22131-79-9, 2214
22204-53-1, 3017
22212-55-1, 3534
22224-92-6, 2923
22248-79-9, 1907
22259-30-9, 2295
22496-45-3, 2712
22781-23-3, 2264
22817-17-0, 2784
22832-87-7, 3527
22862-76-6, 3073
22900-79-4, 2028
22916-47-8, 3524
22936-75-0, 2357
22936-86-3, 1773
23043-88-1, 1742
23103-98-2, 2337
23135-22-0, 1202
23167-99-9, 2710
23184-66-9, 3497
23214-92-8, 4029
23325-78-2, 3326
23409-17-8, 2034
23505-41-1, 2928
23560-59-0, 1732
23576-23-0, 2804
23593-75-1, 3867
23597-82-2, 2689

23767-00-2, 1677
23784-45-4, 2257
23784-51-2, 3070
23784-56-7, 3222
23815-28-3, 637
23945-44-0, 381
23947-60-6, 2338
23950-58-5, 2572
24009-06-1, 1873
24017-47-8, 2672
24151-93-7, 3114
24253-33-6, 248
24270-82-4, 2108
24280-93-1, 3462
24282-51-7, 2134
24353-58-0, 3241
24356-66-9, 2086
24448-94-0, 766
24535-67-9, 2706
24539-94-4, 1356
24606-94-8, 3228
24691-76-7, 2837
24691-80-3, 2576
24759-34-0, 3804
24759-35-1, 3802
24759-37-3, 3869
24759-38-4, 3797
24759-41-9, 3868
24851-65-8, 1048
24851-98-7, 2927
24934-91-6, 506
25006-32-0, 2789
25057-89-0, 1989
25141-27-9, 1281
25210-30-4, 701
25311-71-1, 3253
25323-68-6, 2497
25366-23-8, 736
25429-29-2, 2432
25496-72-4, 3856
25569-80-6, 2515
25687-48-3, 2785
25687-50-7, 1652
25735-67-5, 2321
25786-28-1, 3405
25911-76-6, 608
26002-80-2, 3921
26087-47-8, 2921
26225-79-6, 2903
26258-70-8, 3670
26259-45-0, 2147
26271-72-7, 3966
26328-59-6, 2963
26399-36-0, 3030
26584-42-9, 228
26601-64-9, 2408
26628-97-7, 1594
26644-46-2, 2055
26761-40-0, 4062
26914-33-0, 2459
26952-23-8, 126
26962-26-5, 3397
27134-24-3, 3015
27196-00-5, 3126

27199-40-2, 2777
27282-86-6, 1733
27314-13-2, 2539
27321-61-5, 208
27397-05-3, 2748
27541-88-4, 2457
27554-26-3, 3969
27563-65-1, 1279
27653-63-0, 1163
27760-74-3, 2066
27858-07-7, 2378
27955-87-9, 3828
27970-50-9, 2131
28176-10-5, 229
28189-85-7, 3379
28249-77-6, 2662
28346-65-8, 2925
28396-87-4, 3299
28406-15-7, 2333
28456-54-4, 760
28463-03-8, 3159
28463-05-0, 3297
28655-71-2, 2398
28805-86-9, 2075
28994-41-4, 2819
29091-05-2, 2256
29136-19-4, 4002
29232-93-7, 2343
29446-15-9, 2455
29826-16-2, 4086
29839-52-9, 1539
29839-54-1, 2168
29839-55-2, 2169
29839-58-5, 1821
29839-59-6, 1540
29839-60-9, 1541
29839-61-0, 1792
29839-62-1, 1793
29839-63-2, 3111
29839-64-3, 1823
29839-66-5, 1817
29839-67-6, 1553
29839-68-7, 1800
29839-70-1, 2162
29839-71-2, 2166
29839-72-3, 2170
29839-73-4, 2171
29839-74-5, 1548
29839-76-7, 1825
29839-77-8, 2172
29839-78-9, 2754
29848-44-0, 479
29848-45-1, 2163
29883-15-6, 3767, 3768
29973-13-5, 2291
30003-26-0, 862
30003-27-1, 2165
30007-47-7, 263
30010-08-3, 1221
30010-09-4, 2164
30091-04-4, 3013
30377-37-8, 403
30448-43-2, 3370
30544-61-7, 3536

30560-19-1, 358
30564-38-6, 1044
30565-25-4, 319
30652-11-0, 1146
30665-94-2, 3300
30746-58-8, 2404
30979-48-7, 1518
31218-83-4, 2156
31220-35-6, 3337
31431-39-7, 3283
31482-09-4, 1810
31508-00-6, 2446
31637-97-5, 3537
31647-36-6, 3796
31817-24-0, 3871
31889-35-7, 1831
32060-78-9, 3944
32365-02-9, 1721
32407-99-1, 1719
32598-10-0, 2480
32598-11-1, 2479
32598-12-2, 2464
32598-13-3, 2463
32620-68-1, 2601
32620-70-5, 2834
32620-72-7, 3002
32690-93-0, 2466
32861-85-1, 2780
33025-41-1, 2475
33089-61-1, 3672
33146-45-1, 2514
33213-65-9, 1607
33244-57-4, 3162
33284-50-3, 2519
33284-52-5, 2462
33284-53-6, 2465
33376-25-9, 1774
33419-42-0, 4071
33423-92-6, 2406
33433-95-3, 1284
33439-45-1, 1933
33564-31-7, 4014
33669-70-4, 614
33693-04-8, 2146
33820-53-0, 3249
33857-26-0, 2454
33979-03-2, 2427
34006-76-3, 3065
34014-18-1, 1790
34123-59-6, 2700
34197-26-7, 3451
34244-80-9, 1046
34392-82-0, 2206
34462-96-9, 2487
34645-84-6, 2967
34801-09-7, 1442
34883-39-1, 2520
34883-43-7, 2518
34968-90-6, 675
35065-27-1, 2409
35065-28-2, 2415
35065-29-3, 2387
35065-30-6, 2389
35075-35-5, 1498

35367-38-5, 2954
35693-92-6, 2493
35693-99-3, 2476
35694-06-5, 2416
35694-08-7, 2382
35801-62-8, 1199
35801-67-3, 3224
35822-46-9, 2375
35846-47-0, 2588
35858-24-3, 3230
36322-90-4, 3167
36330-85-5, 3294
36355-01-8, 2402
36370-80-6, 2724
36378-49-1, 3698
36378-71-9, 3614
36559-22-5, 2471
36566-80-0, 1179
36614-38-7, 1267
36653-71-1, 619
36653-82-4, 3403
36729-58-5, 2181
36734-19-7, 2821
36756-79-3, 3391
36765-01-2, 1032
37076-68-9, 1409
37139-21-2, 3393
37517-28-5, 3908
37680-65-2, 2502
37680-66-3, 2501
37680-73-2, 2430
37723-78-7, 3200
37764-25-3, 1476
37809-02-2, 158
37924-13-3, 2991
38026-46-9, 1110
38260-54-7, 2114
38363-40-5, 3599
38379-99-6, 2442
38380-01-7, 2445
38380-02-8, 2439
38380-03-9, 2448
38380-07-3, 2418
38380-08-4, 2412
38411-25-5, 2391
38434-77-4, 149
38444-73-4, 2503
38444-76-7, 2491
38444-78-9, 2500
38444-81-4, 2504
38444-85-8, 2505
38444-86-9, 2496
38444-90-5, 2492
38444-93-8, 2470
38713-56-3, 3389
38765-78-5, 1471
38821-53-3, 3342
38964-22-6, 2456
38973-73-8, 3600
38973-74-9, 3708
38973-75-0, 3786
38973-76-1, 3854
39001-02-0, 2768
39075-90-6, 2080

39196-18-4, 1805
39227-28-6, 2381
39227-53-7, 2489
39227-54-8, 2488
39227-58-2, 2429
39227-61-7, 2385
39588-36-8, 2213
39719-87-4, 2242
39832-36-5, 1744
40038-46-8, 2561
40186-70-7, 2390
40186-72-9, 2376
40326-33-8, 1555
40363-76-6, 3137
40487-42-1, 2908
40596-69-8, 3711
40691-50-7, 3138
41047-52-3, 1425
41198-08-7, 2286
41295-28-7, 3318
41394-05-2, 1938
41464-39-5, 2473
41464-40-8, 2468
41464-41-9, 2477
41464-43-1, 2482
41464-47-5, 2485
41464-51-1, 2437
41483-43-6, 2929
41492-69-7, 2346
41541-11-1, 3568
41541-13-3, 3453
41541-14-4, 3459
41814-78-2, 1630
42175-34-8, 2937
42509-80-8, 1797
42576-02-3, 2955
42782-57-0, 3435
43121-43-3, 3026
45376-90-7, 467
47000-92-0, 1959
50471-44-8, 2542
50512-35-1, 2720
50785-22-3, 2222
51146-56-6, 2900
51146-57-7, 2899
51166-71-3, 4169
51203-19-1, 1972
51203-20-4, 2035
51207-31-9, 2403
51218-45-2, 3240
51235-04-2, 2726
51333-22-3, 3995
51337-71-4, 3655
51338-27-3, 3288
51437-89-9, 3396
51437-90-2, 3602
51437-91-3, 3788
51437-92-4, 3904
51437-95-7, 3855
51453-65-7, 2870
51527-19-6, 2197
51543-29-4, 2616
51543-30-7, 2656
51543-31-8, 2269

51736-24-4, 2691
51908-16-8, 2414
51940-44-4, 3050
51953-05-0, 383
51953-13-0, 235
51953-17-4, 234
52061-82-2, 3122
52222-35-2, 3245
52645-53-1, 3806
52663-58-8, 2484
52663-59-9, 2472
52663-60-2, 2434
52663-62-4, 2431
52663-63-5, 2424
52663-65-7, 2393
52663-67-9, 2394
52663-68-0, 2399
52663-69-1, 2396
52663-70-4, 2388
52663-71-5, 2386
52663-74-8, 2395
52663-77-1, 2377
52704-70-8, 2422
52709-86-1, 3303
52712-04-6, 2425
52712-05-7, 2397
52717-51-8, 1743
52717-52-9, 1997
52744-13-5, 2421
52756-22-6, 3640
52918-63-5, 3870
53162-40-6, 2217
53469-21-9, 2499
53496-15-4, 1218
53535-33-4, 1261
53648-05-8, 2906
53744-47-1, 133
53780-34-0, 2255
53799-78-3, 3918
53982-07-3, 3447
53983-00-9, 423
53983-01-0, 1180
53994-73-3, 3168
54010-85-4, 789
54135-80-7, 995
54612-53-2, 2916
54942-37-9, 3594
54942-38-0, 3783
54942-39-1, 3903
54942-40-4, 3973
54942-41-5, 3380
54942-42-6, 2260
55030-48-3, 3492
55125-24-1, 3851
55205-89-5, 2690
55215-17-3, 2441
55215-18-4, 2410
55283-68-6, 2828
55312-69-1, 2440
55335-06-3, 972
55380-34-2, 843
55441-71-9, 714
55511-98-3, 2102
55648-40-3, 2271

55702-45-9, 2498
55774-17-9, 4082
55774-19-1, 4067
55791-76-9, 2905
56030-56-9, 2411
56070-16-7, 1857
56209-30-4, 1443
56265-21-5, 3765
56265-22-6, 3668
56265-23-7, 3662
56265-24-8, 3756
56265-26-0, 3678
56265-27-1, 3562
56288-27-8, 3207
56529-85-2, 1617
56558-16-8, 2436
56563-18-9, 694
56741-95-8, 1890
56995-20-1, 3194
57045-86-0, 1328
57094-83-4, 4013
57117-31-4, 2384
57117-44-9, 2380
57229-74-0, 301
57381-79-0, 3865
57440-16-1, 2648
57837-19-1, 3229
58182-63-1, 3416
58447-18-0, 3542
58471-47-9, 1106
58522-87-5, 1930
58528-60-2, 3438
58947-88-9, 615
58965-05-2, 2207
59080-33-0, 551
59080-37-4, 2452
59536-65-1, 2401
59602-16-3, 2081
59746-11-1, 1671
59831-97-9, 3044
59831-98-0, 3358
59831-99-1, 3378
59832-00-7, 3481
59832-01-8, 3593
59832-02-9, 3696
59832-03-0, 3845
59832-05-2, 3209
59832-06-3, 3357
59832-07-4, 3482
59832-08-5, 3681
59865-13-3, 4166
59886-40-7, 3901
59886-43-0, 3574
59886-45-2, 3029
59886-47-4, 3386
59886-48-5, 3597
59886-49-6, 3061
59886-50-9, 3083
59886-52-1, 2674
59886-53-2, 3084
59886-54-3, 3086
59886-55-4, 3387
59917-58-7, 2058
60041-48-7, 179

60077-04-5, 178
60102-87-6, 184
60102-88-7, 183
60129-67-1, 3328
60145-20-2, 2433
60166-93-0, 3475
60166-94-1, 3471
60168-88-9, 3411
60208-45-9, 3473
60233-24-1, 2481
60469-53-6, 3124
60631-05-2, 3123
60631-06-3, 3121
60631-07-4, 3119
60631-08-5, 3118
60631-09-6, 3089
60631-10-9, 3088
60730-94-1, 1492
60766-57-6, 788
60766-61-2, 1931
60784-70-5, 2926
60908-29-4, 1105
61251-77-2, 2188
61336-70-7, 3344
61358-28-9, 3989
61358-30-3, 4032
61576-93-0, 3509
61576-97-4, 3506
61798-70-7, 2420
61986-93-4, 1926
62046-37-1, 1648
62113-41-1, 1084
62113-42-2, 2024
62249-37-0, 3255
62433-26-5, 2775
62475-58-5, 1235
62536-86-1, 2280
62570-20-1, 3327
62666-20-0, 3424
63283-80-7, 825
64098-82-4, 1734
64124-14-7, 1536
64124-15-8, 2159
64124-17-0, 2124
64124-20-5, 2751
64124-21-6, 1812
64649-43-0, 1656
64649-57-6, 2259
64649-63-4, 2872
64706-54-3, 3960
64810-90-8, 3110
64810-91-9, 3262
64947-06-4, 3085
65125-87-3, 3437
65174-99-4, 2884
65267-94-9, 2849
65267-95-0, 3688
65267-96-1, 3900
65267-97-2, 3028
65277-42-1, 4009
65445-09-2, 3994
65510-45-4, 2438
65783-69-9, 2877
65882-61-3, 620

65882-62-4, 612
65926-28-5, 2660
65926-30-9, 3077
65926-31-0, 3238
65926-32-1, 3235
65926-33-2, 3236
65926-34-3, 3382
66460-51-3, 3595
66461-73-2, 896
66542-36-7, 1410
66542-37-8, 1699
66542-48-1, 1957
66542-50-5, 3377
66843-01-4, 2062
66947-87-3, 810
66999-97-1, 2190
66999-98-2, 2544
66999-99-3, 2288
67028-18-6, 2405
67051-26-7, 2704
67337-73-9, 1184
67562-39-4, 2374
67877-88-7, 700
67888-96-4, 2428
68066-37-5, 1127
68194-14-9, 2426
68194-16-1, 2392
68325-41-7, 3661
68325-42-8, 3383
68411-44-9, 2046
68659-48-3, 2220
68797-31-9, 3530
69472-19-1, 3660
69477-71-0, 2031
69577-07-7, 1789
69655-05-6, 1999
69679-30-7, 3498
69679-31-8, 3598
69679-32-9, 3784
69698-47-1, 3477
70136-02-6, 3090
70362-47-9, 2467
70362-48-0, 2469
70424-68-9, 2447
70424-70-3, 2435
70648-26-9, 2379
71067-09-9, 3940
71067-10-2, 4001
71119-16-9, 3440
71119-19-2, 2844
71119-20-5, 2226
71119-21-6, 2208
71119-29-4, 2896
71119-30-7, 2897
71119-31-8, 2675
71119-32-9, 2276
71119-34-1, 3035
71119-35-2, 3366
71119-36-3, 3576
71119-37-4, 2244
71119-38-5, 1976
71138-72-2, 2635
71177-53-2, 3853
71720-65-5, 2845

71759-43-8, 640
71759-45-0, 1698
72487-80-0, 1329
72509-76-3, 3546
72762-00-6, 395
73042-04-3, 1085
73239-20-0, 2059
73383-40-1, 1181
73429-89-7, 4033
73632-76-5, 2122
73632-77-6, 2342
73632-78-7, 3213
73632-79-8, 2123
73632-81-2, 2856
73632-82-3, 2100
73632-83-4, 2312
74103-08-5, 3519
74103-09-6, 3406
74103-11-0, 2546
74103-12-1, 2809
74158-10-4, 1720
74472-34-7, 2483
74472-37-0, 2449
74472-42-7, 2417
74472-44-9, 2423
74692-14-1, 15
74713-68-1, 4070
74713-69-2, 4073
74713-70-5, 3946
74973-14-1, 1669
74973-19-6, 3429
74973-22-1, 4054
75121-79-8, 2320
75410-15-0, 639
75410-16-1, 1083
75410-27-4, 1086
75410-28-5, 2189
75922-48-4, 3189
76006-86-5, 1315
76432-30-9, 2250
76432-33-2, 2251
76432-35-4, 2252
76466-16-5, 1643
76748-72-6, 3740
76842-07-4, 2443
77340-50-2, 317
77632-11-2, 719
77658-97-0, 2297
77868-40-7, 3350
77868-41-8, 3351
77868-43-0, 3348
77868-44-1, 3349
77868-45-2, 3474
77868-46-3, 3682
77868-48-5, 3573
77942-93-9, 3476
78002-88-7, 1201
79578-14-6, 3950
80421-90-5, 3132
80496-87-3, 1282
80702-24-5, 3753
81103-11-9, 4124
82310-91-6, 298
82410-32-0, 1770

84043-25-4, 2641
84043-26-5, 2865
84043-27-6, 2864
84043-28-7, 3069
84052-69-7, 3068
84833-58-9, 3175
84869-93-2, 3985
85326-32-5, 1697
85872-51-1, 3454
86357-19-9, 2886
86357-20-2, 3239
87970-03-4, 2659
87970-05-6, 3025
87984-85-8, 3237
88110-71-8, 3496
88110-72-9, 3697
88110-76-3, 3468
89073-47-2, 3190
89073-49-4, 3683
89073-57-4, 3460
89073-58-5, 3836
89324-38-9, 611
90204-40-3, 1517
90870-76-1, 1995
90996-54-6, 4113
91399-12-1, 1667
91399-13-2, 1984
91625-66-0, 2887
91763-77-8, 2709
91969-06-1, 2279
92788-60-8, 2665
96445-34-0, 3467
96884-73-0, 2349
96933-13-0, 2219
97096-67-8, 2983
97108-48-0, 3302
98204-08-1, 2697
98550-33-5, 1052
98827-15-7, 2278
98827-16-8, 3490
98827-17-9, 3489
98827-18-0, 2277
98827-19-1, 3216
98827-20-4, 3217
98827-21-5, 2000
98827-22-6, 2001
98827-24-8, 2245
98827-27-1, 3076
98846-64-1, 1371
98846-65-2, 2799
98846-66-3, 2552
99167-69-8, 2098
100613-21-6, 2861
100888-67-3, 3133
100915-04-6, 3157
101398-19-0, 861
102170-44-5, 1049
102273-25-6, 1965
102892-46-6, 3988
102943-71-5, 3517
103951-39-9, 2649
103951-40-2, 2874
104663-14-1, 3528
105394-75-0, 2298

106231-50-9, 1962
106231-51-0, 2645
106231-53-2, 2236
106231-54-3, 2261
106231-55-4, 3226
106231-56-5, 3225
106231-57-6, 3494
106231-58-7, 3195
106231-59-8, 2650
106231-60-1, 2875
106231-61-2, 2881
106231-62-3, 2625
106231-63-4, 3049
106231-64-5, 2612
106231-67-8, 3047
106231-68-9, 2840
106231-74-7, 3758
106396-19-4, 3319
106415-21-8, 2003
106593-25-3, 2915
106596-34-3, 3055
106873-99-8, 443
107203-72-5, 3215
108030-77-9, 1307
108979-58-4, 3404
109232-73-7, 1411
109232-74-8, 2254
110459-55-7, 3604
111328-65-5, 2733
111440-78-9, 3615
112599-90-3, 1988
113222-29-0, 1996
113321-22-5, 1703
113977-19-8, 3128
114665-08-6, 2263
114665-09-7, 2652
114665-11-1, 3586
114665-16-6, 3813
114665-17-7, 3324
114665-18-8, 3566
114665-19-9, 3671
114665-20-2, 3761
114681-69-5, 3666
114849-58-0, 1983
115178-63-7, 3232
115178-64-8, 3233
115178-66-0, 2607
115178-67-1, 2839
115178-68-2, 3211
115178-69-3, 3360
115178-70-6, 3361
115178-71-7, 3048
115178-74-0, 2823
115178-75-1, 3021
115178-76-2, 3196
115178-77-3, 2644
115178-78-4, 2869
115178-79-5, 3071
115178-80-8, 2609
115193-27-6, 2646
115193-28-7, 2868
115193-29-8, 3227
115193-30-1, 2624
115193-31-2, 2882

115193-33-4, 3493
116482-56-5, 3212
116482-75-8, 3486
116482-76-9, 3485
116482-77-0, 3339
116482-82-7, 3033
116482-84-9, 3814
116529-61-4, 961
118247-01-1, 2841
118247-02-2, 2831
118247-03-3, 3197
118247-04-4, 2842
118247-07-7, 3042
118247-09-9, 3362
118353-05-2, 2270
121032-20-0, 3648
121032-21-1, 3535
121032-22-2, 3369
121032-31-3, 3540
121032-34-6, 3533
121032-35-7, 3650
121032-36-8, 3647
121032-37-9, 3651
121032-38-0, 3059
121032-39-1, 3554
121032-41-5, 3219, 3221
121032-42-6, 3367
121032-43-7, 2860
121032-44-8, 3218, 3220
121032-51-7, 3653
121032-52-8, 3649
121032-55-1, 3553
124485-63-8, 2939
124669-93-8, 3910
125867-92-7, 2930
125867-93-8, 3388
125867-94-9, 2359
126974-79-6, 2732
129618-40-2, 3174
129968-26-9, 3478
130482-28-9, 1568
131287-51-9, 168
132312-81-3, 3185
132312-82-4, 3187
132312-85-7, 3008
132686-75-0, 3180
132686-95-4, 3183
133627-12-0, 3001
134698-27-4, 3012
134698-31-0, 3186
134698-40-1, 2805
134894-45-4, 3004
135794-72-8, 3161
135794-75-1, 3188
135794-77-3, 3316
135794-80-8, 3332
135794-88-6, 3315
135795-08-3, 3345
135812-04-3, 3329
136370-32-6, 3958
142963-77-7, 3368
142963-79-9, 3588
143288-61-3, 2063
143288-62-4, 2308

143288-63-5, 3101
143288-64-6, 3258
143458-55-3, 2370
143458-56-4, 2756
143458-57-5, 3117
143585-01-7, 1444
143585-02-8, 2914
145913-38-8, 3058
145913-39-9, 3491
145913-40-2, 3587
145913-41-3, 3689
145913-42-4, 3776
145913-43-5, 4047
145913-44-6, 3541
145913-46-8, 3652
145913-47-9, 3060
145913-48-0, 2677
145913-49-1, 3557
145913-50-4, 3480
145913-52-6, 4043
145913-53-7, 4080
147318-81-8, 4102
148832-09-1, 114
150454-14-1, 678
151272-78-5, 4178
153474-30-7, 3635
153474-31-8, 3736
155206-46-5, 3346
155206-47-6, 3172

155206-48-7, 3289
155206-49-8, 3274
160348-99-2, 3198
160349-00-8, 3571
160349-01-9, 3810
160349-02-0, 3024
160349-03-1, 3577
171664-62-3, 699
180262-60-6, 3011
181518-40-1, 3173
187610-86-2, 13
188124-94-9, 3872
188124-96-1, 3912
188124-97-2, 3618
188367-19-3, 4158
194715-43-0, 4175
204443-26-5, 1424
204443-27-6, 1045
204443-29-8, 2065
204443-30-1, 1051
207004-30-6, 2413
223674-01-9, 240
245115-83-7, 1327
245652-81-7, 4021
252967-17-2, 2953
290294-31-4, 3777
292870-71-4, 1423
300360-23-0, 2942

Index 3: Names and Synonyms

A 3322, 3074
A-56268, 4124
A-9991, 3417
9AA, 2795
AAQ, 2962
Abaphos, 3355
Abate, 3355
Abboxampam, 3141
Abietic Acid, 3780
Abietin(Dihydrate), 3374
Abitrate, 2640
Abol, 2337
ABPP, 1890
AC 64475, 828
AC 94320, 1857
Acaraben, 3287
Acaralate, 3425
Acarin, 2958
AcDNal-DCpa-Ser-Tyr-Dhai-Leu-Lys(iPr)-Pro-Dala-NH2, 4178
3,4'-Ace-1,2-benzanthracene, 3722
Acenaphthalene, 2511
Acenaphthene, 2553
Acenaphthylene, 2511
Acephate, 358
Acetal, 934
AcetaldehyD-diethylacetal, 934
Acetaldehyde Diethyl Acetal, 934
Acetaldehyde Homopolymer, 1558
Acetaldehyde Tetramer, 1558
Acetamid, 87
Acetamide, 87
Acetamide, N-Acctyl-2-(bcnzoyloxy)-, 2220
Acetamide, N-Acetyl-2-(benzoyloxy)-N-methyl-, 2609
Acetamide, 2-Amino-N-[4-[(4-aminophenyl)sulfonyl]phenyl], 3024
Acetamide, N-[4-[[(Aminoiminomethyl)amino]sulfonyl]phenyl]-, 1749
Acetamide, N-(2-Amino-2-oxoethyl)-2-(benzoyloxy)-, 2236
Acetamide, N-(2-Amino-2-oxoethyl)-2-(benzoyloxy)-N-methyl-, 2625
Acetamide, N-[5-(Aminosulfonyl)-3-methyl-1,3,4-thiadiazol-2(3H)-ylidene]-, 426
Acetamide, 2-(Benzoyloxy)-, 1656
Acetamide, 2-(Benzoyloxy)-N,N-Acetamide, 2-(benzoyloxy)-N,N-dibutyl-, 3494
Acetamide, 2-(Benzoyloxy)-N,N-bis(2-hydroxyethyl)-, 2881
Acetamide, 2-(Benzoyloxy)-N,N-bis(2-hydroxypropyl)-, 3232
Acetamide, 2-(Benzoyloxy)-N,N-bis(2-methoxyethyl)-, 3233
Acetamide, 2-(Benzoyloxy)-N,N-bis(1-methylethyl)-, 3225

Acetamide, 2-(Benzoyloxy)-N,N-bis(2-methylpropyl)-, 3493
Acetamide, 2-(Benzoyloxy)-N-butyl-, 2868
Acetamide, 2-(Benzoyloxy)-N,N-di-Acetamide, 2-(benzoyloxy)-N,N-di-2-propenyl-, 3195
Acetamide, 2-(Benzoyloxy)-N,N-diethyl-, 2872
Acetamide, 2-(Benzoyloxy)-N,N-dimethyl-, 2261
Acetamide, 2-(Benzoyloxy)-N,N-dipropyl-, 3226
Acetamide, 2-(Benzoyloxy)-N-ethyl-, 2259
Acetamide, 2-(Benzoyloxy)-N-ethyl-N-(2-hydroxyethyl)-, 2875
Acetamide, N-[2-(Benzoyloxy)ethyl]-N-methyl-, 2648
Acetamide, 2-(Benzoyloxy)-N-hexyl-, 3227
Acetamide, 2-(Benzoyloxy)-N-[2-hydroxy-1,1-bis(hydroxymethyl)ethyl]-, 2882
Acetamide, 2-(Benzoyloxy)-N-(2-hydroxyethyl)-N-methyl-, 2650
Acetamide, 2-(Benzoyloxy)-N-methyl-, 1962
Acetamide, 2-(Benzoyloxy)-N-(1-methylethyl)-, 2646
Acetamide, N-[4-(Benzoyloxy)phenyl]-, 3165
Acetamide, 2-(Benzoyloxy)-N-propyl-, 2645
Acetamide, N,N'-[Carbonylbis(oxy-4,1-phenylene)]bis-, 3431
Acetamide, 2-chloro-N,N-diethyl-, 824
Acetamide, N-(4-Chlorophenyl)-, 1354
Acetamide, N-[4-[[(2,6-Dimethyl-4-pyrimidinyl)amino]sulfonyl]phenyl]-, 3037
Acetamide, N-[4-(1-ethoxyethoxy)phenyl]-, 2691
Acetamide, N-[4-[[(5-Ethyl-1,3,4-thiadiazol-2-yl)amino]sulfonyl]phenyl]-, 2633
Acetamide, N-(4-Fluorophenyl)-, 1360
Acetamide, N-(4-Formylphenyl)-, 1654
Acetamide, N-[4-[(2-Hydroxy-5-methylphenyl)azo]phenyl]-, 3182
Acetamide, N-(4-methoxyphenyl)-, 1707
Acetamide, N-Methyl-N-phenyl-, 1702
Acetamide, N-{4-Methyl-3-{{(trifluoromethyl)sulfonyl}amino}phenyl}-, 1959
3-Acetamido-5-acetamidomethyl-2,4,6-triiodobenzoic Acid, 2573
p-Acetamidobenzaldehyde, 1654
p-(p-Acetamidobenzamido)phenyl Acetate, 3429
p-(p-Acetamidobenzamido)phenyl Prostaglandin E2, 4111
p-(p-Acetamidobenzamido)phenyl Prostaglandin F2 α, 4112
2-p-Acetamidobenzenesulphonamido-4:6-dimethylpyri-, 3036
4-Acetamido-2,6-diethylphenol, 2690
4-Acetamido-2,6-diisopropylphenol, 3092
4-Acetamido-2,5-dimethylphenol, 2031
4-Acetamido-2,6-dimethylphenol, 2028
4-Acetamidophenol, 1416
p-Acetamidophenyl Decanoate, 3594

4-Acetamidophenyl Iodide, 1362
3-Acetamidophenyl Isothiocyanate, 1635
m-Acetamidophenyl Isothiocyanate, 1635
p-Acetamidophenyl Prostaglandin E2, 4053
2-Acetamidopteridine, 1349
4-Acetamidopteridine, 1348
7-Acetamidopteridine, 1347
5-Acetamido-1,3,4-thiadiazole-2-sulfonamide, 268
Acetamine Rubine B, 3321
Acetaminophen, 1416
p-Acetaminophen, 1416
Acetaminophen Acetate, 1963
Acetaminophen Butyrate, 2647
Acetaminophen Decanoate, 3594
Acetaminophen Dodecanoate, 3783
Acetaminophen Hexadecanoate, 3973
Acetaminophen Hexanoate, 3072
Acetaminophen Laurate, 3783
Acetaminophen Myristate, 3903
Acetaminophen Octadecanoate, 4024
Acetaminophen Octanoate, 3380
1-(p-Acetaminophenoxy)-1-ethoxyethane, 2691
2-(p-Acetaminophenoxy)tetrahydropyran, 2870
Acetaminophen Palmitate, 3973
Acetaminophen Propionate, 2260
Acetaminophen Stearate, 4024
Acetaminophen Tetradecanoate, 3903
p-4-Acetaminophenyl Acetaminophen, 3431
Acetanilid, 1414
Acetanilide, 1414
Acetanilide, 4'-Bromo-, 1353
Acetanilide, 4'-Chloro-, 1354
Acetanilide, 4'-Formyl-, 1654
Acetanilide, 4'-Hydroxy-, Benzoate (Ester), 3165
Acetanilide, 4'-Hydroxy-, Butyl Carbonate (Ester), 2880
Acetanilide, 4'-Hydroxy-, Butyrate, 2647
Acetanilide, 4'-Hydroxy-, Carbonate (2:1) (Ester), 3431
Acetanilide, 4'-Hydroxy-, Chloroacetate (Ester), 1932
Acetanilide, 4'-Hydroxy-, 2-Chloroethyl Carbonate
  (Ester), 2227
Acetanilide, 4'-Hydroxy-, Cinnamate (Ester), 3421
Acetanilide, 4'-Hydroxy-, Crotonate (Ester), 2608
Acetanilide, 4'-Hydroxy-, Ethyl Carbonate (Ester), 2265
Acetanilide, 4'-Hydroxy-, Hexyl Carbonate (Ester), 3231
Acetanilide, 4'-Hydroxy-, Hydrogen Succinate (Ester),
  2613
Acetanilide, 4'-Hydroxy-, Isobutyl Carbonate (Ester),
  2879
Acetanilide, 4'-Hydroxy-, Isopropyl Carbonate, 2651
Acetanilide, 4'-Hydroxy-, Methyl Carbonate (Ester),
  1966
Acetanilide, 4'-Hydroxy-, Octyl Carbonate (Ester), 3495
Acetanilide, 4'-Hydroxy-, Phenyl Carbonate (Ester), 3166
Acetanilide, 4'-Hydroxy-, Pivalate (Ester), 2873
Acetanilide, 4'-Hydroxy-, Stearate (Ester), 4024
Acetanilide, 4'-Hydroxy-, Succinate, 3737
Acetanilide, 4'-Hydroxy-, Succinate (2:1) (Ester), 3737
Acetanilide, 4'-Iodo-, 1362
Acetanilide, 4'-[(5-Methyl-3-isoxazolyl)sulfamoyl]-,
  2618
p-Acetanisidide, 1707
p-Acetanisidine, 1707
9-[5-O-(Acetate-β-D-arabinofuranosyl)]-6-methoxy-9H-
  purine (Hemihydrate), 2860

Acetate Fast Orange R, 3144
Acetate Fast Rubine B, 3321
Acetate, Phenylmercuric, 1386
Acetate Red Violet R, 2971
Acetazolamide, 268
Acetdimethylamide, 342
Acetessigsaeure-aethyl Ester, 797
Acethydroximsaeure-chlorid, 72
Acetic Acid, 81
Acetic Acid, Chloro-, 4-(Acetylamino)phenyl Ester, 1932
Acetic Acid Glacial, 81
Acetic Acid, Iminodi-, Monoethyl Ester, 816
Acetic Acid Isoamyl Ester, 1217
Acetic Acid Isobutyl Ester, 860
Acetic Acid Phenylmethyl Ester, 1684
Acetic Acid, (2,3,4,6-Tetrachlorophenoxy)-, 1274
Acetic Acid, (2,3,4-Trichlorophenoxy)-, 1281
Acetic Acid, (2,3,5-Trichlorophenoxy)-, 1284
Acetic Acid, (2,3,6-Trichlorophenoxy)-, 1286
Acetic Acid, (2,4,5-Trichlorophenoxy)-, 1285
Acetic Acid, (2,4,6-Trichlorophenoxy)-, 1283
Acetic Acid, (3,4,5-Trichlorophenoxy)-, 1282
Acetic Anhydride, 279
Acetoacetic Acid Ethyl Ester, 797
Acetohexamid, 3214
Acetohexamide, 3214
Acetohydroxamic Acid Chloride, 72
Acetonchloroform, 295
Acetone Oxime N-Phenylcarbamate, 1986
Acetone N-(Phenylcarbamoyl)oxime, 1986
Acetonitril, 66
Acetonitrile, 66
Acetonitrile, Diphenyl-, 2984
p-Acetophenetidide, 2025
Acetophenon, 1378
Acetophenone, 1378
Acetoquinone Blue BF, 3539
Acetoquinone Light Heliotrope NL, 2971
Acetoquinone Light Rubine BLZ, 3321
Acetothioamide, 92
p-Acetoxyacetanilide, 1963
p-Acetoxy-acetanilide, 1963
2-(Acetoxy)-benzoic Acid, (Methylsulfinyl)methyl Ester,
  2251
2-(Acetoxy)-benzoic Acid, (Methylsulfonyl)methyl Ester,
  2252
1-(2-Acetoxyethyl)-4-[3-(2-chloro-10-
  phenothiazinyl)propyl]piperazine, 3925
1-Acetoxymethyl Allopurinol, 1371
1-Acetoxymethyl-3-benzoyl-5-fluoro-2,4(1H,3H)-
  pyrimidinedi-one, 2983
1,3-bis(Acetoxymethyl)-5-fluoro-2,4(1H,3H)-
  pyrimidinedi-one, 1957
1,3-bis(Acetoxymethyl)-5-fluorouracil, 1957
1-Acetoxymethyl-3-benzoyl-5-fluorouracil, 2983
1-Acetoxymethyl-5-fluoro-2,4(1H,3H)-pyrimidinedi-one,
  1084
3-Acetoxymethyl-5-fluoro-2,4(1H,3H)-pyrimidinedi-one,
  1085
1-Acetoxymethyl-5-fluorouracil, 1084
3-Acetoxymethyl-5-fluorouracil, 1085
O-Acetoxymethyl Methyl Salicylamide, 1965
O-(acetoxymethyl) Salicylamide, 1965
5-Acetoxy-2-oxindole, 1669

[R]-[-]-α-(Acetoxy)phenylacetic Acid, 1948
m-Acetoxyphenyl Isothiocyanate, 1628
α-Acetoxytoluene, 1684
Acetrizoic Acid, 1608
Aceturic Acid, 297
Acetyl Acetaminophen, 1963
Acetylaceton, 430
Acetylacetone, 430
p-Acetylacetophenone, 1945
3-Acetylamino-5-acetylaminomethyl-2,4,6-triiodobenzoic
    Acid, 2573
2'-Acetylamino-4'-[bis(acetoxyethyl)amino]-6-bromo-2,4-
    dinitro-5'-ethoxyazobenzene, 3947
9-(2-O-Acetyl-β-D-arabinofuranosyl)adenine, 2659
9-[5'-(O-Acetyl)-β-D-arabinofuranosyl]adenine Ester,
    2660
4-Acetylbenzenesulfonamide, 1421
p-Acetylbenzenesulfonamide, 1421
1-(p-Acetylbenzenesulfonyl)-3-cyclohexylurea, 3214
Acetylbenzoylaconine, 4106
(±)-cis-1-Acetyl-4-(4-[(2-[2,4-dichlorophenyl]-2-[1H-
    imidazol-1-ylmethyl]-1,3-dioxolan-4-yl)-
    methoxy]phenyl)piperazine, 4009
N4-Acetyl-2,4-dimethoxy-6-sulfanilamidopyrimidine,
    3040
3-Acetyl-10-(3-dimethylaminopropyl)phenothiazine,
    3665
Acetylen, 50
Acetylene, 50
cis-Acetylene Dichloride, 52
trans-Acetylene Dichloride, 54
Acetylene Tetrabromide, 51
Acetylene Trichloride, 46
Acetyl Ethyl Tetramethyl Tetralin, 3589
3-Acetyl-5-fluoro-2,4(1H,3H)-pyrimidinedi-one, 639
3-Acetyl-5-fluorouracil, 639
N-Acetyl Glycine, 297
Acetylharnstoff, 162
3-Acetyl-2-hexanone, 1502
(R)(-)O-Acetylmandelic Acid, 1948
O-Acetylmandelic Acid, 1948
Acetyl-r-mandelic Acid, 1948
2'-Acetyl-6-methoxypurine Arabinoside (Hemihydrate),
    2860
Acetylmethyl Hexyl Ketone, 2130
Acetylmidicel, 2833
2-[(Acetyloxy)methoxy]-benzamide, 1965
N-Acetyl-L-phenylalanine Ethyl Ester, 2871
3-Acetylphenyl Isothiocyanate, 1626
4-Acetylphenyl Isothiocyanate, 1627
m-Acetylphenyl Isothiocyanate, 1626
p-Acetylphenyl Isothiocyanate, 1627
Acetylphosphoramidothioic Acid O,S-Dimethyl Ester,
    358
Acetylpromazine, 3665
3-Acetyl Propionic Acid, 433
Acetylsalicylic Acid, 1642
Acetylsalicylic Acid, Methyl Ester, 1951
(O-Acetylsalicyloyloxy)acetamide, 2222
Acetyl-salicylsaeure, 1642
Acetyl Sulfacetamide, 1446
Acetyl Sulfadiazine, 2598
N4-Acetyl Sulfadiazine, 2598
N4-Acetylsulfadiazine, 2598

N4-Acetylsulfadimethoxypyrimidine, 3040
Acetylsulfadimethylisoxazole, 2846
Acetyl Sulfaethylthiadiazole, 2633
Acetylsulfamethazine, 3036
Acetylsulfamethoxazole, 2618
Acetyl Sulfamethoxypyridazine, 2833
4'-Acetyl-3-sulfa-5-methylisoxazole, 2618
Acetyl Sulfamethylthiazole, 2242
Acetylsulfapyridine, 2824
Acetyl Sulfathiazole, 2225
Acetyl Sulfisoxazole, 2847
4-N-Acetylsulfisoxazole, 2846
N-Acetylsulfisoxazole, 2846
N1-Acetyl-sulfaisoxazole, 2847
N4-Acetylsulfamerazine, 2832
N4-Acetylsulfamethazine, 3036
N4-Acetyl Sulfamethizole, 2242
N4-Acetylsulfamezathine, 3036
N1-Acetylsulfanilamide, 1446
N4-Acetylsulfanilamide, 1445
3-(N1-Acetylsulfanilamido)-6-methoxypyridazine, 2833
2-N4-Acetylsulfanilamido-4-methylpyrimidine, 2832
2-N4-Acetylsulfanilamidopyrimidine, 2598
2-(N4-Acetylsulfanilylamino)-4-n-amylpyrimidine, 3479
2-(N4-Acetylsulfanilylamino)-4-ethyl-5-
    methylpyrimidine, 3203
2-(N4-Acetylsulfanilylamino)-4-ethylpyrimidine, 3038
2-(N4-Acetylsulfanilylamino)-4-isobutylpyrimidine,
    3353
2-(N4-Acetylsulfanilylamino)-4-phenylpyrimidine, 3532
2-(N4-Acetylsulfanilylamino)-4-n-propylpyrimidine,
    3202
2,5-di-(N4-Acetylsulfanilylamino)pyrimidine, 3738
N4-Acetylsulfanilylguanidine, 1749
N4-Acetylsulfapyrazine, 2599
N4-Acetyl Sulfapyridine, 2824
N4-Acetylsulfathiazole, 2225
N4-Acetylsulfisoxazole, 2846
N4-Acetylsulphacetamide, 1990
N4-Acetylsulphadimethoxine, 3040
N4-Acetylsulphamerazine, 2832
N4-Acetylsulphamethazine, 3036
N4-Acetylsulphamethomidine, 3039
N4-Acetylsulphanilamide, 1445
N4-Acetylsulphapyrazine, 2599
N4-Acetylsulphasomidine, 3037
N4-Acetylsulphathiazole, 2225
17-O-Acetyltestosterone, 3842
Acetyltetraglycine Ethyl Ester, 2727
Acetyl Tributyl Citrate, 3789
N-Acetyl-L-tryptophan Ethyl Ester, 3201
N-Acetyl-L-tyrosinamide Acetate, 2855
N-Acetyl-L-tyrosinamide Prostaglandin E2, 4089
N-Acetyl-L-tyrosinamide Prostaglandin F2 α, 4090
N-Acetyl-L-tyrosine Ethyl Ester, 2878
1-Acetylurea, 162
AcGlyOEt, 816
Achromycin V, 3879
Acide n-Butylmalonique, 1193
Acide Chloromethionique, 15
Acide Dimethylsuccinique-sym, 798
Acide Isoamylmalonique, 1509
Acide Malonique, 136
Acide Methionique, 26

Acide Methylmalonique, 280
Acide Methylsuccinique, 435
Acide Orthotrifluortoluique, 1287
Acide n-Propylmalonique, 803
Acide 2,4,6-Trinitrobenzoique, 960
Acido D-Feniltartrammico Tartranilico, 1967
Acido p-Ossifeniltartrammico, 1968
Acid succinyl Acetaminophen, 2613
Acimetion, 500
Acintene DP Dipentene, 2089
Aconitine, 4106
Acridin, 2783
1-Acridinamine, 2796
2-Acridinamine, 2797
3-Acridinamine, 2798
4-Acridinamine, 2794
Acridine, 2783
Acrolein, 135
Acronine, 3733
Acronize, 3877
Acronize PD, 3877
Acronycine, 3733
Acrylaldehyde, 135
Acrylamide, 147
Acrylanilide, 3',4'-Dichloro-2-methyl-, 1902
Acrylic Acid Methyl Ester, 272
Acrylonitrile, 119
Actactinomycin A IV, 4165
Actidione, 3248
Actinomycin AIV, 4165
Actinomycin D, 4165
Actinomycin I1, 4165
Actispray, 3248
Activated 7-Dehydrocholesterol, 4048
Actol, 2781
Adagio, 1989
Adakane 12, 2757
Adalin, 1198
Adapin, 3657
Adcortyl-A, 3951
Adenin, 400
Adenine, 400
Adenocard, 2043
Adenosin, 2043
Adenosine, 2043
Adenosine, Cyclic 3',5'-(hydrogen phosphate), 2004
Adenosine 3':5'-Monophosphate, 2004
2'-Adenylic Acid, 2069
3'-Adenylic Acid, 2070
2'-Adenylsaeure, 2069
3'-Adenylsaeure, 2070
Adepress, 3743
Adepril, 3743
Adipamide, 831
Adiphenine, 3757
Adipic Acid, 799
Adipinsaeure, 799
Adipinsaeurediamid, 831
Adonit, 523
Adonite, 523
Adonitol, 523
Adrenalin, 1765
Adrenaline, 1765
Adrenosterone, 3680

Adriamycin, 4029
Adriblastin, 4029
Adrona, 850
Advil, 2898
Aeo-antergan, 3487
D(-)-Aepfelsaeure, 285
Aerolate, 1113
Aesculin (Dihydrate), 3193
Aethan, 99
4-Aethoxy-phenylharnstoff, 1738
Aethylamin, 104
Aethylarsin, 107
Aethylbromid, 83
Aethylchlorid, 84
Aethylenimin, 86
Aethyliodid, 85
Aethylmalonsaeure, 436
3-Aethyl-pyridin, 1136
4-Aethyl-pyridin, 1135
Aethyl-vinyl-aether, 322
AETT, 3589
Afaxin, 3778
AFB1, 3414
Afesin, 1692
Aflatoxin B1, 3414
Aflatoxin B2, 3419
Aflatoxin G1, 3415
Aflatoxin G2, 3420
Afugan, 3082
Agovirin, 3896
Agricur, 2330
Agritox, 1985
Agrotect, 3052
AH-42, 3074
Airet, 2068
Ajmalan-17,21-diol, (17R,21α)-, 3763
Ajmaline, 3763
Akaritox, 2486
Alachlor, 3079
L-Ala-Dapsone, 3198
ALANAP-1, 3520
β-Alanin, 197
Alanine, 198
α-Alanine, 198
β-Alanine, 197
D-Alanine, 201
D(-)-Alanine, 201
DL-Alanine, 194
DL-α-Alanine, 194
α-Alanine Hydantoic Acid, 317
β-Alanine Hydantoic Acid, 319
Alanine, 3-(1(4H)-Naphthylidene)-, 2822
Alar, 833
Albalon-A, 3452
Albuterol, 2920
Alclofenac, 2214
Alcohol C-10, 2181
Aldicarb, 1208
Aldifen, 601
Aldocortin, 3832
Aldosterone, 3832
Aldrin, 2522
Aldrite, 2522
Aldrosol, 2522

Alfacet, 3168
Alfacron, 1355
Alfamat, 1477
Algamon, 1091
Algiamida, 1091
Algocor, 3413
Alizarin, 2947
Alizarine, 2947
Alizarine Bordeaux B, 2950
Alkane C(12), 2757
Alkane C(24), 3983
Alkron, 2056
Allantoin, 267
Allantoine, 267
Allicin, 793
Allidochlor, 1483
Allied GC 9160, 3412
Allobarbital, 1987
Alloisoleucine, 898
L-allo-Isoleucine, 898
Allopurinol, 385
DL-Allothreonine, 351
DL-allo-Threonine, 351
Alloxan, 226
Alloxane, 226
Alloxantin, 1314
Alloxantin Hydrate, 1314
All-trans-Retinal, 3770
β-All-trans-Retinoic Acid, 3772
All-trans retinoic Acid, 3772
All-trans Vitamin A Aldehyde, 3770
4-Allylanisole, 2008
Allyl Bromide, 137
Allylbutylbarbituric Acid, 2307
5-Allyl-5-butylbarbituric Acid, 2307
5-Allyl-5-sec-butylbarbituric Acid, 2310
Allyl-sec-butyl-barbituric Acid, 2310
Allyl Chloride, 139
Allyl 2,4-Dichlorophenoxyacetate, 2207
5-Allyl-5-ethylbarbituric Acid, 1741
5-Allyl-5-ethylbutylbarbituric Acid, 2913
Allylguaiacol, 2011
4-Allylguaiacol, 2011
5-Allyl-5-isobutylbarbituric Acid, 2305
5-Allyl-5-isopropylbarbituric Acid, 2061
Allyl Isothiocyanate, 256
1-Allyl-3-methoxy-4-hydroxybenzene, 2011
5-Allyl-5-(1-methylbutyl)barbituric Acid, 2703
5-Allyl-5-(1-methyl-butyl)-barbituric Acid, 2702
5-Allyl-5-(1-methylbutyl)-2-thiobarbituric Acid, 2702
5-Allyl-5-methylhexylcarbinylbarbituric Acid, 3263
Allyl Mustardiol, 256
(4-Allyloxy-3-chlorophenyl)acetic Acid, 2214
1-[o-(Allyloxy)phenoxy]-3-(isopropylamino)-2-propanol, 3247
5-Allyl-5-phenylbarbituric Acid, 2817
Allylsenfoel, 256
Allyl Trichloride, 143
Alophen, 3726
Alphadrol, 3824
Alphamine, 4168
Alrheumat, 3295
Altretamine, 1809
Alvit, 2524

Alzapam, 3135
Amacel Fast Brown 3R, 3169
Amaze, 3253
Ambush, 3806
Amchem 65-81-B, 1334
Ameisensaeure-aethyl Ester, 175
Ameisensaeure-propylester, 328
American Cyanamid 3911, 1266
Ametrex, 1802
Ametryn, 1802
Ametryne, 1802
Amicardine, 3000
m-Amidobenzenesulfonamide, 762
Amidopyrine, 2883
Amidosal, 1091
Amid-Sal, 1091
Amikacin, 3908
Aminacrine, 2795
2-Aminoacetanilide, 1442
3-Aminoacetanilide, 1441
4-Aminoacetanilide, 1440
m-Aminoacetanilide, 1441
o-Aminoacetanilide, 1442
p-Aminoacetanilide, 1440
3'-Aminoacetophenone, 1415
4'-Aminoacetophenone, 1413
m-Aminoacetophenone, 1415
p-Aminoacetophenone, 1413
1-Aminoacridine, 2796
2-Aminoacridine, 2797
3-Aminoacridine, 2798
4-Aminoacridine, 2794
9-Aminoacridine, 2795
α-Aminoadipic Acid, 815
α-Amino-adipinsaeure, 815
6-Amino-8-[[(aminocarbonyl)oxy]methyl]-1,1α,2,8,8α,8β-hexahydro-8α-methoxy-5-methyl,, 3204
6-[D-2-amino-2-(4-aminophenyl)-acetamido]-3,3-dimethyl-7-oxo-4-thia-1-azabicyclo[3,2,0]hept-2-yl-5-t, 3354
2-Amino-N-[4-[(4-aminophenyl)sulfonyl]phenyl]-, (S)-, 3198
2-Amino-N-[4-[(4-aminophenyl)sulfonyl]phenyl]-4-methyl-, (S)-, 3571
2-Aminoanthracene, 2985
β-Aminoanthracene, 2985
1-Amino-9,10-anthracenedione, 2961
2-Amino-9,10-Anthracenedione, 2962
Aminoanthraquinone, 2962
1-Aminoanthraquinone, 2961
1-Amino-9,10-anthraquinone, 2961
2-Aminoanthraquinone, 2962
2-Amino-9,10-anthraquinone, 2962
4-Aminoantipyrine, 2266
10-Amino-5-azaanthracene, 2795
4-Aminoazobenzene, 2578
p-Aminoazobenzene, 2578
4-Amino-azobenzol, 2578
2-Amino-5-azotoluene, 3022
o-Aminoazotoluene, 3022
Aminobenzene, 737
p-Aminobenzenesulfamido-tert-butylthiodiazole, 2676
2-(Aminobenzene-4'-sulfamido)-pyridine, 2224
m-Aminobenzenesulfonamide, 762

o-Aminobenzenesulfonamide, 763
4-Aminobenzenesulfonamide (Monohydrate), 765
2-(p-Aminobenzenesulfonamido)-4-methylthiazole, 1973
3-Aminobenzenesulfonic Acid, 745
4-Aminobenzenesulfonic Acid, 744
3-Aminobenzenesulfonic Acid (Sesquihydrate), 746
1-(p-Aminobenzenesulfonyl)-5,5-diphenyl-hydantoin,
   3804
p-Aminobenzenesulfonylthiourea, 1155
p-Aminobenzenesulphonamide, 764
2-p-Aminobenzenesulphonamido-4,5,6-
   trimethylpyrimidine, 2858
3-Amino-benzoesaeure, 1095
4-Amino-benzoesaeure, 1088
2-Aminobenzoic Acid, 1089
3-Aminobenzoic Acid, 1095
4-Aminobenzoic Acid, 1088
m-Aminobenzoic Acid, 1095
o-Aminobenzoic Acid, 1089
p-Aminobenzoic Acid, 1088
p-Aminobenzoicacid, 1088
4-Aminobenzoic Acid-2-(butyl-amino)ethyl Ester, 2912
4-Aminobenzoic Acid Butyl Ester, 2289
4-Aminobenzoic Acid-2-(diethyl-amino)butyl Ester,
   3256
4-Aminobenzoic Acid-2-(diethyl-amino)propyl Ester,
   3100
p-Aminobenzoic Acid Dodecyl Ester, 3707
4-Aminobenzoic Acid-2-(ethyl-amino)ethyl Ester, 2304
4-Aminobenzoic Acid Ethyl Ester, 1708
4-Aminobenzoic Acid Hexyl Ester, 2905
4-Aminobenzoic Acid Methyl Ester, 1419
4-Aminobenzoic Acid Octyl Ester, 3244
4-Aminobenzoic Acid Pentyl Ester, 2687
2-Aminobenzoic Acid 3-Phenyl-2-propenyl Ester, 3304
4-Aminobenzoic Acid 2-(Propyl-amino)ethyl Ester, 2695
4-Aminobenzoic Acid-2-(Propyl-amino)ethyl Ester, 2695
4-Aminobenzoic Acid Propyl Ester, 2026
2-[Aminobenzol-4'-sulfamid]-pyridin, 2224
9-[5-O-(4-Aminobenzoyl-β-D-arabinofuranosyl)]-6-
   methoxy-9H-purine (0.3 Hydrate), 3554
2'-(o-Aminobenzoyl)-6-methoxypurine Arabinoside,
   3553
Aminobenzylpenicillin, 3341
4-Aminobiphenyl, 2574
5-Amino-1-(bis(dimethylamino)phosphoryl)-3-phenyl-
   1,2,4-triazole, 2723
5-Amino-2-bromobenzenesulfonic Acid, 678
Amino-4-bromo-2-phenyl-3(2H)-pyridazinone, 1891
2-Amino-5-bromo-6-phenyl-py-rimidin-4(3H)-one, 1890
2-Amino-5-bromophenylsulfonic Acid, 679
2-Amino-5-bromophenylsulfonic Acid (Monohydrate),
   680
1-Aminobutane, 372
DL-2-Aminobutanedioic Acid, 302
2-Aminobutanoic Acid, 346
γ-Amino-buttersaeure, 349
4-Amino-N-[(butylamino)carbonyl]-benzenesulfonamide,
   2326
4-Amino-6-tert-butyl-3-(methylthio)-as-triazin-5(4H)-one,
   1499
4-Amino-N-(5-butyl-1,3,4-thiadiazol-2-
   yl)benzenesulfonamide, 2675
α-Aminobutyric Acid, 346

β-Amino-n-butyric Acid, 346
β-Aminobutyric Acid, 348
β-Amino-n-butyric Acid, 348
DL-2-Aminobutyric Acid, 345
DL-α-Aminobutyric Acid, 345
γ-Aminobutyric Acid, 349
γ-Amino-n-butyric Acid, 349
6-Aminocaproic Acid, 897
α-Aminocaproic Acid, 890
D-2-Amino-n-caproic Acid, 886
DL-2-Amino-n-caproic Acid, 893
ε-Aminocaproic Acid, 897
α-Aminocaproic Hydantoic Acid, 1210
ε-Aminocaproic Hydantoic Acid, 1209
ε-Amino-capronsaeure, 897
Aminocarb, 2303
N-(Aminocarbonyl)-2-bromo-2-ethylbutanamide, 1198
1-Amino-4-carboxybenzene, 1088
5-Amino-4-carboxymethylaminopyrimidine, 769
1-Amino-4-chlorobenzene-3-sulfonic Acid, 687
3-Amino-6-chlorobenzenesulfonic Acid, 687
1-Amino-4-chloro-2-benzenesulfonic Acid
   (Monohydrate), 689
1-Amino-4-chlorobenzene-3-sulfonic Acid
   (Monohydrate), 688
p-Amino-p'-chlorobiphenyl, 2555
4-Amino-4'-chlorodiphenyl, 2555
Amino-2-chloro-6-ethylamino-s-triazine, 125
6-Amino-2-chloropurine, 378
6-Aminochrysene, 3518
5-Amino-4-chloro-2-phenyl-3(2H)-pyridazinone, 1892
6-Amino-4-(diallylamino)-1,2-dihydro-1-hydroxy-2-
   imino-s-triazine, 1777
3-Amino-2,5-dichlorobenzoic Acid, 992
2-Amino-6,7-diethylpteridine, 2038
4-Amino-6,7-diethylpteridine, 2037
2-Amino-1,9-dihydro-9-((2-hydroxy-1-
   (hydroxymethyl)ethoxy)methyl)-6H-purin-6-one, 1770
2-Amino-4:6-dihydroxpteridine, 673
4-Amino-4'-(dimethylamino)azobenzene, 3034
Aminodimethylbenzene, 1478
p-Amino-N,N-dimethylbenzenesulfonamide, 1484
4-Amino-N-(3,4-dimethyl-5-
   isoxazolyl)benzenesulfonamide, 2267
6-Amino-2,4-dimethyl-pyrimidin, 780
6-Amino-2,4-dimethylpyrimidine, 780
1-Amino-2,4-dinitrobenzene, 665
4-Amino-2',4'-dinitrodiphenylamine, 2563
2-Amino-4,6-dinitro-phenol, 666
N,N'-bis(2-Aminoethyl)-ethylenediamine, 950
N-(2-Aminoethyl)-N'-(2-((2-aminoethyl)amino)ethyl)-1,2-
   ethanediamine, 1600
4-(2-Aminoethyl)phenol, 1479
2-Aminofluorene, 2807
D-2-Aminoglutaramic Acid, 456
D-2-Aminoglutaric Acid, 447
DL-2-Aminoglutaric Acid, 446
L-2-Aminoglutaric Acid, 445
2-Aminohexanedioic Acid, 815
2-Aminohexanoic Acid, 893
D-2-Aminohexanoic Acid, 886
DL-2-Aminohexanoic Acid, 893
D-α-Aminohydrocinnamic Acid, 1709
1-Amino-4-hydroxyanthraquinone, 2964

2-Amino-4-hydroxy-6,7-diethylpteridine, 2039
2-Amino-4-hydroxy-6:7-diethylpteridine, 2039
4-Amino-2-hydroxy-6,7-diethylpteridine, 2040
4-Amino-2-hydroxy-6:7-diethylpteridine, 2040
2-Amino-2-(hydroxymethyl)-1,3-propanediol, 373
DL-2-Amino-3-(4-hydroxyphenyl)-propanoic Acid, 1716
2-Amino-3-hydroxypropanoic Acid, 206
D-2-Amino-3-hydroxypropanoic Acid, 207
DL-2-Amino-3-hydroxypropanoic Acid, 205
2-Amino-4-hydroxypteridine, 672
4-Amino-2-hydroxypteridine, 670
7-Amino-6-hydroxypteridine, 671
N-(p-(((2-Amino-4-hydroxy-6-
    ptcridinyl)methyl)amino)benzoyl)-L-glutamic Acid,
    3641
2-Amino-6-hydroxypurine, 401
2-Aminohypoxanthine, 401
DL-α-Amino-3-indolepropionic Acid, 2234
α-Amino-isobuttersaeure, 347
α- Aminoisobutyric Acid, 347
α-Aminoisobutyric Acid, 347
4-Amino-N-(5-isopentyl-1,3,4-thiadiazol-2-
    yl)benzenesulfonamide, 2896
2-Amino-4-isopropylamino-6-chloro-s-triazine, 790
4-Amino-N-(5-isopropyl-1,3,4-thiadiazol-2-
    yl)benzenesulfonamide, 2275
β-Amino-isovalerian-saeure, 498
β-Aminoisovaleric Acid, 498
2-Amino-3-(lH-indol-3-yl)-propanoic Acid, 2233
2-Amino-3-mercaptopropanoic Acid, 203
Aminomethane, 28
4-Amino-N-(6-methoxy-3-pyridazinyl)-
    benzenesulfonamide, 2241
1-Amino-4-(N-methylamino)anthraquinone, 3154
1-Amino-2-methylanthraquinone, 3144
4-Amino-N-methylbenzenesulfonamide, 1160
2-Amino-5-methylbenzene Sulfonic Acid, 1152
4-Amino-2-methylbenzene Sulfonic Acid, 1151
4-Amino-3-methylbenzene Sulfonic Acid, 1150
2-Amino-3-methylbutyric Acid, 495
L-2-Amino-3-methylbutyric Acid, 495
2-Amino-4-methyl-5-carboxanilidothiazole, 2223
β-(Aminomethyl)-p-chlorohydrocinnamic Acid, 1979
4-Amino-N-(5 methyl-3-isoxazolyl)benzenesulfonamide,
    1975
2-Amino-4-methylpentanoic Acid, 891
D-2-Amino-4-methylpentanoic Acid, 887
DL-2-Amino-3-methylpentanoic Acid, 895
DL-2-Amino-4-methylpentanoic Acid, 894
L-2-Amino-4-methylpentanoic Acid, 891
4-Amino-3-methyl-6-phenyl-1,2,4-triazin-5-one, 1938
2-Amino-4-(methylthio)butanoic Acid, 499
DL-2-Amino-4-(methylthio)butyric Acid, 500
D-2-Amino-4-methylvaleric Acid, 887
DL-2-Amino-4-methylvaleric Acid, 894
1-Aminonaphthalene, 1908
4-Amino-1-naphthalenesulfonic Acid, 1920
2-Amino-5-naphthol-1-sulfonic Acid, 2144
7-Amino-1-naphthol-3-sulfonic Acid, 1925
7-Amino-naphtol-(1)-sulfosaeure-(3), 1925
4-Amino-4'-nitroazobenzene, 2562
1-Amino-2-nitrobenzene, 704
1-Amino-3-nitrobenzene, 705
4-Amino-nitrobenzene, 706

p-Aminonitrobenzene, 706
1-Amino-2-nitroguanidine, 29
1-Amino-3-nitroguanidine, 29
3-Amino-1-nitroguanidine, 29
1-Aminooctadecane (Dihydrate), 3613
1-Aminooctane, 1587
2-Amino-4-oxopteridine, 672
4-Amino-2-oxopteridine, 670
7-Amino-6-oxopteridine, 671
Aminoparathion, 1481
4-Amino-N-(5-pentyl-1,3,4-thiadiazol-2-
    yl)benzenesulfonamide, 2897
Aminophen, 737
Aminophenazone, 2266
2-Amino-phenol, 740
3-Aminophenol, 741
4-Aminophenol, 739
m-Aminophenol, 741
o-Aminophenol, 740
p-Aminophenol, 739
4-Amino-phenol-N-acetat, 1416
2-Aminophenol-4-sulfonic Acid, 747
4-Aminophenol-2-sulfonic Acid, 748
2-Amino-phenol-sulfosaeure-(4), 747
4-Amino-phenol-sulfosaeure-(2), 748
(2S,5R,6R)-6-[(R)-2-Amino-2-phenylacetamido]-3,3-
    dimethyl-7-oxo-4-thia-1-azabicyclo[3.2.0]heptane-2-
    carboxylic Acid, 3341
2-Aminophenylacetic Acid, 1420
2-Amino-phenyl-essigsaeure, 1420
2-(4-Aminophenyl)-6-methyl-benzothiazole, 2992
2-Amino-3-phenylpropanoic Acid, 1706
D-α-Amino-β-Phenylpropionic Acid, 1709
[(4-Aminophenyl)sulfonyl]dimethylamine, 1484
p-Aminophenylsulfonylthiourea, 1155
2-Aminopropanoic Acid, 198
DL-2-Aminopropionic Acid, 194
L-2-Aminopropionic Acid, 198
4'-Aminopropiophenone, 1704
p-Aminopropiophenone, 1704
4-Amino-N-(5-propyl-1,3,4-thiadiazol-2-
    yl)benzenesulfonamide, 2276
2-Aminopteridine, 668
4-Aminopteridine, 669
7-Aminopteridine, 667
2-Aminopteridin-4-one, 672
2-Amino-4(1H)-pteridinone, 672
2-Amino-4(3H)-pteridinone, 672
4-Aminopteridin-2-one, 670
7-Aminopteridin-6-one, 671
2-Amino-4-pteridone, 672
4-Amino-2-pteridone, 670
7-Amino-6-pteridone, 671
Aminopyrine, 2883
2-Amino-9-β-D-ribofuranosyl-9H-purine-6-(1H)-one,
    2044
4-Aminosalicylic Acid, 1097
p-Aminosalicylic Acid, 1097
4-Amino-salicylsaeure, 1097
4-Aminostilbene, 3020
p-Aminostilbene, 3020
4-Aminosulfonyl-1-bromobenzene, 677
5-Amino-tetrazol, 20
5-Aminotetrazole, 20

2-Amino-1,3,4-thiadiazole-5-sulfonamide, 80
5-Amino-1,3,4-thiadiazole-2-sulfonamide, 80
5-Amino-1,3,4-thiadiazol-2-sulfonamide, 80
4-Amino-N-2-thiazolyl-, 1661
6-Amino-2-thiouracil, 258
2-Amino-toluol-sulfosaeure-(5), 1152
4-Amino-toluol-sulfosaeure-(2), 1151
4-Amino-toluol-sulfosaeure-(3), 1150
Aminotriazole, 79
3-Amino-1,2,4-triazole, 79
3-Amino-s-triazole, 79
4-Amino-3,5,6-trichloropicolinic Acid, 562
2-Amino-4:6:7-trihydroxypteridine, 674
N-(3-Amino-2,4,6-triiodobenzoyl)-N-phenyl-β-alanine,
   3280
N-(3-Amino-2,4,6-triiodophenyl)-3-acetamido-2-
   methylpropionic Acid, 2604
β-(3-Amino-2,4,6-triiodophenyl)-α-ethylpropionic Acid,
   2230
1-Amino-2,4,6-trinitrobenzene, 618
11-Aminoundecanoic Acid, 2372
Amino-11-undecanoique Acide, 2372
5-Aminouracil, 260
5-Amino-uracil, 260
DL-2-Aminovaleric Acid, 488
L-(+)-2-Aminovaleric Acid, 489
δ-Aminovaleric Hydantoic Acid, 834
Amiodarone, 3991
Amitraz, 3672
Amitriptyline, 3743
Amitrole, 79
Ammelide, 134
Ammelin, 151
Ammeline, 151
Ammoidin, 2532
2-Ammoniopropanoate, 198
Amobarbital, 2334
Amoxicillin (Trihydrate), 3344
Amoxone, 3052
Amphicol, 2228
Amphotericin B, 4140
Ampicillin, 3341
Ampicillin (Trihydrate), 3343
Amplivix, 3413
Ampyrone, 2266
Amygdalic Acid, 1392
Amygdalin, 3767
(R)-Amygdalin, 3767, 3768
D-(-)-Amygdalin, 3768
Amygdalin (Trihydrate), 3768
(R)-Amygdaloside, 3767
Amyl Acetate, 1225
iso-Amyl Alcohol, 510
sec-Amyl Acetate, 1218
Amyl α-Acetoxypropionate, 2135
Amylacetylene, 1175
Amyl Alcohol, 516
n-Amyl Alcohol, 516
sec-Amyl Alcohol, 510
tert-Amylalkohol, 515
10-Amyl-1,2-benzanthracene, 3912
Amylbenzene, 2301
n-Amylbenzene, 2301
t-Amylbenzene, 2299

tert-Amylbenzene, 2299
n-Amyl Bromide, 486
n-Amyl n-Butyrate, 1828
n-Amyl Carbamate, 884
tert-Amyl Carbamate, 888
Amyl Carbinol, 914
Amyl 2,4-Dichlorophenoxyacetate, 2848
α-n-Amylene, 450
β-Amylene, 449
β-n-Amylene, 449
Amylene Bromide, 486
n-Amyl β-Ethoxypropionate, 2174
n-Amyl Formate, 864
Amyl Lactate, 1551
n-Amyl β-Methoxypropionate, 1834
Amylmethylcarbinol, 1260
tert-Amyl Methyl Ether, 926
Amylobarbitone, 2334
3-Amyl-2,4-pentanedione, 2131
2-n-Amylphenol, 2318
4-sec-Amylphenol, 2321
o-n-Amylphenol, 2318
p-n-Amylphenol, 2319
p-sec-Amylphenol, 2321
Amyl n-Propanoate, 1550
4-n-Amyl-resorcin, 2322
4-n-Amyl Resorcinol, 2322
Anaxirone, 2297
Ancymidol, 3184
Androgen, 3896
Androsan, 3896
Androstane-3-β,11-β-diol-17-one, 3704
Androstanedione, 3692
5α-Androstane-3,17-dione, 3692
Androstane-17-one, 3698
Androstanolone, 3701
Androstanolone Acetate, 3850
Androstanolone Butyrate, 3939
Androstanolone Formate, 3781
Androstanolone Propionate, 3902
Androstanolone Valerate, 3966
Androstan-3-one, 17-(Acetyloxy)-, (5α,17β)-, 3850
5β-Androstan-17-one, 3α,11-Dihydroxy-, 3706
5α-Androstan-3-one, 17-Hydroxy-, Formate, 3781
Androstan-3-one, 17-(1-Oxobutoxy)-, (5α,17β)-, 3939
Androstan-3-one, 17-[(1-Oxopentyl)oxy]-, (5α,17β)-, 3966
Androstan-3-one, 17-(1-Oxopropoxy)-, (5α,17β)-, 3902
Androst-4-en-3,17-dion, 3685
Androstenedione, 3685
4-Androstene-3,17-dione, 3685
δ-4-Androstene-3-one, 3684
Androstene-3,11,17-trione, 3680
Androst-4-en-17β-ol-3-one Formate, 3774
Androst-4-en-3-one, 17-(Acetyloxy)-, (17β)-, 3842
Androst-5-en-17-one, 3-(Acetyloxy)-, (3β)-, 3840
Androst-4-en-3-one, 17-(Benzoyloxy)-, (17β)-, 4016
17-β-(4-Androsten-3-one)-N-2-(2-desoxyglucosyl), 4020
Androst-5-en-17-one, 3α-Hydroxy-, Formate, 3773
17-β-(4-Androsten-3-one)-N-methyl-N-1-(1-
   desoxyglucosyl) Carbamate, 4046
Androst-5-en-17-one, 3-(1-Oxobutoxy)-, (3β)-, 3936
Androst-4-en-3-one, 17-(1-Oxobutoxy)-, (17bet)-, 3934
Androst-4-en-3-one, 17-[(1-Oxopentyl)oxy]-, (17β)-,
   3964

Androst-5-en-17-one, 3-[(1-Oxopentyl)oxy]-, (3β)-, 3965
Androst-4-en-3-one, 17-(1-Oxo-3-phenylpropoxy)-, (17β)-, 4052
Androst-4-en-3-one, 17-[(Phenylacetyl)oxy]-, (17β)-, 4036
Androsterone, 3703
Anethole, 2007
Anfram 3pb, 1780
α-Angelica-lacton, 411
α-Angelica Lactone, 411
(7α-Angelyloxy-5,6,7,8α-Tetrahydro-3H-pyrrolizin-1-yl)methyl-2,3-dihydroxy-2-(1'-methoxyethyl)-3-methylbutyrate, 3852
Anhydrogitalin, 4127
Anhydroglucochloral, 1477
Anhydrohydroxyprogesterone, 3829
Anhydrous 6-(1-Aminocyclohexanecarboxamido)penicillanic Acid, 3250
ANI, 2192
Anilazine, 1603
Aniline, 737
p-Anilinesulfonamide (Monohydrate), 765
Anilinoazobenzene, 2577
2-Anilinoethanol, 1480
4-Anilino-4'-nitroazobenzene, 3526
Anisaldehyd, 1382
p-Anisaldehyde, 1382
Anise Camphor, 2007
m-Anisic Acid, 1394
o-Anisic Acid, 1389
p-Anisic Acid, 1397
2-Anisidine, 1144
o-Anisidine, 1144
p-Anisidine, 1145
Anisole, 1118
Anisole, 2,3,4,5-Tetrachloro-, 974
Anisomycin, 3073
Anissaeure, 1397
Anisylbutamide, 2706
ANIT, 2192
Annamene, 1351
Ansaid, 3160
Anspor, 3342
Antabuse, 2158
Antadix, 2158
Antarelix, 4178
Antazoline, 3452
Anthisan, 3487
2-Anthracenamine, 2985
Anthracene, 2965
Anthracene Amine, 2985
9-Anthracenecarboxylic Acid, 3136
Anthracene-9-carboxylic Acid, 3136
9,10-Anthracenedione, 2950
9,10-Anthracenedione, 1-Amino-4-hydroxy-2-phenoxy-, 3800
9,10-Anthracenedione, 1,4-bis(Methylamino)-, 3291
9,10-Anthracenedione, 1,5-Dihydroxy-4,8-bis(methylamino)-, 3293
9,10-Anthracenedione, 1,4,5,8-Tetraamino-, 2993
1-Anthracenesulfonic Acid, 5-Chloro-9,10-dihydro-9,10-dioxo-, 2941
2-Anthracenesulfonic Acid, 5-Chloro-9,10-dihydro-9,10-dioxo-, 2942

Anthrachinon-disulfosaeure-(1,8), 2951
Anthracin, 2965
2-Anthramine, 2985
Anthranilsaeure, 1089
Anthranol, 2973
1-Anthranol, 2973
2-Anthranol, 2974
Anthraquinone, 2946
9,10-Anthraquinone, 2946
1,5-Anthraquinone Disulfonic Acid, 2953
Anthraquinone-1,5-Disulfonic Acid, 2953
1,6-Anthraquinone Disulfonic Acid, 2952
Anthraquinone-1,6-Disulfonic Acid, 2952
1,8-Anthraquinone Disulfonic Acid, 2951
Anthraquinone-1,8-disulfonic Acid, 2951
1,5-Anthraquinone Disulfonic Acid Anilide, 4008
1,8-Anthraquinone Disulfonic Acid Anilide, 4007
Anthraxcene, 2965
1-Anthrol, 2973
2-Anthrol, 2974
2-Anthrylamine, 2985
Antibiotic X 465A, 4092
Antibiotic BB-K8, 3908
Antioxidant D, 814
Antipyrin, 2232
Antipyrine, 2232
ANTU, 2212
Apazone, 3352
Aphidan, 1852
Apholate, 2752
Aphosal, 1477
Aphox, 2337
Apirolio 1431C, 2497
Apl-Luster, 1888
APN, 2752
Apocholic Acid, 3972
Apo-Lorazepam, 3135
Apomorphin, 3433
Apomorphine, 3433
Apo-Oxazepam, 3141
Appa, 2231
Aprobarbital, 2061
Aprobarbitone, 2061
Apron, 3229
APZ, 3352
Aquamox, 1982
Aquaphor, 3177
Aquaphor (Diuretic), 3177
DL-Arabinitol, 522
9-β-D-Arabino Furanosyl Adenine, 2086
L-Arabinopyranose, 483
L-Arabinose, 483
Arabinosyladenine, 2086
β-D-Arabinosyladenine, 2086
Arabinosyladenine 5'-Formate, 2271
Arabinosyladenine 5'-O-Formate Ester, 2271
(±)-Arabitol, 522
Aracnol K, 2486
Arasan, 841
Aratac, 3991
Arbutin (Monohydrate), 2684
Aresin, 1692
A-Rest, 3184
Aretit, 2627

L(+)-Arginin, 913
Arginine, 913
(±)-Arginine, 912
DL-Arginine, 912
L-Arginine, 913
Aristocort, 3826
Aristoderm, 3951
Arochlor 1221, 2536
Arochlor 1242, 2499
Arochlor 1248, 2460
Arochlor 1254, 2461
Arochlor 1260, 2419
Aroclor 1221, 2536
Aroclor 1242, 2499
Aroclor 1248, 2460
Aroclor 1254, 2461
Aroclor 1260, 2419
Arosol, 1467
Arresin, 1692
Arsen, 107
Arsine Oxide, Hydroxydimethyl-, 106
Arsonoacetic Acid, 98
Arsono-essigsaeure, 98
Ascabin, 2996
Ascorbic Acid, 772
L-Ascorbic Acid, 772
L-Ascorbinsaeure, 772
Aseptil 2, 1973
L-Asparagin, 316
Asparagine, 316
L-Asparagine, 316
L-Asparagine Monohydrate, 320
Asparagine, Monohydrate, L-, 320
L-(+)-Asparaginic Acid, 300
L(+)-Asparaginsaeure, 300
Asp-arg-val-tyr-ile-his-pro-D-phe-his-leu-phe-val-tyr, 4182
Aspartic Acid, 300
DL-Aspartic Acid, 302
L-Aspartic Acid, 300
L-(+)-Aspartic Acid, 300
Aspidospermidine, 1-Acetyl-17-methoxy-, 3892
Aspidospermine, 3892
Aspirin, 1642
Aspirin Phenylalanine Ethyl Ester, 3740
Asulam, 1448
As-o-xylenol, 1455
AT-7, 2772
ATA, 79
Athylacetat, 333
Ativan, 3135
Atratone, 1801
Atrazine, 1497
Atromid S, 2640
Atropasaeure, 1641
Atropic Acid, 1641
Atropin, 3483
Atropine, 3483
Aurin, 3626
Avical, 952
Avlothane, 111
9-Azafluorene, 2545
1-Azanaphthalene, 1618
5-Azaphenanthrene, 2782

9-Azaphenanthrene, 2782
Azapropazone, 3352
Azathioprine, 1631
Azatioprin, 1631
Azelaic Acid, 1794
Azelainsaeure, 1794
1H-Azepine, 1-[(Benzoyloxy)acetyl]hexahydro-, 3211
Azetidine, 1-[(Benzoyloxy)acetyl]-, 2607
2-Azetidinecarboxylic Acid, 1-[(Benzoyloxy)acetyl]-, 2823
Azidocarbonicdisulfide, 114
2-Azido-4-ethylamino-4-t-butylamino-s-triazine, 1791
2-Azido-4-isopropylamino-6-methylmercapto-s-triazine, 1173
Azidoschwefel-kohlenstoff, 114
Azinepurine, 605
Azinos, 2673
Azinphos-ethyl, 2673
Azinphos-ethyl O-Analog, 2678
Azinphos-methyl, 1993
Aziprotryne, 1173
Aziridine, 86
Aziridine (Dihydrate), 93
Aziridine, 1-(1-Oxo-3-phenyl-2-propenyl)-, 2217
Azobenzene, 2557
Azobenzide, 2557
Azobenzol, 2557
Azole, 252
Ba 2684, 3695
Ba 2750, 1746
Baam, 3672
Baclofen, 1979
Bacteramid (Monohydrate), 765
Bactine, 3843
Bactrim, 1975
Badional, 1155
Badische Acid, 1918
Baldinol, 1155
BAM, 991
Barban, 2198
Barbital, 1486
Barbituric Acid, 241
Barbituric Acid, 5-Allyl-5-ethyl, 1741
Barbituric Acid, 5-Allyl-5-phenyl, 2817
Barbituric-2-14C Acid, 5,5-Diallyl, 1988
Barbituric Acid, 5,5-Diethyl-2-thio, 1485
Barbituric Acid, 5,5-Diisopropyl, 2098
Barbituric Acid, 5,5-Dimethyl, 766
Barbituric Acid, 5,5-Diphenyl, 3275
Barbituric Acid, 5-Ethyl-5-(3-methyl-2-butenyl), 2309
Barbituric Acid, 5-Ethyl-5-(1-methylbutyl)-2-thio, 2332
Barbituric Acid, 5-Methyl-5-phenyl, 2211
Barbitursaeure, 241
Basagran 4E, 1989
Basamaize, 2588
BAS 2103H, 2925
BAS 351H, 1989
Basic Violet 14, 3735
Basudin, 2731
Bay 25141, 2330
Bay 37289, 1985
Bay 49854, 2021
Bay 5712α, 2021
Bay 68138, 2923

Baycid, 2088
Bayer 6159H, 1499
Baygon, 2293
Bayleton, 3026
Bayluscid, 2778
Bayrusil, 2654
Baythion, 2655
B[B]F, 3713
BDAM, 4032
BDCM, 1
BE 419, 3507
Beet-Klean, 1737
Beet Sugar, 2744
BEI 1293, 3177
Belclene 310, 2357
Benazolin, 1604
Bendiocarb, 2264
Bendroflumethiazide, 3170
Benefin, 2852
Benfluralin, 2852
Benodanil, 2793
Benomyl, 3056
Benquinox, 2811
Bensulide, 3108
Bentazon, 1989
Bentranil, 2960
Benylate, 2996
1,2-Benzacenaphthene, 3270
Benzadox, 1657
Benzalacetone, 1944
Benzaldehyd, 1054
Benzaldehyde, 1054
Benzaldehyde, 2-(β-D-Glucopyranosyloxy)-, Hydrate
    (4:3), 2863
Benzamid, 1087
Benzamide, 1087
Benzamide, 2-[(Acetyloxy)methoxy]-, 1965
Benzamide, N-[7-[[3-O-(Aminocarbonyl)-6-deoxy-5-C-
    methyl-4-O-methyl-β-L-lyxo-hexopyranosyl]oxy]-4-
    hydroxy-8-methyl-2-oxo-2H-1-benzopyran-3-yl]-4-
    hydroxy-3-(3-methylbutyl)-, 4086
Benzamide, 2-Amino-3,5-dichloro-, 1032
Benzamide, N,N'-(9,10-Dihydro-3-methyl-9,10-dioxo-1,8-
    anthracenediyl)bis, 4064
Benzamide, N-[(Dimethylamino)methyl]-, 2058
Benzamide, N-[(Ethylamino)methyl]-, 2059
Benzamide, N-(1-Pyrrolidinylmethyl)-, 2665
Benzanthracene, 3512
1,2-Benzanthracene, 3512
2,3-Benzanthracene, 3515
Benz[a]pyrene, 3712
1-Benzazine, 1618
1-Benzazole, 1337
Benz[b]indeno[1,2-d]pyran-9(6H)-one, 6α,7-Dihydro-
    3,4,6α,10-Tetrahydroxy-, 3277
3,4-Benzchrysene, 3861
BenzDAM, 4082
Benzenamine, 706, 1478
Benzenamine, 2,4,6-Trinitro-N-(2,4,6-trinitrophenyl)-,
    2451
Benzene, 676
Benzeneacetamide, N-hydroxy-α-dipropyl, 3089
Benzeneacetic Acid, 3-Benzoyl-α-methyl-, 2-[(2-Amino-
    2-oxoethyl)methylamino]-2-oxoethyl Ester, 3814

Benzeneacetic Acid, 4-Benzoyl-α-methyl-, 2-[(2-Amino-
    2-oxoethyl)methylamino]-2-oxoethyl Ester, 3813
Benzeneacetic Acid, 2-(2,4-Dichlorophenoxy)-, 2967
Benzeneacetic Acid, α-(Hydroxymethyl)-, 8-Methyl-8-
    azabicyclo[3.2.1]oct-3-yl Ester, [3(S)-Endo]-, 3484
Benzeneacetic Acid, α-Hydroxy-α-phenyl-, 2998
Benzeneacetic Acid, α-Methyl-4-(2-methylpropyl)-, 2-[(2-
    Amino-2-oxoethyl)methylamino]-2-oxoethyl Ester,
    3586
Benzeneacetic Acid, 4-[[(Phenylamino)carbonyl]amino]-,
    3173
Benzenearsonic Acid, 755
Benzeneazobenzene, 2557
Benzenecarboxylic Acid, 1055
1,3-Benzenediamine, N1,N1-Diethyl-2,6-dinitro-4-
    (trifluoromethyl)-, 2256
1,4-Benzenedicarbonitrile, 1275
1,2-Benzenedicarboxamide, 1363
1,3-Benzenedicarboxamide, N,N'-bis(2,3-
    Dihydroxypropyl)-5S-[(hydroxyacetyl)amino]-2,4,6-
    triiodo- [RS-(RS*,RS*)]-, 3350
1,3-Benzenedicarboxamide, N,N'-bis(2,3-
    Dihydroxypropyl)-5S-[(2-hydroxy-1-oxopropyl)amino]-
    2,4,6-triiodo-(RS,S)-, 3470
1,3-Benzenedicarboxamide, N,N'-bis(2,3-
    Dihydroxypropyl)-5S-[(2-hydroxy-1-oxopropyl)amino]-
    2,4,6-triiodo-[RS-(RS*,RS*)]-, 3476
1,3-Benzenedicarboxamide, N,N'-bis(2,3-
    Dihydroxypropyl)-5S-[(2-hydroxy-1-oxopropyl)amino]-
    2,4,6-triiodo-[S-(S*,S*)]-, 3469
1,3-Benzenedicarboxamide, N,N'-bis(2,3-
    Dihydroxypropyl)-5RS-[(2-hydroxy-1-
    oxopropyl)amino]-2,4,6-triiodo-[RS-(RS*,RS*)]-,
    3471
1,3-Benzenedicarboxamide, N,N'-bis(2,3-
    Dihydroxypropyl)-5RS-[(2-hydroxy-1-
    oxopropyl)amino]-2,4,6-triiodo-[RS-(RS*,S*)]-, 3472
1,3-Benzenedicarboxamide, N,N'-bis(2,3-
    Dihydroxypropyl)-5RS-[(2-hydroxy-1-
    oxopropyl)amino]-2,4,6-triiodo-[RS-(S*,S*)]-, 3475
1,3-Benzenedicarboxamide, N,N'-bis[2-Hydroxy-1,1-
    bis(hydroxymethyl)ethyl]-5-[(2-hydroxy-1-
    oxopropyl)amino]-2,4,6-triiodo-, (S)-, 3682
1,3-Benzenedicarboxamide, N,N'-bis[2-Hydroxy-1-
    (hydroxymethyl)ethyl]-5-[(2-hydroxy-1-
    oxopropyl)amino]-2,4,6-triiodo-, (S)-, 3475
1,3-Benzenedicarboxamide, N,N'-bis[2-Hydroxy-1-
    (hydroxymethyl)ethyl]-5-[(2-hydroxy-1-
    oxopropyl)amino]-2,4,6-triiodo-, (RS)-, 3473
1,3-Benzenedicarboxamide, 5-[(2,3-Dihydroxy-1-
    oxobutyl)amino]-N,N'-bis[2-hydroxy-1-
    (hydroxymethyl)ethyl]-2,4,6-triiodo-, 3478
1,3-Benzenedicarboxamide, 5-[(2,3-Dihydroxy-1-
    oxopropyl)amino]-N,N'-bis[2-hydroxy-1-
    (hydroxymethyl)ethyl]-2,4,6-triiodo-, 3477
1,3-Benzenedicarboxamide, N-(2,3-Dihydroxypropyl)-N'-
    (2-hydroxyethyl)-5-[(2-hydroxy-1-oxopropyl)amino]-
    2,4,6-triiodo-(RS), 3348
1,3-Benzenedicarboxamide, N-(2,3-Dihydroxypropyl)-N'-
    [2-hydroxy-1-(hydroxymethyl)ethyl]-5-[(2-hydroxy-1-
    oxopropyl)amino]-2,4,6-triiodo-, 3474
1,3-Benzenedicarboxamide, 5-[(Hydroxyacetyl)amino]-
    N,N'-bis[2-hydroxy-1-(hydroxymethyl)ethyl]-2,4,6-
    triiodo-, 3351

1,3-Benzenedicarboxamide, N-(2-Hydroxyethyl)-N'-[2-
    hydroxy-1-(hydroxymethyl)ethyl]-5-[(2-hydroxy-1-
    oxopropyl)amino]-2,4,6-triiodo-, (S)-,   3349
1,3-Benzenedicarboxamide, 5RS-[(2,3-Dihydroxy-1-
    oxobutyl)amino]-N,N'-bis(2,3-dihydroxypropyl)-2,4,6-
    triiodo-[RS-(RS*,S*)]-,   3573
1,2-Benzenedicarboxamide, N,N,N',N'-Tetramethyl-,
    2666
1,3-Benzenedicarboxamide, N,N,N',N'-Tetramethyl-,
    2667
1,4-Benzenedicarboxamide, N,N,N',N'-Tetramethyl-,
    2668
1,2-Benzenedicarboxylic Acid,   1320, 3971
Benzene-1,2-dicarboxylic Acid,   1320
1,3-Benzenedicarboxylic Acid,   1321
1,2-Benzenedicarboxylic Acid Anhydride,   1278
1,2-Benzenedicarboxylic Acid, bis(2-Methylpropyl)
    Esterpalatinol,   3372
1,2-Benzenedicarboxylic Acid 2-Butoxy-2-oxoethyl Butyl
    Ester,   3581
1,2-Benzenedicarboxylic Acid Butyl Phenylmethyl Ester,
    3654
1,2-Benzenedicarboxylic Acid, Dicyclohexyl Ester,   3766
1,2-Benzenedicarboxylic Acid, Didodecyl Ester,   4096
1,2-Benzenedicarboxylic Acid Diisooctyl Ester,   3969
1,2-Benzenedicarboxylic Acid, di(2-Methoxyethyl) Ester,
    3065
1,2-Benzenedicarboxylic Acid, Dimethyl Ester,   1952
1,4-Benzenedicarboxylic Acid Dimethyl Ester,   1949
Benzene, 1,2-Dichloro-4,5-dimethoxy-,   1358
Benzene, 1,3-Dichloro-2-methoxy-,   1035
Benzene, 2-(2,4-Dinitrophenoxy)-1,3,5-trinitro-,   2450
Benzene, 1,1'-(1,2-Ethanediyl)bis-,   3006
Benzene Hexachloride,   691, 953
α-Benzene Hexachloride,   692
β-Benzene Hexachloride,   693
δ-Benzene Hexachloride,   690
Benzenemethanol,   1117
Benzene, 1-Methoxy-2,4-dinitro-,   1039
Benzenenitrile,   1005
Benzene, Nitro-,   651
Benzenepentanoic Acid,   2282
Benzenepropanamide, α-Amino-N-[4-[(4-
    aminophenyl)sulfonyl]phenyl]-, (S)-,   3810
Benzenepropanamide, N-Hydroxy-α2,3-pentamethyl,   3118
Benzenepropanamide, N-Hydroxy-α2,4,6-pentamethyl,
    3088
Benzenepropanoic Acid, Ethyl Ester,   2281
Benzenesulfamide,   764
Benzenesulfonamide,   742
Benzenesulfonamide, 4-Amino-N-(5-amino-2-pyridinyl)-,
    2238
Benzenesulfonamide, 4-Amino-N-(2-amino-5-
    pyrimidinyl)-,   1976
Benzenesulfonamide, 4-Amino-N-(2-bromo-5-pyridinyl)-,
    2204
Benzenesulfonamide, 4-Amino-N-(5-bromo-2-pyridinyl)-,
    2203
Benzenesulfonamide, 4-Amino-N-(2-chloro-5-
    pyrimidinyl)-,   1901
Benzenesulfonamide, 4-Amino-N-(5-chloro-2-
    pyrimidinyl)-,   1900
Benzenesulfonamide, 4-Amino-N-(1,2-dihydro-1-methyl-
    2-pyridinyl)-,   2656

Benzenesulfonamide, 4-Amino-N-(2,3-dihydro-3-methyl-
    2-thiazolyl)-,   2035
Benzenesulfonamide, 4-Amino-N-(4,5-dihydro-2-
    thiazolyl)-,   1721
Benzenesulfonamide, 4-Amino-N-(3-ethoxy-2-pyridinyl)-,
    2844
Benzenesulfonamide, 4-Amino-N-(6-ethoxy-3-pyridinyl)-,
    2845
Benzenesulfonamide, 4-Amino-N-(5-iodo-2-pyridinyl)-,
    2208
Benzenesulfonamide, 4-Amino-N-(2-methoxy-5-
    pyrimidinyl)-,   2244
Benzenesulfonamide, 4-Amino-N-(4-methoxy-2-
    pyrimidinyl)-,   2243
Benzenesulfonamide, 4-Amino-N-[5-(3-methylbutyl)-
    1,3,4-thiadiazol-2-yl]-,   2896
Benzenesulfonamide, 4-Amino-N-(2-methyl-4-
    pyrimidinyl)-,   2237
Benzenesulfonamide, 4-Amino-N-(5-nitro-2-pyridinyl)-,
    2213
Benzenesulfonamide, 4-Amino-N-(5-pentyl-1,3,4-
    thiadiazol-2-yl)-,   2897
Benzenesulfonamide, 4-Amino-N-(1,2,3,4-tetrahydro-2,4-
    dioxo-5-pyrimidinyl)-,   1943
Benzenesulfonamide, 4-Chloro-3-nitro-,   631
Benzenesulfonamide, 3,4-Dichloro-,   637
Benzenesulfonamide, 4-(1,3-Diethyl-2,3,6,7-tetrahydro-
    2,6-dioxo-1H-purin-8-yl)-N-[2-(dimethylamino)ethyl]-,
    3683
Benzenesulfonamide, N-[2-(Dimethylamino)ethyl]-4-
    (2,3,4,5,6,7-hexahydro-2,6-dioxo-1,3-dipropyl-1H-
    purin-8-yl)-,   3836
Benzenesulfonic Acid,   730
Benzenesulfonic Acid, 2-Amino-4-iodo-,   699
Benzenesulfonic Acid, 2-Amino-5-iodo-,   697
Benzenesulfonic Acid, 3-Amino-4-iodo-,   698
Benzenesulfonic Acid, 3-Amino-5-iodo-,   696
Benzenesulfonic Acid, 3-Amino-6-iodo-,   695
Benzenesulfonic Acid, 4-Amino-2-iodo-,   700
Benzenesulfonic Acid, 4-Amino-3-iodo-,   701
Benzenesulfonic Acid, 4-(1,3-Diethyl-2,3,6,7-tetrahydro-
    2,6-dioxo-1H-purin-8-yl)-,   3190
Benzenesulfonic Acid (Dihydrate),   733
Benzene Sulfonic Acid Ethyl Ester,   1470
Benzenesulfonic Acid (2.5 Hydrate),   732
Benzenesulfonic Acid (Monohydrate),   731
Benzenesulfonic Acid, 4-(2,3,6,7-Tetrahydro-2,6-dioxo-
    1,3-dipropyl-1H-purin-8-yl)-,   3460
Benzenesulfonic Acid (Trihydrate),   734
N-Benzenesulfonyl-N'-n-butylurea,   2311
1-Benzenesulfonyl-5,5-diphenyl-hydantoin,   3801
1-Benzenesulfonyl-5-ethyl-5-phenyl-hydantoin,   3430
1,2,4,5-Benzenetetracarboxylic Acid,   1878
Benzene, 1,2,3,4-Tetrachloro-,   539
Benzene, 1,2,3,4-Tetrachloro-5-methoxy-,   974
Benzene, 1,2,3,5-Tetrachloro-4-nitro-,   530
1,2,3-Benzenetricarboxylic Acid,   1613
1,3,5-Benzenetricarboxylic Acid,   1615
Benzene, 1,2,3-Trichloro-,   559
Benzene, 1,2,4-Trichloro-,   560
Benzene, 1,3,5-Trichloro-,   561
1,2,3-Benzenetriol,   729
1,3,5-Benzenetriol,   726
Benzhydrol,   2818

Benzidin, 2589
Benzidin-disulfosaeure-(2,2'), 2596
Benzidine, 2589
3-Benzidine, 2590
m-Benzidine, 2590
Benzidine-2,2'-disulfonic Acid, 2596
Benzilen, 3287
Benzilic Acid, 2998
1H-Benzimidazole-1-carboxylic Acid, 6-Benzoyl-2-
    [(methoxycarbonyl)amino]-, Ethyl Ester, 3635
1H-Benzimidazole-1-carboxylic Acid, 6-Benzoyl-2-
    [(methoxycarbonyl)amino]-, Methyl Ester, 3528
1H-Benzimidazole-1-carboxylic Acid, 6-Benzoyl-2-
    [(methoxycarbonyl)amino]-, Propyl Ester, 3736
2-Benzimidazolinone, 1-[1-[4,4-bis(p-
    Fluorophenyl)butyl]-4-piperidyl]-, 4051
2H-Benzimidazol-2-one, 1-[1-[4-(4-Fluorophenyl)-4-
    oxobutyl]-1,2,3,6-tetrahydro-4-pyridinyl]-1,3-dihydro-,
    3874
1H-Benzimidazol-2-ylcarbamic Acid Methyl Ester, 1660
2,3-Benzindene, 2787
Benziodarone, 3413
Benz[g,h,i]perylene, 3859
3-Benzisothiazolinone 1,1-dioxide, 1011
1,2-Benzisothiazol-3(2H)-one-1,1-dioxide, 1011
Benz[k]acephenanthrene, 3722
Benzo[a]fluorene, 3407
11H-Benzo[a]fluorene, 3407
1,2-Benzoanthracene, 3512
Benzo[a]pyrene, 3712
Benzoate, 2-(2,4-Dichlorophenoxy)ethyl-, 3149
Benzo[b]fluorene, 3408
11H-Benzo[b]fluorene, 3408
Benzo[b]pyridine, 1618
Benzo[b]thiophene, 1326
Benzocaine, 1708
Benzochinhydrone, 622
Benzo[def]phenanthrene, 3271
Benzoesaeure, 1055
Benzoesaeure-aethyl Ester, 1683
Benzo(b)fluoranthene, 3713
Benzo(j)fluoranthene, 3715
Benzo(k)fluoranthene, 3716
10,11-Benzofluoranthene, 3715
11,12-Benzofluoranthene, 3716
Benzo-12,13-fluoranthene, 3715
2,3-Benzofluoranthene, 3713
3,4-Benzofluoranthene, 3713
8,9-Benzofluoranthene, 3716
1,2-Benzofluorene, 3407
2,3-Benzofluorene, 3408
Benzoic Acid, 1055
Benzoic Acid, 2-(Acetyloxy)-, 4-(Acetylamino)phenyl
    Ester, 3422
Benzoic Acid, 2-(Acetyloxy)-, (Acetyloxy)methyl Ester,
    2601
Benzoic Acid, 2-(Acetyloxy)-, 2-[(2-Amino-2-
    oxoethyl)amino]-2-oxoethyl Ester, 2831
Benzoic Acid, 2-(Acetyloxy)-, 2-Amino-2-oxoethyl Ester,
    2222
Benzoic Acid, 2-(Acetyloxy)-, 2-[bis(1-
    Methylethyl)amino]-2-oxoethyl Ester, 3485
Benzoic Acid, 2-(Acetyloxy)-, 2-Carboxyphenyl Ester,
    3278

Benzoic Acid, 2-(Acetyloxy)-, 2-(Diethylamino)-1-
    methyl-2-oxoethyl Ester, 3362
Benzoic Acid, 2-(Acetyloxy)-, 2-(Diethylamino)-2-
    oxoethyl Ester, 3212
Benzoic Acid, 2-(Acetyloxy)-, 2-(Dimethylamino)-2-
    oxoethyl Ester, 2842
Benzoic Acid, 2-(Acetyloxy)-, (2,5-Dioxo-1-
    pyrrolidinyl)methyl Ester, 3002
Benzoic Acid, 2-(Acetyloxy)-, 2-(Dipropylamino)-2-
    oxoethyl Ester, 3486
Benzoic Acid, 2-(Acetyloxy)-, 2-[(2-Ethoxy-2-
    oxoethyl)methylamino]-2-oxoethyl Ester, 3339
Benzoic Acid, 2-(Acetyloxy)-, 2-(Ethylamino)-2-oxoethyl
    Ester, 2841
Benzoic Acid, 2-(Acetyloxy)-, Methyl Ester, 1951
Benzoic Acid, 2-(Acetyloxy)-, (Methylthio)methyl Ester,
    2250
Benzoic Acid, 2-(Acetyloxy)-, (1-Oxobutoxy)methyl
    Ester, 3042
Benzoic Acid, 2-(Acetyloxy)-, Phenyl Ester, 3158
Benzoic Acid Amide, 1087
Benzoic Acid, 2-[bis(4-Hydroxyphenyl)methyl]-, 3731
Benzoic Acid, 3-Bromo-2-nitro-, 961
Benzoic Acid, 2-Chloro-5-[[(5,6-dihydro-2-methyl-1,4-
    oxathiin-3-yl)carnonyl]amino]isopropyl Ester, 3329
Benzoic Acid, 3,4-Dichloro-, 969
Benzoic Acid, 3,5-Dichloro-, 971
Benzoic Acid, 2-Hydroxy-, Butyl Ester, 2284
Benzoic Acid, 2-Hydroxy-, 2-(Diethylamino)-2-oxoethyl
    Ester, 2877
Benzoic Acid, 2-Hydroxy-, 2-[(2-
    Hydroxyethyl)methylamino]-2-oxoethyl Ester, 2652
Benzoic Acid, 4-Hydroxy-, (4-Methylphenyl)methyl
    Ester, 3175
Benzoic Acid, 2-Hydroxy-, 2,2,2-Trichloroethyl Ester,
    1617
Benzoic Acid, 4-(Methylamino)-, 2-(Diethylamino)ethyl
    Ester, 3099
Benzoic Acid, p-(Methylamino)-, 2-(Diethylamino)ethyl
    Ester, 3099
Benzoic Acid Nitrile, 1005
Benzoic Acid Phenylmethyl Ester, 2996
Benzoicacidpropyl Ester, 2009
Benzoin, 2994
Benzo[g,h,i]perylene, 3859
3,4-Benzoisoquinoline, 2782
11,12-Benzo[k]fluoranthene, 3716
Benzo[j,k]fluorene, 3270
Benzol, 676
Benzolene, 676
Benzo[l]fluoranthene, 3715
1-Benzol β Pyrrol, 1337
Benzolsulfosaeure, 730
Benzolsulfosaeure-amid, 742
Benzol-tetracarbonsaeure-(1,2,4,5), 1878
Benzol-tricarbonsaeure-(1,2,3), 1613
Benzol-tricarbonsaeure-(1,3,5), 1615
Benzomarc, 3286
Benzonitril, 1005
Benzonitrile, 1005
Benzonitrile, 2,6-Dichloro-, 956
Benzophenone, 2800
Benzophosphate, 2639
Benzopyran-2-one, 1610

2H-1-Benzopyran-2-one, 1610
4H-1-Benzopyran-4-one, 7-[[2-O-(6-Deoxy-α-L-
   mannopyranosyl)-β-D-glucopyranosyl]oxy]-2,3-
   dihydro-5-hydroxy-2-(4-hydroxyphenyl)-, (S)-, 4034
2H-1-Benzopyran-2-one, 6-(β-D-Glucopyranosyloxy)-7-
   hydroxy-, 3193
2H-1-Benzopyran-2-one, 7-Methoxy-8-(3-methyl-2-
   butenyl)-, 3192
Benzo(a)pyrene, 3712
Benzo(e)pyrene, 3714
1,2-Benzopyrene, 3712
4,5-Benzopyrene, 3714
Benzopyridine, 1618
Benzopyrone, 1610
1,2-Benzopyrone, 1610
Benzopyrrole, 1337
2,3-Benzopyrrole, 1337
1,4-Benzoquinone, 622
Benzoquinone-2,5-bisaziridinyl-3,6-bismethyl Amino,
   2674
2H-1,2,4-Benzothiadiazine -7-sulfonamide, 6-Chloro-3-
   (chloromethyl)-3,4-dihydro-2-methyl-1,1-dioxide,
   1694
2,1,3-Benzothiadiazin-4(3H)-one, 1989
2H-1,2-Benzothiazine-3-carboxamide, 4-Hydroxy-2-
   methyl-N-2-pyridinyl-, 1,1-dioxide, 3167
Benzothiazole, 1315
N-2-Benzothiazolyl-N,N'-dimethylurea, 1969
Benzothiazol-2-yl-3-methylurea, 1659
N-2-Benzothiazolyl-N'-methylurea, 1659
Benzothiofuran, 1326
1-Benzothiophene, 1326
Benzoxydiglycine, 2221
Benzoyl Acetaminophen, 3165
Benzoylaminoacetic Acid, 1655
((Benzoylamino)oxy)acetic Acid, 1657
9-[5'-(O-Benzoyl)-β-D-arabinofuranosyl]adenine Ester,
   3435
9-[5-O-(Benzoyl-β-D-arabinofuranosyl)]-6-methoxy-9H-
   purine, 3540
N-Benzoylbenzamide, 2986
Benzoylbenzene, 2800
Benzoyldiantipyrylmethane, 4082
N-Benzoyl-N-(3,4-dichlorophenyl)-DL-alanine Ethyl
   Ester, 3534
Benzoyl-1-(3,4-dichlorophenyl)-3,3-dimethylurea, 3286
3-Benzoyl-5-fluorouracil, 2188
Benzoylformic Acid, 1319
N-Benzoylglycine, 1655
Benzoylhydrazone of Quinone Oxime, 2811
7-Benzoylindoline, 3162
p-Benzoylmandelic Acid, 3157
Benzoyl-r-mandelic Acid, 3157
2'-Benzoyl-6-methoxypurine Arabinoside (0.75 Hydrate),
   3541
Benzoyl-mitomycin C, 3876
2-(Benzoyloxy)-N,N-dimethylacetamide, 2261
2-(Benzoyloxy)-N-ethylacetamide, 2259
1-Benzoyloxymethyl Allopurinol, 2799
Benzoyl-peroxid, 2976
Benzoyl Peroxide, 2976
Benzoylphenylalanine, 3305
N-Benzoyl-DL-phenylalanine, 3305
Benzoylphenylcarbinol, 2994

4-Benzoyl Phenylisothiocyanate, 2963
2-(meta-Benzoylphenyl) Propionic Acid, 3295
Benzoylprop-ethyl, 3534
Benzoyltryptophan, 3531
N-Benzoyl-DL-tryptophan, 3531
N-Benzoyl-L-tyrosinamide Acetate, 3550
N-Benzoyl-L-tyrosinamide Prostaglandin E2, 4115
N-Benzoyl-L-tyrosinamide Prostaglandin F2 α, 4116
Benzoyltyrosine, 3306
1,2-Benzphenanthrene, 3513
9,10-Benzphenanthrene, 3514
3,4-Benzpyrene, 3712
Benzthiazuron, 1659
Benzulfide, 3108
Benzyl Acetate, 1684
9-[5-O-(Benzyl Acetate-β-D-arabinofuranosyl)]-6-
   methoxy-9H-purine (0.05 Hydrate), 3651
Benzyl Alcohol, 1117
Benzylalkohol, 1117
Benzylbenzene, 2815
Benzyl Benzoate, 2996
Benzylbutyl Phthalate, 3654
Benzyl Carbamate, 1417
O-Benzyl Carbamate, 1417
Benzylcarbonyl-mitomycin C, 3916
Benzyl Cellosolve, 1757
Benzylcellosolve, 1757
Benzylchlorophenol, 2806
6-Benzyl-m-cresol, 3013
S-Benzyl Di-sec-butylthiocarbamate, 3391
2-Benzyl-4,6-dichlorophenol, 2791
Benzyl 2,4-Dichlorophenoxyacetate, 3150
Benzyl 2,2-Diethylmalonurate, 3213
Benzyl-2,2-diethylmalonurate, 3213
S-Benzyl O,O-di-isopropyl Phosphorothioate, 2328
S-Benzyl O,O-Di-isopropyl Phosphorothioate, 2921
2-Benzyl-3,5-dimethyl-4-chloro-phenol, 3178
N-Benzyl-N',N'-dimethyl-N-2-pyridylethylenediamine,
   3363
1-O-Benzylethanediol, 1757
9-[5-O-(Benzyl Formyl-β-D-arabinofuranosyl)]-6-
   methoxy-9H-purine (0.1 Hydrate), 3647
Benzyl-harnstoff, 1439
Benzyl 4-hydroxybenzoate, 2999
Benzyl Isothiocyanate, 1346
Benzylisothiocyanate, 1346
Benzyl-mitomycin C, 3882
Benzyloxycarbonyl Amine, 1417
1-Benzyloxycarbonyl-5-fluorouracil, 2544
Benzyloxycarbonyl Glycine, 1964
Benzyloxycarbonyl-mitomycin C, 3917
N-Benzyloxycarbonyl-L-tyrosine, 3434
Benzylparaben, 2999
2-Benzylphenol, 2819
4-Benzylphenol, 2820
o-Benzylphenol, 2819
p-Benzylphenol, 2820
Benzylurea, 1439
B[E]P, 3714
Bepadin, 3960
Bepridil, 3960
Berberine, 3734
Berberine (Hexahydrate), 3734
Bergamol, 2728

Bermat, 2016
Bernsteinsaeure, 281
Bersteinsaeure-diamid, 314
Bersteinsaeure-dinitril, 233
Betain, 491
Betaine, 491
Betaine (Monohydrate), 502
Betamec, 3108
Betamethasone, 3889
Betamethasone Acetate, 3952
Betamethasone-17-acetate, 3952
Betamix 70 WP, 3312
Betanal-475, 3312
Betanex, 3312
Betasan, 3108
Betula, 1391
Betula Oil, 1391
Bexone, 2253
1,4-Bezenedicarboxylic Acid, 1322
α-BHC, 692
β-BHC, 693
γ-BHC, 691
BH 2,4-D, 3052
Biacetyl Mono(2-pyridyl)hydrazone, 1720
Biacetyl Mono(2-pyridyl)-hydrazone, 1720
Biallyl, 787
Biaxin, 4124
Bibenzyl, 3006
Bicep 6L, 3240
Bichlorendo, 2185
cis-Bicyclo[4.4.0]decane, 2119
Bicyclo[2.2.1]heptan-2-one, 1,3,3-Trimethyl-, (1S)-,
    2104
Bicyclo[2.2.1]heptylcarbinol, 1500
5-Bicyclo[3.2.1]oct-2-en-3-yl-5-ethylbarbituric Acid,
    3054
5-Bicyclo[3.2.1]oct-2-en-3-yl-5-ethyl-2,4,6(1H,3H,5H)-
    pyrimidinetri-one, 3054
Bidisin, 1934
Bifenox, 2955
Biformylchlorazin, 2055
Bigitalin, 4127
Bihexyl, 2757
Bilevon, 2772
Bilijodon, 2230
21H-Biline-8,12-dipropanoic Acid, 2,17-Diethenyl-
    1,10,19,22,23,24-hexahydro-3,7,13,18-tetramethyl-
    1,19-dioxo-, 4100
Bilirubin, 4100
Biliverdic Acid, 1422
Biliverdinsaeure, 1422
2,3,1',8'-Binaphthylene, 3716
Biofanal, 4141
Biotin, 2099
D-Biotin, 2099
Biotin d, 2099
Bi-PC, 2205
Biphenyl, 2554
1,1'-Biphenyl, 2554
 4-Biphenylacetic Acid, 2997
1,1'-Biphenyl, 4-Bromo-, 2534
1,1'-Biphenyl-2,2'-dicarboxylic Acid, 2977
2,2'-Biphenyldicarboxylic Acid, 2977
1,1'-Biphenyl, 2,3'-Dichloro-, 2515

1,1'-Biphenyl, 2,4-Dichloro-, 2519
1,1'-Biphenyl, 2,5-Dichloro-, 2520
1,1'-Biphenyl, 2,6-Dichloro-, 2514
1,1'-Biphenyl, 3,3'-Dichloro-, 2517
1,1'-Biphenyl, 3,4-Dichloro-, 2516
o-Biphenylenemethane, 2787
1,1'-Biphenyl, Heptachloro-, 2398
1,1'-Biphenyl, 2,2',3,3',4,4',5-Heptachloro-, 2389
1,1'-Biphenyl, 2,2',3,3',4,4',6-Heptachloro-, 2386
1,1'-Biphenyl, 2,2',3,3',4,5,5'-Heptachloro-, 2395
1,1'-Biphenyl, 2,2',3,3',4,5,6-Heptachloro-, 2392
1,1'-Biphenyl, 2,2',3,3',4,5',6-Heptachloro-, 2390
1,1'-Biphenyl, 2,2',3,3',4,5,6'-Heptachloro-, 2391
1,1'-Biphenyl, 2,2',3,3',4,5',6'-Heptachloro-, 2388
1,1'-Biphenyl, 2,2',3,3',4,6,6'-Heptachloro-, 2393
1,1'-Biphenyl, 2,2',3,3',5,5',6-Heptachloro-, 2394
1,1'-Biphenyl, 2,2',3,4,4',5,5'-Heptachloro-, 2387
1,1'-Biphenyl, 2,2',3,4,4',5',6-Heptachloro-, 2396
1,1'-Biphenyl, 2,2',3,4',5,5',6-Heptachloro-, 2399
1,1'-Biphenyl, Hexachloro-, 2408
1,1'-Biphenyl, 2,2',3,3',4,4'-Hexachloro-, 2418
1,1'-Biphenyl, 2,2',3,3',5,6'-Hexachloro-, 2421
1,1'-Biphenyl, 2,2',3,3',6,6'-Hexachloro-, (+)-, 2413
1,1'-Biphenyl, 2,2',3,4,4',5-Hexachloro-, 2416
1,1'-Biphenyl, 2,2',3,4,4',5'-Hexachloro-, 2415
1,1'-Biphenyl, 2,2',3,4,4',6-Hexachloro-, 2411
1,1'-Biphenyl, 2,2',3,4,5,5'-Hexachloro-, 2425
1,1'-Biphenyl, 2,2',3,4',5,5'-Hexachloro-, 2414
1,1'-Biphenyl, 2,2',3,4,4',5'-Hexachloro-, 2426
1,1'-Biphenyl, 2,2',3,5,5',6-Hexachloro-, 2424
1,1'-Biphenyl, 2,2',4,4',6,6'-Hexachloro-, 2427
1,1'-Biphenyl, 2,3,3',4,4',6-Hexachloro-, 2417
1,1'-Biphenyl, 2,3,3',4',5,6-Hexachloro-, 2423
3-Biphenyl Isothiocyanate, 2786
4-Biphenyl Isothiocyanate, 2785
m-Biphenyl Isothiocyanate, 2786
p-Biphenyl Isothiocyanate, 2785
o-Biphenylmethane, 2787
1,1'-Biphenyl, 2,2',3,3',4,5,5',6,6'-Nonachloro-, 2377
1,1'-Biphenyl, 2,2',4,5,5'-Pentabromo-, 2428
1,1'-Biphenyl, 2,2',3,3',4-Pentachloro-, 2431
1,1'-Biphenyl, 2,2',3,3',5-Pentachloro-, 2433
1,1'-Biphenyl, 2,2',3,3',6-Pentachloro-, 2434
1,1'-Biphenyl, 2,2',3,4,4'-Pentachloro-, 2438
1,1'-Biphenyl, 2,2',3,4',5-Pentachloro-, 2437
1,1'-Biphenyl, 2,2',3,5',6-Pentachloro-, 2442
1,1'-Biphenyl, 2,2',4,4',5-Pentachloro-, 2445
1,1'-Biphenyl, 2,3,3',4',5-Pentachloro-, 2447
1,1'-Biphenyl, 2,3,3',4,5'-Pentachloro-, 2443
1,1'-Biphenyl, 2,3,3',4',6-Pentachloro-, 2448
1,1'-Biphenyl, 2,3,4,4',5-Pentachloro-, 2449
1,1'-Biphenyl, 2,3',4,4',5-Pentachloro-, 2446
1,1'-Biphenyl, 2,3',4',5,5'-Pentachloro-, 2435
1,1'-Biphenyl, 2,3,4,5,6-Pentachloro-, 2444
1,1'-Biphenyl, Tetrachloro-, 2459
1,1'-Biphenyl, 2,2',3,3'-Tetrachloro-, 2470
1,1'-Biphenyl, 2,2',3,4-Tetrachloro-, 2472
1,1'-Biphenyl, 2,2',3,4'-Tetrachloro-, 2471
1,1'-Biphenyl, 2,2',3,5'-Tetrachloro-, 2473
1,1'-Biphenyl, 2,2',3,6'-Tetrachloro-, 2485
1,1'-Biphenyl, 2,2',4,4'-Tetrachloro-, 2474
1,1'-Biphenyl, 2,2',4,5-Tetrachloro-, 2467
1,1'-Biphenyl, 2,2',5,5'-Tetrachloro-, 2476
1,1'-Biphenyl, 2,2',5,6'-Tetrachloro-, 2477

1,1'-Biphenyl, 2,2',6,6'-Tetrachloro-, 2478
1,1'-Biphenyl, 2,3,3',4'-Tetrachloro-, 2482
1,1'-Biphenyl, 2,3,4,4'-Tetrachloro-, 2475
1,1'-Biphenyl, 2,3',4,4'-Tetrachloro-, 2480
1,1'-Biphenyl, 2,3,4,5-Tetrachloro-, 2465
1,1'-Biphenyl, 2,3,4',5-Tetrachloro-, 2483
1,1'-Biphenyl, 2,3',4',5-Tetrachloro-, 2479
1,1'-Biphenyl, 2,3',4',5'-Tetrachloro-, 2469
1,1'-Biphenyl, 2,3',4,6-Tetrachloro-, 2481
1,1'-Biphenyl, 2,3,4',6-Tetrachloro-, 2484
1,1'-Biphenyl, 2,4,4',5-Tetrachloro-, 2466
1,1'-Biphenyl, 2,4,4',6-Tetrachloro-, 2464
1,1'-Biphenyl, 3,3',5,5'-Tetrachloro-, 2462
1,1'-Biphenyl, 2,4,6-Tribromo-, 551
1,1'-Biphenyl, 2,2',3-Trichloro-, 2500
1,1'-Biphenyl, 2,2',4-Trichloro-, 2501
1,1'-Biphenyl, 2,2',5-Trichloro-, 2502
1,1'-Biphenyl, 2,2',6-Trichloro-, 2503
1,1'-Biphenyl, 2',3,4-Trichloro-, 2496
1,1'-Biphenyl, 2,3,4'-Trichloro-, 2505
1,1'-Biphenyl, 2,3',5-Trichloro-, 2504
1,1'-Biphenyl, 2,3,6-Trichloro-, 2498
1,1'-Biphenyl, 2,3',6-Trichloro-, 2491
1,1'-Biphenyl, 2,4,5-Trichloro-, 2494
1,1'-Biphenyl, 2,4,6-Trichloro-, 2493
3-(4-Biphenylylcarbonyl) Propionic Acid, 3294
2,2'-Bipyridine, 1894
α,α'-Bipyridyl, 1894
2,2'-Bipyridyl, 1894
4,4'-Bipyridyl, 1893
2,2'-Biquinoline, 3516
2,2'-Biquinolyl, 3516
BI-RG 587, 3174
Birlane, 2621
Birlanex, 2621
bis(2-Butoxyethyl) Phthalate, 3782
bis(2-N-Butoxyethyl) Phthalate, 3782
N,N-bis(Carboxymethyl)glycine, 779
N,N'-bis(3-Carboxy-2,4,6-triiodophenyl)-diglycolamide,
    3507
1,2-bis(2-Chloroethoxy)ethane, 826
4-[bis(2-Chloroethyl)amino]-L-Phenylalanine, 2891
3-[bis(2-Chloroethyl)carbamate, 3928
bis(2-Chloro-1-methylethyl) Ether, 825
bis(5-(p-Chlorophenyl)biguanidinio)hexane, 3891
2,2-bis(4-Chlorophenyl)-1,1-dichloroethylene, 2944
N,5-bis(4-Chlorophenyl)-3,4-dihydro-3-((1-
    methylethyl)imino)-2-phenazinamine, 4028
2,2-bis(p-Chlorophenyl)-1,1,1-trichloroethane, 2957
bis-Cyclopentadienyliron, 1929
bis(Di-n-hexyl-phosphinyl)butane, 4063
bis(Di-n-hexyl-phosphinyl)ethane, 4027
bis(Di-n-hexyl-phosphinyl)methane, 4005
bis(Di-n-hexyl-phosphinyl)propane, 4050
bis(Dimethylamino)-(3-amino-5-phenyl-1,2,4-triazol-1-
    yl)-phosphine Oxide, 2723
4,4[-bis(Dimethylamino)benzophenone, 3455
p,p'-bis(N,N-Dimethylamino)benzophenone, 3455
bis[4-(Dimethylamino)phenyl]-methanone, 3455
bis(2-Ethoxyethyl) Phthalate, 3373
bis(Ethylamino)-6-(methylthio)-s-triazine, 1523
4,6-bis(Ethylamino)-s-triazin-2-ol, 1203
bis(2-Ethylhexyl) Adipate, 3907
bis-(2-Ethylhexyl) 1,2-Benzenedicarboxylate, 3967

bis(2-Ethylhexyl) Phthalate, 3967
1,4-bis[(2-Hydroxyethyl)amino]anthraquinone, 3539
4-[bis(2-Hydroxyethyl)amino]azobenzene, 3340
p-[bis(2-Hydroxyethyl)amino]azobenzene, 3340
4-[bis(2-Hydroxyethyl)amino]-2',4'-dinitroazobenzene,
    3328
4-[bis(2-Hydroxyethyl)amino]-2'-methoxy-2-methyl-4'-
    nitroazobenzene, 3568
4-[bis(2-Hydroxyethyl)amino]-2-methylazobenzene,
    3466
2,2-bis(Hydroxymethyl)-1,3-Propanediol, 521
3,4-bis(4-Hydroxyphenyl)-2,4-hexadiene, 3543
2-[bis(4-Hydroxyphenyl)methyl]benzoic Acid, 3726
4-[bis-(p-Hydroxyphenyl)methylene]-2,5-cyclohexadien-
    1-one, 3626
2,2-bis-[4-Hydroxyphenyl]-propan, 3191
2,2-bis(-p-Hydroxyphenyl)-1,1,1-trichloroethylene, 2982
2,2-bis-(4-Hydroxypheny)-propane, 3191
bis(Isooctyl) Phthalate, 3969
bis(Isopropylamino)hydroxy-s-triazine, 2150
bis(2-Methoxyethyl) Phthalate, 3066
2,5-bis(Methylaziridinyl)-3,6-bis(methylamino)-1,4-
    benzoquinone, 3085
bis(1-Methylethyl) 1,3-Dithiolan-2-ylidenepropanedioate,
    2720
N,N'-bis(1-Methylethyl)-6-methylthio-1,3,5-triazine-2,4-
    diamine, 2153
O,O-bis(1-Methylethyl) S-(2-
    ((Phenylsulfonyl)amino)ethyl) Phosphorodithioate,
    3108
bis(1-Methyl-ethyl) Phthalate, 3063
bis(2-Methyl Propyl) Ether, 1580
2,2-bis(p-Methylsulfinylphenyl)-1,1,1-trichloroethane,
    3299
2,2-bis(p-Methylsulfonylphenyl)-1,1,1-trichloroethane,
    3300
2,2-bis-(p-Methylthiophenyl)-1,1,1-trichloroethane, 3301
bis(Methylxanthogen) Disulfide, 278
bis(n-Octyl) Phthalate, 3971
Bisphenol A, 3191
N,N-bis(Phosphonomethyl)glycine, 374
1,3-bis(Pivaloyloxymethyl)-5-fluoro-2,4(1H,3H)-
    pyrimidinedi-one, 3377
1,3-bis(Pivaloyloxymethyl)-5-fluorouracil, 3377
1,2-bis(Propionyloxy)ethane, 1511
bis(1-Propyl) Dixanthogen, 1506
bis(Tereoctyl) Phthalate, 3968
bis(Tributyltin) Oxide, 3987
bis(cis-3,3,5-Trimethylcyclohexyl) Phthalate, 4021
p,p'-Bitoluene, 3005
Biuret, 94
B[K]F, 3716
Bladex, 1763
Blastin, 958
Blattanex, 2293
Blattosep, 2293
BNOA, 2569
Bolasterone, 3847
endo-2-Bornanol, 2126
Borneocamphor, 2126
Borneol, 2129
D-Borneol, 2126
L-Borneol, 2129
BPH, 1720

BPMC, 2685
Brassylic Acid, 2931
Bravo, 1601
Brenzkatechin, 724
Brofene, 1352
Bromacil, 1761
DL-Brombernsteinsaeure, 245
DL-Brombuttersaeure, 292
α-Bromethylpropylaceturea, 1491
Bromfenim, 2773
Bromkal 80, 2378
2-Brom-4-nitro-phenol, 576
4'-Bromoacetanilide, 1353
5-(2-Bromoallyl)-5-sec-butylbarbituric Acid, 2287
p-Bromoaniline-m-sulfonic Acid, 678
p-Bromoaniline-o-sulfonic Acid, 679
p-Bromoaniline-o-sulfonic Acid (Monohydrate), 680
Bromoantifebrin, 1353
Bromobenzene, 625
4-Bromobenzenesulfonamide, 677
p-Bromobenzenesulfonamide, 677
4-Bromobenzenesulfonic Acid, 627
p-Bromobenzenesulfonic Acid, 627
p-Bromobenzenesulfonic Acid (2.5 Hydrate), 629
p-Bromobenzenesulfonic Acid (Monohydrate), 628
3-Bromobenzoic Acid, 987
4-Bromobenzoic Acid, 986
m-Bromobenzoic Acid, 987
p-Bromobenzoic Acid, 986
3-Bromobenzyl Isothiocyanate, 1296
4-Bromobenzyl Isothiocyanate, 1295
m-Bromobenzyl Isothiocyanate, 1296
p-Bromobenzyl Isothiocyanate, 1295
4-Bromobiphenyl, 2534
Bromobutane, 335
1-Bromo-3-butene, 290
4-Bromobutene-1, 290
4-Bromo-1-butene, 290
5-Bromo-3-tert-butyl-6-methyluracil, 1760
α-Bromobutyric Acid, 292
DL-2-Bromobutyric Acid, 292
Bromochlorodifluoromethane, 32
Bromochlorodifluoromethine, 32
1-Bromo-2-chloroethane, 70
Bromochloromethane, 7
Bromo-chloro-methane, 7
O-(4-Bromo-2-chlorophenyl)-O-ethyl-S-propyl
    phosphorothioate, 2286
3-(4-Bromo-3-chlorophenyl)-1-methoxy-1-methylurea,
    1664
N'-(4-Bromo-3-chlorophenyl)-N-methoxy-N-methylurea,
    1664
1-Bromo-3-chloropropane, 154
7-Bromo-5-chloro-8-quinolinyl 2-propenoate, 2487
2-Bromo-2-chloro-1,1,1-trifluoroethane, 45
Bromodichloromethane, 1
O-(4-Bromo-2,5-dichlorophenyl) O,O-Diethyl
    Phosphorothioate, 1978
O-(4-Bromo-2,5-dichlorophenyl) O,O-Dimethyl
    phosphorothioate, 1352
O-(4-Bromo-2,5-dichlorophenyl) O-Methyl
    phenylphosphonate, 2789
Bromodiethylacetylcarbamide, 1198
Bromodiethylacetylurea, 1198

5-Bromo-2,2-dimethyl-5-nitro-1,3-dioxane, 788
Bromodiphenyl, 2534
Bromoethane, 83
1-Bromo-ethyl-butyryl-urea, 1198
5-Bromo-2-ethyl-5-nitro-1,3-dioxane, 789
Bromofenoxim, 2773
1-Bromo-2-fluorobenzene, 575
1-Bromo-3-fluorobenzene, 574
2-Bromofluorobenzene, 575
3-Bromofluorobenzene, 574
Bromoform, 3
1-Bromoheptane, 1236
1-Bromohexane, 882
1-Bromo-3-isothiocyanato-benzene, 962
1-Bromo-4-isothiocyanato-benzene, 963
α-Bromo-isovaleric Ureide, 811
Bromol, 552
Bromomethane, 12
Bromomethionic Acid, 13
α-Bromo-methyl-acetic Ureide, 291
1-Bromo-3-methylbenzene, 1070
3-Bromo-1-methylbenzene, 1070
1-Bromo-3-methylbutane, 485
3-Bromo-2-methyl-butanoic Ureide, 807
DL-N-(2-Bromo-2-methylbutanoyl)urea, 807
5-Bromo-6-methyl-3,5-butyluracil, 1761
α-Bromo-methyl-ethyl-acetate, 440
5-Bromo-2-methyl-5-nitro-1,3-dioxane, 423
1-Bromo-2-methylpropane, 334
1-Bromonaphthalene, 1880
2-Bromonaphthalene, 1879
1-Bromo-2-naphthylisothiocyanate, 2186
2-Bromo-4-nitroacetanilide, 1328
2-Bromo-5-nitroacetanilide, 1327
3-Bromo-2-nitrobenzoic Acid, 961
5-Bromo-5-nitro-1,3-dioxane, 263
2-Bromo-4-nitrophenol, 576
2-Bromo-2-nitropropane-1,3-diol, 155
1-Bromooctane, 1559
1-Bromopentane, 486
4-Bromophenol, 626
p-Bromophenol, 626
5-Bromo-2-p-phenol-5-nitro-1,3-dioxane, 1931
3-Bromophenyl Isothiocyanate, 962
4-Bromophenyl Isothiocyanate, 963
3-(p-Bromophenyl)-1-methoxy-1-methylurea, 1690
N'-(4-Bromophenyl)-N-methoxy-N-methylurea, 1690
5-Bromo-2-p-phenyl-5-nitro-1,3-dioxane, 1930
(4-Bromophenyl)sulfonamide, 677
Bromophos, 1352
Bromophos-ethyl, 1978
Bromo-pivalate Ureide, 1197
Bromopropane, 187
1-Bromopropane, 187
3-Bromo-1-propanol, 188
3-Bromopropene, 137
(2-Bromopropionyl)urea, 291
α-Bromopropionylurea, 291
Bromopropylate, 3423
3-Bromopropyl chloride, 154
5-Bromo-2-propyl-5-nitro-1,3-dioxane, 1180
Bromosuccinic Acid, 245
3-Bromotoluene, 1070
m-Bromotoluene, 1070

11-Bromoundecanoic Acid, 2351
Bromo-11-undecanoique Acide, 2351
α-Bromo-valeric Acid Ureide, 810
β-Bromo-valeric Acid Ureide, 812
γ-Bromo-valeric Acid Ureide, 808
Bromoxynil, 954
3-Brom-propanol-(1), 188
Brompyrazone, 1891
Bronchodid Duracap, 1113
Bronco, 213
Bronidox, 263
Bronidox L, 263
Bronkodyl, 1113
Bronkotabs, 1113
Bronner's Acid, 1922
Bronopol, 155
Brophene, 1352
Bropirimine, 1890
Brown AP, 729
Brucin, 3919
Brucine, 3919
Brucine (Tetrahydrate), 3920
Bryamycin, 4176
Budesonide, 3995
Bu-paraben, 2283
Bupirimate, 2929
Bupivacaine, 3596
DL-Bupivacaine, 3596
Bupivicaine, 3596
Butabarbital, 2096
Butacarb, 3392
Butachlor, 3497
1,3-Butadiene, 262
Butadiyne, 225
Butalbital, 2305
Butallylonal, 2287
Butamben, 2289
n-Butanal, 324
Butanamide, N-(Aminocarbonyl)-2-bromo-3-methyl-,
    811
Butanamide, 4-(Benzoyloxy)-N,N-dimethyl-, 2869
Butanamide, 4-[[(4-Chlorophenyl)(5-fluoro-2-
    hydroxyphenyl)methylene]amino]-, 3424
2-Butanamine, N-Ethyl-, 941
Butanamine, N-Ethyl-N-nitroso-, 910
Butane, 356
n-Butane, 356
Butane, 2,3-Dichloro-, 307
Butanedioic Acid, bis[4-(Acetylamino)phenyl] Ester,
    3737
Butanedioic Acid, Diethyl Ester, 1512
Butanedioic Acid, Mono[4-(acetylamino)phenyl] Ester,
    2613
2,3-Butanedione, 276
Butane, 1-Nitro-, 343
Butanetetracarboxylic Acid, 1474
1,2,3,4-Butanetetracarboxylic Acid, 1474
1,2,3,4,-Butane Tetracarboxylic Acid, 1474
meso-1,2,3,4-Butanetetracarboxylic Acid, 1474
DL-1,2,3,4-Butanetetrol, 368
Butanex, 3497
Butanimide, 254
Butanoic Acid, 346
Butanoic Acid, 4-(Acetylamino)phenyl Ester, 2647

Butanoic Acid, 2-[[2-[3-(Acetylamino)-2,4,6-
    triiodophenoxy]ethoxy]methyl]-, 3200
Butanoic Acid, [2-(Aminocarbonyl)phenoxy]methyl Ester,
    2649
Butanoic Acid, 4-Amino-2-hydroxy-4-oxo-, 298
Butanoic Acid, 4-[(4-Chlorobenzoyl)(4-
    methoxyphenyl)amino]-, 3536
Butanoic Acid, 4-(Diethylamino)-4-oxo-, (4,5-Dihydro-4-
    oxo-1H-pyrazolo[3,4-d]pyrimidin-1-yl)methyl Ester,
    3076
Butanoic Acid, (4,5-Dihydro-4-oxo-1H-pyrazolo[3,4-
    d]pyrimidin-1-yl)methyl Ester, 2000
Butanoic Acid, (4,5-Dihydro-4-oxo-2H-pyrazolo[3,4-
    d]pyrimidin-1-yl)methyl Ester, 2001
Butanoic Acid Ethyl Ester, 856
Butanoic Acid, (5-Fluoro-3,4-dihydro-2,4-dioxo-1(2H)-
    pyrimidinyl)methyl Ester, 1699
Butanoic Acid, 2-(Formylamino)-, 443
Butanoic Acid, 3-Methoxy-3-oxopropyl Ester, 1555
Butanol-(1), 360
1-Butanol, 360
DL-Butanol-(2), 361
n-Butanol, 360
tert-Butanol, 366
Butanon-(2), 325
Butazolidin, 3644
Butazone, 3644
2-Butenamide, N-[(4-Aminophenyl)sulfonyl]-3-methyl-,
    2274
Butene-1, 306
1-Butene, 306
α-Butene, 306
2-Butenediamide, N,N,N',N'-Tetraethyl-, 2733
2-Butenediamide, N,N,N',N'-Tetramethyl-, (Z)-, 1498
2-Butenedioic Acid (Z)-, 3790
2-Butenoic Acid, 277
3-Buten-2-yl Acetate, 794
3-Butenyl Bromide, 290
Butethal, 2096
Buthidazole, 2102
4-Butoxybenzoic Acid 2-(Diethyl-amino)ethyl Ester,
    3500
Butoxydibutylphosphine Oxide, 2762
Butoxyl, 1226
Butoxyl (3-Methoxy-N-butyl Acetate), 1226
N-(Butoxymethyl)-2-chloro-2',6'-diethylacetanilide, 3497
N-(Butoxymethyl)-2-chloro-N-(2,6-
    diethylphenyl)acetamide, 3497
2-Butoxyphenol, 2080
3-Butoxyphenol, 2078
m-Butoxy Phenol, 2078
o-Butoxyphenol, 2080
3-n-Butoxyphenyl N-Methylcarbamate, 2692
3-sec-Butoxyphenyl N-Methylcarbamate, 2693
m-n-Butoxyphenyl N-Methylcarbamate, 2692
m-sec-Butoxyphenyl N-Methylcarbamate, 2693
Buttersaeure, 331
Buttersaeure-methyl Ester, 476
Buttersaeure-propyl Ester, 1219
Buturon, 2602
Butyl Acetaminophen, 2880
n-Butylacetat, 863
Butyl Acetate, 863
1-Butyl Acetate, 863

DL-sec-Butyl Acetate, 858
n-Butyl Acetate, 863
sec-Butyl Acetate, 858
Butyl α-Acetoxypropionate, 1795
9-[5'-(O-tert-Butylacetyl)-β-D-arabinofuranosyl]adenine
  Ester, 3383
Butylacetylene, 786
n-Butylacetylene, 786
n-Butyl Acohol, 360
Butyl Alcohol, 360
DL-sec-Butyl Alcohol, 361
n-Butyl Alcohol, 360
sec-Butyl Alcohol, 361
sec-DL-Butyl Alcohol, 361
tert-Butyl Alcohol, 366
n-Butylallylbarbitone, 2307
n-Butylallylbarbituric Acid, 2307
DL-sec-Butylamin, 371
n-Butylamin, 372
DL-sec-Butylamine, 371·
n-Butylamine, 372
sec-Butylamine, 371
Butyl p-Aminobenzoate, 2289
4-Butylaminobenzoic Acid 2-(Diethyl-amino)ethyl Ester,
  3501
(1-(Butylamino)carbonyl)-1H-benzimidazol-2-yl)carbamic
  Acid Methyl Ester, 3056
2-(Butylamino)ethyl 4-Aminobenzoate, 2912
2-sec-Butylamino-4-ethylamino-6-methoxy-s-triazine,
  2147
Butylate, 2371
Butylated Hydroxytoluene, 3260
10-Butyl-1,2-benzanthracene, 3872
Butylbenzene, 2046, 2052
n-Butylbenzene, 2046, 2052
sec-Butylbenzene, 2045
t-Butylbenzene, 2050
tert-Butylbenzene, 2050
Butyl Benzyl Phthalate, 3654
Butylbenzyl Phthalate, 3654
n-Butyl Bromide, 335
N-Butyl-1-butanamine, 1588
Butyl 3-Butoxypropionate, 2369
n-Butyl β-n-Butoxypropionate, 2369
Butyl Butyrate, 1545
n-Butyl n-Butyrate, 1545
Butyl Carbamate, 496
iso-Butyl Carbamate, 494
n-Butyl Carbamate, 496
O-t-Butyl Carbamate, 497
tert-Butyl Carbamate, 497
1-Butylcarbamoyl-5-fluorouracil, 1734
t-Butyl Carbinol, 511
Butyl Carbitol Acetate, 2176
Butyl Carbobutoxymethyl Phthalate, 3581
Butylcellosolve Acetate, 1552
n-Butyl Cellosolve Acetate, 1552
n-Butyl Chloride, 339
sec-Butyl Chloride, 336
tert-Butyl Chloride, 338
N3-Butyl-N1-p-chlorobenzenesulfonylurea, 2020
3-tert-Butyl-5-chloro-6-methyluracil, 1762
N-Butylcinnamamide, 2866
Butyl Citrate, 3605

5-tert-Butyl-o-cresol, 2317
Butyldiantipyrylmethane, 4032
Butyl Dibutyl Phosphinate, 2762
Butyl Dibutylphosphinate, 2762
1-Butyl-3-(3,4-dichlorophenyl)-1-methylurea, 2663
Butyl Diethyl Phosphate, 1595
2-Butyl-4,5-dimethylphenol, 2715
2-Butyl-4,6-dimethylphenol, 2714
4-Butyl-2,5-dimethylphenol, 2709
4-Butyl-2,6-dimethylphenol, 2711
1-Butyl-3,7-dimethylxanthine, 2314
1-n-Butyl-3,7-dimethylxanthine, 2314
7-Butyl-1,3-dimethylxanthine, 2101
Butyl Diphenyl Phosphate, 3347
1-Butylene, 306
α-Butylene, 306
Butylene Glycol Diacetate, 1510
Butylene Oxide, 326
Butyl Ether, 1579
iso-Butyl Ether, 1580
n-Butyl Ether, 1579
n-Butyl β-Ethoxypropionate, 1830
Butylethylacetic Acid, 1542
sec-Butylethylamine, 941
5-Butyl-2-(ethylamino)-4-hydroxy-6-methylpyrimidine,
  2338
5-Butyl-2-(ethylamino)-6-methyl-4-pyrimidinyl
  dimethylsulfamate, 2929
5-Butyl-5-ethylbarbituric Acid, 2096
4-[[(4-(N-Butyl-N-
  cthylnitrilc)amino)phcnyl]azo]nitrobenzene, 3659
2-Butyl-4-ethylphenol, 2713
2-Butyl-6-ethylphenol, 2712
N-Butyl-5-fluoro-2,4-dioxo pyrimidinecarboxamide,
  1734
Butyl Formate, 472
Butyl Glycol Phthalate, 3782
Butyl 4-Hydroxybenzoate, 2283
n-Butyl-9-hydroxyfluorene-(9)-carboxylate, 3545
Butyl 2-Hydroxypropanoate, 1227
Butyl α-Hydroxypropionate, 1227
16,17-Butylidenebis(oxy)-11-,21-dihydroxypregna-1,4-
  diene-3,20-dione, 3995
n-Butyl Iodide, 340
t-Butyl Isopropyl Ether, 1253
Butyl Lactate, 1227
n-Butylmalonic Acid, 1193
n-Butyl β-Methoxypropionate, 1556
Butyl(methyl)acetylene, 1178
i-Butylmethylcarbinol, 930
n-Butylmethylcarbinol, 919
t-Butylmethylcarbinol, 916
tert-Butyl Methyl Ether, 512
5-tert-Butyl-2-methylphenol, 2317
Butyl Nicotinate, 2030
n-Butyl Nicotinate, 2030
Butyl Nitrate, 352
N-Butyl Nitrate, 352
Butyl Octyl Phenyl Phosphate, 3604
1-Butyloxycarbonyl-5-fluorouracil, 1697
n-Butyl-paba-β-cyclodextrin, 4139
n-Butyl-paba-γ-cyclodextrin, 4150
Butylparaben, 2283
n-Butyl Pentanoate, 1826

2-n-Butylphenol, 2075
4-n-Butylphenol, 2076
4-sec-Butylphenol, 2074
4-t-Butylphenol, 2077
o-n-Butylphenol, 2075
p-n-Butylphenol, 2076
p-sec-Butylphenol, 2074
p-tert-Butylphenol, 2077
2-sec-Butylphenyl Methylcarbamate, 2685
3-tert-Butylphenyl N-Methylcarbamate, 2686
m-tert-Butylphenyl N-Methylcarbamate, 2686
Butyl Phenyl-methyl Phthalate, 3654
N-Butyl-3-phenyl-2-propenamide, 2866
n-Butyl Phthalate, 3371
Butylphthalyl Butyl Glycolate, 3581
Butyl Phthalyl Butyl Glycollate, 3592
Butyl Propionate, 1220
n-Butyl Propionate, 1220
Butyl Salicylate, 2284
n-Butyl Salicylate, 2284
1-Butyl-3-sulfanilyl Urea, 2326
1-Butyl Theobromine, 2314
7-Butyl Theophylline, 2101
1-(5-tert-Butyl-1,3,4-thiadiazol-2-yl)-1,3-dimethylurea,
    1790
3-(5-tert-Butyl-1,3,4-thiadiazol-2-yl)-4-hydroxy-1, 2102
1-Butyl-3-(para-tolylsulfonyl) Urea, 2705
Butyl Valerate, 1826
Butyl Valerianate, 1826
Butylxanthogenic Acid, 471
1-Butyne, 261
Butyraldehyd, 324
Butyraldehyde, 324
2-Butyraldehyde, 324
Butyramide, 341
n-Butyramide, 341
9-[5-O-(Butyrate-β-D-arabinofuranosyl)]-6-methoxy-9H-
    purine, 3219
Butyric Acid, 331
n-Butyric Acid, 331
N-Butyric Acid Isopropyl Ester, 1223
Butyric Acid, p-Isothiocyanatophenyl Ester, 2219
Butyric Acid, 4-Phosphono-, 355
Butyric Ether, 856
β-Butyrolacetone, 275
n-Butyroniitrile, 296
γ-Butyronitrile, 296
n-Butyronitrile, 296
Butyryl Acetaminophen, 2647
9-(2-O-Butyryl-β-D-arabinofuranosyl)adenine, 3025
9-[5'-(O-Butyryl)-β-D-arabinofuranosyl]adenine Ester,
    3077
5'-Butyryl 5-iodo-2'-deoxyuridine, 2865
2'-Butyryl-6-methoxypurine Arabinoside (0.3 Hydrate),
    3221
1-Butyryloxymethyl Allopurinol, 2000
2-Butyryloxymethyl Allopurinol, 2001
1-Butyryloxymethyl-3-benzoyl-5-fluoro-2,4(1H,3H)-
    pyrimidinedi-one, 3302
1-Butyryloxymethyl-3-benzoyl-5-fluorouracil, 3302
1-Butyryloxymethyl-5-fluorouracil, 1699
O-Butyryloxymethyl Salicylamide, 2649
O-(Butyryloxymethyl) Salicylamide, 2649
Buvilan, 2828

Bykahepar, 3536
C-3470, 3331
C-7019, 1173
Cacodylic Acid, 106
Cacodylic Acid, 106
Caffeine, 1449
Caffeine (Monohydrate), 1450
Calcozine Magenta xx, 3735
Calirus, 2793
Camphechlor, 1937
Campheclor, 1937
D-Campher, 2103
D-Camphersaeure, 2110
D-Campholic Acid, 2132
D-Campholsaeure, 2132
Camphor, 2103
D-Camphor, 2103
D-Camphoric Acid, 2110
L-Camphoronic Acid, 1778
L-Camphoronsaeure, 1778
Camphor Tar, 1889
Candeptin, 4167
Candicidin, 4167
Candyl, 3167
Cane Sugar, 2744
Canrenone, 3887
Cantharides, 2014
Cantharidin, 2014
Caparol, 2153
Caparol 80W, 2153
n-Capric Acid, 2167
Capric Acid Methyl Ester, 2365
Capric Alcohol, 2181
Caprinsaeure, 2167
Caproamide, 883
n-Caproic Acid, 859
Caproic Alcohol, 914
Caproic Aldehyde, 845
Caprolactam, 813
ε-Caprolactam, 813
n-Capronsaeure, 859
n-Capronsaeure-amid, 883
9-[5'-(O-Caproyl)-β-D-arabinofuranosyl]adenine Ester,
    3382
Caprylene, 1527
Caprylic Acid, 1543
Caprylic Alcohol, 1581
sec-Caprylic Alcohol, 1583
Caprylic Aldehyde, 1537
Caprylsaeure-amid, 1561
Caprylsaure, 1543
Caprylylamide, 1561
9-[5'-(O-Caprylyl)-β-D-arabinofuranosyl]adenine Ester,
    3595
Captafol, 1905
Captan, 1634
Caragard, 2146
Carbamate, n-Propyl-di-n-propylthio-, 2177
Carbamazepine, 3152
Carbamic Acid, [2-Amino-6-[[(4-
    fluorophenyl)methyl]amino]-3-pyridinyl]-, Ethyl Ester,
    3194
Carbamic Acid, Diethyldithio-2chloroallyl Ester, 1496
Carbamic Acid, Dipropylthio-, S-propyl Ester, 2177

Carbamic Acid Ethyl Ester, 202
Carbamic Acid, N-Methyl-, 3,5-di-tert-Butylphenyl Ester, 3392
Carbamic Acid, Methyl-, 4-(Dimethylamino)-m-tolyl Ester, 2303
Carbamic Acid, Octyl Ester, 1836
Carbamic Acid, N-Phenyl, 3-((Ethoxycarbonyl)amino)phenyl Ester, 3312
Carbamidodiglycylglycine, 1185
Carbamidoglycylglycine, 1186
Carbamidsaeure-aethyl Ester, 202
Carbamidsaeure-methyl Ester, 90
5-Carbamoyl-5H-Dibenz[b,f]azcpinc, 3152
Carbamoylglycine, 166
N-Carbamoylglycine, 166
Carbamult, 2688
Carbamylurea, 94
Carbanilide, 2816
Carbaryl, 2575
Carbazepine, 3152
Carbazole, 2545
9H-Carbazole, 2545
Carbendazim, 1660
Carbenoxolone, 4107
Carbetamide, 2671
4-Carbethoxyphenylisothiocyanate, 1914
Carbinamine, 28
Carbitol, 935
Carbobenzoxydiglycine, 2614
Carbobenzoxyglycine, 1964
Carbobenzoxytyrosine, 3434
N-Carbobenzyloxyglycine, 1964
Carbocain, 3242
Carbocaine, 3242
5-Carboethoxy-2-thiouracil, 1110
Carbofos, 2154
Carbofuran, 2643
Carbolic Acid, 722
Carbon Dioxide, 43
Carbon Disulfide, 44
Carbon Disulphide, 44
Carbonic Acid, 4-(Acetylamino)phenyl 2-chloroethyl Ester, 2227
Carbonic Acid, 4-(Acetylamino)phenyl Ethyl Ester, 2265
Carbonic Acid, 4-(Acetylamino)phenyl Hexyl Ester, 3231
Carbonic Acid, 4-(Acetylamino)phenyl Methyl Ester, 1966
Carbonic Acid, 4-(Acetylamino)phenyl 1-Methylethyl Ester, 2651
Carbonic Acid, 4-(Acetylamino)phenyl Phenyl Ester, 3166
Carbonic Acid, Butyl Ester, Ester with 4'-Hydroxyacetanilide, 2880
Carbonic Acid Gas, 43
Carbonic Acid, Isobutyl Ester, Ester with 4'-hydroxyacetanilide, 2879
Carbonic Acid, Methyl Ester, Ester with Methyl Lactate, 805
Carbonic Acid, Octyl Ester, Ester with 4'-hydroxyacetanilide, 3495
Carbonic Anhydride, 43
Carbon Tetrabromide, 34
Carbon Tetrachloride, 40

Carbon Tetrafluoride, 41
Carbonyl Sulfide, 42
Carbophenothion, 2302
Carbophenothion-methyl, 1731
Carbostyril, 1619
Carbothialate, Ethyl-1-hexa-methylene Imine-, 1798
Carbovir, 2270
Carbox, 2958
Carboxin, 2606
1-Carboxy-3,4-dimethylbenzene, 1681
2-Carboxy-1,4-dimethylbenzene, 1679
4-Carboxy-1,3-dimethylbenzene, 1682
2-Carboxyethylphosphonic Acid, 211
2-Carboxy-5-hydroxy-4-pyrone, 624
3-Carboxylhexylphenylisothiocyanate, 3046
m-Carboxylhexylphenylisothiocyanate, 3046
7-Carboxylic Acid Methyl Ester Canrenone, 3957
3-Carboxyloctylphenylisothiocyanate, 3359
m-Carboxyloctylphenylisothiocyanate, 3359
m-Carboxylpentylphenylisothiocyanate, 2838
5-Carboxymethylhydantoin, 410
N2-Carboxymethyl-N2,N4,N4,N6,N6-pentamethylmelamine, 2124
N-(Carboxymethyl)urea, 166
3-Carboxyphenylisothiocyanate, 1289
4-Carboxyphenylisothiocyanate, 1290
p-Carboxyphenylisothiocyanate, 1290
2-Carboxyphenyl Pivalate, 2637
5-Carboxyuracil, 381
Carbromal, 1198
Carbutamide, 2326
Card-20(22)-enolide, 3-[(6-Deoxy-4-O-β-D-glucopyranosyl-3-O-methyl-β-D-galactopyranosyl)oxy]-14,16-dihydroxy-, (3β,5β,16β)-, 4117
Card-20(22)-enolide, 3-[(2,6-Dideoxy-4-O-β-D-glucopyranosyl-3-O-methyl-β-D-ribo-hexopyranosyl)oxy]-5,14-dihydroxy-, (3β,5β)-, 4084
Card-20(22)-enolide, 3,14-Dihydroxy-, (3β,5β)-, 3937
Cardrase, 1674
Carmine, 3873
Carminic Acid, 3873
Carminsaeure, 3873
Carubinose, 872
Carvacrol, 2072
L-Carvone, 2071
Carvotan-aceton, 2107
Carvotan Acetone, 2107
Cassella's Acid F, 1917
Catanil, 2020
Catechol, 724
CB 10-375, 1812
CB 118, 2446
CB 138, 2415
CB 170, 2389
CBM, 7
N-CBZ-Glycine, 1964
CCRIS 805, 1999
CCRIS 805Didanosine, 1999
Ccucol, 1631
CDAA, 1483
CDBM, 2
CDCA, 3976
CDEA, 824
CDEC, 1496

Ceclor, 3168
Cefaclor, 3168
Cefalexin, 3325
Cefanex, 3325
Cekudifol, 2958
Celfume, 12
Celliton Discharge Scarlet RNL, 3552
Celliton Fast Brown 3R, 3169
Celliton Fast Pink B, 2964
Celliton Fast Scarlet RN, 3552
Cellobiose, 2743
D-(+)-Cellobiose, 2743
Cellosolve Acetate, 866
Celontin, 2605
Centrax, 3631
Cephaclor, 3168
Cephalexin, 3325
Cephalexin (Monohydrate), 3326
Cephaloglycin, 3551
Cephaloridine, 3633
Cephradine, 3342
Cerenox, 2811
Ceresan, 1386
Cerise B, 3735
Certodorm, 1485
Cestocid, 2778
Cetacaine, 3254
Cetane, 3401
Cetyl Alcohol, 3403
Cevadine, 4095
Cevane-3,4,12,14,16,17,20-heptol, 4,9-epoxy-, 3-[(2Z)-2-
    methyl-2-butenoate], (3β,4α,16β)-, 4095
CFC-114, 108
CGA-12223, 1797
Chartreusin, 4092
Chavicyl Methyl Ether, 2008
Chelidonic Acid, 984
Chelidonsaeure, 984
Chemagro 25141, 2330
Chemform, 3298
Chenodeoxycholic Acid, 3976
Chinasaeure, 1196
Chinhydron, 2570
Chinidin, 3747
Chinin, 3748
Chininon, 3742
Chinolin, 1618
Chinonamid, 2457
Chipman Merbam, 1719
(+)-Chiro-Inositol, 878
D-Chiro-Inositol, 878
Chlomethoxyfen, 2780
Chlomethoxynil, 2780
Chloraceton, 140
Chloradracylic, 989
Chloral-hydrat, 47
Chloral (Monohydrate), 47
Chloralose, 1477
Chloramben, 992
Chlorambenamide, 1032
Chloramben Methyl, 1334
Chloramben Methyl Ester, 1334
Chlorambucil, 3067
Chloramphenicol, 2228

Chloramphenicol Palmitate, 4042
α-Chloramphenicol Palmitate, 4042
β-Chloramphenicol Palmitate, 4042
Chloranil, 951
Chloranilic Acid, 536
Chloranilsaeure, 536
Chlorazine, 2341
2-Chlor-benzoesaeure, 988
3-Chlor-benzoesaeure, 990
4-Chlor-benzoesaeure, 989
4-Chlor-benzolsulfosaeure, 635
D(+)-Chlor-bernsteinsaeure, 250
L(-)-Chlor-bernsteinsaeure, 250
3-Chlorbiphenyl, 2538
Chlorbromuron, 1664
Chlorbufam, 2205
Chlorbupham, 2205
α-Chlor-crotonsaeure, 247
β-Chlor-crotonsaeure, 249
Chlorcyan, 35
Chlorcyclizine, 3560
Chlordane, 1874
Chlordecone, 2184
Chlordene, 1870
Chlordene Epoxide, 1871
Chlordene Hydroxide, 1871
Chlordiazepoxide, 3285
Chlordimeform, 2016
4-Chlor-1,3-dinitrobenzol, 553
Chloressigsaeureamid, 72
Chloreton, 295
Chlorfenac, 1280
Chlorfenprop-methyl, 1934
Chlorfenvinphos, 2621
Chlorflurazole, 1271
Chlorflurecol-methyl, 3142
Chlorflurenol, 3142
Chlorhexidin, 3891
Chlorhexidine, 3891
Chlorinated Champhene, 1937
α-Chlor-isocrotonsaeure, 248
β-Chlor-isocrotonsaeure, 246
Chlormephos, 506
Chlormethazole, 1605
Chlormite, 3425
Chlornitrofen, 2458
Chloroacetamide, 72
2-Chloroacetamide, 72
p-Chloroacetanilide, 1354
Chloroacetone, 140
Chloroacetyl Acetaminophen, 1932
2-Chloroadenine, 378
2-Chloroallyl Diethyldithiocarbamate, 1496
(3-Chloro-4-allyloxyphenyl)acetic Acid, 2214
4-Chloro-4'-aminobiphenyl, 2555
2-Chloro-6-aminopurine, 378
2-Chloroaniline, 683
3-Chloroaniline, 681
4-Chloroaniline, 682
m-Chloroaniline, 681
o-Chloroaniline, 683
p-Chloroaniline, 682
p-Chloroaniline-m-sulfonic Acid, 687
p-Chloroaniline-m-sulfonic Acid (Monohydrate), 688

p-Chloroaniline-o-sulfonic Acid (Monohydrate), 689
3-(p-Chloroanilino)-10-(p-chlorophenyl)-2,10-dihydro-2-
    (isopropylimino)phenazine, 4028
1-((p-(2-(5-Chloro-o-anisamido)ethyl)phenyl)-sulfonyl)-3-
    cyclohexylurea, 3926
2-Chloroanisole, 1077
3-Chloroanisole, 1076
4-Chloroanisole, 1075
m-Chloroanisole, 1076
o-Chloroanisole, 1077
p-Chloroanisole, 1075
1,5-Chloroanthraquinone Sulfonic Acid, 2941
1,6-Chloroanthraquinone Sulfonic Acid, 2942
1,7-Chloroanthraquinone Sulfonic Acid, 2943
4-Chloroazobenzene, 2540
Chlorobenzene, 630
2-Chlorobenzenesulfonamide, 686
4-Chlorobenzenesulfonamide, 684
m-Chlorobenzenesulfonamide, 685
o-Chlorobenzenesulfonamide, 686
p-Chlorobenzenesulfonamide, 684
p-Chlorobenzenesulfonic Acid, 635
p-Chlorobenzenesulfonic Acid (2.5 Hydrate), 636
1-(p-Chlorobenzenesulfonyl)-5,5-diphenyl-hydantoin,
    3797
Chlorobenzilate, 3287
2-Chlorobenzoic Acid, 988
3-Chlorobenzoic Acid, 990
4-Chlorobenzoic Acid, 989
m-Chlorobenzoic Acid, 990
meta-Chlorobenzoic Acid, 990
o-Chlorobenzoic Acid, 988
p-Chlorobenzoic Acid, 989
9-[5-O-(4-Chlorobenzoyl-β-D-arabinofuranosyl)] 6
    methoxy-9H-purine (Hemihydrate), 3533
4-Chlorobenzyl N,N-diethylthiocarbamate, 2662
S-4-Chlorobenzyl Diethylthiocarbamate, 2662
3-Chlorobenzyl Isothiocyanate, 1297
4-Chlorobenzyl Isothiocyanate, 1298
m-Chlorobenzyl Isothiocyanate, 1297
p-Chlorobenzyl Isothiocyanate, 1298
7-Chlorobicyclo[3.2.0]hepta-2,6-dien-6-yl Dimethyl
    Phosphate, 1732
2-Chlorobiphenyl, 2535
3-Chlorobiphenyl, 2538
4-Chlorobiphenyl, 2537
2-Chloro-4,6-bis-(diethylamino)-s-triazine Chlorazine,
    2341
2-Chloro-4,6-bis(ethylamino)-s-triazine, 1182
Chlorobromomethane, 7
3-Chloro-1-bromopropane, 154
ω-Chlorobromopropane, 154
1-Chlorobutane, 339
2-Chlorobutane, 336
1-Chloro-2-butene, 293
4-Chloro-2-butynyl-m-chlorocarbanilate, 2198
4-Chloro-2-butynyl-N-(3-chlorophenyl)carbamate, 2198
3-Chlorochlordene, 1861
N-[3-Chloro-4-(chlorodifluoromethylthiol)phenyl]-N',N'-
    dimethylurea, 1933
3-[3-Chloro-4-(chlorodifluoromethylthio)phenyl]-1,1-
    dimethylurea, 1933
N-[3-Chloro-4-(chlorodifluoromethylthio)phenyl]-N',N'-
    dimethylurea, 1933

6-Chloro-3-(chloromethyl)-3,4-dihydro-2-methyl-2H-
    1,2,4-benzothiadiazine -7-sulfonamide 1,1-dioxide,
    1694
7-Chloro-5-(o-chlorophenyl)-1,3-dihydro-3-hydroxy-2H-
    1,4-benzodiazepin-2-one, 3135
4-Chloro-α-(4-chlorophenyl)-α-
    (trichloromethyl)benzenemethanol, 2958
Chlorocresol, 1074
4-Chloro-3-cresol, 1074
4-Chloro-o-cresol, 1079
6-Chloro-o-cresol, 1078
2-Chlorocrotonic Acid, 247
3-Chlorocrotonic Acid, 249
3-Chloro-6-cyanonorbornanone-2-oxime-O,N-
    methylcarbamate, 1981
Chlorocyclizine, 3560
6-Chloro-3-cyclopentylmethyl-3,4-dihydro-2H-1,2,4-
    benzothiadiazine-7-sulphonamide 1,1-dioxide, 2890
2-Chloro-4-cyclopropylamino-6-isopropylamino-1,3,5-
    triazine, 1773
7-Chloro-1-(cyclopropylmethyl)-1,3-dihydro-5-phenyl-
    2H-1,4-benzodiazepin-2-one, 3631
2-Chloro-N,N-diallylacetamide, 1483
1-Chlorodibenzo-p-dioxin, 2489
2-Chlorodibenzo-p-dioxin, 2488
Chlorodibromomethane, 2
1-Chloro-2,3-dibromopropane, 138
5-Chloro-2-(2,4-dichlorophenoxy)-phenol, 2507
2-Chloro-1-(2,4-dichlorophenyl)Ethenyl Phosphoric Acid,
    Diethyl Ester, 2621
2-Chloro-2',3'-dideoxyadenosine, 1983
2-Chloro-4-diethylamino-6-diethylamino-s-triazine, 2341
2-Chloro-4-diethylamino-6-ethylamino-s-triazine, 1786
2-Chloro-4-diethylamino-6-isopropylamino-s-triazine,
    2121
2-Chloro-2',6'-diethyl-N-(methoxymethyl)acetanilide,
    3079
Chlorodifluorobromomethane, 32
N-(3-Chloro-4-difluorochloromethylthiophenyl)-N',N'-
    dimethylurea, 1933
Chlorodifluoromethane, 4
7-Chloro-5,11-dihydrodibenz[b,e][1,4]oxazepine-5-
    carboxamide, 2980
2-Chloro-5-(2,3-dihydro-1-hydroxy-3-oxo-1H-isoindol-1-
    yl)benzenesulfonamide, 2981
7-Chloro-1,3-dihydro-3-hydroxy-5-phenyl-2H-1,4-
    benzodiazepin-2-one, 3141
7-Chloro-2,3-dihydro-1-methyl-5-phenyl-1H-1,4-
    benzodiazepine, 3296
5-Chloro-N-(2-(dimethylamino)ethyl)-N-(2-pyridyl)-2-
    thenylamine, 3051
4-Chloro-5-(dimethylamino)-2-(α,α,α-trifluoro-m-tolyl)-
    3(2H)-pyridazinone, 2804
1-Chloro-1,1-dimethyl-2,2-bis(p-ethoxylphenyl)ethane,
    3756
5-Chloro-3-(1,1-dimethylethyl)-6-methyl-2,4(1H,3H)-
    pyrimidinedione, 1762
4-Chloro-2,5-dimethylphenol, 1406
2-Chloro-N,N-dimethyl-10H-phenothiazine-10-
    butanamine, 3561
2-Chloro-N,N-dimethyl-10H-phenothiazine-10-
    ethanamide, 3320
1-Chloro-N,N-dimethyl-10H-phenothiazine-10-
    propanamide, 3441

3-Chloro-N,N-dimethyl-10H-phenothiazine-10-
   propanamide, 3442
4-Chloro-N,N-dimethyl-10H-phenothiazine-10-
   propanamide, 3443
2-Chloro-N,N-dimethyl-10H-phenothiazine-10-
   propanamine, 3444
2-Chloro-N-(2,6-dimethyl)phenyl-N-
   isopropoxymethylacetamide, 3241
1-Chloro-2,4-dinitrobenzene, 553
4-Chloro-1,3-dinitrobenzene, 553
1-Chloro-2,4-dinitronaphthalene, 1860
1-(o-Chloro-α,α-diphenylbenzyl)imidazole, 3867
p-Chlorodiphenyl Oxide, 2541
Chloroethane, 84
Chloroethyl Acetaminophen, 2227
2-Chloro-4-ethylamino-6-tert-butylamino-s-triazine,
   1788
2-Chloro-4-ethylamino-6-diethylamino-s-triazines, 1786
2-Chloro-4-ethylamino-6-ethylamino-s-triazine, 1182
2-Chloro-4-ethylamino-6-isopropylamino-s-triazine,
   1497
2-[[4-Chloro-6-(ethylamino)-1,3,5-triazin-2-yl]amino]-2-
   methylpropanenitrile, 1763
Chloroethylene, 61
2-Chloro-N-(2-ethyl-6-methylphenyl)-N-(2-methoxy-1-
   methylethyl)acetamide, 3240
7-Chloro-2-ethyl-1,2,3,4-tetrahydro-4-oxo-6-
   quinazolinesulfonamide, 1982
6-Chloro-N-ethyl-1,3,5-triazine-2,4-diamine, 125
1-Chloro-2-fluorobenzene, 580
1-Chloro-3-fluorobenzene, 579
2-Chlorofluorobenzene, 580
3-Chlorofluorobenzene, 579
4'-Chloro-5-fluoro-2-hydroxy Benzophenone, 2775
Chloroform, 5
1-Chloroheptane, 1237
5-Chloro-2-hydroxydiphenylmethane, 2806
3-Chloro-2-hydroxytoluene, 1078
5-Chloro-2-hydroxytoluene, 1079
6-Chloro-3-hydroxytoluene, 1074
Chloro-IPC, 1980
2-Chloroisocrotonic Acid, 248
3-Chloroisocrotonic Acid, 246
2-Chloro-N-isopropylacetanilide, 2272
2-Chloro-4-isopropylamino-6-biethylamino-s-triazines,
   2121
2-Chloro-4-isopropylamino-6-isopropylamino-s-triazine,
   1787
1-Chloro-3-isothiocyanato-benzene, 967
Chloromethane, 14
Chloromethapyrilene, 3051
Chloromethionic Acid, 15
1-Chloro-3-methoxybenzene, 1076
1-Chloro-4-methoxybenzene, 1075
N'-(3-Chloro-4-methoxyphenyl)-N,N-dimethylurea, 2019
2-Chloro-4-methylamino-6-isopropylamino-s-triazine,
   1183
7-Chloro-2-(methylamino)-5-phenyl-3H-1,4-
   benzodiazepine-4-oxide, 3285
2-Chloro-4-methyl Amino-6-propyl Amino-s-triazines,
   1181
4-Chloro-5-(methylamino)-2-(α,α,α-trifluoro-m-tolyl)-
   3(2H)-pyridazinone, 2539
6-Chloro-10-methyl-1,2-benzanthracene, 3618

1-Chloro-3-methylbenzene, 1071
1-Chloro-4-methylbenzene, 1072
2-Chloro-1-methylbenzene, 1073
4-Chloro-1-methyl-benzene, 1072
1-Chloro-1-methyl-2,2-bis(p-ethoxylphenyl)ethane, 3668
2-Chloro-N-(2-methyl-6-t-butylphenyl)acetamide, 2776
2'-Chloro-α-methylene-γ-oxo[1,1'-biphenyl]-4-butanoic
   Acid, 3416
6-Chloro-N-(1-methylethyl)-1,3,5-triazine-2,4-diamine,
   790
1-Chloro-1-methyl-2-(p-methylphenyl)-2-p-
   Ethoxylphenyl)ethane, 3562
2-Chloro-6-methylphenol, 1078
4-Chloro-2-methylphenol, 1079
6-Chloro-2-methylphenol, 1078
(4-Chloro-2-methylphenoxy)acetic Acid, 1645
2-(4-Chloro-2-methylphenoxy)propionic Acid, 1955
7-Chloro-1-methyl-5-phenyl-2H-1,4-benzodiazepin-2-one,
   3279
N'-(4-Chloro-2-methylphenyl)-N,N-
   dimethylmethanimidamide, 2016
N'-(3-Chloro-4-methylphenyl)-N,N-dimethylurea, 2017
2-Chloro-2-methylpropane, 338
1-Chloro-2-methylpropene-2, 293
5-Chloro-4-methyl-2-propionamide-thiazole, 1130
2-Chloro-N-(1-methyl-2-propynyl)acetanilide, 2588
1-Chloronaphthalene, 1881
2-Chloronaphthalene, 1882
α-Chloronaphthalene, 1881
β-Chloronaphthalene, 1882
Chloroneb, 1357
2-Chloro-4-nitroacetanilide, 1330
2-Chloro-5-nitroacetanilide, 1329
1-Chloro-3-nitrobenzene, 583
3-Chloronitrobenzene, 583
m-Chloronitrobenzene, 583
o-Chloronitrobenzene, 585
p-Chloronitrobenzene, 584
4-Chloro-3-nitro-benzenesulfonamide, 631
3-Chloro-2-nitrobenzoic Acid, 964
4-Chloro-3-nitrobenzoic Acid, 965
5-Chloro-2-nitrobenzoic Acid, 966
1-Chloronitroethane, 73
1-Chloro-1-nitroethane, 73
2-Chloro-4-nitrophenylamide-6-chlorosalicylic Acid,
   2778
O-(2-Chloro-4-nitrophenyl) O,O-Dimethyl
   phosphorothioate, 1405
1-Chloro-1-nitropropane, 157
1-Chloro-2-nitropropane, 158
7-Chloro-2-oxo-3(2H)-benzothiazolacetic Acid, 1604
Chlorooxuron, 3176
Chlorophenamidine, 2016
Chlorophenate, 632
Chlorophene, 2806
2-Chlorophenol, 633
3-Chlorophenol, 632
4-Chlorophenol, 634
4-Chloro-phenol-, 634
m-Chlorophenol, 632
o-Chlorophenol, 633
p-Chlorophenol, 634
4-(3-(2-Chlorophenothiazin-10-YL)propyl)-1-
   piperazineethanol, 3820

2-Chlorophenoxyacetic Acid, 1331
3-Chlorophenoxyacetic Acid, 1332
4-Chlorophenoxyacetic Acid, 1333
m-Chlorophenoxyacetic Acid, 1332
o-Chlorophenoxyacetic Acid, 1331
p-Chlorophenoxyacetic Acid, 1333
1-Chloro-4-phenoxybenzene, 2541
4-(4-Chlorophenoxy)butyric Acid, 1954
1-(4-Chlorophenoxy)-3,3-dimethyl-1-(1H-1,2,4-triazol-1-
    yl)-2-butanone, 3026
2-(p-Chlorophenoxy)-2-methylpropionic Acid Ethyl Ester,
    2640
(N'-4-(4-Chlorophenoxy)phenyl-N,N-dimethylurea),
    3176
3-[p-(p'-Chlorophenoxy)phenyl]-1,1-dimethylurea, 3176
N-4-(4'-Chlorophenoxy)phenyl-N',N'-dimethylurea, 3176
2-(o-Chlorophenoxy)propionic Acid, 1643
DL-2-(2-Chlorophenoxy)propionic Acid, 1643
DL-2-(4-Chlorophenoxy)propionic Acid, 1644
p'-Chloro-p-phenylaniline, 2555
1-Chloro-4-phenyl benzene, 2537
1-[α-(2-Chlorophenyl)benzhydryl]imidazole, 3867
1-(2-Chlorophenyl)-1-(4-chlorophenyl)-2,2-
    dichloroethane, 2968
1-(2-Chlorophenyl)-1-(4-chlorophenyl)-2,2-
    dichloroethylene, 2945
α-(4-Chlorophenyl)-α-(1-2-(2-chloro)phenylethenyl)-1H-
    1,2,4-triazole-1-ethanol, 3409
α-(2-Chlorophenyl)-α-(4-chlorophenyl)-5-
    Pyrimidinemethanol, 3411
1-(2-Chlorophenyl)-1-(4-chlorophenyl)-2,2,2-
    trichloroethane, 2956
2-(2-Chlorophenyl)-2-(4-chlorophenyl)-1,1,1-
    trichloroethane, 2956
4'-[Chlorophenyl}-3,4-dichlorophenylbenzene-
    sulphonamide, 2543
N'-(4-Chlorophenyl)-N,N-dimethyl-urea, 1691
N-(4-Chlorophenyl)-2,2-dimethylvaleramide, 2889
N-(p-Chlorophenyl)-o-hydroxybenzamide, 2790
4-[4-(p-Chlorophenyl)-4-hydroxypiperidino]-4'-
    fluorobutyrophenone, 3815
3-Chlorophenyl Isothiocyanate, 967
3-(4-Chlorophenyl)-1-methoxy-1-methylurea, 1692
3-(para-Chlorophenyl)-1-methyl-1-(1-methyl-2-propynyl)
    Urea, 2602
4-Chlorophenyl Phenyl Ether, 2541
4-(p-Chlorophenyl)-2-phenyl-5-thiazoleacetic Acid, 3410
1-(p-Chlorophenyl)-1-(2-pyridyl)-3-
    dimethylaminopropane, 3338
N-(p-Chlorophenyl)salicylamide, 2790
S-p-Chlorophenylthiomethyl O,O-Dimethyl
    Phosphorodithioate, 1731
N'-4-Chlorophenyl-O,N,N-trimethylisourea, 2018
3-Chlorophthalic Acid, 1279
6-Chloropicolinic Acid, 582
Chloropicrin, 39
1-Chloropromazine, 3441
3-Chloropromazine, 3442
4-Chloropromazine, 3443
Chloropropane, 189
1-Chloropropane, 189
2-Chloropropane, 190
3-Chloro-1-propanol, 191
1-Chloro-2-propanone, 140

3-Chloro-1-propene, 139
2-Chloro-4-propylamino-6-isopropylamino-s-triazine,
    1785
Chloropropylate, 3425
7-Chloropteridine, 554
N1-(6-Chloro-3-pyridyl)sulfanilamide, 2206
Chloropyrilene, 3051
N-(7-Chloro-4-quinolyl)anthranilate, 3632
4'-Chloro Salicylanilide, 2790
D-Chlorosuccinic Acid, 250
L-Chlorosuccinic Acid, 250
4-Chloro-3-sulfoaniline, 687
Chlorothalonil, 1601
Chlorothen, 3051
Chlorothiazide, 1030
2-Chlorotoluene, 1073
3-Chlorotoluene, 1071
4-Chlorotoluene, 1072
m-Chlorotoluene, 1071
o-Chlorotoluene, 1073
p-Chlorotoluene, 1072
2-Chloro-6-(trichloromethyl)pyridine, 568
2-Chloro-1-(2,4,5-trichlorophenyl)vinyl Dimethyl
    Phosphate, 1906
2-Chloro-1-(2,4,5-
    trichlorophenyl)vinyldimethylphosphate, 1907
(2S-trans)-7-Chloro-2',4,6-trimethoxy-6'-
    methylspiro[benzofuran-2(3H),1'-[2]cyclohexene]-3,4'-
    dione, 3432
Chlorotrinitromethane, 36
Chlorovinyldichloroarsinc, 55
Chloroxone, 3052
4-Chloro-2,5-xylcnol, 1406
Chloroxylenol, 1407
4-Chloro-2,6-xylenol, 1408
Chlorozide, 1104
Chlorphenethazine, 3320
Chlorpheniramine, 3338
3-Chlor-phthalsaeure, 1279
Chlorpikrin, 39
Chlorpromazine, 3444
Chlorpropamide, 2020
3-Chlor-propanol-(1), 191
Chlorpropham, 1980
Chlorprothixene, 3538
Chlorpyrifos, 1695
Chlorpyrifos-methy, 1081
Chlorpyrifos-methyl, 1081
Chlorpyrifos Oxon, 1696
Chlorpyrifos Oxygen Analog, 1696
Chlorquinox, 1269
Chlortalidone, 2981
Chlortetracycline, 3877
7-Chlortetracycline, 3877
Chlorthalidone, 2981
Chlorthiamid, 993
Chlorthion, 1404
Chlortokem, 2017
Chlortoluron, 2017
2-Chlor-1,3,5-trinitrobenzol, 535
Chlor-trinitro-methan, 36
Chlorure de Picryle, 535
Chlorvinylarsin-dichlorid, 55
(C6H10O5)6, 4119

Choladine, 2230
Cholan-24-oic Acid, 3,12-Dihydroxy-, (3α,5β,12α)-, 3978
Cholanthrene, 3721
Cholebrine, 2604
Cholecalciferol, 4048
Cholest-5-ene-3β-ol, 4049
(3β)-Cholest-5-en-3-ol, 4049
Cholest-5-en-3-ol (3β)-, 4049
3-β-(5-Cholestenyl)-N-2-(2-desoxyglucosyl) Carbamate,
    4108
3-β-(5-Cholestenyl)-N-methyl-N-1-(1-desoxyglucosyl)
    Carbamate, 4114
Cholesterol, 4049
Cholevid, 2230
Cholic Acid, 3981
Cholsaeure, 3981
CHP, 1758
6-Chrysenamine, 3518
Chrysene, 3513
Chrysene, 5,6-Dimethyl-, 3730
CHT, 1101
C.I. 10375, 2563
C.I. 11025, 3034
C.I. 42510, 3735
C.I. 52015, 3356
C.I. 61505, 3427
C.I. Acid Blue 74(Free Acid), 3272
Ciafos, 1670
Ciba 3753, 1973
2-CIDDA, 1983
Cidial, 2698
C.I. Disperse Blue 1, 2993
C.I. Disperse Blue 14, 3291
C.I. Disperse Blue 19, 3724
C.I. Disperse Blue 23, 3539
C.I. Disperse Blue 24, 3800
C.I. Disperse Blue 26, 3293
C.I. Disperse Blue 3, 3427
C.I. Disperse Blue 79, 3947
C.I. Disperse Dye, 3552
C.I. Disperse Orange 11, 3144
C.I. Disperse Orange 3, 2562
C.I. Disperse Orange 5, 3169
C.I. Disperse Organe 1, 3526
C.I. Disperse Red 1, 3335
C.I. Disperse Red 11, 3234
C.I. Disperse Red 13, 3321
C.I. Disperse Red 15, 2964
C.I. Disperse Red 17, 3458
C.I. Disperse Red 19, 3336
C.I. Disperse Red 3, 3281
C.I. Disperse Red 5, 3445
C.I. Disperse Red 7, 3322
C.I. Disperse Red 9, 3143
C.I. Disperse Violet 1, 2971
C.I. Disperse Yellow 1, 2551
C.I. Disperse Yellow 1, 2551
C.I. Disperse Yellow 14, 2550
C.I. Disperse Yellow 3, 3182
C.I. Disperse Yellow 54, 3511
C.I. Disperse Yellow 9, 2563
C.I. Mordant Red 11, 2947
Cinchocaine, 3775
Cinchomeronic Acid, 1019

(8α,9R)-Cinchonan-9-ol, 3664
(9S)-Cinchonan-9-ol, 3663
Cinchonan-9-ol, 3663
Cinchonan-9-ol, 10,11-Dihydro-6'-methoxy-, (8α,9R)-,
    3762
Cinchonan-9-ol, 6'-Methoxy-, (9S)-, 3747
Cinchonan-9-ol, 6'-Methoxy-, Trihydrate, (8α,9R)-, 3749
Cinchonan-9-one, 6'-Methoxy-, (8α)-, 3742
Cinchonidin, 3664
Cinchonidine, 3664
L-Cinchonidine, 3664
Cinchonine, 3663
(+)-Cinchonine, 3663
Cinchoninon, 3643
Cinchoninone, 3643
Cinchophen, 3273
Cinchotine, 3674
Cineol, 2127
Cineole, 2127
1,8-Cineole, 2127
DL-Cineolic Acid, 2111
DL-Cineolsaeure, 2111
Cinmetacin, 3805
Cinnamaldehyde, 1637
Cinnamamide, N,N-Dipropyl-, 3222
Cinnamamide, N-Propyl-, 2642
Cinnamic Acid, 1638
cis-Cinnamic Acid, 1639
trans-Cinnamic Acid, 1640
Cinnamic Acid Dimethylamide, 2258
Cinnamic Acid, Ester with 4'-Hydroxyacetanilide, 3421

1-Cinnamoyl-2-methyl-5-methoxyindolyl-3-acetic Acid,
    3805
1-Cinnamoylpiperidine, 3045
Cinnamyl Acetaminophen, 3421
Cinnamyl Alcohol, 3304
Cinnamyl Anthranilate, 3304
Cinnarizine, 4010
C.I. Oxidation base 1, 737
C.I. Pigment Violet 12, 2948
cis 1,3-Dichloro-propene, 127
C.I. Solvent Yellow 1, 2578
C.I. Solvent Yellow 2, 3023
C.I. Solvent Yellow 58, 3340
C.I. Solvent Yellow 7, 2559
Citraconic Acid, 412
Citraconsaeure, 412
Citral, 2106
Citral A, 2106
Citralite, 774
Citric Acid Anhydrous, 774
Citric Acid (Monohydrate), 775
Citro, 774
Citronellol, 2161
β-Citronellol, 2161
Citronensaeure, 774
CL 11,366, 1373
CL 18,161, 1268
CL 35,024, 1266
CL 36,010, 1982
CL 5,343, 80
CL 64,475, 828
Claigeon, 2927

Clanobutin, 3536
Clarithromycin, 4124
CL 911C, 3744
1,6-Cleve's Acid, 1923
Clofazimine, 4028
Clofibrate, 2640
Clopidol, 1080
Clopyralid, 555
Cloramical, 2228
Clotrimazole, 3867
CMPT, 1130
2-CNB, 585
4-CNB, 584
3 COB, 3303
8 COB, 3819
Cobalamin, Co-(5'-deoxy-5'-adenosyl)-, 4177
Cobamamide, 4177
L-Cocain, 3464
Cocaine, 3464
L-Cocaine, 3464
Cocositol, 881
Codein, 3564
Codeine, 3564
Codeine (Monohydrate), 3565
Coenzyme B12, 4177
Coffein, 1449
Colchicin, 3883
Colchicine, 3883
Colepax, 2230
Colifoam, 3843
Colistatin, 2825
Collunosol, 567
Compa-Z, 3745
Compazine, 3745
Complexon I, 779
Compound 733, 1760
Coniferin (Dihydrate), 3374
Coniferosi(Dihydrate), 3374
D-Coniin, 1560
Coniine, 1560
D-Coniine, 1560
Conoderm, 3524
Contraven, 1850
Cope, 3074
Copsamine, 3487
Corbeton, 3247
Corbit, 2946
Cordarone, 3991
Corlutin, 3838
Corlutina, 3838
Coronen, 3942
Coronene, 3942
Corophyllin-N, 1747
Corozate, 840
Cortaid, 3843
Cortate, 3830
Cortef, 3843
Cortisol Acetate, 3933
Cortisol 21-acetate, 3933
Cortisone, 3830
Cortisone Acetate, 3844
Cortisone Caprylate, 4077
Cortisone-21-hemi-succinate, 3930
Cortisone 17-Propionate, 3958

Cortisone Tricarballylate, 4037
Corzide, 3170
Cotofor, 2356
Cotoran Multi, 3240
Cotranzine, 3745
Cotrim, 1975
Cotrimoxazole, 1975
Coumaphos, 3027
Coumarin, 1610
Counter 15G, 1850
Counter Sulfone, 1857
Coversan, 951
Coyden, 1080
3-CP, 1643
CP 31675, 2776
CP 52223, 3241
CP 6343, 1483
4-CPA, 1333
4-(4-CPB), 1954
Creatine, 354
Creatinine, 304
Credazine, 2210
Cremosuxidine, 2825
2-Cresol, 1119
3-Cresol, 1115
4-Cresol, 1116
m-Cresol, 1115
o-Cresol, 1119
p-Cresol, 1116
o-Cresol, 5-tert-Butyl-, 2317
m-Cresol, 6-Tert-Butyl-2,4-dinitro-, Acetate, 2857
m-Cresol, 4-Chloro-5-ethyl-, 1693
o-Cresol, 4,6-Dichloro-α-phenyl-, 2791
m-Cresol, 5-Ethyl-, 1750
o-Cresol, 4-Hexyl-, 2917
o-Cresol, 6-Hexyl-, 2915
m-Cresol, 2,4,5,6-Tetrachloro-, 973
m-Cresol, 2,4,6-Trichloro-, 994
m-Cresotic Acid, 1399
o-Cresotic Acid, 1387
p-Cresotic Acid, 1388
o-Cresyl-p-toluene Sulfonate, 3018
Crisfolatan, 1905
Crisfuran, 2643
Crodacid, 3116
Croneton, 2291
Crossbow Turflon, 972
Crotonaldehyd, 271
trans-Crotonaldehyde, 271
Crotonic Acid, 277
trans-Crotonic Acid, 273
Crotonic Acid, Ester with 4'-hydroxyacetanilide, 2608
trans-Crotonsaeure, 273
Crotonyl Acetaminophen, 2608
Crotoxyphos, 3078
Crufomate, 2721
Crystamine, 4168
Crystodigin, 4126
Cumarin, 1610
Cumene, 1728
Cumene Hydroperoxide, 1758
Cumol, 1728
Curacron, 2286
Curafume, 71

Curamil, 3082
Curaterr, 2643
Cyamelid, 121
Cyamelide, 121
Cyanamid, 11
Cyanamide, 11
Cyanazine, 1763
Cyanessigsaeure-aethyl Ester, 415
Cyanessigsaeure-amid, 131
Cyanoacetamide, 131
Cyanobenzene, 1005
4-Cyanobenzenesulfonamide, 1036
p-Cyanobenzenesulfonamide, 1036
4-Cyanobenzyl Isothiocyanate, 1609
p-Cyanobenzyl Isothiocyanate, 1609
4-Cyano-2,6-dibromophenol, 954
4-Cyano-2,6-diiodophenol, 959
Cyanofenphos, 3171
Cyanogen, 112
Cyanogen Azidodithiocarbonate, 113
Cyanogen Chloride, 35
2-Cyano-10-[3'-(4"-
    hydroxypiperidino)propyl]phenothiazine, 3816
Cyanoject, 4168
4-Cyano-4'-octyloxybiphenyl, 3819
O-(4-Cyanophenyl) O-Ethyl Phenylphosphonothioate, 3171
3-Cyanophenyl Isothiocyanate, 1276
m-Cyanophenyl Isothiocyanate, 1276
Cyanophos, 1670
1-Cyanopropane, 296
4-Cyano-4'-propyloxybiphenyl, 3303
p-Cyanotoluene, 1338
Cyanthoate, 2145
Cyanuric Acid, 122
Cyanursaeure, 122
Cyanursaeure-triallylaether, 2657
CYAP, 1670
Cyclacillin (Dihydrate), 3252
Cyclacilllin, 3250
Cyclil Decene, 2089
Cycloate, 2352
Cyclobarbital, 2670
α-Cyclodextrin, 4119
β-Cyclodextrin, 4131
γ-Cyclodextrin, 4142
Cyclodextrin Hydrate, 4131
β-Cyclodextrin Hydrate, 4131
β-Cyclodextrin, Tetradeca-O-ethyl-, 4175
β-Cyclodextrin,
    2A,2B,2C,2D,2E,2F,2G,6A,6B,6C,6D,6E,6F,6G-
    Tetradeca-O-methyl-, 4169
β-Cyclodextrin, Tetradeca-O-methyl-, 4158
N-Cyclododecylcinnamamide, 3845
Cyclogeraniolane, 1803
Cycloheptaamylose Hydrate, 4131
Cycloheptane, 1207
Cycloheptatriene, 1101
1,3,5-Cycloheptatriene, 1101
Cycloheptene, 1174
(1Z)-Cycloheptene, 1174
cis-Cycloheptene, 1174
5-(1-Cyclohepten-1-yl)-5-ethylbarbituric Acid, 2893
5-(1-Cyclohepten-1-yl)-5-ethyl-2,4,6(1H,3H,5H)-
    pyrimidinetrione, 2893

N-Cycloheptylcinnamamide, 3358
N-Cycloheptyl-3-phenyl- 2-Propenamide,, 3358
1,4-Cyclohexadiene, 756
2,5-Cyclohexadiene-1,4-dione, 2,5-bis(1-Aziridinyl)-3-
    fluoro-6-(4-morpholinyl)-, 3029
Cyclohexan, 823
Cyclohexan-carbonsaeure, 1189
cis-Cyclohexan-dicarbonsaeure-(1,2), 1489
trans-Cyclohexan-dicarbonsaeure-(1,2), 1490
trans-Cyclohexan-dicarbonsaeure-(1,4), 1488
Cyclohexane, 823
Cyclohexanecarboxylic Acid, 1189
cis-Cyclohexane-1,2-dicarboxylic Acid, 1489
trans-Cyclohexane-1,2-dicarboxylic Acid, 1490
trans-Cyclohexane-1,4-dicarboxylic Acid, 1488
Cyclohexane, Ethyl-, 1530
cis-1,2,3,5-trans-4,6-Cyclohexanehexol, 875
Cyclohexanol, 850
1-Cyclohexanol, 850
Cyclohexanol Acetate, 1501
Cyclohexanol, 5-Methyl-2-(1-methylethyl)-, (1α,2β,5α)-,
    2160
Cyclohexanon, 791
Cyclohexanone, 791
Cyclohexanone Oxime, 814
Cyclohexatriene, 676
Cyclohexen-(1)-dicarbonsaeure-(1,4), 1472
Cyclohexen-(2)-dicarbonsaeure-(1,2), 1471
Cyclohexene, 784
Cyclohexene-1,4-dicarboxylic Acid, 1472
2-Cyclohexene-1,2-dicarboxylic Acid, 1471
5-(1-Cyclohexenyl)-1,5-dimethylbarbituric Acid, 2669
5-(1-Cyclohexen-1-yl)-1,5-dimethylbarbituric Acid,
    2669
Cycloheximide, 3248
Cyclohexyl Acetate, 1501
Cyclohexyl Alcoho, 850
N-Cyclohexylcinnamamide, 3208
Cyclohexyl 2,4-Dichlorophenoxyacetate, 3028
3-Cyclohexyl-6-dimethylamino-1-methyl-1,3,5-triazine-
    2,4-dione, 2726
2-Cyclohexyl-4,6-dinitrophenol, 2626
1-Cyclohexyloxycarbonyl-5-fluorouracil, 2254
2-Cyclohexylphenol, 2680
4-Cyclohexylphenol, 2679
o-Cyclohexylphenol, 2680
p-Cyclohexylphenol, 2679
4-Cyclohexylresorcinol, 2681
p-Cyclohexylresorcinol, 2681
Cyclohexyl-4-toluene Sulfonamide, 3676
Cyclohexyl-p-toluene Sulfonamide, 3676
1-Cyclohexyl-3-para-tolylsulfonylurea, 3081
3-Cyclohexyl-5,6-trimethyleneuracil, 2892
Cyclo(L-leucyl-D-phenylalanyl-L-prolyl-L-valyl-L-
    ornithyl-L-leucyl-D-phenylalanyl-L-prolyl-L-valyl-L-
    ornithyl), 4162
Cyclomaltooctaose, 4142
Cyclonite, 172
Cyclooctaamylose, 4142
Cyclooctane, 1529
N-Cyclooctylcinnamamide, 3481
Cyclooctyl-1,1-dimethylurea, 2361
N'-Cyclooctyl-N,N-dimethylurea, 2361
4H-Cyclopenta[def]phenanthrene, 3134

2-Cyclopentamethylene-4-methoxycarbamyl-1,3-
    dioxolane, 2094
2-Cyclopentamethylene-4-methoxycarbamyl-1,3-
    dithiolane, 2092
2-Cyclopentamethylene-4-methoxycarbamyl-1,3-
    oxathiolane, 2093
Cyclopentane, 448
Cyclopentane, 1,1,3-Trimethyl-, 1525
Cyclopentene, 419
Cyclopenthiazide, 2890
N-Cyclopentylcinnamamide, 3044
2-Cyclopentyl-4-methoxycarbamyl-1,3-dithiolane, 2092
β-Cyclopentylpropionyl Salicylate, 3206
Cyclopropane, 153
Cyclopropane, Ethoxy-, 468
N-Cyclopropylcinnamamide, 2217
α-Cyclopropyl-α-(4-methoxyphenyl)-5-
    pyrimidinemethanol, 3184
N-(Cyclopropylmethyl)-2,6-dinitro-N-propyl-4-
    (trifluoromethyl)benzenamine, 3030
Cyclosporin A, 4166
Cycluron, 2361
Cyhexatin, 3608
Cymel, 171
4-Cymene, 2053
p-Cymene, 2053
Cymetrin, 1523
Cyomin, 4168
Cypreth Ether, 468
Cyprethylene Ether, 428
Cyprozine, 1773
Cysteine, 203
L-Cysteine, S-[(2R)-2-Amino-2-carboxyethyl]-, 836
2-S-Cysteinyl-4,6-bis-(dimethylamino)-s-triazine, 2005
Cystine, 838
D-Cystine, 837
DL-Cystine, 838
L-Cystine, 835
meso-Cystine, 839
Cythion, 2154
Cytisin, 2273
Cytisine, 2273
Cytosine, 257
Cytostatics, 1631
2,3-D, 1301
2,4-D, 1300
2,5-D, 1302
2,6-D, 1303
3,4-D, 1304
3,5-D, 1305
D 263, 3777
D 66, 2381
D 73, 2375
dA, 2042
Dacamox, 1805
DAC PRD, 264
Dahl's Acid, 1915
Dalapon, 130
Dambose, 875
Daminozide, 833
Dantoin, 405
Dapsone, 2592
Darvocet N-100, 4093
Darvocet N-50, 4093

Darvon-N, 4093
Dasanit, 2330
Dasanit Sulfone, 2331
Dasanit Sulphone, 2331
DATC, 2112
Daturine, 3484
Daxtrom, 377
Dazomet, 458
2,4-DB, 1935
DBC, 3755
DBD, 1567
DBDPO, 2767
2,4-DBE, 3150
DBF, 2530
2,7-DCDD, 2454
DCDF, 2453
2,8-DCDF, 2453
DCIP, 825
2,6-DCP, 593
3,4-DCP, 592
3,5-DCP, 591
DCPA, 1869
DDA, 2041
DDD, 2969
o,p'-DDD, 2968
o,p'-DDE, 2945
p,p'-DDE, 2944
2,4'-DDT, 2956
o,p'-DDT, 2956
p,p'-DDT, 2957
DEANAB, 3333
3-Deazauracil, 398
2,4-DEB, 3149
Debenal, 1939
Decabromodiphenyl Ether, 2767
Decabromodiphenyl Oxide, 2767
Decachlorbiphenyl, 2770
Decachlorobiphenyl, 2770
2,2',3,3',4,4',5,5',6,6'-Decachlorobiphenyl, 2770
Decachloroketone, 2184
1,2,3,5,6,7,8,9,10,10-
    Decachloropentacyclo[5.2.1.0(2,6).0(3,9).0(5,8)]decano
    -4-one, 2184
2,4-Decadione, 2130
Deca-Durabolin, 4060
Decahydro-2-methylnaphthalene, 2340
Decahydronaphthalene, 2120
cis-Decahydronaphthalene, 2119
Decalin, 2120
cis-Decalin, 2119
Decanamide, 2,2-Diethyl-N-hydroxy, 3121
Decan-dicarbonsaeure-(1,10), 2736
Decane, 2179
n-Decane, 2179
Decane(S-(carboxymethyl)-L-cysteine)), 4145
Decanediamide, N,N,N',N'-Tetramethyl-, 3115
1,10-Decanedicarboxylic Acid, 2736
Decanedioic Acid Dibutyl Ester, 3609
Decanoic Acid, 2167
Decemthion, 2231
Dechlorane 4070, 2185
Declomycin, 3807
n-Decyl Alcohol, 2181
n-Decylcinnamamide, 3696

Decyl Hydride, 2179
n-Decyl p-hydroxybenzoate, 3498
Decyl p-hydroxybenzoate, 3498
Decyl-p-hydroxybenzoate, 3498
Decyl Lactate, 2937
Deethylatrazine, 790
1,2-Dehydroacenaphthalene, 2511
Dehydrocholic Acid, 3962
Dehydroepiandrosterone, 3691
Dehydroisoandrosterone, 3691
5,6-Dehydroisoandrosterone, 3691
5,6-Dehydroisoandrosterone Acetate, 3840
5,6-Dehydroisoandrosterone Butyrate, 3936
5,6-Dehydroisoandrosterone Formate, 3773
5,6-Dehydroisoandrosterone Propionate, 3897
5,6-Dehydroisoandrosterone Valerate, 3965
Dehydrostilbestrol, 3543
Dehydrothio-N-toluidin, 2992
Dehydrothio-N-toluidine, 2992
Delcortin, 3823
Delmadinone Acetate, 3922
Delphinine, 4103
Deltamethrin, 3870
Demeclocycline, 3807
Demephion, 525
Demerol, 3223
Demethylchlortetracycline, 3807
4'-Demethylepipodophyllotoxin ethylidene-β-D-glucoside,
    4071
Demetonthiol, 1592
Demetonthione, 1593
Demetrin, 3631
Demosan, 1357
Denarin, 2055
Deoxyadenosine, 2042
2'-Deoxyadenosine, 2042
Deoxyadenosylcobalamin, 4177
Deoxycholic Acid, 3978
Deoxycorticosterone, 3841
11-Deoxycorticosterone, 3841
Deoxycorticosterone Acetate, 3932
2'-Deoxy-5-iodouridine, 1700
6-Deoxy-L-mannose, 869
D-2-Deoxy-2-(3-methyl-3-nitrosoureido)glucopyranose,
    1519
9-Deoxy-9-oxocinchonine, 3643
2-Deoxyphenobarbital, 2623
Deprol, 1806
Deptran, 3657
Dermadex, 2772
Dermafos, 1359
Dermaton, 2621
Derris, 3913
Desethyl Simazine, 125
Desfenuron, 1434
Desipramine, 3567
Desmedipham, 3312
Desmethyldoxepin, 3549
Desmetryne, 1522
21-Desoxy-9α-fluoro-6α-methyl-prednisolone, 3888
Desoxyphenobarbitone, 2623
Desphenuron, 1434
Despirol, 3412
Destrol, 3559

Destun, 2991
Detonal, 1799
Devermine, 2778
Devicarb, 2575
Devrinol, 3463
Devrinol 2E, 3463
Devrinol 10G, 3463
Devrinol 50W, 3463
Dexamethasone, 3890
Dexamethasone Acetate, 3953
Dexamethasone Acetate, 3953
Dexamethasone-17-acetate, 3953
Dexamethasone Alcohol, 3890
Dexamethasone TBA, 4019
Dexibuprofen, 2900
Dexoxadrol, 3744
α-Dextrin, 4119
Dextromethorphan HBr, 3583
Dextropropoxyphene Napsylate, 4093
Dextrose, 879
Dexy Pearl γ-100, 4142
DHPG, 1770
Diaalylacetamide, 1492
Diabaril, 2020
Diabeta, 3926
Diabinese, 2020
9-(1,3-Diacetate-2-propoxymethyl)guanine, 2886
1,4-Diacetoxybutane, 1510
2,5-Diacetoxymethyl Allopurinol, 2245
Diacetyl, 276
3,5-Diacetylamino-2,4,6-triiodobenzoic Acid, 2200
3,5-Diacetylamino-2,4,6-triiodobenzoic Acid Methyl-
    glucamide, 3582
Diacetylen, 225
N1,N4-Diacetylsulfanilamide, 1990
Diaethylaether, 362
Diaethylsulfon, 367
Dial, 2287
Dialen 6, 822
Dialifor, 3043
Dialifos, 3043
Diallate, 2112
Diallyl, 787
α,α-Diallylacetamide, 1492
5,5-Diallylbarbituric Acid, 1987
N,N-Diallyl-2-chloroacetamide, 1483
N, N-Diallyl Dichloroacetamide, 1476
N,N-Diallyldichloroacetamide, 1476
Diallyl Phenyl Phosphonate, 2661
Diallyl Phthalate, 3019
1,4-Diamino-9,10-anthraquinone, 2971
p-Diaminobiphenyl, 2589
1,8-Diamino-3,6-diazaoctane, 950
4,4'-Diamino-3,3'-dichlorobiphenyl, 2556
4,4'-Diaminodiphenyl Sulphone, 2592
2,4-Diaminopteridine, 718
2:4-Diaminopteridine, 718
4,6-Diaminopteridine, 721
4,7-Diaminopteridine, 720
4:6-Diaminopteridine, 721
4:7-Diaminopteridine, 720
1,11-Diamino-3,6,9-triazaundecane, 1600
Diamyl Phthalate, 3591
Dianex, 3711

1,4:3,6-Dianhydro-D-glucitol dinitrate, 768
Dianisidine, 3032
o-Dianisidine, 3032
Diantipyrylmethane, 3915
4,4'-Diantipyrylmethane, 3915
Diatrazoic Acid, 2200
3,6-Diazaoctane-1,8-diamine, 950
7,9-Diazaspiro[4.5]decane-6,8,10-trione, 1443
2,4-Diazaspiro[5.6]dodecane-1,3,5-trione, 2063
2,4-Diazaspiro[5.11]heptadecane-1,3,5-trione, 3258
2,4-Diazaspiro[5.10]hexadecane-1,3,5-trione, 3101
6,8-Diazaspiro[3.5]nonane-5,7,9-trione, 1109
5,7-Diazaspiro[2.5]octane-4,6,8-trione, 707
2,4-Diazaspiro[5.7]tridecane-1,3,5-trione, 2308
2,4-Diazaspiro[5.5]undecane-1,5-dione, 3-Thioxo-, 1739
2,4-Diazaspiro[5.5]undecane-1,3,5-trione, 1740
2,4-Diazaspiro[5.5]undecane-1,3,5-trione, 3-Thio, 1739
Diazemuls, 3279
Diazene, (4-Chlorophenyl)phenyl-, (E)-, 2540
Diazene, (p-Nitrophenyl)phenyl-, (E)-, 2548
Diazepam, 3279
Diazinon, 2731
2,5-Diaziridinyl-3,6-bis(butylamino)-1,4-benzoquinone, 3597
2,5-Diaziridinyl-3,6-bis(dimethylamino)-1,4-benzoquinone, 3083
2,5-Diaziridinyl-3,6-bis(ethylamino)-1,4-benzoquinone, 3084
2,5-Diaziridinyl-3,6-bis(glycinamide)-1,4-benzoquinone, 3061
2,5-Diaziridinyl-3,6-bis(hydroxyethylamino)-1,4 benzoquinone, 3086
2,5-Diaziridinyl-3,6-bis(hydroxylethylmethylamino)-1,4-benzoquinone, 3387
2,5-Diaziridinyl-3,6-bis(2'-hydroxyl-3'-hydroxylpropylamino)-1,4-benzoquinone, 3387
2,5-Diaziridinyl-3,6-bis(methylamino)-1,4-benzoquinone, 2674
2,5-Diaziridinyl-3,6-bis(propylamino)-1,4-benzoquinone, 3386
2,5-Diaziridinyl-3,6-di(1'-piperazineethanol)-1,4-benzoquinone, 3901
2,5-Diaziridinyl-3,6-dipyrrolidino-1,4-benzoquinone, 3574
2,5-Diaziridinyl-3-floro-6-morpholino-1,4-benzoquinone, 3029
Diazoaminobenzene, 2577
Dibencozide, 4177
1:2,6:7-Dibenzacridine, 3793
1:2,8:9-Dibenzacridine, 3794
3:4,6:7-Dibenzacridine, 3795
Dibenzamid, 2986
1,2,5,6-Dibenzanthracene, 3863
1,2:3,4-Dibenzanthracene, 3862
1,2:5,6-Dibenzanthracene, 3863
1,2:7,8-Dibenzanthracene, 3864
Dibenz[de,kl]anthracene, 3717
6H-Dibenz[b,e]azepin-6-one, 5,11-Dihydro-5-(2,2,2-trifluoroethyl)-, 3274
6H-Dibenz[b,e]azepin-6-one, 5-(2-Fluoroethyl)-5,11-dihydro-, 3289
5H-Dibenz[b,f]azepine-5-carboxamide, 3152
Dibenz[b,f][1,4]oxazepin-11(10H)-one, 8-Amino-2-methyl-, 3172

Dibenz[b,f][1,4]oxazepin-11(10H)-one, 10-Ethyl-, 3163
Dibenz[a,j]anthracene, 3864
1,2,5,6-Dibenzocarbazole, 3720
1:2,5:6-Dibenzocarbazole, 3720
1:2,7:8-Dibenzocarbazole, 3719
13H-Dibenzo(a,i)carbazole, 3719
3,4,5,6-Dibenzocarbazole, 3718
3:4,5:6-Dibenzocarbazole, 3718
Dibenzo-18-crown-6, 3755
Dibenzo[1,4]dioxin, 2531
Dibenzo-p-dioxin, 2531
Dibenzo[b,d]pyrrole, 2545
Dibenzo[b,e][1,4]dioxin, 1,2,3,4,7-Pentachloro-, 2385
Dibenzo[b,e][1,4]dioxin, 1,2,4-trichloro-, 2429
Dibenzofuran, 2530
Dibenzo[b,k][1,4,7,10,13,16]hexaoxacyclooctadecin, Icosahydro-, 3791
Dibenzopyrazine, 2528
2,3,5,6-Dibenzopyridine, 2783
Dibenzopyrrole, 2545
Dibenzo-1,4-thiazine, 2547
Dibenzothiophene, 2533
1,4-Dibenzoylaminoanthraquinone, 4064
1,2,7,8-Dibenzphenanthrene, 3861
1,2-Dibromaethan, 71
DL-Dibrom-bernsteinsaeure, 230
meso-Dibrom-bernsteinsaeure, 230
Dibrom-methan, 8
1,3-Dibromobenzene, 578
1,4-Dibromobenzene, 577
m-Dibromobenzene, 578
p-Dibromobenzene, 577
4,4'-Dibromobiphenyl, 2512
p,p'-Dibromobiphenyl, 2512
Dibromochloromethane, 2
1,2-Dibromo-3-chloropropane, 138
1,2-Dibromoethane, 71
3-(2,2-Dibromoethenyl)-2,2-dimethylcyclopropanecarboxylic Acid Cyano(3-phenoxyphenyl)methylEster, 3870
3,5-Dibromo-4-hydroxybenzaldehyde 2,4-dinitrophenyloxime, 2773
3,5-Dibromo-4-hydroxybenzonitrile, 954
1,4-Dibromonaphthalene, 1865
2,3-Dibromonaphthalene, 1866
1,3-Dibromopropane, 156
2,3-Dibromo-1-propanol Phosphate (3:1), 1780
2,3-Dibromopropyl Phosphate, 1780
2,6-Dibromoquinone-3-chlorimide, 534
2,6-Dibromoquinonechloroimide, 534
2,6-Dibromoquinone Oxime, 550
DL-2,3-Dibromosuccinic Acid, 230
meso-2,3-Dibromosuccinic Acid, 230
Dibucaine, 3775
Dibutoxybutylphosphine Oxide, 2764
Dibutoxyethyl Phthalate, 3782
Dibutoxyphenylphosphine Oxide, 3107
Dibutoxyphosphine Oxide, 1591
Dibutylaceturethane, 2933
Di-tert-butylacetylene, 2118
n-Dibutylamine, 1588
Di-n-butylamine, 1588
N,N-Dibutylamine, 1588
N,N-Dibutyl-1-butanamine, 2759

Dibutyl Butanephosphonate, 2764
Dibutylbutoxyphosphine Oxide, 2762
Dibutyl Butyl Phosphonate, 2764
Dibutyl Butylphosphonate, 2764
2,6-Di-tert-Butyl-p-Cresol, 3260
Dibutyl Decanedioate, 3609
Dibutyl Decyl Phosphate, 3615
Dibutyl Decyl Phosphonate, 3614
Dibutyl Ether, 1579
Dibutyl Ethoxybutyl Phosphate, 3133
Dibutyl Ethyl Phosphate, 2183
Dibutyl Ethyl Phosphonate, 2182
Di-tert-butylethyne, 2118
Dibutyl Hexadecyl Phosphonate, 3985
Dibutyl Hexyl Phosphate, 3132
Dibutyl Hexyl Phosphonate, 3129
Dibutyl Hydrogen Phosphonate, 1591
2,2-Dibutyl-N-hydroxyhexanamide, 3122
2,6-Di-tert-Butyl-1-Hydroxy-4-Methylbenzene, 3260
Dibutyl Isooctyl Phosphonate, 3404
Dibutyl Ketone, 1815
Dibutyl Maleate, 2729
Di-n-butyl Maleate, 2729
Di-n-butyl Methanephosphonate, 1851
Dibutyl Methyl Phosphate, 1854
Dibutyl Methyl Phosphonate, 1851
Dibutylnitrosamine, 1578
Di-n-butylnitrosamine, 1578
Dibutyl Octyl Phosphate, 3405
3,5-di-tert-Butylphenyl Methylcarbamate, 3392
Dibutyl Phenyl Phosphonate, 3107
Dibutyl Phenylphosphonate, 3107
Di-n-Butyl Phosphite, 1591
Dibutyl Phthalate, 3371
Dibutyl Sebacate, 3609
Di-n-butyl Sebacate, 3609
Dibutyl Succinate, 2735
Dibutyl Tartrate, 2739
(2R,3R)-Di-n-Butyl Tartrate, 2739
2,6-Di-tert-butyl-p-tolyl Methylcarbamate, 3499
9-(1,3-Dibutyrate-2-propoxymethyl)guanine, 3496
1,5-Dibutyryloxymethyl Allopurinol, 3216
2,5-Dibutyryloxymethyl Allopurinol, 3217
Dicamba, 1299
Dicaptan, 1405
Dicapthon, 1405
Dicarboethoxyethyl O,O-Dimethyl Phosphorodithioate, 2154
2,5-Dicarboxyfuran, 623
Di(3-carboxy-2,4,6-triiodoanilido)adipic Acid, 3723
Dichlobenil, 956
Dichlofenthion, 2022
Dichlofluanid-methyl, 2021
Dichlone, 1859
1,1-Dichloraethan, 75
1,2-Dichloraethan, 74
2,4-Dichlor-benzoesaeure, 970
2,6-Dichlor-benzoesaeure, 968
D(+)-Dichlor-bernsteinsaeure, 232
L(-)-Dichlor-bernsteinsaeure, 232
β,β'-Dichlor-diaethylsulfid, 311
β,β'-Dichlor-diaethylsulfon, 310
β,β'-Dichlor-diaethylsulfoxid, 309
cis-Dichlorethylene, 52

trans-Dichlorethylene, 54
Dichlormate, 1647
Dichlor-methan, 9
Dichlormid, 1476
2-(Dichloroacetamido)-3-chloro-1,4-naphthoquinone, 2457
2-[(Dichloroacetyl)amino]-3-chloro-1,4-naphthoquinone, 2457
S-(2,3-Dichloroallyl-N,N-diisopropylthiocarbamate, 2112
2,3-Dichloroanisole, 1033
2,6-Dichloroanisole, 1035
3,5-Dichloroanthranilamide, 1032
Dichlorobenzamide, 991
2,6-Dichlorobenzamide, 991
1,2-Dichlorobenzene, 586
1,3-Dichlorobenzene, 588
1,4-Dichlorobenzene, 587
m-Dichlorobenzene, 588
o-Dichlorobenzene, 586
p-Dichlorobenzene, 587
3,4-Dichloro-benzenesulfonamide, 637
3,3'-Dichlorobenzidine, 2556
o,o'-Dichlorobenzidine, 2556
2,4-Dichlorobenzoic Acid, 970
2,6-Dichlorobenzoic Acid, 968
3,4-Dichlorobenzoic Acid, 969
3,5-Dichlorobenzoic Acid, 971
2,6-Dichlorobenzonitrile, 956
3,4-Dichlorobenzyl N-Methylcarbamate, 1647
2,4,-Dichloro-6-benzyl-phenol, 2791
Dichlorobiphenyl, 2513
2,2'-Dichlorobiphenyl, 2521
2,3'-Dichlorobiphenyl, 2515
2,4-Dichlorobiphenyl, 2519
2,4'-Dichlorobiphenyl, 2518
2,5-Dichlorobiphenyl, 2520
2,6-Dichlorobiphenyl, 2514
3,3'-Dichlorobiphenyl, 2517
3,4-Dichlorobiphenyl, 2516
4,4'-Dichlorobiphenyl, 2513
3,3'-Dichloro-4,4'-biphenyldiamine, 2556
1,1-Dichloro-2,2-bis(p-chlorophenyl)ethane, 2969
1-Dichloro-2,2-bis(p-ethoxylphenyl)ethane, 3556
Dichlorobromomethane, 1
2,3-Dichlorobutane, 307
2,4-Dichloro-6-butyl-phenol, 1984
4,6-Dichloro-N-(2-chlorophenyl)-1,3,5-triazin-2-amine, 1603
2,4-Dichloro-6-(cyclohexylamino)triazine, 1733
2,4-Dichloro-6-cyclohexylamino-1,3,5-triazine, 1733
2,3-Dichlorodibenzodioxin, 2455
2,3-Dichlorodibenzo-p-dioxin, 2455
2,7-Dichlorodibenzo-p-dioxin, 2454
2,8-Dichlorodibenzodioxin, 2454, 2456
2,8-Dichlorodibenzo-p-dioxin, 2456
2,8-Dichlorodibenzofuran, 2453
1-[2,4-Dichloro-β-[(2,4-dichlorobenzyl)oxy]phenethyl]imidazole, 3524
2,2'-Dichlorodiethylether, 308
β,β'-Dichlorodiethylsulfone, 310
β,β'-Dichlorodiethylsulfoxide, 309
Dichlorodifluoromethane, 37
β,β'-Dichlorodiisopropyl Ether, 825

1,4-Dichloro-2,5-dimethoxybenzene, 1357
3,5-Dichloro-2,4-dimethoxy-6-(trichloromethyl), 1307
1,1-Dichloro-N-((dimethylamino)sulfonyl)-1-fluoro-N-(4-
    methylphenyl)methanesulfenamide, 2021
1,3-Dichloro-5,5-dimethylhydantoin, 405
1,3-Dichloro-5,5-Dimethyl-2,4-Imidazolidinedione, 405
3,5-Dichloro-N-(1,1-dimethyl-2-propynyl)benzamide,
    2572
3,5-Dichloro-2,6-Dimethyl-4-Pyridinol, 1080
3,6-Dichloro-9,10-dioxaanthracene, 2456
Dichlorodiphenyldichloroethane, 2969
2,4'-Dichlorodiphenyldichloroethylene, 2945
p,p'-Dichlorodiphenyldichloroethylene, 2944
2,2-Dichloro-N,N-di-2-propenylacetamide, 1476
1,1-Dichloroethane, 75
3-(2,2-Dichloroethenyl)-2,2-
    dimethylcyclopropanecarboxylic Acid (3-
    Phenoxyphenyl)methyl Ester, 3806
N,N-di-(2-Chloroethyl)-γ-(p-aminophenyl)butyric Acid,
    3067
1,1-Dichloroethylene, 53
cis-1,2-Dichloroethylene, 52
trans-1,2-Dichloroethylene, 54
sym-Dichloroethyl Ether, 308
2,4-Dichloro-6-ethyl-phenol, 1356
1,3-Dichlorohydantoin, 115
O-(2,5-Dichloro-4-iodophenyl) O,O-Dimethyl
    Phosphorothioate, 1355
Dichloroisopropyl Ether, 825
Dichloromethane, 9
1,2-Dichloro-3-methoxybenzene, 1033
3-[2,4-Dichloro-5-(1-methylethoxy)phenyl]-5-(1,1-
    dimethylethyl)-1,3,4-oxadiazol-2(3H)-one, 3199
1,3-Dichloro-5-methylhydantoin, 231
2,4-Dichloro-6-methyl-phenol-, 1034
2,6-Dichloro-4-methyl-phenol, 1034
1,4-Dichloronaphthalene, 1867
2,3-Dichloro-1,4-naphthalenedione, 1859
1,2-Dichloro-3-nitrobenzene, 558
1,2-Dichloro-4-nitrobenzene, 556
1,4-Dichloro-2-nitrobenzene, 557
2,3-Dichloronitrobenzene, 558
2,5-Dichloronitrobenzene, 557
3,4-Dichloronitrobenzene, 556
Dichloronitroethane, 62
1,1-Dichloro-1-nitroethane, 62
1,1-Dichloro-1-nitropropane, 142
2',5-Dichloro-4'-nitrosalicylanilide, 2778
Dichlorophen, 2792
2,3-Dichlorophenol, 590
2,4-Dichlorophenol, 589
2,5-Dichlorophenol, 594
2,6-Dichlorophenol, 593
3,4-Dichlorophenol, 592
3,5-Dichlorophenol, 591
4,5-Dichlorophenol, 592
(2,4-Dichlorophenoxy)acetic Acid, 1300
2,3-Dichlorophenoxyacetic Acid, 1301
2,4-Dichlorophenoxyacetic Acid, 1300
2,5-Dichlorophenoxyacetic Acid, 1302
2,6-Dichlorophenoxyacetic Acid, 1303
3,4-Dichlorophenoxyacetic Acid, 1304
3,5-Dichlorophenoxyacetic Acid, 1305
2,4-Dichlorophenoxyacetic Acid Allyl Ester, 2207

2,4-Dichlorophenoxyacetic Acid Benzyl Ester, 3150
2,4-Dichlorophenoxyacetic Acid Butyl Ester, 2620
2,4-Dichlorophenoxyacetic Acid n-Butyl Ester, 2620
2,4-Dichlorophenoxyacetic Acid sec-Butyl Ester, 2619
2,4-Dichlorophenoxyacetic Acid Capryl Ester, 3364
2,4-Dichlorophenoxyacetic Acid Cyclohexyl Ester, 3028
2,4-Dichlorophenoxyacetic Acid Ethyl Ester, 1936
2,4-Dichlorophenoxyacetic Acid 1-Ethylpropyl Ester,
    2849
2,4-Dichlorophenoxyacetic Acid n-Hexyl Ester, 3052
2,4-Dichlorophenoxyacetic Acid Isopropyl Ester, 2229
2,4-Dichlorophenoxyacetic Acid Iso-Propyl Ester, 2229
2,4-Dichlorophenoxyacetic Acid 2-Methylbutyl Ester,
    2850
2,4-Dichlorophenoxyacetic Acid β-Monochloroethyl
    Ester, 1904
2,4-Dichlorophenoxyacetic Acid n-Octyl Ester, 3364
2,4-Dichlorophenoxyacetic Acid n-Pentyl Ester, 2848
2,4-Dichlorophenoxyacetic Acid n-Tetradecyl Ester,
    3900
2,4-Dichlorophenoxyacetic Acid n-Undecyl Ester, 3688
4-(2,4-Dichlorophenoxy)propionic Acid, 1935
α-(2,4-Dichlorophenoxy)propionic Acid, 1632
p,p-Dichlorophenylbutazone, 3637
1-[2-(2,4-Dichlorophenyl)-2-[(2,4-
    dichlorophenyl)methoxy]ethyl]-1H-imidazole, 3524
3-(3,4-Dichlorophenyl)-1,1-dimethylurea, 1665
3-(3,5-Dichlorophenyl)-5-ethenyl-5-methyl-2,4-
    oxazolidinedione, 2542
3-(3,4-Dichlorophenyl)-1-methoxy-1-methylurea, 1666
2,4 Dichlorophenyl 3-Methoxy-4-nitrophenyl Ether,
    2780
O-(2,4-Dichlorophenyl)-O-methyl, 2054
2 (3,4-Dichlorophenyl)-4-methyl-1,2,4-oxadiazolidine-
    3,5-dione, 1605
2,4-Dichlorophenyl-4-nitrophenyl Ether, 2490
3,6-Dichloropicolinic Acid, 555
Dichloroprop, 1632
trans 1,3-Dichloro-propene, 128
Dichloropropane, 160
1,2-Dichloropropane, 160
1,3-Dichloropropane, 159
1,3-Dichloro-2-propanol, 161
(E)-1,3-Dichloro-1-Propene, 128
trans 1,3-Dichloro-propene, 128
(Z)-1,3-Dichloropropene, 127
1,2-Dichloropropene, 126
1,3-Dichloropropene, 129
cis-1,3-Dichloropropene, 127
cis-1,3-Dichloro-1-propene, 127
E-1,3-Dichloropropene, 128
trans-1,3-Dichloropropene, 129
trans-1,3-Dichloro-1-propene, 128
3',4'-Dichloropropionanilide, 1646
1,3-Dichloropropropylene (cis), 127
Dichloropropylene, 126
cis-1,3-Dichloropropylene, 127
trans-1,3-Dichloropropylene, 129
1,3-Dichloropropylene (trans), 129
2,4-Dichloro-6-propyl-phenol, 1667
3,6-Dichloro-2-pyridinecarboxylic Acid, 555
2,4'-Dichloro-α-(5-pyrimidinyl)benzhydryl Alcohol, 3411
2,3-Dichlorosuccinic Acid, 232
D-2,3-Dichlorosuccinic Acid, 232

L-2,3-Dichlorosuccinic Acid, 232
meso-2,3-Dichlorosuccinic Acid, 232
3,4-Dichlorosulfolane, 264
1,2-Dichlorotetrafluoroethane, 108
sym-Dichlorotetrafluoroethane, 108
3,4-Dichlorotetrahydrothiophene Dioxide, 264
3,4-Dichlorotetrahydrothiophene 1,1-dioxide, 264
2,6-Dichlorothiobenzamide, 993
3,4-Dichlorothiolane 1,1-dioxide, 264
4,4'-Dichloro-α-(trichloromethyl)benzhydrol, 2958
Dichloro-2-(trifluoromethyl)benzimidazole, 1271
4,5-Dichloro-2-(trifluoromethyl)-benzimidazole, 1271
4,5-Dichloroveratrole, 1358
2,4-Dichlor-phenol, 589
2,5-Dichlor-phenol, 594
5-(2,4-Dichlorphenoxy)-2-nitro-benzoic Acid Methyl
    Ester, 2955
Dichlorprop, 1632
1,2-Dichlor-propan, 160
1,3-Dichlor-propan, 159
1,3-Dichlor-propanol-(2), 161
α,α-Dichlor-propionsaeure, 130
Dichlorvos, 294
2,4-Dichorophenoxyacetic Acid Methyl Ester, 1633
Diclophenthion, 2022
Diclotop-methyl, 3288
Dicofol, 2958
Dicoumarol, 3617
Dicryl, 1902
Dicumarol, 3617
Dicuran, 2017
Dicyan, 112
Dicyandiamid, 78
Dicyandiamide, 78
1,4-Dicyanobenzene, 1275
Dicyanodiamide, 78
2,3-Dicyano-1,4-dithiaantraquinone, 2940
Dicyclohexano-18-crown-6, 3791
cis-Dicyclohexano-18-crown-6, 3791
N,N-DicyclohexylCinnamamide, 3834
Dicyclohexyl-18-crown-6, 3791
N,N-Dicyclohexyl-3-phenyl2-Propenamide, 3834
Dicyclohexyl Phthalate, 3766
N,N-Dicyclopentylcinnamamide, 3681
Didanosine, 1999
Di-n-decyl Phthalate, 4061
7,8-Didehydro-4,5-epoxy-3-ethoxy-17-methylmorphinan-
    6-ol, 3669
7,8-Didehydro-4,5-α-epoxy-3-methoxy-17-
    methylmorphinan-6-α-ol, 3564
7,8-Didehydro-4,5-epoxy-17-methylmorphinan-3,6-diol,
    3449
2',3'-Dideoxyadenosine, 2041
2',3'-Dideoxyinosine, 1999
(3β,5β)-3-[(O-2,6-Dideoxy-β-D-ribo-hexopyranosyl-(1-
    >4)-O-2,6-dideoxy-β-D-ribo-hexopyranosyl-(1->4)-2,6-
    dideoxy-β-D-ribo-hexopyranosyl)oxy]-14-hydroxycard-
    20(22)-enolide, 4126
3β-((O-2,6-Dideoxy-β-D-Ribo-hexopyranosyl-(1rightarrow4)-
    O-2,6-dideoxy-β-D-Ribo-hexopyranosyl-(1rightarrow4)-
    2,6-dideoxy-β-D-Ribo-hexopyranosyl)oxy)-12β,14-
    dihydroxy-5β-card-20(22)-enolide, 4128
1,5-Di(2,4-dimethylphenyl)-3-methyl-1,3,5-triazapenta-
    1,4-diene, 3672

Didodecyl Phthalate, 4096
Dieldrin, 2524
Dienestrol, 3543
Diethion, 1858
4-Diethoxybenzene, 2079
p-Diethoxybenzene, 2079
1-(3,4-Diethoxybenzylidene)-6,7-diethoxy-1,2,3,4-
    tetrahydroisoquinoline, 3954
Diethoxyethane, 933
1,2-Diethoxyethane, 933
Diethoxyethyl Adipate, 2138
Diethoxyethyl Phthalate, 3373
Diethoxymethane, 520
2-[(Diethoxyphosphinothioyl)oxy]-5-methylpyrazolo[1,5-
    a]pyrimidine-6-carboxylic Acid Ethyl Ester, 3082
Diethquinalphion, 2654
Diethyl, 356
Diethylacetalformaldehyde, 520
2,6-Diethyl-4-acetaminophenol, 2690
Diethylacetic Acid, 857
Diethylaceturethane, 1799
Diethylacetylene, 785
Diethylacetyl Salicylate, 2861
Diethyl Adipate, 2138
2-Diethylamino-2',6'-acetoxylidide, 3098
2-(Diethyl(amino)butyl 4-Aminobenzoate, 3256
2-(Diethylamino)-N-(2,6-dimethylphenyl)acetamide,
    3098
Diethylaminoethyl p-Anisate, 3093
2-Diethylaminoethyl Diphenylacetate, 3757
N-[2-(Diethylamino)ethyl]-2-ethoxyquinoline-4-
    carboxamide, 3584
N-[2-(Diethylamino)ethyl]-2-hexoxyquinoline-4-
    carboxamide, 3899
N-[2-(Diethylamino)ethyl]-2-methoxyquinoline-4-
    carboxamide, 3488
N-[2-(Diethylamino)ethyl]-2-pentoxyquinoline-4-
    carboxamide, 3846
N-[2-(Diethylamino)ethyl]-2-propoxyquinoline-4-
    carboxamide, 3687
4[[(4-Diethylamino)phenyl]azo]nitrobenzene, 3334
2-Diethylamino)propyl 4-Aminobenzoate, 3100
Diethyl Amyl Phosphate, 1855
Diethylaniline, 2082
2,6-Diethylaniline, 2082
5,5-Diethylbarbituric Acid, 1486
1,2-Diethylbenzene, 2047
1,4-Diethylbenzene, 2048
o-Diethylbenzene, 2047
p-Diethylbenzene, 2048
Diethyl Benzenephosphonate, 2117
Diethyl Benzoyl Phosphonate, 2298
O,O-Diethyl S-Benzyl Thiophosphate, 2328
N,N'-Diethyl-bicyclo(2.2.1)hept-5-ene-2,3-trans-
    dicarboxamide, 2911
Diethyl Butyl Phosphate, 1595
Diethylcarbamothioic Acid S-[(4-Chlorophenyl)methyl]
    Ester, 2662
Diethyl Carbinol, 513
Diethyl Carbonate, 481
Diethyl Cellosolve, 933
O,O-Diethyl O-(3-Chloro-4-methylcoumarinyl-7)
    Thiophosphate, 3027
Diethyl S-(Chloromethyl) Dithiophosphate, 506

Diethyl O-(5-Chloro-1-(1-methylethyl)-1H-1,2,4-triazol-3-yl) phosphorothioate, 1797
Diethyl S-((6-chloro-2-oxobenzoxazolin-3-yl)methyl) phosphorodithioate, 2639
O,O-Diethyl S-(4-Chlorophenylthiomethyl) Dithiophosphate, 2302
Diethyl S-(2-Chloro-1-phthalimidoethyl) Phosphorodithioate, 3043
N,N-Diethylcinnamamide, 2867
Diethyl Decyl Phosphate, 3131
Diethyl O-Dichlorophenyl phosphorothioate, 2022
Diethyl O-(2-(Diethylamino)-6-methyl-4-pyrimidinyl) phosphorothioate, 2928
O,O-Diethyl (1,3-Dihydro-1,3-dioxo-2H-isoindol-2-yl) Phosphonothioate, 2622
O,O-Diethyl S-(((1,1-Dimethylethyl)thio)methyl) Phosphorodithioic Acid, 1850
N3,N3-Diethyl-2,4-dinitro-6-(trifluoromethyl)-1,3-phenylenediamine, 2256
N3,N3-Diethyl-2,4-dinitro-6-(trifluoromethyl)-m-phenylenediamine, 2256
Diethyl 1,3-Dithietan-2-ylidenephosphoramidate, 828
Diethylene Glycol Acetate Butyl Ether, 2176
Diethylene Glycol Butyl Ether Acetate, 2176
Diethylene Glycol Dipropionate, 2141
4,4'-(1,2-Diethyl-1,2-ethenediyl)bisphenol, 3559
Diethyl Ether, 362
4,4'-(1,2-Diethylethylene)diphenol, 3569
O,O-Diethyl 2-Ethylmercaptoethyl Thiophosphate, 1593
Diethyl S-(2-Ethylsulfonylethyl) Phosphorodithioate, 1596
O,O'-Diethyl S-Ethylsulfonylmethyl-phosphorodithioate, 1268
O,O-Diethyl-O-(2-(ethylthio)-ethyl)ester Thiophosphoric Acid, 1593
O,O-Diethyl-S-(2-(ethylthio)-ethyl)ester Thiophosphoric Acid, 1592
N,N-Diethyl Glycolamide Salicylate, 2877
N,N-Diethylglycolamide Salicylate, 2877
N,N-Diethylglycyloxymethyl-1-allopurinol, 2697
Di(2-ethylhexyl) Azelate, 4004
Di-2-ethylhexyl Isophthalate, 3970
Diethyl Hexyl Phosphonate, 3130
Di(2-ethylhexyl)phthalate, 3967
Di-(2-ethylhexyl)-phthalate, 3967
N,N-Diethyl-6-hydroxynorbornane-2-carboxamide-3,5-lactone, 2907
Diethyl Isobutyl Phosphate, 1594
O,O-Diethyl S-(N-Isopropylcarbamylmethyl) Dithiophosphate, 1843
O,O-Diethyl O-(2-Isopropyl-6-methyl-4-pyrimidinyl), Phosphorothioate, 2731
Diethyl Ketone, 461
Diethyl Malonate, 1192
Diethylmalonylurea, 1486
N,N-Diethylmethylamine, 524
5,5'-Diethyl-1-methylbarbituric Acid, 1775
Diethylmethylcarbinol, 925
O,O-Diethyl S-(N-Methyl-N-carboethoxycarbamoylmethyl) Dithiophosphate, 2157
3,3-Diethyl-5-methyl-2,4-piperidinedione, 2113
O,O-Diethyl O-(4-(Methylsulfinyl)phenyl) Phosphorothioate, 2330
2,2-Diethyl-5-methyl-tetrahydrofuran-3,4-diol, 1831

O,O-Diethyl O-[p-(Methylthio)phenyl] Phosphorothioate, 2329
N,N'-Diethyl-6-(methylthio)-1,3,5-Triazine-2,4-diamine, 1523
N,N-Diethyl-2-(1-naphthyloxy)propanamide, 3463
Diethyl p-Nitrophenyl Phosphate, 2057
O,O-Diethyl O-p-Nitrophenyl Phosphorothioate, 2056
Diethyl Nitrosamine, 359
Diethyl Octanephosphonate, 2763
Diethyl Octyl Phosphonate, 2763
O,O-Diethyl S-[(4-Oxo-1,2,3-benzotriazin-3(4H)-yl)methyl] Phosphorodithioate, 2673
3,5-Diethylparacetamol, 2690
3,3-Diethylpentane, 1838
Diethyl Pentyl Phosphate, 1855
O,O-Diethyl O-Pentyl Phosphate, 1855
O,O-Diethyl O-5-Phenylisoxazol-3-yl Phosphorothioate, 2853
Diethyl Phenyl Phosphonate, 2117
Diethyl Phenylphosphonate, 2117
N,N-Diethyl-3-phenyl-2-propenamide, 2867
O,O-Diethyl O-(1-Phenyl-1H-1,2,4-triazol-3-yl) Phosphorothioate, 2672
1,3-Diethyl-8-phenylxanthine, 3189
Diethyl Phthalate, 2638
Di-ethyl Phthalate, 2638
6,7-Diethylpteridine, 1994
O,O-Diethyl O-Pyrazinyl Thiophosphate, 1493
Diethyl O-(2-Quinoxalyl) phosphorothioate, 2654
Diethylstilbestrol, 3559
Diethylstilboestrol, 3559
N,N-Diethylsuccinamyloxymethyl-1-allopurinol, 3076
Diethyl Succinate, 1512
Diethyl Sulfone, 367
N6,N6-Diethyl-N2,N2,N4,N4-tetramethylmelamine, 2362
5,5-Diethyl-2-thiobarbituric Acid, 1485
Diethyl Thioether, 370
Diethyl Trichloromethyl Phosphonate, 452
O,O-Diethyl O-3,5,6-Trichloro-2-pyridyl Phosphorothioate, 1695
Difenoxuron, 3331
Diflorasone Diacetate, 4014
Diflubenzuron, 2954
Difluorodichloromethane, 37
1,1-Difluoroethane, 76
6α,9α-Difluoro-16α Hydroxyprednisolone-16,17-acetonide, 3949
6α,9α-Difluoro-16α,17α-isopropylidenedioxy-1,4-pregnadiene-3,20-dione, 3949
Difluorophate, 904
α-(2,4-Difluorophenyl)-α-(1-2-(2-chloro)phenylethenyl)-1H-1,2,4-triazole-1-ethanol, 3417
α-(2,4-Difluorophenyl)-α-(1-2-(2-pyridyl)phenylethenyl)-1H-1,2,4-triazole-1-ethanol, 3910
Difluron, 2954
Difolatan, 1905
Digallic Acid, 2979
m-Digallic Acid, 2979
m-Digallussaeure, 2979
Digifortis, 4126
Digitalin, 4117
Digitalinum Verum, 4117
Digitoxigenin, 3937

Digitoxin, 4126
Diglycine, 318
Diglycine Hydantoic Acid, 1186
Diglycolic Acid, 283
Diglykol-monobutylaether-acetat, 2176
Di-glykolsaeure, 283
Digoxin, 4128
9-(1,3-Dihemisuccinate-2-propoxymethyl)guanine, 3468
Diherbid, 3323
Dihydrate 6-(1-
    Aminocyclohexanecarboxamido)penicillanic Acid,
    3252
1,2-Dihydroacenaphthene, 2553
6-(1,3-Dihydro-7-acetate-5-methoxy-4-methyl-1-
    oxoisobenzofuran-6-yl)-4-methyl-4-hexanoic Solketal
    Ester, 3996
1,4-Dihydrobenzene, 756
2,3-Dihydro-5-carboxanilido-6-methyl-1,4-oxathiin,
    2606
2,3-Dihydro-5-carboxanilido-6-methyl-1,4-oxathiin-4,4-
    dioxide, 2610
L-Dihydro-carvon, 2105
L-Dihydrocarvone, 2105
Dihydrodiethylstilbestrol, 3569
1,4-Dihydro-3,5-diiodo-4-oxopyridine-1-acetic Acid,
    1004
2,3-Dihydro-2,2-dimethyl-7-benzofuranol
    methylcarbamate, 2643
10,11-Dihydro-N,N-dimethyl-5H-Dibenz[b,f]azepine-5-
    propanamine, 3673
1,2-Dihydro-1,5-dimethyl-4-(isopropyl)-2-phenyl-pyrazol-
    3-one, 3053
1,2-Dihydro-1,5-dimethyl-2-phenyl-3H-pyrazol-3-one,
    2232
6-(1,3-Dihydro-7-hydroxy-5-methoxy-4-methyl-1-
    oxoisobenzofuran-6-yl)-4-methyl-4- Hexanoic Acid,
    3462
2,3-Dihydroindene, 1662
2,3-Dihydroindole, 1412
2,3-Dihydro-1H-indole, 1412
3,4-Dihydro-4-keto-3-methylpteridine, 1048
3:4-Dihydro-4-keto-3-methylpteridine, 1048
1-(2,3-Dihydro-5-methoxybenzo[b]furan-2-ylmethyl)-4-
    (o-methoxyphenyl)piperazine, 3821
S-2,3-Dihydro-5-methoxy-2-oxo-1,3,4-thiadiazol-3-
    ylmethyl O,O-dimethylphosphorodithioate, 817
3,12-Dihydro-6-methoxy-3,3,12-trimethyl-7H-pyrano(2,3-
    c)acridin-7-one, 3733
1,2-Dihydro-3-methyl-benz[j]aceanthrylene, 3799
3,4-Dihydro-6-methyl-N-phenyl-2H-pyran-5-
    carboxamide, 2837
1,3-Dihydro-7-nitro-5-phenyl-2H-1,4-benzodiazepin-2-
    one, 3145
Dihydronovobiocin, 4086
S-(3,4-Dihydro-4-oxobenzo[d][1,2,3]triazin-3-ylmethyl)
    O,O-Dimethyl Phosphorodithioate, 1993
Dihydropyridazine-3,6-dione, 238
10,11-Dihydroquinine, 3762
2,8-Dihydroxyadenine, 403
2,4-Dihydroxy-6-amino-1,3,5-triazine, 134
1,4-Dihydroxyanthraquinone, 2948
2,4-Dihydroxyazobenzene, 2560
2,4-Dihydroxy-azobenzol, 2560
3,4-Dihydroxy-benzaldehyd, 1063

3,4-Dihydroxybenzaldehyde Methylene Ketal, 1318
2,5-Dihydroxy-benzeneacetic Acid, 1401
1,2-Dihydroxybenz[j]aceanthrylene, 3721
2,4-Dihydroxy-benzoesaeure, 1066
2,5-Dihydroxy-benzoesaeure, 1064
2,6-Dihydroxy-benzoesaeure, 1067
3,4-Dihydroxy-benzoesaeure, 1065
2,4-Dihydroxybenzoic Acid, 1066
2,4-Dihydroxybenzoicacid, 1066
2,5-Dihydroxybenzoic Acid, 1064
2,5-Dihydroxybenzoicacid, 1064
2,6-Dihydroxybenzoic Acid, 1067
3,4-Dihydroxybenzoic Acid, 1065
3α,12α-Dihydroxy-5β-cholanoic Acid, 3978
3α,6α-Dihydroxy-5β-cholanoic Acid, 3979
6,7-Dihydroxycoumarin 6-glucoside, 3193
2,2'-Dihydroxy-5,5'-dichlorodiphenylmethane, 2792
2,4-Dihydroxy-6,7-diethylpteridine, 1996
2,4-Dihydroxy-6:7-diethylpteridine, 1996
4'-[(N,N-Dihydroxyethyl)amino]-4-nitroazobenzene,
    3336
N,N-di(Hydroxyethyl)aniline, 2085
5,16-β-Dihydroxy-6-β-methyl-3,11-dioxo-5-α-pregn-
    17(20)-ene-cis-20-carboxylic Acid Methyl Ester, 3894
5,16-β-Dihydroxy-6-β-methyl-3,11-dioxo-5-α-pregn-
    17(20)-ene-cis-20-carboxylic Acid Methyl Ester Cycl,
    3999
3,3'-Dihydroxy-4,4'-methylenedi-2-naphthoic Acid, 3909
2,4-Dihydroxy-5-methylpyrimidine, 409
2,3-Dihydroxynaphthalene, 1899
2,6-Dihydroxynaphthalene, 1898
2,3-Dihydroxy-naphthalin, 1899
2,6-Dihydroxy-naphthalin, 1898
11,17-Dihydroxy-21-(1-oxobutoxy)-pregn-4-ene-3,20-
    dione, 3998
2,5-Dihydroxyphenylacetic Acid, 1401
L-3,4-Dihydroxyphenylalanin, 1718
DL-3-(3,4-Dihydroxyphenyl)alanine, 1717
L-1-(3,4-Dihydroxyphenyl)-2-methylaminoethanol, 1765
2,3-Dihydroxypropyl-N-(7-chloro-4-
    quinolinyl)anthranilate, 3632
7-(2,3-Dihydroxypropyl)theophylline, 2068
β,γ-Dihydroxypropyltheophylline, 3011
2,4-Dihydroxypteridine, 610
2,6-Dihydroxypteridine, 611
2,7-Dihydroxypteridine, 612
2:4-Dihydroxypteridine, 610
2:6-Dihydroxypteridine, 611
2:7-Dihydroxypteridine, 612
4,6-Dihydroxypteridine, 613
4,7-Dihydroxypteridine, 614
4:6-Dihydroxypteridine, 613
4:7-Dihydroxypteridine, 614
6,7-Dihydroxypteridine, 614
6:7-Dihydroxypteridine, 614
2,4-Dihydroxypyridine, 398
2,4-Dihydroxypyrimidine, 239
4,6-Dihydroxypyrimidine, 237
2,4-Dihydroxypyrimidine-5-carboxylic Acid, 381
D-(-)-Dihydroxysuccinic Acid, 288
3,3-Dihydroxy-2,2,5,5-tetramethyl-4-
    carbamyltetrahydrofuran, 1800
Diiod-methan, 10
1,4-Diiodobenzene, 596

p-Diiodobenzene, 596
3,5-Diiodo-4-pyridone-1-acetic Acid, 1004
3,5-Diiodo-4-pyridone-N-acetic Acid, 1004
3,5-Diiodosalicylic Acid, 977
3,5-Diiodotyrosine, 1650
3,5-Diiodo-L-tyrosine, 1649
L-3,5-Diiodotyrosine, 1649
3,5-Diiod-pyridon-(4)-N-essigsaeure, 1004
3,5-Diiod-DL-tyrosin, 1650
Diisobutylcarbinol, 1844
Di-isobutyl Ether, 1580
Diisobutyl Ketone, 1813
Diisobutyl Phthalate, 3372
Diisodecyl Phthalate, 4062
Diisooctyl Phenyl Phosphonate, 3906
Diisooctyl Phthalate, 3969
Diisopropyl, 902
Diisopropylacetamide, 1562
2,6-Diisopropyl-4-acetaminophenol, 3092
5,5-Diisopropylbarbituric Acid, 2098
Diisopropyl Carbinol, 1255
4,6-Diisopropyl-1,1-dimethyl-7-propionylindan, 3777
Di-isopropyl 1,3-Dithiolan-2-ylidinemalonate, 2720
Diisopropyl Ether, 928
O,O-Diisopropyl S-[(Ethylsulfinyl)methyl]
    Dithiophosphate, 1852
Diisopropylfluorophosphate, 904
Di-isopropylnitrosamine, 909
3,5-Diisopropylparacetamol, 3092
N,N'-Diisopropylphosphorodiamidic Fluoride, 947
Diisopropyl Phthalate, 3063
Diketopiperazine, 265
Dilactone, 2015
Dilafurane, 3413
Dilantin, 3153
Dilaphyllin, 1747
Dilitursaeure, 229
Dimelin, 3214
Dimelone, 2285
Dimercaprol, 216
2,3-Dimercapto-1-propanol, 216
Dimerin, 2113
Dimetan, 2324
Dimethametryn, 2357
Dimethindene, 3746
Dimethirimol, 2339
Dimethoate, 507
3,5-Dimethoxy-acetophenetide, 2694
1,3-Dimethoxybenzene, 1462
4-Dimethoxybenzene, 1464
m-Dimethoxybenzene, 1462
o-Dimethoxybenzene, 1463
p-Dimethoxybenzene, 1464
3,3'-Dimethoxybenzidine, 3032
9-(1,3-Dimethoxycarbonyl-2-propoxymethyl)guanine,
    2887
3,5-Dimethoxycinnamic Acid, 2249
Dimethoxyethyl Adipate, 2738
Dimethoxyethyl Phthalate, 3065
2,6-Dimethoxyphenol, 1469
N,N-Dimethylacetamide, 342
2,5-Dimethyl-4-acetaminophenol, 2031
2,6-Dimethyl-4-acetaminophenol, 2028
Dimethylaether, 100

4-Dimethylaminoantipyrine, 2883
Dimethylaminoazobenzene, 3023
p-Dimethylaminoazobenzene, 3023
4-Dimethylaminoazobenzol, 3023
Dimethylaminobenzene, 1478
4-Dimethylaminobenzoic Acid, 1713
4-(Dimethylamino)benzoic Acid, 1713
N-Dimethylamino-β-carbamyl Propionic Acid, 833
4-Dimethylamino-3,5-Dimethylphenol Methylcarbamate
    Ester, 2701
2-(Dimethylamino)-5,6-dimethyl-4-pyrimidinyl
    Dimethylcarbamate, 2337
N-(2-Dimethylaminoethyl)-N-2-pyridyl-3-thenylamine,
    3075
2-Dimethylamino-4-hyroxy-5-n-butyl-6-
    methylpyrimidine, 2339
N-(Dimethylaminomethyl)benzamide, 2058
10-(2-Dimethylamino-2-methylethyl)phenothiazine,
    3457
4-[[(4-Dimethylamino)phenyl]azo]nitrobenzene, 3009
p-Dimethylaminophenyldiantipyrylmethane, 4085
3-Dimethylaminophenyl Isothiocyanate, 1676
4-Dimethylaminophenyl Isothiocyanate, 1675
N',N'-Dimethyl-m-aminophenyl Isothiocyanate, 1676
1,3-N,N-Dimethylaminophenyl N-Methylcarbamate,
    2060
m-N,N-Dimethylaminophenyl N-Methylcarbamate, 2060
5-[3-(Dimethylamino)propyl]-10,11-dihydro-5H-
    dibenz[b,f]azepine, 3673
5-(3-Dimethylaminopropylidene)-5H-dibenzo[a,d]-10,11-
    dihydrocycloheptene, 3743
10-(2-Dimethylaminopropyl)phenothiazine, 3457
2-Dimethylaminopteridine, 1425
4-Dimethylaminopteridine, 1426
7-Dimethylaminopteridine, 1424
N,N-Dimethylaniline, 1478
9,10-Dimethylanthracene, 3284
Dimethylarsinsaeure, 106
N,N-Dimethyl-4,4'-azodian, 3034
5,5-Dimethyl Barbituric Acid, 766
5,5-Dimethylbarbituric Acid, 766
5,5-Dimethylbarbitursaeure, 766
7,12-Dimethylbenz[a]anthracene, 3729
9,10-Dimethyl-benz[a]anthracene, 3729
7,12-Dimethyl-1,2-benzanthracene, 3729
9,10-Dimethyl-1,2-benzanthracene, 3729
Dimethylbenzene, 1428
1,2-Dimethylbenzene, 1431
1,4-Dimethylbenzene, 1430
Dimethyl 1,4-Benzenedicarboxylate, 1949
3,3'-Dimethylbenzidine, 3031
2,2-Dimethyl-1,3-benzodioxol-4-ol methylcarbamate,
    2264
2,4-Dimethylbenzoic Acid, 1682
2,5-Dimethylbenzoic Acid, 1679
3,4-Dimethylbenzoic Acid, 1681
1,3-Dimethyl-3-(2-benzothiazolyl)urea, 1969
α,α-Dimethylbernsteinsaeure, 800
DL-α,α'-Dimethylbernsteinsaeure, 801
4,4'-Dimethylbiphenyl, 3005
4,4'-Dimethyl-1,1'-biphenyl, 3005
1,1-Dimethyl-2,2-bis(p-ethoxylphenyl)ethane, 3765
2,2-Dimethyl-5-bromo-5-nitro-1,3-dioxane, 788
1,2-Dimethyl-1,3-butadiene, 795

3,4-Dimethylbutadiene, 795
2,2-Dimethylbutane, 900
2,3-Dimethylbutane, 902
1,1-Dimethyl-1-butanol, 922
2,2-Dimethyl-1-butanol, 915
2,2-Dimethyl-3-butanol, 916
2,3-Dimethyl-1-butanol, 917
3,3-Dimethyl-1-butanol Carbamate, 1241
3,3-Dimethylbutanone-2, 847
3,3-Dimethyl-2-butanone, 847
Dimethylbutylacetamide, 1567
Dimethyl-tert-butylcarbinol, 1258
1-Dimethylcarbamoyl-5-fluoro-2,4(1H,3H)-pyrimidinedi-
    one, 1105
1-(N,N-Dimethylcarbamoyl)-5-fluorouracil, 1105
Dimethyl Carbate, 2285
2,4-Dimethyl-5-carboxanilidothiazole, 2591
2,5-Dimethyl-4-chloro-phenol, 1406
2,6-Dimethyl-4-chloro-phenol, 1408
3,5-Dimethyl-4-chloro-phenol-, 1407
1,1-Dimethyl-3-(p-chlorophenyl)urea, 1691
5,6-Dimethylchrysene, 3730
N,N-Dimethylcinnamide, 2258
Dimethyl O-(p-Cyanophenyl) Phosphorothioate, 1670
Dimethyl-β-cyclodextrin, 4169
1,2-trans-Dimethylcyclohexane, 1532
1,4-Dimethylcyclohexane, 1526
1-cis-2-Dimethylcyclohexane, 1528
cis-1,2-Dimethylcyclohexane, 1528
p-Dimethylcyclohexane, 1526
trans-1,2-Dimethylcyclohexane, 1532
trans-1,4-Dimethylcyclohexane, 1524
Dimethyl Cyclohexyl Oxalate, 2136
3,3-Dimethyl-D-(-)-cysteine, 501
O,O-Dimethyl S-(4,6-Diamino-1,3,5-triazinyl-2-methyl)
    Dithiophosphate, 844
Dimethyl O-2,5-Dichloro-4-iodophenyl Thiophosphate,
    1355
1,1-Dimethyl-3-(3,4-dichlorophenyl)urea, 1665
O,O-Dimethyl O-2-Dichlorovinyl Phosphate, 294
5,5-Dimethyldihydroresorcinyl N,N-Dimethylcarbamate,
    2324
2,6-Dimethyldinitrosopiperazine, 843
N,N-Dimethyldiphenylacetamide, 3323
Dimethyleneimine, 86
Dimethyl-3,6-epoxyperhydrophthalic Anhydride, 2014
1,1-Dimethylethane, 357
Dimethyl Ether, 100
1,3-Dimethyl Ether Pyrogallol, 1469
Dimethyl-S-(α-ethoxycarbonylbenzyl)
    Phosphorodithioate, 2698
1,1-Dimethylethylbenzene, 2050
O,O-Dimethyl S-(N-Ethylcarbamoylmethyl)
    Dithiophosphate, 905
Dimethylethylcarbinol, 515
Dimethyl O-(2-Ethyl-4-ethoxy-pyrimidin-6-
    yl)thionophosphate, 2114
N-(1,1-Dimethylethyl)-N'-ethyl-6-(methylthio)-1,3,5-
    Triazine-2,4-diamine, 2151
O,O-Dimethyl S-(2-Ethylmercaptoethyl) Dithiophosphate,
    943
N,N-Dimethyl Glycolamide Salicylate, 2263
Dimethylglyoxim, 313
Dimethylglyoxime, 313

2,6-Dimethyl-4-heptanol, 1844
2,6-Dimethyl-4-heptanone, 1813
5,5-Dimethyl-2,4-hexadione, 1503
2,2-DimethylHexanamide, 1567
2,3-Dimethylhexane, 1572
2:3-Dimethylhexane, 1572
2,2-Dimethyl-3-hexyne, 1495
5,5-Dimethylhydantoin, 424
5,5'-Dimethylhydantoin, 424
1,2-Dimethyl-3-hydroxy-4-pyridone, 1146
O,O-Dimethyl (1-Hydroxy-2,2,2-
    trichloroethyl)phosphonate, 312
5,5-Dimethyl-2,4-imidazolidinedione, 424
5,5-Dimethylimidazolidine-2,4-dione, 424
Dimethyl-isopropylcarbinol, 917
N1-(3,4-Dimethyl-5-isoxazolyl)-N4-acetylsulfanilamide,
    2846
Dimethylmalonic Acid, 438
Dimethyl-malonsaeure, 438
N,N-Dimethylmethanamine, 220
N,N-Dimethyl-N'-(p-methoxybenzyl)-N'-(2-
    pyrimidyl)ethylenediamine, 3365
2,2-(Dimethyl)-4-(methoxycarbamyl)-1,3-dithiolane,
    1199
2,2-(Dimethyl)-4-(methoxycarbamyl)-1,3-oxathiolane,
    1201
O,O-Dimethyl S-(5-Methoxypyronyl-2-methyl)
    Thiophosphate, 1772
O,O-Dimethyl S-2-(1-N-
    Methylcarbamoylethylmercapto)ethyl Thiophosphate,
    1577
O,O-Dimethyl S-(N-Methylcarbamoylmethyl)
    Dithiophosphate, 507
N,N-Dimethyl-α-methylcarbamoyloxyimino-α-
    (methylthio)acetamide, 1202
N',N'-Dimethyl-N-[(methylcarbamoyl)oxy]-1-
    thiooxamimidic Acid Methyl Ester, 1202
N,N-Dimethyl-N'-(4-(1-methylethyl)phenyl)urea, 2700
O,O-Dimethyl S-(N-Methyl-N-formylcarbamoylmethyl)
    Dithiophosphate, 829
O,O-Dimethyl 2-Methylmercaptoethyl Thiophosphate,
    525
1,1-Dimethyl-2-(p-methylphenyl)-2-p-
    Ethoxylphenyl)ethane, 3678
N,N-Dimethyl-N'-3-methylphenylthiourea, 2064
3,3-Dimethyl-1-(methylthio)-2-butanone O-
    ((methylamino)carbonyl)oxime, 1805
1,3-Dimethylnaphthalene, 2587
1,4-Dimethylnaphthalene, 2586
1,5-Dimethylnaphthalene, 2585
2,3-Dimethylnaphthalene, 2583
2,6-Dimethylnaphthalene, 2581
O,O-Dimethyl O-4-Nitro-3-Chlorophenyl Thiophosphate,
    1404
1,2-Dimethyl-5-nitroimidazole, 418
Dimethylnitromethane, 199
2,6-Dimethylnitrosomorpholine, 832
Dimethyl O-(4-Nitro-m-tolyl) Phosphorothioate, 1736
7α,17-Dimethyl-19-nortestosterone, 3848
trans-3,7-Dimethyl-2,6-octadienal, 2106
2,6-Dimethylocta-2,7-dien-6-ol, 2128
3,7-Dimethyl-1,6-octadien-3-ol, 2128
3,7-Dimethylocta-1,6-dien-3-ol, 2128
3,7-Dimethyl-1,6-octadien-3-yl Acetate, 2728

4,4-Dimethyloctane, 2180
3,7-Dimethyl-6-octen-1-ol, 2161
N-(4,5-Dimethyloxazol-2-yl)sulfanilamide, 2268
3-((R)-2-((1S,3S,5S)-3,5-Dimethyl-2-oxocyclohexyl)-2-
  hydroxyethyl)glutarimide, 3248
2,2-Dimethylpentane, 1242
2,3-Dimethylpentane, 1245
2,4-Dimethylpentane, 1248
3,3-Dimethylpentane, 1243
2,2-Dimethylpentanol-3, 1263
2,2-Dimethyl-3-pentanol, 1263
2,3-Dimethylpentanol-2, 1257
2,3-Dimethyl-2-pentanol, 1257
2,3-Dimethylpentanol-3, 1251
2,3-Dimethyl-3-pentanol, 1251
2,4-Dimethylpentanol-2, 1256
2,4-Dimethyl-2-pentanol, 1256
2,4-Dimethylpentanol-3, 1255
2,4-Dimethyl-3-pentanol, 1255
2,4-Dimethylpentanone-3, 1214
2,4-Dimethyl-3-pentanone, 1214
4,7-Dimethyl-1,10-phenanthroline, 3003
4,7-Dimethyl-o-phenanthroline, 3003
2,3-Dimethylphenol, 1458
2,4-Dimethylphenol, 1452
2,4-Dimethyl-phenol-, 1452
2,5-Dimethylphenol, 1457
2,5-Dimethyl-phenol-, 1457
2,6-Dimethylphenol, 1456
3,4-Dimethylphenol, 1455
3,5-Dimethylphenol, 1454
2',3'-Dimethyl-N-phenyl-anthranilic Acid, 3179
N,N-Dimethyl-α-phenylbenzeneacetamide, 3323
p,p-Dimethylphenylbutazone, 3645
3,4-Dimethylphenyl Methylcarbamate, 2027
3,4-Dimethylphenyl N-Methylcarbamate, 2027
N,N-Dimethyl-3-phenyl-2-propenamide, 2258
2,3-Dimethyl-1-phenyl-3-pyrazolin-5-one, 2232
o,p-Dimethylphenylthioethylphenylbutazone, 3988
N,N-Dimethyl-N-phenylurea, 1737
Dimethylphosphate of α-Methylbenzyl-3-hydroxy-cis-
  crotonate, 3078
Dimethyl Phthalate, 1952
trans-2,5-Dimethyl-piperazin, 906
trans-2,5-Dimethylpiperazine, 906
2,2-Dimethylpropane, 504
Dimethyl-propanedioic Acid, 438
3,3-Dimethylpropene, 451
Dimethyl-i-propylcarbinol, 917
Dimethyl-n-propylcarbinol, 922
N-(1,2-Dimethylpropyl)-N'-ethyl-6-(methylthio)-1,3,5-
  triazine-2,4-diamine, 2357
p-(1,1-Dimethylpropyl)phenol, 2316
p-(α,α-Dimethylpropyl)phenol, 2316
3,7-Dimethyl-1-propyl-xanthine, 2065
3,7-Dimethyl-7-propyl-xanthine, 2066
6,7-Dimethylpteridine, 1368
6:7-Dimethylpteridine, 1368
4,6-Dimethyl-2-pyranone, 1123
4,6-Dimethyl-2H-pyran-2-one, 1123
2,4-Dimethyl-pyridin, 1131
2,5-Dimethyl-pyridin, 1141
2,6-Dimethyl-pyridin, 1142
2,3-Dimethylpyridine, 1140

2,4-Dimethylpyridine, 1131
2,5-Dimethylpyridine, 1141
2,6-Dimethylpyridine, 1142
3,4-Dimethylpyridine, 1138
3,5-Dimethylpyridine, 1137
N,N-Dimethyl-N',2-pyridinyl-N'-(2-thienylmethyl)-1,2-
  ethanediamine, 3074
N,N-Dimethyl-N'-(2-pyridyl)-N'-(5-chloro-2-
  thenyl)ethylenediamine, 3051
1,3-Dimethyl-2,4-pyrimidinedione, 761
2,4-Dimethyl-α-pyrone, 1123
4,6-Dimethyl-1,2-pyrone, 1123
4,6-Dimethyl-α-pyrone, 1123
Dimethylpyruvic Acid, 434
2,4-Dimethylquinoline, 2216
2,7-Dimethylquinoline, 2215
Dimethylresorcinol, 1462
7,8-Dimethyl-10-ribitylisoalloxazine, 3461
2,2-Dimethylsuccinic Acid, 800
DL-2,3-Dimethylsuccinic Acid, 801
sym-Dimethylsuccinic Acid, 798
p-(Dimethylsulfamoyl)aniline, 1484
N1-Dimethylsulfanilamide, 1484
3,4-Dimethyl-5-sulfanilamidoisoxazole, 2267
Dimethyl Sulfate, 103
Dimethyl TerePhthalate, 1949
7α-17-Dimethyltestosterone, 3847
Dimethyl Tetrachloroterephthalate, 1869
O,S-Dimethyl Tetrachlorothioterephthalate, 1868
2,2-Dimethyltetrahydrofuran-3-ol, 861
3,5-Dimethyl-1,2,3,5-tetrahydro-1,3,5-thiadiazinethione-2,
  458
5,6-Dimethylthiouracil, 760
5,6-Dimethyl-2-thiouracil, 760
Dimethyl Trichlorophenylthiophosphate, 1359
Dimethyl 3,5,6-Trichloro-2-pyridinyl Phosphate, 1082
N-(2,4-Dimethyl-5-
  ((((trifluoromethyl)sulfonyl)amino)phenyl)acetamide,
  2255
1,1-Dimethyl-3-(α,α,α-trifluoro-m-tolyl)urea, 1958
3,7-Dimethyl-9-(2,6,6-trimethyl-1-cyclohexen-1-yl)-
  2,4,6,8-nonatetraenoic Acid, 3772
1,3-Dimethyluracil, 761
N,N'-Dimethyluracil, 761
N,N-1,3-Dimethyluracil, 761
N1,N3-Dimethyluracil, 761
m-(3,3-Dimethylureido)phenyl-tert-butylcarbamate, 3095
1,3-Dimethylxanthine, 1113
Dimethylxanthogen Disulfide, 278
Dimetindene, 3746
Dimetridazole, 418
Dimpylate, 2731
Dinaphthanthracene, 3864
Dinex, 2626
Dinicotinic Acid, 1012
Dinitramine, 2256
2,4-Dinitroaminobenzene, 665
2,4-Dinitro-4'-aminodiphenylamine, 2563
2,4-Dinitroaniline, 665
2,6-Dinitroaniline, 664
p-(2,4-Dinitroanilino), 2551
Dinitroanisole, 1039
2,4-Dinitroanisole, 1039
2,4-Dinitrobenzenamine, 665

2,6-Dinitrobenzenamine, 664
1,2-Dinitrobenzene, 597
1,3-Dinitrobenzene, 598
1,4-Dinitrobenzene, 599
m-Dinitrobenzene, 598
o-Dinitrobenzene, 597
p-Dinitrobenzene, 599
2,4-Dinitro-1,3-benzenediol, 603
4,6-Dinitro-1,3-benzenediol, 604
2,4-Dinitrobenzoesaeure, 981
2,6-Dinitrobenzoesaeure, 982
3,4-Dinitrobenzoesaeure, 980
3,5-Dinitrobenzoesaeure, 979
2,4-Dinitrobenzoic Acid, 981
2,6-Dinitrobenzoic Acid, 982
3,4-Dinitrobenzoic Acid, 980
3,5-Dinitrobenzoic Acid, 979
2,4-Dinitro-6-sec-butylphenol, 1992
4,6-Dinitro-2-S-butylphenol, 1992
2,4-Dinitro-1-chlorobenzene, 553
2,4-Dinitrochloronaphthalene, 1860
2,4-Dinitro-1-chloronaphthalene, 1860
Dinitrocresol, 1038
Dinitro-o-cresol, 1038
2,4-Dinitro-6-cyclohexylphenol, 2626
4,6-Dinitro-2-cyclohexylphenol, 2626
2,4-Dinitrodiphenylamin, 2550
2,4-Dinitrodiphenylamine, 2550
2,6-Dinitro-N,N-dipropylcumidene, 3249
2,6-Dinitro-N,N-dipropylcumidine, 3249
1,2-Dinitroglycerol, 168
2,4-Dinitro-4'-hydroxydiphenylamine, 2551
2,4-Dinitro-6-methylphenol, 1038
1,5-Dinitronaphthalene, 1877
1,8-Dinitronaphthalene, 1876
1,5-Dinitronaphthalin, 1877
1,8-Dinitronaphthalin, 1876
2,4-Dinitro-1-naphthyl Chloride, 1860
3,5-Dinitro-N4,N4-dipropylsulfanilamide, 2708
2,4-Dinitrophenol, 601
2,6-Dinitrophenol, 602
3,5-Dinitrophenol, 600
α-Dinitrophenol, 601
β-Dinitrophenol, 602
2,4-Dinitroresorcinol, 603
4,6-Dinitroresorcinol, 604
Dinitrosopentamethylenetetramine, 459
Dinitrosopiperazine, 321
N,N'-Dinitrosopiperazine, 321
3,7-Dinitroso-1,3,5,7-tetraazabicyclo[3.3.1]nonane, 459
2,4-Dinitrotoluene, 1037
2,4-Dinitro-toluol, 1037
Dinoprostone, 3785
Dinoseb, 1992
Dinoseb Acetate, 2627
Dioctyl Adipate, 3907
Dioctyl Azelate, 4004
Dioctyl Ester, 3790
Dioctyl Isophthalate, 3970
Dioctyl Maleate, 3790
Dioctyl Phenyl Phosphonate, 3905
Di-n-Octyl Phenylphosphonate, 3905
Di-n-Octyl Phthalate, 3971
Di-sec-octyl Phthalate, 3967

Dioctyl Sebacate, 4026
Diodon, 1004
Diosgenin, 4044
Dioxabenzofos, 1427
Dioxabenzophos, 1427
Dioxacarb, 2262
1,4-Dioxan, 332
Dioxane, 332
1,4-Dioxane, 332
m-Dioxane, 5-Bromo-2,2-dimethyl-5-nitro-, 788
Dioxane, 5-Bromo-2-methyl-5-nitro-, 423
1,3-Dioxane, 5-Bromo-5-nitro-, 263
m-Dioxane, 5-Bromo-5-nitro-2-phenol-, 1931
1,3-Dioxane, 5-Bromo-5-nitro-2-phenyl-, 1930
m-Dioxane, 5-Bromo-5-nitro-2-phenyl-, 1930
3,6-Dioxaoctane, 933
1,1-Dioxide-1,2-Benzisothiazol-3-(2H)-one, 1011
9,10-Dioxoanthracene, 2946
1,3-Dioxolane-4-methanol, 2-(2-Butoxyethyl)-2-methyl, 2370
1,3-Dioxolane-4-methanol, 2-Butyl-2-methyl, 1833
1,3-Dioxolane-4-methanol, 2-Heptyl-2-methyl, 2755
1,3-Dioxolane-4-methanol, 2-[2-(Heptyloxy)ethyl]-2-methyl, 3117
1,3-Dioxolane-4-methanol, 2-Hexyl-2-methyl, 2368
1,3-Dioxolane-4-methanol, 2-[2-(Hexyloxy)ethyl]-2-methyl, 2939
1,3-Dioxolane-4-methanol, 2-Methyl-2-pentyl, 2175
1,3-Dioxolane-4-methanol, 2-Methyl-2-[2-(pentyloxy)ethyl], 2756
2-(1,3-Dioxolan-2-yl)phenyl Methylcarbamate, 2262
2-(1,3-Dioxolan-2-yl)-phenyl N-methylcarbamate, 2262
[1,3]Dioxolo[4,5-h]-1,3-dioxolo[7,8][2]benzopyrano[3,4-a][3]benzazepine, 5β,6,7,8,13β,15-Hexahydro-15-methoxy-6-methyl-, (5bR,13bR,15S), 3809
1,3-Dioxophthalan, 1278
3,11-Dioxo-4,17(20)-cis-pregnadien-21-oic Acid Methyl Ester, 3754
3,11-Dioxo-4,17(20)-cis-pregnadien-20-oic Acid Methyl Ester 3-Oxime, 3759
2,6-Dioxopurine, 386
2,8-Dioxyadenine, 403
Dioxyethylaminoazobenzene, 3398
Dipane, 3487
Dipentum, 2972
Dipentyl Phthalate, 3591
Diphenamid, 3323
Diphenatrile, 2984
Diphenic Acid, 2977
Diphenyl, 2554
Diphenylacetic Acid, 2995
DL-1,2-Diphenyl-aethanol, 3014
Diphenylamine, 2574
5,5-Diphenylbarbituric Acid, 3275
1,2-Diphenyl Benzene, 3521
1,3-Diphenyl Benzene, 3522
1,4-Diphenyl Benzene, 3523
N,N'-Diphenylbenzidine, 3943
1,2-Diphenyl-4-butyl-3,5-dioxopyrazolidine, 3644
Diphenyldiazene, 2557
Diphenyl Diimide, 2557
(+)-2-(2,2-Diphenyl-1,3-dioxolan-4-yl)piperidine, 3744
Diphenyleneglycollic Acid, 2975
Diphenylenemethane, 2787

Diphenylene Oxide, 2530
Diphenylene Sulfide, 2533
Diphenylenimine, 2545
Diphenyl-essigsaeure, 2995
1,2-Diphenylethane, 3006
DL-1,2-Diphenylethanol, 3014
1,1-Diphenylethene, 2989
1,2-Diphenylethene, 2990
Diphenyl Ether, 2564
1,1-Diphenylethylene, 2989
trans-Diphenylethylene, 2990
trans-1,2-Diphenylethylene, 2990
trans-α, β-Diphenylethylene, 2990
Diphenylglycolic Acid, 2998
5,5-Diphenylhydantoin, 3153
2,2-Diphenyl-2-hydroxyacetic Acid, 2998
5,5-Diphenyl-2,4-imidazolidinedione, 3153
Diphenylmethane, 2815
Diphenylmethanol, 2818
Diphenylmethanone, 2800
2,4-Diphenyl-4-methyl-2-pentene, 3555
Diphenyl Methyl Phosphate, 2826
Diphenylmorpholidophosphate, 3330
Diphenylnitrosamine, 2558
1,2-Diphenyl-4-(2-phenylthioethyl)-3,5-
    pyrazolidinedione, 3911
Diphenyl Phosphate, 2580
Diphenyl Phthalate, 3727
3,3-Diphenyl-phthalid, 3725
3,3-Diphenylphthalide, 3725
N-(3,3-Diphenylpropyl)-α-methylphenylethylamine, 3948
1,3-Diphenyltriazene, 2577
Diphenylurea, 2816
N,N'-Diphenylurea, 2816
Diphonate, 2087
1,2-Di(4-piperazine-2,6-dione)propane, 2315
9-(1,3-Dipivaloate-2-propoxymethyl)guanine, 3697
1,5-Dipivaloyloxymethyl Allopurinol, 3490
2,5-Dipivaloyloxymethyl Allopurinol, 3489
Diplin, 2771
Di-2-propenyl Phthalate, 3019
Dipropetryn, 2356
9-(1,3-Dipropionate-2-propoxymethyl)guanine, 3239
Dipropylaceturethane, 2353
Dipropylaether, 921
Dipropylamine, 942
n-Dipropylamine, 942
1,3-Dipropyl-8-(2-amino-4-chlorophenyl)xanthine,
    3454
5,5-Dipropylbarbituric Acid, 2097
5,5-Dipropylbarbitursaeure, 2097
Dipropyl Carbinol, 1262
N,N-Dipropylcinnamamide, 3222
Dipropyl Dixanthogen, 1506
Dipropyl Ether, 921
Dipropylether, 921
Dipropyl Ketone, 1215
O,O-Dipropyl O-4-Methylthiophenyl Phosphate, 2922
Dipropylnitrosamine, 908
Di-n-propylnitrosamine, 908
2,4-Dipropylphenol, 2710
2,6-Dipropylphenol, 2716
Dipropyl Phthalate, 3062
Di-n-propyl Phthalate, 3062

N,N-Dipropylpropanamine, 1849
N,N-Dipropyl-1-propanamine, 1849
Dipropyl Thioperoxydicarbonate, 1506
Dipropyl Xanthogen Disulfide, 1506
6H-Dipyrido[3,2-b:2',3'-e][1,4]diazepine-6-thione, 11-
    Ethyl-5,11-dihydro-5-methyl, 3012
6H-Dipyrido[3,2-b:2',3'-e][1,4]diazepine-6-nitrile, 11-
    Cyclopropyl-5,11-dihydro-4-methyl, 3307
6H-Dipyrido[3,2-b:2',3'-e][1,4]diazepin-6-one, 2-Chloro-
    11-ethyl-5,11-dihydro-, 2805
6H-Dipyrido[3,2-b:2',3'-e][1,4]diazepin-6-one, 2-Chloro-
    11-ethyl-5,11-dihydro-5-methyl-, 3001
6H-Dipyrido[3,2-b:2',3'-e][1,4]diazepin-6-one, 11-
    Cyclobutyl-5,11-dihydro-5-methyl-, 3315
6H-Dipyrido[3,2-b:2'3'-e][1,4]diazepin-6-one, 11-
    Cyclopropyl-5,11-dihydro-2,4-dimethyl-, 3316
6H-Dipyrido[3,2-b:2',3'-e][1,4]diazepin-6-one, 11-
    cyclopropyl-5,11-dihydro-4-methyl, 3174
6H-Dipyrido[3,2-b:2',3'-e][1,4]diazepin-6-one, 5,11-
    Diethyl-5,11-dihydro-, 3187
6H-Dipyrido[3,2-b:2',3'-e][1,4]diazepin-6-one, 5,11-
    Dihydro-5-methyl-11-propyl-, 3185
6H-Dipyrido[3,2-b:2',3'-e][1,4]diazepin-6-one, 2-
    Dimethylamino)-11-ethyl-5,11-dihydro-4-methyl-,
    3345
6H-Dipyrido[3,2-b:2',3'-e][1,4]diazepin-6-one, 11-(1,1-
    Dimethylethyl)-5,11-dihydro-5-methyl-, 3332
6H-Dipyrido[3,2-b:2',3'-e][1,4]diazepin-6-one, 11-Ethyl
    5,11-dihydro-2,4-dimethyl-, 3186
6H-Dipyrido[3,2-b:2',3'-e][1,4]diazepin-6-one, 11-Ethyl-
    5,11-dihydro-2-[(2 hydroxyethyl)methylamino, 3346
6H-Dipyrido[3,2-b:2',3'-e][1,4]diazepin-6-one, 11-Ethyl-
    5,11-dihydro-2-methoxy-4-methyl-, 3188
6H-Dipyrido[3,2-b:2',3'-e][1,4]diazepin-6-one, 11-Ethyl-
    5,11-dihydro-5-methyl, 3008
6H-Dipyrido[3,2-b:2',3'-e][1,4]diazepin-6-one, 11-Ethyl-
    5,11-dihydro-2-methyl-4-(trifluoromethyl)-, 3161
2,2'-Dipyridyl, 1894
α,α'-Dipyridyl, 1894
γ,γ'-Dipyridyl, 1893
Direx, 1603
Dis. A. 5, 3333
Dis. A. 1, 3453
Dis. A. 10, 3438
Dis. A. 12, 3327
Dis. A. 13, 3459
Dis. A. 14, 3328
Dis. A. 2, 3466
Dis. A. 3, 2549
Dis. A. 6, 3660
Dis. A. 7, 3010
Dis. A. 8, 3437
Dis. A. 9, 3568
Disflamoll DPO, 3769
Disflamoll TOF, 3986
Disperse Black 1, 3314
Disperse Black 3, 3034
Disperse Blue 19, 3724
Disperse Brightner, 3508
Disperse Orange 1, 3526
Disperse Red 1, 3335
Disperse Red 15, 2964
Disperse Red 19, 3336
Disperse Red 3, 3281

Disperse Red 66, 3281
Disperse Violet 4, 3154
Disperse Violet 4K, 3453
Disperse Yellow 23, 3525
Disperse Yellow 54, 3511
Distokal, 111
Distopan, 111
Distopin, 111
2,5-Disulfanilamidopyridine, 3308
2,5-Disulfanilamidopyrimidine, 3317
Disulfiram, 2158
1,8-Disulfonic Acid Anthraquinone, 2951
Disulfoton, 1590
Disulfoton Dioxide, 1596
Disulfoton Sulfone, 1596
Disyston, 1590
Disyston Sulfone, 1596
DIT, 1649
Ditalimfos, 2622
1,4-Dithiaanthraquinone-2,3-dinitrile, 2940
Dithianon, 2940
3,3'-Dithiobis(2-aminopropanoic Acid), 835
D-(+)-3,3'-Dithiobis(2-aminopropanoic Acid), 837
Dithiodiantipyrinylmethane, 3918
N,N'-(Dithiodicarbonothioyl)bis(N-methylmethanamine), 841
1,3-Dithiolane-4-methanol, 2,2-Dimethyl-, Carbamate, 1199
Ditridecyl Phthalate, 4109
Diucardin, 1361
Diulo, 3310
Diuresal, 1030
Diuretic C, 1674
Diurex, 3177
Diuron, 1665
Divinyl Ether, 270
Dixanthogen, 796
Djenkoic Acid, 1211
Djenkolsaeure, 1211
p-DMAPhDAM, 4085
DMASA, 833
DMDNP, 843
DMHP, 1146
DMNM, 832
DMPA, 2054
DNOC, 1038
Dodecanamide, N-Hydroxy-2,2-dimethyl, 3123
Dodecane, 2757
N-Dodecane, 2757
Dodecane(S-(carboxymethyl)-L-cystein)), 4163
1,12-Dodecanedicarboxylic Acid, 3112
Dodecanedioc Acid, 2736
Dodecanoic Acid, 2753
Dodecanoic Acid Methyl Ester, 2935
Dodecanol, 2758
Dodecyl Alcohol, 2758
Dodecylamine (Tetrahydrate), 2760
Dodecyl p-Aminobenzoate, 3707
Dodecyl p-Hydroxybenzoate, 3705
Dodecyl Lactate, 3268
p-(Dodecyloxy)benzoic Acid, 3705
Dolantin, 3223
Dolcymine, 2053
Donalgin, 2781

Donco-163, 568
Dopa, 1717
DL-Dopa, 1717
DOPP, 3905
Dorantamin, 3487
Doriden, 2836
Dormigene, 811
Dormitiv, 1741
Dosaflo, 2019
Dosanex, 2019
Dotan, 506
DOWCO 179, 1695
DOWCO 180, 1696
Dowco 213, 3608
DOWCO 217, 1082
Dowicide 2, 567
Dowicide 25, 563
Dowicide 7, 532
Doxepin, 3657
Doxy-caps, 3880
Doxycycline (Monohydrate), 3880
Doxylamine Ethanamine, 3686
Doxylin, 3880
DPA, 1646
DPNA, 908
DPX 1410, 1202
DPX 3674, 2726
DPX 6774, 2700
DPX-D732, 1762
DRC 1201, 2604
1,1-Drichloro-1-methyl-2,2-bis(p-methoxylphenyl)ethane, 3207
Drihydroxyestrin, 3580
Dronabinol, 3837
Droperidol, 3874
Drotaverine, 3954
DS-15647, 1805
Dual, 3240
Duboisine, 3484
Dulcin, 1738
Dulcit, 936
Dulcitol, 936
Duodecane, 2757
DuP 860, 3409
DuP 991, 3417
Durene, 2049
Durol, 2049
Dursban Oxygen Analog, 1696
2,2-Dwumetylopentan, 1242
2,3-Dwumetylopentan, 1245
2,4-Dwumetylopentan, 1248
3,3-Dwumetylopentan, 1243
Dyamid, 3323
Dybar, 1736
Dye II, 3009
Dye III, 3334
Dye IV, 3335
Dye V, 3336
Dye VI, 3526
Dye VII, 3659
Dyfonate®, 2087
Dymelor, 3214
Dynacin, 3924
Dyphylline, 2068

Dyrene, 1603
E-101, 3461
E-48, 2853
EB, 1432
EBD, 1565
Ebufac, 2898
Ecgonine, 1782
Econazole Nitrate, 3530
Ectiban, 3806
EDAM, 3989
E.E.S, 4122
1-Eicosene, 3792
n-Eicosene, 3792
EID, 1568
Ekalux, 2654
Ekamet G, 2114
Ekamet ULV, 2114
L-Ekgonin, 1782
Ekko, 3153
Ektasolve EB Acetate, 1552
Elcoril, 3067
Electrocortin, 3832
Elocron, 2262
Elsan, 2698
Embark, 2255
Embonic Acid, 3909
Emetan, 6',7',10,11-Tetramethoxy-, 4076
Emetine, 4076
Emulsion 212, 1781
Enaven, 3534
ENC, 149
Endomid, 3502
α-Endosulfan, 1606
β-Endosulfan, 1607
Endosulfan A, 1606
Endosulfan I, 1606
Endosulfan II, 1607
Endothal, 1473
Endothall, 1473
Endothion, 1772
Endrin, 2523
Enide, 3323
ENS, 2611
4-ENS, 2611
ENT 26316, 2752
ENT 396, 2739
Entex, 2088
Enzactin, 1779
Eorthcicle, 952
EP 201, 3390
Ephedrine, 2083
(1R,2S)-(-)-Ephedrine, 2083
Epiandrosterone, 3702
Epichloridrina, 141
Epichlorohydrin, 141
Epihydroxyetioallocholan-17-one, 3703
Epinephrine, 1765
Epipen, 1765
Epitiostanol, 3699
Epitol, 3152
EPN, 3007
1,4-Epoxybutane, 326
1,2-Epoxyethylbenzene, 1375
1-Epoxyethyl-3,4-Epoxycyclohexane, 1487

2,3-Epoxy-4,5,6,7,8,8-hexachloro-3α,4,7,7α-tetrahydro-
   4,7-methanoindene, 1871
3,4-Epoxy-6-Methylcyclohexylmethyl-3,4-Epoxy-6-
   Methylcyclohexane Carboxylate, 3390
Eptam, 1835
Eptapur, 2602
EPTC, 1835
Equilenin, 3544
Equilin, 3558
Equiphen, 3644
ER-5461, 3030
Eriodictyol, 3139
Erypar, 4122
Erythrit, 369
Erythritol, 369
L-Erythro-2-(methylamino)-1-phenylpropan-1-ol, 2083
Erythromycin, 4122
Erythromycin Anhydrate, 4122
Erythromycin (Dihydrate), 4123
Erythromycin Estolate, 4151
Erythromycin Ethyl Succinate, 4133
Erythromycin Lactobionate, 4144
Esculin, 3193
Esestrolo, 3569
Esperal, 2158
Espotabs, 3726
Essigsaure, 81
Essigsaureaethyl Ester, 333
Essigsaure-anhydrid, 279
Essigsaure-n-butyl Ester, 863
Essigsacure N chloramid, 72
Essigsaurefluorid, 65
Essigsaureisoamyl Ester, 1217
Essigsaureisobutyl Ester, 860
Essigsaureisopropyl Ester, 474
Essigsaurepropyl Ester, 477
Essigsaures Methyl, 176
Estone, 3752
Estradiol, 3578
17-α-Estradiol, 3579
17-β-Estradiol, 3578
Estradiol-17β, 3578
α-Estradiol, 3579
Estradiol Benzoate, 3990
7β-Estradiol-3-benzoate, 3990
Estradiol 3-[bis(2-Chloroethyl)carbamate], 3928
Estradiol Monobenzoate, 3990
Estragole, 2008
Estramustine, 3928
1,3,5-10,6,8-Estrapentaen-3-ol-17-one, 3544
1,3,5(10),7-Estratetraen-3-ol-17-one, 3558
1,3,5(10)-Estratrien-3-ol-17-one, 3570
Estra-1,3,5(10)-Trien-17-one, 3-Hydroxy-, 3570
Estr-4-en-3-one, 17-Hydroxy-, (17β), 3590
Estriol, 3580
Estrone, 3570
Ethalfluralin, 2828
1,2-Ethandiol, 101
Ethane, 99
1,2-Ethanediamine, N,N-Dimethyl-N'-2-pyridinyl-N'-(3-
   thienylmethyl)-, 3075
1,2-Ethanediamine, N-[(4-Methoxyphenyl)methyl]-N',N'-
   dimethyl-N-2-pyridinyl-, 3487
Ethanedioic Acid, Dihydrate, 59

1,2-Ethanediol, Dipropanoate, 1511
Ethane, Isocyano-, 145
Ethane Pentachloride, 49
Ethane, 1,1,1,2-Tetrachloro-, 57
1,1,2,2-Ethanetetrol, Tetraacetate, 2081
Ethanethioamide, 92
Ethanol, 2-[[4,6-bis(Dimethylamino)-1,3,5-triazin-2-
    yl]amino]-, 1810
Ethanol, 2-[[4,6-bis(Dimethylamino)-s-triazin-2-
    yl]amino]-, 1810
Ethanol, 2-Chloro-, (2,4-Dichlorophenoxy)acetate, 1904
Ethanol, 2,2'-[[4-[(2-Chloro-4,6-dinitrophenyl)azo]-3-
    methylphenyl]imino]bis-, 3437
Ethanol, 2,2'-[[4-[(2-Chloro-4-nitrophenyl)azo]-3-
    methylphenyl]imino]bis-, 3445
Ethanol, 2,2'-[[3-Chloro-4-[(4-
    nitrophenyl)azo]phenyl]imino]bis-, 3322
Ethanol, 2-[[4-[(2,6-Dichloro-4-
    nitrophenyl)azo]phenyl]methylamino], 3169
Ethanol, 2,2'-[4-(2,6-Dichloro-4-nitrophenylazo)-m-
    tolylimino]di-, 3438
Ethanol, 2-(2,4-Dicholrophenoxy)-, Benzoate, 3149
Ethanol, 2-[[4-[(2,4-
    Dinitrophenyl)azo]phenyl]ethylamino]-, 3327
Ethanol, 2,2'-[4-(2,4-Dinitrophenylazo)-m-tolylimino]di-,
    3453
Ethanol, 2,2'-[1,2-Ethanediylbis(oxy)]bis-, Dipropanoate,
    2740
Ethanol, 2,2'-[[4-[(2-Methoxy-4-nitrophenyl)azo]-3-
    methylphenyl]imino]bis-, 3568
Ethanol, 2,2'-[[4-[(2-Methoxy-4-
    nitrophenyl)azo]phenyl]imino]bis, 3459
Ethanol, 2,2'-[p-(2-Methoxy-4-
    nitrophenylazo)phenylimino]di-, 3459
Ethanol, 2,2'-[[3-Methyl-4-[(4-
    nitrophenyl)azo]phenyl]imino]bis-, 3458
Ethanol, 2,2'-[[3-Methyl-4-(phenylazo)phenyl]imino]bis-,
    3466
Ethanol, 2-(4-Nonylphenoxy)-, 3503
Ethanol, 2-[2-[2-(4-Nonylphenoxy)ethoxy]ethoxy]-,
    3855
Ethanol, 2-[2-[2-[2-(4-
    Octylphenoxy)ethoxy]ethoxy]ethoxy]-, 3904
Ethanol, 2,2'-Oxybis-, Dipropanoate, 2141
Ethanone, 1-(3-Ethyl-5,6,7,8-tetrahydro-5,5,8,8-
    tetramethyl-2-naphthyl)-, 3589
Ethanone, 1,1'-(1,4-Phenylene)bis-, 1945
Ethanox, 1858
Ethene, 69
Ethenylbenzene, 1351
Ethide, 62
Ethinyl Estradiol, 3752
Ethinyloestradiol, 3752
Ethinyl Trichloride, 46
Ethiofencarb, 2291
Ethion, 1858
Ethiophencarp, 2291
Ethirimol, 2338
Ethisterone, 3829
Ethoate-methyl, 905
Ethofumesate, 2903
Ethohexadiol, 1585
Ethonyphenyl Tartramic Acid, 2653
Ethoprop, 1589

Ethoprophos, 1589
Ethosuximide, 1170
p-Ethoxyacetanilide, 2025
Ethoxybenzene, 1453
4-Ethoxybenzoic Acid-2-(diethylamino)ethyl Ester, 3246
6-Ethoxy-2-benzothiazolesulfonamide, 1674
Ethoxycaffeine, 2067
Ethoxycarbonylmethyl Ethyl Phthalate, 3041
Ethoxycyclopropane, 468
2-Ethoxy-N-[2-(diethyl-amino)ethyl]-4-quinoline
    Carboxamide, 3584
2-Ethoxy-2,3-dihydro-3,3-dimethyl-5-benzofuranyl
    Methanesulfonate, 2903
2-(2-Ethoxyethoxy)ethanol, 935
2-Ethoxyethyl Acetate, 866
1-Ethoxyethyl-4-allopurinyl Ether, 1743
2-Ethoxy-2-oxoethyl Methyl Ester, 2835
2-Ethoxyphenol, 1465
3-Ethoxyphenol, 1466
m-Ethoxy Phenol, 1466
o-Ethoxyphenol, 1465
p-Ethoxyphenol, 1468
4-Ethoxy-7-phenyl-3,5-dioxa-6-aza-4-phosphaoct-6-ene-
    8-nitrile 4-sulfide, 2655
3-Ethoxyphenyl Isothiocyanate, 1653
4-Ethoxyphenyl Isothiocyanate, 1652
m-Ethoxyphenyl Isothiocyanate, 1653
p-Ethoxyphenyl Isothiocyanate, 1652
1,2-Ethoxyphenyl N-Methylcarbamate, 2034
1,3-Ethoxyphenyl N-Methylcarbamate, 2033
m-Ethoxyphenyl N-Methylcarbamate, 2033
o-Ethoxyphenyl N-Methylcarbamate, 2034
(4-Ethoxyphenyl)urea, 1738
3-Ethoxypropionic Acid Methyl Ester, 867
Ethoxzolamide, 1674
Ethyl Acetaminophen, 2265
Ethyl Acetate, 333
Ethyl Acetoacetate, 797
Ethyl 2-Acetoxypropionate, 1194
Ethyl α-Acetoxypropionate, 1194
Ethylacetylene, 261
Ethyl 2-(Acetyloxy)propanoate, 1194
Ethyl Acetylthiodiazole, 1099
Ethyl N-Acetyl-L-tyrosinate, 2878
Ethyl Acrylate, 432
5-Ethyl-5-allylbarbituric Acid, 1741
Ethylamine, 104
Ethyl p-Aminobenzoate, 1708
Ethyl p-Aminobenzoic Acid, 1708
4-Ethylaminobenzoic Acid-2-(diethyl-amino)ethyl Ester,
    3257
2-(Ethylamino)butane, 941
2-(Ethylamino)ethyl 4-Aminobenzoate, 2304
4-(Ethylamino)-2-methoxy-6-(tert-butylamino)-s-triazine,
    2146
N-(Ethylaminomethyl)benzamide, 2059
4-(Ethylamino)-6-[(1-methylethyl)amino]-1,3,5-triazin-
    2(1H)-one, 1515
2-(N-Ethylanilino)ethanol, 2084
Ethylarsine, 107
10-Ethylbenz[a]anthracene, 3728
10-Ethyl-1,2-benzanthracene, 3728
Ethylbenzene, 1432
Ethyl Benzenesulfonate, 1470

Ethyl Benzoate, 1683
Ethyl p-Benzoate, 1683
Ethylbenzol, 1432
Ethyl Benzoyl Benzoate, 3319
Ethyl N-Benzoyl-N-(3,4-dichlorophenyl)-2-
  aminopropionate, 3534
Ethyl Biscoumacetate, 3866
Ethyl Bromide, 83
2-Ethyl-5-bromo-5-nitro-1,3-dioxane, 789
Ethyl DL-2-Bromopropionate, 440
Ethyl DL-α-Bromopropionate, 440
α-Ethyl-β-bromo-propionic Ureide, 439
N-Ethylbutan-1-amine, 939
Ethyl Butanoate, 856
2-Ethyl-butanoic Acid, 857
2-Ethylbutanol, 918
2-Ethyl-1-butanol, 918
2-Ethyl-4-butanol, 920, 932
Ethylbutylacetamide, 1565
2-Ethyl Butyl Acetate, 1549
Ethylbutylamine, 939
N-Ethylbutylamine, 939
N-Ethyl-n-butylamine, 939
N Ethyl-sec-butylamine, 941
5-Ethyl-5-n-butylbarbituric Acid, 2096
Ethyl-n-butylnitrosamine, 910
Ethyl Butyrate, 856
Ethylbutyric Acid, 857
2-Ethylbutyric Acid, 857
Ethyl Carbamate, 202
1-Ethylcarbamoyl-5-fluoro-2,4(1H,3H)-pyrimidinedi-one,
  1106
1-Ethylcarbamoyl 5-fluorouracil, 1106
Ethyl Carbonate, 481
Ethyl Chloride, 84
N-Ethylcinnamamide, 2257
Ethyl Cinnamate, 2247
Ethyl (E)-Cinnamate, 2247
4-Ethyl-m-cresol, 1751
Ethyl Cyanoacetate, 415
Ethyl Cyclohexane, 1530
Ethyl Cyclopropyl Ether, 468
Ethyldiantipyrylmethane, 3989
Ethyl Dibutyl Phosphonate, 2182
Ethyl 4,4'-Dichlorobenzilate, 3287
Ethyl (2,4-Dichlorophenoxy)acetate, 1936
Ethyl 2,2-Diethylmalnurate, 2122
Ethyl-2,2-diethylmalonurate, 2122
Ethyl Dihexyl Phosphinate, 3128
1-Ethyl-1,4-dihydro-7-methyl-4-oxo-1,8-naphthyridine-3-
  carboxylic Acid, 2594
5-Ethyldihydro-5-phenyl-4,6(1H,5H)-pyrimidinedione,
  2623
S-Ethyl Diisobutylthiocarbamate, 2371
1-Ethyl-3,7-dimethylxanthine, 1744
7-Ethyl-1,3-dimethylxanthine, 1742
O-Ethyl-S,S-dipropylphosphorodithioate, 1589
Ethyl S,S-Dipropyl phosphorodithioate, 1589
S-Ethyl Dipropylthiocarbamate, 1835
S-Ethyl N,N-di-n-Propylthiocarbamate, 1835
Ethyl N,N'-di-n-Propylthiolcarbamate, 1835
Ethyl Dixanthogen, 796
Ethyle Acetyle Thiodiazolique, 1099
Ethylene, 69

5,5-Ethylenebarbituric Acid, 707
Ethylene Chlorobromide, 70
Ethylene Dibromide, 71
Ethylene Dichloride, 74
Ethylene Glycol, 101
Ethylene Glycol Diacetate, 802
Ethylene Glycol Dibutyrate, 2137
Ethylene Glycol Di-N-Butyrate, 2137
Ethylene Glycol Diethyl Ether, 933
Ethylene Glycol Dipropionate, 1511
Ethylene Glycol Divalerate, 2737
Ethylene Glycol Monobutyl Ether Acetate, 1552
Ethylene Glycol Mono-n-butyl Ether Acetate, 1552
Ethylene Glycol Phenyl Ether, 1467
Ethyleneimine, 86
Ethyleneimine (Dihydrate), 93
1,8-Ethylenenaphthalene, 2553
Ethylene Tetrachloride, 110
Ethylenethiourea, 169
Ethylenimine, 86
Ethylenzene, 1432
Ethyl Ether, 362
Ethyl β-Ethoxypropionate, 1228
n-Ethyl β-Ethoxypropionate, 1228
S-Ethyl N,N-Ethylcyclohexylthiocarbamate, 2352
Ethylethylene, 306
S-Ethyl N-Ethylthiocyclohexanecarbamate, 2352
Ethylethyne, 261
N-Ethyl-5-fluoro-3,4-dihydro-2,4-dioxo-1-
  pyrimidinecarboxamide, 1106
Ethyl Formate, 175
Ethyl Glyme, 933
Ethyl Guthion, 2673
5-Ethyl-5-heptylbarbituric Acid, 2926
5-Ethyl-5-n-heptylbarbituric Acid, 2926
S-Ethyl Hexahydro-1H-azepine-1-carbothioate, 1798
2-Ethylhexanamide, 1565
2-Ethyl-1,3-hexanediol, 1585
2-Ethyl-1-hexanoic Acid, 1542
2-Ethyl Hexanol, 1582
2-Ethylhexanol, 1582
2-Ethyl-1-hexanol, 1582
2-Ethylhexan-1-ol, 1582
2-Ethylhexoic Acid, 1542
5-Ethyl-5-n-hexylbarbituric Acid, 2725
Ethyl n-Hexyl Carbinol, 1847
D-(2-Ethylhexyl) Isophthalate, 3970
5-Ethylhydantoin, 425
Ethyl Hydrocinnamate, 2281
Ethyl 4-hydroxybenzoate, 1687
Ethyl o-Hydroxybenzoate, 1688
Ethyl p-Hydroxybenzoate, 1687
Ethyl m-Hydroxycarbanilate Carbanilate, 3312
1-[N-Ethyl-N-(2-hydroxyethyl)amino]-4-(4-
  nitrophenylazo)benzene, 3335
n-Ethyl-6-hydroxynorbornane-2-carboxamide-3,5-lactone,
  2294
Ethylidene Chloride, 75
Ethylidene Fluoride, 76
Ethyl Iodide, 85
5-Ethyl-5-isoamylbarbituric Acid, 2334
Ethylisobutylacetamide, 1568
Ethyl Isobutyl Phosphate, 1594
Ethyl Isocyanide, 145

1-Ethyl-2-isopropylacetylene, 1179
5-Ethyl-5-isopropylbarbituric Acid, 1776
Ethylisopropyl Ether, 517
Ethyl 4-Isothiocyanatobenzoate, 1914
Ethyl p-Isothiocyanatobenzoate, 1914
Ethyl 3-Isothiocyanobenzoate, 1913
Ethyl m-Isothiocyanobenzoate, 1913
Ethylmalonic Acid, 436
Ethylmercaptomethylphenyl-N-methylcarbamate, 2291
Ethyl Methane Dicarboxylate, 1192
5-Ethyl-5-methylbarbituric Acid, 1163
1-Ethyl-2-methylbenzene, 1727
1-Ethyl-4-methylbenzene, 1724
5-Ethyl-5-(1-methyl-1-butenyl)barbituric Acid, 2306
5-Ethyl-5-(3'-methylbut-2'-enyl)barbituric Acid, 2309
5-Ethyl-5-(1-methyl-butyl)-barbituric Acid, 2335
5-Ethyl-5-(1-methylbutyl)-2-thiobarbituric Acid, 2332
5-Ethyl-5-(1-methyl-butyl)-2-thiobarbituric Acid, 2332
Ethyl 2-Methyl-2-cyclohexenyl-6-methylmalonurate, 3102
Ethyl-2-methyl-2-cyclohexenyl-6-methylmalonurate, 3102
N-Ethyl-N'-(1-methylethyl)-6-(methylthio)-1,3,5-Triazine-2,4-diamine, 1802
p-5-Ethyl-5-methylhexylcarbinylbarbituric Acid, 3109
2-Ethyl-3-methyl-Pentanamide, 1566
2-Ethyl-4-methylPentanamide, 1568
4-Ethyl-3-methylphenol, 1751
5-Ethyl-1-methyl-5-phenylbarbituric Acid, 2830
5-Ethyl-N-methyl-5-phenylbarbituric Acid, 2830
Ethyl 1-Methyl-4-phenylpiperidine-4-carboxylate, 3223
N-Ethyl-N-(2-methyl-2-propenyl)-2,6-dinitro-4-(trifluoromethyl)benzenamine, 2828
5-Ethyl-5-(2-methylpropyl)barbituric Acid, 2095
2-Ethyl-2-methylsuccinimide, 1170
Ethylmethylthiophos, 1735
Ethylmorphine, 3669
1-Ethylnaphthalene, 2584
2-Ethylnaphthalene, 2582
O-Ethyl O-p-Nitrophenyl Benzenephosphonothioate, 3007
Ethyl O-(p-nitrophenyl) benzenethiophosphonate, 3007
Ethyl O-(p-Nitrophenyl) Phenylphosphonothionate, 3007
Ethylnitrosocyanamide, 149
5-Ethyl-5-n-nonylbarbituric Acid, 3262
5-Ethyl-5-n-octylbarbituric Acid, 3110
Ethyl Orotate, 1111
3-Ethyloxycarbonyl-5-fluoro-2,4(1H,3H)-pyrimidinedione, 1086
1-Ethyloxycarbonyl-5-fluorouracil, 1086
3-Ethyloxycarbonyl-5-fluorouracil, 1086
n-Ethyl-paba-β-cyclodextrin, 4134
Ethylparaben, 1687
Ethyl Paraoxon, 2057
1-Ethyl-1-pentanol, 1250
3-Ethyl-pentanol-3, 1249
3-Ethyl-3-pentanol, 1249
5-Ethyl-5-pentylbarbituric Acid, 2336
4-Ethylphenol, 1459
p-Ethylphenol, 1459
Ethyl Phenylacrylate, 2247
N-Ethyl-2-(((phenylamino)carbonyl)oxy)propanamide, 2671
5-Ethyl-5-Phenylbarbituric Acid, 2595

(+)-2-(2-Ethyl-2-phenyl-1,3-dioxolan-4-yl)piperidine, 3379
Ethyl Phenyl Ethanolamine, 2084
Ethyl S-Phenyl Ethylphosphonothiolthionate, 2087
5-Ethyl-5-phenylhydantoin, 2235
5-Ethyl-5phenyl-1(phenylsulfonyl)-2,4-imidazolidinedione, 3430
N-Ethyl-3-phenyl-2-propenamide, 2257
Ethyl 3-Phenyl Propenoate, 2247
Ethyl 3-Phenylpropionate, 2281
Ethyl Phenylsulfonate, 1470
Ethyl Phosphate, 946
Ethyl Phthalate, 2638
Ethylphthalyl Ethyl Glycolate, 3041
Ethylphthalyl Ethylglycolate, 3041
Ethyl Phthalyl Ethyl Glycollate, 3064
Ethyl Propanedioate, 1192
Ethyl Propenoate, 432
Ethyl Propionate, 473
Ethylpropylaceturethane, 2142
5-Ethyl-5-n-propylbarbituric Acid, 1774
N-(1-Ethylpropyl)-2,6-dinitro-3,4-xylidine, 2908
2-Ethylpyridine, 1139
3-Ethylpyridine, 1136
4-Ethylpyridine, 1135
N-[(1-Ethyl-2-pyrrolidinyl)methyl]-2-methoxy-5-sulfamoylbenzamide, 3251
Ethyl Salicylate, 1688
Ethyl Sulfide, 370
4-Ethylsulfonylnaphthalene-1-sulfonamide, 2611
S-Ethylsulphinylmethyl O,O-Di-isopropyl Phosphorodithioate, 1852
1-Ethyl-2-tertbutylacetylene, 1495
1-(3-Ethyl-5,6,7,8-tetrahydro-5,5,8,8-tetramethyl-2-naphthalenyl)ethanone, 3589
Ethyl Tetryl, 1350
1-Ethyl Theobromine, 1744
7-Ethyl Theophylline, 1742
2-(Ethylthio)-4,6-bis(isopropylamino)-s-triazine, 2356
Ethyl Thiodiazole, 1998
2-((Ethylthio)methyl)phenyl Methylcarbamate, 2291
Ethylthiometon, 1590
2-Ethylthiophene, 776
Ethyl 2-Thiouracil-5-carboxylate, 1110
2-Ethyltoluene, 1727
4-Ethyltoluene, 1724
o-Ethyltoluene, 1727
p-Ethyltoluene, 1724
Ethyl O-(2,4,5-Trichlorophenyl) Ethylphosphonothioate, 1985
α-Ethyltryptamine, 2664
Ethyl Vinyl Ether, 322
Ethynodiol Diacetate, 3955
Ethynylbenzene, 1294
Ethynyl Estradiol, 3752
Ethynyl Testosterone, 3829
17α-Ethynyl Testosterone, 3829
Eticol, 2057
Etiocholanolone, 3700
Etiocholanol-11-one, 3695
Etofibrate, 3537
Etofylline, 1747
Etoposide, 4071
Etoxadrol, 3379

Etrafon, 3820
Etrimfos, 2114
Etrimphos, 2114
Etryptamine, 2664
Eucalyptol, 2127
Eugenol, 2011
Euparen M, 2021
Exofene, 2772
Exotherm, 1601
Exporsan, 3108
Exxsol Cyclopentane S, 448
Eythyl Urethan, 202
F 118, 2379
F 121, 2380
F 130α, 57
F 131, 2374
F 1379, 3416
FA, 3270
FAC 20, 1843
Famid, 2262
Faneron, 2773
Fatex, 1934
FDNB, 569
Febrinina, 2883
Febron, 2883
Felbinac, 2997
Feldene, 3167
Felodipine, 3546
Fenac, 1280
Fenamiphos, 2923
Fenarimol, 3411
Fenbufen, 3294
Fenchlorphos, 1359
(+)-Fenchone, 2104
α-Fenchone, 2104
D-Fenchone, 2104
Fenclofenac, 2967
Fenergan, 3457
Fenfosphorin, 1427
Fenfuram, 2576
Fenitrothion, 1736
Fenoprop, 1616
Fenoxyl Carbon N, 601
Fensaid, 3167
Fensulfothion, 2330
Fensulfothion Sulfide, 2329
Fensulfothion Sulfone, 2331
Fenthion, 2088
Fenthoate, 2698
Fentiazac, 3410
Fentin Acetate, 3732
Fenuron, 1737
Ferbam, 1807
Ferbeck, 1807
Fermine, 1952
Fernex, 2928
Fernos, 2337
Ferriamicide, 2185
Ferrocene, 1929
Ferrotsen, 1929
Fezudin, 2731
Figsen, 3726
Filariol, 1978
Fioricet, 2305

Fire Master BP-6 (Hexabromophenyl mixture), 2401
FireMaster FF-1 (Hexabromobiphenyl mixture), 2400
Fitios, 905
Flagyl, 783
Flamex t 23p, 1780
Flamprop-isopropyl, 3640
Flexol TOF, 3986
Flogovital, 2781
Florinef, 3833
Fludrocortisone, 3833
Fludroxycortide, 3959
Flufenamic Acid, 2970
Flufenprop-isopropyl, 3640
Flumethasone, 3885
Flumethasonpivalate, 3885
Flumethiazide, 1308
Fluocinolide, 4015
Fluocinolone Acetonide, 3949
Fluocinolone Acetonide Acetate, 4015
Fluocinonide, 4015
Fluometuron, 1958
Fluopromazine, 3548
Fluoranthene, 3270
3-Fluor-benzoesaeure, 997
4-Fluor-benzoesaeure, 999
Fluorbenzol, 638
2-Fluorenamine, 2807
9H-Fluoren-2-amine, 2807
Fluorene, 2787
Fluoridamid, 1959
4-Fluoroacetanilide, 1360
4'-Fluoroacetanilide, 1360
Fluoroacetic Acid, 65
1-Fluoroanthracene, 2959
Fluorobenzene, 638
p-Fluorobenzenesulfonic Acid (2.5 Hydrate), 645
p-Fluorobenzenesulfonic Acid (Monohydrate), 644
p-Fluorobenzenesulfonic Acid (Tetrahydrate), 647
p-Fluorobenzenesulfonic Acid (Trihydrate), 646
2-Fluorobenzoic Acid, 998
3-Fluorobenzoic Acid, 997
4-Fluorobenzoic Acid, 999
m-Fluorobenzoic Acid, 997
o-Fluorobenzoic Acid, 998
p-Fluorobenzoic Acid, 999
2-Fluorobenzyl Chloride, 1028
3-Fluorobenzyl Chloride, 1029
m-Fluorobenzyl Chloride, 1029
o-Fluorobenzyl Chloride, 1028
5-Fluoro-1-(butoxycarbonyl)uracil, 1697
Fluorocarbon 113, 109
Fluorodifen, 2774
5-Fluoro-3,4-dihydro-N-methyl-2,4-dioxo-
    pyrimidinecarboxamide, 694
9-Fluoro-11β,17-dihydroxy-6α-methylpregna-1,4-diene-
    3,20-dione, 3888
21-(9-α-Fluoro-11β, 17α-Dihydroxy-4-pregnen-3,20-
    dione)-N-2-(2-desoxyglucosyl) Carbamate, 4056
21-(9-α-Fluoro-11β, 17α-Dihydroxy-4-pregnen-3,20-
    dione)-N-methyl-N-1-(1-desoxyglucosyl) Carbamate,
    4078
1-Fluoro-2,4-dinitrobenzene, 569
9α-Fluorohydrocortisone, 3833
9α-Fluorohydrocortisone Acetate, 3929

9α-Fluoro-17-hydroxycorticosterone, 3833
6-Fluoro-16α-hydroxyhydrocortisone-16,17-acetonide, 3959
21-(9-α-Fluoro-11β-hydroxy-16α, 17α-Isopropylidenedioxy-1,4-pregnadien-3,20-dione)-N-1-(1-desoxyglucosyl) Carbamate, 4088
21-(9-α-Fluoro-11β-hydroxy-16α, 17α-Isopropylidenedioxy-1,4-pregnadien-3,20-dione)-N-2-(2-desoxyglucosyl) Carbamate, 4087
21-(9-α-Fluoro-11β-hydroxy-16α, 17α-Isopropylidenedioxy-1,4-pregnadien-3,20-dione)-N-methyl-N-1-(1-desoxyglucosyl) Carbamate, 4094
9α-Fluoro-16α-hydroxyprednisolone, 3826
9α-Fluoro-16α-hydroxyprednisolone acetonide, 3951
1-Fluoro-4-iodobenzene, 595
4-Fluoro-1-iodobenzene, 595
p-Fluoroiodobenzene, 595
Fluoromar, 251
Fluoromethane, 16
Fluorometholone, 3888
3-Fluoro-10-methyl-1,2-benzanthracene, 3620
4-Fluoro-4-methyl-1,2-benzanthracene, 3619
9α-Fluoro-16β-methylprednisolone-21-acetate, 3952
2-Fluorophenol, 641
3-Fluorophenol, 643
4-Fluorophenol, 642
m-Fluorophenol, 643
o-Fluorophenol, 641
p-Fluorophenol, 642
3-Fluoro-4-phenylhydratropic Acid, 3160
p-Fluorophenyl Iodide, 595
5-Fluoro-2,4(1H,3H)-Pyrimidinedione, 227
Fluororuracil, 227
9α-Fluoro-11β,16α,17α,21-tetrahydroxy-1,4-pregnadiene-3,20-dione, 3826
Fluorotribromomethane, 33
Fluorotrichloromethane, 38
6α-Fluoro-11β,17,21-trihydroxypregna-1,4-diene-3,20-dione17,21-trihydroxypregna-1,4-diene-3,20-dione, 3824
Fluorouracil, 227
5-Fluorouracil, 227
5 Fluorouracil, 227
2-Fluor-phenol, 641
Fluothane, 45
Fluotrimazole, 3865
Fluphenazine, 3884
Flupirtine, 3194
Fluprednisolone, 3824
Fluprednisolone (Monohydrate), 3825
Flurandrenolone, 3959
Flurbiprofen, 3160
Flurecol-butyl, 3545
Flurenol-n-butyl Ester, 3545
Fluroblastin, 227
Fluroxene, 251
3-FMBA, 3620
4-FMBA, 3619
Folacin, 3641
Folbex, 3287
Folcid, 1905
Folcysteine, 3641
Folic Acid, 3641
Foliclal, 2056

Folliculin, 3570
Folpan, 1602
Folpel, 1602
Folpet, 1602
Fonofos, 2087
Fontamide, 1155
Formaldehyd-diaethyl-acetal, 520
Formaldehyd-dimethyl-acetal, 217
Formaldehyde Diethyl Acetal, 520
Formaldehyde Oxime, 18
Formaldehyd-oxim, 18
Formetanate, 2295
Formic Acid Butyl Ester, 472
Formic Acid Ethyl Ester, 175
Formothion, 829
Formyl-α-aminobutyric Acid, 443
Formylglycine, 148
N-Formyl Glycine, 148
N-Formylleucine, 1200
N-Formyl-DL-leucine, 1200
1-(4'-Formyl-1-piperizinyl)-3,5-bis(dimethylamino)-s-triazine, 2732
Formyl Trichloride, 5
Forte Mefenamic Acid, 3179
Fortrol, 1763
Forturf, 1601
Fosfacol, 2057
Fosferno, 2056
Fospirate, 1082
Fosthietan, 828
Fostion, 1843
Fourrine 85, 729
Freon 11, 38
Freon 113, 109
Freon 12, 37
Freon 22, 4
Frescon, 3914
Froben, 3160
β-D-Fructofuranosyl-α-D-glucopyranoside, 2744
Fructose, 873
D-Fructose, 873
D-(-)-Fructose, 873
Frusemide, 2571
Ftorafur, 1409
Fuam, 2264
Fubol, 3229
Fuclasin, 840
Fuji-one, 2720
Fuklasin, 840
Fulvicin, 3432
Fumaric Acid, 243
trans-Fumaric Acid, 243
Fumarsaeure, 243
Funacomide, 4177
Fundex, 2016
Fungicidin, 4141
Fungifen, 532
Funginex, 2055
Furadanx, 2643
2-Furaldehyde, 391
3-Furamide, Tetrahydro-4,4-dihydroxy-2,2,5,5-tetramethyl-, 1800
Furan-carbon-saeure-(2), 394
Furan-dicarbon-saeure-(2,5), 623

3,4-Furandiol, 2,2-Diethyltetrahydro-5-methyl-, 1831
3,4-Furandiol, Tetrahydro-2,2,5,5-tetramethyl-, 1553
3-Furanol, 2-Butyltetrahydro-5,5-dimethyl-, 2166
3-Furanol, 5-Butyltetrahydro-5-methyl-, 1820
3-Furanol, 2,5-Diethyltetrahydro-2,5-dimethyl-, 2164
3-Furanol, 5,5-Diethyltetrahydro-2,2-dimethyl-, 2172
3-Furanol, 2,5-Diethyltetrahydro-2-methyl-, 1823
3-Furanol, 5,5-Diethyltetrahydro-2-methyl-, 1827
3-Furanol, 5,5-Diisopropyltetrahydro-, 2169
3-Furanol, 2,5-Dimethyltetrahydro-5-propyl-, 1818
3-Furanol, 2,5-Dipropyltetrahydro-, 2165
3-Furanol, 5,5-Dipropyltetrahydro-, 2168
3-Furanol, 2-Ethyltetrahydro-2,5-dimethyl-, 1541
3-Furanol, 2-Ethyltetrahydro-5,5-dimethyl-, 1540
3-Furanol, 5-Ethyltetrahydro-5-methyl-, 1221
3-Furanol, 2-Ethyltetrahydro-5-methyl-5-propyl-, 2171
3-Furanol, 2-Ethyltetrahydro-5-methyl-2-propyl-, 2170
3-Furanol, 2-Ethyltetrahydro-5-methy-5-propyl-, 1539
3-Furanol, 2-Ethyltetrahydro-2,2,5,5-tetramethyl-, 1548
3-Furanol, 3-Ethyltetrahydro-2,2,5-trimethyl-, 1821
3-Furanol, 3-Ethynyltetrahydro-2,2,5,5-tetramethyl-, 2108
3-Furanol, 5-Isobutyltetrahydro-5-methyl-, 1819
3-Furanol, 5-Methyltetrahydro-2-pentyl-, 2163
3-Furanol, 2,2,5,5-Tetraethyltetrahydro-, 2754
3-Furanol, Tetrahydro-2,2-dimethyl-, 861
3-Furanol, Tetrahydro-2,5-dimethyl-, 862
3-Furanol, Tetrahydro-2-isopropyl-5,5-dimethyl-, 1817
3-Furanol, Tetrahydro-2-methyl-, 479
3-Furanol, Tetrahydro-2,2,4,5,5-pentamethyl-, 1825
3-Furanol, 2,5,5-Triethyltetrahydro-, 2162
Furatoin, 1313
Furfural, 391
Furfurin, 3156
Furfurol, 391
Furloe, 1980
Furo[3,4-b]quinolin-3(1H)-one, 9-Hydroxy-1,6-dimethyl-, 2810
Furo[3,4-b]quinolin-3(1H)-one, 9-Hydroxy-1,7-dimethyl-, 2809
Furo[3,4-b]quinolin-3(1H)-one, 9-Hydroxy-1-methyl-, 2546
Furo[3,4-b]quinolin-3(1H)-one, 9-Hydroxy-6-methyl-1-phenyl-, 3519
Furo[3,4-b]quinolin-3(1H)-one, 9-Hydroxy-1-phenyl-, 3406
4H-Furo[3,2-d]-1,3,2-dioxaphosphorin, Adenosine Deriv, 2004
7H-Furo[3,2-g][1]benzopyran-7-one, 2187
2-Furoic Acid, 394
Furosemide, 2571
β-2-Furylacrylic Acid, 1059
β-Furyl-(2)-acrylsaeure, 1059
β-2-Furyncrylic Acid, 1059
FX 2182, 3534
Fyrol FR-2, 1781
G 30414, 1203
G 30451, 1785
G 31432, 2149
G-1, 3911
G-20, 3637
G-21, 3645
G-23, 3639
G-24163, 3425

G-3, 3944
G-32911, 1523
G-4, 2792
G-696, 2591
G-8, 3988
Gabrene, 3424
Galactaric Acid, 806
D-Galactaric Acid, 806
Galactitol, 936
β-D-Galactopyranoside, (3β)-Solanid-5-en-3-yl O-6-deoxy-α-L-mannopyranosyl-(1®2)-O-[β-D-glucopyranosyl-(1-3)]-, 4136
4-O-B-D-Galactopyranosyl-D-glucose, 2741
4-O-β-D-Galactopyranosyl-D-glucose, 2742
Galactose, 877
(+)-Galactose, 877
D-Galactose, 877
D(+)-Galactose, 877
6G-α-D-Galactosylsucrose, 3606
6G-α-D-Galactosylsucrose (Pentahydrate), 3607
Galecon, 2016
Galipan, 1604
Gallic Acid, 1068
Gallussaeure, 1068
Gallussaeuremethyl Ester, 1403
Ganciclovir, 1770
Gantrisin, 2267
Gardona, 1906, 1907
Garlon, 972
Garrathion, 2302
Garvox, 2264
Gatnon, 1659
Gentianic Acid, 2978
Gentianin, 2978
Gentisic Acid, 1064
Gentisin, 2978
Geofos, 828
Geonter, 1762
Geranial, 2106
Geranialdehyde, 2106
Gesagard, 2153
Gesaran, 2355
Gestoral, 3829
Gitoxin, 4127
Glafenine, 3632
Glaxoridin, 3633
Glenbar, 1868
Glibenclamide, 3926
Glipasol, 2676
Glomycin, 3881
Glucamide, 2020
Glucamine 9-α-Fluorohyfrocortisome (Monohydrate), 4058
GlucamineTriamcinolone Acetonide (Monohydrate), 4088
Glucid, 1011
Gluco-Heptose, 1235
D-α-Glucoheptose, 1235
α-D-Glucopyranosyl β-D-fructofuranoside, 2744
4-O-β-D-Glucopyranosyl-D-glucose, 2743
4-β-D-Glucopyransoyl-D-glucopyranose, 2743
Glucosamine Cholesterol, 4108
Glucosamine 9-α-Fluorohyfrocortisome (Monohydrate), 4056

Glucosamine Testosterone, 4020
Glucosamine Triamcinolone Acetonide (Monohydrate), 4087
Glucose, 874
α-Glucose, 879
α-D-Glucose, 879
D-Glucose, 874
D(+)-Glucose, 874
D-α-Glucose, 879
D-Glucose, 4-O-α-D-glucopyranosyl-, 2745
Glucose (Monohydrate), 880
α-Glucose-penta-acetat, 3376
β-Glucose-penta-acetat, 3375
α-Glucose Pentaacetate, 3376
β-D-Glucose Pentaacetate, 3375
6-O-α-D-Glucosyl-α-cyclodextrin, 4130
6-O-α-D-Glucosyl-γ-cyclodextrin, 4153
Glutamic Acid, 445
D-Glutamic Acid, 447
DL-Glutamic Acid, 446
L-Glutamic Acid, 445
L(+)-Glutamin, 457
Glutamine, 457
D-Glutamine, 456
L-Glutamine, 457
L(+)-Glutamine, 457
L(+) Glutaminic Acid, 445
L(+)-Glutaminsaeure, 445
Glutaric Acid, 437
Glutarsaeure, 437
Glutathion, 2116
Glutathione, 2116
Glutethimide, 2836
Glyburide, 3926
Glybuthiazole, 2676
DL-Glyceraldehyde, 182
Glycerin, 218
DL-Glycerin-aldehyd, 182
Glycerin-α,α'-dinitrat, 167
Glycerin-α-nitrat, 208
Glycerol, 218
Glycerol 1,2-Dinitrate, 168
Glycerol 1,3-Dinitrate, 167
Glycerol-α,α'-dinitrate, 167
Glycerol Monooleate, 3856
Glycerol-α-nitrate, 208
Glycerol Triacetate, 1779
Glycerol Trichlorohydrin, 143
Glyceryl Tributyrate, 3264
Glycin, 88
Glycin-N-acetat, 297
Glycine, 88
Glycine-N-acetate, 297
Glycine, N-Acetylglycylglycylglycyl-, Ethyl Ester, 2727
Glycine, N-[[[2-(Acetyloxy)benzoyl]oxy]acetyl]-, Ethyl Ester, 3197
Glycine, N-(Aminocarbonyl)-N-methyl-, 319
Glycine, N-[(Benzoyloxy)acetyl]-N-methyl-, 2612
Glycine, N-[(Benzoyloxy)acetyl]-N-methyl-, Ethyl Ester, 3049
Glycine, N-(Carboxymethyl)-, 1-Ethyl Ester, 816
Glycine, N,N-Diethyl-, (4,5-Dihydro-4-oxo-1H-pyrazolo[3,4-d]pyrimidin-1-yl)Methyl Ester, 2697
Glycine Dipeptide, 303

Glycine Sulfate, 949
Glycine, N-[(3α,5β,7α,12α)-3,7,12-Trihydroxy-24-oxocholan-24-yl]-, 4025
Glycine Tripeptide, 818
Glycocholic Acid, 4025
Glycocoll, 88
Glycocyamine, 210
Glycol, 101
Glycolamide, 89
Glycolamide, Benzoate, 1656
Glycol Diacetate, 802
Glycolic Acid, 82
Glycolic Acid Amide, 89
Glycolic Acid Phenyl Ether, 1400
Glycolic Amide, 89
Glycoluric Acid, 166
Glycolylglycineamide, 455
Glycophen, 2821
Glycyclamide, 3081
N-Glycylglycine, 318
Gly-Dapsone, 3024
Glykolsaeure, 82
Glyoxal-tetraacetat, 2081
Glyoxal Tetraacetate, 2081
Glypasol, 2676
Glyphosate, 213
Glyphosine, 374
Glyprothiazole, 2275
Goltix, 1938
Gopher Bait, 3812
Gopher Getter, 3812
Gossypose, 3606
Gramicidin, 4162
Gramicidin S-A, 4162
Gramicidin S, 4162
Graslan, 1790
Green Oil, 2965
Grifulvin, 3432
Grisactin, 3432
Griseofulvin, 3432
Griseofulvin-4-carboxy-methoxime, 3642
Griseofulvin-4'-ol, 3446
Griseofulvin-4'-oxime, 3436
Griseostatin, 3432
GS 11526, 2150
GS-14254, 2147
GS-19851, 3423
Guaiacol, 1122
Guanidin-carbonat, 222
Guanidineacetic Acid, 210
Guanidine Carbonate, 222
Guanidin-essigsaeure, 210
p-(Guanidinosulfonyl)acetanilide, 1749
Guanine, 401
Guanine Riboside, 2044
Guanosin, 2044
Guanosine, 2044
Guthion, 1993
Haematein, 3277
Halacrinate, 2487
Haldol, 3815
Halocarbon 113, 109
Halocarbon 22, 4
Halocrinate, 2487

Halon 10001, 17
Halon 1211, 32
Halon 242, 108
Haloperidol, 3815
Halotensin, 3693
Halothane, 45
Haltox, 71
Harnsaeure, 388
Harnstoff, 22
HB 419, 3926
HC 600, 356
HCB, 953
HCC 130α, 57
α-HCH, 692
HDAM, 4072
HDPB, 4063
HDPE, 4027
HDPM, 4005
HDPP, 4050
Hedione, 2927
Helicin (0.75 Hydrate), 2863
Heliotropine, 1318
Hematein, 3277
Hemel, 1809
Hemellitol, 1725
Hemimellitene, 1725
Hemimellitic Acid, 1613
Hemo-sol, 2090
n-Hendecane, 2373
Hendecenoic Acid, 2347
Hepachlor Epoxide, 1862
Heptabarbital, 2893
Heptabarbitone, 2893
Heptachlor, 1861
2,3,4,5,6,2',5'-Heptachlorbiphenyl, 2397
Heptachlor Epoxide, 1862
Heptachlorobiphenyl, 2398
2,2',3,3',4,4',5-Heptachlorobiphenyl, 2389
2,2',3,3',4,4',6-Heptachlorobiphenyl, 2386
2,2',3,3',4,5,5' Heptachlorobiphenyl, 2395
2,2',3,3',4,5,6-Heptachlorobiphenyl, 2392
2,2',3,3',4',5,6-Heptachlorobiphenyl, 2388
2,2',3,3',4,5',6-Heptachlorobiphenyl, 2390
2,2',3,3',4,5,6'-Heptachlorobiphenyl, 2391
2,2',3,3',4,6,6'-Heptachlorobiphenyl, 2393
2,2',3,3',5,5',6-Heptachlorobiphenyl, 2394
2,2',3,4,4',5,5'-Heptachlorobiphenyl, 2387
2,2',3,4,4',5',6-Heptachlorobiphenyl, 2396
2,2',3,4,5,5',6-Heptachlorobiphenyl, 2397
2,2',3,4',5,5',6-Heptachlorobiphenyl, 2399
1,2,3,4,6,7,8-Heptachlorodibenzo-p-dioxin, 2375
Heptachlorodibenzo-p-dioxin, 2375
1,2,3,4,6,7,8-Heptachlorodibenzofuran, 2374
3,4,5,6,7,8,8α-Heptachlorodicyclopentadiene, 1861
Heptachlorodiphenyl, 2398
1,4,5,6,7,8,8-Heptachloro-2,3-epoxy-3α,4,7,7α-
   tetrahydro-4,7-methanoindan, 1862
1,4,5,6,7,8,8-Heptachloro-3α,4,7,7α-tetrahydro-4,7-
   methano-1H-indene, 1861
Heptadecanoic Acid, 3504
Heptadecanol, 3505
1-Heptadecanol, 3505
1,6-Heptadiene, 1177
1,6-Heptadiyne, 1103

Heptakis(2,6-di-O-ethyl)-β-cyclodextrin, 4175
Heptakis(2,6-di-O-methyl)-β-cyclodextrin, 4158, 4169
N,N-Heptamethylenecinnamamide, 3357
Heptanal, 1216
Heptane, 1247
n-Heptane, 1247
3-Heptanecarboxylic Acid, 1542
Heptanedioc Acid, 1190
Heptanoic Acid, 1224
n-Heptanoic Acid, 1224
Heptanol, 1261
(±)-3-Heptanol, 1250
Heptanol-(1), 1259
1-Heptanol, 1259
Heptan-1-ol, 1259
2-Heptanol, 1260
3-Heptanol, 1250
4-Heptanol, 1262
2-Heptanone, 1213
Heptan-2-one, 1213
4-Heptanone, 1215
2-Heptanone, Cyclic (hydroxymethyl)ethylene Acetal,
   2175
2'-(2-Heptanoyl-2-hexanyl-acetyl)-6-methoxypurine
   Arabinoside (0.3 Hydrate), 4043
2'-Heptanyl-6-methoxypurine Arabinoside (Hemihydrate),
   3587
1-Heptene, 1204
1-n-Heptene, 1204
2-Heptene, 1206
n-Hept-1-ene, 1204
Heptenophos, 1732
Heptobarbital, 2211
Heptoic Acid, 1224
4-Heptoxybenzoic Acid-2-(diethyl-amino)ethyl Ester,
   3786
Heptylacetylene, 1783
n-Heptylacetylene, 1783
n-Heptyl Alcohol, 1259
Heptyl Aldehyde, 1216
Heptyl 4-Aminobenzoate, 3091
Heptyl p-Aminobenzoate, 3091
4-Heptylaminobenzoic Acid-2-(diethyl-amino)ethyl Ester,
   3787
9-[5-O-(Heptylate-β-D-arabinofuranosyl)]-6-methoxy-9H-
   purine, 3588
Heptyl Bromide, 1236
Heptyl Carbamate, 1569
n-Heptyl Carbamate, 1569
Heptyl Chloride, 1237
n-Heptylcinnamamide, 3378
n-Heptyl 4-hydroxybenzoate, 3087
Heptyl p-Hydroxybenzoate, 3087
2-Heptyl-4-hydroxymethyl-2-methyl-1,3-dioxolane,
   2755
Heptyl Iodide, 1238
Heptylmethylcarbinol, 1848
2-n-Heptylphenol, 2919
4-n-Heptylphenol, 2918
o-n-Heptylphenol, 2919
p-n-Heptylphenol, 2918
1-Heptyne, 1175
1-n-Heptyne, 1175
2-Heptyne, 1178

Herbazolin, 1604
Hercules 14503, 3043
Herplex, 1700
β-Hexaamylose, 4119
Hexabarital, 2669
Hexabromobiphenyl, 2402
2,2',4,4',6,6'-Hexabromobiphenyl, 2402
2,3,4,2',3',4'-Hexachlorbiphenyl, 2418
2,3,4,5,2',3'-Hexachlorbiphenyl, 2410
2,3,5,6,2',3'-Hexachlorbiphenyl, 2422
Hexachlorobenzene, 953
Hexa-chlorobenzene, 953
Hexachlorobiphenyl, 2408
2,2',3,3',4,4'-Hexachlorobiphenyl, 2418
2,2',3,3',4,5-Hexachlorobiphenyl, 2410
2,2',3,3',4,5'-Hexachlorobiphenyl, 2410
2,2',3,3',4,6-Hexachlorobiphenyl, 2420
2,2',3,3',5,6-Hexachlorobiphenyl, 2422
2,2',3,3',5,6'-Hexachlorobiphenyl, 2421
2,2',3,3',6,6'-Hexachlorobiphenyl, 2413
2,2',3,4,4',5-Hexachlorobiphenyl, 2416
2,2',3,4,4',5'-Hexachlorobiphenyl, 2415
2,2',3,4,4',6-Hexachlorobiphenyl, 2411
2,2',3,4,5,5'-Hexachlorobiphenyl, 2425
2,2',3,4',5,5'-Hexachlorobiphenyl, 2414
2,2',3,4,5',6-Hexachlorobiphenyl, 2426
2,2',3,4',5',6'-Hexachlorobiphenyl, 2420
2,2',3,5,5',6-Hexachlorobiphenyl, 2424
2,2',4,4',5,5'-Hexachlorobiphenyl, 2409
2,2',4,4',6,6'-Hexachlorobiphenyl, 2427
2,3,3',4,4',5-Hexachlorobiphenyl, 2412
2,3,3',4,4',5'-Hexachlorobiphenyl, 2412
2,3,3',4,4',6-Hexachlorobiphenyl, 2417
2,3,3',4',5,6-Hexachlorobiphenyl, 2423
2,4,5,2',4',5'-Hexachlorobiphenyl, 2409
Hexachlorobutadiene, 375
Hexachloro-1,3-butadiene, 375
α-Hexachlorocyclohexane, 692
α-1,2,3,4,5,6-Hexachlorocyclohexane, 692
β-Hexachlorocyclohexane, 693
β-1,2,3,4,5,6-Hexachlorocyclohexane, 693
δ-1,2,3,4,5,6-Hexachlorocyclohexane, 690
Hexachlorocyclopentadiene, 526
1,2,3,4,5,5-Hexachlorocyclopentadiene, 526
1,2,3,4,5,5-Hexachloro-1,3-Cyclopentadiene, 526
Hexachloro-1,3-cyclopentadiene, 526
1,2,3,4,5,5-Hexachloro-1,3-cyclopentadiene dimer, 2185
1,2,3,4,7,8-Hexachlorodibenzo[1,4]dioxin, 2381
1,2,3,4,7,8-Hexachlorodibenzo-p-dioxin, 2381
1,2,3,4,7,8-Hexachlorodibenzo[b,e][1,4]dioxin, 2381
1,2,3,4,7,8-Hexachlorodibenzofuran, 2379
1,2,3,6,7,8-Hexachlorodibenzofuran, 2380
2,3,4,7,8,9-Hexachlorodibenzofuran, 2380
1,2,3,4,10,10-Hexachloro-6,7-epoxy-1,4,4α,5,6,7,8,8α-
  octahydro-1,4-endo-endo-5,8-dimethano-naphthalene,
  2523
Hexachloroethane, 111
1,1,1,2,2,2-Hexachloroethane, 111
1,2,3,4,10,10-Hexachloro-1,4,4α,5,8,8α-hexahydro-
  1,4:5,8-dimethanonaphthalene, 2522
Hexachloro-5-norbornene-2,3-dimethanol, Cyclic Sulfite,
  endo-, 1606
Hexachloro-5-norbornene-2,3-dimethanol, Cyclic Sulfite,
  exo-, 1607

3,4,5,6,9,9-Hexachloro-1α,2,2α,3,6,6α,7,7α-octahydro-
  2,7:3,6-Dimethanonaphth[2,3-b]oxirene, 2524
Hexachlorophene, 2772
4,5,6,7,8,8-Hexachloro-3α,4,7,7α-tetrahydro-4,7-
  methanoindene, 1870
Hexadecane, 3401
n-Hexadecane, 3401
hexadecanoic Acid, 3399
Hexadecanol, 3403
Hexadecyl 4-Hydroxybenzoate, 3940
Hexadecyl p-Hydroxybenzoate, 3940
2'-Hexadecyl-6-methoxypurine Arabinoside, 4047
1,5-Hexadiene, 787
Hexadienoic Acid, 771
2,4-Hexadienoic Acid, 771
Hexaferb, 1807
N-(((Hexahydro-1H-azepin-1-yl)amino)carbonyl)-4-
  methylbenzenesulfonamide, 3096
Hexahydrobenzoic Acid, 1189
Hexahydro-3α,7α-dimethyl-4β,7β-epoxyisobenzofuran-
  1,3-dione, 2014
2,2,4,4,6,6-Hexahydro-2,2,4,4,6,6-hexakis(1-aziridinyl)-
  1,3,5,2,4,6-triazatriphosphorine, 2752
3-(Hexahydro-4,7-methanoindan-5-yl)-1,1-dimethylurea,
  2924
Hexahydro-1-(1-oxo-3-phenyl-2-propenyl)1H-Azepine,
  3209
1,2,3,6,7,8-Hexahydropyrene, 3309
Hexahydrotoluene, 1205
2,2,4,4,6,6-Hexakis(1-aziridinyl)cyclotriphosphaza-1,3,5-
  triene, 2752
Hexaldehyde, 845
Hexalin Acetate, 1501
N,N-Hexamethylenecinnamamide, 3209
Hexamethylenediamine, 948
1-(Hexamethyleneiminel)-3,5-bis(dimethylamino)-s-
  triazine, 2930
Hexamethylen-tetramin, 842
Hexamethylmelamine, 1809
n-Hexanal, 845
Hexanamide, 883
Hexanamide, 2,6-Diamino-N-[4-[(4-
  aminophenyl)sulfonyl]phenyl]-, (S), 3577
Hexanamide, 2,2-Dibutyl-N-hydroxy, 3122
Hexanamide, 2-Hydroxy-, 896
Hexane, 899
n-Hexane, 899
1,6-Hexanediamine, 948
Hexanema, 2022
Hexane, 2,2,5-Trimethyl-, 1841
Hexanitrodiphenylamine, 2451
Hexanoic Acid, 4-(Acetylamino)phenyl Ester, 3072
Hexanoic Acid, Amide, 883
1-Hexanol, 914
2-Hexanol, 919
3-Hexanol, 929
n-Hexanol, 914
2-Hexanone, 848
Hexanone-3, 854
3-Hexanone, 854
2'-(2-Hexanoyl-2-pentanyl-acetyl)-6-methoxypurine
  Arabinoside, 3997
Hexanyl Acetaminophen, 3072
2'-Hexanyl-6-methoxypurine Arabinoside, 3491

Hexaplas DTDP, 4109
Hexastat, 1809
Hexatriacontane, 4121
n-Hexatriacontane, 4121
Hexavin, 2575
Hexazinone, 2726
Hexaziridinocyclotriphosphazene, 2752
Hexene, 822
1-Hexene, 822
1-n-Hexene, 822
Hexen-1-ol-3, 846
1-Hexen-3-ol, 846
Hexen-4-ol-3, 851
4-Hexen-3-ol, 851
2-2-Hexenylphenol, 2320
o-2-Hexenylphenol, 2320
Hexestrol, 3569
Hexethal, 2725
Hexobarbital, 2669
4-Hexoxybenzoic Acid-2-(diethyl-amino)ethyl Ester, 3708
2-Hexoxy-N-[2-(diethyl-amino)ethyl]-4-quinoline Carboxamide, 3899
Hexyl Acetaminophen, 3231
Hexyl Acetate, 1549
sec-Hexyl Acetate, 1546
Hexyl α-Acetoxypropionate, 2349
Hexylacetylene, 1494
n-Hexylacetylene, 1494
n Hexyl Alcohol, 914
tert-Hexyl Alcohol, 929
Hexyl p-Aminobenzoate, 2905
4-Hexylaminobenzoic Acid-2-(diethyl-amino)ethyl Ester, 3709
Hexylbenzene, 2699
n-Hexylbenzene, 2699
Hexyl Bromide, 882
Hexyl Carbamate, 1239
n-Hexyl Carbamate, 1239
tert-Hexyl Carbamate, 1241
2-Hexyl-p-cresol, 2916
Hexyldiantipyrylmethane, 4072
Hexyldithiopyrylmethane, 4073
n-Hexyl β-Ethoxypropionate, 2366
Hexyl Ethyl Carbinol, 1847
n-Hexyl 2-hydroxybenzoate, 2902
Hexyl p-Hydroxybenzoate, 2901
Hexyl Lactate, 1829
Hexyl Methyl Ketone, 1538
2-Hexyl-4-methylphenol, 2916
2-Hexyl-6-methylphenol, 2915
4-Hexyl-2-methylphenol, 2917
Hexyl Nicotinate, 2689
n-Hexyl Nicotinoate, 2689
1-Hexyloxycarbonyl-5-fluorouracil, 2288
2-n-Hexylphenol, 2717
4-n-Hexylphenol, 2718
o-n-Hexylphenol, 2717
p-n-Hexylphenol, 2718
4-n-Hexylresorcin, 2719
4-Hexylresorcinol, 2719
n-Hexyl Salicylate, 2902
1-Hexyne, 786
3-Hexyne, 785

3-Hexyne, 2,2,5-Trimethyl-, 1784
HHDN, 2522
Hiltachlor, 3497
Hippuric Acid, 1655
Hippursaeure, 1655
His-pro-D-phe-his-leu-leu-thr-tyr, 4146
His-pro-phe-his-leu-leu-val-tyr, 4148
His-pro-phe-his-leu-D-leu-val-tyr, 4149
His-pro-D-phe-his-leu-leu-val-tyr-serinol, 4156
His-pro-phe-his-leu-phe-val-tyr, 4155
L-Histidin, 782
Histidine, 782
L-Histidine, 782
HMM, 1809
Hoelite, 2946
HOE 13465OH, 2457
Holdem, 1589
Homoallyl Bromide, 290
DL-Homocystine, 1534
DL-meso-Homocystine, 1534
Homogentisic Acid, 1401
Hostacortin H, 3831
Hostaquick, 1732
Hostathion, 2672
HOX 1901, 2291
1,2,3,4,6,7,8-HpCDD, 2375
1,2,3,4,6,7,8-HpCDF, 2374
o-HPhDAM, 4069
HT 972, 2829
1,2,3,4,7,8-HxCDD, 2381
1,2,3,4,7,8-HxCDF, 2379
1,2,3,6,7,8-HxCDF, 2380
Hydantoic Acid, 166
Hydantoin, 132
Hydantoin, 1,3-Dichloro-5-methyl-, 231
Hydantoin, 5,5-Diphenyl-1-(o-tolylsulfonyl)-, 3868
Hydantoin of α-Aminobutyric Acid, 425
Hydantoin of Aspartic Acid, 410
Hydantoin of DL-Leucine, 1184
Hydracrylic Acid, 180
Hydracrylic Acid, Butyl Ester, Acetate, 1795
Hydracrylic Acid, Pentyl Ester, Acetate, 2135
Hydracrylic Acid, Propyl Ester, Acetate, 1513
Hydracrylsaeure, 180
Hydram, 1798
Hydrastic Acid, 1614
Hydrastin, 3808
Hydrastine, 3808
(1R,9S)-β-Hydrastine, 3808
Hydrastsaeure, 1614
Hydrazinecarboximidamide, N-Nitro-, 29
4-Hydrazinopteridine, 719
Hydrindane, 1662
Hydrindene, 1662
Hydrobenzoin, 3015
DL-Hydrobenzoin, 3015
Hydrobexan, 4168
Hydrochinon, 725
Hydrochlorothiazide, 1104
Hydrocinchonin, 3674
Hydrocinchonine, 3674
Hydrocinnamic Acid, 1680
9-[5'-(O-Hydrocinnamoyl)-β-D-arabinofuranosyl]adenine Ester, 3661

Hydrocortisone, 3843
Hydrocortisone Acetate, 3933
Hydrocortisone-21-Acetate, 3933
Hydrocortisone Butyrate, 3998
Hydrocortisone-21-butyrate, 3998
Hydrocortisone-21-caproate, 4041
Hydrocortisone Heptanoate, 4059
Hydrocortisone-21-heptanoate, 4059
Hydrocortisone-21-hexanoate, 4041
Hydrocortisone Propionate, 3963
Hydrocortisone-21-propionate, 3963
Hydrocortisone Tebutate, 4041
Hydrocortisone Valerate, 4022
Hydrocortisone-21-valerate, 4022
Hydroflumethiazide, 1361
Hydromox, 1982
Hydrophenol, 850
Hydroquinine, 3762
Hydroquinol, 725
Hydroquinone, 725
Hydroquinonecarboxylic Acid, 1064
Hydroquinone-β-D-glucopyranoside Monohydrate, 2684
Hydroquinone Monoethyl Ether, 1468
Hydroquinone Monomethyl Ether, 1124
2-Hydroxyacetamide, 89
p-Hydroxyacetanilide, 1416
4'-Hydroxyacetanilide Hexanoate, 3072
2-Hydroxyacetimidic Acid, 89
4'-Hydroxy-acetophenon, 1380
4-Hydroxyacetophenone, 1380
2-Hydroxyacridine, 2784
o-Hydroxyacridine, 2784
2-Hydroxy-6-aminopurine, 402
3β-Hydroxy-13α-amino-13,17-seco-5α-androstan-17-oic-
    13,17-lactam-4-N,N-bis-(chloroethyl)amino Phenyl-
    acetate, 4074
3α-Hydroxy-5β-androstane-11,17-dione, 3695
3α-Hydroxy-5β-androstane-17-one, 3700
Hydroxy-17-androstanone, 3703
3α-Hydroxy-17-androstanone, 3703
3α-Hydroxy-5α-androstan-17-one, 3703
Hydroxy-5α-androstan-17-one, 3703
17β-Hydroxyandrost-4-en-3-one, 3693
p-Hydroxyanisole, 1124
Hydroxyatrazine, 1515
2-Hydroxy atrazine, 1515
3-Hydroxyazobenzene, 2561
4-Hydroxyazobenzene, 2559
p-Hydroxyazobenzene, 2559
3-Hydroxy-azobenzol, 2561
3-Hydroxy-benzaldehyd, 1057
4-Hydroxy-benzaldehyd, 1058
m-Hydroxybenzaldehyde, 1057
p-Hydroxybenzaldehyde, 1058
2-Hydroxybenzanilide, 2808
Hydroxybenzene, 722
(S)-α-Hydroxybenzeneacetic Acid, 1398
α-Hydroxy-benzeneacetic Acid, 1392
4-Hydroxybenzenecarboxylic Acid, 1062
3-Hydroxy-benzoesaeure, 1061
4-Hydroxy-benzoesaeure, 1062
2-Hydroxybenzoic Acid, 1060
3-Hydroxybenzoic Acid, 1061
4-Hydroxybenzoic Acid, 1062

m-Hydroxybenzoic Acid, 1061
m-Hydroxybenzoicacid, 1061
o-Hydroxybenzoic Acid, 1060
p-Hydroxybenzoic Acid, 1062
p-Hydroxybenzoicacid, 1062
2-Hydroxybenzoicacidamide, 1091
2-Hydroxy-benzoic Acid, Butyl Ester, 2284
2-Hydroxybenzoic Acid, 2-(dimethylamino)-2-oxoethyl
    Ester, 2263
4-Hydroxybenzoic Acid Ethyl Ester, 1687
4-Hydroxybenzoic Acid N-Hexyl Ester, 2901
2-Hydroxybenzoic Acid Phenyl Ester, 2801
p-Hydroxybenzoic Acid Tridecyl Ester, 3784
Hydroxybenzopuridine, 1624
p-Hydroxybiphenyl, 2566
p-Hydroxy-p-bis(azobenzene), 3525
3-Hydroxybutanoic Acid β-Lactone, 275
3-Hydroxy-2-butyl-5,5-methyltetrahydrofuran, 2166
α-Hydroxycaproamide, 896
4-Hydroxy-chinolin, 1621
4-Hydroxy-chinolin-carbonsaeure-(2), 1885
Hydroxychlor, 2982
1-Hydroxychlordene, 1872
1-Hydroxychlordene Epoxide, 1873
4-Hydroxychlorobenze, 634
3-Hydroxychlorobenzene, 632
4-Hydroxychlorobenzene, 634
3α-Hydroxycholanic Acid, 3974
3α-Hydroxy-5β-cholan-24-oic Acid, 3974
3β-Hydroxy-5β-cholanoic Acid, 3975
7α-Hydroxy-5β-cholanoic Acid, 3975
3β-Hydroxycholest-5-ene, 4049
7-Hydroxycoumarin, 1611
3-Hydroxy-p-cymene, 2073
p,p'-Hydroxy-DDT, 2982
17-Hydroxy-11-dehydrocorticosterone, 3830
4-Hydroxy-3,5-Di-tert-Butyltoluene, 3260
2-Hydroxy-6,7-diethylpteridine, 1995
2-Hydroxy-6:7-diethylpteridine, 1995
4-Hydroxy-6,7-diethylpteridine, 1995
4-Hydroxy-6:7-diethylpteridine, 1995
3-Hydroxy-2,2-diethyltetrahydrofuran, 1547
2-Hydroxy-3,5-diiod-benzoesaeure, 977
4-Hydroxy-3,5-diiodobenzonitrile, 959
3-Hydroxy-5,5-diisopropyltetrahydrofuran, 2169
1-Hydroxy-2,4-dimethylbenzene, 1452
3-Hydroxy-2,2-dimethyl-5,5-diethyltetrahydrofuran,
    2172
3-Hydroxy-2,5-dimethyl-2,5-diethyltetrahydrofuran,
    2164
4-Hydroxy-6,7-dimethylpteridine, 1369
4-Hydroxy-6:7-dimethylpteridine, 1369
3-Hydroxy-2,2-dimethyltetrahydrofuran, 861
3-Hydroxy-2,5-dimethyltetrahydrofuran, 862
4-Hydroxy-2',4'-dinitrodiphenylamine, 2551
2-Hydroxy-diphenyl-aether, 2567
2-Hydroxy-2,2-diphenylethanoic Acid, 2998
2-Hydroxy-1,2-diphenylethanone, 2994
2-Hydroxydiphenyl Ether, 2567
Hydroxydiphenylhydantoin, 3155
p-Hydroxydiphenylhydantoin, 3155
4-Hydroxy-6,7-diphenylpteridine, 3517
4-Hydroxy-6:7-diphenylpteridine, 3517
3-Hydroxy-2,5-dipropyltetrahydrofuran, 2165

3-Hydroxy-5,5-dipropyltetrahydrofuran, 2168

3-Hydroxy-2,5-dispirocyclohexyltetrahydrofuran, 3111

1-Hydroxy-2,3-epoxy-4,5,6,7,8,8-hexachloro-3α,4,7,7α-tetrahydro-4,7-methanoindene, 1873

N-(2-Hydroxyethyl)-1-aminoanthraquinone, 3281

1-[(2-Hydroxyethyl)amino]-4-(methylamino)-9,10-anthracenedione, 3427

β-Hydroxyethyl Aniline, 1480

Hydroxyethyl-β-cyclodextrin, 4135

3-Hydroxy-2-ethyl-5,5-dimethyltetrahydrofuran, 1540

3-Hydroxy-5-ethyl-2,5-dimethyltetrahydrofuran, 1541

2-[(2-Hydroxyethyl)[4-(4-nitrophenylazo)phenyl]amino]ethanol, 3336

3-Hydroxy-2-ethyl-5-propyl-5-methyltetrahydrofuran, 2171

7-β-Hydroxyethyltheophylline, 1747

3-Hydroxy-3-ethyl-2,2,5-trimethyltetrahydrofuranol, 1821

3-Hydroxy-3-ethynyl-2,2,5,5-tetramethyltetrahydrofuran, 2108

11-Hydroxyetiocholanolone, 3706

1-Hydroxyheptane, 1259

2-Hydroxyheptane, 1260

3-Hydroxyheptanc, 1250

1-Hydroxy-4,5,6,7,8,8-hexachloro-3α,4,7,7α-tetra-hydro-4,7-methanoindene, 1872

2-Hydroxyhexanamide, 896

(Hydroxyimino)Cyclohexane, 814

Hydroxyisoandrosterone, 3704

2-Hydroxyisophthalic Acid, 1323

4-Hydroxyisophthalic Acid, 1325

5-Hydroxyisophthalic Acid, 1324

2-Hydroxy-iso-phthalsaeure, 1323

5-Hydroxy-iso-phthalsaeure, 1324

4-Hydroxy-iso-phthasaeure, 1325

3-Hydroxy-2-isopropyl-5,5-dimethyltetrahydrofuran, 1817

3-Hydroxy-17-keto-δ(1,3,5-10,6,8)estrapentaene, 3544

3-Hydroxy-17-keto-δ(1,3,5-10,7)estratetraene, 3558

1-(Hydroxylamino)-3,5-bis(dimethylamino)-s-triazine, 1536

1-(p-Hydroxylbenzenesulfonyl)-5,5-diphenyl-hydantoin, 3802

o-Hydroxylphenyldiantipyrylmethane, 4069

17-Hydroxy-7α-mercapto-3-oxo-17α-pregn-4-ene-21-carboxylic Acid γ-Lactone Acetate, 3956

Hydroxymesitylene, 1755

2-Hydroxymesitylene, 1755

4-Hydroxy-3-methoxybenzaldehyde, 1396

4-Hydroxy-3-methoxy-1-(γ-hydroxypropenyl)benzene-4-D-glucoside (Dihydrate), 3374

1-Hydroxy-4(2-methyl-2-butyl)benzene, 2316

3-Hydroxy-5-methyl-5-butyltetrahydrofuran, 1820

4-Hydroxy-2-methyl-chinolin, 1912

9-[4α-(Hydroxymethyl)-cyclopent-2-ene-1α-yl]guanine, 2270

3-Hydroxy-2-methyl-2,5-diethyltetrahydrofuran, 1823

3-Hydroxy-2-methyl-5,5-diethyltetrahydrofuran, 1827

3-Hydroxy-17β-{[(1-methyl-1,4-dihydropyridin-3-yl)-carbonyl]oxy}-estra-1,3,5(10)-triene, 3993

3-(Hydroxymethyl)-5,5-diphenyl-2,4-imidazolidinedione, 3292

3-Hydroxy-5-methyl-5-ethyltetrahydrofuran, 1221

3-Hydroxy-5-methyl-5-isobutyltetrahydrofuran, 1819

3-Hydroxy-5-methyl Isoxazole, 255

17-Hydroxy-6-methyl-16-methylenepregna-4,6-diene-3,20-dione Acetate, 3822

3-(Hydroxymethyl)nitrofurantoin, 1636

3-Hydroxy-4-methylol-2,2,5,5-tetramethyltetrahydrofuran, 1837

(Hydroxymethyl)pentamethylmelamine, 1811

N-(Hydroxymethyl)pentamethylmelamine, 1811

2-(Hydroxymethyl)phenyl-β-D-glucopyranoside, 2904

3-(Hydroxymethyl)phenytoin, 3292

3-Hydroxy-5-methyl-5-propyltetrahydrofuran, 1539

4-Hydroxy-6-methylpteridine, 1047

4-Hydroxy-7-methylpteridine, 1046

8-Hydroxymethylpurine, 713

Hydroxymethylpyrone, 728

3-Hydroxy-2-methyl-4-pyrone, 728

4-Hydroxy-2-methylquinoline, 1912

3-Hydroxy-2-methyl-5-spirocyclopentyltetrahydrofuran, 1793

3-Hydroxy-2-methyltetrahydrofuran, 479

p-Hydroxy-p'-nitroazobenzene, 2549

4-Hydroxy-2-nitrophenol, 657

3-Hydroxyoxolane, 327

N2-Hydroxy-N2,N4,N4,N6,N6-pentamethylmelamine, 1536

17-Hydroxy-3-oxo-17α-pregna-4,6-diene-21-carboxylic Acid Lactone, 3887

3-Hydroxy-2,2,4,5,5-pentamethyltetrahydrofuran, 1825

3-Hydroxy-2,3,4,5,5-pentamethyltetrahydrofuran, 1814

3-Hydroxy-2-pentyl-5-methyltetrahydrofuran, 2163

1-Hydroxy-2-phenoxyethane, 1467

α-Hydroxyphenylacetic Acid, 1392

Hydroxy-2-phenyl Acetophenone, 2994

2-Hydroxy-2-phenylacetophenone, 2994

3-(4-Hydroxyphenyl)-D-alanine, 1715

3-(4-Hydroxyphenyl)-DL-alanine, 1716

3-(4-Hydroxyphenyl)-L-alanine, 1714

2-Hydroxy-N-phenylbenzamide, 2808

p-Hydroxyphenylbutazone, 3646

2-(p Hydroxyphenyl)ethylamine, 1479

4-Hydroxyphenylethylamine, 1479

3-Hydroxyphenyl Isothiocyanate, 1006

4-Hydroxyphenyl Isothiocyanate, 1007

4-Hydroxyphenylisothiocyanate, 1007

m-Hydroxyphenyl Isothiocyanate, 1006

DL-5-(p-Hydroxyphenyl-5-phenylhydantoin, 3155

p-Hydroxyphenytoin, 3155

21-Hydroxypregn-4-ene-3,20-dione, 3841

3β-Hydroxy-5-pregnen-20-one, 3849

3β-Hydroxypregn-5-en-20-one, 3849

17-α-Hydroxyprogesterone, 3839

21-Hydroxyprogesterone, 3841

(4S)-4-Hydroxy-L-proline, 444

L-Hydroxyproline, 444

L-4-Hydroxyproline, 444

trans-4-Hydroxy-L-Proline, 444

Hydroxyprometryne, 2150

3-Hydroxy-1-propanesulfonic Acid γ-sultone, 185

2-Hydroxypropionamide, 195

2-Hydroxypropionic Acid Decyl Ester, 2937

3-Hydroxy-2-propyl-5,5-diethyltetrahydrofuran, 2363

3-Hydroxy-5-propyl-2,5-dimethyltetrahydrofuran, 1818

3-Hydroxy-2-propyl-5-methyl-5-ethyltetrahydrofuran, 2170

β-Hydroxypropyltheophylline, 2036
8-Hydroxypsoralon, 2196
2-Hydroxypteridine, 608
4-Hydroxypteridine, 609
6-Hydroxypteridine, 607
7-Hydroxypteridine, 606
8-Hydroxypurine, 383
2-Hydroxypyridine, 395
3-Hydroxypyridine, 397
4-Hydroxypyridine, 396
5-(α-Hydroxy-α-2-pyridylbenzyl)-7-(α-2-
    pyridylbenzylidene)-5-norbornene-2,3-dicaboximide,
    4097
2-Hydroxypyrimidine, 235
4-Hydroxypyrimidine, 234
8-Hydroxyquinaldine, 1911
2-Hydroxyquinoline, 1619
3-Hydroxyquinoline, 1620
4-Hydroxyquinoline, 1621
5-Hydroxyquinoline, 1622
6-Hydroxyquinoline, 1623
7-Hydroxyquinoline, 1625
8-Hydroxyquinoline, 1624
4-Hydroxysalicylic Acid, 1066
Hydroxysimazine, 1203
2-Hydroxysimazine, 1203
3-Hydroxy-5-spirocyclohexyltetrahydrofuran, 1792
1-Hydroxy-2,3,4,6-tetrachlorobenzene, 543
Hydroxytetracycline, 3881
N-Hydroxytetradecanamide, 3120
3-Hydroxy-2,2,5,5-tetraethyltetrahydrofuran, 2754
(±)-3-Hydroxytetrahydrofuran, 327
3-Hydroxytetrahydrofuran, 327
(RS)-3-Hydroxytetrahydrofuran, 327
3-Hydroxy-2,2,5,5-tetramethyltetrahydrofuran, 1548
α-Hydroxytoluene, 1117
2-Hydroxy-m-toluic Acid, 1387
3-Hydroxy-p-toluic Acid, 1395
4-Hydroxy-m-toluic Acid, 1387
6-Hydroxy-m-toluic Acid, 1388
2-Hydroxy-m-tolylsaeure-(1), 1387
2-Hydroxy-p-tolylsaeure-(1), 1399
3-Hydroxy-p-tolylsaeure-(1), 1395
4-Hydroxy-m-tolylsaeure-(1), 1387
6-Hydroxy-m-tolylsaeure-(1), 1388
2-Hydroxytricarballylic Acid, 774
2-Hydroxytricarballylic Acid (Monohydrate), 775
Hydroxy-3,5,6-trichloropyridine, 376
3-Hydroxy-2,5,5-triethyltetrahydrofuran, 2162
1-Hydroxy-2,3,5-trimethylbenzene, 1752
1-Hydroxy-2,4,6-trimethylbenzene, 1755
3-Hydroxy-2-tropane Carboxylic Acid, 1782
Hydrozimtsaeure, 1680
Hygroton, 2981
Hylemox, 1858
Hymexazol, 255
Hyodeoxycholic Acid, 3979
Hyoscine, 3465
Hyoscyamin, 3484
Hyoscyamine, 3484
L-Hyoscyamine, 3484
Hypoxanthin, 384
Hypoxanthine, 384
Hypoxanthine Ribonucleoside, 2002

IBDAM, 4031
IBIB, 1544
IBP, 2328, 2921
Ibudros, 2906
Ibuprofen, 2898
D-Ibuprofen, 2900
l-Ibuprofen, 2899
r-Ibuprofen, 2899
S-Ibuprofen, 2900
Ibuprofen N-Methyl-N-carbamoyl Methyl Glycolamide
    Ester, 3586
Ibuprofen N-Methyl-N-crabamoyl Methyl Glycolamide
    Ester, 3585
Ibuproxam, 2906
ICI 46638, 2771
Idobutal, 2307
Idoxuridine, 1700
Idryl, 3270
Imdur, 768
Imidan, 2231
Imidazol-di-carbonsaeure-(4,5), 379
4,5-Imidazoledicarboxylic Acid, 379
α,β-Imidazoledicarboxylic Acid, 379
1H-Imidazole, 2-(1-Methylethyl)-4-nitro-, 781
2-Imidazole Sulfonic Acid, 133
1H-Imidazole, 2,4,5-tri-2-furanyl-4,5-dihydro-, 3156
1-Imidazolidinecarboxamide, 3-(3,5-Dichlorophenyl)-N-
    (1-methylethyl)-2,4-dioxo-, 2821
2,4-Imidazolidinedione, 132
2,4-Imidazolidinedione, 1,3-Dichloro-, 115
2,4-Imidazolidinedione, 1,3-Dichloro-5-methyl-, 231
2,4-Imidazolidinedione, 5-Ethyl-5-phenyl-, 2235
2-Imidazolidinone, 3-[5-(1,1-Dimethylethyl)-1,3,4-
    thiadiazol-2-yl]-4-hydroxy-1-methyl-, 2102
2-Imidazolidinone, 1-Nitroso-3-(5-nitro-2-thiazolyl)-,
    675
2-Imidazoline, 2,4,5-Tri-2-furyl-, 3156
Imidazol-sulfosaeure-(2), 133
7-Imidazo(4,5-d)pyrimidine, 382
Imidole, 252
Iminodiacetic Acid, 299
Imino-diessigsaeure, 299
Iminostilbene, 3152
Imipramine, 3673
Imuran, 1631
Indaconitine, N-Deethyl-3-deoxy-N-methyl-, 4103
Indan, 1662
1H-Indene, 2,3-Dihydro-, 1662
Indeno(1,2,3-cd)pyrene, 3860
Indeno[1,2,3-cd]pyrene, 3860
Indigo-disulfosaeure-(5,5'), 3272
Indigotinsulfonic Acid, 3272
Indolacin, 3805
Indole, 1337
1H-Indole-3-alanine, 2234
Indoline, 1412
3-Indol-3-ylalanine, 2233
L-β-3-Indolylalanine, 2233
Indomethacin, 3628
Indomethacine,Form IV, 3628
Inosin, 2002
Inosine, 2002
Inositol, 875
D-Inositol, 878

D(+)-Inositol, 878
Interchem Acetate Violet 6B, 3154
Intramyctin, 2228
Iobenzamic Acid, 3280
Iocetamic Acid, 2604
Iodaethan, 85
Iodamide, 2573
4-Iodanilin-N-acetat, 1362
Iodipamide, 3723
4-Iodoacetanilide, 1362
p-Iodoacetanilide, 1362
p-Iodoaniline-N-acetate, 1362
2-Iodoaniline-4-sulphonic Acid, 700
3-Iodoaniline-4-sulphonic Acid, 701
4-Iodoaniline-2-sulphonic Acid, 699
4-Iodoaniline-3-sulphonic Acid, 698
5-Iodoaniline-2-sulphonic Acid, 697
5-Iodoaniline-3-sulphonic Acid, 696
6-Iodoaniline-3-sulphonic Acid, 695
Iodobenzanilide, 2793
Iodobenzene, 648
p-Iodobenzenesulfonyl Chloride, 581
2-Iodobenzoic Acid, 1002
3-Iodobenzoic Acid, 1001
4-Iodobenzoic Acid, 1003
m-Iodobenzoic Acid, 1001
o-Iodobenzoic Acid, 1002
p-Iodobenzoic Acid, 1003
3-Iodobenzyl Isothiocyanate, 1310
4-Iodobenzyl Isothiocyanate, 1309
m-Iodobenzyl Isothiocyanate, 1310
p-Iodobenzyl Isothiocyanate, 1309
Iodobutane, 340
(+)-5-Iodo-2'-deoxyuridine, 1700
5-Iodo-2'-deoxyuridine 5'-butyrate, 2865
5-Iodo-2'-deoxyuridine 5'-isobutyrate, 2864
5-Iodo-2'-deoxyuridine 5'-pivalate, 3069
5-Iodo-2'-deoxyuridine 5'-propionate, 2641
Iodoethane, 85
Iodofenphos, 1355
p-Iodofluorobenzene, 595
Iodoform, 6
1-Iodoheptane, 1238
Iodol, 224
Iodomethane, 17
1-Iodonaphthalene, 1883
α-Iodonaphthalene, 1883
4-Iodophenol, 649
p-Iodophenol, 649
2-Iodo-N-phenylbenzamide, 2793
4-Iodophenyl Iodide, 596
3-Iodophenyl Isothiocyanate, 975
4-Iodophenyl Isothiocyanate, 976
4-Iodophenylisothiocyanate, 976
m-Iodophenyl Isothiocyanate, 975
Iodopropane, 193
2-Iodopropane, 192
β-Iodopropionic Acid, 144
β-Iod-propionsaeure, 144
Ioglycamic Acid, 3507
DL-Iopamidol, 3475
L-Iopamidol, 3475
Iopanoic Acid, 2230
Iopodic Acid, 2603

Iopronic Acid, 3200
Iothalamic Acid, 2200
Ioxynil, 959
Ipatone, 2354
Ipatryne, 2358
Ipazine, 2121
IPC, 2029
IP Carrier T 40, 630
IPDAM, 4012
IPO 4328, 1434
Ipodic Acid, 2603
Iprobenfos, 2921
Iprodial, 2821
Iprodione, 2821
Ipronidazole, 1172
IPSP, 1852
Irgamide, 2274
Iron bis(Cyclopentadiene), 1929
Isazophos, 1797
Isoamyl Acetate, 1217
Isoamyl Alcohol, 519
tert-Isoamyl Alcohol, 519
Isoamyl Bromide, 485
Isoamyl o-Hydroxybenzoate, 2683
Isoamylmalonic Acid, 1509
Isoamyl Salicylate, 2683
Isoamylurethan, 885
Isoamylurethane, 885
N-Isoamylurethane, 1570
Isoandrosterone, 3702
5-Isoandrosterone, 3700
1,3-Isobenzofurandione, 1278
Isobrenzschleimsaeure, 393
Isobromyl, 811
Isobutane, 357
1-Isobutoxy-2-pyrrolidino-3-N-benzylanilino-propane, 3960
Isobuttersaeure, 330
Isobutyl Acetaminophen, 2879
Isobutyl Acetate, 860
Isobutyl Alcohol, 364
Isobutylbenzene, 2051
Isobutyl Bromide, 334
Isobutyl Carbamate, 494
Isobutylchlorid, 337
Isobutyl Chloride, 337
Isobutyl (+/-)-2-[4-(4-Chlorophenoxy)phenoxy]propionate, 3655
Isobutyldiantipyrylmethane, 4031
Isobutyldithiopyrylmethane, 4033
Isobutylene, 305
5-Isobutylhydantoin, 1184
r-(-)-p-Isobutylhydratropic Acid, 2899
Isobutyl Isobutyrate, 1544
Isobutyl 2-Methylpropanoate, 1544
1-Isobutyloxycarbonyl-5-fluorouracil, 1698
(R)-2-(4-Isobutylphenyl)propanoic Acid, 2899
(S)-(+)-2-(4-Isobutylphenyl)propionic Acid, 2900
2-(4-Isobutylphenyl)propionic Acid, 2898
2-(4-Isobutylphenyl)propionohydroxamic Acid, 2906
9-[5-O-(Isobutyrate-β-D-arabinofuranosyl)]-6-methoxy-9H-purine (0.25 Hydrate), 3218
Isobutylurethan, 1240
Isobutyl Urethane, 1240

Isobutyraldehyde, 323
Isobutyric Acid, 330
5'-Isobutyryl 5-iodo-2'-deoxyuridine, 2864
2'-Isobutyryl-6-methoxypurine Arabinoside, 3220
L-Isocamphersaeure, 2109
L-Isocamphoric Acid, 2109
Isocarbamid, 1518
Isocarboxazid, 2615
Isochlorthion, 1405
Isochrysene, 3514
Isocil, 1475
Isocinchomeronic Acid, 1018
Isocomene, 1730
Isocyanuric Acid, 122
Isocyanursaeure, 122
Isoestragole, 2007
Isofenphos, 3253
Isofluorphate, 904
Isogen, 768
Isoguanine, 402
Isohexene, 821
Isohexyl Alcohol, 923
Isokitazine, 2921
L(+)-Isoleucin, 892
Isoleucine, 892
DL-Isoleucine, 895
L-Isoleucine, 892
Isoniazid, 749
Isonicotinic Acid Hydrazide, 749
Isonoruron, 2925
Isooctane, 1574
Isopentane, 503
Isopentyl Alcohol, 519
Isopentyl Urethane, 885
Isophthalic Acid, 1321
Isoprene, 420
Isopropalin, 3249
Isopropenylbenzene, 1663
3-Isopropoxyphenyl Isothiocyanate, 1961
m-Isopropoxyphenyl Isothiocyanate, 1961
o-Isopropoxyphenyl Methylcarbamate, 2293
Isopropyl Acetaminophen, 2651
Iso-propylacetat, 474
Isopropyl Acetate, 474
Isopropylacetone, 849
Isopropyl Alcohol, 215
5-Isopropyl-5-allylbarbituric Acid, 2061
Isopropylantipyrine, 3053
Isopropylbarbituric Acid, 1164
Isopropylbenzene, 1728
Isopropyl N-Benzoyl-N-(3-chloro-4-fluorophenyl)alanine, 3640
Isopropylbromid, 186
Isopropyl Bromide, 186
Isopropyl tert-Butyl Ether, 1253
Isopropyl Butyrate, 1223
Isopropyl N-Butyrate, 1223
Isopropyl Carbanilate, 2029
Isopropyl Chloride, 190
N-Isopropyl-2-chloroacetanilide, 2272
N-Isopropyl-α-chloroacetanilide, 2272
Isopropyl m-Chlorocarbanilate, 1980
6-Isopropyl-m-cresol, 2073
Isopropyldiantipyrylmethane, 4012

Isopropyl 2,2-Diethylmalonurate, 2342
Isopropyl-2,2-diethylmalonurate, 2342
4-Isopropyl-2,3-dimethyl-5-oxo-1-phenyl-3-pyrazoline, 3053
4-Isopropyl-2,6-dinitro-N,N-dipropylaniline, 3249
2,4-D-isopropyl Ester, 2229
2,4-D Isopropyl Ester, 2229
Isopropyl Ether, 928
Isopropylethylene, 451
α, 3-o-Isopropylidene Pyriridoxine, 2292
Isopropyl Iodide, 192
Isopropyl (2E,4E)-11-Methoxy-3,7,11-trimethyl-2,4-dodecadienoate, 3711
5-Isopropyl-5-(3-methylbut-2-enyl)barbituric Acid, 2704
2-Isopropyl-3-methyl-butyramide, 1562
2-Isopropyl-5-methyl Phenol, 2073
2-Isopropyl-4-nitroimidazole, 781
2-Isopropyl-4(5)-nitroimidazole, 781
2-Isopropyl-5-nitroimidazole, 781
1-Isopropyloxyarbonyl-5-fluorouracil, 1411
Isopropyl-N-phenyl Carbamate, 2029
3-(4-Isopropylphenyl)-1,1-dimethylurea, 2700
3-Isopropylphenyl N-Methylcarbamate, 2290
m-Isopropylphenyl N-Methylcarbamate, 2290
2-(p-Isopropylphenyl)-2-methyl-4-(methoxycarbamyl)-1,3-dioxolane, 3230
2-(p-Isopropylphenyl)-2-methyl-4-(methoxycarbamyl)-1,3-dithiolane, 3224
2-(p-Isopropylphenyl)-2-methyl-4-(methoxycarbamyl)-1,3-oxathiolane, 3228
Isopropylphosphoramidothioate, 2054
13-Isopropylpodocarpa-7,13-dien-15-oic Acid, 3780
2-Isopropyl-2-propylacetamide, 1563
N1-(5-Isopropyl-1,3,4-thiadiazol-2-yl)sulfanilamide, 2275
S-2-Isopropylthioethyl O,O-Dimethyl Phosphorodithioate, 1267
5-Isopropyl-m-tolyl Methylcarbamate, 2688
2-Isopropylvaleramide, 1563
Isoprothiolane, 2720
Isoproturon, 2700
Isopseudocumenol, 1752
Isopyromucic Acid, 393
Isoquinoline, 3-Methyl-, 1910
DL-Isoserin, 204
DL-Isoserine, 204
Isosorbide Dinitrate, 768
Isothioate, 1267
Isothiocyanatobenzene, 1022
3-Isothiocyanato-benzonitrile, 1276
4-Isothiocyanatobenzophenone, 2963
4-Isothiocyanato-N,N-dimethyl-benzenamine, 1675
Isothiocyanatomethane, 68
Isothiocyanatomethylbenzene, 1346
1-Isothiocyanatonaphthalene, 2192
2-Isothiocyanatonaphthalene, 2191
Isothiocyanic Acid, p-Cyanobenzyl Ester, 1609
m-Isothiocyanobenzoic Acid, 1289
3-Isothiocyanophenyl Isothiocyanate, 1277
m-Isothiocyanophenyl Isothiocyanate, 1277
Isovaleriansaeure, 475
Isovaleric Acid, 475
Isovalerylacetone, 1504

9-[5'-(O-Isovaleryl)-β-D-arabinofuranosyl]adenine Ester, 3235
DL-Isovalin, 492
DL-Isovaline, 492
Isoxathion, 2853
Isoxazolol, 5-Methyl-, 255
Isoxylic Acid, 1679
Itaconic Acid, 413
Itaconsaeure, 413
Itamidone, 2883
Itanoxone, 3416
Itobarbital, 2305
Izoacridina, 2795
Izopentan, 503
Jayflex DTDP, 4109
Jodfenphos, 1355
Jodomiron 380, 2573
K 138, 2415
Kabat, 3711
Kaken, 3248
Kakodylsaeure, 106
Kandiset, 1011
Kanekrol 500, 2432
Kaoxidin, 2825
Karagard, 2146
Karbam White, 840
Karbaspray, 2575
Karbutilate, 3095
Karphos, 2853
Kativ-G, 2194
Kayaphos, 2922
Kayphosnac, 2922
Kebuzone, 3638
Keflex, 3325
Keflodin, 3633
Kelevan, 3412
Kemate, 1603
Kepone, 2184
Kerb 50W, 2572
Kesscocide, 2192
Ketoconazole, 4009
11-Ketoetiocholanolone, 3695
α-Ketoisovaleric Acid, 434
Ketone, Methyl Tetrahydro-3-hydroxy-2,2,5,5-tetramethyl-3-furyl, 2134
Ketoprofen, 3295
Ketoprofen, N-methyl-N-carbamoylmethyl Glycolamide Ester, 3814
Khellin, 3000
Kitazin, 2328
Kitazin L, 2921
Kitazin P, 2921
Ortho-Klor, 1874
KNI-272, 4102
Knockmate, 1807
Kohlenoxidsulfid, 42
Komenic Acid, 624
Komensaeure, 624
Kopmite, 3287
Korksaeure, 1508
Kreatin, 354
Kreatinin, 304
m-Kresotinsaeure, 1399
Krystar 300, 873

Kusol, 1701
Kyanmethin, 780
Kyanol, 737
Kynostatin, 4102
Kynurenic Acid, 1885
Kynurensaeure, 1885
Kypfarin, 3630
L 581490, 259
Lactamide, 195
Lactic Acid Butyl Ester, 1227
Lactose, 2741
B-Lactose, 2742
β-Lactose, 2742
(R)-Laenitrile, 3767
Laevulinsaeure, 433
Lambdamycin, 4092
Lamprene, 4028
Landruma, 2781
Ianiazid, 749
Lannate, 454
Lanoxicaps, 4128
Lanoxin, 4128
DL-Lanthionine, 836
Laptran, 2622
Laraflex, 3017
Largactil, 3444
Lasiocarpine, 3852
Laurent's Acid, 1916
Lauric Acid, 2753
Laurostearic Acid, 2753
Lauryl Alcohol, 2758
Laxettes, 3726
Lederfen, 3294
Leguarme, 2671
Lemonene, 2554
Lenacil, 2892
Leostesin, 3098
Leptophos, 2788
Leptophos Oxon, 2789
Lethox, 2302
Leucine, 891
(2S)-α-Leucine, 891
D-Leucine, 887
DL-Leucine, 894
L-Leucine, 891
L(-)-Leucine, 891
Leucopterin, 674
L-Leu-dapsone, 3571
Leukeran, 390
Leukersan, 3067
Leukomycin, 2228
Levatol, 3599
Levo-citronellol, 2161
Levodopa, 1718
Levomepromazine, 3675
Levulinic Acid, 433
D-(-)-Levulose, 873
C-Lexin, 3325
Lexone, 1499
Leymin, 1604
LFA 2043, 2821
Librium, 3285
Lidocaine, 3098
Lignocaine, 3098

Limonene, 2089
(R)-(+)-Limonene, 2090
D-Limonene, 2090
Linalol, 2128
Linalool, 2128
Linalyl, 2728
Linalyl Acetate, 2728
Linarotox, 2960
Lincomycin Hexadecylsulfamate, 4110
Lincomycin Octadecylsulfamate, 4120
Lindane, 691
Linfolizin, 3067
Linfolysin, 3067
Linuron, 1666
Linurotox, 2960
Lioresal, 1979
Liothyronine, 3151
Lipo-Merz, 3537
Liquefied Petroleum Gas, 356
Lithocholic Acid, 3974
Lontrel, 555
Lopurin, 385
Lorazepam, 3135
Loridine, 3633
Lotrimin, 3867
Lucel, 1269
Lufyllin-EPG, 2068
Lumazine, 610
Lutein, 3838
2,3-Lutidine, 1140
2,4-Lutidine, 1131
2,5-Lutidine, 1141
2,6-Lutidine, 1142
3,4-Lutidine, 1138
3,5-Lutidine, 1137
α-Lutidine, 1139
β-Lutidine, 1136
Lutidinic Acid, 1017
L-Lys-Dapsone, 3577
L(+)-Lysin, 911
Lysine, 911
L(+)-Lysine, 911
Lysine Estrone Ester, 3961
Lyxo-2-Hexulose, 876
MAA, 30
Machete, 3497
Maclurin, 2803
Macrobid, 1313
Macrodantin, 1313
L-β-Malamidic Acid, 301
D-β-Malaminsaeure, 298
L-β-Malaminsaeure, 301
R-β-Malaminsaeure, 298
Malathion, 2154
Maleic Acid, 242
Maleic Hydrazide, 238
Maleinsaeure, 242
Malic Acid, 284
D-Malic Acid, 285
DL-Malic Acid, 284
Malonamide, 163
Malonic, 1192
Malonic Acid, 136
Malonic Acid Diamide, 163

Malonic Acid Monoethyl Ester, 436
Malonic Ester, 1192
Malonodiamide, 163
Malononitrile, 116
Malonsaeure, 136
Malonsaeure-diamid, 163
Malonsaeure-dinitril, 116
Malonsaeure-monoaethyl Ester, 436
Maloran, 1664
Maltol, 728
Maltose, 2745
α-Maltose, 2745
6-O-α-Maltosyl-α-cyclodextrin, 4143
6-O-α-Maltosyl-β-cyclodextrin, 4152
6-O-α-D-Maltosyl-α-cyclodextrin, 4143
6-O-α-D-Maltosyl-β-cyclodextrin, 4152
6-O-α-D-Maltosyl-γ-cyclodextrin, 4164
6-O-α-Maltosyl-γ-cyclodextrin, 4164
6-O-α-Maltotriosyl-α-cyclodextrin, 4154
6-O-α-Maltotriosyl-β-cyclodextrin, 4164
6-O-α-D-Maltotriosyl-α-cyclodextrin, 4154
6-O-α-D-Maltotriosyl-β-cyclodextrin, 4164
6-O-α-Maltotriosyl-γ-cyclodextrin, 4171
6-O-α-D-Maltotriosyl-γ-cyclodextrin, 4171
Malt Sugar, 2745
Mandelic Acid, 1392
(R)(-)Mandelic Acid, 1398
(S)-(+)-Mandelic Acid, 1398
D-Mandelic Acid, 1398
DL-Mandelic Acid, 1393
L-Mandelic Acid, 1398
DL-Mandelsaeure, 1393
Mandrax, 3290
D-Mannit, 937
Mannitol, 937
D-Mannitol, 937
D-Manno-α-heptit, 1265
D-Mannoheptose, 1234
L-Mannomethylose, 869
D-Mannose, 872
D-(+)-Mannose, 872
D-Mannose-phenylhydrazon, 2707
D-Mannosephenylhydrazone, 2707
Maralate, 3298
Marcaine, 3596
Marcaine (Hydrochloride Monohydrate), 3596
Margaric Acid, 3504
Marlate, 3298
Marplan, 2615
MATB, 618
MB 8882, 1367
MBN, 509
MBR 12325, 2255
MBR6033, 1959
MBR 8251, 2991
MC 1488, 2857
MCB, 630
MCP, 820
MCPA, 1645
MCPB, 2253
4-(MCPB), 2253
2-(MCPP), 1955
MDAM, 3945
Mebaral, 2830

Mebendazole, 3283
Mec, 417
Mecarbam, 2157
Meconic Acid, 985
Meconin, 1950
Mecoprop, 1955
Medazepam, 3296
Medigesic, 2305
Medinoterb Acetate, 2857
Medrogestone, 3931
Medrol, 3893
Mefenamic Acid, 3179
Mefluidide, 2255
Mekonin, 1950
Mekonsaeure, 985
Melamine, 171
Melitose, 3606
Melitriose, 3606
Melphalan, 2891
Menadione, 2194
Menaphthone, 2194
Menazon, 844
Mendrin, 2523
Menidazole, 259
Menrium, 3285
D-1,8-p-Menthadiene, 2090
p-Mentha-1,8-diene, 2089
p-Mentha-6,8-dien-2-one, 2071
r-(-)-p-Mentha-6,8-dien-2-one, 2071
3-p-Menthanol, 2160
p-Menthan-3-one, 2125
trans-p-Menthan-3-one, 2125
Menthol, 2160
(-)-Menthone, 2125
L-Menthone, 2125
Meobal, 2027
Me-paraben, 1390
Mepazine, 3667
Meperidine, 3223
Mephobarbital, 2830
DL-Mepivacaine, 3242
Mcprobamate, 1806
Meprospan, 1806
Merabitol, 3763
2-Mercapto-4-amino-6-hydroxypyrimidine, 258
2-Mercapto-6-aminouracil, 258
Mercaptofos, 2088
Mercaptoimidazoline, 169
4-Mercapto-7-methylpteridine, 1052
2-Mercaptopteridine, 621
4-Mercaptopteridine, 620
7-Mercaptopteridine, 619
Mercaptopurine, 390
6-Mercaptopurine, 390
Mercaptothion, 2154
D-3-Mercaptovaline, 501
Mercozen, 169
Mercurialin, 28
Merex, 2184
Merfenl 51, 1719
Merpan 90, 1634
Mertect, 1888
Mesaconic Acid, 414
Mesaconsaeure, 414

Mesitelene, 1729
Mesitene Lactone, 1123
Mesityl Alcohol, 1755
Mesitylene, 1729
Mesitylenecarboxylic Acid, 2010
Mesitylene Phosphinous Acid, 1771
Mesityloxid, 792
Mesityl Oxide, 792
Mesocystine, 839
Mesoinosit, 875
Mesooxalsaeure-diaethyl Ester, 1169
Mesoxalic Acid Diethyl Ester, 1169
Metalaxyl, 3229
Metaldehyde, 1558
Metamitron, 1938
Metanilamide, 762
Metanilic Acid, 745
Metanilic Acid (Sesquihydrate), 746
Metathionine, 1736
Metazolamide, 426
Methabenzthiazuron, 1969
Methacrylic Acid Methyl Ester, 431
Methacrylonitrile, 253
Methacycline Base, 3875
Methafluoridamid, 2255
Methan, 21
Mcthane, 21
Methanearsonic Acid, 30
Methane Dichloride, 9
Methanedisulfonic Acid, 26
Methanedisulfonic Acid, Bromo-, 13
Methanesulfonic Acid Methyl Ester, 102
Methane Tetrachloride, 40
Methane, Tribromofluoro-, 33
4,7-Methano-1H-inden-1-ol, 4,5,6,7,8,8-hexachloro-
    3α,4,7,7α-tetrahydro-, 1871
Methanol, 25
Methapyrilene, 3074
Methaqualone, 3290
Metharbital, 1775
Methazolamide, 426
Methazole, 1605
Methenamide, 426
Methenamine, 842
Methiazinic Acid, 3164
Methidathion, 817
Methionic Acid, 26
Methionic Acid (Monohydrate), 27
DL-Methionin, 500
Methionine, 499
DL-Methionine, 500
L-(-)-Methionine, 499
Methiuron, 2064
Methomyl, 454
Methoprene, 3711
Methoprotryne, 2355
Methotrimeprazine, 3675
Methoxalen, 2532
Methoxsalen, 2532
p-Methoxyacetanilide, 1707
9-[5-O-(Methoxyacetate-β-D-arabinofuranosyl)]-6-
    methoxy-9H-purine (Hemihydrate), 3059
2'-Methoxyacetyl-6-methoxypurine Arabinoside (0.9
    Hydrate), 3060

2-Methoxy-4-allylphenol, 2011
2-Methoxy-1-aminobenzene, 1144
4-Methoxy-1-aminobenzene, 1145
1-Methoxy-2-amino-4-nitrobenzene, 1108
o-Methoxyaniline, 1144
p-Methoxyaniline, 1145
p-Methoxybenzaldehyde, 1382
2-Methoxybenzenamine, 1144
4-Methoxybenzenamine, 1145
Methoxybenzene, 1118
4-Methoxybenzenesulfonamide, 1153
p-Methoxybenzenesulfonamide, 1153
2-Methoxy-4H-benzo-1,3,2-dioxaphosphorin-2-thione,
    1427
3-Methoxy-benzoesaeure, 1394
2-Methoxybenzoic Acid, 1389
3-Methoxybenzoic Acid, 1394
3,4-Methoxybenzoic Acid, 1689
4-Methoxybenzoic Acid, 1397
m-Methoxybenzoic Acid, 1394
o-Methoxybenzoic Acid, 1389
p-Methoxybenzoic Acid, 1397
4-Methoxybenzoic Acid-2-(diethylamino)ethyl Ester,
    3093
9-[5-O-(4-Methoxybenzoyl-β-D-arabinofuranosyl)]-6-
    methoxy-9H-purine (0.25 Hydrate), 3650
2'-(p-Methoxybenzoyl)-6-methoxypurine Arabinoside,
    3653
(2R,3R,4R)-2-(4-Methoxybenzyl)-3,4-pyrrolidinediol-3-
    acetate, 3073
2-Methoxy-4,6-bis(ethylamino)-s-triazine, 1521
2-Methoxy-4,6-bis-(isopropyl-amino)-s-triazine, 2148
2-Methoxy-4,6-bis-isopropylamino-s-triazine, 2148
3-Methoxy-1-Butanol Acetate, 1226
3-Methoxy Butyl Acetate, 1226
3-Methoxybutyl Acetate, 1226
8-Methoxycaffeine, 1748
1-Methoxycarbonyl-5-fluorouracil, 640
Methoxycarbonylmethyl 2,2-Diethylmalonurate, 2313
Methoxycarbonylmethyl-2,2-diethylmalonurate, 2313
Methoxychlor, 3298
Methoxy DDT, 3298
2-Methoxy-3,6-dichlorobenzoic Acid, 1299
2-Methoxy-N-[2-(diethyl-amino)ethyl]-4-quinoline
    Carboxamide, 3488
4-Methoxy-3,3'-dimethylbenzophenone, 3318
2-Methoxy-4-ethylamino-6-tert-butylamino-s-triazine,
    2146
2-Methoxy-4-ethylamino-6-diethylamino-s-triazine, 2149
2-Methoxy-4-ethylamino-6-isopropylamino-s-triazine,
    1801
2-Methoxy-4-ethylamino-6-isopropylamino-s-triazines,
    1801
8-Methoxyfuranocoumarin, 2532
8-Methoxy-2',3',6,7-furocoumarin, 2532
3-Methoxy-4-hydroxybenzaldehyde, 1396
1-(p-Methoxylbenzenesulfonyl)-5,5-diphenyl-hydantoin,
    3869
2-Methoxy-4-methylamino-6-isopropylamino-s-triazine,
    1520
6-Methoxy-9-(5-O-[4-methylbenzoyl]-β-D-
    arabinofuranosyl)-9H-purine (Hemihydrate), 3648
8-Methoxy-16-methyl-2,3:10,11-bis[methylenebis(oxy)]-,
    (8β)-, 3809

Methoxymethyl 2,2-Diethylmalonurate, 2123
Methoxymethyl-2,2-diethylmalonurate, 2123
Methoxymethyl-2-methyl-2-cyclohexenyl-6-
    methylmalonurate, 3103
(S)-6-Methoxy-α-methyl-2-naphthaleneacetic Acid, 3017
6-Methoxy-α-methyl-2-naphthaleneacetic Acid, 3017
N-(Methoxymethyl)pentamethylmelamine, 2159
6-Methoxy-9-(5-O-[4-nitrobenzoyl]-β-D-
    arabinofuranosyl)-9H-purine, 3535
S-((5-Methoxy-2-oxo-1,3,4-thiadiazol-3(2H)-yl)methyl)
    O,O-Dimethyl Phosphorodithioate, 817
3-Methoxyphenol, 1120
4-Methoxyphenol, 1124
o-Methoxyphenol, 1122
p-Methoxyphenol, 1120, 1124
N-4-(4'-Methoxyphenoxy)phenyl-N',N'-dimethylurea,
    3331
N-(4-Methoxyphenyl)acetamide, 1707
N-(4-Methoxyphenyl)acetic Acid Amide, 1707
o-Methoxyphenylamine, 1144
p-Methoxyphenylamine, 1145
Methoxyphenylbutazolamide, 2706
2-p-Methoxyphenyl-2-p-hydroxyphenyl-1,1,1-trichloro-
    ethane, 3159
3-Methoxyphenyl Isothiocyanate, 1340
4-Methoxyphenylisothiocyanate, 1339
m-Methoxyphenyl Isothiocyanate, 1340
p-Methoxyphenyl Isothiocyanate, 1339
6-Methoxy-9-(5-O-pivalate-β-D-arabinofuranosyl)]-9H-
    purine (Hemihydrate), 3367
1-Methoxypropane, 363
2-Methoxypropane, 365
1-Methoxy-4-(2-propen-1-yl)benzene, 2008
Methoxy-4-propenylbenzene, 2007
2-Methoxy-4-(2-propenyl)phenol, 2011
2-Methoxypteridine, 1049
4-Methoxypteridine, 1044
7-Methoxypteridine, 1045
6-Methoxypurine Arabinoside, 2279
Methoxytolbutamide, 2706
1-Methoxy-2,4,6-trichlorobenzene, 996
2-Methoxy-1,3,5-trinitro-benzene, 1024
6-Methoxy-9-(5-O-valerate-β-D-arabinofuranosyl)]-6-
    methoxy-9H-purine (Hemihydrate), 3368
Methsuximide, 2605
Methyl Acetaminophen, 1966
2-Methyl-4-acetaminophenol, 1710
N-Methylacetanilide, 1702
Methylacetat, 176
Methyl Acetate, 176
Methyl 2-Acetoxybenzoate, 1951
Methyl 2-Acetoxypropionate, 804
Methyl α-Acetoxypropionate, 804
Methyl, [3-(Acetylamino)phenyl]-, 1703
Methyl Acetylene, 124
Methylacetylene, 124
Methyl O-Acetyllactate, 804
Methyl 2-Acetyloxypropanoate, 804
Methyl Acetyl Sulfathiazole, 2617
Methyl Acetylthiodiazole, 663
α-Methylacrolein, 269
α-Methyl-acrolein, 269
Methyl Acrylate, 272
3-Methylacrylic Acid, 277

3-Methyladipic Acid, 1191
Methylal, 217
2-Methylalanine, 347
Methyl Alcohol, 25
1-Methylallyl Acetate, 794
5-Methyl-5-allylbarbituric Acid, 1444
α-Methylallyl Chloride, 293
Methylamine, 28
1-(Methylamino)-9,10-anthraquinone, 3143
4-Methylaminobenzenesulfonamide, 1161
N-Methyl-4-aminobenzenesulfonamide, 1160
N-Methyl-p-aminobenzenesulfonamide, 1160
p-Methylaminobenzenesulfonamide, 1161
Methyl N-(4-Aminobenzenesulphonyl)carbamate, 1448
Methyl p-Aminobenzoate, 1419
Methyl-p-aminobenzoate, 1419
4-Methylaminobenzoic Acid-2-(diethyl-amino)ethyl Ester, 3099
1-(Methylamino)-3,5-bis(dimethylamino)-s-triazine, 1535
Methyl 3-amino-2,5-dichlorobenzoate, 1334
L-α-(1-Methylaminoethyl)benzyl Alcohol, 2083
Methylamino-4-methylthio-6-isopropylamino-1,3,5-triazine, 1522
3-Methyl-4-amino-6-phenyl-1,2,4-triazin-5(4H)-one, 1938
2-Methylaminopteridine, 1100
Methyl Amyl Acetate, 1546
Methyl Amyl Alcohol, 930
Methyl tert-Amyl Ether, 926
2-Methyl-2-n-amyl-4-hydroxymethyl-1,3-dioxolane, 2175
Methyl-n-amylnitrosamine, 907
Methylaniline, 1133
4-Methylaniline, 1134
N-Methylaniline, 1133
Methyl Anisate, 1686
Methyl Anone, 1187
2-Methylanthracene, 3147
9-Methylanthracene, 3146
N-Methylanthranilic Acid, 1418
N-Methyl-anthranilsaeure, 1418
2-Methyl-1-anthraquinonylamine, 3144
Methylarsin, 31
Methylarsine, 31
Methylarsonsaeure, 30
8-Methyl-8-azabicyclo[3.2.1]octan-3-yl 3-hydroxy-2-phenylpropionate, 3483
4-Methylbenzaldehyde, 1377
p-Methylbenzaldehyde, 1377
10-Methyl-1,2-benzanthracene, 3622
1'-Methyl-1,2-benzanthracene, 3621
9-Methyl-1,2-benzanthracene, 3623
Methylbenzene, 1102
4-Methylbenzenecarbonitrile, 1338
4-Methylbenzenesulfonamide, 1147
m-Methylbenzenesulfonamide, 1149
o-Methylbenzenesulfonamide, 1148
p-Methylbenzenesulfonamide, 1147
Methylbenzenesulfonic Acid, 1125
4-Methylbenzenesulfonic Acid, 1125
2-Methyl-benzenesulfonic Acid (Monohydrate), 1167
1-(o-Methylbenzenesulfonyl)-5,5-diphenyl-hydantoin, 3868

2'-(p-Methylbenzenesulfonyl)-6-methoxypurine Arabinoside, 3557
Methyl Benzoate, 1385
2-Methylbenzoic Acid, 1383
3-Methylbenzoic Acid, 1384
4-Methylbenzoic Acid, 1379
β-Methylbenzoic Acid, 1384
m-Methylbenzoic Acid, 1384
p-Methylbenzonitrile, 1338
Methyl 5-Benzoyl Benzimidazole-2-carbamate, 3283
Methyl 5-Benzoyl-2-benzimidazolecarbamate, 3283
Methyl Benzoyl Benzoate, 3175
5-Methyl-3,4-benzpyrene, 3796
4-Methylbenzyl Isothiocyanate, 1658
p-Methylbenzyl Isothiocyanate, 1658
Methylbenzylnitrosamine, 1438
DL-Methyl-bernsteinsaeure, 434
4-Methylbiphenyl, 2814
5-Methyl-brenzschleimsaeure, 727
Methyl Bromide, 12
3-Methyl-1-bromobenzene, 1070
α-Methyl-γ-bromo-butanoic Ureide, 809
2-Methyl-1,3-butadiene, 420
2-Methylbutane, 503
2-Methyl-1-butanol, 518
2-Methylbutan-1-ol, 518
3-Methyl-1-butanol, 519
3-Methyl-2-butanol, 514
DL-2-Methyl-1-butanol, 518
3-Methylbutanone-2, 465
3-Methyl-2-butanone, 465
1-Methyl-3-butene, 450
2-Methyl-3-butene, 451
3-Methyl-1-butene, 451
Methyl β-n-Butoxypropionate, 1555
1-Methylbutyl Acetate, 1218
1-Methyl-2-butylacetylene, 1178
Methyl-1,3-Butylene Glycol Acetate, 1226
Methyl tert-Butyl Ether, 512
3-Methylbutyl o-Hydroxybenzoate, 2683
Methyl Butyl Ketone, 848
Methyl n-Butyl Ketone, 848
Methyl-n-butylnitrosamine, 509
2-Methyl-5-t-butylphenol, 2317
N-Methyl O-sec-butylPhenylcarbamate, 2685
3-Methylbutyl Salicylate, 2683
Methyl Butyrate, 476
n-Methyl n-Butyrate, 476
8-Methyl Caffeine, 1745
Methyl Caprate, 2365
Methyl Caproate, 1222
Methyl Caprylate, 1822
Methyl Carbamate, 90
Methylcarbamic Acid, 1208
Methylcarbamic Acid, Ester with N'-(m-hydroxyphenyl)-N,N-dimethylformamidine, 2295
1-Methylcarbamoyl-5-fluoro-2,4(1H,3H)-pyrimidinedione, 694
1-Methylcarbamoyl-5-fluorouracil, 694
N-Methyl-N-Carbamoyl Methyl Glycolamide Salicylate, 3813
N-Methyl-N-carbamoylmethyl Glycolamide Salicylate, 2652
N-Methylcarbamoyloxime,2-methyl-2-methylsulfenylpropionaldehyde, 1208

Methyl Chloramben, 1334
Methyl Chloride, 14
1-Methyl-2-chlorobenzene, 1073
2-Methylchlorobenzene, 1073
O-Methyl O-2-Chloro-4-tert-butyphenyl
    N-Methylamidophosphate, 2721
Methyl 2-chloro-3-(p-chlorophenyl)propionate, 1934
Methyl-2-chloro-9-hydroxyfluorene-9-carboxylate, 3142
2-Methyl-4-chloro-phenol, 1079
2-Methyl-6-chloro-phenol, 1078
3-Methyl-4-chlorophenol, 1074
3-Methyl-4-chloro-phenol-, 1074
4-(2-Methyl-4-chlorophenoxy)butyric Acid, 2253
2-(2-Methyl-4-chlorophenoxy)propionic Acid, 1955
Methylchloropindol, 1080
Methylchlorotetracycline, 3807
Methylchlothiazide, 1694
20-Methylcholanthrene, 3799
3-Methylcholanthrene, 3799
5-Methylchrysene, 3624
6-Methylchrysene, 3625
Methyl Cinnamate, 1946
trans-α-Methyl-cinnamic Acid, 1947
N-Methylcinnamide, 1960
α-Methyl-crotonaldehyd, 429
α-Methylcrotonaldehyde, 429
Methyl Cyclohexane, 1205
Methylcyclohexane, 1205
10-Methyl-1,2-cyclohexane Anthracene, 3636
Methyl Cyclohexanone, 1187
2-Methylcyclohexanone, 1187
3-Methylcyclohexanone, 1188
m-Methylcyclohexanone, 1188
1-Methylcyclohexene, 1176
1-Methyl-1-cyclohexene, 1176
1-(2-Methylcyclohexyl)-3-phenylurea, 3080
Methylcyclopentane, 820
o-Methylcyohexanone, 1187
5-Methylcytosine, 417
2-Methyldecalin, 2340
Methyl Decanoate, 2365
Methyl 2,2-Diallylmalonurate, 2100
Methyl-2,2-diallylmalonurate, 2100
Methyldiantipyrylmethane, 3945
Methyl Dibutyl Phosphate, 1854
1-Methyl-1,1-dichloro-2,2-bis(p-ethoxylphenyl)ethane,
    3662
Methyl α-p-Dichlorohydrocinnamate, 1934
Methyl (2,4-Dichlorophenoxy)acetate, 1633
Methyl (+/-)-2-[4-(2,4-Dichlorophenoxy)
    phenoxy]propionate, 3288
N-Methyldiethylamine, 524
Methyl 2,2-Diethylmalonurate, 1789
Methyl-2,2-diethylmalonurate, 1789
Methyl Dihydrojasmonate, 2927
Methyl 4-Dimethylaminobenzoate, 2032
Methyl p-Dimethylaminobenzoic Acid, 2032
Methyl N-(2,6-Dimethyl-phenyl)-N-(2'-methoxyacetyl)-
    DL-alaninate, 3229
Methyl Diphenyl Phosphate, 2826
Methyldithiopyrylmethane, 3946
Methyl Dixanthogen, 278
Methyl Dodecanoate, 2935
18-Methyleicosane, 3858

19-Methyleicosane, 3857
2-Methyleicosane, 3857
3-Methyleicosane, 3858
Methylenblau, 3356
Methylen-citronensaeure, 1129
2,2'-Methylenebiphenyl, 2787
1,1'-Methylenebis-benzene, 2815
4,4'-Methylenebisbenzeneamine, 2829
3,3'-Methylene-bis(4-hydroxycoumarin), 3617
4,4'-Methylenebis[3-hydroxy-2-naphthalenecarboxylic
    Acid], 3909
2,2'-Methylenebis[3,4,6-trichlorophenol], 2772
Methylene Blue, 3356
Methylene Bromide, 8
Methylene Chloride, 9
Methylenecitric Acid, 1129
4,4'-Methylenedianiline, 2829
4,4'-Methylenediantipyrine, 3915
Methylene Dichloride, 9
Methylene, (Diethoxyphosphinyl)phenyl-, 2298
Methylenedioxy Procatechuic Aldehyde, 1318
Methylene Iodide, 10
4,5-Methylenephenanthrene, 3134
Methylester Propanoic Acid, 329
Methyl Ether, 100
Methyl 3-Ethoxypropionate, 867
Methyl β-Ethoxypropionate, 867
Methyl 2-Ethyl-2-allylmalonurate, 2312
Methyl-2-ethyl-2-allylmalonurate, 2312
5-Methyl-5-ethylbarbituric Acid, 1163
1-Methyl-2-ethylbenzene, 1727
1-Methylethyl N-Benzoyl-N-(3-chloro-4-fluorophenyl)-
    DL-alanine, 3640
1-Methylethyl-4-bromo-α-(4-bromophenyl)-α-
    hydroxybenzeneacetate, 3423
1-Methylethyl-4-chloro-α-(4-chlorophenyl)-α-
    hydroxybenzeneacetate, 3425
3-Methyl-5-ethyl-4-chloro-phenol, 1693
Methyl Ethylene, 152
Methyl Ethylene Oxide, 173
Methylethyl 2-((Ethoxy((1-methylethyl)
    amino)phosphinothioyl)oxy)benzoate, 3253
Methylethyl (E)-3-(((Ethylamino)methoxy-
    phosphinothioyl)oxy)-2-butenoate, 2156
1-Methyl-2-ethylethylene, 449
sym-Methylethylethylene, 449
1-(Methylethyl)-O-ethyl-O-(3-methyl-4-
    (methylthio)phenyl)phosphoramidate, 2923
Methyl Ethyl Ketone, 325
2-(1-Methylethyl)-4-nitro-1H-imidazole, 781
O-Methyl O-Ethyl O-4-Nitrophenyl Thiophosphate, 1735
3-Methyl-4-ethylphenol, 1751
3-Methyl-5-ethyl-phenol, 1750
Methyl 2-Ethyl-2-phenylmalonurate, 2856
Methyl-2-ethyl-2-phenylmalonurate, 2856
5-(1-Methylethyl)-5-(2-propenyl)-2,4,6(1H,3H,5H)-
    pyrimidinetrione, 2061
Methylethylthiofos, 1735
Methylethylthiophos, 1735
O-Methyl O-Ethyl O-2,4,5-Trichlorophenyl
    Thiophosphate, 1668
1-Methylfluorene, 2988
1-Methyl-9H-fluorene, 2988
Methylfluoride, 16

Methyl Furoate, 727
5-Methylfuroic Acid, 727
Methyl Gallate, 1403
Methylgelb, 3023
N-Methylglucamine Cholesterol, 4114
N-Methylglucamine 9-α-Fluorohyfrocortisome
    (Monohydrate), 4078
N-Methylglucamine Testosterone, 4046
N-Methylglucamine Triamcinolone Acetonide
    (Monohydrate), 4094
α-Methyl-D-glucosid, 1233
β-Methyl-D-glucosid, 1231
α-D-Methylglucoside, 1233
α-Methyl-D-glucoside, 1233
β-Methyl-D-glucoside, 1231
Methyl Glycol Phthalate, 3066
Methylglyoxim, 164
Methylglyoxime, 164
Methyl Gusathion, 1993
6-Methyl-2,4-heptadione, 1504
2-Methyl-Heptanamide, 1564
2-Methylheptane, 1575
3-Methylheptane, 1573
2-Methyl-4,6-heptanedione, 1504
(1-Methylheptyl)benzene, 3097
Methyl n-Heptyl Carbinol, 1848
2-Methylhexane, 1244
3-Methylhexane, 1246
3-Methylhexanedioic Acid, 1191
Methyl Hexanoate, 1222
2-Methylhexanol-2, 1254
2-Methyl-2-hexanol, 1254
3-Methylhexanol-3, 1252
3-Methyl-3-hexanol, 1252
Methyl Hexyl Carbinol, 1583
2-Methyl-3-hexyne, 1179
1-Methyl-L-histidine, 1171
L-1-Methylhistidine, 1171
Methylhydantoic Acid, 317
3-Methyl-4-hydroxyacetanilide, 1710
Methyl Hydroxybenzoate, 1391
Methyl 4-Hydroxybenzoate, 1390
Methyl p-Hydroxybenzoate, 1385
Methyl p-Hydroxybenzoic Acid, 1390
Methyl m-Hydroxycarbanilate m-Methylcarbanilate,
    3313
3-Methyl-indol, 1651
3-Methylindole, 1651
Methyliodide, 17
Methyl-iodide, 17
Methyl Iodine, 17
Methyl Ionone, 3104
6-Methylionone, 3104
Methyl Isobutyl Ketone, 849
Methyl-isocyanid, 67
Methylisocyanide, 67
1-Methyl-4-isopropenyl-6-cyclohexen-2-one, 2071
1-Methyl-4-isopropylbenzene, 2053
Methylisopropylcarbinol, 514
Methyl Isopropyl Ether, 365
1-Methyl-2-isopropyl-5-nitro-imidazole, 1172
2-Methyl-5-isopropylphenol, 2072
5-Methyl-2-isopropyl-1-phenol, 2073
3-Methyl-isoquinoline, 1910

Methyl Isothiocyanate, 68
Methyl m-Isothiocyanobenzoate, 1628
5-Methyl-3(2H)-isoxazolone, 255
Methyl Laurate, 2935
Methylmalonic Acid, 280
Methyl-malonsaeure, 280
α-Methyl-D-mannosid, 1232
α-Methyl-D-mannoside, 1232
2-Methylmercapto-4-isopropylamino-6-diethylamino-s-
    triazine, 2358
4-Methylmercapto-3-methylphenyl Dimethyl
    Thiophosphate, 2088
Methylmercuridicyanodiamide, 170
Methyl Mesylate, 102
Methyl Methacrylate, 431
Methyl Methanesulphonate, 102
Methyl-4-methoxybenzoate, 1686
Methyl 3-Methoxypropanoate, 482
Methyl 3-Methoxypropionate, 482
Methyl β-Methoxypropionate, 482
Methyl-N'-methyl-N'-(2-benzothiazolyl)urea, 1969
5-Methyl-5-(3-methylbut-2-enyl)barbituric Acid, 2062
Methyl 2-Methyl-2-cyclohexenyl-6-methylmalonurate,
    2895
Methyl-2-methyl-2-cyclohexenyl-6-methylmalonurate,
    2895
(R)-1-Methyl-4-(1-methylethenyl)cyclohexene, 2090
2-Methyl-2-(1-methylethoxy)-propane, 1253
3-Methyl-2-(1-methylethyl)Butanamide, 1562
(-)-5-Methyl-2-(1-methylethyl)cyclohexanone, 2125
N-Methyl-N'-(1-methylethyl)-6-(methylthio)-1,3,5-
    triazine-2,4-diamine, 1522
5-Methyl-2-(1-methylethyl)phenol, 2073
N1-Methyl-N1-(5-methyl-3-isoxazolyl)sulfanilamide,
    2269
N-Methyl-α-methyl-α-phenylsuccinimide, 2605
Methyl 2-Methyl-2-Propenoate, 431
2-Methyl-2-(methylthio)propanal O-
    [(Methylamino)carbonyl]oxime, 1208
Methylmorphin, 3564
1-Methyl Naphthalene, 2202
1-Methylnaphthalene, 2202
2-Methyl Naphthalene, 2201
2-Methylnaphthalene, 2201
α-Methylnaphthalene, 2202
α-Methyl Naphthalenes, 2202
β-Methyl Naphthalenes, 2201
2-Methyl-1,4-naphthoquinone, 2194
1-Methyl-napthalene, 2202
O-Methylnarcotoline, 3878
Methyl Nicotinate, 1093
Methyl N-(4-Nitrobenzenesulphonyl)carbamate, 1367
3-Methyl-2-nitrobenzoic Acid, 1342
3-Methyl-6-nitrobenzoic Acid, 1343
5-Methyl-2-nitrobenzoic Acid, 1343
1-Methyl-1-nitro-2,2-bis(p-ethoxylphenyl)ethane, 3670
1-Methyl-1-nitro-2,2-bis(p-methoxylphenyl)ethane, 3451
2-Methyl-4(5)-nitroimidazole, 259
2-Methyl-5-nitroimidazole, 259
2-Methyl-5-nitro-1-imidazoleethanol, 783
2-Methyl-5-nitroimidazole-1-ethanol, 783
6-(1-Methyl-p-nitro-5-imidazolyl)-thiopurine, 1631
1-Methyl-1-nitro-2-(p-methylphenyl)-2-p-
    Ethoxylphenyl)ethane, 3447

1-Methyl-3-nitro-1-nitrosoguanidine, 96
N-Methyl-N'-Nitro-N-Nitrosoguanidine, 96
3-Methyl-4-nitrophenol, 1096
Methylnitrosoacetamide, 165
Methylnitrosourea, 95
7α-Methyl-19-nortestosterone, 3690
3-Methyloctane, 1842
4-Methyloctane, 1840
Methyl Octanoate, 1822
Methyl-octyl-alcohol, 1848
Methyl Octylate, 1822
N-Methylolpentamethylmelamine, 1811
N-Methylolpentamethylmelamine Methyl Ether, 2159
Methylol-riboflavin, 3835
Methylol Riboflavine, 3835
2'-(2-Methyl-3-one-pentanyl)-6-methoxypurine
    Arabinoside (0.75 Hydrate), 3480
1-Methylorotic Acid, 708
Methyl Oxalate, 282
Methyl Oxolane, 467
2-Methyl Oxolane, 466
Methyl 3-Oxo-2-pentylcyclopentaneacetate, 2927
Methyl Paraben, 1390
Methylparaben, 1390
3-Methylparacetamol, 1710
Methyl Parathion, 1433
Methylparathion, 1433
2-Methylpentadecane, 3402
3-Methylpentadecane, 3400
3-Methyl-1,3-pentadione, 795
2-Methylpentane, 903
3-Methylpentane, 901
1-Methyl Pentanol, 919
2-Methyl-2-pentanol, 922
2-Methyl-3-pentanol, 931
3-Methylpentanol, 920, 932
3-Methyl-1-pentanol, 920
3-Methyl-2-pentanol, 924
3-Methyl-3-pentanol, 925
4-Methyl-1-pentanol, 923
4-Methyl-2-pentanol, 930
3-Methylpentanone-2, 852
3-Methyl-2-pentanone, 852
4-Methyl-2-pentanone, 849
4-Methylpentanone-3, 855
4-Methyl-3-pentanone, 855
2-Methyl-1-pentene, 819
4-Methylpentene, 821
4-Methyl-1-pentene, 821
4-Methyl-4-pentene, 819
2-Methylpenten-4-ol-3, 853
2-Methyl-4-penten-3-ol, 853
Methylpentylacetamide, 1564
3-Methyl-2-pentyl Alcohol, 924
2-Methyl-2-pentyl-1,3-dioxolane-4-methanol, 2175
1-Methylphenanthrene, 3148
1-(Methylphenethylamino)-3,5-bis(dimethylamino)-s-
    triazine, 3388
Methylphenobarbital, 2830
2-Methyl-phenol-, 1119
m-Methylphenol, 1115
o-Methylphenol, 1119
p-Methylphenol, 1116
3-(2-Methylphenoxy)-pyridazine, 2210

3-Methylphenyl Bromide, 1070
Methyl Phenyl Ether, 1118
2-Methyl-N-phenyl-3-furancarboxamide, 2576
3-Methylphenyl Isothiocyanate, 1345
m-Methylphenyl Isothiocyanate, 1345
Methyl Phenyl Ketone, 1378
2-Methyl-1-phenylpropane, 2051
N-Methyl-2-phenyl-succinimide, 2218
Methyl-4-(phenylsulfonyl)trifluoromethanesulfonanilide,
    2991
2-Methylphenyl Tosylate, 3018
1-Methyl-3-phenylurea, 1434
N-Methyl-N'-phenylurea, 1434
Methylphthalyl Ethyl Glycolate, 2835
Methyl Phthalyl Ethyl Glycollate, 2862
Methyl Picrate, 1024
Methyl Picric Acid, 1026
1-Methyl-2',6'-pipecoloxylidide, 3242
2-Methyl-piperazin, 508
2-Methylpiperazine, 508
2-(4-Methyl-1-piperazinyl)-4,6-bis(dimethylamino)-s-
    triazine, 2747
S-(2-(2-Methyl-1-piperidinyl)-2-oxoethyl) O,O-dipropyl
    Phosphorodithioate, 3114
1-(4'-Methyl-1-piperizinyl)-3,5-bis(dimethylamino)-s-
    triazine, 2747
Methylprednisolone, 3893
6α-Methylprednisolone, 3893
2-Methyl Propanal, 323
2-Methylpropane, 357
2-Methyl-1-propanol, 364
2-Methyl-2-propanol, 366
2-Methylpropene, 305
2-Methyl-2-Propenenitrile, 253
Methyl Propionate, 329
Methyl β-n-Propoxypropionate, 1229
(2-Methylpropyl)-benzene, 2051
1-Methylpropylbenzene, 2045
Methyl Propyl Carbinol, 510
Methyl Propyl Ether, 363
Methyl Propyl Ketone, 470
2-Methylpropyl 2-Methylpropanoate, 1544
N-(2-Methylpropyl)-2-oxo-1-imidazolidinecarboxamide,
    1518
2-(1-Methylpropyl)phenol Methylcarbamate, 2685
2-Methyl-2-propyl-1,3-propanediol Dicarbamate, 1806
1-Methylpropyn-2-yl N-(m-Chlorophenyl)carbamate,
    2205
Methylprotocatechuic Aldehyde, 1396
2-Methylpteridine, 1043
4-Methylpteridine, 1041
7-Methylpteridine, 1042
8-Methylpurine, 712
N1-Methyl-N1-(2-pyridyl)sulfanilamide, 2616
N1-Methyl-N1-2-pyridyl-sulfanilamide, 2616
3-(1-Methyl-2-pyrrolidinyl)-indole, 2854
2-Methyl 8-Quinolinol, 1911
Methyl Salicylate, 1391
Methylsecodione, 3753
α-Methylstyrene, 1663
Methyl Styryl Ketone, 1944
Methylsuccinic Acid, 435
N1-Methylsulfamethoxazole, 2269
N1-Methylsulfanilamide, 1160

N1-Methylsulfathiazole, 1972
N-(4-Methyl-2-sulfamoyl-D2-1,3,4-thiadiazolin-5-ylidene)acetamide, 426
1-Methyl-2-sulfanilamide-1,2-dihydropyridine, 2656
3-Methyl-2-sulfanilamide-2,3-dihydrothiazole, 2035
4-Methyl-2-sulfanilamidothiazole, 1973
Methyl Sulfathiazole, 1970
7-Methylsulfinyl-2-xanthonecarboxylic Acid, 3138
4-(Methylsulfonyl)-2,6-dinitro-N,N-dipropylaniline, 2909
2'-Methylsulfonyl-6-methoxypurine Arabinoside, 2677
2-(p-Methylsulfoxidephenyl)-1,1,1-trichloroethane, 3299
Methyltestosterone, 3779
Methyl-testosterone, 3779
17-Methyltestosterone, 3779
17-α-Methyltestosterone, 3779
Methyltestosterone Acetate, 3895
17-α-Methyltestosterone Acetate, 3895
17-α-Methyltestosterone Propionate, 3935
Methyl-2,3,5,6-tetrachloro-N-methoxy-N-methylterephthalamate, 2199
1-Methyl Tetrahydrofuran, 467
2-Methyl Tetrahydrofuran, 466
α-Methyl Tetramethylene Oxide, 467
β-Methyl Tetramethylene Oxide, 466
N-Methyl-N,2,4,5-tetranitroaniline, 1027
2-Methyl-1,3,4-thiadiazole, 228
N1-Methyl-N1-2-thiazolyl-sulfanilamide, 1972
Methylthiobenzothiazole, 1315
Methylthio-4,6-bis(ethylamino)-s-triazine, 1523
2-Methylthio-4-ethylamino-6-tert-butylamino-s-triazine, 2151
(2-Methylthio-4-ethylamino-6-isopropylamino-s-triazine, 1802
2-Methylthio-4-isopropylamino-6-(3-methoxypropylamino)-s-triazine, 2355
Methylthio-4-isopropylamino-6-methylamino-s-triazine, 1522
4-(Methylthio)phenyl Dipropyl Phosphate, 2922
2-(p-Methylthiophenyl)-2-(p-methylsulfinylphenyl)-1,1,1-trichloroethane, 3297
2-Methylthiopteridine, 1053
4-Methylthiopteridine, 1050
7-Methylthiopteridine, 1051
Methylthiouracil, 407
5-Methyl-2-thiouracil, 406
6-Methyl-2-thiouracil, 407
7-Methylthio-2-xanthonecarboxylic Acid, 3137
2-Methyl-3-(o-tolyl)-6-sulfamyl-7-chloro-1,2,3,4-tetrahydro-4-quinazolinone, 3310
5-Methyl-1,2,4-triazolo[3,4-b]benzothiazole, 1630
Methyl-1,2,4-triazolo(3,4-b)benzothiazole, 1630
Methyl Tribromide, 3
Methyl Trichloride, 5
Methyl 2,4,6-Trichlorophenyl Ether, 996
Methyl-3,4,5-trihydroxybenzoate, 1403
3-Methyl-2,4,6-trinitrophenol, 1026
1-Methyluracil, 408
3-Methyluracil, 404
4-Methyl-uracil, 416
5-Methyluracil, 409
6-Methyluracil, 416
Methyl Urethane, 90
1-Methyluric Acid, 715

9-Methyluric Acid, 714
α-Methyluric Acid, 715
Methyl Vinyl Carbinol Acetate, 794
α-Methyl-trans-zimtsaeure, 1947
N1-Methyluracil, 408
N9-Methyluric Acid, 714
Methy Propyl Ketone, 470
Methyprylon, 2113
17-β-Methythiocarbonyl-9α-fluoro-11β, 3950
Metiazinic Acid, 3164
Metizolin, 1604
Metobromuron, 1690
Metocorten, 3823
Metolachlor, 3240
Metolazone, 3310
Metoxuron, 2019
Metribuzin, 1499
Metrisone, 3893
Metronidazole, 783
Metrozine, 783
2-Metyloheksan, 1244
3-Metyloheksan, 1246
3-Metyloheptan, 1573
4-Metylooktan, 1840
2-Metylopentan, 903
3-Metylopentan, 901
Mexacarbate, 2701
Mexacarbole, 2701
Mezopur, 1605
Michler's Ketone, 3455
Miconazole, 3524
Miconazole Nitrate-β Cyclidextrin Complexant, 3527
Micotiazone, 1654
Microcide I, 263
Milchzucker, 2742
Milcurb Super, 2338
Milgo, 2338
Milk Sugar, 2741
Milontin, 2218
Milstem, 2338
Miltown, 1806
Minims, 3483
Minocin, 3924
Minocycline, 3924
Mintezol, 1888
Miotisal, 2057
Mipafox, 947
Mipax, 1952
Miral, 1797
Mirex, 2185
Mitac, 3672
Mitomycin C, 3204
Mitomycinum, 3204
MMC, 3204
MNA, 165
MNNG, 96
MNU, 95
Mobilis, 3167
Modecate, 3884
Modown, 2955
Modown 4 Flowable, 2955
Mogadon, 3145
Molinate, 1798
MON 5783, 685

Monacrin, 2795
Monalide, 2889
Monobromobenzene, 625
Monochlorobenzene, 630
4-Monochloro-biphenyl, 2537
1-Monochlorodibenzodioxin, 2489
2-Monochlorodibenzo-p-dioxin, 2488
Monochloroethane, 84
Monocrotaline, 3381
(-)-Monocrotaline, 3381
Monoctylamine, 1587
Monodox, 3880
Mono-Ethyl Malonate, 436
D-Monofeniltartramide Tartranilamide, 1991
Monolinuron, 1692
1-Monoolein, 3856
α-Monoolein, 3856
Monotrim, 3057
Monuron, 1691
Mordant Violet 26, 2950
Morin, 3140
Morkit, 2946
Morphin, 3449
Morphinan-3,6-diol, 7,8-Didehydro-4,5-epoxy-17-methyl-
     (5α,6α)-, Monohydrate, 3450
Morphinan-6-ol, 7,8-Didehydro-4,5-epoxy-3-methoxy-17-
     methyl-, Monohydrate, (5α,6α), 3565
Morphinan-6-one, 7,8-Didehydro-4-hydroxy-3-methoxy-
     17-methyl-, 3563
Morphinan, 6,7,8,14-Tetrahydro-4,5α-epoxy-3,6-
     dimethoxy-17-methyl-, 3658
Morphine, 3449
Morphine (Monohydrate), 3450
Morpholine, 4-[(Benzoyloxy)acetyl]-, 2840
Morpholine, 4,4',4"-Phosphinothioylidynetris-, 2749
Morpholine, 4,4',4"-Phosphinylidynetris-, 2750
1-(Morpholinyl)-3,5-bis(dimethylamino)-s-triazine, 2345
Mothballs, 1889
Mowilith Red 3B(IG), 4006
MP 620, 2604
MPD, 1564
MPMC, 2027
Muconic Acid, 735
Muconsaeure, 735
Multimet, 2264
Murex, 1477
Murocoll, 3465
Murvin, 2575
Mustard Gas, 311
E-Mycin, 4122
Mycophenolic Acid, 3462
Mycostatin, 4141
Mykrox, 3310
Myristic Acid, 3116
Myristohydroxamic Acid, 3120
Myristyl Alcohol, 3127
NAA, 2568
Nalidic Acid, 2594
Nalidixic Acid, 2594
Nandrolone Butyrate, 3898
Nandrolone Decanoate, 4060
Nandrolone Nonanoate, 4045
Nandrolone Undecanoate, 4081
Naphthacene, 3515

Naphthalene, 1889
1-Naphthaleneacetic Acid, 2568
2-Naphthaleneacetic Acid, 6-Methoxy-α-methyl-, 2-
     Amino-2-oxoethyl Ester, 3324
2-Naphthaleneacetic Acid, 6-Methoxy-α-methyl-, 2-
     Amino-2-oxoethyl Ester, (S), 3324
2-Naphthaleneacetic Acid, 6-Methoxy-α-methyl-, 2-[(2-
     Amino-2-oxoethyl)methylamino]-2-oxoethyl Ester,
     3666
2-Naphthaleneacetic Acid, 6-Methoxy-α-methyl-, 2-
     [bis(2-Hydroxyethyl)amino]-2-oxoethyl Ester, 3761
2-Naphthaleneacetic Acid, 6-Methoxy-α-methyl-, 2-
     (Diethylamino)-2-oxoethyl Ester, 3758
2-Naphthaleneacetic Acid, 6-Methoxy-α-methyl-, 2-
     (Diethylamino)-2-oxoethyl Ester, (S), 3758
2-Naphthaleneacetic Acid, 6-Methoxy-α-methyl-, 2-
     (Dimethylamino)-2-oxoethyl Ester, 3566
2-Naphthaleneacetic Acid, 6-Methoxy-α-methyl-, 2-
     (Dimethylamino)-2-oxoethyl Ester, (S), 3566
2-Naphthaleneacetic Acid, 6-Methoxy-α-methyl-, Ethyl
     Ester, (alphaS)-, 3337
2-Naphthaleneacetic Acid, 6-Methoxy-α-methyl-, 2-[(2-
     Hydroxyethyl)methylamino]-2-oxoethyl Ester, 3671
Naphthalene, 1-Bromo-, 1880
Naphthalene, 2-Bromo-, 1879
2-Naphthalenecarboxylic Acid, 2195
Naphthalene, 1,4-Dibromo-, 1865
Naphthalene, 2,3-Dibromo-, 1866
Naphthalene, 1,4-Dichloro-, 1867
1,2-(1,8-Naphthalenediyl)benzene, 3270
Naphthalene, 1,4,5-Trimethyl-, 2827
1,3,6-Naphthalenetrisulfonic Acid, 4-Amino-, 1926
Naphthionic Acid, 1920
Naphtho[1,2-b]furan-2,8(3H,4H)-dione, 3α,5,5α,9β-
     Tetrahydro-3,5α,9-trimethyl-, (3S,3αS,5αS,9βS)-, 3205
2-Naphthoic Acid, 2195
β-Naphthoic Acid, 2195
1-Naphthol, 1896
2-Naphthol, 1897
α-Naphthol, 1896
β-Naphthol, 1897
(2-Naphthoxy)acetic Acid, 2569
β-Naphthoxyacetic Acid, 2569
α-Naphthoylamine, 1908
β-(α-Naphthyl)-β-alanine, 2822
α-Naphthylamin, 1908
β-Naphthylamin, 1909
Naphthylamine-(2), 1909
1-Naphthylamine, 1908
2-Naphthylamine, 1909
α-Naphthylamine, 1908
β-Naphthylamine, 1909
α-Naphthylamine-o-monosulfonic Acid, 1924
1-Naphthylamine-2-sulfonic Acid, 1921
1-Naphthylamine-4-sulfonic Acid, 1920
1-Naphthylamine-5-sulfonic Acid, 1916
1-Naphthylamine-6-sulfonic Acid, 1923
1-Naphthylamine-8-sulfonic Acid, 1919
2-Naphthylamine-1-sulfonic Acid, 1924
2-Naphthylamine-5-sulfonic Acid, 1915
2-Naphthylamine-6-sulfonic Acid, 1922
2-Naphthylamine-7-sulfonic Acid, 1917
2-Naphthylamine-8-sulfonic Acid, 1918
1-Naphthylamine-2,4,7-trisulfonic Acid, 1926

2-((1-Naphthylamino)carbonyl)benzoic Acid, 3520
Naphthylamin-(1)-sulfosaeure-(2), 1921
Naphthylamin-(1)-sulfosaeure-(4), 1920
Naphthylamin-(1)-sulfosaeure-(5), 1916
Naphthylamin-(1)-sulfosaeure-(6), 1923
Naphthylamin-(1)-sulfosaeure-(8), 1919
Naphthylamin-(2)-sulfosaeure-(5), 1915
Naphthylamin-(2)-sulfosaeure-(6), 1922
Naphthylamin-(2)-sulfosaeure-(7), 1917
Naphthylamin-(2)-sulfosaeure-(8), 1918
1-Naphthyl Chloride, 1881
1-Naphthyl Isothiocyanate, 2192
2-Naphthyl Isothiocyanate, 2191
1-Naphthyl N-Methylcarbamate, 2575
α-Naphthyl Mustard Oil, 2192
β-Naphthyl Mustard Oil, 2191
Naphthylphthalamic Acid, 3520
N-1-Naphthylphthalamic Acid, 3520
1-Naphthylthiourea, 2212
Napropamide, 3463
Naproxen, 3017
Naproxen, N,N-Diethanol Glycolamide Ester, 3761
Naproxen, N,N-Diethyl Glycolamide Ester, 3758
Naproxen, N,N-Diethyl Glycolamide Ester, 3758
Naproxen,N,N-Dihydroxyethyl Glycolamide Ester, 3761
Naproxen, N,N-Dimethyl Glycolamide Ester, 3566
Naproxen N,N-Dimethyl Glycolamidc Ester, 3566
Naproxen Ethyl Ester, 3337
Naproxen, N,N-Glycolamide Ester, 3324
Naproxen, N,N-Glycolamide Ester, 3324
Naproxen, N-Methyl-N-carbamoyl Methyl-glycolamide
     Ester, 3751
Naproxen N-Methyl -N-Carbamoyl Methyl Glycolamide
     Ester, 3666
Naproxen N-Methyl-N-Ethanol Glycolamide Ester,
     3671
Naproxen, N-Methyl-N-hydroxyethyl Glycolamide Ester,
     3760
Naptalam, 3520
Napthalene, 1889
Naramycin, 3248
Narceine, 3923
Narcotine, 3878
Naringin, 4034
Natamycin, 4104
Naturetin, 3170
Navilene Blue GL, 3293
Naxol, 850
NdiPA, 909
NDPA, 908
NEBA, 910
Neburon, 2663
NegGRAM, 2594
Nemacide, 2022
Nemacur, 2923
Nemagon, 138
Nematak, 828
Nendrin, 2523
Neo-Diophen, 3483
Neohetramine, 3365
Neohexane, 900
Neopentane, 504
Neopentyl Alcohol, 511
Neoron, 3423

Neothylline, 2068
Nephocarp, 2302
Neptazaneat, 426
Neral, 2106
Neurolidol, 3874
Nevarapine, 3174
Nevirapine, 3174
Nevulose, 873
Nexagan, 1978
Nexion, 1352
NHMI, 830
Niacin, 650
Niacinamidc, 702
Nialate, 1858
Nibroxane, 423
Niclosamide, 2778
Nicotiamide, 702
Nicotinamide, 702
Nicotinic Acid, 650
Nicotinic Acid n-Hexyl Ester, 2689
Nicotinic Acid n-Octyl Ester, 3090
3-Nicotinoyl-5-fluorouracil, 1875
1-Nicotinoyloxymethyl Allopurinol, 2552
Nicotinsaeure-methyl Ester, 1093
Nifedipine, 3439
Niflumic Acid, 2781
Nifuradene, 1372
Nifurtoinol, 1636
NIH 10760, 3923
Nimrod, 2929
Niridazole, 716
Nirodazole, 716
Nirvanol, 2235
Nisolone, 3831
Nitralin, 2909
Nitramine, 1027
Nitrapyrin, 568
Nitrazepam, 3145
o-Nitroacetanilide, 1365
p-Nitroacetanilide, 1366
2-Nitroacetotoluide, 1672
4-Nitroacetotoluide, 1673
o-Nitroacetotoluide, 1672
p-Nitroacetotoluide, 1673
1-Nitro-4-acetylaminobenzene, 1366
4'-Nitro-4-aminoazobenzene, 2562
m-Nitroaminobenzene, 705
Nitroaminoguanidine, 29
1-Nitro-3-aminoguanidine, 29
3-Nitro-anilin, 705
2-Nitro-anilin-N-acetat, 1365
4-Nitro-anilin-N-acetat, 1366
2-Nitroaniline, 704
2-Nitro-aniline, 704
3-Nitroaniline, 705
4-Nitroaniline, 706
m-Nitroaniline, 705
o-Nitroaniline, 704
p-Nitroaniline, 706
2-Nitroaniline-N-acetate, 1365
4-Nitroaniline-N-acetate, 1366
4-Nitroanisol, 1098
4-Nitro-anisol, 1098
p-Nitroanisol, 1098

4-Nitroazobenzene, 2548
5-Nitrobarbituric Acid, 229
2-Nitro-benzaldehyd, 1010
3-Nitro-benzaldehyd, 1008
2-Nitrobenzaldehyde, 1010
3-Nitrobenzaldehyde, 1008
4-Nitrobenzaldehyde, 1009
m-Nitrobenzaldehyde, 1008
o-Nitrobenzaldehyde, 1010
p-Nitrobenzaldehyde, 1009
3-Nitrobenzenamine, 705
4-Nitrobenzenamine, 706
Nitrobenzene, 651
2-Nitro-1,3-benzenediol, 659
3-Nitro-1,2-benzenediol, 655
4-Nitro-1,2-benzenediol, 656
4-Nitro-1,3-benzenediol, 658
2-Nitrobenzenesulfonamide, 709
3-Nitrobenzenesulfonamide, 711
4-Nitrobenzenesulfonamide, 710
m-Nitrobenzenesulfonamide, 711
o-Nitrobenzenesulfonamide, 709
p-Nitrobenzenesulfonamide, 710
5-p-Nitrobenzenesulfonamidotetrazole, 2813
4-Nitrobenzenesulfonic Acid, 660
p-Nitrobenzenesulfonic Acid, 660
p-Nitrobenzenesulfonic Acid (Dihydrate), 661
p-Nitrobenzenesulfonic Acid (Tetrahydrate), 662
1-(p-Nitrobenzenesulfonyl)-5,5-diphenyl-hydantoin,
    3798
2-Nitrobenzoic Acid, 1015
3-Nitrobenzoic Acid, 1016
4-Nitrobenzoic Acid, 1014
m-Nitrobenzoic Acid, 1016
o-Nitrobenzoic Acid, 1015
p-Nitrobenzoic Acid, 1014
Nitrobenzol, 651
3-Nitrobenzyl Isothiocyanate, 1311
4-Nitrobenzyl Isothiocyanate, 1312
m-Nitrobenzyl Isothiocyanate, 1311
p-Nitrobenzyl Isothiocyanate, 1312
o-Nitro-o-bromacetanilide, 1327
p-Nitro-o-bromacetanilide, 1328
1-Nitrobutane, 343
Nitrocarbol, 19
3-Nitrocatechol, 655
4-Nitrocatechol, 656
o-Nitro-o-chloracetanilide, 1329
p-Nitro-o-chloracetanilide, 1330
2-Nitrochlorobenzene, 585
4-Nitrochlorobenzene, 584
m-Nitrochlorobenzene, 583
2-Nitro-3-chlorobenzoic Acid, 964
2-Nitro-5-chlorobenzoic Acid, 966
3-Nitro-4-chlorobenzoic Acid, 965
4-Nitro-2-chloro-4'-[ethyl(2-
    hydroxyethyl)amino]azobenzene, 3321
3-Nitro-p-cresol, 1096
4-Nitro-4'-diethylaminoazobenzene, 3333
4-Nitro-4'-N,N-diethylaminoazobenzene, 3333
Nitroetan, 91
Nitroethane, 91
4-Nitro-4'-[ethyl(2-hydroxyethyl)amino]azobenzene,
    3335

Nitrofen, 2490
5-Nitro-2-Furaldehyde Semicarbazone, 717
Nitrofurantoin, 1313
Nitrofurazone, 717
1-[(5-Nitrofurfurylidene)amino]hydantoin, 1313
1-[5-Nitrofurfuryllidene)Amino]-2-Imidazolidinone,
    1372
Nitroglycerin, 150
Nitroglycerol, 150
Nitroguanidin, 24
Nitroguanidine, 24
α-Nitroguanidine, 24
Nitrohydroquinone, 657
Nitroisopropane, 199
3-Nitro-p-kresol, 1096
Nitromethane, 19
4-Nitro-2-methoxy-4'-di(β-hydroxyethyl)-
    aminoazobenzene, 3459
2-Nitro-3-methylbenzoic Acid, 1342
2-Nitro-5-methylbenzoic Acid, 1343
6-Nitro-3-methylbenzoic Acid, 1343
4-Nitro-5-methylphenol, 1096
1-Nitronaphthalene, 1884
1-Nitro-naphthalin, 1884
1-Nitro-naphthol-(2), 1886
1-Nitro-2-naphthol, 1886
Nitropentaerythritol, 427
3-Nitropentane, 490
5-Nitro-1,10-phenanthroline, 2508
5-Nitro-o-phenanthroline, 2508
2-Nitrophenol, 652
3-Nitrophenol, 654
4-Nitrophenol, 653
m-Nitrophenol, 654
o-Nitrophenol, 652
p-Nitrophenol, 653
4-(4-Nitrophenylazo)aniline, 2562
4-(4-Nitrophenylazo)diphenylamine, 3526
4-(p-Nitrophenylazo)diphenylamine, 3526
4-[(4-Nitrophenyl)azo]phenol, 2549
p-Nitrophenylazophenol, 2549
m-Nitrophenyldiantipyrylmethane, 4065
o-Nitrophenyldiantipyrylmethane, 4066
p-Nitrophenyldiantipyrylmethane, 4067
3-Nitrophenyl Isothiocyanate, 978
m-Nitrophenylisothiocyanate, 978
4-Nitrophenyl 2,4,6-trichlorophenyl Ether, 2458
p-Nitrophenyl α,α,α-Trifluoro-2-nitro-p-tolyl Ether, 2774
3-Nitrophthalic Acid, 1293
6-Nitro-phthalid, 1291
6-Nitrophthalide, 1291
3-Nitro-phthalsaeure, 1293
1-Nitropropane, 200
2-Nitropropane, 199
n-Nitropropane, 200
2-Nitroquinol, 657
2-Nitroresorcinol, 659
4-Nitroresorcinol, 658
3-Nitrosalicylic Acid, 1021
5-Nitrosalicylic Acid, 1020
3-Nitro-salicylsaeure, 1021
5-Nitrosalicylsaeure, 1020
N-Nitroso(benzyl)methylamine, 1438
N-Nitroso-N-butylethylamine, 910

N-Nitroso-di-n-Butylamine, 1578
N-Nitrosodiethylamine, 359
N-Nitrosodiisopropylamine, 909
Nitrosodipropylamine, 908
N-Nitroso(ethyl)-n-butylamine, 910
Nitroso-N-ethyl-n-butylamine, 910
Nitrosoethylurea, 209
N-Nitroso-N-Ethylurea, 209
N-Nitrosohexamethyleneimine, 830
4-Nitroso-hydrochlorothiazide, 1031
N-Nitroso-N-methylbenzenemethanamine, 1438
N-Nitroso(methyl)benzylamine, 1438
N-Nitroso-N-methylbenzylamine, 1438
N-Nitroso(methyl)pentylamine, 907
Nitrosomethylurea, 95
N-Nitroso-N-methyl-urethan, 315
N-Nitroso-N-methylurethane, 315
3'-Nitrosoniridazole, 675
N-Nitroso-N-Phenylaniline, 2558
N-Nitrosopiperidine, 453
N-Nitroso-N-propyl-1-propanamine, 908
3'-Nitroso-tolbutamide, 2696
1-Nitro-2,3,4,5-tetrachlorobenzene, 528
4-Nitrotoluene, 1090
m-Nitrotoluene, 1094
o-Nitrotoluene, 1092
p-Nitrotoluene, 1090
2-Nitro-m-toluic Acid, 1342
3-Nitro-o-toluidin, 1107
3-Nitro-o-toluidine, 1107
2-Nitro-toluol, 1092
3-Nitro-toluol, 1094
4-Nitro-3-(trifluoromethyl)benzenesulfonamide, 1000
NK 711, 2788
NM, 19
Nobrium, 3296
Nomersan, 841
2,3,4,5,6,2',3',4',5'-Nonachlorbiphenyl, 2376
2,2',3,3',4,4',5,5',6-Nonachlorobiphenyl, 2376
2,2',3,3',4,5,5',6,6'-Nonachlorobiphenyl, 2377
Nonadecylbenzene, 4002
1,8-Nonadiene, 2,8-Dimethyl-5-methylene-, 2724
1,8-Nonadiyne, 1726
n-Nonan, 1839
n-Nonanal, 1816
Nonan-dicarbonsaeure-(1,9), 2348
Nonane, 1839
Nonanecarboxylic Acid, 2167
Nonanediamide, N,N,N',N'-Tetramethyl-, 2934
1,9-Nonanedicarboxylic Acid, 2348
Nonanedioic Acid, 1794
n-Nonanoic Acid, 1824
Nonanol, 1846
2-Nonanol, 1848
3-Nonanol, 1847
5-Nonanone, 1815
1-Nonene, 1804
1-n-Nonene, 1804
α-Nonene, 1804
n-Non-1-ene, 1804
Nonyl Acarbinol, 2181
n-Nonyl Alcohol, 1846
Nonyl Aldehyde, 1816
Nonyl 4-Aminobenzoate, 3393

Nonyl p-Aminobenzoate, 3393
N-Nonylcinnamamide, 3593
Nonyl 4-Hydroxybenzoate, 3389
Nonyl p-Hydroxybenzoate, 3389
Nonylic Acid, 1824
2'-Nonyl-6-methoxypurine Arabinoside, 3776
Nonylol, 1845
Nonyloxycarbonyl-mitomycin C, 3997
4-Nonylphenol, 3259
4-t-Nonylphenol, 3259
4-Nonylphenol Diethoxylate, 3710
4-Nonylphenol Monoethoxylate, 3503
4-Nonylphenol Pentaethoxylate, 4003
4-Nonylphenol Tetraethoxylate, 3941
p-Nonylphenol Tetraethoxylate, 3941
4-Nonylphenol Triethoxylate, 3855
1-Nonyne, 1783
Norandrostenolone Decanoate, 4060
Noratone, 1520
Norazine, 1183
Norbormide, 4097
5-Norbornene-2,3-dimethanol, 1,4,5,6,7,7-Hexachloro-,
    Cyclic Sulfite, endo-, 1606
5-Norbornene-2,3-dimethanol, 1,4,5,6,7,7-Hexachloro-,
    Cyclic Sulfite, exo-, 1607
2-Norcamphanemethanol, 1500
Norea, 2924
Norethindrone, 3764
Norethindrone Acetate, 3886
Norethindrone Benzoate, 4030
Norethindrone Biphenyl 4-carboxylate, 4098
Norethindrone Dimethylpropionate, 3994
Norethindrone Heptanoate, 4039
Norethindrone Pentamethyldisiloxyl Ether, 4000
Norethindrone 4-Phenoxybenzoate, 4099
Norethisterone, 3764
Norethisterone Acetate, 3886
Norethisterone Heptanoate, 4017
Norflurazon, 2539
Norleucine, 890
D-Norleucine, 886
DL-Norleucine, 893
L-Norleucine, 890
Normal Hexane, 899
Normephenytoin, 2235
19-Nor-7α-methyltestosterone, 3690
Norpramine, 3567
19-Norpregn-4-ene-3,20-dione, 3771
19-Norpregn-4-en-20-yn-3-one, 17-(2,2-Dimethyl-1-
    oxopropoxy)-, (17α)-, 3994
19-Norprogesterone, 3771
Nortestosterone, 3590
Nortran, 2903
Noruron, 2924
DL-Norvaline, 488
L-Norvaline, 489
Noscapine, 3878
Notensil, 3665
Novacaine, 2910
Novathion, 1736
Novokain, 2910
Noxyron, 2836
NPA, 3520
m-NPhDAM, 4065

o-NPhDAM, 4066
p-NPhDAM, 4067
NPIP, 453
NSC 171240, 2271
NSC 33669, 4076
NSC 338720, 1307
NSC 5159, 4092
Nucite, 875
Nucofed, 3564
Nudrin, 454
Nuvanol, 1736
Nuvanol-N, 1355
Nystatin, 4141
Nystex, 4141
OBBP, 2378
OCDD, 2769
O8CDD, 2769
OCDF, 2768
O8CDF, 2768
OCS-21693, 2199
OCT, 1073
Octabromobiphenyl, 2378
Octachlor, 1874
2,3,4,5,2',3',4',5'-Octachlorbiphenyl, 2382
2,3,5,6,2',3',5',6'-Octachlorbiphenyl, 2383
2,2',3,3',4,4',5,5'-Octachlorobiphenyl, 2382
2,2',3,3',5,5',6,6'-Octachlorobiphenyl, 2383
1,2,3,4,6,7,8,9-Octachlorodibenzodioxin, 2769
Octachlorodibenzo-p-dioxin, 2769
Octachlorodibenzo[b,e][1,4]dioxin, 2769
Octachlorodibenzofuran, 2768
1,2,3,4,6,7,8,9-Octachlorodibenzofuran, 2768
1,2,4,5,6,7,8,8-Octachloro-4,7-methano-3α,4,7,7α-
    Tetrahydroindane, 1874
Octadecane, 3611
n-Octadecane, 3611
Octadecanoic Acid, 3610
Octadecanoic Acid, 4-(Acetylamino)phenyl Ester, 4024
Octadecanol, 3612
9-Octadecenoic Acid (Z)-, Monoester with 1,2,3-
    Propanetriol, 3856
Octadecyl Alcohol, 3612
Octadecylamine (Dihydrate), 3613
Octadecyl-p-hydroxybenzoate, 4001
2,4-Octadione, 1505
Octahydro-1-(1-oxo-3-phenyl-2-propenyl) Azocine,
    3357
Octahydro-1-(1-oxo-3-phenyl-2-propenyl)1H-Azonine,
    3482
Octaldehyde, 1537
N,N-Octamethylenecinnamamide, 3482
n-Octanal, 1537
Octanamide, N-Hydroxy-2,2-dipropyl, 3118
Octanamide, 2,2,4-triethyl-N-hydroxy, 3119
1-Octanamine, 1587
Octane, 1576
n-Octane, 1576
1-Octanecarboxylic Acid, 1824
Octanediamide, N,N,N',N'-Tetramethyl-, 2748
Octane, 3-Methyl-, 1842
1-Octanesulfonic Acid, 3-Hydroxy-, γ-Sultone, 1557
Octanoic Acid, 4-(Acetylamino)phenyl Ester, 3380
1-Octanol, 1581
2-Octanol, 1583

DL-Octanol-(2), 1584
DL-2-Octanol, 1584
n-Octanol, 1581
2-Octanone, 1538
Octan-2-one, 1538
2-Octanone, Cyclic (hydroxymethyl)ethylene Acetal,
    2368
2'-(2-Octanoyl-2-heptanyl-acetyl)-6-methoxypurine
    Arabinoside (0.4 Hydrate), 4080
2'-Octanyl-6-methoxypurine Arabinoside, 3689
1-Octene, 1527
4-Octoxybenzoic Acid-2-(diethyl-amino)ethyl Ester,
    3854
Octyl Acetaminophen, 3495
Octyl α-Acetoxypropionate, 2932
Octyl Alcohol, 1582
n-Octyl Alcohol, 1581
sec-Octyl Alcohol, 1583
Octylamine, 1587
n-Octylamine, 1587
Octyl 3-Aminobenzoate, 3245
Octyl m-Aminobenzoate, 3245
Octyl p-Aminobenzoate, 3244
2-Octylbenzene, 3097
n-Octyl Bromide, 1559
n-Octyl Carbamate, 1836
N-Octylcinnamamide, 3492
Octyldiphenyl Phosphate, 3769
n-Octyl β-Ethoxypropionate, 2938
Octyl-(2-ethyl Hexyl) Alcohol, 1582
n-Octyl 4-Hydroxybenzoate, 3243
Octyl p-Hydroxybenzoate, 3243
Octyl Lactate, 2367
Octyl Nicotinate, 3090
2-n-Octylphenol, 3105
4-Octylphenol, 3106
o-n-Octylphenol, 3105
p-n-Octylphenol, 3106
4-Octylphenol Diethoxylate, 3602
4-Octylphenol Monoethoxylate, 3396
4-Octylphenol Tetraethoxylate, 3904
4-Octylphenol Triethoxylate, 3788
2-[2-(p-Octylphenoxy)ethoxy]ethanol, 3602
tri-n-Octylphosphine Oxide, 3984
Octyl Phthalate, 3967
1-Octyne, 1494
O,O-Dimethyls-isopropylthioethyl Phosphoroditjioate,
    1267
Oenanthaldehyd, 1216
Oenanthylylurethane, 2143
Oestrin, 3570
Oestriol, 3580
Oestrone, 3570
Oftanol, 3253
Oil Orange SS, 3418
Olean-12-en-29-oic Acid, 3-(3-Carboxy-1-oxopropoxy)-
    11-oxo-, (3β,20β)-, 4107
Oleophosvel, 2788
Oleovitamin d3, 4048
1-Oleoyl-sn-glycerol, 3856
Olsalazine, 2972
Omexan, 1352
OMU, 2361
Ontrack 8E, 3240

Opian, 3878
Opianic Acid, 1953
Opianin, 3878
Opiansaeure, 1953
Oramide, 2705
Orange OT, 3418
Orap, 4051
Orbil, 3280
Oreton, 3693
Orinase, 2705
Ormazine, 3444
Ornalin, 2542
Orotamide, N-Benzyl-, 2579
Orotamide, N-Butyl-, 1767
Orotamide, N-Cyclohexyl-, 2296
Orotamide, N,N-Diethyl-, 1766
Orotamide, N-Methyl-, 750
Orotic Acid, 380
Orotic Acid Allylamide, 1423
Orotic Acid 2-Amide-2-methyl-1,3-propanediol, 1769
Orotic Acid Benzylamide, 2579
Orotic Acid n-Butylamide, 1767
Orotic Acid Choline, 2115
Orotic Acid Cyclohexylamide, 2296
Orotic Acid Diethylamine, 1766
Orotic Acid Ethanol Amide, 1158
Orotic Acid Ethylamide, 1156
Orotic Acid Isobutanolamine, 1768
Orotic Acid Methylamide, 750
Orotic Acid Methylglucamide, 2722
Orotic Acid Morpholine, 1722
Orotic Acid Nicotinmide, 2193
Orotic Acid Pyridine, 1887
Orotic Acid Triethanolamide, 2327
Orotic Acid Triethylamide, 2325
Ortal, 2725
Orthanilamide, 763
Orthanilic Acid, 743
Orthanilsaeure, 743
Orthene, 358
Orthocid-83, 1634
Ortol, 2725
Orudis, 3295
Oruvail, 3295
Oryzalin, 2708
Osbil, 3280
Osbiland, 3280
Osthole, 3192
Ouabain, 4079
Oubain, 4079
Ovasyn, 3672
Ovulen-50, 3955
Oxadiazon, 3199
7-Oxadispiro[5.1.5.2]pentadecan-14-ol, 3111
6-Oxa-5,7-distannaundecane, 5,5,7,7-Tetrabutyl-, 3987
Oxalic Acid, 58
Oxalic Acid Dihydrate, 59
Oxalic Acid Ethyl Ester, 282
Oxalsaeure, 58
Oxalsaeure-diamid, 77
Oxalsaeure-monoaethyl Ester, 282
2-(Oxalylamino)benzoic Acid, 1629
Oxamide, 77
Oxamyl, 1202

Oxanil-carbonsaeure-(2), 1629
Oxanil-o-carboxylic Acid, 1629
Oxanilic Acid, 1341
Oxanilsaure, 1341
Oxanthrene, 2531
1-Oxaspiro[4.5]decan-3-ol, 1792
1-Oxaspiro[4.4]nonan-3-ol, 2-Methyl-, 1793
Oxathiin Carboxanilide, 3329
1,2-Oxathiolane 2,2-dioxide, 185
1,3-Oxathiolane-5-methanol, 2,2-Dimethyl-, Carbamate, 1201
1,2-Oxathiolane, 5-pentyl-, 2,2-dioxide, 1557
Oxazepam, 3141
Oxidized DL-Homocysteine, 1534
18-Oxocorticosterone, 3832
18-Oxo-11β,21-dihydroxy-4-pregnene-3,20-dione, 3832
α-Oxodiphenylmethane, 2800
Oxoditane, 2800
α-Oxoditane, 2800
2-Oxo-5-Indolinyl Acetate, 1669
2-Oxo-5-Indolinyl Prostaglandin F2α, 4054
α-Oxo-β-methylol-γ-butyrolactone Betrachten, 2015
4-Oxopentanoic Acid, 433
1-(1-Oxo-3-phenyl-2-propenyl)-Piperidine, 3045
17-(1-Oxopropoxy)-(17β)-androst-4-en-3-one, 3896
Oxprenolol, 3247
Oxralox, 2524
2-Oxy-4-amino Pyrimidine, 257
2,2'-Oxybis[1-chloropropane], 825
1,1'-Oxybisethene, 270
1-Oxybutylphenylbutazone, 3639
8-Oxy-caffeine, 1451
Oxycarboxin, 2610
Oxyclozanide, 2771
Oxyphenbutazone, 3646
Oxypsoralen, 2532
Oxytetracycline, 3881
Oxytetracycline, 6-Methylene-, 3875
Paarlan, 3249
Paarlan EC, 3249
PACPX, 3454
PAD, 1517
Palatinolic, 3372
Palatinol M, 1952
Palatone, 728
Palmitic Acid, 3399
Pamoic Acid, 3909
Panalgesic, 1391
Panasol, 3823
Panmycin, 3879
Panogen, 170
Pano-ram, 2576
Panosine, 2194
Pantelmin, 3283
Pantocaine, 3254
Pantoyl Taurine, 3739
Papaverine, 3739
Parabanic Acid, 117
Parabansaeure, 117
Parachlorophenol, 634
Paraldehyd, 868
Paraldehyde, 868
Paramorphine, 3658
Paranaphthalene, 2965

Paraoxon, 2057
Parathion, 2056
Parathion-amino, 1481
Parathion-methyl, 1433
Parethoxycaine, 3246
PASIT, 2275
Pathibamate, 1806
Patoran, 1690
Patrovine, 3757
Pattonex, 1690
Paxilon, 1605
Payze, 1763
PBB 101, 2428
PBZ, 3363
(+)-PCB 136, 2413
PCB 10, 2514
PCB 104, 2436
PCB 107, 2447
PCB 110, 2448
PCB 114, 2449
PCB 116, 2444
PCB 118, 2446
PCB 122, 2443
PCB 124, 2435
PCB 128, 2418
PCB 129, 2410
PCB 131, 2420
PCB 135, 2421
PCB 137, 2416
PCB 138, 2415
PCB 139, 2411
PCB 141, 2425
PCB 144, 2426
PCB 146, 2414
PCB 151, 2424
PCB 155, 2427
PCB 158, 2417
PCB 163, 2423
PCB 170, 2389
PCB 171, 2386
PCB 172, 2395
PCB 173, 2392
PCB 174, 2391
PCB 175, 2390
PCB 176, 2393
PCB 177, 2388
PCB 178, 2394
PCB 18, 2502
PCB 180, 2387
PCB 183, 2396
PCB 185, 2397
PCB 187, 2399
PCB 194, 2382
2-PCB, 2535
2,2'-PCB, 2521
2,2',4,5,5'-PCB, 2430
2,4'-PCB, 2518
2,4,4'-PCB, 2495
2,4,5,2',4',5'-PCB, 2409
2,4,5,2',5'-PCB, 2430
PCB 208, 2377
PCB 31, 2506
4,4'-PCB, 2513
PCB 40, 2470

PCB 41, 2472
PCB 42, 2471
PCB 44, 2473
PCB 46, 2485
PCB 47, 2474
PCB 48, 2467
PCB 49, 2468
PCB 52, 2476
PCB 53, 2477
PCB 54, 2478
PCB 56, 2482
PCB 60, 2475
PCB 61, 2465
PCB 63, 2483
PCB 64, 2484
PCB 66, 2480
PCB 69, 2481
PCB 70, 2479
PCB 74, 2466
PCB 75, 2464
PCB 76, 2469
PCB 80, 2462
PCB 82, 2431
PCB 83, 2433
PCB 84, 2434
PCB 85, 2438
PCB 86, 2440
PCB 87, 2437, 2439
PCB 88, 2441
PCB 95, 2442
PCB 99, 2445
PCBA, 958
PCDD 1, 2489
PCDD 10, 2455
PCDD 12, 2456
PCDD 14, 2429
PCDD 2, 2488
PCDD 29, 2405
PCDD 42, 2406
PCDD 50, 2385
PCDD 66, 2381
PCDD 73, 2375
PCDF 118, 2379
PCDF 121, 2380
PCDF 131, 2374
2,3,4,7,8-P5CDF, 2384
PCP, 532
PCT, 1072
PDAM, 4011
PDEA, 2085
PE 200, 521
PEA, 1480
Pebulate, 2178
Pecazine, 3667
PeCDF, 2,3,4,7,8-, 2384
Pelamine, 3363
Pelargonic Acid, 1824
Pelletierin, 1516
Pelletierine, 1516
Penbutolol, 3599
Penclomedine, 1307
Pendecane(S-(carboxymethyl)-L-cysteine)), 4179
Pendimethalin, 2908
Penicillamine, 501

D-Penicillamine, 501
Penta-acetyl-gitoxin, 4147
1,4,7,10,13-Pentaazatridecane, 1600
2,2',4,5,5'-Pentabromobiphenyl, 2428
Pentachlorbenzyl Alcohol, 958
2,3,4,2',5'-Pentachlorbiphenyl, 2439
2,3,4,5,2'-Pentachlorbiphenyl, 2440
2,3,4,6,2'-Pentachlorbiphenyl, 2441
Pentachlorethane, 49
Pentachlorobenzene, 531
Penta-chlorobenzene, 531
Pentachlorobiphenyl, 2432
2,2',3,3',4-Pentachlorobiphenyl, 2431
2,2',3,3',5-Pentachlorobiphenyl, 2433
2,2',3,3',6-Pentachlorobiphenyl, 2434
2,2',3,4,4'-Pentachlorobiphenyl, 2438
2,2',3,4,5-Pentachlorobiphenyl, 2440
2,2',3',4,5-Pentachlorobiphenyl, 2437
2,2',3,4,5'-Pentachlorobiphenyl, 2439
2,2',3,4,6-Pentachlorobiphenyl, 2441
2,2',3,4',6-Pentachlorobiphenyl, 2436
2,2',3,5',6-Pentachlorobiphenyl, 2442
2,2',4,4',5-Pentachlorobiphenyl, 2445
2,2',4,4',6-Pentachlorobiphenyl, 2432
2,2',4,5,5'-Pentachlorobiphenyl, 2430
2,2',4,6,6'-Pentachlorobiphenyl, 2436
2',3,3',4,5-Pentachlorobiphenyl, 2443
2,3,3',4',5-Pentachlorobiphenyl, 2447
2,3,3',4',6-Pentachlorobiphenyl, 2448
2,3,4,4',5-Pentachlorobiphenyl, 2449
2,3',4,4',5-Pentachlorobiphenyl, 2446
2',3,4,4',5'-Pentachlorobiphenyl, 2435
2,3,4,5,6-Pentachlorobiphenyl, 2444
1,2,3,4,7-Pentachlorodibenzo-p-dioxin, 2385
2,3,4,7,8-Pentachlorodibenzofuran, 2384
3,5,6,3',5'-Pentachloro-2,2'-dihydroxybenzanilide, 2771
Pentachloroethane, 49
Pentachloro-ethane, 49
Pentachloronitrobenzene, 952
Pentachlorophenol, 532
2,3,4,5,6-Pentachloro-phenol-, 532
Pentachlorophenol Acctate, 1272
Pentachlorophenoxyacetic Acid, 1273
2,3,4,5,6-Pentachlorophenoxyacetic Acid, 1273
Pentachlorophenyl Acetate, 1272
1,15-Pentadecandioic Acid, 3265
Pentadecanoic Acid, 3267
Pentadecanol, 3269
1-Pentadecanol, 3269
Pentadecan-1-ol, 3269
1-Pentadecene, 3266
Pentadecylic Acid, 3267
1,4-Pentadiene, 422
Penta-1,4-diene, 422
Pentaerythritol, 521
Pentaerythritol Tetranitrate, 427
Pentafluorophenol, 533
2,4,6,3',4'-Penta-hydroxy-benzophenol, 2803
2,4,6,3',4'-Pentahydroxybenzophenon, 2803
3,5,7,2',4',-Penta-hydroxyflavon, 3140
L-1,3,4,5,6-Pentahydroxyhexan-2-one, 871
Pentalin, 49
Pentamethylbenzene, 2300
1,2,3,4,5-Pentamethyl Benzene, 2300

Pentamethylene, 448
N,N-Pentamethylenecinnamamide, 3045
Pentamethylene Oxide, 460
Pentamethylmelamine, 1535
Pentamide, N-(3-Chloro-4-methylphenyl)-2-methyl-,
2888
n-Pentanal, 469
Pentanamide, 487, 3571
Pentanamide, N-(Aminocarbonyl)-2-bromo-, 810
Pentanamide, N-hydroxy-α,α-dipropyl, 3113
Pentanamide, N-hydroxy-4-methyl-2,2-bis(2-
methylpropyl), 3124
Pentane, 505
n-Pentane, 505
2,4-Pentanedione, 430
Pentane, 3-Nitro-, 490
Pentanitrophenylether, 2450
Pentanochlor, 2888
Pentanol, 516
1-Pentanol, 516
2-Pentanol, 510
3-Pentanol, 513
Pentan-3-ol, 513
2-Pentanone, 470
Pentan-2-one, 470
3-Pentanone, 461
1-Pentene, 450
2-Pentene, 449
3-Pentene, 449
Penten-1-ol-3, 462
1-Penten-3-ol, 462
Penten-3-ol-2, 464
3-Penten-2-ol, 464
Penten-4-ol-1, 463
4-Penten-1-ol, 463
o-2-Pentenylphenol, 2280
Pentobarbital, 2335
Pentothiobarbital, 2332
4-Pentoxybenzoic Acid-2-(diethyl-amino)ethyl Ester,
3600
2-Pentoxy-N-[2-(diethyl-amino)ethyl]-4-quinoline
Carboxamide, 3846
3-Pentoxyphenol, 2323
m-Pentoxy Phenol, 2323
Pentyl Acetate, 1225
2-Pentyl Acetate, 1218
Pentylacetylene, 1175
Pentyl Alcohol, 516
tert-Pentyl Alcohol, 515
Pentyl p-Aminobenzoate, 2687
4-Pentylaminobenzoic Acid-2-(diethylamino)ethyl Ester,
3601
Pentylbenzene, 2301
n-Pentylbenzene, 2301
n-Pentylbenzene1-Phenylpentane, 2301
Pentyl Bromide, 486
Pentyl n-Butanoate, 1828
Pentyl Butyrate, 1828
n-Pentyl Carbamate, 884
O-Pentyl Carbamate, 884
tert-Pentyl Carbamate, 888
t-Pentylcarbinol, 915
n-Pentylcinnamamide, 3070
1-Pentylcyclopentane, 2155

n-Pentylcyclopentane, 2155
Pentyl 2,4-Dichlorophenoxyacetate, 2848
2,4-D Pentyl Ester, 2848
5-n-Pentyl-5-ethylbarbituric Acid, 2336
Pentyl Formate, 864
n-Pentyl Lactate, 1551
Pentyl 3-Methoxypropionate, 1834
Pentyloxycarbonyl-mitomycin C, 3827
3-Pentyl-2,4-pentadione, 2131
4-n-Pentylphenol, 2319
p-tert-Pentylphenol, 2316
Pentyl Propionate, 1550
n-Pentyl Propionate, 1550
1-Pentyne, 421
Pent-1-yne, 421
PERC, 110
Perchloroethylene, 110
Perflan, 1790
Perfluidone, 2991
Peri Acid, 1919
Periciazine, 3816
Pericyazine, 3816
peri-Dinaphthalene, 3717
peri-Ethylenenaphthalene, 2553
Periplocin, 4084
Periplocoside, 4084
Permethrin, 3806
Permitil, 3884
Perphenazine, 3820
(+)-Perseitol, 1265
Pertofran, 3567
Perylene, 3717
Pethidine, 3223
Pevaryl, 3530
PF-3, 904
PF 38, 1781
PFP, 533
PH 40-21, 527
Phaltan, 1602
Phalton, 1602
Phanodorm, 2670
Pharorid, 3711
Phaseomannite, 875
PhDAM, 4068
Phe, 1706
D-PHE, 1709
L-Phe-Dapsone, 3810
Phelam DP, 1719
Phenacetin, 2025
PhenAcide, 1937
Phenallymal, 2817
Phenanthracene, 2966
Phenanthrene, 2966
Phenanthridin, 2782
Phenanthridine, 2782
m-Phenanthrolin, 2525
o-Phenanthrolin, 2526
p-Phenanthrolin, 2527
1,10-Phenanthroline, 2526
m-Phenanthroline, 2525
o-Phenanthroline, 2526
p-Phenanthroline, 2527
Phenazine, 2528
Phenazone, 2232

Phenbutamide, 2311
Phene, 676
Phenetole, 1453
Pheniramine Maleate, 3750
Phenisobromolate, 3423
Phenmedipham, 3313
Phennin, 3158
Phenobarbital, 2595
Phenobenzuron, 3286
Phenodioxin, 2531
Phenol, 722
Phenol, 3-Butoxy-, 2078
Phenol, 2-Butyl-4,6-dichloro-, 1984
Phenol, 2-Butyl-4,5-dimethyl-, 2715
Phenol, 4-Butyl-2,6-dimethyl-, 2711
Phenol, 2-Butyl-4-ethyl-, 2713
Phenol, 2-Butyl-6-ethyl-, 2712
Phenol, 4-Chloro-3-methyl-, 1074
Phenol, 2,3-Dichloro-, 590
Phenol, 2,4-Dichloro-6-ethyl-, 1356
Phenol, 2,4-Dimethyl-, 1452
Phenol, 2,5-Dimethyl-, 1457
Phenol, 4-(Dimethylamino)-3-methyl, Methylcarbamate
    (Ester), 2303
Phenol, θ-Dinitro-, 600
Phenol, 4,6-Dinitro-2-sec-butyl-, 1992
Phenol, 2,4-Dipropyl-, 2710
Phenol, 2,6-Dipropyl-, 2716
Phenol, 3-Ethyl-5-methyl-, 1750
Phenol, 2-Methyl-, 1119
Phenol, 5-Methyl-2-(phenylmethyl)-, 3013
Phenol, 2,3,4,5,6-Pentachloro-, 532
Phenol, 2-(2-Pentenyl)-, 2280
Phenol, 3-Pentoxy-, 2323
Phenol, 4-[[4-(Phenylazo)phenyl]azo]-, 3525
Phenolphthalein, 3726
Phenolphthalin, 3731
Phenol, 3-Propoxy-, 1756
Phenolrot, 3627
Phenolsulfonaphthalein, 3627
Phenol, 2,3,4,5-Tetrachloro-, 544
Phenol, 2,3,4,6-Tetrachloro-, 543
Phenol, 2,3,5,6-Tetrachloro-, 545
Phenol, 2,3,4,6-Tetrachloro-5-methyl-, 973
Phenol, 2,4,5-Trichloro-, 567
Phenol, 4-[2,2,2-Trichloro-1-(4-methoxyphenyl)ethyl]-,
    3159
Phenothiazine, 2547
Phenothrin, 3921
Phenoxethol, 1467
Phenoxyacetic Acid, 1400
2'-Phenoxyacetyl-6-methoxypurine Arabinoside
    (Hemihydrate), 3652
3-Phenoxybenzyl D-cis and trans-2,2-dimethyl-3-(2-
    methylpropenyl)cyclopropanecarboxylate, 3921
2-Phenoxyethanol, 1467
Phenoxyethyl Alcohol, 1467
(3-Phenoxylphenyl)methyl 2,2-dimethyl-3-(2-methyl-1-
    propenyl)cyclopropanecarboxylate, 3921
Phensuximide, 2218
Phent, 2698
Phenthoate, 2698
Phentolamine Methanesulfonate, 3572
Phenyglyoxilic Acid, 1319

Phenyl Acetaminophen, 3166
Phenylacetic Acid, 1381
Phenyl 2-Acetoxybenzoate, 3158
Phenylacetylene, 1294
3-(1'-Phenyl-2'-acetylethyl)-4-hydroxycoumarin, 3630
2'-Phenylacetyl-6-methoxypurine Arabinoside, 3649
Phenyl Acetylsalicylate, 3158
Phenylacrolein, 1637
Phenylacrylic Acid, 1638
trans-β-Phenylacrylic Acid, 1640
DL Phenylalanin, 1711
Phenylalanine, 1706
(S)-Phenylalanine, 1706
(S)-(-)-Phenylalanine, 1706
D-Phenylalanine, 1709
DL-Phenylalanine, 1711
L-Phenylalanine, N-[2-(Acetyloxy)benzoyl]-, Ethyl Ester, 3740
1-Phenyl-4-amino-5-bromo-6-pyridazone, 1891
D-β-Phenyl-α-Aminopropionic Acid, 1709
Phenylarsonsaeure, 755
N-(Phenylazo)aniline, 2577
1-Phenylazo-2-naphthylamine, 3282
4-Phenylazophenol, 2559
Phenylbenzene, 2554
2-Phenyl-3,1-benzoxazin-4-one, 2960
Phenylbenzyl, 2815
Phenyl Bromide, 625
1-Phenylbutane, 2046, 2052
Phenylbutazone, 3644
4-Phenyl-3-buten-2-one, 1944
3-Phenyl-n-butyric Acid, 2012
β-Phenylbutyric Acid, 2012
6-Phenylcaproic Acid, 2682
ε-Phenylcaproic Acid, 2682
Phenylcarbinol, 1117
Phenyl Carboxamide, 1087
Phenyl Chloride, 630
2-Phenylcinchoninic Acid, 3273
Phenyl Cyanide, 1005
Phenyldiantipyrylmethane, 4068
Phenyl Diethanolamine, 2085
N-Phenyldiethanolamine, 2085
8-Phenyl-1,3-diethylxanthine, 3189
1-Phenyl-2,3-dimethyl-3-pyrazoline-5-thione, 2597
3-Phenyl-1,1-dimethylurea, 1737
Phenyl(di-morpholido)-phosphate, 3094
Phenyldithiopyrylmethane, 4070
m-Phenylendiamin, 758
o-Phenylendiamin, 757
p-Phenylendiamin-mono-N-acetat, 1437
1,4-Phenylenediamine, 759
m-Phenylenediamine, 758
o-Phenylenediamine, 757
p-Phenylenediamine, 759
p-Phenylenediaminemono-N-acetate, 1437
o-Phenylenepyrene, 3860
Phenylessigsaeure, 1381
Phenylethane, 1432
Phenyl Ethanolamine, 1480
Phenylethanolamine, 1480
N-Phenylethanolamine, 1480
Phenyl Ether, 2564
Phenyl Ethyl Alcohol, 1460

Phenylethylalcohol, 1460
Phenylethylene, 1351
N-Phenyl-N-ethylethanolamine, 2084
5-Phenyl-5-ethylhydantoin, 2235
Phenylethylmalonylurea, 2595
DL-2-Phenylglycine, 1420
O-Phenylglycolic Acid, 1400
1-Phenylhexane, 2699
6-Phenylhexanoic Acid, 2682
4-Phenylhydrazine Sulfonic Acid, 767
Phenylhydrazin-sulfosaeure-(4), 767
Phenyl Hydride, 676
Phenylhydroxylamin, 738
Phenylhydroxylamine, 738
5-Phenyl-5-(p-hydroxy)phenyl-hydantoin, 3155
2,2'-(Phenylimino)diethanol, 2085
Phenyl Isothiocyanate, 1022
Phenylmercuric Acetate, 1386
Phenylmercury Dimethyldithiocarbamate, 1719
Phenylmethanol, 1117
Phenylmethyl Acetate, 1684
Phenylmethylbarbituric Acid, 2211
Phenylmethyl Benzoate, 2996
Phenylmethyl Ester, 2999
3-Phenyl-, Methyl Ester, 1946
Phenyl Methyl Ether, 1118
1-Phenyl-3-methylpyrazolyl-5-dimethylcarbamate, 2843
N-Phenyl-N'-methylurea, 1434
Phenyl Mustard Oil, 1022
1-Phenylnonadecane, 4002
N-Phenyloxalic Acid Monoamide, 1341
3-Phenyloxycarbonyl-5-fluoro-2,4(1H,3H)-pyrimidinedi-one, 2190
1-Phenyloxycarbonyl-5-fluorouracil, 2189
3-Phenyloxycarbonyl-5-fluorouracil, 2190
2-Phenylphenol, 2565
o-Phenylphenol, 2565
p-Phenylphenol, 2566
Phenyl Phosphate, 3529
Phenylphosphinic Acid, 753
Phenyl-phosphinigsaeure, 753
Phenylphosphonic Acid, 754
Phenylphosphonothioic Acid O-(4-Bromo-2,5-dichlorophenyl) O-Methyl Ester, 2788
Phenylphosphonsaeure, 754
Phenyl Phthalate, 3727
1-Phenylpropane, 1730
2-Phenylpropane, 1728
3-Phenyl-2-propenal, 1637
3-Phenyl-2-propenaldehyde, 1637
2-Phenylpropene, 1663
2-Phenyl-1-propene, 1663
β-Phenylpropene, 1663
(E)-3-Phenyl-2-Propenoic Acid, 1640
3-Phenylpropenoic Acid, 1638
trans-3-Phenyl-2-propenoic Acid, 1640
3-Phenyl-2-propenyl 2-Aminobenzoate, 3304
3-Phenyl-2-propen-1-yl Anthranilate, 3304
p-Phenylpropylphenylbutazone, 3944
1-Phenyl-1-(2-pyridyl)-3-dimethylaminopropane Maleate, 3750
2-Phenyl-4-quinolinecarboxylic Acid, 3273
Phenyl Salicylate, 2801
Phenylsemicarbazide, 1154

4-Phenylsemicarbazide, 1154
N-(Phenylsulfonyl)-N'-butylurea, 2311
p-Phenylthioethylphenylbutazone, 3911
Phenylthioharnstoff, 1112
1-Phenyl-2-thiourea, 1112
4-Phenyltoluene, 2814
4-Phenyluracil, 1895
4-Phenyl-uracil, 1895
p-(3-Phenylureido)phenyl Acetate, 3173
5-Phenylvaleric Acid, 2282
δ-Phenylvaleric Acid, 2282
Phenytoin, 3153
Phloral, 1461
Phloridzin (Dihydrate), 3818
Phloroglucinol, 726
Phorate, 1266
Phorate Sulfone, 1268
Phosalone, 2639
Phosmet, 2231
Phosphine Oxide, Trimorpholino-, 2750
Phosphine Sulfide, Trimorpholino-, 2749
Phosphinic Acid, Dihexyl-, Ethyl Ester, 3128
Phosphinic Acid, Hexyl-, Dibutyl Ester, 3129
Phosphinic Acid, Hexyl-, Diethyl Ester, 3130
Phosphinic Acid, (2,4,6-Trimethylphenyl)-, 1771
Phosphoacetic Acid, 97
Phosphonic Acid, (3-Carboxypropyl)-, 355
Phosphonic Acid, Hexadecyl-, Dibutyl Ester, 3985
Phosphonic Acid, Phenyl-, Di-2-Propenyl Ester, 2661
Phosphonic Acid, (Trichloromethyl)-, Diethyl Ester, 452
Phosphonoacetic Acid, 97
4-Phosphonobutyric Acid, 355
γ-Phosphono-n-butyric Acid, 355
N-(Phosphonomethyl)glycine, 213
3-Phosphonopropionic Acid, 211
Phosphoramidothioic Acid, Isopropyl-o-(2,4-
   dichlorophenyl)-o-methyl Ester, 2054
Phosphor Carboxymethyl-phosphonsaeure, 97
Phosphoric Acid, Decyl Ester, 3131
Phosphoric Acid, Dibutyl Hexyl Ester, 3132
Phosphoric Acid, Diethyl 2-Methylpropyl Ester, 1594
Phosphoric Acid, Dimethyl 3,5,6-Trichloro-2-Pyridyl
   Ester, 1082
Phosphoric Acid, Diphenyl Ester, 2580
Phosphoric Acid, Triethyl Ester, 946
Phosphoric Acid Triphenyl Ester, 3529
Phosphorodithioic Acid O,O-Diethyl S-[2-
   (Ethylsulfonyl)ethyl] Ester, 1596
Phosphorodithioic Acid O,O-Diethyl S-
   [(Ethylsulfonyl)methyl] Ester, 1268
Phosphorodithioic Acid O,O-Diethyl S-[2-
   (Ethylthio)ethyl] Ester, 1590
Phosphorodithioic Acid O,O-Diethyl S-
   [(Ethylthio)methyl] Ester, 1266
Phosphorodithioic Acid, S-[[(1,1-Dimethylethyl)
   sulfinyl]methyl] O,O-Diethyl Ester, 1853
Phosphorodithioic Acid, S-[[(1,1-Dimethylethyl)
   sulfonyl]methyl] O,O-Diethyl Ester, 1857
Phosphorodithioic Scid S-[(1,3-Dihydro-1,3-dioxo-2H-
   isoindol-2-yl)methyl] O,O-Dimethyl Ester, 2231
Phosphorofluoridic Acid bis(1-Methylethyl) Ester, 904
Phosphorothioic Acid, S-(2-((1-Cyano-1-
   methylethyl)amino)-2-oxoethyl) O,O-Diethyl Ester,
   2145

Phosphorothioic Acid, O,O-Diethyl O-[p-
   (Methylsulfonyl)phenyl] Ester, 2331
Phosphorothioic Acid, O,O-Diethyl O-[4-
   (Methylthio)phenyl] Ester, 2329
Phosphorsaeure-trimethyl Ester, 221
Phosphoryl Trimorpholide, 2750
Phosphothion, 2154
Phosvel, 2788
Phosvel Oxon, 2789
Phoxim, 2655
Phrenilin, 2305
o-Phthalaldehyd, 1316
Phthalamide, 1363
1,3 Phthalandione, 1278
Phthalate Butyl Benzyl Ester, 3654
Phthalic Acid, 1320
m-Phthalic Acid, 1321
o-Phthalic Acid, 1320
p-Phthalic Acid, 1322
Phthalic Acid Anhydride, 1278
Phthalic Acid Diisobutyl Ester, 3372
Phthalic Acid Ethyl Ester, 2638
Phthalic Anhydride, 1278
Phthalic Dicarboxaldehyde, 1316
Phthalimid, 1288
Phthalimide, 1288
Phthalonic Acid, 1612
Phthalonsaeure, 1612
Phthalsaeure, 1320
Phthalsaeure-diaethyl Ester, 2638
Phygon, 1859
Phygon Paste, 1859
Phygon XL, 1859
Phyomone, 2569
Picene, 3861
Pichloram, 825
Picloram, 562
Picolinic Acid, m-Isothiocyanatophenyl Ester,
   2779
Picramic Acid, 666
Picramine, 618
Picric Acid, 572
Picrolonic Acid, 1928
Picronitric Acid, 572
Picrotoxin, 4083
Picrotoxine, 4083
Picrylaniline, 2529
Picryl Chloride, 535
PID, 1563
Pikrinsaeure, 572
Pikrolonsaeure, 1928
Pillarcap, 1634
Pilocarpic Acid, 2333
Pimafucin, 4104
Pimelic Acid, 1190
Pimozide, 4051
Pinacolone, 847
Pindone, 3016
Pinolhydrat, 2133
Pipemidic Acid, 3050
Pipemidique Acide, 3050
1-Piperazinecarboxaldehyde, 4-[4,6-bis(Dimethylamino)-
   1,3,5-triazin-2-yl]-, 2732
2,5-Piperazinedione, 265

2,6-Piperazinedione, 4,4'-(1-Methyl-1,2-ethanediyl)bis-, 2315
2,6-Piperazinedione, 4,4'-(1-Methyl-1,2-ethanediyl)bis-, (±)-, Polymer with 1,3-Dibromopropane, 2315
N,N'-[1,4-Piperazinediylbis(2,2,2-trichloroethylidene)] Bisformamide, 2055
Piperidine, 1-[5-(1,3-Benzodioxol-5-yl)-1-oxo-2,4-pentadienyl]-, (E,E)-, 3448
Piperidine, 1-[(Benzoyloxy)acetyl]-, 3047
Piperidine, 1-[(Benzoyloxy)acetyl]-2,6-dimethyl-, 3361
Piperidine, 1-[(Benzoyloxy)acetyl]-2-ethyl-, 3360
4-Piperidinol, 1-[(Benzoyloxy)acetyl]-, 3048
N-(Piperidinomethyl)benzamide, 2665
1-(Piperidinyl)-3,5-bis(dimethylamino)-s-triazine, 2734
α-Piperidyl-3,6-bis(trifluoromethyl)-9-phenanthrenemethanol, 3871
Piperine, 3448
1-(1-Piperizinyl)-3,5-bis(dimethylamino)-s-triazine, 2359
Piperonal, 1318
Piperonyl Aldehyde, 1318
Piperophos, 3114
N-[(E,E)-Piperoyl]piperidine, 3448
Pipsyl Chloride, 581
Pirias Acid, 1920
Pirimicarb, 2337
Pirimiphos-ethyl, 2928
Pirimiphosmethyl, 2343
Pirimiphos-methyl, 2343
Pirofos, 1599
Piroxicam, 3167
PITC, 1022
Pivadorm, 811
Pivadorn, 811
Pivalic Acid, 478
Pivaloylacetone, 1503
Pivaloylacetylmethane, 1503
9-[5'-(O-Pivaloyl)-β-D-arabinofuranosyl]adenine Ester, 3236
2-Pivaloylindandione-1,3, 3016
5'-Pivaloyl 5-iodo-2'-deoxyuridine, 3069
1-Pivaloyloxymethyl Allopurinol, 2277
2-Pivaloyloxymethyl Allopurinol, 2278
1-Pivaloyloxymethyl-5-fluoro-2,4(1H,3H)-pyrimidinedione, 2024
1-Pivaloyloxymethyl-5-fluorouracil, 2023, 2024
O-Pivaloyloxymethyl Salicylamide, 2874
O-(Pivaloyloxymethyl) Salicylamide, 2876
Pivaloyl Salicylate, 2659
Pivalyl Acetaminophen, 2873
Plantvax, 2610
Plegicil, 3665
Plendil, 3546
Plexonal, 3465
Plictran, 3608
Plivafen, 3665
Plondrel, 2622
PMA, 1386
PMAC, 1386
Polaris, 374
Polybromilated Biphenyl, 2402
Polycizer 962BPA, 4109
Polycron, 2286
Ponstan, 3179
Ponstel, 3179

Poperidinecarbothioic Acid, S-Ethyl Ester, 1798
Pounce, 3806
Pramitol, 2148
Prasterone, 3691
Prazepam, 3631
Prazina, 399
Precor, 3711
Prednisolone, 3831
Prednisolone Acetate, 4018
Prednisolone 21-Trimethylacetate, 4018
Prednisone, 3823
Prednisone Acetate, 3927
Predominantly Trans Isomer, 2249
Predonin, 3831
Prefix, 993
Pregard, 3030
1,4-Pregnadiene-17α,21-diol-3,11,20-trione, 3823
1,4-Pregnadiene-17x,21-diol-3,11,20-trione, 3823
Pregna-1,4-diene-3,20-dione, 16,21-bis(Acetyloxy)-9-fluoro-11,17-dihydroxy-, (11β,16apha)-, 3992
Pregna-4,6-diene-3,20-dione, 6,17-Dimethyl-, 3931
Pregna-1,4-diene-3,20-dione, 9-Fluoro-11,17,21-trihydroxy-16-methyl-, (11β,16β)-, 3889
Pregna-1,4-diene-3,11,20-trione, 21-(Acetyloxy)-17-hydroxy-, 3927
3,20-Pregnanedione, 3847
Pregna-1,4,6-triene-3,20-dione, 17-(Acetyloxy)-6-chloro-, 3922
Pregn-4-ene-3,20-dione, 3838
Pregn-4-ene-3,20-dione, 21-(Acetyloxy)-, 3932
Pregn-4-ene-3,20-dione, 21-(Acetyloxy)-9-fluoro-11,17 dihydroxy-, (11β)-, 3929
Pregn-4-ene-3,20-dione, 17-Hydroxy-, 3839
δ4-Pregnene-3,20-dione, 3838
4-Pregnene-20-one-3-spiro-2'-(4'-ethoxycarbonyl-1',3'-thiazolidine), 4023
5-Pregnene-20-one-3-spiro-2'-(1',2'-thiazolidine), 3938
Pregn-4-ene-3,11,20-trione, 21-(Acetyloxy)-17-hydroxy-, 3844
Pregn-4-ene-3,11,20-trione, 21-Hydroxy-17-(1-oxopropoxy)-, 3958
Pregneninolone, 3829
4-Pregnen-21-ol-3,20-dione, 3841
Pregnenolone, 3849
5-Pregnen-3β-ol-20-one, 3849
Prenylamine, 3948
Preparation 5633, 1969
Preservastat, 771
Preventol I, 567
Primatene, 1765
Primatol M, 1788
Primatol O, 2148
Primatol Q, 2153
Primatol S, 1182
Primazine, 3456
Primidone, 2623
Primotec, 2928
Probarbital, 1776
Procaine, 2910
Prochlorperazine, 3745
Prodix, 3839
Prodox, 3839
Profenofos, 2286
Profluralin, 3030

Progabide, 3424
Progesterone, 3838
Pro-his-pro-phe-his-leu-leu-val-tyr, 4160
Pro-his-pro-phe-his-leu-D-leu-val-tyr, 4159
Pro-his-pro-phe-his-leu-phe-val-tyr, 4161
DL-Proline, 441
L-Proline, 442
L-Proline, 1-[(Benzoyloxy)acetyl]-, 3021
L-Proline, 1-[(Benzoyloxy)acetyl]-, Methyl Ester, 3196
Prolixin, 3884
Proloprim, 3057
Promacortine, 3893
Promazine, 3456
Promecarb, 2688
Promethazine, 3457
Prometon, 2148
Prometone, 2148
Prometryne, 2153
Prominal, 2830
Pronamide, 2572
Pronone, 2726
Propachlor, 2272
Propaldehyde, 174
Propan, 212
Propanal, 174
Propanamide, 3198
Propanamide, N-(Aminocarbonyl)-2-bromo-, 291
Propanamide, 2-[[(Benzoyloxy)acetyl]amino]-, 2624
Propanamide, 2-(Benzoyloxy)-N,N-diethyl-, 3071
Propanamide, 3-(Benzoyloxy)-N,N-dimethyl-, 2644
1-Propanamine, 3-(2-Chloro-9H-thioxanthen-9-ylidene)-
    N,N-dimethyl-, (3Z)-, 3538
1-Propanamine, 3-Dibenz[b,e]oxepin-11(6H)-ylidene-N-
    methyl-, 3549
2-Propanamine, N-(1-Methylethyl)-N-nitroso-, 909
Propane, 212
Propane, 1-Chloro-1-nitro-, 157
Propane, 1-Chloro-2-nitro-, 158
Propanediamide, 163
Propane, 1,3-Dibromo-, Polymer with (±)-4,4'-(1-Methyl-
    1,2-ethanediyl)bis[2,6-piperazinedione], 2315
1,1-Propanedicarboxylic Acid, 436
1,2-Propanedicarboxylic Acid, 435
1,3-Propanedicarboxylic Acid, 437
Propane, 1,1-Dichloro-1-nitro-, 142
Propanedioic Acid Diethyl Ester, 1192
1,3-Propanediol, 2,2-bis[(nitrooxy)methyl]-, Dinitrate
    (Ester), 427
Propane, 2-Ethoxy-, 517
Propanenitrile, 3-[Butyl[4-[(4-
    nitrophenyl)azo]phenyl]amino]-, 3660
Propanenitrile, 3-[(2-Hydroxyethyl)[3-methyl-4-[(4-
    nitrophenyl)azo]phenyl]amino]-, 3552
1,3-Propane Sultone, 185
1,2,3-Propanetricarboxylic Acid, 773, 774, 3789
1,2,3-Propanetriol 1,2-Dinitrate, 168
1,2,3-Propanetriol, Tripropanoate, 2730
Propane-1,2,3-triyl Triacetate, 1779
1,2,3-Propanetriyl Tributyrate, 3264
1,2,3-Propanetriyl Tripropionate, 2730
Propanil, 1646
Propanilide, 1705
2-Propanoic Acid, 1227
Propanoic Acid, 2-(Acetyloxy)-, Hexyl Ester, 2349

Propanoic Acid, 2-(Acetyloxy)-, Octyl Ester, 2932
Propanoic Acid, 2-[(Amoxycarbonyl)oxy]-, Methyl Ester,
    2140
Propanoic Acid, 2-Bromo-, Ethyl Ester, 440
Propanoic Acid, 3-Butoxy-, Butyl Ester, 2369
Propanoic Acid, 2-[(Butoxycarbonyl)oxy]-, Methyl Ester,
    1796
Propanoic Acid, 2,2-Dimethyl-, 4-(Acetylamino)phenyl
    Ester, 2873
Propanoic Acid, 2,2-Dimethyl-, [2-
    (Aminocarbonyl)phenoxy]methyl Ester, 2874
Propanoic Acid, 2,2-Dimethyl-, (4,5-Dihydro-4-oxo-2H-
    pyrazolo[3,4-d]pyrimidin-1-yl)methyl Ester, 2277
Propanoic Acid, 2,2-Dimethyl-, (4,5-Dihydro-4-oxo-2H-
    pyrazolo[3,4-d]pyrimidin-2-yl)methyl Ester, 2278
Propanoic Acid, 2-[(Ethoxycarbonyl)oxy]-, Butyl Ester,
    2140
Propanoic Acid, 2-[(Ethoxycarbonyl)oxy]-, Methyl Ester,
    1195
Propanoic Acid Ethyl Ester, 473
Propanoic Acid, 2-[(Hexthoxycarbonyl)oxy]-, Methyl
    Ester, 2350
Propanoic Acid, 2-Hydroxy-, Dodecyl Ester, 3268
Propanoic Acid, 2-Hydroxy-, Hexyl Ester, 1829
Propanoic Acid, 2-Hydroxy-, Octyl Ester, 2367
Propanoic Acid, 3-Methoxy-, Butyl Ester, 1556
Propanoic Acid, 2-[(Methoxycarbonyl)oxy]-, Butyl Ester,
    1796
Propanoic Acid, 2-[(Methoxycarbonyl)oxy]-, Methyl
    Ester, 805
Propanoic Acid Pentyl Ester, 1550
Propanoic Acid, 2-[(Propoxycarbonyl)oxy]-, Methyl Ester,
    1514
Propanoic Acid, 3-Propoxy-, Methyl Ester, 1229
Propanoic Acid, 3-Propoxy-, Propyl Ester, 1832
Propanoic Acid, 2-[(Proxycarbonyl)oxy]-, Butyl Ester,
    2360
Propanol, 214
2-Propanol, 215
2-Propanol, 1-(2-Cyclopentylphenoxy)-3-[(1,1-
    dimethylethyl)amino]-, (S)-, 3599
1-Propanol, 3-[(2,3,6-Trichlorobenzyl)oxy]-, 1956
1-Propanone, 1-[2-(β-D-Glucopyranosyloxy)-4,6-
    dihydroxyphenyl]-3-(4-hydroxyphenyl)-, Dihydrate,
    3818
Propaphos, 2922
Propazine, 1787
2-Propenal, 135
2-Propenamide, 147
2-Propenamide, N-Cyclododecyl-3-phenyl, 3845
2-Propenamide, N-Cyclohexyl-3-phenyl-, 3208
2-Propenamide, N-Cyclooctyl-3-phenyl-, 3481
2-Propenamide, N-Cyclopentyl-3-phenyl-, 3044
2-Propenamide, N-Decyl-3-phenyl-, 3696
2-Propenamide, N,N-Dicyclopentyl-3-phenyl-, 3681
2-Propenamide, N-Dodecyl-3-phenyl-, 3851
2-Propenamide, N-Heptyl-3-phenyl-, 3378
2-Propenamide, N-Methyl-3-phenyl-, 1960
2-Propenamide, N-Nonyl-3-phenyl-, 3593
2-Propenamide, N-Octyl-3-phenyl-, 3492
2-Propenamide, N-Pentyl-3-phenyl-, 3070
2-Propenamide, 3-Phenyl-N-propyl-, 2642
Propene, 152
2-Propene-1-sulfinothioic Acid S-2-propenyl Ester, 793

Propenitrile, 119
2-Propenoic Acid, 1946
2-Propenoic Acid Ethyl Ester, 432
2-Propenoic Acid Methyl Ester, 272
2-Propenoic Acid, 3-Phenyl-, 1638
2-Propen-1-ol, 3-Phenyl-, 2-Aminobenzoate, 3304
2-Propenylacrylic Acid, 771
Propenylanisole, 2007
p-Propenylanisole, 2007
2-(2-Propenyl)4-Pentenamide, 1492
Propetamphos, 2156
Propham, 2029
Prophenpyridamine Maleate, 3750
Propionanilide, 1705
Propionic Acid, 177
n-Propionic Acid, 177
Propionic Acid, p-Acetamidophenyl Ester, 2260
Propionic Acid, 3-Ethoxy-, Butyl Ester, 1830
Propionic Acid, 3-Ethoxy-, Hexyl Ester, 2366
Propionic Acid, 3-Ethoxy-, Pentyl Ester, 2174
Propionic Acid, 3-Ethoxy-, Propyl Ester, 1554
Propionic Acid, 3-Methoxy-, Methyl Ester, 482
Propionic Acid, 3-Methoxy-, Propyl Ester, 1230
Propionic Acid N-Propyl Ester, 865
Propionic Acid, 2(2,4,5-Trichlorophenoxy)-, 1616
n-Propioniitrile, 146
Propionitrile, 146
Propionsaeure-anilid, 1705
Propionsaeure-nitril, 146
9-(2-O-Propionyl-β-D-arabinofuranosyl)adenine, 2884
9-[5'-(O-Propionyl)-β-D-arabinofuranosyl]adenine Ester, 2885
3-Propionyl-5-fluoro-2,4(1H,3H)-pyrimidinedi-one, 1083
3-Propionyl-5-fluorouracil, 1083
5'-Propionyl 5-iodo-2'-deoxyuridine, 2641
Propionyl-r-mandelic Acid, 2248
2'-Propionyl-6-methoxypurine Arabinoside (Hemihydrate), 3058
1-Propionyloxymethyl-5-fluoro-2,4(1H,3H)-pyrimidinedi-one, 1410
1-Propionyloxymethyl-5-fluorouracil, 1410
Proponal, 2097
Propoxur, 2293
4-Propoxybenzoic Acid-2-(diethyl-amino)ethyl Ester, 3394
2-Propoxy-N-[2-(diethyl-amino)ethyl]-4-quinoline Carboxamide, 3687
2-Propoxyphenol, 1759
3-Propoxyphenol, 1756
m-Propoxy Phenol, 1756
o-Propoxyphenol, 1759
Pro-pro-pro-his-pro-phe-his-leu-leu-val-tyr, 4173
Pro-pro-pro-his-pro-phe-his-leu-D-leu-val-tyr, 4172
Pro-pro-pro-his-pro-phe-his-leu-phe-val-tyr, 4174
Pro-pro-pro-pro-pro-his-pro-phe-his-leu-leu-val-tyr, 4180
Pro-pro-pro-pro-pro-his-pro-phe-his-leu-phe-val-tyr, 4181
Propycil, 1159
Propyl Acetate, 477
Propyl α-Acetoxypropionate, 1513
n-Propyl Alcohol, 214
Propyl Aldehyde, 174
Propylallylacetamide, 1517
Propylamin, 219

Propylamine, 219
n-Propylamine, 219
Propyl-p-aminobenzoate, 2026
4-Propylaminobenzoic Acid-2-(diethyl-amino)ethyl Ester, 3395
2-(Propylamino)ethyl 4-Aminobenzoate, 2695
Propylbenzene, 1730
n-Propylbenzene, 1730
Propyl Benzoate, 2009
n-Propyl Benzoate, 2009
Propylbromid, 187
Propyl Bromide, 187
2-Propyl-5-bromo-5-nitro-1,3-dioxane, 1180
Propylbutylaceturethane, 2746
S-Propyl Butylethylthiocarbamate, 2178
Propyl Butyrate, 1219
n-Propyl n-Butyrate, 1219
Propyl Carbamate, 344
n-Propyl Carbamate, 344
Propyl Chloride, 189
n-propylcinnamamide, 2642
Propyl cyanide, 296
1-Propylcyclopentane, 1531
n-Propylcyclopentane, 1531
Propyldiantipyrylmethane, 4011
S-Propyl Dipropylthiocarbamate, 2177
Propyldithiopyrylmethane, 4013
Propyl Dixanthogen, 1506
Propylene, 152
Propylene Chloride, 160
Propylene Dichloride, 160
Propylene Oxide, 173
Propyl Ether, 921
n-Propyl β-Ethoxypropionate, 1554
i-Propylethylcarbinol, 931
n-Propylethylcarbinol, 929
Propylethylene, 450
Propyl Formate, 328
Propyl Formate, 328
n-Propyl Formate, 328
Propyl 4-Hydroxybenzoate, 2013
Propyl p-Hydroxybenzoic Acid, 2013
n-Propyl Iodide, 193
Propylisopropylacetamide, 1563
Propyl-isopropyl-aether, 927
Propyl Isopropyl Ether, 927
n-Propylmalonic Acid, 803
Propyl methanoate, 328
n-Propyl β-Methoxypropionate, 1230
Propyloxycarbonyl-mitomycin C, 3677
n-Propyl-paba-β-cyclodextrin, 4137
Propyl Paraben, 2013
Propylparaben, 2013
3-Propyl-2,4-pentadione, 1502
2-Propyl-4-Pentenamide, 1517
2-Propylphenol, 1753
2-n-Propylphenol, 1753
4-Propylphenol, 1754
p-n-Propylphenol, 1754
α-Propylpiperidine, 1560
Propyl Propionate, 865
n-Propyl Propionate, 865
n-Propyl β-n-Propoxypropionate, 1832
1-Propyl Theobromine, 2065

7-Propyl Theophylline, 2066
N1-(5-Propyl-1,3,4-thiadiazol-2-yl)sulfanilamide, 2276
Propylthiouracil, 1159
6-Propyl-2-thiouracil, 1159
Propylurethan, 889
n-Propyl Urethane, 889
N-Propylurethane, 889
Propyl Xanthogen Disulfide, 1506
Propyne, 124
Propyphenazone, 3053
2-Propyphenol, 1753
4-Propyphenol, 1754
Propyzamide, 2572
Prosta-5,13-dien-1-oic Acid, 9,11,15-Trihydroxy-, 2,3-
    Dihydro-2-oxo-1H-indol-5-yl Ester,
    (5Z,9α,11α,13E,15S)-, 4054
Prostaglandin E2, 3785
Protazine, 3457
Prothoate, 1843
Protocatechualdehyde, 1063
Protocatechuic Acid, 1065
Protocatechuic Aldehyde Methylene Ether, 1318
Protoporphyrin IX, 4105
Protoprophyrin IX, 4105
Proxypham, 1986
Prozine, 3456
Pr-paraben, 2013
Pryfon, 3253
Pseudocumene, 1723
Pseudodigitoxin, 4127
Psoralen, 2187
PSP 204, 1852
PTAP, 2316
PTB 31, 3293
2-Pteridinamine, 668
4-Pteridinamine, 669
7-Pteridinamine, 667
2-Pteridinamine, N,N-Dimethyl-, 1425
4-Pteridinamine, N,N-Dimethyl-, 1426
7-Pteridinamine, N,N-Dimethyl-, 1424
Pteridine, 605
Pteridine, 7-Chloro-, 554
Pteridine, 2-Methoxy-, 1049
Pteridine, 4-Methoxy-, 1044
Pteridine, 7-Methoxy-, 1045
Pteridine, 2-Methyl-, 1043
Pteridine, 4-Methyl-, 1041
Pteridine, 7-Methyl-, 1042
Pteridine, 2-(Methylamino)-, 1100
Pteridine, 2-(Methylthio)-, 1053
Pteridine, 4-(Methylthio)-, 1050
Pteridine, 7-(Methylthio)-, 1051
2-Pteridinethiol, 621
4-Pteridinethiol, 620
7-Pteridinethiol, 619
4-Pteridinethiol, 7-Methyl-, 1052
2(1H)-Pteridinethione, 621
4(1H)-Pteridinethione, 620
7(1H)-Pteridinethione, 619
2-Pteridinol, 608
4-Pteridinol, 609
6-Pteridinol, 607
7-Pteridinol, 606
4-Pteridinol, 6-Methyl-, 1047

4-Pteridinol, 7-Methyl-, 1046
4(1H)-Pteridinone, Hydrazone, 719
Pteroylglutamic Acid, 3641
PTSA, 1125
Purifrigor Iso 3.5, 357
9H-Purin-6-amine, 9-[3,5-bis-O-[(1,1-
    Dimethylethyl)dimethylsilyl]-2-O-(1-oxobutyl)-β-D-
    arabinofuranosyl]-, 3025
1H-Purin-6-amine, 2-Chloro-, 378
Purine, 382
9H-Purine, 9-β-D-Arabinofuranosyl-6-methoxy-, 2279
1H-Purine-2,6-dione, 8-(2-Amino-4-chlorophenyl)-3,7-
    dihydro-1,3-dipropyl-, 3454
1H-Purine-2,6-dione, 8-(2-Aminophenyl)-3,7-dihydro-1,3-
    dipropyl-, 3467
1H-Purine-2,6-dione, 7-Butyl-3,7-dihydro-1,3-dimethyl-,
    2101
1H-Purine-2,6-dione, 1,3-Diethyl-3,7-dihydro-8-phenyl-,
    3189
1H-Purine-2,6-dione, 3,7-Dihydro-, 386
1H-Purine-2,6-dione, 3,7-Dihydro-1,3-dimethyl-,
    Monohydrate, 1166
1H-Purine-2,6-dione, 3,7-Dihydro-7-(2-hydroxyethyl)-
    1,3-dimethyl-, 1747
1H-Purine-2,6-dione, 3,7-Dihydro-8-methoxy-1,3,7-
    trimethyl-, 1748
1H-Purine-2,6-dione, 3,7-Dihydro-1,3,7-trimethyl-,
    Monohydrate, 1450
1H-Purine-2,6-dione, 1-Ethyl-3,7-dihydro-3,7-dimethyl-,
    1744
1H-Purine-2,6-dione, 7-Ethyl-3,7-dihydro-1,3-dimethyl-,
    1742
Purine-8-methanol, 713
9H-Purine, 6-Methoxy-9-[2-O-(methylsulfonyl)-β-D-
    arabinofuranosyl]-, 2677
1H-Purine, 8-Methyl-, 712
6-Purinethiol, 390
Purine-6-thiol, 390
1H-Purine-2,6,8(3H)-trione, 7,9-Dihydro-9-methyl-, 714
1H-Purine-2,6,8(3H)-trione, 7,9-Dihydro-1,3,7,9-
    tetramethyl-, 1746
9H-Purin-8-ol, 383
Purivel, 2019
Purpurin, 2949
Pyracarbolid, 2837
Pyralene 1498, 2459
Pyralene 3011, 2497
Pyranol 1499, 2497
Pyrazinamide, 399
Pyrazine-2-carboxamide, 399
Pyrazino[2,3-d]pyrimidine, 605
3H-Pyrazole-3-thione, 4,4'-Butylidenebis[1,2-dihydro-1,5-
    dimethyl-2-phenyl-, 4013
3H-Pyrazole-3-thione, 4,4'-Ethylidenebis[1,2-dihydro-1,5-
    dimethyl-2-phenyl-, 3946
3H-Pyrazole-3-thione, 4,4'-Heptylidenebis[1,2-dihydro-
    1,5-dimethyl-2-phenyl-, 4073
3H-Pyrazole-3-thione, 4,4'-(3-Methylbutylidene)bis[1,2-
    dihydro-1,5-dimethyl-2-phenyl-, 4033
3H-Pyrazole-3-thione, 4,4'-Methylenebis[1,2-dihydro-1,5-
    dimethyl-2-phenyl-, 3918
3H-Pyrazole-3-thione, 4,4'-(Phenylmethylene)bis[1,2-
    dihydro-1,5-dimethyl-2-phenyl-, 4070
3,5-Pyrazolidinedione, 3638

3,5-Pyrazolidinedione, 4-Butyryl-1,2-diphenyl-, 3639

3,5-Pyrazolidinedione, 1,2-Diphenyl-4-(3-phenylpropyl)-, 3944

3,5-Pyrazolidinedione, 1,2-Diphenyl-4-[2-(2,4-xylylthio)ethyl]-, 3988

1H-Pyrazolo[3,4-d]pyrimidine, 4-(1-Ethoxyethoxy)-, 1743

1H-Pyrazolo[3,4-d]pyrimidine, 4-[(Tetrahydro-2H-pyran-2-yl)oxy]-, 1997

4H-Pyrazolo[3,4-d]pyrimidin-4-one, 1-[(Acetyloxy)methyl]-1,5-dihydro-, 1371

4H-Pyrazolo[3,4-d]pyrimidin-4-one, 1-[(Benzoyloxy)methyl]-1,5-dihydro-, 2799

4H-Pyrazolo[3,4-d]pyrimidin-4-one, 2,5-bis[(Acetyloxy)methyl]-2,5-dihydro-, 2245

1H-Pyrazolo(3,4-d)pyrimidin-4-ol, 385

Pyrazon, 1892

Pyrazophos, 3082

Pyrene, 3271

Pyridazinone, 5-Amino-4-bromo-2-phenyl-, 1891

Pyridine, 1307

Pyridinecarboxylic Acid, 6-Chloro-, 582

3-Pyridinecarboxylic Acid, 2-[2-(4-Chlorophenoxy)-2-methyl-1-oxopropoxy]ethyl Ester, 3537

3-Pyridinecarboxylic Acid, (4,5-Dihydro-4-oxo-1H-pyrazolo[3,4-d]pyrimidin-1-yl)methyl Ester, 2552

m-Pyridine Carboxyphenylisothiocyanate, 2779

Pyridine-2,3-Dicarboxylate, 1013

2,3-Pyridinedicarboxylic Acid, 1013

Pyridine-2,3-Dicarboxylic Acid, 1013

2,4-Pyridinedicarboxylic Acid, 1017

2,5-Pyridinedicarboxylic Acid, 1018

Pyridine-2,5-Dicarboxylic Acid, 1018

3,4-Pyridinedicarboxylic Acid, 1019

3,5-Pyridinedicarboxylic Acid, 1012

3,5-Pyridinedicarboxylicacid, 3439

3,5-Pyridinedicarboxylic Acid, 4-(2,3-Dichlorophenyl)-1,4-dihydro-2,6-dimethyl-, Ethyl Methyl Ester, 3546

Pyridine, 2-[1-[2-[2-(Dimethylamino)ethyl]inden-3-yl]ethyl]-, 3746

2,4-Pyridinediol, 398

Pyridine, Hexahydro-N-nitroso, 453

2,3,4-Pyridinetricarboxylic Acid, 1292

2-Pyridinol, 395

3-Pyridinol, 397

4-Pyridinol, 396

Pyridinone, 3,5,6-trichloro-, 376

Pyridin-tricarbonsaeure-(2,3,4), 1292

5H-Pyrido[2,3-b][1,5]benzodiazepine-5-one, 11-Ethyl-6,11-dihydro-6-methyl-, 3180

5H-Pyrido[2,3-b][1,5]benzodiazepine-5-thione, 11-Ethyl-6,11-dihydro-6-methyl-, 3183

Pyrido[2,3-b][1,5]benzoxazepin-5(6H)-one, 3-Amino-6,9-dimethyl-, 3004

Pyrido[2,3-b][1,5]benzoxazepin-5(6H)-one, 3-Amino-6,7,9-trimethyl-, 3181

3,4-Pyridyl-(5)-2-chlorophenyl-1,2,4-oxadiazole, 2777

N-(2-Pyridyl)sulfanilamide, 2224

4-Pyrimidinecarboxamide, 1,2,3,6-Tetrahydro-2,6-dioxo-N-2-propenyl-, 1423

1(2H)-Pyrimidinecarboxylic Acid, 5-Fluoro-3,4-dihydro-2,4-dioxo-, Cyclohexyl Ester, 2254

1(2H)-Pyrimidinecarboxylic Acid, 5-Fluoro-3,4-dihydro-2,4-dioxo-, Hexyl Ester, 2288

1(2H)-Pyrimidinecarboxylic Acid, 5-Fluoro-3,4-dihydro-2,4-dioxo-, Methyl Ester, 640

1(2H)-Pyrimidinecarboxylic Acid, 5-Fluoro-3,4-dihydro-2,4-dioxo-, 1-Methylethyl Ester, 1411

1(2H)-Pyrimidinecarboxylic Acid, 5-Fluoro-3,4-dihydro-2,4-dioxo-, 2-Methylpropyl Ester, 1698

1(2H)-Pyrimidinecarboxylic Acid, 5-Fluoro-3,4-dihydro-2,4-dioxo-, Phenylmethyl Ester, 2544

4-Pyrimidinecarboxylic Acid, 1,2,3,6-Tetrahydro-1-methyl-2,6-dioxo-, 708

4,6-Pyrimidinediol, 237

4,6(1H,5H)-Pyrimidinedione, 5,5-Diethyldihydro-2-thioxo, 1485

4,6(1H,5H)-Pyrimidinedione, 5-Ethyldihydro-5-(1-methylbutyl)-2-thioxo, 2332

2,4(1H,3H)-Pyrimidinedione, 1-Methyl-, 408

2,4(1H,3H)-Pyrimidinedione, 3-Methyl-, 404

2,4,6-Pyrimidinetriol, 240

2,4,6(1H,3H,5H)-Pyrimidinetrione, 5,5-bis(1-Methylethyl), 2098

2,4,6(1H,3H,5H)-Pyrimidinetrione, 5,5-Dimethyl, 766

2,4,6(1H,3H,5H)-Pyrimidinetrione, 5-(1,1-Dimethylethyl)-5-(3-methyl-2-butenyl), 2914

2,4,6(1H,3H,5H)-Pyrimidinetrione, 5,5-Diphenyl, 3275

2,4,6(1H,3H,5H)-Pyrimidinetrione, 5-Ethyl-5-hexyl-, 2725

2,4,6(1H,3H,5H)-Pyrimidinetrione, 5-Ethyl-5-(3-methyl-2-butenyl), 2309

2,4,6(1H,3H,5H)-Pyrimidinetrione, 5-Ethyl-5-(3-methyl-2-butenyl)-, 2309

2,4,6(1H,3H,5H)-Pyrimidinetrione, 5-Ethyl-5 (1-methylethyl), 1776

2,4,6(1H,3H,5H)-Pyrimidinetrione, 5-Ethyl-5-octyl-, 3110

2,4,6(1H,3H,5H)-Pyrimidinetrione, 5-Ethyl-5-propyl-, 1774

2,4,6(1H,3H,5H)-Pyrimidinetrione, 5-(3-Methyl-2-butenyl)-5-(1-methylethyl), 2704

2,4,6(1H,3H,5H)-Pyrimidinetrione, 5-(1-Methylethyl)-, 1164

2,4,6(1H,3H,5H)-Pyrimidinetrione, 5-Methyl-5-(3-methyl-2-butenyl), 2062

2,4,6(1H,3H,5H)-Pyrimidinetrione, 5-Methyl-5-phenyl, 2211

2,4,6(1H,3H,5H)-Pyrimidinetrione, 5-Methyl-5-(2-propenyl), 1444

2,4,6(1H,3H,5H)-Pyrimidinetrione, 5-Phenyl-5-(2-propenyl), 2817

2-Pyrimidinol, 235

4(1H)-Pyrimidinone, 236

2(1H)-Pyrimidinone, 4-Amino-, 257

4(1H)-Pyrimidinone, 2,3-Dihydro-5,6-dimethyl-2-thioxo-, 760

4(1H)-Pyrimidinone, 2,3-Dihydro-5-methyl-2-thioxo-, 406

N1-(2-Pyrimidinyl)-sulfanilamide, 1939

Pyrimido[4,5-b]pyrazine, 605

4(3H)-Pyrimidone, 234

Pyrocatechol, 724

Pyrogallol, 729

Pyrogallol-1,3-dimethylaether, 1469

Pyrolan, 2843

Pyromellitic Acid, 1878

Pyrrobutamine, 3741

Pyrrole, 252
Pyrrolidine, 1-[(Benzoyloxy)acetyl]-, 2839
2-Pyrrolidinecarboxamide, 1-[(Benzoyloxy)acetyl]-,
    3033
2-Pyrrolidinecarboxylic Acid, 442
Pyrrolidine-2-carboxylic Acid, 441
Pyrrolidine, 1-[4-(4-Chlorophenyl)-3-phenyl-2-butenyl]-,
    3741
2,5-Pyrrolidinedione, 254
1-Pyrrolidino-3,5-bis(Dimethylamino)-s-triazine, 2344
1-(Pyrrolidinyl)-3,5-bis(dimethylamino)-s-triazine, 2344
N-(1-Pyrrolidinylmethyl)tetracycline, 4035
Pyrrolo[2,1-b]quinazolin-9(1H)-one, 2,3-Dihydro-3-
    hydroxy-, (3S)-, 2209
Pyrrolylene, 262
Quaalude, 3290
Quabain, 4079
Quercinitol, 881
D-Quercit, 870
D-Quercitol, 870
Quinalizarin, 2950
Quinalphos, 2654
Quinethazone, 1982
Quinhydrone, 2570
Quinic Acid, 1196
D-(-)-Quinic Acid, 1196
Quinidine, 3747
Quinine, 3748
Quinine Alkaloid, 3748
Quinine, Compd. with Valeric Acid (1:1), Hydrate, 3749
Quinine (Trihydrate), 3749
Quininone, 3742
Quinizarin, 2948
Quinoline, 1618
4-Quinolinecarboxamide, 2-Butoxy-N-[2-
    (diethylamino)ethyl]-, 3776
Quinoline, 2,4-Dimethyl-, 2216
Quinoline, 2,7-Dimethyl-, 2215
Quinolinic Acid, 1013
2-Quinolinol, 1619
3-Quinolinol, 1620
4-Quinolinol, 1621
5-Quinolinol, 1622
6-Quinolinol, 1623
7-Quinolinol, 1625
8-Quinolinol, 1624
Quinonamid, 2457
Quinone, 622
p-Quinone, 622
Quintobenzene, 952
Quintox, 2524
Quintozene, 952
R 14827, 3530
R 25788, 1476
R 4749, 3874
R 600α, 357
R 600 (alkane), 356
Rabcon, 1272
Rabon, 1906
Racemethorphan, 3583
Raffinose, 3606
Raffinose (Pentahydrate), 3607
Ragadan, 1732
Rampart, 1266

Randox, 1483
Rapid, 2337
Raugalline, 3763
Rauwilid, 4101
Rauwiloid, 4101
Rauwolfine, 3763
Rauzide, 3170
Ravage, 2102
Razebil, 3280
RDX, 172
Redax, 2558
Redeptin, 251
Regitine Mesylate, 3572
Regitine Methanesulfonate, 3572
Relane, 3744
Rentovet, 3538
Reposal, 3054
Reserpine, 4101
Reserptyl, 2543
Resiren Blue TG, 3293
Resorcin, 723
Resorcinol, 723
Resorcinol Monoethyl Ether, 1466
Resorcinol Monomethylether, 1120
β-Resorcyclic Acid, 1066
β-Resorcylic Acid, 1066
γ-Resorcylic Acid, 1067
Retinal, 3770
Retinene, 3770
Retinoic Acid, 3772
Retinol, 3778
Retinol, Hexadecanoate, 4118
Retinyl Palmitate, 4118
rG, 2044
RH-315, 2572
Rhamnose, 869
α-L-Rhamnose, 869
L-Rhamnose, 869
Rhinocort, 3995
Rhizoxin, 4113
Rhodan, 1040
Rhodanin, 120
Rhodanine, 120
Rhodiachlor, 1861
Rhodiatox, 2056
Rhoeadine, 3809
Riboflavin, 3461
Riboflavine, 3461
9-β-D-Ribofuranosyladenine, 2043
9-B-D-Ribofuranosyl-9H-purin-6-amine Adenine
    Riboside, 2043
Ricinin, 1364
Ricinine, 1364
Ridomil, 3229
Rifampicin, 4132
Rifampin, 4132
Rimidin, 3411
Ringdex C, 4142
Riomitsin, 3881
Risocaine, 2026
RJ-64, 2777
Ro 3-0658, 452
Robiflavine, 3461
Robitet, 3879

Robitussin AC, 3564
Rolitetracycline, 4035
Romate, 1647
Rondomycin, 3875
RO-Neet, 2352
Ronilan, 2542
Ronnel, 1359
Ronstar, 3199
Ropredlone, 3831
Rosaniline, 3735
Rosex, 3630
Rosolic Acid, 3626
Rospin, 3425
Rotenone, 3913
Rotetra, 2486
Rovokil, 1589
Rowmate, 1648
Rozex, 783
RP-17623, 3199
RP 2254, 2275
RP 8532, 259
Rubigan 4AS, 3411
Rubitox, 2639
Rufen, 2898
RX 67408, 2967
Saccharin, 1011
Saccharose, 2744
Safrotin, 2156
Salbutamol, 2920
Salicin, 2904
Salicoside, 2904
Salicyl Alcohol, 1121
Salicylaldehyd, 1056
Salicylaldehyde, 1056
Salicylaldehyde β-D-Glucoside, 2863
Salicylalkohol, 1121
Salicylamide, 1091
Salicylanilide, 2808
Salicylic Acid, 1060
Salicylic Acid Acetate, Ester with N-
   (hydroxymethyl)succinimide, 3002
Salicylic Acid Acetate, Hydroxymethyl Ester Acetate,
   2601
Salicylic Acid Acetate, Hydroxymethyl Ester Propionate,
   2834
Salicylic Acid n-Butyl Ester, 2284
Salicylic Acid, 2-Ethylbutyrate, 2861
Salicylic Acid Methyl Ether, 1389
Salicylic Acid, Pivalate, 2637
Salicylolhydrazone of Picolinealdehyde, 2987
2',6'-Salicyloxylidide, 4-Chloro-5-sulfamoyl-, 3177
Salicylsaeure-methylaether, 1389
Salicylsaeure-methyl Ester, 1391
Salithion, 1427
Salol, 2801
Saluron, 1361
Samaron Violet, 4006
Samaron Yellow, 3510
San 6706, 2804
Sancap, 2356
Sancap 80W, 2356
Sandimmun, 4166
Sandimmun Neoral, 4166
Sang-35, 4166

Santonin, 3205
Saprol, 2055
Sarcosine, 196
1-(Sarcosino)-3,5-bis(dimethylamino)-s-triazine, 2124
Sarkosin, 196
Scabagen, 2996
Schleimsaeure, 806
Schwefelkohlenstoff, 44
Scopolamin, 3465
Scopolamine, 3465
Scotts OH I, 3199
Scyllit, 881
Scyllitol, 881
SD 45418, 1763
SD 8447, 1906
SDZ-OXL 400, 4166
Sebacic Acid, 2139
Sebacic Acid bis(2-Ethylhexyl) Ester, 4026
Sebacil, 2655
Sebacinsaeure, 2139
Secbumeton, 2147
Secobarbital, 2703
Seconal, 2703
1,2-Secopilocarpin-2-oic Acid, 2333
D-Sedoheptose, 1234
Seedrin, 2522
Seedvax, 2223
Selecron, 2286
Semeron, 1522
α-Semicarbazono-p-tolyl Acetate, 1974
α-Semicarbazono-p-tolyl Prostaglandin E2, 4055
α-Semicarbazono-p-tolyl Prostaglandin F2 α, 4057
Semiken, 3074
Seminose, 872
Sencor, 1499
Sencorex, 1499
Septra, 1975
Seractil, 2900
Seraphos, 2156
Serax, 3141
Seredon, 2811
Serenace, 3815
Serilene Fast Pink BT, 3143
Serilene Red 2BL, 3800
L(-)-Serin, 206
Serine, 206
D-Serine, 207
DL-Serine, 205
N-Serve(R), 568
Shikimic Acid, 1168
Shikimisaeure, 1168
Shoxin, 4097
SI-6711, 2853
Sicarol, 2837
Siduron, 3080
Silvex, 1616
Simazine, 1182
Simetone, 1521
Simetryne, 1523
Sinbar 80W, 1762
Sinequan, 3657
Sintodril, 3874
Sirmate, 1648
Sirmate 4E, 1648

Skatole, 1651
Skellysolve B, 899
SKF 962A, 3757
SK-Lygen, 3285
SL 76-002, 3424
SL79.182, 2775
Smidan, 2231
Smoke Red M, 3143
Sobrerol, 2133
Solan, 2888
Solanan, 2828
Solanidane, β-D-Galactopyranoside Deriv, 4136
Solanine, 4136
Solgard, 2928
Solumedrol, 3893
Solvent Red 146, 3800
Solvirex, 1590
Somanil, 817
Somnafac, 3290
Sorbic Acid, 771
Sorbidin, 768
Sorbistat, 771
Sorbitol, 938
D-Sorbitol, 938
Sorbose, 871
L-Sorbose, 871
Spanish Fly, 2014
Sparine, 3456
(-)-Spartein, 3261
Sparteine, 3261
Spasmolysin, 2036
Spectazole, 3530
Spectracide, 2731
Spike, 1790
Spike 20P, 1790
Spiractin, 3956
Spiro[barbituric Acid-5,1'-cyclohexane], 1740
Spiro[benzofuran-2(3H),1'-[2]cyclohexene]-3,4'-dione, 7-
    Chloro-2',4,6-trimethoxy-6'-methyl-, 4'-oxime, 3436
Spiro[benzofuran-2(3H),1'-[2]cyclohexen]-3-one, 7-
    Chloro-4'-hydroxy-2',4,6-trimethoxy-6'-methyl-, 3446
Spirocyclopentabarbituric Acid, 1443
Spirocyclopropane-1',5-barbituric Acid, 707
Spiroform, 3158
Spironolactone, 3956
(25R)-Spirost-5-en-3β-ol, 4044
Spongoadenosine, 2086
SQ 22982, 378
ST 100, 1850
Stadacain, 3500
Staflex DTDP, 4109
Staleydex 111, 874
Staleydex 333, 874
Stanolone, 3701
Stanolone Acetate, 3850
Stauffer N-2790, 2087
Stearic Acid, 3610
Stearinsaeure, 3610
Stearoyl Acetaminophen, 4024
Stearyl Alcohol, 3612
Steladone, 2621
Stelazine, 3817
Steraffine, 3612
α-Sterol, 3778

Stevacin, 3881
trans-Stilbene, 2990
Stinger, 555
Stirofos, 1906
Streptovaricin, 4125
Streptovaricin C, 4125
Streptozocin, 1519
Streptozotocin, 1519
Strophanthin, 4091
γ-Strophanthin, 4079
k-Strophanthin, 4091
Strychnidin-10-one, 3812
Strychnidin-10-one, 2,3-Dimethoxy-, Tetrahydrate, 3920
Strychnine, 3812
L-Strychnine, 3812
Stugeron, 4010
Styphnic Acid, 573
Styphninsaeure, 573
Styrene, 1351
Styrene Oxide, 1375
Styrol, 1351
Styrolene, 1351
Subdue, 3229
Suberic Acid, 1508
Succinamide, 314
Succinic Acid, 281
Succinic Acid di-n-Butyl Ester, 2735
Succinic Acid 2,2-Dimethylhydrazide, 833
Succinimide, 254
Succinonitrile, 233
Succinyl Acetaminophen, 3737
7,7'-Succinylditheophylline, 3542
2-(N(4)-Succinylsulfanilamido)thiazole, 2825
Succinylsulfathiazole, 2825
Sucrose, 2744
Suffix, 3534
Sugordomycin, 3634
Sulerex, 2019
Sulfabenz, 2593
Sulfacetamide, 1446
Sulfadiazine, 1939
Sulfadicramide, 2274
Sulfadimethoxine, 2636
Sulfadimethoxine N4-Acetate, 3040
Sulfadimezine, 2629
Sulfaethidole, 1998
Sulfaethylthiadiazole, 1998
Sulfaguanidin, 1165
Sulfaguanidine, 1165
4-Sulfahydroxymethylnaphthalene, 3428
Sulfaisopropylthiadiazole, 2275
Sulfamerazine, 2239
Sulfameter, 2240
Sulfamethazine, 2629
Sulfamethazine Hemihydrate, 2630
Sulfamethizole, 1678
Sulfamethomidine, 2634
Sulfamethoxazole, 1975
Sulfamethoxypyridazine, 2241
1-Sulfamethylnaphthalene, 3426
Sulfamethylpyrimidine, 2239
Sulfamethylthiadiazole, 1678
Sulfamethylthiazole, 1973
Sulfamide, (4-Methylphenyl)-, 1162

Sulfamoxole, 2268
Sulfanaphthoquinone, 3276
Sulfanilamide, 764
Sulfanilamide (Monohydrate), 765
Sulfanilamide, N1-(5-Butyl-1,3,4-thiadiazol-2-yl)-, 2675
Sulfanilamide, N1-4-Pyrimidinyl-, 1941
Sulfanilamide, N1-5-Pyrimidinyl-, 1942
Sulfanilamide, N1-1,3,4-Thiadiazol-2-yl-, 1370
Sulfanilamide, N1-(4,5,6-Trimethyl-2-pyrimidinyl)-, 2858
2-Sulfanilamido-5-aminopyridine, 2238
5-Sulfanilamido-2-aminopyrimidine, 1976
2-Sulfanilamidobenzothiazole, 2812
2-Sulfanilamidobornylenepyrimidine, 3575
2-Sulfanilamido-5-bromopyridine, 2203
5-Sulfanilamido-2-bromopyridine, 2204
5-Sulfanilamido-2-chloropyridine, 2206
2-Sulfanilamido-5-chloropyrimidine, 1900
5-Sulfanilamido-2-chloropyrimidine, 1901
6-Sulfanilamido-2,4-dimethylpyrimidin, 2628
2-Sulfanilamido-4,5-dimethylpyrimidine, 2632
2-Sulfanilamido-4,6,-dimethylpyrimidine, 2629
6-Sulfanilamido-2,4-dimethylpyrimidine, 2628
2-Sulfanilamido-4-p-diphenylthiazole, 3803
2-Sulfanilamido-3-ethoxypyridine, 2844
5-Sulfanilamido-2-ethoxypyridine, 2845
2-Sulfanilamido-4-ethoxypyrimidine, 2635
5-Sulfanilamido-2-hydroxypyridine, 2226
2-Sulfanilamido-5-iodopyridine, 2208
2-Sulfanilamido-4-methoxypyrimidine, 2243
5-Sulfanilamido-2-methoxypyrimidine, 2244
2-Sulfanilamido-4-methyl-5-n-amylpyrimidine, 3366
4-Sulfanilamido-2-methylpyrimidine, 2237
2-Sulfanilamido-4-methylthiazole, 1973
2-Sulfanilamido-5-nitropyridine, 2213
4-Sulfanilamido-1-phenyl-2,3-dimethyl-5-pyrazolone, 3440
4-Sulfanilamidopyrimidine, 1941
5-Sulfanilamidopyrimidine, 1942
2-Sulfanilamido-5,6,7,8,-tetrahydro-8-isopropyl-5-methyl-quinazoline, 3576
2-Sulfanilamido-5,6,7,8-tetrahydroquinazoline, 3035
2-Sulfanilamido-5,6,7,8,-tetrahydroquinazoline, 3035
2-Sulfanilamido-1,3,4-thiadiazole, 1370
5-Sulfanilamidouracil, 1943
Sulfanilguanidin, 1165
Sulfanilic Acid, 744
m-Sulfanilic Acid, 745
Sulfanilid, 2593
Sulfanilsaeure, 744
Sulfanilsaeure-amid, 764
2-Sulfanilylamino-4-amylpyrimidine, 3215
2-Sulfanilylamino-4-ethyl-5-methylpyrimidine, 2859
2-Sulfanilylamino-4-ethylpyrimidine, 2631
2-Sulfanilylamino-4-isobutylpyrimidine, 3055
N-Sulfanilyl-β,β-dimethylacrylamide, 2274
Sulfanilylguanidine, 1165
Sulfanilylharnstoff, 1157
1-Sulfanilyl-3-methyl-5-pyrazolone, 1927
Sulfanilylurea, 1157
6-Sulfapurine, 2246
Sulfapyrazine, 1940
Sulfapyridine, 2224
2-Sulfapyridine, 2224

Sulfapyridine Acetylee, 2824
4-Sulfapyrimidine, 1941
5-Sulfapyrimidine, 1942
Sulfapyrrole, 1971
Sulfathiadiazole, 1370
Sulfathiazol Acetyle, 2225
Sulfathiazole, 1661
Sulfathiazoline, 1721
Sulfathiazol Methyle, 1970
Sulfathiazol Methyle Acetyle, 2617
Sulfathiourea, 1155
Sulfirgamid, 2274
Sulfirgamide, 2274
Sulfisomidine, 2628
Sulfisoxazole, 2267
Sulfonal, 1264
Sulfonethylmethane, 1586
Sulfonmethane, 1264
Sulfotepp, 1599
Sulfuno, 2268
Sulfure β'-Ethyl Dichlore, 311
Sulfuric Acid Dimethyl Ester, 103
Sulgin ASG, 1749
Sulphadiazine, 1939
Sulphadimethoxine, 2636
Sulphamethazine (Hemihydrate), 2630
Sulphamethomidine, 2634
Sulphamethoxydiazine, 2240
Sulphamethoxypyridazine, 2241
Sulphapyrazine, 1940
Sulphapyridine, 2224
Sulphasomidine, 2628
Sulphathiazole, 1661
Sulpiride, 3251
Sumatra Camphor, 2126
Sumikaron Red E-FBL, 3800
Sumithrin, 3921
Sumycin, 3879
Suncide, 2293
Supracet Brilliant Violet 3R, 2971
Supracide, 817
Supra Light Yellow GGL(IG), 3510
Surecide, 3171
Sustar, 1959
Sylvic Acid, 3780
Symclosene, 223
Synotodecin, 4035
Syntetrin, 4035
Syraprim, 3057
Systox, 1593
2,3,4-T, 1281
2,3,5-T, 1284
2,3,6-T, 1286
2,4,5-T, 1285
2,4,6-T, 1283
3,4,5-T, 1282
T-1703, 904
Tabutrex, 2735
Tachigaren, 255
Tagatose, 876
DL-Tagatose, 876
Talbutal, 2310
D-Talogalactaric Acid, 806
D-Taloschleimsaeure, 806

Tandex, 3095
Taractan, 3538
Tarragon, 2008
D-(-)-Tartaric Acid, 288
DL-Tartaric Acid, 287
L-Tartaric Acid, 286
L(+)-Tartaric Acid, 286
meso-Tartaric Acid, 289
Tartaric Acid (Racemic), 287
Tartran, 2145
Taterpex, 1980
Taurin, 105
Taurine, 105
2,4,5-TB, 1903
4-(2,4,5-TB), 1903
2,3,6-TBA, 957
TBPO, 2761
TC 44, 2764
TCA, 48, 57
TCBC, 541
TCDD, 2407
1,2,3,4-TCDD, 2404
2,3,7,8-T4CDF, 2403
TCP, 543
TCPP, 1781
TDCPP, 1781
p,p'-TDE, 2969
p,p'-TDEE, 2957
TE-031, 4124
Tebulan, 3411
Tebuthiuron, 1790
Tecnazene, 529
Tecto, 1888
Tedion, 2486
TEHP, 3986
Temephos, 3355
Temik, 1208
Temorine, 1746
Temurin, 1746
Tenfidil, 3075
Tenoran, 3176
TEP, 946
Terasil Violet P 4RT, 3453
Terbacil, 1762
Terbufos, 1850
Terbufos Sulfone, 1857
Terbufos Sulfoxide, 1853
Terbuthylazine, 1788
Terbutol, 3499
Terbutrex, 2151
Terbutryn, 2151
Terbutryne, 2151
Terbutylazine, 1788
Terebumeton, 2146
tere-Butyl Phthalate, 3370
Terephthalaldehyd, 1317
Terephthalate Acid Dimethyl Ester, 1949
Terephthaldicarboxaldehyde, 1317
Terephthalic Acid, 1322, 1949
Terephthalonitrile, 1275
Terephthalsaeure-dimethyl Ester, 1949
m-Terphenyl, 3522
o-Terphenyl, 3521
p-Terphenyl, 3523

Terpin-hydrat, 2173
Terpin (Monohydrate), 2173
Terrafungine, 3881
Terraneb, 1357
Terraneb B, 1357
Terraneb SP, 1357
Testex, 3693
Testosterone, 3693
Testosterone Acetate, 3842
Testosterone Benzoate, 4016
Testosterone Butyrate, 3934
Testosterone Formate, 3774
Testosterone 17-Formate, 3774
Testosterone (Monohydrate), 3694
Testosterone Monohydrate -I, 3694
Testosterone Phenylacetate, 4036
Testosterone Phenylbutyrate, 4075
Testosterone Phenyl Propionate, 4052
Testosterone Propionate, 3896
Testosterone-17-Propionate, 3896
Testosterone Valerate, 3964
Testosterone 17-Valerate, 3964
1,4,7,10-Tetraazadecane, 950
1,3,5,8-Tetraazanaphthalene, 605
1,1,2,2-Tetrabrom-aethan, 51
Tetrabromoacetylene, 51
Tetrabromobiphenyl, 2452
2,2',5,5'-Tetrabromobiphenyl, 2452
1,1,2,2-Tetrabromoethane, 51
sym-Tetrabromoethane, 51
Tetrabromomethane, 34
Tetracaine, 3254
Tetracene, 3515
3,4,3',4'-Tetrachlorbiphenyl, 2463
2,3,4,5-Tetrachloroanisole, 974
1,2,3,4-Tetrachlorobenzene, 539
1,2,3,5-Tetrachlorobenzene, 540
1,2,4,5-Tetrachlorobenzene, 542
1,2,4,6-Tetrachlorobenzene, 540
s-Tetrachlorobenzene, 542
2,4,5,6-Tetrachloro-1,3-benzenedicarbonitrile, 1601
Tetrachloro-1,2-Benzenedicarboxylic Acid, 1270
2,3,5,6-Tetrachloro-p-benzoquinone, 951
Tetrachloro-p-benzoquinone, 951
Tetrachlorobiphenyl, 2459
2,2',3,3'-Tetrachlorobiphenyl, 2470
2,2',3,4-Tetrachlorobiphenyl, 2472
2,2',3,4'-Tetrachlorobiphenyl, 2471
2,2',3,5'-Tetrachlorobiphenyl, 2473
2,2',3,6'-Tetrachlorobiphenyl, 2485
2,2',4,4'-Tetrachlorobiphenyl, 2474
2,2',4,5-Tetrachlorobiphenyl, 2467
2,2',4,5'-Tetrachlorobiphenyl, 2468
2,2',4,5'-Tetrachlorobiphenyl, 2468
2,2',5,5'-Tetrachlorobiphenyl, 2476
2,2',5,6'-Tetrachlorobiphenyl, 2477
2,2',6,6'-Tetrachlorobiphenyl, 2478
2,3,3',4'-Tetrachlorobiphenyl, 2482
2,3,4,4'-Tetrachlorobiphenyl, 2475
2,3',4,4'-Tetrachlorobiphenyl, 2480
2,3,4,5-Tetrachlorobiphenyl, 2465
2',3,4,5-Tetrachlorobiphenyl, 2469
2,3,4',5-Tetrachlorobiphenyl, 2483
2,3',4',5-Tetrachlorobiphenyl, 2479

2,3',4,6-Tetrachlorobiphenyl, 2481
2,3,4',6-Tetrachlorobiphenyl, 2484
2,4,4',5-Tetrachlorobiphenyl, 2466
2,4,4',6-Tetrachlorobiphenyl, 2464
3,3',4,4'-Tetrachlorobiphenyl, 2463
3,3',5,5'-Tetrachlorobiphenyl, 2462
2,3,5,6-Tetrachloro-2,5-cyclohexadiene-1,4-dione, 951
1,2,3,4-Tetrachlorodibenzo-p-dioxin, 2404
1,2,3,7-Tetrachlorodibenzo-p-dioxin, 2405
1,3,6,8-Tetrachlorodibenzo[1,4]dioxin, 2406
1,3,6,8-Tetrachlorodibenzo-p-dioxin, 2406
2,3,7,8-Tetrachlorodibenzodioxin, 2407
2,3,7,8-Tetrachlorodibenzo-p-dioxin, 2407
1,2,3,4-Tetrachlorodibenzo[b,e][1,4]dioxin, 2404
2,3,7,8-Tetrachlorodibenzofuran, 2403
3,4,5,6-Tetrachloro-1,2-dimethoxybenzene, 1306
2,4,5,4'-Tetrachlorodiphenyl Sulfone, 2486
1,1,1,2-Tetrachloroethane, 57
1,1,2,2-Tetrachloroethane, 56
sym-Tetrachloroethane, 56
Tetrachloroethene, 110
Tetrachloroethylene, 110
Tetrachloro-ethylene, 110
Tetrachlorohydroquinone, 546
2,3,5,6-Tetrachlorohydroquinone, 546
Tetrachloromethane, 40
2,4,5,6-Tetrachloro-3-methyl-phenol, 973
1,2,3,4-Tetrachloro-5-nitrobenzene, 528
2,3,4,5-Tetrachloronitrobenzene, 528
2,3,4,5-Tetrachloro-1-nitrobenzene, 528
2,3,4,6-Tetrachloronitrobenzene, 530
2,3,5,6-Tetrachloronitrobenzene, 529
2,3,4,5-Tetrachlorophenol, 544
2,3,4,6-Tetrachlorophenol, 543
2,3,5,6-Tetrachlorophenol, 545
2,3,4,6-Tetrachlorophenoxyacetic Acid, 1274
Tetrachlorophthalic Acid, 1270
Tetrachloroquinoxaline, 1269
5,6,7,8-Tetrachloroquinoxaline, 1269
2,4,4''',6-Tetrachloro-p-terphenyl, 3506
2,4,4''',6-Tetrachloro-1,1':4',1''-terphenyl, 3506
Tetrachloroveratrole, 1306
Tetrachlorphthalsaeure, 1270
Tetrachlorvinphos, 1906
Tetracosane, 3983
n-Tetracosane, 3983
Tetracycline, 3879
Tetradecakis-2,6-O-methylcycloheptaamylose, 4169
Tetradeca-O-methyl-β-cyclodextrin, 4169
Tetradecanamide, N-hydroxy, 3120
Tetradecane, 3125
n-Tetradecane, 3125
Tetradecanedioic Acid, 3112
Tetradecanoic Acid, 3116
Tetradecanol, 3126, 3127
Tetradecyl 4-Hydroxybenzoate, 3853
Tetradecyl p-Hydroxybenzoate, 3853
Tetradifon, 2486
N,N,N',N'-Tetraethyl-bicyclo(2.2.1)hept-5-ene-2,3-
    dicarboxamide, 3502
Tetraethyl Dithiopyrophosphate, 1599
Tetraethylenepentamine, 1600
N,N,N',N'-Tetraethylfumaramide, 2733
N,N,N',N'-Tetraethylisophthalamide, 3384

Tetraethylmethane, 1838
O,O,O,O-Tetraethyl S,S-Methylene
    bisPhosphorodithioate, 1858
Tetraethylsilane, 1597
Tetraethylsilicane, 1597
Tetraethylsilicon, 1597
Tetraethylstannane, 1598
N,N,N',N'-Tetraethylterephthalamide, 3385
Tetraethylthioperoxydicarbonothioic Diamide, 2158
Tetraethylthiuram Disulfide, 2158
Tetraethyltin, 1598
Tetrafenphos, 3355
1,2,3,5-Tetrafluorobenzene, 547
1,2,4,5-Tetrafluorobenzene, 548
1,2,4,6-Tetrafluorobenzene, 547
1,3,4,5-Tetrafluorobenzene, 547
2,3,5,6-Tetrafluorobenzene, 548
m-Tetrafluorobenzene, 547
p-Tetrafluorobenzene, 548
1,2,4,5-Tetrafluoro-3-hydroxybenzene, 549
Tetrafluoromethane, 41
2,3,5,6-Tetrafluorophenol, 549
Tetrahydro, 1861
1,2,3,4-Tetrahydrobenzene, 784
Tetrahydrocannabinol, 3837
δ9-Tetrahydrocannabinol, 3837
1,2,3,6-Tetrahydro-2,6 dioxo-4-pyrimidinecarboxylic
    Acid, 380
1,2,3,6-Tetrahydro-2,6-dioxo-4-pyrimidine-carboxylic
    Acid, Ethyl Ester, 1111
Tetrahydrofuran, 326
Tetrahydro-3-furanol, 327
1-(2-Tetrahydrofuryl)-5-fluorouracil, 1409
1,2,12,12α-Tetrahydro-2α-isopropenyl-8,9-
    dimethoxy(1)benzopyrano(3,4-b)furo(2,3-
    h)(1)benzopyran-6(6α H)-one, 3913
1,2,3,4-Tetrahydro-10-methyl-1,2-benzanthracene, 3636
1,2,3,4-Tetrahydronaphthalene, 1977
5,6,7,8-Tetrahydro-naphthol-(2), 2006
5,6,7,8-Tetrahydro-2-naphthol, 2006
2-Tetrahydropuran-4-allopurinyl Ether, 1997
Tetrahydropyran, 460
1,2,3,4-Tetrahydroquinoline, 1701
cis-3α,4,7,7α-Tetrahydro-2-(1,1,2,2-tetrachloroethyl)thio-
    1H-Isoindole-1,3(2H)-dione, 1905
1,2,5,8-Tetrahydroxyanthraquinone, 2950
1,3,4,5-Tetrahydroxycyclohexanecarboxylic Acid, 1196
5,7,3',4'-Tetra-hydroxyflavon, 3139
2,4,6,7-Tetrahydroxypteridine, 617
2,3,4,5-Tetraiodpyrrol, 224
Tetralin, 1977
N,N,N',N'-Tetramethylazelamide, 2934
1,2,4,5-Tetramethylbenzene, 2049
Tetramethyl-bernsteinsaeure, 1507
N,N,N',N'-Tetramethylbutanediamide, 1533
Tetramethyldiaminobenzophenone, 3455
Tetramethylene Acetate, 1510
5,5-Tetramethylenebarbituric Acid, 1443
1,1,2,2-Tetramethylethane, 902
1,1,4,4-Tetramethyl-6-ethyl-7-acetyl-1,2,3,4-
    tetrahydronaphthalene, 3589
cis-N,N,N',N'-Tetramethylfumaramide, 1498
2,2,5,5-Tetramethyl-3-hexyne, 2118
N,N,N',N'-Tetramethylisophthalamide, 2667

Tetramethylmelamine, 1212
N2,N2,N4,N4-Tetramethylmelamine, 1212
Tetramethylolmethane, 521
3,4,7,8-Tetramethyl-1,10-phenanthroline, 3311
N,N,N',N'-Tetramethylphthalamide, 2666
1,1,2,2-Tetramethylpropanol, 1258
1,1,2,2-Tetramethylpropyl alcohol, 1258
1,3,7,9-Tetramethylpyrimido(5,4-γ) Pteridine-
    2,4,6,8(1H,3H,7H,9H)-tetrone, 2600
N,N,N',N'-Tetramethylsebacamide, 3115
N,N,N',N'-Tetramethylsuberamide, 2748
N,N,N',N'-Tetramethylsuccinamide, 1533
Tetramethyl Succinic Acid, 1507
N,N,N',N'-Tetramethylterephthalamide, 2668
2,2,5,5-Tetramethyltetrahydrofuran-3,4-diol, 1553
2,2,5,5-Tetramethyl-tetrahydro-3-hydroxy-3-furanyl
    Methyl Ketone, 2134
Tetramethyl O,O'-Thiodi-p-phenylene Phosphorothioate,
    3355
Tetramethylthioperoxydicarbonothioic Diamine, 841
Tetramethylthiuram Disulfide, 841
1,3,7,9-Tetramethyluric Acid, 1746
1,3,7,8-Tetramethylxanthine, 1745
2,4,6,7-Tetraminopteridine, 770
2:4:6:7-Tetraminopteridine, 770
2,3,5,6-Tetranitroanisol, 983
2,4,2',4-Tetranitro-diphenylamin, 2509
2,4,2',4'-Tetranitrodiphenylamine, 2509
2,4,5,6-Tetranitrodiphenylamine, 2510
Tetra-n-propylgermane, 2766
Tetrapropylgermanium, 2766
Tetraverin, 4035
Tetrethyl, 1350
Tetrosin SP, 630
Tetryl, 1027
α-TGT, 2658
TH 6040, 2954
Thalitone, 2981
1,3,5-THB, 726
THC, 3837
Thebaine, 3658
Thebainon, 3563
Thebainone A, 3563
Thefanil, 3075
Thenfadil, 3075
3-Thenoic Acid, 392
Thenyldiamine, 3075
Theobromin, 1114
Theobromine, 1114
Theophylline, 1113
Theopylline (Monohydrate), 1166
THFFU, 1409
TH 052 H, 527
4-Thia-1-azabicyclo(3,2,0)heptane-2-carboxylic Acid
    (Trihydrate), 3344
5-Thia-1-azabicyclo[4.2.0]oct-2-ene-2-carboxylic Acid,
    3551
5-Thia-1-azabicyclo[4.2.0]oct-2-ene-2-carboxylic Acid, 7-
    [[(2R)-Aminophenylacetyl]amino]-3-chloro-8-oxo-,
    (6R,7R)-, 3168
Thiabendazole, 1888
Thiacetamide, 92
Thiacyclopentadiene, 244
Thiadiazinol, 1989

Thiamylal, 2702
Thianaphthene, 1326
Thiazafluron, 736
4-Thiazolidinecarboxamide, N-(1,1-Dimethylethyl)-3-
    [(2S,3S)-2-hydroxy-3-[[(2R)-2-[[(5-
    isoquinolinyloxy)acetyl]amino]-3-(methylthio)-1-
    oxopropyl]amino]-1-oxo-4-phenylbutyl]-, (4R)-, 4102
2-(Thiazol-4-yl)benzimidazole, 1888
N1-2-Thiazolyl-, 1661
p-2-Thiazolylsulfamoylsuccinanilic Acid, 2825
Thiazon, 458
Thiazone, 458
Thiergan, 3457
Thimet, 1266
Thimet Sulfone, 1268
Thioacetamide, 92
2-Thio-4-amino-6-hydroxypyrimidine, 258
Thiobencarb, 2662
1,1'-Thiobisethane, 370
Thiocarbazil, 3391
Thiochlormethyl, 1933
4-Thiocyanoaniline, 1040
Thiodan I, 1606
Thiodan II, 1607
Thiodemeton, 1590
Thiodemeton Sulfone, 1596
3-Thio-2,4-diazaspiro[5.5]undecane-1,3,5-trione, 1739
Thiodiazolique Ethyle, 1998
Thiodiazolique Methyle, 228
Thiodiazolique Methyle Acetyle, 663
Thiodiphenylamine, 2547
O,O'-(Thiodi-4,1-phenylene)bis(O,O'-
    dimethylphosphorothioate), 3355
Thioessigsaeureamid, 92
Thiofanox, 1805
Thiofuran, 244
Thiolo-Demeton, 1593
Thiolo-Methyl Demeton, 944
Thiolo-Methylmercaptophos, 944
Thiolo-Tinox, 525
Thiometon, 943
1-(Thiomorpholinyl)-3,5-bis(dimethylamino)-s-triazine,
    2346
Thionazin, 1493
Thiono-Methyl Demeton, 945
Thiono-Methylmercaptophos, 945
Thiopental, 2332
Thioperoxydicarbonic Diazide, 114
Thiophanox, 1805
Thiophen-carbonsaeure-(3), 392
Thiophene, 244
2,2'-(2,5-Thiophenediyl)bisbenzoxazole, 3508
Thiophene, 2-Ethyl-, 776
Thiophos, 2088
Thiophosphorsaeure-O,O-diaethyl-O-[2-(aethylthio)-
    aethyl]-ester, 1593
Thiophosphorsaeure-O,O-diaethyl-S-[2-(aethylthio)-
    aethyl]-ester, 1592
Thiophosphoryl Trimorpholide, 2749
Thiopropazate, 3925
Thiopyrine, 2597
Thiostrepton, 4176
2-Thiothymine, 406
Thiouracil, 236

2-Thiouracil, 236
Thiourea, 23
Thiouree, 23
Thioxamyl, 1202
Thiram, 841
Thorazine, 3444
Thor-prom, 3444
THQ, 1701
DL-Threitol, 368
Threonine, 353
(±)-Threonine, 350
DL-Threonine, 350
L-Threonine, 353
D-(-)-Threo-1-(p-nitrophenyl)-2-dichloroacetamido-1,3-
  propanediol, 2228
Thymine, 409
Thymol, 2073
DL-Thyronin, 1650
Tianafac, 2197
Timobesone Acetate, 3950
Tiocarbazil, 3391
Tio-urasin, 80
TMMT, 2199
TMPPT, 2600
TO-2, 1130
Tolazamide, 3096
Tolban, 3030
Tolbutamide, 2705
Tolcyclamide, 3081
o-Tolidine, 3031
Tolinase, 3096
Tolkan, 2700
Toluene, 1102
Toluenecarboxylic Acid, 1379
Toluenesulfamide, 1162
m-Toluenesulfonamide, 1149
o-Toluenesulfonamide, 1148
p-Toluenesulfonamide, 1147
Toluene-4-sulfonic Acid, 1125
p-Toluenesulfonic Acid, 1125
o-Toluenesulfonic Acid (Dihydrate), 1127
o-Toluenesulfonic Acid (Monohydrate), 1167
p-Toluenesulfonic Acid (Monohydrate), 1126
p-Toluenesulfonic Acid (Tetrahydrate), 1128
N-(p-Toluenesulfonyl)-N'-hexamethyleniminourea, 3096
m-Toluic Acid, 1384
o-Toluic Acid, 1383
p-Toluic Acid, 1379
p-Toluidin, 1134
m-Toluidin-N-acetat, 1703
2-Toluidine, 1132
3-Toluidine, 1134
m-Toluidine, 1134
o-Toluidine, 1132
p-Toluidine, 1134
m-Toluidine-N-acetate, 1703
p-Toluonitrile, 1338
3-o-Toluoxypyridazine, 2210
Tolurex, 2017
Toluylene, 2990
o-Toluylic Acid, 1383
2'-(p-Toluylyl)-6-methoxypurine Arabinoside, 3649
1-(o-Tolylazo)-2-naphthol, 3418
1-o-Tolylazo-2-naphthol, 3418

m-Tolyl Chloride, 1071
p-Tolyl Chloride, 1072
Tolylfluanid, 2021
p-Tolylhydroxylamin, 1143
p-Tolylhydroxylamine, 1143
4-Tolylisothiocyanate, 1344
p-Tolyl Isothiocyanate, 1344
3-Tolyl Methylcarbamate, 1712
m-Tolyl Methylcarbamate, 1712
o-Tolylsaeure, 1383
p-Tolylsulfamide, 1162
2-Tolyl Tosylate, 3018
o-Tolyl Tosylate, 3018
1-(2-Tolyl)urea, 1436
1-(4-Tolyl)urea, 1435
o-Tolylurea, 1436
p-Tolylurea, 1435
Tonox, 2829
Tonzilamine, 3365
Topcide, 1657
TOPO, 3984
Topusyn, 1522
Torak, 3043
Torelle, 1082
Tosic Acid, 1125
Tosylcyclopentylurea, 2894
Tosylcyclopentyluree, 2894
Tosylurea, 1447
Tosyluree, 1447
Totacillin, 3341
Toxakil, 1937
Toxaphene, 1937
Toxichlor, 1874
TPP, 3529
Tramat, 2903
Tranid, 1981
Transderm-SCOP, 3465
1,4-Transdimethylcyclohexane, 1524
Tranzetil, 3757
Trestolone, 3690
Triacetin, 1779
Triacetylglycerol, 1779
Triacetyloleandomycin, 4129
Triadimefon, 3026
Triaethylamin, 940
Triallate, 2091
Triallyl Cyanurate, 2657
Triamcinolone, 3826
Triamcinolone Acetonide, 3951
Triamcinolone 16α,17-Acetonide, 3951
Triamcinolone 16,21-Diacetate, 3992
Triamefon, 3026
Triamifos, 2723
2,4,7-Triaminopteridine, 752
2:4:7-Triaminopteridine, 752
4,6,7-Triaminopteridine, 751
4:6:7-Triaminopteridine, 751
Triamiphos, 2723
Triamylose (Tetrahydrate), 3603
Triasyn, 1603
3,6,9-Triaza-1,11-undecanediamine, 1600
3,6,9-Triazaundecane-1,11-diamine, 1600
Triazid, 3672
1,3,5-Triazin-2-amine, 4,6-dichloro-N-cyclohexyl-, 1733

1,3,5-Triazine, 2341
s-Triazine, 2,4-bis(Dimethylamino)-6-morpholino-, 2345
s-Triazine, 2,4-bis(Dimethylamino)-6-piperidino-, 2734
s-Triazine, 2-Chloro-4-methylamino-6-propylamino-,
    1181
1,3,5-Triazine-2,4-diamine, 2341
1,3,5-Triazine-2,4-diamine, 6-Chloro-N-methyl-N'-
    propyl-, 1181
1,3,5-Triazine-2,4-diamine, N,N-Diethyl-6-methoxy-N'-
    (1-methylethyl), 2354
1,3,5-Triazine-2,4-diamine, N-(1,1-Dimethylethyl)-N'-
    ethyl-6-methoxy-, 2146
1,3,5-Triazine-2,4-diamine, 6-(Hexahydro-1H-azepin-1-
    yl)-N,N,N',N'-tetramethyl-, 2930
1,3,5-Triazine-2,4-diamine, N,N,N',N'-Tetramethyl-6-(1-
    piperazinyl)-, 2359
1,3,5-Triazine-2,4-diamine, N,N,N',N'-Tetramethyl-6-(4-
    thiomorpholinyl)-, 2346
1,3,5-Triazine, 2-(Diethylamino)-4-(isopropylamino)-6-
    methoxy, 2354
1,3,5-Triazine-2,4(1H,3H)-dione, 3-Cyclohexyl-6-
    (dimethylamino)-1-methyl-, 2726
1,3,5-Triazine-2,4,6-triamine, 171
1,3,5-Triazine-2,4,6-triamine, N,N-Diethyl-N',N',N'',N''-
    Tetramethyl-, 2362
1,3,5-Triazine-2,4,6-triamine, N,N',N''-Triethyl-, 1808
1,3,5-Triazine-2,4,6-triamine, N,N',N''-Triethyl-N,N',N''-
    trimethyl-, 2751
s-Triazine-2,4,6-trithiol, 123
1,3,5-Triazin-2(1H)-one, 4,6-bis(ethylamino)-, 1203
1,3,5-Triazin-2(1H)-one, 4,6-bis[(1-methylethyl)amino]-,
    2150
s-Triazole, 2,4-bis(Ethylamine)-6-methoxy-, 1521
s-Triazole, 2,4-bis(isopropylamine)-6-methylmercapto-,
    2152
1H-1,2,3-Triazole, 1-[Diphenyl[3-
    (trifluoromethyl)phenyl]methyl]-, 3865
Triazophos, 2672
2,2,2-Tribrom-aethanol, 60
2,4,6-Tribrom-benzoesaeure, 955
2,4,6-Tribromobenzoic Acid, 955
2,4,6-Tribromobiphenyl, 551
2,2,2-Tribromoethanol, 60
Tribromo-fluoro-methane, 33
Tribromomethane, 3
Tribromophenol, 552
2,4,6-Tribromophenol, 552
2,4,6-Tribrom-phenol, 552
Tribunil, 1969
Tributanoylglycerol, 3264
Tributoxyethyl Phosphate, 3616
Tri-n-butylacetohydroxamic Acid, 3122
Tributyl Acetylcitrate, 3789
Tributylamine, 2759
Tributyl Citrate, 3605
Tri-n-butyl Citrate, 3605
Tributyl Phosphate, 2765
Tri-n-butyl Phosphate, 2765
Tributyl Phosphine Oxide, 2761
Tributylphosphine Oxide, 2761
Tributyrin, 3264
Tricarballylic Acid, 773
Tricarballylsaeure, 773
Tricerol, 3537

1,1,1-Trichloethane, 63
3,4,4'-Trichlorbiphenyl, 2492
Trichlorfon, 312
Trichlormetafos-3, 1668
β,β,β-Trichlor-milchsaeure, 118
Trichloroacetic Acid, 48
S-(2,3,3-Trichloroallyl)diisopropylthiocarbamate, 2091
2,3,4-Trichloroanisole, 995
2,4,6-Trichloroanisole, 996
1,2,3-Trichlorobenzene, 559
1,2,4-Trichlorobenzene, 560
1,3,5-Trichlorobenzene, 561
2,3,6-Trichlorobenzoic Acid, 957
4,5,7-Trichloro-2,1,3-benzothiadiazole, 527
Trichlorobenzyl Chloride, 541
2,3,6-Trichlorobenzyloxypropanol, 1956
Trichlorobiphenyl, 2497
2,2',3-Trichlorobiphenyl, 2500
2,2',4-Trichlorobiphenyl, 2501
2,2',5-Trichlorobiphenyl, 2502
2,2',6-Trichlorobiphenyl, 2503
2',3,4-Trichlorobiphenyl, 2496
2,3,4'-Trichlorobiphenyl, 2505
2,3',5-Trichlorobiphenyl, 2504
2,3,6-Trichlorobiphenyl, 2498
2,3',6-Trichlorobiphenyl, 2491
2,4,4'-Trichlorobiphenyl, 2495
2,4,5-Trichlorobiphenyl, 2494
2,4',5-Trichlorobiphenyl, 2506
2,4,6-Trichlorobiphenyl, 2493
2,5,4'-Trichlorobiphenyl, 2506
3,4,4'-Trichlorobiphenyl, 2492
1,1,1-Trichloro-tert-butanol, 295
2,4,6-Trichloro-m-cresol, 994
1,2,4-Trichlorodibenzo-p-dioxin, 2429
2,4,6-Trichloro-3,5-dimethyl-phenol, 1335
Trichloroethane, 63
1,1,1- Trichloroethane, 63
1,1,1-Trichloroethane, 63
1,1,2-Trichloroethane, 64
1,1,2-β-Trichloroethane, 64
Trichloroethene, 46
Trichloroethylene, 46
Trichloro-ethylene, 46
1,1,2-Trichloroethylene, 46
1,1'-(2,2,2-Trichloroethylidene)-bis[4-methoxybenzene],
    3298
1,2-O-(2,2,2-Trichloroethylidene)-α-D-glucofuranose,
    1477
Tri-β-chloroethyl Phosphate, 827
Trichloroethyl Salicylate, 1617
Trichlorofluoromethane, 38
Trichlorohydrin, 143
2,3,5-Trichloro-4-hydroxypyridine, 377
Trichloroisocyanuric Acid, 223
β,β,β-Trichlorolactic Acid, 118
Trichloromethane, 5
1,2,3-Trichloro-4-methoxy-benzene, 995
2,4,6-Trichloro-3-methylphenol, 994
N-Trichloromethylthio-4-cyclohexene-1,2-dicarboximide,
    1634
N-(Trichloromethylthio)phthalimide, 1602
Trichloronat, 1985
Trichloronate, 1985

1,2,3-Trichloro-4-nitrobenzene, 538
1,2,4-Trichloro-5-nitrobenzene, 537
1,4,5-Trichloro-2-nitrobenzene, 537
2,3,4-Trichloronitrobenzene, 538
2,3,4-Trichloro-1-nitrobenzene, 538
2,4,5-Trichloronitrobenzene, 537
2,4,5-Trichloro-1-nitrobenzene, 537
3,4,6-Trichloronitrobenzene, 537
1,3,5-Trichloro-2-(4-nitrophenoxy)benzene, 2458
2,3,4-Trichlorophenol, 564
2,3,5-Trichlorophenol, 565
2,3,6-Trichlorophenol, 566
2,4,5-Trichlorophenol, 567
2,4,5-Trichloro-phenol, 567
2,4,6-Trichlorophenol, 563
(2,4,5-Trichlorophenoxy)acetic Acid, 1285
2,3,4-Trichlorophenoxyacetic Acid, 1281
2,3,5-Trichlorophenoxyacetic Acid, 1284
2,3,6-Trichlorophenoxyacetic Acid, 1286
2,4,5-Trichlorophenoxyacetic Acid, 1285
2,4,6-Trichlorophenoxyacetic Acid, 1283
3,4,5-Trichlorophenoxyacetic Acid, 1282
2,4,5-Trichlorophenoxy-γ-butyric Acid, 1903
4-(2,4,5-Trichlorophenoxy)butyric Acid, 1903
2-(2,4,5-Trichlorophenoxy)propionic Acid, 1616
2,3,6-Trichlorophenylacetic Acid, 1280
1',3',5'-Trichlorophenyl-4-nitrophenyl Ether, 2458
1,2,3-Trichloropropane, 143
3,5,6-Trichloropyridinol, 376
3,5,6-Trichloro-2-pyridinol, 376
(3,5,6-Trichloro-2-pyridinyl)oxyacetic Acid, 972
3,5,6-Trichloro-2-Pyridyl Diethyl Phosphate, 1696
2,4'',5-Trichloro-p-terphenyl, 3509
2,4'',5-Trichloro-1,1':4',1''-terphenyl, 3509
1,1,2-Trichloro-1,2,2-trifluoroethane, 109
3,4,5-Trichloroveratrole, 1336
4,5,6-Trichloroveratrole, 1336
2,3,4-Trichlorphenol, 564
2,3,5-Trichlorphenol, 565
2,3,6-Trichlorphenol, 566
2,4,6-Trichlorphenol, 563
Triclopyr, 972
Triclosan, 2507
Tricresyl Phosphate, 3811
Tri-p-cresyl Phosphate, 3811
Tricuron, 2925
Tricyclazole, 1630
Tricyclohexylhydroxystannane, 3608
Tricyclohexyltin Hydroxide, 3608
1-Tridecanecarboxylic Acid, 3116
Tridecane(S-(carboxymethyl)-L-cyateine)), 4170
1,13-Tridecanedicarboxylic Acid, 3265
1,13-Tridecanedioic Acid, 2931
Tridecanoic Acid, 2936
n-Tridecanoic Acid, 2936
Tridecyl p-Hydroxybenzoate, 3784
Tridione, 778
Trientine, 950
Trietazine, 1786
Triethylamine, 940
N,N,N'-Triethyl-bicyclo(2.2.1)hept-5-ene-2,3-trans-
    dicarboxamide, 3255
Triethyl Carbinol, 1249
Triethylene Glycol Dibutyrate, 3397

Triethylene Glycol Dipropionate, 2740
Triethylenetetramine, 950
N2,N4,N6-Triethylmelamine, 1808
Triethyl Phosphate, 946
N,N',N''-Triethyl-1,3,5-triazine-2,4,6-triamine, 1808
N2,N4,N6-Triethyl-N2,N4,N6-trimethylmelamine, 2751
N2,N4,N6-Trimethyl-N2,N4,N6-trimethylolmelamine,
    1812
Trifenmorph, 3914
Trifluoperazine, 3817
α,α,α-Trifluoro-2,6-dinitro-N,N-dipropyl-p-toluidine,
    2851
2,2,2-(Trifluoroethoxy)ethene, 251
2-[3-(Trifluoromethyl)anilino]nicotinic Acid, 2781
6-(Trifluoromethyl)-2H-1,2,4-benzothiadiazine-7-
    sulfonamide 1,1-dioxide, 1308
2-Trifluoromethyl-4,5-Dichlorobenzimidazole, 1271
2-Trifluoromethyl-N,N-dimethyl-10H-phenothiazine-10-
    propanamide, 3656
4-Trifluoromethyl-N,N-dimethyl-10H-phenothiazine-10-
    propanamide, 3547
3-Trifluoromethyl-4-nitrobenzenesulfonamide, 1000
N-(3-Trifluoromethylphenyl)anthranilic Acid, 2970
1,1,1-Trifluoro-N-(2-methyl-4-
    (phenylsulfonyl)phenyl)methanesulfonamide, 2991
Trifluoro-N-(2-methyl-4-
    (phenylsulfonyl)phenyl)methanesulfonamide, 2991
6-Trifluoromethyl-7-sulfamoyl-4H-1,2,4-benzothiadiazine
    1,1-dioxide, 1308
Trifluoromethylthiazide, 1308
Trifluorophenol, 570
2,3,4-Trifluorophenol, 570
α, α, α-Trifluoro-o-toluic Acid, 1287
Trifluoro-o-toluic Acid, 1287
N-(α,α,α-Trifluoro-m-tolyl)anthranilic Acid, 2970
Triflupromazine, 3548
Trifluralin, 2851
Triforine, 2055
Trifungol, 1807
1,3,5-Triglycidyl-S-triazinetrione, 2658
Triglycidylurazol, 2297
Triglycine, 779
Triglycine Hydantoin Acid, 1185
Triglycine Sulfate, 949
Triglycol Dichloride, 826
1,2,4-Trihydroxy-anthrachinon, 2949
1,2,3-Trihydroxybenzene, 729
1,3,5-Trihydroxybenzene, 726
2,3,4-Trihydroxybenzoesaeure, 1069
3,4,5-Trihydroxybenzoesaeure, 1068
2,3,4-Trihydroxybenzoic Acid, 1069
2,4,6-Trihydroxy-benzophenon, 2802
2,4,6-Trihydroxybenzophenone, 2802
3α, 6α, 7α -Trihydroxy-5β-cholanate, 3982
3α,7β,12α-Trihydroxy-5β-cholanoic Acid, 3980
11β,17α,21-Trihydroxypregna-1,4-diene-3,20-dione, 3831
11β,17,21-Trihydroxypregn-4-ene-3,20-dione, 3843
2,4,6-Trihydroxypteridine, 617
2,4,7-Trihydroxypteridine, 616
2:4:6-Trihydroxypteridine, 617
2:4:7-Trihydroxypteridine, 616
4,6,7-Trihydroxypteridine, 615
4:6:7-Trihydroxypteridine, 615
2,4,6-Trihydroxypyrimidine, 240

2,4,6-Triiodo-3-acetaminobenzoic Acid, 1608
Triiodomethane, 6
3,3',5-Triiodothyronine, 3151
Triisobutene, 2724
Trilafon, 3820
Trimelamol, 1812
Trimercapto-s-triazine, 123
Trimesic Acid, 1615
Tri-methacycline, 3875
Trimethadione, 778
Trimethoprim, 3057
3,4,5-Trimethoxybenzoyl Methyl Reserpate, 4101
5-(3,4,5-Trimethoxybenzyl)-2,4-diaminopyrimidine, 3057
Trimethylacetic Acid, 478
2'-Trimethylacetyl-6-methoxypurine Arabinoside, 3369
Trimethylacetyl Salicylate, 2637
Trimethylamine, 220
α,α',α''-Trimethylaminetricarboxylic Acid, 779
Trimethylammonioacetate (Monohydrate), 502
2,4,5-Trimethylanilin, 1764
2,4,5-Trimethylaniline, 1764
1,2,3-Trimethylbenzene, 1725
1,2,4-Trimethylbenzene, 1723
1,3,5-Trimethylbenzene, 1729
2,4,6-Trimethylbenzoic Acid, 2010
endo-1,7,7-Trimethyl-bicyclo[2.2.1]heptan-2-ol, 2129
1,1,1-Trimethyl-2,2-bis(p-ethoxylphenyl)ethane, 3828
1,1,1-Trimethyl-2,2-bis(p-methyloxylphenyl)ethane, 3679
2,3,3-Trimethyl-2-butanol, 1258
1,1,3-Trimethylcyclohexane, 1803
Trimethylcyclohexyl Phthalate, 4021
1,1,3-Trimethylcyclopentane, 1525
3,5,5-Trimethyl-2,4-diketooxazolidine, 778
1,3,7-Trimethyl-2,6-dioxo-8-ethoxypurine, 2067
Trimethylene, 153
5,5-Trimethylenebarbituric Acid, 1109
Trimethylene Bromide, 156
α-Trimethylene Trisulphide Dioxide, 178
β-Trimethylene Trisulphide Dioxide, 179
α-Trimethylene Trisulphoxide, 184
β-Trimethylene Trisulphoxide, 183
Trimethylessigsaeure, 478
2,2,5-Trimethylhexane, 1841
3,5,5-Trimethylhexanol, 1845
3,5,5-Trimethyl-1-hexanol, 1845
3.,5,5-Trimethyl Hexanol, 1845
2,2,5-Trimethyl-3-hexyne, 1784
Trimethylmethane, 357
1,4,5-Trimethylnaphthalene, 2827
D-1,3,3-Trimethyl-2-norbornanone, 2104
3,5,5-Trimethyl-2,4-oxazolidinedione, 778
2,3,4-Trimethylpentane, 1571
2:2:4-Trimethylpentane, 1574
2,2,3-Trimethylpentanol-3, 1376
2,2,3-Trimethyl-3-pentanol, 1376
2,3,5-Trimethyl-phenol, 1752
2,4,6-Trimethylphenol, 1755
Trimethyl Phosphate, 221
2,6,7-Trimethylpteridine, 1677
2:6:7-Trimethylpteridine, 1677
N,N',N''-Trimethyl-N,N',N''-trimethylolmelamine, 1812
1,3,7-Trimethyluric Acid, 1451

Trimeturon, 2018
Trimorpholin, 777
Trimorpholine, 777
Trimpex, 3057
2,4,6-Trinitroaniline, 618
2,4,6-Trinitroanisole, 1024
1,3,5-Trinitrobenzene, 571
sym-Trinitrobenzene, 571
2,4,6-Trinitrobenzoesaeure, 960
2,4,6-Trinitrobenzoic Acid, 960
1,3,5-Trinitro-benzol, 571
2,4,6-Trinitro-1-chlorobenzene, 535
2,4,6-Trinitro-m-cresol, 1025, 1026
2,4,6-Trinitrodiethylaniline, 2003
2-4-6-Trinitrodiethylaniline, 2003
2,4,6-Trinitrodiphenyllamine, 2529
2,4,6-Trinitroethylaniline, 1374
2,4,6-Trinitro-m-kresol, 1025
2,4,6-Trinitro-3-methylphenol, 1026
2-4-6-Trinitromonoethylaniline, 1374
1,3,8-Trinitronaphthalene, 1863
1,4,5-Trinitronaphthalene, 1864
1,3,8-Trinitronaphthalin, 1863
1,4,5-Trinitronaphthalin, 1864
2,4,6-Trinitrophenol, 572
Trinitrophenylethylnitramine, 1350
2,4,6-Trinitrophenylethylnitramine, 1350
2,4,6-Trinitrotoluene, 1023
Trioctylphosphine oxide, 3984
Trional, 1586
4,6,10-Trioxa-1-azatricyclo[3.3.1.13,7]decane, 777
1,3,5-Trioxan, 181
s-Trioxane, 181
3,7,12-Trioxo-5β-cholanic Acid, 3962
2,4,6-Trioxo-5-methyl-5-phenylhexahydropyrimidine, 2211
Tripelennamine, 3363
Triphenylcarbinol, 3629
Triphenylene, 3514
Triphenylmethanol, 3629
Triphenyl Phosphate, 3529
Triphenyl Phosphoric Acid Ester, 3529
Triphenyltin Hydroxide Acetate, 3732
Tripropanoylglycerol, 2730
Tripropionin, 2730
Tripropionylglycerol, 2730
Tripropylamine, 1849
Tri-n-propylamine, 1849
Tripropyl Phosphate, 1856
Tri-n-propyl Phosphate, 1856
Tris-BP, 1780
Tris-n-butylamine, 2759
Tris-(2-chloroethyl) Phosphate, 827
Tris(2,3-dibromopropyl)Phosphate, 1780
Tris(1,3-dichloroisopropyl) Phosphate, 1781
2,4,6-Tris(dimethylamino)-1,3,5-triazine, 1809
Tris(dimethyldithiocarbamate)iron, 1807
Tris(ethylamino)-1,3,5-triazine, 1808
2,4,6-Tris(ethylamino)-1,3,5-triazine, 1808
2,4,6-Tris(ethylamino)-s-triazine, 1808
Tris-(2-ethylhexyl) Phosphate, 3986
Tris-(hydroxymethyl)-amino-methan, 373
Tris-(hydroxymethyl)-Aminomethane, 373
Tris(hydroxymethyl)methylamine, 373

1,3,5-Trithiane, 1,3-Dioxide, cis-, 179
1,3,5-Trithiane, 1,3-Dioxide, trans-, 178
1,3,5-Trithiane, 1,3,5-Trioxide, (1α,3α,5α)-, 184
1,3,5-Trithiane, 1,3,5-Trioxide, (1α,3α,5β)-, 183
Trithiocyanuric Acid, 123
Trithion, 2302
Tritolyl Phosphate, 3811
N-Tritylmorpholine, 3914
Triumph, 1797
TRI-VC13, 2022
2,3,4-Trojmetylopentan, 1571
Troleandomycin, 4129
Tromethamine, 373
Tromexan, 3866
2,4,6-Tronitrotoluol, 1023
Tropacocaine, 3210
DL-Tropasaeure, 1685
DL-Tropic Acid, 1685
Tropilidene, 1101
Tropium, 3285
Trp, 2233
Truflex DTDP, 4109
Tryptophan, 2233
(S)-(-)-Tryptophan, 2233
DL-Tryptophan, 2234
Tubatoxin, 3913
Tunic, 1605
Tylosin, 4138
Tylosterone, 3559
Tyramin, 1479
Tyramine, 1479
Tyrene, 996
DL-Tyrosin, 1716
L-Tyrosin, 1714
Tyrosine, 1714
(S)-(-)-Tyrosine, 1714
D-Tyrosine, 1715
DL-Tyrosine, 1716
L-Tyrosine, 1714
p-Tyrosine, 1714
U 17835, 3096
U 3096, 3839
U-20235, 3894
U-26,952, 3162
U-2726, 3754
U-34865, 4014
U-5954, 342
UC-10854, 2290
UC 21149, 1208
UC 22463, 1648
UDCA, 3977
Ulfamerazine, 2239
Ultracide, 817
Ultrazine, 3745
Umbelliferone, 1611
Unasyn, 3341
1,4,7,10,13,16,19,22,25,28,31-Undecaazacyclo-
    tritriacontane, Cyclic Peptide Deriv., 4166
Undecane, 2373
n-Undecane, 2373
Undecane(S-(carboxymethyl)-L-cysteine)), 4157
1,11-Undecanedicarboxylic Acid, 2931
Undecanedioic Acid, 2348
Undecanoic Acid, 2364

Undecanoique Acide, 2364
Undecyl Carbinol, 2758
Undecylenic Acid, 2347
10-Undecylenic Acid, 2347
Undecyl 4-Hydroxybenzoate, 3598
Undecyl p-Hydroxybenzoate, 3598
Unimoll DM, 1952
Unisomnia, 3145
Unitex OB, 3508
Uracil, 239
Uracil, 5-Bromo-3-isopropyl-6-methyl-, 1475
Uracil-carbonsaeure-(4), 381
5-Uracilcarboxylic Acid, 381
Uracil, 3-Methyl-, 404
α-Uramidocaproic Acid, 1210
ε-Uramidocaproic Acid, 1209
β-Uramidopropionic Acid, 319
δ-Uramidovaleric Acid, 834
Urea, 22
Urea, N-Benzoyl-N-(3,4-dichlorophenyl)-N',N'-dimethyl-,
    3286
Urea, (2-Bromo-2-methylbutyryl)-, 807
Urea, N'-(4-Chlorophenyl)-N-methyl-N-(1-methyl-2-
    propynyl), 2602
Urea, N,N'-Dimethyl-N-[5-(trifluoromethyl)-1,3,4-
    thiadiazol-2-yl]-, 736
Urea, 3-[Hexahydro 4,7-methanoindan-1(or 2)-yl]-1,1-
    dimethyl-, 2925
Uree, 22
4-Ureidophenyl Acetate, 1671
p-Ureidophenyl Acetate, 1671
p-Ureidophenyl Prostaglandin E2, 4038
p-Ureidophenyl Prostaglandin F2 α, 4040
Urethan, 202
Urethane, 202
Uric Acid, 388
Uric Acid (Dihydrate), 389
Uridine, 2'-Deoxy-5-iodo-, 5'-Butanoate, 2865
Uridine, 2'-Deoxy-5-iodo-, 5'-(2,2-Dimethylpropanoate),
    3069
Uridine, 2'-Deoxy-5-iodo-, 5'-(2-Methylpropanoate),
    2864
Uridine, 2'-Deoxy-5-iodo-, 5'-Pentanoate, 3068
Uridine, 2'-Deoxy-5-iodo-, 5'-Propanoate, 2641
Urisoxin, 2267
Urocanic Acid, 703
Urocaninsaeure, 703
Urogan, 2267
Uromaline, 1392
Uromiro, 2573
Uromiron, 2573
Uroxine, 1314
Ursocholic Acid, 3980
Ursodeoxycholic Acid, 3977
USB 3584, 2256
USR 604, 1859
Uvitex EBF, 3508
Vagestrol, 3559
Valeral, 469
Valeraldehyde, 469
n-Valeraldehyde, 469
Valeramide, 487
Valeric Acid, 480
Valeric Acid, Normal, 480

Valerylacetone, 1505
9-(2-O-Valeryl-β-D-arabinofuranosyl)adenine, 3237
9-[5'-(O-Valeryl)-β-D-arabinofuranosyl]adenine Ester, 3238
5'-Valeryl 5-iodo-2'-deoxyuridine, 3068
2'-Valeryl-6-methoxypurine Arabinoside, 3369
DL-Valin, 493
Valine, 495
D-Valine, 498
DL-Valine, 493
L-Valine, 495
L-(+)-valine, 495
Valium, 3279
Valmethamide, 1566
Valnoctamide, 1566
Valrelease, 3279
Vamidothion, 1577
Vanay, 1779
Vancide 89, 1634
Vanillaldehyde, 1396
Vanillic Acid, 1402
Vanillic Aldehyde, 1396
Vanillin, 1396
Vanillinsaeure, 1402
Vanobid, 4167
Vasicinone, 2209
(-)-Vasicinone, 2209
L-Vasicinone, 2209
Vasomax, 3572
Vazepam, 3279
VCD, 1566
Vegiben 2E, 1334
VEL 5026, 2102
Velosef, 3342
Velsicol 1068, 1874
Velsicol 506, 2788
Ventolin, 2920

Vepesid, 4071
o-Veratric Acid, 6-[[6-[2-(Dimethylamino)ethyl]-2-methoxy-3,4-(methylenedioxy)phenyl]acetyl]-, 3923
Veratrine, 4095
Veratrole, 1463
Veratrumsaeure, 1689
Vernam, 2177
4-Vinylcyclohexene, 1482
4-Vinyl-1-cyclohexene, 1482
DL-Weinsaeure, 287
L(+)-Weinsaeure, 286
meso-Weinsaeure, 289
9H-Xanthen-9-one, 1,7-Dihydroxy-3-methoxy-, 2978
1,2-Xylene, 1431
1,3-Xylene, 1429
1,4-Xylene, 1430
m-Xylene, 1429
o-Xylene, 1431
p-Xylene, 1430
1,3,2-Xylenol, 1456
2,3-Xylenol, 1458
2,4-Xylenol, 1452
2,5-Xylenol, 1457
2,6-Xylenol, 1456
3,4-Xylenol, 1455
3,5-Xylenol, 1454
m-Xylenol, 1452
p-Xylenol, 1457
2,5-Xylenol, 4-Butyl-, 2709
2,6-Xylenol, 2-Butyl-, 2714
2,6-Xylenol, 4-Butyl-, 2711
3,5-Xylenol, 2,4,6-trichloro-, 1335
L-Xylo-2-Hexulose, 871
α-Xylose, 484
D-Xylose, 484
N-(2,3-Xylyl)anthranilic Acid, 3179
3,4-Xylyl Methylcarbamate, 2027
cis-Zimtsaeure, 1639